# Contents

# Metals Handbook®
# Ninth Edition
## Volume 16
## Machining

Prepared under the direction of the
ASM INTERNATIONAL Handbook Committee

**Joseph R. Davis**, Senior Editor
**Steven R. Lampman**, Technical Editor
**Theodore B. Zorc**, Technical Editor
**Heather J. Frissell**, Editorial Supervisor
**George M. Crankovic**, Assistant Editor
**Alice W. Ronke**, Assistant Editor
**Karen Lynn O'Keefe**, Word Processing Specialist
**Jeanne Patitsas**, Word Processing Specialist

**Robert L. Stedfeld**, Director of Reference Publications
**Kathleen M. Mills**, Manager of Editorial Operations

Editorial Assistance
**Lois A. Abel**
**Robert T. Kiepura**
**Penelope Thomas**
**Nikki D. Wheaton**

ASM INTERNATIONAL™

METALS PARK, OH 44073

First printing, March 1989

Metals Handbook is a collective effort involving thousands of technical specialists. It brings together in one book a wealth of information from world-wide sources to help scientists, engineers, and technicians solve current and long-range problems.

Great care is taken in the compilation and production of this Volume, but it should be made clear that no warranties, express or implied, are given in connection with the accuracy or completeness of this publication, and no responsibility can be taken for any claims that may arise.

Nothing contained in the Metals Handbook shall be construed as a grant of any right of manufacture, sale, use, or reproduction, in connection with any method, process, apparatus, product, composition, or system, whether or not covered by letters patent, copyright, or trademark, and nothing contained in the Metals Handbook shall be construed as a defense against any alleged infringement of letters patent, copyright, or trademark, or as a defense against liability for such infringement.

Comments, criticisms, and suggestions are invited, and should be forwarded to ASM INTERNATIONAL.

Library of Congress Cataloging-in-Publication Data

ASM INTERNATIONAL

Metals handbook.

Vol. 16: Prepared under the direction of
the ASM INTERNATIONAL Handbook Committee.
Includes bibliographies and indexes.
Contents: v. 1. Properties and selection—
v. 2. Properties and selection—nonferrous alloys
and pure metals—[etc.]—v. 16. Machining

1. Metals—Handbooks, manuals, etc.
I. ASM Handbook Committee.
II. ASM INTERNATIONAL. Handbook Committee.
TA459.M43 1978 669 78-14934
ISBN 0-87170-007-7 (v. 1)
SAN 204-7586

Printed in the United States of America

# Foreword

In the 22 years since the 8th Edition *Metals Handbook* volume on machining was published, material removal operations have undergone dynamic changes. The mechanics of the cutting process are better understood, new cutting tool materials have been developed, machine controls and computer-aided engineering have rapidly advanced, and nontraditional machining methods continue to be refined. The difficult challenges faced by industry have necessitated these developments. Requirements for high-strength materials and the introduction of difficult-to-machine structural ceramics, composites, and electronic components have placed new and greater demands on machining technology, and have also spurred continued research and development in material removal techniques.

Volume 16 of the 9th Edition describes the evolution of machining technology comprehensively, with great attention to detail and accuracy. In addition to providing valuable information on recent developments, the Handbook devotes exhaustive coverage to more standard, traditional machining methods. This new Volume is also the final step in the fulfillment of ASM's commitment to provide the materials science community with the most authoritative information available on the processing and fabrication of metals. *Machining* completes the coverage of metalworking technology in the 9th Edition, taking its place alongside Volume 6 (*Welding, Brazing, and Soldering*), Volume 7 (*Powder Metallurgy*), Volume 14 (*Forming and Forging*), and Volume 15 (*Casting*).

This enormous undertaking was made possible by the combined efforts of many dedicated and selfless authors and reviewers, the ASM Handbook Committee, and the ASM editorial staff. Special recognition is also due to Metcut Research Associates Inc. and its president, William P. Koster, for permission to use tabulated data published in Volumes 1 and 2 of the *Machining Data Handbook* (3rd Edition). To all the men and women who contributed to the planning and preparation of this Volume, we extend our sincere thanks.

Richard K. Pitler
President,
ASM INTERNATIONAL

Edward L. Langer
Managing Director,
ASM INTERNATIONAL

**The Ninth Edition of Metals Handbook**
**is dedicated to the memory of**
**TAYLOR LYMAN, A.B. (Eng.), S.M., Ph.D.**
**(1917-1973)**
**Editor, Metals Handbook, 1945-1973**

# Preface

Machining is one of the most important of the basic manufacturing processes. Almost every manufactured product contains components that require machining, often to great precision. Yet material removal operations are among the most expensive; in the U.S. alone, more than $100 billion will be spent this year on machining. These high costs put tremendous economic pressures on production managers and engineers as they struggle to find ways to increase productivity. Compounding their problems is the increasing use of more difficult-to-machine materials, such as nickel-base superalloys and titanium-base alloys in aerospace applications, structural ceramics, high-strength polymers, composites (both metal-matrix and resin-matrix), and electronic materials.

The present Volume of *Metals Handbook* has been structured to provide answers to the questions and challenges associated with current machining technology. Following a general introduction to machining processes, 9 major sections containing 78 articles cover all aspects of material removal. Much of this material is new. In fact, 30 articles in this Volume were not included in its 8th Edition predecessor. Noteworthy are the articles that have been added to describe the mechanics of the cutting process and advances in new materials, new processes, new methods of machine control, and computer-aided engineering.

The first Section of the Handbook reviews the fundamentals of the machining process. Included are articles describing the mechanics of chip formation, the forces, stresses, and power at the cutting tool, the principles of tool wear and tool life, and the relationship between cutting and grinding parameters and surface finish and surface integrity.

In the following Section, extensive data are provided on the applications, advantages and limitations, properties, tool geometries, and typical operating parameters for seven classes of tool materials: high-speed tool steels (both conventional wrought and powder metallurgy), cast cobalt alloys, cemented carbides, cermets, ceramics, and ultrahard tool materials (polycrystalline diamond and cubic boron nitride). Recent developments in wear-resistant coatings that are applied on high-speed steel, carbides, and ceramics are also discussed.

The third Section focuses on cutting and grinding fluids—their functions, selection criteria, and application. Coverage of proper maintenance procedures (storage, handling, recycling, and disposal) and the toxicology and biology associated with cutting and grinding fluids is included.

The next Section contains 21 articles that summarize the process capabilities, machines, cutting parameters and variables, and applications of traditional chip removal processes, such as turning, drilling, and milling. Advanced tooling used in multiple-operation machining, proper tool fixturing, and tool condition monitoring systems are also discussed, along with computer numerical controlled machining centers, flexible manufacturing systems, and transfer machines.

Although near net shape technology, including a greater use of precision casting, powder metallurgy, and precision forging, has lessened the need for some traditional machining operations, abrasive machining is being employed to a greater extent than in the past. The fifth Section of the Handbook examines the principles, equipment, and applications of grinding, honing, and lapping as well as recent developments in super-abrasives, used for precision grinding of difficult-to-machine and/or brittle materials.

The sixth Section looks at a variety of nontraditional machining methods that do not produce chips or a lay pattern in the surface. Mechanical, electrical, thermal, and chemical nontraditional techniques are described. Applications of these methods are emphasized, with practical examples involving nontraditional machining of metals, ceramics, glasses, plastics, and electronic components.

The next Section describes high-speed and high removal rate processes that have been developed to dramatically increase productivity. The effects of high-speed processing on chip formation and tool wear are discussed, along with materials that are being machined using these processes.

The eighth Section introduces the reader to two of the most rapidly developing and important areas in machining technology: machine controls and computer applications. Although the basic configurations of many machine tools have not changed significantly, the advent of numerical control and adaptive control has substantially improved manufacturing productivity and workpiece quality. Machine controls and the integration of CAD/CAM technology into machine tools are described in articles written with the engineer, not the software expert, in mind.

The last Section of the Handbook covers specific machining practices for 23 different metal systems, including all structural alloy systems, and relates the latest information on such topics as powder metals, metal-matrix composites, and honeycomb structures. Machining parameters (speeds, feeds, depth-of-cut, etc.) and the influence of microstructure on machinability are described in detail. Coverage includes difficult-to-machine aerospace alloys and high-silicon cast aluminum alloys, as well as materials such as beryllium and uranium that require special considerations during machining. Finally, an article on machinability test methods examines various types of tests used to study cutting tool and workpiece machining characteristics.

Much of the credit for the content and organization of this Handbook must be given to the Steering Committee that worked with the ASM staff during the early stages of the project. This group includes Professor George E. Kane, Lehigh University; Dr. William P. Koster, Metcut Research Associates Inc.; Dr. Ranga Komanduri, National Science Foundation; Dr. Richard P. Lindsay, Norton Company; Mr. Gary F. Benedict, Allied-Signal Aerospace Company, Garrett Engine Division; and Mr. Michael E. Finn, Stelco Inc. We are also indebted to the officers of the Society of Carbide and Tool Engineers for their assistance in the planning of the Volume. Finally, we gratefully acknowledge the countless hours of time and expertise loaned to the project by the nearly 200 authors and reviewers. Without the collective efforts of all these individuals, the successful completion of this Handbook would not have been possible.

The Editors

## Officers and Trustees of ASM INTERNATIONAL



## Members of the ASM Handbook Committee (1988–1989)



## Previous Chairmen of the ASM Handbook Committee

# Policy on Units of Measure

By a resolution of its Board of Trustees, ASM INTERNATIONAL has adopted the practice of publishing data in both metric and customary U.S. units of measure. In preparing this Handbook, the editors have attempted to present data in metric units based primarily on Système International d'Unités (SI), with secondary mention of the corresponding values in customary U.S. units. The decision to use SI as the primary system of units was based on the aforementioned resolution of the Board of Trustees and the widespread use of metric units throughout the world.

For the most part, numerical engineering data in the text and in tables are presented in SI-based units with the customary U.S. equivalents in parentheses (text) or adjoining columns (tables). For example, pressure, stress, and strength are shown both in SI units, which are pascals (Pa) with a suitable prefix, and in customary U.S. units, which are pounds per square inch (psi). To save space, large values of psi have been converted to kips per square inch (ksi), where 1 ksi = 1000 psi. The metric ton (kg × $10^3$) has been shown in megagrams (Mg). Some strictly scientific data are presented in SI units only.

To clarify some illustrations, only one set of units is presented on artwork. References in the accompanying text to data in the illustrations are presented in both SI-based and customary U.S. units. On graphs and charts, grids corresponding to SI-based units appear along the left and bottom edges. Where appropriate, corresponding customary U.S. units appear along the top and right edges.

Data pertaining to a specification published by a specification-writing group may be given in only the units used in that specification or in dual units, depending on the nature of the data. For example, the typical yield strength of aluminum sheet made to a specification written in customary U.S. units would be presented in dual units, but the sheet thickness specified in that specification might be presented only in inches.

Data obtained according to standardized test methods for which the standard recommends a particular system of units are presented in the units of that system. Wherever feasible, equivalent units are also presented.

Conversions and rounding have been done in accordance with ASTM Standard E 380, with attention given to the number of significant digits in the original data. For example, an annealing temperature of 1570 °F contains three significant digits. In this case, the equivalent temperature would be given as 855 °C; the exact conversion to 854.44 °C would not be appropriate. For an invariant physical phenomenon that occurs at a precise temperature (such as the melting of pure silver), it would be appropriate to report the temperature as 961.93 °C or 1763.5 °F. In some instances (especially in tables and data compilations), temperature values in °C and °F are alternatives rather than conversions.

The policy on units of measure in this Handbook contains several exceptions to strict conformance to ASTM E 380; in each instance, the exception has been made in an effort to improve the clarity of the Handbook. The most notable exception is the use of g/$cm^3$ rather than kg/$m^3$ as the unit of measure for density (mass per unit volume).

SI practice requires that only one virgule (diagonal) appear in units formed by combination of several basic units. Therefore, all of the units preceding the virgule are in the numerator and all units following the virgule are in the denominator of the expression; no parentheses are required to prevent ambiguity.

# Authors and Reviewers

**T.E. Aaron**
Anocut, Inc.
**Gary Adams**
Cominco Metals
**John Agapiou**
General Motors Technical Center
**M.S. Ahmed**
Transfer Technology Limited (England)
**G. Albares**
Technical Consultant
**Tom Andrew**
Harper Company
**James A. Aris**
Rockwell International
**William N. Ault**
Norton Company
**A. Bagchi**
Ohio State University
**J. Gary Baldoni**
GTE Laboratories
**Moshe M. Barash**
Purdue University
**Carl Bartholed**
Reishauer Corporation
**Alan M. Bayer**
Teledyne Vasco
**Abdel E. Bayoumi**
Washington State University
**Bruce N. Beauchesne**
Laser Services, Inc.
**Bruce A. Becherer**
Teledyne Vasco
**Guy Bellows**
Metcut Research Associates Inc.
**Gary F. Benedict**
Allied-Signal Aerospace Company
Garrett Engine Division
**R.C. Benn**
Inco Alloys International, Inc.
**E.O. Bennett**
University of Houston
**Michael Bess**
Aluminum Smelting & Refining Company, Inc.
Certified Alloys Company
**Hugh Bettis**
DoAll Company
**J. Binns, Jr.**
Binns Machinery Products
**J. Binns, Sr.**
Binns Machinery Products
**J T. Black**
Auburn University
**Mark Bobert**
Technical Consultant
**J.F. Boland**
Rockwell International
**S.P. Boppana**
GTE Valenite
**F.W. Boulger**
Technical Consultant
**K. Brach**
General Electric Company
**J. Bradley**
Technical Consultant
**José R.T. Branco**
Colorado School of Mines
**R. Bratt**
Technical Consultant
**R.W. Breitzig**
INCO Alloys International
**James Brewer**
Fairfield Manufacturing Company
**Chris Brookes**
The University of Hull (England)
**S.T. Buljan**
GTE Laboratories
**Virgil Buraczynski**
Besley Products Corporation
**Stephen J. Burden**
GTE Valenite Corporation
**John H. Burness**
The Timken Company
**A.C. Carius**
General Electric Company
**Nick Cerwin**
A. Finkl & Sons, Inc.
**Harry E. Chandler**
ASM INTERNATIONAL
**S. Chandrasekar**
Purdue University
**Chao-Hwa Chang**
University of California, Los Angeles
**John D. Christopher**
Metcut Research Associates Inc.
**T.J. Clark**
General Electric Company
**Hilary A. Clouser**
Extrude Hone Corporation
**Joseph W. Coniglio**
Gould & Eberhardt Gear Machinery Corporations
**John Conlon**
Conlon Industries, Inc.
**S. Cook**
LTV Aerospace Company
**David Cunningham**
General Electric Superabrasives
**Richard Dabeck**
Coral Chemical Company
**Dilip Dalal**
The Cross Company
**J. Dalton**
Bardons & Oliver
**Timothy Danielson**
Chem Tronics, Inc.
**C.V. Darragh**
The Timken Company
**D.W. Davies**
BNF Metals Technology Centre (England)
**Warren J. Demery**
Sossner Tap & Tool Company
**Amedeo deRege**
Domfer Metal Powders Limited (Canada)
**Warren R. DeVries**
Rensselaer Polytechnic Institute
**Kurt Dieme**
Reed Rolled Thread Die Company
**J. Dimitrious**
Pfauter-Maag Cutting Tools
**Phil Diskins**
DiCo Corporation
**Charles A. Divine, Jr.**
AL Tech Specialty Steel Corporation
**R. Dixon**
Crucible Specialty Metals
**Stephan Donelson**
Colorado School of Mines
**Carl J. Dorsch**
Crucible Materials Corporation
**Clifford E. Drake**
ENERPAC Group Applied Powers, Inc.
**W. Dresher**
International Copper Research Association
**D. Dykehouse**
Technical Consultant
**Robert P. Eichorst**
United Technologies
**Ahmad K. Elshennawy**
University of Central Florida
**Dana Elza**
Coherent General
**Phil Esserkaln**
Kempsmith Machining Company
**J. Richard Evans**
Dowty Canada Ltd. (Canada)
**John J. Fickers**
Los Alamos National Laboratory
**Michael Field**
Metcut Research Associates Inc.
**M.E. Finn**
Steltech Inc. (Canada)

**Thomas Fisher**
Surftran Division
Robert Bosch Corporation
**Donald G. Flom**
Technical Consultant
**Thomas O. Floyd**
Seco-Carboloy
**John E. Foley**
S. Baird Corporation
**David Fordanick**
The Cross Company
**Paul Frederick**
Dow Chemical Company
**Howard Friedman**
Fotofabrication Corporation
**John E. Fuller**
Rockwell International
**Roland Galipeau**
ThermoBurr Canada
(Canada)
**Douglas V. Gallagher**
Rockwell International
**Ramesh Gandhi**
Alliance Tool & Manufacturing Inc.
**Geoffrey Y. Gill**
Muskegon Tool Industries Inc.
**J. Ginsberg**
Photo Chemical Machining Institute
**M.A. Glandt**
Giddings & Lewis
**Claus G. Goetzel**
Technical Consultant
**F. Gorsler**
General Electric Company
**Leigh Gott**
Kearney & Trecker Corporation
**Dennis Grable**
The Cross Company
**Allan M. Grant**
Allan M. Grant & Associates
**Mikell P. Groover**
Lehigh University
**Walter W. Gruss**
Kyocera Feldmuehle, Inc.
**J. Gurland**
Brown University
**Clarence J. Hagstrom**
Lindmark Machine Works
**T.E. Hale**
Carboloy, Inc.
**James E. Hanafee**
Lawrence Livermore National
Laboratory
**R. Hanson**
Ferranti Sciaky, Inc.
**R.E. Hardesty**
Electrofusion Corporation
**S.M. Harrington**
Thomas & Betts Corporation
**Derek Hartell**
Cleerman Machine Tool, Inc.
**P.J. Heath**
De Beers Industrial Diamond Division
(FTY) Ltd. (England)
**Barry Heller**
Teledyne Firth Sterling
**Gene Herron**
Metem Corporation
**Thomas Hill**
Speedsteel of New Jersey, Inc.
**R.M. Hooper**
University of Exeter (England)
**L. Houman**
Axon EDM, Inc.
**David J. Howell**
Roll-A-Matic, Inc.
**Fred Huscher**
Rockwell International Automotive
Division
**Richard M. Jacobs**
Consultant Services Institute, Inc.
**J. Jackson**
Radian Corporation
**E.C. Jameson**
Transtec, Inc.
**Ernest Jerome**
Zagar Inc.
**Mark Johnson**
Tapmatic Corporation
**C.E. Johnston**
Flow Systems, Inc.
**K. Jones**
Tooling Systems Inc.
**John F. Kahles**
Metcut Research Associates Inc.
**Serope Kalpakjian**
Illinois Institute of Technology
**A. Karl**
Garrett Turbine
**K. Katbi**
GTE Valenite
**L. Alden Kendall**
Washington State University
**B. Klamecki**
University of Minnesota
**J.B. Kohls**
Institute of Advanced Manufacturing
Sciences, Inc.
**Ranga Komanduri**
National Science Foundation
**Yoram Koren**
University of Michigan
**Ted Kosa**
Carpenter Technology Corporation
**William P. Koster**
Metcut Research Associates Inc.
**T. Kozinski**
Precision Art Coordinators
**James E. Krejci**
Keystone Threaded Products
Division
**Theodore J. Krenzer**
The Gleason Works
Gleason Company
**Gerald Kusar**
Ajax Manufacturing Company
**John B. Lambert**
Fansteel
**Eugene M. Langworthy**
Aerochem, Inc.
**L.K. Lauderbaugh**
Rensselaer Polytechnic Institute
**J.A. Laverick**
The Timken Company
**Frank D. Leone**
Pitney Bowes, Inc.
**D. Levinson**
Taussig Associates, Inc.
**Terry L. Lievestro**
Lehr Precision, Inc.
**Richard P. Lindsay**
Norton Company
**Steven Lochmoeller**
Roton Products Inc.
**R. Luke**
DoAll Company
**Pel Lynah**
P. R. Hoffman Machine Products
**Gerald Makuh**
Weldon Tool Company
**Reza A. Maleki**
Moorhead State University
**Stephan Malkin**
University of Massachusetts at Amherst
**O. Masory**
Texas A&M University
**Larry Mayer**
TIMET
**Bob McLemore**
The Marquardt Company
**Alan McMechan**
McDonnell Douglas Canada (Canada)
**Pankaj K. Mehrotra**
Kennametal Inc.
**Fred Meyer**
Precitec Corporation
**Thomas W. McClure**
Balax Inc.
**W. Mihaichuk**
Eastern Alloys, Inc.
**James Millar**
Lapmaster Division of Crane Packing
Company
**Brian Mitchell, Sr.**
General Broach and Engineering
Company
**Walter R. Mohn**
Advanced Composite Materials
Corporation
**Frank Moravcik**
The Cross Company
**Mary Moreland**
Bullen Ultrasonics Inc.
**Jonathon Morey**
Morey Machining Company
**R.A. Morley**
Reynolds Aluminum
**T.O. Morris**
Martin Marietta Energy Systems, Inc.
**David Moskowitz**
Technical Consultant
**Bill Murphy**
Rodeco Company
**Elliot S. Nachtman**
Tower Oil & Technology Company

**Steven J. Neter**
Peterson Precision Engineering Company

**M. Anthony Newton**
NItech, Inc.

**Ronald P. Ney**
Carpenter Technology Corporation

**Roger Nichting**
Colorado School of Mines

**P. Niessen**
University of Waterloo (Canada)

**Bernard North**
Kennametal Inc.

**Raymond J. Novotny**
Technical Consultant

**J. Padgett**
J.R. Padgett Associates

**Ralph Panfil**
Davenport Machine

**Jeffrey T. Paprocki**
Kearney & Trecker Corporation

**W. Neil Peters**
Corning Glass Works

**Robert E. Phillips**
Everite Machine Products Company

**R. Pierce**
Radian Corporation

**Kenneth E. Pinnow**
Crucible Metals Corporation

**Robert A. Powell**
Hoeganaes Corporation

**D. Powers**
Leybold Vacuum Systems, Inc.

**J. Prazniak**
The Timken Company

**Ralph E. Prescott**
Monarch Machine Tool Company

**Allen Queenen**
Kearney & Trecker Corporation

**S. Ramanath**
Norton Company

**V. Rangarajan**
Colorado School of Mines

**M.P. Ranson**
Inco Alloys International, Inc.

**James Reichman**
Kenworth Truck Company

**Lawrence J. Rhoades**
Extrude Hone Corporation

**C.E. Rodaitis**
The Timken Company

**Harvey W. Rohmiller**
Lodge & Shipley Division Manuflex Corporation

**Stuart Salmon**
Advanced Manufacturing Science & Technology

**Shyam K. Samanta**
National Science Foundation

**Ron Sanders**
Laserdyne

**A.T. Santhanam**
Kennametal Inc.

**K. Scheucher**
Modtech Corporation

**Ronald W. Schneider**
MG Industries

**Scott Schneier**
Regal Beloit Corporation

**Michael Shultz**
Wisconsin Drill Head Company

**R. Seely**
Corning Glass Works

**W.R. Sharpe**
Battelle Pacific Northwest Laboratories

**Chi-Hung Shen**
General Motors Technical Center

**T. Slawson**
Ridge Metals Inc.

**Ted A. Slezak**
Armstrong-Blum Manufacturing Company

**William M. Spurgeon**
University of Michigan—Dearborn

**D.R. Stashko**
GTE Valenite Corporation

**William Stasko**
Crucible Materials Corporation

**Larry E. Stockline**
PROMESS, Inc.

**Glenn E. Stork**
S.S. White Industrial Products Division of Pennwalt Corporation

**K. Subramanian**
Norton Company

**Lewis Sylvia**
Morse Cutting Tools

**D. Taylor**
Manufacturing Systems Extension Center

**R.A. Thompson**
General Electric Company

**Thomas Thompson**
Badger Meter Company

**P. Tierney**
Kennametal Inc.

**Jiri Tlusty**
University of Florida

**C. Treadwell**
Sonic-Mill Albuquerque

**J. Tulloch**
Wells Saw Division

**Charles I. Turner**
Kearney & Trecker Corporation

**William R. Tyrell**
Branson Ultrasonics Corporation

**A. Galip Ulsoy**
University of Michigan

**G.L. Van Arsdale**
Battelle Pacific Northwest Laboratories

**M.R. Van den Bergh**
Composites Specialties, Inc.

**Christopher Van De Motter**
The Ohio Broach & Machine Tool Company

**Philip A. Ventura**
The Cross Company

**Don Vick**
Ingersoll Milling Machine Company

**Craig E. Virkus**
Elliott Company

**R.J. von Gutfeld**
Thomas J. Watson Research Center
International Business Machines

**Charles F. Walton**
Technical Consultant

**L. Walton**
Latrobe Steel Company

**I. Weber**
Technical Consultant

**R. Terrence Webster**
Teledyne Wah Chang Albany

**W.R. Welton**
Welton Rolled Thread Corporation

**Robert Werkema**
Technical Consultant

**Robert I. Werner**
R.D. Werner Company Inc.

**Gene White**
Coherent General

**Richard F. Williams**
Natco, Inc.

**M.L.H. Wise**
University of Birmingham (England)

**William Wonnacott**
Thread Grinding Service

**R.E. Wood**
Lockheed Aeronautical Systems Company

**Hiroshi Yaguchi**
Inland Steel Company

**Patrick Yeko**
ENERPAC Group Applied Powers, Inc.

**C. Zimmerman**
GTE Valenite

**Emory W. Zimmers, Jr.**
Lehigh University

# Contents

# Introduction to Machining Processes

J T. Black, Auburn University

MACHINING is a term that covers a large collection of manufacturing processes designed to remove unwanted material, usually in the form of chips, from a workpiece. Machining is used to convert castings, forgings, or preformed blocks of metal into desired shapes, with size and finish specified to fulfill design requirements. Almost every manufactured product has components that require machining, often to great precision. Therefore, this collection of processes is one of the most important of the basic manufacturing processes because of the value added to the final product. By the same token, machining processes are often the most expensive.

The majority of industrial applications of machining are in metals. Although the metal cutting process has resisted theoretical analysis because of its complexity, the application of these processes in the industrial world is widespread.

Machining processes are performed on a wide variety of machine tools. Figure 1 shows an example of a machine tool—a dual-turret numerically controlled (NC) lathe. Workpieces are held in workholding devices, such as a three-jaw chuck. The tools used to cut metal are in the turrets. Other examples of basic machine tools are milling machines, drill presses, grinders, shapers, broaching machines, and saws.

Each of the basic machine tool types has many different configurations. Lathes, for example, may be engine lathes, turret lathes, tracer lathes, or automatic-screw machines. Lathes have followed the trend of other machine tools, and NC lathes can now be routinely purchased.

The primary chip formation processes are listed below, with alternative versions in parentheses. Each process is performed on one or more of the basic machine tools. For example, drilling can be performed on drill presses, milling machines, lathes, and some boring machines:

- Turning (boring, facing, cutoff, taper turning, form cutting, chamfering, recessing, thread cutting)
- Shaping (planing, vertical shaping)
- Milling (hobbing, generating, thread milling)
- Drilling (reaming, tapping, spot facing, counterboring, countersinking)
- Sawing (filing)
- Abrasive machining (grinding, honing, lapping)
- Broaching (internal and surface)

Processes can be combined into multiple-capability machines, known as machining centers. The machining center shown in Fig. 2 is capable of performing the machining processes normally performed on a milling machine, drilling machine, and a boring mill and is numerically controlled. The position and velocity of the tool with respect to the work is under feedback control. Different tools can be automatically inserted into the spindle as needed to do different machining processes. The horizontal spindle machine shown in Fig. 2 was one of the first NC machining centers to be able to change workpiece pallets.

For each of the basic machine tool types, there are many different kinds of workholders, cutting tools, and cutting tool holders, resulting in a rather formidable list of equipment and processes. In this Volume, a Section entitled "Fundamentals of the Machining Process" is presented first, with the intent of putting these processes into perspective and helping the reader to understand the problems associated with using machining processes in the manufacture of products.

**Fig. 1** A dual-turret NC turning center with 16 tool stations. Courtesy of Cincinnati Milacron

## Overview of Machining Process Variables

Metal cutting processes can be viewed as consisting of independent (input) variables, dependent variables, and independent-dependent interactions or relationships. The engineer or machine tool operator has direct control over the input variables and can specify or select them when setting up the machining process. Several input variables are described below. Figure 3 summarizes the input/output relationships associated with metal cutting.

### Independent Input Variables

**Workpiece Material.** The metallurgy and chemistry of the workpiece can either be specified or is already known. Quite often, a material is selected for a particular application chiefly because it machines well. Cast iron and aluminum, for example, are known to machine easily. Other metals, such as stainless steel or titanium, are difficult to machine. They often have large cutting forces or poor surface finishes, which can result in short cutting tool life, yet these metals are selected to meet other functional design criteria. Machining practice for specific workpiece materials are reviewed in the Section "Machining of Specific Metals and Alloys" in this Volume.

**Fig. 2** Numerically controlled machining center that can change workpieces as well as cutting tools. Courtesy of Kearney and Trecker Corporation

**Starting Geometry.** The size and shape of the workpiece may be dictated by preceding processes (casting, forging, forming, and so forth) or may be selected from standard machining stock (for example, bar stock for screw machines). Usually this variable directly influences the machining process or processes that are selected, as well as the depths of cut.

**Specific Machining Processes.** The selection of machining processes required to convert the raw material into a finished product must be based on the geometry of the part (size and shape, rotational or nonrotational), the required finishes and tolerances, and the quantity of the product to be made. Machining processes can be grouped into three broad categories. These include traditional chip formation processes, abrasive machining processes, and nontraditional machining processes.

*Chip Formation Processes.* As described earlier, there are seven basic chip formation processes: turning, shaping, milling, drilling, sawing, broaching, and abrasive machining. The equipment and principles of operation associated with each of these processes (with the exception of abrasive machining, which is treated separately) are described in the Section titled "Traditional Machining Processes" in this Volume.

*Abrasive machining* is the basic process by which chips are formed by very small cutting edges that are integral parts of abrasive particles. The principles of abrasive machining, the fundamental differences between metal cutting and grinding, and the abrasives and equipment used for abrasive machining operations are described in the Section "Grinding, Honing, and Lapping" in this Volume.

*Nontraditional Machining Processes.* Machining processes that involve compression/shear chip formation have a number of inherent disadvantages. These include:

- High costs incurred with chip formation (high energy output and chip removal, disposal, and/or recycling)
- Heat buildup that often results in workpiece distortion
- High forces that create problems in holding the workpiece and which can also cause distortion
- Undesirable cold working and residual stresses in the workpiece that often necessitate further processing to remove the harmful effects
- Limitations as to the size and delicacy of the workpiece

In order to avoid these limitations, nontraditional machining processes are increasingly being used. Nontraditional methods usually do not produce chips or a lay pattern in the surface and often involve new energy modes (see the Section "Nontraditional Machining Processes" in this Volume). Volumetric material removal rates, however, are much lower than with traditional machining processes.

**Tool Materials.** The three most common cutting tool materials currently in use for production machining operations are high-speed steel (HSS), both in wrought and powder metallurgy (P/M) form; carbides; and coated tools. Cubic boron nitride (CBN), ceramics, and diamonds are also being widely employed. Generally speaking, HSS is used for general-purpose tools, for tools of complex design or for tools used when cutting speeds are more modest. Carbide and ceramic tool materials, which can operate at faster cutting speeds, come in a wide variety of grades and geometries. Titanium nitride and titanium carbide coatings for HSS and carbides are now commonplace. Selection of a tool material that provides reliable service while fulfilling the functional requirements is still an art. The harder the tool material, the better it can resist wear at faster cutting speeds. The faster the cutting speed, the higher the cutting temperature and the shorter the tool life. Retention of hardness at elevated temperatures as well as long tool life are desirable characteristics in cutting tools. See the Section "Cutting Tool Materials" in this Volume for descriptions of the processing, properties, and applications associated with the aforementioned materials.

**Cutting Parameters.** For every machining operation, it is necessary to select a cutting speed, a feed, and a depth of cut. Many factors impinge on these decisions because all of the dependent variables are influenced by them. Proper selection of variables also depends on the other input variables that have been selected; that is, the total amount of material to be removed, the workpiece and tool materials, and the machining process or processes. These need to be selected before preliminary choices for speed, feed, and depth of cut can be made.

**Tool Geometry.** Cutting tools are usually designed to accomplish specific operations, and thus the tool geometry (angles) is selected to accomplish specific machining functions. Generally speaking, large rake and clearance angles are preferred, but they are possible only on HSS tools. Tools made from carbides, ceramics, and other very hard materials must be given small tool angles, which keep the tool material in compression during machining and thereby avoid tensile failure and brittle fractures of the tool. The greater the precision required of the process, the better the geometry of the cutting edge itself must be.

**Workholding Devices.** Workpieces are located (held in specific position with respect to the tools) and clamped in workholding devices in or on the machine tools. For every machine tool, there are many different kinds of workholding devices, ranging from general-purpose vises to specifically designed jigs and fixtures (see the article "Proper Fixturing" in this Volume). The workholding devices are the key to precision manufacturing; thus, the selection (or design and construction) of the correct workholding devices is every bit as important as the selection of the right cutting tool and machine tool.

**Cutting Fluids.** The selection of the right cutting fluid for a particular combination of work material and tool material can mean the difference between success and failure in almost every production machining process. Cutting fluids serve to cool the workpiece, tool, and chips; reduce friction by means of lubrication; carry the chips away from the cutting region; help improve the surface finish; and provide surface protection to the workpiece (a more complete discussion may be found in the article

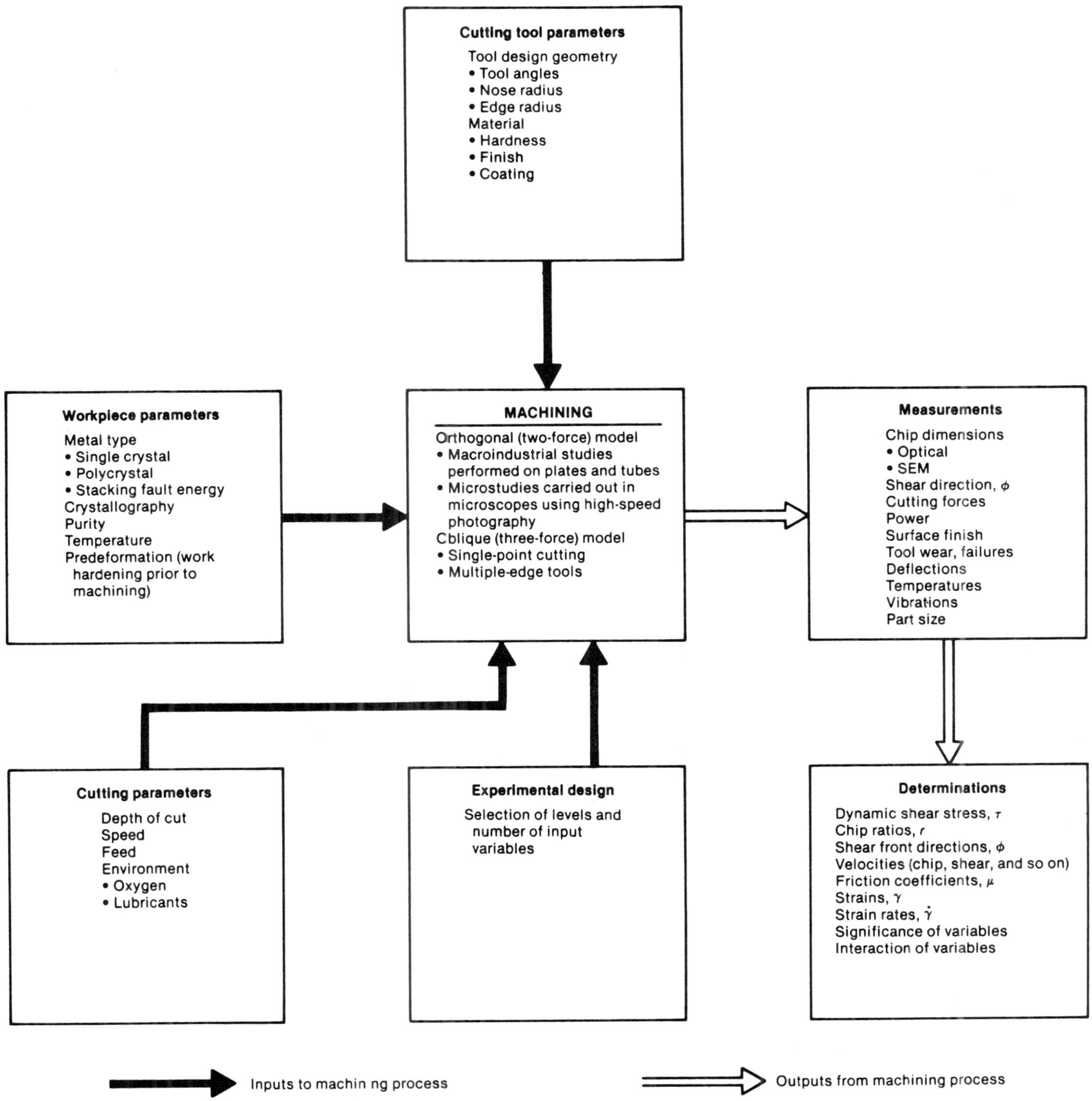


**Fig. 3** Input/output relationships in metal cutting (machining)

"Metal Cutting and Grinding Fluids" in this Volume).

## Dependent Variables

Dependent variables are determined by the process based on the prior selection of the input or independent variables. Thus, the manufacturing engineer's control over these is usually indirect. The important dependent variables are cutting force and power, size and properties of the finished product, surface finish, and tool wear and tool failure.

**Cutting Force and Power.** To machine metal at a specified speed, feed, and depth of cut, with a specified lubricant, cutting tool material, and geometry, generates cutting forces and consumes power. A change in any of the variables alters the forces, but the change is indirect in that the engineer does not specify the forces, only the parameters that generate those forces. Forces are important in that they influence the deflections in the tools, the workpieces, and the workholders, which in turn affect the final part size. Forces also play a roll in chatter and vibration phenomena common in machining. Obviously, the manufacturing engineer would like to be able to predict forces (and power) so that he can safely specify the equipment for a manufacturing operation, including the machine tool, cutting tool, and workholding devices. The basic concepts associated with the modeling and understanding of cutting forces and power are explained in the article "Forces, Power, and Stresses in Machining" in this Volume.

**Size and Properties of the Finished Product.** Ultimately, the objective of machining is to obtain a machined surface of desired size and geometry with the desired mechanical properties. Because machining is a localized, plastic deformation process, every machined surface will have some residual deformation (stresses) left in it. These residual stresses are usually tensile in nature and can interact with surface flaws to produce part failure from fatigue or to cause corrosion. In addition, every process has some inherent process variability (variations about average size) that changes with almost all of the input variables. Thus, the manufacturing engineer must try to select the proper levels of input variables to produce a product that is within the tolerance specified by the designer and has satisfactory surface properties.

**Surface Finish.** The final finish on a machined surface is a function of tool geometry, tool material, workpiece material, machining process, speed, feed, depth of cut, and cutting fluid. Surface finish is also related to the process variability. Rough surfaces have more variability than smooth surfaces. Often it is necessary to specify multiple cuts, that is, roughing and finish cuts, to achieve the desired surface finish, or it may be necessary to specify multiple processes, such as following turning with cylindrical grinding, in order to obtain the desired finish. The effect of various machining processes on surface finish and on the properties of the final products are described in the article "Surface Finish and Surface Integrity" in this Volume.

**Tool Wear and Tool Failure.** The plastic deformation and friction inherent in machining generate considerable heat, which raises the temperature of the tool and lowers its wear resistance. The problem is subtle, but significant. As the tool wears, it changes in both geometry and size. A dull cutting edge and change in geometry can result in increased cutting forces that in turn increase deflections in the workpiece and may create a chatter condition. The increased power consumption causes increased heat generation in the operation, which accelerates the wear rate. The change in the size of the tool changes the size of the workpiece. Again, the engineer has only indirect control over these variables. He can select slow speeds, which produce less heat and lower wear rates, but which decrease the production rates because the metal removal rate is decreased. Alternatively, the feed or depth of cut can be increased to maintain the metal removal rate while reducing the speed. Increasing either the feed or depth of cut directly increases the cutting forces. Therefore, while tool life may be gained, some precision may be lost due to increased deflection and chatter. Wear mechanisms, determination of modes of tool failure, and tool life testing are examined in the article "Tool Wear and Tool Life" in this Volume.

## Relations Between Input Variables and Process Behavior

Understanding the connections between input variables and process behavior is important knowledge for the manufacturing engineer. Unfortunately, this knowledge is difficult to obtain. Machining is a unique plastic deformation process in that it is constrained only by the cutting tool and operates at very large strains and very high strain rates. The tremendous variety in the input variables results in an almost infinite number of different machining combinations. Basically, there are three ways to deal with such a complex situation.

**Experience** requires long-term exposure, because knowledge is basically gained by trial and error, with successful combinations transferred to other, "similar" situations. This activity goes on in manufacturing every time a new material is introduced into the production facility. It took years for industry to learn how to machine titanium. Unfortunately, the knowledge gained through one process may not transfer well to another even though their input variables appear very similar.

**Experiments.** Machining experiments are expensive, time consuming, and difficult to carry out. Tool life experiments, for example, are quite commonly done, yet tool life data for most workpiece/tool material combinations are not available. Even when laboratory data have been published, the results are not necessarily transferable to the particular machine tools and cutting tools on the shop floor. Tool life equations are empirically developed from turning experiments in which all input variables except cutting speed are kept constant. The experimental arrangement may limit the mode of tool failure to wear. Such results are of little value on the shop floor, where tools can and do fail from causes other than wear.

**Theories.** There have been many attempts to build mathematical models of the metal cutting process. Many of the theories are extensions of the mechanics presented in the following Section, "Fundamentals of the Machining Process." These theories try to predict the direction of the shearing process of metal cutting. These models range from crude, first-order approximation to complex, computer-based models using finite-element analysis. Recently, some modest successes have been reported in the literature in which accurate predictions of cutting forces and tool wear were made for certain materials. Clearly such efforts are extremely helpful in understanding how the process behaves. However, the theory of plastic deformation of metals (dislocation theory) has not yet been able to predict values for shear stresses and tool/chip interface from the metallurgy and deformation history of the material. Therefore, it has been necessary to devise two independent experiments to determine the shear strength ($\tau_s$) of the metal at large strains and high strain rates and the sliding friction situation at the interface between the tool and chip (see the article "Mechanics of Chip Formation" in this Volume).

## Future Trends

The metal cutting process will continue to evolve, with improvements in cutting tool materials and machine tools leading the evolution. More refined coatings on cutting tools will improve tool life and reliability, as will more robust, rigid machine tools. The challenge for machining will involve dealing with the new types of materials that will need to be machined, including aluminum and titanium alloys, alloy steels, and superalloys. These materials, because of improved processing techniques, are becoming stronger and harder and therefore more difficult to machine. The objective should be to design and build cutting tools that have less variability in their tool lives rather than longer tool lives. The increasing use of structural ceramics, high-strength polymers, composites, and electronic materials will also necessitate the use of nontraditional methods of machining. In addition, grinding will be employed to a greater extent than in the past, with greater attention to creep feed grinding and the use of superabrasives (diamond and cubic boron nitride).

As the cutting tools improve, the machine tools will become smarter, with on-board computers providing intelligent algorithms interacting with sensory data from the process. Programmable machine tools, if equipped with the proper sensors, are capable of carrying out measurements of the product as it is being produced. These product data will be fed back to the control program, which is then modified to improve the product or corrected for errors. Thus, the machine will be able to make the adjustments necessary to prevent defective products from being produced. The goal of such control programs should be improved quality (designed not to make a defect), rather than optimum speed or lowest cost. Advancements in computer-aided machining processes are discussed in the Section "Machine Controls and Computer Applications in Machining" in this Volume.

Another area in which significant advances will be made is the design of workholders that are capable of holding various parts without any downtime for setups. Included in this search for flexible fixtures will be workholding devices that can be changed over by a robot—the same robot used to load or unload parts from the machine.

# Fundamentals of the Machining Process

# Mechanics of Chip Formation

J T. Black, Auburn University

THE BASIC MECHANISM involved in metal cutting is that of a localized shear deformation on the work material immediately ahead of the cutting edge of the tool. The relative motion between the tool and the workpiece during cutting compresses the work material near the tool and induces a shear deformation (called the primary deformation), which forms the chip. The chip passes over the rake face of the cutting tool and receives additional deformation (called the secondary deformation) because of the shearing and sliding of the chip against the tool.

These two plastic deformation processes have a mutual dependence. The material element that rubs the rake face has been heated and plastically deformed during its passage through the primary shear process; therefore, the secondary process is influenced by the phenomena on the shear plane. At the same time, the shear direction is directly influenced by the rake face deformation and friction processes. The shear direction influences the heating and straining of the chip in the primary process. In terms of metal cutting theory, this means that shear stress and shear direction must be determined simultaneously. Such theoretical analyses are usually based on the mechanics of the process.

This article will review the following:

- The fundamental nature of the deformation process associated with machining
- The principles of the orthogonal cutting model
- The effect of workpiece properties on chip formation
- The mechanics of the machining process

Additional information on the modeling and analysis of chip formation can be found in the article "Forces, Power, and Stresses in Machining," which immediately follows in this Section.

## Fundamental Mechanism of Metal Deformation

**Cutting Models.** Before the mechanics of machining are presented, a brief discussion of the fundamental nature of the deformation processes is helpful in understanding the assumptions that accompany the mechanics. The machining geometry can be simplified from the three-dimensional (oblique) geometry, which typifies most industrial processes, to a two-dimensional (orthogonal) geometry. Figure 1 compares the oblique and the orthogonal cutting geometries. Orthogonal machining can be obtained in practice by:

- End cutting a tube wall by turning (Fig. 1b)
- Machining a plate as shown in Fig. 2

Oblique cutting is obtained when the cutting edge and the cutting motion are not perpendicular to each other. Because the orthogonal case is more easily modeled, it will be used in this article to describe the deformation process.

In the orthogonal cutting of a tube (Fig. 1b), the width of the cut is equal to the thickness of the tube wall, $w$. The direction of shear is specified by $\phi$, the shear angle. The cross-sectional area of the chip is given by $t_c \cdot w_c$, where $t_c$ is chip thickness and $w_c$ is the width of the chip. The cutting edge of the tool is perpendicular to the feed direction. The measured horizontal cutting force, $F_c$, is the force in the direction of the cutting velocity (or cutting speed). The force in the direction of the feed (vertical or tangential) and perpendicular (orthogonal) to $F_c$ is denoted by $F_t$. With this two-dimensional model of chip formation, the influence of the most critical elements of the tool geometry (rake angle, $\alpha$, and the edge radius of the tool) and the interactions that occur between the tool and the chip can be more easily examined.

**Shear Zone.** Basically, the chip is formed by a localized shear process that takes place over very narrow regions. Classically called the shear zone or shear plane, this deforma-

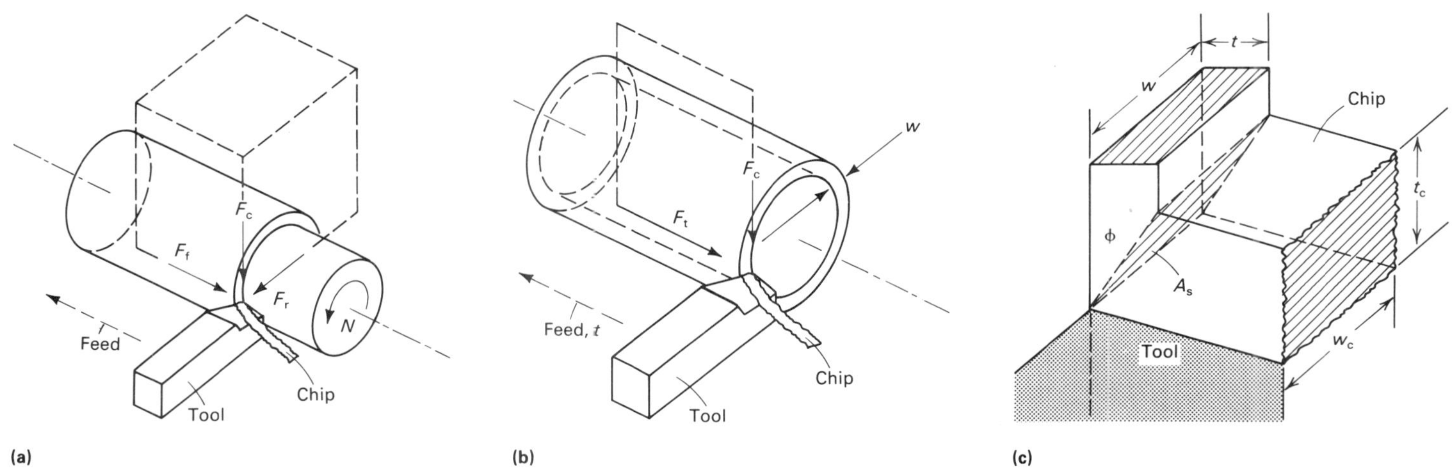


**Fig. 1** Comparison of oblique and orthogonal geometry machining (a) Three-force oblique machining. $F_c$ is the primary cutting force, $F_f$ is the feed force, and $F_r$ is the radial or thrust force. (b) Two-force orthogonal machining. $F_c$ is the measured cutting force, and $F_t$ is the feed (tangential) force. A tube-cutting application is shown; the cutting edge of the tool is perpendicular to the direction of motion. (c) For orthogonal cutting, the shear area, $A_s$, occurs for a shear angle $\phi$, width of cut $w$, and feed $t$.

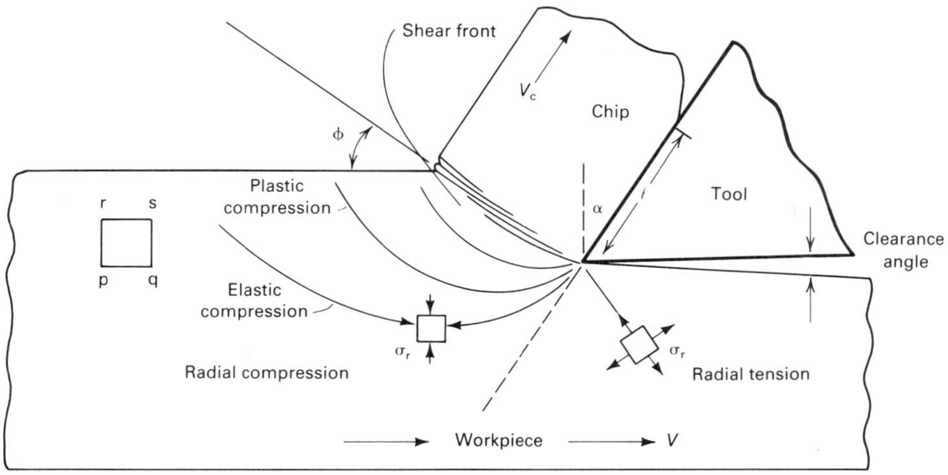


**Fig. 2** Development of the shear front-lamella structure. As shown by this orthogonal geometry, shear deformation evolves from a radial compression zone. See Fig. 5 for an explanation of the effects of shear deformation on area p-q-r-s.

tion evolves out of a radial compression zone that travels ahead of the shear process as the tool passes over the workpiece (Fig. 2). Like all plastic deformations, this radial compression zone has an elastic compression region that converts to a plastic compression region as the material approaches the cutting edge. This plastic compression generates dense dislocation tangles and networks in annealed metals. When this work-hardened material reaches the tool, the material shears in the direction of the free surface.

**Shear Front-Lamella Structure.** The shear process itself is a nonhomogeneous (discontinuous) series of shear fronts (or narrow bands) that produce a lamellar structure in the chips. This fundamental structure occurs on the microscale in all metals when they are machined and accounts for the unique behavior of the machining process.

Individual shear fronts (Fig. 2) coalesce into narrow shear bands. The shear bands are very narrow (20 to 200 nm) compared to the thickness of a lamella (2 to 4 μm) and account for the large strain and high strain rates that typify this process.

These fundamental structures are difficult to observe in normal metal cutting, but can be readily observed in a scanning electron microscope with specially prepared workpieces. Figure 3 shows micrographs from an orthogonal machining experiment performed inside a scanning electron microscope. The fundamental shear front-lamella structure is readily observed. The side of the workpiece has been given a mirror polish so that the shear fronts can be observed. The shear fronts are produced by the activation of many dislocations traveling in waves from the tool tip to the free surface. The lamella represents heavily deformed material that has been segmented by the shear fronts. When machined, all metals deform by this basic mechanism. The shear fronts relieve the applied stress.

Chips are sometimes produced with a sawtooth pattern on the top side—the side that did not rub against the tool. This sawtooth pattern is not produced by the shear front-lamella structure but rather by the unloading of the elastic energy stored in the tool and workpiece, which results in chatter and vibration during cutting. The shear front-lamella structure can and does exist without any vibration of the tool or workpiece. If each sawtooth were to be observed in the scanning electron microscope, many fine shear front-lamella structures would be found in each sawtooth region. The geometry of the sawtooth can be changed (even eliminated) by altering the rigidity of the setup or the machine. The shear front-lamella structure is fundamental to, and characteristic of, the plastic deformation process itself; therefore, it is relatively invariant with respect to cutting parameters and certainly cannot be eliminated.

## Orthogonal Machining Fundamentals

Orthogonal machining setups are used to model oblique machining processes. Processes such as turning, drilling, milling, and shaping are all three-force, or oblique, cutting methods. However, the orthogonal model shown in Fig. 4 is an excellent illustration of the behavior of oblique processes without the complications of the third dimension.

**Chip Ratio.** As described earlier in this article, orthogonal machining can be accomplished by machining a plate or can be approximated by cutting the end of a tube wall in a turning setup. For the purposes of modeling, the following are assumed: The shear process is a plane, the cutting edge is perfectly sharp, and there is no friction contact between the flank of the tool and the workpiece surface. Because plane-strain conditions are assumed, the chips are assumed to have no side flow ($w = w_c$, Fig. 1c), and the cutting velocity is constant. The shear process occurs at angle $\phi$ for a tool with back rake angle $\alpha$. The chip has velocity $V_c$ and makes contact with the rake face of the tool over length $\ell$ (Fig. 2). Defining the ratio of the uncut chip thickness, $t$, to the chip thickness, $t_c$, as the chip ratio, $r$, produces the following:

$$r = \frac{t}{t_c} = \frac{\sin \phi}{\cos (\phi - \alpha)} \quad \text{(Eq 1)}$$

Solving Eq 1 for $\phi$ yields:

$$\tan \phi = \frac{r \cos \alpha}{1 - r \sin \alpha} \quad \text{(Eq 2)}$$

In practical tests, the average chip thickness can be obtained by carefully measuring the length $L$ and the weight $W$ of a piece of a chip. Then:

$$t_c = \frac{W}{\rho t L} \quad \text{(Eq 3)}$$

where $\rho$ is the density of the work material and $t$ is the feed or uncut chip thickness. Chip thickness is usually greater than the depth of cut, $t$, and is constrained by the rake face of the cutting tool.

**Shear Angle.** There are numerous other ways to measure or compute the shear angle, both during (dynamically) the cutting process and after (statically) it has been halted. The shear angle can be measured statically by instantaneously interrupting the cut through the use of quick-stop devices. These devices disengage the cutting tool from the workpiece while cutting is in progress, leaving the chip attached to the workpiece. Optical microscopy and scanning electron microscopy are then used to observe the shear angle. High-speed motion pictures have also been used to observe the process at frame rates as high as 30 000 frames per second. More recently, machining stages have been built that allow the process to be performed inside a scanning electron microscope and recorded on videotape for high-resolution, high-magnification examination of the deformation process. The micrographs shown in Fig. 3 were created in this manner. This technique has been used to measure the velocity of the shear fronts, $V_c$, during cutting, thus verifying experimentally that the vector sum of $V$ and $V_c$ equals $V_s$ (Fig. 4b).

For constancy of volume, it was observed that:

$$\frac{V_c}{V} = \frac{t}{t_c} = r = \frac{\sin \phi}{\cos (\phi - \alpha)} \quad \text{(Eq 4)}$$

Equation 4 indicates that the chip ratio (and therefore the shear angle) can be determined dynamically if a reliable means of measuring chip velocity can be found. Thus, one could determine $\phi$ dynamically for a known tool geometry. Therefore, cut-

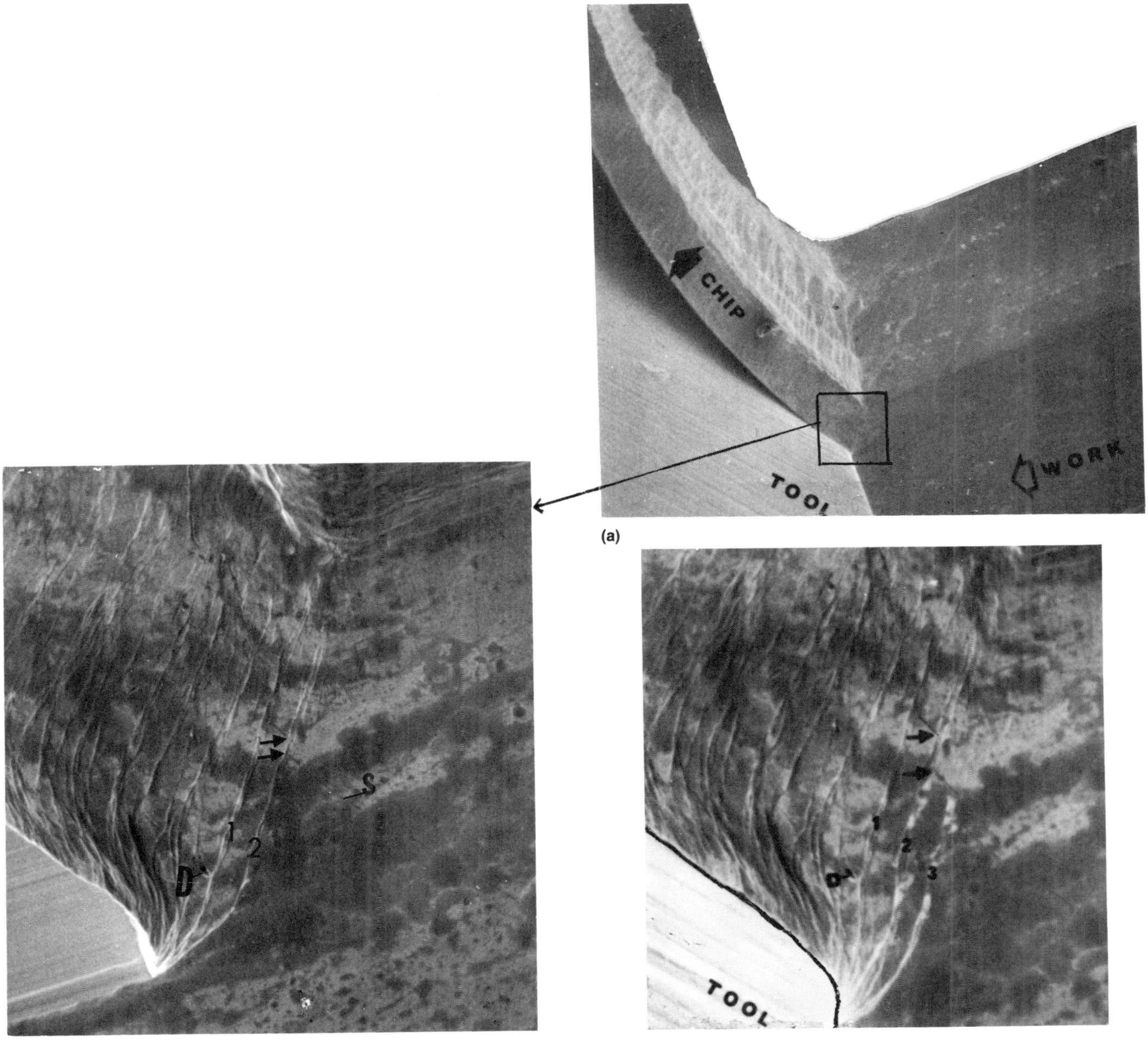


(b) (c)

**Fig. 3** Chip formation process viewed inside a scanning electron microscope. The workpiece is a rectangular plate of high-purity gold that was polished on the sides so that the plastic deformation of the shear process can be readily observed. The boxed area in (a), which is shown at a higher magnification in (b), shows the shear fronts, numbered 1 and 2, advancing from the tool tip region toward the free surface of the workpiece. The letter D indicates a defect on the side of the chip. The arrows indicate a scratch (S) that has been sheared. The tool has been withdrawn from the workpiece. In (c), the tool has been reinserted and slightly advanced. This produced additional shear on shear front No. 2 and new shear front No. 3. Note the movement of defect D. These shear fronts are difficult to observe unless the specimen is polished and examined in a scanning electron microscope.

ting forces can be dynamically predicted, an important consideration in adaptive control machining (see the article "Adaptive Control" in this Volume). Velocities are also important in power calculations, heat determinations, and vibration analyses associated with chip formation.

**Shear Strain.** When an area of metal (for example, area p-q-r-s in Fig. 2) passes through the shear process, it is plastically deformed into a new shape, as shown in Fig. 5. The amount of plastic deformation is related to the shear angle, $\phi$, and the rake angle, $\alpha$.

Therefore, the chip undergoes a shear strain, $\gamma$, of:

$$\gamma = \frac{\cos \alpha}{\sin \phi \cos (\phi - \alpha)} \quad \text{(Eq 5)}$$

The meaning of shear strain, as well as of the units in which it is measured, is shown in the inset diagram in Fig. 5. A unit displacement of one face of a unit cube is a shear strain of 1 ($\gamma = 1$). Figure 5 illustrates the relationship between the shear strain in orthogonal cutting and the shear plane angle for three values of the rake angle. For any rake angle, there is a minimum strain at which the mean chip thickness is equal to the feed ($t_c = t$). For zero rake angle, this occurs at $\phi = 45°$. The change in shape of a unit cube after it passes through the shear plane for different values of the shear plane angle is shown in the lower diagram in Fig. 5 for a tool with a zero rake angle. The minimum strain at $\phi = 45°$ is apparent from the shape change. The shaded region in Fig. 5 shows the typical values of $\phi$ found in practice.

At a zero rake angle, the minimum shear strain is 2. The minimum strain occurs when there is no friction at the tool/chip interface. The minimum strain decreases as the rake

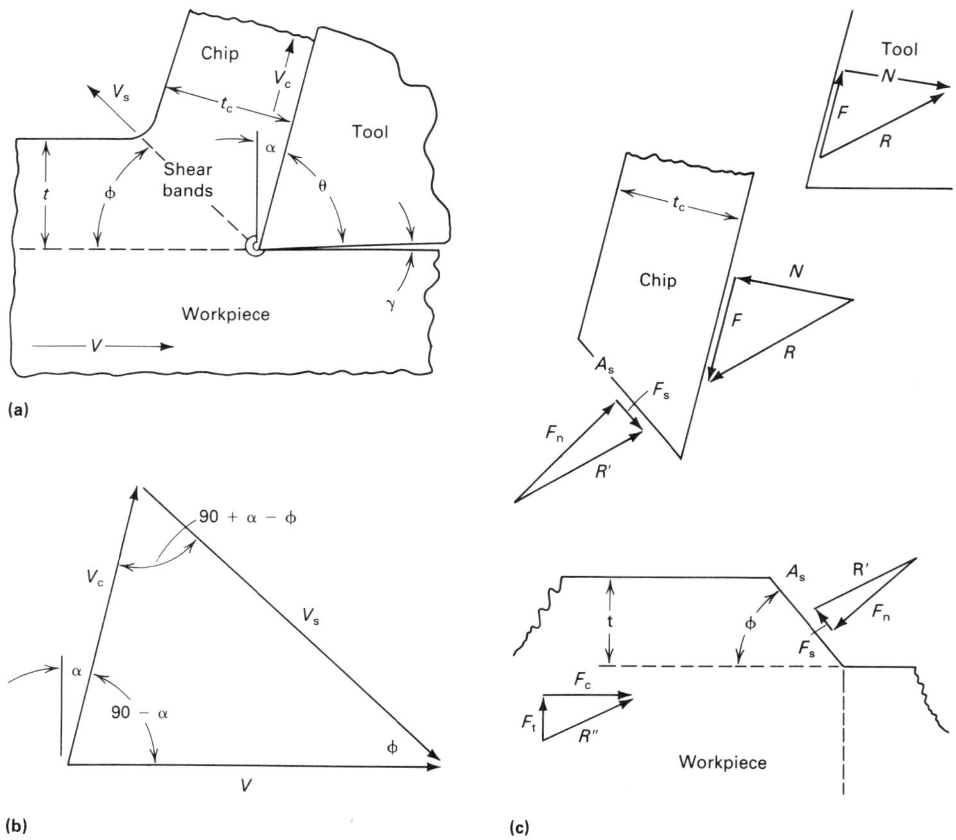


**Fig. 4** Schematics of orthogonal metal cutting mechanics. (a) Orthogonal model. $t$, uncut chip thickness (feed or depth or cut); $t_c$, chip thickness; $\phi$, shear angle; $\alpha$, back rake angle; $\gamma$, clearance angle; $\theta$, edge angle [$\theta = 90 - (\alpha + \gamma)$]. (b) Velocity triangle. $V_s$, shear velocity; $V_c$, chip velocity; $V$, cutting velocity. (c) Chip freebody diagram. $F$, friction force; $N$, normal to friction force; $F_s$, shear force; $F_n$, normal to shear force; $F_c$, cutting force; $F_t$, tangential force; $R$, resultant force

angle increases. If the rake angle is too large, the tool is weak and will fracture. Rake angles larger than 30° are seldom used in industry. With carbides and ceramics, the tendency has been to decrease the rake angle to make the tools more robust, allowing these harder but less tough tool materials to be used. Therefore, even under optimum cutting conditions, chip formation involves very severe plastic deformation, resulting in considerable work hardening and structural change. Metals and alloys lacking in ductility periodically fracture on the shear plane, producing discontinuous chips (see the section "Effect of Work Material Properties" in this article).

In general, metal cutting strains are quite large compared to other plastic deformation processes, being of the order of 2 to 4 mm/mm (2 to 4 in./in.). However, this large strain occurs over very narrow regions (the shear band), which results in extremely high shear strain rates, typically of the order of $10^4$ to $10^8$ mm/mm ($10^4$ to $10^8$ in./in.). This strain rate can be estimated from:

$$\dot{\gamma} = \frac{V}{d} \quad \text{(Eq 6)}$$

where $d$ is the thickness of the shear bands. This combination of large strains and high strain rates operating within a process constrained only by the workpiece and the tool (actually, the deformation interface at the rake face of the tool) causes great difficulties in theoretical analyses of the process.

## Effect of Work Material Properties

**Principal Chip Types.** The properties of the work material control chip formation. Work material properties include yield strength, shear strength under compressive loading, strain-hardening characteristics, friction behavior, hardness, and ductility. As noted in the section "Shear Strain" in this article, work material ductility is an important factor. Highly ductile materials not only permit extensive plastic deformation of the chip during cutting, which increases work, heat generation, and temperature, but also result in longer, continuous chips that remain in contact longer with the tool face, thus causing more frictional heat. Chips of this type are severely deformed and have a characteristic curl. On the other hand, some materials, such as gray cast iron, lack the ductility necessary for appreciable plastic chip formation. Consequently, the compressed material ahead of the tool can fail in a brittle manner anywhere ahead of the tool, producing small fragments. Such chips are termed discontinuous or segmented (Fig. 6).

The cutting parameters also influence chip formation. Cutting parameters include tool materials, tool angles, edge geometries (which change due to wear, cutting speed, feed, and depth of cut), and the cutting environment (machine tool deflections, cutting fluids, and so on). Further complications result from the formation of the built-up edge on the cutting tool.

**A built-up edge** is work material that is deposited on the rake face near the cutting edge (Fig. 6c). It is the product of the localized high temperature and extreme pressure at the tool/chip interface. The work material adheres to the cutting edge of the tool (similar to a dead-metal zone in extrusion). Although this material protects the cutting edge, it also modifies the geometry of the tool. Built-up edges are not stable and will slough off periodically, adhering to the chip or passing under the tool and adhering to the machined surface. Built-up edge formation can often be eliminated or minimized by reducing the depth of the cut, increasing the cutting speed, using positive rake tools, or applying a coolant, but these techniques greatly increase the complexity of the chip formation process analysis.

## Mechanics of Machining

Orthogonal machining has been defined as a two-component force system, while oblique cutting involves a three-force situation. Figure 4(c) shows a free body diagram of a chip that has been separated at the shear plane. The resultant force $R$ consists of the friction force, $F$, and the normal force, $N$, acting on the tool/chip interface contact area (length $\ell$ times width $w$). The resultant force $R'$ consists of a shear force, $F_s$, and a normal force, $F_n$, acting on the shear plane area, $A_s$.

The forces $R$ and $R'$ are assumed to be equal, opposite, and colinear. Determination of these forces necessitates a third set that can be measured. A dynamometer, mounted in the workholder or the toolholder, can be used to measure $F_c$ and $F_t$. This set has resultant $R''$, which is equal in magnitude and colinear to the other resultant forces in the diagram. To express the desired forces ($F_s$, $F_n$, $F$, $N$) in terms of the dynamometer components $F_c$ and $F_t$ and appropriate angles, a circular force diagram is developed in which all six forces are collected in the same force circle. This is shown in Fig. 7. In Fig. 7, $\beta$ is the angle between the normal force, $N$, and the resultant force $R$. It is used to describe the friction coefficient, $\mu$, on the tool/chip interface area, which is defined as $F/N$ so that:

$$\beta = \tan^{-1} \mu = \tan^{-1} \frac{F}{N} \quad \text{(Eq 7)}$$

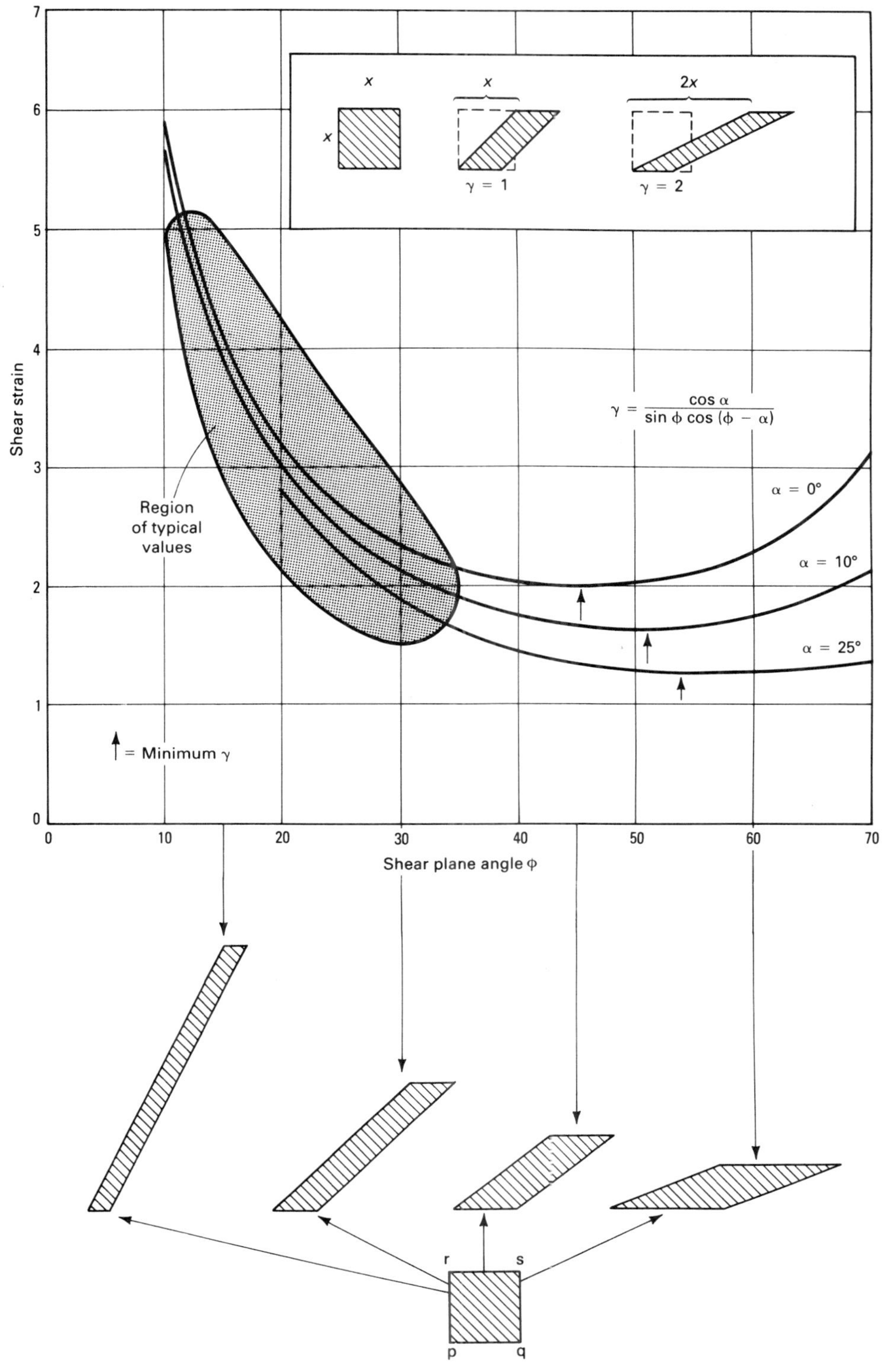


**Fig. 5** Strain on shear plane, γ, versus shear plane angle, φ, for three values of rake angle, α

The friction force, $F$, and its normal force, $N$, can be shown to be:

$$F = F_c \sin \alpha + F_t \cos \alpha \qquad \text{(Eq 8)}$$

$$N = F_c \cos \alpha - F_t \sin \alpha \qquad \text{(Eq 9)}$$

where

$$R = (F_c^2 + F_t^2)^{1/2} \qquad \text{(Eq 10)}$$

When the back rake angle, $\alpha$, is zero, then $F = F_t$ and $N = F_c$.

The forces parallel and perpendicular to the shear plane can be shown (from the force circle diagram) to be:

$$F_s = F_c \cos \phi - F_t \sin \phi \qquad \text{(Eq 11)}$$

$$F_n = F_c \sin \phi - F_t \cos \phi \qquad \text{(Eq 12)}$$

The shear force, $F_s$, is of particular interest because it is used to compute the shear stress on the shear plane. The shear stress, $\tau_s$, is defined as:

$$\tau_s = \frac{F_s}{A_s} \qquad \text{(Eq 13)}$$

where $A_s = tw/\sin \phi$.

Recalling that $t$ is the depth of the cut and $w$ is the width of the workpiece, the shear stress is:

$$\tau_s = \frac{F_c \sin \phi \cos \phi - F_t \sin^2 \phi}{tw} \qquad \text{(Eq 14)}$$

For a given polycrystalline metal, this shear stress is a material constant that is not sensitive to variations in cutting parameters, tool material, or the cutting environment.

Some researchers are attempting to derive (predict) the shear stress, $\tau_s$, and the shear direction from dislocation theory, but this has not yet been accomplished. Correlations of the shear stress with metallurgical measures, such as hardness or dislocation stacking fault energy, have been useful in these efforts.

The cutting force, $F_c$, is the dominant force in this system, and it is important to understand how it varies with changes in the cutting parameters. As shown in Fig. 8, the cutting forces typically double when the feed or depth of cut is doubled, but remain constant when speed is increased. In addition, the forces will increase (and change direction) when the rake angle is reduced. More detailed information on the determination of cutting forces can be found in the article "Forces, Power, and Stresses in Machining" in this Section.

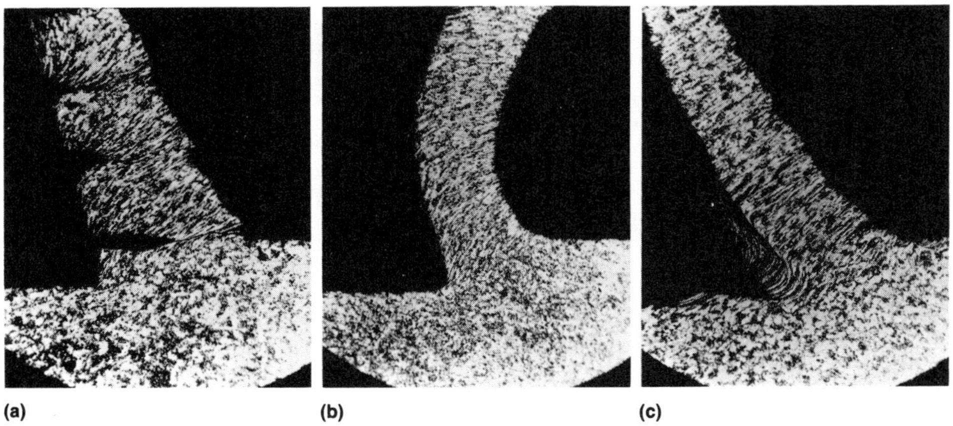


**Fig. 6** Three characteristic types of chips. (a) Discontinuous. (b) Continuous. (c) Continuous with built-up edge

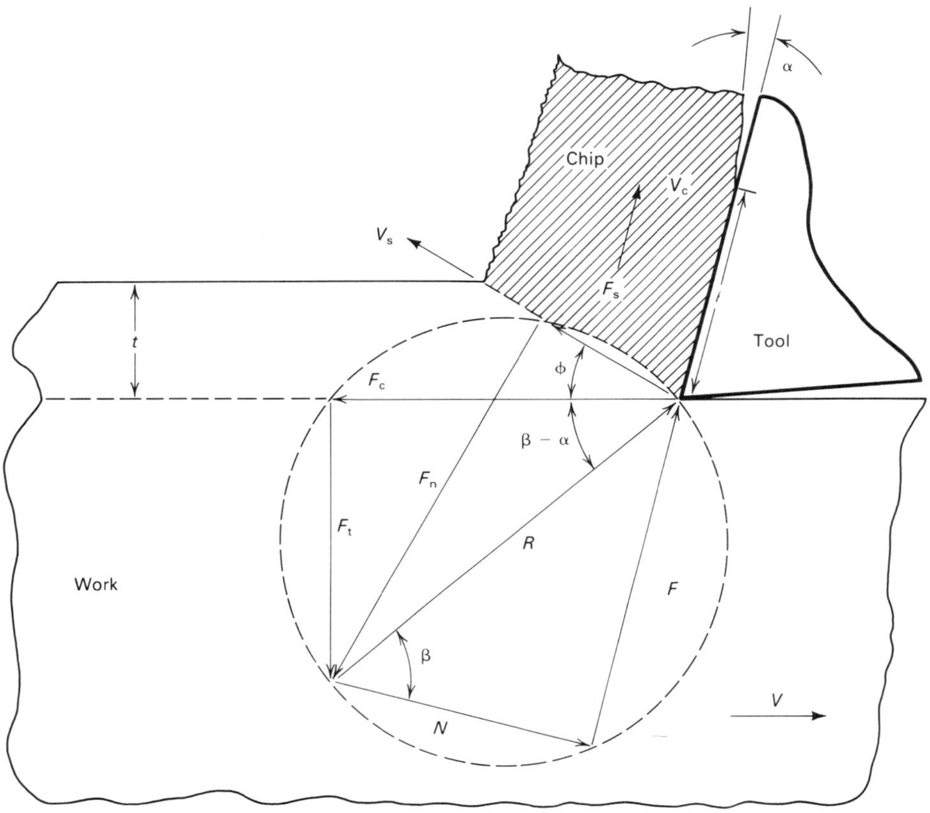


**Fig. 7** Circular force diagram for orthogonal chip formation

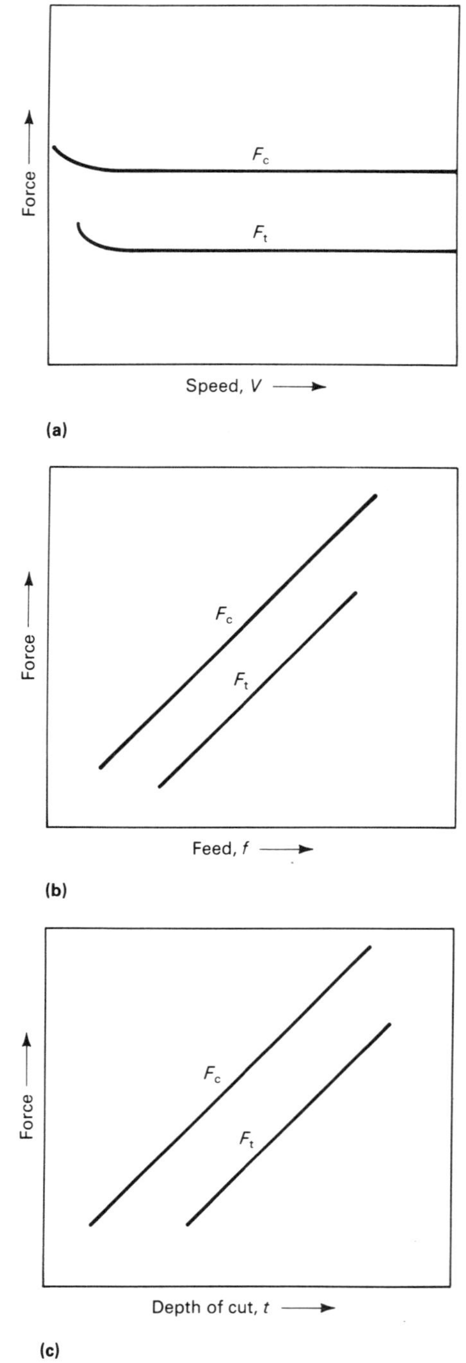


**Fig. 8** General relationship of orthogonal cutting forces to primary cutting parameters speed (a), feed (b), and depth of cut (c)

# Forces, Power, and Stresses in Machining

Paul H. Cohen, The Pennsylvania State University

THE MODELING AND ANALYSIS of chip formation has been a continuing exercise over the past century. The metal cutting process is a unique and complex production process distinguished by:

- Large shear strains, usually of the order of 2 to 5 (Ref 1)
- Exceptionally high shear strain rates, typically from $10^3$ to $10^5$ $s^{-1}$ with local variations as high as $10^7$ $s^{-1}$ (Ref 2, 3)
- The rubbing of the tool flank over a freshly cut surface that is chemically clean and active
- Many process and tooling parameters with a wide range of settings that can drastically alter the cutting process
- A large number of metallurgical parameters in the workpiece that can influence its response to the cutting tool

These factors and others make the modeling of metal cutting a difficult task that continues to evolve over time. The models and the discussion presented in this article will attempt to explain the basic concepts of the many complex factors that influence the forces, power, and stresses in machining.

## Forces and Energy in Orthogonal Machining

Although most production machining processes are oblique (that is, having three component forces), models of the orthogonal (that is, two force) machining of metals are useful for understanding the basic mechanics of machining and can be extended for modeling of the production processes.

**Forces.** The classical thin zone mechanics was developed for materials that yield continuous chips with a planar shear process coupled with the following assumptions (Ref 4, 5):

- The tool tip is sharp, and no rubbing occurs between the tool and the workpiece
- Plane strain conditions prevail (that is, no side spread occurs)
- The stresses on the shear plane are uniformly distributed
- The resultant force, $R$, on the chip is equal, opposite, and colinear to the force $R'$ at the tool/chip interface (Fig. 1)

The modeling of the orthogonal cutting process defines two regions of deformation (primary and secondary), each described by

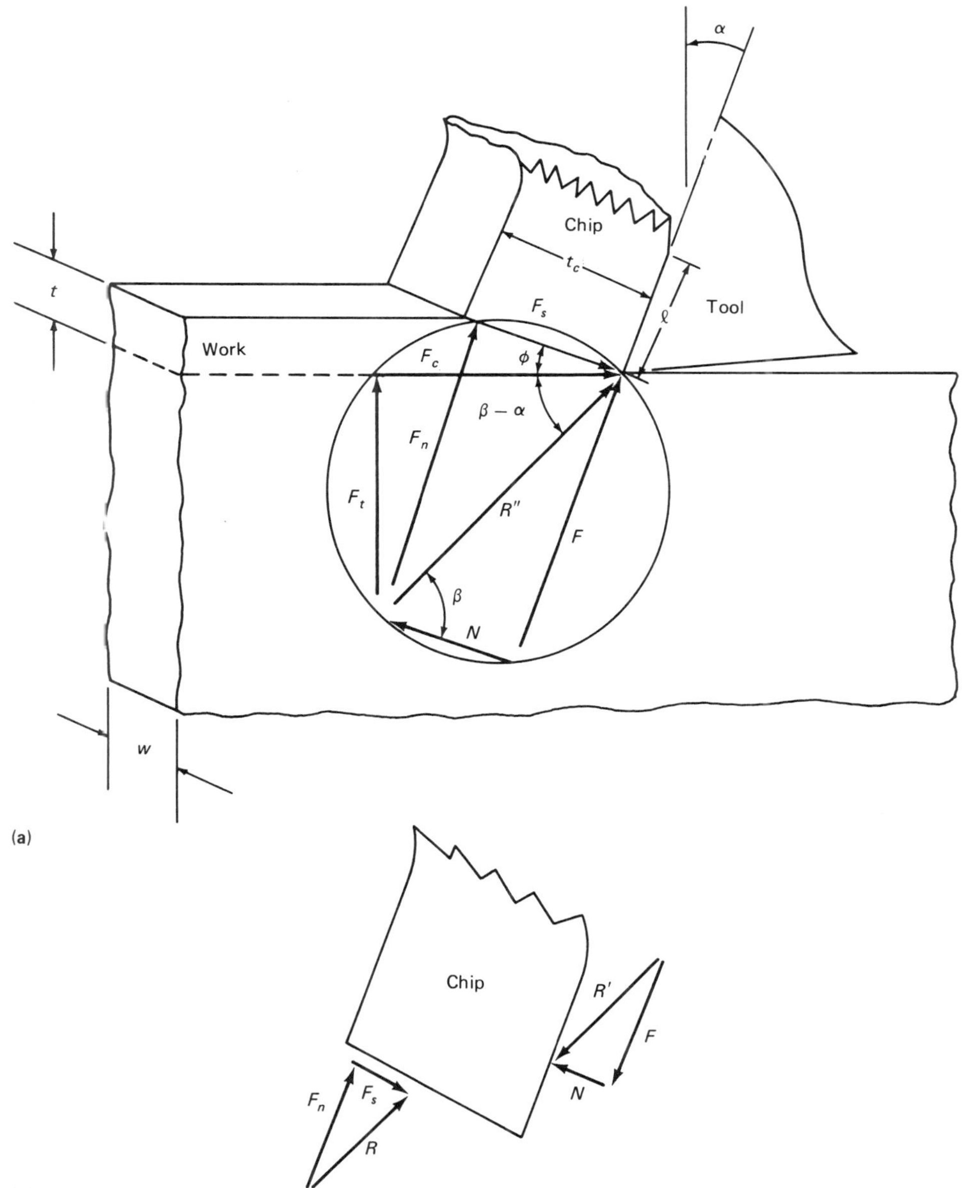


**Fig. 1** The geometry (a) and forces (b) in orthogonal cutting

its own set of orthogonal forces, as shown in Fig. 1(b). Because these force components cannot be directly measured (except for the forces on the rake face of the tool when $\alpha = 0°$), a dynamometer must be used to measure the primary (horizontal) cutting force, $F_c$, and the tangential (vertical) force, $F_t$. Thus, the measured forces can be resolved onto the shear plane through the shear angle, $\phi$, and onto the rake face through the back rake angle, $\alpha$.

The shear angle, $\phi$, is the angle the primary shear plane makes with respect to the horizontal motion of the tool. Although it is possible to observe and measure this angle in special experiments by using high-speed photography or the machining stages within a scanning electron microscope, $\phi$ is typically computed by using a ratio of the depth of cut, $t$, to chip thickness, $t_c$, as follows:

$$\phi = \tan^{-1}\left[\frac{(t/\tau_c)\cos\alpha}{1 - (t/t_c)\sin\alpha}\right] \qquad \text{(Eq 1)}$$

The shear strain necessary to shear the work material at this angle, $\phi$, is:

$$\gamma = \tan(\phi - \alpha) + \cos\phi \qquad \text{(Eq 2)}$$

Analyzing the primary shearing process in Fig. 1, the shear and normal forces on the shear plane can be written as functions of the measured horizontal and vertical (dynamometer) forces and shear angle, as follows:

$$F_s = F_c \cos\phi - F_t \sin\phi \qquad \text{(Eq 3)}$$

$$F_n = F_c \sin\phi + F_t \cos\phi \qquad \text{(Eq 4)}$$

Similarly, the forces on the rake face can be written as functions of the same measured force components and the tool back rake angle as:

$$F = F_c \sin\alpha + F_t \cos\alpha \qquad \text{(Eq 5)}$$

$$N = F_c \cos\alpha - F_t \sin\alpha \qquad \text{(Eq 6)}$$

The resultant force, $R$, which acts on the chip and is shown in Fig. 1(b), can be written as the vector sum of the measured forces, the forces acting on the shear plane, or the forces acting on the rake face of the tool. Therefore:

$$R = (F_c^2) + F_t^2)^{1/2} \qquad \text{(Eq 7)}$$

$$R = (F_s^2 + F_n^2)^{1/2} \qquad \text{(Eq 8)}$$

$$R = (F^2 + N^2)^{1/2} \qquad \text{(Eq 9)}$$

**Energy of Chip Formation.** During the cut, the total energy per unit time (or power) can be calculated simply as the product of the primary cutting force, $F_c$, and the velocity of cut, $V$. However, because many parameters can be varied in the cutting process that change the total energy consumed, this energy value is typically normalized by dividing by the rate at which material is removed. The material removal rate is calculated by multiplying the area being cut ($t \cdot w$ for the case of the plate of width $w$ shown in Fig. 1a), by the velocity perpendicular to that area at which the material is removed ($V$ in this case). Thus, the energy per unit time, or specific energy, $u$, can be calculated as:

$$u = \frac{F_c V}{twV} = \frac{F_c}{tw} \qquad \text{(Eq 10)}$$

The specific energy can be partitioned into four components (Ref 6, 7):

- Shear energy per unit volume, $u_s$
- Friction energy per unit volume, $u_f$
- Kinetic (momentum) energy per unit volume, $u_m$
- Surface energy per unit volume, $u_a$

The shear energy per unit volume can be calculated by substituting the energy per unit time necessary to shear the material in place of the total energy per unit time in Eq 8. Thus:

$$u_s = \frac{F_s V_s}{twV} \qquad \text{(Eq 11)}$$

where $V_s$ is the shear velocity (where $V_s = V\cos\alpha/\cos(\phi - \alpha)$, as defined in the article "Mechanics of Chip Formation" in this Section; see the discussion of shear angle measurement during orthogonal machining). The shear energy per unit volume is the largest of the four components, typically representing more than 75% of the total.

The friction energy per unit volume is consumed as the chip slides on the rake face of the tool. This component is very sensitive to cutting velocity and can be written as:

$$u_f = \frac{FV_c}{twV} \qquad \text{(Eq 12)}$$

where $V_c$ is the velocity of the chip as it flows over the tool ($V_c = V\sin\phi/\cos(\phi - \alpha)$, as defined in Eq 4 of the article "Mechanics of Chip Formation" in this Section).

The kinetic (momentum) energy per unit volume required to accelerate the chip is generally neglected but takes on increasing importance with very high speed machining. It can be written as:

$$u_m = \frac{F_m V_s}{twV} \qquad \text{(Eq 13)}$$

where $F_m$ is the momentum force $= \rho V^2 tw\gamma \sin\phi$, where $\rho$ is the density of the material being cut and $\gamma$ is the shear strain.

Additional energy is required to produce a new uncut surface. The surface energy per unit volume needed to create this new surface can be written as:

$$u_a = \frac{T \cdot 2Vw}{twV} = \frac{2T}{t} \qquad \text{(Eq 14)}$$

where $T$ is the surface energy of the material being cut. This component is also generally neglected.

Therefore, for most machining applications, the specific energy can be accurately estimated as:

$$u \sim u_s + u_f \qquad \text{(Eq 15)}$$

except at high speeds (above ~900 to 1200 m/min, or 3000 to 4000 sfm) for which the kinetic specific energy should be included. Specific energies can be used to calculate the power per unit volume per unit time (specific horsepower) and are readily accessible for most engineering materials. They are a good measure of the difficulty involved in machining a particular material.

## Stress Distributions in Metal Cutting

High shear and normal stresses occur both in the primary shear plane and on the rake face of the tool. This region of friction or secondary shear is critical in understanding the process mechanics and the wear of cutting tools.

**Stresses in the Workpiece.** As discussed in the article "Mechanics of Chip Formation" in this Section, the fundamental mechanism of chip formation requires prior work hardening before the workpiece material reaches the shear plane. Experimental results have shown that the material is elastically deformed at distances sufficiently far from the tool tip. As the material approaches the tool, the compressive stresses will begin to plastically deform the workpiece material as shown in Fig. 2. Behind the tool tip, the stresses will be tensile.

The distance of the elastic-plastic boundary from the tool tip will depend on tooling parameters, cutting parameters, and workpiece material properties. In particular, the amount of prior strain hardening and the ability of the workpiece material to work harden will alter the magnitude of the stresses in the workpiece and will affect the placement of the elastic-plastic boundary. Materials with little prior strain hardening will extend their boundaries farther out from the tool tip.

**Stresses on the Shear Plane.** Consistent with the assumptions in the section "Forces" in this article, the shear plane is generally modeled to have uniform distributions of both shear and normal forces over its entire area. The shear area, $A_s$, is the area of cut ($A_c = t \cdot w$) inclined at the shear angle $\phi$. Thus, the shear area is:

$$A_s = \frac{A_c}{\sin\phi} = \frac{t \cdot w}{\sin\phi} \qquad \text{(Eq 16)}$$

The shear stress on the shear plane can then be calculated as follows:

$$\tau_s = \frac{F_s}{A_s}$$

$$\tau_s = \frac{F_c \cos\phi - F_t \sin\phi}{t \cdot w/\sin\phi}$$

$$\tau_s = \frac{F_c \cos\phi \sin\phi - F_t \sin^2\phi}{t \cdot w} \qquad \text{(Eq 17)}$$

and the normal stress can be computed similarly as:

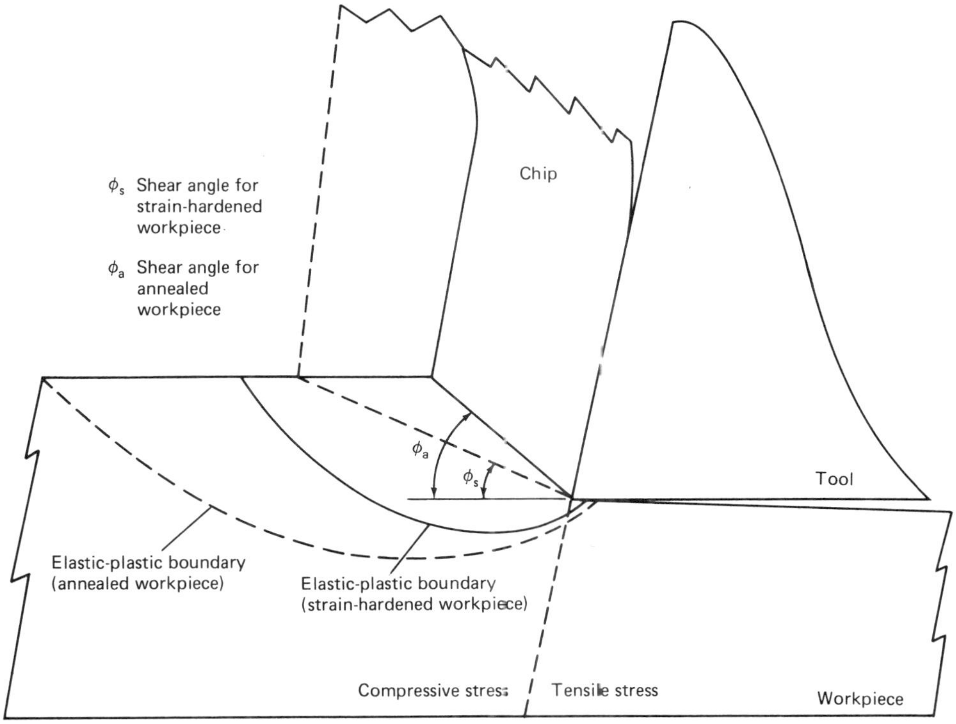


**Fig. 2** Stresses in the workpiece

$$\sigma = \frac{F_n}{A_s}$$

$$\sigma = \frac{F_c \sin \phi + F_t \cos \phi}{t \cdot w/\sin \phi}$$

$$\sigma = \frac{F_c \sin^2 \phi + F_t \sin \phi \cos \phi}{t \cdot w} \qquad \text{(Eq 18)}$$

Therefore, the stresses rely only on measured cutting forces ($F_c$ and $F_t$), the geometry of the cut ($t$ and $w$), and the deformation geometry ($\phi$).

The shear stress, $\tau_s$, takes on a constant value for a particular material. Figure 3 and Table 1 provide typical values for a variety of metals.

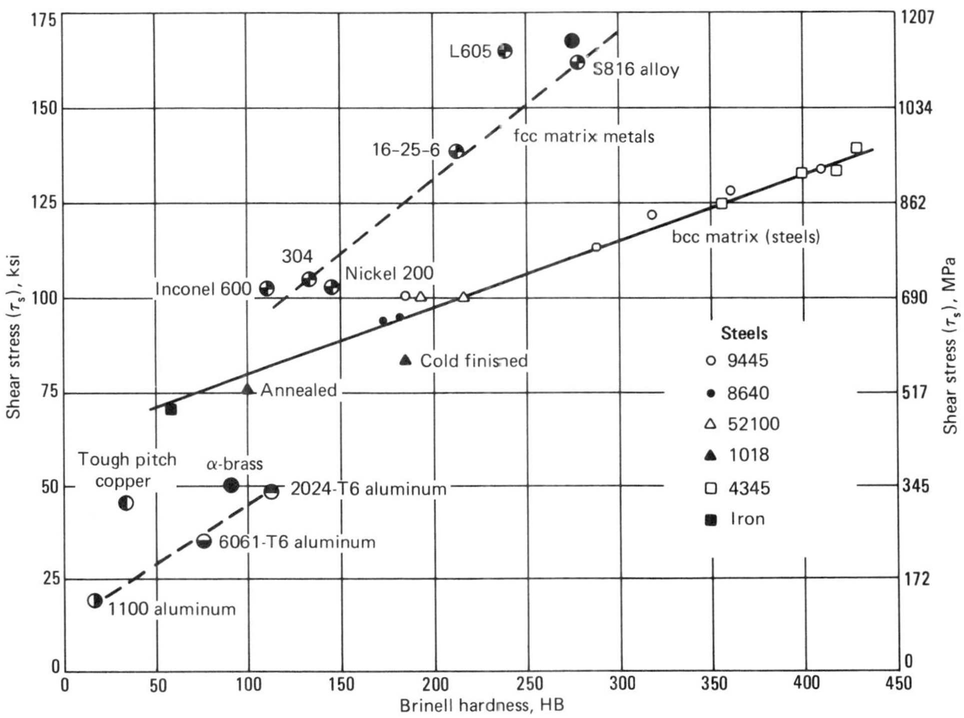


**Fig. 3** Shear stress variation with Brinell hardness for ferrous and nonferrous metals. Source: Ref 9

**Table 1 Shear stresses and specific horsepowers of selected engineering materials**

| Material | Shear stress, psi | Specific horsepower, hp/in.$^3$/min | Hardness, HB |
|---|---|---|---|
| Magnesium | 28 000 | 0.17 | ... |
| 1100 aluminum alloy | 16 700 | ... | ... |
| 6061-T6 aluminum alloy | 35 722 | 0.35 | ... |
| 2024-T4 aluminum alloy | 50 000 | 0.46 | ... |
| Copper | 44 850 | 0.78 | ... |
| 60-40 brass | 47 000 | | |
| 65-35 brass | 50 000 | | |
| 70-30 brass | 56 940 | 0.59 | ... |
| AISI 1020 steel | 61 500 | 0.58 | 150–175 |
| | | 0.67 | 176–200 |
| AISI 1112 steel | 63 500 | 0.5 | 150–175 |
| Type 304 stainless steel | 105 000 | 1.1–1.9 | ... |
| Titanium | 173 500 | 1.9 | ... |

Source: Ref 2, 8

**Stress Distributions on the Rake Face.** The nature of the tool/chip interface and the distribution of the shear and normal stresses are critical in understanding the cutting process and the performance of cutting tools. The high stresses, coupled with the high temperatures and large strains in the chip adjacent to the tool face, make the secondary shearing process difficult to model.

**Uniform Stresses on the Rake Face.** The classical analysis of the forces and stresses on the rake face assumes that Coulombic sliding friction is present and that the stresses are uniformly distributed. Therefore, the coefficient of sliding friction is simply the frictional force, $F$, divided by the normal force, $N$, acting on the rake face. Thus:

$$\mu = \frac{F}{N}$$

$$\mu = \frac{F_c \sin \alpha + F_t \cos \alpha}{F_c \cos \alpha - F_t \sin \alpha} \qquad \text{(Eq 19)}$$

The coefficient of friction is velocity dependent, with increasing speeds yielding lower friction.

The area of contact on the tool/chip interface is the product of the width of cut, $w$, and the length of sliding contact, $\ell$, as illustrated in Fig. 1. Thus, the area of sliding contact on the rake face is:

$$A_f = w \cdot \ell \qquad \text{(Eq 20)}$$

and the shear stress at the interface can be calculated as:

$$\tau_f = \frac{F}{A_f}$$

$$\tau_f = \frac{F_c \sin \alpha + F_n \cos \alpha}{w \cdot \ell} \qquad \text{(Eq 21)}$$

Analogously, the normal stress on the rake face can be written as:

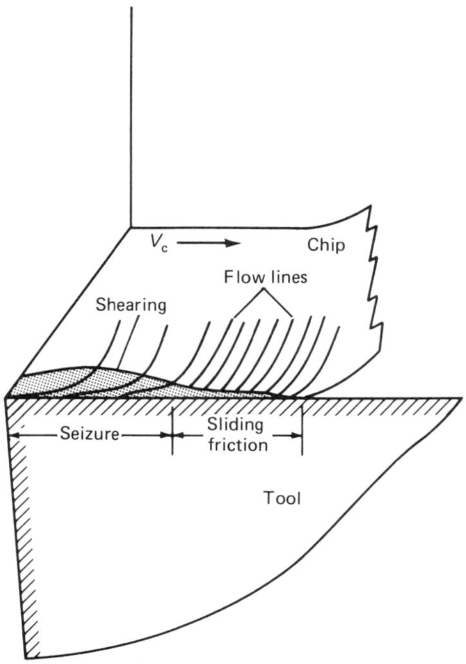


**Fig. 4** Flow lines in a chip

$$\tau_f = \frac{N}{A_f}$$

$$\tau_f = \frac{F_c \cos \alpha - F_n \sin \alpha}{w \cdot \ell} \quad \text{(Eq 22)}$$

These models have been found to be useful approximations of the behavior of the chip as it slides over the tool. However, there is a large body of experimental evidence to suggest that the stresses are not uniformly distributed on the rake face.

**Nonuniform Stress Distributions on the Rake Face.** The body of experimental evidence indicating the nonuniformity of the stresses on the rake face is extensive, using a wide variety of experimental techniques and observations. Perhaps the simplest observation supporting this conclusion is the transfer of workpiece material to the tool as observed by the unaided eye, light microscope, or scanning electron microscope (Ref 10). The transfer of this material does not occur over the entire contact area, but near the tip of the tool. Experiments utilizing photography through transparent sapphire tools (Ref 11), photoelastic tools (Ref 12), quick-stop devices to observe metal flow in the chip (Ref 13), and other techniques have revealed the nonuniformity of the stresses.

Quick-stop devices that separate the tool from the chip freeze the flow pattern of the material in the chip. Such studies have revealed two major regions on the rake face with respect to flow. When polished and etched, it is clear from the flow lines that the material near the tool tip is seized by the tool. This can be shown by the flow lines in Fig. 4, which run parallel to the tool face. The rest of the contact area exhibits sliding contact.

This concept of seizure is quite different from the standard notions of sliding friction. Because of the high interface temperatures and pressures, the material adjacent to the tool surface is almost stationary, and relative shearing takes place in the chip.

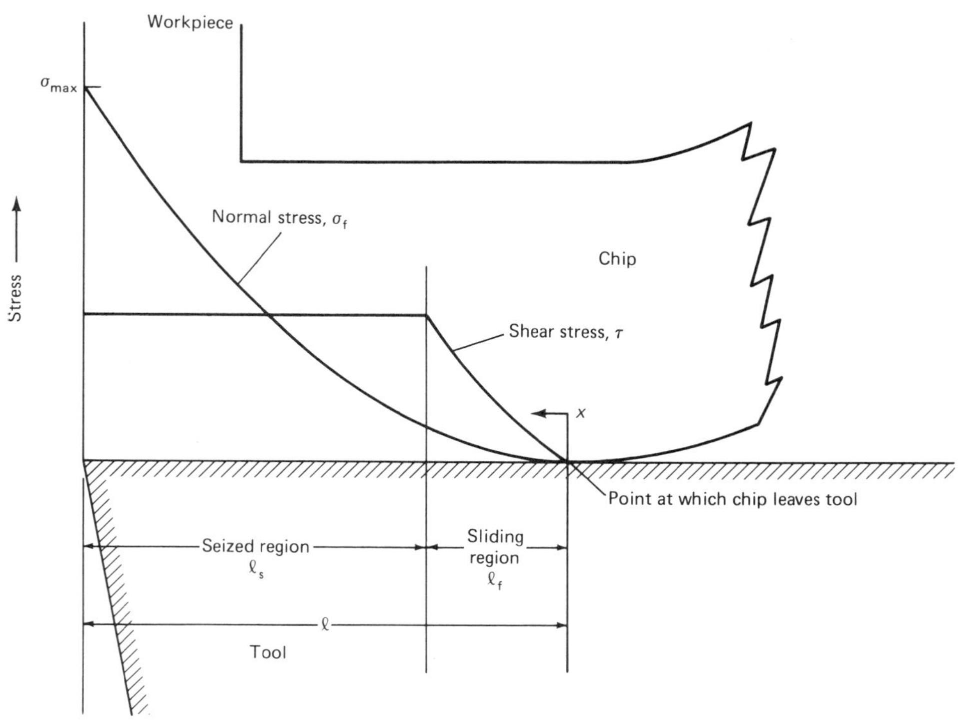


**Fig. 5** Model of stress distribution on tool during cutting. Source: Ref 14

As originally developed by Zorev (Ref 14) and consistent with the empirical results presented, the stresses on the rake face are inherently nonlinear, as shown in Fig. 5. The normal stress is assumed to take on a maximum value, $\sigma_{max}$, at the tip of the tool; the stress then decreases as a power function of the distance from the tool tip to the point at which the chip leaves the tool (Fig. 5). The shear stress is constant in the region of seizure and then decreases as a power function to the point at which the chip leaves the tool.

The normal stress on the rake face is defined by:

$$\sigma_f = \sigma_{max}(x/\ell)^n \quad \text{(Eq 23)}$$

where $\sigma_{max}$ is the maximum normal stress at the tool tip (or $x = \ell$), $\ell$ is the total length of contact of the chip on the tool, $x$ is the distance from the point at which the chip leaves the tool to the point of interest, and $n$ is the exponent. The normal force can be obtained by integrating the normal stress over the area of contact on the tool face:

$$N = \int_0^{\ell} \sigma_f \cdot w \cdot dx$$

$$N = \frac{\sigma_f w \ell}{1 + n} \quad \text{(Eq 24)}$$

The shear stress is more complicated to evaluate because the behavior of the chip material as it passes over the tool varies along the rake face. The region of seizure close to the tool tip must be modeled differently from the region of sliding (Coulombic) friction. Over the region of seizure ($\ell_f \leq x \leq \ell$), the shear stress has a constant value, $\tau$, because the chip material shears internally, as illustrated by the flow lines in Fig. 4.

Over the sliding region, the shear and normal stresses are related by:

$$\tau = \mu\sigma_f$$

$$\tau = \mu\sigma_{max}(x/\ell)^n \quad \text{(Eq 25)}$$

Thus, the shear stress over the entire face can be conveniently expressed as:

$$\tau_f = \begin{cases} \mu\sigma_{max}(x/\ell)^n & 0 \leq x \leq \ell_f \\ \tau & \ell_f \leq x \leq \ell \end{cases} \quad \text{(Eq 26)}$$

To determine the friction force, $F$, on the rake face, the shear stress given in Eq 26 must be integrated over the area of contact. This yields:

$$F = \int_0^{\ell_f} \mu\sigma_{max}(x/\ell)^n w dx + \int_{\ell_f}^{\ell} \tau w dx$$

$$F = \tau w \, [\ell_s + \ell_f/(1 + n)] \quad \text{(Eq 27)}$$

where $\ell_s$ is the length of seizure (that is, $\ell - \ell_f$).

Although such models are useful in understanding the process, it is difficult to determine the lengths associated with seizure and sliding. Typically, a tool is ground

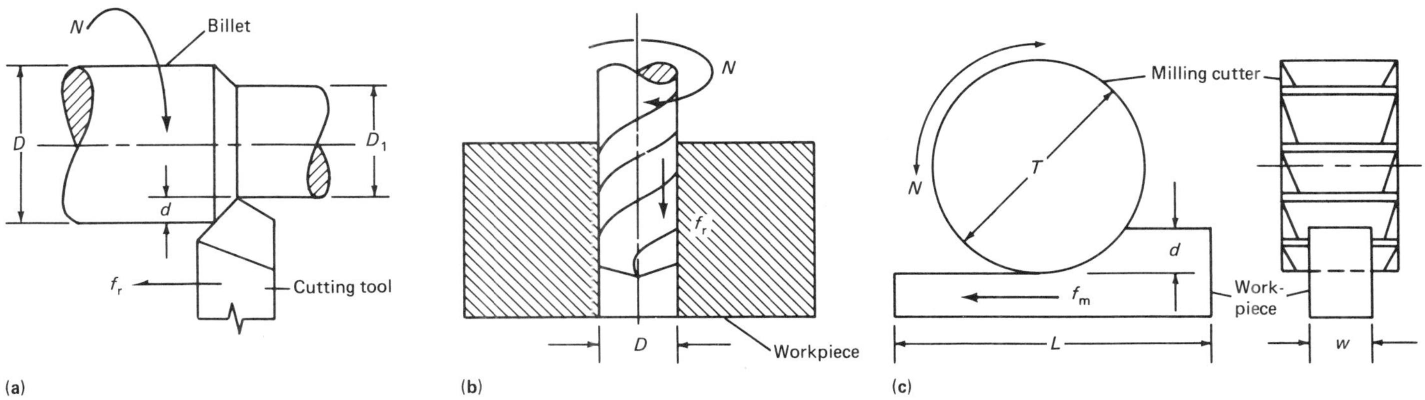


**Fig. 6** Setups for turning (a), drilling (b), and milling (c) operations

to restrict the total length of contact and the length of seizure determined by the flow lines, as detailed previously.

From this simple orthogonal model, it is clear that the shear strength of the chip material, the relative amount of seizure, and other parameters will significantly alter the machining forces. Increasing shear strengths and lengths of seizure (for constant contact length) will increase the cutting forces. Therefore, the use of tool materials with less propensity for chip seizure or the use of lubricants that decrease $\ell_s$ will lower the cutting forces accordingly.

## Power Consumption in Production Processes

Although one may wish to describe the energy per unit volume needed to form the chip, machine tools are typically rated in terms of power. Unit (or specific) power values can be calculated by dividing the power input to the process, $F_cV$, by the volumetric rate at which material is removed and then dividing this quantity by 33 000 to convert to horsepower. The specific power, $P_s$, is a measure of the difficulty involved in machining a particular material and can be used to estimate the total cutting power, $P$. Typical specific horsepower values are given in Table 1.

The specific power is the power required to remove a unit volume per unit time. Therefore, the specific and total powers are related as follows:

$$P = P_s \cdot \text{MRR} \qquad \text{(Eq 28)}$$

where MRR is the material removal rate, or volume of material removed per unit time. The material removal rate can be computed as the uncut area multiplied by the rate at which the tool is moved perpendicular to the uncut area. As previously determined for a plate, the material removal rate is the uncut area, $t \cdot w$, multiplied by the velocity of the tool, $V$. Thus, the cutting parameters and machine tool kinematics define the material removal rate. There are many standard sources for specific power values for a variety of materials.

Unfortunately, machine tools are not completely efficient. Losses due to component wear, friction, and other sources prevent some power from reaching the tool. Therefore, the gross power, $P_g$, needed by the motor can be defined as:

$$P_g = \frac{P}{\eta} \qquad \text{(Eq 29)}$$

where $\eta$ is the efficiency of the machine.

**Power in Turning.** As with the plate, the total power required in a turning operation can be calculated as $P = P_s \cdot \text{MRR}$. However, the material removal rate must be redefined for turning. Consider the turning operation illustrated in Fig. 6(a), in which a billet of diameter $D$ is turned with depth of cut $d$ to diameter $D_1$. The billet is rotated at $N$ revolutions per minute, while the tool is fed at $f_r$ units (millimeters or inches) per revolution, which can be set directly on the machine. Recommended cutting speeds (in meters or feet per minute) are generally available from handbooks and can be converted to rotational speed where $V = \pi DN$. Suggested feeds are also available.

The material removal rate has been defined as the uncut area multiplied by the rate at which the material is removed perpendicular to the area. For turning, the area removed is an annular ring of outside diameter $D$ and inside diameter $D_1$. Thus, the uncut area is $\pi(D^2 - D_1^2)/4$. The rate at which the tool is fed, $f_m$ (in unit distance per minute), is $f_r \cdot N$. Therefore, the material removal rate for turning is:

$$\text{MRR} = \frac{\pi(D^2 - D_1^2)}{4} \cdot f_r \cdot N \qquad \text{(Eq 30)}$$

and the total cutting power is $P_s \cdot \text{MRR}$, which can then be adjusted for machine tool efficiency.

The specific power and the material removal rate can also be used to estimate the main cutting force, $F_c$. The total horsepower can be written as $P = P_s \cdot \text{MRR}$ or as the product of the main cutting force multiplied by the velocity, as described for the plate in the section "Energy of Chip Formation" in this article. Equating these two equivalent expressions yields:

$$F_c = \frac{P_s \cdot \text{MRR}}{V} \qquad \text{(Eq 31)}$$

**Power in Drilling.** In Fig. 6(b), a drill of diameter $D$ is rotated at $N$ revolutions per minute and fed at $f_r$ (unit distance per revolution). The uncut area is the area covered by the drill, or $\pi D^2/4$, while the feed rate perpendicular to the area is $f_m = f_r \cdot N$. As with turning, recommended velocities, $V$, and feeds, $f_r$, are available in handbooks and the literature. The cutting velocity is again related to the rotational velocity by $V = \pi DN$. The material removal rate is $\pi D^2/4 \cdot f_m$, and the power can be calculated as:

$$P = P_s(\pi D^2/4 \cdot f_r \cdot N) \qquad \text{(Eq 32)}$$

Thus, the calculation of the material removal rate and the power is quite analogous to the case of turning.

**Power in Milling.** For the milling operation illustrated in Fig. 6(c), the cutter has diameter $D$ with $T$ teeth. The cutter is rotated at $N$ revolutions per minute. Unlike turning and drilling, the table (tool) is indexed at a feed rate of $f_m$ (millimeters or inches per minute), which is set directly on the machine. However, the feed rate is not given in handbooks. Instead, the feed (in unit distance) per tooth, $f_t$, is specified because too large a cut with any tooth may damage the tooth and because different cutters may contain varying numbers of teeth. The feed rate is related to the feed per tooth by:

$$f_m = f_t \cdot T \cdot N \qquad \text{(Eq 33)}$$

and the material removal rate is specified by:

$$\text{MRR} = \text{Uncut area} \cdot \text{Feed rate}$$
$$\text{MRR} = (d \cdot w)f_m$$
$$\text{MRR} = dwf_tTN \qquad \text{(Eq 34)}$$

The power required is therefore $P_s \cdot \text{MRR}$.

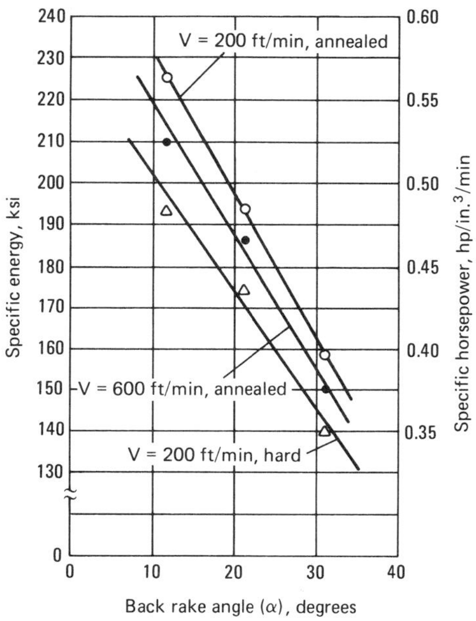


**Fig. 7** Influence of speed, tool geometry, and prior strain hardening on the specific energy of brass

**Factors Affecting Specific Power.** The specific power is a material-related property that can be significantly altered by cutting parameters, prior strain hardening of the workpiece material, and the material and geometry of the cutting tool.

The cutting velocity has a significant effect on the specific power and/or energy because the coefficient of friction on the rake face is speed dependent. Increasing speeds decrease the friction (up to a point), thus decreasing the specific power through the frictional component of specific power (Fig. 7).

The cutting tool material will also change the specific power due to changes in friction. For example, under similar conditions, the coefficient of friction and therefore the frictional component of the specific power will be lower for carbide tools than for high-speed steel tools. Therefore, the specific power will be higher, resulting in higher cutting forces. The tool geometry will also play a role in determining the specific power. In particular, the back rake angle, α, will influence the friction and therefore the specific power. Larger back rake angles will result in lower power and/or energy consumed per unit volume per unit time, provided the tool maintains its integrity. Figure 7 illustates this in the machining of brass. In addition, cutting tools (inserts) are available in a variety of nose radii. The radius of the tool has been shown to change the specific power and cutting forces significantly with larger radii yielding higher specific power.

Research has shown that ductile materials such as aluminum, brass, or copper will behave quite differently during machining, depending on the amount of strain hardening from prior processing. These materials have a great propensity for seizing the tool when in an annealed state. Cutting forces and specific powers are greater than for the same materials with significantly more work hardening. Thus, the processing history of a metal is an important determining factor for the specific power requirements.

## REFERENCES


1. S. Ramalingam and J T. Black, An Electron Microscopy Study of Chip Formation, *Metall. Trans.*, Vol 4, 1973, p 1103
2. J T. Black, Flow Stress Model in Metal Cutting, *J. Eng. Ind.*, Vol 101, 1979, p 403
3. B.F.von Turkovich, Dislocation Theory of Shear Stress and Strain Rate in Metal Cutting, in *Advances in Machine Tool Design and Research*, Pergamon Press, 1967, p 531
4. M.E. Merchant, Mechanics of the Metal Cutting Process—I, *J. Appl. Phys.*, Vol 16 (No. 5), 1945, p 267-275
5. M.E. Merchant, Mechanics of the Metal Cutting Process—II, *J. Appl. Phys.*, Vol. 16 (No. 6), 1945, p 318-325
6. E. Vidosic, *Metal Machining and Forming Technology*, The Ronald Press, 1964
7. N. Cook, *Manufacturing Analysis*, Addison-Wesley, 1966
8. E.P. DeGarmo, J T. Black, and R.A. Kosher, *Materials and Processes in Manufacturing*, 7th ed., Macmillan, 1988
9. S. Ramalingham and K.J. Trigger, in *Advances in Machine Tool Design and Research*, Pergamon Press, 1971
10. E.M. Trent, *Metal Cutting*, Butterworths, 1984, p 24
11. P.K. Wright, Friction Interactions In Machining: Comparisons Between Transparent Sapphire Tools and Steel Cutting Tools, *Met. Technol.*, Vol 8, 1981, p 150
12. E. Amini, *J. Strain Anal.*, Vol 3, 1968, p 206
13. P.K. Wright and J.L. Robinson, Material Behaviour in Deformation Zones of Machining Operation, *Met. Technol.*, May 1977, p 240
14. N.N. Zorev, Interrelationship Between Shear Process Occurring Along Tool Face and on Shear Plane in Metal Cutting, *International Research in Production Engineering*, Presented at International Production Engineering Research Conference (Pittsburgh), 1963, p 42

# Surface Finish and Surface Integrity

Michael Field, John F. Kahles, and William P. Koster, Metcut Research Associates Inc.

A PART SURFACE has two important aspects that must be defined and controlled. The first concerns the geometric irregularities of the surface, and the second concerns the metallurgical alterations of the surface and the surface layer. This second aspect has been termed surface integrity. Both surface finish and surface integrity must be defined, measured, and maintained within specified limits in the processing of any product. Standards have been adopted for surface finish and are available in ANSI/ASME B46.1-1985 (Ref 1). A companion standard for surface texture symbols is ANSI Y14.36-1978 (Ref 2). The standard for surface integrity is ANSI B211.1-1986 (Ref 3).

## Surface Finish or Surface Texture

Surface finish as described in Ref 1 is concerned with only the geometric irregularities of surfaces of solid materials and the characteristics of instruments for measuring roughness. Surface texture is defined in terms of roughness, waviness, lay, and flaws (Fig. 1):

- Surface roughness consists of fine irregularities in the surface texture, usually including those resulting from the inherent action of the production process, such as feed marks produced during machining
- Waviness is a more widely spaced component of surface texture and may result from such factors as machine or work deflections, vibration, or chatter
- Lay is the direction of the predominant surface pattern
- Flaws are unintentional, unexpected, and unwanted interruptions in the surface—for example, cracks, nicks, scratches, and ridges

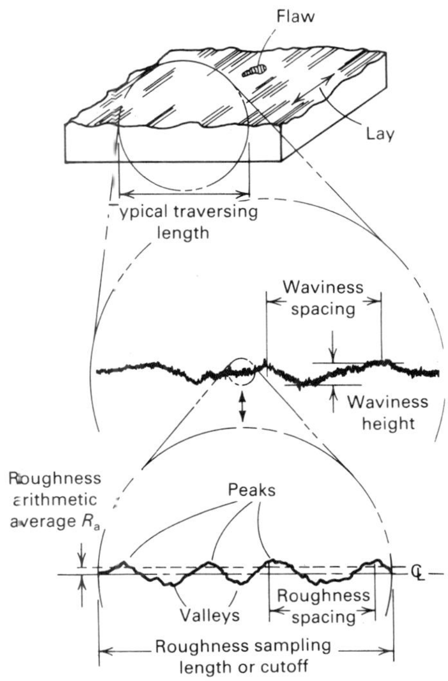


**Fig. 1** Schematic of roughness and waviness on a surface with unidirectional lay and one flaw. See Fig. 2 for definition of $R_a$ and waviness height. Source: Ref 1

Both surface roughness and waviness can be measured by a variety of instruments, including both surface contact and noncontact types. By far the most universal technique is to measure surface roughness with a stylus contact-type instrument that provides a numerical value for surface roughness. Such instruments can usually provide an indication of roughness in terms of the arithmetic average, $R_a$ (Fig. 2a), or the root mean square (rms) value, $R_q$ (Fig. 2b).

**Surface texture symbols** used for illustrations and specifications are shown in Fig. 3. Symbols for defining lay and its direction are shown in Fig. 4.

**Designations of Surface Roughness.** Figure 2 illustrates some of the designations of surface roughness. The most common method of designating surface roughness in the United States is the arithmetical average $R_a$, although the rms value $R_q$ is also used. The ratio between $R_q$ and $R_a$ varies with the manufacturing process producing the surface (Table 1). A preferred series of roughness values is given in Table 2.

**Surface Roughness Produced in Manufacturing Processes.** The predominant method of producing engineering surfaces is by a machining process, although some finished surfaces result from primary techniques such as casting, extruding, or forging. Each surface-producing method has a characteristic surface roughness range, some of which are shown in Fig. 5. The finer finishes are generally produced by machining techniques. Traditional machining techniques include chip removal processes (such as turning, milling, and reaming) and abrasive processes (such as grinding, polishing, buffing, and superfinishing). A variety of surface finishes can also be produced by nontraditional machining techniques such as electrical discharge machining, electrochemical machining, or laser beam machining. Surface finish requirements for representative machine tool components and aircraft engine components are given in Tables 3 and 4, respectively.

**Surface Roughness and Dimensional Tolerances.** Surface roughness is closely tied to the accuracy or tolerance of a machine component (Table 5). A close-tolerance dimension requires a very fine finish, and the finishing of a component to a very low roughness value may require multiple machining operations. For example, a 3.2 μm (125 μin.) surface roughness can be produced by milling or turning, while a very fine (low roughness value) surface would require grinding or additional subsequent operations, such as honing, superfinishing, buffing, or abrasive flow. Therefore, specifying very fine finishes will normally result in increased costs (Table 5.)

**Theoretical Surface Roughness Produced by Milling and Turning Tools.** It is possible to calculate the theoretical surface roughness profile produced by milling cutters and lathe tools. Surface roughness calculations have been made for three of the most common cutting tool shapes. These tool shapes (Fig. 6) are designated as Type A, a sharp-nose milling tooth, and Types B and C, which are a round tool and a tool with a nose radius, respectively. Types B and C can be used for either milling (Fig. 6a) or turning (Fig. 6b). Calculations have been made of the theoretical surface roughness as a function of the feed, the tool radius, the end cutting edge angle (ECEA), and the side cutting edge angle (SCEA) (Fig. 6). The theoretical surface roughness obtained from these calculations represents the best finish

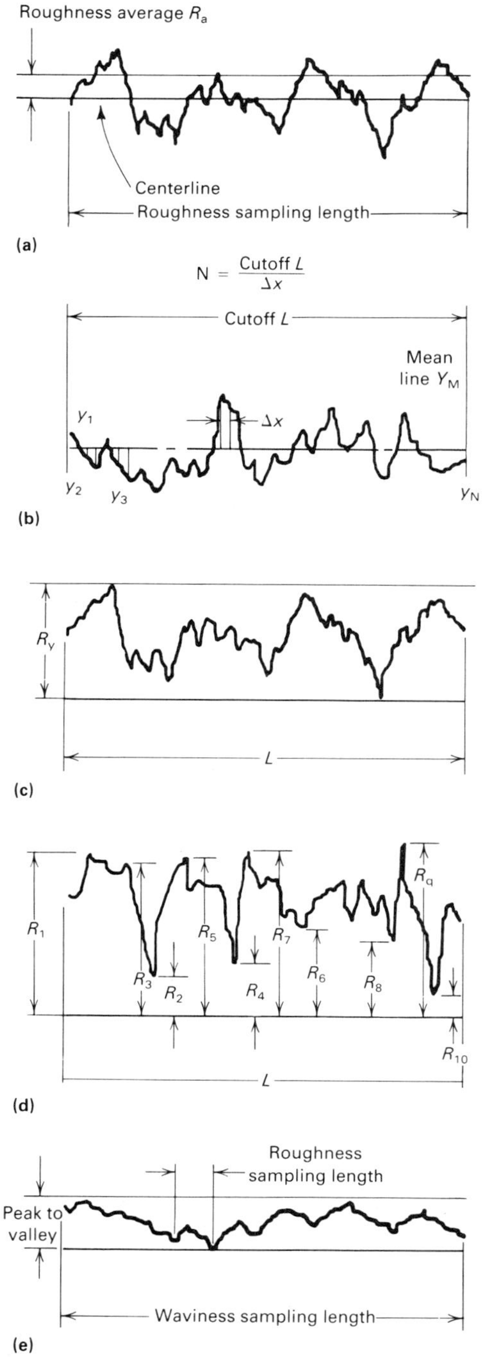


**Roughness average $R_a$, AA, or CLA** is

$$R_a = \frac{1}{L}\int_{x=0}^{x=L} |y|\,dx,$$

where $L$ is the sampling length and $y$ is the ordinate of the profile from the centerline.

$$\textbf{rms roughness } R_q = \left(\frac{1}{L}\int_{x=0}^{x=L} y^2dx\right)^{1/2}$$

which can be approximated as: rms roughness =

$$\sqrt{\frac{(y_1 - Y_M)^2 + (y_2 - Y_M)^2 + \ldots. (y_N - y_M)^2}{N}}$$

**Maximum peak-to-valley roughness height $R_y$ or $R_{max}$** is the distance between two lines parallel to the mean line that contacts the extreme upper and lower points on the profile within the roughness sampling length.

**Ten-point height $R_z$** is the average distance between the five highest peaks and the five deepest valleys within the sampling length measured from a line parallel to the mean line and not crossing the profile. It should be:

$$R_z = \frac{(R_1 + R_3 + R_5 + R_7 + R_9) - (R_2 + R_4 + R_6 + R_8 + R_{10})}{5}$$

**Waviness height $W$** is generally a maximum peak-to-valley measurement of the total profile within a waviness sampling length minus the average peak-to-valley roughness value within that length.

**Fig. 2** Some commonly used designations of surface texture. (a) $R_a$. (b) $R_q$. (c) $R_y$ or $R_{max}$. (d) $R_z$. (e) $W$. Source: Ref 1

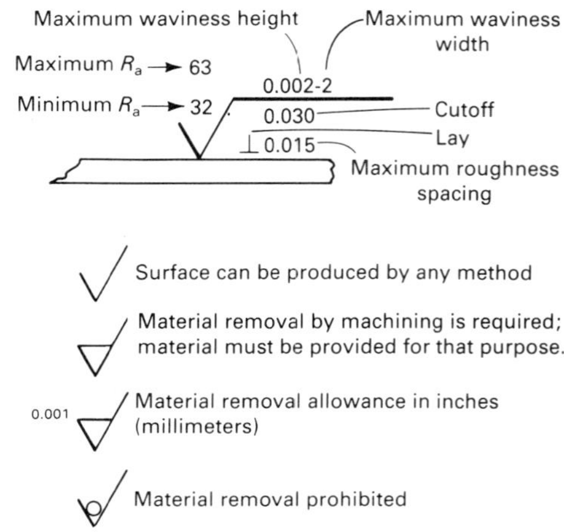


**Fig. 3** Surface texture symbols used for drawings or specifications. In this example, all values are in inches except $R_a$ values, which are in microinches. Metric values (millimeters and micrometers) are used on metric drawings. Source: Ref 2

| Lay symbol | Meaning | Example showing direction of tool marks |
|---|---|---|
| = | Lay approximately parallel to the line representing the surface to which the symbol is applied | |
| ⊥ | Lay approximately perpendicular to the line representing the surface to which the symbol is applied | |
| X | Lay angular in both directions to line representing the surface to which the symbol is applied | |
| M | Lay multidirectional | |
| C | Lay approximately circular relative to the center of the surface to which the symbol is applied | |
| R | Lay approximately radial relative to the center of the surface to which the symbol is applied | |
| P[3] | Lay particulate, nondirectional, or protuberant | |

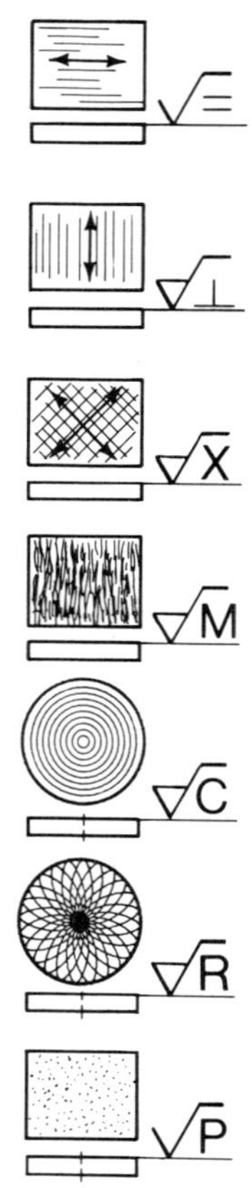

**Fig. 4** Symbols used to define lay and its direction. Source: Ref 2

commonly produced by that particular milling or turning tool and thus provides an indication of the minimum surface roughness possible with a designated tool shape and feed rate. The actual surface roughness may be poorer, because the surface is further degraded by a built-up edge that is usually formed as a characteristic of the machining process. Under some less common occurrences, a finer surface finish than the theoretical is produced because of the wear of the cutting edge that produces the finished surface; the worn tool develops a wearland that provides the tool with a wiping action, which tends to smooth out the theoretical surface irregularities. Surface roughness is sometimes improved in milling by providing the milling cutter with one additional finishing or wiper tooth designed to produce a broad finished machining path following the cutting action of the regular chip producing teeth in the cutter.

The theoretical surface roughness produced by a face milling cutter containing teeth with a zero nose radius is plotted in Fig. 7. The theoretical surface for turning or face milling with round cutting edges is illustrated in Fig. 8, and the theoretical surface roughness for turning or face milling with a radius of 0.396 mm (0.0156 in.) and various end cutting edge angles is shown in Fig. 9. Numerical tables of the theoretical

**Table 1 Ratio of root mean square roughness to arithmetic average roughness**

| | |
|---|---|
| Root mean square roughness | $R_q$ |
| Arithmetic average roughness | $R_a$ |
| Theoretical ratio of sine waves, $R_q/R_a$ | 1.11 |
| Actual ratios of $R_q/R_a$ for various processes | |
| Turning | 1.17 to 1.26 |
| Milling | 1.16 to 1.40 |
| Surface grinding | 1.22 to 1.27 |
| Plunge grinding | 1.26 to 1.28 |
| Soft honing | 1.29 to 1.48 |
| Hard honing | 1.50 to 2.10 |
| Electrical discharge machining | 1.24 to 1.27 |
| Shot peening | 1.24 to 1.28 |
| Practical first approximation of $R_q/R_a$ | |
| For most processes | 1.25 |
| For honing | 1.45 |

Source: Ref 4

roughness produced by milling and turning tools as a function of the feed, end cutting edge angle, and side cutting edge angle are provided in Ref 5. Table 6 gives the arithmetic roughness average, $R_a$, and the maximum peak-to-valley roughness height, $R_y$, for turning and milling.

## Surface Integrity

The specification and manufacture of unimpaired or enhanced surfaces require an understanding of the interrelationship among metallurgy, machinability, and mechanical testing. To satisfy this requirement, an encompassing discipline known as surface integrity was introduced, and it has gained worldwide acceptance. Surface integrity technology describes and controls the many possible alterations produced in a surface layer during manufacture, including their effects on material properties and the performance of the surface in service. Surface integrity is achieved by the selection and control of manufacturing processes, estimating their effects on the significant engineering properties of work materials.

**Table 2 Preferred series of roughness average values ($R_a$)**

| μm | μin. | μm | μin. |
|---|---|---|---|
| 0.012 | 0.5 | 1.25 | 50 |
| 0.025(a) | 1(a) | 1.60(a) | 63(a) |
| 0.050(a) | 2(a) | 2.0 | 80 |
| 0.075 | 3 | 2.5 | 100 |
| 0.10(a) | 4(a) | 3.2(a) | 125(a) |
| 0.125 | 5 | 4.0 | 160 |
| 0.15 | 6 | 5.0 | 200 |
| 0.20(a) | 8(a) | 6.3(a) | 250(a) |
| 0.25 | 10 | 8.0 | 320 |
| 0.32 | 13 | 10.0 | 400 |
| 0.40(a) | 16(a) | 12.5(a) | 500(a) |
| 0.50 | 20 | 15 | 600 |
| 0.63 | 25 | 20 | 800 |
| 0.80(a) | 32(a) | 25(a) | 1000(a) |
| 1.00 | 40 | | |

(a) Recommended. Source: Ref 2

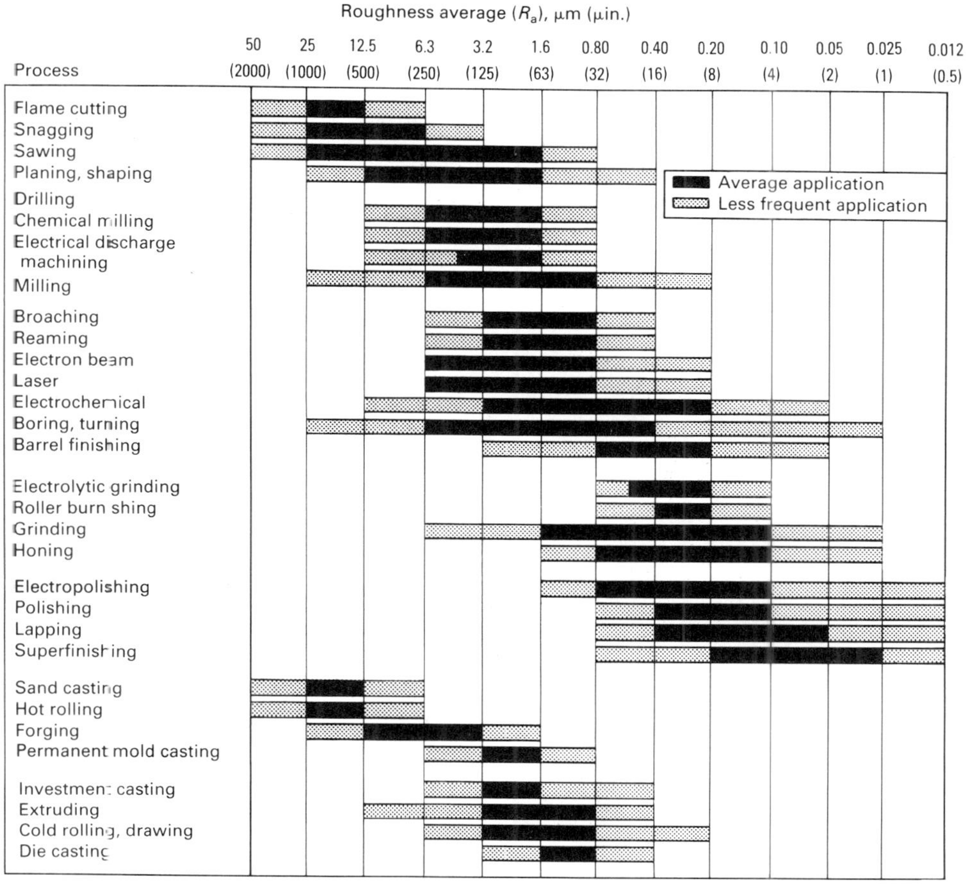


**Fig. 5** Surface roughness produced by common production methods. The ranges shown are typical of the processes listed. Higher or lower values can be obtained under special conditions. Source: Ref 1

Surface integrity involves the study and control of both surface roughness or surface topography, and surface metallurgy. Both of these factors influence the quality of the machined surface and subsurface, and they become extremely significant when manufacturing structural components that have to withstand high static and dynamic stresses. For example, when dynamic loading is a principal factor in a design, useful strength is frequently limited by the fatigue characteristics of materials. Fatigue failures almost always nucleate at or near the surface of a component; similarly, stress corrosion is also a surface phenomena. Therefore, the nature of the surface from both a topographical and a metallurgical point of view is important in the design and manufacture of critical hardware.

**Table 3 Typical surface finish requirements for machine tool components**

| Part name (material) | Machining process | Surface finish required, $R_a$ μm | μin. |
|---|---|---|---|
| Quill (4145 H) | | | |
| End face | Mill | 1.60 | 63 |
| Outside diameter | Lathe | 1.60 | 63 |
| Holes | Drill | 1.60 | 63 |
| Inside diameter | Grind | 0.40 | 16 |
| Cam (1018) | Grind | 0.40–0.80 | 16–32 |
| Key (1018) | Mill | 3.2 | 125 |
| Holder (1018) | Mill | 3.2 | 125 |
| Bracket (1018) | Mill | 3.2 | 125 |
| Plate (1018) | Mill | 3.2 | 125 |
| Block (1018) | Mill | 3.2 | 125 |
| Junction block (1018) | Grind | 1.60 | 63 |
| Ball screw (4150) | | | |
| Keyways | Mill | 3.2 | 125 |
| Outside diameter | Turn | 3.2 | 125 |
| Thread diameter | Grind | 0.80 | 32 |
| Ball nut (8617) | | | |
| Slots | Mill | 3.2 | 125 |
| Diameters | Grind | 1.60 | 63 |
| Holes | Drill | 3.2 | 125 |

**Table 4 Typical surface finish requirements for aircraft engine components**

| Part name and material | Operation | Surface finish requirements, $R_a$ μm | μin. |
|---|---|---|---|
| **Fan disk (Ti-6Al-4V) and turbine disk (Inconel 718)** | | | |
| Ultrasonic envelope | Turned | 1.60 | 63 |
| General surfaces | Turned | 1.60–3.2 | 63–125 |
| Bolt holes | Reamed | 0.80–1.60 | 32–63 |
| Dovetails | Broached | 0.80–1.60 | 32–63 |
| Corner breaks | Mass media finished | 0.80 | 32 |
| **Compressor casing (M-152 stainless steel)** | | | |
| Flowpath (inside diameter) | Turned | 1.60–3.2 | 63–125 |
| Outside (outside diameter) | Milled | 3.2 | 125 |
| Vane bores | Bored | 1.60 | 63 |
| Flange faces | Turned | 1.60 | 63 |
| **High-pressure turbine blade (René·80)** | | | |
| Airfoil | Tumbled | 0.80 | 32 |
| Dovetail form | Ground | 0.80 | 32 |
| General surfaces | Ground | 1.1 | 45 |
| **High-pressure turbine vane X40 (cobalt base)** | | | |
| Airfoil, convex | Tumbled | 0.61–0.80 | 24–32 |
| Airfoil, concave | Tumbled | 1.1 | 45 |
| Flowpath | Tumbled | 0.80 | 32 |
| General surfaces | Ground, tumbled | 0.80 | 32 |
| **Turbine shaft (Inconel 718)** | | | |
| Journals (chromium plated) | Ground | 0.40 | 16 |
| Bolt holes | Reamed | 0.80 | 32 |
| General surfaces | Turned | 1.60–3.2 | 63–125 |
| **Fan blade (Ti-6Al-4V)** | | | |
| Airfoil, convex | Belt ground, tumbled | 0.61–0.80 | 24–32 |
| Airfoil, concave | Belt ground, tumbled | 0.80–1.1 | 32–45 |
| Dovetail | Broached | 0.80 | 32 |

**Table 5 Classification of machined surface finishes**

| Class | Roughness, $R_a$ μm | μin. | Suitable for tolerance of plus or minus mm | in. | Typical method of producing finish | Approximate relative cost to produce |
|---|---|---|---|---|---|---|
| Super finish | 0.10 | 4 | 0.0125 | 0.0005 | Ground, microhoned, lapped | 40 |
| Polish | 0.20 | 8 | 0.0125 | 0.0005 | Ground, honed, lapped | 35 |
| Ground | 0.40 | 16 | 0.025 | 0.001 | Ground, lapped | 25 |
| Smooth | 0.80 | 32 | 0.050 | 0.002 | Ground, milled | 18 |
| Fine | 1.60 | 63 | 0.075 | 0.003 | Milled, ground, reamed, broached | 13 |
| Semifine | 3.2 | 125 | 0.100 | 0.004 | Ground, broached, milled, turned | 9 |
| Medium | 6.3 | 250 | 0.175 | 0.007 | Shaped, milled, turned | 6 |
| Semirough | 12.5 | 500 | 0.330 | 0.013 | Milled, turned | 4 |
| Rough | 25 | 1000 | 0.635 | 0.025 | Turned | 2 |
| Cleanup | 50 | 2000 | 1.25 | 0.050 | Turned | 1 |

The importance of surface integrity is further heightened when high stresses occur in the presence of extreme environments. Heat-resistant, corrosion-resistant, and high-strength alloys are used in a wide variety of such applications. Typical alloys used in these applications include alloy steels with hardnesses of 50 to over 60 HRC and heat-treated alloys with strength levels as high as 2070 MPa (300 ksi). Additional materials include stainless steels, titanium alloys, and high-temperature nickel-base alloys developed for high-temperature and corrosion-resistant applications.

Unfortunately, the alloys suitable for high-strength applications are frequently difficult to machine. The hard steels and high-temperature alloys, for example, must be turned and milled at low speeds, which tend to produce a built-up edge and poor surface finish. The machining of these alloys tends to produce undesirable metallurgical surface alterations, which have been found to reduce fatigue strength. Typical surface integrity problems created in metal removal operations include:

- Grinding burns on high-strength steel aircraft landing gear components
- Untempered martensite in drilled holes in steel components
- Grinding cracks in the root sections of cast nickel-base gas turbine buckets
- Lowering of fatigue strength of parts processed by electrical discharge machining
- Distortion of thin components
- Residual stress induced in machining and its effect on distortion, fatigue, and stress corrosion

## Surface Alterations

The types of surface alterations associated with metal removal practices are (Ref 6):

### Mechanical

- Plastic deformations (as result of hot or cold working)
- Tears and laps and crevicelike defects (associated with built-up edge produced-in-machining)
- Hardness alterations
- Cracks (macroscopic and microscopic)
- Residual stress distribution in surface layer
- Processing inclusions introduced
- Plastically deformed debris as a result of grinding
- Voids, pits, burrs, or foreign material inclusions in surface

### Metallurgical

- Transformation of phases
- Grain size and distribution
- Precipitate size and distribution
- Foreign inclusions in material
- Twinning
- Recrystallization
- Untempered martensite or overtempered martensite
- Resolutioning or austenite reversion

### Chemical

- Intergranular attack
- Intergranular corrosion
- Intergranular oxidation
- Preferential dissolution of microconstituents
- Contamination
- Embrittlement by the chemical absorption of elements such as hydrogen and chlorine
- Pits or selective etch
- Corrosion
- Stress corrosion

### Thermal

- Heat-affected zone
- Recast or redeposited material
- Resolidified material
- Splattered particles or remelted metal deposited on surface

### Electrical

- Conductivity change
- Magnetic change
- Resistive heating or overheating

The principal causes of surface alterations produced by machining processes are:

- High temperature or high-temperature gradients developed in the machining process
- Plastic deformation

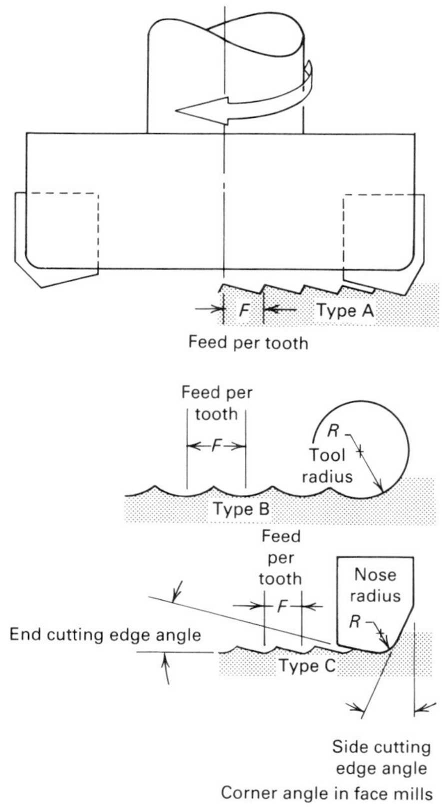


**Fig. 6(a)** Theoretical surfaces produced in models of face milling with a sharp-nose milling tool (Type A), a round tool (Type B), and a round-nose tool (Type C). Source: Ref 5

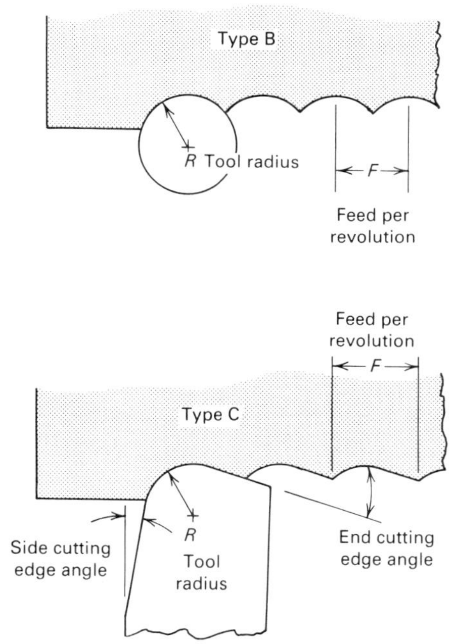


**Fig. 6(b)** Theoretical surfaces produced by turning with a round tool (Type B) and a round-nose tool (Type C). Source: Ref 5

- Chemical reactions and subsequent absorption into the nacent machined surface
- Excessive electrical currents

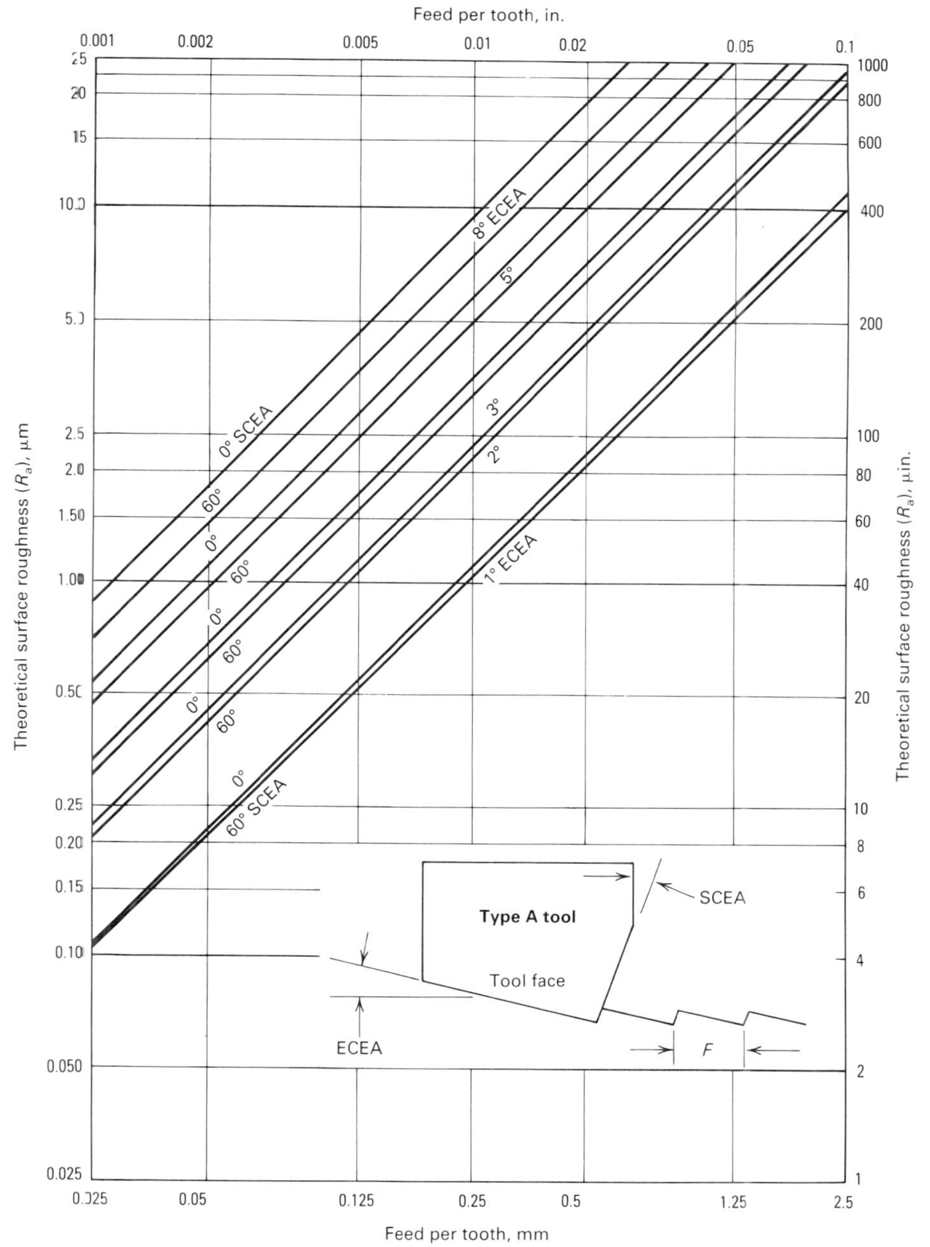


**Fig. 7** Theoretical surface roughness for a face milling cutter containing teeth with a zero nose radius. Source: Ref 5

- Excessive energy densities during processing

Virtually all material removal methods produce altered surface and subsurface conditions. The possible surface alterations resulting from various processes are summarized in Table 7. The mechanically and metallurgically altered zones produced by material removal processes may also extend into the surface to a considerable depth as a function of whether roughing or finishing conditions are used in the material removal process (Table 8).

**Surface Alterations Produced in Traditional Machining Operations.** Traditional machining methods, which are the principal means of metal removal, include chip cutting (such as milling, drilling, turning, broaching, reaming, and tapping) and abrasive machining methods (such as grinding, sanding, and polishing). These machining operations can produce surface layer alterations when abusive machining conditions are used (Fig. 10 and 11). In general, abusive machining promotes higher temperatures and excessive plastic deformation. Gentle machining operations occur when a sharp tool is employed and when machining conditions result in reduced machining forces.

*Surface Effects in Grinding.* In grinding, there are five important parameters that determine gentle or abusive conditions:

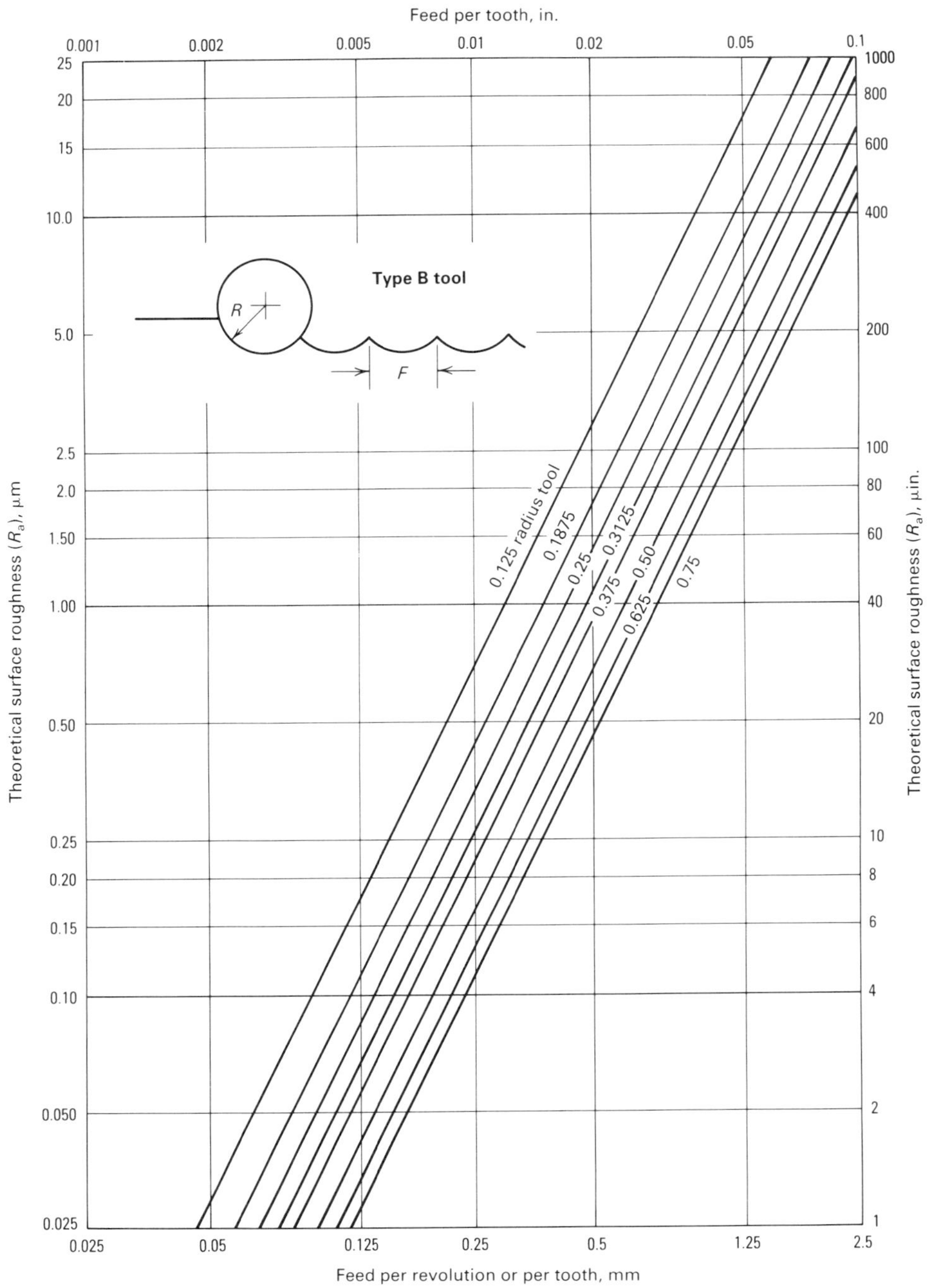


**Fig. 8** Theoretical surface roughness for turning or face milling tools with round cutting edges. Source: Ref 5

wheel grade, wheel speed, downfeed or infeed, wheel dressing, and grinding fluid. As grinding parameters become more aggressive (that is, harder wheels, higher wheel speeds, increased infeed, and so on), the grinding process becomes more abusive and therefore more likely to produce surface damage. Gentle, or low-stress, grinding conditions for a variety of alloys are summarized in Table 9.

Figure 12(a) illustrates the surface characteristics produced by the low-stress grinding of AISI 4340 steel quenched and tempered to 50 HRC. The low-stress condition produced no visible surface alterations, while the abusive grinding condition (Fig. 12b) produced an untempered martensite layer 0.03 to 0.13 mm (0.001 to 0.005 in.) deep with a hardness of 65 HRC. Below this white layer there was an overtempered martensitic zone with a hardness of 46 HRC. The hardness returned to its normal value at a depth of 0.30 mm (0.012 in.) below the surface (Fig. 12c).

Abusive grinding also tends to produce a residual stress within the altered surface layer. A residual stress profile can be readily obtained by using x-ray diffraction techniques both at the surface and then by stepping into the surface with repeated x-ray diffraction readings after successive surface etching. The abusive grinding produced high tensile stresses in the altered zone, while low-stress grinding produced a surface with small compressive stresses (Fig. 12d).

Fatigue tests conducted on flat specimens indicate that abusive grinding may seriously reduce the fatigue strength, as shown in Fig. 12(e). In this example, the abusive grinding depressed the endurance limit from 760 MPa (110 ksi) for low-stress grinding to 520 MPa (75 ksi).

**Surface Alterations Produced in Nontraditional Machining Operations.** Nontraditional machining includes a variety of methods for removing and finishing materials. Examples of nontraditional operations are electrical discharge machining, laser beam machining, electrochemical machining, electropolishing, and chemical machining.

*Electrical discharge machining (EDM)* tends to produce a surface with a layer of recast spattered metal that is usually hard and cracked and porous to some degree (Fig. 13). Below the spattered and recast metal, it is possible to have some of the same types of surface alterations that occur in abusive or traditional machining practices.

The effect is more pronounced when EDM is used under abusive or roughing conditions. Figure 14 shows a surface produced by EDM under finishing and roughing conditions. Under roughing conditions, globs of recast metal are spattered onto a white layer of rehardened martensite. An overtempered zone up to 46 HRC is also found beneath the surface. The surface produced under finishing conditions contains discontinuous patches of recast metal plus a thin layer of rehardened martensite 0.003 mm (0.0001 in.) deep.

*Laser beam machining (LBM)* tends to produce the same types of surface alterations as EDM. Figure 15 illustrates the heat-affected zone produced by LBM on Inconel 718. The intense heat generated by the laser beam resulted in a recast surface layer at the entrance and exit of the hole produced by the laser beam.

*Electrochemical machining (ECM)* is capable of producing a surface that is essentially free of metallurgical surface layer alterations. However, when ECM is not properly controlled, selective etching or intergranular attack may occur (Fig. 16). Abusive ECM conditions also tend to degrade surface roughness (Fig. 17).

*Electropolishing (ELP) and Chemical Machining (CM).* Surface softening is produced on most materials by electrochemical machining as well as by electropolishing and chemical machining, also referred to as chemical milling. Figure 17 illustrates the

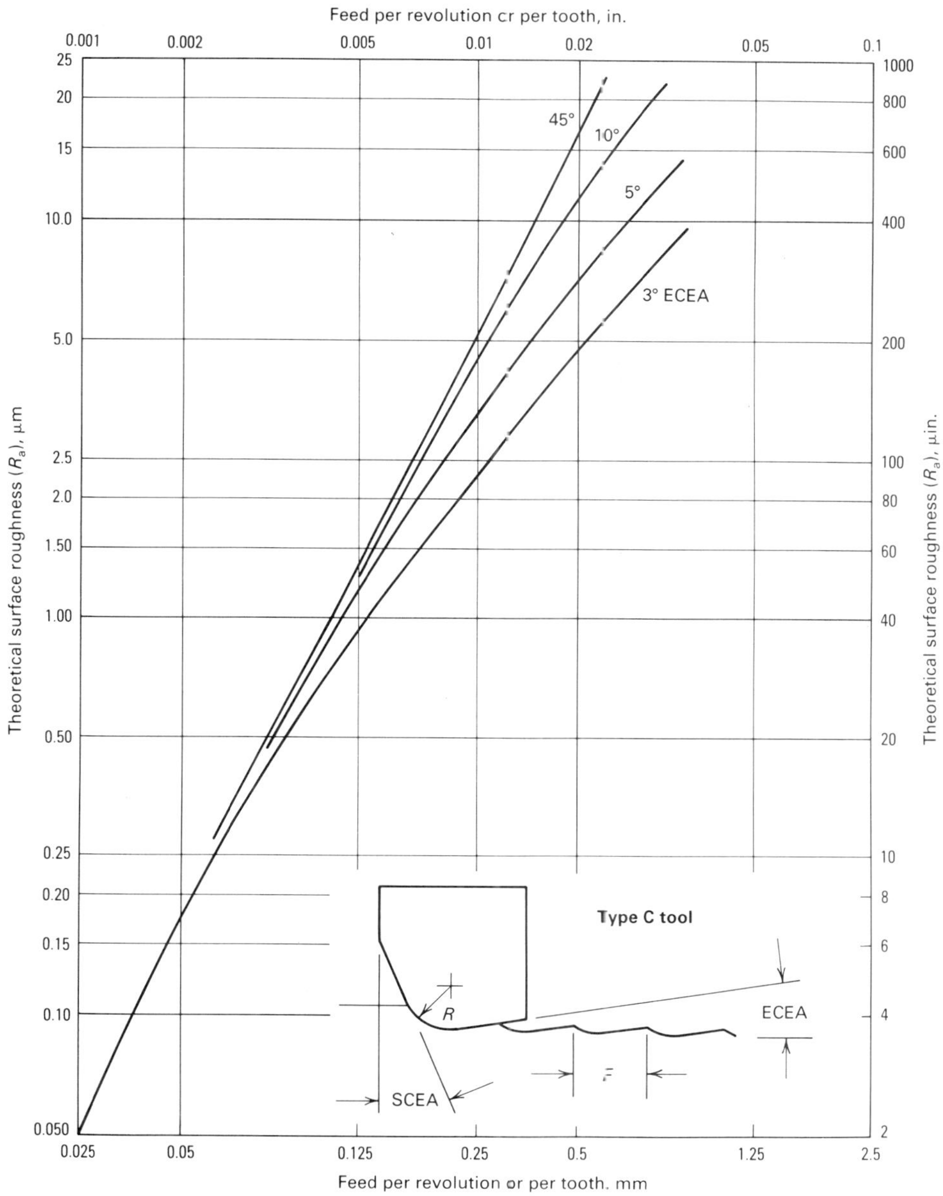


**Fig. 9** Theoretical surface roughness for turning or face milling tools with a radius of 0.39 mm (0.0156 in.) and various ECEAs. Source: Ref 5

surface softening produced by both gentle and abusive ECM on 4340 steel. The surface is about 5 HRC points lower in hardness than the interior to approximately 0.05 mm (0.002 in.) in depth. With CM, the same steel had its surface softened by about 5 HRC points to a depth of about 0.05 mm (0.002 in.) (Fig. 18). This softening may be severe enough and deep enough to affect the fatigue strength and other mechanical properties of metals and may necessitate postprocessing.

**Residual Stress.** Machining processes impart a residual stress in the surface layer. In grinding, the residual stress tends to be tensile when more abusive conditions are used (Fig. 19). By using gentle grinding conditions, the stress can be reduced in magnitude and can even become compressive. In milling, the residual stress tends to be compressive (Fig. 20). In facing milling 4340 steel (Fig. 20), the stresses are tensile at the surface but go into compression below the surface.

*Distortion and residual stress* are two related results from abusive machining conditions. In grinding, for example, a more abusive machining condition (such as an increase in downfeed) increases the distortion of the workpiece (Fig. 21) and creates more residual stress at the surface (Fig. 19). Figures 22 and 20 illustrate a similar relation in milling. As the machining condition becomes more abusive with a duller tool, the distortion (Fig. 22) and residual stress (Fig. 20) become greater. Residual stress and distortion thus exhibit the following relation: The greater the area under the residual stress curve, the greater the distortion of the workpiece.

## Mechanical Property Effects

The surface alterations produced by abusive metal removal conditions generally have a minor effect on the static mechanical properties of materials. A major exception to this, however, has been a marked decrease in ductility and tensile strength on materials that have been processed using EDM, followed by stress-relief heat treatment.

**Ductility and Tensile Strength.** Figure 23 illustrates the change in ductility and tensile strength from an EDM application. It was found that a carbon deposit produced on the surface during the EDM operation was diffused into the grain boundaries during a subsequent stress-relief treatment and caused an excessive grain-boundary precipitation of carbides. These precipitates were responsible for reductions in ductility and strength, the extent of which was found to be a function of surface roughness (Fig. 23). Reductions in ductility as high as 80% were noted after the heat treatment of Inconel specimens that had been EDM-machined to a surface finish of 16.5 µm (650 µin.) $R_a$.

**Fatigue Strength.** Surface alterations produced in machining are known to affect the fatigue and stress-corrosion properties of many materials. Extensive investigations have been performed on high-strength steels, and data illustrating the effect of some machining methods on fatigue strength are given in Table 10. The electropolishing of 4340 steel resulted in a 12% decrease in fatigue strength compared to gentle grinding. Chemical machining of Ti-6Al-4V resulted in an 18% drop in fatigue strength compared to gentle grinding, while the electrochemical machining and electrical discharge machining of Inconel 718 produced a 35% and a 63% drop, respectively, in endurance limit compared to gentle grinding. Generally, ECM, ELP, and CM produce a stress-free surface. These surfaces exhibit reduced fatigue strength when compared to a gently ground surface because the fatigue enhancement by the compressive stress associated with low-stress grinding is not present.

*Effect of Grinding.* Endurance limits vary with selected changes in grinding parameters (Fig. 24). When abusively grinding, there is a tendency to form patches or streaks of untempered martensite (UTM) or overtempered martensite (OTM) on the surface. It has been found that the presence of either one of these two microconstituents is usually associated with a significant drop in fatigue strength. For example, the presence of a depth of UTM as little as 0.01 mm (0.0005 in.) or as large as 0.09 mm (0.0035 in.) produces a decrease in endurance limit from 760 MPa (110 ksi) down to 480 to 520

**Table 6 Theoretical values for arithmetic roughness average, $R_a$, and maximum peak-to-valley roughness height, $R_y$, for turning or milling with a Type C tool**

Tool nose radius = 0.40 mm (0.031 in.)

| Feed/rev (turning) or feed/tooth (milling) mm (in.) | End cutting edge angle | | | | | | | | | | | | | | | |
|---|---|---|---|---|---|---|---|---|---|---|---|---|---|---|---|---|
| | 3° | | 5° | | 6° | | 10° | | 15° | | 30° | | 40° | | 45° | |
| | $R_a$ | $R_y$ | $R_a$ | $R_y$ | $R_a$ | $R_y$ | $R_a$ | $R_y$ | $R_a$ | $R_y$ | $R_a$ | $R_y$ | $R_a$ | $R_y$ | $R_a$ | $R_y$ |
| | Surface roughness, µm (µin.) | | | | | | | | | | | | | | | |
| 0.020 | 0.03 | 0.13 | 0.03 | 0.13 | 0.03 | 0.13 | 0.03 | 0.13 | 0.03 | 0.13 | 0.03 | 0.13 | 0.03 | 0.13 | 0.03 | 0.13 |
| (0.001) | (1.0) | (4.0) | (1.0) | (4.0) | (1.0) | (4.0) | (1.0) | (4.0) | (1.0) | (4.0) | (1.0) | (4.0) | (1.0) | (4.0) | (1.0) | (4.0) |
| 0.040 | 0.13 | 0.50 | 0.13 | 0.50 | 0.13 | 0.50 | 0.13 | 0.50 | 0.13 | 0.50 | 0.13 | 0.50 | 0.13 | 0.50 | 0.13 | 0.50 |
| (0.002) | (4.1) | (16.0) | (4.1) | (16.0) | (4.1) | (16.0) | (4.1) | (16.0) | (4.1) | (16.0) | (4.1) | (16.0) | (4.1) | (16.0) | (4.1) | (16.0) |
| 0.060 | 0.28 | 1.1 | 0.29 | 1.1 | 0.29 | 1.1 | 0.29 | 1.1 | 0.29 | 1.1 | 0.29 | 1.1 | 0.29 | 1.1 | 0.29 | 1.1 |
| (0.003) | (9.3) | (36.0) | (9.3) | (36.0) | (9.3) | (36.0) | (9.3) | (36.0) | (9.3) | (36.0) | (9.3) | (36.0) | (9.3) | (36.0) | (9.3) | (36.0) |
| 0.080 | 0.46 | 1.7 | 0.51 | 2.0 | 0.51 | 2.0 | 0.51 | 2.0 | 0.51 | 2.0 | 0.51 | 2.0 | 0.51 | 2.0 | 0.51 | 2.0 |
| (0.004) | (16.0) | (63.0) | (16.0) | (64.0) | (16.0) | (64.0) | (16.0) | (64.0) | (16.0) | (64.0) | (16.0) | (64.0) | (16.0) | (64.0) | (16.0) | (64.0) |
| 0.100 | 0.65 | 2.4 | 0.79 | 3.0 | 0.80 | 3.1 | 0.80 | 3.1 | 0.80 | 3.1 | 0.80 | 3.1 | 0.80 | 3.1 | 0.80 | 3.1 |
| (0.005) | (25.0) | (93.0) | (26.0) | (100.0) | (26.0) | (100.0) | (26.0) | (100.0) | (26.0) | (100.0) | (26.0) | (100.0) | (26.0) | (100.0) | (26.0) | (100.0) |
| 0.120 | 0.85 | 3.1 | 1.1 | 4.0 | 1.1 | 4.3 | 1.2 | 4.5 | 1.2 | 4.5 | 1.2 | 4.5 | 1.2 | 4.5 | 1.2 | 4.5 |
| (0.006) | (34.0) | (125.0) | (37.0) | (144.0) | (37.0) | (145.0) | (37.0) | (145.0) | (37.0) | (145.0) | (37.0) | (145.0) | (37.0) | (145.0) | (37.0) | (145.0) |
| 0.140 | 1.1 | 3.9 | 1.4 | 5.1 | 1.5 | 5.5 | 1.6 | 6.2 | 1.6 | 6.2 | 1.6 | 6.2 | 1.6 | 6.2 | 1.6 | 6.2 |
| (0.007) | (43.0) | (159.0) | (50.0) | (191.0) | (50.0) | (196.0) | (51.0) | (197.0) | (51.0) | (197.0) | (51.0) | (197.0) | (51.0) | (197.0) | (51.0) | (197.0) |
| 0.160 | 1.3 | 4.6 | 1.7 | 6.3 | 1.9 | 6.9 | 2.1 | 8.0 | 2.1 | 8.1 | 2.1 | 8.1 | 2.1 | 8.1 | 2.1 | 8.1 |
| (0.008) | (53.0) | (194.0) | (64.0) | (242.0) | (66.0) | (253.0) | (66.0) | (257.0) | (66.0) | (257.0) | (66.0) | (257.0) | (66.0) | (257.0) | (66.0) | (257.0) |
| 0.180 | 1.5 | 5.4 | 2.0 | 7.5 | 2.2 | 8.2 | 2.6 | 9.9 | 2.6 | 10.0 | 2.6 | 10.0 | 2.6 | 10.0 | 2.6 | 10.0 |
| (0.009) | (63.0) | (230.0) | (79.0) | (294.0) | (82.0) | (312.0) | (84.0) | (326.0) | (84.0) | (326.0) | (84.0) | (326.0) | (84.0) | (326.0) | (84.0) | (326.0) |
| 0.200 | 1.7 | 6.2 | 2.4 | 8.7 | 2.6 | 9.6 | 3.2 | 12.0 | 3.2 | 13.0 | 3.2 | 13.0 | 3.2 | 13.0 | 3.2 | 13.0 |
| (0.010) | (73.0) | (268.0) | (94.0) | (349.0) | (100.0) | (373.0) | (103.0) | (403.0) | (103.0) | (403.0) | (103.0) | (403.0) | (103.0) | (403.0) | (103.0) | (403.0) |
| 0.250 | 2.2 | 8.3 | 3.2 | 12.0 | 3.6 | 13.0 | 4.7 | 17.0 | 5.1 | 20.0 | 5.1 | 20.0 | 5.1 | 20.0 | 5.1 | 20.0 |
| (0.012) | (94.0) | (344.0) | (126.0) | (462.0) | (136.0) | (503.0) | (149.0) | (579.0) | (149.0) | (582.0) | (149.0) | (582.0) | (149.0) | (582.0) | (149.0) | (582.0) |
| 0.300 | 2.8 | 10.0 | 4.1 | 15.0 | 4.7 | 17.0 | 6.3 | 23.0 | 7.2 | 27.0 | 7.4 | 29.0 | 7.4 | 29.0 | 7.4 | 29.0 |
| (0.014) | (115.0) | (423.0) | (159.0) | (581.0) | (174.0) | (639.0) | (202.0) | (771.0) | (204.0) | (795.0) | (204.0) | (795.0) | (204.0) | (795.0) | (204.0) | (795.0) |
| 0.350 | 3.4 | 13.0 | 5.0 | 19.0 | 5.7 | 21.0 | 7.9 | 29.0 | 9.5 | 36.0 | 10.0 | 40.0 | 10.0 | 40.0 | 10.0 | 40.0 |
| (0.016) | (136.0) | (503.0) | (192.0) | (704.0) | (213.0) | (781.0) | (259.0) | (975.0) | (267.0) | (1043.0) | (267.0) | (1043.0) | (267.0) | (1043.0) | (267.0) | (1043.0) |
| 0.400 | 3.9 | 15.0 | 5.9 | 22.0 | 6.8 | 25.0 | 9.6 | 35.0 | 12.0 | 44.0 | 14.0 | 54.0 | 14.0 | 54.0 | 14.0 | 54.0 |
| (0.018) | (158.0) | (585.0) | (226.0) | (830.0) | (253.0) | (927.0) | (319.0) | (1189.0) | (339.0) | (1318.0) | (339.0) | (1326.0) | (339.0) | (1326.0) | (339.0) | (1326.0) |
| 0.450 | 4.5 | 17.0 | 6.9 | 26.0 | 7.9 | 29.0 | 11.0 | 42.0 | 14.0 | 53.0 | 17.0 | 69.0 | 17.0 | 69.0 | 17.0 | 69.0 |
| (0.020) | (180.0) | (669.0) | (261.0) | (959.0) | (294.0) | (1076.0) | (381.0) | (1410.0) | (417.0) | (1609.0) | (420.0) | (1646.0) | (420.0) | (1646.0) | (420.0) | (1646.0) |
| 0.500 | 5.1 | 19.0 | 7.8 | 29.0 | 9.0 | 34.0 | 13.0 | 48.0 | 17.0 | 62.0 | 22.0 | 85.0 | 22.0 | 88.0 | 22.0 | 88.0 |
| (0.025) | (236.0) | (881.0) | (350.0) | (1291.0) | (398.0) | (1463.0) | (543.0) | (1992.0) | (636.0) | (2389.0) | (665.0) | (2613.0) | (665.0) | (2613.0) | (665.0) | (2613.0) |
| 0.600 | 6.3 | 24.0 | 9.7 | 36.0 | 11.0 | 42.0 | 17.0 | 62.0 | 22.0 | 82.0 | 31.0 | 120.0 | 33.0 | 132.0 | 33.0 | 135.0 |
| (0.030) | (293.0) | (1099.0) | (441.0) | (1634.0) | (506.0) | (1865.0) | (711.0) | (2607.0) | (873.0) | (3232.0) | (972.0) | (3842.0) | (972.0) | (3842.0) | (972.0) | (3842.0) |
| 0.700 | 7.5 | 28.0 | 12.0 | 44.0 | 14.0 | 51.0 | 21.0 | 76.0 | 28.0 | 102.0 | 42.0 | 158.0 | 46.0 | 183.0 | 48.0 | 192.0 |
| (0.035) | (351.0) | (1321.0) | (534.0) | (1985.0) | (615.0) | (2277.0) | (885.0) | (3247.0) | (1120.0) | (4124.0) | (1348.0) | (5320.0) | (1350.0) | (5370.0) | (1350.0) | (5370.0) |
| (0.040) | (409.0) | (1545.0) | (628.0) | (2343.0) | (726.0) | (2698.0) | (1063.0) | (3906.0) | (1377.0) | (5055.0) | (1783.0) | (6940.0) | (1808.0) | (7253.0) | (1808.0) | (7253.0) |
| (0.045) | (468.0) | (1772.0) | (723.0) | (2706.0) | (839.0) | (3127.0) | (1244.0) | (4583.0) | (1640.0) | (6020.0) | (2263.0) | (8684.0) | (2355.0) | (9456.0) | (2360.0) | (9581.0) |
| (0.050) | (527.0) | (2001.0) | (819.0) | (3074.0) | (953.0) | (3562.0) | (1429.0) | (5274.0) | (1910.0) | (7015.0) | (2778.0) | (10542.0) | (2983.0) | (11874.0) | (3018.0) | (12273.0) |
| (0.060) | (646.0) | (2465.0) | (1013.0) | (3821.0) | (1185.0) | (4447.0) | (1805.0) | (6693.0) | (2466.0) | (9080.0) | (3889.0) | (14572.0) | (4447.0) | (17356.0) | (4629.0) | (18554.0) |

Source: Ref 5

MPa (70 or 75 ksi) (Fig. 24). Typically, residual tensile stresses are involved in this condition.

Retempering of the workpiece does not correct the problem. Although it tempers the UTM and reduces its brittleness, it does not restore the softening of the OTM to its prior condition. In addition, the tempering cycle does not reduce the residual tensile stresses formed in the abusive grinding operation.

The effect on fatigue strength from EDM and grinding of aged Inconel 718 is illustrated in Fig. 25. After gentle grinding, the Inconel 718 had an endurance limit of 410 MPa (60 ksi). With either gentle or abusive EDM, the endurance limit dropped to 150 MPa (22 ksi).

The effect of ECM on the fatigue properties of Ti-6Al-4V is shown in Fig. 26. The endurance limit after low-stress grinding was 460 MPa (67 ksi). The tests were conducted on flat specimens that were longitudinally surface ground to a surface finish of 0.35 µm (14 µin.). Specimens that were electrochemically machined by a frontal or pocketing operation had an endurance limit of 410 MPa (60 ksi) for the same surface finish. When the titanium specimens were electrochemically machined by a trepanning operation, the endurance limit was reduced to 280 MPa (40 ksi), a 40% drop with respect to the low-stress grinding conditions.

**Improving or Preserving Mechanical Properties With Shot Peening.** Shot peening provides the most practical technique for enhancing the mechanical properties affected by surface alterations. It is frequently used in industry to improve the fatigue strength and stress-corrosion properties of most structural alloys subjected to harsh environments and high stress. Shot peening not only enhances the properties of gently machined materials but also improves the properties of metals that have been processed by machining techniques that tend to abuse or degrade fatigue strength and other mechanical properties. Shot peening parameters should be selected to meet surface finish specifications.

Fatigue properties can also be improved by other cold-work finishing processes, such as the burnishing of holes and the roll burnishing of fillets. The advantage of shot peening is that it can be applied to large surfaces and to contoured or irregular surfaces. With shot peening, the magnitude and depth of the compressive stress layer can be regulated by controlling the type and size of the shot, the intensity, and the coverage. However, the compressive stress produced by shot peening can be reduced in service by prolonged exposure to elevated temperature.

*Improving Fatigue Strength.* Shot peening can improve the fatigue strength of machined metals (Table 11). On 4340 steel at 50 HRC, for example, shot peening improved the fatigue strength of a gently ground surface by 10% (Table 11). Shot peening after roughing EDM increased the fatigue strength to 130% of the gently ground level. Similar improvements in fatigue strength by shot peening occur on surfaces produced by ECM and ELP (Table 11).

*Retarding Stress-Corrosion Cracking.* It has been shown that the compressive stresses introduced by shot peening retard stress-corrosion cracking (Ref 14). For example, shot peening increased the stress-corrosion life of 4340 steel specimens that had been prestressed to 75% of the yield strength and exposed to 3.5% sodium chloride solution at room temperature. Shot peening increased the life of gently ground specimens from 400 to 850 h (Fig. 27). On abusively ground specimens, shot peening increased the life from 200 to over 1000 h, while on an electropolished specimen, the life was increased from 300 to 400 h by shot peening.

**Abrasive tumbling** is also an effective technique for improving fatigue and surface properties. Abrasive tumbling can be used to reverse unfavorable tensile stress by inducing a compressional stress in the surface layer.

## Measuring Surface Integrity Effects

In a systematic study of surface integrity, there are two recommended sets of data: a minimum data set and a standard data set.

**The minimum data set** is used for initial screening and consists of measurements of surface texture, macrostructure, microstructure, and microhardness alterations. The minimum recommended data set consists of the following (Ref 12):

- Surface roughness measurement according to Ref 1 or a microtopographic map
- Lay designation or photograph
- Macroscopic examination at a magnification of 10× or less
- Macroetch indications from die penetrants or magnetic flux
- Chemical etches
- Microscopic examination of a cross section of the altered surface (cross section examination at 1000× preferred)
- Indications of heat-affected zones and microhardness alterations
- (Optional) A scanning electron microscope can occasionally supplement surface texture measurements with a series of photographs at increasing magnifications (preferably 20, 200, 1000, and 2000×)

**The standard data set** in measuring surface integrity consists of the above minimum data set plus the following additional tests (Ref 12):

- Residual stress profile or distortion measurements
- Fatigue tests (screening only)

Residual stresses in surfaces can be determined with x-ray diffraction techniques. A residual stress profile requires a determination of stress not only at the surface but also into the surface. This can be done by taking multiple x-ray readings after successive layers are removed by etching. Full-reverse bending at room temperature using tapered flat specimens taken to $10^7$ cycles is recommended for fatigue testing.

**Macroscopic examination** involves some type of visual inspection for detecting visible cracks and other surface defects. Die penetrants are generally applied to nonmagnetic materials, while magnetic particle inspection can detect small cracks in magnetic materials. Parts manufactured from martensitic high-strength steels can be visually inspected after a macroetch for evidence of untempered or overtempered martensite resulting from grinding or machining. A dilute solution of nitric acid is usually used to etch the damaged areas so that they can be visually detected. Typically, untempered martensite appears white, and the overtempered areas appear darker than the back-

**Table 7 Summary of possible surface alterations resulting from various material removal processes**

| | | | |
|---|---|---|---|
| R | Roughness of surface | UTM | Untempered martensite |
| PD | Plastic deformation and plastically deformed debris | OTM | Overtempered martensite |
| L & T | Laps and tears and crevicelike defects | OA | Overaging |
| MCK | Microcracks | RS | Resolution or austenite reversion |
| SE | Selective etch | RC | Recast, respattered metal, or vapor-deposited metal |
| IGA | Intergranular attack | HAZ | Heat-affected zone |

| | Process | | | | |
|---|---|---|---|---|---|
| | Conventional | | Nontraditional | | |
| Material | Milling, drilling, or turning | Grinding | Electrical discharge machining | Electrochemical machining | Chemical machining |
| Nonhardenable 1018 steel | R | R | R | R | R |
| | PD | PD | MCK | SE | SE |
| | L & T | | RC | IGA | IGA |
| Hardenable 4340 and D6ac steels | R | R | R | R | R |
| | PD | PD | MCK | SE | SE |
| | L & T | MCK | RC | IGA | IGA |
| | MCK | UTM | UTM | | |
| | UTM | OTM | OTM | OTM | |
| | OTM | | | | |
| D2 tool steel | R | R | R | R | R |
| | PD | PD | MCK | SE | SE |
| | L & T | MCK | RC | IGA | IGA |
| | MCK | UTM | UTM | | |
| | UTM | OTM | OTM | | |
| | OTM | | | | |
| Type 410 stainless steel (martensitic) | R | R | R | R | R |
| | PD | PD | MCK | SE | SE |
| | L & T | MCK | RC | IGA | IGA |
| | MCK | UTM | UTM | | |
| | UTM | OTM | OTM | | |
| | OTM | | | | |
| Type 302 stainless steel (austenitic) | R | R | R | R | R |
| | PD | PD | MCK | SE | SE |
| | L & T | | RC | IGA | IGA |
| 17-4 PH steel | R | R | R | R | R |
| | PD | PD | MCK | SE | SE |
| | L & T | OA | RC | IGA | IGA |
| | OA | | OA | | |
| 250 grade maraging (18% Ni) steel | R | R | R | R | R |
| | PD | PD | RC | SE | SE |
| | L & T | RS | RS | IGA | IGA |
| | RS | OA | OA | | |
| | OA | | | | |
| Nickel and cobalt-base alloys | HAZ | HAZ | | | |
| Inconel alloy 718 | R | R | R | R | R |
| René 41 | PD | PD | MCK | SE | SE |
| HS 31 | L & T | MCK | RC | IGA | IGA |
| IN-100 | MCK | | | | |
| Ti-6Al-4V | HAZ | HAZ | | | |
| | R | R | R | R | R |
| | PD | PD | MCK | SE | SE |
| | L & T | MCK | RC | IGA | |
| Refractory alloy TZM | R | R | R | R | R |
| | L & T | MCK | MCK | SE | SE |
| | MCK | | | IGA | |
| Tungsten (pressed and sintered) | R | R | R | R | R |
| | L & T | MCK | MCK | SE | SE |
| | MCK | | | MCK | MCK |
| | | | | IGA | IGA |

Source: Ref 7

## Table 8 Comparison of depth of surface integrity effects observed in material removal processes

| Property and type of effect | Condition | Maximum observed depth of effect(a): Turning or milling, mm | Turning or milling, in. | Drilling, mm | Drilling, in. | Grinding, mm | Grinding, in. | Chemical machining, mm | Chemical machining, in. | Electrochemical machining, mm | Electrochemical machining, in. | Electrochemical grinding, mm | Electrochemical grinding, in. | Electrical discharge machining, mm | Electrical discharge machining, in. | Laser beam machining, mm | Laser beam machining, in. |
|---|---|---|---|---|---|---|---|---|---|---|---|---|---|---|---|---|---|
| **Mechanically altered material zones** | | | | | | | | | | | | | | | | | |
| Plastic deformation | Finishing(b) | 0.043 | 0.0017 | 0.020 | 0.0008 | 0.008 | 0.0003 | (c) | (c) | (c) | (c) | (c) | (c) | (c) | (c) | (c) | (c) |
| | Roughing(d) | 0.076 | 0.0030 | 0.119 | 0.0047 | 0.089 | 0.0035 | (c) | (c) | (c) | (c) | (c) | (c) | (c) | (c) | (c) | (c) |
| Plastically deformed debris | Finishing | (c) | (c) | (c) | (c) | 0.013 | 0.0005 | (c) | (c) | (c) | (c) | (c) | (c) | (c) | (c) | (c) | (c) |
| | Roughing | (c) | (c) | (c) | (c) | 0.033 | 0.0013 | (c) | (c) | (c) | (c) | (c) | (c) | (c) | (c) | (c) | (c) |
| Hardness alteration(e) | Finishing | 0.013 | 0.0005 | 0.025 | 0.0010 | 0.038 | 0.0015 | 0.025 | 0.0010 | 0.036 | 0.0014 | 0.018 | 0.0007 | 0.025 | 0.0010 | (c) | (c) |
| | Roughing | 0.127 | 0.0050 | 0.508 | 0.0200 | 0.254 | 0.0100 | 0.079 | 0.0031 | 0.051 | 0.0020 | 0.038 | 0.0015 | 0.203 | 0.0080 | (c) | (c) |
| Microcracks or macrocracks | Finishing | 0.013 | 0.0005 | 0.013 | 0.0005 | 0.013 | 0.0005 | (c) | (c) | 0.008 | 0.0003 | 0.000 | 0.0000 | 0.013 | 0.0005 | 0.015 | 0.0006 |
| | Roughing | 0.038 | 0.0015 | 0.038 | 0.0015 | 0.229 | 0.0090 | (c) | (c) | 0.038 | 0.0015 | 0.025 | 0.0010 | 0.178 | 0.0070 | 0.102 | 0.0040 |
| Residual stress(f) | Finishing | 0.152 | 0.0060 | (c) | (c) | 0.013 | 0.0005 | 0.025 | 0.0010 | 0.000 | 0.0000 | 0.000 | 0.0000 | 0.051 | 0.0020 | 0.005 | 0.0002 |
| | Roughing | 0.356 | 0.0140 | (c) | (c) | 0.318 | 0.0125 | 0.025 | 0.0010 | 0.000 | 0.0000 | 0.000 | 0.0000 | 0.076 | 0.0030 | (c) | (c) |
| **Metallurgically altered material zones** | | | | | | | | | | | | | | | | | |
| Recrystallization | Finishing | (c) | (c) | (c) | (c) | 0.013 | 0.0005 | (c) | (c) | (c) | (c) | (c) | (c) | (c) | (c) | (c) | (c) |
| | Roughing | (c) | (c) | (c) | (c) | (c) | (c) | (c) | (c) | (c) | (c) | (c) | (c) | (c) | (c) | (c) | (c) |
| Intergranular attack | Finishing | (c) | (c) | (c) | (c) | (c) | (c) | 0.008 | 0.0003 | 0.008 | 0.0003 | 0.000 | 0.000 | (c) | (c) | (c) | (c) |
| | Roughing | (c) | (c) | (c) | (c) | (c) | (c) | 0.152 | 0.0060 | 0.038 | 0.0015 | (c) | (c) | (c) | (c) | (c) | (c) |
| Selective etch, pits, protuberances | Finishing | 0.010 | 0.0004 | (c) | (c) | 0.005 | 0.0002 | 0.015 | 0.0006 | 0.010 | 0.0004 | 0.003 | 0.0001 | 0.013 | 0.0005 | (c) | (c) |
| | Roughing | 0.025 | 0.0010 | 0.076 | 0.3000 | 0.010 | 0.0004 | 0.038 | 0.0015 | 0.064 | 0.0025 | 0.013 | 0.0005 | 0.041 | 0.0016 | (c) | (c) |
| Metallurgical transformations | Finishing | 0.010 | 0.0004 | 0.038 | 0.0015 | 0.013 | 0.0005 | (c) | (c) | 0.000 | 0.0000 | 0.003 | 0.0001 | 0.015 | 0.0006 | 0.015 | 0.0006 |
| | Roughing | 0.076 | 0.0030 | 0.508 | 0.0200 | 0.152 | 0.0060 | (c) | (c) | 0.005 | 0.0002 | 0.008 | 0.0003 | 0.127 | 0.0050 | 0.038 | 0.0015 |
| Heat-affected zone or recast layers | Finishing | 0.003 | 0.0001 | (c) | (c) | 0.018 | 0.0007 | (c) | (c) | (c) | (c) | (c) | (c) | 0.015 | 0.0006 | 0.015 | 0.0006 |
| | Roughing | 0.025 | 0.0010 | 0.076 | 0.0030 | 0.318 | 0.0125 | (c) | (c) | (c) | (c) | (c) | (c) | 0.127 | 0.0050 | 0.038 | 0.0015 |

(a) Normal to the surface. (b) Finishing, gentle, or low-stress conditions. (c) No occurrences or not expected. (d) Roughing, off-standard, or abusive conditions. (e) Depth to point at which hardness becomes less than ±2 points HRC (or equivalent) of bulk material hardness (hardness converted from Knoop microhardness measurements). (f) Depth to point at which residual stress becomes and remains less than 140 MPa (20 ksi) or 10% of tensile strength, whichever is greater. Source: Ref 8

ground material. A specific etching technique for detecting grinding damage in hardened steel is given in Table 12.

**Microscopic examination** of a cross section of an altered surface is used to detect the following microstructural alterations:

- Microcracks
- Plastic deformation (a cross section parallel to the lay is suggested)
- Phase transformations
- Intergranular attack
- Micro defects such as laps and inclusions
- Built-up edges or deposits of debris
- Recast layers
- Selective etching
- Metallurgical transformations
- Heat-affected zones

These surface alterations can be detected by magnifying a cross section of an altered surface. A metallurgical mount holds the test specimen, and a cross section of the altered surface is ground and polished. The edge of the altered surface is then examined under a magnifying instrument with a magnification factor of 500 to 1000×.

When polishing the cross section of an altered surface, special techniques for mounting and polishing are used to reduce edge rounding. Edge retention is important because surface alterations are only about 0.025 to 0.075 mm (0.001 to 0.003 in.) deep and because depth of field decreases as magnification factors increase. One technique is to vacuum cast the metallurgical mount and to use special procedures for polishing. Polishing units with a horizontal specimen holder, which is rotated with a preselected force against various grinding and polishing surfaces, have also been successfully used for edge retention. An additional mounting and preparation technique for edge retention is as follows (Ref 7):

- Samples are sectioned from the workpiece in a manner that leads to the least possible distortion or burring. Band sawing or hacksawing is preferred. A minimum of 0.75 mm (0.030 in.) is then removed from the cut surface using a 120-grit silicon carbide paper on a low-speed polisher
- Copper molds (or tubes), 31.8 mm (1¼ in.) in inside diameter by 70 mm (2¾ in.) high, are placed on a pallet approximately

## Table 9 Low-stress grinding procedures

| Grinding parameters | Steels and nickel-base high-temperature alloys(a) | Titanium |
|---|---|---|
| **Surface grinding** | | |
| Wheel | A46HV | C60HV |
| Wheel speed, m/s (sfm)(b) | 13–15 (2500–3000) | 10–15 (2000–3000) |
| Downfeed per pass, mm (in.) | 0.005–0.013 (0.0002–0.0005) | 0.005–0.013 (0.0002–0.0005) |
| Table speed, m/min (sfm)(c) | 12–30 (40–100) | 12–30 (40–100) |
| Crossfeed per pass, mm (in.) | 1–1.25 (0.040–0.050) | 1–1.25 (0.040–0.050) |
| Grinding fluid | Highly sulfurized oil | (d) |
| **Traverse cylindrical grinding** | | |
| Wheel | A60IV | C60HV |
| Wheel speed, m/s (sfm)(b) | 13–15 (2500–3000) | 10–15 (2000–3000) |
| Infeed per pass, mm (in.) | 0.005–0.013 (0.0002–0.0005) | 0.005–0.013 (0.0002–0.0005) |
| Work speed, m/min (sfm)(c) | 20–30 (70–100) | 20–30 (70–100) |
| Grinding fluid | Highly sulfurized oil | (d) |

(a) For a wide variety of metals (including high-strength steels, high-temperature alloys, titanium, and refractory alloys), low-stress grinding practices develop very low residual tensile stresses. In some materials, the residual stress produced near the surface is actually in compression instead of tension. (b) Low-stress grinding requires wheel speeds lower than the conventional 30 m/s (6000 sfm). To apply low-stress grinding, it would be preferable to have a variable-speed grinder. Because most grinding machines do not have wheel-speed control, it is necessary to add a variable-speed drive or to make pulley modifications. (c) Increased work speeds even above those indicated are considered to be advantageous for improving surface integrity. (d) Cutting fluids should be nitrate free for health reasons. Some manufacturers also prohibit cutting fluids with chlorine when machining titanium. Source: Ref 10

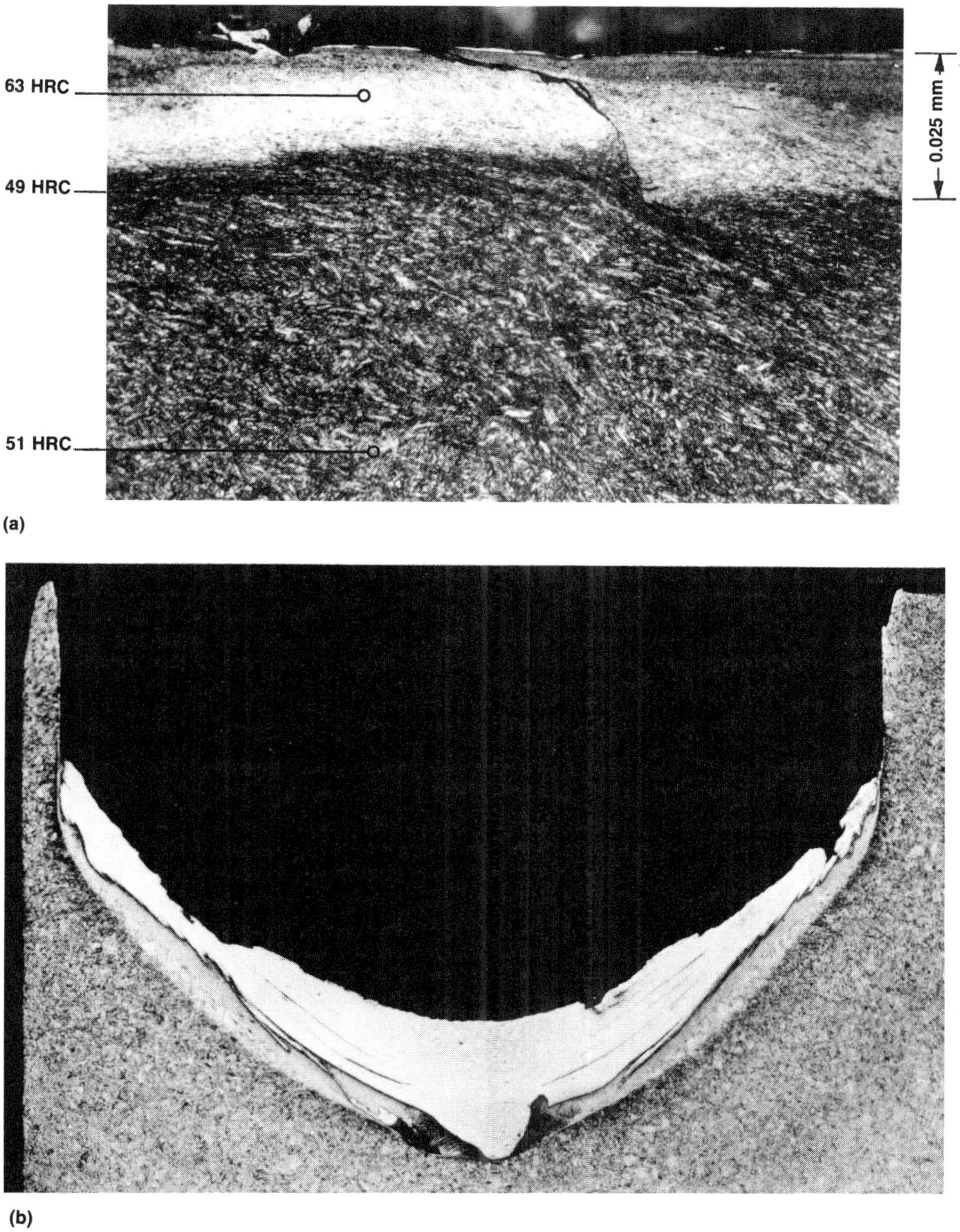


**Fig. 10** Surface alterations produced from drilling with dull tools. (c) Section perpendicular to the drill-hole axis in a 4340 steel (48 HRC). Abusive drilling produced a cracked untempered martensite surface alteration. Also note the softer overtempered zone below the untempered martensite layer. (b) Cross section through a hole drilled in type 410 stainless with a dull drill (1.5 mm, or 0.060 in., wearland). The drill broke down at the corner during the test and friction welded a portion of the high-speed steel drill bit to the workpiece. The base metal exhibits a rehardened and subsequently overtempered zone as a result of the high localized heating. ~20×. Source: Ref 9

125 mm (5 in.) in diameter. The inner surface of the molds and the surface of the pallet are previously sprayed with a silicone releasing agent

- After placing a metallurgical specimen in a mold, a mixture of epoxy resin, hardener, and pelletized aluminum oxide sufficient to produce a layer 6.4 to 9.5 mm (¼ to ⅜ in.) is poured over the specimen. The ratio of resin to hardener is 4 to 1. The amount of pellets added is in the range of 10 to 15 g (0.35 to 0.5 oz). The hardness or abrasive level of the pelletized material used (low, medium, or high fired) is strictly a function of the alloy to be prepared and its hardness characteristics
- The pallet containing the molds is placed in a vacuum chamber at 10 to 100 Pa (1 × $10^{-2}$ to 1 × $10^{-3}$ torr) to degas the mixture, thus improving the adherence of the epoxy and pellets to the surface of the specimen. When vigorous bubbling of the mixture decreases after vacuum impregnation, sufficient resin and hardener (4 to 1 ratio) is added to produce a mount approximately 25 mm (1 in.) high
- The mounts are cured at a temperature not greater than 20 °C (70 °F) for approximately 10 h. Casting of the mounts is accomplished during the latter portion of the laboratory workday so that curing occurs overnight
- After curing, the mounts are placed in an oven at a temperature of 65 °C (150 °F) for 1 h, after which they are removed from the molds
- Approximately 0.50 mm (0.020 in.) of stock is then removed from the as-mounted metal surface on a positive-positioning automatic polishing unit, using the side of a 25 × 330 mm (1 × 13 in.) aluminum oxide 320-grit grinding wheel as the grinding medium. Water is used as a coolant
- Subsequent rough grinding is performed wet on silicon carbide papers or equivalent ranging from 240 to 600 grit
- For steels and nickel- and cobalt-base superalloys, the intermediate polish is performed on an automatic polisher using a polishing cloth with a soft nap texture and 6 μm diamond paste. The final polish is achieved using deep nap or pile cloth similar to billiard cloth with a suspension of 0.1 μm or finer aluminum oxide in water. Titanium and refractory alloys require an etch-polish cycle (using a slurry of hydrogen peroxide, water, and 0.1 μm or finer aluminum oxide), which is accomplished between a diamond polish and a final polish procedure. The final polish for titanium and refractory alloys is accomplished on a vibratory polisher using a deep pile cloth with a suspension of 0.1 μm or finer aluminum oxide in water
- Samples are etched by swabbing. Examples of some typical etchants used are given below:

| Material | Etchant |
|---|---|
| Steels | 2% $HNO_3$ and 98% denatured anhydrous alcohol |
| Nickel-base alloys | 100 mL HCl, 5 g $CuCl_2 \cdot 2H_2O$, and 100 mL denatured anhydrous alcohol |
| Titanium alloys | 2% HF and 98% $H_2O$ or 2% HF, 3% $HNO_3$, and 95% $H_2O$ |

## Guidelines for Material Removal, Postprocessing, and Inspection

The following guidelines are meant to serve only as general or starting recommendations for the machining of structural components. More detailed information on guidelines for surface integrity is given in Ref 16.

Data and experience indicate that these practices will lead to increased surface integrity in some important applications. However, the current state of knowledge of surface alterations is such that general rec-

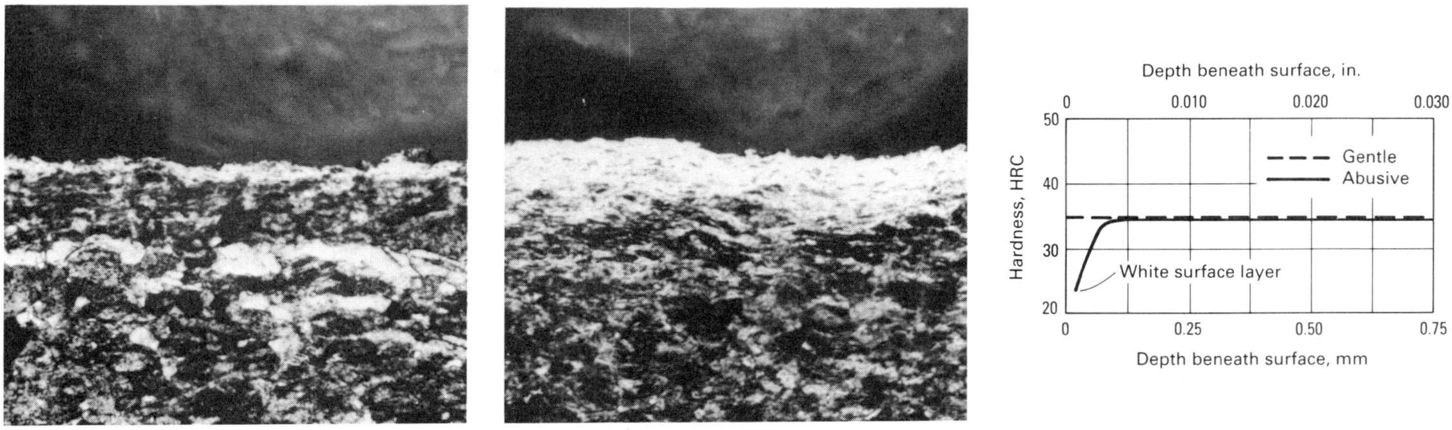


**Fig. 11** Surfaces produced by the face milling of Ti-6Al-4V (aged, 35 HRC). (a) With gentle machining conditions, a slight white layer is visible, but changes in microhardness are undetected. 1000×. (b) With abusive conditions, an overheated white layer about 0.01 mm (0.0004 in.) deep and a plastically deformed layer totalling 0.04 mm (0.0015 in.) deep are visible. 1000×. (c) Microhardness measurements show a total affected zone 0.08 mm (0.003 in.) deep from abusive conditions. Source: Ref 9

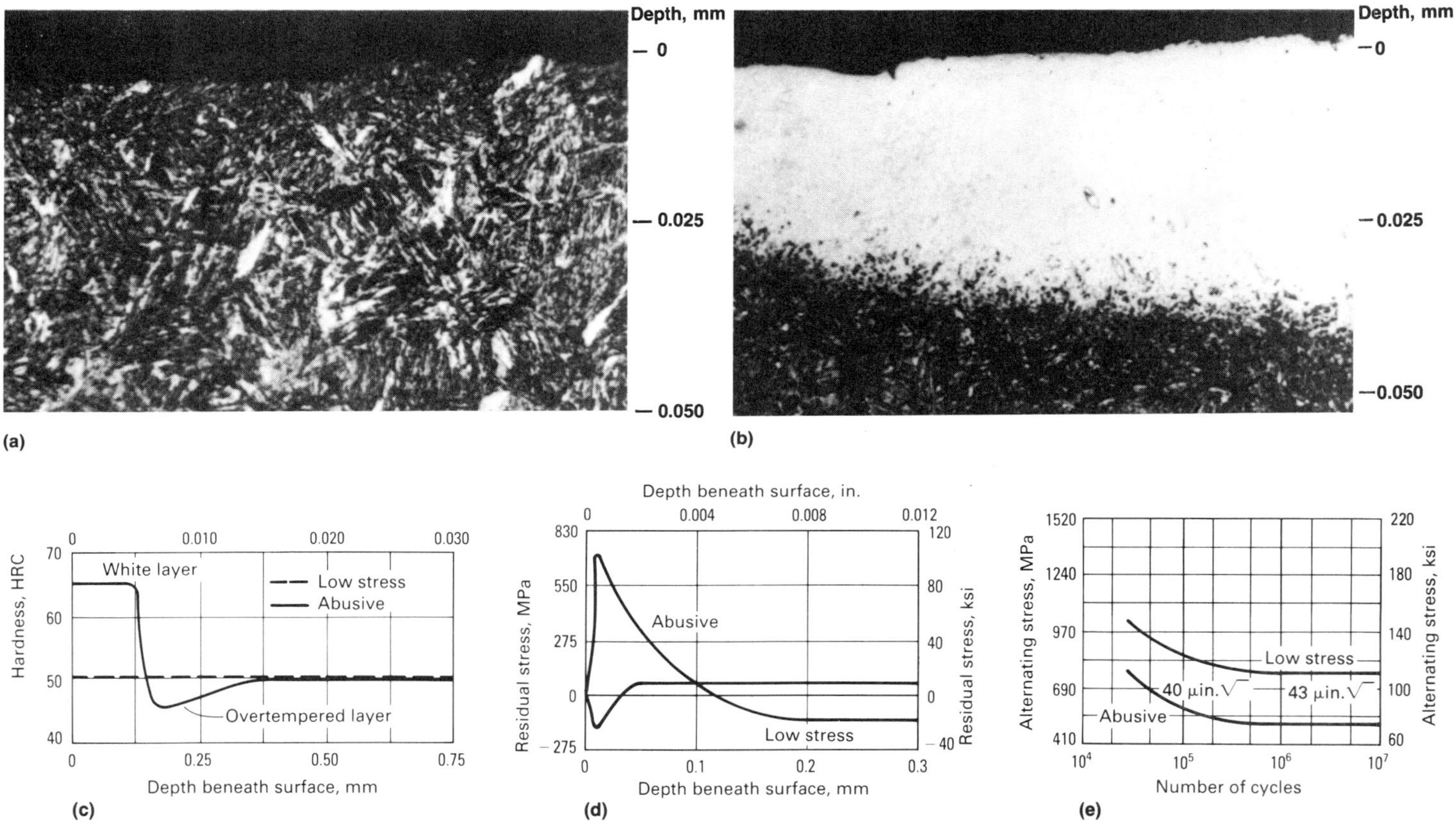


**Fig. 12** Surface characteristics produced by the low-stress and abusive grinding of AISI 4340 steel. (a) Low-stress grinding produced no visible surface alterations. (b) The white layer shown from abusive conditions has a hardness of 65 HRC and is approximately 0.025 to 0.050 mm (0.001 to 0.002 in.) deep. (c) Plot of microhardness alterations showing a total heat-affected zone of 0.33 mm (0.013 in.) from abusive conditions. (d) Plot of residual stress. (e) Effect on fatigue strength. Source: Ref 11

ommendations are not always applicable for all specific surface integrity situations. For highly critical parts, it is mandatory to make individual, specific evaluations. It is also important to note that these guidelines were formulated based on experience with structural surfaces as opposed to mating or bearing surfaces. However, many of these guidelines are also applicable to the mating surfaces of bearings, cams, gears, and other similar parts.

Surface integrity controls often result in increased manufacturing costs and decreased production rates. Therefore, surface integrity practices should not be implemented unless the need exists. Process parameters that provide surface integrity should be applied selectively to critical parts or to critical areas of given parts to help minimize cost increases.

These surface integrity guidelines are primarily intended for application to metal removal processes used for final surface generation rather than roughing cuts. It is important, however, to know the type and depth of surface alterations produced during roughing so that adequate provisions

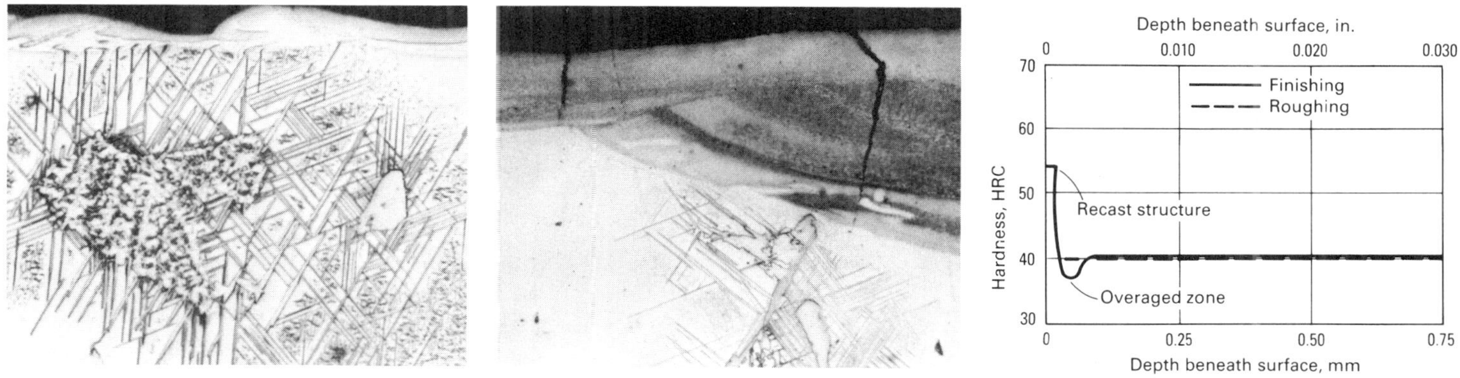


(a) (b) (c)

**Fig. 13** Surface characteristics of cast Inconel 718 (aged, 40 HRC) produced by EDM. (a) Finishing conditions produced a variable recast layer 0.005 mm (0.0002 in.) thick. 860×. (b) Roughing conditions produced an extensively cracked variable recast layer up to 0.05 mm (0.002 in.). The random acicular structure is not related to the surface phenomena. 660×. (c) The recast structure has a hardness of about 53 HRC, and slight overaging due to localized surface overheating was also noted. Source: Ref 9

(a) (b) (c)

**Fig. 14** Surface characteristics of AISI (quenched and tempered, 50 HRC) produced by EDM under finishing and roughing conditions. (a) Finishing conditions produce discontinuous patches of recast metal plus a thin layer (0.0025 mm, or 0.0001 in.) of rehardened martensite. 620×. (b) Roughing conditions produce globs of recast metal and a white layer of rehardened martensite 0.075 mm (0.003 in.) deep. 620×. (c) Microhardness measurements show a total heat-affected zone approaching 0.25 mm (0.010 in.). Globs of recast and the white layer are at 62 HRC. An overtempered zone as soft as 46 HRC is found beneath the surface. Source: Ref 9

**Table 10 Effect of method of machining on fatigue strength**

| Alloy | Machining operation | Endurance limit in bending, $10^7$ cycles MPa | ksi | Gentle grinding, % |
|---|---|---|---|---|
| 4340 steel, 50 HRC | Gentle grinding | 703 | 102 | 100 |
| | Electropolishing | 620 | 90 | 88 |
| | Abusive grinding | 430 | 62 | 61 |
| Ti-6Al-4V, 32 HRC | Gentle milling | 480 | 70 | 113 |
| | Gentle grinding | 430 | 62 | 100 |
| | Chemical milling | 350 | 51 | 82 |
| | Abusive milling | 220 | 32 | 52 |
| | Abusive grinding | 90 | 13 | 21 |
| Inconel 718, aged, 44 HRC | Gentle grinding | 410 | 60 | 100 |
| | Electrochemical machining | 270 | 39 | 65 |
| | Conventional grinding | 165 | 24 | 40 |
| | Electrical discharge machining | 150 | 22 | 37 |

Source: Ref 9

can be made for establishing surface integrity during finishing operations by removing damaged surface layers.

**Abrasive Processes.** Grinding distortion and surface damage can be reduced by using low-stress grinding conditions. Low-stress grinding conditions (Table 9) differ from conventional practices by employing softer-grade grinding wheels, reduced grinding wheel speed, reduced infeed rates, chemically active cutting fluids, and coarse wheel dressing procedures. Low-stress grinding and other guidelines should be used as follows:

- Low-stress grinding should be used for removing the last 0.25 mm (0.010 in.) of material
- If low-stress grinding is required for finish grinding, then conventional grinding can be used to within 0.25 mm (0.010 in.) of finish size if the materials being ground are not sensitive to cracking
- Alloys for high-stress applications, such as titanium and high-temperature nickel and cobalt alloys, should be finished with low-stress grinding instead of conventional grinding
- Frequent dressing of grinding wheels can reduce surface damage by keeping the wheels open and sharp, thus helping to reduce temperatures at the wheel/workpiece interface
- Cutting fluids, properly applied, help promote surface integrity. As a general rule, at least 10 L/kW (2 gal./hp) per minute is needed
- Hand wheel grinding of sensitive alloys should be discouraged
- When abrasive cutoff is used, steps should be taken to determine the extent of the disturbed layer and to make proper stock allowances for subsequent cleanup by suitable machining

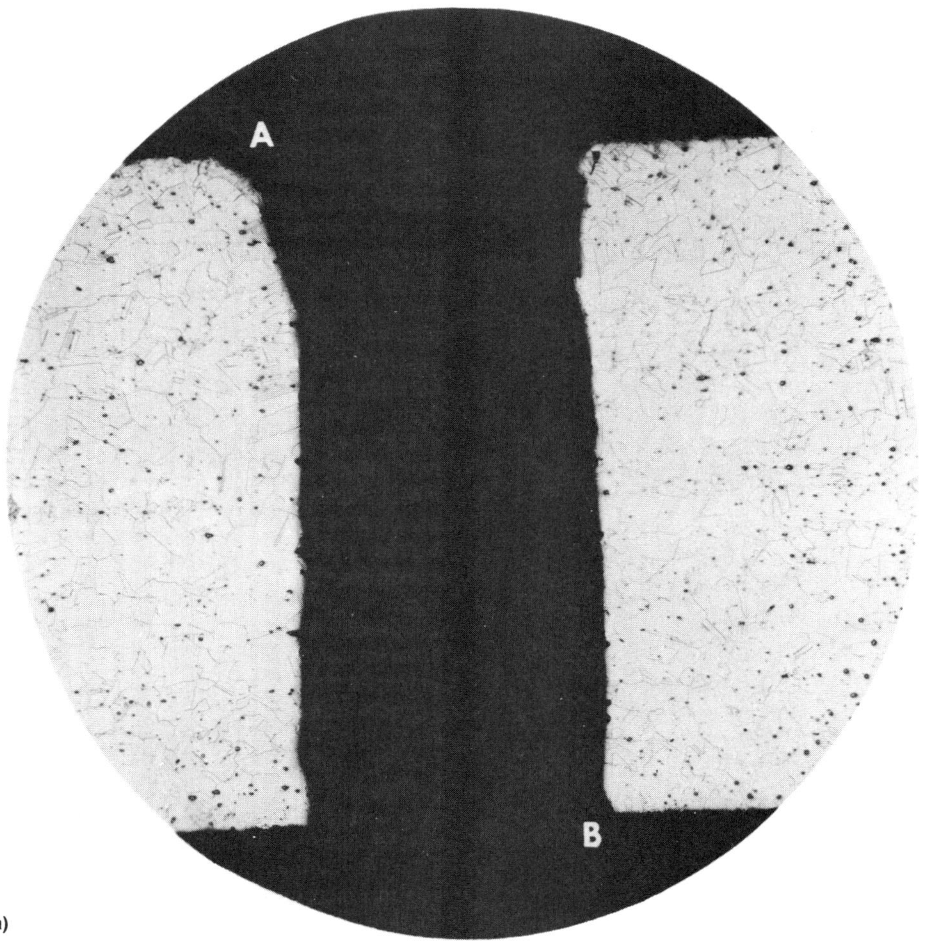


(a)

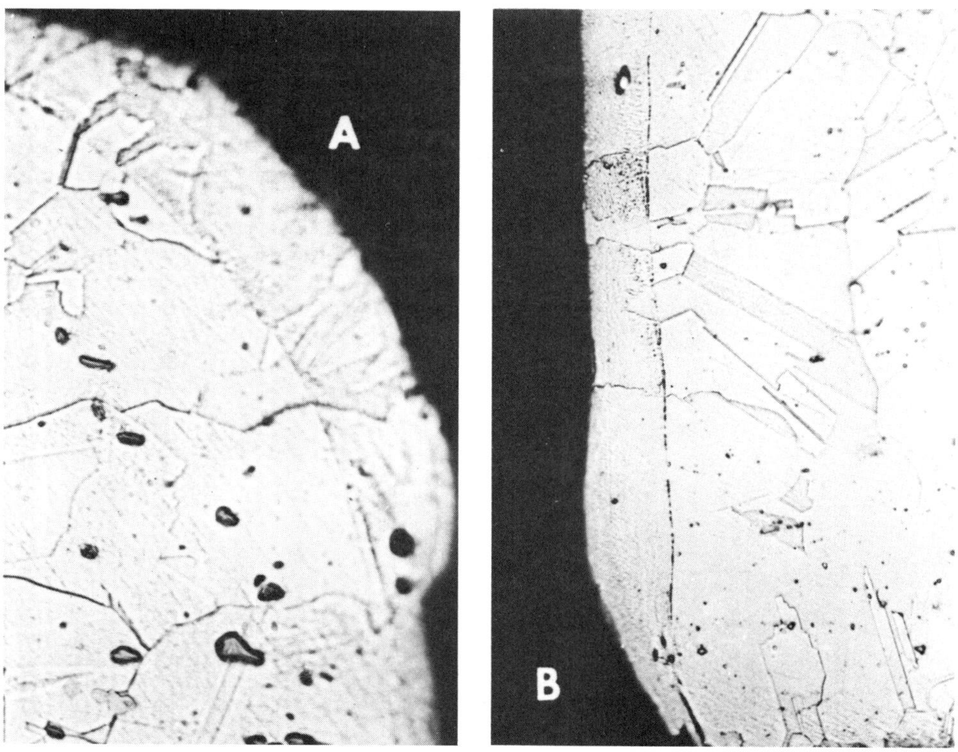


(b)

**Fig. 15** Surfaces from the laser beam drilling of Inconel 718 shown at magnifications of 185× (a) and 1160× (b). Note the grain structure in the heat-affected zones of the entrance (A) and the exit (B). Source: Ref 9

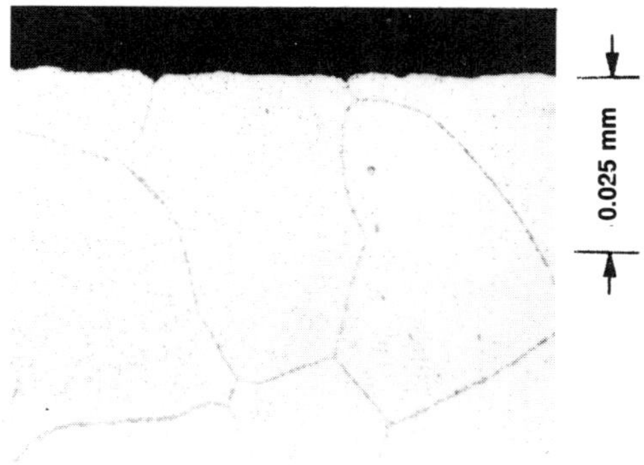


(a)

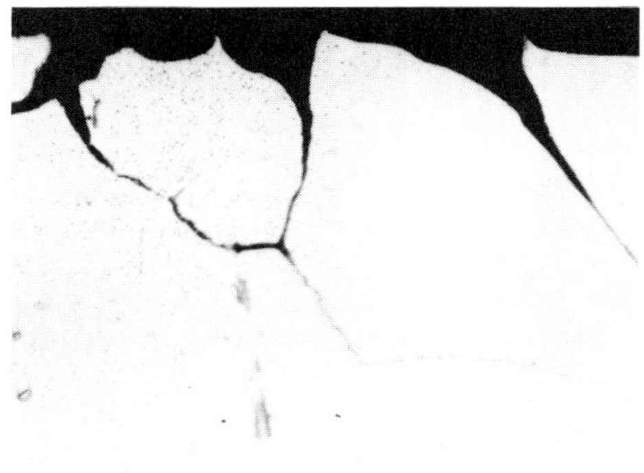

(b)

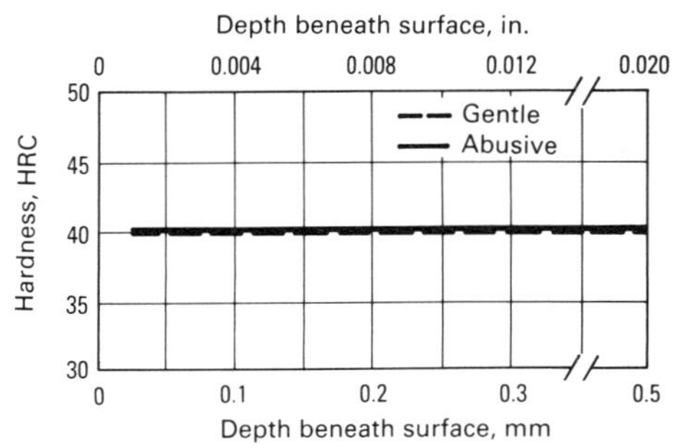


(c)

**Fig. 16** Surface characteristics of Waspaloy (aged, 40 HRC) produced by ECM. (a) Gentle conditions produce a slight roughening of the surface and some intergranular attack. (b) Abusive conditions produce severe intergranular attack. (c) Microhardness is unaffected by the abusive conditions. Source: Ref 9

- Controls for hand power sanders should be maintained
- The relatively new high-speed grinding processes should not be used for finishing highly stressed structural parts unless a standard data set is developed
- Abrasive processing and especially finish grinding must be accomplished under strict process control when employed for the manufacture of aerospace components

**Chip Removal Operations.** For turning and milling, there are at least two very important steps that will improve surface integrity. First, machining conditions should be selected that will provide long

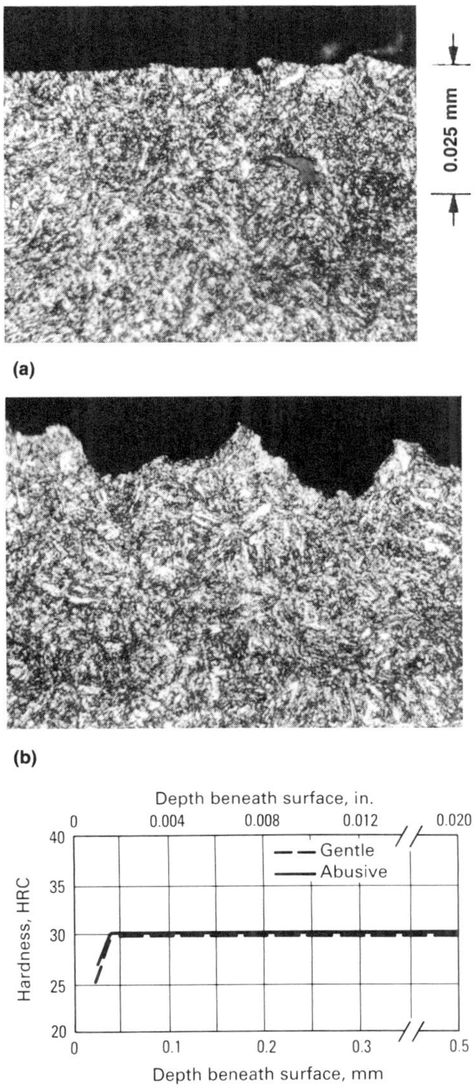


**Fig. 17** Surface characteristics of 4340 steel (quenched and tempered, 30 HRC) produced by ECM. (a) Gentle conditions produce slight surface pitting but no other visible changes. (b) Abusive conditions produce surface roughening but no other visible effect on microstructure. (c) Gentle and abusive conditions both produce a slight hardness loss at the surface. Source: Ref 12

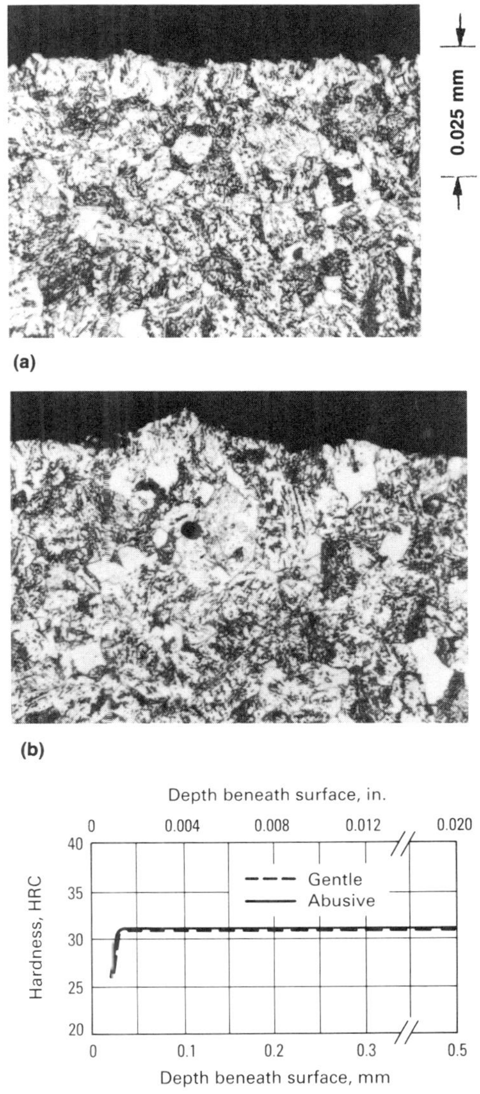


**Fig. 18** Surface characteristics of 4340 steel (annealed, 31 to 36 HRC) produced by CM. (a) Gentle conditions produce no visible surface effects and a surface finish of 0.9 μm (35 μin.). (b) Abusive conditions produce a slight roughening and a surface finish of 3 μm (120 μin.). (c) Gentle and abusive conditions both produce a softening at the surface. Source: Ref 9

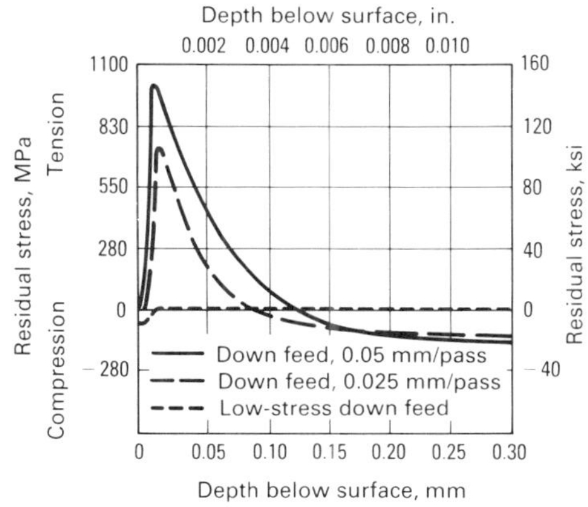


| | |
|---|---|
| Wheel | A46K8V |
| Wheel speed, m/min (ft/min) | 1800 (6000) |
| Cross feed, mm/pass (in./pass) | 1.25 (0.050) |
| Table speed, m/min (ft/min) | 12 (40) |
| Depth of grind, mm (in.) | 0.25 (0.010) |
| Grinding fluid | Soluble oil (1:20) |
| Specimen size, mm (in.) | 1.5 × 19 × 108 (0.060 × 3/4 × 4 1/4) |

**Fig. 19** Residual stress from surface grinding of D6AC steel (56 HRC)

tool life and good surface finish. Second, all machining should be done with sharp tools. Sharp tools minimize distortion and generally lead to better control during machining. The maximum flank wear when turning or milling should be 0.13 to 0.20 mm (0.005 to 0.008 in.). A good rule of thumb is to remove the tool when the wearland becomes visible to the naked eye. Guidelines for other chip removal operations are as follows:

- Rigid, high-quality machine tools are essential
- All hand feeding during drilling should be avoided
- The wearland on drills should be limited to 0.13 to 0.20 mm (0.005 to 0.008 in.)
- The finishing of drilled holes is imperative. The entrance and exit of all holes should be carefully deburred and chamfered or radiused
- Special precautions should be taken when reaming holes in sensitive alloys. The power feeding of power-driven machines should be employed for the reaming of straight holes. On tapered holes (using power-driven machines), hand feeding is permissible, but power feeding is preferred
- All straight holes 7.9 mm (5/16 in.) or larger should be double reamed, with a minimum metal removal of 1.2 mm (3/64 in.) on the diameter. The operator should visually check the reamer for sharpness after each operation. At the first sign of chipping, localized wear, or average flank wear beyond specification, the reamer should be replaced, and the hole should be inspected. In addition, regardless of the hole and reamer condition, a maximum number of holes should be specified for reamer replacement
- Deburring and chamfering or radiusing should be used to remove all sharp edges
- Honing is an excellent finishing operation for developing surface integrity. A multistone head is preferred; heads with steel shoes and/or steel wipers are not recommended
- Boring can be used as a finish machining operation if roughness is within the manufacturing engineering limits. The tool wearland in finish boring should be limited to 0.13 mm (0.005 in.), but it should often be far less than this in order to achieve the desired accuracy and surface finish

**Electrical, Chemical, and Thermal Material Removal Processes.** Guidelines for these processes are as follows:

- Whenever EDM is used in the manufacture of highly stressed structural parts, the heat-affected layer produced should be removed. Generally, during EDM roughing, the layer showing microstructural changes, including a melted and resolidified layer, is less than 0.13 mm (0.005 in.) deep. During EDM finishing, it is less than 0.025 mm (0.001 in.) deep
- Surface integrity evaluations should be made when CM and ECM processes are used for finishing critical parts

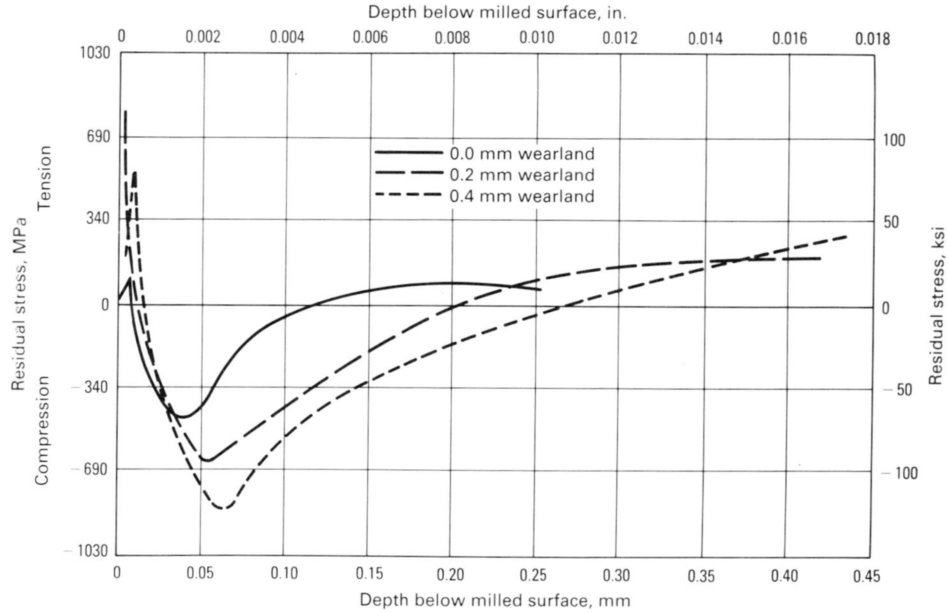


| | |
|---|---|
| Tool | 100 mm (4 in.) diam single-tooth face mill with Carboloy 370 (C-6) carbide |
| End cutting edge angle | 5° |
| Peripheral clearance | 8° |
| Cutting speed, m/min (ft/min) | 55 (180) |
| Feed, mm/tooth (in./tooth) | 0.125 (0.005) |
| Depth of cut, mm (in.) | 1.0 (0.040) |
| Width of cut, mm (in.) | 1.9 (0.750) |
| Cutting fluid | Dry |
| Specimen size, mm (in.) | 1.5 × 19 × 108 (0.060 × ¾ × 4¼) |

**Fig. 20** Residual stress from surface milling 4340 steel (quenched and tempered to 52 HRC)

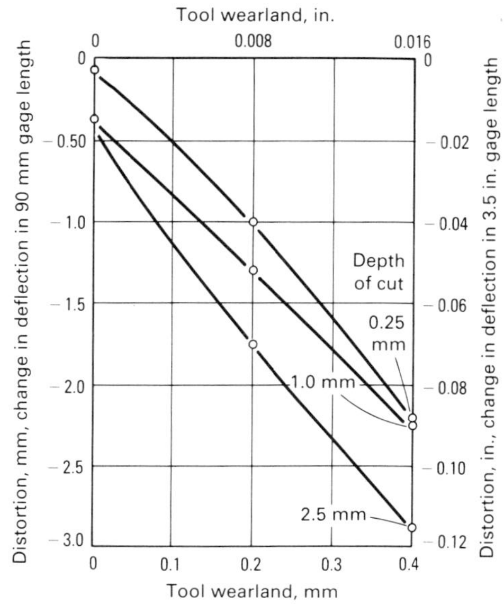


| | |
|---|---|
| Tool | 100 mm (4 in.) diam single-tooth face mill with Carboloy 370 (C-6) carbide |
| End cutting edge angle | 5° |
| Peripheral clearance | 8° |
| Cutting speed, m/min (ft/min) | 55 (180) |
| Feed, mm/tooth (in./tooth) | 0.125 (0.005) |
| Cutting fluid | Dry |

**Fig. 22** Change in deflection versus tool wearland for the face milling of 4340 steel (quenched and tempered to 52 HRC)

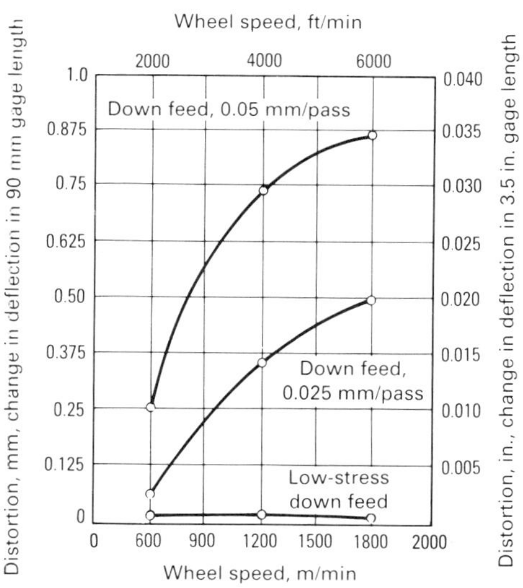


| | |
|---|---|
| Wheel grade | A46K8V |
| Cross feed, mm/pass (in./pass) | 1.25 (0.050) |
| Table speed, m/min (ft/min) | 12 (40) |
| Depth of cut, mm (in.) | 0.25 (0.010) |
| Specimen size, mm (in.) | 1.8 × 19 × 108 (0.070 × ¾ × 4¼) |
| Grinding fluid | Soluble oil (1:20) |

**Fig. 21** Change in deflection versus wheel speed and down feed in the surface grinding of D6AC steel (56 HRC)

- When substituting ECM for other machining processes, it may be necessary to add postprocessing operations such as steel shot or glass bead peening or mechanical polishing. Some companies require the peening of all electrochemically machined surfaces of highly stressed structural parts
- Special cognizance should be taken of the surface softening that occurs in the chemical and electrochemical machining of aerospace materials. Hardness reduction for CM and ECM range from 3 to 6 HRC points to a depth of 0.025 mm (0.001 in.) for CM and 0.05 mm (0.002 in.) for ECM. Shot peening or other suitable postprocessing should be used on such surfaces to restore mechanical properties
- Laser beam machining develops surfaces showing the effects of melting and vaporization. It is suggested that critical parts made by LBM be tested to determine if surface alterations lower the critical mechanical properties

**Postprocessing** guidelines are given below:

- Cracks, heat-affected layers, and other detrimental layers created during material removal processing should be removed from critically stressed parts. In addition to conventional machining processes, chemical

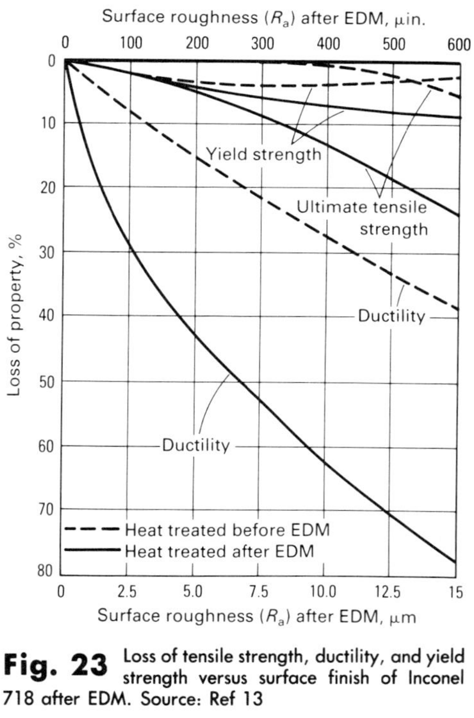


**Fig. 23** Loss of tensile strength, ductility, and yield strength versus surface finish of Inconel 718 after EDM. Source: Ref 13

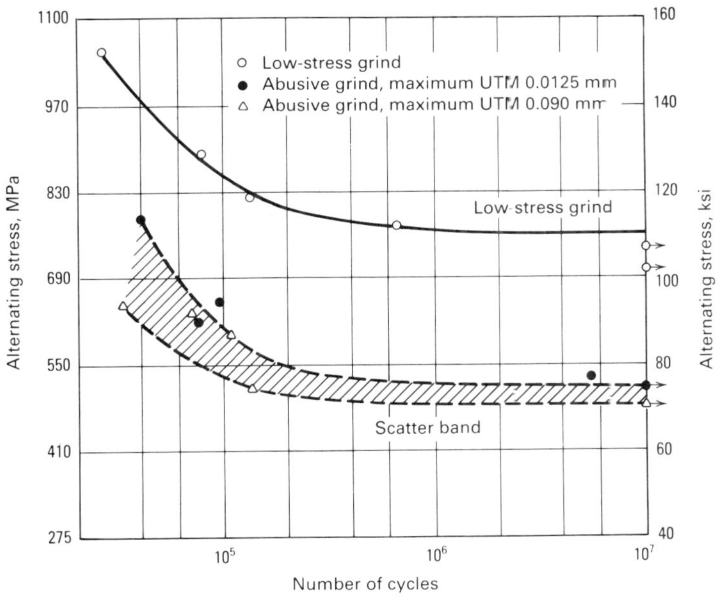


**Fig. 24** Loss of fatigue strength from the abusive grinding of 4340 steel (quenched and tempered to 50 HRC). Fatigue tests involved cantilever bending at room temperature and zero mean stress. Source: Ref 9

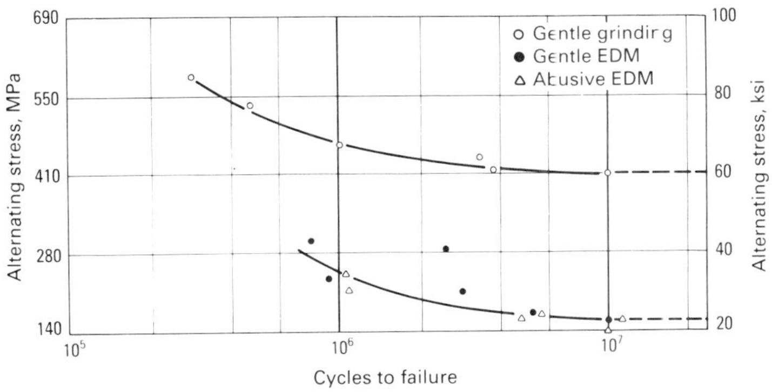


**Fig. 25** Effect of EDM and grinding on the fatigue strength of Inconel 718. Fatigue tests involved cantilever bending at room temperature and zero mean stress. Source: Ref 9

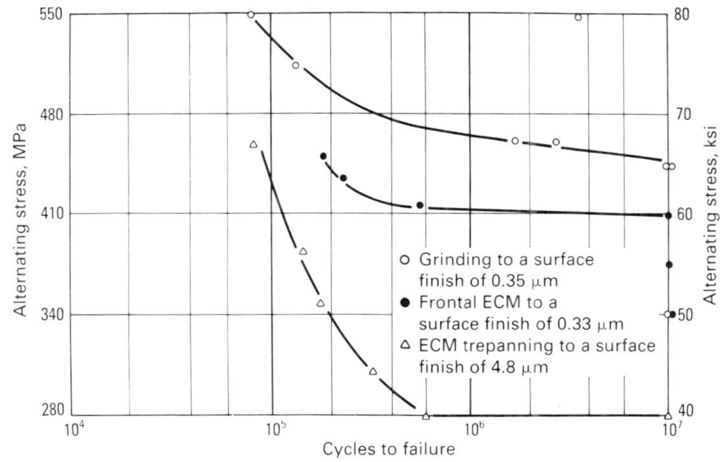


**Fig. 26** Effect of ECM on the fatigue strength of Ti-6Al-4V. Fatigue tests involved cantilever bending at room temperature and zero mean stress. Source: Ref 9

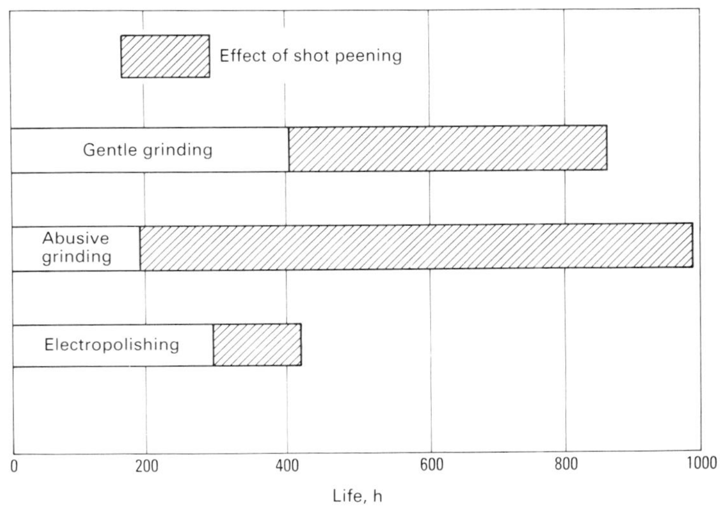


**Fig. 27** Effect of shot peening on the stress-corrosion resistance of AISI 4340 steel (50 HRC). Source: Ref 14

machining and abrasive flow machining are successfully used for this purpose

- Steel shot and glass bead peening as well as burnishing can be used to improve surface integrity
- The use of controlled shot peening practices to restore the fatigue life of components processed by electrical, chemical, and thermal removal processes should be evaluated. Shot peening has been shown to be extremely effective in improving the fatigue life of specimens processed by ECM, EDM, and ELP (Table 11). Component tests are recommended to confirm the favorable trends shown in laboratory tests
- Postheat treatments following material removal are of limited utility. Stress-relief treatments, used to soften the hardened layers produced during the grinding of steels, do not restore the hardness of overtempered layers present immediately below the damaged surface layer. In addition, heat treatment does not repair any cracks produced during material removal
- Abrasive tumbling is an effective technique for improving surface properties, including fatigue. Abrasive tumbling can be used to reverse unfavorable tensile stresses by inducing a compression-stressed surface layer
- Washing procedures should be employed for critical parts and assemblies to remove all traces of cutting fluids, which may cause stress corrosion. Typical suspect compounds are sulfur compounds on aluminum- and nickel-base alloys and chlorine compounds on titanium alloys. Some companies use this precaution only for parts subjected to temperatures over 260 °C (500 °F). For applications at less than 260 °C (500 °F), carefully controlled washing procedures are often used to remove the chlorinated and sulfurized cutting oils

**Inspection practices** should be reviewed and amplified to meet surface integrity requirements. Some inspection practices include the following:

- Microscopic examination, including microhardness testing, is used on a sampling basis to determine the type of surface layer being produced and its depth. This method can be used to check for such defects as microcracks, pits, folds, tears, laps, built-up edge, intergranular attack, and sparking
- The white layer or overtempered martensite produced during the grinding of steels can be detected by immersion etching using a 3 to 5% aqueous nitric acid solution
- Magnetic particle, penetrant inspection, ultrasonic testing, and eddy current techniques are recommended for detecting macrocracks
- X-ray diffraction methods are available for detecting residual surface stresses

## REFERENCES


1. "Surface Texture (Surface Roughness, Waviness and Lay)," ANSI/ASME B46.1-1985, American Society of Mechanical Engineers, 1985
2. "Surface Texture Symbols," ANSI Y14.36-1978, American Society of Mechanical Engineers, 1978
3. "Surface Integrity," ANSI B211.1-1986, Society of Manufacturing Engineers, 1986
4. J. Peters, P. Vanherck, and M. Sastrodinoto, Assessment of Surface Typology Analysis Techniques, *Annals of the CIRP*, Vol 28 (No. 2), 1979
5. *Machining Data Handbook*, Vol 2, 3rd ed., Metcut Research Associates, 1980, p 18-15 to 18-37
6. G. Bellows and D.N. Tishler, "Introduction to Surface Integrity," Technical Report TM 70-974, General Electric Company, 1970, p 3
7. L.R. Gatto and T.D. DiLullo, "Metallographic Techniques for Determining Surface Alterations in Machining," Technical Paper IQ 71-225, Society for Manufacturing Engineers, 1971
8. *Machining Data Handbook*, Vol 2, 3rd ed., Metcut Research Associates, 1980, p 18-58
9. M. Field and J.F. Kahles, Review of Surface Integrity of Machined Components, *Annals of the CIRP*, Vol 20 (No. 2), 1971
10. *Machining Data Handbook*, Vol 2, 3rd ed., Metcut Research Associates, 1980, p 18-87
11. M. Field, J.F. Kahles, and J.T. Cammett, A Review of Measuring Methods for Surface Integrity, *Annals of the CIRP*, Vol 2 (No. 1), 1972
12. W.P. Koster *et al.*, "Surface Integrity of Machined Structural Components," AFML-TR-70-11, Metcut Research Associates, March 1970, p 2
13. A.R. Werner and P.C. Olson, Paper MR 68-710, Society of Manufacturing Engineers, 1968
14. W.P. Koster, L.R. Gatto, and J.T. Cammett, Influence of Shot Peening on Surface Integrity of Some Machined Aerospace Materials, in *Proceedings of the First International Conference on Shot Peening* (Paris, France), Sept 1981, Pergamon Press, p 287-293
15. W.E. Littman, "The Influence of Grinding on Workpiece Technology," Paper MR 67-593, Society of Manufacturing Engineers, 1967
16. J.F. Kahles, G. Bellows, and M. Field, "Surface Integrity Guidelines for Machining," Paper 69-730, Society of Manufacturing Engineers, 1969


**Table 11 Effect of shot peening on fatigue strength**

| Alloy | Machining operation | Endurance limit in bending, $10^7$ cycles MPa | ksi | Gentle grinding, % |
|---|---|---|---|---|
| 4340 steel, 50 HRC | Gentle surface ground | 703 | 102 | 100 |
| | Gentle surface ground and shot peened | 772 | 112 | 110 |
| | Abusive surface ground | 430 | 62 | 61 |
| | Abusive surface ground and shot peened | 630 | 92 | 90 |
| | Electropolished | 620 | 90 | 88 |
| | Electropolished and shot peened | 660 | 96 | 94 |
| Inconel 718, solution treated and aged, 44 HRC | Gentle surface ground | 410 | 60 | 100 |
| | EDM roughing | 170 | 25 | 42 |
| | EDM roughing and shot peened | 540 | 78 | 130 |
| | EDM finishing | 170 | 25 | 42 |
| | EDM finishing and shot peened | 480 | 70 | 117 |
| | ECM | 285 | 41 | 68 |
| | ECM and shot peened | 560 | 81 | 135 |
| | ELP | 290 | 42 | 70 |
| | ELP and shot peened | 540 | 79 | 132 |

Note: Shot size: S110; shot hardness: 50–55 HRC; coverage: 300%. Source: Ref 14

**Table 12 Etching techniques for the detection of grinding injury in hardened steel**

| Operation | Solution used | Description, time, or function |
|---|---|---|
| **Double-etch method** | | |
| 1. Etch No. 1 | 4–5% nitric acid in water | Until black, 5–10 s; do not over etch. |
| 2. Rinse | Warm water | To remove acid |
| 3. Rinse | Methanol or acetone(a) | To remove water |
| 4. Etch No. 2 | 5–10% hydrochloric acid in methanol or acetone(a) | Until black smut is removed, 5–10 s |
| 5. Rinse | Running warm water | To remove acid |
| 6. Neutralize | 2% sodium carbonate + phenolpthalien indicator in water | To neutralize any remaining acid |
| 7. Rinse | Methanol | To remove water |
| 8. Dry | Warm air blast | |
| 9. Oil dip | Low-viscosity mineral oil with rust inhibitor | To enhance contrast, prevent corrosion |
| **Nital etch method** | | |
| 1. Etch | 5–10% nitric acid in ethanol or methanol | Until contrast is evident |
| 2. Repeat steps 5–9 above. Dark areas are overtempered, light areas are rehardened, uniform gray indicates no injury. | | |

(a) 4% nitric acid in water for etch No. 1 used with 2% hydrochloric acid in acetone for etch No. 2 sometimes gives greater sensitivity on high-carbon hardened steel. It is important that appropriate precautions be taken to avoid fire hazards, and good ventilation must be provided. Source: Ref 15

# Tool Wear and Tool Life

L. Alden Kendall, Washington State University

CUTTING TOOLS WEAR because normal loads on the wear surfaces are high and the cutting chips and workpiece that apply these loads are moving rapidly over the wear surfaces. The cutting action and friction at these contact surfaces increase the temperature of the tool material, which further accelerates the physical and chemical processes associated with tool wear. In order to remove the unwanted material as chips, these forces and motions are necessary; therefore, cutting tool wear is a production management problem for manufacturing industries.

To successfully manage machining processes, production engineers and managers need to establish a system that:

- Selects the proper machine tools and cutting tools to produce the geometric features in a part being machined from a particular material
- Ensures that the tool distribution system provides quality tools having the required geometry
- Specifies the correct cutting velocity, tool feed rate, and tool engagements with the workpiece
- Establishes on-line or off-line procedures to monitor the condition of the cutting tool and the quality of the surfaces machined by the tool
- Has maintenance procedures that ensure consistent machine tool operation
- Takes into account the cost of machine tool operation and tool use, permitting a clear idea of the economic objective for the machining system

Such a system can provide the necessary information to determine when a tool should be changed. Unfortunately, there are many variables to consider; thus, it is not surprising that tool wear assessment and tool change decisions are difficult problems.

## The Wear Environment

Cutting tool wear occurs along the cutting edge and on adjacent surfaces. Figure 1 shows a view of the cutting process in which the rake and clearance surfaces intersect to define the cutting edge. Figure 2 in the article "Mechanics of Chip Formation" in this Volume shows a similar view but adds the stress and strain states that are present in the material and chip; Fig. 4 from the same article also shows the machining forces and velocities that produce this state of stress and strain. These are average forces and velocities. Cutting tool wear is localized on specific surfaces where stress, strain, velocity, and temperature are above critical levels. It is important to understand where these critical conditions exist and how they interact to cause tool wear.

**Wear Surfaces.** Figure 2 shows how the sharp tool of Fig. 1 may wear. Along the rake surface, the chip motion and high normal stress have produced a wear scar called crater wear. Along the clearance surface, the tool motion and high normal stress have increased the area of contact between the tool and work, producing flank wear. Lastly, the cutting edge radius has increased. Figure 3 shows the characteristic wear surfaces on a turning tool insert, end mill, form tool, and drill. The cutting edge view shown in Fig. 1 and 2 is identified as section A—A in Fig. 3.

These figures show how the wear process changes the geometry of these different types of cutting tools. Flank wear decreases the diameter of the end mill as well as the

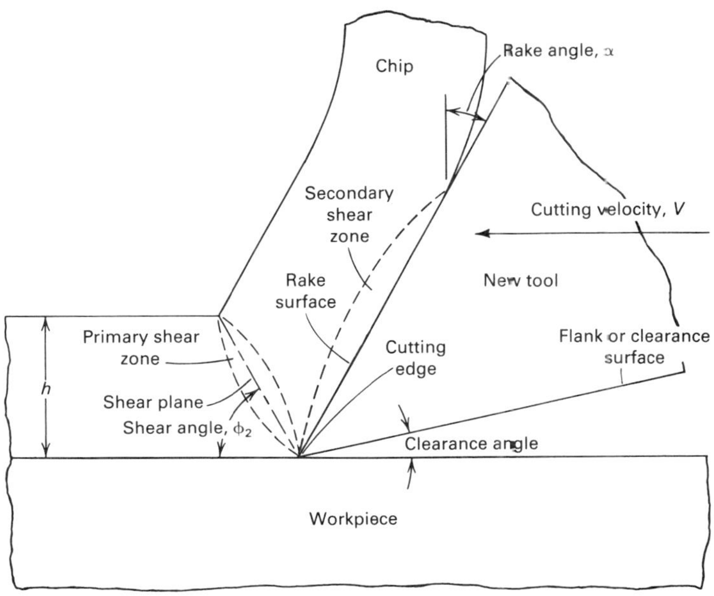


**Fig. 1** Chip, workpiece, and tool relationship

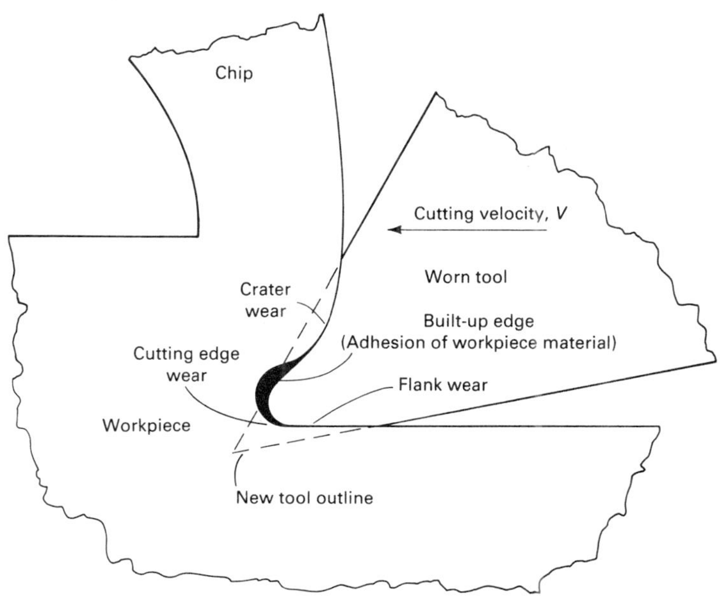


**Fig. 2** Typical wear surfaces

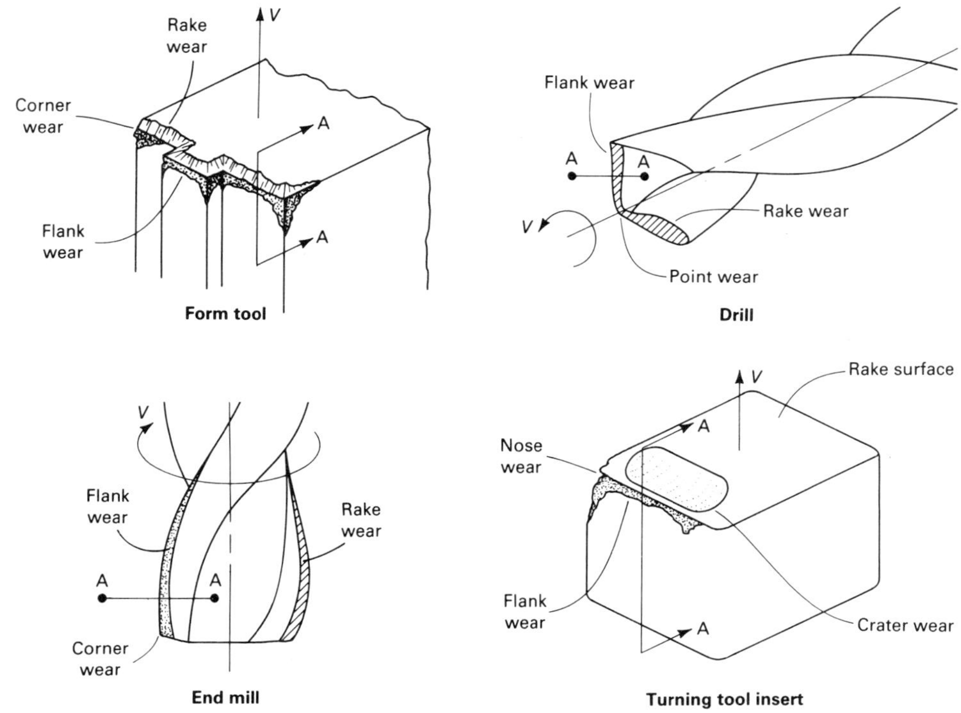


**Fig. 3** Wear surfaces on common tools due to the tool motion, *V*

depth of cut for a lathe tool. Both of these changes in the geometry of the cutting tool could produce out-of-tolerance dimensions on machined parts. The edge wear and crater wear on the rake surface alter the state of stress and strain in the cutting region, thereby changing cutting forces and the mechanics associated with the chip-making process. Severe geometric changes that decrease the angle between the rake and clearance surfaces can weaken the tool so that the edge may suddenly fracture.

It should be apparent that the location and size of these wear surfaces play an important role in determining the useful life of the cutting tool. Localized stresses on cutting tool surfaces are a major contributing factor in regard to location and size of wear surfaces.

**Stresses on Tool Wear Surfaces.** Figure 4 shows the approximate distribution of normal and shear stresses on the tool wear surfaces. The normal stresses, $\sigma_n$, are caused by normal forces acting along the rake surface, the cutting edge surface, and the clearance surface. In addition to the normal stresses, Fig. 4 shows the shear stresses, $\tau$, that act along the surface of the tool and are associated with sticky and sliding shear processes. For the sticky zone, the normal force has a magnitude that results in a shear stress component that equals the shear yield strength, $\tau_y$, of the strain-hardened workpiece material. Rather than sliding along the surface, the chip tends to adhere and periodically separate along shear fracture planes. The existence and size of this sticky zone is dependent upon the magnitude of the normal force and the friction coefficient along these surfaces. The sliding zones have friction forces and associated surface shear stresses that vary according to the normal force and coefficient of friction. The state of strain created by the shear stresses associated with the sticky and sliding zones on the rake surface is generally called the secondary shear zone and is identified in Fig. 1. Tool surface roughness and lubrication conditions affect the magnitude of these surface shear stresses.

The primary shear zone shown in Fig. 1 extends from the cutting edge to the surface and is the zone where the chip material is plastically deformed and sheared from the work material. The complex state of stress along the cutting edge is caused by the strain associated with the chip separating from the work and moving along the rake surface of the tool and the strained material that remains a part of the work material. This conflict between chip motion strain and stationary workpiece strain produces a plowing action by the cutting edge. Normal stresses can become very high and exceed the strength of the tool material, causing plastic deformation or fracture of the cutting edge. A sticky zone may not exist for certain cutting conditions; however, the plowing action along the cutting edge always exists to some degree because it is impossible to create a cutting edge with no radius and with a primary shear zone that is a perfect plane. The magnitude of the state of stress in the cutting region also varies with time and creates a potential fatigue failure environment.

**Motion Along the Wear Surfaces.** One way to increase machining productivity is to increase the volumetric chip removal rate. Volumetric chip removal rate is the product of the engagement area of the tool with the work material times the cutting velocity, $V$; thus, one option the process planner has is to increase the cutting velocity. This productivity gain must be balanced against increased tool wear caused by higher cutting velocities.

In the article "Mechanics of Chip Formation" in this Volume, the chip velocity, $V_c$, and the shear velocity, $V_s$, were shown to

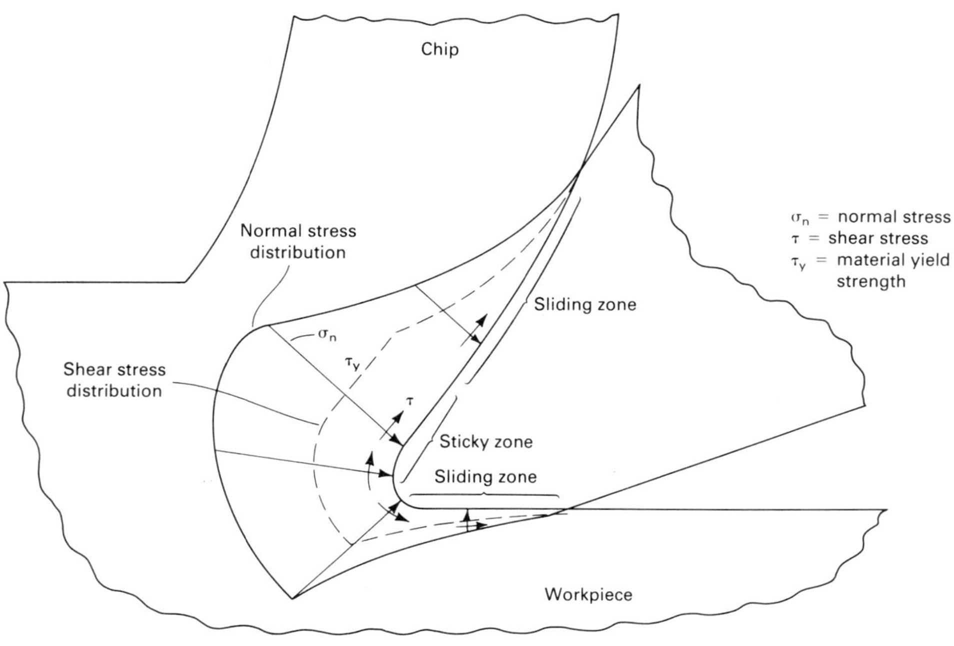


**Fig. 4** Wear surface stresses

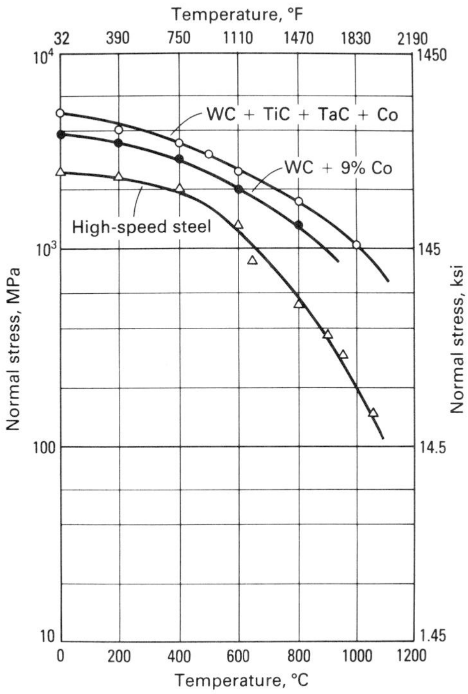


**Fig. 5** Cutting tool materials yield strength as a function of temperature. Lower curve is high-speed steel. Upper two curves are tungsten carbide. Source: Ref 1

be functionally related to the primary shear angle, $\phi$, the rake angle, $\alpha$, and the cutting velocity, $V$. The cutting velocity is the relative velocity between the clearance surface of the tool and the work, while the chip velocity is the relative velocity between the chip and the rake surface of the tool. The magnitude of these two velocities and the related shear stresses at these interfaces determine the amount of thermal energy released per unit of contact area. The magnitude of the shear velocity causes a high strain rate in the primary shear zone and in the sticky zone. The volume of material strained by this strain rate releases additional thermal energy.

This thermal energy is the heat source that causes the temperature of the workpiece, cutting tool, and chip to increase. The productivity gain from increasing the cutting velocity proportionately increases wear surface velocities, strain rates, and release of thermal energy, thus increasing the wear environment temperature.

**Temperatures in the Wear Zones.** The difference between the thermal-energy release rate and the thermal-energy dissipation rate determines the temperature of the material in these wear zones. Thermal-energy dissipation is a function of the thermal-conductivity properties of the tool material and workpiece material. Additionally, the workpiece size and specific heat determine the workpiece heat capacity; to a lesser extent, the surface area plays a role in convective-heat transfer to the surrounding air. If a cutting fluid is used, its conductivity and convective-heat transfer boundary coefficient with the hot tool and workpiece play an important role.

The development and ultimate selection of tool materials are based on the ability of the tool to maintain hardness, toughness, and chemical stability at high temperatures. Even with the best of tools, these properties eventually adversely change with increasing temperatures. Figure 5 shows the change in yield strength as the temperature changes for three common cutting tool materials.

There have been many studies of temperatures in the cutting tool. Most commonly, average temperature conditions are approximated using the following type of relationship (Ref 2):

$$T = u\left(\frac{Vh}{k\rho c}\right)^{1/2}$$

where $T$ is the mean temperature (°F) of tool-chip interface, $u$ is the specific cutting energy, which is the energy used per unit volume of material removed, $V$ is the cutting speed, $h$ is the undeformed chip thickness (see Fig. 1), and $k$, $\rho$, $c$ are the conductivity, density, and specific heat, respectively, of the workpiece material.

Figure 6 shows a typical temperature distribution. The high temperature gradient can cause portions of the tool to be at dangerously high temperature levels, well above the mean temperature. The mean temperature relationship identifies the important variables and their effect on temperature. High-strength work materials have high values of specific cutting energy. Some of these materials, such as titanium, are also poor conductors and are of low density. This combination of conditions results in higher tool temperatures, and the cutting tool material used must maintain its hardness and strength at these temperature levels if it is to have a useful service life.

Coolants are a mixture of lubricating and heat dissipation constituents; however, their primary function is to cool the tool. The lubricating component becomes more important at lower cutting speeds where there is less heat generated but a greater tendency for the chip and work material to adhere to the surfaces of the tool. More detailed information on machining coolants can be found in the article "Metal Cutting and Grinding Fluids" in this Volume.

Combining these temperature effects with the state of stress, strain rates, and material motion makes the wear zones of the tool a complex battleground in which these conditions interact to trigger the mechanisms that cause wear.

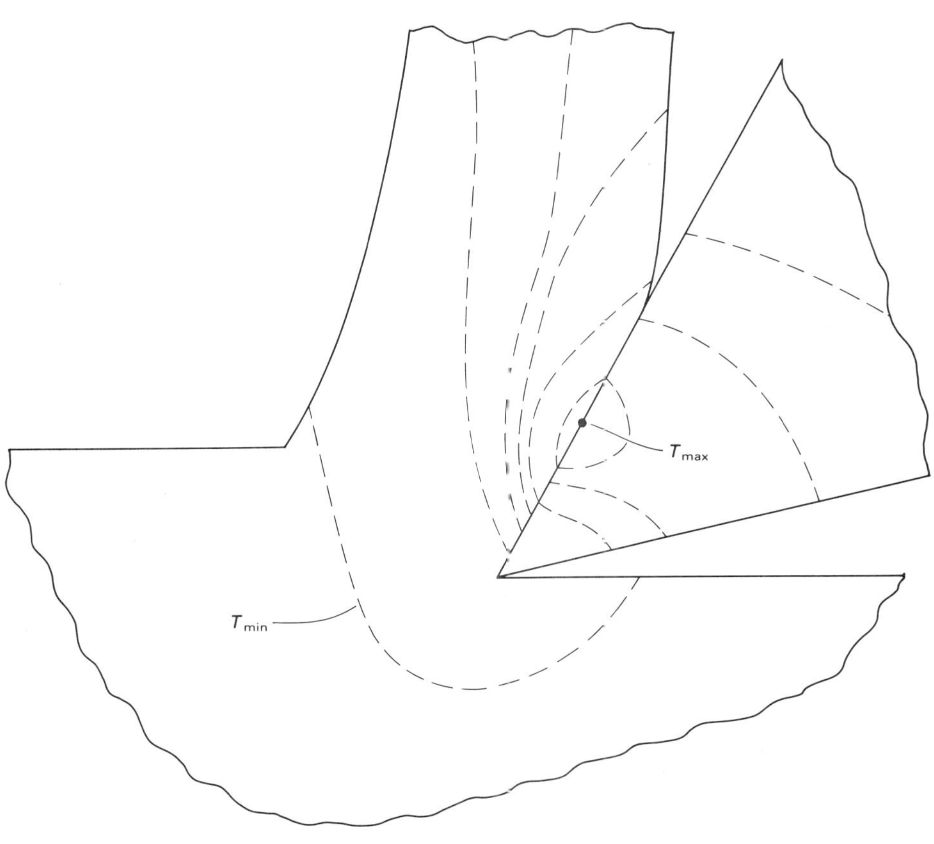


**Fig. 6** Generalized temperature distributions in the cutting region

## Wear Mechanisms

Material scientists and engineers have documented many volumes of information on wear. One important area of concern in these studies has been the identification of wear mechanisms. What has emerged is

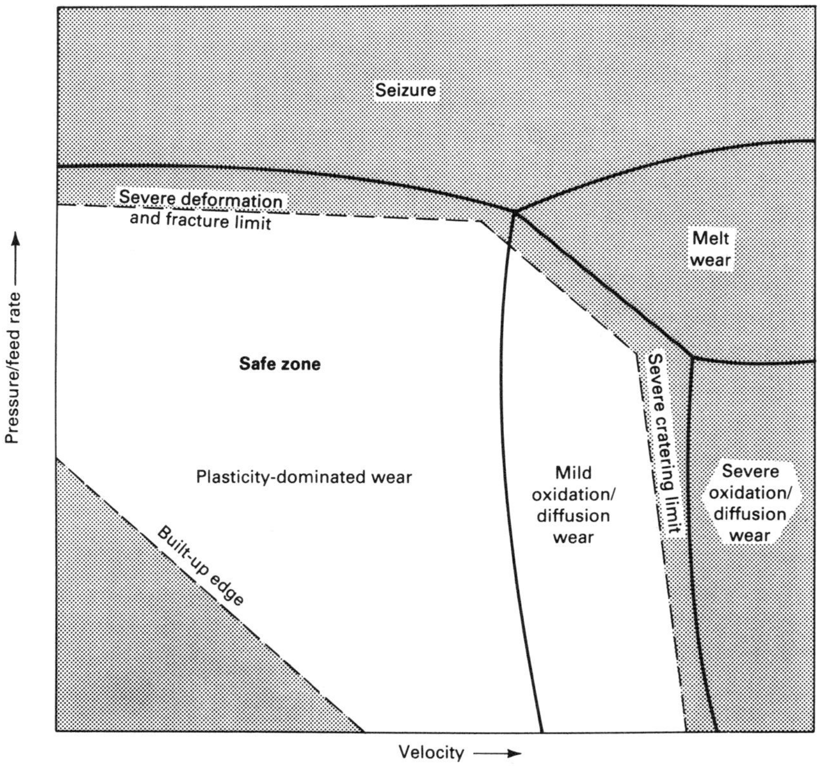


**Fig. 7** Wear mechanism maps and safe operating regions for cutting tools. Source: Ref 3

that a particular wear mechanism is dependent upon the contact stress, relative velocities at the wear interface, temperature, and physical properties of the materials in contact.

For a particular set of contacting materials, wear mechanism maps have been used to identify the ranges of normal pressure and velocity that result in a particular wear mechanism (Ref 3). Figure 7 shows the general structure of such a map when the area of the contacting wear surface remains constant. Temperature is not a mapped variable; however, it is a variable that is dependent upon the pressure (normal stress), velocity, and size of the wear surface, as discussed in the previous section. The four broad classes of mechanisms shown in Fig. 7 are:

- Seizure
- Melt wear
- Oxidation/diffusion-dominated wear
- Plasticity-dominated wear

Cutting tool wear research has established similar maps for feed rate, which is related to pressure for a given engagement condition, and velocity (Ref 1). The dashed lines in Fig. 7 show some possible boundaries that could be used to define a safe operating zone. This overlay of the two maps provides a framework to discuss the progressive wear and failure conditions for cutting tools.

**Initial Wear Mechanisms.** The two materials in contact have surface roughness irregularities in the form of protrusions or asperities. At the interface, asperities from the two materials touch, creating tiny contact areas. The total area from these contact points is a fraction of the projected area of the contact surface. The stresses and heat are intensified in the asperities, and partial removal may occur due to seizure accompanied by fracture of the asperity or melting in the asperity. As these asperities are removed, the initial surface roughness is altered and the contact area increases. If the force conditions remain unchanged, pressure decreases and the active wear mechanisms change to plasticity and/or mild oxidation/diffusion dominated wear. This initial wear period creates small, visible wear surfaces.

**Steady-State Wear Mechanisms.** A velocity and normal stress condition that would continue to cause seizure and melt should be avoided because it would soon cause complete failure of the cutting tool. Assuming that such conditions do not exist, the wear surfaces get progressively larger. If the wear surfaces are plasticity dominated, small particles of material are mechanically deformed and fractured away from the wear surface. Generally called abrasion, this is the most common wear process along the clearance surfaces of most tools.

As discussed earlier, normal stress and temperature vary over the wear surfaces so that a plasticity mechanism that dominates in one wear zone may not dominate in another. Figure 6 indicates that the maximum tool surface temperature occurs on the rake surface a small distance from the cutting edge. This is where the crater wear condition occurs, with diffusion wear often being the dominant mechanism (Ref 4). The high temperature and pressure cause atoms to move between the contacting materials, and this diffusion process locally aids in the removal of tool material to form the crater. In very hard cutting tool materials, such as ceramics, for which high cutting velocities are commonly used, these oxidation and diffusion mechanisms may be responsible for a majority of the wear.

A built-up edge (BUE) condition, which is indicated by a dashed line in Fig. 7, affects the process in two ways. Near the cutting edge, the higher pressures could cause particles of work material to adhere to the cutting tool in the sticky zone. If the shear forces due to chip movement are high enough, the bond will be temporary and the adhered material will fracture away from the tool surface. When it fractures, small particles of tool material may be removed with the previously adhered material. This then is a wear process and would be associated with conditions in the safe zone just outside the built-up edge line. The second effect on the process occurs when the BUE is not fractured away by the chip motion and remains to alter the geometry of the cutting edge. The presence of the BUE changes the shear angle, causing instabilities in the chip-forming process and damage to the machined surface. The lubricating characteristics of cutting fluids are helpful in eliminating this BUE condition. A discussion of the BUE condition can also be found in the article "Mechanics of Chip Formation" in this Volume.

Wear can also occur as chipping along the cutting edge. Such chipping more commonly occurs when the cutting edge intermittently removes chips. This results in cyclic impact and thermal loading of the cutting edge. These two cyclic loading states can initiate small cracks and then propagate these cracks or other residual cracks to form the chips.

These abrasion, oxidation, diffusion, and chipping wear mechanisms that occur at operating conditions within the safe zone shown in Fig. 7 cause the initial wear surfaces to enlarge over time. This surface life period is often referred to as the steady-state wear period.

**Tertiary Wear Mechanisms.** The steady-state period of wear eventually enlarges the wear surfaces to a critical size that triggers accelerated wear. In tools that have a hard, wear-resistant coating such as titanium nitride, wear through this coating or separation of small volumes of the coating (Fig. 8a) exposes the less-resistant core material (Fig. 8b), resulting in accelerated wear. The pressures and velocities on these enlarged surfaces begin increasing the temperature

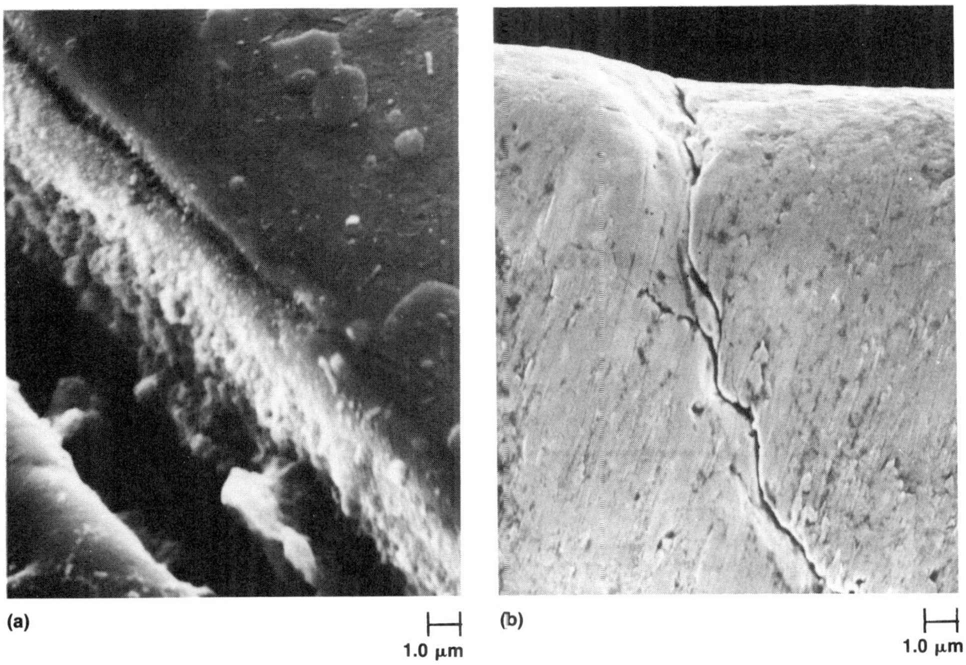


**Fig. 8** Delamination of titanium nitride coating from HSS end mill. (a) 940×. (b) 4700×. Source: Ref 5

so that the rapid oxidation/diffusion and local seizure or melting conditions shown in Fig. 7 exist, causing rapid destruction of the tool. A similar vulnerability exists for the softer, core material of coated tools. A tool change must be made before this point is reached. Figure 9 shows these three zones in terms of amount of wear over time.

As a final factor complicating wear mechanisms, there is a growing trend toward high-speed machining. This means cutting velocities five to ten times normal speeds are used. New tool materials and machine tool designs have made this possible; however, these changes alter the wear environment. Proportionally more thermal energy is removed by the chip due to the decrease in contact time between the tool and the chip. The higher velocities increase absolute temperatures on tool wear surfaces. Abrasive wear may become less important, and the diffusion and oxidation processes dominate in creating and enlarging the wear surfaces. See the Section "High-Productivity Machining" in this Volume for information on high-speed machining practice.

## Machine, Cutting Tool, and Tool Wear Interactions

Tool wear, too, is influenced by the construction of the tool; the rigidity of the tool, workpiece, and machine tool system; and the proper positioning of the tool. Other Sections of this Volume provide information on tool materials, different machining processes, and the mechanics of the cutting process. The information and suggestions given in these Sections must be followed in order to avoid some undesirable tool wear side effects. A number of these effects are emphasized in this article.

Cutting tools may be made of high-speed steel; high-speed steel coated with a thin, hard surface; or cemented carbide materials. As mentioned earlier, the coated material provides good wear resistance until the hard coating is removed, exposing the softer core material. The cemented carbides are very hard, but brittle, and have a lower thermal conductivity than high-speed steel. Inserts may easily fracture if the mechanical attachment to the insert toolholder does not firmly hold the insert. An insert cannot easily dissipate heat through the attachment interfaces to the more massive toolholder; this situation could cause unacceptably high temperatures in the insert. Carbide inserts that are brazed in place are better supported and dissipate heat better; however, they are still susceptible to fracture failures caused by impact or cyclic loads. At times, the application requires small cutting tools and tool inserts. In such instances, thermal overload and fracture failures can be avoided by decreasing cutting velocities and feed rates, reducing the size of cut, and/or increasing the coolant flow rate.

Sometimes tool wear is excessive even when the recommended cutting conditions are used. The cause of the problem may be the rigidity of the machine tool and workholding fixture. An older machine tool or one that has not been properly maintained can cause vibrations at the tool that accelerate the breakdown and wear of the cutting edge.

Even with a well-maintained rigid machine and workpiece system, improper positioning of the tool can cause tool wear problems. Excessive overhang of the tool can cause dynamic instabilities similar to those mentioned above. In the cutting zone, the cutting edge must be positioned so that clearance surfaces do not rub against the machined surface. For a single-point cutting tool, this requires correct alignment of the point of the tool with the rotating workpiece centerline and careful positioning of the rake and flank surfaces so that correct an-

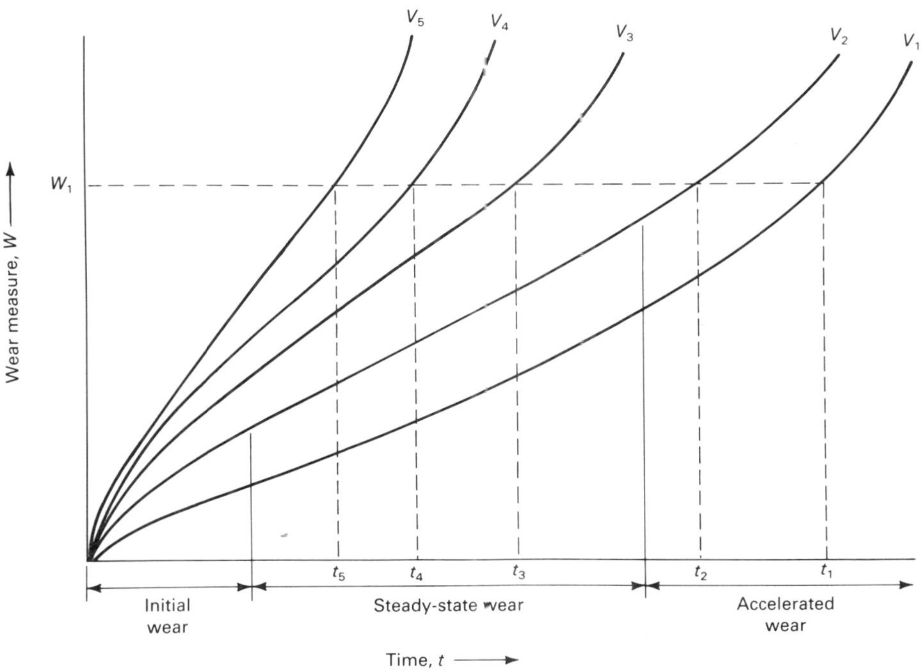


**Fig. 9** Tool wear curves for different cutting velocities

gles exist between the tool surfaces and the workpiece surfaces. Each type of machining operation has such specifications, and the tool engineer should ensure that they are followed when tooling setups are made.

Understanding the tool service environment and the conditions that cause wear is the first step in planning and controlling of tool use in machining. Unfortunately, research and development efforts in tool wear have not progressed to the point that the life of a cutting tool can be predicted using basic tool and work material properties plus fundamental science and engineering principles. Tool replacement decisions are dependent on predicting or sensing the magnitude of the wear damage caused by the active wear mechanisms on critical wear surfaces.

## Tool Replacement

Tools are replaced to minimize the probable consequences of a failure event. The tool wear environment is so complex that even with the utmost care it is possible that the tool may fail in service. A very conservative strategy may replace the tool frequently to reduce the probability of an in-process failure, but, as a consequence, interrupt the process so frequently that productivity decreases and tool costs increase. To establish the strategy objectively, a clear understanding of the failure event must be agreed upon, and suitable data concerning tool performance and operating costs must be available.

**Tool Failure Events.** The steady-state wear process removes material gradually until damage reaches a critical level. Wear beyond this level is considered damaging to the process and represents a failed state for the tool based on wear. This is called a wear failure event.

The wear environment also weakens the tool so that sudden variations in tool thermal and/or mechanical loading can cause rapid destruction of the tool cutting surfaces. These sudden variations can come from interrupted cuts, hard spots in the work material, or sudden loss of coolant in the cutting region. Sudden destruction can also be caused by residual defects in new tools. Even a new tool with no residual defects can be destroyed if the abrupt load changes are large enough. This second failure state is called a catastrophic-failure event.

**Failure event consequences** are dependent on the type of feature being machined, the quality specifications associated with the feature, and the production economics. The type of feature determines the type of machine and tool used for processing. A feature such as a small-diameter hole has a potential risk for a catastrophic failure event that could have severe consequences. A sudden increase in torsional loading could fracture the drill, leaving the fractured part in the hole and requiring expensive rework to remove the broken portion and resize the hole. In contrast, a pocket being milled out to a rough size using a large end mill has less severe consequences. A wear or catastrophic-failure event could mean loss of the tool, but little damage to the part. However, a final finish cut for this same pocket will change the failure event consequences: Excessive tool wear may result in an unacceptable surface finish or out-of-tolerance dimensions.

Economic analysis can be used to establish a tool replacement strategy that optimizes the performance objective associated with the making of each feature. Common objectives are minimum cost, maximum production rate, and maximum profit; however, minimum acceptable reliability or minimum downtime could also be used. The tool replacement strategy is dependent on the mechanism by which wear and catastrophic-failure events influence the objective.

To access failure event consequences and achieve feature production objectives, tool wear must be mathematically modeled and/or measured in a timely way in order to change tools strategically. Real time measurement of tool wear is an emerging but promising technology (Ref 6). Until it matures and is commercially available, mathematical models of tool wear will be primarily used.

**Modeling Tool Wear.** One of the earliest applications of science to production management was done by F.W. Taylor (Ref 7). Recognizing that tool wear was dependent on cutting velocity, he developed the following equation using data from tool life tests:

$$Vt^n = C$$

This became known as Taylor's tool life equation, in which the tool lifetime, $t$, was related to cutting velocity, $V$, by means of the constants n and C. These constants were obtained by testing cutting tools at different cutting velocities and using a "tool life criterion" to establish the point at which the useful life of the cutting tool had ended. This criterion was a wear limit that could not be exceeded if a wear failure event was to be avoided. Figure 9 shows typical tool wear curves for different cutting velocities. The wear limit or failure criterion, $W_1$, shows that the elapsed time before tool replacement increases with a decrease in cutting velocity. Taylor's relationship models this behavior (Fig. 10) and has been used by industry as originally conceived or in modified forms to the present time. In fact, the constant C, which equals the cutting velocity for 1 min of elapsed time before reaching the wear limit of the tool, has been widely used as a measure of machinability for a particular material being machined using a particular cutting tool and cutting condition. The cutting condition in this context is the feed rate and depth of cut (engagement) used to collect the data. Table 1 contains other tool life models that add feed rate and/or depth of cut as variables that influence tool life. In all of these, life testing of tools must be conducted.

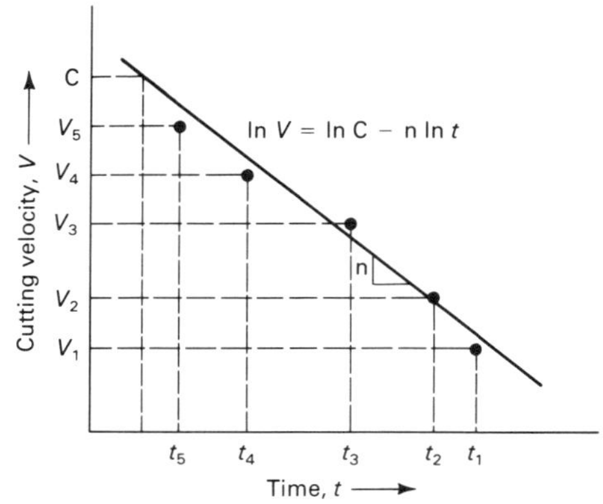


**Fig. 10** Taylor tool life model using data from Fig. 9 and a ln-ln plot

The wear limit or failure criterion is an important decision to be made in the process of developing the tool life model. The size and characteristics of particular wear surfaces are often used. Figure 11 shows the variety of wear and fracture surfaces that may be present on a worn or failed cutting tool. This figure also shows how to measure the amount of wear on one or more of these wear surfaces. Flank wear has been commonly used, and Table 2 presents some typical values for some common tools.

To use these wear limits, the tool life test must be stopped and a measurement made using optical instruments at suitable magnification levels. Generally, a toolholding fixture must be made to position the tool properly and consistently in the field of view. Typical flank wear scars are shown in Fig. 12(a) to (d). Each of these figures shows that the size of the scar varies. The end mill in Fig. 12(a) has enlarged flank wear toward the end where the flank wear from the end cutting edge and flute cutting

**Table 1 Tool life models proposed in the literature**

| | |
|---|---|
| Linearized Taylor model . . | $\ln t = b_0 + b_1x_1 + b_2x_2$ |
| Linearized Taylor model with interaction term . . . . | $\ln t = b_0 + b_1x_1 + b_2x_2 + b_{12}x_1x_2$ |
| Partial second-order model . . . . . . . . . . . . . . . . | $\ln t = b_0 + b_1x_1 + b_2x_2 + b_{12}x_1x_2 + b_{11}x_1^2 + b_{22}x_2^2$ |
| Full second-order model . . . | $\ln t = b_0 + b_1x_1 + b_2x_2 + b_3x_3 + b_{12}x_1x_2 + b_{13}x_1x_3 + b_{23}x_2x_3 + b_{11}x_1^2 + b_{22}x_2^2 + b_{33}x_3^2$ |
| Konig and DePiereux(a) . . . | $\ln t = b_0 + b_1V^{b_2} + b_3f^{b_4}$ |

(a) Ref 8. $t$ = tool life; $x_1 = \ln V$, where $V$ is the cutting velocity; $x_2 = \ln f$, where $f$ is the feed rate; $x_3 = \ln D$, where $D$ is the depth of cut; $b_0, b_1, b_2, \ldots, b_{33}$ are experimentally determined coefficients.

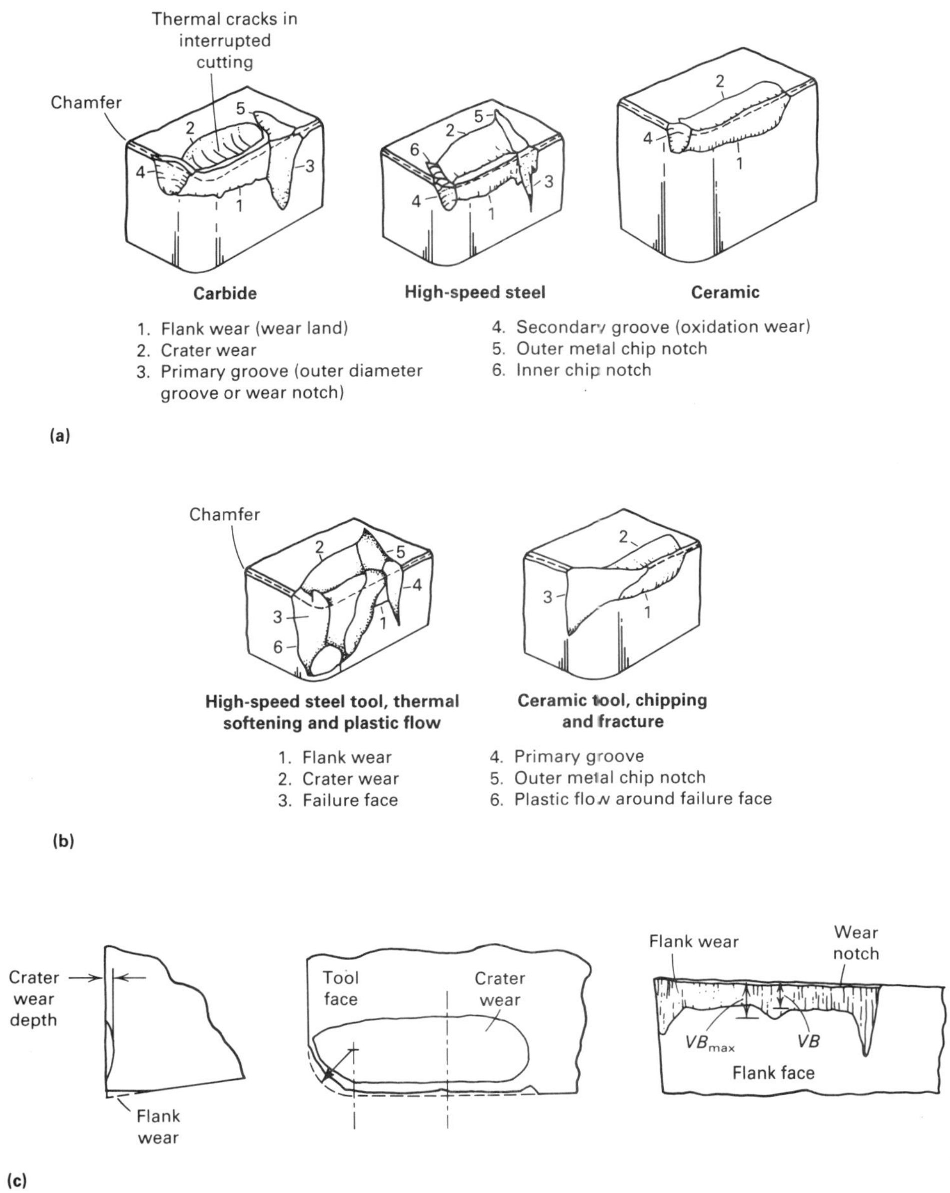


**Fig. 11** Cutting tool failure modes. (a) Characteristic wear and fracture surfaces on cutting tools. (b) Catastrophic failure. (c) Typical wear measurements for a turning tool. $VB$ = flank wear. Source: Ref 9

edge overlap. The drill in Fig. 12(b) shows that the width of the flank wear scar is roughly proportional to the distance from the center of the drill. Velocity increases with distance from the center, and such increased wear is consistent with the dependence of wear on velocity.

Figures 12(c) and (d) show the flank wear on a turning tool that was used for an orthogonal cut on the edge of a ring. In these figures, the nature and size of the scar reflect the difference between coated and uncoated high-speed steel tools. These figures also illustrate the importance of establishing where the flank measurements are to be made. It is apparent, too, that it is a good practice to take multiple measurements along the cutting edge; with multiple-tooth cutters, such measurements should be made on all of the cutting edges.

If crater wear is part of the wear limit criterion, the topography can be analyzed by the depth of the crater or the projected area of the crater. For chipping, the size and number of such chips can be determined during examination. A wear limit criterion for chipping may have to be based on the number and size that detrimentally affect surface finish.

Often it is necessary to use more detailed metallographic examination techniques. For external topographical examination of wear surfaces, a portion of the tool needs to be appropriately prepared for viewing under varying levels of magnification. A scanning electron microscope is sometimes used for this purpose. To evaluate internal damage, sectional views perpendicular to the cutting edge are generally needed. Near-surface thermal softening and phase changes are of interest, and microhardness measurements can be used to quantify the magnitude of these changes. Figure 8 is an example of the way high magnifications can be used to examine a failed tool and provide insight concerning the wear mechanisms that caused failure.

**Table 2 Typical flank wear limits**

| Operation and material | Flank wear land: Average wear, mm | Average wear, in. | Maximum local wear, mm | Maximum local wear, in. |
|---|---|---|---|---|
| Turning | | | | |
| HSS | 1.5 | 0.060 | 1.5 | 0.060 |
| Carbide | 0.45 | 0.015 | 0.9 | 0.030 |
| Face milling | | | | |
| HSS | 1.5 | 0.060 | 1.5 | 0.060 |
| Carbide | 0.45 | 0.015 | 0.9 | 0.030 |
| End milling-slotting | | | | |
| HSS | 0.30 | 0.012 | 0.50 | 0.020 |
| Carbide | 0.30 | 0.012 | 0.50 | 0.020 |
| End milling-peripheral | | | | |
| HSS | 0.30 | 0.012 | 0.50 | 0.020 |
| Carbide | 0.30 | 0.012 | 0.50 | 0.020 |
| Drilling | | | | |
| HSS | 0.45 | 0.015 | 0.45 | 0.015 |
| Carbide | 0.45 | 0.015 | 0.45 | 0.015 |
| Reaming | | | | |
| HSS | 0.15 | 0.006 | 0.15 | 0.006 |
| Carbide | 0.15 | 0.006 | 0.15 | 0.006 |
| Tapping | | | | |
| HSS | Go-no go gage | | Tap fracture | |

Source: Ref 10

## Tool Life Testing

A successful tool life testing program depends on establishing the test plan objective, designing the test, conducting the test, analyzing the results, and applying the results. These points are described below. Additional information on tool life testing can also be found in the article "Machinability Test Methods" in this Volume.

### Testing Program Expertise

Metallurgical, manufacturing-process, and experimentation knowledge is needed for a testing program. The first step is to find an individual or assemble a team that can provide this expertise. Metallurgical assistance is needed to:

- Provide the material information concerning the cutting tool and the workpiece
- Provide metallographic knowledge during examination of worn tools
- Interpret evidence concerning wear mechanisms and tool fracture

Manufacturing-process help is required to:

- Select the machine tool, cutting tool, and geometric feature to be machined for the cutting test
- Ensure that the machine is properly set up and calibrated for the test
- Establish the correct operating procedure for the test equipment

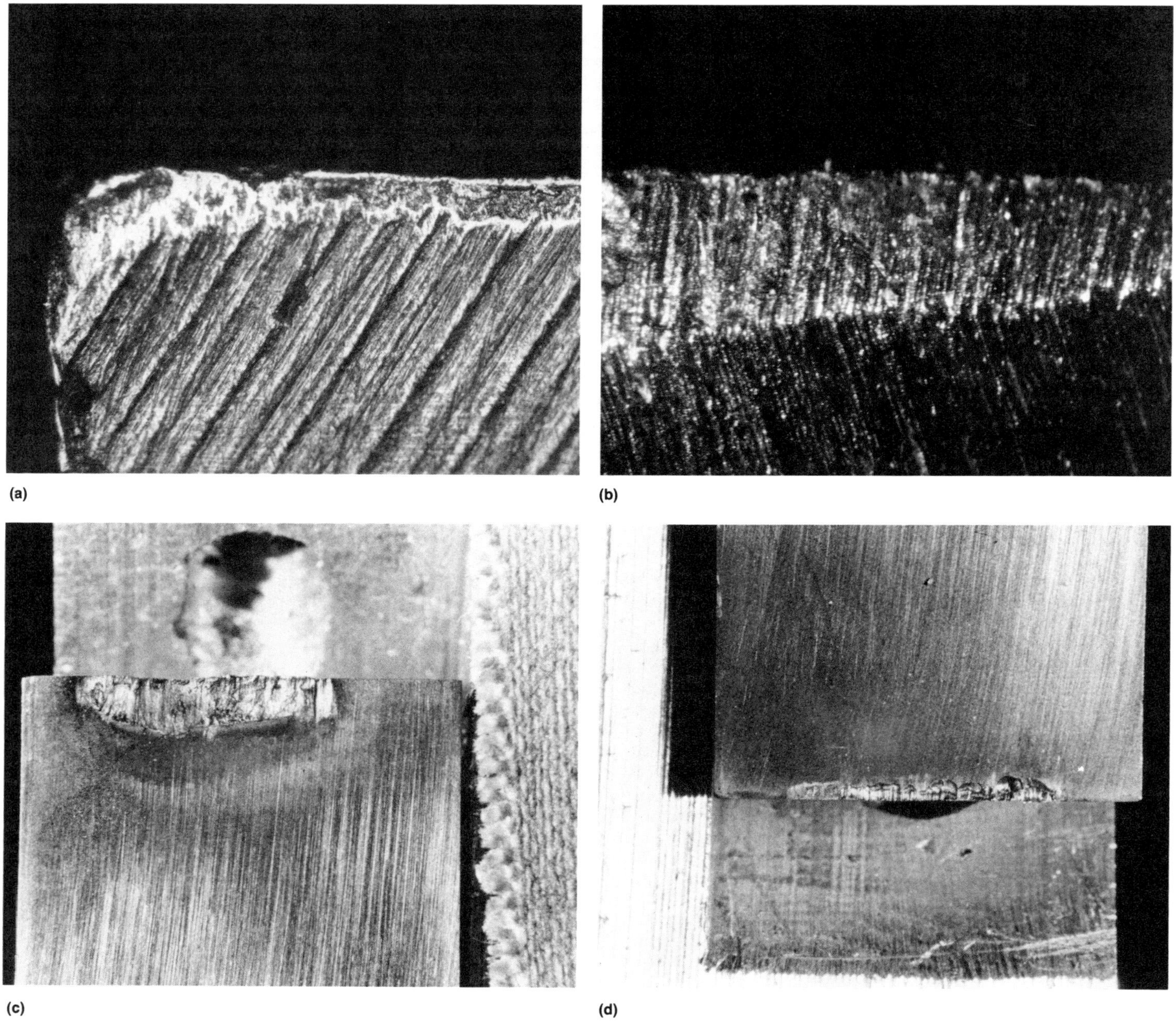

**Fig. 12** Typical flank wear scars. (a) Flank wear on an end mill. (b) Flank wear on the cutting edge of a drill. (c) and (d) Flank wear on a turning tool used for an orthogonal cut on a ring. (c) High-speed steel tool. (d) Coated high-speed steel tool. Courtesy of A.E. Bayoumi, Washington State University

- Help relate the results to productivity and economic gains

Experimentation assistance is needed to:

- Design the test plan including the statistically designed experiment if one is to be used
- Conduct the calibration tests and determine what assignable errors exist in the test equipment
- Conduct the data analysis and compute any parameters that are needed in the wear models

With this expertise in place, the testing program can begin.

## Test Plan Objectives

There are global and local objectives that must be dealt with. The global concern often comes from management, who have decisions to make that require tool wear data. Only two objectives will be defined and expanded upon using examples. (However, the tool wear problems of industry can provide many, diverse reasons to conduct tool wear tests.) The first is the problem of selecting tools to buy. This will be referred to as the *acceptance objective*. The second concerns the development of cutting condition data for a new tool-operation-material combination. This will be referred to as the *operating objective*. Many aspects of the test plan will be the same for these objectives; however, there are a number of important differences, which will be identified. These two objectives might be stated as:

- *The acceptance objective*: To establish which HSS steel end mills to purchase for use in a machining center that machines parts from an AISI 4340 steel
- *The operating objective*: To develop a tool life model for an HSS end mill used in a machining center to cut AISI 4340 alloy steel

The local objectives are the testing conditions and wear criterion that should be used for these global objectives. These target more specifically the intended use of the tool and how its performance will be judged. For example, the following local objectives will be established:

- *Acceptance local objectives*: A sizable rough cut will be used at feed and speed conditions that are at the maximum generally used to machine this material on this particular machine. A limiting flank wear value will be used as a failure criterion
- *Operating local objective*: A statistically designed experiment will be used to develop the tool life model for the range of operating conditions normally used for this tool-machine-material combination. A limiting flank wear value will be used as a failure criterion

It should be noted that these objectives are still broadly stated. During the test plan design stage, specific values will be established that appropriately define the bounds for these objectives.

## Designing the Test

The establishment of test conditions, a sampling plan, and test plans are all integral to the design of the test.

**Establishment of Test Conditions.** A machine that represents the type of machining center that will be used in production must be identified and made available. It is important to examine this machine closely and perform all maintenance that is necessary to ensure that it will operate in a nominal way. The speed and feed settings must be checked to see whether they produce the correct motion values. The runout of the table and spindle must be checked.

It is desirable to add instrumentation to the machine and connect it to an appropriate data acquisition system. Power, spindle torque, and force components on the tool are useful measurements to obtain (Ref 5). Acoustic emission sensors are also proving to be of value (Ref 11). These measurements will not be used in the procedure described in this article; however, they should be obtained if the wear study results are to be used for process monitoring and control using on-line sensors.

Size of the workpiece should be considered next. As discussed earlier, tool temperature influences the type of wear mechanism and the wear rate, and the mass of the workpiece is a variable heat sink as it is cut. For the current example, this temperature condition will be handled as follows for the use of acceptance and operating tests.

The acceptance test part size will be large plates. The tools to be tested will be sequenced so that each vendor's tool will have a balanced exposure to the heat sink mass effects of the plates as they are reduced in size during machining. Coolant flow will be constant and will be measured to verify.

The operating test part size is more difficult to establish. In a production environment, the goal would be to use coolant flow rates and cutting conditions that would maintain a tool temperature level below a safe upper bound. In the test environment, this same upper-bound temperature constraint is desirable, but it is also necessary to allow normal temperature changes below this upper bound caused by changes in the operating conditions. The recommendation is to use the largest workpiece for the machine being used and to monitor the surface temperature of the material a fixed distance from the cut. An infrared sensor could be used. When this temperature exceeds the upper bound value, a larger piece of material is called for.

Cutting conditions for the acceptance objective must severely test each vendor's tools, whereas operating objective tests require conditions that represent the range of values used for the operating environment.

Acceptance cutting conditions will be the same for every test cut. Some analysis will have to be made of the maximum speed and feed conditions normally used on that machine for the rough-cut size selected. Some sample cuts will most likely be necessary in order to establish that these conditions can be adequately handled by the machine tool.

Operating cutting conditions will be systematically varied over the range of values that are anticipated for use in the production environment. A statistically designed experiment will be used to provide data for estimation of the coefficients in the partial second order model described in Ref 11 and given in Table 1. The cutting conditions to be varied will be feed rate and cutting velocity. A representative depth of cut will be used for all tests. A full $2^2$ factorial design will be used with center point replication for a total of six tests. A fractional replication of this full design requiring fewer tests is another option, and more recent optimal designs have been proposed for tool life models (Ref 12).

**Sampling Plan.** The work material for each of these test programs should come from the same manufacturer and carry the same heat number. Verification of chemical composition and condition should be obtained. The tool sampling should be:

- *Acceptance tools*: Randomly selected from a batch of tools from each vendor. In this case, five tools should be used
- *Operating tools*: Randomly selected from a batch of the particular tool type being tested. In this case, six would be needed

**Test Points.** For both the acceptance and operating test plans, the test data are used to generate a wear curve for each tool. This means that the test must be interrupted periodically to measure the flank wear on the tool. Equal intervals of time should be used.

## Conducting the Test

An unneeded tool should be used to machine a starting surface on the edge of the workpiece. The test series can now be started and should be conducted in the same manner for each test plan. The end mills should be removed and placed in the measurement fixture. Each flute of these mills will have multiple measurements made along the length of the flank wear surface. Averages for each mill must be used when the results are analyzed. The test can be interrupted and measurements made until the wear limit has been exceeded. At the end of the test, metallographic examinations are conducted using sectioned portions of tools. These examinations will provide data concerning wear mechanisms and wear surface topography.

## Analyzing the Results

The first step is the same for both testing plans; the wear curves are plotted. If all the test data fall within the steady-state wear region, a linear wear trend may exist, and such a line could be fitted to the data. If one assumes such data exist for the acceptance test, as plotted in Fig. 13, this data would provide a set of five tool life values for each tool vendor. The operating test plan would produce similar plots and six tool life values, one for each test condition. Further analysis will differ for each of the plans.

**Acceptance data analysis** seeks a quality measure for the tool of each vendor. Reliability is the measure and can be computed using the five tool life values for each vendor. The method of median ranks and a graphical technique are used to establish the parameters for the Weibull probability distribution for each vendor (Ref 13). This is shown in Fig. 14 for some hypothetical data. To compare one vendor with another, a tool life value, $t$, was selected, and the reliability computed using the values of $\theta$ (characteristic life parameter) and $\beta$ (shape parameter) for each vendor's tools. For the parameter values $\theta = 100$ and $\beta = 6$ calculated in Fig. 14 for $t = 70$ minutes, $R(t) = 0.889$. This number is the probability that the vendor's tools will survive this service life. Higher reliability values imply that the tools are of higher quality. A statistician can provide other tests that compare tool life values; however, this reliability measure is used to provide background for application of this test information for tool change decisions.

**Operating data analysis** establishes the value of the significant coefficients in the tool life model. Any coefficients that are statistically insignificant will be set to zero, which eliminates the product term in which the coefficient appears. The terms that remain define the tool life equation. Any

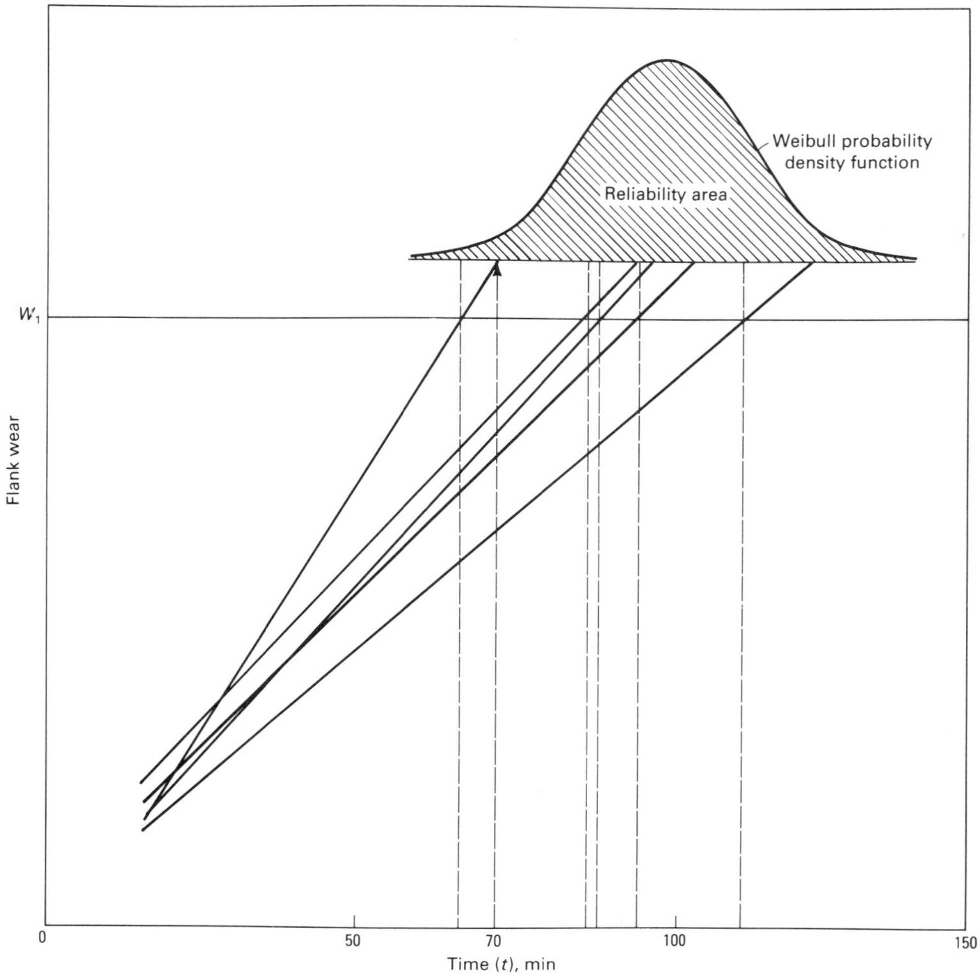


**Fig. 13** Acceptance test linearized wear curves for the tools of one vendor. The relationship of the Weibull probability density function to the tool life variability is shown, as well as the reliability probability.

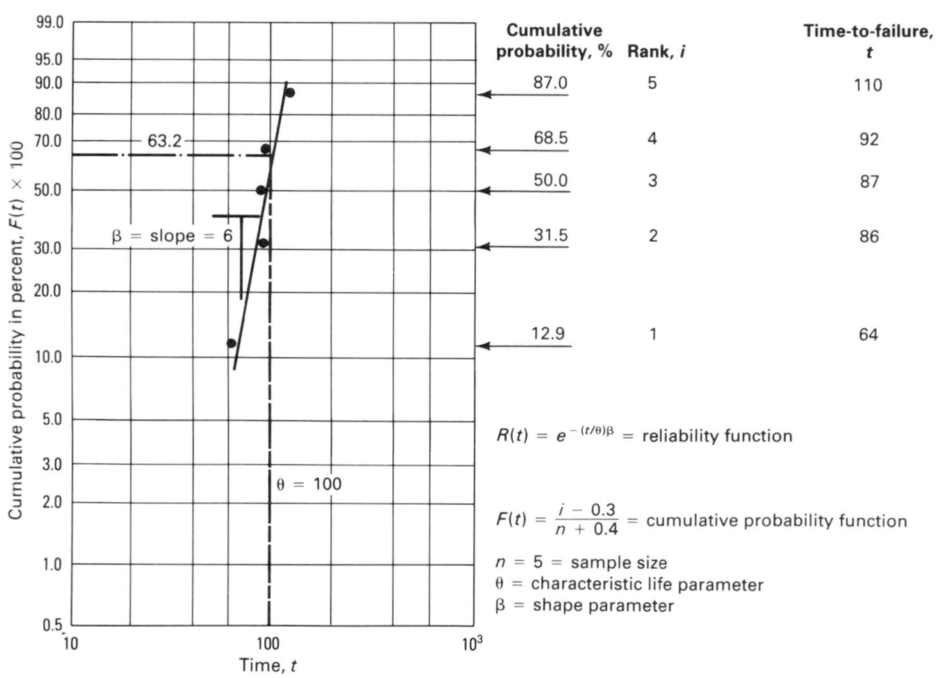


**Fig. 14** Graphical method to develop the Weibull parameters

combination of cutting condition variables within the span of the test cutting conditions can be substituted into the equation, and the tool life can then be computed. The method can also provide a confidence interval for this tool life estimate.

## Applying the Results

The results for these two tests will be used to indicate the economic impact of the wear information on the production process. Before addressing each of these test plans individually, some general background concerning such analyses is presented below.

The primary objective is to select cutting conditions to obtain a tool service life that meets the process management objectives. The decision-making process must take into account the possibility that a wear or catastrophic-failure event may occur before replacement is made. These will be called premature failures.

The tool life models in Table 1 predict average performance. If a statistical technique was used to develop the equation, this is an accurate assumption. A simple way to use these models is to select a replacement interval and then choose cutting conditions that satisfy the model. At these cutting conditions, about half of the tools will prematurely fail. Because this is generally unacceptable, the cutting conditions must be changed or the interval must be shortened to ensure fewer premature failures. The lower confidence bound on a statistically developed model can be used to make this margin of safety more precise.

These approaches ignore some very important economic consequences. First, the decision does not take into account the gains to be made by increased material removal rates and tool usage penalties associated with such productivity gains. Figure 15 shows this trade-off for costs, but such relationships could also be based on production rates or profits. The recommended operating point is obtained by computing the operating condition that minimizes the total cost function shown in Fig. 15. Second, the method incorrectly assumes that replacing premature failures costs the same as scheduled replacements. Premature failure not only stops the process, it can damage the part and add rework or scrapped-part expense to the process.

**Acceptance test plan evaluations** are conducted to measure the impact of poor tool quality. The wear data provide a probability distribution for tool life for each vendor's tools. Table 3 presents two common fixed interval replacement policies that can be used. The scheduled replacement policy is used in this example, and the expected replacement cost per unit of operating time computed, for different scheduled replacement intervals, $t_s$. There are formal optimization approaches that can

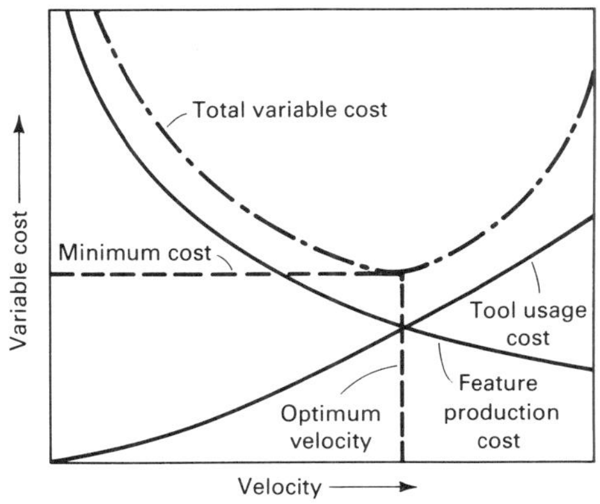


**Fig. 15** Minimum cost operating point for a cutting process

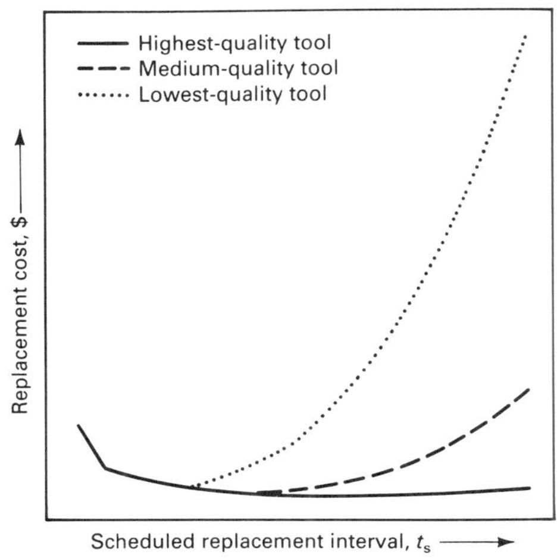


**Fig. 16** The expected replacement cost for three quality levels of tools

compute this interval directly (Ref 14); however, good, useful information can be easily obtained using a spreadsheet program with plotting capabilities. The θ and β parameters of the Weibull distribution are used to obtain renewal function values from statistical tables (Ref 15). For certain conditions, an approximation for this function can be obtained directly by using the reliability function (Ref 16). Figure 16 shows typical plots for three vendors and indicates that the expected cost for the higher-quality tools was less sensitive to the choice of the scheduled replacement interval. This knowledge would certainly be of comfort to a process manager who otherwise would not know that by using the higher-quality tools, more flexibility was available concerning the choice of the replacement interval.

**Operating test plan evaluations** are used to establish operating conditions that minimize the variable costs to machine this particular feature. The tool life model that was developed has two independent variables. It is possible to establish the combination of values for these variables that will minimize this cost; however, there are often other machining considerations that will fix

**Table 3 Expected cost functions for tool replacement policies**

**Preventive planned replacement**

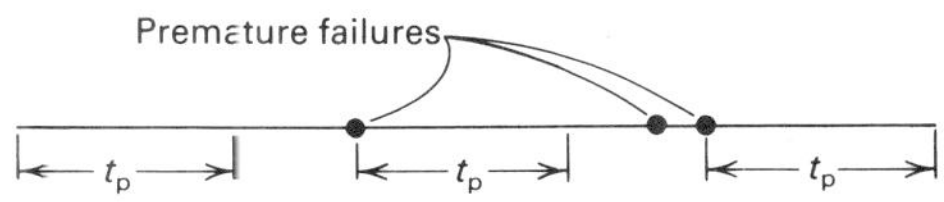


Expected replacement cost when replacing the tool at the end of planned period $t_p$ =

$$\{c_pR(t_p) + c_fF(t_p)\}/\{\int_0^{t_p} R(t)dt\}$$

**Scheduled replacement**

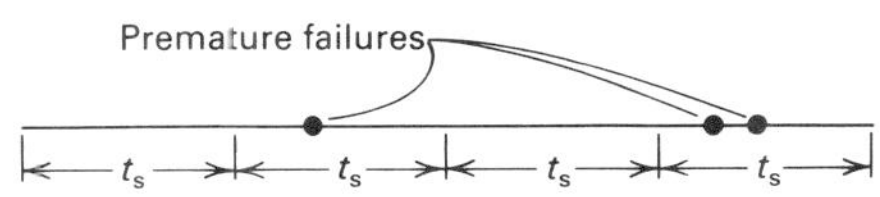


Expected replacement cost when replacing the tool at the end of scheduled period $t_s$ =

$$\{c_s + c_fH(t_s)\}/t_s$$

$c_p$-$c_s$ = cost for planned and scheduled replacements; $c_f$ = cost for a premature failure replacement ($c_f >> c_p$ or $c_s$); $R$, $F$, and $H$ designate the reliability, cumulative, and renewal function, respectively, for the probability density function that models tool life.

one of these. As shown in Fig. 15, this variable cost, $C$, equals the sum of the production costs and the tool usage costs for the feature. The functional expression is:

$$C = c_0t_0 + n_t(c_0t_t + C_t)$$

where $c_0$ is the cost per unit time for machine operation; $t_0$ is the time to produce the feature; $n_t$ is the fraction of tool life used for the feature, which is equal to $t_0/t$; $t_t$ is the time required for a tool change; and $C_t$ is the cost of the cutting tool.

The proper expression for $t_0$ depends on the type of tool and feature machined by the tool. The value of $t$ for any combination of operating conditions comes from the tool life model, and an evaluation of the cost $C$ could be accomplished using a spreadsheet or formal optimization procedures (Ref 17). This procedure works well for a tool that is used to produce this same feature again and again until the tool is replaced. A typical application of this type is a transfer line. When the tool is used to produce a variety of features such as in a machining center, the analysis becomes more complex. A system that estimates the proportion of the life of the tool used for each feature must be used and then these values must be added together to establish when the tool needs to be changed.

# Future Trends

The complex machining wear environment will continue to be studied and better understood by tribologists, material scientists, and manufacturing engineers. Its importance deserves such attention, and the research efforts by these investigators needs continuing support. The use of this wear knowledge for tool change decisions, process monitoring, and control strategies is even more important. Dependence on off-line laboratory testing and model development will eventually become too costly and time consuming for the current and future automated machining systems. The process will become the direct source of wear information through the use of new monitoring techniques and sensors. Local data bases that store performance information concerning the production of each feature will be able to access the progressive wear of the tool more precisely. Some of the motivation to develop such systems will come from changes in controller technology that will include the intelligent decision-making capabilities currently done by human experts.

## REFERENCES

1. P.K. Wright, Physical Models of Tool Wear for Adaptive Control in Flexible Machining Cells, in *Proceedings of the Symposium on Computer Integrated Manufacturing*, Vol 8, American Society of Mechanical Engineers, 1986
2. M.C. Shaw, *Metal Cutting Principles*, Oxford University Press, 1984
3. S.C. Lim and M.F. Ashby, Overview No. 55, Wear-Mechanism Maps, *Acta Metall.*, Vol 35 (No. 1), 1987, p 1-24
4. B.M. Kramer and N.P. Suh, Tool Wear by Solution: A Quantitative Understanding, *J. Eng. Ind. Trans. ASME*, Vol 102, 1980, p 303
5. A.E. Bayoumi and L.A. Kendall, Modeling and Measurement of Wear of Coated and Uncoated High Speed Steel End Mills, *J. Mater. Shap. Technol.*, to be published
6. S. Jetly, Measuring Cutting Tool Wear On-Line: Some Practical Considerations, *Manuf. Eng.*, July 1984, p 55
7. F.W. Taylor, On the Art of Cutting Tools, *Trans. ASME*, Vol 28, 1907
8. B.N. Colding and W. Konig, Validity of the Taylor Equation in Metal Cutting, *Ann. CIRP*, Vol 19 (No. 4), 1971, p 793
9. S. Kalpakjian, *Manufacturing Processes for Engineering Materials*, Addison-Wesley, 1984, p 489
10. D.A. Dornfeld, The Role of Acoustic Emission in Manufacturing Process Monitoring, in *Proceedings of the 13th NAMRC*, Society of Manufacturing Engineers, May 1985, p 69-74
11. S.M. Wu, Tool Life Testing by Response Surface Methodology, Parts I and II, *J. Eng. Ind. Trans. ASME*, Series B, Vol 86, 1964, p 105
12. W.J. Zdeblick and R.E. Devor, An Experimental Strategy for Designing Tool Life Experiments, *J. Eng. Ind. Trans. ASME*, Vol 100, 1978, p 441

13. K.C. Kapur, chapter 11 in *Reliability in Engineering Design*, John Wiley & Sons, 1977
14. A.K. Sheikh, L.A. Kendall, and S.M. Pandit, Probabilistic Optimization of Multitool Operations, *J. Eng. Ind. Trans. ASME*, Vol 102, 1980
15. J.S. White, "Weibull Renewal Analysis," Paper presented at the 3rd Annual Aerospace Reliability Maintainability Conference (Washington, D.C.), June/July 1964
16. S.B. Billatos and L.A. Kendall, Approximate Renewal Functions for Optimization, in *Proceedings of the 40th Meeting of the Mechanical Failures Prevention Group*, T.R. Shives, Ed., Cambridge University Press, 1985
17. G. Boothroyd, chapter 6 in *Fundamentals of Metal Machining and Machine Tools*, McGraw-Hill, 1975

# Cutting Tool Materials

# High-Speed Tool Steels

Alan M. Bayer and Bruce A. Becherer, Teledyne Vasco

HIGH-SPEED TOOL STEELS and their requirements are defined by The American Society for Testing and Materials in Specification A600-79 as follows: High-speed tool steels are so named primarily because of their ability to machine materials at high cutting speeds. They are complex iron-base alloys of carbon, chromium, vanadium, molybdenum, or tungsten, or combinations thereof, and in some cases substantial amounts of cobalt. The carbon and alloy contents are balanced at levels to give high attainable hardening response, high wear resistance, high resistance to the softening effect of heat, and good toughness for effective use in industrial cutting operations. Commercial practice has developed two groups of cutting materials:

- The recognized standard high-speed tool steel, which serves almost all applications under mild to severe metal-cutting conditions
- A smaller group of intermediate steels, which are satisfactory for limited applications under mild to moderate metal-cutting conditions

The minimum requirements that must be met to be classed as a standard high-speed tool steel, and those for an intermediate high-speed tool steel, are listed in Table 1. To be acceptable for either group, an alloy must meet all of the requirements shown for that group.

A chronology of some of the significant developments in high-speed tool steels is given in Table 2. The research work in 1903 on a 14% tungsten alloy led to the development of the first high-speed tool steel, which is now designated T1.

**Table 1 Requirements for high-speed tool steels per ASTM A 600**

| Requirement | Standard | Intermediate |
|---|---|---|
| **Chemical requirements** | | |
| Minimum alloy content by major elements | | |
| Carbon | 0.65 | 0.70 |
| Chromium | 3.50 | 3.25 |
| Vanadium | 0.80 | 0.80 |
| Tungsten + 1.8% molybdenum | 11.75 | 6.50 |
| Minimum total alloy content based on tungsten equivalents (⅓ Cr + 6.2 V + W+ 1.8 Mo) | | |
| Grades containing less than 5% cobalt | 22.50 | 13.00 |
| Grades containing 5% or more cobalt | 21.00 | 12.00 |
| **Hardening response requirements** | | |
| Ability to be austenitized, and tempered at a temperature not less than 510 °C (950 °F) with a fine-grain structure (Snyder-Graff grain size 8 min) to | 63 HRC | 62 HRC |

**Table 2 Significant dates in the development of high-speed tool steels**

| Date | Development |
|---|---|
| 1903 | 0.70% C, 14% W, 4% Cr prototype of modern high-speed tool steels |
| 1904 | 0.30% V addition |
| 1906 | Introduction of electric furnace melting |
| 1910 | Introduction of first 18-4-1 composition (AISI T1) |
| 1912 | 3 to 5% Co addition for improved hot hardness |
| 1923 | 12% Co addition for increased cutting speeds |
| 1939 | Introduction of high-carbon high-vanadium super high-speed tool steels (M4 and T15) |
| 1940–1952 | Increasing substitution of molybdenum for tungsten |
| 1953 | Introduction of sulfurized free-machining high-speed tool steel |
| 1961 | Introduction of high-carbon high-cobalt super hard high-speed tool steels (M40 series) |
| 1970 | Introduction of powdered metal high-speed tool steels |
| 1973 | Addition of higher silicon/nitrogen content to M-7 to increase hardness |
| 1980 | Development of cobalt-free super high-speed tool steels |
| 1982 | Introduction of aluminum-modified high-speed tool steels for cutting tools |

## M and T Classification

There are presently more than 40 individual classifications of high-speed tool steels, according to the American Iron and Steel Institute (AISI). When these are compounded by the number of domestic manufacturers, the total number of individual steels in the high-speed tool steel category exceeds 150. The AISI established a classification system for the high-speed tool steels many years ago. That system consists of a T for those steels that have tungsten as one of their primary alloying elements and an M for those steels that have molybdenum additions as one of their primary alloying elements. In addition, there is a number that follows either the M or the T. Thus, there are high-speed tool steels designated M1, M2, M41, T1, T15, and so on. That number does not have any special significance other than to distinguish one from another. For example, M1 does not mean that it is more highly alloyed than M2 or has greater hardenability or poorer wear resistance, and so on. It merely separates the types and attempts to simplify selection for the user. Table 3 lists the nominal analyses of the common M and T types.

## Effect of Alloying Elements

The T series contains 12 to 20% tungsten, with chromium, vanadium, and cobalt as the other major alloying elements. The M series contains approximately 3.5 to 10% molybdenum, with chromium, vanadium, tungsten, and cobalt as the other alloying elements. All types, whether molybdenum or tungsten, contain about 4% chromium; the carbon and vanadium contents vary. As a general rule, when the vanadium content is increased, the carbon content is usually increased (Ref 1).

The tungsten type T1 does not contain molybdenum or cobalt. Cobalt-base tungsten types range from T4 through T15 and contain various amounts of cobalt.

Molybdenum types M1 through M10 (except M6) contain no cobalt, but most contain some tungsten. The cobalt-base, molybdenum-tungsten, premium types are generally classified in the M30 and M40 series. Super high-speed steels normally range from M40 upward; they are capable of being heat treated to high hardnesses.

The M series steels generally have higher abrasion resistance than the T series steels and less distortion in heat treatment; also, they are less expensive (Ref 2). Tools made of high-speed tool steel can also be coated with titanium nitride, titanium carbide, and numerous other coatings by physical vapor deposition technique for improved performance and increased tool life.

**Table 3 Composition of high-speed tool steels**

| AISI type | UNS designation | C | Si | Cr | V | W | Mo | Co |
|---|---|---|---|---|---|---|---|---|
| **Molybdenum high-speed tool steels** | | | | | | | | |
| M1 | T11301 | 0.83 | 0.35 | 3.75 | 1.18 | 1.75 | 8.70 | ... |
| M2 | | | | | | | | |
| Regular C | T11302 | 0.83 | 0.33 | 4.13 | 1.98 | 6.13 | 5.00 | ... |
| High C | ... | 1.00 | 0.33 | 4.13 | 1.98 | 6.13 | 5.00 | ... |
| M3 | | | | | | | | |
| Class 1 | T11313 | 1.05 | 0.33 | 4.13 | 2.50 | 5.88 | 5.63 | ... |
| Class 2 | T11323 | 1.20 | 0.33 | 4.13 | 3.00 | 5.88 | 5.63 | ... |
| M4 | T11304 | 1.33 | 0.33 | 4.25 | 4.13 | 5.88 | 4.88 | ... |
| M6 | T11306 | 0.80 | 0.33 | 4.13 | 1.50 | 4.25 | 5.00 | 12.00 |
| M7 | T11307 | 1.01 | 0.38 | 3.75 | 2.00 | 1.75 | 8.70 | ... |
| M10 | | | | | | | | |
| Regular C | T11310 | 0.89 | 0.33 | 4.13 | 2.00 | ... | 8.13 | ... |
| High C | ... | 1.00 | 0.33 | 4.13 | 2.00 | ... | 8.13 | ... |
| M15 | T11315 | 1.50 | 0.33 | 4.00 | 5.00 | 6.50 | 3.50 | 5.00 |
| M30 | T11330 | 0.80 | 0.33 | 4.00 | 1.25 | 2.00 | 8.00 | 5.00 |
| M33 | T11333 | 0.89 | 0.33 | 3.75 | 1.18 | 1.70 | 9.50 | 8.25 |
| M34 | T11334 | 0.89 | 0.33 | 3.75 | 2.10 | 1.75 | 8.48 | 8.25 |
| M35 | T11335 | 0.80 | 0.33 | 4.00 | 2.00 | 6.00 | 5.00 | 5.00 |
| M36 | T11336 | 0.85 | 0.33 | 4.13 | 2.00 | 6.00 | 5.00 | 8.25 |
| M41 | T11341 | 1.10 | 0.33 | 4.13 | 2.00 | 6.63 | 3.75 | 8.25 |
| M42 | T11342 | 1.10 | 0.40 | 3.88 | 1.15 | 1.50 | 9.50 | 8.25 |
| M46 | T11346 | 1.26 | 0.53 | 3.95 | 3.15 | 2.05 | 8.25 | 8.30 |
| M48 | T11348 | 1.50 | 0.33 | 3.88 | 3.00 | 10.00 | 5.13 | 9.00 |
| M50(a) | T11350 | 0.80 | 0.40 | 4.13 | 1.00 | ... | 4.25 | ... |
| M52(a) | T11352 | 0.90 | 0.40 | 4.00 | 1.93 | 1.25 | 4.45 | ... |
| M62 | T11362 | 1.30 | 0.28 | 3.88 | 2.00 | 6.25 | 10.50 | ... |
| **Tungsten high-speed tool steels** | | | | | | | | |
| T1 | T12001 | 0.73 | 0.30 | 4.13 | 1.10 | 18.00 | ... | ... |
| T4 | T12004 | 0.75 | 0.30 | 4.13 | 1.00 | 18.25 | 0.70 | 5.00 |
| T5 | T12005 | 0.80 | 0.30 | 4.38 | 2.10 | 18.25 | 0.88 | 8.25 |
| T6 | T12006 | 0.80 | 0.30 | 4.38 | 1.80 | 19.75 | 0.70 | 12.00 |
| T8 | T12008 | 0.80 | 0.30 | 4.13 | 2.10 | 14.00 | 0.70 | 5.00 |
| T15 | T12015 | 1.55 | 0.28 | 4.38 | 4.88 | 12.38 | 1.00 | 5.00 |

(a) Intermediate high-speed tool steel

Various elements are added to M and T series high-speed tool steels to impart certain properties to the tool steels. These elements and their effects are discussed below.

**Carbon** is by far the most important of the elements and is very closely controlled. While the carbon content of any one high-speed tool steel is usually fixed within narrow limits, variations within these limits can cause important changes in the mechanical properties and the cutting ability. As the carbon concentration is increased, the working hardness also rises; the elevated temperature hardness is higher; and the number of hard, stable, complex carbides increases. The latter contribute much to the wear resistance and other properties of the high-speed tool steels.

**Silicon.** The influence of silicon on high-speed tool steel, up to about 1.00%, is slight. Increasing the silicon content from 0.15 to 0.45% gives a slight increase in maximum attainable tempered hardness and has some influence on carbide morphology, although there seems to be a concurrent slight decrease in toughness. Some manufacturers produce at least one grade with silicon up to 0.65%, but this level of silicon content requires a lower maximum austenitizing temperature than does a lower silicon level in the same grade, if overheating is to be avoided. In general, however, the silicon content is kept below 0.45% on most grades.

**Manganese.** Generally, manganese is not high in concentration in high-speed tool steels. This is because of its marked effect in increasing brittleness and the danger of cracking upon quenching.

**Phosphorus** has no effect on any of the desired properties of high-speed tool steels, but because of its well-known effect in causing cold shortness, or room-temperature brittleness, the concentration of phosphorus is kept to a minimum.

**Chromium** is always present in high-speed tool steels in amounts ranging from 3 to 5% and is mainly responsible for the hardenability. Generally, the addition is 4% because it appears that this concentration gives the best compromise between hardness and toughness. In addition, chromium reduces oxidation and scaling during heat treatment.

**Tungsten.** In the high-speed tool steels, tungsten is of vital importance. It is found in all T-type steels and in all but two of the M-type steels. The complex carbide of iron, tungsten, and carbon that is found in high-speed tool steels is very hard and significantly contributes to wear resistance. Tungsten improves hot hardness, causes secondary hardening, and imparts marked resistance to tempering. When the tungsten concentration is lowered in high-speed tool steels, molybdenum is usually added to make up for its loss.

**Molybdenum** forms the same double carbide with iron and carbon as tungsten does but has half the atomic weight of tungsten. As a consequence, molybdenum can be substituted for tungsten on the basis of approximately one part of molybdenum, by weight, for two parts of tungsten.

The melting point of the molybdenum steels is somewhat lower than that of the tungsten grades, and thus they require a lower hardening temperature and have a narrower hardening range. The M-type high-speed tool steels are tougher than the T-type high-speed tool steels, but the hot hardness is slightly lower. Compensation for this reduced hot hardness is partially accomplished by the addition of tungsten and, to a lesser extent, vanadium to the plain molybdenum grades. This is one important reason for the popularity of the tungsten-molybdenum grades (like M2, M3, M4): they afford good hot hardness, which is so desirable in high-speed tool steels.

**Vanadium** was first added to high-speed tool steels as a scavenger to remove slag impurities and to reduce nitrogen levels in the melting operation, but it was soon found that the element materially increased the cutting efficiency of tools. The addition of vanadium promotes the formation of very hard, stable carbides, which significantly increase wear resistance and, to a lesser extent, hot hardness. An increase in vanadium, when properly balanced by carbon additions, has relatively little effect on the toughness. For this reason, vanadium-bearing grades are a very good choice when very fast cutting operations are demanded, as in finishing cuts, or when the surface of the material is hard and scaly. The special characteristics of the high-speed tool steels that are due to high vanadium additions have given rise to several specially developed steels for very severe service requiring high toughness as well as exceptional hot hardness and wear resistance. The T15, M4, and M15 grades are in this category; their vanadium content is 4.88, 4.13, and 5.00%, respectively.

**Cobalt.** The main effect of cobalt in high-speed tool steel is to increase the hot hardness and thus to increase the cutting efficiency when high tool temperatures are attained during the cutting operation. Cobalt raises the heat-treating temperatures because it elevates the melting point. Hardening temperatures for cobalt high-speed tool steels can be 14 to 28 °C (25 to 50 °F) higher than would be normal for similar grades without cobalt. Cobalt additions slightly increase the brittleness of high-speed tool steels.

Cobalt steels are especially effective on rough or hogging cuts, but they are not usually suited to finishing cuts that do not involve high temperatures. Cobalt types usually perform quite well when cutting materials that have discontinuous chips such as cast iron or nonferrous metals.

The necessity of using deep cuts and fast speeds or of cutting hard and scaly materials justifies the use of cobalt high-speed tool steels.

**Sulfur,** in normal concentrations of 0.03% or less, has no effect on the properties of high-speed tool steels. However, sulfur is added to certain high-speed tool steels to contribute free-machining qualities, as it does in low-alloy steels. The amount of free-machining high-speed tool steels is a small but significant percentage of the total consumption of high-speed tool steels. One of the major areas for free-machining high-speed tool steels is in larger-diameter tools such as hobs, broaches, and so on.

Sulfur forms complex sulfides, containing chromium, vanadium, and manganese, which are distributed throughout the steel as stringer-type inclusions. The stringers interrupt the steel structure and act as notches, which aid the metal-removing action of a cutting tool when machining the high-speed steel, because the resulting chip is discontinuous, a characteristic of free-machining steels. Very high sulfur additions (up to 0.30%) are made to some powder metallurgy (P/M) high-speed tool steels for improved machinability/grindability by forming globular sulfides rather than stringers (see the article "P/M High-Speed Tool Steels" in this Volume).

**Nitrogen** is generally present in air-melted high-speed tool steel in amounts varying from approximately 0.02 to 0.03%. The nitrogen content of some high-speed tool steels is deliberately increased to about 0.04 to 0.05%, and this addition, when combined with higher than usual amounts of silicon, results in a slight increase of maximum attainable tempered hardness and some change of carbide morphology.

## Properties of High-Speed Tool Steels

High-speed tool steels, regardless of whether they are an AISI M-type or T-type, have a rather striking similarity in their physical makeup:

- They all possess a high-alloy content
- They usually contain sufficient carbon to permit hardening to 64 HRC
- They harden so deeply that almost any section encountered commercially will have a uniform hardness from center to surface
- They are all hardened at high temperatures, and their rate of transformation is such that small sections can be cooled in still air and be near maximum hardness

All high-speed tool steels possess excess carbide particles, which in the annealed state contain a high proportion of the alloying elements. These carbide particles contribute materially to the wear resistance of hardened high-speed tool steel. By partially dissolving during heat treatment, these carbides provide the matrix of the steel with the necessary alloy and carbon content for hardenability, hot hardness, and resistance to tempering.

While all high-speed tool steels have many similar mechanical and physical characteristics, the properties may vary widely due to changes in chemical composition. Basically, the most important property of a high-speed tool steel is its cutting ability. Cutting ability depends on a combination of properties, the four most important of these being:

- *Hardness*: Resistance to penetration by diamond-hard indenter, measured at room temperature
- *Hot hardness*: The ability to retain high hardness at elevated temperatures
- *Wear resistance*: Resistance to abrasion, often measured by grindability, metal-to-metal, or various other types of tests to indicate a relative rating
- *Toughness*: Ability to absorb (impact) energy

The relative importance of these properties varies with every application. High machining speeds require a composition with a high initial hardness and a maximum resistance to softening at high temperatures. Certain materials may abrade the cutting edge of the tool excessively; hence, the wear resistance of the tool material may well be more important than its resistance to high cutting temperatures.

Hardness is necessary for cutting harder materials and generally gives increased tool life, but it must be balanced against the toughness required for the application.

The desired combination of properties in a high-speed tool steel may be obtained, first, by selection of the proper grade and,

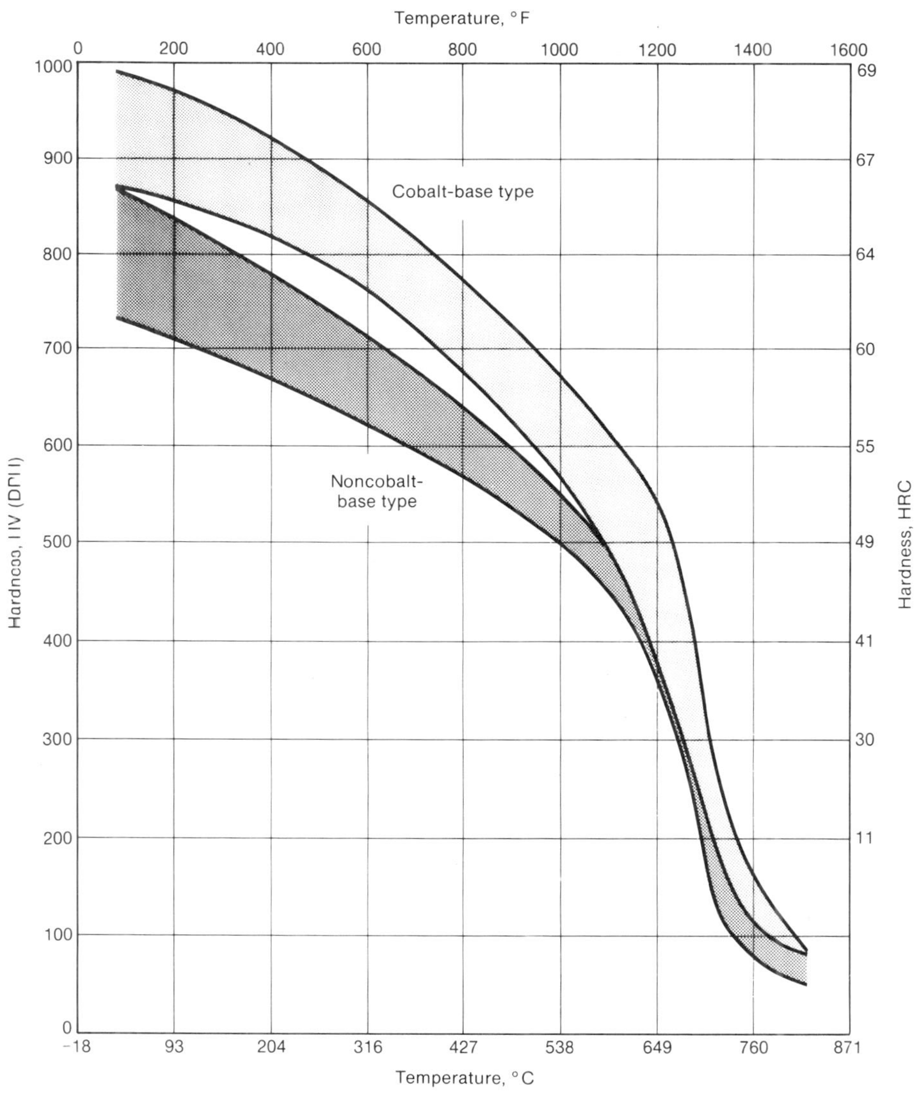


**Fig. 1** Comparison of the hot hardness of cobalt-base (M33, M36, M4, and T15) type versus noncobalt-base (M1 M2, M4, M7, and T1) type high-speed tool steels

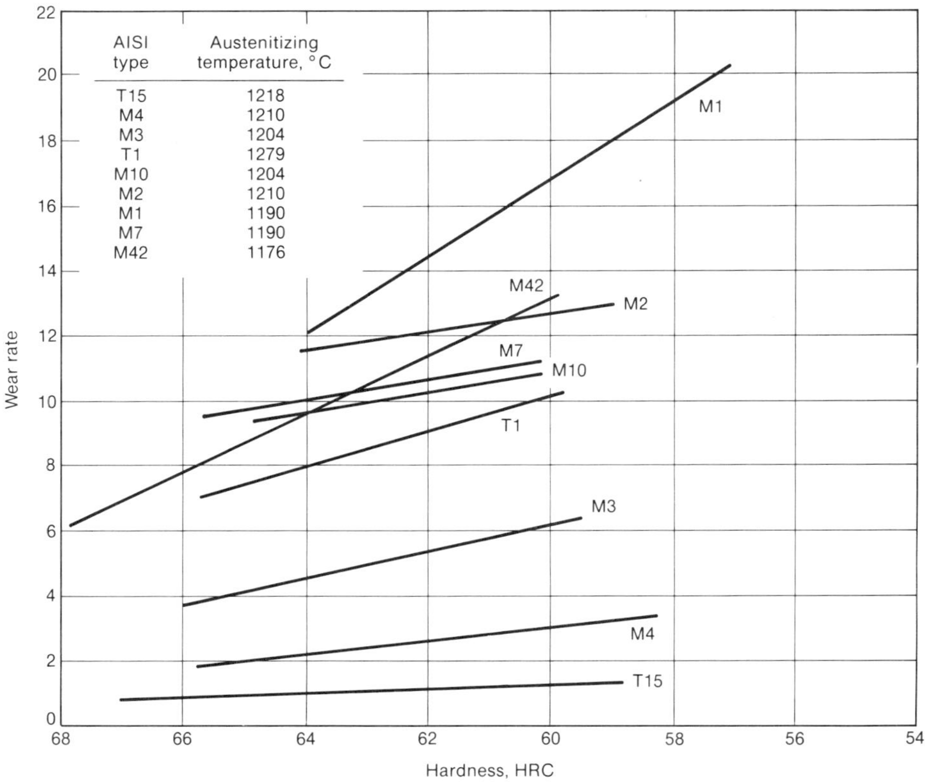


**Fig. 2** Effect of hardness on wear rate for high-speed tool steels, each having been double tempered to the indicated hardness

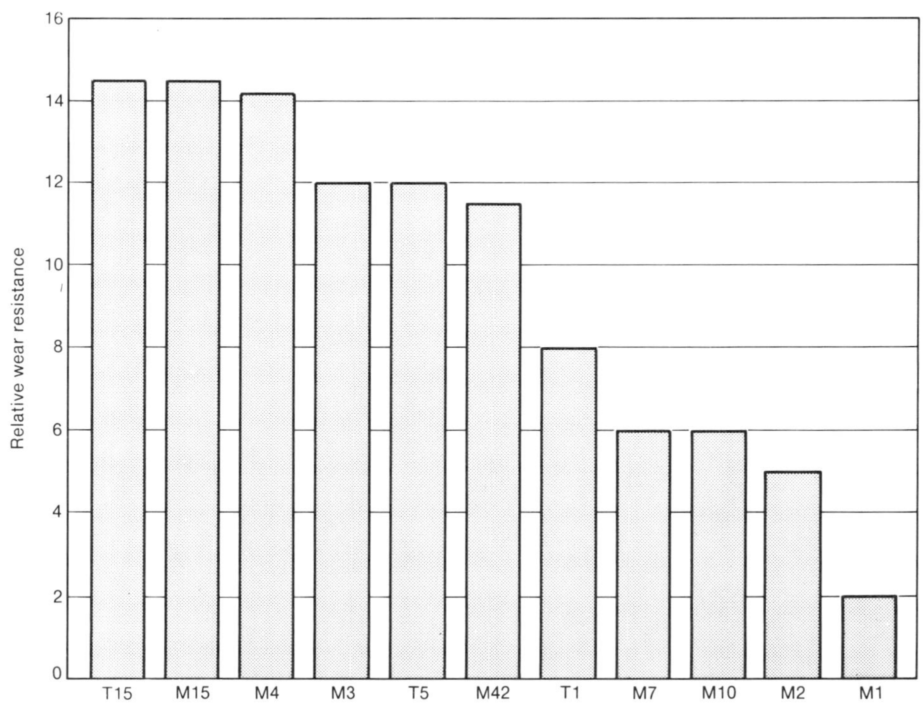


**Fig. 3** Comparison of relative abrasion resistance at typical working hardness for high-speed tool steels

second, by the proper heat treatment, two equally important decisions.

**Hardness.** Hardness is the most commonly stipulated requirement of a high-speed tool steel and is used as an acceptance check of a heat-treated tool. All high-speed tool steels can be hardened to room temperature hardness of 64 HRC, while the M40 series, some of the M30 series, and T15 can reach nearly 69 HRC.

**Hot Hardness.** A related and important component of cutting ability is hot hardness. It is simply the ability to retain hardness at elevated temperatures. This property is important because room temperature hardness values are not the same values that exist at the elevated temperature produced by friction between the tool and workpiece.

Hot hardness values of some representative grades are plotted in Fig. 1. It is noteworthy that the cobalt-base types as a group exhibit higher hot hardness than non-cobalt-base types. For a comparison of the hot hardness of other metals, alloys, carbides, and ceramics to that of high-speed tool steels, see Fig. 1 in the article "Cast Cobalt Alloys" in this Volume.

**Wear Resistance.** The third component of cutting ability is resistance to wear. Wear resistance of high-speed tool steels is affected by the matrix hardness and composition, by precipitated $M_2C$ and MC carbides responsible for secondary hardness, by the volume of excess alloy carbides, and by the nature of these excess carbides. In practically any given high-speed tool steels, wear resistance strongly depends on hardness of the steel, and higher hardness, however achieved, is an aim when highly abrasive cutting conditions are encountered (Fig. 2).

For the ultimate in wear resistance, carbon content has been increased simultaneously with vanadium content, to permit the introduction of a greater quantity of total carbide and a greater percentage of extremely hard vanadium carbide in high-speed tool steel. Examples of this effect are given when discussing the effect of vanadium on the properties of high-speed tool steels. Steels T15, M3 (class 2), M4, and M15 are in this category, and all have extremely high wear resistance.

Laboratory tests for wear resistance are diverse, making comparisons between different procedures difficult. Therefore, production tests on actual tools are used to a great extent. However, laboratory tests can produce valuable data on the relative wear resistance for these steels (Fig. 3). The data given in Fig. 3 were generated by measuring the volume loss of a high-speed tool steel sample against the volume loss of a known vitrified abrasive wheel after a predetermined grinding procedure.

**Toughness.** The fourth component of cutting ability mentioned above is toughness, which is defined as a combination of two factors:

- The ability to deform before breaking (ductility)
- The ability to resist permanent deformation (elastic strength)

If either of these factors is to be used to describe toughness (a practice not to be condoned), the second appears more practical for high-steel tool steel because rarely are large degrees of flow or deformation

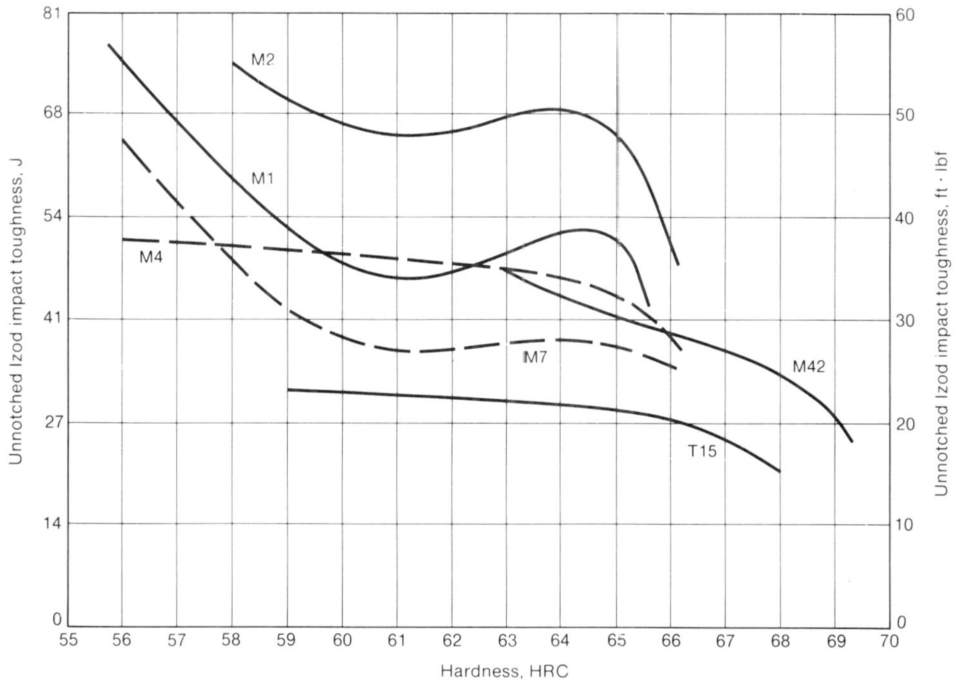


**Fig. 4** Plot of impact toughness versus hardness for high-speed tool steels

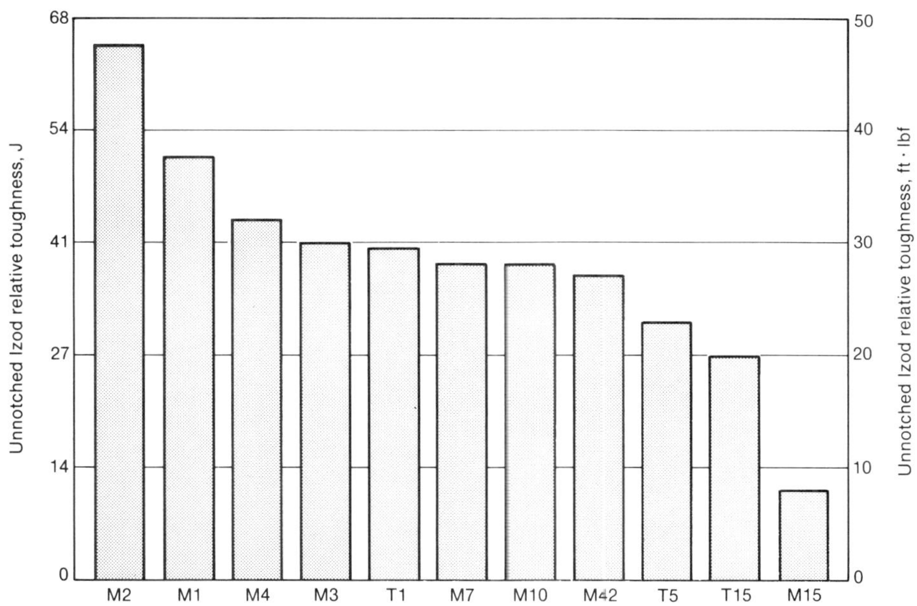


**Fig. 5** Relative toughness of high-speed tool steels at typical working hardness

permissible with fine-edge tools. The first, however, cannot be ignored, as frequently the stress applied to a tool (through overloads, shock, notches, and sharp corners) exceeds the elastic strength.

Toughness tests on high-steel tool steel are usually conducted at room temperature. Tool failures that occur from spalling of the tool edge generally occur during the initial contact of the tool with the work, and once the tool becomes heated, its performance in this respect is much superior. Therefore, room temperature tests are perhaps of greater value when toughness is considered than when hardness is in question. Laboratory tests for the measurement of toughness of hardened high-speed tool steel include bend, unnotched or C-notch impact, static torsion, and torsion impact tests. Figures 4 and 5 compare relative unnotched impact values for representative high-speed tool steels. Modest improvements in toughness (within a grade) can be made by lowering the tempered hardness. Lower austenitizing temperatures enhance the toughness for a given hardness and grade.

## Heat Treatment of High-Speed Tool Steels

Proper heat treatment is as critical to the success of the cutting tool as material selection itself. Often the highest-quality steel made into the most precise tools does not perform because of improper heat treatment.

The object of the heat treating or hardening operation is to transform a fully annealed high-speed tool steel consisting mainly of ferrite (iron) and alloy carbides into a hardened and tempered martensitic structure having carbides that provide the cutting tool properties (see Fig. 6 and 7).

The heat treatment process can be divided into four primary areas, preheating, austenitizing, quenching, and tempering. Figure 8 outlines graphically these four heat treatment steps.

**Preheating.** From a metallurgical standpoint, preheating plays no part in the hardening reaction; however, it performs three important functions. The first of these is to reduce thermal shock, which always results when a cold tool is placed into a warm or hot furnace. Minimizing thermal shock reduces the danger of excessive distortion or cracking. It also relieves some of the stresses developed during machining and/or forming, although conventional stress relieving is more effective.

The second major benefit of preheating is to increase equipment productivity by decreasing the amount of time required in the high-heat furnace. Thirdly, if the high-heat furnace is not neutral to the surface of the tool or part, preheating will reduce the amount of carburization and decarburization that would result if no preheat were employed. In commercial salt bath hardening, a two-step preheat is typically used for high-speed tool steels. The first preheat is carried out between 650 and 760 °C (1200 and 1400 °F); the second preheat cycle is carried out between 815 and 900 °C (1500 and 1650 °F). In atmosphere or vacuum heat treating, the furnace is usually heated slowly to a single preheat of 790 to 845 °C (1450 to 1550 °F). Preheat duration is of little importance as long as the part is heated throughout its cross section.

**Austenitizing (hardening)** is the second step of the heat treatment operation. Austenitizing is a time/temperature dependent reaction. High-speed tool steels depend upon the dissolving of various complex alloy carbides during austenitizing to develop their properties. These alloy carbides do not dissolve to any appreciable extent unless the steel is heated to a temperature within 28 to 56 °C (50 to 100 °F) of their melting point. This temperature is dependent upon the particular high-speed tool steel being treated and is in the range of 1150 to 1290 °C (2100 to 2350 °F). The generally recommended hold time for high-speed tool steel is approximately 2 to 6 min, depending upon

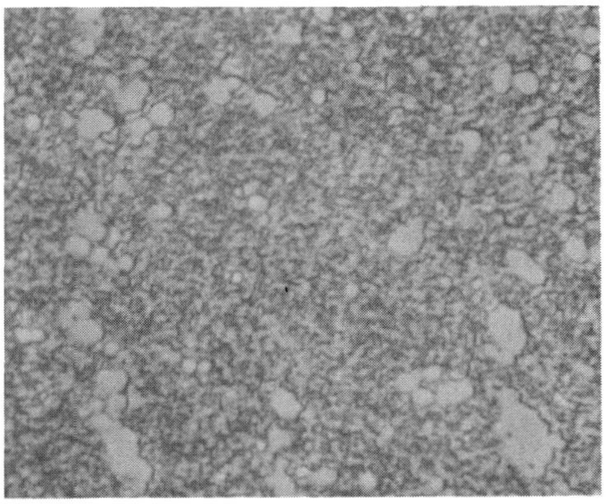

**Fig. 6** Microstructure of fully annealed high-speed tool steel consisting of ferrite (iron) and alloy carbides. 1000×

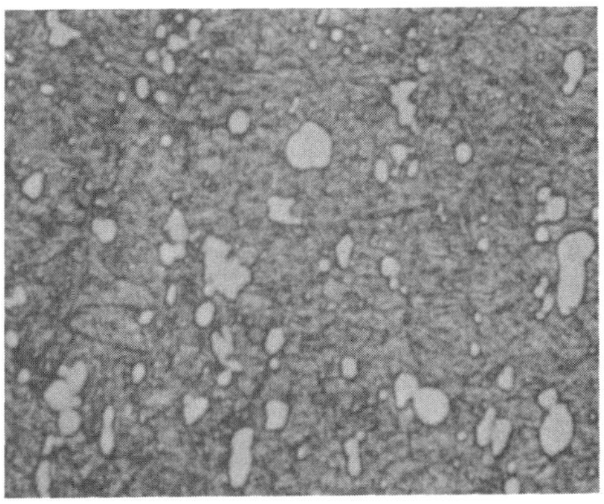

**Fig. 7** Microstructure of hardened, tempered high-speed tool steel having martensitic structure with carbides. 1000×

high-speed tool steel type, tool configuration, and cross-sectional size.

Lowering the hardening temperature (underhardening) generally improves the impact toughness while lowering the hot hardness. Raising the hardening temperature increases heat-treated room-temperature hardness and also increases the hot hardness.

**Quenching.** The quenching or cooling of the workpiece from the austenitizing temperature is designed to transform the austenite that forms at the high temperature to a hard martensitic structure. The rate of cooling, which must be controlled, is dictated by the analysis of the particular steel. Sometimes high-speed steels are two-step quenched, initially in a molten salt bath maintained at approximately 540 to 595 °C (1000 to 1100 °F) or an oil quench, followed by air cooling to near ambient temperature. The least drastic form of quenching is cooling in air, although only in the smaller and/or thinner cross sections would high-speed tool steels air quench rapidly enough to transform the majority of the structure into the desirable martensitic condition. The austenite-martensite transformation is exemplified in Fig. 9 illustrating a time-temperature-transformation curve.

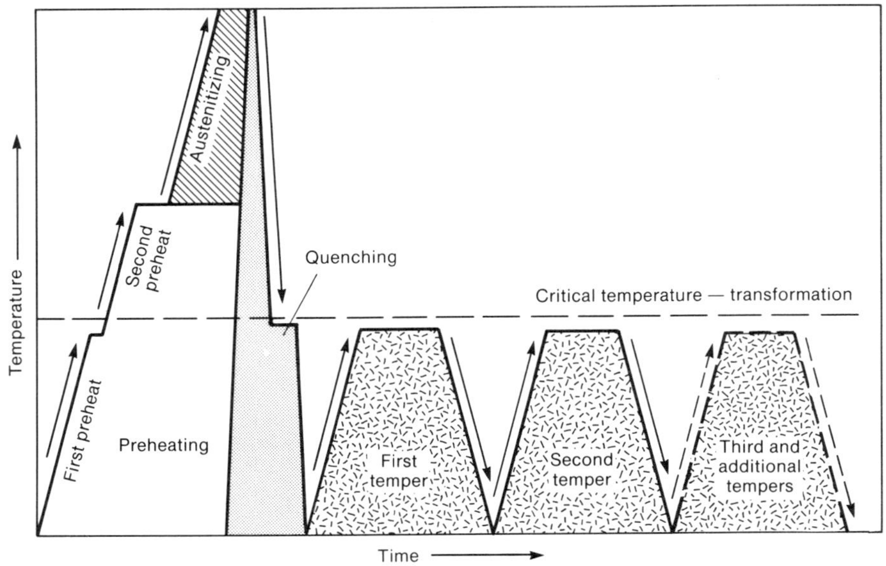


**Fig. 8** Time versus temperature plot illustrating sequences required to properly heat treat high-speed tool steels

**Tempering.** Following austenitizing and quenching, the steel is in a highly stressed state and therefore is very susceptible to cracking. Tempering (or drawing) increases the toughness of the steel and also provides secondary hardness, as illustrated by the peak on the right of the tempering curve in Fig. 10. Tempering involves reheating the steel to an intermediate temperature range (always below the critical transformation temperature), soaking, and air cooling.

Tempering serves to stress relieve and to transform retained austenite from the quenching step to fresh martensite. Some precipitation of complex carbide also occurs, further enhancing secondary hardness. It is this process of transforming retained austenite and tempering of newly formed martensite that dictates a multiple tempering procedure. High-speed tool steels require 2 to 4 tempers at a soak time of 2 to 4 h each. As with austenitizing temperatures and quenching rates, the number of tempers is dictated by the specific grade. High-speed tool steels should be multiple tempered at 540 °C (1000 °F) minimum for most grades. It is essential to

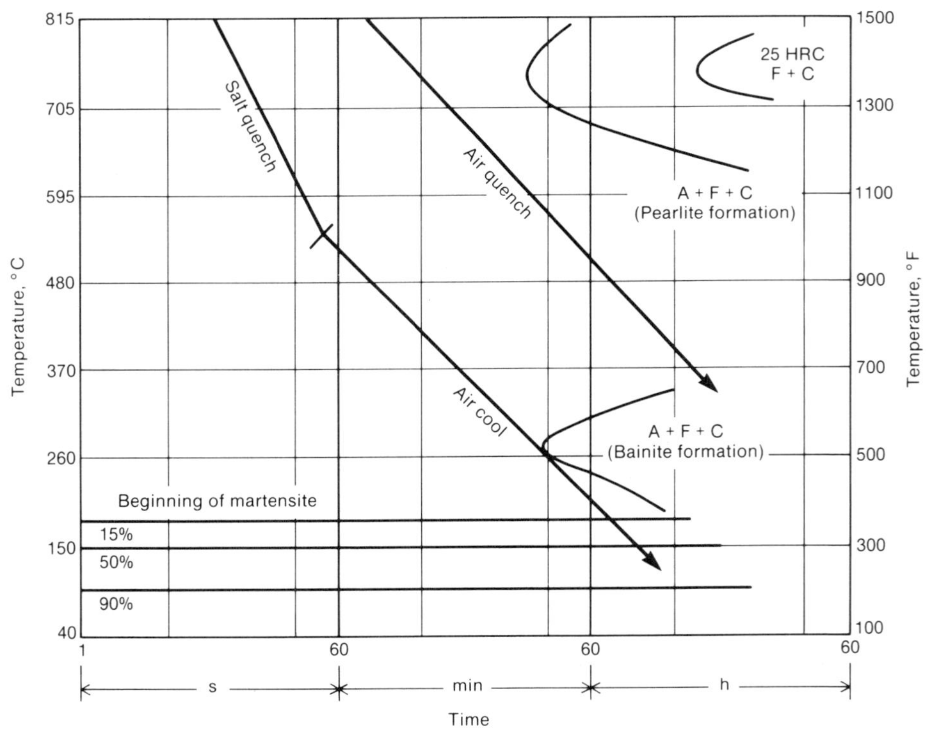


**Fig. 9** Time-temperature-transformation diagram for M2 high-speed tool steel that was annealed prior to quenching. Austenitizing temperature was 1230 °C (2250 °F), and critical temperature was 830 °C (1530 °F).

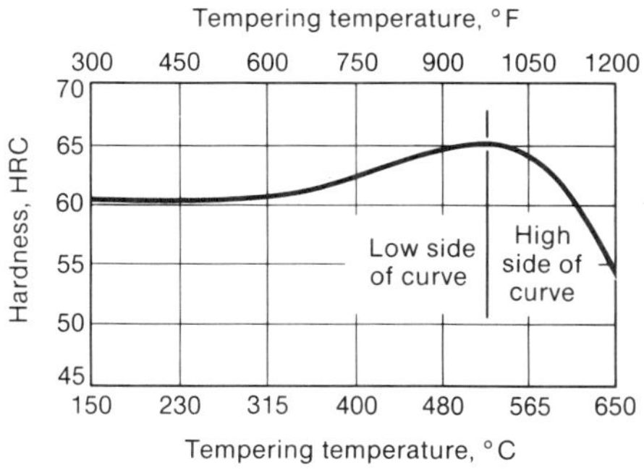


**Fig. 10** Tempering curve for M2 high-speed tool steel. To optimize the transformation of retained austenite to fresh martensite during the tempering sequence, the high (right) side of the secondary hardness peak curve is preferred, and the low (left) side should be avoided.

favor the right (high) side of the secondary hardness peak of the tempering curve in order to optimize the above-described transformations.

Subzero treatments are sometimes used in conjunction with tempering in order to continue the transformation of austenite to martensite. Numerous tests have been run on the effect of cold treatments, and the findings generally prove that cold treatments used after quenching and first temper enhance the transformation to martensite, in much the same way that multiple tempering causes transformation. Cold treatments administered to high-speed tool steels immediately after quenching can result in cracking or distortion because the accompanying size change is not accommodated by the newly formed, brittle martensite. It is generally accepted that subzero treatments are not necessary if the steel is properly hardened and tempered.

## Surface Treatments

Tools made of high-speed tool steel are available with either a bright, black oxide or nitride finish or they can be coated with titanium nitride and other coatings using a vapor disposition process that greatly increases tool life.

**Bright Finish.** Most tools are finished with a ground or mechanically polished surface that would be categorized as a bright finish. Bright finished tools are often preferred to tools with an oxide finish for machining nonferrous work material. The smooth or bright finish tends to resist galling, a type of welding or buildup associated with many nonferrous alloys. However, work materials of ferrous alloys tend to adhere to similar, iron-base tools having a bright finish. This buildup on the cutting edges leads to increased frictional heat, poor surface finish, and increased load at the cutting edge.

**Black Oxide Finish.** This characteristic black finish is typically applied to drills and other cutting tools by oxidizing in a steam atmosphere at approximately 540 °C (1000 °F). The black oxide surface has little or no effect on hardness, but serves as a partial barrier to galling of similar ferrous metals. The surface texture also permits retention of lubricant.

**Nitride Finish.** Nitriding is a method of introducing nitrogen to the surface of high-speed tool steels at a typical temperature of 480 to 595 °C (900 to 1100 °F) and is accomplished either by the dissociation of ammonia gas, exposure to sodium cyanide salt mixtures, or bombardment with nitrogen ions in order to liberate nascent nitrogen, which combines with the steel to form a hard iron nitride. Nitriding improves wear resistance of high-speed steel, at the expense of notch toughness.

**Coated High-Speed Tool Steels.** The addition of wear-resistant coatings to high-speed tool steel cutting tools lagged behind the coating of carbide inserts by approximately 10 years until the development of the low-temperature physical vapor deposition (PVD) process, an innovation, which is much more suitable for coating high-speed tool steels than is the older chemical vapor deposition (CVD) process, and which also eliminates the need for subsequent heat treatment (Ref 3). As described in Ref 4, titanium nitride is the most commonly used and most durable coating available, although substitutes such as other nitrides (hafnium nitride and zirconium nitride) and carbides (titanium carbide, zirconium carbide, and hafnium carbide) are being developed. These other coatings are expected to equal or surpass the desired properties of titanium nitrides in future years. The hard thin (2 to 5 μm, or 80 to 200 μin. thick) deposit of high-density titanium nitride, which has 2500 HV hardness and imparts a characteristic gold color to high-speed tool steels, provides excellent wear resistance, minimizes heat buildup, and prevents welding of the workpiece material, while improving the surface finish of high-speed tool steels (Ref 5).

The initial use, in 1980, of titanium nitride coatings was to coat gear cutting tools. Subsequent applications include the coating of both single-point and multipoint tools such as lathe tools, drills, reamers, taps, milling cutters, end mills, and broaches (Ref 3). Today, titanium nitride coated hobs and shapers dominate high-production applications in the automotive industry to such an extent that 80% of such tools use this coating.

As described in Ref 6, significant cost savings are possible because the titanium nitride coating improves tool life up to 400% and increases feed and speed rates by 30%. This is primarily attributable to the increased lubricity of the coating because its coefficient of friction is one-third that of the bare metal surface of a tool.

Examples of increased tool life obtained when using coated versus uncoated single-point and multipoint cutting tools are listed in Table 4. The increased production obtained with a coated tool justifies the application of the coating despite the resulting 20 to 30% increase in the base price of the tool (Ref 3).

Coated tools can meet close-tolerance requirements and significantly improve the machining of carbon and alloy steels, stainless steels (especially the 300 series, where galling can be a problem), and aluminum alloys (especially aircraft grades). Coated high-speed tool steels are less of a factor in the machining of certain titanium alloys and some high-nickel alloys because of chemical reactions between the coatings and the workpiece materials (Ref 3).

## High-Speed Tool Steel Applications

High-speed tool steels are used for most of the common types of cutting tools including single-point lathe tools, drills, reamers, taps, milling cutters, end mills, hobs, saws, and broaches.

### Single-Point Cutting Tools

The simplest cutting tools are single-point cutting tools, which are often referred to as tool bits, lathe tools, cutoff tools, or inserts. They have only one cutting surface or edge in contact with the work material at any given time. Such tools are used for turning, threading, boring, planing, or shaping, and most are mounted in a toolholder that is made of some type of tough alloy steel. The performance of such tools is dependent on the tool material as well as factors such as the material being cut, the speeds and feeds, the cutting fluid, and fixturing. Following is a discussion of material characteristics and recommendations for the most popular lathe tools.

M1, M2, and T1 are suitable for all-purpose tool bits. They offer excellent strength and toughness and are suitable for both roughing and finishing and can be used for machining wrought steel, cast steel, cast iron, brass, bronze, copper, aluminum, and so on (see the Section "Machining of Specific Metals and Alloys" in this Volume). These are good economical grades for general shop purposes.

M3 class 2 and M4 high-speed tool steels have high-carbon and high-vanadium contents. The wear resistance is several times that of standard high-speed steels. These bits are hard and tough, withstanding intermittent cuts even under heavy feeds. They are useful for general applications and especially recommended for cast steels, cast iron, plastics, brass, and heat-treated steels. On tool bit applications where failure occurs from rapid wearing of the cutting edge, M3 class 2 and M4 will be found to surpass the performance of regular tool bits.

T4, T5, and T8 combine wear resistance resulting from the higher carbon and vanadium contents together with a higher hot hardness, resulting from a cobalt content.

**Table 4 Increased tool life attained with coated cutting tools**

| Cutting tool — Type | High-speed tool steel, AISI type | Coating | Workpiece material | Workpieces machined before resharpening — Uncoated | Coated |
|---|---|---|---|---|---|
| End mill | M7 | TiN | 1022 steel, 35 HRC | 325 | 1 200 |
| End mill | M7 | TiN | 6061-T6 aluminum alloy | 166 | 1 500 |
| End mill | M3 | TiN | 7075T aluminum alloy | 9 | 53 |
| Gear hob | M2 | TiN | 8620 steel | 40 | 80 |
| Broach insert | M3 | TiN | Type 303 stainless steel | 100 000 | 300 000 |
| Broach | M2 | TiN | 48% nickel alloy | 200 | 3 400 |
| Broach | M2 | TiN | Type 410 stainless steel | 10 000–12 000 | 31 000 |
| Pipe tap | M2 | TiN | Gray iron | 3 000 | 9 000 |
| Tap | M2 | TiN | 1050 steel, 30-33 HRC | 60–70 | 750–800 |
| Form tool | T15 | TiC | 1045 steel | 5 000 | 23 000 |
| Form tool | T15 | TiN | Type 303 stainless steel | 1 840 | 5 890 |
| Cutoff tool | M2 | TiC-TiN | Low-carbon steel | 150 | 1 000 |
| Drill | M7 | TiN | Low-carbon steel | 1 000 | 4 000 |
| Drill | M7 | TiN | Titanium alloy 662 layered with D6AC tool steel, 48-50 HRC | 9 | 86 |

Source: Ref 3

Because of the good resistance to abrasion and high hot hardness, these steels should be applied to the cutting of hard, scaly, or gritty materials. They are well adapted for making hogging cuts, for the cutting of hard materials, and for the cutting of materials that throw a discontinuous chip, such as cast iron and nonferrous materials. The high degree of hot hardness permits T4, T5, and T8 to cut at greater speeds and feeds than most high-speed tool steels. They are much more widely used for single-point cutting tools, such as lathe, shaper, and planer tools, than for multiple-edge tools.

Superhard tool bits made from the M40 series offer the highest hardness available for high-speed tool steels. The M40 steels are economical cobalt alloys that can be treated to reach a hardness as high as 69 HRC. Tool bits made from them are easy to grind and offer top efficiency on the difficult-to-machine space-age materials (titanium and nickel-base alloys, for example) and heat-treated high-strength steels requiring high hot hardness.

T15 tool bits are made from a steel capable of being treated to a high hardness, with outstanding hot hardness and wear resistance. The exceptional wear resistance of T15 has made it the most popular high-speed tool steel for lathe tools. It has higher hardness than most other steels, and wear resistance surpassing that of all other conventional high-speed tool steels as well as certain cast cutting tool materials. It has ample toughness for most types of cutting tool applications, and will withstand intermittent cuts. These bits are especially adapted for machining materials of high-tensile strength such as heat-treated steels and for resisting abrasion encountered with hard cast iron, cast steel, brass, aluminum, and plastics. Tool bits of T15 can cut ordinary materials at speeds 15 to 100% higher than average.

Often an engineer will specify a grade that is not necessary for a given application. For example, selecting M42 for a general application that could be satisfied with M2 does not always prove to be beneficial. The logic is that the tool can be run faster and therefore generate a higher production rate. What happens many times is that the M42 will chip because of its lower toughness level, whereas the M2 will not.

## Multipoint Cutting Tools

Applications of high-speed tool steels for other cutting tool applications such as drills, end mills, reamers, taps, threading dies, milling cutters, circular saws, broaches, and hobs are based on the same parameters of hot hardness, wear resistance, toughness, and economics of manufacture. Some of the cutting tools that require extensive grinding have been produced of P/M high-speed tool steels (see the article "P/M High-Speed Tool Steels" in this Volume).

**General-purpose drills,** other than those made from low-alloy steels for low production on wood or soft materials, are made from high-speed tool steels, typically M1, M2, M7, and M10. For lower cost hardware quality drills, intermediate high-speed tool steels M50 and M52 are sometimes used although they cannot be expected to perform as well as standard high-speed tool steels in production work. For high hot hardness required in the drilling of the more difficult-to-machine alloys such as nickel-base or titanium product, M42, M33, or T15 are used.

High-speed tool steel twist drills are not currently being coated as extensively as gear cutting tools because many drills are not used for production applications. Also, the cost of coating (predominantly with titanium nitride) is prohibitive because it represents a higher percentage of the total tool cost.

Drills coated with titanium nitride reduce cutting forces (thrust and torque) and improve the surface finishes to the point that they eliminate the need for prior core drilling and/or subsequent reaming. Coated drills have been found especially suitable for cutting highly abrasive materials, hard nonferrous alloys, and difficult-to-machine materials such as heat-resistant alloys. These tools are not recommended for drilling titanium alloys because of possible chemical bonding of the coating to the workpiece material. When drilling gummy materials (1018 and 1020 steels, for example) with coated tools, it may be necessary to provide for chipbreaking capabilities in the tool design (Ref 3).

**End mills** are produced in a variety of sizes and designs, usually with two, four, or six cutting edges on the periphery. This shank-type milling cutter is typically made from general-purpose high-speed tool steels M1, M2, M7, and M10. For workpieces made from hardened materials (over 300 HB), a grade such as T15, M42, or M33 is more effective. Increased cutting speeds can be used with these cobalt-containing high-speed tool steels because of their improved hot hardness.

One manufacturer realized a fourfold increase in the tool life of end mill wear lands when he switched to a titanium-nitride coated tool (Fig. 11). Titanium nitride coated end mills also outperform uncoated solid carbide tools. When machining valves made from type 304 stainless steel, a switch from solid carbide end mills to titanium nitride coated end mills resulted in a fivefold increase in tool life, that is, 150 parts compared to 30 finished with the carbide tools (Ref 3). Furthermore, the cost of the coated high-speed steel end mills was only one-sixth that of the carbide tools. Both types of 19 mm (3/4 in.) fluted end mills were used to machine a 1.6 mm (1/16 in.) deep slot at a speed of 300 rev/min and a feed of 51 mm/min (2 in./min).

**Reamers** are designed to remove only small amounts of metal and therefore require very little flute depth for the removal of chips. For this reason, reamers are designed as rigid tools, requiring less toughness from the high-speed tool steel than a deeply fluted drill. The general-purpose grades M1, M2, M7, M10, and T1 are typically used at maximum hardness levels. For applications requiring greater wear resistance, grades such as M3, M4, and T15 are appropriate.

**Milling Cutters.** The size, style, configuration, complexity, and capacity of milling cutters is almost limitless. There are staggered-tooth and straight-tooth, form-re-

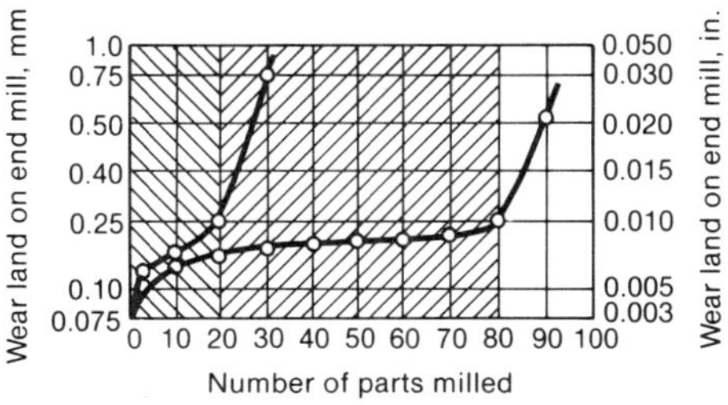


**Fig. 11** Wear lands developed with uncoated and titanium nitride coated end mills show a 4:1 increase in tool life with coated tools. The cross-hatched area at left (extending from 0 to 20 parts) indicates the number of pieces produced by uncoated end mill after 0.25 mm (0.010 in.) wear land on the tool; the crosshatched area at right represents quantity produced by titanium nitride coated end mill after 0.25 mm (0.010 in.) wear land on tool. Source: Ref 3

lieved and formed milling cutters with sizes that range from 51 to 305 mm (2 to 12 in.) and are used to machine slots, grooves, racks, sprockets, gears, splines, and so on. They cut a wide variety of materials, including plastics, aluminum, steel, cast iron, superalloys, titanium, and graphite structures. The general-purpose high-speed tool steel used for more than 70% of milling cutter applications is M2, usually the free-machining type. It has a good balance of wear resistance, hot hardness, toughness, and strength and works well on carbon, alloy, and stainless steels, aluminum, cast iron, and some plastics (generally any material that is under 30 HRC in hardness). When higher hardness materials or more wear-resistant materials need to be milled, M3 or M4 are selected. The higher carbon and vanadium content in those materials improves wear resistance and allows for the machining of materials greater than 35 HRC in hardness. For workpiece hardness levels above that and as high as 50 HRC, either M42 with its high hardness and high hot hardness properties or T15 with its high wear resistance and high hardness characteristics are desirable. The powder metallurgy grades in M4 and T15 are increasing in popularity for milling cutters because of their ease of grinding and regrinding.

**Hobs** are a type of milling cutter that operates by cutting a repeated form about a center, such as gear teeth. The hob cuts by meshing and rotating about the workpiece, forming a helical pattern. This type of metal cutting creates less force at the cutting edge (less chip load on the teeth) than do ordinary milling cutters. Accordingly, less toughness and edge strength is required of hob materials; wear is more commonly a mode of failure. Most hobs are made from a high-carbon version of M2, although normal carbon levels are also used. M2 with a sulfur addition or P/M product for improved machinability and surface finish is often used for hobs.

**Saws** are quite similar to milling cutters in style and application, but they are usually thinner and tend to be smaller in diameter. Sizes range from 0.076 mm (0.003 in.) thick by 13 mm (½ in.) outside diameter to more than 6.4 mm (¼ in.) thick by 203 mm (8 in.) outside diameter. Used for cutting, slitting, and slotting, saws are available with straight-tooth, staggered-tooth, and side-tooth configurations and are made from alloys similar to those used for milling cutters. Again, M2 high-speed tool steel is the general-purpose saw material, but, because of the typical thinness of these products, toughness is optimized with lower hardness. There are relatively few saws that are made from M3 or M4 high-speed tool steel because generally T15 and M42 are the two alternative materials to the standard M2 steel. M42 is often used to machine stainless steels, aluminum, and brass because it increases saw production life and can be run at considerably higher speeds. T15 is used for very specialized applications. Saws made of high-speed tool steel are used to cut, slit, and slot everything from steel, aluminum, brass, pipe, and titanium to gold jewelry, fish, frozen foods, plastics, rubber, and paper.

**Broaches.** M2 high-speed tool steel is the most frequently used material for broaches. This includes the large or circular broaches that are made in large quantities as well as the smaller keyway and shape broaches. Sometimes the higher-carbon material is used, but generally free-machining M2 is used because it results in a better surface finish. P/M products are very popular for broaches in both M2 as well as M3 class 2 and M4 when they are used to improve wear resistance. M4 is probably the second most widely used material for this application. M42 and T15 are often used for difficult-to-machine materials such as the nickel-base alloys and other aerospace-type alloys.

A high-nickel (48%) alloy magnet manufacturer using a 3.2 × 13 × 305 mm (⅛ × ½ × 12 in.) flat broach made of M2 increased tool life from 200 pieces to 3400 pieces when a titanium nitride coating was added, and also obtained a smoother surface finish. Replacing the flat broach with an uncoated 11.99 mm (0.472 in.) diam by 660 mm (26 in.) long round broach increased the production to about 7000 pieces, and coating the round broach with titanium nitride further increased the magnet production to about 19 000 pieces (Ref 3). Thus, going from an uncoated flat broach to a coated round broach increased production by a factor of 95.

## Factors in Selecting High-Speed Tool Steels

No one composition of high-speed tool steel can meet all cutting tool requirements. The general-purpose molybdenum steels such as M1, M2, and M7 and tungsten steel T1 are more commonly used than other high-speed tool steels. They have the highest toughness and good cutting ability, but they possess the lowest hot hardness and wear resistance of all the high-speed tool steels. The addition of vanadium offers the advantage of greater wear resistance and hot hardness, and steels with intermediate vanadium contents are suited for fine and roughing cuts on both hard and soft materials. The 5% V steel (T15) is especially suited for cutting hard metals and alloys or high-strength steels, and is particularly suitable for the machining of aluminum, stainless steels, austenitic alloys, and refractory metals. Wrought high-vanadium high-speed tool steels are more difficult to grind than their P/M product counterparts. The addition of cobalt in various amounts allows still higher hot hardness, the degree of hot hardness being proportional to the cobalt content. Although cobalt steels are more brittle than the noncobalt types, they give better performance on hard, scaly materials that are machined with deep cuts at high speeds.

High-speed tool steels have continued to be of importance in industrial commerce for 70 to 80 years despite the inroads made by competitive cutting tool materials such as cast cobalt alloys, cemented carbides, ceramics, and cermets. The superior toughness of high-speed tool steels guarantees its niche in the cutting tool materials marketplace.

### ACKNOWLEDGMENTS

The authors wish to express their thanks to the following individuals for their useful contributions to this article: Gene Bistrich, National Broach and Machine Division, Lear Siegler, Inc.; George Kiss, The Cleveland Twist Drill Company; Jim Kucynski, The Weldon Tool Company; Ron Oakes, Niagara Cutter Inc.; Robert Phillips, Pfauter Maag; Ira Redinger, Thurston Manufacturing Company; and Scott Schneier, Regal-Beloit Corporation.

### REFERENCES

1. *Machining*, Vol 1, *Tool and Manufacturing Engineers Handbook*, Society of Manufacturing Engineers, 1983, p 3-6
2. S. Kalpakjian, *Manufacturing Processes for Engineering Materials*, Addison-Wesley, 1984, p 524
3. C. Wick, HSS Cutting Tools Gain a Productivity Edge, *Manufacturing Engineering*, May 1987, p 38
4. W.D. Sproul, Turning Tests of High Rate Reactively Sputter-Coated T-15 HSS Inserts, *Surf. Coat. Tech.*, Vol 33, 1987, p 133
5. TiN Coatings Continue to Revolutionize the Metalworking Industry, *Machining Source Book*, ASM INTERNATIONAL, 1988, p 98
6. How Cutting Tools Can Help You Make a Quick Buck, *Machining Source Book*, ASM INTERNATIONAL, 1988, p 20

# P/M High-Speed Tool Steels

Revised by Kenneth E. Pinnow and William Stasko, Crucible Materials Corporation

POWDER METALLURGY (P/M) high-speed tool steels are used extensively for drills, taps, end mills, reamers, broaches, and other cutting tools because of their excellent manufacturing and performance characteristics. For most applications, they offer distinct advantages over conventional high-speed tool steels which, as a result of pronounced ingot segregation, often contain a coarse, nonuniform microstructure, accompanied by poor toughness and grindability, and also present problems of size control and hardness uniformity in heat treatment. Rapid solidification of the atomized powders used in the production of P/M high-speed tool steels eliminates such segregation and produces a very fine microstructure with a uniform distribution of carbides and nonmetallic inclusions. As a result, a number of important end properties of high-speed tool steels have been improved by powder processing, notably toughness, dimensional control during heat treatment, grindability, and cutting performance under difficult conditions when good toughness is essential (Ref 1). Further, powder processing allows the production of high-speed tool steels with much greater alloy contents than are practical or possible by conventional ingot methods. Two examples of such highly alloyed high-speed tool steels are CPM Rex 76 and ASP 60.

Since the early 1970s, several P/M methods for producing high-speed tool steels have been developed, including controlled spray deposition (CSD), the Osprey process, rapid omnidirectional compaction, consolidation at atmospheric pressure (CAP process), the STAMP process, and injection molding. These processes are discussed in Volume 7 of the 9th edition of *Metals Handbook*.

The present discussion describes procedures for producing tool steel powder by inert-gas atomization, followed by compaction by hot isostatic pressing (HIP). These processes include the Anti-Segregation Process (ASP), developed in Sweden by Stora Kopparberg and ASEA, and the Crucible Particle Metallurgy process, developed in the United States by the Crucible Materials Corporation. The FULDENS process, which uses water-atomized powders compacted by vacuum sintering, is also discussed. It was developed in the United States by Consolidated Metallurgical Industries, Inc.

For additional data concerning the classification, composition, heat treatment, and properties of conventionally processed and P/M processed high-speed tool steel materials, see the articles on "Tool Steels" in Volume 3, "High-Speed Tool Steels" in this Volume, and "P/M Tool Steels" in Volume 7 of the 9th Edition of *Metals Handbook*.

## The Anti-Segregation Process

The Anti-Segregation Process, or ASEA-STORA process, is used to produce high-speed tool steels by powder metallurgy. In this process, an alloy steel melt is atomized in an inert gas to form spherical powder particles. These are poured into cylindrical sheet steel capsules (cans), which are vibrated to pack the particles as tightly as possible. A cover is then welded onto the capsule and the air inside is evacuated. The capsule and its contents are cold isostatically pressed at 400 MPa (58 ksi).

The capsule is hot isostatically pressed at 100 MPa (14.5 ksi) at 1150 °C (2100 °F) to full density. After compaction, the steel is conventionally hot worked by forging and rolling to the desired dimensions. Figure 1 compares the processing of conventional (wrought) high-speed tool steels with that of ASP high-speed tool steels.

This processing results in a fine-grain material with a uniform distribution of small carbides. The homogeneous material, free from segregation, has a uniform structure, regardless of bar size and alloy content. Figure 2 compares the microstructures of conventional high-speed tool steel and P/M processed ASP high-speed tool steel.

## Properties of ASP Steels (Ref 2)

The primary benefits of ASP techniques include improved toughness and ultimate strength due to uniform carbide distribution and the absence of metallurgical defects. Improved grindability due to the small carbide size and improved dimensional stability in heat treatment caused by the absence of segregation are also benefits. Additionally, wear resistance can be improved by increasing alloy content, without sacrificing toughness or grindability.

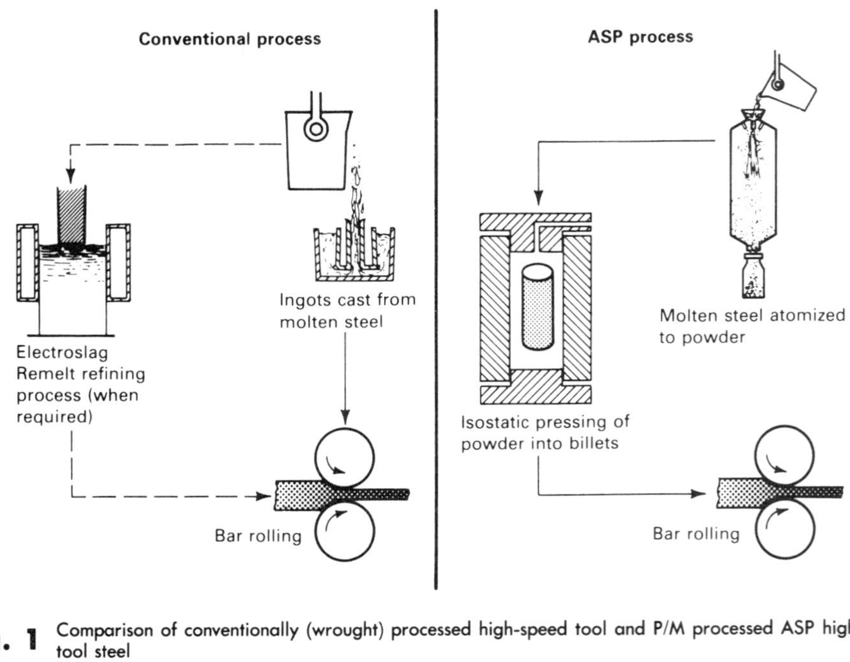


**Fig. 1** Comparison of conventionally (wrought) processed high-speed tool and P/M processed ASP high-speed tool steel

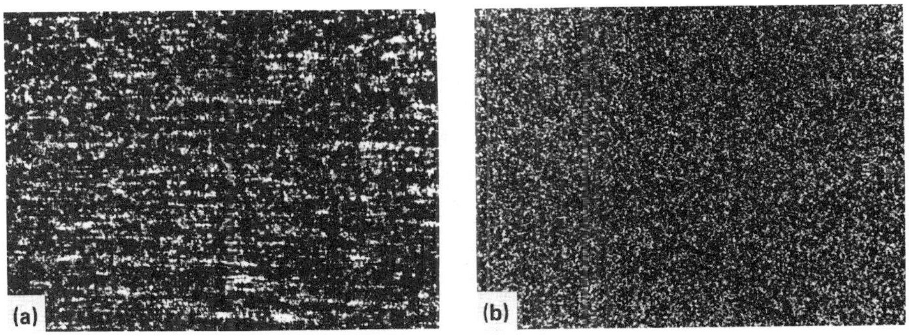

**Fig. 2** Comparison of microstructures of conventional high-speed tool steel and P/M high-speed tool steel. (a) Conventional high-speed tool steel microstructure showing carbide segregation. (b) Microstructure of P/M processed ASP steel showing small, uniformly distributed carbide particles. Courtesy of Speedsteel, Inc.

**Table 1 ASP steel grades, compositions, hardnesses, and applications**

| ASP grade | Composition, % C | Cr | Mo | W | V | Co | Typical hardness, HRC | Recommended applications |
|---|---|---|---|---|---|---|---|---|
| 23 | 1.28 | 4.2 | 5.0 | 6.4 | 3.1 | . . . | 65–57 | For ordinary applications of most cutting tools when hot hardness is not of primary concern. Also for tools used in cold-working applications |
| 30 | 1.28 | 4.2 | 5.0 | 6.4 | 3.1 | 8.5 | 66–58 | For cutting tool applications when hot hardness is important. Suitable for cutting most stainless steels and superalloys, and for cutting at higher speeds. Also for cold work-tools when wear resistance is critical |
| 60 | 2.30 | 4.0 | 7.0 | 6.5 | 6.5 | 10.5 | 67–59 | For cutting tools when wear resistance and hot hardness are critical. Particularly suitable for extratough applications (cutting titanium, high-hardness materials, and iron forgings). Also for cold-work tools requiring highest wear resistance |

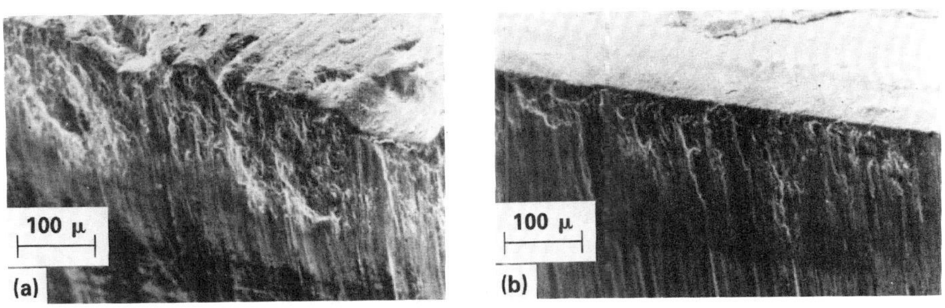

**Fig. 3** Comparison of cutting edge wear of a conventional high-speed tool steel and a P/M high-steel tool steel. (a) Cutting edge of tool made of conventional AISI M2 material, showing severe microchipping. (b) Cutting edge of tool made of P/M-processed ASP 23 material, showing no microchipping under the same service conditions. Courtesy of Speedsteel, Inc.

Currently, ASP high-speed tool steel is available in three grades: ASP 23, 30, and 60 (ASP 60 can be made only by the powder metallurgy process). The compositions and recommended applications of these grades are given in Table 1. Additional information on applications of ASP steels can be found in the section "Applications of P/M High-Speed Tool Steels" in this article.

**Wear resistance** is generally a function of the hardness of the tool and the specific alloy content or type of carbide. The higher hardness that is possible with P/M high-speed tool steels, plus the higher carbon and vanadium contents, promote better wear resistance.

**Toughness** of a tool or high-speed tool steel is usually defined as a combination of strength and ductility or as resistance to breaking or chipping. A tool that deforms from lack of strength is useless, and one that lacks adequate ductility will fail prematurely.

The importance of toughness of high-speed tool steel is illustrated in Fig. 3. A cutting edge may suffer from repeated microchipping. As shown in Fig. 3, the ASP 23 cutting edge shows minimal wear. The M2 cutting edge, however, shows microchipping under the same service conditions. Microchipping blunts the cutting edge, increases stress, and accelerates other wear factors.

One method of measuring toughness of high-speed tool steel after heat treatment is bend testing. Bend yield strength, ultimate bend strength, and deflection are measured on 5 mm (0.2 in.) diam test bars on which a load is exerted. The results of these laboratory tests correlate well with shop experience.

As shown in Fig. 4, toughness and hardness can be controlled by varying the hardening temperature. A low hardening temperature produces good toughness. Raising the hardening temperature increases hardness, but lowers toughness.

**Grindability** of ASP steel is superior to that of conventional high-speed tool steel of the same chemical composition. This is due to the small carbide size and the uniform distribution of carbides, regardless of bar size. Figure 5 compares the grindability of several tool steels. These data are based on laboratory measurements, but results are confirmed by shop experience.

## Heat Treatment of ASP High-Speed Tool Steels

Only with proper heat treatment can optimum mechanical properties of tools and dies be obtained. Improper heat treatment may result in a tool with greatly reduced productivity or even an unusable tool. Heat treatment consists of four stages: preheating, austenitizing, quenching, and tempering. The heat treatment procedure for ASP high-speed tool steels is essentially the same as for wrought high-speed tool steels. Optimum heat-treating temperatures may vary, however, even if chemical compositions are identical.

The following procedures should be used to heat treat ASP high-speed tool steels as well as all P/M high-speed tool steels:

- *Annealing*: Heat to 850 to 900 °C (1560 to 1650 °F). Slow cool 10 °C/h (18 °F/h) to 700 °C (1290 °F). Hardness values are 260 HB maximum for ASP 23, 300 HB for ASP 30, and 340 HB for ASP 60
- *Stress relieving*: Hold for approximately 2 h at 600 to 700 °C (1110 to 1290 °F). Slow cool to 500 °C (930 °F) in furnace
- *Hardening*: Preheat in two steps, first at 450 to 500 °C (840 to 930 °F) and then at 850 to 900 °C (1560 to 1650 °F). Austenitize at 1050 to 1180 °C (1920 to 2155 °F) and quench, preferably in a neutral salt bath. Cool to hand warmth. See Table 2 for recommended temperatures

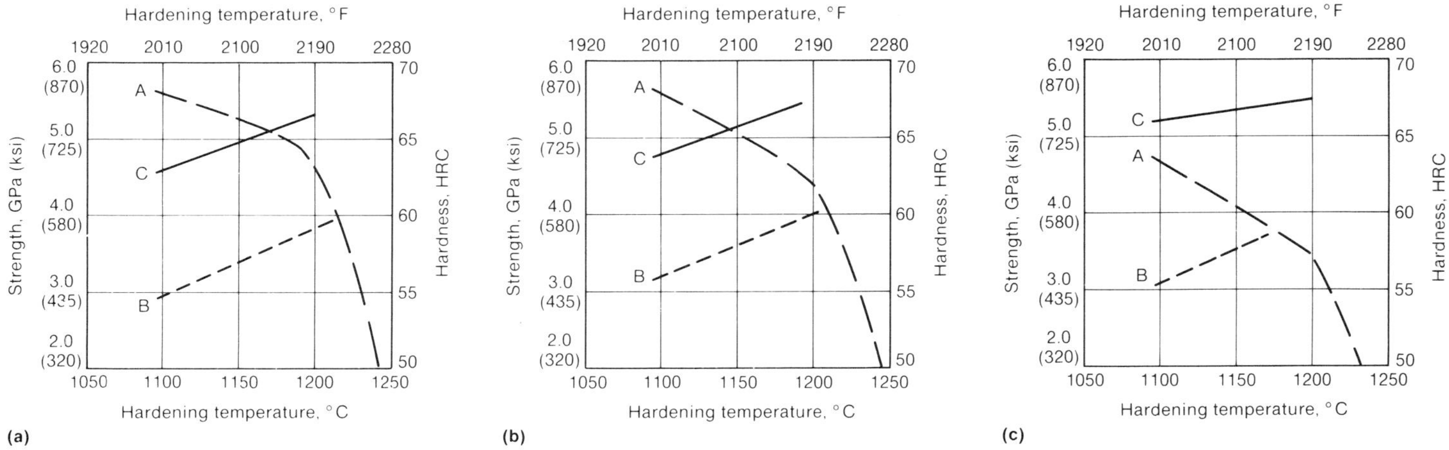


**Fig. 4** Bend test results to determine toughness of PM/processed ASP high-speed tool steels. A, ultimate bend strength; B, bend yield strength; C, hardness (HRC). (a) Bend strength of a test bar of ASP 23 steel after hardening and tempering at 560 °C (1040 °F) (three times for 1 h). (b) Bend strength of a test bar of ASP 30 steel after hardening and tempering at 560 °C (1040 °F) (three times for 1 h). (c) Bend strength of a test bar of ASP 60 steel after hardening and tempering at 560 °C (1040 °F) (three times for 1 h). Ultimate bend strength may vary ±10%; bend yield strength may vary ±5%; hardness values may vary ±1%. Courtesy of Speedsteel, Inc.

- *Tempering*: Raise temperature to 560 °C (1040 °F) or higher three times for at least 1 h at full temperature. Cool to room temperature between tempers

Hardness of ASP high-speed tool steel after hardening and tempering is shown in Fig. 6.

**Dimensional Stability in Heat Treatment.** Three types of distortion are experienced metallurgically during heat treatment:

- Normal volume change due to phase transformations in the steel
- Variations in volume change in different parts of the tool due to the segregation in the steel
- Distortion due to residual stress caused by machining or nonuniform heating and cooling during heat treatment

P/M grades, however, differ significantly from conventionally manufactured high-speed tool steels. Dimensional changes are more uniform in all directions. Because P/M high-speed tool steels are segregation free, variations in dimensional change are smaller. As a result, dimensional change occurring during hardening can be predicted more accurately. Conventionally processed high-speed tool steels go out-of-round in a four-sided pattern. The extent of distortion during heat treatment depends on the type and degree of segregation. In P/M high-speed tool steels, anisotropy is smaller, and out-of-roundness occurs in a close, circular pattern. Figure 7 shows typical results of measuring 102 mm (4 in.) diam disks after hardening and tempering. With P/M high-speed tool steels, cracking and variation of hardness are minimized because of their fine-grain, uniform structure.

The same precautions must be taken to control distortion due to residual stresses during heat treating. Mechanical stresses from rough machining can be eliminated by stress relieving prior to finish machining and heat treating.

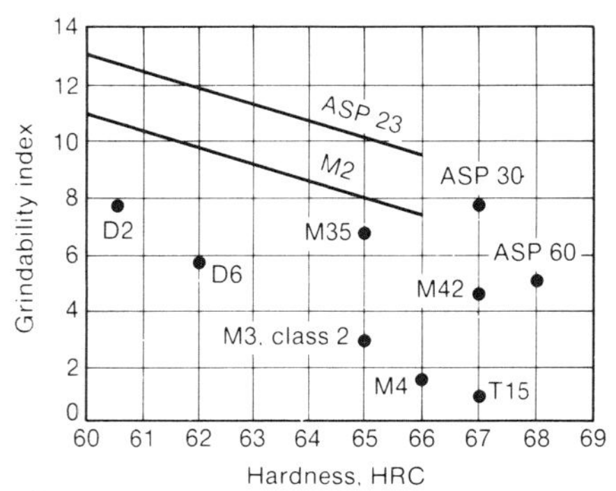


**Fig. 5** Grindability of P/M high-speed tool steel and conventional high-speed tool steel materials. Grindability index is the ratio of the volume of material removed to the volume of grinding wheel wear.

## Crucible Particle Metallurgy Process

Since 1970, Crucible Materials Corporation has been producing powder metal tool steels commercially by the Crucible Particle Metallurgy (CPM) process. The process consists of induction melting and inert-gas atomizing, screening, and containerizing the prealloyed particles, followed by hot isostatic pressing to full density. See Fig. 8 for a schematic of the process elements. The desired chemical composition is melted, and the molten stream is poured into an atomizing chamber where high-pressure gas jets disperse it into spheroidal droplets that are rapidly quenched to ambient temperature. Powder is removed from the atomizing chamber, dried, and screened to obtain the desired size fraction. It is then poured into cylindrical steel cans that are evacuated and sealed. The cans are subsequently heated to a specific temperature and hot isostatically compacted to achieve a fully dense product. Compacts are processed to the desired billet and bar sizes by conventional hot rolling and forging (Ref 3).

As stated earlier, the most detrimental tendency of conventionally produced high-alloy high-speed tool steels is the high degree of alloy and carbide segregation that occurs during ingot solidification. This segregation not only reduces the hot workability and machinability of these alloys, but also results in reduced mechanical properties and tool performance. An increase in the carbon and alloy content results in increased segregation and low product yield after hot working of conventional ingot products.

The CPM process was developed to minimize alloy segregation in standard high-alloy high-speed tool steel grades. Additionally, the CPM process is used to produce more highly alloyed grades than can be made by conventional practices.

**Table 2 Austenitizing temperatures of ASP 23 steel**

| Hardness(a), HRC | Temperature °C | Temperature °F | Salt bath(b) min/mm | Salt bath(b) min/in. | Other furnace(c), min |
|---|---|---|---|---|---|
| 58 | 1000 | 1830 | 0.59 | 15 | 30 |
| 60 | 1050 | 1925 | 0.47 | 12 | 25 |
| 62 | 1100 | 2010 | 0.39 | 10 | 20 |
| 64 | 1140 | 2085 | 0.31 | 8 | 15 |
| 66 | 1180 | 2155 | 0.24 | 6 | 10 |

(a) After triple temper at 560 °C (1040 °F); hardness values may vary by ±1%. (b) Total immersion time after preheating. (c) Holding time in minutes after tool has reached full temperature

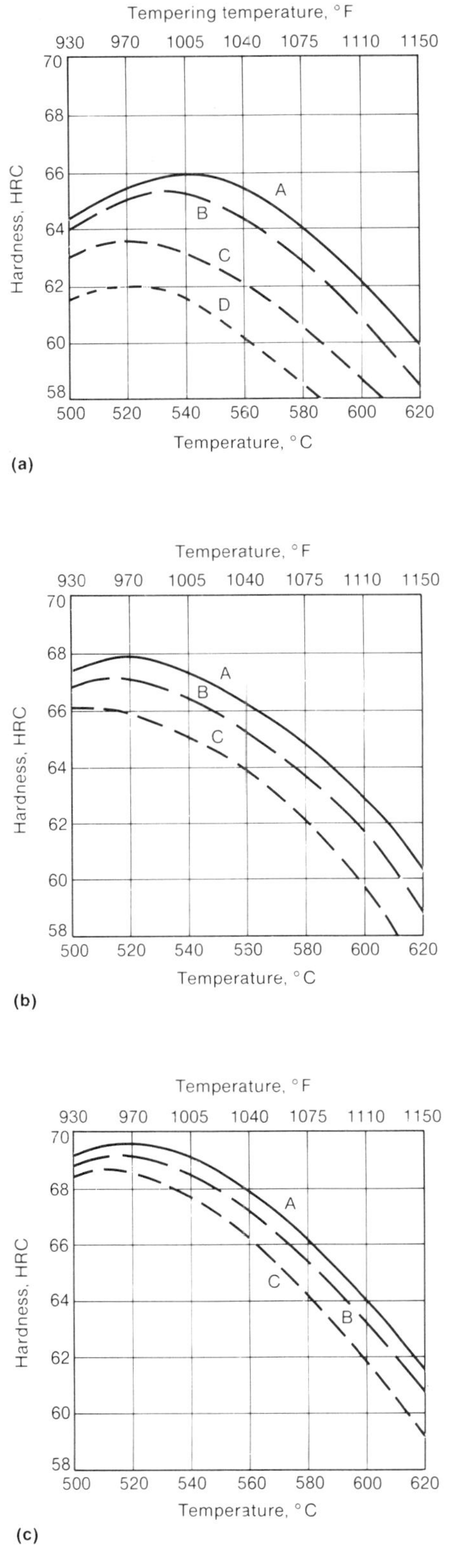


**Fig. 6** Hardness of ASP steels after hardening and tempering a 25 mm (1 in.) diam specimen three times for 1 h. (a) ASP 23. (b) ASP 30. (c) ASP 60, cooled in step bath. Hardening temperature for curves is: A, 1180 °C (2155 °F); B, 1150 °C (2100 °F); C, 1100 °C (2010 °F); D, 1050 °C (1920 °F)

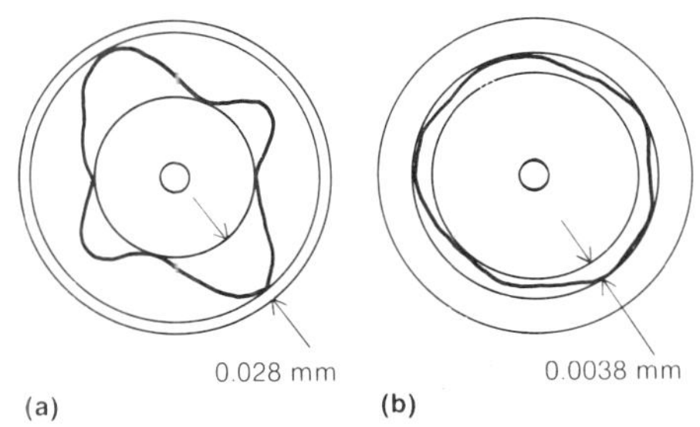


**Fig. 7** Out-of-roundness measurements on test disks after hardening and tempering. Test disks machined from 102 mm (4 in.) diam bars. (a) AISI M2. (b) ASP 30

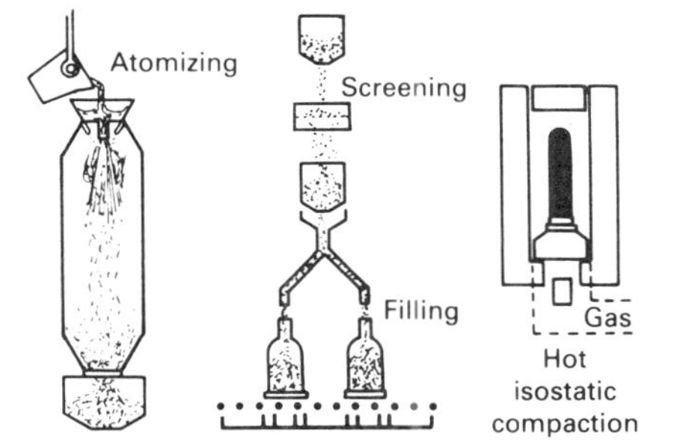


**Fig. 8** Schematic of CPM processing

## Properties of CPM High-Speed Tool Steels

A variety of CPM high-speed tool steels are available, as is shown in Table 3. Some of these grades are standard AISI steels such as M2, M3, M35, and M42, which normally are produced by conventional means, but when made by the CPM process offer notable advantages in toughness, out-of-roundness after heat treatment, and grindability. Others such as M4 and T15 are very difficult to produce by conventional means, but are readily producible by the CPM process with an improvement in properties. Still others, such as CPM Rex 20 and CPM Rex 76, are superhigh-speed tool steels that are very difficult or impossible to produce by conventional means and can only be made by the CPM route.

AISI T15 (Fe-12.25W-5Co-5.0V-4Cr-1.55C) demonstrates the advantages of the CPM process. This high-speed tool steel is one of the most wear- and heat-resistant grades of the standard American Iron and Steel Institute (AISI) high-speed tool steel materials. However, T15 usage has been limited because this highly alloyed, carbide-rich high-speed tool steel is difficult to produce by conventional production methods. The CPM process makes it possible to produce such difficult compositions in high volume.

The size distributions of the primary carbides in CPM and conventional T15 are shown in Fig. 9. Most carbides in CPM high-speed tool steel are less than about 3 μm (120 μin.), whereas those in the conventional product cover the entire size range to approximately 34 μm (1360 μin.) with a median size of 6 μm (240 μin.). The microstructures of CPM and conventionally processed T15 are compared in Fig. 10.

Figure 11 shows a high-speed tool steel graph that summarizes the relative toughness, wear resistance, and red (hot) hardness characteristics of CPM and conventional versions of various AISI high-speed tool steel grades. As shown by the graph, the wear resistance and red hardness of a given grade of high-speed tool steel are equal for both its CPM and conventional versions. However, the wear resistance and red hardness generally increase with increasing alloy content, and it is the very highly alloyed grades, such as CPM M4, CPM T15, CPM Rex 20, CPM Rex 45, and CPM Rex 76, that are best produced or can only be produced by the CPM method. The toughness comparison shows that the CPM version of each grade is notably tougher than the conventional high-speed tool steel.

The alloyed grade CPM Rex 76 is an example of a high-speed tool steel designed for production by the CPM process. This grade is a cobalt-rich high-speed tool steel with exceptional hot hardness and wear resistance and greatly increased tool life in difficult cutting operations. The high alloy content (32.5% compared to 27.8% for T15 and 25% for M42) renders this alloy unforgeable if produced by conventional processing.

Two prominent high-speed steel cutting tool grades used in machining difficult-to-machine superalloys and titanium alloys used by the aircraft industry are T15 and M42, which contain 5 and 8% Co, respec-

**Table 3 Commercial CPM high-speed tool steel compositions**

| Steel designation | | Composition, % | | | | | | | Typical hardness, HRC |
|---|---|---|---|---|---|---|---|---|---|
| Trade name | AISI | C | Cr | W | Mo | V | Co | S | |
| CPM Rex M2HCHS | M2 | 1.00 | 4.15 | 6.40 | 5.00 | 2.00 | . . . | 0.27 | 64–66 |
| CPM Rex M3HCHS | M3 | 1.30 | 4.00 | 6.25 | 5.00 | 3.00 | . . . | 0.27 | 65–67 |
| CPM Rex M4 | M4 | 1.35 | 4.25 | 5.75 | 4.50 | 4.00 | . . . | 0.06 | 64–66 |
| CPM Rex M4 HS | M4 | 1.35 | 4.25 | 5.75 | 4.50 | 4.00 | . . . | 0.22 | 64–66 |
| CPM Rex M35 HCHS | M35 | 1.00 | 4.15 | 6.00 | 5.00 | 2.00 | 5.00 | 0.27 | 65–67 |
| CPM Rex M42 | M42 | 1.10 | 3.75 | 1.50 | 9.50 | 1.15 | 8.00 | . . . | 66–68 |
| CPM Rex 45 | . . . | 1.30 | 4.00 | 6.25 | 5.00 | 3.00 | 8.25 | 0.03 | 66–68 |
| CPM Rex 45 HS | . . . | 1.30 | 4.00 | 6.25 | 5.00 | 3.00 | 8.25 | 0.22 | 66–68 |
| CPM Rex 20 | M62 | 1.30 | 3.75 | 6.25 | 10.50 | 2.00 | . . . | 0.06 | 66–68 |
| CPM Rex T15 | T15 | 1.55 | 4.00 | 12.25 | . . . | 5.00 | 5.00 | 0.06 | 65–67 |
| CPM Rex T15 HS | T15 | 1.55 | 4.00 | 12.25 | . . . | 5.00 | 5.00 | 0.22 | 65–67 |
| CPM Rex 76 | M48 | 1.50 | 3.75 | 10.00 | 5.25 | 3.10 | 9.00 | 0.06 | 67–69 |
| CPM Rex 76 HS | M48 | 1.50 | 3.75 | 10.00 | 5.25 | 3.10 | 9.00 | 0.22 | 67–69 |

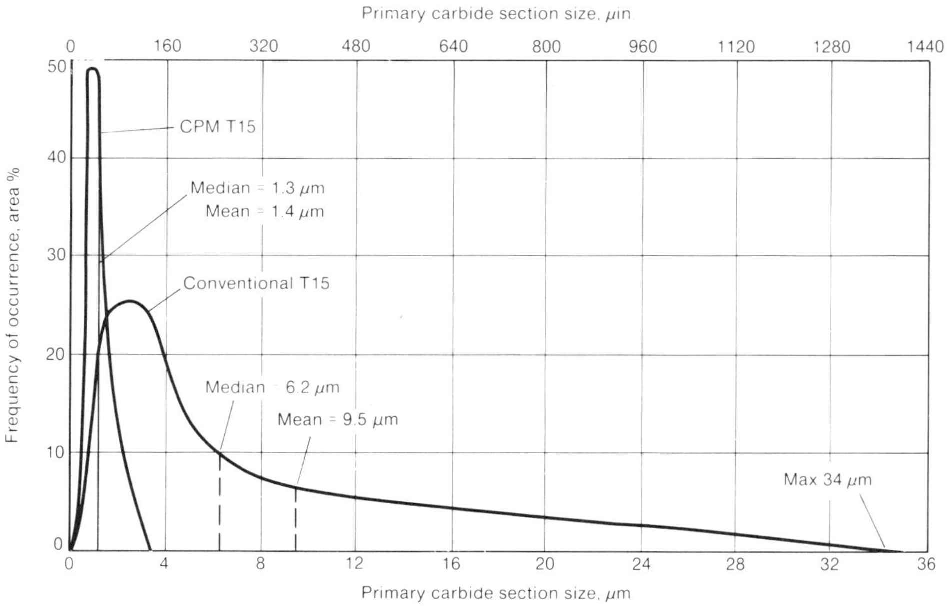


**Fig. 9** Primary carbide size distributions in CPM and conventionally produced T15 high-speed tool steel

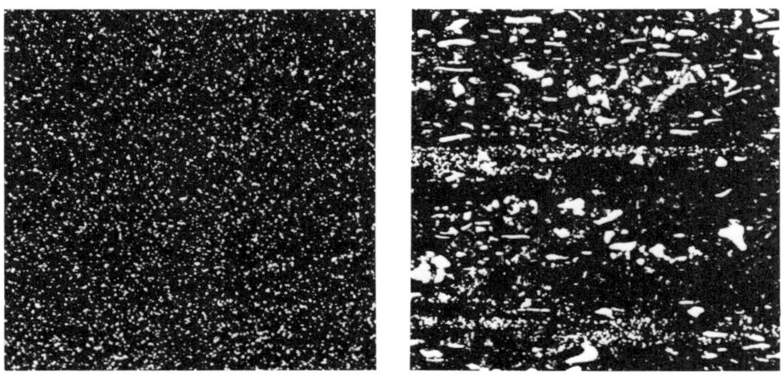

**Fig. 10** Microstructures of high-speed tool steels. Left: CPM T15. Right: Conventional T15. Carbide segregation and its detrimental effects are eliminated with the CPM process, regardless of the size of the products. Courtesy of Crucible Materials Corporation

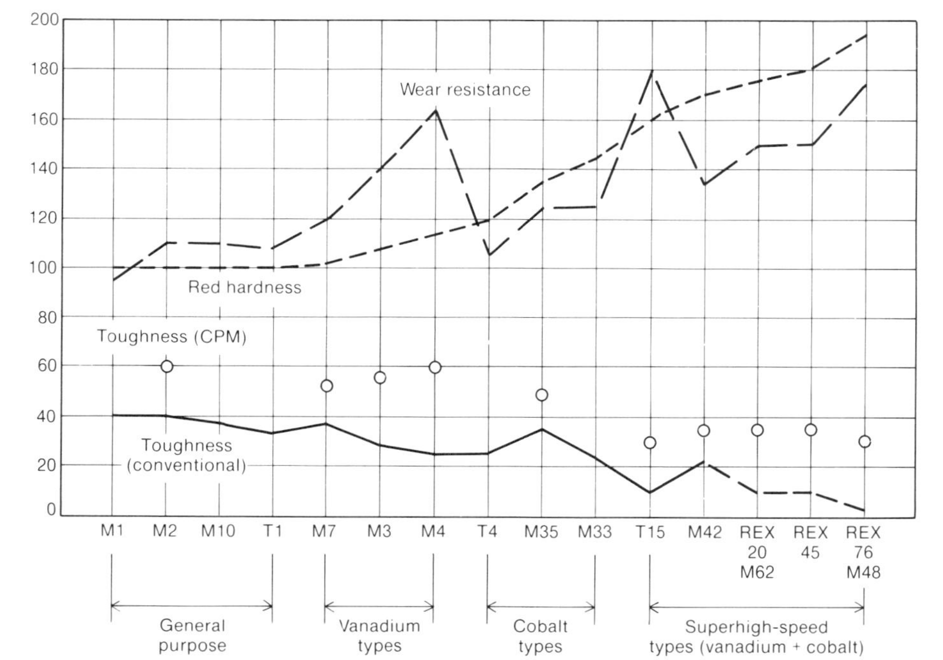


**Fig. 11** Comparagraph showing wear resistance, red (hot) hardness, and toughness of CPM and conventional high-speed tool steels

tively (Table 3). Cobalt increases the solidus temperature in high-speed tool steels, thereby permitting the use of high austenitizing temperatures to achieve greater homogeneous mixing of alloying elements. Cobalt also enhances the secondary hardening reaction, which results in a 1 to 2 HRC hardness advantage in the fully heat-treated condition. It also enhances hot hardness and temper resistance, thus allowing a tool to retain a sharp cutting edge at higher machining speeds that generate heat. Despite the advantages of cobalt additions, the high cost and occasional lack of availability of cobalt have necessitated the development of cobalt-free alternatives.

CPM Rex 20 was developed as a cobalt-free P/M equivalent of M42. The chemical composition of CPM Rex 20 is listed in Table 3, and property comparisons of CPM Rex 20 with those of CPM and conventional M42 are shown in Tables 4 to 6.

Tables 4 and 5 show the results of temper resistance and hot hardness comparisons for specimens that were heat treated to full hardness. These results show that CPM Rex 20 and CPM and conventional M42 are equivalent in both temper resistance and hot hardness. Table 6 compares the Charpy C-notch impact energy and bend fracture strength values obtained for CPM Rex 20 with those obtained for CPM and conventional M42. Both the Charpy C-notch impact energy and the bend fracture strength of CPM Rex 20 are equal to those of CPM M42, but are notably higher than those of conventional M42. Table 7 shows the results of laboratory lathe tool tests in single-point turning on H13 and P/M René 95 superalloy. The overall performance of CPM Rex 20 is comparable to that of CPM M42.

## The FULDENS Process

Another method for the consolidation of P/M high-speed tool steels is the FULDENS process. This process differs from the others discussed in this article in that it uses water-atomized powders compacted either mechanically or by cold isostatic pressing and sintered in a vacuum to full density. For a discussion of mechanical properties of water-atomized powders that have been compacted and sintered, see the article "Mechanical Properties of P/M Materials" in Volume 7 of the 9th Edition of *Metals Handbook*. The FULDENS process allows close-tolerance complex shapes to be made with mechanical properties and performance characteristics comparable to those of equivalent parts made by conventional machining, with considerable material and labor savings.

**Processing Steps.** Figure 12 is a flowchart for the FULDENS process. The powders, usually water atomized, are specially prepared in regard to composition as well as

**Table 4 Temper resistance of CPM alloys**

| Alloy grade | Hardness, HRC: As heat treated at 1190 °C (2175 °F) + 550 °C (1025 °F) 3 times/2 h | As heat treated + 595 °C (1100 °F)/ 2 h | As heat treated + 595 °C (1100 °F)/ 2 + 2h | As heat treated + 650 °C (1200 °F)/ 2 h | As heat treated + 650 °C (1200 °F)/ 2 + 2 h |
|---|---|---|---|---|---|
| CPM Rex 20 | 67.5 | 66 | 65.5 | 60 | 57 |
| CPM M42 | 67.5 | 65.5 | 65 | 59 | 55.5 |
| Conventional M42 | 67.5 | 65 | 65 | 59 | 55 |

Source: Ref 3

**Table 5 Hot hardness of CPM alloys**

| Alloy grade | Hot hardness, HRC: At room temperature before test | At 540 °C (1000 °F) | At 595 °C (1100 °F) | At 650 °C (1200 °F) | At room temperature after test |
|---|---|---|---|---|---|
| CPM Rex 20 | 67.5 | 58.0 | 56 0 | 47.5 | 64.0 |
| CPM M42 | 67.0 | 58.5 | 56 0 | 48.0 | 63.0 |
| Conventional M42 | 66.5 | 58.5 | 56 0 | 48.0 | 62.0 |

Source: Ref 3

**Table 6 Charpy C-notch impact energy and bend fracture strengths of CPM alloys and conventional alloy**

| Alloy | Austenitizing temperature(a) °C | °F | Hardness, HRC | Charpy C-notch impact energy J | ft·lbf | Bend fracture strength MPa | ksi |
|---|---|---|---|---|---|---|---|
| CPM Rex 20 | 1190 | 2175 | 67.5 | 16 | 12 | 4006 | 581 |
| CPM M42 | 1190 | 2175 | 67.5 | 16 | 12 | 4006 | 581 |
| Conventional M42 | 1190 | 2175 | 67.5 | 7 | 5 | 2565 | 372 |

(a) 4 min soak in salt bath and oil quenched. Tempered at 550 °C (1025 °F) three times for 2 h. Source: Ref 3

**Table 7 Lathe tool test results on CPM alloys**

| Alloy grade | Austenitizing temperature °C | °F | Hardness, HRC | Tool life, minutes to 0.038 mm (0.015 in.) flank wear: Intermittent cut on H13 steel at 33 HRC | Continuous cut on H13 steel at 33 HRC | Continuous cut on P/M René 95 at 33 HRC |
|---|---|---|---|---|---|---|
| CPM Rex 20 | 1190 | 2175 | 67.5 | 8.5 | 14 | 31 |
| CPM M42 | 1190 | 2175 | 67 | 8 | 16 | 27 |
| **Test conditions** | | | | | | |
| Speed, m/s (sfm) | | | | 0.20 (40) | 0.20 (40) | 0.06 (12) |
| Feed, mm/rev (in./rev) | | | | 0.10 (0.004) | 0.14 (0.0055) | 0.18 (0.007) |
| Depth of cut, mm (in.) | | | | 1.57 (0.062) | 1.57 (0.062) | 1.57 (0.062) |
| Coolant | | | | None | None | None |

particle shape and size distribution. The powder is then annealed and pressed into green compacts, either by conventional mechanical pressing or by cold isostatic pressing. When part geometry allows, the part is compacted by filling a closed die with annealed powder and compacting with pressure ranging from 414 to 690 MPa (60 to 100 ksi). Cold isostatic pressing is more suitable for parts of relatively low volume, high complexity, and liberal tolerance; mechanical pressing is more suitable for parts of relatively high volume, low complexity, and close tolerance. Examples of isostatically pressed and mechanically pressed parts are shown in Fig. 13. For more information on cold isostatic pressing, see the article "Cold Isostatic Pressing of Metal Powders" in Volume 7 of the 9th Edition of *Metals Handbook*.

The compact is then sintered in a specially built vacuum sintering furnace. After sintering, it emerges from the furnace fully dense. For more information on vacuum sintering, see the article "Production Sintering Practices for P/M Materials" in Volume 7.

**Advantages.** Hardenability, elongation, and impact strength of materials drop significantly if small amounts of porosity (even 1 or 2%) are present. This has often limited the applicability of pressed and sintered powdered metals in demanding applications. Because the FULDENS process provides a finished part of nearly 100% density, good properties and product performance can be expected.

In the FULDENS process, only the net weight of materials in the preform is actually used. In conventional manufacturing, much material is wasted in the form of chips. With higher-alloy high-speed tool steels that contain cobalt, molybdenum, and tungsten (materials often in short supply), savings in scrap reduction by using P/M processing can be substantial. The FULDENS process also eliminates much preheat-treat machining labor.

**Applications.** Components produced by the FULDENS process should be considered for applications in which the following conditions exist:

- Low net-to-gross weight ratios
- Relatively high strength-to-ductility ratios
- High wear environments
- High temperature environments
- High Hertz stresses

Applications in which fully dense P/M high-speed tool steel parts are currently in use include screw machine tooling, gear cutting tools, high-speed tool steel indexable inserts, and forming tools.

## Applications of P/M High-Speed Tool Steels

**Milling.** Milling cutters, such as those shown in Fig. 14, are emerging as a major application of P/M high-speed tool steels. Stock removal rates can usually be increased by raising the cutting speed and/or feed rate. In general, the feed per cutter tooth is increased in roughing operations, and the cutting speed is increased for finishing operations.

The performances of conventionally processed and P/M high-speed tool steel end mills in milling Ti-6Al-4V have been evaluated. In these tests, ASP 30 and ASP 60 were compared to M42. The cutting conditions used for this evaluation are given in Fig. 15, which shows tool life versus cutting speed. Both feed per tooth (0.203 mm, or 0.008 in.) and cutting speeds (>45.7 m/min, or 150 sfm) are higher than those used in production practice for machining aircraft parts. At a constant metal removal rate that corresponds to a cutting speed of 53.3 m/min (175 sfm), ASP 60 and ASP 30 lasted eight times and 4.5 times longer, respectively, than the M42 end mill.

**Other materials** machined by P/M high-speed tool steel milling cutters include tough, hardened steels such as 4340, austenitic stainless steels such as AISI type 316, and nickel-base superalloys such as Nimonic 80.

**Hole Machining.** Reamers, taps, and drills (Fig. 16) are also made from P/M high-speed tool steels. In one application, the tool life of ¼-20 GH3 four-flute plug taps made from CPM M4 and conventional M1, M7, and M42 were compared. The operation consisted of tapping a reamed 5.18 mm (0.204 in.) diam, 12.7 mm (0.500 in.) deep through-hole in AISI 52100 steel at 32 to 34 HRC using a speed of 7.8 m/min (26 sfm) and chlorinated tapping oil. Eight to thirteen taps of each grade were tested. The CPM M4 tap had an average tool life of 157 holes tapped before tool failure, compared to 35 holes for M1, 18 holes for M7, and 32 holes for M42. The tool life of the CPM M4 in this application was about five times the life of conventional M42.

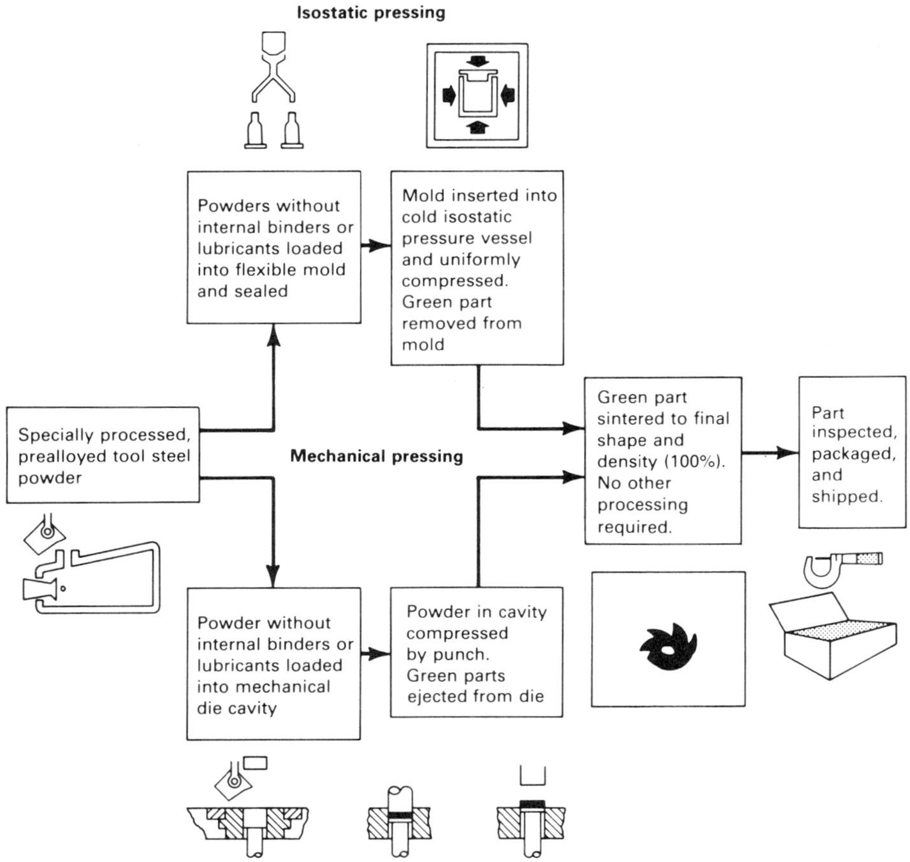


**Fig. 12** The FULDENS process for producing tool steel powder. Source: Ref 4

**Fig. 14** Typical milling cutters made from P/M high-speed tool steels. Courtesy of Speedsteel Inc.

**Broaching.** These tools constitute another major application for P/M high-speed tool steels because tool life is often improved when P/M steels are used to broach difficult-to-cut materials such as case-hardened steels and superalloys. One application required broaching six ball tracks that are used in front-wheel-drive automobiles in constant-velocity joint hubs made of a case-hardening steel. Figure 17 shows the joint hubs and broaching tool used in this application.

In this broaching application, surface finish and form tolerance requirements are high because subsequent machining is not performed on the ball tracks. In broaching with tools 18.0 mm (0.709 in.) in diameter and 185.0 mm (7.283 in.) in length made from low-carbon M35 steel (similar to M41 in chemical composition), the total number of hubs machined per tool was 5600. The M35 tools experienced severe flank wear and developed a large built-up edge, which

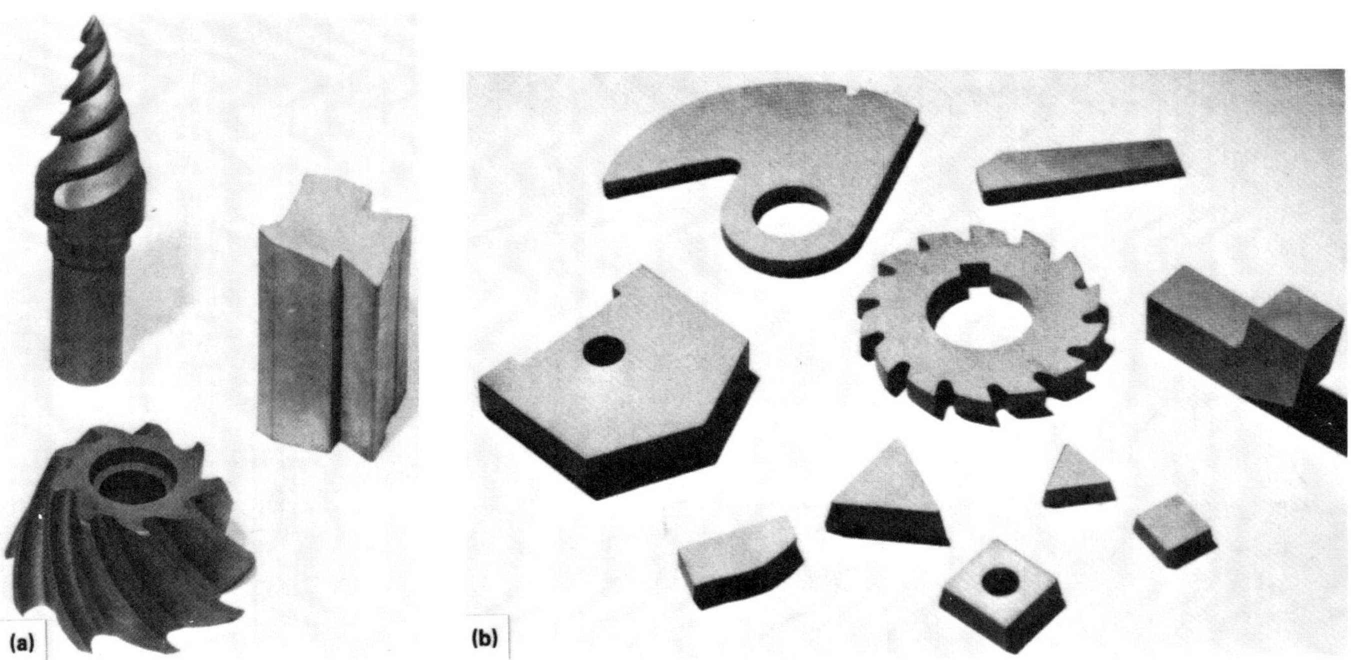


**Fig. 13** Examples of parts manufactured by the FULDENS process. Note the complexity of shapes attainable by this process. (a) Using cold isostatic pressing. (b) Using mechanical pressing

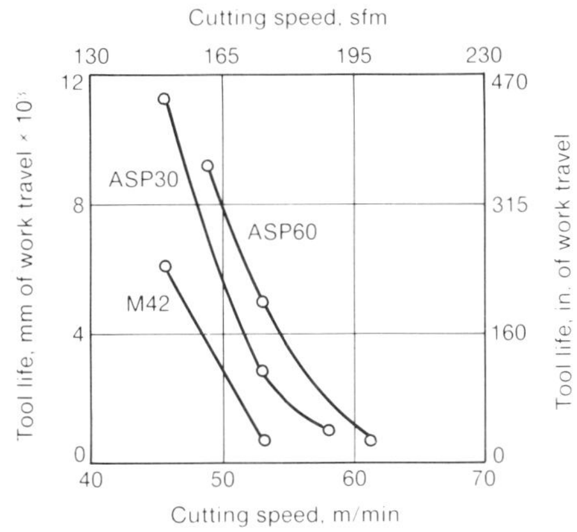


| | |
|---|---|
| Cutter | 25 mm (1 in.) diam end mills |
| Feed | 0.203 mm/tooth (0.008 in./tooth) |
| Radial depth of cut | 6.35 mm (0.250 in.) |
| Axial depth of cut | 25.4 mm (1.000 in.) |
| Cutting fluid | Soluble oil (1:20) |
| Tool life end point | 0.5 mm (0.020 in.) wear |

**Fig. 15** End mill test on Ti-6Al-4V. Hardness: 321 HB

produced poor surface finishes. With an ASP 30 tool, 20 000 parts were produced.

Large broaching tools, such as those shown in Fig. 18, are also being made from P/M high-speed tool steels, such as P/M M3 and M4, to upgrade the broach material. In general, large rounds for broaches are not available in conventional high-speed tool steels in sizes above about 254 mm (10 in.), but larger sizes are available in P/M high-speed tool steels. One application for these tools is the broaching of involute splines in bores of truck transmission gear blanks. Bores up to 305 mm (12 in.) in diameter by 1380 mm (54¼ in.) long have been cut using such tools.

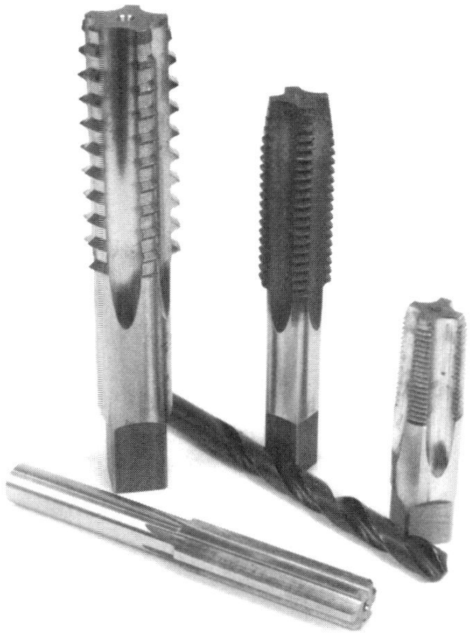

**Fig. 16** Reamers, taps, and drills made from P/M high-speed tool steels. Courtesy of Crucible Materials Corporation

**Gear Manufacturing.** Gear hobs (Fig. 19) made from P/M high-speed tool steels can also provide substantial cost reductions by increasing machining rates. One application called for hobbing of rear axle gears for heavy-duty trucks and tractor differentials. Hobs made of conventionally processed AISI M35 (65 HRC) and ASP 30 (67 HRC) were compared. Test parameters for both materials were:

- *Hob dimensions*: 152 mm (6 in.) diam × 50 mm (2 in.) diam × 205 mm (8 in.) length
- *Work material*: Case-hardening steel (Fe-1.2Ni-1Cr-0.2C-0.12Mo); hardness: 160 to 180 HB
- *Cutting speed*: 70 m/min (230 ft/min)
- *Spindle speed*: 150 rev/min
- *Roughing*: Feed, 4.24 mm (0.167 in.); depth of cut, 15.0 mm (0.591 in.)
- *Finishing*: Feed, 5.92 mm (0.233 in.); depth of cut, 0.81 mm (0.032 in.)
- *Coolant*: Cutting oil
- *Number of parts per resharpening*: 20

Production results showed that the flank wear land on the hobs made of ASP 30 (0.44 mm, or 0.017 in.) was much less than hobs made of M35 (0.71 mm, or 0.028 in.). Chipping of the edges was infrequent on the ASP 30 hobs, while the M35 hobs frequently displayed such damage. ASP 30 is now the standard grade for hobs used by one automotive manufacturer.

**Tool Bits.** Figure 20 shows tool bits that have been produced using P/M high-speed tool steels. Typical applications include machining of turbine blades made from superalloys and turning of hardened steels, such as AISI 4340 (1.9Ni-0.75Cr-0.4C).

## REFERENCES

1. R. Riedl *et al.*, Developments in High Speed Tool Steels, *Steel Res.*, Vol 58 (No. 8), 1987, p 339-352
2. O. Siegwarth, "Higher Productivity with ASP Tooling Material," Technical Paper MF 81-137, Society of Manufacturing Engineers, 1981, p 1-22
3. F.R. Dax, W.T. Haswell, and W. Stasko, Cobalt-Free CPM High Speed

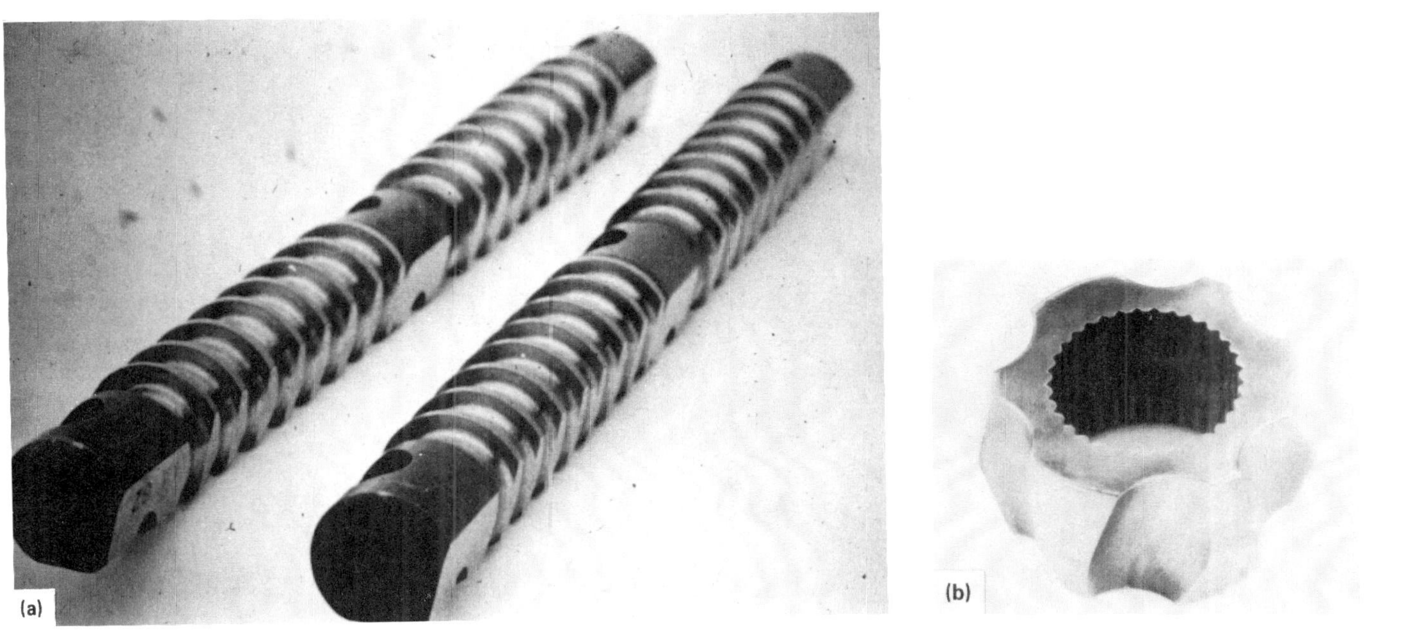


**Fig. 17** Broaching application. (a) Tool made of P/M high-speed tool steel that was used to produce ball tracks on joint hub. (b) ASP 30 tools produced 20 000 parts compared to 5600 parts by tools made from conventional high-speed tool steel. Courtesy of Speedsteel Inc.

**Fig. 18** Large broaching tool made from P/M high-speed tool steel that was used for broaching involute splines in bores of truck transmission gear blanks. Courtesy of Crucible Materials Corporation

Steels, in *Processing and Properties of High Speed Tool Steels*, The Metallurgical Society of the American Institute of Mining, Metallurgical, and Petroleum Engineers, 1980, p 148-158
4. M.T. Podob and R.P. Harvey, Advantages and Applications of CMI's FUL-DENS Process, in *Processing and Properties of High Speed Tool Steels*, The Metallurgical Society of the American Institute of Mining, Metallurgical, and Petroleum Engineers, 1980, p 181-195

## SELECTED REFERENCES


- B. Alvelid, and H. Wisell, Wear of High Speed Steel in Orthogonal Milling, *Scand. J. Metall.*, Vol 9, 1980, p 59-67
- J.T. Berry, "High Performance High Hardness High Speed Steels," Climax Molybdenum Company, 1970
- E.A. Carlson, J.E. Hansen, and J.C. Lynn, Characteristics of Full-Density P/M Tool Steel and Stainless Steel Parts, in *Modern Developments in Powder Metallurgy*, Vol 13, Metal Powder Industries Federation, Princeton, NJ, 1980
- B.-A. Cehlin, "Improving Productivity With High Strength P/M High Speed Steel Cutting Tools," Technical Paper MR82-948, Society of Manufacturing Engineers, presented at Increasing Productivity With Advanced Machining Concepts Clinic (Los Angeles), 1982

**Fig. 19** Gear hobs made from P/M high-speed tool steels. Courtesy of Speedsteel Inc.

**Fig. 20** Tool bits made from P/M high-speed tool steels. Courtesy of Crucible Materials Corporation


- "Crucible CPM Rex-High Speed Steel for Superior Cutting Tools," Colt Industries
- E.J. Dulis and T.A. Neumeyer, Particle-Metallurgy of High-Speed Tool Steel, in *Materials for Metal Cutting*, Publication 126, The Iron and Steel Institute, 1970, p 112-118
- P. Hellman, Wear Mechanism and Cutting Performance of Conventional and High-Strength P/M High-Speed Steels, *Powder Metall.*, Vol 25 (No. 2), 1982
- P. Hellman *et al.*, The ASEA-STORA-Process, *Modern Developments in Powder Metallurgy*, Vol 4, Plenum Press, 1970, p 573-582
- P. Hellman and H. Wisell, "Effect of Structure on Toughness and Grindability of High Speed Steels," Paper presented at Colloquium on High Speed Steels (Saint-Étienne, France), Nov 1975
- W.E. Henderer and B.F. von Turkovich, "The Influence of Heat and Surface Treatments on the Performance of M1 HSS Taps," Paper presented at American Society for Metals Symposium on Hole Making Operations (Boston), May 1977
- W.J. Huppmann and P. Beiss, "Sintering of P/M Tool Steels to Full Density," Paper presented at Fully Dense P/M Materials for High Performance Applications, and Metal Powder Industries Federation Short Course, Sintermetallwerk Krebsoge (New Orleans), Feb 1982
- A. Kasak and E.J. Dulis, Powder-Metallurgy Tool Steels, *Powder Metall.*, Vol 21 (No. 2), 1978, p 114-123
- A. Kasak, G. Steven, and T.A. Neumeyer, "High-Speed Tool Steels by Particle Metallurgy," Paper 720182, Society of Automotive Engineers, 1972
- *Properties and Selection: Stainless Steels, Tool Materials and Special-Purpose Metals*, Vol 3, 9th ed., *Metals Handbook*, American Society for Metals, 1980
- *Powder Metallurgy*, Vol 7, 9th ed., *Metals Handbook*, American Society for Metals, 1984
- G.A. Roberts, J.C. Hamaker, Jr., and A.R. Johnson, *Tool Steels*, American Society for Metals, 1962, p 710-713
- S. Söderberg, S. Hogmark, H. Haag, and H. Wissel, "Wear Resistance of High Speed Steel Milling Tools," Report 821 OR., Uppsala University, Institute of Technology
- "Tool Steels Today Newsletter," Committee on Tool Steel Producers, American Institute of Mining, Metallurgical, and Petroleum Engineers, Washington, D.C., Jan 1976

# Cast Cobalt Alloys

CAST COBALT ALLOYS were developed to bridge the gap between high-speed steels and carbides. Although comparable in room-temperature hardness to high-speed steel tools, cast cobalt alloy tools retain their hardness to a much higher temperature (Fig. 1) and can be used at higher (~20%) cutting speeds than high-speed steel tools. Unlike the high-speed steels that can be heat treated to obtain the desired hardness, cast cobalt alloys are hard in the as-cast condition and cannot be softened or hardened by heat treatment.

## Processing, Properties, and Applications

**Processing.** Cast cobalt alloys are produced by electric or induction melting under a protective atmosphere, and for cutting tool applications they are preferably cast in permanent graphite molds. However, they can be cast in investment, shell, or sand molds to produce special and intricate shapes. Each of the melting and casting processes mentioned above is discussed in Volume 15 of the 9th Edition of *Metals Handbook*.

**Properties and Applications.** Cast cobalt alloys contain a primary phase of cobalt-rich solid solution that is strengthened by chromium and tungsten and is dispersion strengthened by complex, hard, refractory carbides of tungsten and chromium (Ref 2, 3). Nominal compositions for two commercially available grades are as follows:

| Element | Alloy: Tantung G, % | Alloy: Tantung 144, % |
|---|---|---|
| Cobalt | 42–47 | 40–45 |
| Chromium | 27–32 | 25–30 |
| Tungsten | 14–19 | 16–21 |
| Carbon | 2–4 | 2–4 |
| Tantalum or niobium | 2–7 | 3–8 |
| Manganese | 1–3 | 1–3 |
| Iron | 2–5 | 2–5 |
| Nickel | 7(a) | 7(a) |

(a) Maximum

The typical microstructure of a permanent graphite mold cast Tantung G alloy is shown in Fig. 2. Properties of Tantung G are given in Table 1. For comparison, permanent mold cast Tantung 144 has a hardness of 61 to 65 HRC, a transverse strength of 2070 MPa (300 ksi), and an elastic modulus of 295 GPa ($43 \times 10^6$ psi).

Tantung G is recommended for general-purpose cutting tools and parts for wear applications, and it is more widely used than Tantung 144. Tantung 144 has higher hardness than Tantung G and was developed for use where resistance to abrasion is paramount and where there is little or no shock or impact. The good resistance of Tantung G and Tantung 144 to foods, particularly those containing acetic acid, makes them highly suitable for food-processing equipment, especially parts requiring good resistance to abrasion and corrosion.

Water containing chlorine and hypochlorites may produce some corrosion and pitting of alloys G and 144. In addition, these

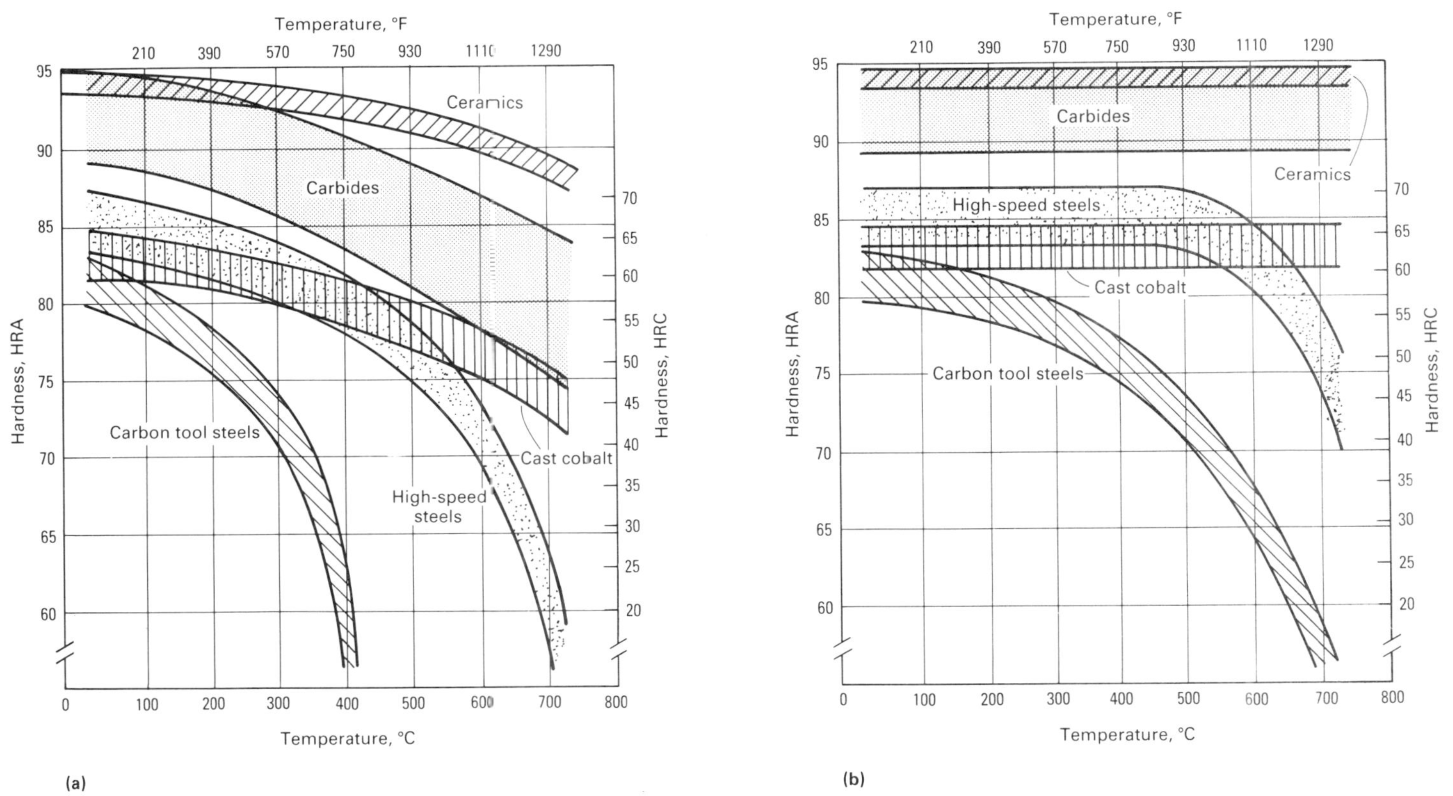


**Fig. 1** Comparison of hot hardness values of cast cobalt alloys with alternate cutting tool materials. (a) Hot hardness as a function of temperature. (b) Recovery hardness as a function of temperature. Source: Ref 1

**Table 1 Typical properties of cast Tantung G**

| Property | Permanent mold cast | Refractory mold cast |
|---|---|---|
| Melting temperature, °C (°F) | 1150–1200 (2100–2200) | 1150–1200 (2100–2200) |
| Casting temperature, °C (°F) | 1370 (2500) | 1370 (2500) |
| Density, g/cm$^3$ (lb/in.$^3$) | 8.3 (0.30) | 8.3 (0.30) |
| Thermal expansion, μm/m · °C (μin./in. · °F) | 4.2 (2.3) | 4.2 (2.3) |
| Thermal conductivity, W/m · K (Btu/ft · h · °F) | 26.8 (15.5) | 26.8 (15.5) |
| Hardness, HRC | 60–63 | 53–58 |
| Transverse strength, MPa (ksi) | 2240 (325) | 1030–1200 (150–175) |
| Modulus of elasticity, GPa ($10^6$ psi) | 265 (41) | . . . |
| Tensile strength, MPa (ksi) | 585–620 (85–90) | 450 (65) |
| Compressive strength, MPa (ksi) | 2760 (400) | 2930 (425) |
| Impact strength, J (ft · lb) | 6.1 (4.5) | 6.1 (4.5) |

alloys are attacked by strong acid solutions, alkalies, and solutions of some heavy-metal salts such as ferric chloride, ferric sulfate, and cupric chloride.

When heated in air, Tantung G and Tantung 144 both tarnish on short-time exposures at 400 °C (750 °F) and lose appreciable weight at 750 °C (1380 °F) or higher. Scaling may be progressive above 1000 °C (1830 °F). Hot hardness data for cast cobalt alloys are shown in Fig. 1.

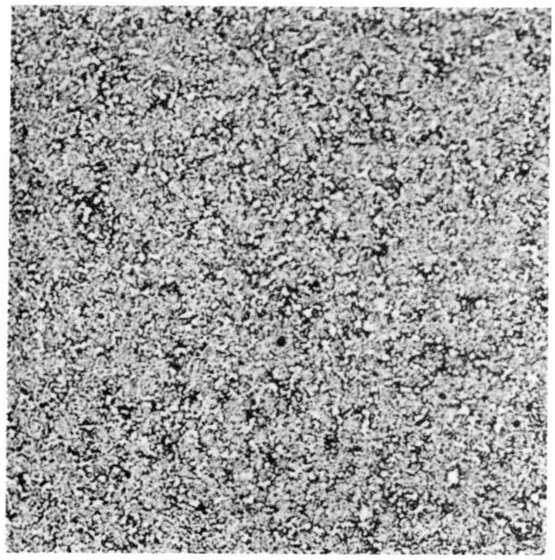

**Fig. 2** Microstructure of cast Tantung G alloy. Etched with Murakami's reagent (standard mix: 10 g sodium hydroxide, 10 g potassium ferricyanide, 100 mL $H_2O$). 400×. Courtesy of G.F. Vander Voort, Carpenter Technology

The use of cast cobalt cutting tools should be considered:

- Where relatively low surface speeds cause buildup with cemented carbides
- Where machines lack the power or rigidity to use cemented carbides effectively
- Where higher production is desired than is possible with high-speed steel tools
- For multiple-tool operations in which the surface speed of one or more operations falls between the recommended speeds for high-speed steel and carbide tools
- For short runs on automatic equipment in which the form grinding of carbide tools is excessively costly
- For machining rough surfaces of castings where the surfaces contain abrasive material, such as residual sand, surface oxides, slag, or refractory particles

Tools made of cast cobalt alloys usually are not recommended for light, very fast finishing cuts. Typical wear applications for these alloys include wear strips for belt sanders; dies for extruding copper, for extruding molybdenum tubing, and for hot swaging tungsten rod; burnishing rolls; internal chuck jaws; drill bushings; and knives for slicing fruits, vegetables, and meat.

## REFERENCES

1. R. Komanduri, Tool Materials, in *Encyclopedia of Chemical Technology*, Vol 23, 3rd ed., John Wiley & Sons, 1983
2. K.J.A. Brooks, *World Directory and Handbook of Hard Metals*, Engineer's Digest Ltd., 1976
3. R.L. Hatschek, *Am. Mach.*, Vol 733 (No. 165), May 1981

# Cemented Carbides

A.T. Santhanam and P. Tierney, Kennametal Inc.

CEMENTED CARBIDES belong to a class of hard, wear-resistant, refractory materials in which the hard carbide particles are bound together, or cemented, by a soft and ductile metal binder. These materials were first developed in Germany in the early 1920s in response to demands for a die material having sufficient wear resistance for drawing tungsten incandescent filament wires to replace the expensive diamond dies then in use. The first cemented carbide to be produced was tungsten carbide (WC) with a cobalt binder.

Tungsten carbide was first synthesized by the French chemist Henri Moissan in the 1890s (Ref 1). There are two types of tungsten carbide: WC, which directly decomposes at 2800 °C (5070 °F), and $W_2C$, which melts at 2750 °C (4980 °F) (Ref 2, 3). Early attempts to produce drawing dies from a eutectic alloy of WC and $W_2C$ were unsuccessful, because the material had many flaws and fractured easily. The use of powder metallurgy techniques by Schroeter in 1923 paved the way for obtaining a fully consolidated product (Ref 4). Schroeter blended fine WC powders with a small amount of iron, nickel, or cobalt powders and pressed the powders into compacts, which were then sintered at approximately 1300 °C (2400 °F). Cobalt was soon found to be the best bonding material. Over the years, the basic WC-Co material has been modified to produce a variety of cemented carbides, which are used in a wide range of applications, including metal cutting, mining, construction, rock drilling, metal forming, structural components, and wear parts. Approximately 50% of all carbide production is used for metal cutting applications. Although the term cemented carbide is widely used in the United States, these materials are better known as hard metals internationally.

This article will discuss the manufacture and composition of cemented carbides and their microstructure, classifications, applications, and physical and mechanical properties. New tool geometries, tailored substrates, and the application of thin, hard coatings to cemented carbides by chemical vapor deposition and physical vapor deposition will also be discussed. This article is limited to tungsten carbide cobalt-base materials. Information on titanium carbide nickel-base materials is given in the article "Cermets" in this Volume. Extensive reviews of the scientific and industrial aspects of cemented carbides are available in Ref 5 to 8.

## Manufacture of Cemented Carbides

Cemented carbides are manufactured by a powder metallurgy process consisting of a sequence of steps in which each step must be carefully controlled to obtain a final product with the desired properties, microstructure, and performance. The steps include:

- Processing of the ore and the preparation of the tungsten carbide powder
- Preparation of the other carbide powders
- Production of the grade powders
- Compacting or powder consolidation
- Sintering
- Postsinter forming

The sintered product can be directly used or can be ground, polished, and coated to suit a given application.

**Preparation of Tungsten Carbide Powder.** There are two methods by which tungsten carbide powders are produced from the tungsten-bearing ores. Traditionally, tungsten ore is chemically processed to ammonium paratungstate and tungsten oxides. These compounds are then hydrogen-reduced to tungsten metal powder. The fine tungsten powders are blended with carbon and heated in a hydrogen atmosphere between 1400 and 1500 °C (2500 and 2700 °F) to produce tungsten carbide particles with sizes varying from 0.5 to 30 μm (Fig. 1). Each particle is composed of numerous tungsten carbide crystals. Small amounts of vanadium, chromium, or tantalum are sometimes added to tungsten and carbon powders before carburization to produce very fine (<1 μm) WC powders.

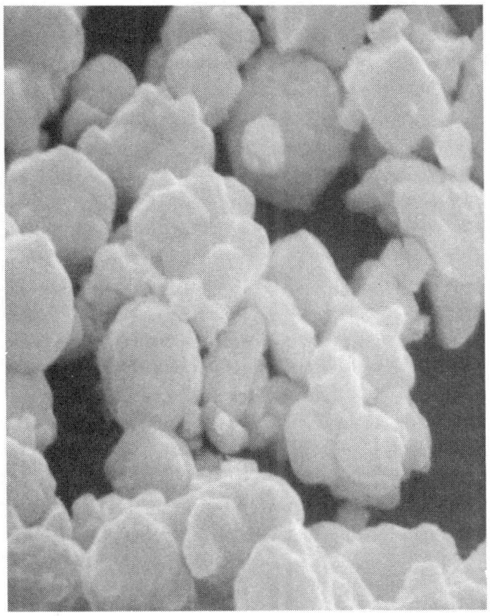

**Fig. 1** Tungsten carbide particles produced by the carburization of tungsten and carbon. 10 000×

In a more recently developed and patented process, tungsten carbide is produced in the form of single crystals through the direct reduction of tungsten ore (sheelite) (Ref 9). The ore is mixed with iron oxide, aluminum, carbon, and calcium carbide. A high-temperature exothermic reaction ($2Al + 3FeO \leftrightarrows Al_2O_3 + 3Fe$) at about 2500 °C (4500 °F) produces a molten mass that, when cooled, consists of tungsten carbide crystals dispersed in iron, and a slag containing impurities. The crystalline WC (Fig. 2) is then chemically separated from the iron matrix.

**Tungsten-titanium-tantalum (niobium) carbides** are used in steel-cutting grades to resist cratering or chemical wear and are produced from metal oxides of titanium, tantalum, and niobium. These oxides are mixed with metallic tungsten powder and carbon. The mixture is heated under a hydrogen atmosphere or vacuum to reduce the oxides and form solid-solution carbides such as WC-TiC, WC-TiC-TaC, or WC-TiC-(Ta, Nb)C. The menstruum method can be used to produce WC-TiC solid solution. In this method, the individual carbides are dissolved in liquid nickel. Solid-solution carbides are then precipitated during cooling (Ref 10).

**Production of Grade Powders.** Cemented carbide grade powders may consist of

**Fig. 2** Tungsten carbide single crystals produced by the direct reduction of tungsten ore. 200×

WC mixed with a finely divided metallic binder (cobalt, nickel, or iron) or with additions of other cubic carbides, such as TiC, TaC, and NbC, depending on the required properties and application of the tool. Intensive milling is necessary to break up the initial carbide crystallites and to blend the various components such that every carbide particle is coated with binder material. This is accomplished in ball mills, vibratory mills, or attritors that use carbide balls. The mills are usually lined with carbide sleeves, although mills lined with low-carbon steel or stainless steel are also used.

Milling is performed under an organic liquid such as heptane or acetone to minimize heating of the powder and to prevent its oxidation. The liquid is distilled off after the milling operation. A solid lubricant such as paraffin wax is added to the powder blend in the final stages of the milling process or later in a blender. The lubricant provides a protective coating to the carbide particles and prevents or greatly reduces the oxidation of the powder. The lubricant also imparts strength to the pressed or consolidated powder mix.

After milling, the organic liquid is removed by drying. In a spray-drying process commonly used in the cemented carbide industry, a hot inert gas such as nitrogen impinges on a stream of carbide particles. This produces free-flowing spherical powder aggregates.

**Powder Consolidation.** A wide variety of techniques are used to compact the cemented carbide grade powders to the desired shape. Carbide tools for mining and construction applications are pill pressed (pressure applied in one direction) in semiautomatic or automatic presses. Metal cutting inserts are also pill pressed, but may require additional shaping after sintering. Cold isostatic pressing, in which the powder is subjected to equal pressure from all directions, followed by green forming, is also a common practice for wear and metal forming tools. Rods and wires are formed by the extrusion process.

Unlike most other metal powders, cemented carbide powders do not deform during the compacting process. Generally, they cannot be compressed to much above 65% of the theoretical upper limit for density. Despite this low green density, carbide manufacturers have developed the technology for achieving good dimensional tolerances in the sintered product.

**Sintering and Postsintering Operations.** The first step in the sintering process is the removal of the lubricant (dewaxing) from the powder compact. The pressed compacts are normally set on graphite trays coated with a graphite paint. The compacts are first heated to about 500 °C (900 °F) in a hydrogen atmosphere or vacuum using either semicontinuous or batch-type graphite furnaces.

After lubricant removal, the compacts are heated in a vacuum (0.1 Pa, or $10^{-3}$ torr) to a final sintering temperature ranging from 1350 to 1600 °C (2460 to 2900 °F), depending on the amount of the cobalt binder and the desired microstructure. The dewaxing and sintering operations can also be performed in a single vacuum cycle using furnaces equipped to condense the lubricant and remove it from the heating chamber.

During the final sintering operation, the cobalt melts and draws the carbide particles together. Shrinkage of the compact ranges from 17 to 25% on a linear scale, producing a virtually pore-free, fully dense product.

In the 1970s, the cemented carbide industry took advantage of hot isostatic pressing (HIP) technology, in which vacuum-sintered material is heated again under a gaseous (argon or helium) pressure of 100 to 150 MPa (15 to 20 ksi) (Ref 11). The temperatures of this additional process are 25 to 50 °C (45 to 90 °F) below the sintering temperature. The high temperatures and pressures employed in the HIP furnace remove any residual internal porosity, pits, or flaws and produce a nearly perfect cemented carbide.

The latest advancement in sintering technology is the sinter-HIP process, which was developed in the early 1980s (Ref 12). In this process, low-pressure hot isostatic pressing (up to about 7 MPa, or 1 ksi) is combined with vacuum sintering, and the pressure is applied at the sintering temperature when the metallic binder is still molten. With this process, void-free products can be produced at costs only slightly higher than those of vacuum sintering.

**Postsinter Forming.** A large number of cemented carbide products are shaped after sintering because of surface finish, tolerance, and geometry requirements. This forming operation is both time consuming and expensive. The sintered material is formed with metal-bonded diamond or silicon carbide wheels, turned with a single-point diamond tool, or lapped with diamond-containing slurries.

## Cemented Carbide Compositions and Microstructures

The performance of carbide cutting tools is strongly dependent on composition and microstructure, and the properties of cemented carbide tools depend not only on the type and amount of carbide but also on the carbide grain size and the amount of binder metal. The basic physical and mechanical properties of refractory metal carbides used in the production of cemented carbide tools are given in Table 1. Tungsten carbide and molybdenum carbide have hexagonal crystal structures, while the carbides of titanium, tantalum, niobium, vanadium, hafnium, and zirconium are cubic. They undergo no structural changes up to their melting points.

**Tungsten Carbide-Cobalt Alloys.** The first commercially available cemented carbides consisted of tungsten carbide particles bonded with cobalt. These are commonly

**Table 1 Properties of refractory metal carbides**

| Carbide | Hardness, HV (50 kg) | Crystal structure | Melting point °C | Melting point °F | Theoretical density, g/cm³ | Modulus of elasticity GPa | Modulus of elasticity psi × $10^6$ | Coefficient of thermal expansion, μm/m · K |
|---|---|---|---|---|---|---|---|---|
| TiC | 3000 | Cubic | 3100 | 5600 | 4.94 | 451 | 65.4 | 7.7 |
| VC | 2900 | Cubic | 2700 | 4900 | 5.71 | 422 | 61.2 | 7.2 |
| HfC | 2600 | Cubic | 3900 | 7050 | 12.76 | 352 | 51.1 | 6.6 |
| ZrC | 2700 | Cubic | 3400 | 6150 | 6.56 | 348 | 50.5 | 6.7 |
| NbC | 2000 | Cubic | 3600 | 6500 | 7.80 | 338 | 49.0 | 6.7 |
| $Cr_3C_2$ | 1400 | Orthorhombic | 1800(a) | 3250 | 6.66 | 373 | 54.1 | 10.3 |
| WC | (0001) 2200<br>(10$\bar{1}$0) 1300 | Hexagonal | ~2800(a) | 5050 | 15.7 | 696 | 101 | (0001) 5.2<br>(10$\bar{1}$0) 7.3 |
| $Mo_2C$ | 1500 | Hexagonal | 2500 | 4550 | 9.18 | 533 | 77.3 | 7.8 |
| TaC | 1800 | Cubic | 3800 | 6850 | 14.50 | 285 | 41.3 | 6.3 |

(a) Not congruently melting, dissociation temperature. Source: Ref 8

**Table 2 Properties of representative cobalt-bonded cemented carbides**

| Nominal composition | Grain size | Hardness, HRA | Density, g/cm³ | Density, oz/in.³ | Transverse strength, MPa | Transverse strength, ksi | Compressive strength, MPa | Compressive strength, ksi | Modulus of elasticity, GPa | Modulus of elasticity, psi × 10⁶ | Relative abrasion resistance(a) | Coefficient of thermal expansion, μm/m · K, at 200 °C (390 °F) | Coefficient of thermal expansion, μm/m · K, at 1000 °C (1830 °F) | Thermal conductivity, W/m · K |
|---|---|---|---|---|---|---|---|---|---|---|---|---|---|---|
| 97WC-3Co | Medium | 92.5–93.2 | 15.3 | 8.85 | 1590 | 230 | 5860 | 850 | 641 | 93 | 100 | 4.0 | ... | 121 |
| 94WC-6Co | Fine | 92.5–93.1 | 15.0 | 8.67 | 1790 | 260 | 5930 | 860 | 614 | 89 | 100 | 4.3 | 5.9 | ... |
| | Medium | 91.7–92.2 | 15.0 | 8.67 | 2000 | 290 | 5450 | 790 | 648 | 94 | 58 | 4.3 | 5.4 | 100 |
| | Coarse | 90.5–91.5 | 15.0 | 8.67 | 2210 | 320 | 5170 | 750 | 641 | 93 | 25 | 4.3 | 5.6 | 121 |
| 90WC-10Co | Fine | 90.7–91.3 | 14.6 | 8.44 | 3100 | 450 | 5170 | 750 | 620 | 90 | 22 | ... | ... | ... |
| | Coarse | 87.4–88.2 | 14.5 | 8.38 | 2760 | 400 | 4000 | 580 | 552 | 80 | 7 | 5.2 | ... | 112 |
| 84WC-16Co | Fine | 89 | 13.9 | 8.04 | 3380 | 490 | 4070 | 590 | 524 | 76 | 5 | ... | ... | ... |
| | Coarse | 86.0–87.5 | 13.9 | 8.04 | 2900 | 420 | 3860 | 560 | 524 | 76 | 5 | 5.8 | 7.0 | 88 |
| 75WC-25Co | Medium | 83–85 | 13.0 | 7.52 | 2550 | 370 | 3100 | 450 | 483 | 70 | 3 | 6.3 | ... | 71 |
| 71WC-12.5TiC-12TaC-4.5Co | Medium | 92.1–92.8 | 12.0 | 6.94 | 1380 | 200 | 5790 | 840 | 565 | 82 | 11 | 5.2 | 6.5 | 35 |
| 72WC-8TiC-11.5TaC-8.5Co | Medium | 90.7–91.5 | 12.6 | 7.29 | 1720 | 250 | 5170 | 750 | 558 | 81 | 13 | 5.8 | 6.8 | 50 |

(a) Based on a value of 100 for the most abrasion-resistant material

referred to as straight grades. These alloys exhibit excellent resistance to simple abrasive wear and thus have many applications in metal cutting. Table 2 lists the representative properties of several straight WC-Co alloys.

The commercially significant alloys contain cobalt in the range of 3 to 25 wt%. For machining purposes, alloys with 3 to 12% Co and carbide grain sizes from 0.5 to more than 5 μm are commonly used.

The ideal microstructure of WC-Co alloys should exhibit only two phases: angular WC grains and cobalt binder phase. Representative microstructures of several straight WC-Co alloys are shown in Fig. 3. The carbon content must be controlled within narrow limits. Too high a carbon content results in the presence of free and finely divided graphite (Fig. 4), which in small amounts has no adverse effects in machining applications. Deficiency in carbon, however, results in the formation of a series of double carbides (for example, $Co_3W_3C$ or $Co_6W_6C$), commonly known as η phase, which causes severe embrittlement. Because the formation of η phase involves the dissolution of the original carbides into the cobalt binder, η phase appears as an irregularly shaped phase in the microstructure (Fig. 5).

**Submicron Tungsten Carbide-Cobalt Alloys.** In recent years, WC-Co alloys with submicron carbide grain sizes (Fig. 6) have been developed for applications requiring more toughness or edge strength. Typical applications include indexable inserts and a wide variety of solid carbide drilling and milling tools. Grain refinement in these alloys is obtained by small additions (0.25 to 3.0 wt%) of tantalum carbide, niobium carbide, vanadium carbide, or chromium carbide. Additions can be made before carburization of the tungsten or later in the powder blend. Vanadium carbide is the

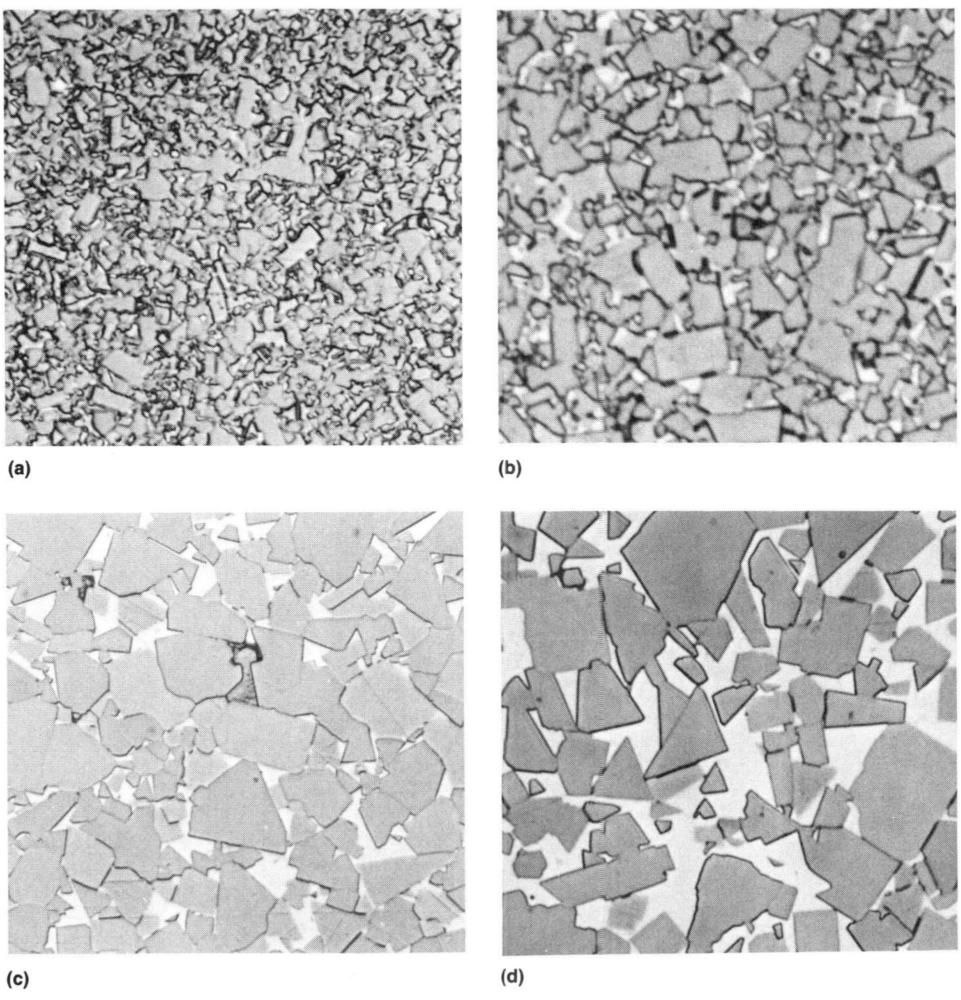

**Fig. 3** Microstructures of straight WC-Co alloys. (a) 97WC-3Co alloy, medium grain size. (b) 94WC-6Co alloy, medium grain. (c) 94WC-6Co alloy, coarse grain. (d) 85WC-15Co alloy, coarse grain. All etched with Murakami's reagent for 2 min. 1500×

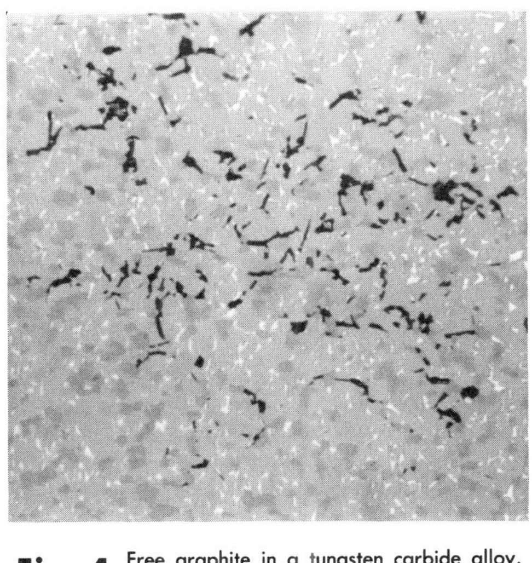

**Fig. 4** Free graphite in a tungsten carbide alloy. Black areas contain graphite and are an example of C-type porosity. Polished 86WC-8(Ta,Ti,Nb)C-6Co alloy. 1005×

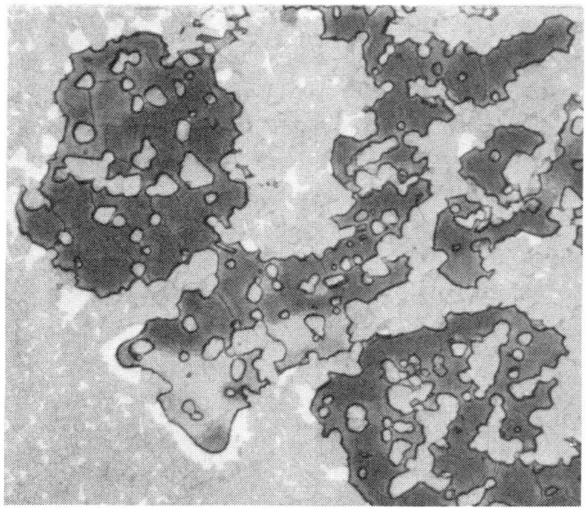

**Fig. 5** η phase microstructure. Micrograph shows a $(Co_3W_3)C$-η phase in detail. η phase appears as various shades of gray with clearly defined grain boundaries. Light gray WC particles surrounded by η phase are rounded because of the solubility of tungsten carbide in the binder. 85WC-8(Ta,Ti,Nb)C-7Co alloy etched with Murakami's reagent for 3 s. 900×

most effective grain growth inhibitor. Chromium carbide, in addition to being an efficient grain growth inhibitor, imparts excellent mechanical properties. Tantalum carbide is not as effective as vanadium carbide or chromium carbide in grain refinement.

**Alloys Containing Tungsten Carbide, Titanium Carbide, and Cobalt.** The tungsten carbide-cobalt alloys, developed in the early 1920s, were successful in the machining of cast iron and nonferrous alloys at much higher speeds than were possible with high-speed steel tools, but were subject to chemical attack or diffusion wear when cutting steel. As a result, the tools failed rapidly at speeds not much higher than those used with high-speed steel. This led to the development of WC-TiC-Co alloys.

Tungsten carbide diffuses readily into the steel chip surface, but the solid solution of

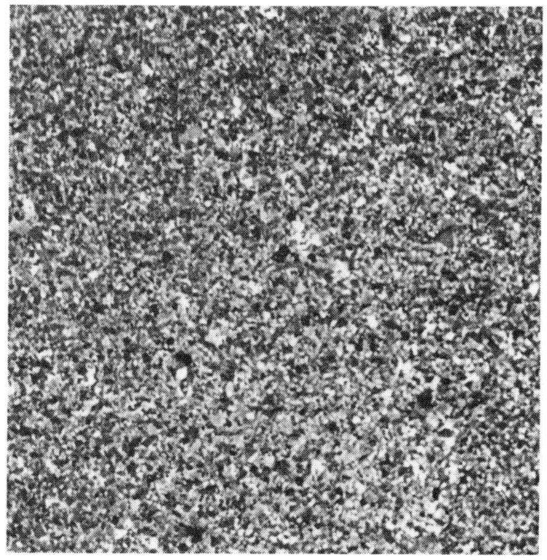

**Fig. 6** Submicron carbide grain size. 94WC-6Co alloy. Etched with Murakami's reagent for 2 min. 1500×

tungsten carbide and titanium carbide resists this type of chemical attack. Unfortunately, titanium carbide and WC-TiC solid solutions are more brittle and less abrasion resistant than tungsten carbide. The amount of titanium carbide added to tungsten carbide-cobalt alloys is therefore kept to a minimum, typically no greater than 15 wt%. The carbon content is less critical in WC-TiC-Co alloys than in WC-Co alloys, and the η phase does not appear in the microstructure unless carbon is grossly inadequate. In addition, free graphite rarely occurs in these alloys.

**Steel-Cutting Grades of Carbide Alloys.** The WC-TiC-Co alloys have given way to alloys of tungsten carbide, cobalt, titanium carbide, tantalum carbide, and niobium carbide. The tungsten carbide-cobalt alloys containing TiC, TaC, and NbC are called complex grades, multigrades, or steel-cutting grades. Adding TaC to WC-TiC-Co alloys partially overcomes the deleterious effect of TiC on the strength of WC-Co alloys. Tantalum carbide also resists cratering and improves thermal shock resistance. The latter property is particularly useful in applications involving interrupted cuts. Tantalum carbide is often added as (Ta,Nb)C because the chemical similarity between TaC and NbC makes their separation expensive. Fortunately, NbC has an effect similar to TaC in most cases. The relative concentrations of tantalum carbide and niobium carbide in these alloys are dependent on the raw material used, the desired composition, the properties, and the microstructure.

Unlike the WC-Co alloys, the microstructure of WC-TiC-(Ta,Nb)C-Co alloys shows three phases: angular WC grains, rounded WC-TiC-(Ta,Nb)C solid-solution grains, and cobalt binder. The solid-solution carbide phase often exhibits a cored structure, indicating incomplete diffusion during the sintering process. Representative microstructures of several WC-TiC-(Ta,Nb)C alloys are shown in Fig. 7. The size and distribution of the phases vary widely, depending on the amounts and grain sizes of the raw materials employed and on the method of manufacture. Similarly, the properties of these complex alloys also vary widely, as indicated in Table 2 for a few representative steel-cutting grades.

## Classification of Cemented Carbides

There is no universally accepted system for classifying cemented carbides. The systems most often employed by producers and users are discussed below. Each system has inherent strengths and weaknesses in describing specific materials, and for this reason close cooperation between user and

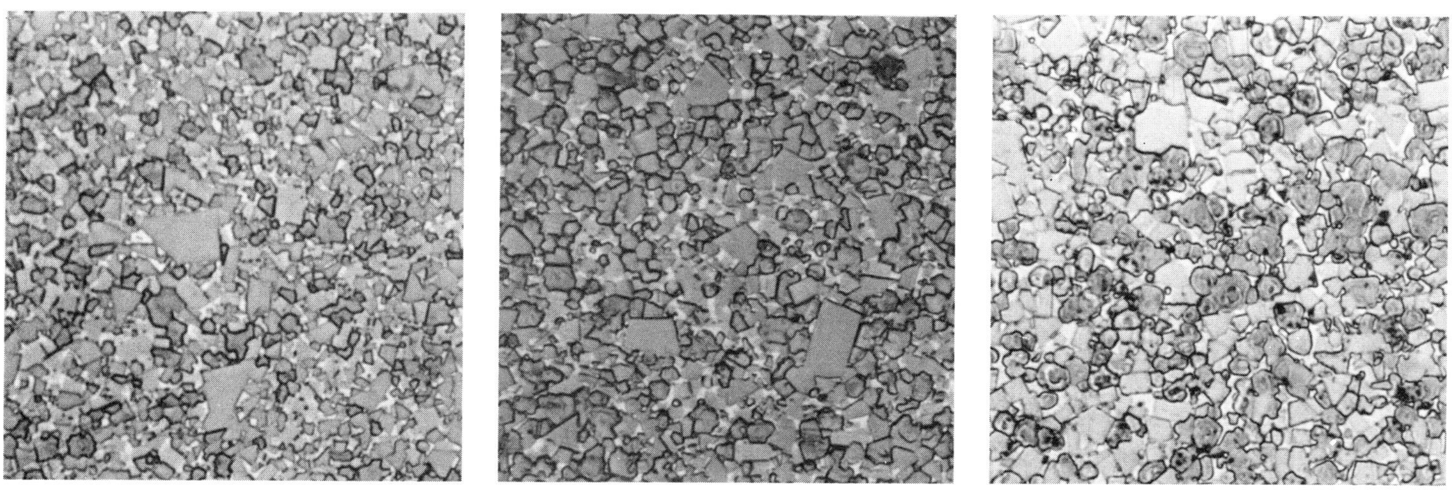

(a) (b) (c)

**Fig. 7** Representative microstructures of steel-cutting grades of tungsten carbide. (a) 85WC-9(Ta,Ti,Nb)C-6Co alloy, medium grain size. (b) 78WC-15(Ta,Ti,Nb)C-7Co alloy, medium grain. (c) 73WC-19(Ta,Ti,Nb)C-8Co alloy, medium grain. The gray, angular particles are WC, and the dark gray, rounded particles are solid-solution carbides. The white areas are cobalt binder. All etched with Murakami's reagent for 2 min. 1500×

**Table 3 C-grade classification of cemented carbides**

| C-grade | Application category |
|---|---|
| **Machining of cast iron, nonferrous, and nonmetallic materials** | |
| C-1 | Roughing |
| C-2 | General-purpose machining |
| C-3 | Finishing |
| C-4 | Precision finishing |
| **Machining of carbon and alloy steels** | |
| C-5 | Roughing |
| C-6 | General-purpose machining |
| C-7 | Finishing |
| C-8 | Precision finishing |

**Table 4 ISO R513 classification of carbides according to use for machining**

| Designation(a) | Groups of application: Material to be machined | Use and working conditions |
|---|---|---|
| P 01 | Steel, steel castings | Finish turning and boring; high cutting speeds, small chip section, accuracy of dimensions and fine finish, vibration-free operation |
| P 10 | Steel, steel castings | Turning, copying, threading, and milling; high cutting speeds, small or medium chip sections |
| P 20 | Steel, steel castings, malleable cast iron with long chips | Turning, copying, milling, medium cutting speeds and chip sections; planing with small chip sections |
| P 30 | Steel, steel castings, malleable cast iron with long chips | Turning, milling, planing, medium or low cutting speeds, medium or large chip sections, and machining in unfavorable conditions(b) |
| P 40 | Steel, steel castings with sand inclusion and cavities | Turning, planing, slotting, low cutting speeds, large chip sections with the possibility of large cutting angles for machining in unfavorable conditions(b) and work on automatic machines |
| P 50 | Steel; steel castings of medium or low tensile strength, with sand inclusion and cavities | For operations demanding very tough carbide: turning, planing, slotting, low cutting speeds, large chip sections, with the possibility of large cutting angles for machining in unfavorable conditions(b) and work on automatic machines |
| M 10 | Steel, steel castings, manganese steel, gray cast iron, alloy cast iron | Turning, medium or high cutting speeds; small or medium chip sections |
| M 20 | Steel, steel castings, austenitic or manganese steel, gray cast iron | Turning, milling; medium cutting speeds and chip sections |
| M 30 | Steel, steel castings, austenitic steel, gray cast iron, high-temperature resistant alloys | Turning, milling, planing; medium cutting speeds, medium or large chip sections |
| M 40 | Mild free-cutting steel, low-tensile steel, nonferrous metals, and light alloys | Turning, parting off, particularly on automatic machines |
| K 01 | Very hard gray cast iron, chilled castings of over 85 scleroscope hardness, high-silicon aluminum alloys, hardened steel, highly abrasive plastics, hard cardboard, ceramics | Turning, finish turning, boring, milling, scraping |
| K 10 | Gray cast iron over 220 HB, malleable cast iron with short chips, hardened steel, silicon aluminum alloys, copper alloys, plastics, glass, hard rubber, hard cardboard, porcelain, stone | Turning, milling, drilling, boring, broaching, scraping |
| K 20 | Gray cast iron up to 220 HB, nonferrous metals: copper, brass, aluminum | Turning, milling, planing, boring, broaching, demanding very tough carbide |
| K 30 | Low-hardness gray cast iron, low-tensile steel, compressed wood | Turning, milling, planing, slotting, for machining in unfavorable conditions(b) and with the possibility of large cutting angles |
| K 40 | Softwood or hardwood, nonferrous metals | Turning, milling, planing, slotting, for machining in unfavorable conditions(b) and with the possibility of large cutting angles |

(a) In each letter category, low designation numbers are for high speeds and light feeds; higher numbers are for slower speeds and/or heavier feeds. Also, increasing designation numbers imply increasing toughness and decreasing wear resistance of the cemented carbide materials. (b) Unfavorable conditions include shapes that are awkward to machine; material having a casting or forging skin; material having variable hardness; and machining that involves variable depth of cut, interrupted cut, or moderate to severe vibrations.

producer is the best means of selecting the proper grade for a given application.

**C-Grade System.** The U.S. carbide industry uses an application-oriented system of classification to assist in the selection of proper grades of cemented carbides. This C-grade system does not require the use of trade names for identifying specific carbide grades (Table 3). Although this classification simplifies tool application, it does not reflect the material properties that significantly influence selection of the proper carbide grade. Additionally, the definitions of work materials involved in this classification scheme are imprecise. There is also no universal agreement on the meanings of the terms used to describe the various application categories. Despite these limitations, the C-grade classification has been successfully used by the manufacturing industry since 1942.

**ISO Classification.** In 1964, the International Organization of Standardization (ISO) issued ISO Recommendation R513 "Application of Carbides for Machining by Chip Removal." The basis for the ISO classification of carbides is summarized in Table 4.

In the ISO system, all machining grades are divided into three color-coded groups:

- Highly alloyed tungsten carbide grades (letter P, blue color) for machining steel
- Alloyed tungsten carbide grades (letter M, yellow color, generally with less TiC than the corresponding P series) for multipurpose use, such as steels, nickel-base superalloys, and ductile cast irons
- Straight tungsten carbide grades (letter K, red color) for cutting gray cast iron, nonferrous metals, and nonmetallic materials

Each grade within a group is assigned a number to represent its position from maximum hardness to maximum toughness. P-grades are rated from 01 to 50, M-grades from 10 to 40, and K-grades from 01 to 40. Typical applications are described for grades at more or less regular numerical intervals. Although the coated grades had not been developed at the time the ISO classification system was prepared, one should be able to classify them as easily as the uncoated grades.

## Tool Wear Mechanisms

The cutting of metals involves extensive plastic deformation of the workpiece ahead of the tool tip, high temperatures, and severe frictional conditions at the interfaces of the tool, chip, and workpiece. Most of the work of plastic deformation and friction is converted into heat. In cutting, about 80% of this heat leaves with the chip, but the other 20% remains at the tool tip, producing high temperatures (≥1000 °C, or 1800 °F) (Ref 13). The stresses on the tool tip are also high; the actual values are dependent on the workpiece material and the machining conditions. In addition, the tool may experience repeated impact loads during interrupted cuts, and the freshly produced chips may chemically interact with the tool material. The cutting tool is thus subjected to a variety of hostile conditions.

The performance of a tool material is dictated by its response to the above conditions existing at the tool tip. High tempera-

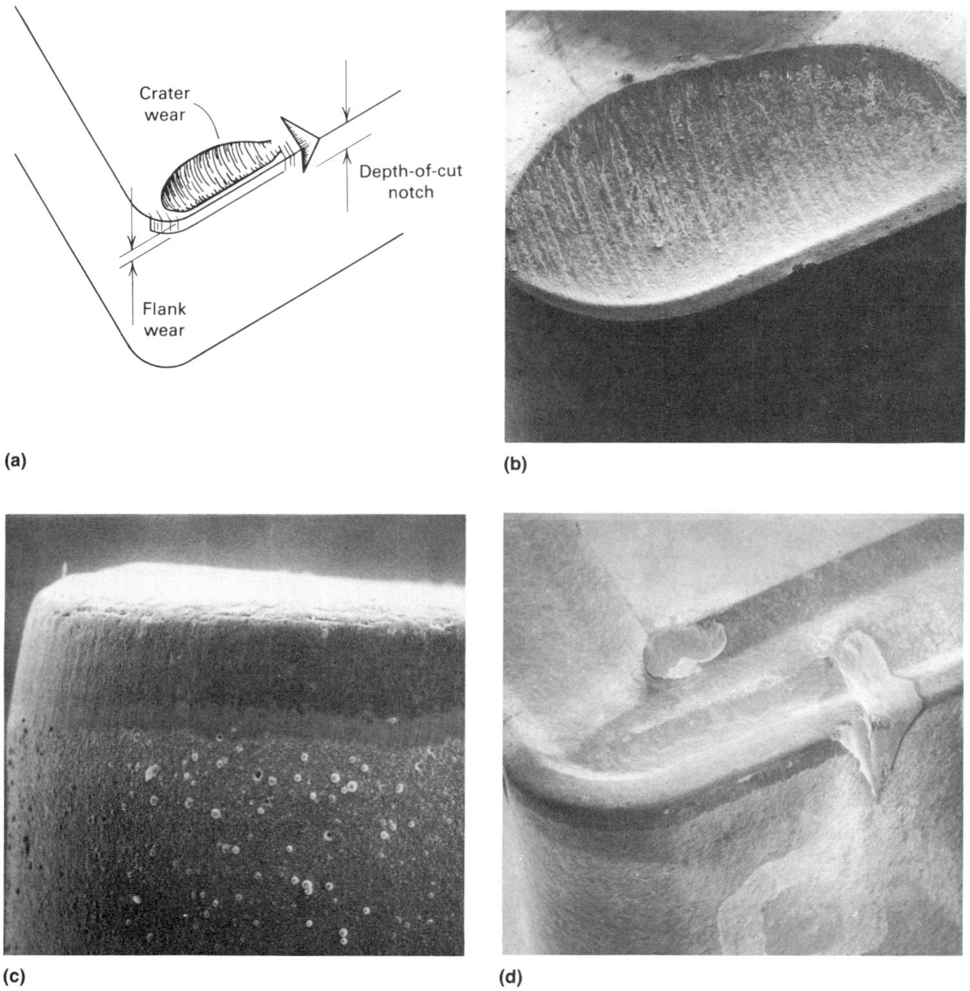


**Fig. 8** Crater wear, flank wear, and depth-of-cut notch wear processes. (a) Schematic of wear mechanisms. (b) Crater wear on a cemented carbide tool produced during the machining of plain carbon steel. 15×. (c) Abrasive wear on the flank face of a cemented carbide tool produced during the machining of gray cast iron. 75×. (d) Depth-of-cut notching on a cemented carbide tool produced during the machining of a nickel-base superalloy. 15×

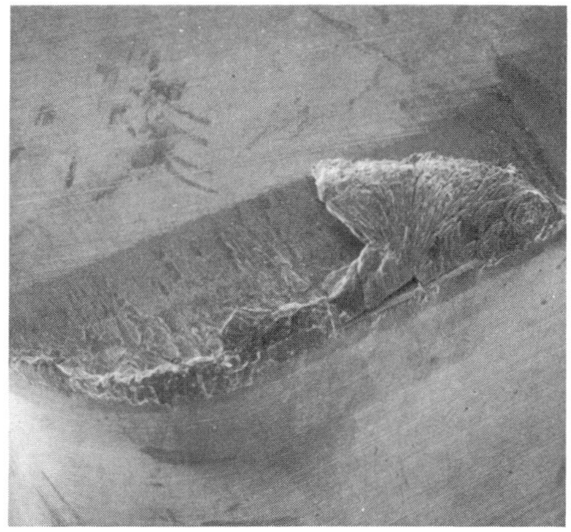

**Fig. 9** Built-up edge on a cemented carbide tool. The built-up edge was produced during the low-speed machining of a nickel-base alloy. 20×

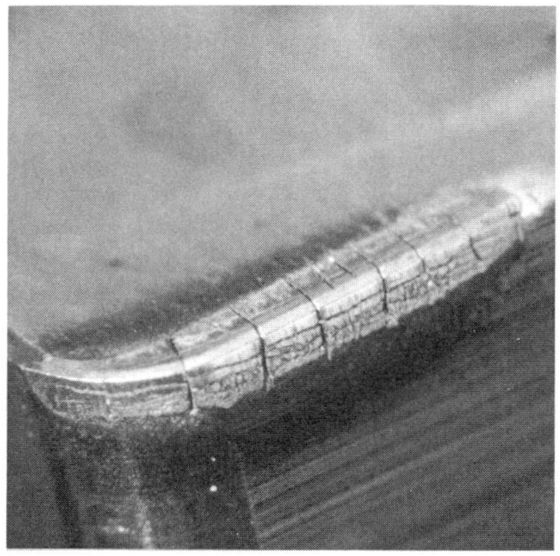

**Fig. 10** Thermal cracks in a cemented carbide insert. The thermal cracks are perpendicular to the cutting edge, and the mechanical cracks are parallel to the cutting edge. 15×

tures and stresses can cause blunting from the plastic deformation of the tool tip, and high stresses may lead to catastrophic fracture. In addition to plastic deformation and fracture, the service life of cutting tools is determined by a number of wear processes, such as crater wear, attrition wear, flank or abrasive wear, thermal fatigue, and depth-of-cut notching.

**Crater wear** (Fig. 8a and b) occurs on the rake face, where the tool temperatures are higher. Crater wear is caused by a chemical interaction between the rake face of the insert and the hot metal chip flowing over the tool. This interaction may involve diffusion or dissolution of the tool material into the chip. The chemical inertness of the tool material (or its coating) relative to the workpiece material is a requisite property for crater resistance.

**Attrition Wear and Built-Up Edge.** If machining is done at relatively low speeds and if the tool tip temperature is not high enough for crater wear or deformation to be significant, attrition may become the dominant wear process. The attrition process may occur when there is an intermittent flow of workpiece material, and this condition is usually associated with a built-up edge (Fig. 9). Built-up edge occurs at low metal cutting speeds and is not a serious problem as long as the edge remains intact on the tool. When machining gray cast iron, built-up edges do not break away from the tool. However, when machining steel at low speeds, built-up edges break off easily. This may result in attrition wear if small fragments of tool material are carried away as the built-up edge breaks off. In some cases, large chunks of tool material may even be carried away. Built-up edge and attrition wear can be minimized by increasing the metal cutting speed, by selecting fine-grain WC-Co alloys, and/or by using positive-rake tools with smooth surface finishes.

**Flank or abrasive wear** (Fig. 8a and c) is often observed on the flank face and is related to the hardness of the tool material or coating. Harder materials provide greater flank and abrasive wear resistance.

Cemented carbide tools may be subjected to abrasive wear when abrasive particles (such as sand on the surface of castings) or hard carbide or alumina inclusions are present in the workpiece materials. Tools with lower binder contents and/or finer carbide grain sizes can resist abrasive wear.

**Thermal Fatigue.** Cemented carbide tools sometimes exhibit a series of cracks perpendicular to the tool edge when applied to interrupted cutting operations such as milling (Fig. 10). These thermal cracks are caused by the alternating expansion and contraction of the tool surface as it is heated during the cut and cooled outside the cut. The cracks initiate on the rake face, then spread across the edge and down the flank face of the tool. With prolonged intermittent cutting, lateral cracks appear parallel with, and close to, the cutting edge. The thermal and lateral cracks may join together and cause small fragments of tool material to break away. The resistance of WC-Co tools to thermal fatigue can generally be improved by TaC additions.

### Table 5 Test methods for determining the properties of cemented carbides

| Property | Test method ASTM/ANSI | CCPA(a) | ISO |
|---|---|---|---|
| Abrasive wear resistance | B 611 | P112 | . . . |
| Apparent grain size | B 390 | M203 | . . . |
| Apparent porosity | B 276 | M201 | 4505 |
| Coercive force | . . . | . . . | 3326 |
| Compressive strength | E 9 | P104 | 4506 |
| Density | B 311 | P101 | 3369 |
| Fracture toughness | . . . | . . . | . . . |
| Hardness, HRA | B 294 | P103 | 3738 |
| Hardness, HV | E 92 | . . . | 3878 |
| Linear thermal expansion | B 95 | P108 | . . . |
| Magnetic permeability | A 342 | P109 | . . . |
| Microstructure | B 657 | M202 | 4499 |
| Poisson's ratio | E 132 | P105 | . . . |
| Transverse rupture strength | B 406 | P102 | 3327 |
| Young's modulus | E 111 | P106 | 3312 |

(a) Cemented Carbides Producers Association

**Depth-of-cut notching** (Fig. 8a and d) consists of a high degree of localized wear on both the rake face and the flank face at the depth of the cut line. Notching is common when machining materials that work harden, such as austenitic stainless steels or high-temperature alloys. This type of wear is attributed to the chemical reaction of the tool material with the atmosphere or to abrasion by the hard, sawtooth outer edge of the chip. Depth-of-cut notching may lead to tool fracture; it can be minimized by:

- Increasing the fracture toughness of the tool material
- Increasing the lead angle (for round inserts) or the side cutting angle (for other insert shapes)
- Chamfering the tool edge
- Varying the depth of cut if multiple passes are made

## Cemented Carbide Properties

Evaluation of the physical and mechanical properties of tool materials is an important prerequisite to the selection of grades for a given metal cutting application and for tool material development. A number of industry, national, and ISO standards have been developed for determining the selected properties of cemented carbides (Table 5).

**Hardness** determines the resistance of a material to abrasion and wear. It is affected not only by composition but also by the level of porosity and microstructure. For straight WC-Co alloys of comparable WC grain size, hardness and abrasion resistance decrease with increasing cobalt content (Fig. 11a and b). However, because both composition and microstructure affect hardness, cobalt content and grain size must be considered. At a given cobalt level, hardness improves with decreasing WC grain size (Fig. 11a).

In cemented carbides, hardness is measured by the Rockwell A-scale diamond cone indentation test (HRA) or by the Vickers diamond pyramid indentation test (HV). Both tests are performed on a finely ground,

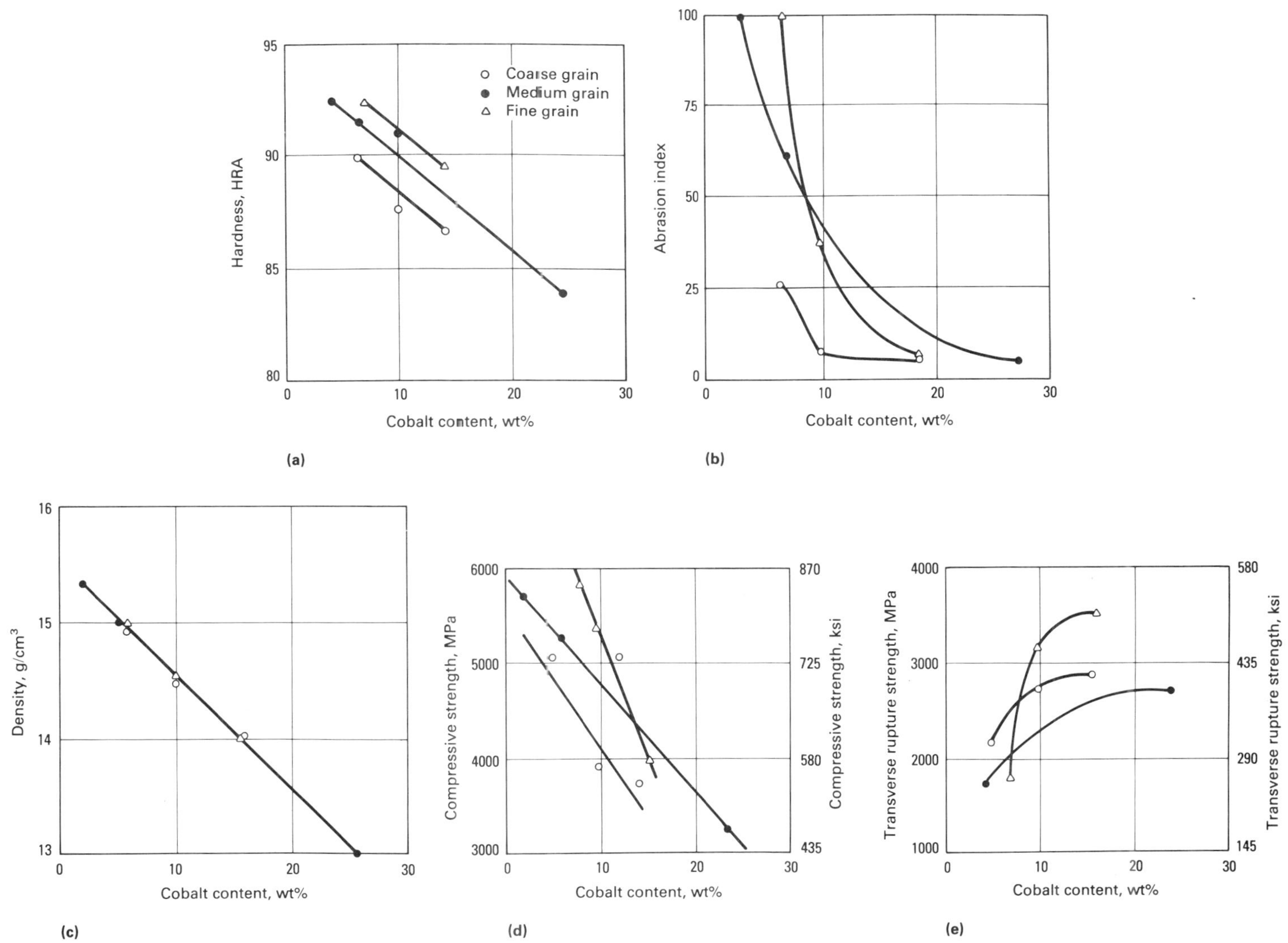


**Fig. 11** Variation in properties with cobalt content and grain size for straight WC-Co alloys. (a) Variation in hardness. (b) Variation in abrasion resistance. (c) Variation in density. (d) Variation in compressive strength (e) Variation in transverse rupture strength

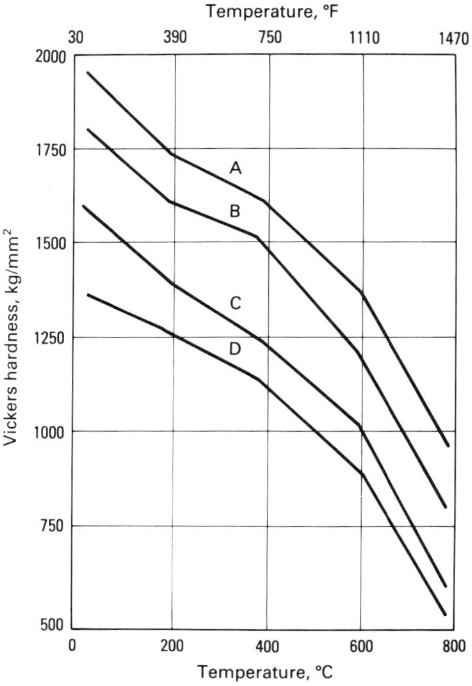


**Fig. 12** Variation in microhardness with temperature. Microhardness is based on a 1 kg load, and all alloys are of medium WC grain size. A, 97WC-3Co alloy; B, 94WC-6Co; C, 80WC-12(Ti,Ta,Nb)C-8Co; D, 86WC-2TaC-12Co

lapped or polished planar surface placed at right angles to the indenter axis. The Rockwell A test employs a load of 60 kg, while a range of loads can be used in the Vickers test. For cemented carbides used in machining applications, hardness values range from 88 to 94 HRA and from 1100 to 2000 HV.

Although the Rockwell scale has been used for decades as a measure of hardness, a true indication of the resistance to plastic deformation in metal cutting operations can be obtained only by measuring hardness at elevated temperatures. Measurements of hardness over a wide range of temperatures are therefore valuable for tool selection.

Hardness testers with high-temperature capability (up to 1200 °C, or 2200 °F) are commercially available and are being increasingly used by the cemented carbide industry. Figure 12 shows hot hardness data for a number of cemented carbides. The hardness of these materials decreases monotonically with increasing temperatures.

**Compressive Properties.** One of the unique properties of cemented carbides is their high compressive strength. Uniaxial compression tests can be performed on straight cylindrical samples or on cylinders having reduced diameters in the middle to localize fracture. The compressive strengths of cemented carbides are greater than those of most other materials. Typical values of compressive strength range from 3.5 to 7.0 GPa (0.5 to 1.0 psi × $10^6$).

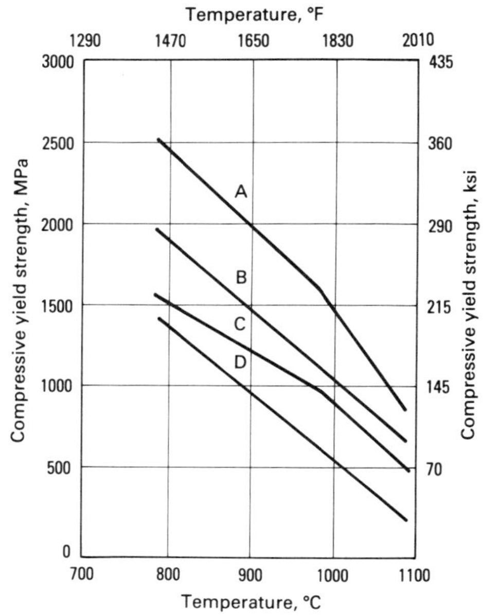


**Fig. 13** Variation in compressive yield strength with temperature. Measured at 0.2% offset strain; all alloys characterized by medium grain size. A, 73WC-22(Ti,Ta,Nb)C-5Co alloy; B, 80WC-12(Ti,Ta,Nb)C-8Co; C, 86WC-8(Ti,Ta,Nb)C-6Co; D, 86WC-2TaC-12Co

The ductility of cemented carbides is generally low at room temperature, so there is little difference between their yield strength and fracture strength. At higher temperatures, however, these materials exhibit a small but finite amount of ductility. Measurement of yield strength is therefore more appropriate at elevated temperatures. High-temperature compressive yield strength is typically measured at 0.2% offset strain. Compression tests are performed in a high-temperature furnace (typically with resistance heating) under a vacuum or in an inert atmosphere. Figure 13 shows yield strength data for selected straight and alloyed WC-Co grades. Like hardness, the compressive yield strengths of cemented carbides decrease monotonically with increasing temperature; the rate of decrease depends on the composition and the microstructure. As in metallic materials, fine-grain alloys tend to lose their yield strengths more rapidly with increasing temperature than coarse-grain grades, although at room temperature the former can exhibit high yield strengths.

**Transverse Rupture Strength.** The most common method of determining the fracture strength of cemented carbides is the transverse rupture test. In this test, a rectangular test bar is placed across two sintered carbide support cylinders, and a gradually increasing load is applied by a third carbide cylinder at the midpoint between the supports. Transverse rupture strength is determined from the dimensions of the test bar, the distance between the supports, and the fracture load. A disadvantage of this test is the large scatter in the experimental data resulting from surface defects introduced

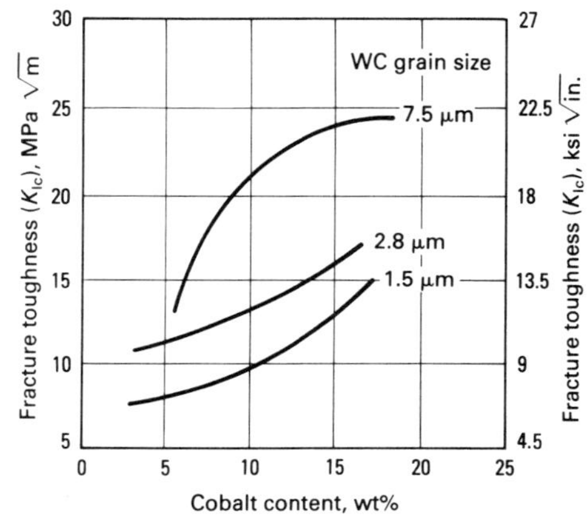


**Fig. 14** Variation in fracture toughness ($K_{Ic}$) with cobalt content for WC-Co alloys with different WC grain sizes. Source: Ref 15

into the test specimens during processing. Nevertheless, it is an excellent quality control test, and it is particularly useful for large carbide components.

Figure 11(e) shows the variation in transverse rupture strength with cobalt content. In metal cutting applications, no clear relationship has been established between transverse rupture strength and turning performance. However, there appears to be a good correlation between transverse rupture strength and milling performance (Ref 14). During milling, the tool is subjected to tensile stresses as it leaves the cut, and a material with high transverse rupture strength should be able to resist fracture under these conditions.

**Fracture toughness** is less sensitive than transverse rupture strength to such extrinsic factors as specimen size, geometry, and surface finish. Fracture toughness is measured by the critical stress intensity factor $K_{Ic}$ (Ref 15-17). This parameter indicates the resistance of a material to fracture in the presence of a sharp crack and thus provides a better measure of the intrinsic strength of the cemented carbide than transverse rupture strength. A variety of specimen geometries have been used in this test, including the single-edge notched beam, the double cantilever beam, the compact tension specimen, and the double torsion specimen. The carbide industry in the United States generally uses commercial equipment for fracture toughness evaluation. The fracture toughness of cemented carbides increases with cobalt content and with WC grain size (Fig. 14). On the other hand, cubic carbide additions lessen the fracture toughness of WC-Co alloys.

As with the other mechanical properties, attention is focused on the development of test techniques to evaluate fracture toughness at elevated temperatures. Figure 15 shows $K_{Ic}$ data for a number of cemented carbides from room temperature to about 1000 °C (1800 °F). Depending on the composition of the cemented carbide, the $K_{Ic}$

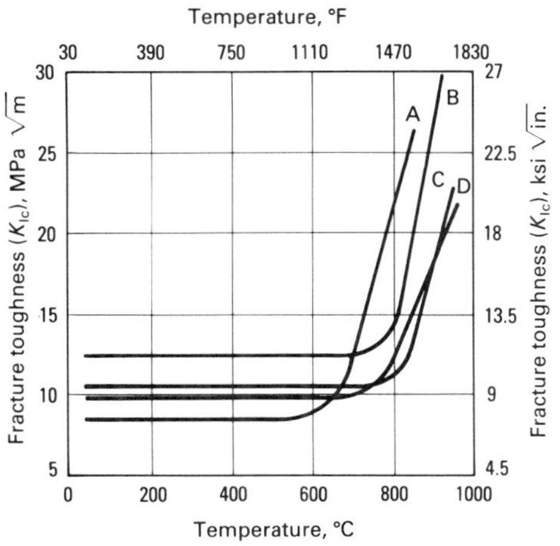


**Fig. 15** Variation in fracture toughness ($K_{Ic}$) with temperature for a number of WC-Co base alloys. A, 86WC-2TaC-12Co; B, 85WC-9(Ti,Ta,Nb)C-6Co; C, 80WC-12(Ti,Ta,Nb)C-8Co; D, 96WC-4Co

parameter is insensitive to temperature up to about 600 °C (1100 °F), but increases rapidly at higher temperatures. This behavior is reminiscent of the ductile-to-brittle transition observed in quenched-and-tempered steels.

**The density,** or specific gravity, of cemented carbides is very sensitive to composition and porosity in the sample and is widely used as a quality control test. Density values of cemented carbides range from 15 g/cm$^3$ for low-cobalt straight WC-Co alloys to about 10 or 12 g/cm$^3$ for highly alloyed carbide grades (Fig. 11c).

**Magnetic Properties.** Tungsten carbide-cobalt alloys lend themselves to the analysis of magnetic properties because cobalt is ferromagnetic. The properties measured are magnetic saturation and coercive force. Both free cobalt and solid solutions of cobalt and tungsten contribute to magnetization. The magnetic saturation of pure cobalt is 201 × $10^{-6}$ T m$^3$/kg. With additions of tungsten to cobalt, the magnetic saturation decreases steadily from 201 × $10^{-6}$ to about 151 × $10^{-6}$ T m$^3$/kg. In this range, the cemented carbide is characterized by two or three phases (WC and Co or WC, a solid-solution carbide, and Co). Values below 151 × $10^{-6}$ T m$^3$/kg indicate the presence of $\eta$ phase. The solubility of tungsten in cobalt is inversely proportional to carbon content. Lower-carbon alloys have more tungsten dissolved in cobalt and are characterized by lower magnetization. Magnetic saturation thus provides an accurate measure of the changes in carbon content in the cemented carbide alloy and is widely used as a quality control test.

The coercive force varies considerably with increasing sintering temperature and indicates the structural changes that take place during sintering. The coercive force of WC-Co alloys reaches a maximum at the optimum sintering temperature and decreases at higher temperatures because of grain growth. Therefore, measurement of the coercive force permits control of the sintering process. The factors influencing the coercive force are complex, varied, and interactive. For a given cobalt and carbon content, the coercive force provides a measure of the degree of distribution of the carbide phase in the microstructure. Commercial units are available for the rapid measurement of magnetic saturation and coercive force in cemented carbides.

**Porosity.** The properties of a cemented carbide are dependent on its density, which in turn is critically dependent on composition and porosity. Porosity is evaluated on the as-polished material. The American Society for Testing and Materials (ASTM) has established a standard procedure (B 276) that rates three types of porosity:

- Type A, covering pore diameters less than 10 μm
- Type B, covering pore diameters between 10 and 25 μm
- Type C (Fig. 4), covering porosity developed by the presence of free carbon

Type A porosity is rated at a magnification of 200×, while types B and C porosity are rated at 100×. The degree of porosity is given by four numbers ranging in value from 02 to 08. The number provides a measure of pore volume as a percentage of total volume of the sample.

**Thermal Shock Resistance.** As discussed earlier in this article, cutting tool materials are subjected to thermal shocks during interrupted cutting operations such as milling. Resistance to thermal shock is therefore an important property that determines tool performance in milling. No laboratory test has yet been developed that can consistently predict the resistance to thermal shock of a tool. However, empirical parameters have been suggested that can be used to evaluate tool materials for their probable resistance to thermal shock (Ref 18). A commonly used parameter is $\sigma k/E\alpha$, where $\sigma$ is the transverse rupture strength, $k$ is the thermal conductivity, $E$ is Young's modulus, and $\alpha$ is the coefficient of thermal expansion. Table 6 lists representative values of this parameter for a number of WC-Co alloys. In general, the higher the value of $\sigma k/E\sigma$, the better the thermal shock resistance.

**Table 6 Thermal shock resistance parameters for cemented carbides**

| Composition, wt% | WC grain size | Transverse rupture strength, $\sigma$ MPa | ksi | Thermal conductivity ($k$), W/m · K | Young's modulus, $E$ GPa | psi × $10^6$ | Coefficient of thermal expansion ($\alpha$), μm/m · K | Thermal shock resistance ($\sigma k/E\alpha$), kW/m |
|---|---|---|---|---|---|---|---|---|
| 97WC-3Co | Medium | 1590 | 230 | 121 | 641 | 93 | 5.0 | 60 |
| 94WC-6Co | Medium | 2000 | 290 | 100 | 648 | 94 | 5.4 | 57 |
| 90WC-10Co | Fine | 3100 | 450 | 80 | 620 | 90 | 6.0 | 67 |
| 71WC-12.5TiC-12TaC-4.5Co | Medium | 1380 | 200 | 35 | 565 | 82 | 6.5 | 13 |
| 72WC-8TiC-11.5TaC-8.5Co | Medium | 1720 | 250 | 50 | 558 | 81 | 6.8 | 23 |

**Abrasive Wear Resistance.** Most producers of cemented carbides use a wet-sand abrasion test to measure abrasion resistance. In this test, a sample is held against a rotating wheel for a fixed number of revolutions while the sample and wheel are immersed in a water slurry containing aluminum oxide particles. Comparative rankings are reported, usually on the basis of a wear rating based on the reciprocal of volume loss. Although standard test procedures are available, carbide producers have not agreed on a single test method, and so the values of abrasion resistance cited in the literature vary widely. Because of this variance, it is almost impossible to make valid comparisons among test results reported by different producers. It is also fallacious to use abrasion resistance as a measure of the wear resistance of cemented carbide materials when they are used for cutting steel or other materials; abrasion resistance in a standard test does not correspond directly to wear resistance in machining operations.

Generally, the abrasion resistance of cemented carbides decreases as cobalt content or grain size is increased (Fig. 11b). Abrasion resistance is also lower for complex carbides than for straight WC grades having the same cobalt content.

## Coated Carbide Tools

One of the challenges in the design of cemented carbide tools is the optimization of toughness associated with straight WC-Co alloys with the superior crater wear resistance of alloyed carbides containing high levels of titanium carbide. This challenge has led to the development of coated carbide tools.

Coated carbides account for a major portion of all commercial metal cutting inserts sold in the United States. The success of coated carbides is based on their proven ability to extend tool life on steels and cast irons by at least a factor of two to three. This is accomplished by a reduction in wear processes, especially at higher cutting speeds.

### Laminated Coatings

An important development in the production of coated carbide tools occurred in the 1960s, when laminated tips consisting of a

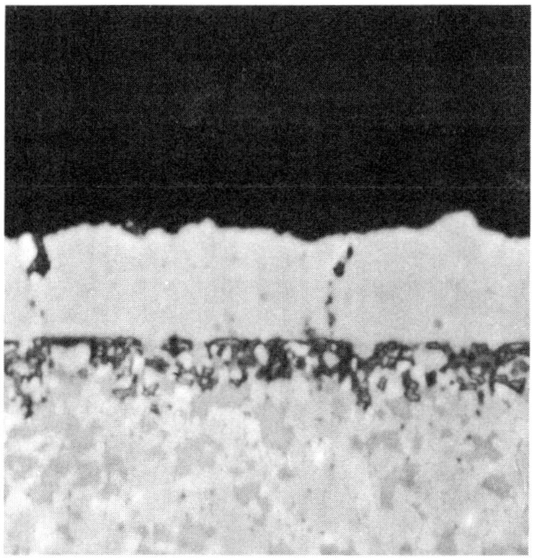

**Fig. 16** Decarburization of a TiC coating. Micrograph shows the η phase at the coating/substrate interface of an 85WC-9(Ti,Ta,Nb)C-6Co alloy with an 8 μm (315 μin.) TiC coating. Etched with Murakami's reagent for 3 s. 1500×

base of WC-Co alloy with a sintered layer of high TiC composition were produced. This development not only enabled higher cutting speeds in steel machining but also reduced crater wear on the tool. Although metal cutting productivity increased with the use of these laminated tools, the thermal expansion mismatch between the substrate and the surface layer caused thermal stresses during metal cutting, and the laminate tended to spall during use.

## Chemical Vapor Deposited Coatings

Further development of laminated tools was superseded in 1969 by the application of a thin layer (~5 μm, or 200 μin.) of hard TiC coating to the cemented carbide tool by chemical vapor deposition (CVD) (Ref 19). The impetus for this development came from the Swiss Watch Research Institute, where vapor-deposited TiC coatings had been used on steel watch parts and cases to combat wear on these components.

**The CVD coating process** consists of heating the tools in a sealed reactor with gaseous hydrogen at atmospheric or lower pressure; volatile compounds are added to the hydrogen to supply the metallic and nonmetallic constituents of the coating. For example, TiC coatings are produced by reacting $TiCl_4$ vapors with methane ($CH_4$) and hydrogen ($H_2$) at 900 to 1100 °C (1650 to 2000 °F). The reaction is:

$$TiCl_4(g) + CH_4(g) + H_2(g) \rightleftharpoons TiC(s) + 4HCl(g) + H_2(g)$$

During the TiC deposition process, a secondary reaction often occurs in which carbon is taken from the cemented carbide substrate:

$$TiCl_4(g) + C(s) + 2H_2(g) \rightleftharpoons TiC(s) + 4HCl(g)$$

The resulting surface decarburization leads to the formation of a brittle η phase at the coating/substrate interface (Fig. 16) and sometimes to premature tool failure due to excessive chipping and insufficient edge strength (Ref 20, 21). These problems are particularly prevalent in severe machining operations involving interrupted cuts.

Performance inconsistencies have been largely eliminated by a number of metallurgical and processing innovations. These include improvements in CVD coating technology, which have resulted in coatings with greater uniformity of thickness, more adherence, and more consistent morphology and microstructure with minimum interfacial η phase and associated porosity (Ref 22).

**Coating compositions** have also evolved from single-layer TiC coatings with narrow application ranges to multilayer hard coatings. Multilayer coatings have a nominal total thickness of about 10 μm (400 μin.) and use various combinations of TiC, TiCN, TiN, $Al_2O_3$, and occasionally HfN (Fig. 17).

Multilayer coatings are intended to suppress both crater wear and flank wear. There

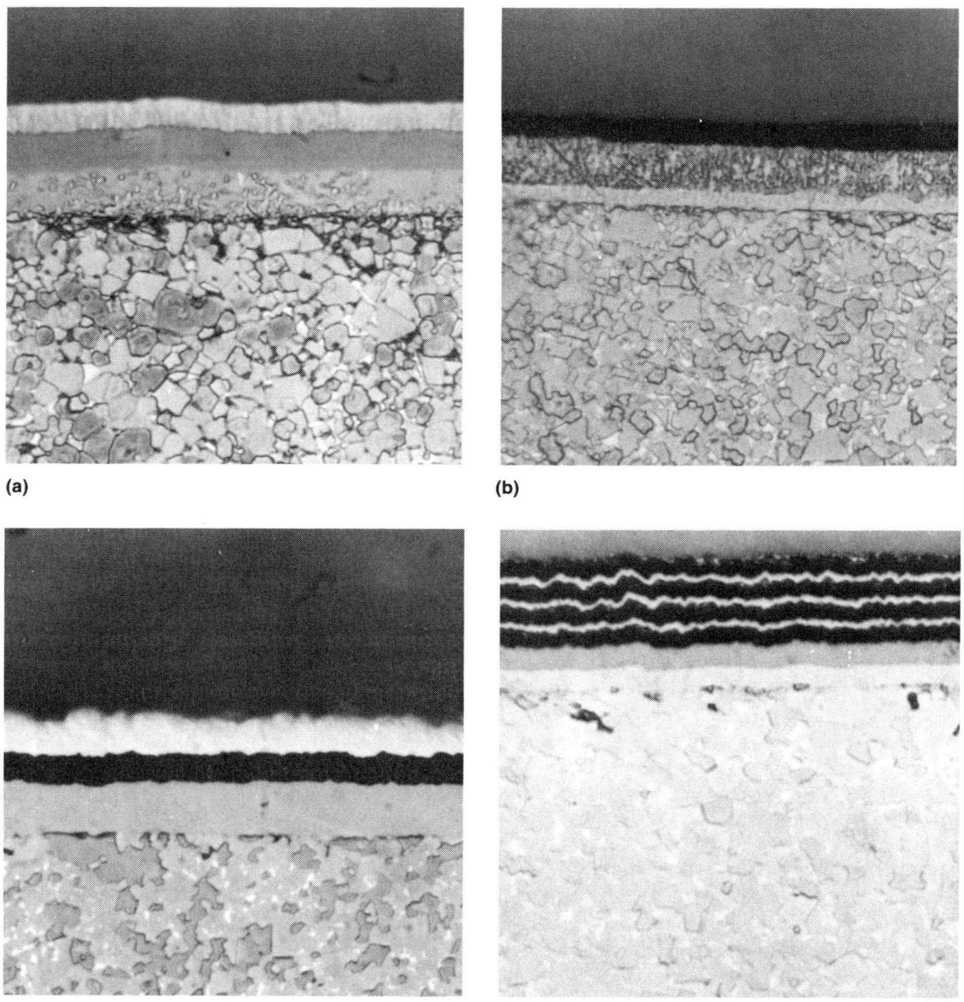


**Fig. 17** Multilayer coatings of carbide substrates. (a) 73WC-19(Ti,Ta,Nb)C-8Co alloy with a TiC/TiCN/TiN coating of about 10 μm (400 μin.) in total thickness. (b) 85WC-9(Ti,Ta,Nb)C-6Co with a TiC/$Al_2O_3$ coating about 9 μm (350 μin.) thick. (c) 85WC-9(Ti,Ta,Nb)C-6Co with a TiC/$Al_2O_3$/TiN coating about 10 μm (400 μin.) thick. (d) 88WC-7(Ti,Ta,Nb)C-5Co with TiC/TiCN coating supporting multiple alternating coating layers of $Al_2O_3$ and TiC. All etched with Murakami's reagent for 3 s. 1500×

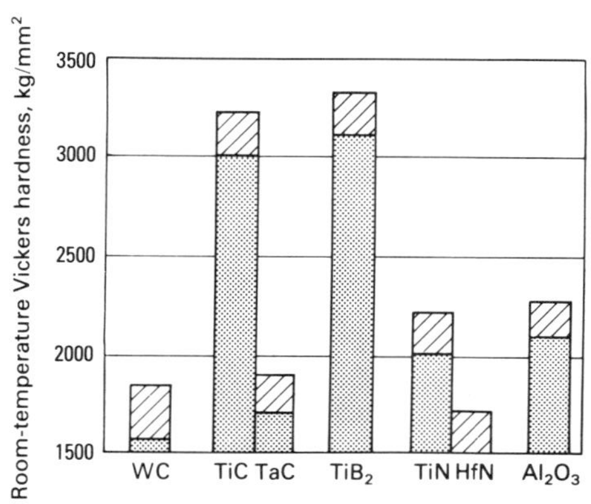


**Fig. 18** Room-temperature microhardness of hard coating materials. The hatched area (unshaded) indicates the range of hardness normally observed in these materials. Source: Ref 24

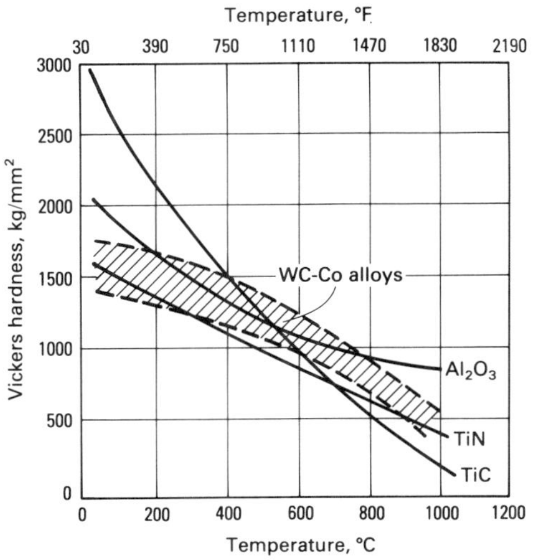


**Fig. 19** Temperature dependence of hardness of TiC, $Al_2O_3$, and TiN. The range of hardness of WC-Co alloys is also shown.

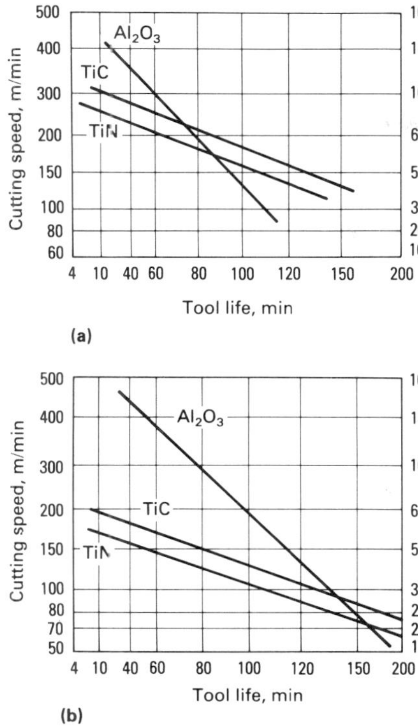


**Fig. 20** Tool life diagrams of coated inserts. Tool life is based on a 0.25 mm (0.01 in.) flank wear criterion. (a) Turning 1045 steel with a 2.5 mm (0.1 in.) depth of cut and a 0.40 mm/rev (0.016 in./rev) feed rate. (b) Turning SAE G4000 gray cast iron with a 2.5 mm (0.1 in.) depth of cut and a 0.25 mm/rev (0.01 in./rev) feed rate. Source: Ref 25

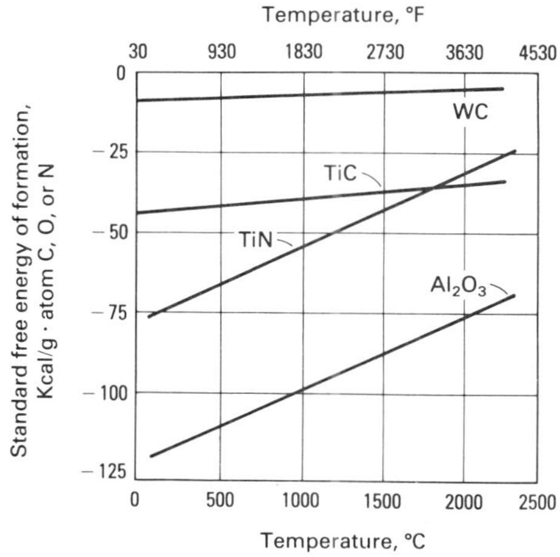


**Fig. 21** Variation in the standard free energy of formation of WC, TiN, TiC, and $Al_2O_3$ with temperature. This parameter gives an indication of the extent to which these materials will undergo diffusion wear.

is also a trend toward the use of multiple alternating coating layers (Fig. 17d), which are believed to produce finer grain sizes and to minimize chipping. These improvements increase tool life and extend the range of application of the coated tool (Ref 23).

**Hardness and Tool Life.** Figure 18 shows the room-temperature Vickers microhardness of various hard coatings deposited on cemented carbide substrates. The range of hardness for WC-Co alloys is shown for reference. The compounds TiC and $TiB_2$ have the highest relative hardness of about 3000 kg/mm$^2$ and offer the best protection against abrasive wear. With increases in temperature, however, TiC loses hardness rapidly, while $Al_2O_3$ retains its hardness to higher temperatures.

At 1000 °C (1800 °F), which is the temperature typically reached on the rake face of the tool during high-speed machining (Ref 13), $Al_2O_3$ has the highest hardness, followed by TiN and then TiC (Fig. 19). For comparison, the hardness of WC-Co substrates is also shown in Fig. 19. The hot hardness data indicate that the TiC coating would be more effective at the flank face, which rarely exceeds 600 °C (1100 °F) during machining. At lower speeds, TiC also offers adequate rake face protection from abrasive wear. On the other hand, $Al_2O_3$ would provide better abrasion resistance at higher speeds, as judged from its high hardness at elevated temperatures.

These concepts are illustrated in tool life plots (Fig. 20) based on flank wear criterion for TiC, TiN, and $Al_2O_3$ coated cemented carbide inserts in turning 1045 steel and gray cast iron workpiece materials. At lower speeds, the TiC coating provides the longest tool life, followed by TiN and $Al_2O_3$. At higher speeds, the ranking is altered. These rankings reflect the varying temperature dependence of their microhardness shown in Fig. 19.

Hard coatings also reduce frictional forces at the chip/tool interface, which in turn reduces the heat generated in the tool and thus results in lower tool tip temperatures. The compound TiN is particularly noted for this lubricity effect (Ref 26).

**Diffusion Wear.** The standard free energy of formation is given in Fig. 21 for WC, TiN, TiC, and $Al_2O_3$ from room temperature to about 2000 °C (3600 °F). This thermodynamic property gives an indication of the extent to which these materials will undergo diffusion wear. The most stable of these materials at all temperatures is $Al_2O_3$.

The dissolution rate of coating materials into the workpiece can also determine the rate of diffusive wear on the tool (Ref 24). The relative dissolution rates of various coating materials into steel at different temperatures are given in Table 7. The dissolution rate of $Al_2O_3$ in steel is an order of magnitude lower than the dissolution rates of other refractory compounds. Therefore, $Al_2O_3$ is the most crater-resistant material and is a very effective coating for cemented carbide tools used in the high-speed machining of steel.

**Thermal Expansion and Coating Adhesion.** The high temperatures employed for CVD coating generally ensure good bonding between the substrate and the coating. However, coating adhesion can be adversely affected by stresses caused by the thermal expansion mismatch between the substrate and the coating. Table 8 lists the thermal properties of various coating mate-

**Table 7 Relative dissolution rates of coating materials into iron**

Rates are relative to TiC.

| Material | Dissolution rate at: 100 °C (212 °F) | 500 °C (930 °F) | 1100 °C (2000 °F) |
|---|---|---|---|
| WC | $1.1 \times 10^{10}$ | $5.4 \times 10^4$ | $3.2 \times 10^2$ |
| TiC | 1.0 | 1.0 | 1.0 |
| TaC | 2.3 | 1.2 | $8.0 \times 10^{-1}$ |
| $TiB_2$ | $9.9 \times 10^1$ | 8.5 | 2.8 |
| TiN | $1.0 \times 10^{-8}$ | $1.8 \times 10^{-3}$ | $2.2 \times 10^{-1}$ |
| HfN | $2.5 \times 10^{-12}$ | $3.8 \times 10^{-5}$ | $2.5 \times 10^{-2}$ |
| $Al_2O_3$ | $1.1 \times 10^{-24}$ | $8.9 \times 10^{-11}$ | $4.1 \times 10^{-5}$ |

Source: Ref 24

**Table 8 Tool material thermal properties**

| Material | Melting point °C | °F | Coefficient of thermal expansion, µm/m · K | Thermal conductivity, W/m · K: 100 °C (212 °F) | 500 °C (930 °F) | 1000 °C (2000 °F) |
|---|---|---|---|---|---|---|
| WC-Co | ... | ... | 5–6 | 38–80 | ... | ... |
| Co | 1492 | 2717 | 12.3 | 70 | ... | ... |
| WC | ~2800 | 5070 | ~5 | 120 | ... | ... |
| TiC | 3100 | 5612 | 7.7 | 33 | 37 | 41 |
| TiN | 2950 | 5342 | 9.4 | 21 | 23 | 26 |
| $Al_2O_3$ | 2050 | 3722 | 8.4 | 28 | 13 | 6 |
| $M_{12}C$ | ... | ... | 7–10 | ... | ... | ... |

(a) (b)

**Fig. 22** Scanning electron macrographs of cobalt-enriched and nonenriched tools after a slotted-bar strength test. Machining parameters: depth of cut, 2.5 mm (0.1 in.); speed, 107 m/min (350 sfm); feed, 0.50 mm/rev (0.02 in./rev). The workpiece was AISI 41L50 steel; average hardness: 26 HRC. (a) Cobalt-enriched tool after 1200 impacts. 10×. (b) Nonenriched tool after 220 impacts. 11×

rials as well as WC-Co substrates. The thermal expansion mismatch is lowest for TiC and highest for TiN, while the thermal expansion coefficients of the coating materials listed are higher than those of the substrates. As a result, hard coatings on cemented carbide substrates are in residual tension at room temperature. Because the stresses are most severe at tool corners, the CVD-coated tools must be honed before coating. Another reason for honing is to minimize the formation of $\eta$ phase, which tends to develop to a greater extent at sharp tool edges.

**Cobalt Enrichment.** Although the early coated tools substantially improved metal cutting productivity, they were prone to catastrophic fracture when applied at high feed rates or in intermittent cutting operations. One solution to the problem of coating-related tool fracture is to improve the fracture toughness of the substrate by increasing its cobalt content and/or by increasing the binder mean free path (mean thickness of the binder between WC grains). Unfortunately, this approach results in decreased deformation resistance, which can cause tool tip blunting.

A major breakthrough in resolving this conflict between fracture toughness and deformation resistance occurred in the late 1970s, when a TiC/TiCN/TiN-coated tool was developed with a peripheral cobalt-enriched zone (10 to 40 μm, or 400 to 1600 μin., thick), which provided superior edge strength while maintaining the edge and crater wear resistance of the coating layers (Ref 27). The cobalt-enriched zone contained essentially a straight WC-Co composition with nearly three times the nominal cobalt level, while the bulk of the tool insert, containing higher levels of solid-solution cubic carbide and less cobalt, provided the necessary deformation resistance. Thus, a single tool combined the fracture resistance of a high-cobalt alloy with the deformation resistance of a lower-cobalt alloy.

The ability of the cobalt-enriched tool to resist edge chipping and catastrophic fracture is illustrated in Fig. 22, which compares a cobalt-enriched insert with a nonenriched tool after an edge strength test. Tool failures such as that shown in Fig. 22(b) can ruin the workpiece and result in excessive production downtime. Such tool failures can be particularly damaging in flexible machining systems and in untended machining centers.

The development of the cobalt-enriched tool permitted the use of heavy interrupted cuts, such as those encountered in scaled forgings and castings. However, optimum performance was obtained only at lower speeds. Further refinements to the cobalt enrichment concept have expanded the application range of this type of tool to higher speeds (Ref 28). Cutting speeds for cobalt-enriched tools range from 60 to 300 m/min (200 to 1000 sfm), with feeds ranging from 0.1 to 1 mm/rev (0.005 to 0.050 in./rev). These capabilities are suitable for most metal removal applications. Two typical cobalt-enriched microstructures that can cover

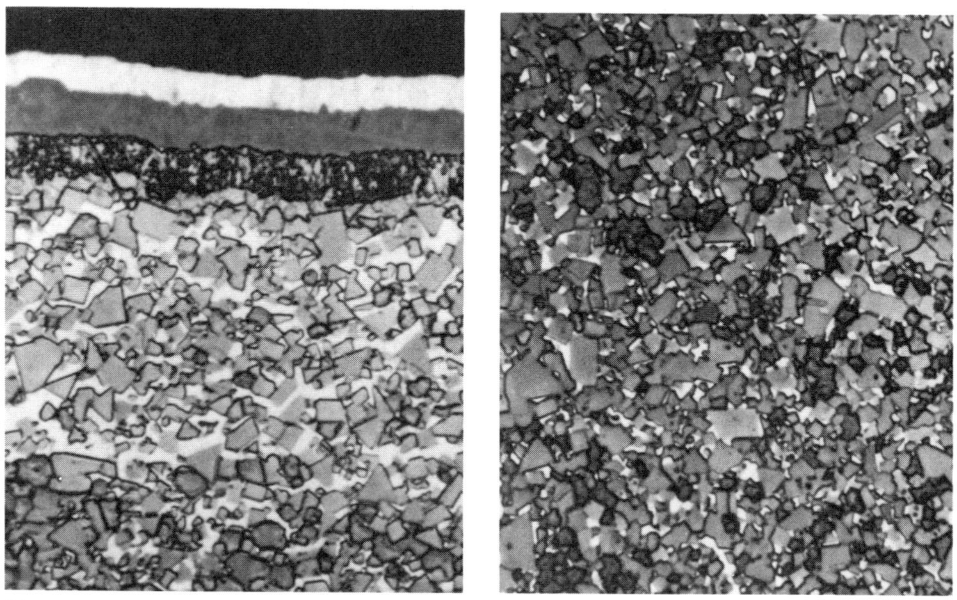

(a) (b)

**Fig. 23** Microstructure of a cobalt-enriched coated tool. 86WC-8(Ti,Ta,Nb)C-6Co tool with a TiC/TiCN/TiN coating. (a) Cobalt-enriched periphery (beneath the coating). (b) Bulk microstructure. Both etched with Murakami's reagent for 2 min. 1500×

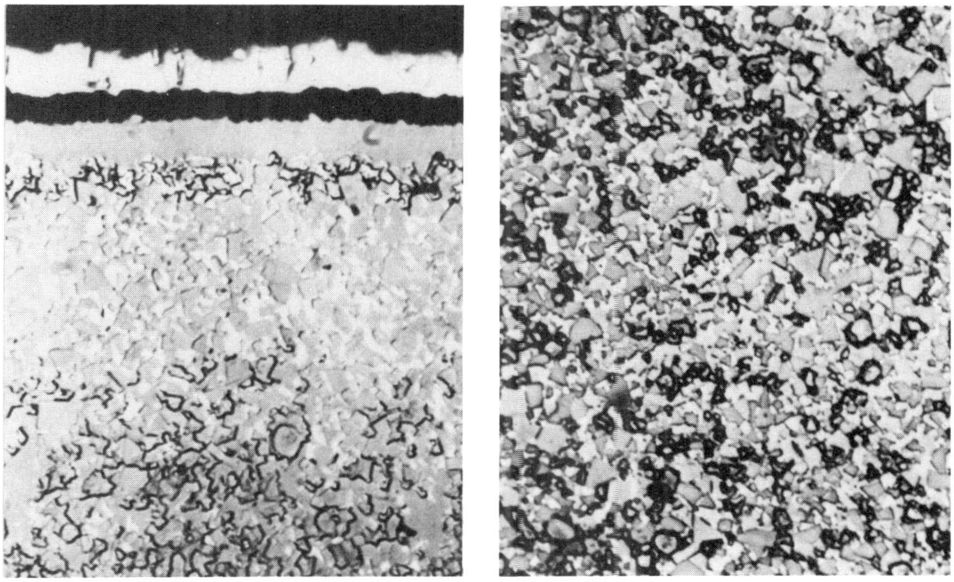


**Fig. 24** Microstructure of a second-generation cobalt-enriched coated tool. 85WC-9(Ti,Ta,Nb)C-6Co tool with a $TiC/Al_2O_3/TiN$ coating. (a) Cobalt-enriched periphery (beneath the coating). (b) Bulk microstructure. Both etched with Murakami's reagent for 2 min. 1500×

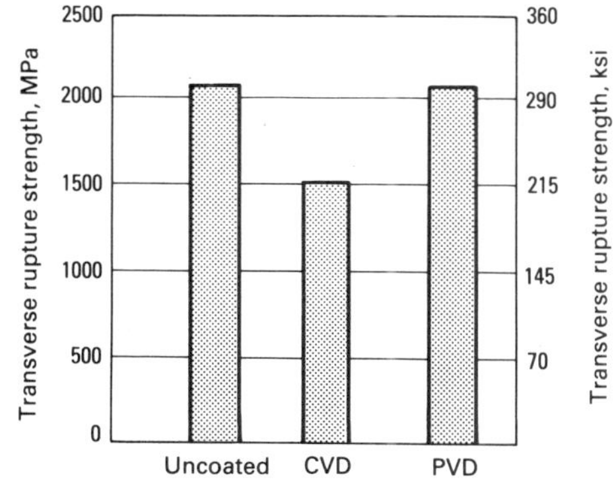


**Fig. 25** Comparison of the transverse rupture strength of uncoated and coated carbide tools. Measured by a three-point bend test on 5 × 5 × 19 mm (0.2 × 0.2 × 0.75 in.) specimens of 73WC-19(Ti,Ta,Nb)C-8Co

such a range of machining speeds and feeds are illustrated in Fig. 23 and 24.

## Physical Vapor Deposited Coatings

Physical vapor deposition (PVD) has recently emerged as a commercially viable process for applying hard TiN coatings onto high-speed steel tools (Ref 29). Physical vapor deposition typically employs lower temperatures (~500 °C, or 900 °F) than chemical vapor deposition. Therefore, the PVD process is attractive for use with cemented carbide tools because the lower deposition temperature prevents η phase formation and provides for refinement of the grain size of the coating layer.

It has been shown by standard three-point bend tests that CVD coatings reduce the transverse rupture strength of cemented carbide tools by as much as 30% because of the presence of interfacial η phase and/or tensile residual stress within the coating (Fig. 25). On the other hand, PVD coatings do not produce η phase. In addition, PVD coatings may induce a compressive residual stress, depending on the deposition technique. Therefore, PVD coatings do not degrade the transverse rupture strength of the carbide tools (Ref 30, 31).

Another advantage of the PVD process is its ability to coat uniformly over sharp cutting edges (Fig. 26). A sharp edge is desirable in a cutting tool because it leads to lower cutting forces, reduced tool tip temperatures, and finer finishes. In many cases, CVD coatings cannot be applied over sharp cutting edges without heavy η phase formation and/or coating buildup, both of which can lead to rapid edge failures. The PVD coating offers the benefit of the abrasive wear resistance of a hard coating layer without the degradation of edge strength.

The above advantages of PVD coatings are especially beneficial in operations such as milling. As noted previously, milling operations can produce severe thermal and mechanical cracks in cutting tools, and these cracks can lead to edge chipping. The CVD coatings often aggravate this situation through stress raisers such as thermally induced cracks within the coating as well as interfacial η phase. These problems can be minimized by honing the cutting edge, by reducing the coating thickness, and by controlling η phase formation, but they can be eliminated by the use of PVD coatings. As a result, PVD-coated milling tools have recently become a commercial reality (Ref 32).

It should be noted that PVD coatings will not replace CVD coatings to any great extent in the near future. Currently, PVD has certain limitations. For example, $Al_2O_3$ coating is not feasible, and multilayer coatings have not yet been commercially developed. Additionally, methods have not yet been developed for coating complex tool geometries.

## Tools and Toolholding

Early carbide metal cutting tools consisted of carbide blanks brazed to steel holders or milling cutters. Tools that became dull were resharpened by grinding. Clearance angles, cutting point radii, and other features could also be ground into the tools to suit particular cutting situations. Special chip-breaker grooves designed to curl and

**Fig. 26** An example of PVD TiN coating on a sharp cemented carbide tool. Etched with Murakami's reagent for 3 s. 1140×

**Fig. 27** Indexable insert secured by pin

**Fig. 28** Pin-and-clamp method of securing an indexable insert to a steel toolholder

**Fig. 29** Screw-on method of securing an indexable insert to a toolholder

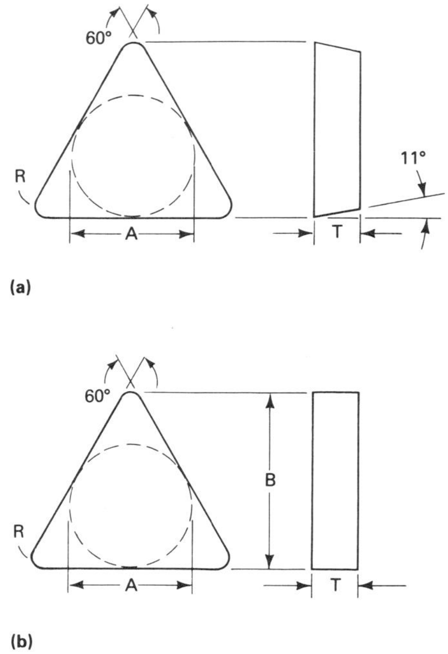


**Fig. 30** Positive- and negative-rake geometries. (a) Triangle positive-rake insert. (b) Triangle negative-rake insert

break the chips generated in metal cutting were also ground into the early tools.

Although these early carbide metal cutting tools provided significant increases in metal cutting productivity, certain disadvantages have become apparent. Regrinding changes the size of the tool; therefore, the cutting tool/workpiece relationship must be readjusted each time the tool is resharpened. Maintaining consistent geometry is difficult with reground tools, and part quality can suffer. Further, because the braze joint can withstand only a limited range of temperature, the selection of usable carbide compositions is restricted, and CVD coating technologies cannot be applied. Nevertheless, brazed tools are still used in applications such as circular saws and small-diameter drills where mechanical clamping is impractical.

**Indexable Carbide Inserts.** The now-familiar prismatically shaped indexable inserts were introduced in the 1950s. These so-called throwaway inserts resembled brazed tools except that the carbide was secured in the holder pocket by a clamp rather than a braze. When a cutting edge wore, a fresh edge was simply rotated or indexed into place. Several other holding methods were subsequently developed.

The most popular holding method employs a pin that passes through a hole in the insert and forces it into the holder pocket (Fig. 27). Clamps are still widely used, often in conjunction with a holding pin (Fig. 28). Another common holding style employs a screw with a tapered head that fits a conical hole in the insert and thus holds the insert securely (Fig. 29).

Consistency and ease of replacement are the main advantages of indexable insert tooling. Consistent positioning of the cutting edge from index to index simplifies machine tool setup and helps ensure uniform product quality. The use of indexable inserts also eliminates labor costs for regrinding and permits CVD-coated tools and a wide variety of carbide compositions to be utilized.

Indexable carbide inserts are available in both positive and negative geometries (Fig. 30). Negative-rake inserts have excellent resistance to breakage and are well suited to difficult operations involving interrupted cuts. Negative-rake inserts can also be used on both sides, effectively doubling the number of cutting edges available per insert. Although positive-rake inserts can be used on only one side, they cut with lower force, reduce the possibility of distorting the workpiece, and help produce better surface finishes. Both negative- and positive-rake inserts are available in a range of tolerances.

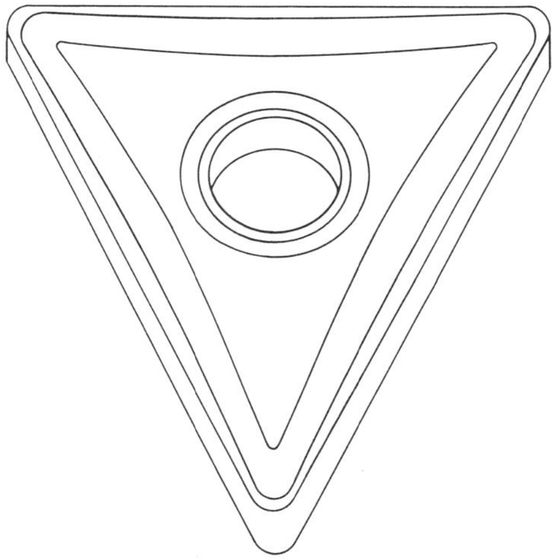

**Fig. 31** Chip-breaker grooves on one side of a triangle insert

*Chip-breaker grooves* (Fig. 31) can also be molded into the surfaces of indexable carbide inserts. In addition to providing chip control, these grooves allow lower cutting forces. On negative-rake inserts, they can be used on both sides for greater economy. Single-sided negative-rake inserts (Fig. 31) can provide the force-reducing capabilities of positive-rake geometries with the strength and chip control properties of a conventional negative-rake design. Specialized chip control geometries have been developed for particular workpiece materials or types of operations; geometries have also been developed to complement particular carbide grades.

**Edge preparation** refers to the practice of modifying the cutting edge itself after the correct overall geometry has been produced on the tool. Edge preparations are applied for two reasons: to prevent the chipping and

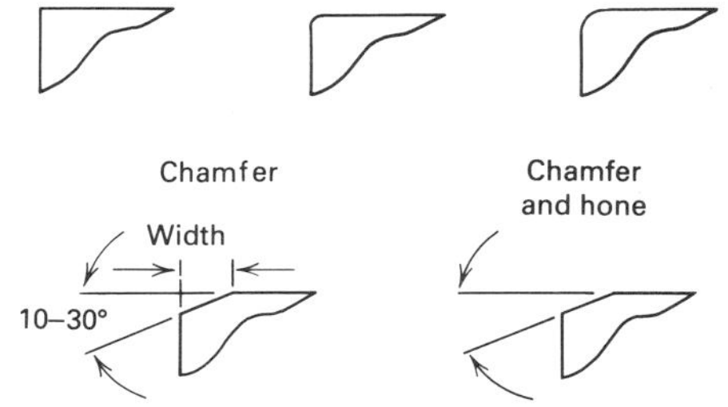


**Fig. 32** Various edge preparations for metal cutting tools

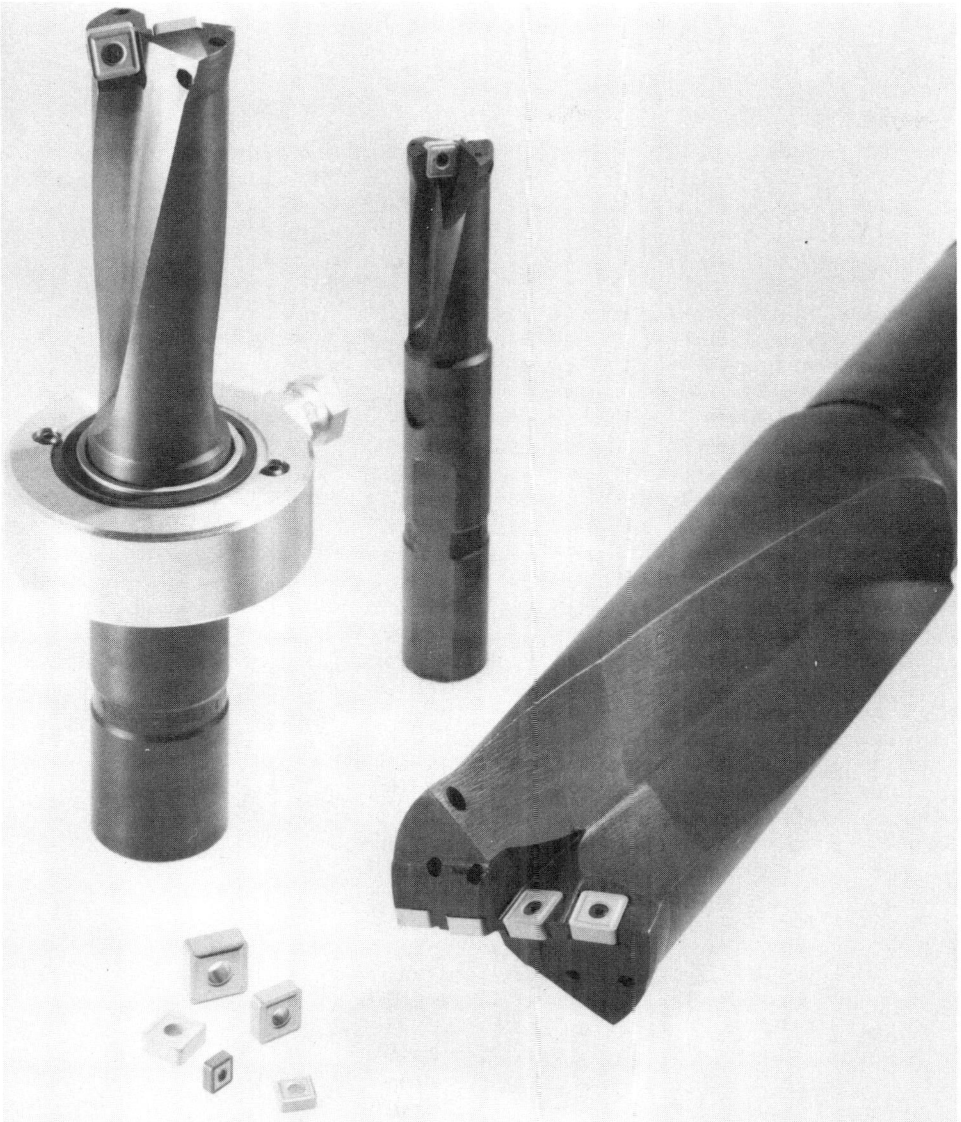

**Fig. 33** Metal cutting drills with indexable carbide inserts

premature failure of a too sharp and therefore weak cutting edge, or to provide a slightly rounded (honed) edge that will optimize the effect of CVD coating. The edge preparations most commonly used are hones or chamfers (Fig. 32).

The edge preparation of brazed tools is most often applied by hand with a stone. Because the radius desired is very small (commonly 0.025 to 0.075 mm, or 0.001 to 0.003 in.), it is very difficult to achieve consistent performance. Too little hone can lead to microchipping and subsequent rapid edge wear; too much hone is very much like a preworn edge and also results in shorter tool life. Indexable carbide inserts, especially coated inserts, are usually equipped with a machine-applied hone of the size the manufacturer considers proper for the insert size, style, grade, and application. In general, only enough hone to prevent chipping should be used.

Cemented carbide tools intended for use in heavy-duty machining applications are often chamfered to provide maximum resistance to edge chipping. Although this may result in the loss of some tool life, the trade-off is reasonable if the mode of failure with honed inserts has been breakage rather than wear. Inserts intended for the machining of nonferrous and some aerospace alloys are generally supplied with sharp edges to reduce cutting forces and thus improve tool life.

**Drills and end mills** are available in both solid cemented carbide and indexable insert versions. The increased production benefits realized from carbide tools can be as high as 6 to 1 compared to high-speed steel products Solid carbide tools are available in the smallest sizes, ranging from circuit board drills of 0.13 mm (0.005 in.) in diameter to end mills with diameters up to 38 mm (1.5 in.). Drills and end mills with indexable inserts are generally available in sizes ranging from 16 mm (⅝ in.) in diameter up to about 75 mm (3 in.) (Fig. 33 and 34). Solid carbide tools have the advantages of greater rigidity, more cutting edges, and greater precision than comparable tools with indexable inserts. Indexable inserts offer the advantage of repeatability, require no resharpening, and permit the use of a wider variety of grades.

**Threading.** A number of methods are available for producing thread forms. These can be divided into two major categories: forming the thread, and machining (or cutting) the thread. A common thread forming process is thread rolling, which utilizes two diametrically opposed dies to cold form a thread into a workpiece. A frequent (and sometimes desirable) by-product of this process in certain materials is work hardening at the root of the thread.

Threads are typically machined by one of two methods: tapping, or single-point machining utilizing an indexable insert/toolholder system. Tapping is used where the diameter of the hole is too small for an indexable carbide/toolholder system and/or the machining speed is too slow for carbide. The three popular insert/toolholder systems used for threading are the laydown triangle, the stand-up (on edge) triangle, and several versions of a proprietary stand-up 55° parallelogram design (Fig. 35).

An undesirable by-product of thread machining is the V-shaped chip removed from the workpiece. Indexable inserts with proprietary chip control geometries molded into the top rake surface of the cutting edge are becoming available, and they control and break the chips with varying levels of success. In the more common thread forms, multitooth thread-chasing inserts are available as a means of reducing the number of passes required to complete a thread, thus improving productivity (Fig. 36). An increasingly popular option available for many thread forms is the cresting insert, which machines the full thread form. Noncresting inserts machine the root and flanks but not the crest of the thread.

*Thread milling*, a thread machining method that is useful when turning is not possible, is performed on multiaxis computer numerical control machines capable of helical interpolation (Fig. 37). A disadvantage of thread milling is that the thread form it produces is slightly imperfect because of the inability of the cutting tool to clear the helical angle of the thread form as it exits the part. However, the threads are sufficiently accurate for all but the most demanding applications.

**Grooving.** There are three different grooving insert styles in common use:

- 90° V-bottom (Fig. 38)
- Proprietary stand-up 55° parallelogram (Fig. 35)
- Stand-up (on-edge) triangle

**Fig. 34** End mills with indexable carbide inserts

**Fig. 35** Proprietary 55° parallelogram carbide insert

**Fig. 36** A multitooth thread-chasing insert

The V-bottom system is the most suitable for deep grooving because the cutting edge of the insert is wider than the body and is directly supported by the toolholder. The compact design and proprietary clamping method of the 55° parallelogram system maximize rigidity in shallow grooving. The on-edge triangle system offers three cutting edges on each insert, as opposed to two in the other common grooving systems. Chip control is a major concern in grooving, and products are becoming available that control and/or break chips with varying levels of success.

**Cutoff.** The early carbide cutoff tools consisted of carbide inserts brazed onto steel shanks. As in the case of carbide turning inserts, efforts to eliminate tool resharpening costs and to improve performance led to the development of mechanically held replaceable cutoff inserts. These inserts are available in a variety of styles, but most have a vee shape in the top or bottom surface, which is gripped by the steel holder for rigidity. Most cutoff inserts have a single cutting edge and are held either by clamping or by wedging directly into the holder. Chip control is available in either molded or ground geometries (Fig. 39).

## Machining Applications

**Uncoated Straight WC-Co Grades.** Despite the advent of coated cemented carbide tools in the late 1970s, uncoated straight WC-Co grades still find a place in many machining operations. Unalloyed gray cast iron is probably the most common workpiece material machined with WC-Co grades, but other materials such as high-temperature alloys, austenitic stainless steels, nonferrous alloys, and nonmetals are often machined with C1 and/or C2 grades. The recommended speed and feed ranges for these materials are listed in Table 9. The higher-cobalt C1 grades are also often used on difficult-to-machine workpieces, such as chilled cast iron and heat-treated steels, where cutting tool strength and shock resistance are important. High-temperature alloys and austenitic stainless steels are machined with C2 grades, typically in positive-rake, sharp-edge geometries at lower surface speeds.

**Uncoated alloyed carbide grades** (C5 to C8 carbide classifications) are primarily used in machining steels in the speed and feed ranges listed in Table 10. Uncoated alloyed carbides are well suited to a number of machining applications. They are widely used when machines do not have sufficient horsepower to utilize the high metal removal capabilities of advanced coated tools. Uncoated carbides also find application in brazed form tools, which are typically made with highly specialized geometries mirroring the part being machined.

Uncoated alloyed carbides are also employed in the machining of special part configurations with thin wall sections and tight tolerances. These parts cannot be subjected to high forces during machining. The CVD-coated tools, which are honed prior to coating, are not effective in these applications, for which high positive, sharp-edged,

**Fig. 37** Thread milling with indexable inserts

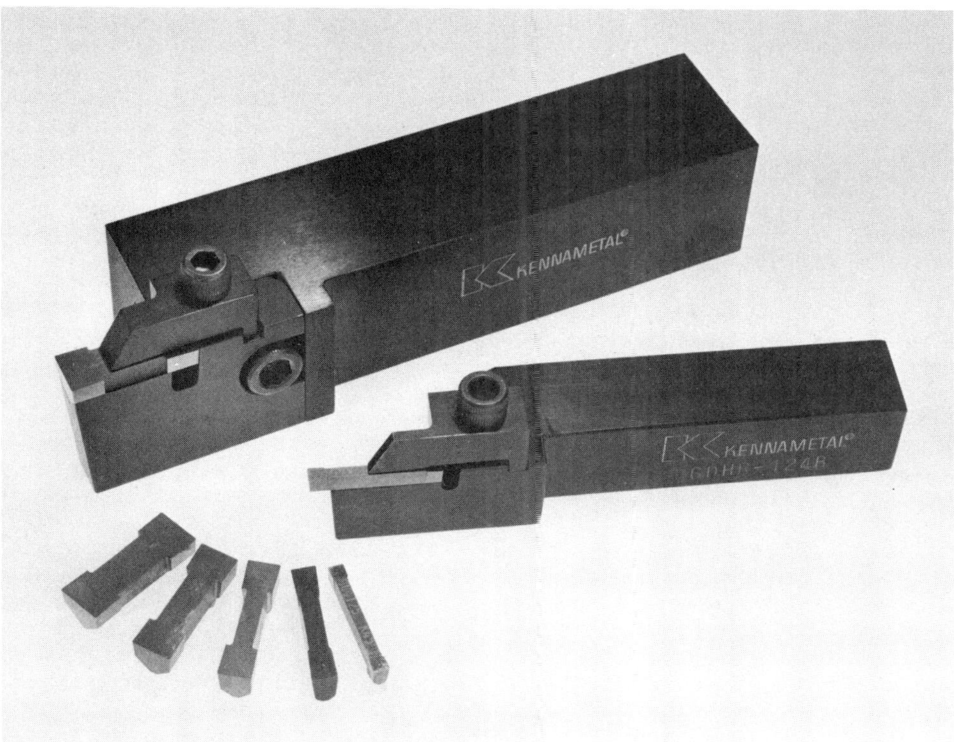


**Fig. 38** 90° V-bottom (dogbone) grooving inserts and toolholders

uncoated alloyed carbides are a better choice. The recently developed PVD-coated inserts with sharp edges may prove effective in such applications.

**Coated carbide grades** provide more abrasion resistance to the tool and permit the use of higher machining speeds and feeds. This is illustrated in the tool life diagram shown in Fig. 40, which compares the performance of a C5 grade with and without TiC-TiCN-TiN coating in turning SAE 1045 steel.

More than 50% of the metal cutting inserts currently sold in the United States are CVD coated. These coatings are used in a wide range of machining applications, including turning, milling, threading, and grooving. Physical vapor deposited coatings can be applied to sharp insert edges without the deleterious effect of $\eta$ phase formation. The sharp, tough PVD inserts are particularly well suited to milling applications; they have also been found to perform well in threading and grooving operations on difficult-to-machine materials such as high-temperature alloys and austenitic stainless steels.

**Submicron Grades for Aerospace Materials.** Although new cutting tool materials such as Sialons and whisker-reinforced ceramics have provided great increases in machining productivity on nickel-base alloys, similar improvements have not occurred in the machining of titanium alloys. However, submicron (fine-grain) carbide grades have shown the capacity to enhance productivity in both titanium alloys and the nickel-base materials. In general, the strength or toughness of a carbide grade is inversely proportional to its abrasion resistance. The microstructures associated with submicron carbide materials provide both strength and abrasion resistance. In titanium turning operations where speed and feed parameters are held constant, tool life improvements of 200% have been recorded after a change to submicron carbide cutting tool materials.

## REFERENCES

1. H. Moissan, *The Electrical Furnace*, V. Lenher, Trans., Chemical Publishing Company, 1904
2. E.K. Storms, *The Refractory Carbides*, Academic Press, 1978
3. M. Hansen and K. Anderko, *Constitution of Binary Alloys*, McGraw-Hill, 1958
4. K. Schroeter, U.S. Patent 1,549,615, 1925
5. E.M. Trent, Cutting Tool Materials, *Metall. Rev.*, Vol 13 (No. 127), 1948, p 129-144
6. K.J.A. Brookes, *World Directory and Handbook of Hardmetals*, 4th ed., International Carbide Data, 1987
7. E. Lardner, *Powder Metall.*, Vol 21, 1978, p 65
8. H.E. Exner, *Int. Met. Rev.*, Vol 24 (No. 4), 1979, p 149-173
9. P.M. McKenna, U.S. Patent 3,379,503, 1968
10. P.M. McKenna, Tool Materials—Cemented Carbides, in *Powder Metallurgy*, J. Wulff, Ed., 1942, p 454-469
11. H.D. Hanes, D.A. Seifert, and C.R. Watts, *Hot Isostatic Processing*, Battelle Press, 1979, p 20-24
12. R.C. Lueth, Advances in Hardmetal Production, in *Proceedings of the Metal Powder Report Conference* (Luzern), Vol 2, MPR Publishing Services Ltd.,

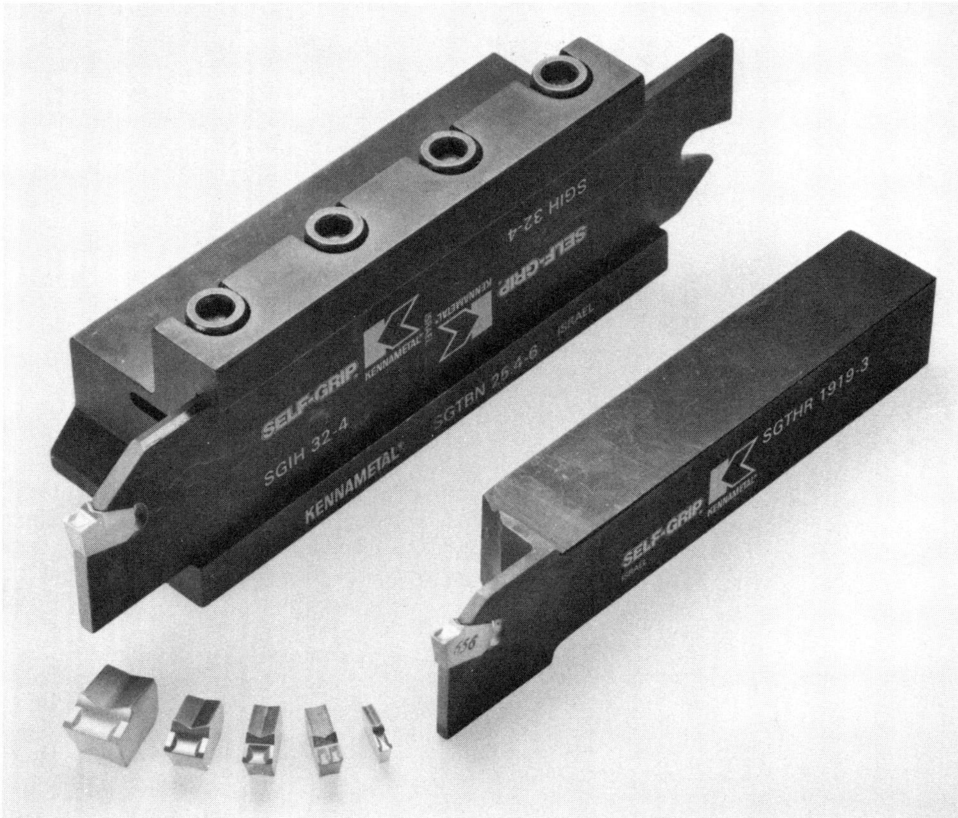


**Fig. 39** Cutoff inserts and toolholders

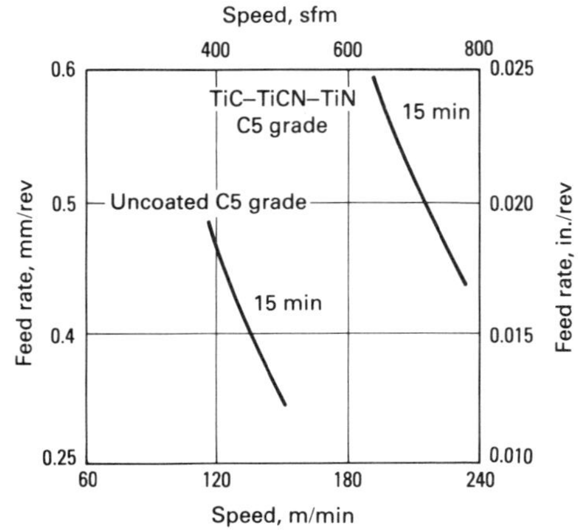


**Fig. 40** Tool life comparison of a coated and an uncoated carbide tool. Constant tool life (15 min) plot for an uncoated and a TiC-TiCN-TiN-coated C5 grade in turning SAE 1045 steel. The depth of cut was 2.5 mm (0.100 in.).

1983
13. P.A. Dearnley, *Met. Technol.*, Vol 10, 1983
14. H. Tanaka, Relationship Between the Thermal, Mechanical Properties and Cutting Performance of TiN-TiC Cermet, in *Cutting Tool Materials*, Conference Proceedings, American Society for Metals, 1981, p 349-361
15. R.C. Lueth, *Fracture Mechanics of Ceramics*, R.C. Bradt *et al.*, Ed, Plenum Press, 1974, p 791-806
16. J.L. Chermant and F. Osterstock, *J. Mater. Sci.*, Vol 11, 1976, p 1939-1951
17. J.R. Pickens and J. Gurland, *Mater. Sci. Eng.*, Vol 33, 1978, p 135-142
18. W.D. Kingery, H.K. Bowen, and D.R. Uhlmann, *Introduction to Ceramics*, 2nd ed., John Wiley & Sons, 1960, p 828
19. C.S. Ekmar, German Patent 2,007,427
20. W. Schintlmeister, O. Pacher, and K. Pfaffinger, in *Chemical Vapor Deposition, Fifth International Conference*, J.M. Blocker, Jr. *et al.*, Ed., The Electrochemical Society Softbound Symposium Series, Electrochemical Society, 1975, p 523

**Table 9 Recommended speed and feed ranges for uncoated straight WC-Co grades**

| | | | Roughing(a) | | | | Finishing(b) | | | |
|---|---|---|---|---|---|---|---|---|---|---|
| | | | Speed | | Feed | | Speed | | Feed | |
| Material | Hardness, HB | Recommended grade | m/min | sfm | mm/rev | in./rev | m/min | sfm | mm/rev | in./rev |
| Gray cast iron | 190–330 | C2 | 99–168 | 325–550 | 0.10–0.36 | 0.004–0.014 | 122–213 | 400–700 | 0.10–0.38 | 0.004–0.015 |
| | 330–450 | C2 | 84–137 | 275–450 | 0.08–0.25 | 0.003–0.010 | 91–183 | 300–600 | 0.08–0.30 | 0.003–0.012 |
| | 450–700 | C1 | 68–106 | 225–350 | 0.05–0.15 | 0.002–0.006 | 91–137 | 300–450 | 0.05–0.25 | 0.002–0.010 |
| Austenitic stainless steel | 120–180 | C2 | 68–106 | 225–350 | 0.15–0.41 | 0.006–0.016 | 84–137 | 275–450 | 0.13–0.51 | 0.005–0.020 |
| | 180–240 | C2 | 61–91 | 200–300 | 0.08–0.30 | 0.003–0.012 | 68–122 | 225–400 | 0.08–0.38 | 0.003–0.015 |
| High-temperature alloys | 150–250 | C2 | 18–43 | 60–140 | 0.13–0.25 | 0.005–0.010 | 23–61 | 75–200 | 0.08–0.30 | 0.003–0.012 |
| | 250–400 | C2 | 15–34 | 50–110 | 0.08–0.20 | 0.003–0.008 | 18–43 | 60–140 | 0.05–0.23 | 0.002–0.009 |
| Nonferrous materials | 80–120 | C2 | 61–244 | 200–800 | 0.20–0.51 | 0.008–0.020 | 91–366 | 300–1200 | 0.13–0.30 | 0.005–0.012 |
| Nonmetals | ··· | C2 | 76–183 | 250–600 | 0.20–0.51 | 0.008–0.020 | 152–244 | 500–800 | 0.13–0.25 | 0.005–0.010 |

(a) Depth of cut greater than 2.5 mm (0.100 in.). (b) Depth of cut less than 2.5 mm (0.100 in.)

**Table 10 Recommended speed and feed ranges for uncoated alloyed carbide grades**

| | | | Roughing(a) | | | | Finishing(b) | | | |
|---|---|---|---|---|---|---|---|---|---|---|
| | | | Speed | | Feed | | Speed | | Feed | |
| Material | Hardness, HB | Recommended grade | m/min | sfm | mm/rev | in./rev | m/min | sfm | mm/rev | in./rev |
| Free-machining steels | 140–190 | C5, C6 | 76–122 | 250–400 | 0.30–0.64 | 0.012–0.025 | 106–168 | 350–550 | 0.13–0.38 | 0.005–0.015 |
| | 190–240 | C5, C6 | 61–106 | 200–350 | 0.20–0.51 | 0.008–0.020 | 91–152 | 300–500 | 0.08–0.30 | 0.003–0.012 |
| Plain carbon | 185–240 | C5, C6 | 61–106 | 200–350 | 0.20–0.51 | 0.008–0.020 | 91–137 | 300–450 | 0.13–0.51 | 0.005–0.020 |
| Alloy steels | 190–330 | C5 | 61–106 | 200–350 | 0.30–0.64 | 0.012–0.025 | 91–122 | 300–400 | 0.13–0.51 | 0.005–0.020 |
| | 330–450 | C5 | 53–91 | 175–300 | 0.25–0.51 | 0.010–0.020 | 76–106 | 250–350 | 0.10–0.41 | 0.004–0.016 |
| | 450–700 | C7 | 38–76 | 125–250 | 0.15–0.38 | 0.006–0.015 | 46–91 | 150–300 | 0.08–0.30 | 0.003–0.012 |
| 400, 500 series stainless steel | 175–210 | C5 | 46–106 | 150–350 | 0.25–0.64 | 0.010–0.025 | 91–152 | 300–500 | 0.10–0.51 | 0.004–0.020 |

(a) Depth of cut greater than 3.2 mm (0.125 in.). (b) Depth of cut less than 3.2 mm (0.125 in.)

21. W. Schintlmeister, O. Pacher, K. Pfaffinger, and T. Raine, *J. Electrochem. Soc.*, Vol 123, 1976, p 924-929
22. V.K. Sarin and J.N. Lindstorm, *J. Electrochem. Soc.*, Vol 126, 1979, p 1281-1287
23. W. Schintlmeister, W. Wallgram, J. Ganz, and K. Gigl, *Wear*, Vol 100, 1984, p 153-169
24. B.N. Kramer and P.K. Judd, *J. Vac. Sci. Technol.*, Vol A3 (No. 6), 1985, p 2439-2444
25. T.E. Hale, Paper presented at the International Machine Tool Show Technical Conference, National Machine Tools Builders Association, 1982
26. H.E. Hintermann, *Wear*, Vol 100, 1984, p 381-397
27. B.J. Nemeth, A.T. Santhanam, and G.P. Grab, in *Proceedings of the Tenth Plansee Seminar* (Reutte/Tyrol), Metallwerk Plansee A.G., 1981, p 613-627
28. A.T. Santhanam, G.P. Grab, G.A. Rolka, and P. Tierney, An Advanced Cobalt-Enriched Grade Designed to Enhance Machining Productivity, in *High Productivity Machining—Materials and Processes*, Conference Proceedings, American Society for Metals, 1985, p 113-121
29. R.F. Bunshah and A.C. Raghuram, *J. Vac. Sci. Technol.*, Vol 9, 1972, p 1385
30. G.J. Wolfe, C.J. Petrosky, and D.T. Quinto, *J. Vac. Sci. Technol.*, Vol A4 (No. 6), 1986, p 2747-2754
31. D.T. Quinto, G.J. Wolfe, and P.C. Jindal, *Thin Solid Films, the International Conference on Metallic Coatings*, Vol 153, 1987, p 19-36
32. D.T. Quinto, C.J. Petrosky, and J.L. Hunt, *Cutting Tool Eng.*, Vol 39, 1987, p 46-52

# Cermets

Walter W. Gruss, Kyocera Feldmuehle, Inc.

CERMETS are a group of powder metallurgy products consisting of ceramic particles bonded with a metal. The ceramic component of cermets provides high hot hardness and oxidation resistance, while the metallic component enhances ductility and thermal shock resistance. The bonding of ceramic components with metals, a complex process, depends largely on the solubility, wetting properties, and phase relations of the selected materials.

## Composition and Microstructure

Ceramics are defined as any class of inorganic or nonmetallic products that are subjected to high temperature during manufacture or use (Ref 1). Typically, but not exclusively, a ceramic is a metallic oxide, boride, or carbide, or a mixture or compound of such materials. By this definition of ceramics, the following materials theoretically fall into the group of cermets:

- WC + Co (Tungsten carbide + cobalt)
- WC/TiC/TaC + Co (Tungsten carbide/titanium carbide/tantalum carbide + cobalt)
- TiC + Ni (Titanium carbide + nickel)
- Ti(C,N) + Ni/Mo (Titanium carbonitride + nickel/molybdenum)

However, the cutting tool industry considers only the titanium carbide and titanium carbonitride base materials to be cermets, while the tungsten carbide based materials are named cemented carbides. Therefore, this article will concentrate on cermets

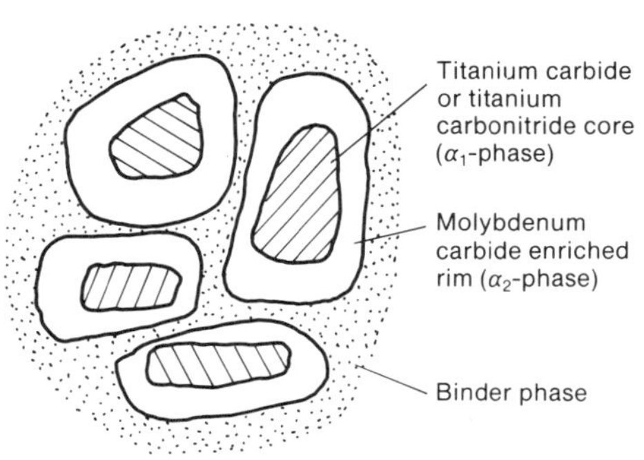


**Fig. 1** Schematic of cermet microstructure

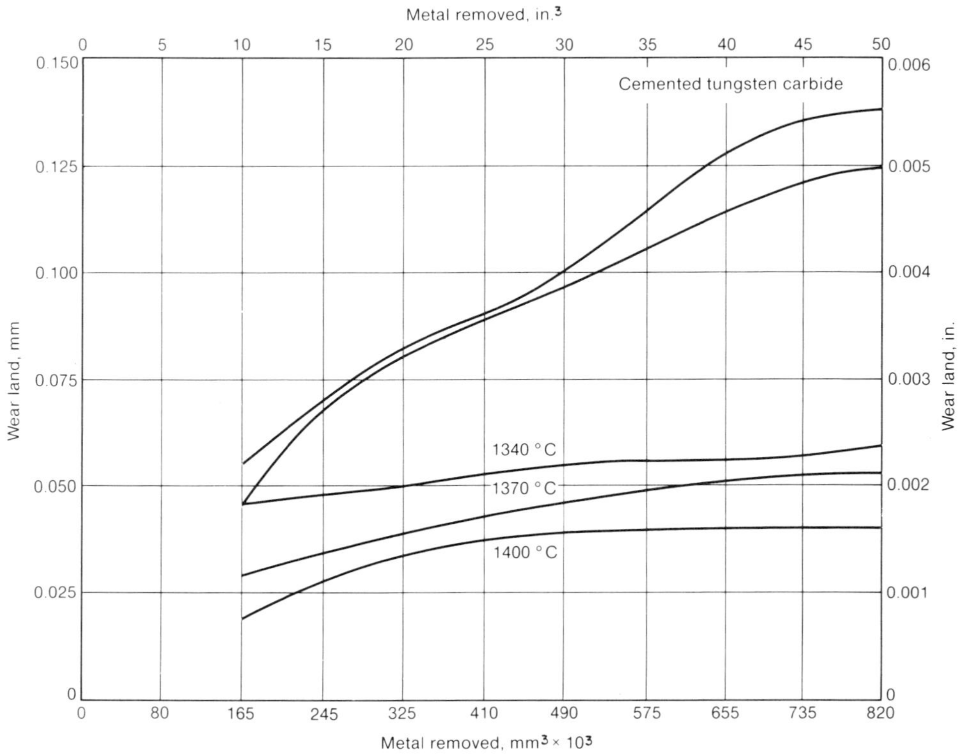


**Fig. 2** Flank wear of titanium carbide cermet sintered at different temperatures. Machining parameters: feed, 0.28 mm/rev (0.011 in./rev); depth of cut, 2.5 mm (0.100 in.); speed, 106 m/min (350 sfm). Workpiece: 1045 steel (163 to 174 HB)

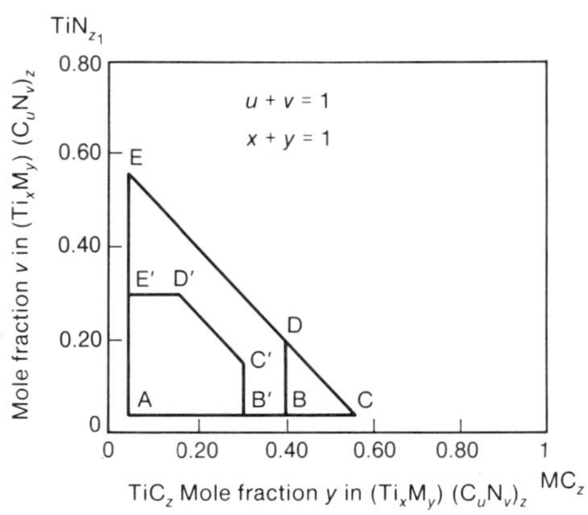


**Fig. 3** Preferred compositions of titanium carbonitride cermets. M, molybdenum and/or tungsten; z, number of moles carbon and nitrogen divided by the number of moles titanium and M; z is variable between the limits 0.80 and 1.07. Source: Ref 6

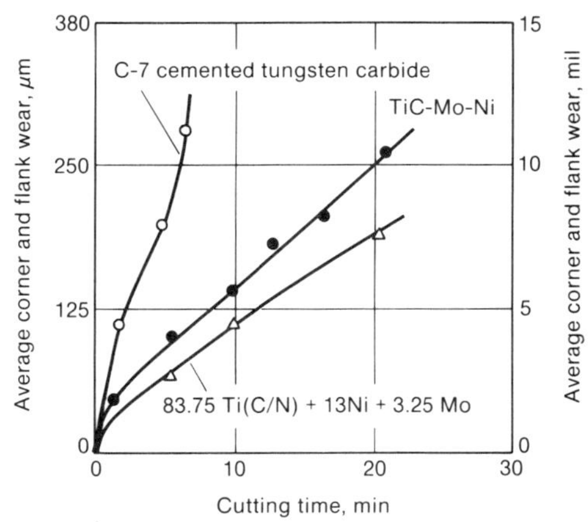


**Fig. 4** Comparison of flank wear for two cermets and a cemented carbide when turning 4340 steel. Source: Ref 6

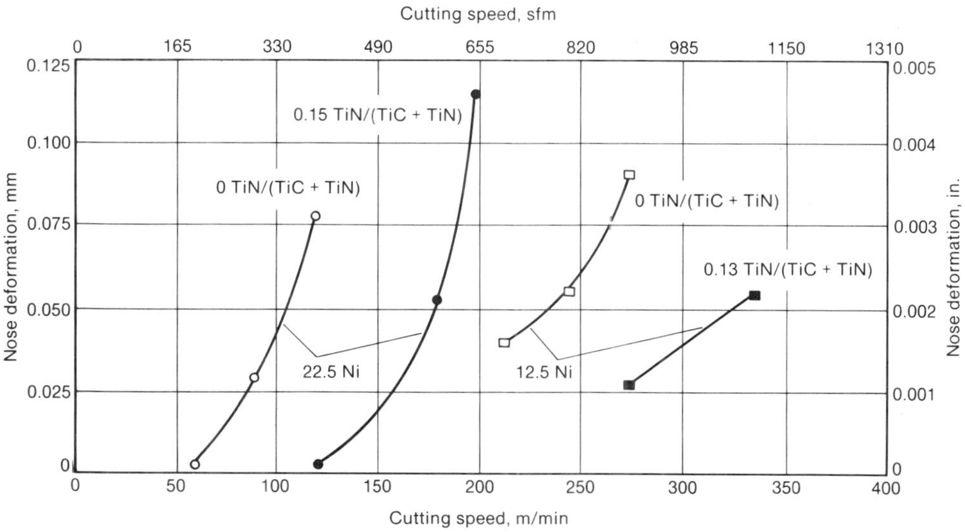


**Fig. 5** Effect of titanium nitride addition on the plastic deformation of a cutting edge. Workpiece, 4340 steel (300 HB). Source: Ref 7

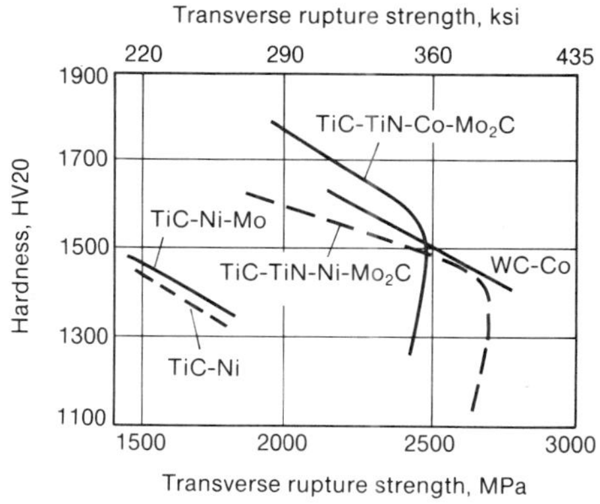


**Fig. 6** Relationship between hardness and modulus of rupture for various cutting materials. Source: Ref 8

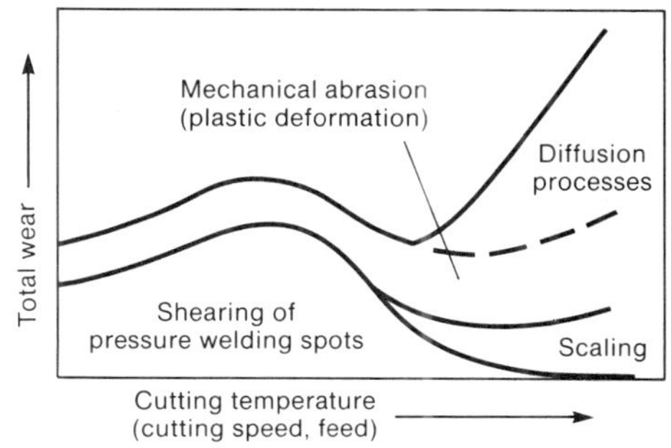


**Fig. 7** Wear mechanisms of cutting tools

**Table 1 Comparison of high-temperature properties of a TiC cermet and a complex carbonitride cermet**

| Composition of cermet | Vickers hardness at 1000 °C (2000 °F), kg/mm² | Transverse rupture strength at 900 °C (1650 °F) MPa | ksi | Oxidation resistance at 1000 °C (2000 °F) weight gain, mg/cm² · h | Thermal conductivity at 1000 °C (2000 °F), W/K⁶ · m |
|---|---|---|---|---|---|
| TiC-16.5Ni-9Mo | 500 | 1050 | 152 | 11.8 | 24.7 |
| TiC-20TiN-15WC-10TaC-5.5Ni-11Co-9Mo | 650 | 1360 | 197 | 1.66 | 42.3 |

**Table 2 Typical properties of titanium carbonitride cermets**

| Tool designation(a) U.S. | ISO | Vickers hardness, kg/mm² | Modulus of rupture MPa | ksi | Fracture toughness ($K_{Ic}$) MPa $\sqrt{m}$ | ksi $\sqrt{in.}$ | Young's modulus of elasticity GPa | psi × $10^6$ |
|---|---|---|---|---|---|---|---|---|
| C8/C7 / C4/C3 (b) | P01/P15 / K01/K10 | 1650 | 1500 | 220 | 8.5 | 7.7 | 450 | 65 |
| C7/C6 / C3/C2 (c) | P10/P20 / K05/K20 | 1500 | 1800 | 260 | 10 | 9 | 410 | 59 |

(a) The ISO and U.S. classification systems for cemented tungsten carbide can be used to designate cermet grades. The ISO and U.S. classification systems are defined in the article "Cemented Carbides" in this Volume. (b) Cermet grades of this group are typically based on the system (Ti,Mo/W) (C,N) and are referred to as molybdenum carbide toughened cermets. (c) Cermet grades of this group are typically based on the system (Ti, Mo/W, Nb, Ta) (C,N) and are referred to as tantalum/niobium carbide toughened cermets.

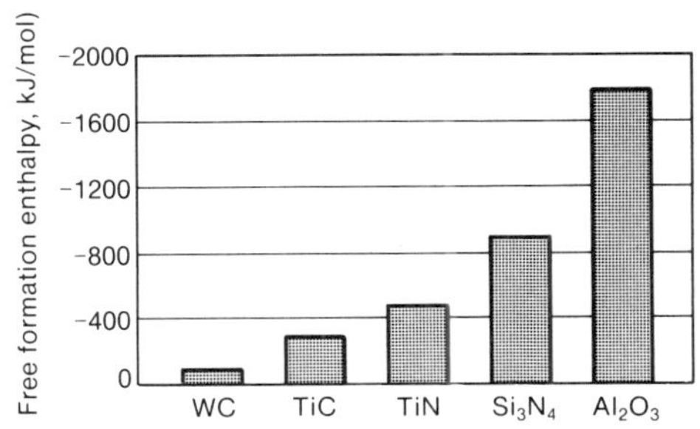


**Fig. 8** Free-formation enthalpy of cutting materials at 260 °C (500 °F)

based on titanium carbide and titanium carbonitride.

**Titanium Carbide Cermets.** The first attempts to apply titanium carbide in sintered, tungsten-free cutting tool materials were made in Germany in 1929, when titanium carbide/molybdenum carbide solid solutions with 15% nickel as binder were manufactured and applied for finish turning of steel (Ref 2). Acceptance was limited because of the low strength and high brittleness of this material. However, the interest in titanium carbide continued, mainly due to the lower cost and availability of its raw material, titanium oxide ($TiO_2$). Also, the higher hardness, melting point, and oxidation resistance of titanium carbide (TiC) compared to that of tungsten carbide (WC) promised greater potential.

The poor wettability of titanium carbide (TiC) with nickel (Ni) was improved drastically with the addition of molybdenum (Mo) or molybdenum carbide ($Mo_2C$) to the nickel binder phase (Ref 3). The microstructure of such a composition is shown schematically in Fig. 1. The core of the carbide phase consists of titanium carbide ($\alpha_1$-phase), while the rim is enriched with molybdenum carbide ($\alpha_2$-phase) (Ref 4 and 5). The molybdenum from the binder phase diffuses into the carbide phase and improves wettability by means of the metal binder. The abrasion resistance of such a composition varies with the sintering temperature (Fig. 2).

**Cermets Based on Metal Carbonitrides.** The development of cermets continued with the introduction of metal carbonitrides. Titanium-molybdenum-carbon-nitrogen and titanium - tungsten - carbon - nitrogen compounds with metal binders, preferably consisting of nickel, molybdenum, cobalt, or a combination thereof, gained specific attention. Considerable improvement was achieved when the carbonitride phase had a composition within the parameters described in Fig. 3.

The microstructure of such a material is shown in Fig. 1. It is possible to observe a core/rim microstructure of the carbonitride phase, with the $\alpha_1$-phase of the core consisting of titanium carbonitride and the $\alpha_2$-phase of the rim being enriched with molybdenum carbide and/or tungsten carbide (Ref 4).

The nitrogen additions to the hard phase resulted in higher wear resistance (Fig. 4) and reduced plastic deformation of the cutting edge (Fig. 5). Additions of cobalt to the binder phase and of tantalum and/or niobium to the hard phase of complex metal carbonitrides also can improve the cutting

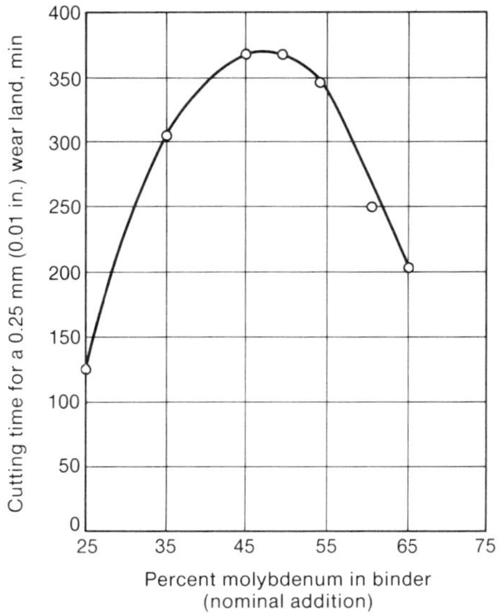


**Fig. 9** Cermet tool life with varying molybdenum content. Machining parameters: feed, 0.28 mm/rev (0.011 in./rev); depth of cut, 2.5 mm (0.100 in.); speed, 180 m/min (600 sfm); coolant workpiece: 1045 steel (163 to 174 HB). Tool: 80TiC-20Ni, Mo. Source: Ref 3

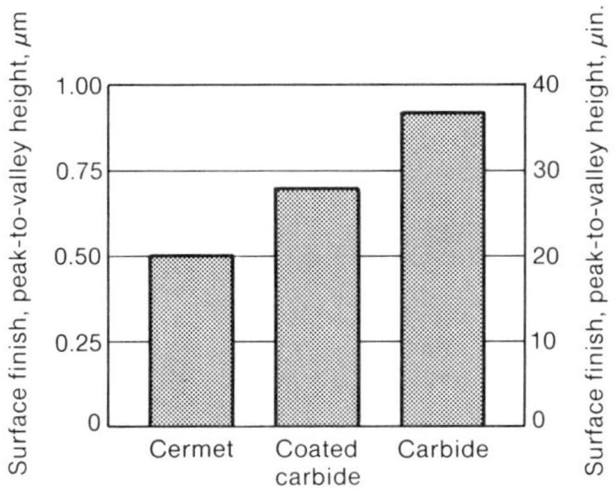


**Fig. 11** Comparison of surface finishes of cermet and cemented tungsten carbide tools. Machining parameters: cutting speed, 250 m/min (825 sfm); feed rate, 0.30 mm/rev (0.012 in./rev); depth of cut, 3.0 mm (0.12 in.); dry, no coolant. Workpiece: 1045 steel

performance of cermets. The high-temperature properties of a complex metal carbonitride is compared with a titanium carbide cermet in Table 1.

## Properties and Grade Selection

The manufacturers of cutting tool materials treat the compositions and properties of their grades as proprietary. The commercially available grades fall into two categories: the titanium carbide base cermets and the titanium carbonitride base cermets. The titanium carbide base cermet grades are in the process of being replaced by the titanium carbonitride base cermet grades because of their higher wear resistance (Fig. 4), hardness, and transverse rupture strength (Fig. 6). Typical properties of tita-

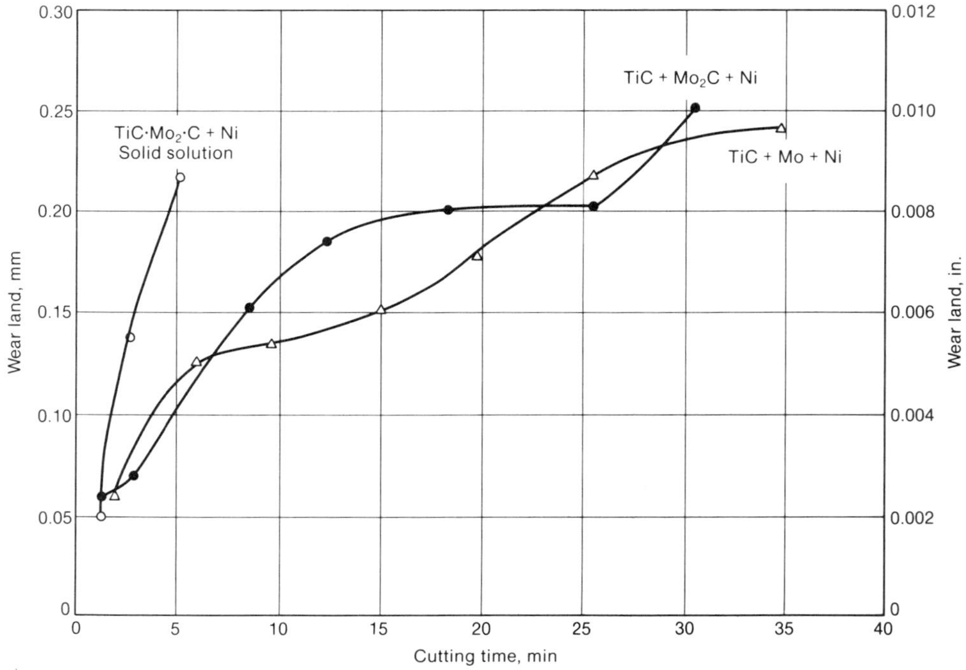


**Fig. 10** Comparison of wear of cermet with molybdenum in three different states: $Mo_2C$ in solid solution with TiC, $Mo_2C$ in the binder phase, and Mo in the binder phase. Machining parameters: feed, 0.28 mm/rev (0.011 in./rev); depth of cut, 1.5 mm (0.060 in.); speed, 365 m/min (1200 sfm); coolant. Workpiece: AISI 1045 (192 to 201 HB). Source: Ref 3

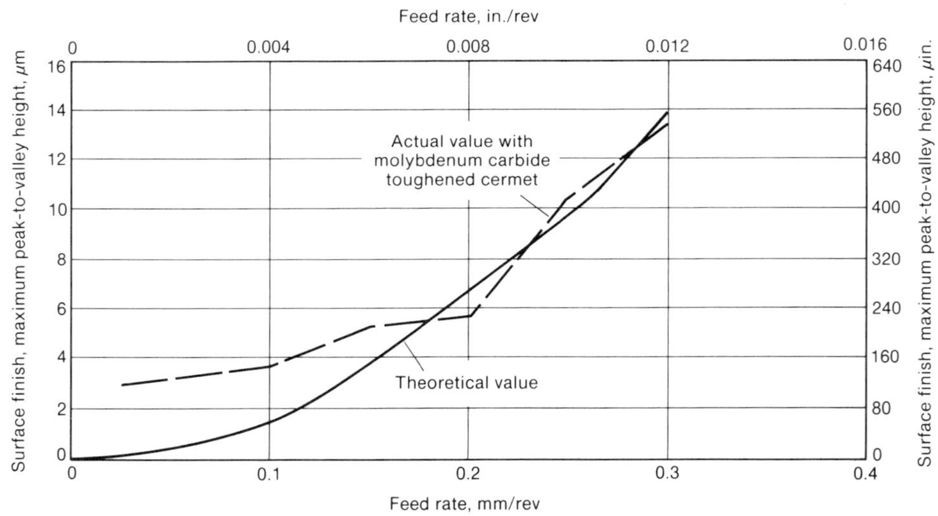


**Fig. 12** Comparison of theoretical surface finish with actual surface produced with cermet tool. Theoretical value = $f^2/8R$, where $f$ = feed rate and $R$ = tool nose radius. Machining parameters: cutting speed, 200 m/min (650 sfm); depth of cut, 2.0 mm (0.080 in.). Workpiece: 4135 steel

**Table 3 Machining applications of cermet cutting tools**

| Tool composition | Operation | Workpiece material |
|---|---|---|
| Molybdenum carbide toughened cermet | Rough and semifinish turning; milling, grooving, and threading | Carbon steels, alloy steels, cast irons; free-machining and nonferrous materials, such as aluminum, copper, brass, and bronze, with hardnesses up to about 38 HRC |
| | Finish turning and milling | Carbon steels, alloy steels, tool steels, stainless steels, and cast irons; free-machining nonferrous materials, such as aluminum, copper, brass, and bronze, with hardnesses up to about 38 HRC |
| Tantalum/niobium carbide toughened cermets | Rough and semifinish turning; milling, grooving, and threading | Carbon steels, alloy steels, cast irons, stainless steels, and tool steels, with hardnesses up to about 38 HRC |

**Table 4 Machining parameters with cermet tools in the turning of various steels**

| Hardness, HB | Operation | Cutting speed, m/min (sfm) Low | Cutting speed, m/min (sfm) Median | Cutting speed, m/min (sfm) High | Feed rate, mm/rev (in./rev) Low | Feed rate, mm/rev (in./rev) Median | Feed rate, mm/rev (in./rev) High | Maximum depth of cut mm | Maximum depth of cut in. |
|---|---|---|---|---|---|---|---|---|---|
| **Carbon steels 1000 series** | | | | | | | | | |
| 100 | Roughing | 73 (240) | 245 (800) | 293 (960) | 0.23 (0.009) | 0.41 (0.016) | 0.46 (0.018) | 7.6 | 0.300 |
| | Semifinishing Finishing | 150 (500) | 300 (1000) | 365 (1200) | 0.20 (0.008) | 0.33 (0.013) | 0.41 (0.016) | 3.8 | 0.150 |
| 300 | Roughing | 50 (160) | 195 (640) | 245 (800) | 0.15 (0.006) | 0.36 (0.014) | 0.41 (0.016) | 5.1 | 0.200 |
| | Semifinishing Finishing | 120 (400) | 245 (800) | 300 (1000) | 0.15 (0.006) | 0.30 (0.012) | 0.38 (0.015) | 3.8 | 0.150 |
| 400 | Roughing | 45 (150) | 120 (400) | 213 (700) | 0.15 (0.006) | 0.30 (0.012) | 0.38 (0.015) | 3.8 | 0.150 |
| | Semifinishing Finishing | 60 (200) | 183 (600) | 245 (800) | 0.10 (0.004) | 0.25 (0.010) | 0.30 (0.012) | 2.5 | 0.100 |
| **Tool steels** | | | | | | | | | |
| 100 | Roughing | 183 (600) | 220 (720) | 275 (900) | 0.25 (0.010) | 0.30 (0.012) | 0.38 (0.015) | 5.1 | 0.200 |
| | Semifinishing Finishing | 183 (600) | 245 (800) | 275 (900) | 0.25 (0.010) | 0.30 (0.012) | 0.38 (0.015) | 2.5 | 0.100 |
| 200 | Roughing | 90 (300) | 165 (540) | 245 (800) | 0.10 (0.004) | 0.28 (0.011) | 0.33 (0.013) | 2.5 | 0.100 |
| | Semifinishing Finishing | 90 (300) | 183 (600) | 245 (800) | 0.10 (0.004) | 0.25 (0.010) | 0.30 (0.012) | 2.0 | 0.080 |
| 260 | Roughing | 90 (300) | 165 (540) | 245 (800) | 0.10 (0.004) | 0.23 (0.009) | 0.30 (0.012) | 2.0 | 0.080 |
| | Semifinishing Finishing | 90 (300) | 183 (600) | 245 (800) | 0.10 (0.004) | 0.23 (0.009) | 0.30 (0.012) | 2.0 | 0.080 |
| 300 | Roughing | 90 (300) | 165 (540) | 245 (800) | 0.10 (0.004) | 0.23 (0.009) | 0.30 (0.012) | 2.0 | 0.080 |
| | Semifinishing Finishing | 90 (300) | 183 (600) | 245 (800) | 0.10 (0.004) | 0.23 (0.009) | 0.30 (0.012) | 2.0 | 0.080 |
| **Alloy steels, 4000, 5000, 6000, 8000 and 9000 series** | | | | | | | | | |
| 150 | Roughing | 45 (150) | 205 (675) | 230 (750) | 0.23 (0.009) | 0.38 (0.015) | 0.46 (0.018) | 5.1 | 0.200 |
| | Semifinishing Finishing | 60 (200) | 275 (900) | 300 (1000) | 0.20 (0.008) | 0.33 (0.013) | 0.41 (0.016) | 3.8 | 0.150 |
| 250 | Roughing | 34 (110) | 183 (600) | 205 (675) | 0.23 (0.009) | 0.36 (0.014) | 0.43 (0.017) | 5.1 | 0.200 |
| | Semifinishing Finishing | 45 (150) | 245 (800) | 275 (900) | 0.20 (0.008) | 0.30 (0.012) | 0.41 (0.016) | 3.8 | 0.150 |
| 350 | Roughing | 34 (110) | 137 (450) | 183 (600) | 0.15 (0.006) | 0.30 (0.012) | 0.41 (0.016) | 4.7 | 0.187 |
| | Semifinishing Finishing | 45 (150) | 183 (600) | 245 (800) | 0.15 (0.006) | 0.28 (0.011) | 0.38 (0.015) | 3.8 | 0.150 |
| 400 | Roughing | 30 (100) | 120 (400) | 183 (600) | 0.15 (0.006) | 0.28 (0.011) | 0.38 (0.015) | 3.8 | 0.150 |
| | Semifinishing Finishing | 45 (150) | 183 (600) | 245 (800) | 0.10 (0.004) | 0.20 (0.008) | 0.25 (0.010) | 2.5 | 0.100 |
| **Stainless steels, 400 series (martensitic, ferritic)** | | | | | | | | | |
| 150 | Roughing | 98 (320) | 170 (560) | 220 (720) | 0.20 (0.008) | 0.33 (0.013) | 0.36 (0.014) | 5.6 | 0.220 |
| | Semifinishing Finishing | 120 (400) | 213 (700) | 275 (900) | 0.20 (0.008) | 0.30 (0.012) | 0.38 (0.015) | 5.1 | 0.200 |
| 200 | Roughing | 73 (240) | 146 (480) | 195 (640) | 0.15 (0.006) | 0.30 (0.012) | 0.33 (0.013) | 4.2 | 0.165 |
| | Semifinishing Finishing | 90 (300) | 183 (600) | 245 (800) | 0.15 (0.006) | 0.25 (0.010) | 0.33 (0.013) | 3.8 | 0.150 |
| 350 | Roughing | 50 (160) | 120 (400) | 170 (560) | 0.10 (0.004) | 0.25 (0.010) | 0.30 (0.012) | 2.8 | 0.110 |
| | Semifinishing Finishing | 60 (200) | 150 (500) | 213 (700) | 0.10 (0.004) | 0.23 (0.009) | 0.30 (0.012) | 2.5 | 0.100 |
| **Stainless steels, 200 and 300 series (austenitic)** | | | | | | | | | |
| 200 | Roughing | 75 (250) | 150 (500) | 183 (600) | 0.15 (0.006) | 0.25 (0.010) | 0.30 (0.012) | 2.8 | 0.110 |
| | Semifinishing Finishing | 90 (300) | 183 (600) | 230 (750) | 0.15 (0.006) | 0.25 (0.010) | 0.30 (0.012) | 2.5 | 0.100 |
| 250 | Roughing | 73 (240) | 137 (450) | 170 (560) | 0.10 (0.004) | 0.23 (0.009) | 0.28 (0.011) | 2.3 | 0.090 |
| | Semifinishing Finishing | 75 (250) | 167 (550) | 213 (700) | 0.10 (0.004) | 0.23 (0.009) | 0.28 (0.011) | 2.0 | 0.080 |
| 350 | Roughing | 50 (160) | 120 (400) | 146 (480) | 0.10 (0.004) | 0.20 (0.008) | 0.28 (0.011) | 2.3 | 0.090 |
| | Semifinishing Finishing | 60 (200) | 150 (500) | 183 (600) | 0.10 (0.004) | 0.20 (0.008) | 0.28 (0.011) | 2.0 | 0.080 |

nium carbonitride cermets are shown in Table 2.

**Cermets Compared With Cemented Carbides.** Titanium carbide and titanium carbonitride cermets are more wear resistant than cemented carbides (Fig. 4) and allow higher cutting speeds than tungsten carbides or coated carbides. The following additional comparisons with cemented tungsten carbides are helpful in the successful application of titanium carbide and titanium carbonitride cermets.

*The hardness* of titanium carbonitride cermets is approximately comparable to that of cemented carbides. The influence of temperature on hardness follows similar patterns because the metal binder system and its volume determine this behavior.

*The strength* of cemented carbides is about 15 to 25% higher than the strength of titanium carbonitride cermets. As a result, feed rates and depth of cut have to be selected more conservatively, specifically

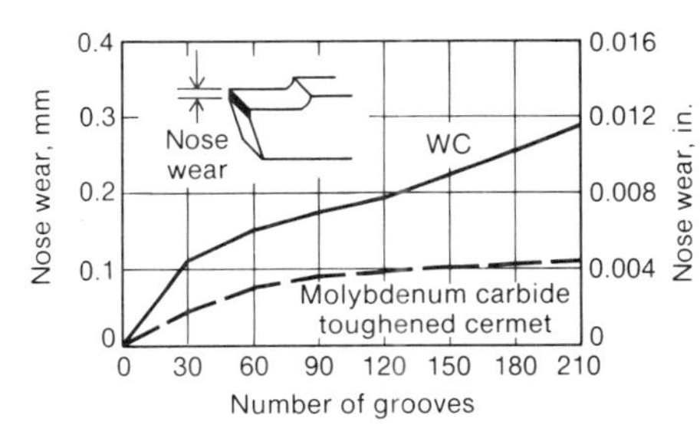


**Fig. 13** Wear comparison between cemented carbide and cermet cutting tools in grooving of 4135 alloy steel

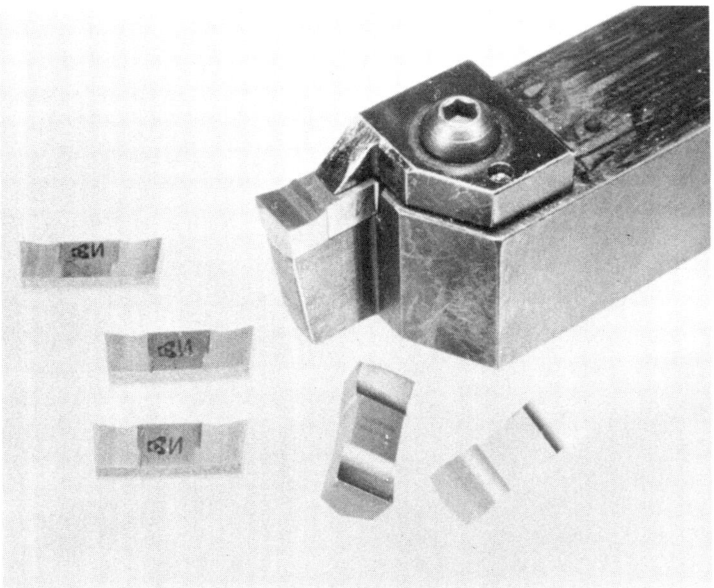

**Fig. 14** Toolholder with cermet indexable inserts for grooving

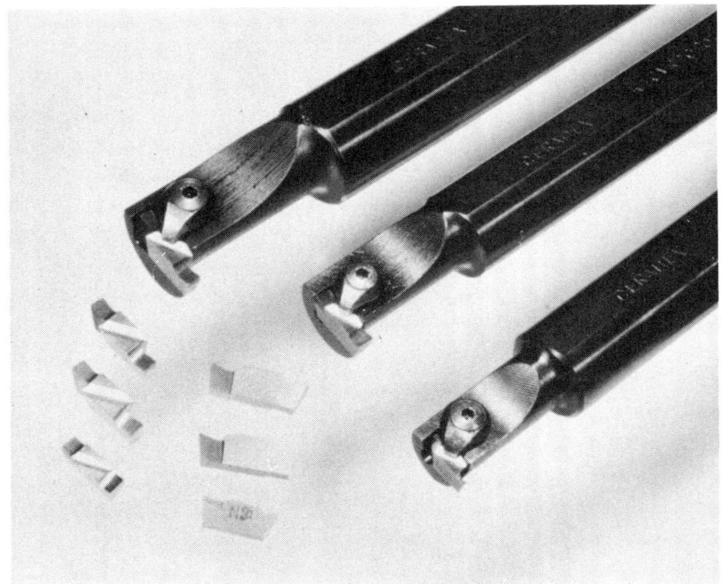

**Fig. 15** Boring bars with cermet indexable inserts for grooving

**Table 5 Machining parameters with cermet tools in the turning of cast iron**

| Hardness, HB | Operation | Cutting speed, m/min (sfm) Low | Median | High | Feed rate, mm/rev (in./rev) Low | Median | High | Depth of cut mm | in. |
|---|---|---|---|---|---|---|---|---|---|
| **Gray cast irons, malleable cast irons** | | | | | | | | | |
| 100 | Roughing | 90 (300) | 150 (500) | 260 (850) | 0.25 (0.010) | 0.41 (0.016) | 0.46 (0.018) | 6.3 | 0.250 |
| | Finishing | 150 (500) | 275 (900) | 365 (1200) | 0.23 (0.009) | 0.38 (0.015) | 0.43 (0.017) | 3.8 | 0.150 |
| 180 | Roughing | 60 (200) | 146 (480) | 245 (800) | 0.23 (0.009) | 0.36 (0.014) | 0.41 (0.016) | 4.7 | 0.187 |
| | Finishing | 150 (500) | 245 (800) | 300 (1000) | 0.20 (0.008) | 0.33 (0.013) | 0.38 (0.015) | 3.2 | 0.125 |
| 250 | Roughing | 60 (200) | 120 (400) | 170 (560) | 0.23 (0.009) | 0.33 (0.013) | 0.38 (0.015) | 4.7 | 0.187 |
| | Finishing | 60 (200) | 150 (500) | 213 (700) | 0.20 (0.008) | 0.30 (0.012) | 0.36 (0.014) | 3.2 | 0.125 |
| 320 | Roughing | 45 (150) | 110 (360) | 146 (480) | 0.18 (0.007) | 0.30 (0.012) | 0.33 (0.013) | 3.8 | 0.150 |
| | Finishing | 45 (150) | 137 (450) | 183 (600) | 0.15 (0.006) | 0.28 (0.011) | 0.30 (0.012) | 2.5 | 0.100 |
| **Nodular cast irons** | | | | | | | | | |
| 140 | Roughing | 75 (250) | 120 (400) | 213 (700) | 0.28 (0.011) | 0.33 (0.013) | 0.38 (0.015) | 6.3 | 0.250 |
| | Finishing | 120 (400) | 213 (700) | 300 (1000) | 0.20 (0.008) | 0.30 (0.012) | 0.38 (0.015) | 3.8 | 0.150 |
| 250 | Roughing | 160 (200) | 110 (360) | 167 (550) | 0.20 (0.008) | 0.28 (0.011) | 0.33 (0.013) | 4.7 | 0.187 |
| | Finishing | 120 (400) | 150 (500) | 245 (800) | 0.18 (0.007) | 0.25 (0.010) | 0.33 (0.013) | 3.2 | 0.125 |
| 320 | Roughing | 45 (150) | 98 (320) | 150 (500) | 0.18 (0.007) | 0.25 (0.010) | 0.30 (0.012) | 2.5 | 0.100 |
| | Finishing | 160 (200) | 120 (400) | 183 (600) | 0.15 (0.006) | 0.20 (0.008) | 0.25 (0.010) | 2.5 | 0.100 |
| 380 | Roughing | 45 (150) | 85 (280) | 137 (450) | 0.15 (0.006) | 0.23 (0.009) | 0.25 (0.010) | 2.5 | 0.100 |
| | Finishing | 160 (200) | 90 (300) | 120 (400) | 0.13 (0.005) | 0.15 (0.006) | 0.20 (0.008) | 2.5 | 0.100 |

**Table 6 Machining parameters with cermet tools in the turning of nonferrous free-machining metals**

| Hardness, HB | Operation | Cutting speed, m/min (sfm) Low | Median | High | Feed rate, mm/rev (in./rev) Low | Median | High | Maximum depth of cut mm | in. |
|---|---|---|---|---|---|---|---|---|---|
| **Nonferrous free-machining alloys (aluminum, brass, bronze, copper, magnesium, zinc)** | | | | | | | | | |
| 100 | Roughing | 440 (1450) | 490 (1600) | 730 (2400) | 0.23 (0.009) | 0.51 (0.020) | 0.56 (0.022) | 7.6 | 0.300 |
| | Semifinishing Finishing | 550 (1800) | 610 (2000) | 915 (3000) | 0.20 (0.008) | 0.46 (0.018) | 0.51 (0.020) | 6.3 | 0.250 |
| 120 | Roughing | 400 (1300) | 425 (1400) | 670 (2200) | 0.23 (0.009) | 0.46 (0.018) | 0.51 (0.020) | 7.6 | 0.300 |
| | Semifinishing Finishing | 490 (1600) | 550 (1800) | 850 (2800) | 0.20 (0.008) | 0.41 (0.016) | 0.46 (0.018) | 6.3 | 0.250 |

in roughing operations, in comparison to cemented carbides.

*The toughness* range of titanium carbonitride cermets is smaller than that of cemented carbides, limiting the use of these cermets in heavy roughing applications. However, the fracture toughness of TiC-base cermets may be in the same range as that of WC-Co, at equivalent binder volumes and carbide grain sizes.

*The temperature shock resistance* of titanium carbonitride cermets is lower than that of cemented carbides and restricts the use of coolant in roughing applications. Coolant is applicable in threading, grooving, and finish turning applications.

**Wear Resistance.** The wear mechanisms of a cutting edge are closely related to cutting temperature (Fig. 7). Low cutting temperatures produce pressure welding, which results in a built-up edge while high

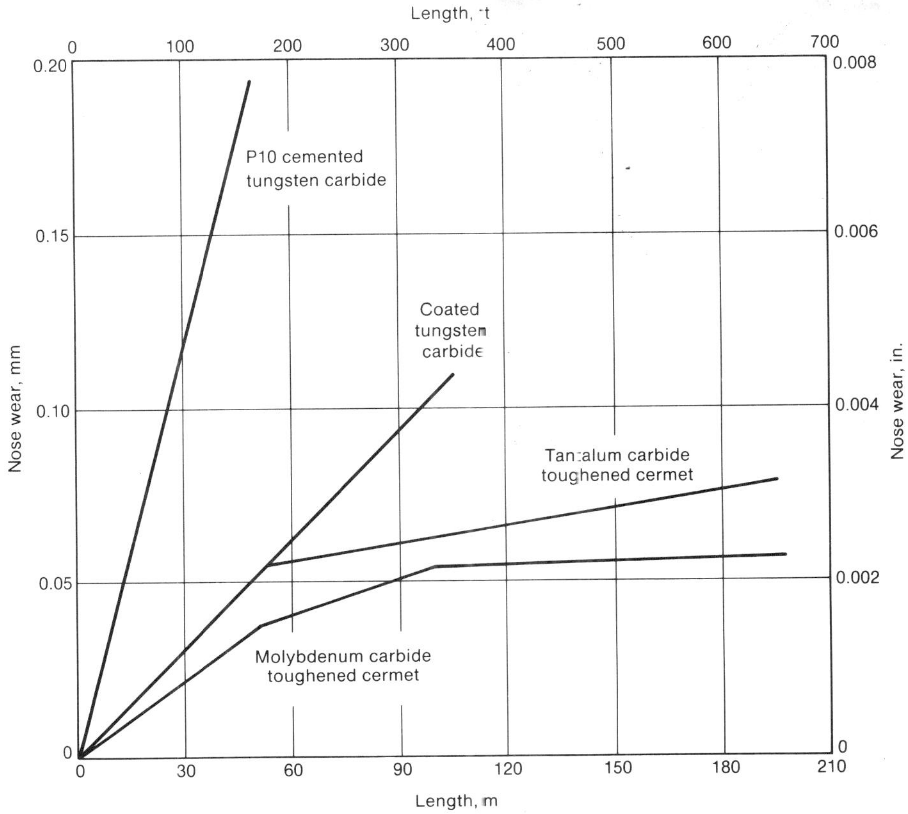


**Fig. 16** Wear comparison between cemented carbide, coated carbide, and cermet cutting tools in the threading of alloy steel. Machining parameters: cutting speed, 130 m/min (430 sfm); six passes with coolant. Workpiece: 4140 steel. Pitch: 6 threads per 25 mm (1 in.)

### Table 7 Machining parameters for the grooving of various steels with cermet tools

Feed rate, 0.05 to 0.13 mm/rev maximum (0.002 to 0.005 in./rev maximum)

| Hardness, HB | Cutting speed, median m/min | sfm |
|---|---|---|
| **Carbon steels** | | |
| 150 | 230 | 750 |
| 200 | 204 | 670 |
| 250 | 183 | 600 |
| 300 | 165 | 540 |
| 350 | 146 | 480 |
| **Alloy steels** | | |
| 150 | 213 | 700 |
| 250 | 174 | 570 |
| 350 | 146 | 480 |
| 400 | 120 | 400 |
| **Stainless steels, 400 series (martensitic, ferritic)** | | |
| 150 | 170 | 560 |
| 200 | 137 | 450 |
| 350 | 106 | 350 |
| **Stainless steels, 300 series (austenitic)** | | |
| 200 | 146 | 480 |
| 250 | 128 | 420 |
| 350 | 106 | 350 |
| **Tool steels** | | |
| 150 | 195 | 640 |
| 200 | 183 | 600 |
| 250 | 165 | 540 |
| 300 | 146 | 480 |

cutting temperatures enhance diffusion and oxidation processes. Diffusion processes between the chip and the top rake surface of the cutting edge result in crater wear, and oxidation reactions with the environment produce scaling of the cutting edge.

Oxidation, diffusion, and pressure welding are thermochemical processes which are roughly correlated to the free-formation enthalpy of the materials involved. Figure 8 shows the free-formation enthalpy of various cutting tool materials. Titanium nitride and titanium carbide, the base materials for cermets, considerably surpass tungsten carbide. Built-up edge, scaling, and crater wear are greatly reduced with cermet cutting tools, and a wider temperature range and higher cutting speeds are possible with cermet cutting tools because of their higher resistance to diffusion and oxidation processes.

*Molybdenum Content.* Abrasion resistance is affected by the molybdenum content in the nickel binder of titanium carbide cermets. Figure 9 shows the flank wear as a function of the molybdenum content in the nickel binder, and Fig. 10 compares the flank wear of a titanium carbide/molybdenum carbide solid solution material and titanium carbide with molybdenum and molybdenum carbide added to the nickel binder.

## Applications

Cermet cutting tools are suitable for the machining of steels, cast irons, cast steels, and nonferrous free-machining alloys, as detailed in Table 3. Cermet cutting tools are capable of operating at higher cutting speeds than cemented or coated carbides, thus allowing better surface finishes with cermet cutting tools. Some comparisons of surface finish are shown in Fig. 11 and Fig. 12.

**Cermet cutting tools** are generally used in the form of indexable inserts. Neither brazed tools nor solid cermet tools have reached any commercial significance. Cermet indexable inserts are manufactured by applying the basic process steps suitable for cemented carbide indexable inserts. The similarity of the manufacturing process results not only in manufacturing costs comparable to uncoated cemented carbides, but also allows the manufacture of inserts in shapes and sizes customary for cemented carbide cutting tools. Cutting tool manufacturers also provide various chip-breaker geometries.

**Turning.** The major reason for the application of cermet cutting tools in turning is their long and consistent tool life over a wide range of cutting speeds. When turning ductile materials, the proper choice of a

### Table 8 Cutting speeds for threading various steels with cermet tools

| Hardness, HB | Cutting speed m/min | sfm |
|---|---|---|
| **Carbon steels** | | |
| 150 | 200 | 650 |
| 200 | 183 | 600 |
| 250 | 174 | 570 |
| 300 | 165 | 540 |
| 350 | 150 | 500 |
| **Alloy steels** | | |
| 150 | 183 | 600 |
| 250 | 150 | 500 |
| 350 | 120 | 400 |
| 400 | 120 | 400 |
| **Stainless steels, 400 series (martensitic, ferritic)** | | |
| 150 | 143 | 470 |
| 200 | 137 | 450 |
| 350 | 120 | 400 |
| **Stainless steels, 300 series (austenitic)** | | |
| 200 | 120 | 400 |
| 250 | 113 | 370 |
| 350 | 107 | 350 |
| **Tool steels** | | |
| 150 | 165 | 540 |
| 200 | 150 | 500 |
| 250 | 137 | 450 |
| 300 | 120 | 400 |

## Table 9 In-feed recommendations for threading with cermet tools

| No. of threads | Corner radius, mm | Corner radius, in. | Total depth of cut, mm | Total depth of cut, in. | Number of passes | Depth of cut per pass, mm (in.)(a): 1 | 2 | 3 | 4 | 5 | 6 | 7 | 8 | 9 | 10 | 11 |
|---|---|---|---|---|---|---|---|---|---|---|---|---|---|---|---|---|
| **External threads per 25 mm (1 in.)** | | | | | | | | | | | | | | | | |
| 28 | 0.10 | 0.004 | 0.63 | 0.025 | 5 | 0.20 (0.008) | 0.18 (0.007) | 0.15 (0.006) | 0.07 (0.003) | 0.03 (0.001) | | | | | | |
| 24 | 0.13 | 0.005 | 0.71 | 0.028 | 6 | 0.20 (0.008) | 0.18 (0.007) | 0.15 (0.006) | 0.10 (0.004) | 0.05 (0.002) | 0.03 (0.001) | | | | | |
| 20 | 0.15 | 0.006 | 0.86 | 0.034 | 7 | 0.20 (0.008) | 0.18 (0.007) | 0.15 (0.006) | 0.15 (0.006) | 0.10 (0.004) | 0.05 (0.002) | 0.03 (0.001) | | | | |
| 18 | 0.18 | 0.007 | 0.94 | 0.037 | 8 | 0.20 (0.008) | 0.18 (0.007) | 0.15 (0.006) | 0.15 (0.006) | 0.10 (0.004) | 0.07 (0.003) | 0.05 (0.002) | 0.03 (0.001) | | | |
| 16 | 0.20 | 0.008 | 1.04 | 0.041 | 9 | 0.20 (0.008) | 0.18 (0.007) | 0.15 (0.006) | 0.13 (0.005) | 0.13 (0.005) | 0.10 (0.004) | 0.07 (0.003) | 0.05 (0.002) | 0.03 (0.001) | | |
| 14 | 0.23 | 0.009 | 1.19 | 0.047 | 10 | 0.23 (0.009) | 0.20 (0.008) | 0.15 (0.006) | 0.13 (0.005) | 0.13 (0.005) | 0.10 (0.004) | 0.10 (0.004) | 0.07 (0.003) | 0.05 (0.002) | 0.03 (0.001) | |
| 12 | 0.25 | 0.010 | 1.40 | 0.055 | 11 | 0.23 (0.009) | 0.20 (0.008) | 0.18 (0.007) | 0.15 (0.006) | 0.15 (0.006) | 0.13 (0.005) | 0.10 (0.004) | 0.10 (0.004) | 0.08 (0.003) | 0.05 (0.002) | 0.03 (0.001) |
| **Internal threads per 25 mm (1 in.)** | | | | | | | | | | | | | | | | |
| 20 | 0.08 | 0.003 | 0.78 | 0.031 | 7 | 0.18 (0.007) | 0.15 (0.006) | 0.13 (0.005) | 0.12 (0.005) | 0.10 (0.004) | 0.07 (0.003) | 0.03 (0.001) | | | | |
| 18 | 0.08 | 0.003 | 0.89 | 0.035 | 8 | 0.18 (0.007) | 0.15 (0.006) | 0.15 (0.006) | 0.13 (0.005) | 0.10 (0.004) | 0.10 (0.004) | 0.05 (0.002) | 0.03 (0.001) | | | |
| 16 | 0.10 | 0.004 | 0.97 | 0.038 | 9 | 0.18 (0.007) | 0.15 (0.006) | 0.15 (0.006) | 0.13 (0.005) | 0.10 (0.004) | 0.10 (0.004) | 0.08 (0.003) | 0.05 (0.002) | 0.03 (0.001) | | |
| 14 | 0.10 | 0.004 | 1.12 | 0.044 | 10 | 0.20 (0.008) | 0.18 (0.007) | 0.15 (0.006) | 0.13 (0.005) | 0.13 (0.005) | 0.10 (0.004) | 0.08 (0.003) | 0.07 (0.003) | 0.05 (0.002) | 0.03 (0.001) | |
| 12 | 0.13 | 0.005 | 1.27 | 0.050 | 11 | 0.20 (0.008) | 0.18 (0.007) | 0.18 (0.007) | 0.15 (0.006) | 0.13 (0.005) | 0.10 (0.004) | 0.10 (0.004) | 0.08 (0.003) | 0.07 (0.003) | 0.05 (0.002) | 0.03 (0.001) |

(a) Based on 4135 steel (180 to 220 HB). Harder workpiece or poor machine condition may require a decrease in depth of cut per pass.

## Table 10 Machining parameters for the milling of various steels with cermet tools

| Hardness, HB | Operation | Cutting speed, m/min (sfm): Low | Median | High | Feed rate, mm/rev (in./rev): Low | Median | High | Maximum depth of cut, mm | in. |
|---|---|---|---|---|---|---|---|---|---|
| **Carbon steels** | | | | | | | | | |
| 150–200 | Roughing | 106 (350) | 230 (750) | 275 (900) | 0.05 (0.002) | 0.15 (0.006) | 0.18 (0.007) | 7.6 | 0.300 |
| | Semifinishing Finishing | 150 (500) | 300 (1000) | 365 (1200) | 0.05 (0.002) | 0.10 (0.004) | 0.13 (0.005) | 6.3 | 0.250 |
| 200–250 | Roughing | 90 (300) | 183 (600) | 230 (750) | 0.05 (0.002) | 0.13 (0.005) | 0.15 (0.006) | 7.6 | 0.300 |
| | Semifinishing Finishing | 137 (450) | 275 (900) | 335 (1100) | 0.05 (0.002) | 0.10 (0.004) | 0.13 (0.005) | 5.1 | 0.200 |
| **Stainless steels** | | | | | | | | | |
| Austenitic 130–180 | Roughing | 75 (250) | 150 (500) | 183 (600) | 0.05 (0.002) | 0.15 (0.006) | 0.18 (0.007) | 7.6 | 0.300 |
| | Semifinishing Finishing | 75 (250) | 150 (500) | 183 (600) | 0.05 (0.002) | 0.10 (0.004) | 0.13 (0.005) | 5.1 | 0.200 |
| Ferritic 130–180 | Roughing | 90 (300) | 183 (600) | 230 (750) | 0.05 (0.002) | 0.13 (0.005) | 0.15 (0.006) | 7.6 | 0.300 |
| | Semifinishing Finishing | 90 (300) | 200 (660) | 245 (800) | 0.05 (0.002) | 0.10 (0.004) | 0.13 (0.005) | 5.1 | 0.200 |
| Martensitic 200–300 | Roughing | 60 (200) | 130 (430) | 150 (500) | 0.05 (0.002) | 0.10 (0.004) | 0.13 (0.005) | 7.6 | 0.300 |
| | Semifinishing Finishing | 60 (200) | 137 (450) | 183 (600) | 0.05 (0.002) | 0.08 (0.003) | 0.10 (0.004) | 5.1 | 0.200 |
| **Alloy steels** | | | | | | | | | |
| 200–250 | Roughing | 90 (300) | 183 (600) | 230 (750) | 0.05 (0.002) | 0.15 (0.006) | 0.18 (0.007) | 7.6 | 0.300 |
| | Semifinishing Finishing | 120 (400) | 245 (800) | 300 (1000) | 0.05 (0.002) | 0.08 (0.003) | 0.10 (0.004) | 5.1 | 0.200 |
| 250–350 | Roughing | 75 (250) | 140 (460) | 183 (600) | 0.05 (0.002) | 0.10 (0.004) | 0.13 (0.005) | 7.6 | 0.300 |
| | Semifinishing Finishing | 90 (300) | 183 (600) | 230 (750) | 0.05 (0.002) | 0.08 (0.003) | 0.10 (0.004) | 5.1 | 0.200 |
| 350–400 | Roughing | 55 (180) | 106 (350) | 150 (500) | 0.05 (0.002) | 0.10 (0.004) | 0.18 (0.005) | 6.3 | 0.250 |
| | Semifinishing Finishing | 75 (250) | 137 (450) | 150 (500) | 0.05 (0.002) | 0.08 (0.003) | 0.10 (0.004) | 3.8 | 0.150 |
| **Cast steels** | | | | | | | | | |
| 100–150 | Roughing | 106 (350) | 213 (700) | 275 (900) | 0.05 (0.002) | 0.13 (0.005) | 0.18 (0.007) | 7.6 | 0.300 |
| | Semifinishing Finishing | 150 (500) | 300 (1000) | 365 (1200) | 0.05 (0.002) | 0.10 (0.004) | 0.13 (0.005) | 5.1 | 0.200 |
| 150–200 | Roughing | 90 (300) | 183 (600) | 245 (800) | 0.05 (0.002) | 0.15 (0.006) | 0.18 (0.007) | 7.6 | 0.300 |
| | Semifinishing Finishing | 120 (400) | 250 (820) | 300 (1000) | 0.05 (0.002) | 0.10 (0.004) | 0.13 (0.005) | 5.1 | 0.200 |
| 200–250 | Roughing | 60 (200) | 120 (400) | 183 (600) | 0.05 (0.002) | 0.13 (0.005) | 0.15 (0.006) | 7.6 | 0.300 |
| | Semifinishing Finishing | 75 (250) | 150 (500) | 183 (600) | 0.05 (0.002) | 0.10 (0.004) | 0.13 (0.005) | 5.1 | 0.200 |

**Table 11 Machining parameters for the milling of cast irons with cermet tools**

| Hardness, HB | Operation | Cutting speed, m/min (sfm) Low | Median | High | Feed rate, mm/rev (in./rev) Low | Median | High | Maximum depth of cut mm | in. |
|---|---|---|---|---|---|---|---|---|---|
| **Gray cast irons** | | | | | | | | | |
| 150 | Roughing | 60 (200) | 150 (500) | 245 (800) | 0.05 (0.002) | 0.13 (0.005) | 0.18 (0.007) | 7.6 | 0.300 |
| | Finishing | 150 (500) | 245 (800) | 300 (1000) | 0.05 (0.002) | 0.10 (0.004) | 0.13 (0.005) | 5.1 | 0.200 |
| 250 | Roughing | 60 (200) | 120 (400) | 170 (560) | 0.05 (0.002) | 0.13 (0.005) | 0.15 (0.006) | 7.6 | 0.300 |
| | Finishing | 60 (200) | 150 (500) | 213 (700) | 0.05 (0.002) | 0.10 (0.004) | 0.13 (0.005) | 5.1 | 0.200 |
| **Malleable cast irons** | | | | | | | | | |
| 150 | Roughing | 60 (200) | 150 (500) | 245 (800) | 0.05 (0.002) | 0.13 (0.005) | 0.18 (0.007) | 7.6 | 0.300 |
| | Finishing | 150 (500) | 245 (800) | 300 (1000) | 0.05 (0.002) | 0.10 (0.004) | 0.13 (0.005) | 5.1 | 0.200 |
| 250 | Roughing | 60 (200) | 120 (400) | 170 (560) | 0.05 (0.002) | 0.13 (0.005) | 0.15 (0.006) | 7.6 | 0.300 |
| | Finishing | 60 (200) | 150 (500) | 213 (700) | 0.05 (0.002) | 0.10 (0.004) | 0.13 (0.005) | 5.1 | 0.200 |
| **Nodular cast irons** | | | | | | | | | |
| 150 | Roughing | 75 (250) | 120 (400) | 213 (700) | 0.05 (0.002) | 0.13 (0.005) | 0.18 (0.007) | 7.6 | 0.300 |
| | Finishing | 120 (400) | 213 (700) | 300 (1000) | 0.05 (0.002) | 0.10 (0.004) | 0.13 (0.005) | 5.1 | 0.200 |
| 250 | Roughing | 60 (200) | 110 (360) | 168 (550) | 0.05 (0.002) | 0.13 (0.005) | 0.15 (0.006) | 7.6 | 0.300 |
| | Finishing | 90 (300) | 150 (500) | 245 (800) | 0.05 (0.002) | 0.10 (0.004) | 0.13 (0.005) | 5.1 | 0.200 |

chip-breaker geometry is essential for a safe and reliable turning operation. Tables 4 to 6 present general machining parameters for turning various metals with cermet cutting tools.

**Grooving** operations are associated with high heat and cutting pressure at the cutting edge. Cermet cutting tools, because of their high resistance to plastic deformation and their thermochemical stability, offer long tool life in comparison to cemented carbides. Typical wear characteristics are shown in Fig. 13.

Cermet indexable inserts are applied in grooving operations to achieve close dimensional control and good surface finish of the machined surfaces. Figures 14 and 15 show cermet indexable inserts mounted on a toolholder and boring bar for grooving. Table 7 contains guidelines for the median cutting speed and feed rate used when various materials are grooved with cermet cutting tools. The use of coolant improves surface finish and tool life.

**Threading.** The development of tougher cermet cutting materials resulted in the economical application of cermet cutting tools in single-point thread turning. Threading requires tougher cutting materials because of the small tool nose radii of most standard thread forms and because of the rapid change of cutting forces at the entry and exit of the cut.

The application of cermet indexable inserts in single-point turning extends tool life, in comparison to carbides and coated carbides (Fig. 16) and allows higher cutting speeds. Table 8 contains guidelines for the cutting speed in single-point turning of threads with cermet cutting tools, and Table 9 gives guidelines on the depth of cut and number of passes. The first pass normally takes a deeper cut, and on each subsequent pass the depth of cut is reduced, as shown in Table 9. Coolant is applicable and results in improved surface finish and longer tool life.

**Milling.** The development of tougher titanium carbonitride compositions widened the application range of cermets in milling. The modern molybdenum carbide and tantalum/niobium carbide grades can endure the mechanical and temperature shock cycling and the variations in chip load associated with milling.

Cermet cutting tools are applied when consistent tool life and good surface finish are required, and the cutting tool industry provides various styles of cermet indexable inserts for the most common milling cutter designs. Tables 10 and 11 contain general guidelines for milling various steels and cast irons with cermet cutting tools.

## REFERENCES


1. E.C. Van Schoick, Ed., *Ceramic Glossary*, The Ceramic Society, 1963
2. R. Kieffer and F. Benesovsky, Hartmetalle, Springer, 1965, p 8, 14, 250
3. M. Humenik and D. Moskowitz, U.S. Patent 2,967,349, June 1959
4. A.H. Heuer, J.S. Sears, and N.J. Zalucec, Analytical Electron Microscopy of Phase Separated Ti/Mo Cemented Carbides and Carbonitrides, chapter 4 in *Institute of Physics Conference Series No. 75*, Adam Hilger, 1986
5. D. Moskowitz and M. Humenik, Jr., *Modern Developments in Powder Metallurgy*, Vol 3, Plenum Press, 1966
6. E. Rudy, U.S. Patent 3,971,656, May 1974
7. D. Moskowitz, L.L. Terner, and M. Humenik, chapter 7 in Some Physical and Metalcutting Properties of Titanium Carbonitride Base Materials, *Institute of Physics Conference Series No. 75*, Adam Hilger, 1986
8. H. Doi, Advanced TiC and TiC-TiN Base Cermets, chapter 6 in *Institute of Physics Conference Series No. 75*, Adam Hilger, 1986

# Ceramics

Ranga Komanduri and Shyam K. Samanta, National Science Foundation

CERAMICS are the newest class of tool materials with potential for a wide range of high-speed finishing operations and for the high removal rate machining of difficult-to-machine materials. The development of current ceramic tool materials was partially based on the advances in high-temperature structural ceramic materials and processing technology developed in the early 1970s for automotive gas turbine and other high-temperature structural applications.

Ceramics are inorganic, nonmetallic materials that are subjected to high temperature during synthesis or use (Ref 1). The use of ceramic tools in metal cutting, after a long period of slow growth, is beginning to increase with the advent of alloyed ceramics and ceramic-matrix composites, as well as with the advances in ceramic processing technology.

## Production Process

The production of ceramic tools involves the consolidation and sintering of powdered material. The sintering provides the necessary densification of the consolidated powder and can be performed with or without the assist of pressure. In pressureless sintering, the powder is first shaped into a green (unsintered) body, which is then sintered to achieve the necessary densification. In hot pressing, the shaping is performed during sintering.

**Hot pressing** involves heating the powder in a die, along with the simultaneous application of a high uniaxial pressure. Although hot-pressed materials are more expensive, they usually have a finer grain size and a higher density and transverse rupture strength than cold-pressed products.

**Hot isostatic pressing** is used to reduce the size of closed pores in high-performance ceramics. This process exposes sintered or unsintered products to a hot pressurized gas. The hot isostatic pressing of unsintered parts requires a gastight container for transmitting the gas pressure to the porous part.

## Compositions

The ceramics currently used in metal cutting are based on either alumina ($Al_2O_3$) or silicon nitride ($Si_3N_4$). Other ceramics (such as magnesia, yttria, zirconia, chromium oxide, and titanium carbide) are used as additives to aid sintering or to form composite ceramics with improved thermo-mechanical properties.

### Alumina-Base Tool Materials

Alumina was considered for certain machining applications as early as 1905, and patents based on this technology were issued in England and in Germany around 1912. However, the strength and toughness of these ceramic tools were inadequate for commercial applications. During World War II, because of the strategic value of tungsten and the potential for increased machining rates, the Germans reconsidered ceramics as cutting tool materials (Ref 2). The pioneering work of Ryschkewitsch on pure oxide ceramics lead to Degussit, the first commercially produced $Al_2O_3$ ceramic tool (Ref 3). A similar effort at the Institute of Chemical Technology in Moscow lead to the development of an $Al_2O_3$ ceramic, Microlith, in 1943. In the United States, the work in this area began as early as 1935, but no progress was made until 1945, after extensive testing of Degussit.

In the 1960s, the development of hot pressing and hot isostatic pressing resulted in stronger and more reliable parts. Ceramic cutting tools were based on sintered or hot-pressed polycrystalline $\alpha$-$Al_2O_3$ (white ceramic) with a variety of sintering aids and compositions (Ref 4). These tools were primarily used for the high-speed finish machining of cast iron and steel for the automotive industry or for the slow-speed machining of extremely hard (and difficult-to-machine) cast or forged alloy steels in the steel roll industry. These ceramics were basically fine-grain (<5 μm) $Al_2O_3$-base materials with magnesia as a sintering aid and grain growth inhibitor. They were also alloyed with suboxides of titanium or chromium to form solid solutions. For example, in 1960, General Electric Company developed an $Al_2O_3$-TiO ceramic made by cold pressing and liquid-phase sintering (Ref 5). This ceramic is characterized by a reasonably fine grain size (3 to 5 μm) and a uniform microstructure. The TiO constitutes about 10%. A hardness of 93 to 94 HRA and a transverse rupture strength of about 550 MPa (80 ksi) were achieved. The low strength and toughness of these $Al_2O_3$-base ceramics limited their use to high-speed finish machining or to the high removal rate machining of chill cast iron rolls on extremely rigid, high-power (<450 kW, or 600 hp), high-precision lathes (Ref 6).

Since the 1960s, the development of alloyed ceramics and ceramic-matrix composites has resulted in ceramic tool materials with improved thermo-mechanical properties. The three main $Al_2O_3$-base tool materials are $Al_2O_3$-TiC, $Al_2O_3$-$ZrO_2$, and $Al_2O_3$ reinforced with silicon carbide (SiC) whiskers. Other $Al_2O_3$-base ceramics have additives of TiN, $TiB_2$, Ti(C,N), and Zr(C,N).

**Alumina and Titanium Carbide.** Improvements in the thermo-mechanical properties occurred in the late 1960s when Japanese researchers added titanium carbide (TiC) to an $Al_2O_3$ matrix. This alloyed ceramic is a dispersion-strengthened ceramic that contains 25 to 40 vol% TiC as a dispersed particulate phase. Alumina-titanium carbide is often called a black ceramic composite due to its color, which results from the presence of titanium carbide. Hot pressing is the typical production process.

Figure 1 shows an electron micrograph of hot-pressed $Al_2O_3$, and Fig. 2 shows an optical micrograph of a hot-pressed $Al_2O_3$-TiC ceramic. The addition of TiC greatly increases thermal conductivity, presumably through the formation of a more conductive intergranular phase (Ref 7). The $Al_2O_3$-TiC ceramics also exhibit a higher hardness and toughness than single-phase $Al_2O_3$ (Ref 8-10).

**Alumina-zirconia** ($Al_2O_3$-$ZrO_2$) is an alloyed ceramic with higher fracture toughness and thermal shock resistance than monolithic $Al_2O_3$. However, the degree of toughening of $Al_2O_3$-$ZrO_2$ ceramics decreases with increasing temperature (Ref 11, 12). The hardness of the ceramic is also reduced with increasing $ZrO_2$.

The toughening of $Al_2O_3$ with $ZrO_2$ exploits a specific crystallographic change (martensitic-type transformation) that is an energy-absorbing mechanism. The presence of metastable tetragonal $ZrO_2$ provides the potential for transformation under stress into a stable monoclinic structure. The

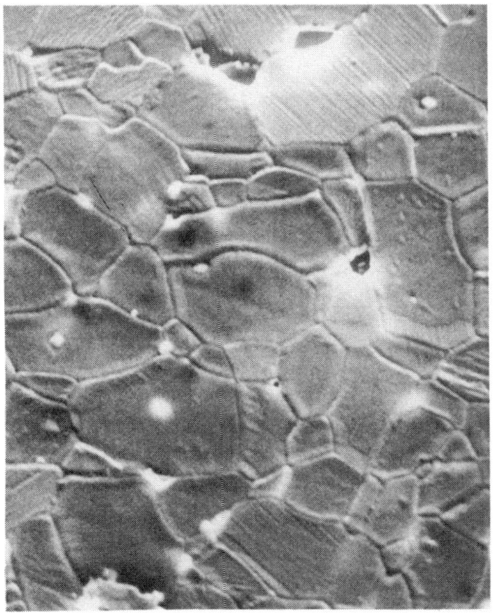

**Fig. 1** Scanning electron micrograph of hot-pressed $Al_2O_3$. Polished and etched. 5000×. Courtesy of Kennametal Inc.

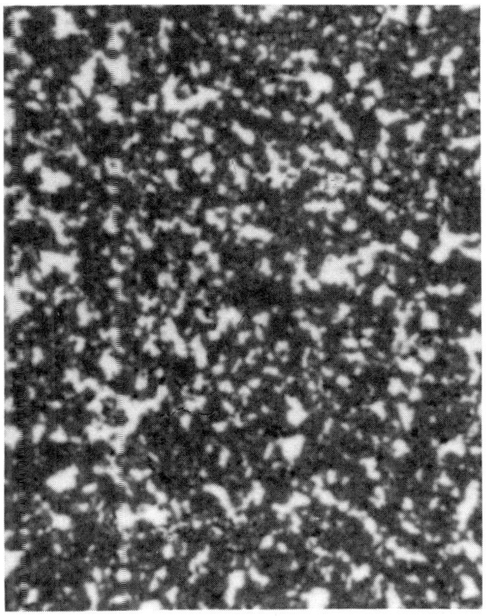

**Fig. 2** Micrograph of a hot-pressed $Al_2O_3$-TiC ceramic tool material. 1500×. Courtesy of Kennametal Inc.

**Fig. 3** Micrograph of a fracture surface of an alumina-zirconia ceramic ($Al_2O_3$ + 8% $ZrO_2$) showing the concentration of zirconia particles (the bright edges) at the alumina grain boundaries. 3000×. Courtesy of Carboloy Inc.

transformation acts as a stress absorber and prevents, even when cracks exist, further cracking by the twinning of the monoclinic phase.

Figure 3 is a micrograph of a fracture surface of an alumina-zirconia alloy. The zirconia particles are concentrated predominantly at the alumina grain boundaries. Although the fracture is intergranular, the presence of these particles is believed to provide additional toughness before failure can occur by fracture.

An $Al_2O_3$-$ZrO_2$ ceramic with traces of tungsten carbide was introduced by the Carboloy Systems Department of General Electric in the early 1980s. The high toughness of this alloy is attributed to rapid freezing from the melt, which results in a dendritic freezing structure and superior grinding performance. The three popular compositions contain 10, 25, and 40% $ZrO_2$; the remainder is $Al_2O_3$. The 40% $ZrO_2$ composition is close to the eutectic. The higher-$ZrO_2$ compositions are less hard, but tougher.

**Silicon carbide whisker reinforced alumina** ($Al_2O_3$-$SiC_w$) is the newest $Al_2O_3$-base tool material. The incorporation of SiC whiskers (25 to 45 vol%) into an $Al_2O_3$ matrix with subsequent hot pressing results in a composite with significantly improved toughness. Whisker reinforcement produces a twofold increase in fracture toughness relative to monolithic $Al_2O_3$ (Ref 13, 14). Figure 4 shows the microstructure of SiC whisker reinforced $Al_2O_3$.

The whiskers, which are small fibers of single-crystal SiC about 0.5 to 1 μm (20 to 40 μin.) in diameter and 10 to 80 μm (400 to 3200 μin.) long, have a higher thermal conductivity and a lower coefficient of thermal expansion than $Al_2O_3$. This improves thermal shock resistance.

The SiC whiskers in the alumina matrix also improve fracture toughness. Although the details of the micromechanisms for improved fracture toughness of $Al_2O_3$-SiC whiskers have not been clearly established, a plausible mechanism for the toughness is based on current knowledge of the general behavior of composite materials. As a crack propagates through the ceramic matrix, bonds between the $Al_2O_3$ matrix and the whiskers are broken. However, the SiC whiskers, because of their inherently high tensile strength (7 GPa, or 1 × $10^6$ psi), remain essentially intact. Consequently, whisker pullout occurs as a result of the separation of the matrix from the whiskers. Interfacial shear stress resisting the whisker pullout absorbs a substantial amount of this fracture energy and inhibits crack propagation. For improved fracture toughness, a strong metallurgical bond between the fibers and the matrix is not desirable; a strong bond would cause the whiskers to fail along with the matrix. This is the case with SiC whisker reinforced $Al_2O_3$ because the bond between the matrix and the fibers is not particularly strong. For optimum results, it is recommended that the fracture energy of

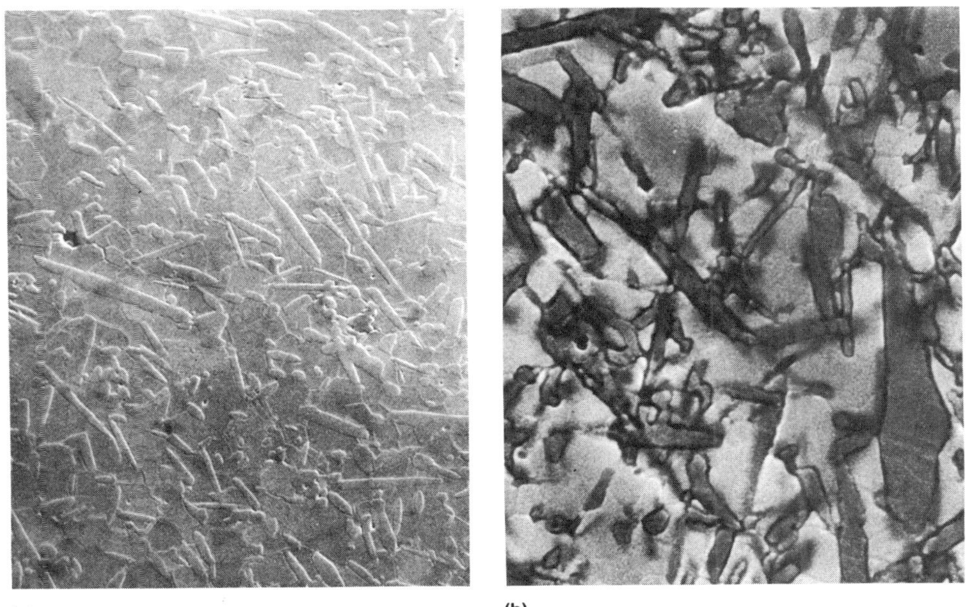

(a) (b)

**Fig. 4** Microstructure of SiC whisker reinforced $Al_2O_3$ composite tool material. (a) 2100×. (b) 5000×. Courtesy of Kennametal Inc.

**Table 1 Sintering additives for $Si_3N_4$**

| Additives | Chemical behavior | Name |
|---|---|---|
| **Systems with extended solid-solution formation** | | |
| $Al_2O_3$, AlN, $SiO_2$ | $Si_{6-x}Al_xO_xN_{8-x}$ | β-SiAlON |
| BeO,$Be_3N_2$, $SiO_2$ | $Si_{6-x}Be_xO_{2x}N_{8-2x}$ | β-SiBeON |
| $Al_2O_3$, AlN, $SiO_2$, BeO, $Be_3N_2$ | $Si_{6-x-y}Be_xAl_yO_{2x+y}$ | β-SiAlBeOn |
| $Al_2O_3$, AlN, metal oxide, metal nitride, M = Li, Ca, Mg, Y | $M_x(Si, Al)_{12}(O,N)_{16}$ | α-SiAlON |
| **Systems without extended solid-solution formation** | | |
| MgO or MgO, $Mg_3N_2$, $SiO_2$ | No quaternary phases | SiMgON |
| $ZrO_2$ | No quaternary phases | SiZrON |
| $Y_2O_3$ or $Y_2O_3$, $SiO_2$ | Quaternary phases | SiYON |
| $Ce_2O_3$ or $CeO_2$ | Quaternary phases | SiCeON |
| **Mixed systems** | | |
| $Al_2O_3$, AlN, $SiO_2$, MgO | β-SiAlON + MgO | SiAlMgON |
| $Al_2O_3$, AlN, $SiO_2$, $Y_2O_3$ | β-SiAlON + $Y_2O_3$ | SiAlYON |
| $Al_2O_3$, AlN, $SiO_2$, $ZrO_2$ | β-SiAlON + $ZrO_2$ | SiAlZrON |

Source: Ref 15

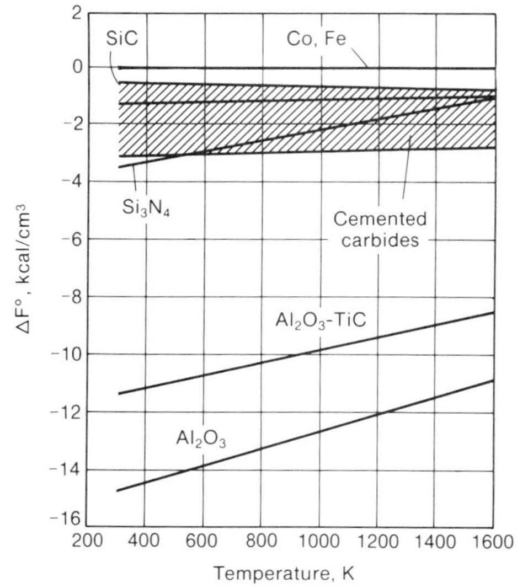


**Fig. 5** Standard free energies of formation (ΔF°) of $Al_2O_3$, $Al_2O_3$-TiC, $Si_3N_4$, SiC, Co, Fe, and cemented carbides at various temperatures. Source: Ref 16

the interface not exceed 10% of the fracture energy of the $Al_2O_3$ matrix.

## Silicon Nitride Base Tool Materials

Silicon nitride recently has attracted much attention as a tool material because of its unique combination of excellent high-temperature mechanical properties and resistance to oxidation and thermal shock. Both transverse rupture strength and fracture toughness of $Si_3N_4$ are higher than for $Al_2O_3$-TiC. Moreover, because poor thermal-shock resistance is a major cause of failure of $Al_2O_3$-TiC tools, the difference in thermal properties is significant. The thermal conductivity of $Si_3N_4$ is approximately double that of $Al_2O_3$-TiC, while the coefficient of thermal expansion is around one-half that of $Al_2O_3$-TiC. These thermal properties result in a much lower sensitivity to temperature changes and improved thermal-shock resistance.

The difficulty of utilizing these properties arises from the problems involved in sintering pure $Si_3N_4$. In $Si_3N_4$, the volume diffusivity is not large enough to offset the densification retardation effects of surface diffusion and volatilization phenomena; as a result, sintering is difficult. The covalent bond is believed to be the reason for such a low volume diffusivity. Therefore, dense $Si_3N_4$-base material can be obtained only by alloying $Si_3N_4$ with additives that promote sintering. The properties of these $Si_3N_4$ alloys, especially at high temperatures, are to a large extent controlled by the additives. As a result, there is not just a single $Si_3N_4$ material, because the properties of $Si_3N_4$ alloys depend on the additives.

Various metal oxides and nitrides have been found to be effective sintering aids. The sintering additives can be divided into three classes. The first class exhibits extended solid-solution formation with β-$Si_3N_4$ or α-$Si_3N_4$; the second class does not. The third class is made up of combinations of the additives from the first and second class and yields a mixed behavior (Table 1). Consequently, by using soluble additives, it is possible to manufacture single-phase alloys; using insoluble additives necessarily yields multiple-phase alloys with either crystalline or amorphous phases.

The most commonly used solid-solution additives to $Si_3N_4$ are $Al_2O_3$, aluminum nitride (AlN), and silica ($SiO_2$). The densification of $Si_3N_4$-base ceramics is obtained by liquid-phase sintering. Dense material of β-$Si_3N_4$ is fabricated by hot pressing α-$Si_3N_4$ powder with a suitable densification additive at 30 MPa (4 ksi) in a graphite die at 1700 to 1800 °C (3100 to 3300 °F). The $Si_3N_4$ powder is generally covered with a thin layer of $SiO_2$. The purpose of the additive is to react with this $SiO_2$ and a small amount of $Si_3N_4$ at the high hot-pressing temperature to form an oxynitride liquid phase in which α-$Si_3N_4$ dissolves and from which β-$Si_3N_4$ is precipitated.

**SiAlON** generally refers to a system made up of $Al_2O_3$ and $Si_3N_4$. The β-SiAlONs are derived from the structure of β-$Si_3N_4$ (hexagonal crystal structure with *ABAB* stacking) having compositions of $Si_{6-x}Al_xO_xN_{8-x}$ ($O \leqq x \leqq 4.2$). The α-SiAlONs are derived from the α-$Si_3N_4$ structure (hexagonal crystal structure with *ABCDABCD* stacking). The β solid-solution formation is purely substitutional; silicon is replaced by aluminum and nitrogen by oxygen. The mechanism of densification of SiAlON is a transient liquid-phase sintering.

Alumina is a desirable additive for cutting tool applications because its chemical inertness limits wear (such as crater wear) by chemical reactions. This benefit associated with $Al_2O_3$ can be demonstrated by considering the free energy of formation, which is a good indicator of chemical inertness. Figure 5 compares the free energy of formation as a function of temperature for various materials. The more negative the free energy, the more chemically inert or corrosion resistant the material should be. In Fig. 5, the free energy of $Si_3N_4$ falls within the range of cemented carbides, while $Al_2O_3$ has the most negative free energy. Therefore, an $Al_2O_3$ additive should improve the wear resistance by making the resultant alloy more chemically inert.

A mixture of two different classes of oxide additives is sometimes used for the densification of $Si_3N_4$. This is done to densify the material at a lower temperature and to yield more refractory-type secondary phases. Alumina is used most often, so the systems studied usually include a mixed alloy of β-SiAlON, such as SiAlMgON or SiAlYON.

The SiAlZrON system also gains interest because of the possibility of the martensitic transformation toughening of $Si_3N_4$ by $ZrO_2$, which has been demonstrated in $Al_2O_3$ ceramics. If only $ZrO_2$ is added to $Si_3N_4$, an undesirable chemical reaction without densification occurs (Ref 17). Further additions of $Al_2O_3$ and AlN yield a dense material consisting of β-SiAlON and $ZrO_2$ (Ref 18).

If yttria ($Y_2O_3$) is used, the material is $Y_2O_3$-stabilized silicon aluminum oxynitride (SiAlYON), which is isostructural with β-$Si_3N_4$. Because of the similarity in crystal structure, β′-SiAlON has physical and mechanical properties similar to those of $Si_3N_4$, and because of its $Al_2O_3$ additive, β′-SiAlON has additional chemical inertness.

To produce β′-SiAlON, a mixture of $Al_2O_3$ (~13%), $Si_3N_4$ (~77%), $Y_2O_3$ (~10%), and AlN is used as the starting material. During sintering, this mixture produces a larger volume of lower-viscosity liquid than in the synthesis of $Y_2O_3$-stabilized $Si_3N_4$ (and its surface silica); therefore, β′-SiAlON can be fully densified by pressureless sintering. The powder mix for β′-SiAlON is first ball milled, then preformed by cold isostatic pressing, and sub-

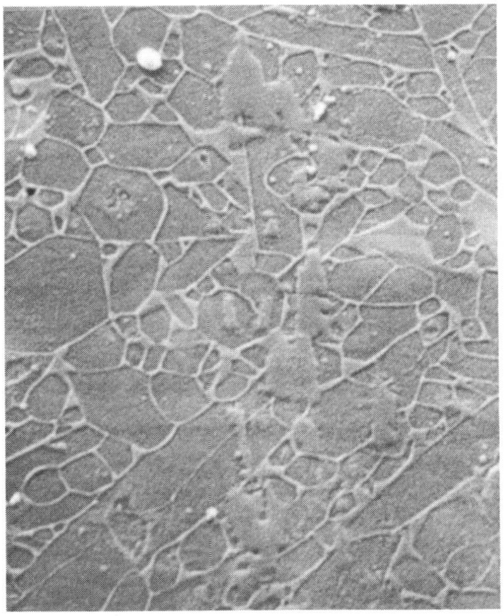

**Fig. 6** Scanning electron micrograph of a SiAlON ceramic tool. 10 000×. Courtesy of Kennametal Inc.

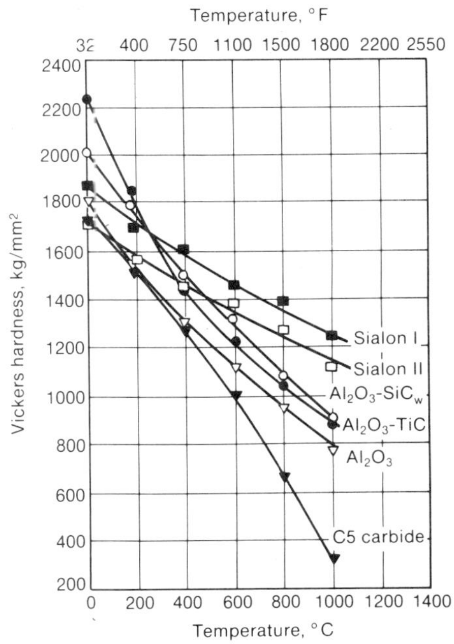


**Fig. 7** Vickers hardness versus temperature. 1 kg load, 10 s indent time. Sialon I and Sialon II are product names of Kennametal Inc. Sialon I is an $\alpha'$/$\beta'$-SiAlYON, and Sialon II is a $\beta'$-SiAlYON. Source: Ref 19

sequently sintered at a maximum temperature of 1800 °C (3300 °F) under isothermal conditions for approximately 1 h before it is allowed to cool slowly. The microstructure of $\beta'$-SiAlON consists of $\beta'$ grains cemented by a glass phase. A scanning electron micrograph of a polished and etched sample is shown in Fig. 6.

Rapid cooling from the processing temperature produces a microstructure of $\beta$-SiAlON grains with an intergranular glassy phase. If $Y_2O_3$ is the sintering aid, a portion of this glassy phase can be converted to crystalline yttrium-aluminum-garnet (YAG) by heat treating and slow cooling. However, most sintered SiAlONs contain some residual glass phase, particularly at grain-boundary triple-points. Like the $Si_3N_4$ sintering aid systems, the properties of the SiAlONs are dependent on the type and amount of sintering aid employed and the processing route followed during part fabrication.

## Properties

Ceramics provide a desirable tool material because their good hot hardness (Fig. 7) and oxidation resistance reduce the amount of tool wear at high cutting temperatures. These properties allow ceramic tools to be used in the high-speed machining (>300 m/min, or 1000 sfm) of even difficult-to-machine metals.

The primary difficulty, however, is that ceramics have a lower fracture toughness and transverse rupture strength than carbides (Table 2). Ceramic tools also have a lower thermal shock resistance, which depends on the fracture toughness ($K_{Ic}$), Young's modulus ($E$), the coefficient of thermal expansion ($\alpha$), and thermal conductivity ($k$). Figure 8 compares some ceramic and carbide tools in terms of two figures-of-merit for thermal shock resistance.

The importance of these properties depends on the machining conditions (Fig. 9). Strength and toughness are important during interrupted cuts and at high feed rates and depths of cut, while hot hardness and abrasion resistance are primarily important as cutting speeds increase. Thermal shock resistance is desired when high mechanical stresses, high cutting speeds, or the use of cutting fluids are involved.

Chemical inertness is also an important property at high cutting speeds. The chemical inertness of ceramics, as discussed earlier, is associated with the free energy of formation (Fig. 5). Alumina is the most chemically inert, while silicon nitride and silicon carbide are more reactive with cobalt and iron. The reactivity of SiC with ferrous materials appears to be the reason why SiC whisker reinforced $Al_2O_3$ is limited to applications with nickel-base superalloy workpieces.

**Table 2 Room-temperature properties of ceramic and tungsten carbide tool materials**

| Tool material | Transverse rupture strength, MPa | Transverse rupture strength, ksi | Hardness, HRA | Fracture toughness, $MPa\sqrt{m}$ | Fracture toughness, $ksi\sqrt{in.}$ |
|---|---|---|---|---|---|
| $Al_2O_3$ | 500–700 | 70–100 | 93–94 | 3.5–4.5 | 3.2–4.1 |
| $Al_2O_3$-$ZrO_2$ | 700–900 | 100–130 | 93–94 | 5.0–8.0 | 4.5–7.3 |
| $Al_2O_3$-TiC | 600–850 | 90–120 | 94–95 | 3.5–4.5 | 3.2–4.1 |
| $Al_2O_3$-$SiC_w$ | 550–750 | 80–110 | 94–95 | 4.5–8.0 | 4.1–7.3 |
| $Si_3N_4$ | 700–1050 | 100–150 | 92–94 | 6.0–8.5 | 5.5–7.7 |
| SiAlON | 700–900 | 100–125 | 93–95 | 4.5–6.0 | 4.1–5.5 |
| WC-Co alloys | 1250–2100 | 180–300 | 91–93 | 10.0–13.5 | 9.1–11.4 |

Source: Ref 20

## Tool Geometries

Because ceramics are relatively brittle, special cutting edge preparation is required to prevent chipping or edge breakage. This consists of honing or grinding a 0.05 to 0.15 mm (0.002 to 0.006 in.) wide flat on the cutting edge, which is made about 30° with respect to the tool face.

Ceramics have primarily used negative rakes with chamfered edges, although improvements in fabrication techniques, mechanical properties, and machine tool rigidity permit the use of positive-rake geometries. Honed edges and a central hole for a locking pin are also possible.

Some simple chip-breaker styles have also been demonstrated. Because hot uniaxial and isostatic pressing are often required to ensure full densification and because the brittleness of ceramics makes grinding difficult, molded chip breakers are difficult to form. However, higher speeds, especially if combined with high feeds, generally provide adequate chip control.

## Applications

Ceramic cutting tools are primarily used in turning and boring operations, and each type of ceramic has different areas of application, as outlined in Fig. 10. Because of their higher hot hardness and chemical inertness, ceramics are used in the high-speed (>300 m/min, or 1000 sfm) finish machining of many hard and difficult-to-machine materials. High removal rates are possible with uninterrupted cuts or more rigid machine setups. The higher thermal shock resistance of the tougher ceramics (SiAlON and SiC whisker reinforced $Al_2O_3$) allows the use of cutting fluids in most applications.

**Cutting Steels and Cast Irons.** With a depth of cut of about 1 mm (0.04 in.), the $Al_2O_3$-TiC ceramics are suitable for the cutting of steels and cast irons at the approximate turning feeds and speeds shown in Fig. 10. The tougher SiAlON ceramics are also suitable for cutting cast irons, but not steels. For more information on the cutting of steels and cast irons with ceramic tools, see the articles "Machining of Cast

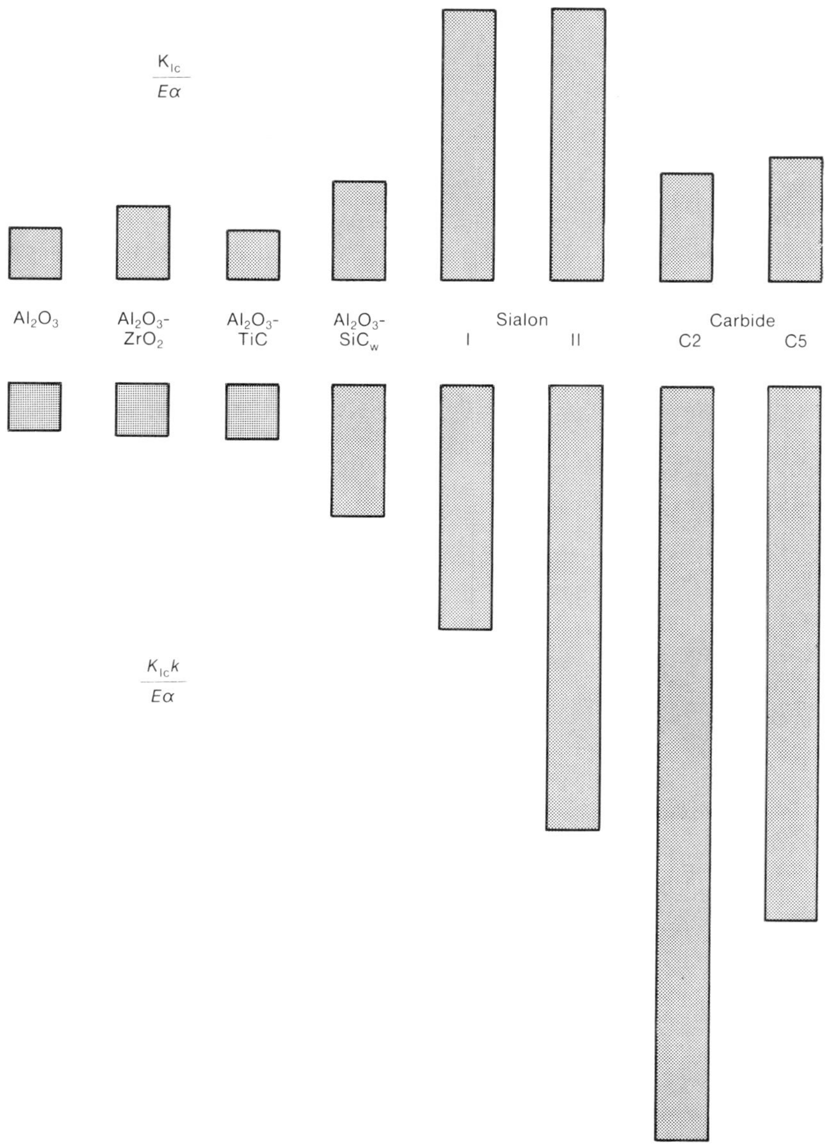


**Fig. 8** Thermal shock figures-of-merit. Thermal conductivity, *k*, at 500 °C (930 °F); α, expansion coefficient, 25 to 870 °C (80 to 1600 °F) range. Sialon I and II are defined in Fig. 7. Source: Ref 19

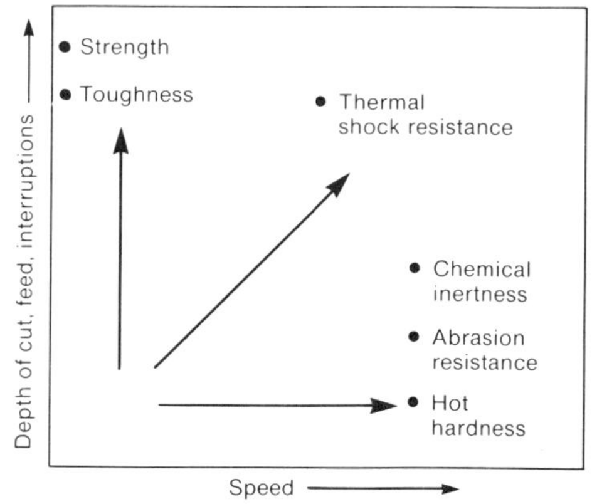


**Fig. 9** Principal tool properties versus machining conditions

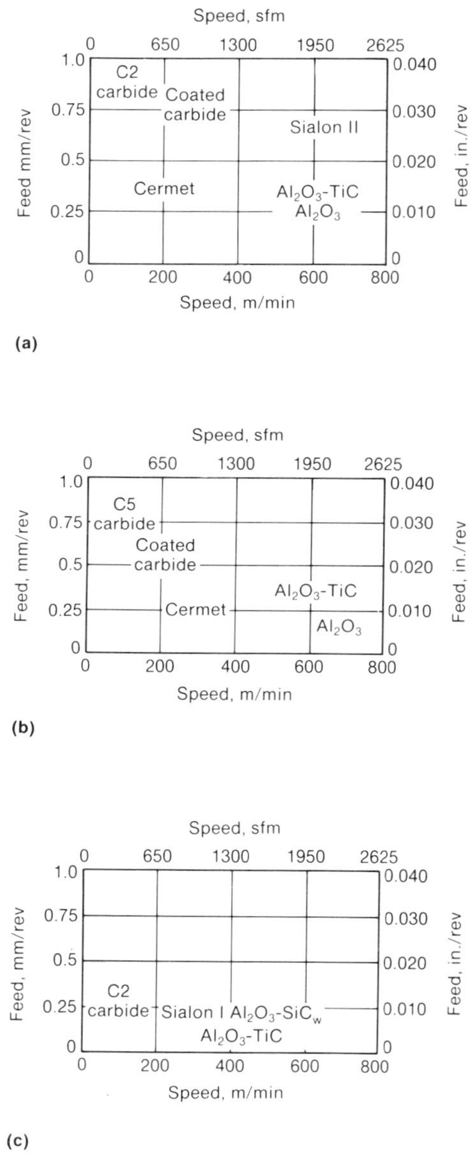


**Fig. 10** Applications of ceramic cutting tools. (a) Cast iron (100 to 300 HB). (b) Alloy steels (190 to 330 HB). (c) Nickel-base alloys (200 to 260 HB). Actual conditions depend on workpiece, cut geometry, machine, and surface requirements. Sialon I and II are defined in Fig. 7. Source: Ref 19

Irons'' and ''Machining of Tool Steels'' in this Volume.

**Cutting Nickel-Base Superalloys.** Depth-of-cut notching is a problem when machining nickel-base superalloys with ceramic tools, and the tougher ceramics (SiAlON and SiC whisker reinforced $Al_2O_3$) provide more resistance against depth-of-cut notching. For more information on cutting nickel-base superalloys with ceramic tools in turning operations, see the article ''Machining of Heat-Resistant Alloys'' in this Volume

**Grinding.** Ceramics are also used in grinding operations. In grinding, polishing, and cutoff operations, ceramics ($Al_2O_3$, $Al_2O_3$-$ZrO_2$, and SiC) and superabrasives (diamond and cubic boron nitride) are invariably used. Alumina and silicon carbide abrasives are discussed in the article ''Grinding Equipment and Processes'' in this Volume. Diamond and cubic boron nitride abrasives are discussed in the article ''Superabrasives.''

**Milling and Drilling.** Ceramics are generally not used in drilling because small-diameter holes and the resultant lower cutting speeds preclude the utility of ceramics. The strength and fracture toughness limitations of current ceramics also limit their application to milling, although some recently introduced ceramics are being successfully applied in the milling of selected work materials. Advances in ceramic tool technology by way of toughened ceramics may enable wider application of ceramics to milling operations.

**Material Limitations.** Not all materials can be machined with ceramics. Ceramics can be successfully used to machine most steels, cast iron, and nickel-base superalloys, even in their hardened condition, as

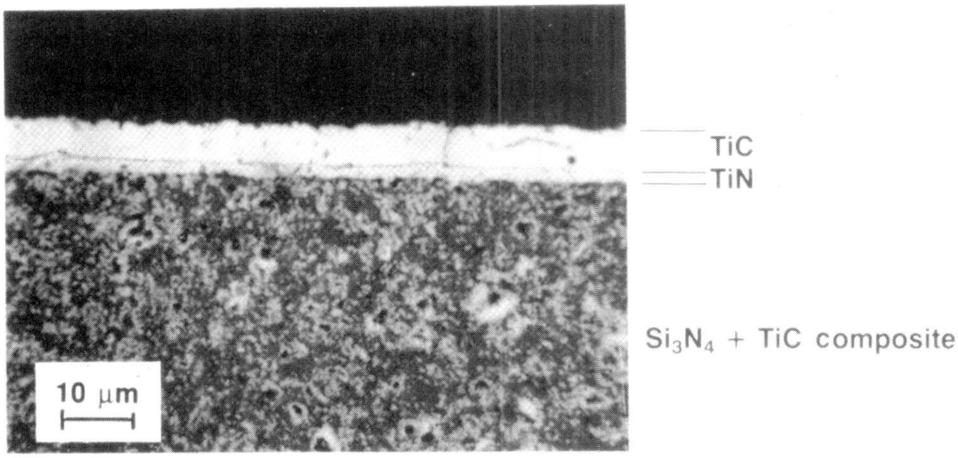


**Fig. 11** Cross section of multiple coatings of TiC on TiN on a $Si_3N_4$-TiC composite tool material

well as many nonferrous alloys and composites. However, the SiAlON and SiC whisker reinforced ceramics are limited in their application because of chemical reactions between the workpiece and ceramic tool. The SiC whisker reinforced ceramics are limited to the cutting of nickel-base superalloys, while SiAlON tools are primarily limited to the cutting of cast iron and nickel-base superalloys.

Titanium alloys, aluminum alloys, and some stainless steels also cannot be machined economically with most ceramic tools, because of chemical interactions between the ceramic tools and these work materials. In contrast, high-speed steel and cemented carbides (for example, WC base and TiC base) are capable of machining a wide range of work materials. This is because the chemistry of both classes of materials can be varied to exhibit a range of metallurgy, hardness, and toughness for a variety of applications.

## Coated Ceramics

Thin coatings (2 to 5 μm) on monolithic ceramic substrates have been developed primarily to limit chemical interactions between the tool and the work material. For example, $Si_3N_4$ tools are successfully used for the high-speed machining of cast iron. However, these tools react significantly with steels and therefore cannot be used for high-speed machining. To take advantage of the high-temperature deformation resistance of this material and to minimize chemical interactions when machining steels at high speeds, multiple coatings of TiC-TiN or $Al_2O_3$-TiC on $Si_3N_4$ and SiAlON substrates were developed that are similar to those for cemented carbides (Ref 21-24). Figure 11 shows a cross section of multiple coatings of TiC on TiN on a $Si_3N_4$-TiC composite tool material. However, the extent to which such coated ceramic tools will be used versus the competing materials for the high-speed machining of steels and other materials depends on the need and the economics of machining. Coated ceramics remain in the experimental stage.

### REFERENCES

1. E.C. Van Schoick, Ed., *Ceramic Glossary*, The Ceramic Society, 1963
2. A.G. King and W.M. Wheildon, *Ceramics in Machining Processes*, Academic Press, 1966
3. E.D. Whitney, "Progress in the Development of Modern Ceramic Cutting Tool Material," Paper 810319, Society of Automotive Engineers, 1981
4. E.D. Whitney, Ceramic Cutting Tools, *Powder Metall. Int.*, Vol 6 (No. 2), 1974, p 73-76
5. E.W. Goliber, "Ceramic and Cermet Compositions Containing TiO," Paper presented at the 66th Annual Meeting, The Ceramic Society, 1960
6. J. Binns, "Cutting Performance on Rough and Hogging Cuts," Paper 633, American Society of Tool and Manufacturing Engineers, 1963, p 1-15
7. J.G. Baldoni and S.T. Buljan, Ceramics for Machining, *Ceram. Bull.*, Vol 67 (No. 2), 1988, p 381-387
8. M. Furakawa *et al.*, Fracture Toughness in the System $Al_2O_3$-TiC Ceramics, *Nippon Tungsten Rev.*, Vol 18, 1985, p 16-22
9. R.P. Wahl and B. Ilschner, Fracture Behavior of Composites Based on $Al_2O_3$-TiC, *J. Mater. Sci.*, Vol 15, 1980, p 875
10. J.G. Baldoni *et al.*, *Particulate Titanium-Carbide Ceramic Matrix Composites*, Institute of Physics Conference Series No. 75, Adam Holger, 1986, p 427
11. F.F. Lange, Transformation Toughening, *J. Mater. Sci.*, Vol 17, 1982, p 225
12. P.F. Becher *et al.*, Relation of Transformation Temperature to Fracture Toughness of Transformation-Toughened Ceramics, *J. Mater. Sci.*, Vol 22, 1987, p 76-83
13. G.C. Wei and P.F. Becher, Development of SiC-Whisker-Reinforced Ceramic Composite, *Am. Ceram. Soc. Bull.*, Vol 64 (No. 2), 1985, p 296-304
14. T.N. Tiegs and P.F. Becher, Whisker-Reinforced Ceramic Composites, in *Tailoring Multiphase and Composite Ceramics*, Plenum Press, 1986, p 639-647
15. J. Lorenz, J. Weiss, and G. Petzow, Dense $Si_3N_4$ Alloys: Phase Relations and Consolidation, Microstructure and Properties, in *Advances in Powder Technology*, American Society for Metals, 1982, p 290-291
16. P.K. Mehrotra, Evaluation of Engineering Ceramics for Wear Applications, in *Wear of Materials*, K.C. Ludema, Ed., The American Society of Mechanical Engineers, 1987
17. J. Weiss, L.J. Gauckler, and T.Y. Yen, *J. Am. Ceram. Soc.*, Vol 62, 1979, p 632
18. J. Weiss, M.S. thesis, University of Stuttgart, 1977
19. B. North, Ceramic Cutting Tools—A Review, *Int. J. High Technol. Ceram.*, Vol 3, 1987, p 113-127
20. S.J. Burden *et al.*, Comparison of Hot-Isostatically-Pressed and Uniaxially Hot Pressed Alumina-Titanium Carbide Cutting Tools, *Ceram. Bull.*, Vol 67 (No. 6), 1988, p 1003
21. V.K. Sarin and S.T. Buljan, Alumina Coated Silicon Nitride Cutting Tools, U.S. Patent 4,440,547, 1984
22. V.K. Sarin and S.T. Buljan, Nitride Coated Silicon Nitride Cutting Tools, U.S. Patent 4,406,668, 1983
23. V.K. Sarin and S.T. Buljan, Carbo-Nitride Coated Composite Modified Silicon Aluminum Oxynitride Cutting Tools, U.S. Patent 4,406,669, 1983
24. V.K. Sarin and S.T. Buljan, Nitride Coated Composite Modified Silicon Aluminum Oxynitride Cutting Tools, U.S. Patent 4,406,670, 1983

### SELECTED REFERENCES

- J.R. Baldoni and S.T. Buljan, Ceramics for Machining, *Ceram. Bull.*, Vol 67 (No. 2), 1988, p 381-387
- S.K. Bhattacharyya and A. Jawaid, Syalon Ceramics in Metal Cutting, *Int. J. Prod. Res.*, Vol 19 (No. 5), 1981, p 589-594
- S.K. Bhattacharyya, A. Jawaid, and J. Wallbank, Wear of Syalon Tooling in the High-Speed Machining of Aerospace Materials, in *Proceedings of the ASME High-*

*Speed Machining Symposium*, Vol 12, R. Komanduri, K. Subramanian, and B.F. von Turkovich, Ed., American Society of Mechanical Engineers, 1984, p 245-262

- E.R. Billman, P.K. Mehrotra, L.F. Shuster, and C.W. Deeghly, Machining With $Al_2O_3$-SiC, Whisker Cutting Tools, *Ceram. Bull.*, Vol 67 (No. 6), 1988, p M16-1019
- S.T. Buljan and S.F. Wayne, Silicon Nitride Based Composite Cutting Tools: Material Design Approach, *Adv. Ceram. Mater.*, Vol 2 (No. 4), 1987, p 813-816
- A. Ezis, S.K. Samanta, and S. Subramanian, Ceramic Cutting Tool Formed From $Si_3N_4Y_2O_3$ $SiO_2$ and Method of Making, U.S. Patent 4,434,238, 1984
- R.L. Hatschek, Take a New Look at Ceramics/Cermets, *Am. Mach.*, May 1981, p 165-176
- K.H. Jack and W.I. Wilson, *Nature*, Vol 238, 1972, p 28
- K.H. Jack, Sialon Tool Materials, *Met. Technol.*, Vol 9, July 1982, p 297-301
- R. Komanduri and R. Desai, Tool Materials, in *Encyclopedia of Chemical Technology*, 3rd ed., Wiley-Interscience, 1983, p 273-309
- R. Komanduri, On the Substitution in the U.S. of Ceramic Tools for Conventional Cutting Tools Containing Strategic Minerals—W, Ta and Co, *Carbide Tool J.* Vol 18 (No. 4), July/Aug 1986, p 33-38
- P.K. Mehrotra, Hot Isostatic Pressing of Ceramic Metal Cutting Tool, *Mater. Process. Rev.*, July-Aug 1987, p 506-510
- Y. Oyama and O. Kamigaito, *Jpn. J. Appl. Phys.*, Vol 10, 1971, p 1637
- S.K. Samanta, L. Wang, and H. Wada, Pressure Sintering of $Si_3N_4$ System for Near-Net Shape Forming, *J. Eng. Mater. Technol.* (*Trans. ASME*), to be published
- S.K. Samanta, S. Subramanian, and A. Ezis, Method of Using $Si_3N_4Y_2O_3$ $SiO_2$ Ceramic System for Machining Cast Iron, U.S. Patent 4,227,842, 1980
- S.K. Samanta and S. Subramanian, Hot Pressed $Si_3N_4$ as a High Performance Cutting Tool Material, in *Proceedings of the North American Manufacturing Research Conference*, University of California, 1985
- V.K. Sarin and S.T. Buljan, Alumina Coated Silicon Nitride Cutting Tools, U.S. Patent 4,440,547, 1984
- K.H. Smith, Whisker Reinforced Ceramic Composite Cutting Tools, *Carbide Tool J.*, Vol 18 (No. 5), Sept-Oct 1986, p 8-10
- N.P. Suh, New Theories of Wear and Their Implications for Tool Materials, *Wear*, Vol 62, 1980, p 1-20
- G.C. Wei, Silicon Carbide Whisker Reinforced Ceramic Composites and Method for Making Same, U.S. Patent 4,543,345, 1985
- F.W. Wilson, Ed., *Machining With Carbides and Oxides*, McGraw-Hill, 1962

# Ultrahard Tool Materials

P.J. Heath, De Beers Industrial Diamond Division (Pty) Ltd. (England)

DIAMOND AND CUBIC BORON NITRIDE (CBN) are the two hardest materials known. They have found numerous applications in industry, both as ultrahard abrasives and as cutting tools. This paper reviews the high-pressure synthesis of these materials, their cutting tool formats, and some examples of their application as cutting tools in the field of machining. For information on the use of these materials for grinding operations, see the article "Superabrasives" in this Volume.

## Comparison of Diamond and Cubic Boron Nitride

Diamond and CBN are very similar in many ways: they are the two hardest materials known, they share the same crystallographic structure, and they exhibit exceptionally high values of thermal conductivity. In other ways they are very different. Diamond, for example, oxidizes in air (Ref 1), reacts with ferrous workpiece material at moderate temperatures, and is subject to graphitization. Cubic boron nitride, on the other hand, is thermally quite stable both in air and in contact with ferrous workpiece materials (Ref 2). As a result of these fundamental property differences, there is a simple split in the basic application of diamond and CBN. Diamond is used in the machining of nonferrous materials such as:

- Aluminum alloys
- Copper alloys
- Abrasive plastics
- Glass and carbon fiber composites
- Green ceramics
- Tungsten carbide
- Abrasive wood/plastic composites
- Natural stone
- Concrete

Cubic boron nitride is used in the machining of ferrous materials such as:

- Tool steels
- Hard irons
- Pearlitic gray cast iron
- Hardfacing alloys
- Surface-hardened steels and irons

**Synthesis of Diamond Grit.** The bulk of diamond used for industrial applications is made synthetically by subjecting carbon, in the form of graphite, to high temperatures and pressures using large special-purpose presses (Fig. 1). By the application of heat and pressure, the hexagonal structure of graphite (Fig. 2a) can be transformed into the cubic structure of diamond (Fig. 2b).

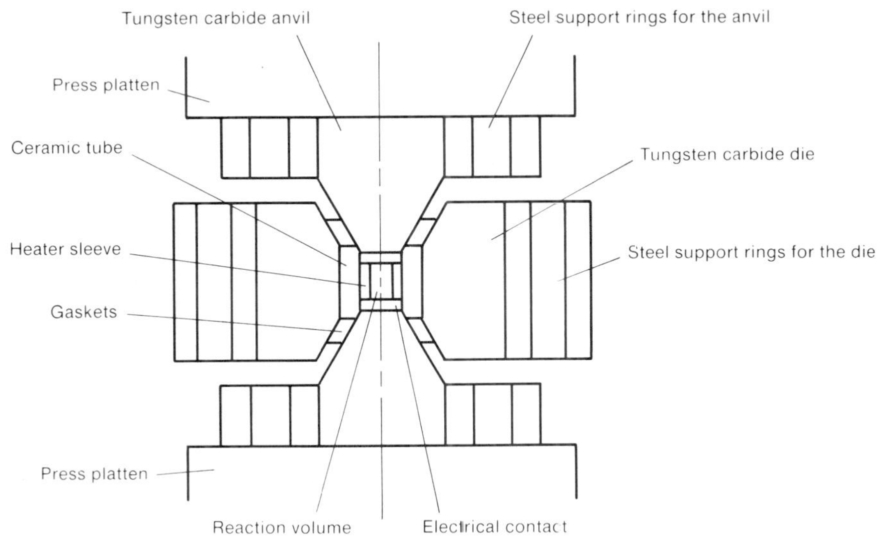


**Fig. 1** The main components of a die set for diamond or CBN synthesis

This transformation does not take place easily, as the immense temperatures and pressures of the graphite-diamond equilibrium diagram show (Fig. 3). In commercial practice, however, the required temperatures and pressures can be reduced by the use of a solvent/catalyst, such as nickel, cobalt, iron, or the alloys thereof. After the transformation process, the reaction mass embedded with diamond crystals (Fig. 4) is extracted from the reaction chamber, and the dissolution of the solvent/catalyst with acid liberates the crystals. In this form, the diamond crystals are recovered for subsequent processing and are used as an abrasive in grinding wheels and stone saws, for example.

**Synthesis of Cubic Boron Nitride Grit.** Boron and nitrogen can form a compound of boron and nitrogen using the reaction:

$$BCl_3 + NH_3 \rightleftharpoons BN + 3HCl$$

where the compound BN has a hexagonal, graphitelike structure with approximately an equal number of boron and nitrogen atoms arranged alternately (Fig. 5a). This compound of hexagonal boron nitride (HBN) is a soft, slippery substance like hexagonal carbon (graphite), and just as hexagonal carbon (graphite) can be transformed into cubic carbon (diamond), HBN can be transformed into CBN as shown in Fig. 5(b). The equilibrium diagram for this transformation is shown in Fig. 6. Here, too, it is normal practice to use a solvent/catalyst. In this case an alkali, an alkali earth metal, or the nitrides thereof, such as lithium nitride, are used.

**Synthesis of Polycrystalline Diamond and Polycrystalline Cubic Boron Nitride.** It is possible also to produce polycrystalline diamond (PCD) or polycrystalline cubic boron nitride (PCBN) by sintering (or binding) many individual crystals of diamond or CBN together to produce a larger polycrystalline mass. It is commercial practice to enhance the rate of polycrystalline formation by the addition of a metal or ceramic second phase (Ref 4). In addition, the whole mass must again be maintained in the cubic region of the respective temperature-pressure phase diagram to prevent the hard cubic crystals from reverting to the soft hexagonal form. By such high-temperature

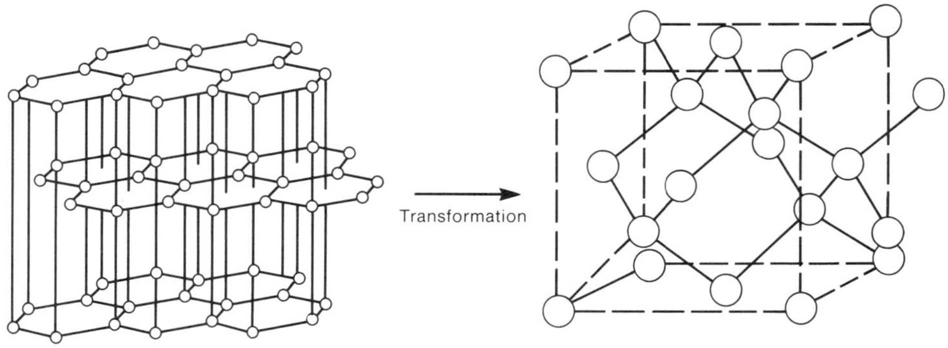


(a) (b)

**Fig. 2** Arrangement of carbon atoms. (a) The hexagonal arrangement of atoms in graphite. (b) The cubic arrangement of atoms in diamond

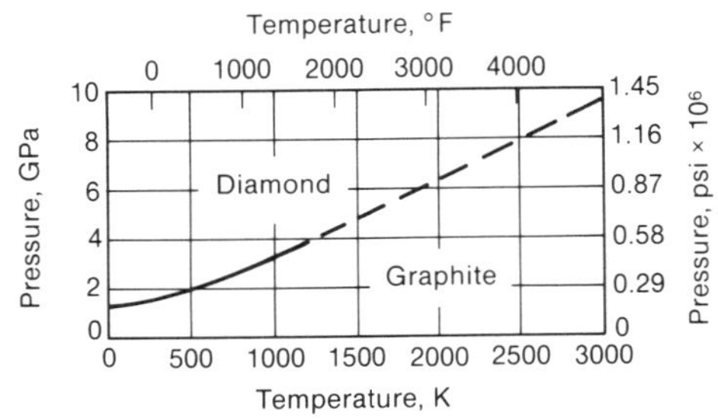


**Fig. 3** Equilibrium diagram of graphite and diamond. Source: Ref 3

**Fig. 4** Crystals of diamond embedded in the reaction mass extracted from a high-temperature press

high-pressure sintering techniques, it is possible to obtain a mass of diamond or CBN in which randomly oriented crystals are combined to produce a large isotropic mass.

An immense range of polycrystalline products can be made of diamond or CBN. Changes in grain size, second phase employed, degree of sintering, particle size distribution, and the presence or absence of inert ceramic, metallic, or nonmetallic fillers, for example, have profound effects on the mechanical, physical, and thermal properties of the final product. By careful formulation it is possible to tailor material properties for particular applications. Figure 7 shows a few of the many microstructures that can be produced in PCD and PCBN products.

**Product Formats.** It is possible not only to change the formulation but also to make PCD and PCBN blanks in many different formats. Figure 8 illustrates some of the possibilities. One of the most commonly used formats for cutting tool applications consists of a layer of PCD or PCBN, between 0.3 to 1.5 mm (0.012 to 0.060 in.) thick, bonded to a cemented tungsten carbide substrate. This format of PCD and PCBN tooling is often used for light cutting of hard materials. Where more arduous machining conditions are encountered, such as in rough machining, solid PCD or PCBN tool inserts can offer advantages over a two-layer product.

**Properties of PCD and PCBN.** Table 1 shows some of the important mechanical and thermal properties of PCD and PCBN, together with properties measured under similar conditions for the conventional cutting tool materials, tungsten carbide, $Al_2O_3$ ceramics, $Al_2O_3$ + TiC mixed ceramics, Sialon, and single-crystal diamonds.

## Polycrystalline Diamond Cutting Tools

Polycrystalline diamond cutting tools are commercially available in either the solid or layered format. Polycrystalline diamond cutting tools with a layered format have a tungsten or tungsten carbide substrate. Standard overall product thicknesses have become established at 1.6 mm (0.063 in.), 3.18 mm (0.125 in.) and 8.0 mm (0.315 in.). The largest cutting tool blank commercially available has a diameter of 71.0 mm (2.75 in.).

The thickness of the diamond layer varies from manufacturer to manufacturer, but is usually between 0.3 mm (0.012 in.) and 1.5 mm (0.060 in.). Over the years, various grades have been developed that are based on different average particle sizes of diamond used in the PCD layer. In the De Beers range, for example, there are three grades based on the average particle sizes of 2 μm (80 μin.), 10 μm (400 μin.), and 25 μm (1000 μin.). A fourth grade, also based on 2 μm (80 μin.) diamond, has recently been developed with enhanced toughness, which finds applications in the woodworking industry, in particular. The etched structures of these four grades of PCD are shown in Fig. 9.

The following discussions refer to layered PCD tools. These tools are brazed to tool shanks or to indexable inserts for use in standard toolholders.

### Tool Fabrication

**Cutting PCD blanks** into smaller pieces is usually done by wire electrical discharge machining (EDM). However, it should be noted that not all available machines have sufficiently powerful generators to cut the PCD layer, which has a relatively low conductivity. If cutting difficulties are experienced, the following factors should be checked:

- Is there good electrical contact to the wire?
- Is there good electrical contact to the workpiece?
- Is the water conductivity less than 20 ohm · cm (7.9 ohm · in.)?
- Is electrolyte flushing adequate?
- Are discharge conditions correctly set for PCD?

**Brazing.** The next step in tool fabrication consists of brazing a suitably shaped blank into a pocket in a steel or cemented carbide substrate. In general, to braze layered PCD

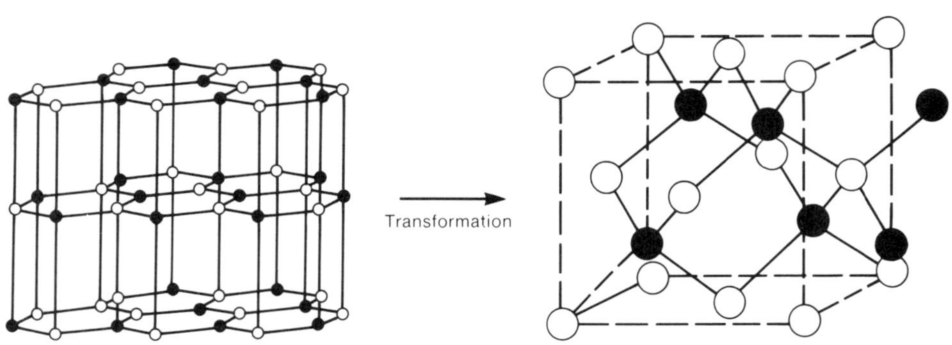


(a) (b)

**Fig. 5** Arrangement of boron and nitrogen atoms. (a) The hexagonal arrangement of boron nitride. (b) The cubic arrangement of boron nitride

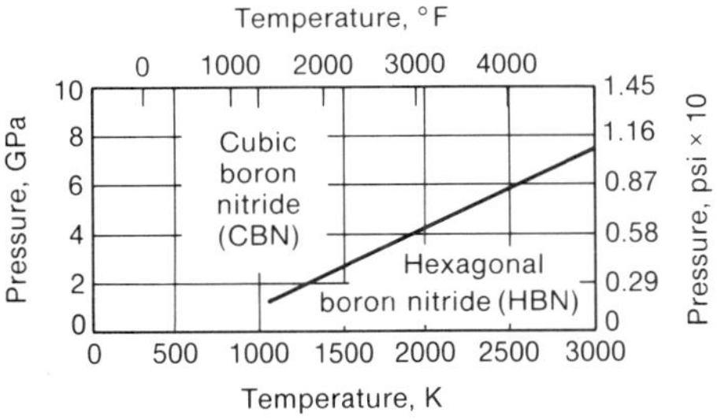


**Fig. 6** Equilibrium diagram for HBN and CBN

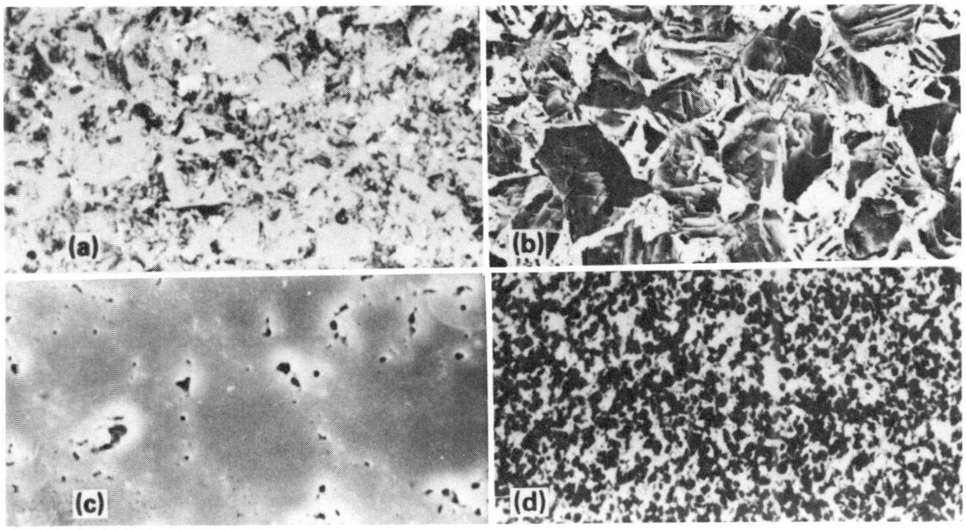


**Fig. 7** Microstructures of four sintered polycrystalline products. (a) Syndite PCD. (b) Syndax3 PCD. (c) Amborite PCBN. (d) DBC50 PCBN

tools, silver alloy brazes with low melting points in the range of 600 to 750 °C (1100 to 1400 °F) are used. Suitable alloys and their manufacturers are listed in Table 2 and should be used in conjunction with the recommended fluxes. Great care must be taken not to overheat the PCD blank; overheating could cause the diamond to revert to graphite (Ref 5). The mating surfaces must be adequately prepared, and cleanliness is of paramount importance. Heating may be by an oxygen-acetylene torch, but the preferred technique is by induction brazing, which gives a greater degree of control.

**Grinding of PCD** is the next step in the tool fabrication process. Grinding and radiusing the tool flank remove the damage produced by wire cutting (typically 0.05 mm, or 0.002 in. deep) and generate the required cutting geometry and edge quality.

Because PCD is extremely hard, grinding can present a problem. For shank-type tools, it is advisable to use rigid tool and cutter grinders with reciprocating wheelheads and facilities to tilt the head so that the necessary clearance angles can be created (Ref 6). Where larger quantities of indexable inserts of a standard size are required, indexable insert grinders suitable for grinding PCD and PCBN are available (Ref 7).

Many grinding wheel manufacturers now produce wheels developed especially for grinding PCD. Vitreous-bond, metal-bond, or resin-bond wheels have all proved successful. There is, however, no consensus on which wheel type is superior. For rough grinding, it is common to use a resin-bonded wheel containing 320/400 U.S. mesh Micron CDA at 100 or 125 concentration. For finishing, vitreous-bonded wheels containing 8 to 16 μm (300 to 600 μin.) or finer CDA grit, or metal-bonded wheels containing MDA at a concentration of 100, are often used (Ref 8).

In grinding PCD, the $G$ ratio (that is, the ratio of the volume of workpiece removed to the volume of wheel lost) is very low because of the hardness of PCD. When diamond wheels are used to grind tungsten carbide, they give typical $G$ ratios on the order of 30 to 500. However, $G$ ratios achieved in the grinding of PCD are often only 0.01 to 0.02; that is, the wheel wears away 50 to 100 times faster than the tool being ground. A typical stock removal rate with a 6A2 wheel grinding small PCD tools is approximately 0.5 mm$^3$/min (0.00003 in.$^3$/min).

**Electrical Discharge Grinding.** A number of manufacturers have introduced machines in which a conventional abrasive

**Table 1 Mechanical, physical, and thermal properties of PCD, PCBN, and other tool materials**

| Tool material | Density, kg/m$^3$ | Density, lb/ft$^3$ | Compressive strength, GPa | Compressive strength, psi × 10$^6$ | Fracture toughness, MPa$\sqrt{m}$ | Fracture toughness, ksi$\sqrt{in.}$ | Knoop hardness, GPa | Knoop hardness, psi × 10$^6$ |
|---|---|---|---|---|---|---|---|---|
| $Al_2O_3$ | 3.91 | 0.244 | 4.00 | 0.580 | 2.33 | 2.12 | 16 | 2.3 |
| $Al_2O_3$ + TiC | 4.28 | 0.267 | 4.50 | 0.653 | 3.31 | 3.01 | 17 | 2.5 |
| Sialon | 3.20 | 0.200 | 3.50 | 0.508 | 5.00 | 4.55 | 13 | 1.9 |
| Tungsten carbide (K10) | 14.70 | 0.918 | 4.50 | 0.653 | 10.80 | 9.83 | 13 | 1.9 |
| PCBN(a) (Amborite) | 3.12 | 0.195 | 3.80 | 0.551 | 6.30 | 5.73 | 28 | 4.1 |
| PCD(b) (Syndax3) | 3.43 | 0.214 | 4.74 | 0.687 | 6.89 | 6.27 | 50 | 7.3 |
| PCD(c) (Syndite 010) | 4.12 | 0.257 | 7.60 | 1.102 | 8.80 | 8.01 | 50 | 7.3 |
| Single-crystal diamond | 3.52 | 0.220 | 8.58 | 1.244 | 3.40 | 3.09 | 57–104 | 8.2–15.1 |
| PCBN(d) (DBC50) | 4.28 | 0.267 | 3.55 | 0.515 | 3.70 | 3.37 | 28 | 4.1 |

| Tool Material | Young's modulus, GPa | Young's modulus, psi × 10$^6$ | Modulus of rigidity, GPa | Modulus of rigidity, psi × 10$^6$ | Bulk modulus, GPa | Bulk modulus, psi × 10$^6$ | Poisson's ratio, $\nu$ | Coefficient of thermal expansion, 10$^{-6}$/K | Thermal conductivity, W/m · K | Thermal conductivity, Btu/ft · h · °F |
|---|---|---|---|---|---|---|---|---|---|---|
| $Al_2O_3$ | 380 | 55 | 153 | 22.2 | 243 | 35.2 | 0.24 | 8.50 | 22.7 | 13.1 |
| $Al_2O_3$ + TiC | 370 | 54 | 160 | 23.2 | 232 | 33.6 | 0.22 | 7.80 | 16.7 | 9.65 |
| Sialon | 300 | 44 | 117 | 16.9 | 227 | 32.9 | 0.28 | 3.20 | 20–25 | 12–14 |
| Tungsten carbide (K10) | 620 | 90 | 258 | 37.4 | 375 | 54.4 | 0.22 | 5.00 | 100 | 58 |
| PCBN(a) (Amborite) | 680 | 99 | 279 | 40.4 | 405 | 58.7 | 0.22 | 4.90 | 100 | 58 |
| PCD(b) (Syndax3) | 925 | 134 | 426 | 61.8 | 372 | 54.0 | 0.086 | 3.80 | 120 | 69 |
| PCD(c) (Syndite 010) | 776 | 112 | 363 | 52.6 | 301 | 43.7 | 0.07 | 4.20 | 560 | 323 |
| Single-crystal diamond | 1141 | 165 | 553 | 80.2 | 442 | 64.1 | 0.07 | 1.50–4.80 | 500–2000(e) | 290–1160(b) |
| PCBN(d) (DBC50) | 587 | 85 | 284 | 41.2 | 254 | 36.8 | 0.15 | 4.70 | 44 | 25 |

(a) Solid PCBN. (b) Solid, thermally stable PCD used mainly in mining applications. (c) Layered PCD. (d) Layered PCD with lowered CBN content. (e) Depending on crystal orientation and perfection

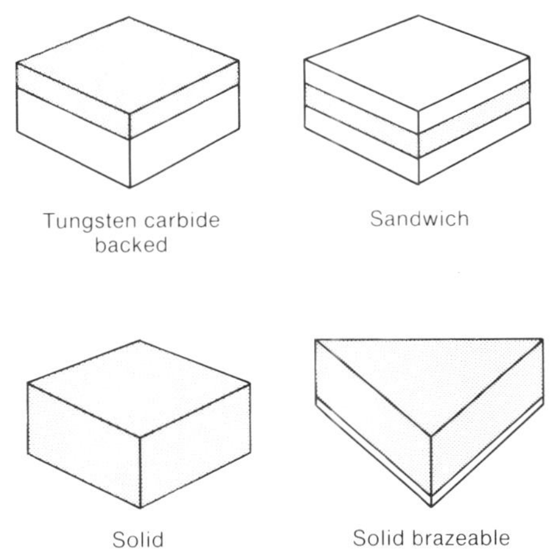


**Fig. 8** Formats for polycrystalline diamond and polycrystalline CBN tools

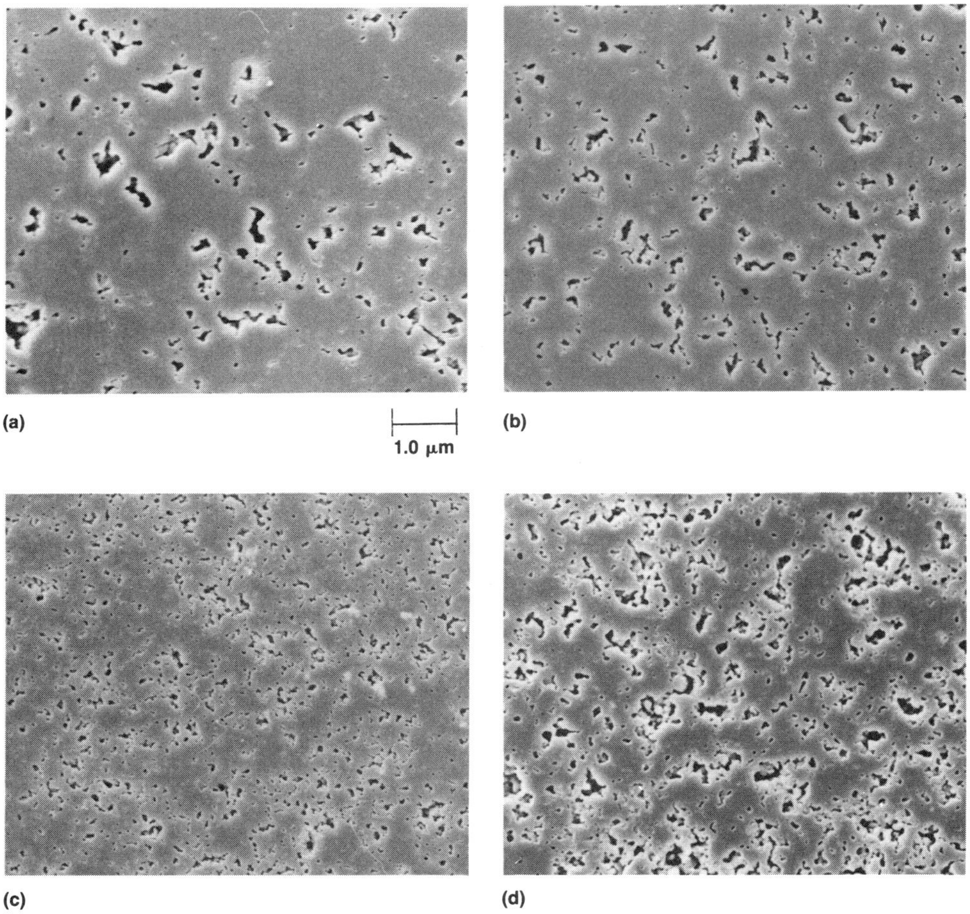


**Fig. 9** The etched microstructures of the four grades of Syndite PCD. (a) Syndite 025. (b) Syndite 010. (c) Syndite 002. (d) Syndite 002C. All 1000×

grinding wheel is replaced by a graphite wheel, and an electrical discharge is struck between the wheel and the tool (see the article "Electrical Discharge Grinding" in this Volume). These rotary EDM, or electrical discharge grinding, machines have the advantage that a profiled graphite wheel can be used to produce similarly profiled tools (Ref 9). Because there is no physical contact and consequently no grinding force, very delicate or complex tool forms can be created. For tool forms in which long cutting edges are required (for example, in woodworking profile cutters), wire EDM cutting is often used to form the cutting edge. To avoid the development of unacceptable stresses, it is recommended that pieces be preshaped before brazing. These are then brazed into the cutter body and finish cut to the required profile. A final trim cut at a low power level and low cutting speed, in which the wire barely penetrates the PCD, produces the final cutting edge with negligible EDM damage.

**Polishing.** The majority of toolmakers in the world now generally supply tools with the top face polished to a mirror finish. This improves the flow of chips across the face of the tool during machining, achieves a sharper and more chip-free edge, and further improves the ability of PCD to resist the formation of a built-up edge. In addition, polished tools are cosmetically attractive.

At present, it is usual practice to lap the table of the smaller individual blanks before tool fabrication takes place (Ref 10). However, it is becoming increasingly common for both toolmakers and blank manufacturers to polish the large blanks before cutting.

## Wear Resistance

In simple abrasive-wear situations, PCD is significantly more wear resistant than any other cutting tool material. PCD is approximately 10 times more wear resistant than PCBN and 100 times more wear resistant than cemented tungsten carbide.

The relative wear resistance of PCD, in comparison with conventional tool materials such as ceramics or tungsten carbide and high-speed steel, for example, can be demonstrated by conducting turning tests on an abrasive workpiece material (Ref 11). The results of cutting tests at various cutting speeds on an abrasive silica flour filled epoxy resin (SFFER) workpiece are given in Fig. 10. It can be seen that, for a tool life of 1.0 min to produce a flank wear of 0.3 mm (0.012 in.), the various tool materials must be used at the cutting speeds summarized in Table 3.

Theoretical relationships have been derived relating wear resistance to fracture toughness, Young's modulus, and hardness (Ref 12). One such equation is:

$$W_r = K_{Ic}^{0.5} \cdot E^{-0.8} \cdot H^{1.43}$$

where $W_r$ is the wear resistance figure of merit, $K_{Ic}$ is the fracture toughness (MPa$\sqrt{m}$, or ksi$\sqrt{in.}$), $E$ is Young's modulus (GPa, or $10^6$ psi), and $H$ is Knoop hardness (GPa, or $10^6$ psi).

Using the mechanical property information given in Table 1, wear resistance figures of merit ($W_r$) can be calculated. These are given in the last column of Table 3.

It can be seen that this ranking is very similar to that of the cutting speeds required to give a 1 min tool life. However, this

**Table 2 Alloys suitable for brazing carbide-backed PCD**

| Commercial designations or AWS(a) brazing alloy | Temperature: Solidus °C | °F | Liquidus °C | °F | Composition, %: Ag | Cu | Zn | Cd | Ni | Comments |
|---|---|---|---|---|---|---|---|---|---|---|
| Easy-flo | 620 | 1148 | 630 | 1166 | 50 | (b) | (b) | (b) | ... | General purpose |
| Easy-flo No. 2 | 608 | 1126 | 617 | 1143 | 42 | (b) | (b) | (b) | ... | General purpose for tungsten carbide and induction brazing |
| Easy-flo No. 3 | 634 | 1173 | 656 | 1213 | 50 | (b) | (b) | (b) | (b) | |
| 5004 Ju | 620 | 1148 | 640 | 1184 | 50 | 15 | 17.5 | 17.5 | ... | Brazing of carbide |
| 5009 | 645 | 1193 | 690 | 1274 | 50 | 15.5 | 15.5 | 16.3 | 3 | |
| Easy-flo No. 45 | 607 | 1124 | 618 | 1144 | 45 | 15 | 16 | 24 | ... | Torch brazing |
| Easy-flo No. 3 | 632 | 1170 | 688 | 1270 | 50 | 15.5 | 15.5 | 16 | 3 | Induction brazing |
| AWS BAg-1 | 605 | 1121 | 620 | 1148 | 45 | 15 | 16 | 24 | ... | ... |
| AWS BAg-3 | 630 | 1166 | 690 | 1274 | 50 | 15.5 | 15.5 | 16 | 3 | ... |

(a) American Welding Society. (b) Proprietary

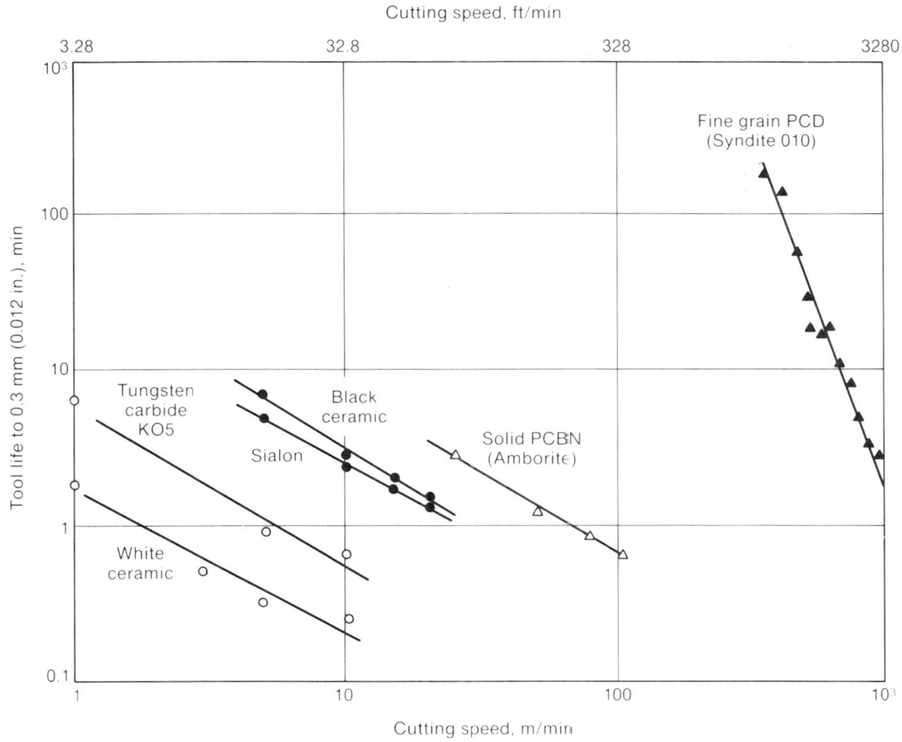


**Fig. 10** Abrasion resistance of PCD, PCBN, and other cutting tool materials. Machining parameters: depth of cut = 1.0 mm (0.040 in.); feed rate = 0.32 mm/rev (0.013 in./rev); approach angle = 45°; top rake = 0°; clearance = 6°; tool nose radius = 0.8 mm (0.030 in.); dry, no coolant. Workpiece: silica flour filled epoxy resin

**Table 3 Cutting speeds required to produce a 1 min tool life on silica flour filled epoxy resin**

| Tool material | Cutting speed m/min | Cutting speed sfm | $W_r$(a) |
|---|---|---|---|
| White ceramic ($Al_2O_3$) | 2 | 6.6 | 0.76 |
| Tungsten carbide (ISO KO5) | 7 | 23 | 0.73 |
| Sialon | 30 | 98 | 0.91 |
| Black ceramic ($Al_2O_3$ + TiC) | 30 | 98 | 0.92 |
| PCBN (Amborite) | 80 | 260 | 2.07 |
| PCD (Syndite grade 010) | 1000 | 3300 | 3.38 |

(a) $W_r$, wear resistance figure of merit

ranking only applies in simple situations of abrasive wear. When cutting metal workpieces, other wear mechanisms, such as chemical or adhesive wear, can become dominant. For example, the sialon materials are particularly good when cutting the nickel-base superalloys, but PCBN or PCD wear rapidly on this class of material, despite the fact that they are significantly harder. It is essential, therefore, to select the correct tool material for each type of workpiece.

**Wear Resistance of Specific Grades.** The relative wear characteristics described below refer to three grades (2 μm, 10 μm, and 25 μm particle size) of Syndite, a commercially produced layered PCD. The discussions do not include the newer chip-resistant grade (Syndite 002C), which is intended only for specific applications in woodworking in which toughness, rather than long tool life, is required. The relative wear resistance of the various grades of PCD can easily be assessed with turning tests on heavily loaded silica flour filled epoxy resin workpieces, as mentioned earlier. Figure 11 shows the progress of flank wear with time under the machining conditions indicated. It can be seen that the rate of wear decreases as the grain size of the PCD increases. A similar trend is observed when machining high-silicon aluminum alloys (Fig. 12). The differences between the grades, however, are less marked.

The coarser grades of PCD give a longer tool life, but the finer grades have a better edge quality and permit a higher quality of surface finish to be produced on homogeneous workpiece materials. If long life is required, coarser grain size PCD should be selected. However, long tool life may have to be sacrificed if a better surface finish is required. For most applications, 10 μm average particle size would be considered the general-purpose tool.

## Tool Geometries

**Metals machined** with PCD tooling are primarily aluminum and copper and their alloys. These are relatively soft, and positive cutting geometries are required. Typically, top rakes of +6° and clearance angles of +6° are common. Nose radii usually lie in the range of 0.2 to 2.0 mm (0.008 to 0.080 in.). Some attempts have been made to generate chipbreakers in the diamond layer, but the difficulties of doing this have prevented their wide acceptance.

**Machining Nonmetals.** Soft, nonfibrous workpieces, including abrasive flour-filled plastics and hard, fibrous materials such as glass fiber reinforced plastics do not require a true cutting action, and 0° top rakes with 5° to 6° clearance angles are used. This geometry gives a stronger cutting edge and improved wear characteristics, in comparison with more positive geometries.

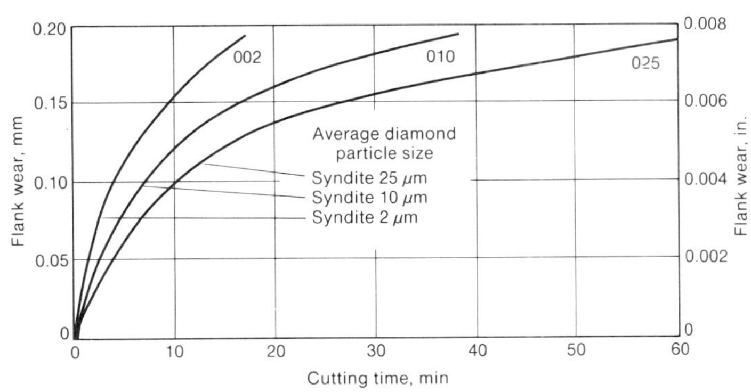


**Fig. 11** Tool life comparison for three PCD grades. Machining parameters: cutting speed = 400 m/min (1300 sfm); feed rate = 0.10 mm/rev (0.004 in./rev); depth of cut = 1.00 mm (0.040 in.); tool nose radius = 0.8 mm (0.030 in.); dry, no coolant. Workpiece: silica flour filled epoxy resin

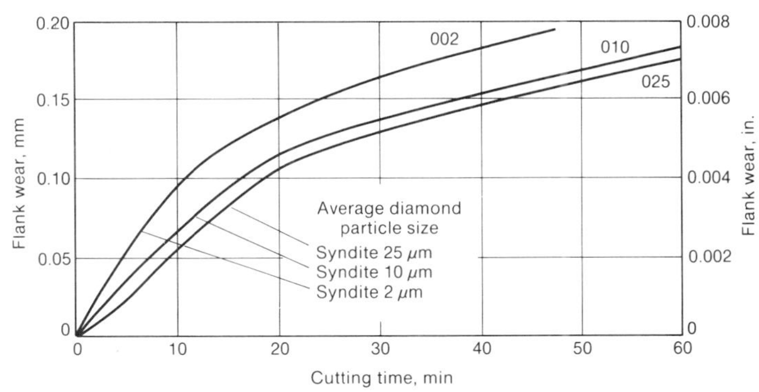


**Fig. 12** Tool life comparison for three PCD grades. Machining parameters: cutting speed = 1000 m/min (3300 sfm); feed rate = 0.10 mm/rev (0.004 in./rev); depth of cut = 0.25 mm (0.010 in.); tool radius = 0.8 mm (0.03 in.); dry, no coolant. Workpiece: Al-18Si

Soft, fibrous materials, such as carbon fiber or Kevlar, for example, require positive-rake tooling (top rake 5° to 6°, clearance 5° to 6°), because it is necessary to cut the fibers cleanly to prevent ripping of the cut surface.

Hard materials, such as stone or sintered tungsten carbide, must be cut with negative-rake cutting geometries (top rake −6°, clearance −6°). Here edge strength is essential to counteract the high cutting forces that develop, and tools with large nose radii should be used.

## Machining Parameters

In many situations, PCD tooling will be a substitute for cemented carbide tooling. PCD cutting tools can provide much longer tool life, better finishes, and higher cutting speeds and feeds than carbide tooling.

**Speed, Feed, and Depth of Cut.** The machining parameters for carbide tools are usually a useful starting point for PCD cutting speeds and feed rates. Under these circumstances, tool lives should increase by a factor of 50 to 200 times. Details of typical machining parameters for turning and milling operations in the major application areas for PCD are given in Table 4.

**Table 4 Typical machining parameters for PCD tooling**

| Workpiece material | Cutting speed | | Feed rate | | Depth of cut | |
|---|---|---|---|---|---|---|
| | m/min | sfm | mm/rev | in./rev | mm | in. |
| **Turning** | | | | | | |
| Aluminum alloys, copper, brass, and their alloys | 300–1000 | 980–3300 | 0.05–0.5 | 0.002–0.020 | ≤10.0 | ≤0.400 |
| Sintered tungsten carbide (using 9.5 mm, or 0.37 in., rounds) | 10–30 | 33–100 | 0.1–0.2 | 0.004–0.008 | ≤2.0 | ≤0.080 |
| Green tungsten carbide | 50–200 | 165–650 | 0.1–0.5 | 0.004–0.020 | ≤5.0 | ≤0.200 |
| Glass and carbon fiber reinforced plastics | 100–600 | 330–2000 | 0.05–0.5 | 0.002–0.020 | ≤2.0 | ≤0.080 |
| Green ceramics | 100–600 | 330–2000 | ≤0.2 | ≤0.008 | ≤2.0 | ≤0.80 |
| **Milling** | | | | | | |
| Aluminum alloys | 500–3000 | 1650–10 000 | 0.1–0.5(a) | 0.004–0.020(b) | ≤5.0 | ≤0.200 |
| Copper, brass, and their alloys | 200–1000 | 650–3300 | 0.1–0.5(a) | 0.004–0.020(b) | ≤2.0 | ≤0.080 |
| Chipboard and fiberboard | 2000–3000 | 6500–10 000 | 1.5–2.0(a) | 0.060–0.080(b) | ≤15.0 | ≤0.600 |

(a) mm/tooth. (b) in./tooth

**Coolants** are generally used in machining metals to aid the cutting action, maintain uniform temperatures, and assist swarf removal. When aluminum alloys are cut with carbide tools, there is a tendency for a built-up edge to form, due to the affinity between aluminum and cemented carbide. Traditional palliatives used with carbide tools have included high tool-rake angles, polished tool tables, and highly polar cutting fluids. With PCD tooling, there is little, if any, tendency for bonding of the aluminum to the tool, and it is, therefore, possible to use simple white water (oil-in-water emulsions) as a cutting fluid. Experiments with lubricants rather than coolants have indicated little, if any, benefit.

Many of the nonmetallic materials on which PCD is used are porous (for example, wood particleboard), and these would be damaged by a cutting fluid. In these situations it is essential to machine dry, although, with these thermally insulating materials, care must be taken to prevent excessive heat buildup in the tools, which could lead to the melting of brazed joints.

## Applications

Polycrystalline diamond cutting tools can be used only on nonferrous workpiece materials, but there are many situations in industry in which cemented tungsten carbide tooling can simply be replaced by PCD. In most situations, the extended tool life more than compensates for the increased tool costs. In others, PCD is often the only tool which can successfully cut the materials to be machined. In mirror or gloss finishing of aluminum or plastic components, however, single-crystal diamond tools may be necessary because it may not be possible to obtain a sufficiently high-quality edge with PCD.

The major applications of PCD tools include the machining of:

- Aluminum and its alloys (Ref 13), particularly the aluminum-silicon alloys used in the automotive industry (Ref 14-16)
- Copper (Ref 17) and brasses (Ref 18, 19)
- Plastics (Ref 20-23), flour-filled resins (Ref 24), and SiC-filled nylons
- Granites (Ref 25, 26), synthetic marble and synthetic stone
- Fiber-reinforced plastics and laminates such as carbon fiber, Kevlar, glass fiber reinforced plastics (Ref 27-32), and printed circuit board laminates
- Wood and wood-base products, particularly particleboards (Ref 33-36), medium density fiberboard (Ref 37), plastic-coated and laminated boards (Ref 38, 39), cement-fiberboards, and abrasive solid hardwoods (Ref 23, 40)
- Ceramics, including green ceramics, unfired clays (Ref 41), and some fired ceramics
- Tungsten carbide, for example, green tungsten carbide and sintered tungsten carbide with high (>12%) cobalt content (Ref 42)

In all of these applications, the tool life of PCD tools is approximately 50 to 100 times greater than that obtained with tungsten carbide cutting tools.

**Woodworking.** PCD use in woodworking has grown rapidly in the last few years. This growth has been stimulated by the increased use of synthetic board materials for modern furniture and encouraged by the recent availability of much larger pieces of PCD, from which tools with long cutting edges can be made.

Modern composite board materials are made from wood particles of a controlled size and character, held together with a resin adhesive. Boards may be surfaced with plastic laminates or natural-wood veneers (bonded with abrasive resins), but the abrasive nature of the composite is often compounded by hard particles of sand or other inclusions in the interior of the board.

The most common cutting techniques used in the woodworking industry are sawing (with circular saws) and routing. Both operations lend themselves readily to the use of cutting tools tipped with PCD inserts. The major benefits of PCD are increased accuracy of cut, fewer tool changes, and overall reduced machining costs.

In volume production situations, large cost savings can be obtained with PCD saws in comparison to carbide tipped saws, even though circular saws tipped with PCD may necessitate a heavy financial investment at the outset. Because PCD saws cut very freely, some tool manufacturers tip only a percentage of the teeth with PCD. Besides reducing tool costs, it is thought that the carbide teeth blunt rapidly during use, resulting in an additional cutting force that aids blade stability.

One problem with PCD saws used in woodworking is that, should a tooth be chipped during use, it becomes necessary to regrind all the teeth on the saw. This can be an expensive and time-consuming operation. For this type of situation, a more chip-resistant grade of PCD should be used. This newly developed grade of Syndite (002C), although slightly less abrasion resistant than the other grades discussed above, has an increased toughness and is easier to cut by electrical discharge machining.

**Wear-Resistant Parts.** An increasingly important use of PCD is for wear-resistant contact faces in measurement probes and bearing support surfaces. An early application in which the properties of PCD provided a solution was in sliding calipers and micrometer spindles. The requirements were for polished surfaces with a surface finish of 0.2 μm (8 μin.) $R_z$ and for the PCD to be plane parallel to less than 2 μm (80 μin.) on opposite measuring faces. Such components are produced not only with flat working faces, but also in shapes contoured to the shape of the component.

A typical application of this type of wear-resistant gage part is for the contact faces of micrometers used to measure the diameter of diamond honing sticks. Because of the abrasive nature of the surfaces being measured, cemented-carbide gage surfaces had to be replaced every few months. Micrometers fitted with PCD contact faces can guarantee accurate readings over periods of more than 1.5 years.

In addition to the benefits of this simple, essentially stationary, contact type of wear surface, PCD has shown significant advantages over tungsten carbide in situations in which precision components are subjected to sliding wear. An example is the use of PCD to form the work guides of centerless grinding rests and V-guides. In this type of application the life of carbide guides between regrinds is typically 18 h, whereas the life obtained with PCD is some 40 times greater, at approximately 700 h. Even greater improvements can be obtained by lapping and polishing the PCD to a flatness of 1 μm (40 μin.) and a surface finish of 0.1 μm (4 μin.) $R_z$. Lives in excess of 2000 h have been reported.

Another application is the use of PCD as steady rests when machining the ends of crankshafts. During precision machining, the crankshaft is supported and spun on four steady rests, each fitted with two halves of a 12.7 mm (0.5 in.) diam disk of PCD. In this way, precision machining can be ensured over long production runs extending over many months.

## Polycrystalline Cubic Boron Nitride Cutting Tools

Polycrystalline diamond cutting tool materials from different manufacturers have essentially the same application areas, although some may perform better than others. Similarly, there are a number of PCBN cutting materials on the market. With PCBN, however, there are a greater number of different formulations than is the case with PCD. Each manufacturer uses different second-phase materials, particle size distributions, and CBN grit concentrations (Ref 11). Therefore, it is not easy to generalize on mechanical properties, tool fabrication techniques, and machining performance of the various grades. However, two broad categories of application can be defined:

- *Rough machining* of hard ferrous materials, particularly hard cast irons with hardnesses in the range of 45 to 65 HRC, and with depths of cut between 0.5 mm (0.020 in.) and 8 mm (0.315 in.)
- *Finish machining* of hardened components, typically tool steels or surface-hardened irons, with depths of cut below 0.5 mm (0.020 in.), and typically 0.2 mm (0.008 in.)

Rough machining of hard ferrous materials requires high CBN content tools, while finishing operations are performed more effectively when PCBN cutting tools have a lowered CBN content. High CBN content tools are not suited when depths of cut are less than 0.5 mm (0.020 in.).

### Tool Fabrication

Like PCD, PCBN is produced in the form of a master blank, which is then cut into smaller blanks before being ground to the desired form. However, cutting solid PCBN into smaller blanks requires laser cutting, not wire EDM cutting, because solid PCBN has a very low electrical conductivity. A layered PCBN format (on WC) is electrically conductive and can be cut by wire EDM.

Cutting tools of PCBN, like those of PCD, are available in either the layered or solid format. PCBN blanks in the layered format have thicknesses up to 4.8 mm (3/16 in.) and are brazed to carbide tool shanks or indexable inserts. The solid PCBN product has a standard thickness of 3.18 mm (0.12 in.) and is ground to form indexable inserts.

### Toolholders for Solid PCBN Inserts

With solid PCBN inserts, it is important that high-quality toolholders be used. In particular, it is recommended that tungsten carbide support shims be used to prevent unnecessary insert breakage. Also, solid PCBN inserts should be lapped so that they are very flat and smooth.

**Fig. 13** Toolholder for solid PCBN inserts. Toolholder consists of a solid ground carbide support shim and a tungsten carbide face clamp.

Adequate mechanical and thermal support can be provided by the shim only if it is flat and clean and offers as large a contact area as possible. Surface-ground rather than lapped shims should be used, because lapping can produce a slightly domed surface. In some toolholder designs, the philosophy of having as large a contact area as possible has been pushed to the limit by using carbide shims that are brazed to steel subshims. The steel subshim is drilled and tapped from beneath, to permit the shim assembly to be fixed to the toolholder from below without the usual central fixing screw.

Toolholders should also have adequate top clamps capable of coping with aggressive service conditions. One particular problem that may occur is erosion of the

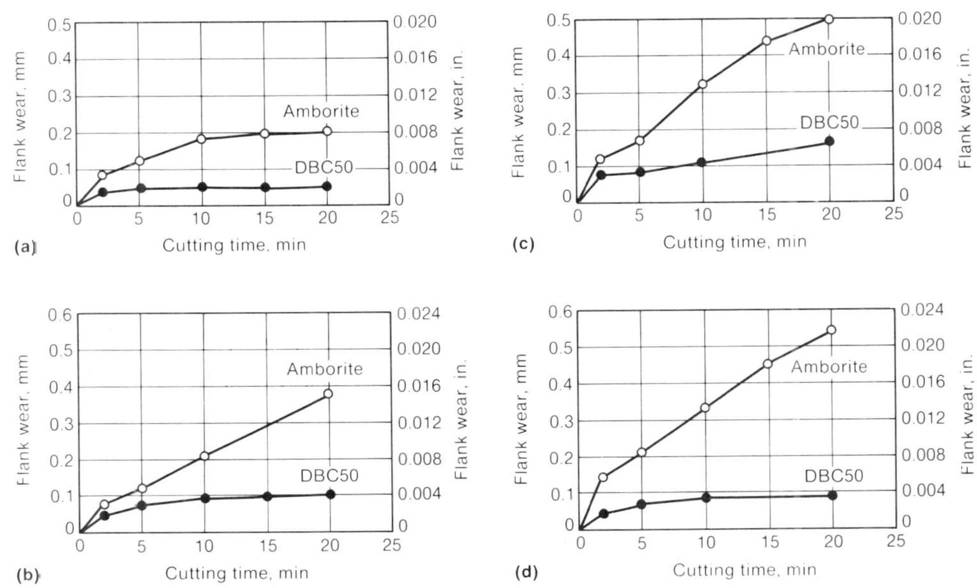


**Fig. 14** Relative wear resistance of Amborite and DBC50. (a) D3 cool-worked tool steel at 60 HRC. (b) Bearing steel at 60 HRC. (c) M2 high-speed steel at 62 HRC. (d) Hot-worked die steel at 50 HRC. Machining parameters: cutting speed = 120 m/min (395 sfm); depth of cut = 0.25 mm (0.010 in.); feed rate = 0.1 mm/rev (0.004 in./rev); dry, no coolant

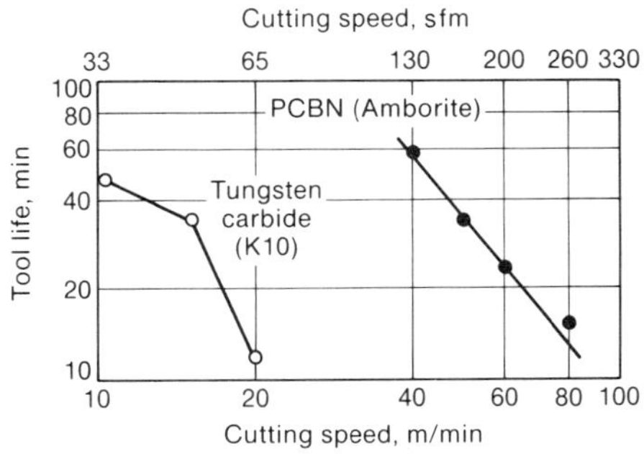


**Fig. 15** Tool life versus cutting speed for PCBN and tungsten carbide. Machining parameters: depth of cut = 2.0 mm (0.080 in.); feed rate = 0.3 mm/rev (0.012 in./rev); dry, no coolant. Workpiece: Ni-HARD 2C

clamp by swarf. This type of problem can be overcome by facing the clamp with tungsten carbide or coating it with a hardfacing alloy. A toolholder that has the features of both a solid, ground carbide support shim and a carbide face clamp is shown in Fig. 13.

## Wear Resistance

The wear resistance of PCBN tools during finishing operations can be increased by reducing the CBN content of the tool. Figure 14, for example, compares the flank wear of a roughing-grade and a finishing-grade PCBN tool. The finishing-grade tool (DBC50) with a lowered CBN content has a wear resistance three to five times higher than that of the roughing grade (Amborite) with a higher CBN content.

For some workpiece materials, increases in cutting speed result in increases in tool life (see the discussion in "Machining of Gray Cast Iron" in this article). For other materials, such as hard martensitic cast irons, increases in cutting speed result in the decrease of tool life (Fig. 15).

**Table 5 Recommended machining conditions for initial trials with finishing grades of PCBN cutting tools (DBC50)**

| Material | Hardness | Speed m/min | Speed sfm | Feed mm/rev | Feed in./min | Depth of cut mm | Depth of cut in. |
|---|---|---|---|---|---|---|---|
| Cold-work tool steels | 58–60 HRC | 100–200 | 330–650 | 0.1 | 0.004 | 0.3 | 0.012 |
| High-speed steels | 60–65 HRC | 100–200 | 330–650 | 0.1 | 0.004 | 0.2 | 0.008 |
| Hot-work die steels | 47–50 HRC | 100–200 | 330–650 | 0.1 | 0.004 | 0.3 | 0.12 |
| Bearing steels | 55–60 HRC | 100–200 | 330–650 | 0.1 | 0.004 | 0.3 | 0.012 |
| Case-hardened steels | 55 HRC | 100–200 | 330–650 | 0.1 | 0.004 | 0.2 | 0.008 |
| Hard facing alloys | 35 HRC | 200–250 | 650–820 | 0.1 | 0.004 | 0.3 | 0.012 |
| Gray cast irons | 200–220 HB | 500–800 | 1650–2600 | 0.1 | 0.004 | 0.3 | 0.012 |
| Sintered irons | 200–220 HB | 300–600 | 980–2000 | 0.1 | 0.004 | 0.2 | 0.008 |
| High tensile steels | 45–50 HRC | 100–200 | 330–650 | 0.1 | 0.004 | 0.3 | 0.012 |

## Tool Geometries

The available shapes of solid PCBN inserts are rounds, squares, triangles, and rhombuses. Various corner radii are available, and edges can be chamfered to increase edge strength. The standard chamfer (k-land) of the Amborite solid PCBN insert is 0.2 mm (0.008 in.). Edges are also honed to 20 μm (800 μin.). The Amborite insert has a thickness to fit standard negative-rake toolholders.

Layered PCBN inserts include 13 mm (½ in.) inscribed-circle triangles in both positive-rake and negative-rake styles. In addition, layered PCBN inserts also include 13 and 16 mm (½ and ⅝ in.) squares and rounds with negative rakes and indexable inserts tipped with 5 and 7 mm (0.20 and 0.28 in.) side length triangles.

## Machining Parameters

When finishing with PCBN, cutting speeds should be higher than when using roughing grades of PCBN tooling. Table 5 suggests some initial conditions for machining various materials with the finishing-grade PCBN tool (DBC50). Table 6 summarizes some machining parameters for the major applications of a roughing-grade PCBN tool (Amborite).

**Coolants** are not generally required for rough machining and are not recommended for milling. Although finishing operations can be carried out dry, use of a flood coolant is recommended. Coolant in finishing operations slightly improves tool life (Fig. 16).

## Rough Machining Applications

**Turning of Martensitic Cast Irons.** Hard martensitic cast irons may be produced by chilling during casting or by alloying with nickel and/or chromium (Ref 44). This class of material typically has hardnesses of 50 to 60 HRC and is used for its abrasion resistance as liners for solids-handling pumps, coal and ore crushing equipment, and mill rolls.

Before the introduction of PCBN cutting tools, this type of material was machined by grinding or cut at very slow cutting speeds using carbide tools. PCBN cutting tools

**Table 6 Typical machining parameters for machining with a roughing grade (Amborite) PCBN tool**

| Workpiece material | Hardness | Speed m/min | Speed sfm | Depth of cut mm | Depth of cut in. | Feed mm/rev | Feed in./rev | Comments |
|---|---|---|---|---|---|---|---|---|
| **Turning** | | | | | | | | |
| Ni-HARD | 58 HRC | 50 | 165 | 2.5 | 0.100 | 0.35 | 0.014 | Round chamfered inserts preferable |
| White/chill cast iron | 55 HRC | 60–100 | 200–330 | 2.0 | 0.080 | 0.4 | 0.016 | Round chamfered inserts used |
| High-speed steel | 62 HRC | 60–120 | 200–390 | 2.0 | 0.080 | 0.2 | 0.008 | Interrupted cuts not possible |
| Cold-work tool steel | 60 HRC | 80 | 260 | 2.0 | 0.080 | 0.25 | 0.010 | ... |
| Nickel/chromium irons | 58 HRC | 55 | 180 | 2.0 | 0.080 | 0.3 | 0.012 | ... |
| Gray cast iron | 220 HB | 600–800 | 2000–2620 | 0.1–2.0 | 0.004–0.080 | 0.1–0.4 | 0.004–0.016 | High cutting speeds and correct workpiece microstructure essential |
| Sintered iron | 200 HB | 300 | 980 | 0.5 | 0.020 | 0.1 | 0.004 | ... |
| Cobalt-base hard facing alloys | ≥35 HRC | 200–250 | 650–820 | 1.0 | 0.040 | 0.25 | 0.010 | Feed and speed must not be too low. Not all alloys can be machined successfully. |
| **Milling** | | | | | | | | |
| Meehanite | 55 HRC | 300 | 980 | 0.5 | 0.020 | 0.5(a) | 0.020(b) | Round chamfered inserts (roughing) used |
| Ni-HARD | 58 HRC | 200 | 650 | 1.0 | 0.040 | 0.25(a) | 0.010(b) | Round chamfered inserts (roughing) used |
| Cold-work tool steel | 60 HRC | 180 | 590 | 1.0 | 0.040 | 0.2(a) | 0.008(b) | ... |
| Nickel-base hard facing alloys | ≥35 HRC | 220 | 720 | 0.5 | 0.020 | 0.3(a) | 0.012(b) | ... |

(a) mm/tooth. (b) in./tooth. Source: Ref 43

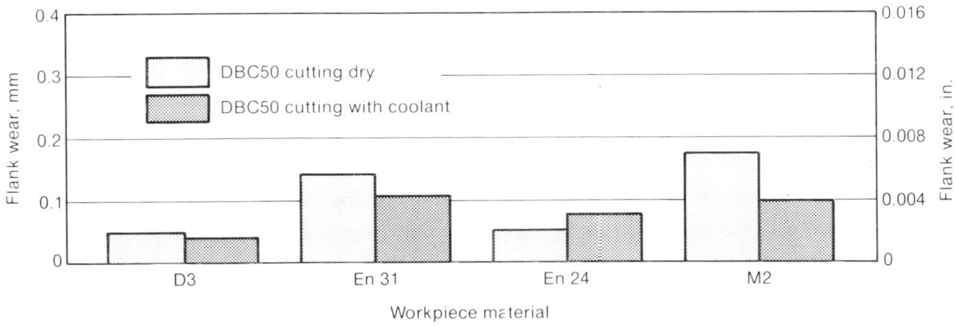


**Fig. 16** Effect of coolant on the flank wear of PCBN tools after 20 min of cutting. Machining parameters: cutting speed = 50 m/min (165 sfm); feed rate = 0.7 mm/rev (0.028 in./rev); depth of cut = 8 mm (0.315 in.)

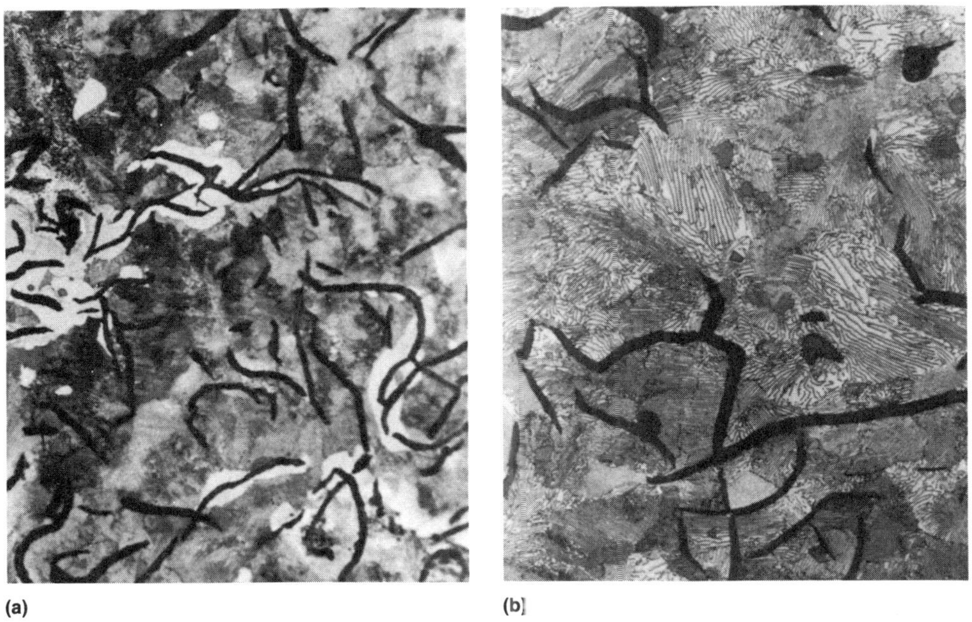

(a) (b)

**Fig. 17** Microstructures of ferritic and pearlitic cast irons. (a) Ferritic gray cast iron. (b) Pearlitic gray cast iron

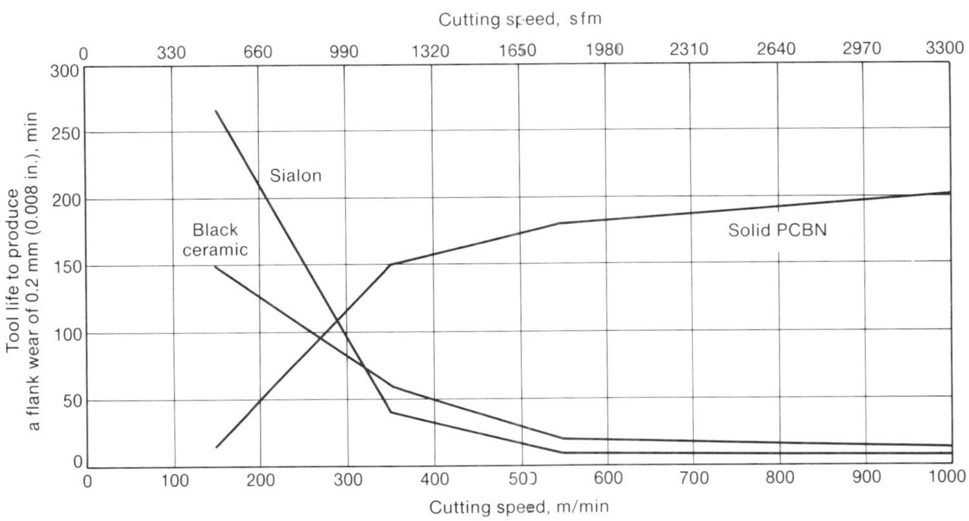


**Fig. 18** Effect of cutting speed on PCBN tool life when machining gray cast iron. Machining parameters: feed rate = 0.1 mm/rev (0.04 in./rev); depth of cut = 1.0 mm (0.040 in.)

**Table 7 The effect of microstructure on machinability when using conventional cutting tool materials**

| Matrix microstructure | Brinell hardness, HB | Tool life index |
|---|---|---|
| Ferrite | 120 | 20 |
| 50% ferrite + 50% pearlite | 150 | 10 |
| Coarse pearlite | 195 | 2 |
| Medium pearlite | 215 | 1.5 |
| Fine pearlite | 218 | 1 |
| Fine pearlite + 50% free ferrite | 240 | 0.3 |

Source: Ref 63

permit significant improvements in stock removal rates and production rates (Ref 45).

Because of their greater geometric strength, round chamfered inserts should be used whenever possible to machine hard cast irons. This is particularly important when "black" castings are being machined, that is, those requiring machining through the casting skin.

A comparison of tool life for Amborite PCBN (RNMN120300T inserts) and tungsten carbide (K10) tools when machining Ni-HARD 2C at a feed rate of 0.3 mm/rev (0.012 in./rev) and a depth of cut of 2.0 mm (0.080 in.) is shown in Fig. 15. It can be seen that while Amborite gives a 30 to 60 min tool life over a speed range of 55 to 40 m/min (180 to 130 sfm), the cemented carbide gives a similar useful life over the much lower speed range of 10 to 20 m/min (33 to 65 sfm). Above 20 m/min (65 sfm), the carbide tool life decreases rapidly, as further increases in cutting speed cause rapid softening of the tool tip.

There are many examples of the use of Amborite for general stock removal on white cast iron cited in the literature (Ref 25, 46-56). With RNMN120300T inserts, stock removal rates of the order of 40 to 50 $mm^3 \times 10^3$/min (2.5 to 3.0 $in.^3$/min) can be achieved without difficulty on stress-relieved castings. This stock removal rate is approximately five times faster than can be obtained with carbide tooling, and the resultant increase in production rate is the major factor that compensates for the increased tool costs. Secondary benefits are longer machine tool life and closer tolerances, because cutting forces are significantly lower than with carbide tooling. It has also been found possible to thread Ni-HARD, and one particular example relates to internal threading of pump impellers (Ref 57).

**Machining of Hardened Tool Steels.** The rough machining of hardened tool steels (45 to 65 HRC) is not a common application for PCBN cutting tools, because tool steels are usually rough machined in the soft state before hardening. In certain situations, however, it is convenient to be able to rough machine in the hardened state such as when it is necessary to refurbish components or correct a mistake (Ref 58).

**Fig. 19** The appearance of a worn PCBN cutting edge after machining 2600 gray cast iron engine block bores

The remachining of rolls in the steel or paper industries or the removal of worn threads from thread rolling dies (Ref 59) are other examples. More recently, there has been some work carried out with PCBN cutting tools, particularly in the bearing industry, to assess production economics of both rough and finish machining in the hardened state (Ref 60, 61).

As with rough machining of hard irons, round, chamfered, indexable inserts should be used whenever possible. Cutting speeds will generally be in the range of 50 to 100 m/min (165 to 330 sfm), and, with round inserts, depths of cut and feed rates of up to 3 mm (0.120 in.) and 0.3 mm/rev (0.012 in./rev), respectively, are possible on 60 HRC tool steels.

**Machining of Hardfacing Alloys.** Higher cutting speeds than those used for cutting hard irons or steels are required for machining hardfacing alloys. This is particularly true for the more temperature-resistant cobalt-base materials. For these alloys, cutting speeds in the range of 200 to 250 m/min (650 to 820 sfm) are generally recommended. Because of the work-hardening nature of these materials, feed rates should not be less than 0.2 mm/rev (0.008 in./rev). Round, chamfered PCBN inserts should be used whenever possible, and the depth of cut should be sufficient to enable the tool to penetrate the solid underlying material, thereby avoiding the interruptions and damaging abrasiveness of the as-cast or as-deposited alloy skin (Ref 62).

Little success has been achieved with the machining of the iron-base hardfacing alloys. The reactivity of PCBN with the soft (nonmartensitic) phases of iron should be borne in mind, even when machining nickel-base and cobalt-base grades applied to steel parent bodies. Often the machinability of the deposited layer can be reduced by the diffusion of iron from the base material.

Interrupted cuts can usually be tolerated in the nickel-base materials, given both a substantial component and a rigid machine tool. However, the milling of cobalt grades is more difficult because these alloys are particularly heat resistant. Rough machining of hardfacing alloys is typically done dry.

**Machining of Gray Cast Iron.** With hard ferrous materials, hardness is generally regarded as a reasonable guide to machinability. With gray cast irons, however, microstructure is a more fundamental indicator. A reduction in the free ferrite content or a refining of the pearlite lamellar spacing reduces the machinability of gray cast iron (Table 7).

Unfortunately, the specifications for critical components in the automotive industry, such as brake disks, usually require irons with a minimal ferrite content and a relatively fine pearlite spacing. Machinability, therefore, can be a problem, and high machining costs and production losses due to frequent tool changes can be significant.

PCBN cutting tools have been found to work well on gray cast iron, provided two important criteria are fulfilled:

- The microstructure should contain little, if any, free ferrite; that is, it should be fully pearlitic
- The cutting speed should be in excess of 500 m/min (1650 sfm)

Comparative microstructures of a ferritic and fully pearlitic gray cast iron are shown in Fig. 17. The ferrite is seen as white areas often associated with the graphite flakes. This phase is highly reactive and produces rapid wear on PCBN tools.

The formation of a fully pearlitic microstructure is mainly dependent on chemical composition and cooling rate. Chromium and copper are pearlite formers, while molybdenum tends to decrease lamellar spacing. Cooling rates depend on both the component cross section and the casting technique. As a general rule, castings which cool quickly and have a small cross section show less tendency for the formation of free ferrite. However, if cooling rates are too rapid, chilling can occur, and hard white iron and intercellular carbides can form. With conventional tools this can be catastrophic, but the high toughness and abrasion resistance of PCBN tooling permit it to cope easily.

The importance of cutting speed in machining gray iron with PCBN cutting tools is

**Fig. 20** Rough machining with a solid PCBN insert. Machining parameters: cutting speed = 50 m/min (165 sfm); feed rate = 0.7 mm/rev (0.028 in./rev); depth of cut = 8 mm (0.315 in.). Workpiece: 58 HRC white iron roll

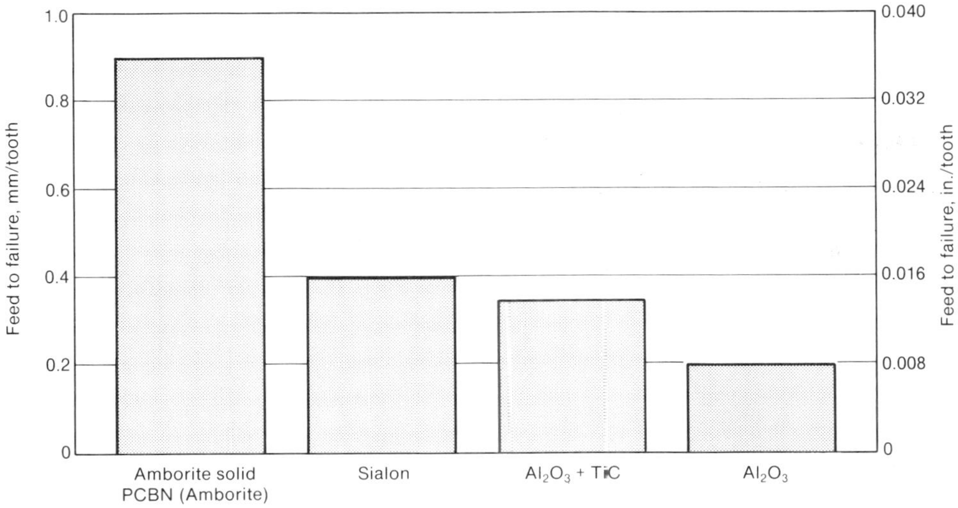


**Fig. 21** Feed-to-failure for solid PCBN and other tool materials when milling D3 steel (58 HRC)

shown in Fig. 18. It can be seen that with Amborite, tool life increases with increasing cutting speed. This differs from conventional ceramic cutting tools (such as mixed ceramics and Sialons), which show a reduction in tool life with increases in cutting speed. This surprising effect has been confirmed both in independent laboratory trials (Ref 64) and in industrial situations. For rough machining, cutting speeds therefore, should be in excess of 500 m/min (1650 sfm) to obtain satisfactory tool lives. When PCBN tools are used for finish machining, speeds of 700 to 800 m/min (2300 to 2600 sfm) are usually employed. Above 1000 m/min (3300 sfm), solid PCBN begins to behave like a conventional tool material in that its tool life decreases with further increases in cutting speed.

(a) (b) (c) (d)

**Fig. 22** Typical applications of PCBN tools with a lowered CBN content. Use of PCBN inserts (DBC50) in hardened steels (55 to 62 HRC). (a) Facing. (b) Copy turning. (c) Threading. (d) Grooving

**Example 1: Cylinder Block Boring of Gray Cast Iron With PCBN.** The bores of automotive cylinder blocks are generally rough-machined with carbide tools at relatively slow cutting speeds and then finish machined with conventional ceramic tooling before honing. The usual criteria for selecting the tool-changing interval in this operation is the surface finish of the bores and their deviation from roundness. Typically, a maximum peak-to-valley height ($R_y$) of less than 20 μm (800 μin.) is required, and the out-of-roundness should not be greater than 30 μm (1200 μin.).

The change from ceramic tools to Amborite for cylinder block boring increased tool lives from 15 to 2600 bores/edge (Ref 65). The appearance of the Amborite cutting edge after machining 2600 bores is shown in Fig. 19. Similar results have also been obtained with brake machining (Ref 66). In this case, the finish and accuracy obtained on the major brake faces of the disk are often of such high quality that the final grinding operation can be eliminated.

Because of the extreme wear resistance of PCBN cutting tools, secondary benefits also arise. Tool wear compensations to maintain tolerances need be made only infrequently, and dimensional variability of the components being supplied to subsequent work stations is reduced considerably. In transfer-line situations, savings can result from reductions in tool changing or stoppages for tool wear compensation.

**Rough machining of large workpieces,** such as mill rolls, often requires stock removal rates that are greater than the rates achieved with the usual PCBN inserts. Multiple insert toolholders have been designed; this enables two or even three inserts to be presented to the workpiece in a staggered formation. More recently, larger pieces of Amborite PCBN have become available. These large (equilateral) triangles permit cuts up to 8 mm (0.315 in.) deep. By using the long cutting edges of these larger inserts at shallow approach angles of 15° to 30°, high feed rates can be achieved even through chromium iron rolls or the black casting skin of white cast iron (55 to 60 HRC). Figure 20 shows the machining of a Ni-HARD iron roll with a solid PCBN insert. The stock removal rate is 280 mm$^3$ × 10$^3$/min (17 in.$^3$/min).

**Milling.** The fracture toughness of solid PCBN inserts (shown in Table 1) enables PCBN tooling to be used successfully for rough milling operations on hardened irons (Ref 50, 67-69) and on some hardened steels. The relative toughness of various tool materials has been evaluated by running fly-cutting tests on a block of D3 cold-work tool steel hardened to 60 HRC. The results of flycutting the face of a block 100 mm (4 in.) long by 75 mm (3 in.) wide with a single 0.8 mm (0.03 in.) nose radius tool set in a 200 mm (8 in.) diam milling cutter head are shown in Fig. 21. In this test, the feed rate is increased by 0.1 mm/tooth (0.004 in./tooth) at each successive pass across the block, and the toughness of the cutting edge is determined by monitoring the feed rate at which fracture of the cutting edge occurs. It can be seen from Fig. 21 that the feed-to-failure for a solid PCBN insert is more than twice as high as that obtained from conventional ceramic cutting tool materials.

## Finish Machining of Hardened Steels and Irons

There are a great many situations that require finish machining operations to correct the size and surface finish of hard ferrous materials. This is traditionally carried out by grinding. Typical examples are the grinding of tool steels, bearing steels, die steels, and high tensile steels after hardening or the finish machining of induction-hardened irons, surface-hardened irons, and some hardfacing alloys.

**Finishing grades** of PCBN cutting tools, which have a lower CBN content than do the roughing grades, are produced by substituting a portion of the CBN with a ceramic. The slightly reduced toughness is not a disadvantage in finishing operations, and these composite grades provide a viable alternative to grinding.

An example of a finishing grade PCBN cutting tool is DBC50. Table 1 gives the mechanical and thermal properties of DBC50, and Fig. 22 shows its use in facing, copy turning, threading, and grooving. Like other layered PCBN products, DBC50 is brazed to carbide tools or indexable inserts and then is ground to the required (usually negative) geometry. Chamfered edges (k-lands) can also be used to improve edge strength. Typically, these are 0.05 to 0.2 mm (0.002 to 0.008 in.) wide at an angle of 20° to 30°.

### REFERENCES


1. C. Phaal, Surface Studies of Diamond, *Ind. Diamond Rev.*, No. 1965, p 486-489
2. A. Sawaoka, Boron Nitride: Structural Changes by Shock Compression and Preparation of Superhard Compacts by Very High Pressure Sintering, *Ceram. Bull.*, Vol 62 (No. 12), 1983, p 1379
3. R. Berman and F. Simon, *Z. Elektrochem.*, Vol 49 (No. 333), 1955
4. R.J. Wedlake, Technology of Diamond Growth, in *The Properties of Diamond*, J. Field, Ed., Academic Press, 1979
5. P.A. Bex and G.R. Shafto, The Influence of Temperature and Heating Time on PCD Performance, *Ind. Diamond Rev.*, Vol 3, 1984, p 128-132
6. P. Herzig, Grinding Polycrystalline Diamond Tools, *Ind. Diamond Rev.*, Vol 4, 1982, p 212-214
7. J.A. Pfluger, Automatic Grinding of PCD/PCBN Blanks, *Ind. Diamond Rev.*, Vol 3, 1986, p 128-130
8. P.J. Heath and M.E. Aytacoglu, Edge Preparation of SYNDITE PCD and Parameters, *Ind. Diamond Rev.*, Vol 3, 1984, p 133
9. P. Silveri, Shaping PCD Tools by Rotary EDM, *Ind. Diamond Rev.*, Vol 3, 1986, p 108-109
10. P. Herzig, New Machine for Lapping the Table of PCD Inserts, *Ind. Diamond Rev.*, Vol 3, 1983, p 134
11. P.J. Heath, "Structure Properties and Applications of Polycrystalline Cubic Boron Nitride," Paper presented at Superabrasives '85, New Developments in Diamond and CBN (Chicago), April 1985
12. A.G. Evans and D.B. Marshall, Wear Mechanisms in Ceramics, in *Fundamentals of Friction and Wear of Materials*, American Society for Metals, 1980, p 439-452
13. S. Herbert, SYNDITE Tools Reduce Downtime by 80:1, *Ind. Diamond Rev.*, Vol 3, 1983, p 117-120
14. M. Wolf and R. Dreher, Machining of Aluminium Engine Parts for the Porsche 928, *Ind. Diamond Rev.*, Vol 5, 1981, p 254-257
15. S. Herbert, Jaguar's Long Distance Runner, *Ind. Diamond Rev.*, Vol 3, 1987, p 100-102
16. S. Herbert, Austin Rover's PCD Switch Pays Dividends, *Ind. Diamond Rev.*, Vol 6, 1985, p 278-281
17. W. Hück, New Techniques in the Machining of Commutators, *Ind. Diamond Rev.*, Vol 2, 1983, p 69-74
18. Diamond Composites Hit the Right Note, *Ind. Diamond Rev.*, Vol 2, 1981, p 67
19. W. Stief, Tumpets Depend on Precision, *Ind. Diamond Rev.*, Vol 3, 1987, p 112
20. E. Quadri, Finish Machining of Plastics, *Ind. Diamond Rev.*, Vol 6, 1980, p 222
21. A Better Way to Machine Phenolics, *Ind. Diamond Rev.*, Vol 4, 1982, p 215
22. G. Biastoch, Contact Lenses Machined With SYNDITE and Natural Diamond, *Ind. Diamond Rev.*, Vol 2, 1983, p 66-68
23. F. Waltz, Machining Wood and Plastics With SYNDITE Tools, *Ind. Diamond Rev.*, Vol 6, 1982, p 339-340
24. S. Herbert, Modern Insulation Materials—A New Machining Approach, *Ind. Diamond Rev.*, Vol 3, 1983, p 144-147
25. Roll Turning Using SYNDITE and AMBORITE Tools, *Ind. Diamond Rev.*, Vol 6, 1984, p 334-340
26. S. Herbert, A Granite Machining Challenge Answered, *Ind. Diamond Rev.*, Vol 4, 1984, p 199-201
27. G. Spur and V.E. Wunsch, Turning FRP With SYNDITE—Test Results, *Ind. Diamond Rev.*, Vol 4, 1985, p 195-199
28. W. Koglmeir, SYNDITE for Machining Fibre Glass Reinforced Epoxy Resin, *Ind. Diamond Rev.*, Vol 11, 1978, p 395-397
29. A Better Way to Machine GRP Pipes, *Ind. Diamond Rev.*, Vol 1, 1980, p 17
30. P. Schimmel, Economic Machining of GRP Laminates, *Ind. Diamond Rev.*, Vol 6, 1982, p 348
31. B. Cullingworth, Syndite Helps Clinch Order, *Ind. Diamond Rev.*, Vol 4, 1985, p 185
32. E. Heimbrand, Machining Composite Metals With PCD, *Ind. Diamond Rev.*, Vol 4, 1985, p 187-190
33. Wood Products Latch Onto PCD, *Ind. Diamond Rev.*, Vol 3, 1984, p 159-162
34. H. Lach, PCD Tools for Woodworking, *Ind. Diamond Rev.*, Vol 4, 1985, p 166-167
35. H. Moitzi, PCD for Chipboard Machining, *Ind. Diamond Rev.*, Vol 4, 1985, p 175-181
36. H. Schulz, Machining Wood Products With PCD, *Ind. Diamond Rev.*, Vol 5, 1984, p 263-265
37. Elizabeth Ann's 100:1 Favourite, *Ind. Diamond Rev.*, Vol 4, 1985, p 163-165
38. Diamond Tools in the Wood Products Industry, *Ind. Diamond Rev.*, Vol 6, 1980, p 214-221
39. H. Prekwinkel, Woodworking With SYNDITE Tools, *Ind. Diamond Rev.*, Vol 3, 1983, p 148-150
40. New Tools Ensure a Good Smoke, *Ind. Diamond Rev.*, Vol 6, 1980, p 211
41. B. Cullingworth, SYNDITE's Potential in Tile Cutting, *Ind. Diamond Rev.*, Vol 5, 1985, p 243
42. G. Ottevanger, Boring Sintered Carbide Rolls With SYNDITE PCD Tools, *Ind. Diamond Rev.*, Vol 3, 1984, p 154-156
43. T.A. Notter and P.J. Heath, "The Selection of Machining Parameters Using AMBORITE," Paper presented at CSIR Second Seminar, Efficient Metal Forming and Machining (Pretoria, South Africa), Nov 1980
44. British Standard B.S. 4844: Part 2: 1972

Abrasion Resisting White Cast Irons, Part 2: Nickel-Chromium Grades
45. "AMBORITE Machining of Ni-HARD 2C (57HRC)," Leaflet T14.1, De Beers Industrial Diamond Division (Pty) Ltd.
46. S. Herbert and P.J. Heath, AMBORITE; an Answer to the Ni-HARD Machining Problem, *Ind. Diamond Rev.*, Vol 2, 1981, p 53-56
47. C. Stevens, A Bold Tooling Investment Brings 40% Productivity Increase, *Ind. Diamond Rev.*, Vol 3, 1982, p 130-134
48. H. Muller and K. Steinmetz, Machining Mineral Crushing Rings With AMBORITE, *Ind. Diamond Rev.*, Vol 1, 1983, p 30-33
49. G. Johnson, Machining in Half the Time, *Ind. Diamond Rev.*, Vol 6, 1985, p 309-310
50. Milling of Ni-HARD 2C, *Ind. Diamond Rev.*, Vol 6, 1983, p 305
51. M.O. Nicolls, Machining Ni-HARD Augers with AMBORITE, *Ind. Diamond Rev.*, Vol 1, 1984, p 28-29
52. S. Herbert, New Potential in the Heavy Machine Shop, *Ind. Diamond Rev.*, Vol 6, 1984, p 307-309
53. P. Silveri, Pump Producer Goes for AMBORITE, *Ind. Diamond Rev.*, Vol 6, 1985, p 287-288
54. F.W. Mansfeld, Turning of Chill Cast Rolls on CNC Production Machines, *Ind. Prod. Eng.*, Vol 1, 1982, p 69
55. S. Herbert, AMBORITE Cuts Machining Time in Mill Roll Reclamation, *Ind. Diamond Rev.*, Vol 5, 1981, p 258-260
56. S. Herbert, Roll Machining Costs Reduced, *Ind. Diamond Rev.*, Vol 3, 1982, p 140-141
57. S. Herbert, High Speed Threading in Parent Hard Metal, *Ind. Diamond Rev.*, Vol 1, 1986, p 19-21
58. B. Cullingworth, Seven Times Cheaper With AMBORITE, *Ind. Diamond Rev.*, Vol 4, 1984, p 195
59. S. Herbert, Cycle Times Reduced by 90% in Re-machining of Thread Rolls, *Ind. Diamond Rev.*, Vol 4, 1981, p 176-178
60. G. Werner and W. Knappert, Machining Hardened Bearing Races With PCBN, *Ind. Diamond Rev.*, Vol 3, 1985, p 117-120
61. W. Konig and T. Wand, Turning Bearing Steel With AMBORITE and Ceramic, *Ind. Diamond Rev.*, Vol 3, 1987, p 117-120
62. P. Fitton, Machining Hard-Faced Components—Better Economics Are Possible, *Ind. Diamond Rev.*, Vol 6, 1984, p 310-312
63. Machining of Cast Iron, in *Machining*, Vol 3, 8th ed., *Metals Handbook*, American Society for Metals, 1967, p 333-352
64. D.K. Aspinwall and W. Chen, Machining of Grey Cast Iron Using Advanced Ceramic Tool Materials, to be published
65. E. Lurger, J. Steber, V. Frommberz, *et al.*, Machining of Cylinder Bores at BMW, *Ind. Diamond Rev.*, Vol 3, 1982, p 135
66. K. Sanger and K. Steinmetz, No Putting the Brakes on AMBORITE, *Ind. Diamond Rev.*, Vol 6, 1982, p 350-355
67. J. James, A Bed-Time Story From Harrison, *Ind. Diamond Rev.*, Vol 3, 1985, p 113-115
68. J. Shanks, AMBORITE Saves the Day, *Ind. Diamond Rev.*, Vol 3, 1983, p 133
69. Milling Meehanite With AMBORITE, *Ind. Diamond Rev.*, Vol 4, 1981, p 179

# Cutting Fluids

# Metal Cutting and Grinding Fluids

Elliot S. Nachtman, Tower Oil & Technology Company

METAL CUTTING AND GRINDING OPERATIONS involve a complex set of operating parameters, and the choice and effectiveness of a cutting or grinding fluid are determined by:

- The design, rigidity, and operating condition of the machine tool
- The speed, feed, and depth of cut
- The composition, finish, and geometry of the cutting tool
- The mode of fluid application
- The geometry of the material to be machined
- Surface coatings
- The composition, microstructure, and residual stress distribution in the workpiece

When properly applied, cutting fluids can increase productivity and reduce costs by making possible the use of higher cutting speeds, higher feed rates, and greater depths of cut. The effective application of cutting fluids can also lengthen tool life, decrease surface roughness, increase dimensional accuracy, and decrease the amount of power consumed as compared to cutting dry.

Knowledge of cutting fluid functions, types, physical limitations, and composition plays an important role in the selection and application of the proper fluid for a specific machining situation. The functions, chemistry, control, application, recycling, and disposal of cutting fluids will be discussed in this article. The health implications and biology of cutting fluids will also be discussed.

## Functions of Cutting and Grinding Fluids

Depending on the machining operation being performed, a cutting or grinding fluid has one or more of the following functions:

- Cooling the tool, workpiece, and chip
- Lubricating (reducing friction and minimizing erosion on the tool)
- Controlling built-up edge on the tool
- Flushing away chips
- Protecting the workpiece tooling and machine from corrosion

The relative importance of each of these functions depends on the work material, the cutting or grinding tool, the machining conditions, and the finish required on the part.

**Grinding fluids** perform several of the same functions as cutting fluids. Grinding fluids lubricate the grit/workpiece interface, thus reducing the generated heat and the power requirements for a given material removal rate.

The primary difference between the functions of grinding and cutting fluids is that lubrication is more important in grinding than in cutting. In metal cutting, most of the heat generated during the cutting operation is carried away in the chip. Relatively less heat is generated in the workpiece and the tool. In the case of grinding, however, most of the heat is retained in the workpiece. Therefore, lubrication becomes more important for grinding fluids than for cutting fluids.

**Cutting Fluids.** Two functions of cutting fluids include lubrication and cooling so that the frictional forces and temperature are reduced at the tool/workpiece interface. In high-speed cutting operations, the cooling provided by the cutting fluid is its most important function. At moderate cutting speeds both cooling and lubrication are important, but at low speeds, lubrication becomes the dominant function of a cutting fluid.

**Chip formation and built-up edges** are related to the frictional effects of metal cutting. Figure 1 shows a schematic of the cutting operation with a single-point tool. A tool moving with a velocity $V$ and a depth of cut $t_o$ creates a chip of thickness $t_c$ that is greater than $t_o$. The chip is generated at a shear plane that makes an angle $\phi$ with the direction of cut. This angle, known as the rake angle, is an important variable in the mechanics of chip formation. The relief, or clearance, angle is also important because it provides potential access to the cutting zone for lubrication.

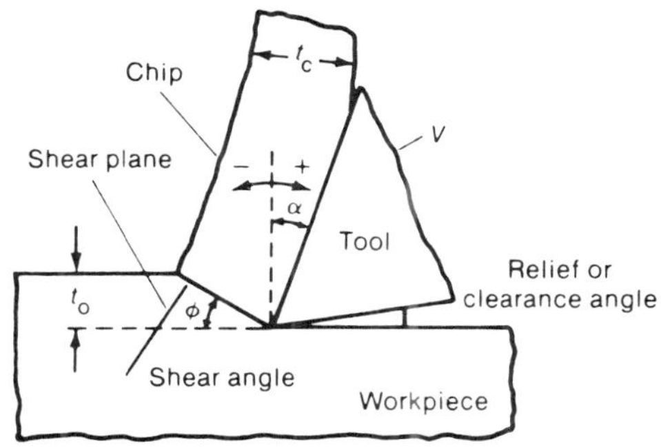


**Fig. 1** Schematic of the cutting process with a single-point tool. Source: Ref 1

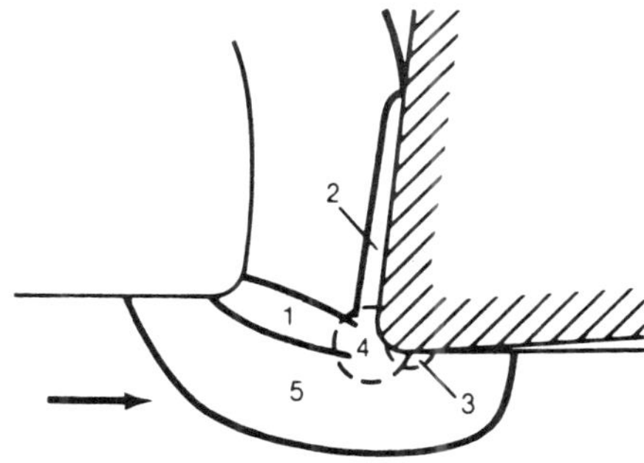


**Fig. 2** Zones of deformation and friction in chip formation. Source: Ref 2

In considering the potential for improving the cutting process with a cutting fluid, Fig. 2 illustrates the major areas of deformation and friction that occur during the generation of chips. In Fig. 2, zone 1 indicates the area of strain hardening that forms in the material being cut ahead of the tool. Microcracking can take place in the zone, and relatively high temperatures result from the deformation and resultant strain hardening. In zone 2, the deformed chip moves out of the shear zone and flows up the surface of the tool. As the chip slides up the face of the rake of the tool, it generates more heat as a result of friction between the chip and the tool. In zone 3, as the tool traverses the freshly cut surface, further rubbing of the tool against the workpiece material takes place, thus generating friction and additional deformation. As chip formation proceeds, the tool edge forms a built-up edge (zone 4), which creates more local plastic deformation and friction. In zone 5, below the area of primary metal removal, additional plastic deformation takes place, along with some strain hardening. The geometry of the chips varies with the workpiece material and the cutting conditions. The various types of chips are illustrated in Fig. 3.

**Lubrication.** Cutting fluids improve tool life and allow higher cutting speeds by reducing the amount of friction that occurs

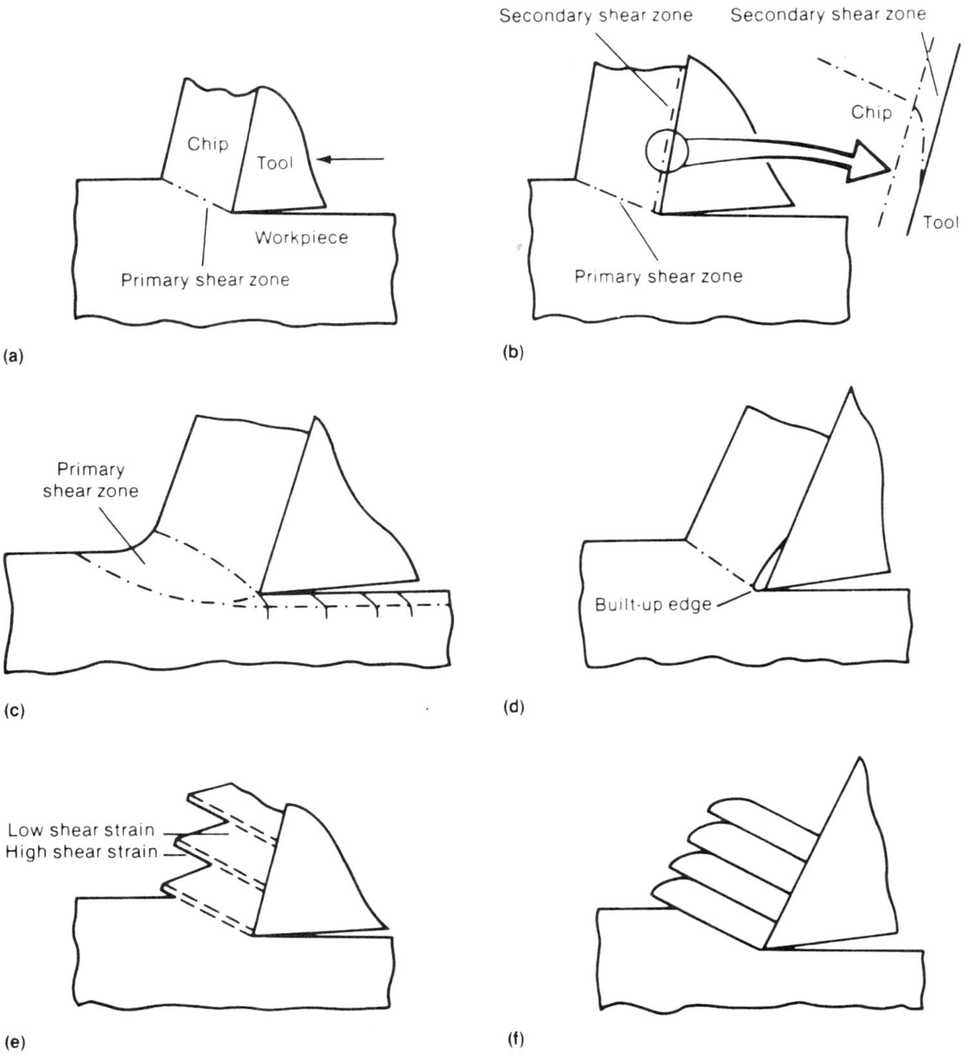


**Fig. 3** Types of chips obtained in metal cutting. (a) Continuous chip. (b) Continuous chip with a secondary shear zone. (c) Continuous chip with a large primary shear zone. (d) Built-up edge in a continuous chip. (e) Inhomogeneous (serrated) continuous chip with regions of low and high shear in the primary zone. (f) Discontinuous chip. Source: Ref 1

during the cutting process. Cutting fluids with good lubricating qualities can also:

- Allow the formation of a continuous chip when low cutting speeds result in the formation of discontinuous chips or serrated continuous chips (Fig. 3)
- Reduce the frictional forces between the tool and the rake face
- Reduce the size of the built-up edge or in some cases eliminate built-up edge formation. Total elimination of the built-up edge will produce a superior finish on the part being machined and will result in less frictional drag on the flank face
- Reduce adhesive wear by reducing adhesion between the tool and the chip or the workpiece
- Produce insignificant lubricant effects in the case of extremely brittle materials that yield very small discontinuous chips

**Cooling.** Cutting fluids reduce the temperature of the metal cutting operation by transferring heat away from the workpiece and the tool. Some of the factors involved in cooling are as follows:

- Cooling effects due to the application of the cutting fluid increase the shear strength of the material being cut, thus increasing the forces required for metal cutting. Generally, this effect is small for most metals
- The cooling effects of cutting fluids may be deleterious if the change in temperature caused in the cutting tool is abrupt and discontinuous. Abrupt changes in temperature may cause fracture and spalling of the tool; ceramic tooling is particularly sensitive in this regard
- The cooling from cutting fluids is generally related to their thermal properties. In general, cooling efficiency is less for an oil than for an emulsion and is greatest with a water solution. Frictional effects, however, may complicate this relationship because the lubricating properties influence the amount of heat generated
- Cooling efficiency can be reduced by the heat transfer characteristics of high-viscosity fluids. High cutting speeds can initially improve the cooling because the viscosity of the cutting fluid decreases with temperature, but beyond a certain temperature this beneficial effect on cooling is no longer present
- The effectiveness of cooling depends on the amount of surface wetting, fluid viscosity, chemical reactivity and molecular size, and the physical characteristics of fluid flow

## Chemistry of Cutting and Grinding Fluids

Metal cutting and grinding fluids are of two general types: solutions and emulsions. Solutions consists of a base fluid such as petroleum oil, a petroleum solvent, a synthetic fluid, or water. These base fluids can then be formulated with various additives that are soluble in the fluid. Emulsions, on the other hand, are composed of two phases: a continuous phase consisting of water, and a discontinuous phase consisting of small particles of oil, petroleum, or synthetic fluid suspended in the water. These emulsions are commonly called soluble oils.

Oil or synthetic solutions generally have the highest lubricating capabilities and the lowest cooling efficiencies. Water-base solutions, on the other hand, have the highest cooling efficiencies and lower lubrication effectiveness. In general, emulsions tend to have moderate properties for both cooling and lubrication.

**Solutions.** Some cutting and grinding solutions are described below with regard to three types of base fluid. In all metal cutting and grinding solutions, additives are selected that are soluble or in some cases dispersible in the fluid. In combination, the fluid and the additive display the desired properties.

*Cutting Oils.* The so-called cutting or grinding oils have either a naphthenic or paraffinic oil or a petroleum solvent as the primary ingredient. Paraffinic mineral oils differ from naphthenic base oils in two ways. First, paraffinic oils have a much higher concentration of straight-chain carbon atoms varying greatly in length, and second, they have a smaller concentration of ring compounds, such as naphthenic and polycyclic aromatic compounds. Paraffinic oils exhibit greater oxidation resistance than naphthenic oils and tend to maintain their viscosity over a wider temperature range. On the other hand, naphthenic oils tend to form more stable solutions of additives than the paraffinic oils. In naphthenic oils, the aromatics in the oil are surface active and therefore provide improved load-carrying capacity. All mineral oils contain a large variety of ring and straight-chain com-

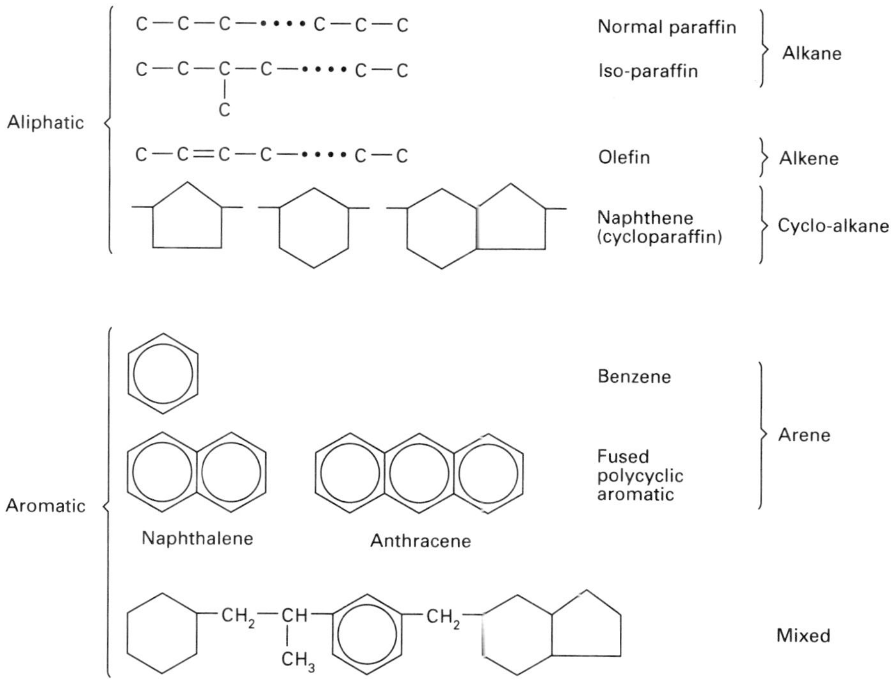


**Fig. 4** Some components of mineral oil

pounds of varying molecular weight. Some of these typical structures are illustrated in Fig. 4.

*Synthetic fluid lubricants* have a controlled molecular structure with predictable properties. Synthesized hydrocarbons such as the polyalphaolefins have been used to replace mineral oil in some special cutting fluid applications. Long-chain alcohols have also been used for special applications. However, a significant market has not developed for these synthetic cutting fluids.

*Water-base solutions* are often termed "synthetics" in industry. Water-base solutions have excellent heat transfer characteristics because of the high heat capacity of water. However, the purity of the water can significantly affect the performance of water-base cutting fluid solutions. Biological attack, foaming, additive displacement, and deposit formation are some consequences of mineral concentrations or other impurities that may be present in the water.

**Emulsions** consist of immiscible fluids that form a relatively stable mixture because of emulsifiers or surface-active chemicals (which are soluble in the fluids). Emulsions for metal cutting and grinding fluids consist primarily of a continuous phase of water containing suspended mineral oil or synthetic fluid. The particle size of the suspended fluid varies and depends on the effectiveness of the surfactant chemicals used to emulsify the system. Clear emulsions can be produced when the suspended phase consists of sufficiently fine particle sizes, but in most cases emulsions have a milky-white or blue-white color, depending on the chemistry of the additives. In these cases, the particle size is larger.

In general, when suitable surfactants are added to a pair of immiscible liquids such as oil and water, the surfactants will be absorbed at the interface between the two liquids. The hydrophilic group, that is, the part of the molecule of the surfactant that is water soluble, will orient itself so as to become part of the water phase, and the lipophilic or oil-miscible portion of the molecule will orient itself so as to become part of the oil phase. A large number of surfactants are available for use in promoting the development of a stable emulsion. In general, combinations of emulsifiers or surfactants are used. Soaps of long-chain fatty acids, phosphate esters, sulfonates, and ethoxylated alcohols are frequently used in appropriate ratios as components of the emulsion-surfactant system.

*The stability of emulsions* is important in metal cutting and grinding operations, and the destabilization of an emulsion is to be avoided. Destabilization often occurs because of the buildup of minerals from frequent additions of impure water and the buildup of swarf from the machining operation. The inadvertent addition of cleaners or dirt may also break the emulsion.

Splitting of the emulsion, when it occurs, generally results in the formation of two distinct liquids: water and the oil floating at the top. However, a phenomenon known as creaming may also occur, which produces a thick cream layer that floats on the surface. This layer is not so much a result of the breaking of the emulsion but rather is the product of two separate emulsions being created. The emulsion at the top has a much higher concentration of suspended oil particles. The presence of the cream may indicate that a process of breaking of the emulsion is about to begin. On the other hand, such mixtures may be advantageous in some metal cutting operations. Typical stages in the breaking of emulsions are illustrated in Fig. 5.

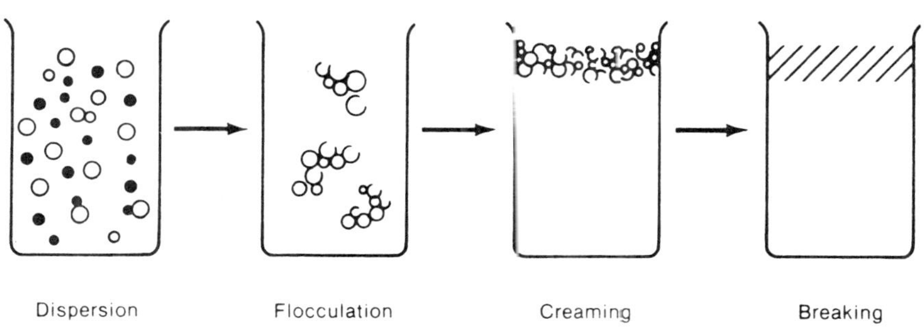


**Fig. 5** Instability in emulsions. Source: Ref 3

**Additives.** Some of the important classes of additives used in both solutions and emulsions are described below.

*Extreme-Pressure (EP) Additives.* These chemical compounds vary in structure and composition and are sufficiently reactive with the metals being machined to form relatively weak compounds at the tool/workpiece interface. Extreme-pressure additives serve as solid lubricants with low binding energy and therefore reduce the friction between the tool and the workpiece. They are primarily sulfurous additives (such as sulfurized esters of fatty acids), chloride additives (such as chlorinated hydrocarbons or chlorinated esters), or phosphorous additives (such as phosphoric acid esters). Solid lubricants such as molybdenum disulfide have also been used in small amounts. These solid lubricants deposit on the metallic surface and reduce the friction between the tool and the workpiece. Borates have been added for the same purpose. However, organic molecules with sulfur and chlorine are by far the most widely used EP additives. The chemical reactions that occur during the formation of sulfides and/or chlorides as a result of workpiece-additive interaction are very complex and not clearly understood.

**Table 1 Biocides for use in cutting and grinding fluids**

| Manufacturer | Trade name and active ingredients | Recommended dose (% of formulated biocide) | EPA registration number | Comments |
|---|---|---|---|---|
| IMC Chemical Group, Inc. | Bioban P-1487: 4-(2-nitrobutyl)-morpholine 70%; 4,4′-(2-ethyl-2-nitrotrimethylene) dimorpholine 20% | 0.01–0.3 | 271–30 | Good oil solubility; stable in oil-water emulsion concentrates |
| | Tris Nitro: tris(hydroxymethyl) nitromethane; aqueous 50%, powder 100% | 0.2 aqueous<br>0.1 powder | 271–26<br>271–18 | Poorly compatible with most concentrates; recommended for tankside use |
| Lehn & Fink Industrial Products, Division of Sterling Drug Inc. | Grotan BK(a): hexahydro-1,3,5-tris (2-hydroxyethyl)-*s*-trizane 78% | 0.15 | 10,000–1 | Stable in many concentrates; less effective against fungi at low dose levels |
| Olin Corporation | Zinc omadine: zinc 2-pyridinethiol-1-oxide; powder 95%, aqueous dispersion 48% | 0.0079<br>0.015 | 1258–840<br>1258–841 | Not soluble in concentrates |
| | Sodium omadine: sodium 2-pyridinethiol-1-oxide; powder 90%, aqueous solution 40% | 0.005<br>0.0115 | 1258–842<br>1258–843 | Soluble in most concentrates except with difficulty in oil-water emulsion concentrates; more effective against fungi and yeasts than bacteria |
| | Triadine 10: hexahydro-1,3,5-tris (2-hydroxyethyl)-*s*-triazine 63.6%, sodium 2-pyridinethiol-1-oxide 6.4% | 0.07 in synthetic fluids or 0.1 in oil-containing fluids | 1258–990 | Used in oil-water emulsion concentrates with care; compatible with most other concentrates |
| Rohm and Haas Company | Kathon 886 MW: 5-chloro-2-methyl-4 isothiazolin-3-one 8.6%; 2-methyl-4-isothiazolin-3-one 2.6% | 0.0025–0.0125 | 707–129 | Stable in some concentrates; not particularly in oil-water emulsion concentrates—possibly inactivated by some amines and sulfides |
| R. T. Vanderbilt Company, Inc. | Vancide TH: hexahydro-1,3,5-triethyl-*s*-triazine 95% | 0.05–0.1 | 1965–55 | Good oil and water solubility; stable in most concentrates; strong amine odor |
| | Vancide 51: sodium dimethyldithiocarbamate 27.6% sodium 2-mercaptobenzothiazole 2.4% | 4% in water as bactericide; 2% in water as fungicide | 1965–8 | Stable in some concentrates but recommended at-use dilution by user |

(a) This product is marketed as Grotan in the United States.

*Detergents.* Compounds such as long-chain alcohols, substituted benzene sulfonic acid, and petroleum sulfonic acids can reduce or prevent deposit formation on the workpiece.

*Antimisting Additives.* Airborne contamination by the metal cutting fluid in the plant is a long-standing problem that occurs when oil-base solutions are used. The addition of small quantities of acrylates or polybutanes will reduce mist formation by encouraging the buildup of larger particle sizes, which are heavier and much less readily airborne.

*Antifoaming Additives.* Foaming generally occurs when agitation from either the cutting operation or fluid handling introduces air into the fluid. To prevent or minimize the formation of foam, the free energy of the film surface must be reduced. Antifoaming agents have been developed for this purpose. Polyalkoxysiloxanes, fumed silica, high molecular weight amides, and polyglycols are effective in specific metal cutting fluids.

*Odor Masks.* High temperatures at the tool/workpiece interface heat the fluid and often result in odors that are disagreeable to the operator. Pine oil, cedar oil, and sassafras essence have been used to mask these odors, thus making the fluid more acceptable in long-term operations.

*Corrosion Inhibitors.* The corrosion of machine parts and the machine tool can be a problem, particularly with water-base fluids. Sulfonates, borates, and benzotriazoles have been used as additives in cutting fluids to help prevent corrosion. Many organic amines and sulfonates, which provide corrosion protection, may serve another purpose—that of providing effective surfactant characteristics as well as corrosion protection. In the case of copper alloys, toluyltriazole is an effective inhibitor.

*Dyes.* Both oil- and water-soluble dyes are used to assist in the identification of the metalworking fluid and to help in identifying the location of the fluid with respect to the application technique.

*Antimicrobial Agents.* Microbial growth will take place in cutting fluids that intentionally or inadvertently contain water. Bacteria, fungi, and/or mold will grow, depending on the growth conditions and the competition for nourishment among these organisms. Sulfate-reducing bacteria produce the well-known "Monday morning stink." These microbes do not require air, and they grow at the bottoms of sumps, attacking the sulfur in EP additives or the sulfur incidentally present in a cutting oil. This produces malodorous complex sulfur compounds. A number of biocides are available for use in attacking and killing bacteria, mold, and fungi. Table 1 lists biocide manufacturers, the active ingredients of the biocides, and other pertinent information.

## Selection of a Cutting or Grinding Fluid

Metal cutting or grinding fluid selection depends on an evaluation of a large number of interrelated factors. Some of the pertinent factors have nothing to do with the particular metal cutting or grinding operation in question, but rather concern the ease of cleaning the part after production, the cost of recycling the fluid, the cost of fluid disposal, the possibility of adverse effects on operator health and safety, and the cost of the fluid itself. Nevertheless, the technical criteria of the machining process must also underlie the choice of a particular metal cutting or grinding fluid. These criteria include the desired tolerances, tool life, surface finish, and energy consumption. Fluids must also be noncorrosive to the equipment and to the part being machined.

## Table 2 Selection guide for cutting fluids

These recommendations are general guidelines. They are influenced by tool material and composition, workpiece material composition and treatment, and the machine tool. Machining with tungsten carbide and ceramic tools can often be carried out more effectively without a cutting fluid on aluminum alloys, copper alloys, plastics, and steels, depending on operating conditions.

| | Operation(a) | | | | | | | |
|---|---|---|---|---|---|---|---|---|
| Material | Broaching (internal and external) | Tapping, threading, deep-hole drilling | Screw machining | Milling, drilling, shaping | Turning | Sawing | Surface, cylindrical, and centerless grinding | Crush and form grinding |
| Carbon steels | E2c, S2c | E2d, S2c, O5d | E2d, O4d, O5d | E2c, S2c, O4c, O5c | E2d, S2c, O4d, O5d | E2c, S2c, O3c | E2c, d; S2c, d; O5c | O3c, O5c |
| Free-machining low-carbon steels | E2d, S2c | E2c, S2c, O5d | E2d, O4d | E2d, S2d, O4d, O5d | E2d, S2c, d; O4d | E2c, S2c, O3c | E2c, d; S2c, d; O5c | E2c, d; O3d, O5d |
| Alloy steels | E2c, O5c, d; O7c | O4c, O5c | E2c, O5c | E2c, S2c, O3c, O5c | E2c, S2c, O3c, O5c | E2c, S2c, O5c | E2c, d; S2c, d; O5c | O4d, O5d |
| Stainless steels | S2c, O3c, O5c | O5c | E2c, O5c, O6d | E2c; O5c, d | E2c, S2c, O3c | E2c, S2c, O5c | E1c, d; S1c, d; O4d | O4d, O5d |
| Tool steels | E2c, S2c, O4c, O5c | O4c, O5c | E2c, O5c | E2c, S2c, O3c, O4c | E2c, S2c, O4c, O5c | E2c, S2c, O5c | E2c, d; S2c, d; O5c | O4d, O5d |
| Cast irons | E1c, E2c, S1c, S2c | E2c, S2c, O5c | E2d, O4d | E1c, E2d, S1c, d; S2c, d | E2c, O4c | E2c, d | E2c, d; S2c, d | E2c, O5d |
| Aluminum and alloys | E2d, S2d, O6d, O7d | E2d, S2d, O5d, O7d | E2c, d; O4d | E2c, d; S2c, d; O4d | E2c, d; S2c, d | E2c, d; S2c, d; O6d | E2d, S2d | E2c, d; O6d, O7d |
| Copper and alloys | E1d, S1d, O6d | E2d, S2d, O5d, O7d | E2d, O7d | E2d, S2c, d; O7d | E2c, d; S2c, d; O4d | E2c, d; S2c, d; O6d | E2d, S2d | E2d, S2d |
| Titanium and alloys | E2c, S2c | O3d, O5c | E2c, O4c | O4c, O5c | E2c, S2c | E2c, S2c, O5c, O7c | E2d, S2d | E2d, O4d |
| Refractory alloys | O5c | O5c | E2c, O4c | O4c, O5c | E2c, S2c | S2c, O5c | E2d, S2d | E2c, d; S2c, d |
| High-temperature alloys (iron, nickel, and cobalt base) | E2c, S2c, O5c | O5c | E2c; O5c, d | S2c, O5c, O7c | E2c, O3c | E2c, S2c, O5c | E1c, d; S1c, d; O4d | O3c, O5c |
| Plastics | E1d, E2d, S1d, S2d, O6d | E2c, S2c, O6d | E2d, O1, O4d | E2d, S2d, O1 | E2d, S2d | E1d, E2d, S1c, S2c | E1d, E2d, S1d, S2d | E1d, E2d, S1d, S2d |

(a) E, emulsions: 1, surface-active; 2, extreme-pressure. S, solutions: 1, surface active; 2, extreme-pressure. O, mineral oils: 1, straight mineral oil; 2, mineral oil + fat; 3, mineral oil + fat + sulfur; 4, mineral oil + fat + chlorine; 5, mineral oil + fat + sulfur + chlorine; 6, mineral oil + fat + inhibited sulfur; 7, mineral oil + fat + inhibited chlorine. c, concentrated; d, diluted. Source: Ref 4

The choice of a cutting or grinding fluid is influenced by the fluid characteristics, the workpiece material, and the machining operation. Table 2 provides general guidelines for the selection of fluids based on the material to be machined and the cutting operation involved. At best, these recommendations provide a starting point for evaluating the preferred cutting or grinding fluid in a given manufacturing environment.

**Fluid Characteristics.** Oil-base solutions possess superior lubricating characteristics, resist bacterial attack, protect surfaces from corrosion, and can be readily recycled with appropriate filtration. Water-base solutions have superior cooling and penetration capabilities, are generally lower in operating costs, are less dependent on mineral oil supply, and may cost less to recycle because of the ease of settling of swarf and chips. Oil and water emulsions tend to have moderate cooling and lubrication characteristics compared to those of oil and water solutions.

**Workpiece Material.** The intrinsic machinability of metals can vary considerably within a specified composition because of variations in structure and homogeneity. Nevertheless, there are some general preferences in the selection of a cutting fluid for a given workpiece material, as follows.

*Free-Machining Steels.* The addition of lead, sulfur, and bismuth to steels improves their machinability. Water-base cutting fluids are most effective with these materials, particularly those containing sulfur-base additives as well as fatty esters. Chlorinated EP additives are not generally effective.

*Low-carbon steels* in the hot-rolled condition tend to be somewhat gummy. Emulsions and low-viscosity oils can be used effectively in machining these materials. Medium- and high-carbon steels as well as alloy compositions in the same carbon range are effectively machined with emulsions and water-base solutions, particularly if machining rates are high.

*Cast Iron.* Because swarf buildup must be avoided, water-base emulsions and solutions are effective. The greater the cutting speed, the more effective the water-base coolant.

*Stainless Steel.* Low-viscosity oils with chlorine, as well as sulfur EP additives, are effective. Emulsions containing sulfur and chlorine are effective fluids at higher cutting speeds.

*Copper Alloys.* Because of the formation of stringy chips and the ease of staining with active sulfur, fatty esters are used in oils, emulsions, and water-base solutions. Soap-base solutions have been used effectively in the machining of copper alloys.

*Aluminum Alloys.* Water-base solutions containing fatty esters and amides are effective. Lightweight oils containing a fatty ester are also effective, particularly for turning and milling operations.

*Titanium Alloys.* Lightweight oils containing chlorinated EP additives are effective. Emulsions containing chlorine have been found to be effective when grinding with a silicon carbide wheel.

**Machining Operation.** Each of the metal cutting operations has characteristics that often influence the effectiveness of a particular cutting fluid. The basic metal removal methods are turning, milling, drilling, and grinding.

*Turning.* Because the cutting tool is in continuous contact with the workpiece, access to the cutting area is restricted. Therefore, the cutting fluids of choice are those with a base fluid and additives of low molecular weight. In general, water-base solutions and emulsions are preferable for most turning operations.

*Milling.* Lubrication is generally more important than cooling in this operation because of the relatively low cutting speeds involved and the easy access to the cutting tool. Therefore, compounded oils and emulsions are frequently preferred.

*Drilling.* Because of the constant engagement of the tool and workpiece and the difficulty of gaining access to the cutting area, drills with cutting fluid access ports should be used when possible. Solutions based on oil and water can be successfully used with sulfur and/or chlorine additives, although the specifics of the fluid chemistry are heavily influenced by the composition of the material being machined. Chlorine, for example, does not seem to be effective in improving the drilling of free-machining steels.

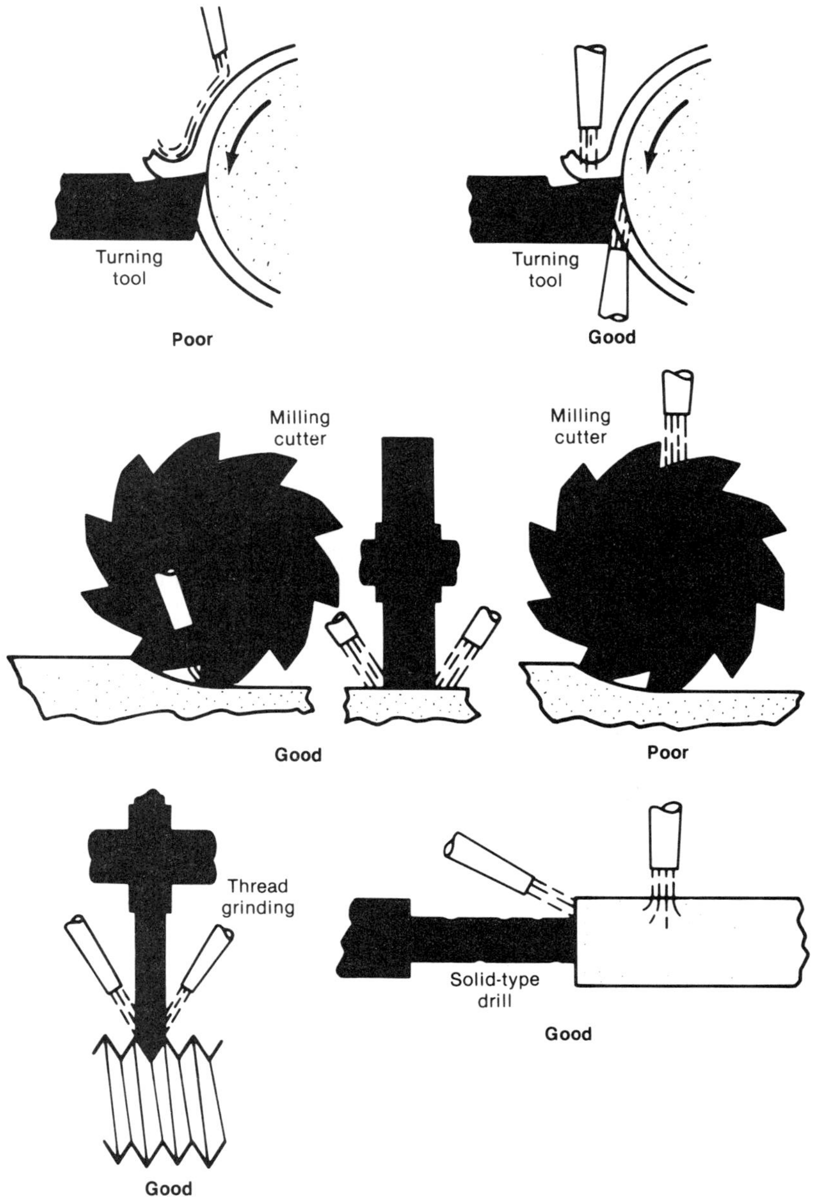


**Fig. 6** Proper and improper methods of applying cutting fluids. Source: Ref 5

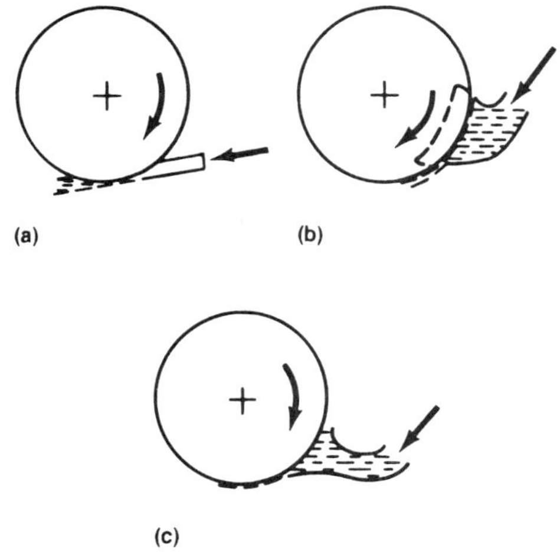


**Fig. 7** Methods of applying grinding fluids. (a) A fan-shaped nozzle covers the width of the wheel and is shaped to break the air film generated by the rotating wheel. (b) A nozzle with a large orifice extending over the sides of the wheel allows gradual acceleration of the fluid. (c) A nozzle that directs the fluid almost perpendicular to the wheel surface allows the fluid to penetrate the air film generated by the wheel. Source: Ref 3

*Grinding*. Because of the high rotational speeds of the grinding wheels, the application of a fluid is extremely important to ensure fluid contact with the wheel and the workpiece. Furthermore, the relationship between the chemistry of the grinding wheel and that of the workpiece is also important. These interactions must be evaluated in choosing an appropriate cutting fluid for a specific grinding wheel material in a production situation. Generally, emulsions and water-base solutions are the fluids of choice, with a wide array of esters, amides, sulfur compounds, and chlorine compounds successfully used in the fluid formulation. Oil-base solutions are chosen when lubrication of the wheel is the critical criterion.

## Application Methods

Correct application of the cutting fluid at the tool/workpiece interface is fundamental to the effective use of the fluid, and the method of application affects not only lubrication and cooling but also the efficiency in removing swarf and chips from the cutting operation. Some recommendations on the placement of the fluid stream are illustrated in Fig. 6 and 7. Frequently, more than one nozzle per tool should be used to optimize chip removal as well as cooling and lubrication. Manual application of a cutting fluid is effective only for very low volume production or toolroom use. Fluid and mist application also affect fluid effectiveness.

**Flooding** of the cutting area is the most widely used method of promoting lubrication, cooling, chip removal, and access to the cutting operation. The volume of fluid per unit time that is applied is critical in achieving optimum results. Volume recommendations vary from less than 1 L/min (0.25 gal./min) to more than 2000 L/min (500 gal./min), depending on feed, speed, and cutting tool material and geometry (Table 3). The optimum pressure varies with operation.

In the case of drilling, access holes in the body of the drill will improve the access of the fluid. The optimum volume and pressure must be developed for each application. Pressures and volumes in excess of the optimum may be harmful. The creation of excessive mist must be guarded against by using an appropriate enclosure for the cutting area.

**Misting.** A particularly effective method of applying a cutting fluid in drilling and cutoff operations involves the creation and application of the lubricant as a mist. The size of the mist droplets can be controlled, depending on the particular effects desired. In addition, more efficient use of the cutting fluid can also be achieved, particularly in the case of water-base solutions and emulsions. Vaporization of the small particles may improve both cooling and lubrication during machining. Care must be taken when misting cutting fluids to prevent excessive buildup in the air and in the workplace in general.

## Control and Test Methods

Control of cutting and grinding fluids depends on the adoption of appropriate test procedures. Table 4 lists some of the Amer-

**Table 3 Cutting fluid flow recommendations**

| Operation | Fluid flow L/min | Fluid flow gal./min |
|---|---|---|
| Turning | 20 | 5 |
| Screw machining | | |
| 25 mm (1 in.) diameter | 132 | 35 |
| 50 mm (2 in.) diameter | 170 | 45 |
| 75 mm (3 in.) diameter | 227 | 60 |
| Milling | | |
| Small cutters | 19/tool | 5/tool |
| Large cutters | Up to 227/tool | Up to 60/tool |
| Drilling, reaming | | |
| 25 mm (1 in.) diameter | 7–11 | 2–3 |
| Drilling, large | 0.3 to 0.43 × diam, mm | 2–3 × diam, in. |
| Gundrilling | | |
| External chip removal type | | |
| 4.6–9.4 mm (0.18–0.37 in.) diameter | 7–23 | 2–6 |
| 9.4–19 mm (0.37–0.75 in.) diameter | 20–65 | 5–17 |
| 19–32 mm (0.75–1.25 in.) diameter | 40–150 | 10–40 |
| 32–38 mm (1.25–1.50 in.) diameter | 64–190 | 17–50 |
| Internal chip removal type | | |
| 7.9–9.4 mm (0.31–0.37 in.) diameter | 20–30 | 5–8 |
| 9.4–19 mm (0.37–0.75 in.) diameter | 30–98 | 8–26 |
| 19–30 mm (0.75–1.18 in.) diameter | 98–250 | 26–66 |
| 30–60 mm (1.18–2.38 in.) diameter | 250–490 | 66–130 |
| Trepanning | | |
| External chip removal heads | | |
| 50–90 mm (2–3.5 in.) diameter | 30–180 | 8–48 |
| 90–150 mm (3.5–6 in.) diameter | 61–300 | 16–80 |
| 150–200 mm (6.8 in.) diameter | 120–394 | 32–104 |
| Internal chip removal heads | | |
| 60–150 mm (2.37–6 in.) diameter | 415–814 | 110–215 |
| 150–300 mm (6–12 in.) diameter | 814–1300 | 215–340 |
| 300–460 mm (12–18 in.) diameter | 1300–1740 | 340–460 |
| 460–610 mm (18–24 in.) diameter | 1740–2160 | 460–570 |
| Honing | | |
| Small | 10/hole | 3/hole |
| Large | 20/hole | 5/hole |
| Broaching | | |
| Small | 40/stroke | 10/stroke |
| Large | 0.45/stroke × length of cut in mm | 3/stroke × length of cut in in. |
| Centerless grinding | | |
| Small | 76 | 20 |
| Large | 150 | 40 |
| Other grinding | 0.75/mm of wheel width | 5/in. of wheel width |

Source: Ref 6

**Table 4 ASTM standards for control of metal cutting and grinding fluids**

| Specification | Topic |
|---|---|
| D 88 | Saybolt viscosity |
| D 92 | Flash and fire points by Cleveland open cup |
| D 94 | Saponification number of petroleum products |
| D 129 | Sulfur in petroleum products (general bomb method) |
| D 130 | Detection of copper corrosion from petroleum products by the copper strip tarnish test |
| D 808 | Chlorine in new and used petroleum products (bomb method) |
| D 811 | Chemical analysis for metals in new and used lubricating oils |
| D 892 | Foaming characteristics of lubricating oils |
| D 893 | Insolubles in used lubricating oils |
| D 1317 | Chlorine in new and used lubricants (sodium alcoholate method) |
| D 1479 | Emulsion stability of soluble cutting oils |
| D 1662 | Active sulfur in cutting fluids |
| D 1748 | Rust protection by metal preservatives in the humidity cabinet |
| D 3601 | Foam in aqueous media (bottle test) |
| D 3705 | Misting properties of lubricating fluids |
| D 3946 | Evaluating the bioresistance of water-soluble metal working fluids |

ican Society for Testing and Materials (ASTM) standard test procedures used in establishing control, and the following sections describe some of the procedures.

**Viscosity** is the most important property of a lubricant. The viscosity of a fluid determines its characteristics of flow, penetration, and oil film thickness. In general, there are two basic measurements of viscosity: absolute or dynamic viscosity and kinematic viscosity. Dynamic viscosity represents the force required to overcome fluid friction. Kinematic viscosity measures viscosity in relation to the density of the fluid. Kinematic viscosity is usually used to characterize lubricants.

Kinematic viscosity is generally expressed in either centistokes (cSt) or Saybolt universal seconds (SUS). Specification ASTM D 445-446 describes the instrumentation and technique for measuring kinematic viscosity. Viscometers of various designs permit flow under controlled conditions, and this flow is translated into viscosity at a specified temperature. Specification ASTM D 88 describes a method that uses a Saybolt universal viscometer. The method also measures fluid flow at a specific temperature. Although this technique is not as precise as the one described in ASTM D 445-446, it is adequate for most applications.

**Concentration.** During processing, various additives will be depleted at different rates, depending on many of the variables observed during machining, such as material chemistry, salt concentration, temperature, and the rate of part production. This change in composition and concentration can result in unsatisfactory tool life, poor tolerances, degradation of surface finish, the production of rusty parts, and the growth of bacteria and/or mold and fungus. The selective addition of appropriate additives may be necessary if the depletion of specific components of the lubricant is not compensated for by periodic additions of the concentrate. Overall concentration can be controlled by the appropriate use of a refractometer, by splitting the emulsion, or by the titration of water-base solutions.

The most significant operating parameter of a water-base emulsion or solution is its concentration of active ingredients and its concentration of the concentrate relative to the water diluent. Control of these concentrations is imperative if the operating characteristics of the cutting fluid are to be consistent. Because the chemical composition of the water to be used in diluting the concentrate is often an uncontrollable variable, determination of the concentration can be an operating problem. Generally, the use of a refractometer for measuring concentration is adequate during the initial preparation of the fluid. However, it frequently becomes inadequate as contaminants in the fluid accumulate and the salt concentration of the water increases.

Frequently, the most effective and reliable method of determining the concentration of a given emulsion is to break the emulsion by adding ionic salts and/or acids to a small portion of it. A sample of the emulsion is selected, sediment and tramp oil are removed, and a strong mineral acid ($H_2SO_4$) is carefully added. Splitting occurs after the mixture is heated for approximately 1 h, and two separate layers are created: one of water and the other of additives and oil. This splitting technique does distort the concentration of oil and additives in the emulsion. Nevertheless, useful information on concentration can be obtained with this technique or its variations.

The titration of emulsions is generally based on the colorimetric titration of the anionic emulsifiers frequently used in producing emulsions. If a petroleum sulfonate has been used as an emulsifier, it can be titrated with hyamine by using a cationic indicator (methylene blue). Titration can also be used as a control method for water-base solutions because these solutions generally vary in pH from 8.5 to 9.5. Titration with a dilute mineral acid to a predetermined pH value can be compared to a standard to establish an alkaline equivalent that rises with a change in pH of the cutting fluid. Experience over time of the relationship of this equivalent to concentration can be useful in controlling solution concentrations.

**Emulsion Stability.** Excessive cold, high operating temperatures, contamination of the cutting fluid by both metal fines and tramp oil, and formulation inadequacies can affect the stability of emulsions. Particle size and distribution are often taken as indications of stability. Larger particle size, which may be desirable in some metal cutting operations often indicates incipient breaking of the emulsion. Size and distribution can be observed under low magnifications in an optical microscope. Regular observation and the creation of a data base of information on particle size and distribution can serve as an adequate control technique. Excessive heat or cold should be avoided because they can lead to an unstable emulsion.

Determination of the stability of fresh emulsions is important in qualifying a new emulsion cutting fluid and for establishing a baseline against which subsequent test procedures can be related. A depletion test based on ASTM D 1479 can be useful in this regard. After letting a sample stand for 24 h, 20 mL (0.7 oz) of the prepared emulsion is drawn from the bottom of the sample. This emulsion is then broken, and its concentration is compared to that obtained by breaking a sample taken prior to the 24-h settling period.

**Foaming** occurs as a result of agitation produced by the machining operation or the transfer of fluid. Foam production can reduce effective film strength, complicate the settling of metal fines, and slow heat transfer. Specification ASTM D 892 is a method for evaluating the relative tendency toward foaming of a given fluid. An air diffuser is immersed in a fluid sample, and a predetermined air flow is initiated for a given period of time. The ensuing volume of foam is measured. The sample is allowed to stand, and the volume of foam that remains is recorded. The foam that remains after the 10-min settling time is usually a fair reflection of the stability of the foam. The test is then repeated at a higher temperature. The relative volume of foam is a measure of the foaming tendency of the cutting fluid. The effects of contaminants during processing can also be evaluated in this manner.

**Particulate Contaminants.** As a result of metal cutting operations, metal particles and organic contaminants from various sources build up in the fluid. These particulates influence tool life and surface finish and may facilitate the chemical breakdown of molecular species in the cutting fluid. Specification ASTM D 273 describes a filtering procedure that can be used to measure particulate concentration under controlled conditions.

**Hydrogen ion concentration** can be measured by using pH paper or a standard pH meter. Because of the relative simplicity of the measuring methods, regular monitoring of pH in a cutting fluid can be performed to measure the overall change in acidity or alkalinity as operations proceed. A change in the pH may reflect chemical degradation or degradation due to biological growth. In plant operations, systematic measurement of pH is a relatively simple procedure for controlling cutting fluid quality.

**Corrosion.** One of the more important characteristics of water-base cutting or grinding fluids is the potential for corrosion. A number of relatively simple test procedures have been developed to measure the tendency toward corrosion or staining. Cast iron chips that have been cleaned and coated with the test fluid can be monitored under controlled conditions of temperature and time to observe a tendency toward staining or corrosion. A similar test can be carried out on coupons made of copper, aluminum, or ferrous alloys. Humidity cabinets of varying designs are also frequently used to evaluate the tendency toward staining and corrosion of selected fluids on selected surfaces. Some cabinets allow for the application of controlled salt concentrations and vapor condensation.

**Biological Tests.** Microbes such as bacteria, mold, and fungi can promote corrosion, dermatitis, and emulsion destabilization. Therefore, techniques are used to evaluate the presence and concentration of microorganisms in cutting fluids. Three biological test procedures are discussed below. They are relatively simple to carry out, and they adequately monitor the presence of microbial infestation of the fluid.

*Oxygen liberation* is a technique for measuring the oxygen released from microbiological infection. Equipment has been developed for measuring the oxygen liberated from the enzymatic conversion of a peroxide substrate into oxygen and water. Oxygen is liberated from the microbes in the cutting fluid and is measured with instrumentation. The method yields rapid, reproducible results and is satisfactory for monitoring the presence of microorganisms in the fluid.

*Dip-Slide Technique.* Small plastic slides coated with a nutritive gel are dipped into the cutting fluid and allowed to drain. Any microbes present will grow in the gel. The gel can be selected to be responsive to either bacteria or yeast and mold. A variation is the recent development of a single slide that has different media on opposite sides. One side is responsive to bacteria, the other side to yeast and mold. Although not quantitative, the method is adequate for monitoring a cutting fluid in the manufacturing environment.

*Ammonia concentration* can signal excessive levels of microbes and metallic particles in a cutting fluid. A straightforward method of evaluating ammonia concentration is by the use of a pH meter equipped with a specific ion electrode designed to measure ammonia.

**Water Quality.** In the case of water-base cutting fluids, the importance of controlling water chemistry cannot be overemphasized. High hardness and low pH will adversely affect the stability of emulsions. Control of the concentration of anions, cations, and pH reflects the control (or lack thereof) of water quality. Cations such as magnesium, aluminum, and calcium are particularly important because they influence the formation of hard water soaps and the complexing of the surfactants used in compounding emulsions and solutions. If the surfactants are complexed, they are no longer effective. High salt concentrations in the water reduce the stability of emulsions and solutions. Total hardness is generally expressed in terms of the calcium carbonate ($CaCO_3$) that is titrated.

**The residues** that remain after the water has evaporated from water-base fluids may cause problems with long-term corrosion, slide movement, and tooling alignment. In general, residues should be soft and nonsticky to maintain their lubricating effectiveness. A simple method of evaluating residue characteristics is to evaporate a measured quantity of the fluid in a flat container such as a petri dish and then examine this residue for tackiness, fluidity, color, crystallization, and concentration.

**The electrical conductivity** of a water-base fluid can be used as a measure of the buildup of salts or metallic impurities. A series of conductivity measurements taken over time can be used to determine the appropriate cleaning and disposal cycles for a fluid. Many other simple metering devices are available for testing the intrinsic electrical conductivity of a cutting fluid.

## Storage, Cleaning, and Disposal of Cutting and Grinding Fluids

The contamination of coolants and lubricants is a constant problem, and cutting and grinding fluids are often recycled and reused. This requires careful attention to the storage, distribution, cleaning, and disposal of cutting and grinding fluids.

### Storage and Distribution

**Fluids** should be stored in a manner that minimizes potential contamination. Water contamination is the concern with oil-base fluids, and oil and particulate contamination is the concern with water-base fluids. In the case of tank storage, the water contamination of oils is of particular importance. Storage tanks require appropriate venting and periodic cleaning to prevent contamination problems.

All fluid containers should be properly labeled for compliance with regulations governing the presence of hazardous materials in the workplace. Material safety data sheets must be available in the workplace for each material being used. Similar precautions should be instituted for sampling in-process fluids.

**Design Considerations.** The design of all parts of the system that will be in contact with the cutting or grinding fluid should take into account the following:

- All surfaces in contact with the fluid should be as smooth as possible to minimize the deposit buildup of metallic or nonmetallic materials as well as microbial agglomeration
- If flumes are constructed in the floor, they should be covered to prevent access of any outside waste material and should be designed to maximize fluid flow in order to minimize possible microbial growth or the buildup of fines and metal chips
- All piping should be sized to maximize fluid flow and should contain as few bends as possible to facilitate cleaning. Further, the pipes should be sized so that they are full during operation, thus preventing the buildup of slime on the walls
- Reservoirs should be constructed of materials that are not subject to chemical attack. The concrete tank is a particularly poor choice because the lime in the concrete can be eroded, subsequently increasing particulate contamination and possibly altering lubricant chemistry
- Equipment of any kind that is likely to be in contact with a fluid should have continuous drainage to prevent the growth of microbes in static pools of lubricant

**System Cleaning.** The reservoir, auxiliary piping, and application devices should be cleaned before the reservoir is filled with machining fluid. This cleaning of the system is important before the initial fill and is even important in subsequent filling cycles. Good cleaning practice consists of the following series of steps:

- Drain fluid from all lines, application devices, sumps, and/or reservoirs
- Remove as much of the solids collected in the system (filters, lines, sump, reservoir, and so on) as possible
- Charge with a suitable cleaner diluted with an appropriate fluid (water, solvent, or oil)
- Circulate the cleaning solution for a sufficient length of time to remove residual machining fluid and accumulations (solids, liquid contaminants) in the total system. The use of a brush after a period of time will often help loosen accumulated organic and inorganic debris
- Drain the system as completely as possible and flush the system with the appropriate light oil, solvent (in the case of oil-base fluids), or water (in the case of water-base fluids)
- Proceed immediately to the next step in the cleaning cycle if the system is flushed with water
- Rinse with a solution containing a biocide and a fungicide for at least 2 h when water-base solutions or emulsions are the machining fluids of choice
- Drain the system once again if the rinse appears to be heavily contaminated. If not, approximately a 1% concentration of the chosen machinery fluid can be added, using the biocidal rinse as part of the makeup. This mixture should be circulated for approximately 30 min prior to final draining
- Mix the desired concentration of concentrate and water in a clean mixing vessel, using deionized water if available. The mixing should be carried out so as to ensure intimate contact of concentrate and water. The concentrate should always be added to the water to facilitate approximate mixing. The system should then be charged

## Recycling and Fluid Cleaning

Cutting and grinding fluids are often collected in a holding tank and recycled many times during their service lives. This important aspect of fluid management often requires cleaning of the lubricant or coolant.

Appropriate recycling procedures and equipment depend on an analysis of the lubricant characteristics and the potential contaminants during the fluid life cycle. Metal cutting and grinding fluids that are to be recycled may contain a wide diversity of liquid or semiliquid contaminants. The chemistry, particle size, geometry, and concentration of these contaminants will influence the equipment and disposal technology selected.

Contamination can be more effectively controlled when the source and frequency of contamination can be predicted. Contaminants that result from a breakdown of the tool or workpiece can be analyzed for appropriate cleaning procedures. Similarly, insoluble precipitates from the water component can be analyzed for appropriate cleaning procedures. Contamination often occurs as a result of such random events as floor sweeping or the disposal of food in reservoirs. These occurrences are difficult to handle on a systematic basis, but they should be kept in mind when process and equipment decisions are made with respect to recycling.

Fluid cleaning equipment is illustrated in Fig. 8 to 10. Settling tanks, flotation tanks, magnetic separators, and centrifuges are used singly or in combination.

**Settling tanks** (Fig. 8a) are often used to remove particulate matter, chips, and swarf. The tank is separated with two partitions. A baffle at the first partition prevents the flow of tramp oil, which can be subsequently removed. A weir at the second partition isolates the clean fluid. A settling tank is often the first stage of the fluid cleaning process.

**In flotation tanks** (Fig. 8b), air bubbles are created by a stirring action or by the introduction of compressed air. Fines attach themselves to the bubbles (especially in the presence of surfactants) as the bubbles rise to the surface. A foam develops on the surface and is subsequently removed.

Flotation tanks are also used to remove oil from water-base fluids. Oil that enters a water-base cutting fluid will increase the growth of anaerobic bacteria, cause corrosion and staining, plug the filters, destabilize the emulsions, and increase part cleaning costs.

Generally, the oil will float on the top. However, in some systems in which mixing of the oil and the water base occurs, partial emulsification of the oil may also take place. Aeration devices are used to promote the formation of tramp oil, which then floats to the surface. Rotating disks, which are wetted by the oil and/or belts of stainless steel, neoprene, or other materials, collect oil as they move through the oil layer.

**Hydrocyclones** (Fig. 8c) separate suspended particles from the fluid by imposed acceleration. The fluid, pumped at high velocity, is fed tangentially into a conical vessel, where the heavier particles are forced to the wall. The contaminants exit at the bottom, while the back pressure resulting from the conical shape of the vessel causes the clean fluid to discharge at the top. Hydrocyclones can be used effectively after the settling tanks to remove residual contaminants. In general, the equipment is effective on light-viscosity fluids of up to approximately 100 SUS at 38 °C (100 °F). If low levels of contamination are present, the use of hydrocyclones can be cost effective.

**Centrifuges** (Fig. 8d) impose higher accelerations than hydrocyclones. The centrifuge is effective over a wide viscosity range and can be designed for specific applications. Low-speed centrifuges are effective in removing particulates, while high-speed units can be used to remove particulates and tramp oil. Bacteria and mold, if agglomerated, can also be removed with the tramp oil. Basket centrifuges are used to remove chips from cutting fluids.

**Magnetic separators** are used to remove metallic fines from lubricants. A magne-

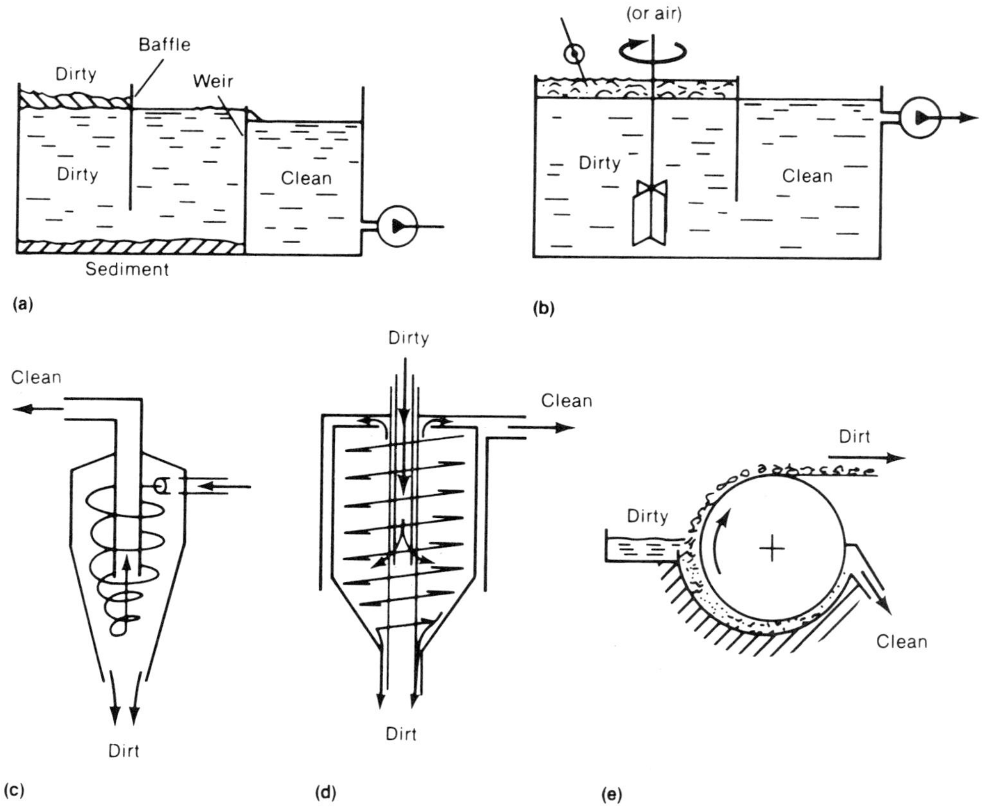


**Fig. 8** Removal of fines from lubricants by settling (a), by flotation (b), in a hydrocyclone (c), in a centrifuge (d), and in a magnetic drum (e)

tized revolving drum (Fig. 8e) can often be used for the desired particulate separation. In general, such devices are more effective in water-base fluids because of their low viscosity. Magnetized conveyors have also been successfully used to remove chips from the cutting fluid. Again, settling tanks, magnetic separators, and centrifugally driven devices and skimmers can be used singly or in combination to promote fluid cleanliness.

**Filtration.** Depending on the size of the particles, various filtration media are used, such as cloth, paper, polymers, and wire screens. The driving force can be gravity, pressure, or vacuum. Filter configurations range from flat, stationary beds to moving belts, rotating drums, tubes (socks), or flattened tubes (leaves). Accumulated solids can be removed by scrapping, shaking, or reverse flow.

*Tube and leaf filters* are used to remove small particles and to promote fine finishes in the machining and/or grinding operation. A compartment containing filter tubes or leaves made of materials such as nylon or woven wire serves as the filter unit, through which the fluid is pumped either by the application of pressure or a vacuum. When fine filtration is required, the filter surface may require recoating to initiate each filtration cycle. Diatomaceous earth improves filtration, but may also remove lubricant components. In this process, as in all filtration procedures, the chemistry of the fluid being filtered must be matched to the filtration system so that the components of the fluid are not filtered out while contaminants are removed.

*Ultrafiltration* (Fig. 9) differs from filtration in that a higher pressure (150 to 200 kPa, or 20 to 30 psi) is maintained over a filter membrane with very small pores (from 25 to 100 Å in diameter). Ultrafiltration performs molecular filtration and is used when high-quality fluids are required. Prior filtration and settling techniques are required to make this procedure effective. In ultrafiltration, molecular size determines filtration efficiency; therefore, the components of the fluids must be chosen selectively with this method in mind.

*Reverse osmosis* (Fig. 10) uses a membrane to separate ions. The pore size of the membrane is approximately 5 to 25 Å. With reverse osmosis, hard water contaminants such as sodium chloride are effectively removed from aqueous solutions. Like ultrafiltration, reverse osmosis requires prior filtration.

**Water Treatment.** Water quality is important in water-base emulsions and solutions because various cations and anions can promote bacterial growth, corrosion, and emulsion instability. A number of processes are used in water treatment.

*Water Softening.* Sodium ion exchange resins in an appropriate container are used to replace calcium, aluminum, iron, and other cations with sodium. The ion exchange resin can be regenerated by flushing with a saturated sodium chloride solution. The total amount of solids present in softened water is not appreciably affected by this procedure, but the hardness is significantly reduced. Although softened water may be beneficial in some water-base fluids, it may lead to excessive foaming.

*Deionization.* Cations and anions present in the water can be completely removed by

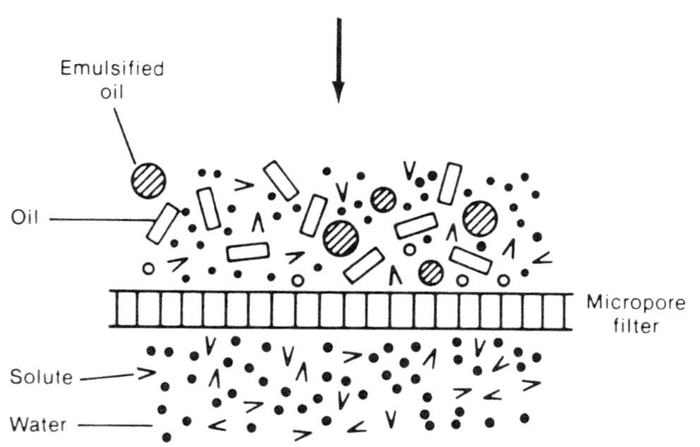


**Fig. 9** Schematic of ultrafiltration

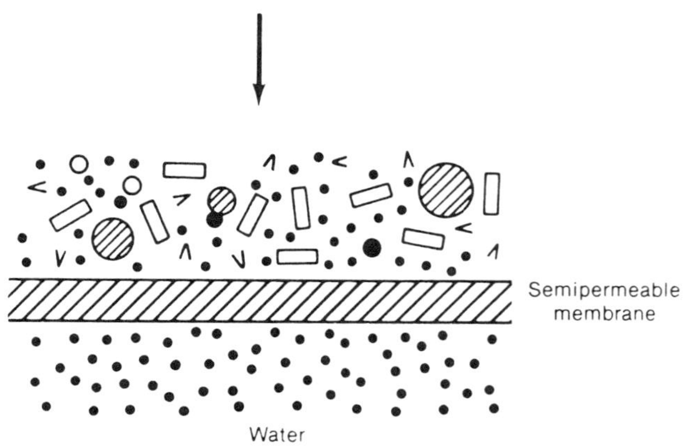


**Fig. 10** Schematic of reverse osmosis

the use of ion exchange resins. A system consisting of two resin columns is used. One is a cation exchanger that replaces cations with hydrogen ions. The second column contains an anion exchanger resin that replaces sulfates, chlorides, and carbonates with hydroxyl ions. These exchange units can be regenerated by appropriate washing with hydrochloric acid and sodium hydroxide, respectively. This cost-effective method is preferred for water treatment, and the water quality is comparable to that from distillation.

*Distillation*. If water is treated by distillation, essentially all the salts precipitate, and the condensed water is very pure. Commercial distillation devices are available that use both higher pressures and temperatures to accelerate the process. Boiler water and rainwater are satisfactory substitutes for distilled water.

**Microbial Control.** Because microbes are so prevalent in the environment, they will grow in cutting and grinding fluids, particularly in water-base emulsions and solutions. Therefore, it is imperative to control their growth.

Generally, the most cost-effective treatment is heating or the addition of biocides. Selective biocides can be effective in the particular operating circumstances. Heating the metalworking fluid to temperatures of approximately 70 °C (160 °F) will kill the microbes present in the cutting fluid. Heating also facilitates the separation of tramp oil and solid contaminants.

**Disposal** is required when the recycling of a fluid is no longer cost effective. Frequently, in-plant reprocessing of the oil and water is not cost effective, therefore, disposal and removal must be carried out by licensed waste treatment companies.

Depending on the composition of the fluid and its contaminants, the fluid can be separated into oil and water layers by heating. Generally, however, this is not sufficient for adequate separation. Chemical treatments are often needed to destabilize the emulsion. A practical cycle used in some facilities consists of the following steps:

- Separate the tramp oil and solid particulates using the methods previously described
- Using the appropriate safety procedures, add the concentrated sulfuric acid to the emulsion until the pH is approximately 3.5. The solution should be mixed while the addition is being made
- Add the aluminum sulfate during mixing. The concentration of the aluminum sulfate depends on volume, chemistry, and pH. The concentration must be high enough to cause precipitation of the aluminum hydroxide
- Using adequate safety precautions, recycle the emulsion back to a pH between 6.5 and 7.0 by adding 50% sodium hydroxide to the emulsion
- Allow the mixture to stand long enough to permit the formation of the aluminum hydroxide floc

The clear water layer that results should be pure enough for disposal in the sewer system. No fluid should be added to a municipal sewer system without prior approval and evaluation by that facility. The aluminum hydroxide floc can be removed and treated with concentrated sulfuric acid to regenerate the aluminum sulfate for the next splitting cycle.

## Biological Effects of Cutting and Grinding Fluids

Although most cutting and grinding fluids generally have a low order of toxicity, some compounds that are normally used as components of cutting fluids have been identified as having a greater potential for toxicity than others. Reduced contact with these chemicals is an important part of good manufacturing procedure when metal removal processes are involved. Contact with the following components of some cutting fluids should be minimized, even though direct evidence of any toxic effects on humans is at present inconclusive:

- *Bactericides*: Formaldehyde donors, halogenated alicylanilides, and mercaptobenzothiazoles
- *EP additives*: Chlorinated compounds, emulsifiers and detergents, soaps, and petroleum sulfonates
- *Antioxidants*: Diphenylamine, hydroxy compounds, alkyl sulfides, and disulfides
- *Corrosion inhibitors*: Hydroxylamines, inorganic and organic nitrites and nitrates, and petroleum sulfonates
- *Dyes*: Azo dyes and fluorescein
- *Water conditions and antiwear agents*: Phosphates and borates

The health effects of fluids have been tested by using animal experiments defined in the Code of Federal Regulations (CFR) under the U.S. Federal Hazardous Substances Acts. These animal tests consist of:

- *Acute oral toxicity*: 16 CFR 1500.3 (C) (1 and 2)
- *Acute inhalation toxicity*: 16 CFR 1500.3 (C) (1 and 2)
- *Acute dermal toxicity*: 16 CFR 1500.40
- *Primary skin irritation*: 16 CFR 1500.41
- *Acute eye irritation*: 16 CFR 1500.42

There is a great deal of uncertainty about the relationship between these test results and human health.

**Skin Effects.** By far the most common effects from cutting fluids are skin disorders resulting from prolonged contact. The four major types of disorders that have been studied are contact dermatitis, folliculitis and acne, pigmentary changes and benign and malignant tumors.

*Contact dermatitis* is primarily caused by the removal of the natural oils in the skin due to the presence of water, solvents, emulsifiers, and/or soaps in the cutting fluid. A fluid with a pH of over 9.0 can also accelerate the occurrence of contact dermatitis.

*Folliculitis*. Prolonged exposure to oil-base cutting fluids can block the hair follicles in the skin. Unless care is taken to keep the skin clean, bacteria may contribute to the formation of folliculitis and oil boils.

*Pigmentary changes* and skin thickening can also be associated with the formation of folliculitis.

*Tumors*. The petroleum-base oils refined from crude oil contain varying concentrations of polycyclic aromatic hydrocarbons. These polycyclic compounds seem to be correlated with the formation of scaly overgrowth of skin and benign and malignant tumors. Appropriate hydrotesting or solvent refining removes most of the objectionable compounds.

**Health Practices.** The potential health hazards from cutting fluids can be reduced by observing good health practices in the workplace. The following guidelines are especially effective:

- Hands and arms, if exposed to cutting fluids, should be coated with a vegetable oil or petroleum jelly before a work shift
- The skin should be washed with a mild soap that does not contain abrasives
- Clothing that has been impregnated with the cutting fluid should be discarded as quickly as possible and replaced
- Impervious armlets and aprons should be used to reduce contact with the fluids
- Towels and white cloths should not be exchanged among workers and should be discarded after one use
- Any abrasion or cut in the skin necessitates that precautions be taken to prevent contact with the fluid
- The equipment and surroundings should be kept clean, and accumulations of debris or biological matter should be removed from reservoirs
- A continuing program of education and training that stresses the importance of personal cleanliness and hygiene should be conducted in the plant
- Solvents of all types that may be used in cleaning the equipment should not be used to clean the skin
- Metal fines should be removed on a regular basis (preferably continuously)
- Foreign matter or contaminants, such as hydraulic fluids or other machine lubricants, should be prevented from entering the fluid, if possible
- The pH and biocide content should be controlled. In water-base fluids, the pH should be held under 9.5

- Biocides, due to their active nature and wide variation in chemistry, should be handled with extreme care. Only government-approved biocides should be used

**Microbes,** which may be bacteria, mold, or yeast can grow in water-base cutting fluids at extremely high rates, depending on the conditions and the microbe type. In general, the population will double every 15 to 30 min. Even in oil-base fluids, contamination with water can cause unwanted microbial infestation.

Bacteria are most commonly associated with emulsions, and mold is most commonly associated with water solutions (synthetics). The microbes in question are frequently subclassified as bacteria and fungi, with fungi being divided into molds and yeast. In general, there is a natural antagonism between bacteria and fungi. In controlling the growth of bacteria, fungi can often get out of control and flourish. Therefore, when using biocides, it is important to use those that suppress the growth of both bacteria and fungi.

Microbes cause a number of unwanted effects because they feed on components of the cutting fluid as well as on organic and inorganic contaminants. Oxidation reduction reactions initiated by microbial attack often result in the removal of side chains from complex molecules, the opening of aromatic rings, the reduction of carbon chain lengths, and the unsaturation of saturated bonds. In addition, acid is produced by the activity of bacteria, although this effect can be masked if the growing bacteria degrade nitrogen additives and liberate ammonia.

The three main types of microbes present in cutting fluids are: aerobic bacteria (primarily the pseudomonas group), anaerobic bacteria (particularly the *Desulfovibrio desulfuricans*), and fungi (principally *Fusarium* and *Cephalosporium*). Candida (yeast) is also prevalent in fungi-containing cutting fluids.

Anaerobic bacteria grow in the absence of oxygen. They usually grow more slowly than aerobic bacteria, but their growth can be very objectionable. Generally, they do not grow in a fresh, clean fluid, but will grow once the fluid has been attacked by the aerobic bacteria. When makeup is added to an odoriferous coolant, momentary relief occurs until the aerobic bacteria break down the additive concentration; this results in the formation of hydrogen sulfide (and the so-called Monday morning stink). Aerobic bacteria, the most aggressive and prevalent type found in cutting fluids, grow in the presence of air (oxygen). They are the principal cause of the biological deterioration of the metal cutting fluid. The pseudomonas bacteria attack oil and multiply rapidly in machines that leak lubricating and hydraulic oils. It is important to remove this oil if the leakage cannot be stopped.

## REFERENCES


1. S. Kalpakjian, *Manufacturing Processes for Engineering Materials*, Addison-Wesley, 1984, p 463, 467
2. G. Warnecke, in *Proceedings of the Fifth North American Research and Manufacturing Conference*, Society of Manufacturing Engineers, 1977, p 229-236
3. J.A. Schey, *Tribology in Metalworking*, American Society for Metals, 1983
4. E.S. Nachtman and S. Kalpakjian, *Lubricants and Lubrication in Metalworking Operations*, Marcel Dekker, 1985
5. H.F. Weindel, *Tool. Prod.*, Nov 1981
6. C. Wick, J.T. Benedict, and R.F. Veilleux, *Tool and Manufacturing Engineers Handbook*, Vol I, 4th ed., *Machining*, Society of Manufacturing Engineers, 1983

# Traditional Machining Processes

# Turning

TURNING is a machining process for generating external surfaces of revolution by the action of a cutting tool on a rotating workpiece, usually in a lathe. Boring is this same action applied to internal surfaces of revolution. In many instances, turning and boring are performed simultaneously or consecutively in the same setup. This article discusses applications in which turning is the sole or major operation in a machining sequence.

## Process Capabilities

Often other machining operations are performed in conjunction with turning. These include facing, longitudinal drilling, boring, reaming, tapping, threading, chamfering, and knurling. Common cutting tool modes used on turning equipment are shown in Fig. 1. Turning operations may be divided into two classes: those in which the workpiece is situated between centers, and those in which the workpiece is chucked or gripped at one end with or without support at the other end. Also, accessories can be obtained for milling, grinding, and cross drilling, although these operations are less frequently combined with turning. When more than two or three different operations are performed on identical parts, it is usually more practical to employ processes that use a single tool with the capability of performing two or more operations simultaneously or consecutively.

**Size and Shape of Workpiece.** Availability of equipment that can hold and rotate the workpiece is the major restriction on the size of the workpiece that can be turned. Turning is done on parts ranging in size from those used in watches to steel propeller shafts more than 25 m (80 ft) long. Aluminum parts (about one-third the density of steel or brass) over 3.0 m (10 ft) in diameter have been successfully turned. In actuality, the weight of the work metal per unit of volume may restrict the size of the workpiece that is practical to turn. Problems in holding and handling increase as weight and size increase. Some large parts are turned in vertical boring mills, some of which are capable of machining up to a 54 Mg (60 ton) workpiece.

Sometimes the entire workpiece is so unwieldy that rotating is virtually impossible. A notable example is in the turning of crankpin diameters on large crankshafts. This condition, however, usually can be overcome, and an acceptable degree of dynamic balance obtained, by counterweighting. Counterweights may be attached either to the spindle of the machine or to the work.

**Torque and Horsepower Requirements.** Engagement of the cutting tool with the rotating work results in a tangential force that, for a specific work metal, tool shape, and feed rate, generally is independent of the cutting speed and directly proportional to the depth of cut. That force, multiplied by the surface speed of the workpiece, serves as a basis for calculating the net horsepower required to remove metal from the piece being turned. Power required to move the tool longitudinally is usually negligible, with the exception of spade-drilling operations.

Capacities of lathes range from fractional horsepower to more than 150 kW (200 hp) for vertical boring mills (see the article "Boring" in this Volume).

The effects of composition and hardness of the work metal on power requirements for turning are illustrated in Fig. 2 for a tool setup which uses the identical feed and depth of cut on each work metal. Power requirements differ greatly for the different families of alloys, averaging about $4.55 \times 10^{-6}$, $1.14 \times 10^{-5}$, and $3.64 \times 10^{-5}$ kW/mm$^3$/min (0.1, 0.25, and 0.8 hp/in.$^3$/min) for magnesium alloys, copper alloys, and steels, respectively. As also shown in Fig. 2, the power values increase with increasing hardness within each family of alloys, the rate of increase being greatest for cast irons and steels harder than 350 HB. Besides having significance for the design of lathes, these data on power requirements provide an indication of the relative ease and cost of turning various metals.

The effect of cutting-tool design on power requirements is discussed in the section "Design of Single-Point Tools" in this article. The use of cutting fluids can have an indirect effect on power requirements through its effect on speed and feed.

**Depth of Cut, Feed, and Speed.** To minimize the number of cuts required, the

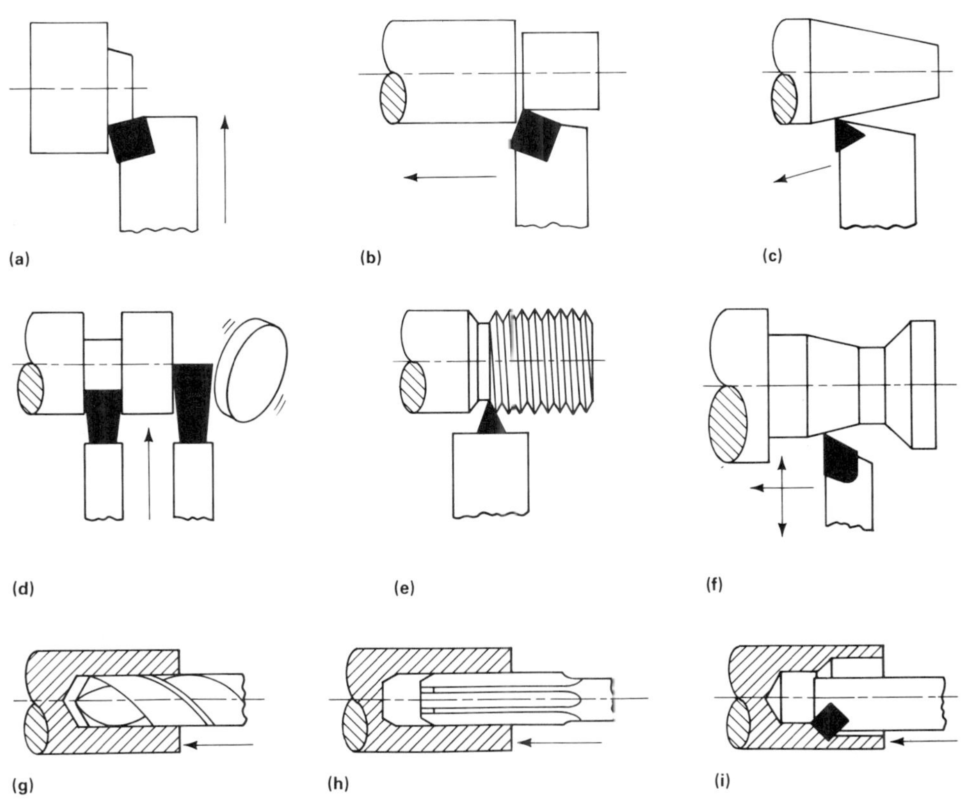


**Fig. 1** Basic operations performed on turning equipment. (a) Facing. (b) Straight turning. (c) Taper turning. (d) Grooving and cutoff. (e) Threading. (f) Tracer turning. (g) Drilling. (h) Reaming. (i) Boring

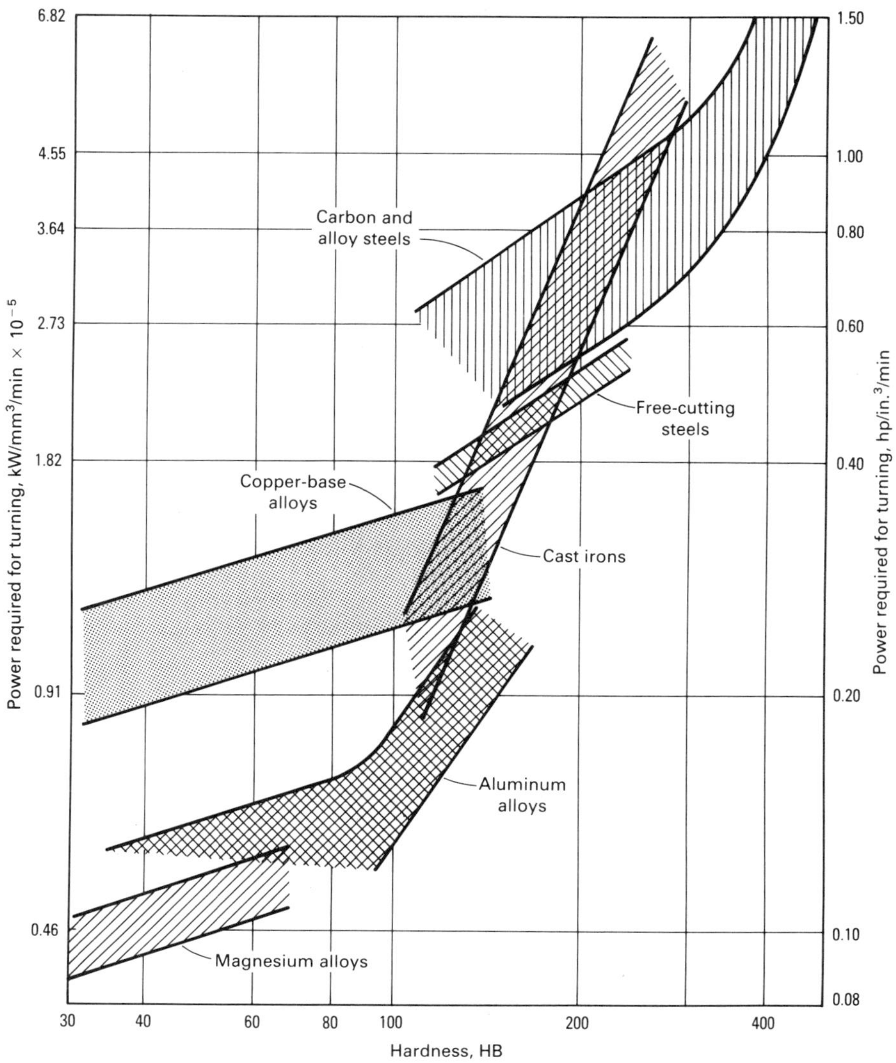


**Fig. 2** Effect of composition and hardness of work metal on horsepower requirements. Data plotted are for 2.54 mm (0.100 in.) depth of cut at a feed of 0.318 mm/rev (0.0125 in./rev) with tools of the following geometry: 8°, 14°, 6°, 6°, 0°, 3/64 in., except on aluminum, for which tool geometry was 20°, 40°, 10°, 10°, 10°, 15°, 0 in.

depth of cut should be as great as is consistent with the strength of the part and the chucking equipment, the power of the machine tool, the strength and size of any cutting tool or carbide inserts when used, and the amount of stock to be removed. As the depth of cut is increased, the cutting force becomes larger. This must be limited to a value that will not distort the part, pull it from the chuck, or overload the machine. Depth of cut ordinarily ranges from a few mils to about 6.4 mm (1/4 in.), but can be 25 mm (1 in.) or more in roughing cuts on large pieces.

The feed depends on the finish desired and the strength and rigidity of the part and the machine. Feed in most applications varies from 0.13 to 0.51 mm/rev (0.005 to 0.020 in./rev). Finishing cuts require a light feed and may be at 0.025 mm/rev (0.001 in./rev) or less; roughing cuts on large workpieces are often taken at a feed rate of 6.35 mm/rev (0.25 in./rev) or more. Cutting speed depends primarily on workpiece hardness and tool material.

## Basic Lathe Components

An engine lathe basically consists of a bed (including two ways), a headstock, a tailstock, a compound slide (carriage), a cross slide, a tool holder mounted on the cross slide or top slide, and a source of power for rotating the workpiece (see Fig. 3). Most engine lathes also incorporate a lead screw that moves the compound slide uniformly along the bed. When used for thread-cutting operations, a rack and pinion drive is typically used for normal feeding of the workpiece.

Engine lathes are available with an almost infinite number of modifications and also have served as the basis for another group of machine tools that includes turret lathes and single-spindle or multiple-spindle bar and chucking machines.

Regardless of other modifications, engine lathes are provided with only two types of headstocks or work drivers, that is, chucking and centering. The chucking type of headstock incorporates a work-holding device that grips the workpiece, thus providing centering and clamping for positive rotation. This type is most often used for turning castings or forgings or for the cutoff of short sections of bar stock without using centers.

In a lathe having the centering type of headstock, the workpiece is secured between a pointed center in the headstock and a counterpart center in the tailstock. The workpiece is gripped for rotation by means of a drive dog, which locks into the face plate or a compensating work driver. The centering type is most commonly used for turning long, symmetrical workpieces, such as shafts. In common practice, the workpiece is turned for most of its length, after which the drive dog is removed, the workpiece reversed and the operation completed.

### Lathe Size

It is common practice to specify the size of an engine lathe by giving the swing (maximum diameter of workpiece over the ways) and distance between centers when the tailstock is flush with the end of the bed. The maximum swing over the ways is usually greater than the nominal swing. The length of the bed is frequently given to specify the overall length of the bed. A lathe size is indicated thus: 356 mm (14 in.) (swing) by 762 mm (30 in.) (between centers) by 1.8 m (6 ft) (length of bed). Lathes are made for light, medium, or heavy-duty work.

### Methods of Control

Regardless of size and design, lathes are classified by method of control in operation, that is, manual, semiautomatic, or automatic.

**Manual lathes** have no automatic controls; all operations are controlled by the operator. Manual lathes are most commonly used for three kinds of applications:

- Machining one or, at most, fewer than a dozen small parts
- Producing prototypes in model shops or toolrooms
- Turning large components for which one turning cut may require one or more hours

For turning the same workpieces, setup time for manual operation is low compared to that for semiautomatic or automatic operation. Axis digital readouts, both in English and metric units, are now being used to improve the accuracy and efficiency of engine lathes.

**Semiautomatic lathes** may vary considerably in the degree to which the operation has been automated. However, all semiautomatic machines require some operator

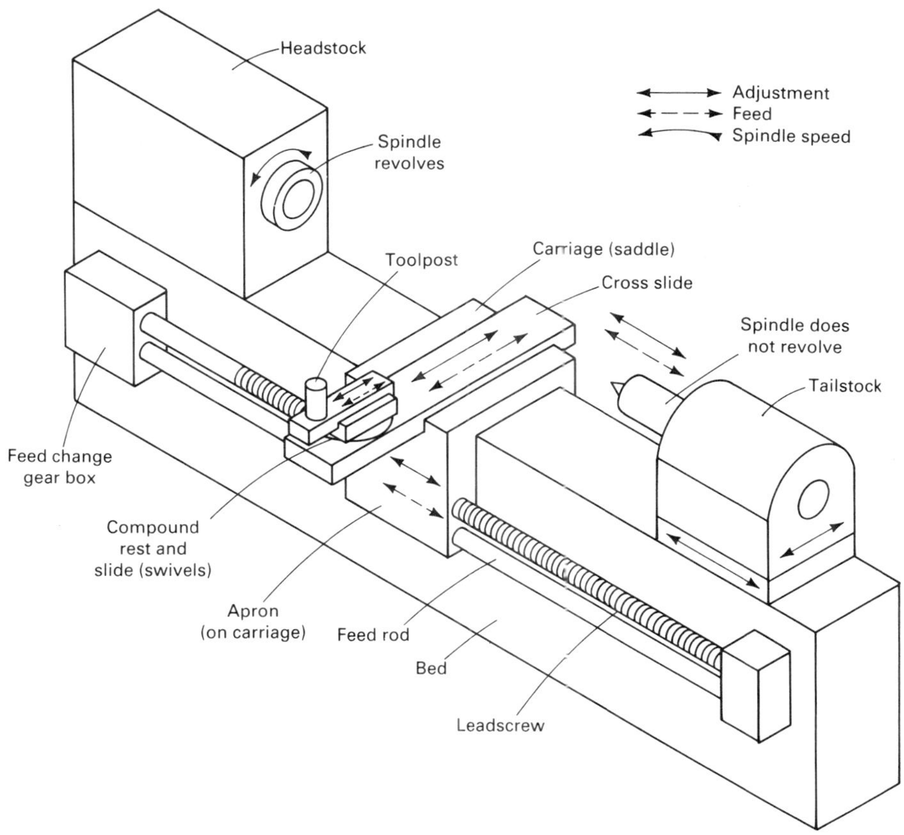


**Fig. 3** Principal components and movements of a lathe

attention during the machining cycle (in addition to loading and unloading). In some instances, the operator may be required only to initiate the action of a cross-slide tool at a specific point in the operation. In other instances, the operator may have to stop the cycle, remove the drive dog, and reverse the workpiece for the next operation in a sequence. Setup time for a semiautomatic lathe is greater than for a manual lathe but is less than for an automatic lathe. Many semiautomatic lathes are now being equipped with tracing attachments.

**Automatic lathes** require no operator attention during the machining cycle. With most types, however, workpieces are loaded and unloaded manually. Also, the cycle is started manually, but the operations are completed (and the machine is stopped) automatically.

For turning several diameters on a workpiece that cannot be placed between the lathe centers but can be secured on an arbor, a logical procedure is:

- Operator assembles workpiece on arbor
- Operator places loaded arbor in lathe and touches start button
- Machining operations take place in a predetermined sequence, and lathe stops automatically

During the machining cycle, the operator prepares another workpiece-and-arbor assembly to permit immediate reloading. It is often possible (depending on the time consumed in the machining cycle) for one operator to run two machines. Minimizing the operator's interaction with the machine allows the operator to perform other important functions as the machine is cycling.

For some high-production operations, automatic lathes that incorporate automatic loading and unloading devices are available, thereby allowing one operator to handle two or more machines. In all instances, setup time is greater for automatic lathes than for manual or semiautomatic types.

## Lathe Classification

A suitable classification of these machines is difficult because there are so many variations in the size, design, method of drive, and application. Most lathes are designated according to some outstanding design characteristic: speed lathes, engine lathes, bench lathes, toolroom lathes, special-purpose lathes, turret lathes, automatic-turning machines, and modifications of these types.

### Speed Lathes

The speed lathe, the simplest of all lathes, consists of a bed, a headstock, a tailstock, and an adjustable slide for supporting the tool. Usually it is driven by a variable-speed motor built into the headstock, although the drive may be a belt to a step-cone pulley. Because hand tools are used and the cuts are small, the lathe is driven at high speed (spindle speeds up to 4000 rev/min are typical), with the work either held between centers on a chuck or attached to a face plate on the headstock.

The speed lathe is principally used for turning wood for small cabinet work or for patterns, and for centering metal cylindrical parts prior to further work on the engine lathe. In the latter operation, the center drill is held in a small chuck fastened to the headstock, and the work is guided to the center drill either by a fixed center rest or by a movable center in the tailstock. Metal spinning is done on lathes of this type by rapidly revolving a stamped or deep-drawn piece of thin, ductile metal and pressing it against a form by means of blunt hand tools or rollers.

### Engine Lathes

The engine lathe derives its name from the early lathes, which were powered by engines. It differs from a speed lathe in that it has additional features for controlling the spindle speed and for supporting and controlling the feed of the fixed cutting tool. There are several variations in the design of the headstock through which the power is supplied to the machine.

**Step-Cone Pulley Drive.** Light- or medium-duty lathes receive their power through a short belt from the motor or from a small cone-pulley countershaft driven by the motor. The headstock is equipped with a four-step-cone pulley, which provides four different spindle speeds when connected directly from the motor countershaft. In addition, these lathes are equipped with back gears which, when connected with the cone pulley, provide four additional speeds.

**Geared-Head Drive.** The spindle speeds of this lathe are varied by a gear transmission, the different speeds being obtained by changing the gear trains through the positioning of levers on the headstock. Such lathes are usually driven by a constant-speed motor mounted on the lathe, but in a few cases variable-speed motors are used. A geared-head lathe has the advantage of a positive drive and has a greater number of spindle speeds available (typically 8, 12, 16, and 24) than are found on a step-cone driven lathe.

Heavy-duty models range in size from 305 to 610 mm (12 to 24 in.) swing and from 610 to 1220 mm (24 to 48 in.) center distances, with swings up to 1270 mm (50 in.) and center distances up to 3.7 m (12 ft) being common.

### Bench Lathes

The name bench lathe is given to a small engine lathe that is mounted on a work bench. In design it has the same features as speed or engine lathes and differs from

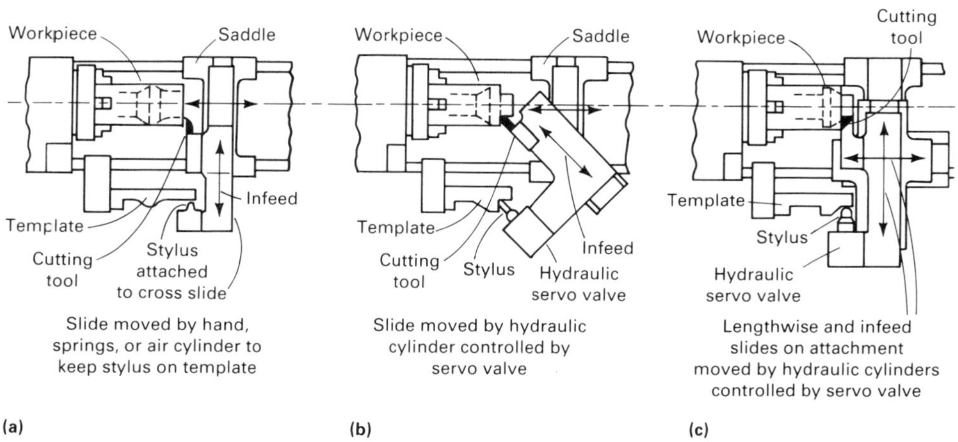


**Fig. 4** Details of the main types of tracer or duplicating attachments. (a) and (b) are motor tracer types, which use lengthwise feed from machine saddle movement. (c) Slide tracer type

these lathes only in size and mounting. It is adapted to small work, having a maximum swing capacity of 255 mm (10 in.) at the face plate. Many lathes of this type are used for precision work on small parts.

## Toolroom Lathes

This lathe, the most modern engine lathe, is equipped with all the accessories necessary for accurate tool work, being an individually driven geared-head lathe with a considerable range in spindle speeds. It is often equipped with center steady rest, quick-change gears, lead screw, feed rod, taper attachment, thread dial, chuck, indicator, draw-in collet attachment, and a pump for a coolant. All toolroom lathes are carefully tested for accuracy and, as the name implies, are especially adapted for making small tools, test gages, dies, and other precision parts. Their beds frequently are shorter than ordinary engine lathes having comparable swing dimensions because they are usually used for machining relatively small parts.

## Special-Purpose Lathes

Several types of special-purpose lathes are made to accommodate specific types of work. These include wheel lathes, hollow-spindle lathes, and gap-frame lathes.

**Wheel lathes** permit the turning of journals and wheel treads of railroad car wheel and axle assemblies. A special headstock drives the assembly at a point between the two wheels.

**Hollow-spindle lathes** permit the loading of tubular or shaftlike workpieces through a hollow spindle in the rear of the headstock. Workpieces may then be extended as required through the headstock into the work area, gripped by jaw-type holders. With this type of lathe, the ends of long members can be turned without the need for a lathe long enough to accommodate the entire length of the workpiece between centers. The distance that the workpiece may extend from the rear of the headstock is virtually unlimited. However, for safety reasons, the workpiece must be supported or contained because as the unsupported length becomes unwieldy and dangerous, steady rests or some other means of outboard support/containment must be provided. Hollow-spindle lathes often are referred to as oil-country lathes, because of their extensive use for turning sections of long members such as drill pipes and sucker rods.

**Gap-frame lathes,** sometimes called gap-bed lathes, have been modified to provide greater diametral clearance adjacent to the headstock. With a gap-frame lathe, workpieces that require off-center mounting or that have irregular protrusions can be turned by using this additional clearance. Otherwise, a larger lathe would be required to rotate the same workpiece. However, this does not mean that the gap-frame lathe can turn the extremes of any workpiece that can be accommodated by the larger clearance. Travel of the compound slide is restricted to the length of the ways. Therefore, as the slide approaches the headstock, it is stopped at the gap, which may not be as far as desired. This disadvantage can be partly overcome by the use of accessories that permit further travel of the tool from the compound slide. However, this involves tool overhang and decreases rigidity.

**Duplicating lathes** include copying, tracer, profiling, numerical-control, and continuous-path turning lathes. All are versatile production machines, and all are adaptable to the turning of external and internal contours. Tracer and numerical-control lathes are the most commonly used types of duplicating lathes. Each employs a different actuating system to guide the single-point cutting tool that generates the desired profile as the workpiece is rotated.

Specially designed lathes known as T lathes are used to turn large diameters. These lathes have their carriage ways running parallel to the face of the chuck or at right angles to the headstock spindle centerline.

*Tracer Lathes.* With tracer equipment, the tool slide is controlled by means of a sensitive stylus, usually servo-actuated hydraulically, which follows an accurate template (see Fig. 4). The two- or three-dimensional template may be a finish-turned workpiece or a profile cut from a flat, thin steel plate.

Tracer equipment is available either as units to be attached to standard lathes or as complete, specially engineered machine tools. The combinations of feeds, speeds, and number of cuts that can be accomplished automatically are almost unlimited. For instance, some tracer lathes can be set up to make six or more passes to permit roughing, semifinishing, and finishing. These cuts may be coordinated with three or more automatic changes of speed or feed, or both, during one machining cycle.

Tracer lathes can be equipped with two tracer units to operate from both ends or both sides of a workpiece, for completely turning the part on one machine. Often, however, limitations caused by the physical angle of approach of the tool do not allow one tool or tracer unit to completely machine the workpiece. A fully equipped tracer lathe incorporates one or more cross slides for use in rough or finish facing, back chamfering, grooving, or undercutting. Optional equipment includes automatic-positioning steady rests and an automatic-indexing tool head that contains both roughing and finishing tools.

*Numerical-control (NC) lathes* differ from hydraulically controlled tracer lathes primarily in that their cutting tools are controlled electronically. All tool movements during cuts, as well as all tool indexing, cross-slide operations, and changes of speed and feed, are pre-engineered and programmed on a punched tape for electronic control of all machine movements. Advantages of the NC lathe include low inventory of tooling (no templates are required, for example) and fully controlled cutting conditions (because speed and feed changes, for example, can be made only by reprogramming and/or punching a new tape accordingly).

For copying work, NC lathes have the capability of machining any cylindrical form without the need for a template. However, compared to tracer lathes, NC machines can be quite expensive.

## Turret Lathes

A turret lathe is a manual lathe having a hexagonal tool-holding turret in place of the tailstock of an engine lathe. The headstock in most cases is geared with provision for 6 to 16 spindle speeds and may double this capability with a two-speed motor. Some turret lathes are designed and equipped for

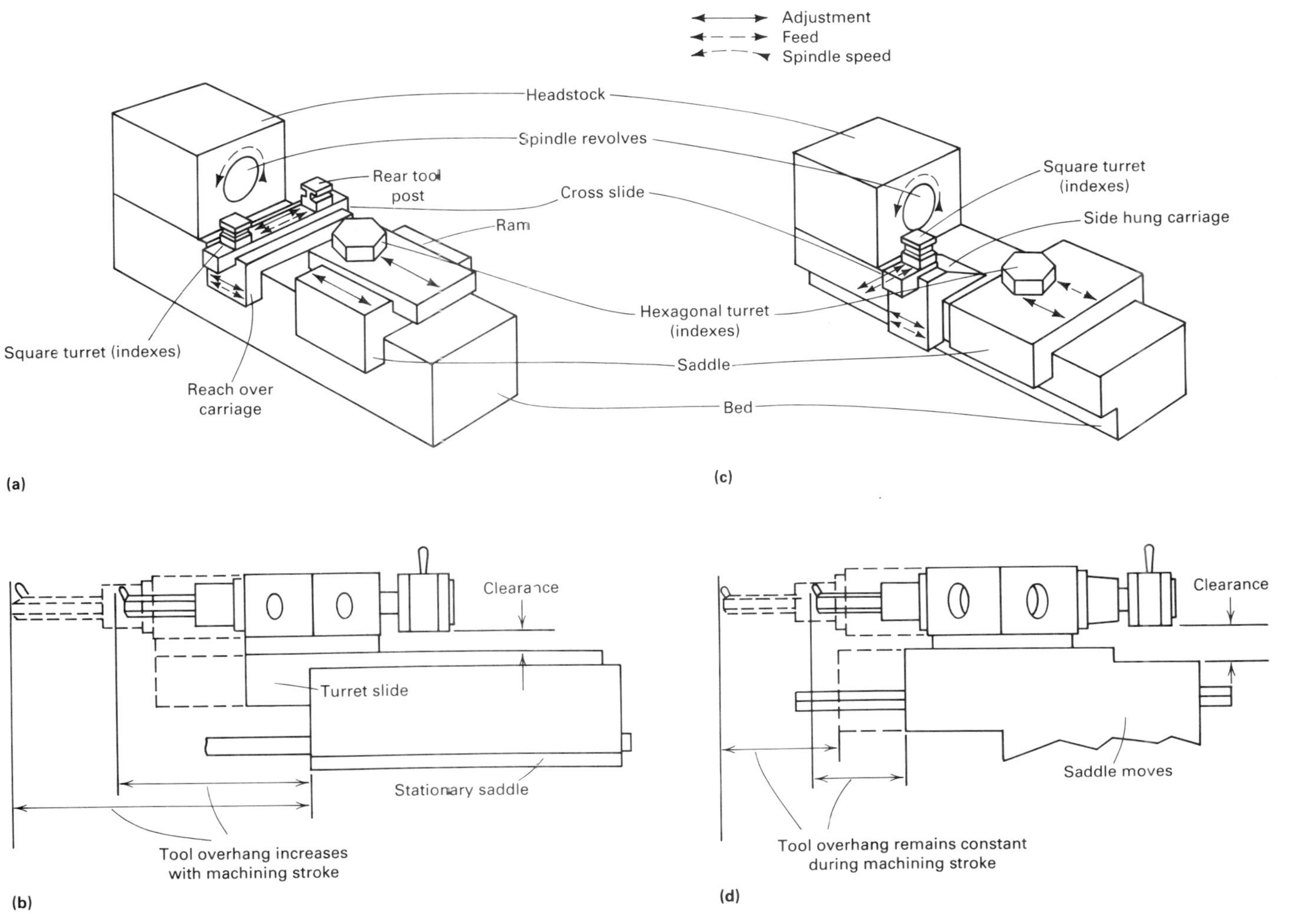


**Fig. 5** Principal components and movements of horizontal turret lathes. (a) Ram type. (b) Detail of turret movement in ram type. (c) Saddle type. (d) Detail of turret movement in saddle type

working on barstock and are called bar-type machines. The name screw machine or hand screw machine has been used for such machines, particularly in the smaller sizes. Other turret lathes are equipped for chuck work.

**Horizontal Turret Lathes.** The name turret lathe alone ordinarily implies a horizontal spindle machine. This type of lathe is made in two general designs, known as ram and saddle.

*A ram-type* turret lathe carries the turret on a ram (Fig. 5a). The ram slides longitudinally on a saddle positioned and clamped on the ways of the bed. Tools in their holders are mounted on the faces of the turret, and the tools on the face toward the headstock are fed to the work when the ram is moved to the left. When the ram is withdrawn, the turret indexes, and the next face, called a station, is positioned to face the headstock.

A ram is lighter and can be moved more quickly than a saddle, but lacks some rigidity. Because of convenience and speed, the ram-type construction is favored for small- and medium-size turret lathes for which the ram does not have to overhang excessively.

*Saddle-Type.* The hexagonal turret of a saddle-type turret lathe is carried directly on a saddle that slides lengthwise on the bed (Fig. 5b). This construction is favored for large turret lathes because it provides good support for the tools and the means to move the tools a long distance when necessary. The saddle is moved toward the headstock by hand or power to feed the tools to the work and is withdrawn to index the turret.

The turret is fixed in the center of the saddle on some machines. On others it may be moved crosswise. This helps reduce tool overhang for machining large diameters and is helpful for taper or contour boring and turning.

Most saddle-type turret lathes have a side hung carriage that does not extend across the entire top of the bed and allows larger pieces to be swung. As a result, the rear tool station on the cross slide is lost.

**Workpiece Capacity.** Turret lathes of the ram type are built in a wide range of sizes, and saddle-type machines are built in still larger sizes. Generally, ram-type turret lathes will produce bar parts up to 76 mm (3 in.) in diameter and chucking work up to approximately 508 mm (20 in.) in diameter; ram slide stroke lengths range from 102 to 330 mm (4 to 13 in.). Standard saddle-type turret lathes can handle bar work up to 305 mm (12 in.) in diameter and chucking work up to approximately 915 mm (36 in.) in diameter.

The newer machines not only are made more rigid and powerful, they embody more automatic features. Some have speed and feed calculators, rapid traverse on longitudinal and cross feed, higher turning speeds, and automatic pressure lubrication.

Figure 6 illustrates the complex configurations which can be machined on a turret lathe. More information on turret lathes can be obtained in the article "Multiple-Operation Machining" in this Volume.

**Vertical Turret Lathes.** With these lathes, the work is mounted on a rotating table, and the tools are mounted on vertical rams or turrets. Vertical turret lathes are used primarily for short, heavy, large-diameter workpieces. They are also ideal for lightweight but bulky parts. Controls are usually

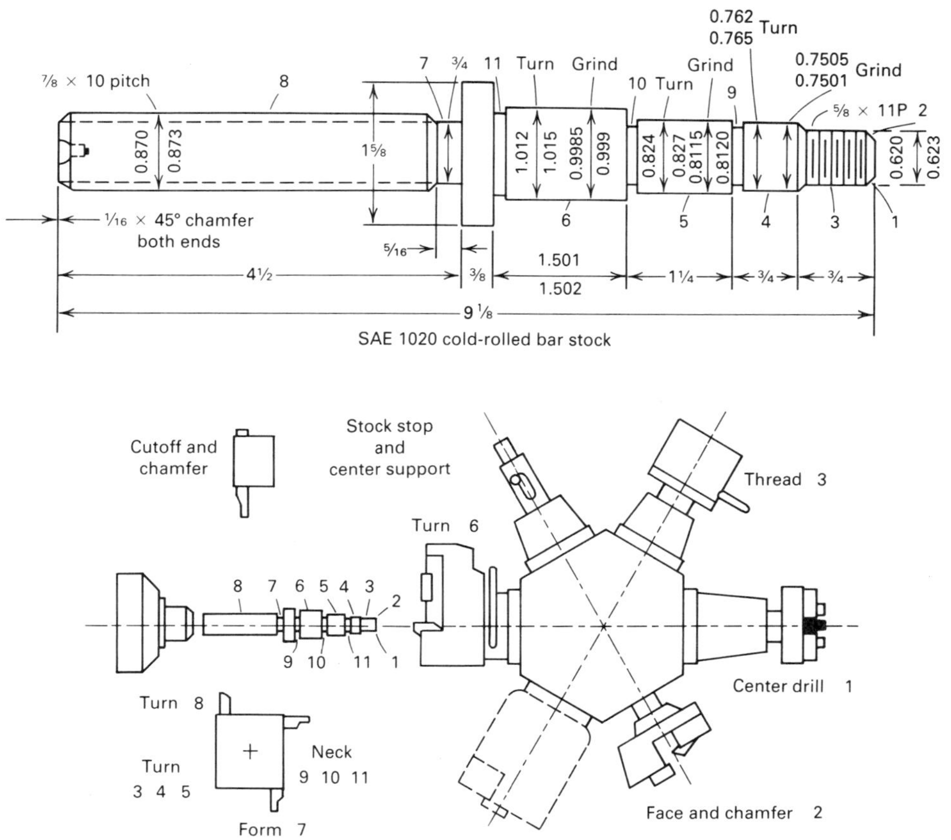


**Fig. 6** A typical part made on a turret lathe and the cutting components on the turret. After one operation is completed, the turret rotates and performs the next operation. Note the various operations performed and the tolerances obtained. Dimensions given in inches

automatic (plugboard or cam) but can also be manual or numerical control.

Vertical turret lathes are more convenient for short, heavy pieces over about 457 mm (18 in.) in diameter because the work can be laid upon the table and fastened more easily than hung on the end of the spindle of a horizontal turret lathe. A turret on a vertical turret lathe is carried on a ram on a crossrail above the worktable and can be fed up or down or crosswise. There is also a side head, essentially a carriage and cross slide, alongside the table with a square turret that can be fed radially and parallel to the axis of the table.

The line of machines of one leading manufacturer is typical; indexing turrets are available with table capacities to 3.7 m (12 ft) in diameter. Larger sizes, and smaller sizes, too, if desired, have single-tool rams; there are normally two rams on the crossrail. These larger machines are called vertical boring mills (see the article "Boring" in this Volume) and are available in sizes up to 18 m (60 ft) in diameter. They perform turning, facing, boring, and grooving operations of huge, fairly round, and symmetrical pieces such as reduction gear housings and turbine casings.

A typical vertical turret lathe has a 1.4 m (55 in.) diam swing and a 65 kW (87 hp) drive. Typical vertical boring mill specifications include 2.3 m (90 in.) diam capacity, 1.7 m (67 in.) maximum work height, 15 Mg (33 000 lb) maximum workpiece weight, and a 90 kW (120 hp) power rating.

## Automatic Turning Machines

A machine that moves the work and tools at the proper rates and sequences through a cycle without the attention of an operator to perform an operation on one piece is commonly called an automatic. Strictly speaking, the machine is a semiautomatic if an operator is required to load and unload the machine and start each cycle. Often an operator can do this for several machines in a group. Workpieces may come to a fully automatic machine on a conveyor or an operator may load a magazine or hopper at intervals. Automatic machines are widely used for drilling, boring, milling, broaching, grinding, and other operations (see the Section "Traditional Machining Processes" in this Volume). Automatic turning machines are made massive, rigid, and powerful to drive cutting tools at their highest speed and get the most from gangs of tools and multiple and combined tooling.

Automatic machines intended for large lots of pieces and infrequent changeover are not usually designed for quick setup. For instance, speeds and feeds are not changed by sliding levers or turning dials, as on an engine lathe, but by removing and replacing pick-off gears. The multiplicity of tools must be set with respect to each other as well as to the machine. In some cases, common practice is to preset the tools in blocks off the machine. The blocks then go into position quickly on the machine at setup time, and machine downtime is saved. Setup time may be less than an hour in favorable cases in which few changes are needed between jobs, but normally changeover takes several hours, and not infrequently up to a day or more, compared to a few minutes on an engine lathe. Careful scheduling to run similar jobs in succession can save changeover time. Although setup time is long, an automatic machine has the advantage when cutting because it is fast and often able to make up the time lost in setup in less than 100 pieces.

A basic mechanical device for driving tool slides and other units is the cam. Some machines require a set of cams for each job. The shape of each cam programs one or more movements. Because the operation is costly, only large production runs are warranted. However, each set of cams can be made to do the job in the shortest time (other machines that do not require cam changes for each job are called camless automatics). One kind has a permanent set of cams that act through adjustable linkages. Without cams, activities are carried out through adjustable electro- and hydromechanical devices.

A different kind of automatic operation is numerical control (NC), in which the program is expressed in numbers that are fed into the controller to make the machine go through the steps of a desired operation. Leading kinds of automatic turning machines are presented here under the classifications of automatic lathes, and single-spindle and multiple-spindle automatic bar and chucking machines.

**Automatic lathes** have the basic units of simple lathes: bed, headstock, tool slides, and sometimes a tailstock. In addition, an automatic lathe drives the tools through all the steps of a cycle without operator attention once the machine has been set up. The workpiece is rotated between centers. The tools are carried on blocks on the front and rear slides. The front slide is traversed along the bed, and the tools in this case make straight cuts along the workpiece, retract at the end of the cut, and are withdrawn to the starting position. The rear tool slide typically feeds the tools toward the center of the workpiece for facing, necking, grooving, and forming, but can be given a sideways movement to relieve the tools at the end of a cut.

All automatic lathes perform basically similar functions, but they appear in a variety of forms. Many have no tailstock. Some models have one slide, and others have two

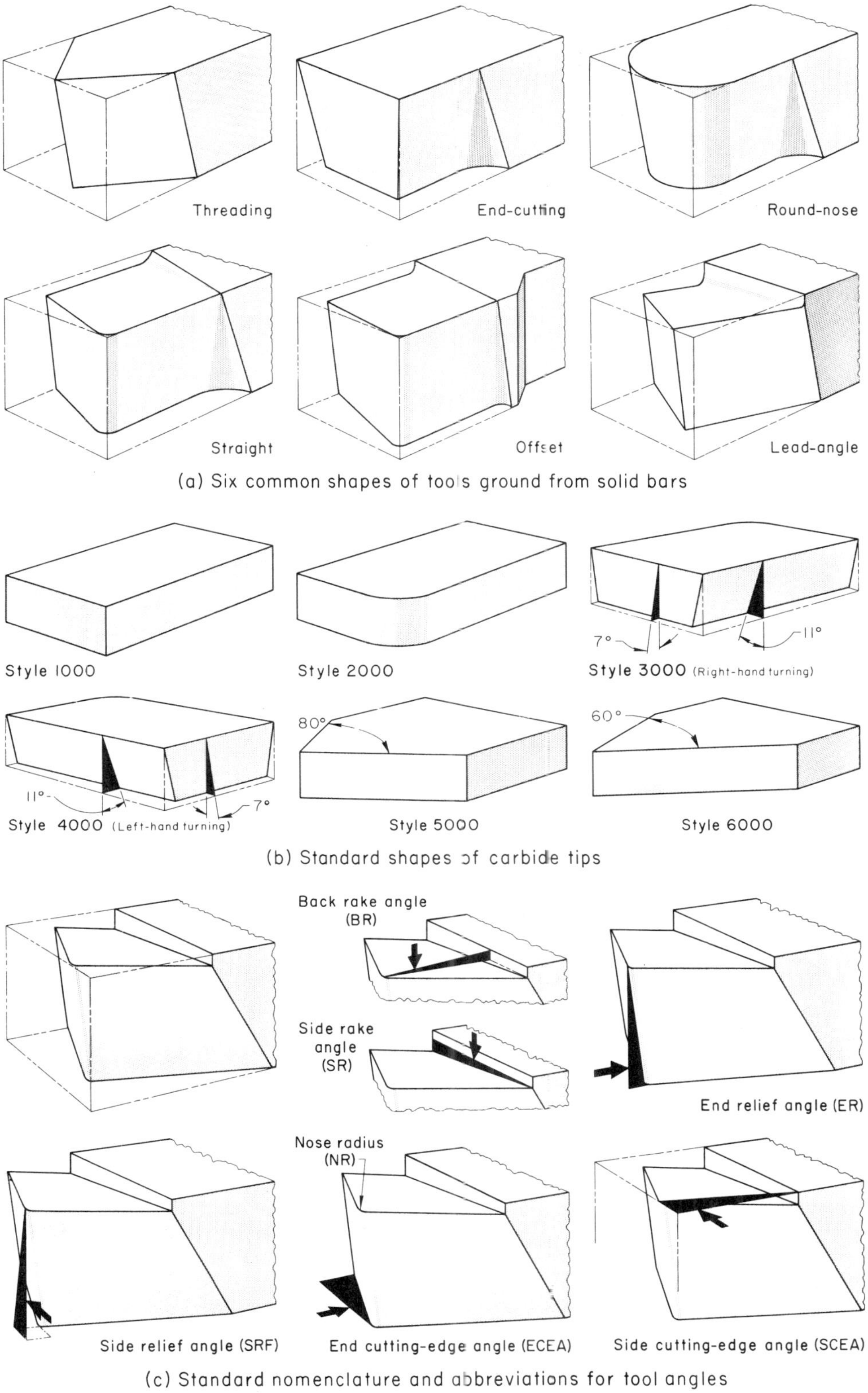


**Fig. 7** Common shapes and standard angles of single-point tools

or three. Some slides move in one direction only; others move in two directions. Some machines have level tool slide ways; others have sloping ways, and some have overhead slides for additional tools. Some slides are fed by screws, others by hydraulic or air-hydraulic means, and still others entirely by cams and templates. There are horizontal and vertical models according to spindle position. Some automatic lathes have two or three work spindles so that two or three sets of tools can perform the same cuts on two or three workpieces at once.

Automatic lathes, particularly vertical machines, without tailstocks are commonly called chucking machines, or chuckers. The distinguishing feature of this automatic lathe type is that all the tools cut the workpiece substantially at the same time. On the other types also called chucking machines, a series of groups of tools or single tools are applied to each workpiece. On the other hand, a type called precision production boring machines, like automatic lathes, makes a number of cuts at one time. The distinction is generally that its movements are in one direction, while the movements on automatic lathes are in two or more directions. Detailed information on automatic lathes is available in the article "Multiple-Operation Machining" in this Volume.

**Automatic Bar and Chucking Machines.** Automatic machines for internal and external operations on barstock have been called automatic screw machines for years, but the term automatic bar machines is now preferred because screws are seldom made on them any more. Their counterparts for individual pieces are called automatic chucking machines. They can be further categorized as single-spindle, Swiss-type screw machine, or multiple-spindle automatics.

*Single-spindle automatics* are often referred to as single-spindle automatic bar and chucking machines. In essence, they are small cam-operated turret lathes. The distinguishing features of these machines are the positions of the turret and tool mountings. The automatic screw machine has front, top, and rear slides that are fed at predetermined rates by means of disk cams. These are designed to handle work ranging from watch staffs to bars approximately 200 mm (7¾ in.) in diameter and chuck work over 305 mm (12 in.) in diameter.

*Swiss-type automatic screw machines* have five radially mounted tools that are cam-controlled. The stock can be fed as it is being cut, making any desired cylindrical shape possible to generate. This type of lathe is especially good for turning out small-diameter instrument parts. The accuracy on small parts can be maintained to ±0.013 mm (±0.0005 in.).

*Multispindle Automatics.* Instead of having only one spindle, as on the engine lathe, four, six, or eight spindles may be used. These are high-production machines. Production quantities may range from 5000 to over 100 000 parts. Stock may be in bar (up to 150 mm, or 6 in., diam) or piece form. In the latter case, the lathes are called chucking machines.

## Lathe Attachments and Accessories

Attachments that increase the capabilities of engine lathes include those below.

**Table 1 Conventional order for listing the angles and the nose radius of single-point tools in the tool signature**

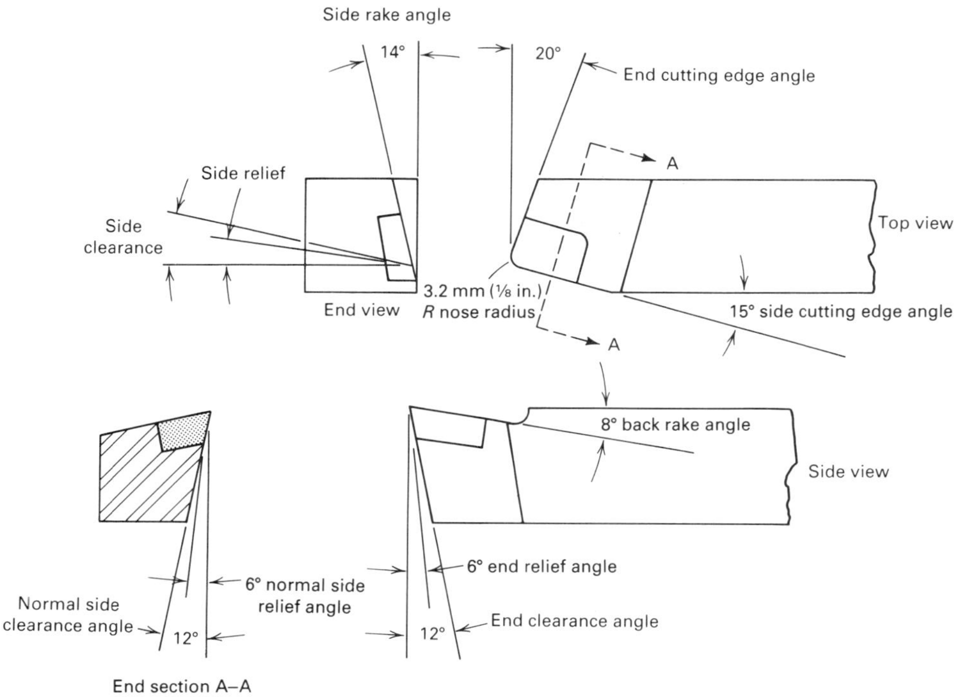


**Tool signature: 8, 14, 6(12), 6(12), 20, 15, ⅛**

| Order listed | Dimension | Abbreviation |
|---|---|---|
| 1 | Back rake angle | BR |
| 2 | Side rake angle | SR |
| 3 | End relief angle | ER |
| 4 | End clearance angle | . . . |
| 5 | Side relief angle | SRF |
| 6 | Side clearance angle | . . . |
| 7 | End cutting-edge angle | ECEA |
| 8 | Side cutting-edge angle | SCEA |
| 9 | Nose radius | NR |

**Grinding attachments** of several types and sizes make it possible to perform many grinding operations without removing the workpiece from the lathe. These attachments, however, do not provide the rigidity of grinding machines, and light grinding cuts therefore are advisable with attachments.

**Milling attachments** permit keyway cutting, T-slotting, dovetailing, angle milling, and thread milling. With one type, the cutter is held in the headstock, and the workpiece is held and correctly positioned by a device attached to the cross slide. With another, the cutter and driving arrangement are mounted on the cross slide, and the workpiece is held between the lathe centers.

**Taper attachments** are controlled by a guide to which the cross slide is attached. This guide is adjustable to various angles to permit turning or boring of tapers on the workpiece.

**Turret attachments** of several types may be added to engine lathes to provide them with some of the multiple-operation versatility of turret lathes.

**Ball-turning rests** replace the compound slide and enable turning or boring of spherical shapes.

**Contouring attachments** follow a flat or three-dimensional template or sample part by means of a servomechanism that positions the tool for the turning of shafts, radii, or irregular shapes.

## Single-Point Cutting Tools

Although form tools ground to specific shapes are frequently used to produce contours on workpieces when tool travel is limited to straight-line movement, most metal removal in lathe turning is accomplished with single-point tools.

Single-point tools may be produced from a solid bar of tool steel by grinding the appropriate cutting edge on one end (Fig. 7a), or they may be made of less-costly stock and provided with a tip, or insert, of carbide or other cutting material (Fig. 7b). The insert may be held in place mechanically or by brazing, soldering, or welding. Brazed, soldered, or welded inserts are resharpened after becoming dull through use; inserts held in place mechanically are usually of the disposable type.

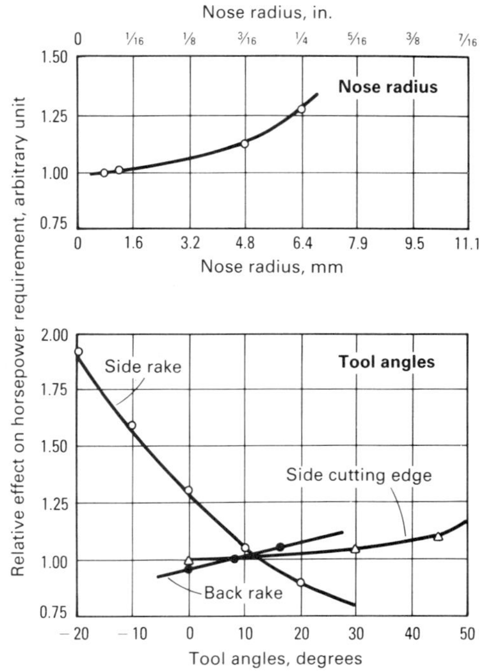


**Fig. 8** Effect of tool angles and nose radius on horsepower required for turning. Data show the effects of deviations from standard single-point tool shape of 8°, 14°, 6°, 6°, 6°, 0°, 1/16. Depth of cut 2.54 mm (0.100 in.); feed, 0.318 mm/rev (0.0125 in./rev)

### Design of Single-Point Tools

Standard angles for single-point tools are illustrated and named in Fig. 7c and in the figure which accompanies Table 1. Figure 7 shows how back rake angle, side rake angle, end relief angle, side relief angle, end cutting-edge angle, side cutting-edge angle, and nose radius are related in order to aid engineers in designing single-point cutting tools.

Back rake angle (BR) is the angle between the cutting face of the tool and the shank or holder, measured parallel to the side of the shank or holder. The angle is positive if, as shown in Fig. 7, it slopes from the cutting point downward toward the shank, and negative if it slopes upward toward the shank.

Side rake angle (SR) is the angle between the cutting face of the tool and the shank or holder, measured perpendicular to the side of the shank or holder. The angle is positive if, as shown in Fig. 7(c), it slopes downward away from the cutting edge to the opposite side of the shank, and negative if it slopes upward.

End relief angle (ER) is the angle between the end face of the tool and a line drawn from the cutting edge perpendicular to the base of the shank or holder and usually is measured at right angles to the end cutting edge.

Side relief angle (SRF) is the angle between the side flank immediately below the side cutting edge and a line drawn through the side cutting edge perpendicular to the base of the tool or tool holder and usually is measured at right angles to the side flank.

End cutting-edge angle (ECEA) is the angle between the end cutting edge of the tool and a line perpendicular to the side of the shank.

Side cutting-edge angle (SCEA), also called lead angle, is the angle between the side cutting edge and the projected side of the shank or holder.

Nose radius (NR) is the radius on the tool between the end and the side cutting edges.

The conventional order in which these angles are listed in specific identifications (the tool signature) is given in Table 1. It should be noted that the nine-digit tool signature shown in Table 1 applies to brazed and inserted tools; a 7-digit tool signature in which both the end clearance and side clearance angles are omitted is used to describe tool geometry for solid tool bits.

On a single-point tool that feeds to the side, the back rake (sometimes called top rake) turns the chip away from the finished work and gives the tool a slicing action. Zero back rake makes the chip spiral more tightly, whereas a positive back rack stretches the spiral chip into a longer helix.

Side rake controls the thickness of the tool behind the cutting edge. A thick tool associated with a small rake angle provides maximum strength, but the small angle produces higher cutting forces than a larger one (Fig. 8).

The end relief angle (also called front clearance) provides clearance between the tool and the finished surface of the work. Wear reduces the angle. If the angle is too small, the tool rubs on the surface of the work and mars the finish. If the angle is too large, the tool may dig into the work and chatter, or show weakness and fail through chipping. If the tool is set above the center of rotation, effective end relief angle is reduced; this must be considered in choosing the value.

The side relief angle provides clearance between the cut surface of the work and the flank of the tool. Tool wear reduces the effective portion of the angle close to the work. If this angle is too small, the cutter rubs and heats. If it is too large, the cutting edge is weak, and the tool may dig in, particularly with a large side rake angle or with play in the feed mechanism. When threads or other helices are being cut, the effective side relief angle is reduced by the helix angle, which should be added to the recommended angle.

The end cutting-edge angle provides clearance between the cutter and the finished surface of the work. An angle too close to zero may cause chatter with heavy feeds, but for a smooth finish the angle on light finishing cuts should be small.

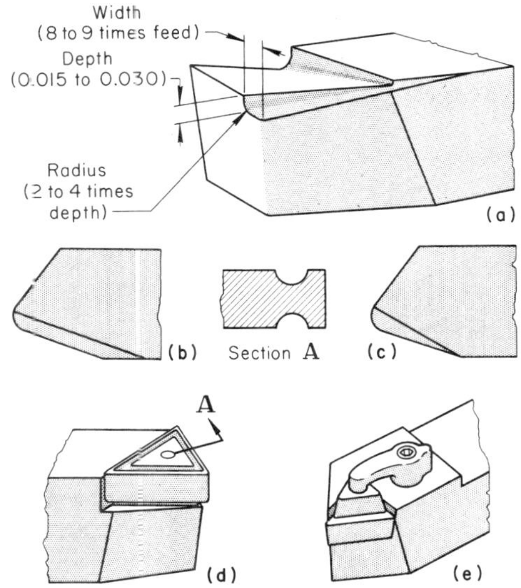


**Fig. 9** Typical chip breakers for turning tools. (a) Details of a typical ground-in chip breaker. (b) Parallel ground-in chip breaker. (c) Angular ground-in chip breaker. (d) Grooved chip breaker on carbide. (e) Mechanically held chip breaker for carbide insert. Dimensions given in inches

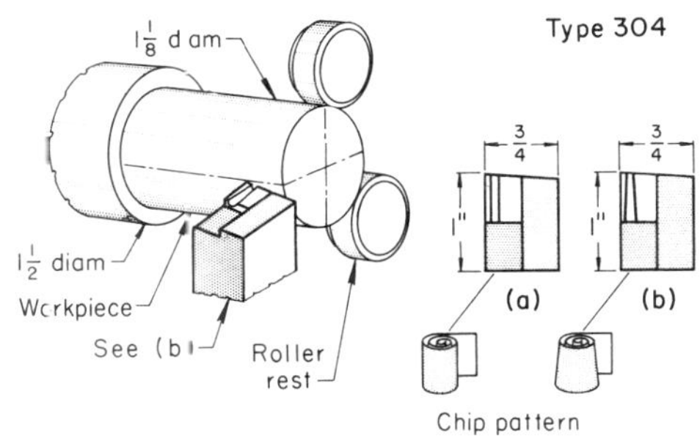


| | |
|---|---|
| Speed, m/min (sfm) | 64 (250) |
| Feed, mm/rev (in./rev) | 0.68 (0.011) |
| Metal removed per cut, mm (in.) | 4.8 (3/16) |
| Cutting fluid | Mixture of soluble oil and water |
| Workpiece hardness, HB | 140–170 |

Note: Bar stock was turned on a turret lathe with spindle capacity of 51 mm (2 in.). A standard carbide-tipped box tool was mounted in the turret.

**Fig. 10** Box tool and (a) improperly designed chip-curling groove and (b) improved design. Dimensions in figure given in inches

The side cutting-edge angle turns the chip away from the finished surface. If this angle is enlarged, the width of the chip is increased inversely with the cosine of the angle, and the chip thickness is reduced proportionately.

The nose contour, normally specified as a radius, removes the fragile corner of the tool, prolongs tool life, and improves finish. The radius may be large for maximum strength or for rough-cutting tools, but it may be reduced for light feeds. A radius too large may cause chatter, but the larger it is without chattering, the better the finish.

Tool design influences the power required at the tool point for turning. The effects of three tool angles and the nose radius are plotted in Fig. 8; the side rake angle has the greatest effect.

Tool design is particularly important in tracer lathe turning, for which only one tool is used to make all cuts. Because the minimum number of passes is desirable, so is the fastest rate of metal removal that is compatible with maintenance of dimensional and finish requirements. The use of only one cutting tool imposes severe restrictions on the shape of the tool. A tool often must be provided with a $-3°$ side cutting-edge angle and a 30° minimum end cutting-edge angle, resulting in a weak structure at the nose of the cutting edge. For this reason, heavy feed forces often must be avoided, if they have not already been ruled out because of required finish. Thus, surface speed becomes the criterion for determining maximum rate of metal removal, and the design of the cutting tool becomes vitally important in obtaining maximum metal removal and tool life. Increasing the surface cutting speed increases tool temperatures and results in premature tool wear.

**Cutoff Tools.** Successive pieces of turned, faced, or bored work frequently are separated or cut away by a radially plunging tool. Providing a front angle of about 15° on parting tools minimizes the size of the teat on the piece as it breaks off; this is not done with thin cutoff blades because the resulting lateral force might cause the parting tool to lead sidewise.

**Chip Breakers.** In most turning, it is preferable to break chips into small pieces rather than permit them to form a continuous chip because:

- Broken chips can be flushed away more readily by the cutting fluid, thus avoiding tangling of stringy chips in the machine
- Small, broken chips are more easily handled and stored
- Broken chips permit better lubrication and cooling by the cutting fluid, thereby increasing tool life

Chips are broken by one of two methods: The chip may be bent to the limit of its ductility, or it may be jammed into a tight corner as it leaves the workpiece. The second method is seldom recommended.

In lathe turning, chips ordinarily curl across the face of the cutting tool. In some instances, particularly with brittle metals, chips break as they are deflected by the tool face. When heavier cuts are taken by increasing the feed, the thicker chip may break as it impinges against the tool face. Where the cut may not be made heavier, or the chip does not break by deflection, the tool must be provided with a chip breaker. Chip breakers may either be ground into the tool or they may be separate additions to

**Table 2 Nominal speeds and feeds for turning a variety of steels and cast irons with high-speed steel (HSS) single-point and box tools**

| | | | | | High-speed steel tool | | | | | |
|---|---|---|---|---|---|---|---|---|---|---|
| | Hardness, | | Depth of cut | | Speed | | Feed | | Tool material | |
| Material | HB | Condition | mm | in. | m/min | sfm | mm/rev | in./rev | ISO | AISI |
| **Carbon steels, wrought** | | | | | | | | | | |
| **Low carbon** | | | | | | | | | | |
| 1005 1010 1020 | 125–175 | Hot rolled, normalized, | 1 | 0.040 | 46 | 150 | 0.18 | 0.007 | S4, S5 | M2, M3 |
| 1006 1012 1023 | | annealed, or cold drawn | 4 | 0.150 | 38 | 125 | 0.40 | 0.015 | S4, S5 | M2, M3 |
| 1008 1015 1025 | | | 8 | 0.300 | 30 | 100 | 0.50 | 0.020 | S4, S5 | M2, M3 |
| 1009 1017 | | | 16 | 0.625 | 24 | 80 | 0.75 | 0.030 | S4, S5 | M2, M3 |
| **Medium carbon** | | | | | | | | | | |
| 1030 1042 1053 | 125–175 | Hot rolled, normalized, | 1 | 0.040 | 43 | 140 | 0.18 | 0.007 | S4, S5 | M2, M3 |
| 1033 1043 1055 | | annealed, or cold drawn | 4 | 0.150 | 35 | 115 | 0.40 | 0.015 | S4, S5 | M2, M3 |
| 1035 1044 1525 | | | 8 | 0.300 | 27 | 90 | 0.50 | 0.020 | S4, S5 | M2, M3 |
| 1037 1045 1526 | | | 16 | 0.625 | 11 | 70 | 0.75 | 0.030 | S4, S5 | M2, M3 |
| 1038 1046 1527 | | | | | | | | | | |
| 1039 1049 | | | | | | | | | | |
| 1040 1050 | | | | | | | | | | |
| **Alloy steels, wrought** | | | | | | | | | | |
| **Medium carbon** | | | | | | | | | | |
| 1330 4135 8625 | 275–325 | Normalized or quenched | 1 | 0.040 | 29 | 95 | 0.18 | 0.007 | S9, S11(a) | T15, M42(a) |
| 1335 4137 8627 | | and tempered | 4 | 0.150 | 23 | 75 | 0.40 | 0.015 | S9, S11(a) | T15, M42(a) |
| 4027 4427 8630 | | | 8 | 0.300 | 18 | 60 | 0.50 | 0.020 | S9, S11(a) | T15, M42(a) |
| 4028 4626 8637 | | | | | | | | | | |
| 4032 5130 94B30 | | | | | | | | | | |
| 4037 5132 | | | | | | | | | | |
| 4130 5135 | | | | | | | | | | |
| **Medium carbon** | | | | | | | | | | |
| 1340 4340 81B45 | 175–225 | Hot rolled, annealed, or | 1 | 0.040 | 41 | 135 | 0.18 | 0.007 | S4, S5 | M2, M3 |
| 1345 50B40 8640 | | cold drawn | 4 | 0.150 | 32 | 105 | 0.40 | 0.015 | S4, S5 | M2, M3 |
| 4042 50B44 8642 | | | 8 | 0.300 | 24 | 80 | 0.50 | 0.020 | S4, S5 | M2, M3 |
| 4047 5046 8645 | | | 16 | 0.625 | 20 | 65 | 0.75 | 0.030 | S4, S5 | M2, M3 |
| 4140 50B46 86B45 | | | | | | | | | | |
| 4142 5140 8740 | | | | | | | | | | |
| 4145 5145 8742 | | | | | | | | | | |
| 4147 5147 | | | | | | | | | | |
| **Low carbon** | | | | | | | | | | |
| 4012 4621 8115 | 225–275 | Hot rolled, normalized, | 1 | 0.040 | 34 | 110 | 0.18 | 0.007 | S4, S5 | M2, M3 |
| 4023 4718 8617 | | annealed, or cold drawn | 4 | 0.150 | 26 | 85 | 0.40 | 0.015 | S4, S5 | M2, M3 |
| 4024 4720 8620 | | | 8 | 0.300 | 20 | 65 | 0.50 | 0.020 | S4, S5 | M2, M3 |
| 4118 4815 8622 | | | 16 | 0.625 | 15 | 50 | 0.75 | 0.030 | S4, S5 | M2, M3 |
| 4320 4817 8822 | | | | | | | | | | |
| 4419 4820 9310 | | | | | | | | | | |
| 4422 5015 94B15 | | | | | | | | | | |
| 4615 5115 94B17 | | | | | | | | | | |
| 4617 5120 | | | | | | | | | | |
| 4620 6118 | | | | | | | | | | |
| **High-strength steels, wrought** | | | | | | | | | | |
| 300M 98BV40 | 225–300 | Annealed | 1 | 0.040 | 26 | 85 | 0.18 | 0.007 | S4, S5 | M2, M3 |
| 4330V D6ac | | | 4 | 0.150 | 20 | 65 | 0.40 | 0.015 | S4, S5 | M2, M3 |
| 4340 H11 | | | 8 | 0.300 | 15 | 50 | 0.50 | 0.020 | S4, S5 | M2, M3 |
| 4340Si H13 | | | 16 | 0.625 | 12 | 40 | 0.75 | 0.030 | S4, S5 | M2, M3 |
| **Stainless steels, wrought** | | | | | | | | | | |
| **Austenitic** | | | | | | | | | | |
| 201 304L 348 | 135–185 | Annealed | 1 | 0.040 | 34 | 110 | 0.18 | 0.007 | S4, S5 | M2, M3 |
| 202 305 384 | | | 4 | 0.150 | 27 | 90 | 0.40 | 0.015 | S4, S5 | M2, M3 |
| 301 308 385 | | | 8 | 0.300 | 21 | 70 | 0.50 | 0.020 | S4, S5 | M2, M3 |
| 302 321 | | | 16 | 0.625 | 17 | 55 | 0.75 | 0.030 | S4, S5 | M2, M3 |
| 304 347 | | | | | | | | | | |
| **Martensitic** | | | | | | | | | | |
| 403 422 | 135–175 | Annealed | 1 | 0.040 | 47 | 155 | 0.18 | 0.007 | S4, S5 | M2, M3 |
| 410 501 | | | 4 | 0.150 | 38 | 125 | 0.40 | 0.015 | S4, S5 | M2, M3 |
| 420 502 | | | 8 | 0.300 | 30 | 100 | 0.50 | 0.020 | S4, S5 | M2, M3 |
| | | | 16 | 0.625 | 24 | 80 | 0.75 | 0.030 | S4, S5 | M2, M3 |
| **Gray cast irons** | | | | | | | | | | |
| **Pearlitic** | | | | | | | | | | |
| ASTM A48: classes 30, 35, 40 | 190–220 | As-cast | 1 | 0.040 | 37 | 120 | 0.18 | 0.007 | S9, S11(a) | T15, M42(a) |
| SAE J431c: grade G3000 | | | 4 | 0.150 | 24 | 80 | 0.40 | 0.015 | S9, S11(a) | T15, M42(a) |
| | | | 8 | 0.300 | 20 | 65 | 0.50 | 0.020 | S9, S11(a) | T15, M42(a) |
| | | | 16 | 0.625 | 17 | 55 | 0.75 | 0.030 | S9, S11(a) | T15, M42(a) |

(a) Any premium HSS, (T15, M33, M41–M47) or (S9–S12). sfm, surface feet per minute. Source: Metcut Research Associates Inc.

**Table 3 Nominal speeds and feeds for turning a variety of steels and cast irons with carbide-tipped single-point and box tools**

| Material | Hardness, HB | Condition | Carbide tool: Uncoated: Speed: Brazed m/min | Brazed sfm | Indexable m/min | Indexable sfm | Feed mm/rev | Feed in./rev | Tool material grade ISO | Tool material grade C | Carbide tool: Coated: Speed m/min | Speed sfm | Feed mm/rev | Feed in./rev | Tool material grade ISO | Tool material grade C |
|---|---|---|---|---|---|---|---|---|---|---|---|---|---|---|---|---|
| **Carbon steels, wrought** | | | | | | | | | | | | | | | | |
| **Low carbon** | | | | | | | | | | | | | | | | |
| 1005 1010 1020 | 125–175 | Hot rolled, | 150 | 485 | 195 | 640 | 0.18 | 0.007 | P10 | C-7 | 290 | 950 | 0.18 | 0.007 | CP10 | CC-7 |
| 1006 1012 1023 | | normalized, | 125 | 410 | 150 | 500 | 0.50 | 0.020 | P20 | C-6 | 190 | 625 | 0.40 | 0.015 | CP20 | CC-6 |
| 1008 1015 1025 | | annealed, | 100 | 320 | 120 | 390 | 0.75 | 0.030 | P30 | C-6 | 150 | 500 | 0.50 | 0.020 | CP30 | CC-6 |
| 1009 1017 | | or cold drawn | 75 | 245 | 95 | 305 | 1.0 | 0.040 | P40 | C-6 | ... | ... | ... | ... | ... | ... |
| **Medium carbon** | | | | | | | | | | | | | | | | |
| 1030 1042 1053 | 125–175 | Hot rolled, | 140 | 465 | 130 | 590 | 0.18 | 0.007 | P10 | C-7 | 280 | 925 | 0.18 | 0.007 | CP10 | CC-7 |
| 1033 1043 1055 | | normalized, | 110 | 360 | 140 | 460 | 0.50 | 0.020 | P20 | C-6 | 185 | 600 | 0.40 | 0.015 | CP20 | CC-6 |
| 1035 1044 1525 | | annealed, | 85 | 280 | 110 | 360 | 0.75 | 0.030 | P30 | C-6 | 145 | 475 | 0.50 | 0.020 | CP30 | CC-6 |
| 1037 1045 1526 | | or cold | 67 | 220 | 85 | 280 | 1.0 | 0.040 | P40 | C-6 | ... | ... | ... | ... | ... | ... |
| 1038 1046 1527 | | drawn | | | | | | | | | | | | | | |
| 1039 1049 | | | | | | | | | | | | | | | | |
| 1040 1050 | | | | | | | | | | | | | | | | |
| **Alloy steels, wrought** | | | | | | | | | | | | | | | | |
| **Medium carbon** | | | | | | | | | | | | | | | | |
| 1330 4135 8625 | 275–325 | Normalized or | 105 | 340 | 130 | 420 | 0.18 | 0.007 | P10 | C-7 | 170 | 550 | 0.18 | 0.007 | CP10 | CC-7 |
| 1335 4137 8627 | | quenched | 79 | 260 | 100 | 330 | 0.40 | 0.015 | P20 | C-6 | 130 | 425 | 0.40 | 0.015 | CP20 | CC-6 |
| 4027 4427 8630 | | and | 60 | 200 | 79 | 260 | 0.50 | 0.020 | P30 | C-6 | 105 | 350 | 0.50 | 0.020 | CP30 | CC-6 |
| 4028 4626 8637 | | tempered | ... | ... | ... | ... | ... | ... | ... | ... | ... | ... | ... | ... | ... | ... |
| 4032 5130 94B30 | | | | | | | | | | | | | | | | |
| 4037 5132 | | | | | | | | | | | | | | | | |
| 4130 5135 | | | | | | | | | | | | | | | | |
| **Medium carbon** | | | | | | | | | | | | | | | | |
| 1340 4340 81B45 | 175–225 | Hot rolled, | 115 | 375 | 150 | 500 | 0.18 | 0.007 | P10 | C-7 | 200 | 650 | 0.18 | 0.007 | CP10 | CC-7 |
| 1345 50B40 8640 | | annealed, | 90 | 300 | 120 | 400 | 0.50 | 0.020 | P20 | C-6 | 160 | 525 | 0.40 | 0.015 | CP20 | CC-6 |
| 4042 50B44 8642 | | or cold | 73 | 240 | 95 | 315 | 0.75 | 0.030 | P30 | C-6 | 120 | 400 | 0.50 | 0.020 | CP30 | CC-6 |
| 4047 5046 8645 | | drawn | 58 | 190 | 76 | 250 | 1.0 | 0.040 | ... | ... | ... | ... | ... | ... | ... | ... |
| 4140 50B46 86B45 | | | | | | | | | | | | | | | | |
| 4142 5140 8740 | | | | | | | | | | | | | | | | |
| 4145 5145 8742 | | | | | | | | | | | | | | | | |
| 4147 5147 | | | | | | | | | | | | | | | | |
| **Low carbon** | | | | | | | | | | | | | | | | |
| 4012 4621 8115 | 225–275 | Hot rolled, | 120 | 400 | 150 | 500 | 0.18 | 0.007 | P10 | C-7 | 200 | 650 | 0.18 | 0.007 | CP10 | CC-7 |
| 4023 4718 8617 | | normalized, | 95 | 310 | 120 | 390 | 0.50 | 0.020 | P20 | C-6 | 150 | 500 | 0.40 | 0.015 | CP20 | CC-6 |
| 4024 4720 8620 | | annealed, | 73 | 240 | 95 | 310 | 0.75 | 0.030 | P30 | C-6 | 120 | 400 | 0.50 | 0.020 | CP30 | CC-6 |
| 4118 4815 8622 | | or cold | 58 | 190 | 73 | 240 | 1.0 | 0.040 | P40 | C-6 | ... | ... | ... | ... | ... | ... |
| 4320 4817 8822 | | drawn | | | | | | | | | | | | | | |
| 4419 4820 9310 | | | | | | | | | | | | | | | | |
| 4422 5015 94B15 | | | | | | | | | | | | | | | | |
| 4615 5115 94B17 | | | | | | | | | | | | | | | | |
| 4617 5120 | | | | | | | | | | | | | | | | |
| 4620 6118 | | | | | | | | | | | | | | | | |
| **High-strength steels, wrought** | | | | | | | | | | | | | | | | |
| 300M 98BV40 | 225–300 | Annealed | 105 | 350 | 135 | 450 | 0.18 | 0.007 | P10 | C-7 | 185 | 600 | 0.18 | 0.007 | CP10 | CC-7 |
| 4330V D6ac | | | 84 | 275 | 105 | 350 | 0.40 | 0.015 | P20 | C-6 | 135 | 450 | 0.40 | 0.015 | CP20 | CC-6 |
| 4340 H11 | | | 66 | 215 | 84 | 275 | 0.50 | 0.020 | P30 | C-6 | 105 | 350 | 0.50 | 0.020 | CP30 | CC-6 |
| 4340Si H13 | | | 52 | 170 | 69 | 225 | 0.75 | 0.030 | P40 | C-6 | ... | ... | ... | ... | ... | ... |
| **Stainless steels, wrought** | | | | | | | | | | | | | | | | |
| **Austenitic** | | | | | | | | | | | | | | | | |
| 201 304L 348 | 135–185 | Annealed | 105 | 350 | 120 | 400 | 0.18 | 0.007 | K01, M10 | C-3 | 160 | 525 | 0.18 | 0.007 | CK01, CM10 | CC-3 |
| 202 305 384 | | | 100 | 325 | 105 | 350 | 0.40 | 0.015 | K10, M10 | C-3 | 135 | 450 | 0.40 | 0.015 | CK10, CM10 | CC-3 |
| 301 308 385 | | | 76 | 250 | 84 | 275 | 0.50 | 0.020 | K10, M10 | C-2 | 105 | 350 | 0.50 | 0.020 | CK10, CM10 | CC-2 |
| 302 321 | | | 60 | 200 | 64 | 210 | 0.75 | 0.030 | K20, M20 | C-2 | ... | ... | ... | ... | ... | ... |
| 304 347 | | | | | | | | | | | | | | | | |
| **Martensitic** | | | | | | | | | | | | | | | | |
| 403 422 | 135–175 | Annealed | 145 | 475 | 190 | 620 | 0.18 | 0.007 | M10, P10 | C-7 | 245 | 800 | 0.18 | 0.007 | CM10, CP10 | CC-7 |
| 410 501 | | | 120 | 400 | 145 | 480 | 0.40 | 0.015 | M10, P10 | C-6 | 190 | 625 | 0.40 | 0.015 | CM10, CP10 | CC-6 |
| 420 502 | | | 100 | 320 | 115 | 380 | 0.75 | 0.030 | M20, P20 | C-6 | 150 | 500 | 0.50 | 0.020 | CM20, CP20 | CC-6 |
| | | | 73 | 240 | 90 | 300 | 1.0 | 0.040 | M30, P30 | C-6 | ... | ... | ... | ... | ... | ... |
| **Gray cast irons** | | | | | | | | | | | | | | | | |
| **Pearlitic** | | | | | | | | | | | | | | | | |
| ASTM A48: classes | 190–220 | As-cast | 115 | 370 | 125 | 410 | 0.18 | 0.007 | K20, M20 | C-3 | 160 | 525 | 0.18 | 0.007 | CK20, CM20 | CC-3 |
| 30, 35, 40 | | | 90 | 300 | 105 | 340 | 0.40 | 0.015 | K20, M20 | C-2 | 135 | 450 | 0.40 | 0.015 | CK20, CM20 | CC-2 |
| Sae J431c: grade | | | 76 | 250 | 84 | 275 | 0.75 | 0.030 | K20, M20 | C-2 | 105 | 350 | 0.50 | 0.020 | CK20, CM20 | CC-2 |
| G3000 | | | 60 | 200 | 69 | 225 | 1.0 | 0.040 | K20, M20 | C-2 | ... | ... | ... | ... | ... | ... |

Source: Metcut Research Associates Inc.

## Table 4 Nominal speeds and feeds for turning a variety of steels and cast irons with ceramic tools

| Material | Hardness, HB | Condition | Depth of cut | | Speed | | Feed | | Type of ceramic |
|---|---|---|---|---|---|---|---|---|---|
| | | | mm | in. | m/min | sfm | mm/rev | in./rev | |
| **Carbon steels, wrought** | | | | | | | | | |
| **Low carbon** | | | | | | | | | |
| 1005 1010 1020 | 85–125 | Hot rolled, normalized, | 1 | 0.040 | 610 | 2000 | 0.13 | 0.005 | CPA |
| 1006 1012 1023 | | annealed, or cold | 4 | 0.150 | 460 | 1500 | 0.25 | 0.010 | CPA |
| 1008 1015 1025 | | drawn | 8 | 0.300 | 335 | 1100 | 0.40 | 0.015 | CPA |
| 1009 1017 | | | | | | | | | |
| **Medium carbon** | | | | | | | | | |
| 1030 1042 1053 | 125–175 | Hot rolled, normalized, | 1 | 0.040 | 580 | 1900 | 0.13 | 0.005 | CPA |
| 1033 1043 1055 | | annealed, or cold | 4 | 0.150 | 395 | 1300 | 0.25 | 0.010 | CPA |
| 1035 1044 1525 | | drawn | 8 | 0.300 | 335 | 1100 | 0.40 | 0.015 | CPA |
| 1037 1045 1526 | | | | | | | | | |
| 1038 1046 1527 | | | | | | | | | |
| 1039 1049 | | | | | | | | | |
| 1040 1050 | | | | | | | | | |
| **Alloy steels, wrought** | | | | | | | | | |
| **Medium carbon** | | | | | | | | | |
| 1330 4135 8625 | 275–325 | Normalized or quenched | 1 | 0.040 | 395 | 1300 | 0.13 | 0.005 | HPC |
| 1335 4137 8627 | | and tempered | 4 | 0.150 | 235 | 775 | 0.25 | 0.010 | HPC |
| 4027 4427 8630 | | | 8 | 0.300 | 175 | 575 | 0.40 | 0.015 | HPC |
| 4028 4626 8637 | | | | | | | | | |
| 4032 5130 94B30 | | | | | | | | | |
| 4037 5132 | | | | | | | | | |
| 4130 5135 | | | | | | | | | |
| **Medium carbon** | | | | | | | | | |
| 1340 4340 81B45 | 175–225 | Hot rolled, annealed, or | 1 | 0.040 | 520 | 1700 | 0.13 | 0.005 | CPA |
| 1345 50B40 8640 | | cold drawn | 4 | 0.150 | 350 | 1150 | 0.25 | 0.010 | CPA |
| 4042 50B44 8642 | | | 8 | 0.300 | 260 | 850 | 0.40 | 0.015 | HPC |
| 4047 5046 8645 | | | | | | | | | |
| 4140 50B46 86B45 | | | | | | | | | |
| 4142 5140 8740 | | | | | | | | | |
| 4145 5145 8742 | | | | | | | | | |
| 4147 5147 | | | | | | | | | |
| **Low carbon** | | | | | | | | | |
| 4012 4621 8115 | 225–275 | Hot rolled, normalized, | 1 | 0.040 | 460 | 1500 | 0.13 | 0.005 | HPC |
| 4023 4718 8617 | | annealed, or cold | 4 | 0.150 | 305 | 1000 | 0.25 | 0.010 | HPC |
| 4024 4720 8620 | | drawn | 8 | 0.300 | 245 | 800 | 0.40 | 0.015 | HPC |
| 4118 4815 8622 | | | | | | | | | |
| 4320 4817 8822 | | | | | | | | | |
| 4419 4820 9310 | | | | | | | | | |
| 4422 5015 94B15 | | | | | | | | | |
| 4615 5115 94B17 | | | | | | | | | |
| 4617 5120 | | | | | | | | | |
| 4620 6118 | | | | | | | | | |
| **High-strength steels, wrought** | | | | | | | | | |
| 300M 98BV40 | 225–300 | Annealed | 1 | 0.040 | 440 | 1450 | 0.13 | 0.005 | HPC |
| 4330V D6ac | | | 4 | 0.150 | 265 | 875 | 0.25 | 0.010 | HPC |
| 4340 H11 | | | 8 | 0.300 | 205 | 675 | 0.40 | 0.015 | HPC |
| 4340Si H13 | | | | | | | | | |
| **Stainless steels, wrought** | | | | | | | | | |
| **Austenitic** | | | | | | | | | |
| 201 304L 348 | 135–185 | Annealed | 1 | 0.040 | 425 | 1400 | 0.13 | 0.005 | CPA |
| 202 305 384 | | | 4 | 0.150 | 275 | 900 | 0.25 | 0.010 | CPA |
| 301 308 385 | | | 8 | 0.300 | 185 | 600 | 0.40 | 0.015 | CPA |
| 302 321 | | | | | | | | | |
| 304 347 | | | | | | | | | |
| **Martensitic** | | | | | | | | | |
| 403 422 | 135–175 | Annealed | 1 | 0.040 | 550 | 1800 | 0.13 | 0.005 | CPA |
| 410 501 | | | 4 | 0.150 | 395 | 1300 | 0.25 | 0.010 | CPA |
| 420 502 | | | 8 | 0.300 | 305 | 1000 | 0.40 | 0.015 | CPA |
| **Gray cast irons** | | | | | | | | | |
| **Pearlitic** | | | | | | | | | |
| ASTM A48: classes 30, | 190–220 | As-cast | 1 | 0.040 | 460 | 1500 | 0.25 | 0.010 | HPC |
| 35,40 | | | 4 | 0.150 | 305 | 1000 | 0.40 | 0.015 | HPC |
| SAE J431c: grade G3000 | | | 8 | 0.300 | 215 | 700 | 0.50 | 0.020 | HPC |

CPA, cold-pressed alumina; HPC, hot-pressed cermet. Source: Metcut Research Associates Inc.

**Table 5 Nominal speeds and feeds for turning a variety of steels and cast irons with high-speed steel and carbide cutoff and form tools**

| Material | Hardness, HB | Condition | Speed m/min | Speed sfm | Feed, mm/rev (in./rev) — Cutoff tool width 1.5 mm (0.062 in.) | Cutoff tool width 3 mm (0.125 in.) | Cutoff tool width 6 mm (0.250 in.) | Form tool width 12 mm (0.500 in.) | Form tool width 18 mm (0.750 in.) | Form tool width 25 mm (1.00 in.) | Form tool width 35 mm (1.50 in.) | Form tool width 50 mm (2.00 in.) | Tool material grade(a) ISO | Tool material grade(a) AISI or C |
|---|---|---|---|---|---|---|---|---|---|---|---|---|---|---|
| **Carbon steels, wrought** | | | | | | | | | | | | | | |
| **Low carbon** | | | | | | | | | | | | | | |
| 1005 1010 1020 | 125–175 | Hot rolled, | 37 | 120 | 0.038 | 0.050 | 0.061 | 0.046 | 0.041 | 0.036 | 0.033 | 0.028 | S4, S5 | M2, M3 |
| 1006 1012 1023 | | normalized, | 120 | 390 | (0.0015) | (0.002) | (0.0024) | (0.0018) | (0.0016) | (0.0014) | (0.0013) | (0.0011) | P40, M40 | C-6 |
| 1008 1015 1025 | | annealed, or | | | 0.038 | 0.050 | 0.061 | 0.046 | 0.041 | 0.036 | 0.033 | 0.028 | | |
| 1009 1017 | | cold drawn | | | (0.0015) | (0.002) | (0.0024) | (0.0018) | (0.0016) | (0.0014) | (0.0013) | (0.0011) | | |
| **Medium carbon** | | | | | | | | | | | | | | |
| 1030 1042 1053 | 125–175 | Hot rolled, | 35 | 115 | 0.036 | 0.043 | 0.056 | 0.043 | 0.038 | 0.036 | 0.033 | 0.028 | S4, S5 | M2, M3 |
| 1033 1043 1055 | | normalized, | 115 | 370 | (0.0014) | (0.0017) | (0.0022) | (0.0017) | (0.0015) | (0.0014) | (0.0013) | (0.0011) | P40, M40 | C-6 |
| 1035 1044 1525 | | annealed, or | | | 0.036 | 0.043 | 0.056 | 0.043 | 0.038 | 0.036 | 0.033 | 0.028 | | |
| 1037 1045 1526 | | cold drawn | | | (0.0014) | (0.0017) | (0.0022) | (0.0017) | (0.0015) | (0.0014) | (0.0013) | (0.0011) | | |
| 1038 1046 1527 | | | | | | | | | | | | | | |
| 1039 1049 | | | | | | | | | | | | | | |
| 1040 1050 | | | | | | | | | | | | | | |
| **Alloy steels, wrought** | | | | | | | | | | | | | | |
| **Medium carbon** | | | | | | | | | | | | | | |
| 1330 4135 8625 | 275–325 | Normalized or | 17 | 55 | 0.033 | 0.041 | 0.050 | 0.041 | 0.036 | 0.033 | 0.028 | 0.023 | S4, S5 | M2, M3 |
| 1335 4137 8627 | | quenched | 53 | 175 | (0.0013) | (0.0016) | (0.002) | (0.0016) | (0.0014) | (0.0013) | (0.0011) | (0.0009) | P40, M40 | C-6 |
| 4027 4427 8630 | | and tempered | | | 0.033 | 0.041 | 0.050 | 0.041 | 0.036 | 0.033 | 0.028 | 0.023 | | |
| 4028 4626 8637 | | | | | (0.0013) | (0.0016) | (0.002) | (0.0016) | (0.0014) | (0.0013) | (0.0011) | (0.0009) | | |
| 4032 5130 94B30 | | | | | | | | | | | | | | |
| 4037 5132 | | | | | | | | | | | | | | |
| 4130 5135 | | | | | | | | | | | | | | |
| **Medium carbon** | | | | | | | | | | | | | | |
| 1340 4340 81B45 | 175–225 | Hot rolled, | 29 | 95 | 0.038 | 0.046 | 0.056 | 0.046 | 0.041 | 0.038 | 0.033 | 0.028 | S4, S5 | M2, M3 |
| 1345 50B40 8640 | | annealed, or | 95 | 305 | (0.0015) | (0.0018) | (0.0022) | (0.0018) | (0.0016) | (0.0015) | (0.0013) | (0.0011) | P40, M40 | C-6 |
| 4042 50B44 8642 | | cold drawn | | | 0.038 | 0.046 | 0.056 | 0.046 | 0.041 | 0.038 | 0.033 | 0.028 | | |
| 4047 5046 8645 | | | | | (0.0015) | (0.0018) | (0.0022) | (0.0018) | (0.0016) | (0.0015) | (0.0013) | (0.0011) | | |
| 4140 50B46 86B45 | | | | | | | | | | | | | | |
| 4142 5140 8740 | | | | | | | | | | | | | | |
| 4145 5145 8742 | | | | | | | | | | | | | | |
| 4147 5147 | | | | | | | | | | | | | | |
| **Low carbon** | | | | | | | | | | | | | | |
| 4012 4621 8115 | 225–275 | Hot rolled, | 24 | 80 | 0.036 | 0.043 | 0.050 | 0.043 | 0.038 | 0.036 | 0.030 | 0.025 | S4, S5 | M2, M3 |
| 4023 4718 8617 | | normalized, | 78 | 255 | (0.0014) | (0.0017) | (0.002) | (0.0017) | (0.0015) | (0.0014) | (0.0012) | (0.001) | P40, M40 | C-6 |
| 4024 4720 8620 | | annealed, or | | | 0.036 | 0.043 | 0.050 | 0.043 | 0.038 | 0.036 | 0.030 | 0.025 | | |
| 4118 4815 8622 | | cold drawn | | | (0.0014) | (0.0017) | (0.002) | (0.0017) | (0.0015) | (0.0014) | (0.0012) | (0.001) | | |
| 4320 4817 8822 | | | | | | | | | | | | | | |
| 4419 4820 9310 | | | | | | | | | | | | | | |
| 4422 5015 94B15 | | | | | | | | | | | | | | |
| 4615 5115 95B17 | | | | | | | | | | | | | | |
| 4617 5120 | | | | | | | | | | | | | | |
| 4620 6118 | | | | | | | | | | | | | | |
| **High-strength steels, wrought** | | | | | | | | | | | | | | |
| 300M 4340Si H11 | 225–300 | Annealed | 18 | 60 | 0.036 | 0.043 | 0.053 | 0.043 | 0.038 | 0.036 | 0.030 | 0.025 | S4, S5 | M2, M3 |
| 4330V 98BV40 H13 | | | 58 | 190 | (0.0014) | (0.0017) | (0.0021) | (0.0017) | (0.0015) | (0.0014) | (0.0012) | (0.001) | P40, M40 | C-6 |
| 4340 D6ac | | | | | 30.036 | 0.043 | 0.053 | 0.043 | 0.038 | 0.036 | 0.030 | 0.025 | | |
| | | | | | (0.0014) | (0.0017) | (0.0021) | (0.0017) | (0.0015) | (0.0014) | (0.0012) | (0.001) | | |
| **Stainless steels, wrought** | | | | | | | | | | | | | | |
| **Austenitic** | | | | | | | | | | | | | | |
| 201 304L 348 | 135–185 | Annealed | 26 | 85 | 0 033 | 0.041 | 0.050 | 0.050 | 0.041 | 0.036 | 0.030 | 0.025 | S4, S5 | M2, M3 |
| 202 305 384 | | | 84 | 275 | (0 0013) | (0.0016) | (0.002) | (0.002) | (0.0016) | (0.0014) | (0.0012) | (0.001) | K40, M40 | C-2 |
| 301 308 385 | | | | | 0 033 | 0.041 | 0.050 | 0.050 | 0.041 | 0.036 | 0.030 | 0.025 | | |
| 302 321 | | | | | (0 0013) | (0.0016) | (0.002) | (0.002) | (0.0016) | (0.0014) | (0.0012) | (0.001) | | |
| 304 347 | | | | | | | | | | | | | | |
| **Martensitic** | | | | | | | | | | | | | | |
| 403 422 | 135–175 | Annealed | 29 | 95 | 0.038 | 0.050 | 0.063 | 0.050 | 0.046 | 0.038 | 0.030 | 0.025 | S4, S5 | M2, M3 |
| 410 501 | | | 100 | 325 | (0.0015) | (0.002) | (0.0025) | (0.002) | (0.0018) | (0.0015) | (0.0012) | (0.001) | K40, M40 | C-6 |
| 420 502 | | | | | 0.038 | 0.050 | 0.063 | 0.050 | 0.046 | 0.038 | 0.030 | 0.025 | | |
| | | | | | (0.0015) | (0.002) | (0.0025) | (0.002) | (0.0018) | (0.0015) | (0.0012) | (0.001) | | |
| **Gray cast irons** | | | | | | | | | | | | | | |
| **Pearlitic** | | | | | | | | | | | | | | |
| ASTM A48: | 190–220 | As-cast | 18 | 60 | 0.075 | 0.13 | 0.18 | 0.18 | 0.13 | 0.13 | 0.102 | 0.075 | S4, S5 | M2, M3 |
| classes 30, 35, 40 | | | 64 | 210 | (0.003) | (0.005) | (0.007) | (0.007) | (0.005) | (0.005) | (0.004) | (0.003) | K40, M40 | C-2 |
| SAE J431c: | | | | | 0.075 | 0.13 | 0.18 | 0.18 | 0.13 | 0.13 | 0.102 | 0.075 | | |
| grade G3000 | | | | | (0.003) | (0.005) | (0.007) | (0.007) | (0.005) | (0.005) | (0.004) | (0.003) | | |

(a) Use submicron carbide when the recommended carbide grade chips excessively. Source: Metcut Research Associates Inc.

**Table 6 Comparison of actual and nominal speeds and feeds for the turning of carbon and low-alloy steels**

| Steel | Average hardness, HB | Comparable steel | Operation | Tool: Material | Tool: Type | Speed: Actual m/min | Speed: Actual sfm | Speed: Nominal(a) m/min | Speed: Nominal(a) sfm | Feed: Actual mm/rev | Feed: Actual in./rev | Feed: Nominal(a) mm/rev | Feed: Nominal(a) in./rev |
|---|---|---|---|---|---|---|---|---|---|---|---|---|---|
| 1037 | 186 | 1045 | Roughing | Carbide | Brazed | 220 | 725 | 102 | 335 | 0.41 | 0.016 | 0.38 | 0.015 |
| 6118 | 207 | 8620 | Roughing | Carbide | Disposable | 165 | 540 | 128 | 420 | 0.76 | 0.030 | 0.38 | 0.015 |
| 5120 | 167 | 8620 | Finishing | Carbide | Brazed | 312 | 1025 | 137 | 450 | 0.30 | 0.012 | 0.18 | 0.007 |
| | | | Forming(b) | Carbide | Brazed | 419 | 1375 | 85 | 280 | 0.30 | 0.012 | 0.13 | 0.005 |
| 1045 | 194 | 1045 | Roughing | Carbide | Brazed | 131 | 430 | 102 | 335 | 0.41 | 0.016 | 0.38 | 0.015 |
| | | | Roughing | Carbide | Brazed | 161 | 530 | 102 | 335 | 0.46 | 0.018 | 0.38 | 0.015 |
| 4130 | 318 | 4140 | Finishing | Carbide | Brazed | 58 | 190 | 107 | 350 | 0.10 | 0.004 | 0.18 | 0.007 |
| | | | Finishing | Carbide | Brazed | 82 | 270 | 107 | 350 | 0.10 | 0.004 | 0.18 | 0.007 |
| 4820 | 228 | 8620 | Roughing | HSS | . . . | 25 | 83 | 23 | 75 | 0.25 | 0.010 | 0.38 | 0.015 |
| | | | Roughing | Carbide | Disposable | 123 | 405 | 120 | 395 | 0.36 | 0.014 | 0.38 | 0.015 |
| 8622 | 187 | 8620 | Roughing | HSS | . . . | 35 | 115 | 27 | 90 | 0.30 | 0.012 | 0.38 | 0.015 |
| | | | Roughing | Carbide | Disposable | 200 | 655 | 128 | 420 | 0.25 | 0.010 | 0.38 | 0.015 |
| 4320 | 227 | 8620 | Roughing | Carbide | Brazed | 115 | 392 | 94 | 310 | 0.38 | 0.015 | 0.38 | 0.015 |
| 1035 | 150 | 1045 | Roughing | Carbide | Brazed | 151 | 495 | 116 | 380 | 0.51 | 0.020 | 0.38 | 0.015 |
| 8620 | 169 | 8620 | Finishing | Carbide | Brazed | 195 | 639 | 137 | 450 | 0.20 | 0.008 | 0.18 | 0.007 |
| | | | Finishing | Carbide | Brazed | 343 | 1125 | 137 | 450 | 0.20 | 0.008 | 0.18 | 0.007 |

(a) Comparable steel. (b) Tool width, 12.7 mm (0.500 in.)

the tool or tool holder. Several types of chip breakers are illustrated and described in Fig. 9.

The efficiency of chip breakers is markedly affected by their design. A step that is too wide or too shallow will permit chips of the more ductile metals to flow over the intended obstruction without breaking. A step that is too narrow will cause the chips to crowd between the tool and the workpiece, resulting in galling of the turned surface and reduced tool life. Generally, the optimum width and depth of a chip breaker are governed by depth of cut and rate of feed. Usually, the depth is maintained between 0.38 and 0.76 mm (0.015 and 0.030 in.). The fillet radius should be two to four times the depth of the chip breaker, and the width should be eight to nine times the feed per revolution. All cutting angles of the tools should be maintained.

**Chip Curlers.** With ductile metals, and especially with metals that combine strength and ductility (such as annealed stainless steel), breaking of chips is often not feasible, and particular attention must be directed to the handling of continuous curls. The characteristics of the spiral chip and its direction of release can be regulated to some extent by adjusting rake and side cutting-edge angles. Further control can be obtained through chip-curler grooves in the tool, as illustrated in the example given below.

**Example 1: Design for Chip-Curler Groove.** A box tool with a chip-curler groove of the design shown in Fig. 10(a) was not satisfactory for reducing the diameter of annealed type 304 stainless steel bar stock from 38 to 29 mm (1½ to 1⅛ in.) over a length of 76 mm (3 in.). The groove design impeded chip flow and caused crowding of the chip and breakage of tools. Adequate chip elimination was obtained when the design of the chip-curler groove was changed to that shown in Fig. 10(b). Other processing details are given in the table accompanying Fig. 10.

## Speed and Feed

Tables 2, 3, and 4 present nominal speeds and feeds for turning numerous steels and cast irons used in the examples that follow at specific hardness levels, using high-speed steel, carbide, or ceramic tools. These nominal values are based on a tool life of 1 to 2

**Table 7 Comparison of cutting speeds for 60 min tool life in turning three different hot-rolled steels with T1 high-speed steel tools(a)**

| Depth of cut mm | Depth of cut in. | 0.05 mm/rev (0.002 in./rev) m/min | sfm | 0.10 mm/rev (0.004 in./rev) m/min | sfm | 0.20 mm/rev (0.008 in./rev) m/min | sfm | 0.4 mm/rev (1/64 in./rev) m/min | sfm | 0.8 mm/rev (1/32 in./rev) m/min | sfm | 1.6 mm/rev (1/16 in./rev) m/min | sfm | 3.2 mm/rev (1/8 in./rev) m/min | sfm |
|---|---|---|---|---|---|---|---|---|---|---|---|---|---|---|---|
| **1112 steel (130 HB)** | | | | | | | | | | | | | | | |
| 0.8 | 1/32 | 611 | 2004 | 393 | 1290 | 249 | 816 | 165 | 540 | 110 | 363 | 78 | 255 | 67 | 219 |
| 1.6 | 1/16 | 427 | 1401 | 283 | 930 | 182 | 597 | 119 | 390 | 79 | 258 | 57 | 186 | 46 | 150 |
| 3.2 | 1/8 | 370 | 1215 | 236 | 774 | 163 | 534 | 100 | 327 | 66 | 216 | 46 | 150 | 35 | 114 |
| 6.4 | 1/4 | 337 | 1107 | 210 | 690 | 136 | 444 | 88 | 288 | 58 | 189 | 39 | 129 | 28 | 93 |
| 13 | 1/2 | 316 | 1038 | 199 | 654 | 125 | 411 | 81 | 267 | 52 | 171 | 34 | 111 | 23 | 75 |
| 25 | 1 | 304 | 999 | 191 | 627 | 121 | 396 | 77 | 252 | 48 | 159 | 30 | 99 | 19 | 63 |
| **1020 steel (127 HB)** | | | | | | | | | | | | | | | |
| 0.8 | 1/32 | 288 | 944 | 185 | 607 | 117 | 384 | 77 | 254 | 52 | 171 | 37 | 120 | 31 | 103 |
| 1.6 | 1/16 | 201 | 660 | 134 | 438 | 86 | 281 | 56 | 184 | 35 | 114 | 27 | 88 | 22 | 71 |
| 3.2 | 1/8 | 174 | 572 | 111 | 364 | 77 | 251 | 47 | 154 | 31 | 102 | 22 | 71 | 16 | 54 |
| 6.4 | 1/4 | 159 | 521 | 99 | 325 | 64 | 209 | 41 | 136 | 27 | 89 | 19 | 61 | 13 | 44 |
| 13 | 1/2 | 149 | 489 | 94 | 308 | 59 | 194 | 38 | 126 | 24 | 80 | 16 | 52 | 11 | 35 |
| 25 | 1 | 144 | 471 | 90 | 295 | 57 | 187 | 36 | 119 | 23 | 75 | 14 | 47 | 9 | 30 |
| **1050 steel (201 HB)** | | | | | | | | | | | | | | | |
| 0.8 | 1/32 | 141 | 464 | 91 | 298 | 58 | 189 | 38 | 125 | 26 | 84 | 18 | 59 | 15 | 51 |
| 1.6 | 1/16 | 99 | 324 | 66 | 215 | 42 | 138 | 27 | 90 | 18 | 60 | 13 | 43 | 11 | 35 |
| 3.2 | 1/8 | 86 | 281 | 55 | 179 | 38 | 124 | 23 | 76 | 15 | 50 | 11 | 35 | 8 | 26 |
| 6.4 | 1/4 | 77 | 254 | 49 | 160 | 31 | 103 | 20 | 67 | 13 | 44 | 9 | 30 | 7 | 22 |
| 13 | 1/2 | 73 | 240 | 46 | 151 | 29 | 95 | 19 | 62 | 12 | 40 | 8 | 26 | 5 | 17 |
| 25 | 1 | 70 | 231 | 44 | 145 | 28 | 92 | 18 | 58 | 11 | 37 | 7 | 23 | 4.5 | 15 |

(a) Tool angles: BR, 8°; SR, 14°; ER, 6°; SRF, 6°; ECEA, 6°; SCEA, 30°; NR, 3.2 mm (1/8 in.). No cutting fluid used

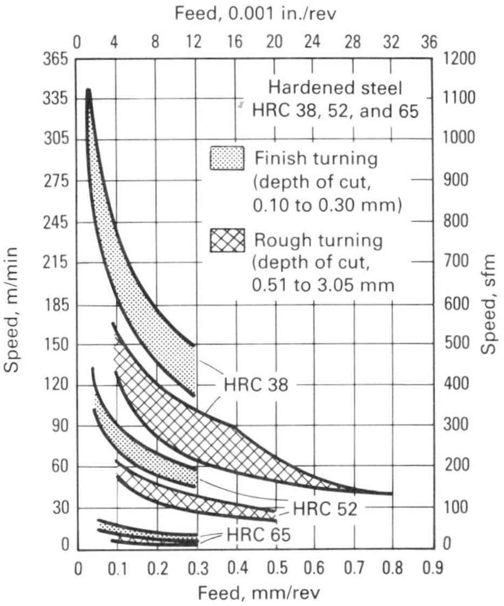


**Fig. 11** Effect of hardness on speed and feed used in rough and finish turning. Speed and feed combinations for 60 min tool life. Data are based on turning with carbide tools (79WC-15TiC-6Co) lapped with boron carbide. Tools for steels of all hardnesses had 45° side cutting-edge angle, 15° end cutting-edge angle, 1.02 mm (0.040 in.) nose radius, 0° side rake, and relief angle of either 15° (for 0.020 mm/rev, or 0.008 in./rev or less) or 10° (for feeds over 0.020 mm/rev, or 0.008 in./rev). Back rake varied with hardness, being 0° for HRC 38, −5° for HRC 52, and −10° for HRC 65.

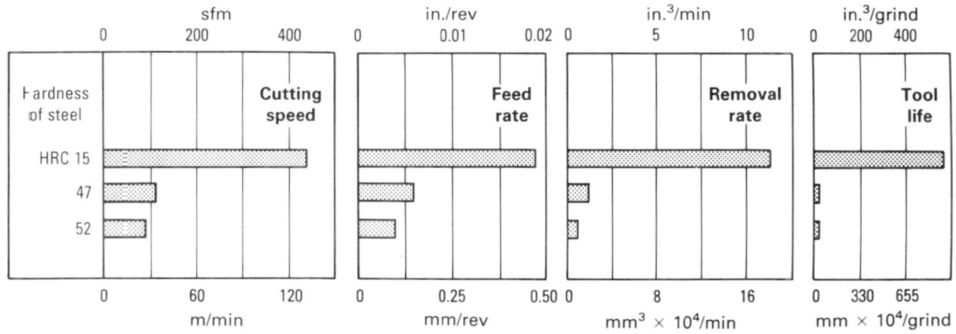


| Machining conditions | Hardness of steel turned(a) 15 HRC | 47 HRC | 52 HRC |
|---|---|---|---|
| Speed, rev/min | 1090 | 350 | 226 |
| Speed, m/min (sfm) | 136 (445) | 35 (115) | 27 (90) |
| Feed, mm/rev (in./rev) | 0.48 (0.019) | 0.15 (0.006) | 0.10 (0.004) |
| Depth of cut, mm (in.) | 3.05 (0.120) | 3.05 (0.120) | 3.05 (0.120) |
| Cutting fluid | None | (b) | (b) |

(a) Steels were 4130 at HRC 15, 4330 at HRC 47, and 4340 at HRC 52. All three steels were turned on a 406 mm (16 in.) lathe with C-6 carbide (77W-8Ti-7C-8Co) tool bits 13 mm (1/2 in.) inscribed circle (IC) by 4.8 mm (3/16 in.). All tools had −7° back and side rake angles, +7° end and side relief angles, 15° side cutting-edge angle, and 0.76 mm (0.030 in.) nose radius; height above center was 0.13 mm (0.005 in.). (b) One-to-one mixture of sulfurized oil and mineral oil

**Fig. 12** Effect of hardness on speed, feed, and results in turning

h for high-speed steel or brazed-carbide tools. Table 5 lists comparable data for turning of these same materials using high-speed and carbide form tools. A tool life of 30 to 60 min would be typical of indexable-insert carbides when used to turn 25 to 76 mm (1 to 3 in.) diam workpieces. The tabulated values are intended to serve as starting points for the selection of optimum machining conditions. More exact settings can be made with experience. In commercial practice, speeds and feeds often differ widely from the values given in Tables 2, 3, and 4. The magnitude of these differences is illustrated in Table 6, which compares nominal values of speed and feed of wrought low- and medium-carbon steels and wrought low-alloy steels listed in Tables 2 to 5 and also compares actual values with nominal values used in turning these materials. The factors responsible for large deviations from nominal values are considered in the individual examples.

**Selection of Speed and Feed.** The nominal speed and feed shown in Table 2, 3, 4, or 5 for a given application reflect only the effect of steel group (general chemical type), hardness, type of operation, and tool material. The final selection of speed and feed is usually made on the basis of experience with similar parts or from production trials, and is influenced by:

- Work metal
- Type of operation
- Tool material
- Tool design
- Shape of workpiece
- Tool life desired
- Machine condition
- Power of available machines
- Surface finish required

**Work Metal.** With the exception of free-cutting additions, the direct effect of chemical composition on optimum speed and feed is small and of little significance in comparison with the normal variation in properties of commercial grades of steel or the larger changes that can be obtained by heat treatment or by cold reduction.

**Free-cutting grades** are usually produced by the addition of 0.10 to 0.30% S or 0.20 to 0.30% Pb (or both) to carbon or low-alloy steel. Except at high hardness, the nominal turning speeds for resulfurized or leaded carbon steels are substantially higher than the values for corresponding plain carbon steels. A similar but somewhat smaller increase in permissible speed results from additions of sulfur or lead to low-alloy steels, compared with corresponding plain low-alloy steels.

Table 7 compares speed-feed-depth relationships for a resulfurized free-cutting steel (1112) and a plain carbon steel (1020), using high-speed steel tools for a (60 min) tool life. Over a range of feed rates and depths of cut, turning speed for the 1112 steel was about twice that for a comparable operation on 1020 steel.

**Carbon and other alloying elements** in carbon and low-alloy steels have only a small direct effect on speed and feed in turning. Plain carbon steels containing less than about 0.15% C machine to give soft, gummy chips that are likely to adhere to the cutting tool, and these steels require relatively low speed and feed. Up to about 0.25% C, optimum speed and feed increase with increasing carbon content, because of the corresponding decrease in chip ductility. A further rise in carbon content leads to the use of lower speed and feed, because of greatly increased strength, hardness, and abrasive characteristics of the metal. The addition of other alloying ele-ments lowers optimum speed and feed by increasing the strength and hardness of the metal. These effects of chemical composition on optimum speed and feed are too small to affect the nominal values listed in Tables 2, 3, 4, and 5.

**Hardness** exerts a major effect on the selection of speed and feed. In fact, as a first approximation, speed and feed for turning carbon and low-alloy steels (in the absence of free-cutting additives) can be considered as functions of hardness only; nominal speed and feed values are identical at equal hardness.

Figure 11 shows the effect of hardness on speeds and feeds for rough and finish turning of hardened steel (quenched and tempered to HRC 38, 52, and 65) regardless of variations in composition. As hardness increases, speed and feed rates for rough turning decrease, speed for finish turning decreases, and the ranges of speed and feed become progressively narrower.

Figure 12 shows a decrease in speed and feed rates with increasing hardness (HRC 15, 47 and 52) and also shows the resulting decrease in metal removal rate and tool life.

The effect of a difference in hardness associated with carbon content is illustrated by the comparison in Table 7 of turning speeds for (60 min) tool life for hot-rolled 1020 steel (127 HB) and hot-rolled 1050 steel (201 HB). Over a wide range of feed and depth of cut,

## Table 8 Effect of variables on cutting speed for 60 min tool life in turning hot-rolled 1020 steel with T1 high-speed steel tools(a)

| Depth of cut | | Speed for feeds of: 0.05 mm/rev (0.002 in./rev) | | 0.10 mm/rev (0.004 in./rev) | | 0.20 mm/rev (0.008 in./rev) | | 0.4 mm/rev (1/64 in./rev) | | 0.8 mm/rev (1/32 in./rev) | | 1.6 mm/rev (1/16 in./rev) | | 3.2 mm/rev (1/8 in./rev) | |
|---|---|---|---|---|---|---|---|---|---|---|---|---|---|---|---|
| mm | in. | m/min | sfm | m/min | sfm | m/min | sfm | m/min | sfm | m/min | sfm | m/min | sfm | m/min | sfm |
| **Tool 1 (SCEA, 0°; NR, 0)(b)** | | | | | | | | | | | | | | | |
| 0.8 | 1/32 | 128 | 420 | 81 | 267 | 53 | 174 | 36 | 117 | 25 | 82 | 25 | 82 | 25 | 82 |
| 1.6 | 1/16 | 127 | 418 | 80 | 264 | 53 | 173 | 35 | 116 | 24 | 78 | 17 | 57 | 16 | 54 |
| 3.2 | 1/8 | 127 | 418 | 80 | 263 | 52 | 171 | 34 | 112 | 23 | 75 | 16 | 52 | 12 | 40 |
| 6.4 | 1/4 | 127 | 416 | 80 | 263 | 51 | 167 | 34 | 110 | 22 | 72 | 15 | 49 | 10 | 34 |
| 13 | 1/2 | 126 | 412 | 80 | 261 | 51 | 167 | 33 | 107 | 21 | 69 | 14 | 45 | 9 | 30 |
| 25 | 1 | 126 | 412 | 79 | 260 | 50 | 164 | 32 | 105 | 20 | 65 | 12 | 41 | 8 | 25 |
| **Tool 2 (SCEA, 0°; NR, 1.6 mm, or 1/16 in.)(b)** | | | | | | | | | | | | | | | |
| 0.8 | 1/32 | 209 | 688 | 134 | 439 | 87 | 285 | 58 | 191 | 43 | 140 | 33 | 109 | 28 | 93 |
| 1.6 | 1/16 | 171 | 562 | 109 | 359 | 71 | 232 | 48 | 157 | 34 | 110 | 25 | 82 | 20 | 66 |
| 3.2 | 1/8 | 150 | 493 | 96 | 315 | 62 | 202 | 41 | 134 | 28 | 93 | 20 | 66 | 16 | 51 |
| 6.4 | 1/4 | 138 | 454 | 88 | 288 | 56 | 185 | 37 | 122 | 25 | 82 | 17 | 57 | 12 | 41 |
| 13 | 1/2 | 131 | 432 | 84 | 276 | 53 | 175 | 34 | 113 | 22 | 73 | 15 | 49 | 10 | 34 |
| 25 | 1 | 129 | 424 | 80 | 261 | 52 | 170 | 33 | 107 | 21 | 68 | 13 | 42 | 8 | 27 |
| **Tool 3 (SCEA, 30°; NR, 3.2 mm, or 1/8 in.)(b)** | | | | | | | | | | | | | | | |
| 0.8 | 1/32 | 257 | 944 | 185 | 607 | 117 | 384 | 77 | 254 | 52 | 171 | 37 | 120 | 28 | 103 |
| 1.6 | 1/16 | 201 | 660 | 134 | 438 | 86 | 281 | 56 | 184 | 35 | 114 | 27 | 88 | 22 | 71 |
| 3.2 | 1/8 | 174 | 572 | 111 | 364 | 77 | 251 | 47 | 154 | 31 | 102 | 22 | 71 | 16 | 54 |
| 6.4 | 1/4 | 159 | 521 | 99 | 325 | 64 | 209 | 41 | 136 | 27 | 89 | 19 | 61 | 13 | 44 |
| 13 | 1/2 | 149 | 489 | 94 | 308 | 59 | 194 | 38 | 126 | 24 | 80 | 16 | 52 | 11 | 35 |
| 25 | 1 | 144 | 471 | 90 | 295 | 57 | 187 | 36 | 119 | 23 | 75 | 14 | 47 | 9 | 30 |
| **Tool 4 (SCEA, 30°; NR, 6.4 mm, or 1/4 in.)(b)** | | | | | | | | | | | | | | | |
| 0.8 | 1/32 | 325 | 1065 | 207 | 680 | 131 | 430 | 86 | 283 | 57 | 188 | 40 | 130 | 31 | 103 |
| 1.6 | 1/16 | 287 | 941 | 182 | 596 | 116 | 379 | 75 | 246 | 49 | 161 | 33 | 109 | 25 | 81 |
| 3.2 | 1/8 | 203 | 667 | 132 | 434 | 84 | 275 | 54 | 178 | 36 | 117 | 24 | 79 | 18 | 58 |
| 6.4 | 1/4 | 175 | 574 | 111 | 365 | 70 | 229 | 45 | 148 | 30 | 97 | 20 | 65 | 14 | 45 |
| 13 | 1/2 | 158 | 517 | 99 | 326 | 63 | 206 | 40 | 131 | 26 | 85 | 17 | 55 | 11 | 37 |
| 25 | 1 | 147 | 482 | 93 | 305 | 58 | 191 | 37 | 123 | 23 | 76 | 15 | 48 | 9 | 30 |
| **Finishing tool(c)** | | | | | | | | | | | | | | | |
| 0.13 | 0.005 | 397 | 1303 | 232 | 762 | 180 | 590 | 131 | 431 | 105 | 343 | 87 | 287 | ... | ... |
| 0.25 | 0.010 | 310 | 1018 | 204 | 668 | 135 | 442 | 97 | 318 | 74 | 243 | 58 | 191 | ... | ... |
| 0.38 | 0.015 | 273 | 896 | 175 | 574 | 117 | 383 | 83 | 271 | 61 | 199 | 49 | 161 | ... | ... |

(a) No cutting fluid used. (b) Tool angles: BR, 8°; SR, 14°; ER, 6°; SRF, 6°; ECEA, 6°. (c) Tool angles: BR, 20°; SR, 0°; ER, 6°; SRF, 6°; ECEA, 6°; SCEA, 0°; NR, 3.2 mm (1/8 in.); flat, 3.2 mm (1/8 in.)

the speed shown for 1050 is about half that for a comparable operation on 1020.

**Other Metallurgical Considerations.** The need for low speed and feed in turning annealed carbon steel of low-carbon content results principally from microstructural considerations. Higher speed and feed can be employed after normalizing or oil quenching such steels. Cold reduction of low-carbon steel also permits higher speed and feed by lowering chip ductility.

Medium-carbon and high-carbon steels also can be heat treated to allow higher speed and feed; the most machinable microstructures for various carbon contents are listed below:

| Carbon, % | Optimum microstructure |
|---|---|
| 0.06–0.20 | As rolled (most economical) |
| 0.20–0.30 | Under 76 mm (3 in.) diam, normalized; 76 mm (3 in.) diam and over, as rolled |
| 0.30–0.40 | Annealed to give coarse pearlite, minimum ferrite |
| 0.40–0.60 | Coarse lamellar pearlite to coarse spheroidite |
| 0.60–1.00 | 100% spheroidite, coarse to fine |

The presence of manganese, chromium, nickel, or molybdenum in the carbide phase of low-alloy steel affects machining behavior in the same way as does the presence of carbon; in solid solution alloying elements toughen and strengthen the ferrite phase, thus reducing permissible speed and feed rates.

**Type of Operation.** For a given work metal, speed is highest for turning with single-point or box tools. Feed is also relatively high for this type of operation. Higher speed and lower feed rates are used for shallow finishing cuts with these tools than is the case with roughing cuts. Brazed carbide tools are used at lower speeds than are disposable carbide tools, primarily to increase tool life and thereby minimize tool changing.

Speeds for turning with form tools or cutoff tools are usually about 40 to 60% of the finish turning speed with single-point tools. Feed rate in form turning varies inversely with tool width, as would be expected, but feed rate for cutoff turning varies directly with tool width, for the range of widths shown. Feed rate is usually lower for these tools than for single-point or box tools. As turning speeds increase, the problems of chip containment and the loss of gripping force on the chuck jaws due to centrifugal force become increasingly severe.

**Tool Material.** Tables 2, 3, 4, and 5 show ranges of nominal speed for the three basic tool materials to be: high-speed steel, 6.1 to 107 m/min (20 to 350 sfm); carbide, 18 to 411 m/min (60 to 1350 sfm); and ceramic, 30 to 792 m/min (100 to 2600 sfm). Nominal feeds for single-point and box tools are given below:

| Nominal speed | Feed rate mm/rev | Feed rate in./rev |
|---|---|---|
| **For single-point and box tools** | | |
| High-speed steel | 0.13–0.38 | 0.005–0.015 |
| Carbide | 0.13–0.38 | 0.005–0.015 |
| Ceramic | 0.08–0.51 | 0.003–0.020 |
| **For form tools** | | |
| High-speed steel | 0.013–0.08 | 0.0005–0.003 |
| Carbide | 0.025–0.18 | 0.001–0.007 |
| **For cutoff tools** | | |
| High-speed steel | 0.025–0.08 | 0.001–0.003 |
| Carbide | 0.051–0.23 | 0.002–0.009 |

High-speed steel cutting tools, although versatile, shock resistant, and readily forged and machined to a wide variety of forms, are limited to relatively low turning speeds. Carbide tools are more economical than high-speed steel tools in most high-production turning applications, but they require a rigid setup and no chatter to avoid chipping or breaking, and they do not per-

**Table 9 Effect of variables on cutting speed for 60 min tool life in turning hot-rolled 1020 steel with carbide tools(a)**

| Depth of cut | | Speed for feeds of: 0.05 mm/rev (0.002 in./rev) | | 0.10 mm/rev (0.004 in./rev) | | 0.20 mm/rev (0.008 in./rev) | | 0.4 mm/rev (1/64 in./rev) | | 0.8 mm/rev (1/32 in./rev) | | 1.6 mm/rev (1/16 in./rev) | | 3.2 mm/rev (1/8 in./rev) | |
|---|---|---|---|---|---|---|---|---|---|---|---|---|---|---|---|
| mm | in. | m/min | sfm | m/min | sfm | m/min | sfm | m/min | sfm | m/min | sfm | m/min | sfm | m/min | sfm |
| **Tool 1 (SCEA, 0°; NR, 0)(b)** | | | | | | | | | | | | | | | |
| 0.8 | 1/32 | 438 | 1438 | 277 | 908 | 178 | 583 | 116 | 379 | 78 | 256 | 78 | 256 | 78 | 256 |
| 1.6 | 1/16 | 438 | 1436 | 276 | 906 | 178 | 583 | 115 | 376 | 75 | 245 | 51 | 169 | 50 | 163 |
| 3.2 | 1/8 | 437 | 1433 | 276 | 905 | 176 | 577 | 114 | 373 | 73 | 241 | 48 | 159 | 34 | 110 |
| 6.4 | 1/4 | 437 | 1433 | 275 | 902 | 175 | 575 | 112 | 369 | 72 | 237 | 47 | 154 | 31 | 102 |
| 13 | 1/2 | 435 | 1428 | 275 | 901 | 174 | 571 | 111 | 365 | 71 | 232 | 45 | 148 | 29 | 95 |
| 25 | 1 | 435 | 1427 | 274 | 898 | 173 | 568 | 110 | 362 | 69 | 227 | 43 | 142 | 27 | 89 |
| **Tool 2 (SCEA, 0°; NR, 1.6 mm, or 1/16 in.)(b)** | | | | | | | | | | | | | | | |
| 0.8 | 1/32 | 725 | 2378 | 461 | 1513 | 298 | 978 | 198 | 651 | 141 | 464 | 109 | 358 | 95 | 311 |
| 1.6 | 1/16 | 590 | 1935 | 376 | 1233 | 241 | 792 | 160 | 526 | 111 | 363 | 80 | 262 | 64 | 210 |
| 3.2 | 1/8 | 519 | 1702 | 330 | 1084 | 210 | 690 | 138 | 452 | 93 | 306 | 64 | 211 | 49 | 161 |
| 6.4 | 1/4 | 478 | 1568 | 310 | 1016 | 194 | 635 | 125 | 411 | 83 | 271 | 55 | 182 | 39 | 129 |
| 13 | 1/2 | 456 | 1495 | 290 | 951 | 183 | 599 | 117 | 385 | 76 | 249 | 50 | 163 | 34 | 110 |
| 25 | 1 | 439 | 1439 | 276 | 903 | 179 | 586 | 113 | 372 | 72 | 237 | 46 | 150 | 28 | 93 |
| **Tool 3 (SCEA, 30°; NR, 3.2 mm, or 1/8 in.)(b)** | | | | | | | | | | | | | | | |
| 0.8 | 1/32 | 1009 | 3309 | 640 | 2101 | 403 | 1323 | 264 | 867 | 177 | 581 | 123 | 403 | 104 | 340 |
| 1.6 | 1/16 | 695 | 2281 | 461 | 1512 | 294 | 964 | 191 | 625 | 125 | 411 | 87 | 286 | 70 | 229 |
| 3.2 | 1/8 | 603 | 1978 | 383 | 1256 | 264 | 865 | 160 | 524 | 104 | 340 | 70 | 230 | 52 | 170 |
| 6.4 | 1/4 | 545 | 1788 | 343 | 1124 | 216 | 718 | 141 | 463 | 91 | 299 | 60 | 198 | 45 | 147 |
| 13 | 1/2 | 510 | 1672 | 323 | 1061 | 204 | 668 | 131 | 430 | 84 | 274 | 54 | 177 | 36 | 119 |
| 25 | 1 | 497 | 1631 | 313 | 1026 | 197 | 646 | 126 | 414 | 79 | 261 | 49 | 163 | 32 | 105 |
| **Tool 4 (SCEA, 30°; NR, 6.4 mm, or 1/4 in.)(b)** | | | | | | | | | | | | | | | |
| 0.8 | 1/32 | 1123 | 3684 | 715 | 2346 | 451 | 1481 | 295 | 969 | 194 | 638 | 133 | 435 | 104 | 342 |
| 1.6 | 1/16 | 992 | 3254 | 627 | 2058 | 398 | 1305 | 257 | 844 | 166 | 545 | 111 | 363 | 81 | 267 |
| 3.2 | 1/8 | 701 | 2301 | 456 | 1497 | 288 | 946 | 186 | 611 | 121 | 396 | 80 | 263 | 57 | 187 |
| 6.4 | 1/4 | 606 | 1988 | 381 | 1251 | 240 | 787 | 155 | 510 | 100 | 327 | 65 | 214 | 45 | 147 |
| 13 | 1/2 | 545 | 1788 | 340 | 1116 | 217 | 711 | 139 | 456 | 89 | 290 | 57 | 186 | 37 | 123 |
| 25 | 1 | 510 | 1674 | 323 | 1061 | 202 | 664 | 130 | 426 | 82 | 269 | 52 | 169 | 33 | 107 |
| **Finishing tool(c)** | | | | | | | | | | | | | | | |
| 0.13 | 0.005 | 1050 | 3444 | 611 | 2007 | 474 | 1556 | 345 | 1131 | 275 | 903 | 229 | 752 | 211 | 691 |
| 0.25 | 0.010 | 821 | 2694 | 536 | 1760 | 354 | 1162 | 253 | 831 | 192 | 630 | 157 | 516 | 139 | 456 |
| 0.38 | 0.015 | 719 | 2360 | 462 | 1515 | 307 | 1007 | 215 | 704 | 157 | 516 | 127 | 416 | 110 | 362 |

(a) No cutting fluid used. (b) Tool angles: BR, 8°; SR, 14°; ER, 6°; SRF 6°; ECEA, 6°. (c) Tool angles: BR, 20°; SR, 0°; ER, 6°; SRF, 6°; ECEA, 6°; SCEA, 0°; NR, 3.2 mm (1/8 in.); flat, 3.2 mm (1/8 in.)

form well at low turning speed. Ceramic tools are used at still higher speeds, but they are subject to the limitations of carbide tools to a greater degree. Cast cobalt-base alloy tools are usually operated at speeds between those used with high-speed steel tools and those used with carbide tools, but they have only a narrow range of application, chiefly when tool temperatures are high and cooling is not feasible.

In multiple-tool setups, it is sometimes possible to use to advantage the relationship between optimum speed and tool material in machining two or more diameters at the same spindle speed. For instance, on a workpiece having a 25 mm (1 in.) diameter and a 102 mm (4 in.) diameter, both diameters could not be turned efficiently with the same tool material at the same time. If the speed were selected for the larger diameter, it would be too slow for the smaller diameter, and vice versa. However, both diameters can be turned efficiently at the same time when a high-speed steel tool is used for the 25 mm (1 in.) diameter and a carbide tool is used for the 102 mm (4 in.) diameter.

**Tool Design.** Single-point tools have already been described in a general way. Figure 7 shows the common shapes and defines the standard angles for these tools, and the function of each tool angle is explained in the section "Design of Single-Point Tools" in this article. The effects of tool angles and nose radius on power requirements are illustrated in Fig. 8. Permissible speed becomes greater as nose radius is increased, up to the radius at which chatter begins. The effect on speed is most pronounced for shallow cuts, and speed is inversely related to feed.

The same relationship holds for side cutting-edge angle, up to the angle at which chatter occurs. This limiting angle is usually above 30° (tool shank perpendicular to work surface) and is lowest for deep cuts and low feeds.

As side rake angle (or back rake angle in end-cutting applications) is increased, speed increases at first and then decreases. The side rake angle for which speed is at a maximum varies with the operation, but usually is in the range of 8 to 22°. Side cutting-edge angle and side rake angle for a given speed are likely to be lower in turning hard steel than in turning soft steel. The remaining standard angles of single-point tools have little or no effect on speed and feed.

The effect of nose radius on speed and feed can be seen in Tables 8 and 9, in which speeds corresponding to 60 min tool life are tabulated over a range of feed and depth of cut for nose radii of 0, 1.6, 3.2, and 6.4 mm (0, 1/16, 1/8, and 1/4 in.). As mentioned previously, speed can be increased as nose radius is increased, and speeds for carbide tools are about three times those for high-speed steel tools. Speeds for the finishing tool are relatively high, in spite of the 0° side rake and side cutting edge angles, in order to produce a smooth surface.

**Configuration of the workpiece,** or of that portion of the workpiece being machined, sometimes influences optimum speed or feed, or both.

When large and small diameters are to be machined within the same cycle, a change in speed during the cycle may be necessary to provide the most suitable speed for each diameter. This can be accomplished by the use of a two-speed drive motor or, more effectively, by the use of an electronically controlled variable-speed unit.

**Example 2: Variable-Speed Control Reduces Machining Time.** The gray iron casting shown in Fig. 13 was consecutively turned on the 149.1 mm (5.87 in.) diameter

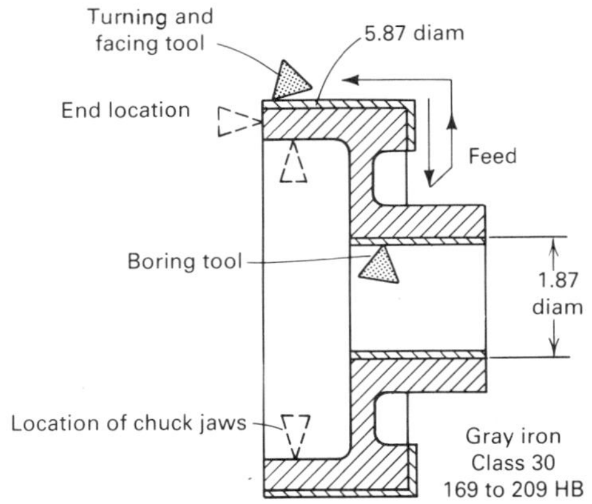


| Machining conditions | Two-speed motor 14.9 mm (5.87 in.) OD | Two-speed motor 4.75 mm (1.87 in.) ID | Variable-speed control 14.9 mm (5.87 in.) OD | Variable-speed control 4.75 mm (1.87 in.) ID |
|---|---|---|---|---|
| Speed, rev/min | 280 | 560 | 280 | 775 |
| Speed, m/min (sfm) | 130 (430) | 85 (275) | 130 (430) | 95 (380) |
| Feed, mm/rev (in./rev) | 0.30 (0.012) | 0.36 (0.014) | 0.30 (0.012) | 0.46 (0.018) |
| Depth of cut, mm (in.) | 0.51 (0.020) | 0.51 (0.020) | 0.51 (0.020) | 0.51 (0.020) |
| Machine cycle, s | 105 | 105 | 80 | 80 |

**Fig. 13** Comparison of machining conditions and cycle time using two-speed motor versus variable-speed control. OD, outside diameter; ID, inside diameter. Dimensions in figure given in inches

and bored on the 47.5 mm (1.87 in.) diameter. Originally, the lathe on which these operations were performed was powered by a two-speed motor rated at 1800 and 900 rpm. By providing the lathe with a variable-speed control, a more nearly optimum speed could be used for the 47.5 mm (1.87 in.) diameter, resulting in about a 25% reduction in machine cycle time. Comparative data for the two methods are presented in the table that accompanies Fig. 13. In some applications of this kind, simultaneous turning with a carbide tool and boring with a high-speed steel tool have proved to be efficient.

**Tool Life Desired.** Analyses are often made with specific setups to determine the optimum feed and speed for maximum tool life, because tool life is many times more sensitive to changes in cutting speed than to any other single factor. However, it is common practice to sacrifice some tool life by purposely increasing speeds (often by as much as 50%) to shorten cycle time and increase productivity. This practice is most often used when tools can be changed readily, with a minimum of downtime.

In any specific application, however, overall cost must be examined to determine whether the gains in productivity outweigh the added cost of sharpening or replacing tools.

In an effort to determine optimum conditions such as cutting speed, specific cutting force, and net power when encountering varying material hardness conditions, tables of compensating factors such as those shown in the example below may be used.

**Example 3: Use of Compensating Factors to Determine Optimum Conditions in the Machining of Grade H1P Material at a 0.30 mm/rev (0.012 in./rev) Feed Rate.** The original material had a 250 HB hardness, and the cutting tool had a 15 min tool life, when the material was machined at 130 m/min (425 sfm/min) to a depth of 4.06 mm (0.160 in.). The next shipment of material is 230 HB, and the goal is to increase the tool life to 40 min, while maintaining the 0.30 mm/rev (0.012 in./rev) feed rate and 4.06 mm (0.160 in.) depth of cut.

The following table relates cutting speed to material hardness:

| Hardness, HB | Compensating factor |
|---|---|
| 190 | 1.07 |
| 210 | 1.04 |
| 230 | 1.02 |
| 250 | 1.00 |
| 270 | 0.98 |
| 290 | 0.95 |
| 310 | 0.94 |
| 330 | 0.93 |

Thus, at 230 HB, the cutting speed should be 1.02 × 425 sfm = 435 sfm.

The compensating factor for cutting speed in regard to the required tool life is obtained from:

| Tool life, min | Compensating factor |
|---|---|
| 5 | 1.40 |
| 10 | 1.15 |
| 15 | 1.00 |
| 20 | 0.92 |
| 30 | 0.81 |
| 40 | 0.75 |
| 60 | 0.66 |

Thus, if a 40 min, rather than a 15 min, tool life is required, the actual cutting speed should be 0.75 × 435 sfm = 326 sfm ($\alpha$ = 0.30).

Calculation of the power required is obtained using the following table, in which the specific cutting force (in tons/in.$^2$, or tsi) is shown at various feed rates and cutting-edge angles:

| Cutting-edge angle | Specific cutting force (tsi) at feed rate, in./rev, of: 0.004 | 0.008 | 0.012 | 0.016 | 0.024 | 0.032 | 0.040 | 0.060 |
|---|---|---|---|---|---|---|---|---|
| 90° | 171 | 140 | 123 | 111 | 98 | 93 | 85 | 79 |
| 75° | 174 | 143 | 127 | 114 | 101 | 95 | 90 | 82 |
| 60° | 181 | 146 | 130 | 117 | 104 | 98 | 90 | 85 |
| 45° | 190 | 155 | 140 | 127 | 111 | 101 | 95 | 90 |
| 30° | 209 | 171 | 152 | 140 | 123 | 114 | 104 | 98 |

Thus, at a feed rate of 0.012 in./rev, the specific cutting force is 123 tsi for a 90° insert entry angle.

The compensating factor for the specific cutting force as a function of material hardness is obtained from:

| Hardness, HB | Compensating factor |
|---|---|
| 190 | 0.73 |
| 210 | 0.82 |
| 230 | 0.91 |
| 250 | 1.00 |
| 270 | 1.08 |
| 290 | 1.17 |
| 310 | 1.26 |

Thus, the specific cutting force required to cut this material having a hardness of 230 HB is 0.91 × 123 tsi = 112 tsi.

Finally, with the cutting speed at 326 sfm, the feed rate at 0.012 in./rev, and the depth of cut at 0.160 in., the net power consumption is:

$$326 \cdot 0.012 \cdot 0.160 \cdot 112 \times 0.067 = 4.7 \text{ hp}$$

**Machine Condition.** Lathes that are worn may require the use of lower than normal speeds and feeds, mainly because they will develop chatter more readily than machines in good condition. It is not good practice to use worn machines, but when there is no alternative, processing must be modified to accommodate the condition of the equipment.

**Horsepower** of available machines may limit the speeds and feeds to be used. It may be necessary to turn a part at less than the optimum speed or reduce the feed rate because of inadequate available power. If purchase of an adequately powered machine is not economically practical, compromises must be made. By reducing the feed, speed may be maintained, but the penalty, of course, is a longer machining cycle.

**Surface finish** is influenced by feed rate per revolution and by the nose radius of the tool. For parts that require a tool with a small nose radius (for example, to maintain a small radius between a shaft diameter and a shoulder), compensation must be made by reducing the feed rate to obtain the required surface finish. The feed rate is usually the compromise value that allows the attainment of maximum possible production consistent with the specified finish.

Surface finishes of 0.50 to 1.25 μm (20 to 50 μin.) are the practical limits that can be expected from turning operations when using well-maintained lathes and tools. Smoother surface finishes, to 0.025 μm (1 μin.) or less, however, can be produced, particularly with precision machines and diamond cutting tools (for nonferrous metals), but generally several cuts are required, resulting in increased manufacturing costs.

**Tolerance Requirements.** Dimensional tolerances that can be maintained in turning vary, depending on the machine and operating parameters, the workpiece, the setup rigidity, and other variables. Practical limits for production applications, with machines and tools in good condition, range from ±0.025 mm (±0.001 in.) for workpieces having diameters of about 6.4 mm (1/4 in.) or less to ±0.08 mm (±0.003 in.) for diameters of 102 mm (4 in.) or more. Closer tolerances to ±0.00127 mm (±0.000050 in.) are often maintained, but maintaining these tolerances generally requires the use of more precise machines and results in higher manufacturing costs.

**Cost Considerations.** Speeds and feeds that are too low consume excessive time, which usually results in an increase in workpiece cost. However, optimum speeds and feeds are not necessarily the maximum that the workpiece and the machine can tolerate. Excessively high speeds and feeds result in shorter tool life and therefore in increased tool cost.

In turning difficult-to-machine alloys, it is especially important that speed and feed be carefully selected and coordinated for optimum results at minimum cost.

## Choice of Equipment and Procedure

A 1983 investigation of more than 13 million workpieces (including cubic and flat parts) machined in 650 plants in the most important industrial nations of the world revealed that:

- 70% of all plants carrying out metal-cutting operations produce batch size of less than 50 pieces
- Rota-symmetrical parts predominate, comprising 75% of the parts produced
- Of all these rota-symmetrical parts (in cumulative terms) 90% are smaller than 200 mm (8 in.) in diameter, 70% are smaller than 42 mm (1 21/32 in.) in diameter, and 70% are shorter than 200 mm (8 in.) in length
- On average, all rota-symmetrical parts (from blank to finished workpiece) require 6 chuckings and 5 min for cutting and setup time
- The initial turning operation was followed by a secondary operation or operations such as turning, with 25% of parts; drilling, with 14% of parts; milling, with 10% of parts; and grinding, with 4% of parts
- The most frequently voiced requirement calls for reductions in setting, idle, transportation, and storage times

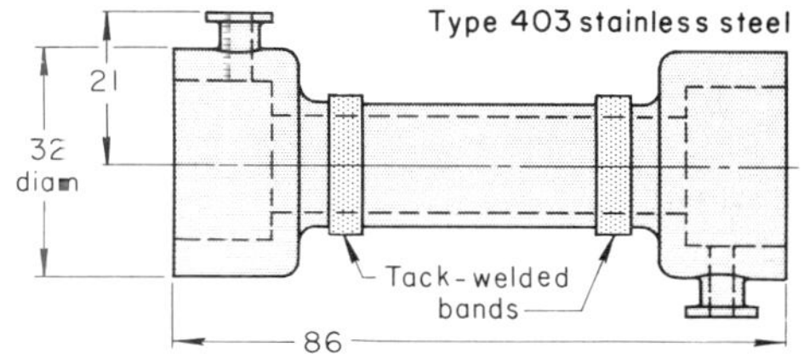


**Fig. 14** Tack-welded bands solved problem of chucking this large fabricated part for two-stage machining of both ends in an engine lathe. Dimensions given in inches

The selection of equipment and machining procedure for a specific part depends largely on:

- Size of workpiece
- Configuration of workpiece
- Equipment capacity (speed, feed, and horsepower range)
- Production quantity
- Dimensional accuracy
- Number of operations
- Surface finish

The following section of this article discusses the influence of these factors and presents examples that describe or compare equipment and techniques for production applications.

### Size of Workpiece

In addition to physically accommodating the work, a suitable lathe provides the workpiece with firm support, rigidly supports the cutting tools and feeds them into the work at the desired rate, and has enough power to maintain the selected rate of metal removal. Thus, size of the workpiece is usually the first consideration in selecting the most appropriate lathe for a specific job.

Small parts requiring average to close tolerances, such as components of instruments, are commonly produced in watchmaker's lathes, bench lathes, or toolroom lathes.

Average-size parts such as automotive spindles and shafts with a length-to-diameter ratio of not more than 10:1, axles and drive shafts long enough to require one or more steady rests to prevent flexing, and similar parts turned between centers comprise a substantial percentage of the parts produced in engine lathes. Average-size parts of relatively short length and large diameter, such as gear blanks, are usually chucked on the outside or inside diameter and are turned on regular engine lathes, or on gap-frame, automatic, stub, and copying (or tracer) lathes.

Large or extremely heavy parts are usually turned on lathes designed specifically for one type of work. Examples are oil-drilling tools, large gun barrels, large steel mill rolls, press columns, and missile parts. Lathes appropriate for parts of this type are heavy-duty long-bed lathes, hollow-spindle lathes, special roll-turning lathes, and missile lathes.

### Workpiece Configuration

Workpiece configurations can be separated into two basic categories, regular and irregular.

**Regular-shape workpieces** are those on which all turned faces are either parallel or perpendicular to the centerline of rotation. Examples include gear blanks, shafts, flanged axles or other parts, cylinder liners, pistons, bearings on camshafts, and ring-shaped parts. Workpieces of this type have no significant angles or radii except normal corner breaks or angular chamfers, which are easily cut by means of tools having corresponding shapes.

Workpieces with cuts only parallel or perpendicular to the centerline of rotation represent a large percentage of lathe work and are machinable on a wide range of standard engine lathes having tool carriages or cross slides that operate either parallel or at 90° to the centerline of rotation. For turning workpieces of this class, size is the major factor in choosing the most suitable equipment.

Regular lathes can be altered to generate more complex shapes (such as angles and large radii) by the use of cams and angular slides. For small production quantities, these auxiliary devices may be impractical because of the setup time required for changing from one shape to another. Large production quantities, however, may justify the use of these modified machines, particularly the duplicating lathes (copying, tracer, profiling, numerical-control, or continuous-path machines).

**Irregular-shape workpieces** are those that require the use of a specific type of lathe in order to be turned satisfactorily. Crankshaft lathes, a notable example, use special center drives to turn and face main bearing sections and use double-end drives to turn and face rod-bearing sections.

Many irregular-shape parts are out of balance when rotated, which may require the application of counterbalances to the spindle, chuck, or workpiece. The need for counterbalancing is influenced by the degree to which parts are out-of-balance, speed of rotation, available power, or a combination of these variables.

Workpieces that have L-shape or T-shape sections and those that have large flanges may require a swing diameter greatly out of proportion to the stem diameter. Often gap-frame lathes are the best choice for these types of workpiece and for others that require large swing clearances in specific locations along the lathe bed.

Workpieces with extremely high length-to-diameter ratios, such as long sections of

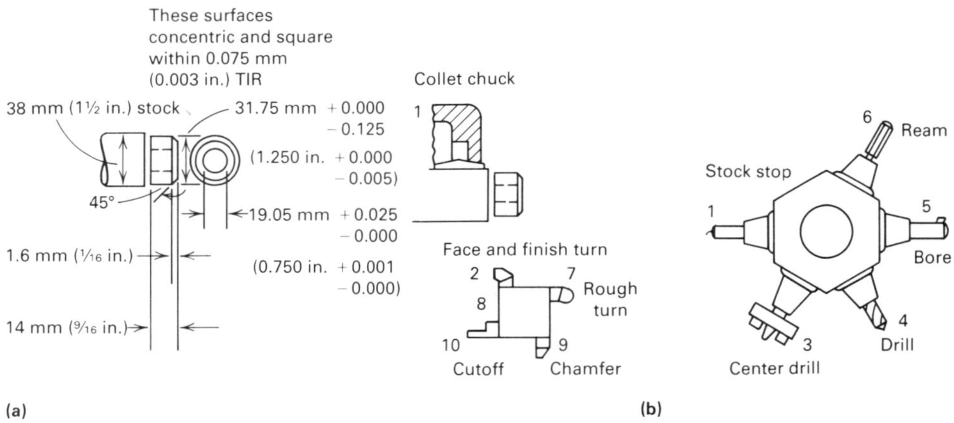


**Fig. 15** (a) Workpiece. (b) Turret lathe tool layout for making a collar of 200 HB mild steel on a No. 4 turret lathe. TIR, total indicator reading

pipe or shafting, are also considered irregular. A stabilizing rest is a common lathe accessory employed in this situation. In many instances, such parts require turning only on the ends and can be machined efficiently in a hollow-spindle or center-drive lathe.

Special lathes can be obtained to machine almost any configuration, but usually their cost is justified only when large quantities of similar parts must be produced. The example that follows describes a method devised for adapting an irregular-shape part for machining in a large engine lathe.

**Example 4: Special Method of Holding a Large, Irregular Part.** The large 770 kg (1700 lb) fabricated stainless steel part shown in Fig. 14 required boring, facing, and threading at each end. Chucking this part presented a problem because during machining each end had to be free of centers of chucks to prevent restriction of the tool. The problem was solved by tack welding two bands to the body of the part, as shown in Fig. 14. Each band was rolled of 13 mm (½ in.) plate and was tack welded in two halves.

The part was chucked externally at the headstock, using a four-jaw chuck, and internally at the tailstock, using a revolving three-jaw chuck. The bands were turned true and to the same diameter. The part was then held and driven by the four-jaw chuck at the headstock. A steady rest was used at the farther band to support the part. After one end had been bored, faced, and threaded, the part was reversed, and the other end was machined. The bands were then removed by grinding away the tack welds.

## Equipment Capacity

The capacity of lathes has been continually increased in terms of power, speed, feed, and thread range.

**Horsepower rating** must be considered when selecting a lathe, because power consumption is in direct ratio to the rate of metal removal, which in turn is related to production rate. With carbide and ceramic cutting tools, it is practical to use surface speeds ranging from 3 to 610 m/min (10 to 2000 sfm). With high-speed steel tools, feed rates up to 1.5 mm/rev (0.060 in./rev) on cuts up to, or beyond, 25 mm (1 in.) in depth are commonly used.

Power requirements for one carbide tool, on average work and operating at optimum speeds, can range from 3.7 to 22 kW (5 to 30 hp); such a range is typical in tracer-lathe turning. In applications involving extremely deep cuts (for example, 64 mm, or 2½ in., cuts in turning steel rolls), more than 225 kW (300 hp) may be required.

**Spindle-Speed Range.** A lathe must be able to rotate a given size of workpiece fast enough to produce the surface speed proper for the tool material, the workpiece material, and the workpiece hardness. For turning carbon and alloy steels, approximate ranges of surface speeds for various tool materials are:

| Tool material | Surface speed, m/min | Surface speed, sfm |
|---|---|---|
| High-speed steel | 3–60 | 10–200 |
| Cast cobalt-base alloy | 15–90 | 50–300 |
| Tungsten carbide | 75–245 | 250–800 |
| Titanium carbide | 75–460 | 250–1500 |
| Ceramic | 150–610 | 500–2000 |

## Production Quantity

The quantity of parts to be machined has a direct bearing on the type of lathe selected, primarily because of cost. Although setup time for an engine lathe is usually less than for a comparable job on any other lathe, the engine lathe requires more operator supervision, and production time per piece is usually greater; consequently, for a majority of parts, it is economical to produce only small quantities on engine lathes. Production volumes above the break-even point favor machinery such as turret lathes and single-spindle and multiple-spindle automatic lathes. This is demonstrated in the three examples that follow.

**Example 5: Comparison of Time Required to Produce a Part on an Engine Lathe Versus a Ram-Type Turret Lathe.** To produce five pieces of the part shown in

### Table 10 Effect of quantity on choice of machine(a)

| Production quantity, pieces | Most suitable machine | Pieces per hour, Per machine | Pieces per hour, Per man | Setup time, hour |
|---|---|---|---|---|
| Under 10 | Engine lathe | 2 | 2 | ½ |
| 10–100 | Turret lathe | 4 | 4 | 4 |
| 100–10 000 | Single-spindle automatic | 7 | 28 | 6 |
| 10 000–100 000 | Six-spindle automatic(b) | 15 | 45 | 12 |

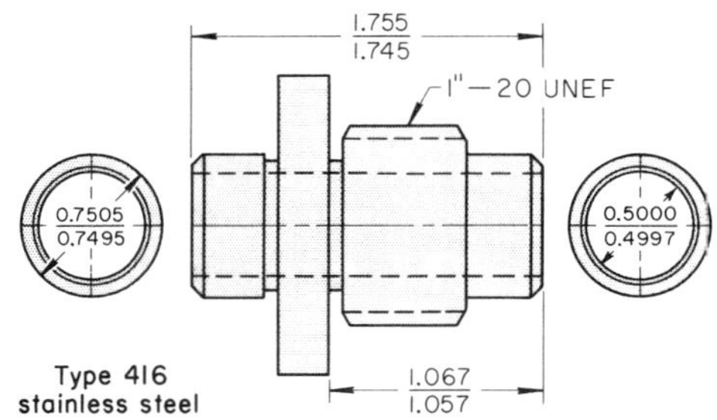


(a) Data are based on turning (at 113 m/min, or 370 sfm), drilling (at 46 m/min, or 150 sfm), and threading (at 7.6 m/min, or 25 sfm) the part illustrated above. (b) Estimated; the part was not actually machined on a six-spindle automatic machine. Dimensions in figure given in inches

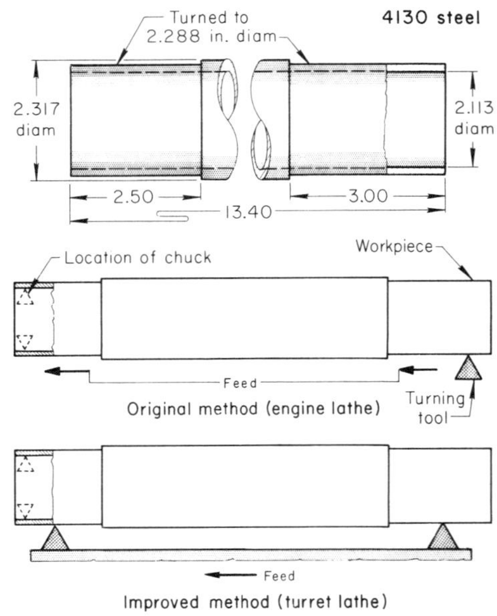


| Machining conditions | Method(a), Engine lathe | Method(a), Turret lathe |
|---|---|---|
| Speed, rev/min | 320 | 450 |
| Speed, m/min (sfm) | 58 (190) | 82 (270) |
| Feed, mm/rev (in./rev) | 0.10 (0.004) | 0.10 (0.004) |
| Setup time, min | 45.6 | 68 |
| Production/h, piece | 9 | 20 |
| Workpiece hardness, HRC | 34 | 34 |

(a) For both methods, carbide tools were used, and each produced about 50 pieces per grind; tool change times were equal, and water-soluble oil was used as the cutting fluid.

**Fig. 16** Comparison of setups and operating conditions for turning a tubular part in an engine lathe and in a turret lathe. Dimensions given in inches

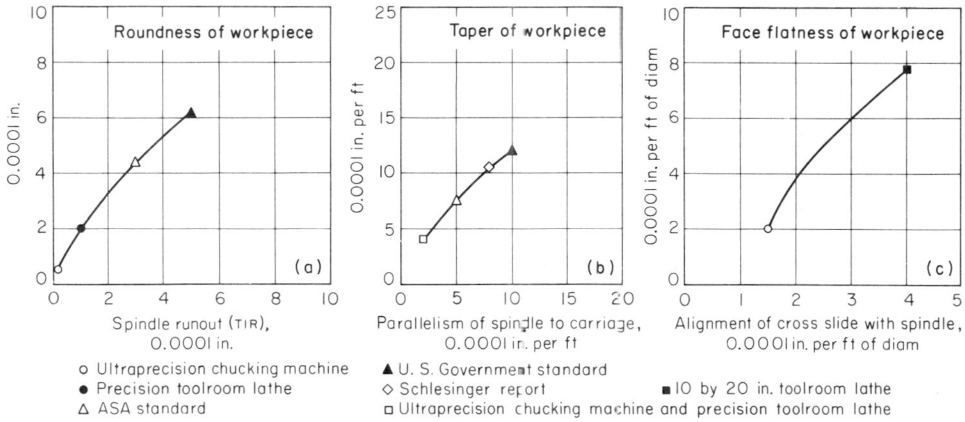


**Fig. 17** Effect of machine variables on roundness, taper, and face flatness of workpieces

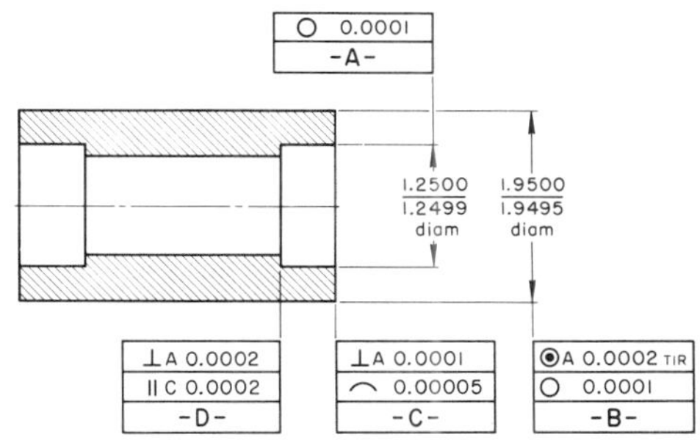


**Fig. 18** Typical tolerance relationships that can be held in precision turning. Dimensions given in inches

Fig. 15(a) takes a total of 41.70 min, or 8.34 min per piece, on the engine lathe, compared with a total of 93.40 min, or 18.68 min per piece, on the turret lathe setup shown in Fig. 15(b). For 100 pieces, the total time is 454 min, or 4.54 min per piece, on the engine lathe, but a total time of only 348 min, or 3.48 min per piece, on the turret lathe. If the labor and overhead rates and capital costs are nearly the same on both machines, and they usually are for machines such as these, a comparison of the operation times indicates which machine is more economical for a required number of pieces. Such an analysis bears out the principle that the engine lathe is usually economical for a few pieces, but for larger numbers, the turret lathe shows lower costs.

**Example 6: Engine Lathe Versus Turret Lathe and Single-Spindle and Six-Spindle Automatics.** The data in Table 10 compare productivity and setup times for four different types of machines, for turning, drilling, and threading a small stainless steel part (inset sketch in Table 10). From this comparison it is evident that for machining this part in quantities of more than ten, machines other than an engine lathe are more efficient. For the engine and turret lathes, one operator was required for each machine, whereas one operator could handle four single-spindle automatics or (it was estimated) three six-spindle automatics.

**Example 7: Engine Lathe Versus Turret Lathe.** The tubular part shown in Fig. 16 was produced in annual quantities of about 1600, in lots of 200 pieces. Originally, a 3.7 kW (5 hp) engine lathe was used (center sketch in Fig. 16). Speed and feed were restricted to values substantially below the nominal (see comparison in Table 6), because of problems in maintaining rigidity. Changing to a more rugged 11 kW (15 hp) turret lathe (bottom sketch in Fig. 16) permitted turning at a 50% increase in speed, and also allowed both ends of the part to be turned simultaneously. Although the turret lathe cost 150% more to set up than the engine lathe, the turret lathe produced parts at 53% of the cost per unit piece of the engine lathe.

Because both tools were operated from the cross slide, an engine lathe rated at 11 kW (15 hp) could have been used. However, a turret lathe was preferred because the turret could be manipulated in less time than a tailstock, thus reducing loading and unloading time.

## Dimensional Accuracy

In turning a simple cylinder with a single-point tool cutting perpendicular to the axis of rotation, straightness of cut depends on:

- Squareness of the cross slide to the spindle centerline
- Axial movement of the spindle
- Vibration of the cutting tool at the tip and tool wear

Assuming inaccuracy to be 0.0025 mm (0.0001 in.) for each of these three fac-

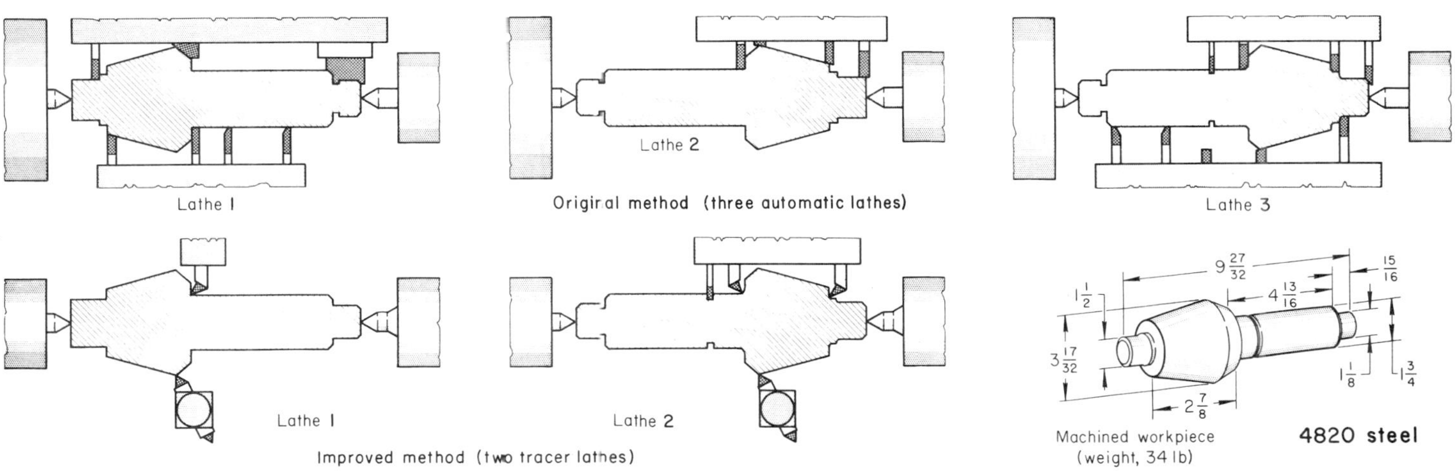


| Method | Speed, rev/min | Speed, m/min (sfm) | Feed, mm/rev (in./rev) | Cutting fluid | Tool material | Tool life, pieces | Tool changing, min/8 h | Machining time per piece, min |
|---|---|---|---|---|---|---|---|---|
| Original (three automatic lathes) | 90 | 25 (83) max | 0.25 (0.010) | Soluble-oil:water (1:25) | High-speed steel | 100 per grind | 62 | 4.124 |
| Improved (two tracer lathes) | 440 880 | 123 (405) | 0.36 (0.014) | Soluble-oil:water (1:25) | Disposable carbide | 50 per insert | 90 | 2.477 |

**Fig. 19** Change from three automatic lathes to two tracer lathes that provided rigidity required for obtaining dimensional accuracy in machining a forged pinion-gear blank. Workpiece hardness, 217 to 241 HB. Dimensions in figure given in inches

tors, minimum variation in machined dimension could not be less than 0.0025 mm (0.0001 in.) and might equal the sum of these three inaccuracies, or 0.008 mm (0.0003 in.).

**Diametral roundness** is related directly to the lathe spindle run-out. Most spindles rotate in a preloaded angular bearing to eliminate side and end movement. Lathes are available with spindles that do not exceed 0.0005 mm (0.000020 in.) run-out, total indicator reading. The relationship of workpiece roundness to spindle run-out is shown in Fig. 17(a).

**Diameter variation (taper)** depends on the relationship of the axis of rotation of the workpiece, in both the vertical and the horizontal planes, to the longitudinal travel of the tool and carriage. Any relationship other than true parallelism will result in taper of the workpiece. The relationship is shown in Fig. 17(b).

**Face flatness** is influenced by the alignment of the cross slide with the axis of rotation of the workpiece. Any variation from a true perpendicular relationship results in either a high or low center, depending on the direction of misalignment. Alignment is checked by facing a surface on a dummy blank approximately 102 mm (4 in.) in diameter. An indicator is then mounted on the cross slide and traveled completely across the machined face. The total indicator movement registers face flatness resulting from both vibration and misalignment of the cross slide due to cam action. Cam action, or end camming, is the amount of movement of the spindle along the spin axis under dynamic conditions; cam action of the spindle on a precision lathe should be less than 0.0025 mm (0.0001 in.). The effect of cross-slide alignment on face flatness of the workpiece is shown in Fig. 17(c).

**Diameter accuracy** depends to some degree on operator skill. The cross-slide positioning dial on precision lathes is graduated in 0.025 mm (0.001 in.) increments, but these increments are about 3.2 mm (1/8 in.) apart on the dial. A capable operator can control turned diameters within 0.008 mm (0.0003 in.) by estimating settings between dial markings. With precision gaging, an experienced operator can sometimes hold diameters within 0.0025 mm (0.0001 in.). Digital readouts can help a less experienced operator. Repeatability of the lathe should be within half of the lowest divisional value on the cross-slide dial.

**Length dimensions** are measured parallel to the axis of rotation. Accuracy depends on positioning of the longitudinal slide. Precision lathes are capable of holding a tolerance of 0.013 mm (0.0005 in.) on length. On manually operated lathes, however, an experienced operator can usually improve this capability by means of dial gages on machine slides for reference positioning.

**The relationship of tolerances** to each other are often more meaningful than tolerances on individual dimensions. Tolerance relationships that can be maintained in precision turning are given in Fig. 18. The data in the charts in Fig. 17 show that tolerances of this order of magnitude can be maintained with precision lathes.

**Lathe selection** on the basis of dimensional accuracy required is illustrated in the example that follows. It describes an application in which reselection of machinery was necessary to meet a more typical tolerance of 0.13 mm (0.005 in.).

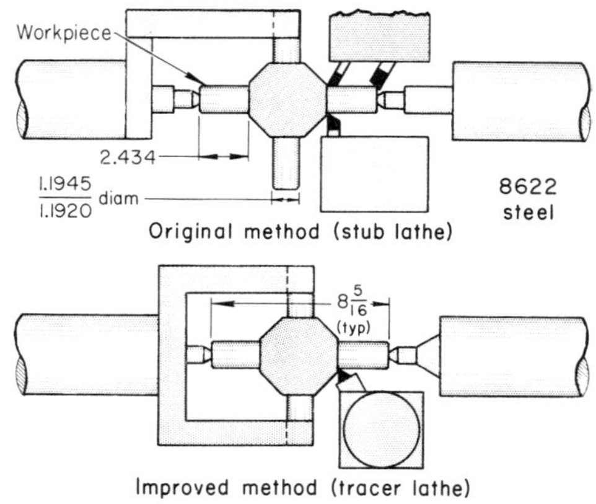


| Machining conditions | Stub lathe | Tracer lathe |
|---|---|---|
| Speed, rev/min | 366 | 2100 |
| Speed, m/min (sfm) | 35 (115) | 200 (655) |
| Feed, mm/rev (in./rev) | 0.300 (0.0118) | 0.25 (0.010) |
| Depth (turning), mm (in.) | 2.4 (3/32) | 2.4 (3/32) |
| Cutting fluid, soluble oil; water | 1:25 | 1:25 |
| Tool material | HSS | Carbide |
| Tool life, piece | 75(a) | 78(b) |
| Setup time, min | 24 | 24 |
| Downtime for tool change, min(c) | 8 | 15 |
| Machining time, min | 3.029 | 1.476 |
| Workpiece hardness, HB | 179–197 | 179–197 |

(a) Per grind. (b) Per tip (disposable type). (c) Per 8 h shift

**Fig. 20** Comparison of setup and processing details for machining differential-cross arms in a stub lathe and in a tracer lathe. Dimensions in figure given in inches

**Example 8: Change From Automatic to Tracer Lathes for Rigidity.** A tolerance of 0.13 mm (0.005 in.) was specified for all finish-machined dimensions on the forged 4820 steel pinion-gear blank shown in Fig. 19. Machining operations included turning, chamfering, facing, and grooving; stock re-

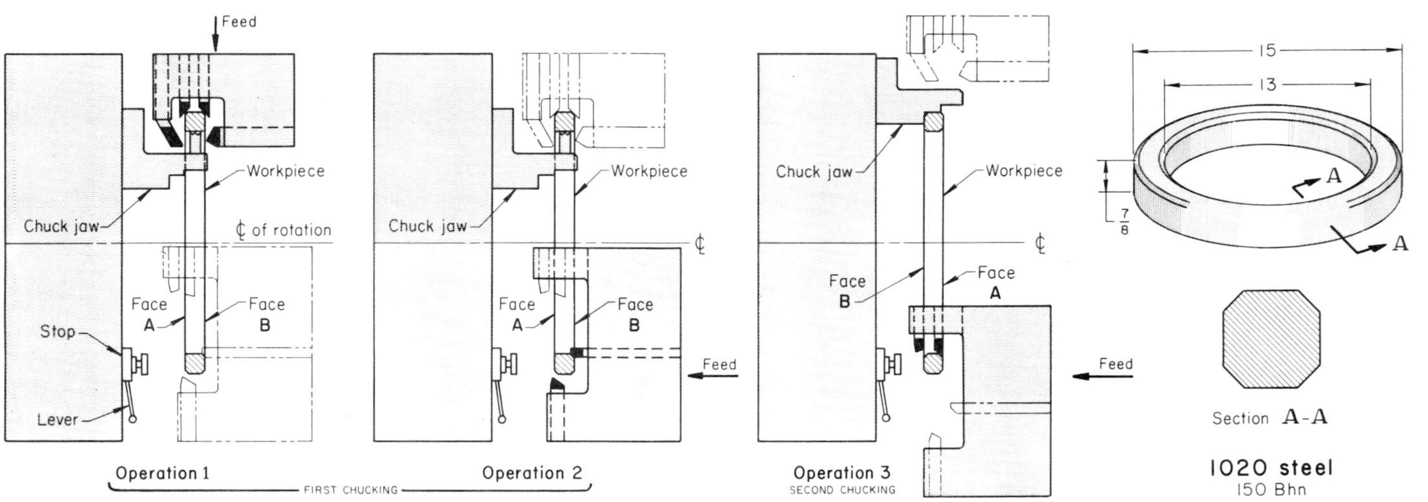


**Operation 1:** Face A is positioned against three retractable stops; then three chuck jaws grip workpiece on inside diameter, and stops are retracted by lever. In this position, both sides are faced and outside-diameter edges are chamfered.

**Operation 2:** Outside diameter is turned, edge of face B chamfered. **Operation 3:** Workpiece is reversed, and face B is placed against stops; then chuck jaws grip workpiece on outside diameter, and stops are retracted. Inside diameter is bored, and edge of face A is chamfered.

**Fig. 21** Setup using specially designed chuck, showing sequence of operations for machining ring-gear blanks in two chuckings. Dimensions given in inches

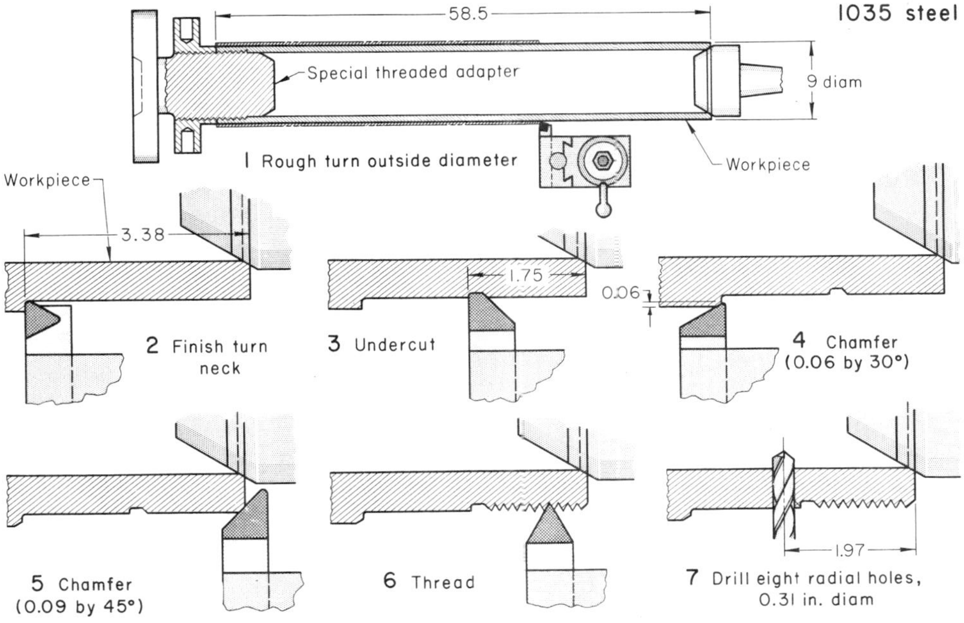


| | | | |
|---|---|---|---|
| Speed, at 210 rev/min, m/min (sfm) | 151 (495) | Tool material | Carbide |
| Feed, turning, mm/rev (in./rev) | 0.51 (0.020) | Tool life per grind, piece | 5 |
| Depth of cut, mm (in.) | | Downtime for tool change, min | 5 |
| Rough turning OD | 4.8 (3⁄16) | Setup time, h | 2 |
| Finish turning neck | 0.8 (1⁄32) | Cycle time per piece, h | 1.178 |
| Cutting fluid | None | Production/h, piece | 0.85 |
| | | Hardness, HB | 140–160 |

**Fig. 22** Setup and conditions for performing seven machining operations in one chucking in an engine lathe, by the use of a special threaded adapter at headstock. Dimensions in figure given in inches

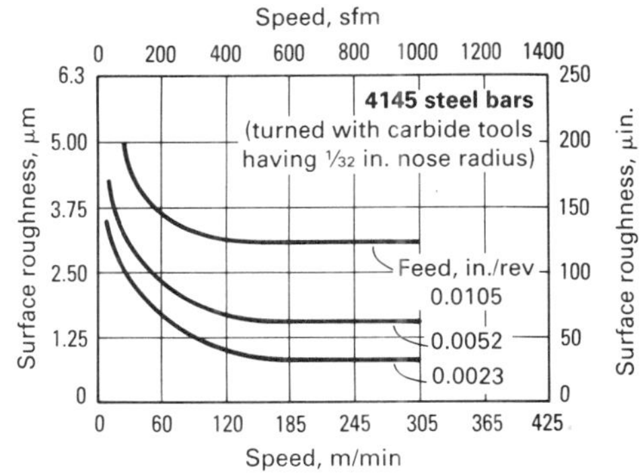


**Fig. 23** Influence of speed on surface finish, at constant feed

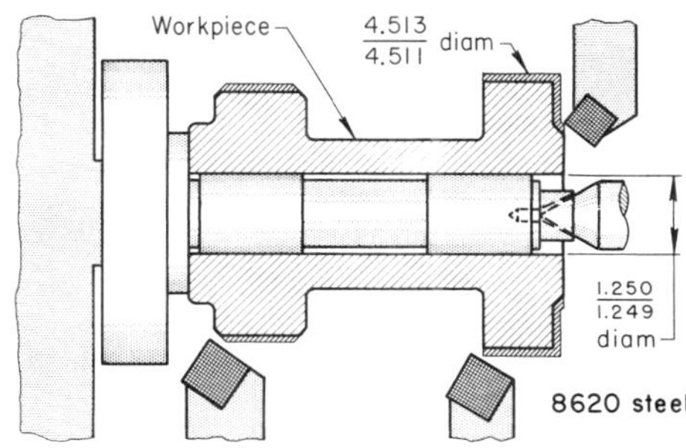


| Machining conditions | Type of carbide tool: Steel-cutting grade (C-8) | Type of carbide tool: Special TiC grade |
|---|---|---|
| Speed, rev/min | 575 | 1040 |
| Speed, m/min (sfm) | 195 (639) | 343 (1125) |
| Feed, mm/min (in./min) | 0.20 (0.008) | 0.20 (0.008) |
| Depth of cut, mm (in.) | 0.51 (0.020) | 0.51 (0.020) |
| Finish obtained, μm (μin.) | 4.5 (180) | 2.00 (80) |
| Cycle time per piece, s | 18 | 10 |
| Workpiece hardness, HB | 149–189 | 149–189 |

**Fig. 24** Improvement in surface finish and machining efficiency that resulted from change in composition of carbide tools used for turning, facing, and chamfering cluster-gear blanks. Dimensions in figure given in inches

moval ranged from 2.4 to 6.4 mm (3⁄32 to 1⁄4 in.) per side.

Originally, three 11 kW (15 hp) automatic lathes and a total of 20 tools were used to complete the sequence of operations (top row of sketches in Fig. 19). These machines, however, were not rigid enough to withstand the forces from the heavy stock removal. Consequently, size variation of the workpieces and tool breakage were constant problems, despite the use of a feed rate two-thirds the nominal value (see comparison in Table 6).

To improve dimensional accuracy and decrease tool breakage, as well as to improve production efficiency, the job was transferred to two 30 kW (40 hp) tracer lathes using only eight tools (Lathe 1 and 2, bottom row, Fig. 19). The superior rigidity of these machines eliminated tolerance and tool breakage problems and made it possible to change from high speed steel tools to carbide tools, thereby increasing speed and feed. The change to the tracer lathes reduced machining time per piece by 40%.

## Number of Operations

The number of operations that can be performed on an engine lathe is almost limitless. The unmodified engine lathe, however, using one single-point tool, can generate only one surface at a time, and the tool or tool position, or both, must be changed before another surface can be generated. There are two methods for increasing the efficiency of an engine lathe:

- By the use of a duplicating attachment, which can do several operations consecutively by means of program control
- By the use of multiple-operation accessories and attachments, such as a turret, to permit simultaneous or consecutive operations by two or more different tools

The three examples in the remainder of this section describe applications in which lathe efficiency was increased by various methods that made it possible to perform several operations in one setup.

**Example 9: Three Operations With One Tool in Tracer Lathe.** Originally, a 3.7 kW (5 hp) stub lathe was used for machining the four arms of a differential cross forged from 8622 steel. By this method (see upper sketch in Fig. 20), three high-speed steel tools were required for turning, chamfering, and facing each of the four sections in separate chuckings.

The job was transferred to an 11 kW (15 hp) tracer lathe, which performed all three operations with one tool, and which also used a double, rather than a single, driving arm (see lower sketch in Fig. 20). The changes in machine and driving mechanism increased rigidity enough to permit the use of a carbide tool, which in turn allowed a higher speed and shorter machining time. Operations at a feed rate below nominal with carbide tools made possible a speed 50% greater than nominal (see Table 6), and five times greater than that used with the high-speed steel tools in the stub lathe. Processing details for the two methods are compared in the table in Fig. 20.

**Example 10: Special Chuck for Ring-Gear Blanks.** Ring-gear blanks that required machining all over were formerly machined by means of two chucks and two setups. To reduce handling and permit machining of all faces in one setup, a special hydraulically operated chuck having 76 mm (3 in.) of jaw travel was designed. This chuck permitted the machining of all surfaces in two chuckings. Setup and sequence of operations are given in Fig. 21.

**Example 11: Seven Operations in One Chucking, Using Threaded Adapter.** By using a threaded adapter at the headstock end of an engine lathe, and a male center in the tailstock, it was feasible to perform seven operations on tubular parts in a single chucking (Fig. 22). The workpiece had pre-

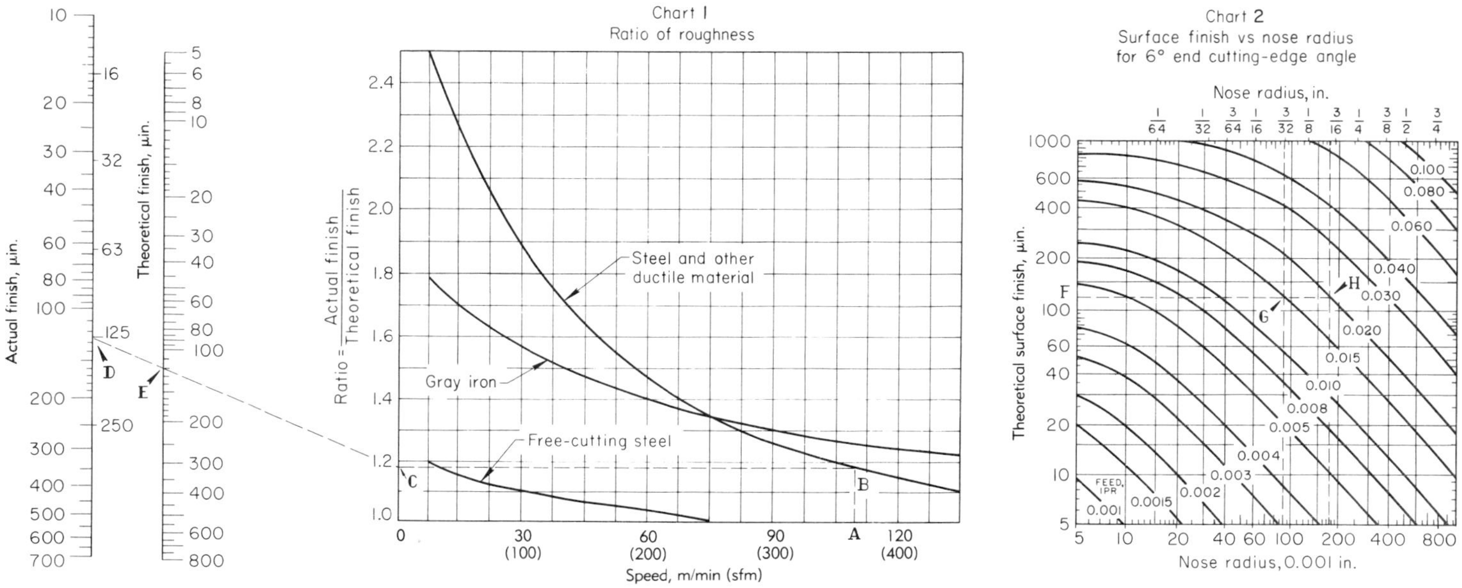


**Example of Use.** To determine the nose radius required for obtaining a finish of 125 µin. when turning 1095 steel at a speed of 111 m/min (365 sfm) and a feed of 0.015 in./rev.

1. On chart 1, locate 365 sfm (point A). From point A, follow a vertical line to its intersection with the "Steel and other ductile material" curve (point B). Follow a horizontal line to determine the ratio of actual to theoretical finish (point C).
2. Locate the specified 125 µin. finish on the "Actual finish" scale (point D), then draw a line from point D to point C. This line crosses the "Theoretical finish" scale at 120 µin. (point E).
3. On chart 2, locate the theoretical finish of 120 µin. (point F). Follow a horizontal line to its intersection with the 0.015 in./rev feed curve (point G). Follow a vertical line and find that required nose radius is 0.090 in.; the nearest standard radius is 3⁄32 in.
4. If machine and work conditions are such that a heavier feed rate could be used, extend line F—G to intersect the 0.020 in./rev feed curve (point H). From point H, the vertical line indicates a required nose radius of 0.175 in.; the standard radius is 3⁄16 in.

**Fig. 25** Nomograph for estimating nose radius required for obtaining specified surface finish

viously been bored and internally threaded to permit the use of the threaded adapter. Machining details are given in the table accompanying Fig. 22.

The first six operations included turning, undercutting, chamfering, and threading. After completion of these operations, rotation was stopped and eight cross holes, equally spaced around the periphery, were drilled through the cylinder wall. The cross-hole drill was powered separately and mounted on a quick-change holder. Adequate accuracy for radial spacing of the holes was obtained from marks on the adapter and a fixed pointer on the headstock. Speed and feed for the turning operation were about one-third higher than the nominal values shown in Table 6.

## Surface Finish

The surface roughness obtained in lathe turning, aside from being dependent on workpiece material and hardness, is influenced by tool material and its relation to speed and feed rate, by tool design (particularly, tool nose radius), by the rigidity of the machine and the tool, and by the type and effectiveness of the cutting fluid used.

**Tool Material, and Speed and Feed.** Welding of chips to cutting tool (edge buildup) is the major cause of surface roughness. Because the speed at which buildup occurs varies for different tool material, selection of tool material for obtaining the smoothest finishes depends on the surface speed to be used.

With high-speed steel tools, a built-up edge forms more readily as speed increases. The smoothest surfaces possible using high-speed steel tools are obtained at speeds of 1.5 to 3 m/min (5 to 10 sfm), which are too low to be practical for most production applications.

Conversely, edge buildup on carbide tools is minimized by using a surface speed high enough to cause plastic flow of the chip (for ductile metals). A satisfactory guide for determining whether speeds are high enough in machining steel is the color of the chip after cooling. The absence of any heat color, even straw color, on steel chips indicates their removal at a surface speed too low for the material being turned, a practice that usually results in poor finish and short tool life.

Figure 23 shows the relationship between finish and surface speed. At any specific feed rate, surface finish stabilizes at 120 to 150 m/min (400 to 500 sfm). These data also show the pronounced influence of feed rate on surface finish.

For any given workpiece material, the finish obtained in turning with carbide tools is influenced also by the composition of the carbide. Straight tungsten carbide is suitable for brittle metals like cast iron, but tools that contain titanium carbide give much better results, in terms of finish, allowable speed, and tool life, for turning steel and other ductile metals.

In addition to conventional steel-cutting grades of carbide (all of which contain titanium carbide), special grades are available that more readily resist adhering to steel workpieces over a wide range of speeds (75 to 365 m/min, or 250 to 1200 sfm), compared to conventional grades. As a result, these special grades are sometimes used for the primary purpose of obtaining better finishes, the higher speeds they can withstand being an added benefit in some instances. These special grades, however, are more brittle than conventional grades, and consequently they must be used for relatively light feeds under conditions of maximum rigidity.

The following example describes an application in which a special grade of titanium carbide proved to be advantageous.

**Example 12: Improved Finish and Higher Speeds With Special Carbide.** The tools originally used for turning, facing, and chamfering cluster-gear blanks in an automatic lathe (setup shown in Fig. 24) were made of a conventional steel-cutting grade of carbide. As shown by the comparison of operating details in the table in Fig. 24, changing to tools made of a special grade of titanium carbide not only allowed higher speed and decreased cycle time, it also resulted in an improvement in surface finish from 4.50 to 2.00 µm (180 to 80 µin.).

Depth of cut (within any reasonable operating range) has little influence on the finish obtained when carbide tools are used. However, as the depth of cut increases, chip control becomes more critical. The chip breaker must provide a uniform movement of the chips away from the turned surface, because chips that are directed onto the turned surface will scratch the workpiece, and particles of these chips will weld to the workpiece surface.

**Nose radius** of the tool exerts an important influence on the surface finish obtainable. The nose radius required for obtaining a specified surface finish may be estimated by means of the nomograph in the example below (see Fig. 25).

**Example 13: Effect of Tool Nose on Finish.** A shaft made of 4130 steel (hardness, HRC 34) was turned in an engine lathe with carbide tools. Specified finish was 1.25 μm (50 μin.). At a speed of 712 rev/min (82 m/min, or 270 sfm), feed of 0.19 mm/rev (0.0075 in./rev) and depth of cut of 0.64 mm (0.025 in.), tools ground with a nose radius of 3.2 mm (1/8 in.) could produce only an unacceptable 2.50 μm (100 μin.) finish. This was reduced to 1.88 μm (75 μin.) by grinding the tool nose to a 6.4 mm (1/4 in.) radius. The specified 1.25 μm (50 μin.) finish was finally obtained by using a 3.2 mm (1/8 in.) radius tool on which a small flat had been ground. The flat, 0.13 to 0.20 mm (0.005 to 0.008 in.) wide to correspond with feed rate, produced a skiving action on the work surface.

**Rigidity** of machines and tools has a large influence on surface finish, other factors remaining constant. Chatter develops at lower speeds in machines that have loose bearings or other vital parts that need repair than in well-maintained machines. Nonrigid tools and holders also allow chatter to develop at lower speeds than when rigidity is good. The immediate result of chatter is roughness of machined surfaces, and the eventual result is short tool life.

Machines having preloaded ball-bearing (or roller-bearing) spindles provide the rigidity necessary for meeting stringent requirements on finish and tolerance.

## Cutting Fluid

Although for some applications of lathe turning cutting fluids are neither needed nor desired (for example, most cast iron parts, and some steel parts, are machined dry) in most applications, some type of cutting fluid is used. Cutting fluids serve the same purposes in lathe turning as in other metal-cutting operations: to cool workpieces and tools, to cool and flush away chips, to promote cutting action by minimizing adherence of tool and workpiece, and to protect the workpiece from corrosion.

### Fluid Classification

Surface finish and tolerance requirements, work metal composition, and the specific types of operation influence the choice of cutting fluid.

**Soluble oil** (in mixtures at various concentrations with water) is the most widely used cutting fluid, both because it is the least expensive and because it is unexcelled by any other fluid in ability to cool and to flush away chips. Soluble oils are nonflammable and nontoxic, and are safe to use with virtually all metals without fear of staining. Most soluble oils contain inhibitors that prevent them from causing ferrous metals to rust. The usual mixture is about 1 part oil to 20 parts water; proportions are not critical, however; in some instances 40 parts water may be used to dilute 1 part oil without significant change in results.

However, soluble-oil emulsions are far less effective than many other cutting fluids for promoting cutting action and preventing edge buildup. As required smoothness and dimensional accuracy increase, some oil or nonaqueous oil mixture is needed.

**Straight mineral oils** are often used when soluble oils do not meet requirements, particularly when the work material is not free-machining or when specified finish exceeds the capability of soluble oil. Mineral oils with a viscosity of about 100 Saybolt Universal Seconds (SUS) (at 40 °C, or 100 °F) are most commonly used, although oils with a viscosity of only about 40 SUS (such as mineral seal oil) are used in many applications.

**Blended cutting oils** of various viscosities are readily available as proprietary compositions. Most of them are basically mineral oils, blended with sulfur compounds, animal fats, and other materials. They may be used as-purchased or cut with mineral oils, depending on prior experience with similar jobs.

Although any straight oils are less effective than soluble oils (oil-water emulsions) for cooling and washing away chips, all oils (particularly those containing sulfur compounds or other special additives) are more effective for improving cutting action. When chatter develops as the result of vibration or other causes, unacceptable surface finish and short tool life are inevitable. Cutting oils are more effective than soluble-oil emulsions for preventing chatter.

**Special Oils and Mixtures.** In applications that demand maximum performance of cutting fluids, high-viscosity thread-cutting oils or lard-oil mixtures are preferred. These special cutting fluids are especially effective for cutting threads that require smooth surfaces.

Lard oil is one of the best for promoting cutting action, but because of its high viscosity, it is impractical for high-production applications. However, it is often used in small lathes for toolroom or pilot production applications.

Both thread-cutting oil and lard oil are often mixed with mineral oil to reduce viscosity to a practical level. These mixtures still retain some of the advantages of the undiluted oils. Neither of these special oils, however, is equal to a soluble-oil emulsion in capability for cooling or for washing away chips.

### Compatibility With Metals

All of the cutting fluids discussed above can be used for machining ferrous metals without danger of staining or corroding the work. However, not all of these cutting fluids are compatible with all metals. For instance, sulfurized oils are likely to stain copper-base alloys, beryllium alloys, and nickel-base alloys. The Section "Machining of Specific Metals and Alloys" in this Volume contains information concerning the cutting fluids that are compatible with both ferrous and nonferrous metals.

**Contamination** of cutting fluids may cause excessive variation in workpiece finish or dimensions, short tool life, or corrosion of the workpieces. Common contaminants include: tramp oil, usually from hydraulic systems; water, from any of several sources; fine chips; and bacteria, which cause rancidity and breakdown of cutting fluids by organic decomposition.

Normal preventive maintenance will usually forestall serious contamination from tramp oil or water. At a minimum, the circulating system should include a screen (50 to 100 mesh) to prevent chips from being returned to the machining area. For more stringent tolerance and finish requirements, cutting fluids should be circulated through filters. Contamination by bacteria can be prevented by purchasing oils that contain microbe inhibitors. Water-oil emulsions (soluble oils) used for machining ferrous metals should contain a rust inhibitor.

# Boring

BORING is a machining process in which internal diameters are generated in true relation to the centerline of the spindle by means of single-point cutting tools, and it is the most commonly used process for enlarging or finishing holes or other circular contours. Although most boring operations are done on simple, straight-through holes (ranging upward in diameter from about 6 mm, or ¼ in.), the process is also applied to a variety of other configurations. Tooling can be designed for the boring of blind holes, holes with bottle configurations, circular-contoured cavities, and bores with numerous steps, undercuts, and counterbores. The process is not limited by length-to-diameter ratio of holes; with the workpiece properly supported, holes having diameters that exceed length (or vice versa) by a factor of 50 or more have been successfully bored.

Boring is sometimes used after drilling to provide drilled holes with greater dimensional accuracy or improved finish. It is more widely used, however, for finishing holes too large to be produced economically by drilling, such as large cored holes in castings or large pierced holes in forgings. In many applications, boring is done in conjunction with turning, facing, or other machining operations. The scope of this article is limited to applications in which boring is the sole operation or in which it is the major operation in a machining sequence.

## Machines

Metal workpieces have been bored on almost every type of machine that has facilities for rotating a spindle or a workpiece. Most boring, however, is done on the machines (or modifications of them) discussed in the following paragraphs. Electronic (numerical) control can be used with many of these machines.

**Engine lathes** are versatile and are used for a variety of boring operations—usually for single-tool jobs. Lathes provide maximum rigidity, because of their massive, single-unit construction, and permit the use of supporting members such as steady rests or boring-bar supports.

In most operations, the workpiece is clamped to the face plate or chuck and is rotated by the spindle in the headstock, the boring tool is secured to a bracket mounted on the tool-post carriage, and power is supplied to the tool from the carriage. Occasionally, however, the workpiece is mounted on the lathe compound and is fed into the rotating boring tool, which is mounted between the headstock and the tailstock and is powered by the headstock spindle.

The use of engine lathes for boring is usually restricted to the machining of a single part (or, at most, a few identical parts), because setups are cumbersome and expensive, and because only one hole can be bored at a time. Other limitations on the use of engine lathes for boring are:

- The swing of the lathe limits the maximum projection from center of the workpiece
- Bed length limits maximum length of carriage feed
- Workpieces must be symmetrical, or very nearly so, because off-center configurations cause a serious out-of-balance condition

**Turret lathes** are modifications of engine lathes and are used extensively for boring. The use of turret lathes, however, is subject to the same limitations with respect to workpiece size and configuration that apply to engine lathes (see list above).

Turret lathes are better adapted to high production than engine lathes. The main advantage of a turret lathe is that the rotating turret can be tooled for performing as many as eight different operations in a continuous sequence. This sequence often includes turning, facing, drilling, reaming, tapping, and other machining operations, in addition to boring.

**Bar machines** (screw machines), which in turn are modifications of turret lathes, enable a further increase in production of parts that are made from bars or tubes. Production can be still further increased by the use of a multiple-spindle automatic bar machine, designed so that every tool is in operation at the same time, but on a different piece of material. This principle is also used on chucking machines.

**Vertical boring mills** embody the fundamental elements of the lathe. At times the choice between these two types of machines depends on availability, although the vertical boring mill has its own area of application. Vertical machines are more appropriate for workpieces that are so large, heavy, or seriously out of balance that they are easier to lay down on a table than to hang on the face plate of a lathe. Hence, vertical boring mills are commonly used for boring and turning operations on heavy workpieces, such as large rings and short cylinders. The weight of a heavy workpiece is distributed uniformly over the table of the boring mill and can easily be supported by the machine base.

Vertical boring mills are especially suited for heavy workpieces that require indicating during setup. With these machines a workpiece can be placed on the horizontal table, set up, leveled, and given a trial cut with temporary clamping. Counterbalance can be applied to the top of the work table to compensate for off-center loads. Two or more tools can be operated simultaneously, thus permitting two or more boring operations, or boring and turning operations, to be done at the same time. Another advantage of the vertical mill is that it requires less floor space than an engine lathe of equivalent capacity.

**Vertical turret lathes** include features of the vertical boring mill. In addition, they are equipped with a turret on the main head and a turret toolholder on the side head. A second vertical head may be mounted on the crossrail, and a second side head may be mounted on the opposite side of the machine; these modifications provide the machine with greater flexibility and increase its capacity for simultaneous multiple cutting on a variety of work.

**Horizontal boring mills** are preferred for a wide variety of production work. In these machines, the workpiece remains stationary, and the tool rotates. In some setups, the work is fed toward the tool; in others, the tool is fed toward the work.

Horizontal boring mills are of three principal types: table, planer, and floor. The table type feeds horizontally on saddle ways, both parallel with and at right angles to the spindle axis. The headstock can be moved vertically on the column, and the spindle is fed horizontally. Because of its flexibility, this type of machine is especially well-suited to work in which other machining operations are performed in conjunction with boring.

The planer type of machine is similar to the table type, except that the supporting table can be moved only at right angles to the spindle. On some planer-type machines, the housing can be fed in and out on a slide, in the same direction as the spindle.

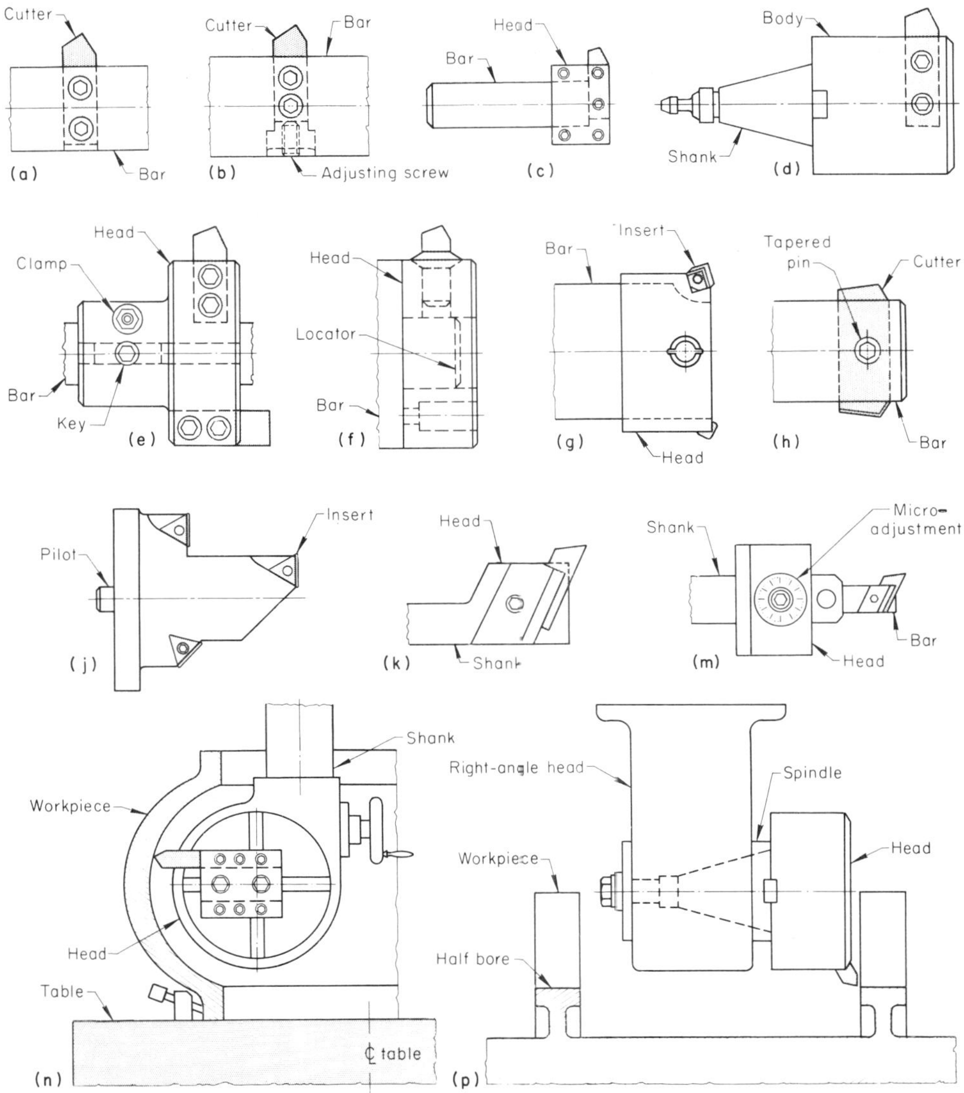


**Fig. 1** Thirteen types of boring tools. (a) Single-point cutter mechanically secured to boring bar, with no screw for adjustment. (b) Similar to (a), except for adjusting screw, which permits advancement of cutter to compensate for wear. (c) Universal head, or box tool. (d) Stub boring bar (e) Detachable head. (f) Detachable head suited to mounting on end of stub or line boring bar. (g) Blade-type tool with two identical cutting inserts 180° apart. (h) Blade-type tool in which cutter is inserted through the body to provide two cutting edges. (j) Multiple-diameter head with indexable inserts. (k) Offset head. (m) Offset head with microadjustment. (n) Head for generating a radius. (p) Head for boring at right angle to axis of boring bar

The floor-type machine uses a stationary, T-slotted floor plate, instead of a table, for supporting workpieces. This type of machine is used for machining workpieces that are too large or heavy for reciprocating tables. Horizontal feeds perpendicular to the spindle axis are obtained by movement of the column along the baseways, rather than by movement of the workpiece.

**Drill presses,** especially of the radial type, are sometimes used for boring, usually when only a few parts require boring. The difficulty of holding tolerances because of lack of rigidity is the main disadvantage in the use of drill presses for boring. This can be partly overcome by clamping workpieces in fixtures that allow the boring-bar extension to enter a bushing in the fixture on the side of the workpiece opposite the spindle.

**Precision boring machines** are required for boring to tolerances of thousandths of a millimeter. These machines are available in either vertical or horizontal models with one or more working spindles.

Precision boring machines are of two basic types:

- Those in which the spindle is mounted on a fixed bridge and the workpiece is mounted on a reciprocating table
- Those in which the spindle is mounted on a reciprocating table and the work holding fixture is mounted on a fixed bridge

With either type, the workpiece may be mounted on the spindle and rotated, while the tool is mounted on a nonrotating table or fixed bridge. Precision boring machines are frequently used in tool making.

**Special machines** include features on the conventional machines discussed above, or are modifications of these machines. Usually, they are "single-purpose" machines, expressly designed either for large-quantity, continuous production of identical parts or for boring work that is too large or unwieldy to be handled in "standard" equipment. An example of the latter is a specially constructed boring mill capable of accommodating workpieces up to 15 m (50½ ft) in diameter and boring them to a tolerance of ±0.05 mm (±0.002 in.). This machine has two toolheads mounted on a 730 kN (82 tonf) crossrail, which is supported by two 6.4 m (21 ft) high columns that span a 10 m (35 ft) diam rotary worktable. Table speeds range from 0.005 to 0.5 rev/min.

## Tools

The simplest form of boring tool, shown in Fig. 1(a), consists of a single-point cutter mechanically secured directly to a straight length of the boring bar. The bar can be rotated and fed into the workpiece, or the workpiece can be rotated and moved while the bar remains stationary. However, adjustment is difficult; when the tool becomes worn, it must be removed for sharpening and must be reset when returned. Resetting requires a fair degree of skill and is sometimes tedious. With the boring tool shown in Fig. 1(b), the cutter can be advanced to compensate for wear by loosening the securing screws and turning the adjusting screw forward.

Increased versatility of operation is provided by a universal boring head (sometimes called a box tool), shown in Fig. 1(c). A head of this type, which is attached to the end of the bar, is designed to hold left-hand or right-hand cutters of a variety of configurations. It can also hold more than one cutter for multiple-diameter work.

Figure 1(d) shows a type of head known as a stub boring bar. This head has a fixed cutter and can be used for only a small range of bore sizes. However, it is simple and widely used.

Detachable heads of the type illustrated in Fig. 1(e) are widely used because of their flexibility. These heads can be located at any desired point along the bar and can hold two or more cutters.

The type of detachable head shown in Fig. 1(f) is mounted at the end of a boring bar. These heads can be designed to hold more than one cutter, and their interchangeability permits the boring bar to be used for a range of bore sizes.

The assembly illustrated in Fig. 1(g) is a blade-type tool using two identical cutting inserts 180° apart. The inserts can be either brazed or secured mechanically. The main advantage of this type of tool is that it equalizes the forces imposed on the bar during operation. It is thus possible to maintain closer tolerances with bars having maximum unsupported length than when using a boring tool that has only one cutting edge. Its disadvantage is that the blades cannot be adjusted to compensate for wear and therefore must be removed for grinding and then be reset. This disadvantage is lessened by the use of mechanically held inserts that can be indexed to maintain size.

Figure 1(h) illustrates another style of blade-type tool. The cutter is inserted through the body, thus providing two cutting edges. This tool is sometimes known as a reaming-type boring tool and may be used without support or with a pilot. The two cutting edges often enable a substantial increase in feed rate over that which is possible when only one cutting edge is used. Advantages and disadvantages of this tool are similar to those described for the tool illustrated in Fig. 1(g).

Numerous modifications of the tool illustrated in Fig. 1(j) are used. This multiple-diameter head may be used with two cutting edges for the same diameter or with two or more cutting edges performing several operations simultaneously or consecutively. Mechanically secured disposable carbide inserts are usually used. These inserts can be indexed, thus using all cutting edges before they are replaced. This is a single-purpose tool and is best suited to high-production boring.

An offset boring head, particularly well suited to the boring of small holes, is illustrated in Fig. 1(k). This type has no means for fine adjustment. An offset boring head with a microadjustment is shown in Fig. 1(m). Adjustment is quickly performed by unlocking the dial, turning it to attain the required tool setting, and then relocking it. This head is useful in low-production or toolroom boring in which frequent changes of diameter are required.

A head for generating a radius is shown in Fig. 1(n). This type of head is used to generate an internal or external torus on a workpiece by means of a lathe. The head illustrated in Fig. 1(n) is hand fed and used for low-production boring. For high production, a power feed can be applied.

Figure 1(p) illustrates a right-angle head, which is often used with stub boring bars on fine bores that normally require piloted boring bars. Using a right-angle head helps to minimize bearing and vibration problems often encountered with long boring bars. Right-angle heads are especially suited to machining half bores.

## Tool Design

Cutting angles for boring are more critical than for operations such as turning or planing, for at least two reasons:

- Boring is more frequently a final machining operation
- Chip flow is of greater concern in boring

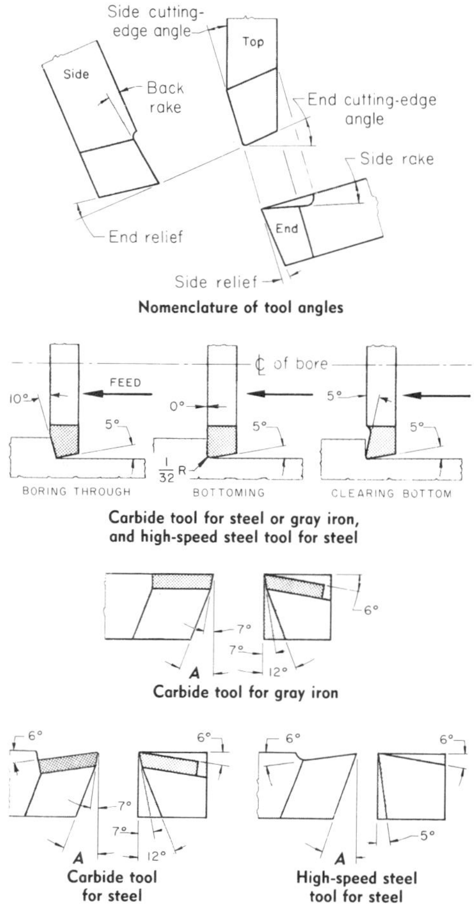


**Fig. 2** Nomenclature and typical configurations of boring tools. End relief angle *A* in lower sketches varies inversely with bore diameter.

Nomenclature of the angles for boring tools is shown in Fig. 2.

The type and the size of the hole being bored are major factors influencing requirements of tool angles. As noted in the center portion of Fig. 2, the side cutting edge angle must be varied for through-boring, bottoming, or clearing bottom. The end relief angle denoted as angle *A* in the lower sketches of Fig. 2 must be sufficient to clear the bore surface. Therefore, this angle must be increased as the bore size is decreased. Excessive end relief is not recommended, however, because it weakens the cutting edge.

Back and side rake angles, in addition to providing cutting action, must act in combination to direct chip flow properly. Chips must flow away from the cut surface toward the center of the bore. If chips are directed toward the side of the bore, they may wrap themselves around the tool in heavy cuts, or mar the finish in a final cut. Avoidance of chip congestion is of particular importance in boring lead-base bearings. If a lead alloy chip becomes entrapped, it is likely to fuse and promote further congestion, damaging the surface of the workpiece or the tool, or both.

Typical values for boring-tool angles are shown in the lower portion of Fig. 2. These tool angles generally give free-cutting action with minimum resistance to the cutter, thus minimizing the likelihood of chatter. Tool angles may be varied considerably, however.

As boring speeds are increased, and as closer dimensional control is required, it becomes increasingly important that cutting angles be duplicated in regrinding. Random grinding of boring tools can cause variation in surface finish, subnormal tool life, and excessive variation in dimensions.

## Tool Materials

High-speed steel is generally more suitable than carbide for slow-speed boring of large workpieces. Carbide cutting edges are used almost exclusively for precision boring, in which speeds are high, depth of cut is low, and maximum rigidity is maintained in the setup. Carbide is less suitable for slow speeds and heavier cuts, especially if rigidity cannot be maintained.

Ceramic tools are being increasingly applied for precision boring applications. Advantages of ceramic inserts include higher cutting speeds, reduced tool wear, better size control, production of smoother surface finishes, and the ability to bore hard materials. Only the newest precision-boring machines are capable of operating at the high speeds for which ceramic inserts are best suited.

Ceramic inserts have proved to be ideal for the accurate boring of cast iron parts. These tools have also been found to be good for precision boring steel parts having a hardness of 60 to 62 HRC, sometimes eliminating the need for subsequent grinding. Ceramic inserts are generally not recommended for heavy, interrupted cuts or for boring refractory metals and certain aluminum alloys because they develop built-up edges.

## Pilots and Supports

Figure 3 illustrates a number of methods that are used for piloting and supporting tools in applications in which long boring bars must be used or close tolerances must be met.

In many setups for which it is impractical to use bearing supports, the bar can be supported by the workpiece. Figure 3(a) shows a piloted bar, often known as a pack head that uses the workpiece for support. Four or more inserts (usually of nylon or bronze) form the bearing surfaces for the pilot. Another type of head that is piloted by the workpiece is illustrated in Fig. 3(b). This trepanning head is most commonly used for boring large-diameter holes from the solid. Wear pads (usually of carbide) located on the diameter 180° from the cutting edge are

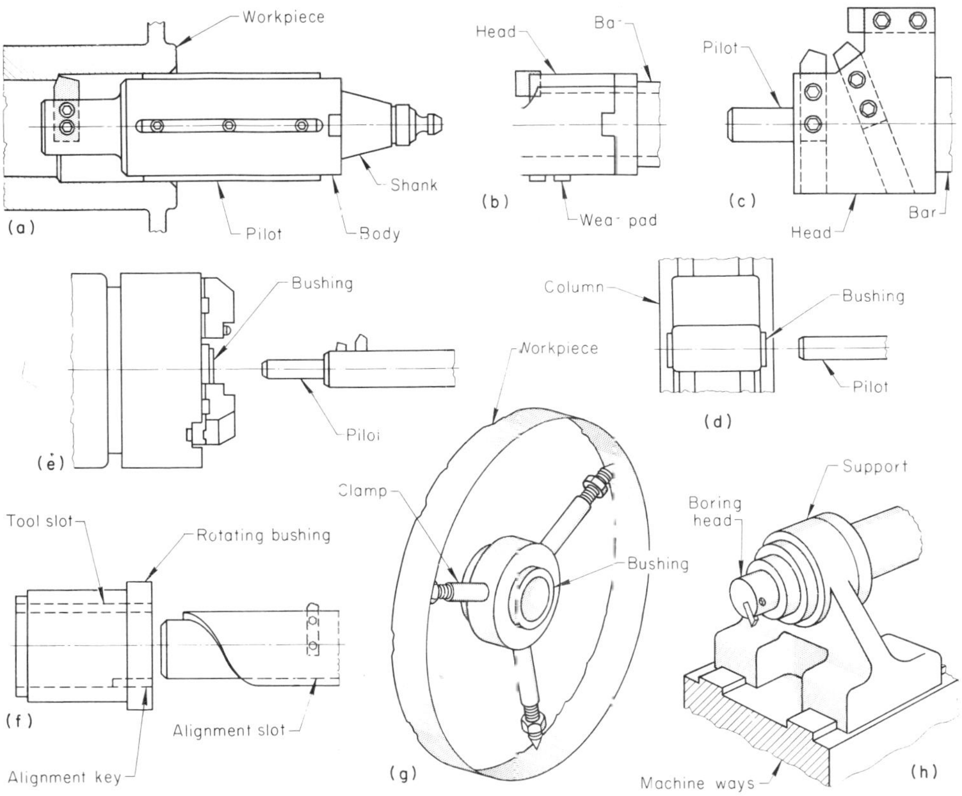


**Fig. 3** Methods of piloting or supporting boring tools to maintain alignment. (a) Pack head type of pilot. (b) Wear-pad support of trepanning head used for boring large diameter holes from solid stock. (c) Piloted head capable of using several cutting edges. (d) Bushing mounted on auxiliary column guides pilot on boring bar. (e) Rotating bushing mounted on front of lathe chuck to receive piloted bar. (f) Rotating bushing to align boring tool with tool slot in pilot bushing. (g) Three-leg, adjustable spider support. (h) Boring-bar support mounted on machine ways in line with centerline of spindle

used with this type of head to aid in maintaining alignment.

Many modifications of the tool assembly illustrated in Fig. 3(c) are used in production boring applications in which it is practical to precede the cutting edges with a pilot. This type of pilot is usually hardened and ground (and may even be chromium plated) to resist wear.

When it is not feasible to use a portion of the workpiece for the pilot to enter, a guide bushing may be established on an auxiliary column. This principle is illustrated in Fig. 3(d). The type of support shown must be adjusted so that the center of the bushing is accurately aligned with the centerline of the spindle.

Another means of pilot support, for use when machining relatively short bores, is illustrated in Fig. 3(e). As seen in the illustration, the pilot must enter the rotating bushing before the cutting edges begin boring the workpiece, which is held on the front of the rotating chuck.

A rotating bushing used with a line boring bar is illustrated in Fig. 3(f). The spiral on the nose of the bar contacts the alignment key upon entry of the bar and rotates the tool slot into alignment with the cutting tools, thus providing support.

In some applications, workpiece configuration allows the use of a spider support. A three-leg, adjustable support of this type is illustrated in Fig. 3(g). This type of support, however, is difficult to align and therefore is ordinarily used only for extremely low-production or toolroom operations.

Supports mounted on machine ways (Fig. 3h) are in common use, particularly when the boring tools are fed into the rotating workpiece, as on a lathe or similar machine tool.

**Maximum unsupported length** of a boring bar depends to a great extent on tolerance requirements. Other governing factors are hardness of the metal being bored, speed, depth of cut, and cutting-tool material.

Steel boring bars have a modulus of elasticity of approximately 200 GPa (30 million psi), and although most steel boring bars are heat treated for additional strength or wear resistance, the modulus is unchanged by heat treatment. For maintaining tolerances on the order of ±0.025 mm (±0.001 in.), an unsupported length of four times diameter approaches the practical limit for steel boring bars, and with this much unsupported bar length, even minor increases in speed or feed are likely to cause chatter. Less stringent tolerances of ±0.1 mm (±0.005 in.) have been maintained with steel bars having an unsupported length several times diameter.

When boring-bar length must be greater than four times diameter, and additional support is not feasible, boring bars made from carbide are sometimes used; the modulus of elasticity of carbide is about three times greater than that of steel, and hence rigidity can be maintained for a greater unsupported length. The use of carbide boring bars is extremely limited, however, because only small sizes are available, and pound for pound, carbide costs about 100 times as much as alloy steel.

## Speed and Feed

Speed and feed affect power requirements, tool life, rate of metal removal, machining cost, and tolerances and surface finish obtainable in boring operations. In most applications, speed and feed are selected on the basis of minimum overall cost, but sometimes a compromise must be made because of the hardness of the work material, special product requirements, or production quantity and schedule. For equal depth of cut and rate of feed, boring speeds are substantially lower than turning speeds.

Table 1 presents nominal speeds and feeds for the rough boring and finish boring of carbon and low-alloy steels (grouped in Table 2) at various hardness levels, using high-speed steel or carbide tools. These rates apply primarily to the boring of holes 50 to 100 mm (2 to 4 in.) in diameter and are useful as a starting point for the selection of efficient and economical machining conditions. As is evident from many of the examples presented later in this article, speeds and feeds used in commercial practice often differ widely from the nominal values given in Table 1. Some of the reasons for this difference are discussed below.

**Bore Size and Diameter.** If rugged equipment with adequate power is available, large diameters can be bored at much higher surface speeds and feed rates than those listed in Table 1. Metal can be removed at extremely rapid rates in heavy roughing cuts, and the resulting rough finish and dimensional variation can be corrected by light finishing cuts at slower speeds and higher feeds than are normally recommended for finish boring.

These techniques are illustrated in Example 3, which describes rough and finish boring of a 160 × $10^3$ kg (180 ton) rolling-mill housing made of 1030 steel. Holes 710, 810, and 890 mm (28, 32, and 35 in.) in diameter were bored with high-speed steel tools. The high surface speed for roughing this large workpiece (98 m/min, or 320 sfm) was obtained with a relatively slow rotational speed (35 rev/min). The high speed and feed removed metal rapidly in a rough-

**Table 1 Nominal speeds and feeds for boring of carbon and low-alloy steels with high-speed steel and carbide tools**

The rates shown here are most applicable to the boring of holes 50 to 100 mm (2 to 4 in.) in diameter, and are useful as starting points for other applications with comparable steels.

| | | Rough boring (depth of cut, 2.5 mm, or 0.100 in.) | | | | Finish boring (depth of cut, 0.25 mm, or 0.010 in.) | | | |
|---|---|---|---|---|---|---|---|---|---|
| | | Speed, m/min (sfm) | | Feed, mm/rev (in./rev) | | Speed, m/min (sfm) | | Feed, mm/rev (in./rev) | |
| Typical steel(a) | Hardness, HB | HSS(b) | Carbide(c) | HSS(b) | Carbide(c) | HSS(b) | Carbide(c) | HSS(b) | Carbide(c) |
| **Carbon and low-alloy steels (except free-cutting grades)** | | | | | | | | | |
| 1020, 1045, 4140, 7140 and 8620, at hardness ranges listed at right | 85–125 | 37 (120) | 126 (415) | 0.25 (0.010) | 0.38 (0.015) | 41 (135) | 140 (460) | 0.13 (0.005) | 0.15 (0.006) |
| | 125–175 | 30 (100) | 104 (340) | 0.25 (0.010) | 0.38 (0.015) | 34 (110) | 116 (380) | 0.13 (0.005) | 0.15 (0.006) |
| | 175–225 | 24 (80) | 93 (305) | 0.225 (0.009) | 0.33 (0.013) | 27 (90) | 102 (335) | 0.13 (0.005) | 0.15 (0.006) |
| | 225–275 | 20 (65) | 85 (280) | 0.175 (0.007) | 0.23 (0.009) | 23 (75) | 94 (310) | 0.10 (0.004) | 0.13 (0.005) |
| | 275–325 | 18 (60) | 75 (245) | 0.175 (0.007) | 0.23 (0.009) | 20 (65) | 82 (270) | 0.10 (0.004) | 0.13 (0.005) |
| | 325–375 | 15 (50) | 62 (205) | 0.175 (0.007) | 0.23 (0.009) | 17 (55) | 69 (225) | 0.075 (0.003) | 0.10 (0.004) |
| | 375–425 | 11 (35) | 49 (160) | 0.15 (0.006) | 0.175 (0.007) | 12 (40) | 53 (175) | 0.075 (0.003) | 0.10 (0.004) |
| | HRC, 50–52 | 6 (20) | 27 (90) | 0.15 (0.006) | 0.20 (0.008) | 6 (20) | 30 (100) | 0.075 (0.003) | 0.10 (0.004) |
| | HRC, 54–56 | ... | 17 (55) | ... | 0.15 (0.006) | ... | 18 (60) | ... | 0.075 (0.003) |
| **Free-cutting carbon and low-alloy steels** | | | | | | | | | |
| 1112 and 1117 ........ | 100–150 | 41 (135) | 154 (505) | 0.25 (0.010) | 0.38 (0.015) | 46 (150) | 170 (560) | 0.13 (0.005) | 0.15 (0.006) |
| | 150–200 | 44 (145) | 170 (560) | 0.25 (0.010) | 0.38 (0.015) | 50 (165) | 189 (620) | 0.13 (0.005) | 0.15 (0.006) |
| 1137 and 12L14....... | 100–150 | 41 (135) | 155 (510) | 0.25 (0.010) | 0.38 (0.015) | 46 (150) | 172 (565) | 0.13 (0.005) | 0.15 (0.006) |
| | 150–200 | 35 (115) | 131 (430) | 0.25 (0.010) | 0.38 (0.015) | 38 (125) | 146 (480) | 0.13 (0.005) | 0.15 (0.006) |
| | 200–250 | 29 (95) | 99 (325) | 0.20 (0.008) | 0.30 (0.012) | 32 (105) | 110 (360) | 0.13 (0.005) | 0.15 (0.006) |
| | 275–325 | 24 (80) | 93 (305) | 0.20 (0.008) | 0.25 (0.010) | 27 (90) | 104 (340) | 0.10 (0.004) | 0.13 (0.005) |
| | 325–375 | 15 (50) | 62 (205) | 0.175 (0.007) | 0.23 (0.009) | 17 (55) | 69 (225) | 0.075 (0.003) | 0.10 (0.004) |
| | 375–425 | 11 (35) | 47 (155) | 0.15 (0.006) | 0.20 (0.008) | 12 (40) | 53 (175) | 0.075 (0.003) | 0.10 (0.004) |
| 4140+S and 41L40.... | 150–200 | 32 (105) | 120 (395) | 0.25 (0.010) | 0.38 (0.015) | 35 (115) | 134 (440) | 0.13 (0.005) | 0.15 (0.006) |
| | 275–325 | 18 (60) | 82 (270) | 0.20 (0.008) | 0.25 (0.010) | 21 (70) | 91 (300) | 0.10 (0.004) | 0.13 (0.005) |
| | 375–425 | 11 (35) | 47 (155) | 0.15 (0.006) | 0.20 (0.008) | 12 (40) | 53 (175) | 0.075 (0.003) | 0.10 (0.004) |
| | HRC, 50–52 | 6 (20) | 27 (90) | 0.15 (0.006) | 0.20 (0.008) | 6 (20) | 30 (100) | 0.075 (0.003) | 0.10 (0.004) |
| | HRC, 54–56 | ... | 17 (55) | ... | 0.15 (0.006) | ... | 18 (60) | ... | 0.075 (0.003) |

(a) Each steel listed is a frequently used grade in a group of similar steels. Table 2 lists the steels in the various groups. (b) High-speed steels M2 and T5, except M2, T5, and T15 for boring steels at hardnesses of 225 to 375 HB, and except T15, M41, M42, M43, and M44 for hardnesses above 375 HB. (c) Carbide grade C-7, except C-8 for hardnesses above 425 HB. Source: Data are adapted from tables compiled by Metcut Research Associates, Inc.

ing cut of 9.5 to 13 mm (3/8 to 1/2 in.). A single finishing pass at low feed and high speed removed 0.175 mm (0.007 in.) for a tolerance of +0.075, −0.00 mm (+0.003, −0.000 in.) on the diameter.

When a workpiece is rotated, its size sometimes limits speeds to below normal values, as in boring an 810 mm (32 in.) diameter on the 6800 kg (15 000 lb) 1045 steel forging of Example 2 with carbide tools. Speeds were limited by the size of the workpiece to less than half the nominal values, but because of the large bore diameter, feed rates could be substantially higher than the nominal values.

**Chip control** is a serious problem with boring cuts, and chips must be broken to have success with this process. In boring tubes of small diameter, chip control is critical and often governs the selection of feed and speed.

**Tolerances and Bore Length.** Close-tolerance bores of high length-to-diameter ratio require slower speed than normal, even though the tool is well supported. When unsupported boring bars of high length-to-diameter ratio are used in close-tolerance operations, speed and feed are limited by the degree of rigidity of the bar.

**Other conditions** may also demand the use of feeds and speeds substantially different from those shown in Table 1. Tool design can have major effects on feed and speed. In an example discussed later in the article, for instance, a change from a single-point to a blade-type cutter for finish boring allowed the feed rate to be increased by 50%.

Surface finish can often be improved by using higher speeds and lower feed rates than normal. With large parts, it may be possible to improve finish by decreasing feed rate without changing speed, but at a sacrifice of production rate.

Production rates in close-tolerance operations can sometimes be increased by using different types of machines for roughing and for finishing to permit optimum speed and feed for each operation.

## Workpiece Size

In selecting the equipment and processing procedure for a specific boring operation, size of workpiece is usually the first factor to consider, because all machines are limited with respect to swing, speed, height, and other capabilities. A small part can be bored on a greater variety of machines than can a large part. However, it would be impractical to process a small part on a machine such as a large boring mill, because it would rotate too slowly.

For extremely large workpieces, large machines would be needed if the part, rather than the tools, had to be rotated. For these parts, it is more practical to line up the workpiece and rotate the tools. Part size may also dictate the choice between horizontal and vertical machines. Extremely heavy workpieces impose so great an overhang on spindle bearings in a horizontal machine that chucking or clamping is impractical.

Size may also dictate machine type for parts that, although relatively light in weight, are too long to be bored in machines having short travel, like turret lathes. For these parts, a horizontal machine may be required because of limitations in building height.

The four examples that follow describe specific applications in which size of workpiece was the major factor in choice of equipment and procedure.

**Example 1: Boring, Turning, and Facing a Pressure Vessel in One Setup.** A 75 kW (100 hp) vertical boring machine with a 6 m (20 ft) diam table and 3.7 m (12 ft) clearance under the crossrail was used for boring, turning, and facing a 3600 mm (140 in.) high pressure vessel made from steel plate and weighing $55 \times 10^3$ kg (60 tons). It was feasible to rotate this part, despite its size and weight, because it was symmetrical. The use of four tools on the crossrail, as shown in Fig. 4, made it possible to use the same setup for facing the ends and turning the flanges as for boring the 1420 mm (56 in.) inside diameter (ID).

The workpiece was set up and centered on the 6 m (20 ft) diam table and was held in place by four jaws at table level and four other jaws mounted on 1900 mm (75 in.) high supports. The higher, supported jaws

**Table 2 Groups of carbon and low-alloy steels referred to in tables of nominal speeds and feeds**

| Typical steel | Other steels in the group |
|---|---|
| **Carbon steels (low carbon)** | |
| 1020 | 1006, 1008, 1009, 1010, 1012, 1015, 1016, 1017, 1018, 1019, 1021, 1022, 1023, 1024, 1025, 1026 |
| **Carbon steels (medium and high carbon)** | |
| 1045 | 1027, 1030, 1033, 1034, 1035, 1036, 1037, 1038, 1039, 1040, 1041, 1042, 1043, 1046, 1049, 1050, 1052, 1055, 1060, 1062, 1064, 1065, 1066, 1070, 1074, 1078, 1080, 1084, 1085, 1086, 1090, 1095 |
| **Resulfurized (free-cutting) low-carbon steels** | |
| 1112 | 1111, 1113, 1119, 1212, 1213, 1215+Te |
| 1117 | 1108, 1109, 1115, 1118, 1120, 1126, 1144, 1211 |
| **Resulfurized (free-cutting) medium-carbon steels** | |
| 1137 | 1132, 1138, 1139, 1140, 1141, 1145, 1146, 1151 |
| **Leaded (free-cutting) low-carbon steels** | |
| 12L14 | 10L18, 10L20, 12L13 |
| **Low-alloy steels (medium and high carbon)** | |
| 4140 | 1330, 1332, 1335, 1340, 1345, 2330, 2335, 2340, 2345, 3130, 3135, 3140, 3141, 3145, 3150, 4030, 4032, 4037, 4042, 4047, 4063, 4130, 4135, 4137, 4142, 4145, 4147, 4150, 4337, 4340, 4640, 50B40, 50B44, 5046, 50B46, 50B50, 50B60, 5075, 5080, 5130, 5132, 5135, 5140, 5145, 5147, 5150, 5155, 5160, 51B60, 50100, 51100, 52100, 6145, 6150, 5180, 6240, 6250, 6260, 6270, 6290, 6342, 6382, 6440, 6475, 81B45, 8630, 8637, 8640, 8642, 8645, 86B45, 8650, 8655, 8660, 8740, 8742, 9255, 9260, 9262, 94B30, 94B40, 9445, 9840, 9845, 9850 |
| **Resulfurized (free-cutting) low-alloy steels** | |
| 4140+S | Same steels as in group following 4140, but with sulfur added |
| **Leaded (free-cutting) low-alloy steels** | |
| 41L40 | 41L30, 41L47, 41L50 43L47, 51L32, 86L20, 86L40, 52L100 |
| **Low-alloy nitriding steels** | |
| 7140 | All grades |
| **Low-alloy carburizing steels** | |
| 8620 | 1320, 2317, 2512, 2515, 2517, 3115, 3120, 3125, 3310, 3316, 4012, 4017, 4023, 4024, 4027, 4028, 4118, 4125, 4128, 4317, 4320, 4608, 4615, 4617, 4620, 4621, 4720, 4815, 4817, 4820, 5015, 5020, 5024, 5120, 6118, 6120, 6317, 6325, 6415, 8115, 8615, 8617, 8622, 8625, 8627, 8720, 8822, 9310, 9315, 94B15, 94B17 |

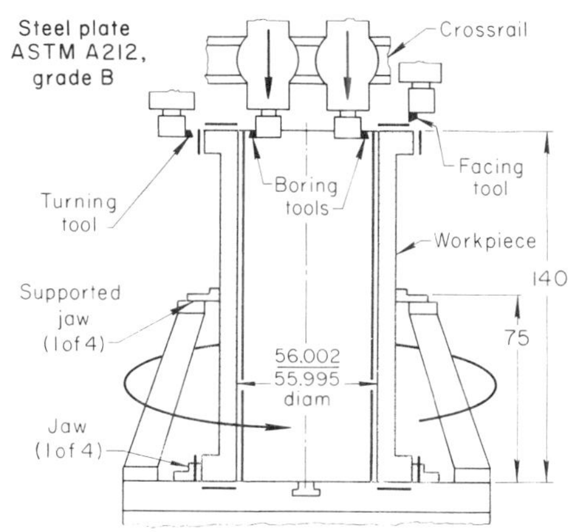


**Operating conditions for boring**

| | |
|---|---|
| Workpiece hardness, HB | 165–170 |
| Speed, roughing and finishing, at 20 rev/min, m/min (sfm) | 89.3 (293) |
| Feed, mm/rev (in./rev) | 0.5 (0.020) |
| Depth of roughing cut, each head, mm (in.) | 1.6 (0.0625) |
| Total depth of cut, mm (in.) | 3.2 (0.125) |
| Tool material | Carbide (brazed) |
| Cutting fluid | Soluble oil |
| Setup time, h/piece | 9 |
| Total time, h/piece | 62 |

**Fig. 4** Boring, turning, and facing a 55 × $10^3$ kg (60 ton) steel pressure vessel. Dimensions in figure given in inches

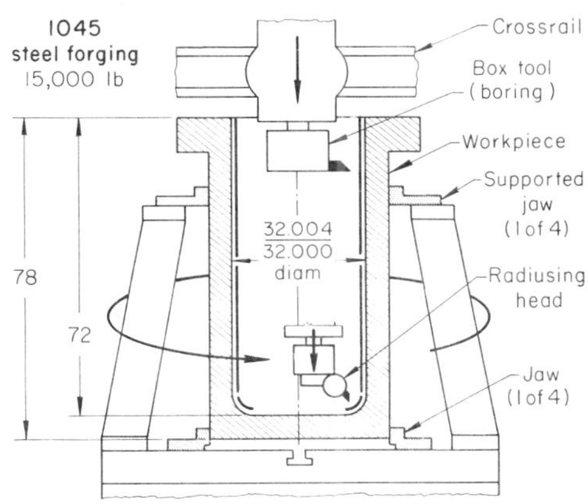


**Operating conditions for boring**

| | |
|---|---|
| Speed, roughing and semifinishing, at 18 rev/min, m/min (sfm) | 46 (150) |
| Feed, roughing, mm/rev (in./rev) | 0.64 (0.025) |
| Feed, semifinishing, mm/rev (in./rev) | 0.5 (0.020) |
| Depth of each roughing cut (four cuts made), mm (in.) | 4.75 (0.187) |
| Depth of semifinishing cut, mm (in.) | 1.5 (0.060) |
| Tool material | Carbide (brazed) |
| Cutting fluid | Soluble oil |
| Setup time, h/piece | 3 |
| Total time, h/piece | 25 |

**Fig. 5** Boring and radiusing a 6800 kg (7½ ton) steel forging. Dimensions in figure given in inches

were locked against the part, and indicators at these jaws ensured that the part was not moved or twisted in machining.

For boring, the workpiece was rotated around the two heads, which were fed downward into it. When 1900 mm (75 in.) of the 3600 mm (140 in.) length had been machined, the workpiece was turned end for end, and the procedure was repeated. Additional processing details for the boring operation are listed in Fig. 4.

**Example 2: Boring and Radiusing a Steel Forging.** A 37 kW (50 hp) vertical boring mill with a 3 m (10 ft) diam table and 2.5 m (8 ft) of clearance under the crossrail was used for boring a 6800 kg (15 000 lb) forging of 1045 steel. The size, weight, and symmetry of this part made it well suited to the setup used (see Fig. 5), in which the workpiece was rotated and tools were fed downward from the crossrail.

The workpiece was secured to the table by four jaws at table level, and additional bracing was provided by four other jaws mounted on supporting columns about three-fourths as high as the workpiece (see Fig. 4 for a generally similar setup).

A box tool was used for rough and semifinish boring, after which the 810 mm (32 in.) inside diameter was polished with a sanding belt to a maximum roughness of 0.80 μm (32 μin.). A radius at the bottom of the cylindrical cavity was formed with a separate head (Fig. 5) in a subsequent operation. The swivel radiusing tool, made of high-speed steel, was operated with an extension handle, which permitted rotation from above the workpiece and was used with a rotating head to eliminate tool grooves. Processing conditions and time per piece for the boring operation are listed in the tabulation accompanying Fig. 5.

**Example 3: Three Diameters in a Rolling Mill Housing.** A horizontal, floor-type mill was used for boring rolling-mill housings made of 1030 steel that weighed up to 160 × $10^3$ kg (180 tons) each. Figure 6 shows the setup used for boring three different diameters (nut, clearance, and wormwheel bores) in a housing 10 m (33 ft) long, 5 m (17 ft) wide overall, and up to 1000 mm (40 in.) in section thickness. With this setup, a 180 mm (7 in.) spindle drove a 200 mm (8 in.) boring bar, which was supported by an outboard bearing suspended from the

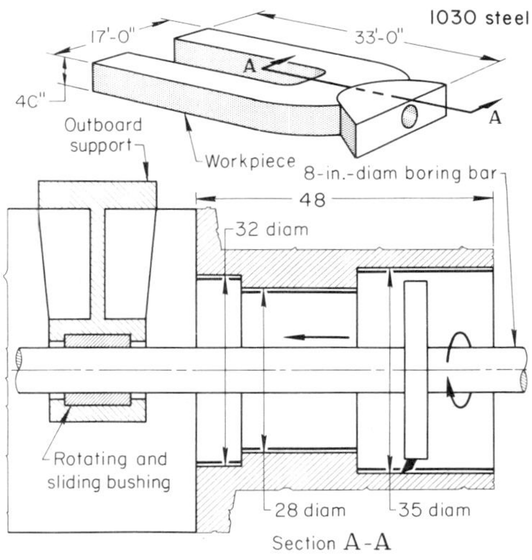


| | |
|---|---|
| Speed, roughing, at 35 rev/min, m/min (sfm) | 98 (320) max |
| Speed, finishing, at 5 rev/min, m/min (sfm) | 14 (46) max |
| Feed, roughing, mm/rev (in./rev) | 6.4 (0.25) |
| Feed, finishing, mm/rev (in./rev) | 1.0 (0.04) |
| Depth of cut, roughing, mm (in.) | 9.5–13 (3/8–1/2) |
| Depth of cut, finishing, mm (in.) | 0.175 (0.007) |
| Tool material | High-speed steel |
| Setup time, h/piece | 8 |
| Total time, h/piece | 48 |

**Fig. 6** Boring a 160 × $10^3$ kg (180 ton) rolling-mill housing. Dimensions in figure given in inches

housing face through a window opening in the housing; a bushing in the outboard bearing rotated with the bar, but slid on the bar for feed. One roughing pass and one finishing pass were made to complete each bore; tolerances of +0.075, −0.00 mm (+0.003, −0.000 in.) were maintained. Additional operating conditions are tabulated with Fig. 6.

**Example 4: Boring a Hole 18½ Diameters Long in a Large Piston.** A long-bed horizontal lathe was used to bore the 4330 steel piston shown in Fig. 7. This part posed a problem in boring because of the extreme length (2071 mm, or 81.535 in.) of the bore to be machined in relation to its diameter (112.6 mm, or 4.434 in.). Concentricity had to be held within 0.5 mm (0.020 in.) total indicator reading (TIR) and diameter had to be within 0.13 mm (0.005 in.), because the bore would subsequently be honed.

To achieve the critical lineup required for a bore of this length, the soft chuck jaws had to be bored to hold concentricity, and the steady rest and guide bushing had to be in line to prevent the bore from leading off. Boring tools were held on a true centerline by the use of a pack head containing four nylon inserts (Fig. 7). This head was of greater than usual length to provide more packing area and thus hold pack wear to a minimum. The packs were ground 0.075 mm (0.003 in.) below tool size in order to allow for thermal expansion with increasing temperature during operation.

The bore was started by holding the pack head in the guide bushing, and the packs entered the bore before leaving the bushing. Soluble oil was pumped through the boring-bar stem to flush out chips and keep the nylon packs cool.

The tool block held two carbide cutters, 180° apart, that were ground in assembly for the diameter to be bored. Each cutter had a 30° lead angle and a 0.6 mm (0.025 in.) deep, 1.5 mm (0.060 in.) wide chip breaker with a 1.5 mm (0.06 in.) base radius and a nose radius of 0.0 to 0.13 mm (0.000 to 0.005 in.). Additional processing details are listed in the table that accompanies Fig. 7.

## Workpiece Configuration

The configuration of a workpiece, irrespective of its size, may be a major factor in determining the most practical boring method. Configuration may affect the rigidity with which a part can be held while being machined and thus may create problems even in otherwise simple operations. Other conditions related to configuration that may pose problems in boring are:

- The maintenance of specific relationships between two or more holes in a single part
- Boring bottle and other complex configurations
- Boring difficult-to-reach areas

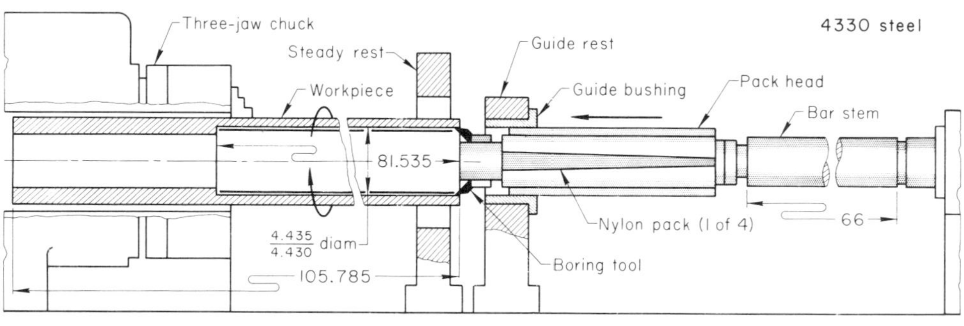


| | |
|---|---|
| Workpiece hardness, HRC | 46–49 |
| Speed, at 100 rev/min, m/min (sfm) | 35 (116) |
| Feed, mm/rev (in./rev) | 3.2 (0.125) |
| Tool material | Carbide |
| Cutting fluid | Soluble oil |
| Setup time, h | 3.5 |
| Handling time, min/piece(a) | 10 |
| Total time, h/piece | 1.4 |

(a) Loading and unloading, positioning of guide bushing, and tool approach to cutting position

**Fig. 7** Boring a long hole to a concentricity of 0.5 mm (0.020 in.) TIR (total indicator reading). Dimensions in figure given in inches

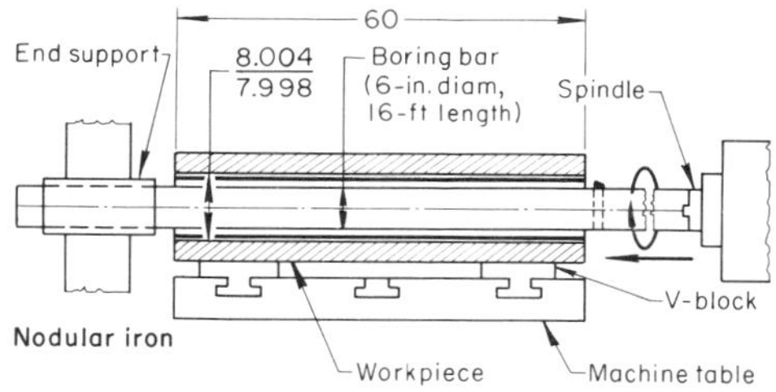


| | |
|---|---|
| Workpiece hardness, HB | 280 |
| Speed, at 70 rev/min, m/min (sfm) | 45 (147) |
| Feed, roughing and semifinishing, mm/rev (in./rev) | 1.0 (0.040) |
| Feed, finishing, mm/rev (in./rev) | 0.41 (0.016) |
| Depth of roughing cut, mm (in.) | 4.75 (0.187) |
| Depth of semifinishing cut, mm (in.) | 0.79 (0.031) |
| Depth of finishing cut, mm (in.) | 0.13 (0.005) |
| Tool material | |
| Roughing and semifinishing | Carbide |
| Finishing | High-speed steel |
| Cutting fluid | Soluble oil |
| Setup time, h/piece | 3.5 |
| Total time, h/piece | 8 |

**Fig. 8** Straight-through boring of a ductile (nodular) iron cylinder. Dimensions in figure given in inches

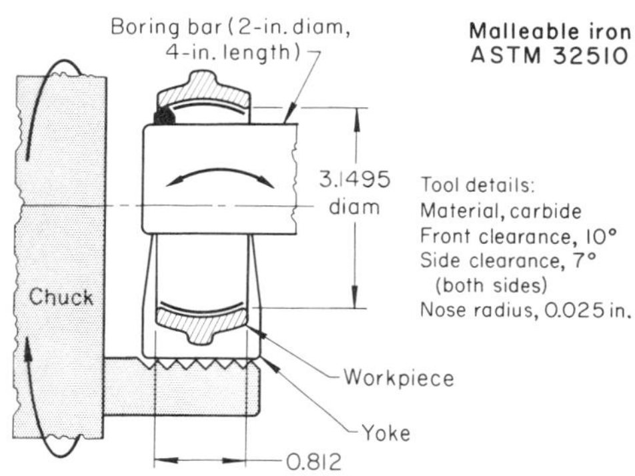


| | |
|---|---|
| Workpiece hardness, HB | 140–150 |
| Tolerance specified, mm (in.) | +0.000, −0.02 (+0.0000, −0.0008) |
| Finish specified, μm (μin.) | 3.20 (125) |

**Operating conditions**

| | |
|---|---|
| Speed, at 545 rev/min, m/min (sfm) | 137 (450) |
| Feed, mm/rev (in./rev) | 0.20 (0.008) |
| Depth of cut, mm (in.) | 2.4 (0 to 3/32) |
| Cutting fluid | Air-mist soluble oil |
| Setup time, h | 1½ |
| Downtime for changing tools, h | ½ |
| Production rate, pieces/h | 130 |
| Tool life, pieces/grind | 900 |

**Fig. 9** Close-tolerance contour boring in a spherical boring machine. Dimensions in figure given in inches

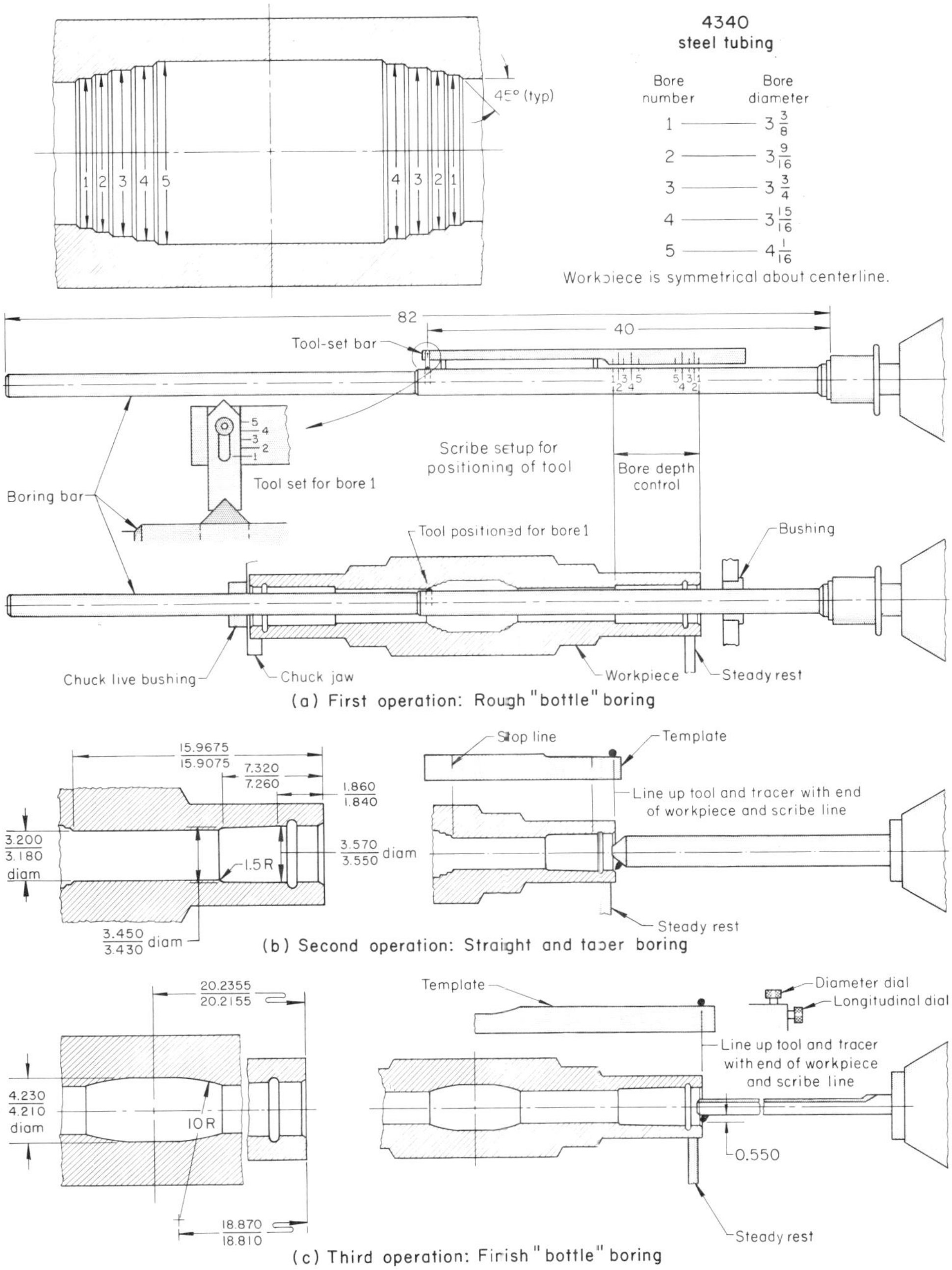


| Item | Operation: First | Second | Third |
|---|---|---|---|
| **Tool details** | | | |
| Length of boring bar, mm (in.) | 2080 (82) | 610 (24) | 775 (30½) |
| Diameter of boring bar, mm (in.) | 71 (2¹³⁄₁₆) | 70 (2¾) | 68 (2¹¹⁄₁₆) |
| Size of tool bit, mm (in.) | 19 × 19 (¾ × ¾) | 19 × 19 (¾ × ¾) | 13 × 13 (½ × ½) |
| Lead angle | 45° | 15° | 15° |
| Tool material | Carbide | Carbide | Carbide |
| Chip breaker dimensions | | | |
| Depth, mm (in.) | 0.6 (0.025) | 0.5 (0.020) | 0.6 (0.025) |
| Width, mm (in.) | 1.6 (¹⁄₁₆) | 1.6 (¹⁄₁₆) | 2.0 (⁵⁄₆₄) |
| Radius at base, mm (in.) | 0.8 (¹⁄₃₂) | 0.8 (¹⁄₃₂) | 0.8 (¹⁄₃₂) |
| Nose radius, mm (in.) | 0.13 (0.005) | 0.4 (¹⁄₆₄) | 0.25–0.38 (0.010–0.015) |
| **Operating conditions(a)** | | | |
| Speed, rev/min | 307 | 183 | 183 |
| Speed, max, m/min (sfm) | 99.4 (326) | 52.1 (171) | 61.6 (202) |
| Feed, mm/rev (in./rev) | 0.23 (0.009) | 0.23 (0.009)(b) | 0.13 (0.005) |
| Number of cuts | 5 | 2 | 4 |
| Production rate, pieces per hour | 1¼ | 1 | 1 |

(a) In all operations, a soluble-oil cutting fluid was fed through the center of the bar to the cutting tool. (b) For roughing. Feed for finishing was 0.175 mm/rev (0.007 in./rev).

**Fig. 10** "Bottle" boring alloy steel tubing in three operations on a tracer lathe. Dimensions in figure given in inches

The following examples describe equipment and techniques used in specific production applications to overcome difficulties presented by workpiece configuration.

**Example 5: Preventing Distortion in Boring a Long Cylinder.** A 30 kW (40 hp) horizontal boring mill with a 150 mm (6 in.) spindle and a 1.8 × 3 m (6 × 10 ft) table was used to bore ductile iron cylinders (Fig. 8). Despite its simple configuration, a part of this type, especially when close tolerances are required, poses problems because of the danger of distortion while it is held in place for machining. The problem was solved here by releasing the hold-down clamps after roughing. The part was then reclamped to the supporting V-blocks, but under a much lighter force, which was adequate for semifinishing and finishing because less stock was being removed. Additional processing details are given in Fig. 8.

**Example 6: Close-Tolerance Contour Boring.** A precision spherical boring machine was used for contour boring small malleable iron castings to extremely close tolerances (Fig. 9). Workpieces were clamped to the fixture with an air-operated yoke clamping base attached to the rotating air chuck. Processing details are tabulated with Fig. 9.

**Example 7: Three-Operation Boring of Bottle-Shape Cavity.** Production of a complex bottle configuration in 4340 steel tubing (Fig. 10) was a problem because of the amount of stock to be removed, the depth of the bore, and the necessity for using a small-diameter boring bar. The configuration was produced using a tracer lathe, in three separate operations and with three different tooling setups (Fig. 10).

In the first operation (Fig. 10a), the boring bar used was equipped with a built-in tool-set bar for raising and lowering the tool bit. This permitted roughing-out of stock so that profile boring could be done in the third operation. Roughing was accomplished in five cuts, as shown in the top sketch of Fig. 10(a).

In the second operation (Fig. 10b), a taper was bored to a distance of 185 mm (7.29 in.), and a straight bore was bored to 405 mm (15¹⁵⁄₁₆ in.), from each end of the workpiece. This was done by removing the cross-slide screw from the lathe and then controlling movement of the cross slide by holding the tracer stylus on the template with hydraulic pressure. Longitudinal feed was controlled by the screw of the lathe.

Profile boring of the roughed-out center section was performed in the third operation (Fig. 10c). The diameter of the boring bar was limited to 68 mm (2¹¹⁄₁₆ in.), because of the depth of the bore and the distance the bar was required to move from the centerline. To obtain maximum rigidity, the bar was bolted to the turret face. This bar was 775 mm (30½ in.) long and had a 560 mm (22 in.) overhang and a 30° angle

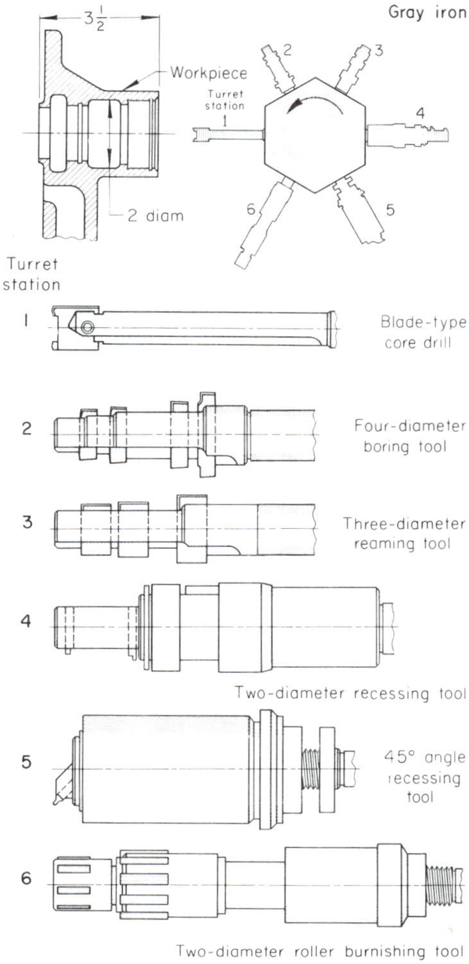


| | |
|---|---|
| Tolerance on reamed bores, mm (in.) | ±0.025 (±0.001) |
| Tolerance on burnished bores, mm (in.) | ±0.013 (±0.0005) |
| **Operating conditions** | |
| Speed | |
| Stations 1, 2, and 3, at 389 rev/min, m/min (sfm) | 62 (205) |
| Stations 4, 5, and 6, at 200 rev/min, m/min (sfm) | 32 (105) |
| Feed | |
| Stations 1 and 2, mm/rev (in./rev) | 0.43 (0.017) |
| Station 3, mm/rev (in./rev) | 0.284 (0.0112) |
| Stations 4 and 5, mm/rev (in./rev) | 0.155 (0.0061) |
| Station 6, mm/rev (in./rev) | 0.866 (0.0341) |
| Depth of cut, mm (in.) | 0.38–3.2 (0.015–0.125) |
| Tool material | Carbide |
| Cutting fluid | None |
| Setup time, h | 3 |
| Downtime for changing tools, min | 10 |
| Production rate, pieces/h | 8 |
| Tool life, pieces/grind | 75 |

**Fig. 11** Use of multiple-diameter tools on a horizontal turret lathe for machining a complex bore in a cast water pump housing. Dimensions in figure given in inches

tool slot. Half of the center section was bored from each end; mismatch had to be kept within the amount that could be removed by a subsequent polishing operation. The sequence of steps in the third operation was:

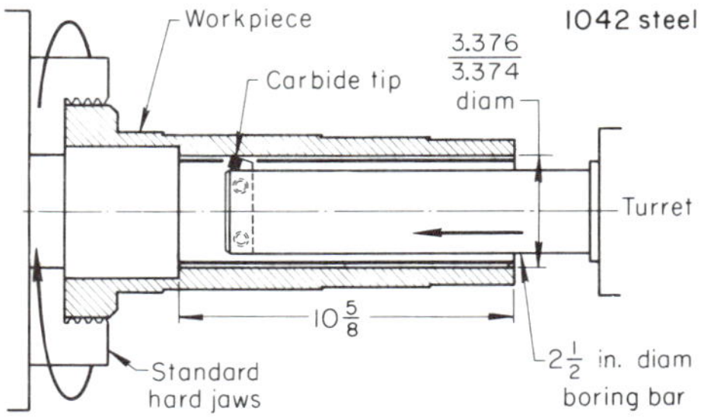


| | |
|---|---|
| **Rough boring (two cuts)** | |
| Speed, at 348 rev/min, m/min (sfm) | 76 and 84.7 (250 and 278) |
| Feed, mm/rev (in./rev) | 0.315 (0.0124) |
| Depth of first cut, mm (in.) | 3.2 (0.125) |
| Depth of second cut, mm (in.) | 3.8 (0.150) |
| **Semifinish boring** | |
| Speed, at 456 rev/min, m/min (sfm) | 120 (400) |
| Feed, mm/rev (in./rev) | 0.23 (0.009) |
| Depth of cut, mm (in.) | 3.8 (0.150) |
| **Finish boring** | |
| Speed, at 456 rev/min, m/min (sfm) | 123 (403) |
| Feed, mm/rev (in./rev) | 0.23 (0.009) |
| Depth of cut | 0.32 (0.0125) |
| **All operations** | |
| Workpiece hardness, HB | 187 |
| Cutting fluid | Soluble oil:water (1:25) |
| Setup time, h | 2 |
| Time for grinding and resetting tools, h | 0.1 |
| Production rate, pieces/h | 1 |
| Tool life, pieces/grind | 12 |

**Fig. 12** Four-cut boring of a drive hub, in the second of three chucking for 17 operations on a horizontal turret lathe. Dimensions in figure given in inches

- Set the tracer pin to the "start" line and the tool bit to the end of the part, with the tool bit set to finish (approximately) the diameter
- Turn the longitudinal dial clockwise four full turns, turn the diameter dial counterclockwise for a 2.5 mm (0.100 in.) depth of cut, then make the first cut 70 mm (2¾ in.) long
- Turn the longitudinal dial one turn counterclockwise, then make a second cut 70 mm (2¾ in.) long
- Turn the longitudinal dial one turn counterclockwise, and turn the diameter dial clockwise to leave approximately 1.5 mm (0.060 in.) of stock. Take a full cut of template
- Turn both dials to the finish setting and make the final cut

Tool details and operating conditions for the three operations are given in the table accompanying Fig. 10.

## Number of Operations

In production boring, cost is reduced by combining as many machining operations as possible. Thus, the same machine setup is often used for two or more boring operations, or for boring and other machining

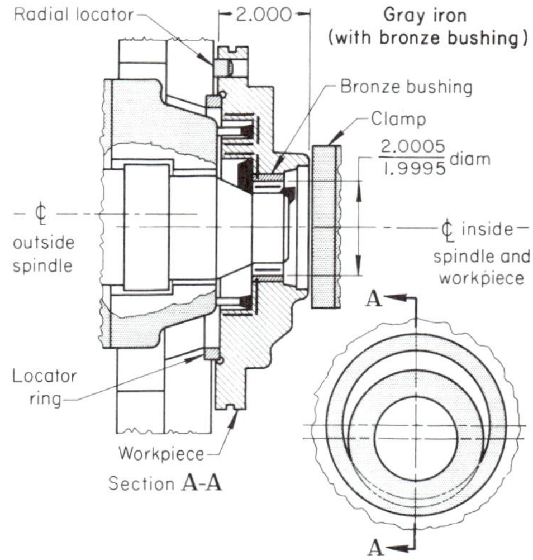


| | |
|---|---|
| Speed, at 420 rev/min, m/min (sfm) | |
| Boring | 67 (220) |
| Counterboring | 104 (340) |
| Turning and facing | 125 (410) |
| Feed, mm/rev (in./rev) | |
| First (all tools) | 0.13 (0.005) |
| Second (counterboring, turning, and facing) | 0.025 (0.001) |
| Depth of cut, on diameter, mm (in.) | 0.3 (0.012) |
| Cutting fluid | None |
| Tool material | Carbide(a) |
| Tool life, h | 8 |
| Production rate, pieces/h | 56 |

(a) Tools on outside spindle had steel shanks, carbide tips; inside-spindle tools were solid carbide.

**Fig. 13** Use of a two-in-one spindle for combining boring and counterboring with eccentric turning, boring, and facing in a single-end precision boring machine. Dimensions in figure given in inches

operations such as turning, facing, drilling, chamfering, and broaching. This may be accomplished either by having two or more boring or other tools on the same mounting so that they may work simultaneously or consecutively, or by indexing the tools or the workpieces so that a number of operations can follow in a cycle.

To achieve these ends, machines such as turret lathes, double-end boring machines, and automatic chucking machines are given primary consideration in equipment selection. The following examples describe some equipment and techniques used in specific applications in which two or more boring operations, or boring and other machining operations, were combined.

**Example 8: Multiple-Diameter Tools on a Six-Station Turret.** A six-station turret lathe was used for machining a complex bore in a gray iron water pump housing. Operations included core drilling, rough boring of four diameters, reaming of three diameters, cutting of a 45° angle recess, and roller burnishing of two diameters. The workpiece and tools at the six-turret stations are shown in Fig. 11.

The multiple-slot bars (stations 2 and 3) permitted combined cuts and aided in holding bore concentricity. The combined recessing tool and the special 45° angle recessing tool (stations 4 and 5) eliminated cross-slide operations. Roller burnishing provided control of diameter within ±0.013 mm (±0.0005 in.) and resulted in desirable work-hardened surfaces. Additional processing details are included with Fig. 11.

**Example 9: Seventeen Operations in Three Chuckings.** Drive hubs were machined from 162 mm (6⅜ in.) diam cold drawn 1042 steel bar slugs in three chuckings on a horizontal turret lathe. The setup and conditions used for boring, which was done in the second chucking, are presented in Fig. 12. The lathe had a two-speed motor; 24 different spindle speeds, ranging from 12 to 628 rev/min; and 32 different feeds, ranging from 0.06 to 4.24 mm/rev (0.0025 to 0.167 in./rev). The operations performed in the three chuckings were as follows:

- Face and turn flange end
- Chuck on flange (Fig. 12), using a stepped face on the jaws for ensuring parallelism of the bar with the centerline of the spindle. Drill through with 64 mm (2½ in.) diam spade drill, rough turn outside diameters, bore (two roughing cuts, one semifinishing cut, and one finishing cut), finish turn outside diameters, groove, face hub and flange, and break corners
- Chuck on hub, finish turn and face flange, counterbore, and break corners

This bore represents about the maximum length that is practical to run without a piloted bar when such a close tolerance (±0.025 mm, or ±0.001 in.) is required. Had speeds and feeds been increased by even as little as 10% more than those used (see tabulation with Fig. 12), chatter would probably have occurred.

**Example 10: Combining Operations on a Precision Boring Machine With a Two-in-One Spindle.** On a single-end precision boring machine, in the setup shown in Fig. 13, boring and counterboring were combined with eccentric turning, boring, and facing to form an eccentric lug in a gray iron casting containing a bronze bushing. This was accomplished by the use of a special two-in-one spindle, that is, an assembly of two spindles, one mounted within the other, that rotated on different centerlines (Fig. 13).

The inside spindle carried a boring and a counterboring tool, and its centerline coincided with the centerline of the workpiece. The outside spindle carried two tools, spaced 180° apart, swinging eccentric to the workpiece centerline. One of these tools bored the outer wall of the eccentric diameter, while the other turned the inner wall. The tools overlapped slightly to provide a smooth bottom. The concentric counterbore on the inside, in combination with the eccentric diameter on the outside, gave the lug a crescent shape. The turning tool and the counterboring tool also overlapped. This required that both spindles be driven at the same speed (420 rev/min) to prevent interference between the tools.

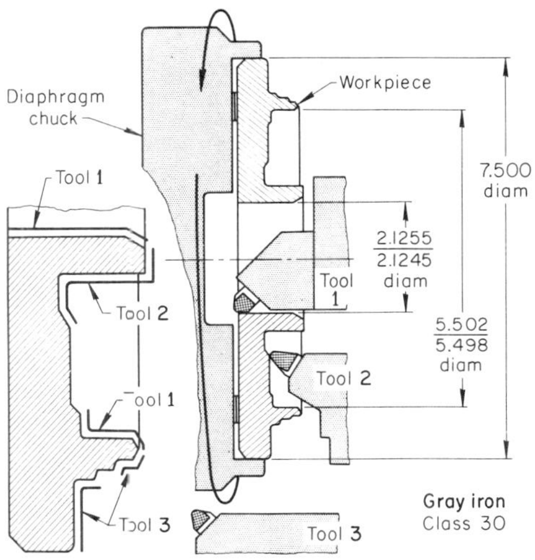


| | |
|---|---|
| Workpiece hardness, HB | 169–207 |
| Speed, at 260 rev/min, m/min (sfm) | 155 (510) max |
| Feed, mm/rev (in./rev) | 0.22–0.23 (0.0088–0.0091) |
| Depth of cut on faces, mm (in.) | 0.25–0.38 (0.010–0.015) |
| Depth of cut on diameters, mm (in.) | 0.5–0.75 (0.020–0.030) |
| Cutting fluid | Soluble oil |
| Tool material | Carbide tips, steel shanks |
| Production rate, pieces/h | 100 |

**Fig. 14** Boring, turning, facing, and chamfering on a cam-operated precision boring machine. Dimensions in figure given in inches

To machine a part, the operator placed it in the fixture over a locator ring that fit in a previously machined, but larger, counterbore. A radial locating pin fit in a drilled hole to orient the cast lug with the eccentricity in the outer spindle. After clamping, the machine was cycled, and all operations were performed simultaneously. Operating conditions are tabulated in Fig. 13.

**Example 11: Four Operations With Three Tools on a Cam-Operated Machine.** Boring, turning, facing, and chamfering were combined on a cam-operated single-end precision boring machine; the three-tool setup used and the gray iron casting being machined are shown in Fig. 14. All three tools were in a fixed mounting and moved as a unit to machine the part. After the casting had been secured in the diaphragm chuck, the machining cycle was begun. The sequence of operations in this cycle (see Fig. 14 for correlation of tools with surfaces machined, and for operating conditions) was:

- Tool 1 moved to the larger (140 mm, or 5½ in. diam) bore, faced the end, and backed off. Tool 1 reapproached the larger bore at an angle, chamfered the opening, then moved "down" the bore, through a radius, and along the "bottom" of the bore. The three-tool unit then backed off
- Tool 1 moved to the smaller (54 mm, or 2⅛ in. diam) bore, chamfered the opening, then continued in to finish this bore to size
- As tool 1 completed the smaller bore, tool 2 contacted the part in the position shown in Fig. 14. Tool 2 generated the angle and continued along the "bottom" to machine a radius. Tool 2 then turned the outside diameter of the central boss, doubled back, and faced the end of this boss. The tools again backed away and moved aside
- Tool 3 approached the part at its outer edge, machined that face, and turned a portion of an outside diameter. Tool 3 continued to move away from the part and was cammed slightly toward the center. The tool reapproached the part and generated two chamfers and a short turned length to complete the operation

Although the combination of operations described in this example could be performed on a turret lathe, the assigned tolerances of ±0.013 and ±0.05 mm (±0.0005 and ±0.002 in.) would be difficult to attain, because repeatability would depend on operator skill rather than on the machine. Production rate would also be lower than on the precision boring mill used.

## Composition and Hardness of Workpiece Metal

In boring, as in turning, the hardness of steels has a greater effect on their machinability than does composition. In general, the harder the steel, the lower the permissible speed and feed for a given operation. Table 1 shows the nominal effect of steel hardness on speed and feed.

The composition and the hardness of the metal being machined are more likely to influence the selection of tool design, tool material, and cutting fluid than they are to determine the type of machine used for a given boring operation. However, metal composition or condition, or both, may dictate not the type but the size of machine required, because more power is needed as hardness increases. The ease with which a metal can be machined also has a direct influence on costs.

Rigidity of setup is needed for virtually any metal-cutting operation, but its importance increases as workpiece hardness increases. For example, an annealed alloy steel might be bored without difficulty in a setup having questionable rigidity, whereas if the steel were heat treated to 350 or 400 HB, it would probably be impossible to machine in such a setup without tool chatter. If chatter occurs, tolerance, finish, and tool life are subsequently impaired.

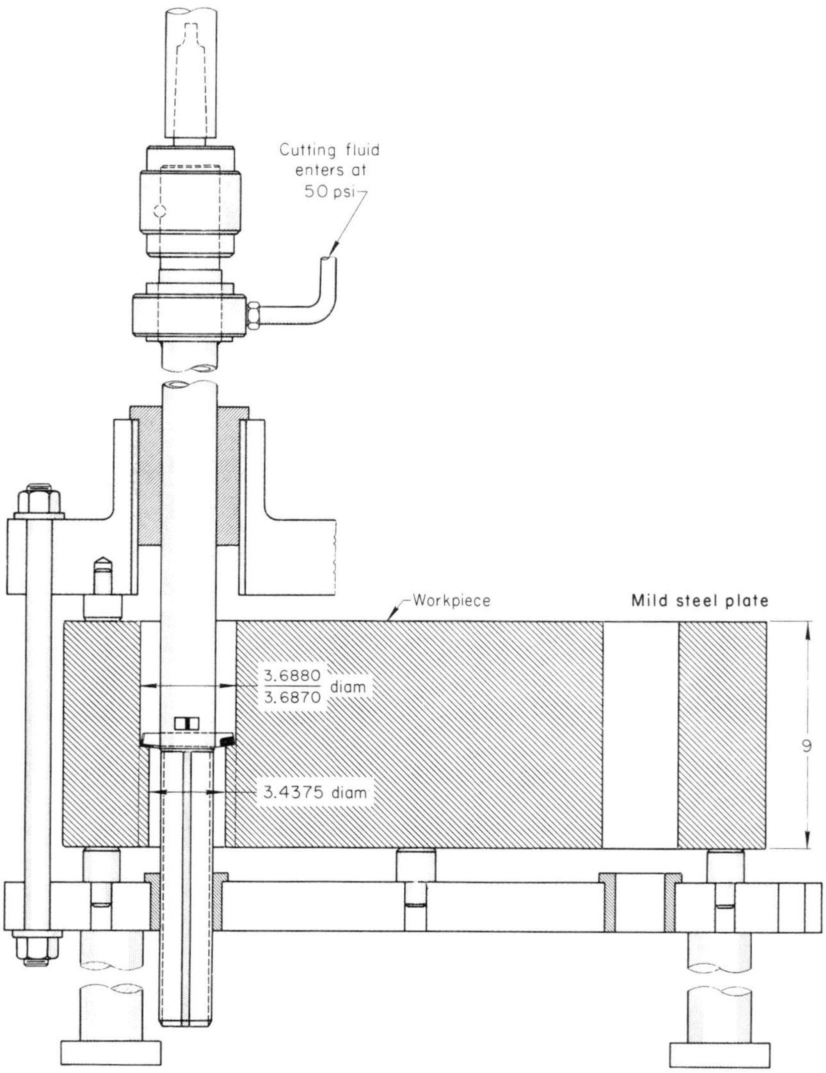


**Fig. 15** Setup involving the use of cutting fluid under pressure for improved finish and dimensional accuracy in drilling and boring a press platen. Dimensions in figure given in inches

## Cutting Fluid

When difficult-to-machine metals are processed, desired results sometimes can be obtained by changing the grade or type of cutting fluid or by changing the method of delivering the cutting fluid to critical areas. The following example focuses on one particular application.

**Example 12: Pressurized Cutting Fluid for Boring Mild Steel Plate.** Mineral oil under no pressure was used as the cutting fluid in drilling and boring press platens flame cut to shape from 230 mm (9 in.) thick mild steel plate (setup shown in Fig. 15). The finish obtained in boring (2.00 to 3.00 μm, or 80 to 120 μin.) was unacceptably rough, and the "ball-up" of the stringy chips characteristic of mild steel adversely affected size control.

The problem was solved by the use of a coolant inductor (a commercially available accessory). This forced the mineral oil, at 345 kPa (50 psi) pressure, through the boring bar, providing a copious supply at the cutting area to flush away chips. With the pressurized cutting fluid and no change in tool design, surface roughness was reduced to 1.00 to 1.5 μm (40 to 60 μin.), and size control was no longer a problem.

## Production Quantity

The number of identical parts to be bored in a production run is a major factor in the selection of the equipment and procedure to be used. When only one part or only a few identical parts are to be produced and no repeat order is anticipated, the simplest suitable machine available (engine lathe, drill press, or boring mill) is generally used, with standard holding devices and boring tools, even though machining time may greatly exceed that on production machines.

As production quantity increases, the need for raising production speed must be weighed against the increased cost of more elaborate tooling. The next step is usually a machine (such as a turret lathe, assuming that part size and shape permit) capable of performing a number of operations consecutively. In many instances in which small to intermediate quantities are required, however, radial drill presses may be the most economical machines for boring. This is particularly true when the hole is small in relation to the size of the workpiece or when the configuration of the hole poses problems for machining in lathes or similar machines.

As requirements are further increased to continuous, high production, progressively more complicated machines may be used, up to and including automatically controlled chuckers that are completely programmed for one specific job. Many large-volume applications employ special machines built for efficient boring of identical workpieces in continuous, high production; a notable example is the high-production boring of automobile engine blocks.

## Accuracy of Form and Diameter

Accuracy of boring relates to size, taper, roundness, concentricity, squareness, and parallelism, and various factors (such as vibration, heat variation, and boring bar deflections) affect these specifications of boring accuracy.

**Diameter** of the bore can be affected by heat generated at the cut, although an adequate flow of cutting fluid can control heat.

**Taper** is an error in a cylindrical bore and may be caused by deflection of the boring bar and by tool wear.

To decrease deflection, pilots and bushings are commonly used to reduce the unsupported length of boring bars. Such a setup changes the boring bar from a cantilever beam with end loading to a beam supported at each end with center loading. Although a bar with end supports may be up to twice as long as a cantilever bar, it deflects only about one-tenth as much as the cantilever bar. The deflection is at a maximum at the start of the cut and decreases as the cutting tool moves closer to the end support bushing. An end support bushing at both ends of a workpiece, as in Example 7 (Fig. 10a), may further reduce deflection of the boring bar during cutting. Deflection can also be reduced by changing the material of the boring bar to one with a higher modulus of elasticity.

Cutting on the return stroke often removes stock left in a hole by a deflected bar. The success of this technique depends on the greatly reduced cutting pressure and the proper radial advance of the tool to remove a very slight amount of metal. Return-stroke cutting increases cycle time, because the return stroke is at a low feed rate rather than at the usual rapid traverse.

**Roundness** of a hole is determined by the variation in radius about a fixed axis. Out-of-roundness (eccentricity) may result when the finish cut is not concentric with the previous cut. Depending on the amount of eccentricity, the depth of cut will vary, which may deflect the tool more during heavy cutting than in a lighter cut. With these conditions it may be necessary to use a semifinish cut and a finish cut and sometimes to cut on the return stroke.

An out-of-balance rotating workpiece or toolholder can produce an out-of-round hole. Careful balancing of the setup can eliminate most of this type of out-of roundness.

**Concentricity** of one surface with another is a common specification that can cause production problems. When concentricity must be held to close limits, it is usually desirable to rotate the work. Almost all types of rotating fixtures are precision boring machines.

Clamping of the workpiece must be such that distortions will not affect hole size or shape. In Example 5 the work was reclamped with a lighter force after the roughing cut. Distortions resulting from a heavy clamping force are released before semifinish and finish cutting and usually are not reinstated by a lighter clamping force.

Multiple-spindle machines or double-end machines can be set up accurately to produce closely concentric surfaces. If the results are not acceptable, it may be possible to plan the sequence of operations so that all close-tolerance surfaces can be machined with tools in the same boring bar.

**Squareness and parallelism** of holes in relation to other features of the work can be affected by changes in room temperature, in the temperature of control system oil, and in the temperature of the cutting fluid.

A change in room temperature can cause expansion or contraction of the machine components, with resulting misalignment of the spindle and workpiece centerlines. The temperature of oil in the hydraulic control system can cause deflection of the feed cylinders or the machine slide. Cooling coils in the oil storage tank can maintain a constant oil temperature.

The heat produced at the tip of the cutting tool may cause a rise in temperature of the cutting fluid. The flow of a warm cutting fluid may cause dimensional changes that affect accuracy of operation. Maintaining the cutting fluid at a constant temperature helps to maintain consistent temperature in both the workpiece and the machine tool. The actual temperature of a system is usually less important than the magnitude of temperature variation.

## Close-Tolerance Boring

As dimensional tolerances become more stringent, machining costs increase. This increase may be caused by the need for more accurate or more rigid tooling and setups. The need for taking lighter cuts (which may involve more passes of the tool) may also slow production in close-tolerance work, although in some applications calling for close-tolerance boring, production rate can be increased by the use of two different types of machines, one for roughing, and the other for finishing to required dimensions. The decreased number of parts that can be machined between tool grinds is another important cost factor, because of increased tool maintenance costs and loss of machine productivity.

**Conventional Equipment.** Boring to close tolerances with conventional equipment is common practice. The various means of guiding the boring tools, flushing away chips, and other refinements of technique contribute to the maintenance of diametral tolerances of ±0.025 mm (±0.001 in.) or even closer in production operation.

**Precision boring machines** are required for meeting tolerances of ±0.13 mm (±0.0005 in.) or less. In many instances such machines are used even when tolerances are no closer than the best obtainable with conventional machines. For example, in high-production boring of relatively small parts to tolerances up to ±0.075 mm (±0.003 in.), the precision type of machine usually is preferred, because it offers increased productivity and decreased cost per piece bored. Precision boring is closely related to and frequently overlaps with, gun drilling.

**Methods.** When close tolerances must be held on complex workpieces, the preferred practice for achieving them at minimum cost is to use a multiple-spindle machine, performing all operations in one setup. Also, some complex parts can be efficiently machined to close tolerances by using double-end machines and combination tools. In precision boring, when workpiece size and configuration permit, it is usually preferable to rotate the part rather than the tools, because problems involving concentricity and parallelism are minimized.

In addition, higher cutting speeds are desirable, and, to maintain sufficient tool rigidity when the ratio of the length of the boring bar to the hole diameter exceeds 4:1, it is generally mandatory that the bar be made of carbide.

## Control of Vibration and Chatter

Vibrations between a tool and a workpiece are either forced or self-excited. In forced vibrations, the alternating force that sustains the motion is created or controlled by the motion itself; when the motion stops, the alternating force disappears. In self-excited vibrations, the sustaining alternating force exists independent of the motion and persists even when the vibratory motion stops.

If an external, periodically varying (cyclic) force is applied to a system such as a machine tool, the mass oscillates at the frequency of the applied cyclic force. Such forced vibrations can be caused by a passing truck or an overhead crane; by machine sources such as faulty bearings or troublesome belts; by loose, worn, or vibrating machine parts; or by interrupted cuts in machining. Such difficulties can be remedied by isolation, by repair or maintenance, or by changing the feed or tool position.

The control of self-excited vibration that builds up in the tool itself is more difficult. In boring bars and quills of unfavorable length-to-diameter ratio, the means of controlling self-excited vibration include the use of a boring bar made of material that has a high modulus of elasticity, the use of an inertia-damped quill, or both.

The static stiffness of a carbide boring bar is three times that of an alloy steel bar because of the difference in modulus of elasticity of the two materials. With its greater static stiffness, a carbide bar can make three times as heavy a cut with the same deflection, or it can make an equal cut with one-third as much deflection and consequently with greater accuracy.

The following data permit comparison of the performance of a steel and a carbide bar:

| Item | Steel bar | Carbide bar |
|---|---|---|
| **Rough boring** | | |
| Feed, mm/rev (in./rev) | Oil 0.1 (0.005) | 0.2 (0.007) |
| Speed, rev/min | 180 | 350 |
| Depth of cut, mm (in.) | 4.8–6.4 (3/16–1/4) | 4.8–6.4 (3/16–1/4) |
| **Finish boring** | | |
| Number of cuts | 2 | 1 |
| Feed, mm/rev (in./rev) | 0.1 (0.005) | 0.1 (0.005) |
| Speed, rev/min | 150 | 400 |
| Depth of cut, mm (in.) | 0.8 (0.030) | 0.2 (0.007) |
| Finish, μm (μin.) | 3.8 (150) | 2.3 (90) |

Each bar had a 16 mm (5/8 in.) diam head, cutting edges with 30° included angles, and a minimum bore diameter of 25 mm (1 in.).

The strength and stiffness of a quill are factors in the deflection caused by cutting forces and have an effect on the frequency of natural vibration. Hence, changing the tool material from steel to carbide increases stiffness and the frequency of natural vibration, but cannot avoid the effects of self-excited vibration.

**Use of a Plug Damper.** One approach to the problem of self-excited vibration is to bore the unsupported end of the quill and insert a damper (a plug of heavy material) that fits the hole, but is 0.05 to 0.075 mm (0.002 to 0.003 in.) smaller in diameter so that it floats in the quill. The end clearance

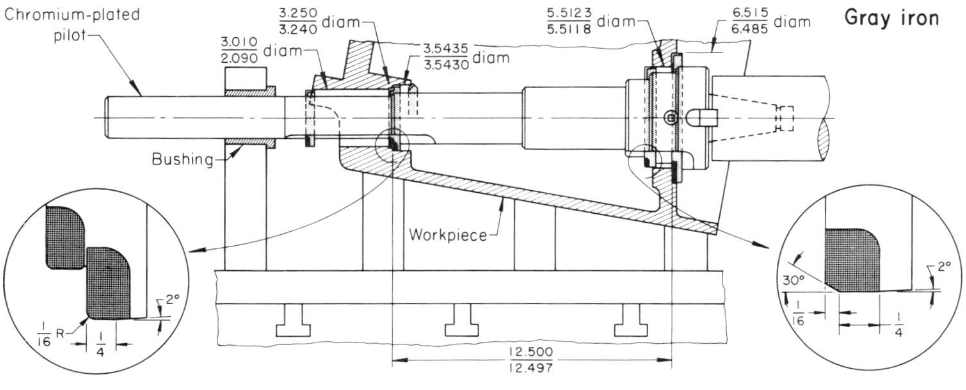


| | |
|---|---|
| Finish specified, μm (μin.) | 1.60 (63) |
| Finish obtained, μm (μin.) | 0.75–0.9 (30–35) |
| **Operating conditions** | |
| Speed, roughing, at 120 rev/min, m/min (sfm) | 28.7–62.2 (94–204) |
| Speed, finishing, at 150 rev/min, m/min (sfm) | 36.0–77.7 (118–255) |
| Feed, roughing, mm/rev (in./rev) | 0.5 (0.020) |
| Feed, finishing, mm/rev (in./rev) | 0.3 (0.012) |
| Depth of roughing cut, mm (in.) | 6.3 (0.250) |
| Depth of finishing cut, mm (in.) | 0.38 (0.015) |
| Cutting fluid | None |
| Tool material | Carbide |
| Setup time, man-hours | 2 |
| Downtime for changing tools, min | 6 |
| Machining time, min/piece | 27 |
| Tool life, pieces/grind | 50 |

**Fig. 16** Two-pass rough and finish boring of five in-line holes, in which the use of specially designed finishing tools (insets) produced better-than-specified surfaces. Dimensions in figure given in inches

of the damper is approximately the same as the clearance on the diameter.

When the quill starts to vibrate, the plug tends to remain stationary because of its inertia. Thus, the air surrounding the plug is caused to flow from one side of the plug to the other. This air flow tends to dissipate and damp vibration before it can build up or resonate. Any tendency of the quill to vibrate is resisted by the plug in the same manner as a spinning gyroscope resists a change, with the air film helping to dissipate the energy that promotes vibration. The vibrations in a quill without a damper die out slowly; in a quill with a damper, the vibrations die out rapidly.

The heavier the weight, the greater the damping effect. Hence, the use of a material such as a tungsten alloy (70% heavier than lead) is indicated. The weight should be made as large as possible without appreciably reducing the stiffness of the quill. It is equally important to place the weight at the location at which the greatest motion is expected. Both of these considerations indicate that the weight should be placed at the free end of the quill, not at the root, where the quill is secured to the spindle.

The damping effect is limited by the amount of weight at the effective damping point. The effective damping force of a given combination of weight and quill increases in direct proportion to the increase in the natural vibration frequency of the quill. By making a portion of the quill from a material having a modulus of elasticity higher than steel, such as tungsten carbide or molybdenum, the natural frequency of vibration is increased, and therefore the same size weight provides a greater damping effect.

In horizontal rotating quills, it is advisable to keep the radial clearance of the damper plug small or to fill the clearance space around the cylindrical inertia weight with oil for equal dashpot effect even though there is a much larger clearance than in air. This is advantageous because of the effect of centrifugal force on the damper plug. The damper plug tends to center itself in the quill at low speeds because of the combined action of gravity and centrifugal force. Above a certain speed, centrifugal force overbalances gravity, and the weight is pulled to one side and held there, thus reducing its damping effectiveness.

The relation between rotational speed and maximum plug clearance that separates the stable region at which gravity keeps the plug centered and the unstable region at which centrifugal forces overbalance gravity is indicated by the following approximate values for speed and radial clearance: 10 000 rev/min, 0.0075 mm (0.0003 in.); 6000 rev/min, 0.025 mm (0.001 in.); 5000 rev/min, 0.036 mm (0.0014 in.); 4000 rev/min, 0.05 mm (0.002 in.); and 3000 rev/min, 0.075 mm (0.003 in.). The smaller the clearance, the greater the permissible speed; hence, the preference is for air as a damping medium rather than oil.

The approximate useful range of the damper appears to be for quill length-to-diameter ratios of 3.5:1 to 8:1. The tendency of the plug to remain at rest decreases with a decrease in the natural vibration frequency of the quill. Making the quill longer or increasing the weight of the free end of the quill decreases the natural vibration frequency. Thus, the damper loses its effectiveness for long (and consequently low-frequency) quills.

**Inertia-Disk Damper.** Another device for dissipating vibration consists of a stack of inertia disks of slightly different diameters inside a cavity in the boring bar. Any vibration in the bar causes the disks to slide against the bar in random timing and thereby reduce the vibration below the magnitude that causes chatter.

The assembly of disks is held together by a draw bar that extends through the shank and screws into a threaded hole in the cutting head. This arrangement puts the bar under compression, for maximum strength, and permits the shop to change heads for different jobs.

An operation in which a damped boring bar was substituted for a steel bar was the rough and finish boring of a 165 mm (6½ in.) diam hole 1300 mm (51 in.) long in a heat-treated 1045 steel compressor body. Two 125 mm (5 in.) diam openings in the wall presented an interrupted-cutting condition. Boring was done from both ends with a maximum bar overhang of 760 mm (30 in.). A standard square-shank brazed boring tool was used as the cutter. The steel boring bar was run at a speed of 30 rev/min and a feed rate of 0.43 mm/rev (0.017 in./rev) for both rough and finish boring.

The roughing operation required three cuts, each 3.2 mm (0.125 in.) deep. Surface finish was not critical, and performance was satisfactory but slow because of the 30 rev/min spindle speed.

Excessive chatter and tool wear were experienced during the finish boring cut of 0.38 mm (0.015 in.) depth. Tool wear was so severe that the tool had to be changed in the middle of the cut. Deflection of the steel bar made it difficult to match the cut with the second tool. The resulting surface irregularities required that the hole be honed to meet size and finish requirements.

Using the damped bar permitted the spindle speed to be increased from 30 to 90 rev/min. At this faster speed but with the same feed of 0.43 mm/rev (0.017 in./rev), two cuts, each 4.75 mm (0.187 in.) deep, were required to rough bore the forging. Finish boring was done in a single cut 0.38 mm (0.015 in.) deep at the same speed and feed as those used for rough boring. The damped bar eliminated chatter and provided a surface finish well within requirements, thereby eliminating the honing operation.

**Other Methods.** Another method of overcoming vibration is by the use of a combination steel and carbide quill. Such a unit consists of a sintered carbide quill press-fitted into a steel mounting flange. On the end of the carbide is brazed a steel extension, which

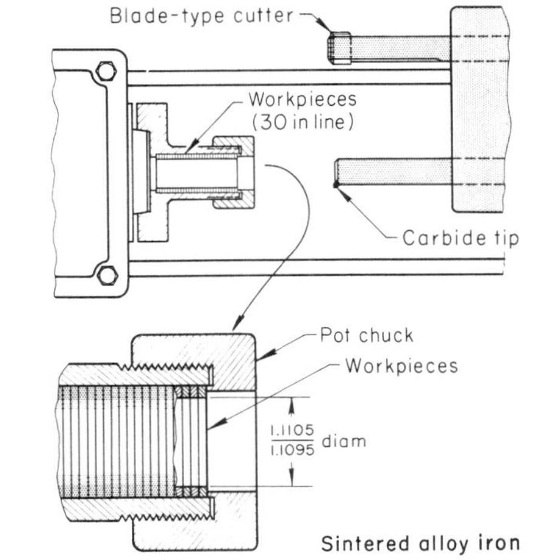


| | |
|---|---|
| Speed, at 700 rev/min, m/min (sfm) | 60 (200) |
| Feed, mm/rev (in./rev) | 0.3 (0.012) |
| Depth of roughing cut, mm (in.) | 0.90 (0.035) |
| Depth of finishing cut, mm (in.) | 0.38 (0.015) |
| Cutting fluid | None |
| Production rate, pieces/h | 1600 |
| Tool life, pieces/grind | 800 |

**Fig. 17** Boring 30 piston rings at a time, using a single-point carbide tool for roughing, and a blade-type cutter for finishing to a specified maximum surface roughness of 0.75 μm (30 μin.). Dimensions in figure given in inches

contains the cutting tool and damper. When possible, the cutting tool is placed between the damper and the carbide quill.

If it is not possible to use dampers in boring quills, oil or cutting fluid under pressure in the support bushing can act as a vibration damper. The liquid provides a film that dissipates some of the vibrational energy.

Another method of damping vibration is to make a quill having the same diameter and length on both sides of the mounting flange. The end without the tool is damped. This design requires that the center portion of the mounting flange be relieved, to act as a diaphragm, and that the spindle of the machine be hollow to accommodate the quill extension.

Design of the workpiece may promote vibration and chatter; for example, a long, small-diameter hub attached to a thin web or to spokes radiating to a rim makes a typical unstable arrangement. The change in direction of the cutting forces as the tool or workpiece rotates can cause deflection and vibration in the hub. If a heavy, ringlike weight is properly clamped to the free end of the hub, it will damp the vibration and improve the finish of the bore.

## Surface Finish

Aside from the inherent machining characteristics of the work material, factors that may singly or collectively affect surface finish are speed, feed, machine condition, tool design, tool support, and cutting fluid. Rigid machines and tool supports are mandatory when low surface roughness must be achieved.

In many instances, finishes are improved by using higher speeds and lighter feeds. In boring larger parts, finish can often be improved by decreasing feed without changing speed.

Often tool design is a major factor in the control of finish, and on many parts finish can be improved by the use of specially designed tools. Sometimes, tool changes improve productivity as well as finish, by permitting higher feed rate, increasing tool life, and decreasing downtime for tool change. The two examples that follow consider the effect of tool design on finish.

**Example 13: Specially Designed Cutters for Low Surface Roughness.** A horizontal boring mill was used for two-pass roughing and finishing of five in-line bores in gray iron machine-tool work heads (see Fig. 16). Maintenance of size control and concentricity dictated the use of a multiple-tool bar.

Achievement of the specified 1.60 μm (63 μin.) finish necessitated the development of specially designed finishing cutters (see insets in Fig. 16), which were inserted in the bar for the second pass. With these special tools, a better-than-specified finish of 0.75 to 0.90 μm (30 to 35 μin.) was obtained. Success resulted from the 30° lead chamfer on the 1.6 mm (1/16 in.) radius lead, and from the 6.4 mm (1/4 in.) long, circle-ground, 0° back taper land, which burnished the bearing bore as it passed through the cut. Operating conditions for roughing and finishing are tabulated in Fig. 16.

**Example 14: Tool Design Change That Improved Productivity as Well as Finish.** A special boring machine equipped with a rocking-type toolholder was used for the rough and finish boring of 30 sintered alloy iron piston rings at a time, using the setup and chucking method shown in Fig. 17. These rings, 35.18/35.05 mm (1.385/1.380 in.) outside diameter, 28.21/28.18 mm (1.1105/1.1095 in.) inside diameter, and 3.15/3.12 mm (0.124/0.123 in.) wide, were required to have a maximum surface roughness of 0.75 μm (30 μin.).

Originally, single-point tools had been used for both rough and finish boring. However, because of the tool configuration required for obtaining the specified finish, the single-point finishing tool deflected excessively under pressure and adversely affected dimensional accuracy.

When a blade-type cutter was substituted as the finishing tool (Fig. 17), tool deflection was eliminated, accurate dimensions and acceptable finish were obtained, and downtime for changing tools was decreased to 2 min (from the 15 min formerly required). Changing from a single-point to a blade-type finishing tool also increased allowable feed rate from 0.2 mm/rev (0.008 in./rev) to 0.3 mm/rev (0.012 in./rev), and increased the number of pieces per tool grind from 400 to 800. Additional operating conditions are given in the table accompanying Fig. 17.

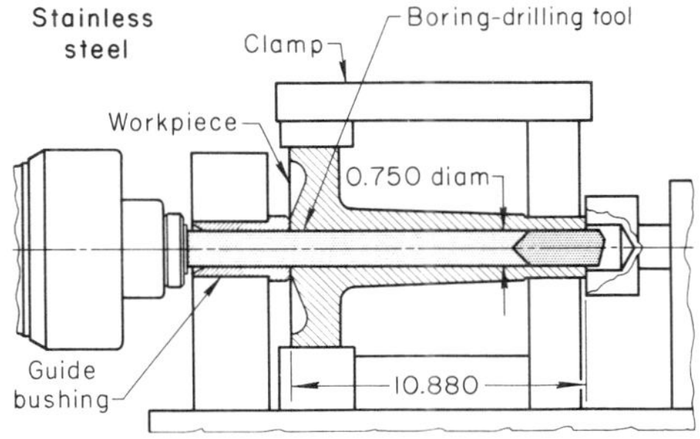


| | |
|---|---|
| **Tolerances** | |
| Diameter, mm (in.) | ±0.02 (±0.00075) |
| Straightness, mm (in.) | Within 0.013 (0.0005) |
| Finish specified, μm (μin.) | 1.60 (63) |
| **Operating conditions** | |
| Speed, at 1150 rev/min, m/min (sfm) | 69 (225) |
| Feed, mm/rev (in./rev) | 0.013 (0.0005) |
| Depth of cut per side, mm (in.) | 0.95 (0.0375) |
| Cutting fluid | Oil |
| Tool material | Carbide tip, steel shank |
| Production rate, pieces/h | 3.1 |

**Fig. 18** Use of a boring-drilling tool in a precision boring machine for simultaneously cutting from solid and finishing a long hole. The work material here was a heat-resisting stainless steel containing approximately 12 Cr, 1 Mn, 3 Mo, 0.25 C, and 0.08 N. Dimensions in figure given in inches

## Use of Boring Equipment for Machining Operations Other Than Boring

Because of their rigidity, adaptability, and accuracy, boring machines are often used for machining operations other than those strictly classified as boring in certain applications. Three such applications are described below.

**Example 15: Deep-Hole Drilling in a Precision Boring Machine.** Because they are rigid and have easily adjustable feed controls, precision boring machines are well suited to driving gun-drilling tools. A single-end precision machine fitted with a special type of gun-drilling tool (for combination drilling and boring) was used for simultaneously drilling from solid and boring a hole 14½ diameters long in a heat-resisting stainless steel workpiece (Fig. 18). The tool was a 378 mm (14⅞ in.) long, single-flute steel shank with a brazed-on carbide cutting edge. Although in many gun-drilling operations the work is rotated while the drill is fed into it, in this application the workpiece was held in a fixture, and the tool was rotated. As in all gun drilling, however, a cutting fluid was used to cool the tool and to flush chips out of the hole along the tool flute.

As shown in the tabulation of processing details that accompanies Fig. 18, this boring-drilling tool produced the hole to a

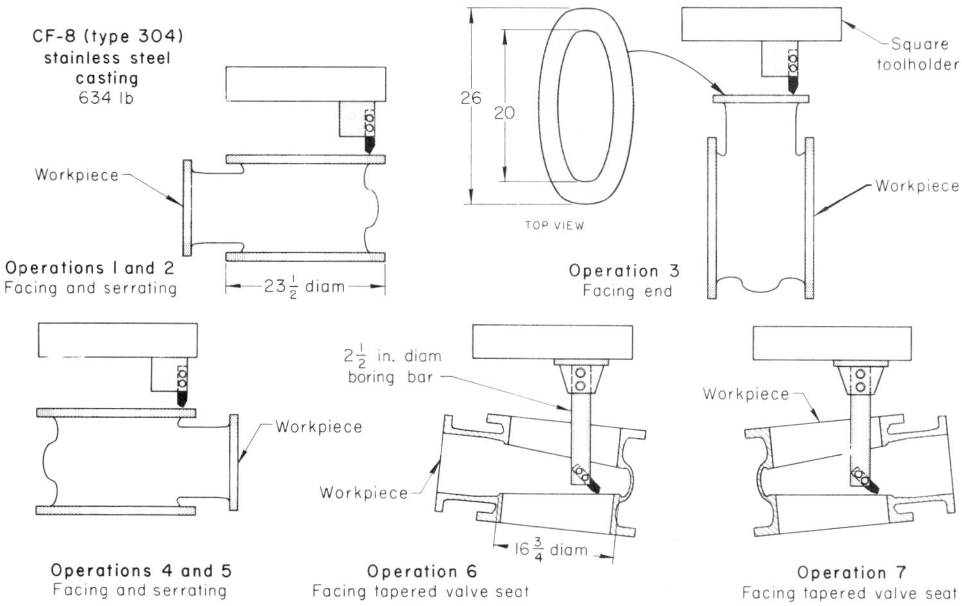


| | |
|---|---|
| **Tolerances** | |
| Inside seat, mm (in.) | +0.13, −0.00 (+0.005, −0.00) |
| All other surfaces, mm (in.) | ±0.4 (±1/64) |
| **Operating conditions** | |
| Speed, m/min (sfm) | |
| Operations 1 and 4, at 22 rev/min | 60 (197) |
| Operations 2 and 5, at 25 rev/min | 46.9 (154) |
| Operation 3, at 32 rev/min | 66.4 (218) |
| Operations 6 and 7, at 25 rev/min | 33.5 (110) |
| Feed, mm/rev (in./rev) | |
| Operations 1, 3, and 4 | 0.20 (0.008) |
| Operations 2 and 5 | 0.84 (0.033) |
| Operations 6 and 7 | 0.15 (0.006) |
| Depth of cut, mm (in.) | |
| Facing, all operations | 6.35 (0.250) max |
| Serrating, 32 cuts per 25 mm | 0.38 (0.015) |
| Tool material | Carbide |
| Setup time, h | 3 |
| Total machining time, h | 12 |

**Fig. 19** Sequence of facing and serrating operations on a gate valve, using a horizontal boring mill with an indexing table. Dimensions in figure given in inches

diametral tolerance of ±0.02 mm (±0.00075 in.) and to straightness within 0.01 mm (0.0005 in.). This type of tool, however, can drift as much as 0.001 mm/mm (0.001 in./in.) and therefore cannot be used to eliminate a separate boring operation when the tolerance on concentricity is extremely tight.

**Example 16: Machining Pockets From Solid in a Contour Boring Machine.** A special vertical contour boring machine was used for cutting contoured pockets in the solid walls of ring-shape parts made from 1015 steel and stress relieved to 159 to 170 HB. The cutting tool used, a special two-flute end mill with one flute acting as a boring tool, removed 19 mm (3/4 in.) of stock from the ring wall in a straight cut, then made a contour finishing cut of 0.1 to 0.79 mm (0.005 to 0.031 in.) to a specified tolerance of ±0.13 mm (±0.005 in.) and a finish of 0.50 to 0.63 μm (20 to 25 μin.).

The carbide cutter, which was ground with 10° front clearance, 15° side clearance, and a negative rake angle of −2°, was operated at 150 m/min (500 sfm) and 0.13 mm/rev (0.005 in./rev); water-soluble oil was used as a cutting fluid.

Workpieces were supported on an indexing fixture plate. Setup time was 2 h, production rate was 16 pieces per hour, tools machined 80 pieces per grind, and downtime for changing tools was 6 min.

**Example 17: Facing and Serrating in a Horizontal Boring Mill.** A horizontal boring mill was used for seven facing or serrating operations on a 288 kg (634 lb) gate valve cast from CF-8 (type 304) stainless steel. The valve and the sequence of operations are shown in Fig. 19. The boring mill was equipped with a 1.2 m (4 ft) square, 360° indexing table on a cross slide. Facing or turning capacity was 1070 mm (42 in.) with 1.8 mm (6 ft) vertical travel. The workpiece was held on the machine bed with standard tie-down clamps. Facing and serrating were done with the same tool, but at different feeds and speeds, as shown in the table of operating conditions that accompanies Fig. 19.

# Trepanning

TREPANNING is a machining process for producing a circular hole or groove in solid stock, or for producing a disk, cylinder, or tube from solid stock, by the action of a tool containing one or more cutters (usually single-point) revolving around a center. The process is used in at least four distinct production applications:

- Round disks
- Large shallow through holes
- Circular grooves
- Deep holes

A hole diameter tolerance of ±0.25 mm (±0.010 in.) can easily be obtained by trepanning, and in some cases, a tolerance of ±0.025 mm (±0.001 in.) has been held. Hole runout can be held to ±0.25 mm/m (±0.003 in./ft) and at times to ±0.08 mm/m (±0.001 in./ft). On heat-treated metal, a surface finish of 3.2 to 3.8 μm (125 to 150 μin.) $R_a$ can be obtained, and on annealed metals 2.5 to 6.3 μm (100 to 250 μin.) $R_a$ is common.

## Round Disks

One common use of trepanning is the production of round disks from flat stock. The process is particularly applicable when production quantities are too small to warrant the use of blanking dies and when flame-cutting equipment is unavailable. Hand-fed drill presses are the machines most used for this application of trepanning.

When a center hole is unobjectionable, the trepanning tool consists of an adjustable fly cutter mounted on a twist drill (Fig. 1); the drill serves as both driver and pilot, and the single-point tool can be positioned as desired to change the size of the circle cut. If a center hole cannot be permitted, a tool of the type shown in Fig. 1 can be used without a drill pilot. By this technique, however, rigidity is more difficult to maintain, and there is greater likelihood of tool chatter and loss of dimensional control.

There is no established maximum thickness or diameter of disks that can be cut by this method of trepanning. Because of the load imposed on the tool, however, the process is seldom used for cutting material more than about 6.4 mm (¼ in.) thick. In addition, because rigidity decreases as diameter increases, disks cut by this method are usually less than 150 mm (6 in.) in diameter. Larger cuts by trepanning are not made if other methods can be used. Smaller trepanning cuts may compete with press-working (if the means are available), and larger cuts may compete with other methods, such as flame cutting.

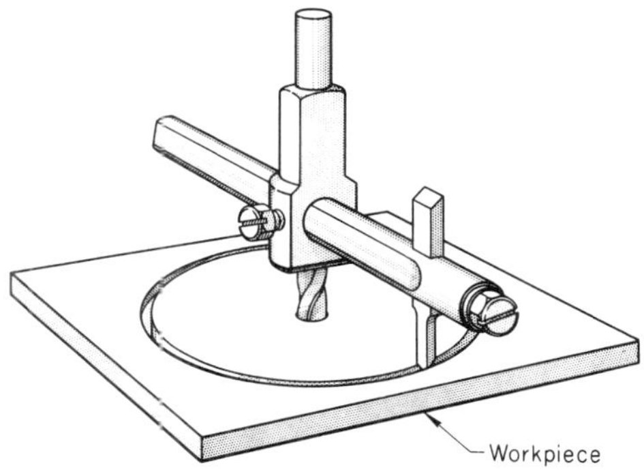


**Fig. 1** Drill-mounted adjustable fly cutter used for trepanning various sizes of disks from flat stock, or grooves around centers

In this type of trepanning, slow speeds (ranging downward from about 10 m/min, or 35 sfm) are ordinarily used and feed is controlled manually. Cutting fluids are seldom employed.

## Large Shallow Through Holes

Round through holes having diameters that exceed depth by a factor of about five or more can often be efficiently and accurately produced by trepanning. Workpiece configuration often dictates the trepanning technique used because special fixturing may be necessary to ensure adequate rigidity. The tools and techniques employed in one application are described in the following example, in which trepanning was preferred to drilling.

**Example 1: Nine Shallow Holes in Web of a Steel Gear.** Trepanning proved more practical than drilling for producing nine weight-reducing holes in the web section of an aircraft accessory drive gear made of 9310 steel (Fig. 2). Depending on the size of these gears, the holes ranged from 19 to 32 mm (¾ to 1¼ in.) in diameter, and web-section thickness ranged from 3.56 to 6.35 mm (0.140 to 0.250 in.).

Trepanning was done with carbide-insert single-point tools at a speed of 91 m/min (300 sfm) and a feed of 0.05 mm/rev (0.002 in./rev). Originally, a top-brazed insert (bottom left, Fig. 2) was used, but tool life was only 30 holes per grind. By changing to side-brazed inserts (bottom right, Fig. 2), tool life was increased to 75 holes per grind. With either tool, maximum tool life was obtained when the tool was not allowed to cut completely through the web. A thin section was left to hold the plug, and this section was easily knocked out during indexing of the gear.

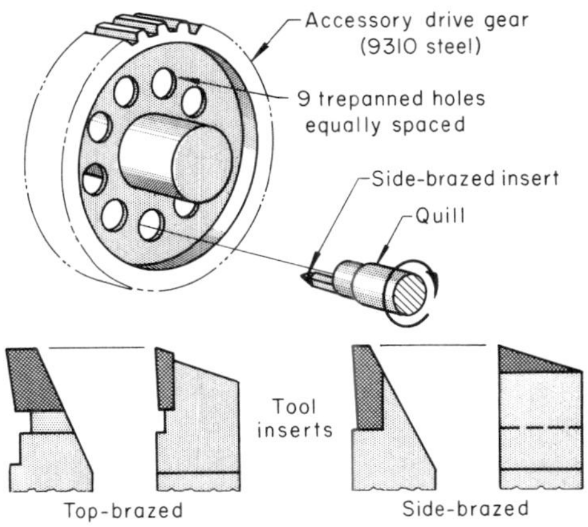


**Fig. 2** Shallow holes produced by trepanning in aircraft gear shown at top left. Top-brazed carbide inserts cut only 40% as many holes per grind as side-brazed inserts.

## Circular Grooves

The tools and techniques used for producing round disks or large shallow through holes can also be used to provide metal parts with circular grooves for accommodating O-rings or for other purposes. The only difference is that the cutter must be shaped to form the desired cross-sectional shape of the groove.

When a groove to be cut is only slightly larger in diameter than a concentric pilot hole, optimum results are obtained by the use of a combination drilling-and-trepanning tool that resembles a hollow mill (Fig. 3). With this tool, a twist drill is inserted into a hollow cutter and held by a setscrew. This type of trepanning cutter usually has two or more cutting edges to provide balance, which assists in maintaining dimensional control. Cutting edges have a back rake angle of about 20° for most applications.

Any machines normally used for drilling are suitable for driving combination tools of the type shown in Fig. 3. Speeds up to 30 m/min (100 sfm) are ordinarily used, in con-

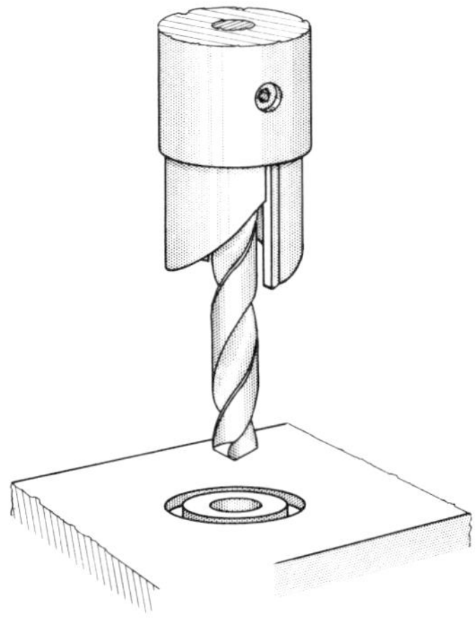

**Fig. 3** Combination tool for producing a groove close to a concentric pilot hole

junction with feeds of about 0.1 mm/rev (0.004 in./rev). Soluble oils are usually satisfactory as cutting fluids, although sulfurized oils are preferred when tolerance and finish requirements are critical. The following example describes an application in which trepanning was more efficient and economical than the use of single-point tools for producing a circular face groove.

**Example 2: Trepanning Versus Single-Point Plunge Cutting.** Originally, the 73 mm (2⅞ in.) groove in the 4140 steel part shown in Fig. 4 was produced in roughing and finishing operations on a lathe, using single-point carbide-insert tools. Chip buildup and tool breakage were continual problems. Tool life per sharpening was only 25 pieces, and production time per piece was 3.9 min.

Tool breakage and chip buildup were eliminated (and grooving was reduced to a one-operation job) by substituting trepanning for the original method. The trepanning tool used had two carbide cutters mounted on an 1141 steel body; tool design is shown in Fig. 4. Trepanning also reduced time per piece to 1.8 min (less than half, compared with the original method) and doubled tool life to 50 pieces per sharpening.

## Deep Holes

Trepanning is often the most practical method of machining deep holes or tubes from the solid. Deep-hole trepanning is similar to gun drilling (see the article "Drilling" in this Volume) in that both processes require a pressurized cutting fluid system and employ self-piloting cutting action. The two main differences are:

- Trepanning is practical only for larger holes (more than about 50 mm, or 2 in., in diameter)
- Trepanning produces a solid core, while gun drilling forms only chips

As a means of producing holes 50 mm (2 in.) in diameter or larger (especially holes

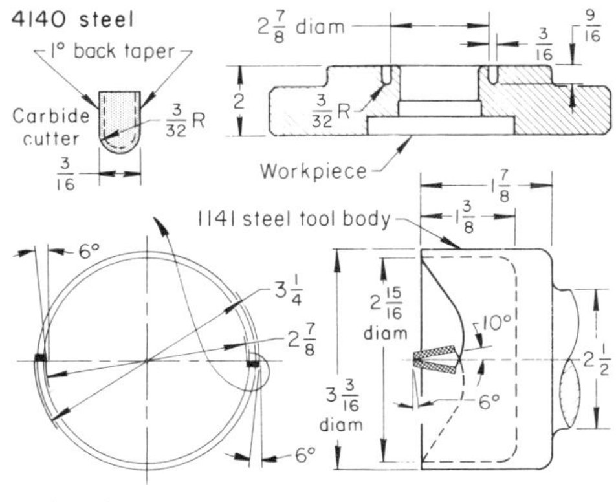


**Fig. 4** Tool used in trepanning operation (shown at top left and bottom). The part is shown at top right. Dimensions given in inches

whose depth is eight or more diameters), trepanning offers the following advantages over spade or twist drilling, with their allied operations:

- Closer tolerances can be met on diameter and straightness
- Drilling of deeper holes is feasible
- Rate of metal removal is higher
- In machining costly work materials, cores are more valuable than chips

Trepanning can also be used to produce a tube from a cylindrical billet when machining space-age metals such as beryllium. The trepanning of a cylindrical core from the center of a solid cylinder of metal is not ordinarily done in regular mass production. Beryllium, however, is a problem metal that needs special methods.

Misalignment is probably the most frequent single source of difficulty in deep-hole trepanning. The misalignment may be caused by insufficient rigidity in the tooling and the setup. Accurate alignment and rigidity are essential for control of dimensions and finish and for satisfactory tool life at high depth-to-diameter ratios.

## Machines for Deep-Hole Trepanning

For trepanning holes less than about five diameters deep, simple vertical drill presses are usually satisfactory. However, as the depth of the hole exceeds five times diameter, any type of vertical equipment becomes progressively more impractical. In addition, as the depth-to-diameter ratio increases, accuracy is lost more rapidly in equipment in which the tool is rotated and the work is held stationary. Therefore, engine lathes, turret lathes, or horizontal drilling machines are preferred for trepanning deep holes. In all of these machines, the workpiece is rotated while the tool remains stationary. This technique results in greater accuracy, other conditions being constant.

Regardless of the type of machine used, it must be rigid and capable of speeds up to 185 m/min (600 sfm) to accommodate carbide tooling. It should also have variable feed control.

**Engine Lathes.** Figure 5 shows a basic engine lathe setup, in which a cylindrical workpiece is rotated, the tool is fed into it, and an inside-diameter-exhaust trepanning head is used (see the section "Tools for Deep-Hole Trepanning" in this article). Such a setup is used for holes about 50 to 115 mm (2 to 4½ in.) in diameter. One end of the workpiece is held in a three-jaw chuck, and the other end in a roller steady rest (rollers are about 150 mm, or 6 in., in diameter). A relatively long workpiece requires an additional steady rest midway between the chucking headstock and the roller steady rest.

The end of the workpiece next to the steady rest must be faced at right angles to the spindle centerline. The facing cut, made by a tool on a cross slide, provides a flat surface for the fluid seal on the guide bushing and prevents the runout that could make the fluid seal leak. The guide bushing should be mounted on ball bearings which are the most economical means of support at the speeds involved.

Cutting fluid under pressure enters the leakproof rotary joint behind the bushing and flows into the annular space around the hollow boring bar. The fluid then flows to the cut between the bushing and the periphery of the tool head, picks up the chips, and flushes them through the head between the cutter and the core and then out along the space between the core and the inner wall of

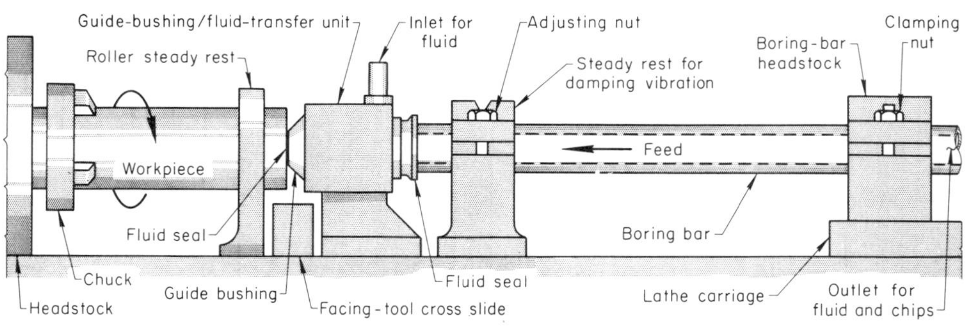


**Fig. 5** Engine lathe setup for trepanning deep holes

the boring bar. An additional fluid seal is necessary at the rear of the guide-bushing/fluid-transfer unit. The tailstock end of the boring bar is mounted in a headstock on the lathe carriage and is rigidly clamped in position by means of a bearing cap and clamping nuts.

To serve as a vibration damper, a steady rest is located directly behind the guide-bushing/fluid-transfer unit. This damper is of the same construction as the boring-bar headstock, except that a two-piece bronze bushing is used to damp vibration and to allow the boring bar to slide.

Alignment is critical. The spindle, chuck, steady rests, guide bushing, and boring-bar headstock must be as nearly in line as possible. In addition, the machine ways must be aligned, and the boring bar must be ground to uniform diameter and straightness.

The bores in the guide-bushing/fluid-transfer unit, the vibration-damper unit, and the boring-bar headstock should be large enough to accommodate the boring bar for the largest-diameter hole to be trepanned on the machine. For smaller holes, smaller boring bars can be used with appropriate bushings.

For holes more than about 115 mm (4½ in.) in diameter, other setups are sometimes more economical and can be used with an outside-diameter-exhaust head for trepanning relatively deep holes (see the section "Tools for Deep-Hole Trepanning" in this article). One of these adaptations is the use of a three-roll support on the bed of an engine lathe. A three-roll support used in conjunction with an outside-diameter-exhaust head effects a savings by eliminating the need for a starting or guide bushing.

**Turret lathes** are also suitable for trepanning relatively deep holes, as indicated in the following example.

**Example 3: Use of a Turret Lathe for Trepanning.** Holes 127 mm (5 in.) in diameter and 915 mm (36 in.) deep were trepanned in vacuum-melted 4340 steel using an outside-diameter-exhaust trepanning head with a 19 mm (¾ in.) wide carbide cutting tip. A turret lathe of 11 kW (15 hp) capacity was used to rotate the workpieces at 190 rev/min (76 m/min, or 250 sfm). Water-soluble oil, under pressure of 275 kPa (40 psi) was pumped through the cutting area at 190 L/min (50 gal./min). At a feed rate of 0.18 mm/rev (0.007 in./rev), two pieces per hour were produced.

## Tools for Deep-Hole Trepanning

Boring bars for trepanning are hollow tubes that allow the workpiece core to enter with enough clearance for cutting fluid to flow to the cutter or for fluid and chips to be forcibly exhausted from the cutter. The bar is usually made from 52100 bearing steel or a similar steel. Wall thickness ranges upward from about 7.9 mm (5⁄16 in.) depending

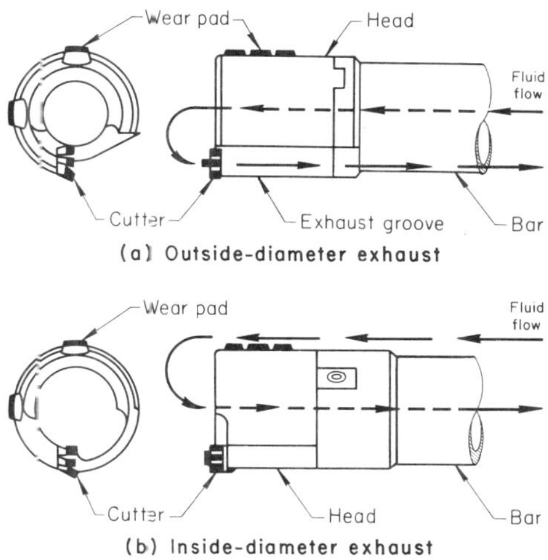


**Fig. 6** Two types of single-cutter trepanning heads

on the length of the bar and the required resistance to torsional forces.

Trepanning heads are cylindrical and usually employ a single solid-carbide or carbide tip cutter. Multiple-cutter heads, despite their desirable chip-breaking action, are used to a lesser extent because they pose problems in attaining balanced cutting action—without which hole accuracy may be sacrificed.

**Single-cutter heads** (Fig. 6) are self-piloting; they are supported and guided by wear pads located about 90 and 180° behind the cutter. The head fits onto the boring bar by means of a pilot diameter and can be driven by three lugs. With this mating design, the head is locked to the bar screws. Some heads are secured to the bar by Acme threads around the inner circumference of the head, but high torsional forces can cause thread seizure. Ahead of the cutting edge on the outside diameter of the head is a relief for intake of cutting fluid or for exhaust of cutting fluid and chips.

One type of head (Fig. 6a), usually for holes up to 115 mm (4½ in.) in diameter and with depths of 12 to 15 diameters, accommodates cutting fluid flow from the inside diameter of the bar and exhausts the fluid on the outside diameter. Recommended maximum depths for holes trepanned with outside-diameter-exhaust heads are as follows:

| Diameter, mm | Diameter, in. | Depth, mm | Depth, in. |
|---|---|---|---|
| 50–64 | 2–2½ | 610–762 | 24–30 |
| 65–89 | 2 9⁄16–3½ | 762–1065 | 30–42 |
| 90–115 | 3 9⁄16–4½ | 1065–1780 | 42–70 |

For holes 116 mm (4 9⁄16 in.) in diameter and larger, maximum depth is limited only by machine design.

With this type of head, chips and fluid are exhausted along a longitudinal groove milled on the outside diameter of the head. The clearance between the core and the inside wall of the head must be controlled so that the volume of cutting fluid is restricted. As a result, there is a high-velocity centrifugal action that forces the chips away from the cutting edge.

In production trepanning, one plant found that a fluid-inlet area of 645 mm$^2$ (1 in.$^2$) produces a pressure of about 345 kPa (50 psi) on a 178 mm (7 in.) diam head. This size of inlet area allows full pump flow through the head and provides sufficient velocity for chip disposal. As the hole size decreases and the inlet area remains constant, the pressure increases. Increased pressure is desirable as hole diameter decreases or as depth increases. For holes less than 100 mm (4 in.) in diameter, however, an inlet area of 645 mm$^2$ (1 in.$^2$) is not possible; for these holes, the inlet area should be made as large as possible to provide adequate volume without weakening the head.

The outside diameter of an outside-diameter-exhaust head is only 0.50 to 0.64 mm (0.020 to 0.025 in.) smaller than the diameter of the hole being cut; this prevents chips from escaping between the head and the wall of the hole. As a result, the possibility of a marred finish is lessened, and channeling of the fluid and chips through the exhaust groove is facilitated.

To minimize the problem of chip disposal in holes more than about 15 diameters deep, an inside-diameter-exhaust head (Fig. 6b) is used. With this type of head, fluid under pressure flows to the cutting edge over the outside diameter of the bar and head. The fluid and the chips are exhausted through the inside diameter of the boring bar.

**Wear pads** (Fig. 6) are essential components of single-cutter heads. Wear pads balance cutting force, control hole size, and provide a burnishing action that may improve finish.

Ordinarily, wear pads have steel bodies and brazed-on carbide wearing surfaces. The wear pad body may have two angular sides that result in a dovetail fit in the head. The pads are circle ground so that when they are positioned in the head they will clear the bore wall by about 0.05 mm (0.002 in.) and will project about 0.25 mm (0.010 in.) from the head.

One pad is located approximately 90° behind the cutting edge; this pad steadies the head against the bore and balances the cutting force. The other pad is about 180° behind the cutter and automatically controls the size of the hole. If the cut is oversize, the bore is large, and the pressure on this pad is immediately decreased. As a result, the head moves away from the surface being cut, reducing the bore size until equilibrium is again established. Similarly, if the cut is undersize, pressure on the pad is increased, and the head moves toward the surface being cut.

**Initial cutting action** is controlled by a guide bushing (Fig. 5), by a counterbore, or

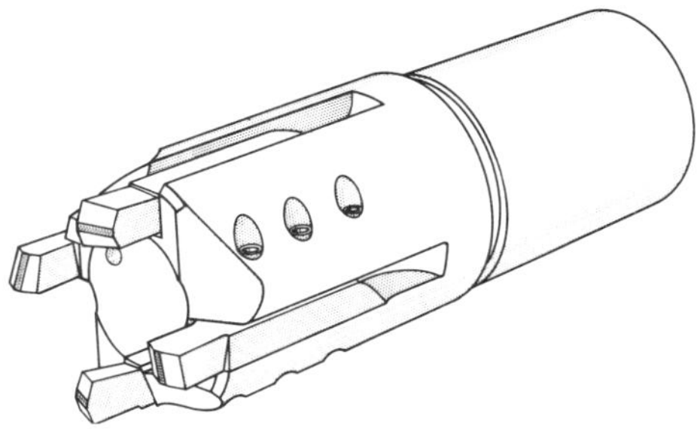

**Fig. 7** Starting tool for trepanning

by a starting tool that cuts a groove (Fig. 7). If a bushing is used, starting feeds are relatively light and characteristically produce stringy chips, which must be removed at intervals. The counterbored hole or starting groove should be deep enough to ensure self-piloting by the trepanning head.

**Conventional cutters** (Fig. 8) have a carbide tip brazed on a tool steel body. (Because index positions are limited and chip flow is obstructed, disposable-insert tooling has not been widely used in trepanning.) The brazed cutter is designed so that the single edge has three steps to break the chip into three equal widths. Each step includes a parallel chip breaker. The cutter is commonly 19 mm (¾ in.) wide, but wider cutters have been successfully used for holes more than about 100 mm (4 in.) in diameter.

The radial position of the cutter when used in a stationary head is important. Viewed from the cutting end of the head, the cutter should be approximately at the 2 o'clock position; thus the cutting fluid will

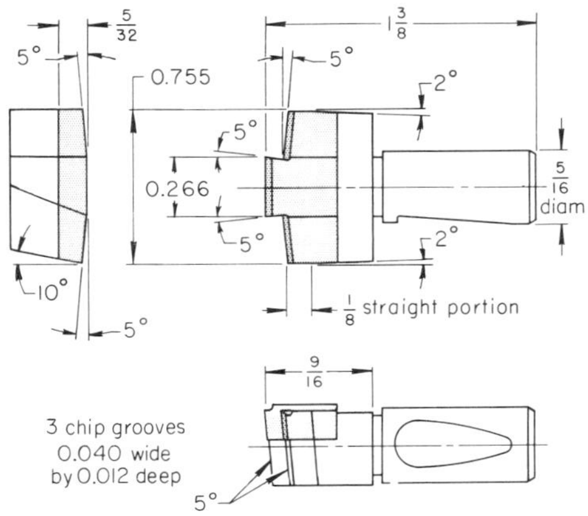


**Fig. 8** Typical design of a trepanning cutter for producing deep holes. Dimensions given in inches

wash the chips away from the cutter through the relief on the outside diameter of the head, and the core, when cut, will fall away from the cutter.

**Multiple-cutter heads** are sometimes more appropriate than those with single cutters. This is especially true when hole starting is difficult or for minimizing shock on the bar or other components in the driving mechanism.

**Example 4: Use of a Three-Cutter Head for Trepanning Transversely Through a Steel Cylinder.** A 64 mm (2½ in.) diam hole was trepanned through the 100 mm (4 in.) diameter of a solid cylinder of 8615 steel. A specially designed three-cutter head (Fig. 9) of 4140 steel was used in order to minimize spindle shock. This head, of the outside-diameter-exhaust type, included three 150 mm (6 in.) long bronze conventional wear strips, spaced 120° apart and 30° behind each cutter. To provide maximum stability and accuracy, an additional set of three cylindrical carbide wear strips (7.9 mm, or 5/16 in. in diameter and 32 mm, or 1¼ in., long) was incorporated. A guide bushing was used in conjunction with a modified drill press that provided 22 kW (30 hp) at the spindle. Cutting fluid (active mineral oil) was pumped through the bar at 345 kPa (50 psi) and 380 L/min (100 gal./min).

*Four-Cutter Heads.* Production rate can be increased with only a slight increase in feed rate or chip thickness, by the use of a four-cutter head. The first pair of cutters produces a narrow groove; this is then widened to final dimensions by the second pair of cutters. Initial and maintenance costs, however, are greater for this type of tool than for a single-cutter head. Figure 10 illustrates the recommended coolant pressure and flow rates for multiple-cutter internal chip removal trepanning tools.

## Speed and Feed in Deep-Hole Trepanning

Table 1 lists speeds and feeds for trepanning deep holes in a variety of carbon, alloy, and stainless steels. These values are useful

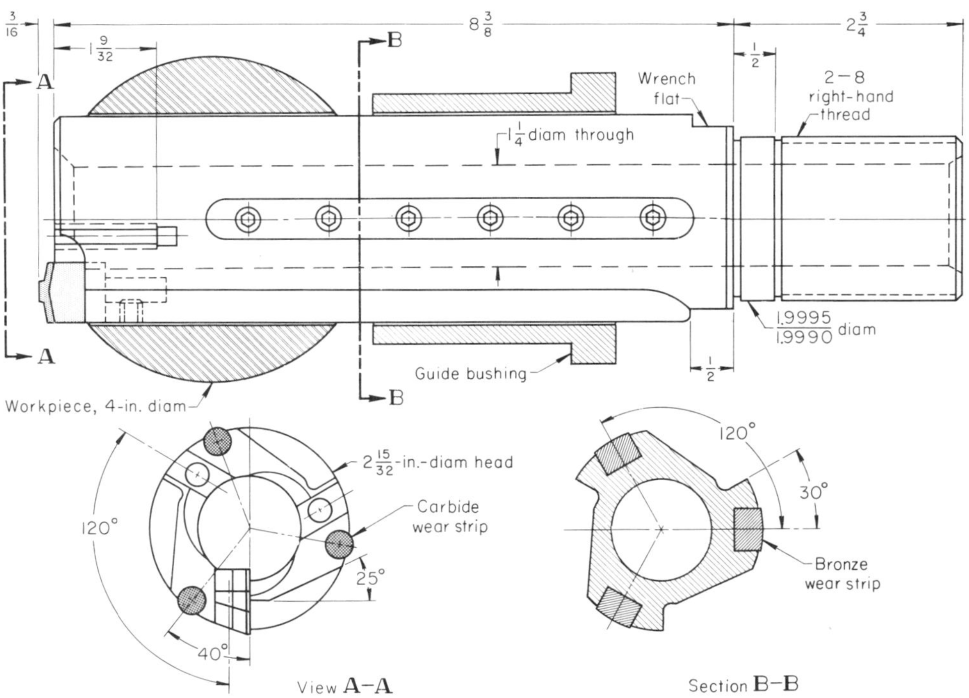


**Fig. 9** Three-cutter trepanning head used for cutting a 64 mm (2½ in.) diam hole through a solid cylinder of 8615 steel. Dimensions given in inches

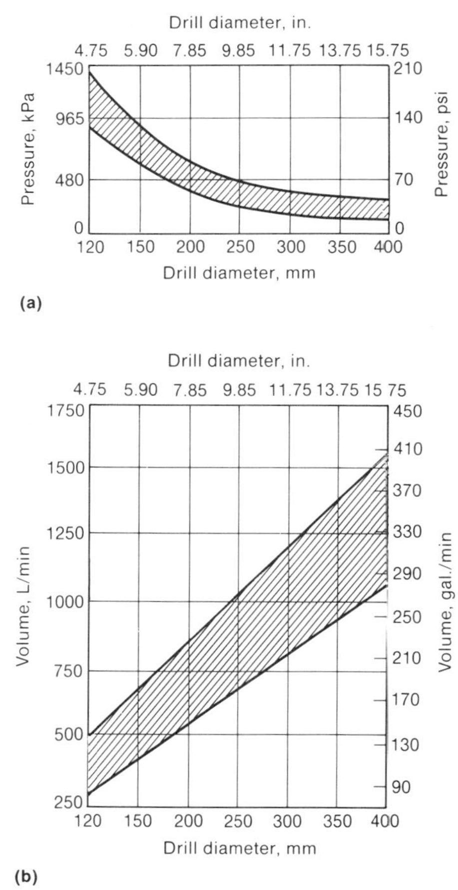


**Fig. 10** Recommended coolant pressures (a) and volumes (b) for multiple-cutter internal chip removal trepanning tools

**Table 1 Speeds and feeds for the deep-hole trepanning of various steels with high-speed tool steels and carbide tools**

| Material | Hardness, HB | Condition | Speed m/min | Speed sfm | Feed(a) mm/rev | Feed(a) in./rev | Tool material ISO | Tool material AISI or C |
|---|---|---|---|---|---|---|---|---|
| **Wrought free-machining carbon steels** | | | | | | | | |
| **Low-carbon resulfurized** | | | | | | | | |
| 1116 1118 1211 | 100–150 | Hot rolled or annealed | 35 | 115 | 0.20 | 0.008 | S4, S5 | M2, M3 |
| 1117 1119 1212 | | | 150 | 500 | 0.20 | 0.008 | P30 | C-6 |
| | 150–200 | Cold drawn | 37 | 120 | 0.20 | 0.008 | S4, S5 | M2, M3 |
| | | | 185 | 600 | 0.20 | 0.008 | P30 | C-6 |
| **Medium-carbon resulfurized** | | | | | | | | |
| 1132 1140 1145 | 175–225 | Hot rolled, normalized, | 32 | 105 | 0.18 | 0.007 | S4, S5 | M2, M3 |
| 1137 1141 1146 | | annealed, or cold drawn | 150 | 500 | 0.18 | 0.007 | P30 | C-6 |
| 1139 1144 1151 | | | | | | | | |
| | 325–375 | Quenched and tempered | 15 | 50 | 0.15 | 0.006 | S9, S11(b) | T15, M42(b) |
| | | | 120 | 400 | 0.18 | 0.007 | P30 | C-6 |
| **Low-carbon leaded** | | | | | | | | |
| 12L13 | 100–150 | Hot rolled, normalized, | 40 | 130 | 0.20 | 0.008 | S4, S5 | M2, M3 |
| 12L14 | | annealed, or cold drawn | 185 | 600 | 0.20 | 0.008 | P30 | C-6 |
| 12L15 | | | | | | | | |
| | 200–250 | Hot rolled, normalized, | 32 | 105 | 0.18 | 0.007 | S4, S5 | M2, M3 |
| | | annealed, or cold drawn | 150 | 500 | 0.18 | 0.007 | P30 | C-6 |
| **Wrought carbon steels** | | | | | | | | |
| **Low carbon** | | | | | | | | |
| 1005 1009 1015 1023 | 85–125 | Hot rolled, normalized, | 34 | 110 | 0.20 | 0.008 | S4, S5 | M2, M3 |
| 1006 1010 1017 1025 | | annealed, or cold drawn | 150 | 500 | 0.20 | 0.008 | P30 | C-6 |
| 1008 1012 1020 | | | | | | | | |
| | 225–275 | Annealed or cold drawn | 23 | 75 | 0.18 | 0.007 | S4, S5 | M2, M3 |
| | | | 120 | 400 | 0.18 | 0.007 | P30 | C-6 |
| **Medium carbon** | | | | | | | | |
| 1030 1039 1045 1055 | 125–175 | Hot rolled, normalized, | 26 | 85 | 0.20 | 0.008 | S4, S5 | M2, M3 |
| 1033 1040 1046 1525 | | annealed, or cold drawn | 135 | 450 | 0.20 | 0.008 | P30 | C-6 |
| 1035 1042 1049 1526 | | | | | | | | |
| 1037 1043 1050 1527 | 375–425 | Quenched and tempered | 9 | 30 | 0.10 | 0.004 | S9, S11(b) | T15, M42(b) |
| 1038 1044 1053 | | | 100 | 325 | 0.13 | 0.005 | P30 | C-6 |
| **Wrought alloy steels** | | | | | | | | |
| **Low carbon** | | | | | | | | |
| 4012 4615 4817 8617 | 125–175 | Hot rolled, annealed, or | 27 | 90 | 0.18 | 0.007 | S4, S5 | M2, M3 |
| 4023 4617 4820 8620 | | cold drawn | 145 | 475 | 0.18 | 0.007 | P30 | C-6 |
| 4024 4620 5015 8622 | | | | | | | | |
| 4118 4621 5115 8822 | 325–375 | Normalized or quenched | 14 | 45 | 0.10 | 0.004 | S9, S11(b) | T15, M42(b) |
| 4320 4718 5120 9310 | | and tempered | 90 | 300 | 0.13 | 0.005 | P30 | C-6 |
| 4419 4720 6118 94B15 | | | | | | | | |
| 4422 4815 8115 94B17 | | | | | | | | |
| **Medium carbon** | | | | | | | | |
| 1340 4145 50B46 8642 | 175–225 | Hot rolled, annealed, or | 18 | 60 | 0.15 | 0.006 | S4, S5 | M2, M3 |
| 1345 4147 5140 8645 | | cold drawn | 115 | 375 | 0.15 | 0.006 | P30 | C-6 |
| 4042 4340 5145 86B45 | | | | | | | | |
| 4047 50B40 5147 8740 | 325–375 | Normalized or quenched | 9 | 30 | 0.10 | 0.004 | S9, S11(b) | T15, M42(b) |
| 4140 50B44 81B45 8742 | | and tempered | 90 | 300 | 0.13 | 0.005 | P30 | C-6 |
| 4142 5046 8640 | | | | | | | | |
| **High carbon** | | | | | | | | |
| 50100 | 175–225 | Hot rolled, annealed, or | 17 | 55 | 0.13 | 0.005 | S4, S5 | M2, M3 |
| 51100 | | cold drawn | 105 | 350 | 0.15 | 0.006 | P30 | C-6 |
| 52100 | | | | | | | | |
| M-50 | | | | | | | | |
| **Wrought stainless steels** | | | | | | | | |
| **Ferritic** | | | | | | | | |
| 405 430 442 | 135–185 | Annealed | 27 | 90 | 0.075 | 0.003 | S4, S5 | M2, M3 |
| 409 434 446 | | | 84 | 275 | 0.15 | 0.006 | P30 | C-6 |
| 429 436 | | | | | | | | |
| **Austenitic** | | | | | | | | |
| 201 304 321 385 | 135–185 | Annealed | 20 | 65 | 0.075 | 0.003 | S4, S5 | M2, M3 |
| 202 304L 347 | | | 84 | 275 | 0.15 | 0.006 | K20 | C-2 |
| 301 305 348 | | | | | | | | |
| 302 308 384 | 225–275 | Cold drawn | 18 | 60 | 0.075 | 0.003 | S4, S5 | M2, M3 |
| | | | 69 | 225 | 0.10 | 0.004 | K20 | C-2 |
| **Martensitic** | | | | | | | | |
| 403 422 | 135–175 | Annealed | 27 | 90 | 0.075 | 0.003 | S4, S5 | M2, M3 |
| 410 501 | | | 76 | 250 | 0.15 | 0.006 | P30 | C-6 |
| 420 502 | 275–325 | Quenched and tempered | 15 | 50 | 0.075 | 0.003 | S9, S11(b) | T15, M42(b) |
| | | | 60 | 200 | 0.10 | 0.004 | P30 | C-6 |

(a) Based on standard 19 mm (¾ in.) cutting edge, single blade. (b) Any premium high-speed tool steel (T15, M33, M41–M47) or (S9, S10, S11, S12). Source: Metcut Research Associates as a starting point for selecting efficient and economical rates.

Processing conditions for a given application are best determined by production trials. The nature and amount of machining to be done, the size and shape of the part, the work material, and the equipment and tooling may necessitate substantial deviations from the nominal speed and feed.

In deep-hole trepanning, the ability to use optimum speed and feed largely depends on the rigidity of the boring bar and on the effectiveness with which the cutting fluid cools, flushes away chips, and prevents chips from forcibly contacting the newly machined surface. The speed and feed values given in Table 1 assume effective use of a suitable cutting fluid.

## Cutting Fluids for Deep-Hole Trepanning

Cutting fluids often represent the difference between success and failure in a trepanning operation. In general, the cutting fluids recommended for trepanning are similar to those recommended for gun drilling, that is, straight oils. Extreme pressure (EP) additives significantly improve tool life, particularly the life of the wear pads.

In applications in which high-nickel or high-cobalt alloys are trepanned, sulfurized or other EP-enhanced cutting oils are critical to the life of the wear pads. If the lubricating film between the wear pad and the workpiece breaks down, the cobalt binder in the carbide is quickly attacked, and the wear pads disintegrate. In critical applications for nuclear materials in which chlorine and sulfur are not allowed because of stress corrosion problems, synthetic coolants used at ratios as low as 5:1 with heavy EP additives can result in a successful application of trepanning.

The volume of the cutting fluid is also critical to the success of a trepanning operation. Trepanning produces a large number of chips at a very high rate of speed in an extremely confined area. Assuming the trepanning bit produces consistent, small chips, there is an immediate need to evacuate these chips as quickly as possible. High volumes of coolant with increased viscosity will normally accomplish this task. The power consumption of the coolant pump often equals the power consumption of the machine spindle driving the tool.

In an application involving the vertical trepanning of platens for plastic injection molding presses, a 45 kW (60 hp) vertical mill was used. The platens were 610 mm (24 in.) thick, and a 318 mm (12½ in.) hole had to be trepanned through 1020 steel. The cutting fluid was an EP-sulfurized mineral oil. Vertical application necessitated the use of large volumes of cutting fluid at high pressure to remove the chips from the hole. A 45 kW (60 HP) coolant pump was successfully used in the operation. The starting bushing was combined with a chip evacuation box surrounding the workpiece to reduce the flow of coolant to the bed of the machine and to direct the waste chips and coolant to a settling tank. Outside-diameter-exhaust trepanning was successfully used in this application.

# Planing

PLANING is a machining process for removing metal from surfaces in horizontal, vertical, or angular planes. In this process, the workpiece is reciprocated in a linear motion against one or more single-point tools. Although planing is most widely used for producing flat, straight surfaces on large workpieces, the process can also be used to produce contours and a variety of irregular configurations, such as deep slots in large rotors, helical grooves in large rolls, and internal guide surfaces in large valves. It is often possible to produce one or two parts on a planer in less time than is required merely to set up for machining by an alternative method; therefore, planing is often used for machining parts to meet production emergencies.

## Process Capabilities

As the hardness of the workpiece increases above about 25 HRC, metal removal rate and tool life decrease. However, metals hardened to 46 HRC or higher can be planed. For example, planing is often used to produce flat surfaces on large, heat-treated die blocks.

**The size of the workpiece** that can be planed is limited only by the capacity of the available equipment. Standard equipment is available that can make cuts as long as 15 m (50 ft), and still larger machines have been built to special order.

Although planing is most widely used for machining large areas, it is also used for machining smaller parts or areas. For example, jig-frame weldments can be squared and tooling-pad surfaces leveled on areas of less than 0.09 $m^2$ (1 $ft^2$). However, 305 mm (12 in.) is about the minimum planing stroke. Another common practice is tandem, or gang, planing, in which a number of relatively small but identical workpieces are lined up on the table and planed at the same time.

**Tolerance and Finish.** Even though the equipment used for planing is relatively large and rugged, close tolerances in both flatness and parallelism can be achieved with planers. For this reason, planing is sometimes selected over competitive processes.

Flatness variations can be held within 0.013 mm (0.0005 in.) total indicator reading on workpieces up to 0.4 $m^2$ (4 $ft^2$) in area or up to about 1.2 m (4 ft) long. Cast iron workpieces can be planed to a finish of about 1.60 μm (63 μin.) and steel workpieces to about 0.80 μm (32 μin.).

## Planers

The planer develops its cutting action from the straight-line reciprocating motion between the tool and the workpiece. On a planer, the work is reciprocated longitudinally while the tools are fed sideways into the work. The feed in planing is intermittent and represents width of cut. Planer tables are reciprocated by either mechanical or hydraulic drives. Most planers, however, are mechanically driven by such means as variable-voltage, constant-torque drives.

The speed at which a mechanical-drive planer operates depends on the speed of the driving motor and on the gear ratio. In hydraulic planers, table speed depends on the effective area of the piston and on the volume of oil pumped against the piston area per unit of time. Regardless of whether the drive is mechanical or hydraulic, the efficiency of planers can be greatly increased by incorporating a means of increasing table speed on the return stroke, during which time no cutting is done.

**Double-housing planers** (Fig. 1) incorporate two vertical uprights that support the crossrail. Double-housing planers are more rigid than open-side planers, but they are limited as to the width of workpiece they can accommodate.

**Open-side planers** differ from double-housing planers mainly in that they have only one upright column, from which the crossrail is cantilevered over the table. Open-side planers can accommodate wider workpieces than the double-housing machines because the workpieces can overhang from the side of the table without interfering with the planer mechanism. When the workpiece is considerably wider than the table, an outboard rolling table can be used to support the overhanging section of the workpiece.

The main disadvantage of open-side planers is that they are less rigid than double-housing planers. Open-side planers are also limited to three tool heads: two on the crossrail and one on the single upright column. However, an open-side planer can be converted into a double-housing planer by using an outboard detachable housing. Outboard housings are available as optional equipment and can be mounted or dismounted in about 1 h. This added support increases the rigidity of the machine and permits the use of an additional tool head.

*Double-housing planer with a tool head on one of the two vertical support uprights in addition to the two tool heads on the crossrail. Another tool head can be supported on the other upright in some models. The two uprights are immovable and limit the width of workpiece that can be planed in this machine.*

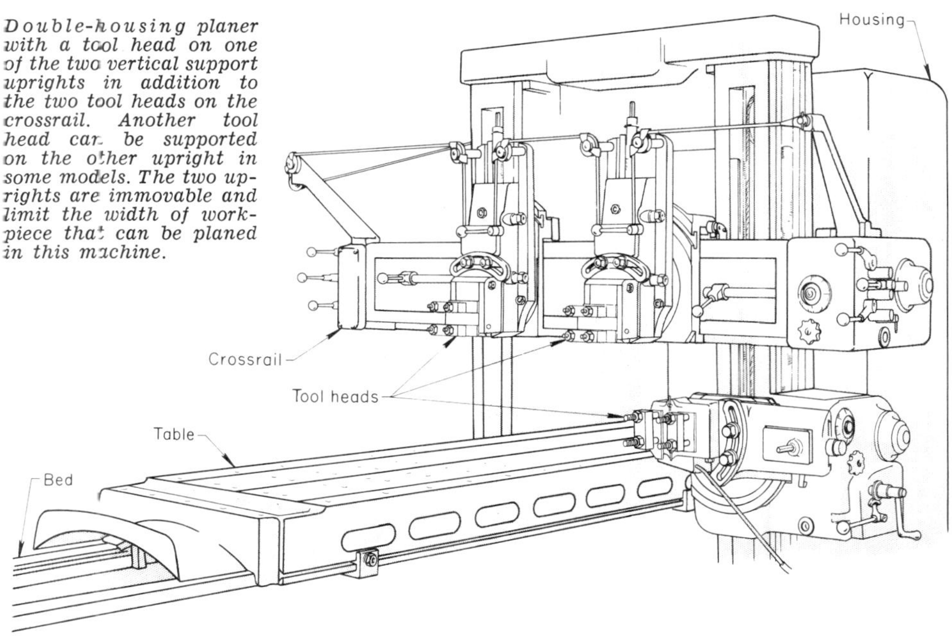


**Fig. 1** Significant components of a double-housing planer

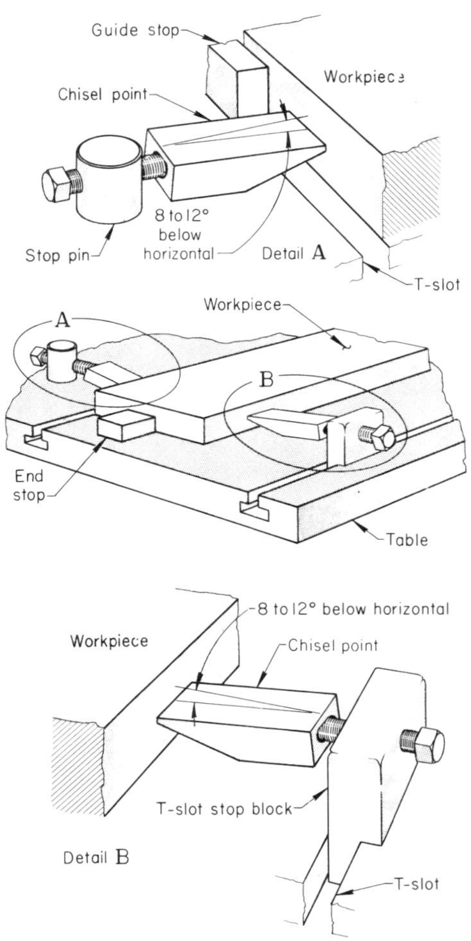


**Fig. 2** A common method for securing a long, flat plate to a planer table

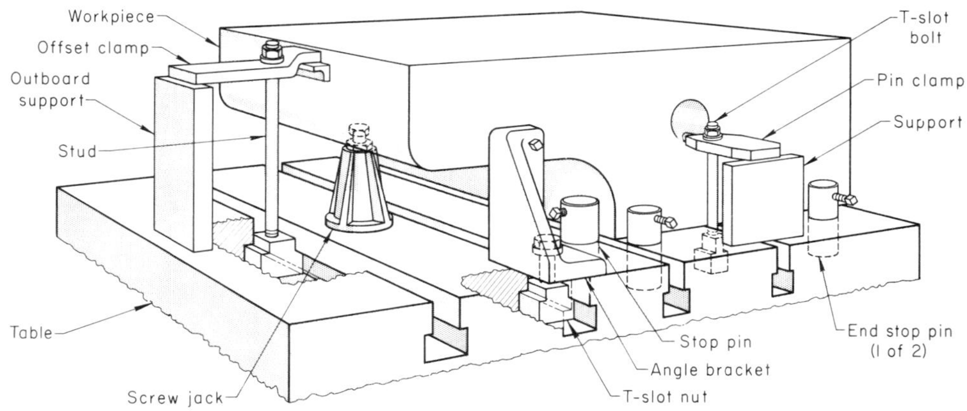


**Fig. 3** Typical arrangement of securing and supporting devices for mounting a large, irregular workpiece on a planer table

**Planer mills (adjustable rail mills)** have been developed from the use of accessory milling and boring heads on planers. These machines have various combinations of planing, milling, and boring heads, and they can perform multiple operations on a single workpiece setup.

**Workpiece Capacity.** Both double-housing and open-side planers are available in a wide range of sizes. Double-housing planers are rated by width, height, and length, in that order. Width refers to the maximum width of workpiece that can pass between the upright housings. Height refers to the maximum height of workpiece that can pass under the crossrail when supported on the table. Length refers to the maximum stroke. Open-side planers are similarly rated with regard to size, except that the width is table width, not maximum workpiece width.

**Tool Capacity.** Although only three tool heads can be used on an open-side planer, and four on a double-housing planer, tool capacity can be increased through the use of special holders that accommodate more than one single-point tool. It is common practice to use a roughing tool and a finishing tool in the same head, thus completing a planing operation in one pass instead of two (usually with some sacrifice in surface finish).

For straight planing, the efficiency of the planer can be greatly increased by using double-cutting tools, in special holders on special heads, which permit planing on both the forward and return strokes of the planer. When these tools are used, the speed of the normally noncutting return stroke—which on most modern planers is several times greater than that of the forward stroke—is of course reduced.

**Special Accessories.** Automatic controls and tracer attachments can be used to improve planing capabilities. Many planers are equipped for full control of speed and feed ranges from one suspended station and for automatic rail positioning and leveling. Tracer attachments permit the contour planing of such parts as large propeller blades, steam chests, and rocker arms for bridge supports.

Grinding and milling accessories can also be mounted in place of one of the rail or side heads. At slow feed rates, milling heads provide an increased stock removal rate. The use of grinding heads during slow feed rates can improve surface finish.

Planing is a rugged cutting operation, and setting up the work is probably the most important aspect. This is particularly true for the tandem planing of small or medium-size pieces. Serious damage and operator injury can also occur if the workpiece moves during planing.

**Clamping Hardware.** Setup hardware includes planer vises, jackscrews, clamps, and T-slot devices. An ample supply of setup hardware in a wide variety of types and sizes can greatly decrease the time required for setting up large and irregularly shaped workpieces, thus decreasing downtime. Inexpensive and seemingly inconsequential items of hardware can often reduce floor-to-floor time. For example, special T-slot nuts can be dropped into a 100 mm (4 in.) long cleared section of the slot at the desired location and given a quarter turn, thus eliminating the need to clear chips from the entire slot for loading another workpiece.

Cleaning chips from T-slots and stop-pin holes consumes operator time and machine time. For example, an operator might spend 40 min planing a workpiece and 30 min picking chips from slots and pin holes before the next workpiece can be set up. This loss of time can be prevented by placing specially cut lengths of hardwood strips in the T-slots, and metal covers on the stop-pin holes. These protective devices can be quickly removed to permit loading of the next workpiece.

**Magnetic Chucks.** Cast iron or steel workpieces are often held to planer tables with magnetic chucks. Magnetic chucking, however, is reliable only for holding workpieces that have a large table-contact area in relation to height. For this reason, holding by means of magnetic chucks is generally restricted to the planing of platelike workpieces. To ensure against movement of a workpiece held by magnetic chucking, an end stop should be placed against it counter to the cutting stroke (if double cutting is used, an end stop should be at each end of the work).

**Setup Plates.** Planing productivity can often be increased through the use of setup plates. A setup plate resembles a planer table in length and width, and it has T-slots for mounting the work. In production planing, the work is secured to the plate away from the planer, and the assembly is then carried to the planer table and located on it by keys at each end of the setup plate that fit into the center T-slot in the table. When two setup plates are used, one load can be planed while another is being set up—a procedure comparable to the use of a duplex table (see below).

**Tandem (Gang) Planing.** Specially designed fixtures that allow quick loading and unloading can greatly increase productivity in tandem planing. However, the cost of such fixtures can rarely be justified, except for continued production of identical or closed similar workpieces.

**Duplex Tables.** Planers with duplex, or divided, tables allow planing on one half of the table while the other half is being unloaded or reloaded. Duplex tables are split crosswise into two sections. The two sections can be fastened together and used as one long table, or they can be split for planing shorter workpieces. When split, one section of the table can be loaded and set up while work is being planed on the other section. This practice greatly increases the productivity of a planer, because in many applications setup time exceeds machining time.

## Workpiece Setup

Once the workpiece is placed on the table, the first step is to ensure that it rests solidly. If there is any wobbling, shim jacking must be used. For large workpieces with relatively slender sections, such as some weldments, jacks or blocks should be used to damp vibrations. A strap or pin clamp should be used over each jack or block to ensure that the workpiece rests solidly.

**Platelike Workpieces.** When magnetic chucks are not available or when a platelike workpiece does not lend itself to this method of mounting, various clamping devices are used. Figure 2 illustrates a common method of securing a long, flat plate to a planer table by using chisel points, T-slot stop blocks, stop pins, guide stops for initial alignment, and an end stop. Two precautions must be observed with this method. First, the adjusting screws contacting the chisel points must not extend beyond the stop blocks or stop pins by more than the distances suggested in Fig. 2 (one or two diameters), or the screws may bend and allow the workpiece to loosen. Second, the chisel points must be set at an angle of 8 to 12° (Fig. 2).

**Irregularly Shaped Workpieces.** A typical method of securing a large, irregular casting to a planer table is illustrated in Fig. 3. This setup uses an angle bracket and an outboard support. If double-cutting practice were employed, an angle bracket and two end stop pins would be used at each end of the workpiece. For workpieces of this type that are more than 1 or 1.5 m (4 or 5 ft) long, the use of two outboard supports and screw jacks is recommended.

**Workpieces for tandem planing** must either be tightly butted together (so that the tool cuts continuously for the entire stroke) or separated by 150 to 200 mm (6 to 8 in.) (for a fully interrupted cut). If the workpieces are separated by only 25 to 50 mm (1 to 2 in.), the tool is likely to break, because of deflection by chips lodged between the workpieces or because the sudden release of pressure on the tool shank as the tool emerges from the cut will cause a vibration and allow the tool to enter the next workpiece before becoming fully seated.

**Other Workpieces.** Bases for large presses or diesel engines are usually large rough castings, forgings, or weldments produced one at a time, and planing is often the first operation in machining them. Setup of these pieces on a planer is extremely critical because the planed surfaces will be used as a means of locating for subsequent operations.

## Tool Materials

High-speed steels, cast cobalt-chromium-tungsten alloys, and carbides are used as materials for planer tools. High-speed steels and cast alloys are often interchanged. In some heavy-duty planing applications, cast alloy tools have proved superior to high-speed steel tools. However, high-speed steel tools are used more often for planing.

Most planer tools have relatively large cross sections because they are made for maximum rigidity (note the typical planer tool and mounting shown in Fig. 4). Because of the size of these tools, it is common practice to make the shank from an alloy steel (such as 4140 or 4340) heat treated for high strength, then use inserts of high-speed steel (brazed or mechanically secured).

**High-Speed Steels.** General-purpose high-speed steels such as T1 and M2 have proved satisfactory for many planer tools or cutting edges. However, in applications involving hard work metals or heavy cuts, the cobalt types of high-speed steel, such as T6, T15, M6, or M44, will give better tool life. These more highly alloyed high-speed steels are generally used as inserts.

**Carbides.** Under conditions of maximum rigidity of machine, tools, and workpiece, carbide tools are more efficient than high-speed steel or cast alloy tools. Planing time is often reduced by 50% or more through the use of carbide instead of high-speed steel cutting tools, because cutting stroke speed can be increased up to 90 m/min (300 sfm).

In selecting tool material for planing, however, overall time is a more important consideration than machining time alone. In many applications, setup time is far greater than planing time, and under these conditions, large reductions in planing time have a relatively small effect on total processing time.

Before considering the use of carbide tools for planing, it must be ensured that the planer is rigid and in good condition and that it is capable of the high speeds required (90 m/min, or 300 sfm). The rail should be securely clamped to the housing or column, and the head should fit snugly onto the rail. The tool box and apron should be free of spring and excessive wear to permit the apron to seat in its clapper box. The planer should also be equipped with tool lifters to prevent the carbide tools from dragging or bouncing on the return stroke. Sufficient power must be available; in the rough planing of steel with one carbide tool, as much as 70 kW (90 hp) is sometimes required.

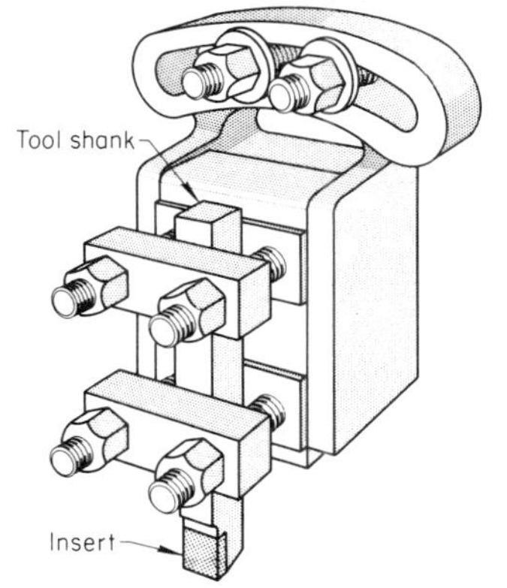


**Fig. 4** Planer tool and method of holding for maximum rigidity

A shock-resistant type of carbide must be used, regardless of whether it is a straight-tungsten grade for planing gray iron or a steel-cutting grade. Carbides suitable for planing usually contain up to about 16% Co.

Carbide planer tools should not be considered for the following:

- Workpieces that limit tool runout to about 75 mm (3 in.); the higher speeds used in carbide planing require a greater runout margin
- Workpieces that require the use of a longitudinal extension on the tool holder for reaching into blind areas (a practice known as poke planing)
- Workpieces that require excessive tool overhang in a vertical plane; as the tool is lowered in the holder, rigidity decreases
- Weldments in which metal hardness at junctions may vary considerably; for this condition, high-speed steel tools operated at slow speed will generally give better results
- Planing of metals that are harder than about 40 HRC; these must be planed at slow speeds, with which the inherent advantages of carbide tools cannot be realized

## Tool Design

Recommended design details for the cutting portions of high-speed steel planer tools are shown in Fig. 5. Tools are available in a variety of configurations suited to the undercutting, slotting, and straight planing of either horizontal or vertical surfaces. Tools having the small nose radii shown in Fig. 5 are preferred for roughing cuts; broad (often as wide as 38 mm, or 1½ in.), flat-nose tools are better for the finishing of most metals.

**Carbide Roughing Tools.** The back rake angle for carbide planer tools ranges from 0

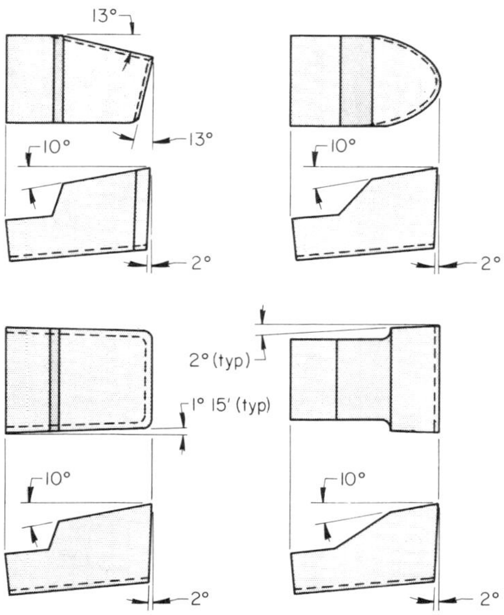


**Fig. 5** Recommended designs of high-speed steel planer tools

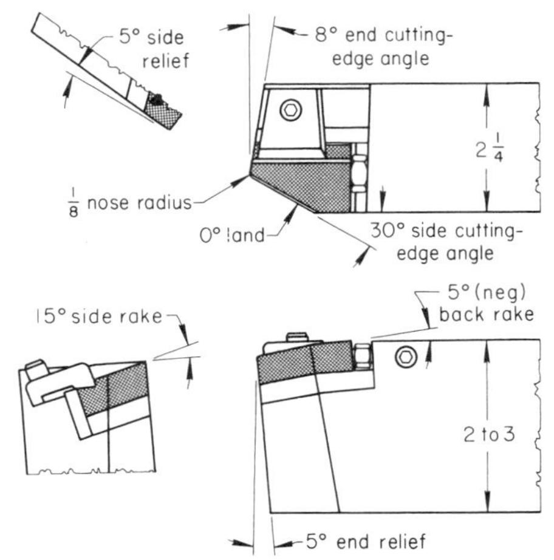


**Fig. 6** Carbide-insert tool for the rough planing of cast iron

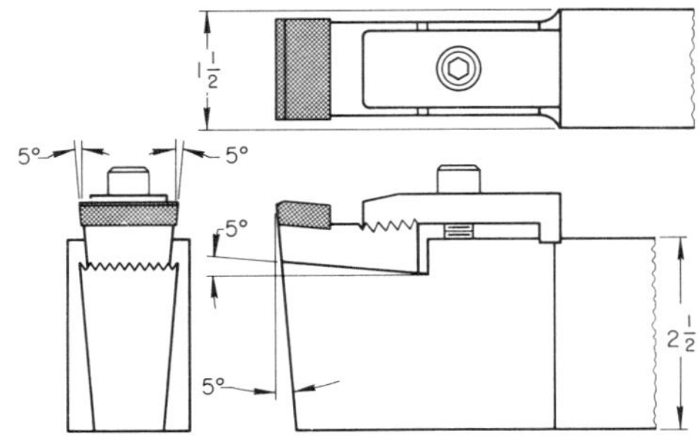


**Fig. 7** Typical insert tool for semifinish and finish planing or for slotting

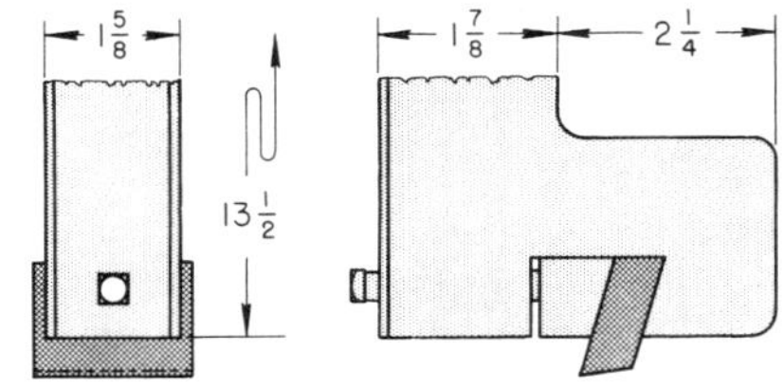


**Fig. 8** Gooseneck-holder tool used for light cuts in finish planing. Dimensions given in inches

to −15°. In general, the more difficult the work metal is to machine, the more negative is the back rake angle. A 0° back rake is used for soft metals; −3 to −5° for cast iron, low-carbon steel, and medium-carbon steels; and −5 to −15° for difficult-to-machine steels (such as 4340 alloy steel at 40 HRC). Negative back rake is also sometimes used when planing gummy metals because it aids chip flow and reduces face wear.

Side rake angles range from 3 to −15°, also depending on the machinability of the work metal. A side rake angle of 3 to 0° is usually suitable for free-cutting metals, 0° for medium-carbon steels, −3° for gummy metals, and up to −15° for difficult-to-machine steels. Excessive negative rake angles should be avoided because they are likely to cause chatter.

Side rake, or the land of the tool, is the largest single factor in controlling chip flow. The use of optimum side rake eliminates the need for chip breakers. This is an advantage because mechanical chip breakers are readily knocked off by the heavy chip produced in planing, and ground-in chip breakers weaken the cutting edge of the tool. Chip control is aided by grinding the land into a triangular shape. This causes the edge of the chip at the surface line of the cut to curl before the edge of the chip at the tool point, thus flowing the chip toward the side of the tool and back onto the workpiece.

The lead, or side cutting edge, angle directly controls chip thickness in relation to feed. A 45° lead angle will produce a thin chip, a 10° lead will produce a thicker chip, and a 0° lead will produce a chip equivalent in thickness to the feed used. The lead angle should be such that the chip produced is thick enough to take the initial shock away from the cutting edge and yet thin enough to curl properly. In planing low-carbon steel, a lead angle less than 15° will cause excessive shock, and a lead angle greater than 35° will produce a long, straight chip. A lead angle of 23 to 25° is near optimum for many metals, allowing the chip to flow onto the workpiece or against the tool holder and break into small segments. However, a lead angle of about 30° is most commonly used to obtain an acceptable chip thickness in roughing.

Carbide tools require a comparatively small nose radius to prevent machining stress from focusing at the radial point of the radius and rupturing the carbide. A small radius permits heat and stress to flow straight through the tip, causing less chatter. A nose radius of 1.6 to 3.2 mm (1/16 to 1/8 in.) is suitable for cast iron, and a 0.8 mm (1/32 in.) radius for steel.

Relief (clearance) angles should vary with the type of material being planed. Side and end relief angles of 5° are usually suitable for cast iron. For the planing of most steels, it is advisable to strengthen the cutting tool as much as possible by reducing the side clearance angle, thus providing a larger area of carbide for dissipating heat and absorbing shock.

Figure 6 shows a type of carbide-insert tool widely used for the rough planing of cast iron and steel. When ground to the configuration shown in Fig. 6, the tool is best suited for planing cast iron, although tools of this design have been successfully used on brass and aluminum. For planing carbon and alloy steels, the design of the tool should be modified in accordance with the recommendations discussed above.

Round or square button-type carbide inserts are also used for rough planing. Between regrinds for sharpening, the inserts can be rotated to present new cutting edges, thus allowing greater total tool life. Square button inserts can be ground in position. Round button inserts 25 mm (1 in.) in diameter and 13 mm (1/2 in.) thick are suitable for depths of cut to 9.5 mm (3/8 in.); for greater depths of cut, inserts 32 mm (1 1/4 in.) in diameter and 19 mm (3/4 in.) thick should be used. For planing cast iron, 0° back and side rakes and a 6° side relief are satisfactory. For planing steel, a −5°, 0.8 mm (1/32 in.) wide land on the cutting face has proved successful.

**Finishing Tools.** Figure 7 shows a typical broad-nose finishing tool that can also be used for slotting. Either a carbide or a high-speed steel insert can be used in this type of tool. As Fig. 7 shows, the bottom of the insert is serrated lengthwise, and the top of the insert is serrated across the width. This design permits offsetting the insert to the left or right as needed and provides a rigid lock. The cross serration at the top prevents vertical movement of the insert.

The insert is held in place mechanically by a serrated clamp, which is secured by a socket screw. When the cutting edge becomes dull, a new insert can be installed without removing the shank from the tool holder.

**Gooseneck-holder finishing tools** (Fig. 8) are primarily intended for use on cast iron, but have been successfully used for the finish planing of other metals. Gooseneck tools carry the cutting edge behind center so that the cutter is dragged in cutting. It does not dig in or chatter as readily as a cutter that is ahead of center.

Inserts for gooseneck holders can be made of high-speed steel or carbide, but greater accuracy is obtained with high-speed steel. High-speed steel inserts, which are available in widths up to 44 mm (1 3/4 in.), should be ground with a slight positive back rake; cutting edges should be honed with an oilstone to obtain maximum sharpness. For surfaces that require extreme accuracy (as for fitting to template gages), it is possible with high-speed steel inserts to take cuts of less than 0.025 mm (0.001 in.) on cast iron.

Carbide inserts, up to 32 mm (1¼ in.) wide, are also used in gooseneck holders for producing flat cuts. Edges of carbide inserts are not as keen as those of high-speed steel inserts; therefore, carbide inserts should not be used for depths of cut less than 0.05 mm (0.002 in.).

**Double-Cutting Tools.** Double-cut planing uses both strokes of the planer table for cutting and requires special tool holders mounted to a spindle in the planer head (Fig. 9). This spindle oscillates 13° clockwise and places a tool in cutting position for the normal cutting stroke. At the completion of the normal cutting stroke, the spindle oscillates 13° counterclockwise and positions a second tool for cutting on the return stroke of the planer table. In addition, a conventional finishing tool can be mounted in the clapper box of the head to semifinish a workpiece simultaneously with the roughing cut. This technique is known as triple planing.

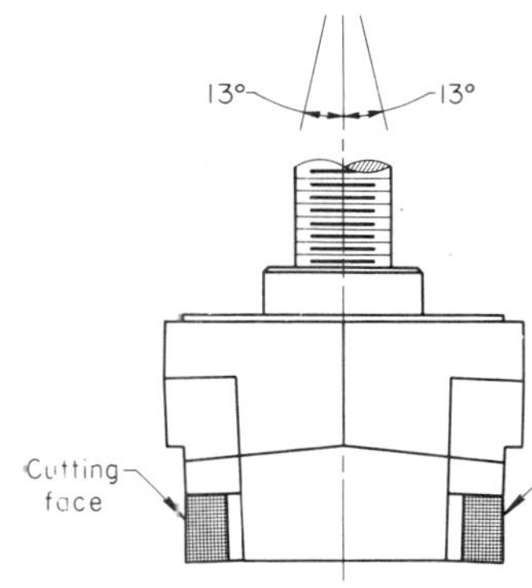


**Fig. 9** Tool for double-cut planing

## Speed, Feed, and Depth of Cut

Table 1 lists recommended speeds and feeds for various depths of cut in the rough and finish planing of several different metals with high-speed steel or carbide tools. The values given in Table 1 are not maximum, but are typical values that have proved consistent with optimum tool life and efficiency of metal removal.

Minor variables in work metal, tool design, or machine conditions often indicate the need for adjustments in speed, feed, or both. On new jobs, it is often necessary to inspect the tool after a few passes for the following conditions:

- Edge wear, which indicates too much speed or not enough feed
- Face wear, which indicates too much feed or not enough speed
- Redness at the tool point, which may occur in extreme cases of interrupted cutting and which should be kept to a minimum by adjusting speed, feed, or both

In general, it is advisable to plane steel with as heavy a feed and as high a speed as possible, because this promotes good chip flow without the aid of mechanical or ground-in chip breakers. A heavy chip will curl and break into small segments better than a thin chip. Increased cutting speed maximizes heating of the chip and is the best means to prevent overheating of the workpiece and the tool. Using carbide tools, thin plates have been more successfully planed by increasing the cutting speed from 70 to 90 m/min (225 to 300 sfm), thus transmitting more heat to the chips.

Uniform cutting speed and feed should be maintained throughout the entire stroke of the planer. It is not necessary to start the cut with a slow speed and a low feed, nor is it advantageous to preheat the tool before beginning a cut.

When carbide tools are used for planing medium grades of gray iron (~200 HB), common practice is to operate at 46 to 53 m/min (150 to 175 sfm) and 2.0 to 2.4 mm (5/64 to 3/32 in.) feed per stroke with a 13 to 19 mm (½ to ¾ in.) depth of cut. By decreasing the

**Table 1 Recommended speeds and feeds for planing with high-speed steel or carbide tools**

| Work metal | Hardness, HB | Depth of cut: 3.2 mm (⅛ in.) Speed, m/min (sfm) | Depth of cut: 3.2 mm (⅛ in.) Feed, mm (in.) per stroke | Depth of cut: 6.4 mm (¼ in.) Speed, m/min (sfm) | Depth of cut: 6.4 mm (¼ in.) Feed, mm (in.) per stroke | Depth of cut: 13 mm (½ in.) Speed, m/min (sfm) | Depth of cut: 13 mm (½ in.) Feed, mm (in.) per stroke | Depth of cut: 25 mm (1 in.) Speed, m/min (sfm) | Depth of cut: 25 mm (1 in.) Feed, mm (in.) per stroke | Finishing speed(a), m/min (sfm) |
|---|---|---|---|---|---|---|---|---|---|---|
| **High-speed steel tools** | | | | | | | | | | |
| Cast iron | 230 | 15 (50) | 2.3–3.2 (0.090–0.125) | 15 (50) | 1.5–2.3 (0.060–0.090) | 12 (40) | 1.1–1.5 (0.045–0.060) | 11 (35) | 0.8–1.1 (0.030–0.045) | 12 (40) |
| Cast iron | 175 | 21 (70) | 2.3–3.2 (0.090–0.125) | 18 (60) | 1.5–2.3 (0.060–0.090) | 15 (50) | 1.1–1.5 (0.045–0.060) | 12 (40) | 1.1–1.5 (0.045–0.060) | 18 (60) |
| Steel | 270 | 11 (35) | 1.5–2.3 (0.060–0.090) | 9 (30) | 1.5–2.3 (0.060–0.090) | 7.6 (25) | 1.1–1.5 (0.045–0.060) | 6 (20) | 1.1–1.5 (0.045–0.060) | 6 (20) |
| Steel | 200 | 11 (35) | 2.3–3.2 (0.090–0.125) | 11 (35) | 1.5–2.3 (0.060–0.090) | 9 (30) | 1.5–2.3 (0.060–0.090) | 8 (25) | 1.1–1.5 (0.045–0.060) | 9 (30) |
| Steel | 130 | 18 (60) | 2.3–3.2 (0.090–0.125) | 15 (50) | 2.3–3.2 (0.090–0.125) | 12 (40) | 1.5–2.3 (0.060–0.090) | 9 (30) | 1.1–1.5 (0.045–0.060) | 15 (50) |
| Bronze | Hard | 18 (60) | 2.3–3.2 (0.090–0.125) | 18 (60) | 2.3–3.2 (0.090–0.125) | 15 (50) | 2.3–3.2 (0.090–0.125) | 12 (40) | 1.5–2.3 (0.060–0.090) | 18 (60) |
| Bronze | Soft | 45 (140) | 4.0–4.8 (0.156–0.188) | 35 (120) | 3.2–4.0 (0.125–0.156) | 30 (100) | 3.2–4.0 (0.125–0.156) | 30 (100) | 2.3–3.2 (0.090–0.125) | 45 (140) |
| **Carbide tools** | | | | | | | | | | |
| Cast iron | 230 | 60 (200) | 2.3–3.2 (0.090–0.125) | 60 (200) | 1.9–2.3 (0.075–0.090) | 60 (200) | 1.5–1.9 (0.060–0.075) | 60 (200) | 1.1–1.5 (0.045–0.060) | 55 (180) |
| Cast iron | 175 | 90 (300) | 2.5–3.2 (0.100–0.125) | 90 (300) | 2.3–2.5 (0.090–0.100) | 90 (300) | 1.9–2.3 (0.075–0.090) | 90 (300) | 1.5–1.9 (0.060–0.075) | 65 (220) |
| Steel | 270 | 75 (250) | 1.5–1.9 (0.060–0.075) | 75 (250) | 1.5–1.9 (0.060–0.075) | 75 (250) | 1.1–1.5 (0.045–0.060) | ... | ... | 75 (250) |
| Steel | 200 | max | 1.5–2.3 (0.060–0.090) | max | 1.5–1.9 (0.060–0.075) | 90 (300) | 1.5–1.9 (0.060–0.075) | ... | ... | 90 (300) |
| Steel | 130 | max | 1.5–2.3 (0.060–0.090) | max | 1.5–2.3 (0.060–0.090) | max | 1.5–2.3 (0.060–0.090) | ... | ... | 90 (300) |
| Bronze | Hard | max | 2.3–3.2 (0.090–0.125) | max | 2.3–3.2 (0.090–0.125) | max | 2.3–2.5 (0.090–0.100) | max | 1.5–2.3 (0.060–0.090) | max |
| Bronze | Soft | max | 2.3–3.8 (0.090–0.150) | max | 2.3–3.2 (0.090–0.125) | max | 2.3–3.2 (0.090–0.125) | max | 2.3–2.5 (0.090–0.100) | max |

(a) For a depth of cut ranging from 0.08–0.38 mm (0.003–0.015 in.). Finishing feeds at these speeds depend on the type of tool used. Flat-nose tools are used for cast iron and bronze at feeds of 13–25 mm (½–1 in.) per stroke; variations of flat-nose tools are used for steel at feeds of 3.2–13 mm (⅛–½ in.) per stroke. Round-nose tools are sometimes used at feeds of 1.1–1.5 mm (0.045–0.060 in.), depending on the nose radius and on the finish desired.

depth of cut to 9.5 mm (3⁄8 in.), speed can safely be increased to 55 to 60 m/min (180 to 200 sfm), and feed can be increased to 2.4 to 3.2 mm (3⁄32 to 1⁄8 in.) per stroke.

Low-carbon steel can be planed at higher speed than gray iron with carbide tools; 90 m/min (300 sfm) and a feed of 2.4 to 3.2 mm (3⁄32 to 1⁄8 in.) per stroke with 9.5 mm (3⁄8 in.) depth of cut is common practice. Medium-carbon low-alloy steel such as annealed 4130 can be planed at 53 to 60 m/min (175 to 200 sfm), 2.0 mm (5⁄64 in.) feed, and 9.5 mm (3⁄8 in.) depth of cut. For higher-carbon higher-alloy steels such as 4350, common practice is 30 to 45 m/min (100 to 150 sfm) at 1.2 to 1.6 mm (3⁄64 to 1⁄16 in.) feed and 9.5 mm (3⁄8 in.) depth of cut. Die blocks at about 36 HRC have been planed with carbide tools at 30 m/min (100 sfm), 1.2 mm (3⁄64 in.) feed per stroke, and 13 mm (1⁄2 in.) depth of cut. Heat-treated die blocks at 46 HRC have been planed with carbide at 4.5 to 6.0 m/min (15 to 20 sfm), 0.8 mm (1⁄32 in.) feed per stroke, and 6.4 mm (1⁄4 in.) depth of cut.

A carbide tool may break because of side pressure as it makes the last one or two passes on a workpiece. To minimize this possibility, on cuts of more than 6.4 mm (1⁄4 in.) depth it is advisable to reduce the feed to about 1 mm (3⁄64 or 1⁄32 in.) and to decrease the speed by 50% for four or five strokes before the tool emerges from a cut. This will often prevent the breakage caused by pressure on the side clearance portion of the tool from the elastic behavior of the work metal. That is, too much feed will deflect the thin flange or remaining portion of the work at the cutting edge of the tool, thus preventing the tool from removing the full amount of metal for which the feed is set. As the tool continues through the stroke, the metal flange, with the deflected amount of stock still intact, will spring back to its normal position and rub against the side clearance edge of the tool on the return stroke, thus generating heat by friction. Even if this does not cause immediate tool failure, breakout of the steel may be caused when the tool, under automatic feed, overtakes the deflected metal and plows through an oversize flange equivalent to a double or triple feed. This will set up intense chatter and cause the tool to chip.

## Special Applications

Planing is primarily used for removing metal in a straight line from flat surfaces. However, a variety of contour operations can be accomplished by template control (Fig. 10). An example of contouring by planing is the cutting of helical grooves in large rolls. For this operation, the roll is set up longitudinally on the planer table with special holding fixtures that include a drive mechanism to rotate the roll at a predetermined speed as it advances into the fixed cutting tool, thus cutting a helical groove.

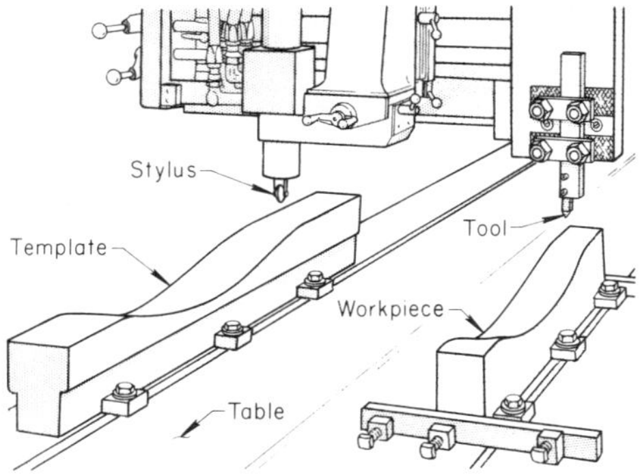


**Fig. 10** Contour planing with a template

## Planing Versus Alternative Processes

Broaching or even sawing can be used as an alternative to planing, but milling and grinding are most often the competitive processes. The selection of planing in preference to an alternative process is usually based on the following considerations:

- The initial cost of a planer is about half that of a planer-type milling machine for performing the same job
- Because of the lower initial cost of the planer, the burden rate is lower, regardless of whether the machine is used part time or full time
- Certain shapes, such as dovetails and V-sections, are particularly suited to planing
- For some parts, such as large machine tool components, the dimensional accuracy required is obtainable only by planing
- Planing is preferred for bearing surfaces that must be finished by scraping because the surface condition that results is more suitable than that produced by milling
- Because of its versatility, low tooling cost, and short tooling-up time, planing is more economical for low production

**Planing Versus Sawing.** In some applications, planing and band sawing are competitive operations. The following describes an application in which either planing or band sawing could have been used to obtain acceptable results, but planing was used because it was less expensive.

**Example 1: Planing Versus Gas Cutting Versus Band Sawing for Slitting Steel Plate.** Figure 11 shows the setup of a planer for slitting 9.5 mm (3⁄8 in.) alloy steel plates that were 925 mm (36½ in.) wide by 1.8 m (6 ft) long into three 300 mm (12 in.) wide strips. Strips were originally produced by band sawing, but this method was slow and costly. Gas cutting had been tried, but was rejected because an additional finishing operation was needed to smooth the cut edges. As indicated in the table with Fig. 11, planing produced the strips in less time than band sawing.

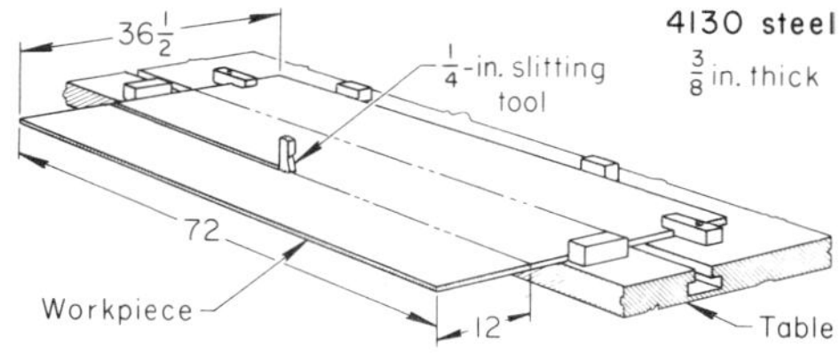


**Fig. 11** Slitting steel plate by planing and comparing with band sawing. Dimensions given in inches

**Planing Versus Milling.** Compared with milling on the basis of volume of metal removed per unit of time, planing is relatively inefficient; metal can be removed about twice as fast by milling as by the most efficient planing method (double-cut). However, the longer setup time required for milling and the more expensive equipment and tooling (for one job, tooling for milling costs fifty times as much as tooling for planing) can be justified only when large quantities of similar parts are to be produced. For extremely large workpieces, production is usually low; therefore, planing is the less costly method.

Planing is often the most practical approach for machining several surfaces to a given absolute level, as in constructing jig-frame components or machining large, irregular workpieces. For such applications, milling would seldom be practical, mainly because only unit production is involved.

## Cutting Fluids

A flood of cutting fluid is seldom used for planing operations because two of the three functions of cutting fluid (chip disposal and cooling) are less important in planing than in operations such as turning. In most planing operations, chips are relatively thick and are thrown clear. Therefore, they seldom interfere with successive cutting strokes. As chips begin to pile up on the table or fixtures they should be brushed away. The use of compressed air is not recommended, because chips are likely to lodge in the mechanism.

In planing, the tool is seldom engaged more than 75% of the time. Therefore, cooling of tools and workpieces usually presents no problem. In some planing operations, however, cutting fluids will improve dimensional accuracy, finish, and tool life by minimizing the adherence of work metal to the tool. When cutting fluids are used, a common practice is to apply the fluid directly to the cutting area with a swab. A spray mist of cutting oil diluted with a lower-viscosity oil such as mineral seal oil is sometimes effective. A film of oil mist aids the cutting action and has a mild cooling effect on the tools in planing operations.

# Shaping and Slotting

SHAPING AND SLOTTING are machining processes that remove metal from surfaces through the use of a single-point tool supported by a ram that reciprocates the tool in a linear motion against the workpiece. Shaping can be done on machines with either vertical or horizontal rams, while slotting is commonly done on machines with vertical rams.

## Process Capabilities

Shaping and slotting, although versatile processes with short setup times and relatively inexpensive tools, are comparatively inefficient means of metal removal. The cost per cubic inch of metal removed by shaping or slotting may be as much as five times that for removal by milling or broaching in a given job. For this reason, shaping and slotting are generally confined to unit or small-quantity production, as in toolrooms or model shops.

As the hardness of the workpiece increases above about 25 HRC, metal removal rate and tool life decrease. On the other hand, when the occasion demands, pieces much harder than this can be cut on a shaper. Steel hardened to 46 HRC, or even higher (heat-treated die blocks, for example), has been successfully machined by shaping and slotting.

**Size of the workpiece** that can be shaped or slotted is limited by the maximum length of stroke, which for standard shapers or slotters is about 915 mm (36 in.). When surfaces longer than 915 mm (36 in.) must be machined, planing or some other suitable process is used.

The usual range of cutting stroke is even less—from 150 to 610 mm (6 to 24 in.). Although machines with longer strokes could be built, they would be impractical for most purposes because dimensional accuracy decreases as stroke length increases. On even the best-maintained shapers, deviation from dimensional accuracy is about 0.04 mm/m (0.0005 in./ft) of ram travel.

**Configuration of Workpiece.** Although shaping is most commonly used for machining flat surfaces, the process can also be used to produce contours and a variety of irregular configurations. Shaping is sometimes used to machine contours because production quantities are too low to justify the expense of the tooling required for producing the same configurations by milling or broaching. In addition, some complex configurations are machined on a shaper because they would be difficult or impossible to produce by milling or broaching—for example, deep internal slots and tortuous contours and configurations in blind holes.

Because of its versatility and short setup time, shaping is often used for the emergency production of gears, splined shafts, racks, or similar parts. It is often possible to produce one or two such parts in a shaper in less time than is required merely to set up for production on other, alternative equipment with a higher output rate.

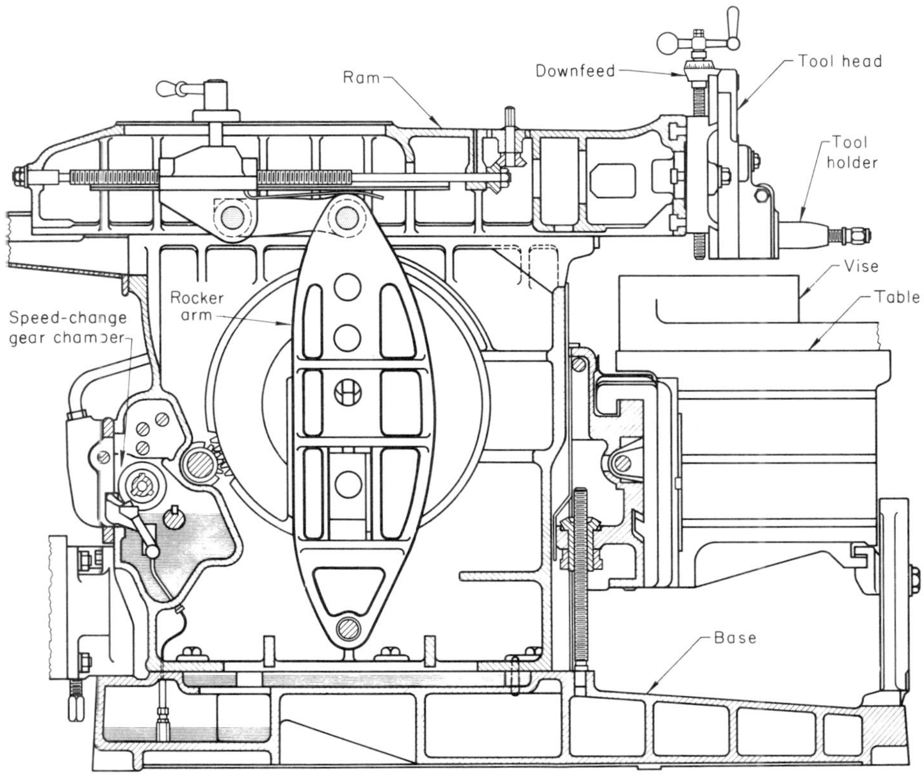


**Fig. 1** Sectional view of a crank-driven horizontal shaper

## Machines

Shaping and slotting machines develop cutting action from a straight-line reciprocating motion between the tool and the work. The tool is driven forward and recovered by a sliding ram. The work is fed at right angles to the direction of the ram stroke in small increments. Most shapers have rams that drive the cutting tool in a horizontal direction, but a few shapers have rams that drive the cutting tool in a vertical direction. Slotters also operate in a vertical direction. In either type of shaper, the workpiece rests on a flat bed, which advances it toward the cutting tool.

**Horizontal shapers** can either be crank driven or operated hydraulically. A hydraulic shaper uses a piston and cylinder to operate the ram. However, because of the higher cost of a hydraulic unit, the comparatively low efficiency of the hydraulic drive, and the difficulty in obtaining stroke length accuracy with hydraulic shapers, most horizontal shapers are crank driven.

Figure 1 shows a sectional view of a crank-driven horizontal shaper and identifies its important working components. The rocker arm is reciprocated by a crankpin mounted on the crank gear. The crank mechanism is an application of a Whitworth quick-return mechanism.

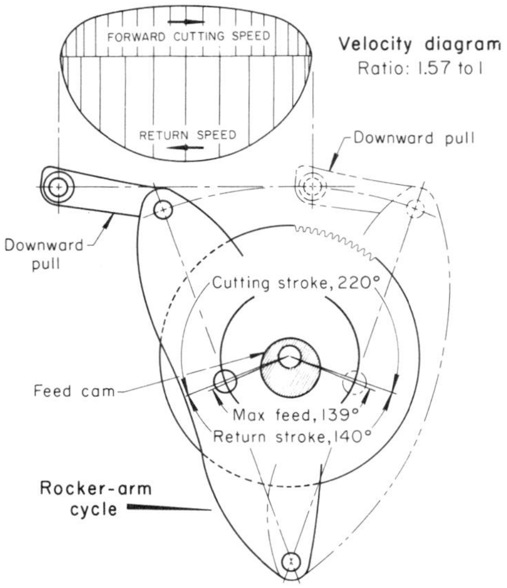


**Fig. 2** Velocity diagram and rocker-arm cycle for a crank-driven horizontal shaper. See text for discussion.

For increased efficiency, shapers are built so that the ram speed in the return stroke is greater than that in the forward, or cutting, stroke. This is accomplished as indicated in the rocker-arm cycle illustrated in Fig. 2. As shown in Fig. 2, 220° of the circle is used for the cutting stroke and only 140° for the return stroke—a ratio of approximately 1.6 to 1. This ratio between forward and return strokes is not a fixed ratio; it varies among different designs of machines and with the length of the stroke.

Horizontal shapers are furnished with either plain or universal tables. The universal table, in conjunction with a swivel vise, provides rotation on all three axes. Graduations on all three movements allow angular setups to be made quickly.

Horizontal shapers range in size from small bench models with a maximum stroke length of less than 150 mm (6 in.) to large, rugged machines with a maximum stroke of as much as 915 mm (36 in.). On a machine of any size, however, the length of stroke can be varied from its maximum to slightly less than 25 mm (1 in.) for the largest machine and to about 3.2 mm (1/8 in.) for the smallest.

Horizontal shapers are commonly provided with powered table feeds ranging from about 0.25 mm (0.010 in.) per stroke (for a machine with a 150 mm, or 6 in., maximum stroke) to 5.00 mm (0.200 in.) per stroke (for a machine with a 915 mm, or 36 in., maximum stroke). Over the same range of machine sizes, vertical power feeds on the tool head will range from about 0.125 to 2.50 mm (0.005 to 0.100 in.) per stroke.

**Slotters and vertical shapers** are very similar and are also much like the horizontal shaper except that the ram operates vertically, rather than horizontally, cutting on the downstroke. The slotter, as the name implies, was first developed for cutting slots or keyways. The vertical shaper, usually a much smaller version of the slotter, was developed for toolroom work. Slotters can have ram strokes up to 1800 mm (72 in.) long. Most vertical shapers have strokes of 150 to 300 mm (6 to 12 in.). In these vertical machines, because the ram must be pulled against the force of gravity on its upward stroke, a counterweight is added to equalize the power requirements on the up and down strokes and to enable a smoother action of the machine.

Most vertical shapers have a means of adjusting the ram and its guides so that it can be set at an angle as great as 15° to the vertical. This permits the cutting of proper clearances in dies and similar work. Many slotters also have this ram adjustability.

Tables on vertical machines can be rotated and can be moved longitudinally or transversely. With this degree of flexibility in direction of feed, a vertical machine can cut almost any type of groove, slot, or keyway.

Vertical shaping machines have longitudinal and transverse power feeds ranging from about 0.05 to 2.50 mm (0.002 to 0.100 in.) per stroke for short-stroke machines and up to 3.80 mm (0.150 in.) per stroke for a 915 mm (36 in.) stroke machine. Rotary feeds usually range from 0.10 to 4.45 mm (0.004 to 0.175 in.) on a 500 mm (20 in.) circle.

**The ram speeds** of most shapers can be adjusted to provide incremental increases in surface speeds from about 1.5 to 90 m/min (5 to 300 sfm). Speeds are changed by positioning the range and speed levers. Typical shapers have four speeds available in each of four ranges; this permits adjustment in order to obtain 16 different speeds within the total speed range of a given machine. Cutting speed in meters per minute is a function of both length of stroke and the number of strokes per minute. This relationship is shown quantitatively in Table 1.

**Tools.** Although most shaping is done with only one single-point tool, there are notable exceptions. For example, form tools are often used in slotters or shapers for machining two or more surfaces simultaneously or for machining conjugate surfaces (such as machining a spur gear with a tool in the form of a gear lock). In other applications, as in cutting closely spaced parallel grooves, a conventional box tool holder is mounted on the tool head, and two or three cutting tools are secured in the holder so that more than one cut can be made simultaneously.

**Automatic Controls.** Because shapers are seldom used for volume production, the degree of automation with which they are provided is generally less than that for other types of machines. On the simplest type of shaper, the operator first sets controls for depth of cut, ram speed, and stroke length and then feeds the work sideways into the tool by turning a handle that has a graduated dial. The work is then advanced into the tool at the end of each return stroke. Most shapers, however, are equipped with power feeds. Therefore, a more common practice is to preset the controls for the desired feed per stroke. When the cut is finished, the operator stops the machine, returns the table to its starting position, resets the downfeed, and starts another cycle.

The degree of automation and process efficiency of shapers can be progressively increased by the addition of:

- Power downfeed, operated either with or independently of the table feed
- Various stops that control the end of a desired cycle
- Complete controls that allow recycling according to a preestablished program
- Automatic indexing devices for cutting slots or grooves according to an established spacing
- Automatic duplicating equipment that guides the cutting tools by means of templates for shaping contours

## Workholding Devices

Shaper tables are usually provided with standard T-slots, which permit the direct mounting of some workpieces by means of bolts, blocks, and clamp plates (Fig. 3a). However, most workpieces are placed in vises, which are secured to the table by T-bolt mounting. It is usually preferred that the direction of clamping in the vise be at right angles to the direction of tool travel. Clamping must be tight. If there is danger of marring the workpiece surfaces by clamping, shims made of soft metal (or, for more positive gripping, shims made of emery paper) can be used.

For workpieces that are too thin to be held firmly in a standard vise, a special clamping device with toe dogs and a stop can be used (Fig. 3b). In this arrangement, clamping forces are parallel with tool travel, and the stop prevents the workpiece from being forced out of the vise. Outboard support by jacks is often utilized for obtaining rigidity in the shaping of irregular configurations such as angles (Fig. 3c).

In contouring and slotting operations, rugged vises (Fig. 3d) often must be used because workpieces are usually of irregular shape and because cutting forces are likely to vary during the operation. If special indexing fixtures are used, they must be designed for maximum rigidity.

## Tool Heads

The principal components of a shaper tool-head assembly are illustrated in Fig. 4. The clapper box holds the tool rigidly on the forward (cutting) stroke and allows it freedom to tilt upward on the return stroke after force is removed from the tool face. There-

## Table 1 Surface speeds in shaping as related to frequency and length of stroke

Surface speeds given in m/min (sfm)

| Strokes per min | 25 (1) | 50 (2) | 75 (3) | 100 (4) | 125 (5) | 150 (6) | 175 (7) | 200 (8) | 225 (9) | 250 (10) | 275 (11) | 300 (12) | 325 (13) | 350 (14) | 375 (15) | 400 (16) | 425 (17) | 450 (18) | 475 (19) | 500 (20) | 550 (22) | 600 (24) | 650 (26) | 700 (28) | 750 (30) | 800 (32) | 850 (34) | 900 (36) |
|---|---|---|---|---|---|---|---|---|---|---|---|---|---|---|---|---|---|---|---|---|---|---|---|---|---|---|---|---|
| | Length of stroke, mm (in.) | | | | | | | | | | | | | | | | | | | | | | | | | | | |
| 500 mm (20 in.) standard shaper (25–500); 400 mm (16 in.) heavy-duty shaper (25–425) | | | | | | | | | | | | | | | | | | | | | | | | | | | | |
| 12.... | 0.9 (3) | 1.2 (4) | 1.8 (6) | 2.4 (8) | 2.7 (9) | 3.0 (10) | 4.0 (13) | 4.6 (15) | 4.9 (16) | 5.5 (18) | 6.1 (20) | 6.4 (21) | 7.0 (23) | 7.3 (24) | 7.9 (26) | 8.2 (27) | 8.5 (28) | 9.1 (30) | 9.5 (31) | 9.8 (32) | ... | ... | ... | ... | ... | ... | ... | ... |
| 17.... | 1.2 (4) | 1.8 (6) | 2.4 (8) | 3.4 (11) | 4.0 (13) | 4.9 (16) | 5.5 (18) | 6.4 (21) | 7.0 (23) | 7.6 (25) | 8.5 (28) | 9.1 (30) | 9.8 (32) | 10.4 (34) | 11 (36) | 11.6 (38) | 12.2 (40) | 12.8 (42) | 13.4 (44) | 14.0 (46) | ... | ... | ... | ... | ... | ... | ... | ... |
| 26.... | 1.5 (5) | 2.4 (8) | 4.0 (13) | 5.2 (17) | 6.1 (20) | 7.3 (24) | 8.5 (28) | 9.8 (32) | 10.7 (35) | 11.9 (39) | 12.8 (42) | 14 (46) | 14.9 (49) | 15.8 (52) | 16.8 (55) | 17.7 (58) | 18.6 (61) | 19.5 (64) | 20.4 (67) | 21.3 (70) | ... | ... | ... | ... | ... | ... | ... | ... |
| 36.... | 1.8 (6) | 3.7 (12) | 5.2 (17) | 7.0 (23) | 8.5 (28) | 10.4 (34) | 11.9 (39) | 13.4 (44) | 14.9 (49) | 16.5 (54) | 18 (59) | 19.2 (63) | 20.7 68 | 21.9 (72) | 23.5 (77) | 24.7 (81) | 25.9 (85) | 27.4 (90) | 28.3 (93) | 29.6 (97) | ... | ... | ... | ... | ... | ... | ... | ... |
| 54.... | 2.7 (9) | 5.5 (18) | 7.9 (26) | 10.4 (34) | 13.1 (43) | 15.2 (50) | 17.7 (58) | 20.1 (66) | 22.3 (73) | 24.7 (81) | 26.8 (88) | 29 (95) | 31.1 (102) | 32.9 (108) | 35 (115) | 36.9 (121) | 39 (128) | 40.5 (133) | 42.7 (140) | 44.5 (146) | ... | ... | ... | ... | ... | ... | ... | ... |
| 77.... | 4.0 (13) | 7.6 (25) | 11.3 (37) | 14.9 (49) | 18.6 (61) | 21.9 (72) | 25.3 (83) | 28.7 (94) | 32 (105) | 35.0 (115) | 38.1 (125) | 41.1 (135) | 44.2 (145) | 46.9 (154) | 50 (164) | 52.4 (172) | 54.9 (180) | 57.6 (189) | 60.6 (199) | 63.4 (208) | ... | ... | ... | ... | ... | ... | ... | ... |
| 115.... | 5.8 (19) | 11.3 (37) | 17.1 (56) | 22.3 (73) | 27.7 (91) | 32.9 (108) | 37.8 (124) | 42.7 (140) | 47.5 (156) | 52.4 (172) | 57.3 (188) | 61.6 (202) | 65.8 (216) | 70.1 (230) | ... | ... | ... | ... | ... | ... | ... | ... | ... | ... | ... | ... | ... | ... |
| 157.... | 7.9 (26) | 15.5 (51) | 23.2 (76) | 30.5 (100) | 38.7 (127) | 44.8 (147) | 51.8 (170) | 58.5 (192) | 65.2 (214) | 71.9 (236) | ... | ... | ... | ... | ... | ... | ... | ... | ... | ... | ... | ... | ... | ... | ... | ... | ... | ... |
| 600 mm (24 in.) standard shaper (25–600); 500 mm (20 in.) heavy-duty shaper (25–500) | | | | | | | | | | | | | | | | | | | | | | | | | | | | |
| 10.... | ... | 0.9 (3) | 1.5 (5) | 2.1 (7) | 2.7 (9) | 3.0 (10) | 3.4 (11) | 4.0 (13) | 4.6 (15) | 4.9 (16) | 5.2 (17) | 5.5 (18) | 5.8 (19) | 6.1 (20) | 6.4 (21) | 6.7 (22) | 7.0 (23) | 7.6 (25) | ... | 8.2 (27) | 8.8 (29) | 9.8 (32) | ... | ... | ... | ... | ... | ... |
| 14.... | ... | 1.5 (5) | 2.1 (7) | 2.7 (9) | 3.4 (11) | 4.0 (13) | 4.6 (15) | 5.2 (17) | 5.8 (19) | 6.4 (21) | 7.0 (23) | 7.3 (24) | 7.9 (26) | 8.5 (28) | 9.1 (30) | 9.8 (32) | 10.1 (33) | 10.7 (35) | ... | 11.9 (39) | 12.8 (42) | 13.7 (45) | ... | ... | ... | ... | ... | ... |
| 21.... | ... | 2.1 (7) | 3.0 (10) | 4.0 (13) | 4.9 (16) | 5.8 (19) | 6.7 (22) | 7.6 (25) | 8.5 (28) | 9.4 (31) | 10.4 (34) | 11.3 (37) | 12.2 (40) | 13.1 (43) | 14.0 (46) | 14.6 (48) | 15.5 (51) | 16.1 (53) | ... | 17.7 (58) | 18.9 (62) | 20.4 (67) | ... | ... | ... | ... | ... | ... |
| 29.... | ... | 3.0 (10) | 4.3 (14) | 5.8 (19) | 7.0 (23) | 8.2 (27) | 9.8 (32) | 11 (36) | 12.2 (40) | 13.4 (44) | 14.6 (48) | 15.5 (51) | 16.8 (55) | 18 (59) | 19.2 (63) | 20.4 (67) | 21.3 (70) | 22.3 (73) | ... | 24.1 (79) | 26.2 (86) | 28 (92) | ... | ... | ... | ... | ... | ... |
| 44.... | ... | 4.6 (15) | 6.7 (22) | 8.5 (28) | 10.7 (35) | 12.5 (41) | 14.3 (47) | 16.5 (54) | 18.6 (61) | 20.4 (67) | 21.9 (72) | 23.5 (77) | 25.3 (83) | 27.1 (89) | 29 (95) | 30.8 (101) | 32.3 (106) | 33.8 (111) | ... | 36.9 (121) | 39.9 (131) | 42.7 (140) | ... | ... | ... | ... | ... | ... |
| 64.... | ... | 6.4 (21) | 9.4 (31) | 12.5 (41) | 15.5 (51) | 18.3 (60) | 21.3 (70) | 24.1 (79) | 26.8 (88) | 29.6 (97) | 32 (105) | 34.4 (113) | 36.9 (121) | 39.3 (129) | 42.1 (138) | 44.8 (147) | 46.9 (154) | 49.1 (161) | ... | ... | ... | ... | ... | ... | ... | ... | ... | ... |
| 94.... | ... | 9.1 (30) | 13.7 (45) | 183 (60) | 22.6 (74) | 26.8 (88) | 31.1 (102) | 35.4 (116) | 39.3 (129) | 43.3 (142) | 46.9 (154) | 50.3 (165) | ... | ... | ... | ... | ... | ... | ... | ... | ... | ... | ... | ... | ... | ... | ... | ... |
| 129.... | ... | 12.8 (42) | 18.9 (62) | 25 (82) | 30.8 (101) | 36.6 (120) | 39.6 (130) | 42.7 (140) | 48.5 (159) | ... | ... | ... | ... | ... | ... | ... | ... | ... | ... | ... | ... | ... | ... | ... | ... | ... | ... | ... |
| 700 mm (28 in.) heavy-duty shaper (25–700); 600 mm (24 in.) heavy-duty shaper (25–600) | | | | | | | | | | | | | | | | | | | | | | | | | | | | |
| 9 | ... | 0.9 (3) | 1.2 (4) | 1.5 (5) | 2.1 (7) | 2.4 (8) | 2.7 (9) | 3.4 (11) | 3.7 (12) | 4.0 (13) | 4.6 (15) | 4.9 (16) | 5.2 (17) | 5.5 (18) | ... | 6.4 (21) | ... | 7.0 (23) | ... | 7.6 (25) | 8.2 (27) | 8.8 (29) | 9.4 (31) | 10.1 (33) | ... | ... | ... | ... |
| 13 | ... | 1.5 (5) | 2.1 (7) | 2.4 (8) | 3.0 (10) | 3.7 (12) | 4.3 (14) | 4.9 (16) | 5.5 (18) | 6.1 (20) | 6.4 (21) | 7.0 (23) | 7.6 (25) | 8.2 (27) | ... | 9.1 (30) | ... | 10.4 (34) | ... | 11.3 (37) | 12.2 (40) | 13.1 (43) | 14 (46) | 14.9 (49) | ... | ... | ... | ... |
| 20 | ... | 2.1 (7) | 3.0 (10) | 4.0 (13) | 4.9 (16) | 5.8 (19) | 6.7 (22) | 7.6 (25) | 8.5 (28) | 9.4 (31) | 10.1 (33) | 11 (36) | 11.9 (39) | 12.8 (42) | ... | 14.3 (47) | ... | 15.8 (52) | ... | 17.4 (57) | 18.6 (61) | 20.1 (66) | 21.3 (70) | 22.6 (74) | ... | ... | ... | ... |
| 27 | ... | 2.7 (9) | 4.0 (13) | 5.2 (17) | 6.4 (21) | 7.6 (25) | 8.8 (29) | 10.1 (33) | 11.3 (37) | 12.5 (41) | 13.7 (45) | 14.6 (48) | 15.8 (52) | 17.1 (56) | ... | 18.6 (61) | ... | 21.0 (69) | ... | 23.5 (77) | 25.3 (83) | 27.1 (89) | 29 (95) | 30.8 (101) | ... | ... | ... | ... |
| 41 | ... | 4.0 (13) | 6.1 (20) | 7.9 (26) | 9.8 (32) | 11.6 (38) | 13.4 (44) | 15.2 (50) | 17.1 (56) | 18.9 (62) | 20.7 (68) | 22.3 (73) | 24.1 (79) | 25.5 (84) | ... | 28.7 (94) | ... | 32 (105) | ... | 35.1 (115) | 38.1 (125) | 41.1 (135) | 43.9 (144) | 46.3 (152) | ... | ... | ... | ... |
| 59 | ... | 5.8 (19) | 8.5 (28) | 11.3 (37) | 14 (46) | 16.8 (55) | 19.5 (64) | 22.3 (73) | 25 (82) | 27.4 (90) | 29.9 (98) | 32.3 (106) | 34.7 (114) | 37.2 (122) | ... | 42.1 (138) | ... | 46.3 (152) | ... | 35.4 (116) | ... | ... | ... | ... | ... | ... | ... | ... |
| 87 | ... | 8.5 (28) | 12.5 (41) | 16.5 (54) | 20.4 (67) | 24.4 (80) | 29 (95) | 32.3 (106) | 36.3 (119) | 40.2 (132) | 43.9 (144) | 47.5 (156) | 51.2 (168) | ... | ... | ... | ... | ... | ... | ... | ... | ... | ... | ... | ... | ... | ... | ... |
| 119 | ... | 11.9 (39) | 17.4 (57) | 22.9 (75) | 28.3 (93) | 33.8 (111) | 39.3 (129) | 44.8 (147) | 50.3 (165) | ... | ... | ... | ... | ... | ... | ... | ... | ... | ... | ... | ... | ... | ... | ... | ... | ... | ... | ... |
| 900 mm (36 in.) heavy-duty shaper (25–900); 800 mm (32 in.) heavy-duty shaper (25–800) | | | | | | | | | | | | | | | | | | | | | | | | | | | | |
| 8 | ... | 0.9 (3) | ... | 1.5 (5) | ... | 2.4 (8) | ... | 3.0 (10) | ... | 3.7 (12) | ... | 4.3 (14) | ... | 4.9 (16) | ... | 5.5 (18) | ... | 6.1 (20) | ... | 7.0 (23) | 7.6 (25) | 8.2 (27) | 8.8 (29) | 9.4 (31) | 10.1 (33) | 10.7 (35) | 11.6 (37) | 11.9 (39) |
| 11 | ... | 1.5 (5) | ... | 2.1 (7) | ... | 3.4 (11) | ... | 4.3 (14) | ... | 5.5 (18) | ... | 6.1 (20) | ... | 7.3 (24) | ... | 8.2 (27) | ... | 9.1 (30) | ... | 10.1 (33) | 10.7 (35) | 11.3 (37) | 11.9 (39) | 12.5 (41) | 13.1 (43) | 13.7 (45) | 14.6 (48) | 15.2 (50) |
| 17 | ... | 1.8 (6) | ... | 3.4 (11) | ... | 4.9 (16) | ... | 6.4 (21) | ... | 7.9 (26) | ... | 9.4 (31) | ... | 11 (36) | ... | 12.2 (40) | ... | 13.4 (44) | ... | 14.6 (48) | 15.8 (52) | 17.1 (56) | 18.3 (60) | 19.5 (64) | 20.7 (68) | 21.6 (71) | 22.9 (75) | 23.8 (78) |
| 23 | ... | 2.4 (8) | ... | 4.6 (15) | ... | 6.7 (22) | ... | 8.8 (29) | ... | 11 (36) | ... | 13.1 (43) | ... | 14.9 (49) | ... | 16.8 (55) | ... | 18.6 (61) | ... | 20.4 (67) | 21.9 (72) | 23.5 (77) | 25 (82) | 26.5 (87) | 28 (92) | 29.3 (96) | 30.8 (101) | 32.0 (105) |
| 35 | ... | 3.7 (12) | ... | 7.0 (23) | ... | 10.4 (34) | ... | 13.7 (45) | ... | 16.8 (55) | ... | 19.8 (65) | ... | 22.6 (74) | ... | 25.3 (83) | ... | 28 (92) | ... | 30.8 (101) | 33.5 (110) | 36 (118) | 38.1 (125) | 40.5 (133) | 42.7 (140) | 44.8 (147) | 46.6 (153) | 48.5 (159) |
| 50 | ... | 5.2 (17) | ... | 9.8 (32) | ... | 14.6 (48) | ... | 19.5 (64) | ... | 23.8 (78) | ... | 28 (92) | ... | 32.3 (106) | ... | 42.5 (119) | ... | 39.9 (131) | ... | 44.2 (145) | 47.9 (157) | 51.2 (168) | ... | ... | ... | ... | ... | ... |
| 74 | ... | 7.6 (25) | ... | 14.9 (49) | ... | 21.9 (72) | ... | 29 (95) | ... | 35.4 (116) | ... | 48.8 (136) | ... | 47.5 (156) | ... | ... | ... | ... | ... | ... | ... | ... | ... | ... | ... | ... | ... | ... |
| 102 | ... | 10.1 (33) | ... | 19.8 (65) | ... | 29.6 (97) | ... | 39.3 (129) | ... | 48.8 (160) | ... | ... | ... | ... | ... | ... | ... | ... | ... | ... | ... | ... | ... | ... | ... | ... | ... | ... |

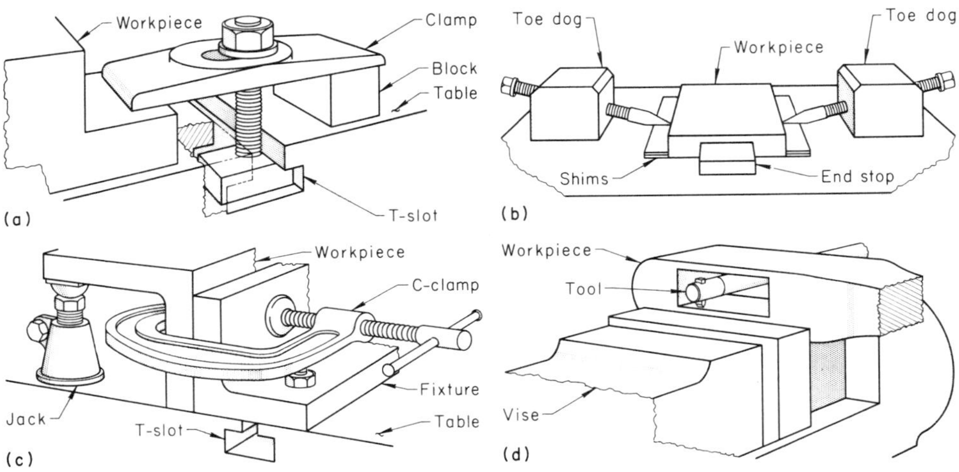


**Fig. 3** Four working methods used in shaping. (a) Standard T-slot with bolts, blocks, and clamp plates. (b) Toe dogs for thin workpieces. (c) Outboard support with jacks for irregular workpieces. (d) Vise

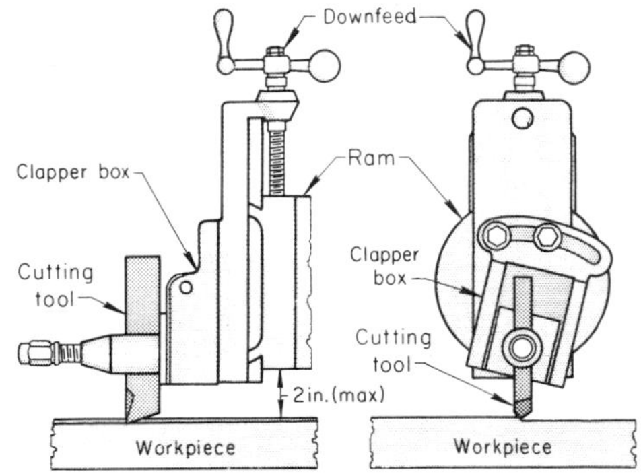


**Fig. 4** Shaper tool-head assembly showing principal components. Side view (left) shows suggested maximum clearance between ram and workpiece. Front view (right) shows typical setting of clapper box for horizontal shaping.

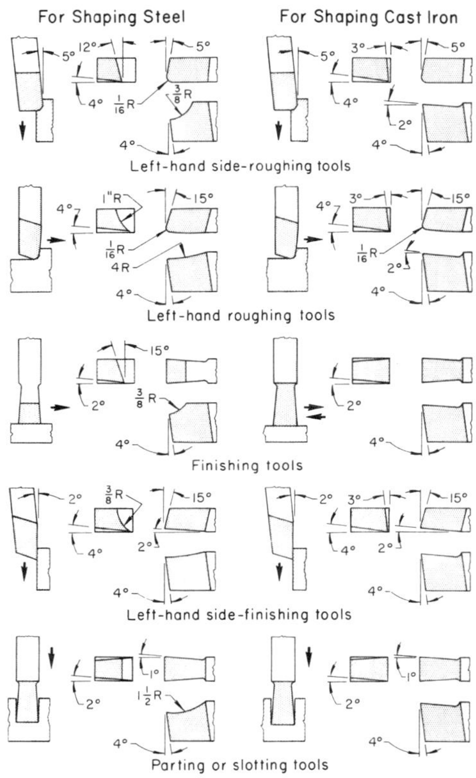


**Fig. 5** Recommended designs of high-speed steel tools for shaping steel and cast iron. Dimensions given in inches

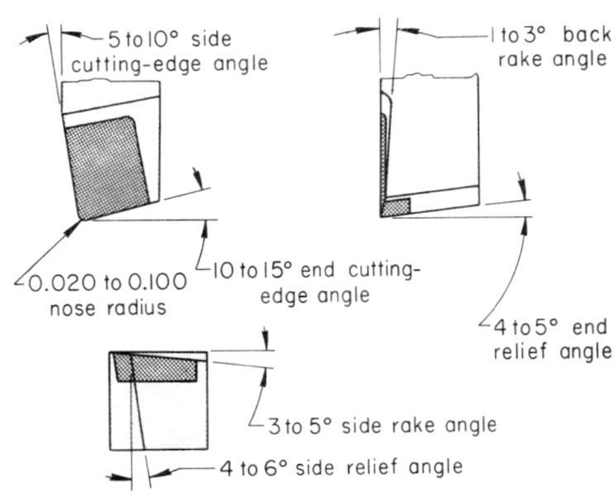


**Fig. 6** Details of a carbide-tipped tool for shaping heat-resistant alloys. Dimension given in inches

fore, on the return stroke, only the weight of the tool bears on the work, which minimizes marring of the surface.

Another means of preventing the tool from marring the workpiece surface on the return stroke consists of keeping the tool in a vertical position and setting the clapper box at an angle away from the direction of feed. With this arrangement (at right, Fig. 4), the tool will swing away from the work during the return stroke.

For applications in which surface finish is so critical that tool drag against the work is unacceptable, devices are available that positively lift the tool clear of the work on the return stroke. Automatic tool lifters are always recommended when carbide cutting tools are used.

## Tool Material and Design

Cutting tools used for shaping can be made of high-speed steel or carbide.

**High-speed steel tools** are used for most shaping, mainly because:

- Shaping generally is not a high-production operation; therefore, the cutting speeds at which carbide functions best are not needed
- Shaping is an interrupted-cutting operation in which the cutting tool is subjected to impact forces at the beginning of each stroke. Therefore, carbide is susceptible to chipping in this operation

General-purpose types of high-speed steel, such as M2 or T1, are satisfactory for most applications. For work made of highly abrasive materials (such as high-alloy tool steels or refractory metals) or for work of high hardness (more than about 35 HRC), one of the highly alloyed types of high-speed steel, such as M6, M36, or T15, will result in longer tool life and usually in lower cost per unit volume of metal removed.

**Carbide tools,** of either the brazed or disposable-insert type, are used mainly:

- In applications in which maximum cutting speeds are being used and carbide has previously been satisfactory
- For making finishing cuts on workpieces such as die blocks (because carbide tools will result in better surface finish)
- For shaping difficult-to-machine metals, such as heat-resistant alloys

For any shaping operation in which carbide tools are used, two recommendations should be followed. The first recommendation is to use a shock-resistant grade of carbide, and the second is to use an automatic tool lifter on the machine to prevent the tool from dragging on the workpiece during the return stroke.

**Design.** Recommended designs of high-speed steel tools for use on steel and cast iron in a variety of shaping operations are shown in Fig. 5. Figure 6 shows the details of a carbide-tipped tool for use on heat-resistant alloys.

The use of lathe (single-point turning) tools for shaping is not recommended, because the angles used on lathe tools cause deflection and digging in, ultimately resulting in chatter and dimensional inaccuracy. The grinding of shaper tools to a large nose radius is another common error. Tools having a nose radius of 0.8 to 1.6 mm (1/32 to 1/16 in.) usually perform best in shapers.

## Ram Stroke and Clearance

It is common practice to set the ram stroke for a distance 16 mm (5/8 in.) greater than the actual length of cut to be made, allowing about 13 mm (1/2 in.) of this excess as clearance at the start of the stroke and 3.2 mm (1/8 in.) at the end. However, allowing more than 13 mm (1/2 in.) of clearance at the beginning of the stroke sometimes proves advantageous, as in the application described below.

**Example 1: Increase in Ram Stroke Clearance for Improved Tool Life.** A vertical shaper was used to cut ten grooves 6.4

**Table 2 Typical speeds used in shaping**

| Work metal | Speed: Roughing m/min | Roughing sfm | Finishing m/min | Finishing sfm |
|---|---|---|---|---|
| Aluminum | 45 | 150 | 60 | 200 |
| Brass and bronze | 45 | 150 | 60 | 200 |
| Gray iron | 18 | 60 | 12 and 30 | (40 and 100)(a) |
| Low-carbon steel | 15 | 50 | 10 and 25 | (35 and 80)(a) |
| Tool steel | 12 | 40 | 18 | 60 |
| Heat-resistant alloys | 3–5 | 10–15 | 6–9 | 20–30 |

(a) Lower speed is used for broad-nose finishing tools and higher speed for conventional or radius-nose tools.

**Table 3 Typical feeds, depths of cut, and metal removal rates for shaping steel and gray iron**

Data are based on use of a 340 mm (13½ in.) stroke on a 3.7 kW (5 hp) shaper having a maximum stroke length of 400 mm (16 in.).

| Feed mm (in.) per stroke | Depth of cut mm | Depth of cut in. | Removal rate $mm^3 \times 10^4$/min | Removal rate $in.^3$/min |
|---|---|---|---|---|
| **1045 steel (annealed)** | | | | |
| At 21 cutting strokes/min (~13 m/min, or 44 sfm): | | | | |
| 1.6 (0.062) | 4.75 | 0.187 | 5.4 | 3.3 |
| 1.9 (0.075) | 4.75 | 0.187 | 6.6 | 4.0 |
| 1.3 (0.050) | 6.35 | 0.250 | 5.7 | 3.5 |
| At 15 cutting strokes/min (~9.8 m/min, or 32 sfm): | | | | |
| 1.9 (0.075) | 6.35 | 0.250 | 6.2 | 3.8 |
| **Gray iron** | | | | |
| At 30 cutting strokes/min (~20 m/min, or 65 sfm): | | | | |
| 0.94 (0.037) | 12.7 | 0.500 | 12.3 | 7.5 |
| 1.27 (0.050) | 12.7 | 0.500 | 16.4 | 10.0 |
| 0.64 (0.025) | 19.1 | 0.750 | 12.5 | 7.6 |
| 0.94 (0.037) | 19.1 | 0.750 | 18.5 | 11.3 |

mm (¼ in.) wide, 4.8 mm (3⁄16 in.) deep, and 38 mm (1½ in.) long on the outer periphery of a cylindrical workpiece made of D2 tool steel at 270 HB (a difficult-to-machine metal). Originally, the machine was adjusted for 6.4 mm (¼ in.) clearance at the beginning of the ram stroke. The service life of the high-speed steel cutting tools used was poor; resharpening was often required before one piece (ten grooves) could be completed. Varying the speed from 12 to 3 m/min (40 to 10 sfm) and varying the depth of cut from 0.635 to 0.125 mm (0.025 to 0.005 in.) per stroke resulted in negligible improvement.

The problem was solved by increasing the clearance at the beginning of the stroke to 19 mm (¾ in.). With this change, one tool cut an average of ten pieces (100 grooves) before requiring resharpening. The tool life improved largely because the increased clearance gave the tool a running start so that it did not hesitate at the instant it contacted the work. However, the greater time allowed for the tool to cool also contributed, because no cutting fluid was used.

**Clearance between ram and workpiece** (Fig. 4) should be kept to a minimum, not exceeding 50 mm (2 in.) whenever practicable. As the clearance between the front of the ram and the workpiece is increased, cutting action must be obtained either by lowering the tool post or by lengthening the cutting tool. In either case, rigidity is decreased as the distance increases, and chatter will occur, causing loss of accuracy and finish and a probable decrease in tool life.

## Speed, Feed, and Depth of Cut

**Typical ram speeds** used in rough and finish shaping are given in Table 2. As these data indicate, speeds are closely related to workpiece composition, microstructure, and hardness, increasing or decreasing with the machinability of the work metal.

**Feed per stroke** in shaping is also influenced by the composition, microstructure, and hardness of the work metal. However, ram speed, depth of cut, and required dimensional accuracy and finish also affect choice of feed.

Typical feed rates for shaping annealed carbon steel and gray iron are given in Table 3. These feed rates typify those used for roughing to remove metal as fast as possible; the metal removal rates given in Table 3 are relatively high.

Common practice in shaping is to make roughing (hogging) cuts at maximum feed but at slow speed and then to reverse this procedure for finishing cuts. Sometimes it is not practical to increase the speed for finishing, but lighter feeds are almost always used.

For almost all shaping operations, regardless of other conditions, there is a practical minimum feed rate. Feed rates that are too light do not allow the average tool to bite sufficiently and will result in chatter. Feed rates less than 0.125 mm (0.005 in.) per stroke are seldom recommended.

**Depth of cut** is usually not less than about 0.38 mm (0.015 in.). Shallower cuts, like extremely light feeds, do not allow sufficient tool bite and result in chatter and glazing. The selection of depth of cut above this minimum is influenced by the following, singly or in combination:

- Composition, microstructure, and hardness of the workpiece metal
- Amount of metal to be removed
- Workpiece size
- Workpiece configuration
- Rigidity of the setup
- Available power

The workpiece configuration is often a factor that affects the allowable depth of cut. For example, a depth of cut as great as 25 mm (1 in.) is often used for machining a solid piece. For machining an intricately shaped workpiece, however, the maximum allowable depth of cut may be less than 3.2 mm (⅛ in.) because the forces for deeper cuts would deform the workpiece.

The rigidity of the setup and the allowable depth of cut are directly related. Under conditions of insufficient rigidity, depth of cut must be reduced to avoid chatter. Depths of cut used at various speeds and feeds in shaping several metals are given in Tables 3 to 5.

## Cutting Fluids

A flood of cutting fluid is seldom used for shaping operations, because two of the three functions of cutting fluids (chip disposal and cooling) are not needed. In most operations, chips are thrown forward and seldom interfere with successive cutting strokes. (As chips begin to pile up on the table or fixtures, however, they should be brushed away by the operator. The use of compressed air for removal of chips is not recommended, because chips are likely to lodge in the shaper mechanism.) Because the rate of metal removal in shaping is relatively low and the tool is seldom engaged in cutting more than about 60% of the time, cooling of tools and workpieces usually presents no problem.

In many applications, however, the use of cutting fluids will result in improved dimensional accuracy, finish, and tool life by minimizing the adherence of work metal to the tool. When cutting fluids are used, the most common practice is to apply the undiluted fluid with a brush or swab.

Sulfurized cutting oils are commonly used for steels. However, many nonferrous metals, such as copper-base alloys, are susceptible to staining from sulfurized oils; when a cutting fluid is needed for shaping these metals, kerosene with an addition of about 10% lard oil is recommended.

**Table 4 Shaping of W2 and D2 tool steel die sections**

| | |
|---|---|
| **Tool details for both steels** | |
| Tool material | M3 high-speed steel |
| Tool size | 13 × 13 × 100 mm (½ × ½ × 4 in.) |
| Type of tool holder | Straight-shank(a) |
| Tool angles | |
| Back rake | 10° |
| Normal side rake | 10° |
| End relief (on tool bit) | 20° |
| Effective end relief (in holder)(a) | 7° |
| Normal side relief | 25° |
| End cutting edge | 50° |
| Side cutting edge | 30° (left-hand) |
| Nose radius | 1.6 mm (1/16 in.) |
| **Machining conditions for W2(b)** | |
| Ram speed, roughing and finishing | 18.0 m/min (59 sfm) |
| Feed, roughing | 0.5 mm (0.020 in.) per stroke |
| Feed, finishing | 0.25 mm (0.010 in.) per stroke |
| Depth of each cut, roughing | 6.35 mm (0.250 in.) |
| Depth of cut, finishing | 1.52 mm (0.060 in.) |
| Length of cut, rough and finish | 203 mm (8.000 in.) |
| Cutting fluid | None |
| Metal removal rate | |
| Roughing | 4.1 × $10^4$ $mm^3$/min (2.5 $in.^3$/min) |
| Finishing | 0.5 × $10^4$ $mm^3$/min (0.3 $in.^3$/min) |
| Machining time per cut, roughing | 2.3 min |
| Machining time per cut, finishing | 4.4 min |
| **Machining conditions for D2(b)** | |
| Ram speed, roughing and finishing | 12.5 m/min (41 sfm) |
| Feed, rough and finish | 0.25 mm (0.010 in.) per stroke |
| Depth of each cut, roughing | 6.35 mm (0.250 in.) |
| Depth of cut, finishing | 1.52 mm (0.060 in.) |
| Length of cut, rough and finish | 178 mm (7.000 in.) |
| Cutting fluid | None |
| Metal removal rate | |
| Roughing | 1.3 × $10^4$ $mm^3$/min (0.77 $in.^3$/min) |
| Finishing | 0.3 × $10^4$ $mm^3$/min (0.18 $in.^3$/min) |
| Machining time per cut | |
| Roughing | 6.1 min |
| Finishing | 5.7 min |

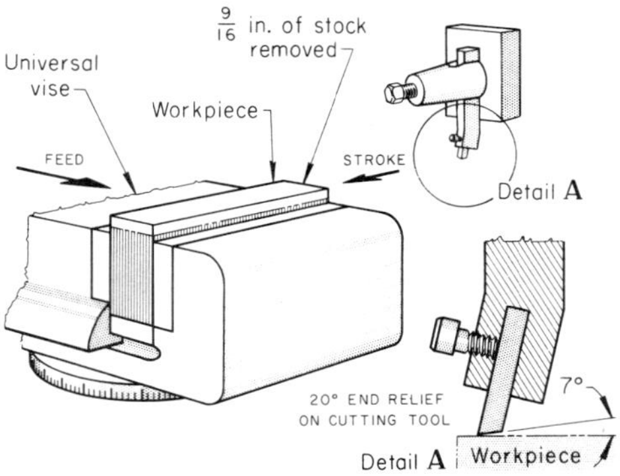


(a) Tool holder angle was 13°. (b) Two roughing cuts, one finishing cut

**Table 5 Shaping of heat-resistant alloys: Hastelloy alloy X, René 41, and HS-25**

| | |
|---|---|
| **Tool details(a)** | |
| Tool material | M30 or M34 high-speed steel |
| Tool angles | |
| Back rake | 0–3° |
| Side rake | 8° |
| End and side relief | 4–6° |
| Nose radius | 1.1–2.2 mm (0.045–0.090 in.) |
| **Machining conditions** | |
| Ram speed | 2.1–4.0 m/min (7–13 sfm) |
| Feed, roughing | 0.5–0.75 mm (0.020–0.030 in.) per stroke |
| Feed, finishing | 0.25–0.380 mm (0.010–0.015 in.) per stroke |
| Depth of cut, roughing | 1.25–2.50 mm (0.050–0.100 in.) |
| Depth of cut, finishing | 0.380–0.750 mm (0.015–0.030 in.) |
| Cutting fluid | Sulfur-free chlorinated oil(b) |

(a) Honing of cutting edges proved beneficial. Optimum cutting efficiency was obtained by keeping edge wear below 0.25 mm (0.010 in.) wear land. (b) Applied by a brush

## Flat Surfaces

Shaping is usually selected as the method for machining flat surfaces when:

- The required flatness is greater than can be obtained by other methods
- The production quantity is too low to justify the cost of machines and tooling for milling or broaching
- A few pieces are needed in an emergency and no production setup is available

**Dimensional Control.** When workpieces such as dies, molds, and fixtures must be matched, thus requiring the highest possible degree of flatness, shapers are preferred. Under good shop conditions (capable operator, machine in good repair, rigid setup), a heavy-duty shaper can machine a surface 460 mm (18 in.) square to a flatness within 0.025 mm (0.001 in.) and can hold variation in flatness among identical parts within 0.05 mm (0.002 in.). Under optimum conditions of control, including a highly skilled operator, the flatness of a 460 mm (18 in.) square surface can be held to within about 0.013 mm (0.0005 in.).

When shaping workpieces that are relatively thin compared to their length and width—for example, a 13 mm (½ in.) thick plate having 6.5 × $10^4$ $mm^2$ (100 $in.^2$) per side—warpage is often a problem. If the two sides are flat and parallel, warping is best avoided by successively removing equal amounts of stock from the two sides in both roughing and finishing; about 0.25 to 0.375 mm (0.010 to 0.015 in.) of material per side is left for the finishing cuts. For thin workpieces from which all the stock must be removed from one side, the best practice is to rough machine to within about 0.380 mm (0.015 in.) of finished size, stress relieve, and then finish machine. Tool details, techniques, and operating conditions in the shaping of flat surfaces are discussed in the following example.

**Example 2: W2 and D2 Tool Steel Die Sections.** A heavy-duty 610 mm (24 in.) machine with 7.5 kW (10 hp) drive was used for the rough and finish shaping of die sections made of either W2 hot-rolled bar (202 HB maximum) or D2 tool steel (255 HB). Before shaping, each section made of W2 measured 65 × 150 × 200 mm (2½ × 6 × 8 in.) and weighed 15 kg (34 lb). Each D2 section measured 55 × 150 × 175 mm (2¼ × 6 × 7 in.) and weighed 12.1 kg (26.7 lb). Shaping reduced the 150 mm (6 in.) dimension on sections of both steels to 138 mm (5 7/16 in.), and the weight to 13.9 and 11.0 kg (30.7 and 24.2 lb) for the W2 and D2 sections. Tool details and machining conditions are listed in Table 4.

Each workpiece was clamped in the shaper vise with the longest dimension parallel to the ram stroke, as shown in the illustration in Table 4, after which it was machined with two 6.4 mm (¼ in.) roughing cuts and a 1.6 mm (1/16 in.) finishing cut. The tools were hand sharpened to the configuration detailed in Table 4.

A comparison of the data in Table 4 for the preceding example demonstrates the influence of machinability of workpiece metal on the machining conditions employed and results obtained. For both the W2 and the D2 die sections, tools of the same design were used in identical machines for removing the same amounts of metal. For the more difficult-to-machine D2, however, speed had to be reduced by about 30% and feed by 50%. The metal removal rate in roughing was only about one-third that for W2.

Another group of difficult-to-machine materials comprises the nickel-base and cobalt-base heat-resistant alloys. As indicated in Table 5, the ram speed used in shaping Hastelloy alloy X (at 20 HRC), René 41 (at 28 HRC), and HS-25 (at 25 HRC) alloys must be much slower, even for shallower cuts, than that used for the highly alloyed D2 tool steel.

## External or Internal Contours

The use of a shaper for cutting external or internal contours is usually limited to unit or low production. When identical workpieces must be produced in large quantities, cost per piece can be greatly reduced by milling or broaching these contours.

However, for cutting internal contours on workpieces such as the one shown in Fig. 7, shaping is often the most satisfactory and economical method for producing up to medium quantities. The contour in this part could be produced in one cut with a broach, but a large and expensive machine would be required. Such an investment could be jus-

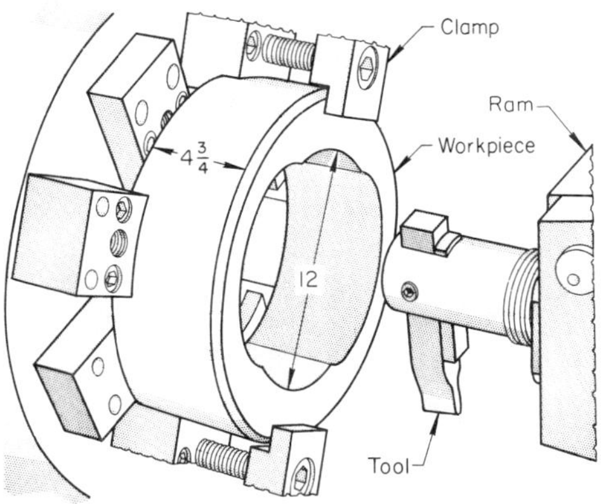


**Fig. 7** Cutting an internal contour with a shaper. Dimension given in inches

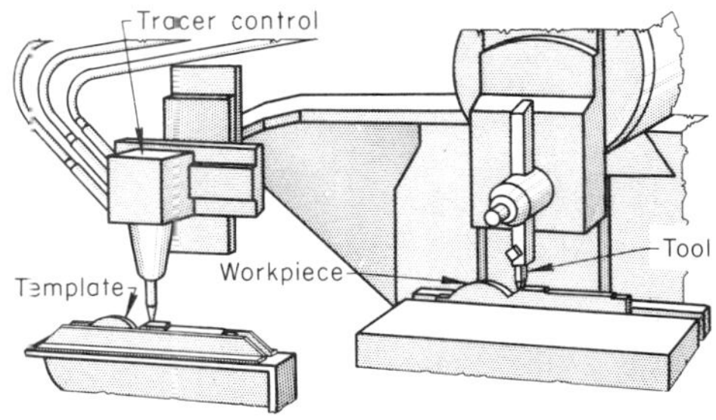


**Fig. 8** Use of duplicating equipment on a shaper for cutting the external contour of a trimming die

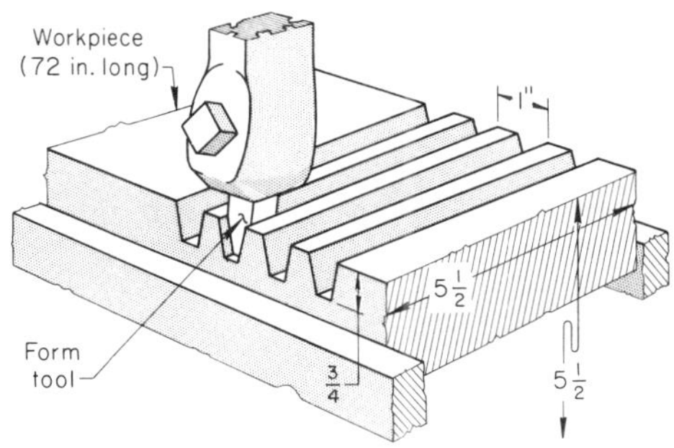


**Fig. 9** Form shaping of teeth in arbor-press racks. Dimensions given in inches

tified only if the parts were to be produced in large quantities. In the contour cutting of the workpiece shown in Fig. 7, the operation was manually controlled. However, contours can be cut by means of automatic duplicating equipment (Fig. 8).

Automatic duplicating makes use of a template followed by a tracer that controls and directs the cutting path of the tool. Setup time is short, and tooling cost is low.

## Form Cutting

In most form cutting on a shaper, the workpiece remains stationary while the form tool is advanced into the work by increasing the downfeed on each stroke. A typical form-cutting operation, that of cutting teeth in racks for emergency use in unit or low-production quantities, is described in the following example.

**Example 3: Form Cutting Heat-Treated Tool Steel Racks.** A 7.5 kW (10 hp) shaper having a 910 mm (36 in.) maximum stroke was used to form-cut teeth in 1830 mm (72 in.) long arbor-press racks made of 0.50% C, low-alloy tool steel and heat treated to a hardness of 300 to 320 HB. The setup is shown in Fig. 9.

The shaper was operated with a 200 mm (8 in.) stroke and at 41 strokes per minute (ram speed: 15 m/min, or 50 sfm). The form tool, which was ground from 25 mm (1 in.) square high-speed steel, was downfed at 0.5 mm (0.020 in.) per stroke. The workpiece was manually indexed and located for each succeeding tooth space with the aid of crossfeed, rapid-traverse, and micrometer dials that were standard equipment on the shaper used. No cutting fluid was used.

## Grooves, Slots, and Keyways

Vertical machines are usually used in preference to horizontal shapers for cutting external or internal grooves. This is particularly true when two or more grooves must be kept in a specific relationship to each other, because indexing for cutting in a vertical plane is usually simpler.

**External grooves** can be machined on shapers to about the same degree of dimensional accuracy as in milling. Holding close dimensions on external grooves cut on a shaper depends largely on the accuracy of the indexing fixture, because there is seldom any reason why rigid tools and tool mountings cannot be used. On most parts, two or more grooves can be held within 0.05 mm (0.002 in.) in relationship to each other.

Shapers are used for cutting external grooves when quantities are too low to justify the cost of tooling up for milling or when workpiece configuration does not permit the use of a milling cutter. As an example of the latter condition, it is often necessary to cut grooves that terminate adjacent to a flange or other obstruction, thus precluding the use of a rotating cutter. With a shaper, it is usually possible to cut grooves or slots to within about 6.4 mm (¼ in.) of an enlarged section. When it is feasible to use milling and when production quantities warrant its use, external grooves are seldom cut by a shaper, because milling costs less under these conditions.

**Internal grooves** are machined on a shaper for the same reasons as noted above for external grooves, namely, low quantity and restrictive configuration. For internal grooves, however, broaching is the alternative process.

Shaping is generally used for cutting internal grooves or slots in tooling components such as dies for hot extrusion. This is because quantities are usually too low to warrant the cost of tooling for broaching and because the production of slots that are several times as deep as they are wide would be impractical for broaching.

The hardness of the work material can also determine whether shaping or broaching is used for cutting intricately shaped internal grooves in dies. Heat-treated tool steels as hard as 46 HRC are commonly machined on a shaper, while the broaching a steel of this hardness would be risky because of the likelihood of broach breakage. Shaping is also virtually the only process by which grooves can be cut to about 6.4 mm (¼ in.) of the end of a blind hole.

The length of internal slots or grooves in relationship to their cross-sectional dimensions limits the use of a shaper. By using holders that extend the cutting tool out from the ram head, it is possible to remove metal from holes as small as about 25 mm (1 in.) in initial cross section. However, as the ratio of length to diameter (or to other cross-sectional dimension) exceeds about 4 to 1, the decrease in accuracy is likely to become intolerable.

## Gear Shaping

Gear shaping is more complex than general-purpose shaping. Gear shaping is a form-generating operation; ordinary shaping is not. A gear shaping cutter is a conjugate of the workpiece and meshes with it; a shaping cutter is a single-point tool. Gear shaping is usually performed on five-axis vertical slotting machines equipped with rack or rotary cutters and a rotary table.

# Broaching

BROACHING is a machining process in which a cutting tool that has multiple transverse cutting edges is pushed or pulled through a hole or over a surface to remove metal by axial cutting. Because broaches are multitoothed cutting tools produced to close tolerances, they are expensive; consequently, the process is usually employed only for high production. Some part configurations, however, are practical to produce only by broaching; for parts of this type, the broaching of low production quantities may be warranted. Because several edges of a broach are cutting at once, forces are much greater than in other machining methods, and broaching is rated as the most severe of all machining operations.

Broaching is similar to shaping, it competes economically with milling and boring, and it is capable of producing precision machined surfaces. The heart of this process is the broaching tool, in which roughing, semifinishing, and finishing teeth are combined into one tool. Broaching is unique in that it is the only one of the basic machining processes in which the feed of the cutting edges into the workpiece, determining the chip thickness, is built into the tool, called a broach. The machined surface is always the inverse of the profile of the broach, and in most cases it is produced with a single, linear stroke of the tool across the workpiece (or the workpiece across the broach).

Broaching was originally developed for machining internal keyways. However, its obvious advantages quickly led to its development for the mass-production machining of various surfaces, such as flat, interior and exterior cylindrical and semicylindrical, and many irregular surfaces. Because there are few limitations regarding the contour form that broach teeth may have, there is almost no limitation on the shape of surfaces that can be produced by broaching. The only physical limitations are that there must be no obstruction to interfere with the passage of the entire tool over the surface to be machined and that the workpiece must be strong enough to withstand the forces involved. In internal broaching, a hole must exist in the workpiece into which the broach may enter. Such a hole can be made by drilling, boring, or coring.

Broaching usually produces better accuracy and finish than can be obtained by milling or reaming. Although the relative motion between the broaching tool and the work is usually a single linear one, a rotational motion can be added to permit the broaching of spiral grooves, as in spiral splines or in gun-barrel rifling.

## Fundamentals of Broaching

Each tooth on a broach is essentially a single-edge cutting tool, arranged much like the teeth on a saw except for the step (rise per tooth), which determines the depth cut by each tooth. The depth of cut can vary from approximately 0.15 mm (0.006 in.) for roughing teeth in machining free-cutting steel to a minimum of 0.025 mm (0.001 in.) for finishing teeth.

**Broach Tooth Terminology.** Usually, a broach is a tapered bar into which teeth have been cut so as to produce a desired contour in a workpiece by a single pass of the tool. A typical broach for cutting a round hole is shown schematically in Fig. 1. This broach has three basic sections of cutting teeth: roughing teeth, intermediate (semifinishing) teeth, and finishing teeth. The broach tapers from the first roughing tooth to the last intermediate tooth with the outside diameter of each tooth being slightly larger than the tooth that precedes it. The difference in height between each tool is termed the tooth rise. The diametral differences of the roughing teeth are usually greater than for the intermediate teeth. Ordinarily, all finishing teeth are of the same diameter.

A standard broach tooth is illustrated in Fig. 2. Individual teeth have a land and face that intersect to form a cutting edge. The face is ground with a face (hook) angle that is determined by the workpiece material. Soft steel workpieces usually require greater hook angles; hard or brittle materials, smaller hook angles.

The land supports the cutting edge against stresses. A slight clearance or backoff angle is ground onto the lands to reduce friction. On roughing and semifinishing teeth, the entire land is relieved with a backoff angle. On finishing teeth, part of the land immediately behind the cutting edge is often left straight so that repeated sharpening (by grinding the face of the tooth) will not alter tooth size.

The distance between teeth, or pitch, is determined by the length of cut and is influenced by the type of workpiece material. A relatively large pitch may be required for roughing teeth to accommodate a greater chip load. Tooth pitch may be smaller on semifinishing and finishing teeth to reduce the overall length of the broach tool. Pitch is calculated so that, preferably, two or more teeth cut simultaneously. This prevents the tool from drifting or chattering.

A broach tool will sometimes vibrate when a heavy cut is taken, especially when the cutting load is not evenly distributed. Vibration may also occur when tooth engagement is irregular. The greatest contributing factors to vibration are poor tooth engagement and extremely hard workpieces. Such problems must be anticipated by the broach designer.

Tooth rise is calculated so that the thickness of the chip does not impose too great a strain on individual teeth. A large tooth rise increases power requirements. When all teeth are simultaneously engaged in the workpiece, too large a tooth rise could cause an increase in power requirements beyond the rated tonnage of the machine. If the rise is too small to permit the teeth to bite into the workpiece, a glazed or galled finish will result.

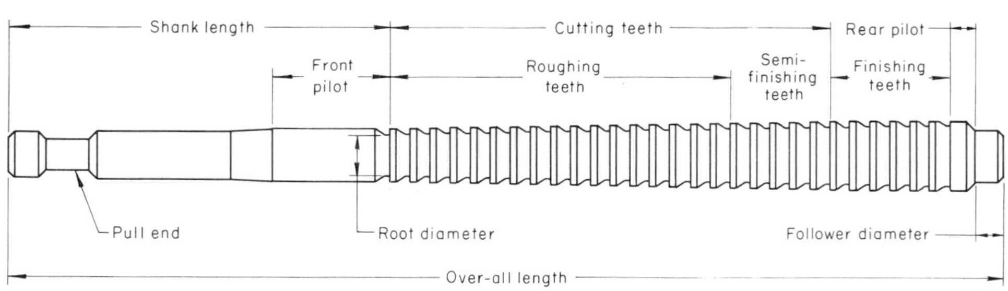


**Fig. 1** Essential features and nomenclature of broaches as typified by an internal pull broach for cutting round holes

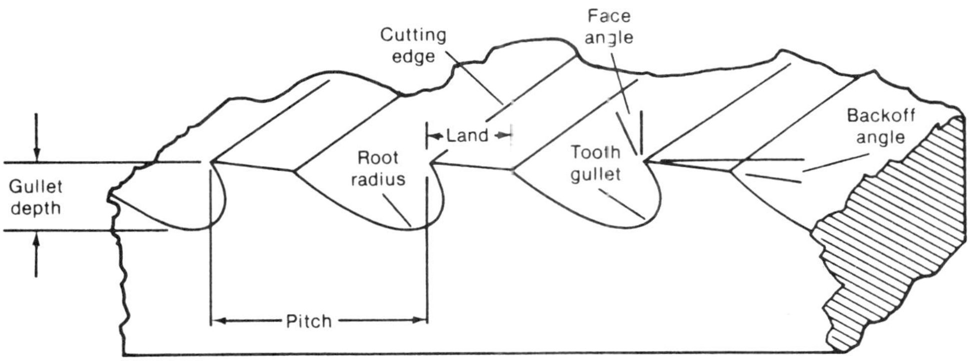


**Fig. 2** Teeth in a typical surface broaching tool. Nomenclature corresponds to that of an internal broach. Gullets shown are average size.

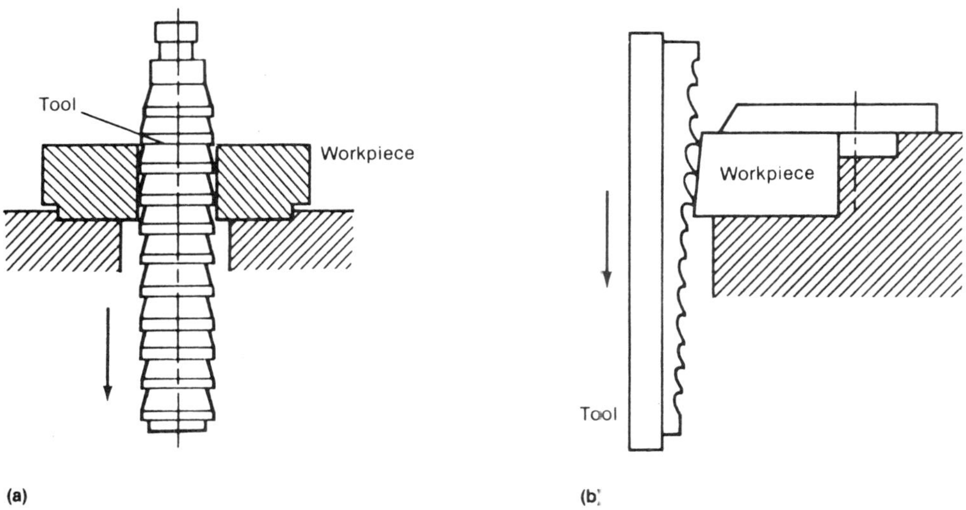


**Fig. 3** Relationship of workpiece to tool for forming chips during broaching. (a) Internal broaching. (b) External broaching

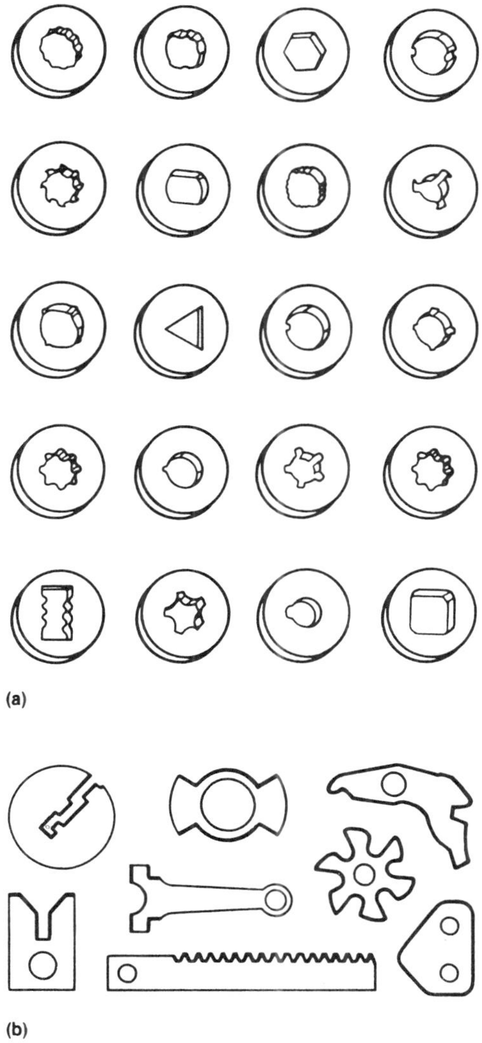


**Fig. 4** Typical configurations produced by internal (a) and external (b) broaching. Heavy lines in (b) indicate surfaces that are broached.

The depth of the tooth gullet is related to the tooth rise, pitch, and workpiece material. The tooth root radius is usually designed so that chips curl tightly within themselves, occupying as little space as possible.

Roughing and finishing operations are usually combined in one broach, and the broached surface normally requires no subsequent finishing operations. Many cuts can be made simultaneously in close dimensional relationship with either internal or external (or surface) broaches.

Internal broaches (Fig. 3a) can be rotated during passage through the work to produce internal helical keyways, splines, or gear teeth. Broached holes can be produced from holes that are cored, forged, punched, drilled, bored, or reamed.

Broaches do not ordinarily remove large amounts of stock, because of machine tonnage and stroke limitations. However, as much as 13 mm (½ in.) of stock has been economically removed in some surface broaching applications (Fig. 3b). Depths of 100 mm (4 in.) have been achieved with broaches. The final configuration obtained with a broach may be a smoother and flatter surface, a larger hole, or a complex splined, toothed, notched, curved, helical, or other irregularly shaped section, as shown in Fig. 4.

**Broach Cutting Action.** A workpiece to be internally broached must be provided with a starting hole through which the broach is pulled or pushed. This hole is just large enough to permit the front pilot section of the broach to enter freely. As the broach progresses through the part, cutting begins gradually, and as each succeeding tooth engages the work, it removes a small amount of metal (Fig. 5). If the first finishing tooth becomes dull or nicked, the work metal is likely to be torn.

A broach is normally sharpened by grinding the cutting edge. Each tool incorporates a backoff or relief angle (usually 1½°, except for finishing, where it may be as small as ¼°). Tooth diameter is changed only slightly when the broach is sharpened on the face because only a small amount of metal is removed. Because the first finishing tooth does the work, only this tooth is ground at resharpening. As this tooth is decreased in size, the next tooth does the finish cutting; this continues until all the finishing teeth have been reground to a point at which size tolerance can no longer be held.

**Broach Size.** Broaches can be as small as 1.3 mm (0.050 in.) in diameter or as large as 380 to 500 mm (15 to 20 in.) in diameter. The internal push broach is limited in length to 25 times the diameter of the finishing teeth, and the internal pull broach is limited in length to 75 times the diameter of the finishing teeth.

## Applicability

Broaching can be employed in a wide range of applications and, where applicable, offers several advantages over other machining processes. Because both roughing and finishing can be done in a single pass of the broach, broaching is rapid and efficient. Moreover, because close tolerances can be held and smooth surfaces provided, subsequent operations are seldom required. Almost any irregular shape can be broached if it is in the same plane as the direction of broach travel. Large surfaces can be broached flat in one pass.

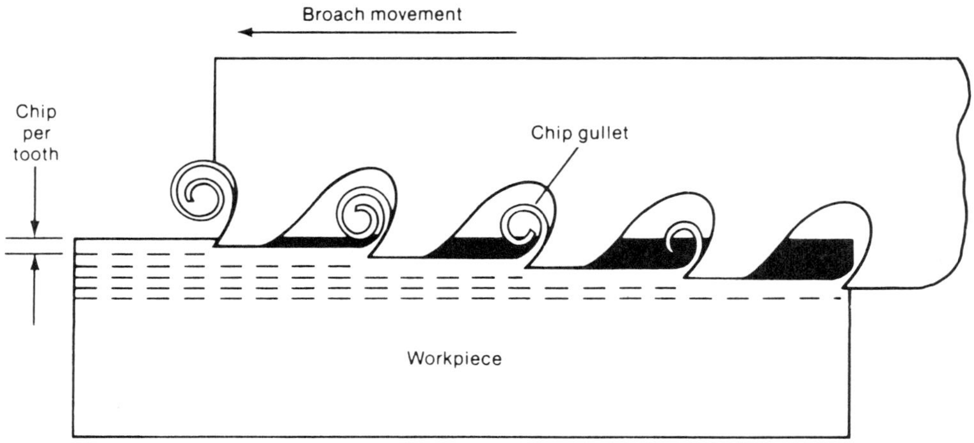


**Fig. 5** Ideal chip formation by a broaching tool

For many applications, broaching is selected because it is less expensive or faster, even though acceptable results could be obtained by other machining methods, such as milling, boring, shaping, or reaming (see the section "Broaching Versus Alternative Processes" in this article). For other applications, the required configuration is such that broaching is the only feasible method—for example, for fir-tree or dovetail slots in turbine wheels.

The limitations of broaching stem from the fundamental characteristics of the process. With the exception that it can be rotated (which permits the broaching of internal or external helical teeth on gear blanks), a broach always moves forward in a straight line. Consequently, all elements of broached surfaces must be parallel to the axis of travel. It is impossible to broach the entire surface of a tapered hole. A blind hole can be broached, but a recess must be provided that is larger in diameter than the broach and deep enough to permit full travel of the broach. A series of push broaches is usually used for a blind hole (Fig. 6).

On external surfaces, it is impossible to broach to a shoulder that is perpendicular to the direction of broach movement. If the shoulder is parallel to the motion, however, broaching to the shoulder is possible and is frequently done.

It is also impractical to change the direction of travel of the broach (except for a spiral twist). Thus, surfaces having curves that lie in two or more planes cannot be broached in a single operation, although such surfaces are often machined by multiple broaching.

Because of the length of stroke and machine tonnage that would be required, broaching is seldom recommended for the removal of large amounts of metal. Where heavy stock must be removed, it is generally better to use a preliminary roughing operation other than broaching or to use two or more roughing broaching passes.

Where size is concerned, the size of the broach required, rather than the size of the workpiece, determines the applicability of the process. Broaching of a hole 150 mm (6 in.) in diameter is commonplace. For a hole 300 mm (12 in.) in diameter, a much heavier broach is needed, and a much larger machine is required. Broaching of holes larger than 300 mm (12 in.) in diameter is seldom practical because of the size and cost of the broach.

The size of the workpiece has little influence on broaching conditions, provided the broach sections are similar in shape, workpiece sizes are in a range that is easily handled, and the materials are comparable in machinability. However, workpiece size does affect production rate per hour.

For the production of die cavities, where the shape permits and where several similar cavities are required, broaching can provide a substantial savings. In the usual toolroom method of die production, labor costs per die are approximately the same whether one cavity or ten cavities are produced. In broaching, the major cost is that of the broach or broaches required, and this cost is reduced in proportion to the number of cavities produced.

## Broaching Machines

The type of broach cutting tool required for a given job is the single most important factor in determining the type of broaching machine to be used. Second in importance is the production requirement. Taken together, these factors usually determine the specific type of machine for the job.

The type of broach tool (internal or external) immediately narrows down the types of machines that could be used. The number of pieces required per hour or over the entire production run will further narrow the field.

For example, a dual-ram machine with one operator may be chosen over two single-ram machines requiring two operators in order to provide higher output per man-hour. The single operator can load one table of a dual-ram unit while the other ram is cutting. Production requirements that are even higher may dictate a continuous horizontal machine. The machine size in a particular model is a function of tool size, workpiece size, broaching power requirements, and available production space.

For internal broaching, the length of a broach in relation to its diameter may determine whether it must be pulled rather

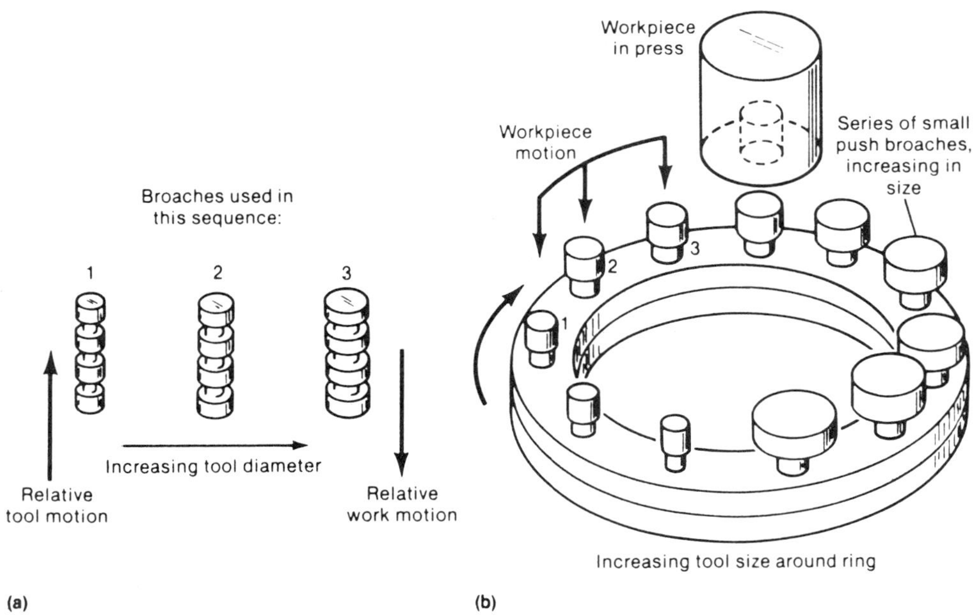


**Fig. 6** Schematic of blind-hole broaching on a press. This operation involves a series of short push broaches, each slightly larger in diameter than the preceding tool (a). The broaches are mounted on a circular indexing table that rotates under the workpiece (b). As each tool stops under the workpiece, the broaching machine pushes the workpiece down over the tool, withdraws it, and then waits for the next broaching tool to index into position.

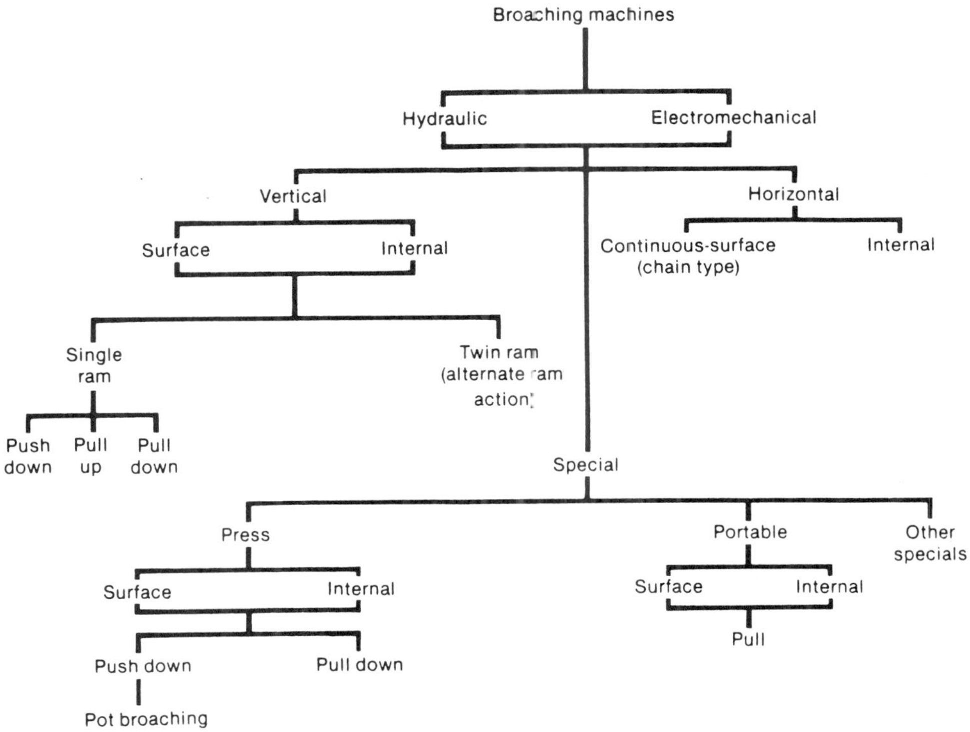


**Fig. 7** Classification of broaching machines

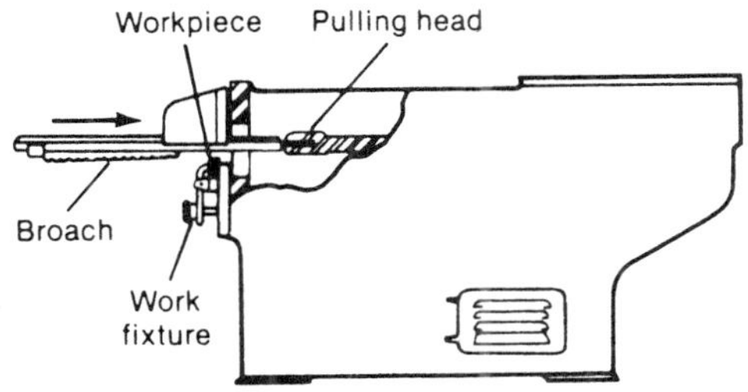


**Fig. 8** Schematic of a typical pull-type horizontal broaching machine

than pushed through the workpiece, because a broach tool is stronger in tension than in compression. This in turn helps determine the type of machine for the job. A short push broach is often handled in a press instead of an expensive ram-type broaching machine. Presses can of course be converted to pull short broaches by the addition of a pull-down adaptor, which converts push strokes to pull strokes.

Lubrication, workpiece size, chip-handling characteristics, and surface finish help determine whether a pull-up or a pull-down broach should be used. Pull-down machines are used most often because gravity helps feed lubricant to the cutting teeth. Large workpieces are more easily handled in a pull-down machine than in a pull-up machine.

The type of drive—hydraulic or electromechanical—is another important factor in machine selection. Convertibility and automation are also significant considerations. Some machine designs allow for conversion from internal to surface work. Some designs are fully automated; others are limited in scope and operate only with close operator supervision.

The two major types of broaching machines—horizontal and vertical—are classified in terms of the direction of broach travel. Smaller portable machines and some sophisticated larger machines are classified as special broaching machines. Figure 7 shows a breakdown of various categories of broaching machines.

**Broaching machine designations** follow a simple letter-number code in which the type and size (capacity and maximum length of stroke) of the machines are identified. Three examples of the use of this code are:

- VPU-5-54—Vertical pull-up machine, 45 kN (5 tonf) capacity, 1370 mm (54 in.) stroke
- VPD-10-60—Vertical pull-down machine, 90 kN (10 tonf) capacity, 1525 mm (60 in.) stroke
- H-5-48—Horizontal machine, 45 kN (5 tonf) capacity, 1220 mm (48 in.) stroke

## Horizontal Broaching Machines

The preferred configuration for broaching machines seems to have come full circle. The original gear- or screw-driven machines were designed as horizontal units. Gradually, the vertical machines evolved as it became apparent that floor space could be utilized more efficiently with vertical units. Currently, the horizontal machine, both hydraulically and mechanically driven, is again finding increasing favor among users because of its very long strokes and the limitation that ceiling height places on vertical machines. About 47% of all broaching machines are horizontal units. Horizontal machines are used exclusively for some types of work, such as the roughing and finishing of automotive engine blocks.

**Horizontal internal or combination machines,** among the first used after the advent of powered broaching, have been driven hydraulically for many years. Hydraulic drives, developed in the early 1920s, offered such pronounced advantages over the various early mechanical driving methods that only within recent years has any other method been used.

By far the greatest amount of horizontal internal broaching is done on hydraulic pull-type machines (Fig. 8), for which configurations have become somewhat standardized over the years. Fully one-third of the broaching machines in existence are of this type, and of these, nearly one-fourth are over 20 years old. They find their heaviest application in the production of general industrial equipment, but can be found in nearly every type of industry.

Hydraulically driven horizontal internal machines are built with pulling capacities from 22 to 670 kN (2½ to 75 tonf). The former value represents machines only about 2.4 m (8 ft) long; the latter, machines over 11 m (35 ft) long. Strokes up to 3050 mm (120 in.) are available, with cutting speeds generally limited to less than 12 m/min (40 sfm).

**Horizontal surface broaching machines** account for only about 10% of existing broaching machines, but this is not indicative of the percentage of the total investment they represent or of the volume of work they produce. Horizontal surface broaching machines belong in a class by themselves in terms of size and productivity. Only the large continuous horizontal units can match or exceed them in productivity. Horizontal surface units are manufactured in both hydraulically and electromechanically driven models, with the latter type becoming dominant.

The older hydraulically driven horizontal surface machines are produced with capacities up to 355 kN (40 tonf), strokes up to 4570 mm (180 in.), and normal cutting speeds of 30 m/min (100 sfm). These machines, a major factor in the automotive industry for nearly 30 years, produce a great variety of cast iron parts. They use standard carbide cutting tools and some of the highest cutting speeds found in broaching.

However, electromechanically driven horizontal surface machines are being used at an ever-increasing rate for certain applications, despite their generally higher cost. Because of their smooth ram motion and the resultant improvements in surface finish and part tolerances, these machines have

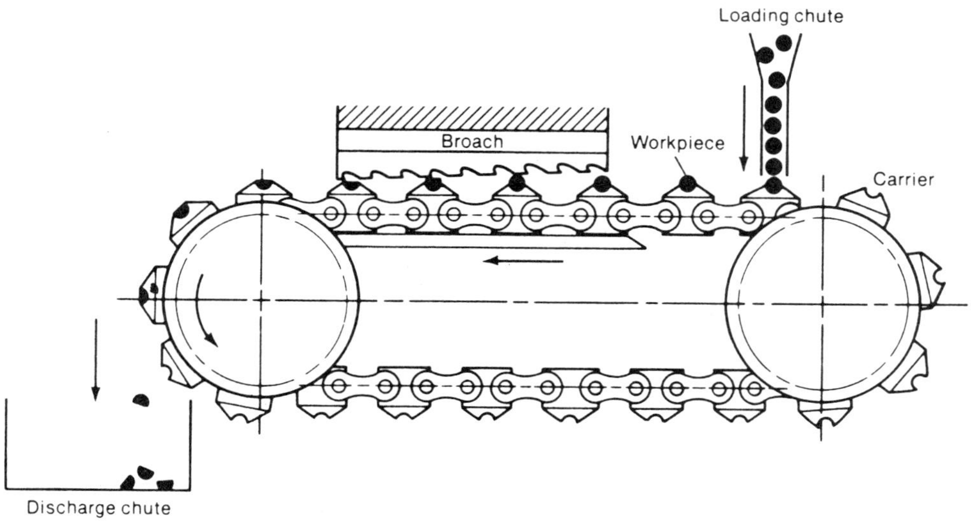


**Fig. 9** Schematic of a continuous surface horizontal broaching machine

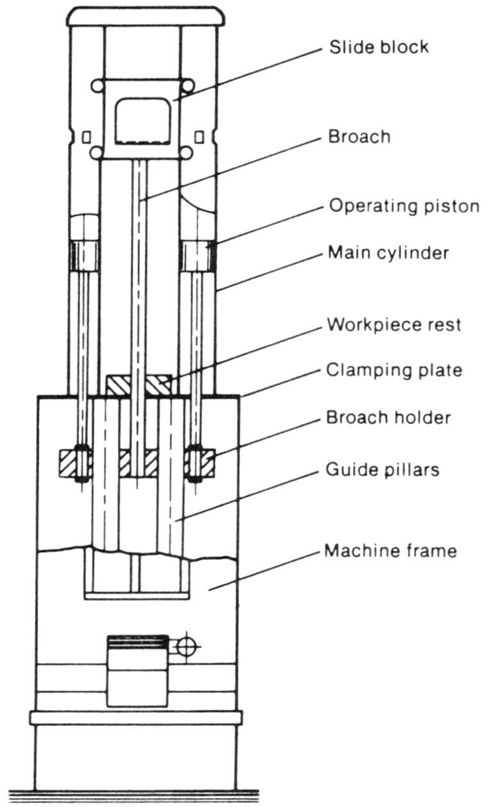


**Fig. 10** Vertical internal broaching machine of the pull-down type

become the largest class of horizontal surface broaching unit built. They are available with pulling capacities in excess of 890 kN (100 tonf), strokes up to 9 m (30 ft), and cutting speeds, in some cases, of more than 90 m/min (300 sfm).

Larger machines have fully stress-relieved welded steel frames rather than gray iron castings. Two sets of cutting tools are often attached to the ram so that parts can be broached on both the forward and return strokes. A common operation on automobile engine blocks consists of broaching head surfaces on one stroke of the ram, followed by the pan rail and bearing surfaces on the return stroke.

These machines can also be equipped with dual-speed controls, by which the ram is driven at one preselected speed during one portion of the stroke and then changed to a second preselected speed during another portion of the stroke. A typical application is the use of the high speed for the initial roughing cut on pine-tree slots in turbine wheels, followed by the slower speed while the finishing teeth are cutting.

**Continuous surface horizontal broaching machines** are rapidly becoming the most widely used type of machine produced for high-production surface broaching. A large continuous machine can be a 38.5 Mg (42.5 ton) giant with an 8.8 m (29 ft) long bed, a 5600 mm (220 in.) stroke, and a 355 kN (40 tonf) broaching capacity. It is capable of executing nine separate operations on 3.4 kg (7½ lb), 318 mm (12½ in.) long connecting rod-and-cap sets for farm machinery engines. One of the smallest continuous machines is a 7.7 Mg (8.5 ton) chain broach with a 22 kN (2.5 tonf) capacity and a 500 mm (20 in.) stroke. It broaches 0.14 kg (5 oz), 64 mm (2.5 in.) long manual transmission shaft shifters in four different configurations for automobiles.

The key to the productivity of a continuous horizontal broaching machine is elimination of the return stroke by mounting the workpieces, or the tools, on a continuous chain. Most frequently, the tools remain stationary, mounted in a tunnel in the top half of the machine, and the chain-mounted workpieces pass underneath them (Fig. 9).

## Vertical Broaching Machines

About 45% of all the broaching machines in existence are verticals, almost equally divided between vertical internal and vertical external or combination machines. Vertical broaching machines, used in every major area of metalworking, are almost all hydraulically driven. One of the essential features that promoted their development, however, is becoming a limitation. The cutting strokes currently in use often exceed existing factory ceiling clearances. When machines reach heights of 6 m (20 ft) or more, expensive pits must be dug for the machine so that the operator can work at factory floor level.

**Vertical internal broaching machines** are pull-up, pull-down, or push-down units, depending on their mode of operation.

*The pull-up type*, in which the workpiece is placed below the worktable, was the first to be introduced. Its principal use is in broaching round and irregularly shaped holes. Pull-up machines have pulling capacities of 55 to 445 kN (6 to 50 tonf), strokes up to 1830 mm (72 in.), and broaching speeds of 9 m (30 sfm). Larger machines are available; some have electromechanical drives for greater broaching speed and higher productivity.

*Pull-Down Machines*. The more sophisticated pull-down machines, in which the work is placed on top of the table (Fig. 10), were developed later than the pull-up type. Pull-down machines are capable of holding internal shapes to closer tolerances by means of locating fixtures on top of the worktable. These machines are available with pulling capacities of 18 to 445 kN (2 to 50 tonf), 380 to 2290 mm (15 to 90 in.) strokes, and speeds of up to 24 m/min (80 sfm).

*Push-down machines* are often nothing more than general-purpose hydraulic presses with special fixtures. They are available with capacities of 18 to 220 kN (2 to 25 tonf), strokes up to 915 mm (36 in.), and speeds as high as 12 m/min (40 sfm). In some cases, universal machines have been designed that combine as many as three different broaching operations, such as push, pull, and surface, simply through the addition of special fixtures.

**Vertical surface or combination broaching machines** are found mainly in the automotive industry. These machines, produced in single- and double-ram versions (and even more rams occasionally), are hydraulically powered, with a few notable exceptions. Capacities range from 27 to 445 kN (3 to 50 tonf), with up to 3050 mm (120 in.) strokes, and speeds of 37 m/min (120 sfm).

Electromechanically driven vertical surface broaching machines are available with either single or double rams and with strokes up to 3050 mm (120 in.), capacities of 220 kN (25 tonf), and speeds of 24 m/min (80 sfm).

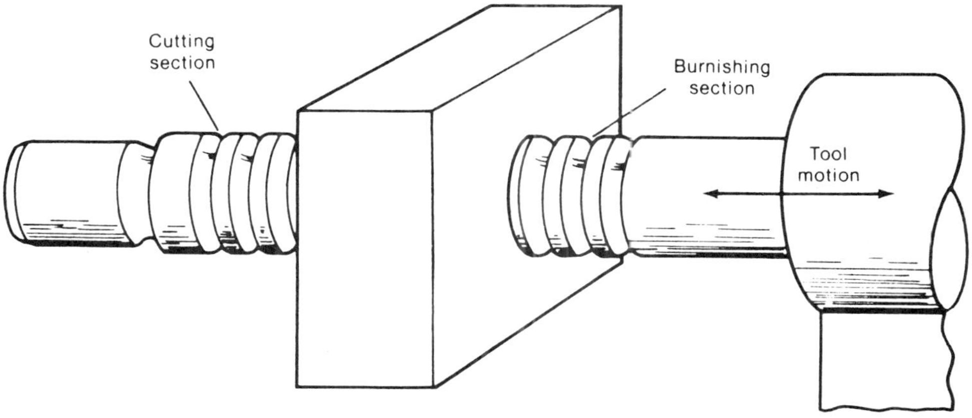


**Fig. 11** Schematic of strip broaching. In this method, the tool is returned over the already-cut work surfaces; this reduces tool life and limits tool precision.

## Special Broaching Machines

Special broaching machines also fall under the general categories of internal or external use, but beyond that it is difficult to classify the wide and often unique variety of special machines. It is sometimes impossible to bring the workpiece to the machine. This is particularly true in the marine, power generation, construction, and airframe industries. Therefore, broaching machine builders have designed portable machines that can be brought to the work.

A form of internal broaching called strip broaching is occasionally used to effect large gains in productivity per machine and man-hour through reduced broaching time cycles. In strip broaching, the broach is returned directly through the hole just broached, immediately after the cutting stroke, eliminating the necessity for disengaging the broach tool from its pulling fixture (Fig. 11). Because the cutting edges rub against the work on the return stroke, broach life is reduced, but not to the extent where the overall savings derived from this technique are lost.

Internal broaching of helicopter rotor spar sections is an unusual special broaching application. In one case, a 7.3 m (24 ft) long workpiece had about 3.2 mm (⅛ in.) of 4153 aluminum removed around the periphery of the irregularly shaped internal form by 35 progressively stepped broach sections. These were pulled through the workpiece one at a time by a special electromechanical horizontal machine with a 19.5 m (64 ft) long bed. Broach sections were semiautomatically loaded and unloaded from the pulling bar at the beginning and end of each stroke. One operator handled the entire job, riding from loading to unloading stations in an electric cart.

**Pot broaching machines** are arranged for machining outside surfaces such as gear teeth in one pass by means of cutter teeth inside a hollow broach. Most pot broaching machines push the work upward as indicated in Fig. 12 because such movement allows the chips to escape most easily and the arrangement is amenable to automatic loading and unloading. However, if the workpiece is small or if the stroke must be long, the push rod would be slender and subject to buckling. For such cases, pull-up broaching machines are available.

Pot broaches produce external forms on parts that are passed through them. They are applied on special vertical pot broaching machines that normally have the broaching tool stationary. The workpieces are fed one or more at a time either up or down through the broach.

A high-production process that is ideally adapted to automation concepts, pot broaching is widely applied to the production of external involute spur running gears, involute splines, cam shapes, and special tooth shapes. There are three basic types of pot broaches: ring-type, stick-type, and ring- and stick-type combination.

*Ring-type pot broaches* produce precision involute tooth forms. Finishing rings on the final portion of ring-type broach sections ensure maximum tooth profile and spacing accuracy with excellent surface finish.

*Stick-type pot broaches* are applied where workpiece accuracy permits. They are particularly adapted to producing profiles with a length of cut under 15 mm (⅝ in.). Cam shapes, straight-sided teeth, interrupted tooth shapes, and special tooth profiles are usually pot broached with stick-type broaches.

*Ring- and stick-type combination pot broaches* have wafers applied in combination with supported sticks to simultaneously produce precision involute teeth with protuberance forms and tip chamfers or slots between the workpiece involute tool sections. The rings form the involute teeth, and the sticks produce the slots or protuberance form.

## Determination of Power Requirements

Power requirements are influenced by the material to be broached, the axial length of the broached section, the perimeter of the cut to be taken, the amount of material to be removed, and the condition of the cutting edges of the broach. To assist in evaluating the broachability of commonly encountered metals, the empirical constants given in Table 1 have been generally accepted. These constants are used in Eq 1 to 3 to determine the force required for broaching. In these computations, it is assumed that the broach is sharp and in good condition.

**Surface Broaching.** The force in pounds, $F$, required for surface broaching is determined by:

$$F = TRC \qquad \text{(Eq 1)}$$

where $T$ is the total length (in inches) of all teeth engaged, $R$ is the rise per tooth (or chip thickness), and $C$ is the broachability constant for the material being broached (Table 1). Total length $T$, could be stated as $N \cdot L$, where $N$ is the number of teeth engaged and $L$ is the effective length of tooth or width of broach.

**Broaching Round Holes.** The force required for broaching round holes is determined by:

$$F = N\pi DRC \qquad \text{(Eq 2)}$$

where $N$ is the number of broach teeth engaged, $D$ is the hole diameter, $R$ is the rise per tooth, and $C$ is the broachability constant from Table 1.

**Broaching Splined Holes.** For broaching splined holes of which the inside diameter has been previously sized, the force required is determined by:

$$F = NSWRC \qquad \text{(Eq 3)}$$

where $N$ is the number of broach teeth engaged, $S$ is the number of splines, $W$ is the width of the spline, $R$ is the rise per tooth, and $C$ is the broachability constant (Table 1). In applying Eq 3 to a ten-spline hole (inside diameter previously sized) in 5140 steel 41.28 mm (1.625 in.) thick, with splines 6.35 mm (0.25 in.) wide, using a broach with rise per tooth of 0.064 mm (0.0025 in.) and a pitch of 12.7 mm (0.5 in.), the force for broaching the hole is calculated as 65 kN (15 000 lbf).

## Types of Broaches

There are three general categories of broaches: solid, shell, and insert-type. Within these categories, broaches can be further classified by the type of cut (internal or external) they are designed to make and by the method by which they are actuated (push or pull). Of the various types of broaches, the solid pull broach (Fig. 13) is probably the most commonly used.

**Solid broaches** are one-piece broaches produced from bar stock (Fig. 1 and 13). They can be provided with greater dimensional accuracy and concentricity than shell broaches. The primary disadvantages of solid broaches lie in the difficulty of repair-

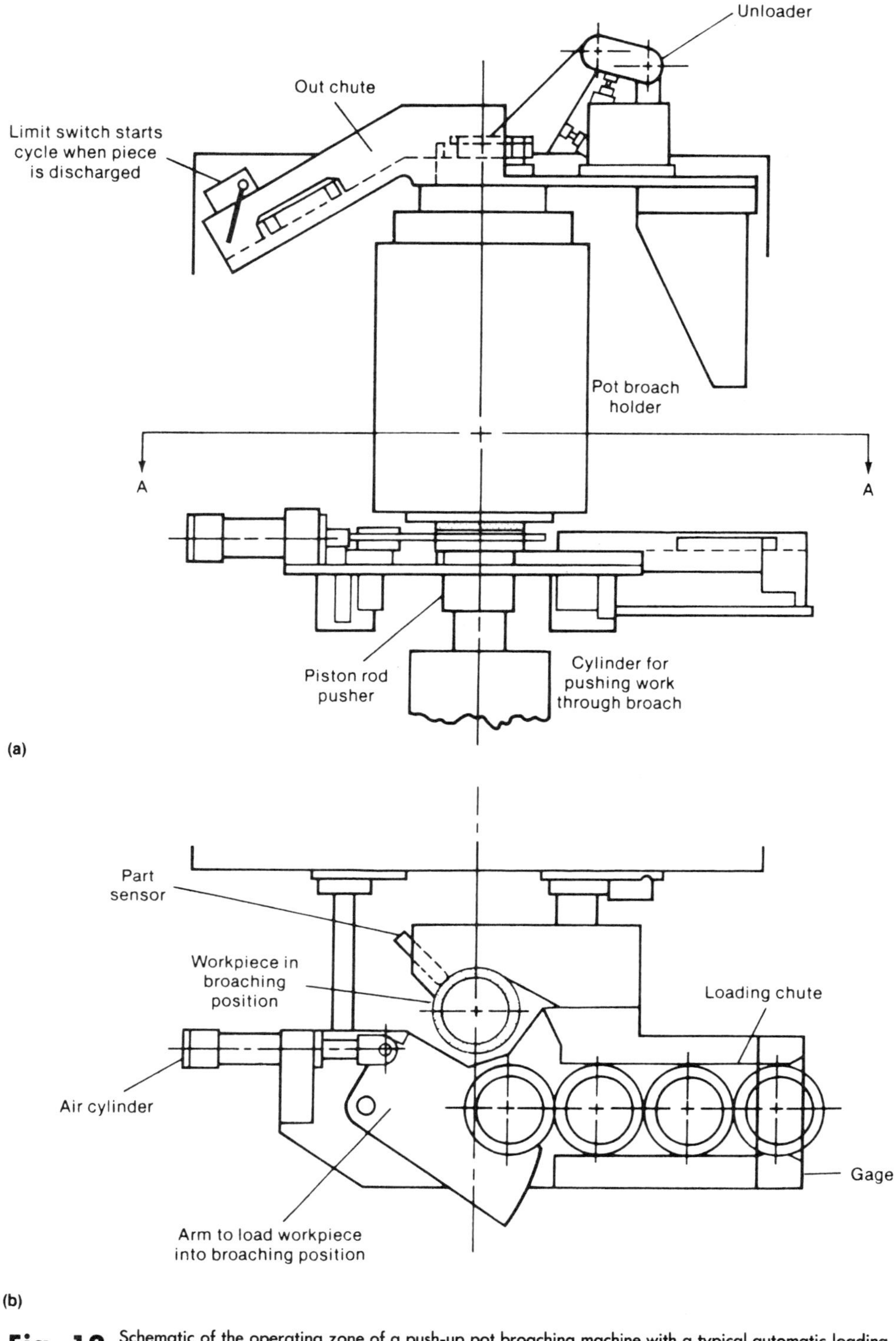


**Fig. 12** Schematic of the operating zone of a push-up pot broaching machine with a typical automatic loading and unloading system. (a) Front view of machine. (b) Section A–A

ing broken teeth and in the cost of replacement.

**Shell broaches** consist of a main broach body (usually the roughing and intermediate broach sections), an arbor section over which a removable shell fits, and the removable shell that is the finishing section. A broach can have more than one removable shell. Shell broaches can be used for internal broaching or for some external operations such as broaching half bores. Shell broaches are superior to solid broaches in that worn or broken sections can be replaced without discarding other sections. The disadvantages of shell broaches are that some accuracy and concentricity are sacrificed as normal variations are multiplied by the number of pieces involved. A typical pull-down removable-shell broach is shown in Fig. 14.

**Insert-type broaches** consist of a toolholder and inserts of high-speed tool steel or carbide, which do the cutting. Although these broaches can be used for broaching contoured shapes, both internal and external, their main use is for broaching large flat surfaces, such as the machined faces of automotive engine blocks. Insert broaches are also used on a wide variety of form broaching operations, such as plier halves or steering gear racks. Large flat areas are only a portion of insert-type broach forms. Dovetail forms, T-slots, and various other configurations can be broached with insert-type broaches.

Carbide inserts can be brazed to the toolholder, as shown in Fig. 15(a). However, the practice of using brazed inserts has been largely replaced by the use of disposable inserts for broaches of this type. Disposable inserts are either triangular or square, thus having six or eight cutting edges, respectively. A square insert secured to a holder is shown in Fig. 15(b). Square carbide inserts with negative rake angles provide eight cutting edges and can typically be used to broach 25 000 parts per edge, totaling 200 000 parts per insert life (with indexing of the part) in broaching cast iron engine blocks. Instead of being sharpened, these inserts are discarded after the cutting edges have been used. The initial cost of a broach with disposable carbide inserts is about the same as that of a solid broaching tool.

Round (button-type) indexable carbide inserts are also used for broaching castings. They offer the same advantages as square or rectangular inserts with the additional benefit, in some applications, of longer tool life. A tool life of 85 000 workpieces per insert index has been reported. Less cutting force is required with round inserts because they have a reduced depth-of-cut capability, which may require lengthening the broaching stroke. On a unit-load basis, a round insert requires greater force than a square one. A half-round broach with double-sided button inserts retained by center screws is shown in Fig. 15(c). Inserts can be overlapped and mounted at a shear angle along the length of the broach to produce the required size progression and to balance the cutting load.

Provided the holder is manufactured to close tolerances, particularly in those sections that determine the height of the cutting edge, gaging is not required when cutters

**Table 1 Constants for the broachability of selected annealed metals**

| Metal | Constant |
|---|---|
| Aluminum | 50 000 |
| Copper | 250 000 |
| Cast iron; bronze | 350 000 |
| Low-carbon steel; steel castings | 450 000 |
| Alloy steels 3115–4615 | 550 000 |
| Alloy steels 5120–6195 | 600 000 |
| Titanium and A-286 alloy | 650 000 |

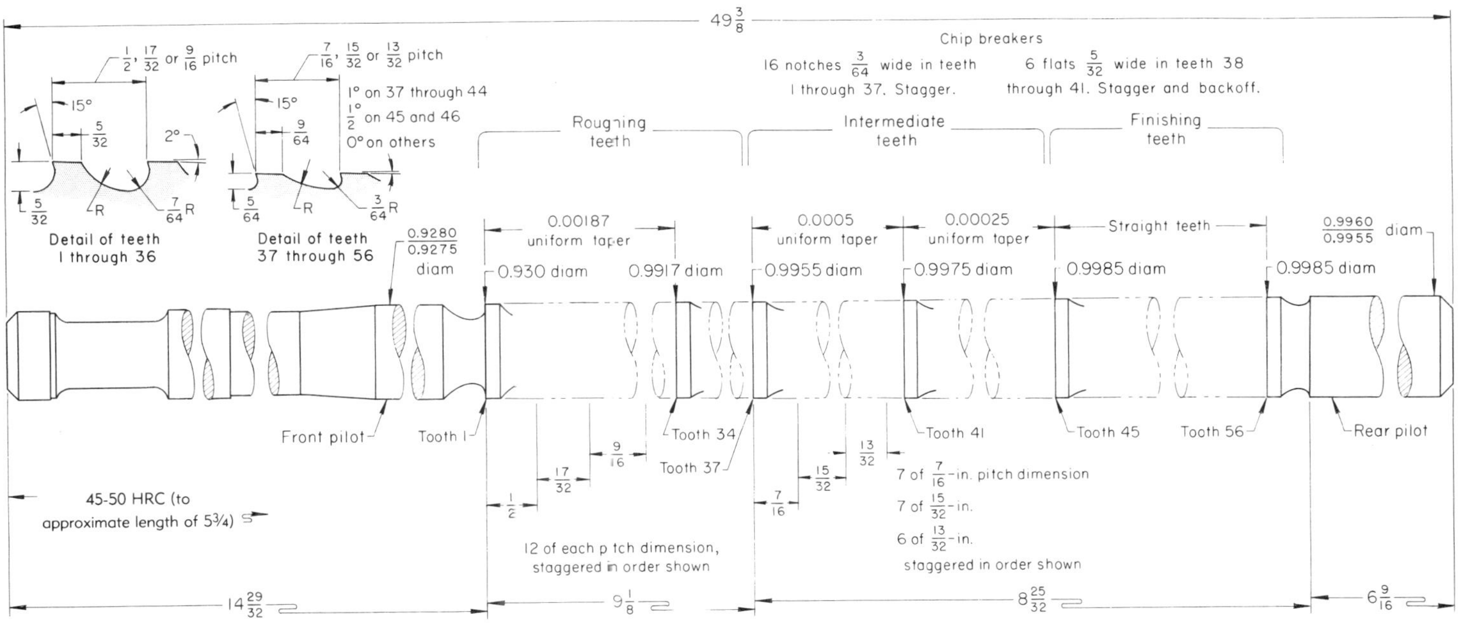


**Fig. 13** Dimensional details of a typical round pull broach made of M2 high-speed tool steel with a shank hardness of 45 to 50 HRC and a main body hardness of 64 to 66 HRC. Dimensions given in inches

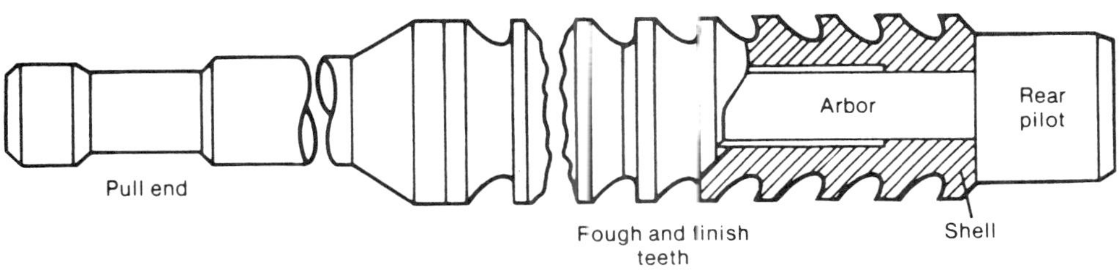


**Fig. 14** Schematic of a shell broach

are changed to present a new cutting edge. Disposable carbide cutters can be inserted in any position without removing the broach from the machine. To provide the required clearance behind the cutting edges, inserts must have a negative rake. Inserts for contour cutting can be sharpened on the top and face, and shimmed to the correct position. A considerable savings in tool cost is made possible by changing from brazed carbide tools to disposable carbide inserts for the same broaching operation.

**Internal and External Broach Configurations.** A number of typical broaches and the operations for which they are intended are shown in Fig. 16.

*A square broach* (Fig. 16a) produces a round-cornered, square hole. Prior to broaching square holes, it is common to drill a round hole having a diameter somewhat larger than the width of the square. Thus, the sides are not completely finished, but this unfinished part is not objectionable in most cases. In fact, this clearance space is an advantage during broaching in that it serves as a channel for the broaching lubricant; moreover, the broach has less metal to remove.

*A round broach* (Fig. 16b) is for finishing round holes. Broaching is superior to reaming for some classes of work because the broach will hold its size for a much longer period, thus ensuring greater accuracy.

*Keyway broaches* (Fig. 16c and d) are for cutting single and double keyways. The single keyway broach is of rectangular section and, when in use, slides through a guiding bushing inserted in the hole.

*The four-spline broach* (Fig. 16e) is for forming four integral splines in a hub.

*The hexagon broach* (Fig. 16f) is for producing hexagonal holes.

*A rectangular broach* (Fig. 16g) is used for finishing rectangular holes. The teeth on the sides of this broach are inclined in opposite directions; this has the following advantages:

- The broach is stronger than it would be if the teeth were opposite and parallel to each other
- Thin work cannot drop between the inclined teeth, as it tends to do when the teeth are at right angles, because at least two teeth are always cutting
- The inclination in opposite directions neutralizes the lateral thrust

The teeth on the edges are staggered, the teeth on one side being midway between the teeth on the other edge, as shown by the dotted line.

*A double-cut broach* (Fig. 16h) is for finishing, simultaneously, both sides of a slot and for similar work.

*An internal gear broach* (Fig. 16i) is the style used for forming the teeth in internal gears. It is a series of gear-shaped cutters, the outside diameters of which gradually increase toward the finishing end of the broach.

*A round broach* (Fig. 16j) is for round holes, but differs from the broach shown in Fig. 16(b) in that it has a continuous helical cutting edge. The broach shown in Fig. 16(j) produces a shearing cut.

*A helical groove broach* is for cutting a series of helical grooves in a hub or bushing. In helical broaching, either the work or the broach is rotated to form the helical grooves as the broach is pulled through.

*Rotary-Cut Broaches.* Rough forgings, malleable iron castings with a hard skin, and sand castings with abrasive surface inclusions are cut with one of three types of rotary-cut broaches (Fig. 17). The design concept is similar to that of a chip-breaking slot; but the cutting edge has been drastically reduced, and the slots between the teeth have become much deeper. Rotary-cut broaching teeth are heavier to withstand the heavy cutting load and are spaced in staggered fashion along the axis of the broach to generate the entire circumference of the hole. The tools are designed to take deep cuts underneath a poor-quality surface. Once this surface has been penetrated, the balance of the broaching tool proceeds to semifinish and finish the underlying metal in the normal manner.

The hexagonal rotary-cut broach (Fig. 17a), used for small-diameter holes, re-

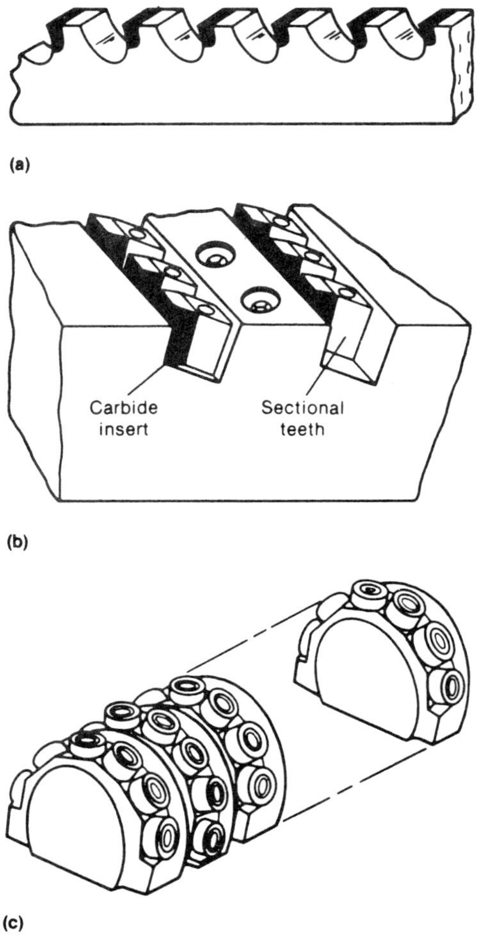


**Fig. 15** Three types of carbide inserts. (a) Brazed inserts. (b) Square indexable throwaway inserts. (c) Round (button-type) indexable inserts on a half-round broach

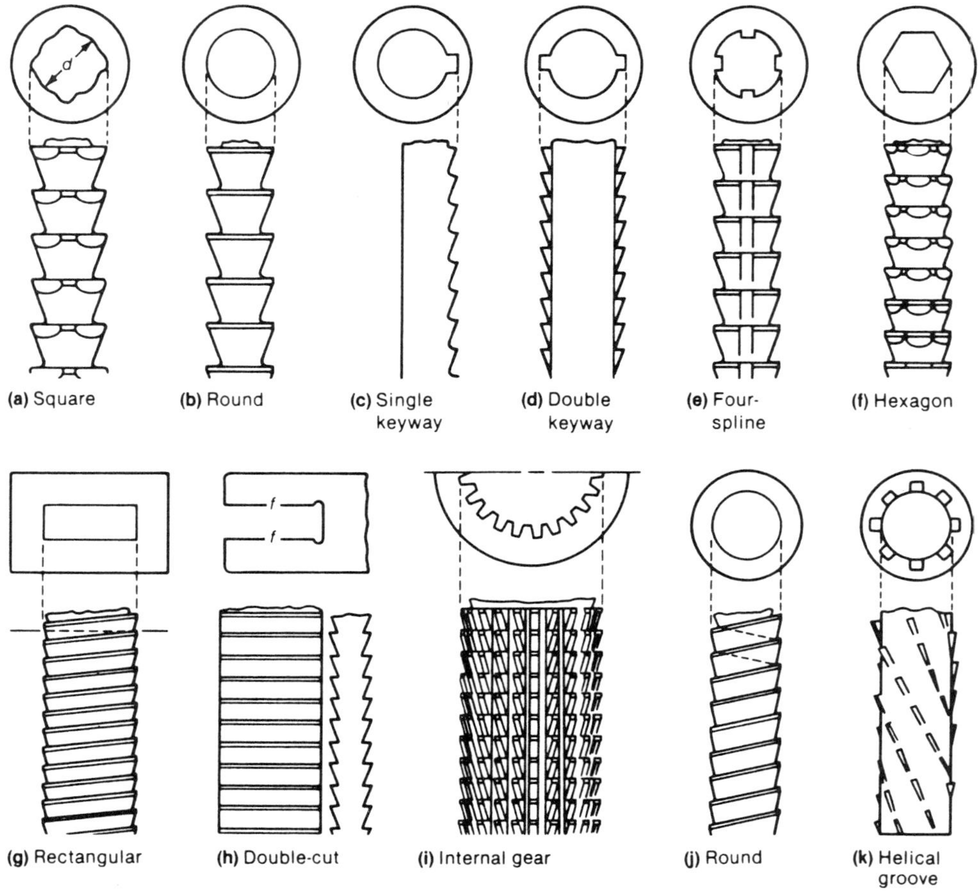


**Fig. 16** Typical broaches and the configurations they generate. See text for discussion.

moves little stock. Depth of cut is limited to the distance across the flats.

The radial rotary-cut broach (Fig. 17b) removes more stock than the hexagonal tool because the cutting portions of the teeth are connected by arcs rather than by flats.

Spline rotary-cut broaches (Fig. 17c) offer a greater degree of flexibility than either of the other tool types and also permit maximum stock removal. The amount of stock removal is primarily governed by the capacity of the broaching machine rather than by any tooling limitations. Rise per tooth may be as much as 1.3 mm (0.050 in.) on such broaches.

In addition to the typical broaches shown in Fig. 16 and 17, many special designs are used for performing more complex operations. Two surfaces on opposite sides of a casting or forging are sometimes machined simultaneously by twin broaches, and in other cases, three or four broaches are drawn through a part at the same time for finishing as many duplicate holes or surfaces.

*Progressive Broaches.* Notable developments have been made in the design of broaches for external broaching. One of these developments is the progressive broach (Fig. 18). Employed primarily for broaching wide, flat surfaces, the first few teeth in progressive sectional broaches completely machine the center, while succeeding teeth are offset in two groups to complete the remainder of the surface.

One-piece progressive broaches (Fig. 18a) have two sets of narrow roughing teeth, with each set positioned at an angle with respect to the centerline of the broach holder, thus forming an inverted vee. Each tooth or insert takes a shear cut, generally to full depth, but covers only a small portion of the workpiece surface. This is similar to a single-point tool on a shaper or planer progressively generating a flat surface on the workpiece.

Full-width teeth for semifinishing and finishing are located behind the roughing teeth on progressive broaches so that the entire surface is cut in one pass. For narrow surfaces, the teeth or inserts at the starting end are V-shaped. On subsequent teeth, the vees gradually widen until the full required width of the surface is cut. The final teeth are flat, similar to those on a slab broach.

**Pull broaches,** as the name implies, are pulled through or against the surface of the workpiece. Most internal broaching is done with pull broaches. Because there is no problem of bending, pull broaches can be longer than push broaches for the same size of hole, and they can also remove more stock in one pass. Pull broaches can be made to long lengths, but cost usually limits the length of solid pull broaches to approximately 2.1 m (7 ft). Broaches longer than 2.1 m (7 ft) are usually made up of sections similar to shell broaches because the cost is less for replacing a damaged or worn section than for replacing the entire broach.

**Push broaches,** for internal broaching, are necessarily shorter than pull broaches because of the problem of bending under load. Push broaches are used for broaching blind holes (Fig. 6) or for multiple-station broaching machines in which several short broaches, rather than a single long broach, are used to reduce the time required for a given operation.

## Broach Tool Materials

Hardened high-speed tool steel is by far the most widely used material for solid broaches or for the cutting teeth of other types of broaches. The tools are usually ground to final dimensions after hardening. The grade of high-speed tool steel is normally chosen on the basis of minimum overall

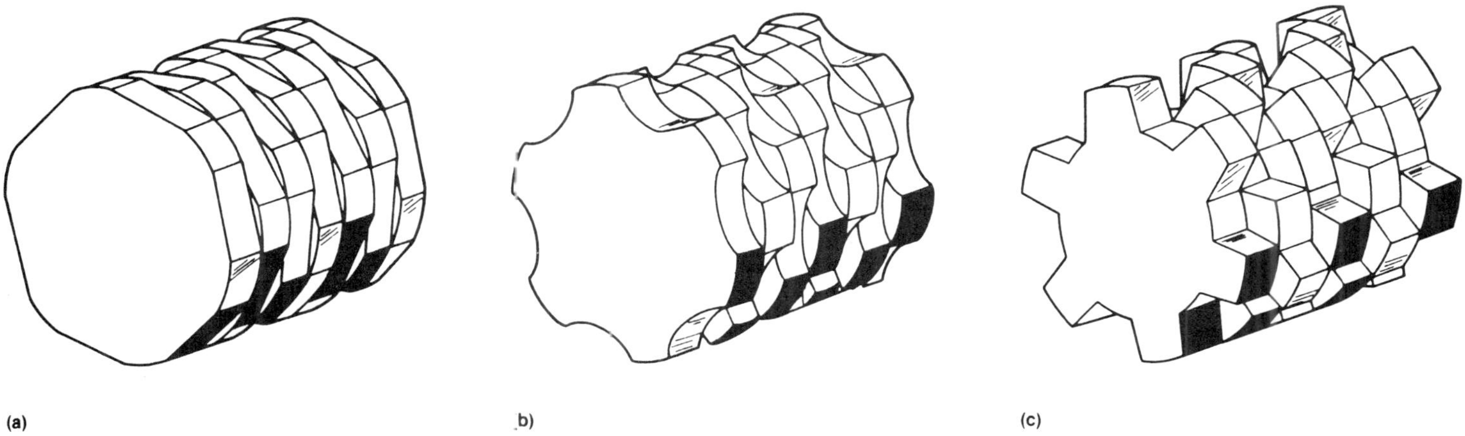


**Fig. 17** Three types of rotary-cut broaching tools designed to penetrate rough skins, as on castings and forgings, without exceeding the power ratings of a broaching machine. (a) Hexagonal rotary cut. (b) Radial rotary cut. (c) Spline rotary cut

cost, balancing tool life and production rate against tool cost (material, heat treatment, fabrication, and regrinding for reuse).

**High-Speed Tool Steels.** In the early stages of broaching technology, broach tools were made from water-hardening tool steels. These tools were used on slow, screw-type broaching machines. With the introduction of new machines with higher speeds and greater production rates, high-speed tool steels became the principal materials for broach tooling. The following is a list of typical tool steels and the materials that are commonly broached with these steels:

- *M2 tool steel*: General use, including brass, aluminum, magnesium, and the following steels: 1020, 1063, 1112, 1340, 1345, B-1113, 4140, 4340, 5140, 8620 (26 HRC), type 347 stainless steel, and type 416 stainless steel (35 to 40 HRC)
- *M3 tool steel*: Aluminum castings, cast irons, A-286 (32 to 38 HRC), Greek Ascoloy (32 to 38 HRC), M-252, D-979 (40 HRC), and the following steels: 4140 (32 HRC), 4337 (29 to 23 HRC), 4340 (32 to 38 HRC), 8617 (30 to 36 HRC), 8620 (32 HRC), 9310 (36 to 38 HRC), 9840 (32 to 36 HRC), type 403 stainless (37 to 40 HRC)
- *M4 (or T5) tool steel*: Cast irons
- *T2 tool steel*: Steels: 1112, 4340 (35 to 40 HRC), type 403 stainless (30 to 35 HRC); titanium alloys, PWA-682 Ti (36 HRC), Lapelloy (30 to 35 HRC), Greek Ascoloy (32 to 38 HRC), 19-9DL (20 to 27 HRC), and Discalloy (23 to 32 HRC)
- *T5 tool steel*: A-286 (29 HRC), Chromalloy (30 to 35 HRC), Incoloy 901, and PWA-682 Ti (34 to 36 HRC)
- *T15 tool steel*: Aluminum 2219, A-286 (32 to 36 HRC), Stellite, 17-22 A(S) (29 to 34 HRC), N-155 (30 to 40 HRC), AMS 4925 titanium (32 to 40 HRC), Waspaloy, Incoloy 901 (32 to 36 HRC), and the following steels: heat-resistant steels, conventional alloy steel forgings, 4340 (35 to 40 HRC), 52100, 9310 (26 to 30 HRC), and 17-4PH

**Carbide-Tip Cutters.** Most of the carbide cutters used to broach cast iron are used in flat surface broaching applications, although contoured cast iron surfaces have been broached successfully. Surface broaching of pine-tree slots has been attempted with carbides on high-temperature alloy turbine wheels, but with little success. The carbide edges tend to chip on the first stroke.

Carbide-tip broaches are seldom used on conventional steel parts and forgings. One reason is that good performance is obtained from high-speed steel tools; another is that the low cutting speeds of most broaching operations (from 3.7 to 9 m/min, or 12 to 30 sfm) do not lend themselves to the advantages of carbide tooling. The success of carbide tooling on cast irons is due to the resistance of carbide to abrasion on the tool flank below the cutting edge.

Another problem with carbide-tip tools is that a broaching machine work fixture must be exceptionally rigid to prevent chipping of the cutting edge. Experimental work with extra-rigid tools and workpiece fixtures, however, has shown that tool life and surface finish can be greatly improved with carbide-tip tools, even when used on alloy steel forgings.

Cast high-speed tool steels are seldom used in broaches. One property of the cast tool materials that prohibits their use in monolithic internal pull broaches is low tensile strength. Most cast alloys that can attain a hardness of 60 HRC or higher do not have ultimate tensile strengths much in excess of 585 MPa (85 ksi).

To provide the optimum combination of abrasion resistance and toughness, broach cutting teeth are normally hardened to 64 to 66 HRC for the general-purpose grade, ranging upward for the more highly alloyed grades to a maximum of 66 to 68 HRC for T15. For longer tool life, surface treatments such as nitriding or oxidizing are sometimes employed. Nitriding increases superficial hardness, and both nitriding and oxidizing minimize sticking or welding of the tool to the work material. Chromium plating will also minimize sticking, although this plating is prone to chipping.

Carbide inserts or rings are used to a limited extent in internal broaching, primarily on small parts made of free-machining materials such as gray iron, usually in applications requiring extremely close tolerances at high production rates. Broaching of steel castings, in which a carbide tool can cut through local hard spots with less tool damage than high-speed tool steel, is another application of carbide inserts.

## Broach Design

Basic broach design is shown in Fig. 13, which presents the dimensional details of a typical pull broach for producing a round hole. This broach has been used for the production broaching of a hole 25.36/25.31 mm (0.9985/0.9965 in.) in diameter in a

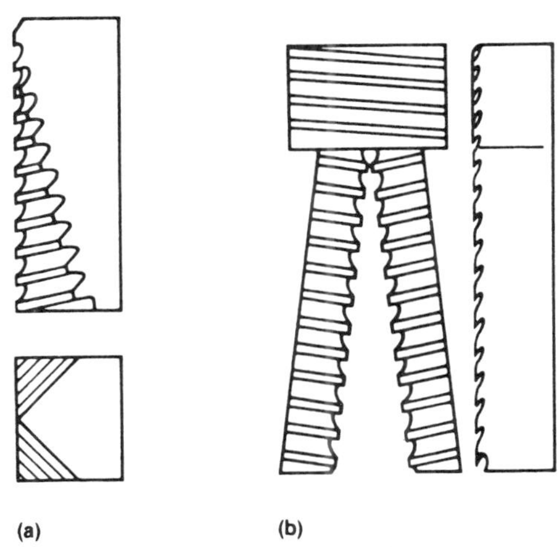


**Fig. 18** One-piece (a) and sectional (b) progressive broaches. Top and side views of both types are shown.

normalized forged steel steering knuckle. A starting hole 23.8 mm (15⁄16 in.) in diameter was drilled through the forging to accommodate the broach. As shown in Fig. 13, the first cutting tooth of this broach is 23.6 mm (0.930 in.) in diameter, and each tooth in the roughing section increases 0.0475 mm (0.00187 in.) in diameter over the one preceding it (the first three or four teeth may cut little or nothing, depending on the exact size of the drilled hole). Thus, as the broach is pulled through the workpiece, cutting begins gradually, and as each succeeding tooth engages the work, it removes a small amount of metal. The progressive increase in tool diameter is usually greatest in the roughing section. In this broach, the increase is 0.0475 mm (0.00187 in.) for the roughing teeth, 0.013 mm (0.0005 in.) for the first four teeth in the intermediate section, and 0.0065 mm (0.00025 in.) for the remaining teeth in the intermediate section. In the finishing section, all teeth are 25.36 mm (0.9985 in.) in diameter—the maximum permissible diameter of the broached hole.

Tooth contours are shown in the upper left corner of Fig. 13. The greater depth of the gullet and the greater pitch for teeth 1 through 36 permit better chip accommodation. This is essential, because these teeth make the greatest advance and therefore remove the most metal. The pitch length (distance between teeth) in both the roughing and the intermediate sections is staggered to prevent chatter as the broach is pulled through the work.

Chip breakers (discussed in the section "Chip Breakers" in this article) are incorporated in the roughing teeth and the first four intermediate teeth, as specified in the notes in Fig. 13. The chip breakers are staggered from tooth to tooth so that the ridges left in the workpiece surface by the discontinuities in any one tooth are removed by the tooth that follows. (Note that the last three intermediate teeth have no chip breakers, so that these teeth can remove all traces of irregularities left by chipbreakers on preceding teeth before the first finishing tooth starts cutting.)

**Face (hook) angles** for broach teeth are selected on the basis of hardness and ductility of the work metal. Metals that yield brittle chips, such as cast iron or leaded brasses, are usually cut most efficiently by teeth with a narrow face angle. Ductile materials, such as annealed or normalized steels, usually respond better to wider face angles. Face angles on broaches are similar to top rake angles on single-point tools. Recommended face angles, along with backoff angles, for various metals are given in Table 2.

When broaching similar parts of different metals, a different face angle must often be used for each of the metals cut.

**Gullets.** The shape of the gullets (chip spaces) of broach teeth influences the efficiency of the broaching operation and has a marked effect on broach life. To obtain maximum efficiency, it may be necessary when sharpening a broach to deviate from conventional gullet shape, as in the following example:

**Table 2 Typical broach face and backoff angles**

| Material | Face angle, degrees | Backoff angle, degrees |
|---|---|---|
| Aluminum | 6–10 | . . . |
| Babbitt | 8–10 | . . . |
| Brass | −5 to 5 | 2–3 |
| Bronze | 0 | ½–2 |
| Cast iron | 6–10 | 2–5 |
| Copper | 15 | 2–3 |
| Zinc | 6 | . . . |
| Aluminum bronze | 15 | 2–3 |
| SAE 1037 | 15 | 1–2 |
| SAE 1112 | 15 | 2½ |
| SAE B-1113 | 15 | 2–3 |
| SAE 1340 | 12 | 1–2 |
| SAE 4140 | 8–15 | 1–3 |
| SAE 4337 | 8–15 | 1–3 |
| SAE 5140 | 15 | 1–2 |
| SAE 5140 (Type 410 stainless) | 18 (roughing) | 2 |
| | 20 (finishing) | 2 |
| SAE 9310 | 18 (roughing) | 2 |
| | 20 (finishing) | 1–2 |
| Type 303 stainless steel | 15 | ½–2 |
| Type 304 stainless steel | 15 | ½–2 |
| Type 403 stainless steel | 15–20 (roughing) | 3 |
| | 30 (finishing) | 5 |
| Type 431 stainless steel | Up to 28 | . . . |
| M-308 | 15 | 3 |
| N-155 | 20 | 2 |
| Greek Ascoloy | 15 | 2–3 |
| Chromalloy | 15 | 2 |
| Lapelloy | 12–15 | 2 |
| A-286 | 10–15 (roughing) | 2–3 |
| | 15–18 (finishing) | |
| René 41 | 15 | 3 |
| Incoloy 901 | 15 (roughing) | 3 |
| | 18 (finishing) | |
| Titanium 140A | 5–15 | 2–4 |
| Titanium 150A | 5–9 | 2–5 |
| Titanium PWA A682 | 12–15 (roughing) | 3 |
| | 15 (finishing) | 3 |

Source: *Metalcutting: Today's Techniques for Engineers and Shop Personnel*, McGraw-Hill, 1979

**Example 1: Gullet Redesign for Increased Broach Life.** Figure 19 shows original and revised designs of the gullet in teeth of a broach used for cutting fir-tree slots in a turbine wheel of Incoloy 901. The original full-radius design encourages the packing of chips in the gullet so tightly that they were almost impossible to remove by wire brushing. The broach became overheated because of the transfer of heat from the packed-in chips. At times, the packing of chips in the gullet caused galling and tears in the broached surface of the workpiece.

When the gullet was ground to the two-angle configuration shown as the improved design in Fig. 19, broach life increased from one piece per grind to three pieces per grind, and galling and tearing of the broached surface were eliminated. The change of gullet had no effect on the number of resharpenings; broaches could still be reground seven or eight times.

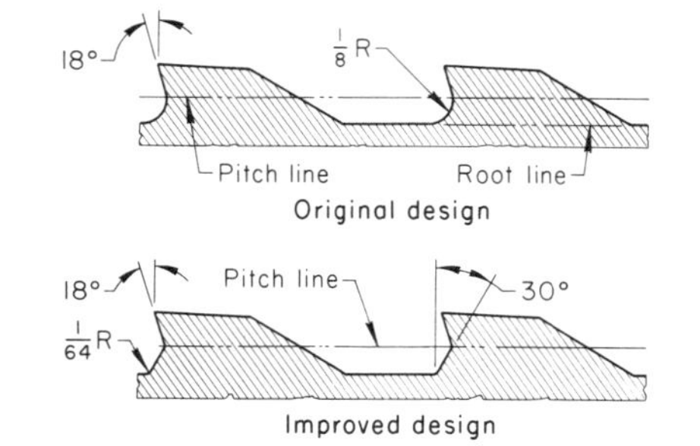


**Fig. 19** Revision of gullet configuration to eliminate problems in broaching Incoloy 901. Dimensions given in inches

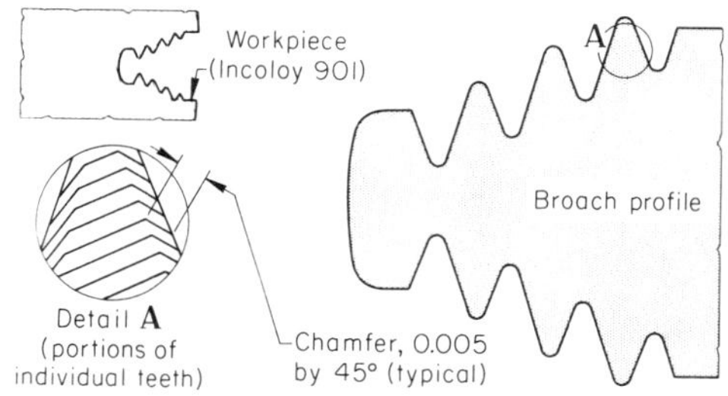


**Fig. 20** Addition of chamfer to corners of broach teeth for increased tool life in the broaching of fir-tree slots in a turbine wheel

**Chamfered Edges.** The sides of broach teeth used for forming configurations such as keyways or fir-tree slots are usually chamfered to prolong tool life. The need for chamfered teeth increases as the machinability of the work metal decreases.

The amount of chamfer may be restricted by the shape being broached, but even as little as 0.13 mm (0.005 in.) chamfer is helpful. The following example demonstrates the beneficial effect of tooth chamfer in the broaching of heat-resistant alloys.

**Example 2: Addition of Chamfer to Prolong Tool Life When in Turbine Wheels.** A cross section of one of 102 fir-tree slots that were broached around the periphery of an aircraft engine turbine wheel made of Incoloy 901 is shown in Fig. 20. The broach was made of T15 high-speed tool steel and was heat treated to 66 to 68 HRC. The wheels were broached in a 760 kN (85 tonf) vertical machine with a 2290 mm (90 in.) stroke at a speed of 3050 mm/min (120 in./min). A broaching oil having a viscosity of 155 Saybolt Universal seconds (SUS) at 40 °C (100 °F), 2% fat content, 0.8% active sulfur, and 2.1% Cl was used.

Before a chamfer was added to the broach teeth (shown at bottom, Fig. 20), broach life was 19.2 wheels, and an average of two wheels could be broached between sharpenings. The addition of this 0.13 mm (0.005 in.) by 45° chamfer parallel to the broaching axis on the sharp corners of the teeth increased broach life to an average of 24.8 wheels and increased to three the number of wheels that could be broached between sharpenings. The improvement in broach life was assisted by improved grinding and resharpening procedures and by more careful handling of the broach.

**Effect of Tooth Design on Chatter.** When chatter develops in broaching, loss of accuracy, poor surface finish on the workpiece, and excessive broach wear are probable results. With extreme chatter, the broach is likely to break. In broaching, cuts are often interrupted, depending on the length of the section to be broached and the distance between successive cutting teeth. In general, the likelihood of chatter increases as the severity of the interruptions in cutting increases.

Conventional broaches having circular teeth such as those shown in Fig. 1 and 13 are more susceptible to chatter than specially designed broaches because there is a complete interruption after each cutting tooth. Either of two approaches is frequently used in broach design to minimize interrupted cutting and chatter:

- In broaching flat surfaces or several internal splines spaced around a periphery, teeth staggered longitudinally provide a more uniform cutting action
- When broaching round holes, a broach having helical teeth is an effective means of eliminating chatter

Specially designed broaches are more expensive than conventional types, but the increased broach cost is often justified.

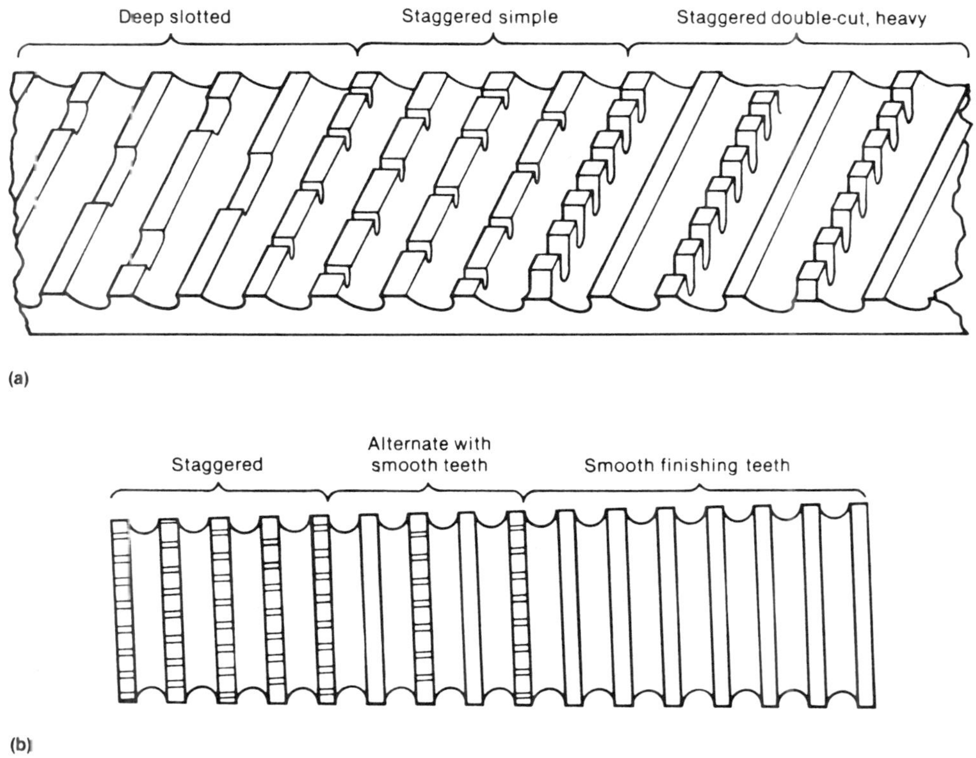


**Fig. 21** Chip breakers on a flat broach (a) and a round broach (b). Notches that split the heavy chips can be either U-shaped or V-shaped breakers.

## Chip Breakers

Chip control is essential in all machining operations, but especially in broaching. In single-point machining such as lathe turning, the chip leaves the cutter as soon as it is formed. In broaching, however, the chip stays in the gullet, or chip space, behind each tooth until that tooth clears the workpiece. Chip space is limited by the pitch of the teeth, and in small broaches by the root diameter of the broach. Thus, it is seldom possible to provide enough space for chips, particularly in small-diameter broaches.

In broaching, therefore, chips must be controlled to keep chip-space requirements to a minimum and to facilitate chip removal. Broaches are provided with chip breakers, which are small grooves or notches approximately 0.8 to 2.4 mm (1/32 to 3/32 in.) wide that transversely break the cutting edge and land of each roughing and semifinishing tooth (Fig. 21). These grooves break the continuity of the width of the chip. Thus, instead of one wide chip, several narrow chips are formed. These narrow chips are easily washed out by the cutting fluid, are brushed out, or fall away when the gullet clears the workpiece.

Broaches that cut around holes especially require chip breakers. Without chip breakers, the chips would be continuous rings that would be difficult to remove from the broach. In addition, the shape of the chips must permit them to drop into the chip space behind each tooth. Chip breakers accomplish this by making smaller chips.

Chip breakers are staggered from tooth to tooth so that the ridge left by the chip breaker in one tooth is removed by the tooth immediately following. Generally, finishing teeth have no chip breakers, although some broaches require them in the first few teeth.

## Fixtures

In broaching, as in other machining operations, fixturing of the workpiece is required. Broaching fixtures have several functions. They must locate the workpiece accurately and hold it securely. For certain applications, such as broaching helical gears, the fixture must position the part accurately and firmly, yet permit the part to rotate as required. Fixtures can be used to carry workpieces in and out of the broaching position or to carry them from one broaching position to another where more than one broach is required. Fixtures can also be used to guide the broach when it moves over or through the workpiece.

Fixtures used in broaching must not obstruct the removal of chips. Broaching fixtures must provide more holding force and more rigidity and support than fixtures for most other machining operations because more teeth are cutting at any one time than is typical for other operations.

Broaching fixtures can be manual, semiautomatic, or automatic. Cost considerations and the quality of parts to be broached will largely determine the type of fixture used. Semiautomatic and automatic fixtures can be operated pneumatically, hydraulically, or mechanically.

**Types of Fixtures.** The fixture used can be as simple as a concave backup plate for positioning the piece. The pressure of the plate on the workpiece holds it firmly as the broach is pulled through.

A simple fixture known as a work horn is a special type of faceplate used when broaching keyways (Fig. 22). A ground diameter on the back of the horn fits snugly into the machine platen or into a reducing bushing, depending on the size of the workpiece. A round pilot on the opposite end is ground approximately 0.025 mm (0.001 in.) smaller than the hole in the work for accurate positioning of the work. The horn is slotted to provide a guide for the broach. The depth of the slot in the horn and the height of the last tooth on the keyway broach determine the depth of the broached keyway. The horn is hardened, and all functioning surfaces are ground to size. All faces of the horn are accurately produced to hold the workpiece square and parallel to the axis of broach travel. The force applied to the workpiece by the broach as it is being pulled through the cut holds the workpiece firmly in position.

**Fig. 22** Work horn that positions part and guides broach during the broaching of a keyway

A major trend in fixture design is the automation of fixture action to assist in the integration of broaching machines into transfer lines and other automatic machining systems. A second trend is toward universal fixtures that can hold similar, but not necessarily identical, workpieces.

## Selection of Broach Length

Length of the broach is usually determined by three interrelated factors:

- Material to be broached
- Type of cut
- Dimensions of the cut

**The material to be cut** has a double influence on the required length of the broach or broach holder. In the broaching of hard metals, chips must be thinner to avoid damage to the broach teeth and to reduce power requirements; therefore, more teeth are required to remove hard metal than to remove the same amount of soft metal that permits deeper cuts.

The type of work metal determines the type of chip. Brass and cast iron produce chips that break up readily and can be packed into a smaller volume than chips from malleable iron, steel, aluminum, and titanium, which produce continuous chips that coil and fill more space. Larger gullets (chip spaces) must be provided for the metals that form coiled chips (Fig. 23). In some broaches, the gullet can be made deeper to accommodate the greater chip volume. In other broaches, however, deepening the gullet would make the broach too weak for service, and as a result the pitch of the teeth is increased, permitting a longer gullet and increasing the length of the broach and the stroke.

**Type of cut**—that is, whether the cut is internal or external or whether it is a simple cut or one with a keyway, spline, or dovetail shape—influences the length of the broach. For example, if a dovetail form is to be broached, the broach must be almost twice as long as required for the keyway slot of comparable size. This is because the general procedure is first to broach a simple slot and then to follow with cutters that will shape and size the dovetail form.

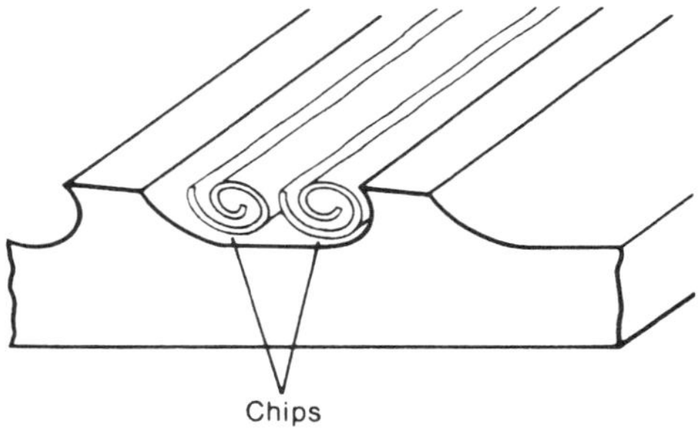


**Fig. 23** Schematic of a flat-bottom gullet, which provides an extended space between teeth to hold more chips

**The length and depth of cut** may have the greatest influence on the length of the broach, particularly on internal cuts. Chips accumulate in the gullets of the teeth. As the cut increases in length, more chip capacity is required for the same amount of tooth advance. Adequate space must be provided to prevent damage to the broach or the surface being broached. It is desirable that chips be loose enough in the gullet so that they fall away (or can be washed away) from the broach when the gullet clears the work. Therefore, for increased length of cut, the size of the gullet is increased, the amount of step per tooth is decreased, or both. For similar materials and the removal of similar amounts of metal, the broach must increase in length proportionately as the surface to be broached increases in length.

## Selection of Stroke Speed

The major consideration in the selection of optimum broaching speed is the need to operate at the lowest overall cost. Related considerations are type and hardness of the work metal, rigidity of the workpiece, and length of the cut.

**Cost.** Up to a point, increasing the broach speed will increase the number of pieces that can be produced per hour. Beyond that point, however, as broach speed is increased, the number of parts that can be produced between sharpenings (and therefore the total parts produced by a broach) will decrease. Maximum efficiency and minimum cost are achieved when that point is determined.

Initial broach cost, setup costs, and broach maintenance costs are relatively high when compared to cost per man-hour of machining time. Therefore, the maximum number of acceptable pieces that can be produced between sharpenings has a greater influence on maximum efficiency and minimum cost than does the total number of pieces produced per hour.

**Work Metal.** In general, steels are broached at speeds of 18 to 24 m/min (60 to 80 sfm). The speed will vary with the hardness of the steel. Hard or tough steels are broached at lower speeds than the free-machining types. Some steels that are relatively soft are difficult to broach without galling or tearing, which results in unacceptable surface finish. Although this can often be corrected by changing the tooth angle or the cutting fluid, an increase in cutter speed will also provide a better finish. Nominal speeds and feeds per tooth for broaching carbon and alloy steels are given in Table 3.

Stainless steels are broached at speeds ranging from 1.5 m/min (5 sfm), for the harder types, to 7.6 m/min (25 sfm), for the free-machining types. Cast iron and malleable iron are broached at speeds up to 9.1 m/min (30 sfm) with high-speed steel broaches and at 37 m/min (120 sfm) with carbide broaches.

Brass and bronze are broached at speeds of 7.5 to 9 m/min (25 to 30 sfm). Aluminum and magnesium can usually be broached at the highest speed of which a machine is capable. With tough alloys such as A-286, René 41, and AM-355, optimum broach life is obtained at speeds below 6 m/min (20 sfm) and sometimes as low as 1.5 m/min (5 sfm). Additional information on speeds and feeds for broaching metals other than steel is available in the Section "Machining of Specific Metals and Alloys" in this Volume.

**The rigidity** of the part and the fixturing employed affects the most efficient cutting speed of the broach. Slight movement of the part, or vibration set up from cutter contact, can cause fracture of the cutting edge of the broach. High cutting speed usually increases the occurrence of fractured edges.

**Length of Cut.** All other factors being equal, surface speed must be decreased as the length of material to be broached increases. Broach life can be shortened if high speeds are employed in the cutting of long areas because of the heat generated. Trapped chips and restricted cutting fluid exposure (especially in horizontal broaching) usually generate more heat than in a similar operation with a shorter length to broach.

## Cutting Fluids

Oils that are relatively high in viscosity and contain substantial amounts of fat, plus compounds of sulfur or chlorine or both, are most commonly used as cutting fluids for broaching steel, stainless steel, and cast iron. These oils are sometimes used for broaching other metals. These specially prepared oil-base compounds are proprietary, but are readily available. They are frequently referred to as broaching oils although they are also used for other machining operations.

Steels of several compositions and hardness levels, as well as cast iron and heat-resistant alloys, are work metals that typically use oils having a viscosity of

**Table 3 Feeds and speeds for broaching various steels with high-speed tool steels and carbide tools**

| Material | Hardness, HB | Condition | Speed m/min | Speed sfm | Chip load mm/tooth | Chip load in./tooth | Tool material grade ISO | Tool material grade AISI or C |
|---|---|---|---|---|---|---|---|---|
| **Wrought free-machining carbon steels** | | | | | | | | |
| **Low-carbon resulfurized** | | | | | | | | |
| 1108 1115 1118 1212 | 100–150 | Hot rolled or annealed | 11 | 35 | 0.10 | 0.004 | S4, S2 | M2, M7 |
| 1109 1116 1119 1213 | 150–200 | Cold drawn | 9 | 30 | 0.10 | 0.004 | S4, S2 | M2, M7 |
| 1110 1117 1211 1215 | | | | | | | | |
| **Medium-carbon resulfurized** | | | | | | | | |
| 1132 1140 1145 | 175–225 | Hot rolled, normalized, annealed, or cold drawn | 7.5 | 25 | 0.075 | 0.003 | S4, S2 | M2, M7 |
| 1137 1141 1146 | | | | | | | | |
| 1139 1144 1151 | 325–375 | Quenched and tempered | 5 | 15 | 0.05 | 0.002 | S9, S11(a) | T15, M42(a) |
| **Low-carbon leaded** | | | | | | | | |
| 10L18 12L14 | 100–150 | Hot rolled, normalized, annealed, or cold drawn | 12 | 40 | 0.10 | 0.004 | S4, S2 | M2, M7 |
| 11L17 12L15 | | | | | | | | |
| 12L13 | 200–250 | Hot rolled, normalized, annealed, or cold drawn | 11 | 35 | 0.075 | 0.003 | S4, S2 | M2, M7 |
| **Wrought carbon steels** | | | | | | | | |
| **Low carbon** | | | | | | | | |
| 1005 1012 1019 1026 | 85–125 | Hot rolled, normalized, annealed, or cold drawn | 9 | 30 | 0.10 | 0.004 | S4, S2 | M2, M7 |
| 1006 1013 1020 1029 | | | | | | | | |
| 1008 1015 1021 1513 | 225–275 | Annealed or cold drawn | 5 | 15 | 0.075 | 0.003 | S4, S2 | M2, M7 |
| 1009 1016 1022 1518 | | | | | | | | |
| 1010 1017 1023 1522 | | | | | | | | |
| 1011 1018 1025 | | | | | | | | |
| **Medium carbon** | | | | | | | | |
| 1030 1042 1053 1541 | 125–175 | Hot rolled, normalized, annealed, or cold drawn | 7.5 | 25 | 0.075 | 0.003 | S4, S2 | M2, M7 |
| 1033 1043 1055 1547 | | | | | | | | |
| 1035 1044 1524 1548 | 325–375 | Quenched and tempered | 3 | 10 | 0.05 | 0.002 | S9, S11(a) | T15, M42(a) |
| 1037 1045 1525 1551 | | | | | | | | |
| 1038 1046 1526 1552 | | | | | | | | |
| 1039 1049 1527 | | | | | | | | |
| 1040 1050 1536 | | | | | | | | |
| **Wrought alloy steels** | | | | | | | | |
| **Low carbon** | | | | | | | | |
| 4012 4615 4817 8617 | 125–175 | Hot rolled, annealed, or cold drawn | 7.5 | 25 | 0.075 | 0.003 | S4, S2 | M2, M7 |
| 4023 4617 4820 8620 | 325–375 | Normalized, or quenched and tempered | 3 | 10 | 0.05 | 0.002 | S9, S11(a) | T15, M42(a) |
| 4024 4620 5015 8622 | | | | | | | | |
| 4118 4621 5115 8822 | | | | | | | | |
| 4320 4718 5120 9310 | | | | | | | | |
| 4419 4720 6118 94B15 | | | | | | | | |
| 4422 4815 8115 94B17 | | | | | | | | |
| **Medium carbon** | | | | | | | | |
| 1330 4047 4161 5060 | 175–225 | Hot rolled, annealed, or cold drawn | 6 | 20 | 0.10 | 0.004 | S4, S2 | M2, M7 |
| 1335 4130 4340 50B60 | 325–375 | Normalized, or quenched and tempered | 3 | 10 | 0.05 | 0.002 | S9, S11(a) | T15, M42(a) |
| 1340 4135 4427 5130 | | | | | | | | |
| 1345 4137 4626 5132 | | | | | | | | |
| 4027 4140 50B40 5135 | | | | | | | | |
| 4028 4142 50B44 5140 | | | | | | | | |
| 4032 4145 5046 5145 | | | | | | | | |
| 4037 4147 50B46 5147 | | | | | | | | |
| 4042 4150 50B50 5150 | | | | | | | | |
| 5155 8627 86B45 9254 | | | | | | | | |
| 5160 8630 8650 9255 | | | | | | | | |
| 51B60 8637 8655 9260 | | | | | | | | |
| 6150 8640 8660 94B30 | | | | | | | | |
| 81B45 8642 8740 | | | | | | | | |
| 8625 8645 8742 | | | | | | | | |
| **High carbon** | | | | | | | | |
| 50100 52100 | 175–225 | Hot rolled, annealed, or cold drawn | 6 | 20 | 0.10 | 0.004 | S4, S2 | M2, M7 |
| 51100 M-50 | 325–375 | Normalized, or quenched and tempered | 3 | 10 | 0.05 | 0.002 | S9, S11(a) | T15, M42(a) |
| **Wrought stainless steels** | | | | | | | | |
| **Ferritic** | | | | | | | | |
| 405 429 434 442 | 135–185 | Annealed | 6 | 20 | 0.075 | 0.003 | S4, S2 | M2, M7 |
| 409 430 436 446 | | | | | | | | |
| **Austenitic** | | | | | | | | |
| 201 304 321 385 | 135–185 | Annealed | 6 | 20 | 0.075 | 0.003 | S4, S2 | M2, M7 |
| 202 304L 347 | 225–275 | Cold drawn | 5 | 15 | 0.075 | 0.003 | S9, S11(a) | T15, M42(a) |
| 301 305 348 | | | | | | | | |
| 302 308 384 | | | | | | | | |
| **Martensitic** | | | | | | | | |
| 403 422 | 135–175 | Annealed | 7.5 | 25 | 0.10 | 0.004 | S4, S2 | M2, M7 |
| 410 501 | 275–325 | Quenched and tempered | 5 | 15 | 0.05 | 0.002 | S9, S11(a) | T15, M42(a) |
| 420 502 | | | | | | | | |

(a) Any premium high-speed steel (T15, M33, M41-M47) or (S9, S10, S11, S12). Source: Metcut Research Associates Inc.

**Table 4 Effect of workpiece hardness on broach life in internal broaching of steel**

For all operations, broaches were of M2 high-speed tool steel, and the cutting fluid was a chlorinated sulfur-base oil with lard oil added to obtain a suitable viscosity.

| Steel and hardness | Length of cut mm | in. | Shape broached | Pieces broached per hour | Pieces per sharpening | Total pieces per broach |
|---|---|---|---|---|---|---|
| 1020, 3–16 HRC | 34.92 | 1.375 | Square | 50–70 | 2500 | 48 000 |
| 1026, 16 HRC | 15.37 | 0.605 | Spline | 90–120 | 4000 | 36 000 |
| 8620, 25–30 HRC | 38.1 | 1.50 | Serrations | 60–90 | 1500 | 28 500 |
| 4140, 27–33 HRC | 24.46 | 0.963 | Spline | 90–120 | 1000 | 24 000 |

approximately 155 SUS and containing 2% fat, 0.8% active sulfur, and 2.1% Cl.

Broaching oils are considered most effective in preventing the adherence of work metal to the broach, thus producing the best finishes, dimensional accuracy, and broach life, other conditions being equal. Discriminating selection among the various proprietary broaching oils is usually based on experience with similar applications, and if tool life or finish on the part is poor, a change of broaching oil should be tried. The disadvantages of broaching oils are:

- Higher initial cost than for common cutting oils
- Staining of some metals, such as copper alloys
- Necessity for 100% removal from heat-resisting alloys before they are heat treated or placed in high-temperature service
- Poorer ability to cool and remove chips than some other cutting fluids, such as soluble-oil emulsions, because of their higher viscosity

Water-soluble oils (1 part oil mixed with 15 to 20 parts water) are being increasingly used. They provide acceptable results in the broaching of steel and other metals. Water-soluble oils are more effective for cooling and flushing chips away, but are far less effective for preventing adherence of the work metal to the broach than are the broaching oils.

Aluminum and some other soft nonferrous metals are sometimes broached without cutting fluid, although there are light oils that are specially prepared for broaching aluminum. Copper alloys are susceptible to staining from oils containing sulfur and chlorine. When staining cannot be tolerated, kerosene containing 10 to 20% lard oil is often used.

Cast iron is sometimes broached without cutting fluid. Whether or not a cutting fluid is used depends to a large extent on the need for cooling and chip removal because cast iron is a free-cutting metal.

Regardless of which cutting fluid is used, it is of utmost importance to obtain an adequate supply at the cutting edges. This is more easily accomplished in vertical broaching than in horizontal broaching. In some horizontal broaching, the problem of supply at the critical areas can be solved only by coating the workpieces with an extremely viscous cutting fluid before broaching. However, pumps supplied with the machines will usually force the cutting fluid into critical areas. Additional information on cutting fluids is available in the Section "Machining of Specific Metals and Alloys" and the article "Metal Cutting and Grinding Fluids" in this Volume.

## Effect of Work Metal and Hardness on Broach Life

Some metals such as René 41, are inherently difficult to broach and have a detrimental effect on broach life. Two factors that influence broach life are work metal hardness and carbon content.

**Hardness.** The effect of the hardness of the workpiece on broach life, and therefore on broach cost per piece broached, is indicated in Table 4, which compares results obtained in broaching internal shapes in four different production parts, each of a different steel and hardness. Although these data indicate that total pieces per broach decreased as workpiece hardness increased, it should not be assumed that soft metals are always easily broached. Some metals (steels in particular) are too soft to be broached successfully because they readily become welded to the cutting edges of the broach. This results in decreased production, shorter tool life, and increased operating costs. Welding can be minimized by the use of highly chlorinated or sulfurized cutting oils and by providing broach teeth with extremely high face and relief angles. High relief angles, however, reduce the number of regrinds possible before the broach loses its ability to hold dimensions. This reduces the effective life of the broach and can add substantially to overall production cost. When practical, hardening of steel workpieces to 22 to 28 HRC is often a better solution. Much of the welding will be eliminated.

**Effect of Carbon Content.** Low-carbon steels (<0.25% C) in the annealed condition are often more difficult to broach than medium-carbon steels. This is because low-carbon steels have a greater tendency to weld to the broach teeth. Heat treatment is not always practical.

## Dimensional Accuracy

Broaching is capable of providing and maintaining close tolerances during a long production run. Table 5 lists commonly broached work metals and typically obtained tolerance and surface finish results. This is inherent in the process for several reasons. Although broaching combines roughing and finishing cuts in a single broach, no single broach tooth performs both functions. Each successive tooth removes only a predetermined amount of metal and is in cutting contact only as long as it takes to pass over the work one time; therefore, a minimum amount of heat is developed. In addition, the finishing teeth are shielded from the heavier cuts by both the roughing and the intermediate teeth, thus giving the finishing teeth maximum production life.

The shape of the broach tooth permits repeated sharpening without loss of accuracy. Finishing teeth may have flat lands 0.13 to 0.76 mm (0.005 to 0.030 in.) in width on each tooth. Thus, a tooth can be sharpened without sacrificing any dimension. However, this land must be held to a practical minimum. A straight land increases friction, and the resulting heat may cause the broach to expand enough to cause galling of the broached surface or even to exceed the tolerance.

It is unnecessary to sharpen all finishing teeth each time the tool is sharpened. Common practice is to sharpen only the first, or the first and second finishing teeth until they have become undersize. Then the second, or the third and fourth teeth are sharpened (depending on whether a single tooth or two teeth are sharpened each time) when the broach has again become dull. This practice permits the broach to hold tolerance for a long production run. Typical variations in dimensions obtained in production applications of broaching are given in the following example.

**Example 3: Variations Found in Broaching Keyways.** The nodular iron casting shown in Fig. 24 required a keywayed through hole broached to the dimensions shown. Parts were broached at the rate of 152 per hour. Approximately 1000 pieces could be broached between regrinds, and the broaches usually could be reground eight times. Figure 24 plots keyway width at entrance and exit ends of every fifth piece in a 250-piece batch.

Because the part shown in Fig. 24 is a rigid casting, the difference in spread between the entering and the exiting dimensions (particularly for dimension B) points out a condition that must be considered in broaching parts with relatively thin walls. The shape and sharpness of the broach teeth strongly influence the generated forces that tend to push the part walls outward. To maintain tolerances on some thin-wall parts, it may be necessary to revise the shape of the broach teeth or to increase the frequency of sharpening. Differences in the

**Table 5 Commonly broached materials and typical results**

| Metal | Heat treatment(a) | Hardness, HRC | Tolerance, mm | Tolerance, in. | Finish, μm | Finish, μin. |
|---|---|---|---|---|---|---|
| 2618-T61 Al | G | 70 HRB | 0.05 | 0.002 | 0.80–1.15 | 32–45 |
| 2014-T6 Al | G | 70 HRB | 0.058 | 0.0023 | 0.80 | 32 |
| Ti-6Al-4V | E | 36–38 | 0.019 | 0.00075 | 0.61–0.80 | 24–32 |
| Stellite 31 | B | 32 | 0.05 | 0.002 | 2.00 | 80 |
| SAE 51410 (type 410SS) | H | 32–36 | 0.05 | 0.002 | 1.60 | 63 |
| Greek Ascoloy | I | 32–38 | 0.025 | 0.001 | 0.89–1.07 | 35–42 |
| Inconel | A | 85 HRB | 0.13 | 0.005 | 2.00 | 80 |
| Inconel X | H | 29 | 0.025 | 0.001 | 0.80 | 32 |
| Timken 16-25-6 | F | 20–28 | 0.025 | 0.001 | 0.80–1.60 | 32–63 |
| A-286 | G | 28–30 | 0.060 | 0.0024 | 0.80 | 32 |
| | | 30–35 | 0.025 | 0.001 | 0.89 | 35 |
| | | 32–38 | 0.015 | 0.0006 | 0.80 | 32 |
| S-816 | G | 23–30 | 0.025 | 0.001 | 0.80–1.00 | 32–40 |
| SAE 3310 | E | 20 | 0.25 | 0.010 | 1.60 | 63 |
| SAE 9310 | I | 36–38 | 0.05 | 0.002 | 1.60 | 63 |
| 17-22A(S) | H | 29–34 | 0.025 | 0.001 | 1.50 | 60 |
| 17-22A | H | 35–40 | 0.075 | 0.003 | ... | ... |
| SAE 9840 | I | 32–36 | 0.025 | 0.001 | 1.25 | 50 |
| SAE 4130 | I | 32 | 0.013 | 0.0005 | 1.60 | 63 |
| SAE 4140 | I | 25–29 | 0.05 | 0.002 | 0.80–1.60 | 32–63 |
| SAE 4340 | I | 38 | 0.05 | 0.002 | 1.14–1.60 | 45–63 |
| M2 tool steel | A | 24–28 | 0.02 | 0.0008 | 0.80 | 32 |
| EMS 544 | ... | 40–47 | 0.025 | 0.001 | 0.75 | 30 |
| Inconel 901 | I | 32–36 | 0.038 | 0.0015 | 1.60 | 63 |
| René 41 | G | 40–42 | 0.060 | 0.0024 | 0.80 | 32 |
| WAD 7823A | ... | 28 | 0.0076 | 0.0003 | 1.0–1.5 | 40–60 |
| D-979 | I | 38–40 | 0.013 | 0.0005 | 1.50 | 60 |
| EMS 73030 | ... | 32–36 | 0.071 | 0.0028 | 1.60 | 63 |
| M-308 | ... | 36–38 | 0.060 | 0.0024 | 0.80 | 32 |
| Chromoloy | ... | 31–32 | 0.10 | 0.004 | 0.80 | 32 |
| PWA-682 (Ti) | ... | 34–36 | 0.025 | 0.001 | 0.80 | 32 |
| Lapelloy | J | 30–37 | 0.20 | 0.008 | 0.80 | 32 |
| Type 303 SS | A | 85 HRB | 0.025 | 0.001 | 1.60 | 63 |
| Type 304 SS | A | 80–85 HRB | 0.05 | 0.002 | 1.60 | 63 |
| Type 403 SS | I | 37–40 | 0.015 | 0.006 | 1.60 | 63 |
| SAE 1010 | D | 60 | 0.025 | 0.001 | 0.75 | 30 |
| SAE 1020 | D | 3–12 | 0.05 | 0.002 | 1.55–2.05 | 60–80 |
| SAE 1037 | I | 15–20 | 0.0076 | 0.0003 | 1.60 | 30 |
| SAE 1045 | I | 24–31 | 0.013 | 0.0005 | ... | ... |
| SAE 1063 | E | 12–18 | 0.10 | 0.004 | 0.63–1.5 | 25–60 |
| SAE 1070 | E | 5–10 | 0.05 | 0.002 | 0.71–1.5 | 28–60 |
| SAE 1112 | ... | 87 HRB | 0.025 | 0.001 | 1.0–1.15 | 40–45 |
| SAE 1145 | C | 13–18 | ... | ... | 1.25–2.5 | 50–100 |
| SAE 1340 | C | 15–20 | 0.075 | 0.003 | ... | ... |
| SAE 4047 | C | 8–15 | 0.05 | 0.002 | 1.5–2.0 | 60–80 |
| SAE 5140 | C | 8–15 | 0.05 | 0.002 | 1.5–2.0 | 60–80 |
| SAE 52100 | D | 25 | 0.013 | 0.0005 | 0.75 | 30 |
| Gray cast iron | B | 90 HRB | 0.075 | 0.003 | 2.0–2.5 | 80–100 |
| KP-7 cast iron | B | ... | 0.013 | 0.0005 | 3.20 | 125 |

SS, stainless steel. (a) Treatment or condition. A, annealed; B, as-cast; C, as-forged; D, cold finished; E, hot finished; F, stress relieved; G, solution and precipitation treated; H, air quench, furnace temper; I, oil quench, furnace temper; J, salt quench, furnace temper. Source: *Metal Cutting: Today's Techniques for Engineers and Shop Personnel*, McGraw-Hill, 1979

depth of broached keyways at the entrance and exit ends are common. To compensate for the differences, the slot in the horn is often tapered.

## Broaching Versus Alternative Processes

Many parts can be produced by more than one process. Selecting the most efficient process requires consideration of available equipment, quantity to be produced, dimensional accuracy, surface finish, and workpiece material.

**The main advantage of broaching** is that it is fast; it commonly takes only seconds to accomplish a task that would require minutes with any other method. Little skill is needed to operate a broaching machine (all the skill is built in the tooling), and automation is easily arranged. Good finish and accuracy are obtained over the life of a broach because roughing and finishing are done by separate teeth.

**Limitations of Broaching.** The main disadvantage of broaching is that broaches are costly to make and sharpen. Standard broaches are available, but most broaches are made especially for and can do only one job.

Special precautions may be necessary in founding or forging to control variations in stock, and additional operations for removing excess stock may have to be added before broaching to protect the broach. These add to the overall cost of manufacturing.

Some of the limitations of broaching make it impracticable for certain work. A surface cannot be broached if it has an obstruction across the path of broach travel. Blind holes and pockets normally are not broached. Frail workpieces are not good subjects for broaching, because they cannot withstand the large forces imposed by the process without distortion or breakage. For example, automobile engines have been redesigned to reduce weight and to save fuel. The engine blocks had previously been broached, but it has been found that the thinner walls of some of the engine blocks cannot withstand the broaching forces and that more costly milling is necessary. Surfaces that run in the same general direction can often be broached at the same time, but surfaces not so related generally must be broached separately. A hole and a perpendicular face can be machined in one operation on a lathe or boring machine, but require two passes in broaching. The lines left on a surface lie in the direction of broach travel. Broaching is not capable of producing a circular pattern in a hole if such a finish is required.

**Compared with reaming,** broaching can hold a closer tolerance for a longer production run. The cutting edges of a reamer are constantly in contact with the work as long as the reamer is cutting; therefore, the reamer travels many times around the circumference of the hole before the hole is complete. Although a reamer has a margin conforming to the circumference of the hole being reamed (comparable to the flat land on a broach tooth), which permits some sharpening of the reamer without loss of tolerance, considerably more grinding is done on a reamer than on a broach. This is because, in a reamer, fewer teeth are required to remove the same amount of metal removed by a greater number of teeth in a broach.

**Broaching Versus Milling.** With respect to tolerance capabilities and tool life, milling may also suffer in comparison with broaching. Although a milling cutter usually has more cutting edges than a reamer, in milling as in reaming, each cutting tooth is required to do more work than is performed by any one tooth on a broach.

**Broaching Versus Gear-Shaping Operations.** Broaches are expensive tools, and any revision in the design of a gear (such as a change in the number of teeth or the pitch diameter) would require a new broach. Furthermore, broaching costs are directly affected by the hardness of the material to be broached and by the tolerances applied to the broached shape. The flexibility of gear shapers minimizes these problems; the same shaping cutter can be used for gears having the same diametral pitch and pressure angle but having a different number of teeth and modified pitch diameters. Gear shapers can also be adjusted to compensate for cutter wear.

**Combination Broaching and Boring.** In many applications where holes to be machined are no larger than approximately 150 mm (6 in.) in diameter (boring is invariably

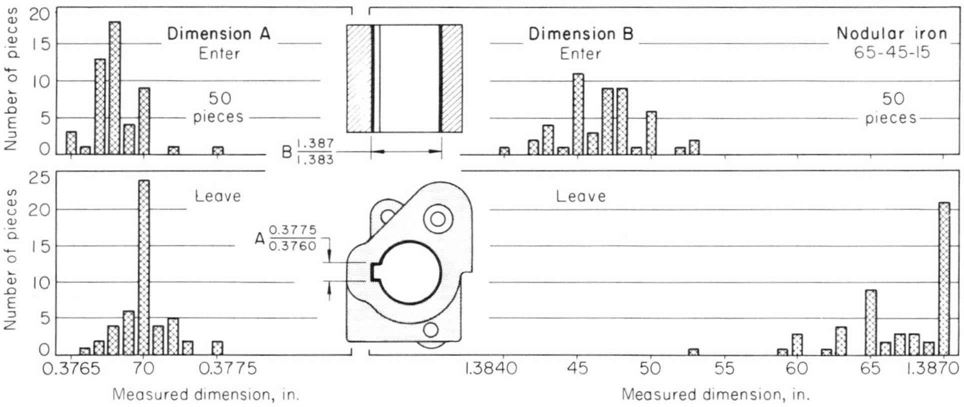


**Fig. 24** Dimensional variations of a broached keyway showing the difference in variation between the dimensions where the broach entered the part and where it left. Dimensions given in inches

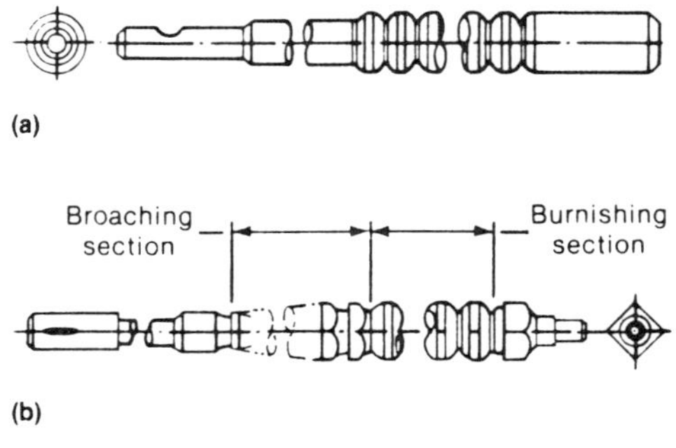


**Fig. 25** Two types of broaches used for burnishing the walls of broached holes. (a) Broach for burnishing only. (b) Broach for cutting and burnishing

more practical for larger holes), broaching and boring are alternative processes. When keyways or other internal forms are required, it is common practice to bore the hole and then to broach the keyway. However, for production quantities, the hole can be broached with the key slot in one operation, using a broach of proper design.

## Burnishing

Burnishing is often done in conjunction with broaching for one or more of the following reasons:

- To provide a smoother surface than can be obtained by broaching
- To provide greater accuracy in the diameter of the hole
- To provide a hole with a better wearing surface

Burnishing is accomplished by the use of either a broach designed especially for burnishing (Fig. 25a) or a broach in which burnishing buttons have been incorporated following the finishing teeth (Fig. 25b). A burnishing broach produces a glazed surface in a steel, cast iron, or nonferrous hole. Burnishing teeth are rounded. They do not cut the surface metal, but compress and cold work it.

The total change in diameter produced by a burnishing operation may be no more than 0.013 to 0.025 mm (0.0005 to 0.001 in.). Burnishing tools, used when surface finish and accuracy are critical, are relatively short and are generally designed as push broaches.

Burnishing buttons are sometimes included behind the finishing-tooth section of a conventional broaching tool. The burnishing section can be added as a special attachment or an easily replaced shell. These replacement shells are commonly used to reduce tooling costs when high wear or tool breakage is expected.

Although burnishing is usually done in a broaching machine, it can be done in conjunction with other machining operations in other machines. A burnishing operation similar to ball sizing can be incorporated as one of the stations of a multiple-spindle automobile machine. This operation differs from ball sizing in two important respects. First, ball sizing is a separate operation in which neither the ball nor the workpiece rotates, while in a multiple-operation machine, both the burnishing tool and the workpiece rotate. Second, in ball sizing, the ball is pressed through the workpiece, while in a multiple operation machine, the burnishing tool is pressed into or through the workpiece and withdrawn. Therefore, ball sizing is limited to through holes, but both through holes and blind holes can be finished by a burnishing tool. Detailed information on burnishing is available in the article ''Roller Burnishing'' in this Volume.

## Causes and Prevention of Broach Breakage

Broaches that are properly designed and well made seldom break in normal use. Broach breakage is usually the result of poor processing practices (such as failure to maintain tooth sharpness, overloading, or improper sharpening), careless handling of the broach, or improper preparation or excessive hardness of the workpiece.

**Poor Processing Practices.** More force is required for pulling a dull broach through the workpiece than is required for a sharp broach. If a broach is sufficiently dull, force requirements may be increased by as much as 50%. Overloading can cause an increase in tensile load that results in broach breakage. For example, a broach designed to machine one piece at each pass may fail if two or more pieces are stacked and broached simultaneously. In addition, a broach designed to machine a soft, free-cutting material may break if the material is changed to one that is harder and tougher.

**Sharpening** in a manner that changes the type of chip may cause breakage if, after sharpening, the chips pack in the gullets so tightly that more force is required for moving the broach through the workpiece.

**Careless Handling.** Because broaches are extremely hard, and consequently brittle, they may break, rather than merely become nicked, if dropped. Another cause of breakage is returning the broach through the completed workpiece. Because there is a small amount of springback of the metal surrounding a broached hole, movement of the broach back through the hole would be restricted.

Broaches can also be broken by being permitted to pass completely through the guide hole in the fixture. On the return stroke, the end of the broach can miss the guide hole and can jam between the platen and the fixture; this usually causes the broach to buckle and break. The stroke length should be set so that the trailing end of the broach remains in the guide hole at the end of the stroke. A rear pilot is sometimes incorporated to guide the broach at the end of the stroke.

**Removing a stuck broach** is an emergency action that requires care to save the costly broach. Hand methods should be used because the broach may be broken if the machine is reversed to pull out the broach. The broach and the work should be removed from the machine. Light hammer blows, carefully placed, may remove the work from the broach. If not, the workpiece can be cut with a hacksaw (which will not cut the hard broach if it contacts it). A round workpiece with a stuck broach can be placed in a lathe, and the work can be cut away from the broach by turning.

**Improper Preparation of the Workpiece.** Misalignment of piercing punches entering from opposite sides of a forging, misalignment of cores from opposite sides for a through hole in a casting, or a bent hole made by a drifting drill may result in overloading of the teeth on one side of the broach. If this overloading exceeds the strength of the tool, teeth can break off, or the broach may break. In the simultaneous broaching of parallel holes, broach breakage can occur if the holes actually are not parallel or if their center-to-center distance is incorrect. Improper fixturing that establishes the centerline of a hole to be broached in a position that is not the centerline of broach travel will cause uneven loading of the broach, which may cause breakage.

**Excessive workpiece hardness** may cause broach breakage. This hardness may result from faulty heat treatment or from work hardening as a result of some previous processing operation. Machining operations such as drilling or boring, especially if the tools are dull, can increase the hardness of the part. The heat developed in poor grinding procedures can raise the hardness of some steels past the safe broaching range and can result in broach breakage.

## Broach Repair

Because of the high initial cost of broaches and the amount of time required for producing them, it is often advisable (and is common practice) to repair a damaged broach rather than to replace it. Usually, before a broach is repaired, the cost of a new broach is weighed against the amount of life left in the old broach and the cost of repairing it. Generally, a broach can be repaired in less time than is required for producing a new one. This is an important consideration if production lines require minimum downtime to meet output schedules.

Repaired teeth usually last as long as the remaining original teeth. The repairs are made by building up the damaged section by welding, using the correct rod. Difficulty is seldom encountered after the sections are properly welded; however, there is some chance that the broach will crack during welding.

**Welding Methods.** Broken broaches have been successfully welded by several different methods. However, welding inevitably produces heat-affected zones in high-speed steel. Tempering can minimize, but not eliminate, the rehardened areas, and it will not eliminate the over-tempered regions. That requires a full anneal cycle and leads to distortion and rework problems. At the very least a retemper is required, but this will not fully restore the properties of the high-speed steel. The risk of a cracked broach or one with soft spots remains.

The method selected depends largely on the type of break and the welding equipment available. When a broach is broken into two pieces, butt welding or butt brazing (addition of filler metal) is commonly used after the surfaces to be joined have been faced. Often, a new section must be inserted, thus requiring two butt-welded or butt-brazed joints.

Broken teeth are usually repaired by arc welding. Gas welding has been used, but is generally less desirable than electric arc welding because the heat-affected areas are much larger in gas welding. Welding rods having a composition similar to that of high-speed tool steel are available for electric arc and gas welding. The weld metal hardens as it is deposited and cooled so that heat treatment of the repaired broach is not required after welding.

# Drilling

A DRILL for cutting metal is a rotary end cutting tool with one or more cutting lips and usually one or more flutes for the passage of chips and the admission of cutting fluid. Drilling is usually the most efficient and economical method of cutting a hole in solid metal.

Drilling is often done in conjunction with other machining operations. This article will discuss only those applications in which drilling is the sole or the major operation in a machining sequence. Additional information on drilling is presented in the Section "Machining of Specific Metals and Alloys" and in the article "Multiple-Operation Machining" in this Volume.

## Process Capabilities

Although most metals drilled are softer than 30 HRC, it is common practice to drill holes in metal as hard as 50 HRC, and steel at 60 HRC has been successfully drilled.

**Brittle material** can be drilled by using backing material or special feed control to prevent damage at breakthrough.

**Hole Size.** Most drilled holes are 3.2 to 38 mm (1/8 to 1 1/2 in.) in diameter. Drills are commercially available, however, for drilling holes as small as 0.025 mm (0.001 in.) in diameter (see the section "Small-Hole Drilling (Microdrilling)" in this article), and special drills are available as large as 152 mm (6 in.) in diameter.

**The length-to-diameter ratio** of holes that can be successfully drilled depends on the method of driving the drill and on straightness requirements. In the simplest form of drilling (feeding a rotating twist drill into a fixed workpiece), optimum results are obtained when hole length is less than three times the diameter. However, by using special tools, equipment, and techniques, straight holes can be drilled in which length is more than eight times the diameter.

## Types of Drilling Machines

The basic work and tool motions that are required for drilling—relative rotation between the workpiece and the tool, with relative longitudinal feeding—also occur in a number of other machining operations. Consequently, drilling can be done on a variety of machine tools, such as lathes, milling machines, and boring machines. This section will focus on machines that are designed, constructed, and used primarily for drilling.

Drilling machines, called drill presses, consist of a base, a column that supports a powerhead, a spindle, and a worktable, as depicted in Fig. 1. On small machines, the base rests on a bench, but on larger machines, it rests on the floor. A column on a base carries a table for the workpiece and a spindle head. The table is raised or lowered manually, often by an elevating screw, and can be clamped to the column for rigidity. Some tables are round and can be swiveled. On vertical drill presses with round columns, the tables can generally be swung out from under the spindle so that the workpieces can be mounted on the base.

The heart of any drilling machine is its spindle. To drill satisfactorily, the spindle must rotate accurately and must resist whatever side forces result from the drilling. In virtually all machines, the spindle rotates in preloaded ball or tapered-roller bearings. The spindle that drives the cutting tool revolves in the nonrevolving quill that is fed up or down. Some machines have only hand feed; others have power feeds. Machines of this type can be equipped with a positive leadscrew for tapping and a spindle-reversing mechanism. As a rule, an adjustable stop is provided to limit the depth of travel of the quill and, with power feed, to disengage the feed or reverse a tapping spindle at a definite depth. Most machines have a Morse taper hole in the end of the spindle, but small machines often have a drill chuck attached to the end of the spindle. The spindles of fractional-horsepower drill presses are usually driven by V-belts; larger spindles have gear transmissions, and some have multiple-speed motors. Numerical control is also utilized.

Drilling machines can usually be classified as follows: bench, upright, radial, gang, multiple-spindle, deep-hole, transfer, and special-purpose. Although the term drill press is often used, particularly in referring to small, light-duty drilling machines, the latter term is used more frequently because of the power features that now are common.

### Bench-Type Drilling Machines

Bench-type drilling machines are classified as either plain or sensitive drilling machines.

**Plain bench-type machines** are very common. The spindle rotates on ball bearings within a nonrotating quill that can be moved up and down in the machine head to provide feed to the drill. The vertical motion is imparted by a hand-operated capstan wheel through a pinion that meshes with a rack on the quill. A spring raises the quill-and-spindle assembly to the highest position when the hand lever is released. The spindle is driven by a step-cone pulley that rides on a splined shaft, thus imparting rotation regardless of the vertical position of the spindle.

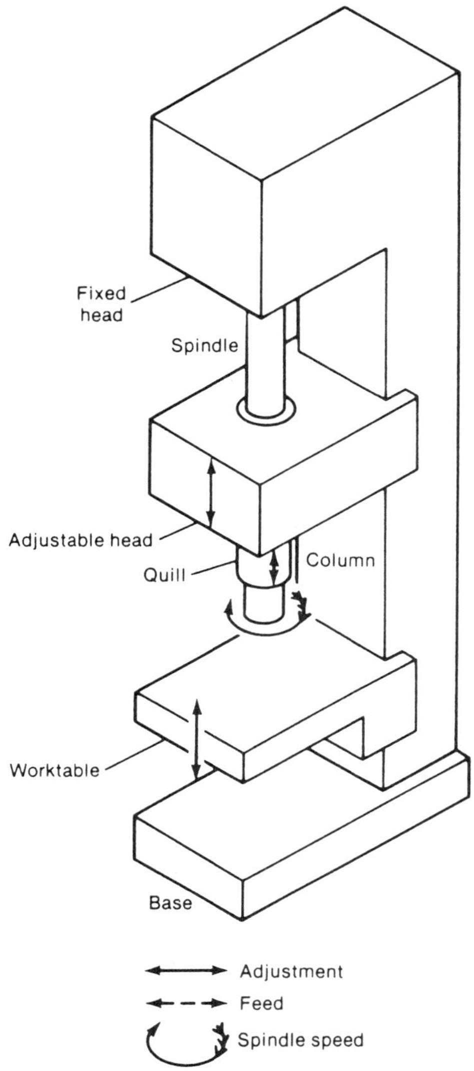


**Fig. 1** Principal parts and movements of a single-spindle upright drill press

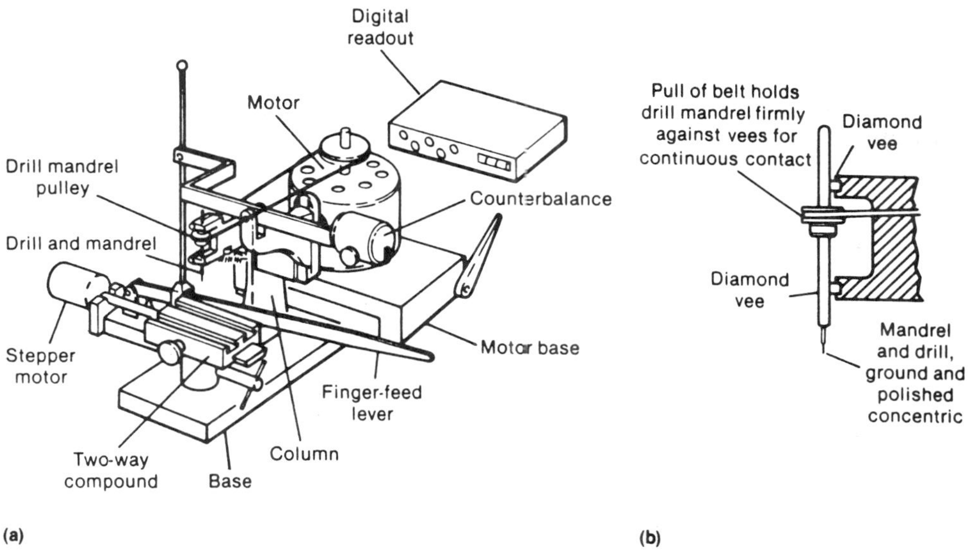


**Fig. 2** Ultrasensitive drilling machine (a) for producing small holes utilizes digital readout and a stereoscopic microscope for accuracy. (b) Schematic of V-bearing mounting assembly, which minimizes tool breakage

Presses of this type can usually drill holes up to 13 mm (½ in.) in diameter and typically offer a selection of eight spindle speeds varying from 600 to 3000 rev/min. Worktables often contain holes for use in clamping work. The same type of machine can be obtained with a long column so that it can stand on the floor instead of on a bench.

The size of bench and upright drilling machines is designated by twice the distance from the centerline of the spindle to the nearest point on the column; therefore, this is an indication of the maximum size of the work that can be drilled in the machine. For example, a 380 mm (15 in.) drill press will permit a hole to be drilled at the center of a workpiece 380 mm (15 in.) in diameter.

**Sensitive bench-type machines** are essentially the same as plain bench-type machines except that they are usually smaller, have more accurate spindles and bearings, and operate at higher speeds—up to 30 000 rev/min. Very sensitive, hand-operated feeding mechanisms are provided for use in drilling small holes. Such machines are advantageous for feeding small drills to avoid breakage.

Very small holes must be drilled on ultrasensitive presses with spindles running quite true at high speeds; V-bearings are used in one design (Fig. 2). The action of the drill is observed through a microscope. Holes less than 25 μm (0.0010 in.) in diameter have been drilled, and diameters of 125 μm (0.005 in.) are common in the instrument and timepiece industries.

## Upright Drilling Machines

The term upright or vertical is applied to drilling machines that stand on the floor and have power spindle feed. Upright machines can be either single-spindle or turret-type units.

**Single-spindle machines** are essentially the same as bench-type units with respect to spindle design, but they are heavier. Upright drilling machines are the most widely used type for heavy-duty drilling. Both round and box-column designs are manufactured, but box-column machines are more common because of their rigidity.

Upright drilling machines usually have spindle speed ranges from 60 to 3500 rev/min and power feed rates, in 4 to 12 steps, from 0.10 to 0.64 mm/rev (0.004 to 0.025 in./rev). Most modern machines use a single-speed motor and geared transmissions to provide the range of speeds and feeds. Some units use multispeed electric motors to obtain at least some of the various spindle speeds.

*Types of Spindle Drives*. Various spindle-drive systems, all suitable for drilling, are used on upright machines. These systems include the following:

- *Four-speed motor drive*: The motor is mounted above and in line with the spindle or is belted to the spindle. Speeds are quickly and easily changed by the operator, who can stand at floor level. This is a good system for a tapping or reversing drive, using the lower motor speeds, and is suitable for toolroom or medium-duty manufacturing. The horsepower available is limited for heavy-duty drilling
- *Multiple V-belt drive*: This is a single-speed drive, generally used on single-purpose machines. It is a good reversing drive and can transmit high horsepower
- *Step-pulley V-belt drive*: Speed changes with this drive are slow because the operator has to climb a ladder to make a change. It is, however, a good general-purpose and reversing drive and can transmit high horsepower
- *Ten-step, quick-change V-belt drive*: This drive system provides ten speeds, arranged in a geometric progression, that can be quickly and easily changed from floor level. It is a good general-purpose drive for drilling and tapping and can transmit high horsepower
- *Variable-speed drive*: With this drive system, speeds are quickly and easily changed at floor level. This system provides a good general-purpose drive for drilling. However, it is not a good reversing drive, because of the inertia of the belts and pulleys and because only moderate horsepowers can be transmitted
- *Gearbox drive*: With this general-purpose drive, four or more speeds can be belted or geared to the spindle, and speed changes can be made at floor level. Low-speed torque is excellent, high horsepowers can be transmitted, and it can be used for moderate-duty reversing
- *Back-gear drive*: This system, usually with a reduction from 4:1 to 6:1, doubles the number of speeds available from the various spindle drives. It also extends their range to lower speeds for large slow-running tools that require increased torque

The feed clutch usually is designed to disengage automatically when the spindle reaches a preset depth or when it reaches the limits of its travel.

The worktables on most upright drilling machines contain holes and slots for use in clamping work and nearly always have a channel around the edges to collect cutting fluid when it is used. On box-column machines, the table is mounted on vertical ways on the front of the column and can be raised or lowered by means of a crank-operated elevating screw.

**Turret-type upright drilling machines,** such as the unit illustrated in Fig. 3, are used where a series of holes of different size or a series of operations (such as center drilling, drilling, reaming, and spot facing) must be done repeatedly in succession. The turrets can be either radial or offset and have six, eight, or ten tools ready for use. After the selected tools are set in the turret, each can be quickly brought into position to be driven by the power spindle merely by rotating the turret. Such machines are particularly adaptable for numerical control (Fig. 4), and the turret is typically mounted on a traveling column having 760 by 1525 mm (30 by 60 in.) movement. Turret-type machines are also ideal for use as machining centers (see the section "Machining Centers" in the article "Multiple-Operation Machining" in this Volume).

## Radial Drilling Machines

When several holes must be drilled at different locations on a large workpiece, radial drilling machines make it possible to move the drill spindle to the desired location instead of having to move and reclamp

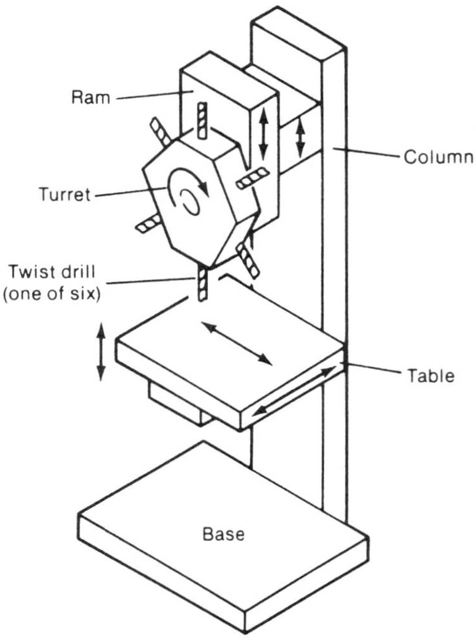


**Fig. 3** Principal components of a turret-type drilling machine and their movements

the workpiece, as would be required on an ordinary upright drilling machine (Fig. 5). Thus, much nonproductive work is eliminated, and larger workpieces can readily be accommodated.

Radial drilling machines have a large, heavy, round, vertical column supported on a large base. The column supports a radial arm that can be raised and lowered by power, and the entire column can rotate on the base. The spindle head, with its speed- and feed-changing mechanism, is mounted on the radial arm so that it can be moved horizontally to any desired position on the arm. Thus, by the combined movements of raising or lowering and swinging the radial arm, along with the horizontal movement of the spindle assembly, the spindle can be quickly brought into proper position for drilling holes at any desired point on a large workpiece mounted either on the base of the machine or on the floor.

Radial drilling machines can be categorized as:

- Plain radial drilling machines, which provide only a vertical spindle motion
- Semiuniversal machines, in which the spindle head can be swung about a horizontal axis normal to the arm to permit the drilling of holes at an angle in a vertical plane
- Universal machines, in which an additional angular adjustment is provided by the rotation of the radial arm about a horizontal axis, thus permitting holes to be drilled at any desired angle

The size of a radial drilling machine is designated by the radius of the largest disk in which a center hole can be drilled when

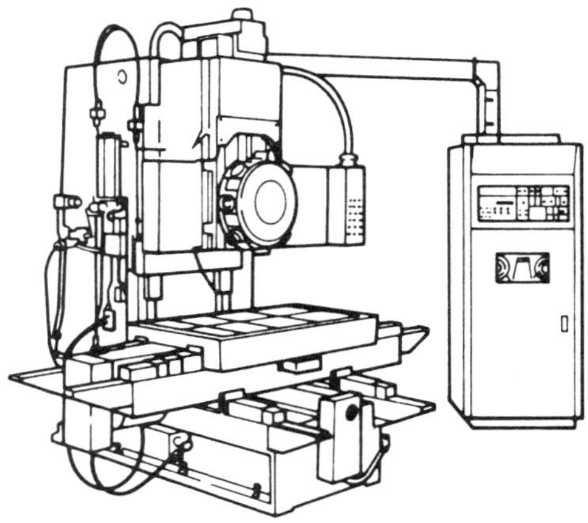

**Fig. 4** Solid-bed, sliding-head drilling machine with eight-station tooling turret and two/three-axis numerical control

the spindle head is at its outermost position. Radial drilling machines are available in sizes from 0.9 to 3.7 m (3 to 12 ft). In addition, it is customary to give the diameter of the column when specifying size, but this is not always done. Column sizes range from 229 to 660 mm (9 to 26 in.). Most radial drilling machines have a wide range of feeds and speeds. For example, one typical machine has 32 spindle speeds from 20 to 1600 rev/min and 16 feeds from 0.076 to 3.18 mm/rev (0.003 to 0.125 in./rev). These also include leads for tapping 8, 11½, 14, and 18 threads per 25 mm (1 in.).

Work can be mounted directly on the base of a radial drill press, using the T-slots that are provided for bolting. Another procedure is to use a worktable that is fastened to the base. When such a machine is used in mass-production work, special jigs or fixtures can be attached to the base to hold the work.

Special types of radial drilling machines are also available:

- Track-type machines rest on rails so that they can be moved and clamped in any desired position along the rails
- Sliding-base machines have columns that can be moved along the base on ways to permit drilling over a wider area
- Portable modification of the sliding-base type can be readily picked up by an overhead crane and set down at a desired location

Most radial drilling machines are equipped with adequately heavy spindle bearings so that they can also be used for boring (see the article "Boring" in this Volume).

## Gang Drilling Machines

A gang drill press is the equivalent of two, three, four, six, or more upright or production drill presses in a row with a common base or table (Fig. 6). A gang drill can be set up so that work can be passed from spindle to spindle to undergo two or more operations. In another mode, the same operation

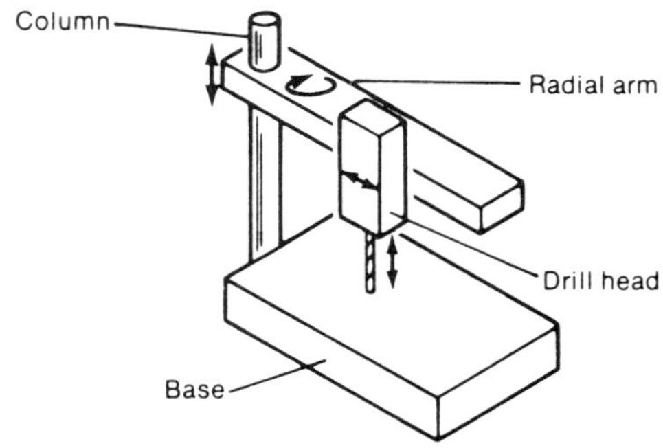


**Fig. 5** Principal components of a radial drilling machine and their movements

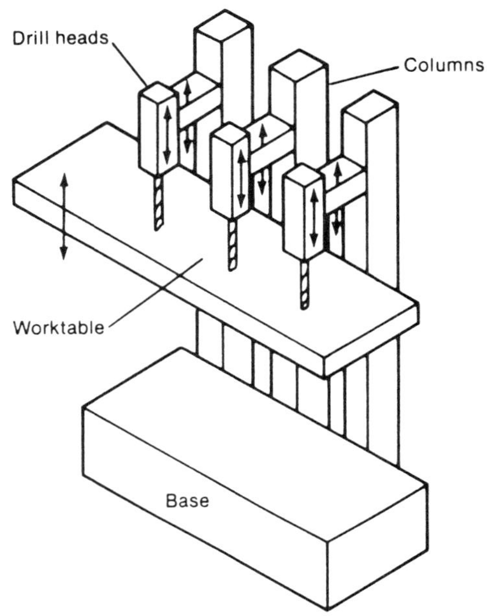


**Fig. 6** Principal components of a gang drilling machine and their movements

can be performed at all spindles; the operator unloads and loads the jig at each spindle in turn while the other spindles are cutting with automatic feed.

On some gang drilling machines, the columns can be moved longitudinally along the base. Such machines are very useful in mass-production work, in which several related operations (such as drilling, reaming, or counterboring holes of varying size) are done on a single part. The work is slid along the table into position for the operation at each spindle.

Gang drill presses are available with and without power feed. One or several operators can be used.

## Multiple-Spindle Drilling Machines

Multiple-spindle drilling machines have 2 to 100 or more spindles per head and are driven with 75 kW (100 hp) motors. These machines can be used where a number of parallel holes must be drilled in a part in high-production applications. The several

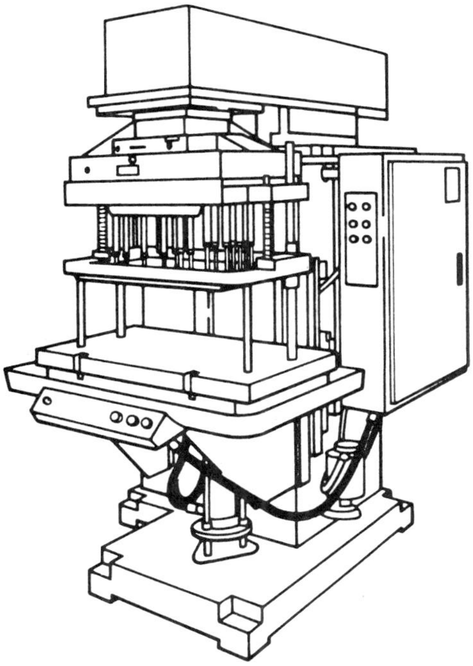

**Fig. 7** Multiple-spindle drilling machine of the open-side type incorporating a way design

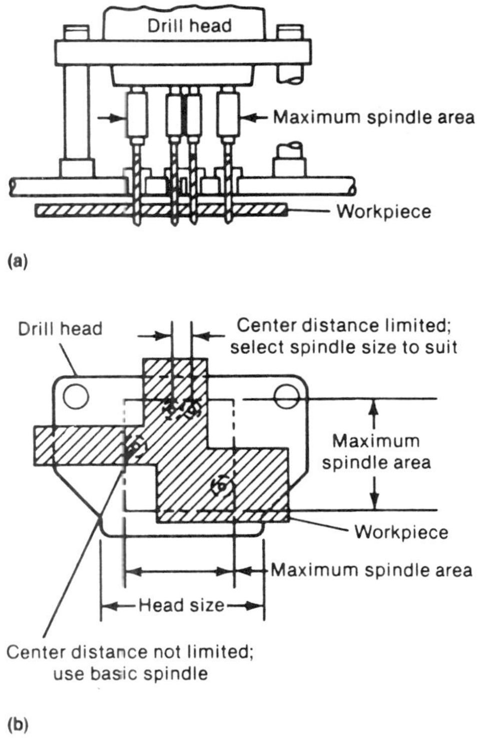


**Fig. 8** Side view (a) and top view (b) of a multiple-spindle drill head showing the relative positions of the spindles, drill head, and workpiece and the dimensions that must be considered in choosing a drill head

spindles are driven by a single powerhead and are fed simultaneously into the work (Fig. 7). The spindles are driven and can be adjusted over a limited area by means of sliding-bar guides. Because the areas that can be covered by adjacent spindles overlap, the machine can be set up to drill holes at any location within its overall capacity. For example, a typical machine having 20 spindles can drill holes at any location within a 760 mm (30 in.) diam circle. A special drill jig is made for each job to provide accurate guidance for each drill (Fig. 8). Although such machines are costly, they can be readily converted for use on different jobs where the quantity to be produced will justify the small setup cost and the cost of the jig.

Multiple-spindle drilling machines are available with a wide range of numbers of spindles in a single head, and two or more heads are frequently combined in a single machine for mass-production work. In some cases, such machines perform drilling operations simultaneously on two or more sides of a workpiece.

**Applications.** Multiple-spindle machines are primarily used for three general types of production operations:

- Multiple operations (drilling, reaming, chamfering, spotfacing, and so on) in a single hole. Machines used for these applications are often equipped with hand-positioned tables, shuttle tables, or rotary indexing tables
- One operation in multiple holes that are the same size or different sizes and are on the same or different planes. Machines used for these operations may require a rotary indexing table if hole center distances are close. Multiple-plane operations are often performed with multiple-position workholding fixtures
- Multiple operations in multiple holes that generally require the machine to be equipped with a rotary indexing table or other type of table, especially when tapping is one of the operations to be performed

**Multiple-spindle drill heads** are designed for various applications and have geared or gearless (crank-type) drives. All tools on a head are fed into the workpiece together, but tool lengths are sometimes staggered so that cutting loads are applied progressively. There are three major types of geared drill heads, as follows.

*Adjustable-arm, universal-joint drill heads* are generally used for frequent changeover or limited production requirements. They provide maximum flexibility because the spindles can be adjusted to any desired location within the head housing. Spindles can be removed or added, up to the capacity of the head. The adjustable-arm spindles are rotated by universally jointed connecting rods that transmit power from a fixed-center, gear-driven, or crank-driven head. All spindles rotate at the same speed, which is normally limited to a maximum of 2000 rev/min primarily because of joint wear problems and the method of lubrication. Some standard machines come equipped with universal-joint heads having a high, neutral, and low-speed shifter range for each spindle. Special heads can be designed to incorporate this speed selector feature.

*Slip-plate drill heads* are often used for repetitive hole patterns and longer production runs. The same type of universal rods as those used on adjustable-arm heads drive slip spindles mounted in a slip plate that can be changed for different hole patterns. The capability of producing various hole patterns on fixed centers can make slip-plate heads more desirable than fixed-center heads; however, greater center distances are sometimes required between spindles, and the locational accuracy of the spindles is not as close. The spindles normally rotate at the same speed unless the head is of special design. These heads are generally used for drilling holes 1.6 to 57 mm (1/16 to 2 1/4 in.) or more in diameter in cast irons and steels.

*Fixed-center drill heads* are used for dedicated, high-production machines. The number of spindles can vary from 2 to 2000 or more, depending on the number of holes required in the workpieces or the ease with which the workpieces can be handled. Gearless heads are required for large numbers of spindles, and all spindles operate at the same speed for light-duty operations. These heads are stronger and more rigid than either the adjustable-arm or slip-plate head because the spindles are permanently fixed for required hole patterns.

Gearless, fixed-center, crank-driven drill heads are used when the need for close-center holes makes the use of gears and universal joints impractical or impossible. Some heads of this type are capable of drilling holes with center distances as close as 4.8 mm (0.19 in.) and diameters of 0.51 to 19.05 mm (0.020 to 0.750 in.). In one application, a gearless drill head having 1800 spindles was mounted on the ram of a punch press to drill acoustical tiles. A 38 kW (50 hp) electric drive motor was mounted on top of the press.

One design of a gearless drill head is illustrated in Fig. 9. The driver (A), in the drilling machine spindle rotates the drive crank (B) in the drill head. This crank moves the oscillator (C) in a short oscillating motion. The oscillating motion rotates the individual drill spindles (D) in the same direction and at the same speed as the drive crank (B).

## Deep-Hole Drilling Machines

Several problems not encountered in ordinary drilling operations arise in the drilling of long holes in rifle barrels, long spindles, connecting rods, and certain oil-well drilling equipment. As the hole length increases, it becomes more and more difficult to support the work and the drill properly.

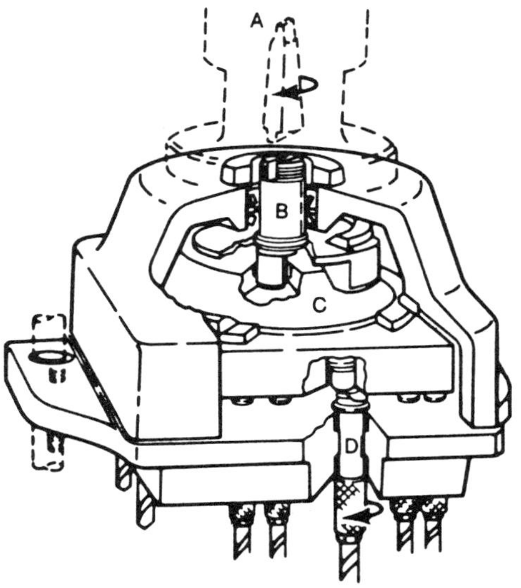


**Fig. 9** Gearless drill head in which an oscillator (C) rotates the drill spindles (D)

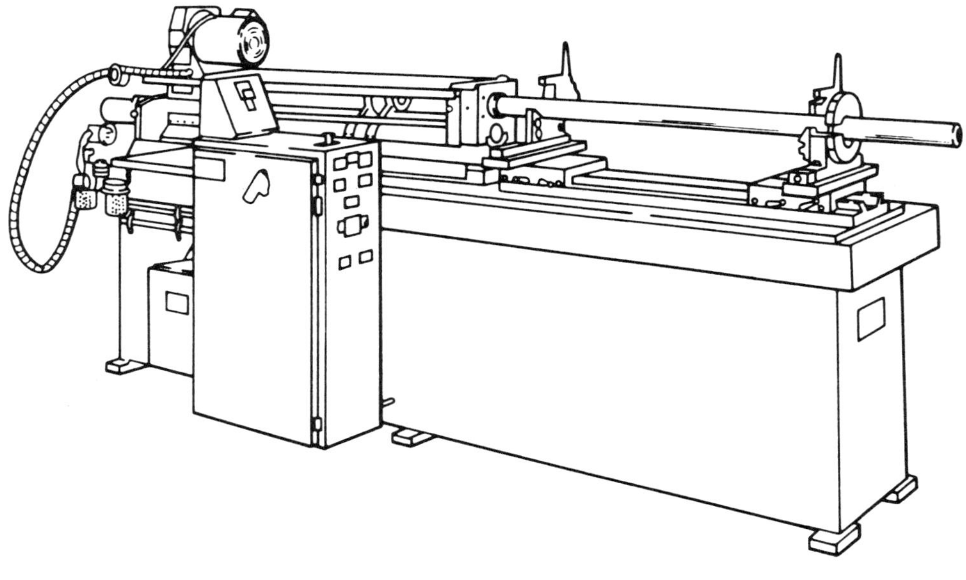

**Fig. 10** Fixed-table, advancing-spindle machine designed for deep-hole drilling

The rapid removal of chips from the drilling operation becomes necessary to ensure the proper operation and accuracy of the drill. Rotational speeds (up to 20 000 rev/min) and feeds must be carefully determined because there is a greater possibility of deflection than when a drill of ordinary size is used.

To overcome these problems, deep-hole drilling machines have been developed (Fig. 10). These machines may be of either the horizontal or the vertical type, with either single-spindle or multiple-spindle construction. In addition, either the workpiece or the drill may revolve. Most machines are of horizontal construction, using a center-cut gun drill having a single cutting edge with a straight flute running throughout its length. Oil, under high pressure, is brought to the cutting edge through a hole in the drill. Normally, the drill is stationary and the work is rotated in a chuck with steady rests providing support along its length, but where it is difficult to rotate the work, the situation is reversed (Fig. 11). In gun drilling, the feed must be light to avoid deflecting the drill.

Because of its long stroke and ease of chip removal, a gun-drilling machine is also suitable for boring, trepanning, and chamfering. The revolving drill is fed into the workpiece resting on the platform. Equipment of this type requires an efficient cooling system to pump the coolant in large volume and high pressure for cooling the tool and flushing the chips. The system also filters the coolant and keeps it at a uniform temperature. Vertical deep-hole drilling machines are also available for work that is not relatively long.

## Transfer Drilling Machines

Frequently designated as automated machines, transfer drilling machines complete a series of machining operations at successive stations and transfer the work from one station to the next. They are, in effect, a production line of connected machines that are synchronized in their operation so that the workpiece, after being loaded at the first station, progresses automatically through the various stations to its completion. Automatic drilling machines are of the indexing table or the in-line transfer type, and they are cost effective only for volume production.

**Indexing Table Transfer Machines.** Parts requiring only a few operations are adapted to indexing table machines, which are made with either vertical or horizontal units spaced around the periphery of an indexing table. A special vertical machine with a six-station indexing table can drill, ream, chamfer, and mill connecting rods at

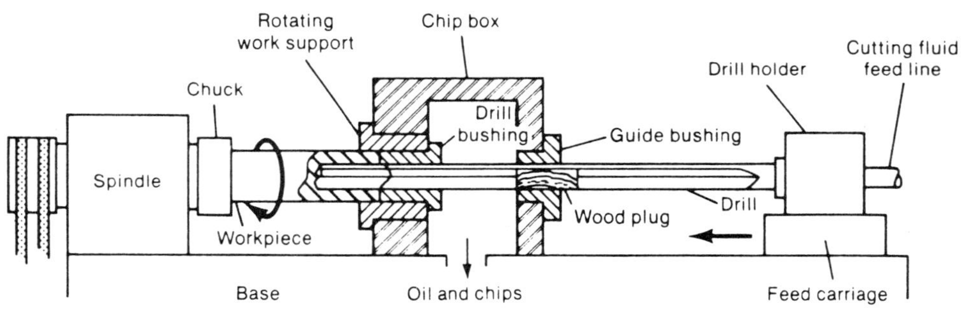


(a)

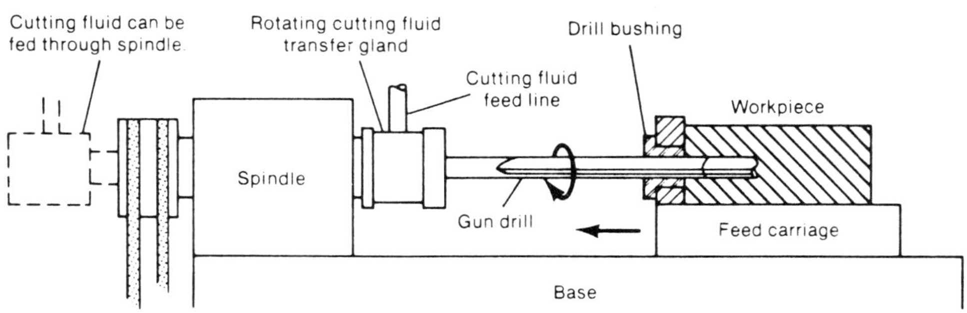


(b)

**Fig. 11** Two methods of gun drilling. (a) Workpiece rotates while the tool is held stationary. Cutting fluid is fed directly through the drill from the back end. (b) Tool rotates while the workpiece is held stationary. Cutting fluid is fed through a rotating oil gland from the fluid feed line or through the machine spindle.

**Table 1 Typical drilling machine specifications**

| Type | Workpiece diameter, mm (in.) | Drill diameter, mm (in.) | Morse taper (No.) | Number of speeds | Power, kW (hp) | Weight, kg (lb) |
|---|---|---|---|---|---|---|
| Hand feed | 395 (15½) | 13 (½) | Chuck | 12 | 0.4 (½) | 100 (220) |
| Upright power feed | 660 (26) | 50 (2) | 5 | 4 | 3.7 (5) | 950 (2100) |
| Upright four-spindle gang type | 660 (26) | . . | 2 | 8 | 1.5 (2)(d) | 2200 (4800) |
| Multiple-spindle (12)(a) | 305 × 510 (12 × 20) | . . | . . . | . . . | 5.6 (7½)(d) | . . . |
| Programmable multispindle(b) | 610 × 1170 (24 × 46) | . . | . . . | Infinitely variable | 18.75 (25)(d) | . . . |
| Radial 330 mm (13 in.)(c) | 1220 (48) | . . | 5 | 16 | 4.5 (6)(d) | 2720 (6000) |

(a) Diameter. (b) Work area. (c) Arm length. (d) Each spindle. Source: Ref 1

the rate of 450 per hour. All operations are performed simultaneously, and when all are complete, the table is automatically indexed to the next station.

**In-Line Transfer Machines.** The principal feature of in-line machines is that they are provided with suitable handling or transfer means between stations. The simplest and most economical method of handling parts is to move them on a rail or a conveyor between stations. When this is impossible because of the shape of the part, it is necessary to provide a holding fixture or pallet on which the work is clamped.

A typical application would be an eight-station automatic transfer machine performing a variety of operations on the center portions of axle housings. Palletized work holding fixtures secure the axles during all operations.

Transfer machines range from comparatively small units having only two or three stations to long straight-line machines with over 100 stations. Transfer machines are primarily used in the automobile industry, in which, because of full production schedules, it is possible to offset their high initial cost by savings in labor. Products processed by these machines include cylinder blocks, cylinder heads, refrigeration compressor bodies, and similar parts.

## Special-Purpose Drilling Machines

Many drilling machines are built in a wide variety of designs and configurations for special-purpose applications. Special-purpose dedicated machines are often used when large quantities of parts require multiple operations. Dedicated units include shuttle-transfer, dial-index, and trunnion-index machines.

**In shuttle-transfer machines,** the workpiece is loaded in a fixture and indexed in a straight line to several machining stations between single or opposed machining heads. After the workpiece is machined, it is indexed to stations down the line for further machining until it reaches the unloading station. These machines are designed for the medium-to-low production of heavy or oddly shaped parts and offer more flexibility in terms of fixture changes than transfer machines.

**Dial-index machines** maximize floor-space use by mounting fixtures on a horizontal-index table and moving the workpiece (held by the fixture) to the machining station, where it is surrounded by an array of radial, horizontal, and vertical machining units. Multiple-spindle heads are extensively used in dial-index machines.

**Trunnion-index machines** provide simultaneous machining operations at each station, along with the production of a completed part each time the trunnion (vertical table) is indexed. Five of the six sides of a rectangular workpiece can be machined in this unit in one cycle. Loading and unloading of the workpiece are done at a single station. Interchangeable heads and fixtures for similar parts allow some degree of flexibility in the production of parts.

Various factors influence the selection of a machine for a specific application. Some of these factors are the size and shape of the workpiece; the diameter and length of the hole to be drilled; the relationship of drilling to the other machining operations; the quantity of pieces to be drilled per hour, per day, or over a longer period; and cost.

## Drilling Machine Selection

**The sizes of drilling machines** are designated in several ways. The most common designation for a vertical drill press is the diameter in millimeters (inches) of the largest disk or workpiece in which a hole can be drilled at the center on the drill press. Other designations are the diameter of the largest drill the press is designed to drive in cast iron or steel and the Morse taper size in the spindle hole. The size of a radial drill press is given in meters (feet) and designates the radius of the largest disk in which a center hole can be drilled with the head at its outermost position on the arm. Some drilling machine manufacturers also specify the diameter of the column in millimeters (inches) (Table 1).

**Machine Selection for Economy of Production.** In selecting a machine, the following options should be considered for their suitability to the specific job:

- *Machine type*: Manual control, numerical control; single spindle or multiple spindle
- *Machine calibrations*: Speed and feed range, manual or automatic machine positioning, and automatic preselect
- *Tooling*: Complex or simple fixturing; high-speed steel or carbide tools

Once these have been decided on, a machine is selected for greatest economy on the basis of size of production lot. Small lots (<100 pieces) seldom justify large expenditures for jigs and fixtures, but numerically controlled machines are being increasingly used for low production quantities. Large production quantities may require more expensive tooling, and single-purpose machine tools often provide the greatest economy.

**Control Systems.** Drilling machines are available for manual, semiautomatic, and automatic operation. Numerical control (NC) and computer numerical control (CNC) are used on many drilling machines, and these types of control are particularly suitable for producing patterns of holes in various workpieces. Numerical control or computer numerical control is usually standard on circuit board, tube sheet, and other special-purpose drilling machines. Programmable controllers are also used on some drilling machines and drill heads.

Tape-controlled radial, turret, or horizontal drilling machines simplify tooling because a means of holding and positioning the work is the only additional requirement. Tape control also minimizes the possibility of operator error and ensures repeatability of operation. The main disadvantages are the cost and complexity of the equipment and the time and skill required for preparing the tape.

The simplest type of numerically controlled machine has a two-position indexing table that automatically places the workpiece under a single vertical drill spindle. The movement of the drill spindle can also be numerically controlled, and this procedure, in conjunction with an automatic tool changer, can produce a completed part. The operator's principal function is to load and unload the workpiece.

The NC/CNC requirements for drilling machines vary, depending on the application. Simple two-axis ($x$ and $y$) positioning systems are used only for table movement on some machines, with the drilling depth ($z$-axis) controlled manually, electrically, electromechanically, or mechanically. Many drilling machines are equipped with more complex systems that can control table positioning, drilling depths, spindle speeds, feed rates, and other functions. Some systems compensate for varying tool lengths.

One manufacturer offers a continuous monitoring control system for its NC machining centers and special drilling ma-

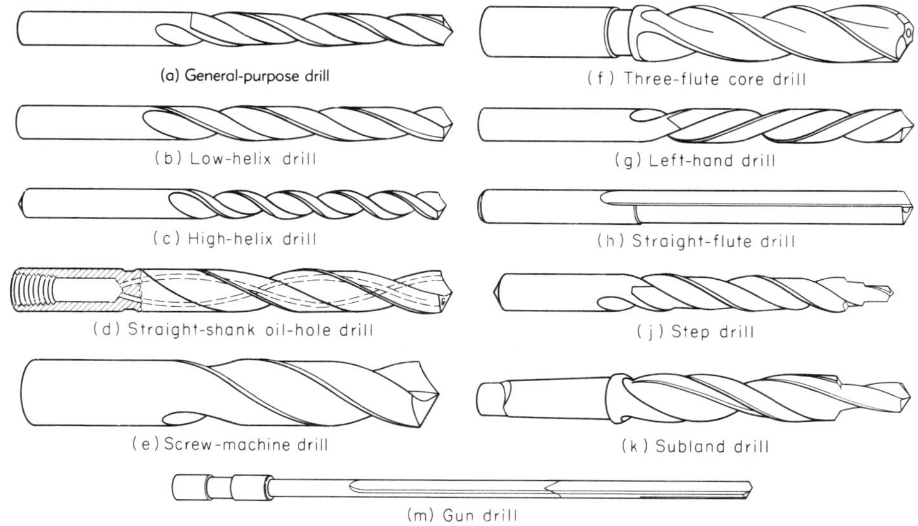


**Fig. 12** Commonly used types of drills. Applications are described in text.

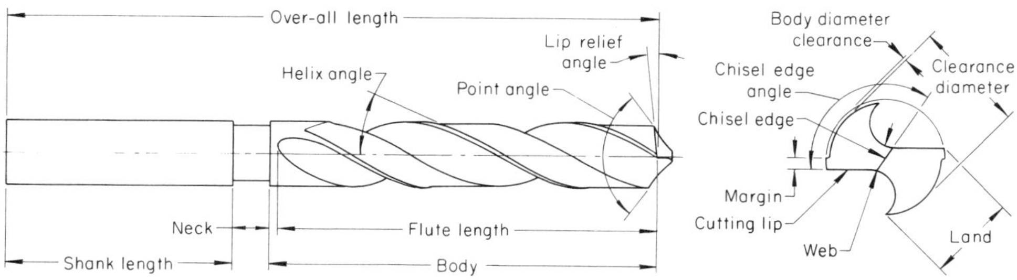


**Fig. 13** Design features of a typical straight-shank twist drill

**Table 2 Lip relief angles for twist drills**

| Drill diameter, mm | Drill diameter, in. | Lip relief angle, degrees: Cast iron or annealed steel | Hard or tough work metal | Soft or free-cutting metal |
|---|---|---|---|---|
| 0.343–0.991 | 0.0135–0.0390 | 24 | 20 | 26 |
| 1.02–2.44 | 0.0400–0.0960 | 21 | 18 | 24 |
| 2.49–3.05 | 0.0980–0.1200 | 18 | 16 | 22 |
| 3.26–6.35 | 0.1285–¼ | 16 | 14 | 20 |
| 6.53–8.7 | 0.2570–11/32 | 14 | 12 | 18 |
| 8.83–12.7 | 0.3480–½ | 12 | 10 | 16 |
| 13.1–19.1 | 33/64–¾ | 10 | 8 | 14 |
| 19.4 and larger | 49/64 and larger | 8 | 7 | 12 |

chines. This system automatically lowers the feed rate when torque on the drill reaches a preset limit or when motor power exceeds the rated amount. Advantages include faster hole production with the same tool life, reduced tool breakage, improved chip-breaking action, and protection of the motor from overloads.

This closed-loop control system continually derives torque at the drill from measured horsepower and spindle speed. The calculated cutting torque is compared to the breaking torque punched into the NC tape. Whenever the cutting torque exceeds one-half the preset limit, the feed rate is automatically reduced. The feed rate is also lowered whenever the horsepower required exceeds the rated motor capacity.

Retrofittable adaptive control for NC drilling machines and machining centers is also available. This is a microprocessor-based system with the capability of combining soft-wired modules for drilling, milling, boring, and tapping in the same unit. The drilling unit uses torque as a feedback.

## Common Types of Drills

Drills are available in a variety of types, and several of these types are illustrated in Fig. 12. Size ranges vary; some drills can be obtained as standard tools up to 90 mm (3½ in.) in diameter, and as special-order tools in even larger sizes (spade drills are usually cost effective for sizes above 38 mm, or 1½ in., and have reportedly been used in sizes up to 380 mm, or 15 in.). The type of shank (straight or taper), the proportionate dimensions of flutes and shanks, and the helix angle also vary widely.

Drills with helical flutes are called twist drills. The features of a typical twist drill are illustrated in Fig. 13.

**Type of Shank.** Taper-shank drills offer these advantages over straight-shank drills:

- Greater tool rigidity
- Closer control of concentricity
- Closer spacing in gang drilling (because screw chucks are not used)
- Drive without slip

For shallow holes, however, straight-shank drills should be considered because they cost less. One disadvantage of straight shanks, particularly in drilling deep holes, is that they sometimes pull out of the chuck on the return stroke of the spindle. When this happens, the operation must be stopped to replace the drill. This pullout is more likely in an operation in which the spindle retracts automatically at intervals to clear chips.

**Drill Design.** The included angle of the point on a conventional twist drill is 118°. The lip relief angle varies with drill diameter and the composition and condition of the work metal (Table 2).

Drill design for specific operations depends on the application in which the drill is used. Drill life can sometimes be greatly improved by the use of a design that varies from the standards of the drill manufacturer. Drill design and rigidity of setup usually receive more attention than selection of the drill material. This is because in most drilling, improvement of design and rigidity offers a greater potential for increasing drill life and reducing cost than does a change in drill material. Special means of clamping may be needed to ensure rigidity of setup.

**Drill Materials.** Most drills are made of common grades of high-speed tool steel (M1, M2, M10, and occasionally T1). At comparatively low cost, these grades provide strength, toughness, and high-temperature hardness suitable for most drilling applications. For excessively hard or abrasive metals, drills can be made of higher-alloy high-speed tool steels such as M3, M4, M6, M33, or T15; these steels, however, are used to make only a small percentage of the drills produced. Additional information on high-speed tool steels is available in the article "High-Speed Tool Steels" and "P/M High-Speed Tool Steels" in this Volume and the article "Selection of Materials for Cutting Tools" in Volume 3 of the 9th Edition of *Metals Handbook*.

Carbide-tip drills are used for special applications, notably, for drilling abrasive metals of low tensile strength (such as cast iron and castings of high-silicon aluminum alloys) or heat-resistant alloys. Special carbide-tip drills with longitudinal passages for conducting coolant to the drill tip are used on steel in gun drilling, for which close tolerances on hole straightness, diameter, and parallelism must be maintained. Another special application of carbide-tip drills is the drilling of steel harder than 48 HRC. Solid-carbide drills (for example, spade drills) are used for extreme rigidity and drilling accuracy.

*Heat Treatment*. The hardening, quenching, and tempering of the high-speed tool steel are critical to the quality and performance of twist drills. A particular high-speed tool steel of the same analysis can be heat treated differently by different manufacturers to develop desired properties for the specific application of a certain style of drill.

Twist drills made from general-purpose high-speed tool steels such as M1 and M7 are usually heat treated to a hardness range of 64 to 66 HRC. Drills made of high-speed tool steels containing cobalt generally have a hardness range of 65 to 67 HRC, although a higher hardness is sometimes used for unusual applications.

*Surface Treatments*. Various surface treatments are applied to high-speed tool steel drills to increase the hardness of the outer layer of material or to reduce the friction between the drill and the workpiece or chips in the flute. These treatments are normally applied after the drills have been finish ground. The treatments that produce a thin hard layer in the outer layers include nitriding, cyaniding, and less frequently, carbonitriding and carburizing. These processes are carried out in either a liquid or gaseous media at elevated temperatures to accelerate the absorption of the element(s) into the outer layers. Because these layers are hard but brittle, they are kept thin to reduce the probability of chipping. Nitride layers range from 0.013 to 0.05 mm (0.0005 to 0.002 in.) in thickness, but carburized layers are several times thicker. Carburizing and carbonitriding usually require temperatures above the tempering range, while gas nitriding or cyaniding can be done at about 540 °C (1000 °F).

In addition to the hard surfaces produced in the outer layers of the drill material by the processes mentioned, there is some small usage of the processes for producing surfaces on the exterior of drills for the same purpose. These surfaces are applied by electroplating, chemical plating, chemical and physical vapor deposition, and flame or arc deposition. Included in such platings are hard chromium, nickel, tungsten, tungsten carbide, titanium nitride, titanium carbide, and numerous other wear-resistant compounds. Considerable care is required during the processing to ensure that the deposition is uniform, continuous, and tightly adherent and that the tempering temperature of the high-speed tool steels is not exceeded.

Surface treatments for reducing friction or improving lubrication include the limited penetration of oxygen or sulfur into tool surfaces in a controlled-atmosphere furnace at an elevated temperature, and vapor or liquid processing. The most widely used surface treatment for drills consists of the development of a thin surface oxide, which has been found to act as a solid lubricant and to prevent welding of chips to the drill. Although various oxide films can be produced in air at temperatures above 205 °C (400 °F), oxide films most advantageous for improved performance are formed in a dry or superheated steam atmosphere at temperatures of 425 to 565 °C (800 to 1050 °F). Oxides created in salt baths are not as effective as those produced by steam. Tools having the dark gray to blue-black surface oxide layers produced in this way are often used for drilling ferrous materials.

Although surface treatments have been discussed as separate entities, they are often used in combination. The most widely used combination treatment for drills is a surface oxide over a nitride layer.

## Selection of Drills

The type of drill selected for a given application depends on:

- Composition and hardness of the work metal
- Rigidity of the tooling setup
- Dimensions of the hole to be drilled
- Type of machine used to rotate the drill or the workpiece
- Whether the drill is used for originating or enlarging holes
- Tolerances on the hole to be drilled
- Whether related operations, such as countersinking or spotfacing, must be performed with drilling
- Cost

**The material drilled** affects drill selection mainly because the need for a drill of rugged design increases with the hardness and strength of the work material. The need for rigidity in the setup also increases with work metal hardness and strength.

Drilling into hard materials such as heat-treated alloy steels, hard steel castings, partly chilled cast iron, and 18-8 stainless steels with high-speed tool steels requires carefully sharpened tools, adequate and uniform feeds, moderate speeds, and coolants suited to the type of material being machined. In an emergency, it is possible to drill holes in materials that would otherwise be considered unmachinable with high-speed tool steels. This is done by making sure the tool cuts continuously once it is started because the material will work harden to the point at which it is unmachinable if the drill is allowed to idle and rub on the cut surface.

Armor plate presents some difficult problems because of its hardness and extreme toughness, but it can be satisfactorily drilled with high-speed tool steels if the work-hardening and flame-hardening properties of armor plate are compensated for. If the material has been flame cut, the areas adjacent to the region to be drilled should be annealed. If the armor plate is of the laminar type (that is, one side of plate is harder than the other), drilling should proceed from the harder side to the softer side because it is virtually impossible to drill successfully from the soft side. Speeds as low as 4.6 m/min (15 sfm) may be required, along with a 135 to 140° included point angle on the drill cutting point.

Molded plastics (resins, phenolics, cellulose acetates, polystyrenes, acrylics, and so on) require quick ejection of chips. If not ejected quickly, the chips will stick and pack in the drill flutes and heat the comparatively low-temperature plastic composition. Overheating is detrimental to both the tool and the work. These materials are drilled at speeds of 30 to 90 m/min (100 to 300 sfm), with tools having 60 to 90° included angle points for small sizes and 90 to 120° included angle points for larger hole sizes. Detailed information on the machining of a variety of materials is available in the Section "Machining of Specific Metals and Alloys" in this Volume.

**Rigidity of Setup.** In setups that are not rigid, drills with short flutes and tapered shanks are usually required.

**Hole Dimensions.** The diameter, length, and length-to-diameter ratio of holes to be drilled also influence selection of the drill. For maximum rigidity and minimum chatter, the shortest possible drill should be used. In drilling blind holes several times greater in length than in diameter, it may be necessary to use oil-hole drills or other special types of drills.

**The type of machine** to be used to drive (or hold) the drill is more likely to determine the design of the drill shank than of the fluted portion. However, some drills are designed for use with specific machines (such as screw-machine drills and left-hand drills).

**Use of Drill.** Whether a drill will be used for originating or enlarging holes also governs its selection. Holes can be enlarged more efficiently by drills designed for this purpose (core drills).

**Tolerances.** Although drilling is generally a roughing operation, with accurate surfaces depending on subsequent operations such as reaming or boring, tolerances do influence drill selection. Tolerances on roundness, diameter, and straightness are often affected by the ability of the drill to remove chips.

**Number of Operations.** When other operations, such as countersinking or spotfacing, are to be done with drilling, the use of combination tools should be considered. Dual-purpose tools are available that can combine the following operations: drilling two different diameters, drilling and reaming, drilling and countersinking, drilling and spotfacing, and drilling and counterdrilling or counterboring.

**Cost.** Carbon tool steel drills have a low initial cost and are acceptable for occasional use, but they must be run slowly. High-speed tool steel drills are the most popular and have good strength. Drills tipped with cemented carbide are economical for high production, but are expensive and must be handled carefully to avoid breakage.

## General Drill Classification

Twist drills are manufactured in a wide variety of types and in many different sizes. To produce a hole of a given diameter, twist drills are commercially available with variations in length, flute and shank configuration, point geometry, and web thickness. In some cases, a dozen or more drills may be available to produce the same size hole.

Drills are made in many different diameter sizes—fractional, number (wire gage), letter, and metric—ranging from 0.150 to 89 mm (0.0059 to 3½ in.):

- *Millimeter series*: 0.01 to 0.50 mm increments, according to size, in diameters from 0.015 mm
- *Numerical series*: No. 97 to No. 1 (0.15 to 5.79 mm, or 0.0059 to 0.228 in.)
- *Lettered series*: A to Z (5.95 to 10.49 mm, or 0.234 to 0.413 in.)
- *Fractional series*: 1/64 to 4 in. (and larger) by 64ths

Many drill styles can be grouped into two categories: general-purpose and heavy-duty. General-purpose drills are the most widely used. Slight alterations of the original point angles sometimes improve performance for given speeds and feeds. Because of their heavier webs, heavy-duty drills are designed to provide greater torsional strength and rigidity than general-purpose drills. Heavy-duty drills can be used to drill steel forgings, hard castings, and high-hardness ferrous alloys.

**Straight-shank twist drills** are made in three series, based on their range of lengths: jobbers length (a medium-length range), taper length (a long-length range), and screw-machine length (a short-length range). Thus, a 6.4 mm (¼ in.) diam drill would have a 70 mm (2¾ in.) flute length and a 102 mm (4 in.) overall length in a jobbers length version. The flute length and the overall length of the same tool would be 95 mm (3¾ in.) and 156 mm (6⅛ in.) for the taper length version and 35 mm (1⅜ in.) and 64 mm (2½ in.) for the screw-machine length version. Heavy-duty drills are also available in screw-machine lengths with 135° split points.

Screw-machine length straight-shank twist drills are used in screw machines of all types when conditions require the use of short drills for maximum rigidity. Other applications for these short drills include portable and sheet metal drilling, body work, and the drilling of tougher, harder steels such as stainless and high-manganese steels.

Tangs can be added to these tools when they are used with ASA drill drivers.

**Taper-shank twist drills** are used for general-purpose drilling in all types of machines for which an American National Standards Institute taper shank is required. These drills are available with standard diameters, flute lengths, and overall lengths, as well as corresponding taper-shank sizes. Some sizes of these drills are available with shorter and longer shanks than standard, which decreases and increases their overall lengths. The high-speed tool steel drill body and the alloy steel shank (when used) on these drills are usually joined by electrical butt or friction welding. In some cases, the drill body can also be press fitted (crimped) into the shank end on smaller sizes.

## Specific Applications by Drill Type

Most drills, particularly those discussed in this section, are specially designed for specific applications. Altering the drill point or drill hardness can greatly increase the range of applications for which these tools can be used. In addition, carbide inserts or indexable-insert cutting edges can be added.

Conventional drills have helix angles of 25 to 33°. Low-helix drills have 15 to 20° helix angles and high-helix drills typically have 35 to 40° helix angles.

**General-purpose drills** are of conventional two-flute design. Because of their versatility, low cost, and availability in a wide range of sizes, these drills are used for the general-purpose drilling of cast iron, steel, and nonferrous metals. They are suited to high-production applications and can be operated under a variety of conditions. A typical general-purpose drill is illustrated in Fig. 12(a). These jobbers length drills are available only with a straight shank. Their short flutes and relatively short length-to-diameter ratio aid in maintaining rigidity during drilling.

**Low-helix drills** (Fig. 12b), also known as slow-spiral drills, have a comparatively thin web, which facilitates the penetration of brass, plastics, and other soft materials and provides maximum space for chips. Low-helix drills are designed to break chips into small pieces and are well-suited for applications in which a large volume of chips is generated. Straight-shank low-helix drills are more rigid than drills with a standard twist and therefore permit greater torque. Low-helix drills are also less likely than standard-helix drills to grab the edge of a hole and hang up or pull through at drill breakthrough. Although developed primarily for drilling brass, these drills are often used successfully for drilling shallow holes in steel and in aluminum and magnesium alloys. Solid carbide or carbide-tip versions of low-helix drills are used to produce holes in cast iron.

**High-helix drills** (Fig. 12c), also known as fast-spiral drills, are made with wide flutes and narrow lands. They are particularly recommended for drilling relatively deep holes in nonferrous metals and other low tensile strength materials. However, carbon, alloy, and stainless steels have been successfully drilled with these drills. The wide flutes and high helix facilitate chip removal. The high helix also creates more land surface per linear inch of drill, thus increasing the bearing surface of the drill. It is advantageous to replace a standard-helix drill with a high-helix drill mainly because a high-helix drill can remove chips from deep holes more efficiently than most other drills—especially in applications requiring relatively high feed rates.

**Oil-Hole Drills.** An oil-hole (or coolant-feeding) drill has one or more continuous holes through its body and shank to permit the passage of a cutting fluid under high pressure. The cutting fluid enters the shank and upon emerging at the drill point, assists in ejecting chips and dirt, which may clog or gall the hole being drilled. These drills can be used in a fixed position, as in a turret lathe or other type of multiple-operation machine, or they can be used with power drive in special sockets.

*Straight-shank oil-hole drills* (Fig. 12d) are well suited to drilling deep holes in hard metals. They are particularly useful when horizontal or inverted drilling is required or when it is difficult to get cutting fluid to the work or to the cutting edges of the tool. An oil-hole drill provides the necessary fluid at the cutting edges, thus permitting a greater feed rate (and therefore a greater volume of chips).

*Taper-shank oil-hole drills* have the following advantages over the straight-shank types:

- They allow closer concentricity
- Speed and feed can be higher
- The tool is more rigid

The following example describes an application for which a taper-shank oil-hole drill was selected in preference to a standard twist drill.

**Example 1: Two Deep Holes.** A taper-shank oil-hole drill was superior to a standard twist drill for drilling the two deep through holes (length-to-diameter ratio: nearly 9:1) in the 4130 steel guide block shown in Fig. 14. The oil-hole drill not only maintained the required straightness but also eliminated a subsequent reaming operation which is needed when using a standard twist drill.

Placed in a box-type drill jig, the guide block was drilled in a single-spindle drill press (19 mm, ¾ in., capacity; 430 mm, or

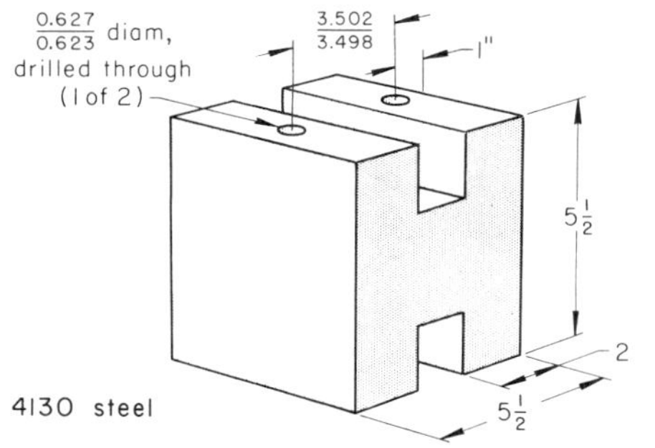


| | |
|---|---|
| Speed, at 730 rev/min, m/min (sfm) | 37 (120) |
| Feed, mm/rev (in./rev) | 0.15 (0.006) |
| Cutting fluid | Soluble oil:water (1:20) |
| Setup time, min | 30 |
| Production rate, pieces (holes)/h | 25 (50) |
| Drill life per grind, pieces | 125 |
| Downtime for tool change, min | 2 |
| Workpiece hardness, HRC | 20-27 |

**Fig. 14** Guide block drilled with a taper-shank oil-hole drill under conditions listed in table. The oil-hole drill produced the deep through holes to the specified finish without subsequent reaming, which was required when a standard twist drill was used. Dimensions in figure given in inches

17 in., swing). Speed, feed, and other operating data are given with Fig. 14.

**Screw machine drills** (Fig. 12e), sometimes called stub drills, are made with short flutes and short overall length for maximum rigidity without loss of cutting ability. They are designed primarily for use in multiple-operation machines and in portable equipment, and they are useful for drilling tough or hard materials. The short flute length (about half that of a standard-length twist drill) eliminates the excessive overhang of standard-length drills used without bushings. When an irregular, angled, or nonflat surface is drilled, excessive overhang causes walking or weaving of the drill, which results in oversize or tapered holes and in reduced tool life.

**Core drills,** which have three to six flutes, are used for enlarging cored, forged, or previously drilled holes. Core drills are sometimes considered as roughing cutters when used primarily for heavy stock removal in holes that are finally sized by boring or reaming.

Three-flute core drills (Fig. 12f) are normally used for holes up to 13 mm (½ in.) in diameter. These drills are designed to allow greater chip clearance for heavy cuts. Four-flute core drills are more commonly used for holes larger than 13 mm (½ in.) in diameter. They are designed to combine maximum flute space with maximum rigidity. Because of their greater number of cutting edges, core drills can be used at higher feed rates than other drills or boring tools without increasing chip load per cutting edge.

Operations such as counterboring and chamfering are often combined with core drilling by step grinding the drill. Core drills normally have a relatively heavy web (up to 50% of the diameter) because they are not designed for center cutting. They are available with three or four flutes, straight or taper shanks, and in fractional sizes only. Straight-shank core drills are made with diameters from 6.4 to 32 mm (¼ to 1¼ in.), flute lengths from 95 to 200 mm (3¾ to 7⅞ in.), and overall lengths from 156 to 318 mm (6⅛ to 12½ in.). Taper-shank core drills are made with diameters from 6.4 to 64 mm (¼ to 2½ in.), flute lengths from 73 to 286 mm (2⅞ to 11¼ in.), and overall lengths from 156 to 476 mm (6⅛ to 18¾ in.).

Larger-diameter core drills usually embody a fluted cutter interchangeably mounted on the end of a fluted body. Such drills often incorporate pilots to ensure accuracy of hole alignment. Four-flute core drills are generally preferred unless the greater chip room of a three-flute tool is needed.

**Left-hand drills** (Fig. 12g) have the same flute length and overall length as general-purpose drills. The principal use for left-hand drills is in multiple-operation machines when the spindle is rotated in the reverse of normal direction. This often occurs when tapping follows drilling or when other machining operations are combined with drilling.

**Straight-flute drills** (Fig. 12h) are especially adapted for drilling brass and other soft materials and for drilling thin sheets of a wide variety of metals. Drills of this type do not run ahead or grab the work during drilling.

Single-fluted drills, also known as bobbin bits, are designed for drilling deep holes in wood. Because of their chip-clearing ability, they are used for the drilling of low-melting thermoset plastics, the chips of which will stick in the flutes of an ordinary drill. These drills are not recommended for use in materials of much higher tensile strength than wood, fiberboard, and soft plastics. Overall and flute lengths are the same as for general-purpose jobbers length drills.

**Half-Round Drills.** The cutting end of this drill is semicircular in cross section. Half-round drills are designed for use in brass and, in some cases, aluminum and die cast materials. They have found wide application in the screw-machine industry in the production of holes of exceptional straightness and high finish.

**Double-margin drills** are often used when holes must be straighter and to closer tolerances with respect to size than can normally be obtained with a single-margin drill. They have a second pair of margins at the trailing edges of the lands, about midway between the primary margins (Fig. 15), thus providing added bearing in bushings and holes with four guide surfaces. This construction also helps to eliminate the effects of errors in point grinding. Double margins are also used on some step drills for producing accurate holes when guide bushings cannot be used.

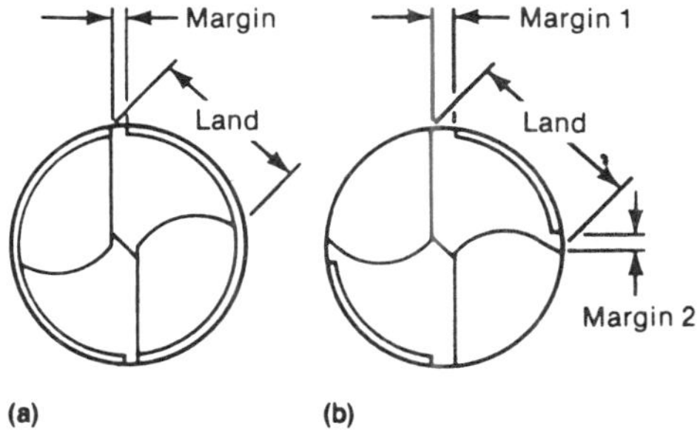


**Fig. 15** Two types of margin configurations used on twist drills. (a) Conventional single margin. (b) Double margin

**Crankshaft drills** are designed for drilling holes in crankshafts, but are also used to produce holes deeper than five or six times the drill diameter in any tough or hard material. They have heavy webs, a higher-than-normal helix angle, a split or crankshaft point, and a surface treatment for increased abrasion resistance.

**Automotive series drills** are straight-shank twist drills of general-purpose construction. They have the same flute and overall lengths as regular jobbers length drills and are furnished with tang drive. They are also available in a long series with taper lengths and tang drive.

**Cotter-pin drills** are heavy-duty jobbers length drills designed for drilling cotter-pin holes in bolts, steering knuckles, pins, and similar parts, as well as cross holes in the heads of bolts. Heavy construction resists the high strains from such applications and makes them suitable for drilling some of the hard alloy steels, including the harder stainless steels. Such drills are generally surface treated.

**Center Drills.** These straight-shank twist drills are often used to center the ends of shaft-type workpieces. Their short flute and overall lengths, as well as their oversize shank diameters, provide good rigidity, thus reducing runout. Larger size center drills are sometimes used for starting operations when drilling is done on screw machines and other machine tools to increase the location accuracies of holes to be produced. When the design of the workpiece permits, the countersink portion of the center-drilled hole should be 0.08 to 0.15 mm (0.003 to 0.006 in.) larger in diameter than the finished hole size.

**Rail Drills.** These heavy-duty cobalt high-speed tool steel tools, originally designed for drilling high-manganese railroad rails, are used to drill a variety of hard, tough, high tensile strength and high-temperature resistant materials used by the aerospace industry. They are made with a low helix angle, heavy web, special web thinning for ease of penetration, and short flute and overall lengths, and they are surface treated.

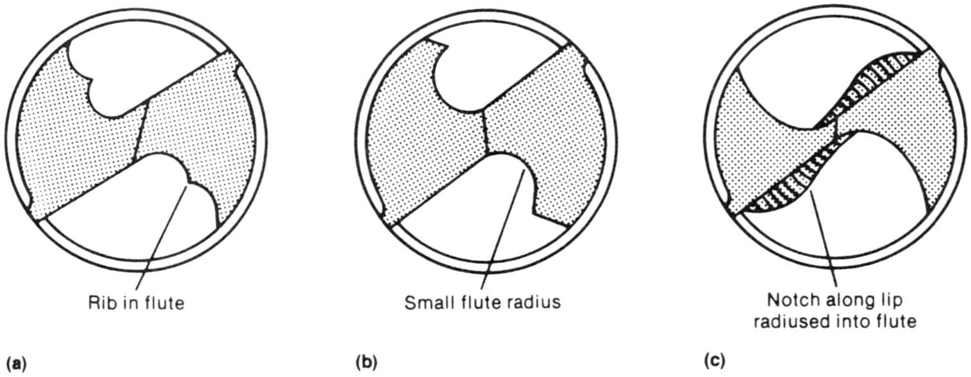


**Fig. 16** Flute geometry of drills modified to provide chip-breaking action through the addition of a rib (a), a radius (b), or a notch (c) to the drill flute

**Aircraft drills,** which are extensively used by the aircraft and aerospace industries, are available as jobbers length drills and as extension drills. These tools have a flute construction conforming to National Aerospace Standard (NAS) 907 prepared by the Aerospace Industries Association, spring-tempered shanks to prevent permanent bending, and overall lengths of 152 to 305 mm (6 to 12 in.). They are available with either 118 or 135° split or offset drill points. These drills are principally used for drilling holes in inaccessible places that cannot be reached with regular-length drills; they are generally operated in portable drilling units.

Straight-shank, jobbers length aircraft drills conforming to NAS 907 are available in two types. Type A drills, with 118° split points, are used for drilling aluminum, low-carbon steels, magnesium, and other soft materials. Type B drills, with 135° split points, are of heavier design and are surface treated for drilling hard steels, stainless steels, titanium, and other hard materials.

Short-flute aircraft drills with 118° points are desirable for tough, portable applications in titanium, heat-treated steels, and aluminum. These drills are surface treated and have no body clearance. Type C drills, with 135° split points, are made in screw-machine lengths for use in machines or portable units to drill hard, tough sheet metals of the heat-resistant stainless and titanium alloy types.

**Chip-breaker drills** have flute modifications or other design features to curl and break the chips produced in most materials in which long, stringy chips are normally produced, thus minimizing clogging problems. Most of the various chip-breaker designs involve a flute configuration that forces long chips to curl and bend sufficiently so that small broken chips are produced.

Until recently, the flute of the drill was thought of only as a path for the chips to follow as they emerge from the hole. However, with industry becoming more safety conscious, stringy chips are now considered to be a significant hazard to the operator. Many approaches are used to modify the flutes to provide a chip-breaking action.

Drill points modified to enhance the chip-breaking action are illustrated in Fig. 16. Such flute designs are built into the drills, and no additional notches or grooves are required when regrinding. For some applications, particularly with larger size twist drills, chip-curling grooves are ground in the rake faces of the cutting lips, or chip-splitting grooves are ground in either the clearance or rake faces. Chip-breaker drills sometimes require more torque than standard drills, and there is a minimum feed rate below which they do not operate effectively.

Chip-breaker drills cost about 10% more than conventional drills, but the added cost is easily offset by the longer tool life and the time saved from not having the problem of long chips. Tool life increases because of the efficient chip ejection and better penetration of the coolant to the drill tip.

**Spotting and centering drills** are straight-shank twist drills designed to produce accurate and true centers. They have short flute and overall lengths and no body clearance, which allows them to be chucked close to the cutting point for accurate hole starting or workpiece centering. A constant web thickness eliminates the need for thinning these drills when repointing.

Holes that are to be drilled and chamfered in workpieces having a hard or scaled surface can be produced more economically by spot drilling (to form the chamfer) prior to drilling. With this method, the spot drill cuts only a short time through stock that has a detrimental effect on tool life, while the hole drill cuts for a longer time through easier-to-machine material. Processing in this way allows the drills to be rapidly advanced closer to the workpiece, thus decreasing the feed stroke and allowing a decrease in the feed rate to increase tool life.

**Step drills** (Fig. 12j) have two or more diameters, produced by grinding various steps on the diameter of the drill. The steps usually consist of a square or angular cutting edge. Step drills are extensively used in applications requiring multiple-diameter holes. They can be made by grinding steps on common drills; usually, some thinning of the web is required. The simplest form of a step drill is a combination drilling and countersinking tool.

**Subland drills** (Fig. 12k) are combination tools having separate lands or margins, which extend the full length of the flutes, for each of the two or more diameters. Subland drills perform two or more operations in one pass of the tool. The different diameters or lands can be ground to permit drilling and countersinking for flathead screws, or drilling and counterboring for socket-head screws. Extensive use is made of subland drills in high-production work.

Subland drills with as many as four diameters have been successfully used. For optimum results, however, the largest diameter should be no greater than twice the smallest diameter because of the variation in cutting speeds and feeds.

Because the lands or margins of subland drills are ground to the correct size along the full length of the flute, each cutting edge can be sharpened separately and repeatedly without affecting the others. When correctly sharpened and operated, subland drills can readily produce accurate holes.

**Gun drills** (Fig. 12m), which are used in horizontal deep-hole drilling machines (see the section "Deep-Hole Drilling Machines" in this article), have a single V-shaped flute; the single cutting face is sharpened offset to form two cutting angles. When drilling, these angles form two chips broken into short lengths for easier expulsion. The drill body is usually a steel tube through which a cutting fluid passes under pressure and in sufficient volume to cool and lubricate the cutting area and flush out chips. To counter the unbalanced cutting force of the single cutting edge, wear pads (usually carbide) are provided in the drill opposite the cutting face to keep the drill on center. Gun drills are used to drill soft as well as hard material.

For accurate holes, gun drills must be used in rigid equipment with a positive feed. With such equipment, gun drilling eliminates the need for subsequent rough and finish reaming. A properly started gun drill guided by a drill bushing will produce a hole that does not drift from the centerline more than 0.013 mm (0.0005 in.) in 50 mm (2 in.) of drill travel. A spotface or a precision center drill can be used for more accurate starting of a gun drill. Gun drills are increasing in use because they can frequently produce holes as accurate as reamed holes in one pass. Tolerances as small as ±0.025 mm (±0.001 in.) and surface finishes up to 0.50 to 0.75 μm (20 to 30 μin.) are not uncommon for some gun drilling applications. For maximum efficiency in gun drilling, the cutting fluid should provide a fluid film between wear pads and work material,

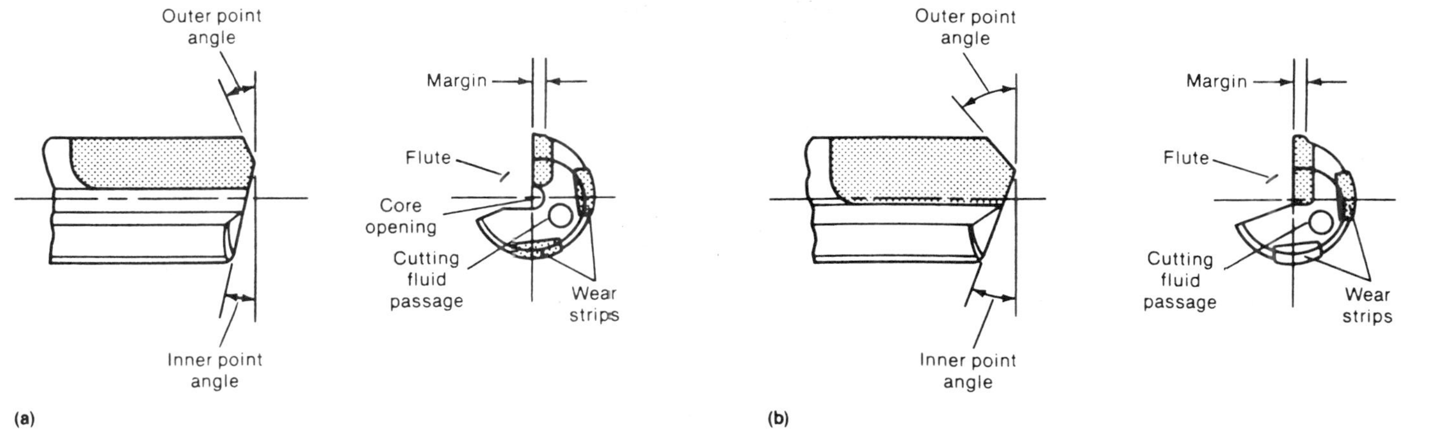


**Fig. 17** Two types of straight-flute gun drills used in deep-hole drilling. (a) Trepanning drill. (b) Center-cut drill

prevent welding of chips to the cutting edge of the drill, cool the drill and the work, and dispose of chips by flushing out the hole.

Two types of straight-flute gun drills are used for deep-hole drilling. One type, known as a trepanning drill or a pin-cutting gun drill (Fig. 17a), has no dead center and leaves a solid core of metal. As the drill advances, the core acts as a continuous center guide at the point where the cutting is done. This prevents the drill from running to one side and the hole accuracy is easier to maintain. The other type, known as a center-cut gun drill (Fig. 17b), has been the conventional drill for many years. It is still used for much deep-hole drilling, such as the drilling of blind holes where a core-type drill cannot be used. Both trepanning and center-cut drills are generally carbide tipped (either solid carbide or a carbide insert).

Gun drills operate at much smaller feeds than conventional twist drills, but the cutting speeds are higher. Each type of drill has only a single cutting edge with a straight flute running throughout its length. The chips are carried out of the hole along the flute of the drill as rapidly as they are formed. Greater accuracy and finish can be obtained in deep holes by the subsequent use of special reamers or broaching tools.

Gun drilling is especially well suited to producing long holes to close tolerances. In some applications, such as the drilling of deep blind holes, gun drilling may be the only practical method of drilling a hole to concentricity tolerances. Hole straightness, roundness, and size can be held within closer limits with a gun drill than with a twist drill, as demonstrated in the following example.

**Example 2: Twist Drill Versus Gun Drill.** A gun drill was compared with a standard twist drill for drilling 9.5 mm (3/8 in.) diam through holes 127 mm (5 in.) deep in gray iron and type 304 stainless steel specimens. The twist drill was used in a general-purpose 1.5 kW (2 hp) vertical drill press; the gun drill, in a 3.8 kW (5 hp) modified boring machine with hydraulic feed. Processing details and results for each material drilled are given in Table 3. As these data show, the gun drill not only produced markedly better finish and dimensional accuracy than the twist drill but also drilled twice as many pieces per grind, under production conditions.

**Table 3 Gun drill versus standard twist drill for through holes in gray iron and stainless steel**

| Condition or result | Gray iron(a) | | Type 304 stainless steel(a) | |
|---|---|---|---|---|
| | Gun drill(b) | Twist drill(c) | Gun drill(d) | Twist drill(e) |
| **Operating conditions** | | | | |
| Speed, rev/min | 2000 | 600 | 2000 | 500 |
| Speed, m/min (sfm) | 60 (200) | 18 (60) | 60 (200) | 15 (50) |
| Feed, mm/rev (in./rev) | 0.025 (0.001) | 0.15 (0.006) | 0.015 (0.0006) | 0.13 (0.005) |
| Feed, mm/min (in./min) | 50 (2) | 91 (3.6) | 30 (1.2) | 64 (2.5) |
| Cutting fluid | Cutting oil(f) | Soluble oil(g) | Cutting oil(f) | Soluble oil(g) |
| **Results** | | | | |
| Surface finish, μm (μin.) | 0.25–0.38 (10–15) | 2.50–3.75 (100–150) | 0.25–0.38 (10–15) | 2.50 (100) |
| Hole straightness, mm (in.) | 0.025 (0.001) | 0.64 (0.025) | 0.025 (0.001) | 0.76–0.89 (0.030–0.035) |
| Out-of-roundness, mm (in.) | 0.005 (0.0002) | 0.05 (0.002) | 0.008 (0.0003) | 0.05 (0.002) |
| Diametral tolerance, mm (in.) | 0.013 (0.0005) | 0.05 (0.002) | 0.025 (0.001) | 0.08 (0.003) |
| Drill life per grind, pieces | 160 | 80 | 50 | 25 |

(a) Specimens of both materials were 22 mm (7/8 in.) in diameter and 125 mm (5 in.) long. Through holes drilled were 9.5 mm (3/8 in.) in diameter and 125 mm (5 in.) deep. (b) Carbide-tip center-cut drill, 9.5 mm (3/8 in.) diameter. (c) Taper-length twist drill 9.5 mm (3/8 in.) diameter. (d) Solid-carbide center-cut drill, 9.5 mm (3/8 in.) diameter. (e) Heavy-web jobbers drill, 9.5 mm (3/8 in.) diameter. (f) Chlorinated mineral oil. (g) Mixed 1 part to 15 parts water

**A spade drill** is generally made up of a spade bit inserted in a holder (Fig. 18a) and fastened securely in a slot with a screw. This type of drill is most widely used for drilling holes 25 to 152 mm (1 to 6 in.) in diameter. (Solid spade drills, usually of carbide, are used for smaller holes.)

Spade drills (size range of 38 to 381 mm, or 1½ to 15 in. in diameter) have risen rapidly in use in the last decade because of their efficiency in making holes and their low cost compared to twist drills. They are the only drills made as stock items in diameters over 89 mm (3.5 in.).

A dull blade can be replaced on the machine, as on indexable throwaway tools, without the necessity of refinding size, resetting stops, breaking down setups, and increasing or decreasing the length of a drilling setup. Because of this feature, spade drills can be easily preset for use on automatic and NC machine tools. These inserted-blade tools, also available with indexable double-edge blades, are a type of flat end cutting drill used to produce large-diameter holes. Spade drilling is done with either the tool or the workpiece rotating.

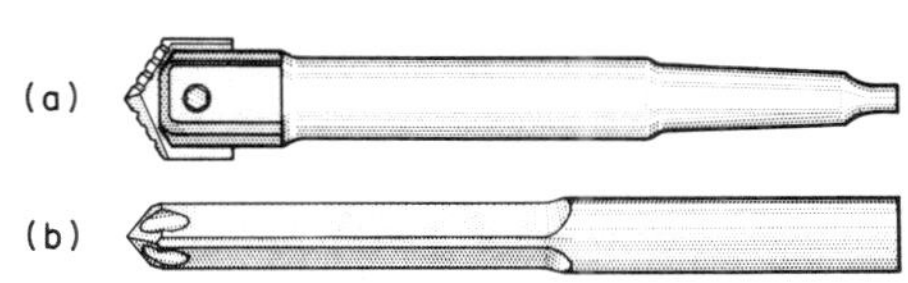


**Fig. 18** Two types of spade drills. (a) Spade drill with interchangeable tip. (b) Special drill for hard steels

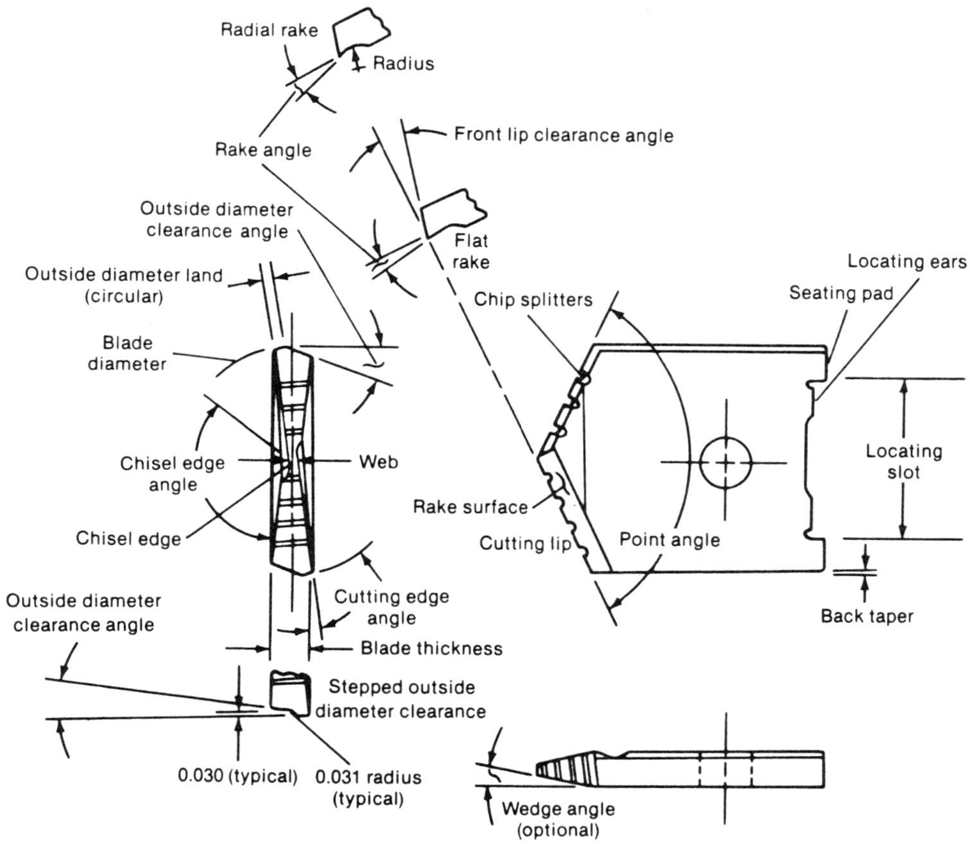


**Fig. 19** Basic spade drill geometry and nomenclature. Dimensions given in inches

Spade-drill holders are made in various sizes with straight or tapered shanks. For deep-hole drilling, the holder has an axial hole through the shank to allow cutting fluid to be applied under pressure to the point of the drill.

Spade drills have heavier cross sections than comparable twist drills (Fig. 19). The additional strength is concentrated along a direct line from the point to the shank of a spade drill. This gives a spade drill greater resistance to end thrust and greater ability to withstand torque. The increased strength minimizes vibration, chipping of cutting edges, and drill breakage. Standard drills are likely to wear into a forward taper, which causes binding. Because the cutting edges of spade drills are shorter and have a greater back taper, binding is less likely.

*Tool Advantages*. One advantage of the type of drill shown in Fig. 18(a) is that the bits are replaceable. By having more than one bit for the holder, the operation can continue while others are being sharpened. Bits are relatively inexpensive. Figure 20 shows various blade configuration available for spade drills.

Another advantage is that they are available in larger diameters than twist drills and can drill deeper holes. Depth-to-diameter ratios have exceeded 120:1 in some horizontal applications of spade drills before the drills needed to be reground. For vertical drilling, however, the depth-to-diameter ratio can seldom exceed 10:1 and is even less for smaller-diameter tools because of chip ejection problems.

*Tool Limitations*. Spade drills are not precision tools, and they should not be used for finishing operations requiring tolerances less than about ±0.25 mm (±0.010 in.). Subsequent finishing holes may be required. Multidiameter blades, however, provide close tolerances with respect to the concentricity of different diameters and radii or chamfers.

The production of holes with spade drills requires high torque and thrust forces. Thrust requirements are sometimes more than those for comparable sizes of twist drills. As a result, rigid and powerful machines are required for spade drilling.

Work surfaces that are cylindrical, spherical, or sloping and that have rough surfaces can create problems when spade drills are used, as can be the case with twist drills. Fragile workpieces are also generally difficult to drill with these tools.

*Blade Materials*. The most common cutting tool materials used in spade drill blades are types M2, M3, and M4 high-speed tool steels. Toughness is an important factor in selecting the blade material, but a compromise between toughness and abrasion resistance generally gives the best results. Premium grades of high-speed tool steel such as T15 are also extensively used.

High-speed tool steels made by powder metallurgy processes are being increasingly applied for spade drill blades. The advantages of these materials include a uniform structure with fine carbide particles and no segregation, faster and more uniform response to heat treatment, improved toughness and tool life, and superior grindability.

Spade drill blades made from cast cobalt-base alloys and carbides are also available. Both solid carbide and carbide-tip blades are used. Carbide-tip blades must be operated at high cutting speeds and with light feed rates to avoid premature wear and crushing at the chisel edge. Solid carbide

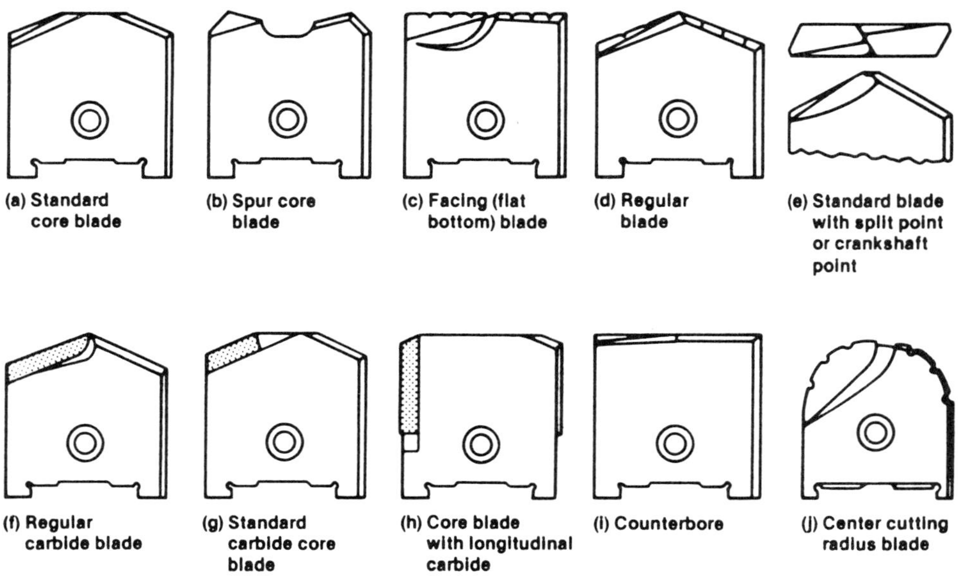


**Fig. 20** Typical spade drill blades

**Table 4 Speed and feed ranges for the spade drilling of ferrous and nonferrous materials**

| | Drill diameter, mm (in.) | | | | | | | | | |
|---|---|---|---|---|---|---|---|---|---|---|
| | 25 (1) | | 50 (2) | | 75 (3) | | 100 (4) | | 125 (5) | |
| Material drilled | Speed, rev/min | Feed rate, mm/rev (in./rev) | Speed, rev/min | Feed rate, mm/rev (in./rev) | Speed, rev/min | Feed rate, mm/rev (in./rev) | Speed, rev/min | Feed rate, mm/rev (in./rev) | Speed, rev/min | Feed rate, mm/rev (in./rev) |
| Aluminum | 1000–1500 | 0.38–0.63 (0.015–0.025) | 550–750 | 0.51–0.76 (0.020–0.030) | 350–500 | 0.76–1.02 (0.030–0.040) | 275–375 | 0.89–1.27 (0.035–0.050) | 225–300 | 1.02–1.52 (0.040–0.060) |
| Brass | | | | | | | | | | |
| Leaded | 400–700 | 0.38–0.51 (0.015–0.020) | 200–350 | 0.51–0.63 (0.020–0.025) | 125–250 | 0.51–0.76 (0.020–0.030) | 100–200 | 0.63–1.02 (0.025–0.040) | 80–150 | 1.02–1.52 (0.040–0.060) |
| Soft | 250–350 | 0.38–0.51 (0.015–0.020) | 125–175 | 0.51–0.63 (0.020–0.025) | 90–125 | 0.51–0.76 (0.020–0.030) | 70–85 | 0.63–1.02 (0.025–0.040) | 50–70 | 1.02–1.52 (0.040–0.060) |
| Hard | 200–250 | 0.30–0.38 (0.012–0.015) | 95–125 | 0.38–0.51 (0.015–0.020) | 60–80 | 0.51–0.63 (0.020–0.025) | 45–60 | 0.51–0.76 (0.020–0.030) | 37–50 | 0.76–1.27 (0.030–0.050) |
| Bronze | | | | | | | | | | |
| Soft | 300–375 | 0.38–0.51 (0.015–0.020) | 150–200 | 0.51–0.63 (0.020–0.025) | 100–125 | 0.51–0.76 (0.020–0.030) | 75–95 | 0.63–1.02 (0.025–0.040) | 60–75 | 1.02–1.52 (0.040–0.060) |
| High-strength | 225–300 | 0.20–0.30 (0.008–0.012) | 100–150 | 0.25–0.38 (0.010–0.015) | 75–95 | 0.38–0.51 (0.015–0.020) | 55–70 | 0.51–0.63 (0.020–0.025) | 45–55 | 0.76–1.02 (0.030–0.040) |
| Cast copper | 175–225 | 0.13–0.20 (0.005–0.008) | 85–125 | 0.18–0.25 (0.007–0.010) | 60–80 | 0.25–0.38 (0.010–0.015) | 40–55 | 0.38–0.51 (0.015–0.020) | 35–45 | 0.51–0.63 (0.020–0.025) |
| Cast iron | | | | | | | | | | |
| Soft | 450–550 | 0.38–0.51 (0.015–0.020) | 225–300 | 0.51–0.63 (0.020–0.025) | 150–200 | 0.51–0.76 (0.020–0.030) | 110–140 | 0.63–0.76 (0.025–0.030) | 90–120 | 0.76–1.02 (0.030–0.040) |
| Medium | 250–375 | 0.20–0.30 (0.008–0.012) | 125–200 | 0.25–0.38 (0.010–0.015) | 90–125 | 0.38–0.51 (0.015–0.020) | 65–95 | 0.51–0.63 (0.020–0.025) | 50–75 | 0.76–1.02 (0.030–0.040) |
| Hard | 150–225 | 0.13–0.20 (0.005–0.008) | 75–110 | 0.20–0.30 (0.008–0.012) | 50–75 | 0.25–0.38 (0.010–0.015) | 35–55 | 0.38–0.51 (0.015–0.020) | 30–45 | 0.51–0.76 (0.020–0.030) |
| Cast steel | 200–275 | 0.15–0.25 (0.006–0.010) | 95–125 | 0.25–0.38 (0.010–0.015) | 65–90 | 0.38–0.51 (0.015–0.020) | 45–65 | 0.51–0.63 (0.020–0.025) | 35–55 | 0.63–0.76 (0.025–0.030) |
| Stainless | | | | | | | | | | |
| Types 201–302 | 150–190 | 0.13–0.20 (0.005–0.008) | 75–95 | 0.20–0.30 (0.008–0.012) | 50–65 | 0.25–0.38 (0.010–0.015) | 35–47 | 0.38–0.51 (0.015–0.020) | 30–37 | 0.51–0.76 (0.020–0.030) |
| Types 303 and 303 Se | 225–275 | 0.15–0.25 (0.006–0.010) | 120–125 | 0.25–0.38 (0.010–0.015) | 75–90 | 0.38–0.51 (0.015–0.020) | 55–65 | 0.51–0.63 (0.020–0.025) | 45–55 | 0.63–0.76 (0.025–0.030) |
| Types 304–316 | 130–170 | 0.13–0.18 (0.005–0.007) | 65–85 | 0.18–0.25 (0.007–0.010) | 45–55 | 0.25–0.38 (0.010–0.015) | 35–45 | 0.38–0.51 (0.015–0.020) | 25–35 | 0.51–0.63 (0.020–0.025) |
| Types 316F and 316 Se | 275–350 | 0.15–0.25 (0.006–0.010) | 140–170 | 0.25–0.38 (0.010–0.015) | 95–110 | 0.38–0.51 (0.015–0.020) | 70–85 | 0.51–0.63 (0.020–0.025) | 55–70 | 0.63–0.76 (0.025–0.030) |
| Types 317–414 | 130–170 | 0.13–0.18 (0.005–0.007) | 65–85 | 0.18–0.25 (0.007–0.010) | 45–55 | 0.25–0.38 (0.010–0.015) | 35–45 | 0.38–0.51 (0.015–0.020) | 25–35 | 0.51–0.63 (0.020–0.025) |
| Types 416, 430F, 430 Se | 300–375 | 0.15–0.25 (0.006–0.010) | 150–190 | 0.25–0.38 (0.010–0.015) | 100–125 | 0.38–0.51 (0.015–0.020) | 75–95 | 0.51–0.63 (0.020–0.025) | 60–75 | 0.63–0.76 (0.025–0.030) |
| Types 420–430 | 150–225 | 0.13–0.18 (0.005–0.007) | 75–110 | 0.18–0.25 (0.007–0.010) | 50–75 | 0.25–0.38 (0.010–0.015) | 35–55 | 0.38–0.51 (0.015–0.020) | 30–45 | 0.51–0.63 (0.020–0.025) |
| Types 431–446 | 150–190 | 0.15–0.20 (0.006–0.008) | 75–95 | 0.20–0.25 (0.008–0.010) | 50–65 | 0.20–0.38 (0.008–0.015) | 35–47 | 0.25–0.51 (0.010–0.020) | 30–37 | 0.38–0.63 (0.015–0.025) |
| Steels | | | | | | | | | | |
| 1018–1095 | 190–275 | 0.13–0.20 (0.005–0.008) | 95–140 | 0.20–0.30 (0.008–0.012) | 65–95 | 0.25–0.38 (0.010–0.015) | 47–70 | 0.38–0.51 (0.015–0.020) | 37–55 | 0.51–0.76 (0.020–0.030) |
| B1112, B1113 | 300–400 | 0.15–0.25 (0.006–0.010) | 200–250 | 0.25–0.38 (0.010–0.015) | 125–160 | 0.38–0.51 (0.015–0.020) | 95–125 | 0.51–0.63 (0.020–0.025) | 75–100 | 0.63–0.76 (0.025–0.030) |
| 1117–1213 | 350–450 | 0.13–0.20 (0.005–0.008) | 170–225 | 0.20–0.30 (0.008–0.012) | 110–150 | 0.25–0.38 (0.010–0.015) | 85–110 | 0.38–0.51 (0.015–0.020) | 70–90 | 0.51–0.76 (0.020–0.030) |
| 1330–3150 | 250–290 | 0.13–0.18 (0.005–0.007) | 125–140 | 0.18–0.25 (0.007–0.010) | 80–95 | 0.25–0.38 (0.010–0.015) | 60–70 | 0.38–0.51 (0.015–0.020) | 50–55 | 0.51–0.63 (0.020–0.025) |
| 4017–4140 | 290–320 | 0.13–0.18 (0.005–0.007) | 140–160 | 0.18–0.25 (0.007–0.010) | 95–110 | 0.25–0.38 (0.010–0.015) | 70–80 | 0.38–0.51 (0.015–0.020) | 55–65 | 0.51–0.76 (0.020–0.030) |
| 4145–4640 | 225–300 | 0.13–0.18 (0.005–0.007) | 120–150 | 0.18–0.25 (0.007–0.010) | 75–100 | 0.25–0.38 (0.010–0.015) | 55–75 | 0.38–0.51 (0.015–0.020) | 45–60 | 0.51–0.76 (0.020–0.030) |
| 5120–8750 | 250–320 | 0.13–0.18 (0.005–0.007) | 125–160 | 0.18–0.25 (0.007–0.010) | 80–110 | 0.25–0.38 (0.010–0.015) | 60–80 | 0.38–0.51 (0.015–0.020) | 50–65 | 0.51–0.76 (0.020–0.030) |
| 52100 | 200–250 | 0.08–0.13 (0.003–0.005) | 100–125 | 0.13–0.20 (0.005–0.008) | 70–80 | 0.15–0.25 (0.006–0.010) | 50–60 | 0.25–0.38 (0.010–0.015) | 40–50 | 0.38–0.51 (0.015–0.020) |
| Tool steels | 110–150 | 0.10–0.15 (0.004–0.006) | 55–75 | 0.15–0.20 (0.006–0.008) | 37–50 | 0.18–0.25 (0.007–0.010) | 27–35 | 0.25–0.38 (0.010–0.015) | 22–30 | 0.38–0.51 (0.015–0.020) |
| Titanium alloys | 110–130 | 0.15–0.25 (0.006–0.010) | 55–65 | 0.25–0.38 (0.010–0.015) | 37–45 | 0.38–0.51 (0.015–0.020) | 27–33 | 0.51–0.63 (0.020–0.025) | 22–27 | 0.63–0.76 (0.025–0.030) |
| Zinc alloys | 750–900 | 0.20–0.30 (0.008–0.012) | 375–475 | 0.25–0.38 (0.010–0.015) | 250–320 | 0.30–0.43 (0.012–0.017) | 190–230 | 0.38–0.63 (0.015–0.025) | 150–190 | 0.63–0.76 (0.025–0.030) |

Source: Ref 2

blades permit faster penetration rates, provide longer life, and are often preferred for drilling very hard or abrasive materials, as well as for some soft materials (except aluminum and magnesium) and for low-carbon or low-alloy steels. Rigid setup conditions are essential for the use of carbide blades.

Spade drills can be used in any machine in which feed and speed can be controlled. Table 4 lists typical feed rates for the spade drilling cf a variety of metals and alloys.

## Special Drills for Hard Steel Applications

Tool steels and other steels with hardness to 60 HRC or higher can be drilled by a special type of drill (Fig. 18b). Blanks for

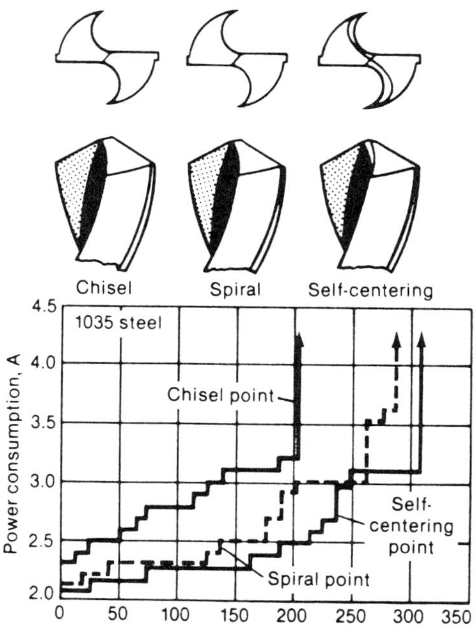


**Fig. 21** Effect of drill point on efficiency in drilling 50 mm (2 in.) deep through holes. Six drills of each type were tested. Each 12 mm (15⁄32 in.) diam drill was operated at a speed of 723 rev/min and a feed of 0.21 mm/rev (0.0083 in./rev).

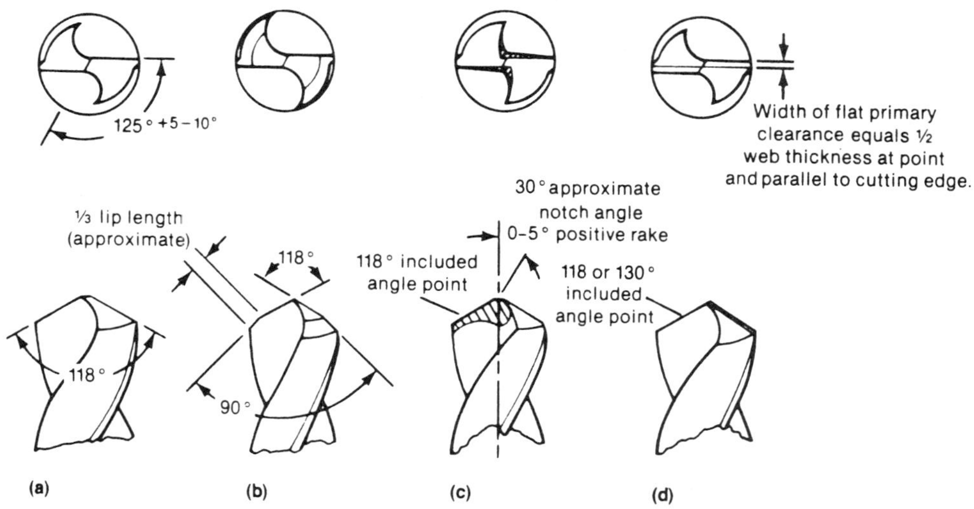


**Fig. 22** Front (top) and side views (bottom) of four types of drill point geometries used for different applications. (a) Single angle. (b) Double angle. (c) Reduced rake. (d) Four facet

these drills are cast from a heat-resistant alloy. They have a round shank (straight or tapered) and a triangular fluted section. The fluted end is ground to a three-sided-pyramid tip and is then notched to provide chip clearance.

These drills work by heating the metal beneath the drill point by friction and then wiping out the softened metal as a chip. They are used in conventional drilling machines at speeds of 30 m/min (100 sfm), for 3.2 mm (⅛ in.) diam drills, to 45 m/min (150 sfm), for 25 mm (1 in.) diam drills. These drills are sometimes used in production drilling, but their greatest use is for salvaging parts when holes were accidentally omitted until after hardening, and for making changes in hardened dies.

In drilling case-hardened parts, these drills should be used only for penetrating the case. Standard drills should be used on the softer core of case-hardened parts.

These drills are often used for drilling under water, which allows a 25% higher speed than dry drilling. Wet drilling also minimizes workpiece distortion and maintains full hardness to the edge of the drilled hole.

The special drill shown in Fig. 18(b) can also be used for reaming. A minimum of 1.6 mm (1⁄16 in.) should be left on the hole diameter for reaming, and speed should be reduced to half that used for drilling. Hole diameters can then be reamed within a total tolerance of 0.08 to 0.13 mm (0.003 to 0.005 in.). Additional information on the drilling of hardened steel is available in the section "Drilling Steel Having 48 to 55 HRC Hardness" in this article.

## Drill Point Modifications

The proper use of controlled drill pointing, which consists of adjustments in the included point angle, cutting relief angle, and clearance angle, can result in substantially lower drilling costs for a particular operation. The effect of drill point design on drill efficiency is demonstrated in the following example.

**Example 3: Chisel Versus Spiral and Self-Centering Drill Points** (Fig. 21). Three different types of points on 12 mm (15⁄32 in.) drills were evaluated for endurance, power consumption, and size retention. A tape-controlled drill press powered the drills as they cut through 50 mm (2 in.) thick 1035 steel plate. The results are shown graphically in Fig. 21. As indicated, the spiral and self-centering points were nearly equal in endurance and power consumption, and both produced better results than the chisel points.

### Material Considerations

The proper point angle (the angle between the two cutting lips of the drill) depends on the material being drilled. An included point angle of 118° is usually satisfactory for drilling all carbon and alloy steels in the annealed or normalized condition. For drilling heat-treated steels as hard as 40 HRC, or for drilling stainless steels, split-point drills are preferred, with the point angle increased to 125 to 135°. For drilling steels or other metals at 40 to 52 HRC, drills with split points of 140 to 150° are preferred.

For drilling cast iron to a depth up to three diameters, the standard 118° point is satisfactory; for deeper holes in cast iron, a point angle of 90 to 100° is preferred. Drills with point angles of 100 to 118° are generally used for drilling copper alloys and aluminum alloys. More detailed information on drill points for drilling metals other than steel is available in the Section "Machining of Specific Metals and Alloys" in this Volume.

### Drill Point Alterations on Stock Tools (Ref 2)

Because drill points form the cutting edges, their geometries are critical to tool performance. A variety of point styles are currently being used; some of the more common ones are described in this section. Proper selection, control, and use of drill points can result in substantial savings in drilling costs. Single-point angles on stock drills cannot only be increased or decreased but also modified to form double-angle points, reduced-rake points, four- or six-facet points, split points, helical points, rounded-edge points, or combined helical/rounded-edge points.

**Single-Angle Points.** Standard twist drills having conventional points with a 118° included angle are the most commonly used because they provide satisfactory results in drilling a wide variety of materials. The cutting lips on these drills are essentially straight lines, with the heel side of each land a smooth curve (Fig. 22a).

As the hardness of the workpiece material decreases, drill performance can be improved by reducing the included angle of the drill point from 118° to between 60 and 90°. Drills having these more acute point angles produce thinner chips for a given feed rate and are commonly used, with low-helix flutes, for producing holes in soft plastics and nonferrous materials. Drills having points with a 90° included angle are occasionally used for drilling soft cast irons and certain woods.

On the other hand, as the hardness of the workpiece material or the depth of hole

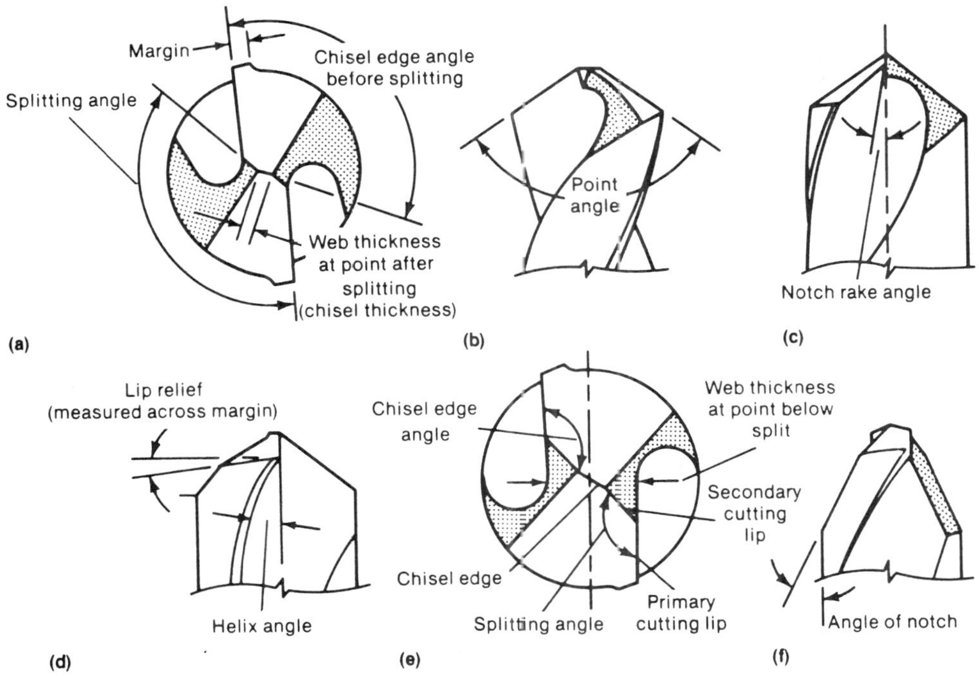


**Fig. 23** Geometry of a split-point twist drill. (a) and (e) Front views. (b), (c), (d), and (f) Side views

increases, the included angle of the drill point is increased from 118° to between 135 and 140°. These larger point angles produce thicker and narrower chips for a given feed rate. Drills with these flatter points are generally used to produce holes in harder, tougher materials, and they usually minimize burring. It is especially important to use guide bushings with drill points having higher angles because there is a tendency for the points to skid or walk on the workpiece surfaces when starting holes.

**Double-Angle Points.** Twist drills with double-angle points (Fig. 22b) are generated by first grinding a larger included angle (118 or 135°) and then a smaller included angle (typically 90°) on the corners. This provides the effect of chamfers and reduces abrasive wear on the corners.

Double-angle points were originally used in drilling medium and hard cast irons as well as other very abrasive materials to reduce corner wear on the drills. More recent applications include improving hole sizes and finishes and drilling very hard materials to reduce chipping of the corners of the lips. Twist drills with double-angle points are often used for the same applications as drills with rounded-edge (radiused lip) points (discussed later in this section).

**Reduced-Rake Points.** A common and easily applied point variation is the flatted cutting lip. Both cutting edges are flatted on their flute faces (called dubbing) from the cutting lip corners to the chisel edge, as illustrated in Fig. 22c. This type of point reduces the effective axial rake to 0 to 5° positive, causing a pushing or plowing of metal rather than a shearing action. Reduced shearing action is an effective method of preventing the tools from digging in when drilling is performed on low tensile strength materials, such as many types of brass, bronze, and some of the harder acrylic plastics (for example, Plexiglas). Reducing the rake also strengthens the cutting lips, and this type of point is often used in operations in which chipping of the lips has been a problem.

**Four- and Six-Facet Points.** The geometry of a four-facet point (Fig. 22d) is generated by grinding flat primary relief (10 to 18°) and secondary clearance angles (25 to 35°) on the end of each flute. The width of the primary relief flat is equal to one-half the web thickness, resulting in four facets on the end of the drill that subtend at a point on the drill axis and entirely remove the chisel edge. Six-facet points are produced by adding two cutting edges at the web of four-facet points.

Because these points are exactly in the middle of the drills, the tools are self-centering, and accurate, straight holes can be produced. They also require less power and thrust and permit increased feed rates. Drills with these points, however, are subject to more wear on their margins, and they cannot be modified to suit the drilling of various materials. Another disadvantage is the cost of resharpening with a special machine.

Four- and six-facet points have found their greatest application in solid-carbide drills used to produce holes in printed circuit board materials such as fiberglass-epoxy. The points can also be used on small-diameter high-speed tool steel drills that do not lend themselves to normal point-splitting techniques.

One major automotive manufacturer has doubled the feed rate over conventionally pointed carbide drills in drilling high-silicon-aluminum engine heads on a transfer line by using six-facet point drills. The six-facet drill points hold 0.05 mm (0.002 in.) or better on location without a bushing plate. According to one major gear manufacturer, the average drill life of the six-facet point is three times that of drills with a standard chisel point in a variety of materials. This longer service life is accompanied by reduced thrust, better hole-size accuracy, and increased production.

**Split Points.** This type of point, also called crankshaft point, was originally developed for use on drills designed for producing small-diameter, deep holes in automotive crankshafts. Since then it has gained widespread acceptance for drilling a wide variety of hard and soft materials. Heavy-duty types with thicker webs are used for drilling stainless steels, titanium, tough alloys, and high-temperature resistant alloys. Drills with this type of point are also extensively used for applications in which guide bushings cannot be used, as well as for portable drilling applications.

In generating split points on drills, the clearance face of each cutting edge is given a sharp (55° typical) secondary relief to the center of the chisel edge (Fig. 23), thus creating a secondary cutting lip on the opposite cutting edge. The angle between these lip segments acts as a chip breaker when drilling is done on many materials; this produces smaller chips, which are readily ejected through the flutes. More important, however, the additional cutting edges produced and the reduction in width of the original chisel edge reduce thrust requirements (typically 25 to 30% compared to conventional 118° points) and improve the centering capability. A disadvantage is the need for a point-splitting grinding machine.

**Helical (Spiral) Points.** This type of point is generated by reducing the drill point from a chisel edge to a helical (spiral) point, as illustrated in Fig. 24. This produces an S-shaped chisel with a radiused crown effect that has its highest point at the center of the drill axis. This S-shaped chisel creates a continuous edge extending from margin to margin across the web.

The advantages of drills with a helical point include a self-centering capability and some reduction of thrust. Their use also results in better hole geometry and improved hole size.

A possible disadvantage of this type of point is that burrs are sometimes produced at hole breakthrough. In addition, the S-shaped chisel is weaker than straight chisel points, resulting in faster dulling when hard materials are drilled. Special machines are required for grinding these points.

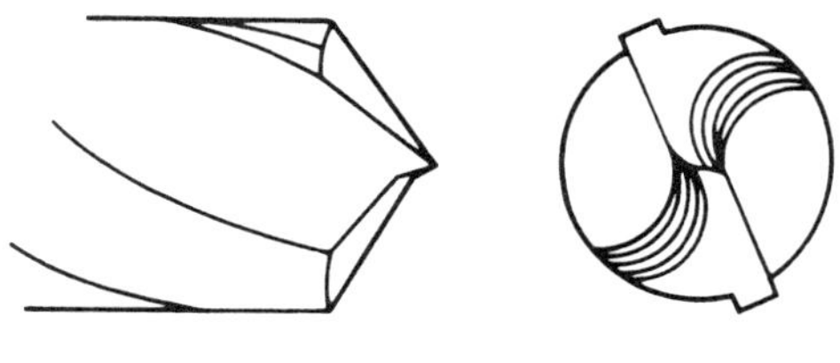

(a) (b)

**Fig. 24** Helical (spiral) drill point that consists of an S-shaped point rather than a straight-line chisel edge. (a) Side view. (b) Front view

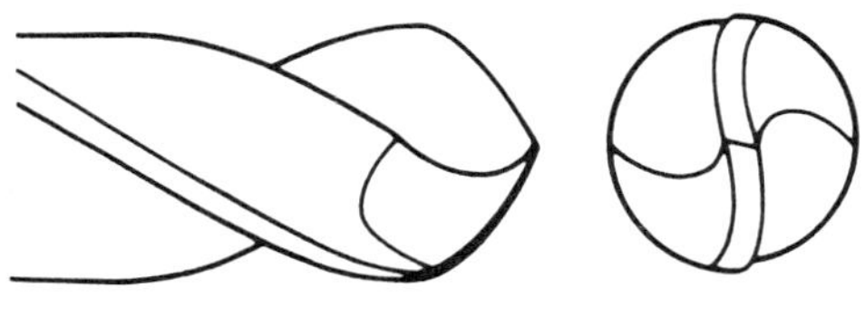

(a) (b)

**Fig. 25** Rounded-edge drill point that contains lips and margins blended by smooth curves. (a) Side view. (b) Front view

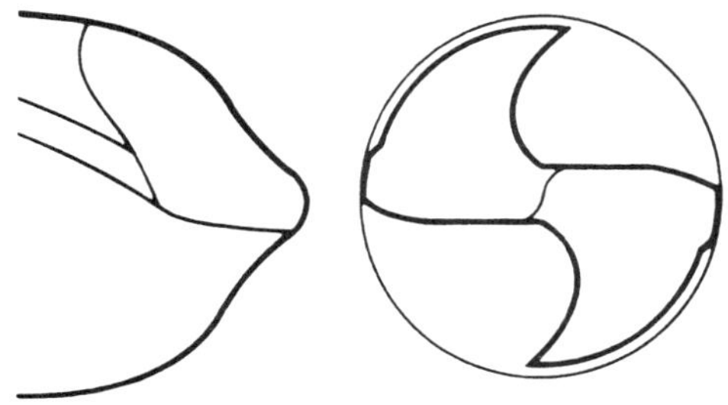

(a) (b)

**Fig. 26** Twist drill point that combines the features of both helical and rounded-edge points. (a) Side view. (b) Front view

**Rounded-edge (radiused-lip) points** are generated by grinding a blended, rounded edge (radiused corner or lip) on conventional points (Fig. 25). Points such as these provide a continuously varying point angle, with the lips and margins blended by a smooth curve. Because the drill (also known as Racon or radiused conventional point) cuts on long, curved lips, there is less load per unit area and less heat generated. Elimination of the corner reduces margin wear. Breakthrough burrs are eliminated, and tool life can be lengthened by a factor of eight to ten times compared to conventionally pointed drills when cast iron is drilled. Feed rates can also be increased because of the improved heat dissipation.

Twist drills with rounded-edge points are used when drill life is most important. Drills with these points are not self-centering and are best applied where guide bushings are used. When they are used on NC machines, prior center drilling is required. The time required for center drilling, however, may be more than offset by longer tool life. A possible limitation is that special grinding machines are required for producing these points. In addition, when steel is drilled, these points cut closer to size, which may reduce drill life compared to that possible with conventional points because of greater corner and margin wear.

**Combined Helical/Rounded-Edge Points.** Drill grinding machines are available that combine the features of both the helical and rounded-edge points. The point produced (Fig. 26) provides the self-centering capability of helical points and the long life, burr-free breakthrough, and higher feed capacity of rounded-edge points. These features make the drills capable of producing accurate holes on NC machines without the need for prior center drilling.

## Radical Drill Point Modification

Figure 27 shows the extensive reworking done on a 17.9 mm (45⁄64 in.) drill to convert it into an end-mill type tool containing a pilot, as discussed in the following example.

**Example 4: Forming Sharp Corners in Counterbores.** The two sharp-cornered counterbores in the turbine-governor case shown at the left in Fig. 27 were produced with a specially ground piloted drill. This tool (at right, Fig. 27) resembled a two-lip flat-bottom end mill and had a primary relief of 8 to 10° and a secondary relief of 30 to 40°.

The complex hole in this part was produced by the following procedure:

- The part was drilled through with a 7.9 mm (5⁄16 in.) diam standard twist drill
- The through hole was enlarged, for about 41.3 mm (1⅝ in.) from each end, with a 17.9 mm (45⁄64 in.) diam standard two-lip taper-shank twist drill
- The specially ground drill (Fig. 27) removed the excess stock from the 17.9 mm (45⁄64 in.) diam counterbores and formed the sharp corners
- The counterbores were reamed to 18.29/18.19 mm (0.720/0.716 in.) in diameter

These parts were drilled in a radial drill press, using a box jig and bushings to ensure alignment of the cutting tools with the spindles. Processing details are given with Fig. 27.

## Dimensional Accuracy of Holes

In most drilling operations, the drill is expected to cut oversize. Because a standard twist drill is primarily a roughing tool, some variation in hole size is expected even under the best conditions. When great accuracy is required (for example, meeting tolerances of less than 0.013 mm, or 0.0005 in., on a 13 mm, or ½ in., diam hole), it is usually necessary to drill undersize and then ream to size or to use a special drill or technique; sometimes both approaches are used.

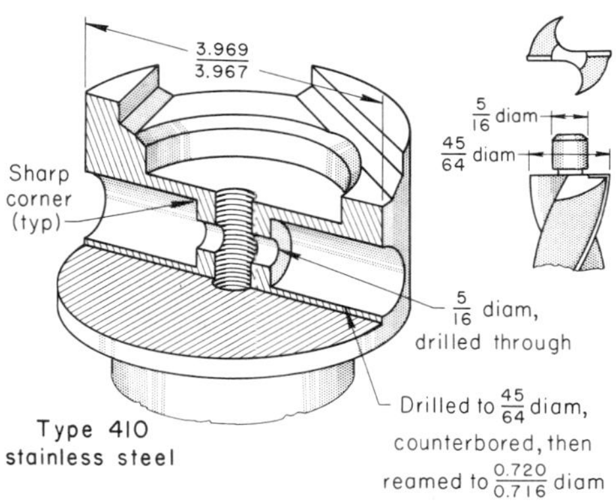


| | |
|---|---|
| Speed, at 428 rev/min, m/min (sfm) | 25 (81) |
| Feed, mm/rev (in./rev) | 0.13 (0.005) |
| Cutting fluid | Soluble oil:water (1:20) |
| Production rate, min/piece | 4.2 |
| Drill life per grind (average), pieces | 80 |
| Workpiece hardness, HB | 156 |

**Fig. 27** Sharp-cornered counterbores that were produced by piloted drill ground as shown at right. Dimensions in figure given in inches

Misalignment occasionally causes a drill to cut oversize, but more often, imprecise sharpening is responsible. If the drill is sharpened with unequal angles (Fig. 28a), the lip with the greater angle does most of the work, and the reaction forces the opposite margin into the wall of the hole. If the drill is sharpened with unequal lip lengths (Fig. 28b), the two lips balance so that their reactions are equal, but the drill sweeps with one margin rubbing against the hole.

Sometimes both conditions exist (Fig. 28c), resulting in holes that vary widely in size and shape. This dual error is a typical result of sharpening drills by hand. Sharpening drills by machine is recommended. If a machine-sharpened drill does not cut within reasonable tolerance, the web may be off-center, or the flute spacing may be inaccurate. Small variations in web concentricity and flute spacing seldom have a significant effect on accuracy.

The amount of oversize cutting that can be expected from machine-sharpened drills has been determined from data compiled by several drill manufacturers. The results of drilling steel or cast iron with drills of six different diameters are shown graphically in Fig. 29. These results were obtained by drilling 468 holes with each size of drill.

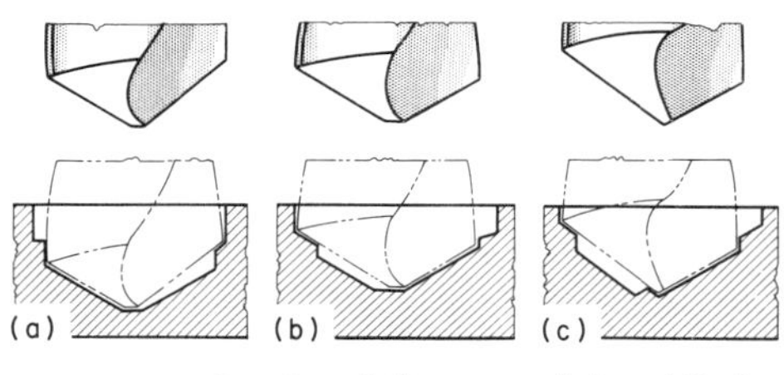


**Fig. 28** Oversize drilling caused by drill lips ground to unequal angles (a), unequal lengths (b), and unequal angles and lengths (c)

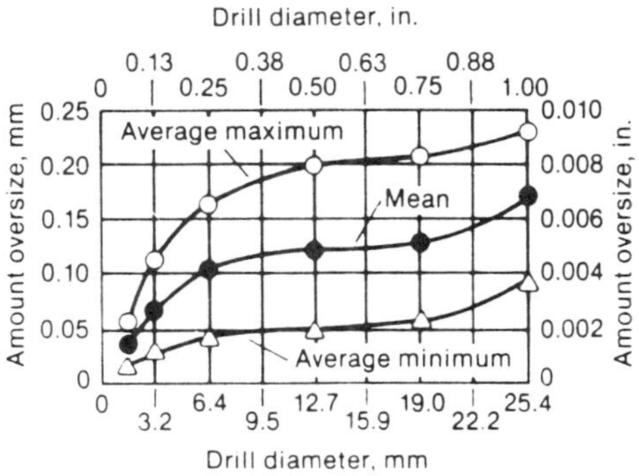


**Fig. 29** Hole oversize in drilling steel or cast iron with machine-sharpened drills. Data are based on drilling 468 holes with drills of each of six different diameters in a series of tests conducted by several drill manufacturers. Drills were used without bushings.

Although this drilling was done with normal good practice, no special equipment (such as bushings) or special technique was used.

As indicated in Fig. 29, for holes larger than 19 mm (3/4 in.) in diameter, oversize increases rapidly with diameter. For close tolerance and uniformity, these holes must always be reamed.

When hole tolerances are closer than those shown in Fig. 29, it may be necessary to do one or more of the following:

- Modify the drill design, or use a different type of drill
- Predrill with a center drill
- Drill in steps
- Use bushings
- Rotate the work instead of the drill (this may require the use of a different drilling machine)
- Change to special drills, such as oil-hole or gun drills
- Use a precision drill collet or chuck to minimize drill runout

Table 5 summarizes the expected accuracy obtained in drilling on the basis of three size ranges and three drilling conditions.

**Drill Design.** Many holes can be drilled with more than one type of drill, but discriminating selection may allow closer tolerances. For example, if a screw-machine drill can be used, somewhat closer tolerances can be met than with a longer drill, because of the decrease in drill overhang. In some applications, a screw-machine drill is used instead of a standard-length drill, thus eliminating the need for bushings. Modification of the drill point often results in greater accuracy.

**Predrilling** with a center drill often improves accuracy because the final drill has no chance to walk at the start. Walking causes out-of-line drilling, which results in holes that are less accurate in size and roundness.

**Drilling in steps** can improve accuracy, although with some types of equipment it can also reduce productivity. For example, to drill a 75 mm (3 in.) deep, 19 mm (3/4 in.) diam hole, it may be advantageous first to drill a 25 mm (1 in.) deep hole with an 18.6 mm (47/64 in.) diam twist drill and then switch to an 18.2 mm (23/32 in.) diam twist drill to drill a 50 mm (2 in.) deep hole. This is followed by a 17.9 mm (45/64 in.) diam twist drill to reach the final 75 mm (3 in.) depth desired. Finally, a 19 mm (3/4 in.) diam four-flute core drill is used as a roughing reamer to obtain an accurate hole.

**Drill bushings** in fixtures ensure uniformly centered starts, which result in better alignment, and thus improve hole roundness, straightness, and accuracy of location.

**Rotating the workpiece** (provided its size and shape permit) will invariably result in more accurate drilling than rotating the drill while the work remains stationary. Machines in which the work is rotated include lathes, turret lathes, and horizontal multiple-station drilling machines. A turret lathe also makes it possible to drill a hole in steps, completing it with a core drill, and then to ream it for still greater accuracy.

## Speed and Feed

Optimum speed and feed for drilling depend on workpiece material, tool material, depth of hole, design of drill, rigidity of setup, tolerance, and cutting fluid. Consequently, it is impossible to recommend speeds and feeds that are applicable under all conditions. The nominal speeds and feeds given in Table 6 are useful as starting points in selecting an optimum combination for a specific job.

When unfavorable conditions prevail, such as less-than-normal rigidity or restrictions on the use of cutting fluid, slower speeds and lower feed rates than those given in Table 6 may be required. Oversize drilling (Fig. 28 and 29), increases as speed increases. Therefore, drill grinding practice, drilling rigidity, and drill design are closely related to the accuracy that can be obtained at a given speed. Feed rate must not exceed that at which chips can be flushed away. Clogging of chips decreases accuracy and eventually leads to drill breakage.

**Gun drills** ordinarily have carbide tips, which can withstand greater speeds. The usual practice is to operate them at considerably higher speeds but lower feeds than those for high-speed tool steel twist drills. This allows gun drills to form thin chips that are more readily flushed away by cutting fluid under pressure. Nominal speeds and feeds for the gun drilling of ferrous materials with carbide-tip drills are given in Table 7.

## Cutting Fluids

Cutting fluids serve the same purposes in drilling as in other metal cutting operations: to cool workpieces and tools, to flush away chips, and to minimize adherence of tool and work metal.

It is suggested that lubrication problems be referred to a reputable manufacturer of cutting oils. The following list of lubricants should be used as suggestions only:

- *Aluminum and aluminum alloys*: Kerosene, kerosene and lard oil, and soluble oil
- *Brass and bronze*: Dry. Deep holes: kerosene and mineral oil, lard oil, and soluble oil
- *Magnesium and magnesium alloys*: Mineral lard oil, kerosene, or dry
- *Copper*: Mineral lard oil and kerosene, soluble oil, or dry
- *Monel metal*: Mineral lard oil
- *Low-carbon steels*: Mineral lard oil
- *Tough alloy steels*: Sulfurized oil
- *Steel forgings*: Sulfurized oil
- *Cast steel*: Soluble oil
- *Wrought iron*: Soluble oil
- *High-tensile steels*: Soluble oil
- *Manganese steel*: Dry
- *Cast iron*: Mineral oil
- *Malleable iron*: Soluble oil or dry
- *Stainless steel*: Soluble oil
- *Titanium alloys*: Soluble oil
- *Tool steel*: Mineral lard oil
- *Abrasives, plastics*: Dry
- *Fiber, asbestos, wood*: Dry
- *Hard rubber*: Dry

Detailed information on types of cutting fluids and principles of selection and use are described in the article "Metal Cutting and Grinding Fluids" in this Volume.

Cutting fluids are used in most drilling applications, except on such materials as

**Table 5 Average accuracy of holes produced with twist drills**

| Drilling condition | Hole diameter, mm (in.): 3.2–6.4 (1/8–1/4) Oversize, mm (in.) | 3.2–6.4 (1/8–1/4) Location, mm (in.) | 6.4–19 (1/4–3/4) Oversize, mm (in.) | 6.4–19 (1/4–3/4) Location, mm (in.) | 19–38 (3/4–1 1/2) Oversize, mm (in.) | 19–38 (3/4–1 1/2) Location, mm (in.) |
|---|---|---|---|---|---|---|
| No center-drilled hole or bushing | 0.08 (0.003) | ±0.18 (±0.007) | 0.15 (0.006) | ±0.20 (±0.008) | 0.20 (0.008) | ±0.23 (±0.009) |
| Center-drilled hole, no bushing | 0.08 (0.003) | ±0.10 (±0.004) | 0.08 (0.003) | ±0.10 (±0.004) | 0.10 (0.004) | ±0.13 (±0.005) |
| With drill bushing | 0.05 (0.002) | ±0.05 (±0.002) | 0.08 (0.003) | ±0.05 (±0.002) | 0.10 (0.004) | ±0.08 (±0.003) |

Source: Ref 2

**Table 6(a) Recommended operating parameters for producing holes in a variety of materials with twist drills**

| Material drilled | Hardness, HB | Hardness, HR | Cutting tool material(a) | Peripheral speed, m/min | Peripheral speed, (sfm) | Feed rate(b) | Helix angle, degrees | Point angle, degrees | Point style |
|---|---|---|---|---|---|---|---|---|---|
| Aluminum and aluminum alloys | 45–105 | to 62 HRB | HSS | 107 | (350) | Z | 32–42 | 90–118 | . . . |
| Asbestos | . . . | . . . | WC | 17 | (55) | Y | 17–20 | 90 | . . . |
| Bakelite | . . . | . . . | WC | 24 | (80) | Y | 30–40 | 90–118 | . . . |
| Carbon | . . . | . . . | HSS | 18–21 | (60–70) | W | 25–35 | 90–118 | . . . |
| Copper and copper alloys | | | | | | | | | |
| High machinability | to 124 | 10–70 HRB | HSS | 61 | (200) | Z | 15–40 | 118 | . . . |
| Low machinability | to 124 | 10–70 HRB | HSS | 21 | (70) | Z | 0–25 | 118 | . . . |
| Fiberglass-epoxy | . . . | . . . | WC | 198 | (650) | 0.063–0.127 mm (0.0025–0.0050 in.) | 35–40 | 118–130 | Four-facet |
| Glass | . . . | . . . | WC | 4.6–7.6 | (15–25) | Light hand | 0 | . . . | Spear |
| High-temperature alloys | | | | | | | | | |
| Cobalt-base | 180–230 | 89–99 HRB | HSS-Co | 6.1 | (20) | W | 28–35 | 118–135 | Split |
| Iron-base | 180–230 | 89–99 HRB | HSS-Co | 7.6 | (25) | X | 28–35 | 118–135 | Split |
| Nickel-base | 150–300 | to 32 HRC | HSS-Co | 6.1 | (20) | W | 28–35 | 118–135 | Split |
| Iron | | | | | | | | | |
| Cast (soft) | 120–150 | to 80 HRB | HSS | 43–46 | (140–150) | Z | 20–30 | 90–118 | . . . |
| | | | WC | 27–50 | (90–165) | Y | 14–25 | 90–118 | . . . |
| Cast (medium hard) | 160–220 | 80–97 HRB | HSS | 24–34 | (80–110) | Y | 20–30 | 90–118 | . . . |
| | | | WC | 27–50 | (90–165) | X | 14–25 | 90–118 | . . . |
| Hard chilled | 400 | 41 HRC | WC | 9 | (30) | X | 0–25 | 130–140 | Notched |
| Malleable | 112–126 | to 71 HRB | HSS | 27–37 | (90–120) | Y | 20–30 | 90–118 | . . . |
| | | | WC | 30–46 | (100–150) | X | 14–25 | 118 | . . . |
| Ductile | 190–225 | to 98 HRB | HSS | 18 | (60) | Y | 20–30 | 118 | . . . |
| | | | WC | 24–30 | (80–100) | X | 14–25 | 118 | . . . |
| Magnesium and magnesium alloys | 50–90 | to 52 HRB | HSS | 46–122 | (150–400) | Z | 25–35 | 118 | . . . |
| Marble | . . . | . . . | WC | 4.6–7.6 | (15–25) | Light hand | 0–10 | 90–130 | . . . |
| Plastics | . . . | . . . | HSS | 30 | (100) | Y | 15–25 | 118 | . . . |
| | | | WC | 30–61 | (100–200) | X | 15–25 | 118 | . . . |
| Rubber (hard) | . . . | . . . | HSS | 30–91 | (100–300) | X | 10–20 | 90–118 | . . . |
| | | | WC | 61–91 | (200–300) | W | 15–25 | 118 | . . . |
| Steel | | | | | | | | | |
| Plain carbon | | | | | | | | | |
| to 0.25 C | 125–175 | 71–88 HRB | HSS | 24 | (80) | Y | 25–35 | 118 | . . . |
| to 0.50 C | 175–225 | 88–98 HRB | HSS | 20 | (65) | Y | 25–35 | 118 | . . . |
| to 0.90 C | 175–225 | 88–98 HRB | HSS | 17 | (55) | Y | 25–35 | 118 | . . . |
| Alloy steel | | | | | | | | | |
| Low carbon (0.12–0.25) | 175–225 | 88–98 HRB | HSS | 21 | (70) | Z | 25–35 | 118 | . . . |
| Medium carbon (0.30–0.65) | 175–225 | 88–98 HRB | HSS | 15–18 | (50–60) | X | 25–35 | 118 | . . . |
| | 488+ | 50+ HRC | WC | 23–30 | (75–100) | 0.013–0.038 mm (0.0005–0.0015 in.) | 25–35 | 135 | Notched |
| Maraging | 275–325 | 28–35 HRC | HSS | 17 | (55) | Y | 25–32 | 118–135 | Split |
| Stainless steel | | | | | | | | | |
| Austenitic | 135–185 | 75–90 HRB | HSS-Co | 17 | (55) | X | 25–35 | 118–135 | Split |
| Ferritic | 135–185 | 75–90 HRB | HSS | 20 | (65) | X | 25–35 | 118–135 | Split |
| Martensitic | 135–175 | 75–88 HRB | HSS-Co | 20 | (65) | Z | 25–35 | 118–135 | Split |
| Precipitation-hardened | 150–200 | 82–94 HRB | HSS-Co | 15 | (50) | X | 25–35 | 118–135 | Split |
| Tool | 196 | 94 HRB | HSS | 18 | (60) | Y | 25–35 | 118 | . . . |
| | 241 | 24 HRC | HSS | 15 | (50) | Y | 25–35 | 118 | . . . |
| Titanium | | | | | | | | | |
| Pure | 110–200 | to 94 HRB | HSS | 30 | (100) | X | 30–38 | 135 | Split |
| α and α-β | 300–360 | 31–39 HRC | HSS-Co | 12 | (40) | Y | 30–38 | 135 | Split |
| β | 275–350 | 29–38 HRC | HSS-Co | 7.6 | (25) | W | 30–38 | 135 | Split |
| Zinc alloys | 80–100 | 41–62 HRB | HSS | 76 | (250) | Z | 32–42 | 118 | . . . |

(a) HSS, high-speed tool steel; HSS-Co, high-speed tool steel with cobalt; WC, tungsten carbide. (b) See Table 6(b). Source: Ref 2

**Table 6(b) Feed rates for materials listed in Table 6(a)**

| Code | Feed rate, mm/rev (in./rev), at a drill diameter of: 3.2 mm (1/8 in.) | 6.4 mm (1/4 in.) | 12.7 mm (1/2 in.) | 19.1 mm (3/4 in.) | 25.4 mm (1 in.) |
|---|---|---|---|---|---|
| W | 0.038 (0.0015) | 0.08 (0.003) | 0.089 (0.0035) | 0.114 (0.0045) | 0.13 (0.005) |
| X | 0.05 (0.002) | 0.089 (0.0035) | 0.15 (0.006) | 0.216 (0.0085) | 0.267 (0.0105) |
| Y | 0.08 (0.003) | 0.13 (0.005) | 0.20 (0.008) | 0.267 (0.0105) | 0.317 (0.0125) |
| Z | 0.08 (0.003) | 0.15 (0.006) | 0.25 (0.010) | 0.394 (0.0155) | 0.483 (0.0190) |

cast iron (for which an air jet may be used instead) or when the use of fluids is incompatible with subsequent operations or the end use of the part. The use of sulfurized cutting oils is almost mandatory for gun drilling operations because of the more accurate dimensions and smoother finishes that are usually required.

## Variables Affecting Drill Life

Apart from drill design and material, and rigidity of the setup, the variables that most affect the service lives of drills are speed and feed and the hardness and composition of the work metal.

**Speed and Feed.** Figure 30 relates speed and feed to drill life. The curves shown in Fig. 30(a) for drilling 4130 and 4340 steels at 341 HB (37 HRC) indicate that:

- Drill life decreased rapidly as cutting speed was increased
- At a given cutting speed, drill life was shortened when the feed rate was increased from 0.05 to 0.13 mm/rev (0.002 to 0.005 in./rev)

Tests on 4340 steel at 514 HB (52 HRC), as shown in the bar graphs of Fig. 30(b), (c), and (d), indicate trends somewhat different from those observed for the softer steels.

## Table 7 Recommended starting conditions for gun drilling

| Gun drill diameter mm | in. | Coolant pressure kPa | psi | Low- and medium-carbon steels to 200 HB rev/min | mm/min | in./min | mm | in.(a) | Alloy steel to 240 HB rev/min | mm/min | in./min | mm | in.(a) | Tool steel to 200 HB rev/min | mm/min | in./min | mm | in.(a) |
|---|---|---|---|---|---|---|---|---|---|---|---|---|---|---|---|---|---|---|
| 1.98 | 0.0781.......... | 10 342 | 1500 | 10 000 | 36 | 1.4 | 114 | 4.5 | 10 000 | 41 | 1.6 | 114 | 4.5 | 8560 | 33 | 1.3 | 132 | 5.2 |
| 2.38 | 0.0938.......... | 10 342 | 1500 | 10 000 | 46 | 1.8 | 127 | 5.0 | 10 000 | 51 | 2.0 | 127 | 5.0 | 7135 | 36 | 1.4 | 150 | 5.9 |
| 3.18 | 0.1250.......... | 10 342 | 1500 | 10 000 | 61 | 2.4 | 147 | 5.8 | 9 165 | 64 | 2.5 | 155 | 6.1 | 5345 | 38 | 1.5 | 203 | 8.0 |
| 3.97 | 0.1562.......... | 8 963 | 1300 | 10 000 | 76 | 3.0 | 165 | 6.5 | 7 335 | 64 | 2.5 | 193 | 7.6 | 4280 | 38 | 1.5 | 254 | 10.0 |
| 4.76 | 0.1875.......... | 7 929 | 1150 | 9 170 | 81 | 3.2 | 190 | 7.5 | 6 110 | 64 | 2.5 | 234 | 9.2 | 3565 | 36 | 1.4 | 305 | 12.0 |
| 5.55 | 0.2188.......... | 7 240 | 1050 | 7 860 | 79 | 3.1 | 218 | 8.6 | 5 240 | 61 | 2.4 | 269 | 10.6 | 3055 | 36 | 1.4 | 353 | 13.9 |
| 6.35 | 0.2500.......... | 6 378 | 925 | 6 875 | 79 | 3.1 | 254 | 10.0 | 4 585 | 61 | 2.4 | 312 | 12.3 | 2675 | 36 | 1.4 | 409 | 16.1 |
| 7.14 | 0.2812.......... | 5 861 | 850 | 6 110 | 76 | 3.0 | 287 | 11.3 | 4 075 | 58 | 2.3 | 351 | 13.8 | 2375 | 33 | 1.3 | 460 | 18.1 |
| 7.94 | 0.3125.......... | 5 344 | 775 | 5 500 | 74 | 2.9 | 320 | 12.6 | 3 665 | 56 | 2.2 | 391 | 15.4 | 2140 | 33 | 1.3 | 511 | 20.1 |
| 8.73 | 0.3438.......... | 4 999 | 725 | 5 000 | 71 | 2.8 | 351 | 13.8 | 3 335 | 56 | 2.2 | 429 | 16.9 | 1945 | 33 | 1.3 | 564 | 22.2 |
| 9.53 | 0.3750.......... | 4 654 | 675 | 4 585 | 71 | 2.8 | 384 | 15.1 | 3 055 | 53 | 2.1 | 470 | 18.5 | 1780 | 30 | 1.2 | 615 | 24.2 |
| 10.32 | 0.4062.......... | 4 309 | 625 | 4 230 | 69 | 2.7 | 417 | 16.4 | 2 820 | 53 | 2.1 | 508 | 20.0 | 1645 | 30 | 1.2 | 665 | 26.2 |
| 11.11 | 0.4375.......... | 4 137 | 600 | 3 930 | 66 | 2.6 | 447 | 17.6 | 2 620 | 51 | 2.0 | 549 | 21.6 | 1530 | 30 | 1.2 | 716 | 28.2 |
| 11.91 | 0.4688.......... | 3 792 | 550 | 3 670 | 66 | 2.6 | 472 | 18.6 | 2 445 | 51 | 2.0 | 587 | 23.1 | 1425 | 30 | 1.2 | 770 | 30.3 |
| 12.70 | 0.5000.......... | 3 620 | 525 | 3 440 | 64 | 2.5 | 511 | 20.1 | 2 290 | 48 | 1.9 | 627 | 24.7 | 1335 | 28 | 1.1 | 820 | 32.3 |
| 13.49 | 0.5312.......... | 3 447 | 500 | 3 235 | 64 | 2.5 | 544 | 21.4 | 2 155 | 48 | 1.9 | 665 | 26.2 | 1260 | 28 | 1.1 | 871 | 34.3 |
| 14.29 | 0.5625.......... | 3 447 | 500 | 3 055 | 61 | 2.4 | 577 | 22.7 | 2 035 | 48 | 1.9 | 706 | 27.8 | 1190 | 28 | 1.1 | 922 | 36.3 |
| 15.08 | 0.5938.......... | 3 275 | 475 | 2 895 | 61 | 2.4 | 512 | 24.1 | 1 930 | 46 | 1.8 | 749 | 29.5 | 1125 | 28 | 1.1 | 980 | 38.6 |
| 15.88 | 0.6250.......... | 3 103 | 450 | 2 750 | 58 | 2.3 | 543 | 25.3 | 1 835 | 46 | 1.8 | 787 | 31.0 | 1070 | 25 | 1.0 | 1031 | 40.6 |
| 16.67 | 0.6562.......... | 2 930 | 450 | 2 620 | 58 | 2.3 | 576 | 26.6 | 1 745 | 46 | 1.8 | 828 | 32.6 | 1020 | 25 | 1.0 | 1082 | 42.6 |
| 17.46 | 0.6875.......... | 2 930 | 425 | 2 500 | 56 | 2.2 | 709 | 27.9 | 1 665 | 43 | 1.7 | 866 | 34.1 | 970 | 25 | 1.0 | 1135 | 44.7 |
| 18.25 | 0.7188.......... | 2 758 | 400 | 2 390 | 56 | 2.2 | 734 | 28.9 | 1 595 | 43 | 1.7 | 899 | 35.4 | 930 | 25 | 1.0 | 1179 | 46.4 |
| 19.05 | 0.7500.......... | 2 758 | 400 | 2 290 | 56 | 2.2 | 772 | 30.4 | 1 530 | 43 | 1.7 | 945 | 37.2 | 890 | 25 | 1.0 | 1240 | 48.8 |

| Gun drill diameter mm | in. | Coolant pressure kPa | psi | Type 300 stainless steel to 200 HB rev/min | mm/min | in./min | mm | in.(a) | Type 400 stainless steel to 200 HB rev/min | mm/min | in./min | mm | in.(a) | Bronze rev/min | mm/min | in./min | mm | in.(a) |
|---|---|---|---|---|---|---|---|---|---|---|---|---|---|---|---|---|---|---|
| 1.98 | 0.0781.......... | 10 342 | 1500 | 9 780 | 30 | 1.2 | 117 | 4.6 | 10 000 | 36 | 1.4 | 114 | 4.5 | 10 000 | 41 | 1.6 | 114 | 4.5 |
| 2.38 | 0.0938.......... | 10 342 | 1500 | 8 150 | 33 | 1.3 | 140 | 5.5 | 10 000 | 46 | 1.8 | 127 | 5.0 | 10 000 | 51 | 2.0 | 127 | 5.0 |
| 3.18 | 0.1250.......... | 10 342 | 1500 | 6 110 | 33 | 1.3 | 188 | 7.4 | 7 640 | 46 | 1.8 | 170 | 6.7 | 10 000 | 71 | 2.8 | 147 | 5.8 |
| 3.97 | 0.1562.......... | 8 963 | 1300 | 4 890 | 33 | 1.3 | 236 | 9.3 | 6 115 | 46 | 1.8 | 211 | 8.3 | 10 000 | 86 | 3.4 | 165 | 6.5 |
| 4.76 | 0.1875.......... | 7 929 | 1150 | 4 075 | 33 | 1.3 | 284 | 11.2 | 5 090 | 46 | 1.8 | 257 | 10.1 | 9 170 | 94 | 3.7 | 190 | 7.5 |
| 5.55 | 0.2188.......... | 7 240 | 1050 | 3 490 | 30 | 1.2 | 330 | 13.0 | 4 360 | 43 | 1.7 | 295 | 11.6 | 7 860 | 91 | 3.6 | 218 | 8.6 |
| 6.35 | 0.2500.......... | 6 378 | 925 | 3 055 | 30 | 1.2 | 381 | 15.0 | 3 820 | 43 | 1.7 | 340 | 13.4 | 6 875 | 89 | 3.5 | 254 | 10.0 |
| 7.14 | 0.2812.......... | 5 861 | 850 | 2 715 | 28 | 1.1 | 429 | 16.9 | 3 395 | 43 | 1.7 | 384 | 15.1 | 6 110 | 89 | 3.5 | 287 | 11.3 |
| 7.94 | 0.3125.......... | 5 344 | 775 | 2 445 | 28 | 1.1 | 472 | 18.8 | 3 055 | 41 | 1.6 | 427 | 16.8 | 5 500 | 86 | 3.4 | 320 | 12.6 |
| 8.73 | 0.3438.......... | 4 999 | 725 | 2 220 | 28 | 1.1 | 526 | 20.7 | 2 780 | 41 | 1.6 | 470 | 18.5 | 5 000 | 84 | 3.3 | 351 | 13.8 |
| 9.53 | 0.3750.......... | 4 654 | 675 | 2 040 | 28 | 1.1 | 574 | 22.6 | 2 545 | 38 | 1.5 | 513 | 20.2 | 4 585 | 81 | 3.2 | 384 | 15.1 |
| 10.32 | 0.4062.......... | 4 309 | 625 | 1 880 | 25 | 1.0 | 622 | 24.5 | 2 350 | 38 | 1.5 | 556 | 21.9 | 4 230 | 81 | 3.2 | 417 | 16.4 |
| 11.11 | 0.4375.......... | 4 137 | 600 | 1 745 | 25 | 1.0 | 671 | 26.4 | 2 180 | 38 | 1.5 | 602 | 23.7 | 3 930 | 76 | 3.0 | 447 | 17.6 |
| 11.91 | 0.4688.......... | 3 792 | 550 | 1 630 | 25 | 1.0 | 719 | 28.3 | 2 035 | 36 | 1.4 | 643 | 25.3 | 3 670 | 76 | 3.0 | 480 | 18.9 |
| 12.70 | 0.5000.......... | 3 620 | 525 | 1 530 | 25 | 1.0 | 767 | 30.2 | 1 910 | 36 | 1.4 | 686 | 27.0 | 3 440 | 74 | 2.9 | 511 | 20.1 |
| 13.49 | 0.5312.......... | 3 447 | 500 | 1 440 | 23 | 0.9 | 815 | 32.1 | 1 800 | 36 | 1.4 | 729 | 28.7 | 3 235 | 74 | 2.9 | 544 | 21.4 |
| 14.29 | 0.5625.......... | 3 447 | 500 | 1 360 | 23 | 0.9 | 864 | 34.0 | 1 700 | 33 | 1.3 | 772 | 30.4 | 3 055 | 71 | 2.8 | 577 | 22.7 |
| 15.08 | 0.5938.......... | 3 275 | 475 | 1 290 | 23 | 0.9 | 914 | 36.0 | 1 610 | 33 | 1.3 | 820 | 32.3 | 2 895 | 71 | 2.8 | 612 | 24.1 |
| 15.88 | 0.6250.......... | 3 103 | 450 | 1 220 | 23 | 0.9 | 965 | 38.0 | 1 530 | 33 | 1.3 | 861 | 33.9 | 2 750 | 69 | 2.7 | 643 | 25.3 |
| 16.67 | 0.6562.......... | 2 930 | 450 | 1 165 | 23 | 0.9 | 1013 | 39.9 | 1 455 | 33 | 1.3 | 907 | 35.7 | 2 620 | 66 | 2.6 | 676 | 26.6 |
| 17.46 | 0.6875.......... | 2 930 | 425 | 1 110 | 23 | 0.9 | 1062 | 41.8 | 1 390 | 33 | 1.3 | 950 | 37.4 | 2 500 | 66 | 2.6 | 709 | 27.9 |
| 18.25 | 0.7188.......... | 2 758 | 400 | 1 060 | 20 | 0.8 | 1102 | 43.4 | 1 330 | 30 | 1.2 | 986 | 38.8 | 2 390 | 66 | 2.6 | 734 | 28.9 |
| 19.05 | 0.7500.......... | 2 758 | 400 | 1 020 | 20 | 0.8 | 1156 | 45.5 | 1 275 | 30 | 1.2 | 1034 | 40.7 | 2 290 | 66 | 2.6 | 772 | 30.4 |

| Gun drill diameter mm | in. | Coolant pressure kPa | psi | A uminum rev/min | mm/min | in./min | mm | in.(a) | Gray cast iron to 180 HB rev/min | mm/min | in./min | mm | in.(a) | Ductile cast iron to 200 HB rev/min | mm/min | in./min | mm | in.(a) |
|---|---|---|---|---|---|---|---|---|---|---|---|---|---|---|---|---|---|---|
| 1.98 | 0.0781.......... | 10 342 | 1500 | 10 000 | 64 | 2.5 | 114 | 4.5 | 10 000 | 91 | 3.6 | 114 | 4.5 | 10 000 | 71 | 2.8 | 114 | 4.5 |
| 2.38 | 0.0938.......... | 10 342 | 1500 | 10 000 | 84 | 3.3 | 127 | 5.0 | 10 000 | 124 | 4.9 | 127 | 5.0 | 10 000 | 97 | 3.8 | 127 | 5.0 |
| 3.18 | 0.1250.......... | 10 342 | 1500 | 10 000 | 122 | 4.8 | 147 | 5.8 | 10 000 | 180 | 7.1 | 147 | 5.8 | 9 165 | 127 | 5.0 | 155 | 6.1 |
| 3.97 | 0.1562.......... | 8 963 | 1300 | 10 000 | 152 | 6.0 | 165 | 6.5 | 9 780 | 226 | 8.9 | 168 | 6.6 | 7 335 | 130 | 5.1 | 193 | 7.6 |
| 4.76 | 0.1875.......... | 7 929 | 1150 | 10 000 | 180 | 7.1 | 183 | 7.2 | 8 150 | 226 | 8.9 | 201 | 7.9 | 6 110 | 127 | 5.0 | 234 | 9.2 |
| 5.55 | 0.2188.......... | 7 240 | 1050 | 10 000 | 208 | 8.2 | 196 | 7.7 | 6 985 | 224 | 8.8 | 234 | 9.2 | 5 240 | 124 | 4.9 | 269 | 10.6 |
| 6.35 | 0.2500.......... | 6 378 | 925 | 10 000 | 231 | 9.1 | 211 | 8.3 | 6 110 | 218 | 8.6 | 269 | 10.6 | 4 585 | 122 | 4.8 | 312 | 12.3 |
| 7.14 | 0.2812.......... | 5 861 | 850 | 10 000 | 257 | 10.1 | 224 | 8.8 | 5 435 | 213 | 8.4 | 305 | 12.0 | 4 075 | 119 | 4.7 | 351 | 13.8 |
| 7.94 | 0.3125.......... | 5 344 | 775 | 9 780 | 272 | 10.7 | 239 | 9.4 | 4 890 | 211 | 8.3 | 338 | 13.3 | 3 665 | 117 | 4.6 | 391 | 15.4 |
| 8.73 | 0.3437.......... | 4 999 | 725 | 8 890 | 264 | 10.4 | 264 | 10.4 | 4 445 | 206 | 8.1 | 373 | 14.7 | 3 335 | 114 | 4.5 | 429 | 16.9 |
| 9.53 | 0.3750.......... | 4 654 | 675 | 8 150 | 259 | 10.2 | 287 | 11.3 | 4 075 | 201 | 7.9 | 406 | 16.0 | 3 055 | 112 | 4.4 | 470 | 18.5 |
| 10.32 | 0.4062.......... | 4 309 | 625 | 7 520 | 254 | 10.0 | 312 | 12.3 | 3 760 | 198 | 7.8 | 439 | 17.3 | 2 820 | 109 | 4.3 | 508 | 20.0 |
| 11.11 | 0.4375.......... | 4 137 | 600 | 6 895 | 249 | 9.8 | 335 | 13.2 | 3 490 | 193 | 7.6 | 475 | 18.7 | 2 620 | 107 | 4.2 | 549 | 21.6 |
| 11.91 | 0.4688.......... | 3 792 | 550 | 6 520 | 244 | 9.6 | 361 | 14.2 | 3 260 | 190 | 7.5 | 508 | 20.0 | 2 445 | 107 | 4.2 | 587 | 23.1 |
| 12.70 | 0.5000.......... | 3 620 | 525 | 6 110 | 239 | 9.4 | 384 | 15.1 | 3 055 | 185 | 7.3 | 544 | 21.4 | 2 290 | 104 | 4.1 | 627 | 24.7 |
| 13.49 | 0.5312.......... | 3 447 | 500 | 5 750 | 234 | 9.2 | 409 | 16.1 | 2 875 | 183 | 7.2 | 577 | 22.7 | 2 155 | 102 | 4.0 | 655 | 26.2 |
| 14.29 | 0.5625.......... | 3 447 | 500 | 5 430 | 229 | 9.0 | 432 | 17.0 | 2 715 | 180 | 7.1 | 612 | 24.1 | 2 035 | 99 | 3.9 | 706 | 27.8 |
| 15.08 | 0.5938.......... | 3 275 | 475 | 5 145 | 226 | 8.9 | 457 | 18.0 | 2 575 | 175 | 6.9 | 648 | 25.5 | 1 930 | 99 | 3.9 | 749 | 29.5 |
| 15.88 | 0.6250.......... | 3 103 | 450 | 4 890 | 221 | 8.7 | 483 | 19.0 | 2 445 | 173 | 6.8 | 683 | 26.9 | 1 835 | 97 | 3.8 | 787 | 31.0 |
| 16.67 | 0.6562.......... | 2 930 | 450 | 4 655 | 218 | 8.6 | 505 | 19.9 | 2 330 | 170 | 6.7 | 716 | 28.2 | 1 745 | 94 | 3.7 | 828 | 32.6 |
| 17.46 | 0.6875.......... | 2 930 | 425 | 4 445 | 216 | 8.5 | 531 | 20.9 | 2 220 | 168 | 6.6 | 752 | 29.6 | 1 665 | 94 | 3.7 | 866 | 34.1 |
| 18.25 | 0.7188.......... | 2 758 | 400 | 4 250 | 211 | 8.3 | 51 | 21.7 | 2 125 | 165 | 6.5 | 780 | 30.7 | 1 595 | 91 | 3.6 | 899 | 35.4 |
| 19.05 | 0.7500.......... | 2 758 | 400 | 4 075 | 208 | 8.2 | 579 | 22.8 | 2 040 | 163 | 6.4 | 818 | 32.2 | 1 530 | 91 | 3.6 | 945 | 37.2 |

(a) Maximum allowable unsupported gun drill shank length. Length in excess of these values will cause shank to whip (balloon). Source: Eldorado Tool & Manufacturing Corporation

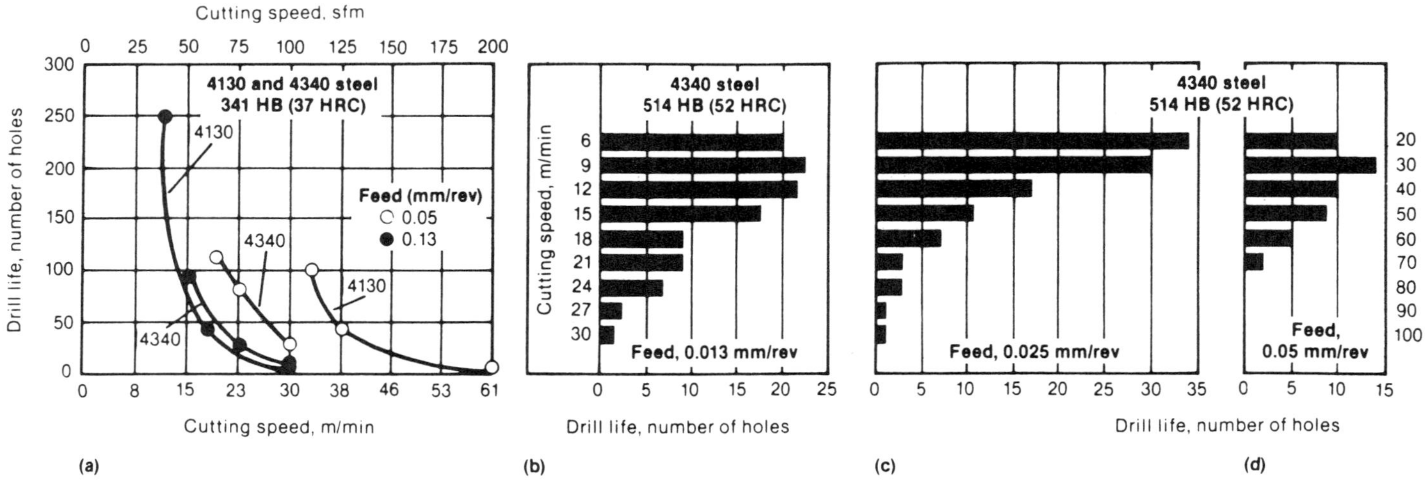


**Fig. 30** Effect of speed and feed on drill life. Holes for both series of tests were drilled through 13 mm (½ in.) thick specimens 75 to 102 mm (3 to 4 in.) in diameter, with 6.4 mm (¼ in.) diam drills (118° included point angle, 29° helix angle, 7° clearance), using a 1:1 mixture of thread-cutting oil and light machine oil as cutting fluid. Drills for data in (a) were 102 mm (4 in.) long and made of M2 high-speed tool steel. Drills for data in (b), (c), and (d) were 70 mm (2¾ in.) long, of T15 high-speed tool steel, and ground with crankshaft points. The end point of drill life was the breakdown of the drill or a 0.38 mm (0.015 in.) wear land on the drill margin, whichever occurred first.

Maximum drill life was found at the intermediate feed rate of 0.025 mm/rev (0.001 in./rev), with shorter drill life observed at either a higher or a lower feed rate. Only at 0.025 mm/rev (0.001 in./rev) feed did drill life decrease progressively as speed was increased; at both the higher and the lower feed rates, maximum drill life was obtained at a cutting speed of 9.1 m/min (30 sfm).

The use of a light feed reduces cutting temperature and cutting force. However, if the feed rate is reduced by half, the area of chip passing over the cutting edge is doubled. As a result, tool wear is likely to increase. Therefore, whether or not reducing the feed rate is advantageous depends on whether the lower cutting temperature and lighter cutting force offset the increased area of chip passing over the cutting edge. At high cutting speed, for which cutting temperature is a critical factor in drill life, reducing the feed is advantageous. At low cutting speed, for which cutting temperature is less critical, the longer chip and the increased rubbing over the cutting edge are likely to offset the advantages of lower feed, and drill life is likely to decrease.

**Hardness and Composition of Work Metal.** The effect of workpiece material on tool life in drilling carbon and low-alloy steels similar in composition and microstructure can be interpreted in terms of hardness. The results of one comparison of this type are given in Fig. 31. Although these results were affected to some degree by the use of a different tool material for drilling the softest steel, drill life (the amount of metal removed before the development of a predetermined wear land on the drill) was progressively shorter for increasing hardness of the steel being drilled.

The composition of carbon and low-alloy steels is usually of only secondary importance in its direct effects on drill life. Effects of practical significance include those of free-cutting additives, differences of 0.10% or more in carbon content, and substantial differences in the content of alloying elements. Some of these differences in composition also produce changes in hardness. Composition is of primary importance when it is necessary to improve drill life by heat treatment or cold reduction of the work material, or by selection of a more suitable work material.

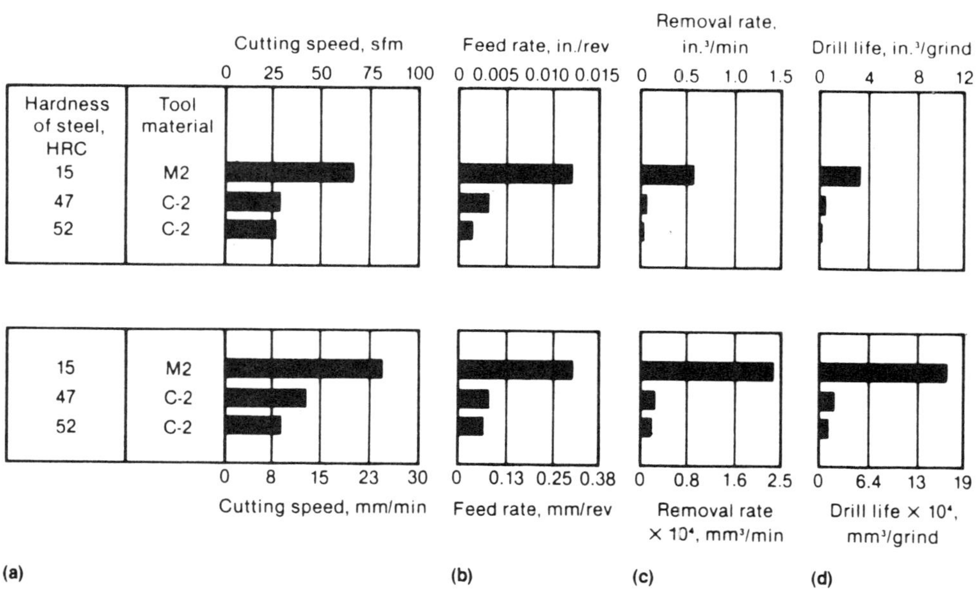


**Fig. 31** Effect of workpiece hardness on drill life for 6.4 mm (¼ in.) (top row) and 13 mm (½ in.) (bottom row) diam holes. Factors considered were cutting speed (a), feed rate (b), removal rate (c), and drill life (d). Comparative drilling tests were conducted on three similar steels, each at a different hardness. On each steel, the tools and machining conditions used had been previously determined to provide maximum economy in drilling that material. Drill life was based on the development of a predetermined amount of wear on the edge of the drill.

## Effect of Workpiece Hardness on Cost

The hardness of the workpiece material exerts a major influence on drilling cost. Meaningful comparisons of the effect of hardness on drilling cost can be made for steels that are similar in composition and microstructure.

**Drilling Cost.** Although drilling cost almost always increases with increasing hardness in the range above 35 HRC (330 HB), the reverse effect is observed for low-carbon steel at lower hardness. For low-carbon steel, there is usually an optimum hardness range for lowest drilling cost; this range varies considerably according to composition. For some low-carbon steels, a heat-treating operation is warranted to increase hardness before drilling.

**Overall Cost.** In considering overall economy of manufacture, cost-per-pound differences among mechanically equivalent steels should be evaluated, along with the differences in cost of machining the steels. Even when machining costs for a grade of steel that is difficult to drill are much greater than for a similar grade that is not, the machining costs may be outweighed by the difference in cost per pound of the two steels. The heavier the part and the less machining done on it, the more important material cost is as a factor in total cost.

## Determination of Optimum Speed and Feed

A wide range of speeds and feeds can be employed to yield acceptable results in the drilling of most materials, particularly those that present no unusual difficulty in machining. For materials that are unusually hard, soft and gummy, abrasive, or otherwise difficult to machine, the range of economical operating conditions is narrow. For these materials, an optimum speed and feed can be determined by making drilling tests to establish the conditions for minimum cost per cubic inch of metal removed.

Optimum drilling conditions are determined by changing one variable (speed or feed) while holding the other constant. The first step is to select a near-optimum feed. Speed is then adjusted in increments to find the most economical speed. Selection of the near-optimum feed is based on previous experience. After this first series of operations (the speed search), speed is held constant, and feed is varied in increments to verify or correct the initially selected feed rate (the feed search).

The development of a predetermined wear land on the edge of the drill is the criterion for tool life. Different speeds and feeds cause wear at different rates. Attainment of a predetermined amount of wear, which is ascertained by the use of a hand microscope, marks the point at which each test is stopped.

**Cost Factors.** In such a study, three factors are included in the determination of machining costs: tool-use cost, tool-change cost, and operating cost. Operating cost is based on a standard cost per minute for all operations. Tool-change cost is obtained by the use of a specific rate of tool changing assigned to the operation. Tool-use cost includes the cost of new tools and the cost of regrinding.

Total cost per hole drilled for each test condition is the sum of the following:

- Operating cost times the number of minutes required to drill one hole
- Tool-change cost divided by tool life in holes per tool
- Tool-use cost divided by tool life in holes per tool

Dividing this total by the volume of metal removed per hole gives the drilling cost in dollars per cubic inch of metal removed for each test condition.

## Testing to Evaluate Drilling Conditions

Laboratory tests are usually conducted to develop and evaluate drilling techniques and drill designs and materials, without concern for machines and fixtures. This requires close control over test conditions (especially rigidity) and the provision of adequate, smooth power to the drill. Production tests are used to evaluate drill performance in combination with drilling machines and fixtures. Accelerated tests may lead to erroneous conclusions because the wear or failure may be different in practice.

**Variation Among Drills.** There is considerable variation in performance among drills of the same type and even on resharpening a single drill. Therefore, enough drills should be tested to allow for this variation, and drills should be sharpened often enough to represent shop practice. Sharpening should reproduce the original point of the drill and should include removal of the part of the drill that has metal pickup on the margin or has reverse back taper. The web should be thinned to its original dimensions, and the resharpened drill should be free from burns and checks.

**Criteria for Drill Life.** Preliminary testing is often necessary to determine realistic criteria for drill life. Workpiece material, cutting fluid, and operating conditions influence the type of failure. Useful criteria include drill noise, drill wear, inability to cut, total failure of drill, inaccurate hole size, poor hole finish, burrs, drill breakage, and increase in the amount of torque required for a given drilling operation.

**Interpretation of Results.** Because of variations in drills, test conditions, and work material, a substantial number of tests usually must be made. As an illustration, differences of 30% or more between groups of identical drills can be detected reliably with about six drills of each type run through several sharpenings. Statistical analysis can indicate whether observed differences are significant.

Caution must be exercised in interpreting test results because of the wide discrepancies shown by machining tests in general. For example, consider the data shown in Fig. 30(a). Results were essentially the same in drilling 4130 and 4340 steel at 0.13 mm/rev (0.005 in./rev) feed, but speed for a given drill life was 40 to 50% higher for 4130 at 0.051 mm/rev (0.002 in./rev) feed. It could validly be concluded from these results that this lot of 4130 could be drilled at higher speeds than 4340 at a feed of about 0.051 mm/rev (0.002 in./rev). However, no conclusion could be drawn about 4130 and 4340 steels in general without additional testing.

## Drilling Steel Having 48 to 55 HRC Hardness

Successful drilling of steel that has been heat treated to a hardness of 48 to 55 HRC depends mainly on the design of the drill, the rigidity of the machine setup, and the choice of tool material. As a rule, improvements in drill design and in machine and workpiece rigidity are of more direct benefit to drilling performance than is a change in tool material. A number of other conditions may also affect drill performance, and in certain cases, they may become critical. Among these conditions are feed and speed, the efficiency and adaptability of power equipment (portable or stationary), and the accuracy of working surfaces of the drill. The following three examples illustrate the effects of operating variables in drilling hardened steel.

**Example 5: Effect of Nitriding on the Drill Life of a High-Speed Tool Steel Drill.** Blind holes, 6.91 mm (0.272 in.) in diameter and 15.75 mm (0.620 in.) deep, were drilled in 4335 steel hardened to 48 to 50 HRC. The drills had a life of only one to four holes per sharpening, and a breakage rate of 1 drill for every 16 to 20 holes. When the drills were nitrided, drill life was increased to 18 to 20 holes per sharpening, and drill breakage was reduced to 1 drill for every 50 to 60 holes.

All drills were standard heavy-duty drills ground from solid heat-treated high-speed steel. Each had a split point ground with a positive axial rake of 3 to 4° on the split portion, an included point angle of 135° and a lip relief angle of 5 to 7°, ground flat 0.8 mm (1/32 in.).

Drilling was done in a radial-arm drill press with multiple speeds and automatic feed. Speed was 147 rev/min (3.20 m/min, or 10.5 sfm), and feed was 0.05 mm/rev (0.002 in./rev).

**Example 6: High-Speed Tool Steel Versus Carbide Drills.** For H11 steel at 50 HRC or below, M33, M34, and M36 high-speed tool steel drills of heavy-duty construction gave good results. Positive-drive equipment was used, with drill speeds of 7.6 to 9.1 m/min (25 to 30 sfm) and feeds of 0.013 to 0.018 mm/rev (0.0005 to 0.0007 in./rev).

Material harder than 50 HRC required solid-carbide drills with short flutes. Speeds of 12 to 15 m/min (40 to 50 sfm) and feeds of 0.013 to 0.025 mm/rev (0.0005 to 0.001 in./rev) produced satisfactory results. The design of the solid-carbide twist drills used for holes of less than 6.4 mm (1/4 in.) diameter in this material is illustrated in Fig. 32(a) and (b); the design of straight-flute solid-carbide drills, for holes 6.4 mm (1/4 in.) in diameter or larger, is shown in Fig. 32(c).

**Example 7: Selection of Carbide Grade.** Three grades of carbide (C-1, C-2, and C-3) were tested to determine which

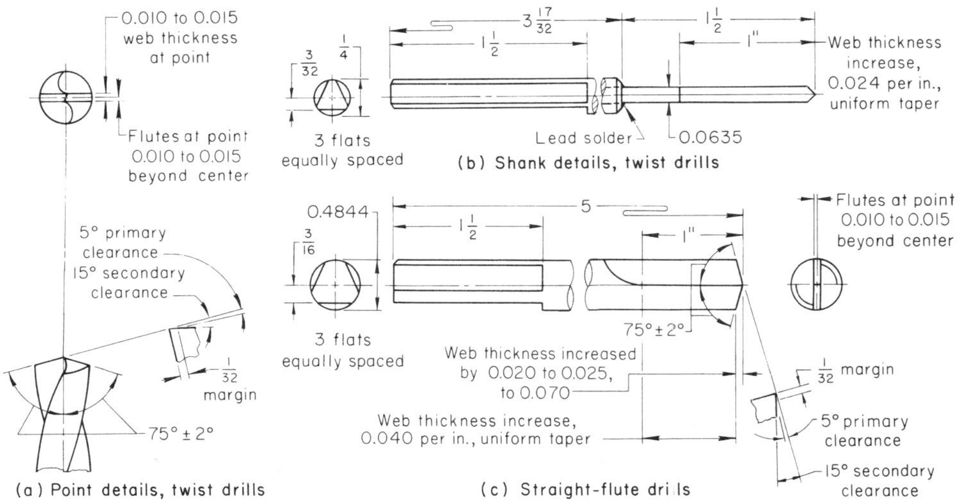


**Fig. 32** Details of solid-carbide drills used successfully on H11 steel harder than 50 HRC. (a) and (b) Twist drills, for holes less than 6.4 mm (¼ in.) in diameter. (c) Straight-flute drills, for holes 6.4 mm (¼ in.) in diameter or larger. Dimensions given in inches

was most suitable for drilling H11 steel sheet (5% Cr, 1.5% Mo) heat treated to 54 HRC. Two sheets of this material, each 2.10 mm (0.083 in.) thick, were drilled simultaneously with 6.35 mm (0.250 in.) diam carbide-tip drills, without a cutting fluid.

The softer 6% Co grade (C-1) failed because of abrasion, excessive burning, and chipping—all of which ultimately resulted in breakage. The hardest (and therefore most brittle) grade, C-3, which contained only 3% Co, failed by excessive chipping and breaking during drilling. The C-3 also chipped and broke in handling.

The most satisfactory carbide was the harder of the two 6% Co grades, C-2, which showed only slight wear after testing; this grade was selected for machining on a production basis. (Although the C-1 and C-2 grades tested both contained 6% Co, the C-2 was harder because of its finer grain size, which decreased the size of the softer cobalt lakes between grains.)

The design of these drills is shown in the upper portion of Fig. 33; the tool life for each grade of carbide is plotted in the lower part. Composition, hardness, and grain size of the three grades of carbide were as follows:

| Grade | Composition, % W | Ta | C | Co | Hardness, HRA | Grain size, μm (μin.) |
|---|---|---|---|---|---|---|
| C-1 | 88.25 | ··· | 5.75 | 6.0 | 91.2 | 3 (120) |
| C-2 | 88.25 | ··· | 5.75 | 6.0 | 91.8 | 1–2 (40–80) |
| C-3 | 87.00 | 4.0 | 6.00 | 3.0 | 92.2 | 1–3 (40–120) |

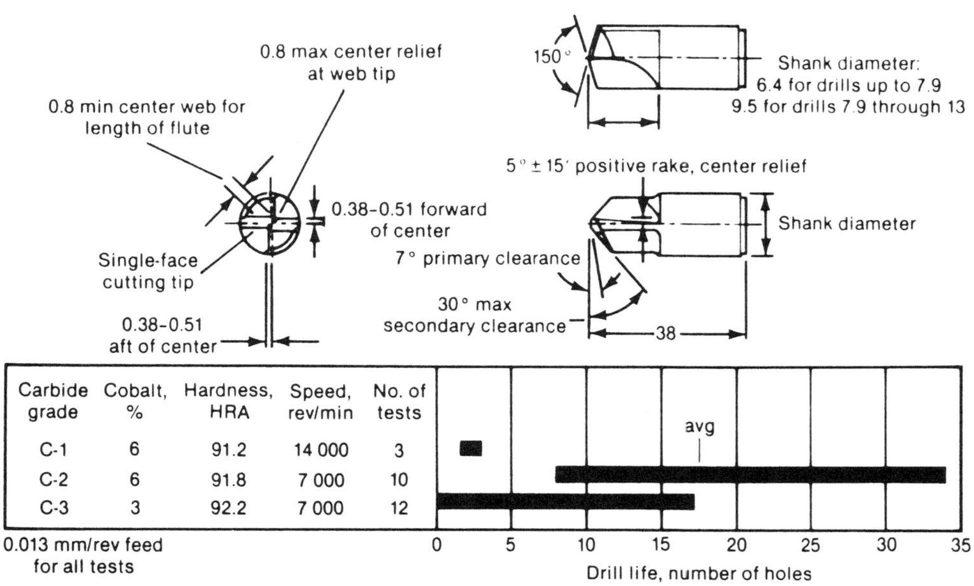


**Fig. 33** Top: design of carbide-tip drills used to drill H11 steel sheet at 54 HRC. Bottom: comparison of tool life for the three grades of carbide tested on drills of the above design for performance in drilling 6.35 mm (0.250 in.) diam holes through two 2.11 mm (0.083 in.) thick sheets of this steel. Dimensions in figure given in millimeters

## Indexable-Insert Drills (Ref 2)

The most important recent advance in drilling technology was the development in the early 1970s of drills with indexable carbide inserts. These tools can produce relatively shallow holes from the solid at faster rates and lower cost than high-speed steel twist drills in many applications.

Studies have shown that about 60% of all drilling applications in industry are considered to be short holes having depths up to about three diameters. Many of these holes, as well as others up to five times the drill diameter, can be drilled with indexable-insert drills. Others are not practical to produce with these tools, because the holes are too small in diameter or because inadequate machines (with respect to speed, power, and rigidity) are employed.

The hole diameters that can be produced with indexable-insert drills vary with the tools available from different manufacturers. A common range of diameters offered by some manufacturers is 19 to 76 mm (¾ to 3 in.). The smallest tool commercially available drills holes 16 mm (⅝ in.) and the largest, 127 mm (5 in.) in diameter. Some of these tools contain up to six inserts. Figure 34 shows two- and four-insert versions.

### Advantages

The major advantages of indexable-insert drills are increased productivity, reduced costs, and better versatility.

**Increased Productivity.** The use of carbide inserts brings drilling close to the machining rates possible with turning and milling. The higher cutting speeds possible permit holes to be drilled substantially faster than with high-speed tool steel twist drills and even faster than with carbide spade drills. The cycle time on transfer machines and other high-production applications is often determined by the capabilities of the high-speed tool steel drills used. With the faster penetration rates of indexable-insert drills often matching the rates of other operations, cycle times can generally be reduced. Potential productivity is also increased because the almost flat lead angle of indexable-insert drills results in a shorter feed stroke before cutting, compared to twist or spade drills that have point angles (Fig. 35). For example, the feed stroke for a twist drill may be 13 mm (½ in.) compared to 2 mm (5/64 in.) for an indexable-insert carbide drill in the case of a 41.4 mm (1.63 in.) diam hole being drilled to a 6 mm (15/64 in.) depth in high-carbon steel; at a 260 m/min (860 sfm) speed and a 500 mm/min (19.7 in./min) feed rate, the carbide drill speed is eleven times faster than the high-speed tool steel twist drill (45 mm/min, or 1.8 in./min).

**Reduced Costs.** The use of low-cost inserts with multiple cutting edges eliminates regrinding costs. The multiple cutting edges

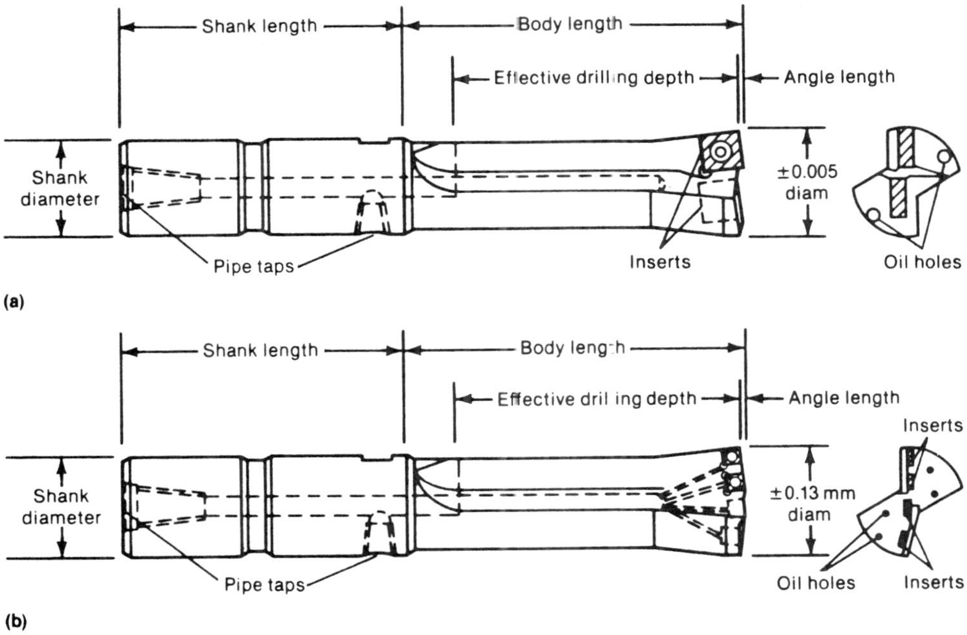


**Fig. 34** Two types of indexable carbide insert drills. (a) Two-insert drill. (b) Four-insert drill

available provide savings from not having to replace the entire tool. Indexing the inserts does not change their positions and the tool length, thus any tool resetting costs are eliminated.

**Versatility.** Indexable-insert drills can be used as nonrotating tools for applications on lathes or other machines or as rotating tools on drilling machines, machining centers, and other machine tools. The machines used, however, must be rigid, must be in good condition, and must have ample speed and power capabilities. Some of the tools have the capability, when used on suitable machines, to perform boring as well as drilling operations. For example, mounted on the cross slide of an NC lathe, some tools can be moved radially outward to drill holes larger than the tool diameter or make a boring pass, thus improving the accuracy and finish of the hole. Other tools can perform turning, contouring, and facing operations, as well as multiple operations such as drilling, chamfering, and spotfacing.

### Limitations

The smallest-diameter hole that can be produced with the indexable-insert drills that are commercially available is 16 mm (5/8 in.). The maximum drilling depth is generally two to three times the hole diameter. Some tools, however, can drill to four times the diameter, and a few specials can drill to depths over five times the diameter. Indexable-insert drills are not precision hole-producing tools, and subsequent operations may be required for improved accuracies and smoother finishes. Small pilot holes are not useful and can be detrimental, and the tools cannot be used to enlarge existing holes.

Indexable-insert drills require less thrust than twist drills because they have no webs or chisel edges, but they do require more power because of increased metal removal rates. Rigid machines in good condition, with adequate speed and power capabilities, and cutting fluid under pressure are necessary to take full advantage of the productive capabilities of these tools. Horsepower requirements increase proportionately with the drill diameter (Fig. 36). Bench, upright, and radial drilling machines are generally not suitable for use with these tools, because they lack sufficient speed, power, or rigidity. Safety guards are required on any machine used for drilling through holes with the workpiece rotating and the drill stationary because the slugs produced can be thrown outward at high velocity.

Indexable-insert drills can be used to drill many materials, but most are not suitable for laminated or stacked materials. This is because the disks or slugs produced would be pressed into or welded to the next layer of material and because increased pressures can damage the inserts and possibly the drill body. The surfaces of workpieces to be drilled should preferably be flat. When using negative-rake inserts, convex surfaces can present problems, and concave surfaces are not generally recommended, because they might throw the drill out of balance. Angular surfaces rising more than 1 mm (0.040 in.) in a 50 mm (2 in.) distance and interrupted cuts are also generally not recommended for use with most of these tools. There are, however, successful applications with angular starting surfaces and interrupted cuts when positive-rake inserts are used.

### Tool Design Considerations

Indexable-insert drills of slightly different designs are available from various tool manufacturers. Most consist of a hardened alloy steel body (with a straight or a taper shank) held by an adapter attached to the machine spindle. Some tools are of one-piece construction and others two-piece, with one of the two pieces being an interchangeable cutting head or nosepiece.

The drills have straight or helical flutes or grooves, and internal coolant holes. Flute or groove design is critical; sufficient space must be provided for the rapid removal of a large volume of chips, but an adequate body

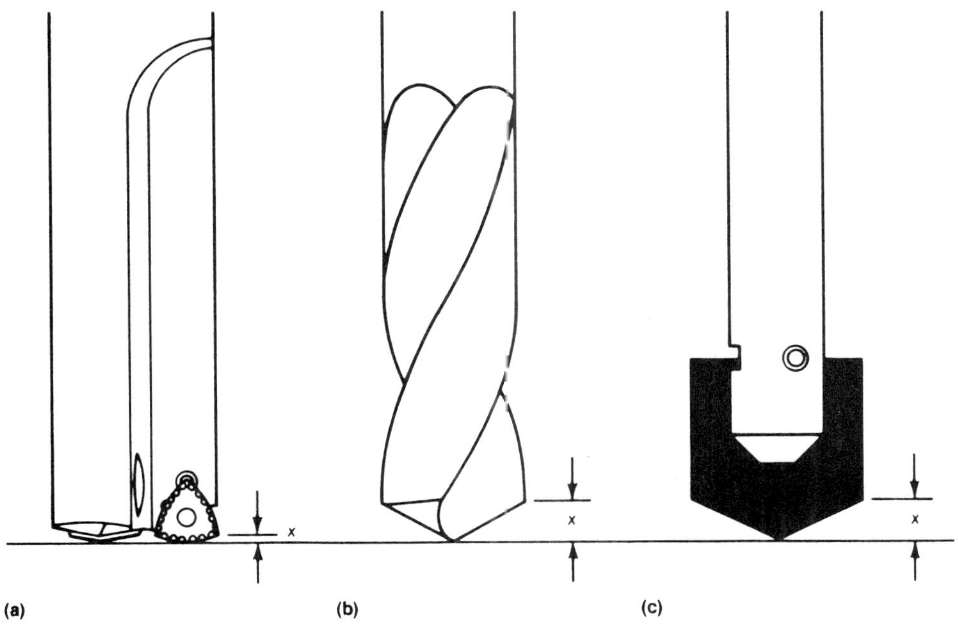


**Fig. 35** Feed strokes (dimension $x$) of three drills. The indexable-insert carbide drill (a), shown here with a trigon insert, has a much shorter feed distance than a standard twist drill (b) or a spade drill (c).

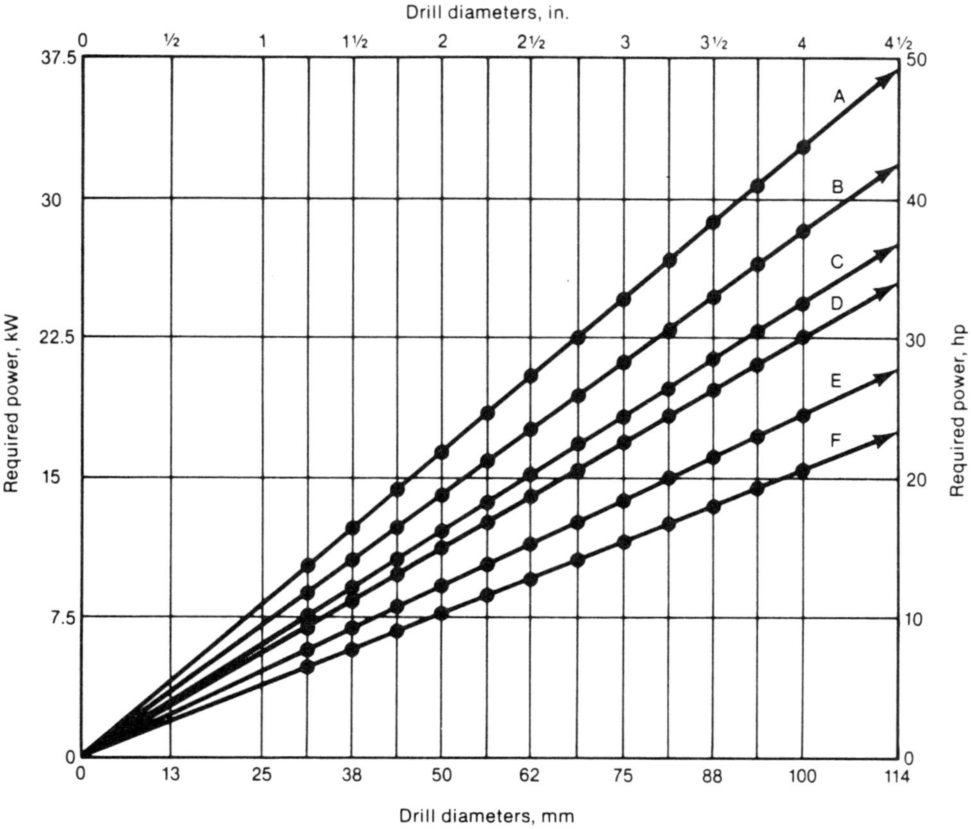


**Fig. 36** Spindle power requirements for various diameters of indexable-insert carbide drills in three materials (4140 steel, hard cast iron, and Type 316 stainless) under varying conditions of speed and feed. A, 4140 steel, 105 m/min (350 sfm), and 0.18 mm/rev (0.007 in./rev); B, 4140 steel, 90 m/min (300 sfm), and 0.18 mm/rev (0.007 in./rev); C, hard cast iron, 105 m/min (350 sfm), and 0.30 mm/rev (0.012 in./rev); D, type 316 stainless steel, 75 m/min (250 sfm), and 0.13 mm/rev (0.005 in./rev); E, 4140 steel, 60 m/min (200 sfm), and 0.18 mm/rev (0.007 in./rev); F, type 316 stainless steel, 55 m/min (175 sfm), and 0.13 mm/rev (0.005 in./rev)

cross section must be maintained for strength and rigidity. Coolant enters the tools, usually through tapped openings, and reaches the cutting zones from orifices near the inserts. Chips and coolant exit through the external flutes or grooves on the tools. In applications in which the tool rotates, an inducer or coolant collar (as with other coolant-fed tools) is required unless coolant is supplied through a hollow spindle on the machine.

At the cutting end of each flute or groove, recessed pockets are provided to locate the indexable inserts. Depending on the drill diameter and design of the tool, one to four inserts are generally used. A few drills, using toolholding cartridges, hold as many as six inserts. Some have a centrally located insert, positioned slightly ahead of the others, for self-centering purposes. The inserts are mounted in positions and at attitudes to counteract each other's lateral cutting forces, thus minimizing side loads. This is necessary because the tools are not guided by the holes being drilled and because guide bushings are not used.

## Insert Configurations

Round, square, triangular, diamond, trigon, trochoid, parallelogram, hexagonal, and octagonal carbide inserts are used by different tool manufacturers (Fig. 37). The inserts may or may not be the same size, depending on the drill design and diameter. Some inserts have dimples, grooves, or special geometries for chip-breaking purposes.

The geometry and positioning of the inserts are important to the performance and efficiency of the drills. Continuing improvements are being made in insert geometries to ensure constant chip control. Depending on the insert shape and application, some drills are equipped with negative-rake inserts, requiring negative-rake placement. Others are designed for positive-rake inserts that are placed to provide either a neutral or a positive axial rake angle. Positive geometries generally develop less cutting and axial thrust forces and require less horsepower than negative geometries. Effective rake in a negative insert, however, can be higher and can develop lower forces than a positive-rake insert. Replaceable anvils under the inserts can help prevent damage to the tools in case of insert breakage.

The grade of carbide used for the inserts varies with the application. Classification C-5 is extensively used for drilling many steels; C-2 is used for cast irons and nonferrous metals. Coated inserts (especially titanium nitride) are also widely employed because they permit higher cutting speeds (to 305 m/min, or 1000 sfm, or more) and are

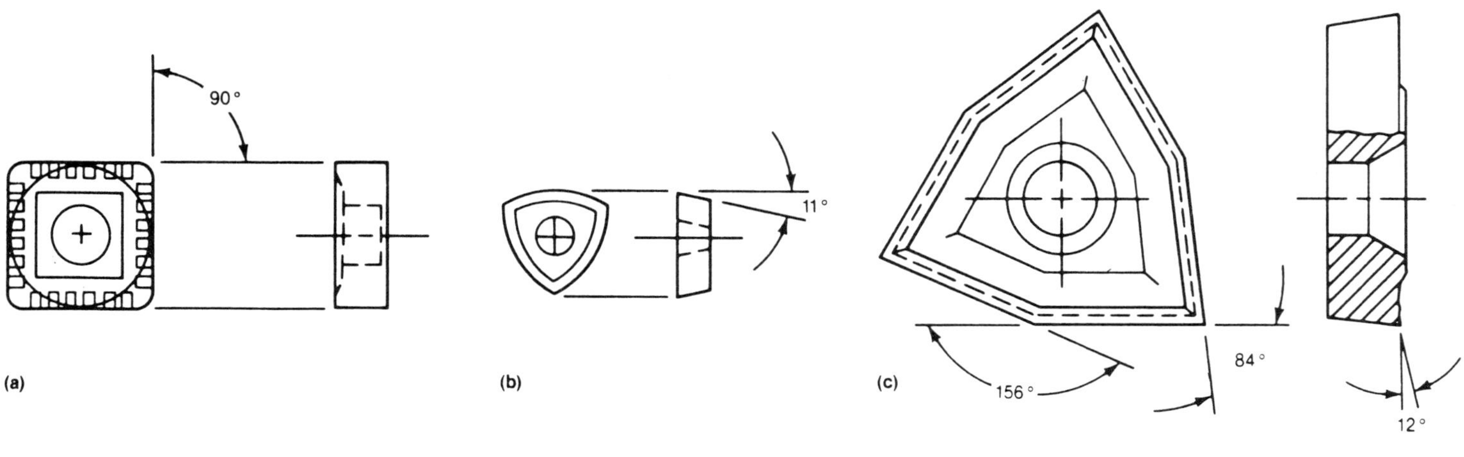


**Fig. 37** Typical insert configurations used in indexable-insert carbide drills. (a) Square. (b) Trochoid. (c) Trigon

**Table 8 Recommended cutting parameters for indexable drills with uncoated and titanium-coated carbide inserts**

| Material | Drill size, mm | Drill size, in. | Speed, Standard length, m/min | Speed, Standard length, sfm | Speed, Stub length, m/min | Speed, Stub length, sfm | Feed, mm/rev | Feed, in./rev | Insert style(a) | Grade |
|---|---|---|---|---|---|---|---|---|---|---|
| Carbon steel (1000 series) | 19–30 | 3/4–1 3/16 | 90–200 | 300–650 | 90–245 | 300–800 | 0.08–0.13 | 0.003–0.005 | SPGM | C-5(b) |
| | 32–36.5 | 1 1/4–1 7/16 | 90–200 | 300–650 | 90–245 | 300–800 | 0.08–0.15 | 0.003–0.006 | SNMG | C-5(b) |
| | 38–47.5 | 1 1/2–1 7/8 | 90–200 | 300–650 | 90–245 | 300–800 | 0.10–0.20 | 0.004–0.008 | SNMM-ND | C-5(b) |
| | 49–75 | 1 15/16–3 | 90–200 | 300–650 | 90–245 | 300–800 | 0.13–0.25 | 0.005–0.010 | SNMM-ND | C-5(b) |
| Free-machining steels (1100 and 1200 series) | 19–30 | 3/4–1 3/16 | 120–215 | 400–700 | 120–275 | 400–900 | 0.08–0.13 | 0.003–0.005 | SPGM | C-5(b) |
| | 32–36.5 | 1 1/4–1 7/16 | 120–215 | 400–700 | 120–275 | 400–900 | 0.10–0.15 | 0.004–0.006 | SNMG | C-5(b) |
| | 38–47.5 | 1 1/2–1 7/8 | 120–215 | 400–700 | 120–275 | 400–900 | 0.08–0.20 | 0.005–0.008 | SNMM-ND | C-5(b) |
| | 49–75 | 1 15/16–3 | 120–215 | 400–700 | 120–275 | 400–900 | 0.15–0.30 | 0.006–0.012 | SNMM-ND | C-5(b) |
| Alloy steels (4000, 5000, 8000 and 9000 series) | 19–30 | 3/4–1 3/16 | 60–185 | 200–600 | 60–200 | 200–650 | 0.08–0.13 | 0.003–0.005 | SPGM | C-5(b) |
| | 32–36.5 | 1 1/4–1 7/16 | 60–185 | 200–600 | 60–200 | 200–650 | 0.10–0.15 | 0.004–0.006 | SNMG | C-5(b) |
| | 38–47.5 | 1 1/2–1 7/8 | 60–185 | 200–600 | 60–200 | 200–650 | 0.13–0.20 | 0.005–0.008 | SNMM-ND | C-5(b) |
| | 49–75 | 1 15/16–3 | 60–185 | 200–600 | 60–200 | 200–650 | 0.15–0.25 | 0.006–0.010 | SNMM-ND | C-5(b) |
| Tool steel | 19–30 | 3/4–1 3/16 | 60–120 | 200–400 | 90–185 | 300–600 | 0.08–0.13 | 0.003–0.005 | SPGM | C-5(b) |
| | 32–36.5 | 1 1/4–1 7/16 | 60–120 | 200–400 | 90–185 | 300–600 | 0.10–0.15 | 0.004–0.006 | SNMG | C-5(b) |
| | 38–47.5 | 1 1/2–1 7/8 | 60–120 | 200–400 | 90–185 | 300–600 | 0.13–0.20 | 0.005–0.008 | SNMM-ND | C-5(b) |
| | 49–75 | 1 15/16–3 | 60–120 | 200–400 | 90–185 | 300–600 | 0.15–0.25 | 0.006–0.010 | SNMM-ND | C-5(b) |
| Cast iron | 19–30 | 3/4–1 3/16 | 120–245 | 400–800 | 120–275 | 400–900 | 0.10–0.20 | 0.004–0.008 | SPGM | C-2(c) |
| | 32–36.5 | 1 1/4–1 7/16 | 120–245 | 400–800 | 120–275 | 400–900 | 0.13–0.25 | 0.005–0.010 | SNMG | C-2(c) |
| | 38–47.5 | 1 1/2–1 7/8 | 120–245 | 400–800 | 120–275 | 400–900 | 0.15–0.30 | 0.006–0.012 | SNMM-ND | C-2(c) |
| | 49–75 | 1 15/16–3 | 120–245 | 400–800 | 120–275 | 400–900 | 0.20–0.38 | 0.008–0.015 | SNMM-ND | C-2(c) |
| Stainless steels (Types 303, 304 and 316, 416) | 19–30 | 3/4–1 3/16 | 45–105 | 150–350 | 45–150 | 150–500 | 0.08–0.10 | 0.003–0.004 | SPGM | C-2(c) |
| | 32–36.5 | 1 1/4–1 7/16 | 45–105 | 150–350 | 45–150 | 150–500 | 0.10–0.13 | 0.004–0.005 | SNMG | C-5(b) |
| | 38–47.5 | 1 1/2–1 7/8 | 45–105 | 150–350 | 45–150 | 150–500 | 0.10–0.15 | 0.004–0.006 | SNMM-ND | C-5(b) |
| | 49–75 | 1 15/16–3 | 45–105 | 150–350 | 45–150 | 150–500 | 0.13–0.20 | 0.005–0.008 | SNMM-ND | C-5(b) |
| Nitralloy 135 | 19–30 | 3/4–1 3/16 | 45–120 | 150–200 | 45–120 | 150–200 | 0.13–0.18 | 0.005–0.007 | SPGM | C-5(b) |
| | 32–36.5 | 1 1/4–1 7/16 | 45–120 | 150–200 | 45–120 | 150–200 | 0.13–0.18 | 0.005–0.007 | SNMG | C-5(b) |
| | 38–47.5 | 1 1/2–1 7/8 | 45–120 | 150–200 | 45–120 | 150–200 | 0.13–0.18 | 0.005–0.007 | SNMG | C-5(b) |
| | 49–75 | 1 15/16–3 | 45–120 | 150–200 | 45–120 | 150–200 | 0.13–0.18 | 0.005–0.007 | SNMG | C-5(b) |
| Inconel 100 | 19–30 | 3/4–1 3/16 | 25–35 | 75–110 | 25–35 | 75–110 | 0.08–0.10 | 0.003–0.004 | SPGM | C-2(c) |
| | 32–36.5 | 1 1/4–1 7/16 | 25–35 | 75–110 | 25–35 | 75–110 | 0.08–0.10 | 0.003–0.004 | SNMG | C-2(c) |
| | 38–47.5 | 1 1/2–1 7/8 | 25–35 | 75–110 | 25–35 | 75–110 | 0.08–0.10 | 0.003–0.004 | SNMG | C-2(c) |
| | 49–75 | 1 15/16–3 | 25–35 | 75–110 | 25–35 | 75–110 | 0.08–0.10 | 0.003–0.004 | SNMG | C-2(c) |
| Aluminum: 2017T and 2024T, 6061T | 19–30 | 3/4–1 3/16 | 185–365 | 600–1200 | 185–490 | 600–1600 | 0.20–0.25 | 0.008–0.010 | SPGM | C-2(c) |
| | 32–36.5 | 1 1/4–1 7/16 | 185–365 | 600–1200 | 185–490 | 600–1600 | 0.20–0.30 | 0.008–0.012 | SNMM-ND | C-2(c) |
| | 38–47.5 | 1 1/2–1 7/8 | 185–365 | 600–1200 | 185–490 | 600–1600 | 0.25–0.38 | 0.010–0.015 | SNMP | C-2(c) |
| | 49–75 | 1 15/16–3 | 185–365 | 600–1200 | 185–490 | 600–1600 | 0.30–0.51 | 0.012–0.020 | SNMP | C-2(c) |

(a) Inserts are designated by a letter code indicating such specifications as shape, clearance angle, tolerances, and hole size. (b) Titanium coated. (c) Uncoated. Source: Metal Cutting Tools, Inc.

less prone to the formation of built-up edges. Inserts made from different grades of carbide are sometimes used in the same tool, depending on insert position and cutting speed.

Inserts are held in the drill pockets by screws or cam-locking pins, eliminating the need for clamps, which would obstruct chip flow. It is important that the screws or pins be securely tightened, but not overtightened. Loose inserts will cause chatter and possible tool breakage. Some indexable-insert drills, especially larger-diameter tools, are equipped with cartridges that reduce costs in case of insert failures because they provide protection for the main body of the tool.

## Indexable-Insert Drill Applications

Indexable-insert drills are primarily used for producing holes in steels and irons. More ductile materials such as aluminum and copper are also drilled with these tools, but chip ejection may be a problem for some applications. When soft, ductile, and gummy materials are being drilled, chip control can be a problem. The thicker chips produced with neutral or negative-rake inserts tend to pack in the flutes or grooves. The tools are not suitable for drilling soft materials such as rubber or plastics.

These tools are especially advantageous for medium- and high-production applications on NC and transfer machines. They are also used for many low-production applications on manual machines, lathes, and other machine tools if the machines have the required power, speed, and rigidity. Table 8 lists suggested cutting for uncoated and titanium nitride coated inserts.

## Small-Hole Drilling (Microdrilling)

The drilling of small holes (diameters of 0.025 to 3.2 mm, or 0.001 to 1/8 in.) requires machines, drills, techniques, and operator skills different from those used in conventional drilling.

**Machines** used for drilling small holes are usually bench-mounted and resemble a jeweler's drill press (Fig. 2). Alignment of spindle and table at any position should be within 0.013 mm in 152 mm (0.0005 in. in 6 in.), or at a radius equal to half the length of the table. Runout of the spindle should not exceed 0.0025 mm (0.0001 in.) in any position. Spindle speeds are high, ranging from 3000 rev/min, for drilling holes near 1.6 mm (1/16 in.) in diameter to 20 000 rev/min or higher for drilling holes 0.25 mm (0.010 in.) in diameter or smaller.

To minimize vibration, the machine is belt-driven by a balanced, vibration-damped motor mounted on a separate stand. The required sensitive feed is provided by a balanced crossarm or by a rack and pinion controlled by a knoblike wheel.

Larger machines for microdrilling have precision chucks or collets to hold the drill. However, for drills smaller than 0.38 mm (0.015 in.) in diameter, a one-piece spindle and drill that revolves in a jeweled V-block is used.

**Drills** larger than 0.20 to 0.25 mm (0.008 to 0.010 in.) in diameter may be of either the twist or the spade type (the latter are also known as pivot-type drills). Drills smaller than this are usually of the spade type, although twist drills have been made in diameters as small as 0.099 mm (0.0039 in.).

Drills smaller than 0.38 mm (0.015 in.) in diameter have their own mandrels. When drills are broken, mandrels are returned to the factory for the insertion of new drills. Incorporating the drill in a mandrel makes it possible to grind the drill point concentric with the mandrel. Drills are available in three classes of tolerance:

- +0.0000, −0.0025 mm (+0.0000, −0.0001 in.)
- ±0.0025 mm (±0.0001 in.)
- ±0.0050 mm (±0.0002 in.)

An included point angle of 135° with an 8° clearance angle is usual for drilling steel; a 118° point angle with a 15° clearance angle is usual for drilling soft nonferrous metals. The point angle influences hole size because small angles improve centering.

Carbon-tungsten special-purpose tool steels such as F2 or F3 are extensively used as materials for small-diameter drills. High-speed tool steels are used less often for these drills than for larger drills.

**Techniques.** To produce small holes that are accurate in size and finish, extreme care must be taken in grinding the drill and in the drilling technique (centering and feeding) because:

- Space for chip removal is limited by the greater ratio of web thickness to diameter of small-hole drills
- Pressure on the end of the drill is greater
- Longitudinal and torsional deflections are greater because of the high ratio of length to diameter

In small-hole drilling, the formation of metal powder rather than chips may result. This condition may lead to packing, causing the drills to break. Packing can be alleviated by frequent clearing of the hole by complete removal of the drill and application of a lubricant. Frequent tool withdrawal to clear chips is known as peck drilling. One successful technique is the following:

- The drill is withdrawn for chip clearance after the initial penetration has reached a depth no greater than three times the drill diameter (a lesser depth for initial penetration may be required for some materials)
- Another withdrawal is made after penetration has progressed for 1½ times the drill diameter beyond the first cut
- The drill is again withdrawn following each succeeding cut to a depth of three-fourths the drill diameter

**Speed and Feed.** The spindle speeds used for drilling small holes are high compared to those used for ordinary drilling, but surface speeds, for holes down to about 0.20 to 0.25 mm (0.008 to 0.010 in.) in diameter, are not necessarily different. The objective is a speed-and-feed combination that produces a true chip. If excessive speed is used, it becomes impossible to obtain a feed great enough to form a chip.

For drilling holes smaller than about 0.20 mm (0.008 in.) in diameter, the smaller the drill, the lower the speed that is permitted. Applications have been reported in which speeds as low as 50 rev/min (0.003 to 0.006 m/min, or 0.01 to 0.02 sfm) were used for drills approaching 0.025 mm (0.001 in.) in diameter.

Feeds as great as 0.025 mm/rev (0.001 in./rev) are often used. However, hand feeding is normal for drilling extremely small holes, and the rate of feed will vary widely among operators. Successful drilling of small holes requires considerable operator skill and feel for feeding the drill. In drilling extremely small holes, even an experienced operator will often break one or two drills when starting a work period before this feel is regained.

**Cutting fluid** requirements for drilling small holes differ somewhat from those for large holes, mainly because intermittent withdrawal of the drill disposes of chips and also cools the drill. Therefore, in small-hole drilling, the main requirement of a cutting fluid is lubrication. The usual practice is to coat the drill with lard oil (or a similar lubricant) at each withdrawal, either by hand brushing or with a lubricant dispenser.

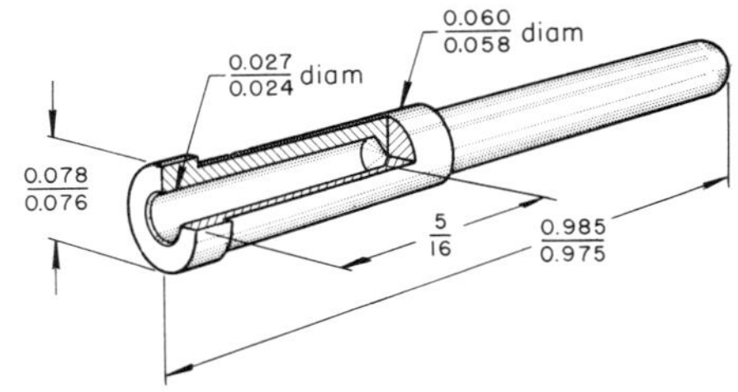


| | |
|---|---|
| Speed, at 7875 rev/min, m/min (sfm) | 15 (52) |
| Feed, mm/rev (in./rev) | 0.25 (0.001) |
| Cutting fluid | Sulfochlorinated oil (plus fat) |
| Drilling time per hole, s | 18 |
| Drill life per grind, pieces | 1200 |

**Fig. 38** The ratio (12.5:1) of depth to diameter of hole in this part was the maximum capability of the standard twist drill used to drill a copper-nickel-tellurium alloy. Dimensions in figure given in inches

**Example 8: Microdrilling of a Copper-Nickel-Tellurium Alloy.** The 0.635 mm (0.025 in.) hole in the copper-nickel-tellurium alloy part shown in Fig. 38 was made with a standard twist drill 0.635 mm (0.025 in.) in diameter, with flute length of 7.9 mm (5⁄16 in.). The part was drilled in a multiple-operation machine. The drill was withdrawn six times to clear the chips. Drilling was more difficult than normal for this type of operation because the hole depth and flute length were the same. This drilled hole represents about the maximum depth that a standard twist drill of 0.635 mm (0.025 in.) diameter can produce within the 0.076 mm (0.003 in.) total tolerance specified. Processing details are given with Fig. 38.

## REFERENCES

1. L.E. Doyle, C.A. Keyser, J.L. Leach, G.F. Schrader, and M.B. Singer, *Manufacturing Processes and Materials for Engineers*, Prentice-Hall, 1985, p 600
2. *Machining*, Vol 1, *Tool and Manufacturing Engineers Handbook*, Society of Manufacturing Engineers, 1983

# Reaming

REAMING is a machining operation in which a rotary tool takes a light cut to improve the accuracy of a round hole and to reduce the roughness of the hole surface. Most reamers have two or more flutes, either parallel to the tool axis or in a helix, which provide teeth for cutting and grooves for chips.

Reaming and boring are related processes, and sometimes their applications overlap. (There are even tools identified as reamers by some and as boring tools by others.) Hole diameter and length, or required straightness or tolerance, usually indicate whether reaming or boring is to be used.

## Process Capabilities

Although steels that range in hardness from about 15 to 30 HRC are the metals most often reamed, the process is widely used for finishing holes in cast iron. Reaming is also used for the softest nonferrous metals as well as for steels with a hardness of 52 HRC or higher.

**Hole Diameter.** Most holes reamed are 3.2 to 32 mm (⅛ to 1¼ in.) in diameter. Reamers are commercially available for holes as small as 0.35 mm (0.0135 in.) in diameter, and specially designed reamers are available for 0.1 mm (0.005 in.) diam holes. Solid reamers are available for holes up to 50 mm (2 in.) in diameter and are specially made for holes up to 75 mm (3 in.) in diameter. Reamers of other types are available for holes up to 150 mm (6 in.) in diameter.

**The hole length** that can be successfully reamed depends on reamer diameter, method of holding and driving the reamer, and required dimensional accuracy. Reamer diameter determines the maximum length of the reamer cutting edge, which affects the length of hole that can be reamed accurately. For example, the cutting edge of a 0.35 mm (0.0135 in.) diam reamer may be as long as 19 mm (¾ in.) (a length-to-diameter ratio of 55:1), but that of a 150 mm (6 in.) diam shell reamer is usually no longer than 150 mm (6 in.). In most applications with standard reamers, the length of the hole being reamed ranges from only slightly longer to considerably shorter than the cutting edge of the reamer. However, the length of the cutting edge does not necessarily limit the length of hole that can be reamed. Shank length can be extended to permit the reaming of holes several times longer than the cutting edge of the reamer, but this makes it difficult to guide the reamer and to hold dimensional tolerances.

In the horizontal reaming of holes several times longer than the cutting edge of the reamer, the difficulty of maintaining finish and dimensions is sometimes increased by misalignment in the machine. This can be minimized by the use of reamers with shorter cutting edges. With special tools, such as gun reamers, accuracy can be attained in reaming holes many times longer than the cutting edge.

**Stock Removal.** Most reaming operations are not intended for the removal of large amounts of stock. This can usually be done more economically by other processes, such as drilling, boring, or core drilling. When more than 0.5 mm (0.020 in.) on diameter must be removed from a hole less than 50 mm (2 in.) in diameter, special reaming methods or boring is usually considered. Special reaming methods may include the use of gun reamers or rough and finish reaming with different types of tools.

The practical minimum stock for reaming is greatly influenced by workpiece composition and hardness. Because reaming is a cutting operation, chip formation is required for efficient operation. If too little stock is being removed, the reamer will burnish the work rather than cut it; this will result in damage to the reamer and the work surface. For soft metals, the removal of 0.2 mm (0.008 in.) on diameter per pass is near the minimum, depending on hole length and tool rigidity. For harder metals, because of the difference in chip formation, this amount can be reduced to 0.13 mm (0.005 in.). For the removal of less than 0.13 mm (0.005 in.) of stock, another machining process, such as honing, is usually preferable.

**Tolerances.** Reamers are ground to size to eliminate tool adjustments during production runs. Tolerances of 0.025 to 0.075 mm (0.001 to 0.003 in.) on diameter are practical in production reaming. Tolerances of less than 0.025 mm (0.001 in.) can be maintained, but this requires closer-than-normal control of reamer dimensions, reaming feed and speed, and all other operating variables. For reaming to extremely close tolerances, it is sometimes helpful to reduce the back taper of the reamer slightly and to match the guide bushing with the reamer so that minimum clearance may be obtained.

**The finish** of a reamed hole depends on workpiece hardness, condition of cutting edges, feed, and speed. Under optimum conditions, it is possible to obtain finishes of 1.00 μm (40 μin.) or less. However, in the production reaming of annealed steel, 2.50 to 3.20 μm (100 to 125 μin.) is more common. When extremely smooth surfaces are required, methods such as honing or burnishing should be considered.

## Workpiece Material and Hardness

The hardness of carbon and low-alloy steels has a greater effect than composition on reamability. The results of tests to determine the effect of workpiece hardness in reaming are shown in Fig. 1. Speed, feed, metal removal rate, and reamer life were compared for reaming holes of 6.4 and 13 mm (¼ and ½ in.) diameter in three similar low-alloy steels (4130, 4330, and 4340) at 15, 47, and 52 HRC, respectively. With increasing hardness of the steel workpiece, the metal removal rate was progressively lower because of the reduced cutting speed and feed rate. Life of the M10 high-speed steel reamers used also decreased progressively despite the reduction in speed and feed.

Soft metals such as aluminum and brass can be reamed at speeds five to ten times greater than those for annealed steel. However, free-cutting low-carbon steel, such as 1113, can be reamed to a smooth finish at nearly the highest speeds. On the other hand, low-carbon steel that does not contain additives for free cutting, such as 1015, produces stringier chips, thus yielding a rougher finish and requiring slower speeds. As carbon content is increased, even in free-cutting steel, abrasiveness is increased, and this shortens tool life.

## Machines

Probably every type of machine capable of rotating a tool or a workpiece has been used for reaming. Reamers are sometimes driven by hand-held air or electric motors, especially when only a few parts require reaming or when the equipment must be taken to the workpiece. Relatively large workpieces are rotated in an engine lathe, and the reamers are fed from the compound rest or the tailstock of the lathe. Most production reaming, however, is done in drilling ma-

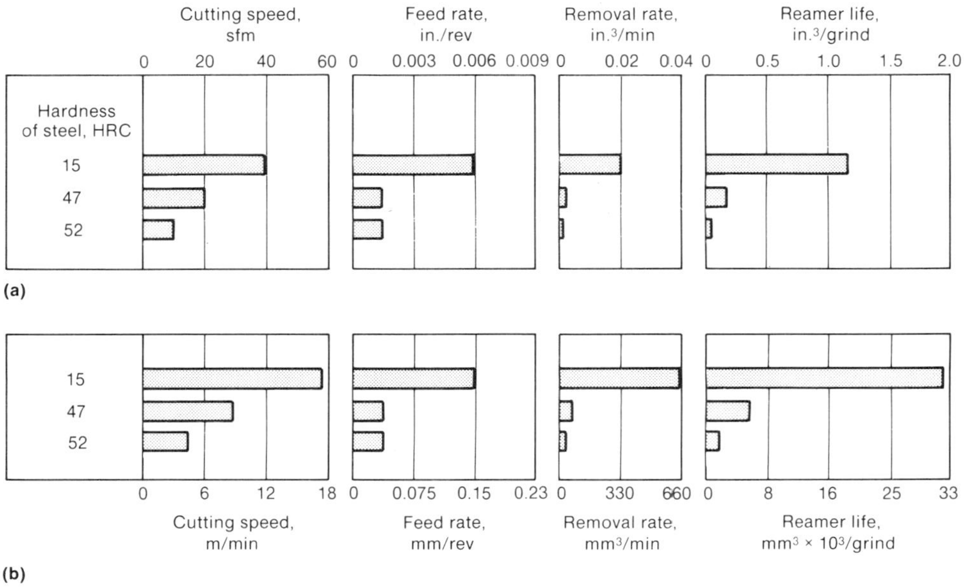


| Reaming conditions (holes of both diameters) | |
|---|---|
| Type of machine | Vertical four spindle |
| Depth of through holes, mm (in.) | 32 (1¼) |
| Stock reamed on diameter, mm (in.) | 0.4 (1/64) |
| Cutting fluid | Sulfurized oil:mineral oil (1:1) |
| Criterion for tool life | Surface finish of 1.0–1.2 μm (40–46 μin.) |

| Reamer details | 6.4 mm (¼ in.) holes | 13 mm (½ in.) holes |
|---|---|---|
| Number of flutes | 4 | 8 |
| Land width, mm (in.) | 0.175 (0.007) | 0.23 (0.009) |
| Helix angle | 0° | 0° |
| Radial rake angle | 3° | 2° |
| Radial relief angle | 17° | 12° |
| Chamfer angle | 42° | 42° |
| Chamfer relief angle | 6° | 5–6° |

**Fig. 1** Effect of the hardness of a low-alloy steel workpiece on reaming conditions and reamer life (M10 high-speed steel reamers). (a) 6.4 mm (¼ in.) diam holes. (b) 13 mm (½ in.) diam holes

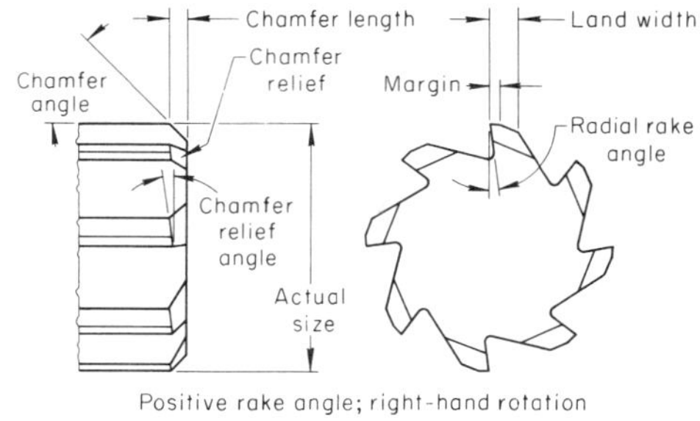


| | |
|---|---|
| Margin, mm (in.) | 1.6 (1/16) |
| Chamfer length, mm (in.) | 1.6 (1/16) |
| Chamfer angle | 45° |
| Chamfer relief angle | 6–9° |
| Radial rake angle | 7° |

**Fig. 2** Typical design of a straight-flute solid reamer. Details in the table are for reamers 13 to 50 mm (½ to 2 in.) in diameter. "Actual size" in illustration refers to the actual measured diameter of a reamer, which is usually slightly greater than the nominal size to allow for wear.

chines or as one machining step in a turret lathe or other multiple-operation machine. Machines having automatic feed can hold close tolerances more consistently than hand-fed machines. The various machines used for production reaming are described in the article "Drilling" in this Volume.

**Selection.** In many applications, reaming is supplementary to, and performed in the same sequence with, other operations, such as drilling. Under these conditions, the machine is selected mainly for the primary operation.

For the production reaming of holes less than 32 mm (1¼ in.) in diameter, machines that rotate the tool and hold the workpiece stationary (drill presses, for example) are usually the most practical and economical. Guide bushings can be used in these machines to maintain tolerances of 0.075 mm (0.003 in.) or less for holes that are several diameters long.

For maximum accuracy, however, it is preferable, if size and shape of the workpiece permit, to rotate the work and hold the tool stationary. This also applies to the reaming of holes that are considerably larger than 32 mm (1¼ in.) in diameter or that have a length-to-diameter ratio of more than about 8:1.

Size and shape of the workpiece are often major factors in the choice of a machine for reaming and for drilling or other operations that precede reaming. In many cases, although only relatively small holes are to be reamed, the workpiece is too large or heavy to be rotated in machines such as turret lathes. In other cases, workpiece size or weight may allow rotation, but asymmetric shape makes rotation impractical. For either condition, the reamer must be rotated in machines such as drill presses or boring mills. Relatively small holes in extremely large workpieces are often reamed by portable, hand-operated machines.

## Reamer Materials

Hand reamers are usually made of a carbon or low-alloy tool steel, such as W1 or O1, hardened to 62 HRC or higher. Reamers for machine operation are made of either high-speed steel or a lower-alloy tool steel for the shank, with carbide inserts for the cutting edges. The high-speed steels most used for reamers include T1, M2, M7, and M10. For reaming especially hard or abrasive metals, high-speed steels with a higher vanadium content, such as T3, T15, M3, or M4, are often used because they give longer life than the lower-alloy steels.

Because the load imposed on the tool in reaming is usually far less than in drilling, reamers require less toughness than drills. Instead, reamers should be of maximum hardness (65 HRC or higher) to obtain optimum surface finish and tool life.

Many standard and special reamers are made of solid carbide or contain carbide inserts. Although more expensive than high-speed steel, carbide will often outlast it ten times or more when reaming steel at near-optimum hardness (~20 HRC). Because of the longer life of carbide reamers, they are preferred for use on steel harder than 40 HRC.

For the efficient use of carbide, maximum rigidity is essential in the machine, reamer, and workpiece. Even with a machine in first-class condition, if the unguided or unsupported length of the reamer is more than six times its diameter, the use of carbide becomes questionable. If chatter develops, the life of carbide tools will be markedly shortened. If chatter is likely to occur, high-speed steel reamers should be used.

## Reamer Design

The design of a typical straight-flute solid reamer is shown in Fig. 2. The cutting angles and other tool details given in the accompanying table usually are not appreciably modified for reaming different work metals, except that the chamfer relief angle is generally increased to 12 to 15° for soft metals such as aluminum.

Sometimes, however, minor changes from the typical values shown in Fig. 2 can improve results. For example, the normal 45° chamfer angle at the lead end of a reamer is sometimes modified to obtain a better finish on the reamed surface.

The eight flutes shown in Fig. 2 are typical of a 25 mm (1 in.) diam reamer. Fewer flutes are used in smaller reamers, and more are used for larger ones (50 mm, or 2 in., diam reamers usually have 12 or more flutes).

If a reamer has too many flutes, it will not provide enough space for chips. If it has too few, it is likely to chatter, especially if it is a straight-flute reamer. Other possible considerations in choosing the number of flutes are discussed in the following two examples.

**Example 1: Six Versus Ten Flutes.** Changing from a six-flute chucking reamer to one with ten flutes increased tool life by nearly 35%. Comparative results from 12 production runs with each type of reamer were:

| | Six-flute | Ten-flute |
|---|---|---|
| Pieces per grind (average) | 1 052 | 1 417 |
| Pieces per tool | 12 624 | 17 004 |

Both types of reamers were operated at a speed of 30 m/min (100 sfm) and a feed of 0.74 mm/rev (0.029 in./rev).

**Example 2: Revised Reamer Design to Meet Close Tolerances.** Close tolerance (±0.01 mm, or ±0.0005 in.) was specified on the diameter of the flat-bottom blind hole reamed to a depth of 2.00/1.97 mm (0.786/0.776 in.) in the alloy 377 forging shown at the top of Fig. 3. Parts were rejected because the eight-flute reamer originally used (lower left, Fig. 3) was forced off center in bottoming out, causing out-of-roundness at the base of the hole.

Changing to a single-flute reamer (lower right, Fig. 3) and incorporating two wear strips to prevent the tool from being forced off center when bottoming out resulted in parts meeting required tolerance. Both reamers had carbide cutting edges and the same speed and feed.

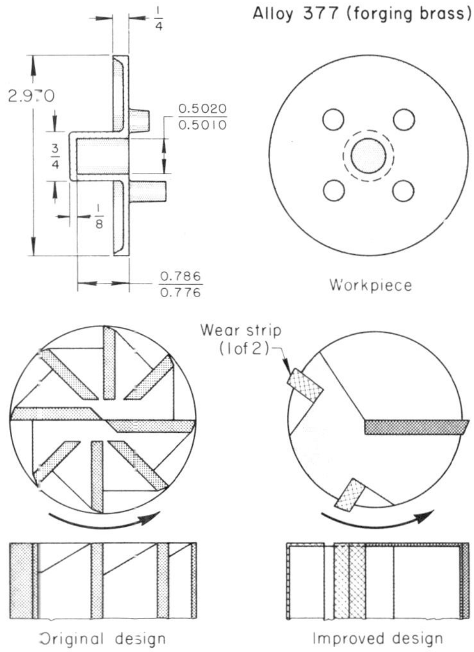


**Fig. 3** Change from eight-flute reamer (bottom left) to single-flute reamer with wear strips (bottom right) to eliminate out-of-roundness in reaming the blind hole in the forging shown at top. Dimensions given in inches. Workpiece hardness: 45 HRB

## Selection of Reamer

Any reamer must of course be compatible with the machine in which it will be used. Apart from this, reamer selection is primarily governed by one or more factors:

- Composition and hardness of workpiece
- Hole diameter
- Hole configuration
- Hole length
- Amount of stock removed
- Type of fixturing, when used
- Accuracy and finish requirements
- Production quantity
- Cost (initial and maintenance)
- Salvage value

When more than one type of reamer can produce acceptable results, the choice may depend on tool availability or established shop practice regarding the standardization of tools.

**Composition and hardness of the workpiece** are more likely to affect the choice of reamer material than type of reamer. However, there are notable exceptions; for example, shell reamers have been less satisfactory than solid reamers for reaming hard and tough metals.

**Hole Diameter.** For holes more than 50 mm (2 in.) in diameter, solid reamers made of high-speed steel are seldom used, mainly because they would be too expensive. Shell reamers are usually a better choice for holes of this size.

**Hole Configuration.** When holes to be reamed have keyways or other irregularities, spiral-flute reamers are preferred. This is because straight flutes may fail to bridge these irregularities, thus causing chatter.

**Hole Length.** When hole length is not more than two diameters, several types of reamers are suitable. However, as the length-to-diameter ratio increases, so does the problem of maintaining accuracy. Guide bushings and pilots often solve the problem, but when accuracy must be maintained in long holes, special reamers are required.

**Amount of Stock Removed.** When large amounts of stock are to be removed, a solid or a special reamer is usually preferred. Shell reamers are not suited to the removal of large amounts of stock. In some heavy-removal applications, reaming is done in two stages, using a shell reamer for the second stage.

**Type of fixturing** (when used) determines whether or not a piloted reamer is the best choice (see the section "Bushings and Fixtures" in this article).

**Accuracy and finish requirements** are related to several other factors. For greater accuracy, the workpiece should be rotated, if possible. Therefore, the reamer shank must fit the chosen machine. Accuracy depends on the rigidity of the setup and on the method of guiding the reamer. Self-guiding reamers are usually chosen for greater accuracy and finer finish in reaming long holes.

**Production Quantity.** For reaming a few pieces that can be produced with a single sharpening of the reamer, the simplest type is the logical choice. For long production runs, adjustable types should be considered to minimize sharpening and downtime.

**Cost.** Reamer cost must be considered from three related standpoints: initial cost, maintenance cost, and salvage value. High initial cost is often warranted because of low maintenance cost; for example, reamers with inserted carbide blades cost more than solid steel reamers, but may finish ten times as many holes per grind. In addition, large reamers made of solid high-speed steel may seem excessively expensive, but their salvage value is high.

**Salvage Value.** The amount realized from the sale of a worn-out high-speed steel reamer is small compared to its original cost. It is common practice, therefore, to rework a worn-out solid reamer to a smaller size rather than sell it for scrap. The economy of reworking depends mainly on whether or not the necessary equipment is available; the advisability of buying such equipment depends on the amount of tool rework to be done. Worn-out reamers can sometimes be sold to tool shops where they will be reworked for resale. If this is done, the salvage value will be greater than if the reamers are sold for scrap.

The extent to which a reamer can be reworked is also governed by the reamer design. A large reamer can be reworked progressively to smaller and smaller sizes, but this practice is limited by reamer design. The limit of decrease in reamer diameter by reworking is usually governed by depth of flute. When the diameter is so small that a further decrease would require deepening of the flutes, the reamer is usually scrapped.

## Types of Reamers

Standard reamers include these principal types:

- Straight-flute chucking reamers
- Spiral-flute chucking reamers
- End-cutting reamers
- Adjustable reamers
- Shell reamers (and expandable reamers)
- Floating-blade reamers
- Gun reamers

There are also numerous types of special-purpose reamers, including diemaker's reamers, taper reamers, bridge reamers, pipe reamers, drill-and-reamer combinations, and multiple-diameter reamers.

In the following eight sections of this article, the various types of reamers named above are described, and their areas of

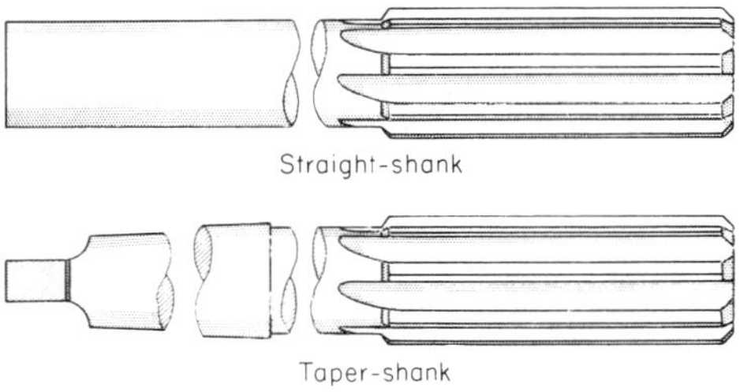


**Fig. 4** Straight-flute chucking reamers

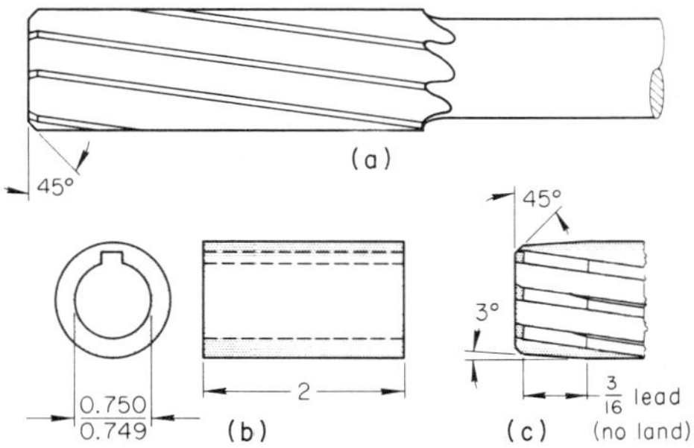


**Fig. 5** Redesign of spiral-flute reamer for keywayed hole. Surface roughness of 2.3 μm (90 μin.), obtained with conventional spiral-flute reamer (a) in reaming keywayed hole (b), was reduced to 1.25 μm (50 μin.) by modifying lead angle as in (c). Dimensions given in inches

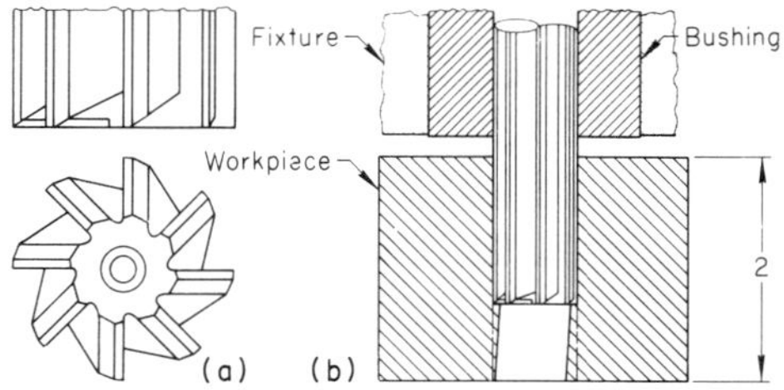


**Fig. 6** End-cutting reamers. (a) A common type of end-cutting reamer used for finishing blind holes. (b) When guided in a bushing, an end-cutting reamer can correct dimensional deviations in through holes. Dimensions given in inches

applicability are discussed. Many of these sections also cite examples of production applications for specific types of reamers.

## Applications of Straight-Flute Chucking Reamers

Straight-flute chucking reamers are general-purpose solid reamers designed for use in various machines, such as drill presses, turret lathes, and automatic bar or chucking machines. They are available with taper or straight shanks (Fig. 4) and therefore can be held in collets or split bushings, or by setscrews. These reamers are most readily available in diameters of 1.2 to 38 mm (3/64 to 1½ in.), in steps of 0.4 mm (1/64 in.); they are seldom made in other sizes.

Straight-flute chucking reamers are normally pointed with a 45° chamfer and are suited to reaming almost all metals. However, keyways or other irregularities in holes, tolerance requirements, blind holes, amount of stock that must be removed, or workpiece hardness may demand the use of reamers other than this general-purpose type.

A jobber's reamer is a modified straight-flute chucking reamer. The primary difference between the two is that the flutes of the jobber's reamer are about twice as long in proportion to overall length.

## Applications of Spiral-Flute Chucking Reamers

Spiral-flute chucking reamers differ from straight-flute reamers only in that their flutes are milled in a helix. They are used in the same machines and are available (with straight or taper shanks) in the same sizes as straight-flute reamers. Spiral-flute reamers cut with a free-reaming action and are used for the more difficult-to-ream materials. A typical spiral-flute chucking reamer is shown in Fig. 5(a).

Spiral-flute reamers are better than straight-flute reamers for reaming holes that have irregularities such as keyways (Fig. 5b). The spiral cutting edges bridge these irregularities, and this minimizes chatter, surface roughness, and size variation and prolongs reamer life.

## Applications of End-Cutting Reamers

An end-cutting reamer, which may have either straight or spiral flutes, has no chamfer on the end for use as a lead; instead, the end has cutting edges at right angles to the reamer axis (Fig. 6a). In this respect, end-cutting reamers resemble end mills.

End-cutting reamers are used for finishing blind holes that must have little or no radius at the bottom. A more important application is the correction of deviations from parallelism in drilled through holes. A reamer having a chamfered end for a lead will usually follow the hole already formed. An end-cutting reamer, when guided by a bushing (Fig. 6b), can correct out-of-parallelism by several hundredths of a millimeter.

The main disadvantage of end-cutting reamers is that they produce comparatively rough surfaces. When these reamers must be used (as for correcting hole deviations), they are usually used as roughing reamers, and a conventional reamer is used for finishing.

## Applications of Adjustable Reamers

Although a number of different types of reamers are adjustable (including floating-blade reamers and expandable shell reamers), the term adjustable reamer is generally used to refer only to a limited number of types. Two of the more common types are inserted-blade and expanding-pin adjustable reamers.

**Inserted-blade reamers,** which are made with and without adjustment for size, are toolholders in which slots are milled to receive inserted flat blades. In the adjustable type (Fig. 7), the blades are slid in angled slots by an adjusting nut to change cutting diameter. The adjusted blades are secured by a locknut and setscrews.

Inserted-blade reamers are available with either straight flutes (Fig. 7) or spiral flutes in diameters of 16 to 150 mm (5/8 to 6 in.). They are especially suited to high-volume applications in which variations in workpiece material or temperature, or in the stability of fixtures or machines, make it difficult to maintain hole size.

Between regrinds to the adjustable type of inserted-blade reamers, minor adjustments

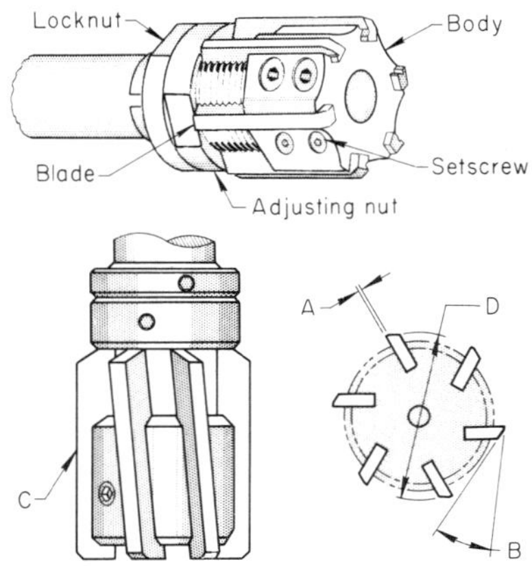


**Fig. 7** Inserted-blade adjustable reamer and typical details of design

(0.0025 to 0.005 mm, or 0.0001 to 0.0002 in.) can be made to compensate for tool wear; greater adjustments between regrinds may result in eccentricity. Recommended major adjustments before regrinding vary with reamer diameter. For example, a 16 mm (5/8 in.) diam reamer should not be adjusted more than 0.8 mm (1/32 in.), but a 90 mm (3½ in.) diam reamer can be adjusted 11 mm (7/16 in.). These adjustments can be exceeded by inserting wider blades. This is usually inadvisable, because the added overhang can cause breakage and chatter. On the other hand, it is sometimes practical in low-volume production where maximum edge sharpness can be maintained through controlled tool changes. The following example describes the procedures used and the results obtained in a high-production application of reaming with inserted-blade adjustable reamers.

**Example 3: Stud Holes to a Tolerance of ±0.025 mm (±0.001 in.).** A 2 kW (3 hp) vertical drill press was used to ream six drilled stud holes in a truck-wheel hub made of annealed 1035 steel (Fig. 8). The hub was clamped in a six-station rotating fixture. The holes had been drilled to allow 0.2 to 0.25 mm (0.008 to 0.010 in.) of stock on the diameter for removal in reaming.

Inserted-blade adjustable reamers with straight flutes were used. Each reamer held six M2 high-speed steel blades and was

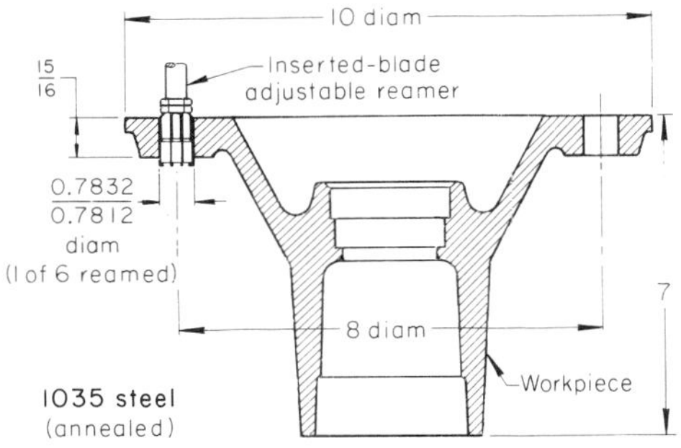


**Reamer details**

| | |
|---|---|
| Number of blade inserts | 6 |
| Insert material | M2 high-speed steel |
| Land width, mm (in.) | 0.2 (0.008) |
| Back taper, mm/mm (in./in.) | 0.002 (0.002) |
| **Operating conditions** | |
| Speed, at 146 rev/min, m/min (sfm) | 9 (30) |
| Feed, mm/rev (in./rev) | 0.89 (0.035) |
| Stock reamed on diameter, mm (in.) | 0.2–0.25 (0.008–0.01) |
| Cutting fluid | Soluble oil:water (1:20) |
| Reaming time per hole | 0.19 min |
| Reamer-blade life per grind | 800 holes |

**Fig. 8** Reaming of six stud holes in a wheel hub with a straight-flute inserted-blade adjustable reamer. Dimensions in figure given in inches

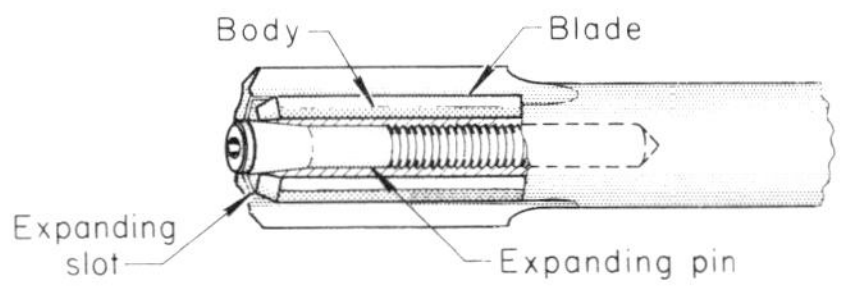


**Fig. 9** Expanding-pin adjustable reamer

ground to a concentricity of 0.005 mm (0.0002 in.) total indicator reading (TIR). Blades were circle ground, backed off to a margin of 0.2 mm (0.008 in.), and back tapered to 0.002 mm/mm (0.002 in./in.). The holes were reamed to ±0.025 mm (±0.001 in.). Additional details are given in the table accompanying Fig. 8.

**Expanding-pin reamers** make use of a tapered, screw-type expanding pin to move blades and change reamer size (Fig. 9). These reamers are available with straight flutes only, and in diameters of 11 to 63.5 mm (7/16 to 2½ in.).

The maximum adjustment for expanding-pin reamers is 0.8 mm (1/32 in.) for diameters of 11 to 6.5 mm (7/16 to ¼ in.), and about 6.5 mm (¼ in.) for diameters of 32 to 65 mm (1¼ to 2½ in.).

**Advantages.** In many high-production operations, inserted-blade and expanding-pin reamers are more economical than solid reamers because blades can be reground many times before they need replacing (body life is unlimited). In addition, the minor adjustments that can be made prolong reamer life between grinds. The major adjustments that can be made are particularly useful in low-production operations because the same reamer can be adjusted to ream holes of slightly different diameter, thus lowering tool inventory. In addition to size adjustment, inserted-blade reamers have two other advantages over solid reamers:

- Blade materials can be changed as required, using the same body
- In regrinding, tool design, including such details as angle of back taper, width of margins and lands, and radial clearance, is more easily modified for an inserted-blade reamer than for a solid reamer

## Applications of Shell Reamers

Shell reamers are used on arbors (taper or straight shank) with driving lugs and have either straight or spiral flutes (Fig. 10). The hole in a shell reamer is ground with a taper to allow the reamer to be firmly seated on the tapered arbor. Usually, hole and arbor diameters are related so that there is enough space between the end of the reamer and the shoulder of the arbor to permit removal of the reamer with minimum danger of damage.

Shell reamers are generally used only when the tool remains stationary and the work rotates, as in a turret lathe. Sometimes, however, shell reamers are used in special machines in which both work and tools are rotated. Because shell reamers, being mounted on arbors, are two-piece assemblies, they are less rigid than solid reamers. Shell reamers are more suitable for finishing operations than for removing larger amounts of stock.

**Expandable Shell Reamers.** As shown in Fig. 11, the shell portion of the expandable shell reamer is a thin-wall fluted body with carbide cutting edges, and it is forced on a hardened steel arbor that has a slight forward taper. The shell has a matching back taper so that it expands in outside diameter as it is forced farther onto the arbor. Expansion is achieved by striking, with a soft hammer, a soft metal sleeve that is inserted over the protruding end of the arbor.

Expandable shell reamers larger than 22 mm (7/8 in.) in diameter are usually of the type shown in Fig. 11(a). This reamer has a short cutting section, followed by a hardened steel flute section (for guiding the cutting section). Reamers 22 mm (7/8 in.) or less in diameter are made with full-length flutes (Fig. 11b). There are also expandable shell reamers with special plugs flush with the end to permit the reaming of blind holes.

Expandable shell reamers are not designed for adjustment to different hole sizes, but only for adjustment to compensate for tool wear. As the reamer wears out of tolerance, it is forced farther back on the arbor until the limit of expansion is reached. The shell is then removed and replaced with a new one.

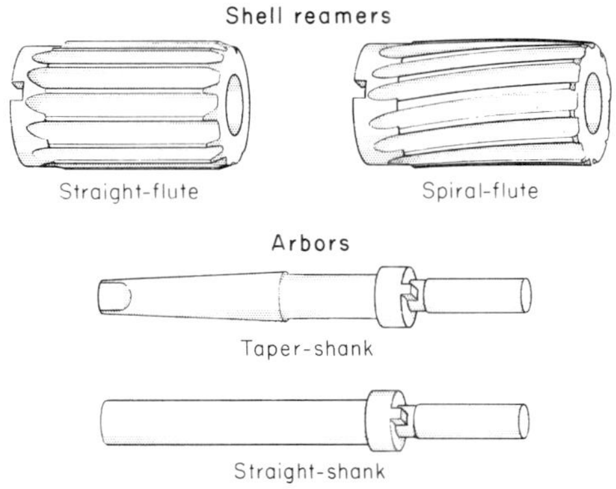


**Fig. 10** Typical shell reamers and accommodating arbors

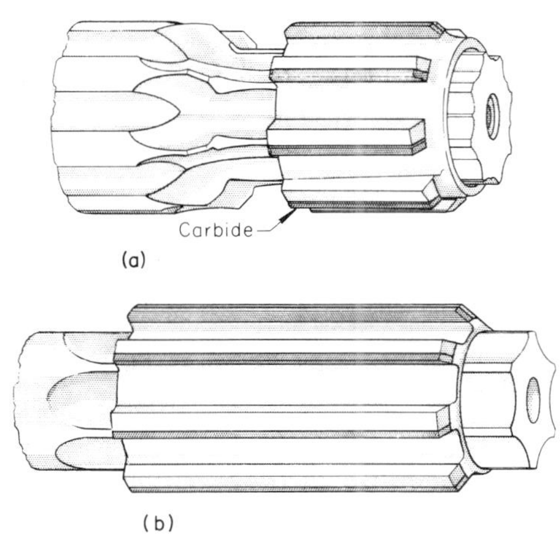


**Fig. 11** Two types of expandable shell reamers. (a) Diameters of more than 22 mm (7/8 in.). (b) Diameters of 22 mm (7/8 in.) or less

The maximum amount a shell reamer can be expanded depends on diameter: approximately 0.18 mm (0.007 in.) for a 13 mm (½ in.) diameter, 0.25 mm (0.010 in.) for a 25 mm (1 in.) diameter, and 0.5 mm (0.020 in.) for a 50 mm (2 in.) diameter. Shells expand evenly and need not be ground for roundness or straightness after each expansion. The following example describes an application in which shell reamers were superior to solid reamers in tool life and tool cost per piece.

**Example 4: Shell Versus Solid Reamers for Gray Iron.** In one plant, solid reamers were compared with shell reamers for reaming 29.97/29.95 mm (1.1798/1.1790 in.) diam shaft holes in gray iron water-pump bodies (131 to 207 HB). Specifications called for maximum out-of-roundness of 0.0075 mm (0.0003 in.), maximum taper of 0.005 mm (0.0002 in.), and finish of 1.5 μm (60 μin.).

Figure 12 shows the workpiece and the two types of reamers used. The solid reamer had a 4.67 mm (1.84 in.) long cutting tip of solid tungsten carbide on a shank of high-speed steel, and the shell reamer had brazed carbide cutting edges.

Both reamers were used in a vertical two-spindle machine with the same operat-

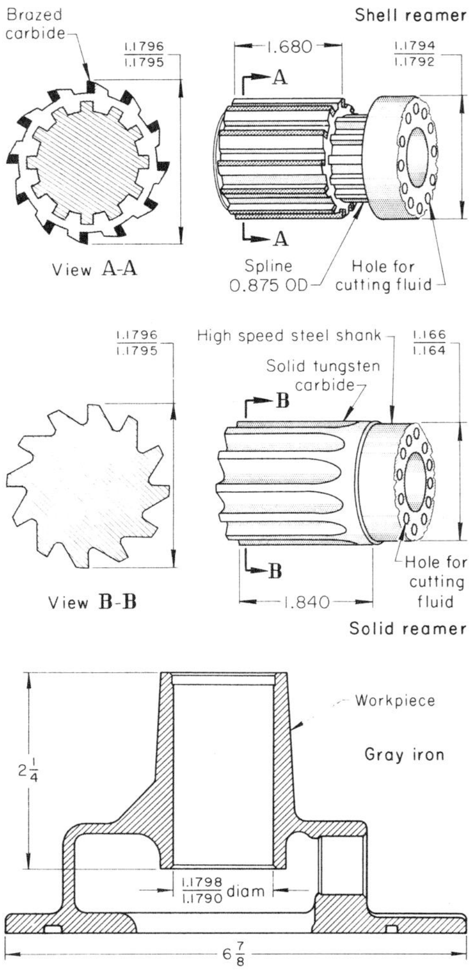


**Operating conditions for both reamers**

| | |
|---|---|
| Type of machine | Vertical two-spindle |
| Speed, at 875 rev/min, m/min (sfm) | 82 (270) |
| Feed, mm/rev (in./rev) | 0.46 (0.018) |
| Depth of cut, mm (in.) | 0.48 (0.019) |
| Reaming time per hole | 13.7 s |
| Setup time | 5 min |
| Cutting fluid | Soluble oil:water (1:20) |

**Comparison of results**

| | Solid | Shell |
|---|---|---|
| Pieces per grind (average) | 2 748 | 3 295 |
| Number of runs | 9 | 13 |
| Number of regrinds | 8 | 12 |
| Regrinding time, min | 30 | 28 |
| Total pieces per tool | 24 732 | 42 835 |

**Fig. 12** Shell versus solid reamers. Shell reamer with brazed carbide cutting edges was superior to solid carbide reamer with high-speed steel shank for reaming the shaft hole in this gray iron water-pump body. Dimensions in figure given in inches. Workpiece hardness: 131 to 207 HB

ing conditions (see table accompanying Fig. 12). In each setup, a guide bushing was used, and cutting fluid was supplied under pressure to the tool.

The results obtained in 9 runs for the solid reamer and in 13 for the shell are tabulated with Fig. 12. As this comparison shows, tool life (number of pieces per grind) for the shell reamers was 20% greater than that for the solid reamers, and tool cost per piece

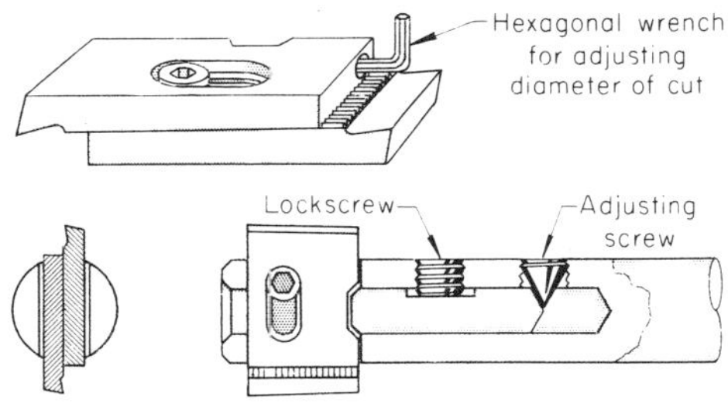


**Fig. 13** Details of a typical floating-blade reamer

was 50% less. The shorter life of the solid reamer was due largely to the difference in the diameters of its shank and head. Because of this difference, the solid reamer was guided through the bushing for only a portion of the cut, while the shell reamer was guided during the entire cut.

## Applications of Floating-Blade Reamers

Floating-blade reamers have replaceable and adjustable cutting edges (Fig. 13). The blades, of either high-speed steel or carbide, can be adjusted for wear (or, within narrow limits, for different hole sizes) by turning an adjusting screw in the blade assembly with a hexagonal wrench. The two-piece blade assembly is held in a slot in the bar (as shown in Fig. 13), in which it can float. The amount of float is controlled by the lockscrew and adjusting screw in the toolholder. By floating, the cutters can follow the surface of the bore being reamed, maintaining tolerances as close as 0.01 mm (0.0005 in.). Floating also permits compensation for machine or holder errors.

Floating-blade reamers provide liberal clearance for chips and in special circumstances may cut faster than solid reamers. Other advantages of floating-blade reamers are:

- Cutters can be replaced, or removed for adjustment, without removing the holder from the machine
- The holder is not expendable and can be used for many cutters
- Tool cost is usually less than that of comparable-size solid reamers. In one application, for example, the initial cost of a carbide-tip floating-blade reamer for 25 mm (1 in.) holes was only about two-thirds that of a solid straight-flute reamer with carbide cutting edges

The use of floating-blade reamers is usually restricted to applications in which the workpiece is rotated, as in turret lathes. An application in which productivity was increased because higher speeds could be used for floating-blade carbide reamers than for solid high-speed steel reamers is described in the following example.

**Example 5: Floating-Blade Carbide Versus Solid High-Speed Steel Reamers.** Holes 50.5 mm (1.988 in.) in diameter and

### Table 1 Solid high-speed steel reamers versus floating-blade carbide reamers

| Reaming condition or result | Type of reamer: Solid | Floating-blade |
|---|---|---|
| Speed, rev/min | 132 | 215 |
| Speed, m/min (sfm) | 21 (69) | 34 (112) |
| Feed, mm/rev (in./rev) | 0.234 (0.0092) | 0.234 (0.0092) |
| Stock reamed on diameter, mm (in.) | 0.3 (0.012) | 0.15 (0.006) |
| Time per piece, min | 0.46 | 0.21 |
| Reamer life/grind, pieces | 70 | 300 |
| Total reamer life, pieces | 1700 | 5500 |

Data are for reaming 50.5 mm (1.988 in.) diam holes 1.3 mm (0.050 in.) deep in gears made of 1118 steel at 160 HB. Reaming was done in a turret lathe using a 1:20 mixture of soluble oil and water as cutting fluid. Both types of reamers produced a 3.20 μm (125 μin.) finish.

1.3 mm (0.050 in.) deep were reamed in gears made of 1118 steel (160 HB) to a total tolerance of 0.025 mm (0.001 in.) using eight-flute solid reamers made of high-speed steel. As Table 1 shows, a change to floating-blade reamers (two carbide blades) allowed a 63% increase in speed at the same feed with a proportionate decrease in reaming time per piece. Reamer life between grinds was more than four times as long for the floating-blade reamer as for the solid reamer; total life was three times as long.

Some of the increased life for the floating-blade reamer was due to the decreased amount of stock on the hole diameter. Less stock was required because the increased accuracy obtainable with the floating-blade reamer permitted drilling closer to final size.

## Applications of Gun Reamers

Gun reamers consist of a hollow shank with a cutting edge (usually carbide) fastened to the end. Cutting fluid is fed through the stem under pressures of 170 to 700 kPa (25 to 100 psi) to enhance cutting, to cool the work and tools, and to flush away chips. Gun reamers are used in machines in which the work is rotated and the tool remains stationary.

Blind-hole gun reamers have full-length flutes to permit backward ejection of chips. Through-hole gun reamers, with fixed or replaceable heads (Fig. 14), have a full round stem to the beginning of the cutter and from this point have a cutaway portion to allow cutting fluid to flow.

Fixed-head (conventional) gun reamers (Fig. 14a) are available in diameters of 3.2 to 50 mm (1/8 to 2 in.), and sometimes larger. Replaceable-head types (Fig. 14b) are available in diameters of 25 to 50 mm (1 to 2 in.).

A gun reamer resembles a gun drill in that it operates on the self-piloting principle and yields similar straightness, tolerance, and finish. Speeds for gun reaming are usually approximately 60 m/min (200 sfm) (about the same as for gun drilling), but feeds are generally 50 to 100% greater than those

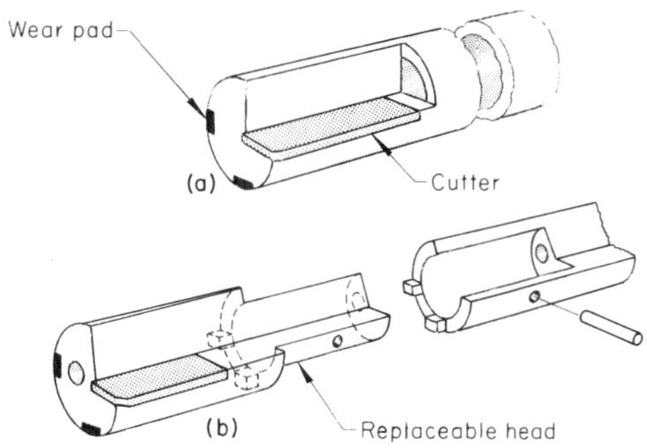


**Fig. 14** Typical gun reamers for through holes. (a) Fixed-head (conventional) type. (b) Replaceable-head type

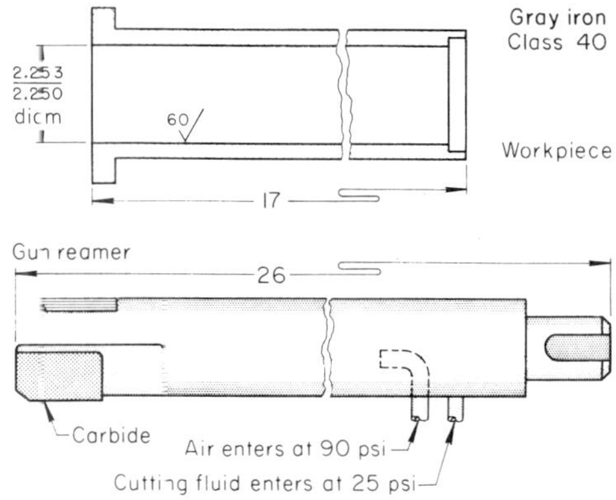


**Reaming conditions**

| | |
|---|---|
| Machine | Turret lathe |
| Speed, at 329 rev/min, m/min (sfm) | 58 (190) |
| Feed, mm/rev (in./rev) | 0.1 (0.004) |
| Stock reamed on diameter, mm (in.) | 4.8–9.5 (3/16–3/8) |
| Cutting fluid | Soluble oil:water (1:15)(a) |
| Setup time | 1 h |
| Reamer life per grind | 41 pieces |
| Maximum permissible runout, mm/m (in./ft) | 0.08 (0.001) |

**Results**

| | |
|---|---|
| Tolerance held on diameter, mm (in.) | Within 0.036 (0.0014) |
| Surface finish obtained, μm (μin.) | 1.0–1.5 (40–60) |

(a) Introduction with air at shank end of tool, as shown in illustration

**Fig. 15** Gun reaming of a taper-cored hole. Dimensions in figure given in inches

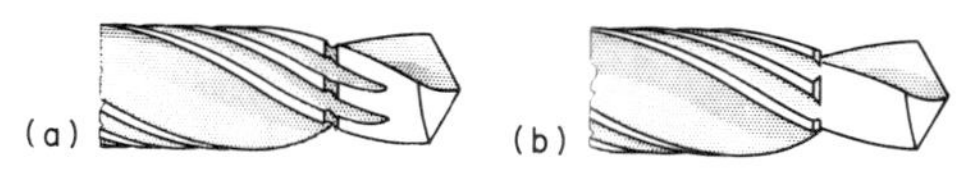


**Fig. 16** Two types of drill-reamer combination tools

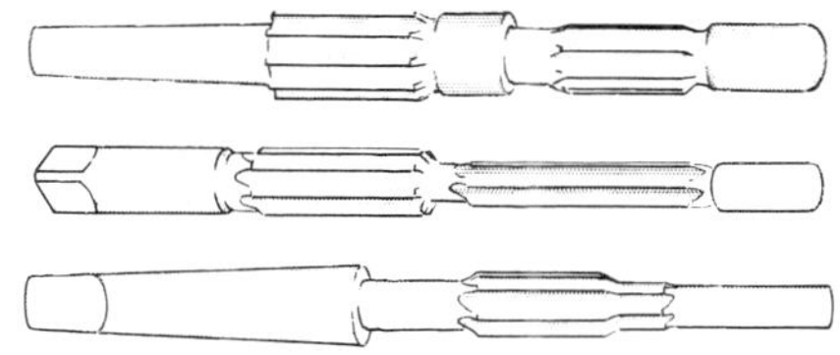

**Fig. 17** Three types of multiple-diameter reamers

used in gun drilling. In the gun reaming of soft metals, feeds as great as 0.25 mm/rev (0.010 in./rev) are often used. In reaming steel, the usual stock allowance is 0.25 to 1.0 mm (0.010 to 0.040 in.) on the diameter, depending on reamer size and workpiece hardness. In reaming softer metals, and sometimes in reaming annealed steel or cast iron, as much as 13 mm (½ in.) of stock can be removed from the diameter. The following example illustrates a typical application of gun reaming.

**Example 6: Accuracy and Finish in Reaming Holes Seven Diameters Long.** Taper-cored through holes in 430 mm (17 in.) long hydraulic-press quills made of class 40 gray iron were gun reamed in a single pass in a turret lathe. During machining, the workpiece (Fig. 15) was held in a three-jaw face plate fixture. Before reaming, a 57.15/57.10 mm (2.250/2.248 in.) diam starting hole was bored to a depth of 13 mm (½ in.). The carbide-tip gun reamer (Fig. 15) was then fed into the workpiece, and it removed stock ranging from about 4.8 mm (3/16 in.) on the diameter at the entrance end (beyond the starting hole) to about 9.5 mm (3/8 in.) at the exit end of the tapered hole.

Although specifications called for tolerance within 0.075 mm (0.003 in.) and a surface finish of 1.5 μm (60 μin.), the gun reamers held tolerance within 0.036 mm (0.0014 in.) and produced an average finish of 1.25 μm (50 μin.). Additional processing details are provided in the table with Fig. 15.

## Applications of Special-Purpose Reamers

Several types of reamers designed for special applications are readily available. Among these are oil-hole reamers, diemaker's reamers, taper reamers, bridge reamers, pipe reamers, drill-and-reamer combinations, and multiple-diameter reamers.

**Oil-hole reamers** have a hollow core through which cutting fluid is forced to flush the chips out through the flutes. They are available as solid reamers in diameters up to about 38 mm (1½ in.) or as inserted-blade reamers in larger sizes. Oil-hole reamers can produce good finish in horizontal reaming.

**Diemaker's reamers** have a high helix angle and left-hand high-spiral flutes and are tapered about 0.013 mm/mm (0.013 in./in.). They are used for reaming dowel-pin holes in die parts.

**Taper reamers** produce tapered holes that receive taper shanks and taper pins.

**Bridge reamers** are used for reaming holes in products such as ship plate and structural members. They are tapered at the end so they can enlarge an out-of-line hole until the entire body of the reamer can enter and enlarge the hole to the required size. They are not intended for reaming tapers. They are available with either spiral or straight flutes and in diameters of 10.3 to 38 mm (13/32 to 1½ in.).

**Pipe reamers** are short, stubby tools used to taper-ream pipe fittings for tapping. They have a taper of 62 mm/m (¾ in./ft) and are of 3.2 to 50 mm (1/8 to 2 in.) pipe size.

**Drill-and-Reamer Combinations.** When producing shallow holes, it is sometimes possible to eliminate a second operation by using a combination drill and reamer. Two types are shown in Fig. 16. The tool shown in Fig. 16(a) is made by milling two spiral flutes the entire length of the tool. The drill section is slightly smaller in diameter than the reamer section, and the drill margins extend only the length of the drill section. The reamer section consists of one to three reamer flutes milled in each land left by the drill-flute milling cutter. This type of tool is subject to the same difficulties in regrinding as those encountered with step drills. These difficulties can be overcome by using the tool shown in Fig. 16(b), which is of subland construction. With this tool, the two drill flutes and cutting edges extend back through the reamer section; reamer cutting edges are in addition to these and begin at a suitable distance from the drill point.

**Multiple-Diameter Reamers.** When two or more concentric holes of different diameters must be reamed, if a separate reamer is used for each hole, several operations are necessary, and it is difficult to keep the holes in line. The use of a multiple-diameter reamer (Fig. 17) permits the operations to be performed together and ensures alignment of the holes.

In many multiple-diameter reamers, each reaming section is preceded by a pilot section to bear in the drilled holes or in bushings in the fixture. Generally, all reaming sections start to cut at the same time.

## Speed and Feed

Nominal speeds and feeds for the reaming of carbon and low-alloy steels at various hardnesses are given in Tables 2 and 3. The rates in Table 2 are for six-flute reamers made of high-speed steel; those in Table 3 are for six-flute reamers with solid steel bodies and carbide cutting edges.

**Speed.** Because of their design, most reamers are more easily damaged than drills. Therefore, it is usual practice to ream a hole at about two-thirds the speed at which it was drilled.

From Tables 2 and 3 it is evident that speed depends mainly on the composition and hardness of the work metal and on tool material. For reaming under similar conditions, recommended speeds for carbide reamers are three to four times those for high-speed steel reamers.

Although the composition of the steel being reamed has some effect on optimum speed, hardness has a much greater effect. The nominal speed in Table 2 for reaming steel at 400 HB is 6 m/min (20 sfm), and it is

**Table 2 Nominal speeds and feeds for the reaming of carbon and low-alloy steels with high-speed steel reamers**

Based on the removal of 0.38–0.5 mm (0.015–0.020 in.) from the hole diameter, with six-flute reamers made of high-speed steels M1, M2, and M7, except T15 for hardness above 375 HB. All speeds and feeds are for use with ample cutting fluid.

| Typical steel(a) | Hardness, HB | Speed, m/min | Speed, sfm | Feed, mm/rev (in./rev), for reamer diameter of: 3.2 mm (1/8 in.) | 6.4 mm (1/4 in.) | 13 mm (1/2 in.) | 25 mm (1 in.) | 38 mm (1 1/2 in.) | 50 mm (2 in.) |
|---|---|---|---|---|---|---|---|---|---|
| **Carbon and low-alloy steels (except free-cutting grades)** | | | | | | | | | |
| 1020, 1045, 4140, 7140, and 8620, at hardness ranges listed at right | 85–125 | 20 | 65 | 0.10 (0.004) | 0.18 (0.007) | 0.25 (0.010) | 0.38 (0.015) | 0.50 (0.020) | 0.64 (0.025) |
| | 125–175 | 17 | 55 | 0.075 (0.003) | 0.13 (0.005) | 0.25 (0.010) | 0.38 (0.015) | 0.50 (0.020) | 0.64 (0.025) |
| | 175–225 | 14 | 45 | 0.075 (0.003) | 0.10 (0.004) | 0.20 (0.008) | 0.30 (0.012) | 0.38 (0.015) | 0.50 (0.020) |
| | 225–275 | 12 | 40 | 0.064 (0.0025) | 0.10 (0.004) | 0.20 (0.008) | 0.30 (0.012) | 0.38 (0.015) | 0.50 (0.020) |
| | 275–325 | 9 | 30 | 0.064 (0.0025) | 0.10 (0.004) | 0.20 (0.008) | 0.30 (0.012) | 0.38 (0.015) | 0.50 (0.020) |
| | 325–375 | 7.5 | 25 | 0.05 (0.002) | 0.075 (0.003) | 0.13 (0.005) | 0.23 (0.009) | 0.30 (0.012) | 0.38 (0.015) |
| | 375–425 | 6 | 20 | 0.05 (0.002) | 0.075 (0.003) | 0.10 (0.004) | 0.18 (0.007) | 0.25 (0.010) | 0.30 (0.012) |
| **Free-cutting carbon and low-alloy steels** | | | | | | | | | |
| 1112 and 12L14 | 100–150 | 23 | 75 | 0.10 (0.004) | 0.15 (0.006) | 0.25 (0.010) | 0.38 (0.015) | 0.50 (0.020) | 0.64 (0.025) |
| | 150–200 | 24 | 80 | 0.10 (0.004) | 0.15 (0.006) | 0.25 (0.010) | 0.38 (0.015) | 0.50 (0.020) | 0.64 (0.025) |
| | 200–250 | 15 | 50 | 0.05 (0.002) | 0.10 (0.004) | 0.20 (0.008) | 0.30 (0.012) | 0.38 (0.015) | 0.50 (0.020) |
| 1117 and 1137 | 100–150 | 21 | 70 | 0.10 (0.004) | 0.15 (0.006) | 0.25 (0.010) | 0.38 (0.015) | 0.50 (0.020) | 0.64 (0.025) |
| | 150–200 | 20 | 65 | 0.075 (0.003) | 0.13 (0.005) | 0.25 (0.010) | 0.38 (0.015) | 0.50 (0.020) | 0.64 (0.025) |
| | 275–325 | 14 | 45 | 0.05 (0.002) | 0.10 (0.004) | 0.20 (0.008) | 0.30 (0.012) | 0.38 (0.015) | 0.50 (0.020) |
| | 325–375 | 9 | 30 | 0.05 (0.002) | 0.10 (0.004) | 0.20 (0.008) | 0.30 (0.012) | 0.38 (0.015) | 0.50 (0.020) |
| | 375–425 | 6 | 20 | 0.05 (0.002) | 0.075 (0.003) | 0.10 (0.004) | 0.18 (0.007) | 0.25 (0.010) | 0.30 (0.012) |
| 4140+S and 41L40 | 150–200 | 18 | 60 | 0.10 (0.004) | 0.15 (0.006) | 0.25 (0.010) | 0.38 (0.015) | 0.50 (0.020) | 0.64 (0.025) |
| | 200–250 | 15 | 50 | 0.075 (0.003) | 0.13 (0.005) | 0.23 (0.009) | 0.33 (0.013) | 0.46 (0.018) | 0.56 (0.022) |
| | 275–325 | 11 | 35 | 0.05 (0.002) | 0.10 (0.004) | 0.20 (0.008) | 0.30 (0.012) | 0.38 (0.015) | 0.50 (0.020) |
| | 325–375 | 7.5 | 25 | 0.05 (0.002) | 0.075 (0.003) | 0.13 (0.005) | 0.23 (0.009) | 0.30 (0.012) | 0.38 (0.015) |
| | 375–425 | 6 | 20 | 0.05 (0.002) | 0.075 (0.003) | 0.10 (0.004) | 0.175 (0.007) | 0.25 (0.010) | 0.30 (0.012) |

(a) Each steel listed is a frequently used grade in a group of similar steels. A listing of the steels in the various groups is provided in Table 2 in the article "Boring" in this Volume. Source: Metcut Research Associates

17 to 24 m/min (55 to 80 sfm) for steel at 160 HB, regardless of differences in composition. As Tables 2 and 3 show, greater differences in nominal speed are related to the differences in hardness of a steel in any of the groups.

Because other variables, such as reamer design, rigidity of the setup, and depth of hole, influence optimum speed, the nominal speeds listed in Tables 2 and 3 can serve only as starting points in the selection of optimum speed for given conditions. Speeds used in Fig. 1 and in the applications described in some examples in this article are summarized in Table 4. Although some speeds shown in Table 4 agree closely with the nominal speed for the same work metal and hardness in Tables 2 and 3, there are several evident differences, but examination of all the details of the applications summarized in Table 4 usually explains the differences. Figure 1, for example, shows that two reaming speeds (12 and 17.7 m/min, or 40 and 58 sfm) were used for reaming 4130 steel at 15 HRC. In Table 2, the nearest nominal speed is shown as 14 m/min (45 sfm) (4140 group at 175 to 225 HB). However, in Fig. 1, a four-flute reamer was used for the 6.4 mm (1/4 in.) diam holes, and an eight-flute reamer was used for the 13 mm (1/2 in.) diam holes, while the speeds in Table 2 are based on the use of six-flute reamers.

In Example 3, annealed 1035 steel was reamed at 14 to 9 m/min (45 and 30 sfm), while Table 2 suggests 17 m/min (55 sfm) for comparable steel (1045 at 125 to 175 HB). However, in Example 3, inserted-blade reamers were used (the data in Table 2 are for solid reamers). The shape of the workpiece in Example 3 also suggests that less-than-normal rigidity prevailed.

In Example 5, the speed of 21.0 m/min (69 sfm) used for reaming annealed 1118 steel at 160 HB with a high-speed steel reamer is very near the 20 m/min (65 sfm) speed Table 2 shows for the 1117 group at 150 to 200 HB, although the reamer used in Example 5 had eight flutes. When a carbide reamer was substituted for this application, it was used at a speed of 34 m/min (112 sfm), or less than half the 79 m/min (260 sfm) speed suggested in Table 3. The carbide reamer, however, was a floating-blade type, and values in Table 3 are based on the use of solid reamers.

In Example 7, a hole in a stack of silicon steel laminations was reamed at 15 m/min (50 sfm). This speed is less than the 20 m/min (65 sfm) speed shown in Table 2 for reaming 1020 in its softest condition. However, even though silicon steels are low in carbon content, the increased silicon content affects reamability. In addition, in Example 7, no cutting fluid was used, while speeds listed in Table 2 are based on the use of cutting fluid.

**Feed** is influenced not only by the variables that govern speed but also by hole size and amount of metal being reamed (depth of cut). The nominal feed rates given in Tables 2 and 3 for six different hole diameters are based on the enlargement of all diameters ranging from 0.38 to 0.5 mm (0.015 to 0.020 in.).

The feed rates used in specific applications are likely to vary more from the nominal rates in Tables 2 and 3 than are speeds. A comparison of the feeds actually used in examples in this article (Table 4) with nominal feeds (Tables 2 and 3) shows that some agree closely and that others differ by a factor of three or more.

In Fig. 1, which involves reaming a total of 0.4 mm (1/64 in.) from 6.4 and 13 mm (1/4 and 1/2 in.) diam holes in 4130 steel at 15 HRC, a feed of 0.15 mm/rev (0.006 in./rev) was used for both sizes. Table 2 suggests 0.1 and 0.2 mm/rev (0.004 and 0.008 in./rev) for reaming 6.4 and 13 mm (1/4 and 1/2 in.) diam holes under similar conditions (4140 group at 175 to 225 HB). In reaming the piece shown in Fig. 5, the feed (Table 4) was only 0.13 mm/rev (0.005 in./rev), less than half the rate suggested in Table 2. Lower feeds were used because the hole had a keyway and because surface finish was a major objective.

It is sometimes advisable to use a higher feed and a slower speed. This is indicated in Example 3, in which the speed was about 45% less than nominal (Table 2 for 1045 steel) and the feed was about 200% greater than nominal.

In Example 2, a feed rate of 0.23 mm/rev (0.0092 in./rev) was used for both a solid high-speed steel reamer and a floating-blade carbide reamer. This is much less than the nominal 0.64 mm/rev (0.025 in./rev) in Tables 2 and 3 for a 50 mm (2 in.) diam hole in 1117 steel.

**Selection of Speed and Feed.** Reaming speed and feed have important effects on results (mainly surface finish and tool life) and on cost. When setting up for a reaming

**Table 3 Nominal speeds and feeds for the reaming of carbon and low-alloy steels with carbide reamers**

Based on the removal of 0.38–0.5 mm (0.015–0.020 in.) from the hole diameter, with six-flute reamers made of grade C-2 carbide. All speeds and feeds are for use when cutting fluid is amply supplied to the cutting edges of the reamer.

| Typical steel(a) | Hardness, HB | Speed m/min | Speed sfm | Feed, mm/rev (in./rev), for reamer diameter of: 3.2 mm (1/8 in.) | 6.4 mm (1/4 in.) | 13 mm (1/2 in.) | 25 mm (1 in.) | 38 mm (1 1/2 in.) | 50 mm (2 in.) |
|---|---|---|---|---|---|---|---|---|---|
| **Carbon and low-alloy steels (except free-cutting grades)** | | | | | | | | | |
| 1020 and 1045, at hardness ranges listed at right | 85–125 | 79 | 260 | 0.10 (0.004) | 0.18 (0.007) | 0.25 (0.010) | 0.38 (0.015) | 0.50 (0.020) | 0.64 (0.025) |
| | 125–175 | 75 | 250 | 0.10 (0.004) | 0.18 (0.007) | 0.25 (0.010) | 0.38 (0.015) | 0.50 (0.020) | 0.64 (0.025) |
| | 175–225 | 58 | 190 | 0.06 (0.0025) | 0.10 (0.004) | 0.20 (0.008) | 0.30 (0.012) | 0.38 (0.015) | 0.50 (0.020) |
| | 225–275 | 50 | 165 | 0.06 (0.0025) | 0.10 (0.004) | 0.20 (0.008) | 0.30 (0.012) | 0.38 (0.015) | 0.50 (0.020) |
| | 275–325 | 30 | 100 | 0.06 (0.0025) | 0.10 (0.004) | 0.20 (0.008) | 0.30 (0.012) | 0.38 (0.015) | 0.50 (0.020) |
| | 325–375 | 27 | 90 | 0.05 (0.002) | 0.075 (0.003) | 0.13 (0.005) | 0.23 (0.009) | 0.30 (0.012) | 0.38 (0.015) |
| | 375–425 | 18 | 60 | 0.05 (0.002) | 0.075 (0.003) | 0.10 (0.004) | 0.18 (0.007) | 0.25 (0.010) | 0.30 (0.012) |
| 4140, 7140, and 8620 | 125–175 | 60 | 200 | 0.075 (0.003) | 0.13 (0.005) | 0.25 (0.010) | 0.38 (0.015) | 0.50 (0.020) | 0.64 (0.025) |
| | 175–225 | 55 | 180 | 0.075 (0.003) | 0.10 (0.004) | 0.20 (0.008) | 0.30 (0.012) | 0.38 (0.015) | 0.50 (0.020) |
| | 225–275 | 49 | 160 | 0.075 (0.003) | 0.10 (0.004) | 0.20 (0.008) | 0.30 (0.012) | 0.38 (0.015) | 0.50 (0.020) |
| | 275–325 | 37 | 120 | 0.06 (0.0025) | 0.10 (0.004) | 0.20 (0.008) | 0.30 (0.012) | 0.38 (0.015) | 0.50 (0.020) |
| | 325–375 | 26 | 85 | 0.05 (0.002) | 0.075 (0.003) | 0.13 (0.005) | 0.23 (0.009) | 0.30 (0.012) | 0.38 (0.015) |
| | 375–425 | 18 | 60 | 0.05 (0.002) | 0.075 (0.003) | 0.10 (0.004) | 0.175 (0.007) | 0.25 (0.010) | 0.30 (0.012) |
| | 45–48 HRC | 12 | 40 | 0.05 (0.002) | 0.075 (0.003) | 0.10 (0.004) | 0.15 (0.006) | 0.20 (0.008) | 0.25 (0.010) |
| | 50–52 HRC | 6 | 20 | 0.05 (0.002) | 0.075 (0.003) | 0.10 (0.004) | 0.15 (0.006) | 0.20 (0.008) | 0.25 (0.010) |
| **Free-cutting carbon and low-alloy steels** | | | | | | | | | |
| 1112 and 12L14 | 100–150 | 90 | 300 | 0.10 (0.004) | 0.15 (0.006) | 0.25 (0.010) | 0.38 (0.015) | 0.50 (0.020) | 0.64 (0.025) |
| | 150–200 | 102 | 335 | 0.10 (0.004) | 0.15 (0.006) | 0.25 (0.010) | 0.38 (0.015) | 0.50 (0.020) | 0.64 (0.025) |
| | 200–250 | 60 | 200 | 0.05 (0.002) | 0.10 (0.004) | 0.20 (0.008) | 0.30 (0.012) | 0.38 (0.015) | 0.50 (0.020) |
| 1117 and 1137 | 100–150 | 90 | 300 | 0.10 (0.004) | 0.15 (0.006) | 0.25 (0.010) | 0.38 (0.015) | 0.50 (0.020) | 0.64 (0.025) |
| | 150–200 | 79 | 260 | 0.075 (0.003) | 0.13 (0.005) | 0.25 (0.010) | 0.38 (0.015) | 0.50 (0.020) | 0.64 (0.025) |
| | 275–325 | 55 | 180 | 0.05 (0.002) | 0.10 (0.004) | 0.20 (0.008) | 0.30 (0.012) | 0.38 (0.015) | 0.50 (0.020) |
| | 325–375 | 30 | 100 | 0.05 (0.002) | 0.10 (0.004) | 0.20 (0.008) | 0.30 (0.012) | 0.38 (0.015) | 0.50 (0.020) |
| | 375–425 | 18 | 60 | 0.05 (0.002) | 0.075 (0.003) | 0.10 (0.004) | 0.175 (0.007) | 0.25 (0.010) | 0.30 (0.012) |
| 4140+S and 41L40 | 150–200 | 75 | 245 | 0.10 (0.004) | 0.15 (0.006) | 0.25 (0.010) | 0.38 (0.015) | 0.50 (0.020) | 0.64 (0.025) |
| | 200–250 | 62.5 | 205 | 0.075 (0.003) | 0.13 (0.005) | 0.20 (0.008) | 0.30 (0.012) | 0.38 (0.015) | 0.50 (0.020) |
| | 275–325 | 46 | 150 | 0.05 (0.002) | 0.10 (0.004) | 0.20 (0.008) | 0.30 (0.012) | 0.38 (0.015) | 0.50 (0.020) |
| | 325–375 | 29 | 95 | 0.05 (0.002) | 0.075 (0.003) | 0.13 (0.005) | 0.23 (0.009) | 0.30 (0.012) | 0.38 (0.015) |
| | 375–425 | 18 | 60 | 0.05 (0.002) | 0.075 (0.003) | 0.10 (0.004) | 0.18 (0.007) | 0.25 (0.010) | 0.30 (0.012) |
| | 45–48 HRC | 12 | 40 | 0.05 (0.002) | 0.075 (0.003) | 0.10 (0.004) | 0.15 (0.006) | 0.20 (0.008) | 0.25 (0.010) |

(a) Each steel listed is a frequently used grade in a group of similar steels. A listing of the steels in the various groups is provided in Table 2 in the article "Boring" in this Volume. Source: Metcut Research Associates

operation without prior experience, it is advisable to use a conservative speed and feed combination first. For example, in reaming holes of 3.2 to 50 mm (1/8 to 2 in.) diameter in low-carbon steel, it would be well to start with a speed of 8 m/min (60 sfm) and a feed of 0.1 to 0.64 mm/rev (0.004 to 0.025 in./rev). Speed, feed, or both can then be increased until the first signs of tool chatter appear. When this point has been determined, it is advisable to keep speeds and feeds below it for optimum finish and tool life; often a change in speed of 10% or less can cause or eliminate chatter.

For reaming soft brass, a speed of 75 m/min (250 sfm) is a good starting point. About twice that, or 150 m/min (500 sfm), is appropriate for a first setting in reaming aluminum. Much slower speeds must be used in steel harder than 20 HRC. As hardness increases, possible speed decreases. The maximum workable speed for steel at 52 HRC is 4.5 m/min (15 sfm).

## Bushings and Fixtures

When a hole to be reamed must be an exact distance from some point or from another hole, reaming in jigs or fixtures is preferred. These devices hold the workpiece securely, and the reamer is guided in bushings set in exact relation to locating points on the workpiece (Fig. 18).

If the hole is more than one diameter long, it is best to guide the reamer in bushings on both sides of the work (Fig. 18a). For this, a special piloted reamer is required. Guide bushings should fit the pilots, but not so tightly that they seize and bind. Pilots should be grooved throughout their length. The grooves serve the double purpose of allowing cutting fluid to lubricate the pilots and of providing an escape for chips that could become wedged between the pilots and bushings. If the hole is not more than one diameter long, the reamer may be guided only at the hole-entry side of the work, in a guide bushing made to fit the reamer (Fig. 18b).

When hardened bushings are used as guides, the machine spindle must be accurately aligned with the bushing. Otherwise, the reamer will hit the bushing and may be damaged. When using sliding fixtures in machines such as gang drill presses, alignment is especially critical. Bronze or fiber caps are sometimes placed over the bushings to assist lead-in, but these caps will not completely protect the reamer if it is not aligned with the bushing.

For either setup shown in Fig. 18, a rigid drive (reamer shank held directly and rigid-

**Table 4 Speeds and feeds used in examples in this article**

| Example or figure | Material reamed | Speed rev/min | Speed m/min (sfm) | Feed mm/rev | Feed in./rev |
|---|---|---|---|---|---|
| Fig. 1(a) | 4130 steel (15 HRC) | 610, 450 | 12, 17.7 (40, 58) | 0.15 | 0.006 |
| | 4330 modified steel (47 HRC) | 325, 225 | 6.5, 9.0 (21.3, 29.5) | 0.038 | 0.0015 |
| | 4340 steel (52 HRC) | 160, 120 | 3.2, 4.8 (10.5, 15.7) | 0.038 | 0.0015 |
| Example 1 | Low-carbon steel | ... | 30 (100) | 0.74 | 0.029 |
| Fig. 5 | Low-carbon steel | 355 | 21.3 (70) | 0.13 | 0.005 |
| Example 3 | 1035 steel, annealed | 146 | 9 (30) | 0.89 | 0.035 |
| Example 4 | Gray iron (131–207 HB) | 875 | 82 (270) | 0.46 | 0.018 |
| Example 5(b) | 1118 steel (160 HB) | 132, 215 | 21, 34 (69, 112) | 0.234 | 0.0092 |
| Example 6 | Gray iron, class 40 | 320 | 58 (190) | 0.10 | 0.004 |
| Example 7 | Low-carbon steel | 90 | 15 (50) | 1.40 | 0.055 |

(a) Figure 1 deals with the reaming of holes of two different diameters; first values listed in "speed" columns here were used for reaming 6.4 mm (1/4 in.) diam holes, second values for 13 mm (1/2 in.) diam holes. (b) Example 5 compares results from tools of different designs and materials; first values in "speed" columns here were used for conventional eight-flute reamer made of high-speed steel; second values were used for a floating-blade reamer with two carbide blades.

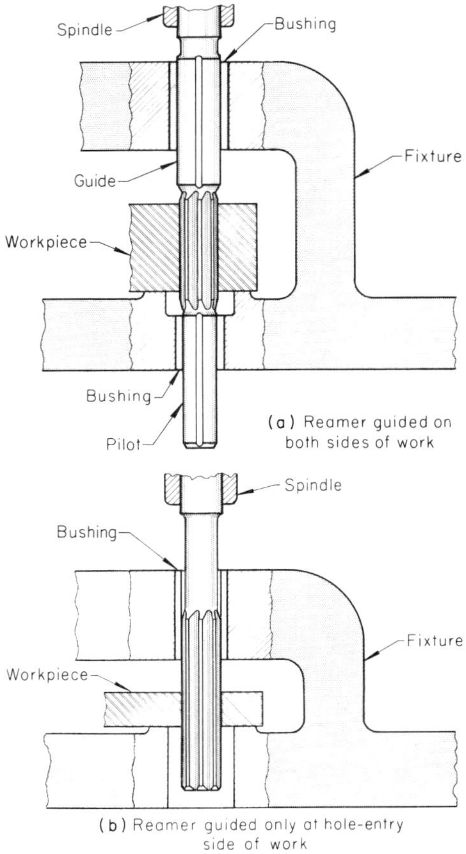


**Fig. 18** Two setups employing fixtures and bushings for the accurate guiding of reamers

ly in the machine spindle) is satisfactory because any slight misalignment between the machine spindle and the work will be corrected by the guide bushings. However, if the reamer is to guide itself into a previously made hole, a rigid drive is not satisfactory because any misalignment of the machine spindle with the work will result in reamed holes that are bell mouthed, tapered, or out-of-round. Some type of floating drive or special reamer is used for these applications.

**Floating drivers** permit some limited angular misalignment of the reamer, some limited parallel misalignment, or both (Fig. 19). The ideal floating driver permits both kinds of float in amounts that can be adjusted for the greatest allowable misalignment in the setup.

## Cutting Fluids

Cutting fluids used in reaming serve the same purposes as in other machining operations:

- Preventing temperature rise in the workpiece as well as in the tool. Excessive temperature rise can adversely affect the dimensions of a reamed hole
- Improving cutting action by minimizing the adherence of the tool to the surface of the workpiece. This adherence occurs most often when reaming soft, gummy metals. When no cutting fluid (or an ineffective one) is used, the finish obtained is often rougher by more than 0.25 to 0.5 μm (10 to 20 μin.)
- Flushing away chips. In finish reaming, if chips are not continuously flushed away, they may be picked up by the cutters and may scratch or gouge the surface of the work

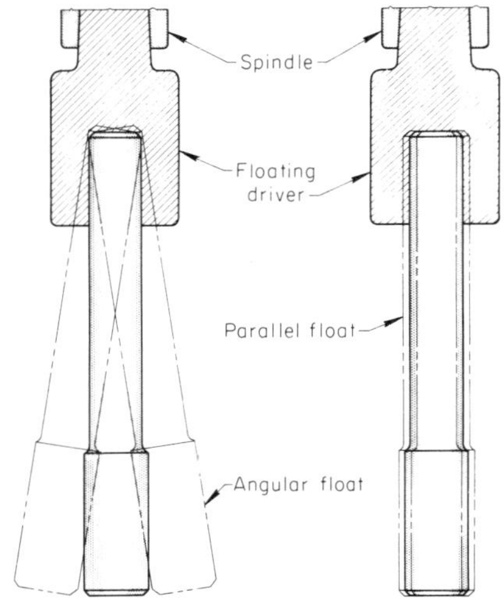


**Fig. 19** Two types of float provided by floating drivers

The cutting fluids most widely used are noted in Table 5. In setups of borderline rigidity, chatter can be minimized or prevented by the use of mineral seal oil or kerosene containing 5 to 25% lard oil or other buffers.

The amount of fluid used is often more important than the type. A copious supply maintains a nearly constant temperature and disperses chips.

**Dry Reaming.** Although cutting fluids are beneficial in most reaming, some applications require no fluid. Gray iron is usually reamed dry. When no cutting fluid is used, it is advisable to direct an air jet on the work for cooling and assistance in chip removal. The use of a cutting fluid is sometimes impractical because of workpiece design, as in the following example.

**Example 7: Dry Reaming of Laminated Steel Rotors.** In reaming shaft holes in rotors for fractional-horsepower electric motors (Fig. 20), cutting fluid was not used, because it would have penetrated between the laminated flat stampings and would have been nearly impossible to remove. The rotors were reamed with ten-flute spiral reamers (Fig. 20) made of M4 high-speed steel. (High-speed steel is preferred for reaming workpieces of this type because they promote tool chatter.) The rotors were held on simple pins (to prevent them from rotating) in a two-station hand-indexing fixture; reamers were held in radial floating holders. Other details are tabulated with Fig. 20.

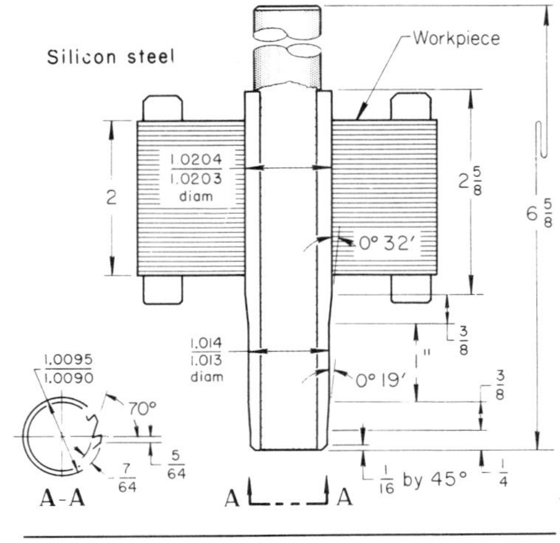


**Reamer details**

| | |
|---|---|
| Type | Spiral, ten-flute(a) |
| Material | M4 high-speed steel |
| Margin width, mm (in.) | 0.05 (0.002) |
| Back taper, mm/mm (in./in.) | 0.0003 (0.0003) |
| Radial hook angle in flutes | 0° |
| Concentricity, mm (in.) | 0.005 (0.0002) TIR |

**Operating conditions**

| | |
|---|---|
| Machine | 2 kW (3 hp) vertical drill press |
| Speed, at 190 rev/min, m/min (sfm) | 15 (50) |
| Feed, mm/rev (in./rev) | 1.4 (0.055) |
| Cutting fluid | None |
| Tool life per grind | 200–250 pieces |

(a) 10° left-hand spiral; right-hand cut. Flute spacing was staggered for elimination of chatter.

**Fig. 20** Dry reaming of a shaft hole in a rotor. Dimensions in figure given in inches

### Table 5 Cutting fluids used in reaming various metals

| Metal | Soluble oil | Sulfurized oil | Kerosene plus lard oil | Non-viscous neutral oil | Dry(a) |
|---|---|---|---|---|---|
| Steel | X | X | | | |
| Stainless steel | X | X | | | |
| Gray iron | | | | | X |
| Malleable or nodular iron | X | | | X | |
| Aluminum | X | | X | | |
| Copper | X | | | | |
| Brass | X | | X | | X |

(a) When no cutting fluid is used, an air jet is advisable for cooling or chip removal.

# Countersinking, Counterboring, and Spotfacing

COUNTERSINKING, COUNTERBORING, AND SPOTFACING are three machining operations used to enlarge the opening of a hole. In countersinking, a conical, reamer-like tool is used to cut a tapered enlargement at the opening of a hole for receiving the head of a fastener, for receiving a center, or for deburring. The surface cut by the tool is concentric with the hole and at an angle of less than 90° to it. Counterboring and spotfacing do not produce a tapered hole enlargement; instead, these techniques enlarge the hole to a given diameter. The difference between counterboring and spotfacing is that a counterbored surface usually has a shoulder at the bottom of the enlarged hole, while a spotfaced surface is flat and always at right angles with the axis of the hole. In addition, the depth of cut in spotfacing is usually shallower than that in counterboring.

Countersinking, counterboring, and spotfacing operations can be carried out on drilling machines and usually follow (or are combined with) drilling operations. Spotfacing, however, sometimes precedes drilling. This is done to provide a contoured workpiece with a flat surface in order to facilitate centering and starting of the drilled hole.

## Countersinking

Although countersinking tools are often ground from drills, tools designed especially for countersinking are more rigid and produce holes with greater accuracy and better finish. Many standard types of these tools are available in various sizes and angles. Five common types of countersinking tools, or countersinks, are shown in Fig. 1 and discussed below.

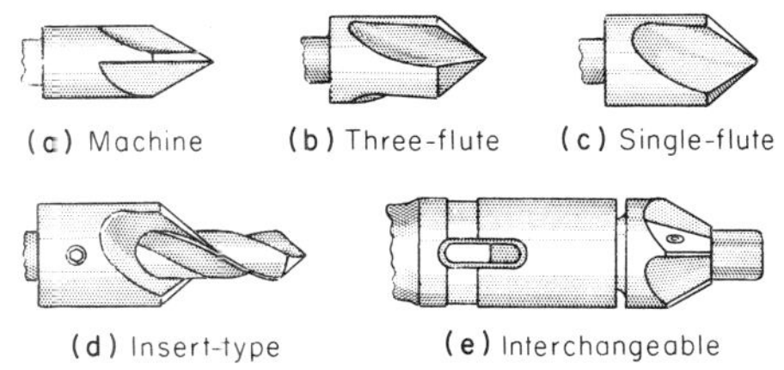


**Fig. 1** Five common types of countersinking tools. See text for discussion.

**Machine countersinks** (Fig. 1a) are made with a radial relief and have four flutes. They are most readily available in angles of 60° for centering, 82° for countersinking flathead screws, and 90° for chamfering or deburring. These are the most common applications. Other angles can be produced as needed.

**Three-flute countersinks** (Fig. 1b) are designed to minimize chatter in countersinking and deburring, and they can be used in either fixed or portable equipment. They are especially suitable for use in portable equipment because they can center readily in drilled holes. Three-flute countersinks are available in many angles and body diameters, and they provide better tool life than single-flute countersinks.

**Single-flute countersinks** (Fig. 1c) are designed for use in machine countersinking and in light-duty countersinking with portable equipment. Much smaller holes can be countersunk with single-flute tools than with multiple-flute tools. However, the diameter of the drilled hole should be at least 10% of the diameter of the countersink. The single-flute tool should be chosen for applications in which multiple-flute tools chatter or when the drilled hole is too small for a multiple-flute tool.

**Insert-type countersinks** (Fig. 1d) are attached to the body of a drill and locked at the desired depth. The body of the drill serves as a pilot and provides rigidity. Near the bottom of the spindle stroke, the cutting edge of the insert engages the surface and countersinks the drilled hole.

An insert-type countersink can also be attached to a tap, but the countersink should then be spring-loaded instead of locked in place. This is required because the point at which the tap reverses is not accurately held and because the lead of a tap is too coarse for a countersinking feed unless the countersink is spring loaded.

**Interchangeable countersinks,** equipped with pilots and adapted to tool holders as shown in Fig. 1(e), are recommended for heavy-duty operations.

**Chatterless countersinks** (Fig. 2a) typically have an odd number of flutes with a radial design. One design uses three main flutes with three intermediate flutes (Fig. 2b). Another chatterless countersink has a hole that passes diagonally through the tool (Fig. 3). The hole generates a single cutting edge and allows a free flow of chips away from the cutting edge.

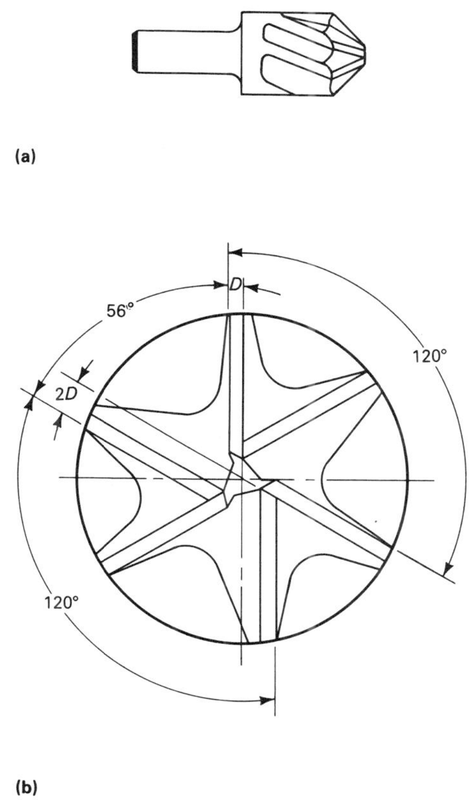


**Fig. 2** Chatterless countersinks. (a) Standard chatterless countersink. (b) Chatterless countersink with three main flutes and three intermediate flutes

**Combination tools,** such as a reamer or a step drill combined with a countersink, are often used. Multiple-flute subland drills can also be adapted to countersinking.

Spot drilling and countersinking are often combined through the use of a drill of countersink diameter that has been ground at the point end to the countersink angle. The drill, guided by a drill bushing, first spot drills the fixtured workpiece and then countersinks to the desired depth. The hole is then drilled using a drill of appropriate smaller diameter—usually through the same drill bushing and without removing the workpiece from the fixture.

**Speed and Feed.** The recommended speed for countersinking is normally one-half to two-thirds the speed used for drilling

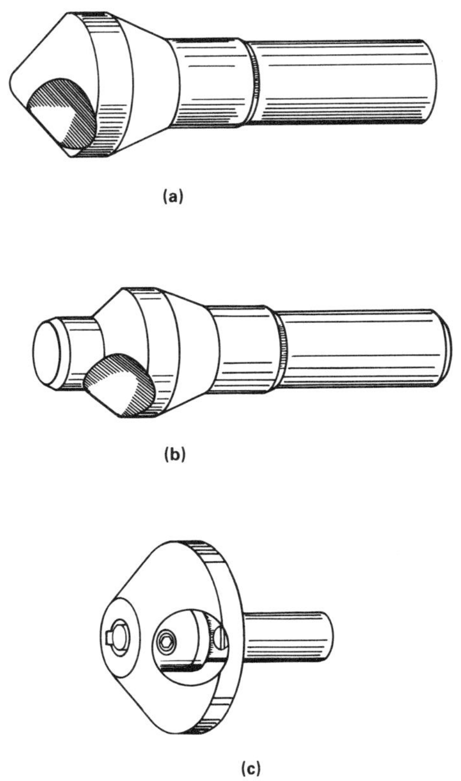


**Fig. 3** Free-cutting, chatterless countersinks with a single cutting edge. (a) Pilotless. (b) Piloted. (c) Removable shank

the same material. When drilling and countersinking are done with a single tool, drilling speed is often penalized as a result of the slower speed required for the countersinking operation.

Feed rates are largely governed by workpiece composition and by the angle, diameter, and depth of the countersink. For example, although a feed of 0.10 to 0.15 mm/rev (0.004 to 0.006 in./rev) is normal for countersinking a 90° angle in a 25 mm (1 in.) diam drilled hole, the above factors may restrict the feed to 0.05 mm/rev (0.002 in./rev) or may permit the use of feeds of 0.30 mm/rev (0.012 in./rev) or more.

## Spotfacing

Many tools used for spotfacing are the same as those used for counterboring. However, tools designed primarily for spotfacing are seldom used for counterboring. Examples are back spotfacers and double-end spotfacers (Fig. 4).

**Back and Double-End Spotfacers.** Back spotfacers (Fig. 4a) are used when the shape of the part makes it impractical to use the down stroke of the spindle. The tool driver is held in the spindle and inserted through the predrilled hole. The spotfacer is then loaded and locked on the driver shaft, and the spindle is rotated and fed upward into the workpiece. Double-end spotfacers (Fig. 4b) permit spotfacing and back spotfacing to

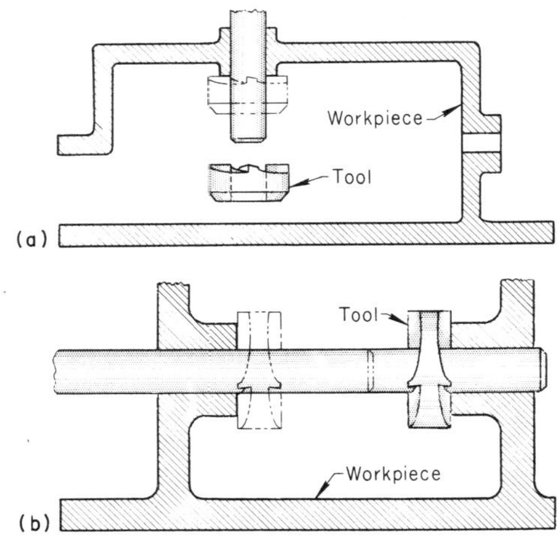


**Fig. 4** Functions of a back spotfacer (a) and double-end spotfacer (b)

be done on two internal surfaces without removing the cutter.

Back or double-end spotfacers can be secured to the driver shaft with one or more setscrews. This makes it easy to reposition the spotfacer on the shaft. A disadvantage of this method is that setscrews are likely to loosen. At least two proprietary devices are available for securing spotfacers to driver shafts. One of these devices employs the principle of a locking pin, and the other is designed with a cam lock.

**Automatic back spotfacers** have a wing that folds into a recess in the tool. The wing carries interchangeable cutters. During insertion, the wing is folded in, and upon spindle reversal, it swings out because of centrifugal force or with the aid of a spring.

## Counterboring

Most tools for counterboring are either of two general types:

- Those that have pilots and can therefore use the predrilled hole to stabilize the cutting action and to minimize chatter
- Those that do not have pilots and therefore usually must be guided by drill bushings

**Pilots for counterbores** are either integral with the cutter or are separate, interchangeable parts selected to fit the guide hole. Interchangeable pilots are either solid or roller style. A large counterbore should not be used with too small a pilot. The pilot diameter should be at least as large as the root diameter of the counterbore, so that chips will not be trapped between pilot and cutter. Chip packing can cause cutter breakage.

The pilot must not be too tight nor too loose in the guide hole. There should be enough clearance so that the pilot can enter the guide hole and rotate without binding. Too much clearance may cause chatter. For solid pilots, clearance of 0.025 mm (0.001 in.) on the diameter is enough for a pilot 3.2 mm (⅛ in.) in diameter; a pilot 75 mm (3 in.)

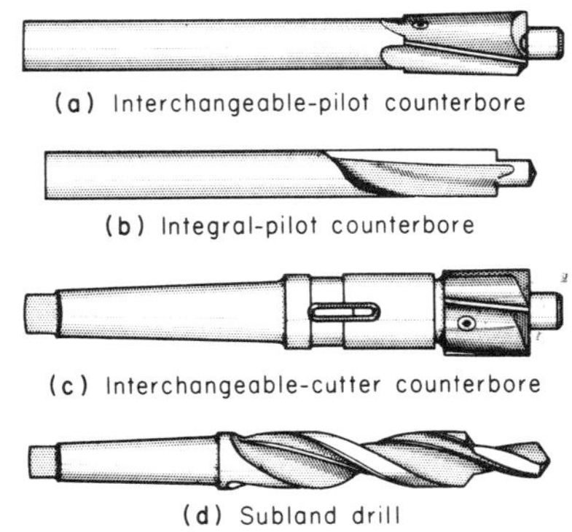


**Fig. 5** Four types of tools used for counterboring

in diameter requires clearance of 0.18 mm (0.007 in.) for comparable action. For greater accuracy, some guide holes are reamed to finish or semifinish size before counterboring. With roller pilots, only 0.0125 to 0.025 mm (0.0005 to 0.001 in.) of running clearance is necessary. Roller pilots provide greater performance while reducing possible hole damage. Four common types of counterboring tools are illustrated in Fig. 5 and discussed below.

**Interchangeable-pilot counterbores** (Fig. 5a) have three, four, or five flutes, depending on the size of the counterbore. They are available in long or short lengths and with either straight or tapered shanks. Pilots of different diameters can be used with a counterbore of specific size. It is usually impractical to use a counterbore with an interchangeable pilot in holes smaller than 6.4 mm (¼ in.) in diameter.

**Integral-pilot counterbores** (Fig. 5b) are designed with two spiral flutes to produce holes for small fillister-head and flathead screws, and they are used for holes up to 9.5 mm (⅜ in.) in diameter. The cutting action of integral-pilot counterbores produces accurate holes without causing chatter. These counterbores are available for fillister-head screws in sizes 0 through 12, and for flathead screws (82° angle) in sizes up to 9.5 mm (⅜ in.) and in some larger sizes.

**Interchangeable-cutter counterbores** (Fig. 4c) are designed so that cutters of various diameters can be inserted interchangeably into a holder designated for a specific range. The inserted cutter is driven by a spline, a key, or a pin. The short, rugged holders, which are available with straight or tapered shanks, make the tool especially well suited for heavy-duty operations. The holder may have an integral or an interchangeable pilot. Interchangeable-cutter counterbores also allow the use of carbide cutters without the expense of solid carbide bodies.

**Subland drills** (Fig. 4d) can be used to produce counterbored holes in conjunction with drilling operations. These multiple-flute tools are of various designs for different applications. Combination tools other than subland drills are also used.

**Table 1 Nominal speeds and feeds for counterboring and spotfacing of carbon and low-alloy steels with high-speed steel and carbide tools**

| | | High-speed steel tools(a) | | | | | | | Carbide tools(b) | | | | | | |
|---|---|---|---|---|---|---|---|---|---|---|---|---|---|---|---|
| | | Speed, | Feed, mm/rev (in./rev), for tool diameter of: | | | | | | Speed, | Feed, mm/rev (in./rev), for tool diameter of: | | | | | |
| Typical steel | Hardness, HB | m/min (sfm) | 6.4 mm (¼ in.) | 13 mm (½ in.) | 25 mm (1 in.) | 38 mm (1½ in.) | 50 mm (2 in.) | 75 mm (3 in.) | m/min (sfm) | 6.4 mm (¼ in.) | 13 mm (½ in.) | 25 mm (1 in.) | 38 mm (1½ in.) | 50 mm (2 in.) | 75 mm (3 in.) |
| **Carbon and low-alloy steels (except free-cutting grades)** | | | | | | | | | | | | | | | |
| 1020, 1045, 4140, 7140, and 8620 at hardness ranges listed at right | 85–125 | 34 (110) | 0.08 (0.003) | 0.10 (0.004) | 0.15 (0.006) | 0.18 (0.007) | 0.18 (0.007) | 0.23 (0.009) | 115 (370) | 0.15 (0.006) | 0.20 (0.008) | 0.28 (0.011) | 0.33 (0.013) | 0.36 (0.014) | 0.41 (0.016) |
| | 125–175 | 27 (90) | 0.08 (0.003) | 0.10 (0.004) | 0.15 (0.006) | 0.18 (0.007) | 0.18 (0.007) | 0.23 (0.009) | 93 (305) | 0.15 (0.006) | 0.20 (0.008) | 0.28 (0.011) | 0.33 (0.013) | 0.36 (0.014) | 0.41 (0.016) |
| | 175–225 | 21 (70) | 0.08 (0.003) | 0.09 (0.0035) | 0.13 (0.005) | 0.13 (0.005) | 0.15 (0.006) | 0.20 (0.008) | 81 (265) | 0.15 (0.006) | 0.18 (0.007) | 0.25 (0.010) | 0.28 (0.011) | 0.30 (0.012) | 0.38 (0.015) |
| | 225–275 | 18 (60) | 0.05 (0.002) | 0.06 (0.0025) | 0.08 (0.003) | 0.10 (0.004) | 0.10 (0.004) | 0.15 (0.006) | 76 (250) | 0.10 (0.004) | 0.15 (0.006) | 0.20 (0.008) | 0.23 (0.009) | 0.25 (0.010) | 0.30 (0.012) |
| | 275–325 | 15 (50) | 0.03 (0.001) | 0.05 (0.002) | 0.08 (0.003) | 0.08 (0.003) | 0.09 (0.0035) | 0.10 (0.004) | 66 (215) | 0.08 (0.003) | 0.10 (0.004) | 0.15 (0.006) | 0.18 (0.007) | 0.18 (0.007) | 0.20 (0.008) |
| | 325–375 | 14 (45) | 0.03 (0.001) | 0.05 (0.002) | 0.08 (0.003) | 0.08 (0.003) | 0.09 (0.0035) | 0.10 (0.004) | 55 (180) | 0.05 (0.002) | 0.10 (0.004) | 0.15 (0.006) | 0.18 (0.007) | 0.18 (0.007) | 0.20 (0.008) |
| | 375–425 | 9 (30) | 0.03 (0.001) | 0.05 (0.002) | 0.08 (0.003) | 0.08 (0.003) | 0.09 (0.0035) | 0.10 (0.004) | 43 (140) | 0.05 (0.002) | 0.10 (0.004) | 0.15 (0.006) | 0.18 (0.007) | 0.18 (0.007) | 0.20 (0.008) |
| **Free-cutting carbon and low-alloy steels** | | | | | | | | | | | | | | | |
| 1112, 1117, and 12L14 | 100–150 | 37 (120) | 0.10 (0.004) | 0.13 (0.005) | 0.18 (0.007) | 0.20 (0.008) | 0.23 (0.009) | 0.28 (0.011) | 135 (440) | 0.20 (0.008) | 0.25 (0.010) | 0.30 (0.012) | 0.36 (0.014) | 0.38 (0.015) | 0.51 (0.020) |
| | 150–200 | 40 (130) | 0.10 (0.004) | 0.13 (0.005) | 0.15 (0.006) | 0.18 (0.007) | 0.20 (0.008) | 0.25 (0.010) | 150 (490) | 0.20 (0.008) | 0.25 (0.010) | 0.30 (0.012) | 0.36 (0.014) | 0.38 (0.015) | 0.51 (0.020) |
| | 200–250 | 26 (85) | 0.08 (0.003) | 0.08 (0.003) | 0.10 (0.004) | 0.13 (0.005) | 0.13 (0.005) | 0.20 (0.008) | 88 (290) | 0.15 (0.006) | 0.18 (0.007) | 0.25 (0.010) | 0.28 (0.011) | 0.30 (0.012) | 0.41 (0.016) |
| 1137 | 175–225 | 32 (105) | 0.08 (0.003) | 0.10 (0.004) | 0.15 (0.006) | 0.18 (0.007) | 0.18 (0.007) | 0.23 (0.009) | 120 (395) | 0.15 (0.006) | 0.20 (0.008) | 0.28 (0.011) | 0.30 (0.012) | 0.33 (0.013) | 0.38 (0.015) |
| | 275–325 | 21 (70) | 0.04 (0.0015) | 0.05 (0.002) | 0.08 (0.003) | 0.08 (0.003) | 0.10 (0.004) | 0.13 (0.005) | 82 (270) | 0.08 (0.003) | 0.10 (0.004) | 0.15 (0.006) | 0.18 (0.007) | 0.20 (0.008) | 0.25 (0.010) |
| | 325–375 | 14 (45) | 0.03 (0.001) | 0.05 (0.002) | 0.08 (0.003) | 0.08 (0.003) | 0.09 (0.0035) | 0.10 (0.004) | 55 (180) | 0.05 (0.002) | 0.08 (0.003) | 0.13 (0.005) | 0.15 (0.006) | 0.18 (0.007) | 0.20 (0.008) |
| | 375–425 | 9 (30) | 0.03 (0.001) | 0.05 (0.002) | 0.08 (0.003) | 0.08 (0.003) | 0.09 (0.0035) | 0.10 (0.004) | 43 (140) | 0.05 (0.002) | 0.08 (0.003) | 0.13 (0.005) | 0.15 (0.006) | 0.18 (0.007) | 0.20 (0.008) |
| 4140+S | 150–200 | 27 (90) | 0.10 (0.004) | 0.13 (0.005) | 0.15 (0.006) | 0.18 (0.007) | 0.20 (0.008) | 0.25 (0.010) | 98 (320) | 0.20 (0.008) | 0.25 (0.010) | 0.30 (0.012) | 0.36 (0.014) | 0.41 (0.016) | 0.51 (0.020) |
| | 200–250 | 21 (70) | 0.08 (0.003) | 0.09 (0.0035) | 0.13 (0.005) | 0.13 (0.005) | 0.15 (0.006) | 0.20 (0.008) | 88 (290) | 0.15 (0.006) | 0.18 (0.007) | 0.25 (0.010) | 0.28 (0.011) | 0.30 (0.012) | 0.36 (0.014) |
| | 275–325 | 15 (50) | 0.04 (0.0015) | 0.05 (0.002) | 0.08 (0.003) | 0.08 (0.003) | 0.10 (0.004) | 0.13 (0.005) | 69 (225) | 0.08 (0.003) | 0.10 (0.004) | 0.15 (0.006) | 0.18 (0.007) | 0.20 (0.008) | 0.25 (0.010) |
| | 375–425 | 9 (30) | 0.03 (0.001) | 0.05 (0.002) | 0.08 (0.003) | 0.08 (0.003) | 0.09 (0.0035) | 0.10 (0.004) | 43 (140) | 0.05 (0.002) | 0.10 (0.004) | 0.15 (0.006) | 0.18 (0.007) | 0.18 (0.007) | 0.20 (0.008) |
| | 50–52 HRC | 5 (15) | 0.03 (0.001) | 0.05 (0.002) | 0.06 (0.0025) | 0.06 (0.0025) | 0.08 (0.003) | 0.08 (0.003) | 24 (80) | 0.05 (0.002) | 0.08 (0.003) | 0.10 (0.004) | 0.10 (0.004) | 0.13 (0.005) | 0.13 (0.005) |
| 41L40 | 150–200 | 38 (125) | 0.10 (0.004) | 0.13 (0.005) | 0.18 (0.007) | 0.20 (0.008) | 0.23 (0.009) | 0.28 (0.011) | 135 (440) | 0.20 (0.008) | 0.25 (0.010) | 0.30 (0.012) | 0.36 (0.014) | 0.41 (0.016) | 0.51 (0.020) |
| | 200–250 | 26 (85) | 0.08 (0.003) | 0.08 (0.003) | 0.10 (0.004) | 0.13 (0.005) | 0.13 (0.005) | 0.20 (0.008) | 88 (290) | 0.15 (0.006) | 0.18 (0.007) | 0.25 (0.010) | 0.28 (0.011) | 0.30 (0.012) | 0.41 (0.016) |
| | 275–325 | 18 (60) | 0.04 (0.0015) | 0.05 (0.002) | 0.08 (0.003) | 0.09 (0.0035) | 0.10 (0.004) | 0.13 (0.005) | 73 (240) | 0.08 (0.003) | 0.10 (0.004) | 0.15 (0.006) | 0.18 (0.007) | 0.20 (0.008) | 0.25 (0.010) |
| | 375–425 | 9 (30) | 0.03 (0.001) | 0.05 (0.002) | 0.08 (0.003) | 0.08 (0.003) | 0.09 (0.0035) | 0.10 (0.004) | 43 (140) | 0.05 (0.002) | 0.08 (0.003) | 0.13 (0.005) | 0.15 (0.006) | 0.18 (0.007) | 0.20 (0.008) |
| | 50–52 HRC | 5 (15) | 0.03 (0.001) | 0.05 (0.002) | 0.06 (0.0025) | 0.08 (0.003) | 0.08 (0.003) | 0.08 (0.003) | 24 (80) | 0.05 (0.002) | 0.08 (0.003) | 0.10 (0.004) | 0.10 (0.004) | 0.13 (0.005) | 0.13 (0.005) |

(a) High-speed steel M2 for carbon and low-alloy steels other than free-cutting grades at hardnesses to 225 HB and for free-cutting grades to 275 HB; then high-speed steels M2, M42, and T15 to 375 HB; above 375 HB, high-speed steels M42 and T15. (b) Carbide grade C-2. Source: Metcut Research Associates, Inc.

**Speed and feed** are the same for counterboring and spotfacing and are influenced by the same factors as for drilling (see the article "Drilling" in this Volume). Speed and feed for spotfacing, however, are often less critical than for counterboring because the depth of cut is less in spotfacing.

Table 1 lists nominal speeds and feeds for counterboring carbon and low-alloy steels with high-speed steel and carbide tools. These rates can serve as starting points for the selection of optimum machining conditions.

Speeds for counterboring and spotfacing are somewhat lower than those for drilling, especially at low work metal hardness, and feeds are lower in most applications. Speed is primarily governed by the tool material used and by the hardness of the work metal; the presence of free-cutting additives in the work metal exerts a secondary effect. As indicated in Table 1, increasing work metal hardness necessitates the use of progressively lower speeds, except for certain free-cutting grades at hardnesses below 200 HB. Speeds for carbide tools are four to five times those for high-speed steel tools in similar operations, and almost all counterboring of steels harder than 52 HRC is done with carbide tools.

Feed is influenced by the same factors as speed and also by tool diameter. Lower feeds are used on steels of higher hardness and with tools of smaller diameter. Feeds for carbide tools are about twice those for high-speed steel tools in similar operations.

When drilling and counterboring are done simultaneously with a combination tool (as with a subland drill), the diameter of the counterbore must be given primary consideration when selecting speed and feed. This often results in the use of subnormal speed and feed for the smaller-diameter drill portion of the combination tool.

# Roller Burnishing

ROLLER BURNISHING is a method of working a metal surface to improve its finish and dimensional accuracy and can provide some degree of work hardening. This is a cold-working process that uses pressure rolling techniques and does not involve metal removal. Roller burnishing is applicable to the finishing of tapered holes, inside and outside cylindrical surfaces, circular flat surfaces, and continuously curved surfaces. By this process, hole diameters can be increased by 0.013 to 0.051 mm (0.0005 to 0.002 in.) without damage to the surface. However, size increase is usually a secondary objective. The primary aim is improved accuracy and finish as well as work-hardened surfaces.

Machines suited to roller burnishing include drill presses, lathes, boring machines, and automatic bar or chucking machines, depending on the degree of quality and accuracy required. The process is applicable to metals softer than about 40 HRC, although diamond burnishing machines are available for finishing materials harder than 40 HRC. However, metals that work harden rapidly must be at lower hardness before roller burnishing. Roller burnishing increases surface hardness to a depth of 0.13 to 0.76 mm (0.005 to 0.030 in.).

## Workpiece Requirements

The workpiece must have a wall thickness great enough to withstand the pressure exerted by the rollers. If the wall is too thin, the workpiece will expand, defeating the purpose of the process. In addition, if thin-wall parts vary significantly in wall thickness, roller burnishing may produce a wavy surface and taper or out-of-roundness.

Workpieces with walls as thin as 2 mm (1/16 in.) can be successfully roller burnished, but for walls this thin, supporting fixtures should be used and size increase should not exceed 0.0076 mm (0.0003 in.) on the diameter.

For optimum results in roller burnishing, the hole must be round and straight. The burnishing tool will not correct deviations from roundness or straightness to any degree.

**Stock Allowance.** Typical stock allowances for holes to be roller burnished are given in Table 1. Stock allowance depends on hole diameter and varies by a factor of five or more as hole diameter increases from 4.8 to 60 mm (3/16 to 2½ in.). These values will vary with different materials, preliminary hardnesses, and preparation processes.

## Tools

Tools for roller burnishing are illustrated in Fig. 1 to 5. The tool shown in Fig. 1 can be adjusted to accommodate a range of diameters in increments as small as 0.013 mm (0.0005 in.). Outside cylindrical surfaces are roller burnished by means of a tool such as the one shown in Fig. 2. Such external tools also have incremental adjustments, but in the opposite direction. For example, an external tool with a 13 mm (½ in.) diameter can be adjusted 0.076 mm (0.003 in.) below or 0.43 mm (0.017 in.) above each 0.8 mm (1/32 in.) size.

Internal tools of the type illustrated in Fig. 1 that are 11.9 mm (0.468 in.) or smaller in diameter can be adjusted to 0.076 mm (0.003 in.) above or 0.43 mm (0.017 in.) below each 0.4 mm (1/64 in.) size. Tools that are 12.7 mm (0.500 in.) or larger in diameter can be adjusted to 0.10 mm (0.004 in.) above or 0.94 mm (0.037 in.) below each 0.8 mm (1/32 in.) size.

The tool shown in Fig. 3 is used for roller burnishing tapered holes. It is not adjustable or self-feeding, and pressure is exerted by the machine spindle. When using this tool with a mechanical feed unit, an overtravel adaptor consisting of stacked Belleville washers is required to control burnishing pressure.

To roller burnish circular flat surfaces as small as 9.5 mm (3/8 in.) in diameter, a tool such as that shown in Fig. 4 is used. This tool is not adjustable, and pressure is exerted by the machine spindle. An overtravel adaptor is required when using this type of tool with mechanical feed.

**Table 1 Typical stock allowances for roller burnishing**

| Hole diameter, mm (in.) | Stock allowance on diameter, mm (in.) |
|---|---|
| 4.75–12.7 (0.187–0.500) | 0.017 (0.00067) |
| 13.48–24.59 (0.531–0.968) | 0.025 (0.001) |
| 25.40–44.45 (1.000–1.750) | 0.038 (0.0015) |
| 45.24–63.50 (1.781–2.500) | 0.051 (0.002) |
| 64.29 and larger (2.531 and larger) | 0.076–0.152 (0.003–0.006) |

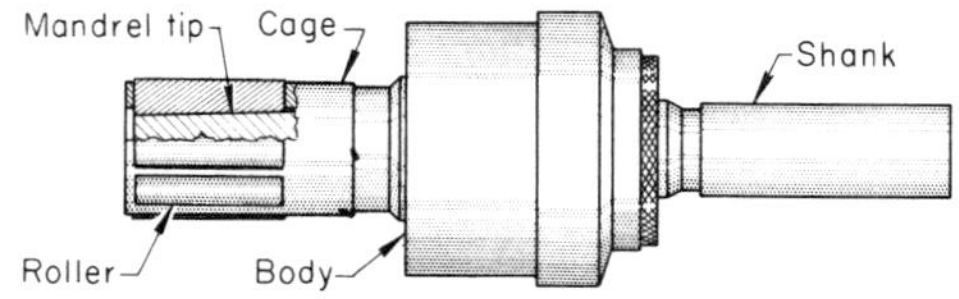


**Fig. 1** Internal tool for the roller burnishing of inside cylindrical surfaces

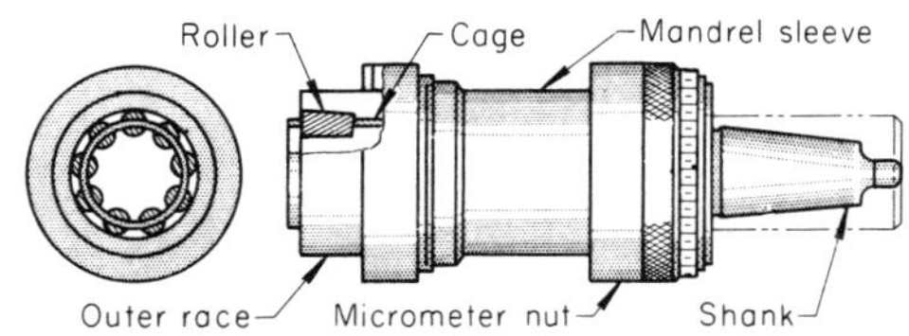


**Fig. 2** External tool for the roller burnishing of outside cylindrical surfaces

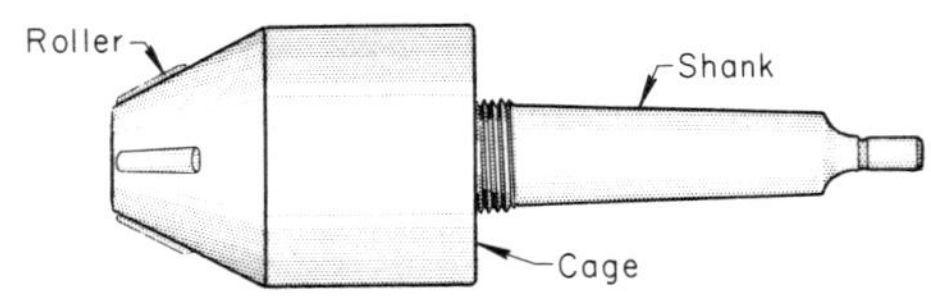


**Fig. 3** Tool for the roller burnishing of tapered holes

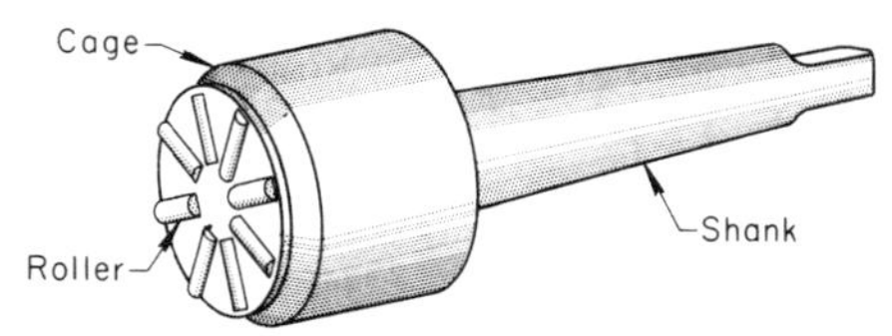


**Fig. 4** Tool for the roller burnishing of circular flat surfaces

A single roller burnishing tool is illustrated in Fig. 5. This type of tool is primarily designed for numerical-control machines, and one tool can often finish the inside and outside surfaces of cylinders with several different diameters. With conventional machines, single rollers are a cost-effective alternative when typical burnishing tools cannot accommodate large-diameter workpieces. Single rollers can also be specially designed for the finishing of curved surfaces.

Because of the single point contact of single rollers, deflection may be a problem with smaller workpieces. However, a smaller radius tool can reduce the force required for burnishing. This reduces the amount of

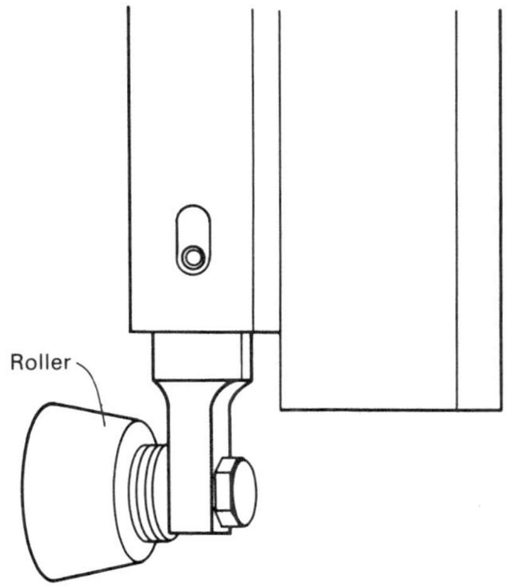


**Fig. 5** Schematic illustrating the components of a single roller burnishing tool

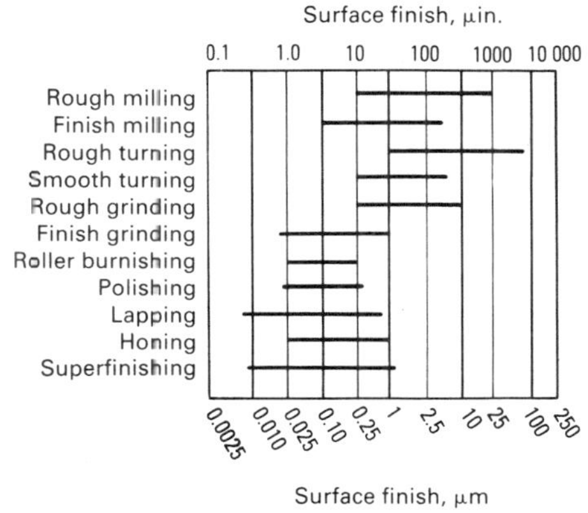


**Fig. 6** Typical surface finishes produced by various metalworking processes

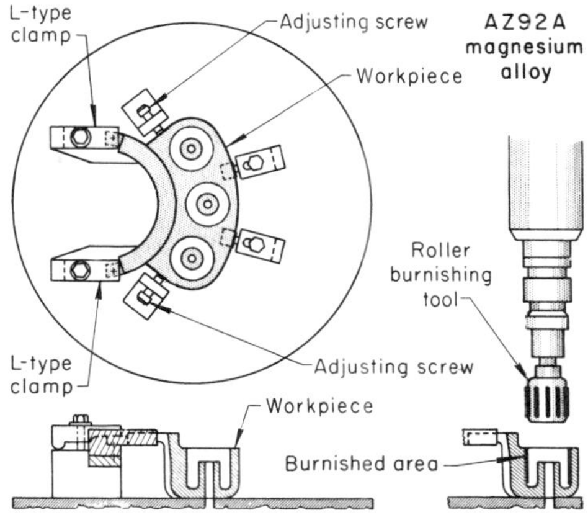


| | |
|---|---|
| Speed | 435 rev/min (82 m/min, or 270 sfm) |
| Feed | 2.5 mm/rev (0.100 in./rev) |
| Cutting fluid | SAE 20 oil:kerosene (1:1) |
| Downtime for changing tools | 12 min |
| Average tool life | 10 000 bores |

**Fig. 7** Roller burnishing cylindrical bores in a magnesium brake housing

deflection; feed rates would also have to be reduced.

## Tolerance and Finish

Roller burnishing can usually work to tolerances with a total variation of 0.013 mm (0.0005 in.). Even closer tolerances can be obtained under closely controlled conditions.

Bores having an initial finish of 2.50 to 3.20 μm (100 to 125 μin.) have been roller burnished in a single pass to a finish of 0.050 to 0.38 μm (2 to 15 μin.), depending on the workpiece metal. Results obtained on specific metals are as follows: 0.050 to 0.20 μm (2 to 8 μin.) on bronze and aluminum, 0.10 to 0.20 μm (4 to 8 μin.) on steel, and 0.25 to 0.38 μm (10 to 15 μin.) on cast iron. In some cases, special care can improve on these finishes (for example, bores finished and sized by pressing a hardened ball through the bore). The finish typically obtained in boring or from an end-cutting reamer is preferred for roller burnishing.

Figure 6 compares the surface finish obtained with roller burnishing to those obtained with other metalworking processes. Additional information on these alternative methods is available in the articles "Milling," "Turning," "Principles of Grinding," "Honing," and "Lapping" in this Volume.

**Table 2 Typical inside and outside diameter speed and feed rates for roller burnishing tools**

| Hole diameter, mm | Hole diameter, in. | Speed, rev/min | Feed, mm/rev | Feed, in./rev |
|---|---|---|---|---|
| 4.8 | 3/16 | 1500–3600 | 0.18–0.30 | 0.007–0.012 |
| 6.4 | 1/4 | 1500–3600 | 0.18–0.30 | 0.007–0.012 |
| 7.9 | 5/16 | 1200–3600 | 0.25–0.38 | 0.010–0.015 |
| 9.5 | 3/8 | 1000–3000 | 0.38–0.51 | 0.015–0.020 |
| 11 | 7/16 | 875–2600 | 0.51–0.64 | 0.015–0.025 |
| 13 | 1/2 | 800–2250 | 0.51–0.64 | 0.015–0.025 |
| 14 | 9/16 | 660–2000 | 0.51–0.64 | 0.015–0.025 |
| 16 | 5/8 | 620–1850 | 0.64–0.89 | 0.025–0.035 |
| 17 | 11/16 | 560–1700 | 0.64–0.89 | 0.025–0.035 |
| 19 | 3/4 | 520–1550 | 0.76–1.02 | 0.030–0.040 |
| 21 | 13/16 | 480–1400 | 0.76–1.02 | 0.030–0.040 |
| 22 | 7/8 | 440–1300 | 0.8–1.14 | 0.035–0.045 |
| 24 | 15/16 | 400–1200 | 1.02–1.27 | 0.040–0.050 |
| 25 | 1 | 375–1150 | 1.14–1.40 | 0.045–0.055 |
| 28.5 | 1 1/8 | 340–1000 | 1.27–1.52 | 0.050–0.060 |
| 32 | 1 1/4 | 300–900 | 1.52–1.78 | 0.060–0.070 |
| 35 | 1 3/8 | 275–850 | 1.78–2.03 | 0.070–0.080 |
| 38 | 1 1/2 | 250–750 | 2.03–2.29 | 0.080–0.090 |
| 41 | 1 5/8 | 240–700 | 2.16–2.41 | 0.085–0.095 |
| 44 | 1 3/4 | 220–650 | 2.29–2.54 | 0.090–0.100 |
| 48 | 1 7/8 | 200–600 | 2.67–2.92 | 0.105–0.115 |
| 51 | 2 | 190–575 | 3.05–3.30 | 0.120–0.130 |
| 54 | 2 1/8 | 180–540 | 3.30–3.56 | 0.130–0.140 |
| 57 | 2 1/4 | 170–510 | 3.68–3.94 | 0.145–0.155 |
| 60 | 2 3/8 | 160–480 | 3.94–4.19 | 0.155–0.165 |
| 63 | 2 1/2 | 150–450 | 4.32–4.57 | 0.170–0.180 |
| 67 | 2 5/8 | 145–420 | 2.16–2.29 | 0.085–0.090 |
| 70 | 2 3/4 | 140–410 | 2.41–2.54 | 0.095–0.100 |
| 73 | 2 7/8 | 130–400 | 2.54–2.67 | 0.100–0.105 |
| 75 | 3 | 125–375 | 2.54–2.67 | 0.100–0.105 |
| 89 | 3 1/2 | 110–325 | 3.18–3.30 | 0.125–0.130 |
| 102 | 4 | 95–285 | 3.81–3.94 | 0.150–0.155 |

## Speed, Feed, and Lubrication

Speed and feed for roller burnishing vary with the diameter being finished, but are not critical. Satisfactory results are obtained at speeds of 18 to 46 m/min (60 to 150 sfm) and feeds of 0.13 mm/rev (0.005 in./rev) for small diameters to 5.1 mm/rev (0.200 in./rev) for large diameters (<127 mm, or 5 in.). Typical spindle speeds and feeds for various inside and outside diameters are given in Table 2. Pressures vary with the application and yield strength of the workpiece.

**Lubrication.** Little heat is generated in roller burnishing. Enough lubrication is provided by applying a few drops (as from a squirt can) of a light spindle oil or of a relatively rich soluble oil (mixed 1 to 10 with water) to each roller before the tool burnishes a hole. The following example describes typical techniques and conditions used in the roller burnishing of magnesium alloy castings.

**Example 1: Three 60.35 mm (2.376 in.) Diam Cylinder Bores.** Figure 7 shows a brake housing, cast from AZ92A magnesium alloy, in which the walls of three 60.35 mm (2.376 in.) diam cylinder bores were roller burnished to a diametral tolerance of ±0.020 mm (±0.0008 in.) and a 0.38 μm (15 μin.) surface finish. These bores had been machined undersize to a diametral tolerance of 0.025 to 0.038 mm (0.001 to 0.0015 in.) and a 1.50 μm (60 μin.) surface finish.

For roller burnishing, the housing was clamped, bore side up, to the fixture base with L-type clamps, as shown in Fig. 7. The burnishing tool contained a series of hardened tapered rollers that rode on a mandrel tapered inversely to the taper of the rollers.

## Fillet Rolling

Fillet rolling is done to improve fatigue resistance. It is a specialized operation that uses a narrow roller of required shape. The rolling and pressing cause combined rolling and sliding (lubrication is used). Forces are not large; for example, a fillet of 0.8 mm (1/32 in.) radius can usually be rolled with a force of 440 N (100 lbf) or less in ten revolutions (passes) around the fillet. A plain roller of oil-hardening tool steel at 62 to 65 HRC can roll fillets on several thousand pieces at low cost.

## Bearingizing

In a modification of roller burnishing known as bearingizing, metal surfaces are finished by a combined rolling and peening action. In this process, hardened rollers rotating around and bearing on cams (Fig. 8) rise and fall rapidly, delivering as many as 200 000 blows per minute. This action produces a smooth surface, improves roundness and straightness, and increases surface hardness to a depth of 0.13 to 0.38 mm (0.005 to 0.015 in.). The inside surfaces of tubes 3 to 6 m (10 to 20 ft) long have been successfully processed by this method. In most applications, however, bore length is relatively short (less than three times diameter). Copper alloys, because of their low hardness and high ductility, are especially suitable for this process.

**Tools.** Three basic types of tools for the finishing of bores are illustrated in Fig. 9. Selection depends on whether the hole is through or blind and, for blind holes, on how close to the bottom that finishing is required. The bottoming tool shown in Fig. 9 can finish to 0.76 mm (0.030 in.) from the bottom of a blind hole. Two styles of cams are also available (Fig. 8); the steep-rise cam is used in thin-wall parts.

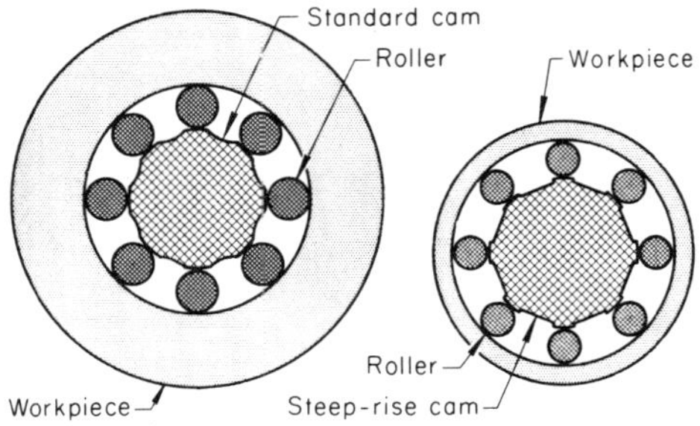


**Fig. 8** Cross sections of bearingizing tools positioned in workpieces showing rollers riding over cams

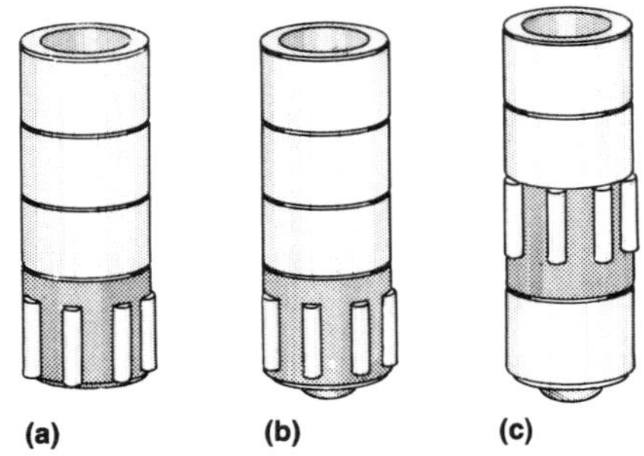


**Fig. 9** Bearingizing tools for finishing through and blind holes. (a) Bottoming. (b) Semibottoming. (c) Through-hole

Taper-shank tools are available from stock in diameters of 4.8 to 25 mm (3/16 to 1 in.) in 0.8 mm (1/32 in.) increments. Overall length and working length (distance from the leading end of the tool to the beginning of the taper shank) vary proportionately, from 75 mm (2 15/16 in.) for the 4.8 mm (3/16 in.) diam tool to 152 mm (6 in.) for the 25 mm (1 in.) diam tool. The number and size of rollers also vary with diameter; for example, the 4.8 mm (3/16 in.) diam tool has 6 rollers, while the 25 mm (1 in.) diam tool has 12.

Tools are available in diameters up to 305 mm (12 in.), but sizes larger than 203 mm (8 in.) in diameter are seldom used. A tool can be adjusted in a total range of about 0.10 mm (0.004 in.) by changing rollers. Rollers are available in increments of 0.0025 mm (0.0001 in.). Because the rollers are diametrically opposed, the tool can be adjusted in increments of 0.0051 mm (0.0002 in.).

**Tolerance and Finish.** Bores are commonly finished to tolerances of ±0.0001 mm/mm (±0.0001 in./in.) of diameter by the bearingizing process. Notable examples are piston-pin bores, which are finished to a total tolerance of 0.0051 mm (0.0002 in.). However, such parts must be finished to close tolerances in a prior operation, because the total expansion for holes less than 25 mm (1 in.) in diameter is only about 0.013 mm (0.0005 in.) (except holes in thin-wall tubing, which can be expanded several hundredths of a millimeter). For holes larger than 25 mm (1 in.) in diameter, a total expansion of 0.025 mm (0.001 in.) is normal. Hole straightness depends largely on prior operations.

The preferred prior surface is one produced by a single-point tool (as in boring) to a roughness of 2.00 to 3.00 μm (80 to 120 μin.). Starting with this roughness, most metals can be finished to about 0.125 μm (5 μin.). For porous metals (such as sintered bronze), however, the resulting finish is likely to be closer to 0.30 μm (12 μin.).

**Speed and Feed.** Neither speed nor feed is critical. Speeds of 90 to 150 m/min (300 to 500 sfm) have proved satisfactory. Because the operation is done in one quick pass, feed rate is difficult to measure, but 3800 to 6350 mm/min (150 to 250 in./min) is normal.

Little time is required per hole; for example, a hole 75 mm (3 in.) deep is finished in slightly over 1 s. Indexing of the workpiece usually requires more time than the actual operation.

**Lubrication** is seldom necessary, but tools should be cleaned frequently with a light oil having a viscosity of about 100 SUS (Saybolt universal second). Spindle oil is often used. Only a few drops applied with a squirt can or a brush are necessary to get oil into the roller cage. Centrifugal force then washes out metal particles or dust.

# Tapping

Revised by Mark Johnson, Tapmatic Corporation

TAPPING is a machining process for producing internal threads. A tap is a cylindrical or conical thread-cutting tool having threads of a desired form on the periphery. Combining rotary motion with axial motion, the tap cuts or forms the internal thread.

Most metals that can be machined with single-point tools can be tapped, but the cost of tapping usually rises sharply as the hardness of the work metal increases beyond 25 HRC. Although steel as hard as 52 HRC can be tapped, efficiency is low and cost is high.

Threads as fine as 360 threads per 25 mm (1 in.) in 0.335 mm (0.0132 in.) diam holes or as coarse as 3 threads per 25 mm (1 in.) in 610 mm (24 in.) diam pipe fittings are routinely tapped.

## Machines and Accessories

The machines most commonly used for tapping are drill presses, tapping machines, gang machines, manual or automatic turret lathes, and other multiple-operation machines. Tapping machines are basically drill presses equipped with lead screws, tap holders, and reversing mechanisms.

**Lead screws** or lead-control devices provide a means of regulating the desired feed rate during tapping. The amount of feed per revolution determines the pitch of the tapped thread. Lead screws convert rotary motion into linear motion so that the axial motion of the tap into the hole conforms with the desired pitch of the thread. Lead-screw control is often used in high-volume applications or with larger tap sizes to ensure quality threads.

A typical lead control is shown in Fig. 1. A driving pinion, keyed to the main drive shaft, drives a gear keyed to the lead-screw shaft. The lead screw passes through the lead-screw nut, which is fastened to the stationary housing. When the main drive shaft rotates, it turns the lead-screw shaft. The rotation of the lead screw within the lead-screw nut causes the assembly to travel up or down at a speed regulated by the lead screw. This controlled action, transmitted to the chuck, drives the tap into the workpiece at a controlled rate. At the end of the stroke, when the tap has penetrated the workpiece to the desired depth, the direction of shaft rotation is reversed, and the lead-control mechanism backs the tap out of the tapped hole.

There are two disadvantages of lead control in tapping. First, the need to return to the starting point to begin each cycle, and to stop rotation between cycles, may lengthen the tapping cycle. For example, when using a collapsible tap, it would be possible to retract the tap more rapidly from the workpiece without lead control. Second, changing taps for different thread sizes consumes more time when lead control is used because the feed-controlling members in the mechanism also must be changed. The additional time required may increase cost in short-run tapping, in which the thread pitch is frequently changed.

**Tension/compression tapping spindles and attachments** provide axial float and compensate for any difference between machine feed and correct tap feed. The axial float of tapping attachments with tension/compression mechanisms also makes it possible to tap several different thread pitches at the same time with a single machine feed rate.

**Self-reversing tapping attachments** eliminate the need for reversing motors for tap retraction and provide precision tapping for machines without reversing motors. These attachments can be used on all machines that provide rotation to the reversing mechanism inside the attachment.

Compact reversing attachments are the most popular type of self-reversing tapping attachment. These devices have a self-feed, or axial float, that permits the tap to act as its own lead screw, and their compact size provides versatility. Other types of self-reversing attachments utilize built-in lead screws or require a higher degree of operator skill in maintaining proper feed rates.

Ball drive provides a reduction in friction during tapping operations. A spring-biased rolling ball transmits rotational power to the tap and compensates for any operator or machine feed error during tapping. The symmetrical design of ball-driven tools has made it possible to tap right- or left-hand threads with a self-reversing tool.

**Nonreversing tapping attachments** must be used on machines equipped with reversing motors. These devices are extensively used on radial drills, milling machines, lathes, and numerical-controlled equipment. The main advantages of nonreversing attachments are their simplicity and compact design.

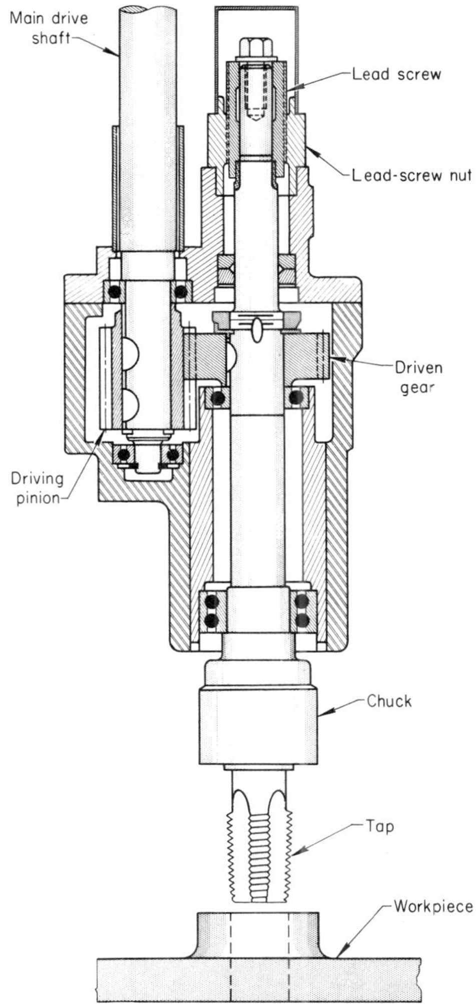


**Fig. 1** Lead-screw mechanism for control of tap feed. See text for discussion.

**Drill Presses.** When no other machining operations are involved, drill presses are often used for tapping because they are easy to set up and simple to operate. Drill presses can be provided with lead-control devices to regulate tap feed rates. However, when lead control is required, tapping machines are ordinarily used rather than drill presses.

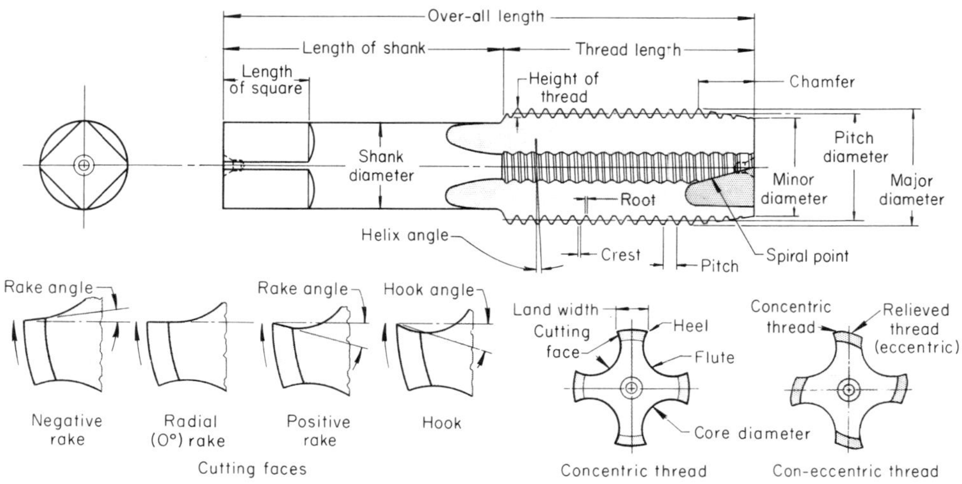


**Fig. 2** Standard nomenclature for design details of solid taps

When a solid tap is used, the drill press must be provided with a tapping attachment or a reversing motor having a tension/compression tap holder. This additional equipment is necessary because the spindle cannot be stopped quickly or precisely enough to hold specified depth tolerance and is not easily reversed to allow the tap to be removed from the workpiece. With a tapping attachment, movement of the feeding lever is stopped at a predetermined point; the tapping attachment then automatically stops the rotation of the tap. Upward movement of the control lever causes the tapping attachment to reverse and spin the tap out of the hole.

With a collapsible tap, a tapping attachment is not required. The tap penetrates the work to a predetermined point at which it automatically collapses and retracts from the work, letting the spindle return without stopping or reversing.

**Single-spindle tapping machines** are generally used for small-to-medium production lots. The simpler models have no lead control but depend on the screw action of the tap in the hole to govern feed.

In a hydraulically driven machine without lead control, axially floated spindles or holders compensate for differences between the feed of the machine and the lead of the tap. An uncontrolled hydraulic feed will not maintain a stable feed rate and might tear the threads.

**Multiple-spindle tapping machines** are for high-volume production. All spindles (some machines have 25 or more) are rotated by a common power source. With these machines, holes of different sizes can be tapped simultaneously. Spindles having axial float can compensate for differences between the lead of the tap and the feed of the spindle. Thus, different thread pitches can be tapped simultaneously in the same machine.

**Gang machines** permit in-line drilling, reaming, and tapping operations, much as in multiple-spindle drilling. Gang machines are intended and used primarily for low production.

**Manual turret lathes** are used for tapping small production lots. Turret lathes are generally more accurate than machines that rotate the tap instead of the workpiece. Moreover, in a turret lathe, tapping can be combined with other operations; therefore, on the same machine the holes can be drilled, bored, reamed, and tapped. This permits the use of higher tapping speed and results in longer tap life than is possible when the holes are less accurate. A lead-control device is almost mandatory when tapping on a turret lathe because the mass of the turret decreases the feel the operator needs to control the feed by hand.

**Automatic Turret Lathes and Bar or Chucking Machines.** Tapping can often be included in a sequence of operations in an automatic turret lathe or in a single-spindle or multiple-spindle bar or chucking machine. However, because of the relatively long setup time required for these machines, they are usually efficient only for large production lots. Moreover, tapping efficiency may depend on the number of other operations that can be incorporated into the automatic sequence; some machines can perform 25 operations per piece. All automatic turret lathes, bar machines, and chucking machines use lead-control devices for regulating the feed.

**Machine Selection.** Selection of the appropriate machine for a tapping operation is based on:

- Size of the workpiece
- Shape of the workpiece
- Production quantity
- Tolerance
- Specified finish
- Number of related operations
- Cost

Many machines tap a wide range of thread sizes. Somewhere within its range, however, a machine is most effective in producing quality threads. Small-diameter, fine-pitch threads should be cut on machines of relatively low power, while larger threads and harder materials require heavier machines with more power. Pipe threads require considerably more torque than straight threads.

## Tap Classification

On the basis of their construction, taps are classified into seven categories:

- Solid taps
- Shell taps
- Sectional taps
- Expansion taps
- Inserted-chaser taps
- Adjustable taps
- Collapsible taps

The following sections describe the design and function of taps in these categories.

### Solid Taps

Solid taps are one-piece taps, usually made of high-speed steel but sometimes of carbon tool steel or of carbide. Solid taps are of two basic types: straight thread and taper thread. Straight-thread taps make threads that do not vary in pitch diameter; taper-thread taps make threads with a uniform reduction in pitch diameter from thread to thread (pipe threads). Standard nomenclature for details of solid taps is given in Fig. 2. Most solid taps have flutes and chamfer.

**Flutes.** Taps have flutes for three reasons: to provide cutting edges, to provide chip clearance and a means of chip control, and to conduct fluid to the cutting sections of the tap. Taps may have straight flutes, spiral flutes, or a combination of both. Taps with straight flutes are the most commonly used because they are more easily made and sharpened than spiral-flute taps and because they perform satisfactorily under many conditions.

**Chamfer.** Solid taps have three types of chamfer:

- Taper chamfer (7 to 9 threads)
- Plug chamfer (3 to 5 threads)
- Bottoming chamfer (1 to 1½ threads)

*Taper chamfer* (Fig. 3a) distributes the cutting load over the greatest number of threads and permits easiest starting of the tap into the workpiece. Therefore, taper-chamfer taps are especially suited for tapping difficult-to-machine metals. However, taper-chamfer taps are seldom suitable for blind holes (too much of the hole is left unthreaded). They also require longer travel than other types to produce full threads in through holes.

*Plug chamfer* (Fig. 3b) is the most commonly used type. Taps with plug chamfer are not required to penetrate a hole as deeply as taps with taper chamfer to pro-

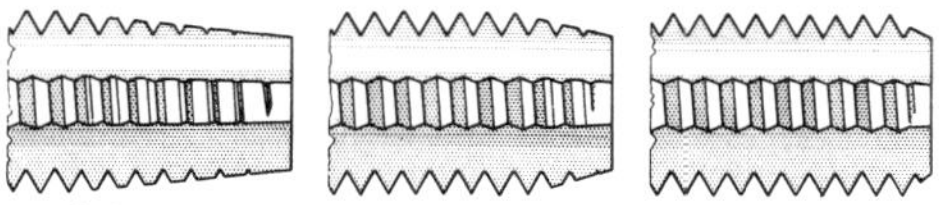

(a) Taper (b) Plug (c) Bottoming

**Fig. 3** Chamfers for solid taps

duce a given length of thread; therefore, they produce at a slightly higher rate. Except in the difficult-to-machine metals, a plug-chamfer tap enters the hole with reasonable ease. If sufficient clearance can be provided, plug-chamfer taps are used successfully in blind-hole tapping.

*Bottoming chamfer* (Fig. 3c) is usually used only for blind holes. When tapping a blind hole in a difficult-to-machine metal, it is common practice to tap as deeply as possible with a taper-chamfer tap or plug-chamfer tap (or sometimes with both, successively) and then to use a bottoming-chamfer tap to finish tapping to the required depth. This practice reduces the time that the 1 to 1½ cutting threads in a bottoming chamfer are under maximum stress.

**Basic Styles of Solid Taps.** The solid taps generally in use are standard hand taps, spiral-point taps, and spiral-flute taps. Each style is described below.

*Hand taps* (Fig. 4a) are the most common type. They were originally used for tapping by hand, and although most are now used in machines, the name has persisted.

Hand taps are produced with either straight or spiral flutes. Many have four flutes, but for tapping metals that produce soft, stringy chips or for ease of chip removal in deep-hole tapping, three or even two flutes can be used.

*Spiral-point taps* (Fig. 4b) have straight flutes supplemented by left-hand angular flutes near the point. The purpose of the spiral point is to push the chips ahead of the tap as tapping progresses.

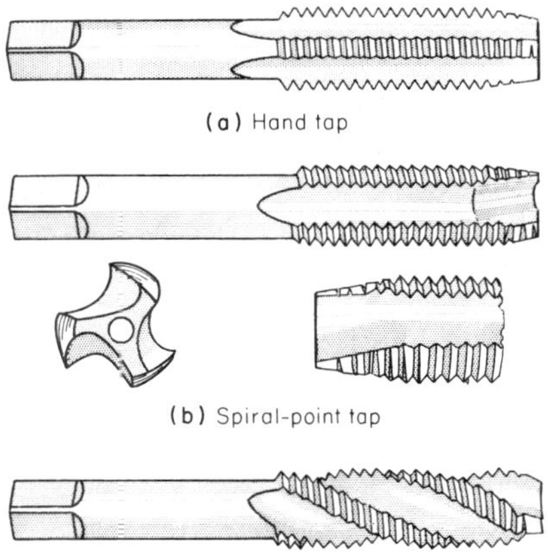

(a) Hand tap

(b) Spiral-point tap

(c) Spiral-flute tap

**Fig. 4** Three basic styles of solid taps

Spiral-point taps are best suited for through holes. However, spiral-point taps with plug chamfer can be used for blind holes, provided there is enough clearance beyond the tapped section to accommodate the chips.

Because the flutes in spiral-point taps are less needed for chip passage, they can be shallower than in standard hand taps, and the tap body can have a stronger cross section. The angular cutting edges generated by the spiral points cut with a shearing action, producing a fine finish on the threads. Furthermore, with the flutes clear of chips, the cutting fluid can move more freely along the flutes to the cutting edges.

*Spiral-flute taps* (Fig. 4c) can have right-hand or left-hand flutes; right-hand flutes are more common. The spiral produces a lifting action that forces the chips along the flute. The spiral of the flutes may be regular (about 25 to 35°) or fast (about 50 to 65°); a fast spiral accelerates chip removal.

Spiral-flute taps are also used to advantage when tapping holes having keyways or other interruptions. The cutting edges meet the interruption progressively, thus cutting more smoothly and being less subject to shock.

Taps cutting regular right-hand threads can be furnished with left-hand spirals. These spirals will push the chips ahead of the tap through the hole for disposal. Keeping the chips out of the flutes minimizes tap breakage and thread damage when the tap is reversed for removal. The shearing action resulting from the angle of the cutting edge on spiral-flute taps produces a better thread finish on difficult-to-machine metals.

**Modified Styles of Solid Taps.** Six of the many modifications of the three basic styles of solid taps are shown in Fig. 5.

*Bent-shank tapper taps* (Fig. 5a) are used for tapping nuts in an automatic tapping machine. The nuts are usually fed from a hopper, and as they are tapped, they pass over the shank and are ejected automatically over the bent end. Consequently, it is unnecessary to reverse the tap as it would be with a conventional tap.

*Combination roughing-and finishing taps* (Fig. 5b) have two stages; the first cuts to rough dimensions, and the second cuts the finished thread. The main disadvantage of these taps is the distance they extend beyond the workpiece at the completion of tapping; this limits their use to the tapping of through holes.

*Step taps* (Fig. 5c) are used for simultaneously cutting threads of the same pitch but of two different diameters.

*Short-flute spiral-point taps* (Fig. 5d) are designed for through holes in thin sections such as webs and sheet metal. Because they are fluted only at the spiral point, these taps are durable. They are limited to tapped holes that are not deeper than one diameter.

*Pulley taps* (Fig. 5e) have the same thread dimensions as hand taps, but they have shanks as large in diameter as the major diameter of the tap thread and much longer than the shanks of hand taps. When the taps thread holes for oil cups or setscrews in pulley hubs, the oversize shanks act as guides in the access holes in the pulley rims to keep the taps aligned.

*Piloted ground-thread taps* (Fig. 5f) are used to maintain the concentricity of threads in workpiece holes. The pilot can be guided by the work hole, a special pilot hole, or a bushing. Piloted taps are specially made to order.

**Sizing Solid Taps.** With solid taps, the relationship between tap size and solid hole size is varied to meet specific conditions. For example, a tapped hole in a part to be electroplated must be oversize by a minimum of four times plate thickness to compensate for the electrodeposited metal. Another condition for which compensation is required is the outward force exerted by the tap when cutting. This force, particularly in thin-wall parts, may be sufficient to influ-

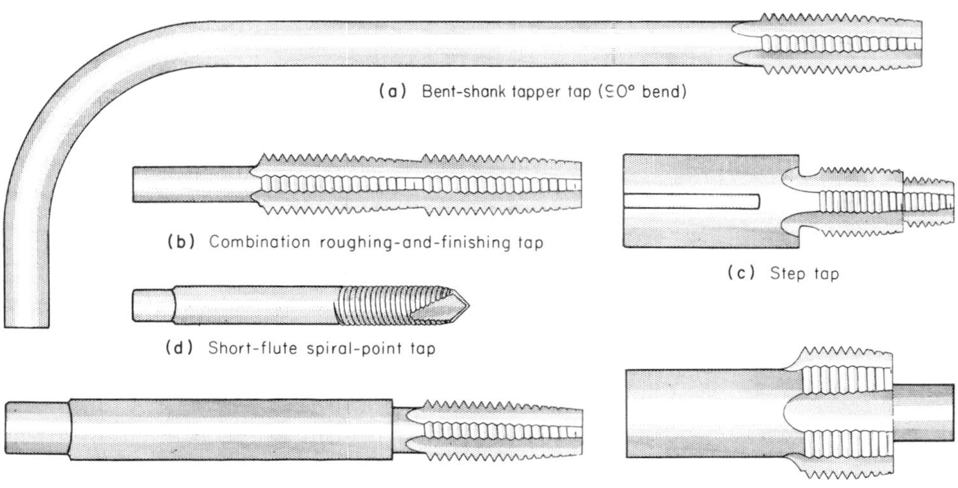

(a) Bent-shank tapper tap (90° bend)

(b) Combination roughing-and-finishing tap

(c) Step tap

(d) Short-flute spiral-point tap

(e) Pulley tap

(f) Piloted ground-thread tap

**Fig. 5** Six modifications of solid taps

**Table 1 Tolerance on pitch diameter of standard taps ground oversize**

| Code(a) | Tolerance |
|---|---|
| **Taps 25 mm (1 in.) or less in diameter** | |
| GH1, mm (in.) | Basic, to basic plus 0.0125 (0.0005) |
| GH2, mm (in.) | Basic plus 0.0125–0.025 (0.0005–0.001) |
| GH3, mm (in.) | Basic plus 0.025–0.0375 (0.001–0.0015) |
| GH4, mm (in.) | Basic plus 0.0375–0.05 (0.0015–0.002) |
| GH5, mm (in.) | Basic plus 0.05–0.0635 (0.002–0.0025) |
| GH6, mm (in.) | Basic plus 0.0635–0.075 (0.0025–0.003) |
| GH7(b), mm (in.) | Basic plus 0.075–0.087 (0.003–0.0035) |
| **Taps over 25 mm (1 in.), and through 38 mm (1½ in.) in diameter** | |
| GH2, mm (in.) | Basic, to basic plus 0.025 (0.001) |
| GH4, mm (in.) | Basic plus 0.025–0.05 (0.001–0.002) |
| GH6, mm (in.) | Basic plus 0.05–0.075 (0.002–0.003) |
| GH8, mm (in.) | Basic plus 0.075–0.10 (0.003–0.004) |

(a) GH, Ground high (oversize). (b) Nonstandard; for tapping workpieces that are susceptible to extreme distortion or to which extremely heavy electrodeposits will be applied. Some manufacturers stock GH7 taps in a special series for tapping zinc die castings.

ence the final results because the part will expand during tapping and will contract when the tap is removed. To meet these conditions, standard taps are ground oversize in established increments and then coded for identification, as indicated in Table 1.

Standard taps ground undersize, identified as GL (ground low), are also available. These taps are used when the tapped hole will be retapped after some process such as heat treating has caused distortion. They are also used when the mating part is undersize. Tolerances for three GL taps are:

| Code | Tolerance |
|---|---|
| GL1 | Basic, to basic minus 0.0125 mm (0.0005 in.) |
| GL2 | Basic minus 0.0125–0.025 mm (0.0005–0.0010 in.) |
| GL3 | Basic minus 0.025–0.0375 mm (0.0010–0.0015 in.) |

## Shell and Sectional Taps

Shell taps are generally made of high-speed steel, without shanks, and are threaded to, or nearly to, their full length. As shown in Fig. 6(a), this type of tap has a full-length hole to accommodate an arbor or shank for driving purposes. The arbor hole is often made with a keyway for a positive drive. Other shell taps may have a tongue or groove across the middle of the back of the tap to fit a specific driver or shank.

Shell taps are made to order, usually 25 mm (1 in.) or larger in diameter and 13 mm (½ in.) or more in length. These taps are best suited to large-diameter holes where thread pitch is eight to the inch or finer and where a short tap is required.

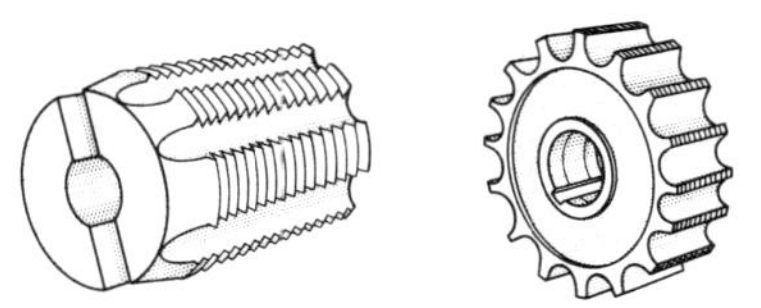

(a) Shell tap (b) Sectional tap

**Fig. 6** Two styles of taps used with inserted shanks

Shell taps can be resharpened in a gang by placing two or more on an arbor. In the large diameters, they are sometimes cheaper than solid taps. However, shell taps of the smaller diameters are generally more expensive than their solid tap counterparts.

Sectional taps (Fig. 6b) are similar to shell taps in that they have inserted shanks. The cutting section is generally shorter for sectional taps than for shell taps. Like shell taps, they can be resharpened in gang, with two or more on an arbor.

## Expansion Taps

Of the various kinds of expansion taps available on special order, all have two features in common:

- An axial hole is drilled from the front end to beyond the thread length of the tap
- One or more radial slots or saw cuts are made from the surface of the tap to this hole between pairs of flutes. The slots, which extend the full working length of the tap, provide springiness to the threaded portion

Two types of expansion taps are illustrated in Fig. 7.

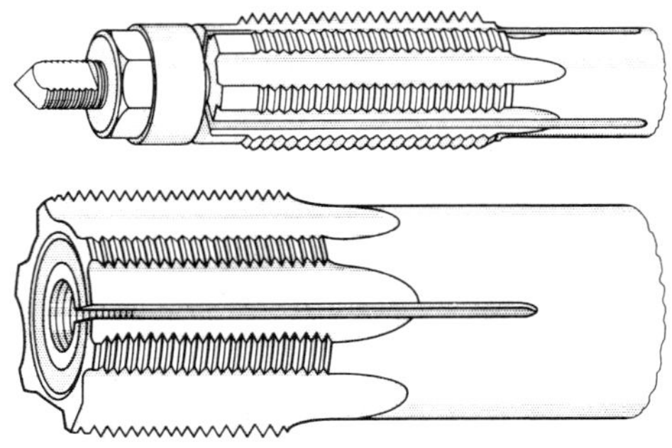

**Fig. 7** Two styles of expansion taps

These taps are available in sizes 9.5 mm (⅜ in.) in diameter and larger. They are ordinarily used only for finishing work or for tapping free-machining metals. They are comparatively high in cost.

## Inserted-Chaser Taps

The body of an inserted-chaser tap has slots that accept sets of four or more chasers held in place either by wedges, screws, and grooves, or by a combination of screws and serrations cut into the chaser body. A typical fixed-chaser type is shown in Fig. 8.

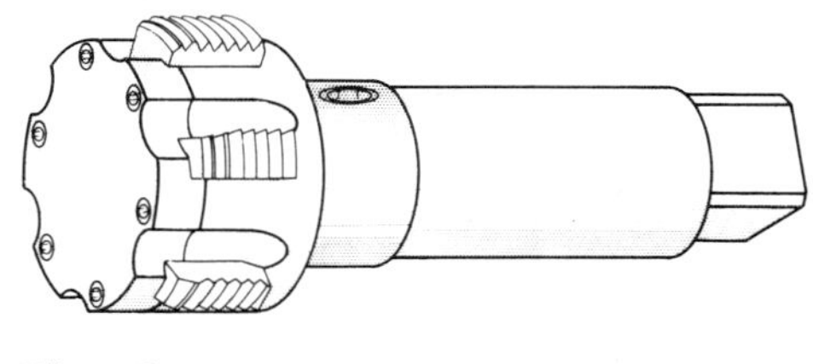

**Fig. 8** Typical inserted-chaser tap

In one common inserted-chaser type, each chaser is wedge shaped (wider at the base than at the cutting surface). The ends of the chasers stop against a shoulder on the body of the tap and are held securely in position by hardened and ground wedges (Fig. 9). These taps, with a cut thread, are standard with some tap manufacturers and are available in sizes ranging from 38 to 150 mm (1½ to 6 in.) in diameter. The initial cost is relatively high, but for long production runs the replacement cost of a set of chasers is considerably lower than that of a solid tap. Ground-thread chasers for tapping taper pipe threads are also available, but are special and designed to fit the user's tap bodies. Oversize holes can be tapped with inserted-chaser pipe taps by placing shims of uniform thickness under the bases of the blades.

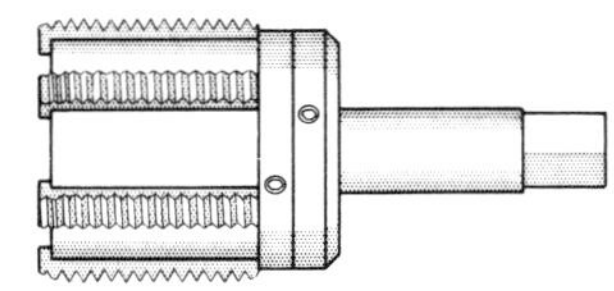

**Fig. 9** Inserted-chaser tap in which chasers are held in place by wedges

**Adjustable Taps.** Another type of inserted-chaser tap is adjustable; that is, it can be made to cut smaller or larger. This tap uses ground-thread chasers with serrations on the backs. The chasers are held in position by locking cams, two for each chaser. The backs of the chasers stop against two hardened and ground nuts on the tap body. The taps are adjusted by loosening the locking cams and turning the nut nearer the chaser. The other nut locks the adjusting nut. The tap will cut larger when the chasers are pushed forward. A still larger cut can be made by removing the chasers from the body and replacing them one serration farther away from the axis of the tap. A smaller cut is made by placing the chasers one serration closer to the axis of the tap.

These taps are used almost exclusively as finishing taps where extreme accuracy is required. They are made only on special order, and the cost is relatively high.

## Collapsible Taps

A collapsible tap has chasers that are set to retract radially after the thread is cut, so that the tap can be withdrawn without reverse rotation. There are two basic types of collapsible taps: rotary and stationary. Rotary taps are used in machines in which the tool rotates, such as drill presses and tapping machines. Stationary taps are used in machines that rotate the workpiece, such as turret lathes. Both types are controlled by a yoke or ring arrangement that automatically withdraws the cutting edges when pressure is applied to the yoke. Stationary collapsible taps (Fig. 10) have a hand lever to reset the tap. Collapsible taps can be used in the same equipment as solid taps, provided there is enough room to accommodate the collapsing head.

Collapsible taps are generally designed with flat or blade-type chasers, but for tapping large diameters, taps with circular chasers are frequently used. With a demountable nosepiece and blocks, either blade or circular chasers can be used in the same collapsible tap.

Taps with blade chasers are generally made for diameters of 30 to 100 mm (1¼ to 4 in.). Circular-chaser taps are made for 90 to 130 mm (3½ to 5 in.) diameters.

**Advantages.** Collapsible taps have two advantages over solid taps: They can be adjusted, and they can be removed from the workpiece without reversing the spindle. The adjustability of collapsible taps is advantageous on first-run jobs in which the reaction to the tapping forces is unknown, because it permits variation in the amount of compensation for distortion of the metal being tapped. Collapsible taps can also be adjusted to allow for electroplated finishes or for growth resulting from heat treatment. Therefore, collapsible taps are adaptable to both short runs and high production.

Collapsing the tap eliminates the wear that would be caused by friction between the cut threads and the tap when the tap is reversed for removal. In addition, a collapsible tap minimizes the danger of tap breakage or thread damage caused by trapped chips as the tap is removed from the tapped hole. This is of particular benefit when tapping metals that make stringy chips.

**Disadvantages.** Compared to solid taps, collapsible taps have four disadvantages:

- Because of the greater overhung inertia of a collapsible tap, a sturdier machine is required than for a solid tap
- Because collapsible taps have a number of moving parts, they require more maintenance; worn or loose components cause excessive variation in tapped holes
- They can seldom be used for holes smaller than 30 mm (1¼ in.) in diameter
- Collapsible taps are expensive

## Tap Materials

Most taps are made of high-speed steel. Carbide blades (chasers) are sometimes

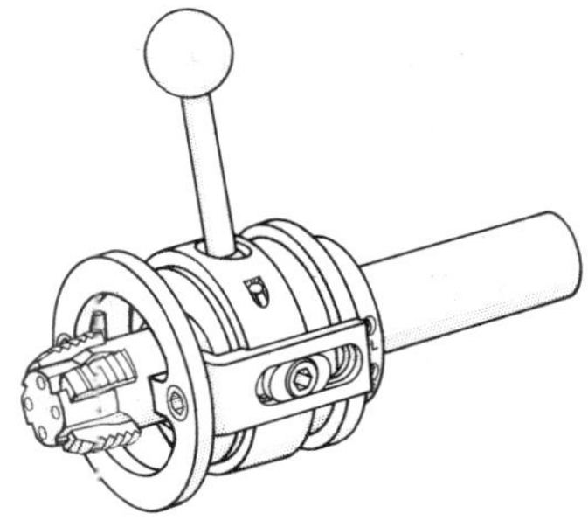

**Fig. 10** Stationary (nonrotating) collapsible tap

used in adjustable or inserted-chaser taps. However, the use of carbide is largely restricted to special applications—for example, the tapping of especially abrasive grades of cast iron or abrasive nonmetals such as fiberglass and filled plastics.

General-purpose high-speed steels, such as M1, M2, M7, and M10, are most widely used for taps and have proved satisfactory for most applications. For tapping difficult-to-machine metals (such as heat-resistant alloys, or steels harder than about 35 HRC), taps made of one of the more highly alloyed high-speed steels, such as M15 or T15, are often justified. However, taps made of the more highly alloyed high-speed steels cost two to three times as much as similar taps made of the general-purpose types; the more highly alloyed types are more difficult to grind and are more susceptible to grinding burn.

For special tap designs that require greater resistance to abrasion, type M3 (class 2) or type M4 is generally used. Where requirements include high abrasion resistance and/or very high resistance to softening at elevated temperatures, types T15 and M42 are typically used.

**Cost.** To the user, tap cost and tap expendability are directly related, especially in smaller sizes. Most users will sacrifice tap life rather than pay for maintenance and repair. A tap chipped along its chamfer or in the leading threads can be salvaged, but replacement may be more economical. In addition, cheaper taps, which are discarded after short runs, may be more economical than costly taps with a longer production life, depending on the cost of downtime for changing taps. In multiple-spindle setups in which tap operating conditions are poor, a sacrifice in tap life may prove less expensive than a second tapping operation.

## Surface Treatment of Taps

Many taps are given one of the following surface treatments: nitriding, titanium nitride coatings, chromium plating, or oxidation (black oxide) by steam. Titanium nitride coatings are the most common treatments used in production shops. These treatments increase tap life and improve thread finish by:

- Reducing adherence of the tap to the work metal
- Reducing chip buildup on cutting edges
- Minimizing wear caused by erosion from chips

Surface treatments can increase tap life by 500% or more. The greatest benefit of surface treatment is realized in metals that form stringy chips or in hard metals such as heat-treated steel. The relative merits of the three types of treatment are debatable; all have proved helpful. However, steam-treated taps are not recommended for use in soft metals such as brass, because they are likely to tap oversize.

The main disadvantages of surface treatments are cost and inconvenience. Regardless of the treatment used, the penetration (or buildup, in the case of plating) is so small that the taps must be re-treated each time they are ground. This is sometimes a great inconvenience, especially when the taps must be sent out for treatment.

## Selection of Tap Features

The main influences on the selection of tap design features are work metal composition and hardness, class of thread required, and cost. The following example describes an application in which changes in the design of the tap proved beneficial.

**Example 1: Tap Redesign to Produce Acceptable Threads.** The tapping of 1-14 UNS threads in free-cutting brass was one of a sequence of operations in a multiple-spindle machine. The taps originally used were standard four-flute taps with a 12° positive rake and relatively wide lands, and they had been subjected to a steam oxide surface treatment. These taps failed to pick up the lead; consequently, the threads were reamed to the extent that a no-go gage would enter the full length of the threads. In addition, the steam oxide treatment caused the taps to cut oversize because the pitted surface from the oxide treatment, combined with the use of soluble oil as a cutting fluid, caused the thread flanks to load.

The following revisions were made:

- The land width of the tap was reduced from 7.6 to 4.2 mm (0.300 to 0.165 in.)
- The rake was reduced to 3°
- The surface treatment of the taps was changed to nitriding
- A light mineral oil (paraffin base) was substituted for the soluble oil

These changes produced acceptable parts. Tap life between grinds was 32 000 pieces. Each tap could be reground three times; therefore, total tap life averaged 128 000 pieces.

**Flute Selection.** Selecting between three-flute and four-flute taps often involves a difficult decision. Four-flute taps have less cutting load per tooth, but three-flute taps in

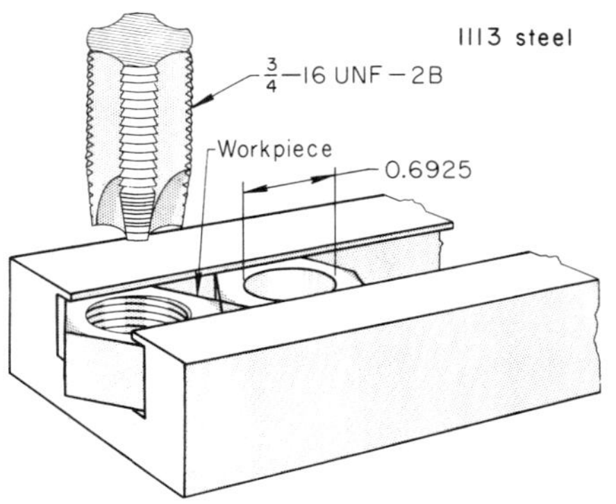


**Tap details**

| | |
|---|---|
| Material | M1 high-speed steel |
| Surface treatment | Steam oxide |
| Hook angle, degrees | 8 |
| Number of chamfered threads | 6 |
| Chamfer angle | 5° 40′ |
| Chamfer relief, mm (in.) | 0.23–0.028 (0.009–0.0011) |
| Spiral-point angle | 7° 45′ |
| Spiral-point length | 8 threads |

**Operating conditions**

| | |
|---|---|
| Speed, at 407 rev/min, m/min (sfm) | 24 (80) |
| Cutting fluid | Sulfurized oil |
| Thread length, mm (in.) | 19 (¾) (through) |
| Percentage of thread | 71% |
| Tap life per grind, pieces | 10 000 |
| Production rate, pieces/h | 500 |

**Fig. 11** Use of a four-flute spiral-point tap for threading nuts in an automatic nut tapper. A three-flute tap was preferred for ¾-10 UNC-2B threads in these nuts.

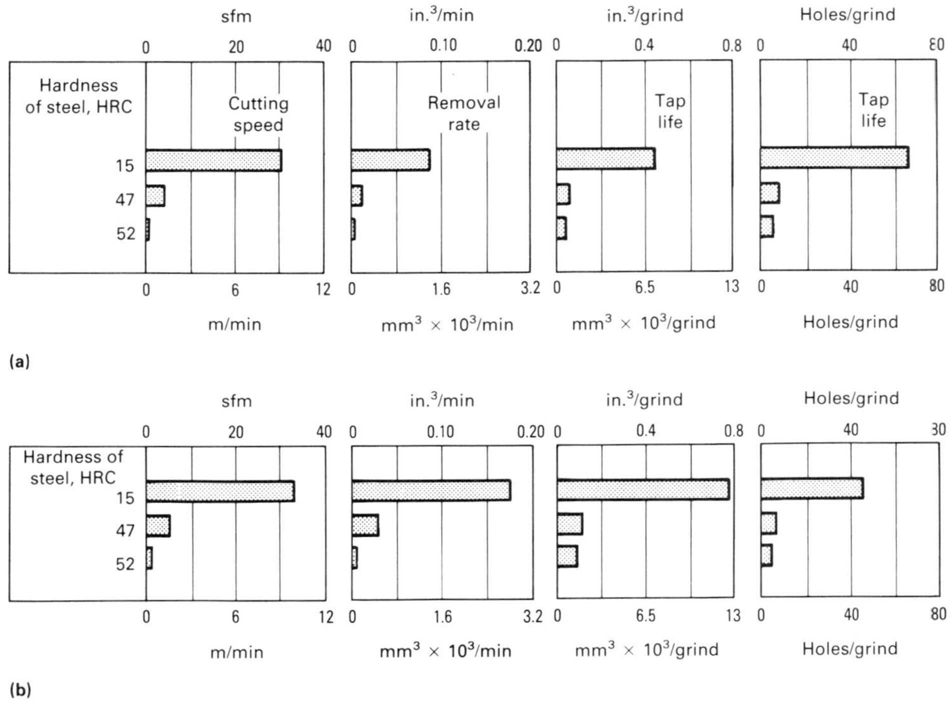


**Tapping conditions (holes of both diameters)**

| | |
|---|---|
| Type of machine | Automatic tapping |
| Depth of through holes, mm (in.) | 32 (1¼) |
| Tap drill for ¼-28, mm (in.) | No. 3, 5.41 (0.213) |
| Tap drill for ½-20, mm (in.) | 11.5 (²⁹⁄₆₄) |
| Cutting fluid | Lithopone-pigmented wax and fatty ester in lard-mineral oil |
| Tap material | M10 high-speed steel |

| Tap details | ¼–28 15 HRC | ¼–28 47, 52 HRC | ½–20 15 HRC | ½–20 47, 52 HRC |
|---|---|---|---|---|
| Flute helix angle, degrees | 0 | 0 | 0 | 0 |
| Hook angle, degrees | 2 | 5 | 2 | 5 |
| Chamfer angle, degrees | 30 | 42 | 30 | 42 |
| Chamfer relief angles, degrees | 5 | 5 | 5 | 5 |
| Number of lead threads | 3 | 3 | 4 | 4 |
| Number of flutes | 4 | 4 | 3 | 3 |

**Fig. 12** Effect of hardness on the tapping of carbon and low-alloy steel. (a) ¼-28 UNF through holes. (b) ½-20 UNF through holes

the same size have greater strength and more chip room. Thread pitch sometimes influences the decision, as indicated in the following example.

**Example 2: Three-Flute Versus Four-Flute Taps.** A four-flute spiral-point tap proved superior to one with three flutes for cutting ¾-16 UNF-2B threads in steel nuts, using an automatic nut tapper (Fig. 11). The greater number of cutting edges on the four-flute tap, together with the smaller chips it produced, resulted in smoother finish on the threads and longer tap life. Tap details and operating conditions are given with Fig. 11. However, for tapping ¾-10 UNC-2B threads in these nuts, a three-flute tap was preferred because of the greater strength in the tap and increased clearance to pass the larger chips.

## Factors That Influence Procedures and Results

The principal factors that influence the selection of equipment and procedure for tapping and that affect thread quality, productivity, and cost are:

- Composition and hardness of the metal being tapped
- Size and shape of the workpiece
- Thread size and depth
- Tolerance and finish specified
- Whether blind or through holes are being tapped
- Speed
- Use or nonuse of lead control
- Cutting fluid

The following sections discuss the influence and control of these variables (except for lead control, which is discussed in the section "Machines and Accessories" in this article).

### Metal Composition and Hardness

A metal is seldom selected for its machinability alone. However, a change of material may increase productivity and decrease cost without sacrificing the required properties. The simplest change is that from a specific alloy to a free-machining counterpart. Invariably, the tapping of free-machining grades results in more accurate threads of better finish at higher production rates and lower cost than the tapping of nonfree-machining grades that are otherwise similar in composition and hardness.

Except for low-carbon steels that do not contain free-cutting additives, as well as certain other soft metals that form gummy, adherent chips in machining, hard metals are less readily tapped than soft metals. Figure 12 illustrates the usual effect of differences in work metal hardness on the tapping of carbon and low-alloy steels. In this test on the tapping of ¼-28 and ½-20 through holes in similar low-alloy steels with M10 high-speed steel tools, the cost for tapping at a hardness of 52 HRC was about 30 times that at 15 HRC; for tapping at a hardness of 47 HRC, about 6 times.

Under the conditions of this test, tap life increased with an increase in hole diameter on the basis of volume of metal removed, but decreased on the basis of number of holes per grind. Tapping cost for the 6.4 mm (¼ in.) diam holes was three times that for the 13 mm (½ in.) diam holes on the basis of volume of metal removed, but was only

**Table 2 Recommended practice for tapping steels in three ranges of hardness**

| Process variable | Hardness of steel tapped, HRC 5–30 | 30–40 | 40–55 |
|---|---|---|---|
| Tap design | Standard | Modified standard, or special(a) | Special(a) |
| Tap material(b) | T1, M1, M7, M10 | M2, M3 | T15, M3, M33, M42 |
| Tap hardness, minimum | 62 HRC | 64 HRC | 66 HRC |
| Surface treatment of tap | Steam oxide(c) | Nitride plus steam oxide | Nitride plus steam oxide |
| Percentage of full thread | 75% | 60% | 55% |
| Speed, m/min (sfm) | 30 max (100 max) | 6–14 (20–45) | 0.15–3.0 (½–10) |
| Cutting oil | Sulfurized | Sulfurized and chlorinated | Sulfurized and highly chlorinated |

(a) Special tap should have four flutes, 3° (positive) hook angle, thread relief, and 3 to 4 thread chamfer. (b) High-speed steels similar in total alloy content to steels listed can be substituted. (c) Or other oxidizing process, see the section "Surface Treatment of Taps" in this article.

slightly greater on a per-hole basis. Machining and tool details are given in Fig. 12.

A tap manufacturer conducted a series of tests on 11 steels that ranged in hardness from 5 to 55 HRC to determine the effect of hardness on tapping. Data obtained in these tests were used as the basis for the recommendations given in Table 2.

As indicated in Table 2, taps made of general-purpose high-speed steels are usually the most economical when the hardness of the metal being tapped does not exceed 30 or 32 HRC; as workpiece hardness increases above this level, one of the more highly alloyed high-speed steels is usually selected. Table 2 also indicates that tap design is more critical when high-hardness steels are being tapped.

A difference in work metal composition can affect requirements for hole preparation before tapping. For example, when tapping aluminum, accuracy and thread finish can usually be improved by reaming after drilling, but in tapping carbon, alloy, or stainless steel, reaming is seldom needed.

## Workpiece Size and Shape

The size and shape of the workpiece must be considered in establishing tapping procedures. Tapping problems are more likely to be encountered with workpieces that are too weak or flimsy to withstand normal tapping forces. This inability can result in a loss of dimensional control or even in damage to the workpiece. The example that follows describes procedures that have been successful for tapping tubular workpieces with a wall 6.0 mm (15/64 in.) thick.

**Example 3: Tapping Seam-Welded Tubing.** An eight-station vertical tapping machine and six-flute solid taps were used for cutting 2-11½ NPSC threads through 54 mm (2⅛ in.) lengths of seam-welded 1020 steel tubing (wall thickness: 6.0 mm, or 15/64 in.) for use as steam-pipe couplings. As shown in Fig. 13, the workpieces were held in air chucks. After each piece had been tapped, it was automatically pushed up the long shank of the tap by the next piece, until the shank was filled, at which time the workpieces were unloaded. The eight stations were synchronized to permit one-man operation.

Solid taps were used in preference to collapsible taps because the variation from true roundness of the inside diameter, together with the seam, required that the tap provide some reaming action in addition to cutting the threads. Tap details and operating conditions are included in Fig. 13.

## Size, Pitch, and Percentage of Thread

Thread size and pitch and the percentage of full depth to which the threads are cut determine the metal removed in any tapping operation and have a large effect on efficiency and tap life. As the size of thread increases, the amount of metal removed increases in approximate proportion to the hole diameter (assuming that other factors remain constant). As the thread becomes coarser, the amount of metal removed in tapping increases for any thread diameter and percentage of thread.

Percentage of thread is half the difference between the basic major diameter and the actual minor diameter of an internal thread, divided by the basic thread height and expressed in percent. As the percentage of thread increases, the amount of metal removed will increase for any thread pitch and size.

**Example 4: Effect of Increase in Percentage of Thread on Speed, Production Rate, and Tap Life.** Two-flute spiral-point taps with a 15° chamfer were used in a two-spindle tapping machine for cutting 8-36 UNF-2B threads in through holes in low-carbon steel plate 2.84 mm (0.112 in.) thick. Speed, production rate, and tap life for tapping these holes to 55 and 78% of full thread depth were as follows:

| | Percentage of full thread depth 55 | 78 |
|---|---|---|
| Speed, rev/min | 1863 | 1398 |
| Speed, m/min (sfm) | 24 (80) | 18 (60) |
| Holes tapped per hour | 2195 | 1925 |
| Tap life, holes | 12 500 | 9000 |

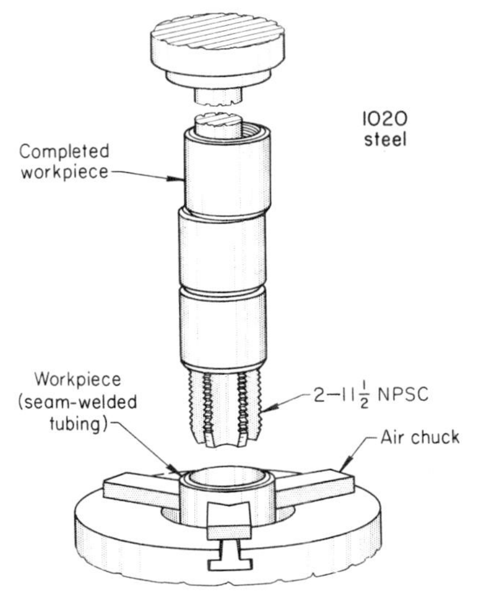


| | |
|---|---|
| **Tap details** | |
| Material | M1 high-speed steel |
| Number of flutes | 6 |
| Overall length, mm (in.) | 380 (15) |
| Chamfer length, mm (in.) | 27.4 (1 5/64) |
| Chamfer angles (double) | 10° 15′ and 4° 36′ |
| **Operating conditions** | |
| Speed, at 80 rev/min, m/min (sfm) | 15 (50) |
| Cutting fluid | Sulfurized oil |
| Tap life per grind, pieces | 1200 |
| Production rate, pieces/h | 80 |
| Workpiece hardness, HRB | 75–80 |

**Fig. 13** Tapping of seam-welded couplings

## Accuracy and Finish

**Conditions** that cause dimensional variation in tapped threads also cause rough finish on the threads. Among these conditions are lack of concentricity between the tap holder and the spindle, worn taps, entrapment of chips in the tapped hole, and chip buildup on cutting edges and flanks of the threads.

**Concentricity of Holder and Spindle.** New taps are seldom responsible for poor threads. However, even a new tap is no more accurate than the combined eccentricity of the holder and rotating spindle. The eccentricity of these rotating members will cause the tapped holes to be oversize.

**Condition of Taps.** Worn taps are a major source of dimensional variation in tapped holes or of poor finish on threads or both. As the cutting edges of a tap become dull, the chips produced are torn, rather than cut or shaved. Improperly sharpened taps, or taps with the wrong cutting-edge angles, will have a similar adverse effect on thread finish.

**Chip Entrapment.** Most taps have relief ground on the trailing part of the lands, which provides space between the heel of the tap and the cut threads. When the tap is backed out of the threaded hole, this space admits fine particles or chips between the tap and the workpiece, and they press against the thread flanks as the tap reverses, resulting in galled thread flanks, broken taps, or both.

Collapsible taps eliminate the entrapment of chips or particles, but they are impractical for threads smaller than 30 mm (1¼ in.) in diameter and sometimes result in excessive variation of threads. In through holes, damage from chips during reversal can be substantially reduced by running the tap far enough to clear all threads in the chamfer and then flushing cutting fluid through the flutes to wash out all the chips before reversing the tap for disengagement.

Blind holes, however, do not permit this through-flushing. Therefore, taps used for threading blind holes should be designed so that they either lift the chips out of the hole, by the action of the flutes, or (if enough clearance can be provided at the bottom of the hole) push the chips ahead of the tap (see the section "Selection of Tap" in this article). When taps are used that lift chips out of the hole by the action of the flutes, it is sometimes helpful to direct a jet of cutting fluid down one flute at the end of the tapping cycle after the tap has stopped rotating. This forces the chips out the remaining flutes. However, when using taps that force the chips down into a clearance, this practice is not recommended. Another aid to chip removal when tapping blind holes is the insertion of a soft wax plug in the hole before tapping. As the tap progresses down the hole, the wax picks up chips and particles and carries them away as it is forced up the flutes of the tap. The wax also lubricates the tap.

**Chip buildup** on the tap (adherence of work metal to the cutting edges) is a major cause of surface roughness. The buildup reduces the cutting efficiency of the tap and produces results similar to those obtained when tapping with a dull tap. Surface treatments of the tap greatly assist in the prevention of chip buildup. Buildup can also be prevented or minimized by changing the cutting fluid or its method of application (see the section "Cutting Fluids" in this article).

**Lead-control tapping** usually produces greater accuracy in tapped holes than manual control of the tap feed. Sometimes, however, the required control of dimensions or finish (or both) can be obtained only by reaming drilled holes before tapping.

## Tapping Blind Holes

In tapping blind holes, as the distance, or clearance, between the last full thread and the bottom of the hole decreases, the chamfer length of the tap used must be reduced. With shorter-chamfer taps, however, the cutting load is borne by fewer teeth, and the chips produced are larger and coarser and, consequently, more difficult to expel. Moreover, as the clearance decreases, there is less room for chips that precede the tap. If more chips enter the clearance space than can be accommodated, they will be compressed by the advancing tap, which may cause tap breakage, hole damage, or both.

**Selection of Tap.** Standard hand taps with straight flutes are seldom used for blind-hole tapping. When there is enough clearance for the chips at the bottom of the hole, spiral-point straight-flute taps (Fig. 4b) are commonly used because they drive the chips ahead of the tap. When clearance at the bottom of the hole is insufficient for the chips, spiral-flute taps (Fig. 4c), with a flute lead angle of 30 to 55°, are used.

Taper-chamfer taps (Fig. 3a) are seldom used for the complete tapping of blind holes, because of the large clearance required for their 7 to 9 thread chamfer. Plug-chamfer taps (Fig. 3b) are more commonly used because they offer a compromise between the ideal load distribution of the taper chamfer and the high load concentration of bottoming-chamfer taps (Fig. 3c). The most common practice is to drill the hole deeper by a distance of at least five threads more than must be tapped and then tap with a plug-chamfer tap.

Tapping to the bottom of a blind hole is slow and expensive because it invariably involves some hand tapping. However, when blind holes must be threaded to the bottom (for example, when creating seals for pressures of 28 MPa, or 4 ksi, or more), the common practice in high-volume production involves the use of a lead-screw tapping unit with a precision lead screw and nut. Depth can be held to ±0.13 mm (±0.005 in.) with this method, and proper tool setting is essential to ensure proper holes. For low-volume production, the practice most commonly employed is as follows:

- Machine-tap as far as possible, using a plug-chamfer tap. In many cases, it is more economical to start the hole with a taper-chamfer tap, thus minimizing the amount of metal that will later be removed by taps having shorter chamfers
- Hand tap, using a standard bottoming-chamfer tap, which can produce threads within 1 to 1½ threads from the bottom
- If required, hand tap with a special minimum-chamfer (~0.25 mm, or 0.010 in.) tap, which can tap one thread or less from the bottom

## Speed

Table 3 lists nominal speeds for tapping carbon and low-alloy steels. It shows that work metal hardness has a great effect on the speed used. Similar tables for tapping other metals will be found in separate articles in this Volume that deal with the machining of specific metals. The direction in which to modify tapping speed for various conditions is as follows:

- As hole length increases, speed must decrease because of chip accumulation
- In short holes, taps with long chamfers can run faster than taps with short chamfers
- Taps with bottoming chamfers must run slower than taps with plug chamfers
- As the percentage of thread being tapped increases, speed must decrease
- As pitch becomes finer for a given hole size, tapping speed can increase
- With all other factors remaining equal, the cutting fluid, together with the amount used and the effectiveness of application, greatly influences optimum speed

In tapping, feed is governed by the pitch of the thread being tapped; therefore, only speed can be adjusted. Optimum tapping speed is usually based on minimum cost per hole and is often a compromise between maximum tap life and maximum productivity. Low speeds result in longer tap life, but if the speed is too low, work hardening of some alloys may occur, causing the tap to break or fuse to the part. Productivity will also be lower.

A wide range of speeds can be successfully used in tapping most metals. In tapping hard and otherwise difficult-to-machine alloys, however, it is especially important to select cutting speed carefully for optimum results because the range of economical operating conditions is relatively narrow.

**Example 5: Effect of Speed on Productivity and Tap Life.** To determine the effect of speed on tap life and production rate, three different speeds were used for tapping 4-40 threads in through holes, 2.4 mm (3/32 in.) deep and 2.26 mm (0.089 in.) in diameter, in 1015 and 1018 steels at 100 to 125 HB. The taps used at all speeds were straight two-flute taps made of general-purpose high-speed steel, with 5° hook, 18° chamfer, and 5° chamfer relief. Sulfurized mineral oil was used as the cutting fluid. Results were as follows:

| Speed | | | |
|---|---|---|---|
| rev/min | m/min (sfm) | Holes per hour | Tap life, holes |
| 4002 | 36 (118) | 1200 | 7200 |
| 2728 | 24 (80) | 1155 | 10 800 |
| 1394 | 12 (41) | 1080 | 9350 |

The greater tap life at 24 m/min (80 sfm) than at 12 m/min (41 sfm) may have resulted from slight variations in grinding the taps, or the slower speed of 12 m/min (41 sfm) may have permitted chips to be trapped.

## Cutting Fluids

A cutting fluid is more important in tapping than in most other machining operations because tap teeth are more susceptible

**Table 3 Nominal speeds for tapping carbon and low-alloy steels with high-speed steel taps**

M1, M7, and M10 high-speed steels are suitable for tapping carbon and low-alloy steel no harder than 375 HB; M3 and M40, for steel of 375 HB and higher.

| Typical steel(a) | Condition | Hardness, HB | Speed m/min | Speed sfm |
|---|---|---|---|---|
| 1020 | Annealed | 85–125 | 17 | 55 |
| | Annealed | 125–175 | 14 | 45 |
| | Annealed | 175–225 | 12 | 40 |
| | Annealed | 225–275 | 9 | 30 |
| 1045 | Annealed | 125–175 | 14 | 45 |
| | Annealed | 175–225 | 12 | 40 |
| | Annealed | 225–275 | 11 | 35 |
| | Annealed | 275–325 | 8 | 25 |
| | Quenched and tempered | 325–375 | 6 | 20 |
| | Quenched and tempered | 375–425 | 3 | 10 |
| 1112 | Annealed | 100–150 | 18 | 60 |
| | Cold drawn | 150–200 | 20 | 65 |
| 1117 | Annealed | 100–150 | 15 | 50 |
| | Cold drawn | 150–200 | 17 | 55 |
| 1137 | Annealed | 175–225 | 15 | 50 |
| | Quenched and tempered | 275–325 | 11 | 35 |
| | Quenched and tempered | 325–375 | 6 | 20 |
| | Quenched and tempered | 375–425 | 3 | 10 |
| 12L14 | Annealed | 100–150 | 18 | 60 |
| | Annealed | 150–200 | 17 | 55 |
| | Annealed | 200–250 | 14 | 45 |
| 4140 | Annealed | 175–225 | 11 | 35 |
| | Quenched and tempered | 275–325 | 8 | 25 |
| | Quenched and tempered | 325–375 | 5 | 15 |
| | Quenched and tempered | 375–425 | 3 | 10 |
| 4140+S | Annealed | 150–200 | 14 | 45 |
| | Quenched and tempered | 275–325 | 8 | 25 |
| | Quenched and tempered | 375–425 | 3 | 10 |
| | Quenched and tempered | 45–48 HRC | 2 | 7 |
| 41L40 | Annealed | 150–200 | 14 | 45 |
| | Quenched and tempered | 275–325 | 6 | 20 |
| | Quenched and tempered | 325–375 | 5 | 15 |
| | Quenched and tempered | 45–48 HRC | 2 | 7 |
| 8620 | Annealed | 125–175 | 14 | 45 |
| | Annealed | 175–225 | 12 | 40 |
| | Cold drawn | 225–275 | 11 | 35 |
| | Cold drawn | 275–325 | 8 | 25 |
| | Quenched and tempered | 325–375 | 6 | 20 |
| | Quenched and tempered | 375–425 | 3 | 10 |

(a) Each steel listed is a common grade in a group of similar steels. Source: Metcut Research Associates, Inc.

to damage from heat than are most other cutting tool surfaces and because chips are more likely to become congested in tapping than in operations in which the cutters are not surrounded by the work material. Cutting fluids are generally used in tapping all metals except cast iron.

For tapping holes longer than about twice the diameter, or blind holes, in cast iron, a cutting fluid or an air blast is recommended. A weak emulsion of soluble oil in water (1 part oil to 40 parts water) or plain water with a rust inhibitor have both been successful.

The cutting fluids most commonly used for tapping various metals are listed in Table 4. Regardless of the type of cutting fluid used, application (conveying the fluid to the cutting areas) is important and is usually more difficult than in operations such as turning or milling. To ensure maximum effectiveness, the cutting fluid should be directed at the tap with sufficient pressure to force it down the flutes of the tap. Under extreme conditions, as in the tapping of deep blind holes, the cutting fluid should be directed at the tap in two streams, one on each side of the tap and as nearly parallel as possible to the axis of the tap.

Another important consideration is the removal of fine metal particles from recirculated cutting fluid. This swarf, as it is called, is highly abrasive. Recirculated fluids should be screened or filtered.

**Sulfurized or chlorinated oils,** used individually or diluted with mineral oil, have proved satisfactory for many tapping applications and are especially desirable for tapping most steels (including stainless steels) and difficult-to-machine metals such as heat-resistant alloys. Sulfurized or chlorinated oil is usually lower in viscosity than a lard oil mixture. Therefore, waste by carryout is reduced. Another advantage of sulfurized or chlorinated cutting oil is adjustability; that is, the oil can be used straight (for full effectiveness) or diluted with mineral oil to lower the viscosity and cost without sacrificing the desirable effects of the additive-containing oil. When prior experience in a similar application is not available, it is advisable to begin a tapping operation with straight sulfurized oil and then to dilute it gradually until results approach unacceptability in accuracy, finish, or tool life.

The main disadvantage in the use of sulfurized or chlorinated oil is that the fluid will stain some metals—for example, many copper-base and nickel-base alloys. When staining is objectionable, a test should be made to determine the compatibility of the cutting oil and the work metal.

**Soluble-oil emulsions** are inexpensive and serve adequately for cooling and flushing away chips, but are less effective than straight oils for preventing the adherence of tools to workpieces and preventing built-up edges. Soluble oil can be used without fear of staining metals such as copper and aluminum. Most soluble oils contain an inhibitor to prevent the rusting of steel workpieces.

**Mineral oil** blended with lard oil or other animal fats (usually 10 to 20%) is effective in preventing adherence of tool to work metal. Mineral oil can also be used on all metals without causing staining or rusting. However, it is less effective for cooling and flushing away chips than soluble oil emulsions. A marked disadvantage of mineral and lard oil mixtures is that, because of their relatively high viscosity and wetting characteristics, they cling tenaciously to chips. Thus, waste by carryout is higher than with less viscous fluids.

**Example 6: Reselection of Extreme Pressure (EP) Additive to Improve Tap Life.** Annealed 1045 steel was tapped with ½-13 UNC-2B threads, 16 mm (⅝ in.) deep. A soluble oil with an EP additive was used

**Table 4 Cutting fluids commonly used in tapping various metals**

| Metal | Sulfurized oil | Sulfurized and chlorinated oil | Soluble oil | Kerosene plus lard oil | Light mineral oil | Dry |
|---|---|---|---|---|---|---|
| Steel, 5–30 HRC | X | ... | ... | ... | ... | ... |
| Steel, 30–40 HRC | ... | X | ... | ... | ... | ... |
| Steel, 40–55 HRC | ... | X(a) | ... | ... | ... | ... |
| Stainless steel | X | X | ... | ... | ... | ... |
| Gray iron | ... | ... | X(b) | ... | ... | X(b) |
| Malleable or nodular iron | X | ... | X | ... | ... | ... |
| Aluminum and alloys | ... | ... | X | X | ... | ... |
| Copper and alloys | ... | ... | X | X | X | ... |
| Heat-resistant alloys(c) | X(d) | X(d) | ... | X(d) | ... | ... |

(a) Highly chlorinated. (b) Dry tapping may cause chip congestion in deep or blind holes; for these, an air blast or a weak soluble-oil: water emulsion (1 part oil to 40 parts water) is recommended. (c) Nickel or cobalt base. (d) Some staining of the work may result from the use of sulfurized oils; when this is objectionable, kerosene plus lard oil can be used.

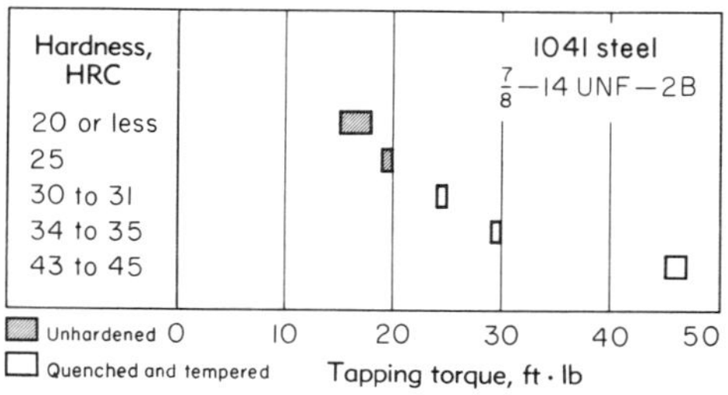


**Fig. 14** Effect of hardness of steel workpiece on torque

for the cutting fluid; however, tap life was not satisfactory. A change was made in the cutting fluid to a more active EP lubricant. Tap life increased 80%, and surface finish showed a substantial improvement.

## Torque for Tapping

Torque demand for tapping generally dictates the size of machine, rigidity of tool holder, and type of workholding devices. Some of the factors that determine torque demand are:

- Workpiece material and hardness
- Tap design
- Surface speed
- Percentage of full thread
- Method of grinding the tap
- Cutting fluid

**Workpiece Hardness.** One company producing threaded nuts in large quantities conducted a series of tests to evaluate the effect of hardness of the steel workpiece on torque. The procedure and results are given in the following example.

**Example 7: Effect of Hardness of Steel Workpiece on Torque.** An automatic nut tapper was used to produce ⅞-14 UNF-2B threads, 73% of full depth, in pierced nuts of 1041 steel. The diameter of the pierced hole was 20.52/20.47 mm (0.808/0.806 in.). Nuts were sorted to obtain one lot that had hardness values of 20 HRC or less and a second lot that was 25 HRC or slightly less. Three additional lots were prepared by heat treating nuts to obtain hardness ranges of 30 to 31, 34 to 35, and 43 to 45 HRC. All five lots of nuts were tapped in the same way. Torque was measured and recorded; results are plotted in Fig. 14.

**Workpiece Composition.** Figures 15 to 18 show that work metal composition also has a marked effect on torque. Gray iron shows the lowest torque for all conditions, and 1020 steel shows the highest of the three ferrous metals tested. This suggests that the more stringy chip characteristic of 1020 steel has a significant adverse effect on torque.

**Tap Design.** The data in Fig. 15 indicate only a small difference in torque between spiral-point and straight-flute taps for tapping O6 tool steel. However, when tapping 1020 steel, torque for the straight flute, plug-chamfer tap exceeds that for the spiral point by a factor of about three.

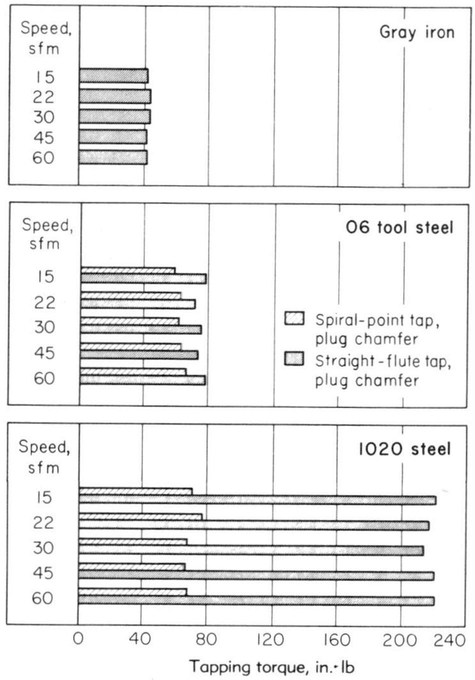


**Fig. 15** Effect of work metal, tap design, and speed on torque. Data were obtained in cutting ⅜-16 UNC-2B threads, to 75% of full depth, through 14 mm (⁹⁄₁₆ in.) stock, with high-speed steel taps (hook angle: 4° 30′), using sulfurized oil as cutting fluid.

Figure 17 shows that there is essentially no difference in torque between 4½ and 11° hook angles when tapping gray iron and that there is only a small difference in tapping O6 tool steel. However, when 1020 steel is being tapped, the 11° hook angle shows less torque than the 4½° hook angle by amounts varying from 10 to 50%, other factors remaining unchanged.

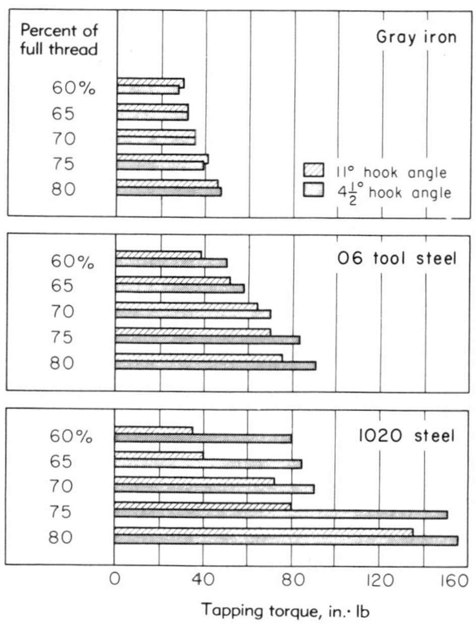


**Fig. 17** Effect of work metal, hook angle of tap, and percentage of full thread on torque. Speed: 18 m/min (60 sfm); other conditions, same as for Fig. 15.

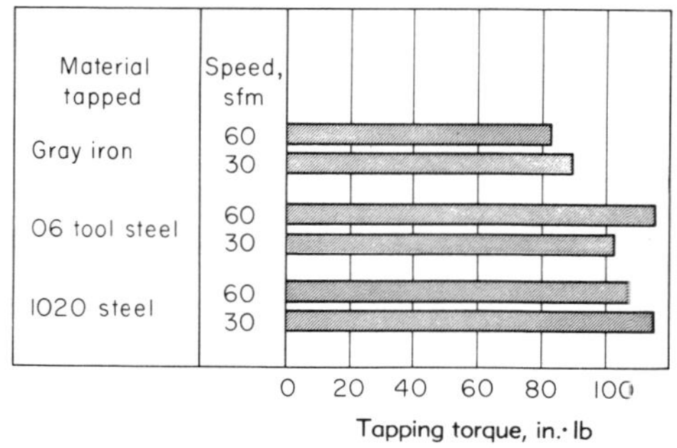


**Fig. 16** Effect of work metal and speed on torque in taper tapping. Data were obtained with ⅛-27 NPT high-speed steel taps for cutting threads in ⁸⁄₆₁ mm (0.339 in.) deep straight reamed holes to gage line. Sulfurized oil was used as cutting fluid.

**Surface speed** has only a minor effect on torque for tapping any of the three materials, as indicated by the data shown in Fig. 15 and 16.

**Percentage of full thread tapped** has a marked influence on torque requirements (Fig. 17) because more metal is removed in producing full threads.

**Grinding the Taps.** Small differences in the grinding of tap chamfers have often proved to be a major variable that causes large differences in tap life and other results. Three methods of grinding tap chamfers were included as part of the investigation of factors influencing torque. Results (Fig. 18) show that taps ground in an automatic chamfer grinder require less torque, regardless of the metal being tapped.

**Cutting fluid** influences torque, as indicated in Fig. 19. Even with cast iron, which is commonly tapped without a cutting fluid, it is apparent that torque is reduced a small amount by using an air blast and more by using liquids.

**Rigidity of the tap** was suspected of affecting torque, and gray iron was tapped using both full-floating and rigid types of

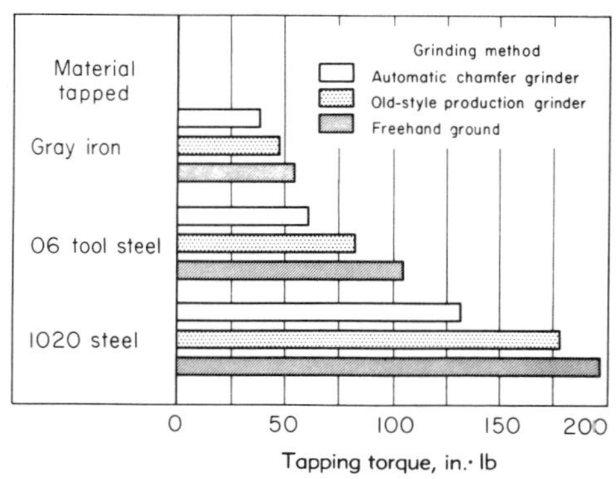


**Fig. 18** Effect of method of grinding tap chamfer on torque for tapping workpieces of three different metals. Speed: 18 m/min (60 sfm); other conditions, same as for Fig. 15.

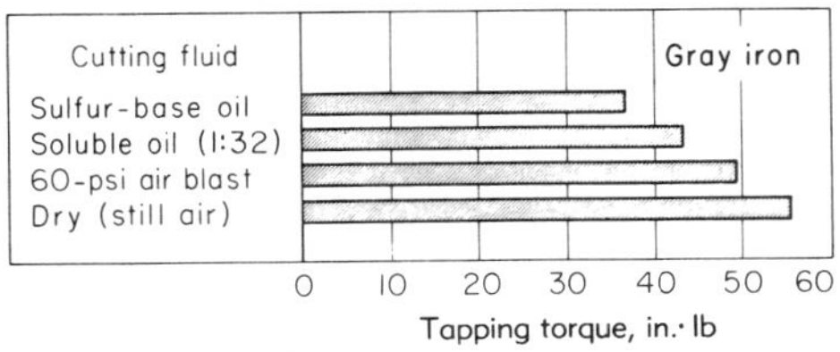


**Fig. 19** Effect of type of cutting fluid on torque required for tapping gray iron. Speed: 18 m/min (60 sfm); other conditions, same as for Fig. 15.

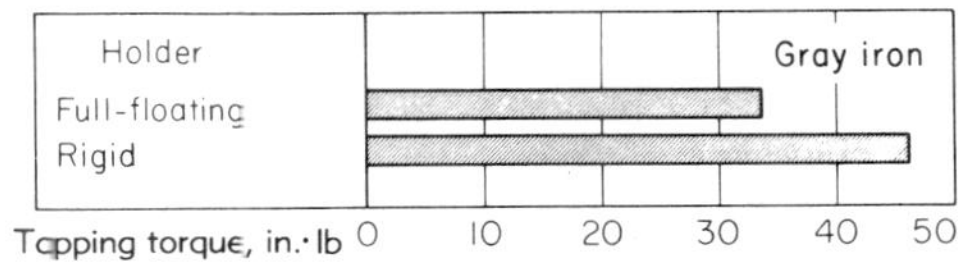


**Fig. 20** Effect of method of holding tap on torque required for tapping gray iron. Speed: 18 m/min (60 sfm); no cutting fluid; other conditions, same as for Fig. 15.

tool holders. Results (Fig. 20) indicate that misalignment of the tap, which cannot be entirely avoided when a rigid holder is used, does increase tapping torque.

## Tapping of Taper Pipe Threads

Cutting taper pipe threads with taps does not differ greatly from tapping machine threads. However, because standard taper pipe threads (NPT) are designed to provide pressure-tight seals, they require closer control in tapping than do thread forms for which 75% of full thread is acceptable. In addition, in tapping taper pipe threads, more teeth are cutting at one time than in tapping straight threads; consequently, pipe thread tapping requires more power than straight thread tapping, size for size. In one case, it was determined by tests that 1.44 kW (1.93 hp) was required to tap a 3/8-18 NPS straight thread, while 1.60 kW (2.14 hp) was required to tap a 3/8-18 NPT taper pipe thread.

Machines used for cutting taper pipe threads are usually the same as those used for other types of tapping. Positive lead control is desirable for optimum results in taper thread tapping because all tap teeth are cutting at one time.

**Tools.** Solid taps for cutting taper pipe threads are of the same two general types as for tapping straight threads: nonadjustable and adjustable. Solid, nonadjustable taps are used for the smaller size where adjustable or collapsible taps would not be practical.

Adjustable solid taps have removable chasers that can be adjusted for pitch diameter and reground or separately replaced when worn. When close control of pitch line taper is required with heavy stock removal, the adjustable solid tap is preferred. Manufacturers of pipe fittings use adjustable solid taps to cut taper pipe threads in gray iron and malleable iron fittings without prior machining of the as-cast diameters.

Collapsible taps permit the individual cutting components to be retracted from the work automatically when the thread is completed and withdrawn without reversing the direction of rotation. Automatic collapsible taps permit shorter machine cycles by eliminating the backing-out time, and they also increase tool life because cutting edges are not dragged back over the previously cut threads. Disadvantages of collapsible taps in comparison with solid adjustable taps are higher initial cost, greater susceptibility to fouling from swarf-contaminated cutting fluids, and increased maintenance.

**Stop Lines.** The prescribed taper in pipe threads is 62 mm/m (3/4 in./ft). Therefore, each successive tooth cuts a progressively larger diameter in the hole as the tap advances. Because each tooth is cutting at the instant of tap reversal, a stop line remains on the thread surface.

Some thread specifications do not permit stop lines. In addition, when solid taps are used, cutting edges must be reversed over these stop lines, which creates an abrasive action that causes excessive tap wear. The severity of the stop lines can be reduced by increasing the number of flutes in the tap. However, for the complete elimination of stop lines, a collapsible tap is required.

In general, stop lines are more pronounced the more ductile the work metal. As stop lines increase in severity, problems of thread quality and tool damage also increase.

**Speed** is usually 20 to 30% lower for pipe tapping than for the cutting of machine threads because more teeth are cutting during pipe tapping. However, in some shops, little or no distinction is made between machine and pipe threads in determining tapping speed. Also, in some shops, the same speed is used regardless of thread pitch. Figure 16 shows the influence of speed of tapping on torque.

Thread quality does not necessarily increase as speed is decreased. Rough and torn threads often result from speeds that are too slow. Excessive speeds always decrease tool life.

**Cutting fluids** recommended for pipe tapping are the same as those used for the cutting of machine threads. A constant supply of fluid to the cutting area is generally more important in pipe threading than in cutting machine threads because more cutting teeth are engaged at one time in pipe threading. In addition, because of the accuracy required in pipe threads and the greater amount of cutting that is being done at one time, susceptibility to thread damage from swarf-contaminated cutting fluids is greater in pipe threading. Damage from this source can be prevented by filtering the cutting fluid.

# Appendix: Cold Form Tapping

Thomas W. McClure, Balax Inc.

Forming taps are gaining in popularity throughout industry because of their elimination of chips, reduced tap breakage, and improvement in part quality. Forming taps and cutting taps both create 60° threads that are interchangeable.

Cold forming taps produce internal threads by an action similar to thread rolling; metal is displaced rather than removed. The thread form is produced by a tool such as the one shown in Fig. 21. This tool has neither flutes nor cutting edges, but the end view (at right, Fig. 21) shows that both the major and minor diameters have lobes that displace the work metal as either the tap or the workpiece is rotated.

These lobes are smaller in diameter at the chamfer, or lead portion, of the tap. The diameter at the chamfer of the tap is sized so that the tap freely enters the pretap hole and starts on center. From there, the lobes gradually increase in diameter until they reach the major diameter of the thread that is to be produced by the tap.

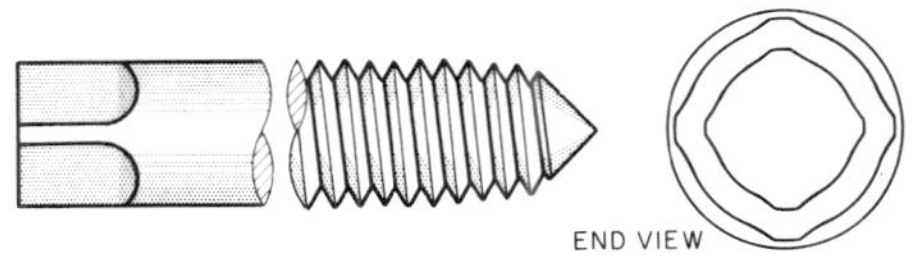


**Fig. 21** Cold forming tap showing the irregular contour by which the tap displaces the work metal when rotated

Because of the taper leading to the reduced diameter at the chamfer portion of a forming tap, it is impossible to have both the tap crests and tap roots exactly on pitch without distorting the 60° thread form (Fig. 22). Taps that are ground with leading crest accuracy are referred to as balanced axially. This process has a major advantage: It reduces tapping torque because the crests of the tap body follow exactly in the path of the lead crests. A noncorrected tap must perform double duty. The off-pitch lead crests on the chamfer form the thread. These lead crests must then be moved axi-

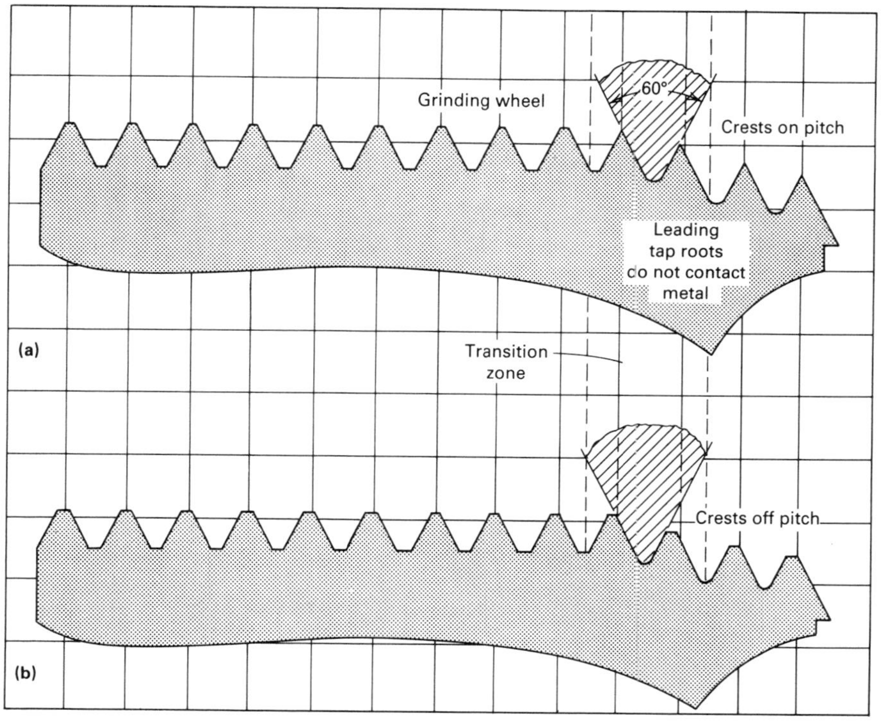


**Fig. 22** Cold forming taps. (a) Ground with leading crest accuracy (balanced axially). Crests are on pitch. (b) Other cold forming taps produce crests off pitch.

ally after they are completely formed to get into pitch with the thread on the cylinder portion of the tap. This requires as much as 30% more torque to accomplish and causes premature wear on the tap flanks, which reduces tap life.

Machine-screw-size and fine-pitch small fractional taps with lead geometry correction do not usually require significantly larger tapping torques than cutting taps. Larger fractional size taps with coarse pitches may require substantially more torque than a cutting tap, depending on the material being tapped.

Most forming taps have a narrow longitudinal groove that performs two separate functions, neither of which has to do with removing material. Primarily, the groove allows trapped air and lubricant to escape to prevent hydraulic locking and bottom blowout when tapping blind holes. Secondarily, the groove provides a path for the distribution of lubrication.

In general, materials that produce stringy chips are ideal for thread forming. Materials that produce powdery chips are usually too brittle. More specific guidelines are that any metal having an elongation of at least 12% in 50 mm (2 in.) and a yield strength below 480 MPa (70 ksi) is readily cold form tapped. If the thread pitch is 20 threads per 25 mm (20 threads per 1 in.) or finer, the metal can have an elongation below 12% in 50 mm (2 in.) or a yield strength of more than 480 MPa (70 ksi) and still be cold formed.

**Table 5 Surface treatments and lubrication for cold forming taps**

| Material category | Material type | Tap treatment | Recommended lubrication |
|---|---|---|---|
| Soft | Aluminum (plate or wrought), diecast, zinc, brass | Bright finish for most applications, or add hard chromium for tap wear and lubricity | Water soluble 5:1 or light tapping oil |
| Soft and abrasive | Diecast aluminum | Nitride | Water soluble 5:1 or light tapping oil |
| | Copper | Balwear or nitride/balwear | Water soluble 5:1 or light tapping oil |
| Intermediate hardness | Low-carbon steel | Nitride or super titanium nitride | EP-rated tapping oil with high sulfur and high chlorine |
| | 300-series stainless steel | Nitride/steam oxide or super titanium nitride | EP-rated tapping oil with high sulfur and high chlorine |
| Hard materials | Alloyed steels, 400-series stainless steel | Nitride/steam oxide or super titanium nitride | EP-rated tapping oil with high sulfur and high chlorine |

Examples of very good materials for cold forming include aluminum, brass, copper, lead, and zinc. Forming taps have typically been limited to small or medium-size blind holes in these very ductile materials. However, recent acceptance of forming tap technology has shown a number of extremely successful applications in larger holes and harder materials. Forming taps are suitable for both blind and through holes in low-carbon steel, leaded steel, and 300-series stainless steel. The most common example of a material a forming tap cannot handle is cast iron.

Surface treatments can prolong tap life by improving tap hardness and lubricity. A number of treatments are available:

- Nitriding hardens the tap surface to resist wear and pickup; it is commonly used for ferrous alloys and aluminum die castings containing silicon
- Steam oxide assists in lubrication; it is good for ferrous alloys, particularly when long or deep holes are involved
- Baked-on chromium plating resists abrasion and prevents pickup; it is well suited to high-copper alloys
- Hard chromium plating resists abrasion and pickup; it is used for low-copper alloys
- Soft chromium functions as a lubricant for ferrous alloys
- Titanium-nitride improves tap lubricity and wear resistance, especially in applications where water-soluble coolants must be used

Recommended surface treatments and tap lubricants for a variety of metals are listed in Table 5.

Plug-style or bottom-style cold forming taps are available. A plug-style tap has 3½ to 4 threads in the tapered lead, and a bottom-style tap has 1¾ to 2 threads in the tapered lead.

The plug style is often used for through holes and will generally outlast a bottom-style tap because it distributes the cold working process over a greater number of forming lobes. The bottom-style tap is used for bottoming or blind holes. Because it forms a thread more rapidly and in a shorter distance than a plug-style tap, it is better suited to work-hardening metals such as stainless steel. A special single-thread lead-style tap is also available and should be used when the clearance at the bottom of the hole is limited.

It is best to specify the largest practical H-number for a forming tap (the H-number describes the amount by which the tap pitch diameter exceeds the basic pitch diameter of the thread). This will provide the maximum allowance for wear and the best (loosest) thread gaging. Suggested H-numbers for forming taps are normally two or three H-numbers larger than H-numbers for cutting taps because of spring-back and also because forming taps, unlike cutting taps, are not likely to produce oversize threads. The H-number recommended for a forming tap is different for each size of thread and class of fit, and it can be determined by consulting the tap manufacturer.

**The advantages** of cold form tapping are chipless tapping, longer tap life, stronger taps, faster tapping speeds, stronger threads, and better thread gaging.

*Chipless Tapping*. The primary advantage of cold form tapping is that no chips are produced. This is especially desirable in the tapping of blind holes.

In one case, during the conventional tapping of aluminum, chips in the bottom of 13 mm (½ in.) deep holes (4-40 UNC-2B threads to 9.5 mm, or ⅜ in., depth) caused rejection of 30 000 pieces. A secondary operation was needed to remove the chips. Acceptable threads were obtained by the cold form tapping of subsequent lots.

In another application, blind holes were being tapped in aluminum die castings, using a two-flute spiral-point tap. Removing the chips required 16 man-hours per 1000 parts. A change to cold forming eliminated the need for chip removal. In this operation, tool life averaged 9750 holes per tap.

*Longer Tap Life*. Cutting taps have cutting edges that dull with use. The lobes on forming taps have much more wear surface to produce the thread because their function is to displace metal rather than cut it. This difference in function and design enables forming taps to last up to 15 times longer than cutting taps.

*Stronger Taps*. A forming tap is stronger than the same size cutting tap because it has no flutes to weaken the overall structure. This becomes especially important in small taps in which flutes substantially reduce the core diameter of a cutting tap.

*Faster Tapping Speeds*. A forming tap uses lobes and does not concentrate all of its threading action on several narrow cutting edges. In addition, the absence of chips minimizes drag, enhances lubrication, and leads to a lower heat buildup, which makes faster tapping speeds possible. Recommended forming tap speeds are 1½ to 2 times faster than cutting tap speeds, if the work is done in pitches of 20 threads per 25 mm (20 threads per 1 in.) or finer.

*Stronger Threads*. In work-hardening materials such as steel and stainless steel, the grain flow of formed threads follows the thread contour, resulting in greater strength. In stampings, in which the thread section is very thin, the added strength of the work-hardened material can be a major improvement to the integrity of the finished part. Pull-test studies and fatigue testing have verified the superiority of formed threads in materials that work harden.

*Better Thread Gaging*. Because a forming tap merely rearranges metal instead of cutting and removing it, it is much less likely to form oversize threads. The flanks of the finished thread are burnished and provide very smooth thread gaging.

Forming taps have very successfully tapped small fine-pitch threads such as 0-80, 2-56, and 4-40 in materials as hard as 25 to 30 HRC. The results have been superior to those for cutting taps for deep blind holes, in which chip binding, combined with the inherent weakness of cutting taps, has resulted in poor cutting tap life. Forming taps with 10 to 20 threads per 25 mm (10 to 20 threads per 1 in.) are effective in 25 to 38 mm (1 to 1½ in.) thread sizes in materials with hardnesses of 16 to 20 HRC. Top performance can be enhanced by additional lubrication grooves, surface treatments, axial oil holes, and the use of special tool steels.

**Processing.** Cold form tapping can be done in almost any machine used for conventional tapping. Because the lead-forming lobes create their own lead, a forming tap does not necessarily require a lead screw. If a lead screw is used and is slightly loose or inaccurate, it may cause detrimental axial forces on the tap unless a tension/compression tap holder is used. Floating tapping heads are also recommended to eliminate tap-to-hole misalignment.

Because the metal is displaced as in thread rolling, the tap drill must be larger than for conventional tapping. Pretap hole size control is more critical with a cold forming tap than with a cutting tap. This can be seen in smaller sizes, such as a 2-56 tap in which 0.05 mm (0.002 in.) change in pretap hole size will cause the finished thread to vary from 55 to 75% thread height. After tapping, minor diameters should be checked with pin gages to verify conformance with specifications.

**Lubricants.** A forming tap does not generate heat the way a cutting tap does. The forming tap dissipates heat into the tap body and into the part being tapped. As a result, the primary consideration in using a forming tap is to reduce the friction between the tap and the part.

Sulfurized and chlorinated oils are preferred for cold form tapping. Other fluids commonly used for metal cutting operations cause galling and tearing of the threads.

The preferred lubrication is an EP-rated heavy-base oil with high sulfur and chlorine levels. Oil-base drawing oils for the stamping industry are designed for lubricity and may also be excellent tap lubricants.

Water-soluble coolants can be used in softer materials such as aluminum and copper, but need to be increased in strength or concentration to maximize lubricity. Water-soluble drawing compounds have provided excellent results in nonferrous applications.

A common mistake is to lubricate the tap but not the hole being tapped. If the hole is dry, the initial threads being formed tend to wipe off the tap, causing the tap to run dry at the bottom of the hole. The hole being tapped should be flooded with lubricant.

# Thread Milling

THREAD MILLING is a method of cutting screw threads with a milling cutter in a thread mill. Thread milling cutters are either single form (Fig. 1) or multiple form (Fig. 2) and are used in either conventional or planetary thread mills. Screw threads cut by thread milling can have pitch diameters that are accurate within 0.025 mm (0.001 in.), surfaces smooth within 1.40 μm (55 μin.) and accurate within 0.8 mm/m (0.001 in./ft), and spacing (of multiple-start threads) accurate within 0.010 mm (0.0004 in.).

Thread mills are usually ramped at a 20° angle when entering the cut (Fig. 3). This reduces the chance of leaving tool marks on the thread flanks, while also ensuring full thread depth around the entire part.

Applications of thread milling are varied. A thread mill can cut internal or external threads and, in some work, can cut both internal and external threads at the same time. The workpiece may be fixed or rotating, making thread mills competitive with lathes and with machines that use taps and dies to cut threads.

The main reasons for choosing thread milling over other thread-cutting methods are:

- Thread milling makes a smoother and more accurate thread than a tap or die
- Thread milling is more efficient than using a single-point tool in a lathe
- The desired thread has too coarse a pitch for cutting with a die
- The desired thread is near a shoulder or other interference, making thread milling the most practical method

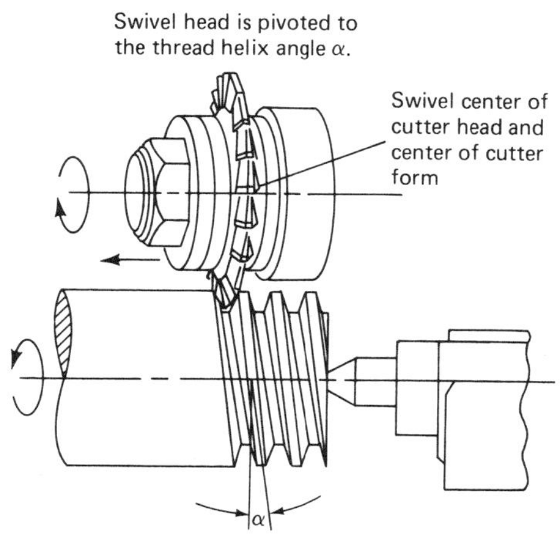


**Fig. 1** Milling a worm thread using a single-form cutter. Relative rotation of the cutter and workpiece illustrate conventional milling.

## Thread Milling Machines

Conventional (up-thread) milling can be done on all types of thread millers, but some models are also designed for climb (down-thread) milling. In climb milling, the relative rotation of the work and the cutter is similar to that of a pair of mating gears. Figure 1 illustrates conventional milling with a single-form cutter.

Thread milling machines can be either planetary or conventional machines. In planetary machines the workpiece does not rotate, while in conventional machines the workpiece does rotate. Conventional milling machines are either universal or production machines.

**Universal thread mills** have a lead screw and can cut all internal or external threads, with the exception of square threads. Change gears permit the milling of threads with leads of 0.8 to 1520 mm (1/32 to 60 in.). Pick-off gears in the cutter drive provide a wide range of speeds. The cutter head, on the cross slide, can be set at the proper angle for any right-hand or left-hand thread helix. A single-form cutter must be set at such an angle, and then must traverse the length of the thread.

**Production thread mills** are semiautomatic; the work is plunged into the nontraveling, rotating cutter by a cam. The workpiece rotates 1.10 revolutions, then backs out of the cut, and the machine stops automatically for unloading and loading.

Production thread mills use a master screw and stationary segment instead of a lead screw. The master screw and segment must be changed for each change in pitch or lead. A no-lead attachment is used for milling annular grooves.

**Planetary thread mills** are used to thread odd-shape parts difficult to chuck (but are not used to thread long pieces, such as lead screws). The work is clamped in a fixture, and the workpiece does not rotate during cutting. Only the cutter moves, rotating on its axis, revolving around the

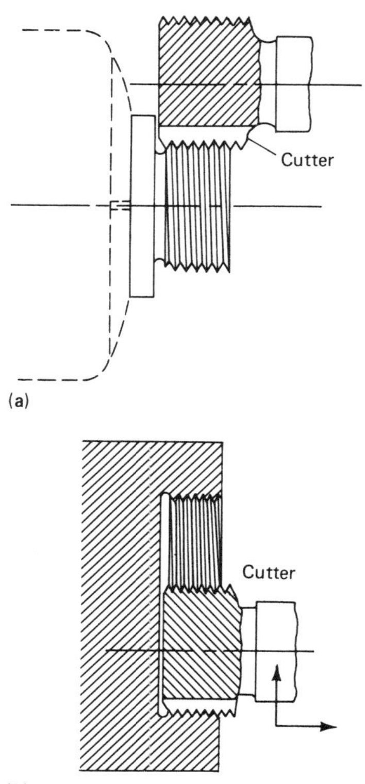


**Fig. 2** (a) External and (b) internal thread milling using a multiple-form cutter

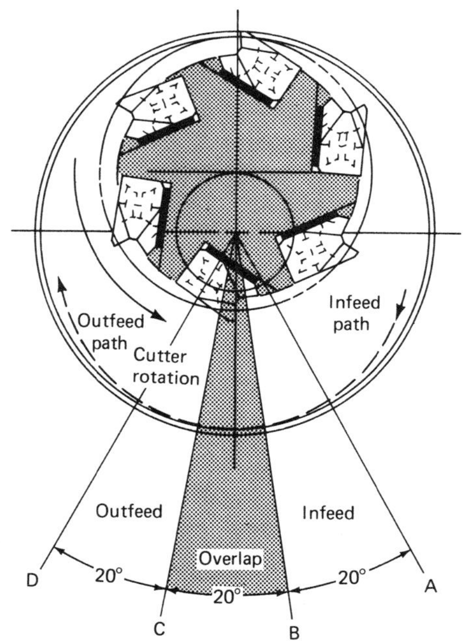


**Fig. 3** Ramping of thread mills. A is the starting point of infeed contact, B is the end of infeed with the cutter at depth, B to C is the overlap cut after a full 360° revolution, C is the starting point of outfeed, and D is the end of cutter engagement with the thread.

**Table 1 Nominal speeds and feeds for thread milling with high-speed steel tools**

| Work material | Hardness, HB | Speed m/min | Speed sfm | Feed mm/tooth | Feed in./tooth |
|---|---|---|---|---|---|
| Plain carbon steels such as 1020 .. | 85–125 | 30 | 100 | 0.05 | 0.002 |
| | 125–175 | 37 | 120 | 0.05 | 0.002 |
| | 175–225 | 30 | 100 | 0.05 | 0.002 |
| | 225–275 | 24 | 80 | 0.04 | 0.0015 |
| | 275–325 | 21 | 70 | 0.025 | 0.001 |
| Free-machining carbon steels such as 1112 .................. | 100–150 | 52 | 170 | 0.05 | 0.002 |
| | 150–200 | 55 | 180 | 0.05 | 0.002 |
| | 200–250 | 38 | 125 | 0.05 | 0.002 |
| | 275–325 | 30 | 100 | 0.025 | 0.001 |
| Alloy steels such as 4140......... | 125–175 | 34 | 110 | 0.05 | 0.002 |
| | 175–225 | 27 | 90 | 0.05 | 0.002 |
| | 225–275 | 21 | 70 | 0.04 | 0.0015 |
| | 275–325 | 20 | 65 | 0.025 | 0.001 |
| Free-machining alloy steels such as 41L40 ................ | 150–200 | 49 | 160 | 0.05 | 0.002 |
| | 200–250 | 40 | 130 | 0.05 | 0.002 |
| | 275–325 | 24 | 80 | 0.025 | 0.001 |
| Ferritic stainless ................ | 135–185 | 34 | 110 | 0.05 | 0.002 |
| Austenitic stainless (type 304) .... | 135–185 | 27 | 90 | 0.05 | 0.002 |
| | 225–275 | 24 | 80 | 0.04 | 0.0015 |
| Free-machining austenitic (type 303) .................... | 135–185 | 30 | 100 | 0.05 | 0.002 |
| | 225–275 | 27 | 90 | 0.04 | 0.0015 |
| Martensitic stainless such as type 403...................... | 135–185 | 34 | 110 | 0.05 | 0.002 |
| | 185–225 | 30 | 100 | 0.05 | 0.002 |
| | 275–325 | 18 | 60 | 0.025 | 0.001 |
| Free-machining martensitic such as type 416 ............. | 135–185 | 46 | 150 | 0.05 | 0.002 |
| | 185–240 | 41 | 135 | 0.05 | 0.002 |
| | 275–325 | 23 | 75 | 0.025 | 0.001 |
| Precipitation-hardening stainless (17-4 PH) ............ | 150–200 | 21 | 70 | 0.05 | 0.002 |
| | 275–325 | 18 | 60 | 0.025 | 0.001 |
| | 325–375 | 14 | 45 | 0.025 | 0.001 |

**Table 2 Suggested cutting speeds for high-speed steel thread milling cutters**

| Material milled | Conventional milling m/min | Conventional milling sfm | Climb milling m/min | Climb milling sfm |
|---|---|---|---|---|
| Cast irons(a) .......................... | 33.5–53.3 | 110–175 | 38.1–61 | 125–200 |
| Steels, HB | | | | |
| Up to 240 ........................ | 33.5–45.7 | 110–150 | 76–91 | 250–300 |
| 240–300 .......................... | 24.4–27.4 | 80–90 | 48.8–53.3 | 160–175 |
| 300–375 .......................... | 12.2–16.8 | 40–55 | 27.4–42.7 | 90–140 |
| 375–450(b) ....................... | 9.1–12.2 | 30–40 | | |
| Stainless steels ..................... | 15.2–30.5 | 50–100 | 15.2–18.3 | 50–60 |
| Brass | | | | |
| Soft ............................. | ≥152 | ≥500 | ... | ... |
| Hard ............................ | ≥61 | ≥200 | ... | ... |
| Aluminum and magnesium ........... | ≥183 | ≥600 | ... | ... |

(a) Using carbide cutters, cast irons can be thread milled at cutting speeds of 76–137 m/min (250–450 sfm). (b) Cutters for higher-hardness steels should be premium high-speed steel or carbide.

work. Double heads can be used to cut both ends of a part, and internal and external threads can be cut at the same time.

**Numerical control (NC) machines** are also used for thread milling, together with other operations, in a single fixturing. Long cutter life and high-quality threads are among the advantages.

## Thread Milling Cutters

Either single- or multiple-form cutters are used in thread milling. Multiple-form cutters are used for milling comparatively short leads and medium or fine pitches, while single-form cutters are used for coarse pitches and long threads.

**Single-form cutters** must traverse the workpiece during cutting, or the workpiece must traverse the cutter. When a single-form cutter is used, the axis of the cutter is pivoted an amount equal to the helix angle of the thread (Fig. 1).

**Multiple-form cutters** have annular rows of teeth of the right form and pitch but no lead. The cutters must be two to three pitches longer than the thread to be cut, or the work must be rotated more than the minimum 1.1 turns.

Most multiple-form cutters are nontopping; that is, they cut only the sides and the root of the thread. The top is usually machined by turning or boring. Topping cutters, not usually recommended, machine all diameters for concentricity, usually removing not more than 0.25 mm (0.010 in.) from the diameter.

## Speed and Feed

**Speeds and feeds** for thread milling depend on a number of variables, such as the material being milled, the tool material, and the rigidity of the machine and workpiece. The choice of either climb or conventional milling is also a factor. Table 1 contains some suggested speeds and feeds with high speed steel tools.

**Climb milling** also permits the use of faster cutting speeds, which improve surface finishes and increase production and cutter life. Table 2 compares cutting speeds for climb and conventional milling. The suggested speeds should be considered a general starting point; increases or decreases may be necessary for desired results.

# Thread Grinding

THREADS are sometimes ground instead of being cut or rolled, for the following reasons:

- If the work is harder than 36 HRC, grinding may be the only feasible production method
- If the work is softer than 17 HRC, grinding may give a better finish
- If a high degree of dimensional accuracy is required, grinding is almost invariably the method preferred

Most threading processes apply methods other than grinding—for example, cutting in a turning or milling operation, or chipless forming carried out by one of the several methods of thread rolling. Thread grinding is applied only when the advantages of the ground threads exceed the disadvantages of the generally higher cost of equipment or when, for technological reasons, the production or quality requirements cannot be met by the generally more productive common methods.

## Thread Grinding Processes

**Grinding Methods.** Threads are ground by contact between a rotating workpiece and a rotating grinding wheel that has been shaped to the desired thread form. In addition to rotation, there is relative axial motion between the wheel and the workpiece to match the pitch of the thread being ground. Thread grinding can produce either external or internal threads.

Methods for grinding threads are:

- Cylindrical grinding, in which the workpiece is held between centers or by means of a chuck
- Centerless grinding, in which the workpiece is supported against the grinding wheel by a rest (work-support blade) and a regulating wheel

When workpiece shape permits and when production quantities are large, centerless grinding is the best choice because productivity is much greater than in cylindrical grinding.

In center-type grinding, the material specifications and the form, length, and quality of the threads determine the number of passes required. The number of passes can vary from one to five to six. With centerless grinding, the part is normally finished in one pass through the machine. As the work moves across the wheel, it is sized to the correct diameter and then the threads are formed.

**Machines for thread grinding** are distinguished by three features:

- Means for imparting a precise axial motion between the wheel and the workpiece to match the thread being ground (the leadscrew principle is commonly used)
- Devices for truing or dressing the grinding wheel to generate the required form in the workpiece
- Inclinable plane of rotation of the grinding wheel to cut the required helix

Thread grinding machines differ in the type of grinding wheel used (single ribbed or multiribbed), the method of supporting the workpiece, and the method of restoring the cutting contour of the grinding wheel (diamond dressing or truing, or crush truing). Thread grinding machines are sometimes classified as external, internal, or universal. The universal type is capable of grinding both external and internal threads.

## Thread Grinding Applications

Threaded parts that are ground include those that are too difficult to cut, mill, or roll when a fine finish is required or when precision form, lead, and pitch requirements must be held before and especially after hardening. Forms that are produced include API, NPT, and other taper pipe threads; 60° unified and metric; 55° Whitworth; 29 and 40° worm; 47°30′ British Association; 53°8′ Lowenhertz; and Buttress. Centerless-ground threaded parts include continuously threaded parts such as setscrews, studs, threaded bushings, threaded size-adjusting bushings for boring heads, thread gages, worm gears, powdered iron screws, and self-threading insert bushings.

Materials that can be thread ground include hardened and annealed screw stock, alloyed high-speed tool and stainless steels, and sintered iron. Sintered iron is widely used for continuously threaded screws.

Threads can be cut and then finish ground after heat treatment, or they can be ground from solid stock. The process of forming threads entirely by grinding, or without preliminary cutting, is applied both in the manufacture of certain classes of threaded parts and in the production of precision tools, such as taps and thread gages. For example, in airplane engine manufacture, certain parts are heat treated and then the threads are ground from the solid. This eliminates distortion and the minute cracks formerly found at the roots of threads that were cut and then hardened. In some cases, steel threads of coarse pitch, which are surface hardened, can be rough threaded by cutting, then hardened and finally corrected by grinding. Many ground thread taps are produced by grinding from the solid after heat treatment. Hardening the high-speed steel taps before the thread is formed eliminates the narrow or delicate crests that would interfere with the application of the high temperature required for uniform hardness and optimum steel structure.

Thread grinding is often specified in highly stressed parts subject to failure by fatigue, such as aircraft, tank, gun, and motor parts. Threading such parts by grinding avoids the minute torn areas that usually exist at the root of cut threads and cause these threads to become the focus of stress concentrations that contribute to progressive rupture by fatigue.

Typically, the upper limits of the thread grinding process can be stated as a 405 mm (16 in.) diameter and a 4.1 m (160 in.) thread length. At the other extreme, that is, the lower limit for manufacturing threads for miniature parts, the finest thread pitch considered practical to produce by using thread grinding is 0.125 mm (0.004921 in.), which is equivalent to 203⅕ threads per 25 mm (1 in.) in the Unified Miniature Screw Thread system.

## Tolerances

Threads are ground for accuracy and finishes unobtainable by other threading methods. Tolerances are held for size to ±0.0001 mm/mm (±0.0001 in./in.) of pitch diameter and for lead within 7.5 μm in 500 mm (0.0003 in. in 20 in.) of length. Hard materials can be threaded more economically by grinding than by other methods. Aluminum and other comparable soft materials can also be ground, but with great difficulty because of their tendency to load the wheel and cause burning.

Threads to be ground should not be specified to have sharp corners at the root.

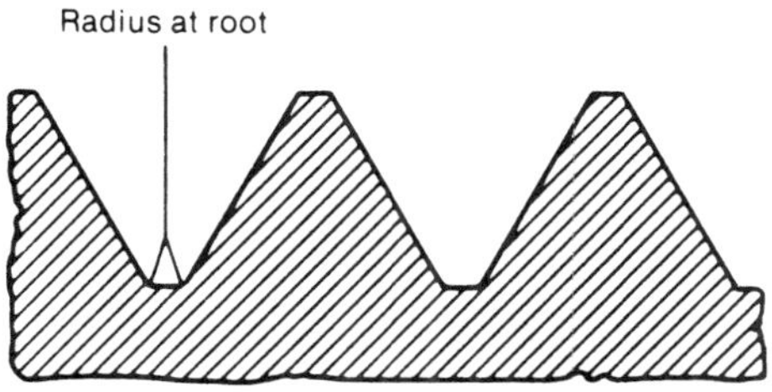


**Fig. 1** Generous root radius necessary in ground threads. A root radius of 0.25 mm (0.010 in.) is preferred, although a minimum root radius of 0.08 mm (0.003 in.) is readily obtained when threads are ground.

Normally, a radius of 0.08 mm (0.003 in.) is the very minimum that can be expected, and much larger radii of the order of 0.25 mm (0.010 in.) are preferred (Fig. 1).

Centerless-ground threads should have a length-to-diameter ratio of at least 1:1; preferably, the length should be longer than the diameter. In addition, parts to be centerless-thread ground should not have large burrs, be flattened or egg-shaped from shearing, or be bent or crooked. Taper and flatness should also be avoided, because they will not be removed by thread grinding.

The number of wheel passes or cuts required for grinding from the solid depends on the type of wheel and the accuracy required. In general, threads of 12 or 14 per 25 mm (1 in.) and finer can be ground in one pass of a single-edge wheel unless the unwrapped thread length is much greater than normal. The unwrapped length is approximately equal to the pitch circumference multiplied by the total number of thread turns. For example, a thread gage 32 mm (1¼ in.) long with 24 threads per 25 mm (1 in.) would have an unwrapped length equal to 30 times the pitch circumference. (If more convenient, outside circumference can be used instead of pitch circumference.) If there is 1.8 or 2.1 m (6 or 7 ft) of unwrapped length on a screw thread having 12 threads per 25 mm (1 in.), one pass might be sufficient for a class 3 fit, but two passes might be recommended for a class 4 fit. When two passes are required, too deep a roughing cut may break down the narrow edge of the wheel. This can be prevented by using a roughing cut depth equal to about two-thirds the total thread depth, thus leaving one-third for the finishing cut.

## Wheel Selection

Typical wheel specifications for the thread grinding of various metals are given in Tables 1 and 2. The selection of abrasive depends on work metal composition and hardness and the number of threads per 25 mm (1 in.). With few exceptions, threads are ground with aluminum oxide wheels, designated by an A in the specification. Silicon carbide wheels, marked C, are used for grinding titanium, and diamond wheels, marked D, for grinding sintered carbide and ceramic materials. The semifriable varieties of aluminum oxide are usually preferred, but in some applications a friable type gives better results, and in a few cases a tough variety is best. Additional information on wheel classification and marking is available in the article "Grinding Equipment and Processes" in this Volume.

**Grit size** is primarily determined by the pitch of the thread being ground; the finer the thread, the finer the grit. Fine grit improves surface finish, but coarse grit removes metal faster. The relationship between thread pitch and grinding wheel grain size is indicated in Table 3.

**Resinoid and Vitrified Wheels.** Either vitrified or resinoid bond is used for thread grinding. Resinoid wheels remove stock more rapidly, but they are less rigid than vitrified wheels and deflect more readily. Therefore, resinoid wheels are usually used in high-production applications, while vitri-

**Table 1 Grinding wheel specifications for thread grinding**

| Application | Type of thread | Threads per 25 mm (1 in.) (from solid) | Specifications for: Precision work | Specifications for: Commercial work |
|---|---|---|---|---|
| **External grinding** | | | | |
| Precision screws | Fine pitch | 32–40 | 32A220-M9VG | ... |
| | | Finer than 40 | A320-09B | ... |
| Screws and studs (heat-treated alloy steel) | American National form threads | 8–12 | ... | 23A100-R9BH |
| | | 14–20 | ... | 23A120-R9BH |
| | | 24 and finer | ... | 23A150-S9BH |
| | | | | 23A150-T9BH |
| | | | | 23A180-T9BH |
| Taps (high-speed steel) | Acme threads | 8–18 | ... | 23A100-R9BH |
| | | 20–26 | ... | 23A150-S9BH |
| | | 27 and finer | ... | 23A150-T9BH |
| | | | | 23A180-T9BH |
| | American National form threads | 4–12 | 32A100-K8VG | 23A100-R9BH |
| | | 14–20 | 32A120-L9VG | 23A120-R9BH |
| | | 24–36 | 32A180-N9VG | 23A180-T9BH |
| | | | | A220-U9BH |
| | American Standard pipe threads | 8–18 | ... | 23A100-R9BH |
| | | 20–26 | ... | 23A150-S9BH |
| | | 27 and finer | ... | 23A150-T9BH |
| | | | | 23A180-T9BH |
| Whitworth form threads | Whitworth form threads | 8–16 | 32A120-K9VG | ... |
| | | 18–24 | 32A150-L9VG | ... |
| Worms and leadscrews | Acme and worm threads | 4–6(a) | 32A80-J8VG | ... |
| | | 8–12(a) | 32A100-K8VG | ... |
| | | 12–16 | 32A120-K9VG | 23A100-R9BH |
| **Internal grinding** | | | | |
| Hardened tool and alloy steel | Acme threads | 2–8(a) | 38A80-J8VG | ... |
| | | 10–20(a) | 38A120-K9VG | ... |
| | American National form threads | 6–16(a) | 38A120-L9VG | 23A100-V9BH |
| | | 20–24(a) | 38A220-M9VG | 23A150-V9BH |
| | | 6–20 | ... | 23A100-V9BH |
| | | 24 and finer | ... | 23A150-V9BH |

(a) Precut threads. Source: Ref 1

**Table 2 Grinding wheel specifications for the multirib grinding of hardened steel**

| Thread pitch | Grinding process | Wheel diameter: <356 mm (14 in.) | Wheel diameter: >356 mm (14 in.) |
|---|---|---|---|
| 5–10 | Centerless | ... | 32A120-Q9VG |
| | Cylindrical | 38A120-N9VG | 32A120-N9VG |
| | Surfacing | 38A120-K9VG | 32A120-K9VG |
| 11–14 | Centerless | 38A150-N9VG | 38A150-Q10VG |
| | Cylindrical | 38A150-N9VG | 32A150-N9VG |
| | Surfacing | 38A150-K9VG | 32A150-K9VG |
| 16–24 | Centerless | 38A2203-N9VG | 38A220-Q10VG |
| | Cylindrical | 38A2203-N9VG | 38A220-N9VG |
| | Surfacing | 38A2203-K9VG | 32A220-K9VG |
| 24–32 | Centerless | 38A320-N10VG | A320-Q11VG |
| | Cylindrical | 38A320-N10VG | A320-N10VG |
| | Surfacing | 38A320-K9VG | A320-K9VG |

Source: Ref 1

**Table 3 Relationship between thread pitch and grinding wheel grain size for vitrified wheels**

| Threads per 25 mm (1 in.) | Approximate equivalent pitch, mm | Grit size for vitrified wheels: Coarsest allowable grit for single-rib wheels | Grit size for vitrified wheels: Recommended grit for multirib wheels |
|---|---|---|---|
| 80–72 | 0.3–0.35 | 240 | ... |
| 64–40 | 0.4–0.6 | 220 | ... |
| 36–32 | 0.7–0.75 | 180 | ... |
| 32–24 | 0.8–1.0 | ... | 320 |
| 28–24 | 0.8–1.0 | 150 | ... |
| 20–16 | 1.25–1.5 | 120 | 220 |
| 14 and 13 | 1.75 and 2.0 | 100 | ... |
| 14–11 | 1.75–2 | ... | 150 |
| 12 and 9 | 2 and 2.5 | 90 | ... |
| 10–5 | 2.5–5 | ... | 120 |
| 8–4½ | 3–5.5 | 80 | ... |
| 4 | 6 | 70 | ... |

Source: Ref 2

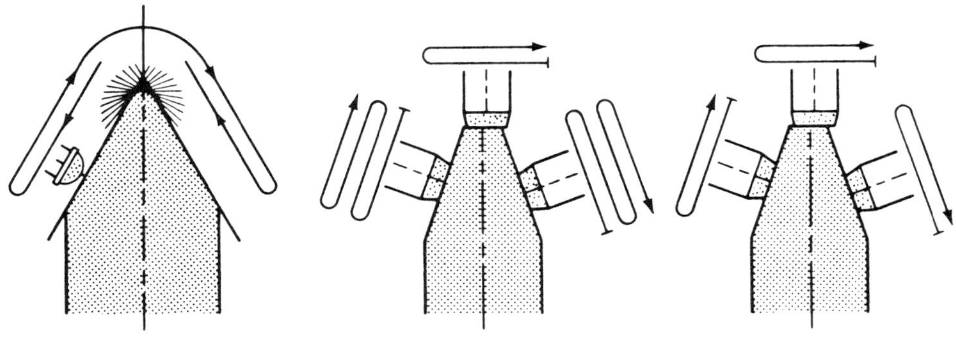


**Fig. 2** Wheel truing of rounded-crest (a) and straight-line (b) thread profiles using diamonds as a truing device. The profile in (a) is generated by a single tool in which the flank surfaces are produced by straight-line motions parallel with the trued profile sections, while the radius results from a large number of tangent-line travels in consecutive positions. The profile in (b) is generated by using three truing heads, each holding a diamond and moving in the paths shown, to dress the wheel for fine (left) and coarse (right) thread grinding.

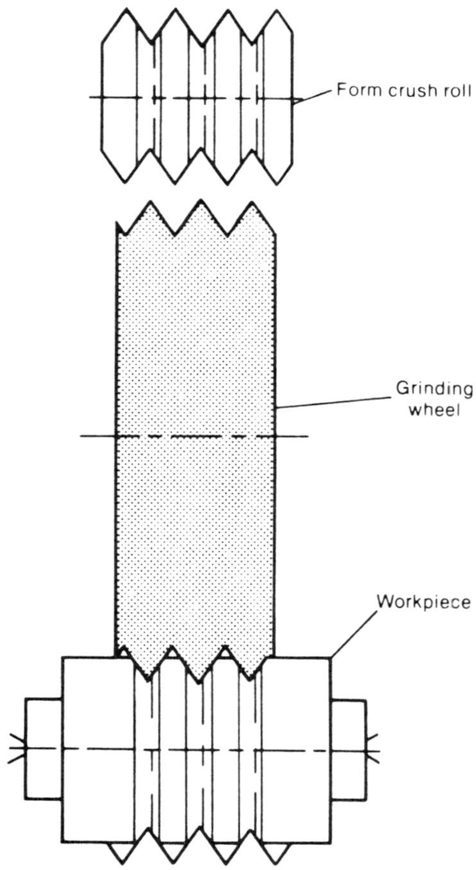


**Fig. 3** Principles of wheel thread shaping by form crush dressing in a plunge-type, profile-grinding process. The relative positions of the form crush roll, the contour-shaped grinding wheel, and the finish-ground workpiece are shown.

fied wheels are used where accuracy is the primary objective, as in grinding threads on gages. Vitrified wheels are also used for correcting lead error in precut threads. The more flexible resinoid wheels tend to follow lead error rather than to correct it. Vitrified bond is also used for wheels dressed by crushing, a method not applicable to the more resilient resinoid-bonded type.

**Diamond wheels** set in a rubber or plastic bond are also used for thread grinding, especially for grinding threads in carbide materials and in other hardened alloys. Thread grinding is done successfully on a commercial basis on both taps and gages made from carbides. Diamond wheels are dressed by means of silicon carbide grinding wheels that travel past the diamond wheel thread form at the angle required for the flanks of the thread to be ground. The action of the dressing wheels is perhaps best described as a scrubbing of the bond that holds the diamond grits. Obviously, the silicon carbide wheels do not dress the diamonds, but they loosen the bond until the diamonds not wanted drop out.

**Wheel grade** depends mainly on the grit size; the finer the grit, the harder the wheel. The work metal, grinding method, thread type and pitch, and wheel size are among the factors that determine which grade will be used.

Wheel hardness or grade selection is based on a compromise between efficient cutting and the durability of the grinding edge. Vitrified wheels usually range from J to M, and resinoid wheels from R to U. For heat-treated screws or studs and the Unified Standard Thread, the following recommendations are helpful:

- For 8 to 12 threads per 25 mm (1 in.), grade S resinoid wheel
- For 14 to 20 threads per 25 mm (1 in.), grade T resinoid
- For 24 threads per 25 mm (1 in.) and finer, grades T or U resinoid

For high-speed steel taps, the recommendations given below are useful:

- For 4 to 12 threads per 25 mm (1 in.), grade J vitrified or S resinoid
- For 14 to 20 threads per 25 mm (1 in.), grade K vitrified or T resinoid
- For 24 to 36 threads per 25 mm (1 in.), grade M vitrified or T resinoid

## Truing Grinding Wheels

Thread grinding wheels are trued to maintain the required thread form and an efficient grinding surface. Thread grinders are usually equipped with precision truing devices that function automatically. One type automatically dresses the wheel and compensates for the slight amount removed in dressing, thus automatically maintaining size control of the work.

**Diamond Dressing.** In most cases, suitably mounted diamond points are the cutting tools used to impart the desired form to the grinding wheel surface. The dressing device imparts to diamond dressing points the predetermined path of movement for generating the desired wheel form (Fig. 2).

In the relatively simpler cases in which the form to be dressed on the wheel is bounded by straight lines, separate diamond tools are either pivotally mounted or slide mounted so that the several tools can traverse the respective surfaces of the wheel. These separate diamond points can be moved in a straight line when supported on respective slides, or they can be pivotally mounted to traverse accurately the wheel surfaces to be dressed by them.

For mechanical reasons and to avoid the necessity of using a diamond point so small that it would have a prohibitively short life, it is preferable that separate diamonds be provided for dressing opposite flanks of the wheel rib. To avoid chipping of the wheel rib in dressing, it is desirable that the diamond points be in cutting contact only while traversing the rib from the apex toward its base and that they return out of contact with the wheel.

Rotating rolls having diamonds plated on their peripheries are a more recent method of dressing wheels for thread grinding. These tools are particularly well suited to the high-production thread grinding of a single design of workpiece. Two-axis numerical control dressers are available for producing various thread forms.

While truing the wheel, a small amount of grinding oil should be used to reduce diamond wear. Light truing cuts are advisable, especially in truing resinoid wheels that can be deflected by excessive truing pressure. A master former for controlling the path followed by the truing diamond may require a modified profile to prevent distortion of the thread form, especially when the lead angles are comparatively large. Such modification usually is not required for 60° threads when the pitches for a given diameter are standard, because then the resulting lead angles are less than 4½°. In grinding Acme threads or 29° worm threads having lead angles greater than 4 or 5°, modified formers may be necessary in order to prevent a bulge in the thread profile. The highest point of this bulge is approximately at the pitch line. A bulge of about 0.025 mm (0.001 in.) may be within allowable limits on some commercial worms, but precision worms for gear hobbers, and so on, require straight flanks in the axial plane.

**Crush Dressing.** Thread grinding wheels are also dressed or formed by the crushing method. This method is used in conjunction with some types of thread grinding machines.

In crush dressing, the brittleness of the wheel bond is utilized by subjecting selected areas of the wheel face to a pressure in excess of the strength of the bond. This causes the bond material to disintegrate in the areas of the applied pressure. The pressure is maintained until the breakdown of the bond reaches the required depth (Fig. 3). When this method is used, the annular ridge or ridges on the wheel are formed by a hardened steel cylindrical dresser or crusher. The crusher has a series of smooth annular ridges that are shaped and spaced like the thread that is to be ground. During the wheel dressing operation, the crusher is positively driven instead of the grinding wheel, and the ridges on the wheel face are formed by the rotating crusher being forced inward.

## Grinding Speed

Table 4 lists the nominal speeds used for the thread grinding of various alloys, based on work metal composition and hardness and the number of threads per inch. Vitrified thread grinding wheels are run at considerably higher speeds than are normal in other operations with vitrified wheels. This is permissible under the safety code only when the wheel has been tested by the wheel manufacturer for operation at such speeds and when the machine is properly guarded to protect the operator against accidental wheel breakage. Although the maximum speed for vitrified wheels on a standard cylindrical grinder is 2000 m/min (6500 sfm), speeds as high as 3700 m/min (12 000 sfm) can be used in thread grinding with vitrified wheels that have been certified by the manufacturer for that speed.

As indicated in Table 4, in most thread grinding, the recommended wheel speed increases as the number of threads per 25 mm (1 in.) increases. Grinding of heat-resistant alloys, however, is an exception; the recommended speed is the same regardless of thread pitch.

## Grinding Fluids

Grinding oils are always used in thread grinding. Water-base fluids have never been successful, although many attempts have been made to replace oils with water-base fluids. Mineral-base sulfochlorinated grinding oils such as G1 or G2 are the most widely used, especially when finish and accuracy of the threads are important:

| | |
|---|---|
| G1 . . . . . . . . | Transparent sulfochlorinated grinding oil containing fats, 4% S, and 2% Cl (both active); viscosity: 230 SUS at 40 °C (100 °F) |
| G2. . . . . . . . . . . . | Dark sulfochlorinated grinding oil containing fats, 3% S, and 0.5% Cl (both active); viscosity: 190 SUS at 40 °C (100 °F) |

Such oils are generally satisfactory for all steels, as well as for other work metals, and many shops have standardized on them for all thread grinding. Other formulations prepared especially for thread grinding are available for specific applications where the conventional sulfochlorinated oils have proved unsatisfactory.

## Cylindrical Grinding of Threads

Five basic methods, identified by the design of the wheel, are employed for cylindrical thread grinding. Four of these methods use multiribbed wheels:

- Single-rib wheel traverse grinding
- Multirib wheel traverse grinding
- Multirib wheel plunge grinding
- Multirib wheel skip-rib, or alternate-rib, grinding
- Multirib wheel three-rib grinding

### Single-Rib Wheel Traverse Grinding

**Machining Process.** In this method (Fig. 4a), the edge of the wheel is trued to the form of the thread to be ground; the profile of the cutting edge works like that of a single-point metal tool for cutting threads on a lathe. Because the grinding wheel is frequently 457 to 508 mm (18 to 20 in.) in diameter, the arc of contact at full depth is appreciable. Thus, for accuracy, the wheel should be inclined to the helix angle; this adjustment is imperative for helices of 4° or more.

Single-rib wheel traverse grinding is the original and most versatile method of thread grinding, and it is still applied in the majority of operations. It can also be considered the method by which the highest degree of dimensional accuracy of the ground thread can be attained. The capacity limits of single-rib wheel thread grinding are the widest available, extending from the finest to the coarsest thread types. The single-rib wheel is also adaptable, with the aid of appropriate truing devices, to many different profile configurations, including those that differ from the conventional thread shapes, such as hobs, worms, and translation screws with special ball tracks.

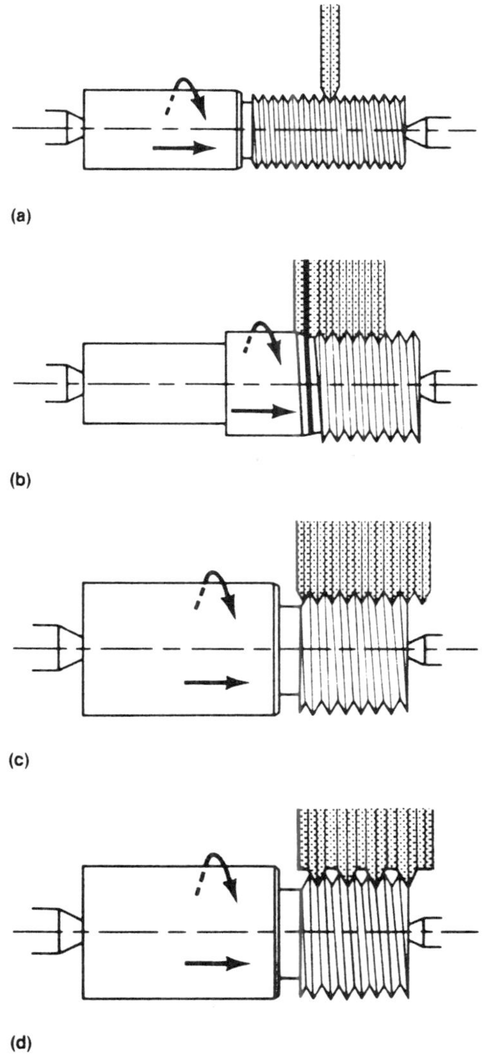


**Fig. 4** Cylindrical grinding of threads using single-rib (a), traverse (b), plunge (c), and skip-rib (d) multiribbed grinding wheels

**Recent Developments to Increase Productivity.** One respect in which single-rib wheel grinding is inferior to methods employing multirib wheels is productivity. Several design improvements, however, have been incorporated into different models of thread grinding machines for increasing the productivity of the single-rib wheel method. These design characteristics, although found only in certain models of thread grinders, include the following:

- The wheel diameter has been increased to 508 mm (20 in.) to extend the length of the operating periphery, thus removing a higher volume of stock before retruing is necessary
- The peripheral speed of the grinding wheel has been increased, often to a multiple of the conventional speed. This preceded by many years the application of high-speed grinding, which more recently is used in other precision grinding methods

## Table 4 Nominal speeds for the thread grinding of ferrous alloys and carbides

| Material | Hardness, HRC | Condition | Thread designation: Pitch, mm | Thread designation: Threads per 25 mm (1 in.) | Wheel identification, ANSI | Wheel speed m/s | Wheel speed sfm | Work speed m/min | Work speed sfm |
|---|---|---|---|---|---|---|---|---|---|
| **Wrought alloy steels** | | | | | | | | | |
| **Low carbon** | | | | | | | | | |
| 4012 4615 4817 8617 | 50 max | Hot rolled, | 6–2.5 | 4–12 | A90RB | 43 | 8 500 | | |
| 4023 4617 4820 8620 | | normalized, | 2–1 | 13–27 | A120SB | 46 | 9 000 | 1.2 | 4.0 |
| 4024 4620 5015 8622 | | annealed, cold | 0.8–0.45 | 28–63 | A180TB | 48 | 9 500 | | |
| 4118 4621 5115 8822 | | drawn, or quenched | 0.4–0.35 | 64–80 | A220UB | 51 | 10 000 | | |
| 4320 4718 5120 9310 | | and tempered | | | | | | | |
| 4419 4720 6118 94B15 | | | | | | | | | |
| 4422 4815 8115 94B17 | >50 | Carburized and/or | 6–2.5 | 4–12 | A100KV | 36 | 7 000 | | |
| **Medium carbon** | | quenched and | 2–1 | 13–27 | A150JV | 38 | 7 500 | 1.8 | 6.0 |
| 1330 4145 5132 8640 | | tempered | 0.8–0.45 | 28–63 | A220MV | 43 | 8 500 | | |
| 1335 4147 5135 8642 | | | 0.4–0.35 | 64–80 | A240PV | 46 | 9 000 | | |
| 1340 4150 5140 8645 | | | | | | | | | |
| 1345 4161 5145 86B45 | | | | | | | | | |
| 4027 4340 5147 8650 | | | | | | | | | |
| 4028 4427 5150 8655 | | | | | | | | | |
| 4032 4626 5155 8660 | | | | | | | | | |
| 4037 50B40 5160 8740 | | | | | | | | | |
| 4042 50B44 51B60 8742 | | | | | | | | | |
| 4047 5046 6150 9254 | | | | | | | | | |
| 4130 50B46 81B45 9255 | | | | | | | | | |
| 4135 50B50 8625 9260 | | | | | | | | | |
| 4137 5060 8627 94B30 | | | | | | | | | |
| 4140 50B60 8630 | | | | | | | | | |
| 4142 5130 8637 | | | | | | | | | |
| **High carbon** | | | | | | | | | |
| 50100 51100 52100 M-50 | | | | | | | | | |
| **Wrought carbon steels** | | | | | | | | | |
| **Low carbon** | | | | | | | | | |
| 1005 1012 1019 1026 | 50 max | Hot rolled, | 6–2.5 | 4–12 | A90RB | 43 | 8 500 | | |
| 1006 1013 1020 1029 | | normalized, | 2–1 | 13–27 | A120SB | 46 | 9 000 | 1.2 | 4.0 |
| 1008 1015 1021 1513 | | annealed, cold | 0.8–0.45 | 28–63 | A180TB | 48 | 9 500 | | |
| 1009 1016 1022 1518 | | drawn, or quenched | 0.4–0.35 | 64–80 | A220UB | 51 | 10 000 | | |
| 1010 1017 1023 1522 | | and tempered | | | | | | | |
| 1011 1018 1025 | | | | | | | | | |
| **Medium carbon** | | | | | | | | | |
| 1030 1042 1053 1541 | >50 | Carburized and/or | 6–2.5 | 4–12 | A100KV | 36 | 7 000 | | |
| 1033 1043 1055 1547 | | quenched and | 2–1 | 13–27 | A150JV | 38 | 7 500 | 1.8 | 6.0 |
| 1035 1044 1524 1548 | | tempered | 0.8–0.45 | 28–63 | A220MV | 43 | 8 500 | | |
| 1037 1045 1525 1551 | | | 0.4–0.35 | 64–80 | A240PV | 46 | 9 000 | | |
| 1038 1046 1526 1552 | | | | | | | | | |
| 1039 1049 1527 | | | | | | | | | |
| 1040 1050 1536 | | | | | | | | | |
| **High carbon** | | | | | | | | | |
| 1060 1074 1085 1566 | | | | | | | | | |
| 1064 1075 1086 1572 | | | | | | | | | |
| 1065 1078 1090 | | | | | | | | | |
| 1069 1080 1095 | | | | | | | | | |
| 1070 1084 1561 | | | | | | | | | |
| **Wrought tool steels** | | | | | | | | | |
| A2 H13 L6 P20 | 50 max | Annealed or quenched | 6–2.5 | 4–12 | A90RB | 43 | 8 500 | | |
| A3 H14 L7 P21 | | and tempered | 2–1 | 13–27 | A120SB | 46 | 9 000 | 1.2 | 4.0 |
| A4 H19 O1 S1 | | | 0.8–0.45 | 28–63 | A180TB | 48 | 9 500 | | |
| A6 H21 O2 S2 | | | 0.4–0.35 | 64–80 | A220UB | 51 | 10 000 | | |
| A8 H22 O6 S5 | | | | | | | | | |
| A9 H23 O7 S6 | >50 | Quenched and | 6–2.5 | 4–12 | A100KV | 36 | 7 000 | | |
| A10 H24 P2 S7 | | tempered | 2–1 | 13–27 | A150JV | 38 | 7 500 | 1.8 | 6.0 |
| H10 H24 P4 W1 | | | 0.8–0.45 | 28–63 | A220MV | 43 | 8 500 | | |
| H11 H26 P5 W2 | | | 0.4–0.35 | 64–80 | A240PV | 46 | 9 000 | | |
| H12 L2 P6 W5 | | | | | | | | | |
| SAE J438b: types W108, W109, W110, W112, W209, W210, W310 | | | | | | | | | |
| **Wrought austenitic stainless steels** | | | | | | | | | |
| 201 304L 310S 330 | 135–275 HB | Annealed or cold | 6–2.5 | 4–12 | A90RB | 43 | 8 500 | | |
| 202 305 314 347 | | drawn | 2–1 | 13–27 | A120SB | 46 | 9 000 | 1.2 | 4.0 |
| 301 308 316 348 | | | 0.8–0.45 | 28–63 | A180TB | 48 | 9 500 | | |
| 302 309 316L 384 | | | 0.4–0.35 | 64–80 | A220UB | 51 | 10 000 | | |
| 302B 309S 317 385 | | | | | | | | | |
| 304 310 321 | | | | | | | | | |

(continued)

**Table 4 (continued)**

| Material | Hardness, HRC | Condition | Thread designation Pitch, mm | Threads per 25 mm (1 in.) | Wheel identification, ANSI | Wheel speed m/s | sfm | Work speed m/min | sfm |
|---|---|---|---|---|---|---|---|---|---|
| **Gray cast irons** | | | | | | | | | |
| **Ferritic** | | | | | | | | | |
| ASTM A48: class 20 | 45 max | As-cast, annealed, | 6–2.5 | 4–12 | A90RB | 43 | 8 500 | 1.2 | 4.0 |
| SAE J431c: grade G1800 | | or quenched and | 2–1 | 13–27 | A120SB | 46 | 9 000 | | |
| **Pearlitic-ferritic** | | tempered | 0.8–0.45 | 28–63 | A180TB | 48 | 9 500 | | |
| ASTM A48: class 25 | | | 0.4–0.35 | 64–80 | A220UB | 51 | 10 000 | | |
| SAE J431c: grade G2500 | | | | | | | | | |
| **Pearlitic** | 45–52 | As-cast, annealed, | 6–2.5 | 4–12 | A90RB | 43 | 8 500 | 1.2 | 4.0 |
| ASTM A48: classes 30, 35, 40 | | or quenched and | 2–1 | 13–27 | A120SB | 46 | 9 000 | | |
| SAE J431c: grade G3000 | | tempered | 0.8–0.45 | 28–63 | A180TB | 48 | 9 500 | | |
| **Pearlitic + free carbides** | | | 0.4–0.35 | 64–80 | A220UB | 51 | 10 000 | | |
| ASTM A48: classes 45, 50 | | | | | | | | | |
| SAE J431c: grades G3500, G4000 | | | | | | | | | |
| **Pearlitic or acicular + free carbides** | | | | | | | | | |
| ASTM A48: classes 55, 60 | | | | | | | | | |
| **Wrought high-temperature nickel-base alloys** | | | | | | | | | |
| AF2-1DA / Inconel 751 | 200–390 HB | Annealed or | 6–2.5 | 4–12 | A90SB | 46 | 9 000 | 0.45–1.2 | 1.5–4.0 |
| Astroloy / M252 | | solution treated | 2–1 | 13–27 | A100TB | | | | |
| Haynes 263 / Nimonic 75 | | | 0.8–0.45 | 28–63 | A180TB | | | | |
| IN-102 / Nimonic 80 | | | 0.4–0.35 | 64–80 | A280TB | | | | |
| Incoloy 901 / Nimonic 90 | | | | | | | | | |
| Incoloy 903 / Nimonic 95 | 300–475 HB | Solution treated and | 6–2.5 | 4–12 | A90SB | 46 | 9 000 | 0.45–1.2 | 1.5–4.0 |
| Inconel 617 / René 41 | | aged | 2–1 | 13–27 | A100TB | | | | |
| Inconel 625 / René 63 | | | 0.8–0.45 | 28–63 | A180TB | | | | |
| Inconel 700 / René 77 | | | 0.4–0.35 | 64–80 | A280TB | | | | |
| Inconel 702 / René 95 | | | | | | | | | |
| Inconel 706 / Udimet 500 | | | | | | | | | |
| Inconel 718 / Udimet 700 | | | | | | | | | |
| Inconel 721 / Udimet 710 | | | | | | | | | |
| Inconel 722 / Unitemp 1753 | | | | | | | | | |
| Inconel X-750 / Waspaloy | | | | | | | | | |
| **Carbides** | | | | | | | | | |
| Titanium carbide | 89–94 HRA | . . . | 6–2.5 | 4–12 | D240R100B1/8 | 23 | 4 500 | 0.45 | 1.5 |
| Tungsten carbide | | | 2–1 | 13–27 | D320T100B1/8 | 25 | 5 000 | | |
| | | | 0.8–0.45 | 28–63 | D400V100B1/8 | 28 | 5 500 | | |
| | | | 0.4–0.35 | 64–80 | D400V100B1/8 | 30 | 6 000 | | |

Source: Ref 3

- Wheel truing systems and devices are used that substantially reduce or even eliminate the loss in productive grinding time due to wheel truing
- For multipass grinding—a procedure that is generally needed for the single-rib wheel method—the accelerated table return has long been provided, but in certain models of modern production grinders it is being superseded by the even more productive two-way grinding. In this latter system, also referred to as two-directional, the grinding is carried out during the table movements in both directions, but it requires specially designed grinders equipped with dependable backlash-eliminator devices. Some manufacturers are building thread grinders that can operate in an automatic cycle using both methods subsequently: the two-way grinding (in roughing) for increased productivity and the one-way grinding (in finishing) for higher work accuracy
- The wheel and work speeds are varied automatically in different phases of the processes. High wheel speeds are used during the roughing passes for increased rate of stock removal. The wheel speed is then decreased, and the work speed increased, during the finishing passes, thus obtaining a better finish on the ground surface

## Multirib Wheel Grinding

Multiribbed wheels have two or more parallel grooves or ribs around the periphery of the wheel. Each rib is trued to the form of the thread to be ground. The thread form is imparted to the wheel by diamond or crush truing.

**Traverse Grinding.** If the grinding wheel is as thick as, or thicker than, the required length of the thread (Fig. 4b), the thread can be completed in one revolution of the work plus a half revolution for feeding-in of the wheel. Traverse grinding operates with a wheel that is wide enough to contain ribs corresponding to six to eight threads of the workpiece. The leading edge of the wheel is usually tapered, with the first two or three ribs truncated in order to cause the wheel to penetrate gradually into the work material as it is traversed axially. The wheel is advanced to the intended radial position of the cut, which may be the full depth of the thread for single-pass grinding or only a part of the total thread depth for multipass grinding. After the wheel has reached its advance position, the traversing and rotating workpiece contacts the leading edge of the wheel and continues the traverse movement across the face of the wheel until the end of the threaded part section has passed the wheel. In the case of multipass grinding, the process of wheel retraction and work return is the same as in single-rib wheel grinding, but a longer return distance must be covered because of the greater wheel width.

Traverse grinding can also be operated in the two-way system. In this case, both sides of the wheel function alternately as leading edges, and the starting ribs are truncated at both ends of the wheel face.

Multirib wheel traverse grinding is more productive than single-rib wheel grinding because of the higher rate of stock removal per pass, and it is frequently used on thread grinding machines that are adapted to this method by the use of wider wheels and the appropriate multirib wheel-truing devices. The application of multirib wheel grinding, however, has certain limitations:

- Thread pitch should not exceed one-eighth of the available wheel width
- There should be no shoulder on the workpiece that could interfere with its passing across the total width of the wheel face

Regarding the accuracy attainable, multirib wheel traverse grinding is successfully applied to threads even with tight tolerances, except for such extreme accuracy requirements as specified for thread gages.

**Plunge Grinding.** For this method of grinding, wheels that are about 1½ thread pitches wider than the entire threaded section of the workpiece must be used, but the complete thread is produced in less than 1½ turns of the workpiece. The wheel, which has ring-shaped thread profile ribs of equal depth around its entire periphery, is advanced gradually into the rotating and traversing workpiece during about ⅓ turn of the latter. The wheel then stays in this position while the workpiece continues making somewhat more than an additional full turn (Fig. 4c). At this position, the entire thread has been ground in a single plunged pass; the wheel retracts and the machine table returns into its starting position, ready to accept the next workpiece.

Multirib wheel plunge grinding is the most productive thread grinding method, and it is successfully used where the dimensions of the workpiece and of the thread permit its application. The accuracy attainable satisfies most general requirements, with the exception of parts having very critical thread tolerances.

The truing of the multirib wheel is commonly performed by crushing, a truing method that was adopted only for certain types of thread grinding machines, including the special machine types built expressly for plunge grinding with crush-dressed wheels. For high-volume production, diamond-impregnated form rolls are also used for the very efficient truing of multirib thread grinding wheels. Because of the relatively high cost of the required special tooling, the crushing rolls, and the diamond rolls (even much more expensive), multirib wheel thread grinding by plunge advance must be used for the production of parts in substantial quantities in order to be economical in comparison to other thread grinding methods.

**Skip-Rib Grinding.** A type of multirib wheel plunge grinding, designated as skip-rib, with rib spacing on the wheel twice that of the thread pitch, is occasionally used for threads of very fine pitch or for resinoid-bonded wheels where the truing of ribs corresponding to the actual thread pitch would be difficult to carry out. The grinding wheel can be formed by crush or diamond truing. Diamond truing is used on resinoid wheels or to dress fine, accurate pitches. In the first pass, the wheel grinds every other thread of the workpiece; in a subsequent pass, after the work is traversed by a distance equal to the thread pitch, the previously untouched threads are ground over a distance comprising a small circumferential overlap for bending (Fig. 4d). This method requires about 2½ total turns of the workpiece for the entire operation.

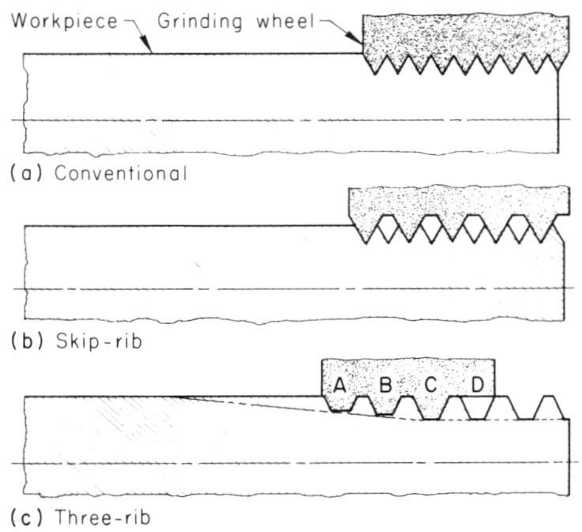


**Fig. 5** Schematic of a three-rib thread grinding wheel. A, roughing rib; B, intermediate rib; C, finishing rib. The flattened area (D) is optional and can be used to finish grind the crest of the thread.

**Three-Rib Grinding.** Another variation of the multirib thread grinding wheel is the three-rib wheel (Fig. 5). This wheel has a roughing rib (A, Fig. 5) that removes about two-thirds of the material, and an intermediate rib (B, Fig. 5) that takes the remainder of the material except for about 0.13 mm (0.005 in.), which is left to be cleaned up by the finishing rib (C, Fig. 5). If desired, the wheel can be dressed so that a flattened area (D, Fig. 5) will finish grind the crest of the thread.

The three-rib wheel will produce threads whose accuracy equals or exceeds that of threads produced by the single-rib wheel. If necessary, this type of wheel can be inclined to the helix angle, provided an allowance is made for the radius of curvature of the work. The three-rib wheel is designed to traverse the work rather than to plunge cut.

## Infeed

When a single-rib wheel is used, a roughing cut of 0.51 to 1.02 mm (0.020 to 0.040 in.) per pass can be made without burning the work metal. An infeed of 0.038 to 0.10 mm (0.0015 to 0.004 in.) per pass is used for close dimensional accuracy, while infeed for larger dimensional tolerances common to many commercial products ranges from 0.10 to 0.25 mm (0.004 to 0.010 in.) per pass.

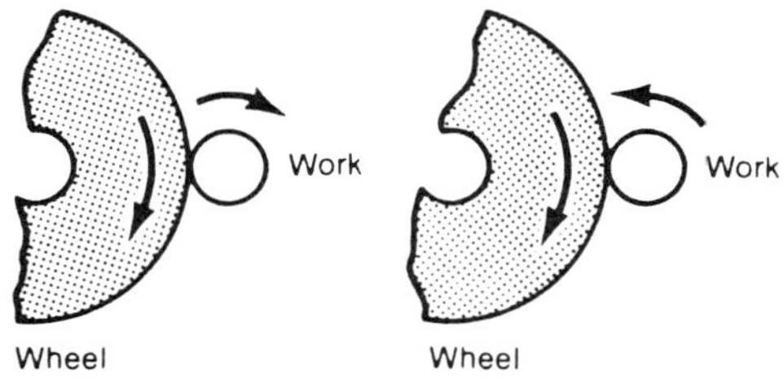


**Fig. 6** Direction of cut in grinding threads. (a) Up cut. (b) Climb cut

The work can be ground to final size in one cut when the depth of cut is less than 1.02 mm (0.040 in.). Work speed for finish grinding in one cut is slow; for example, a work speed of 0.46 to 0.61 m/min (1½ to 2 sfm) is used for grinding 20 threads per 25 mm (1 in.) with a thread depth of 0.81 mm (0.032 in.). For depths greater than 1.02 mm (0.040 in.), roughing cuts of 0.51 to 1.02 mm (0.020 to 0.040 in.) or more per pass at work speeds of 0.9 to 1.2 m/min (3 to 4 sfm) are required. Work speed for the finish cut of 0.038 to 0.10 mm (0.0015 to 0.004 in.) is 1.8 to 2.4 m/min (6 to 8 sfm).

In multicut operations, infeed should be controlled so that the cutting edge of the grinding wheel will not break down before the finishing cut. For example, an infeed of 0.91 mm (0.036 in.) per pass for roughing and 0.46 mm (0.018 in.) per pass for finishing is satisfactory for grinding 20 screw threads per 25 mm (1 in.).

Thread grinding can be done on the up cut, usually with a better finish than that possible when grinding is done on the climb cut (Fig. 6). With up cut, the materials removed from the work and the wheel do not have to pass over the finished surface. Figure 7 shows a tube held in a chuck while being threaded internally using a cylindrical grinding setup.

## Centerless Grinding of Threads

Centerless grinding is the most productive method of grinding screw threads. This method is capable of grinding threads of classes 2, 2A, 3, 3A, and 7 at high production rates; threads of class 4 and 5A are produced at slower rates.

Wheels for centerless thread grinding can also be either single-ribbed or multiribbed. Grain size ranges from 100 to 320, depending on the number of threads per 25 mm (1 in.), as indicated in Tables 1 and 2. Harder wheels are used in centerless thread grinding than for other centerless grinding operations on the same materials. Grade (hardness) ranges from M to R. Wheels of grade Q are generally used for such products as headless setscrews made of steel at 40 HRC. Wheel structure is usually fairly open, ranging from 8 to 11, depending on the wheel manufacturer.

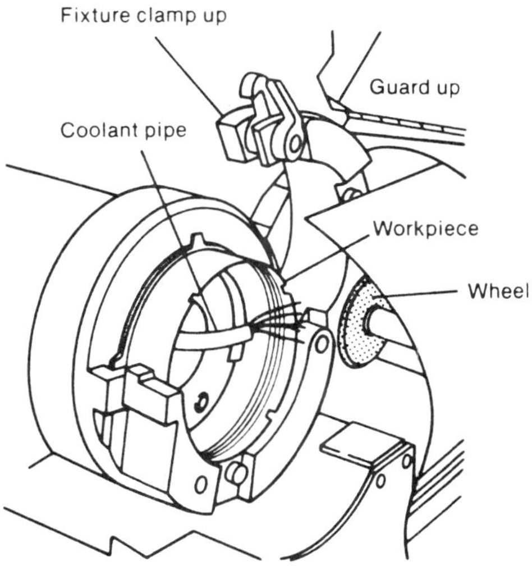


**Fig. 7** Schematic of a cylindrical grinding setup in which a grinding wheel is generating internal threads on a piece of tubing

Wheel life depends on the type and hardness of the material being ground, thread tolerance, and production rate. Because maximum wear occurs at the thread crests of the grinding wheel, the wheel must be dressed to maintain the tolerance of the root of the product thread. Centerless thread grinding wheels are dressed by crushing.

**Regulating wheels** for centerless thread grinding are similar to those used for other centerless grinding operations. For precision thread grinding, however, regulating wheels of finer grain size are used.

Regulating wheels usually revolve in the same direction as the grinding wheel (down grinding, Fig. 8a). However, for coarse threads (1 to 8 coarser), the regulating wheel rotates in the direction opposite to that of the grinding wheel (up grinding, Fig. 8b).

Regulating wheels are dressed with a single-point diamond dressing tool. The frequency of dressing depends on the depth of the grooves that develop in the wheel face during thread grinding. In general, the regulating wheel is dressed each time the grinding wheel is dressed, but in some operations the regulating wheel is dressed only once for every two or three dressings of the grinding wheel.

**Screw Thread Production.** Screw threads can be ground from the solid by the centerless method. Adjustments are provided to accommodate work of different sizes and for varying the rates of feed. The grinding wheel is a multirib type, which consists of a series of annular ridges across the face. These ridges conform in pitch and profile with the thread to be ground. The grinding wheel is inclined to suit the helix or lead angle of the thread. In grinding threads on such work as socket-type setscrews, the blanks are fed automatically and passed between the grinding and regulating wheels

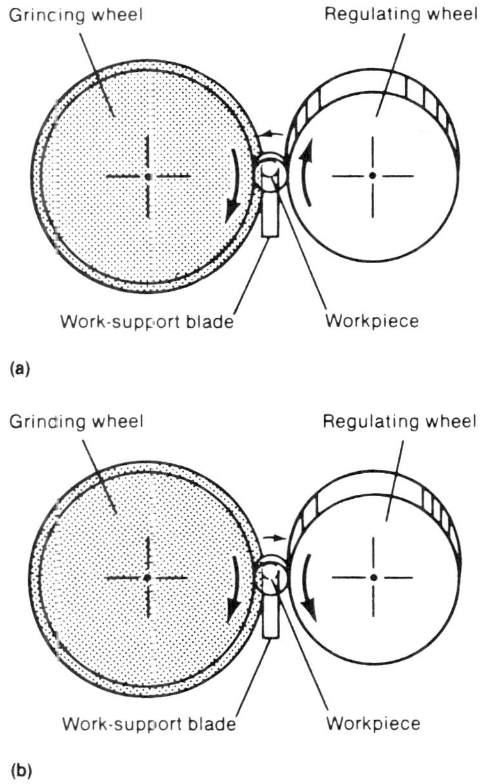


**Fig. 8** Two methods of centerless thread grinding. (a) Down grinding. (b) Up grinding

in a continuous stream. To illustrate production possibilities, hardened socket setscrews of ¼-20 size can be ground from the solid at the rate of 60 to 70 per min and with the wheel operating continuously for 8 h without redressing (Fig. 9). The lead errors of centerless ground screw threads can be limited to 0.0005 mm/mm (0.0005 in./in.) or even less by reducing the production rate. The pitch diameter tolerances are within 0.0051 to 0.0076 mm (0.0002 to 0.0003 in.) of the basic size. The grain size for the wheel is selected with reference to the pitch of the thread. The following sizes are recommended:

- For 11 to 13 threads per 25 mm (1 in.), 150
- For 16 threads per 25 mm (1 in.), 180
- For 18 to 20 threads per 25 mm (1 in.), 220
- For 24 to 28 threads per 25 mm (1 in.), 320
- For 40 threads per 25 mm (1 in.), 400

## Production Practice

Pitches as fine as 80 threads per 25 mm (1 in.) can be ground with conventional thread grinding procedures. Special techniques have been developed for grinding threads as fine as 400 per 25 mm (1 in.). Most threads ground in regular production are considerably coarser than 80 per 25 mm (1 in.).

A hard thread can be obtained by thread grinding because the workpiece can be ground after it has been fully hardened. This eliminates distortion, which is commonly

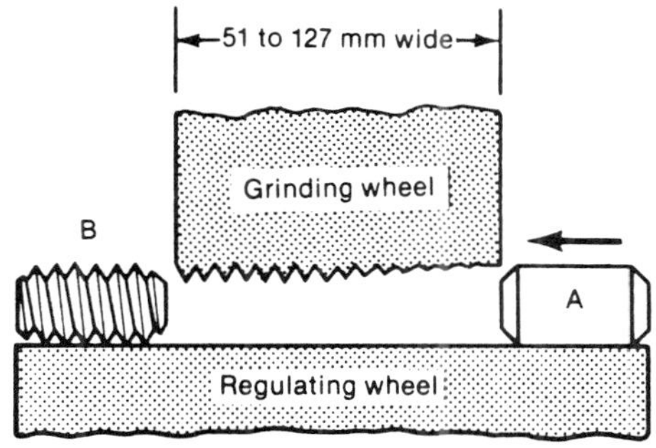


**Fig. 9** Principle of centerless thread grinding. This method is used to produce headless setscrews. The blanks are hopperfed to position A. The regulating wheel causes them to traverse the grinding wheel face, from which they emerge at position B in completed form. A production rate of 60 to 70 screws of 12.7 mm (½ in.) length per min is possible.

found in workpieces threaded before heat treating, and produces true thread forms with the lead and pitch diameter held to close tolerances. Limits of 0.005 mm (0.0002 in.) or less can be held on pitch diameter, and less than 0.0002 mm/mm (0.0002 in./in.) on lead tolerances.

Thread grinding can be used on surfaces containing drilled holes, keyways, or other irregularities. Shell-type parts are more amenable to thread grinding than to other methods of producing threads, in which pressure from the tool can cause distortion. Small worms can be roughed faster by thread grinding than by conventional thread milling.

Thread grinding is applicable to virtually all materials and all hardnesses. On the other hand, thread cutting is superior from the standpoint of material handling and production rate, especially when the thread tolerances are broad. However, these advantages of thread cutting are quickly offset when close tolerances are required. When threads of fine finish and close tolerances are needed, thread grinding is more economical than other methods, including thread cutting and thread milling. Internal threads can also be produced by grinding, using a machine especially designed for internal threading or a universal machine.

### High-Volume Applications of Thread Grinding

In production quantities, thread grinding can be used to produce diverse types of threaded parts, as well as other types of parts that have helical profiles of particular shapes. Table 5 lists commonly used mechanical parts whose threading and similar features are often produced by thread grinding.

Thread grinding as a metalworking method is not limited to parts whose quality specifications warrant or even require its use. In many cases, thread grinding can be the most productive method of producing certain part configurations. In such applica-

**Table 5 Mechanical parts often produced with ground threads or similar features**

| Designation | Discussion |
|---|---|
| **Cutting tools** | |
| Thread cutting taps | Taps were the first types of workpieces manufactured in high-volume production with ground threads, and they still represent one of the principal applications of the method. Thread grinding ensures superior accuracy of all the important parameters and can produce relieved thread profiles. By grinding threads in solid blanks, thread grinding also has a productivity rate exceeding that of only approximately equivalent alternative processes. |
| Thread cutting dies | Universal thread grinding machines with internal grinding attachments are commonly used for grinding the threaded surfaces of these tools and will produce threads with excellent dimensional accuracy and the unique advantage of relieved thread profile. |
| Thread cutting chasers | Both the radial and the tangential types, used in threading heads, can be produced on thread grinding machines equipped with special devices imparting a straight-line reciprocating motion to the workpieces, which are usually mounted in sets. Similar methods can be used for multirib thread chasers used in lathe work. Round thread chasers are ground on basic thread grinding machine equipment. |
| Thread milling cutters | Multirib thread milling cutters are generally manufactured with ground threads that have relieved form on the entire peripheral sections between the gashes. Producing these relieved profiles by thread grinding results in superior accuracy, and thread grinding is also faster than the earlier applied relief-form turning. The relief grinding of thread milling cutters with helical gashes generally requires a special attachment. |
| Gear-generating hobs | Hobs used as the tools of continuously operating gear milling machines generally have ground profiles, particularly for ensuring a high accuracy of tooth form and spacing, but also for higher cutting ability of tool edges that are free of decarburization. Special attachments or thread grinding machines, operating with small-diameter wheels, are needed. |
| **Forming tools** | |
| Crush rolls for thread forms | Multirib wheels are the means of very efficient thread grinding methods, and crush dressing is the most commonly used process for forming the wheels. The crush rolls are ground on thread grinding machines equipped with devices to grind no-lead (annular) thread forms. |
| Thread rolling dies | Both types, the flat dies and the round dies, also known as thread forming rolls, are ground on thread grinding machines. The flat dies require a reciprocating attachment. Higher thread accuracy and better service life of the dies are obtained by thread grinding than by milling followed by heat treatment. |
| **Measuring instruments** | |
| Thread plug and ring gages | The accuracy level that thread gages for internal and external surfaces must satisfy, with respect to all the significant parameters, is most effectively and dependably obtained by thread grinding. Plug gages and (with the aid of an internal grinding attachment) ring gages are ground with all types of profiles, both straight and tapered. |
| Micrometer screws | Screw threads are a reliable and sensitive mechanical means of imparting accurately controlled linear displacement, combined with the ability to display such motions at a high rate of dependably proportional amplification. Micrometer screws are commonly manufactured with ground threads for the highest degree of lead accuracy. |
| Gage and instrument racks | Racks, in combination with pinions or worms, have the potential to produce very sensitively specific linear displacements, whose accuracy is controlled by the shape and spacing of the meshing teeth. Thread grinders with special attachments for work reciprocation and incremental table traverse are suitable for grinding such precise racks. |
| **Translation elements** | |
| Leadscrews and feed screws | The potential advantages of screws serving as translation elements, such as the favorable ratio of input torque to the resulting translational force, with sensitive and precise motion control, can best be implemented by screws with ground threads that can be produced even on very long lead screws. |
| Ball screws and ball nuts | Screws and nuts connected by recirculating balls represent translation elements with excellent force efficiency; they operate with practically no play and have very long service lives. Precision-type ball screws and nuts, with a circular or gothic-arch ball-track profile, are produced with high form and lead accuracy on thread grinding machines. |
| Worms | The efficiency of worm gear drives is particularly sensitive to the accuracy of the driving member. Therefore, ground worms, with flank contours correct either with respect to the axis or the helix, are often required and produced on thread grinding machines adapted to those operations that may need special wheel profiles and very high helix angles. |
| **Fasteners** | |
| Screws and bolts | High-strength threaded fasteners are preferably made by thread grinding because of the favorable stress distribution that results from the maximum flank contact ensured by accurate flank angles and lead. High production rates can be achieved on conventional or centerless automatic thread grinders. |
| Machine elements with threaded sections | Hardened steel machine elements having screw threads only on limited sections but in a very precise relation to other ground portions of the surface are best finished by thread grinding. Preferably, the part is held on the same mounting surfaces that were used in the preceding operations for producing the locationally related surface elements. |
| **Special helical profiles** | |
| Machine elements with helical features | When these must be produced with specific contours, precise lead, and excellent surface texture, thread grinding may be the best suited or only method. Standard machines or special equipment is used. Examples include metering feed screws and steel regulating wheels for special centerless grinding processes. |

Source: Ref 2

tions, the accuracy, finish, and other quality-related properties of the ground surface constitute additional benefits. With respect to productivity, the advantages of thread grinding over those of alternative methods may arise from various conditions, as follows.

**Increased Rate of Stock Removal.** The limiting conditions of chip disposal from a confined work area can be overcome more easily in grinding than in other metalworking methods.

**Wheel Truing.** Instead of frequent tool changes, only the regular truing of the installed grinding wheel needs to be applied, often in a manner not interfering with the productive cycle time of the operation. This property of thread grinding may prove particularly valuable in the production of long threaded parts, such as leadscrews, where the volume of stock removed may exceed the edge-holding limits of a single-point cutting tool.

**Roughing and Finishing Operations.** Roughing, even from the solid, and subsequent finishing can be combined into a single operation, using the same tool, by

**Table 6 Principal categories of thread grinding machines**

| Category | Field of application | General characteristics |
|---|---|---|
| Universal thread grinding machines | Toolroom work for single pieces or limited volumes; general production of a wide range of workpieces, including special helical, annular, or even straight-line element shapes | A particularly adaptable machine type, several models also operating with multirib wheels; wide tilt range of the wheelhead; capable of relief grinding. Can be supplied with many different accessories for extending the scope of application to include various profile types and work surface configurations |
| Production-type thread grinding machines | Construction parts for precision machinery with threads or worms; thread plug gages and thread cutting tools; high-strength threaded fasteners, and so on, produced in medium lots or large volumes | This category comprises several models with different work-size capacities, generally ranging from about 150 to 300 mm (6 to 12 in.) maximum diameter and 200 to 1000 mm (8 to 40 in.) maximum length. Some of these models have change gears; others have interchangeable thread-generating elements, considering the infrequent pitch change needed in high-volume production. Many models can operate with single-rib or multirib wheels, and depending on the design, most or all phases of the grinding cycle are automated, with the variables adjustable to suit the operational requirements. |
| Fully automatic tap grinders | Taps are typical, but not exclusive, examples of small-size threaded parts that, by their overall dimensions, permit automatic work handling and are required in volumes warranting equipment for completely automatic operation. | The thread grinders for this type of application are manufactured in models using single-rib wheels in one-way or two-way grinding at multiple cuts and high work speeds and in models using multirib wheels usually operating in a plunge grinding system, with wheel dressing by crush rolls or diamond rolls. In either case, the operation is continuously automatic, including all the elements of the grinding cycle as well as the loading and positioning of the workpiece. |
| Thread grinding machines for long workpieces | Single-setup grinding of very long, threaded machine components, particularly leadscrews for machine tools, feed and actuator screws, and ball screws with gothic-arch thread profiles | Manufactured in different sizes for maximum grinding lengths of the order of 4 m (12 ft). Some models have hollow work spindles to accommodate parts longer than the grinding capacity; usually operated in a two-way grinding cycle; designed to reduce the extent of, or to compensate for, thermal effects; equipped with various types of wheel-dressing devices; accept special steady rests for providing intermediate support to long and thin workpieces which are prone to deflection |
| Thread grinding machines for internal threads | Production grinding of threads on internal surfaces; principal applications are thread ring gages, nuts of precision-type translation elements (including those with ball tracks and Acme threads), threading dies, and so on | Usually designed with entirely automatic grinding cycle, operating in different modes, such as one-way or two-way grinding. Lead generation by interchangeable master screws and nuts; supplied with interchangeable grinding spindles for various speed ranges. Grinds right- or left-hand threads, as well as annular profiles, and can be used for multistart threads |
| Special-purpose thread grinding machines | Workpieces that, because of extreme size or a particular configuration, cannot be accommodated on standard thread grinding machines. Typical examples are parts requiring the grinding of special profiles or particular feature-generating movements. | Overall design and capacities adapted to the special purpose, although the operating principles and the methods of wheel dressing are often similar to those of general-purpose thread grinders. This category also includes such unconventional designs as the centerless thread grinding machines. |

Source: Ref 2

changing the infeed rate. The condition of the work produced in a single pass is usually adequate, thus eliminating the need for finishing in a second operation.

**Workpiece Configuration.** The repetitive shape of the screw thread frequently permits the use of one-pass plunge grinding, producing (by an essentially single turn of the workpiece) the entire threaded section. In principle, this is similar to the operation of the multirib thread milling cutter, but it is generally accomplished with a much higher productivity and with superior accuracy.

**Subsequent Machining of the Workpiece.** Threads or similar profiles can be produced along a helical path on a basically cylindrical part with ribs having eccentric surfaces. The resulting surfaces are thus relieved with respect to the leading edges, which are located along the individual rib boundaries, and the configuration produced is highly beneficial to the effective operation of thread cutting tools, especially taps and multirib thread milling cutters.

## Thread Grinding Machines

The classification of thread grinding machines can be based on the degree of automation, distinguishing manually controlled, semiautomatic, automatic, and fully automatic thread grinders. The fully automatic classification comprises equipment that can operate essentially without attendance, performing even the loading and unloading of the work by automatic controls. Although the degree of automation is an important consideration in the selection of equipment intended for particular production purposes, thread grinding machines of widely different systems, capacities, and application purposes are available in designs providing different degrees of automation.

A better method of classification is based on the principal application of the machine. The principal application is usually the governing aspect in the design of the various types of thread grinders. A survey of this classification is presented in Table 6. This classification is also a flexible one; overlaps will occasionally occur because of the expanded adaptability of the basic machine and particularly because of the use of optional accessories.

## REFERENCES


1. *Machining*, Vol 1, 4th ed., *Tool and Manufacturing Engineers Handbook*, Society of Manufacturing Engineers, 1983
2. F.T. Farago, *Abrasive Methods Engineering*, Vol 2, Industrial Press, 1980
3. *Machining Data Handbook*, Vol 2, 3rd ed., Metcut Research Associates, 1980

# Thread Rolling

THREAD ROLLING (also known as roll threading) is a cold forming process for producing threads or other helical or annular forms by rolling the impression of hardened steel dies into the surface of a cylindrical or conical blank. Polygonal blanks are also thread rolled for the purpose of fabricating thread-forming and self-locking screws. The preferred polygonal shape is trilobular and is produced in flat die machines.

In contrast to thread cutting and thread grinding, thread rolling does not remove metal from the work blank. Rather, thread rolling dies displace the surface metal of the blank to form the roots and crests of a thread.

Dies for threading rolling may be either flat or cylindrical (Fig. 1). Flat dies operate by a traversing motion. Methods that use cylindrical dies are classified as radial infeed, tangential feed, through feed, planetary, and internal. Each method is discussed in a separate section of this article.

## Capabilities and Limitations

Most thread rolling is done on blanks having a hardness of 32 HRC or less. However, threads on fasteners used for high-temperature service are rolled in metal as hard as 52 HRC. Some metal products such as gray iron castings and sintered metal pieces cannot be thread rolled because of their low ductility. These materials crumble rather than conform to the contour of the die.

All of the commonly used straight and tapered thread forms can be rolled. These include Unified, International Standard (the same as UNR), metric, Whitworth, Acme, worm, buttress, screw shell, wood screw, tapping screw, lag screw, and drive screw. Thread diameters vary from less than 1.25 mm (0.050 in.), for instrument threads, to 380 mm (15 in.); the larger threads can be as long as 6 m (20 ft).

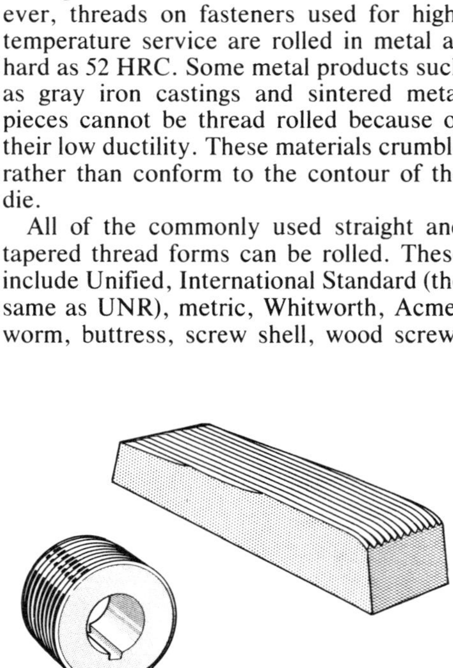

**Fig. 1** Two common types of thread rolling dies, flat and cylindrical

Thread forms of 60° roll easily. In more blunt forms, metal flows with greater difficulty. Threads with fully rounded roots are more easily rolled than threads with wide, flat roots. Flank angles of less than 10° included angle and thread depths exceeding one-sixth of the major diameter should be avoided, except with the most ductile metals. For multiple threads, a thread depth of one-fourth the major diameter for double and quadruple lead threads, and one-fifth the major diameter for triple lead threads, is generally acceptable.

The rolling process can also accomplish many nonthreading operations, such as the rolling of splines, helical and annular grooves, knurls, and involute teeth. Rolling may also be used for burnishing and for displacing metal to form flanges and similar cylindrical shapes.

**Surface Finish.** When properly made, thread rolling dies impart smooth, burnished roots and flanks to threads. Rolled threads are free of tears, chatter marks, or cutting-tool marks common to cut threads. Such imperfections nucleate wear and can serve as starting points for fatigue failure. Surface roughness on rolled threads is usually from 0.20 to 0.60 μm (8 to 24 μin.), whereas on cut threads it is often 1.63 to 3.18 μm (64 to 125 μin.) with small ridges or unevenness along the flanks of the thread. However, because the surface finish of thread flanks is extremely difficult to check, it is rarely specified on drawings.

In general, the coefficient of friction of a rolled thread surface in sliding contact is considerably lower than that of a comparable cut thread surface. The coefficient of friction between the thread and its mating nut determines how effectively a bolt can be tightened or a moving screw can transmit power. Therefore, the relatively low coefficient of friction of rolled thread surfaces provides more uniform and consistent tightening of fasteners and less loss of power in overcoming friction when a load is moved by a screw. The reduced and more uniform friction of rolled threads also contributes to the torque control of threaded connections with self-locking features. The smoother finish of rolled threads also retards corrosion.

**Strength of Rolled Threads.** Thread rolling deforms the blank plastically as it is forced to flow along the contour imposed by the dies. The worked metal is appreciably harder and stronger than the blank prior to rolling. Thus, fasteners with rolled threads are harder and stronger than those with cut threads, as indicated in Table 1.

Heating thread rolled fasteners made of steel above the transformation range during a heat-treating process completely relieves the favorable compressive stresses induced by rolling. Therefore, steel fasteners produced by the thread rolling of blanks already quenched and tempered to a given hardness usually have higher fatigue strength than fasteners quenched and tempered to the same hardness after thread rolling. Figure 2 shows this difference for the most common range of hardness. The different effects of hardness on fatigue strength of bolts that were roll threaded before and after heat treatment is illustrated in Fig. 3.

## Evaluation of Metals for Thread Rolling

The three characteristics that are important in evaluating and selecting metals for thread rolling are rollability, flaking, and seaming.

**Rollability** involves ductility and the resistance of a metal to flow when subjected to cold forming in thread rolling dies. Rollability indexes for 17 steels and for 6 nonferrous alloys commonly threaded are given in Table 2. The power required to form a given thread shape at a given rate in various metals is inversely proportional to the rollability indexes of the metals. If material in an operation is changed for one of lower rollability index, the production rate per horsepower for a rolled form in that operation decreases. For example, if a through-feed machine using its full 7.5 kW (10 hp) output produces ½-13 UNC-2A threads at a rate of 11.5 m/min (450 in./min) in a solid bar of steel with an index of 1.00, the rate will be only 6.9 m/min (270 in./min) when a steel of index 0.60 is threaded in the same machine under the same operating conditions.

The rollability index also provides a means of comparing radial die loads and expected die life in rolling two materials under identical operating conditions. The

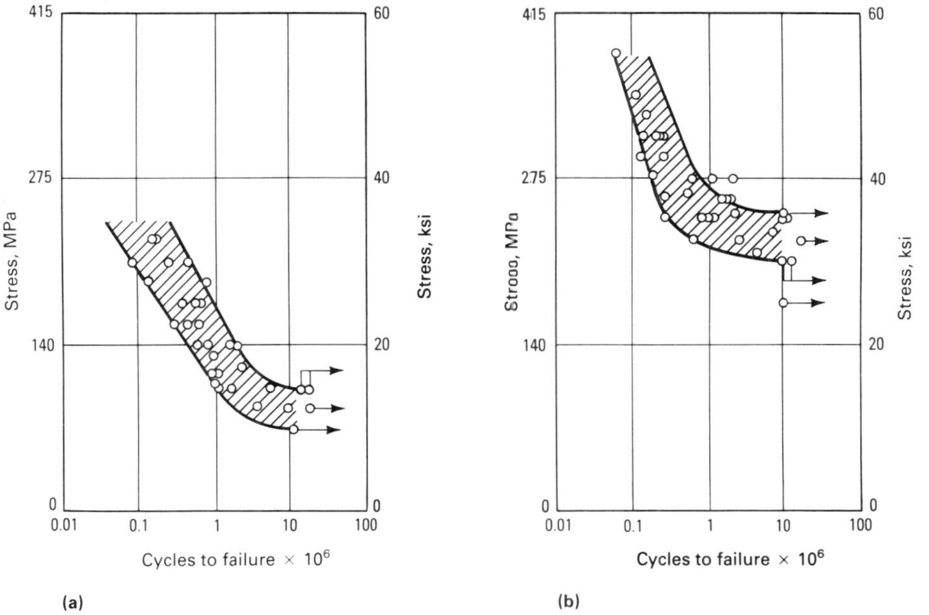


**Fig. 2** Effect of rolling/heat-treating sequence on the fatigue strength of 16 mm (5/8 in.) diam 50B40 steel bolts. (a) Four different lots that were rolled before being tempered to an average hardness of 22.7, 26.6, 27.6, and 32.6 HRC. (b) Five different lots that were rolled after being tempered to an average hardness of 23.3, 27.4, 29.6, 31.7, and 33 HRC. Harder bolts in both (a) and (b) had fatigue strength toward the high side of the ranges shown.

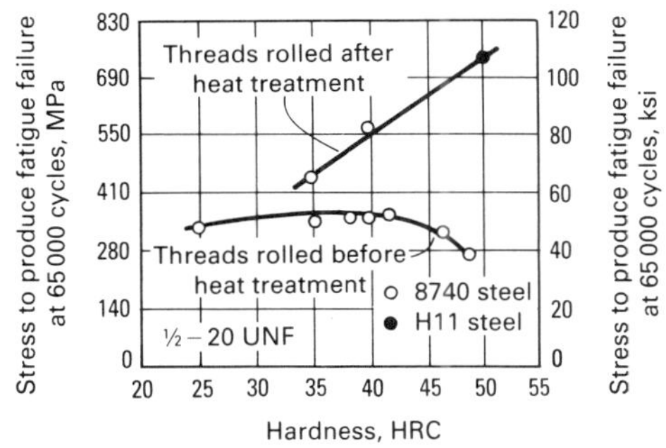


**Fig. 3** Effect of hardness on fatigue strength for bolts with threads rolled before or after heat treatment

### Table 1 Average mechanical properties of hexagonal-head capscrews with rolled and cut threads

Data are based on 15 pieces of each size and each threading method, made of 4027 steel.

| Screw size and pitch | Type of thread | Core hardness(a), HRB Shank | Core hardness(a), HRB Thread area | Tensile strength MPa | Tensile strength ksi | Fatigue life, cycles to failure(b), ×$10^3$ |
|---|---|---|---|---|---|---|
| 7/8–9 | Rolled | 82 | 92 | 631.2 | 91.55 | 71.8 |
| | Cut | 82 | 82 | 489.2 | 70.95 | 14.3 |
| 1–8 | Rolled | 91 | 94.5 | 678.1 | 98.35 | 51.8 |
| | Cut | 91 | 91 | 630.5 | 91.45 | 21.3 |
| 1 1/8–7 | Rolled | 91 | 96.5 | 710.9 | 103.1 | 68.5 |
| | Cut | 91 | 91 | 629.9 | 91.35 | 49.3 |

(a) Core hardness refers to the hardness at the area of the centerline of the thread below the pitch diameter. Converted from Knoop readings. (b) Fatigue test was tension-tension, at 415 MPa (60 ksi) using a preload equal to 10% of maximum.

radial die loads required for roll threading various metals are approximately inversely proportional to the rollability indexes of the metals. Die life is approximately proportional to the third or fourth power of the indexes if final die failure is due to crumbling of thread crests of the die. Thus, if a die life of 30 000 m (100 000 ft) of threaded rod is obtained when rolling steel having an index of 1.00, a die life of about 1800 to 3700 m (6000 to 12 000 ft) can be expected when steel with an index of 0.5 is threaded under the same conditions.

**Flaking** is related to the shear strength of the metal being rolled. Lead and sulfur in brass and steel increase susceptibility to flaking during rolling. An increase in the carbon content of steel decreases susceptibility to flaking. In general, flaking increases directly with the amount of previous cold working of the blank material. This is true of almost all rollable metals, and especially of the work-hardening alloys such as series 300 stainless steel, copper, and some aluminum alloys. Annealing prior to rolling reduces flaking.

Work metals may be classified into four groups with respect to susceptibility to flaking:

- *Group A*: Little or no susceptibility, regardless of whether or not the material was previously cold worked, or regardless of the bluntness of form to be rolled
- *Group B*: Minor susceptibility
- *Group C*: Strong susceptibility
- *Group D*: Excessive susceptibility, which precludes the rolling of all but the most simple, shallow forms

Table 2 indicates susceptibility to flaking for metals most commonly thread rolled. As indicated, all metals listed in Table 2 have either minor or strong susceptibility to flaking. Not many metals can be classified as having little or no susceptibility to flaking. Copper and some of the extremely ductile copper alloys when in the annealed condition are sometimes given the A rating. Metals that contain excessive amounts of free-machining additives, such as the specially prepared screw-machine steels, are likely to fall within the D class. Also, some metals that work harden at an excessive rate (some of the stainless steels and the less-ductile heat-resisting alloys) are likely to fall in the D class.

**Seaming.** If, during rolling, the work metal flows up the flanks of the die threads faster than it does at the center of the thread form, the displaced metal may fold together to form a seam as the metal fills the full crest of the thread form, as shown in Fig. 4. The formation of seams, or folds, depends first on the metal being rolled, and second on the shape of the thread form.

Open seams in the thread crests may occur when undersize blanks are rolled. The open seams can shorten the service life of the thread in a corrosive environment, although they are not usually detrimental in normal service, in which corrosion is less important.

The softer, more ductile metals usually form deeper seams than the harder, less ductile metals. Figure 5 shows the types of metal flow associated with four degrees of seaming. Table 2 indicates susceptibility to seaming for specific alloys commonly used for thread rolling stock.

## Preparation and Feeding of Work Blanks

The diameter of the blank to be threaded is between the major and minor diameters of the thread to be rolled, as shown in Fig. 6. It is common practice to produce blanks for rolling Unified Standard threads with tolerances greater than 0.05 mm (0.002 in.) by extruding, by cold heading, or by shaving on automatic machines. Most of the class 3A threads in the sizes generally used have pitch diameter tolerances greater than 0.05 mm (0.002 in.). Therefore, they can be rolled on extruded, cold-headed, or shaved blanks. Some of the smaller class 3A threads have tolerances closer than 0.05 mm (0.002 in.). Blanks for these threads should be ground.

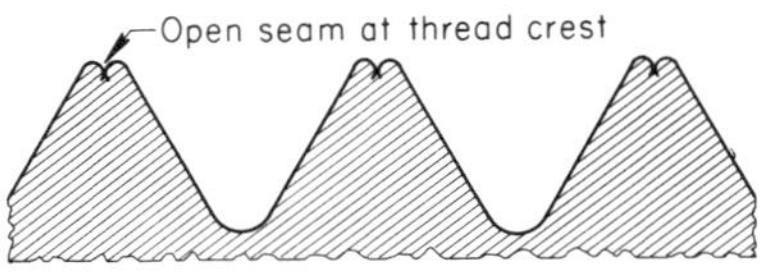


**Fig. 4** Seam at crest of thread caused by faster metal flow along flanks of die thread

To produce threads having a pitch diameter within a tolerance of 0.05 mm (0.002 in.), the tolerance of the blank diameter should be within 0.013 mm (0.0005 in.). Closer thread tolerances can increase the cost of blank preparation, sometimes far beyond the usual costs.

Blanks with close-tolerance diameters are ground. Blanks of material not suitable for extruding or cold heading, such as titanium and some stainless steels, are also ground.

Blank diameter must be within the tolerance required for the particular size and class of thread specified. It is not practical to roll threads to a close tolerance except on blanks held to appropriate diameter. Overrolling an undersize blank to provide a screw of correct size causes premature die failure. Maximum die life is obtained when the crest is not rolled full. When rolling a class 2A thread, for example, the most economical procedure is to use a blank with a diameter such that the thread can be rolled to the mean class 2A pitch diameter (halfway between the high and low limit of pitch diameter) and to maintain the major diameter of the thread just above the lower tolerance limit.

**Relation of Blank and Pitch Diameters.** The thread form is said to be balanced when the volume of cavity below the pitch line (Fig. 7) is equal to the volume of metal above the pitch line. (Pitch line is defined as the location at which the widths of the thread ridge and the thread groove are equal.) In such threads, the correct blank diameter is substantially the same as the pitch diameter. With metals that are commonly cold headed, the maximum blank diameter is generally equal to the mean pitch diameter. The optimum blank diameter varies with different work metals; some adjustment may be needed to get the desired crest formation. Additional dimensional allowance is necessary if subsequent plating is planned. If the form to be rolled is unbalanced (Fig. 7), the blank diameter is not the same as the pitch diameter.

**Bevel on Blanks.** Ground and extruded blanks are made with a bevel at the end or ends of the section to be threaded. This bevel ranges from 15° to 45°. Blanks for the majority of thread rolled products have a bevel angle of 30°. Thread rolling, however, increases the bevel by 15° to 30°. Therefore, the bevel angle on the blank must be less than the angle desired on the finished part.

In addition to the bevel angle, some blanks have a bevel in the area that connects the section to be threaded with the section that will remain unthreaded. This bevel, sometimes called the extruding angle, is usually 30°.

**Feeding.** The various thread rolling methods employ three basic techniques for feeding the blank into the dies:

- *Radial infeed*: The die, usually cylindrical, moves in a radial line directly toward the axis of rotation of the workpiece
- *Tangential feed*: The die, either cylindrical or flat, moves past the workpiece on a path that brings the pitch line of the thread form tangent to the work surface

**Table 2 Rollability of alloys**

| Metal | Hardness, HB | Rollability index(a) | Flaking tendency(b) | Seaming tendency(c) |
|---|---|---|---|---|
| **Carbon and low-alloy steels** | | | | |
| 1010 | 137 | 1.11 | B | C |
| 1018 | 148 | 1.08 | B | C |
| 1020 | 156 | 0.96 | B | C |
| 1095 | 260 | 0.47 | B | B |
| 1095 | 320 | 0.42 | B | B |
| 1112 | 198 | 1.00 | C | C |
| 1117 | 173 | 1.03 | C | C |
| 1144 | 225 | 0.78 | B | C |
| 4140 | 205 | 0.93 | B | C |
| 4140 | 234 | 0.57 | B | C |
| 4140 | 300 | 0.42 | B | B |
| 4340 | 235 | 0.45 | B | B |
| 8620 | 215 | 0.60 | B | C |
| **Stainless steels** | | | | |
| 303 | 174 | 0.46 | C | B |
| 316 | 150 | 0.45 | B | B |
| 416 | 221 | 0.58 | C | B |
| 430 | 225 | 0.56 | C | B |
| **Nonferrous alloys** | | | | |
| Aluminum, 2017 and 2024 | 135 | 1.40 | B | C |
| Brass | | | | |
| Cartridge | 190 | 1.55 | B | B |
| Naval | 155 | 1.00 | C | B |
| Phosphor bronze | 130 | 1.28 | C | B |
| Monel | 235 | 0.93 | B | B |

(a) Index applies to metals rolled at room temperature. (b) B, minor susceptibility; C, strong susceptibility. (c) B, negligible susceptibility; C, moderate susceptibility. See Fig. 5, which also indicates the two extremes of seaming tendency (A and D).

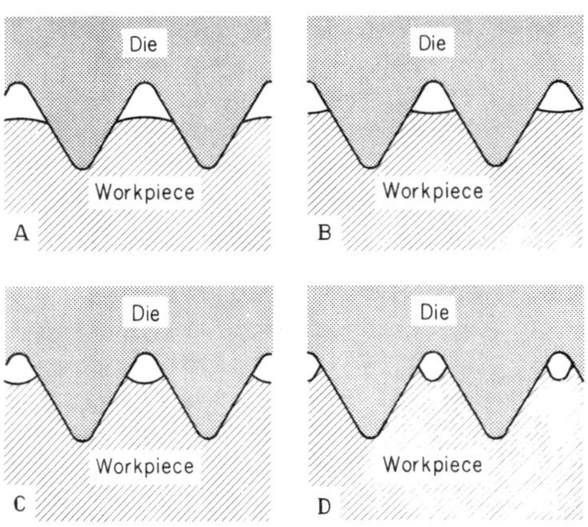


**Fig. 5** Four degrees of susceptibility to seaming during thread rolling. (A) Negative susceptibility to form seams. Metal flow adjacent to the die surface is slower than in the middle of the roll form. This is characteristic of metals having a relatively high coefficient of friction with the die steel. (B) Negligible susceptibility to form seams. Metal flows up with an almost flat top during rolling in conventional thread forms. (C) Moderate susceptibility to form seams, typical of low-carbon steels. (D) Excessive susceptibility to form seams. Cavity is likely to be formed under crest of thread.

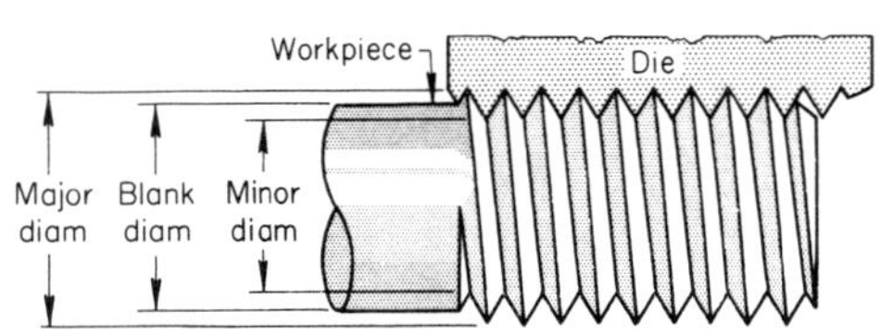


**Fig. 6** Relation of blank diameter to major and minor diameters of threads

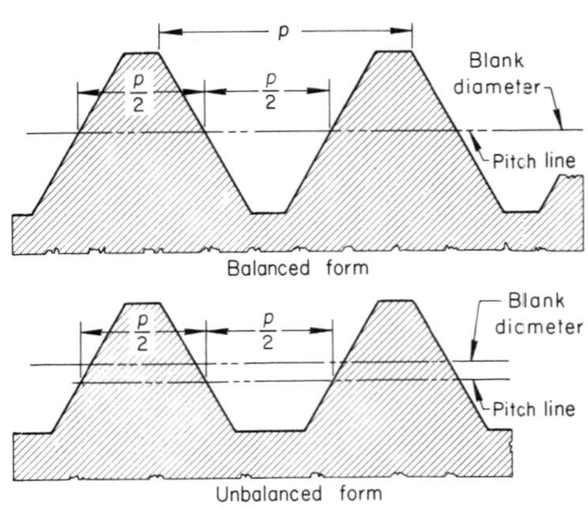


**Fig. 7** Relation of blank diameter to pitch line for balanced and unbalanced thread forms

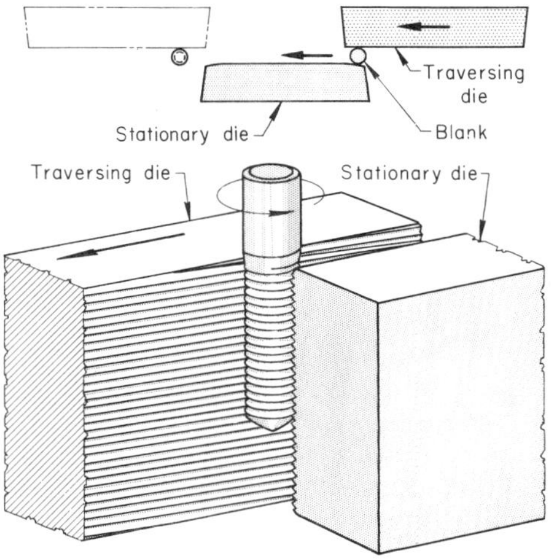


**Fig. 8** Operating principle of flat traversing die thread rolling

- *End feed or through feed*: The cylindrical die "tracks" on the workpiece causing the workpiece to move axially as it rotates

## Flat Die Rolling

Flat traversing dies are the type most commonly used for rolling threads in commercial fasteners and similar parts. One technique involves the use of two flat, rectangular dies; one is stationary, and the other traverses in a plane parallel to the stationary die and separated from it by a distance equal to the minor diameter of the thread to be rolled.

In another technique, both dies traverse the workpiece. Figure 8 shows the most common arrangement of dies and workpiece. As the blank is forced into the space between the dies by a feed finger, it is engaged by the forward motion of the traversing die and caused to roll between the threaded faces of the dies. This action forms the thread.

A thread is rolled on one blank at a time during the forward stroke of the machine. There is no appreciable axial movement of the blank during rolling. The diameter of the finished thread is controlled by the diameter of the blank and the distance between the faces of the dies at the end of the stroke.

**Machines.** The generic term flat die rollers encompasses a large and varied family of machines that are made in a number of sizes, each for a limited diameter range and with a specified die length. As a general rule, the correct machine size accommodates a die length that allows the blank to complete from six to eight full revolutions.

Most flat die rollers have the dies in the side-by-side position shown in Fig. 8. Most of the newer machines also have dies that traverse in the horizontal plane, but with the faces of the dies at an angle to the vertical. The feed track is at an angle, thereby permitting gravity feed down an incline into the inserting mechanism.

Specialized types of vertical and horizontal rolling machines also are available. One horizontal traversing unit has dies placed one above the other rather than in the usual side-by-side position. It is used to roll splines and related forms as well as threads. Another variation of the horizontal traversing machine uses an inclined feed chute at right angles to, rather than parallel with, the axis of traverse. This type of machine incorporates a different method of die alignment. Dies are adjusted longitudinally to match the threads of the two dies, rather than vertically by shims, as in most other machines.

**Dies** used in flat die rollers consist of matching pairs of rectangular plates, with each of the opposed faces having a reverse image of the form to be produced on the part. Dies are made in various widths and are used to roll screws of any thread length up to the maximum die capacity.

The lead angle of flat dies can theoretically vary from 0° to 45°, but for producing most standard screw threads, it is less than 5°. (Lead angle of the die is defined as the angle between the thread form and the longitudinal axis of the die.)

In flat die design, penetration rate is primarily governed by the length of the die. Best practice calls for complete penetration prior to the last revolution of the workpiece in the die. The last complete revolution of the workpiece should only iron out small irregularities. It is important that the die be long enough to prevent an excessive penetration rate.

The above principles can be used to roll more than one form on a part by use of a multiple stack of dies, which are inserted one atop the other in the machine and held together with a clamp. Thus, many combinations of forms that would be impossible to generate on a one-piece die can be easily produced on a multiple die, provided that the spread between diameters of the individual forms to be rolled on the same part is minimal. In addition to the dies, secondary tooling is required for feeding, sorting, orienting, and inserting the parts between the dies.

**Capabilities.** The flat-die process is commonly used for all types of straight- and taper-threaded commercial fasteners. Flat die rolling can produce more than one form in one operation, such as two entirely different types of threads at opposite ends of a part, knurling and a thread, or knurling and an annular groove, on the same part.

Duplex face dies can be used for rolling straight threads. Such dies have threads on both the front and back sides so that they provide two rolling surfaces. When the screw length is less than half the die width, the die can be reversed, top for bottom, so that four rolling edges are available for still greater economy.

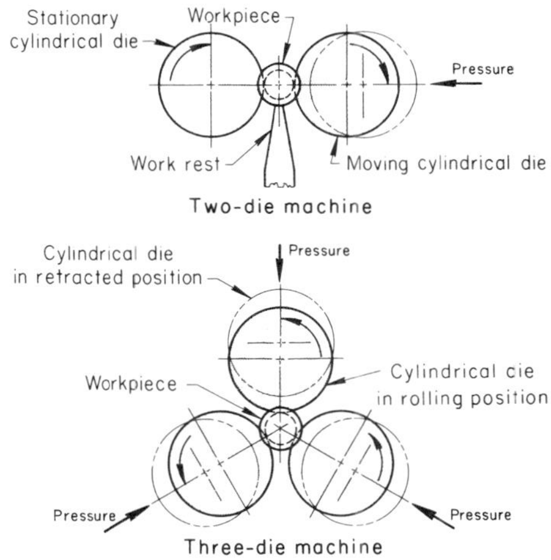


**Fig. 9** Operating principle of radial-infeed cylindrical-die thread rolling

Production rates vary widely and usually are inversely proportional to the size of the product. The small machines are capable of producing parts at a rate of 10 000 to 36 000 per hour. Larger units, producing 9.5 or 13 mm (3⁄8 or 1⁄2 in.) bolts, can roll 3000 to 12 000 pieces per hour. Products such as 32 mm (1 1⁄4 in.) diam bolts are thread rolled at much slower rates, ranging from 900 to 3000 pieces per hour.

**Limitations.** In general, flat dies are used for threading metals no harder than 32 HRC before rolling, although steel as hard as 52 HRC can be roll threaded. Thread diameters are commonly limited to 25 mm (1 in.), although a few machines can roll up to 38 mm (1 1⁄2 in.) diam threads. Thread lengths up to 265 mm (10 1⁄2 in.) are rolled. These limitations—hardness, diameter, and length—are interrelated so that a workpiece having more than one or two of these measurements near maximum value may not be rollable.

The flat-die method is also limited to parts of an overall size that can be accommodated in the machine. Because of interference between the part and elements of the machine and die, part size as well as thread dimensions must be considered before the flat die method is selected for a specific piece.

## Radial-Infeed Rolling

Radial-infeed thread rolling consists of moving a rotating cylindrical die or dies radially toward the center of the rotating workpiece. The operating principle of this method is shown in Fig. 9.

A minimum amount of axial movement between the dies and the workpiece occurs during the rolling cycle. This characteristic distinguishes infeed rolling from the through-feed method of cylindrical-die thread rolling. Axial movement is canceled by designing the die with an effective lead

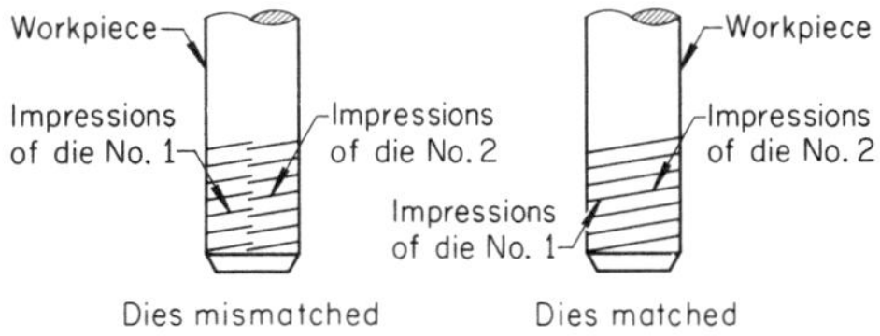


**Fig. 10** Effect of mismatched and correctly matched dies on thread impression made by cylindrical dies

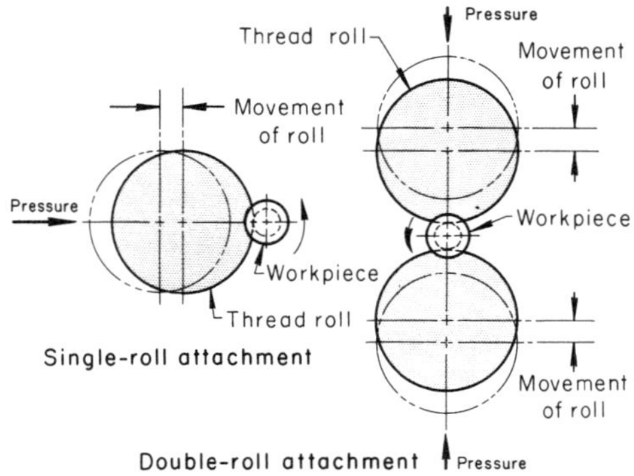


**Fig. 11** Operating principle of two types of radial attachments for thread rolling on lathes and automatic bar machines

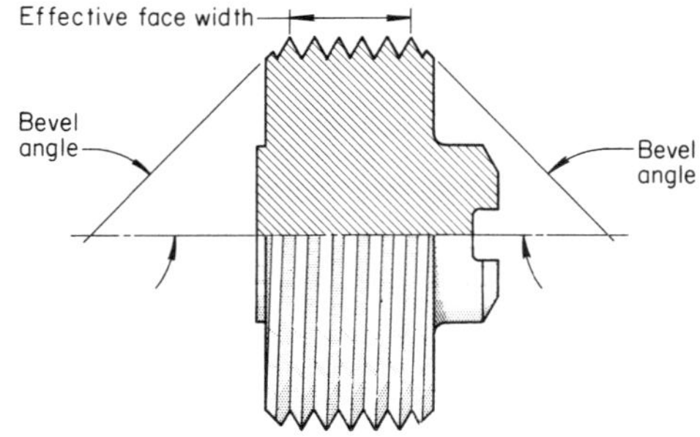


**Fig. 12** Bevel on a cylindrical thread rolling die

angle equal in magnitude but opposite in direction to that on the work. In rolling a 9/32-32 double-lead worm thread, movement would be as much as 4.8 mm (3/16 in.) during the rolling cycle if there were no compensating lead on the die. Axial movement does not affect thread quality, but it may restrict the ability to produce a full thread close to a shoulder and will reduce the amount of full thread that can be produced with a special die face.

The effective lead angle varies slightly during die penetration so that some axial movement does occur. The amount of movement is usually insignificant when rolling standard threads, but can be considerable when rolling blunt or very deep thread forms, or those with large lead angles.

Dies can be designed to give slight axial movement to the blank to increase die life or to simplify regrinding of the dies. Movement of 3.2 to 6.4 mm (1/8 to 1/4 in.) is common; movement of up to 13 mm (1/2 in.) has advantages in some applications.

When two or three dies are used, they must be matched so that the helical path produced by one die is a continuation of that produced by the other die or dies; otherwise, there will be steps in the product thread (Fig. 10). Dies are matched by rotating one or more dies in relation to the others, or by moving one or more dies axially, to produce a continuous helix on the work.

**Cylindrical-die machines** capable of infeed thread rolling are equipped with either two or three dies (Fig. 9). Two-die machines are usually of the horizontal type; that is, the workpiece is horizontal during rolling. Three-die machines are available in both horizontal and vertical models.

In two-die machines, either one or both dies can move radially. If one die does not move radially, the work must move radially during die penetration. If both dies move equally, the work can stay in place.

Most three-die machines provide equal radial movement of all dies so that the work position does not change during rolling. A few three-die machines have one or two stationary dies, and radial movement of the work must be allowed for.

Two-die machines require a work support to position the centerline of the work slightly below (usually about 0.25 mm, or 0.010 in.) the same plane as the centerline of the dies. This offset prevents the workpiece from rising out of the dies. A work rest, as shown in the two-die machine of Fig. 9, can be used for short, manually loaded parts. Larger parts may require additional supports. In many instances, the parts can be inserted in a tube or bushing, or held between centers, to provide proper positioning for rolling. A spring-loaded work stop can be used for positioning the part in proper axial location.

The dies in a three-die machine serve to locate the part so that often the only fixture required for manual loading is a spring-loaded work stop to provide correct axial position.

**Lathes and Automatic Machines.** Radial infeed rolling can be done in lathes and automatic bar machines equipped with single-roll or double-roll radial threading attachments. In these attachments, the cylindrical dies are commonly called thread rolls. Figure 11 shows the operating principle. The single-roll attachment is a simple roll or knurl holder mounted on a cross slide. As the cross slide advances, the roll is pressed into the workpiece so that it rotates with the work and thus forms the thread. The travel of the cross slide is controlled so that the thread roll in its final position produces a thread of correct size. After the cross slide completes its full length of travel, it is rapidly retracted.

Double-roll radial attachments operate by means of a toggle arrangement that causes the rolls to close approximately radially to contact the rotating work, at which time the rolls begin to turn to form the thread. After reaching full depth, the rolls and attachment are rapidly retracted.

**Dies (Rolls).** A little axial movement occurs between the work and the cylindrical dies, and it is necessary that the threaded length of the dies be about two to three threads per product lead longer than the length of the thread to be rolled. In practice, it is desirable to provide a bevel at each end of the die (Fig. 12); thus, the width of the die face must exceed the length of the thread to be rolled by twice the width of a bevel.

Long bevels with a small bevel angle increase die life by reducing breakage at the edge of the die. Usually 30° bevels are recommended, although workpiece requirements may necessitate 45°. For the harder workpieces, bevel angles of less than 30° are desirable; they may need to be as small as 15° for the hardest rollable metals.

**Die Size.** For infeed rolling, the pitch diameter of the die must be a multiple of the pitch diameter of the finished workpiece. For single-lead threads, the number of thread starts is equal to the ratio of die-to-work pitch diameters. For rolling multiple-lead threads, the number of die thread starts equals the number of thread starts on the work multiplied by the ratio of die-to-work pitch diameters.

The diameter of dies for two-die machines is limited by the size of the machine and fixtures and is approximately constant regardless of work diameter. For instance, a typical two-die machine having the capacity for 1.6 to 38 mm (1/16 to 1 1/2 in.) workpieces uses dies that are 125 to 150 mm (5 to 6 in.) in diameter for all sizes of work. In general, but not necessarily, two-die machines use dies of larger diameter than three-die machines for threading the same size of workpiece.

Three-die machines generally cannot use dies larger than about five times the work diameter, because larger dies will contact each other before reaching full thread depth in the work. Slightly larger dies can be used for rolling multiple-lead threads. The size of thread rolls for radial infeed lathe attachments is determined by the dimensions of the attachment being used, rather than by the work diameter.

**Supports for Die Spindles.** Because of the limitation on the die-to-work diameter, three-die machines require a series of spindles and spindle supports, graduated in size, to accommodate the entire work-diameter capacity range. For small-diameter work, the spindles are necessarily quite slender and may not be strong enough to roll hard alloys or long thread lengths.

**Capabilities.** The minimum practical diameter of a workpiece for rolling in two-die machines or attachments is 1.3 mm (0.050 in.). The maximum diameter is limited only by the capacity of available equipment. Two-die machines capable of rolling threads 380 mm (15 in.) in diameter and 400 mm (16 in.) long are in use.

Three-die machines roll threads from 6.4 to 115 mm (¼ to 4½ in.) in diameter and up to 125 mm (5 in.) long. They are seldom practical for rolling threads smaller than 6.4 mm (¼ in.) in diameter.

The rate of die penetration into the work is adjustable; thus metals of various hardnesses can be threaded. Most metals threaded by cylindrical dies or attachments are no harder than about 32 HRC. However, work metals as hard as 52 HRC have been thread rolled on cylindrical dies. Die life does deteriorate rapidly when material harder than 32 HRC is rolled.

The versatility of the radial infeed method makes possible the rolling of thin-wall parts such as tubing or stampings and also some metals harder than 48 HRC that would be difficult or impossible on other types of machine. Rolling of thin-wall parts is discussed in a subsequent section of this article.

The radial infeed method permits rolling of threads close to shoulders, with a minimum of imperfect threads. Also, threads can be rolled between two sections of larger diameter, as is common when rolling worms on large transmission shafts. Threading close to shoulders is considered in the section "Rolling Threads Close to Shoulders" in this article.

The three-point support provided by three-die machines is advantageous for rolling parts with irregular or unbalanced overhangs and parts requiring a thread length considerably shorter than one diameter. Often such parts can be rolled while being supported only by the three dies, whereas cylindrical two-die machines or other types of threading equipment may require expensive or unwieldy fixturing.

Radial-feeding single-roll or double-roll attachments are used to best advantage for

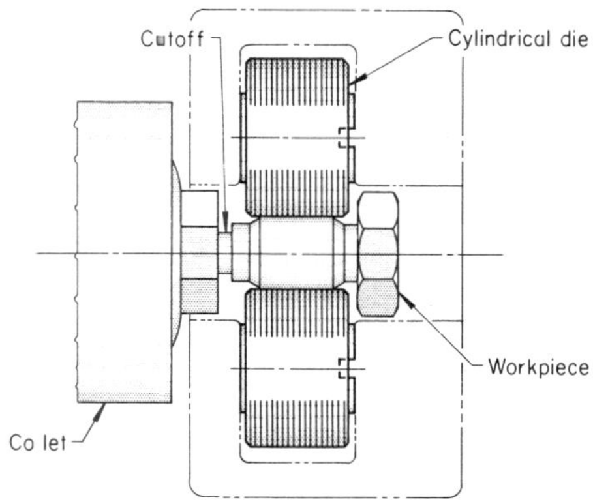


**Fig. 13** Use of a double-roll attachment for thread rolling near the collet of an automatic bar machine

rolling threads at the collet end of pieces being machined in a lathe or automatic bar machine (Fig. 13). Such threads are usually behind a shoulder making radial infeed and tangential feed rolling (discussed below) the only practical thread rolling methods.

Because double-roll attachments exert a minimum of transverse pressure on the work, small diameters of considerable length can be rolled, and at a much greater distance from the collet than with the single-roll type of attachment. For example, a double-roll attachment can roll a ½-20 UNF, 19 mm (¾ in.) long thread on a 13 mm (½ in.) diam bar at a distance of 25 mm (1 in.) from the collet to the first thread. With a single-roll attachment, maximum distance from the collet is 6.4 mm (¼ in.).

Production capabilities of cylindrical-die infeed machines vary according to the method and equipment for feeding the work and the type of work involved. In general, hand feeding is not practical at rates above 25 pieces per minute. Automatic feeding equipment can be installed on two-die or three-die machines to produce up to about 90 pieces per minute. Typical applications of cylindrical-die infeed thread rolling are shown in Table 3.

**Limitations.** The axial travel caused by thread rolling machines when they are producing deep, blunt thread forms or high lead angles is not objectionable except when it causes interference during rolling of a thread close to a shoulder. In single-roll or double-roll lathe attachments, the work cannot move axially, and roll movement is usually limited to approximately one-third of the pitch so that rolls must be designed to cause a minimum of axial travel. The rolling of deep, blunt forms or high lead angles requires careful attention to the roll design and also to the rate of penetration of the rolls, the accuracy of the setup, and the condition of the attachment.

The accuracy of threads rolled by single-roll infeed attachments depends on the accuracy of the cross-slide travel. Thread accuracy is limited also because of the bending action developed by the force of radial-infeed rolling. In some applications, back-up rolls, bearing on a plain cylindrical surface outside the threaded area, can be used to reduce the bending. If a plain surface is not available or machine tooling does not permit the use of backup rolls, single-roll attachments are restricted to thread rolling near the collet and to short thread lengths in the softer materials.

## Tangential Rolling

Tangential thread rolling is similar to infeed rolling except that the dies (rolls) are fed past the blank on a path parallel to the

**Table 3 Typical applications of cylindrical-die thread rolling**

| Product | Steel | Thread | Length, mm | Length, in. | Insertion | Die speed, rev/min | Rate, pieces/min | Die life, pieces |
|---|---|---|---|---|---|---|---|---|
| **Infeed rolling** | | | | | | | | |
| Adjusting worm(a) | 8620 | 9/16–10 | 29 | 1⅛ | . . . | 85 | 8–10 | . . . |
| Armature shaft (worm)(b) | 1045 | 1.7 mm (0.065-in.) pitch | 19 | ¾ | Manual | 45 | . . . | 6 000 |
| Armature shaft (worm)(c) | 1040 | 0.2805–32 | . . . | . . . | Manual | . . . | 6 | . . . |
| Double-end stud | 1018 | 1½–8 UN-2A | 51 | 2 | Manual | . . . | 4(d) | . . . |
| Feed screw(e) | 410 | 0.330–56 Acme | 95 | 3¾ | . . . | 170 | 8–10 | . . . |
| Feed screw(a) | 410 | ⅜–33⅓ Acme LH | 56 | 2 7/32 | . . . | . . . | 10 | . . . |
| Worm | 4140 | Buttress(f) | . . . | . . . | Manual | 85 | 8–10 | . . . |
| Worm | 8620 | ¾–5; 4 starts | 38 | 1½ | Manual | . . . | 8–10 | 20 000 |
| **Through-feed rolling** | | | | | | | | |
| Threaded rod | 1018 | ⅜–16 UNC | 1800 | 72 | Automatic | 500 | 6 | 260 000 |
| Jackscrew(g) | 1018 | ½–10 Acme stub | 210 | 8¼ | Manual | 260 | . . . | 25 000 |
| Setscrew | (h) | ⅜–19 BSP F(h) | 30 | 1.2 | Hopper | . . . | 300 | . . . |
| **Automatic continuous rolling** | | | | | | | | |
| Automotive stud | 8115 | ½–20 UNF-2A | 16.3 | 0.640 | Hopper | . . . | 125 | 500 000 |
| Double-end stud | 1335 or 1041(j) | 5/16–24 UNF-3A(k) | | (k) | Hopper | . . . | 200(d) | . . . |
| Double-end stud | 3135(j) | ⅝–11 UNC(m) | | (m) | Hopper | . . . | 80(d) | . . . |

(a) Pressure angle, 14½°. (b) Pressure angle, 20°. (c) Pressure angle, 25°; two leads on thread. (d) Both ends. (e) Pressure angle, 29°. (f) Modified buttress; 2 starts; 11½ threads per inch; flanks 10° and 30°. (g) Infeed and through-feed rolling. (h) British Standard Parallel Fastener; made of resulfurized screw stock. (j) Cold drawn, cut to length, and ends extruded to blank diameter. (k) Thread length, 14 mm (9/16 in.) on each end of 117 mm (4⅝ in.) long stud. (m) One end class 3A, 29 mm (1⅛ in.) long; the other end, class 5A, 25 mm (1 in.) long; total length of stud, 70 mm (2¾ in.)

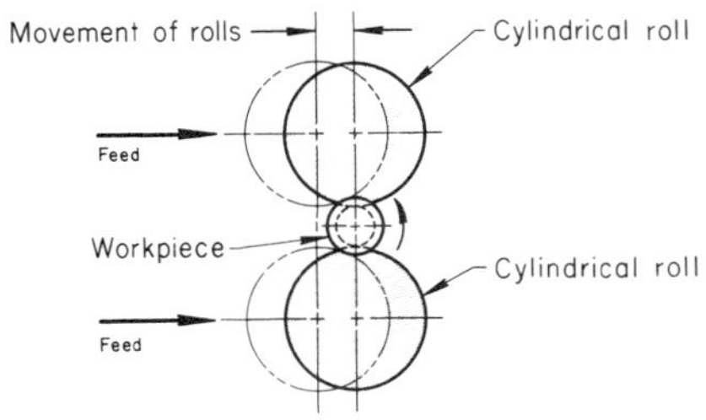


**Fig. 14** Operating principle of tangential-feed thread rolling

radial path at a distance such that when the axis of the roll is opposite the axis of the blank the pitchline of the thread form is tangent with the surface of the blank. Figure 14 shows the operating principle of this process. As the rolls advance, they reach maximum penetration when the centerline of the rolls is directly opposite the centerline of the work. The total depth of penetration is determined by the amount the rolls are offset in relation to the work. As in radial-infeed rolling, only a slight axial movement occurs between the rolls and the work.

**Machines.** Tangential rolling is done in lathes or automatic bar machines equipped with one-roll or two-roll attachments mounted on a cross slide of the machine. The rolls are rotated by their contact with the rotating work. Two-roll attachments are the most common for tangential rolling. They are available in various sizes; each size has a capacity for a range of work diameters. Capacity up to a work diameter of 65 mm (2½ in.) is commonly available, and larger sizes are obtainable for special applications.

**Capabilities.** Tangential feeding attachments have essentially the same capabilities as radial feeding attachments. Some advantage is gained with tangential attachments because the adjustments and control for a given size of work are made within the attachment rather than by the travel of the machine slide. With a two-roll tangential attachment, no radial movement occurs between the roll spindles during rolling, and rolling pressures are greater than in radial-infeed rolling.

Threads can be rolled at spindle speeds compatible with other machining operations; therefore, speed changes for threading are unnecessary. Two-roll tangential rolling produces bending loads somewhat higher than the two-roll radial-feeding attachment, but low enough to allow the rolling of threads on relatively hard work metals at a considerable distance from the collet. For instance, a 19 mm (¾ in.) long ½-20 UNF thread can be rolled on a 13 mm (½ in.) diam bar hardened to 30 HRC, at a distance of 16 mm (⅝ in.) from the collet to the first thread.

**Limitations.** Thread rolling with a two-roll tangential attachment is limited primarily by the size and capacity of the equipment. Rolling threads with high lead angles or with deep, blunt forms can be troublesome because of axial travel. However, with proper attention to roll design and setup, using only attachments in very good condition, the problem can usually be overcome.

Single-roll tangential rolling produces transverse loads capable of bending the workpiece. Therefore, this procedure is limited to rolling threads of short length near the collet in soft materials such as nonferrous alloys or soft steel (generally no greater than 197 HB). In rolling a ½-20 UNF thread next to the collet on a 13 mm (½ in.) diam bar of 1112 steel, the maximum practical thread length would be approximately 22 mm (⅞ in.). As with radial single-roll attachments, support rolls can sometimes be used to increase this capacity.

## Through-Feed Rolling

In through-feed rolling, the work moves axially through the dies. Through-feed dies are designed with a lead angle generally different from that of the work, so that the part can feed. The dies are made with a starting taper, so that the thread is formed progressively as the blank feeds through the dies. The finish end of the dies also is tapered slightly so that rolling pressure is released gradually without marking the work.

**Feed rate,** in terms of feed per work revolution, is proportional to the ratio of the difference in lead angles of the dies to the lead angle of the work. The die lead angle can be either greater or less than that of the work and may be zero (annular form dies). Also, the die lead angle can be the same hand as the work (right-hand lead dies to produce right-hand threads).

**Machines.** Any cylindrical-die thread rolling machine is capable of through-feed rolling, but the capacity may be restricted. Vertical three-die machines, for example, can feed only short lengths because the gearbox or other equipment located a short distance below the dies prevents passage of longer work. In three-die horizontal machines, a passage through the gearbox usually allows unlimited length, but restricts the diameter of long work. Obstructions behind the dies of some two-die machines limit the length or diameter of the work.

Machines having no provision for skewing the spindles are limited to parallel-axis through-feed rolling, for which the feed rates are low. Most machines, of both two-die and three-die types, have infinitely variable die tilt adjustments up to 10°. This permits the use of annular dies, which are less expensive and more productive than helical dies.

Most conventional cylindrical thread rolling machines have one or all dies mounted on slides or pivoting arms for infeed rolling. For through-feed roll threading of the full length of a blank, the dies are held in the closed position by a hydraulic or mechanical system. Such machines can also be used for rolling threads on only a portion of the blank, either by feeding through until the desired length has been threaded, and then opening the dies, or by inserting the work between the open dies to the correct position, and then closing the dies so that the work feeds out of the dies. Some machines designed specifically for through-feed rolling of the entire workpiece are not equipped with die advance-retract mechanisms that permit partial-length rolling.

**End-Feeding Attachments.** Cylindrical through-feed rolling can be done with two-die and three-die end-feeding heads. Two-die heads are used for small work. The rolling heads can be mounted on the rotating headstock of a bolt-threading machine in which the blank is clamped to a slide that advances the work toward the head. When used in an automatic bar or chucking machine, the head is mounted on the tool slide, which advances it toward the rotating workpiece.

The dies are made with annular grooves, and the axes of the dies are set at an angle with the work axis equal to the required lead angle of the product thread. The rate of feed per revolution is equal to the lead of the thread being rolled. Some heads have interchangeable frontplate units that can accommodate a range of lead angles. In other heads, lead angles are varied by interchangeable bushings. End-feeding attachments are adjustable to produce correct thread size.

With the dies locked in the closed position, they engage the blank and roll the thread as they pass over the blank or the blank passes between the dies. When the desired length of thread has been rolled, the head opens, and the work is withdrawn from the head. When used for continuous rolling, the head remains in the closed position at all times.

**Dies** for through-feed rolling are usually relieved at both ends. Through-feed dies for Acme, worm, or other wide threads often have a modified, pointed thread form at the starting end for efficient penetration into the blank.

The number of thread starts, which, together with the diameter of the die, determines the lead angle, is different for through-feed rolling on cylindrical machines than for infeed rolling a similar size thread. Compared to infeed dies of similar diameter for a specific thread, the through-feed dies may have more or fewer starts for parallel-axis rolling, or no thread starts (annular form) for skewed-axis rolling.

**Capabilities.** Through-feed rolling is applicable to threading the full length of a cylindrical part of uniform diameter and to threading one or more sections of the largest diameter of a multiple-diameter cylindrical part. End-feeding heads are used for straight and tapered threads. Annular rings can be produced by through-feed rolling in skewed-axis dies or rolls. Typical examples of parts made by through-feed rolling are commercially threaded rod, high-strength studs, headless set screws, threaded mounting tubes for electrical fittings, pole line hardware, recirculating ball screws, and jackscrews of all types. Threads are through-feed rolled to partial length on parts such as compressor studs, large-diameter cap screws, clamp and jackscrews, finned heat-exchanger tubing, and reinforcing rods. Three specific applications of through-feed rolling are given in Table 3.

Most thread rolling machines and heads except the three-die vertical types are virtually unrestricted in the length of bar that can be threaded. Mill length bars or tubes 3 to 5 m (10 to 16 ft) long are commonly threaded. Heat exchanger tubing is through-feed rolled to produce integral fins in lengths up to 15 m (50 ft). Sometimes it is economical to through feed parts of short length. For example, blanks for socket set screws and various types of studs are thread rolled at high production rates.

Rods or parts up to 6.4 mm (1/4 in.) in diameter can be through feed rolled at speeds up to 20 m/min (800 in./min) depending on the type of machine, available horsepower, and hardness of the blanks. Larger sizes will feed slower; 300 mm/min (12 in./min) is a typical speed for 75 mm (3 in.) diam steel parts at a hardness of 22 HRC.

In many applications, a larger thread can be rolled in a given machine by partial-length through-feed rolling than by infeed rolling. Threads of very small diameter, such as 0-80 UNF, can be through-feed rolled on two-die machines or end-feeding heads. Three-die machines usually cannot roll threads smaller than about 6.4 mm (1/4 in.) in diameter because of die interference. Machines are available for 115 mm (4½ in.) diam work up to about 0.6 m (2 ft) long.

Standard end-feeding attachments are available for threads up to 230 mm (9 in.) in diameter. Feed rates depend on the maximum speed at which the head or work can be rotated with the available horsepower. Feed rates of 7.6 m/min (300 in./min) are common for thread sizes up to 16 mm (5/8 in.) in diameter. Most standard heads have clearance holes through the center so that long thread lengths can be rolled. The maximum length depends on the ability of the equipment to grip the workpiece to prevent rotation or excessive torsional windup of the piece.

All of the common thread forms can be through-feed rolled, including blunt forms such as Acme and worm threads and ball-screw forms. Threads of very blunt form are produced with less difficulty by through-feed rolling than by infeed rolling.

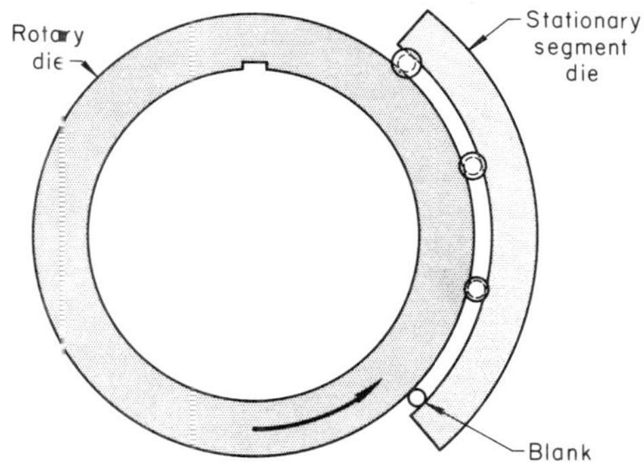


**Fig. 15** Operating principle of planetary thread rolling. More than one segment die may be used.

**Limitations.** Like other types of cylindrical rolling, the process is limited primarily by the characteristics of the equipment being used.

## Planetary Thread Rolling

Planetary thread rolling machines have one central rotating die on a fixed axis and one or more stationary concave segment dies located near the outside of the rotary die, as shown in Fig. 15. One or several blanks may be rolled on a segment die at one time, depending on the gearing of the starting mechanism.

The starting end of the segment is adjusted so that the blank contacts both dies. As the rotary die revolves, the blank is rolled between the dies until it traverses the full arc of the segment die, after which it drops out of the threading area.

For most applications, the finish end of the segment die is adjusted to produce the desired thread size. However, when rolling easily work-hardened parts, the dies may have to be adjusted so that the starting end of the die does most of the work and the finishing end does little, the part being completely formed by the time it is halfway through the die. Final thread size, however, does not depend entirely on correct adjustment of the die; hardness and size of the blank can cause variations. The effect of blank hardness is illustrated by the example that follows.

**Example 1: Effect of Hardness of Blank on Thread Dimensions.** The effect of blank hardness on thread dimensions (primarily pitch diameter and major diameter) was investigated in an effort to reduce the rejection rate of 4-40 UNC-2A machine screws. (Rejection was for undersize pitch diameter and major diameter.) The screws, which were made of 1038 steel cold heading wire, were produced in an automated planetary die threader. Blank diameter was 2.35 to 2.36 mm (0.0925 to 0.0930 in.). A rejection rate of not more than 0.1% was accepted.

Blanks of four different hardness levels were threaded: 60 HRB, 95 HRB, 28 HRC, and 32 HRC. Blank diameter was held within specified limits, and the same type of lubricant was used for all tests. Results are shown in Fig. 16. Some endwise stretching in the softer blanks was observed, probably because the part was short and of small diameter. The harder blanks work hardened perceptibly, resulting in undersize thread dimensions and shorter die life. Other details of the threading operation are listed in the table that accompanies Fig. 16.

**Machines.** Planetary machines are made in several sizes, each having a different maximum rolling capacity. Although machines are usually rated on the basis of nominal work diameter, the blank hardness and length of thread rolled have considerable influence on the practical capacity of a machine.

The basic planetary machine is comparatively simple, consisting of a spindle for the rotary die; a mounting block for the segment die, which also provides size adjustment; and a starting mechanism for inserting the blanks. The starting finger is adjusted by means of gearing or an adjustable cam, so that the workpiece is inserted at the exact point on the rotary die where it is in match with the segment. There is one such match point for each thread start on the rotary die. The number of thread starts varies from 10 to more than 100; however, it is seldom possible to feed a blank into the die at every thread start. Planetary machines usually can feed from three to eight parts per die revolution, depending on size; five pieces per revolution is common for this mass-production process.

To make use of the high production capacity of planetary machines, automatic feeders are essential and are generally supplied as an integral part of the machine.

**Dies.** Lead angles are similar to those of radial infeed and flat dies; therefore, axial travel is at a minimum. Bevels are similar to those used on other types of dies.

Planetary dies vary in diameter from 100 to 350 mm (4 to 14 in.); the most commonly used machines have dies approximately 180 mm (7 in.) in diameter. The segment die has an inside radius (IR) equal to or slightly greater than the sum of the rotary die radius and the minor diameter of the threaded part. The width of the dies can be the maximum accepted by the machine or can be much narrower when short-length threads are rolled. Dies for straight threads, such as those on machine screws can be reversed so that both the upper and lower portions can be used for screws that have a thread length of less than half the die width.

The maximum length of the segment die is limited by the length of the pocket in which it is held. Shorter dies are used for small-diameter work so that the work receives the proper number of revolutions during rolling. Because the surface of the rotary die is longer than that of the segment die, the rotary die generally has a longer life.

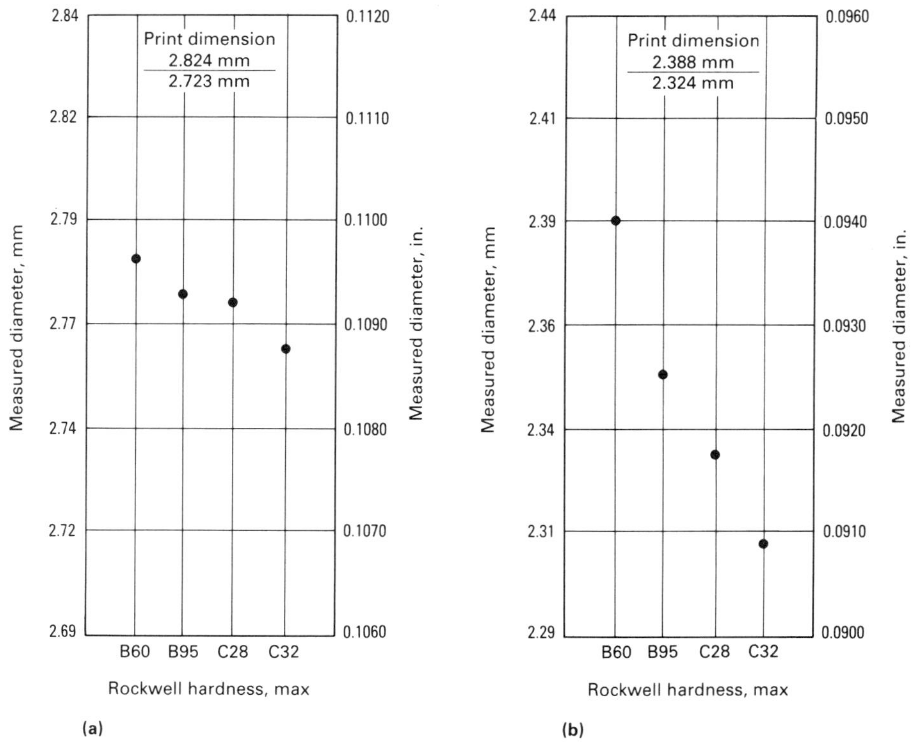


| | |
|---|---|
| Thread | 4-40 UNC-2A by 6.4 mm (¼ in.) |
| Blank diameter | 2.35–2.36 mm (0.0925–0.0930 in.) |
| Machine | No. 10 planetary die threader |
| **Operating conditions** | |
| Die speed, rev/min | 90 |
| Feed rate(a), pieces/rev | 4 |
| Cutting fluid | Soluble oil |
| Production rate, pieces/min | 360 |
| **Cost breakdown** | |
| Material, % | 15 |
| Manufacture(b), % | 65 |
| Finish(c), % | 20 |

(a) Material was fed into dies by a four-lobe cam. (b) Includes overhead. (c) Zinc chromate finish by the barrel method

**Fig. 16** Effect of blank hardness on (a) major and (b) pitch diameters of machine screws made of 1038 steel

Planetary dies for gimlet-pointed screws, such as type A and AB sheet metal screws, are very similar to flat dies for the same screws. The length of the segment die is equivalent to one flat die, and the circumference of the rotary die is usually equivalent to five segments and is capable of threading at least five pieces during each revolution of the die.

**Capabilities.** Planetary thread rolling machines can roll most of the smaller parts that can be rolled on flat die machines. Production rates of planetary rollers are higher than those of other types of thread rolling equipment. Rates of 3000 pieces per minute can be reached on small pieces. The practical limit of production rate depends on the ability to feed the blanks rather than on speed of rolling. Planetary rollers lose their economic advantage over other equipment when it is necessary to reduce the production rate in order to feed difficult parts properly, or when quantities are too low (for instance, less than 1 million 6.4 mm (¼ in.) screws).

Because they are produced in large quantities and can be fed easily, headed parts comprise most of the production on planetary machines. Typical products are machine screws; types A, AB, and B sheet metal screws; and drive screws. Size of product ranges in the popular machine sizes are No. 4 machine screws to 16 mm (⅝ in.) diam screws, although machines with capacities up to 29 mm (1⅛ in.) have been built.

**Limitations.** The cost of rotary dies is in proportion to die size. Die cost per piece produced is generally competitive with other types of dies, but unless volume is high enough to use up the available die life, large inventories of dies will accumulate.

## Continuous Rolling

Continuous rolling is a high-production method suited to cylindrical-die machines. The method uses two cam-type segmental dies, maintained at a predetermined center distance, as required for the desired pitch diameter.

Workpieces are fed from a hopper or magazine to a revolving cage-type workrest that indexes them into and away from the rolling position. Depending on the number of die segments, one, two, or three workpieces are threaded for each revolution of the dies.

Continuous rolling in a two-roll cylindrical-die machine provides the highest rate of production for headed workpieces. This method is applicable to the threading of double-end studs and similar parts. Threads of different diameters, pitches, and tolerances, as well as those of identical specifications, can be produced in a single pass. Parts can be rolled to produce threads on one end and knurling or splines on the other.

When thread rolling double-end studs, two sets of cam-type dies are used. Each spindle contains two dies maintained a fixed distance apart, depending on stud size. Both ends of the stud can be rolled simultaneously when thread diameters are the same. Except for studs larger than about 19 mm (¾ in.) in diameter, segmented dies can be used to roll different pitches on the two ends of a stud in one spindle revolution.

Studs and similar parts up to 19 mm (¾ in.) in diameter and 345 mm (13½ in.) long have been produced by this method. Production rate varies with size and shape of the part. Three applications of continuous rolling are shown in Table 3.

Threads on double-end studs 8.0 mm (5/16 in.) in diameter and 120 mm (4¾ in.) long can be produced at a rate of 240 complete parts per minute by this method.

## Internal Thread Rolling

The rolling of internal threads requires dies of comparatively small diameter, which greatly limits die life and the load-carrying capabilities of die bearings and spindles. As a result, internal thread rolling is limited in its use. However, it has been used successfully for forming helical fins on the internal surfaces of heat-exchanger tubes and for rolling internal threads in pipe couplings.

Internal threads can be rolled by impressing the inside surface of a workpiece shell of suitable wall thickness onto a close-fitting threaded mandrel, as shown in Fig. 17(a). Pressure is provided by three or four rotating plain external dies. The workpiece and mandrel may be clamped in a stationary position with the dies mounted in a rotating die head, or the dies may be stationary while the work and mandrel rotate. It is necessary to unscrew the part from the mandrel after threading has been completed.

To be thread rolled internally, the workpiece must be made of highly ductile metal such as aluminum, brass, or low-carbon steel. The wall must be thick enough to

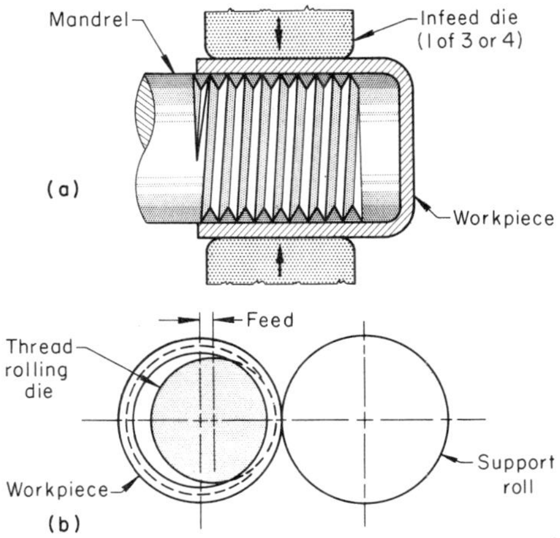


**Fig. 17** Internal thread rolling. (a) With a close-fitting threaded mandrel. (b) With a threading die that is considerably smaller than the inside diameter being rolled

provide adequate material to fill the die thread, but not so thick that the external dies are prevented from creating sufficient load to cause the threaded mandrel to penetrate the workpiece.

Parts have also been threaded internally by using a smaller, threaded mandrel or die and a single, plain external die or support roll, as shown in Fig. 17(b). In this procedure, the part feeds axially when a single-start thread is being rolled. Dies can roll a multiple-start thread with a minimum of axial movement. Cold form tapping, discussed in the article "Tapping" in this Volume, is another method of producing internal threads by metal displacement without the production of chips.

## Selection of Rolling Method

Table 4 shows approximate ranges of production rates for different types of thread rolling equipment. Actual production rates depend on the condition of the machine being used, the work-handling equipment, the type of workpiece, and the metal being rolled. The quantity of pieces to be rolled is also an important factor in the selection of a machine.

In general, a low volume of identical parts (up to several thousand pieces) can be produced most economically on a hand-fed flat die machine. Hopper feeding is usually most economical for more than 10 000 pieces. Depending on the size of part, a production run of more than 100 000 pieces can often be produced most economically on a planetary-die machine.

In one operation involving standard hexagon-head screws ranging from 6.4 to 9.5 mm (¼ to ⅜ in.) in diameter, the optimum run for a hand-fed flat die machine was approximately 10 000 pieces. The hopper-fed flat die machine was best for quantities between 10 000 and 400 000 pieces, and the planetary-die machine proved to be the most practical for runs of more than 400 000 pieces.

The quantity of pieces economically producible on cylindrical-die machines is difficult to assess. These machines can roll special thread forms, and the parts produced are frequently of superior quality, including greater precision. They also can roll a greater range of diameters than either flat die or planetary-die machines. Selection of a single machine that is suitable for a variety of applications is discussed in the example that follows.

**Example 2: Equipment for Varied Product Mix in Quantities from 50 to 1000 Pieces.** Various products produced in one plant were originally threaded by cutting tools in secondary-operation equipment. Thread sizes ranged from 6.4 to 25 mm (¼ to 1 in.) in diameter and from two diameters to 300 mm (12 in.) long. Production quantities ranged from 50 to 1000 pieces. Although thread cutting produced a quality product, the company decided to purchase thread rolling equipment capable of rolling class 2A and 3A threads and of handling the varied product mix. The various types of thread rolling equipment were considered, and selection was based on the factors discussed below.

*Traversing Flat Die Machine.* Die and setup costs for this equipment were favorable, and a high production rate could have been obtained for most of the thread lengths involved. The limited diameter capacity, however, would have necessitated the purchase of more than one size machine in order to include the total range of thread diameters of the product mix, and this in turn would have been unsuitable for the long thread lengths required and the variety of workpiece configurations.

*Planetary-Die Machine.* The quantity of parts to be threaded was insufficient to warrant the purchase of this high-production type of equipment. Also, die cost per piece would have been excessive, and the machine would have been unsuitable for the long thread lengths required.

*Cylindrical-Die Machines.* This type of equipment would have been satisfactory from the aspect of versatility in application to the product mix. Infeed rolling could have been used for the short thread lengths, and through-feed rolling could have been used for the long thread lengths. Two-die and three-die machines were available, and setup time would have been about the same as for other machines of similar capacity. However, because these machines rotate the workpiece, it was doubtful that the equipment could have been used for all sizes and shapes of product.

*End-Feeding Heads.* Thread-rolling machines equipped with end-feeding heads (thread rolls) could have rolled both short and long threads. Because each head could have rolled only a limited range of thread diameters, several sizes would have been required for the product mix involved. Setup of the machine and changing of the head would have been done easily and quickly. Cost of thread rolls was low because of their small size. Because end-feeding heads rotate around the workpiece, simple work holding fixtures could have been employed.

*Selection.* The equipment selected was a thread rolling machine equipped with end-feeding heads. Although a cylindrical-die machine would have been suitable for the work required in this plant, it was more expensive. Final selection of the machine with end-feeding heads was based on the low cost of heads and thread rolling dies and the simplicity of fixtures needed for holding the work.

**Table 4 Approximate range of production rates of thread rolling equipment**

| Thread diameter | | Infeed rolling, threads per minute | | | | Through-feed or end-feed rolling, threads per minute | |
|---|---|---|---|---|---|---|---|
| | | | Cylindrical die | | | | |
| mm | in. | Flat-die traversing | Single revolution(a) | Multiple revolution | Rotary planetary | Parallel-axis dies | Skewed-axis dies |
| 3.2 | ⅛ | 40–500 | 75–300 | 20–90 | 450–2000 | 20–40 | 140–280 |
| 6.4 | ¼ | 40–400 | 60–150 | 20–90 | 250–1200 | 20–40 | 200–450 |
| 13 | ½ | 25–90 | 50–100 | 15–70 | 100–400 | 25–55 | 110–300 |
| 19 | ¾ | 20–60 | ... | 10–50 | ... | 25–65 | 80–300 |
| 25 | 1 | 15–50 | ... | 8–40 | ... | 20–50 | 70–300 |
| 38 | 1½ | ... | ... | 6–30 | ... | 15–30 | 50–200 |
| 51 | 2 | ... | ... | 4–25 | ... | 10–20 | 30–140 |
| 64 | 2½ | ... | ... | 3–20 | ... | 6–15 | 20–90 |
| 76 | 3 | ... | ... | 2–15 | ... | 4–10 | 15–40 |
| 100 | 4 | ... | ... | 1–5 | ... | 1–3 | 5–10 |
| 125 | 5 | ... | ... | ½–1 | ... | ... | ... |
| 150 | 6 | ... | ... | ¼–½ | ... | ... | ... |

(a) Two threads can be rolled on double-end studs in one die revolution in some machines.

## Factors Affecting Die Life

The life of thread rolling dies is determined primarily by the rate of deterioration of the profile of the die threads. Rolling imposes severe stress on the dies from pressure and bending and sliding action.

Dies usually fail by spalling and crumbling of the thread crests, which roughens the minor diameter of the product thread and causes the screw thread to go out of

**Table 5 Nominal die life for three thread rolling methods**

| Thread diameter, mm | Thread diameter, in. | Die life: Flat die pieces | Cylindrical dies: Infeed, pieces | Through-feed, km | Through-feed, ft × 10³ |
|---|---|---|---|---|---|
| **Threading steel at 85 HRB (550 MPa, or 80 ksi, tensile strength)** | | | | | |
| 6.4 | 1/4 | 700 000 | 350 000 | 21 | 70 |
| 13 | 1/2 | 625 000 | 315 000 | 19 | 63 |
| 25 | 1 | 450 000 | 245 000 | 15 | 50 |
| 38 | 1 1/2 | ... | 190 000 | 9.1 | 30 |
| 64 | 2 1/2 | ... | 125 000 | 5.3 | 17.5 |
| 89 | 3 1/2 | ... | 70 000 | 2.1 | 7 |
| 115 | 4 1/2 | ... | 18 000 | ... | ... |
| **Threading steel at 30 HRC (965 MPa, or 140 ksi, tensile strength)** | | | | | |
| 6.4 | 1/4 | 35 000 | 20 000 | 4.3 | 14 |
| 13 | 1/2 | 32 000 | 18 000 | 3.8 | 12.5 |
| 25 | 1 | 24 000 | 14 000 | 3.4 | 11 |
| 38 | 1 1/2 | ... | 11 000 | 1.8 | 6 |
| 64 | 2 1/2 | ... | 7 000 | 1.1 | 3.5 |
| 89 | 3 1/2 | ... | 4 000 | 0.4 | 1.4 |
| 115 | 4 1/2 | ... | 1 000 | ... | ... |

tolerance. Failure is probably caused by fatigue from the stresses imposed in rolling. The best products and maximum die life can be obtained only when the dies are properly set up and the correct die speed and number of blank revolutions are used. The surface of the material being worked should be relatively free of oxide and scale.

Spalling or chipping may be the direct result of endwise extrusion or stretching of the blank during rolling, over-rolling, improper blank design, or inferior design or quality of tools. When spalling occurs, relatively large pieces are broken out of the threads of the tools. This failure usually occurs near the edges of the tools or in the tool area where the ends of the blanks are rolled.

Spalling is frequently the result of having improper bevels on the blanks and tools, or of excessive tool hardness. Excessive variation or irregularity of blank diameter and hardness increases susceptibility to spalling, as does contact between the edge thread on the tools and the shoulders and fillets on the workpiece. Mismatching of the tools during rolling imposes transverse loading on the tool threads and causes spalling.

When crumbling takes place on the crests of die threads, failure usually starts in the more highly stressed rolling areas and gradually spreads over other portions of the die threads. Excessively sharp die crests are subject to greater initial crumbling action.

Useful tool life is greatly affected by the hardness and work hardening characteristics of the metal being rolled. Premature failures may be minimized by preventing over-rolling and by using clean blanks of correct, uniform size and hardness.

The wear of thread rolling dies seldom need be considered. When wear does occur, it is primarily associated with the abrasive action of scale or dirt on the surface of the blanks, or the use of contaminated coolant. Deep threads are generally subject to the most rapid abrasive wear because of the greater amount of sliding action between workpiece thread and die thread. The rolling of deep threads of comparatively short length and small minor diameter may result in excessive extrusion of the blank, which causes permanent bending or chipping of the die thread profile at the end of the blank.

Die life ranges from millions of pieces for soft work metals to a few hundred pieces for hardened steels. Table 5 shows the approximate relationship between die life and method of rolling different thread sizes in steel at 85 HRB and 30 HRC.

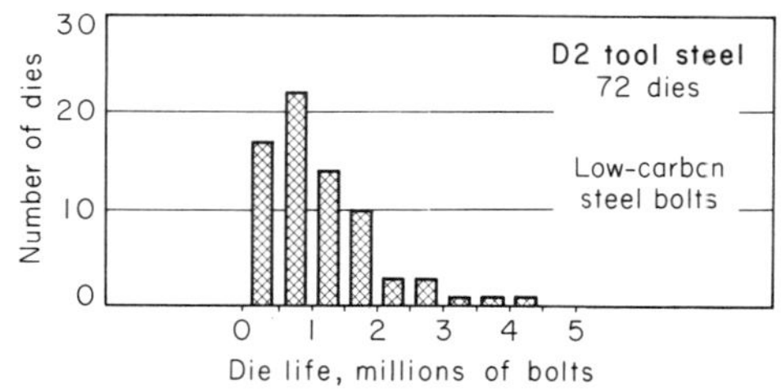


**Fig. 18** Distribution of die life of 72 dies used for threading low-carbon steel bolts in a 13 mm (1/2 in.) boltmaker machine

Even when production conditions are as nearly constant as is feasible, die life can vary greatly, as illustrated in Fig. 18. Seventy-two identical thread rolling dies, made of D2 steel by hobbing before hardening, failed with the die life distribution shown in Fig. 18. These data for dies of the same design and tool steel, all used for rolling the same threads on the same blank material, represented about the narrowest variation of factors affecting die life possible in normal production.

**Fine Threads.** Dies for rolling 40 or more threads per inch, especially with class 3A fits, require greater precision than do dies for rolling coarse threads, but they impose less cold work on the blanks during rolling and have longer life. Fasteners with the closest tolerances and finest pitches require grinding of dies after hardening to achieve required accuracy.

**Die Life Versus Hardness of Blank.** Increasing hardness of the blanks being rolled shortens die life. Figure 19(a) shows the approximate average life of a number of flat dies made of D2 tool steel in rolling 1/4-20 threads on 1022 steel blanks of various hardnesses.

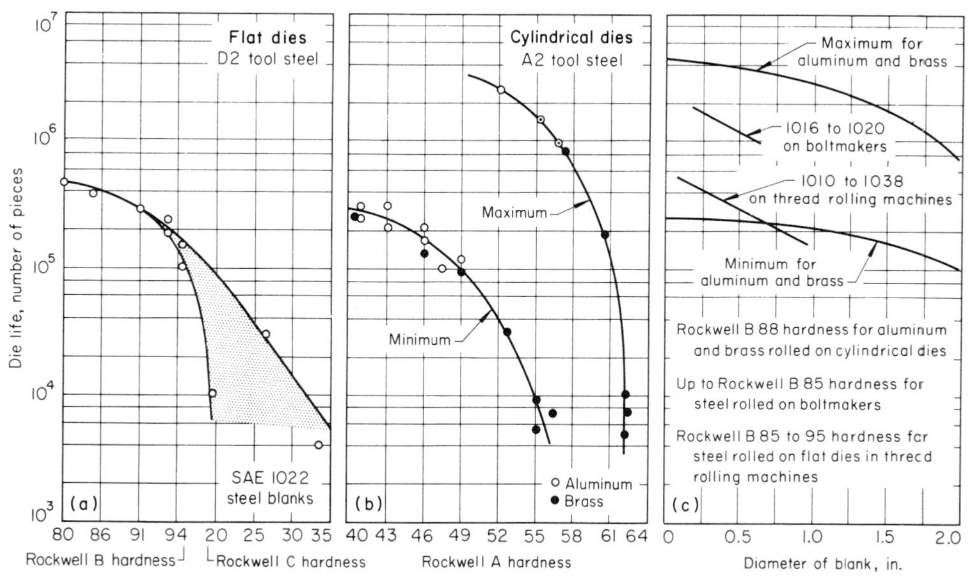


**Fig. 19** Relation of die life to hardness and diameter of blank. (a) Die life versus hardness of blanks of 1022 steel threaded with flat dies of D2 tool steel. (b) Die life versus hardness of blanks of aluminum or brass threaded with cylindrical dies of A2 tool steel. (c) Die life versus diameter of steel blanks threaded with standard flat dies or with special-contour dies in boltmakers, and of aluminum and brass blanks threaded with cylindrical dies

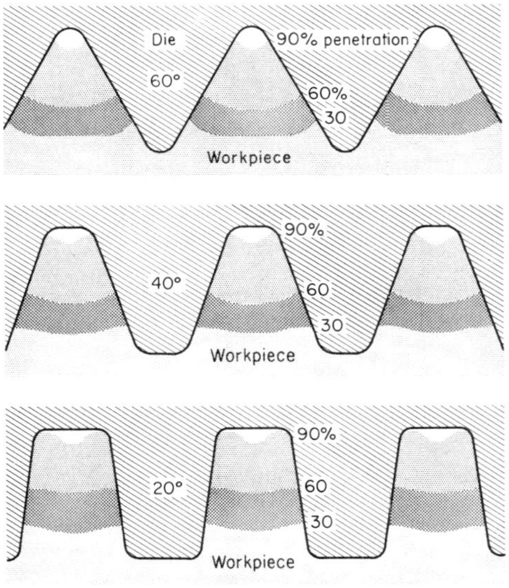


**Fig. 20** Effect of three different thread forms on seam formation at thread crest, and the rolled shape of the workpiece at three levels of die penetration (30, 60, and 90%)

**Table 6 Specifications for high-quality rolled threads on ground and extruded blanks**

| Thread size | p (pitch) mm | p (pitch) in. | 0.25p (min) mm | 0.25p (min) in. | R (min runout radius) mm | R (min runout radius) in. | T (min root radius)(a) mm | T (min root radius)(a) in. |
|---|---|---|---|---|---|---|---|---|
| 10–32 | 0.7938 | 0.03125 | 0.20 | 0.008 | 0.30 | 0.012 | 0.08 | 0.003 |
| ¼–28 | 0.9070 | 0.03571 | 0.23 | 0.009 | 0.33 | 0.013 | 0.10 | 0.004 |
| 5/16–24 | 1.0584 | 0.04167 | 0.25 | 0.010 | 0.38 | 0.015 | 0.13 | 0.005 |
| ⅜–24 | 1.0584 | 0.04167 | 0.25 | 0.010 | 0.38 | 0.015 | 0.13 | 0.005 |
| 7/16–20 | 1.2700 | 0.05000 | 0.30 | 0.012 | 0.46 | 0.018 | 0.15 | 0.006 |
| ½–20 | 1.2700 | 0.05000 | 0.30 | 0.012 | 0.46 | 0.018 | 0.15 | 0.006 |
| 9/16–18 | 1.4112 | 0.05556 | 0.36 | 0.014 | 0.51 | 0.020 | 0.15 | 0.006 |
| ⅝–18 | 1.4112 | 0.05556 | 0.36 | 0.014 | 0.51 | 0.020 | 0.15 | 0.006 |
| ¾–16 | 1.5875 | 0.06250 | 0.41 | 0.016 | 0.58 | 0.023 | 0.18 | 0.007 |

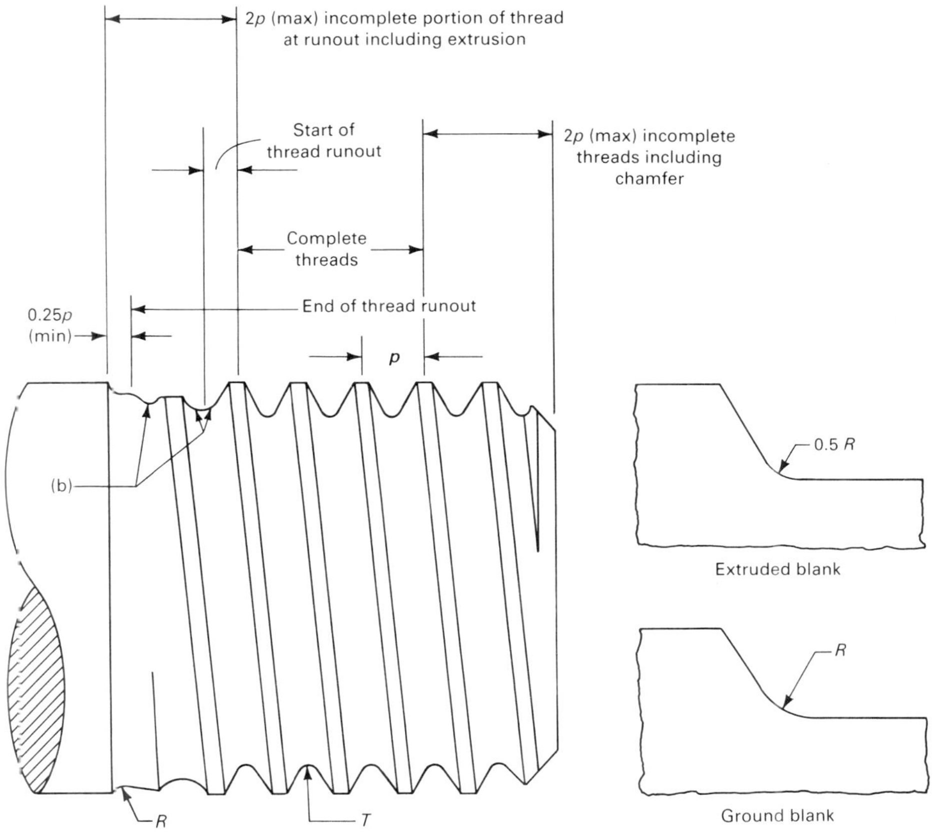


(a) Maximum root radius is limited by requirement that threads accept a "go" thread gage. (b) Radius at root of runout thread must not be less than minimum radius at root of full thread.

Some of the tests were run on hardened blanks, including the two hardest specimens plotted. On blanks harder than 94 HRB, die life was low and inconsistent. The shaded portion of the curve indicates both the spread and inconsistency in die life. A blank hardness of about 32 HRC is the limit for normal thread rolling.

Figure 19(b) shows minimum and maximum die life versus hardness of the screw blank for a large number of circular dies made of A2 tool steel. These dies were used mainly in rolling fine-pitch threads on parts made mostly from free-cutting brass and aluminum alloys. The data include some high-production runs in which conditions of setup and blank material were nearly ideal. Therefore, the average life of cylindrical dies would be less than the mean between the two curves shown. Also, dies that roll fine threads on small parts have longer average life than those that roll coarser threads on larger parts.

Figure 19(c) shows the relationship between die life and diameter of threads rolled on aluminum and brass with circular dies and on steel with flat dies. The upper and lower curves relate to A2 circular dies, most of which were ground after hardening and used primarily for rolling fine-pitch threads, the majority of them being special threads. The original data indicated no detectable difference in die life between aluminum and brass at a hardness of about 88 HRB.

The curve relating to 1016 and 1020 steel gives the average life in a single setting of several hundred D2 dies with special contour on entering threads and used as hardened after machining. The other curve gives similar data for 1010 and 1038 steels on machines of the traversing type.

Unusual variables can alter die life considerably. For example, rolling 16 mm (⅝ in.) diam hot-rolled 1040 steel blanks that were nonuniform in diameter, scaled, and unannealed, has reduced die life to only 45 000 pieces.

## Effect of Thread Form on Processing

The form of the thread affects the radial die load, seam formation, surface finish, and thread dimensions.

**Radial Die Load.** Greater loads are required as the form becomes more blunt. For example, if a radial load of 45 kN (10 000 lbf) is required to produce a ½-13 UNC thread 50 mm (2 in.) long, a radial load 40 to 70% greater would be required to roll a ½-10 Acme thread under the same conditions. The amount of increase in load is determined by the blend radii on the crest of the die thread and by the degree of fullness of thread required.

**Seam formation** at the crest of the thread is greatly affected by the shape of the thread form, as indicated in Fig. 20. Sharp crests on the dies cause sharper seams than those produced when a wider form is rolled more deeply.

**Surface finish** is usually independent of the form for 60° threads. However, as the thread form becomes more blunt, or if it has sharp root corners, the normal, smooth flow pattern of the metal being rolled is altered. Restriction of metal flow may cause localized subsurface shear failures. During subsequent die contacts, these shear failures

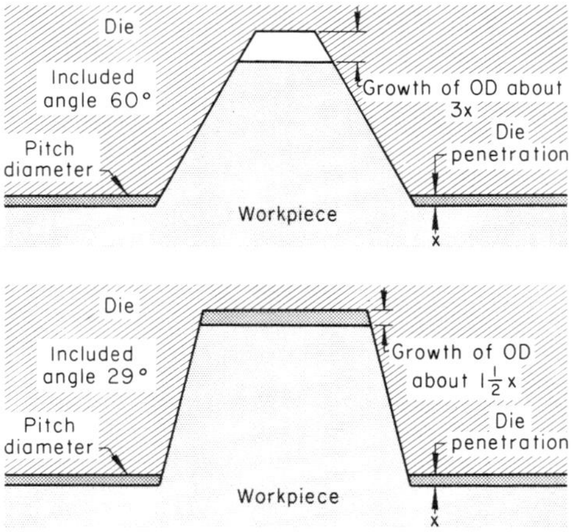


**Fig. 21** Effect of thread form and die penetration on the relation between pitch diameter and outside diameter of the thread

form small flakes, which downgrade the surface finish.

**Thread Dimensions.** Table 6 shows proper specifications for thread rolling high-quality threads in ground and extruded blanks. It is worth noting that no specification for chamfer angle on the end thread is required. In addition to the dimensional details tabulated and shown in the inset illustration, principal dimensions such as the major and minor diameters should, of course, be specified.

Changes in the thread form affect the relationship between pitch ($p$) diameter and outside diameter of the thread. Figure 21 shows that if a standard 60° thread form with a typical sharp thread crest is not quite fully rolled, an additional die penetration of 0.025 mm (0.001 in.) on the diameter, with a commensurate reduction in the pitch diameter, results in an increase of 0.08 mm (0.003 in.) in the outside diameter until the thread form is filled.

For the more blunt Acme thread, the outside diameter increases at only about 1½ times the rate of decrease of the pitch diameter until the material fills the form through the crest radii.

The class of thread affects the size of the blank diameter and the major and minor diameters for a given thread. Classes 1A, 2A, and 3A threads of the same size and pitch in the American Standard series can be rolled with the same die. However, a different die may be required for rolling a given thread of a class other than 1A, 2A, or 3A.

**Table 7 Operating speeds of thread rolling and thread cutting tools**

| Tool | Surface speed m/min | sfm |
|---|---|---|
| **Thread rolling tools** | | |
| Flat traversing die | 30–100 | 100–325 |
| Cylindrical die | 20–180 | 70–600 |
| Lathe attachments | 20–90 | 70–300 |
| **Thread cutting tools** | | |
| High speed steel tools | 3–45 | 10–150(a) |
| Carbide tools | 75 | 250 |

(a) Speed range for threading aluminum, copper, steel, and Monel

## Surface Speed

Permissible surface speeds for thread rolling are governed by the mechanical power limitations of the threading equipment and, when the workpiece is rotated, by the speed (rev/min) of the workpiece or of the holding equipment used with end-feeding heads.

Table 7 compares die surface speeds in modern thread rolling equipment with the surface speeds of thread cutting tools. Table 8 shows the approximate thread rolling time for different spindle speeds used with tangential-infeed double-roll attachments.

## Penetration Rate and Load Requirements

The total die penetration per revolution of the blank varies with different machines, the kind of work, and the type and hardness of the metal being rolled. Low penetration rates are necessary for hard metals, hollow workpieces, and workpieces of nonrigid cross section. Higher penetration rates are used for rolling metals that work harden at an excessive rate.

The rate of die penetration is normally limited by the rigidity of the workpiece and the machine and by the hardness of the metal being rolled. Penetration rates for infeed rolling range from 0.0125 to 0.15 mm (0.0005 to 0.0060 in.) per revolution of the workpiece. Penetration rates for through-feed and end-feed rolling are governed by the lead and pitch of the thread and the length of the entrance threads on the dies.

If the pitch and other rolling variables are held constant and only the diameter of the work is increased, the rolling load will increase about one-third as much as the increase in diameter. For example, if a load of 17 kN (3800 lbf) is required to roll a 25 mm (1 in.) length of ¼-20 thread, the initial die load for rolling a ½-20 thread under the same conditions would be about 22 kN (5000 lbf). However, because the thread pitch normally increases proportionately with the diameter, a proportional change in the rolling conditions causes the die load to increase about half to three-fourths as much as the increase in the workpiece diameter.

Many rolling machines have insufficient power to provide proportionately rapid penetration rates on larger work; thus, an increase in workpiece size presents a problem when rolling threads of large diameter in a metal that work hardens rapidly. In such applications, the inability of the equipment to produce adequate absolute penetration rates increases the number of die contacts necessary and causes too rapid an increase in the hardness of the work metal. As a result, disproportionately large radial die loads are required to complete the thread form. This is hard on machines and dies. More powerful machinery must be used if work hardening of the metal being rolled causes this kind of load escalation.

Rated motor horsepower of flat die and cylindrical die infeed thread rolling machines is not a direct measure of peak power consumption on specific jobs. The motor horsepower of these intermittently loaded machines is supplemented by the stored energy of the drive system momentum. Motor horsepower of through-feed machines can be rated more directly, because power is applied continuously. Each unit of power delivered to the through-feed die rolls threads in material with a rollability index of 1.00 at approximately the rates shown in Table 9.

In addition to the power used for threading the workpiece, the rolling process develops various power losses. Because of these losses, power available for actual threading ranges from 30 to 90% of rated power of the machine.

Energy delivered to the thread rolling dies is dissipated in forming the metal and also because of friction between the work and the die. On shallow threads, die friction accounts for about 10% of the energy delivered. For deep threads, losses due to die friction may be as much as 25% of the energy delivered; the remainder is used to flow the metal.

## Warm Rolling

Low-strength materials, such as aluminum, brass, and low-carbon steel, are al-

**Table 8 Thread rolling time for tangential infeed double-roll attachments**

| Threads per 25 mm (per inch) | Approximate work revolutions(a) | Thread rolling time (s) for spindle speed (rev/min) of: 500 | 1000 | 1500 | 2500 | 5000 |
|---|---|---|---|---|---|---|
| 32 | 11–27 | 1.3–3.2 | 0.7–1.6 | 0.4–1.1 | 0.3–0.6 | 0.1–0.3 |
| 24 | 14–31 | 1.7–3.7 | 0.8–1.9 | 0.6–1.2 | 0.3–0.7 | 0.2–0.4 |
| 18 | 17–35 | 2.0–4.2 | 1.0–2.1 | 0.7–1.4 | 0.4–0.8 | 0.2–0.4 |
| 14 | 20–39 | 2.4–4.7 | 1.2–2.3 | 0.8–1.6 | 0.5–0.9 | 0.2–0.5 |
| 10 | 23–43 | 2.8–5.2 | 1.4–2.6 | 0.9–1.7 | 0.6–1.0 | 0.3–0.5 |
| 8 | 26–47 | 3.1–5.6 | 1.6–2.8 | 1.0–1.9 | 0.6–1.1 | 0.3–0.6 |

(a) The actual number of work revolutions used within the ranges shown depends on the material rolled and the size of the thread rolling attachment.

**Table 9 Approximate rates of through-feed thread rolling for metals with a rollability index of 1.00**

| Thread diameter mm | in. | Threads per 25 mm (per inch) | Rolling rate(a) mm/min/kW | in./min/hp |
|---|---|---|---|---|
| 6.4 | 1/4 | 28 | 5930 | 174 |
| | | 20 | 5650 | 166 |
| 7.9 | 5/16 | 24 | 5310 | 156 |
| | | 18 | 4120 | 121 |
| 11 | 7/16 | 20 | 2180 | 64 |
| | | 14 | 2080 | 61 |
| 13 | 1/2 | 20 | 1600 | 47 |
| | | 13 | 1500 | 44 |
| 14 | 9/16 | 18 | 1460 | 43 |
| | | 12 | 1330 | 39 |
| 16 | 5/8 | 18 | 1360 | 40 |
| | | 11 | 1190 | 35 |
| 19 | 3/4 | 16 | 1160 | 34 |
| | | 10 | 950 | 28 |
| 22 | 7/8 | 14 | 890 | 26 |
| | | 9 | 750 | 22 |
| 25 | 1 | 14 | 610 | 18 |
| | | 8 | 540 | 16 |

Note: To determine the through-feed rate for a metal of different rollability index, multiply the rolling rate by the rollability index. (a) Power is actual power delivered to the dies.

ways thread rolled at room temperature. Because of their relatively low yield strength, these metals are easily penetrated by the threading die. However, some metals, such as high-strength steels and heat-resisting alloys, offer considerable resistance to die penetration. To facilitate thread rolling, blanks of these materials are often heated and rolled while warm.

For example, warm rolling is used to thread fasteners made of 5% Cr-Mo-V steel (H11) that has been quenched and tempered. The H11 steel is induction heated to about 480 °C (900 °F) immediately before threading. Heating temporarily decreases the yield strength so that threads can be formed without great difficulty. However, for optimum life, the dies should not be allowed to reach too high a temperature. An ideal ambient die temperature is 90 °C (200 °F). Dies can be cooled by spraying, as needed, with a conventional die-cooling oil or a soluble-oil emulsion.

Workpieces that have been heated under controlled conditions and warm rolled on dies with a positive lead have a tensile strength equal to, and a fatigue strength greater than, similar parts rolled cold. In addition, die life in warm rolling is greater than in cold rolling of high-strength materials.

## Threading of Thin-Wall Parts

Three-die machines and end-feeding heads are better suited to the threading of hollow work than are two-die machines, because the application of rolling forces at three points on the circumference has much less tendency to collapse the work.

Table 10 shows the minimum wall thickness for thread rolling tubular workpieces on three-die machines. For satisfactory thread rolling of tubular workpieces in a two-die machine, a minimum wall thickness of twice the thickness shown in Table 10 for any given set of conditions is the recommended practice.

When the wall thickness of the tubular blank is too thin to be rolled as a finished part, solid metal or heavy-wall tubular stock is used for the thread rolling operation, and the threaded workpiece is then drilled or bored to the desired wall thickness in a secondary operation.

## Threading Work-Hardening Materials

The crests of threads rolled in metals such as austenitic stainless steel that work harden rapidly usually have a more pronounced seam or fold (Fig. 4) than metals that work harden at lower rates. Although seams do not affect thread strength significantly and are not detrimental for most applications, some thread specifications do not permit seams. One of the main objections to seams at the crest of threads, particularly in stainless steel, is that they are likely to become focal points for corrosive attack of the material. To avoid them requires either the use of a metal that does not form seams, or specially constructed dies that minimize the seams.

Other factors that can lead to difficulty when rolling threads in metal that has a high rate of work hardening are:

- Lead contraction
- Out-of-round threads
- Rough surface finish
- Increased power requirement
- Reduced production rate
- Decreased die life

All of these effects are undesirable; some may be tolerable; most can be minimized by combining high penetration rates, heavy-duty equipment, and special extra-hard dies with expanded leads. Dies as hard as 68 HRC are available for thread rolling work-hardening metals.

Too many revolutions of the blank can work harden some materials, resulting in reduced die life and failure of the threads to meet dimensional tolerances, as in the example below.

**Example 3: Die Setting and Blank Diameter for Threading 1038 Carbon Steel.** The available planetary-die equipment in one plant could not produce machine screws of 1038 steel to the required dimensions when blanks of standard diameter were used. The screws were out of tolerance because the length of the segment die required too many revolutions of the blank. This problem was not encountered with similar parts made on flat dies of shorter length. The use of the planetary-die machine was necessary, however, because of the large volume of screws to be made and the high production rate required.

Satisfactory screws were made by experimentally adjusting the blank size and by setting the dies so that the thread was nearly all formed in the first few revolutions of the blank, after which the threaded blank dwelled in the remaining portion of the die. This procedure produced adequate die life (about 1 500 000 pieces per grind) and the required production rate (20 000 pieces per hour).

Figure 22 shows the variation of major and pitch diameters obtained with standard and oversize blanks. A variation in blank diameter of about 1% resulted in loss of control over these dimensions.

Two characteristic types of thread rolling phenomena were observed when too large a blank diameter was used. A large number of "roll-ups" occurred. A roll-up is a part that rides up in the dies as it rotates, resulting in no helix angle but merely a number of disconnected annular rings. Also, oversize blanks caused overloading of the dies, and when the parts were released from the dies, considerable expansion occurred; some pieces "exploded" upon being released, resulting in hollow parts.

**Table 10 Preferred minimum wall thickness for thread rolling tubular sections**

| Threads per 25 mm (per inch) | Minimum wall thickness, mm (in.), for thread pitch diameter, 25 mm (in.), of: <13 (<1/2) | 13–25 (1/2–1) | 25–50 (1–2) | 50–75 (2–3) | 75–100 (3–4) | 100–125 (4–5) |
|---|---|---|---|---|---|---|
| 32 | 1.0 (0.040) | 1.3 (0.050) | 1.8 (0.070) | 2.4 (0.095) | 2.8 (0.110) | 3.3 (0.130) |
| 24 | 1.4 (0.055) | 1.8 (0.070) | 2.4 (0.095) | 3.0 (0.120) | 3.8 (0.150) | 4.4 (0.175) |
| 20 | 1.7 (0.065) | 2.0 (0.080) | 2.9 (0.115) | 3.7 (0.145) | 4.6 (0.180) | 5.3 (0.210) |
| 18 | 1.8 (0.070) | 2.3 (0.090) | 3.3 (0.130) | 4.1 (0.160) | 5.0 (0.195) | 5.8 (0.230) |
| 16 | 2.0 (0.080) | 2.5 (0.100) | 3.6 (0.140) | 4.6 (0.180) | 5.6 (0.220) | 6.7 (0.265) |
| 14 | 2.4 (0.095) | 2.9 (0.115) | 4.2 (0.165) | 5.3 (0.210) | 6.3 (0.250) | 7.6 (0.300) |
| 12 | 2.8 (0.110) | 3.4 (0.135) | 4.8 (0.190) | 6.1 (0.240) | 7.6 (0.300) | 8.9 (0.350) |
| 10 | ... | 4.1 (0.160) | 5.8 (0.230) | 7.4 (0.290) | 9.1 (0.360) | 10.7 (0.420) |
| 8 | ... | ... | 7.2 (0.285) | 9.1 (0.360) | 11.4 (0.450) | 13.4 (0.530) |

Note: Data apply to a thread length of one diameter or less, and are based on 1010 steel in the soft condition, rolled in a three-die machine. For rolling the same threads in a two-die machine, minimum wall thicknesses twice those shown here are recommended.

## Rolling Threads Close to Shoulders

The minimum dimension between a rolled thread and a shoulder depends on the rolling method, the type and size of thread, the

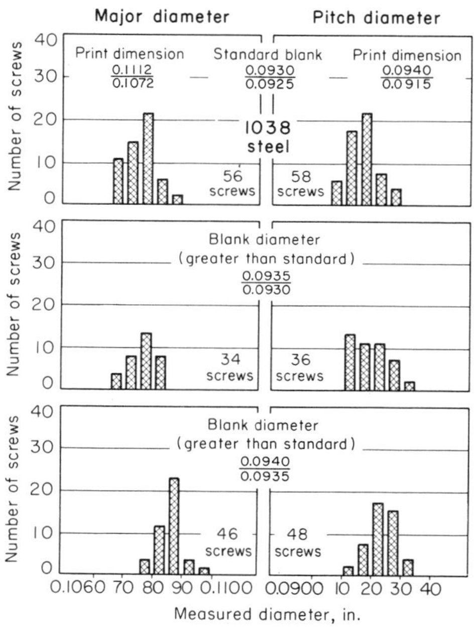


**Fig. 22** Effect of blank size on the major and pitch diameters of 1038 steel screws with thread length of 6.4 mm (¼ in.). See the table accompanying Fig. 16 for screw details and threading conditions. All dimensions in figure given in inches

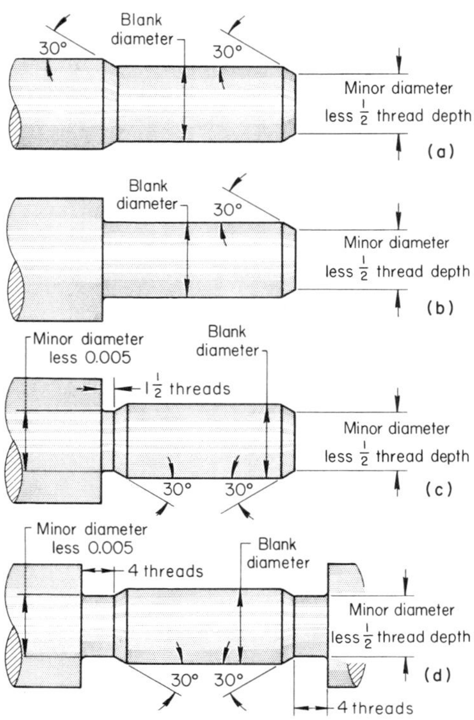


**Fig. 23** Shape of blanks for rolling threads close to a shoulder. See text for description of each type and its application. Dimensions given in inches

diameter of the shoulder in relation to the thread diameter, and the metal being rolled.

Infeed rolling dies usually are beveled, as shown in Fig. 12. The most common bevel angle is 30°. When unified threads are being rolled with a die of this type, a distance of 1½ threads from the shoulder to the root of the first full thread usually allows sufficient clearance between the edge of the die and the shoulder. When longer bevels are required on the dies, the distance from the shoulder to the root of the first full thread should be increased proportionately. For infeed dies, the distance from the shoulder to the root of the first full thread usually runs from 1½ to 2 threads per lead.

The rolls of end-feeding heads have imperfect starting threads similar to those on the throats of die chasers. In general, the length of imperfect thread on heads with nonreversible rolls for rolling unified threads is 1½ threads. For heads with reversible rolls, the length of the imperfect starting threads varies from about 1½ to 2½ threads. In addition to the length of the imperfect starting threads, the distance from the shoulder to the root of the first full thread varies with the lead angle of the thread and the clearance from the shoulder required to trip and open the head for withdrawal at the end of the rolling cycle. Therefore, depending on the design of the head, the approximate distance for rolling threads close to a shoulder using end-feeding heads ranges from 2 to 3 threads per lead. Specially designed rolls can thread closer.

Threads can also be rolled fairly close to a shoulder in cylindrical-die machines with the through-feed rolling method. Because of the longer imperfect starting threads on dies used in through-feed rolling, they can roll threads no closer to an obstruction than 5 or 10 threads, depending on the type of equipment.

The length of starting thread discussed in the preceding paragraphs is applicable to rolling threads in soft and medium-hard materials. These distances must be increased by 1 to 2 threads when threads are being rolled close to a shoulder in steel harder than about 25 HRC.

**Blank Preparation.** Four different blanks that involve rolling threads close to a shoulder are shown in Fig. 23. The blank shown in Fig. 23(a) is a type used extensively for cap screws and bolts. The 30° bevel at the junction of the large and small diameters matches the bevel on the thread rolling die, thereby allowing a thread to be rolled close to the bevel. After rolling a blank of this proportion, the large diameter shown and the major diameter of the rolled thread are usually the same.

A blank like that shown in Fig. 23(b) is used when the thread runs out on the blank without relation to the shoulder. Use of an undercut, as illustrated in Fig. 23(c), is the most practical approach for threading to a shoulder. Under these conditions, the blank can be threaded to within 1½ threads of the shoulder; the unthreaded portion is undercut so that a nut can be tightened against the shoulder.

There is likely to be end movement of the blank between the dies; thus, rolling between two shoulders, as in Fig. 23(d), requires a wider undercut, as shown. Under these conditions, rolling to within 4 threads of the shoulder is about as close as is practical. Worm threads are often rolled on the type of blank illustrated in Fig. 23(d).

**Dies Without Bevels.** Sometimes it is necessary to roll a full thread closer to a shoulder than can be done with a beveled die. If this is the case, a die is used that has no bevel on the leading end. Although the omission of bevel shortens the life of the die, it eliminates the imperfect threads and permits forming of a full thread closer to a shoulder.

## Fluids for Thread Rolling

There are two reasons for using fluids in thread rolling: to cool the dies and the work, and to improve the finish on the rolled products.

Low-carbon steel or nonferrous screws under 6.4 mm (¼ in.) in diameter are often rolled dry on flat dies at rates as high as 24 000 pieces per hour. This is especially true when rolling pointed screws on which the lubrication supplied by even a low-concentration soluble-oil emulsion may cause detrimental slippage. Soft steel or nonferrous metal screws have been rolled in larger sizes (13 mm, or ½ in. in diameter) without fluid, but only at speeds no greater than 300 pieces per hour.

At higher production rates, a fluid is required to control temperature. For cooling only, a low concentration of soluble oil in water (as weak as 1 part oil to 40 parts water) is satisfactory. However, as surface finish requirements become more stringent, a fluid with better lubricating qualities is required. In many applications, more concentrated mixtures of soluble oil in water (1 part oil to 8 parts water) will suffice, but, for the best results, lubricating oil (mineral oil) is recommended. A wide variety of oils have been successfully used. Sometimes, for extremely severe rolling (deep threads or hard work metal, or both), extreme pressure (EP) lubricants are used. However, for most applications, a low-viscosity mineral oil is sufficient. Any oil selected, in addition to having good lubricating properties, must be nontoxic and free from additives that would stain the work metal.

Cylindrical dies are more susceptible to heat buildup than are flat dies; therefore, some type of fluid is usually used for rolling in cylindrical dies. Soluble oils are preferred for maximum die life and optimum machine performance, but some high-tensile fasteners and other high-quality threads require the use of lubricating oil to provide the quality of finish required. Many three-die machines use the same oil on the work and dies as is used for lubrication of the rolling spindles.

Lubricating oils do not dissipate heat as readily as the soluble-oil emulsions. Thus, in high-production rolling in cylindrical-die machines, it is often necessary to provide an oil cooler in order to control temperatures. Generally, such cooling equipment is required only in the most extreme applications when using soluble-oil emulsions.

## Thread Rolling Versus Alternative Processes

For commercial products, the types of equipment available, size and shape of workpiece, number of pieces to be made, and the accuracy required are the principal considerations in determining the method that will be used for producing threads to the required specifications.

**Rolling Versus Cutting.** As production quantity decreases, it becomes more economical to cut the thread rather than roll it. For instance, 500 19 mm (¾ in.) screws 150 mm (6 in.) long with 50 mm (2 in.) threaded length could be made more economically by die threading at about 300 per hour than by setting up a roller that could produce 3000 screws per hour. Thread rolling would be worthwhile, however, for 7500 or more such screws.

Production quantity is not a criterion for determining the method of threading when special products are involved. For example, manufacturers of aircraft products have employed thread rolling for a single piece in order to obtain a product that would meet service requirements.

Thread milling, another method of cutting threads, is more expensive than thread rolling or die threading. Thread milling is seldom used except when extreme accuracy is needed or the workpiece is of such a size or shape that roll or die threading is impractical. In thread milling, threads can be produced with pitch diameter held within 0.025 mm (0.001 in.) and lead error held to 0.08 mm/m (0.001 in./ft).

**Rolling Versus Grinding.** Thread grinding is much more expensive than rolling and is used in preference to rolling only when:

- Work metal is too hard to be rolled or cut
- Work metal is extremely soft, under which condition grinding may be the only way to hold required dimensions
- A high degree of dimensional accuracy is required

Before selecting thread grinding rather than rolling, however, the various cutting procedures should be considered.

# Die Threading

DIE THREADING is a machining process for cutting external threads in cylindrical or tapered surfaces by the use of solid or self-opening dies. Die threading is a slower method of producing external threads than thread rolling, but it is faster than single-point threading in a lathe.

Hardness of the work metal limits the application of die threading. Attempting to die-thread metals harder than HRC 36 usually causes excessive wear or breakage of tools. Single-point threading or thread grinding is recommended for metals harder than HRC 36.

Size of workpiece seldom limits application of the process. Die threading has been used to cut 11 threads per millimeter (280 per inch) in rods 0.43 mm (0.017 in.) in diameter. At the other extreme three threads per 25 mm (three threads per inch) have been die cut into 610 mm (24 in.) pipe for the petroleum industry.

## Machines

The machines most used for die threading are drill presses, manual turret lathes, single-spindle and multiple-spindle automatic machines, and special threading machines.

**Drill Presses.** When workpiece shape permits, and if no other operations are involved, drill presses are often preferred for die threading, because they are easy to set up and simple to operate. A part that can be held in a fast-acting chuck or similar workholding device can be threaded in a manually operated drill press at least as fast as in any other machine.

Lead-control devices (see the article "Tapping" in this Volume) can be incorporated in a drill press. However, if lead control is required, a more rigid type of machine, such as a lathe, is preferred.

**Manual turret lathes** are often used for threading small to medium quantities, particularly when threading is one of a sequence of required machining operations. When threading is the only operation, the productivity of a turret lathe is lower than that of a drill press, because manipulating a turret is slower than moving a drill-press spindle.

Turret lathes can be equipped to handle either bar stock or semifinished parts. They can also thread pieces that are too large for practical manipulation in a drill press.

Starting the die onto the workpiece by hand feeding is more difficult in a turret lathe than in a drill press, because the weight of the turret makes the operator less sensitive to the resistance of an improperly chased thread. For this reason, many turret lathes used for threading are equipped with lead control devices.

**Automatic Machines.** Because of the longer setup time and higher running cost involved, automatic machines (automatic turret lathes, single-spindle or multiple-spindle automatic bar or chucking machines, or other automatic equipment) are used only for medium or high production. On these machines, several operations in addition to threading can be done in one cycle.

**Special threading machines** are available that perform only die threading, on either cylindrical or irregular-shape parts. Workpieces can be loaded and unloaded by hand, or the machines can be hopper fed and fully automatic. These machines usually incorporate lead control devices and are capable of producing class 5 threads. Collet-type machines that can handle long workpieces are commonly used to thread rod, shafts, and pipe.

**Selection** of the most appropriate machine is influenced by the same factors discussed in the article "Tapping" in this Volume. Machines that can produce the various thread classes are selected according to class, as follows:

- *Class 1A and 2A threads*: Drill presses or turret lathes using manual feed
- *Class 3A threads*: Drill presses, turret lathes, or threading machines equipped with lead control
- *Class 5 (interference fit) threads*: Precision threading machines equipped with devices for accurate lead control

## Solid Dies

Solid threading dies may be of one-piece construction, with integral cutting edges, or they may have replaceable cutting edges (chasers). One-piece solid dies can be either nonadjustable or adjustable; insert-type solid dies are adjustable. One-piece dies are usually discarded when worn out of tolerance; inserted chasers can be resharpened.

**Nonadjustable one-piece dies** have all cutting sections in rigidly fixed relationship (Fig. 1). They are available in standard sizes to fit various types of holders.

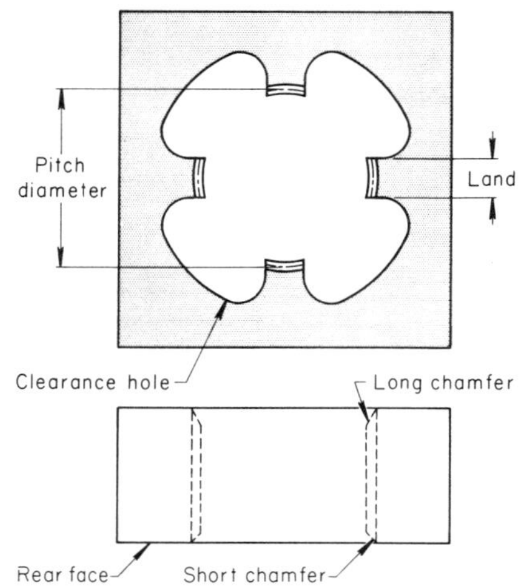


**Fig. 1** Nonadjustable one-piece solid die for threading

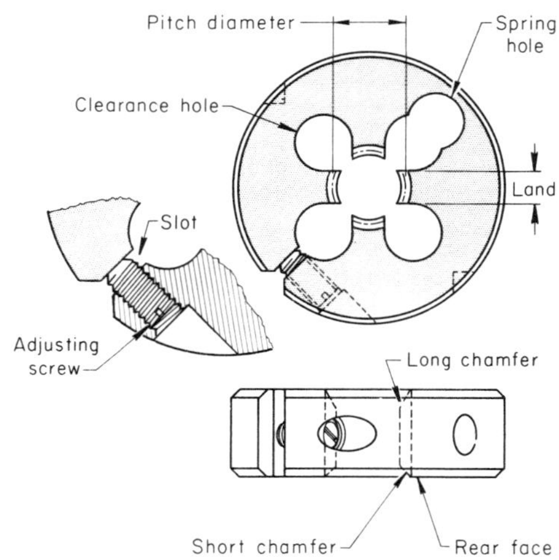


**Fig. 2** Adjusting-screw type of solid die for threading

**Adjustable one-piece dies** are of two types. One type (Fig. 2) is similar to the nonadjustable type except that the body has been slotted, a spring (relief) hole has been added, and an adjusting screw has been incorporated. The adjusting screw permits a small amount of adjustment to compensate for wear and to retain greater accuracy than is possible with the nonadjustable type.

The other type of adjustable one-piece die is the spring type (Fig. 3). Adjustability is provided by the design of the body and the holder, which function like a collet chuck.

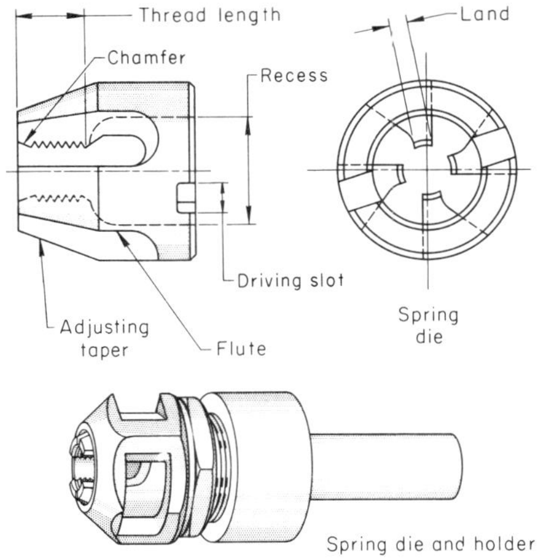


**Fig. 3** Spring-type adjustable die and holder for threading

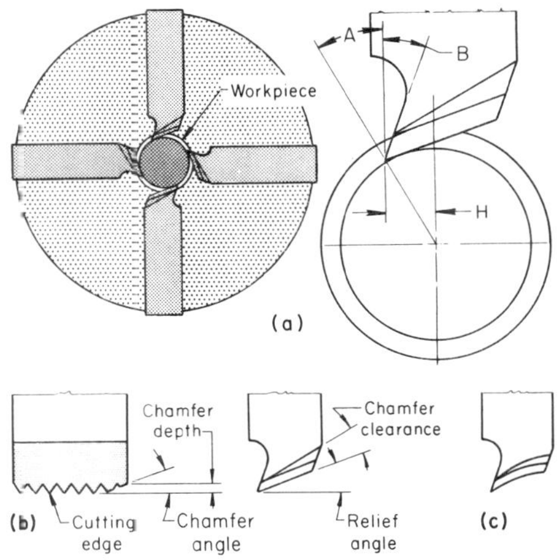


**Fig. 4** Radial chasers and their positions in a self-opening die

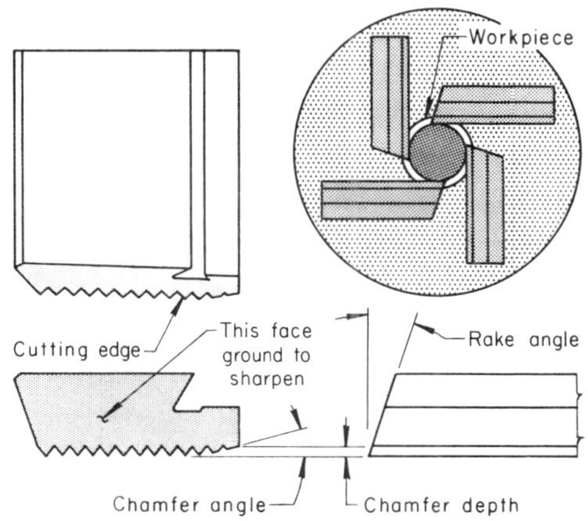


**Fig. 5** Tangential chasers and their positions in a self-opening die

As the outer nut on the holder is tightened, it applies pressure to the tapered section of the die, forcing the jaws inward. The nut can be moved and locked in position as desired, permitting adjustment of the cutting sections and greater accuracy than is possible with nonadjustable dies.

**An inserted-chaser solid die** consists of a holder and three or more inserted cutting edges (chasers). The inserts can be adjusted to compensate for wear, and can be removed for sharpening and replaced as required. Consequently, in long production runs, tool cost is lower than with one-piece dies.

Like one-piece dies, however, inserted-chaser solid dies can be removed from the work only by being back-tracked over the cut threads. Moreover, size for size, they cost more than one-piece dies.

**Circular chasers,** too, are used in solid dies, in sets of four or five. With five chasers, the die does less burnishing of the work in following the thread than it does using four chasers. The reduction in rubbing helps to make smoother threads and makes the die last longer. Besides, with five chasers, the die cuts better with less torque.

## Self-Opening Dies

There are two basic types of self-opening dies, revolving and stationary. The cutting elements of both types are called chasers, and each type is further identified by chaser design or position, or both. Chasers are designated as radial, tangential, and circular.

**Revolving** self-opening dies have a fixed workpiece and a rotating tool, as in a drill press. These dies are usually opened by a yoke arrangement in the die. As the die advances, the yoke meets a stop. The pressure against the yoke opens the die, retracting the chasers from the workpiece. The die can then return to its starting position for the next cycle.

**Stationary** self-opening dies, in contrast, have a rotating workpiece and a fixed tool, as in a turret lathe. Stationary dies are opened at a predetermined point by any of several mechanical devices, and the die returns to its starting position. Although stationary dies do not rotate, they may feed axially as the threading progresses, and thus do not remain fixed.

**Radial Chasers.** Figure 4(a) shows the positions of four radial chasers in a self-opening die and the relation of each chaser to the workpiece. The true top rake angle of a radial chaser is determined by adding the visible rake angle B (if positive, as in Fig. 4a) to the point-height angle A. If angle B is negative, the true rake angle is determined by subtracting angle B from the point-height angle A.

Radial chasers are intended to operate at a specific distance from the centerline of the work being threaded. This distance (H, in Fig. 4a) is called the point height. It varies according to the diameter of the work, and is determined by point height angle constants supplied for specific models of dies. If distance H is reduced, the amount of bearing surface (chaser against workpiece) and the top rake angle are also reduced.

Two designs of cutting edge for radial chasers made of high-speed steel are illustrated in Fig. 4(b) and (c). A straight cutting edge (Fig. 4b) can be generated by a milling machine or shaper. A circular cutting edge (Fig. 4c) is generated by a rotary tool such as a hob.

Radial chasers can be effectively sharpened only on the chamfer; the removal of more than about 0.025 mm (0.001 in.) from the area of the rake angle will adversely affect the chaser-workpiece relationship (Fig. 4a). Consequently, the total life of radial chasers is relatively short, so their use is sometimes restricted to the threading of soft, easy-to-machine metals such as aluminum or free-cutting brass.

**Tangential Chasers.** The position of four tangential chasers in a self-opening die is shown in Fig. 5. These chasers are made in sets, with their cutting edges extending along one face, as shown.

The chamfer angle is permanent for the life of the chaser and does not require regrinding. The chamfer section (including the first full tooth) not only cuts the initial thread form, but also removes excess material from oversize stock.

The rake angle of tangential chasers (Fig. 5) is varied according to the machining characteristics of the work metal; typical values are given in Table 1. Tangential chasers are especially suited to threading steel and other metals that are relatively hard, because of their long total life (including resharpenings). Repeated sharpening is permissible as long as a sufficient length of chaser remains to be held for sharpening or to be secured in the die.

**Circular Chasers.** The positions of circular chasers in a die are shown in Fig. 6. Circular chasers are made in sets of four or five chasers with annular thread form. A single flute, cut axially across the chaser, provides the cutting edge. The bore of each chaser is serrated for positioning on a holder in the die.

Face and hook angles (Fig. 6) can be modified as required for different work metals. Face angle is usually about 1°30′ to 2°; hook angle may vary from 5° negative, for threading soft free-cutting brass, to 20° pos-

**Table 1 Typical rake angles used on tangential chasers**

| Work metal | Rake angle(a) |
|---|---|
| Gray iron | 15°–25° |
| Malleable iron | 18° |
| Steel, 160 to 200 HB | 25° |
| Steel, 200 to 300 HB | 18°–22° |
| Steel, low carbon: | |
| Free cutting | 15°–25° |
| Other than free cutting | 25°–35° |

(a) All angles are positive.

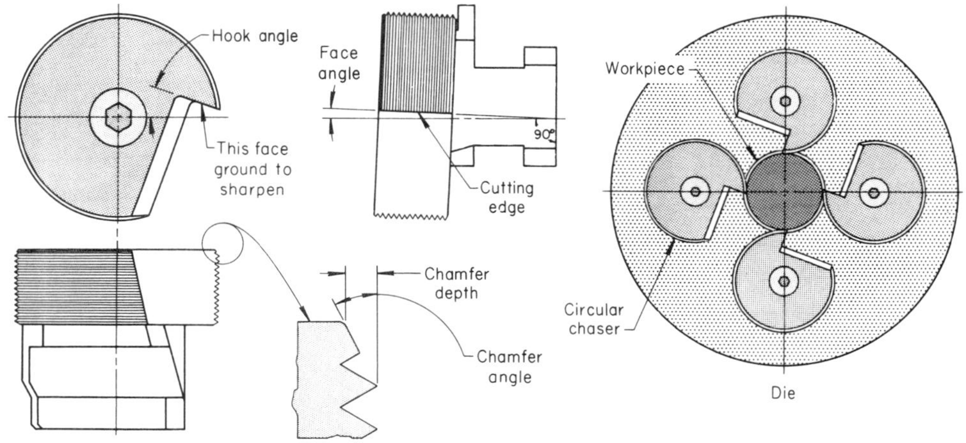


**Fig. 6** Circular chasers and their positions in a self-opening die

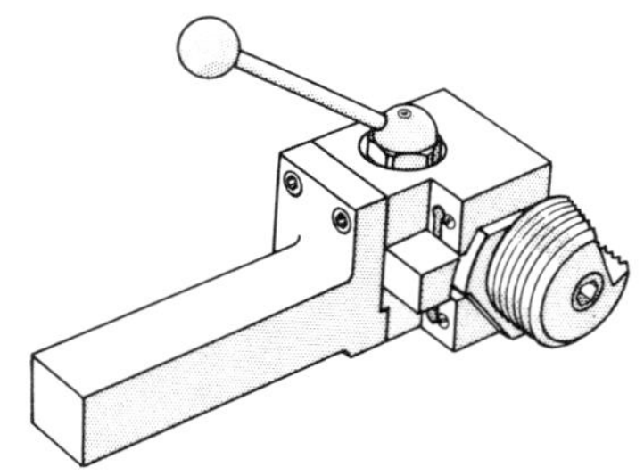

**Fig. 7** Circular chaser mounted in a self-opening holder, for single-chaser threading

itive, for metals that form stringy chips, such as mild steel or copper. Typical face and hook angles for ferrous metals are given in Table 2.

Circular chasers are used in high production for all metals that are threaded. They have a long total life, because they can be resharpened as long as there is enough of the chaser remaining to retain the cutting edge and the serrated hole for mounting.

## Solid Versus Self-Opening Dies

The use of solid dies is generally restricted to small-quantity threading in manually operated machines, or to machines (such as multiple-spindle automatics) that provide insufficient space to permit a self-opening die to operate. The principal reason that solid dies are not preferred for high-production threading is that the spindle must be reversed for die removal after the threads are cut. Backing the die over the threads is time consuming and increases die wear; it can also trap chips between the threads, causing damaged threads or a broken die, or both.

Circumstances sometimes dictate the use of solid dies, despite these disadvantages. For example, when metal beyond the outside diameter of the thread must be removed in the threading operation, solid dies are often used because they are stronger than self-opening dies. A manufacturer of cast iron fittings used self-opening dies to increase production rate, but because excessive metal had to be removed, the dies frequently jammed and failed to open, thus breaking chasers. Breakage was eliminated by changing to solid dies.

Although self-opening dies cost more initially than one-piece solid dies, replacement (or sharpening) of the chasers costs much less than replacing a solid die, and therefore self-opening dies have a lower overall cost. Also, because self-opening dies have a greater range of adjustability than solid dies, they can be used for threading to any degree of accuracy. In addition, self-opening dies eliminate the necessity of reversing the spindle to remove the die from the workpiece, provide easier chip clearance than solid dies, and eliminate stop lines.

**Die Removal.** Ease of removing the die from the workpiece is usually the most important advantage of self-opening dies. The use of a die that returns in the open position eliminates the need for stopping the spindle, regardless of whether the tool or the work is rotating. In some applications, this feature alone increases productivity by 50% or more. In addition to saving time, self-opening dies greatly reduce the possibility of damage to threads from entrapped chips. Furthermore, because the chasers make only one trip over the work, chaser wear is reduced and tool life is increased.

**Chip clearance** is a chronic problem in thread cutting. Adequate clearance must be provided, so that the chips can curl, break up, and drop away from the cutting edges. When the chip is forced into a pocket, or cannot clear the cutting edge, the threads being cut may be scored, and the tool edge can be damaged.

Solid dies made of expensive tool steel are produced only in small sizes. Practice in designing solid dies limits the length of threading fingers to an absolute minimum, to minimize chatter and premature die failure. This practice reduces the area available for chip clearance.

**Table 2 Typical face and hook angles used on circular chasers**

| Work metal | Face angle(a) | Hook angle |
|---|---|---|
| Gray iron | 2° | 10° |
| Malleable iron | 1°30′ | 10° |
| Steel | 1°30′ | 20° |

(a) For fewer than three threads, double the angle. Decrease the angle with higher-than-normal surface speeds, or if threads are excessively rough. Increase the angle for subnormal speeds. In general, the larger the angle, the rougher the finish.

Although self-opening dies inherently provide more space for chips to enter and drop away from the work, available chip clearance in any die is influenced by the diameter of the threads being cut. As thread diameter decreases, the problem of chip congestion increases.

**Stop lines** are a problem in die-threaded pieces when solid dies are used. As the die advances on the workpiece, metal is removed by the cutting section. When the cutter stops at the end of the cycle, metal removal stops abruptly, and a stop line, or step, is left on the work at the farthest advance of the cutting edges. There is no practical way to avoid this line when solid dies are being used. Self-opening dies can be opened just ahead of the point of full advance, causing the cutters to leave the work gradually with a forward rotating motion, thus avoiding cut marks.

## Single-Chaser Threading

Thread chasers can be used singly, rather than arranged in groups in a die head. Cutting threads with a single chaser is often advantageous when a part is rotated on centers and a die or holder cannot surround the part, or when there is not enough tool space to permit the use of multiple-chaser holder. The chaser can be held rigidly in a standard tool holder, or, as shown in Fig. 7, in a holder that opens and closes automatically by means of bumper stops on the machine.

Single-chaser threading often reduces threading time by 50%, compared to single-point threading, particularly if a relatively flat chamfer angle (10°, for example) can be used. A flat chamfer angle allows more threads to participate in the progressive removal of metal to the final thread form. It also increases tool life, although even with a normal chamfer angle of 15 to 20°, tool life of a single chaser is usually greater than that of a single-point tool.

## Modification of Chasers

True thread forms are usually cut into the chaser for reproduction on the workpiece. Sometimes, however, the first three or four teeth on the chaser are modified to distrib-

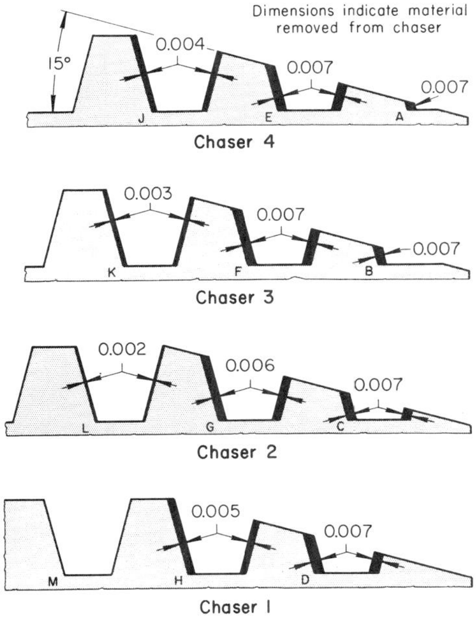


**Fig. 8** Modification of starting threads on chasers in self-opening dies to produce acceptable thread finish on Monel shafts. Dimensions given in inches

**Table 3 Effect of hardness on speed and tool life in die threading 4350 steel, gray iron, and free-cutting brass**

Dimensions in figure given in inches

| Item | 4350 steel | | Gray iron | | Free-cutting brass | |
|---|---|---|---|---|---|---|
| | 200 HB | 225 HB | 174 HB | 200 HB | Half hard | Full hard |
| Speed, rev/min | 190 | 155 | 535 | 535 | 690 | 460 |
| Speed, m/min (sfm) | 8 (25) | 6 (20) | 20 (70) | 20 (70) | 25 (90) | 18 (60) |
| Removal rate, $mm^3$/min × $10^3$ ($in.^3$/min) | 6.30 (0.384) | 4.98 (0.304) | 17.2 (1.05) | 17.2 (1.05) | 22.9 (1.4) | 14.4 (0.880) |
| Cutting fluid | Sulfurized oil | Sulfurized oil | Soluble oil | Soluble oil | Mineral oil | Mineral oil |
| Total tool life, $mm^3$ × $10^3$ ($in.^3$) of metal removed | 65 (4.0) | 52 (3.2) | 105 (6.4) | 85 (5.2) | 262 (16.0) | 184 (11.2) |
| Tool life per grind, pieces | 250 | 200 | 400 | 325 | 1000 | 700 |

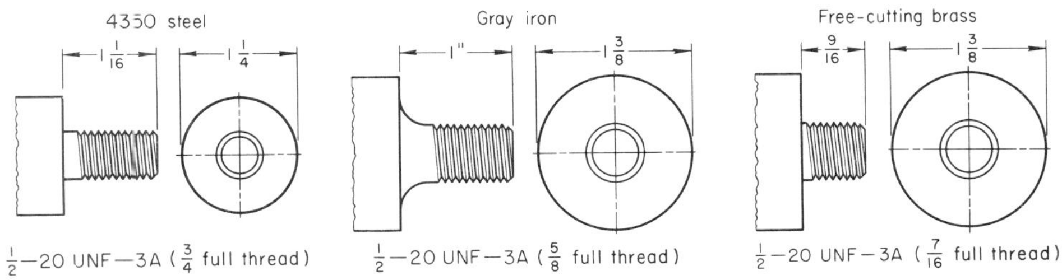


ute the cutting forces over a larger area. This technique improves surface finish and prolongs chaser life when hard materials are being threaded. Chasers can be modified by grinding them so that the first thread is cut oversize and succeeding threads are progressively shaved to the correct size, as illustrated in the following example.

**Example 1: Modified Chasers for Threading Monel.** When chasers having true-form lead threads were used for cutting ¾-6 Acme threads in a Monel shaft, the threads were torn and rough along both flanks. The shafts were threaded with a four-chaser, self-opening die operated at 2.4 m/min (8 sfm) and flooded with sulfurized oil.

To obtain a better finish when removing the relatively large amount of metal required by the Acme form, chasers were modified as shown in Fig. 8. The profiles of the 15° chamfered threads and of one or two full threads were changed by removing material, as indicated by the black areas lettered A to L in Fig. 8. The sequence of reduction in cutting width of the chamfered threads was then: Surface A on chaser 4 was the starting point for a thread flank; surfaces B to F were consecutive steps along the thread flank progressively cut by each chaser; at surface G, and continuing to surface M, each chaser tooth removed progressively smaller amounts of metal from the thread flanks until the desired thread form was finished by the unmodified chaser teeth.

## Factors That Influence Output

The principal factors that influence thread quality, production rate, and cost in die threading are:

- Composition and hardness of work metal
- Accuracy and finish
- Thread size
- Obstacles, such as shoulders or steps
- Speed
- Lead control
- Cutting fluid

These factors are usually related, and their relative importance varies greatly in different applications.

## Composition and Hardness of Work Metal

When a choice is possible, the free-cutting grades of any metal will result in more accurate threads of better finish, at higher production rate and lower cost, than the nonfree-cutting grades. When free-cutting metal cannot be used, comparable alloys for a given part will cause different amounts of tool wear, and the permissible speed of threading will usually vary somewhat.

**Hardness.** For threading metals softer than about 24 HRC, standard chasers, made of general-purpose high-speed steel and not surface treated, are usually satisfactory. For metals in the range of HRC 24 to 31, chasers that have been given a surface treatment, such as liquid nitriding, or a coating, such as titanium nitride, are recommended.

It is possible to die-thread metals as hard as about 36 HRC, but chasers made from the more highly alloyed grades of high-speed steel (such as M3, M4, M42, or T15) are necessary, and chaser life between grinds is short. Metals harder than 36 HRC are usually threaded by a single-point tool or by grinding.

As work-metal hardness increases, more power is required, and rigidity in machines and tools becomes more critical. Also, the advantages of lead control are more pronounced when harder metals are being threaded.

Soft metals, especially nonfree-cutting grades, sometimes cause difficulty in die threading, because they form stringy chips that adhere to the chasers.

The hardness for best results in threading stainless steel and other austenitic steels is more critical than that for carbon or low-alloy steel. The preferred range for threading austenitic steels is 200 to 225 HB. When these metals are significantly softer than 200 HB, they are gummy and adhere to the chasers, causing dimensional variation and poor finish, and sometimes damage to the chasers. When these metals are significantly harder than 225 HB, excessive chaser wear results.

The following example demonstrates the effect of hardness in die threading:

**Example 2: 4350 Steel Versus Gray Iron Versus Brass.** Table 3 compares operating conditions and tool life for cutting ½-20 UNF (Unified fine-thread series) threads in alloy steel, gray iron, and free-cutting brass parts (illustrated in the table), each at two different hardnesses. All threading was done in turret lathes using four-chaser dies having 30° chamfer angles.

As indicated, greater speed could be used for threading the softer alloy steel and brass parts, but the same speed was used for the gray iron part regardless of the hardness. For all three metals, however, tool life showed a substantial increase when the softer parts were threaded.

## Accuracy and Finish

In die threading, dimensional accuracy and surface finish are usually closely related; excessive dimensional variation is commonly accompanied by rough surfaces on the threads.

In addition to the composition and hardness of the metal being threaded (see the preceding section), the major variables affecting accuracy and finish are the type and condition of the machine and tool used (see discussion below), and the type of cutting fluid used (see Example 5 in the section "Cutting Fluids" in this article).

**Machines.** Other conditions being constant, closer dimensional control is invariably obtained in a machine that rotates the workpiece rather than the tool. For example, better dimensional control can be obtained in a turret lathe than in a drill press. Regardless of which type is used, the machine must be maintained at maximum capability by adjustments that compensate for wear in bearings and other moving parts.

**Tools.** To obtain a condition of balance, thread-cutting dies contain a minimum of three chasers (or, in solid dies, cutting lands). Increasing the number of chasers is helpful in maintaining close tolerances and good finish, because the greater the number of cutting edges, the lower the chip load per tooth. Five-chaser dies are often used for improving accuracy and finish.

Chasers with short chamfers (35 to 45° angles) may produce rough or substandard threads, especially on metals that are difficult to machine, because cutting forces on the leading teeth increase as chamfer length decreases. The relation between chamfer angle and the amount of metal removed in threading is tabulated with Fig. 9. These data show the desirability of using chasers having the longest possible chamfer, because as the length of chamfer increases, chips become thinner and chip load per tooth is reduced.

The cutting edges of chasers must be kept sharp by grinding or replacement, depending on the type of die. With self-opening dies, it is essential that all moving parts be hardened and that necessary compensation (by replacement if necessary) be made for wear. If dies having worn components are used, loss of accuracy will result.

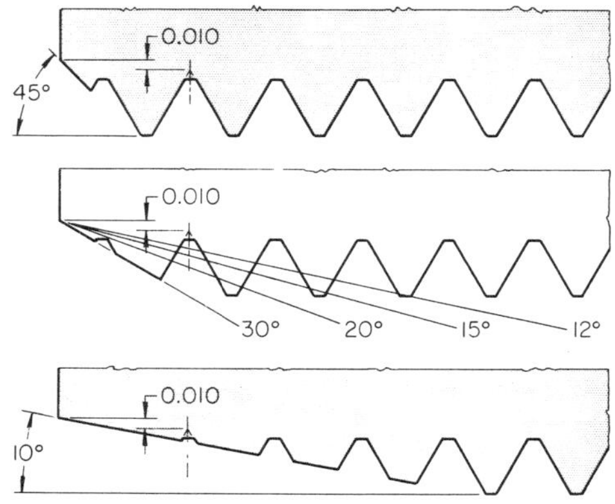


| Chamfer angle | Chip thickness mm | in. |
|---|---|---|
| 45° | 0.450 | 0.0177 |
| 30° | 0.318 | 0.0125 |
| 20° | 0.218 | 0.0086 |
| 15° | 0.165 | 0.0065 |
| 12° | 0.132 | 0.0052 |
| 10° | 0.109 | 0.0043 |

**Fig. 9** Relationship between chamfer angle and number of chamfered threads (in sketch) or thickness of chip cut per thread (in table) for 10-pitch Unified standard threads

## Thread Size

Other factors being equal, diameter of the part being threaded has a significant effect on procedure, production rate, and cost per piece threaded. In the following example, the initial cost for chasers was the same for two different diameters, but tool cost per piece threaded increased almost in proportion to diameter.

**Example 3: Threading Two Sizes of Carbon Steel Tubing.** The workpieces illustrated in Fig. 10 were threaded in a manually operated turret lathe of appropriate size, using self-opening dies. The workpieces, designated part A and part B in Fig. 10, were threaded at the same surface speed. Tool cost per piece threaded was 250% higher for the 70 mm (2¾ in.) diam than for the 22 mm (⅞ in.) diam thread, even though initial cost for chasers was the same.

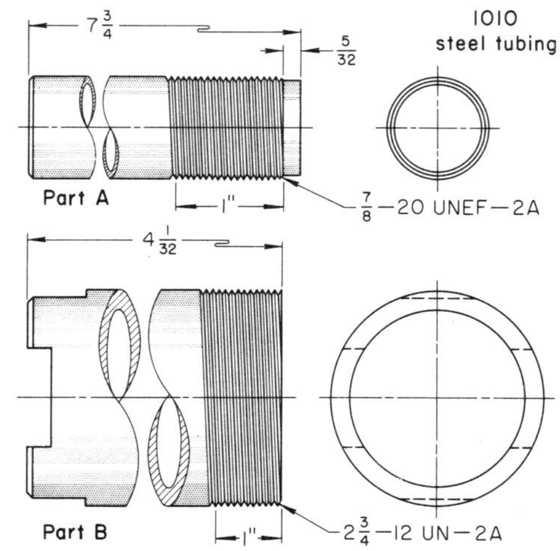


| Condition | Part A (⅞–20) | Part B (2¾–12) |
|---|---|---|
| **Chaser details** | | |
| Number per die | 4 | 4 |
| Pieces threaded per set | 800 | 320 |
| **Operating details(a)** | | |
| Speed, rev/min | 85 | 30 |
| Production, pieces/h | 120 | 41 |
| Tool life per grind, pieces | 200 | 80 |

(a) Soluble-oil emulsion used for both parts

**Fig. 10** Influence of thread size on productivity. Dimensions given in inches

## Threading to a Shoulder

A workpiece that is to be threaded close to a shoulder should have a relief groove wide enough to admit the full chamfer of the chaser and the first full thread, and to provide extra clearance for overtravel to allow the die to open without bumping the shoulder. Recommended minimum widths of relief grooves for chasers ranging in pitch from 32 to 4 and with four different chamfer angles are given in Table 4.

Workpiece design often limits the width of relief groove. When possible, the relief width should be revised to allow the use of chasers having maximum length of chamfer, to provide maximum efficiency of operation (see Fig. 9 and the previous discussion "Accuracy and Finish").

Aside from the danger of die breakage from hitting the shoulder, the two main disadvantages of threading to a shoulder are increased tool cost because of shorter tool life, and decreased productivity because of lower speeds and more downtime for changing tools.

## Speed

Cuts that are made too slowly increase machining time and raise production costs. Slow cuts across keyways can also produce rough, irregular surfaces. On the other hand, cuts that are made too quickly generate excessive heat, shortening tool life, and adversely affect thread accuracy.

However, higher-than-average speed can sometimes be a help in maintaining the accuracy of a cut. When cutting across keyways, holes, flats, parting lines, or other irregularities, a threading die makes a better surface if it is cutting faster than normal. Of course, tool life will suffer.

Table 5 gives nominal threading speeds for various materials. The effect of hardness on threading speed for two steels is shown in Table 6.

## Lead Control

Lead control is control of the axial advance of the chasers in die threading. Lead screws or other devices used for lead control are adjusted to advance the chasers at a rate suited to the pitch of the thread being cut.

The common types of commercially available lead control are those with:

- A hydraulic actuator
- An air actuator
- A cam actuator
- A hydraulic (or air) and cam combination
- Feed gears
- A lead screw and nut
- A lead screw with hydraulic control

**Table 4 Recommended minimum widths of relief grooves for cutting UN threads close to a shoulder**

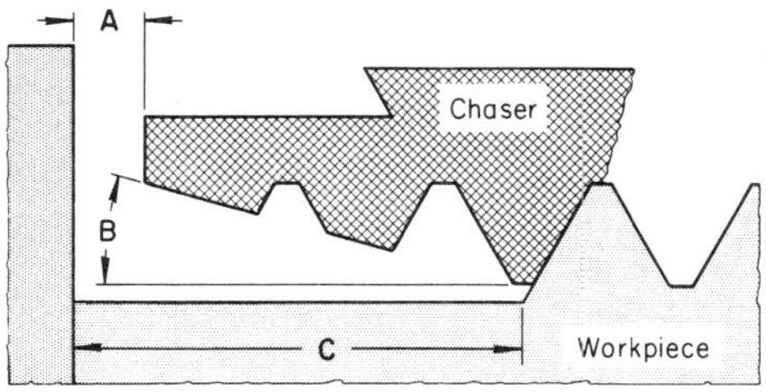


A = Extra clearance (for overtravel)
B = Chamfer angle (15° in sketch)
C = Total width required for relief

| Pitch | Extra clearance (A in drawing) | | Total width for relief (C in drawing) for chamfer angle (B in drawing) of: 45° | | 30° | | 20° | | 15° | |
|---|---|---|---|---|---|---|---|---|---|---|
| | mm | in. | mm | in. | mm | in. | mm | in. | mm | in. |
| 32 | 0.8 | 1/32 | 1.6 | 1/16 | 1.6 | 1/16 | 2.4 | 3/32 | 2.8 | 7/64 |
| 28 | 0.8 | 1/32 | 1.6 | 1/16 | 2.0 | 5/64 | 2.4 | 3/32 | 3.2 | 1/8 |
| 24 | 0.8 | 1/32 | 1.6 | 1/16 | 2.0 | 5/64 | 2.8 | 7/64 | 3.6 | 9/64 |
| 20 | 0.8 | 1/32 | 2.0 | 5/64 | 2.4 | 3/32 | 3.2 | 1/8 | 4.0 | 5/32 |
| 18 | 0.8 | 1/32 | 2.0 | 5/64 | 2.4 | 3/32 | 3.2 | 1/8 | 4.4 | 11/64 |
| 16 | 0.8 | 1/32 | 2.0 | 5/64 | 2.8 | 7/64 | 3.6 | 9/64 | 4.8 | 3/16 |
| 14 | 0.8 | 1/32 | 2.0 | 5/64 | 2.8 | 7/64 | 4.0 | 5/32 | 5.2 | 13/64 |
| 13 | 1.2 | 3/64 | 2.4 | 3/32 | 3.6 | 9/64 | 4.8 | 3/16 | 6.0 | 15/64 |
| 12 | 1.2 | 3/64 | 2.4 | 3/32 | 3.6 | 9/64 | 5.2 | 13/64 | 6.4 | 1/4 |
| 11½ | 1.2 | 3/64 | 2.4 | 3/32 | 3.6 | 9/64 | 5.2 | 13/64 | 6.7 | 17/64 |
| 11 | 1.2 | 3/64 | 2.4 | 3/32 | 4.0 | 5/32 | 5.6 | 7/32 | 6.7 | 17/64 |
| 10 | 1.2 | 3/64 | 2.8 | 7/64 | 4.0 | 5/32 | 6.0 | 15/64 | 7.5 | 19/64 |
| 9 | 1.2 | 3/64 | 2.8 | 7/64 | 4.4 | 11/64 | 6.4 | 1/4 | 7.9 | 5/16 |
| 8 | 1.2 | 3/64 | 3.2 | 1/8 | 4.8 | 3/16 | 6.7 | 17/64 | 9.1 | 23/64 |
| 7 | 1.6 | 1/16 | 4.0 | 5/32 | 5.6 | 7/32 | 7.9 | 5/16 | 10.3 | 13/32 |
| 6 | 1.6 | 1/16 | 4.4 | 11/64 | 6.4 | 1/4 | 9.1 | 23/64 | 11.9 | 15/32 |
| 5 | 1.6 | 1/16 | 5.2 | 13/64 | 7.1 | 9/32 | 10.7 | 27/64 | 13.9 | 35/64 |
| 4½ | 1.6 | 1/16 | 5.6 | 7/32 | 7.9 | 5/16 | 11.5 | 29/64 | 15.5 | 39/64 |
| 4 | 1.6 | 1/16 | 6.0 | 15/64 | 8.7 | 11/32 | 12.7 | 1/2 | 17.0 | 43/64 |

Without external lead control, accuracy of the thread depends partly on the skill of the operator and partly on the ability of the chasers to develop a "nut action." This nut action is influenced by the weight of any machine components that drag on the chasers.

Manual control is usually satisfactory for threading diameters up to 6.4 mm (1/4 in.) because the machine components that move with a threading die of this size are easily handled by the operator. As thread diameter or work metal hardness, or both, increase, the weight of the moving machine components is also increased, because heavier and more powerful equipment is required. Thus, accurate manual control becomes more difficult.

Under conditions of inadequate control, the leading threads will be poorly formed, and the remaining threads will be shaved undersize. Too much force applied by the operator or too much drag from moving components produces similar effects.

**Disadvantages** of lead control are:

- Initial cost of the lead control unit
- Susceptibility of the unit to wear, with resulting requirements for servicing and replacement
- The need for continuing maintenance
- The time required for resetting the unit to the starting point after each cycle, which usually lengthens the total time for performing a threading operation

## Cutting Fluids

Selection of a cutting fluid for die threading is essentially the same as for tapping. Sulfurized cutting oil is effective for most die threading applications. The two examples that follow describe applications in which results were improved when soluble-oil emulsions were replaced by sulfurized oil.

**Example 4: Improved Tool Life.** A soluble-oil emulsion (1 part oil to 20 parts water) was used as a cutting fluid for die threading fittings made of 1018 steel. The quality of the threads (3 3/16-12) was acceptable, but chaser life was only 180 pieces per grind. Changing the cutting fluid to straight sulfurized oil increased chaser life per grind to 225 pieces.

**Example 5: Improved Accuracy and Finish.** Class 3A fit and finish were required for 1½-12 threads on malleable iron castings die threaded at 9 m/min (30 sfm) using a 1-to-20 mixture of soluble oil and water as a cutting fluid. Thread flanks were excessively rough, and some pieces did not meet class 3A requirements. When straight sulfurized cutting oil was substituted, all parts met class 3A, and acceptable surface finish was maintained.

**Table 5 Nominal speeds for die cutting of screw threads**

| Work metal | Speed for pitch of: ≤7 | | 8–15 | | 16–24 | | ≥25 | |
|---|---|---|---|---|---|---|---|---|
| | m/min | sfm | m/min | sfm | m/min | sfm | m/min | sfm |
| **Ferrous metals** | | | | | | | | |
| Gray iron (class 30) | 7.5 | 25 | 12 | 40 | 15 | 50 | 24 | 80 |
| Malleable iron | 6 | 20 | 9 | 30 | 12 | 40 | 15 | 50 |
| Stainless steel | | | | | | | | |
| Types 303, 410, 420F, 440F | 3 | 10 | 4.5 | 15 | 6 | 20 | 7.5 | 25 |
| Types 416, 430F | 3.5 | 12 | 6 | 20 | 9 | 30 | 12 | 40 |
| Types 302, 304, 321 | 2.5 | 8 | 3 | 10 | 4.5 | 15 | 6 | 20 |
| Steel, free-cutting | 4.5 | 15 | 7.5 | 25 | 12 | 40 | 18 | 60 |
| Steel, other than free-cutting | | | | | | | | |
| Medium-carbon | 2.5 | 8 | 3 | 10 | 4.5 | 15 | 6 | 20 |
| Low-carbon | 3.5 | 12 | 6 | 20 | 9 | 30 | 12 | 40 |
| **Nonferrous metals** | | | | | | | | |
| Aluminum | 15 | 50 | 34 | 110 | 55 | 180 | 76 | 250 |
| Brass, free-cutting | 15 | 50 | 34 | 110 | 55 | 180 | 76 | 250 |
| Bronze, manganese | 9 | 30 | 18 | 60 | 30 | 100 | 46 | 150 |
| Bronze, aluminum(a) | 4.5 | 15 | 7.5 | 25 | 12 | 40 | 18 | 60 |
| Copper, free-cutting | 4.5 | 15 | 7.5 | 25 | 12 | 40 | 18 | 60 |
| Magnesium | 15 | 50 | 34 | 110 | 55 | 180 | 76 | 250 |
| Monel | 2.5 | 8 | 3 | 10 | 4.5 | 15 | 6 | 20 |
| Zinc | 15 | 50 | 34 | 110 | 55 | 180 | 76 | 250 |

(a) Also, nickel bronze or phosphor bronze

## Taper Threading of Pipe

The cutting of taper pipe threads with dies does not differ greatly from cutting standard screw threads. However, because American National standard taper pipe threads (NPT) are designed to provide pressure-tight seals, they require closer control in chasing than do thread forms for which 75% of full thread is acceptable. Also, in chasing taper pipe threads, more teeth are cutting at one time than in chasing standard screw threads; thus, taper pipe threading requires more power than screw threading, size for size. Machines used for threading pipe are usually the same as those used for other types of die threading. Positive lead control is essential in taper threading, because all chaser teeth are cutting at one time.

**Table 6 Effect of hardness on speed in threading 1040 steel and type 440A stainless**

| HRC | Speed(a) m/min | Speed(a) sfm |
|---|---|---|
| **1040 steel** | | |
| 6 | 6 | 20 |
| 20 | 4.5 | 15 |
| 30 | 3 | 10 |
| **440A stainless** | | |
| 20 | 3 | 10 |
| 30 | 2.5 | 8 |
| 40(b) | 1.5 | 5 |

(a) Based on cutting 8 to 25 threads per 25 mm (8 to 15 threads per inch). (b) Harder than the limit recommended for die threading

## Tools for Pipe Threading

**Solid dies** for cutting threads on pipe are of the same two general types as for cutting machine threads, that is, adjustable and nonadjustable. With either type, the die must back over the threads to be removed from the work when threading is complete.

Adjustable solid dies have removable chasers that can be adjusted for pitch diameter, and that can be reground or separately replaced when worn. When close control of pitch line taper is required under conditions of heavy stock removal, the adjustable solid die is preferred. Manufacturers of pipe fittings use adjustable solid dies to cut taper pipe threads on gray iron and malleable iron fittings without prior machining of the as-cast diameters.

Nonadjustable solid dies are only used for hand operations or for low-production threading in a simple machine. They are seldom resharpened when worn, and cannot be adjusted for pitch diameter.

**Self-opening dies** permit the individual cutting components to be automatically retracted from the work when threading is complete, so that the die can be withdrawn without the need to reverse rotation. Self-opening dies permit shorter machine cycles than do solid dies, by eliminating the backing-off time. Self-opening dies also increase chaser life because cutting edges are not dragged back over the previously cut threads. Disadvantages of self-opening dies in comparison with adjustable solid dies are higher initial and maintenance costs and greater susceptibility to becoming fouled from swarf-contaminated cutting fluids.

Self-opening dies are also used to cut straight pipe threads. Straight pipe threads, the same size as taper pipe threads (at the large end of the taper) are needed for tank fittings and other places where nuts must be used along with standard taper pipe connections. Straight pipe threads are also known as locknut threads (pipe). If a short pipe goes through a wall, as in tank fittings, a full-length straight pipe thread joins the taper pipe threads at each end. A locknut, with matching internal straight pipe thread, screws onto the thread at the face of the wall, while fittings screw on at the ends.

**Chasers.** Radial, tangential, and circular chasers, in addition to being used to produce screw threads (see the section "Self-Opening Dies" and Fig. 4 to 6 in this article), are also used for threading pipe. For high-production pipe threading, standard (stock) chasers are not always the best choice, regardless of type. Specially designed chasers should be considered for long production runs. For instance, in threading malleable iron pipe fittings that are threaded without prior machining, maximum rigidity is required because of the large and varying amount of stock removal. For such purposes, a specific type of solid, adjustable die equipped with tangential chasers has been designed. These chasers are ground with a 30° throat angle starting 0.25 mm (0.010 in.) below the root of the thread. Tungsten carbide chasers are long lasting in this application if they do not break, but they are susceptible to breakage because some workpiece castings are not well annealed. In general, nitrided high speed steel chasers are the most practical for this application.

## Stop Lines in Pipe Threading

The prescribed taper in pipe threads is 1.6 mm (1/16 in.) in diameter for each inch of length. Thus, each chaser tooth cuts a progressively smaller diameter on the workpiece as the die advances. Because each chaser tooth is cutting at the instant of die opening or reversal, a stop line remains on the thread surface. When solid dies are used, the reversal of the chasers over the stop lines creates an abrasive action that causes excessive wear of the cutting edges.

The kind of metal being threaded has a marked effect on the severity of stop lines. In general, stop lines become more pronounced as ductility of the work metal increases. With solid dies, the severity of stop lines can be reduced by increasing the number of chasers in the die. However, when specifications call for complete elimination of stop lines, a full-receding type of self-opening die is required; otherwise the piece must be threaded with a single-point tool.

In one plant in which alloy steel pipe was being taper threaded with solid dies, stop lines were severe. When the dies were reversed for removal, fine chips were trapped between the cutting teeth and the threads, resulting in damage to both. Rejection rate because of scarred threads was excessive. Changing to a self-opening die and frequently resharpening the chaser teeth solved the problem. The die was adjusted so that it tripped just prior to full advance of the spindle; chasers retracted smoothly, and stop lines were eliminated.

## Speed in Pipe Threading

Speed, in meters per minute or surface feet per minute, is usually 20 to 30% slower for pipe threading than it is for the cutting of screw threads (see Table 5), because more teeth are cutting during pipe threading. In some shops, however, little or no distinction is made between straight threads and taper threads in selecting threading speed. Also, in some shops the same threading speed is used regardless of thread pitch. For instance, one manufacturer taper threads malleable iron as-cast surfaces that range in pipe size from 3.2 to 102 mm (1/8 to 4 in.) at the same surface speeds, using the same type of dies and cutting fluid. Straight pipe threads may be cut at faster speeds, speeds that equal those used in cutting unified threads of equal size.

Thread quality does not necessarily increase as threading speed decreases; rough and torn threads often result from threading speeds that are too low. Excessive speeds, however, always decrease tool life.

**Cutting fluids used for pipe threading** are the same as for the cutting of screw threads. Flooding the cutting area is generally more important in pipe threading because more cutting teeth are engaged at one time.

In addition, because of the accuracy required in pipe threads and the amount of cutting that is being done at one time, susceptibility to thread damage from swarf-contaminated cutting fluids is greater in pipe threading than in screw threading. Damage from this source can be prevented by filtering the cutting fluid.

# Milling

MILLING is a machining process in which metal is removed by a rotating multiple-tooth cutter; each tooth removes a small amount of metal with each revolution of the spindle. Because both workpiece and cutter can be moved in more than one direction at the same time, surfaces having almost any orientation can be machined.

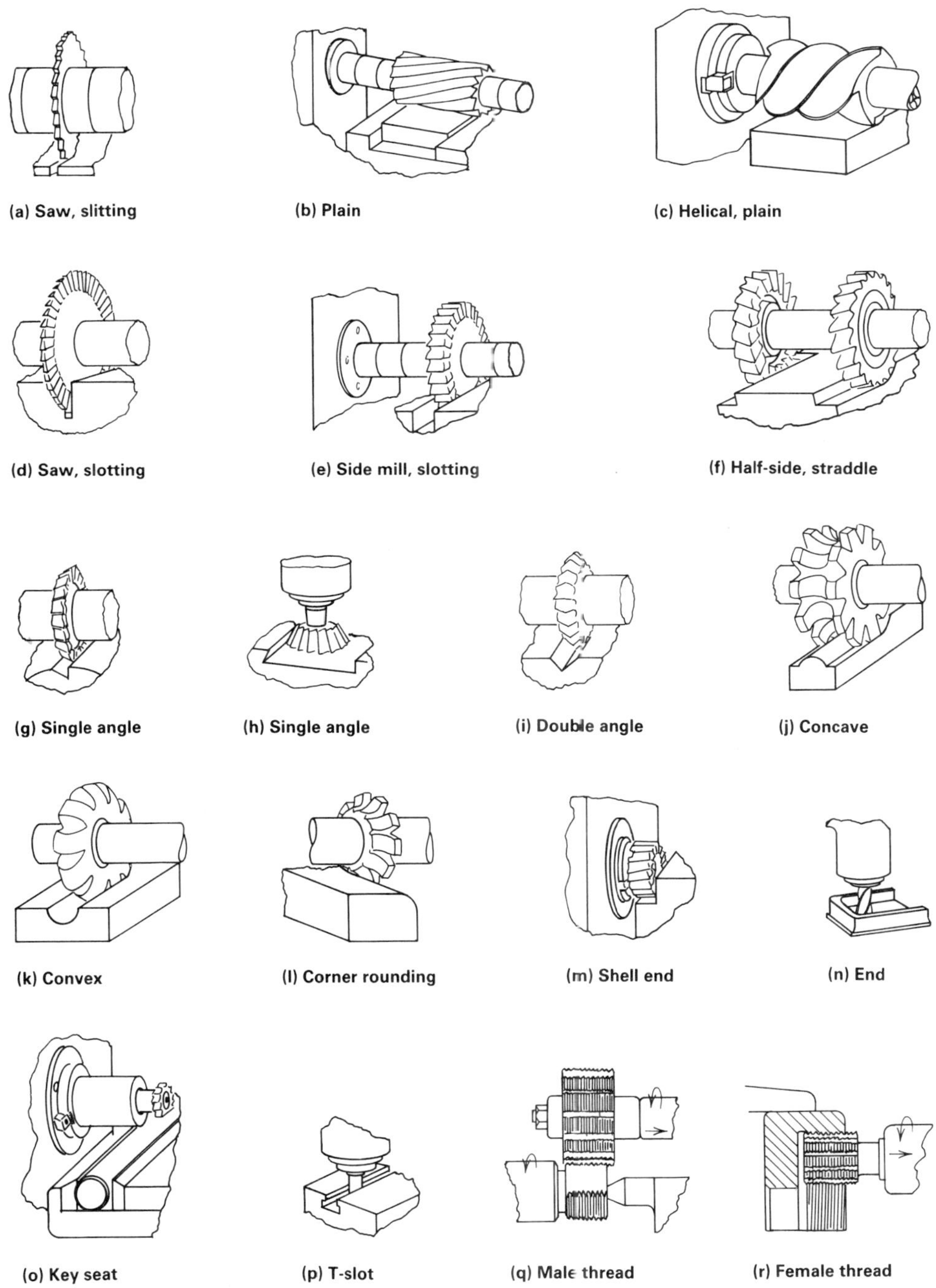


**Fig. 1** Basic milling operations and cutters illustrating the variety of surfaces and surface combinations that can be generated

The principal differences between milling and other machining processes are:

- The interruptions in cutting that occur as the teeth of the milling cutter alternately engage and leave the workpiece
- The relatively small size of the chips in milling
- The variation in thickness within each chip

Chip thickness varies during the cut of any individual tooth because feed is measured in the direction of table motion (workpiece moving into cutter), while chip thickness is measured along the radius of the cutter.

Milling is most efficient when the work is no harder than 25 HRC. However, steel at 35 HRC is commonly milled, and steel as hard as 56 HRC has been successfully milled. The effect of high hardness of the work metal on cutter design is discussed later in this article.

## Milling Machines

Most milling is done in machines designed primarily for milling, but milling has been done in almost every type of machine that can rigidly hold and rotate a cutter while allowing a workpiece to be fed into the cutter, or vice versa. Figure 1 illustrates the wide range of cutter configurations available for use in milling operations. For example, milling can be done in single-spindle or multiple-spindle bar and chucking machines in sequence with other machining operations (see the section "Automatic Lathes" in the article "Multiple-Operation Machining" in this Volume). Conversely, some specially designed milling machines include provisions for other machining operations. Machines used for production milling should incorporate:

- Variable-speed motors
- Ball-screw or hydrostatic worm precision gears and bearings
- Hydrostatic liners
- Numerical control (NC) or computer numerical control (CNC)
- Automatic drives
- Automatic tool changer

Spindle orientation and construction type are two designations used to classify milling machines. Machines that drive cutters with horizontal spindles are the most common, although vertical spindles are also widely

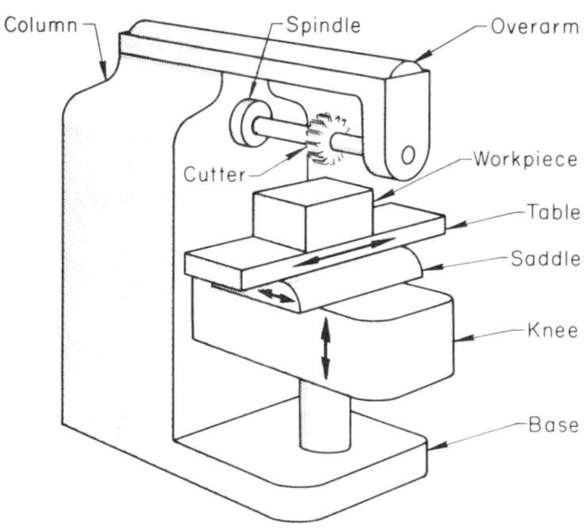


**Fig. 2** Principal components of a plain knee-and-column milling machine with a horizontal spindle

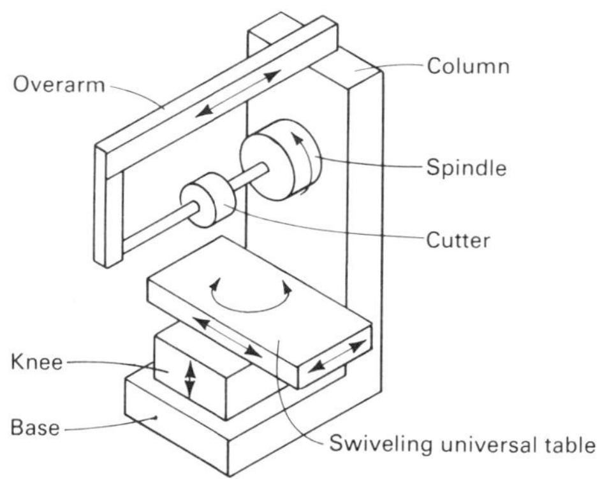


**Fig. 3** Principal components of a universal horizontal knee-and-column milling machine

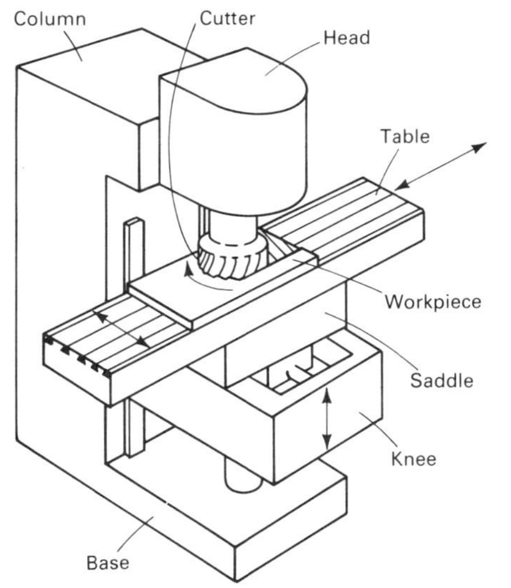


**Fig. 4** Principal components of a vertical-spindle knee-and-column milling machine

used. Some special-purpose machines have horizontal, vertical, and angular spindles that operate consecutively, simultaneously, or both. Tape-controlled milling machines can have up to five axes of motion: standard $x$-, $y$-, and $z$- axes plus two degrees of rotary motion. The A-axis is the rotary motion of the spindle carrier in the vertical plane, while the B-axis is the rotary motion of the column on its base in the horizontal plane.

The type of construction is the basis for classifying machines as:

- Knee-and-column type
- Bed-type
- Planer-type
- Special types

The first two types are available in a wide range of capacities and with many modifications and accessories. Special machines are individually designed and constructed for specific requirements.

## Knee-and-Column Machines

Knee-and-column machines (Fig. 2), in which the milling spindle is fixed, consist of six basic components:

- The base, which supports the other components
- The column, which contains the spindle and its driving mechanism
- The overarm, which provides support for arbor-mounted cutters
- The knee, a separate section attached to the column and movable vertically on the column
- The saddle, which is supported by and moves horizontally (in and out) on the knee
- The table, which supports the workpiece and can move horizontally at right angles to the saddle movement

Knee-and-column milling machines are designed for general-purpose work and can perform not only the straight milling of plane and curved surfaces but also gear and thread cutting, drilling, boring, and slotting (when suitably equipped). These machines can be classified as follows:

- Plain
- Universal
- Vertical
- Turret-type universal
- Ram-type universal

**The plain knee-and-column machine** permits three-dimensional movement of the workpiece by the up-and-down motion of the knee and by the two-direction horizontal motion of the saddle and table. This machine has a horizontal hollow spindle in the column to drive the cutters. Arbors and adapters are held by a quick-change clamping sleeve on the nose of this spindle, but on many machines a draw bar from the rear is screwed into arbors to hold them in the spindle. The overarm extending from the top of the column carries supports for cutter arbors. Braces are available to tie the arbor supports and overarm to the knee for added rigidity. The knee, saddle, and table are positioned and fed (by hand or power) in the three coordinate directions shown in Fig. 2. One model of a general-purpose milling machine has the saddle and the table on a fixed base instead of a knee with the cutter spindle in a housing sliding on vertical ways on the column. This design provides rigid support for heavy workpieces and retains the flexibility needed for general-purpose milling.

**A universal knee-and-column machine** is similar to the plain type except that the table is mounted on a housing that can be swiveled in a horizontal plane (Fig. 3). This machine is especially useful for milling angles and helices, as found in twist drills, milling cutters, and helical gear teeth. The saddle is made in two parts. The upper part swivels on the lower and carries the table with it—to a limit of about 45°. The leadscrew that feeds the table is arranged so that a geared drive can be taken from it to drive a dividing head on the table. Some universal-type machines have additional features. On one unit, the knee can be swiveled to tilt the table end over end, and an auxiliary spindle head on the side of the column can be swiveled to almost any angle.

Most knee-and-column machines have a removable head that is attached to the column and converts the machine into a vertical-spindle machine. Some of these heads can be swiveled 180°, thus creating a universal attachment.

**Vertical-spindle knee-and-column machines** are also available (Fig. 4). Large vertical-spindle machines have the spindle and driving mechanism mounted on a ram attached to an overhang on the column. On small machines, the vertical spindle and driving components are attached to a horizontal slide at the top of the column. Vertical-spindle machines are especially well suited to face milling and end milling operations.

**Turret-type knee-and-column machines** (Fig. 5) have dual heads that can be swiveled about a horizontal axis on the end of a horizontally adjustable ram. This permits the milling to be done horizontally, vertically, or at any angle. This added flexibility is advantageous where a variety of work has to be done, as in tool and die or experimental shops. Turret-type machines are available with either plain or universal tables.

**Ram-Type Universal Machine.** Another modification of the knee-and-column principle is the ram-type, or ram-head, milling machine (Fig. 6). This design has two distinguishing features. First, the horizontal- and vertical-spindle knee-and-column milling machine principles are combined into one design; its milling head can be positioned for horizontal, vertical, or angular milling by simple adjustment, without the addition or removal of any attachments or accessories. The milling head, situated at or near the front end of the ram, is a single- or double-swivel-mounted spindle in fixed or quill mounting.

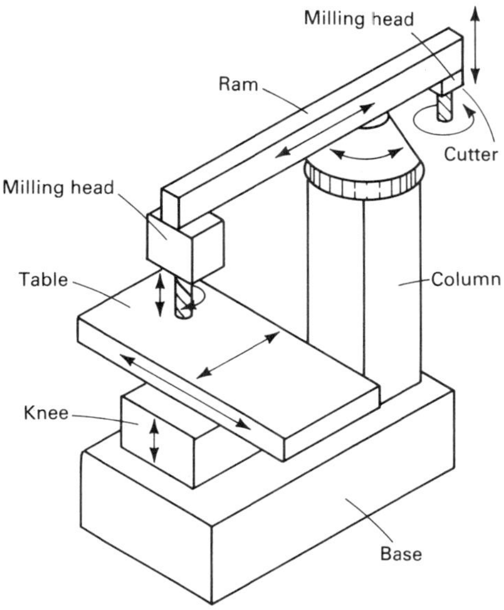


**Fig. 5** Principal components of a turret-type knee-and-column milling machine

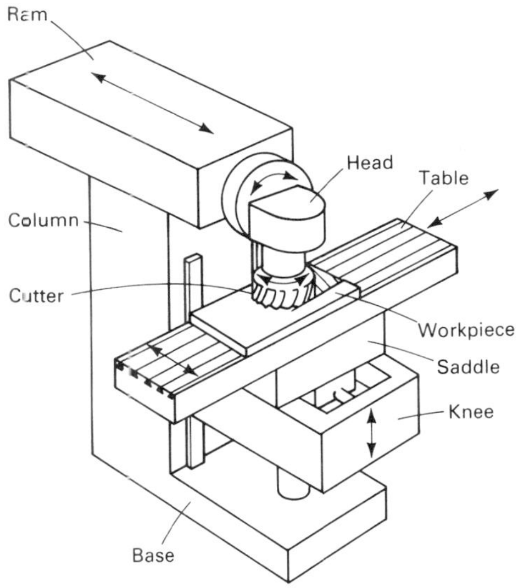


**Fig. 6** Principal components of the ram-type knee-and-column milling machine. This ram can be moved in a direction parallel to the saddle movement.

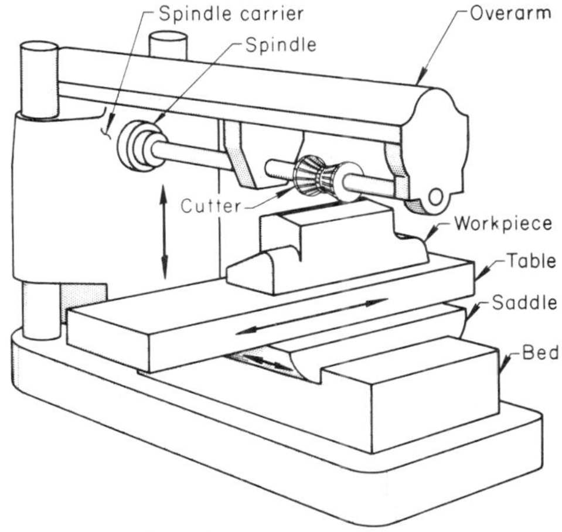


**Fig. 7** Principal components of a horizontal-spindle bed-type milling machine

The second feature is that the ram, which carries the milling head, can be moved horizontally outward or inward with respect to the table, thus bringing the milling cutter to the workpiece for maximum rigidity in setup. The table is equipped with cross and longitudinal power feeds, and the saddle assembly has vertical travel; these means afford tridirectional table movement.

**Milling Machine Size.** The usual power range available for knee-and-column machines is 0.75 to 37 kW (1 to 50 hp). These machines are also rated according to the size of workpiece they can handle. A numbering system is used to designate milling machines in accordance with the largest workpiece that can be milled; the higher the number, the larger the maximum workpiece size. Milling machine size is designated by numbers from 1 through 6, which are approximate indicators of the longitudinal table travel, as follows:

| Size | Table travel mm | Table travel in. |
|---|---|---|
| 1 | 559 | 22 |
| 2 | 711 | 28 |
| 3 | 813 | 32 |
| 4 | 1015 | 40 |
| 5 | 1270 | 50 |
| 6 | 1525 | 60 |

These relationships are not standardized, and a particular machine may vary considerably from these values. A typical No. 3 machine has maximum table, saddle, and knee movement of 812, 305, and 508 mm (32, 12, and 20 in.), respectively. A No. 6 machine has a table travel of 1525 mm (60 in.), cross travel of 457 mm (18 in.), and a vertical travel of 508 mm (20 in.).

**Advantages and Limitations.** The method of operating knee-and-column machines varies from manual control of all movements to power control of all movements. In addition, through the use of stops and other control devices, the machine can be adapted for automatic cycles. Because of their versatility in machining workpieces of different sizes and shapes, these machines are widely used for low-production milling.

The main disadvantage of the knee-and-column machine is its inherent lack of rigidity. The machine has three or four joints with sliding fits, each requiring a minimum of 0.051 mm (0.002 in.) clearance. The joint between the knee and the column bears the combined weight of the knee, saddle, table, fixture, and workpiece. Deflection is therefore inevitable under milling stress. To add some rigidity and to reduce deflection, an outboard support can be placed between the overarm and the knee. To prevent chatter, limitations must be imposed on feed, speed, and depth of cut in milling. Chatter will cause a decrease in dimensional accuracy, unacceptable surface finish, and reduced cutter life.

## Bed-Type Machines

Bed-type machines, also known as fixed-bed machines, are characterized by the extremely rigid construction afforded by a rectangular bed casting or weldment. These units are supported and leveled along their entire length, which can vary from 0.9 to 9 m (3 to 30 ft) or more.

The principal components of a bed-type machine are shown in Fig. 7. These machines are almost as versatile as knee-and-column machines, and they have at least 50% greater rigidity. In a bed-type machine, the table and saddle are mounted on a bed of fixed vertical position, and vertical movement is obtained from the spindle carrier. Consequently, less weight is moved on guideways, which are about 50% longer for a given size of machine as compared to the movement of the knee. In addition, spindle overhang is reduced by as much as 75% because the cutter can be located closer to the vertical way. Bed-type machines are available in horizontal-, vertical-, and multiple-spindle versions.

**Horizontal-Spindle Bed-Type Machines.** The horizontal machine consists basically of a headstock or column bolted to, or integral with, a fixed bed. On this member, a spindle carrier, head, or block is mounted, carrying a spindle that is horizontally positioned. The axis of this spindle is parallel to the table surface and is at right angles to the axis of table movement. The carrier usually moves (manually or automatically) in a vertical direction on the headstock, and the spindle is adjustable axially (in and out) through a spindle quill or ram. Because of the more rugged construction of this type of machine, high-power cuts are made more easily. In addition, better surface finishes and closer machining tolerances can generally be obtained than are possible with knee-type machines.

Horizontal-spindle machines are also available with tracer control to the vertical axis. This permits various shapes and contours to be generated by causing the spindle carrier or head to raise or lower at a specified time, as dictated by a cam or template mounted to the table of the machine. These units are commonly called rise-and-fall milling machines.

Machining cycles can be manual or automatic and are actuated by preset dogs and electric switches or hydraulic plungers. Numerical control systems and similar systems are also frequently used, with all machine functions being programmed.

Horizontal-spindle machines are designated by sizes 10, 14, 18, 22, and 30. Each size may have a range of dimensions. For each dimensional size, machines are usually made for light, medium, and heavy duty, with corresponding strength and power.

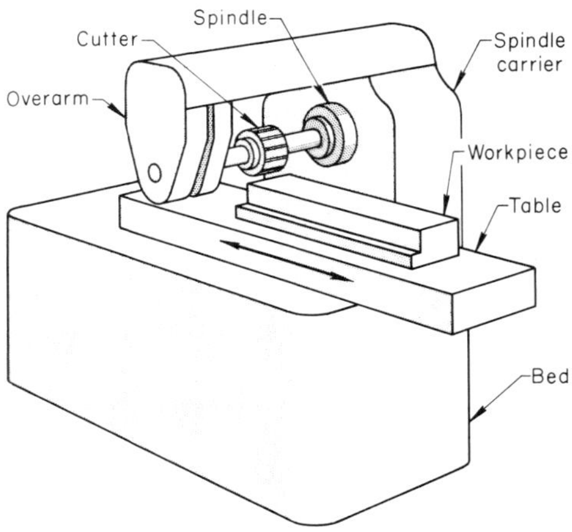


**Fig. 8** Principal components of a manufacturing-type milling machine with one spindle carrier

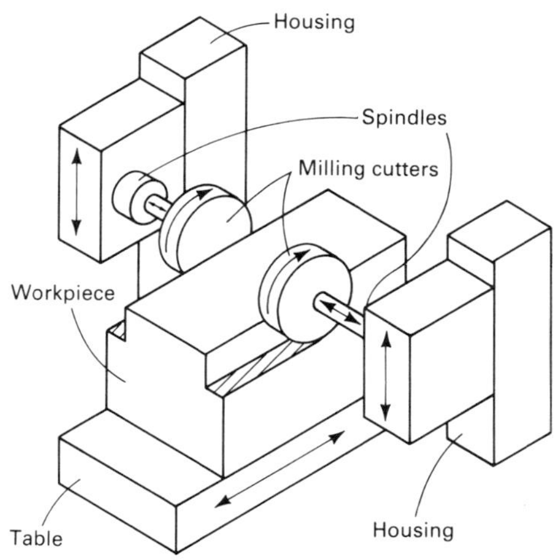


**Fig. 9** Principal components of a duplex horizontal milling machine with two horizontal spindles

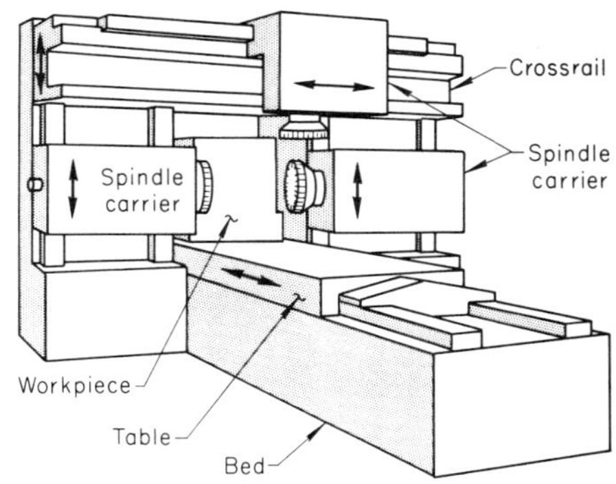


**Fig. 10** Principal components of a triplex planer-type milling machine

**Vertical-spindle bed-type machines** incorporate a rigid headstock on which the entire spindle carrier can slide up and down as required. The vertical milling machine differs from the horizontal machine principally by the position of the carrier and spindle axis. A ram or spindle carrier is mounted on a rear base, which is normally fastened to the fixed bed. This carrier houses the vertical spindle, whose axis is perpendicular to the top of the table and is adjustable vertically along its own axis through a quill. The carrier assembly is sometimes fixed (with no movement) and in some cases is adjustable manually or by power to travel laterally, so that the spindle can be positioned crosswise to the table (forward or to the rear of the table centerline).

The vertical machine is well suited to face milling operations and can handle heavy cuts with close tolerance and finish requirements. Although face milling and end milling are the most common operations performed on this type of machine, arbor-mounted cuts can also be made; however, these are the exception rather than the rule. As a result, the vertical machine is more limited and less versatile than the horizontal machine.

For close tolerances, good finish requirements, and large-lot production with minimum time per part, the vertical milling machine, properly tooled, is a valuable asset. Tracer and numerical control can be adapted to vertical machines. These machines are available in a wide range of sizes and variations. Automatic cycles and numerous accessories are offered as options.

Specific advantages of the bed-type machine in comparison with the knee-and-column machine are:

- Greater rigidity, permitting heavier cuts and closer dimensional control
- Constant reloading level of the table
- Controls at a uniform level
- Greater range of vertical movement
- Versatility that can be increased by providing a longer saddle and outboard support

The main disadvantage of the bed-type milling machine is its high initial cost.

**Simplex, Duplex, and Triplex Models.** Milling machines of the bed type are designed for production and therefore are also termed manufacturing or production millers. They are available as standard models with up to 225 kW (300 hp) and can be modified or equipped as needed for mass production. On some units, the spindle carrier is a ram slide mounted on a vertical column.

The machine illustrated in Fig. 8 is a simplex model (one spindle carrier). Duplex models have a headstock and spindle carrier on each side of the bed. This configuration allows the simultaneous operation of opposing cutters in a single pass, thus doubling productivity with a relatively low (about 30%) increase in investment (Fig. 9). All carrier features, speeds, feeds, and ranges are duplicated. Maintaining accurate parallelism by the two-spindle setup is one of the advantages of the duplex machine. A logical extension of the duplex model is the triplex model, which has three spindles to mill three surfaces (engine blocks, for example) simultaneously in one pass.

## Planer-Type Machines

Planer-type machines are so termed because of their structural resemblance to a planer. Known also as adjustable-rail or fixed-bridge milling machines, they can provide almost any combination of vertical, horizontal, or angular spindles for driving milling cutters and boring bars. Planer-type milling machine spindle drives generally range from 22 to 75 kW (30 to 100 hp). These machines can also perform several milling and boring operations simultaneously. The quill, carrying the spindle, is fed in and out for drilling and boring.

**Fixed-Bridge Milling Machines.** The machine illustrated schematically in Fig. 10 is a triplex model (three spindle carriers). On a triplex machine, workpieces are secured to the table, which carries them back and forth between the two vertical columns and under the crossrail. Two horizontal spindle carriers move vertically on the columns, and a vertical spindle carrier moves horizontally on the crossrail. The crossrail can move vertically on the two columns.

Figure 11 shows that the planer-type milling machine can utilize one to four milling spindle heads, with the spindles either parallel or perpendicular to the movement of the table. These machines overcome the inherent disadvantages of ordinary planers, which can use only single-point cutting tools and cannot be reciprocated rapidly (see the article "Planing" in this Volume). Planer-type milling machines utilize several milling heads to remove large amounts of metal while permitting the table and workpiece to move quite slowly, but they often require only a single pass of the cutters. This is a great advantage where heavy workpieces such as machine tool bases, aircraft wing spars, and missile bodies are being machined.

**Moving-Bridge or Gantry-Type Milling Machines.** Some large milling machines, similar to planer-type mills, are built on the gantry principle. The workpiece is mounted on a stationary table, and the milling heads are mounted on a traveling column, also called a gantry, driven by a ball screw or a rack-and-pinion drive system (Fig. 12). The maximum working capacities of such a machine are as follows: length, 9000 mm (354 in.); width, 4900 mm (193 in.); and height, 4000 mm (157 in.).

The spindle carriers mount to a saddle that travels in the cross, or *y*-axis, direction on a way system. They can also be moved in the *z*-axis, or vertical, direction, either independently or by means of a saddle that rides on a set of vertical ways. As with planer or bridge-type machines, the number of spindle carriers is optional, and they can be mounted on either side of the crossrail. Additional axes of motions are available if

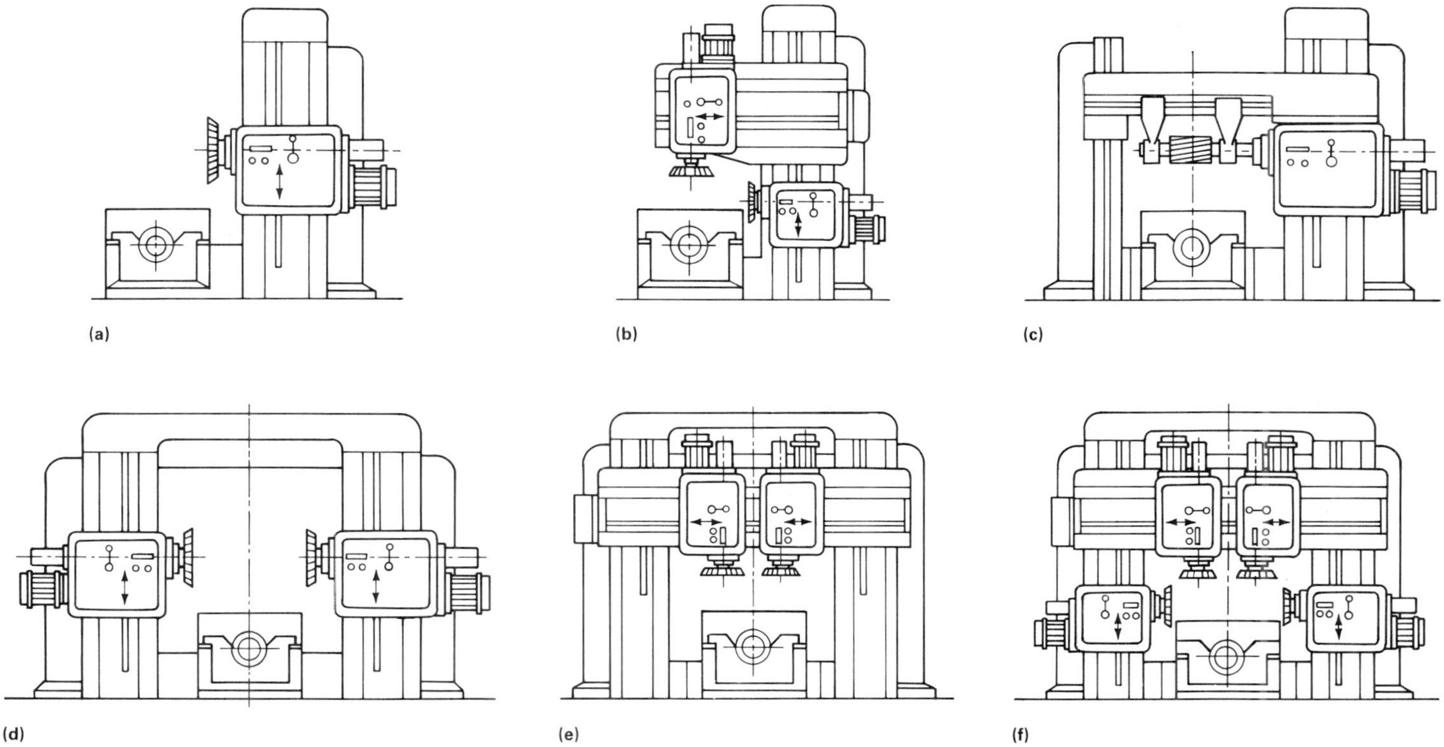


**Fig. 11** End views of planer-type milling machine configurations in addition to that shown in Fig. 10. Top row: single-column machines with one milling spindle head (a); overhanging beam and two milling spindle heads (b); and cross slide, support stand, and one milling spindle head (c). Bottom row: double-column machines with cross beam and two milling heads (d); cross beam, cross slide, and two milling spindle heads (e); cross beam, cross slide, and four milling spindle heads (f)

required; one is usually a swivel motion in the *yz*-plane, and the other a swivel motion in the *xz*-plane. These motions can be either full contouring or positioning only, making possible the machining of complicated geometric shapes such as marine propeller blades. The crossrail is usually designed so that it can be raised for additional clearance over the work surface.

*Skin Mills*. Gantry-type machines are a common style of structure for a variety of special machines used by the aerospace industry. These machines are used to manufacture large wing, missile, and rocket skins. The increased size of aircraft and missiles has dictated machine sizes with up to 6.1 m (20 ft) or more of work width and up to about 60 m (200 ft) of longitudinal travel. Most skins are aluminum, but in recent years skins of steel and titanium have also been manufactured by these machines.

*Template and honeycomb mills* are another common gantry-type machine. These machines are usually smaller than a skin mill and are of much lighter construction because template and honeycomb work requires high-speed, low-torque spindle carriers. Honeycomb milling often requires a multiple-axis machine. In addition to the usual three axes (*x*, *y*, and *z*), one or two swivel axes are generally supplied.

*A spar mill* is also a gantry-type machine, usually with a long longitudinal travel but a very narrow work surface (610 to 914 mm, or 24 to 36 in.). Spar mills usually have horizontal spindles with a swivel axis of motion and/or a vertical spindle, also with a swivel motion. This swivel motion is essential because of the warps inherent in aircraft spars. Recent spar mills have been designed to mill aluminum, steel, or titanium spars. Some seven-axis CNC spar mills are capable of milling workpieces that are 33 m (108 ft) long, and they can machine a pair of opposite-hand (mirror-image) workpieces simultaneously with two spindles.

The gantry-type machine offers several advantages over the planer or fixed-bridge construction. The gantry-type machines maintain the inherent stiffness of the closed-type construction of a bridge machine, thus allowing for heavy cuts with minimum deflection. With *x*-axis travels exceeding the upright length, the gantry machine offers floor-space savings over a similar bridge type. The gantry machine also presents a stationary work surface that facilitates double loading and offers the option of extending the beds for additional *x*-axis travel. This adding of bed sections can take place later if desired. Many installations of gantry machines have several gantries with overlapping travels on one long continuous bed.

The chip removal and coolant problems of gantry-type machines are similar to those of the fixed-bridge machines. However, because the gantry-type machines are usually longer and have a way system running along each side of the bed, the problems are somewhat magnified. Again, as with the fixed-bridge machines, the operator must be positioned on the work surface, or the spindles must be moved to an extreme *y*-axis position to replace or exchange cutters.

Planer-type machines are large and represent a major capital investment. Therefore, their use is generally restricted to removing large amounts of metal from massive workpieces such as mill or power-plant components or to mass-production milling in which identical workpieces are arranged in a row to be milled together. Tables are often split to permit simultaneous setup and milling of the work.

## Special-Purpose Machines

Innumerable special-purpose milling machines have been designed and built, sometimes for use on special workpiece configurations, but more often for high-volume production of a specific part. Most special-purpose machines are combinations of two or more of the basic machines described earlier in this article.

The cost of special-purpose milling machines can be justified only for the continuous high production of identical workpieces or if several operations can be combined in one handling to reduce manufacturing costs. Specially designed machines are usually more readily adapted to automatic control than are standard machines, although standard machines can be automated to various degrees. Special-purpose machines include the following:

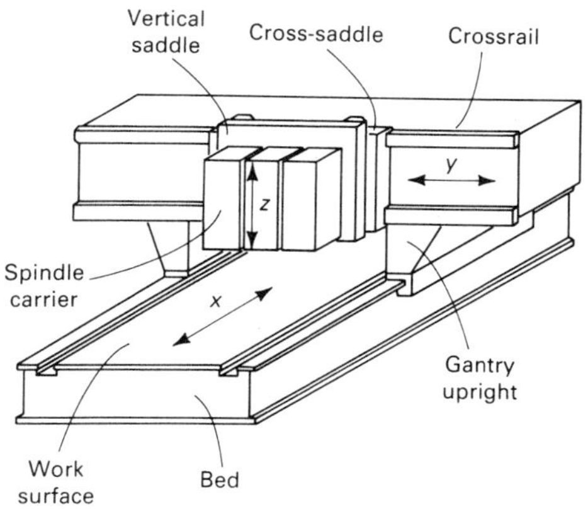


**Fig. 12** Principal components of a moving-bridge or gantry-type milling machine

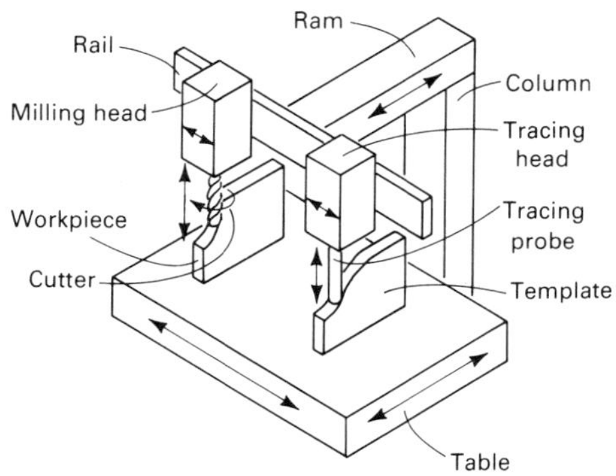


**Fig. 13** Principal components of a vertical-spindle profiler

- Profilers
- Duplicators (die sinking machines)
- Rotary millers
- Planetary millers

**Profilers.** Milling machines that can duplicate external or internal profiles in two dimensions are called profilers or tracer-controlled contouring machines. As shown in Fig. 13, a tracing probe follows a two-dimensional template and, through electronic or air-actuated mechanisms, controls the cutting spindles in two mutually perpendicular directions. The spindles (usually more than one) are set manually in the third dimension.

**Duplicators** produce forms in three dimensions. A tracing probe follows a three-dimensional master. Often the probe does not actually contact the master; a variation in the length of a spark between the probe and the master controls the drives to the quill and the table, thus avoiding wear on the master or possible deflection of the probe. On some machines, the ratio between the movements of the probe and cutter can be varied.

Duplicators are widely used to machine molds and dies. They are extensively used in the aerospace industry to machine parts from wrought plate or bar stock as substitutes for forgings; in such cases, the small number of parts required would make the cost of forging dies uneconomical. To a great extent, profilers and duplicators have been replaced by NC machines, in which a punched tape or a computer input eliminates the need to make a template or a master mold. Die shops are also using double-column, table-type vertical/horizontal portal-type machines in place of profilers and duplicators.

**Rotary millers,** usually of the vertical-spindle type, are used for such operations as contour and slab milling, channeling, milling tongues and jaws, trepanning, and end milling. These machines, however, are suited primarily to the production of large, heavy components in small or medium quantities. Where mass production is contemplated on parts with surfaces that can be face milled in a single pass, a single-purpose type of rotary miller is employed. Jigs or fixtures mounted on the table provide for virtually continuous cutting; parts are loaded, passed first beneath the roughing and then the finishing spindle, and automatically released for replacement with a rough part.

*Rotary-Table Milling Machines*. Some types of face milling in mass-production manufacturing are often done on rotary-table milling machines. These machines are adaptations of the vertical milling machine to a specialized use. In this case, there are two vertical spindles, each equipped with a facing mill. Cylinder heads, for example, are roughed at the first station and then finish milled as they pass the second station, while the workpiece is held in fixtures on the rotating table. The operation is continuous, and there is ample time for the operator to load and unload the machine during the milling. This machine is fast but is limited to the milling of flat surfaces.

*The rotary-head milling machine* is an unusual departure from conventional rotary machines. Suited primarily to tool, die, and small-quantity production, this miller adds to the versatility of the rotary-table machine but is designed for much smaller parts. Intricate radial cam work can be readily produced from drawings without the use of templates.

*Rotary-Drum Milling Machine*. Having as many as five horizontal spindles, this machine is designed for the mass production of large parts such as motor blocks, gear cases, and clutch housings. On these machines, parts are carried in fixtures mounted on a drum, which rotates continuously, carrying the parts between adjustable face mills.

*The rotary vertical-offset milling machine* also uses a rotating table with fixtures. Cutters are carried on a short vertical spindle, which rotates on an axis eccentric to the table axis. Suited primarily to the production of small machine parts, the offset miller can perform facing, slotting, sawing, straddle milling, and some simple form milling.

**Planetary milling machines,** also known as eccentric-drum millers, employ eccentric drums that carry the milling spindle. These units are so named because of the automatic planetary action obtained (Fig. 14). Adapted to internal or external thread milling where high production is contemplated, the planetary machine simplifies the manufacture of parts that are difficult to hold on swing. It also simplifies the production of concentric bores or diameters, either plain or threaded. Both cutter rotation and feed are provided by the eccentric drums; the part being milled is held stationary during the processing. For thread milling with multiple-thread mills, the drums are provided with a lead screw feed, and they finish an entire thread in one and a small fraction of a revolution. The work possible ranges from about 7.9 to 508 mm (5/16 to 20 in.) diameter and from about 16 to 508 mm (5/8 to 20 in.) in inside diameters with thread leads up to 28.58 mm (1.125 in.). Length of milled surfaces in most cases should not exceed 1 to 1½ times the diameter. For special operations, many single-purpose variations of the planetary principle in both horizontal and vertical models are employed. However, in most cases, these machines require special cutters and fixtures and are therefore suited primarily to mass-production output.

**Machining Centers.** In terms of construction, many machining centers are similar to the basic types of manual milling machines discussed earlier (see the section "Machining Centers" in the article "Multiple-Operation Machining" in this Volume). Some small machining centers are patterned closely after the ram-type milling machine. Vertical-spindle machining centers in the medium size category are constructed similarly to fixed-bed vertical milling machines. Both fixed-bed and gantry-type planer milling machines are commonly fitted with numerical controls, although this type of machine does not meet the definition of a machining center.

Medium-size horizontal-spindle machining centers, on the other hand, do not have a common manual twin; the bed is fixed, but the machine is capable of more motions than all but a few fixed-bed horizontal machines. A table and saddle combination provides feed motion in the longitudinal direction and to and from the column face. Vertical-spindle feed is provided by mounting the spindle carrier on column ways or by mounting the spindle in a quill.

The machining center offers a number of basic advantages relative to manual milling machines. Productivity is generally much higher, even though metal removal rates may vary only slightly. The reason is that the machining center generally spends about 70 to 80% of its time cutting chips, compared to about 20 to 30% for a manual milling machine. The reasons for this include the following:

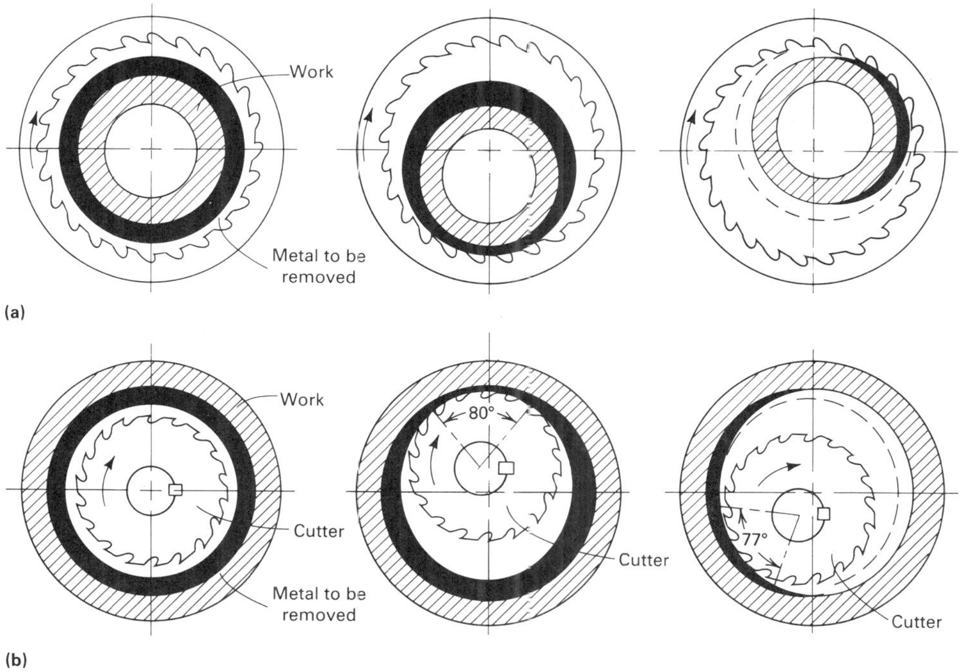


**Fig. 14** Planetary action and cutter path of an eccentric-drum miller. (a) Internal work. (b) External work. Cutter is shown in neutral position for loading (left), in radial feed to depth (center), and in planetary feed around work (right).

- The machining center can perform many different operations in addition to milling; this often eliminates the need to move the part to another machine
- It can work on several sides of the part without a new setup
- It can often accurately hold close tolerances and relationships without special jigs or fixtures

The machining center also provides for greater repeatability than a manual milling machine. Although a highly skilled machinist can hold very close tolerances on a manual machine, a machining center can hold close tolerances day after day, without depending on the skill and attentiveness of the machinist. Furthermore, a tape can be put away and reused several months later; it will still produce identical parts.

The configuration of the part is extremely important in determining whether it would be most efficiently produced on a manual milling machine or on a machining center. The following types of parts are generally well suited to machining centers:

- Parts machined on several faces
- Parts requiring a large number of operations
- Parts with close tolerances
- Parts in which design changes are anticipated
- Parts that are very expensive (because a machining center rarely produces a scrap part

It should be noted that nearly all the advantages attributed to machining centers also apply to automated batch-manufacturing systems and transfer lines. Machining centers are most useful in small production runs in which each batch differs significantly from the others.

## Optimization of Milling Machine Setup

Setup often accounts for most of the time required in milling one or a few pieces. When a variety of work is to be milled, time can be saved by planning and scheduling the jobs so that similar parts are milled in succession.

A milling machine can be tooled in different ways to obtain the lowest cost with different quantities of pieces. Some common milling arrangements are illustrated in Fig. 15.

Plain, or simple, milling (Fig. 15a) involves the loading and milling of one piece at a time and is the usual arrangement for one or a few pieces. Cutting time is saved in string, or line, milling (Fig. 15b) by having two or more pieces arranged in a row because the cutter can enter one piece as it leaves another. Efficiency is achieved by arranging for the operator to load at one station while the cut is taken at another, as in reciprocating milling (Fig. 15d). In this arrangement, the cutter cuts upward in one direction and downward in the other. If this is undesirable, it can be eliminated by index base milling (Fig. 15f). An index base consists of two plates mounted one above the other on the machine table. The upper plate carries a fixture on each end and is indexed 180° when retracted from the cutters. One fixture is then advanced to the cut while the other is unloaded and loaded. Fixtures on a rotating round table are loaded with pieces and fed continuously to the cutter(s) in circular, or rotary, milling (Fig. 15h). This can be done on a continuous miller with an integral rotary table or on an auxiliary circular table mounted on a conventional milling machine.

Cutting time is saved when two or more cutters are set to work at the same time. Duplex milling is shown schematically in Fig. 15(c). Gang milling (Fig. 15e) utilizes two or more cutters on one arbor. If side milling cutters machine two sides of a workpiece at the same time, the operation is called straddle milling. If the purpose is to machine two or more parts side by side, the term is abreast milling (Fig. 15g).

## Automatic Feed Mechanisms

Automatic feed can be applied to any or all sides on simplex, duplex, or triplex manufacturing machines. The most critical feeding device on manufacturing machines is the one that drives the table. Four drives are in common use: leadscrew and nut box (ball screws), hydraulic cylinder, half nut and screw, and rack and pinion.

The leadscrew drive is suitable only for short table feeds because of play in the components. A backlash eliminator is needed for climb milling when a leadscrew drive is used. Ground screws with recirculating ball nuts are used to eliminate backlash in modern machines.

Hydraulic-cylinder drives are usually limited to a table feed of about 3 m (10 ft) because of compressibility in a long column of oil. Hydraulic-cylinder drives are not suitable for climb milling because it is difficult to control cutter pressure variations.

Half-nut-and-screw or rack-and-pinion drives are both suitable for long table feeds. These drives also eliminate backlash, permit variable speed for workpieces on which the amount of stock removal varies, and can be used with a tracer roller for contour milling. Larger planer-type machine tables are fed with worm-and-rack mechanisms.

## Automatic Controls

Mechanical-electric, mechanical-hydraulic, mechanical-electric-hydraulic, and numerical control are the four types of automatic controls used on milling machines. Because of its high degree of sophistication, numerical control has gained wide acceptance.

**Mechanical-electric controls** operate by means of limit switches that actuate solenoids to engage or disengage mechanical components of the machine. Controls of this type are used only for single-cycle operations that involve few variations.

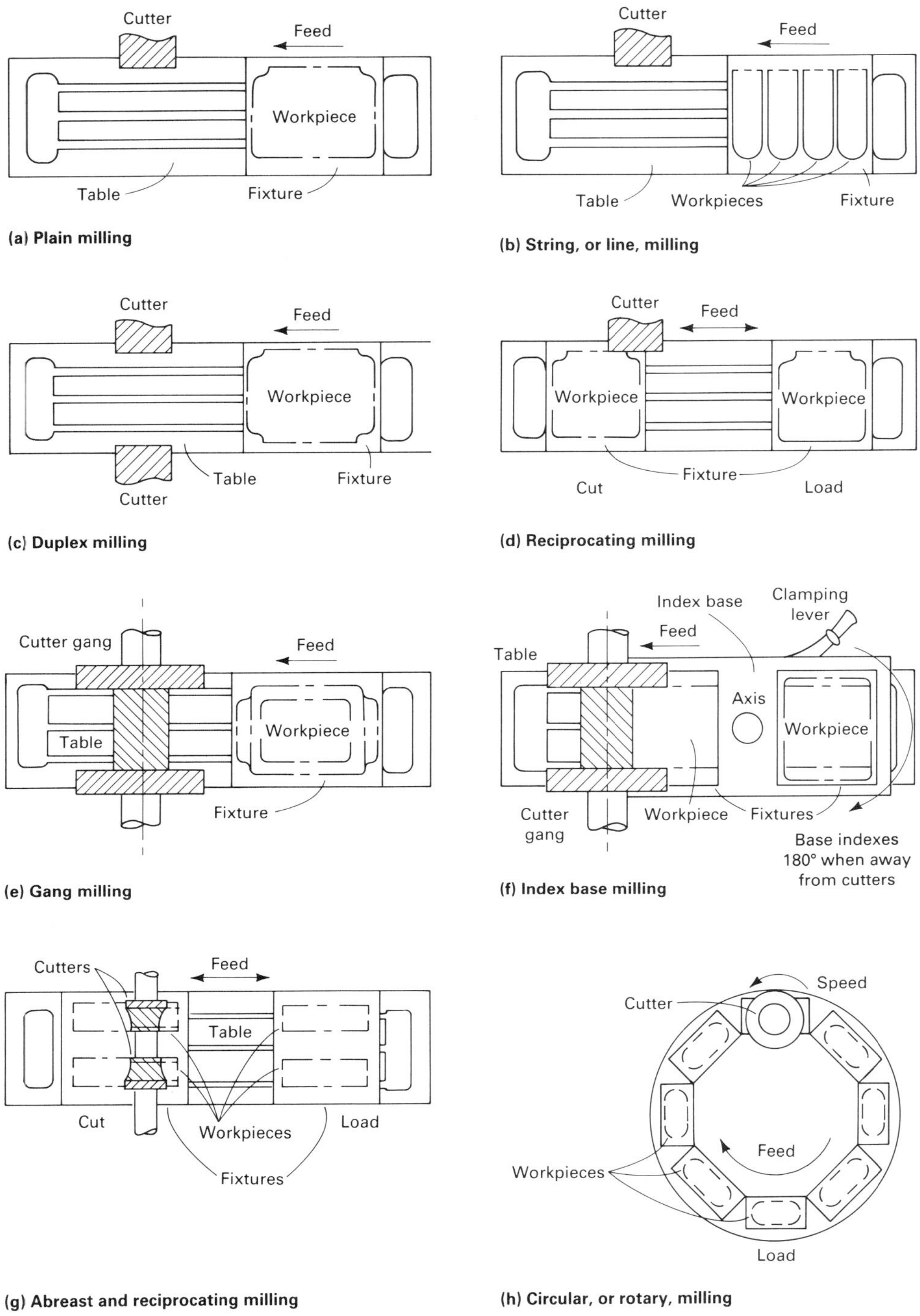


**Fig. 15** Common forms of production milling operations

**Mechanical-hydraulic controls** are actuated by means of a trip plunger having radial and axial movement for positioning a hydraulic valve. Mechanical components of the machine are engaged or disengaged by oil emerging under pressure from ports in the hydraulic valve. These controls are used only for single-cycle operations involving few variations. The addition of a sequence valve and a cycle selector will provide control for many variables. The cycle selectors are arranged with variable-depth cams that depress plungers in a valve block to select the direction of all slides, to start and stop the spindle, and to control indexing and reloading operations.

**Mechanical-electric-hydraulic controls** are an improvement over the types described above for controlling complex machine cycles. These controls utilize an electric stepping switch and connectors or on/off switches. By this means, an automatic cycle is changed by repositioning the connectors or resetting the switches. Because this system lacks automatic positioning, the operator must set stops for length of traverse for each sliding component.

When connectors are used, the tool designer can punch out the cycle in advance on plastic sheets. This punched sheet is then forwarded to the shop with the fixture and tooling, and the operator has only to insert the connectors in the punched-out spaces to control the operation automatically.

**Numerical Control.** The ultimate in equipment for production milling is a machine with numerical control and an automatic tool changer. A punched tape or card establishes the programming, selects the cutter for the tool changer, and determines milling speed and feed. The machine operator inserts the tape (or card) in the control, positions the workpiece on the table with reference to the starting point (known as the zero joint), and then starts the cycle. All operations are performed automatically with one handling of the workpiece. Modern computer-driven numerical control eliminates the tape and can be downloaded from the mainframe.

The advantages of numerical control are usually related closely to cost reduction, better reproducibility, or both. The more important advantages are:

- Multiple cuts are possible with one handling of the workpiece
- All cuts are taken from the same location points
- Feed and speed are computed, rather than being dependent on operator judgment
- Results are more consistent because operator variables are eliminated
- Changes in the machining cycle can be made without detailed instructions to the operator
- Operator duties are decreased, thus allowing more time for housekeeping around the machine
- Closer control of operations permits more accurate cost estimates for the milling of similar workpieces
- Irregular contours can be produced efficiently

## Adapters and Attachments

Longitudinal T-slots are standard on milling machine tables; but cross-table T-slots are gaining acceptance. Many workpieces can be secured directly to the table by means of simple clamps. Workpieces that cannot be fastened directly to the table are held in standard vises or specially designed work holders secured to the table with T-bolts.

All standard milling machines are supplied with an American Standard taper hole in each spindle. Standard tool holders have this taper and are available in at least eight

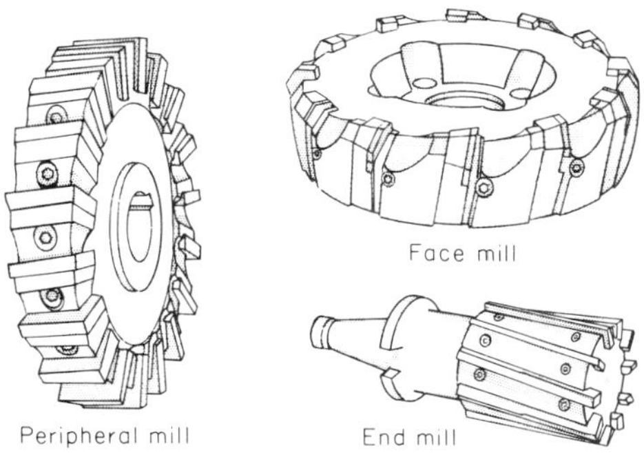


**Fig. 16** Three types of milling cutters

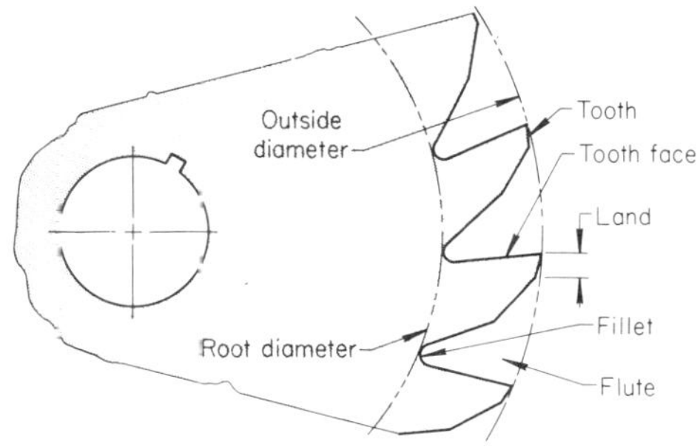


**Fig. 17** Nomenclature for essential features of milling cutters

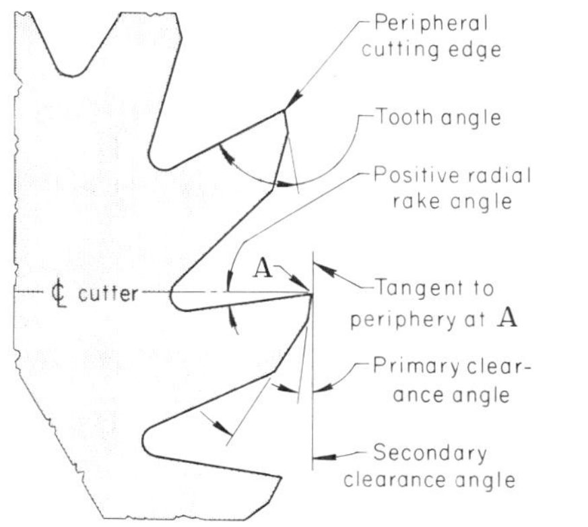


**Fig. 18** Nomenclature of tooth design of milling cutters

designs to accommodate all common types of cutters.

**Attachments.** The number of special attachments that can be used to extend the use of a milling machine is almost unlimited. Some widely used attachments are:

- Rotary tables, which are fixtures designed to rotate the workpiece accurately through angles for the graduating, indexing, or milling of helices. Rotary tables can be operated manually or by power
- Independent or dependent overhead spindles that have as many as eight speeds; these attachments can be applied to any machine having a slide in the top of the column
- Dividing heads, for accurate indexing to the desired angle; these are usually secured directly to the machine table
- Low- or high-lead attachments, which are attached to the end of the table. They are driven off the table leadscrew through various change gears and into the dividing head. They are used to create a helix angle in a workpiece being held by the dividing head
- Some vertical-spindle machines have a provision that permits the use of a keywaying or slotting attachment

## Milling Cutters

Milling cutters vary widely in type and size and are broadly classified as peripheral mills, face mills, end mills, and special mills. Cutters can be of the solid, tipped, or inserted tooth types and have the same materials as single-point tools. Large cutters commonly have teeth of expensive material that are inserted and locked in place in a soft steel or cast iron body.

**Peripheral mills** (Fig. 16) are so named because cutting is mainly done by teeth on the periphery to produce the finished surface. A peripheral mill is usually mounted on an arbor with the axis parallel to the surface being milled. Peripheral mills are further classified as slab mills, slotting cutters, half-side mills, angle cutters, and form cutters. Circular saws are also a type of peripheral mill.

Peripheral mills are available in a wide range of diameters, widths, and width-to-diameter ratios. Straight or spiral cutting edges are also available. The most widely available range of diameters for peripheral mills is from about 102 to 457 mm (4 to 18 in.).

Slotting cutters as thin as 4.8 mm (3/16 in.) can be obtained in diameters up to about 127 mm (5 in.). The minimum available width increases with diameter. For example, in a 305 mm (12 in.) diam cutter, the minimum thickness is usually 11 mm (7/16 in.). Slab mills represent the opposite extreme in width-to-diameter ratio and are often four or five diameters wide.

Thin slotting cutters or complicated form cutters are usually of the solid type, regardless of diameter. Plain peripheral milling cutters less than about 76 mm (3 in.) in diameter are usually of the solid type; larger ones have inserted blades. Most peripheral cutters have zero to positive radial rake, with either positive or negative axial rake.

**Face mills** (Fig. 16) are so named because the workpiece surface is produced by the face of the cutter, although the outside diameter or bevel-cutting edge removes most of the stock. The cutter is driven by a spindle on an axis perpendicular to the surface being milled.

Face mills can be made to almost any size, but usually range in diameter from 70 to 508 mm (2¾ to 20 in.). Face mills smaller than 70 mm (2¾ in.) are seldom needed, and machine power and rigidity restrict the use of cutters larger than 508 mm (20 in.). Face mills less than 76 mm (3 in.) in diameter are usually machined from solid; large face mills characteristically have inserted blades. The tooth design on face mills is usually one of the following:

- Double positive (positive axial rake and positive radial rake)
- Double negative (negative axial rake and negative radial rake)
- Negative radial rake, positive axial rake, and a corner angle

**End mills** (Fig. 16) have cutting edges on both the face end and the periphery; therefore, they can be used for facing cuts, peripheral cuts, or both. As peripheral cutters, they can be used to form either plane or irregular surfaces on the workpiece.

End mills are generally available in diameters of 1.6 to 102 mm (1/16 to 4 in.), although larger ones have been used. End mills less than 51 mm (2 in.) in diameter are usually made of solid tool steel; those larger than 51 mm (2 in.) have inserted blades of either high-speed steel or carbide.

End mills normally have zero to positive radial rake and positive axial rake. Axial rake angles in the form of true spiral or helix angles will produce smoother surfaces. For a combination of end and peripheral cutting, a cutter with a right-hand cut and a right-hand helix is used. For peripheral milling only, most cutters have a right-hand cut and a left-hand helix, but they can also have a left-hand helix and a right-hand cut.

**Special milling cutters** can be made to almost any design. For example, the teeth can be of either the profile-sharpened or the form-relieved type and can be integral with the body or inserted. The cutter can be designed for mounting on an arbor or directly on the spindle nose, or it can have a shank integral with the body. Special cutters can be used singly, or they can be made in sections and combined to mill the desired contour. Material and angle considerations that apply to standard cutters generally also apply to the special designs. The higher cost of special cutters can often be justified by combining operations, thus reducing machine time and machining cost.

**The nomenclature** of milling cutters is illustrated in Fig. 17 and defined below. These terms apply primarily to plain milling cutters.

*Outside diameter* is the diameter of a circle passing around the peripheral cutting edges, and it is used to calculate surface speed from spindle speed.

*Root diameter* is the diameter of a circle tangent to the roots of the teeth fillets.

*The tooth* is the part of the cutter starting at the body and ending with the peripheral cutting edge. Inserted teeth are called blades.

*Tooth face* is the surface of the tooth between the fillet and the cutting edge, where the chip slides during cutting.

*Land* is the area behind the cutting edge on the tooth that is relieved to avoid interference.

*Flute* is the space for chip flow between the teeth.

*Fillet* is the radius at the bottom of the flute to promote chip flow and curling.

The following terms refer to tooth design (Fig. 18).

*Peripheral cutting edge* is the edge aligned principally in the direction of the cutter axis. In peripheral milling, it is the edge that removes metal.

*Face cutting edge* (not shown in Fig. 18) is the metal-removing edge on a face mill that travels in a plane perpendicular to the axis. It is the edge that sweeps the milled surface in normal face milling.

*Tooth angle* is the angle included between the face and the land of the cutter tooth. This angle should be as large as possible to provide maximum tooth strength and better dissipation of heat.

*Radial rake angle* (shown as positive in Fig. 18) is the angle between the tooth face and a radial line passing through the cutting edge in a plane perpendicular to the cutter axis.

*Clearance angle* is the angle included between the land on the back or flank of the milling cutter tooth, and the tangent to the periphery of the cutter at the cutting edge. Clearance angles are always positive and are usually divided into primary and secondary angles (Fig. 18).

*Axial rake angle* (not shown in Fig. 18) is measured between the peripheral cutting edge and the axis of the cutter, when looking radially at the point of intersection. For

**Table 1 Tool geometry used on arbor-mounted side and slot milling cutters with high-speed steel or carbide teeth for milling ferrous and nonferrous alloys**

| Material | Hardness, HB | High-speed steel: Axial rake angle, degrees | Radial rake angle, degrees | Axial relief angle, degrees | Radial relief angle, degrees | Carbide: Axial rake angle, degrees | Radial rake angle, degrees | Axial relief angle, degrees | Radial relief angle, degrees |
|---|---|---|---|---|---|---|---|---|---|
| Wrought free-machining carbon steels, wrought and cast carbon steels, wrought free-machining alloy steels | 85–325 | 10–15 | 10–15 | 3–5 | 4–8 | 0 to −5 | −5 to 5 | 2–4 | 5–8 |
| | 325–425 | 10–12 | 5–12 | 3–5 | 4–8 | 0 to −5 | −5 to 5 | 2–4 | 5–8 |
| | 45–52 HRC | 10–12 | 5–12 | 2–4 | 3–7 | −5 to −10 | 0 to −10 | 2–4 | 5–8 |
| Wrought and cast alloy steels | 125–425 | 10–12 | 5–12 | 3–5 | 4–8 | −5 to −10 | 0 to −10 | 2–5 | 5–8 |
| | 45–52 HRC | 10–12 | 5–12 | 2–4 | 3–7 | −5 to −10 | 0 to −10 | 2–4 | 3–6 |
| Wrought high-strength steels, maraging steels, and tool steels | 100 HB–52 HRC | 10–12 | 5–12 | 2–4 | 3–7 | −5 to −10 | 0 to −10 | 2–4 | 5–8 |
| Wrought nitriding steels | 200–350 | 10–12 | 5–12 | 2–4 | 3–7 | −5 to −10 | 0 to −10 | 2–4 | 3–6 |
| Wrought armor plate | 250–320 | 0–5 | 0–5 | 2–4 | 3–7 | −5 to −10 | −5 to −10 | 2–4 | 3–6 |
| Wrought structural steels | 100 HB–50 HRC | 10–12 | 5–12 | 3–5 | 4–8 | 0 to −5 | 0 to −10 | 2–4 | 5–8 |
| Wrought free-machining stainless steels | 135–425 | 10–12 | 5–12 | 3–5 | 4–8 | 0–5 | −5 to 5 | 2–4 | 5–8 |
| Wrought and cast ferritic and austenitic stainless steels | 135 HB–52 HRC | 10–12 | 5–12 | 3–5 | 4–8 | 0–5 | −5 to 5 | 2–4 | 5–8 |
| Wrought and cast martensitic stainless steels | 135 HB–52 HRC | 10–12 | 5–12 | 2–4 | 3–7 | −5 to −10 | 0 to −10 | 2–4 | 5–8 |
| Wrought and cast precipitation-hardening stainless steels | 150–450 | 10–12 | 5–12 | 2–4 | 4–8 | 0 to −5 | 0 to −10 | 2–4 | 5–8 |
| Gray, ductile, malleable cast irons | 100–400 | 10–12 | 10–12 | 2–4 | 3–7 | 0 to −10 | 5 to −10 | 3–5 | 5–8 |
| Wrought and cast aluminum alloys | 30–150 (500 kg) | 12–25 | 10–20 | 5–7 | 5–11 | 10–20 | 5–15 | 5–7 | 7–10 |
| Wrought and cast magnesium alloys | 40–90 (500 kg) | 12–25 | 10–20 | 5–7 | 5–11 | 10–20 | 5–15 | 5–7 | 7–10 |
| Wrought titanium alloys | 110–440 | 10–15 | 5–10 | 5–7 | 5–11 | 0 to −10 | 0 to −10 | 5–7 | 5–8 |
| Wrought and cast copper alloys | 40–200 (500 kg) | 12–25 | 10–20 | 5–7 | 5–11 | 10–20 | 5–10 | 4–7 | 5–8 |
| Wrought and cast nickel alloys | 80–360 | 10–20 | 10–15 | 3–5 | 4–8 | −5 to −10 | 0 to −10 | 3–5 | 5–8 |
| Wrought and cast high-temperature alloys | 140–300 | 10–15 | 10–15 | 1–5 | 5–10 | −5 to −10 | 0 to −10 | 3–5 | 5–8 |
| | 300–475 | 10–12 | 5–12 | 1–5 | 4–8 | −5 to −10 | 0 to −10 | 3–5 | 5–8 |
| Wrought, cast, and P/M niobium and molybdenum alloys | 170–290 | 0 | 15–20 | 3–5 | 5–10 | 0 | 5–15 | 7–10 | 7–10 |
| Wrought, cast, and P/M tantalum alloys | 200–250 | 0 | 15–20 | 3–5 | 5–10 | 0 | 5–15 | 7–10 | 7–10 |
| Wrought, cast, and P/M tungsten alloys | 180–320 | . . . | . . . | . . . | . . . | −10 to −15 | 5–15 | 10–15 | 10–15 |
| Cast zinc alloys | 80–100 | 10–20 | 10–20 | 5–7 | 8–11 | 10–15 | 10–15 | 7–10 | 7–10 |

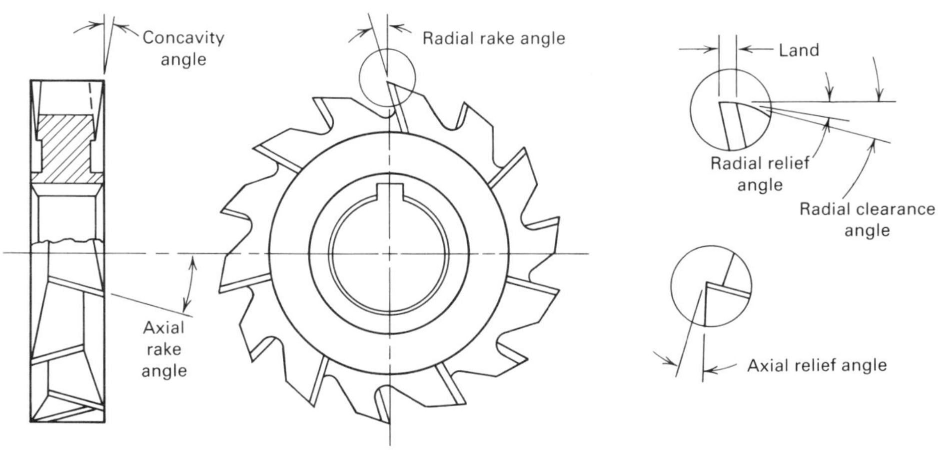


Source: Metcut Research Associates Inc.

**Table 2 Tool geometry used on face milling cutters with high-speed steel, indexable carbide, and brazed carbide teeth for milling ferrous and nonferrous alloys**

| | | High-speed steel | | Indexable carbide | | Brazed carbide | | | | | |
|---|---|---|---|---|---|---|---|---|---|---|---|
| Material | Hardness, HB | Axial rake angle, degrees | Radial rake angle, degrees | Axial rake angle, degrees | Radial rake angle, degrees | Axial rake angle, degrees | Radial rake angle, degrees | Corner angle, degrees | End cutting edge angle, degrees | Axial relief angle, degrees | Radial relief angle, degrees |
| Wrought and cast free-machining carbon steels and carbon steels | 85–270 | 10–15 | 10–15 | 5–7 | −5 to −14 | 0 to −7 | 0 to −7 | 30 | 5–10 | 5–7 | 3–7 |
| Wrought and cast free-machining alloy steels and alloy steels | 270–325 | 10–15 | 10–15 | −4 to −8 | −3 to −11 | 0 to −7 | 0 to −7 | 30 | 5–10 | 5–7 | 3–7 |
| Wrought maraging steels | 325–425 | 10–12 | 10–12 | −4 to −8 | −3 to −11 | 0 to −10 | 0 to −10 | 30 | 5–10 | 5–7 | 3–7 |
| Wrought tool steels | 43–50 HRC | 5–10 | 5–10 | −4 to −8 | −3 to −11 | −5 to −15 | −5 to −15 | 45 | 4–7 | 5–7 | 3–7 |
| Wrought structural steels | 50–56 HRC | . . . | . . . | −4 to −8 | −3 to −11 | −5 to −15 | −5 to −15 | 45 | 4–7 | 8 | 8 |
| Wrought high-strength steels | 225–425 | 5–10 | 0–10 | −4 to −8 | −3 to −11 | −5 to −15 | −5 to −15 | 45 | 4–7 | 5–7 | 3–7 |
| | 45–58 HRC | . . . | . . . | −4 to −8 | −3 to −11 | −5 to −15 | −5 to −15 | 45 | 4–7 | 8 | 8 |
| Wrought nitriding steels | 200–350 | 5–10 | 5–10 | −4 to −8 | −3 to −11 | 0 to −10 | −5 to −15 | 45 | 5–10 | 5–7 | 3–5 |
| Wrought armor plate | 250–320 | 0–5 | 0–5 | −4 to −8 | −3 to −11 | 0 to −10 | −5 to −15 | 45 | 4–7 | 5–7 | 3–5 |
| Wrought free-machining stainless steels | 135–275 | 10–15 | 10–12 | 5–11 | −5 to −11 | 0 | 0–5 | 45 | 5 | 8–10 | 8–10 |
| | 275–425 | 5–10 | 5–10 | 5–11 | −5 to −11 | 0 | 0 to −5 | 45 | 5 | 8–10 | 8–10 |
| Wrought and cast ferritic and austenitic stainless steels | 135–275 | 10–15 | 10–12 | 5–11 | −5 to −11 | 0–5 | 0 to −5 | 45 | 5 | 8–10 | 8–10 |
| Wrought and cast martensitic stainless steels | 135–425 | 5–10 | 5–10 | 5–11 | −5 to −11 | 0 | 0 to −5 | 45 | 5 | 8–10 | 8–10 |
| Wrought and cast precipitation-hardening stainless steels | 150–450 | 5–10 | 5–10 | 5–7 | 0–5 | 0 | 0 | 45 | 5 | 8–10 | 8–10 |
| Gray, ductile, malleable, compacted graphite, and white cast irons | 100–400 | 20–30 | −5 to −10 | 5–11 | −5 to −11 | 5–10 | 5 to −10 | 45 | 5–10 | 4–7 | 4–7 |
| Wrought and cast aluminum alloys | 30–150 (500 kg) | 20–35 | 20–35 | 5–7 | 0–5 | 10–20 | 10–20 | 45 | 7–12 | 3–5 | 10–12 |
| Wrought and cast magnesium alloys | 40–90 (500 kg) | 20–35 | 20–35 | 5–7 | 0–5 | 10–20 | 10–20 | 45 | 7–12 | 3–5 | 10–12 |
| Wrought and cast titanium alloys | 110–440 | 5 | 5 | 0 to −5 | 0 to −5 | 0 to −5 | −10 | 45 | 6–12 | 10–12 | 10–12 |
| Wrought and cast copper alloys | 40–200 (500 kg) | 12–25 | 10–12 | 5–7 | 0–5 | 3–10 | 3–10 | 45 | 7–12 | 3–5 | 5–10 |
| Wrought and cast nickel alloys, chromium-nickel, and beryllium-nickel alloys | 80–360 | 7 | 15 | 5–11 | −5 to −14 | 5–10 | 0 to −5 | 45 | 5 | 7–9 | 7–9 |
| Wrought nitinol alloys | 210–340 (48–60 HRC) | . . . | . . . | 5–7 | 0–5 | 0 | 0 | 45 | 10 | 12 | 12 |
| Wrought and cast high-temperature alloys | 200–475 | 5–10 | 5–10 | 0–5 | 0 to −5 | 0–5 | 0 to −5 | 45 | 5 | 7–10 | 7–10 |
| Wrought, cast, and P/M niobium alloys | 170–225 | 0 | 20 | 5–7 | 0–5 | 0 | 10 | 45 | 5–10 | 10 | 10 |
| Wrought, cast, and P/M molybdenum alloys | 220–290 | 0 | 20 | 5–7 | 0–5 | 0 | 0 | 45 | 5–10 | 10 | 10 |
| Wrought, cast, and P/M tantalum alloys | 200–250 | 0 | 20 | 5–7 | 0–5 | 0 | 0 | 45 | 5–10 | 10 | 10 |
| Wrought, cast, and P/M tungsten alloys | 180–320 | . . . | . . . | −4 to −8 | −3 to −11 | −15 | 0 | 45 | 5–10 | 15 | 15 |
| Cast zinc alloys | 80–100 | 10–15 | 10–15 | 5–7 | 0–5 | 10–12 | 10–12 | 45 | 7–12 | 10 | 10–12 |

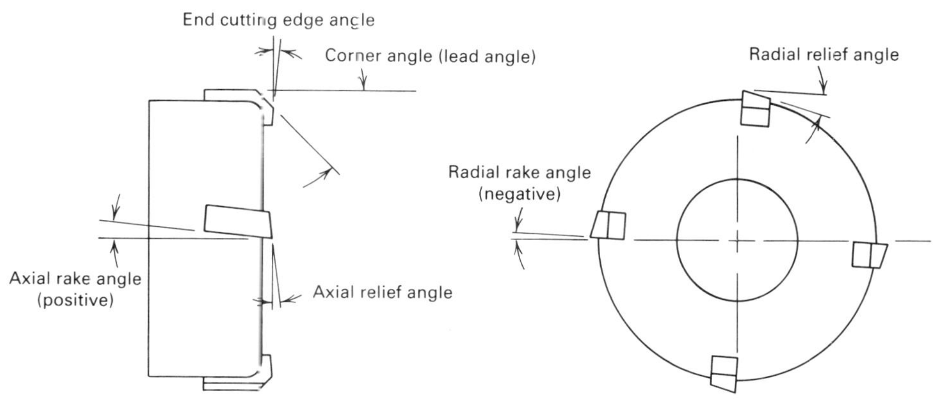


Source: Metcut Research Associates Inc.

a cutter with helical teeth, the axial rake angle equals the helix angle.

*Blade setting angle*, or cone angle (not shown in Fig. 18), is the nominal angle with the cutter axis, along which the blades are moved for adjustments. The cutter angle combinations given in Tables 1 to 3 have proved satisfactory for milling a wide range of metals and alloys using high-speed steel and carbide (both brazed and indexable) cutters for peripheral, face, and end mills.

## Material for Milling Cutters

There are three basic types of construction for milling cutters:

- Solid
- Inserted blade
- Indexable insert

The solid and blade types can be resharpened and are often referred to as grind-type cutters. Although indexable-insert cutters are widely used, grind-type cutters are still required for certain applications. One advantage of grind-type cutters is that their geometry can be tailored to suit a wide variety of milling applications; in addition, runout can be controlled more closely than with other types of cutters.

### Solid Milling Cutters

Solid milling cutters are made from one piece of steel (usually high-speed steel) or

**Table 3 Tool geometry for high-speed steel end mills for the peripheral and slot milling of ferrous and nonferrous alloys**

| Nominal cutter diameter, mm | Nominal cutter diameter, in. | General purpose—30–35° helix: steels, cast irons, copper alloys, titanium alloys, nickel alloys, high-temperature alloys, and zinc alloys: Radial primary relief angle, degrees | Primary land width, mm | Primary land width, in. | Radial secondary clearance angle, degrees | 35–45° helix: aluminum and magnesium alloys: Radial primary relief angle, degrees | Primary land width, mm | Primary land width, in. | Radial secondary clearance angle, degrees |
|---|---|---|---|---|---|---|---|---|---|
| 1.6 | 1/16 | 20–21 | 0.18–0.25 | 0.007–0.010 | 30–35 | 20–22 | 0.18–0.25 | 0.007–0.010 | 30–35 |
| 3 | 1/8 | 12–13 | 0.25–0.38 | 0.010–0.015 | 22–28 | 14–18 | 0.25–0.38 | 0.010–0.015 | 25–30 |
| 4 | 3/16 | 12–13 | 0.25–0.51 | 0.010–0.020 | 20–25 | 14–18 | 0.25–0.51 | 0.010–0.020 | 25–30 |
| 6 | 1/4 | 10–11 | 0.25–0.51 | 0.010–0.020 | 20–25 | 12–15 | 0.25–0.51 | 0.010–0.020 | 22–28 |
| 7 | 5/16 | 10–11 | 0.38–0.64 | 0.015–0.025 | 20–25 | 12–14 | 0.38–0.64 | 0.015–0.025 | 21–28 |
| 8 | 3/8 | 10–11 | 0.38–0.64 | 0.015–0.025 | 17–20 | 12–14 | 0.38–0.64 | 0.015–0.025 | 19–26 |
| 10 | 7/16 | 9–10 | 0.51–0.76 | 0.020–0.030 | 17–20 | 11–13 | 0.51–0.76 | 0.020–0.030 | 18–25 |
| 12 | 1/2 | 9–10 | 0.51–0.76 | 0.020–0.030 | 17–20 | 11–13 | 0.51–0.76 | 0.020–0.030 | 18–25 |
| 14 | 5/8 | 9–10 | 0.64–0.89 | 0.025–0.035 | 17–20 | 11–13 | 0.64–0.89 | 0.025–0.035 | 18–25 |
| 16 | 3/4 | 8–9 | 0.76–1.00 | 0.030–0.040 | 15–18 | 10–12 | 0.76–1.00 | 0.030–0.040 | 17–24 |
| 20 | 7/8 | 8–9 | 0.76–1.00 | 0.030–0.040 | 15–18 | 10–12 | 0.76–1.00 | 0.030–0.040 | 17–24 |
| 22 | 1 | 8–9 | 0.89–1.27 | 0.035–0.050 | 15–18 | 10–12 | 0.89–1.27 | 0.035–0.050 | 16–23 |
| 25 | 1 1/4 | 7–8 | 1.00–1.52 | 0.040–0.060 | 13–18 | 9–11 | 1.00–1.52 | 0.040–0.060 | 14–22 |
| 32 | 1 1/2 | 7–8 | 1.00–1.52 | 0.040–0.060 | 11–17 | 9–11 | 1.00–1.52 | 0.040–0.060 | 13–21 |
| 40 | 1 3/4 | 7–8 | 1.00–1.52 | 0.040–0.060 | 10–16 | 8–10 | 1.00–1.52 | 0.040–0.060 | 12–20 |
| 45 | 2 | 6–7 | 1.00–1.52 | 0.040–0.060 | 9–15 | 8–10 | 1.00–1.52 | 0.040–0.060 | 12–20 |

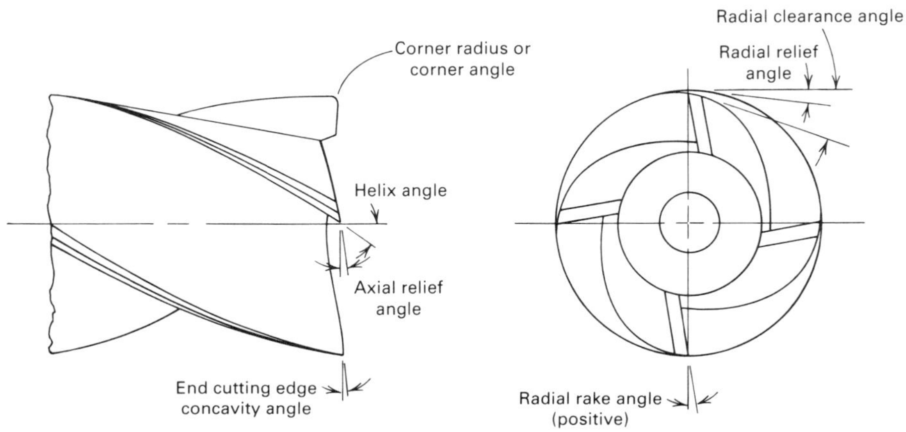


Source: Metcut Research Associates Inc.

carbide by machining the required shape and number of teeth, with the specified cutting and relief angles ground on the teeth. The cutters can be tipped with cast alloy or carbide cutting tool materials if this is warranted by either the nature of the work material or the production requirements. Solid cutters are usually the lowest in initial cost, depending on size, and therefore may be best for short runs or general toolroom use.

High-speed steel is widely used as a tool material for all types of milling cutters. With high-speed steel, the cutter body and teeth can be integral. The required shapes are machined from annealed high-speed steel and then hardened and ground to a finished cutter. Many large cutters of simple design consist of high-speed steel blades inserted in bodies made of heat-treated alloy steel. However, as cutter shape increases in complexity, the use of separate blades becomes more difficult. Consequently, complex cutters are often made from solid high-speed steel, regardless of size.

Cutters made from the more highly alloyed grades of high-speed steel (especially those containing cobalt) have higher hot hardness than the general-purpose grades. The higher hot hardness permits faster cutting speeds. As illustrated in Fig. 19(a), in end milling 4340 steel quenched and tempered to 49 HRC, the cutting speed for a tool life of 1270 mm (50 in.) of workpiece machined was approximately 14 m/min (45 sfm) for M7 high-speed steel and 21 m/min (70 sfm) for T15 high-speed steel, an increase of more than 50%.

The life of milling cutters of T1, M1, and M10 high-speed steels decreases quickly as the hardness of the workpiece rises beyond 35 HRC. The more highly alloyed high-speed steels can be used to cut work at 35 to 50 HRC. For work that is harder than 50 HRC, cutters of high-speed steel would have limited service lives, as shown in Fig. 19(b).

Solid carbide cutters are used for work that is harder than 50 HRC. Carbide cutters are better, in general, because they are used at cutting speeds three to six times those for high-speed steel. The smallest cutters (especially end mills) are sometimes made of solid carbide, but high cost and high susceptibility to breakage limit the use of solid carbide to specialized applications.

On many of the nickel or cobalt-base high-temperature alloys, such as U-500 and HS-25, carbide milling cutters are not recommended. Tool life is usually poor in milling these alloys with carbide, because of edge chipping. Cutters made of cast cobalt-chromium-tungsten alloys have been unsatisfactory for the milling of nickel or cobalt high-temperature alloys. A combination of low cutting speed and a cobalt high-speed cutter is required for milling these alloys. Figure 19(c) indicates the advantage of T15 over T1 high-speed steel, as well as the effect of cutting speed on the life of both tool materials, in the face milling of U-500. The performance of all high-speed cutters, including the high-cobalt types, is greatly improved by the addition of a titanium-nitride coating.

## Inserted-Blade Cutters

Many milling operations are performed with cutters that use a high-speed steel or carbide cutting material locked by mechanical means in a steel body. Some cutters have the material brazed to the body. An advantage of cutters that use a mechanically locked cutting material is that the body of the cutter does not have to be replaced every time the cutting material wears out.

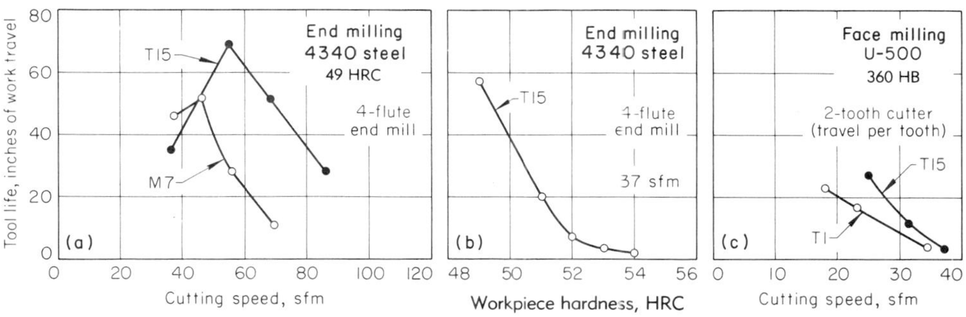


| Graph | Cutter diameter mm | Cutter diameter in. | Feed mm/tooth | Feed in./tooth | Depth of cut mm | Depth of cut in. | Width of cut mm | Width of cut in. | Ratio of soluble oil to water | Wear land mm | Wear land in. |
|---|---|---|---|---|---|---|---|---|---|---|---|
| **Milling conditions** | | | | | | | | | | | |
| (a) | 19 | ¾ | 0.025 | 0.0010 | 6.4 | ¼ | 19 | ¾ | 1:20 | 0.41 | 0.016 |
| (b) | 19 | ¾ | 0.013 | 0.0005 | 6.4 | ¼ | 19 | ¾ | 1:20 | 0.41 | 0.016 |
| (c) | 102 | 4 | 0.13 | 0.0050 | 1.52 | 0.060 | 51 | 2 | 1:20 | 0.41 | 0.016 |

| Graph | Helix angle | Radial rake | Corner angle | Side clearance | End clearance |
|---|---|---|---|---|---|
| **Cutter angles** | | | | | |
| (a) | 35° rh, rh cut | 15° | 45°, 1.52 mm (0.060 in.) wide | 5° pr, 10° s | 5° pr, 20° s |
| (b) | 35° rh, rh cut | 15° | 45°, 1.52 mm (0.060 in.) wide | 5° pr, 10° s | 5° pr, 20° s |
| (c)* | . . . | 10° | 45° | . . . | . . . |

*For graph (c), other cutter angles were: axial rake, 10°; true rake, 14°; inclination, 0°; end cutting edge angle, 5°; peripheral clearance, 10°.

**Fig. 19** Comparative tool life of milling cutters made of different grades of high-speed steel when used under the conditions indicated

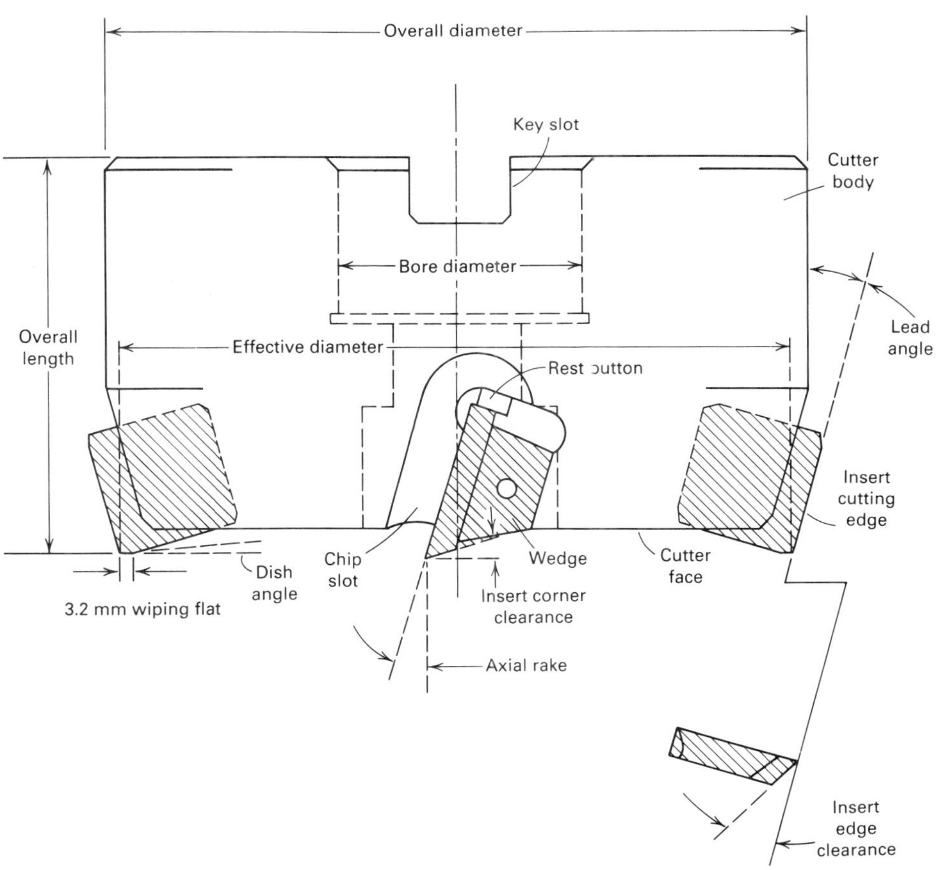


**Fig. 20** Nomenclature for an indexable-insert milling cutter

These cutters are normally higher in initial cost than solid cutters, but generally prove to be lowest in unit cost per piece on production runs.

The first milling cutter to use nonintegral cutting edges was called an inserted-blade cutter. The bodies of inserted-blade milling cutters are generally made of carbon steel, forged alloy steel, or cast steel, depending on the strength and hardness required. Small cutters with brazed tips are frequently made of high-strength cast iron. When a hard cutter body is desired, a carbon steel such as SAE 1095, 4140, or other suitable steel, heat treated to 42 to 48 HRC, can be used.

Blades for inserted-blade cutters are usually made of high-speed steel, solid carbide, carbide-tipped material, or cast alloy, and they are brazed or wedged into place. These cutters are ideal for close-tolerance finishing operations because the blades can be reground while in the cutter body, thus eliminating the cumulative tolerance error associated with some types of indexable-insert cutters.

Solid and brazed-tip cutters are used less frequently for the production of simple flat surfaces because of the production economies offered by indexable-insert cutters. However, because they can be produced relatively inexpensively in a wide variety of configurations, they are still commonly used for many more complex shapes, such as gear teeth, concave or convex surfaces, radiused corners, keyways, and especially small-diameter end-milled surfaces.

## Indexable-Insert Cutters

The indexable-insert cutter is a recently introduced style of construction for milling cutters that is being extensively used (Fig. 20). These cutters use a small cutting insert normally made of carbide, coated carbide, ceramic, or diamond implanted on a carbide substrate. The insert has one or more cutting edges located at various corners and often on both sides. In the case of round inserts, the cutting edges are located around the circumference. The inserts are locked into place with mechanical clamping devices such as wedges or clamps. When they become dull, the inserts are indexed or repositioned so that new cutting edges contact the workpiece.

The role of insert configuration in milling is illustrated in Fig. 21, which shows typical throwaway inserts for rough and finish face milling. Inserts used for rough face milling can leave serrated grooves on the machined surface. The flat corner used for finish face milling produces a smoother machined surface. However, because of the increased width of contact and lack of clearance, finish face milling cuts must be taken at much lower feed rates than those used for rough milling.

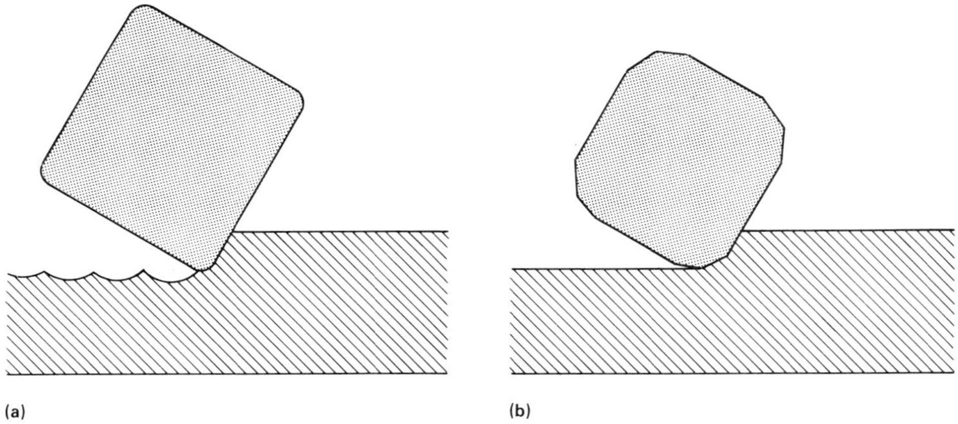


**Fig. 21** Insert configuration used for roughing (a) and finishing (b) milling operations. A throwaway insert for rough face milling (a) leaves a serrated finish on the workpiece, while the flat edge used for finish milling (b) will leave a smoother finish on the workpiece.

The advantages of indexable-insert cutters include the elimination of regrinding, low cost per cutting edge, and the availability of inserts in a wide variety of cutting materials and geometries. Disadvantages include higher cutting forces and the possibility of rougher surface finishes being produced.

**Indexable inserts for milling cutters** can be located or clamped by the use of nonadjustable insert pockets, adjustable pockets, open slots and rails, and cartridges.

*The nonadjustable insert pocket* design (Fig. 22a) locates the insert axially and radially on two locating rails, which are an integral part of the cutter body. This pocket design is strong and allows the cutter to be indexed without a setting fixture. One disadvantage of the design is that, over extended periods of time, the pockets may lose their accuracy and require costly refurbishment. In some welded anvil designs, the anvil seat can be replaced.

A variation of the nonadjustable pocket design overcomes this problem by using a small threaded button within the pocket to provide axial location. By using a button of a different thickness, axial insert location can be changed whenever necessary. Another method of providing axial location is to use rest pads under the inserts.

*Adjustable Insert Pockets.* The adjustable-pocket type of cutter body design (Fig. 22b) uses an adjustable locating support that allows variable positioning of the insert axially with respect to the cutter body. Face runout due to body inaccuracy is largely eliminated by this design—at a cost of some reduction in strength. Inserts in a cutter with this style of pocket can be indexed on the machine, although occasional pocket adjustment on a setting fixture is required.

*The open slot and rail design* is the third basic design for cutter body pockets (Fig. 22c). Strictly speaking, there is no pocket; the insert is radially located in a precision rail and is wedge locked. A setting fixture is used to locate the insert axially during indexing. This design provides the possibility of minimizing face runout because the inserts are set to a common face plane.

*The cartridge style* of holding indexable inserts allows axial adjustment and changes in geometries and lead angles. In case of damage, the cartridges are easily replaced.

**Milling Inserts.** Inserts used for milling are usually different from those employed for turning. Heavier-duty inserts of different grades, precision ground and having distinct corner geometries and edge preparation, are generally used for milling, and chipbreakers are seldom required. Scalloped-edge inserts are unique to milling; extensive use of on-edge inserts as opposed to conventional insert mounting is also a feature of milling.

*Scalloped-Edge Inserts.* Square, rectangular, and round inserts with scalloped or serrated cutting edges are used exclusively for milling cutters. The inserts are arranged in the cutter so that each insert removes the ridges left on the workpiece by the preceding insert. Two-phase square inserts break the cut into two smaller cuts, three-phase triangular inserts break it into three cuts, and four-phase square or rectangular inserts break it into four cuts. Scalloped-edge inserts reduce cutting forces, thus allowing deeper and faster milling.

*On-Edge Inserts.* Larger milling cutters with the inserts mounted for edge cutting are available (Fig. 23). For some applications, faster stock removal rates and longer cutter life are obtained. With the on-edge design (Fig. 23b), the maximum cross section of each insert is behind its cutting edge, thus increasing rigidity. Tangential cutting forces, the key factor in power consumption, are directed along the plane of each insert to make optimum use of the high transverse rupture strength of the carbide, and bending loads are reduced.

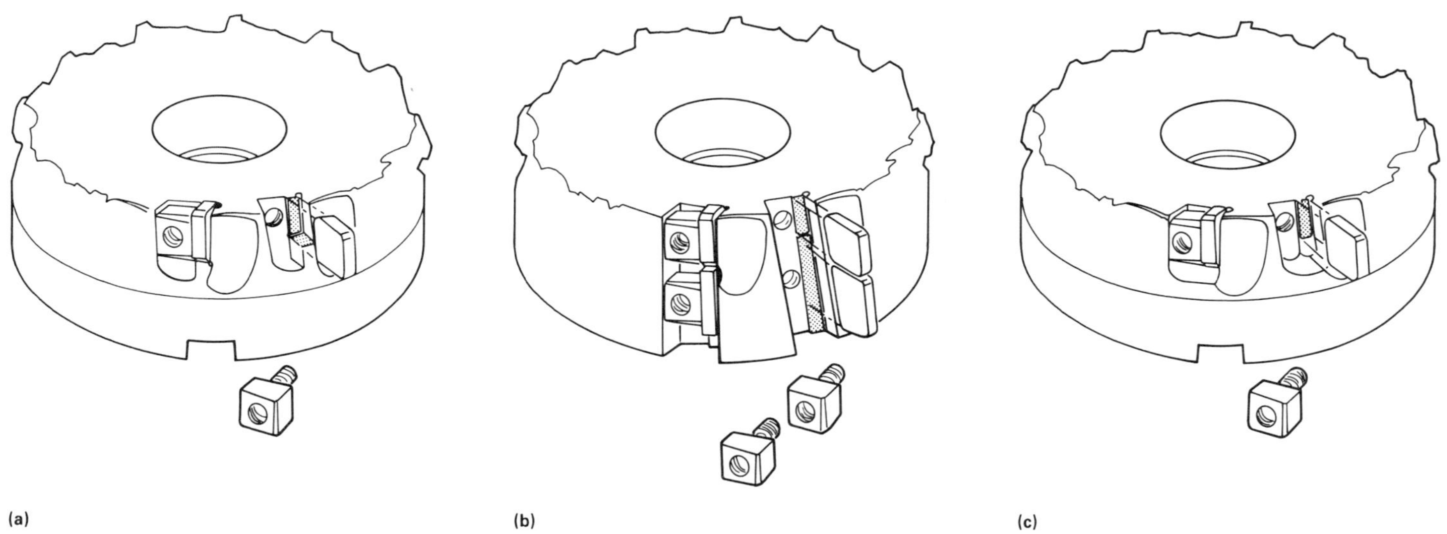


**Fig. 22** Three methods of clamping indexable inserts in milling cutters. (a) Nonadjustable pocket. (b) Adjustable pocket. (c) Open slot and rail. Shading indicates locating surfaces.

## Secondary Factors Affecting Choice of Tooling Material

Other factors to be considered in selecting the tool material for milling cutters are:

- Rigidity of setup
- Available horsepower
- Cutter-sharpening facilities

**Rigidity of setup** is particularly important when carbide milling cutters are used. Looseness in the bearings or slides of the machine can result in vibrations that lead to poor cutter life. If the setup cannot be made rigid, high-speed steel cutters should be used.

**Available Horsepower.** Carbide cutters give optimum results when used at higher speeds. As speed is increased, more power is required. Therefore, when considering the use of carbide cutters, it should be made certain that enough power is available.

**Sharpening Facilities.** Carbide cutters are more difficult to sharpen than high-speed steel cutters. Diamond wheels are commonly used, although some grinding machines use silicon carbide wheels for grinding carbide cutters.

## Geometrical Relation of Cutter to Work

Axial rake, radial rake, and corner angle in face milling are respectively equivalent to back rake, side rake, and side cutting edge angle in turning. These are the angles by which the relationship of cutting edge to work is usually defined (see the article "Turning" in this Volume for the identification of the basic elements and nomenclature of a single-point tool).

The most significant angle in any machining operation is the true rake angle. True rake angle directly affects the shear angle in the chip-forming process and therefore greatly affects tool force, power requirement, and temperature. The larger the positive value of the true rake angle, the lower the force. It is limited in magnitude, however, by the strength required of the tool for a given machining operation. Negative rake or geometry is used when cutting harder materials such as Hy 80 and Hy 100 armor plate and titanium steels for aircraft applications. In milling, the true rake angle is the resultant of the axial rake, radial rake, and corner angle.

Figure 24 shows the geometric arrangement of the various rake angles. Each rake angle is measured with respect to a reference plane, which, in milling, passes through the axis of cutter rotation and the point of the tool. True rake (by definition) is measured in a plane perpendicular to the projection of cutting edge on reference plane.

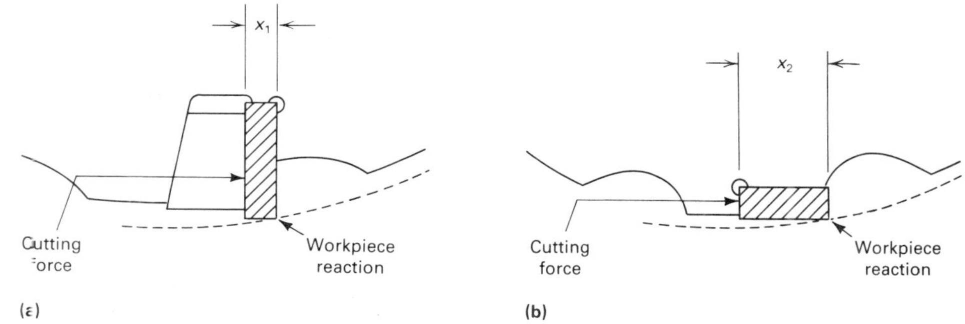


**Fig. 23** Schematics of milling inserts. (a) Conventional wedge-held insert. (b) On-edge insert. Cutting forces are transmitted through the thin section ($x_1$) of the insert in (a) and through the heavy section ($x_2$) in (b).

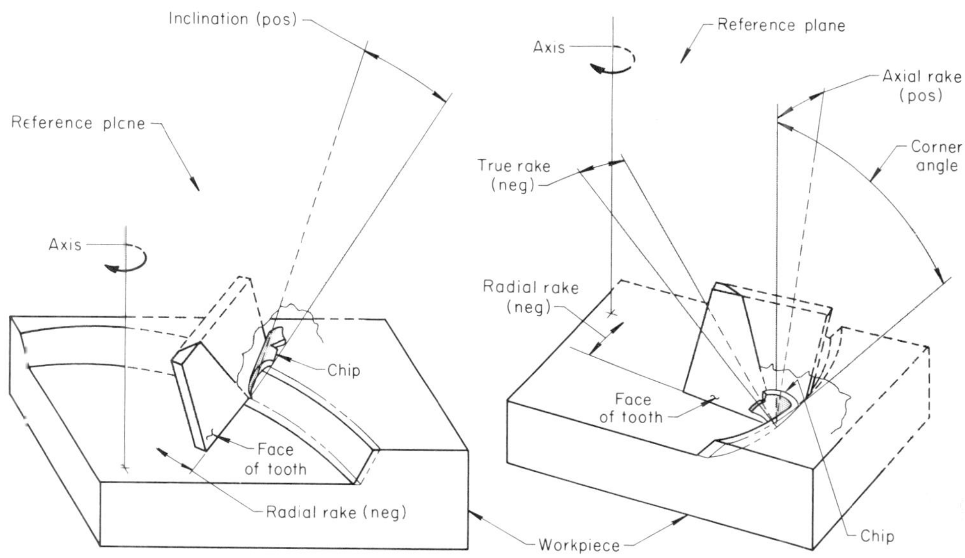


**Fig. 24** Geometric relations of tool angles and reference plane

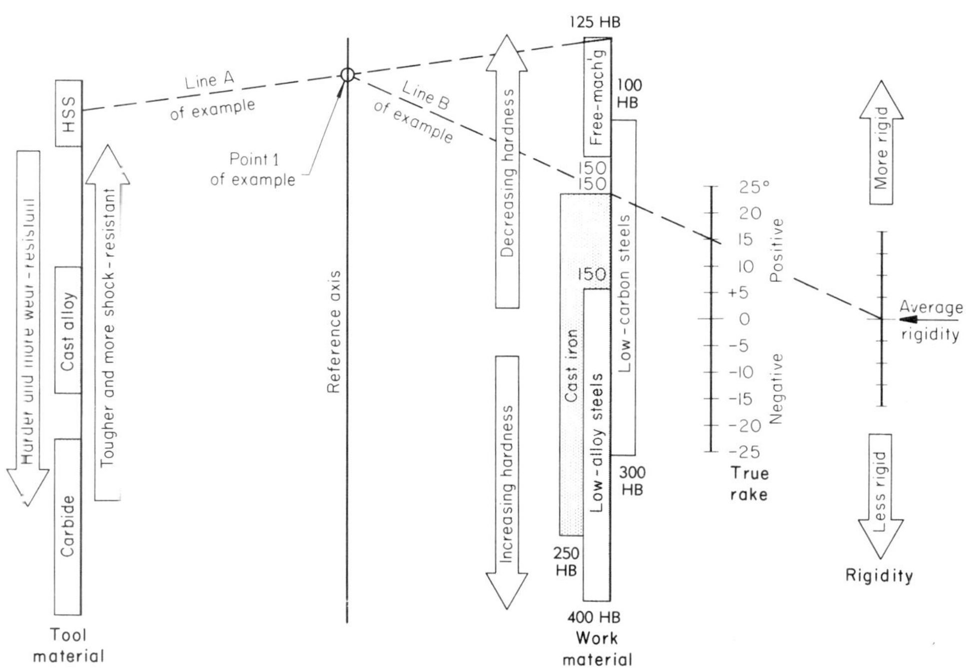


**Fig. 25** Alignment chart for the approximation of true rake

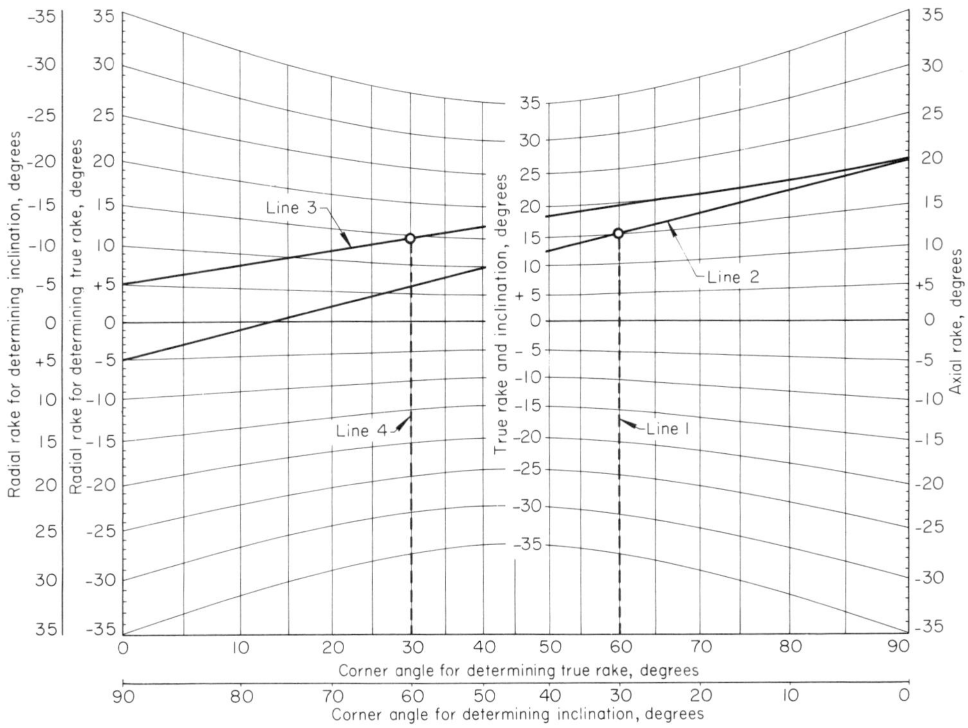


**Fig. 26** Chart for determining true rake components and inclination of teeth for milling

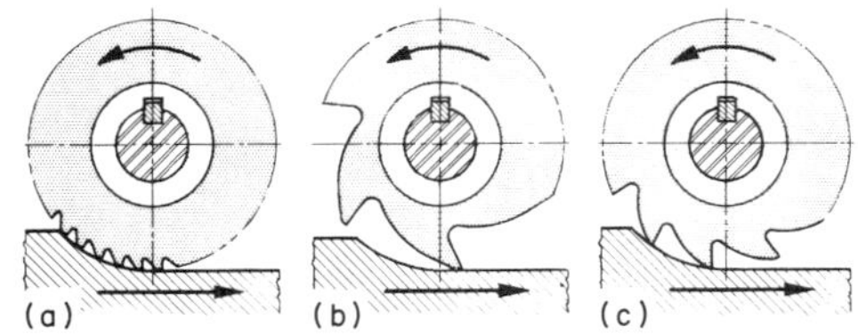


**Fig. 27** Effect of number of teeth on a milling cutter. (a) Too many teeth, resulting in chip crowding and interference. (b) Too few teeth, resulting in intermittent contact. (c) Compromise for satisfactory milling

The second angle of importance in machining is inclination. This is the angle that the cutting edge makes with the reference plane, which, by definition, is perpendicular to the direction of tool travel. Inclination determines the direction of chip curling. When the inclination is zero, chip flow is virtually in the plane of true rake.

Many combinations of axial rake, radial rake, and corner angle other than those listed in Tables 1 to 3 have been successfully used in practice. The requirements are:

- To select a true rake angle large enough for the particular cutting conditions to provide effective chip formation and yet leave an included angle for the tool large enough to provide the required strength and heat conduction
- To select an inclination for the cutting edge that will provide the desired direction of chip flow

In face milling, positive inclination directs the chip outward, and negative inclination directs the chip toward the center of the cutter. A positive inclination is therefore generally desired.

The alignment chart shown in Fig. 25 is a semiquantitative guide for the selection of true rake angles for various combinations of tool materials and common ferrous metals. The rigidity scale is a qualitative modifier of tool-work combination to allow for various degrees of rigidity in work, fixture, or machine. Increased rigidity permits a positive increase in true rake.

After the suggested initial value for true rake has been selected from Fig. 25, it can be resolved into a tentative combination of rake and corner angles by means of the chart shown in Fig. 26. The inclination provided by this combination can then be checked on the same chart.

**Example 1: Use of Charts in Fig. 25 and 26.** Assume a face milling operation on free-machining 1112 steel with a high-speed steel cutter and a setup and machine of average rigidity. On the alignment chart (Fig. 25), connect the high-speed steel index with the upper end of the free-machining steel block as indicated by line A, and mark its intersection with the reference axis at point 1. Then connect point 1 with the average rigidity point by the line B, and note its intersection on the true rake line at the recommended value of +15°.

Next, consult Fig. 26. Usually one can begin with a desired or assumed value of corner angle—for example, 60°. (For long tool life and maximum production, the corner angle should be as large as possible. If chatter or heavy scale is encountered, the corner angle should be decreased.) Note the intersection of the true rake corner angle line 1 for 60° and the true rake line for 15°. Through this point draw line 2 connecting some tentative value for axial rake (for example, 20°) with the radial rake scale, and read the value (−5°) for radial rake.

Next check this combination to find the corresponding inclination. Connect the 20° axial rake point with the −5° inclination point by line 3. (Note that the scale for radial rake is inverted for determining inclination.) Read inclination (−14°) at the intersection of this line with the vertical line 4 for 60° corner angle. If some other value of inclination is desired (so as to provide a different direction of chip flow), a different orientation of the line 2 through the 15° true rake line must be tried. Many different combinations of axial and radial rake angles will provide a given true rake with a given corner angle, but only one combination will provide both a given true rake and a given inclination with a given corner angle.

## Effect of Cutter Design on Efficiency

The angular relations of the cutting edge greatly affect cutting efficiency. The important elements in these relations, as well as their effects, are discussed below.

**Radial rake** has a major effect on the power efficiency of metal removal and on cutter life. Generally, zero to positive radial rakes are used on high-speed steel cutters, and negative radial rakes on carbide cutters. Negative rakes are less efficient, but are usually necessary for a satisfactory service life for the carbide cutters because carbide edges are brittle.

**Axial rake** controls chip flow, thrust force of cut, and the strength of the cutting edges. The axial rake of high-speed steel cutters is usually positive, except for end mills that cut only on the periphery, which often have negative axial rake to transfer cutter thrust back against the spindle bearings while simultaneously applying downward thrust to the workpiece. Carbide cutters can have either positive or negative axial rake, depending on the workpiece material and hardness and on the type of cutter.

**Corner Angle.** Cutters may have a corner angle of 90° to cut a square shoulder, or (as in many face mills) may have a corner angle ranging from 30 to 60° to reduce the chip thickness and to ease the blades into the cut with less shock.

**Clearance.** A clearance angle is necessary behind all cutting edges to prevent the cutter teeth from dragging or rubbing across the workpiece. The primary clearance angle (Fig.

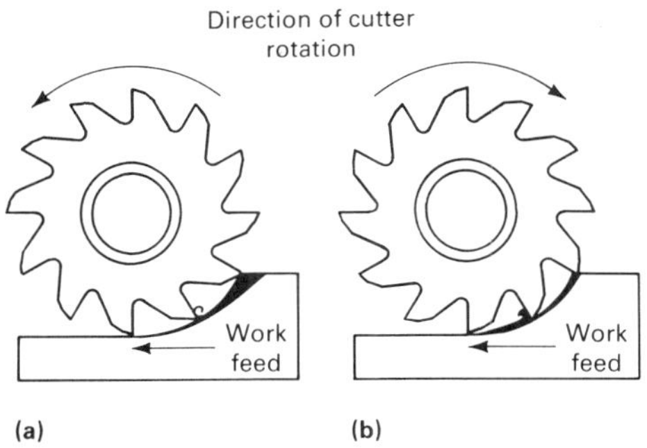


**Fig. 28** Schematics of conventional (a) and climb (b) milling

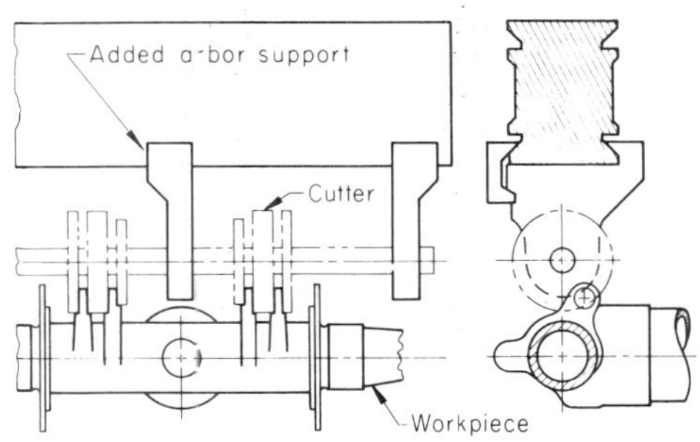


**Fig. 29** Added arbor support that increased the rigidity of gang milling setup

18) usually ranges from 3 to 7°, depending on the type of cut and the workpiece material.

**Number of Teeth.** A milling cutter should have enough teeth to ensure uninterrupted contact with the work metal, and yet not so many as to provide too little space between teeth for chip disposal. Figure 27(a) shows cutter teeth so close together that chip interference is likely. This will increase power consumption and may result in damage to the cutter, the workpiece, or both. Figure 27(b) shows cutter teeth that are too far apart; one tooth will leave the work before the next is engaged. This will cause vibration and chatter, resulting in poor finish, dimensional inaccuracy, and excessive tool wear. A nearly optimum compromise is shown in Fig. 27(c).

The characteristics of the metal being milled also influence the number of teeth in a milling cutter. The tooth spacing shown in Fig. 27(a) would be satisfactory for milling a brittle metal such as cast iron because chips are fine and therefore less likely to interfere. The same tooth spacing would cause chip problems in the milling of steel or other metals that yield stringy chips.

When other cutting conditions are constant, an increase in the number of teeth causes a finer feed per tooth, resulting in a higher specific energy of metal removal and causing greater rubbing on the land. This condition requires that the cutter be ground to closer tolerances to ensure uniform chip size. On the other hand, surface finish is better with finer feeds. Coarser feed per tooth results in a rougher finish and a higher force on each tooth; however, the specific energy of metal removal is lower.

There must be chip space between the teeth on the cutter, although excessively large flutes between closely spaced teeth will weaken the teeth. Obtaining adequate chip space does not depend entirely on the number of teeth in the cutter. Other design features can also contribute. For example, in peripheral milling, it is often possible to use a helical-tooth cutter when a minimum number of teeth are needed for chip space. Helical teeth allow a longer period of contact, thus avoiding the condition that would occur when using a straight-tooth cutter, as shown in Fig. 27(b). Lower peak forces on the teeth of helical cutters reduce vibration and susceptibility to chatter.

Limitations in power and table speed also affect the number of teeth on the cutter. Doubling the number of teeth on the cutter requires that the table speed also be doubled in order to maintain a constant feed per tooth. This doubles the metal removal rate and power consumption. However, if power is available and chips can be disposed of, increasing the number of teeth in the cutter is one means of increasing productivity.

In face milling, the tooth-contact arc is usually longer than in peripheral milling. Therefore, fewer teeth can be used successfully in face milling when chip space is a problem or when available power is limited.

Spindle speed is established for the most economical operation for each cutter diameter and workpiece and cutter material. The number of chips per minute can be varied only by changing the number of teeth on the cutter. The product of feed per tooth and chips per minute (number of teeth times spindle revolutions per minute) is the table speed. Thus, to remain within specific power limitations when milling at high speed, the number of teeth on the cutter can be reduced. This reduces the number of chips per minute, which in turn reduces table speed.

## Power Requirements

The power required for a milling operation is usually computed from the metal removal rate, as follows:

$$P_n = uvdw \qquad \text{(Eq 1)}$$

where $P_n$ is the power required at the cutter, $u$ is the specific energy (in hp/in.$^3$/min), $v$ is the table speed (in in./min), $d$ is the depth of cut (in inches), and $w$ is the width of cut (in inches).

If the metal removal factor is known, $vdw$ can be divided by the factor to find the power required. Equation 1 gives the average power consumption. The instantaneous rate of doing work will vary with the chip thickness and may be considerably higher than the value calculated from Eq 1. However, the rotating parts of the machine and the cutter act like a flywheel to smooth out power drawn from the motor. Therefore, metal removal rates for milling can be safely based on calculations of average power consumption. In general, the milling of most irons and steels requires $4.6 \times 10^{-5}$ kW/mm$^3$ (1 hp/in.$^3$) of metal removal.

## Climb Milling Versus Conventional Milling

The relationships between cutter rotation and feed direction in climb milling and in conventional milling are shown in Fig. 28. In climb milling, as implied by the name, the milling cutter attempts to climb the workpiece. Climb milling is also called down milling. Conventional milling is often referred to as up milling.

In climb milling, chips are cut to maximum thickness at the initial engagement of cutter teeth with the work and then decrease to zero thickness at the end of engagement. In conventional milling, the reverse occurs: The chips start with no thickness and increase in size as the teeth progress through the cut.

The climb technique can be used for most milling applications. Its widespread use has been prevented by the lack of rigid machines with backlash eliminators, which are essential for climb milling. With such equipment, climb milling has several advantages over conventional milling:

- Fixtures and holding devices are simpler and less costly because climb milling exerts a downward force on the workpiece
- Cutters with higher rake angles can be used, decreasing power requirements
- Chips are less likely to be carried by the tooth, reducing the possibility of marring the machined surface
- Chip disposal is easier because chips pile up behind the cutter rather than in front of it
- Cutter wear is less because chip thickness is maximum at the start of the cut
- Finishes are generally improved because the rubbing action in starting a chip is eliminated

The main advantage of conventional milling is the lower impact encountered at initial tooth-workpiece engagement (zero chip thickness). Furthermore, the direction of milling force compensates for the backlash of the feed mechanism. Conventional milling is preferred over climb milling in the following cases:

- When backlash or clearance exists in the feeding mechanism
- The milling of surfaces on which depth of cut varies excessively (for example, by 20%)
- The milling of castings or forgings with very rough surfaces due to sand or scale

## Rigidity of Setup

Optimum results in milling depend greatly on good rigidity of the tooling-and-workpiece setup. Tool deflection and chatter resulting from lack of rigidity cause excessive tool wear and breakage, damage to workpieces, dimensional inaccuracy, and unacceptable surface finish. Merely increasing the size of the arbor holding the cutter(s) can add greater rigidity, which results in higher feed and in lower production costs. Rigid setups minimize or eliminate these adverse conditions, thus increasing production and lowering production cost. The selection of more efficient cutting tools and the increased feed and speed possible with a rigid setup further increase production rates. One method of increasing the rigidity of a tool setup is demonstrated in the following example.

**Example 2: Added Support Between Gang Cutters for Increased Rigidity.** In milling landing-gear main inner cylinders made of 300M steel at 46 to 49 HRC, the use of only one arbor support contributed to a low production rate and an erratic surface finish. Milling speed was only 12 rev/min, and feed was only 6.35 mm/min (0.250 in./min). Surface finish varied from 3.20 to 5.0 μm (125 to 200 μin.).

Placing a second arbor support midway between the six cutters, as shown in Fig. 29, permitted the cutting speed to be increased to 22 rev/min and feed to 22.23 mm/min (0.875 in./min). A uniform finish of less than 3.20 μm (125 μin.) was consistently obtained, and cutter life was increased by 65%.

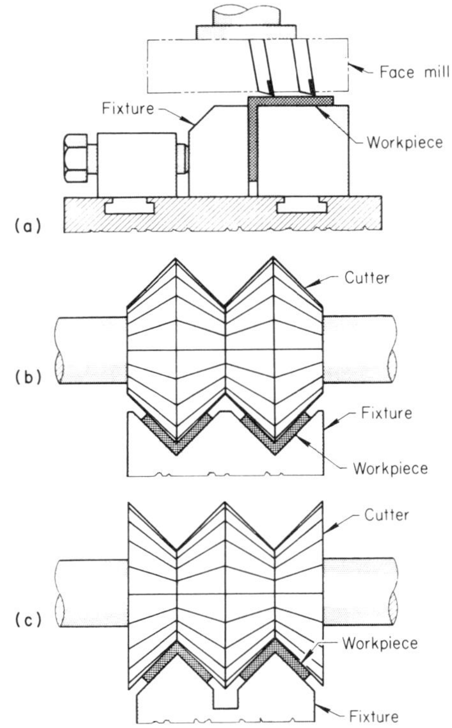


| | |
|---|---|
| **Cutter details** | |
| Type | 45° angle peripheral (matched sets of four) |
| Size | 152 mm (6 in.) diameter |
| Number of teeth | 20 |
| Material | M3 high-speed steel |
| **Machining conditions** | |
| Speed, at 22 rev/min, m/min (sfm) | 11 (35) |
| Feed | 0.089 mm/tooth (0.0035 in./tooth) or 38 mm/min (1.5 in./min) |
| Depth of cut, mm (in.) | 2.54 (0.100) |
| Cutting fluid | Sulfurized oil |
| Finish obtained (approx), μm (μin.) | 3.2 (125) |
| Production rate (approx) | 1 piece/h |
| Cutter life per grind, linear m (linear ft) | 23–30 (75–100) |
| Ultimate tensile strength, MPa (ksi) | |
| H11 steel | 862 (125) |
| Ti-6Al-4V | 1000 (145) |

**Fig. 30** Original (a) and improved (b, c) methods of milling angle extrusions. Extrusions were 2.4 m (8 ft) long; each leg of L was 51 mm (2 in.) long and (before being milled) 7.9 mm (5⁄16 in.) thick. (a) In original method, each surface was milled separately—outside surfaces by face milling, as shown, and inside surfaces by peripheral milling. (b) and (c) Improved method, in which two pairs of peripheral cutters milled both inside or outside surfaces of two extrusions simultaneously (details in table)

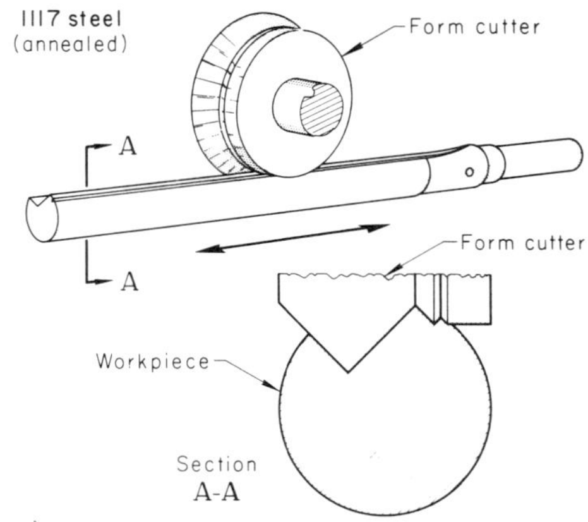


| | |
|---|---|
| **Cutter details** | |
| Type | Special form |
| Size, mm (in.) | 89 (3.5) max diameter |
| Number of teeth | 12 |
| Material | High-speed steel |
| **Machining conditions** | |
| Speed, at 225 rev/min, m/min (sfm) | 63 (206) max |
| Feed | 0.056 mm/tooth (0.0022 in./tooth) or 150.9 mm/min (5.94 in./min) |
| Depth of cut, mm (in.) | 6.4 (1⁄4) max |

**Fig. 31** Use of a special form milling cutter for producing a six-surface contour

## Milling Methods

The principal methods of milling are classified as peripheral, face, and end. These terms refer to the type of cutter used (Fig. 16) and to the relationship of the spindle to the surface being milled. In some cases, the differences between the three methods are clearly defined, but more often a given milling operation is a combination of two methods. This is particularly true of end milling, which is almost invariably a combination of peripheral and face milling.

The milling method selected for a specific application depends largely on the amount of metal to be removed, workpiece size and shape, and the configuration to be milled. Total quantity to be produced, production rate (quantity per unit of time), work metal hardness, and cost are more likely to influence modifications of procedure within a given method than to determine selection of the method itself.

Details of specific milling applications are given in examples in the following sections. These descriptions of practice are presented under headings of peripheral, face, or end milling, although many of the examples represent modifications of these broad classifications.

### Peripheral Milling

Peripheral milling is a method of generating a machined surface by cutting with the teeth on the periphery of a cutter whose axis is parallel to the milled surface. The operation is usually performed on horizontal milling machines in which one or more cutters are mounted by means of keys on arbors with outboard support.

In the simplest type of peripheral milling, cutting is done only by teeth on the periphery of the cutter. However, many applications, such as slotting, require cutting action from both the periphery and the sides. Some side cutting action is often needed for gang, slab, and straddle milling applications.

**Capabilities.** Peripheral milling can be used for removing metal from simple flat surfaces, for cutting keyways and deep slots, and for milling contoured surfaces and surfaces having two or more angles or complex forms. Long keyways can be milled faster by peripheral milling with a keyway cutter or a slotting cutter than by end milling. In addition, the width of the keyway is maintained more accurately, because in peripheral milling cutter wear is distributed over a larger number of teeth than in end milling.

Deep slots, including T-slots, are usually generated by peripheral milling because greater rigidity can be maintained than in end milling. Staggered-tooth slotting cutters are generally used for the peripheral milling of deep slots.

Some configurations of contoured surfaces can be produced by peripheral milling; cutter direction is controlled by rise-and-fall templates and tracers. Forms that can be milled in a single direction by the peripheral method are limited only by cutter design. Several flat and angular surfaces can be produced simultaneously by combining peripheral and face milling.

**Limitations.** In terms of metal removed per unit of time, peripheral milling is less efficient than face milling for producing simple flat surfaces. Other considerations, however, sometimes make peripheral milling the preferable method. Peripheral milling is more limited than end milling for cutting complex configurations; therefore, pockets and intricate recesses are usually cut by end milling.

**Peripheral Versus Face Milling.** In many applications, either peripheral milling or face milling can be used for machining a specific area of the workpiece. Selection of the milling method will then be based on the surface finish required, rigidity of setup, length of feed stroke, and simplicity of fixturing. Production applications that exemplify the advantages of peripheral milling over face milling for specific applications are described in the following example.

**Example 3: Increased Accuracy and Efficiency in Milling Angle Extrusions.** In milling both sides of each leg of right-angle extrusions of H11 steel (annealed) and titanium alloy (solution treated), deformed extrusions with unacceptable surface finish were being produced at low production rates. Originally, complicated fixturing and clamping were required to face mill the outside surfaces (as shown in Fig. 30a) and then to peripheral mill the inside surfaces, one at a time.

In the revised method, two pairs of 45° angle peripheral cutters were used to mill the inside surfaces of two extrusions, as shown in Fig. 30(b). Cutters were reversed for milling the outside surfaces, as shown in Fig. 30(c). The cutters were made in matched sets of four to ensure that all diameters were equal and that keyways aligned the teeth in correct relation.

In addition to doubling production rates, the revised method produced straighter extrusions. The accurate positioning possible with the simplified fixtures equalized cutter forces, minimizing vibration, and thus improved surface finish. Cutter details and machining conditions are tabulated with Fig. 30.

**Milling of Keyways.** The greater accuracy possible with peripheral milling makes it preferable to end milling in the machining of keyways, particularly of long keyways such as those found on machine arbors.

**Slotting.** The accuracy and efficiency of peripheral slotting cutters (which essentially are circular saws) make them preferable to power saws for the machining of slots.

*T-slots* are produced in two separate milling operations. Removal of the largest possible amount of metal, proportional to slot dimensions, is accomplished in an initial cut with a peripheral mill or an end mill. A second cut is then made, using a staggered-tooth T-slotting cutter (solid or tipped) at a

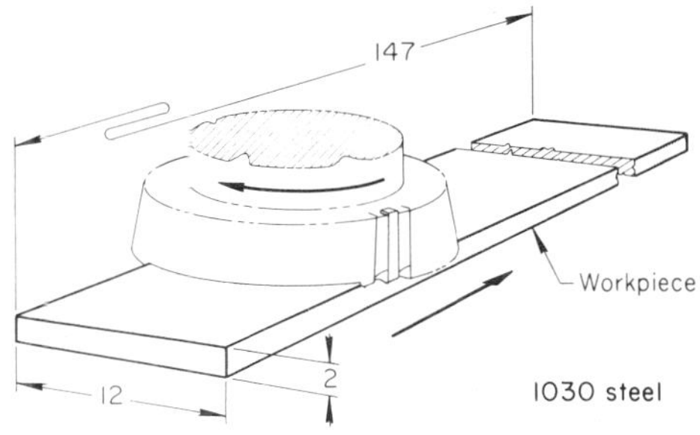


| | |
|---|---|
| **Cutter details** | |
| Type | Face mill, inserted blade |
| Size | 406 mm (16 in.) in diameter, 95 mm (3¾ in.) thick |
| Number of teeth | 32 |
| Material | Carbide-tipped blades |
| **Machining conditions** | |
| Speed, at 72 rev/min, m/min (sfm) | 92 (302) |
| Feed | 0.69 mm/tooth (0.027 in./tooth) or 1575 mm/min (62 in./min) |
| Depth of cut, mm (in.) | 3.2 (⅛) |
| Cutting fluid | None |
| Tolerances | |
| Parallelism error, mm (in.) | 0.076 (0.003) max |
| Depth of cutter marks, mm (in.) | 0.076 (0.003) max |
| Power required, kW (hp) | 71 (95) |
| Metal removal rate, $mm^3$/min ($in.^3$/min) | $2.00 \times 10^6$ (122) |
| Cutter life per grind (approx), $m^2$ ($in.^2$) | 19 (30 000) |
| Workpiece hardness, HB | 165–185 |

**Fig. 32** Face milling of a steel plate in one pass per side. Dimensions in figure given in inches

constant feed rate. To prevent damage to the cutter or the workpiece, cutting fluids (or air blasts in dry milling) are used copiously to flush chips out of the slot.

**Gang milling** refers to milling with two or more peripheral cutters mounted on the same arbor. The cutters can be of the same size or of different sizes, and the surfaces being milled can be adjacent or separated. Gang milling is commonly used to produce several different steps simultaneously in a workpiece or to produce sections of the same thickness from bars or extrusions.

**Form Milling.** The number of parallel surfaces and angular relationships that can be machined by peripheral milling is limited almost solely by cutter design. Form cutters are expensive, but often there is no other satisfactory means of producing complex contours such as the one discussed in the following example.

**Example 4: Form Milling of Six Surfaces With One Cutter.** For small job lots, a knee-and-column horizontal milling machine with a special form cutter was used in milling a six-surface contour in 1117 steel shafts, as illustrated and detailed in Fig. 31. A vise mounted on the machine table held the workpiece, and the cutter was positioned manually. For production lots of more than 1000 pieces, the cutter was used on a bed-type machine with rise and fall of the spindle, and fixtures with air-clamping attachments.

## Face Milling

Face milling is used for machining flat surfaces by means of cutters having teeth on their peripheral faces and driven by spindles whose axes are perpendicular to the surface being milled. In face milling, the tooth path is generated by a combined action of the tooth similar to actions obtained in climb and conventional milling with peripheral cutters. The chip thickness is minimum at the points where the tooth enters and leaves the work and is maximum at the transition point between climb and conventional milling.

Face milling cuts are deep radially and narrow axially, while peripheral cuts are shallow radially and wide axially. Therefore, face milling removes a given amount of metal with less power than peripheral milling.

**Capabilities.** In addition to being a more efficient method of removing metal, other advantages of face milling over peripheral milling are:

- The cutter has greater rigidity because it is attached directly to the spindle nose
- Large areas can be milled with little protrusion of the spindle
- Outboard bearings are not required, which allows greater flexibility in workpiece size
- Cutting forces are more evenly distributed
- Less time is needed for tool changes
- Cutter grinding costs are less than for peripheral cutters of comparable size or are zero if indexable inserts are used in the cutter

The efficiency of face milling in removing large amounts of metal and in milling to close tolerances is described in the following example.

**Example 5: Face Milling 305 mm (12 in.) Wide Plates in One Pass per Side.** Face milling was selected for machining both sides of the 305 mm (12 in.) wide steel plates shown in Fig. 32, mainly because it was more efficient than peripheral milling. The operation was performed on a fixed-rail bed-type milling machine. The plates were secured horizontally by means of a magnetic table, and the vertical spindle was driven by a 75 kW (100 hp) head.

If this operation had been performed on a slab miller (peripheral milling), a helical cutter at least 152 mm (6 in.) in diameter and having about ten rows of inserted blades would have been required. Carbide-tipped blades would have been prohibitively expensive for a peripheral cutter of this size; therefore, high-speed steel inserts would have been used. The allowable feed and

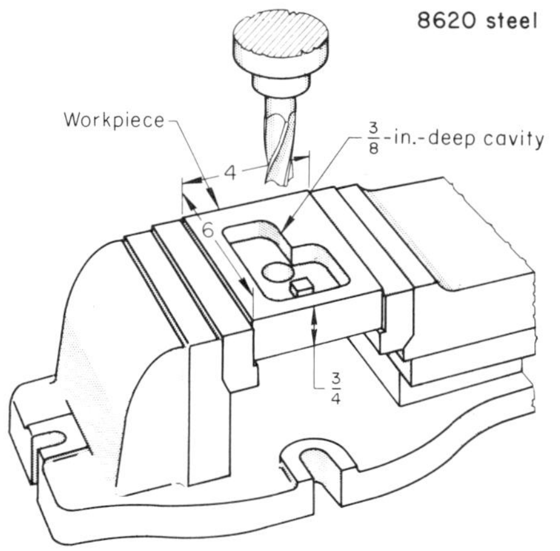


**Cutter details**

| | |
|---|---|
| Type | End mill |
| Size | 19 mm (¾ in.) in diameter |
| Number of teeth | 4 |
| Material | High-speed steel |

**Machining conditions**

| | |
|---|---|
| Speed at 400 rev/min, m/min (sfm) | 24 (79) |
| Feed per tooth, mm (in.) | 0.064 (0.0025) |
| Cutting fluid | Soluble oil |
| Tolerances | |
| Horizontal cuts, mm (in.) | ±0.051 (±0.002) |
| Vertical cuts, mm (in.) | ±0.025 (±0.001) |
| Finish obtained, μm (μin.) | 3.20 (125) |
| Setup time(a) | 30 min |
| Cutter life per grind (approx) | 10 pieces |
| Workpiece hardness, HB | 190–210 |

(a) Using prepared plastic control sheet

**Fig. 33** Milling of a complex cavity with a single cutter under three-axis control. Dimensions in figure given in inches

speed for peripheral milling with high-speed steel would have resulted in a metal removal rate of only $2.0 \times 10^5$ mm³/min (12 in.³/min), compared with $2.00 \times 10^6$ mm³/min (122 in.³/min) by face milling with carbide cutters. Cutter and machining details are tabulated with Fig. 32.

**Limitations** of face milling compared with peripheral milling are:

- Face milling is restricted to the machining of flat surfaces
- As the width of the cut increases in relation to the diameter of the face mill, the larger number of teeth engaged in cutting exerts a greater force at right angles to the direction of feed. Therefore, more rugged fixturing and clamping of the workpiece are required, particularly if the part configuration is sensitive to tangential forces imposed by the cutter

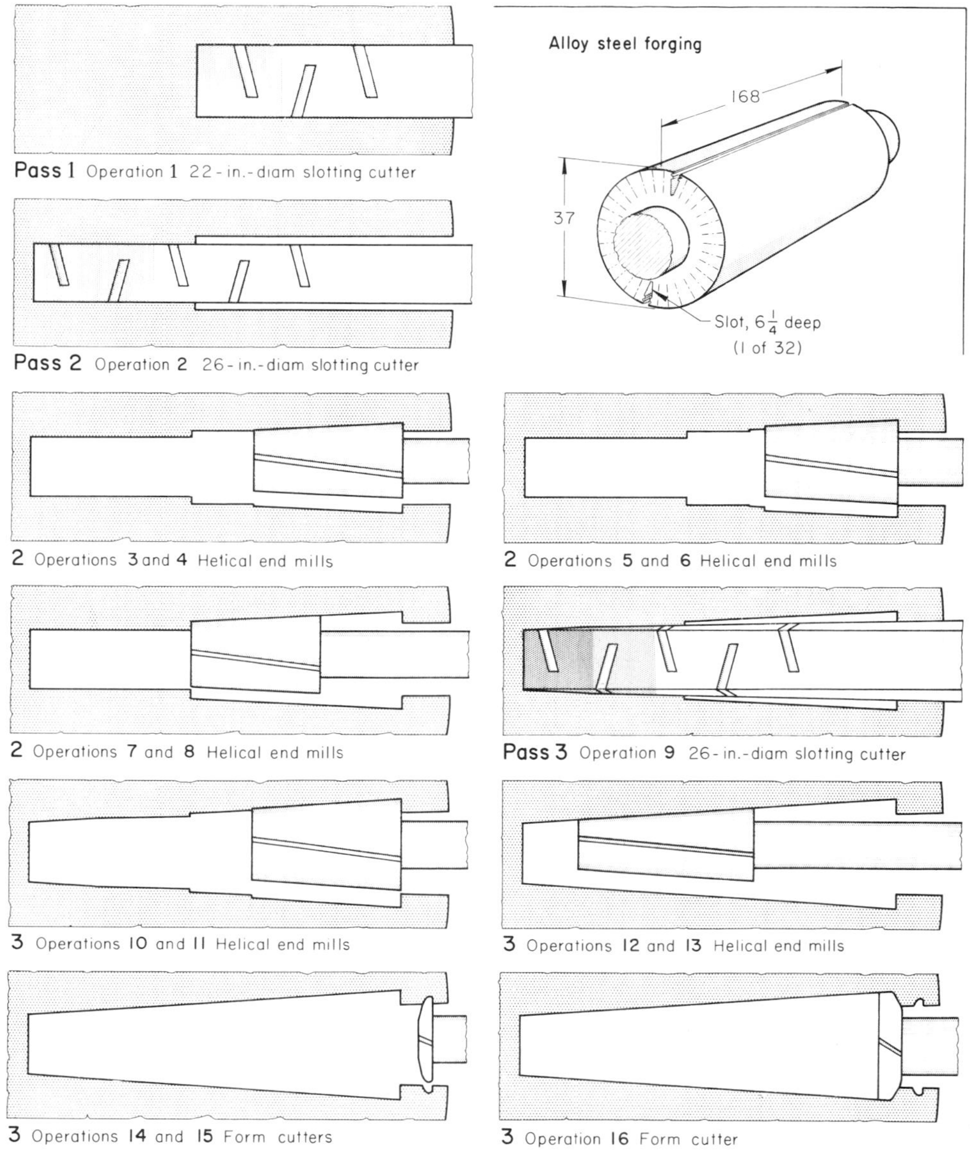


| Pass | Operation(a) | Speed m/min | Speed sfm | Feed mm/rev | Feed in/rev |
|---|---|---|---|---|---|
| 1 | 1, slotting | 84 | 275 | 190 | 7½ |
| 2 | 2, slotting | 84 | 275 | 127 | 5 |
| | 3 through 8, end milling | 18 | 60 | 127 | 5 |
| 3 | 9, slotting | 61 | 200 | 254 | 10 |
| | 10 through 13, end milling | 30 | 100 | 254 | 10 |
| | 10 through 16, form cutting | 30 | 100 | 254 | 10 |

(a) Slotting cutters, carbide tipped; end mills and form tools, solid high-speed steel

**Other machining conditions**

| | |
|---|---|
| Cutting fluid | Soluble oil:water (1:20), air mist |
| Setup time (approx), h | 4 |
| Downtime for tool change, min | |
| Slotting cutters | 30 |
| End mills | 5 |
| Production rate, h/rotor | 65 |
| Cutter life per grind | |
| Slotting cutters, linear m (linear ft) | 30 (100) |
| End mills, roughing, linear m (linear ft) | 23 (75) |
| End mills, finishing, linear m (linear ft) | 38 (125) |
| Workpiece hardness, HB | 250 |

**Fig. 34** Sequence and conditions for milling tapered slots in a large rotor in 16 operations on a special machine. Dimensions in figure given in inches

## End Milling

As shown in Fig. 16, end mills have cutting edges on the face end as well as on the periphery. Therefore, cuts that vary widely in configuration can be made with an end mill by using the side and end consecutively or simultaneously. End milling is usually differentiated from peripheral milling or face milling by application rather than by definition.

End milling is less efficient than other milling methods in removing metal because the end of the cutter is not supported and the length-to-diameter ratio is usually high. Consequently, heavy cuts are not feasible. But despite its low efficiency in removing metal, end milling is often preferred for facing, profiling, slotting, recessing, and box milling.

**Unsupported Length.** End milling is characterized by the unsupported length of the cutter or its driving member or both. As this unsupported length increases, the difficulty in maintaining dimensional accuracy also increases because of cutter deflection. The length-to-diameter ratio that can be tolerated depends largely on the dimension-

## Table 4 Recommended speeds and feeds for the face milling of ferrous and nonferrous alloys with both high-speed steel and carbide cutters

| | | | High-speed steel tool | | | Carbide tool | | | | | | |
|---|---|---|---|---|---|---|---|---|---|---|---|---|
| | | | | | | Uncoated | | | | Coated | | |
| | | | | | | Speed | | | | | | |
| Material (hardness) | Condition | Depth of cut(a), mm (in.) | Speed, m/min (sfm) | Feed per tooth, mm (in.) | Tool material, ISO (AISI) | Brazed, m/min (sfm) | Indexable, m/min (sfm) | Feed per tooth, mm (in.) | Tool material grade, ISO (C) | Speed, m/min (sfm) | Feed per tooth, mm (in.) | Tool material grade, ISO (C) |
| **Wrought aluminum alloys (30–80 HB, 500 kg)** | | | | | | | | | | | | |
| EC 3005 6066 | Cold drawn | 1 (0.040) | 365 (1200) | 0.25 (0.010) | S4, S2 (M2, M7) | 610 (2000) | Max | 0.25 (0.010) | K10, M20 (C-2) | . . . | . . . | . . . |
| 1060 4032 6070 | | 4 (0.150) | 245 (800) | 0.40 (0.015) | S4, S2 (M2, M7) | 550 (1800) | Max | 0.50 (0.020) | K10, M20 (C-2) | . . . | . . . | . . . |
| 1100 5005 6101 | | 8 (0.300) | 200 (650) | 0.50 (0.020) | S4, S2 (M2, M7) | 365 (1200) | Max | 0.65 (0.025) | K20, M30 (C-2) | . . . | . . . | . . . |
| 1145 5050 6151 | | | | | | | | | | | | |
| 1175 5052 6253 | | | | | | | | | | | | |
| 1235 5056 6262 | | | | | | | | | | | | |
| 2011 5083 6463 | | | | | | | | | | | | |
| 2014 5086 6951 | | | | | | | | | | | | |
| 2017 5154 7001 | | | | | | | | | | | | |
| 2018 5252 7004 | | | | | | | | | | | | |
| 2021 5254 7005 | | | | | | | | | | | | |
| 2024 5454 7039 | | | | | | | | | | | | |
| 2025 5456 7049 | | | | | | | | | | | | |
| 2117 5457 7050 | | | | | | | | | | | | |
| 2218 5652 7075 | | | | | | | | | | | | |
| 2219 5657 7079 | | | | | | | | | | | | |
| 2618 6053 7175 | | | | | | | | | | | | |
| 3003 6061 7178 | | | | | | | | | | | | |
| 3004 6063 | | | | | | | | | | | | |
| **Wrought copper alloys (60–100 HRB)** | | | | | | | | | | | | |
| 145 342 377 | Cold drawn | 1 (0.040) | 185 (600) | 0.25 (0.010) | S4, S2 (M2, M7) | 395 (1300) | 436 (1430) | 0.25 (0.010) | K10, M20 (C-2) | . . . | . . . | . . . |
| 147 349 385 | | 4 (0.150) | 150 (500) | 0.36 (0.014) | S4, S2 (M2, M7) | 305 (1000) | 335 (1100) | 0.40 (0.015) | K20, M30 (C-2) | . . . | . . . | . . . |
| 173 350 482 | | 8 (0.300) | 120 (400) | 0.45 (0.018) | S4, S2 (M2, M7) | 215 (700) | 260 (850) | 0.50 (0.020) | K30, M40 (C-2) | . . . | . . . | . . . |
| 187 353 485 | | | | | | | | | | | | |
| 191 356 544 | | | | | | | | | | | | |
| 314 360 623 | | | | | | | | | | | | |
| 316 365 624 | | | | | | | | | | | | |
| 330 366 638 | | | | | | | | | | | | |
| 332 367 642 | | | | | | | | | | | | |
| 335 368 782 | | | | | | | | | | | | |
| 340 370 | | | | | | | | | | | | |
| **Gray cast irons (pearlitic + free carbides) (220–260 HB)** | | | | | | | | | | | | |
| ASTM A48: Classes | As-cast | 1 (0.040) | 26 (85) | 0.15 (0.006) | S4, S2 (M2, M7) | 120 (400) | 135 (440) | 0.18 (0.007) | K10, M20 (C-2) | 200 (650) | 0.18 (0.007) | CK10, CM20 (CC-2) |
| 45, 50; SAE J431c: | | 4 (0.150) | 20 (65) | 0.25 (0.010) | S4, S2 (M2, M7) | 90 (300) | 100 (330) | 0.25 (0.010) | K20, M30 (C-2) | 130 (425) | 0.25 (0.010) | CK20, CM30 (CC-2) |
| Grades G3500, | | 8 (0.300) | 15 (50) | 0.36 (0.014) | S4, S2 (M2, M7) | 62 (205) | 76 (250) | 0.36 (0.014) | K30, M40 (C-2) | 100 (325) | 0.36 (0.014) | CK30, CM40 (CC-2) |
| G4000 | | | | | | | | | | | | |
| **Wrought carbon steels (225–275 HB)** | | | | | | | | | | | | |
| **Low carbon** | | | | | | | | | | | | |
| 1005 1010 1020 | Annealed or | 1 (0.040) | 49 (160) | 0.15 (0.006) | S4, S2 (M2, M7) | 145 (475) | 160 (525) | 0.18 (0.007) | P20 (C-6) | 240 (790) | 0.18 (0.007) | CP20 (CC-6) |
| 1006 1012 1023 | cold drawn | 4 (0.150) | 35 (115) | 0.25 (0.010) | S4, S2 (M2, M7) | 120 (400) | 135 (440) | 0.25 (0.010) | P30 (C-6) | 175 (570) | 0.25 (0.010) | CP30 (CC-6) |
| 1008 1015 1025 | | 8 (0.300) | 27 (90) | 0.36 (0.014) | S4, S2 (M2, M7) | 85 (280) | 105 (345) | 0.36 (0.014) | P40 (C-5) | 135 (450) | 0.36 (0.014) | CP40 (CC-5) |
| 1009 1017 | | | | | | | | | | | | |
| **Medium carbon** | | | | | | | | | | | | |
| 1030 1042 1053 | Hot rolled, | 1 (0.040) | 38 (125) | 0.15 (0.006) | S4, S2 (M2, M7) | 135 (450) | 150 (500) | 0.18 (0.007) | P20 (C-6) | 230 (750) | 0.18 (0.007) | CP20 (CC-6) |
| 1033 1043 1055 | normalized, | 4 (0.150) | 30 (100) | 0.25 (0.010) | S4, S2 (M2, M7) | 115 (375) | 125 (415) | 0.25 (0.010) | P30 (C-6) | 165 (540) | 0.25 (0.010) | CP30 (CC-6) |
| 1035 1044 1525 | annealed, | 8 (0.300) | 24 (80) | 0.36 (0.014) | S4, S2 (M2, M7) | 81 (265) | 100 (325) | 0.36 (0.014) | P40 (C-5) | 130 (420) | 0.36 (0.014) | CP40 (CC-5) |
| 1037 1045 1526 | cold drawn, | | | | | | | | | | | |
| 1038 1046 1527 | or quenched | | | | | | | | | | | |
| 1039 1049 | and tempered | | | | | | | | | | | |
| 1040 1050 | | | | | | | | | | | | |
| **Wrought medium-carbon alloy steels (275–325 HB)** | | | | | | | | | | | | |
| 1330 4135 8625 | Normalized, or | 1 (0.040) | 30 (100) | 0.15 (0.006) | S9, S11(b) (T15, M42)(b) | 115 (375) | 140 (460) | 0.15 (0.006) | P20 (C-6) | 215 (700) | 0.13 (0.005) | CP20 (CC-6) |
| 1335 4137 8627 | quenched and | 4 (0.150) | 24 (80) | 0.23 (0.009) | S9, S11(b) (T15, M42)(b) | 90 (300) | 110 (365) | 0.20 (0.008) | P30 (C-6) | 145 (475) | 0.18 (0.007) | CP30 (CC-6) |
| 4027 4427 8630 | tempered | 8 (0.300) | 18 (60) | 0.30 (0.012) | S9, S11(b) (T15, M42)(b) | 72 (235) | 87 (285) | 0.25 (0.010) | P40 (C-5) | 115 (375) | 0.23 (0.009) | CP40 (CC-5) |
| 4028 4626 8637 | | | | | | | | | | | | |
| 4032 5130 94B30 | | | | | | | | | | | | |
| 4037 5132 | | | | | | | | | | | | |
| 4130 5135 | | | | | | | | | | | | |
| **Wrought austenitic stainless steels (135–185 HB)** | | | | | | | | | | | | |
| 201 304L 348 | Annealed | 1 (0.040) | 40 (130) | 0.20 (0.008) | S4, S2 (M2, M7) | 130 (430) | 145 (475) | 0.20 (0.008) | K10, M20 (C-2) | 215 (700) | 0.20 (0.008) | CK10, CM20 (CC-2) |
| 202 305 384 | | 4 (0.150) | 30 (100) | 0.30 (0.012) | S4, S2 (M2, M7) | 100 (325) | 110 (360) | 0.30 (0.012) | K20, M30 (C-2) | 145 (475) | 0.30 (0.012) | CK20, CM30 (CC-2) |
| 301 308 385 | | 8 (0.300) | 24 (80) | 0.40 (0.016) | S4, S2 (M2, M7) | 70 (230) | 85 (280) | 0.40 (0.016) | K30, M40 (C-2) | 105 (350) | 0.40 (0.016) | CK30, CM40 (CC-2) |
| 302 321 | | | | | | | | | | | | |
| 304 347 | | | | | | | | | | | | |
| **Wrought α + α−β titanium alloys (320–380 HB)** | | | | | | | | | | | | |
| Ti-6Al-4V | Solution | 1 (0.040) | 17 (55) | 0.075 (0.003) | S9, S11(b) (T15, M42)(b) | 44 (145) | 49 (160) | 0.102 (0.004) | K10, M20 (C-2) | . . . | . . . | . . . |
| Ti-6Al-4V ELI | treated and | 4 (0.150) | 15 (50) | 0.13 (0.005) | S9, S11(b) (T15, M42)(b) | 34 (110) | 37 (120) | 0.15 (0.006) | K20, M30 (C-2) | . . . | . . . | . . . |
| Ti-6Al-2Sn-4Zr-2Mo | aged | 8 (0.300) | 12 (40) | 0.18 (0.007) | S9, S11(b) (T15, M42)(b) | 24 (80) | 29 (95) | 0.20 (0.008) | K20, M30 (C-2) | . . . | . . . | . . . |
| Ti-6Al-2Sn-4Zr-2Mo-0.25Si | | | | | | | | | | | | |
| Ti-6Al-2Sn-4Zr-6Mo | | | | | | | | | | | | |

(a) Depth of cut is measured parallel to axis of cutter. (b) Any premium high-speed steel (T15, M33, M41–M47) or (S9, S10, S11, S12). Source: Metcut Research Associates Inc.

**Table 5 Recommended speeds and feeds for the slab (or plain) milling of ferrous and nonferrous alloys with high-speed steel cutters**

| Material | Hardness, HB | Condition | Depth of cut(a) mm | Depth of cut(a) in. | Speed m/min | Speed sfm | Feed per tooth mm | Feed per tooth in. | High-speed steel tool material ISO | High-speed steel tool material AISI |
|---|---|---|---|---|---|---|---|---|---|---|
| **Wrought aluminum alloys** | | | | | | | | | | |
| EC 3005 6066 | 30–80 | Cold drawn | 1 | 0.040 | 365 | 1200 | 0.30 | 0.012 | S4, S2 | M2, M7 |
| 1060 4032 6070 | 500 kg | | 4 | 0.150 | 320 | 1050 | 0.36 | 0.014 | S4, S2 | M2, M7 |
| 1100 5005 6101 | | | 8 | 0.300 | 260 | 850 | 0.40 | 0.016 | S4, S2 | M2, M7 |
| 1145 5050 6151 | | | | | | | | | | |
| 1175 5052 6253 | | | | | | | | | | |
| 1235 5056 6262 | | | | | | | | | | |
| 2011 5083 6463 | | | | | | | | | | |
| 2014 5086 6951 | | | | | | | | | | |
| 2017 5154 7001 | | | | | | | | | | |
| 2018 5252 7004 | | | | | | | | | | |
| 2021 5254 7005 | | | | | | | | | | |
| 2024 5454 7039 | | | | | | | | | | |
| 2025 5456 7049 | | | | | | | | | | |
| 2117 5457 7050 | | | | | | | | | | |
| 2218 5652 7075 | | | | | | | | | | |
| 2219 5657 7079 | | | | | | | | | | |
| 2618 6053 7175 | | | | | | | | | | |
| 3003 6061 7178 | | | | | | | | | | |
| 3004 6063 | | | | | | | | | | |
| **Wrought copper alloys** | | | | | | | | | | |
| 145 342 377 | 60–100 HRB | Cold drawn | 1 | 0.040 | 130 | 425 | 0.18 | 0.007 | S4, S2 | M2, M7 |
| 147 349 385 | | | 4 | 0.150 | 105 | 350 | 0.23 | 0.009 | S4, S2 | M2, M7 |
| 173 350 482 | | | 8 | 0.300 | 84 | 275 | 0.28 | 0.011 | S4, S2 | M2, M7 |
| 187 353 485 | | | | | | | | | | |
| 191 356 544 | | | | | | | | | | |
| 314 360 623 | | | | | | | | | | |
| 316 365 624 | | | | | | | | | | |
| 330 366 638 | | | | | | | | | | |
| 332 367 642 | | | | | | | | | | |
| 335 368 782 | | | | | | | | | | |
| 340 370 | | | | | | | | | | |
| **Gray cast irons (pearlitic + free carbides)** | | | | | | | | | | |
| ASTM A48: Classes 45, 50; SAE J431c: Grades G3500, G4000 | 220–260 | As-cast | 1 | 0.040 | 24 | 80 | 0.15 | 0.006 | S4, S2 | M2, M7 |
| | | | 4 | 0.150 | 18 | 60 | 0.20 | 0.008 | S4, S2 | M2, M7 |
| | | | 8 | 0.300 | 14 | 45 | 0.25 | 0.010 | S4, S2 | M2, M7 |
| **Wrought carbon steels** | | | | | | | | | | |
| **Low carbon** | | | | | | | | | | |
| 1005 1010 1020 | 225–275 | Annealed or cold drawn | 1 | 0.040 | 40 | 130 | 0.13 | 0.005 | S4, S2 | M2, M7 |
| 1006 1012 1023 | | | 4 | 0.150 | 30 | 100 | 0.18 | 0.007 | S4, S2 | M2, M7 |
| 1008 1015 1025 | | | 8 | 0.300 | 24 | 80 | 0.23 | 0.009 | S4, S2 | M2, M7 |
| 1009 1017 | | | | | | | | | | |
| **Medium carbon** | | | | | | | | | | |
| 1030 1042 1053 | 225–275 | Hot rolled, normalized, annealed, cold drawn, or quenched and tempered | 1 | 0.040 | 37 | 120 | 0.13 | 0.005 | S4, S2 | M2, M7 |
| 1033 1043 1055 | | | 4 | 0.150 | 27 | 90 | 0.18 | 0.007 | S4, S2 | M2, M7 |
| 1035 1044 1525 | | | 8 | 0.300 | 21 | 70 | 0.23 | 0.009 | S4, S2 | M2, M7 |
| 1037 1045 1526 | | | | | | | | | | |
| 1038 1046 1527 | | | | | | | | | | |
| 1039 1049 | | | | | | | | | | |
| 1040 1050 | | | | | | | | | | |
| **Wrought medium-carbon alloy steels** | | | | | | | | | | |
| 1330 4135 8625 | 275–325 | Normalized, or quenched and tempered | 1 | 0.040 | 27 | 90 | 0.13 | 0.005 | S9, S11(b) | T15, M42(b) |
| 1335 4137 8627 | | | 4 | 0.150 | 20 | 65 | 0.15 | 0.006 | S9, S11(b) | T15, M42(b) |
| 4027 4427 8630 | | | 8 | 0.300 | 15 | 50 | 0.18 | 0.007 | S9, S11(b) | T15, M42(b) |
| 4028 4626 8637 | | | | | | | | | | |
| 4032 5130 94B30 | | | | | | | | | | |
| 4037 5132 | | | | | | | | | | |
| 4130 5135 | | | | | | | | | | |
| **Wrought austenitic stainless steels** | | | | | | | | | | |
| 201 304L 348 | 135–185 | Annealed | 1 | 0.040 | 55 | 180 | 0.15 | 0.006 | S4, S2 | M2, M7 |
| 202 305 384 | | | 4 | 0.150 | 43 | 140 | 0.20 | 0.008 | S4, S2 | M2, M7 |
| 301 308 385 | | | 8 | 0.300 | 34 | 110 | 0.25 | 0.010 | S4, S2 | M2, M7 |
| 302 321 | | | | | | | | | | |
| 304 347 | | | | | | | | | | |
| **Wrought α + α−β titanium alloys** | | | | | | | | | | |
| Ti-6Al-4V Ti-6Al-4V ELI | 320–380 | Solution treated and aged | 1 | 0.040 | 14 | 45 | 0.102 | 0.004 | S9, S11(b) | T15, M42(b) |
| Ti-6Al-2Sn-4Zr-2Mo | | | 4 | 0.150 | 12 | 40 | 0.13 | 0.005 | S9, S11(b) | T15, M42(b) |
| Ti-6Al-2Sn-4Zr-2Mo-0.25Si | | | 8 | 0.300 | 9 | 30 | 0.15 | 0.006 | S9, S11(b) | T15, M42(b) |
| Ti-6Al-2Sn-4Zr-6Mo | | | | | | | | | | |

(a) Depth of cut is measured perpendicular to axis of cutter. (b) Any premium high-speed steel (T15, M33, M41–M47) or (S9, S10, S11, S12). Source: Metcut Research Associates Inc.

**Table 6 Recommended speeds and feeds for the end milling (peripheral) of ferrous and nonferrous alloys with both high-speed steel and carbide cutters**

| | | | High-speed steel tool | | | | | | Carbide tool | | | | | |
|---|---|---|---|---|---|---|---|---|---|---|---|---|---|---|
| | | | | Feed, mm/tooth (in./tooth), at cutter diameter of: | | | | | | Feed, mm/tooth (in./tooth), at cutter diameter of: | | | | |
| Material (hardness) | Condition | Radial depth of cut(a), mm (in.) | Speed, m/min (sfm) | 10 mm (3/8 in.) | 12 mm (1/2 in.) | 18 mm (3/4 in.) | 25–50 mm (1–2 in.) | Tool material, ISO (AISI) | Speed, m/min (sfm) | 10 mm (3/8 in.) | 12 mm (1/2 in.) | 18 mm (3/4 in.) | 25–50 mm (1–2 in.) | Tool material grade, ISO (C) |
| **Wrought aluminum alloys (30–80 HB, 500 kg)** | | | | | | | | | | | | | | |
| EC, 1060, 1100, 1145, 1175, 1235, 2011, 2014, 2017, 2018, 2021, 2024, 2025, 2117, 2218, 2219, 2618, 3003, 3004, 3005, 4032, 5005, 5050, 5052, 5056, 5083, 5086, 5154, 5252, 5254, 5454, 5456, 5457, 5652, 5657, 6053, 6061, 6063, 6066, 6070, 6101, 6151, 6253, 6262, 6463, 6951, 7001, 7004, 7005, 7039, 7049, 7050, 7075, 7079, 7175, 7178 | Cold drawn | 0.5 (0.020) | 245 (800) | 0.075 (0.003) | 0.102 (0.004) | 0.13 (0.005) | 0.18 (0.007) | S4, S5, S2 (M2, M3, M7) | 395 (1300) | 0.075 (0.003) | 0.102 (0.004) | 0.13 (0.005) | 0.18 (0.007) | K20, M20 (C-2) |
| | | 1.5 (0.060) | 185 (600) | 0.102 (0.004) | 0.15 (0.006) | 0.20 (0.008) | 0.25 (0.010) | S4, S5, S2 (M2, M3, M7) | 305 (1000) | 0.102 (0.004) | 0.15 (0.006) | 0.20 (0.008) | 0.25 (0.010) | K20, M20 (C-2) |
| | | diam/4 diam/4 | 150 (500) | 0.075 (0.003) | 0.102 (0.004) | 0.15 (0.006) | 0.20 (0.008) | S4, S5, S2 (M2, M3, M7) | 275 (900) | 0.075 (0.003) | 0.13 (0.005) | 0.15 (0.006) | 0.20 90.008) | K20, M20 (C-2) |
| | | diam/2 diam/2 | 120 (400) | 0.050 (0.002) | 0.075 (0.003) | 0.13 (0.005) | 0.15 (0.006) | S4, S5, S2 (M2, M3, M7) | 245 (800) | 0.050 (0.002) | 0.102 (0.004) | 0.13 (0.005) | 0.15 (0.006) | K20. M20 (C-2) |
| **Wrought copper alloys (60–100 HB)** | | | | | | | | | | | | | | |
| 145, 147, 173, 187, 191, 314, 316, 330, 332, 335, 0340, 342, 349, 350, 353, 356, 360, 365, 366, 367, 368, 360, 377, 385, 482, 485, 544, 623, 624, 638, 642, 782 | Cold drawn | 0.5 (0.020) | 150 (500) | 0.050 (0.002) | 0.075 (0.003) | 0.13 (0.005) | 0.18 (0.007) | S4, S5, S2 (M2, M3, M7) | 290 (950) | 0.050 (0.002) | 0.075 (0.003) | 0.13 (0.005) | 0.18 (0.007) | K20, M20 (C-2) |
| | | 1.5 (0.060) | 120 (400) | 0.075 (0.003) | 0.13 (0.005) | 0.20 (0.008) | 0.25 (0.010) | S4, S5, S2 (M2, M3, M7) | 225 (730) | 0.075 (0.003) | 0.13 (0.005) | 0.20 (0.008) | 0.25 (0.010) | K20, M20 (C-2) |
| | | diam/4 diam/4 | 105 (350) | 0.050 (0.002) | 0.075 (0.003) | 0.13 (0.005) | 0.15 (0.006) | S4, S5, S2, (M2, M3, M7) | 185 (610) | 0.050 (0.002) | 0.075 (0.003) | 0.13 (0.005) | 0.18 (0.007) | K20, M20 (C-2) |
| | | diam/2 diam/2 | 90 (300) | 0.038 (0.0015) | 0.050 (0.002) | 0.102 (0.004) | 0.13 (0.005) | S4, S5, S2 (M2, M3, M7) | 170 (560) | 0.038 (0.0015) | 0.050 (0.002) | 0.102 (0.004) | 0.15 (0.006) | K20, M20 (C-2) |
| **Gray cast irons (pearlitic + free carbides) (220–260 HB)** | | | | | | | | | | | | | | |
| ASTM A48: Classes 45, 50; SAE J431c: Grades G3500, G4000 | As cast | 0.5 (0.020) | 27 (90) | 0.025 (0.001) | 0.038 (0.0015) | 0.050 (0.002) | 0.075 (0.003) | S4, S5, S2 (M2, M3, M7) | 90 (300) | 0.025 (0.001) | 0.050 (0.002) | 0.075 (0.003) | 0.13 (0.005) | K20, M20 (C-2) |
| | | 1.5 (0.060) | 21 (70) | 0.050 (0.002) | 0.063 (0.0025) | 0.075 (0.003) | 0.102 (0.004) | S4, S5, S2 (M2, M3, M7) | 70 (230) | 0.050 (0.002) | 0.075 (0.003) | 0.102 (0.004) | 0.15 (0.006) | K20, M20 (C-2) |
| | | diam/4 diam/4 | 18 (60) | 0.038 (0.0015) | 0.050 (0.002) | 0.063 (0.0025) | 0.089 (0.0035) | S4, S5, S2 (M2, M3, M7) | 60 (200) | 0.038 (0.0015) | 0.050 (0.002) | 0.075 (0.003) | 0.102 (0.004) | K20, M20 (C-2) |
| | | diam/2 diam/2 | 15 (50) | 0.025 (0.001) | 0.038 (0.0015) | 0.050 (0.002) | 0.075 (0.003) | S4, S5, S2 (M2, M3, M7) | 55 (180) | 0.025 (0.001) | 0.038 (0.0015) | 0.050 (0.002) | 0.075 (0.003) | K20, M20 (C-2) |
| **Wrought carbon steels (225–275 HB)** | | | | | | | | | | | | | | |
| **Low carbon** 1005, 1006, 1008, 1009, 1010, 1012, 1015, 1017, 1020, 1023, 1025 | Annealed or cold drawn | 0.5 (0.020) | 37 (120) | 0.025 (0.001) | 0.050 (0.002) | 0.075 (0.003) | 0.102 (0.004) | S4, S5, S2 (M2, M3, M7) | 140 (455) | 0.025 (0.001) | 0.050 (0.002) | 0.102 (0.004) | 0.13 (0.005) | P20, M20 (C-5) |
| | | 1.5 (0.060) | 27 (90) | 0.050 (0.002) | 0.075 (0.003) | 0.102 (0.004) | 0.13 (0.005) | S4, S5, S2 (M2, M3, M7) | 105 (350) | 0.050 (0.002) | 0.075 (0.003) | 0.13 (0.005) | 0.18 (0.007) | P20, M20 (C-5) |
| | | diam/4 diam/4 | 24 (80) | 0.025 (0.001) | 0.050 (0.002) | 0.075 (0.003) | 0.102 (0.004) | S4, S5, S2 (M2, M3, M7) | 90 (300) | 0.038 (0.0015) | 0.063 (0.0025) | 0.102 (0.004) | 0.13 (0.005) | P20, M20 (C-5) |
| | | diam/2 diam/2 | 21 (70) | 0.018 (0.0007) | 0.025 (0.001) | 0.050 (0.002) | 0.075 (0.003) | S4, S5, S2 (M2, M3, M7) | 85 (280) | 0.025 (0.001) | 0.050 (0.002) | 0.075 (0.003) | 0.102 (0.004) | P20, M20 (C-5) |
| **Medium carbon** 1030, 1033, 1035, 1037, 1038, 1039, 1040, 1042, 1043, 1044, 1045, 1046, 1049, 1050, 1053, 1055, 1525, 1526, 1527 | Hot rolled, normalized, annealed, cold drawn or quenched and tempered | 0.5 (0.020) | 34 (110) | 0.025 (0.001) | 0.050 (0.002) | 0.075 (0.003) | 0.102 (0.004) | S4, S5, S2 (M2, M3, M7) | 130 (430) | 0.025 (0.001) | 0.050 (0.002) | 0.102 (0.004) | 0.13 (0.005) | P20, (C-5) M20 |
| | | 1.5 (0.060) | 26 (85) | 0.050 (0.002) | 0.075 (0.003) | 0.102 (0.004) | 0.13 (0.005) | S4, S5, S2 (M2, M3, M7) | 100 (330) | 0.050 (0.002) | 0.075 (0.003) | 0.13 (0.005) | 0.18 (0.007) | P20, M20 (C-5) |
| | | diam/4 diam/4 | 23 (75) | 0.025 (0.001) | 0.050 (0.002) | 0.075 (0.003) | 0.102 (0.004) | S4, S5, S2 (M2, M3, M7) | 85 (280) | 0.038 (0.0015) | 0.063 (0.0025) | 0.102 (0.004) | 0.13 (0.005) | P20, M20 (C-5) |
| | | diam/2 diam/2 | 20 (65) | 0.018 (0.0007) | 0.025 (0.001) | 0.050 (0.002) | 0.075 (0.003) | S4, S5, S2 (M2, M3, M7) | 81 (265) | 0.025 (0.001) | 0.050 (0.002) | 0.075 (0.003) | 0.102 (0.004) | P20, M20 (C-5) |
| **Wrought medium-carbon alloy steels (275–325 HB)** | | | | | | | | | | | | | | |
| 1330, 1335, 4027, 4028, 4032, 4037, 4130, 4135, 4137, 4427, 4626, 5130, 5132, 5135, 8625, 8627, 8630, 8637, 94B30 | Normalized or quenched and tempered | 0.5 (0.020) | 26 (85) | 0.018 (0.0007) | 0.038 (0.0015) | 0.075 (0.003) | 0.102 (0.004) | S4, S5, S2 (M2, M3, M7) | 95 (310) | 0.025 (0.001) | 0.050 (0.002) | 0.102 (0.004) | 0.13 (0.005) | P20, M20 (C-5) |
| | | 1.5 (0.060) | 20 (65) | 0.025 (0.001) | 0.050 (0.002) | 0.102 (0.004) | 0.13 (0.005) | S4, S5, S2 (M2, M3, M7) | 72 (235) | 0.050 (0.002) | 0.075 (0.003) | 0.13 (0.005) | 0.15 (0.006) | P20, M20 (C-5) |
| | | diam/4 diam/4 | 7 (55) | 0.018 (0.0007) | 0.038 (0.0015) | 0.075 (0.003) | 0.102 (0.004) | S4, S5, S2 (M2, M3, M7) | 62 (205) | 0.038 (0.0015) | 0.050 (0.002) | 0.102 (0.004) | 0.13 (0.005) | P20, M20 (C-5) |
| | | diam/2 diam/2 | 15 (50) | 0.013 (0.0005) | 0.025 (0.001) | 0.050 (0.002) | 0.075 (0.003) | S4, S5, S2 (M2, M3, M7) | 58 (190) | 0.025 (0.001) | 0.038 (0.0015) | 0.075 (0.003) | 0.102 (0.004) | P20, M20 (C-5) |

(a) For standard length end mills, max axial depth can be up to 1.5 × the cutter diameter. (b) Any premium high-speed steel (T15, M33, M41-M47) or (S9, S10, S11, S12). Source: Metcut Research Associates Inc.

(continued)

**Table 6 (continued)**

| Material (hardness) | Condition | Radial depth of cut(a), mm (in.) | High-speed steel tool: Speed, m/min (sfm) | Feed, mm/tooth (in./tooth), at cutter diameter of: 10 mm (3/8 in.) | 12 mm (1/2 in.) | 18 mm (3/4 in.) | 25–50 mm (1–2 in.) | Tool material, ISO (AISI) | Carbide tool: Speed, m/min (sfm) | Feed, mm/tooth (in./tooth), at cutter diameter of: 10 mm (3/8 in.) | 12 mm (1/2 in.) | 18 mm (3/4 in.) | 25–50 mm (1–2 in.) | Tool material grade, ISO (C) |
|---|---|---|---|---|---|---|---|---|---|---|---|---|---|---|
| **Wrought austenitic stainless steels (135–185 HB)** | | | | | | | | | | | | | | |
| 201 304L 348<br>202 305 384 | Annealed | 0.5 (0.020) | 34 (110) | 0.025 (0.001) | 0.050 (0.002) | 0.102 (0.004) | 0.13 (0.005) | S4, S5, S2 (M2, M3, M7) | 110 (360) | 0.013 (0.0005) | 0.025 (0.001) | 0.050 (0.002) | 0.102 (0.004) | K20, M20 (C-2) |
| 301 308 385<br>302 321 | | 1.5 (0.060) | 24 (80) | 0.050 (0.002) | 0.075 (0.003) | 0.13 (0.005) | 0.15 (0.006) | S4, S5, S2 (M2, M3, M7) | 82 (270) | 0.025 (0.001) | 0.050 (0.002) | 0.075 (0.003) | 0.13 (0.005) | K20, M20 (C-2) |
| 304 347 | | diam/4 diam/4 | 21 (70) | 0.025 (0.001) | 0.050 (0.002) | 0.102 (0.004) | 0.13 (0.005) | S4, S5, S2 (M2, M3, M7) | 72 (235) | 0.025 (0.001) | 0.038 (0.0015) | 0.063 (0.0025) | 0.102 (0.004) | K20, M20 (C-2) |
| | | diam/2 diam/2 | 18 (60) | 0.025 (0.001) | 0.038 (0.0015) | 0.075 (0.003) | 0.102 (0.004) | S4, S5, S2 (M2, M3, M7) | 67 (220) | 0.013 (0.0005) | 0.025 (0.001) | 0.050 (0.002) | 0.075 (0.003) | K20, M20 (C-2) |
| **Wrought α + α−β titanium alloys (320–380 HB)** | | | | | | | | | | | | | | |
| Ti-6Al-4V<br>Ti-6Al-4V ELI | Solution treated and aged | 0.5 (0.020) | 26 (85) | 0.025 (0.001) | 0.050 (0.002) | 0.075 (0.003) | 0.13 (0.005) | S9(b) (T15)(b) | 69 (225) | 0.025 (0.001) | 0.050 (0.002) | 0.102 (0.004) | 0.15 (0.006) | K20, M20 (C-2) |
| Ti-6Al-2Sn-4Zr-2Mo | | 1.5 (0.060) | 23 (75) | 0.050 (0.002) | 0.075 (0.003) | 0.102 (0.004) | 0.15 (0.006) | S9(b) (T15)(b) | 60 (200) | 0.050 (0.002) | 0.075 (0.003) | 0.13 (0.005) | 0.18 (0.007) | K20, M20 (C-2) |
| Ti-6Al-2Sn-4Zr-2Mo-0.25Si | | diam/4 diam/4 | 12 (40) | 0.025 (0.001) | 0.038 (0.0015) | 0.050 (0.002) | 0.075 (0.003) | S9(b) (T15)(b) | 38 (125) | 0.038 (0.0015) | 0.050 (0.002) | 0.13 (0.005) | 0.15 (0.006) | K20, M20 (C-2) |
| Ti-6Al-2Sn-4Zr-6Mo | | diam/2 diam/2 | 9 (30) | 0.025 (0.001) | 0.025 (0.001) | 0.038 (0.0015) | 0.050 (0.002) | S9(b) (T15)(b) | 30 (100) | 0.025 (0.001) | 0.025 (0.001) | 0.102 (0.004) | 0.13 (0.005) | K20, M20 (C-2) |

(a) For standard length end mills, max axial depth can be up to 1.5 × the cutter diameter. (b) Any premium high-speed steel (T15, M33, M41-M47) or (S9, S10, S11, S12). Source: Metcut Research Associates Inc.

al accuracy required, method of advancing the cutter (manual or automatic), workmetal hardness, and rate of cutter feed. Excessive unsupported length can sometimes be tolerated when soft metals are being milled with light feeds. For most applications, however, an unsupported length greater than about five diameters is excessive (assuming the shank and cutter body are made of steel).

When workpiece shape or other conditions demand that milling be done with cutters having excessive unsupported length, cutter deflection can be minimized by the use of solid carbide cutters (for greater modulus of rigidity), electronically controlled feed, extremely light cuts, and collet-type adapters for securing the end mill. In addition, the exact amount of runout for each specific off-the-shelf cutter should be determined by an indicator. When the runout is known, it can be corrected before the cutter is used. Minimizing runout will invariably prolong cutter life and is usually helpful in producing a better surface finish.

**Complex contours** are regularly machined by end milling, although some type of automatic control is almost mandatory for achieving good tool life, dimensional control, surface finish, reproducibility, and efficiency of metal removal. End milling with automatic control has been particularly effective in machining complex configurations for aircraft and aerospace vehicles and in machining small complex components for firearms or instruments. The following example describes applications of end milling with automatic control.

**Example 6: Use of Automatic Control for Producing a Complex Cavity With One End Mill.** The need for consistent reproduction of the L-shaped cavity in the 8620 steel part shown in Fig. 33 justified the use of a vertical-spindle milling machine equipped with automatic controls for positioning the end milling cutter along the $x$-, $y$-, and $z$-axes. The workpiece was secured in a vise having step jaws. Starting at a predrilled hole, the cutter was moved by automatic control (punched plastic sheet) to form the cavity, leaving a square island. Then the spindle was raised, and the island was milled to the required height. Cutter and machining details are presented with Fig. 33. The setup time given in Fig. 33 would be much less with NC/CNC five-axes machining than with the tracer milling machine discussed above.

### Combination Processes

For production quantities ranging from 1000 to 10 000 or for steady high production, specially designed equipment for machining several surfaces consecutively or simultaneously is sometimes economical. With this equipment, two or more methods of milling are often combined with other machining operations, as described in the following example.

**Example 7: Sixteen-Operation Milling of Slots in Large Rotors.** Thirty-two taper form slots 159 mm (6¼ in.) deep and 4270 mm (168 in.) long were milled in a large forged steel rotor in 16 operations, as illustrated and detailed in Fig. 34. Milling was done in a special machine of rugged construction to withstand the cutting force required for removing the large volume of steel.

Two vertical spindles and 14 horizontal spindles on each side of the machine made it possible to cut two opposed slots simultaneously and, by taking successive cuts, to complete each pair of slots in only three passes. Carbide-tipped slotting cutters in the vertical spindles removed most of the stock from the slots, thus minimizing adverse effects in subsequent end milling with cutters of high length-to-diameter ratio.

## Speed, Feed, and Depth of Cut

Nominal speeds for milling ferrous and nonferrous alloys are given in Table 4 for face milling, in Table 5 for slab (also known as plain) milling, and in Table 6 for end milling. The speeds in Tables 4 to 6 reflect differences in cutter material and depth of cut, but not in rigidity of setup, surface finish required, and some other variables. For this reason, speeds used in practice are likely to differ considerably from the nominal speeds given in the tables.

Speeds as low as 6.1 m/min (20 sfm) are used in milling some difficult-to-machine metals with cutters made of high-speed steel. At the other extreme, speeds as high as 6100 m/min (20 000 sfm) have been used to mill aluminum or magnesium with carbide cutters (see the section "High-Speed Milling" in this article).

When rigidity of the setup permits, carbide-tipped cutters can usually be operated three to ten times faster than high-speed steel cutters for a given application. Rigidity of setup has a marked effect on the maximum speed that will not cause chatter. Chatter is a condition in which the cutter and the workpiece vibrate in resonance at a frequency usually determined by the natural frequency of one or more elements of the machine. Chatter adversely affects tolerance, finish, and tool life. Carbide tools are especially susceptible to chatter and usually fail by chipping when chatter occurs. When it is impossible to increase rigidity, lower speeds and high-speed steel cutters must be used.

Speed is also influenced by feed rate. If the established feed rate is high, it is often necessary to lower the speed because of inadequate power. Conversely, if a high speed is established, it may be necessary to decrease feed per tooth or to use a cutter having fewer teeth in order to stay within the limits of available power. The speed used for a specific application is ordinarily a compromise between a high speed that will yield a high rate of metal removal and produce the best finish and a low speed that will result in desirable tool life.

**Feed** is expressed as the rate at which the work moves past the cutter (or vice versa) in millimeters per minute (inches per minute) or in millimeters per tooth (inches per tooth). Feed per tooth is the linear distance of tooth advance for each revolution of the cutter; therefore, feed per revolution of the cutter is the product of feed per tooth and the number of teeth on the cutter. Feed in inches per minute can be converted to feed per tooth by dividing inches per minute by the product of the cutter revolutions per minute and the number of teeth in the cutter.

For the highest efficiency of metal removal and the least susceptibility to chatter, the feed per tooth should be as high as possible in any milling operation. However, several factors influence or limit the rate of feed: type of cutter, number of teeth, cutter material, work metal composition and hardness, depth of cut, width of cut, speed, rigidity of the setup, and available power.

Nominal feed rates for the face, slab, and end milling of ferrous and nonferrous alloys are given in Tables 4 to 6. Because Tables 4 to 6 do not reflect all the pertinent variables, feeds used in practice may differ from the nominal rates by a factor of 20 or more.

**Determination of Optimum Speed and Feed.** A wide range of speeds and feeds can be successfully used. However, for optimum results in milling hard or otherwise difficult-to-machine alloys, speed and feed must be carefully selected because the range of workable and economical operating conditions is relatively narrow.

**Depth of cut** in rough milling is usually 3.2 mm (1/8 in.) or more. In finish milling, depth of cut may vary from several hundredths of a millimeter to 1.6 mm (1/16 in.). For the optimum surface finish (particularly when using high-speed cutters), roughing cuts followed by finishing cuts are generally required. When using carbide-tipped cutters, it is often possible to attain the required surface finish in a single cut.

## Surface Finish

The surface finish obtainable by milling depends on work metal composition and condition, speed, feed, tool material, tool design, and cutting fluid. Peripheral and face milling produce different types of surfaces because of the different relations of cutter rotation to the workpiece in the two methods.

**Peripheral milling** produces a surface characterized by lines approximately parallel to the cutter axis. Sometimes these lines lie parallel and normal to the helix angle of the cutter teeth and are spaced at a distance equal to the feed per tooth. Finish in peripheral milling is also affected as follows:

- A high or low tooth produces a mark on the surface
- A bent arbor results in surface waviness
- An arbor too small in diameter causes chatter, resulting in a rough surface
- Uneven spacing of the cutter teeth causes vibrations that result in rough surfaces
- Climb milling produces a different surface texture than conventional milling because in climb milling maximum chip thickness occurs at the beginning of the cut, while in conventional milling maximum chip thickness occurs at the end of the cut. Tooth and revolution marks are usually less noticeable than in climb milling
- Climb milling causes more shock to inserted blades and indexable inserts than conventional milling

**Face Milling.** The surface texture produced by face milling is characterized by a series of arcs when the heel or following edge of the cutter is slightly elevated during milling. If the work surface is parallel to the cutter face, a lattice-type texture is produced.

**Practical Limits.** A finish of 3.20 μm (125 μin.) can usually be obtained with either carbide or high-speed steel cutters without stringent control of process variables. Finishes of 1.60 μm (63 μin.) or less are often obtained in milling with carbide tools. Under good conditions—which include the use of carbide cutters operated at high speed, a free-cutting grade of work metal, and a sulfurized (or sulfurized and chlorinated) cutting fluid—a finish of 0.80 μm (32 μin.) can be achieved. With special cutters used under closely controlled conditions 0.25 μm (10 μin.) finishes and steam-tight joint surfaces have been obtained by milling.

**Improvement** of surface finish has been obtained by altering the heat treated structure of the work metal, by modifying cutter design, and by increasing cutting speed.

## Cutting Fluids

Some type of cutting fluid is ordinarily used with high-speed steel cutters in milling steel, stainless steel, copper-base alloys, and heat-resisting alloys. Aluminum and magnesium alloys and gray iron are usually milled dry. Cutting fluid is used more for malleable and nodular iron than for gray iron.

Regardless of the work metal, cutting fluids are used less often with carbide-tipped cutters than with high-speed steel cutters because carbide withstands higher temperature and is less susceptible to build-up at the cutting edges. With carbide-tipped cutters, the functions of the cutting fluid are to keep tool-workpiece temperature low, thus reducing the possibility of cracking in the carbide tips, and to flush away chips.

Cutting fluids are not recommended or needed when ceramic milling cutters are used. This is because the ceramic remains relatively cool due to the low thermal conductivity of the material, and most of the heat generated in milling is carried away in the chips. When diamond milling cutters are used, a cutting fluid is generally recommended as excessive heat can burn or crack the tool material; however, there are successful applications that are performed dry.

When heavy roughing cuts are taken on annealed steel (high rate of metal removal), cooling of the tools and flushing away of chips are important, and finish (influenced largely by built-up edges) is of lesser importance. Conversely, in light finishing cuts (low rate of metal removal), surface finish may be far more important than cooling and flushing away chips. When milling difficult-to-machine metals, cooling is the most important function of the cutting fluid; chip disposal is less important because the rate of metal removal is low. In climb milling, chip disposal is less of a problem than in conventional milling because the chips are forced out behind the cutter in climb milling but ahead of the cutter in conventional milling.

**Types of Fluids Used in Milling Applications.** Characteristics of the principal types of cutting fluid are discussed below.

*Soluble oils* have superior cooling and flushing ability, but are not effective in preventing adhesion of the tool to the workpiece. When surface finish is critical, it may be necessary to sacrifice some cooling and flushing and to select a cutting fluid that is more effective in preventing adhesion.

*Cutting oils* prevent tool-workpiece adhesion, thus allowing closer dimensional control and better finish. Many cutting oils are dark in color, and this makes it difficult for operators to observe the operation. Viscosity of cutting oils should be kept below 100 Saybolt Universal seconds (SUS) at 40 °C (100 °F) to prevent excessive amounts being carried away with the chips.

*Compressed air*, supplied through jets, is often used in place of a cutting fluid and is effective as a coolant and in keeping

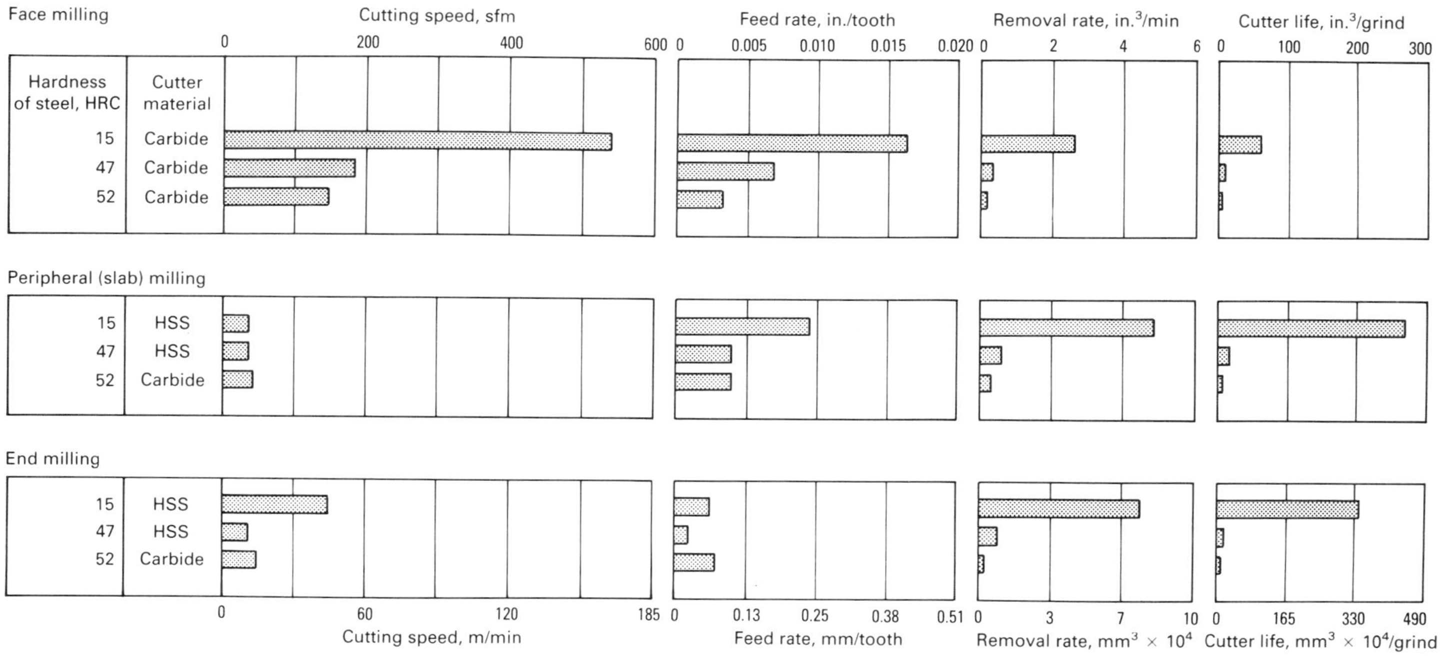


| Item | Face milling(a) 15 HRC | Face milling(a) 47 HRC | Face milling(a) 52 HRC | Peripheral (slab) milling(b) 15 HRC | Peripheral (slab) milling(b) 47 HRC | Peripheral (slab) milling(b) 52 HRC | End milling(c) 15 HRC | End milling(c) 47 HRC | End milling(c) 52 HRC |
|---|---|---|---|---|---|---|---|---|---|
| **Machining conditions** | | | | | | | | | |
| Speed, rev/min | 344 | 117 | 94 | 19 | 17 | 17 | 527 | 137 | 180 |
| Speed, m/min (sfm) | 166 (543) | 56 (185) | 45 (148) | 11.5 (37.8) | 10.8 (35.6) | 13 (42) | 46 (150) | 11 (36) | 14 (47) |
| Feed, mm/min (in./min) | 142.8 (5.623) | 22.2 (0.875) | 7.938 (0.3125) | 82.6 (3.25) | 20.64 (0.8125) | 20.64 (0.8125) | 133 (5.25) | 14.29 (0.5625) | 19.1 (0.75) |
| Feed, mm/tooth (in./tooth) | 0.414 (0.0163) | 0.172 (0.0068) | 0.084 (0.0033) | 0.241 (0.0095) | 0.10 (0.004) | 0.10 (0.004) | 0.064 (0.0025) | 0.025 (0.001) | 0.071 (0.0028) |
| Depth of cut, mm (in.) | 3.18 (0.125) | 3.18 (0.125) | 3.18 (0.125) | 25.4 (1.0) | 25.4 (1.0) | 25.4 (1.0) | 24.64 (0.970) | 24.64 (0.970) | 2.54 (0.100) |
| Width of cut, mm (in.) | 102 (4.0) | 102 (4.0) | 102 (4.0) | 19.0 (0.75) | 10.2 (0.4) | 5.1 (0.2) | 21.59 (0.850) | 21.59 (0.850) | 44.45 (1.75) |
| **Cutter details** | | | | | | | | | |
| Material(d) | Carbide | Carbide | Carbide | T15 HSS | T15 HSS | Carbide | M10 HSS | M10 HSS | Carbide |
| Diameter, mm (in.) | 152 (6) | 152 (6) | 152 (6) | 203 (8) | 203 (8) | 203 (8) | 25.4 (1.0) | 25.4 (1.0) | 44.45 (1.750) |
| Axial rake angle | −7° | −7° | −7° | 10° | 10° | 10° | 30° (e) | 30° (e) | 3° 30′ (e) |
| Radial rake angle | −7° | −7° | −7° | 10° | 10° | 10° | 6° | 6° | −1° 30′ |
| Axial relief angle | 7° | 7° | 7° | 5° | 5° | 2° | 3° | 3° | 3° 30′ |
| Radial relief angle | 7° | 7° | 7° | 5° | 5° | 7° | 10° | 10° | 1° 30′ |
| Corner radius, mm (in.) | 0.76 (0.030) | 0.76 (0.030) | 0.76 (0.030) | 1.78 (0.070) | 1.78 (0.070) | ... | 3.05 (0.120) | 3.05 (0.120) | 0.76 (0.030) |

(a) Steels were milled in a vertical No. 5 machine, using climb cut, and with no cutting fluid. Single-blade cutter (inserts: 19 × 19 × 4.8 mm, or ¾ × ¾ × 3⁄16 in.) had a 45° lead angle. Cutter life was based on the development of 0.46 mm (0.018 in.) wear land. (b) Steels were milled in a vertical No. 4 machine, using conventional cut, and with a 1:1 mixture of sulfurized oil and mineral oil as the cutting fluid. Cutters had 38 mm (1½ in.) diam arbor holes and 16 teeth (for steels at 15 and 47 HRC) or 18 teeth (for 52 HRC). Cutter life was based on the development of a 0.51 mm (0.020 in.) wear land. (c) Steels were milled in a horizontal No. 4 machine, with a 1:1 mixture of sulfurized oil and mineral oil as the cutting fluid. Cutters were 114 mm (4½ in.) long, with four flutes (for steels at 15 and 47 HRC) or three flutes (for 52 HRC). Cutter life was based on the development of a 0.38 mm (0.015 in.) wear land. (d) Approximate composition of carbide used was 67.5W-10.5Ta-8.5Co-7.5C-6.0Ti (grade C-5). (e) Helix angle

**Fig. 35** Effect of work metal hardness on machining conditions, metal removal rate, and cutter life in the milling of low-alloy steels

chips cleared from the cutter path. Air is used when milling cast iron or for operations when using liquids is not practical.

**Methods of Application.** In milling, cutting fluids are usually distributed by nozzles to the cutter and workpiece. A pressure of 35 kPa (5 psi) or less is usually sufficient. Higher pressure results in loss of fluid by splashing and serves no useful purpose.

Nozzles vary in shape from a plain round tube to a manifold type, depending mainly on width of the milling cut. The width of the nozzle should approach the width of the cut to maintain uniform distribution of the fluid.

In peripheral milling, the nozzle should be at the outgoing side of the cutter so that the stream of cutting fluid hits the outgoing teeth and outer surface of the chip. A nozzle in this position allows the most effective use of the cutting fluid because, as a tooth leaves the work after cutting a chip, it will be wetted and cooled by the cutting fluid. Thus, a tooth will carry a film of fluid as it revolves and begins another cut.

In addition to directing cutting fluid less effectively, a nozzle at the ingoing side can be dragged under the cutter, damaging the equipment and the work. Also, the cutting fluid may wash small chips under the ongoing teeth, resulting in gouging of the workpiece surface.

In face milling, optimum results are obtained by applying the cutting fluid on the side where the cutting is done rather than on the idle side. Another effective method of application consists of spraying the fluid with compressed air and directing the jets of vapor at the area being milled. Soluble-oil or proprietary chemical emulsions are readily sprayed. This method is especially effective in large milling installations in which it is otherwise difficult to supply liquid to, and collect it from, the cutting area.

A recent innovation involves applying the coolant through the spindle and subsequently through the tool. This is the best and most effective method yet devised.

## Effect of Workpiece Hardness

The production milling of steel at hardnesses of 40 to 55 HRC is not unusual.

However, the difficulty of milling increases as hardness increases and is manifested mainly in reduced tool life (Fig. 19b), even at the reduced speeds and feeds necessary. Speed, feed, cutter material and design, rigidity, and cutting fluid all require more careful consideration as workpiece hardness increases. Control of these variables will influence the magnitude of reduction in tool life, but even under the best conditions, cycle time will increase and tool life will decrease as hardness increases. The net result is increased machining cost, particularly for roughing operations.

Figure 35 summarizes the results of tests made to determine the influence of hardness on conditions, results, and costs in rough milling. In these tests, three similar low-alloy steels (4130, 4330, and 4340) at 15, 47, and 52 HRC, respectively, were machined by face, peripheral, and end milling under the conditions given in the table accompanying Fig. 35. The steels were milled under average shop conditions and in sufficient quantity to ensure that the data obtained would approximate production practice. Tests were halted when cutter edges developed a predetermined wear land.

As shown in Fig. 35, as work metal hardness increased, it was necessary to use lower speed and feed or (for peripheral and end milling) to change from high-speed steel to carbide tools. The product of depth times width of cut was also lower for milling the harder materials, except in face milling. The metal removal rate was much lower in milling at the higher hardnesses, and—in spite of the changes in feed, speed, and tool material—so was cutter life.

**Cutter Design and Material for Milling Steel Harder Than 40 HRC.** The strength of milling cutters is more critical when high-hardness, high-strength steel is being milled. The number of cutter teeth affects strength, feed rate, and elapsed time per cut. A design that incorporates the maximum number of teeth (short of interfering with necessary chip clearance or exceeding power capacity) is the most economical. Other operating factors, however, may require that the number of cutting teeth be reduced (to as few as one, under certain extreme conditions).

Cutter design must permit the maximum conduction of heat away from the heat-sensitive edges of cutting teeth. Inadequate heat conduction will inevitably shorten tool life. The effects of high cutting temperature can be offset to some extent by using a more heat-resistant material for the cutter.

*Tooth angles* must ensure adequate and properly directed chip flow. When chips are accommodated adequately, there is no crowding or jamming in the spaces between teeth. The direction of chip flow is determined by the inclination angle of the cutting edge, which is a combination of axial rake, radial rake, and corner angles.

*Proper relief angles* minimize built-up edge and rubbing action between cutter surfaces and the workpiece. This rubbing action, in addition to accelerating tool wear, produces work hardening in many of the heat-resistant alloys. Work hardening reduces cutting efficiency and adversely affects surface finish.

*Cutter design and material* have major effects on cutter life and productivity. These factors are usually interrelated with speed and feed.

*Rigidity*. Both the cutter and the setup must be as rigid as possible. Body sections and tool sections must be able to support the cutting load without chatter or excessive vibration. Arbor holes should be as large as possible, and the workpiece must be firmly seated and anchored.

## High-Speed Milling

Considerable developmental work is being performed on the use of high-speed milling, which is defined as milling at elevated speeds, feeds, and depths of cut. The normal speed rate might be 120 to 185 m/min (400 to 600 sfm), but high-speed milling involves speeds 200 times greater.

**Advantages.** A major advantage of high-speed milling is increased productivity. Investigations have shown that the rate of temperature increase tends to decrease at higher cutting speeds. There is no detrimental effect on the finishes produced. In fact, in most instances, smoother surface finishes are obtained.

**Limitations.** A possible limitation to high-speed milling is the need for rigid, adequately powered machines with high speed and feed capabilities, strong fixturing, secure workholding, and ample guarding, which add to capital equipment costs. In addition, because machining time is only a small percentage of the total production time for any workpiece, the added cost for high-speed milling may not be economically feasible. Work is being done with gas hydrodynamic, hydrostatic, and active magnetic bearings to increase the high-speed capabilities of milling machines.

**Applications.** Much of the work in high-speed milling has been done with small end mills for cutting aluminum alloys and honeycomb materials that are easy to machine. One major defense contractor found that a cutting speed increase of 500% resulted in a 300% improvement in cutter efficiency (cubic inches of metal removed per minute per horsepower). This firm is milling 7075 aluminum alloy at cutting speeds to 5500 m/min (18 000 sfm) with feeds to 4600 mm/min (180 in./min), using solid carbide, two-flute end mills.

During tests at another aerospace manufacturer, 25 mm (1 in.) diam, two-flute end mills (both high-speed steel and carbide) were operated at 20 000 rev/min and at feed rates of 5100 mm/min (200 in./min) or more with good cutter life. Axial depths of cut were 6.35 mm (0.250 in.), and radial depths were 13.97 mm (0.550 in.). The optimum feed was 0.20 to 0.25 mm/tooth (0.008 to 0.010 in./tooth), depending on the aluminum alloy being milled. More detailed information on high-speed milling methods is available in the article "High-Speed Machining" in this Volume.

## Milling Compared With Alternative Processes

Any surface that is accessible can be milled. This means that milling machines are to some extent competitive with all other machine tools. However, when the work is to be revolved, a milling machine can be used, but is seldom selected for the job because machines of the lathe family are inherently more efficient for such purposes.

Milling machines are capable of machining holes and locating them with a fair degree of accuracy; tolerances range from 25 to 50 μm (0.001 to 0.002 in.). A milling machine is economical for doing such work in small quantities without additional equipment. If the holes do not need to be located accurately, a drill press will perform the task more quickly and easily. For large quantities, the milling machine is usually slower and unable to compete with jigs on drilling machines or with production boring machines. Jig boring machines are necessary where holes must be located more precisely than can be done with milling machines. Large pieces require the capacity and range of horizontal boring machines. Flat, straight, and many curved and irregular surfaces can be shaped, planed, or broached as well as milled.

**Broaching** is more economical than milling in many cases for large quantities, but is disadvantageous in other cases. Milling often is best for moderate quantities.

**Planing or Shaping.** The initial cost of a milling machine is considerably greater than that of a planer or a shaper that can machine workpieces of similar size. In addition, the tooling for milling usually costs up to 50 times as much, and setup time is usually longer. However, milling is far more efficient than planing or shaping in removing metal within a given time.

**Grinding** is capable of producing parts to closer tolerances and finer finishes and can remove harder materials than milling. Some grinding is done entirely from rough stock in a process called abrasive machining, but many parts are milled before grinding.

Grinding is often preferred to milling when the amount of metal to be removed is small and dimensional accuracy and surface finish are critical. Milling and grinding are frequently used in combination, but when grinding can perform the entire machining operation, the time and expense of an extra setup are avoided.

# Gear Manufacture

Revised by Theodore J. Krenzer, The Gleason Works, Division of Gleason Corporation
and Joseph W. Coniglio, Gould & Eberhardt Gear Machinery Corporation

GEARS are machine elements that transmit rotary motion and power by the successive engagement of teeth on their periphery. They constitute an economical method for such transmission, particularly if power levels or accuracy requirements are high. Gears are basically modifications of friction disks; teeth are added to prevent slipping and to ensure that their relative motions are constant. However, it should be noted that the addition of teeth does not change the relative velocities of the disks and shafts; the velocity ratio is determined by the diameters of the disks. Although wooden teeth or pegs were attached to disks to make gears in ancient times, the teeth of modern gears are produced by making cuts into disks that are sufficiently large to contain the outer portions of the teeth, or by forming processes that cause the metal in the teeth to plastically flow outward from a disk.

Gears can be produced commercially by sand casting, die casting, stamping, and powder metallurgy. All these processes are suited to gears for low wear requirements, low power transmission (except large sand cast gears), and relatively low accuracy of transmitted motion. When the application involves higher values of one or more of these characteristics, cut or machined gears are used.

The methods of gear cutting can be divided broadly into form cutting and generating processes. The tooth profile in the former is obtained by using a form cutting tool. This may be a multiple-point cutter used in a milling machine or a broaching machine, or a single-point tool for use in a shaper. A variation of the form cutter method is based on transferring the profile from a templet. This method is used in the shaperlike cutting of large tooth forms and in cutting bevel gear teeth on a bevel gear planer. In all these processes, the workpiece is held stationary until a tooth is finished; the piece is then indexed for successive teeth.

In the generating process, the tooth profile is obtained by a tool that simulates one or more teeth of an imaginary generating gear. A relative rolling motion of the tool with the workpiece generates the tooth surface. This method is used in the hobbing, shaping, and milling processes for manufacturing spur and helical gears and in the face milling and face hobbing of bevel gears.

A wide variety of machines are used to cut gear teeth. As shown in Fig. 1, there are four more or less distinct ways to cut material from a gear blank so as to leave a toothed wheel after cutting. First, the cutting tool can be threaded and gashed. If so, it is a hob, and the method of cutting is called hobbing. Second, when the cutting tool is shaped like a pinion or a section of a rack, it will be used in a cutting method called shaping. Third, in the milling process, the cutting tool is a toothed disk with a gear tooth contour ground into the sides of the teeth. The fourth general method uses a tool (or a series of tools) that wraps around the gear and cuts all teeth at the same time. Methods of this type are termed broaching, punching, or shear cutting.

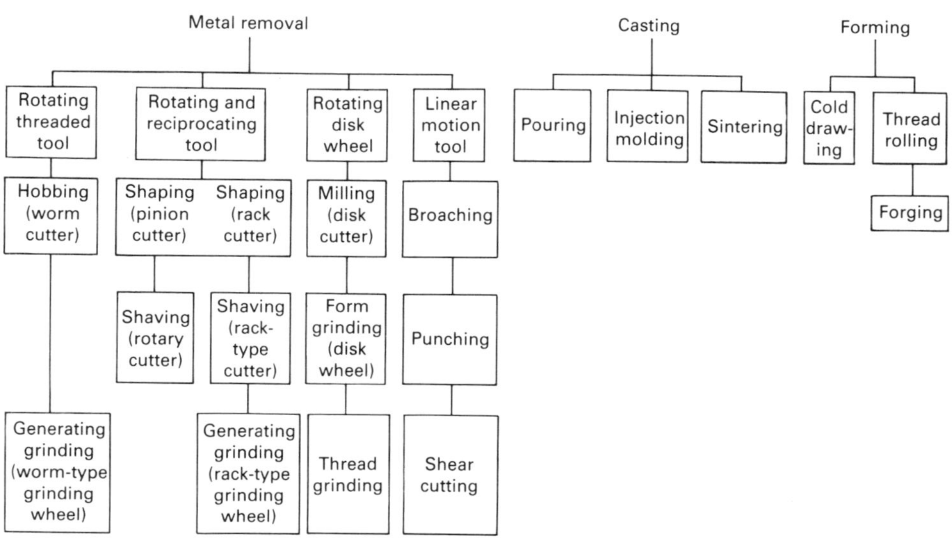


**Fig. 1** Outline of methods of producing gear teeth

Gears are useful when the following kinds of power or motion transmission are required:

- A change in the speed of rotation
- A multiplication or division of the torque or magnitude of rotation
- A change in the direction of rotation
- A conversion from rotational to linear motion or vice versa (rack gears)
- A change in the angular orientation of the rotational motion (bevel gears)
- An offset or change in the location of the rotating motion

Power transmission gears are normally of coarse pitch and can be very large. Power plant gears as large as 7.3 m (24 ft) have been made. They can transmit tens of thousands of horsepower. Contrasted with these are fine-pitch miniature instrument and watch gears as small as 2 mm (0.080 in.) in pitch diameter. Normally, these small gears transmit motion only; the power they transmit is negligible.

Gears of 20 diametral pitch or coarser are classified as coarse-pitch gears. Fine-pitch gears are those with a diametral pitch greater than 20. The usual maximum fineness is 120 diametral pitch. However, involute-tooth gears can be fabricated with diametral pitches as fine as 200, and cycloidal-tooth gears with diametral pitches to 350.

## Terminology

Several terms commonly used in describing gears are defined below:

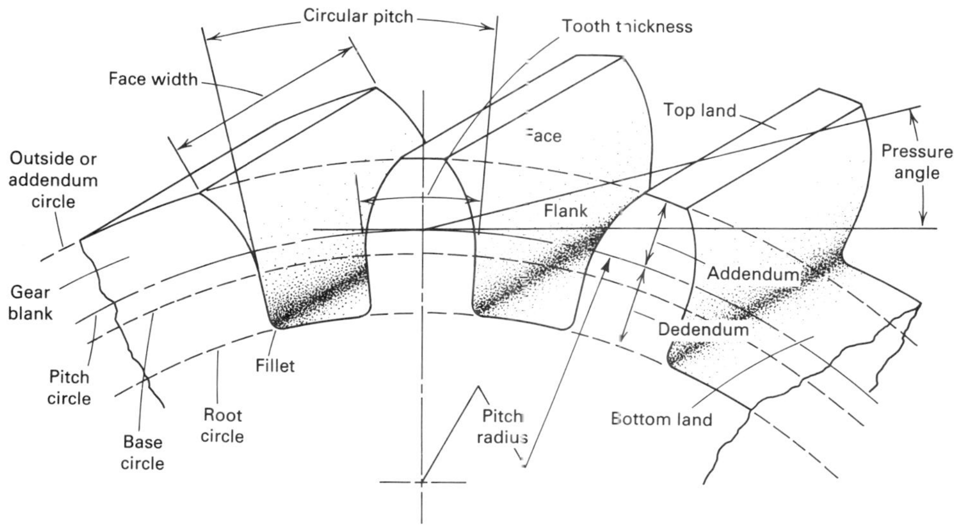


**Fig. 2** Schematic of typical gear tooth nomenclature

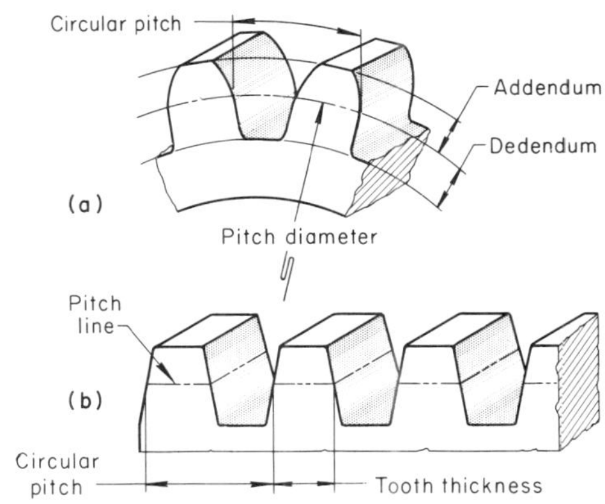


**Fig. 3** Sections of a spur gear (a) and a spur rack (b)

- *Addendum*: Height of tooth above pitch circle
- *Base circle*: Circle from which the involute tooth profiles are generated
- *Circular pitch*: Length of arc of the pitch circle between corresponding points on adjacent teeth
- *Dedendum*: Depth of a tooth space below the pitch circle
- *Diametral pitch*: A measure of tooth size in the English system. In units, it is the number of teeth per inch of pitch diameter. As the tooth size increases, the diametral pitch decreases. Diametral pitches usually range from 25 to 1
- *Face width*: Length of the gear teeth in the axial plane
- *Module*: A measure of tooth size in the metric system. In units, it is millimeters of pitch diameter per tooth. As the tooth size increases, the module also increases. Modules usually range from 1 to 25
- *Pitch circle*: A circle of radius equal to the distance form the gear axis to the pitch point
- *Pitch diameter*: Diameter of the pitch circle
- *Pitch point*: The tangency point of the pitch circles of two mating gears
- *Pressure angle*: The angle between a tooth profile and a radial line at its pitch point

These terms and others are shown schematically in Fig. 2.

## Types of Gears

Gears can be classified as being of the spur, helical, herringbone, crossed-axes helical, worm, internal, rack, bevel, or face type. Each is discussed below.

**Spur gears** (Fig. 3a), the most common type, are used to transmit power or motion between parallel shafts or between a shaft and a rack. Spur gears have radial teeth uniformly spaced around the outer periphery of their circular shapes and parallel to the shafts on which they are mounted. Tooth contact between mating spur gears is in a straight line tangent to the pitch circles.

**Helical gears** (Fig. 4a) are used to transmit motion between parallel or crossed shafts or between a shaft and a rack by meshing teeth that lie along a helix at an angle to the shaft. Because of this angle, mating of the teeth occurs in such a way that more than one tooth of each gear is always in mesh. This condition permits smoother action than with spur gears. However, unlike spur gears, some end thrust is inevitable in helical gears, causing loss of power and requiring thrust bearings.

**Herringbone gears** (Fig. 5), sometimes called double helical gears, are used to transmit motion between parallel shafts. Herringbone gears combine the principal advantages of spur and helical gears because their tooth engagement is progressive, with two or more teeth sharing the load at the same time. Because they have equal right-hand and left-hand helixes, end thrust is eliminated. Herringbone gears can be operated at higher pitch-line velocities than spur gears.

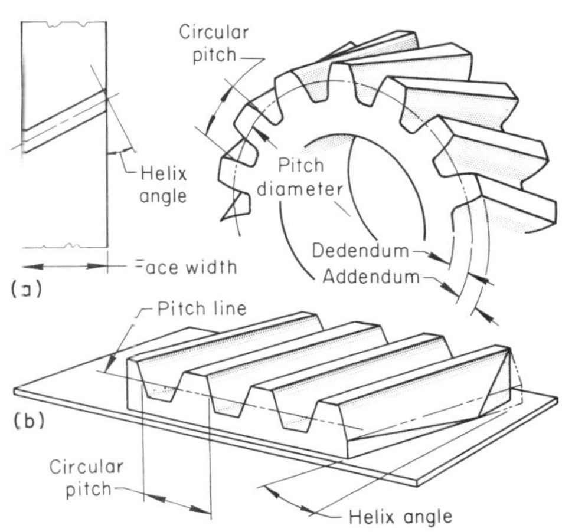


**Fig. 4** Sections of a helical gear (a) and a helical rack (b)

**Crossed-axes helical gears** (Fig. 6) operate with shafts that are nonparallel and nonintersecting. Crossed-helical gears are essentially nonenveloping worm gears, that is, both members are cylindrical. The action between mating teeth has a wedging effect, which results in sliding on tooth flanks. These gears have low load-carrying capacity, but are useful where shafts must rotate at an angle to each other.

**Worm gear sets** are usually right-angle drives consisting of a worm gear (or worm wheel) and a worm. A single-enveloping worm gear set has a cylindrical worm, but the gear is throated (that is, the gear blank has a smaller diameter in the center than at the ends of the cylinder, the concave shape increasing the area of contact between them) so that it tends to wrap around the worm. In the double-enveloping worm gear set, both members are throated, and both members wrap around each other. A double-enveloping worm gear set is shown in Fig. 7. Worm gear sets are used where the ratio of the speed of the driving member to

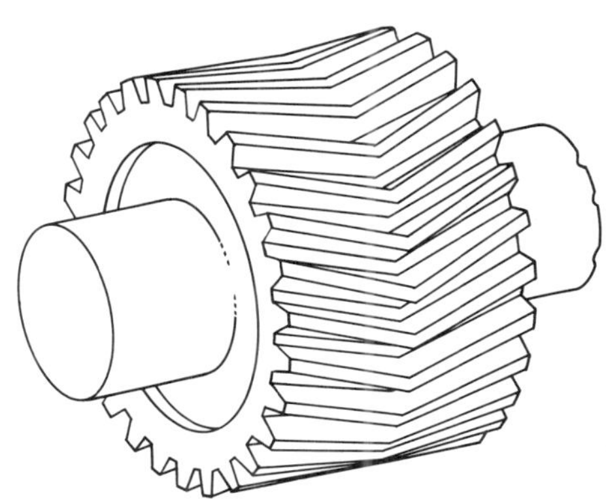

**Fig. 5** A typical one-piece herringbone gear. The opposed helixes permit multiple-tooth engagement and eliminate end thrust.

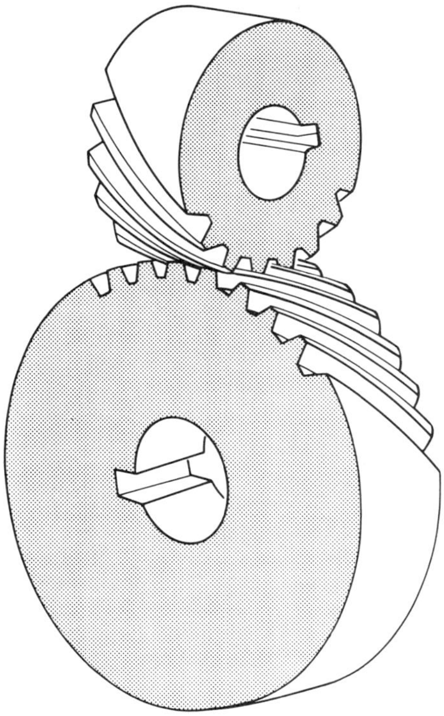

**Fig. 6** Mating crossed-axes helical gears

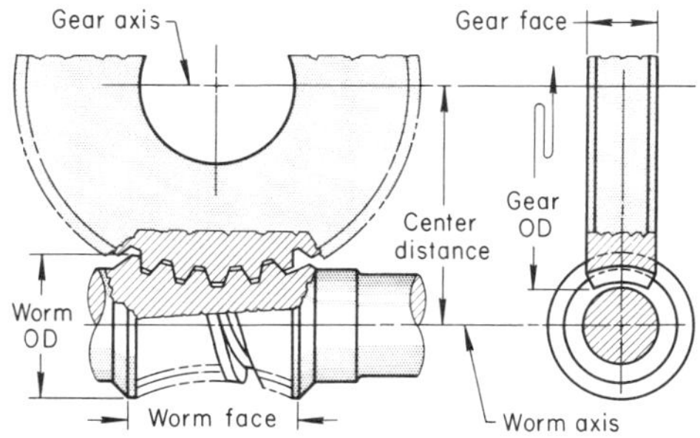


**Fig. 7** Mating of worm gear (worm wheel) and worm in a double-enveloping worm gear set

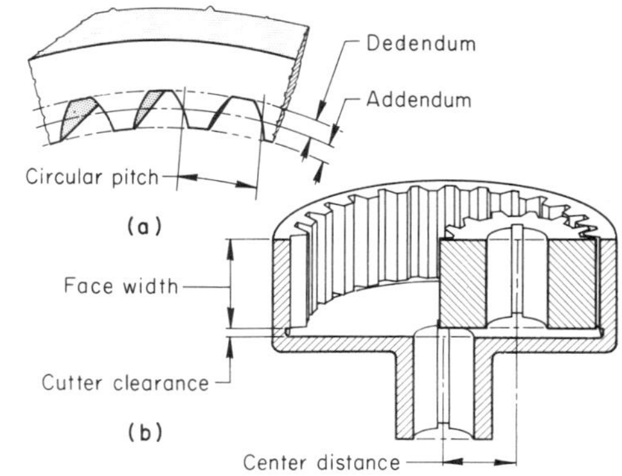


**Fig. 8** Section of a spur-type internal gear (a) and relation of internal gear with mating pinion (b)

the speed of the driven member is large, and for a compact right-angle drive.

**Internal gears** are used to transmit motion between parallel shafts. Their tooth forms are similar to those of spur and helical gears except that the teeth point inward toward the center of the gear. Common applications for internal gears include rear drives for heavy vehicles, planetary gears, and speed-reducing devices. Internal gears are sometimes used in compact designs because the center distance between the internal gear and its mating pinion is much smaller than that required for two external gears. A typical relation between an internal gear and a mating pinion is shown in Fig. 8.

**Racks.** A rack is a gear having a pitch circle of infinite radius. Its teeth lie along a straight line on a plane. The teeth may be at right angles to the edge of the rack and mesh with a spur gear (Fig. 3b), or the teeth on the rack may be at some other angle and engage a helical gear (Fig. 4b).

**Bevel gears** transmit rotary motion between two nonparallel shafts. These shafts are usually at 90° to each other.

*Straight bevel gears* (Fig. 9a) have straight teeth that, if extended inward, would intersect at the axis of the gear. Thus, the action between mating teeth resembles that of two cones rolling on each other (see Fig. 10 for angles and terminology). The use of straight bevel gears is generally limited to drives that operate at low speeds and where noise is not important.

*Spiral bevel gears* (Fig. 9b) have teeth that are curved and oblique. The inclination of the teeth results in gradual engagement and continuous line contact or overlapping action; that is, more than one tooth will be in contact at all times. Because of this continuous engagement, the load is transmitted more smoothly from the driving to the driven gear than with straight bevel gears. Spiral bevel gears also have greater load-carrying capacity than their straight counterparts. Spiral bevel gears are usually preferred to straight bevel gears when speeds are greater than 300 m/min (1000 sfm), and particularly for very small gears.

*Zerol bevel gears* (Fig. 9c) are curved-tooth bevel gears with zero spiral angle. They differ from spiral bevel gears in that the teeth are not oblique. They are used in the same way as spiral bevel gears, and they have somewhat greater tooth strength than straight bevel gears.

*Hypoid gears* (Fig. 9d) are similar to spiral bevel gears in general appearance. The important difference is that the pinion axis of the hypoid pair of gears is offset somewhat from the gear axis. This feature provides many design advantages. In operation, hypoid gears run even more smoothly and quietly than spiral bevel gears and are somewhat stronger.

Spiral bevel, Zerol bevel, and hypoid gears are of two types—generated and nongenerated. In appearance, the two types are nearly identical, the only difference being a slight variation in the profile shape of the teeth. In a generated pair, the teeth of both the gear and pinion are cut on a generating-type machine, while in a nongenerated pair, only the pinion member is generated, the teeth of the gear being straight-sided. In generating a pinion to operate with a nongenerated gear, the tooth profile is modified to compensate for the lack of profile curvature in the gear tooth. For reasons of tooth design, nongenerated gears are usually limited to ratios of at least 2.5:1.

Nongenerated gears are used primarily for economy. Because no generating roll is required when cutting the gear member,

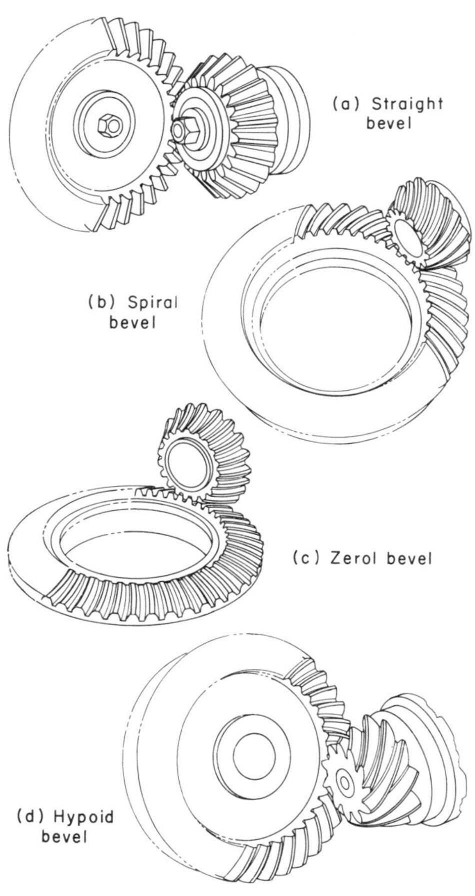


**Fig. 9** Four types of bevel gears

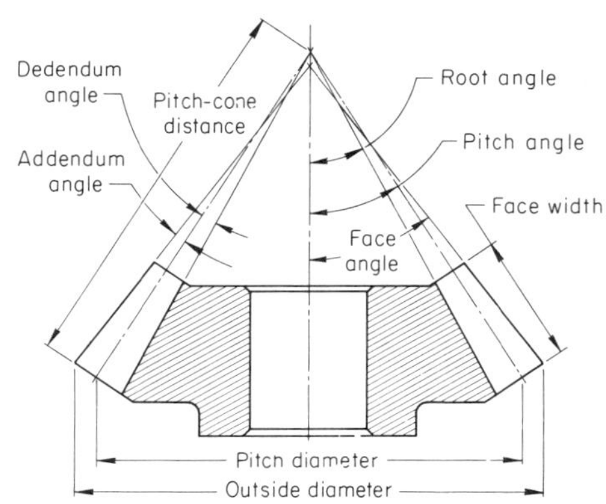


**Fig. 10** Angles and terminology for straight bevel gears

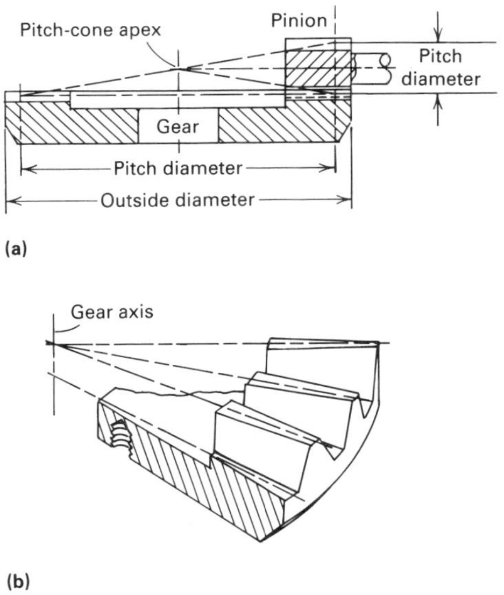


**Fig. 11** Face gear terminology. (a) Cross-sectional view showing gear and pinion positions. (b) Relationship of gear teeth to gear axis

machining is several times faster than for a generated counterpart. For this reason, nongenerated bevel gears are widely used when mass production is required.

**Face gears** have teeth cut on the end face of a gear, as the term face gear implies. They are not ordinarily thought of as bevel gears, but functionally they are more akin to bevel gears than to any other type.

A spur pinion and a face gear are mounted (like bevel gears) on shafts that intersect and have a shaft angle (usually 90°). The pinion bearings carry mostly radial load, while the gear bearings have both thrust and radial load. The mounting distance of the pinion from the pitch-cone apex is not critical, as it is in bevel or hypoid gears. Figure 11 shows the terminology used with face gears.

The pinion that goes with a face gear is usually made spur, but it can be made helical if necessary. The formulas for determining the dimensions of a pinion to run with a face gear are no different from those for the dimensions of a pinion to run with a mating gear on parallel axes. The pressure angles and pitches used are similar to spur gear (or helical gear) practice.

The gear must be finished with a shaper-cutter that is almost the same size as the pinion. Equipment for grinding face gears is not available. The teeth can be lapped, and they can be shaved without too much difficulty, although ordinarily shaving is not used.

The face gear tooth changes shape from one end of the tooth to the other. The face width of the gear is limited at the outside end by the radius at which the tooth becomes pointed. At the inside end, the limit is the radius at which undercut becomes excessive. Due to practical considerations, it is usually desirable to make the face width somewhat short of these limits. The pinion to go with a face gear is usually made with a 20° pressure angle.

**Table 1 Types of gears in common use**

| |
|---|
| **Parallel axes** |
| Spur external |
| Spur internal |
| Helical external |
| Helical internal |
| **Intersecting axes** |
| Straight bevel |
| Zerol bevel |
| Spiral bevel |
| Face gear |
| **Nonintersecting and nonparallel axes** |
| Crossed helical |
| Single-enveloping worm |
| Double-enveloping worm |
| Hypoid |
| Spiroid |

## Proper Gear Selection

The first step in designing a set of gears is to select the correct type. In many cases, the geometric arrangement of the apparatus that needs a gear drive will considerably affect the selection. If the gears must be on parallel axes, then spur or helical gears are appropriate. Bevel and worm gears can be used if the axes are at right angles, but they are not feasible with parallel axes. If the axes are nonintersecting and nonparallel, then crossed-helical gears, hypoid gears, worm gears, or Spiroid gears can be used. Worm gears, though, are seldom used if the axes are not at right angles to each other. Table 1 lists the principal types of gears and how they are mounted.

There are no dogmatic rules that tell the designer which gear to use. The choice is often made after weighing the advantages and disadvantages of two or three types of gears. Some generalizations, though, can be made about gear selection.

**External helical gears** are generally used when both high speeds and high horsepowers are involved. External helical gears have been built to carry as much as 45 000 kW (60 000 hp) of power on a single pinion and gear. Larger helical gears could also be designed and built. It is doubtful if any other type of gear could be built and used successfully to carry this much power on a single mesh.

**Bevel gears** are ordinarily used on right-angle drives when high efficiency is needed. These gears can usually be designed to operate with 98% or better efficiency. Worm gears seldom operate at efficiencies above 90%. Hypoid gears do not have as good efficiency as bevel gears, but hypoid gears can carry more power in the same space, provided the speeds are not too high.

**Worm gears** are ordinarily used on right-angle drives when very high ratios (single-thread worm and gear) are needed. They are also widely used in low-to-medium ratios (multiple-thread worm and gear) as packaged speed reducers. Single-thread worms and worm gears are used to provide the mechanical indexing accuracy on many machine tools. The critical function of indexing hobbing machines and gear shapers is nearly always done by worm gear drive.

**Spur gears** are relatively simple in design and in the machinery used to manufacture and check them. Most designers prefer to use them wherever design requirements permit.

Spur gears are ordinarily thought of as slow-speed gears, while helical gears are thought of as high-speed gears. If noise is not a serious design problem, spur gears can be used at almost any speed that can be handled by other types of gears. Aircraft gas-turbine precision spur gears sometimes operate at pitch-line speeds above 50 m/s (10 000 sfm). In general, though, spur gears are not used much above 20 m/s (4000 sfm).

## Machining Processes for Gears

Simple gear tooth configurations can be produced by basic processes such as milling, broaching, and form tooling. Complex gear tooth configurations require more sophisticated processes designed especially for the manufacture of gears.

### Processes for Gears Other Than Bevel Gears

The methods used to cut the teeth of gears other than bevel gears are milling, broaching, shear cutting, hobbing, shaping, and rack cutting. In any method, a fixture must hold the gear blank in correct relation to the cutter, and the setup must be rigid.

**Milling** produces gear teeth by means of a form cutter. The usual practice is to mill one tooth space at a time. After each space is milled, the gear blank is indexed to the next cutting position.

Peripheral milling can be used for the roughing of teeth in spur and helical gears. Figure 12 shows teeth in a spur gear being cut by peripheral milling with a form cutter. End milling can also be used for cutting teeth in spur or helical gears and is often used for cutting coarse-pitch teeth in herringbone gears.

In practice, gear milling is usually confined to one-of-a-kind replacement gears, small-lot production, the roughing and finishing of coarse-pitch gears, and the finish milling of gears having special tooth forms. Although high-quality gears can be produced by milling, the accuracy of tooth spacing on older gear milling machines was limited by the inherent accuracy of the

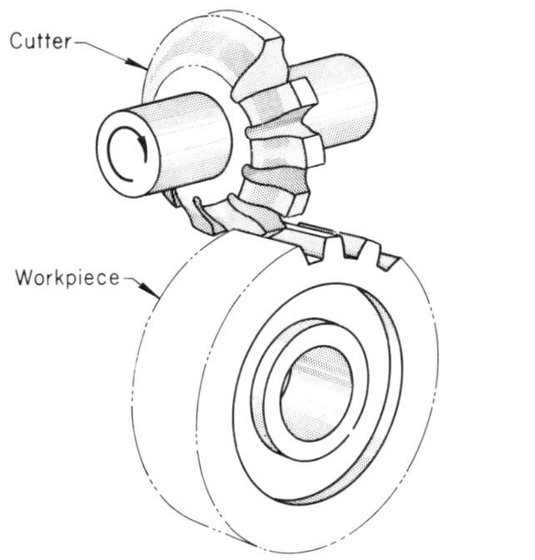


**Fig. 12** Relation of cutter and workpiece when milling teeth in a spur gear

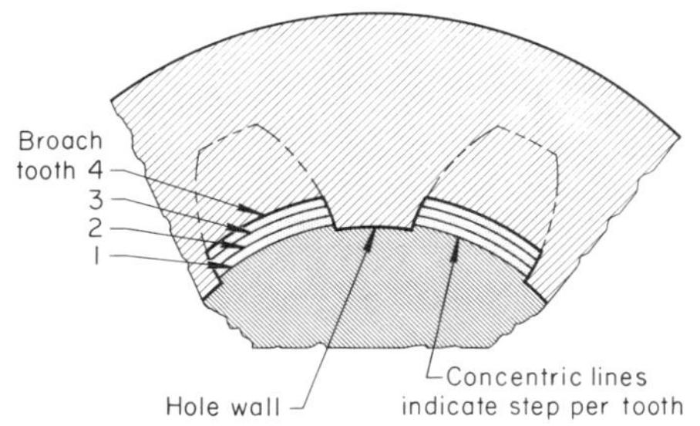


**Fig. 13** Progressive action of broach teeth in cutting teeth of an internal spur gear

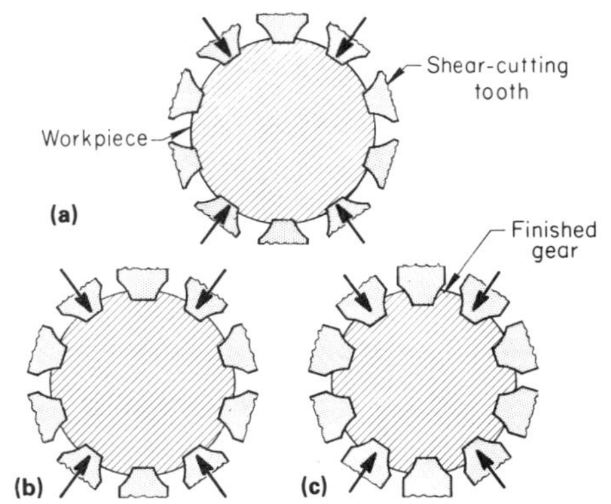


**Fig. 14** Progressive action in shear cutting teeth of an external spur gear. Shear cutting operation proceeds from roughing (a) to intermediate (b) to finishing (c) operations.

indexing mechanism. Most indexing techniques used on modern gear milling machines incorporate numerical control or computer numerical control, and the accuracy can rival that of hobbing machines.

**Broaching.** Both external and internal gear teeth, spur or helical, can be broached, but conventional broaching is usually confined to cutting teeth in internal gears. Figure 13 shows progressive broach steps in cutting an internal spur gear. The form of the space between broached gear teeth corresponds to the form of the broach teeth. The cutting action of any single broach tooth is similar to that of a single form tool. Each cross section of the broach has as many teeth as there are tooth spaces on the gear. The diameter of each section increases progressively to the major diameter that completes the tooth form on the workpiece. Broaching is fast and accurate, but the cost of tooling is high. Therefore, broaching of gear teeth is best suited to large production runs.

**Shear cutting** is a high-production method for producing teeth in external spur gears. The process is not applicable to helical gears. In shear cutting, as in broaching, all tooth spaces are cut simultaneously and progressively (Fig. 14). Cutting speeds in shear cutting are similar to those for broaching the same work metal. Machines are available for cutting gears up to 508 mm (20 in.) in diameter, with face width up to 152 mm (6 in.).

The shear cutting head is mounted in a fixed position, and the gear blanks are pushed through the head. Cutting tools are fed radially into the head, a predetermined amount for each stroke, until the required depth of tooth space is reached. In shear cutting, some space is required for overtravel, although most workpieces with integral shoulders or flanges (such as cluster gears) do have enough clearance between sections to allow shear cutting to be used. Therefore, this process is best suited to large production runs.

**Hobbing** is a practical method for cutting teeth in spur gears, helical gears, worms, worm gears, and many special forms. Conventional hobbing machines are not applicable to cutting bevel and internal gears. Tooling costs for hobbing are lower than those for broaching or shear cutting. Therefore, hobbing is used in low-quantity production or even for a few pieces. On the other hand, hobbing is a fast and accurate method (compared to milling, for example) and is therefore suitable for medium and high production quantities.

Hobbing is a generating process in which both the cutting tool and the workpiece revolve in a constant relation as the hob is being fed across the face width of the gear blank. The hob is a fluted worm with form-relieved teeth that cut into the gear blank in succession, each in a slightly different position. Instead of being formed in one profile cut, as in milling, the gear teeth are generated progressively by a series of cuts (Fig. 15). The hobbing of a spur gear is shown in Fig. 16.

**Gear shaping** is the most versatile of all gear cutting processes. Although shaping is most commonly used for cutting teeth in spur and helical gears, this process is also applicable to cutting herringbone teeth, internal gear teeth (or splines), chain sprockets, ratchets, elliptical gears, face gears, worm gears, and racks. Shaping cannot be used to cut teeth in bevel gears. Because tooling costs are relatively low, shaping is practical for any quantity of production.

Workpiece design often prevents the use of milling cutters or hobs (notably, for cluster gears), and shaping is the most practical method for cutting the teeth.

Gear shaping is a generating process that uses a toothed disk cutter mounted on a spindle that moves in axial strokes as it rotates. The workpiece is carried on a second spindle. The workpiece spindle is synchronized with the cutter spindle and rotates as the tool cuts while it is being fed gradually into the work. The action between a shaping cutter and a gear blank is illustrated in Fig. 17. Shaping applied to the cutting of a worm (Fig. 18) involves no axial stroke of the cutter spindle.

**Rack cutting** is done with a cutter in the form of a rack with straight teeth (usually three to five). This cutter reciprocates parallel to the gear axis when cutting spur gears and parallel to the helix angle when cutting helical gears. Metal is removed by a shaper-like stroke similar to the cutting action in gear shaping. In addition to the reciprocating action of the rack cutter, there is synchronized rotation of the gear blank with each stroke of the cutter, with a corre-

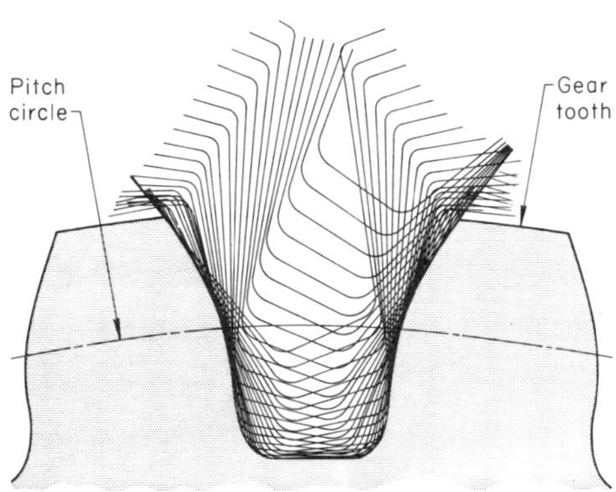


**Fig. 15** Schematic of hobbing action. Gear tooth is generated progressively by hob teeth.

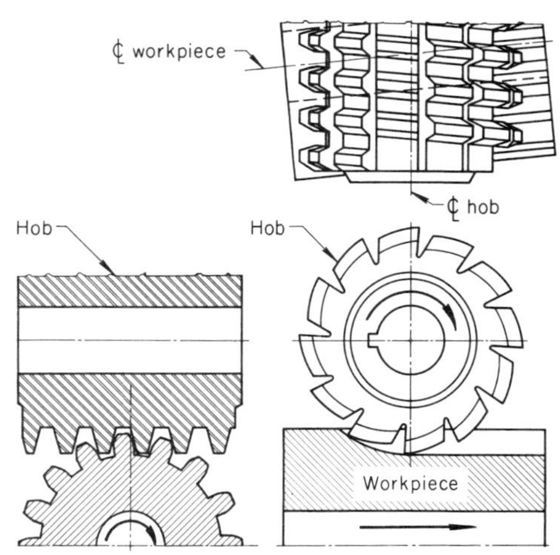


**Fig. 16** Hobbing of a spur gear

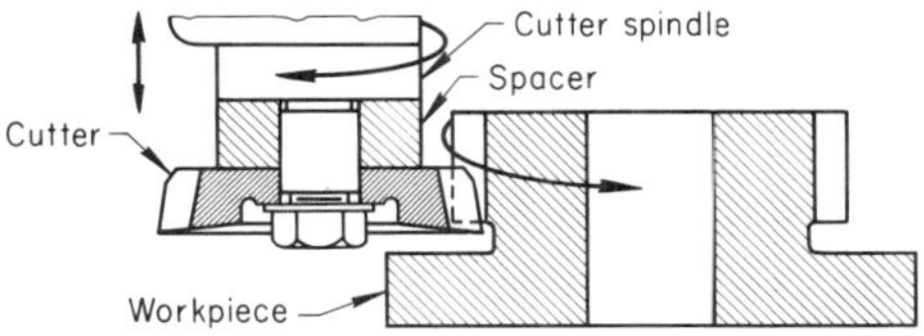


**Fig. 17** Relation of cutter and workpiece in shaping gear teeth

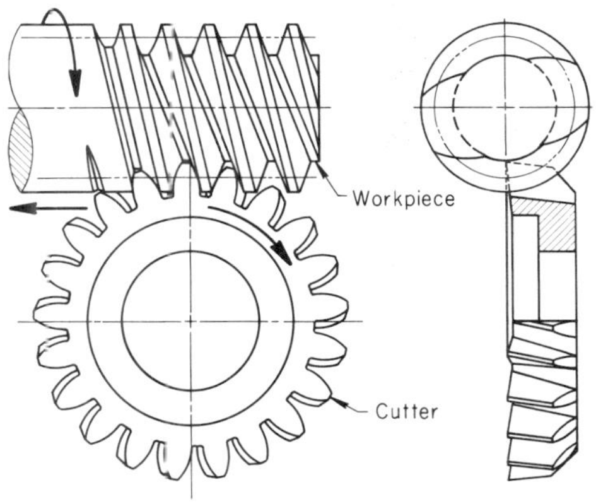


**Fig. 18** Relation of cutter and workpiece in generating a worm by shaping

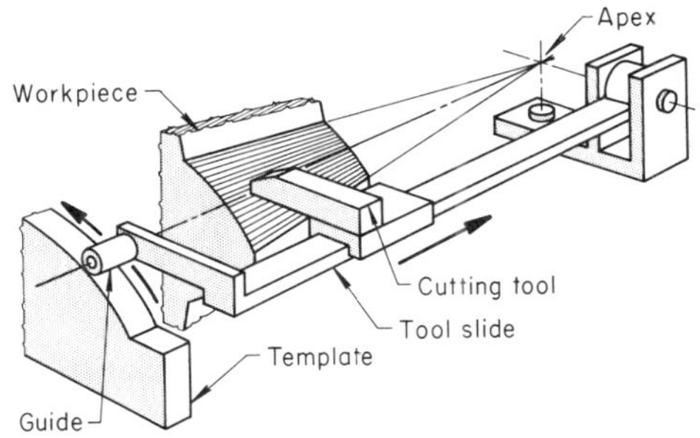


**Fig. 19** Cutting teeth in a large straight bevel gear by template machining in a bevel gear planer

sponding advance of the cutter as in the meshing of a gear and rack. By these combined actions, the true involute curve of the gear tooth is developed. Several gear cutting machines use this principle. Rack cutters are less expensive than hobs. Rack cutting is especially adapted to the cutting of large gears or gears of coarse pitch or both. Gears with diametral pitch as coarse as ⅜ are commonly cut by the rack method.

## Processes for Bevel Gears

The machining of bevel gears is treated as a separate subject because most bevel gears are cut in special machines with special cutters. However, the action of these cutters bears a close resemblance to one or more of the basic processes—milling, broaching, or shaping. Both generating and nongenerating processes are discussed below.

**Milling** is not widely used for cutting bevel gears, because of the accuracy limits of indexing devices and because the operation is time consuming (as many as five cuts around the gear may be required for completing one gear). Straight bevel gears are sometimes roughed by milling and then finished by another method. This two-operation procedure is more common when the availability of special gear cutting machines is limited.

**Template machining** is a low-production, nongenerating method used to cut the tooth profiles of large bevel gears using a bevel gear planer (Fig. 19). Because the setup can be made with a minimum of tooling, template machining is useful when a wide variety of coarse-pitch gears are required. Template machining uses a simple, single-point cutting tool guided by a template several times as large as the gear tooth to be cut. Under these conditions, high accuracy in tooth forms is possible.

The necessary equipment is unique. The setup utilizes two templates; one for each side of the gear tooth. In theory, a pair of templates would be required for each gear ratio, but in practice a pair is designed for a small range of ratios. A set of 25 pairs of templates encompasses all 90° shaft angle ratios from 1:1 to 8:1 for either 14½ or 20° pressure angles. This system of templates is based on the use of equal-addendum tooth proportions for all ratios. The tooth bearing localization is produced by a slight motion of the tool arm as the cutting tool moves along the tooth. The length and position of the cut can be controlled by the machine operator.

To produce a finished gear, five or six cutting operations are required. The first is a roughing operation, made by feeding a slotting tool or a corrugated V-tool to full depth. If the first cut is made with a slotting tool, a second cut is required with a V-tool or a corrugated V-tool. Cuts are made with the template follower resting on a straight guide. After roughing, the templates are set up, and the teeth are finished by making two cuts on each side. Slotting tools are specified by point width and depth of cut; corrugated roughing tools, by point width, depth of cut, and pressure angle; finishing tools, by point width only. To set up a straight bevel gear planer for template machining, the operator need know only the tooth proportions of the gear to be cut, plus the template list and the index gear list furnished with the machine.

**Formate cutting and Helixform cutting** are nongenerating methods for cutting spiral bevel and hypoid gears. Nongenerating methods can be used for cutting the gear member of spiral bevel and hypoid pairs when the gear-to-pinion ratio is 2.5:1 or greater. The two principal methods are Formate cutting and Helixform cutting.

*Formate cutting* is applicable to both the roughing and single-cycle finishing of gears with pitch diameters up to 2540 mm (100 in.). Roughing and finishing are sometimes both done by one cutter in the same machine. More often, for greater efficiency, roughing and finishing operations are done in different machines.

The Formate method can be used with the two-cut roughing and finishing method, in which the gear is roughed to depth and then single-cycle finished, or with a one-cut roughing and finishing method known as completing. In the completing method, the cutter is plunged to approximately 0.25 mm (0.010 in.) of whole depth, the cutting speed is doubled, and the cutter is fed to whole depth, taking a light chip. This final portion of the cycle removes the built-up edge on the cutter and produces an acceptable surface finish. The completing method is used for both face milling and face hobbing.

In the Formate single-cycle finishing method, one tooth space is finish cut in one revolution of the cutter. Stock removal is accomplished by cutting blades mounted in a circular cutter that resembles a face milling cutter. Each blade in the cutter is slightly longer and wider than the preceding blade; thus, the cutting action is, in effect, that of a circular broach. A gap between the first and last cutting blades permits indexing of the workpiece as each tooth space is completed. The relative positions of cutter and gear during Formate single-cycle cutting are illustrated in Fig. 20.

*Helixform cutting*, another nongenerating method of cutting spiral bevel and hypoid gears, is generally similar to Formate cutting. However, there is one significant difference in the method and in the finishing operation for the gear member.

One turn of the Helixform cutter finishes both sides of a tooth space. The cutter has both rotational and reciprocating motion, and this combination makes the path of the cutter-blade tips tangent to the root plane of the gear. Because the cutting edge is a straight line, the surface cut by a Helixform gear cutter is a true helical surface. The principal advantage of Helixform cutting compared to Formate cutting is that the gear produced by Helixform is conjugate to

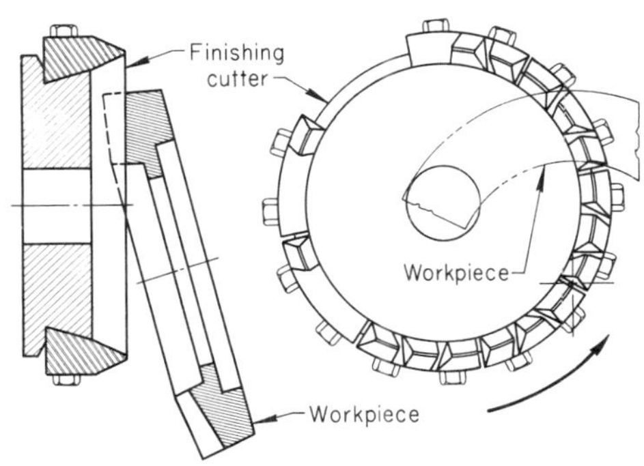


**Fig. 20** Relative positions of cutter and workpiece in Formate single-cycle cutting

the mating pinion, and the resulting contact pattern has little or no bias.

**The Cyclex method** is also a nongenerating method, and the result is the same as that for the Formate method. The Cyclex method was developed for the rough and finish cutting of gears in one operation and is particularly suitable when production quantities are not great enough to warrant separate Formate roughing and finishing machines for the gear member. Cyclex machines of the generator type can cut a wide range of gear sizes and can be used for cutting both gears and pinions.

In the Cyclex method, the gear is roughed and finished in one chucking from the solid blank. The finishing blades of the cutter are set below the roughing blades and do not contact the work during the roughing cycle. Several revolutions of the cutter may be necessary for roughing, the number required depending on the pitch of the gear. In the final revolution, after the last roughing blade has passed through the cut, the work is rapidly advanced, permitting the finishing blades to make contact and finish the tooth space to size. After the finishing blades have passed through the cut, the work is rapidly withdrawn and indexed, and the cycle is repeated until all teeth are completed.

**Face mill cutting** machines are used to finish cut teeth in spiral bevel, Zerol, and hypoid gears. Machines and cutters are available for cutting gears ranging from small instrument gears up to about 2540 mm (100 in.) in diameter.

The three types of face mill cutters are identified by the design of their cutting blades: integral, segmental, and inserted. All of these can be used for both roughing and finishing. Solid or integral-blade cutters are made from a single piece of tool steel and are usually less than 152 mm (6 in.) in diameter. They are used for fine-pitch gears. Segmental cutters are made up of sections, each having two or more blades. The segments are bolted to the cutter head around the periphery. Inserted-blade cutters are of two types. The first type has blades that are bolted to the slotted head. This cutter usually has parallels for changing diameter, and adjusting wedges for truing individual blades. The blades are sharpened in a radial or near-radial plane. The second type of inserted-blade cutter has blades that are clamped in the slot. The blades are sharpened by topping the blades down and repositioning them in the head.

For finishing, the three types of cutters can be furnished with all outside blades, all inside blades, or alternate outside and inside blades. Roughing cutters and completing cutters can have either alternate inside and outside blades or have end-cutting or bottom-cutting blades alternately spaced with inside and outside blades.

There are four basic cutting methods: completing, single-side, fixed-setting, and single-setting. In each case, the rotating cutting edges of a face mill cutter represent the imaginary gear surface.

*The completing method* is the generation of the part by a circular face mill or face hob cutter with alternate inside and outside blades that cut the tooth surfaces on both sides of a tooth space at the same time. With this method, each member is finished in one operation.

*In the single-side method*, the part is finished by a circular face mill cutter with alternate inside and outside blades that cut the tooth surface on each side of a tooth space in separate operations with independent machine settings.

*In the fixed-setting method*, the part is finished by two circular face mill cutters: one with inside blades only for cutting the convex side of the tooth, and the other with outside blades only for cutting the concave side. The two sides of the tooth are produced separately in two entirely different machine setups. For large production runs, a pair of machines is used. One machine is for cutting one side of the tooth, and the other is for the other side of the tooth.

*The single-setting method* is a variation of the completing method and is used when the available cutters have point widths too small for completing cutting. Both sides are cut with the same machine settings, and the blank is rotated on its axis to remove the amount of stock necessary to produce the correct tooth thickness. After the first cut, only one side of the cutter is cutting in the tooth slot. Figure 21 illustrates the action of a face mill when generating pinion teeth by the fixed-setting method (inside blades).

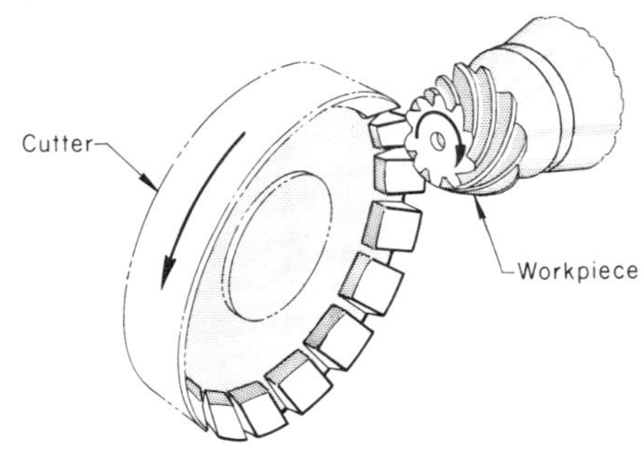


**Fig. 21** Face mill cutter shown in position to generate a pinion by the fixed-setting method

**Face hob cutting** is similar to face milling except that the indexing is superimposed on the generating cycle. Cutters are arranged with blade groupings. As each blade group passes through the cut, the work is being indexed one pitch (Fig. 22). Gears cut by this method are generally completed in one operation. Most face hob cutters are of the inserted-blade type. Integral and segmental systems are available.

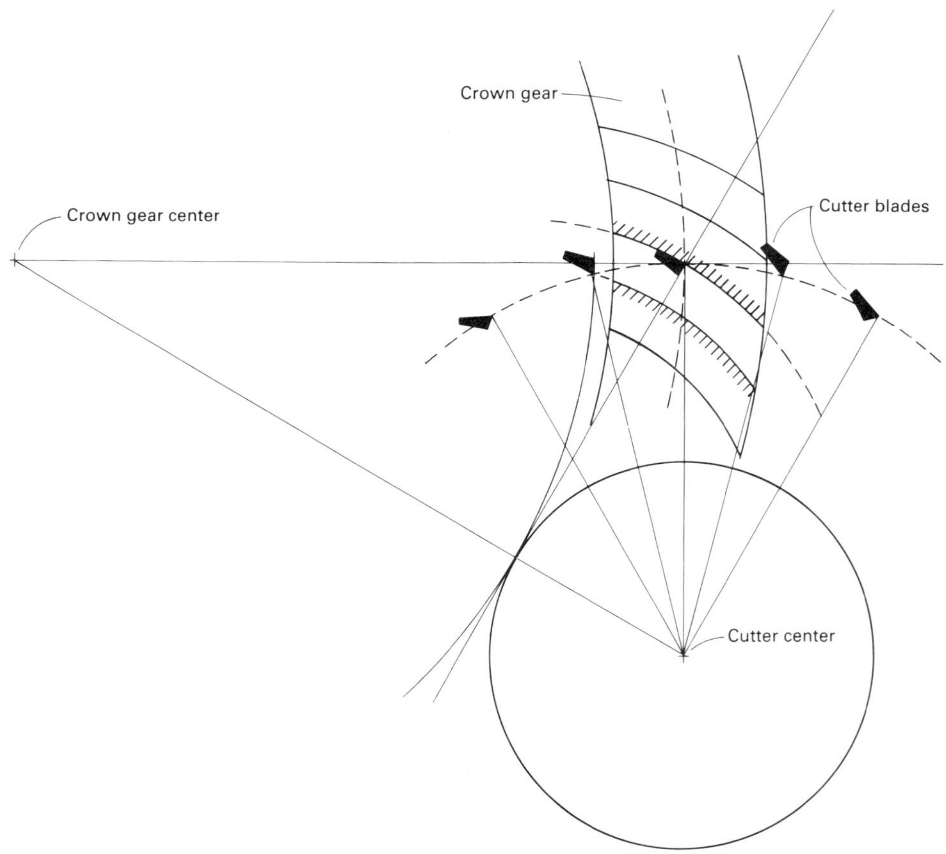


**Fig. 22** Schematic of the face hob cutting method

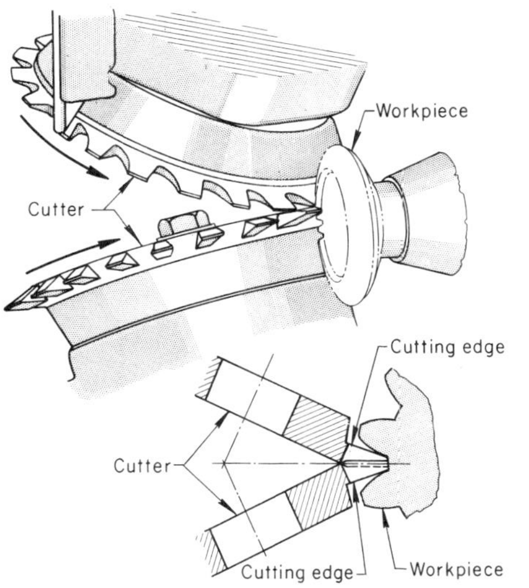


**Fig. 23** Relation of interlocking cutters (completing generators) with the bevel gear being cut

**Interlocking cutters,** known also as completing generators, generate the teeth on straight bevel gears or pinions from a solid blank in one operation.

In this method, two interlocking disk-type cutters rotate on axes inclined to the face of the mounting cradle, and both cut in the same tooth space (Fig. 23). The cutting edges present a concave cutting surface that removes more metal at the ends of the teeth, giving localized tooth contact. The gear blank is held in a work spindle that rotates in timed relation with the cradle on which the cutters are mounted. A feed-cam cycle begins with the workhead and blank moving into position for rough cutting, without generating roll, and cutting proceeds until the cut is just short of full depth. After a rough generating roll, the work is fed in to full depth, and a fast up-roll finish generates the tooth sides. At the top of the roll, the work backs out, and the cradle and work spindle roll down again into roughing position. During this short down-roll, the blank is indexed.

**Revacycle** is a generating process used for cutting straight bevel gears up to about 255 mm (10 in.) pitch diameter in large production runs. This is the fastest method for producing straight bevel gears of commercial quality. Initial tooling cost is greater for the Revacycle process than for other processes for cutting straight bevel gears, but the high production rate results in the lowest cost for mass production.

Most gears produced by the Revacycle method are completed in one operation, using cutters 406, 457, 533, or 635 mm (16, 18, 21, or 25 in.) in diameter that rotate in a horizontal plane at a uniform speed (Fig. 24). The cutter blades, which extend radially outward from the cutter head, have concave edges that produce convex profiles on the gear teeth. During cutting, the work-

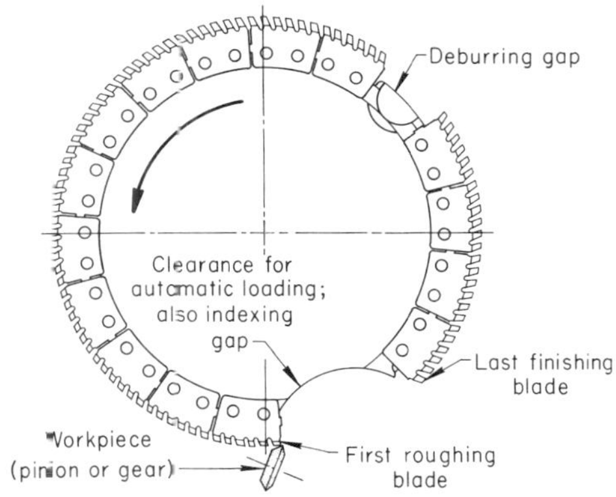


**Fig. 24** Revacycle cutter in position to cut a bevel gear

piece is held motionless while the cutter is moved by means of a cam in a straight line across the face of the gear and parallel to its root line. This motion produces a straight tooth bottom while the desired tooth shape is being produced by the combined effect of the motion of the cutter and the shapes of the cutter blades. The cutter is actually a circular broach in that each successive cutting tooth is larger than the one that precedes it along the circumference of the cutter. The cutter makes only one revolution per tooth space.

Feed is obtained by making cutter blades progressively longer, rather than by moving the entire cutter into the work. The completing cutter contains three kinds of blades: roughing, semifinishing, and finishing. One revolution of the cutter completes each tooth space, and the work is indexed in the gap between the last finishing blade and the first roughing blade. For the small amount of Revacycle work that is too deep to be completed in one cut, separate roughing and finishing operations are used. Under these conditions, separate cutters and setups are required for each operation. The cutters and machine cycles are similar to those for completing cutters, except that the roughing cutters have no semifinishing or finishing blades. A second cutter has only semifinishing and finishing blades.

**Two-tool generators** are used for cutting straight bevel gears by means of two reciprocating tools that cut on opposite sides of a tooth (Fig. 25). Tooling cost is low for two-tool generators, but production rates are lower than those for other straight bevel generators, such as interlocking cutters and Revacycle machines. Two-tool generators are usually used when:

- The gears are beyond the practical size range (larger than about 254 mm, or 10 in., pitch diameter) of other types of generators
- Gears have integral hubs or flanges that project above the root line, thus preventing the use of other generators

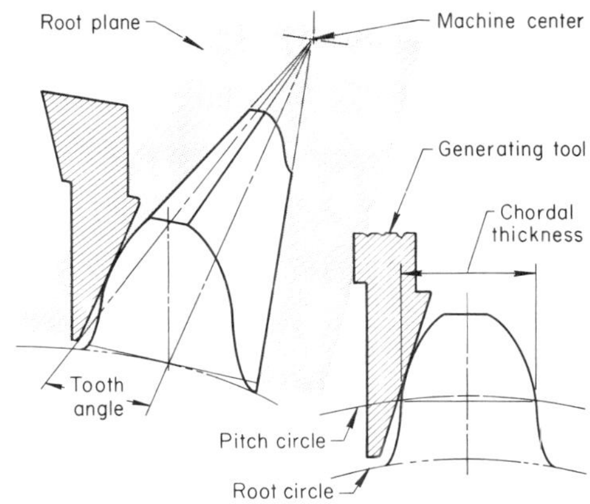


**Fig. 25** Angle of straight bevel gear tooth and sections of tools used for two-tool generating

- A small production quantity or a variety of gear sizes cannot be accommodated by other types of machines used for cutting straight bevel gears

Two-tool generators are used for both rough and finish cutting. When warranted by production quantities, roughing is done in separate machines, which are the same as the generators except that the machines used only for roughing have no generating roll. For small production quantities, both roughing and finishing cuts are made in the generators, the roughing cut being made without generating roll.

To make the machine setup for producing a gear that will operate at right angles to its mating gear, the operator must have the gear specifications and one calculated machine setting called the tooth angle (Fig. 25). The remaining setup data are taken from tables furnished with the machine. When the shaft angle of the two gears is not a right angle, the ratio of roll, as well as the data required for checking the roll, must be calculated.

Most two-tool generators can produce straight bevel gears with teeth crowned lengthwise to localize tooth contact. Crowned teeth are produced by means of two angularly adjustable guides on the back of each slide. The guides ride on a pair of fixed rollers (Fig. 26). When the guides are in line with each other, the tool stroke is a straight line, and when they are out of line, the tool is stroked along a curved path. The amount of curvature is controlled by setting the two guides. A table with each machine lists the guide settings for making the tooth contact approximately one-half the face width. The machine settings can be varied to shorten or lengthen the tooth contact.

**Planing generators** are unique because they can cut both straight-tooth and curved-tooth bevel gears. However, the use of planing generators is ordinarily restricted to cutting gears about 889 mm (35 in.) in diameter or larger or to diametral pitch coarser

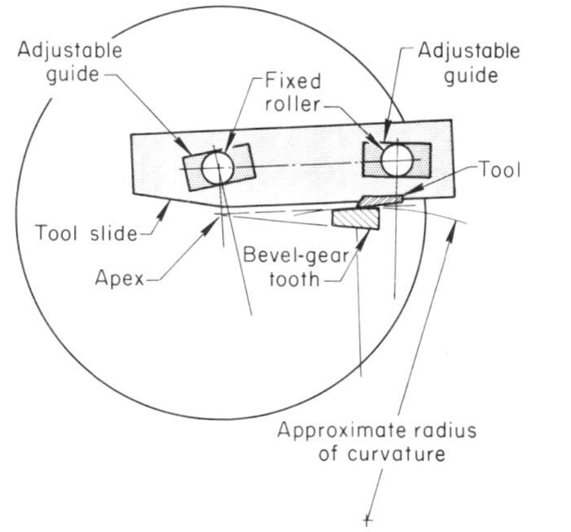


**Fig. 26** Provision for crowning gear teeth by means of adjustable guides in two-tool generators

than 1½. Standard machines can generate straight, Zerol, and spiral bevel gears. Special heads can be added to standard machines to permit the cutting of hypoid gears.

Tools have straight cutting edges and are mounted on a reciprocating slide that is carried on the face of the cradle and connected to a rotating crank by a connecting rod (Fig. 27). Tooth profiles are made by rolling the work with the generating gear. The lengthwise shape of the teeth is formed by a combination of three motions:

- Stroke of the tool
- Continuous, uniform rotation of the work
- An angular oscillation of the work produced by the eccentric shown in Fig. 27

The eccentric motion modifies the effect that the first two motions have on the shape of the teeth. The eccentric is timed for the correct tooth relief.

The spiral angle of the teeth is controlled by the angular offset of the tool slide from the angle of the cradle axis. Continuous rotation of the work is principally for indexing. In effect, the tool makes a cut on all teeth in succession in one generating position, and then the cradle and the work roll together a slight distance before another cut is taken on all teeth in the new generating position. Actually, however, rolling is continuous and occurs gradually until all teeth are completely generated in the last pass around the gear.

Several passes are required to complete a gear, the number depending on tooth depth and shape. Flat gear blanks are usually roughed without a roll, first using a corrugated tool and then using a single cut with a V-roughing tool. This is followed by at least two side-cutting operations on each side of the tooth, including generating cuts with roll. A similar sequence is used for cutting pinions except that roughing is done with roll.

Tools for use in planing generators are simple and inexpensive. Corrugated tools are furnished with a 14½° pressure angle,

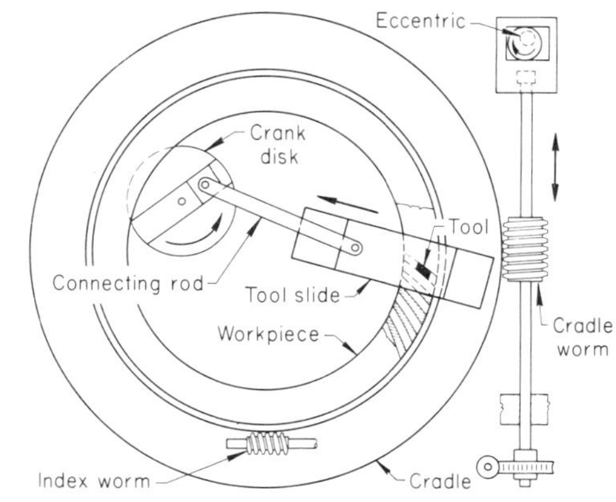


**Fig. 27** Components of a planing generator

regardless of the pressure angle of the gear being cut, but point width and depth of cut must be specified when the tools are ordered. The operator must know the specifications for the gear being cut. A table of settings for specific requirements is supplied with the machine.

## Selection of Machining Process

Each gear cutting process discussed in the preceding sections has a field of application to which it is best adapted. These fields overlap, however, so that many gears can be produced satisfactorily by two or more processes. In such cases, the availability of equipment often determines which machining process will be used.

The type of gear being machined (spur, helical, bevel, or other) is usually the major factor in the selection of machining process, although one or more of the following factors usually must be considered in the final choice of the method:

- Size of the gear
- Configuration of integral sections (flanges or other)
- Quantity requirements
- Accuracy requirements
- Gear-to-pinion ratio
- Cost

The following sections consider the type of gear as the major variable and discuss the machining methods best suited to specific conditions.

### Machining of Spur Gears

Milling, shear cutting, hobbing, and shaping are the methods most commonly used for cutting teeth in spur gears.

**Form milling,** with the cutter ground to the desired shape of the tooth space (Fig. 12), is a simple means of cutting teeth in spur gears. Tooling cost is low, and the process requires only a conventional milling machine, a form cutter, and an indexing mechanism. Except for low-quality production, milling is seldom used for cutting spur gears. The main disadvantage in the form milling of spur gears is the lack of accuracy in tooth spacing, which depends on the accuracy of the indexing mechanism. In addition, form milling is much slower than shear cutting or hobbing.

One milling cutter is not universal for all numbers of teeth, as are hobs and shaper cutters. To produce theoretically correct gear teeth, the tooth form of the cutter must be designed for the exact number of teeth. However, if a small departure in tooth form is acceptable, cutters have been standardized for a range of teeth, the form being correct for the lowest number of teeth in that particular range. Thus, all teeth within the range are provided with sufficient tip relief. The same form is produced in all tooth spaces within that range. For reasonably accurate gear cutting, eight standard involute gear cutters are required to cut all sizes of gears of a given pitch:

| Cutter No. | Gear tooth range |
|---|---|
| 1 | 135 to rack |
| 2 | 55–134 |
| 3 | 35–54 |
| 4 | 26–34 |
| 5 | 21–25 |
| 6 | 17–20 |
| 7 | 14–16 |
| 8 | 12–13 |

If a more accurate tooth form is desired for gears near the higher part of the ranges, seven half-number cutters can be obtained. A No. 1½ cutter, for example, will cut gears with 80 to 134 teeth; a No. 2½ cutter, 42 to 54 teeth; and a No. 3½ cutter, 30 to 34 teeth. For still greater accuracy, cutters of the proper shape for an exact number of teeth can be furnished by most tool manufacturers on short notice.

The tooth form of a single cutter is centered with the gear axis so that a symmetrical tooth space is produced. By the use of gang cutters, portions of adjacent tooth spaces can be rough machined simultaneously. Normally, a roughing and a finishing cutter are ganged, the finishing cutter being centered with the gear axis. Gang cutters and multiple-tooth cutters are specially designed for the specific application.

**Shear cutting** is faster and more accurate than milling for cutting teeth of almost any involute modification in spur gears. Total cutting time is often less than 1 min for gears up to about 152 mm (6 in.) in diameter. However, tooling cost is high, and shear cutting is therefore practical only for large-scale production.

**Hobbing** is the process most widely used for cutting teeth in spur gears, usually for one or more of the following reasons:

- High accuracy in a wide range of gear sizes
- Flexibility in quantity production

- Low cost
- Adaptability to work metals having higher-than-normal hardness

The shape of the workpiece sometimes limits the use of hobbing—for example, if the teeth to be cut are close to another portion of the workpiece having a diameter larger than the root diameter of the gear. The axial distance between the two sections must be large enough to allow for hob overtravel at the end of the cut. This overtravel is about one-half the hob diameter. The clearance required between the gear being cut and any flange or other projecting portion of the workpiece is, therefore, about one-half the hob diameter plus additional clearance to allow for the hob thread angle.

In many cases, it is necessary to cut teeth on heat-treated gear blanks to avoid the difficulties caused by distortion in heat treating. Hobbing is especially suitable for cutting gear teeth in hardened steel (sometimes as hard as 48 HRC). Although hob wear increases rapidly as workpiece hardness increases, normal practice should produce an acceptable number of parts per hob sharpening. Success in hobbing gears at high hardness depends greatly on maintaining minimum backlash in the machine and on rigid mounting of both the hob and the workpiece.

The ability to cut teeth in two or more identical spur gears in one setup can also justify use of the hobbing method. Inexpensive fixturing is often utilized for cutting two or more gears at one time when the ratio of face width to pitch diameter is small.

**Shaping** can produce high accuracy in cutting spur gears because shaping is a generating process. Although seldom as fast as hobbing, shaping is used for a wide range of production quantities. Many types of gears can be produced to requirements by either shaping or hobbing, and the availability of equipment determines which of the two processes is used. However, if the workpiece configuration cannot be hobbed, shaping is often the only practical method.

Cutting of teeth in cluster gears that must meet close tolerances is sometimes a problem because the method used must frequently be restricted to shaping and shaving. The same quality requirements cannot be met by shaping and shaving as by hobbing and grinding. When tolerances for cluster gears are closer than can be met by shaping and shaving, a corrective procedure sometimes must be employed. A cluster gear can be hobbed to greater precision by separating it into a two-piece assembly that is rigidly attached using threaded fasteners.

## Machining of Helical Gears

Milling, hobbing, shaping, and rack cutting are methods most used for producing teeth in helical gears. Rack cutting is most often used for large gears. Identical machining methods are applicable to conventional helical and crossed-axes helical gears.

**Milling** is used less than any other method because of the difficulty in obtaining accuracy and productivity. However, for some low production requirements, milling is the most satisfactory method because the tooling cost is low. In low-volume production, milling is sometimes used for roughing only, and the gear is finished by hobbing or shaping.

The milling of helical gears usually requires cutters specially designed for the specific gear. In milling helical teeth, the cutter travels along the helix angle of the gear. At this setting, the cutter axis of rotation is in the normal plane through the center of the gear tooth space. Under these conditions, only one point on the finished profile is produced in this normal plane. All the others are produced in different planes. Therefore, the form of the cutter teeth is not reproduced in the gear. In addition to the setting angle, the diameter of the cutter affects the gear tooth form and must be considered in designing the cutter.

**Hobbing** is extensively used for generating the teeth of helical gears for any production volume. With the exception mentioned below, procedures for hobbing helical gears are the same as those for spur gears: When hobbing spur gears with a single-thread hob, the blank rotates one tooth space for each rotation of the hob, the rotation being synchronized by means of change gears. When hobbing helical gears, the rotation of the work is retarded or advanced, through the action of the machine differential, in relation to the rotation of the hob, and the feed is also held in definite relation to the work and the hob. The decision to advance or retard the workpiece rotation depends mainly on whether the hob is a right-hand or left-hand type or whether the helix angle is of right-hand or left-hand configuration. The amount by which the workpiece is retarded or advanced depends on the helix angle. In medium-to-high production, it is common to use fixtures that allow hobbing of two or more identical gears in one loading of the machine.

Gears with integral shanks can usually be hobbed without difficulty. The shanks can assist in fixturing and handling for loading and unloading. When warranted by high-volume production, hobbing can be done in automatic machines utilizing automatic unloading and loading.

Although hob life decreases as workpiece hardness increases, helical gears of hardness as high as 48 HRC are sometimes hobbed. When hardnesses of 48 HRC or lower can be tolerated, the sequence of rough hobbing, heat treating, and finish hobbing is likely to cost less than grinding after heat treatment.

**Shaping** is a practical process for generating teeth of helical gears having helix angles up to 45°. The only difference between shaping helical gears and spur gears is that the machines used for cutting helical gears must impart additional rotary motion to the cutter spindle as it reciprocates. The amount of rotation per stroke is controlled by a helix guide. The lead of the guide must be the same as the lead of the cutter.

Workpiece configuration is often the main factor in the selection of shaping as a process for cutting helical gears. For example, rotary cutters such as hobs cannot be used when the teeth being cut are too close to the flange.

## Machining of Herringbone Gears

Milling, hobbing, and shaping are the methods most often used for cutting herringbone gears. Selection of method depends largely on whether the gear is designed with a gap between the two helixes or whether the herringbone is continuous.

Rotary cutters such as form milling cutters and hobs can be used to cut herringbone teeth only when there is a gap wide enough to permit cutter runout between the right-hand and left-hand helixes. Hobbing machines have been built that can cut herringbone teeth in gears up to 5590 mm (220 in.) in diameter.

**End milling** can also be used for machining teeth in herringbone gears, regardless of whether the gears have center slots. The end mills for cutting herringbone gears are used in special machines. Many large-diameter herringbone gears are cut by end milling.

**Shaping** is also a suitable method for cutting teeth on herringbone gears; those designed with a center slot as well as the continuous herringbone can be shaped. The type of shaper used for cutting herringbone gears is similar in principle to the type used for helical gears, except that for herringbone gears two cutters, one for each helix, are operated simultaneously. Both cutters reciprocate, one cutting in one direction to the center of the gear blank and the other cutting to the same point from the opposite direction when the motion is reversed. The cutters not only reciprocate, but also rotate. Both the gear blank and the cutters turn slowly, thus generating the teeth the same way as in a conventional shaper.

## Machining of Internal Gears

Broaching, shear cutting, and shaping are the methods most frequently used for cutting internal parts. Milling is seldom used, except for some very large gears.

**The broaching** of internal spur gear teeth is restricted to workpieces having configurations that permit the broach to pass completely through the piece. The action of a broach in cutting internal gear teeth is shown in Fig. 13. The broaching of gear teeth, which is similar to other types of broaching, is discussed in the section "Broaching" in this article. Broaching is an extremely fast and accurate means of machining internal gear teeth, but tooling cost is high; therefore, broaching is practical only for high-volume production.

**Shear cutting** is applicable to internal gear teeth. The principle involved is essentially the same as that illustrated in Fig. 14 for cutting external spur gears, except that the cutting edges and direction of radial feed are reversed. Unlike broaching, shear cutting is not restricted to parts where the tool must pass completely through the piece. Shear cutting can be used with no more than a 3.2 mm (1/8 in.) relief groove between the end of the cut and a shoulder. Because tooling cost is high for shear cutting, the process is practical only for high-volume production.

**Shaping** is applicable to cutting internal gear teeth. Tooling cost is lower than that for broaching and shear cutting; therefore, shaping is applicable to low-volume production. Shaping is less restricted to specific configurations than broaching is, because gear teeth can be cut to within 3.2 mm (1/8 in.) of a shoulder; however, for adequate chip clearance and to avoid the danger of striking the shoulder, it is better to have ample clearance. Shaping is often the only practical method for cutting teeth in large internal gears because the cost of large broaches or shear cutting tools is prohibitive.

## Machining of Worms

Several types of worms are used for power transmission, among them the double-enveloping type (Fig. 7), also known as the hour-glass worm because of its shape. Milling, hobbing, and shaping are used to machine the various types of worms.

**Milling.** For double-thread worms of low lead angle and commercial accuracy, a duplex cutter can be used. Each milling cutter is specially designed for cutting a specific worm. Another technique for machining worms utilizes the multiple-thread cutter. The cutter is set with its axis parallel to the work axis and is fed to depth. The work then makes one revolution for completion. The infeed can be made automatic, and because no indexing is required, this method is adaptable to volume production.

**Hobbing** produces the highest-grade worm at the lowest machining cost, but hobbing can be used only when production quantities are large enough to justify the tooling cost. Because of the large helix angle at which most worm hobs operate, the teeth at the entering end are chamfered to reduce the cutting load. Hobs for cutting worms are made to the same tolerance standards as those for cutting spur gears. When it is necessary to increase the hob diameter to provide more flutes, the tolerances are increased proportionately. The number of flutes in a worm hob is increased to improve surface finish, because the greater the number of flutes, the smaller the feed marks.

**Shaping.** Worms can also be generated by a shaper-cutter. In this technique, a helical gear cutter is used in a special machine similar to a hobber. Both the work and the cutter rotate, and the cutter is rolled axially along the worm, providing true generating action (Fig. 18).

## Machining of Racks

Milling and shaping are used to cut teeth in spur and helical racks.

**Milling** can be used to produce teeth in both spur and helical racks. The milling cutter must have the exact tooth space form. Racks can be cut in any standard milling machine; requirements are essentially the same as those for a conventional milling operation, and the rack to be cut must be rigidly clamped. Either manual or automatic indexing mechanisms are available for milling all sizes of racks in high-volume production.

**The shaping** of spur and helical racks involves the rolling action of the operating pitch circle of the generating shaper-cutter along the corresponding pitch line of the rack. In cutting racks on a gear shaper, the machine is equipped with a special fixture to hold the work. Several arrangements are used for imparting a transverse indexing movement to the member carrying the rack. One method employs a face gear secured on the work spindle that meshes with a pinion. The latter, by means of change gears, drives a lead screw, which operates the slide carrying the rack. Another method is to attach a pinion to the work spindle, which meshes with a master rack attached directly to the slide carrying the rack being cut. The first method is necessary when high ratios are involved in the drive; the second method needs only the regular work change gears.

## Machining of Bevel Gears

Face mill cutting, face hob cutting, Formate cutting, Helixform cutting, the Cyclex method, interlocking cutters, Revacycle, two-tool generators, planing generators, and template machining are used to cut teeth in straight and spiral bevel gears. The fundamentals of these processes are discussed in the section "Processes for Bevel Gears" in this article. Choice of method depends mainly on the type of gear being cut (straight or spiral bevel), size, configuration, accuracy requirements, and production quantities.

Template machining and planing in a gear generator are more often used for cutting teeth in gears larger than 813 mm (32 in.) outside diameter.

**Straight Bevel Gears.** The two-tool generator is widely used for cutting straight bevel gears and is especially well adapted to the cutting of gears in a wide range of sizes (up to about 889 mm, or 35 in., outside diameter) in low-to-medium production quantities because tool cost is low. Two-tool generating is also adaptable to the cutting of gears that have protruding portions (such as front hubs) that preclude the use of some processes.

Interlocking cutters provide a means of completing straight bevel gears in one operation. The interlocking-cutter method is faster than two-tool generating, but more costly. Gears without front hubs and less than 406 mm (16 in.) in outside diameter are best adapted to machining with interlocking cutters.

The Revacycle process is the fastest method for cutting straight bevel gears. Gear teeth are often completed at the rate of 1.8 s per tooth. This method was primarily designed for cutting gears having up to 254 mm (10 in.) pitch diameter at 4:1 ratio with the pinion and having a maximum face width of 29 mm (1 1/8 in.). In the Revacycle method, the cutter must have an uninterrupted path; therefore, the process cannot cut gears that have front hubs. Because of the high tooling cost, Revacycle cutting is economical only for high-volume production.

*Cost Versus Quantity (Straight Bevel Gears).* The two-tool generator, because it has the lowest tooling cost, is the most economical method for producing up to approximately 150 pairs of gears, at which point the two-tool generator and interlocking-cutter methods are equivalent. For large production runs, the two-tool generating method becomes prohibitively expensive. The Revacycle and interlocking-cutter methods are equivalent in cost at about 1200 pairs of gears, beyond which the Revacycle method is cheaper.

**Spiral Bevel Gears.** Spiral, Zerol, and hypoid bevel gears (Fig. 9) are cut in the same type of equipment and by the same general procedures. (Hypoid gears are by far the most numerous, being used in quantities exceeding those of spiral and Zerol gears combined.) Pinions are generally cut by some type of generator. Gears may or may not be cut in generators. When the gear-to-pinion ratio is greater than about 2.5:1, it is common practice to cut the gears without generating roll. Therefore, the Formate completing, Formate single-cycle, Helixform, and Cyclex methods are extensively used for cutting spiral bevel gears having a ratio of 2.5:1 or greater.

Nongenerated gears are less expensive than their generated counterparts, although there is a smaller difference in cost between the various methods for cutting spiral bevel gears (of less than 813 mm, or 32 in., in outside diameter) than between the various methods for cutting straight bevel gears.

## Machining of Large Gears

There is no one dimension that defines a large or a small gear. As various sizes are reached, some methods of manufacture become impractical, and other methods must be used.

**Herringbone Gears.** The same conditions previously discussed in the section "Machining of Herringbone Gears" in this article also apply to the machining of large herringbone gear components.

**Spur and Helical Gears.** Milling, hobbing, and rack cutting are the methods most commonly used for cutting large spur and helical gears.

*Milling* is the least expensive and the least accurate of these three methods; therefore, the accuracy required in the gear will determine whether or not milling can be used.

*Hobbing* is more costly than milling, but produces more accurate gears. Hobs are available for cutting gears well over 2540 mm (100 in.) in outside diameter, provided the diametral pitch is finer than 1. Because of the size of the hob required and the limitations of hobbing machines, it is difficult to hob gears of 1 diametral pitch and coarser.

Once it has been decided that hobbing will be used, it must be determined whether a ground hob will be required or whether an unground hob will provide the required degree of accuracy. The two types vary greatly in cost. However, if a ground hob is selected, extreme care must be used in resharpening; if this is not done, the original accuracy will not be maintained, and errors may occur in the gear tooth form.

*Rack Cutting.* For machining large gears that have large teeth (coarser than 1 diametral pitch), rack cutting (Fig. 28) is usually the most practical method. Rack cutting may be less expensive than hobbing, even when teeth are finer than 1 diametral pitch.

**Bevel Gears.** Face milling and face hobbing, two-tool generating, and planing generating are the methods most commonly used for cutting large straight and spiral bevel gears.

*Large face mill generators* that can cut gears up to 2540 mm (100 in.) in outside diameter are the fastest and most accurate machines for cutting large bevel gears.

*Two-tool generators* offer a practical means for cutting straight bevel gears having diameters up to about 889 mm (35 in.), face widths up to 152 mm (6 in.), and teeth as coarse as 1 diametral pitch. Tooling cost for the two-tool generating method is also low.

*Planing generators* can be used for cutting straight bevel gears, but they are most widely used for cutting spiral bevel gears ranging from 889 to 1830 mm (35 to 72 in.) in outside diameter, with up to 254 mm (10 in.) face width and teeth as coarse as ¾ diametral pitch.

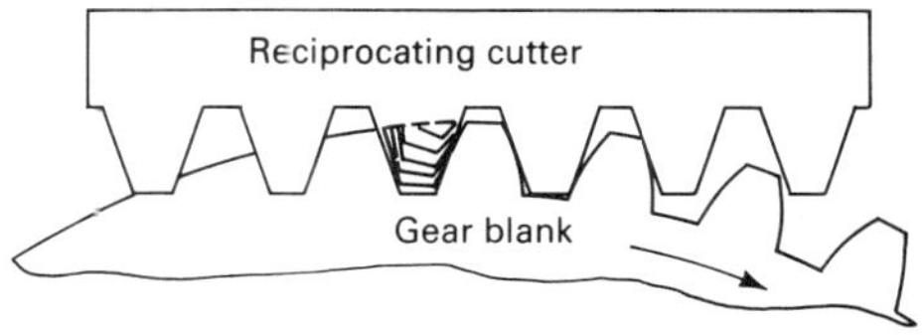


**Fig. 28** Rack-type cutter generating the teeth of a spur gear

## Shaving of Spur and Helical Gears

Gear shaving is a finishing operation that removes small amounts of metal from the flanks of gear teeth. It is not intended to salvage gears that have been carelessly cut, although it can correct small errors in tooth spacing, helix angle, tooth profile, and concentricity. Shaving improves the finish on tooth surfaces and can eliminate tooth-end load concentration, reduce gear noise, and increase load-carrying capacity. Shaving has been successfully used in finishing gears of diametral pitches from 180 to 2. Standard machines and cutters are available for shaving gears that range in size from 6.4 to 5590 mm (¼ to 220 in.) pitch diameter.

Leaving excessive stock for shaving will impair the final quality of the shaved gear. For maximum accuracy in the shaved gear and maximum cutter life, a minimum of stock should be allowed for removal by shaving; the amount depends largely on pitch. As little as 0.008 to 0.025 mm (0.0003 to 0.001 in.) of stock should be left on gears having diametral pitch as fine as 48; 0.08 to 0.13 mm (0.003 to 0.005 in.) is allowable for gears having diametral pitch of 2.

**Operating Principles.** The shaving operation is done with cutter and gear at crossed axes; helical cutters are used for spur gears, and vice versa. The action between gear and cutter is a combination of rolling and sliding. Vertical serrations in the cutter teeth take fine cuts from the profiles of the gear teeth.

During operation, the tip of the shaving cutter must not contact the root fillet, or uncontrolled, inaccurate involute profiles will result. For gears to be shaved, protuberance-type hobs that provide a small undercut at the flank of the tooth may be preferred. This type of hob avoids the initial tip loading of the shaving cutter.

**Shaving Cutters.** A typical rotary gear-shaving cutter is shown in Fig. 29(b). This cutter is serrated on the profile to form the cutting edges. The depth of the serrations governs total cutter life in terms of the number of sharpenings permitted. A shaving cutter is sharpened by regrinding the tooth profiles, thus reducing the tooth thickness. This causes a reduction in operating center distance for the same backlash and in turn changes the operating pressure angle. These changes are compensated for by a change in addendum after resharpening.

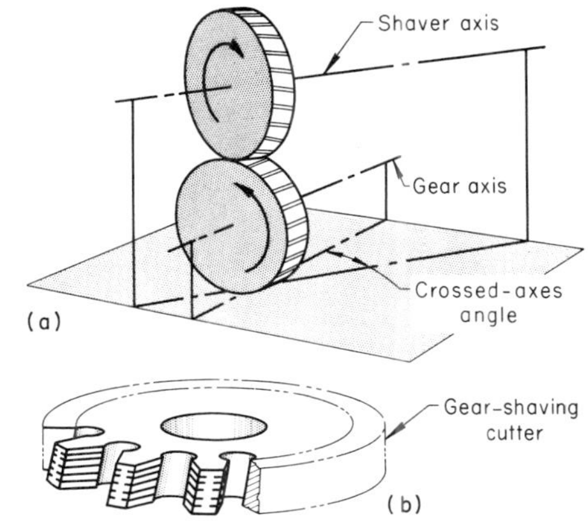


**Fig. 29** Shaving of gears. (a) Work gear in mesh with shaving cutter. (b) Serrated gear-shaving cutter

Tolerance is an important consideration in original purchase and resharpening. Shaving cutters are manufactured to standardized tolerances, not unlike those of master gears. For example, cumulative tooth spacing error can be held to 0.008 mm (0.0003 in.) and profile to 0.00064 mm (0.000025 in.). Because the engineering and facilities necessary to produce such accuracy are not available in most gear manufacturing plants, cutters are ordinarily returned to a tool manufacturer for resharpening.

**Shaving Methods.** Shaving is done by two basic methods: rack and rotary.

*In rack shaving*, the rack is reciprocated under the gear, and infeed takes place at the end of each stroke. Because racks longer than 508 mm (20 in.) are impractical, 152 mm (6 in.) is the maximum diameter of gear that can be shaved by the rack method.

*Rotary Shaving.* The several applications of rotary shaving include underpass, modified underpass, transverse, axial traverse, and angular traverse. Crown shaving can be incorporated in all of these modifications.

In rotary shaving, the cutter has the approximate form of a gear (Fig. 29). The size of gear that can be shaved is limited by the machine rather than by the cutter. Rotary shaving can be any of three types: underpass, modified underpass, and transverse.

Underpass shaving is used on cluster gears or gears with shoulders. To avoid interference with the adjacent gear or shoulder, the cross-axes angle is usually 4 to 6°. The face of the tool must be wider than the face of the shaved gear. Because underpass shaving is a one-cycle, short-stroke process, it is the fastest method of shaving. Disadvantages include relatively short tool life and light stock removal, thus requiring precise size control of the preshaved gear.

Modified underpass shaving is the most widely used method because it is a rapid one-cycle process. Tool cost is moderate because the cutter need be no wider than the gear and may be narrower. The high

**Table 2 Feeds and speeds for the shaving of carbon and low-alloy steel gears with high-speed steel tools**

| Material | Hardness, HB | Condition | Gear tooth size: Module | Gear tooth size: Diametral pitch | Feed per revolution of gear(a): mm | Feed per revolution of gear(a): in. | Cutter pitch line speed: m/min | Cutter pitch line speed: sfm | High-speed steel tool material: ISO | High-speed steel tool material: AISI |
|---|---|---|---|---|---|---|---|---|---|---|
| **Wrought free-machining carbon steels** | | | | | | | | | | |
| **Low-carbon resulfurized** | 100–150 | Hot rolled or annealed | 25–6 | 1–4 | 0.3 | 0.012 | 185 | 610 | S4, S2 | M2, M7 |
| 1116 1119 | | | 5–3 | 5–10 | 0.2 | 0.008 | | | | |
| 1117 1211 | | | 2–1.5 | 11–19 | 0.12 | 0.005 | | | | |
| 1118 1212 | | | 1 and finer | 20 and finer | 0.07 | 0.003 | | | | |
| | 150–200 | Cold drawn | 25–6 | 1–4 | 0.3 | 0.012 | 205 | 675 | S4, S2 | M2, M7 |
| | | | 5–3 | 5–10 | 0.2 | 0.008 | | | | |
| | | | 2–1.5 | 11–19 | 0.12 | 0.005 | | | | |
| | | | 1 and finer | 20 and finer | 0.07 | 0.003 | | | | |
| **Medium-carbon resulfurized** | 175–225 | Hot rolled, normalized, | 25–6 | 1–4 | 0.3 | 0.012 | 150 | 500 | S4, S2 | M2, M7 |
| 1132 1144 | | annealed, or cold drawn | 5–3 | 5–10 | 0.2 | 0.008 | | | | |
| 1137 1145 | | | 2–1.5 | 11–19 | 0.12 | 0.005 | | | | |
| 1139 1146 | | | 1 and finer | 20 and finer | 0.07 | 0.003 | | | | |
| 1140 1151 | 325–375 | Quenched and tempered | 25–6 | 1–4 | 0.3 | 0.012 | 84 | 275 | S4, S2 | M2, M7 |
| 1141 | | | 5–3 | 5–10 | 0.2 | 0.008 | | | | |
| | | | 2–1.5 | 11–19 | 0.12 | 0.005 | | | | |
| | | | 1 and finer | 20 and finer | 0.07 | 0.003 | | | | |
| **Low-carbon leaded** | 100–150 | Hot rolled, normalized, | 25–6 | 1–4 | 0.3 | 0.012 | 215 | 700 | S4, S2 | M2, M7 |
| 12L13 | | annealed, or cold drawn | 5–3 | 5–10 | 0.2 | 0.008 | | | | |
| 12L14 | | | 2–1.5 | 11–19 | 0.12 | 0.005 | | | | |
| 12L15 | | | 1 and finer | 20 and finer | 0.07 | 0.003 | | | | |
| | 200–250 | Hot rolled, normalized, | 25–6 | 1–4 | 0.3 | 0.012 | 185 | 600 | S4, S2 | M2, M7 |
| | | annealed, or cold drawn | 5–3 | 5–10 | 0.2 | 0.008 | | | | |
| | | | 2–1.5 | 11–19 | 0.12 | 0.005 | | | | |
| | | | 1 and finer | 20 and finer | 0.07 | 0.003 | | | | |
| **Wrought carbon steels** | | | | | | | | | | |
| **Low carbon** | 85–125 | Hot rolled, normalized, | 25–6 | 1–4 | 0.3 | 0.012 | 160 | 525 | S4, S2 | M2, M7 |
| 1005 1015 1023 | | annealed, or cold drawn | 5–3 | 5–10 | 0.2 | 0.008 | | | | |
| 1006 1016 1025 | | | 2–1.5 | 11–19 | 0.12 | 0.005 | | | | |
| 1008 1017 1026 | | | 1 and finer | 20 and finer | 0.07 | 0.003 | | | | |
| 1009 1018 1029 | 225–275 | Annealed or cold drawn | 25–6 | 1–4 | 0.3 | 0.012 | 115 | 375 | S4, S2 | M2, M7 |
| 1010 1019 1513 | | | 5–3 | 5–10 | 0.2 | 0.008 | | | | |
| 1011 1020 1518 | | | 2–1.5 | 11–19 | 0.12 | 0.005 | | | | |
| 1012 1021 1522 | | | 1 and finer | 20 and finer | 0.07 | 0.003 | | | | |
| 1013 1022 | | | | | | | | | | |
| **Medium carbon** | 125–175 | Hot rolled, normalized, | 25–6 | 1–4 | 0.3 | 0.012 | 135 | 450 | S4, S2 | M2, M7 |
| 1030 1042 1053 1541 | | annealed, or cold drawn | 5–3 | 5–10 | 0.2 | 0.008 | | | | |
| 1033 1043 1055 1547 | | | 2–1.5 | 11–19 | 0.12 | 0.005 | | | | |
| 1035 1044 1524 1548 | | | 1 and finer | 20 and finer | 0.07 | 0.003 | | | | |
| 1037 1045 1525 1551 | 325–375 | Quenched and tempered | 25–6 | 1–4 | 0.3 | 0.012 | 84 | 275 | S5 | M3 |
| 1038 1046 1526 1552 | | | 5–3 | 5–10 | 0.2 | 0.008 | | | | |
| 1039 1049 1527 | | | 2–1.5 | 11–19 | 0.12 | 0.005 | | | | |
| 1040 1050 1536 | | | 1 and finer | 20 and finer | 0.07 | 0.003 | | | | |
| **Wrought free-machining alloy steels** | | | | | | | | | | |
| **Medium-carbon resulfurized** | 150–200 | Hot rolled, normalized, | 25–6 | 1–4 | 0.3 | 0.012 | 150 | 500 | S4, S2 | M2, M7 |
| 4140 4145Se | | annealed, or cold drawn | 5–3 | 5–10 | 0.2 | 0.008 | | | | |
| 4140Se 4147Te | | | 2–1.5 | 11–19 | 0.12 | 0.005 | | | | |
| 4142Te 4150 | | | 1 and finer | 20 and finer | 0.07 | 0.003 | | | | |
| | 325–375 | Quenched and tempered | 25–6 | 1–4 | 0.3 | 0.012 | 76 | 250 | S5 | M3 |
| | | | 5–3 | 5–10 | 0.2 | 0.008 | | | | |
| | | | 2–1.5 | 11–19 | 0.12 | 0.005 | | | | |
| | | | 1 and finer | 20 and finer | 0.07 | 0.003 | | | | |
| **Medium- and high-carbon leaded** | 150–200 | Hot rolled, normalized, | 25–6 | 1–4 | 0.3 | 0.012 | 160 | 525 | S4, S2 | M2, M7 |
| 41L30 41L50 86L20 | | annealed, or cold drawn | 5–3 | 5–10 | 0.2 | 0.008 | | | | |
| 41L40 43L40 86L40 | | | 2–1.5 | 11–19 | 0.12 | 0.005 | | | | |
| 41L45 51L32 | | | 1 and finer | 20 and finer | 0.07 | 0.003 | | | | |
| 41L47 52L100 | 325–375 | Quenched and tempered | 25–6 | 1–4 | 0.3 | 0.012 | 84 | 275 | S5 | M3 |
| | | | 5–3 | 5–10 | 0.2 | 0.008 | | | | |
| | | | 2–1.5 | 11–19 | 0.12 | 0.005 | | | | |
| | | | 1 and finer | 20 and finer | 0.07 | 0.003 | | | | |
| **Wrought alloy steels** | | | | | | | | | | |
| **Low carbon** | 125–175 | Hot rolled, annealed, or | 25–6 | 1–4 | 0.3 | 0.012 | 145 | 475 | S4, S2 | M2, M7 |
| 4012 4615 4817 8617 | | cold drawn | 5–3 | 5–10 | 0.2 | 0.008 | | | | |
| 4023 4617 4820 8620 | | | 2–1.5 | 11–19 | 0.12 | 0.005 | | | | |
| 4024 4620 5015 8622 | | | 1 and finer | 20 and finer | 0.07 | 0.003 | | | | |
| 4118 4621 5115 8822 | 325–375 | Normalized or quenched | 25–6 | 1–4 | 0.3 | 0.012 | 76 | 250 | S5 | M3 |
| 4320 4718 5120 9310 | | and tempered | 5–3 | 5–10 | 0.2 | 0.008 | | | | |
| 4419 4720 6118 94B15 | | | 2–1.5 | 11–19 | 0.12 | 0.005 | | | | |
| 4422 4815 8115 94B17 | | | 1 and finer | 20 and finer | 0.07 | 0.003 | | | | |

(a) Feed recommendations apply to conventional (axial-transverse) gear shaving. Feeds should be increased 100% for gears shaved by the diagonal (angular-transverse) method. Source: Metcut Research Associates Inc.

(continued)

**Table 2 (continued)**

| Material | | | | Hardness, HB | Condition | Gear tooth size: Module | Gear tooth size: Diametral pitch | Feed per revolution of gear(a): mm | Feed per revolution of gear(a): in. | Cutter pitch line speed: m/min | Cutter pitch line speed: sfm | High-speed steel tool material: ISO | High-speed steel tool material: AISI |
|---|---|---|---|---|---|---|---|---|---|---|---|---|---|
| **Medium carbon** | | | | 175–225 | Hot rolled, annealed, or cold drawn | 25–6 | 1–4 | 0.3 | 0.012 | 120 | 400 | S4, S2 | M2, M7 |
| 1330 | 4145 | 5132 | 8640 | | | 5–3 | 5–10 | 0.2 | 0.008 | | | | |
| 1335 | 4147 | 5135 | 8642 | | | 2–1.5 | 11–19 | 0.12 | 0.005 | | | | |
| 1340 | 4150 | 5140 | 8645 | | | 1 and finer | 20 and finer | 0.07 | 0.003 | | | | |
| 1345 | 4161 | 5145 | 86B45 | 325–375 | Normalized or quenched and tempered | 25–6 | 1–4 | 0.3 | 0.012 | 69 | 225 | S5 | M3 |
| 4027 | 4340 | 5147 | 8650 | | | 5–3 | 5–10 | 0.2 | 0.008 | | | | |
| 4028 | 4427 | 5150 | 8655 | | | 2–1.5 | 11–19 | 0.12 | 0.005 | | | | |
| 4032 | 4626 | 5155 | 8660 | | | 1 and finer | 20 and finer | 0.07 | 0.003 | | | | |
| 4037 | 50B40 | 5160 | 8740 | | | | | | | | | | |
| 4042 | 50B44 | 51B60 | 8742 | | | | | | | | | | |
| 4047 | 5046 | 6150 | 9254 | | | | | | | | | | |
| 4130 | 50B46 | 81B45 | 9255 | | | | | | | | | | |
| 4135 | 50B50 | 8625 | 9260 | | | | | | | | | | |
| 4137 | 5060 | 8627 | 94B30 | | | | | | | | | | |
| 4140 | 50B60 | 8630 | | | | | | | | | | | |
| 4142 | 5130 | 8637 | | | | | | | | | | | |

(a) Feed recommendations apply to conventional (axial-transverse) gear shaving. Feeds should be increased 100% for gears shaved by the diagonal (angular-transverse) method. Source: Metcut Research Associates Inc.

cross-axes angle of 30 to 60° promotes rapid stock removal and smoother surface finish.

Transverse shaving is the slowest shaving method because multiple passes are required. It is a method of handling gears much wider than the cutter; therefore, cutter cost is moderate for gears with wide faces.

Crown shaving is used to relieve load concentration at the ends of gear teeth caused by the misalignment of axes in operation. Crowning is a modification of the tooth profile in both the radial and axial planes. In the axial traverse method of shaping, crowning is done by rocking the worktable as it is reciprocated. In the higher-production angular traverse method, the cutter is modified to provide crowning. The amount of crown varies, but usually 0.0003 to 0.0005 mm/mm (0.0003 to 0.0005 in./in.) of face width is sufficient.

**Speed and Feed.** Although cutting speeds are always high, the optimum speed of rotation for gear shaving varies considerably with work metal hardness and composition. Speeds and feeds for several steels and hardness ranges are given in Table 2.

## Honing and Lapping of Gears

The teeth of hardened steel gears are sometimes honed to remove nicks and burrs, to improve finish, and to make minor corrections in tooth shape. This process is discussed in the article "Honing" in this Volume.

Lapping is sometimes required for sets of hardened steel gears that must run quietly. Lapping is discussed in the article so titled in this Volume.

# Cutter Material and Construction

High-speed tool steel is used almost exclusively as the material for cutting edges of gear cutting tools. The steels most widely used are the general-purpose grades such as M2 or M7. Grade M3 (higher in carbon and vanadium than general-purpose grades) is also used in many gear cutting applications and is often preferred to M2 and M7 for cutting quenched and tempered alloy steels. The more highly alloyed grades of high-speed tool steel such as T15 or M30 are recommended only for conditions where greater red hardness is necessary. Such conditions include hard work metal, inadequate supply of cutting fluid, or high cutting speeds. Cutters are made from these highly alloyed grades only when the general-purpose grades (or M3) have proved inadequate.

Carbide cutters are used to hand finish gears cut by the face mill and face hob methods when small quantities of high-quality parts are needed. For most applications, carbide cutters are not economical. However, one application in which they are a cost-effective tool is the Tangear generator method.

**Tangear Generator.** Single-point cemented carbide cutting tools are rotated in opposite directions on the peripheries of two cutter heads with horizontal and parallel axes on the Tangear generator as shown in Fig. 30. The workpiece, with its axis vertical, is rotated as it is fed horizontally (typically a short distance of 8 mm, or 5/16 in.) between the cutter heads. The motions are synchronized so that the cutters act as though in mesh with the gear and progressively generate the teeth and form cut the root fillets. The teeth are rough cut as the gear moves to the cutters and are finish cut on the back stroke. The relationship set between workpiece and cutter head velocities determines the helix angle.

Conventional and crowned helical gears can be cut with diameters from 19 to 102 mm (3/4 to 4 in.), face widths up to 25 mm (1 in.), and helix angles from 10 to 40°. Gears can be generated at the rate of about 600 per hour, six to ten times as fast as hobbing. The time required for changing cutting tools is about 30 min, but this needs to be done only every 30 to 50 h. Although fast, the process is applicable only for the large-quantity production of a limited variety of gears.

**Construction.** Most reciprocating tools such as those used in gear shaping and planing operations are made of solid high-speed tool steel. Rotary cutters (hobs and milling cutters) for spur and helical gears can be either solid or inserted blade. Economy is the governing factor; the two methods of construction are equally satisfactory in terms of producing acceptable gears. Cutters less than about 75 mm (3 in.) in diameter are invariably solid. As cutters increase in size, the practice of using high-speed tool steel cutting edges (blades) as inserts in alloy steel bodies is usually more

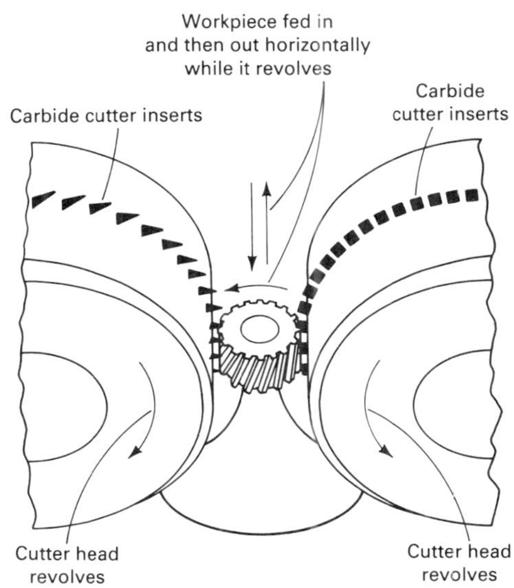


**Fig. 30** Schematic of operation of the Tangear gear cutter

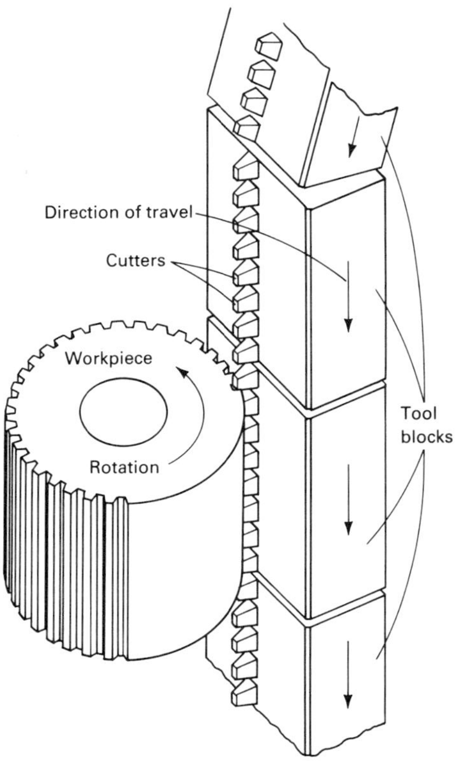


**Fig. 31** Method of operation of a G-Trac gear generator

economical. Inserts are normally held by mechanical fasteners.

Rotary cutters (face mill, rotary broach, and other) for bevel gears are usually designed to use inserts. In rotary cutters, feed rate is a direct function of the number of cutting faces. Therefore, it is desirable to have as many cutting edges as possible to increase the amount of metal removed during each revolution. However, tooth strength and chip clearance must be considered when selecting the number of blades (cutting edges). Despite the limitations imposed by chip clearance and tooth strength, gear cutters are often redesigned for greater efficiency.

*The G-Trac gear generator* is a special machine that uses large numbers of single-point high-speed tool steel cutting tools carried on an endless chain of tool blocks. One model for moderate quantities of gears has a single row of cutting tools (Fig. 31). As these are sped around the track, the work is fed into the cutters and rolled as though in mesh with a single tooth simulated by the cutters. Thus, one tooth space is generated. The work is then withdrawn, indexed to the next tooth space, fed in again, and so on, around the gear.

Another G-Trac model for large quantities has a number of rows of teeth (typically 14) around the chain of blocks. The work is fed into the cutters and revolved continuously as though in mesh with a simulated rack, until all tooth spaces are finished. On both models, spur and helical gears (up to 45° helix angle) as large as 356 mm (14 in.) in diameter can be cut and crowned and can be stacked to a height of 190 mm (7½ in.) times the cosine of the helix angle.

G-Trac generator cutter life is 6 to 175 times as long as that for hobs. Special machines are available for sharpening the cutters, and the cutters can also be sharpened all at one time in a special fixture on a standard surface grinder. A G-Trac machine has a 19 kW (25 hp) motor. It is six to ten times faster than hobbing, but the machine is expensive.

## Speed and Feed

In addition to the type of gear being machined and the method of cutting, other factors that influence the choice of cutting speed are work metal composition and hardness, diametral pitch of the work gear, rigidity of the setup, tolerance and finish requirements, and cutting fluid used.

**Hobbing Speed.** Nominal speeds and feeds for gear hobbing that take into account several of the above variables are listed in Table 3. As carbon content, alloy content, and hardness of the steel workpiece increase, recommended cutter speed decreases; for any given work metal, cutting speed increases as diametral pitch becomes finer. For the numerical values given in Table 3, it is assumed that an additive-type cutting fluid will be used. Low cutting speeds prolong tool life and eliminate the difficulties involved in changing tools and matching cuts.

**Shaping speed** is influenced by variables similar to those that affect hobbing speeds, with the exception of diametral pitch. Nominal speeds and feeds for gear shaping under a variety of conditions are given in Table 4. Cutter speed varies with the composition and hardness of the work metal but not with diametral pitch. The speeds listed in Table 4 are generally conservative.

**Speeds for Interlocking and Face Mill Cutters.** Nominal speeds for cutting straight and spiral bevel gears with two common types of cutters are given in Table 5. As in shaping, speed is not varied with diametral pitch when other conditions are the same. For both the interlocking cutters and the face mill cutters, speeds are increased for finishing when roughing and finishing are done separately. The speeds used for cutting straight and spiral bevel gears with rotating cutters are close to those suggested in Table 5. When straight bevel gears are being cut with reciprocating cutters, much lower speeds are required.

**Feeds.** Nominal feeds for the hobbing of steel gears are included in Table 3. Values are given in millimeters per revolution (inches per revolution) of the gear being hobbed and are based on the assumption that the gears will be ground after hobbing. Hobbing feeds should be decreased when gears are to be finished by shaving. As indicated in Table 3, hobbing feeds are sensitive to differences in diametral pitch, but feeds are not ordinarily changed for differences in the composition or hardness of the gear steel when diametral pitch is the same, except for the coarser-pitch heat-treated gears.

Nominal feeds for cutting gears by shaping are given in Table 4. These values are based on the assumption that the gears will be finished by grinding; a reduction in speed is recommended when gears are to be finished by shaving. Feeds for shaping vary with diametral pitch and vary comparatively little with work metal composition and hardness. Conditions for a specific operation may require feeds higher or lower than those given in Table 4.

Feeds for cutting straight and spiral bevel gears by rotary cutters are suggested in Table 5. The optimum feed is greatly influenced by diametral pitch; work metal composition and hardness have a minor influence. The same rates of feed are commonly used for both roughing and finishing, although the cutting edges of the teeth are altered for finishing, as indicated in Table 5.

## Cutting Fluids

Cutting fluids are recommended for all gear cutting, although for some applications (notably the cutting of large gears) the use of a cutting fluid is impractical. The three cutting fluids most commonly used for gear cutting are:

- Mineral oil (without additives) having a viscosity of 140 to 220 SUS at 40 °C (100 °F)
- Cutting oil, sulfurized or chlorinated (or both), usually diluted with mineral oil to a viscosity of 180 to 220 SUS at 40 °C (100 °F)
- Motor oil (SAE 20 grade), either detergent or nondetergent type

Water-soluble oils, which are widely used in many metal cutting operations, are used to a lesser extent for gear cutting because the oils referred to above are more effective for producing the surface finishes desired on gear teeth and for prolonged cutter life. Nevertheless, soluble oils are sometimes used in gear cutting. One notable application for soluble oil (20 parts water to 1 part oil) is the finish broaching of fine-pitch gears.

The relative advantages and disadvantages of the three oils mentioned above are matters of opinion among different plants producing steel gears. Most data in speed and feed tables are based on the use of cutting oils with additives (the second of the cutting fluids listed above). However, in some plants, mineral oils (including motor oils) are used when records have proved that the use of these oils resulted in longer

## Table 3 Feeds and speeds for the hobbing of carbon and low-alloy steel gears with high-speed steel tools

For hobbing class 9 (per AGMA 390.03) or better gears, it may be necessary to reduce speeds and feeds by 50% and/or take two cuts. To meet finish requirements, it may be necessary to try both conventional and climb cutting.

| Material | Hardness, HB | Condition | Gear tooth size: Module | Gear tooth size: Diametral pitch | Number of cuts | Feed per revolution of workpiece(a): mm | Feed per revolution of workpiece(a): in. | Hob speed: m/min | Hob speed: sfm | High-speed steel tool material: ISO | High-speed steel tool material: AISI |
|---|---|---|---|---|---|---|---|---|---|---|---|
| **Wrought free-machining carbon steels** | | | | | | | | | | | |
| **Low-carbon resulfurized** | 100–150 | Hot rolled or annealed | 25–13 | 1–2 | 2 | 1.50 | 0.060 | 67 | 220 | S4, S2 | M2, M7 |
| 1116 | | | 12–2.5 | 3–10 | 1 | 1.50 | 0.060 | 70 | 230 | | |
| 1117 | | | 2–1.5 | 11–19 | 1 | 1.50 | 0.060 | 73 | 240 | | |
| 1118 | | | 1–0.5 | 20–48 | 1 | 1.25 | 0.050 | 76 | 250 | | |
| 1119 | | | 0.5 and finer | 48 and finer | 1 | 0.75 | 0.030 | 85 | 280 | | |
| 1211 | 150–200 | Cold drawn | 25–13 | 1–2 | 2 | 1.50 | 0.060 | 70 | 230 | S4, S2 | M2, M7 |
| 1212 | | | 12–2.5 | 3–10 | 1 | 1.50 | 0.060 | 73 | 240 | | |
| | | | 2–1.5 | 11–19 | 1 | 1.50 | 0.060 | 81 | 265 | | |
| | | | 1–0.5 | 20–48 | 1 | 1.25 | 0.050 | 84 | 275 | | |
| | | | 0.5 and finer | 48 and finer | 1 | 0.75 | 0.030 | 90 | 300 | | |
| **Medium-carbon resulfurized** | 175–225 | Hot rolled, normalized, annealed, or cold drawn | 25–13 | 1–2 | 2 | 1.50 | 0.060 | 52 | 170 | S4, S2 | M2, M7 |
| 1132 1144 | | | 12–2.5 | 3–10 | 1 | 1.50 | 0.060 | 55 | 180 | | |
| 1137 1145 | | | 2–1.5 | 11–19 | 1 | 1.50 | 0.060 | 58 | 190 | | |
| 1139 1146 | | | 1–0.5 | 20–48 | 1 | 1.25 | 0.050 | 60 | 200 | | |
| 1140 1151 | | | 0.5 and finer | 48 and finer | 1 | 0.75 | 0.030 | 67 | 220 | | |
| 1141 | 325–375 | Quenched and tempered | 25–13 | 1–2 | 2 | 1.50 | 0.060 | 21 | 70 | S5, S11 | M3, M42 |
| | | | 12–2.5 | 3–10 | 1 | 1.50 | 0.060 | 27 | 90 | | |
| | | | 2–1.5 | 11–19 | 1 | 1.50 | 0.060 | 34 | 110 | | |
| | | | 1–0.5 | 20–48 | 1 | 1.25 | 0.050 | 37 | 120 | | |
| | | | 0.5 and finer | 48 and finer | 1 | 0.75 | 0.030 | 40 | 130 | | |
| **Low-carbon leaded** | 100–150 | Hot rolled, normalized, annealed, or cold drawn | 25–13 | 1–2 | 2 | 1.65 | 0.065 | 78 | 255 | S4, S2 | M2, M7 |
| 12L13 | | | 12–2.5 | 3–10 | 1 | 1.65 | 0.065 | 81 | 265 | | |
| 12L14 | | | 2–1.5 | 11–19 | 1 | 1.50 | 0.060 | 90 | 295 | | |
| 12L15 | | | 1–0.5 | 20–48 | 1 | 1.25 | 0.050 | 95 | 305 | | |
| | | | 0.5 and finer | 48 and finer | 1 | 0.75 | 0.030 | 105 | 340 | | |
| | 200–250 | Hot rolled, normalized, annealed, or cold drawn | 25–13 | 1–2 | 2 | 1.50 | 0.060 | 50 | 165 | S4, S2 | M2, M7 |
| | | | 12–2.5 | 3–10 | 1 | 1.50 | 0.060 | 58 | 190 | | |
| | | | 2–1.5 | 11–19 | 1 | 1.50 | 0.060 | 69 | 225 | | |
| | | | 1–0.5 | 20–48 | 1 | 1.25 | 0.050 | 73 | 240 | | |
| | | | 0.5 and finer | 48 and finer | 1 | 0.75 | 0.030 | 84 | 275 | | |
| **Wrought carbon steels** | | | | | | | | | | | |
| **Low carbon** | 85–125 | Hot rolled, normalized, annealed, or cold drawn | 25–13 | 1–2 | 2 | 1.80 | 0.070 | 62 | 205 | S4, S2 | M2, M7 |
| 1005 1016 1026 | | | 12–2.5 | 3–10 | 1 | 1.80 | 0.070 | 75 | 245 | | |
| 1006 1017 1029 | | | 2–1.5 | 11–19 | 1 | 1.50 | 0.060 | 81 | 265 | | |
| 1008 1018 1513 | | | 1–0.5 | 20–48 | 1 | 1.25 | 0.050 | 85 | 280 | | |
| 1009 1019 1518 | | | 0.5 and finer | 48 and finer | 1 | 0.75 | 0.030 | 88 | 290 | | |
| 1010 1020 1522 | 225–275 | Annealed or cold drawn | 25–13 | 1–2 | 2 | 1.50 | 0.060 | 37 | 120 | S4, S2 | M2, M7 |
| 1011 1021 | | | 12–2.5 | 3–10 | 1 | 1.50 | 0.060 | 49 | 160 | | |
| 1012 1022 | | | 2–1.5 | 11–19 | 1 | 1.50 | 0.060 | 56 | 185 | | |
| 1013 1023 | | | 1–0.5 | 20–48 | 1 | 1.25 | 0.050 | 59 | 195 | | |
| 1015 1025 | | | 0.5 and finer | 48 and finer | 1 | 0.75 | 0.030 | 60 | 200 | | |
| **Medium carbon** | 125–175 | Hot rolled, normalized, annealed, or cold drawn | 25–13 | 1–2 | 2 | 1.80 | 0.070 | 50 | 165 | S4, S2 | M2, M7 |
| 1030 1044 1526 | | | 12–2.5 | 3–10 | 1 | 1.80 | 0.070 | 53 | 175 | | |
| 1033 1045 1527 | | | 2–1.5 | 11–19 | 1 | 1.50 | 0.060 | 56 | 185 | | |
| 1035 1046 1536 | | | 1–0.5 | 20–48 | 1 | 1.25 | 0.050 | 60 | 200 | | |
| 1037 1049 1541 | | | 0.5 and finer | 48 and finer | 1 | 0.75 | 0.030 | 64 | 210 | | |
| 1038 1050 1547 | 325–375 | Quenched and tempered | 25–13 | 1–2 | 2 | 1.15 | 0.045 | 21 | 70 | S5, S11 | M3, M42 |
| 1039 1053 1548 | | | 12–2.5 | 3–10 | 1 | 1.15 | 0.045 | 24 | 80 | | |
| 1040 1055 1551 | | | 2–1.5 | 11–19 | 1 | 0.65 | 0.025 | 30 | 100 | | |
| 1042 1524 1552 | | | 1–0.5 | 20–48 | 1 | 0.65 | 0.025 | 32 | 105 | | |
| 1043 1525 | | | 0.5 and finer | 48 and finer | 1 | 0.65 | 0.025 | 37 | 120 | | |
| **Wrought free-machining alloy steels** | | | | | | | | | | | |
| **Medium-carbon resulfurized** | 150–200 | Hot rolled, normalized, annealed, or cold drawn | 25–13 | 1–2 | 2 | 1.50 | 0.060 | 44 | 145 | S4, S2 | M2, M7 |
| 4140 | | | 12–2.5 | 3–10 | 1 | 1.50 | 0.060 | 47 | 155 | | |
| 4140Se | | | 2–1.5 | 11–19 | 1 | 1.50 | 0.060 | 50 | 165 | | |
| 4142Te | | | 1–0.5 | 20–48 | 1 | 1.25 | 0.050 | 55 | 180 | | |
| 4145Se | | | 0.5 and finer | 48 and finer | 1 | 0.75 | 0.030 | 60 | 200 | | |
| 4147Te | 325–375 | Quenched and tempered | 25–13 | 1–2 | 2 | 1.15 | 0.045 | 20 | 65 | S5, S11 | M3, M42 |
| 4150 | | | 12–2.5 | 3–10 | 1 | 1.15 | 0.045 | 23 | 75 | | |
| | | | 2–1.5 | 11–19 | 1 | 0.65 | 0.025 | 29 | 95 | | |
| | | | 1–0.5 | 20–48 | 1 | 0.65 | 0.025 | 30 | 100 | | |
| | | | 0.5 and finer | 48 and finer | 1 | 0.65 | 0.025 | 41 | 135 | | |
| **Medium- and high-carbon leaded** | 150–200 | Hot rolled, normalized, annealed, or cold drawn | 25–13 | 1–2 | 2 | 1.50 | 0.060 | 52 | 170 | S4, S2 | M2, M7 |
| 41L30 43L40 | | | 12–2.5 | 3–10 | 1 | 1.50 | 0.060 | 55 | 180 | | |
| 41L40 51L32 | | | 2–1.5 | 11–19 | 1 | 1.50 | 0.060 | 58 | 190 | | |
| 41L45 52L100 | | | 1–0.5 | 20–48 | 1 | 1.25 | 0.050 | 60 | 200 | | |
| 41L47 86L20 | | | 0.5 and finer | 48 and finer | 1 | 0.75 | 0.030 | 67 | 220 | | |
| 41L50 86L4 | 325–375 | Quenched and tempered | 25–13 | 1–2 | 2 | 1.15 | 0.045 | 21 | 70 | S5, S11 | M3, M42 |
| | | | 12–2.5 | 3–10 | 1 | 1.15 | 0.045 | 27 | 90 | | |
| | | | 2–1.5 | 11–19 | 1 | 1.15 | 0.045 | 34 | 110 | | |
| | | | 1–0.5 | 20–48 | 1 | 1.15 | 0.045 | 37 | 120 | | |
| | | | 0.5 and finer | 48 and finer | 1 | 0.75 | 0.030 | 40 | 130 | | |

(a) Feeds are based on the largest standard recommended hob diameter. When using a smaller hob diameter, the feed must be reduced proportionally. Source: Metcut Research Associates Inc.

(continued)

**Table 3 (continued)**

| Material | | | | Hardness, HB | Condition | Gear tooth size: Module | Gear tooth size: Diametral pitch | Number of cuts | Feed per revolution of workpiece(a): mm | Feed per revolution of workpiece(a): in. | Hob speed: m/min | Hob speed: sfm | High-speed steel tool material: ISO | High-speed steel tool material: AISI |
|---|---|---|---|---|---|---|---|---|---|---|---|---|---|---|
| **Wrought alloy steels** | | | | | | | | | | | | | | |
| **Low carbon** | | | | 125–175 | Hot rolled, annealed, or cold drawn | 25–13 | 1–2 | 2 | 1.50 | 0.060 | 49 | 160 | S4, S2 | M2, M7 |
| 4012 | 4615 | 4817 | 8617 | | | 12–2.5 | 3–10 | 1 | 1.50 | 0.060 | 52 | 170 | | |
| 4023 | 4617 | 4820 | 8620 | | | 2–1.5 | 11–19 | 1 | 1.50 | 0.060 | 55 | 180 | | |
| 4024 | 4620 | 5015 | 8622 | | | 1–0.5 | 20–48 | 1 | 1.25 | 0.050 | 58 | 190 | | |
| 4118 | 4621 | 5115 | 8822 | | | 0.5 and finer | 48 and finer | 1 | 0.75 | 0.030 | 64 | 210 | | |
| 4320 | 4718 | 5120 | 9310 | 275–325 | Normalized or quenched and tempered | 25–13 | 1–2 | 2 | 1.15 | 0.045 | 27 | 90 | S4, S2 | M2, M7 |
| 4419 | 4720 | 6118 | 94B15 | | | 12–2.5 | 3–10 | 1 | 1.15 | 0.045 | 30 | 100 | | |
| 4422 | 4815 | 8115 | 94B17 | | | 2–1.5 | 11–19 | 1 | 1.15 | 0.045 | 34 | 110 | | |
| | | | | | | 1–0.5 | 20–48 | 1 | 1.15 | 0.045 | 37 | 120 | | |
| | | | | | | 0.5 and finer | 48 and finer | 1 | 0.75 | 0.030 | 40 | 130 | | |
| **Medium carbon** | | | | 175–225 | Hot rolled, annealed, or cold drawn | 25–13 | 1–2 | 2 | 1.50 | 0.060 | 38 | 125 | S4, S2 | M2, M7 |
| 1330 | 4427 | 81B45 | | | | 12–2.5 | 3–10 | 1 | 1.50 | 0.060 | 41 | 135 | | |
| 1335 | 4626 | 8625 | | | | 2–1.5 | 11–19 | 1 | 1.25 | 0.060 | 44 | 145 | | |
| 1340 | 50B40 | 8627 | | | | 1–0.5 | 20–48 | 1 | 1.25 | 0.050 | 49 | 160 | | |
| 1345 | 50B44 | 8630 | | | | 0.5 and finer | 48 and finer | 1 | 0.75 | 0.030 | 55 | 180 | | |
| 4027 | 5046 | 8637 | | 325–375 | Normalized or quenched and tempered | 25–13 | 1–2 | 2 | 1.15 | 0.045 | 18 | 60 | S5, S11 | M3, M42 |
| 4028 | 50B46 | 8640 | | | | 12–2.5 | 3–10 | 1 | 1.15 | 0.045 | 20 | 65 | | |
| 4032 | 50B50 | 8642 | | | | 2–1.5 | 11–19 | 1 | 0.65 | 0.025 | 23 | 75 | | |
| 4037 | 5060 | 8645 | | | | 1–0.5 | 20–48 | 1 | 0.65 | 0.025 | 24 | 80 | | |
| 4042 | 50B60 | 86B45 | | | | 0.5 and finer | 48 and finer | 1 | 0.65 | 0.025 | 29 | 95 | | |
| 4047 | 5130 | 8650 | | | | | | | | | | | | |
| 4130 | 5132 | 8655 | | | | | | | | | | | | |
| 4135 | 5135 | 8660 | | | | | | | | | | | | |
| 4137 | 5140 | 8740 | | | | | | | | | | | | |
| 4140 | 5145 | 8742 | | | | | | | | | | | | |
| 4142 | 5147 | 9254 | | | | | | | | | | | | |
| 4145 | 5150 | 9255 | | | | | | | | | | | | |
| 4147 | 5155 | 9260 | | | | | | | | | | | | |
| 4150 | 5160 | 94B30 | | | | | | | | | | | | |
| 4161 | 51B60 | | | | | | | | | | | | | |
| 4340 | 6150 | | | | | | | | | | | | | |

(a) Feeds are based on the largest standard recommended hob diameter. When using a smaller hob diameter, the feed must be reduced proportionally. Source: Metcut Research Associates Inc.

cutter life than when prepared cutting oils were used.

In some plants, additive cutting oils are used for roughing operations and mineral oils for finishing when the operations are done separately in different machines. An adequate supply (a flood) of fluid under slight pressure (about 35 kPa, or 5 psi) at the cutting area is extremely important—usually more important than the composition of the oil.

## Comparison of Steels for Gear Cutting

End use is the main factor in the selection of steel for a specific gear. However, two or more steels will often serve equally well. Under these conditions, the cost of the steel and the cost of processing it are significant factors in making a final selection, and in all selections both these factors must be considered. When the cost of machining standard steels is compared with the cost of free-machining counterparts, the use of a free-machining grade will almost always result in lower machining costs. However, when changing from one standard grade of a given steel to its free-machining counterpart, the additional cost of the free-machining grade must be considered in the cost comparison. When changes in steel composition that result in lower machining costs require changes in the method of heat treating, the cost of heat treating must also be considered.

A change in microstructure, independent of steel composition, also has a significant effect on tool life or surface finish or both. Altering the heat treatment can produce a more machinable microstructure that results in increased cutter life.

## Rolling of Gears

Spur and helical gears, like splines, are roll formed. Millions of high-quality gears are produced annually by this process; many of the gears in automobile transmissions are made this way. As indicated in Fig. 32, the process is basically the same as that by which screw threads are roll formed, except that in most cases the teeth cannot be formed in a single rotation of the forming rolls; the rolls are gradually fed inward during several revolutions.

Because of the metal flow that occurs, the top lands of roll-formed teeth are not smooth and perfect in shape; a depressed line between two slight protrusions can often be seen. However, because the top land plays no part in gear tooth action, if there is sufficient clearance in the mating gear, this causes no difficulty. Where desired, a light turning cut is used to provide a smooth top land and correct addendum diameter.

Rolling produces gears 50 times as fast as gear cutting and with surfaces as smooth as 0.10 μm (4 μin.). Not only does rolling usually need no finish operation, but rolling refines the microstructure of the workpiece.

Production setup usually takes only a set of rolling dies and the proper fixture to equip the rolling machine. By either the infeed (plunge) method or the throughfeed method, the rolling dies drive the workpiece

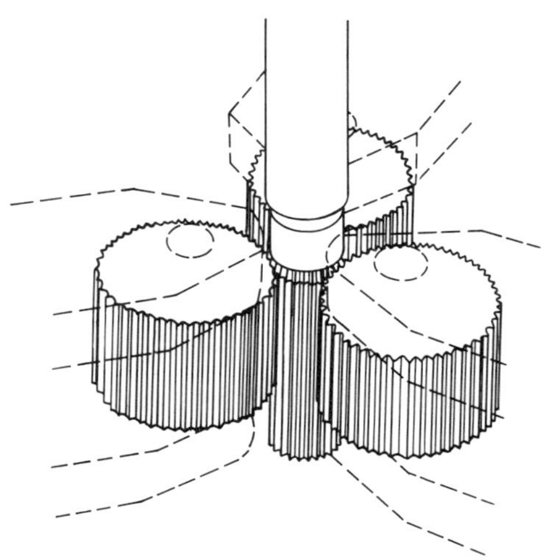

**Fig. 32** Method for forming gear teeth and splines by cold forming

**Table 4 Feeds and speeds for the shaping of carbon and low-alloy steel gears with high-speed steel tools**

| Material | Hardness, HB | Condition | Number of cuts(a) | Gear tooth size: Module | Gear tooth size: Diametral pitch | Rotary feed per cutter stroke (102 mm, or 4 in., pitch diameter cutter)(a): mm | in. | Cutter speed: m/min | sfm | High-speed steel tool material: ISO | AISI |
|---|---|---|---|---|---|---|---|---|---|---|---|
| **Wrought free-machining carbon steels** | | | | | | | | | | | |
| **Low-carbon resulfurized** | 100–150 | Hot rolled or annealed | 4 | 25–6 | 1–4 | 0.55 | 0.022 | 26 | 85 | S4, S2 | M2, M7 |
| 1116 1119 | | | 2 | 5.5–2.5 | 5–10 | 0.40 | 0.016 | | | | |
| 1117 1211 | | | 2 | 2–1.5 | 11–19 | 0.28 | 0.011 | | | | |
| 1118 1212 | | | 2 | 1–0.5 | 20–48 | 0.20 | 0.008 | | | | |
| | 150–200 | Cold drawn | 4 | 25–6 | 1–4 | 0.55 | 0.022 | 27 | 90 | S4, S2 | M2, M7 |
| | | | 2 | 5.5–2.5 | 5–10 | 0.40 | 0.016 | | | | |
| | | | 2 | 2–1.5 | 11–19 | 0.28 | 0.011 | | | | |
| | | | 2 | 1–0.5 | 20–48 | 0.20 | 0.008 | | | | |
| **Medium-carbon resulfurized** | 175–225 | Hot rolled, normalized, | 4 | 25–6 | 1–4 | 0.55 | 0.022 | 24 | 80 | S4, S2 | M2, M7 |
| 1132 1144 | | annealed, or cold drawn | 2 | 5.5–2.5 | 5–10 | 0.40 | 0.016 | | | | |
| 1137 1145 | | | 2 | 2–1.5 | 11–19 | 0.28 | 0.011 | | | | |
| 1139 1146 | | | 2 | 1–0.5 | 20–48 | 0.20 | 0.008 | | | | |
| 1140 1151 | 325–375 | Quenched and tempered | 4 | 25–6 | 1–4 | 0.30 | 0.012 | 12 | 40 | S5, S11 | M3, M42 |
| 1141 | | | 2 | 5.5–2.5 | 5–10 | 0.25 | 0.010 | | | | |
| | | | 2 | 2–1.5 | 11–19 | 0.20 | 0.008 | | | | |
| | | | 2 | 1–0.5 | 20–48 | 0.20 | 0.008 | | | | |
| **Low-carbon leaded** | 100–150 | Hot rolled, normalized, | 4 | 25–6 | 1–4 | 0.55 | 0.022 | 38 | 125 | S4, S2 | M2, M7 |
| 12L13 | | annealed, or cold drawn | 2 | 5.5–2.5 | 5–10 | 0.40 | 0.016 | | | | |
| 12L14 | | | 2 | 2–1.5 | 11–19 | 0.28 | 0.011 | | | | |
| 12L15 | | | 2 | 1–0.5 | 20–48 | 0.20 | 0.008 | | | | |
| | 200–250 | Hot rolled, normalized, | 4 | 25–6 | 1–4 | 0.55 | 0.022 | 27 | 90 | S4, S2 | M2, M7 |
| | | annealed, or cold drawn | 2 | 5.5–2.5 | 5–10 | 0.40 | 0.016 | | | | |
| | | | 2 | 2–1.5 | 11–19 | 0.28 | 0.011 | | | | |
| | | | 2 | 1–0.5 | 20–48 | 0.20 | 0.008 | | | | |
| **Wrought carbon steels** | | | | | | | | | | | |
| **Low carbon** | 85–125 | Hot rolled, normalized, | 4 | 25–6 | 1–4 | 0.45 | 0.018 | 27 | 90 | S4, S2 | M2, M7 |
| 1005 1013 1021 1518 | | annealed, or cold drawn | 2 | 5.5–2.5 | 5–10 | 0.30 | 0.012 | | | | |
| 1006 1015 1022 1522 | | | 2 | 2–1.5 | 11–19 | 0.25 | 0.010 | | | | |
| 1008 1016 1023 | | | 2 | 1–0.5 | 20–48 | 0.20 | 0.008 | | | | |
| 1009 1017 1025 | 225–275 | Annealed or cold drawn | 4 | 25–6 | 1–4 | 0.40 | 0.016 | 18 | 60 | S4, S2 | M2, M7 |
| 1010 1018 1026 | | | 2 | 5.5–2.5 | 5–10 | 0.30 | 0.012 | | | | |
| 1011 1019 1029 | | | 2 | 2–1.5 | 11–19 | 0.25 | 0.010 | | | | |
| 1012 1020 1513 | | | 2 | 1–0.5 | 20–48 | 0.20 | 0.008 | | | | |
| **Medium carbon** | 125–175 | Hot rolled, normalized, | 4 | 25–6 | 1–4 | 0.45 | 0.018 | 23 | 75 | S4, S2 | M2, M7 |
| 1030 1042 1053 1541 | | annealed, or cold drawn | 2 | 5.5–2.5 | 5–10 | 0.30 | 0.012 | | | | |
| 1033 1043 1055 1547 | | | 2 | 2–1.5 | 11–19 | 0.25 | 0.010 | | | | |
| 1035 1044 1524 1548 | | | 2 | 1–0.5 | 20–48 | 0.20 | 0.008 | | | | |
| 1037 1045 1525 1551 | 325–375 | Quenched and tempered | 4 | 25–6 | 1–4 | 0.30 | 0.012 | 12 | 40 | S5, S11 | M3, M42 |
| 1038 1046 1526 1552 | | | 2 | 5.5–2.5 | 5–10 | 0.25 | 0.010 | | | | |
| 1039 1049 1527 | | | 2 | 2–1.5 | 11–19 | 0.20 | 0.008 | | | | |
| 1040 1050 1536 | | | 2 | 1–0.5 | 20–48 | 0.20 | 0.008 | | | | |
| **Wrought free-machining alloy steels** | | | | | | | | | | | |
| **Medium-carbon resulfurized** | 150–200 | Hot rolled, normalized | 4 | 25–6 | 1–4 | 0.55 | 0.022 | 24 | 80 | S4, S2 | M2, M7 |
| 4140 4145Se | | annealed, or cold drawn | 2 | 5.5–2.5 | 5–10 | 0.45 | 0.018 | | | | |
| 4140Se 4147Te | | | 2 | 2–1.5 | 11–19 | 0.28 | 0.011 | | | | |
| 4142Te 4150 | | | 2 | 1–0.5 | 20–48 | 0.20 | 0.008 | | | | |
| | 325–375 | Quenched and tempered | 4 | 25–6 | 1–4 | 0.30 | 0.012 | 11 | 35 | S5, S11 | M3, M42 |
| | | | 2 | 5.5–2.5 | 5–10 | 0.25 | 0.010 | | | | |
| | | | 2 | 2–1.5 | 11–19 | 0.20 | 0.008 | | | | |
| | | | 2 | 1–0.5 | 20–48 | 0.20 | 0.008 | | | | |
| **Medium- and high-carbon leaded** | 150–200 | Hot rolled, normalized, | 4 | 25–6 | 1–4 | 0.55 | 0.022 | 24 | 80 | S4, S2 | M2, M7 |
| 41L30 41L47 51L32 86L40 | | annealed or cold drawn | 2 | 5.5–2.5 | 5–10 | 0.40 | 0.016 | | | | |
| 41L40 41L50 52L100 | | | 2 | 2–1.5 | 11–19 | 0.28 | 0.011 | | | | |
| 41L45 43L40 86L20 | | | 2 | 1–0.5 | 20–48 | 0.20 | 0.008 | | | | |
| | 325–375 | Quenched and tempered | 4 | 25–6 | 1–4 | 0.30 | 0.012 | 11 | 35 | S5, S11 | M3, M42 |
| | | | 2 | 5.5–2.5 | 5–10 | 0.25 | 0.010 | | | | |
| | | | 2 | 2–1.5 | 11–19 | 0.20 | 0.008 | | | | |
| | | | 2 | 1–0.5 | 20–48 | 0.20 | 0.008 | | | | |
| **Wrought alloy steels** | | | | | | | | | | | |
| **Low carbon** | 125–175 | Hot rolled, annealed, or | 4 | 25–6 | 1–4 | 0.45 | 0.018 | 21 | 70 | S4, S2 | M2, M7 |
| 4012 4615 4817 8617 | | cold drawn | 2 | 5.5–2.5 | 5–10 | 0.30 | 0.012 | | | | |
| 4023 4617 4820 8620 | | | 2 | 2–1.5 | 11–19 | 0.25 | 0.010 | | | | |
| 4024 4620 5015 8622 | | | 2 | 1–0.5 | 20–48 | 0.20 | 0.008 | | | | |
| 4118 4621 5115 8822 | 325–375 | Normalized or quenched | 4 | 25–6 | 1–4 | 0.30 | 0.012 | 12 | 40 | S5, S11 | M3, M42 |
| 4320 4718 5120 9310 | | and tempered | 2 | 5.5–2.5 | 5–10 | 0.25 | 0.010 | | | | |
| 4419 4720 6118 94B15 | | | 2 | 2–1.5 | 11–19 | 0.20 | 0.008 | | | | |
| 4422 4815 8115 94B17 | | | 2 | 1–0.5 | 20–48 | 0.20 | 0.008 | | | | |

(a) For cutting gears of class 9 (AGMA 390.03) or better, the rotary feed should be reduced, and the number of cuts should be increased. Source: Metcut Research Associates Inc.

(continued)

**Table 4 (continued)**

| Material | | | Hardness, HB | Condition | Number of cuts(a) | Gear tooth size: Module | Gear tooth size: Diametral pitch | Rotary feed per cutter stroke (102 mm, or 4 in., pitch diameter cutter)(a): mm | Rotary feed per cutter stroke: in. | Cutter speed: m/min | Cutter speed: sfm | High-speed steel tool material: ISO | High-speed steel tool material: AISI |
|---|---|---|---|---|---|---|---|---|---|---|---|---|---|
| **Medium carbon** | | | 175–225 | Hot rolled, annealed, or cold drawn | 4 | 25–6 | 1–4 | 0.45 | 0.018 | 18 | 60 | S4, S2 | M2, M7 |
| 1330 | 4427 | 81B45 | | | 2 | 5.5–2.5 | 5–10 | 0.30 | 0.012 | | | | |
| 1335 | 4626 | 8625 | | | 2 | 2–1.5 | 11–19 | 0.25 | 0.010 | | | | |
| 1340 | 50B40 | 8627 | | | 2 | 1–0.5 | 20–48 | 0.20 | 0.008 | | | | |
| 1345 | 50B44 | 8630 | 325–375 | Normalized or quenched and tempered | 4 | 25–6 | 1–4 | 0.30 | 0.012 | 11 | 35 | S5, S11 | M3, M42 |
| 4027 | 5046 | 8637 | | | 2 | 5.5–2.5 | 5–10 | 0.25 | 0.010 | | | | |
| 4028 | 50B46 | 8640 | | | 2 | 2–1.5 | 11–19 | 0.20 | 0.008 | | | | |
| 4032 | 50B50 | 8642 | | | 2 | 1–0.5 | 20–48 | 0.20 | 0.008 | | | | |
| 4037 | 5060 | 8645 | | | | | | | | | | | |
| 4042 | 50B60 | 86B45 | | | | | | | | | | | |
| 4047 | 5130 | 8650 | | | | | | | | | | | |
| 4130 | 5132 | 8655 | | | | | | | | | | | |
| 4135 | 5135 | 8660 | | | | | | | | | | | |
| 4137 | 5140 | 8740 | | | | | | | | | | | |
| 4140 | 5145 | 8742 | | | | | | | | | | | |
| 4142 | 5147 | 9254 | | | | | | | | | | | |
| 4145 | 5150 | 9255 | | | | | | | | | | | |
| 4147 | 5155 | 9620 | | | | | | | | | | | |
| 4150 | 5160 | 94B30 | | | | | | | | | | | |
| 4161 | 51B60 | | | | | | | | | | | | |
| 4340 | 6150 | | | | | | | | | | | | |

(a) For cutting gears of class 9 (AGMA 390.03) or better, the rotary feed should be reduced, and the number of cuts should be increased. Source: Metcut Research Associates Inc.

between them, forming the teeth by pressure.

**Limits.** Spur gears can be rolled if they have 18 teeth or more. Fewer teeth cause the work to roll poorly. Helical gears can be rolled with fewer teeth if the helix angle is great enough.

It is usually impractical to roll teeth with pressure angle less than 20°. Lower angles have wide flats at root and crest that need more pressure in rolling. Lower angles also hinder metal flow. Although 0.13 mm (0.005 in.) radius fillets can be rolled, 0.25 mm (0.010 in.) is a better minimum. For greater accuracy, gear blanks are ground before rolling. Chamfers should be 30° or less.

Steels for gear rolling should not have more than 0.13% S and preferably no lead. Blanks should not be harder than 28 HRC.

## Computer Numerical Control (CNC) Milling and Hobbing Machines

In practice, gear milling is usually confined to replacement gears or to the small-lot production, roughing, and finishing of coarse-pitch gears. The milling of low-production, special tooth forms is also done. Milling is often used as the first operation prior to finish hobbing or shaping, especially in large diametral pitch spur, helical, and straight bevel gears. The gear is then finished to final size by another method selected by the manufacturer.

Special high-power (56 to 75 kW, or 75 to 100 hp) CNC machines, called gashers (Fig. 33), are used for the roughing of gears from 6 to ⅝ diametral pitch. These machines

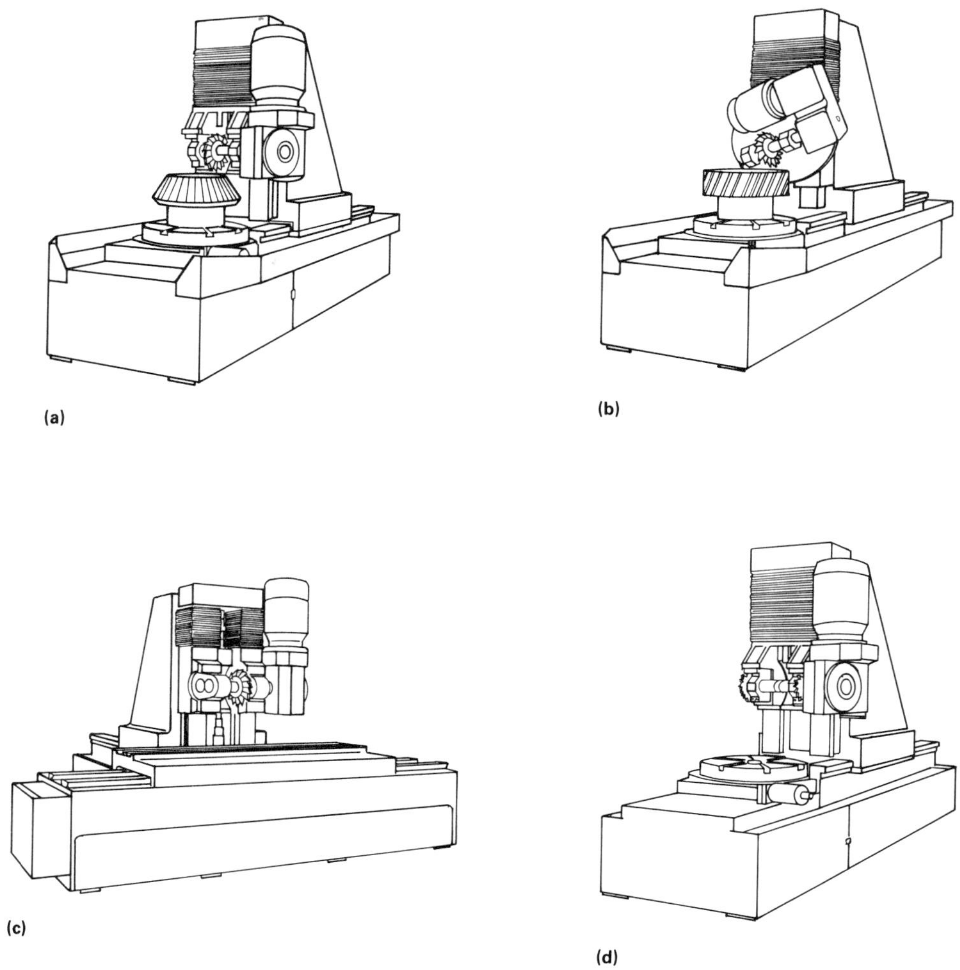


**Fig. 33** Four types of gashers. (a) Duo-axis. (b) Helical machine. (c) Rack machine. (d) Slitting head

**Table 5 Gear cutting speeds for the rough and finish cutting of carbon and low-alloy steel straight and spiral bevel gears with high-speed steel tools**

| Material | Hardness, HB | Condition | Cutting speed(a) Roughing m/min | Roughing sfm | Finishing m/min | Finishing sfm | High-speed steel tool material ISO | AISI |
|---|---|---|---|---|---|---|---|---|
| **Wrought free-machining carbon steels** | | | | | | | | |
| **Low-carbon resulfurized** | 100–150 | Hot rolled or annealed | 38 | 125 | 76 | 250 | S4, S2 | M2, M7 |
| 1116 1118 1211 | 150–200 | Cold drawn | 43 | 140 | 84 | 275 | S4, S2 | M2, M7 |
| 1117 1119 1212 | | | | | | | | |
| **Medium-carbon resulfurized** | 175–225 | Hot rolled, normalized, annealed, or cold drawn | 26 | 85 | 53 | 175 | S4, S2 | M2, M7 |
| 1132 1140 1145 | | | | | | | | |
| 1137 1141 1146 | 325–375 | Quenched and tempered | 15 | 50 | 30 | 100 | S5 | M3 |
| 1139 1144 1151 | | | | | | | | |
| **Low-carbon leaded** | 100–150 | Hot rolled, normalized, annealed, or cold drawn | 44 | 145 | 88 | 290 | S4, S2 | M2, M7 |
| 12L13 | | | | | | | | |
| 12L14 | 200–250 | Hot rolled, normalized, annealed, or cold drawn | 34 | 110 | 69 | 225 | S4, S2 | M2, M7 |
| 12L15 | | | | | | | | |
| **Wrought carbon steels** | | | | | | | | |
| **Low carbon** | 85–125 | Hot rolled, normalized, annealed, or cold drawn | 27 | 90 | 58 | 190 | S4, S2 | M2, M7 |
| 1005 1012 1019 1026 | | | | | | | | |
| 1006 1013 1020 1029 | 225–275 | Annealed or cold drawn | 21 | 70 | 44 | 145 | S4, S2 | M2, M7 |
| 1008 1015 1021 1513 | | | | | | | | |
| 1009 1016 1022 1518 | | | | | | | | |
| 1010 1017 1023 1522 | | | | | | | | |
| 1011 1018 1025 | | | | | | | | |
| **Medium carbon** | 125–175 | Hot rolled, normalized, annealed, or cold drawn | 26 | 85 | 53 | 175 | S4, S2 | M2, M7 |
| 1030 1042 1053 1541 | | | | | | | | |
| 1033 1043 1055 1547 | 325–375 | Quenched and tempered | 15 | 50 | 30 | 100 | S5 | M3 |
| 1035 1044 1524 1548 | | | | | | | | |
| 1037 1045 1525 1551 | | | | | | | | |
| 1038 1046 1526 1552 | | | | | | | | |
| 1039 1049 1527 | | | | | | | | |
| 1040 1050 1536 | | | | | | | | |
| **Wrought free-machining alloy steels** | | | | | | | | |
| **Medium-carbon resulfurized** | 150–200 | Hot rolled, normalized, annealed, or cold drawn | 26 | 85 | 53 | 175 | S4, S2 | M2, M7 |
| 4140 4142Te 4147Te | | | | | | | | |
| 4140Se 4145Se 4150 | 325–375 | Quenched and tempered | 14 | 45 | 29 | 95 | S5 | M3 |
| **Medium- and high-carbon leaded** | 150–200 | Hot rolled, normalized, annealed, or cold drawn | 26 | 85 | 53 | 175 | S4, S2 | M2, M7 |
| 41L30 41L50 86L20 | | | | | | | | |
| 41L40 43L40 86L40 | 325–375 | Quenched and tempered | 14 | 45 | 27 | 90 | S5 | M3 |
| 41L45 51L32 | | | | | | | | |
| 41L47 52L100 | | | | | | | | |
| **Wrought alloy steels** | | | | | | | | |
| **Low carbon** | 125–175 | Hot rolled, annealed, or cold drawn | 26 | 85 | 53 | 175 | S4, S2 | M2, M7 |
| 4012 4615 4817 8617 | | | | | | | | |
| 4023 4617 4820 8620 | | | | | | | | |
| 4024 4620 5015 8622 | 325–375 | Normalized or quenched and tempered | 15 | 50 | 30 | 100 | S5 | M3 |
| 4118 4621 5115 8822 | | | | | | | | |
| 4320 4718 5120 9310 | | | | | | | | |
| 4419 4720 6118 94B15 | | | | | | | | |
| 4422 4815 8115 94B17 | | | | | | | | |
| **Medium carbon** | 175–225 | Hot rolled, annealed, or cold drawn | 20 | 65 | 41 | 135 | S4, S2 | M2, M7 |
| 1330 4145 5132 8640 | | | | | | | | |
| 1335 4147 5135 8642 | 325–375 | Normalized or quenched and tempered | 15 | 50 | 30 | 100 | S5 | M3 |
| 1340 4150 5140 8645 | | | | | | | | |
| 1345 4161 5145 86B45 | | | | | | | | |
| 4027 4340 5147 8650 | | | | | | | | |
| 4028 4427 5150 8655 | | | | | | | | |
| 4032 4626 5155 8660 | | | | | | | | |
| 4037 50B40 5160 8740 | | | | | | | | |
| 4042 50B44 51B60 8742 | | | | | | | | |
| 4047 5046 6150 9254 | | | | | | | | |
| 4130 50B46 81B45 9255 | | | | | | | | |
| 4135 50B50 8625 9260 | | | | | | | | |
| 4137 5060 8627 94B30 | | | | | | | | |
| 4140 50B60 8630 | | | | | | | | |
| 4142 5130 8637 | | | | | | | | |

(a) Cutting speed recommendations are for use with alternate-tooth milling cutters or gear generators. For planer-type generators, use the recommended cutting speeds in Table 4. Source: Metcut Research Associates Inc.

employ strong, rigid machine components and are extremely powerful. They operate at peak efficiency with large inserted carbide cutters. The ability to remove large amounts of metal at substantial feed rates makes this process quite economical for coarse pitches. The gears can be finished on the same machines with high-speed tool steel formed cutters. The accuracy of index on these machines is more than adequate to produce tooth spacing requirements.

The advent of the CNC-controlled index mechanism of the gasher allows the manufacture of large, coarse-pitch sprockets

**Fig. 34** CNC gear hobbing machine producing a gear

for roller chains of all sizes. The face capacity of these machines permits the stacking of many plate sprockets at one time with good quality and reliable production.

The power of these machines and their ability for two-axis interpolation make coarse-tooth straight bevel gears a natural application. The time savings can be significant; two-tool generators can be used to finish the gashed bevel blank.

Computer numerical control gear hobbing machines are available in 305, 508, 914, 1220, and 1625 mm (12, 20, 36, 48, and 64 in.) capacities. Figure 34 shows a CNC machine hobbing a gear.

## Grinding of Gears

Gear teeth can be produced entirely by grinding, entirely by cutting, or by first cutting and then grinding to the required dimensions. Usually, gear grinding removes only several hundredths of a millimeter of metal from precut gears to make accurate teeth for critical applications. Teeth made entirely by grinding are usually only those of fine pitch, for which the total amount of metal removed is small. The grinding of fine-pitch gear teeth from uncut blanks may be less costly than the two-step procedure if there is not much metal to be removed.

The two basic methods for the grinding of gear teeth are form grinding (nongenerating) and generation grinding. Many varieties of machines have been built specifically to grind gears and pinions.

**Advantages and Disadvantages.** A gear is ground when it is so hard that it cannot be finished by other methods or when the required accuracy is greater than can be obtained by other methods. If the hardness of the gear does not exceed 40 HRC and if its size is within the capacity of the machine, shaving will produce teeth of an accuracy close to that produced by grinding. Gears made of steel harder than 40 HRC are finished by grinding when dimensional accuracy is required or when correction beyond that feasible by honing is required.

The major disadvantage of grinding gears is the cost. It is usually more expensive to grind gears than to cut them because material is removed in small increments in grinding. Ground gears are usually subjected to more inspection than cut gears. Magnetic particle inspection of the finished teeth detects fine cracks, and macroetching with dilute nitic acid detects grinding burn. Both inspection procedures are advised in gear grinding.

A carbide network in carburized steels, or the presence of sufficient retained austenite or of hydrogen in any hardened steel, can increase the susceptibility of the steel to grinding cracks. Therefore, it is necessary to keep these conditions under control. Reducing the heat generated in grinding (by decreasing the rate of stock removal or by the use of a grinding fluid) can help to eliminate cracking and burning of the ground surface.

**Stock Allowance.** When workpieces are prepared for grinding by cutting, the least amount of stock possible should be left to be removed by grinding. This stock should be evenly distributed to reduce the wheel dressings and operating adjustments needed to maintain wheel contour when using conventional (aluminum oxide or silicon carbide) dressable wheels. With cubic boron nitride (CBN) wheels, where stock is removed in one pass, evenly distributed stock results in more uniform grinding pressure.

In general, it is not practical to remove more than 0.15 mm (0.006 in.) of stock from each face of the teeth of spur gears up to 178 mm (7 in.) in diameter or 0.20 mm (0.008 in.) from gears between 178 and 254 mm (7 and 10 in.) in diameter. These amounts of stock can usually be removed in three to five passes with conventional dressable wheels or in one pass with CBN wheels. However, the number of passes depends on number of teeth, pressure angle, and required finish. Sometimes nine or ten passes are needed with conventional wheels.

The recommended stock allowance for grinding straight bevel teeth is 0.08 mm (0.003 in.) at the root of the tooth and 0.08 to 0.13 mm (0.003 to 0.005 in.) on the face of each tooth. Grinding time depends on the amount of stock to be removed. For example, removing 0.076 to 0.089 mm (0.0030 to 0.0035 in.) of stock from a gear of 4 diametral pitch and 32 teeth takes 30 s per tooth; a gear of 10 diametral pitch and 32 teeth takes about 20 s per tooth. Figure 35, which shows the cross sections of gear teeth of various diametral pitches, can be used to gage the relative sizes of the 4 and 10 diametral pitch gears.

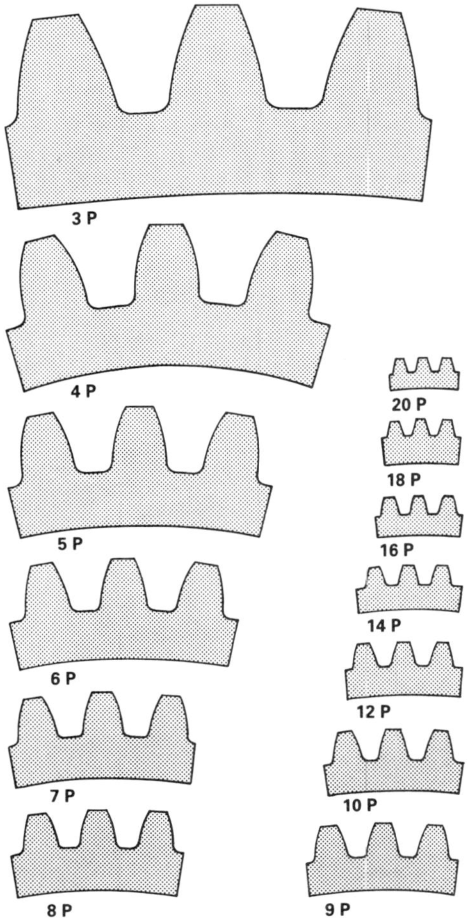


**Fig. 35** Gear teeth of various diametral pitch

Grinding stock allowance per tooth side for spiral bevel and hypoid gears should be 0.10 to 0.13 mm (0.004 to 0.005 in.) for gears coarser than 10 diametral pitch and 0.08 to 0.10 mm (0.003 to 0.004 in.) for gears of 10 to 20 diametral pitch. Stock allowance at the root of the tooth should be about 0.08 mm (0.003 in.) for gears up to 20 diametral pitch.

**Multiple-Piece Loads.** Because gear grinding is an expensive operation, it is important to control the amount of metal to be removed by grinding, to reduce runout and spacing errors, and to try to grind multiple-piece loads. Multiple-start hobs can cause enough spacing error to make extra grinding passes necessary. Therefore, when a hobbed gear is to be ground, it is better to use a single-thread hob. The multiple-start hob will not save enough to pay for the extra grinding passes. Single-thread hobs are available with nonground class-C tooth forms.

**Wheel Specifications.** Aluminum oxide wheels are preferred for the grinding of teeth in through-hardened and case-hardened steel gears. For nitrided gears, silicon carbide wheels are sometimes used. Grit size, which depends mainly on the type of grinding and the pitch of the gear, ranges from 46 to 100, although somewhat finer grits can be used for form grinding. A grit size of 60 is generally recommended for diametral pitch in the range of 1 to 10; 80 for the range of 8 to 12; and 100 for gears of pitch finer than 14. With crush-formed wheels, grit sizes as small as 400 are used in grinding gears of diametral pitch as fine as 200. For grade (hardness), a range of H to M is generally recommended.

Vitrified bond wheels are used because of their rigidity and ability to be diamond dressed with a minimum of diamond wear. Wheel rigidity is particularly important for dish-shaped wheels of the large sizes used in generation grinding. Resinoid bond cup wheels are satisfactory for generating bevel gear teeth.

Vitrified bond wheels are used for forming. They are generally crush trued. The vitrified bond is friable and is rigid enough to support the force of crush truing without deflection.

Steel-core wheels plated with CBN are commonly used because their open grain texture allows for slow traverse across the tooth surface and rapid stock removal without burning. Surfaces are generally finished in one pass. Because of the hardness of the material, several thousand parts can be ground before wheel contours need to be replated. Wheel classification and factors in selection are discussed in the articles "Grinding Equipment and Processes" and "Superabrasives" in this Volume.

**Speed.** Most gears ground with conventional wheels are ground with wheel surface speeds of 1700 to 2000 m/min (5500 to 6500 sfm). When very small amounts of stock are being removed, speeds higher than 2000 m/min (6500 sfm) can be used. When grinding gears with cup-shaped or dish-shaped wheels, slower speeds (140 m/min, or 450 sfm, or less) are common. Wheel speeds up to 3000 m/min (10 000 sfm) are used with CBN wheels.

**Infeed.** When using conventional wheels, pitch of the gear being ground is the main factor influencing optimum infeed. For example, in the plunge grinding of case-hardened or hardened-and-tempered steels (50 to 62 HRC), an infeed of 0.038 mm/pass (0.0015 in./pass) is recommended for gears having diametral pitches of 1 to 4. Infeed is reduced to 0.030 mm/pass (0.0012 in./pass) for pitches of 5 to 8 and to 0.013 mm/pass (0.0005 in./pass) for pitches finer than 8. With CBN wheels, 0.13 to 0.25 mm (0.005 to 0.010 in.) of stock is removed in one pass.

## Correcting for Distortion in Heat Treating

Most steel gears are heat treated. Because some diametral change occurs during heat treatment, gear teeth are cut prior to heat treating and ground to finish dimensions after heat treating. The magnitude of the dimensional change depends on the type of heat treatment and the shape of the part. Distortion is greatest when the entire part is heated and quenched. Induction hardening and flame hardening cause small changes because only the tooth area of the part is heated and quenched.

Usually, nitrided gears do not need to be ground to correct for distortion. When gears are properly nitrided, dimensional change is so small that any correction required can be accomplished by honing or lapping.

When the entire gear is heat treated, dimensional change is less in gears of nearly uniform cross section than in those of intricate design or large variation in cross section. For example, short, stubby pinions are least likely to lose accuracy because of unequal cooling rates during quenching, while thin-web gears with heavy-section teeth or heavy rims change the most.

The use of a heat-treating procedure that allows maximum dimensional control, such as placing the heated part in a fixture for quenching, helps reduce the amount of grinding and sometimes has eliminated the need for grinding. However, some dimensional changes occur during heat treatment regardless of special techniques. The following example is typical of a gear that required grinding even though it had been quenched in a die to control distortion.

**Example 1: Grinding Die-Quenched Gears.** The gear teeth for the part illustrated in Fig. 36 were cut by hobbing, after which they were carburized and quenched in a die to minimize distortion. However, this practice did not prevent distortion beyond tolerance, and lightening holes caused a scalloping of the pitch line by approximately 0.10 mm (0.004 in.) total indicator reading. Therefore, grinding was necessary. No grinding burns or cracks could be accepted.

Conditions of the grinding operation are given in the table with Fig. 36. Damper plates prevented excessive vibration of the gear during grinding, thus reducing the likelihood of both burns and cracks.

## Form Grinding

Usually, form grinding consists of passing a formed wheel through a tooth space to grind to root depth the left side of one tooth and the right side of the next tooth at the same time (see modifications in Fig. 37). When straight teeth are ground, as in spur gears, the workpiece is held in a fixed radial position during a pass. For helical gears, however, the workpiece must be rotated so that the helix will be followed as the wheel passes through the tooth space. A lead bar having the same lead as the helix controls rotation during grinding.

When the wheel pass through a tooth space is completed, the workpiece is indexed to the next position, and the procedure is repeated. The indexing mechanism usually consists of an index plate with the

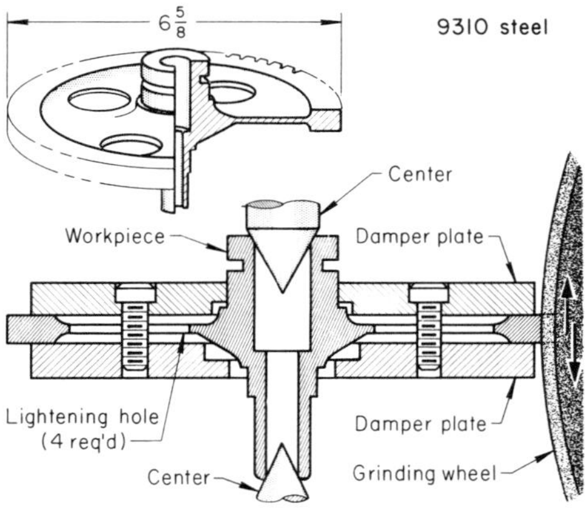


| | |
|---|---|
| **Gear details** | |
| Type | Involute spur |
| Number of teeth | 63 |
| Diametral pitch | 12 |
| Pressure angle | 20° |
| Tolerance | |
| Involute error, mm (in.) | 0.005 (0.0002) |
| Tooth-to-tooth spacing, mm (in.) | Within 0.005 (0.0002) |
| Spacing between any three adjacent teeth, mm (in.) | Within 0.013 (0.0005) |
| Finish required, involute sides and root fillet, μm (μin.) | 0.80 (32) max |
| Hardness, HRC | 58–63 |
| **Grinding wheel** | |
| Classification | A-180-H9-V |
| Size, mm (in.) | 349 (13¾) OD |
| Form | Diamond dressed |
| **Operating conditions** | |
| Machine | Gear grinding, rack type |
| Speed, at 1850 rev/min, m/min (sfm) | 2030 (6650) |
| Infeed per pass | |
| Roughing, mm (in.) | 0.030 (0.0012) |
| Finishing, mm (in.) | 0.020 (0.0008) |
| Axial feed | |
| Roughing, mm (in.) | 0.89 (0.035) |
| Finishing, mm (in.) | 0.41 (0.016) |
| Grinding fluid | Sulfurized grinding oil |
| Time per piece, min | 12 |

**Fig. 36** Setup for grinding a thin-web gear. Damper plates prevent excessive vibration. Dimensions in figure given in inches

same number of spaces at the workpiece. Control of tooth spacing errors depends primarily on the accuracy of this plate.

Spur gears are ground with a single straight wheel with a periphery that has been formed to produce the space between teeth (Fig. 37a), a single wheel with several ribs that grind more than one tooth space during each wheel pass (Fig. 37b), or two single-ribbed wheels (Fig. 37c). Helical gears are ground with a formed, single straight wheel. Bevel gears are form ground with a cup-shaped wheel. Form grinding can also be used to finish grind precut threads of worms.

Wheels for form grinding are usually less than 152 mm (6 in.) in diameter; wheel thickness depends on the size of the teeth to be ground. Wheels of larger diameter can be used, but the smaller wheels are more common because they need less clearance in

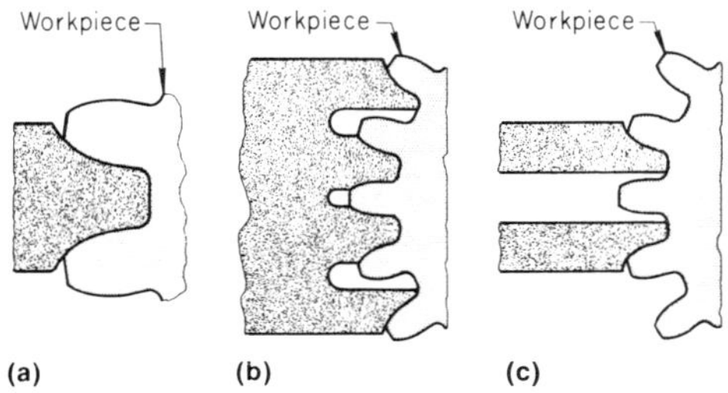


**Fig. 37** Relations of wheel and workpiece in the form grinding of spur gears (top views). (a) Single-ribbed grinding wheel. (b) Multiribbed grinding wheel. (c) Two single-ribbed wheels, also known as a straddle wheel

finishing a gear that is close to a larger gear or to a shoulder. In addition, smaller wheels are better for grinding internal gears.

Standard gear grinding machines are available for form grinding spur gears ranging from a minimum root diameter of 19 mm (¾ in.) to an outside diameter of 914 mm (36 in.). Special machines have been built to grind spur gears 2.4 m (8 ft) in diameter. Machines are available that are capable of form grinding external helical gears up to 457 mm (18 in.) in diameter.

## Generation Grinding

There are four methods of generating gear teeth by grinding. Each method is identified by the type of wheel used: straight, cup-shaped, dish-shaped, and rack-tooth worm wheels.

**Straight Wheels.** In this method, a straight wheel beveled on both sides reciprocates across the periphery of the workpiece as the workpiece rolls under it in a direction perpendicular to its reciprocating motion. The reciprocating action of the wheel is similar to that of a reciprocating gear cutter. This grinding method can be used to generate the teeth of spur gears.

**Cup-Shaped Wheels.** In this method, the gear tooth is rolled against a cup-shaped wheel, the sides of which are beveled to an angle equal to the pressure angle required on the teeth. During grinding, the wheel moves in a straight line along the length of the tooth. The sides of the wheel simultaneously grind adjacent sides of two teeth. A number of mechanical methods are used to control the generating motion and the indexing of the workpiece. Electronic controls are used on newer machines.

Teeth can be generated by grinding with a cup-shaped wheel on spur and helical gears up to 457 mm (18 in.) in pitch diameter. However, when helical gears are being ground, the generating motion must match the pitch in the plane of rotation of the gear to be ground. With this grinding method, it is impractical to generate tip and root reliefs.

**Dish-Shaped Wheels.** This method uses two dish-shaped grinding wheels, which may be inclined 15 or 20° or may be parallel in a vertical position (0°), as shown in Table 6.

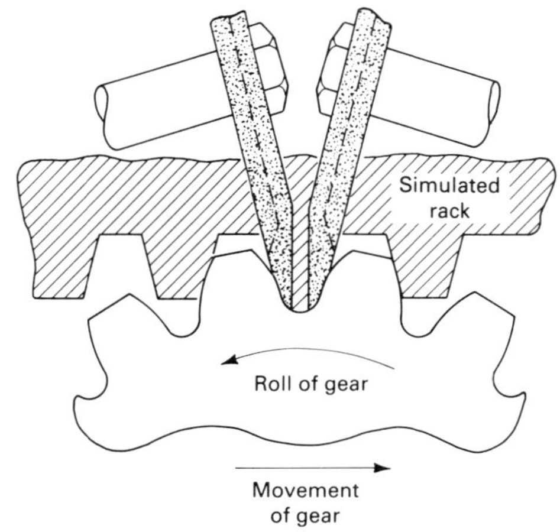


**Fig. 38** Schematic of dish-shaped wheels simulating the motion of a rack. The gear being machined rolls and moves. Wheels may or may not be reciprocated.

The generating motion is sometimes controlled by steel bands that are fastened to a pitch block of the same diameter as the pitch circle of the workpiece minus the thickness of the rolling bands. The shape of the tip and root of the teeth can be modified by using a cam to change the normal generating motion.

Usually an index plate similar to that used in form grinding indexes the workpiece if a pitch block is used. Large machines use worm gearing for indexing. Gears are ground dry, and machines incorporate a device depending on a feeler diamond to compensate for wheel wear. Machines are available for grinding spur and helical gears ranging from 25 to 3610 mm (1 to 142 in.) in diameter, with diametral pitch 2 to 17 by this method.

When the wheels are in the 15 or 20° position, as in illustration (a) in Table 6, their active faces rotate in planes forming an acute angle. In this position, the wheels simulate the tooth flanks of a rack that the workpiece engages during the rolling-generating motion (Fig. 38). The inclination of the grinding wheels to the vertical in the pitch plane generally equals the pressure angle of the gear to be ground (usually 15 or 20°). Under these conditions, the generating circle coincides with the pitch circle of the gear. This method produced a criss-cross pattern of grooves (lay of surface roughness) and has been widely used because machine settings are simple and the desired pressure angle is accurately and simply set by adjusting the inclination angles of the wheels.

The 15 or 20° wheel position is not suitable for making longitudinal modifications. Although the two simultaneous points of contact between the grinding wheels and the respective tooth flanks move along the tooth-form involutes during each transverse rolling-generating motion, the locations of these points relative to each other change constantly.

**Table 6 Grinding of gears with dish-shaped wheels in the 0° position**

Details for seven different spur and helical gears that had been carburized and hardened

| Parameter | Gear details | | | | | | |
|---|---|---|---|---|---|---|---|
| Diametral pitch | 10 | 5 | 4 | 3 | 5 | 3 | 2.75 |
| Module | 2.5 | 5 | 6 | 8 | 5 | 8 | 9.5 |
| Number of teeth | 23 | 44 | 80 | 80 | 120 | 100 | 379 |
| Face width, mm | 55.9 | 61.0 | 119 | 119 | 300 | 180 | 419 |
| (in.) | (2.2) | (2.4) | (4.7) | (4.7) | (11.8) | (7.1) | (16.5) |
| Tip diameter, mm | 71.1 | 234 | 503 | 653 | 630 | 828 | 3350 |
| (in.) | (2.8) | (9.2) | (19.8) | (25.7) | (24.8) | (32.6) | (132) |
| Pressure angle, degree | 15 | 20 | 20 | 20 | 20 | 15 | 15 |
| Helix angle, degree | 26 | 20 | 10 | 10 | 15 | 0 | 7 |
| Grinding allowance per flank, mm | 0.15 | 0.18 | 0.20 | 0.23 | 0.18 | 0.23 | 0.25 |
| (in.) | (0.006) | (0.007) | (0.008) | (0.009) | (0.007) | (0.009) | (0.010) |

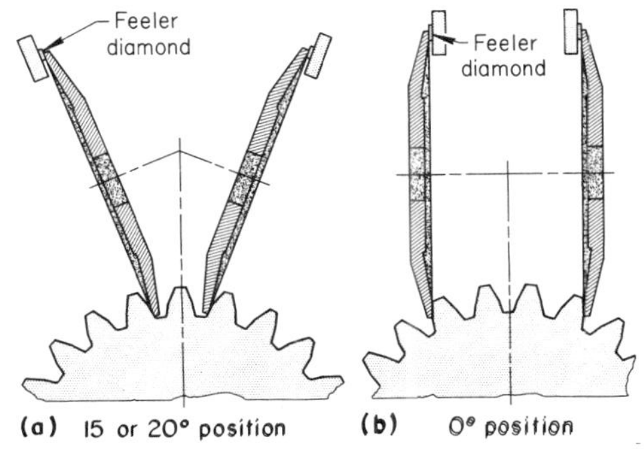


For helical gears, this differential position applies not only along the involutes but also to the position in the longitudinal tooth direction. Therefore, it would be impractical to try to coordinate modification impulses of the grinding wheel and the generating or feed motion to make longitudinal modifications.

In a newer method, the angle between the two wheels is 0°, as shown in illustration (b) in Table 6. The active surfaces of the wheels are parallel and face each other. Because the grinding pressure angle is 0° and corresponds to the pressure angle on the base circle, the generating motion is obtained by rolling on the base circle.

The advantages of grinding gears in the 0° position over the 15 or 20° position are:

- Greater production without sacrifice in quality
- Ability to grind more exacting flank profiles and to make longitudinal modifications

The main disadvantage is the lack of uniform transition between the flank and the tooth root, which is objectionable in some gearing applications.

**Rack-Tooth Worm Wheels.** The wheel used in this method of generating gear teeth is crush trued into the shape of a conventional rack-tooth worm or a CBN-plated rack-tooth worm. Grinding a gear by this method is illustrated in Fig. 39.

The workpiece is mounted vertically on an arbor or in a fixture that fits between the centers of the machine. For grinding helical gears, the workholding slide of the machine is set at an angle equal to the helix angle required on the gear. The workpiece is rotated by a motor that is synchronized with the wheel motor.

The gear can be ground with a small portion of the length of the grinding worm wheel. Generation takes place as a result of rotating the worm wheel and gear in mesh as the wheel is traversed axially across the gear face.

A soft, conventional aluminum oxide or silicon carbide wheel is used to prevent burning the hardened teeth, but this results in greater wheel breakdown and loss of correct tooth profile. To continue using a fresh, unbroken part of the wheel, the work is fed in a direction parallel to the axis of the worm using the setup shown in Fig. 40.

Success in gear grinding with a conventional grinding worm wheel depends greatly on careful dressing of the wheel to the required contour and on synchronization of the workpiece rotation with the rotation of the worm wheel. Special diamond wheel dressers are required. The accuracy of the diamond and of the dressing operation determine to a great extent the quality of the gear teeth produced.

Wheels are about 50 mm (2 in.) wide, and they can be rough formed by crushing and can be finished to the proper pressure angle by diamond dressing. Some machines use wheels that have been finish dressed by crushing only. Because a considerable amount of time is required to crush true a new wheel, it is advisable to stock wheels for each pitch and pressure angle required. In diamond dressing, two diamonds are needed to dress both sides of the thread at the same time. Gear tooth design can be easily modified by lapping the tip of the dressing diamond.

Burning the hardened steel is much less of a problem when using CBN-plated wheels because of the open texture of the wheel surface. The hardness of the CBN material and the stiffness of the steel core allow for several thousand pieces to be ground before the worm needs replating and returning. Modifications of the gear tooth design are manufactured into the core before it is plated.

Spur gears and helical gears up to 508 mm (20 in.) in diameter can be ground in machines using a threaded wheel. Although fine-pitch gears can be ground from a solid gear blank, gear teeth with diametral pitch of 24 or coarser should be cut prior to grinding.

The advantages of gear grinding with a threaded wheel are accuracy, speed, and ease of incorporating changes in tooth design. Machines are also available that crown the gear teeth during grinding. Because mechanical grinders are complicated and sensitive, a considerable amount of operator skill is required. However, with electronic CNC machines, crowning is a standard feature. Gear grinding with a threaded wheel is six to thirty times as fast as the other methods discussed in this section.

## Grinding of Bevel Gears

When accuracy is required, grinding is the most economical method for finishing teeth as coarse as 1.5 diametral pitch on straight bevel, spiral bevel, and hypoid

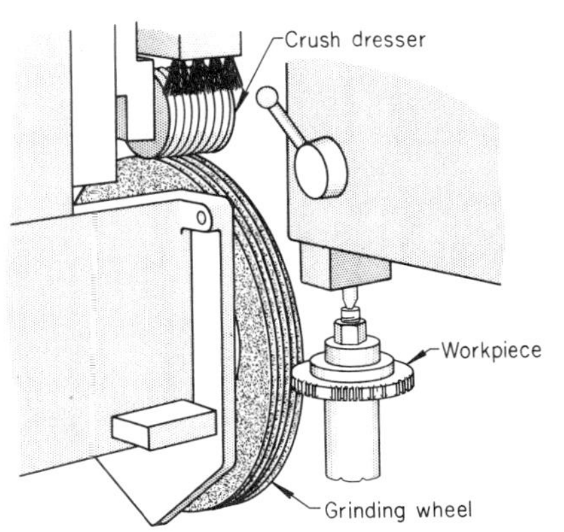


**Fig. 39** Generation grinding with a rack-tooth worm wheel

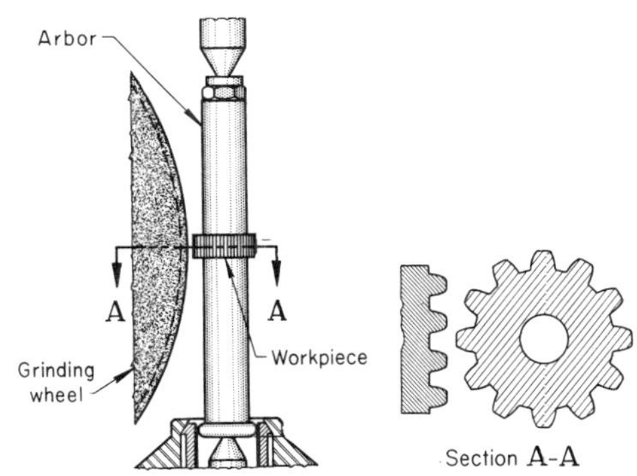


**Fig. 40** Setup used to grind an aircraft pinion by the worm wheel method

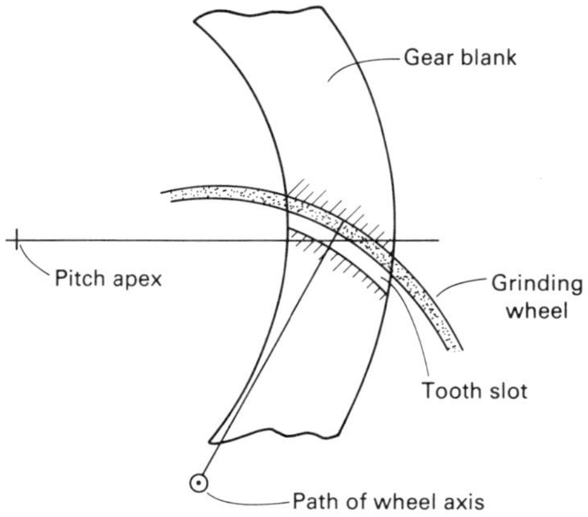


**Fig. 41** Waguri grinding method for nongenerated spiral bevel and hypoid gears

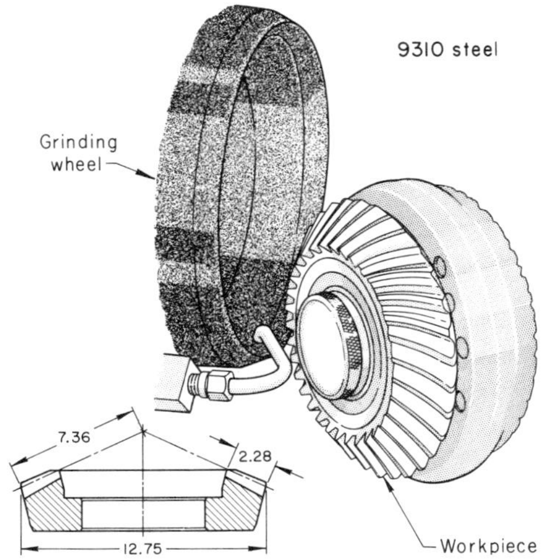


| | |
|---|---|
| **Gear details** | |
| Type | Spiral bevel (octoid) |
| Number of teeth | 51 |
| Diametral pitch | 4 |
| Helix angle | 30° |
| Whole depth, mm (in.) | 12.07(0.475) |
| Tolerance, tooth-to-tooth, mm (in.) | 0.008 (0.0003) |
| Surface finish, μm (μin.) | 0.63–0.76 (25–30) |
| Hardness, HRC | 58–60 |
| **Grinding wheel** | |
| Type | Cup(a) |
| Classification | A-54-J8-V |
| Size, mm (in.) | 305 (12) OD |
| **Operating conditions** | |
| Wheel speed, at 1200 rev/min, m/min (sfm) | 1150 (3770) |
| Feed | 3.2 s/tooth(b) |
| Grinding fluid | Straight mineral oil (no active sulfur) |
| Production rate, pieces/h | 1.7(c) |

(a) Straight-angled grinding area. (b) 12 passes per tooth. (c) At 100% efficiency

**Fig. 42** Grinding of a spiral bevel gear. Dimensions in figure given in inches

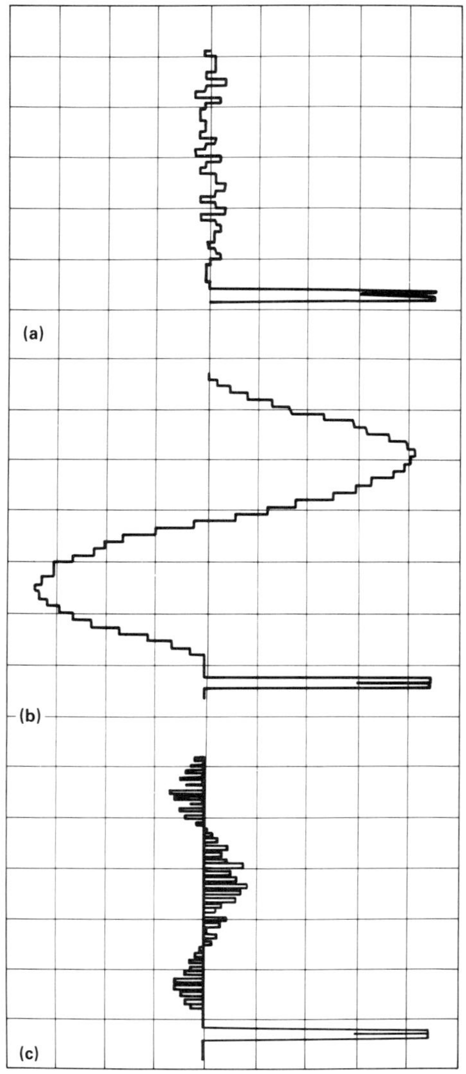


**Fig. 43** Typical data obtained on charts generated by automated gear-checking machines. (a) Tooth-to-tooth pitch error. (b) Accumulated pitch error. (c) Spacing error

gears. With good grinding practice, bevel and hypoid gears can be finished with a tooth-to-tooth spacing accurate within 0.005 mm (0.0002 in.), eccentricity with bore or shaft within 0.0064 mm (0.00025 in.), as little backlash as 0.025 to 0.051 mm (0.001 to 0.002 in.), and a surface finish of 0.38 to 0.76 μm (15 to 30 μin.). In general, any bevel or hypoid gear that is cut with a circular cutter can be ground with a wheel of similar shape and grinding action, except nongenerated gears that are cut with a face mill.

**Nongenerated (Formed) Gears.** The gear teeth are ground by two methods. The first method involves a cup-shaped wheel that simultaneously grinds the facing sides and root of adjacent teeth. There is no generating motion, and the wheel produces a curved tooth whose sides have a straight profile. Three or four revolutions of the workpiece are required for the complete finish grinding of a bevel gear using conventional dressable wheels. A predetermined amount of stock is removed with each revolution of the workpiece. With CBN-plated wheels, the entire finishing stock is removed in one revolution of the workpiece. The grinding wheel makes line contact with the tooth being ground. This method is applicable for finishing hypoid and spiral gears and for special spiral gears that have a 0° mean spiral angle.

The second method is called the Waguri method. The grinding wheel is circular and has an average point width of 0.25 to 0.50 mm (0.010 to 0.020 in.) less than the slot width of the workpiece. The wheel axis follows a circular path with a diameter equal to the difference between its point width and its slot width. Thus, the desired surfaces are ground with line contact (Fig. 41). Conventional dressable wheels can be used with several revolutions of the workpiece, or CBN-plated wheels with one revolution of the workpiece.

An A-46-J-V wheel is the conventional wheel commonly used to grind gears ranging from 2 to 10 diametral pitch. Wheel speed ranges from 1200 to 1400 m/min (3800 to 4500 sfm). The wheel is dressed automatically at predetermined intervals between grinds.

Cubic boron nitride-plated wheels cover the same range of diametral pitch. Wheel speeds range from 1400 to 2000 m/min (4500 to 6500 sfm).

**Generated Gears.** Spiral bevel and hypoid gears are ground with a cup-shaped wheel. The conventional grinding wheel is dressed to produce a smooth blend of flank and root profile. The core of the CBN wheel is machined to produce a smooth blend of the flank and tooth profile. The curved profile of the tooth is generated by relative rolling motion between the grinding wheel and the workpiece.

Vitrified bond and resinoid bond aluminum oxide cup-shaped wheels are the conventional wheels used for spiral bevel and hypoid gears. Typical wheel classifications are A-60-J-V and A-60-M-B. Wheel speed ranges from 1200 to 1400 m/min (3800 to 4500 sfm). Cubic boron nitride-plated wheels are also used.

Generating gear teeth with a cup-shaped wheel produces spiral bevel gears with highly accurate tooth spacing, concentricity, and profile shape. The following is an example of a typical application of this grinding method.

**Example 2: Grinding Spiral Bevel Gears With a Cup-Shaped Wheel.** The spiral bevel gear illustrated in Fig. 42 was carburized, hardened, and tempered before being ground by the generation method with a cup-shaped wheel. An automatic wet-type 3.7 kW (5 hp) machine designed for grinding spiral bevel, Zerol, and hypoid gears was used for this application. Additional manufacturing information is given in the table with Fig. 42.

## Grinding Fluids

The use of a grinding fluid is generally recommended for the grinding of steel gears. A grinding fluid prolongs wheel life between dressings, flushes away chips, and

**Table 7 Comparison of gear classification systems**

| AGMA quality No.(a) | Approximate DIN quality No. | Quality classification | Typical application |
|---|---|---|---|
| (1) | ... | Commercial | Appliances, hand tools, pumps, clocks, farm machines, fishing reels, hoists, slow-speed machinery |
| (2) | ... | | |
| 3–4 | 12 | | |
| 4–5 | 11 | | |
| 5–6 | 10 | | |
| 7 | ... | | |
| 8 | 9–10 | Precision | Instruments, aircraft engines, turbines, professional cameras, machine-tool speed drives, automotive transmissions, high-speed machinery |
| 9 | 8–9 | | |
| 10 | 7–8 | | |
| 11 | 6–7 | | |
| 12 | 5–6 | | |
| 13 | 4–5 | | |
| 14 | 3–4 | Ultraprecision | Military navigation, precision instruments, computer equipment |
| 15 | 2–3 | | |
| 16 | 2 | | |
| (17) | 1 | | |
| (18) | ... | | |
| (19) | ... | | |

(a) Nos. 1, 2, 17, 18, and 19 are derived by extrapolating AGMA tolerance curves for quality Nos. 3–16. Source: J. G. Bralla, *Handbook of Product Design for Manufacturing*, McGraw-Hill, 1986

improves gear tooth finish. Flooding the work surface minimizes the possibility of burning the surfaces of gear teeth. The complete absence of grinding burn is essential for high-quality gears. Virtually all the grinding fluids mentioned in the section "Cutting Fluids" in this article have been used successfully for the grinding of gears.

In many critical gear-grinding applications, mineral-base sulfochlorinated or sulfurized oils are used. Plain soluble-oil emulsions have proved satisfactory for many gear grinding applications. Chemical solutions have also been satisfactory. Additional information on cutting fluids is available in the article "Metal Cutting and Grinding Fluids" in this Volume.

## Surface Finish

Grinding produces a finish unlike that produced by any other process. Whether this finish is more desirable than some other finish depends on the service requirements of the gear. In some applications, a gear with a ground finish operates more quietly, even though grinding has not improved the dimensional accuracy of the product. The effect of a ground finish, in comparison with other finishes, on the lubrication of gears has not been precisely evaluated.

The surface finish of ground gears usually ranges from 0.38 to 0.80 μm (15 to 32 μin.). Under conditions of unusually good control, a surface finish of 0.25 μm (10 μin.) or better can be produced. Surface finish sometimes becomes the major consideration in determining whether a gear will be ground. Grinding can have the dual purpose of correcting dimensional change from heat treating and of producing a fine surface finish.

# Gear Inspection

As with all manufactured products, gears must be checked to determine whether the resulting product meets design specifications and requirements. Because of the irregular shape of gears and the number of factors that must be measured, such inspection is somewhat difficult. Among the factors to be checked are the linear tooth dimensions (thickness, spacing, depth, and so on), tooth profile, surface roughness, and noise. Several special devices, most of them automatic or semiautomatic, are used for this inspection.

Gear tooth vernier calipers can be used to measure the thickness of gear teeth on the pitch circle. However, inspection is usually done by special machines, which in one or a series of operations check several factors, including eccentricity, variations in circular pitch, variations in pressure angle, fillet interference, and lack of continuous action. The gear is usually mounted and moved in contact with a master gear. The movement of the latter is amplified and recorded on moving charts, as shown in Fig. 43.

Noise level is important in many applications, not only from the standpoint of noise pollution but also as an indicator of probable gear life. Therefore, special equipment for its measurement is quite widely used, sometimes integrated into mass-production assembly lines.

Dimensional variations in gears result in noise, vibration, operational problems, reduced carrying ability, and reduced life. These problems are compounded at higher gear-operating speeds. The American Gear Manufacturers Association (AGMA) and its German counterpart Deutsche Industrie-Normen (DIN) have incorporated the critical dimensions of pitch, concentricity, tooth profile, tooth thickness, and tooth surface finish into a comprehensive accuracy classification system. Table 7 compares AGMA and DIN quality numbers for typical applications. Table 8 lists the tolerances in terms of AGMA quality numbers for various gear-making processes.

**Table 8 Recommended tolerances in terms of AGMA quality numbers for various gear manufacturing processes**

| Process | Highest quality number: Normal | Highest quality number: With extra care |
|---|---|---|
| Sand mold casting | 1 | 3 |
| Plaster mold casting | 3 | 5 |
| Permanent mold casting | 3 | 5 |
| Investment casting | 4 | 6 |
| Die casting | 5 | 8 |
| Injection and compression molding | 4 | 8 |
| Powder metallurgy | 6 | 9 |
| Stamping | 6 | 9 |
| Extrusion | 4 | 6 |
| Cold drawing | 6 | 9 |
| Milling | 6 | 9 |
| Hobbing | 8 | 13 |
| Shaping | 8 | 13 |
| Broaching | 7 | 12 |
| Grinding | 10 | 15 |
| Shaving | 9 | 14 |
| Honing | 10 | 15 |
| Lapping | 10 | 15 |

Source: J. G. Bralla, *Handbook of Product Design for Manufacturing*, McGraw-Hill, 1986

# Sawing

Revised by Ted A. Slezak, Armstrong-Blum Manufacturing Company

SAWING is the process of cutting a workpiece with power band saws, hacksaws, and circular saws. Each of these methods is used in cutoff operations (cutting pieces to a required length), although band sawing also provides a method of cutting contours. The most satisfactory range of material hardness for sawing is about 180 to 250 HB, but steels up to 400 HB and some free-machining steels up to 450 HB can be sawed at reduced cutting rates.

Table 1 compares the general characteristics of power band sawing, hacksawing, and circular sawing. In general, hacksawing and band sawing involve lower-cost machines with greater versatility and a larger cutting range. Circular sawing produces the smoothest finishes and the closest dimensional tolerances, especially on nonferrous materials. The tool life and tool cost ratings in Table 1 are only general and may change completely for different applications.

## Band Sawing

Band sawing involves a long, endless saw band traveling in one direction over two or more wheels. One wheel provides drive, and the others are idler wheels. The saw band must be thin enough to bend over driving and idler wheels, yet must have sufficient thickness and beam strength to withstand the pressures exerted during cutting. In all types of band sawing, the power rating of the machine limits the cutting rate and the thickness and hardness of the metal to be cut.

**Table 1 Cost and performance comparisons of hacksawing, band sawing, and circular sawing**

| Characteristics | Type of machine(a): Lowest | | Highest |
|---|---|---|---|
| Machine cost | H → | B → | C |
| Power requirements | B → | H → | C |
| Productivity | H → | C → | B |
| Versatility | H → | C → | B |
| Tool cost | C → | H → | B |
| Tool life | H → | B → | C |
| Accuracy and finish | H → | B → | C |
| Kerf loss | B → | H → | C |

(a) H, hacksawing machine; B, band sawing machine; C, circular sawing machine

Band saws have a thinner cutting tool than the other sawing machines. Band sawing produces a kerf of about 1.6 mm (1/16 in.); hacksawing and circular sawing produce kerfs of 7.9 and 9.5 mm (5/16 and 3/8 in.), respectively. Because the saw band is thinner than the other sawing tools, less power is required for cutting, and binding in the kerf is less likely. Another advantage of band sawing is that the uniform, continuous cutting action produces even wear over the entire length of the saw band. The blades for band saws are also changed less frequently because they are longer than the blades in other types of saws.

### Capabilities and Limitations

Band sawing differs from the other sawing methods in that its blade and cutting action allow the cutting edge to follow a contoured path during cutting. When compared to other machining methods (milling, for example), contour cutting with a band saw has advantages such as:

- Unwanted material is removed in sections instead of chips
- Downward cutting action (vertical band saws only) holds work to the table, thus simplifying fixturing
- Narrower tooth kerf minimizes power requirements for cutting and the amount of material reduced to chips

**Contour band sawing** is performed on a vertical band saw having a C-shaped, open-yoke frame. Because of the open-yoke frame, clearance between the workpiece and the frame imposes a size limitation in contour band sawing. Workpiece height or thickness can be as much as 1400 mm (55 in.), which is the maximum capacity between the guides of standard machines. However, special machines have been built with yoke heights up to 3000 mm (120 in.).

Convex radii of less than 1.6 mm (1/16 in.) can be cut in a single pass using commercially available bands, thus making it possible to produce complex contours in one straightforward machining operation. To produce internal contours, the ends of the saw band are welded together after the band has been inserted through a hole provided in the workpiece for this purpose.

**The dimensional tolerances** that can be maintained in contour and cutoff band sawing depend greatly on the dexterity of the operator, the suitability of the setup, tooling and machining conditions, and the availability of accessories, such as servo controls. Automated band saws have error control devices that can assure the accuracy of cut. If the blade cuts beyond the given tolerance of the programmed shape of the cut, then the machine shuts off, indicating the need for a blade change. In cutoff band sawing, cutting accuracy (straightness of cut) is usually within 0.002 mm/mm (0.002 in./in.).

*Contouring.* A servo-controlled contour sawing attachment maintains a constant feed force and, by lessening the effort required, permits the operator to concentrate more fully on following the line to be cut, thus increasing overall accuracy. Under optimum conditions, a skillful operator with the aid of a magnifying glass can follow a contour to within ±0.25 or ±0.38 mm (±0.010 or ±0.015 in.). A tolerance of ±0.8 mm (±1/32 in.) is more typical of production work. When using a power table for ordinary work thicknesses, the flatness of the cut surface can be held to 0.004 mm/mm (0.004 in./in.) of work thickness or per 25 mm (1 in.) of cut length.

**Surface finish** also varies with operator skill, equipment, and operating conditions. A surface roughness of 5.0 to 7.6 μm (200 to 300 μin.) results under ordinary production conditions. With the use of a fine-pitch blade, high band speed, and low feed force, a finish of 1.5 to 5.0 μm (60 to 200 μin.) can be produced, and a surface roughness as low as 0.63 μm (25 μin.) has been obtained under specially controlled conditions.

### Types of Machines

Most band saws are designed for either vertical or horizontal movement of the saw band, although some manufacturers offer combination vertical-horizontal band saws for light to medium-duty cutting. The band saws available include contour band saws, cutoff band saws, tilt-frame universal band saws, and plate band saws.

**Contour band saws** are vertical machines with C-shaped, open-yoke frames. Although this equipment can perform cutoff operations, it is seldom used for this pur-

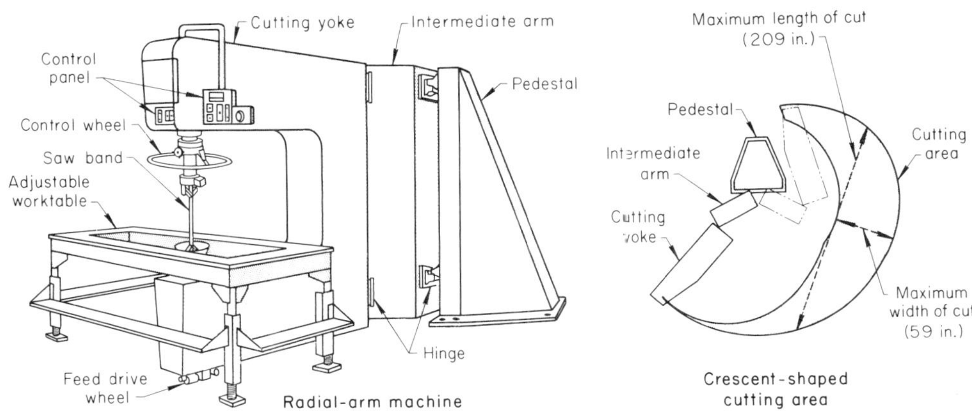


**Fig. 1** Radial-arm contour band sawing machine and shaded crescent showing the total area within which the cutting yoke can move. The workpiece, mounted on the adjustable worktable, remains stationary.

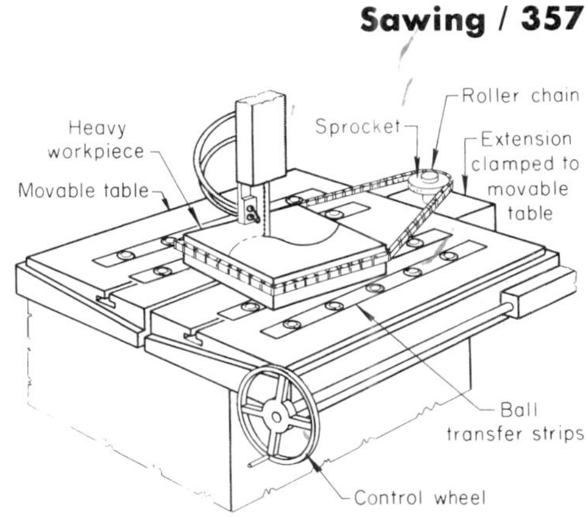


**Fig. 2** Worktable setup for the contour band sawing of heavy workpieces

pose. Cutoff operations are usually done on a horizontal machine.

Contour band saws are available in a wide range of sizes and modifications. There are three general types: fixed table, power table, and radial arm.

*Fixed-Table and Power-Table Machines.* With a fixed-table machine, the work must be fed by hand. Power-table machines, which are usually heavier than fixed-table machines, are equipped with a worktable that pushes the work into the saw band, thus relieving the operator of pushing or manual feeding. These machines have enough power to use high-speed steel bands, while fixed-table machines usually employ a lower cutting rate and carbon steel bands.

*Radial-arm machines* are designed for handling large, heavy workpieces. The articulated structure of the equipment provides the capability for unlimited cutting within a crescent-shaped area. The machine shown in Fig. 1 has a cutting crescent area of 9.2 $m^2$ (99 $ft^2$) and consists of three major members, of which two are movable and the third is stationary. The two moving members—an intermediate arm and a cutting yoke—permit the cutting edge of the saw frame to move anywhere within the prescribed area, while the workpiece—mounted on a worktable that can be raised or lowered—remains stationary. The longest straight cuts that can be made on this machine are 530 mm (209 in.) across the crescent and 1500 mm (59 in.) to the depth of the crescent, as shown by the shaped portion of the cutting-area diagram in Fig. 1.

**Cutoff band saws** cut horizontally or vertically, but in a straight line only. Cutting angle, however, is adjustable. In a cutoff band sawing machine, the saw band is twisted through carbide guides to bring the blade perpendicular to the surface of the worktable.

Cutoff band saws range from machines used for light, intermittent toolroom work to automatic production machines of high capacity. There are also machines for angular cutoff. Unlike contour band sawing machines, cutoff machines have no welders; prewelded bands are used (no internal sawing is done).

Cutoff band saws can accommodate workpieces as large as 2000 × 2000 mm (80 × 80 in.), and they have cutting rates up to 19 × $10^3$ $mm^2$/min (30 in.$^2$/min) in machine steels using welded-edge high-speed steel band saw blades. Cutting rates as high as 46 × $10^3$ mm/min (72 in.$^2$/min) are possible when using blades with triple-chip tungsten carbide inserts. Special machines are also available for sawing aluminum alloys at rates of 250 × $10^3$ $mm^2$/min (400 in.$^2$/min).

**Tilt-frame universal band saws** are widely used for angle-cutting operations and for producing compound miters. On these machines, the sawing head is mounted with pivot bearings on a moving carriage.

**Plate band sawing machines** are vertical band saws used for cutting plate stock. These plate saws are gaining acceptance in steel service centers and in the steel-producing industry. Instead of stocking many sizes of bars, the service center can slice a bar from a plate by using this type of saw. The saw blade is thin and produces little waste. These machines have work heights up to 1300 mm (51 in.) and throats up to 1525 mm (60 in.). They are capable of handling lengths up to 6000 mm (236 in.).

## Fixtures and Attachments

Much of the work done on band saws requires a device to hold or guide the workpiece. In contour band sawing, the downward cutting force of the saw band can assist in holding the workpiece to the table and simple, standard attachments are usually adequate. When they are not, special fixtures must be employed. Contour band sawing may also require devices for guiding the workpiece.

**A work-squaring bar** is a simple attachment that serves as a guide in making straight-line cuts. It consists of a movable workstop that is held securely to a backup bar by means of a cam lock. The backup bar acts as the prime locator and is attached to T-slots in the worktable by means of T-nuts and socket-head screws. The movable workstop slides along the calibrated backup bar and can be clamped to it at any point with the cam-locking lever.

**Contour sawing attachments** provide additional capability for holding and rotating the work and for work or table feed. Heavy workpieces are usually handled with table feed and, to minimize friction, are supported on ball transfer strips on the movable table (Fig. 2). The sprocket is mounted on an extension arm that is clamped to the movable table, and the roller chain feeds the workpiece into the saw band. Servo control on the hydraulic feed system maintains a constant feed force at the value selected for the job, regardless of variation in radius of cut, work thickness, work hardness, or other factors. Turning the hand control wheel rotates the sprocket and pulls the chain to rotate the workpiece as needed to follow the contour of the cut. Three positions of a foot switch give forward or reverse feed or stop.

Light workpieces rest directly on the table and can be manipulated and fed without the use of table feed. Instead, the workpiece is fed into the saw band by a roller chain partly wrapped around the workpiece (or around a work-holding jaw containing the workpiece). In this arrangement, the table-feed piston is disconnected from the table and exerts the feed force against a movable extension arm that holds the sprocket. Servo control of hydraulic feed pressure can be used, as described above, but it is needed less often than for cutting heavy workpieces. Turning the hand control wheel rotates the workpiece as desired. On fixed-table machines, the feed force can be supplied by weights attached to the chain or by other means, and the work-holding jaw can be rotated manually by handles attached to each end.

**A servo-feed attachment** facilitates the handling of heavy work and improves productivity, resulting in more accurate work handling and the avoidance of underfeed or overfeed. A servo-feed attachment can also result in an improved surface finish, as described in the following example.

**Example 1: Sawing Aluminum Honeycomb Sections.** In the contour band sawing of aluminum hobe blanks (unexpanded honeycomb sections) to obtain a surface finish of 2.8 to 3.8 μm (110 to 150 μin.), optimum results were obtained with the following tooling and operating conditions: an 8-pitch, regular-form blade; a band speed of 915 m/min (3000 sfm); and a constant, hydraulically controlled feed of 1900 mm$^2$/min (3 in.$^2$/min) (for a 50 mm, or 2 in., section thickness). These conditions provided a surface finish of 3.3 to 4.0 μm (130 to 160 μin.). The surfaces obtained by manual feeding were poor.

**Welders.** Most contour band sawing machines are equipped with built-in resistance-type butt welders to make possible the cutting of internal contours. The saw band is cut to length, threaded through a hole drilled in the workpiece for this purpose, and welded into a continuous band. The weld is annealed and ground, and the band is placed on the machine.

To obtain optimum cutting performance and maximum life from a saw band, the weld area should be identical in strength and flexibility to the remainder of the band. Welds in carbon steel and welded-edge bands approach this ideal more closely than those in solid high-speed steel or intermediate-alloy tool steel bands, because welds in the latter two materials are somewhat brittle, as a result of the short welding and annealing cycle used. (The solid high-speed steel and intermediate-alloy bands are obsolete.)

**Vises and Nesting Fixtures.** Workpieces must be held securely during cutoff operations. The work is clamped in either a vise or a nesting fixture, depending on the shape, size, and quantity of pieces to be held. Rectangular and square bars can be readily stacked and held firmly in a vise; small and medium-size rounds can also be clamped two abreast and held firmly in a vise. However, holding a larger number of stacked rounds requires the use of a nesting fixture such as that shown in Fig. 3. This type of fixture is widely used for stacking pipe and structural shapes. Stack sawing with the aid of a nesting fixture is most effective when the total area to be sawed is roughly half the capacity of the nesting vise and when the nest is higher than it is wide. Special precautions must be taken in the stack sawing of round pieces to ensure the positive clamping of all pieces because the rotation of a piece during cutting can cause premature band failure. Instead of being stacked as shown in Fig. 3, a number of round, hexagonal, or irregularly shaped bars can be held by special jaws in standard vises.

**Worktables.** The cutoff band saw is usually equipped with at least two work-

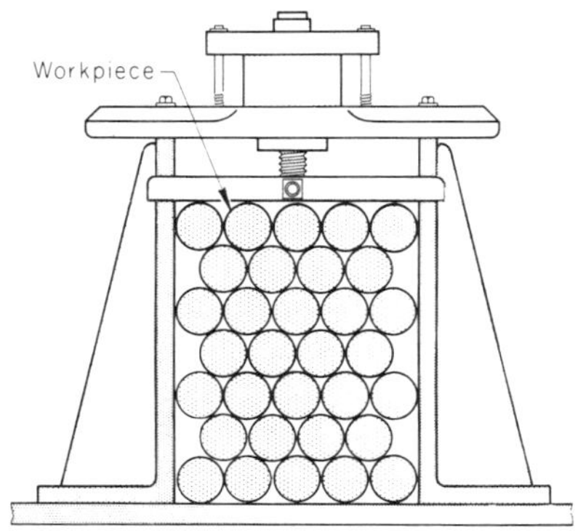


**Fig. 3** Nesting fixture used with a standard vise in cutoff band sawing

tables—a stack feeding table on which are mounted one or more vises for gripping and indexing the work to be cut, and a discharge table that provides continuous support for the workpiece and the stack from which it is cut. These tables are made in various lengths to suit operating requirements, and additional tables can be added to accommodate the longest stack length being handled.

**Cutting fluid systems** are essential for the effective performance of cutoff band saws. They consist of a reservoir and pump, a screening system for chips, draining elements, and a chip drawer or automatic chip remover. The system must be drained and cleaned when changing from one type of cutting fluid to another or when replacing contaminated fluid. Mist or spray systems are sometimes used to apply the cutting fluid.

## Band Construction and Materials

Bands are made of carbon steel or are a bimetallic type. The bimetallic, or composite, types are made with high-speed steel cutting edges that are electron beam welded to a high tensile (AISI 6150) steel back or with tungsten carbide inserts brazed or welded to a high tensile, alloy steel back. These welded-edge composite bands have replaced the carbon and the solid high-speed steel bands.

**Carbon steel** bands are seldom used for the contour band sawing of metals as they have been replaced by the composite welded-edge bands. Fixed-table machines seldom have adequate power, feed mechanisms, and cutting fluid distribution systems for other types of bands. Satisfactory cutting rates and tool life are obtained in sawing carbon and low-alloy steels, tool steels, and the more readily machinable nonferrous alloys.

Carbon steel bands come in two types: the flexible-back band and the hard-back band. The flexible-back band is not heat treated across its entire width. Only the teeth are heat treated to increase their hardness and wear resistance. The hard-back band is first heat treated across its entire width to a spring temper. This allows the blade to be tensioned on the band saw to a higher degree for increased beam strength. After spring tempering, the teeth are heat treated to full hardness.

A typical nominal composition for carbon steel bands is:

| Element | Composition, % |
|---|---|
| Carbon | 1.3 |
| Manganese | 0.3 |
| Silicon | 0.2 |
| Chromium | 0.2 |

**Welded-edge high-speed steel** bands are used on heavy-duty machines equipped with systems for circulating cutting fluid. These are usually power-table machines rated at 1.1 kW (1½ hp) or more and designed for continuous, high-volume production. The welded-edge bands have replaced the solid high-speed steel bands and are used under the same operating conditions as the solid high-speed steel bands.

Welded-edge high-speed steel bands give higher cutting rates and longer tool life than carbon steel bands in cutting the same materials, and they are required for contour sawing the more difficult-to-cut metals, such as stainless steels, heat-resistant alloys, the more highly alloyed tool steels, and some nonferrous alloys. The cutting edge of these bands are usually made of M2, Matrix 2 or M42 high-speed steel. (See Table 2 for material composition.) Welded-edge bands coated with titanium nitride are also available for difficult applications and increased tool life.

**Carbide Inserts.** Bands with tungsten carbide cutting edges brazed or welded to an alloy steel back are used for cutting the most difficult-to-machine alloys, such as nickel-base and cobalt-base heat-resistant alloys, and for sawing sections thicker than about 150 mm (6 in.) of common metals. Compared to high-speed steel, these blades have relatively low shock resistance, but they provide maximum hot hardness and wear resistance.

**Hardening of Bands.** Heat-treating procedures vary with band material and manufacturer. Hardness of the teeth for carbon steel bands is usually 63 to 65 HRC after tempering.

To improve service life and cutting performance, special procedures are employed to impart increased hardness and strength to the body of the band by increasing back hardness from 25 to 32 HRC to 40 to 47 HRC. Most manufacturers flame harden or induction harden the cutting edges to about 63 or 65 HRC.

The components of welded-edge high-speed steel bands are selected so that optimum properties for the teeth and the back of the band are developed in a single-temperature heat treatment. The composite

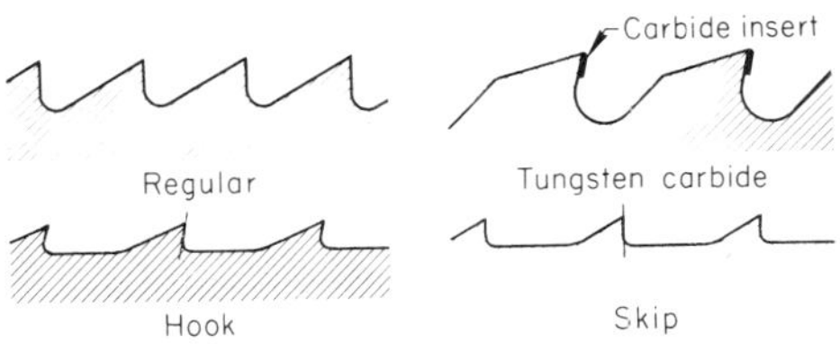


**Fig. 4** Standard tooth forms for steel and carbide-tip band saw blades

bands combine the welding characteristics and fatigue properties of carbon steel with the heat resistance and wear resistance of high-speed steel.

## Tooth Form

As shown in Fig. 4, steel bands are available in three tooth forms: regular, skip, and hook; bands with carbide inserts are also available in a special form. Individual manufacturers of saw bands have referred to the tooth shapes by various names; the terminology followed in this article is based on "Simplified Practice Recommendation R214-55" (U.S. Department of Commerce).

The regular form is the only form available for saws that are finer than 6-pitch in straight teeth or uniform number of teeth per 25 mm (1 in.). For 6-pitch and coarser, the hook form provides the best tool life and the fastest cutting rate. For optimum surface finish, either a regular or a skip tooth form is usually recommended.

**The regular tooth form** is most frequently used in contour band sawing. It has a deep gullet with a smooth radius at the bottom. The rake angle is normally 0°, although positive rake angles applied to regular tooth forms allow faster cutting and are now available from most manufacturers. The back clearance angle is about 30° (see Fig. 5 for an explanation of the nomenclature applied to blade angles). This tooth form produces fine-finish cuts accurately and has ample chip capacity for most sawing operations. The largest selection of widths is available in this form.

**The skip tooth form** is similar to the regular tooth form except that the teeth are more widely spaced to provide greater chip clearance. The skip tooth form has a special gullet design, but rake angle and back clearance angle are the same as in the regular tooth form. Because of its shallow gullet, the skip tooth form may have a coarser pitch on a narrow band. This tooth form is recommended for making deep cuts in soft metals.

**The hook tooth form** has a positive rake angle that permits faster cutting rates, reduced feeding pressures, and longer tool life. The back clearance angle is slightly less than that of the regular and skip tooth forms, and the wide gullet is of a special design.

## Blade Design

Pitch, width and thickness of the blade, and type of set and set dimension are important factors in the selection of a blade for a particular application.

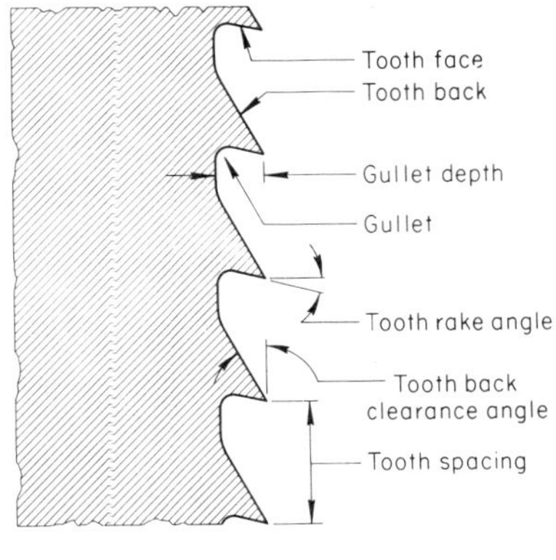


**Fig. 5** Standard nomenclature for saw blade teeth

**The pitch** of a saw blade is the number of teeth per 25 mm (1 in.) of blade. Each of the tooth forms previously discussed is available in various pitches. The pitch of a blade is primarily selected on the basis of the thickness and shape of the cross section to be cut; the type of material to be cut is of minor importance. Therefore, a given cross-sectional thickness of aluminum, low-carbon steel, or tool steel would be cut with blades having identical pitch, although speed and feed would vary.

At least two teeth must remain in contact with the workpiece at all times; it is preferable to have more teeth in constant contact, thus reducing proportionately the load on each tooth and increasing tool life. Therefore, thin sections are usually sawed with a blade of 10-pitch or finer, while heavier sections employ a coarser pitch. A pitch as coarse as 3 or 4 is used to cut thick sections, and bands with a pitch less than 1 tooth per 25 mm (1 in.) have been developed for sawing very thick sections.

Aside from the basic relationship between tooth pitch and the thickness of the workpiece to be cut, a tooth that is too small for a given application will cut at a slow rate and will bind and load up. If a tooth is too large for the application, tooth breakage and stripping are likely.

The noise from band sawing can be reduced by using blades with different pitch combinations. These blades have a variable pitch pattern, where the spacing between the teeth varies and repeats itself about every 25 to 50 mm (1 to 2 in.). This variable spacing of the teeth causes interference in the sound patterns thus reducing the amplitude of the resultant noise. Variable tooth spacing also reduces the amplitude of vibrations, which is particularly important when sawing thin workpieces.

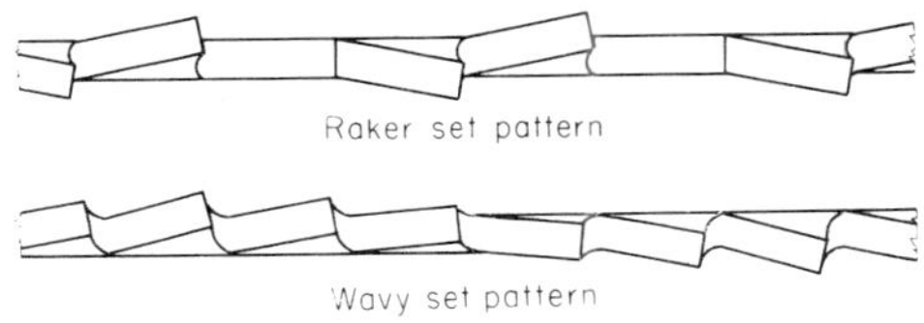


**Fig. 6** Set patterns for saw blades

**Blade Width.** Beam strength increases in proportion to the cube of the blade width, thus permitting the use of higher feed force. In addition, the accuracy of cutting along a straight line is greater for wider blades. Instead of increasing blade width when greater beam strength is needed for difficult straight cuts, the band is sometimes supported by a carbide-faced backup plate of the same thickness.

**The thickness (or gage)** of a saw blade is usually not open to choice; it has been standardized. Therefore, blades that are 13 mm (½ in.) in width or less are generally 0.64 mm (0.025 in.) thick, 16 and 19 mm (⅝ and ¾ in.) widths are generally 0.81 mm (0.032 in.) thick, and a 25 mm (1 in.) width is 0.89 mm (0.035 in.) thick. Blades that are 32 mm (1¼ in.) wide are generally 1.1 mm (0.042 in.) thick, a 38 mm (1½ in.) width is generally 1.3 mm (0.050 in.) thick, and widths from 50 to 120 mm (2 to 4¾ in.) are 1.6 mm (0.063 in.) thick. Beam strength increases linearly with thickness.

In general, a blade of standard gage is adequate for all applications except those involving large workpieces and requiring extreme accuracy. For these applications, a heavier gage is recommended because it will offer increased resistance to side displacement. Similarly, in cutting the more difficult-to-machine alloys, a thicker blade will cut more efficiently up to the full capacity of the machine.

**Set.** The teeth of a saw band are intentionally offset to provide clearance for the back of the band and to permit the cutting of contours. The set dimension is the distance from the extreme corner of one tooth to the extreme corner of the tooth set to the opposite direction. The maneuverability of the

**Table 2 Band saw blade composition**

| Product | Composition, % C | Si | Mn | Cr | V | W | M | Co | Heat resistance, °C (°F) | Hardness, HRC Teeth | Body |
|---|---|---|---|---|---|---|---|---|---|---|---|
| M-2 high-speed welded-edge band saw | 0.79–0.86 | 0.25 | 0.35 | 4.25 | 1.95 | 6.50 | 5.00 | ... | 540 (1000) | 64–66 | 40–47 |
| Matrix band saw | 0.70–0.78 | 0.30 | 0.25 | 4.25 | 1.00 | 1.00 | 5.00 | 8.00 | 590 (1100) | 65–67 | 40–47 |
| Cobalt M-42 band saw | 1.05–1.10 | 0.25 | 0.25 | 3.75 | 1.15 | 1.50 | 9.50 | 8.00 | 700 (1300) | 67–69 | 40–47 |

**Table 3 Recommended band width for the contour sawing of various radii**

| Radius to be cut, mm | Radius to be cut, in. | Band width, mm | Band width, in. |
|---|---|---|---|
| <1.6 | 1/16 | 1.6 | 1/16 |
| 1.6 | 1/16 | 2.4 | 3/32 |
| 3.2 | 1/8 | 3.2 | 1/8 |
| 8 | 5/16 | 4.8 | 3/16 |
| 16 | 5/8 | 6.4 | 1/4 |
| 36.5 | 1 7/16 | 9.5 | 3/8 |
| 65 | 2 1/2 | 13 | 1/2 |
| 95 | 3 3/4 | 16 | 5/8 |
| 140 | 5 7/16 | 19 | 3/4 |
| 180 | 7 1/4 | 25 | 1 |
| 300 | 12 | 32 | 1 1/4 |
| 530 | 21 | 38 | 1 1/2 |
| 710 | 28 | 50 | 2 |

band increases as band width decreases (Table 3) and as the set dimension increases.

Both the raker and wavy set patterns are used for sawing metals (Fig. 6). The raker pattern is developed by a series of three consecutive teeth: one set to the left, one on center and one to the right. This pattern is repeated for each successive group of three teeth. (Raker set is also available in 5/7 pitch combination patterns.) In contrast, the wavy pattern consists of series of teeth that are gradually offset, first to the right and then to the left, to form a wavelike pattern.

Raker set blades are recommended for all sawing applications except those involving workpieces with marked changes in cross section, such as tubing, pipe, and structural shapes, or in thin cross sections. The wavy set performs better than the raker set in thin cross sections because they cut into the work more gradually and uniformly, thus minimizing shock loading of the cutting teeth.

**Special Saw Blades.** In addition to the types of blades already described, three special types—spiral tooth, diamond edge, and aluminum oxide edge—are available. The spiral tooth blade is capable of cutting accurate contours to a minimum radius of 0.25 mm (0.010 in.). Because it has an effective cutting edge of 360°, it is well adapted to cutting intricate patterns in lightgage metal. Diamond-edge and aluminum-oxide-edge blades can be used to cut metals that are extremely tough, such as nickel-base and cobalt-base heat-resistant alloys and steel that has been heat treated to high hardness. Both types of blade generate a great deal of heat, and the use of a cutting fluid is mandatory.

**Table 4 Nominal speed, cutting rate, and band life for the cutoff band sawing of steel bars**

| Steel being cut | Hardness, HB | Band speed(a), m/min | Band speed(a), sfm | Cutting rate(a), mm² × 10³/min | Cutting rate(a), in.²/min | Band life in terms of total area cut(b), mm² × 10⁶ | Band life in terms of total area cut(b), in.² × 10³ |
|---|---|---|---|---|---|---|---|
| **Carbon and low-alloy steels** | | | | | | | |
| 1008–1013 | 150–175 | 100–84 | 325–275 | 9.0–6.4 | 14–10 | 4.3 | 6.7 |
| 1015–1035 | 160–175 | 106–90 | 350–300 | 9.7–7.1 | 15–11 | 4.8 | 7.5 |
| 1036–1064 | 160–180 | 68–58 | 225–190 | 5.8–4.5 | 9–7 | 2.7 | 4.2 |
| 1065–1095 | 180–205 | 52–44 | 170–145 | 5.2–3.9 | 8–6 | 1.9 | 3.0 |
| 1108–1132 | 125–175 | 106–84 | 350–275 | 9.7–7.7 | 15–12 | 5.2 | 8.0 |
| 1137–1151 | 155–180 | 80–68 | 260–225 | 6.4–5.2 | 10–8 | 3.5 | 5.4 |
| 1212–1213 | 150–175 | 106–90 | 350–300 | 9.7–7.7 | 15–12 | 5.5 | 8.5 |
| 1330–1345 | 200–220 | 65–58 | 210–190 | 5.2–3.9 | 8–6 | 2.3 | 3.5 |
| 4023–4047 | 170–220 | 80–70 | 260–230 | 5.2–3.9 | 8–6 | 2.4 | 3.7 |
| 4130–4140 | 190–215 | 75–67 | 250–220 | 5.8–4.5 | 9–7 | 2.1 | 3.3 |
| 4320–4340 | 200–250 | 70–55 | 230–180 | 4.5–3.2 | 7–5 | 1.9 | 3.0 |
| 4815–4820 | 220–240 | 58–53 | 190–175 | 3.9–2.9 | 6–4.5 | 1.6 | 2.5 |
| 5046 | 170–190 | 75–67 | 250–220 | 5.8–4.5 | 9–7 | 2.1 | 3.3 |
| 5140–5160 | 200–220 | 70–60 | 230–200 | 4.2–3.2 | 6.5–5 | 1.6 | 2.5 |
| 50100–52100 | 210–230 | 50–37 | 170–120 | 3.9–2.6 | 6–4 | 1.6 | 2.5 |
| 6118–6150 | 180–220 | 68–45 | 225–150 | 4.8–2.6 | 7.5–4 | 1.7 | 2.6 |
| 8615–8645 | 160–220 | 70–53 | 230–175 | 4.5–3.2 | 7–5 | 2.5 | 3.8 |
| 8720–8740 | 180–215 | 68–53 | 225–175 | 4.5–3.2 | 7–5 | 2.05 | 3.2 |
| 9310 | 210–240 | 53–45 | 175–150 | 2.6–1.9 | 4–3 | 1.3 | 2.0 |
| **Tool steels** | | | | | | | |
| W1 | 155–195 | 67–55 | 220–180 | 3.9–3.2 | 6–5 | 1.9 | 3.0 |
| S2, S5 | 175–230 | 45–33 | 150–110 | 2.6–1.9 | 4–3 | 1.0 | 1.5 |
| O1, O2 | 190–205 | 65–55 | 210–180 | 3.9–2.6 | 6–4 | 1.6 | 2.5 |
| A2 | 215–240 | 60–52 | 200–170 | 2.6–1.9 | 4–3 | 1.5 | 2.3 |
| D2, D3 | 215–240 | 37–27 | 120–90 | 1.9–1.3 | 3–2 | 1.0 | 1.5 |
| D7 | 230–255 | 27–18 | 90–60 | 1.3–0.6 | 2–1 | 0.55 | 0.85 |
| H12, H13, H21 | 205–230 | 58–49 | 190–160 | 3.2–2.6 | 5–4 | 1.3 | 2.0 |
| T1, T2 | 215–250 | 40–30 | 130–100 | 2.3–1.3 | 3.5–2 | 1.1 | 1.7 |
| T6, T8 | 220–295 | 30–21 | 100–70 | 1.6–0.6 | 2.5–1 | 0.7 | 1.2 |
| T15 | 230–255 | 23–15 | 75–50 | 1.3–0.6 | 2–1 | 0.6 | 1.0 |
| M1 | 215–230 | 45–37 | 150–120 | 3.2–1.9 | 5–3 | 1.1 | 1.7 |
| M2, M3 | 215–240 | 33–24 | 110–80 | 2.6–1.3 | 4–2 | 1.0 | 1.5 |
| M4, M10, M15 | 215–230 | 27–18 | 90–60 | 1.6–0.6 | 2.5–1 | 0.7 | 1.2 |
| L6 | 190–230 | 55–49 | 180–160 | 3.9–2.6 | 6–4 | 1.6 | 2.5 |
| **Stainless steels** | | | | | | | |
| 201, 202, 302, 304 | 130–190 | 37–24 | 120–80 | 2.6–1.3 | 4–2 | 1.9 | 3.0 |
| 303, 303F | 150–200 | 40–27 | 130–90 | 3.2–1.3 | 5–2 | 2.1 | 3.3 |
| 308, 309, 310, 330 | 160–220 | 24–18 | 80–60 | 1.3–0.6 | 2–1 | 0.85 | 1.3 |
| 314, 316, 317 | 160–220 | 23–15 | 75–50 | 1.3–0.6 | 2–1 | 0.75 | 1.2 |
| 321, 347 | 165–200 | 37–27 | 120–90 | 2.6–1.3 | 4–2 | 1.5 | 2.4 |
| 410, 420, 420F | 140–185 | 43–30 | 140–100 | 2.6–1.3 | 4–2 | 1.0 | 1.5 |
| 416, 430F | 155–195 | 55–43 | 180–140 | 4.5–3.2 | 7–5 | 1.6 | 2.5 |
| 430, 446 | 170–215 | 27–18 | 90–60 | 2.6–1.9 | 4–3 | 1.1 | 1.7 |
| 440A, B, C | 160–190 | 33–21 | 110–70 | 2.6–1.3 | 4–2 | 1.0 | 1.5 |
| 440F, 443 | 175–215 | 40–30 | 130–100 | 2.6–1.3 | 4–2 | 0.8 | 1.3 |
| 17-7 PH, 17-4 PH | 150–360 | 27–15 | 90–50 | 2.6–1.3 | 4–2 | 1.1 | 1.7 |

(a) Based on the use of a 25 mm (1 in.) wide high-speed steel band, regular tooth form (except hook tooth form for metal thicker than about 250 mm, or 10 in.), raker set, to cut scale-free, solid bar stock up to 460 mm (18 in.) thick; based on the use of a cutting fluid, except for D2, D3, and D7 tool steels, which are cut dry. (b) For 3 m (10 ft) band; proportionate life for other band lengths

## Band Selection

The most widely used welded-edge high-speed steel bands are available in at least three metallurgical grades: M2, Matrix 2, and M42 (see Table 2 for material composition). For the production sawing of a very thin walled steel tubing (<1.5 mm, or 0.060 in.) or a very small bar size, M2 with a tooth tip hardness of 64 HRC is recommended. For sawing carbon and alloy steels, a cutting edge of Matrix 2 with a tooth tip hardness of 67 HRC or variable pitch blades with Matrix 2 edges are recommended. For sawing high-temperature alloys, heat-treated steels, stainless steels (such as type 304, type 316, type 347, and 17-4PH), and superalloys such as A-286, an M42 tooth tip hardness of 69 HRC is advised. When sawing diameters of 100 mm (4 in.) or greater, the M42 blades are most suitable. In addition, the M42 blades should have a positive rake, rather than the standard 90° rake design, for easier and greater penetration. Finally, when sawing superalloys such as Inconel 718, Waspaloy, Astroloy, and 6/4 alloy titanium in large sizes, improved efficiency may be achieved by using carbide-tip band blades. Special welded-edge blades are also available for these applications. They utilize changing width patterns to force the teeth to cut below any work-hardened layers. Welded-edge blades are generally more economical than carbide-tipped blades for these applications.

**Band Width.** To maintain accuracy and a high cutting rate, the widest band capable of cutting the desired radius should be used during contour band sawing (Table 3). Wider bands are also used in cutoff operations because only straight-line cuts are made. Wider bands provide greater beam strength and permit higher loading. Bands 25 to 125 mm (1 to 5 in.) wide are preferred in cutoff operations.

**Table 5 Nominal speed, cutting rate, and band life for the cutoff band sawing of nonferrous alloys**

| Work metal | Hardness, HB | Band speed(a) m/min | Band speed(a) sfm | Cutting rate(a) $mm^2 \times 10^3$/min | Cutting rate(a) $in.^2$/min | Band life in terms of total area cut(b) $mm^2 \times 10^6$ | Band life in terms of total area cut(b) $in.^2 \times 10^3$ |
|---|---|---|---|---|---|---|---|
| **Copper alloys** | | | | | | | |
| 170, beryllium copper | 100–120 | 84–60 | 275–200 | 5.2–3.9 | 8–6 | 2.5 | 3.8 |
| | 220–250 | 68–53 | 225–175 | 3.9–2.6 | 6–4 | 1.7 | 2.7 |
| | 310–340 | 43–27 | 140–90 | 1.9–1.3 | 3–2 | 1.1 | 1.7 |
| 510, phosphor bronze 5% A | 60–100 | 90–75 | 300–250 | 6.4–5.2 | 10–8 | 3.7 | 5.8 |
| | 180–210 | 53–38 | 175–125 | 3.2–1.9 | 5–3 | 1.6 | 2.5 |
| 614, aluminum bronze D | 70–90 | 106–90 | 350–300 | 9.0–6.4 | 14–10 | 4.3 | 6.7 |
| | 190–220 | 53–38 | 175–125 | 3.2–1.9 | 5–3 | 1.6 | 2.5 |
| 656, high-silicon bronze | 70–100 | 100–384 | 325–275 | 9.7–7.7 | 15–12 | 4.8 | 7.5 |
| | 180–210 | 53–38 | 175–125 | 3.9–1.9 | 6–3 | 1.6 | 2.5 |
| 675, manganese bronze A | 95–120 | 100–584 | 325–275 | 9.7–7.7 | 15–12 | 4.8 | 7.5 |
| | 180–190 | 60–45 | 200–150 | 3.9–2.6 | 6–4 | 1.7 | 2.7 |
| **Nickel alloys** | | | | | | | |
| Inconel | 150–200 | 30–18 | 100–60 | 1.9–1.3 | 3–2 | 0.4 | 0.65 |
| Inconel X-750 | 200–300 | 24–18 | 80–60 | 1.0–0.3 | 1.5–0.5 | 0.25 | 0.4 |
| Monel 400 | 125–200 | 30–18 | 100–60 | 1.9–0.6 | 3–1 | 0.55 | 0.85 |
| Monel R-405 | 145–180 | 45–23 | 150–75 | 2.6–1.3 | 4–2 | 0.6 | 1.0 |
| Monel K-500 | 160–210 | 24–18 | 80–60 | 1.3–0.3 | 2–0.5 | 0.3 | 0.5 |
| Monel 501 | 160–210 | 30–18 | 100–60 | 1.9–0.6 | 3–1 | 0.5 | 0.85 |
| Hastelloy A | 210–260 | 37–23 | 120–75 | 1.9–1.0 | 3–1.5 | 0.6 | 1.0 |
| Hastelloy B | 230–270 | 30–23 | 100–75 | 1.6–0.6 | 2.5–1 | 0.55 | 0.85 |
| Hastelloy C | 185–250 | 27–18 | 90–60 | 1.0–0.45 | 1.5–0.7 | 0.38 | 0.6 |
| **Titanium alloys** | | | | | | | |
| Ti; Ti–1.5 Fe–2.5 Cr | 270–350 | 27–18 | 90–60 | 0.6–0.2 | 1–0.3 | 0.25 | 0.4 |
| Ti–4 Al–4 Mn; Ti–6 Al–4 V | 290–360 | 33–21 | 110–70 | 1.3–3.9 | 2–6 | 1.5 | 2.4 |
| Ti–2 Fe–2 Cr–2 Mo | 300–330 | 27–18 | 90–60 | 1.0–0.3 | 1.5–0.5 | 0.48 | 0.75 |

(a) Based on the use of a 25 mm (1 in.) wide high-speed steel band, regular tooth form (except hook tooth form for metal thicker than about 250 mm, or 10 in.), raker set, to cut scale-free, solid bar stock up to 460 mm ( 8 in.) thick; based on the use of a suitable cutting fluid. (b) For 3 m (10 ft) long band; proportionate life for other band lengths

**Noise Reduction.** When the pitch of the blade is varied (or modulated), a smoother, quieter operation results. At recommended speeds, the noise from a pitch-modulated blade seldom exceeds 80 dB.

## Machining Variables

Speed, feed, and thickness of work are variables in cutoff and contour band sawing operations. Metal composition, hardness, structural homogeneity, and work-hardening potential are also important variables.

**Cutoff Operations.** Nominal speed, cutting rate, and band life for the cutoff band sawing of various metals are given in Tables 4 and 5. Tables 4 and 5 are based on the cutting of scale-free rounds, 75 to 125 mm (3 to 5 in.) in diameter, with 25 mm (1 in.) wide high-speed steel bands, and are applicable to solid bar stock up to 450 mm (18 in.) thick. Substantial amounts of scale on the work will necessitate the use of lower band speed and cutting rate and will shorten band life, particularly in cutting thin material.

**Cutting of Hollow Shapes.** Optimum cutting rates are obtained in sawing solid materials because many teeth are uniformly loaded at all times. In sawing pipe, tubing, and structural sections, only a fraction of the total cross section through which the saw band must pass is metal, and the cutting rate must therefore be lower to keep the feed force per tooth at an acceptable level.

The following factors, which depend on the minimum wall thickness to be cut, should be applied to the cutting rates of Tables 4 and 5 to estimate rates for sawing pipe, tubing, and structural shapes:

| Minimum wall thickness mm | Minimum wall thickness in. | Factor |
|---|---|---|
| <4.8 | <3/16 | 0.4 |
| 4.8–9.5 | 3/16–3/8 | 0.5 |
| 9.5–16 | 3/8–5/8 | 0.6 |
| >16 | >5/8 | 0.7 |

The pitch of the saw band should be selected on the basis of the minimum wall thickness to be sawed.

**Contour Operations.** Tables 6 through 11 list nominal speeds for the contour band sawing of various metals in different ranges of thickness. These data are intended to serve as starting points for the selection of optimum sawing conditions for specific applications.

Feed rate is given in the footnote of Tables 6 to 11 only in a generalized way, that is, as a minimum, average, or maximum feed force. Feed rate is usually controlled by adjusting feed force (9 to 35 N/tooth, or 2 to 8 lbf/tooth) to obtain the proper cutting action, as gaged by the formation of a clean, tightly curled chip or by obtaining a cutting rate determined by experience to be optimum for the work material, stock thickness, equipment, and saw band. Linear cutting rates are summarized in Table 12.

**Work Metal Composition and Hardness.** The general effects of work metal composition and hardness are reflected in Tables 4 to 11. Tables 4 to 11 apply directly only to the hardness ranges shown, which represent the conditions in which the metals listed are most frequently sawed (as rolled or annealed). Speed and feed should be adjusted when metals in other ranges of hardness are sawed. Lower band speed (and feed force) is usually required for the higher hardnesses of a given work metal (see data for carbon and low-allow steels in Table 13). There are, however, exceptions in which the reverse is true; for example, higher speeds are required for sawing free-

**Table 6 Speeds for the contour band sawing of carbon and low-alloy steels**

Data are based on the use of a suitable cutting fluid.

| Steel being cut | Hardness, HB | Speed, m/min (sfm), for stock thickness of: Carbon steel bands 6.4–13 mm (1/4–1/2 in.)(b) | Carbon steel bands 25–75 mm (1–3 in.)(c) | Carbon steel bands 150–300 mm (6–12 in.)(d) | High-speed steel bands(a) 6.4–13 mm (1/4–1/2 in.)(e) | High-speed steel bands(a) 25–75 mm (1–3 in.)(c) | High-speed steel bands(a) 150–300 mm (6–12 in.)(d) |
|---|---|---|---|---|---|---|---|
| 1008–1013 | 150–175 | 60 (200) | 45 (150) | 38 (125) | 106 (350) | 81 (265) | 53 (175) |
| 1015–1035 | 160–175 | 68 (225) | 53 (175) | 38 (125) | 106 (350) | 85 (280) | 60 (200) |
| 1036–1064 | 160–180 | 45 (150) | 30 (100) | 23 (75) | 76 (250) | 55 (180) | 37 (120) |
| 1065–1095 | 180–205 | 45 (150) | 30 (100) | 20 (65) | 68 (225) | 49 (160) | 30 (100) |
| 1108–1132 | 125–175 | 80 (260) | 65 (210) | 41 (135) | 130 (425) | 94 (310) | 60 (200) |
| 1137–1151 | 155–180 | 68 (225) | 53 (175) | 33 (110) | 98 (320) | 70 (230) | 45 (150) |
| 1212–1213 | 150–175 | 84 (275) | 68 (225) | 44 (145) | 137 (450) | 98 (320) | 60 (200) |
| 1330–1345 | 200–220 | 45 (150) | 30 (100) | 21 (70) | 75 (245) | 55 (180) | 38 (125) |
| 4023–4047 | 170–220 | 45 (150) | 38 (125) | 27 (90) | 105 (345) | 76 (250) | 45 (150) |
| 4130–4140 | 190–215 | 38 (125) | 23 (75) | 15 (50) | 98 (320) | 67 (220) | 38 (125) |
| 4320–4340 | 200–250 | 37 (120) | 20 (65) | 15 (50) | 84 (275) | 58 (190) | 33 (110) |
| 4815–4820 | 220–240 | 38 (125) | 23 (75) | 15 (50) | 73 (240) | 52 (170) | 30 (100) |
| 5046 | 170–190 | 38 (125) | 23 (75) | 15 (50) | 91 (300) | 70 (230) | 45 (150) |
| 5140–5160 | 200–220 | 37 (120) | 24 (80) | 15 (50) | 85 (280) | 60 (200) | 38 (125) |
| 50100–52100 | 210–230 | 45 (150) | 30 (100) | 15 (50) | 64 (210) | 41 (135) | 21 (70) |
| 6118–6150 | 180–220 | 38 (125) | 23 (75) | 15 (50) | 79 (260) | 53 (175) | 29 (95) |
| 8615–8645 | 160–220 | 33 (110) | 20 (65) | 15 (50) | 87 (285) | 52 (170) | 32 (105) |
| 8720–8740 | 180–215 | 33 (110) | 20 (65) | 15 (50) | 85 (280) | 49 (160) | 30 (100) |
| 9310 | 210–240 | 30 (100) | 15 (50) | 15 (50) | 68 (225) | 46 (150) | 27 (90) |

(a) Data are also suitable for welded-edge high-speed steel bands. (b) Regular tooth form; 14-pitch; minimum feed force. (c) Regular tooth form; 6- to 8-pitch; average feed force. (d) Hook tooth form; 3-pitch; maximum feed force. (e) Regular tooth form; 10-pitch; minimum feed force. Source: Data are adapted from tables in "Fundamentals of Band Machining," Wilkie Brothers Foundation.

## Table 7 Speeds for the contour band sawing of tool steels

Based on the use of a suitable cutting fluid, except for D2, D3, and D7, which are sawed dry

| | | Speed, m/min (sfm), for stock thickness of: | | | | | |
|---|---|---|---|---|---|---|---|
| | | Carbon steel bands | | | High-speed steel bands(a) | | |
| Steel being cut | Hardness, HB | 6.4–13 mm (¼–½ in.)(b) | 25–75 mm (1–3 in.)(c) | 150–300 mm (6–12 in.)(d) | 6.4–13 mm (¼–½ in.)(e) | 25–75 mm (1–3 in.)(f) | 150–300 mm (6–12 in.)(d) |
| W1 | 155–195 | 45 (150) | 30 (100) | 15 (50) | 82 (270) | 55 (180) | 30 (100) |
| S2, S5 | 175–230 | 24 (80) | 15 (50) | 15 (50) | 58 (190) | 37 (120) | 18 (60) |
| O1, O2 | 185–205 | 45 (150) | 30 (100) | 15 (50) | 80 (260) | 55 (180) | 30 (100) |
| A2 | 215–240 | 50 (170) | 30 (100) | 20 (65) | 76 (250) | 53 (175) | 30 (100) |
| D2(g), D3(g) | 215–240 | 30 (100) | 18 (60) | 15 (50) | 45 (150) | 30 (100) | 15 (50) |
| D7(g) | 230–255 | 24 (80) | 15 (50) | 15 (50) | 33 (110) | 21 (70) | 15 (50) |
| H12, H13, H21 | 205–230 | 30 (100) | 18 (60) | 15 (50) | 73 (240) | 49 (160) | 27 (90) |
| T6(g), T8(g) | 220–295 | 27 (90) | 17 (55) | 15 (50) | 38 (125) | 24 (80) | 15 (50) |
| M4(g), M10(g), M15(g) | 215–240 | 27 (90) | 15 (50) | 15 (50) | 33 (110) | 21 (70) | 15 (50) |
| L6 | 190–230 | 30 (100) | 20 (65) | 15 (50) | 70 (230) | 49 (160) | 27 (90) |

(a) Data are also suitable for welded-edge high-speed steel bands. (b) Regular tooth form; 14-pitch; minimum feed force, except average feed force for steels that are footnoted. (c) Regular tooth form; 6- to 8-pitch; average feed force, except maximum feed force for steels that are footnoted. (d) Hook tooth form; 3-pitch; maximum feed force. (e) Regular tooth form; 10-pitch; minimum feed force, except average feed force for steels that are footnoted. (f) Regular tooth form; 6-pitch; average feed force, except maximum feed force for steels that are footnoted. (g) Operating conditions for these tool steels differ slightly from those for the other tool steels; differences are indicated in footnotes (b), (c), (e), and (f). Source: Data are adapted from tables in "Fundamentals of Band Machining," Wilkie Brothers Foundation.

## Table 8 Speeds for the contour band sawing of cast iron (125–250 HB) with high-speed steel and carbon steel saw bands

Data are based on dry sawing.

| Work metal (ASTM grade) | Speed, m/min (sfm), for stock thickness of: 6.4–13 mm (¼–½ in.)(a) | 25–75 mm (1–3 in.)(b) | 150–300 mm (6–12 in.)(c) |
|---|---|---|---|
| **High-speed steel bands(d)** | | | |
| Gray iron | | | |
| Class 30 | 76 (250) | 56 (185) | 33 (110) |
| Class 35; Class 40 | 60 (200) | 44 (145) | 50 (165) |
| Nodular iron | | | |
| 60-45-10; 65-45-12 | 120 (400) | 90 (300) | 60 (200) |
| 80-60-03; 80-55-06 | 73 (240) | 55 (180) | 33 (110) |
| 100-70-03 | 60 (200) | 44 (145) | 49 (160) |
| Malleable iron | | | |
| 32150; 35018 | 106 (350) | 75 (250) | 49 (160) |
| 53004; 60003 | 70 (230) | 49 (160) | 49 (160) |
| **Carbon steel bands(e)** | | | |
| Nodular iron | | | |
| 60-45-10; 65-45-12 | 76 (250) | 60 (200) | NR |
| Malleable iron | | | |
| 32510; 35018 | 64 (210) | NR | NR |

(a) Regular tooth form; 10-pitch for high-speed steel bands, 14-pitch for carbon steel bands; minimum feed force. (b) Regular tooth form; 6-pitch for high-speed steel bands, 8-pitch for carbon steel; average feed force. (c) Hook tooth form for class 30 gray iron, 60-45-10 and 80-60-03 nodular, and 32510 malleable, and carbide tooth form for the other cast irons; 2.5- to 3-pitch; maximum feed force. (d) Data are also suitable for welded-edge high-speed steel bands. (e) Use of carbon steel bands is recommended only for the cast irons and conditions shown; NR, not recommended. Source: Data are adapted from tables in "Fundamentals of Band Machining," Wilkie Brothers Foundation.

## Table 9 Speeds for the contour band sawing of stainless steels with M42 welded-edge high-speed steel saw bands

Data are based on the use of a suitable cutting fluid.

| Steel being cut | Hardness, HB | Speed, m/min (sfm), for stock thickness of: 6.4–13 mm (¼–½ in.)(a) | 25–75 mm (1–3 in.)(b) | 150–300 mm (6–12 in.)(c) |
|---|---|---|---|---|
| 201, 202, 302, 304 | 130–190 | 45 (150) | 30 (100) | 21 (70) |
| 303, 303F | 150–200 | 49 (160) | 40 (130) | 30 (100) |
| 308(d), 309(d), 310(d), 330(d) | 160–220 | 33 (110) | 24 (80) | 15 (50) |
| 314(d), 316(d), 317(d) | 160–220 | 29 (95) | 18 (60) | 12 (40) |
| 321, 347 | 165–200 | 45 (150) | 30 (100) | 21 (70) |
| 410, 420, 420F | 140–185 | 53 (175) | 33 (110) | 26 (85) |
| 416, 430F | 155–195 | 70 (230) | 49 (160) | 24 (80) |
| 430, 446 | 170–215 | 43 (140) | 27 (90) | 21 (70) |
| 440 A(d), B(d), C(d) | 160–190 | 33 (110) | 27 (90) | 15 (50) |
| 440F, 443 | 175–215 | 43 (140) | 35 (115) | 18 (60) |
| 17-7 PH(d), 17-4 PH(d) | 150–360 | 33 (110) | 21 (70) | 15 (50) |

(a) Regular tooth form; 10-pitch; minimum feed force, except average force for steels that are footnoted. (b) Regular tooth form; 6-pitch; average feed force, except maximum force for steels that are footnoted. (c) Hook tooth form; 3-pitch; maximum feed force. (d) Operating conditions for these stainless steels are slightly different from those for the others, as defined in footnotes(a) and (b). Source: Data are adapted from tables in "Fundamentals of Band Machining," Wilkie Brothers Foundation.

cutting steels similar to 1112 to 1117 in the hardness range of 150 to 200 HB than in the range of 100 to 150 HB, as indicated in Table 13.

**Effect of Work Thickness.** The pitch of the blade must be greater for thin material. The recommended tooth form may be different when thickness exceeds about 75 mm (3 in.). For example, in contour sawing 1015 steel with a high-speed steel band, a section thickness of 25 mm (1 in.) indicates the use of a 6-pitch band with regular tooth form and a band speed of 85 m/min (280 sfm). However, a 180 mm (7 in.) section calls for a 3-pitch, hook tooth band and a band speed of 60 m/min (200 sfm). In cutting 316 stainless steel of the same thickness, blade requirements remain unchanged, but cutting speed is markedly reduced. When sawing irregular shapes, blade selection and band speed should be based on the thinnest section to be cut in traversing the workpiece.

**Effect of Stacking.** When sheet materials are stacked for contour band sawing, selection of the saw band should not be based on the total thickness to be cut, but more nearly on the thickness of an individual sheet. Therefore, if a 10-pitch blade would normally be required to cut a single sheet, this would be slightly modified when cutting a stack of sheets, and an 8-pitch blade would be used. Band speed should be reduced to that called for by the total thickness of the stack.

## Cutting Fluids

Cutting fluids are used to provide lubrication and to prevent chip weldment and overheating of the saw band and the workpiece. Sometimes the use of fluids can be harmful in sawing certain work-hardening alloys if the fluid interferes with the cutting action and creates a rubbing or sliding action instead.

**Types of Cutting Fluid.** Three general types of cutting fluid are used in contour band sawing: mineral oils that contain no water, emulsions of soluble oil and water, and chemical solutions. For some recommendations on the cutting fluids to be used in the sawing of various metals, see Table 2 in the article "Metal Cutting and Grinding Fluids" in this Volume.

*Straight mineral oils* provide maximum lubricity and are particularly useful on difficult-to-machine alloys that are cut at low band speeds (≤53 m/min, or 175 sfm), where lubricity is more important than cooling power. These oils cannot be used at high cutting rates, because they overheat and smoke. They also have the disadvantage of leaving an oily film on the workpiece, which requires thorough cleaning for removal. Copper, brass, and bronze will tarnish if exposed to straight mineral oil containing sulfur additives.

*Soluble-oil emulsions* are suited to both the free-machining and difficult-to-machine

**Table 10 Speeds for the contour band sawing of heat-resistant alloys with welded-edge high-speed steel saw bands**

Data are based on the use of a suitable cutting fluid.

| | | Speed, m/min (sfm), for stock thickness of: | | |
|---|---|---|---|---|
| Work metal | Hardness, HB | 6.4–13 mm (¼–½ in.)(a) | 25–75 mm (1–3 in.)(b) | 150–300 mm (6–12 in.)(c) |
| A-286 | ... | 23 (75) | 15 (50) | 15 (50) |
| Discaloy | ... | 24 (80) | 20 (65) | 20 (65) |
| Hastelloy A | 210–260 | 24 (80) | 17 (55) | 15 (50) |
| Hastelloy B | 230–270 | 27 (90) | 18 (60) | 17 (55) |
| Hastelloy C | 185–250 | 24 (80) | 17 (55) | 15 (50) |
| Inconel | 150–200 | 30 (100) | 20 (65) | 15 (50) |
| Inconel 700 | ... | 24 (80) | 17 (55) | 15 (50) |
| Inconel X-750 | ... | 18 (60) | 15 (50) | 15 (50) |
| Waspaloy | ... | 24 (80) | 17 (55) | 15 (50) |
| U-500 | ... | 27 (90) | 17 (55) | 15 (50) |
| René 41 | ... | 24 (80) | 17 (55) | 15 (50) |
| Refractaloy 26 | ... | 17 (55) | 11 (35) | 14 (45) |

(a) Regular tooth form; 8-pitch; average feed force. (b) Regular tooth form; 4- to 6-pitch; maximum feed force. (c) Carbide tooth form: 2.5-pitch; maximum feed force. Source: Data are adapted from tables in "Fundamentals of Band Machining," Wilkie Brothers Foundation.

**Table 11 Speeds for the contour band sawing of nonferrous alloys**

Based on the use of a suitable cutting fluid, except for copper alloys 314, 360, and 544, which are sawed dry; based on sawing with carbon steel bands, except for tooth forms marked C, where carbide is used, and except for footnoted speed entries, where high-speed steel bands are used

| | 6.4–13 mm (¼–½ in.) thick | | | | 25–75 mm (1–3 in.) thick | | | | 150–300 mm (6–12 in.) thick | | | |
|---|---|---|---|---|---|---|---|---|---|---|---|---|
| Work metal | Tooth form(a) | Pitch | Speed(b) m/min | Speed(b) sfm | Tooth form(a) | Pitch | Speed(c) m/min | Speed(c) sfm | Tooth form(a) | Pitch | Speed(d) m/min | Speed(d) sfm |
| **Aluminum alloys** | | | | | | | | | | | | |
| 1100, 2011, 2017, 3003, 3004 | R | 10 | 2130 | 7000 | R | 6 | 1520 | 5000 | H | 3 | 1220 | 4000 |
| 2014, 2018, 2025, 6053, 7075 | R | 14 | 850 | 2800 | R | 2 | 730 | 2400 | H | 3 | 610 | 2000 |
| 2024, 5052 | R | 14 | 2590 | 8500 | R | 6 | 1980 | 6500 | H | 3 | 1525 | 5000 |
| 4032, 6151 | R | 14 | 460 | 1500 | R | 6 | 240 | 800 | H | 3 | 90 | 300 |
| 6061, 6063 | R | 14 | 1520 | 5000 | R | 6 | 1070 | 3500 | H | 3 | 760 | 2500 |
| 13, 43, 85 | R | 14 | 460 | 1500 | R | 6 | 240 | 800 | H | 3 | 90 | 300 |
| 108, A108, A132 | R | 14 | 300 | 1000 | R | 6 | 180 | 600 | H | 3 | 60 | 200 |
| 113, 152, B214, 312, 333 | R | 14 | 430 | 1400 | R | 6 | 275 | 900 | H | 3 | 90 | 300 |
| 122, 214, 218, 220 | R | 14 | 1520 | 5000 | R | 6 | 1220 | 4000 | H | 3 | 910 | 3000 |
| 138, B195, 212, 355, 356 | R | 14 | 430 | 1400 | R | 6 | 275 | 900 | H | 3 | 90 | 300 |
| 142, 195, 750 | R | 14 | 850 | 2800 | R | 6 | 730 | 2400 | H | 3 | 610 | 2000 |
| 360, 380 | R | 14 | 430 | 1400 | R | 6 | 275 | 900 | H | 3 | 90 | 300 |
| **Copper alloys** | | | | | | | | | | | | |
| 102, oxygen-free copper | R | 10 | 90 | 300(e) | R | 6 | 60 | 200(e) | S | 3 | 30 | 100(e) |
| 170, beryllium copper | R | 10 | 84 | 275 | R | 6 | 99 | 325(e) | S | 3 | 60 | 200(e) |
| 220, commercial bronze, 90% | R | 14 | 120 | 400 | R | 6 | 120 | 400(e) | S | 3 | 60 | 200(e) |
| 240, low brass, 80% | R | 10 | 150 | 500(e) | R | 6 | 110 | 360(e) | S | 3 | 60 | 200(e) |
| 314, leaded commercial bronze | R | 14 | 760 | 2500 | S | 3 | 460 | 1500 | H | 3 | 150 | 500 |
| 360, free-cutting brass | R | 14 | 1040 | 3400 | R | 6 | 610 | 2000 | S | 3 | 305 | 1000 |
| 544, phosphor bronze B-2 | R | 14 | 610 | 2000 | R | 6 | 370 | 1200 | S | 3 | 120 | 400 |
| 694, silicon red brass | R | 14 | 180 | 600 | R | 6 | 99 | 325(e) | C | 2.5 | 76 | 250 |
| 757, nickel silver 65-12 | R | 10 | 150 | 500(e) | R | 6 | 110 | 360(e) | S | 3 | 60 | 200(e) |
| 614, aluminum bronze D | R | 14 | 60 | 200 | R | 6 | 68 | 225(e) | C | 2.5 | 41 | 135 |
| **Other nonferrous metals and alloys** | | | | | | | | | | | | |
| Magnesium | R | 8 | 1370 | 4500 | H | 4 | 1070 | 3500 | H | 3 | 915 | 3000 |
| AM100A, AZ63A, AZ80A, AZ91A, AZ92A, M1A | R | 10 | 1070 | 3500 | H | 3 | 760 | 2500 | H | 3 | 610 | 2000 |
| AZ31B, AZ61A | R | 6 | 1520 | 5000 | H | 3 | 1220 | 4000 | H | 3 | 610 | 2000 |
| Monel 400, Monel K-500 | R | 8 | 20 | 65(e) | R | 4 | 15 | 50(e) | C | 2.5 | 15 | 50 |
| Monel 501, Monel R-405 | R | 8 | 32 | 105(e) | R | 4 | 20 | 65(e) | C | 2.5 | 17 | 55 |
| Titanium, 99% | R | 10 | 37 | 120(e) | R | 6 | 27 | 90(e) | C | 2.5 | 30 | 100 |
| Ti–6 Al–4 V | R | 10 | 24 | 80(e) | R | 6 | 20 | 65(e) | C | 2.5 | 15 | 50 |
| Ti–4 Al–4 Mn; Ti–2 Fe–2 Cr–2 Mo | R | 10 | 33 | 110(e) | R | 6 | 23 | 75(e) | C | 2.5 | 21 | 70 |
| Zircaloy-2; Zirconium (2.5 Hf) | R | 10 | 30 | 100(e) | R | 4 | 27 | 90(e) | C | 2.5 | 30 | 100 |

(a) Tooth form codes: R, regular; S, skip; H, hook; C, carbide tooth form. (b) Minimum feed force. (c) Average feed force, except maximum feed force for footnoted speed entries under "Other Nonferrous Metals and Alloys." (d) Maximum feed force. (e) Band material for this speed and metal is high-speed steel. Source: Data are adapted from tables in "Fundamentals of Band Machining," Wilkie Brothers Foundation.

materials and cover the operating speed range from 45 to 90 m/min (150 to 300 sfm). The ratio of soluble oil to water varies with specific job requirements. Therefore, a concentrated mixture (1:3 to 1:5) is selected for maximum lubricity, while a dilute mixture (1:10 to 1:15) is selected for maximum cooling capacity.

*Chemical solutions*, usually solutions containing wetting agents, are formulated to provide excellent cooling qualities and are employed at speeds above 75 m/min (250 sfm).

## Applications

The examples presented in this section compare contour band sawing with milling and shaping in various applications.

**Example 2: Saving in Material by Contour Band Sawing Instead of Milling.** Figure 7 shows layouts for two methods of producing 100 kg (220 lb) parts from low-alloy (chromium-molybdenum) steel—by contour band sawing from billets (Fig. 7a) and milling from individual blocks (Fig. 7b). Because contour band sawing permitted nesting of the parts, 20 kg (44 lb) less steel was required for each part than in milling.

Sawing was done on a radial-arm band saw, which permitted following the contour by guiding the saw band at a fixed feed without moving the billet. Cutting rate was 1600 mm$^2$/min (2.5 in.$^2$/min), and the area of the cut was 155 × 10$^3$ mm$^2$ (240 in.$^2$). Total time for sawing each piece was 96 min.

**Example 3: Cutting Relief Grooves.** The machining of two grinding reliefs in the hot-rolled 1035 steel slide base shown in Fig. 8(a) was originally done on a milling machine. A change to a contour band saw resulted in a reduction in setup time from 30 to 10 min and an increase of 100% in production rate. The 1.6 mm (1⁄16 in.) reliefs were sawed with a 19 mm (¾ in.) wide, 6-pitch, high-speed steel saw band at a speed of 60 m/min (200 sfm).

**Example 4: Cutting of Slots.** The overall cost of producing the 3.2 mm (⅛ in.) slot in the gray iron finger holder shown in Fig. 8(b) in lots of 200 to 300 pieces was substantially reduced by changing from a milling machine to a contour band saw. The milling operation required special fixtures and a large-diameter slitting saw. Setup time was 45 min, and machining time was 6.8 min per piece. No special fixtures were required on the contour band saw; setup time was 20 min, and a standard high-speed steel band sawed the slot in 2.5 min, a reduction of 64% in machining time. The slot area was sawed with a 6-pitch 19 mm (¾ in.) wide saw band at a speed of 84 m/min (275 sfm). No cutting fluid was used.

**Example 5: Contour Band Sawing Versus Shaping of 1020 Steel.** Lever brackets of 1020 steel (20 HRC), one size of which is shown in Fig. 8(c), were machined

### Table 12 Linear cutting rates for contour band sawing

| Work metal | Cutting rate, mm/min (in./min), for stock thickness of: Carbon steel bands, 6.4 mm (¼ in.) | 13 mm (½ in.) | 25 mm (1 in.) | 38 mm (1½ in.) | 50 mm (2 in.) | 75 mm (3 in.) | 100 mm (4 in.) | 150 mm (6 in.) | 200 mm (8 in.) | High-speed steel bands, 6.4 mm (¼ in.) | 13 mm (½ in.) | 25 mm (1 in.) | 38 mm (1½ in.) | 75 mm (3 in.) | 150 mm (6 in.) | Carbide, 75 mm (3 in.) | 150 mm (6 in.) | 300 mm (12 in.) |
|---|---|---|---|---|---|---|---|---|---|---|---|---|---|---|---|---|---|---|
| Carbon and low-alloy steels (low carbon) | 229 (9.00) | 101 (4.00) | 44 (1.75) | 41 (1.62) | 22 (0.87) | 13 (0.50) | 11 (0.44) | 7 (0.29) | 6 (0.22) | 533 (21.00) | 274 (10.80) | 203 (8.00) | 152 (6.00) | 76 (3.00) | 38 (1.50) | 114 (4.50) | 64 (2.50) | 38 (1.50) |
| Alloy steel (high-carbon) | 57 (2.25) | 25 (1.00) | 13 (0.50) | 9 (0.37) | 6 (0.25) | 4.5 (0.17) | 3 (0.12) | 2 (0.08) | 1.5 (0.06) | 183 (7.20) | 120 (4.72) | 67 (2.62) | 41 (1.62) | 17 (0.66) | 7.5 (0.30) | 19 (0.75) | 13 (0.50) | 7.5 (0.30) |
| Gray iron | 406 (16.00) | 190 (7.50) | 83 (3.25) | 54 (2.12) | 41 (1.62) | 25 (1.00) | 20.6 (0.81) | 13 (0.50) | 10 (0.40) | 813 (32.00) | 406 (16.00) | 254 (10.00) | 203 (8.00) | 102 (4.00) | 51 (2.00) | 140 (5.50) | 79 (3.10) | 51 (2.00) |
| Tool steel | 114 (4.50) | 54 (2.12) | 25 (1.00) | 16 (0.62) | 13 (0.50) | 7.8 (0.31) | 6.4 (0.25) | 4.1 (0.16) | 3 (0.12) | 298 (11.75) | 154 (6.06) | 79 (3.12) | 49 (1.92) | 21 (0.82) | 10 (0.39) | 30 (1.20) | 20 (0.80) | 13 (0.50) |
| Titanium | ... ... | ... ... | ... ... | ... ... | ... ... | ... ... | ... ... | ... ... | ... ... | 58 (2.30) | 38 (1.50) | 15 (0.60) | 7.5 (0.30) | 5 (0.20) | 3.8 (0.15) | 9 (0.35) | 6.4 (0.25) | 3.8 (0.15) |

### Table 13 Effect of hardness of steel cut on band speed for contour band sawing

For work up to 75 mm (3 in.) thick, with M2 bands

| Typical steel(a) | Hardness, HB | Speed m/min | Speed sfm |
|---|---|---|---|
| **Carbon and low-alloy steels (except free-cutting steels)** | | | |
| 1020, 1045, 4140, 7140, and 8620 | 85–125 | 85 | 280 |
| | 125–175 | 82 | 270 |
| | 175–225 | 70 | 230 |
| | 225–275 | 49 | 160 |
| | 275–325 | 40 | 130 |
| | 325–375 | 30 | 100 |
| **Free-cutting steels** | | | |
| 1112 and 1117 | 100–150 | 87 | 285 |
| | 150–200 | 90 | 295 |
| 1137, 12L14, 4140+S, and 41L40 | 100–150 | 98 | 320 |
| | 150–200 | 85 | 280 |
| | 200–250 | 69 | 225 |
| | 275–325 | 47 | 155 |
| | 325–375 | 30 | 100 |

(a) Each steel represents a group of similar steels. Source: Metcut Research Associates Inc.

on a shaper in quantities of three to five in a series of sizes. Setup time for special tooling and holding fixtures was 30 min; production rate was 180 min per piece.

Substantial savings resulted from switching the operation to a contour band saw; setup time was essentially zero, operator costs were lower, and machining time was reduced from 180 min per piece to 58 min per piece. The piece was sawed at a rate of 1100 mm²/min (1.7 in.²/min), using a 6-pitch, 19 mm (¾ in.) wide high-speed steel saw band at a speed of 76 m/min (250 sfm). Total area of the cut surface was $64 \times 10^3$ mm² (99 in.²).

**Example 6: Contour Band Sawing Versus Milling of 1020 Steel Nuts.** A productivity increase of 300%, in addition to a reduction in setup time from 45 to 10 min, was realized in changing from a milling operation to contour band sawing in making large quantities of the hot-rolled 1020 steel split nut shown in Fig. 8(d). The cuts of 240 mm² (⅜ in.²) total area were made at a band speed of 84 m/min (275 sfm), using an 8-pitch, 19 mm (¾ in.) wide, regular tooth high-speed steel band.

## Safety

Several important safety features have been incorporated into the design of standard band saws, particularly the more recent models. Older machines may or may not be similarly equipped, but they can usually be modified to include many of the latest safety features.

**Machine Safety Features.** Because the saw band is frequently required to travel at high speed, a most important safety feature is an automatic wheel brake that instantaneously stops the drive wheel when a saw band breaks, thus minimizing damage that might be caused by the broken band. Another important device is the safety interlock; this should be installed on all doors, hatches, and drawers that permit access to compartments containing moving parts, such as belts, wheels, and gears. All corner surfaces of the machine—inside or outside—should be rounded to avoid snagging. Safety guards should always be in place and should be kept operable at all times. Limit switches and automatic devices should also be kept in good operating condition and should be inspected periodically.

**Dust and Fire Hazards.** The sawing of beryllium and magnesium is potentially hazardous. The dust and fumes generated by the sawing of beryllium are extremely toxic. To guard against this hazard, the operator should wear respiratory, protective equipment, and the band saw should be equipped with an effective exhaust system.

The sawing of magnesium presents a fire hazard. Therefore, band saws that are to be used for cutting magnesium should be identified by a bright color and should not be used for sawing other materials, because of the possibility of spark generation. Band saws used to cut magnesium should be

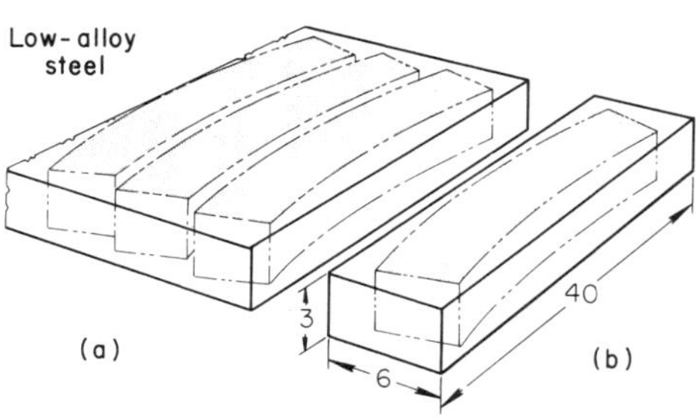


**Fig. 7** Layouts for producing 100 kg (220 lb) parts (a) by contour band sawing several from a single billet and (b) by milling each piece from an individual block. Dimensions given in inches

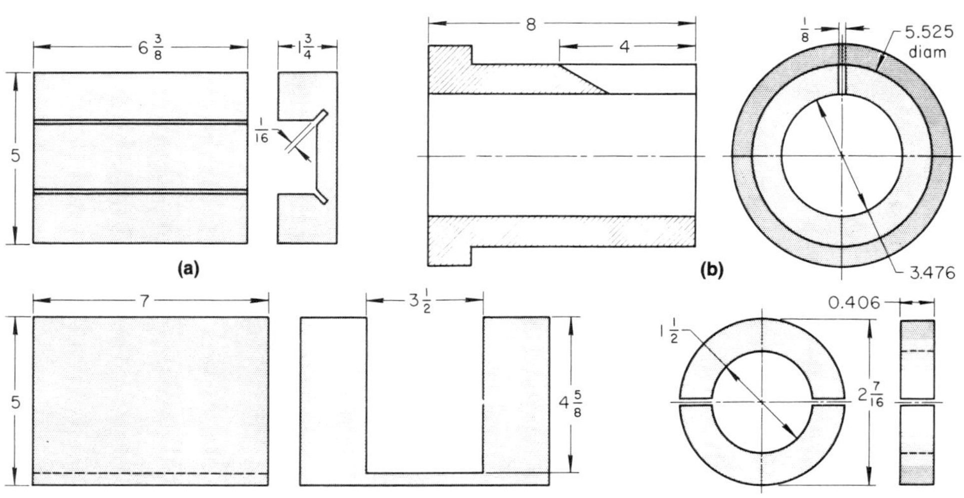


**Fig. 8** Production parts that were produced more economically by contour band sawing than by other machining processes. (a) Example 3. (b) Example 4. (c) Example 5. (d) Example 6. Dimensions given in inches

cleaned before starting and after completing the operation. Pumps and cutting fluid lines should be inspected before and during operation. When in operation, the machine should not be left unattended. The area in which the band saw is located should be equipped with fire-control implements intended for magnesium fires, such as a dispenser for graphite-base powder.

### Friction Band Sawing

In friction band sawing, the work metal is softened (or melted) just ahead of the saw band by frictional heat from dry cutting at high band speeds (1800 to 4600 m/min, or 6000 to 15 000 sfm). The saw teeth remove the softened metal. These band speeds are about 50 to 100 times those for conventional band sawing of the same metals with a cutting fluid. The saw band is not overheated, because only a small part of the rapidly moving band is in the work, and heat is readily dissipated from the rest.

**Applicability.** Friction band sawing is primarily applied to ferrous metals that are harder than 42 HRC or that work harden rapidly. It is also used for the distortion-free cutting of thin material and can produce complex contours as well as straight cuts; in these applications, however, abrasive waterjet machining may also be recommended as a competitive process. For a description of this alternative, see the article "Waterjet/Abrasive Waterjet Machining" in this Volume.

Friction band sawing gives high cutting rates, long band life, and low machining cost on material up to 25 mm (1 in.) thick. Most steels can be cut efficiently because of their low thermal conductivity and wide softening range; alloys of copper or aluminum, as well as most cast irons, are not suitable for friction sawing. Compared to thermal conductivity, the hardness of the work metal is of minor importance.

**Saw bands** for friction sawing are made of special silicon-carbon steel. They are thicker than standard bands and have wider set; the teeth are designed for heavy shear loads, and the gullets are shaped for plastic flow of chips. Sharp teeth are not required; saw bands are used until they break from flexing. Standard saw bands can also be used, but are likely to shatter at the high band speeds used.

## Circular Sawing

Circular sawing involves the use of a rotating cutting blade that can be fed horizontally, vertically, or at an angle into the material. Circular sawing is highly accurate because of the rigidity of the machines and the cutting blade. The length tolerance for most feed stock systems is ±0.10 mm (±0.004 in.), and the accuracy of cut is generally ±0.001 mm/mm (±0.001 in./in.) in the direction of blade travel.

Circular sawing can produce burr-free surfaces and can reduce the need for secondary finishing operations. For blades up to 400 mm (16 in.) in diameter, finishes typically range from 1.5 to 3.2 μm (60 to 125 μin.). Finishes as smooth as 0.2 and 0.8 μm (8 and 32 μin.) have been produced in the circular sawing of aluminum and steel, respectively. Generally, the harder the material, the smoother the finish.

Circular saws represent a larger capital investment than band saws or hacksaws. Circular sawing produces a larger kerf than band sawing, although circular saws as thin as 1.5 mm (0.060 in.) are available. These thin blades, however, cannot maintain the tolerances and high cutting forces for which circular sawing is noted.

## Hacksawing

Hacksawing is characterized by the reciprocating action of a relatively short, straight blade that is drawn back and forth over the workpiece. Hacksawing differs from the other sawing methods in that the back-and-forth motion produces a discontinuous cut.

**Capabilities and Limitations.** Power hacksaws constitute a relatively lower capital investment, and they can handle a wide range of stock sizes within their capacities. However, band saws are generally easier to set up and are able to cut a wider variety of metals (with higher hardnesses). Power hacksawing is essentially a roughing operation and at least 0.05 mm (0.002 in.) should be left on the cut surfaces for subsequent finishing.

# Multiple-Operation Machining

MULTIPLE-OPERATION MACHINING refers to the machining of identical parts in high volume when the operations are performed consecutively or simultaneously to permit complete machining of the workpiece in one setup. Turning, cutoff, facing, drilling, boring, tapping, threading, and other machining operations typically performed on separate machines for low-volume production requirements can be executed on multifunction machines when relatively high-volume production requirements make them cost effective.

## Multifunctional Systems

**Automatic Lathes.** Increasing labor costs, the competitive nature of the business, the shortage of skilled labor, and varying customer demands necessitate the use of multifunction machines. These include single-spindle automatic lathes, manual turret lathes, single-spindle automatic bar and chucking machines, Swiss-type automatic bar machines, multiple-spindle automatic bar and chucking machines, and multiple-spindle vertical chucking machines.

**Machining Centers.** In the late 1950s, the introduction of a single-spindle horizontal machine that could automatically change tools revolutionized the machine tool industry. An arm on the machine could remove and exchange tools in the spindle of the machine. The machine could perform milling, drilling, tapping, reaming, and boring operations automatically under numerical control. This was a significant development because numerical control had become commercially available on a limited basis only a few years earlier.

Thus, traditional machining entered the age of machining centers. Machining centers make it possible for a complete range of machining operations (drilling, milling, reaming, screw cutting, and so on) to be carried out in one setting of the workpiece. Cutting tools are normally exchanged automatically. Machines are usually built in knee-type, bed-plate, or boring-mill construction forms. The cutting tool can be guided in a minimum of three coordinated axes, while the work is clamped during machining on interchangeable tables, rotary tables, or platens. Machining centers usually have the following features:

- Numerical control
- Tool magazines with automatic tool changers
- Rotary tables enabling all sides of the work to be machined
- Within flexible production systems, automatic loading and unloading devices

Machining centers are primarily intended for small to medium-size production quantities because they can be reset for different production requirements in a relatively short time, owing to the high degree of automation. A machining center normally consists of 1 spindle (although multiple-spindle centers have been built) and 30 cutting tools (some have up to 60 tools), which are stored in a tool matrix.

**Transfer machines,** or transfer lines, consist of many machines, including an array of machining centers. Some mechanism must be provided for moving workpieces from station to station on production machines and from one machine to another. Machines that are provided with such automatic mechanisms are called transfer machines. Transfer machines consist of a number of workstations (turning, drilling, milling, and so on) arranged behind each other and linked with an automatic work transportation unit, which governs their positions and the timing cycle. Transfer machines are rigid production systems that are primarily used for mass production. The work advances in discrete steps; that is, when all individual workstations have completed their work cycles, all the components are advanced to the next workstation.

If the entire machine had to be shut down each time a single tool became dull and had to be replaced, productivity would be very low. This is avoided by designing the tooling so that certain groups have similar lives and by utilizing control panels that record tool wear in each group and shut down the machine before the tooling has deteriorated. This avoids shutting down the entire machine in the event that only one or two stations become inoperative. This is usually accomplished by arranging the individual units in groups, or sections with 10 to 12 stations per section, and by providing a small amount of buffer storage (banking) of workpieces between the sections. Thus, production can continue on all remaining sections for a short time while one is shut down for tool changing or repair. All the tools in the affected group can then be changed so that repeated shutdowns are unnecessary.

Transfer machines are a major capital investment. They are custom designed for a specific process and are economical only for large-scale production. A typical application for such a system would be the machining of V-8 engine blocks in which 688 separate machining processes are grouped into 50 machining stations.

**Flexible Manufacturing Systems.** As a result of the demands for production systems that satisfy the contradictory requirements of high productivity and high flexibility, so-called flexible production units have been developed. A computer-controlled component transportation system links a number of individual, interchangeable numerically controlled machines, normally machining centers. In addition, storage magazines for the unmachined and machined components can be connected to the transportation system. The cutting tools can also be automatically moved from the machines to the appropriate tool maintenance units for regrinding and resetting. A process computer controls the complete installation. In such installations, all available automation techniques in production technology are used. Such flexible linked lines are designed to handle single-unit production, small-batch production, spare part production, and component families production. Differing components can be machined on such a system in any order desired.

Flexible manufacturing systems are automated, with some of the units tooled for specific operations and the rest capable of random operation. Sequencing of operations can be varied as required. Thus, the system is a hybrid between a fixed-program, such as a transfer machine, and an assembly of general-purpose stand-alone machines. It has advantages over both alternatives—for example, more flexibility than the fixed-program system and less operation time than the stand-alone machines.

Flexible manufacturing systems are suitable for small to medium-size production runs and provide a compromise between situations in which productivity is the primary concern and situations requiring a high degree of flexibility in the manufacturing process. Typically, a flexible manufacturing system consists of 5 to 12 machine tools, with each machine capable of chang-

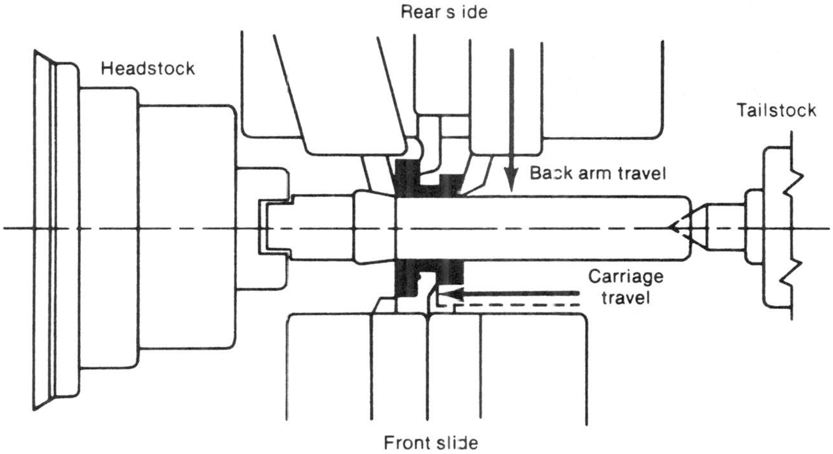


**Fig. 1** Typical setup on an automatic lathe showing three tool bits on the rear slide and one tool bit on the front slide. The workpiece is held in the headstock using a chuck, while the other end is held by the tailstock with a lathe center.

ing cutting tools to perform different tasks, such as drilling, boring, or milling.

A flexible manufacturing system can also be comprised of numerous manufacturing cells. The manufacturing cell is a system in which computer numerically controlled machines and an industrial robot are used to manufacture a specific part or several parts of similar geometry. A manufacturing philosophy that identifies and exploits the similarities of parts and manufacturing processes is termed group technology. A more complete definition of a manufacturing cell is a small unit within a manufacturing system, consisting of one to several workstations and having some degree of capability for automatic:

- Loading and unloading of the workstations with workpieces
- Tool changing at workstations
- Transfer of workpieces (and sometimes tools) between workstations
- Scheduling and control of workstation loading

All these operations are under computer control to produce a family of parts with similar features. Central to these activities in a manufacturing cell is an industrial robot. The workstations around the robot can be several machine tools, each performing a different operation (milling, drilling, turning, and so on) on the part.

# Automatic Lathes

All machines used for multiple operations are modifications of the basic engine lathe; some have undergone several stages of evolution. This first multiple-operation machine was built by replacing the tailstock of an engine lathe with an indexing turret that contained several tools, thus permitting a sequence of operations to be completed without removing the workpiece from the rotating chuck. The need for duplication of repeated motions led to the addition of cams, levers, and other powered devices. The resulting machine was a single-spindle automatic unit. This development was followed by the introduction of the multiple-spindle automatic machine to satisfy the need for greater productivity.

The terminology for different machines used for essentially the same purposes varies among different manufacturers. Some terms that have been extensively used are inaccurate, misleading, or inadequate. The term screw machine, for example, has become inappropriate because nearly all screws are now made by other methods. More descriptive terms are used in this article.

Although multiple-operation machines are almost limitless in variety, they can be classified as either chucking machines or bar machines. Machines in either category are further classified as:

- Manual turret lathes
- Automatic turret lathes (single-spindle automatics)
- Multiple-spindle automatics

Bar machines are available only as horizontal models. Chucking machines are either horizontal or vertical, although the use of vertical models is generally confined to workpieces that are too large to be conveniently placed in horizontal chucks. Aside from the number of spindles, the principal differences among the various machines are:

- Method of loading, holding, or feeding the work material
- Method of holding and moving the cutting tools into and out of position

When forgings, castings, extrusions, or bar slugs are being machined, they are loaded into chucks at the front face of the headstock. When bars or tubes are being machined, hollow headstocks and collets are needed to permit bringing the work material through from the rear of the machine.

There are three basic methods of bringing tools into and out of position:

- A revolving turret that can hold one or more tools on each of its index positions
- Cross slides that feed tools to workpieces at right angles to the machine spindles
- Tool carriers that advance or move back on a main slide parallel to the machine spindle

Single-spindle bar or chucking machines use turrets and cross slides for moving tools. In multiple-spindle bar or chucking machines, tools are moved by cross slides and a sliding tool carrier.

## Single-Spindle Automatic Lathes

Plain turning operations constitute one of the primary machining functions employed in manufacturing. The capabilities of the engine lathe—which generally have been supplanted by those of the more cost-effective turret lathe and the automatic bar or chucking machine—are available in a wide variety of automatic lathes. All the advantages of single-point tooling for maximum metal removal, finish accuracy, center turning, and so on, are at the designer's disposal, offering production speeds comparable to those of the fastest processing equipment available.

In automatic lathes, the tools are automatically fed to the work and withdrawn after the cycle is complete. Automatic lathes have the basic components of simple lathes: bed, headstock, tool slides, and sometimes a tailstock. Most single-spindle automatic lathes are arranged to turn workpieces held between centers, but some are designed for chucking work, with the workpiece held in a chuck, collet, or special workholding fixture. Figure 1 shows a typical setup on an automatic lathe. Most machines are of horizontal configuration, but some are built with vertical spindles. Depending on the manufacturer, feed of the slides is accomplished with hydraulics, air/hydraulics, leadscrews, or cams. Figure 2(a) illustrates a machine with leadscrew feed for the front slide, and Fig. 2(b) shows a machine with a cam-fed front slide.

Because most lathes of this type require that the operator manually place the part to

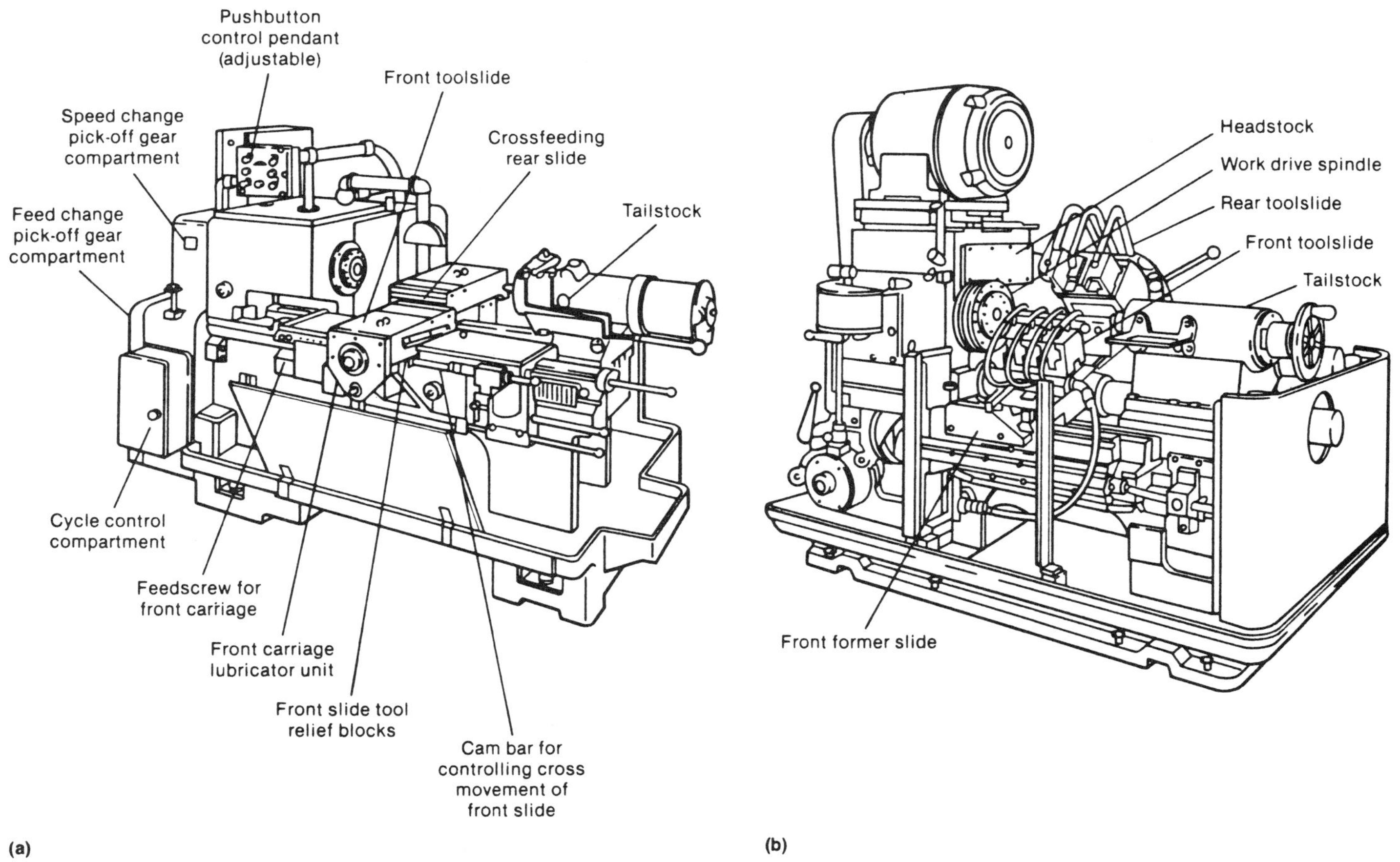


**Fig. 2** Single-spindle automatic lathes. (a) Cam-fed front slide. (b) Leadscrew-fed front slide

be machined in the lathe and remove it after the work is completed, they are perhaps incorrectly called automatic lathes. The machines can be single-cycle automatic, with manual loading and unloading, or fully automatic, with mechanical loading and unloading (Fig. 3). Machines in this group differ principally in the manner of feeding the tools to the work. Most machines, especially those holding the work between centers, have front and rear tool slides. Others, adapted for chucking jobs, have an end tool slide located in the same position as the turret on the turret lathe. These machines can also have the two side tool slides. Another configuration employs a flat table located in front of the chucking spindle on which tool slides can be mounted at any angle or in any position. Each tool slide has individual feed and receives its power from individual drive shafts at the end of the machine.

## Machine Classifications

Turning equipment can be classified into several groups. Fully automatic types with magazine feeds and semiautomatic hand-loaded types constitute the first category. Fully automatic numerical control (NC) lathes and semiautomatic tracers or shape turners constitute the second category. Numerical control lathes are available for chucking and center work with automatic bar feeds, and some models are virtually machining centers. Automatic lathes in the first group are available in a variety of standard or special models with mechanical, electrical, or hydraulic actuation. Tracer lathes that use either a flat template for reproduction or an actual pilot model workpiece are available.

**Automatic Cycle Lathes.** Employing primarily only simple, single-point turning operations, automatic cycle lathes are ideally suited to the production of a wide variety of machine parts, such as automotive and aircraft pistons, aircraft engine cylinders, transmission cluster gear blanks and shafts, camshafts, bearing races, and spur and bevel gear blanks. Unlike the turret lathe and the automatic bar or chucking machine, automatic cycle lathes are well adapted to the machining of forgings, castings, and bars held between centers; work supported on fixtures; and ordinary chuck work. Plain bores on many parts, preferably for accuracy in chucking, are best finished in a preliminary operation on a turret lathe or by drilling and broaching. Special bores or tapers, however, can often be more readily produced on automatic lathes, owing to the advantage of pivoted tool posts.

**Automatic tracers,** like automatic cycle lathes, use single-point tooling but, ordinarily, only one tool. The primary limitation to design is that all details must be accessible with a sharp-point tool. In general, parts that are considered suitable for obtaining the advantages of the automatic tracer are somewhat similar to those mentioned previously, but usually contain a complicated form, taper, or other details that cannot be reproduced except with complicated banks of tools or special attachments. Characteristic parts are valve plugs, valve bodies, nozzles, orifices, stepped shafts, and contoured rolls.

Employing low-cost templates for reproduction, automatic tracers are ideally adapted to production quantities ranging from only a few pieces to large-quantity output. Compressor crankshafts, for example, having five stepped diameters, two tapers, radii, and shoulders can be turned in 24 s from the rough castings. Automatic tracer lathes range in sizes paralleling those of the automatic cycle lathes and offer a broad coverage of machine parts.

**Automatic shape turners** offer possibilities for design that cannot be obtained by any other means. Although originally developed for the manufacture of hubs, dies, and molds of compound shapes, these machines

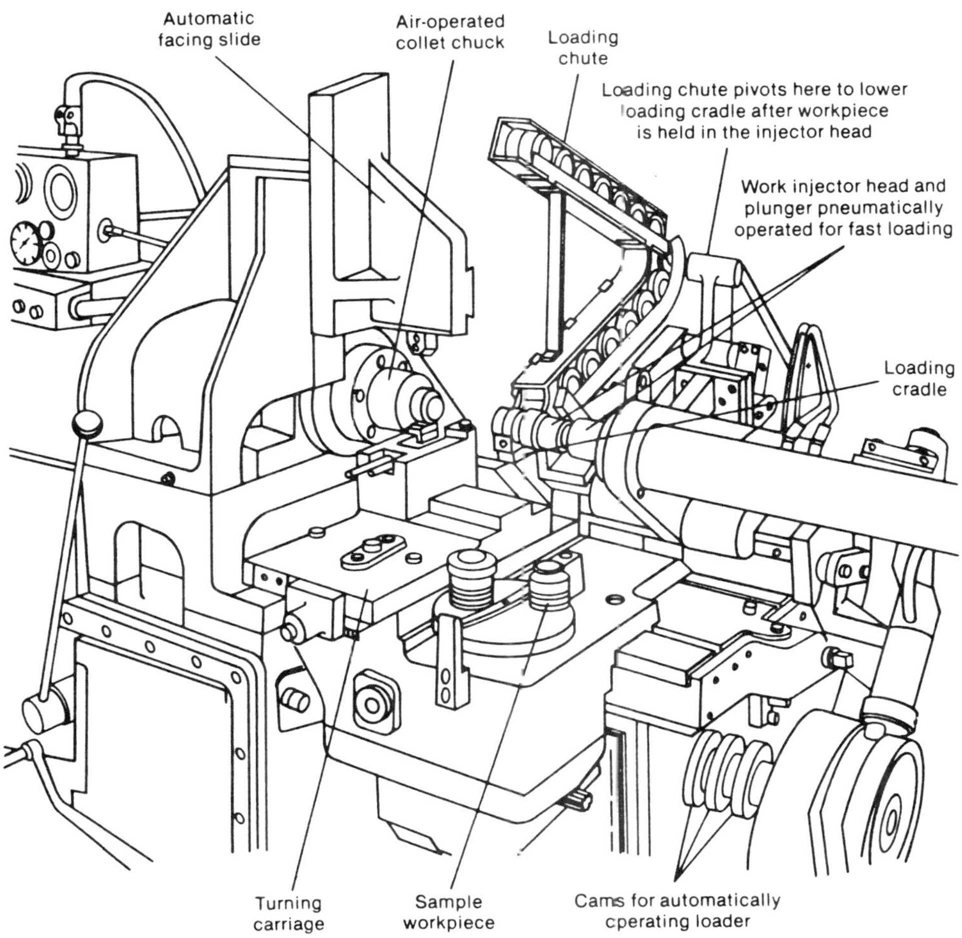


**Fig. 3** Automatic loading device for a single-spindle automatic lathe

often find use in the efficient, economical production of parts for some of the newer, more complex mechanisms, such as helical pump rotors and body sections.

Automatic shape turners are capable of handling parts up to about 508 mm (20 in.) in diameter and, like standard engine lathes, can be adapted to either center or chucking work. Bored contours can be as readily produced as external ones and to a high degree of accuracy. Similarly, various facing designs can also be produced, ranging from squares to extremely complicated contours. Accurate and closely fitted mating components of complicated design can be readily produced in production. Speed, however, is somewhat lower than with the other lathes.

## Process Capabilities

High-speed production is a major advantage of single-spindle automatic lathes, and it results from having a number of tools cutting simultaneously, with no time lost for tool indexing or positioning. Customarily, as many tools as possible are used without exceeding the power of the machine and the rigidity constraints of the workpiece.

Single-spindle automatic lathes are generally best suited to high production requirements in turning and facing multiple-diameter or long workpieces. The more complex the setup, the longer the production run should be for minimum cost because setup and changeover are time consuming.

**Workpiece Requirements.** In general, parts that are suitable for machining in automatic lathes include:

- Those that can only be turned satisfactorily on centers because of length or special requirements for accuracy and finish. This includes work turned in preparation for grinding or other machining requiring the use of centers
- Those that, because of having been forged, cast, or welded, cannot be cut from a bar or fed through a hollow spindle
- Forgings, castings, and so on, that, because of an irregular shape, cannot be readily held in chucking machines
- Those that require machining all over or at both ends
- Those that cannot be chucked and lend themselves to stacking for magazine feed
- Those parts required in small quantities not within the economical range of automatic screw machines

Where parts are long and require machining at both ends, a center drive attachment can be used that acts both as a steady rest and a driver. Large, long parts that require machining all over are usually driven by spur drives.

**Size Range of Machines.** Automatic lathes range in size from small units designed to handle work 13 to 35 mm (½ to 1⅜ in.) in diameter by 64 to 457 mm (2½ to 18 in.) in length to units capable of handling diameters up to 2.1 m (84 in.) and lengths up to 3.1 m (10 ft) or more. These machines are powered by motors ranging in size from 3.7 kW (5 hp) on the small units up to 93 kW (125 hp) or more on the larger machines, with the full advantage of carbide tooling wherever practicable and the maximum possible production speed. Automatic tool relief at the end of each tool stroke avoids scraping of the tools, increasing tool life and providing better finish. Single-spindle automatic lathes are available in a range of capacities. Table 1 lists the capacities of four standard machines offered by one manufacturer. Although the lengths that can be turned or shaped are necessarily limited by the carriage travel of specific models of any of these machines, long lengths of turn can be handled by special machines or by multiple tooling. Truck steering knuckles, for example, are produced in 0.9 min, or at a rate of 66 per hour (Fig. 4).

**Table 1 Typical capacities of single-spindle automatic lathes**

| Maximum power | | Maximum length between centers | | Swing | | | |
|---|---|---|---|---|---|---|---|
| | | | | Over ways | | Over slides | |
| kW | hp | mm | in. | mm | in. | mm | in. |
| 5.6 | 7 | 305 | 12 | 279 | 11 | 114 | 4½ |
| 18.6 | 25 | 1015 | 40 | 279 | 11 | 140 | 5½ |
| 44.8 | 60 | 2540 | 100 | 406 | 16 | 178 | 7 |
| 112 | 150 | 3050 | 120 | 597 | 23½ | 356 | 14 |

## Manual Turret Lathes

The essential components and operating principles of a horizontal turret lathe are illustrated schematically in Fig. 5. The workpiece shown could be either a short-length part (such as a casting or a forging) held by a chuck, or a long bar or tube fed through a hollow headstock from the rear and held by a collet.

The turret, mounted on a ram or a saddle, advances to the workpiece in an axial direction on the latheways. The turret shown in Fig. 1 has six tool-holding faces or index positions and is therefore called a six-station turret.

A smaller turret on the cross slide may have four tool-indexing positions, obtained

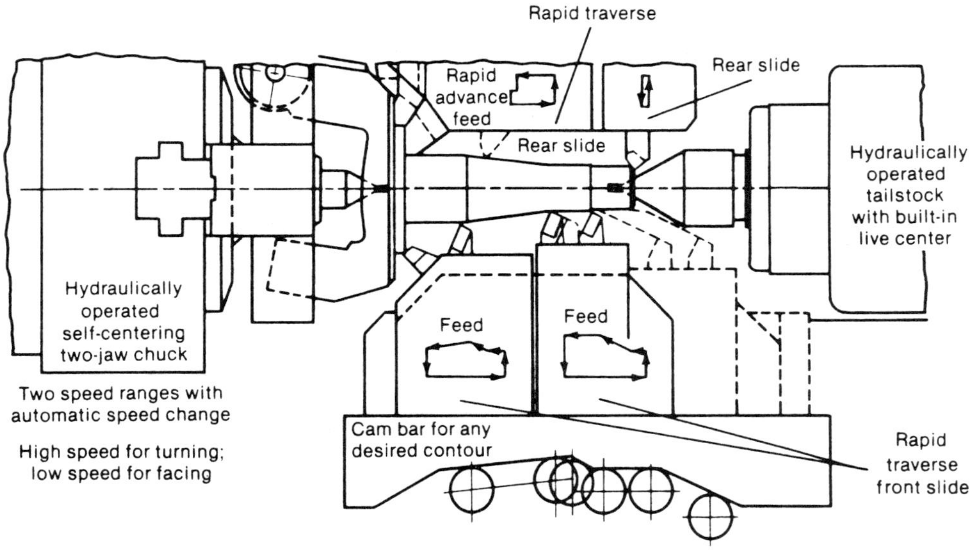


**Fig. 4** Layout of tooling for truck steering knuckles. Production speed is 0.9 min floor to floor, with 0.6 min machining time.

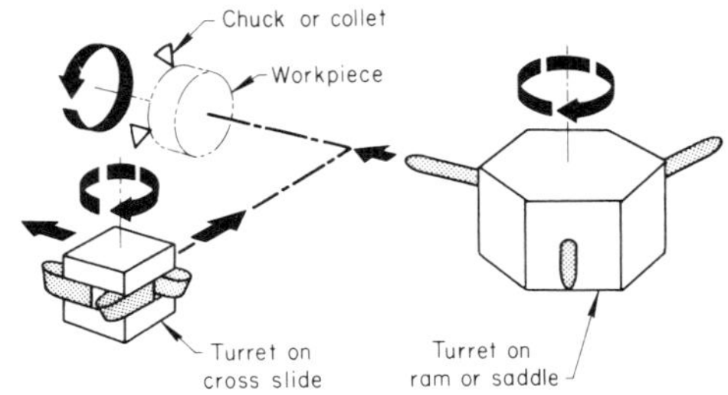


**Fig. 5** Basic components and their motion in a horizontal turret lathe

by clockwise rotation. The turret and cross slide are supported on a saddle and can therefore be moved axially to perform operations such as turning. This turret can also move on the cross slide at right angles to the longitudinal axis of the workpiece to perform such operations as form turning, facing, or grooving. Figure 6 illustrates the multifunction capabilities of turret lathes.

The number of stations that can be used in the main turret of a machine of this type is primarily governed by the tool clearance needed for a given job. If tools extend only a short distance from the turret or if the workpiece is small in diameter compared with the size of the turret, more stations can be used than when the above conditions are reversed. In addition, the workpiece must clear the tools extending from stations next to the one from which the operating tools extend. Turrets having as many as nine stations have been built, but turrets with four, five, or six stations are more usual.

## Ram-Type Versus Saddle-Type Lathes

A difference in construction of the main turret unit separates turret lathes into two general classes: ram-type and saddle-type. The article "Turning" in this Volume contains more information on the principal components of both turret lathe types.

**Ram-Type Turret Lathes.** In this type of machine, the main turret is mounted on a separate slide, or ram, which moves along the lathe axis on a saddle supported by the ways. During setup, the saddle containing the ram and turret is positioned along the ways as close as possible to the workpiece to provide maximum support for the ram and turret, which in operation will move axially on the saddle. When the setup is completed, the saddle is locked in position and remains fixed during machining.

The main advantage of the ram type over the saddle type is that the operator has less mass to move during operation, which results in easier and faster handling. The main disadvantage of the ram-type machine is that, as the turret and ram move forward, overhang of the ram from the saddle increases, and some sag of the turret is inevitable. This sag causes tools to cut tapers—to the extent that the use of a ram-type machine may be precluded for making long turning or boring cuts, particularly when the cutting tools are in a vertical position.

**Saddle-Type Turret Lathes.** In this type of machine, the turret is mounted directly on the saddle, which moves axially during operation. This type of construction provides greater rigidity than that of the ram-type machine because it overcomes the problem of ram overhang and turret sag. The main disadvantage of the saddle type is that greater operator effort is required than with the ram type, but this can be overcome by adding a power attachment for moving the turret-and-saddle unit.

Manually operated turret lathes are normally used for producing parts in small-to-medium production runs for which an engine lathe is too slow and an automatic machine is unwarranted by production demands. Manually operated means that an operator must be in attendance during the entire machining cycle, but the exact duties of an operator may vary, depending on how the machine is equipped.

When the operation is entirely manual, the operator loads the workpiece in the chuck, starts the machine, and then manually performs all operations, including starting and stopping the headstock, indexing the turret, advancing the turret, moving and retracting the cross slide, and unloading the workpiece. This procedure is generally applicable only when fewer than about 12 identical parts are being made and when highly skilled operators are available.

## Accessories

The productivity and versatility of a manual turret lathe can be greatly increased, and the need for operator skill decreased, by the addition of one or more of the following accessories:

- Automatically controlled headstock
- Power feed on turret slide (axial)
- Power feed on cross slide
- Cross slide turret
- Automatic turret indexing

**Automatic headstock control** permits starting, stopping, speed changing, and reversing of the spindle by indexing and axial movement of the turret. It can substantially increase productivity, especially where handling time constitutes one-half or more of the floor-to-floor time or where a number of changes of spindle speed or direction, or both, are required during a short machining cycle. In many cases, when the machine is equipped with headstock control, the operator can complete cycles by handling only the turret.

**Power feed on the turret** reduces operator fatigue and often saves enough operator time to permit an additional machine to be tended by one person. In addition, a powered turret results in greater accuracy in some machining operations, especially tapping or die threading.

**Power Feed on the Cross Slide.** In simple operations, a power-fed cross slide is seldom worthwhile, but in operations such as the facing of large areas, the power feed can be used to increase efficiency and to produce more uniform results.

**The cross slide turret,** with its ability to feed in four directions, adds greatly to the versatility of the manual turret lathe. This unit is available as standard equipment, but its use is generally restricted to large turret lathes having saddle-type construction.

A cross slide turret is especially advantageous for machining small lots of workpieces because it often permits the machining of several surfaces (such as inside diameters) with the same tool. A graduated dial for cross motion allows the use of the

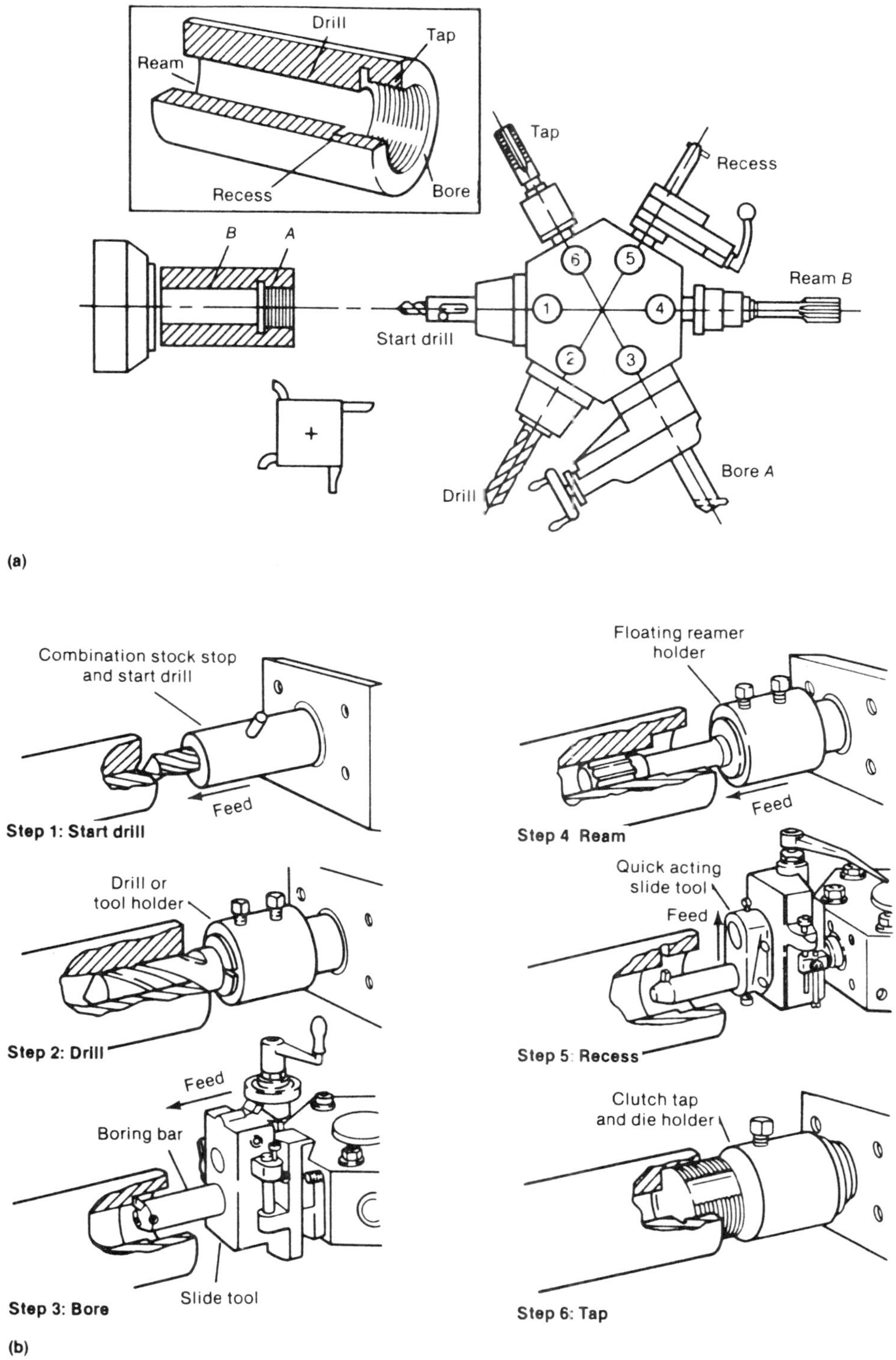


**Fig. 6** Schematic of a basic hexagon-turret setup (a) illustrating the correct sequence of operations (b) to handle the required internal cuts on the threaded adapter shown

turret for turning different diameters on parts in low-quantity production, for which the cost and setup time for more elaborate tooling would be impractical. In addition to its advantages for low-quantity production, the cross slide turret is often advantageous for machining the following types of work:

- Large bores, which can be machined more easily using standard boring bars because cutter overhang is less than on a fixed-center machine
- Deep internal faces and recesses
- Large-diameter work, which limits or prohibits the use of a conventional turret

**Automatic turret indexing** can be applied to virtually any type of manual turret lathe. This feature, as well as the other accessories described in this section, increases setup time, but will also increase productivity and decrease operator effort. The total quantity to be produced and the parts required per unit of time usually determine whether automatic turret indexing and other accessories should be considered.

The use of one or more of the accessories discussed above often makes it possible for one operator to attend more than one machine, provided they are properly grouped. In some plants, manual turret lathes that have been made partially automatic are preferred for large-quantity production, primarily because of their versatility.

## Single-Spindle Automatic Bar and Chucking Machines

The terms automatic turret lathe and single-spindle automatic bar or chucking machine are frequently synonymous. In operating an automatic turret lathe for chucking applications, the operator loads the workpiece and pushes the start button. All machining operations are then conducted automatically by a preestablished sequence, after which the machine stops automatically and is unloaded by the operator. If the workpiece is a bar or a tube, automatic stock feeding can be incorporated, thus permitting continuously repeated cycles.

Automatic turret lathes do not vary greatly from manual turret lathes equipped with the accessories discussed in the preceding section. The main difference between a manual turret lathe with added accessories and one designed and built for automatic operation is that the latter usually has both back and front cross slides, thus increasing its tooling capacity (Fig. 7).

Automatic turret lathes are most often used where production requirements are too high for efficient production on manually operated machines but too low to warrant the use of multiple-spindle equipment. The setup time is lower, and less elaborate tooling is needed, for automatic turret lathes than for multiple-spindle machines.

Automatic turret lathes are generally of two types. In the first type, the turret rotates in a horizontal plane and is attached to a saddle that slides axially on the latheways. In the second type, the turret indexes by rotating around an axis parallel to the machine spindle (thus, the turret actually rotates in the vertical plane) and is supported by a main slide, or turret bar.

**Turret in Horizontal Plane.** Essential components and operating principles of the first of these types are shown in Fig. 5 and discussed in the section "Manual Turret Lathes" in this article. Machines incorporating these principles are available as chucking or bar machines and in a wide range of sizes. The maximum diameter of workpiece that can be machined is usually 305 mm (12 in.).

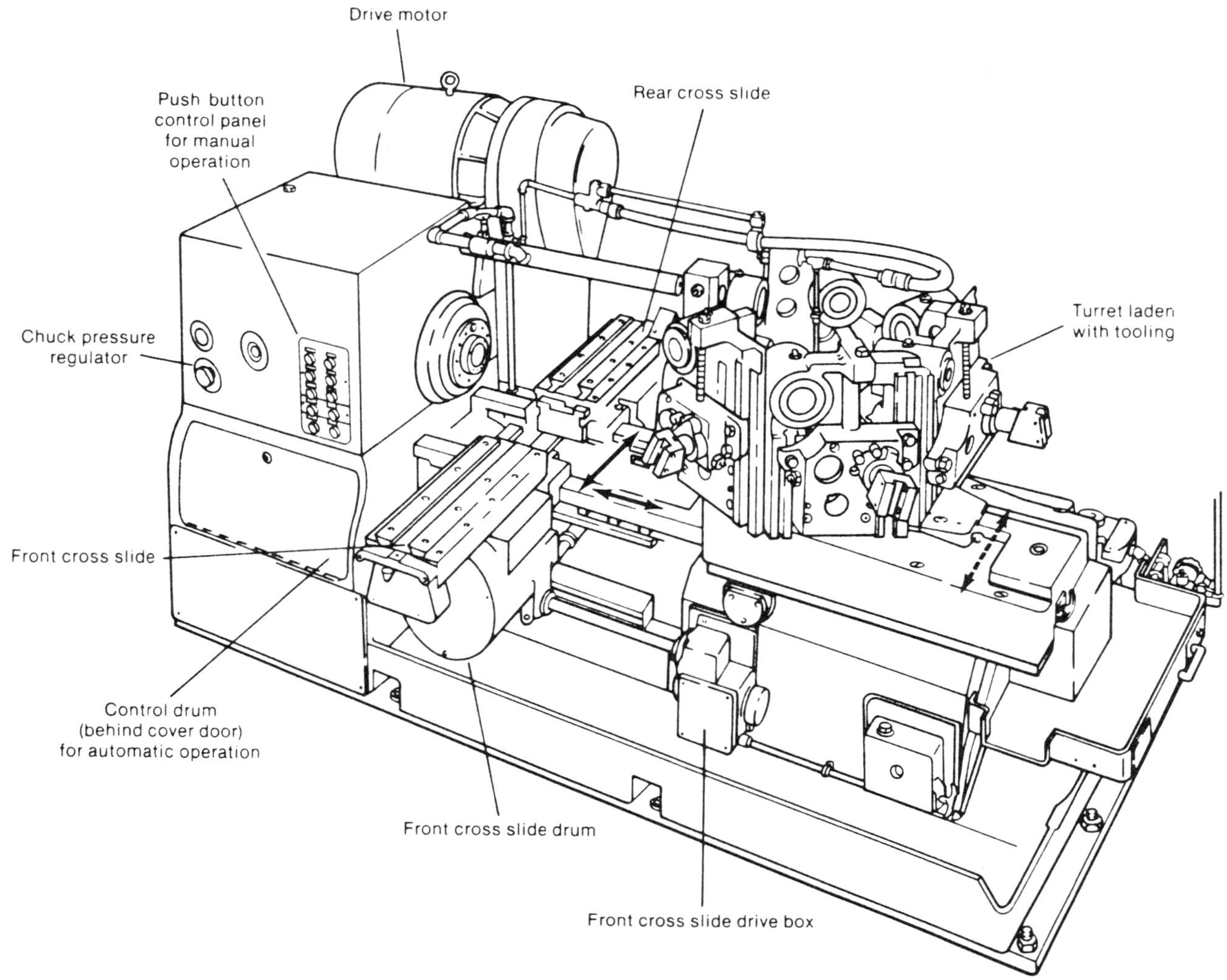


**Fig. 7** Automatic horizontal turret lathe capable of manual or automatic operation

**Turret in Vertical Plane.** The essential components and operating principles of the second type indicated above are shown schematically in Fig. 8.

*Chucking Machine Configuration.* This type is available only as a chucking machine and cannot be used for producing parts directly from random lengths of bar or tube. It is made in sizes large enough to accommodate workpieces somewhat greater than 305 mm (12 in.) in diameter. As indicated by the arrows in Fig. 8, the tools are advanced to the workpiece from three directions and three sources. All axial cutting operations are performed by tools mounted in the turret. The turret is rigidly supported by, and moves axially on, a main slide, and it can be indexed to any of its operating positions (usually five, as indicated in Fig. 8) by rotating in a counterclockwise direction. When the tool is in operating position, its centerline coincides with the centerline of the workpiece, as indicated. Turret tools are secured in standard adapters, which are fastened into the turret by dovetails. Machines of this type are equipped with one front and one rear cross slide. Tools from the cross slides move only in directions at right angles to the axis of spindle rotation. Consequently, cross slide tools are restricted to operations such as facing, form turning, grooving, or knurling.

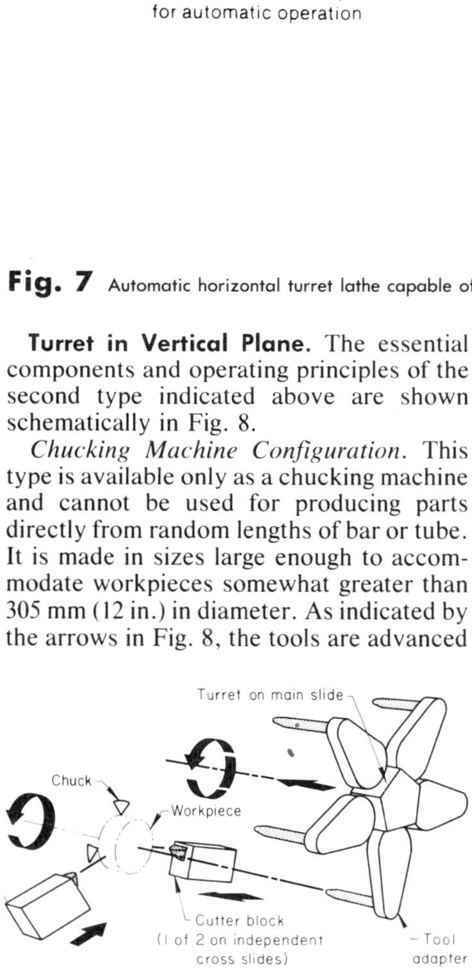


**Fig. 8** Fundamentals of a single-spindle automatic chucking machine (one type of automatic turret lathe)

*Combined Chucking and Bar Configuration.* The working end of a different design of single-spindle automatic chucking machine is shown in Fig. 9. In this design, the turret rotates around a horizontal axis parallel to and directly above the spindle centerline while located in the headstock instead of the tailstock. The main tool turret on this machine has five faces for toolholders and tools. Each face is indexed in turn to working position. The turret then moves lengthwise to feed the tools at the station to and from the workpiece. The two cross slides below the work spindle can be made to feed at the same time as any turret station. The tool turret is carried on one end of a large round slide. Standard holders are available to adapt commonly used turret lathe tooling (Fig. 10). On the other end is a drum from which adjustable dogs set stroke lengths and select speeds and feeds. The machine can also be controlled manually from push buttons and switches on a control panel. Other models are completely numerically controlled. A typical medium-size automatic chucking machine has a 305 mm (12 in.) diam chuck and a 19 kW (25 hp) motor.

The above design can also be used as a bar machine. Single-spindle bar and chucking automatic lathes are almost identical in construction except for the normal differences in work-holding devices and the cutoff slide on the bar machines. Basically, both types of machines are built on a verti-

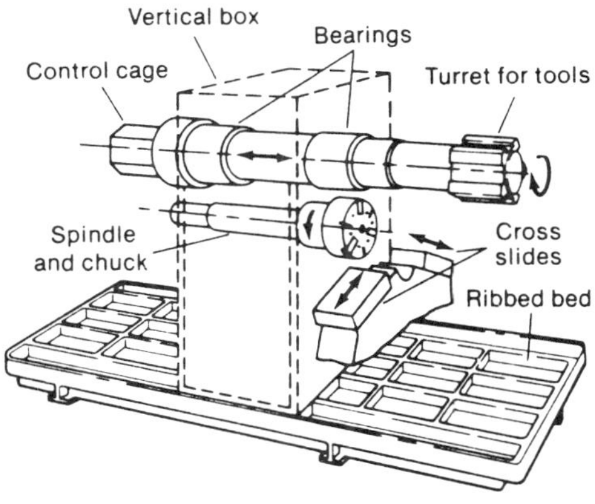


**Fig. 9** Schematic of bar and chucking automatic lathe illustrating the basic relationship among turret, work spindle, and cross slides. The column casting also houses transmissions and other operating subassemblies.

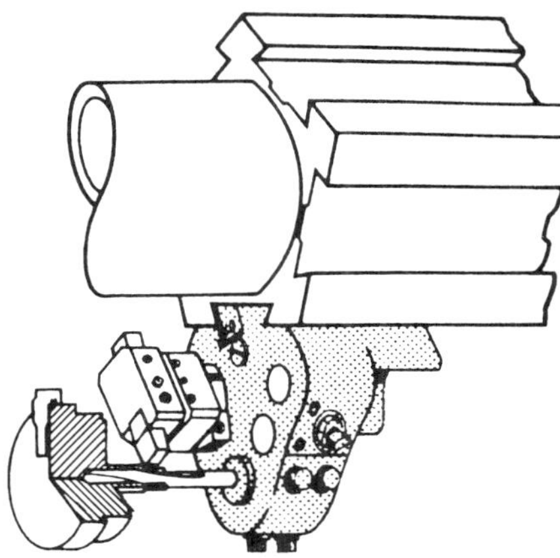

**Fig. 10** Multiple turning head used in a single-spindle automatic chucking machine and mounted on the pentagon turret. The workpiece is turned and drilled simultaneously.

cal, massive box column fastened to a heavy ribbed base. Alignment problems are eliminated because the spindle and turret are housed in the headstock instead of being housed separately in the headstock and tailstock, as in a typical lathe. At operating height, a one-piece forged steel support bar and five-position tool turret extend through the column, parallel to the base. This turret bar feeds in and out and is indexed by an electric motor driven in a plane at right angles to the axis of the bar. Large sleeve bearings support the turret in the column. On the back end of the machine, a cage for mounting speed and feed dogs is fastened to the end of the turret bar. Other setup controls are also located in this area.

The machine control system does not use removable cams for setup purposes. Machine action is controlled through a system of dogs and pins as well as standard electrical relays and switches. The controls are designed so that the machine can operate automatically, manually (through hand-controlled switches), or by tape.

The pentagon turret bar is actuated by a barrel cam, and the cross slides are actuated

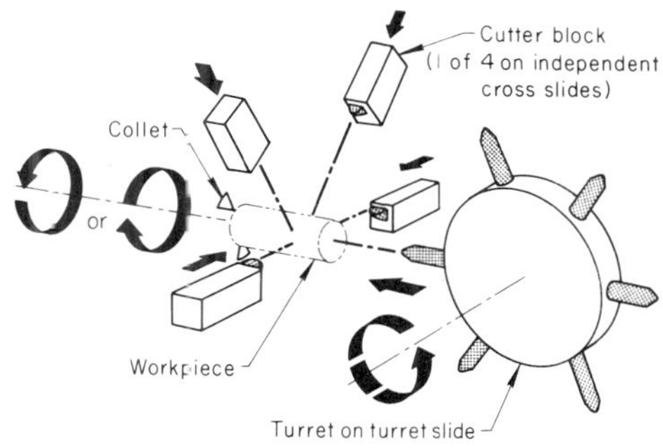


**Fig. 11** Fundamentals of one type of single-spindle automatic bar machine

either by barrel cams or a face cam, depending on the size of the machine. The main drive motor operates speed and feed transmissions. Rate changes in these constant-mesh transmissions are made by hydraulic clutches acting through limit-switch-controlled valving. Transmissions, indexing units, and rapid traverse units are removable individually without disassembling the entire machine.

A work spindle is mounted in antifriction bearings in the column just beneath the overhead bar. Air chucks, fixtures, or hand chucks are mounted to the flanged spindle nose of the chucker. Hydraulically operated collet chucks and bar feeds are standard on bar automatic lathes.

## Basics of Single-Spindle Automatic Bar Machines

Manually operated single-spindle bar machines are the counterparts of manually operated turret lathes (Fig. 5). Manual turret lathes for machining bars are sometimes called hand screw machines because they were originally used to produce screws. The term automatic bar machine is preferable because, today, screws are seldom produced using these machines. However, the term automatic screw machine persists. The essential difference between the manual turret lathe and the chucking type of turret lathe is the construction of the headstock end. To accommodate bars or coil stock, this portion of the machine must be constructed so as to permit the bars to be fed through from the rear and to be held by collets rather than chucks.

Special attachments can also be incorporated on bar machines to increase productivity, to decrease operator effort, or both (see the section "Accessories" in this article). Machines that can handle bars up to about 203 mm (8 in.) in diameter are available, but it is seldom economical to use this type of machine for solid bars this large, because of the time needed for cutoff.

Single-spindle automatic bar machines are classified as cam controlled and programmable (camless) machines. Cam-controlled automatic bar machines, requiring cams and speed-change gears for specific applications, are used more extensively for the high-production requirements of a single part or family of parts. Camless automatic machines, programmed by toggle switches, pegboards, or numerical control/computer numerical control (CNC) units, provide simpler setup, faster changeover, and increased flexibility for shorter-run production.

Single-spindle automatic bar machines are the bar machine counterparts of automatic turret lathes with regard to the operations performed. In essence, they are small cam-operated turret lathes.

**Spindle-Turret Configuration.** There are several different types of single-spindle automatic machines that produce parts directly from lengths of bar or tube, but all types utilize cross slides and some form of turret to manipulate the tools. Some types are constructed so as to operate by the principles of a horizontal turret lathe (Fig. 5).

The essential components and operating principles of another type of single-spindle automatic bar machine are illustrated schematically in Fig. 11. In this type of machine, the turret indexes by revolving on a horizontal axis at right angles to the longitudinal axis of the machine spindle and workpiece. The turret, which may have up to eight index stations, advances to the workpiece on a turret slide supported by the ways. All axial operations are performed by tools mounted in the turret. Tools mounted on the four cross slides can perform, consecutively or simultaneously, operations such as facing, form turning, grooving, or knurling.

For any of these machines, work metal is supplied from a bar or tube held firmly in the spindle by a spring collet. After each piece has been completed, the bar is repositioned for machining of the next piece by being automatically moved forward and butted against an established stop. Some means must be provided for supporting the bar section extending out the rear of the headstock in order to minimize the whipping action. Whipping will cause excessive vibration in the machine.

Bars less than 50 mm (2 in.) in diameter represent the majority of those machined in this type of equipment. Single-spindle automatic bar machines are available to handle work ranging from minute parts such as watch staffs to bars approximately 203 mm (8 in.) in diameter. A large number of standard tools and holders similar to turret lathe attachments are available for use on automatic bar machines.

**Cam-Controlled Automatic Bar Machines.** The components of a cam-controlled, single-spindle automatic bar machine are illustrated in Fig. 12. A bar of stock is held firmly in the spindle by a spring collet and is rotated. Two sprockets on the spindle are driven by roller chains from driveshafts in the base. These sprockets run free, and either one or the other is engaged to drive the spindle by means of a

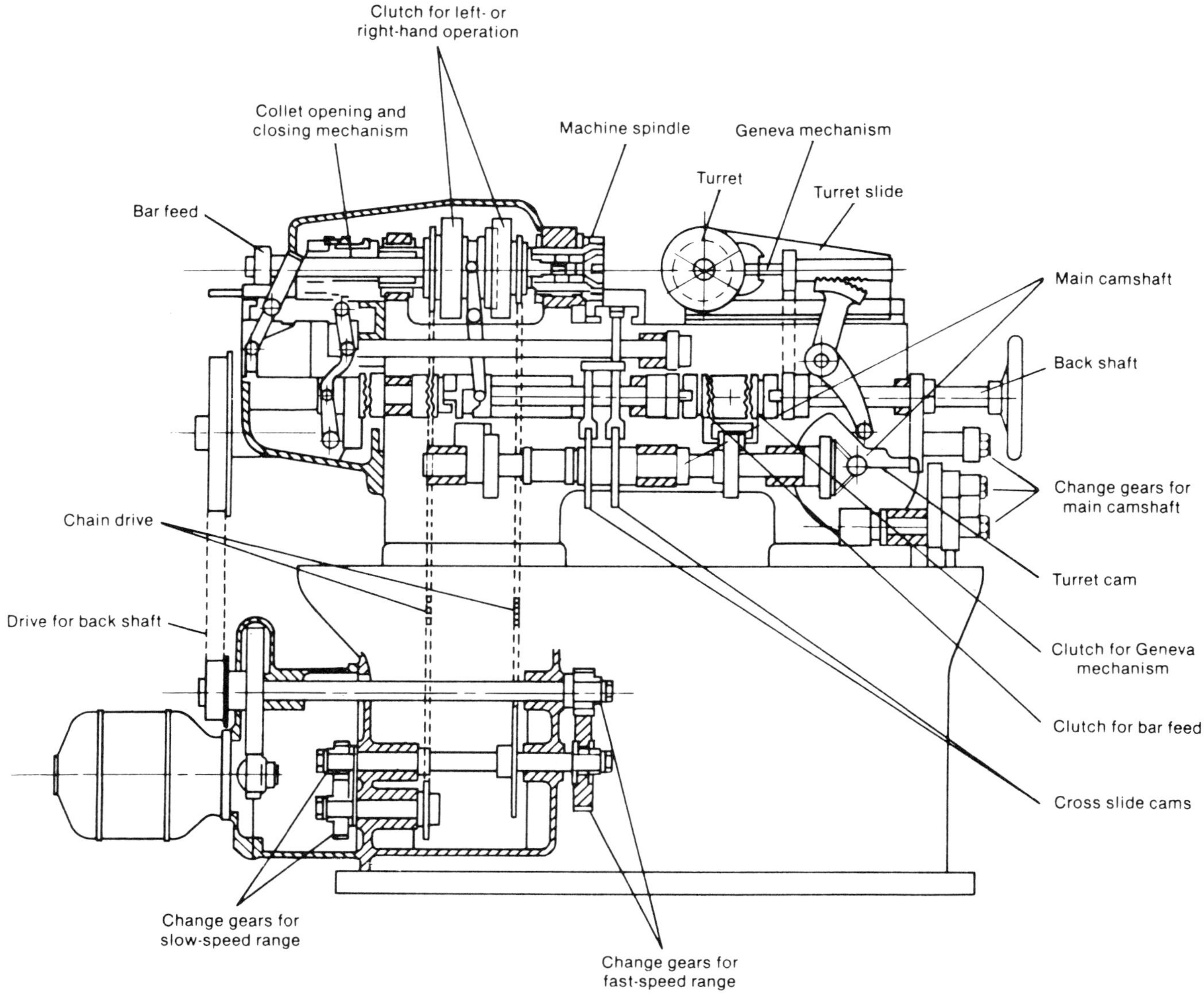


**Fig. 12** Cross-sectional view of the drive and control mechanisms of a single-spindle automatic bar machine

friction clutch between the sprockets. This type of drive allows the direction of rotation to be reversed.

Cutting operations are performed on the bar of stock by cutting tools held in a turret and on the cross slides. The regular turret has six or eight tool positions and is automatically indexed to bring each tool into position. Each operation is rapidly completed, and it typically requires ½ s to index the turret to proceed with the next operation. A variety of standard and special attachments enable the machine to perform screw slotting and burring, cross drilling, and milling. The cross slides are two independent slides at right angles to the spindle. Both the turret and the cross slides are moved to the work through levers actuated by plate cams designed to give the proper feed to each tool. The slides are withdrawn from the work by springs. Adjustable stops for the cross slides permit the accurate repetition of formed diameters within close tolerances. The completed workpiece is cut off by the cross slides or by one of the upper slides actuated by a plate cam and lever.

The bar of stock is automatically fed forward to the new cutting position after each piece has been completed. It is located by being butted up against a swing stop or turret stop. Coupled driving shafts, at the rear of the machine, are driven at a constant speed through transposing gears, the positioning of which determines the direction of rotation and driveshaft speed.

The timing of idle movements of the machine, such as indexing, feeding stock, and reversing the spindle, is governed by clutches mounted on the driving shaft. These clutches are actuated by trip levers and trip dogs, which are adjustable on the dog carriers that rotate with the camshafts (Fig. 13).

The main driveshaft supplies power through feed change gears to the lead wormshaft (in the machine bed), which operates the two front camshafts through separate worms and adjustable wormwheels. The turret is indexed from the driveshaft through an intermediate gear and a sliding gear, then through a simple Geneva movement (Fig. 14) to index the turret.

**Programmable Automatic Bar Machines.** Camless single-spindle automatic bar machines are programmed by toggle switches, pegboards, control drums, punched cards, or NC/CNC units. They are simple to set up, permit faster changeover, and provide considerable flexibility. As a result, they are used more extensively than cam-controlled automatic machines for shorter-run production requirements.

## Swiss-Type Automatic Bar Machine

The principle of the Swiss-type automatic bar machine was originally developed more than 50 years ago for the clock and watch manufacturing trades in Switzerland. The American version of the Swiss high-precision automatic bar machine is particularly well adapted to the turning of pinion blanks,

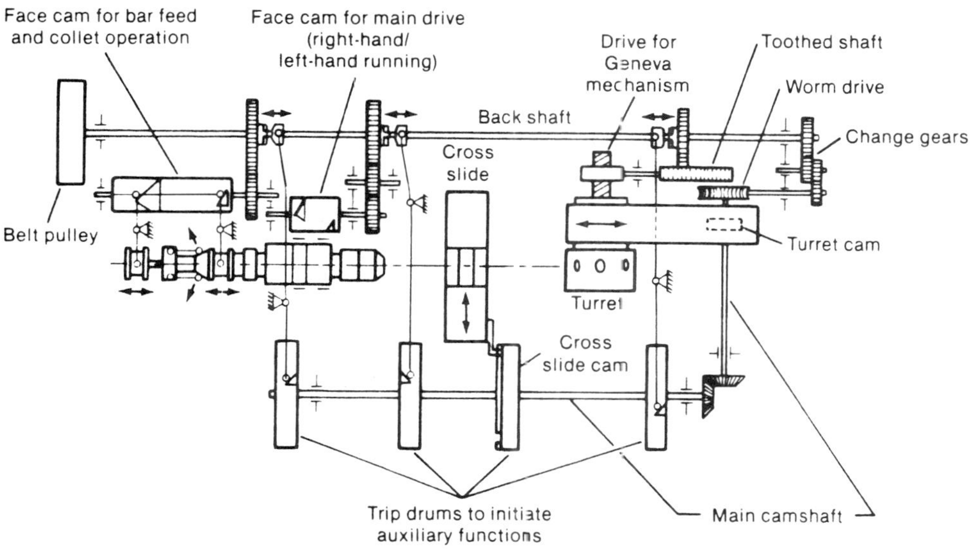


**Fig. 13** Schematic of the drive system employed on a single-spindle bar machine showing the automatic auxiliary control draft and the main cam shaft

studs, worm gears, indicator staffs, shafts, and other minute or slender parts used in clocks, meters, radio or electronic equipment, calculating machines, and a wide variety of industrial and laboratory instruments.

The Swiss-type automatic bar machine works somewhat differently from other bar machines. The turning method employed in these machines makes them ideally suited to the production of small and slender, finely finished parts, and it consists of feeding a revolving piece of stock through a guide bushing past radially fed tools (Fig. 15a). Unlike the turret-type single-spindle automatic screw machine, only single-point turning tools and narrow form tools are used. A more accurate designation for this type of bar machine would be precision sliding-headstock automatic lathe.

## Cam-Controlled Machines

Figure 15(b) shows an end view of a Swiss-type bar machine. The single-point tools used on this machine are placed radially around the carbide-lined guide bushing, through which the stock is advanced during machining operations. Turning of the diameter is primarily done by the two horizontal tool slides, while the other three tool slides in the upper half of the semicircle are principally used for such operations as knurling, chamfering, cutting off, and recessing. The stock is held by a rotating collet in the headstock back of the tools, and all longitudinal feeds are accomplished by a cam that moves the headstock forward as a unit. This forward motion advances the stock through the guide bushing and to the single-point tools, which are controlled and positioned by cams. By coordinating their movement with the forward movement of the stock, any desired shape can be turned.

Sliding headstock and the five radial tools are traversed by individual cam action. Independent cam control of each of these units eliminates the need for wide form tools because the proper combination of headstock and tool movement generates pivot points, back shoulders, multiple diameters, taper, and so on. The complexity of the form-generating, back-shoulder, and back-recessing work that can be accomplished within the capacity of the machines is shown in Fig. 16. Many parts incorporating pivot points, extension spherical ends, and so on, can be completed and finished to final form in the cutoff operation with one automatic cycle of the machine, thus eliminating secondary machining operations and the attendant loss of close concentricity.

**Workpiece Size Limitations.** American machines range in chuck capacity from 4 to 13 mm (5⁄32 to ½ in.). Four machine sizes within this range are available for production and have 4, 9.5, 11, and 13 mm (5⁄32, 3⁄8, 7⁄16, and ½ in.) maximum chuck capacities. The principal advantages of the smallest machines are higher spindle speeds and proportionally smaller working parts, which permit closer running fits and consequently the greatest possible accuracy.

Parts produced on the 4 mm (5⁄32 in.) machine range from 0.8 to 40 mm (1⁄32 to 1 9⁄16 in.). The largest of the machines, having a 13 mm (½ in.) maximum chuck capacity, can handle the greatest range of sizes. Using the standard headstock-feed plate cam, parts up to 70 mm (2¾ in.) in length can be produced. By substituting a special bell cam, the maximum length can be increased to 102 mm (4 in.). However, with the addition of a special steady rest to provide adequate support and a special stock stop to ensure accurate length, parts up to 229 mm (9 in.) can be produced by double feeding the stock.

**Production Speeds.** Suited primarily to the production of parts that are too small, slender, or intricate for a turret-type automatic machine, the Swiss-type automatic bar machine reduces the production time and improves the accuracy, compared to the usual secondary operations and precision bench-lathe work. Most standard single-spindle turret-type automatic machines produce small screw machine parts of simple design in 3 s to approximately 6 min. Cycle times on Swiss-type automatic bar machines average about 1 min for most parts. Depending on the size, design intricacy, and finish required, however, cycle times may vary from as little as 2 s to as much as 4 or 5 min per workpiece. Reported scrap losses as high as 50% using other machining methods have been reduced to less than 10% using the Swiss-type automatic bar machine. Output can often be considerably increased by designing parts so that as many as three or four completed pieces can be produced and cut off before the nonproductive time of retracting the headstock takes place.

**Tolerances.** The cutting action on the Swiss-type machine is confined close to the support bushing. In this way, small tolerances can be held and good finishes obtained.

*Diameter tolerances* of ±0.005 to ±0.013 mm (±0.0002 to ±0.0005 in.) are common. The machine performs step, straight, taper, back, and form cutting and with an attachment can perform centering, drilling, and reaming. Tools are simple and adjustments easy. All tools that can, work at once, and a piece is finished in one pass, making the time short for most work.

*Lengths.* Turned lengths to shoulders or other locations can normally be held to ±0.013 mm (±0.0005 in.). Normal tolerances specified are ±0.05 or ±0.08 mm (±0.002 or ±0.003 in.), with ±0.13 to ±0.25 mm (±0.005 to ±0.010 in.) wherever design permits for maximum economy.

*Surface Roughness.* With proper feeds and speeds, surface finishes from approximately 1.25 μm (50 μin.) to as fine as 0.125 μm (5 μin.) can be obtained. In the manufacture of instrument parts, surface accuracy of 0.30 to 0.40 μm (12 to 16 μin.) is an average obtained in regular production.

## Computer Numerical Control Machines

Swiss-type automatic bar machines with computer numerical control that eliminate the need for cams and minimize both setup times and necessary operator skill are extensively used for lower-volume production requirements. Another important advantage of CNC-equipped machines is that the restriction to 360° layouts on cam-controlled machines is eliminated, permitting the incorporation of machining operations that are difficult or impossible to perform on cam-controlled machines. These operations include deep-hole drilling, single-point threading (both left and right hand), taper cutting, and contouring.

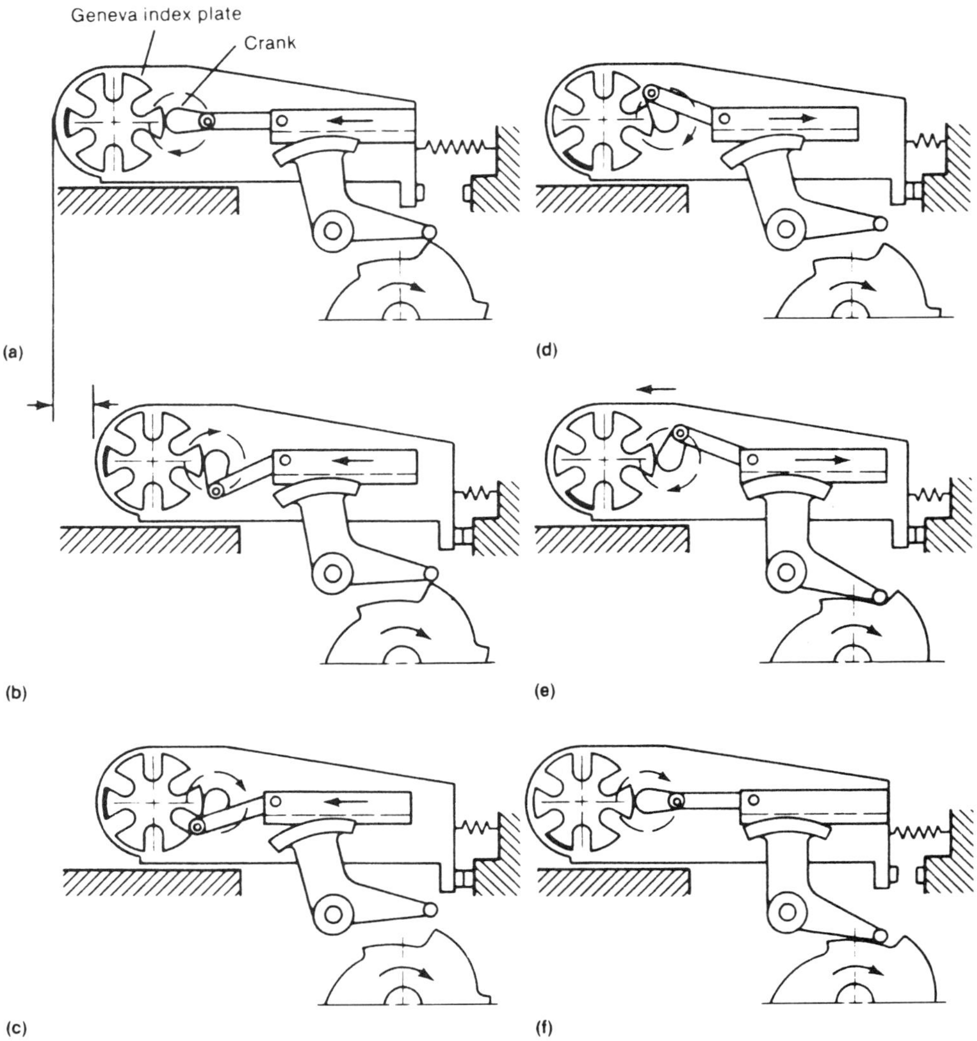


**Fig. 14** Schematic of the rapid indexing mechanism of the turret on the single-spindle automatic bar machine illustrated in Fig. 12. A trip mounted on a drum on the cam shaft (Fig. 13) engages the Geneva mechanism dog clutch, which is mounted on the back shaft. The individual steps that occur during indexing are illustrated above. Before the crank moves (a), the components of the drive are in the horizontal position and the turret head is locked; thus, the slide movement is entirely controlled by the movement of the cam follower on the contour of the cam, and the quadrant is engaged with the rack. A strong spring ensures that the cam follower is in positive contact with the cam. When the crack begins to rotate (b), the distance between the rack and the Geneva indexing plate shortens, so that the spring can pull the slide to its extreme rear position against the positive stop. In this position, the pin on the crack indexes the Geneva plate by one division (c). Because the slide is hard against the dead stop, the movement of the crank also causes the cam follower to lift from the cam plate. In this way, changes in the cam profile are overridden. After indexing (d), the crank will return the cam follower to the cam profile (e). In the meantime, the cam—without participating in the indexing process—has continued to revolve a small amount to the start of the next profile. When the cam follower is again in positive contact with the cam, the crank is provided with the necessary reaction to push the slide forward during the final phase (f) against the spring force. With the crank back in its horizontal position, the turret is then locked with a locking pin. In the indexed condition, the slide will again move entirely under the control of the cam profile for the next operation.

A typical CNC Swiss-type automatic bar machine has simultaneous two-axis control for the $x$-axis toolholder and the $z$-axis headstock. This permits straight, taper, and circular turning, as well as threading. Five toolholders, arranged in a radial configuration on these machines, are controlled by a servodisk (Fig. 17). Programmed signals from the CNC unit command the disk to select the predetermined toolholders, which travel the required distances by rotation of the disk.

The sliding headstock on these machines has a hydraulic chucking device that carries the bar stock longitudinally. Advancement and retraction of the headstock are controlled by a servomotor, which rotates a ballscrew for travel in increments of 0.0010 mm (0.00004 in.). The machine spindle is driven by an ac motor and provides infinitely variable, programmable speeds.

## Multiple-Spindle Automatic Chucking and Bar Machines

Although single-spindle screw machines eliminate the need for constant operator attendance, only one or two of the tooling positions are used at any given time. Thus, the total cycle time per workpiece is the sum of the individual machining and tool-positioning times of the several cutting tools. On multiple-spindle chucking and bar machines, sufficient spindles, usually four, six, or eight, are provided so that all tools cut simultaneously (Fig. 18). Therefore, the cycle time per piece is equal to the maximum cutting time of a single tool position plus the time required to index the spindles from one position to the next.

Multiple-spindle chucking and bar machines have two distinctive features. First, the multiple spindles are carried in a rotatable drum that indexes in order to bring each spindle into a different working position. Second, a nonrotating tool slide contains the same number of toolholders as there are spindles and thus provides and positions a cutting tool (or tools) for each spindle and imparts feed to these tools by longitudinal reciprocating motion. In addition, most machines have a cross (or side) slide at each spindle position so that an additional tool can be fed from the side for facing, grooving, knurling, beveling, and cutoff operations. These slides are shown in Fig. 18. All motions are controlled automatically.

The side slides are mounted radially around the spindles. There are four standard slides on the four-spindle machine and five standard slides on the six-spindle machine, with an additional optional slide available on each. The eight-spindle machine has six standard slides. These slides hold tools that will cut from the side of the workpiece. Each side slide is cammed individually and is driven by its own cam drum. Therefore, the feed rate can be different for each slide. The side slides feed slowly into the work to perform their cutting operations and then return to clear the spindles for indexing.

In general, the two slides are used for making heavy roughing and forming cuts. The other slides are used to complete subsequent machining operations to the required accuracy.

With a tool position available on the end tool slide for each spindle (except for a stock-feed stop at one position), when the slide moves forward, these tools cut essentially simultaneously. At the same time, the tools in the cross slides move inward and make their cuts. These operations are illustrated schematically in Fig. 19. When the forward cutting motion of the end tool slide is completed, it moves away from the work, accompanied by the outward movement of the radial slides. The spindles are indexed one position, by rotation of the spindle carrier, to position each part for the next operation to be performed. At one spindle position, finished pieces are cut off, and the bar stock or workpiece is fed to the correct length for the beginning of the next operation. Therefore, a piece is completed each time the tool slide moves forward and backward. The principal components that regulate the motion of a

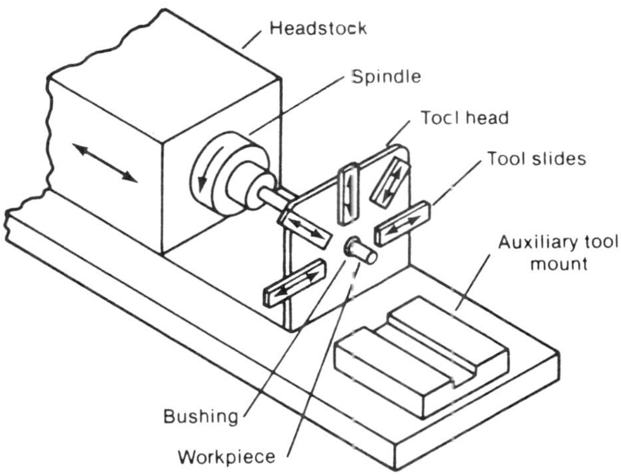


(a)

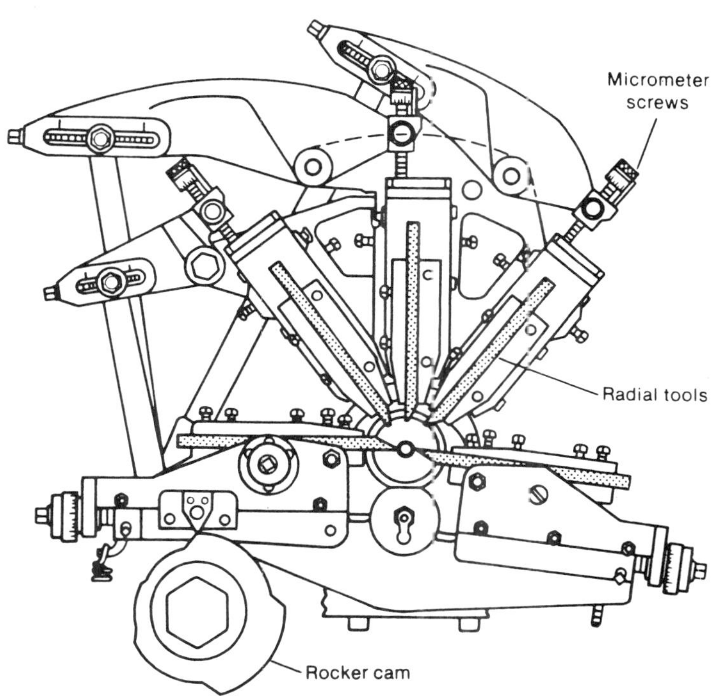


(b)

**Fig. 15** Schematic of a Swiss-type automatic bar machine. (a) Primary components and their motion. (b) End view showing rocker cam and tool control mechanism

spindle turret in a multispindle automatic machine are shown in Fig. 20.

Multiple-spindle chucking and bar machines are made in a considerable range of sizes. The sizes are determined by the diameter of the stock that can be accommodated in the spindles (Table 2).

In tooling a multiple-spindle chucking or bar machine, it should be remembered that, because all cutting operations occur simultaneously, the operating cycle of the main tool slide is determined by the operation that requires the longest time. Consequently, the operations should be balanced so that each requires the same amount of time. Although this ideal frequently cannot be achieved, careful planning permits it to be approached. Proper sequencing is important, and long operations can be broken up so as to be completed using two or more positions. Where a simple part is to be made, it is sometimes possible to arrange the tooling on a multiple-spindle automatic machine so that two pieces are produced in each revolution of the spindle carrier.

The only attention a multiple-spindle chucking or bar machine requires is to keep the bar stock feed rack supplied and to check the finished products occasionally to make sure they are within the desired tolerances. One operator can usually handle several machines.

Most multiple-spindle chucking or bar machines use cams composed of specially shaped segments that are bolted onto a drum to control the motions, thus reducing the need for special cams. Setting these cams and the tooling for a given job may require 2 to 20 h. However, once such a machine is placed in operation, the productivity is very great. A piece can often be completed every 10 s. Typically, 2000 to 5000 parts are required in a lot to justify tooling a multiple-spindle automatic.

The precision of multiple-spindle chucking or bar machines is good, but seldom as good as that of single-spindle machines. However, tolerances of 0.013 to 0.025 mm (0.0005 to 0.001 in.) on the diameter are common.

## Multiple-Spindle Automatic Chucking Machines

Multiple-spindle chucking machines are available in models that can accommodate chuck sizes up to 305 mm (12 in.) and that have four, five, six, or eight spindles. These spindles, which rotate the chucks that hold the workpieces, are arranged radially around a main slide, which does not rotate but moves longitudinally.

The spindle carrier is supported by, and indexes in, the frame of the headstock. The carrier is rotated by the index arm from the main camshaft and by the index plate, which is geared to the carrier.

Cutting tools are secured in adapters, which are fastened to the cross slides or to the tool carriage by means of T-slots. In some machines, the tool carriage is actuated by means of a drum cam on the main camshaft so that all tools moving axially are controlled simultaneously. Other types of machines have independent slides for each axial tool, controlled by individual cams. This permits greater versatility for tools working from the carriage.

Multiple-spindle chucking machines are equipped with devices that permit some spindles to be stopped while machining continues on others. Stopping specific spindles allows loading and unloading as well as operations such as cross drilling, milling, or broaching. Loading and unloading of the workpiece are usually done manually, but if the production quantity justifies the cost, special equipment can be designed for the magazine loading of many kinds of workpieces.

**Advantages.** Multiple-spindle chucking machines are intended to accomplish the same machining operations as single-spindle machines (automatic turret lathes). The main advantage of the multiple-spindle over the single-spindle machine is a reduction in machining time per piece. In contrast to the single-spindle machine, in which tools from only one turret position are working at one time, the multiple-spindle machine can be set up so that all tools are working at one time. In the single-spindle machine, the time required for completing one piece is the sum of the

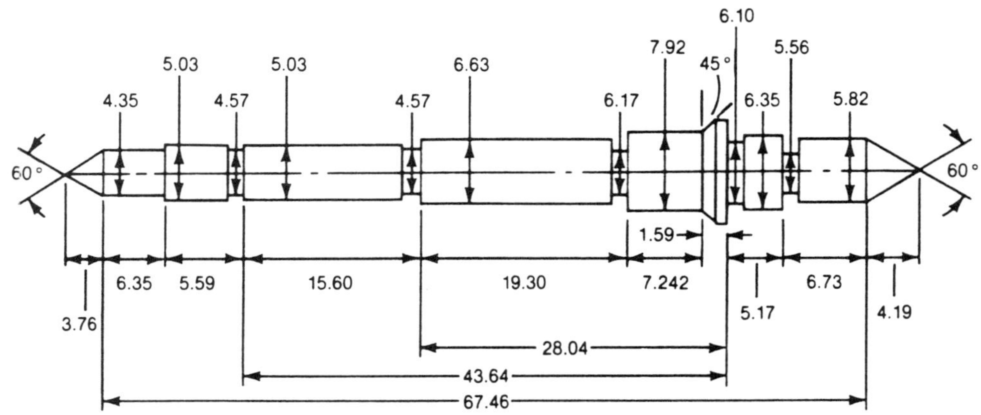


**Fig. 16** Typical intricate designs, which can be produced in one cycle of a Swiss-type automatic bar machine. Machining this shaft required 4 min, 36 s. Dimensions given in millimeters

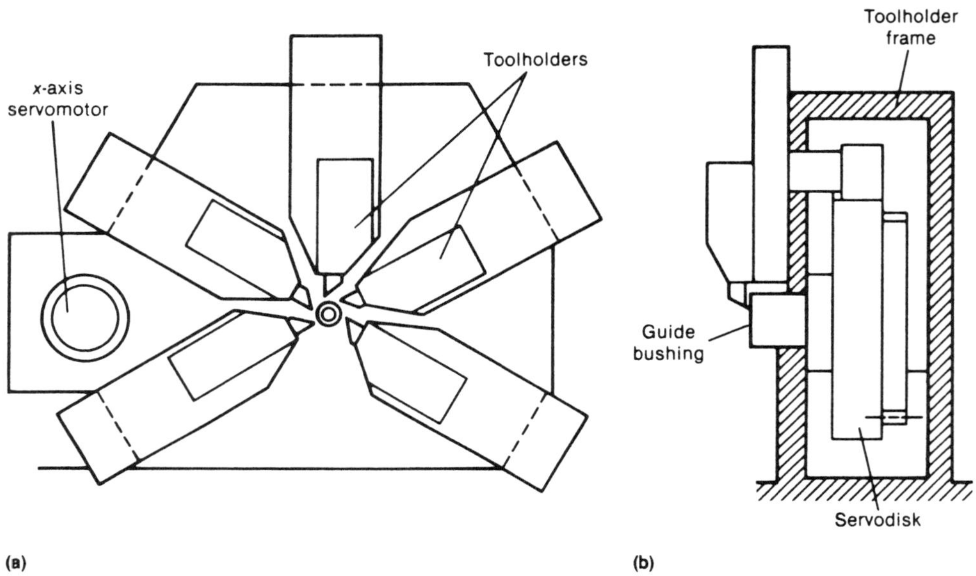


**Fig. 17** End view (a) and front view (b) of a CNC Swiss-type automatic bar machine in which a servodisk control mechanism regulates the movement of the five toolholders

times required for the individual turret operations (not considering cross slide operations that may be conducted simultaneously with turret operations). In multiple-spindle operation, the time required for completing one piece is usually that required for the most time-consuming single operation.

**Limitations.** The disadvantages of multiple-spindle chucking machines are high initial cost, increased setup time, and increased cost of tooling. Therefore, these machines are intended for long production runs.

## Multiple-Spindle Automatic Bar Machines

With respect to indexing of the spindles and tool-positioning methods, multiple-spindle automatic bar machines operate on principles essentially the same as those described in the section "Multiple-Spindle Automatic Chucking Machines" in this article and illustrated schematically in Fig. 19. However, the construction of bar machines is entirely different from that of chucking machines with regard to spindles and related devices for holding and feeding the work metal. In multiple-spindle bar machines, the bars are supported in a stock reel attached to the rear of the machine (Fig. 21). The bars are moved automatically from the stock reel through hollow spindles and spring collets by means of feed fingers.

Six-spindle machines are the most common, although these machines are also available with four, five, or eight spindles. Some models can handle bars up to 203 mm (8 in.) in diameter, but few are used to machine bars larger than 127 mm (5 in.) in diameter, mainly because the cutoff operation consumes too much time to be compatible with other operations.

As an indication of the capability of these machines, the spark plug shell illustrated in Fig. 22 was manufactured on a 32 mm (1¼ in.) capacity automatic bar machine at a rate of 800 pieces per hour, or 4.5 s per piece. Eighteen separate operations were

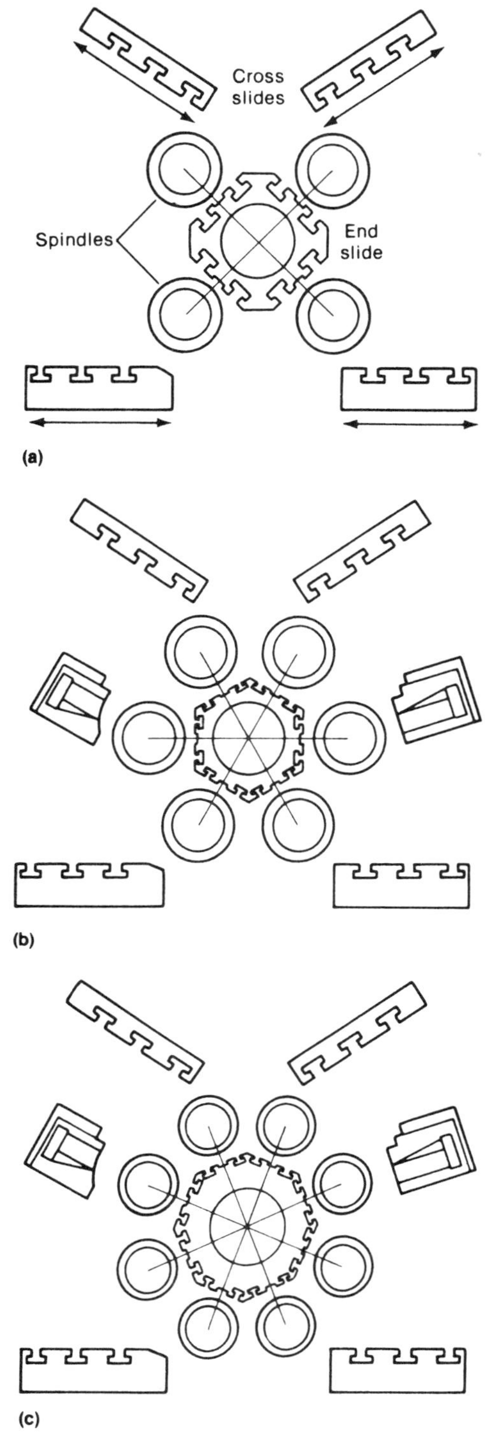


**Fig. 18** End views of the cross slides, end slide, and spindle arrangements for four-spindle (a), six-spindle (b), and eight-spindle (c) machines

required to machine this component using 21 mm (13/16 in.) hex-shaped steel.

## Vertical Multiple-Spindle Automatic Chucking Machines

Machines of this type, as illustrated in Fig. 23, are designed for high production and are usually provided with either five or

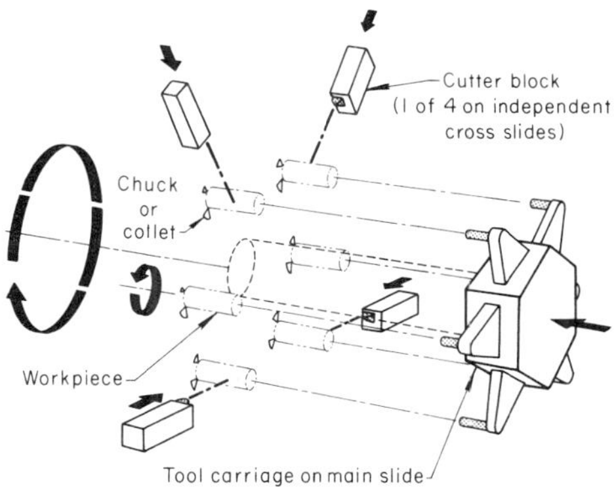


**Fig. 19** Principal components and their operation in a multiple-spindle automatic chucking or bar machine

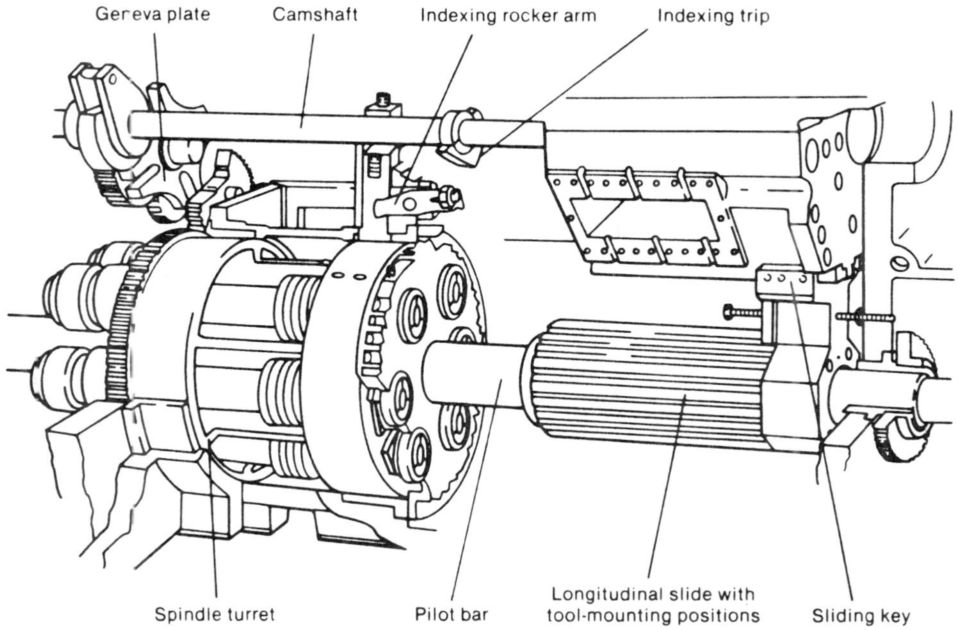


**Fig. 20** Spindle-turret mechanism of a multispindle automatic chucking or bar machine

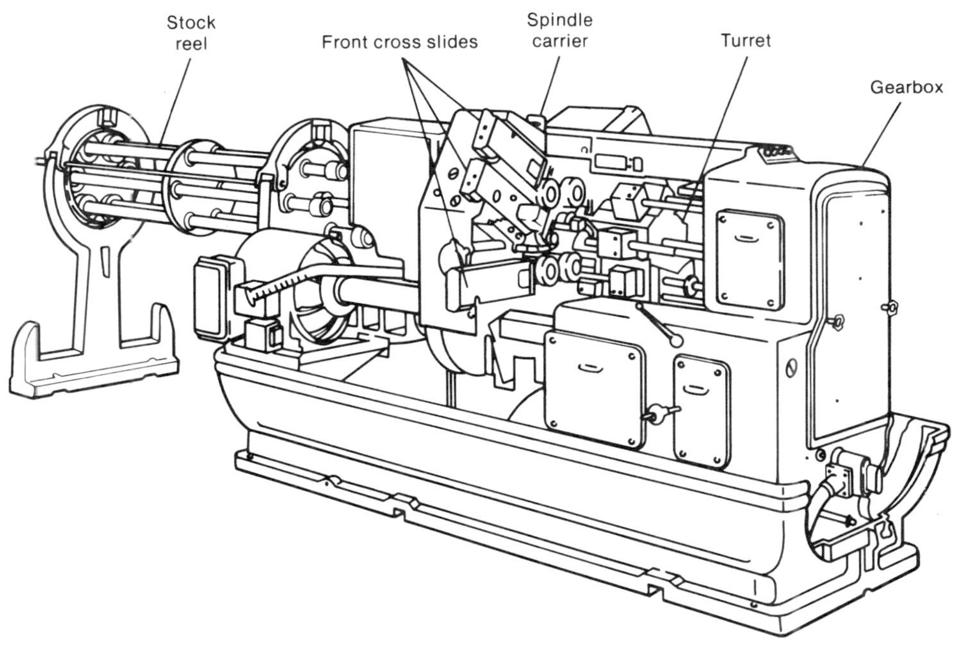


**Fig. 21** Basic components of a multiple-spindle automatic bar machine

seven workstations and a loading position. Thus, a machine may have either a six- or eight-spindle configuration. In some machines, two spindles are provided at each station, doubling the capacity for small-diameter work. The work to be machined is mounted in chucks; the larger machines have chucking capacities up to 457 mm (18 in.) in diameter. Both plain and universal heads can be used; the latter provides for tool feed in any direction. All varieties of machining operations can be performed, including milling, drilling, threading, tapping, reaming, and boring. The advantage of this machine is that all operations can be done simultaneously and in the proper sequence. At each station, except the loading one, an operation is performed that leads to the completion of the part when it has been indexed through all the working stations. The work cannot be indexed to the succeeding positions until the operation requiring the longest time is completed. Each time when all operations are completed, the work spindles stop, the tools are retracted, and the work table holding the spindles is indexed to the next position.

**Advantages.** A vertical multiple-spindle automatic chucking machine occupies a minimum of floor space for its capacity, and each station is readily accessible to facilitate setup and adjustments. A typical machine would have a bed measuring 2200 mm (87½ in.) in diameter and 3600 mm (140 in.) in height, excluding the motor assembly on the very top of the machine. The tools are carried and fed on slides around a central column (Fig. 24). Slides are available for vertical, horizontal, angular, and combination movements, and each slide feeds independently of the other stations.

**Applications.** Components normally machined on vertical multiple-spindle chucking machines are round, have short lengths in relation to their diameters, and are too cumbersome for conventional horizontal multiple-spindle machines. These units have been extensively used to machine automotive components, such as torque converter housings and 410 mm (16⅛ in.) diam cast iron transmission cases. Use of the vertical multiple-spindle machine is especially advantageous in the latter case; scrap due to workpiece breakage is lessened because of reduced handling of the cast iron workpiece.

## Standard Cutting Tools

Most cutting tools used in multiple-operation machining are similar to those used for the same operations conducted individually. For example, a drill made of the same cutting angles will ordinarily be used for drilling a given work metal regardless of whether the drilling is done as a single operation in a drill press or as one of several operations in a sequence on a multiple-spindle automatic machine. The same is usually true for reamers, taps, threading dies, boring tools, and single-point turning tools.

There are sometimes exceptions for the tools listed above. In a sequence of operations in multiple-operation machining, one tool may be unavoidably abused in order to obtain better efficiency from other tools that operate in sequence. Under these circumstances, tools such as drills or reamers

**Table 2 Typical automatic bar and chucking machines and capacities**

| Machine type | Maximum stock diameter or chucking capacity, mm | in. | Maximum length of turn with main slide, mm | in. | Maximum stock feed in one movement, mm | in. |
|---|---|---|---|---|---|---|
| **Bar machines** | | | | | | |
| Single spindle | 51 | 2 | 127 | 5 | 152 | 6 |
| | 67 | 2⅝ | 152 | 6 | 305 | 12 |
| | 203 (241 tubing) | 8 (9½ tubing) | 229 | 9 | 330 | 13 |
| | 203 | 8 | 178 | 7 | 457 | 18 |
| | 25 | 1 | ... | ... | 102 | 4 |
| Four spindle | 121 | 4¾ | 203 | 8 | 254 | 10 |
| | 178 | 7 | 200 | 7⅞ | 203 | 8 |
| | 57 | 2¼ | 152 | 6 | 152 | 6 |
| Six spindle | 89 | 3½ | 203 | 8 | 254 | 10 |
| | 89 | 3½ | 254 | 10 | 305 | 12 |
| | 64 | 2¼ | 152 | 6 | 152 | 6 |
| Eight spindle | 67 | 2⅝ | 203 | 8 | 254 | 10 |
| | 67 | 2⅝ | 254 | 10 | 203 | 8 |
| **Chucking machines** | | | | | | |
| Four spindle | 254(a) | 10(a) | 152 | 6 | ... | ... |
| | 305 (298)(a) | 12 (11¾)(a) | 203 | 8 | ... | ... |
| Five spindle | 222(a) | 8¾(a) | 152 | 6 | ... | ... |
| Six spindle | 305(a) | 12(a) | 203 | 8 | ... | ... |
| | 229 (260)(a) | 9 (10¼)(a) | 203 | 8 | ... | ... |
| | 178 (184)(a) | 7 (7¼)(a) | 152 | 6 | ... | ... |
| Eight spindle | 152(a) | 6(a) | 152 | 6 | ... | ... |
| | 203(a) | 8(a) | 203 | 8 | ... | ... |

(a) Maximum swing

are often made from a more highly alloyed high-speed steel, such as M33 or T15, than they would be otherwise.

Tool design is sometimes modified from that normally used for single-operation machining when experience with a specific application indicates the need. Information on tool material and design for specific operations is available in the Section "Traditional Machining Processes" in this Volume. Features of cutting tools that are unique to the machining of specific metals are covered in the Section "Machining of Specific Metals and Alloys" in this Volume.

**Radial Tangential Turning.** Some multiple-operation machines (notably turret lathes) have a saddle-type, positively driven tool carrier supported on the ways. With this type of equipment, turning tools are mounted on the carrier, thus permitting turning of the workpiece by the same procedures used with an engine lathe (see the article "Turning" in this Volume).

When using machines that do not have a saddle-type tool carrier, turning must be done from the turret or the tool carriage supported by a main slide (depending on the type of machine). Under these circumstances, more elaborate toolholders are necessary to maintain rigidity in the tools and sometimes to prevent bending of the workpiece (see the section "Tool Adapters and Mountings" in this article). However, mounting the tools for turning from turrets or tool carriages does permit greater flexibility in locating the tools radially at various positions than when such tools are mounted on a carrier supported by the ways.

In tangential cutting, the turning tool is set so that the bottom of the side cutting edge is on the vertical centerline of the work (Fig. 25). In radial cutting, the top of the side cutting edge is set on the horizontal centerline (Fig. 25). In many cases, the choice between tangential and radial cutting (where it is convenient to use either one) is arbitrary. Tangential cutting is often preferred for finish turning, particularly of brass and aluminum.

**Tool material, design, and rigidity** are all major influences on productivity, tool life, tolerance, and finish. Experience with specific applications often indicates that changes in tool material or design, or both, will improve results.

Although the composition of the metal being machined is usually the most important factor affecting the selection of tool material and design, the rigidity of the equipment and the specific conditions of the workpiece must also be considered. For example, carbide tools are not recommended for applications in which chatter is likely to develop or for workpieces requiring interrupted cutting.

Lack of rigidity in toolholders can erase all benefit gained from careful selection of tool material and design. Vibration allows chatter, and chatter results in dimensional inaccuracy and poor finish, as well as in subnormal tool life. For drills, reamers, boring bars, or other shank-mounted tools, the ratio of the unsupported length of the tool to its diameter is the main factor influencing rigidity. For holding tolerances of ±0.025 mm (±0.001 in.), an unsupported length 3½ times the tool diameter is about the limit that can be machined. This ratio is sometimes exceeded. In applications in which tolerances are not close and all cutting tools are made of high-speed steel, it is unlikely that carbide cutters would survive this lack of rigidity. When a high length-to-diameter ratio is mandatory for a specific application, solid carbide tools or boring bars are sometimes used, because carbide has a modulus of elasticity about three times that of steel. Methods of holding and supporting tools are discussed in the section "Tool Adapters and Mounting" in this article.

## Form Tools

Tools used for form turning are more important in multiple-operation machining than in engine lathe work. Form tools are used for turning a wide variety of tapers, grooves, and special shapes—often, several simultaneously.

Although there is no limit to the complexity of shape that can be incorporated in form tools and transferred to the workpiece, there is no limit to the total width of forming cut that is practical in a machine of a given horsepower. Form tools are moved in at right angles to the axis of the workpiece, thus making plunge cuts. Therefore, forming cuts are seldom wider than 83 mm (3¼ in.), and most are less than 64 mm (2½ in.) wide, mainly because of the power that would be required for forcing wider tools into the work.

Relief and back rake angles for form tools are essentially the same as for single-point tools, which vary with the metal being machined (see the article "Turning" and the Section "Machining of Specific Metals and Alloys" in this Volume). A major difference between form tools and single-point tools, however, is that single-point tools are usually ground with a side rake angle, while form tools are not. Because form tools are used for plunge cutting, by movement of the cross slide, the cutting edge is kept parallel to the longitudinal axis of the workpiece, and no side rake is used. This simplifies the sharpening of form tools because only the top of the cutting edge need be ground; thus, the established back rake angle is retained.

Regardless of size or design, form tools are of two general types: flat and circular (Fig. 26). Each has advantages and disadvantages over the other, as noted in the following discussion.

**Flat form tools** are most commonly of the dovetail type (which refers to the method of mounting), shown in Fig. 26(a), although simple flat form tools for mounting on standard tool posts are also available. Rigidity is especially difficult to maintain in form turning, and the dovetail method of mounting the cutting tool in the holder or adapter is preferred because it provides greater rigidity than tool post mounting.

Most flat form tools are made from high-speed steel, mainly because high-speed steel is less sensitive than carbide to lack of

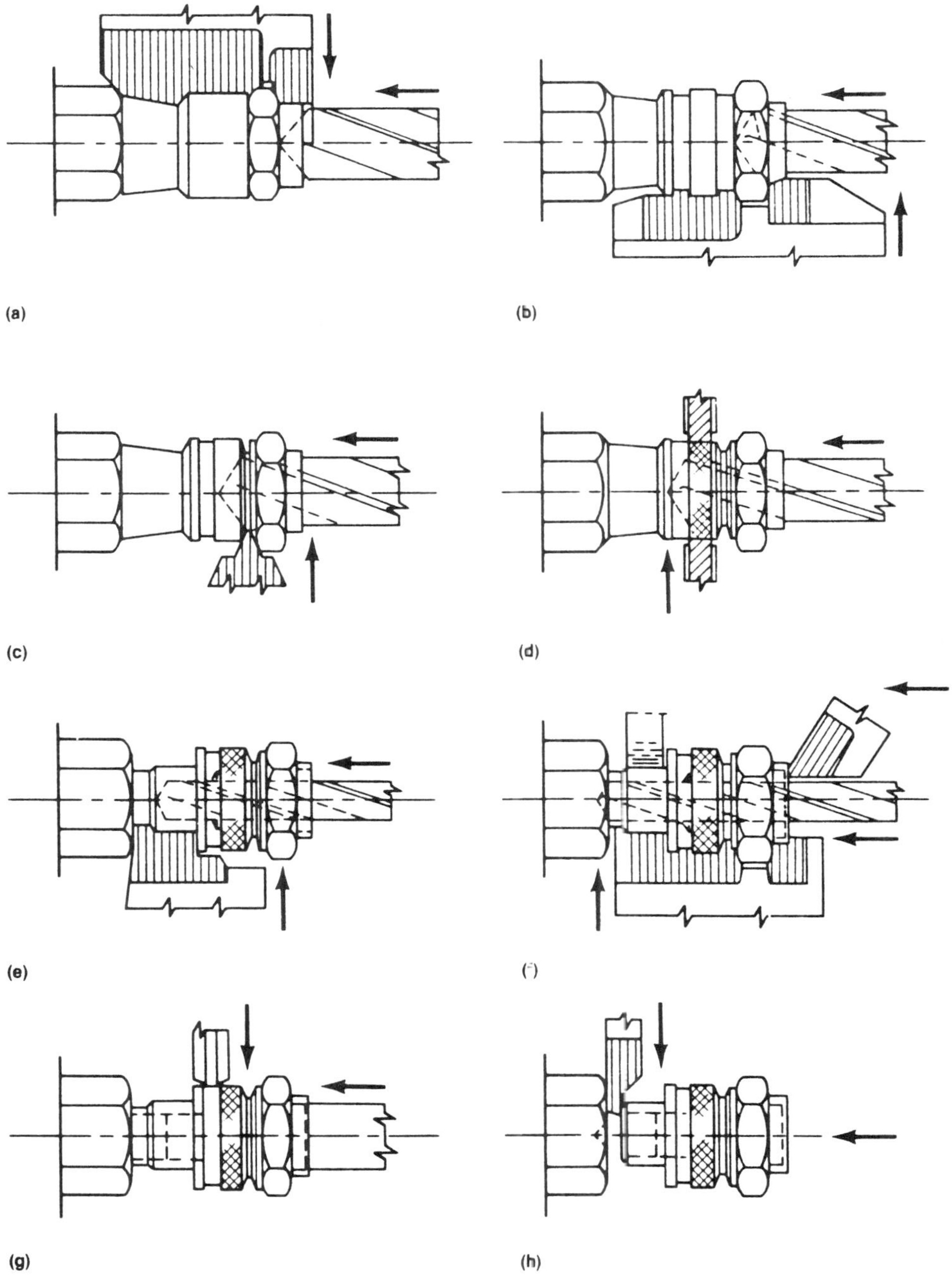


**Fig. 22** Tool layout for a 21 mm ($^{13}/_{16}$ in.) hex-shaped by 32 mm (1¼ in.) long spark plug shell machined on an eight-spindle automatic bar machine. It required 4.5 s, at a production rate of 800 pieces per hour, to machine the part of open-hearth Grade A steel using the following sequence of operations. (a) Position 8: feed out rough form, spot drill. (b) Position 1: finish form portion, drill partway. (c) Position 2: finish form portion, drill partway. (d) Position 3: knurl, drill partway. (e) Position 4: finish form remainder, mark for cutoff, drill partway. (f) Position 5: shave, chamfer, drill remainder. (g) Position 6: stencil, accelerate ream. (h) Position 7: cutoff

rigidity. Some tools have an alloy steel body and inserted cutting edges of high-speed steel. The inserting practice is most often used for larger tools or in applications in which the general-purpose high-speed steels have proved inadequate; in such applications, special grades such as T15, M4, or M44 are required. It is seldom practical to make the entire form tool from these more highly alloyed grades of high-speed steel, because they cost two or three times as much as the general-purpose types. In addition, the highly alloyed grades are extremely difficult to machine, which further increases cost.

Carbide inserts are also used for flat form tools and are satisfactory if the setup is sufficiently rigid. Carbide cutting edges are especially well adapted to the form turning of soft metals such as aluminum and brass at high speeds (often up to 180 m/min, or 600 sfm).

Rigidity is the main advantage of flat form tools of the dovetail type over circular form tools. The main disadvantage of flat form tools is higher cost. Flat form tools (particularly of the dovetail type) are more expensive to produce because of the numerous machining operations involved. In addition, the total life of a flat form tool is usually less than that of a circular tool because a flat tool cannot be ground as many times.

**Circular form tools** (Fig. 26b) are simpler to make than flat tools because the required contour can be produced by turning. After turning, only a milling operation is needed to provide the cutting edges. As indicated in Fig. 26(b), the tool is often locked in the adapter by means of serrations on one face that engage matching serrations in the adapter.

Circular form tools are seldom made with inserts, because this would destroy their two main advantages over flat form tools; namely, simplicity of production and longer tool life. When made from solid material, circular form tools have extremely long life because they can be reground repeatedly as long as there is enough metal remaining for mounting. Most solid circular form tools are made from a general-purpose high-speed steel, such as M2, although a more highly alloyed grade can be used when necessary. Some circular form tools are made of solid carbide. The primary disadvantage of circular form tools is the difficulty of mounting to obtain the same degree of rigidity as with dovetail tools.

## Tool Adapters and Mountings

The T-slot principle of mounting, shown on the two form tool adapters in Fig. 26, is used for most tools that operate from cross slides, regardless of how the adapter is designed. T-slot mounting is also used for securing tool adapters to carriers that move axially on the main slides of multiple-spindle machines (Fig. 19). The tees used for mounting are similar to those used for cross slide mounting (Fig. 26), except that they are parallel to the axial cutting direction of the tool instead of being at right angles to the direction of cutting.

**Turret Tool Configurations.** Turret tools are usually mounted by one of three methods, depending on the type of turret. The most common method of mounting tools on the main turret of a conventional horizontal turret lathe (Fig. 5) is the use of a standard type of adapter that is fastened directly to the turret face by means of four bolts (Fig. 27). For single shank-type tools such as drills or reamers, the adapter can be a simple holder that is modified for various sizes of tool shanks by means of bushings. The adapter shown in Fig. 27 is of a more complex design that holds one turning tool and one boring tool, thus enabling two operations to be performed at one position of the turret. Special adapters are available

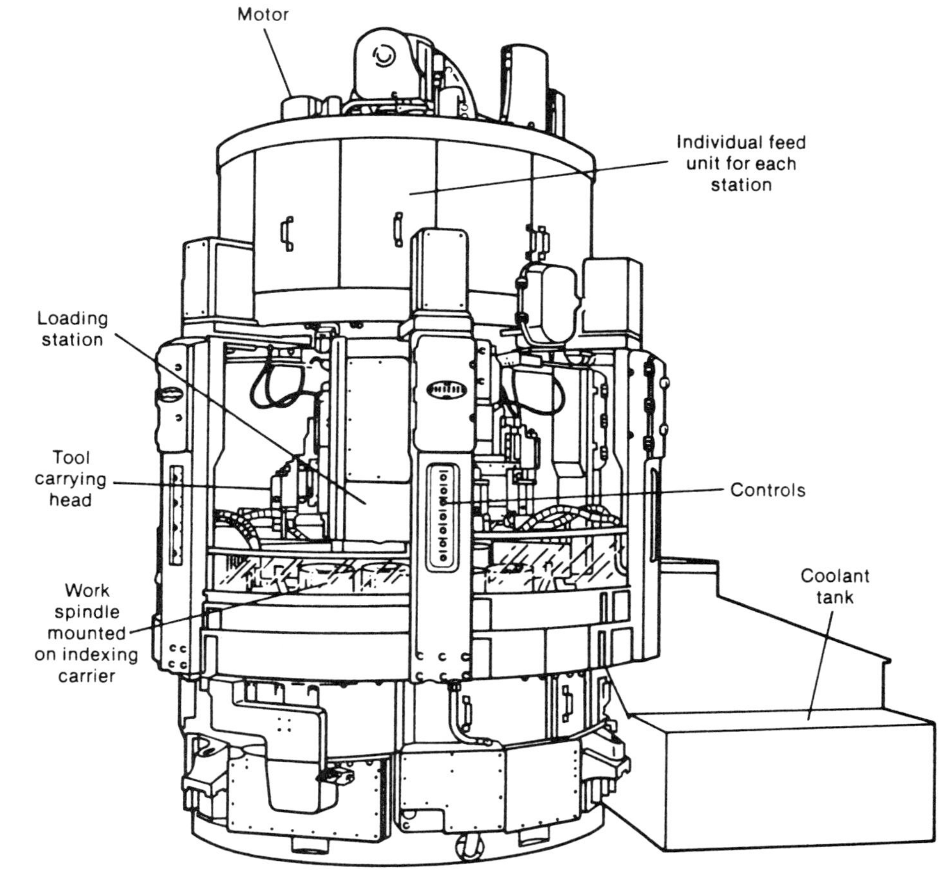


**Fig. 23** Basic components of a vertical multiple-spindle automatic chucking machine

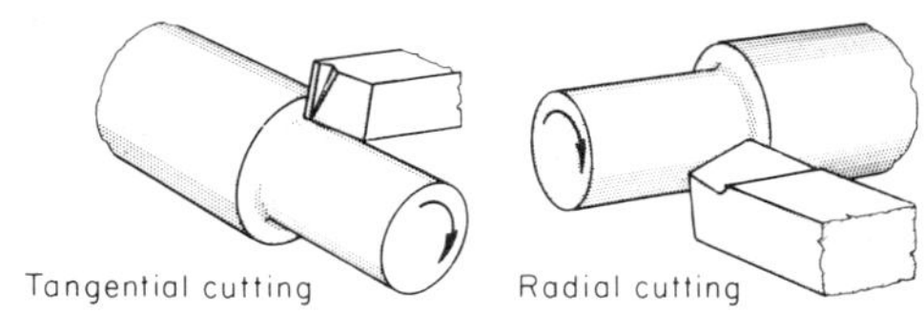


**Fig. 25** Positions of tool and workpiece for tangential and radial cutting

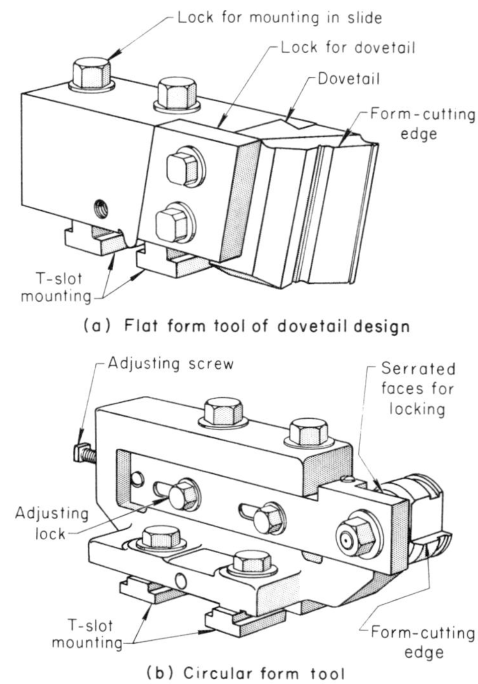


**Fig. 26** Two principal types of form tools shown in adapters for mounting on the cross slide

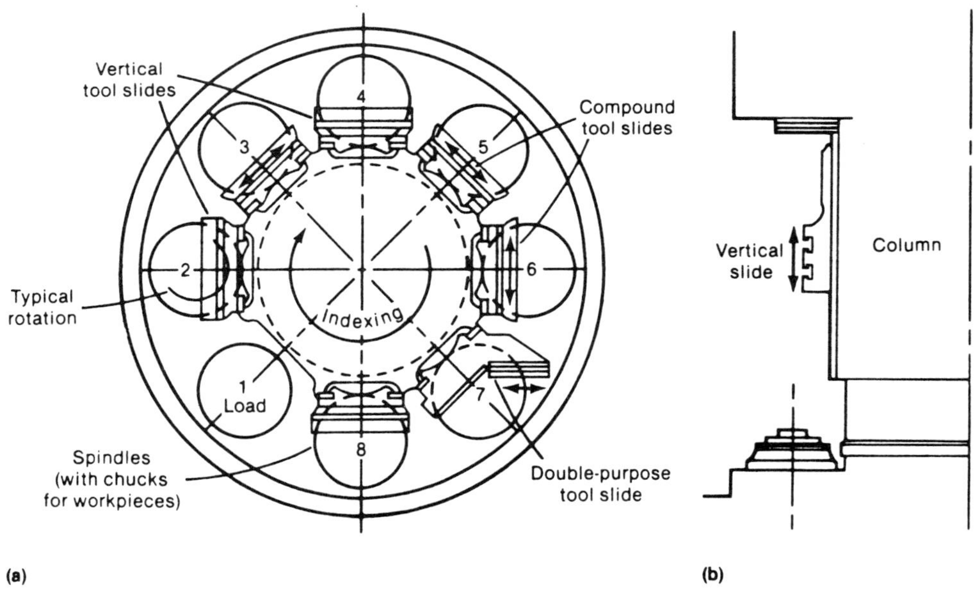


**Fig. 24** Schematic showing a horizontal view (a) and a vertical view (b) of an eight-spindle vertical automatic chucking machine illustrating seven workstations plus one loading station

that hold three or four tools for performing three or four operations simultaneously, particularly when the turret is capable of moving horizontally in two directions.

Turrets that rotate on a horizontal axis parallel to that of the workpiece (Fig. 8) require a rugged mounting between the turret and the tool to maintain rigidity. Consequently, the tool adapters (Fig. 8) are secured to the turret with dovetails. Coarse adjustment of the tool is obtained by loosening the entire adapter and sliding it along the dovetail slot in the turret. Finer adjustment is obtained by moving the tool within the adapter. Turrets that rotate on a horizontal axis at right angles to the axis of the workpiece (Fig. 11) have bored holes in which tool shanks can be inserted directly, if of the same diameter as the holes, or (more usually) in which round shanks of standard tool adapters are inserted and locked.

## Workpiece Supports

One feature of construction common to all types of machines discussed in this article is the absence of a tailstock. Consequently, workpiece overhang can become a problem. As unsupported length increases, the difficulty in controlling dimensions increases, along with the possibility that the workpiece will climb on the tool, which has often been responsible for a costly wreck. The amount of unsupported length that can be tolerated depends greatly on the operations being performed. For cross slide operations such as form turning, an unsupported length of about 3½ times the diameter is usually the recommended maximum.

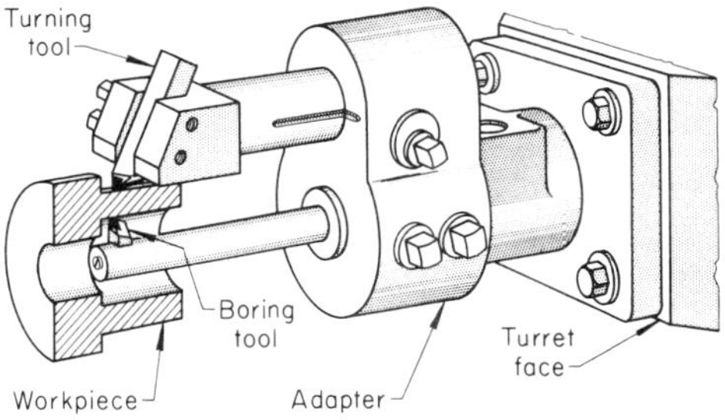


**Fig. 27** Method of mounting tools to the face of the main turret on a horizontal turret lathe

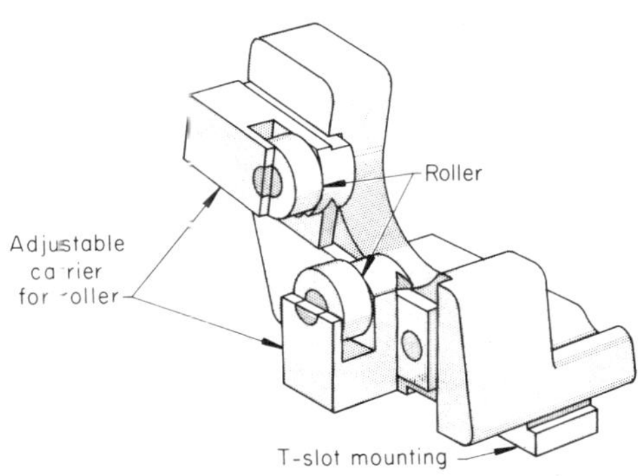


**Fig. 28** Roller support used to prevent the workpiece from bending during machining

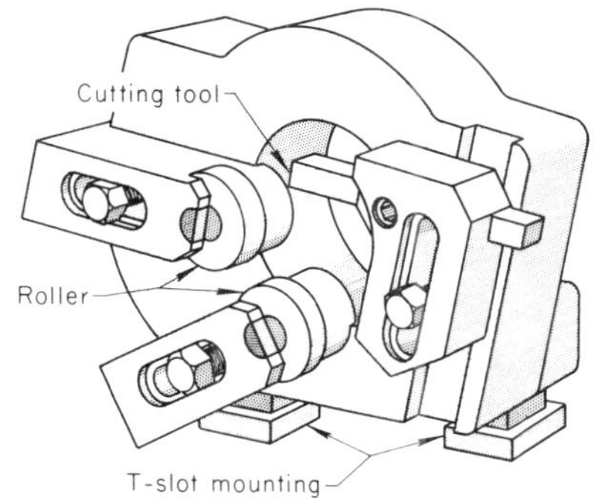


**Fig. 29** Box tool with rollers for support of the workpiece during axial turning

**Natural Support.** Sometimes, a sequence of operations to be performed can be planned such that the workpiece is provided with natural support from the tools. For example, it may be possible to perform simultaneous operations with axially and radially mounted tools; in this case, a drill provides a measure of support while form turning is accomplished from the cross slide. Cutting simultaneously from two or more cross slides is another method by which sufficient support is achieved by natural means. Usually, however, where unsupported workpiece length is equal to several diameters, some special means of support must be provided when the outside diameter of the workpiece is machined.

**Roller supports** (Fig. 28) are frequently used where workpiece overhang is excessive. Usual practice is to mount such a device on a cross slide opposite the direction of the cutting forces; this prevents bending of the workpiece during machining. As seen in Fig. 28, the rollers can be adjusted to accommodate a range of workpiece diameters. Supports similar to the type shown in Fig. 28 but in which the work is supported in a vee instead of by rollers have also been successfully used, but supports that use rollers are preferred.

**Box Tools.** Roller (or vee) support can also be used for axial turning, as from a turret. Tools used for this procedure combine support with one or more cutting tools. Tools that incorporate both support and cutters are often called box tools and are made in a variety of designs and sizes. A box tool that incorporates two supporting rollers and one cutting edge is illustrated in Fig. 29.

In turning with this type of tool, the rotating workpiece passes between the rollers and cutter. The cutter is set slightly ahead of the rollers (usually about 0.8 mm, or 1/32 in.), which allows the rollers to begin their function of providing support as turning progresses. As indicated in Fig. 29, both the rollers and the cutter can be adjusted, permitting considerable size variation (commonly about 100%) for any given tool.

**Die Head Support.** The use of adjustable die heads is another means of providing support for workpieces while performing axial cutting. Die heads used for this purpose are similar in construction to those used for die threading (see the article "Die Threading" in this Volume) and can be mounted on the face of a turret or to a sliding tool carriage. The die head serves as a holder for four identical cutting tools. The cutting tools can be used for plain turning, contour turning, taper turning, turning of multiple diameters, and some types of form turning. The four tools are mounted in the die head 90° apart; therefore, each tool is directly opposite another tool, and the cutting forces during operation are balanced.

**End Support.** In some cases, particularly for extremely small workpieces, end support is more practical than side support. Either male or female supports can be mounted in a turret or a tool carrier that slides axially. In effect, this type of support serves the same purpose as a tailstock on an engine lathe.

## Selection of Equipment and Procedure

In choosing the most appropriate machine for processing a given workpiece at the lowest cost, one or more of the following factors must be considered:

- Workpiece shape or product form
- Workpiece size
- Type and number of operations required
- Production quantity
- Production rate
- Special requirements
- Cost

The following sections discuss the influence of these factors and present examples that describe equipment and techniques selected for specific production applications.

### Workpiece Shape and Product Form

In most cases, the product form of the work material is known before the machine and procedure are selected; therefore, the choice between a bar machine and a chucking machine is established. It is common practice to cut bars into slugs for processing in a chucking machine, particularly when the bars are large and the time required for the cutoff operation on a bar machine would impair the efficiency of the operation sequence.

The shape of a casting or a forging may determine whether a single-spindle or a multiple-spindle chucking machine is the better choice. Workpieces that are symmetrical and can be placed directly in a two-jaw or three-jaw chuck present less of a problem than irregularly shaped workpieces that require special holders. Work holders add substantially to the cost of tooling, and to justify the cost of tooling multiple spindles, total production quantities of parts requiring work holders must be larger than for parts that can be chucked directly. Large, irregularly shaped parts that must be fastened to a face plate are invariably machined on a single-spindle machine. When none of the multiple-operation machines uses a tailstock, the amount of overhang of the workpiece (or its length-to-diameter ratio) may affect the selection of machine and procedure.

Some workpieces are so complex, fragile, or asymmetrical (or all three) that they are not well suited to machining on any of the conventional multiple-operation machines. One alternative is to machine small quantities of such parts on an engine lathe and to machine large quantities on units specially built for the specific application. Another alternative is to use a combination of machines, as in the following example, in which both an engine lathe and turret lathes were used.

**Example 1: Use of Turret and Engine Lathes for Machining a Complex, Nonrigid Casting.** The malleable iron casting shown in Fig. 30 presented problems in machining because a lack of rigidity resulted from the thin web connecting the hub and the rim and because the asymmetrical shape caused severely interrupted cutting. All lathe operations on these parts were done in three separate steps: two in manually operated turret lathes and one in an engine lathe.

As shown in Fig. 30 (see operation 1), each casting was first located in the chuck of a turret lathe by means of specially designed spring-loaded bushings. Two

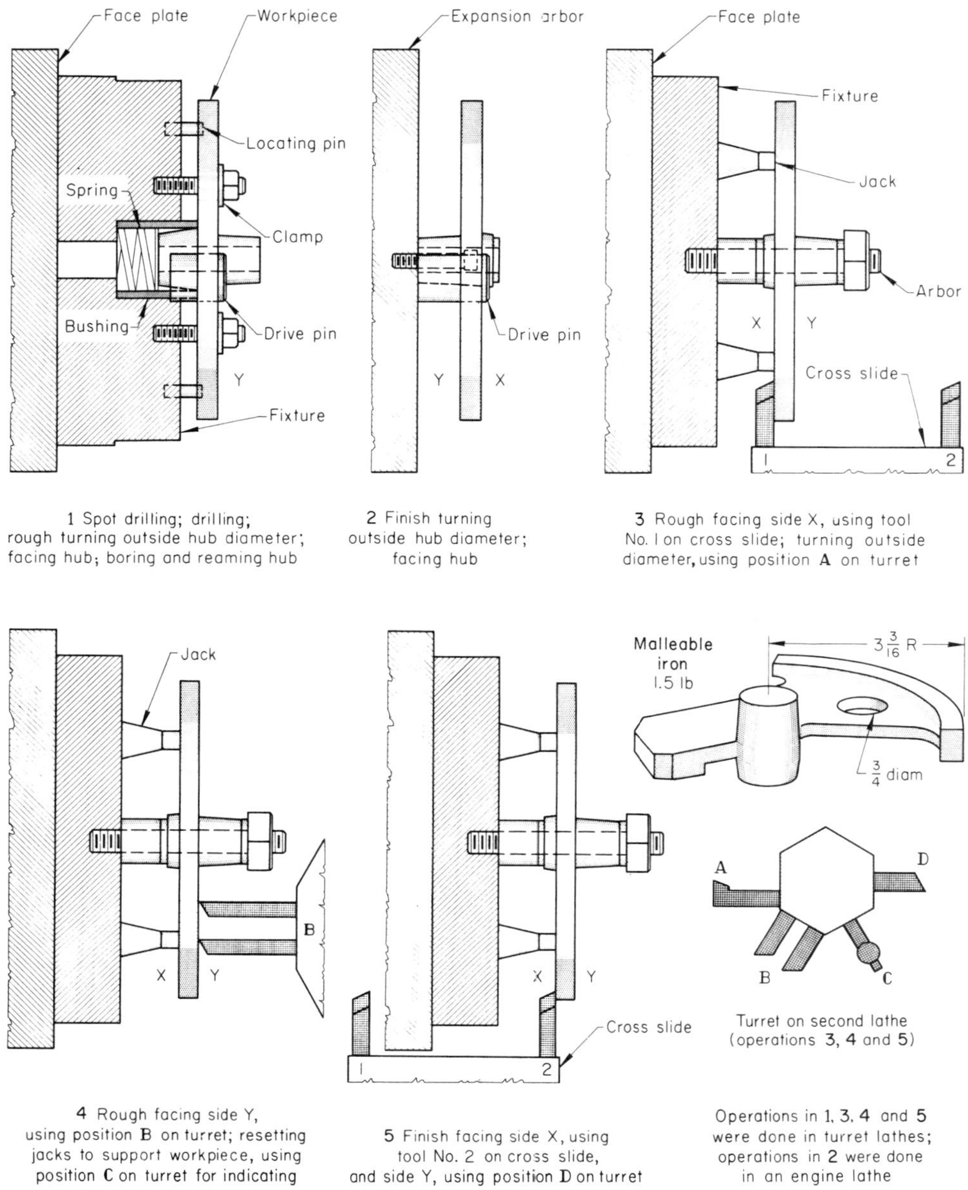


**Fig. 30** Machining an asymmetrical, nonrigid casting. Dimensions given in inches

**Table 3 Guide for the selection of chucking machines on the basis of workpiece weight and production quantity**

| Lot size, pieces | Machine for processing forgings or castings weighing: 2.3–11 kg (5–25 lb) | 11–23 kg (25–50 lb) | Over 23 kg (50 lb) |
|---|---|---|---|
| Up to 10 | Horizontal turret lathe or engine lathe | Vertical turret lathe | Vertical turret lathe |
| 10–100 | Horizontal automatic lathe | Vertical turret lathe (automatic or with tape control) | Vertical turret lathe (automatic or with tape control) |
| 100–1000 | Horizontal multiple-spindle automatic | Vertical multiple-spindle automatic | Vertical automatic turret lathe |
| 1000–100 000 | Horizontal multiple-spindle automatic (possibly with automatic loading) | Vertical multiple-spindle automatic | One or two double-spindle vertical automatic turret lathes |
| Continuous high production | Horizontal multiple-spindle automatic; automatic loading | Vertical multiple-spindle automatic; automatic loading | Special machine for particular part and operation |

clamps held the workpiece rigidly against the locating pins. Driving was accomplished with a pin that extended through the 19 mm (¾ in.) diam hole in the web. After being drilled, turned, faced, bored, and reamed, the workpiece was transferred to an engine lathe, where it was mounted in a reverse position on an expansion arbor and again driven with a pin extending through the 19 mm (¾ in.) diam hole for machining as indicated in operation 2 in Fig. 30.

The workpiece was then reversed again, mounted on a face plate, and chucked in a second turret lathe. The short side of the hub (side X, Fig. 30) was held against a shoulder on a threaded arbor, and two jacks were set in position against the web section to keep the part from springing while it was being machined. With this setup, lathe operations were completed in three stages, utilizing the cross slide and turret tools as indicated in operations 3, 4, and 5 in Fig. 30. The cross slide tools (1 and 2) were spaced far enough apart to straddle the face plate as it revolved.

As Fig. 30 indicates, only four of the six positions on the main turret were used in operations 3, 4, and 5, and only three of these were used for cutting. Position C on the turret was used for indicating to reset the supporting jacks.

All operations were done without cutting fluid at a spindle speed of 255 rev/min. Feeds were manually controlled at 0.089 mm/rev (0.0035 in./rev). Cutting tools were made from high-speed steel. This casting, which weighed 0.68 kg (1.5 lb), was not large or heavy enough to require counterbalancing while being rotated, but workpieces of similar shape that weighed more than 1.4 kg (3 lb) would require counterbalanced chucks.

## Workpiece Size

Workpiece size, alone or in conjunction with shape, sometimes restricts the choice of machine that would be suitable. Some plants have developed guides to the selection of machines on the basis of workpiece size or weight and other variables. Table 3, for example, was developed in one plant to be used in selecting chucking machines for processing forgings and castings on the basis of weight and quantity of workpieces. Although useful as guides, such tables are often deficient because they do not consider all the factors by which machine selection is determined. For example, in Table 3, a multiple-spindle machine is recommended for machining workpieces weighing 2.3 to 11 kg (5 to 25 lb) in lots of 1000 to 100 000. This recommendation does not take into account that some workpieces are of such shape that a multiple-spindle machine could not be used or that some requirements, such as special tolerances, might preclude the consideration of multiple-spindle equipment.

Although most bar machines (particularly multiple-spindle bar machines) are built for handling bars or tubes less than about 76 mm (3 in.) in diameter, single-spindle machines, as well as multiple-spindle machines with up to four spindles, are available that can handle bars up to 203 mm (8 in.) in diameter. The maximum diameter of work material that can be handled in machines having more than four spindles decreases as the number of spindles increases.

Parts are seldom machined from solid bar stock larger than about 76 mm (3 in.) in diameter because:

- The cutoff operation is too lengthy to be compatible with other operations
- The initial cost of large multiple-spindle machines often becomes prohibitive because the high burden rate is reflected in the cost of the finished part

Consequently, when making parts from the larger sizes of bars, it is usually more practical to cut them into slugs in a preliminary operation and then use a chucking machine. This practice, however, has a disadvantage when the metal enclosed by the chuck also requires machining. Under these conditions, either a preliminary or a secondary operation is required.

The size of the workpiece alone seldom influences a choice between a hand-operated and an automatic machine, but workpiece size is likely to influence the choice between single-spindle and multiple-spindle machines. This is particularly true for chucking machines. For example, some workpieces are of such shape that they can be held in relatively small chucks, but other portions would require clearance between workpieces that would be out of proportion to the chuck size, thus precluding the use of a multiple-spindle machine. An example of such a workpiece is a wheel-like part having a relatively small hub that can be used for chucking.

In other applications, the chucks required for a given size of workpiece are too large for practical consideration of multiple-spindle equipment. For example, if a multiple-spindle machine were to be used for machining 34 mm (1 7/16 in.) diam castings, the spindles would necessarily be located on a 1015 mm (40 in.) diam circle (minimum) and would require a work space at least 1.5 m (5 ft) in diameter. Although such a large machine could be obtained, the large capital investment could not be justified unless quantity requirements were extremely high.

## Type and Number of Operations

The number of operations frequently refers to the number of dimensions obtained

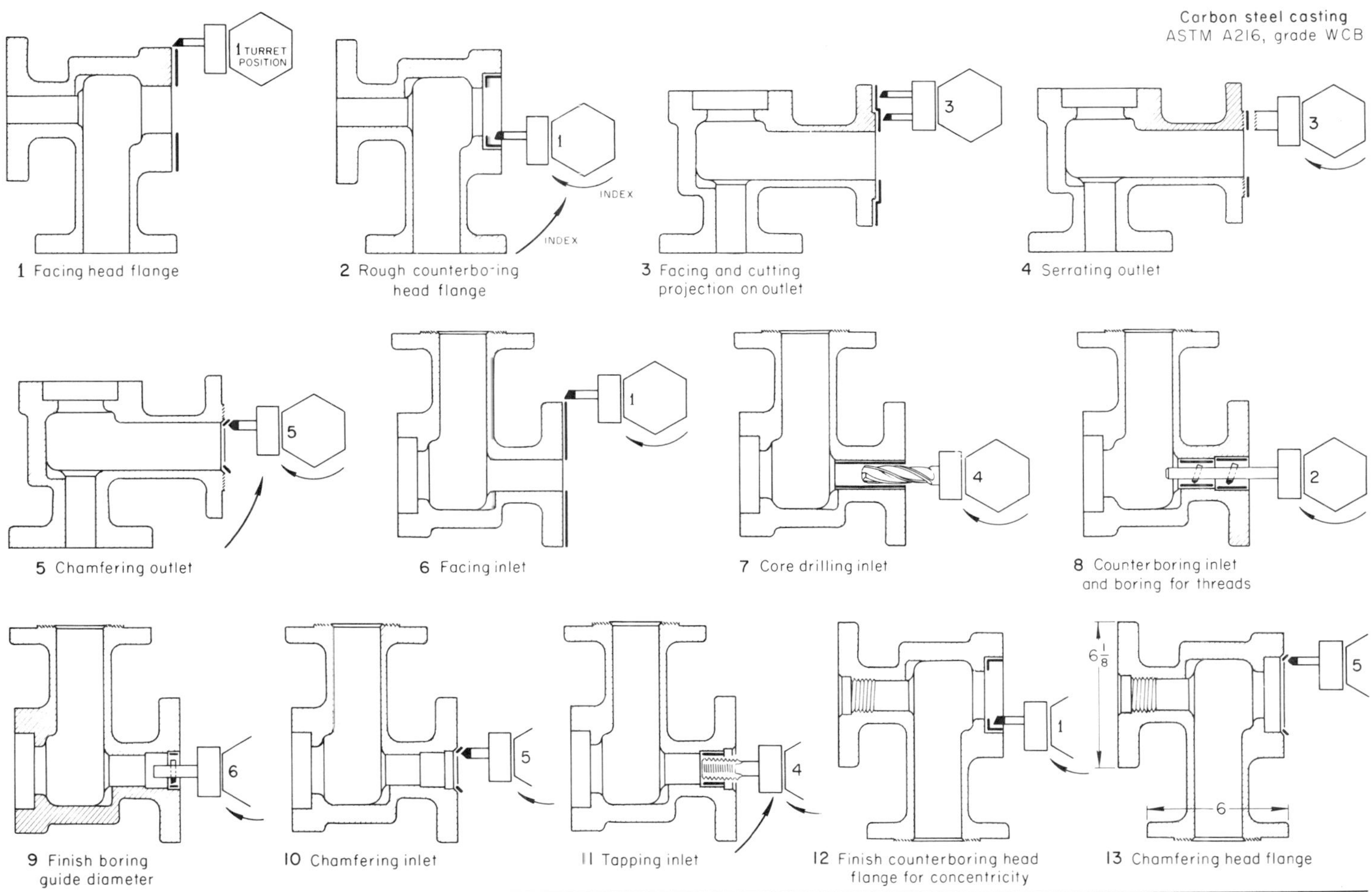


| Parameter | Operation(a) 1 | 2 | 3 | 4 | 5 | 6 | 7 | 8 | 9 | 10 | 11 | 12 | 13 |
|---|---|---|---|---|---|---|---|---|---|---|---|---|---|
| Speed, rev/min | 211 | 211 | 211 | 47 | 211 | 211 | 89 | 211 | 211 | 211 | 32 | 211 | 211 |
| Speed, m/min (sfm) | 76 (250) | 49 (160) | 91 (300) | Contact | Contact | 94 (310) | 10 (33) | 23 (75) | 24 (80) | Contact | 4 (13) | 49 (160) | Contact |
| Feed, mm/rev (in./rev) | 0.58 (0.023) | 0.58 (0.023) | 0.58 (0.023) | Contact | Hand | 0.58 (0.023) | 0.38 (0.015) | 0.58 (0.023) | 0.58 (0.023) | Hand | Lead | 0.58 (0.023) | Hand |
| Length of cut, mm (in.) | 41 (1 5/8) | 16 (5/8) | 51 (2) | 20.6 (13/16) | Contact | 51 (2) | 76 (3) | 76 (3) | 13 (1/2) | 3.2 (1/8) | 25 (1) | 41 (1 5/8) | 3.2 (1/8) |
| Machining time, h | 0.0434 | 0.0200 | 0.0566 | 0.0166 | 0.0066 | 0.0566 | 0.0834 | 0.0354 | 0.0066 | 0.0132 | 0.0134 | 0.0216 | 0.0066 |
| Handling time, h | 0.0334 | 0.0200 | 0.0200 | 0.0200 | 0.0200 | 0.0250 | 0.0250 | 0.0200 | 0.0200 | 0.0200 | 0.0268 | 0.0334 | 0.0200 |

(a) Serrating tool, core drill, and tap were made of high-speed steel; all other tools were of carbide. All machining was done without cutting fluid.

**Fig. 31** Sequence of operations for machining steel castings, held and indexed in a box chuck, in a horizontal turret lathe. Dimensions given in inches on a workpiece by machining. This number may differ from the number of tools because in many applications one tool performs two or more operations. Seldom should any type of turret lathe or automatic machine be considered for parts that can be machined with fewer than three different tools.

Because of the way they are constructed, turret lathes generally have less tool capacity than machines having a sliding tool carriage. For example, although turret lathes having as many as 9 turret stations have been used in production for performing as many as 25 operations, most turret lathes have no more than 6 turret stations and a cross slide. The tool clearance usually limits the number of turret stations. Therefore, when more than eight or nine tools are required, machines having cross and longitudinal slides should be considered.

Although it is possible to mount as many different axial tools in a single-spindle machine as in a multiple-spindle machine, there are limitations on the number that can be cross slide mounted in single-spindle machines. Therefore, when some of the tools are not adaptable to turret mounting, a single-spindle machine will have less capacity than a multiple-spindle machine. Some tools (for example, thread rolling tools) must be mounted accurately in relation to other tools; under these conditions, turret mounting is inadequate. Consequently, there are many parts for which the types of operations or the number of operations, or both, indicate the most practical machine and procedure.

In making some parts, spindle rotation must be stopped to permit cross-machining operations. In multiple-spindle machines, one or more spindles can be stopped while the others continue working. Consequently, for applications in which operations such as cross drilling or milling are incorporated in the sequence, a multiple-spindle machine is usually the most practical.

When workpiece shape is such that all surfaces being machined can be indexed around a common center, a box chuck can be used, thus extending the number of possible operations. One such part is described in the following example.

**Example 2: Thirteen Operations in a Turret Lathe.** A horizontal turret lathe was used to perform 13 operations on carbon steel castings, as shown in Fig. 31. The shape of these castings made it possible for all machined surfaces to be indexed around a common center; therefore, they were held in a box chuck that could be indexed through 360° around a vertical axis.

As shown in Fig. 31, from the first index position of the box chuck, two operations were performed using one turret position. In the second index position of the workpiece, two turret positions were utilized to perform three operations. Position 3 of the turret was equipped with a multiple toolholder, which held the facing tools used in operation 3 as well as the serrating tool used in operation 4.

Six operations were performed in a third index position of the workpiece, utilizing five turret positions and six tools. Turret position 4 was equipped with taper holders that allow quick changes of shank-type tools (operations 7 and 11). The workpiece was then returned to the same index position used for performing operations 1 and 2. In this final position of the workpiece, two more operations were performed from two different positions of the turret. Additional processing details are included with Fig. 31.

In cases in which quantities are low, this same general type of part is machined on vertical turret lathes; however, production quantities of the part in this example were sufficient to warrant the cost of fixtures and tooling, which resulted in lower processing costs and improved reproducibility.

**Single-Spindle Versus Multiple-Spindle Automatic Machines.** For some parts, it is impractical to complete all operations on a single-spindle machine because of insufficient cross slide capacity, and it is necessary to use secondary operations or to

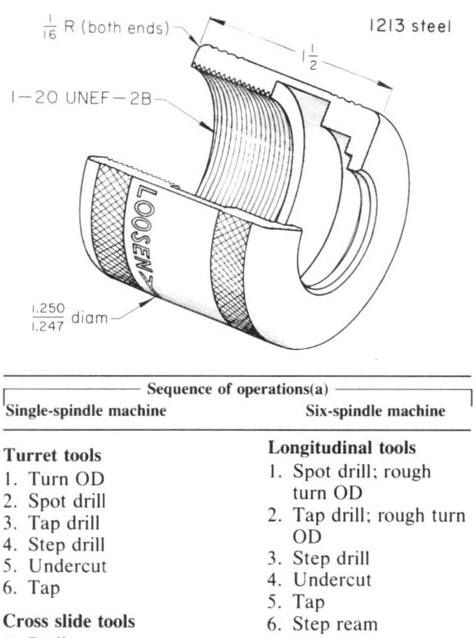


| Sequence of operations(a) | |
|---|---|
| **Single-spindle machine** | **Six-spindle machine** |
| **Turret tools** | **Longitudinal tools** |
| 1. Turn OD | 1. Spot drill; rough turn OD |
| 2. Spot drill | 2. Tap drill; rough turn OD |
| 3. Tap drill | 3. Step drill |
| 4. Step drill | 4. Undercut |
| 5. Undercut | 5. Tap |
| 6. Tap | 6. Step ream |
| **Cross slide tools** | **Cross slide tools** |
| 1. Radius corners | 1. Radius back corner |
| 2. Cut off | 2. Finish form OD |
| **Secondary operations** | 3. Radius front corner |
| Step ream | 4. Stamp letters |
| Stamp | 5. Knurl |
| Knurl | 6. Cut off |

| Results(a) | Single-spindle | Six-spindle |
|---|---|---|
| Setup time, h | 4 | 14 |
| Tool life per grind, pieces | 100 | 208 |
| Production, pieces per hour | 12(b) | 35 |

(a) For both machines, spindle speed was 645 rev/min (64 m/min, or 210 sfm, max), feed was 0.079 mm/rev (0.0031 in./rev), carbide tools were used, and cutting fluid was a blend of lard and sulfurized oils mixed 4 to 1 with mineral oil having a viscosity of 100 Saybolt Universal seconds (SUS). (b) Parts still required secondary machining operations, which had been completed on the six-spindle machine.

**Fig. 32** Comparison of single-spindle and six-spindle automatic bar machines in machining the part shown. Dimensions given in inches

**Sequence of operations**

| | |
|---|---|
| 1. Feed stock | 5. Bend end |
| 2. Rough turn | 6. Mill flats |
| 3. Rough point | 7. Form |
| 4. Finish point | 8. Cut off |

**Processing details**

| | |
|---|---|
| Speed | 2260 rev/min (59 m/min, or 192 sfm) |
| Feed | 0.025 mm/rev (0.001 in./rev) |
| Cutting fluid | Thread-grinding oil(a) |
| Production rate, pieces/h | 65 |

(a) One part heavy thread-grinding oil (75% fat, 15% sulfur) mixed with four parts mineral oil having a viscosity of 100 SUS

**Fig. 33** Complex part machined to close tolerances from profile stock in a single-spindle automatic bar machine. Dimensions given in inches

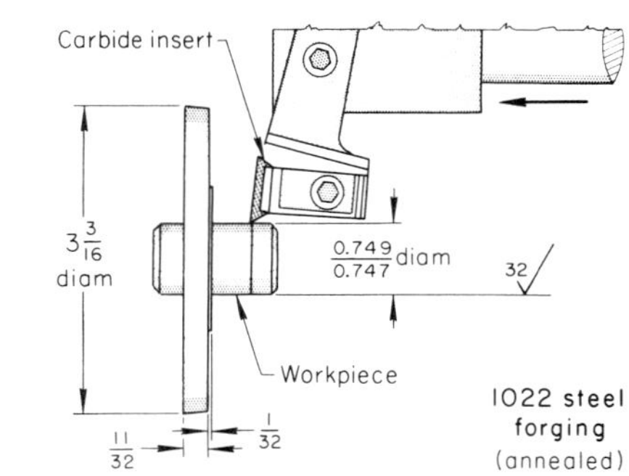


**Insert material(a)**

| | |
|---|---|
| Original | Conventional tungsten carbide |
| Revised | Special titanium carbide |

**Processing details (for inserts of both materials)**

| | |
|---|---|
| Speed | 1564 rev/min (94 m/min, or 307 sfm) |
| Feed | 0.15 mm/rev (0.006 in./rev) |
| Cutting fluid | Soluble-oil:water (1:20) |
| Production rate, pieces/h | 27.6 |
| Tool life per edge | 100 pieces |

(a) Inserts of both materials were triangular (6.4 mm, or 1/4 in., inscribed circle) with negative rake. Original insert had 0.8 mm (1/32 in.) nose radius; revised insert, 0.4 mm (1/64 in.). A 0.4 mm (1/64 in.) nose radius was tried with the original insert, but the finish was unsatisfactory.

**Fig. 34** Turning operation in which score marks were eliminated by changing from a tungsten carbide to a titanium carbide insert. Dimensions given in inches change to multiple-spindle machines. The following example describes an application in which a multiple-spindle machine was preferred over a single-spindle machine because it proved to be more efficient in machining the workpiece.

**Example 3: Single-Spindle Versus Six-Spindle Bar Machines.** Figure 32 compares procedures and results for single-spindle and six-spindle bar machines in producing a small part from 32 mm (1¼ in.) diam 1213 steel bar stock. As noted, with the single-spindle machine, it was necessary to stamp, knurl, and step ream in secondary operations because of inadequate cross slide capacity. With the six-spindle machine, it was possible to produce complete parts at a rate three times as fast as the single-spindle machine could produce parts that still required three operations.

## Production Quantity

For lots of fewer than about 12 pieces, none of the machines discussed in this article is economical. Regardless of the operations required, a lower-cost machine is more practical for machining a few pieces. Manual turret lathes are the most logical choice for lots of about 12 to 200 pieces and are sometimes used for much larger quantities.

When lot size exceeds about 200 pieces, an automatic turret lathe (single-spindle chucking or bar machine) should be considered, especially if repeat orders for the same or similar parts are expected. Labor is markedly reduced by the use of automatic machines because:

- Machining time per part is reduced to a fraction of that required on a hand-operated machine
- Even though the automatic machines are manually loaded and unloaded, one operator can handle two or more units if they are ganged (placed for efficiency)

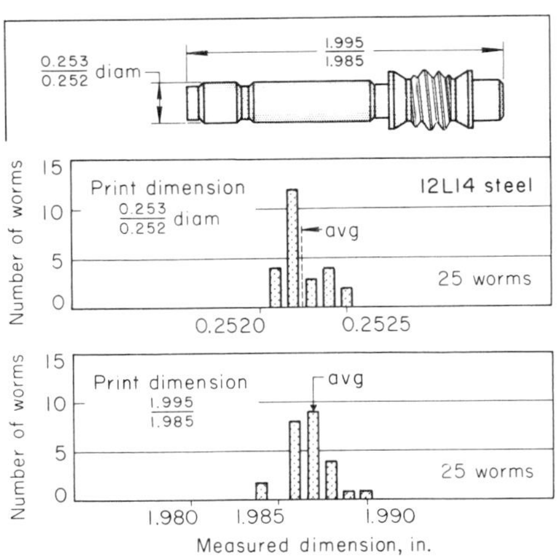


**Fig. 35** Variation in diameter and length of worms consecutively machined in an eight-spindle automatic. Dimensions given in inches

Manual machines require one operator for each machine.

The lot size that warrants selection of multiple-spindle machines is often influenced more by the required rate of production than by total quantity requirements, because parts can usually be produced on multiple-spindle machines two to five times as fast as on single-spindle machines. Some plants use a minimum of 100 000 parts (total) as the production quantity for which the use of multiple-spindle machines is warranted. In other plants, however, it has been found that multiple-spindle machines are justified for lot sizes of fewer than 50 000 pieces, and in some cases for even fewer than 5000 pieces.

## Production Rate

Quantity and productivity are often of equal importance in the selection of methods for machining. In many applications, however, the rate at which a given part must be produced has a greater influence on the selection of machine and method than total quantity requirements. For example, assuming a total lot size of 100 000 specific pieces, if only 1000 per day were needed and a manual turret lathe were available, it may be advisable to use existing equipment. On the other hand, if 5000 pieces per day were needed and could not be produced on existing equipment, single-spindle automatic chucking or bar machines (as required by product form) would be considered, and for higher productivity, multiple-spindle machines would be the logical choice.

In some plants, the availability of floor space is a major factor, thus emphasizing the selection of equipment that can produce the greatest number of parts in the least space. Productivity requirements are sometimes based directly on shipments or on other manufacturing operations, which are often interdependent.

## Special Requirements

Special requirements or unique workpieces often govern the selection of a specific type of machine. For example, for 1.8 m (6 ft) long tubular parts that require several identical machining operations on each end, a bar machine is a logical choice because the workpiece can be loaded through the back of the headstock and machined on one end, then reversed and machined on the opposite end—a two-stage operation. The obvious economies made possible by this type of procedure are often important enough to warrant designing (or redesigning) the parts to be compatible with such processing.

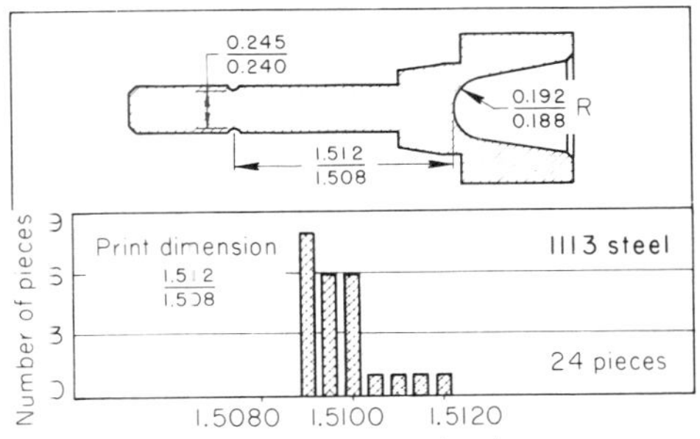


**Fig. 36** Distribution of dimensions, although all are within tolerance, indicating that the six-spindle bar machine used for producing this part was outside the $R_a$ limit. See Table 5. Dimensions given in inches

Required dimensional accuracy often influences the choice between a single-spindle and a multiple-spindle bar or chucking machine. Machine condition and other factors being equal, closer tolerances can be held in a single-spindle machine because it is inherently more rigid. This results mainly from the absence of a spindle carrier, which is a source of inaccuracy in the multiple-spindle machines. In multiple-spindle bar machines, moreover, the weight of heavy bar stock fed into the spindles can be an additional source of inaccuracy.

A choice between a manual and an automatic machine is usually influenced by considerations other than the type of operations each can perform. There are, however, applications in which one type of machine is preferable to the other because it is better suited to a specific operation. For example, automatic turret lathes are preferable for workpieces that require taper boring. Although taper boring attachments can be obtained for manual turret lathes, they are seldom available with standard tooling. However, on a single-spindle automatic machine, tapers can be formed with a quill-mounted single-point boring tool.

Machining to close tolerances or to obtain close relations between dimensions may require careful selection of machine and the use of special procedures. For machining small parts that require concentricity of two or more diameters, a single-spindle machine is usually preferred to a multiple-spindle machine regardless of the quantity to be produced, because of the inaccuracies that inherently result from performing successive operations at different index positions of the spindles.

Special tooling, together with careful choice of machines and procedures, often makes possible the production of special shapes or machining to unusual dimensional accuracy, or both, as in the following example.

**Example 4: Milling and Bending in an Automatic Bar Machine.** The workpiece illustrated in Fig. 33 was completely machined from profile stock on a single-spindle automatic bar machine. The sequence of operations is also given in Fig. 33. It was found that the close tolerances specified for this part were easier to hold by completing the operations on one machine than by the use of separate machines for turning, milling, and bending.

**Table 4 Typical tolerance capabilities for multiple-spindle automatic bar machines**

| Tolerance | | R(a) | | $R_{ar}$(a) | |
|---|---|---|---|---|---|
| mm | in. | mm | in. | mm | in. |
| **Four-spindle machines** | | | | | |
| 0.05 | 0.002 | 0.03 | 0.001 | 0.010 | 0.0004 |
| 0.08 | 0.003 | 0.04 | 0.0015 | 0.015 | 0.0006 |
| 0.10 | 0.004 | 0.05 | 0.002 | 0.020 | 0.0008 |
| 0.13 | 0.005 | 0.06 | 0.0025 | 0.028 | 0.0011 |
| 0.25 | 0.010 | 0.13 | 0.005 | 0.056 | 0.0022 |
| 0.38 | 0.015 | 0.19 | 0.0075 | 0.084 | 0.0033 |
| 0.51 | 0.020 | 0.25 | 0.010 | 0.112 | 0.0044 |
| 0.76 | 0.030 | 0.38 | 0.015 | 0.168 | 0.0066 |
| **Six-spindle machines** | | | | | |
| 0.05 | 0.002 | 0.04 | 0.0014 | 0.018 | 0.0007 |
| 0.08 | 0.003 | 0.05 | 0.0021 | 0.027 | 0.00105 |
| 0.10 | 0.004 | 0.07 | 0.0028 | 0.036 | 0.0014 |
| 0.13 | 0.005 | 0.09 | 0.0035 | 0.043 | 0.0017 |
| 0.25 | 0.010 | 0.18 | 0.007 | 0.089 | 0.0035 |
| 0.38 | 0.015 | 0.27 | 0.0105 | 0.132 | 0.0052 |
| 0.51 | 0.020 | 0.36 | 0.014 | 0.178 | 0.007 |
| 0.76 | 0.030 | 0.53 | 0.021 | 0.264 | 0.0104 |
| **Eight-spindle machines** | | | | | |
| 0.05 | 0.002 | 0.04 | 0.0016 | 0.020 | 0.0008 |
| 0.08 | 0.003 | 0.06 | 0.0024 | 0.030 | 0.0012 |
| 0.10 | 0.004 | 0.08 | 0.0032 | 0.042 | 0.00165 |
| 0.13 | 0.005 | 0.10 | 0.0039 | 0.053 | 0.0021 |
| 0.25 | 0.010 | 0.20 | 0.0078 | 0.106 | 0.0042 |
| 0.38 | 0.015 | 0.30 | 0.0117 | 0.160 | 0.0063 |
| 0.51 | 0.020 | 0.40 | 0.0156 | 0.213 | 0.0084 |
| 0.76 | 0.030 | 0.59 | 0.0234 | 0.320 | 0.0126 |

(a) $R$ is the spread between high and low ratings on pieces produced in a round (one piece from each spindle, in succession) or on consecutive pieces from a spindle; $R_{ar}$ is the average range of four consecutive rounds. See Table 5 for actual dimensional data that illustrate these definitions for a particular part.

**Table 5 Measurements of 1.512/1.508 in. dimension on part shown in Fig. 36**

| Item | Round 1 | Round 2 | Round 3 | Round 4 | Sum of readings | Spindle average | Spindle $R$ |
|---|---|---|---|---|---|---|---|
| Spindle 1 | 1.509 | 1.509 | 1.510 | 1.5105 | 6.0385 | 1.5096 | 0.0015 |
| Spindle 2 | 1.5095 | 1.5095 | 1.5095 | 1.509 | 6.0375 | 1.5094 | 0.0005 |
| Spindle 3 | 1.5115 | 1.512 | 1.511 | 1.510 | 6.0445 | 1.5111 | 0.002 |
| Spindle 4 | 1.5095 | 1.509 | 1.510 | 1.509 | 6.0375 | 1.5094 | 0.001 |
| Spindle 5 | 1.510 | 1.5095 | 1.509 | 1.510 | 6.0385 | 1.5096 | 0.001 |
| Spindle 6 | 1.510 | 1.509 | 1.5095 | 1.509 | 6.0375 | 1.5094 | 0.001 |
| High | 1.5115 | 1.512 | 1.511 | 1.5105 | ... | ... | ... |
| Low | 1.509 | 1.509 | 1.509 | 1.509 | ... | ... | ... |
| Range $R$ | 0.0025 | 0.003 | 0.002 | 0.0015 | | $R_{ar}$ = 0.00225 | |

To produce this part on an automatic bar machine, three special devices were required:

- A spindle-stopping device to locate and hold the part in the collet in definite relation to the centerline of the gear teeth (profile-stock shape)
- A special bending tool, on the turret, to reshape the tapered end after turning
- One open cross slide position on which to mount a special milling attachment that must clear all turret tools

A special spindle-stopping attachment was developed that utilized the longitudinal gear teeth of the profile stock to position the stock in the collet and to stop spindle rotation in a specific position with reference to the gear teeth.

The bending fixture was a two-piece forming die that had taper on the outer edges. Forward movement of the die halves was provided by the turret cam. As the dies moved forward, they were forced inward as they moved along the taper on their outer edges, thus forming the end section of the part.

For the milling operation, an air cylinder was linked with the rear slide holding a special milling attachment. The cylinder provided the necessary movement to retract the milling attachment far enough to clear all turret tools. Because both cross slides were occupied for milling and forming, cutoff was done by a tool attached to the vertical slide.

## Cost

When two or more types of machines can produce acceptable parts, cost per piece usually determines the final choice. Machining cost per piece is influenced most by cost of setup, cost of labor, and burden (amortization) of machine and tools. High-production machines always cost more to set up than simpler machines, but for long runs the cost of setup is rapidly absorbed. As the number of pieces required increases, the cost per piece decreases. Labor costs are also lower when one operator can attend two or more machines.

As the production capabilities of machines increase, however, so do the initial cost of the machines and the cost of tooling. Therefore, prior calculations that indicate that specific parts can be produced at lower cost on expensive, high-production equipment may prove erroneous unless machine burden is taken into account. This is most likely to occur when the quantity of one specific part available for a run at one time is borderline and more frequent setups are required, thus involving more setup time and downtime of the machine. The full value of highly automatic multiple-operation machines can be realized only when they are operating for a major portion of the time. Cost studies are often made to determine break points that indicate which of two or more machines and procedures will prove the most economical for specific lot sizes.

## Tool Material and Design

In multiple-operation machining, changes are often made in tool material or design that improve tool life, workpiece finish, dimensional accuracy, or productivity. When operations are done in a sequence, it is common for one tool to be abused by being run at excessive speed or feed, or both, in order to obtain greater efficiency from other tools that operate in the sequence.

Because form tools are usually subjected to rigorous service, they are often made from some material that is more highly alloyed than a general-purpose type of high-speed steel. In some plants, the established practice is to use general-purpose types of high-speed steel for form tools, changing to a cobalt-containing grade only when specific experience indicates the necessity. The metal being machined is a major factor in selecting form tool material. For cutting difficult-to-machine metals, such as stainless steels, form tools made from the more highly alloyed types of high-speed steel are nearly always superior.

Soft nonfree-machining steels such as 1008 or 1010 are not preferred for high-production, multiple-operation machining, because they are susceptible to tearing and because excessive amounts of burr are probable. When it is necessary to machine these steels, special attention must be paid to the design of the cutting tools used. Minor changes in tool design are often effective.

**Carbide** is used for cutting edges on many tools for multiple-operation machining. Changing the design and the method of mounting carbide tools often results in increased productivity or increased tool life. Sometimes, reselection of carbide grade has a significant effect on results, as in the following example.

**Example 5: Surface Finish Improved by Changing Grades of Carbide.** The 19.02/18.97 mm (0.749/0.747 in.) diameter of the 1022 steel forging illustrated in Fig. 34 required a turned finish that could be burnished to a maximum roughness of 0.80 μm (32 μin.). Turning was done in a single-spindle chucker using triangular indexable carbide inserts.

Inserts made from a conventional grade of tungsten carbide formed built-up edges, resulting in score marks that made it impossible to meet surface finish requirements. Changing to inserts made from titanium carbide corrected this difficulty. Both types of inserts were used under identical conditions, as detailed in the tabulation with Fig. 34, except for the difference in nose radius. However, both 0.8 and 0.4 mm (1/32 and 1/64 in.) nose radii were tried with the conventional carbide, but the finish was unsatisfactory.

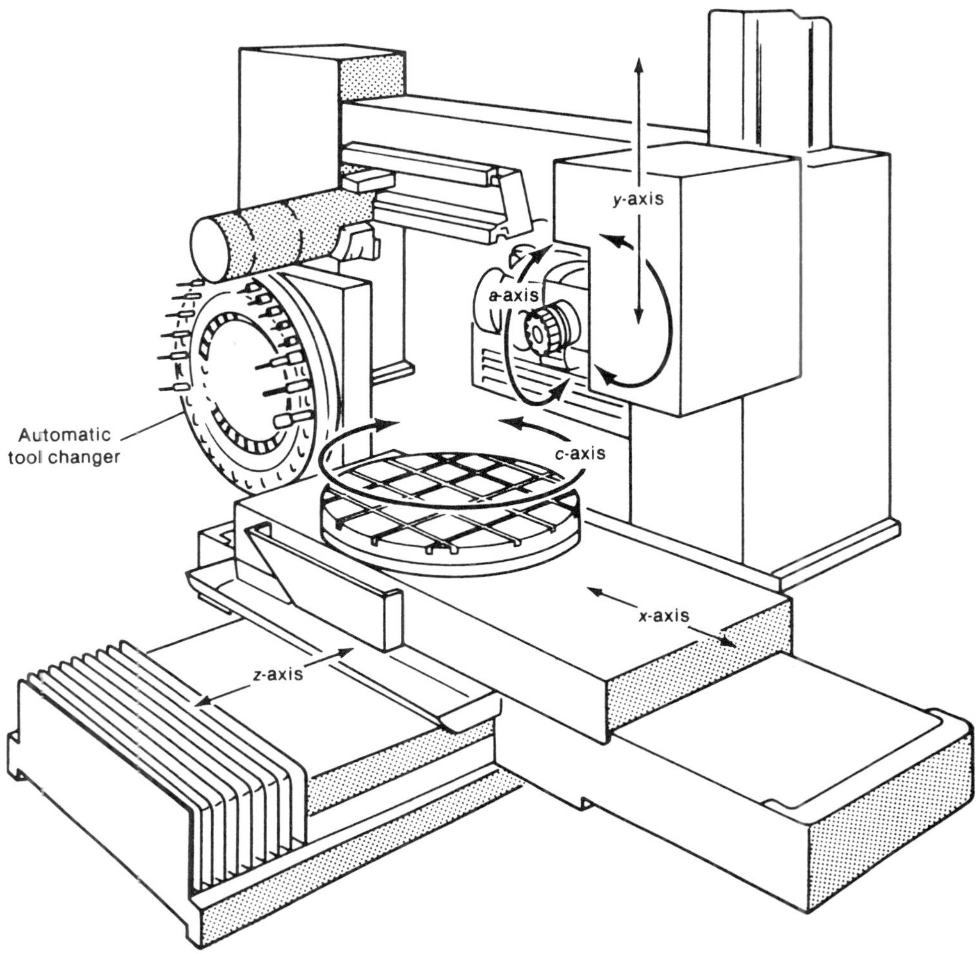


**Fig. 37** Schematic of a five-axis horizontal-spindle machining center

## Machinability and Physical Condition of Work Metal

Metals are usually selected because of required mechanical properties rather than for their machinability. In many applications, however, minor changes can be made in metal selection that will favorably affect machinability without impairing service performance.

### Free-Machining Metals

Many steels and some nonferrous metals are available as free-machining grades. These are similar in composition to the nonfree-machining grades except that they contain small, measured additions of sulfur, phosphorus, lead, tellurium, selenium, or bismuth, singly or in combination. The purpose of these additions is to promote better chip breakage and to minimize friction between tools and work metal, thus permitting the use of higher speeds and contributing to improved tool life, better surface finish, and greater control of dimensions.

The free-machining grades are especially desirable in multiple-operation machining because of the numerous tools involved. In most cases, each type of tool has its own optimum speed, but when several operations are performed simultaneously, some tools are operated at excessive speeds, while others are operated too slowly for efficient performance. Free-machining work metals contribute markedly to minimizing the sensitivity of tools to higher cutting speeds.

Free-machining metals are more expensive than their nonfree-machining counterparts, and because cost per part is significant, it is essential to balance the added cost of the metal against the decreased cost of machining before deciding to use a free-machining grade. In many applications, the amount of machining to be done (number of operations or weight of metal removed or both) has a large influence on whether or not the higher cost for free-machining grades is warranted. For example, if only two or three operations were used to reduce the volume of a steel bar by 10% in producing a finished part, the additional cost for a free-machining steel probably would not be warranted. On the other hand, if ten or more operations were utilized to reduce the volume of a bar by 50% in producing a finished part, a free-machining grade probably would be warranted. Therefore, each workpiece design requires individual consideration, even when there are no specific restrictions on the use of free-machining grades. Metal cost, machining cost, and machine capacity must all be considered simultaneously.

Maximum return on the large investment that a multiple-operation machine represents can be realized only by keeping the machine in operation at or near its full capacity. For this reason, in most plants, the addition of new machines lags behind the demand for machine time. Under conditions of peak demand, it is therefore most important to operate machines at maximum speed. As a result, demand for machine time is often an impetus to the selection of a free-machining work metal.

Paying a higher price for a steel with better machinability may result in lower manufacturing cost solely on the basis of increased tool life. This is most often true when the machine is already being operated at maximum speed; under these circumstances, cycle time per part cannot be decreased, regardless of work metal machinability. However, longer tool life results in lower tool cost and usually in some increase in productivity because downtime for changing tools is decreased.

### Physical Conditions

Workpiece variations other than composition that affect productivity, dimensional accuracy, surface finish, machine wear, tool life, and cost are:

- Dimensions of cross sections (bars, tubes, or extrusions)
- Surface finish

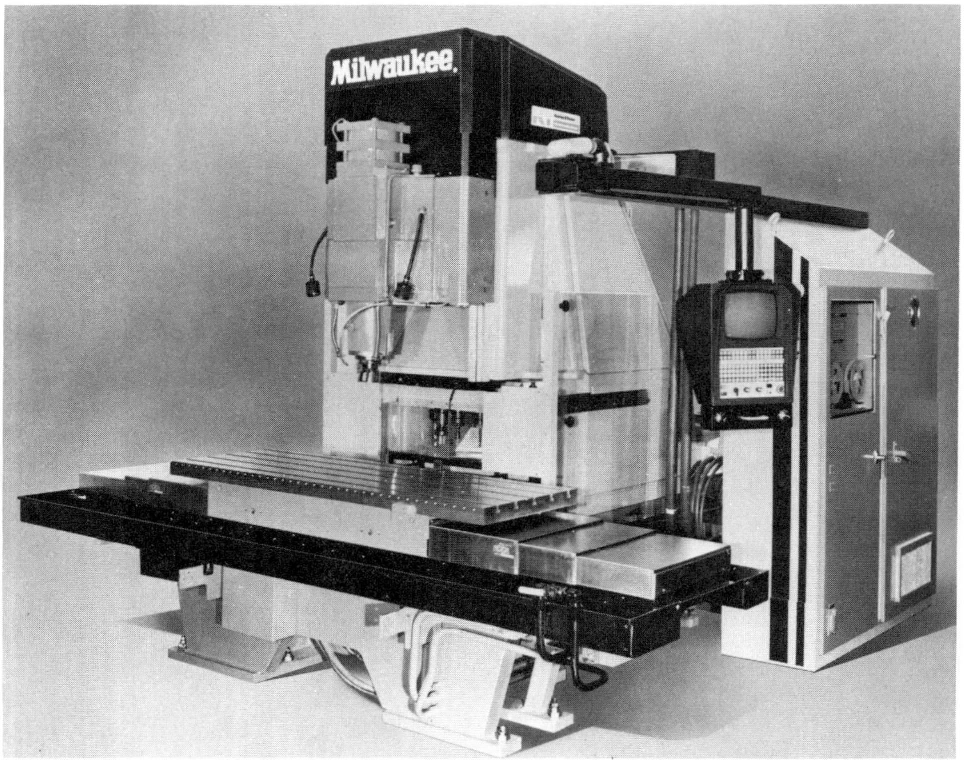


**Fig. 38** Vertical-spindle machining center that incorporates a 56 kW (75 hp) motor yet can be used at minimum spindle speeds of 50 rev/min. Courtesy of Kearney & Trecker Corporation

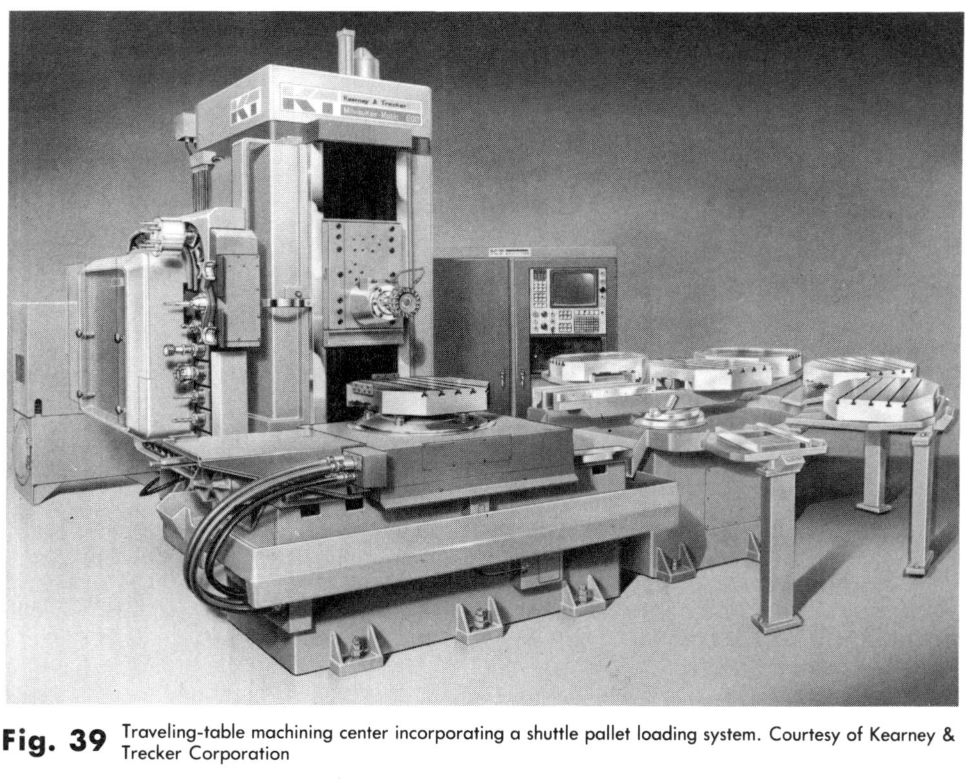

**Fig. 39** Traveling-table machining center incorporating a shuttle pallet loading system. Courtesy of Kearney & Trecker Corporation

- Straightness (bars, tubes, or extrusions)
- Length (bars, tubes, or extrusions)

**Dimensions of Cross Sections.** Cold-drawn products are usually preferred for multiple-operation machining, mainly because of their close dimensional tolerance and smooth finish. Hot-rolled and pickled bars are sometimes used because they cost less than their cold-drawn counterparts, but not all machines will operate successfully with hot-rolled bars, because of dimensional variation. Therefore, the capability of the feeding mechanism to be used must be considered and the tolerance of the hot-rolled bars must be known (and, if possible, the bars tried out on location) before a decision is made to substitute hot-rolled and pickled bars for cold-drawn bars.

Not only do significant variations in cross section (including out-of-roundness) affect feeding mechanisms, but they also affect tool life. Dimensional variations cause nonuniform amounts of stock removal on initial cuts, which may damage tools.

Some automatic machines are so sensitive to dimensional variations of the work material that they may require adjustment for different shipments of material purchased to the same specifications. Consequently, for machines that are extremely sensitive, it is best to separate shipments from different suppliers or to sort the stock.

**Surface Finish.** Bars or tubes that have mill scale are often unsuitable for processing in an automatic machine, because the abrasive scale impairs the action of the feeding mechanism. In addition, taking initial cuts on scaly surfaces damages tools and shortens their life. If hot-rolled bars or tubes are processed, they usually must be descaled, by pickling or blasting, before being machined.

**Straightness.** Regardless of whether cold- or hot-finished bars or tubes are used, they must be machine straightened. Bars or tubes that have been straightened to commercial limits are straight enough for operation in most machines. Additional straightness can be obtained at a premium price, but is seldom needed.

Bars or tubes that are not commercially straight develop whip in the stock reels, which causes abnormal vibration throughout the machine. This eventually results in excessive wear on machine parts and tools, tool chatter, and loss of dimensional control.

**Length.** For single-spindle machines, uniform length in bars or tubes seldom provides any benefit, except for ease of handling and storage. In a multiple-spindle machine, however, extreme nonuniformity of stock length decreases productivity because, as shorter lengths are fed through, some tools will be idle until the longest bar or tube is consumed. In addition, significant differences in the length of large, heavy bars due to the variation in weight can adversely affect the action of the stock reel. Random-length bars are usually segregated as they are loaded into stock reels so that the bars in a reel at any one time do not vary in length by more than a few inches.

## Dimensional Control

Quality control by statistical methods can help to establish practical limits within

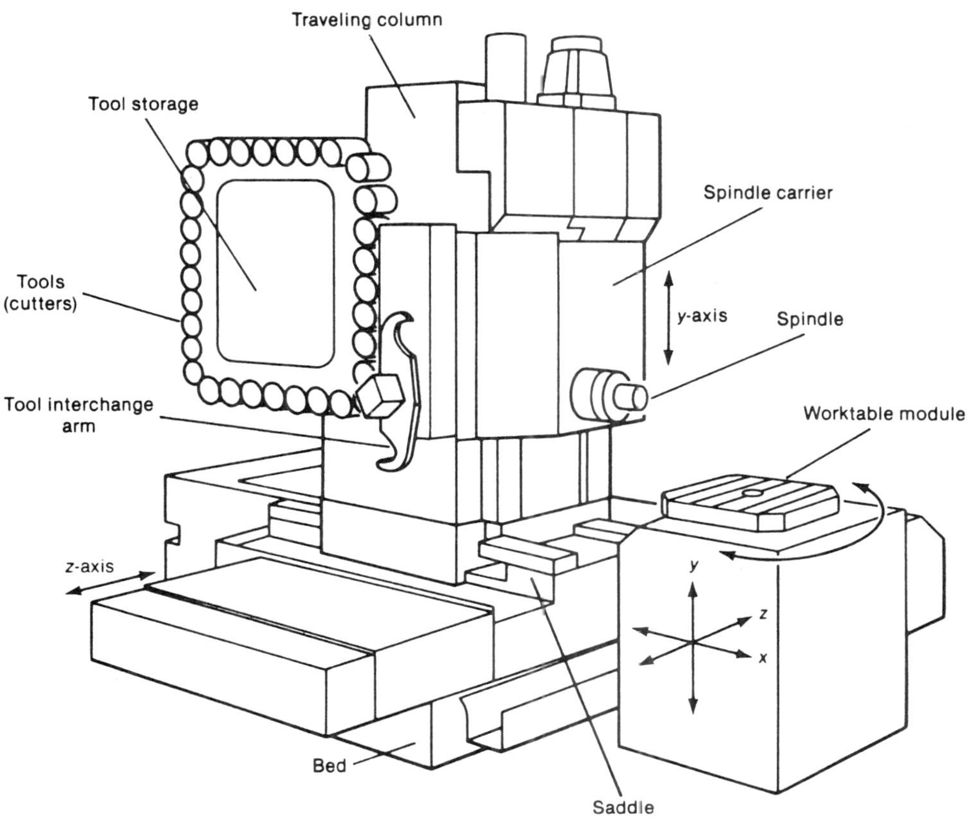


**Fig. 40** Schematic of a horizontal-spindle traveling column machining center and its accompanying worktable module

which dimensional variations may be permitted to differ before the process should be corrected. The dimensions of most pieces produced in a given operation are usually in an average group or cluster close to the nominal, but fliers (pieces departing to an unusual degree from the nominal) are often observed. Statistical quality control determines the trend of the average or majority group in relation to specifications or tolerances. The following example illustrates typical variations in dimensions obtained with a multiple-spindle machine.

**Example 6: Diameter and Length of Worms Produced in an Eight-Spindle Automatic.** Variations in the bearing diameter and the cutoff length of worms produced from 12L14 steel bar stock in an eight-spindle automatic machine are shown in Fig. 35. These data represent measurements made on 25 consecutively produced worms taken as a sample from a continuous production run using four machines. The bearing diameter was form turned, hollow milled, and shaved. When the sample was taken, the shaving tool had been operated for 5 h, and the other tools (including the cutoff tool) for 7 h, since being sharpened.

The spread of dimensions for the bearing diameter (upper chart, Fig. 35) was within allowable tolerance, but the location of the spread with reference to the nominal dimension indicates that slight adjustment of the tools or stops would bring the average dimension closer to nominal. This condition is more pronounced for the cutoff length (lower chart, Fig. 35); two parts were under the minimum, and corrective action was needed.

**Control of Machine Adjustment.** Parts cannot be produced on any multiple-operation machine to tolerances closer than variations existing in the machine. Loose bearings and poorly adjusted slides or stops are the most common conditions that influence dimensional variation. Among other conditions are broken collet-lockup pins, worn cams, and broken connectors in slide-linkage mechanisms.

Periodic checking of the machine and a method for controlling its condition are mandatory for maintaining close and uniform tolerances. In many plants, methods of statistical quality control have been successfully applied to monitor the condition of multiple-spindle automatics. The use of these methods involves the following terms:

- *R*. The spread between the high and low reading on a round, or of consecutive pieces from a spindle
- $R_{ar}$. The average range of four consecutive rounds. The symbol $\overline{R}$ is widely used instead of $R_{ar}$
- *Round*. One piece from each spindle, in succession
- *Spindle round*. Four pieces, in succession, from a given spindle
- *Spindle average*. The average dimensional reading of consecutive pieces from a given spindle

It is first required to have a table of established constants of $R$ and $R_{ar}$. Typical values that serve as a guide to expected tolerance capabilities of automatic bar machines with four, six, or eight spindles are given in Table 4.

A typical procedure used for practical application of quality control principles to the checking of machine condition is described in the following example. As Example 7 demonstrates, data from a simple study can be used to keep machines adjusted and to rectify difficulties before unacceptable parts are produced.

**Example 7: $R_{ar}$ Study of a Six-Spindle Bar Machine.** The 1113 steel part shown in Fig. 36 was produced on a six-spindle automatic bar machine. Maintaining the 38.40/38.30 mm (1.512/1.508 in.) dimension within the 0.10 mm (0.004 in.) tolerance range was a chronic problem, and statistical quality control methods were applied to try to solve it.

Experience proved that for an $R_{ar}$ study, data for four rounds, kept separate by round and spindle, would provide adequate information on the operating condition of the machine. The procedure used was as follows:

- The operation range limit $R$, 0.071 mm (0.0028 in.), was determined from the data for six-spindle machines in Table 4 using the 0.10 mm (0.004 in.) tolerance allowed by specifications
- The machine study range limit $R_{ar}$, 0.036 mm (0.0014 in.), was also determined from Table 4
- Daily checks were made to the operational $R$ (one round)
- Weekly checks were made to $R_{ar}$
- If, on the daily check, $R$ was out of limits, an immediate study was made to determine the cause and to correct it

Details of one $R_{ar}$ study are given in Table 5 and are summarized in the histogram shown in Fig. 36. These readings show that, although no out-of-tolerance pieces were found, the machine was operating at an $R_{ar}$ of 0.0572 mm (0.00225 in.)—well outside the 0.036 mm (0.0014 in.) $R_{ar}$ limit. The "Spindle average" column in Table 5 shows that spindle 3 was cutting 0.038 mm (0.0015 in.) longer than the next-longest-cutting spindle. The end play of spindle 3 was reduced to nominal by adjustment of the bearings, and the average length of the dimension produced by spindle 3 was thus reduced to 38.34 mm (1.5095 in.).

**Loading the machine** before it is set up is the best practice because the accuracy of the machine can be affected by setting up an almost empty machine and then filling it with heavy bars. Bar weight increases rapidly with diameter. For example, ten 3.0 m (10 ft) long steel bars 25 mm (1 in.) in diameter weigh about 120 kg (265 lb), while ten 127 mm (5 in.) diam steel bars of the same length weigh about 3025 kg (6670 lb). When the total weight in the stock reel is too great for the equipment, shorter bars must be used.

## Speed and Feed

Speeds used in multiple-operation machining are not necessarily different from those recommended and used for specific single operations. Additional information is available in the Section "Traditional Machining Processes" in this Volume.

Feeds used in multiple-operation machining are essentially the same as those used in the specific operations when performed singly (see the Section "Traditional Machining Processes" in this Volume).

## Cutting Fluids

The main factors influencing the choice of cutting fluid are the specified finish and tolerance, composition of the work metal, and the specific type of operations to be performed.

**Soluble oil** is the most widely used cutting fluid because it is the least expensive and is equal to, if not better than, any other fluid for cooling and for flushing away chips.

**Straight mineral oils** are often used where soluble oils do not meet requirements, particularly when the work metal is nonfree-machining or where better surface finishes are necessary. Mineral oil is necessary for obtaining acceptable finishes and tool life when low-carbon steel extrusions are being machined. In many cases, the type of machining operation performed, not the raw material used, dictates the type of cutting fluid required to obtain acceptable results.

**Proprietary cutting oils** are more effective than soluble-oil emulsions for preventing chatter under abnormal conditions of unsupported length or vibration.

**Special Oils and Mixtures.** Special cutting fluids are often prepared for unusual applications. They are especially desirable for use in tapping operations.

Lard oil is often used in small single-spindle machines for such special purposes. Thread-cutting or lard oils are often mixed with mineral oils to improve viscosity.

**Dry Cutting.** Materials such as cast iron, malleable iron, and some copper-base metals can be machined dry. However, compressed air instead of a fluid can be used to cool the workpiece and to remove the metal chips created during the machining process.

**Fig. 41** Machining of a rocket fuel blade on the pallet in the background using a traveling-table machine. Empty pallet in the foreground is ready to be loaded. Courtesy of Kearney & Trecker Corporation

## Safety and Protection

One of the dangers involved in operating the machines described in this article is that of being caught by the revolving stock or by the revolving chuck of the machine. Machine tool manufacturers have taken the precaution to provide substantial enclosure guards around the chucks or face plates of the machines.

Machine operators are frequently injured because of displacement or breakage of the tool or the stock being processed. Insecurely or improperly placed tools often work loose and cause problems in other ways, such as digging into the work. Unless the machine is stopped immediately, the tool may break or be forced out of the holder, the workpiece may be thrown out of the chuck, or both of these mishaps may occur simultaneously. In addition, machine tools should never be changed or adjusted while the machine is in motion.

A hazard is present in the operation of turret lathes and automatic bar and chucking machines when revolving stock is permitted to project beyond the ends of the machine. The work on these machines should itself be safeguarded, and long bars of stock should be supported. A good practice is to install sections of piping through which bar stock can be fed to the machine. If the stock ends cannot be prevented from protruding, guard rails can be installed around revolving stock to protect persons standing close to the machines from direct contact with the workpiece when it is in motion.

Cutting oil thrown from the tools of automatic machines makes the floor area around the machines dangerously slippery to operators and other plant personnel. In the interest of safety, various practices are recommended, among which the following are widely endorsed:

- Installation of sheet metal screens for interception of the oil
- Floor mats of finely meshed metal
- Frequent and systematic cleaning of the floor and frequent changing of the sawdust spread on the floor

A safety measure of wide adoption is the provision of a nonslip and nonflammable floor on which to rest the machines.

Skin disorders among machine operators using cutting oils and compounds can be prevented if operators wash their arms and hands thoroughly at proper intervals. Work clothes should be changed often; clothes that are worn too long become saturated with the cutting mixture.

# Machining Centers

A machining center is a single-station computer numerically controlled (CNC) machine tool capable of milling, drilling, boring, reaming, and tapping. This machine tool is usually equipped with one automatic tool changer and is designed to perform operations on different surfaces of a workpiece as the workpiece rests on a rotating table. Therefore, after a particular operation, the workpiece does not have to be removed and taken to another machine for further processing. Instead, the tools and the workpiece to be machined are brought to the machine. The tool changer stores idle tools and interchanges them with those in the spindle according to program commands that ensure optimum speeds and feeds as well as the correct space coordinates necessary to machine the workpiece.

A machining center may be equipped with 100 or more tools. It is capable of manufacturing large and complex parts efficiently and accurately.

**Types of Machining Centers.** Machining centers with a horizontal spindle (Fig. 37) are generally more flexible than vertical-spindle (Fig. 38) models and are available in a wider range of sizes. Machining centers can be further categorized as being of the traveling-table or the traveling-column type. Traveling-table designs are used to machine small parts, and they incorporate a shuttle pallet system to carry the workpiece, as shown in Fig. 39. Traveling-column designs are extensively used for machining large parts. These machining centers usually have one workholding module containing a 60, 360, or 720 programmable position rotary index table (Fig. 40) or a 360 000 programmable position full contouring table.

The machine motions are governed by numerical control in all three major axes and often in additional secondary axes. Machining center programs can be stored and also provide input to the machining center CNC control via punched eight-channel Mylar tape, 89 mm (3½ in.) floppy disks, or directly from an external computer. Many machining centers have an internal minicomputer that can store several programs, eliminating the need to input data each time a part is run.

**Typical Operating Sequence of a Machining Center.** Because a machining center represents such a large investment, it must be operated continuously and efficiently to be profitable. Each cutting tool is preset for diameter and depth in a quick-change holder; therefore, when the tool is placed in the spindle, it cuts its programmed dimensions without any adjustment on the machine. Tools are placed in holes in a rotary drum on top of the machine column. When the program calls for a cutting tool, it is identified by a sensor from the coded rings around the tool shank and then transferred by a mechanical tool changer in a few seconds to the machine spindle on the front of the column. The spindle, which is powered by a 1.5 to 56 kW (2 to 75 hp) motor, typically has a speed range of 50 to 4000 rev/min to accommodate many sizes of tools. Figure 41 shows a workpiece fixture on a pallet in the workstation (on the machine table) being machined by a tool on the machine spindle. At right is another pallet on which work can be set up while a piece is being machined in the workstation. When one piece is done, it is slid off mechanically, another is moved on its pallet into the workstation, and cutting is immediately resumed, with no loss of machine time.

A further advancement is the unmanned machining center. In this unit, 15 to 20 or more parts are loaded on pallets in a repository next to the machine. The parts are transferred into and from the workstation automatically in turn. The CNC control has a memory that is large enough to store programs for all the work, and each program is called out to suit the part in the workstation. The machining center can be used conventionally, and the pallets and programs can be loaded during the day. The machine can then operate all night or over a weekend with little or no supervision.

During the second and third shifts, when labor costs are higher and electrical power costs are lower, the pallets automatically move to the machining center, which selects the appropriate program for the specific workpiece to be machined. Similar operations should be performed on the various workpieces to keep tooling requirements within the capacity of the machine.

The control unit monitors spindle speeds and torque, and it automatically changes the feed rates to maintain the programmed values. The machining centers are also equipped with automatic devices that sense tool wear. Tools are brought into contact with a gage on the machine that detects the wear on lands or teeth. If wear exceeds the programmed limit, the tool is rejected and an alternate chosen, or a warning light flashes. A gaging post on the machine detects broken tools, and the spindle can orient the tools in 1° increments to measure tool diameters.

**Advantages of Machining Centers.** Increased productivity and versatility are the principal advantages of machining centers. The capability of performing drilling, turning, reaming, boring, milling, contouring, and threading operations on a single machine eliminates the need for a number of individual machine tools, thus reducing capital equipment and labor requirements. One relatively unskilled operator can often attend two machining centers and sometimes more. Most workpieces can be completed on a single machining center, often with one setup.

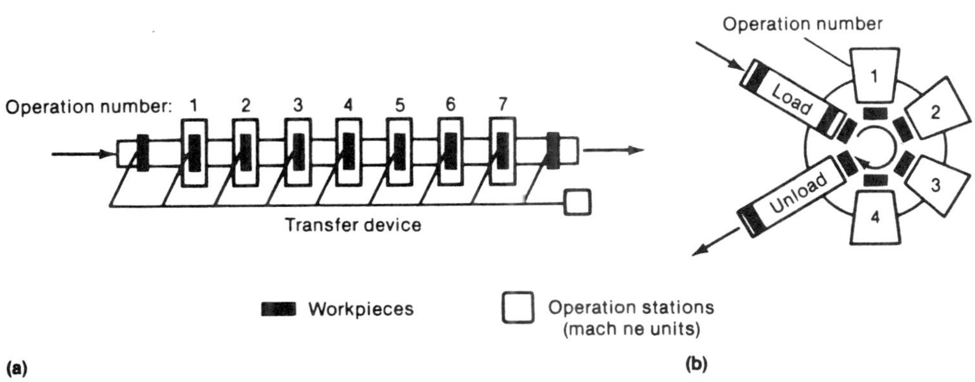


**Fig. 42** Schematic of two types of synchronous system transfer machines. (a) In-line transfer machine. (b) Rotary index transfer machine

Additional savings result from reduced material handling, fixture costs, and floor space requirements. The substantial amount of time conventionally spent transporting work from machine to machine is saved, and throughput is much faster. In addition, in-process inventory, represented by skids of workpieces normally seen at several machines, is replaced by work at only one machine.

Most machining centers maintain close, consistently repetitive tolerances, resulting in higher-quality parts and reduced inspection costs and scrap. In particular, the relationship of machined features on the several faces of a workpiece are more easily held within tolerances. Changeover from the production of one workpiece to another can be done quickly.

Actual machining time on machining centers can be two or more times that of single-purpose, manually operated machine tools. Estimates of increases in productivity per man-hour range from 300 to 500% or more, especially on applications requiring many tools and frequent changeover.

Machining centers have a higher initial cost than many other machine tools, but the annual return on investment has been conservatively estimated to be about 30%. Smaller, compact models are available that make these machines affordable even to small job shops. The accuracies that can be maintained and the reliability of the machines and their controls have been continuously improved since the introduction of machining centers in the late 1950s.

**Applications.** Machining centers have traditionally been used primarily for automating the manufacture of small lots of a wide variety of workpiece shapes and sizes that require multiple operations. Currently, machining centers are being increasingly used for medium-lot requirements, and some have been tooled for long production runs of a single workpiece.

Careful work scheduling is necessary to keep machining centers operating and therefore to realize their full profit potential. Most shops try to operate their machining centers, as well as other NC machines, at least two shifts per day to obtain a more rapid return on investment. Another application for machining centers is the integration of these centers with other NC machines to form flexible machining systems.

The configuration of the part is extremely important in determining whether or not it would be most efficiently produced on a machining center. The following types of parts are generally well suited to machining centers:

- Parts machined on several faces
- Parts with a large number of operations
- Parts with close tolerances
- Parts in which design changes are anticipated
- Parts that are very expensive (because a machining center very rarely makes a scrap part)

Nearly all the advantages attributed to machining centers also apply to automated batch manufacturing systems and transfer lines. Machining centers are most useful in small production runs in which each batch differs significantly from the others.

**Disadvantages of Machining Centers.** The typical machining center has one major disadvantage for large-quantity production: It applies only one cutting tool at a time to a workpiece. Machining centers do have a place in flexible manufacturing systems, along with multiple-tool machines. Efforts have been made to make machining centers more competitive for large-quantity production. A few units have been built with multiple spindles, but a major development has been the addition of multiple-toolhead storage and transfer devices. Multiple toolheads, such as drill heads, can be stored in a rack or on an endless conveyor, taken out on call by a tool changer, and then affixed to the machine spindle. Multiple toolheads can also be intermixed with single tools as required.

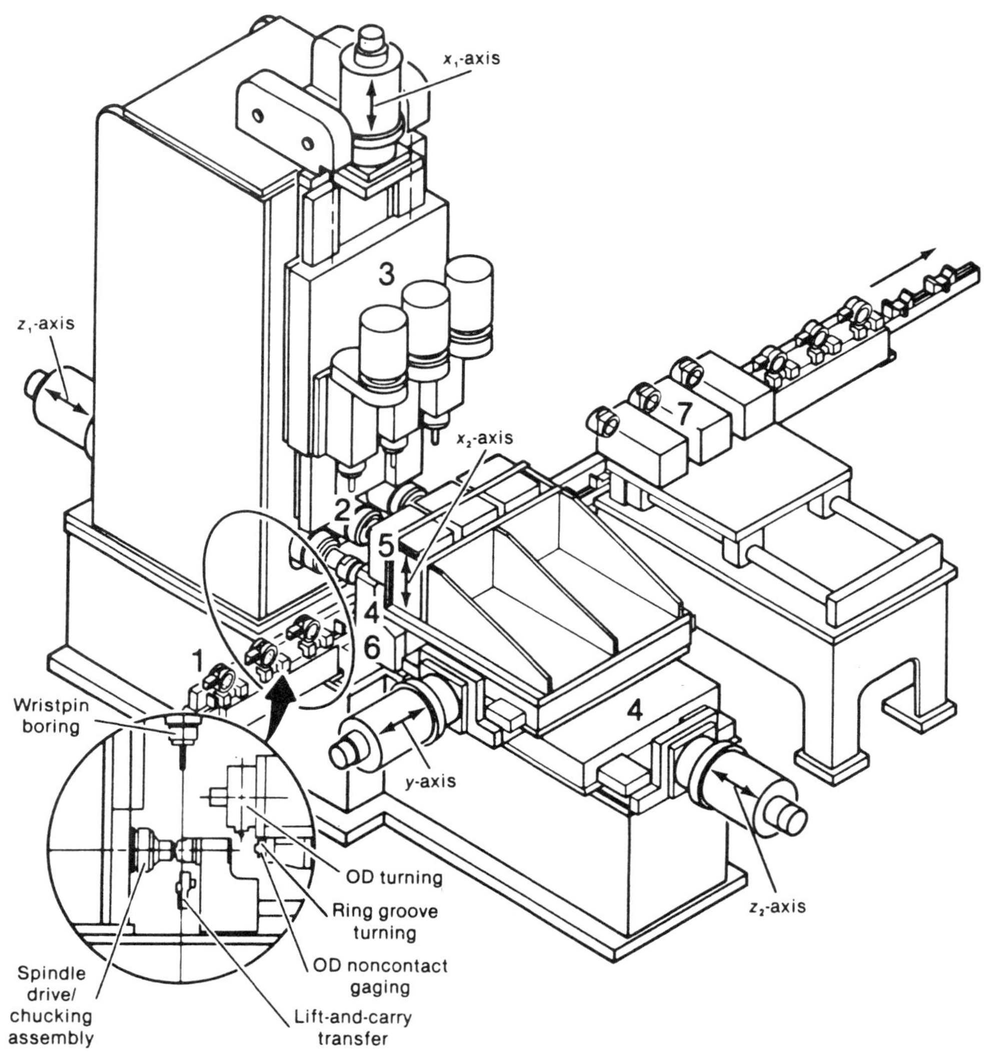


**Fig. 43** Schematic of a machining and inspection station in an in-line transfer machine. This camless piston-manufacturing system is capable of producing 1800 automotive engine pistons per hour using diamond turning tools operating at up to 4000 rev/min. Sequence of operations is as follows: 1, Parts are transferred in individual transfer nests to eliminate part-to-part contact. A coolant-controlled environment maintains constant part temperature; 2, Servomotor-driven $z_1$-axis positions the spindle drive/chucking assemblies. Optical encoders enable the computer to determine exact position of the part during rotation. Similar feedback mechanisms for all other axes provide the computer with exact velocity and position data; 3, Servomotor-driven $x_1$-axis feeds the wrist pin bore semifinish and finish boring tools; 4, Servomotor-driven $z_2$-axis feeds the ringland and skirt diameter turning tools and positions the $y$-axis slide for rough and finish ring groove turning; 5, Motors ($x_2$-axis) replace cams and spring-actuated cam followers to position the outside diameter turning tools. The stroke is computer-controlled in relationship to the rotating part. Variations to outside diameter geometry are made automatically by modifying this relationship; 6, After finish turning, noncontact gages measure the outside diameter during part rotation to provide the computer with data for automatic compensation of out-of-tolerance conditions; 7, In-process gages measure the wrist pin bore diameter, groove width, and minor diameters to provide automatic compensation for out-of-tolerance conditions.

# Transfer Machines

Transfer machines, also referred to as transfer lines, are rigid production systems used mainly for mass production. A transfer machine consists of a series of highly specialized single- or multiple-operation machine tools linked by a material handling system and controlled by a programmable controller or relay network. It is designed to allow continuous processing of a single part or a family of parts having minor variations. A transfer machine is basically an integrated handling device and is applicable to such operations as welding, molding, assembly, and inspection in addition to machining.

Transfer lines are built after the product is designed and are custom made to machine one specific component. If the part becomes obsolete, the transfer line must be scrapped or renovated because the cost of retooling for another part is approximately 90% of the original total cost of the machine.

Transfer lines can be classified as being either synchronous or nonsynchronous.

**A synchronous system** index-type transfer line, also known as a dependent unit system, may have a series of machine or operation units arranged in-line (Fig. 42a) or around a rotary or circular index (Fig. 42b). When the work is completed at all stations, all parts in the circuit are moved simultaneously to succeeding stations, and another cycle is started. The time required for a cycle depends on the slowest operation, and the highest efficiency is obtained when all operation times are the same and are minimal. For tool changes and breakdowns, all stations are shut down, and efficiency decreases as the number of stations increases.

*In-Line.* The most common type of synchronous transfer line is the in-line transfer machine consisting of a series of single-purpose machine tools through which a stream of parts is slid or lifted and carried from station to station (Fig. 43). This system has been widely used for large-quantity production and even for moderate-quantity production (with modifications). The straight-line workflow conforms naturally to continuous production and fits readily into desirable plant layouts. Considerable flexibility is permitted in planning for the number, type, and position of stations; floor space and cost are the only limiting factors. Workpieces can be reoriented readily at desired stations in the line, and the minimum restriction is imposed on the approach of the tools to the part from any direction. Vibration, heat, and strain at any one station can be isolated from the others.

There are also some disadvantages associated with the use of in-line transfer machines. Some parts can be handled only on pallets or in individual fixtures for each piece, and this adds about 20% to production costs. All pallets or parts moved from station to station must be relocated and reclamped at each station. Gaging and automatic tool compensation usually require additional stations, and this also adds to production costs.

Transfer machines vary widely in size. The smaller units have only a few stations and may cost less than $100 000, while the largest units, having over 100 stations, may cost several million dollars. Each is specially designed and constructed.

There are three major types of in-line transfer machines:

- Sliding or free transfer
- Walking-beam (lift-and-carry) transfer
- Palletized transfer

The type of transfer mechanism used to move workpieces from station to station is usually dictated by the shape, size, rigidity, and material composition of the workpieces, as well as their locating surfaces or points. Some transfer machines even use combinations of these mechanisms.

When the shape of the workpieces is suitable, the most economical method of moving them from station to station is with a synchronous free-transfer machine using a transfer or shuttle bar, without any fixture or workholding means during transfer. With synchronous (intermittent) motion, all workpieces are transferred at the same time. A method of transfer sometimes used when workpieces have flat surfaces is the sliding of parts from station to station on skid rails by means of a hydraulically or mechanically operated transfer-bar mechanism.

The walking-beam (lift-and-carry) system picks the part up at one station, carries it to the next station, and lowers it into position without damaging the work surfaces. This method is frequently used for parts that do not have sliding surfaces or that are unsuitable for sliding. Advantages of the walking-beam system, compared to the sliding method, include better control (thus eliminating the possibility of workpiece damage because of sliding) and the ability to use fixed pins for workpiece location. For heavy

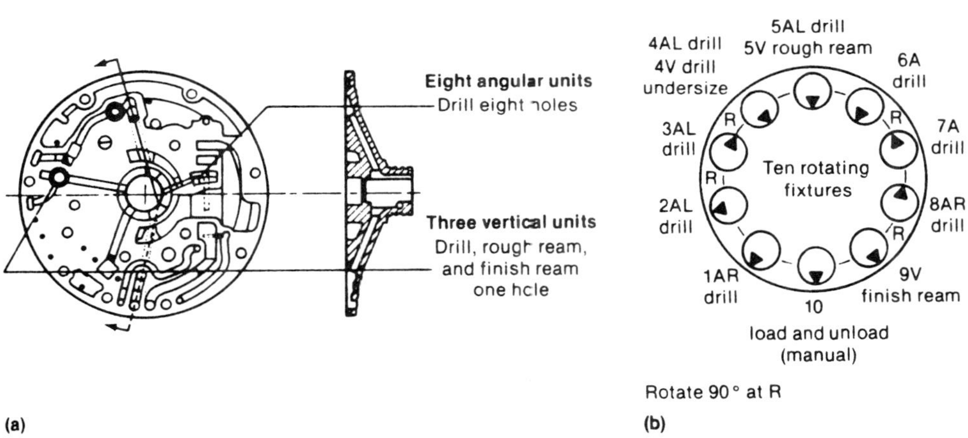


**Fig. 44** Schematic of a cast iron transmission pump cover (a) in which eight holes were drilled at an angle and one hole was drilled and reamed vertically using a ten-station rotary index transfer machine (b). Workpiece is being rotated 90° between stations with R designation to ensure alignment.

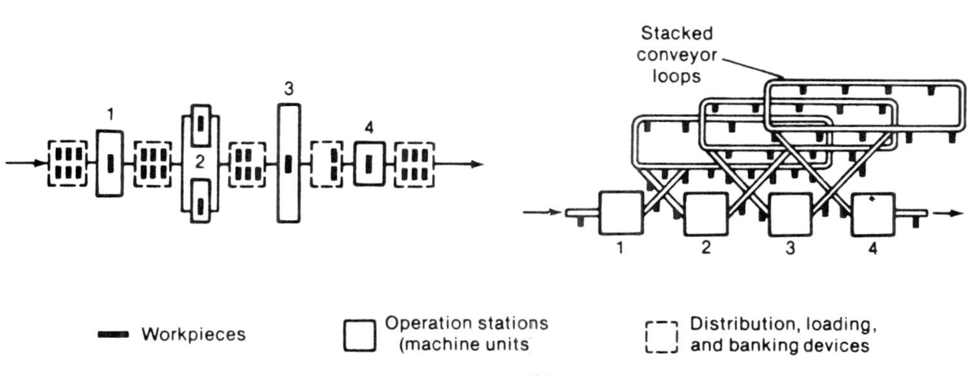


**Fig. 45** Schematic of two types of nonsynchronous system transfer machines. (a) Straight-line flow. (b) Parallel flow

**Fig. 46** Free-sliding, transfer bar engaged in-line transfer machine for processing V-6 automotive engine blocks by drilling the sides and bottom of the block (shown in foreground). Courtesy of The Cross Company

parts such as railroad-car axles, an overhead lift-and-carry transfer system is sometimes used. Such systems have arms that lower to grip all workpieces and then raise, traverse, lower, unclamp, and retract. An advantage of the overhead lift-and-carry system is that its arms aid in keeping the transfer line free of metal chips.

If workpieces are of a shape that is difficult if not impossible to handle in free-transfer systems, if they are not provided with suitable locating holes or surfaces, or if the nature of the operations does not permit other transfer means, workpieces are often placed in individual pallets. Pallets are holding fixtures in which the workpieces are located and securely clamped throughout all operations. Rotating transfer bars are sometimes used to traverse pallets to successive stations. Palletized transfer machines are available with walking-beam systems. Pallets can also be linked together and pushed or pulled; for small parts, a continuous chain is sometimes used.

*Rotary Index.* A moderate-size rotary index machine is illustrated in Fig. 44. The workpieces, located and clamped in floating, pallet-type fixtures, are carried on a circular table with a vertical axis. The centers of the machines are open, permitting horizontal, vertical, and angular machining units to be mounted inside as well as outside the pallet-carrying rails. This type of system normally requires less floor space and is less expensive to construct than an equivalent in-line system, but is limited to a few stations. Typically, on machines with more stations (and for larger workpieces) on larger-diameter circles, the workpieces are carried on pallets that are moved from station to station on a circular track. An 11-station carousel rotary index machine for machining gate valve bodies is 10.4 m (34 ft) in diameter.

Another less common design has a small table, called a trunnion, with a horizontal axis. The toolheads are arrayed on both sides, and the trunnion indexes the work in a vertical plane from station to station.

**A nonsynchronous system,** also called an independent unit system or power-free system, provides for a bank of parts to supply each machine and permits it to operate at its fastest rate. Any machine in the system can be shut down temporarily without stopping the others.

On most nonsynchronous transfer machines, pallets are freely supported and are moved from station to station by a continuously running endless chain. Each self-contained station is independent. Pallets are accumulated ahead of each station and are supplied to the station on demand. When one or more stations are down for any reason, others can continue to operate from the float of parts, thus minimizing production interruptions. Manual or automatic stations can be added or removed and spaced at random to accommodate changes in product design or to increase or decrease the automation of the system. For operations requiring longer cycles, parts can be indexed into identical stations. Parts can also be transferred through several short-cycle stations while a long-cycle operation is being performed on another part.

Such a system has a high overall efficiency and can be planned for the best combination of the machines. For straight-line flow (Fig. 45a), the banks of parts between machines can be contained in hoppers, magazines, or elevators. Parallel flow (Fig. 45b) is an arrangement of conveyor-stored banks between machines. The conveyors can be stacked one above the other in the least possible space adjacent to the line.

A disadvantage of the nonsynchronous system is its size. It is not compact, and some floor space is wasted. In addition, it is not suitable for some sizes and shapes of parts, such as large engine blocks.

**Transfer Machine Design Considerations.** Certain provisions can be made in the design of a transfer machine for efficiency. Every provision must be made for the safety of the equipment and personnel. This includes foolproofing and interlocking to prevent wrecks and personnel injuries, preinspection of rough workpieces, and a means to stop the machine when it misses. The machine must take over everything done by an operator, including clearing and disposing of chips, rough checking and orienting of workpieces, and watching for trouble. Probe mechanisms can be used to verify that a particular operation has been performed (for example, a hole drilled) and to detect broken tools or tools that have broken off during the machining process and may still be attached to the workpiece itself (see the article "Tool Condition Monitoring Systems" in this Volume). Idle stations are commonly included for space to add operations if workpiece design changes occur or for unloading and loading; therefore, entire sections of the machine can be eliminated and workpieces routed partly over conventional machines if breakdowns occur. Some sections can be made to be replaceable so that several sizes of workpieces can be processed. For example, the transfer devices between stations were changeable on a large transfer line for pressing sheet metal panels so that several sizes of panels could be accommodated.

Tool problems must be solved before a transfer machine is designed; this ensures as much dependability as possible. Even so, tools become dull and breakdowns do occur. To minimize interruptions, provisions are made to preset sharp tools before they are put on the machine, to mount them in quick-change holders, for tool racks to make the preset tools readily available when needed, for signals to alert the attendants to the times to change tools, and for schedules for changing the tools in batches. The best of components and adequate lubrication help prevent breakdowns.

Although each transfer machine is a special project, many of its units, such as toolheads, hydraulic drive units, hydraulic or pneumatic probes, controls, and slides, can and should be standard items. This will result in a considerable savings in construction cost and facility repair.

**Applications.** A transfer line is a highly specialized piece of machinery that is usually able to produce only one part. Where production volume is high enough to justify its use, the transfer line offers the ultimate in productivity and repeatability. The automotive, farm equipment, and construction equipment industries are major users of transfer machines.

Gundrilling or reaming on transfer machines is an ideal application if the proper machining units are employed and good bushing practices are followed. Contour boring and turning of spherical seats and other surfaces can be done with tracer-controlled single-point inserts, thus eliminating the need for costly special form tools. In-process gaging of reamed or bored holes and automatic tool setting are done on transfer machines to maintain close tolerances. Less conventional operations sometimes performed on transfer machines include grinding, induction heating of ring gears for shrink-fit pressing on flywheels, induction hardening of valve seats, deep rolling to apply compressive preloads, and burnishing.

Transfer machines have long been used in the automotive industry for producing identical components at high production rates with a minimum of manual part handling (Fig. 46). In addition to decreasing labor requirements, such machines ensure consistently uniform, high-quality parts at lower cost. They are no longer confined to rough machining and often eliminate the need for subsequent operations such as grinding and honing.

A common trend in transfer machine design is to make systems more versatile so that they run varieties of parts and/or smaller batches economically. This can be done by using interchangeable toolheads for different parts. For example, a tractor manufacturer has installed a transfer line to machine (randomly and without shutdown for changeover) five different families of front- and rear-end axle housing components weighing about 225 to 375 kg (about 500 to 825 lb).

# Flexible Manufacturing Systems

Jeffrey T. Paprocki, Kearney & Trecker Corporation

Today's manufacturing systems for discrete metal parts production successfully combine machine tools, material-handling systems, and computers to meet a wide variety of manufacturing requirements.

In the past, flexible manufacturing systems (FMS) have only been applied to midvolume, midvariety manufacturing, but with the advent of increased sophistication of all components, flexible manufacturing systems are becoming a viable solution to high-production applications and high-variety conditions.

## Goals and Objectives of Flexible Manufacturing Systems

The successful development of a manufacturing system requires careful planning. The design team should follow sound guidelines based on goals and objectives established at the very start of the planning process. This is perhaps the most important step in the entire project. Once the objectives and ultimate goal have been thoroughly discussed and organized, a manufacturing system can be custom tailored to meet the specific requirements of the customer.

While flexible manufacturing systems are not the answer to all manufacturing problems, they can help users achieve many cost-effective measures as well as increased flexibility in the production process.

## Cost-Effective Measures

Cost-effective measures to be considered when choosing flexible manufacturing systems include reduced labor costs, improved equipment use, reduced inventory, maximized cutting efficiency, and increased management control.

**Reduced Labor Costs.** Labor-related costs and problems have become a major concern in today's production world. Improved use of direct and indirect labor has become an important goal of manufacturing systems.

The amount of direct manpower required to operate a manufacturing system is relatively low compared to that of other manufacturing alternatives. The integration of material handling, computerized numerical control (CNC) machine tools, and an overall control system has reduced the proportion of direct labor involved in the manufacturing process. Table 6 indicates the approximate ratios of manpower to function for different technologies.

**Improved Equipment Utilization.** Machine tools in a flexible manufacturing system operate at a relatively high rate of equipment utilization, compared to alternative methods. In fact, manufacturing systems are usually designed to operate at 85% efficiency, with efficiency being defined as the time a system is available to receive work. These systems maintain a high level of equipment utilization by the elimination of many areas that normally cause poor utilization.

The material-handling capability of flexible manufacturing systems has eliminated much of the time machine tools normally spend idle while workpieces are loaded or unloaded. The material-handling system (MHS), under computer direction, maintains a constant supply of workpieces to the machine tools.

Improved maintenance techniques and machine tool reliability have lowered previous levels of machine downtime due to ineffective maintenance. One major advancement in this area has been successful application of computer-aided diagnostic troubleshooting of machine tool problems.

Computer operation has allowed the systems to be operated with minimum manpower requirements during the second and third shifts of plant operation, allowing the system to be fully utilized 24 h a day, thereby improving productivity.

Increased capacity of each machine tool station in a system is directly related to its utilization, also known as the cycle-on mode. Table 7 compares the utilization of stations in various manufacturing environments.

**Reduced Inventory.** These systems operate with minimum work-in-process invento-

**Table 6 Labor involvement in various automated manufacturing systems**

| | Ratio of labor to workstations or machines | | |
|---|---|---|---|
| Work function | Transfer line (dedicated technology), people:stations | Stand-alone technology, people:stations | Flexible manufacturing system technology, people:stations |
| Loading and operating | 1:15 | 1:2 | 1:15 |
| Setup and tooling | 1:20 | 1:6 | 1:10 |
| Materials movement | 1:25 | 1:8 | 1:12 |
| Inspection | 1:20 | 1:6 | 1:8 |
| Maintenance | | | |
| Mechanical | 1:30 | 1:8 | 1:10 |
| Electrical | 1:50 | 1:6 | 1:6 |

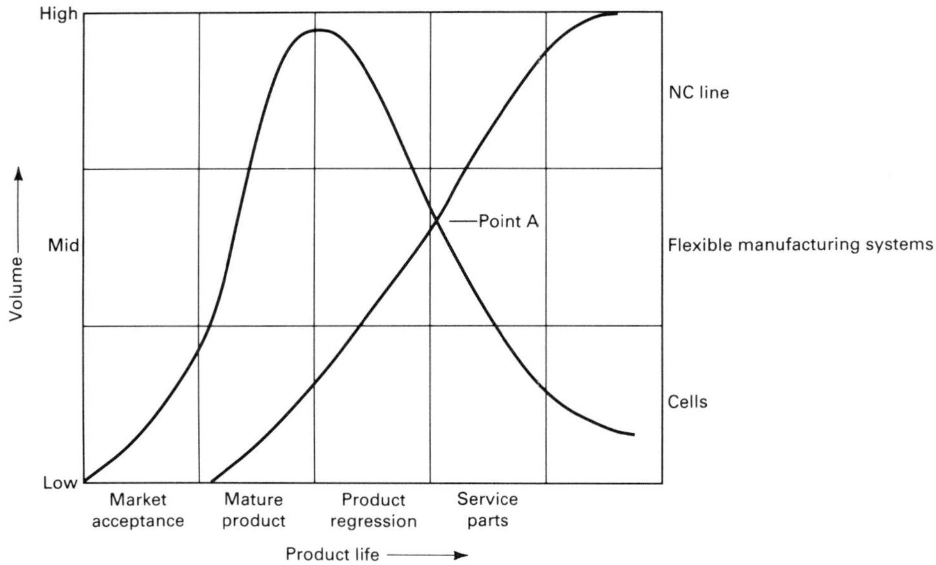


**Fig. 47** Volume plotted against product life to show which flexible-automation method should be used. See text for discussion.

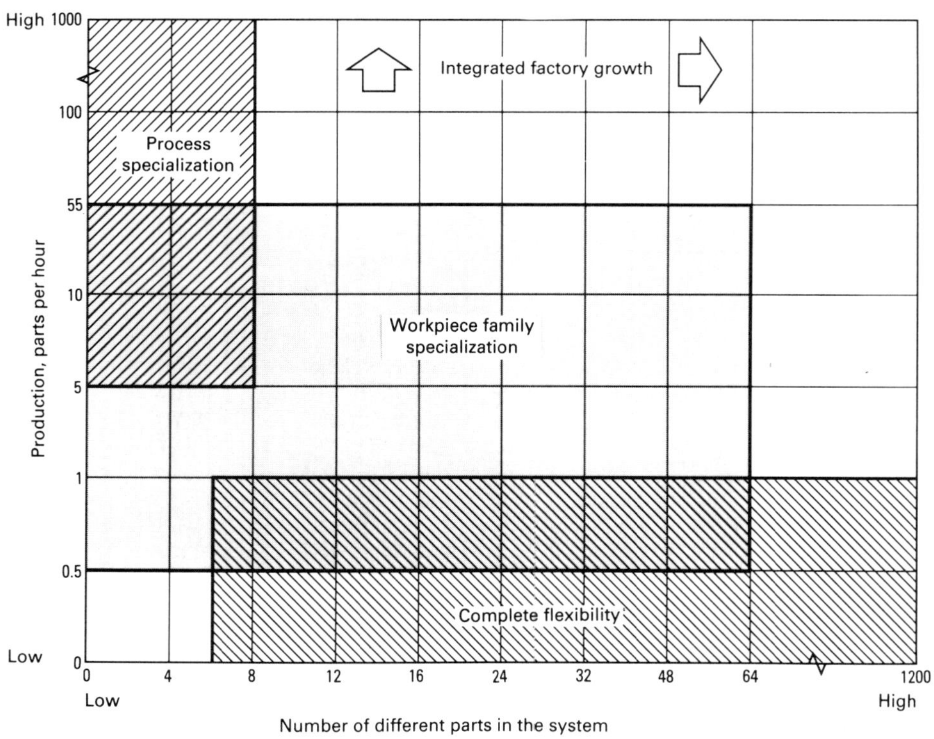


**Fig. 48** Three flexible automation concepts: total flexibility, workpiece family specialization, and process specialization

ry, even though in many systems the workpieces are processed across a number of machine tools. The main reason for this capability of the system is the palletized material-handling systems, which automatically delivers and removes workpieces from the work stations. This eliminates the manual handling of the workpieces at each of the machine tools, which in turn decreases the work-in-process time for loading and unloading. In most systems, the workpieces are manually handled only once for each orientation of the workpiece, and only at specified loading/unloading areas.

This feature has definite advantages. It:

- Releases floor space normally used for work-in-process inventory
- Reduces the number of workpieces a machine shop needs to have on hand to produce required production levels (these workpieces, depending on the amount of work done to them, represent varying accumulated costs)
- Reduces the potential of scrap and rework. Once a problem is discovered, very few pieces are made that do not meet tolerance specifications

**Table 7 Station utilization**

| Type | Utilization, % | Comments |
|---|---|---|
| Stand-alone NC machine | 35–50 | Average industry utilization can be higher with extraordinary means. |
| Transfer line | 50–60 | Buy twice as much capacity than needed. |
| Flexible manufacturing systems | | |
| Machining center | 85–95 | A typical FMS station will outproduce a stand-alone machine tool by a 2:1 margin. |
| Special station(a) | 90–95 | |
| Total flexible manufacturing system | 85–95 | |

(a) Head indexers, head changers, milling and boring modules

**Maximum Cutting Efficiency.** The use of process-specialized manufacturing technologies introduces some of the advantages of production equipment to the system. Such equipment can be successfully implemented in a system either for cost advantages or to meet critical tolerance requirements.

To be effective, like all machine tools in a system, specialized tools need to be flexible. This flexibility is attained through the use of numerical control. A good example of this type of machine tool is a multiple-spindle head changer. This type of machine tool features complete NC control of its three linear axes and its index table. It also has the capability of automatically and randomly changing large multiple-spindle heads. By its design, this machine tool, when applied to a system, does not come dedicated to one part number, but can be used to machine various hole patterns in various numbers of parts.

**Increased Management Control.** Flexible manufacturing systems have been developed to provide management with increased control of the manufacturing process. In sophisticated system designs, a central control system generates a series of reports at desired intervals about production schedules, system component utilization, tooling wear, and maintenance schedules. These records establish a historical data base from which logical decisions can be made.

## Increased Flexibility

Increased flexibility is an outgrowth of innovations in computer technology and tooling.

**Quick and Uncomplicated Reaction to Engineering Changes.** Manufacturing sys-

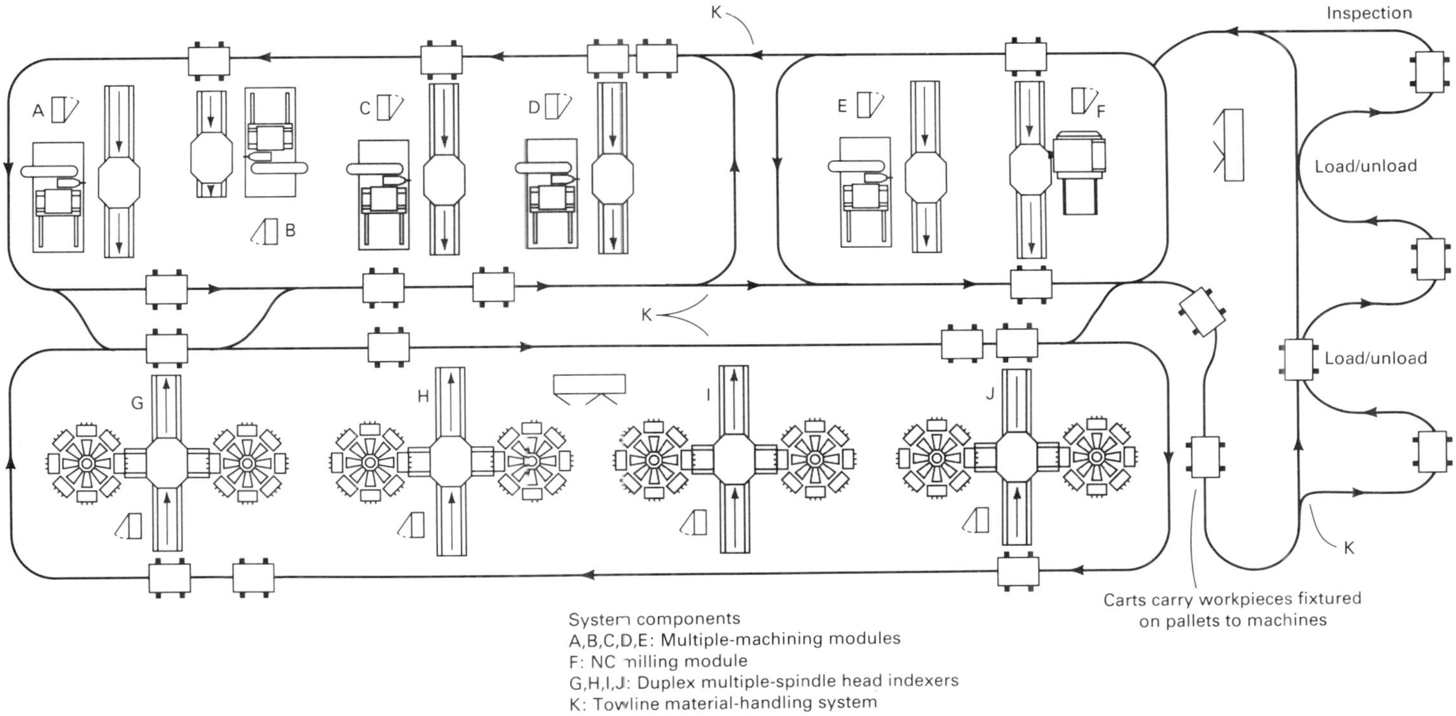


**Fig. 49** Flexible manufacturing system for producing agricultural equipment components. This system measures 84 × 34 m (275 × 110 ft). Shaded components represent FMS components added after initial installation. A central computer routes workpieces and directs machine operations.

**Fig. 50** Manufacturing cell consisting of a machining center and a vertical turning center linked by a pallet shuttle

tems with the flexibility of numerical control can quickly, in an uncomplicated fashion, react to engineering changes. Fixturing, tooling, and NC part programs are involved.

## Capability to Respond to Market Changes

Not only must the flexible manufacturing system work efficiently to produce the product for which it was initially designed, it must also allow product alterations affecting the production process.

**Adaptability to New Workpiece Introduction.** Manufacturing systems, by their very nature, can quickly adapt to the production of new workpieces. In most cases, workpiece changeover can be accomplished with a change in tooling and NC part programming.

The passage of time has shown that there are four phases to any new product, and flexible automation can address any phase of a product life cycle. Figure 47 shows the relationship between volume and product life and how it can be addressed with flexible automation. Point A is the intersection of the curves of two different product life cycles that could represent simply a change in tooling, fixturing, or NC part programming.

**Incremental Phased Installation to Match Production Requirements.** Many manufacturing systems are designed to phase in equipment over a period of time. In this approach, machine tools are purchased and installed as production needs dictate. Depending on the number of distinct market levels, there may be a number of types of equipment to be designed, built, and installed in a system with minimal production disturbance.

The phased-in approach gives the system the unique characteristic of being able to develop with the market. Should the market increase, additional machine tools can be implemented. On the other hand, should the market not develop to its maximum projected level, the user is in the position of having

**Fig. 51** Manufacturing cell consisting of two machining centers and a horizontal lathe, both of which are serviced by an overhead gantry robot

**Fig. 52** Manufacturing cell that uses a rail-guided linear transporter

purchased only the equipment needed to produce for the actual market level, thereby realizing a definite savings in capital investments.

## Production Techniques

To fulfill the production requirements of potential users, system design always starts with the study and analysis of parts to be manufactured. This encompasses the selection of workpieces and the formation of part families.

**Workpiece Selection.** Choosing the workpieces to be machined by a manufacturing system defines the scope of the project. The FMS designer has the unique opportunity of beginning his design with a blank piece of paper. His workpiece selection directs the system design because the design and its components are individually tailored to the user's goals and workpiece selection. This is a unique opportunity, when compared with established methods of developing a manufacturing process, that is, producing parts with existing facilities and machine tools.

Like a good tailor who allows for future alteration in his suits, the system designer must plan for the future of his system, and the planning stage is the time to look into the future and forcast production requirements. Careful planning at the outset determines the success of the system in the future.

**Families of Parts.** Manufacturing systems are designed to machine families of parts. Dividing parts into families provides the user with the opportunity to look at them as an entity and determine if they truly are candidates for a manufacturing system. A definition of families is difficult to isolate. One classification system for three basic workpiece families is:

- *By assembly*: A common method of grouping workpieces that, when assembled, represent a unit. The major components of a transmission or engine are good examples. A system designed to machine workpieces of an assembly family would conform to a production schedule, rather than batch order quantities. Assembly families tend to be directed by market demand for product shipment by model
- *By type*: A method of grouping workpieces into generic families. Engine cylinder heads and industrial butterfly valves are two groups of workpieces that can be categorized into type families
- *By size or similar operation*: Workpieces that share common machining operations or physical size can be placed into this family. A variety of part numbers using the same tooling or fixturing (for example, shafts, spools, and so on) would fit very well into this family

## Choosing a Production Technique

Once objectives have been established and workpieces selected, a production technique to meet specific requirements can be chosen. Figure 48 divides production flexibility into three categories:

- Total flexibility
- Workpiece family specialization
- Process specialization

As the flexibility of a particular manufacturing concept increases, its productivity decreases. On the other hand, if productivity is the main concern, the cost per workpiece and product life cycle must once again be reviewed to determine the extent of flexibility that will be integrated into the system design.

**Flexible Automation Concepts.** Concepts designed to produce workpieces on a low-volume but high-variety basis fit into the category of complete flexibility. They also typically have low productivity.

Midvolume, midvariety manufacturing concepts are included in the category of workpiece family specialization. These concepts have a moderate degree of flexibility and productivity.

High-volume, low-variety manufacturing concepts are included in the category of process specialization. A degree of flexibility is sacrificed for the sake of high production output.

**Flexible Manufacturing Alternatives.** The flexible automation concepts shown in Fig. 48 can be addressed with one or more combinations of the three flexible manufacturing alternatives. The alternatives are manufacturing cells, classic flexible manufacturing systems, and flexible NC lines. The complexity and scope of any flexible manufacturing system is readily demonstrated by the sophistication of the machines used and the amount of floor space required (Fig. 49).

*Manufacturing cells* are the bridge between stand-alone NC machining centers and full-scale flexible manufacturing systems. A cell can be treated as a station in a large-scale flexible manufacturing system, or it can be an entity unto itself that can produce a high variety of service or prototype parts on a limited scale.

Manufacturing cells can actually address two automation concepts, the first being complete flexibility, in which the cell would machine as many as 300 or more various part numbers per year. The total yearly production per part number would probably not exceed 50. The other automation concept that a manufacturing cell could address is workpiece family specialization, in which it could produce a low variety of four to six workpieces with an annual production volume of approximately 50 000 to 150 000.

**Fig. 53** Flexible manufacturing system incorporating automated guided vehicle (AGV) material-handling devices and multiple-machining modules

**Fig. 54** Flexible manufacturing system featuring AGV material-handling and multiple-machining modules

**Fig. 55** A typical flexible NC line

Manufacturing cells are very similar to a classic flexible manufacturing system, but fall short in their expansion capabilities in the areas of computer control and material handling. Gantry robots and linear transporters are both examples of common material-handling devices used in cells. Manufacturing cells can be a subset of a larger system in which computer control can operate a number of cells, and each cell can be treated as an FMS station. Figures 50, 51, and 52 show three examples of manufacturing cells.

*A classic flexible manufacturing system* is a grouping of machine modules or cells designed to combine the labor savings benefits and productivity of traditional transfer lines with the flexibility of the NC machining centers. Flexible manufacturing systems are used to produce midvolume, midvariety workpieces and are included in the workpiece family specialization category. The midvolume, midvariety classification refers to approximately 75% of today's production needs. It includes yearly production volumes of 15 to 15 000 per part number, combined with part varieties ranging from 2 to 800.

Classic flexible manufacturing systems are particularly suited to midvolume, midvariety manufacturing. Addressing the midpoint concept of our production techniques, classic flexible manufacturing systems possess advantages from both high- and low-production concepts. Figures 53 and 54 illustrate two typical systems. The designs shown incorporate a central computer control with real time and decision support logic, automated guided vehicle (AGV) material handling, and multiple machining modules.

In a classic flexible manufacturing system, under automatic computer control, palletized workpieces of different types are randomly and simultaneously transported between and processed at different machining modules and other workstations (for example, wash stations, inspection modules, and storage devices), according to individual processing and production requirements. A system designer can tailor a flexible manufacturing system to meet almost any production volume-variety mix, while taking into account user business objectives of cost-per-workpiece and product life cycle.

*Flexible NC lines* are included in the process specialization category because they are designed to produce low-variety high-volume workpieces. A group of NC modules that were designed to manufacture a limited variety of workpieces, with changeover being internal to the machine modules, would be used to produce as few as 2000 workpieces of one part number per year.

NC lines are the most production oriented but least flexible of the three alternatives. Typically they produce only a few part numbers at a relatively high production rate. NC lines incorporate many advantages of transfer line techniques while maintaining a degree of flexibility.

An NC line gets its flexibility from the use of NC machine modules. Numerical control allows random, automatic tool or head changing; variable spindle speeds; variable feed rates; and multiple-axis movement without manual intervention.

The major feature that separates an NC line from the other flexible manufacturing alternatives is its mode of operation. The NC line is capable of randomly and simultaneously machining any of several part numbers. However, each of the workpieces is machined in a sequential order as it moves from station to station along a predetermined path on the material-handling system. Figure 55 is a good example of a flexible NC line.

**The Volume-Variety Relationship of Flexible Automation Alternatives.** The three flexible manufacturing alternatives have now been defined. Figure 56 illustrates the approximate areas these three technologies will fill on our volume-variety graph. The degree of flexibility and production output are the major differences between the three alternatives.

The three flexible automation alternatives do overlap in their applications. Therefore, one particular type of system may not be the only answer to a production problem. This flexibility of design and overlapping of types permits custom-tailored solutions to a multitude of production problems. As previously discussed, the proper system selection is also influenced by the operating goals and objectives of the user.

## System Manpower Requirements

FMS technology will have a widespread effect on any operation, from accounting and quality control to almost every function of the manufacturer. An awareness of the new requirements that are placed on the organization, in addition to the technical changes, is a must. The skills of personnel, as well as the systems and procedures, must be reoriented if maximum benefits are to be realized.

An integrated teamwork concept is needed because a typical flexible manufacturing system can produce considerable quantities of a mix of parts in a production-oriented environment. Because machines are not individually manned, a particular malfunction can be due to many factors. Tooling interface areas (tool/spindle, part/fixture, fixture/pallet, pallet/machine) and human communications create the need for constant vigilance to maintain FMS quality output.

The workforce required to operate a flexible manufacturing system is relatively small, compared to other NC manufacturing alternatives. The major components of the flexible manufacturing system, computers and material handling, minimize manual-intervention requirements.

Personnel are primarily administrators of the system. For example, labor is used to perform the loading and unloading of workpieces under computer direction at the loading/unloading area. The machine tool operators have been replaced by direct numerical control (DNC), which monitors and controls the machine and operations. Rover operations are used to perform manual checks and operations at the machines. Workers normally used to deliver and remove workpieces to the machine tools in more conventional manufacturing methods have been replaced by an automated material-handling system.

Maintenance of a manufacturing system is critical to its success, and quick response to its needs becomes important. Systems are designed to be as low maintenance and failsafe as possible. Routine maintenance and minor repairs are the responsibility of the cell attendant. However, skilled maintenance people are needed to service the system in the categories of electrical, mechanical, and hydraulical maintenance.

Manpower requirements also tend to be greater at the start-up of the system and until all personnel involved are familiar with the operating characteristics of the system. The manpower functions for most manufacturing systems are system manager, FMS operators, electrical and mechanical technicians, and quality assurance personnel.

*Implementation.* Flexible manufacturing systems are an innovative approach to manufacturing. It is a significant change for many plant operations. Personnel on all levels of responsibility, from the shop personnel and supervisors through the ranks of management. will have different areas of concern on how to use flexible manufacturing systems. Workers should be informed early of the decision to go to flexible manufacturing and should have input in the design of the system. Key individuals should be selected early to receive training to operate the system even before it is operational. The solution of phase one planning will fail if enthusiastic input is not received by personnel at all levels. Likewise, it is important to establish that responsibility for the system lies with the user, not the builder.

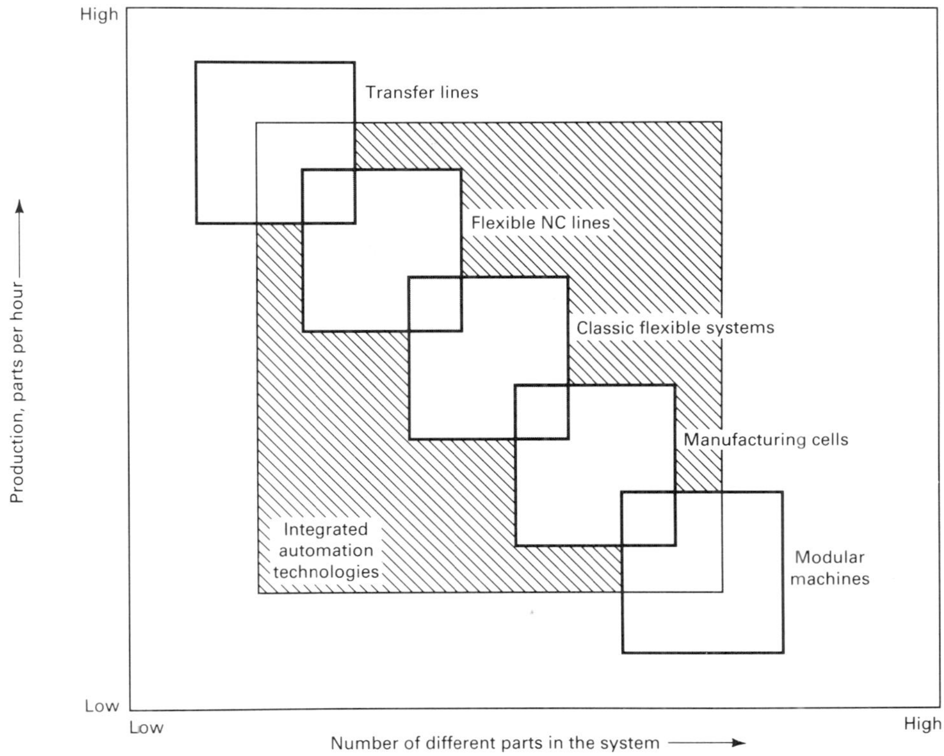


**Fig. 56** Relationship of manufacturing cell, flexible manufacturing system, and flexible NC line technologies on a volume-variety graph

# Proper Fixturing

Clifford E. Drake, ENERPAC Group Applied Power, Inc.

A FIXTURE is a special workholding and supporting device that is securely held or fixed to a machine. The key characteristic is that it is designed and built for a particular part or shape. A general-purpose device, such as a vise or clamp, is not a fixture. The specific purpose of a fixture is to locate and hold a workpiece against the machining forces. A jig differs from a fixture in that it not only locates and holds a workpiece, but also guides the tool to ensure accurate positioning. Thus, fixtures are more open in design than jigs to facilitate the placement and removal of the workpiece. The primary function of fixtures is to:

- Enable a variety of part configurations to be machined to specification
- Produce quality products
- Reduce cost of manufacturing
- Maximize machine utilization

Fixtures are used mostly on machining centers, milling, drilling, and boring machines. They are also used in welding, assembly, and test applications.

Fixtures to hold a particular part are made from gray cast iron or from steel plate by welding or bolting. It is fixed, that is, bolted, clamped, or set with a low-melting alloy, to the machine. A fixture has locating pins or machined blocks against which the workpiece is tightly held by clamping or bolting. In order to ensure part location over time, the locating devices are made from hardened steel. Many fixtures are massive because they, like a machine frame, may have to withstand large, dynamic forces. Because all fixtures are between the workpiece and the machine, their rigidity and the rigidity of their attachment to the machine are paramount. Duplex fixtures have been built to allow for the loading or unloading of one side of a fixture while the machining operation is taking place on a part clamped on the other side.

Fixtures can be made by using standardized components, such as drill bushings, locating buttons, and clamping devices. Fixtures can be expensive, however, even with standard components. Obviously, their cost is a part of the total cost of production, and one must determine whether fixtures can be justified economically by the savings in labor and machine cost that will result from their use.

Fixturing costs are justified when the need for frequent changes in machining setup is balanced by the volume of required work. This goal coincides with the objectives of both flexible manufacturing systems (FMS) and just-in-time (JIT) production. A typical FMS operation, for example, may contain 5 to 12 machine tools, with each machine capable of changing cutting tools to perform different tasks, such as drilling, boring, or milling, for small-to-medium lot sizes. Fixtures make the changeover cost-effective by eliminating the problem of setup time for workholding. JIT production, with its goal of minimizing inventory by providing small lot sizes on an as-needed basis, also greatly benefits from the elimination of setup time provided by fixturing.

## Optimum Fixture Design

When designing fixtures, many things need to be taken into consideration. It is important to have specific information on:

- Type and size of machine
- Specifications of the machine tool
- Process to be applied
- Part size
- Tooling
- Number of parts per fixture
- Size and number of pallets and size of machine bed
- Clearances in the machine enclosure
- Feeds and speeds
- Number of different parts to be produced
- Production forecasts and scheduling needs
- Constraints on pallet change system
- Extent of automation (if any)
- Tolerances
- Part rigidity
- Tool change time
- Pallet change time

Proper fixture design is based on sound basic workholding principles:

- The part should use the 3-2-1 approach to determine part location in the three planes (Fig. 1). Different operations require varying numbers of locating planes, depending on the complexity of the operation (Fig. 2). An alternative approach may be used, depending on circumstances
- The fixture should be designed so that the clamping and cutting forces are directed against the points of location and work support (Fig. 3a and 3b)
- The fixtures should be kept as simple as possible within the constraints laid out at the beginning of the project
- Hydraulic and pneumatic fixtures should be designed so that piping is kept out of the way to prevent possible damage or chip buildup
- Chip traps should be kept to a minimum or guards should be installed to prevent buildup. Correct and incorrect methods of locating the fixture in relation to the workpiece for chip removal are shown in Fig. 4
- Hydraulic and pneumatic fixtures should be designed so that parts stay located before being clamped without the operator holding them in place

The challenge of fixturing is tailoring the design to the production needs and to the characteristics of the machine. Production demand may dictate a design that maxi-

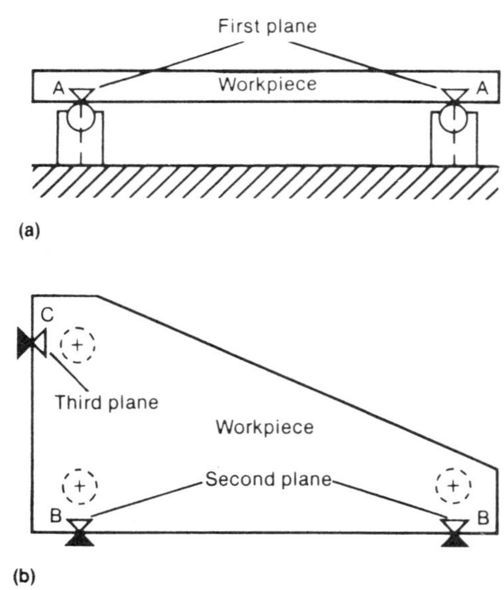


**Fig. 1** Exact positioning of the workpiece, which lacks locating holes, in three planes. (a) The first plane is located by placing the workpiece on three support balls at location A. (b) The second and third planes are located by butting the workpiece against two locating points at location B and flush against the third locating point, C, respectively.

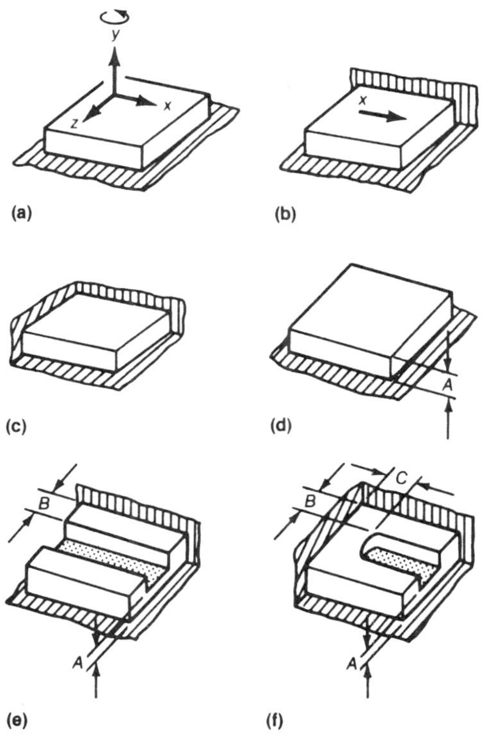


**Fig. 2** Degrees of freedom and number of locating planes for differing production requirements: (a) and (d) at partial location, (b) and (e) at location, and (c) and (f) at full location. Top row shows degrees of freedom. (a) Using one locating plane leaves the component with three degrees of freedom. (b) Using two locating planes leaves component with one degree of freedom. (c) Using three locating planes removes all degrees of freedom from component. Bottom row shows typical production application. (d) Adequate for planing a workpiece with dimension *A*. (e) Adequate for milling an open slot with dimensions *A* and *B*. (f) Necessary for milling a blind slot with dimensions *A*, *B*, and *C*

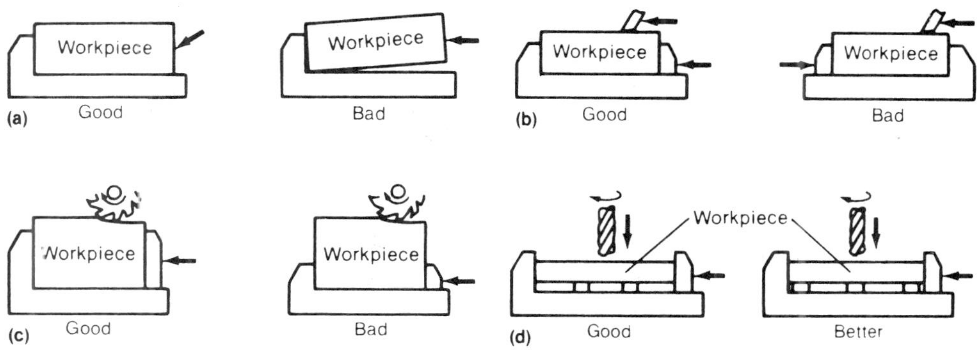


**Fig. 3** Good and bad methods for workpiece support and clamping: (a) Clamping forces should direct the work against the points of location and work support. (b) Whenever possible, cutting forces should act against the fixed portion of a jig or fixture. (c) The points of clamping should be in line with the cutting forces as much as possible, so the workpiece is less likely to be pulled from the clamps. This also allows lower clamping stresses, thus reducing distortion. (d) Location points should be as far apart as possible and positioned so that cutting forces do not distort the workpiece. On the right, three buttons, rather than two, are used to form a triangular support to reduce distortion. Compliance with these methods also aids in reducing workpiece vibration and tool chatter.

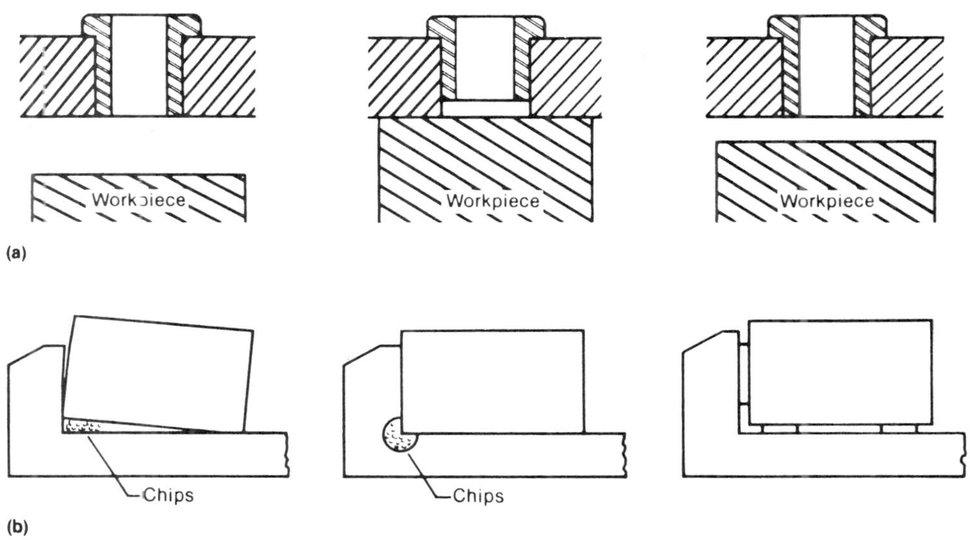


**Fig. 4** Incorrect and preferred methods of chip clearance when machining with (a) a drill bushing and (b) a right-angle workpiece. (a) Drill bushing setup at left permits tool drift due to excessive clearance between bushing and workpiece; at center, bushing retains chips because work abuts bushing; at right, proper fixturing. (b) Right-angle workpiece progresses from improper fixturing at left to better setup at center to proper fixturing at right.

mizes or minimizes the number of parts per fixture. A large part family may require the use of pickoff plates to accommodate all the parts without having to build individual fixtures (Fig. 5). This reduces costs and required storage space and permits a quick changeover. Part geometry may lend itself to a column- or angle-type fixture so that more sides are accessible with fewer setups.

Hydraulic or pneumatic clamping may be directly or indirectly applied, depending on the number of parts per fixture and the constraints of space. This approach can provide flexibility and ease of setup for a family of parts by simply changing clamp location, clamp arm, or clamping linkage.

## Clamping Methods

Clamping of the work is closely related to support of the work. However, there are certain principles that relate specifically to clamping. First, clamping stresses should be kept low. Clamping induces some elastic stresses that tend to cause some distortion of the workpiece. If this distortion is

**Fig. 5** An array of pickoff plates combined to form one fixture equivalent to several standard fixtures, thereby permitting rapid changeover

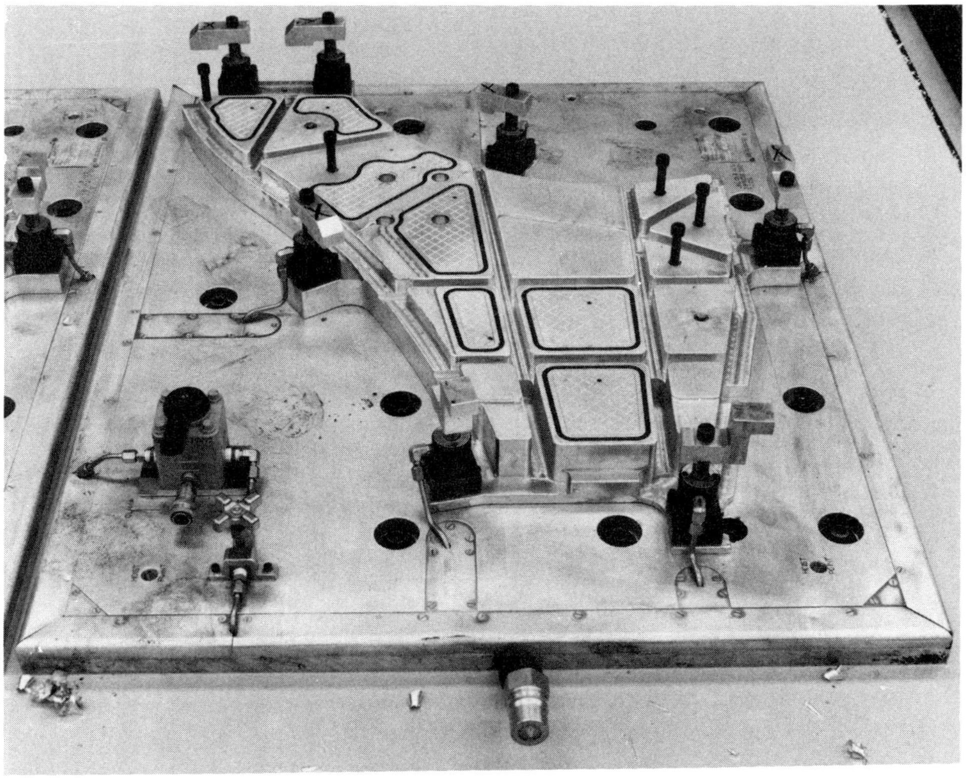

**Fig. 6** Typical hydraulic clamping setup with the vacuum clamping form of fixturing

measurable, it will cause some inaccuracy in final dimensions. The obvious solution is to spread the clamping forces over a sufficient area to reduce the stresses to a magnitude that will not produce appreciable distortion.

Second, the clamping forces should direct the work against the points of location and work support (Fig. 3). It is not safe to assume that if there is no designed clamping-force component either upward or downward, the work will be held properly and will not tend to rise from the supports. Clamped surfaces often have some irregularities that may produce force components in an undesired direction. Consequently, clamping forces should be applied in directions that will ensure that the work remains in the desired position.

An important corollary to the second principle is that, whenever possible, fixtures should be designed so that the forces induced by the cutting process act to hold the workpiece in position against the supports. These forces are predictable, and proper utilization of them can materially aid in reducing the magnitude of the required clamping stresses.

The third principle is that as many operations as practicable should be performed with each clamping of the workpiece. This principle has both physical and economic aspects. Because some stresses result from each clamping, with the possibility of accompanying distortion, greater accuracy is probable if as many operations as are practicable are performed with one clamping. Thus a fixture should be designed to serve several operations when possible. From an economic viewpoint, it is obvious that if the number of fixtures is reduced, a smaller investment of capital is usually required and fewer man-hours are necessary for positioning and removing the workpiece from the fixtures.

## Manual Clamping

Manual clamping is not only the most popular approach to designing fixtures, it is also one of the simplest and most flexible. One of the most popular manual clamps is the machinist vise, although strap clamps, edge clamps, and toggle clamps are also frequently selected. Most manual clamps use some kind of mechanical advantage in applying the clamp force. The compact size, ease of use, great flexibility, and low cost make manual clamps very attractive for fixture design.

While manual clamping offers many benefits, its inherent characteristics can also create problems when manufacturing with more advanced fixturing. It relies on the operator to position and tighten it properly. Parts can be deformed if clamped too tightly, or able to move if not clamped tightly enough. Clamping force may vary from operator to operator, which can subsequently alter the quality of the finished workpiece. Manual clamping and unclamping usually requires more time than pneumatic or hydraulic approaches. Manual clamping is also more difficult to automate as more unmanned systems are installed.

## Pneumatic Clamping

Other forms of clamping were developed as manufacturing needs changed. Increased productivity and increased utilization of machine tools required faster clamping methods. Faster clamping would allow quicker loading and greater production.

Pneumatics was the next form of clamping to be pursued. It was easy to automate and to use. Pneumatic clamping is now used extensively for assembly and light-welding fixtures. These applications require automation and consistent clamping forces. Pneumatic clamps may be applied directly to hold parts in fixturing or indirectly using toggles or linkages.

Pneumatic clamps are relatively large for the amount of force they apply. In addition, the compressibility of pneumatics makes them more difficult to control in a smooth, progressive manner. However, the use of pneumatics allows simplified piping because compressed air can be returned to the atmosphere without environmental concerns or pollution disposal problems.

**Vacuum clamping/fixturing,** another form of pneumatic clamping, is used extensively by the aerospace industry and by other manufacturers of large thin-section parts (Fig. 6). Fixtures are designed with O-ring seals near the perimeter of the part. A vacuum is then pulled inside the O-ring area to draw the part into the fixture. Because of the large surface area of the part, high clamp force can be generated. Manual, pneumatic, or hydraulic clamps are usually used in conjunction with vacuum to ensure that parts are adequately seated to the O-ring.

## Hydraulic Clamping

Hydraulic clamping, another alternative, improves the overall speed of the clamping process, enhances both the repeatability and consistency of clamping, and provides high clamping forces from a compact clamp package. Hydraulics is very controllable and gives a smooth clamp action for improved life and quality of parts. With more conventional machine tools, the primary benefit of hydraulics is the speed of clamping and unclamping, allowing the operator to load and unload the part rapidly and get the machine back into operation as soon as possible. If a part is a thin section one, prone to distortion, the consistency and repeatability of hydraulic clamping make it an attractive alternative to manual clamping. As awareness of quality has grown in recent years, hydraulic clamping has found more widespread use in industry.

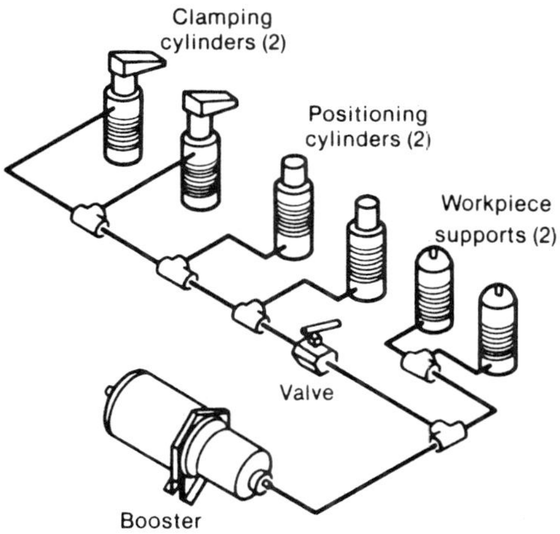


**Fig. 7** Typical hydraulic clamping system, consisting of miniature threaded cylinders, swing clamps, and work supports, and powered by the booster. The sequence of cylinder actions for positioning, supporting, and clamping are actuated by the sequence valve.

Basic hydraulic clamp systems incorporate miniature threaded cylinders or swing clamps, often used in conjunction with work supports (Fig. 7). Work supports replace the mechanical screw jacks required to prevent distortion of the part when initially clamped or during the machining process. Power sources used to operate the systems may be air/hydraulic or electric. They build and maintain pressure to ensure that parts are held throughout the machining cycle.

The high clamping forces generated by the relatively compact hydraulic devices are becoming the accepted method of clamping today's high horsepower machining centers. Hydraulic clamping systems have the added advantage of being able to absorb much of the vibration associated with the machining process.

With nonpalletized machining operations, a power source is usually located with the machine tool and connected to the fixturing to provide clamping force. The power source ensures that hydraulic pressure is maintained throughout the machining cycle. With palletized or flexible manufacturing system (FMS) operations, it is usually not possible to maintain hydraulic connections while the fixture is moving in and out of the machine environment. For such situations, alternative approaches have been developed to ensure that the clamping force remains constant.

Three approaches are currently available for palletized-fixturing applications:

- Conventional hydraulics
- Positive-lock clamping
- Hydraulic/mechanical locking

**Conventional hydraulics** is most often applied and used in conjunction with an accumulator and a shutoff valve to maintain pressure. The accumulator is a stored oil charge that compensates for any clamp settle-in or temperature variation in the system (Fig. 5). This approach requires a good, tight hydraulic system free from leaks. A periodic maintenance program must be instituted to ensure that the system stays leak free. When set up and properly maintained, it provides trouble-free service.

**Positive-Lock Clamping.** The second hydraulic approach uses high-force springs to provide clamping and hydraulics to release clamps for part loading and unloading. This positive-lock type of clamping does not rely on the maintenance of hydraulic pressure.

There are several characteristics associated with spring clamping that need to be considered when designing the fixture. Strokes associated with high-force springs are generally relatively short, and it is therefore important to take into consideration possible casting and forging variations in these cylinders. The other primary characteristic associated with spring clamping is the rate change with deflection. As spring deflection rate is increased, the clamping rate increases. Clamps should be sized so that they are used within the middle ranges of the clamps.

One of the big advantages of spring clamps is the simple piping associated with the system. Positive-lock clamping ensures uninterrupted clamping throughout the machining cycle.

**Hydraulic/Mechanical Locking Clamps.** The third form of hydraulic clamping for palletized operations is a combination of hydraulic and mechanical locking clamps (Fig. 8). The hydraulic portion gives all the advantages of longer strokes, consistent and repeatable clamping, and consistent clamping anywhere in the stroke, and when combined with a locking collet, the system provides positive locking without the necessity for maintaining hydraulic pressure. The part cannot come loose even if stored for extended periods of time.

The hydraulic positive-locking type of clamp is larger than its conventional hydraulic counterpart and requires more complex plumbing to actuate the hydraulic cylinder and the locking mechanism. These clamps are available as swing clamps, push clamps, and work supports.

When selecting the proper mechanism, it is necessary to consider several factors, that is, space, clamp forces, stroke requirements, the need (or absence thereof) for positive locking, the number of parts per fixture, and the ability to route the necessary piping.

## Automation

As automation is applied to more machine tools, it is necessary to add equipment to clamping systems to compensate for the absence of the operator. More sensing is required, and the computer numerical control (CNC) or programmable logic controller (PLC) actuates solenoid valves to control the process. Automation is usually applied to the load/unload station with robot load and automatic clamping systems for moving clamps from the cutter path and then reclamping once the cutter has passed. The computer numerical control is programmed to energize solenoid valves which, in turn, move the clamps. Clamps may be equipped with sensors that provide a feedback signal, providing the controller with verification of clamp position. Automatic couplers are used to connect hydraulics, air, vacuum, and electricity to palletized fixturing (Fig. 9 and 10).

## Modular Fixturing for Limited Production

It is usually quite easy to justify dedicated fixturing for high-run production parts but it becomes increasingly more difficult with limited production runs. This applies to short and medium run parts or those for which uncertainties exist regarding fixture production and scheduling. Modular fixturing is gaining acceptance as a solution to short- and medium-run needs, especially when frequent setups are required. While not a new concept, reusable fixturing provides greater flexibility while reducing fixture cost (Table 1). John Warton has been credited with developing the first modular fixturing system in the early 1940s, for the British war effort. This system was brought to the United States in the late 1940s. There are several systems currently available.

The flexibility of modular fixturing lends itself to today's systems. Reusable fixturing is particularly suited to short- and medium-production runs for which justification may be a problem. Modular fixturing provides the ability to:

- Manufacture prototype parts during the product development process, when changes are frequent and the part configuration has not yet been finalized
- Get into production much quicker once the design has been finalized
- Introduce new products when sales volumes would not otherwise justify the use of dedicated fixtures
- Produce pilot runs before designing and building dedicated fixtures
- Build service parts on an as-required basis without having to store and maintain the old production fixtures
- Assemble back-up fixturing in the event of damage to conventional fixtures or when peak customer demand requires additional manufacturing capacity

It is feasible to run production on a modular fixture because its flexible and reusable features are a more cost-effective way to

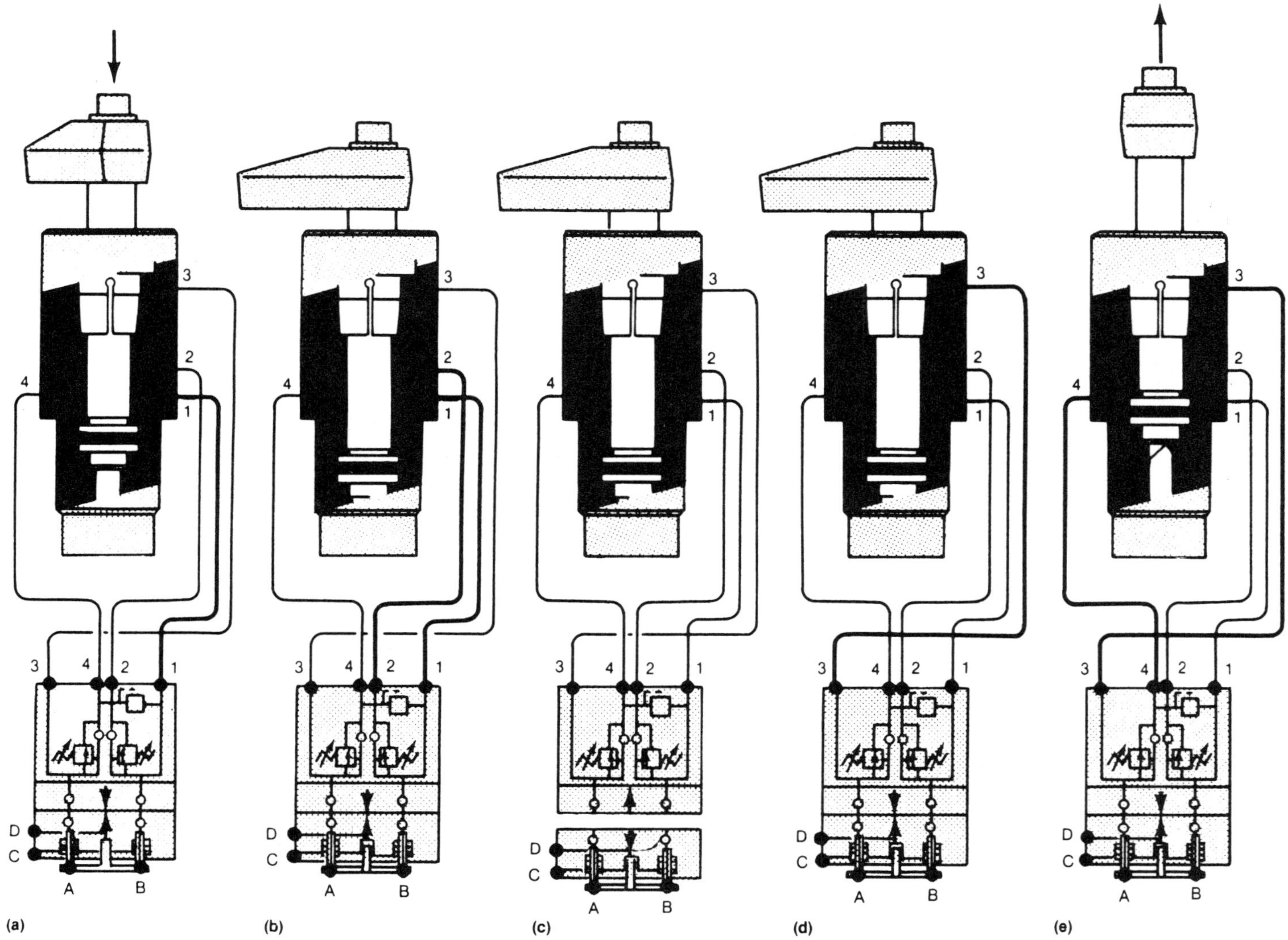


**Fig. 8** Sequence of operations for a positive lock swing clamp: (a) The port C of the autocoupler is pressurized to engage the autocoupler. Second port B of the autocoupler is pressurized, which connects to port 1 of the clamp. The clamp will then clamp the workpiece. (b) System pressure builds as the workpiece is clamped, and the sequence valve then opens. This, in turn, pressurizes port 2 of the cylinder, closing the self-locking collet. (c) System pressure to the cylinder can then be released without loss of clamp force. The autocoupler is disengaged, and the palletized fixture can be shuttled into the machine. (d) Once the pallet is returned to the load station, the autocoupler is engaged, and port A is pressurized; this in turn pressurizes port 3, and unlocks the locking mechanism. (e) System pressure builds up, the sequence valve to port 4 opens, and the cylinder unclamps.

manufacture a broad range of products than by building dedicated fixturing.

A modular fixturing system consists of a series of standard locating and clamping units (Fig. 11) that are mounted to fixturing bases machined with a grid pattern (Fig. 12). The locating clamps are assembled in conjunction with one another to create a fixture. The fixture can be designed in a tool design department using conventional drafting techniques or on computer-aided design (CAD), using the CAD data base. The drawing includes a complete bill of material and all the coordinate locations so that the setup room can assemble the fixtures as required.

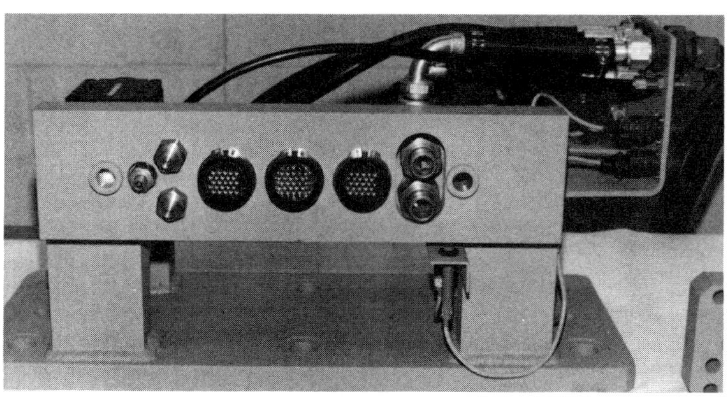

**Fig. 9** Automatic couplers used in conjunction with a network of hydraulic, vacuum, air, and electric lines to provide palletized fixturing

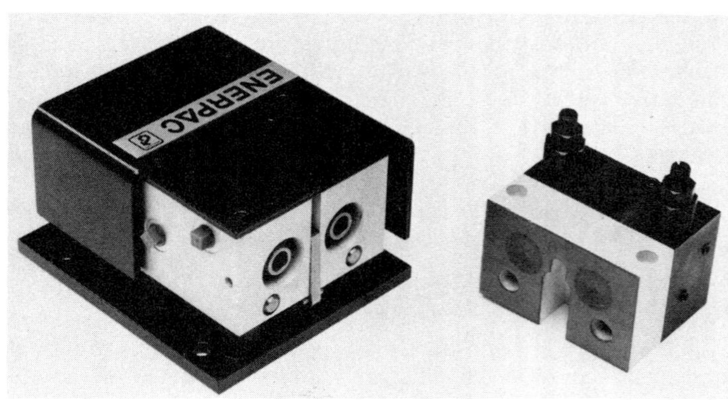


**Fig. 10** Compact automatic couplers used to control the movements of a palletized fixture

**Fig. 11** Standard locating and clamping units used in modular fixturing systems

Another approach is to have the setup room assemble the fixture around the part. It can then be documented by photographs or sketches for later duplication. As the setup room becomes proficient, this approach can be a very time-effective way of building modular fixtures.

**Table 1 Long-term operating costs for modular versus dedicated fixtures**

| | Dedicated fixture | Modular fixture |
|---|---|---|
| Design, $50/h | $800.00 | $100.00 |
| Material cost | 275.00 | . . . |
| Fabrication, $50/h | 1800.00 | . . . |
| Assembly, $50/h | . . . | 150.00 |
| Amortization(a) | . . . | 480.00 |
| Inspection/test, $50/h | 100.00 | 50.00 |
| Storage space per year | 50.00 | . . . |
| Disassembly, $50/h | . . . | 50.00 |
| **Total** | **$3025.00** | **$830.00** |

(a) Cost of base and elements, $4800, divided by ten different usages

To be successful with modular fixturing, all of the components must be cataloged and inventoried in an orderly fashion. This allows ease of use and ensures that parts do not get lost and are always in good repair. The drawings for setting up the fixtures can be put together in a setup book, which may include photographs of the fixture. The setup book should have all the information necessary to put the fixture together, including any adjustable height dimensions or settings that may be required.

**Self-Adapting Fixture Element (SAFE) Modular Fixturing System.** In most systems, a grid base is used with a series of bushed holes to locate the components accurately. Threaded mounting holes may be located directly under the bushings or adjacent to the bushings. To accommodate a variety of different part shapes that may or may not lend themselves to the exact grid locations, grid adjusters are available to ensure clamping and location anywhere within the pattern. Typical accuracy for the hole pattern on a grid plate is ±0.010 mm (0.0004 in.), relative to hole spacing over a 500 mm (20 in.) span. Flatness and perpendicularity follow the same tolerances. High-accuracy systems are often necessary to ensure the same part tolerances even with second and third operations.

Another feature of the SAFE system is the use of self-adapting fixture elements, which consist of balls fitted in an O-ring-sealed, lubricated, precision socket that self-aligns and allows clamping and locating on irregular surfaces without distorting the part when clamped (Fig. 13). This distortion-free clamping ensures improved part quality during the manufacturing process.

## Cost Factors in Fixturing

Cost considerations usually play a large part in fixturing decisions. With added complexity, such as more parts per fixture and more fixtures per machine, fixturing costs rise considerably. In modular fixturing, however, other factors must be included in the justification, and a value must be placed on each. First, fixturing should help improve overall machine utilization. To this end, feeds and speeds should be optimum for the application, and tool and pallet changes (if a palletized system is used) should be minimized. Frequent indexing is usually faster than tool or pallet changes. A faster loading-unloading cycle for nonpalletized operations is desirable. Also, fixture components that are reusable for a number of different parts improve machine utilization.

Other factors to consider in justifying the cost of fixturing include:

- *Amount and value of scrap*: Reduction of scrap is a factor. Second- and third-operation scrap is far more expensive
- *Operator utilization*: More complex fixturing results in less operator time for setup and handling
- *Mode of operation*: This influences the fixturing approach (batched versus just-in-time, or JIT, for example)
- *Production lot size*: Lot size may affect the mode of operation
- *Flexible versus dedicated fixturing*: Choice depends on lot size and mode of operation
- *Level of automation*: This dictates the use of automated fixtures and affects cost dramatically; fixturing should be considered in the decision to automate

When budgeting for a machine tool purchase, fixturing should be considered early in the project. On conventional vertical machining centers without pallet changers, fixturing is 5 to 10% of the machine cost. With horizontal machining centers with pallet changers, it is closer to 25 to 35% of the machine tool cost. If not properly budgeted, the success of the project is compromised by under-utilization of expensive equipment.

The cost of modular fixturing was initially higher than that of dedicated fixturing, but reusability for different fixtures makes the cost of modular fixturing lower because costs are amortized over a number of setups. Modular fixturing component life can exceed 10 years if maintained properly.

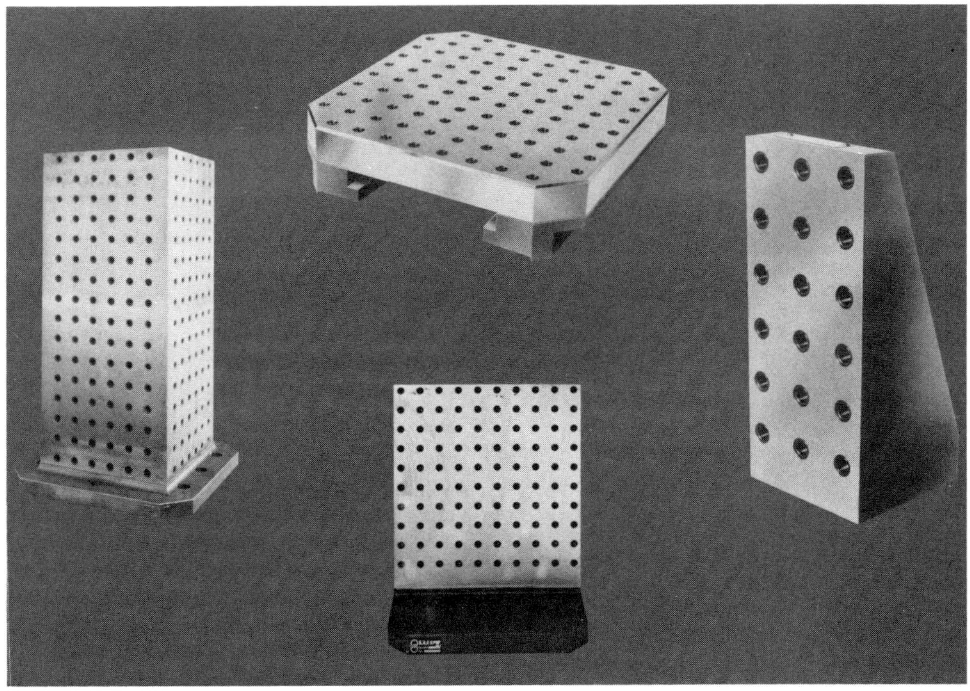

**Fig. 12** Grid pattern configurations available for mounting the locating and clamping units of Fig. 11 to construct a modular fixturing system

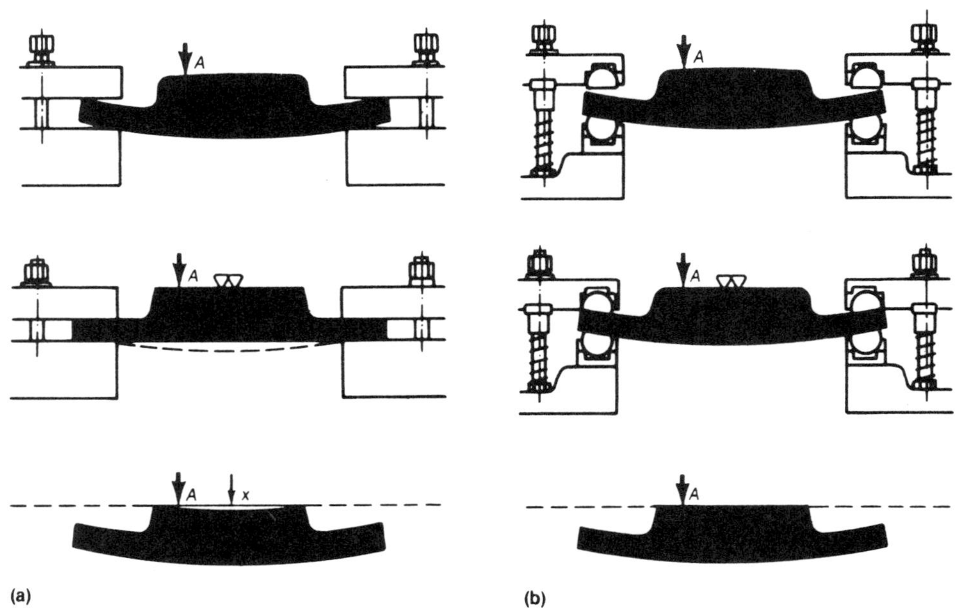


**Fig. 13** Replacing conventional clamping method (a) with SAFE clamping method (b) in a modular fixturing system eliminates costly rework or part rejection due to warpage normally encountered when clamping castings and forgings. (a) In conventional clamping, workpiece is normally clamped at three points by strap clamps, and the warpage progresses as follows: warped undersurface of workpiece rests unevenly at clamping points (top); clamping pressure produces severe strains (center); and the machined surface springs out of flat when the clamps are released (bottom). (b) In the SAFE clamping method, the workpiece floats into position: flatted balls of upper clamps float to secure workpiece (top); the lower arms of the clamps are similarly fitted with identical ball and socket components to avoid strain and provide distortion-free clamping (center); and the machined surface stays true when the workpiece is removed (bottom).

When justifying any fixture purchase, there are short-term costs and long-term costs. Short-term is usually the cost of initial purchase and setup, while long-term is the operating cost of the system that has been chosen. For example, low initial cost may result in higher operating cost of the system, but higher initial cost of better fixturing may be the most cost-effective approach in the long run.

# Tool Condition Monitoring Systems

Larry E. Stockline, Promess, Inc.

IN-PROCESS TOOL MONITORING SYSTEMS can electronically detect excessive tool wear or warn of impending tool failure to lessen machine downtime and prevent the production of out-of-tolerance parts. The degree of sophistication of this technology becomes evident when a monitored machine achieves 90% of available uptime when only two years earlier it would rarely have exceeded 60% uptime. This article discusses the sensing technology currently available for manufacturing applications. The advantages, disadvantages, and applications of this technology, as well as several case studies, are included in this overview.

## Historical Evolution

Early metal removal machining operations employed slow, belt-driven equipment, using tool steels as the cutting material. The metal removal industry followed the example of the wood machining industry in both technique and shape of tooling. Machines were typically manned full time. Sensing of the machining process was primarily done by sight, sound, and touch of the human operator.

In time, production demands grew. With the advent of the Industrial Revolution, more efficient machines and cutting materials were introduced into the workplace. The development of heat-resistant cutting tools in the 1930s and 1940s increased life; however, many of the tools introduced during that period did not have the strength of tool steel cutting tools and were brittle.

Speeds increased as bearings that could handle heavier machining loads were developed, and machining output became even greater. However, as machine efficiency increased, the ability of the operator to respond to unexpected events became less likely, and damage to equipment and injuries to individuals increased significantly.

In the early 1930s, monitoring of the electrical motor became one of the first sensing devices to mimic the human senses and help stop uncontrolled catastrophic events. Machining torque was generally quite high with very low revolutions per minute (rpm) (speeds less than 30 m/min, or 100 sfm), making this overload sensing technique quite effective for the 1930s and 1940s.

With the mass production of the automobile and the high volume demand for consumer products in general, speeds continued to climb, and torque became less important in the newer machines. Measuring motor current became less and less meaningful in detecting potential machine breakdowns.

The rapid rise in speed is demonstrated by the following table, which lists machining speeds over the past decades and those projected for the future:

| Decade | Speed, m/min | Speed, sfm |
|---|---|---|
| 1930s | 15 | 50 |
| 1940s | 30 | 100 |
| 1950s | 60 | 200 |
| 1960s | 90 | 300 |
| 1970s | 120–185 | 400–600 |
| 1980s | 185–915 | 600–3000 |
| 1990s | >915 | >3000 |

State-of-the-art equipment has spindles that machine materials at 100 000 rev/min with point positioning of the tool in excess of 425 mm/s (1000 in./min). The capability of tooling in the 1980s and the potential capability of equipment in the 1990s is:

| Variable | 1980s average | 1990s requirements |
|---|---|---|
| Available machine uptime | 60% | Must be >80% |
| Average cost of collision | $100/s | $1000/s |
| Quantity shipped | 80–90% | 98–100% |
| Machine warranty | 1 year | 5 years |
| Maximum utilization of cutting process | 60% | 90% |
| Human impact | Operator-attended manufacturing | Limited or no operator attendance |

The need for sensing, therefore, becomes clear. Machines must be designed to protect themselves so that all of the advantages mentioned previously can be achieved. Without sensing techniques, these objectives cannot be reached. It is a known fact that every metal removal machine will eventually fail (spindle collides with the machine bed and causes metal-to-metal impact). The machine must be designed to detect failure or catastrophic machining events so they can be prevented.

## Variables Monitored

There are three basic elements to any modern sensing system, a sensing source, a signal amplifier, and a microprocessor or translator to interpret information.

### Sensing Source

The sensing tasks for a manufacturing operation are substantial. In addition to being cost effective, sensors must be nonintrusive, environmentally safe for the operator, abuse resistant, sheltered from the manufacturing environment, uninfluenced by outside phenomena (other than that which is being monitored), repeatable, and highly accurate. The sensor should also have self-calibrating or zeroing capabilities and on-line diagnostics.

Adding to the complexity of the sensor are the multitude of machine types for which the sensor is expected to perform, such as numerical control (NC) machines, milling machines, multispindle screw machines, lathes, machining centers, single-purpose dedicated machines, transfer lines, grinders, tapping and threading machines, broaching machines, and so on.

The type of sensor used depends on the specific application and the results desired by the users. Clearly identifying the goals to be accomplished is a primary objective with every monitoring application. These include:

- Machine overload detection only
- Tool breakage monitoring, using either one nuisance tool only or all tools, such as those on a lathe or machining center
- Tool wear detection
- Detection of missing tool that acts as a probe to sense when the process has not been performed
- Monitoring of machine conditions such as machine being up to temperature, preloads approval, lubrication approval, bearings approval, and detection of water in spindle
- Automatic tool wear compensation

The consensus is that 70 to 80% of all metal removal and metal-forming applications can be monitored while in process.

**Fig. 1** Typical plate transducer consisting of four piezoelectric quartz crystal sensors

Various sensing devices have been developed to measure specific machining parameters such as force, tension, feed, and torque, and the advantages and disadvantages of these devices must be considered on an individual basis for each application to provide optimum utilization of each sensor (transducer). Sensors used for tool-monitoring applications can be classified by what they measure, that is, plate, bearing, tension, feed force, and torque.

**Measuring Plate Transducers.** The sensors in this device are integrated into a plate (Fig. 1) that is then mounted between the turret and the carriage. The actual sensing element is a piezoelectric quartz crystal, which is very durable and capable of accurately measuring relatively heavy loads because of its sensitivity. Among the disadvantages of this transducer is the fact that it is not retrofitable and calibration is difficult to stabilize.

**Bearing transducers** are instrumented on standard spindle bearings to measure existing radial and axial forces (Fig. 2). The advantages of this device are its retrofitability; sensitivity to tool wear, tool breakage, and missing-tool condition; self-calibration; and ease of installation. A major disadvantage is that the system operates only when the spindle is actually rotating and has high sensitivity.

**Tension-measuring transducers** measure forces that cause deformation (Fig. 3). Sensitive transducers can measure these deformations even on large structures such as ram head vertical turret lathes. These sensing devices are designed for ease of installation and can be retrofitted to upgrade machinery. Their use is limited to overload applications.

**Feed force transducers** are used in multiple spindle applications that require monitoring of each individual spindle. The transducers (Fig. 4) are mounted on the circumference of a spindle shaft, as shown in the two-spindle setup of Fig. 5. These transducers offer advantages such as monitoring of individual tool information and high sensitivity to tool wear, tool breakage, and missing-tool conditions. Unfortunately, the units are not easily retrofitted and may require some changes in machine design.

**Tapping transducers** measure in-process torque generated during tapping operations (Fig. 6). Measuring torque is complex, but it has been accomplished in production applications. Tapping transducers are retrofitable, usually measure tapping force in both directions, measure missing force, and can be interfaced with analog or infrared systems. Some systems, however, are limited to measuring torque in one direction only.

## Signal Amplifier

With a self-calibrating sensor established to measure a machining parameter, this information must be relayed to a preamplifier, where the signal is intensified and subsequently sent to a microprocessor. The reaction time for a state-of-the-art sensing system is significantly improved over now outdated motor current monitoring systems.

**Motor current monitoring systems** detect a signal within 100 ms and transmit the signal back to the motor to confirm that a problem exists. These two intervals and the additional time needed to transmit the information to the computer numerically controlled (CNC) system total 1000 to 3000 ms of elapsed time to bring the machine to a complete stop.

**Modern Sensing Technology.** If the signal is reliable, no confirmation of the signal is required, as in motor current monitoring. Problem detection expends 10 ms, and communication time expends 50 to 250 ms before inertia is used to stop the machine, for a total elapsed time of 310 ms. This is one-tenth to one-third of the elapsed time required with the motor current monitoring system. If the total response time is kept under 400 ms, any major damage to the cutting edge will be avoided.

## The Role of Microprocessing

After a sensor has detected a potential problem and communicated it to the operator, microprocessor technology is used to transfer the information to the controls of the machine.

While most sensing technologies work under optimum laboratory conditions, few survive the rigors of the manufacturing environment. The microprocessor has overcome this problem and allowed sensors to be introduced into the hostile environment of the shop floor and to stay permanently.

**Microprocessor Capabilities.** All machine operators will readily admit that there is a difference in machining results when operating a cold machine rather than a hot

**Fig. 2** Bearing transducer mounts on standard spindle bearings to measure radial and axial forces

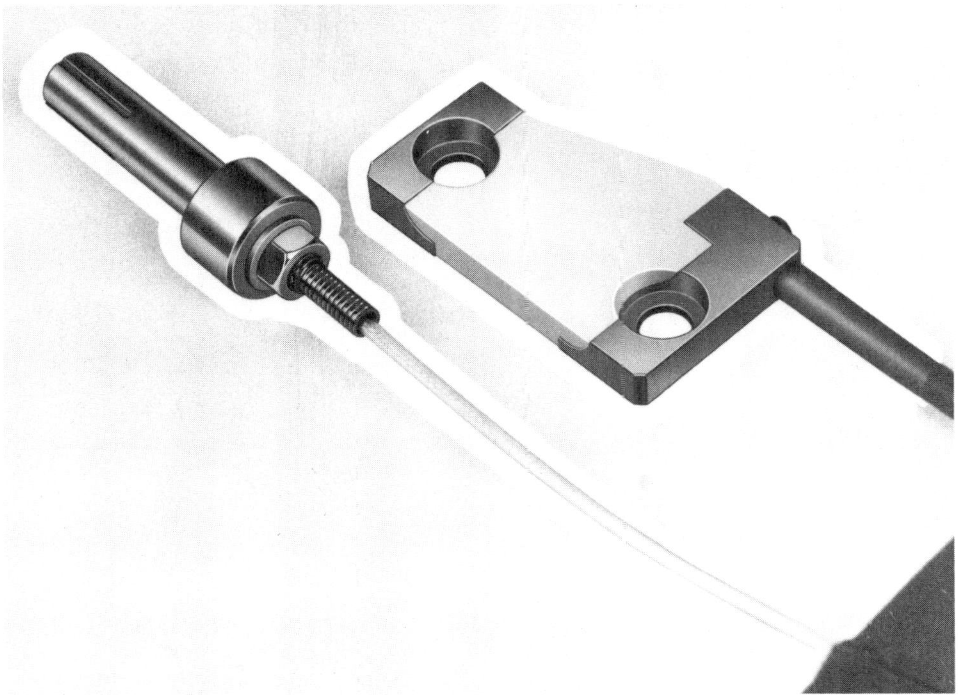

**Fig. 3** Tension-measuring transducer used to measure deformation forces

machine. The sensor detects the difference and must adjust precisely to this condition, as well as to hundreds of other variables that occur in the real world of manufacturing. The microprocessor absorbs all variables in milliseconds and is equipped to make decisions based on this information. For example, if the operator inadvertently hits the machine with a hammer, the sensor feels the shock. In milliseconds the microprocessor must decide whether the shock was from the cutting edge, or some other source. The choice made must be the correct one, otherwise the operator will receive a nuisance trip, which will eventually lead to a loss of confidence in the system.

The microprocessor must be user friendly, and the information on the faceplate must be informative and meaningful to the operator (Fig. 7). Furthermore, the microprocessor should be self-teaching with adjustable limits so that limit values are established within the first hours after installation.

**Example 1: Single Tool Monitoring System Satisfies Objective of Four Users.** Microprocessing provides numerous options in choosing strategy depending on user objectives. The case of four users with four separate objectives demonstrates the options available.

- *User 1*: Machine overload detection for multitool setup
- *User 2*: Tool overload detection on multitool setup
- *User 3*: Tool wear, tool breakage, and missing-tool detection on multitool setup
- *User 4*: Drilling with a single tool, tool breakage, and missing tool

*User 1.* Figure 8(a) illustrates different force values obtained for four machine tools using the sensor layout shown in Fig. 8(b). Within minutes, forces from each tool are identified, and a single overload limit is put over the largest value and becomes analogous to an electronic shear pin. When the machine is overloaded beyond the set limit, the machine is put into feed hold. The sensor should be located close to the cutting action. A surface-tension measuring transducer mounted to the turret is a low cost solution.

The electronic unit consists of an amplifier and microprocessor to monitor the limit. The output signal overload is sent directly to the feed drive and stops the slide movement in the shortest possible time.

The load on the turret can be divided in two sections: the normal (standard) load range and overload range. The standard load range is defined by the dimensions of the machine. Major damage that requires expensive repairs on the turret or spindle should not occur within this range. But overloading the turret shifts the tool-changing mechanism and can damage the spindle and feed drives. The overload monitoring system shuts the machine down immediately if the maximum load limit is reached.

*User 2* is also using four separate tools, but wants additional protection, such as machine overload protection, a single limit over each tool (limit 2), and a limit between tool changes (limit 1). These variables are illustrated in Fig. 9. At intervals $t_1$, $t_3$, $t_5$, $t_7$,

**Fig. 4** Bushing transducer that mounts on circumference of a machine spindle shaft

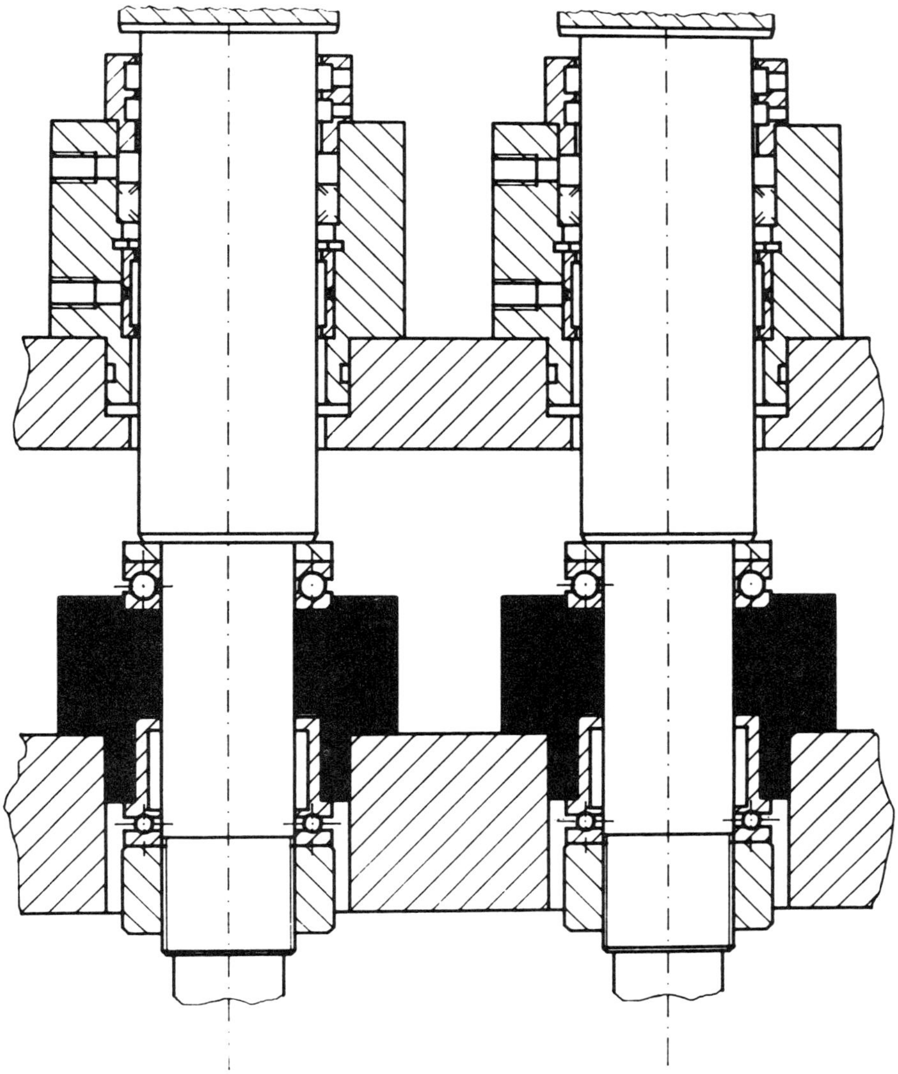

**Fig. 5** Multispindle bearing transducers positioned on two adjacent machine spindle shafts

and $t_9$, the tools mounted on the rotating machine spindle are in a no-load or air-cutting mode. At interval $t_2$, the force, $F_1$, generated by the facing tool is midway between the lower limit (1) and upper limit (2). At $t_4$, the roughing tool generates a cutting force, $F_2$, which approaches but does not surpass the upper limit. At $t_6$, the first finishing tool generates a cutting force, $F_3$, that just barely exceeds the lower limit. At $t_8$, the second finishing tool generates a cutting force, $F_4$, that is less than the lower limit. The actual cutting force is adjustable, but the two fixed limits are fixed in terms of the lower limit being a fixed percentage of the upper limit. At no time did the cutting force approach the overload limit, which is the maximum allowable force capable of being sustained by the cutting tool. Tool wear is a function of the constantly increasing cutting force as the tool is dulled. The sensor also monitors when the tool is missing (no part in fixture due to loading error) or broken should the force exceed the overload limit.

*User 3* wants to know when a part is out of position and when casting is oversize or undersize. He also needs to be aware of tool wear for each tool, tool breakage, and missing tool on three different tools. The monitoring necessary to achieve these goals is illustrated in Fig. 10.

At the beginning of each lot, the device is switched to the teach-in mode. The microprocessor stores the actual value for each section and sets limits for tool wear and breakage with the aid of adjustable factors.

In the automatic mode, the actual force is then monitored. Forces exceeding the limits are reported to the machine control.

At intervals $t_1$, $t_3$, $t_5$, and $t_7$, the tools mounted on the machine spindle are in a no-load or air-cutting mode. For the facing operation (interval $t_2$), the force is below the upper limit of $F_1$. For the turning operation (interval $t_4$), the force is below the upper limit of $F_2$, where $F_2 < F_1$. For the boring operation (interval $t_6$), the force is below the upper limit of $F_3$, where $F_3 > F_1 > F_2$. The lower limit, $F_4$, remains constant throughout the entire machining cycle.

*User 4* was experiencing tap breakage because upstream drills were breaking and were unable to provide the necessary holes for the tapping station. The user wanted to ensure that holes were always present. The strategy to eliminate the problem required monitoring the variables shown in Fig. 11.

In tapping applications, the assurance that the hole is tapped before the part goes to the customer is of the utmost importance. Figure 11 describes a fail-safe method with sensing. Figure 11(a) shows a normal tapping force as it completes its cycle. Figure 11(b) shows the reaction to a broken tap. However, what happens when the operator fails to put the tool in the spindle or when the tap breakage is not seen? Figure 11(c) describes a fail-safe technique, in which a missing force is sensed by the transducer to confirm that the cycle is completed without a force. The machine tool is put into a feed-hold mode until the hole is tapped and a force is felt.

The complete tool monitoring system that would meet the needs of User 4 as well as the previous three users is shown in Fig. 12.

## Design Considerations

Two case studies from two different ends of the metal removal spectrum, broaching and drilling, have been selected to emphasize the cost effectiveness of using a tool condition monitoring system.

### Monitoring System for a Broaching Machine

**The Need for Monitoring.** The broaching process is characterized by a varying force that is applied during the machining cycle. The monitoring system must, therefore, use a floating limit to control the tool. The forces, however, increase and decrease with the position of the broaching tool, because the tool has a heavy-duty part at the beginning and a finishing part at its end. Therefore, the pattern strategy is a reasonable solution.

**Sensing Element.** A load-sensitive transducer is used to measure the forces. This transducer is designed to measure the resistance to cut versus the cutting forces. The cutting forces are generally not very informative about the actual tool condition because of the variables around the cutting edge.

**Tool Condition Monitoring Strategy.** A monitoring system consists of three components: the transducer, the preamplifier, and the microprocessor. The system functions as follows:

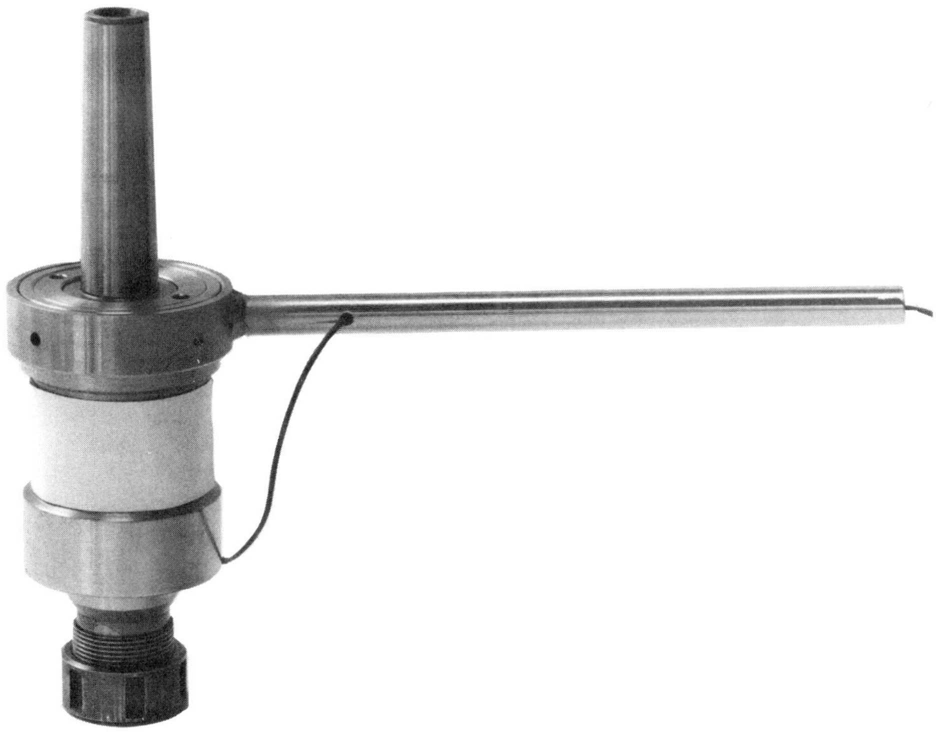

**Fig. 6** Torque transducer used to monitor in-process torque generated in tapping operations

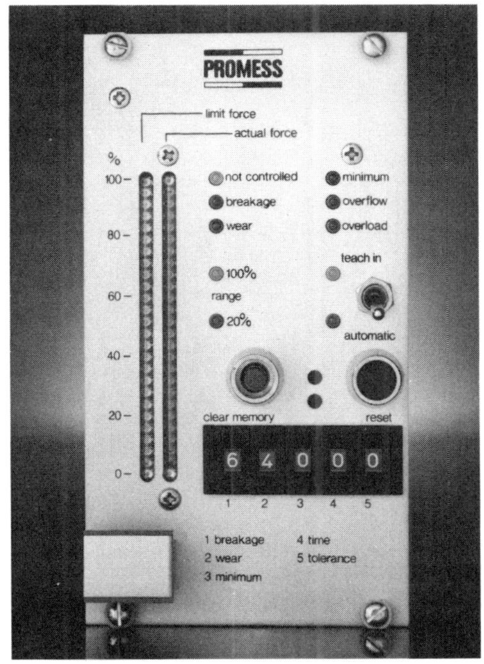


**Fig. 7** Typical user-friendly microprocessor that provides instantaneous monitoring of tool systems

- In the home position of the machine, the force transducer is automatically zeroed
- When the broaching operation starts, the position of the tool and the resistance to cut are measured
- In long broaching strokes, it may be necessary to look only at a portion (high loads for example) of the broach, or it may be necessary to look at the entire stroke under a pattern recognition system that identifies time and resistance and then automatically duplicates this data on subsequent broaching strokes

The cut limit strategy being used at this automotive manufacturer involves four individual broaching positions (see Fig. 13) with four transducers monitoring approximately 30 kN (7000 lbf) (sharp tool) per guided ram of the broach machine.

The component being produced is an automotive steering yoke. The components are positioned side by side (four to a machine) with a broach being pulled through each one of the components. In this application, the broaches are 1220 mm (48 in.) long and are pulled through the workpiece in about 4 s.

**Customer Problem.** The need for tool condition monitoring resulted because of the following tooling problems encountered in the production of an automotive front-end steering gear linkage used in the steering wheel shaft:

- Six broken broaches per week
- Badly ground broaches not being caught until broken
- Irregular tool life between 400 and 4000 parts
- Part misalignment causing broach damage
- All of the above contributed to less than 50% available uptime on machine

**Results.** A tool conditioning monitoring system produced an increase in tool life and productivity.

- Broach breakage reduced from six broken broaches per week to 0.166 per week, or two per quarter
- Badly ground broaches exceeded wear limit almost immediately and were withdrawn to eliminate breakage. A new regrind program was instituted because cutter grind could see forces from badly ground broach
- Part misalignment gave immediate overload indication; no broken broaches resulted from this condition
- Tool life averaged approximately 5000 parts after breakage problem was corrected
- Machine uptime increased more than 20%
- System payback was achieved in 8 months

## Monitoring System for Drilling an Automotive Support Arm

**Operating Parameters.** An automotive parts supplier used a drill with disposable

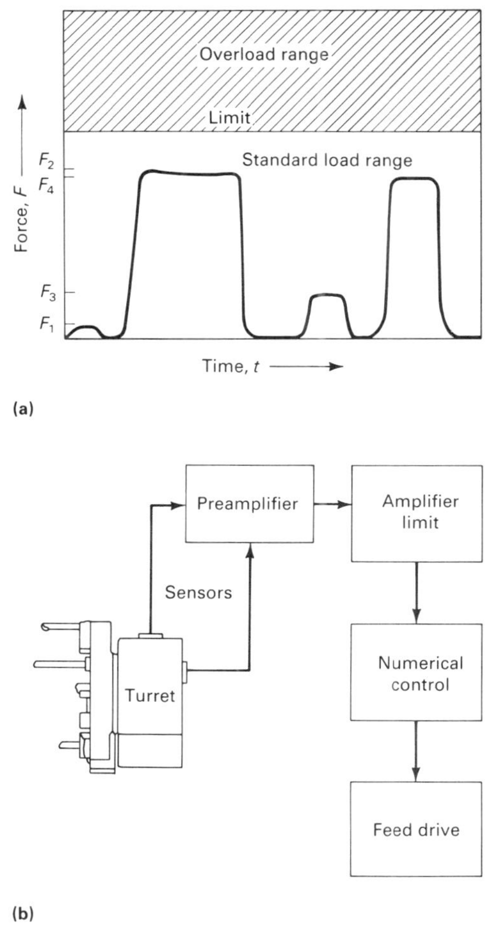


**Fig. 8** Screen display (a) of four forces being monitored for overload in a turret assembly using the electronic sensor setup shown in (b). See text for discussion.

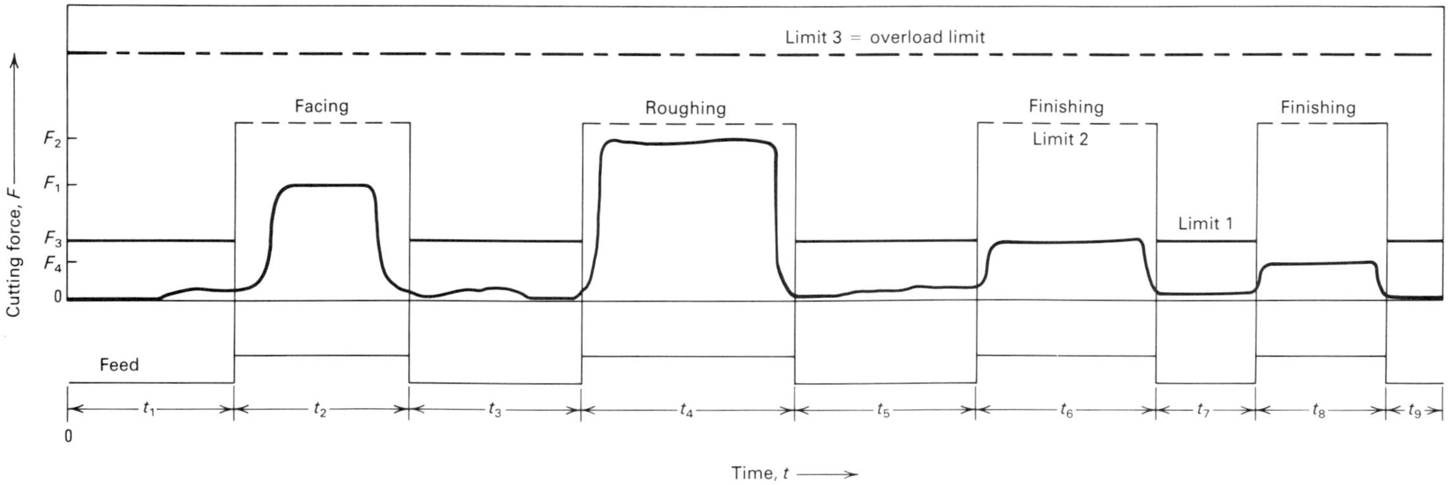


**Fig. 9** Plot of cutting force against time for four separate tooling operations: facing, roughing, and finishing (2). Limit 1 is monitored during rapid traverse to reduce reaction time. Limit 2 is calculated from the highest load and is used to monitor the roughing tool. Limit 3 provides overload protection and is activated during the whole machining process. A large number of software variations is available for special tasks and machines. See text for discussion.

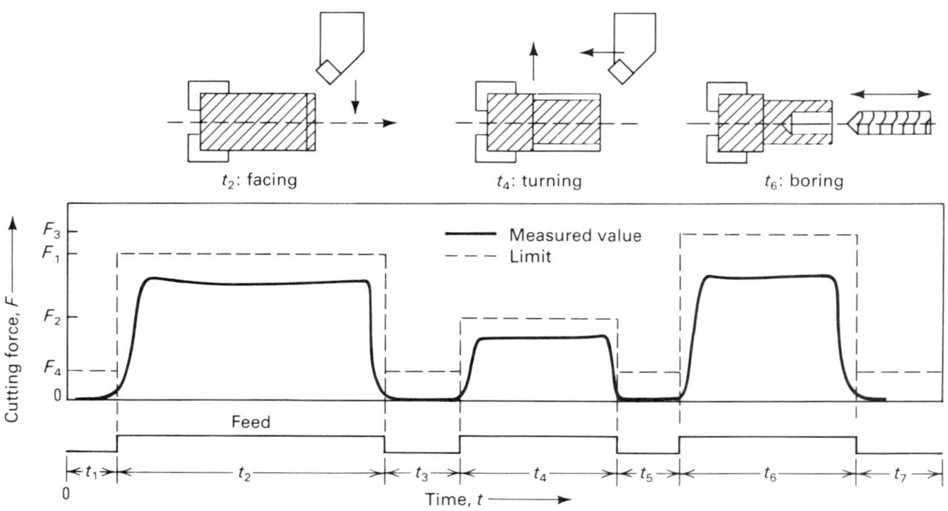


**Fig. 10** Plot of cutting force against time for a force that has a variable upper limit at each of the three machining stations: facing, turning, and boring. See text for discussion.

carbide-tipped inserts to machine a 27.43 mm (1.080 in.) hole in a forged-steel support arm using a four-spindle drilling machine at 2000 rev/min, at a speed of 172 m/min (565 sfm), and at a feed of 255 mm/min (10 in./min).

**Results.** Table 1 compares the cost and total production figures obtained over a span of 24 months (3 shifts, and 616 working days) for production without tool condition monitoring as well as the production and cost figures obtained when motor current, ball screw, and strain gage monitoring systems were used to track tool condition. Payback from tool condition monitoring systems was achieved within nine months of being introduced into plant operations.

After one year, the operator had given up piece count and was using only force reading to gage tool wear. The operator normally replaced drills every 300 to 400 pieces.

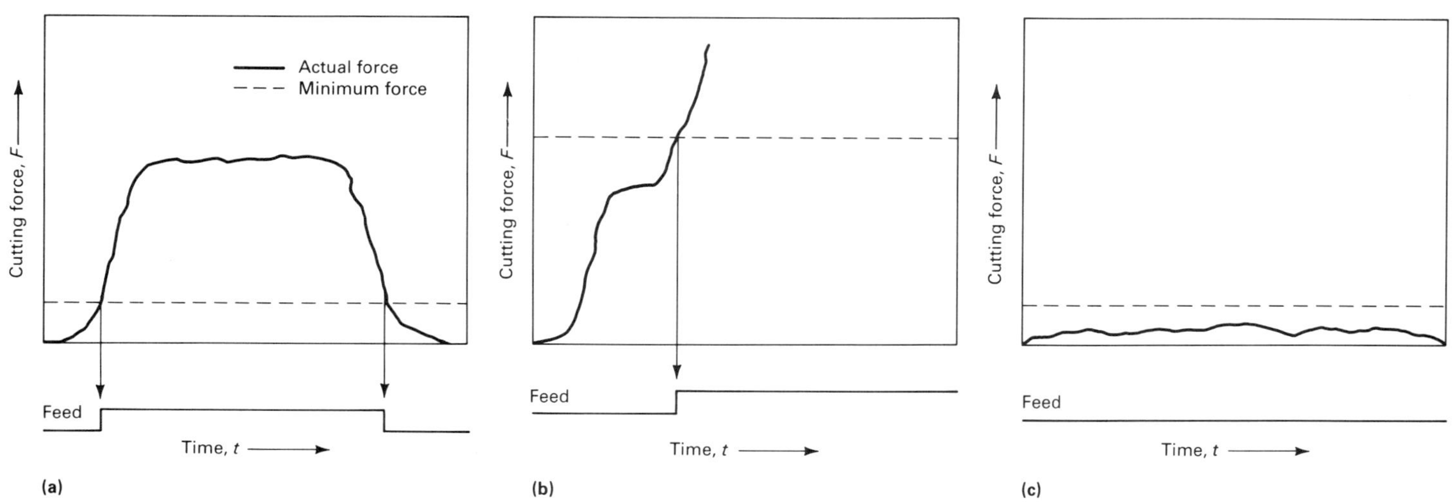


**Fig. 11** Plot of tapping force against time for three conditions: (a) normal cutting, (b) tool breakage, and (c) tool missing. See text for discussion.

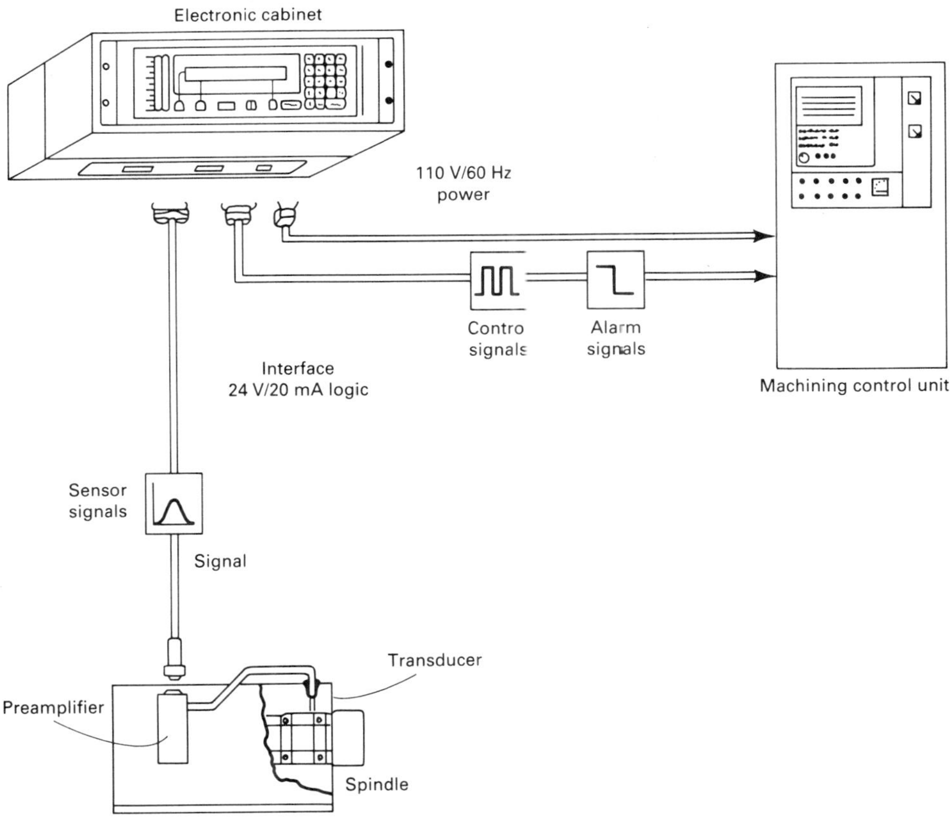


**Fig. 12** A complete tool monitoring system incorporating strain-gage instrumentation

**Table 1 Savings and productivity increases obtained with three different tool condition monitoring systems, compared to tool costs and production figures obtained with no monitoring of tool condition, when drilling an automotive support arm**

| | No tool monitoring | Motor monitoring | Ball screw monitoring(a) | Bearing monitoring |
|---|---|---|---|---|
| Holes produced per holder, average number | 5 000 | 7 000 | 12 000 | 22 000 |
| Tool cost, four spindles(b) | $91 300.00 | $87 760.00 | $63 420.00 | $41 316.00 |
| Hours of tool change per year, h(c) | 832 | 582 | 216 | 83 |
| Cost per hour | $32.00 | $32.00 | $32.00 | $32.00 |
| Cost for tool change | $26 624.00 | $28 624.00 | $6 912.00 | $2 656.00 |
| **Total cost** | **$117 924.00** | **$106 384.00** | **$70 332.00** | **$43 972.00** |

(a) Electronic problems in ball screw monitoring system necessitated the use of backup power monitoring. (b) Tool cost includes scrapped parts, broken fixtures, tool holders, and indexable insert use. (c) Tool change hours include downtime caused by both broken and worn tools.

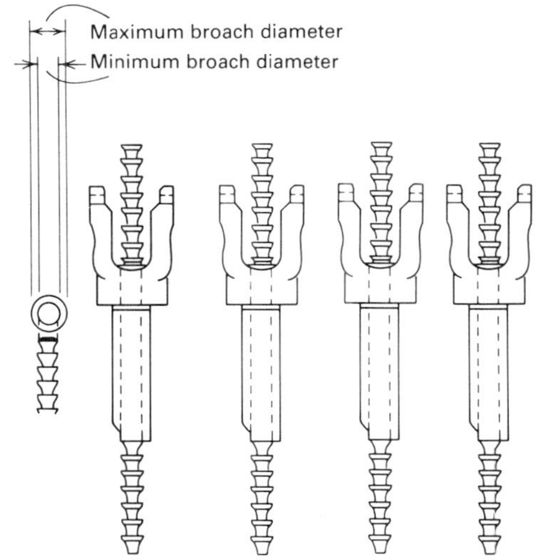


**Fig. 13** Pull broaching of automotive steering wheel yoke components on a broaching machine ram. Each of the four broaching stations is equipped with a separate transducer to monitor part production.

With the force count, he changed tools every 1000 pieces with the tools of one manufacturer and 900 pieces with the tools of another manufacturer. The operator relied exclusively on the tool condition monitoring system to determine the need for tool replacement. An additional benefit was that bearing life was increased by 30% due to the elimination of spindle overload.

## Future Developments

Many other sensing techniques in addition to the ones mentioned are being explored, including acoustics, infrared light, fiber optics, and vibration monitoring. Development has been slow because of the environmental sensitivity of many of the newer sensors. Some new developments will have limited, very select applications. Whatever sensor is selected, it must not be affected by anything other than the element it is monitoring, or the system will be useless in a manufacturing process.

# Grinding, Honing, and Lapping

# Principles of Grinding

Richard P. Lindsay, Norton Company

GRINDING MODES are all similar. An abrasive surface is pressed against a work surface with a force perpendicular to the contact zone (normal direction), and material is removed from the work and wheel. This is true for grinding wheels or belts, using conventional abrasives such as aluminum oxide or silicon carbide, superabrasives such as cubic boron nitride (CBN) or diamond, or the newest abrasive, seeded gel (Ref 1). Three factors determine how much material will be mutually removed:

- The sharpness of the abrasive surface
- The magnitude of the normal force
- The durability of the abrasive

In addition, a second force, tangent to the wheel-work contact, when multiplied by wheel speed and a constant, determines the power used by the operation. The tangential and normal forces are related by a coefficient of friction; therefore, in production grinding, in which the normal force is almost never known (without a dynamometer of some type), the tangential force can be calculated from power measurements and used as a proxy for the normal force.

## Grinding Wheel and Workpiece Parameters

Figure 1 shows an external grinding system, but this discussion pertains to any mode. The subscript W refers to workpiece parameters, and the subscript S is used for wheel factors. The slide infeed rate $\overline{V}_F$, compressed the wheel and work springs, generating a normal force, $F_N$, which causes the work to be ground at a radial rate, $\overline{V}_W$, and the wheel to wear at a radial rate, $\overline{V}_S$. The tangential force, $F_T$, develops as a fraction (from 0.25 to ~0.4) of the normal force. Volumetric removal rates—either total, $Z$ (in units of in.$^3$/min), or on a unit-width basis, $Z'$ (in units of in.$^3$/min, in.)—can be calculated using the equations given in Fig. 1. The grinding width, $B$, is given in inches. In the steady state, when $F_N$ and $F_T$ have reached a constant value after some grinding time:

$$\overline{V}_F = \overline{V}_W + \overline{V}_S \quad \text{(Eq 1)}$$

Figure 2 shows a surface grinding mode in which some wheel depth of cut, $a$, is usually set on the machine. Again, this forced deflection compresses the system springs, developing the normal force $F_N$. The volumetric removal rates are calculated using the depth of cut. For cylindrical grinding, the wheel depth of cut can be expressed as:

$$a = \frac{\overline{V}_W}{N_W} \quad \text{(Eq 2)}$$

where $N_W$ is the workspeed in (in./s)/(rev/s).

Figure 3 shows a typical wheel-work characteristic chart relating surface finish, wheel wear rate, metal removal rate, and power to the normal force. Slight variations in work material, wheel specification, wheel speed, coolant, and grinding wheel-work conformity will all affect the slopes of these lines, but not the basic relationships. Such charts can be used to plan complete production grinding cycles for short-term operations. However, if extended grinding is done (grinding many parts without wheel dressing), the wheel may dull or sharpen, and the slopes will change again. Therefore, time can be thought of as a third axis going into the paper where the slopes of the lines in Fig. 3 may change, depending on the elapsed time. (Superabrasives are less sen-

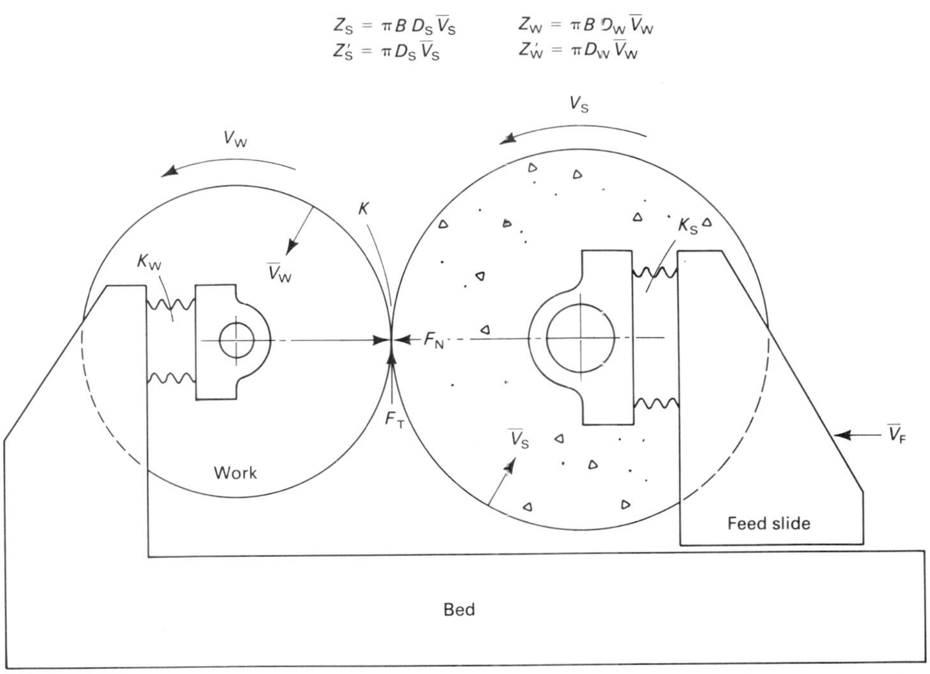


**Fig. 1** Schematic of a cylindrical grinding system. The equations are used to determine total volumetric removal rates ($Z_S$ or $Z_W$) and unit-width volumetric removal rates ($Z'_S$ or $Z'_W$). See text for discussion. Source: Ref 2

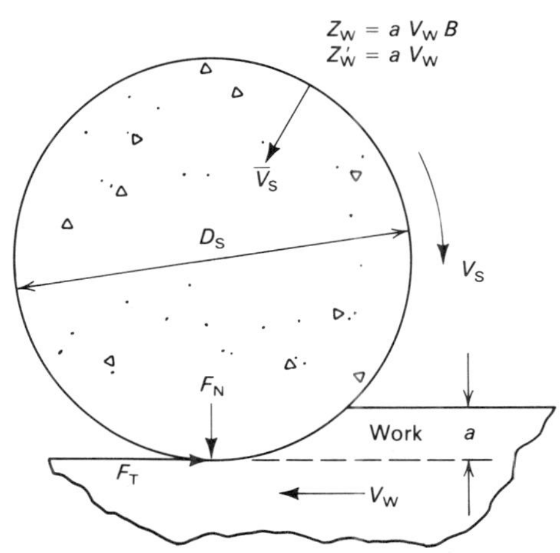


**Fig. 2** Schematic of a surface grinding system. The equations given are used to determine removal rates $Z_W$ (in.$^3$/min) and $Z'_W$ (in.$^3$/min, in.).

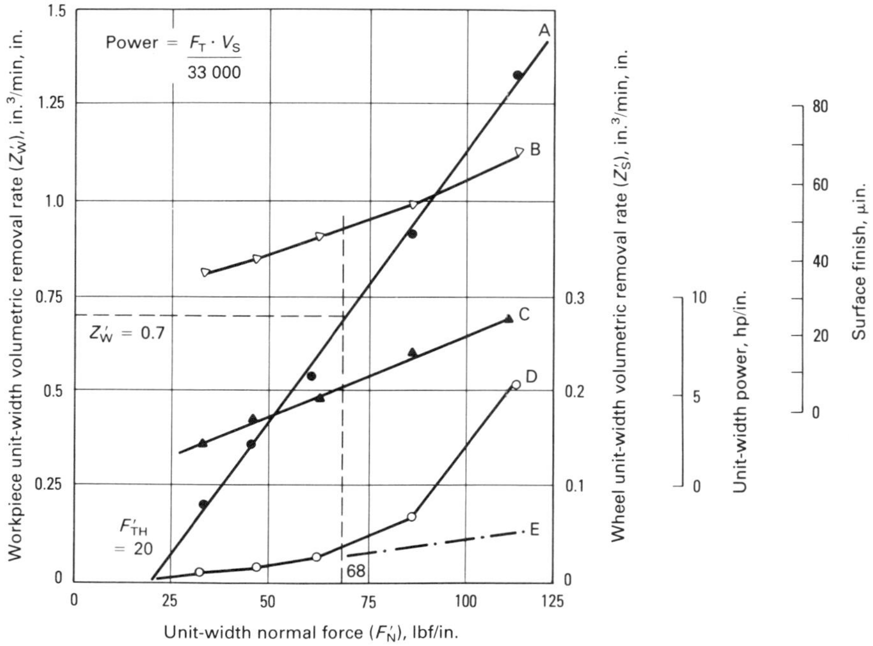


**Fig. 3** System characteristic chart for 52100 bearing steel at 60 HRC that was ground using a 32A5418VS wheel. Other variables were $V_S$ = 52.3 m/s (10 300 sfm), $V_W$ = 1.0 m/s (200 sfm), $B$ = 0.375 in., and work removal parameter = 0.014 in.$^3$/min, lbf. A, workpiece unit-width volumetric removal rate; B, surface finish; C, unit-width power; D, wheel unit-width volumetric removal rate; E, wheel metal removal parameter (slope = 0.000555 in.$^3$/min/lbf). $F'_N$ = 20 lbf/in. is the threshold force, or the force at $Z'_W$ = 0.

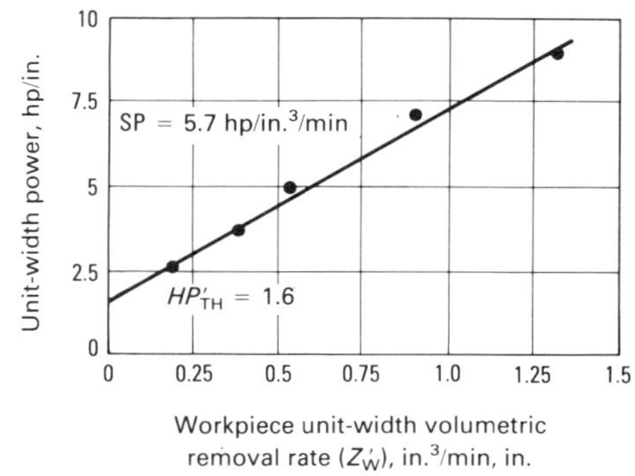


**Fig. 4** Specific power (slope = 5.7 hp/in.$^3$/min) obtained when unit-width power is plotted against unit-width metal removal rate for the data used in Fig. 3. Threshold power $HP'_{TH}$, or power at $Z'_W$ = 0, is 1.6 hp/in.

sitive to time and are used for many parts before reshaping of the wheel face, usually with a rotary diamond; see the article "Superabrasives" in this Volume.) The equation for power (see Fig. 3) is valid for horsepower, using pounds of force and feet per minute for $F_T$ and wheel speed, respectively.

In Fig. 4, the power is plotted against the metal removal rate; both are plotted on a unit-width basis. The slope, that is, the specific power (SP), is given in hp/in.$^3$/min. It is analogous to the metal removal rate versus the normal force slope of Fig. 3, shown as the work removal parameter (WRP), which is given in in.$^3$/min, lbf. When the work removal parameter is larger, the corresponding specific power will be small (sharp-acting wheels); when the specific power is larger, the corresponding work removal parameter is smaller (dull-acting wheels).

Other researchers have constructed different data presentation methods. Figure 5 shows external grinding data from a European university. An equivalent chip thickness, $H_{eq}$, where $H_{eq}$ = metal removal rate/(wheel speed), has been used as an independent variable (horizontal axis), with $G$ ratio (the volume of metal removed/volume of wheel used), finish, and $F'_N$ and $F'_T$ plotted against it. If a wheel speed is chosen, measurements from Fig. 5 can be plotted as in Fig. 6. Figure 6 shows the same relationships as Fig. 3 and 4. Therefore, even though the data in Fig. 3 and 4 were obtained from a controlled-force machine and the data in Fig. 5 from an infeed-rate machine, the data are related exactly the same in Fig. 3, 4, and 6.

## Conformity or Equivalent Diameter

Figure 7 shows three cylindrical grinding systems: one internal (Fig. 7(a), at left) and two external (Fig. 7(b), at right). The equations for the equivalent diameter, $D_E$, are given in Fig. 7 for both modes. The equivalent diameter is the size of wheel that would be used for surface grinding (at right, Fig. 7) to represent the way the original wheel-work pair fit together. The internal system produced a 914 mm (36.0 in.) equivalent diameter, while the two external systems produced equivalent diameter values of 83.8 mm (3.3 in.) for the 102 mm (4 in.) part and 24.1 mm (0.95 in.) for the 25.4 mm (1.0 in.) part. Therefore, for some applied normal force, for example, 450 N (100 lbf), the external systems would generate smaller contact lengths than the internal systems. This means that the internal mode would have a smaller pressure (pressure = $F_N$/contact length) than the external mode. A softer wheel grade (J or K, for example) would be used to ensure that dulled abrasive grains are released properly under this lower pressure. In contrast, the higher-pressure external system could use a harder grade (M to O, for example); otherwise, its high pressure would cause a soft grade to wear too rapidly. Wheel grades should be chosen to balance applied pressure. Detailed information on grinding wheels is available in the articles "Grinding Equipment and Processes" and "Superabrasives" in this Volume.

Figure 8 shows the effects of $D_E$ on cutting and power on two grinding systems. In Fig. 8(a), the external system ($D_E$ of 61.0 mm, or 2.4 in.) needed only 14 lbf/in. of threshold force to begin cutting and then had a 0.0153 WRP slope. The internal system ($D_E$ of 28.96 mm, or 11.4 in.) needed 35 lbf/in. to begin cutting and then had a shallower WRP slope of 0.0129. This means that for any removal rate (any horizontal line up the $Z'_W$-axis) the internal system will use higher forces. In Fig. 8(b), at any removal rate (any vertical line along the $Z'_W$-axis), the internal system would use higher power because of its larger threshold power (4.4 versus 2.6 hp/in.) and steeper SP slope (11.3 versus 8.59 hp/in.$^3$/min). Therefore, the high-conformity internal system needs more force to cut at the same removal rate because of its low pressure due to its longer contact length.

## Wheel Dressing

Wheel truing or dressing is done with single-point or multipoint stationary diamond tools or rotary diamonds. For stationary tools, the dressing lead and the depth of dress are two important factors, as illustrated in Fig. 9. The lead is calculated by the axial velocity per wheel-rotary-speed ratio. Figure 10 shows the effects of harsh (lead: 0.51 mm/rev, or 0.020 in./rev; diametral depth: 0.05 mm, or 0.002 in.) and gentle (lead: 0.10 mm/rev, or 0.004 in./rev; diametral depth: 0.025 mm, or 0.001 in.) dressing on cutting, power requirements, and $G$ ratio. The harsh dressing generated a steeper WRP slope in Fig. 10(a) (meaning it would take lower forces at all $Z'_W$ rates), lower power in Fig. 10(b) because of its shallower specific power slope, but poorer $G$ ratios at all removal rates in Fig. 10(c). Therefore, the harsh dressing sharpened the wheel, thus reducing forces and power, but made it wear faster at all $Z'_W$ rates.

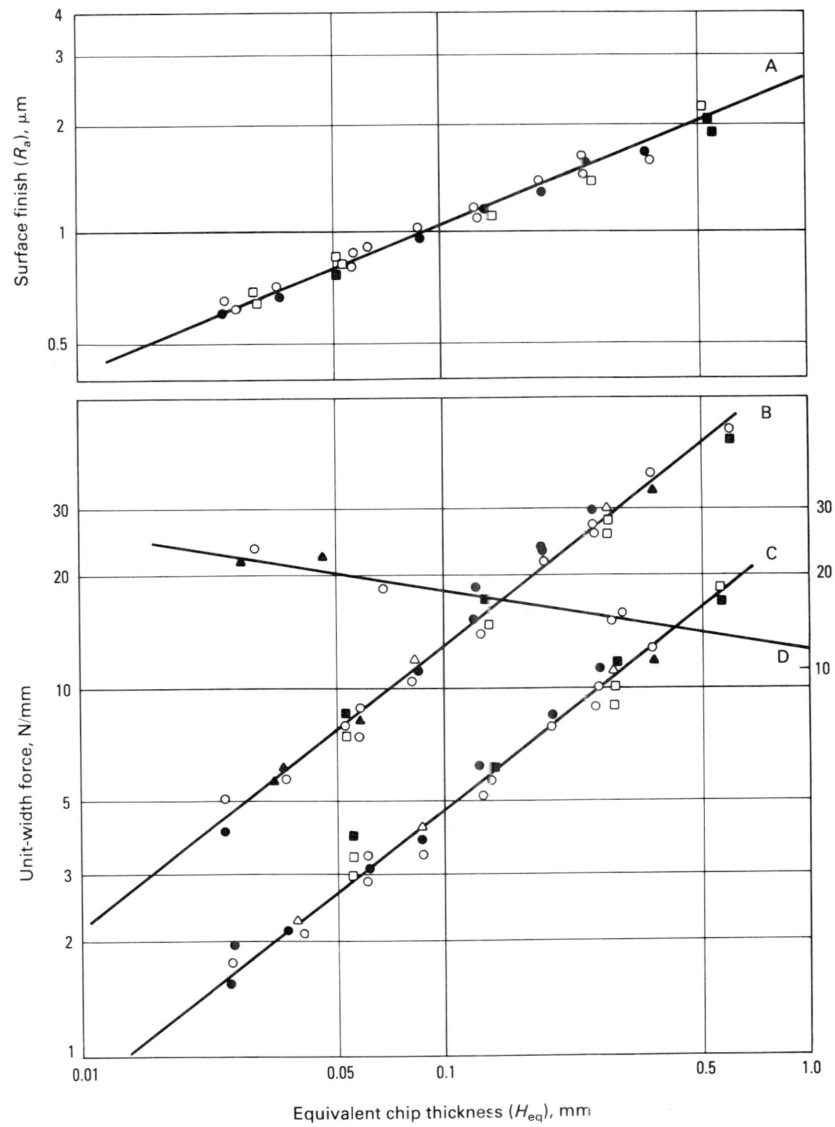


**Fig. 5** Effect of equivalent chip thickness, $H_{eq}$. A, surface finish, $R_a$; B, unit-width normal force, $F'_N$; C, unit-width tangential force, $F'_T$; D, $G$ ratio. Source: Ref 3

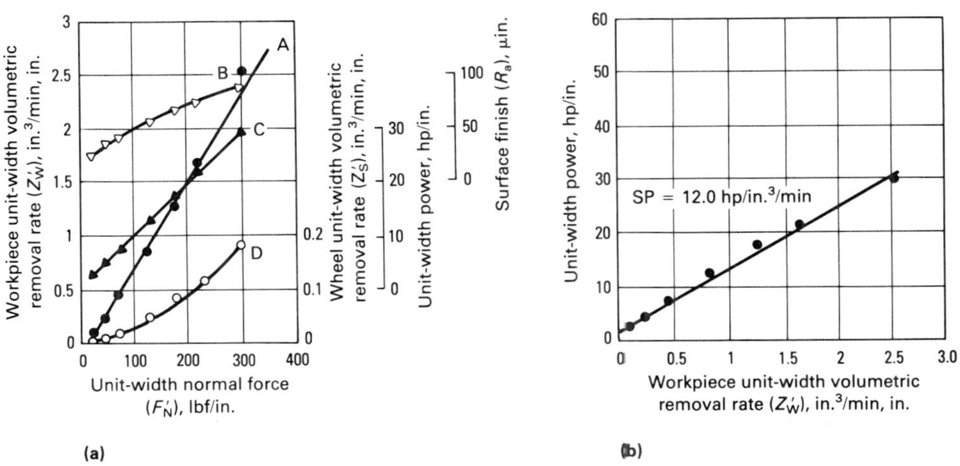


**Fig. 6** System characteristic charts derived from the $H_{eq}$ chart in Fig. 5 with WRP = 0.008 in.$^3$/min, lbf. (a) Workpiece unit-width volumetric removal rate plotted against unit-width normal force. A, workpiece unit-width volumetric removal rate, $Z'_W$; B, surface finish, $R_a$; C, unit-width power; D, wheel unit-width volumetric removal rate, $Z'_S$. (b) Specific power obtained when unit-width power is plotted against workpiece unit-width volumetric removal rate, $Z'_W$

Figure 11 shows a famous graph developed at the Technical University of Braunschweig, West Germany. The horizontal axis is the speed of a parallel-axis diamond roll divided by the wheel speed. A positive value (to the left of the zero axis) indicates that the roll and the wheel are traveling in the same direction at contact. To the right of the zero axis is the negative region, where the directions are opposite at contact. Figure 11(a) shows wheel sharpness after truing as determined by tracing a stylus instrument over the trued wheel. Peak values of roughness (the sharpest wheel) were obtained at +1.0 (wheel and roll traveling at the same speed in the same direction) and at zero (roll not moving). Both peak conditions are not recommended, because the +1.0 condition reportedly rips diamonds from the roll and because the zero position would, of course, wear a flat on the roll. However, any positive value generates a sharper wheel than any negative ratio because the negative-direction curves decrease continuously (the wheel is becoming duller). In addition, the radial infeed of the roll per wheel revolution can also increase or decrease wheel roughness, as shown by the four parallel curves in Fig. 11. Therefore, forces and power can be reduced by using a larger infeed per revolution and a positive direction (+0.8 is good). Better finishes can be obtained by using a shallow infeed per revolution and negative ratios.

Figure 12 shows the results of recent tests conducted on a creep-feed grinder. The speed ratio was varied from +0.85 to −0.85, and normal force and power (not shown) were measured while grinding 4340 at four metal removal rates. The forces (and power) were lowest at the +0.85 ratio and increased steadily until about −0.2, then remained essentially constant. The curves shown in Fig. 12 follow the sharpness trends of Fig. 11, confirming those directions.

## Coolant Effects

Coolants are described in the article "Metal Cutting and Grinding Fluids" in this Volume, but a brief discussion of coolant effects is appropriate in this article. Figure 13 shows the effects of three fluids: a sulfur-chlorinated oil, a water-soluble oil at 10% in water, and a synthetic oil at 5% in water. As shown in Fig. 13(a), the sulfur-chlorinated oil required the least power at any removal rate. Sulfur-chlorinated oil also produced the best $G$ ratios (60 to 400% higher than the water-soluble oil) (Fig. 13b) and, generally, the smoothest surface finishes (Fig. 13c). The synthetic cutting fluid was the poorest-performing fluid, giving the lowest $G$ ratios and the roughest surface finishes.

Figure 14 shows a water-soluble oil at 6% in water versus a rust inhibitor at 3% in water. Figure 14(a) shows that water-solu-

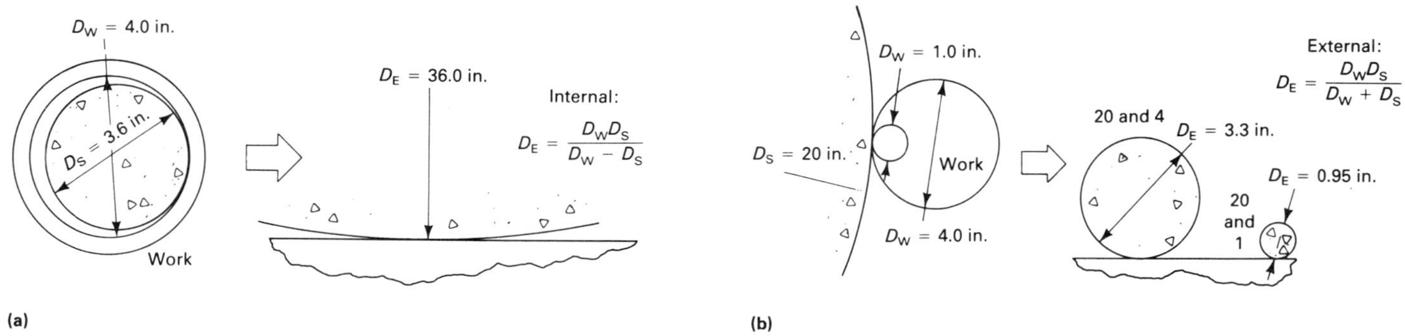


**Fig. 7** Relating internal and external cylindrical grinding to surface grinding using the equivalent diameter. (a) For internal grinding with wheel diameter, $D_S$, of 91.4 mm (3.6 in.) and workpiece diameter, $D_W$, of 102 mm (4.0 in.), the equivalent diameter, $D_E$ obtained is 914 mm (36.0 in.). (b) For external grinding with $D_W$ of 25.4 mm (1.0 in.) and $D_S$ of 508 mm (20 in.), $D_E$ is 24.1 mm (0.95 in.). For $D_W$ of 102 mm (4.0 in.) and $D_S$ of 508 mm (20 in.), $D_E$ is 83.8 mm (3.3 in.).

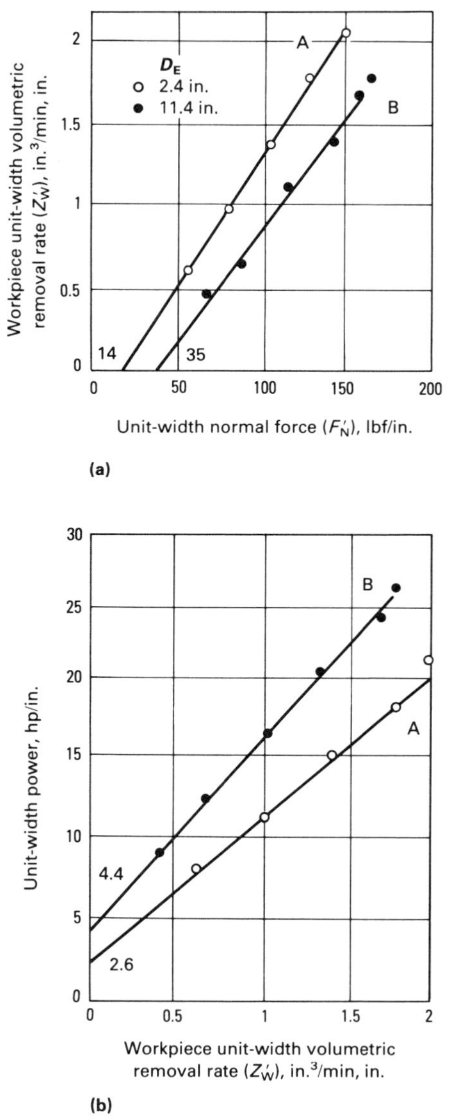


**Fig. 8** Effect of equivalent diameter at high wheel speed. The material, 52100 bearing steel at 60 HRC, is ground by a 53A120K8V127 grinding wheel at $V_S$ = 60 m/s (12 000 sfm). (a) Workpiece unit-width volumetric removal rate plotted against unit-width normal force. A, WRP = 0.0153 in.$^3$/min, lbf; B, WRP = 0.0129 in.$^3$/min, lbf. (b) Unit-width power plotted against workpiece unit-width volumetric removal rate. A, specific power = 8.59 hp/in.$^3$/min; B, specific power = 11.3 hp/in.$^3$/min

ble oil used less power at all $Z'_W$ values. As shown in Fig. 14(b) and (c), the water-soluble oil gave 70 to 80% higher $G$ ratios and 15% better finishes, respectively. There are numerous types of grinding oils (heavy-duty, light, and so on), and there are many more types of water-base fluids. The three grinding fluids discussed in this section demonstrate that these fluids can be significant factors in determining optimum grinding parameters.

## Effect of Grinding Time

Figure 15 shows the effects of grinding time (or metal volume removed) on $G$ ratio, power and surface finish with the same wheel and water-soluble oil (these are the individual grinds that produced the average values shown in Fig. 14). As more metal was removed, the wheel was progressively dulled, which under these controlled-force conditions was shown as slower $Z'_W$ rates at constant power (Fig. 15b). This means that at some $Z'_W$ rate, for example, 1.0, the power increased from 10.5 hp/in. (curve A) to 13.0 hp/in. (curve B) to 15.8 hp/in. (curve C) as the wheel was used without dressing. Figure 15(a) shows that the $G$ ratio steadily increased from 23 to 52 after grind A to 189 to 638 during grind C. (The average values for all grinds were 113 to 188, as indicated by the open circles in Fig. 14b.) In Fig. 15(c), the surface finishes were 44 to 51 μin. $R_a$ after the first grinds (curve A), 35 to 49 μin. $R_a$ after three grinds (curve B), and 38 to 57 μin. $R_a$ after six grinds (curve C). Therefore, even though the $G$ ratios increased with time, the surface finishes remained the same for curves A and B, but increased after six grinds at curve C. As a result, absolute wheel wear slowed with time (as shown by increasing $G$ ratios), but the ability of the wheel to produce the desired surface finish deteriorated.

## Surface Integrity

The surface integrity of the ground surface is extremely important. In steel parts, damage may be hidden because a clean, nondiscolored surface may have subsurface tensile stresses induced by grinding. These stresses can cause premature part failure. In the bearing industry, a common practice is to soak the parts in a hot hydrochloric acid bath after grinding. If tensile stresses exist, the surface of the part, which is made brittle by the acid bath, will crack. If no tensile stresses exist, the surface will not be cracked. However, in either case, the part will be destroyed.

Figure 16 shows the force necessary to cause etch cracks in bearing steel plotted against work speed. Curves A and B show the boundary lines for safe, uncracked parts when the work removal parameter slope is steep (sharp wheel, WRP = 0.0064 in.$^3$/min, lbf) and shallow (dull wheel, WRP = 0.0024 in.$^3$/min, lbf). The safe region is the area beneath each curve. However, dulling can occur with extended grinding time. Therefore, a force of, for example, 200 lbf/in., which at 2.5 m/s (500 sfm) work speed would be safe for a sharp wheel (beneath the upper curve), would cause cracking if the wheel dulled down to the lower sharpness because 200 lbf/in. is above curve B. As a result, the first few parts in a skip-dress sequence would show no cracks when etched, but cracks would suddenly occur and propagate unless the wheel were dressed. As previously noted, aggressive

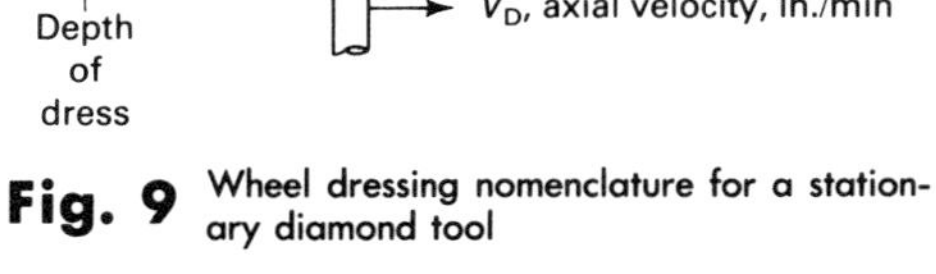


**Fig. 9** Wheel dressing nomenclature for a stationary diamond tool

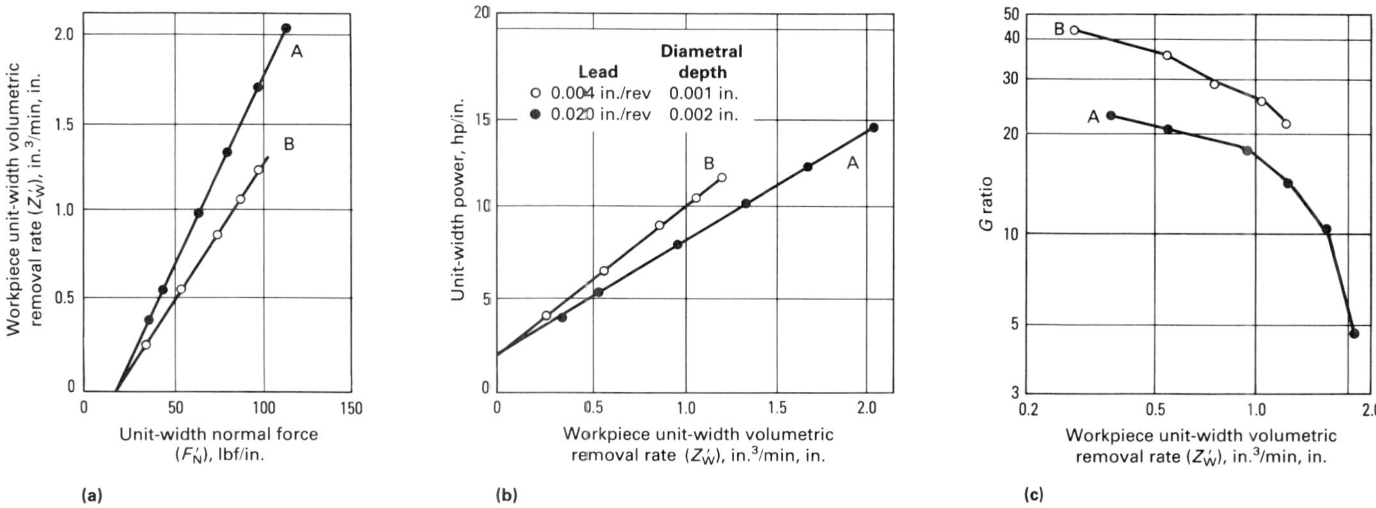


**Fig. 10** Effect of wheel dressing on cutting, power, and $G$ ratio. Material being ground with 38A100M8VBE abrasive wheel is 52100 bearing steel at 60 HRC with $V_S$ of 57.9 m/s (11 400 sfm), $V_W$ of 0.5 m/s (100 sfm), and $D_E$ of 45.7 mm (1.8 in.). Wheel was dressed with a single-point diamond. (a) Cutting: A, WRP = 0.022 in.$^3$/min, lbf; B, WRP = 0.016 in.$^3$/min, lbf. (b) Power: A, SP = 6.0 hp/in.$^3$/min; B, SP = 8.0 hp/in.$^3$/min. (c) $G$ ratio plotted versus workpiece unit-width volumetric removal rate for the two lead and diametral depth conditions mentioned above

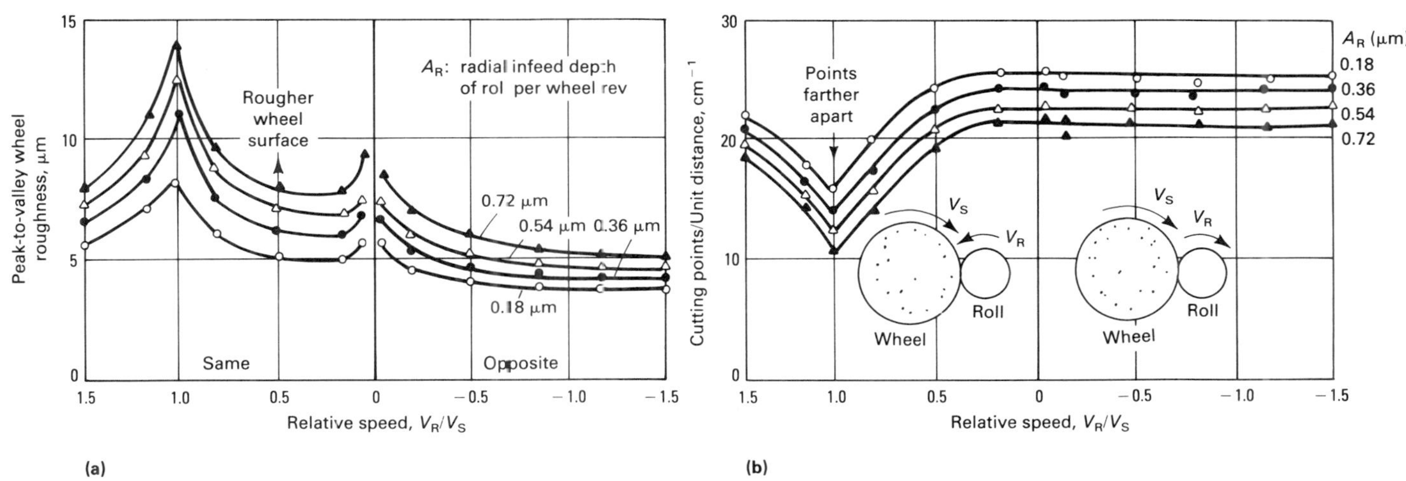


**Fig. 11** Dressing nomenclature for a parallel-axis rotating diamond roll. (a) Wheel sharpness after truing as determined by tracing a stylus device over the trued wheel. (b) Plot of cutting points/unit distance versus relative speed ratio. Source: Ref 4

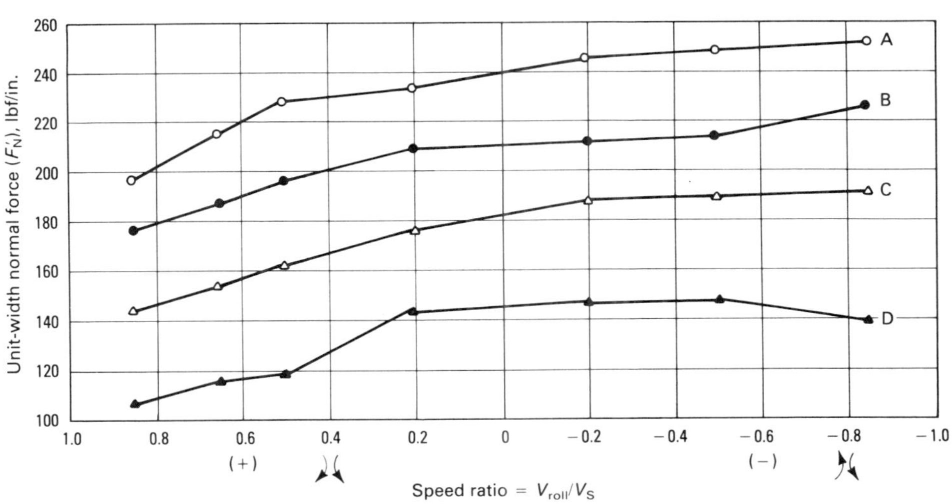


**Fig. 12** Effect of speed ratio on a creep-feed grinder. Going from positive to negative roll per wheel speed direction increases the normal force and the power. Workpiece was 4340 steel at 50 HRC. It was ground with a 38A801-D25VX514 abrasive wheel dressed with a 30/40 mesh reverse-plated diamond roll having a 0.0020 mm/rev (0.000080 in./rev) feed rate and 5% water-soluble oil coolant. Wheel speed, $V_S$, was 28 m/s (5500 sfm). Curves are plotted for four values of $Z'_W$, workpiece unit-width volumetric removal rate (in.$^3$/min, in.). A, 1.03; B, 0.78; C, 0.53; D, 0.27

dressing and softer wheel grades will sharpen wheels to help prevent this dulling. Grinding at work speeds of 0.25 to 0.50 m/s (50 to 100 sfm) will not reveal this wheel sharpness factor and will allow only very low forces to be used, with the resultant low productivity.

Figure 17 shows the maximum residual stress in a finish ground part plotted against the average metal removal rate. The stresses were obtained by stepwise etching away thin layers from the specimen and measuring the resultant change in distortion. The maximum tensile stress (worst condition) was produced at condition 2 using a 0.25 m/s (50 sfm) work speed and a dull wheel, which was created by very gentle dressing. Then, still with the dull-dressed wheel, the work speed was increased to 2.5 m/s (500 sfm), which reduced the maximum tensile stress to 415 MPa (60 ksi) at condition 3. By dressing more aggressively, the wheel was sharpened to WRP = 0.0058 in.$^3$/min,

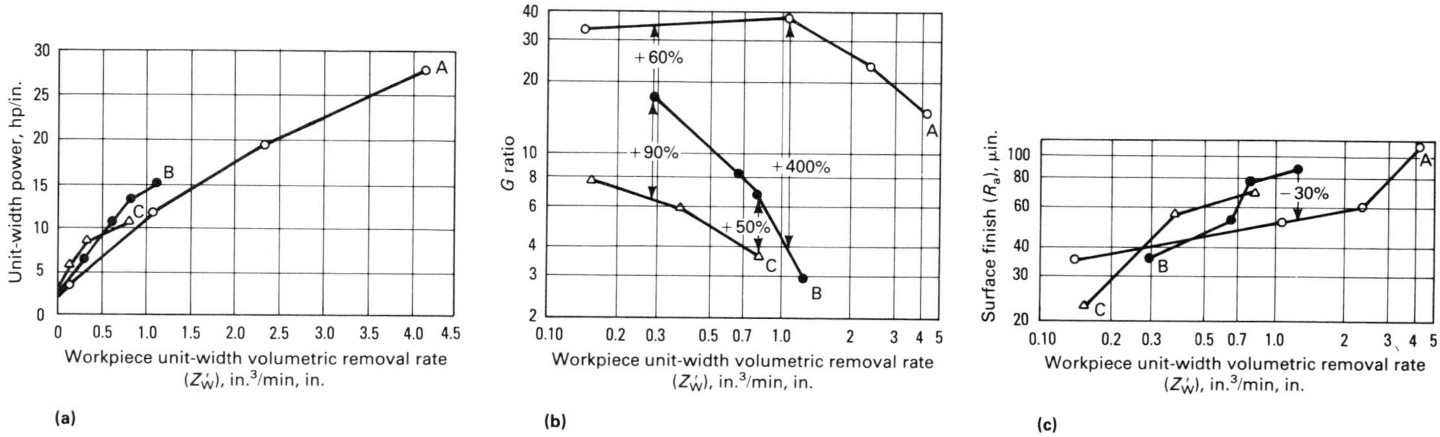


**Fig. 13** Effect of coolants on power (a), *G* ratio (b), and surface finish (c). Material machined is M7 steel at 61 HRC, which was externally ground at $V_W$ = 15 m/min (50 sfm) using a 53A80N8V128 abrasive wheel with a speed of $V_S$ = 43 m/s (8500 sfm) and a $D_E$ of 53.3 mm (2.1 in.). A, sulfur-chlorinated oil; B, water-soluble oil at 10%; C, synthetic fluid at 5% in water

lbf (up from 0.00288 in.³/min, lbf) and produced a desirable compressive stress at condition 7. Finally, the sharp wheel was used at a work speed of 2.5 m/s (500 sfm), followed by a 10-s sparkout, which cold worked the surface to produce a still higher compressive stress at condition 6.

As each step was taken, successively faster metal removal rates were used. Therefore, the final condition (condition 6) produced the same desirable compressive stress as the low-stress condition usually recommended, but at a much higher productivity. Low forces, low power (which sometimes means low wheel speed), sharp-acting wheels (usually soft grades), high work speed, and a lubricating fluid (oil) rather than a cooling fluid (water based) will improve surface integrity. Reference 6 provides more detailed information on surface integrity.

# Appendix: Comparison of Cutting and Grinding

Material cut with tools or ground with abrasives can be described as a system whose behavior is determined by such variables as forces, power, wear, surface finish, and metal removal rates. Different methods of data presentation have been used for cutting with tools (milling, turning, boring, and so on) as opposed to grinding with abrasives. In this section, one method is used for both in order to compare forces, power, and metal removal rates on an identical basis.

## Systems

Figure 18(a) shows a turning operation on a lathe. The work rotates at a certain speed, $V_C$, and the tool has a depth of cut (DOC) and axial feed per revolution along the part. Figure 18(b) shows how these are related. The three forces acting on the tool are shown in Fig. 18(a):

- Cutting force, $F_C$
- The force radial or perpendicular to the work surface, $F_R$
- Axially, the feed force, $F_F$

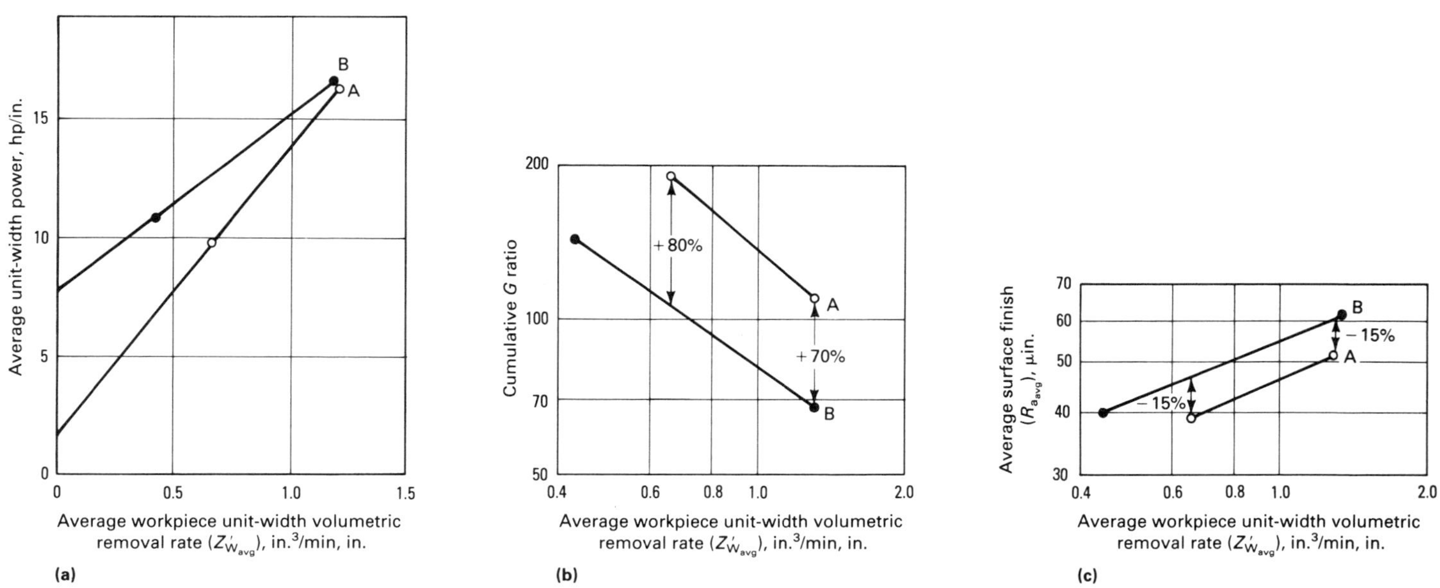


**Fig. 14** Effect of water-mixed coolants on average power (a), cumulative *G* ratio (b), and average surface finish (c). Material machined is 52100 bearing steel at 60 HRC that was ground with a 32A80M8VS abrasive wheel having $V_S$ of 43 m/s (8500 sfm), $V_W$ of 46 m/min (150 sfm), and $D_E$ of 213 mm (8.4 in.). A, water-soluble oil at 6%; B, rust inhibitor at 3%

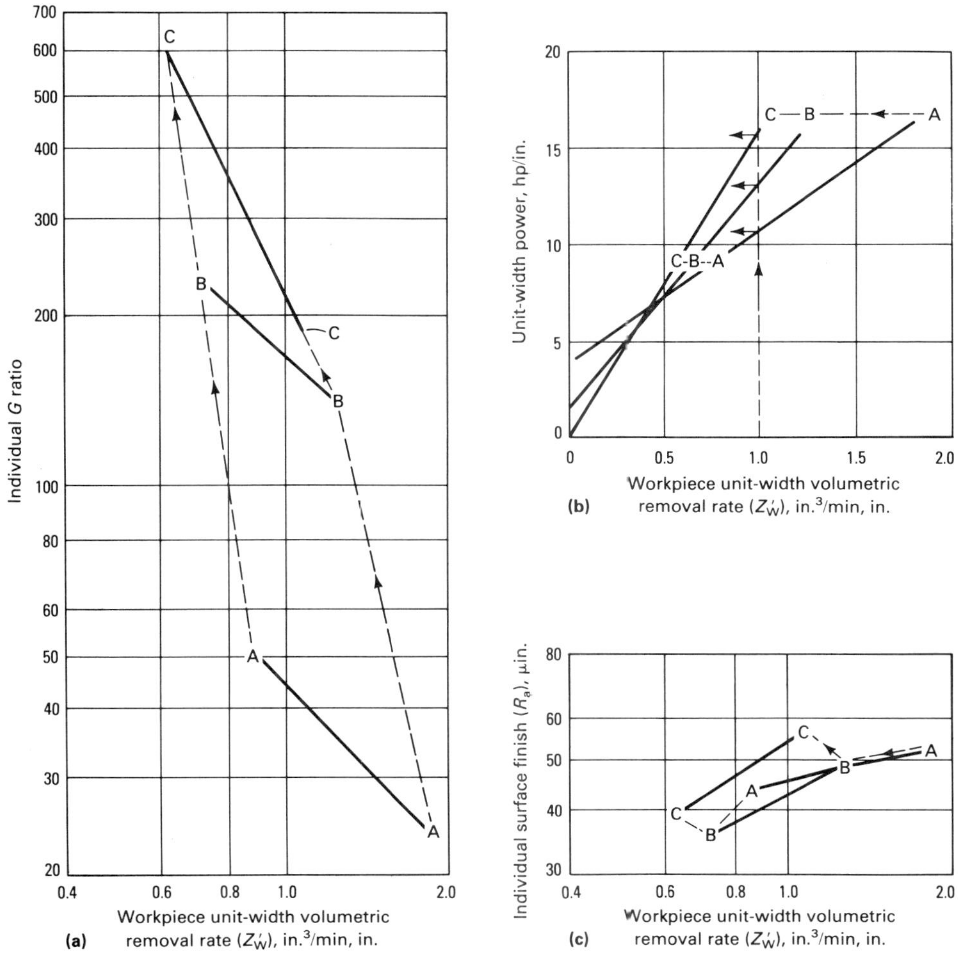


**Fig. 15** Effects of grinding time (or metal removed) on power, individual G ratio, and individual surface finish for water-soluble oil at 6%. Same system as in Fig. 14. (a) Individual *G* ratio plotted against workpiece unit-width volumetric removal rate. Metal removed (in.$^3$/in.): A–A, 0.150; B–B, 0.420; C–C, 0.860. (b) Specific power obtained (hp/in.$^3$/min): A–A, 7.05; B–B, 12.13; C–C, 14.3. (c) Individual surface finish plotted against workpiece unit-width volumetric removal rate

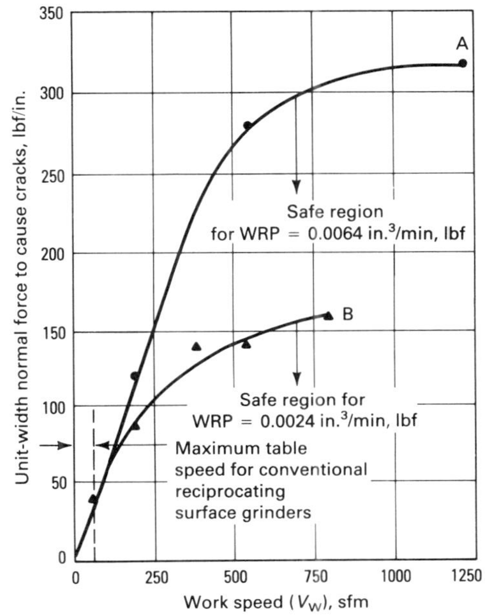


**Fig. 16** Reducing the etch cracking of bearing steel with sharp wheels and high work speeds. Material machined is 52100 bearing steel at 60 HRC (soaked in 65 °C, or 150 °F, hydrochloric acid for 10 min) ground with a __A100L6V__ abrasive wheel having $V_S$ of 39 m/s (7700 sfm). A, WRP = 0.0064 in.$^3$/min, lbf; B, WRP = 0.0024 in.$^3$/min, lbf

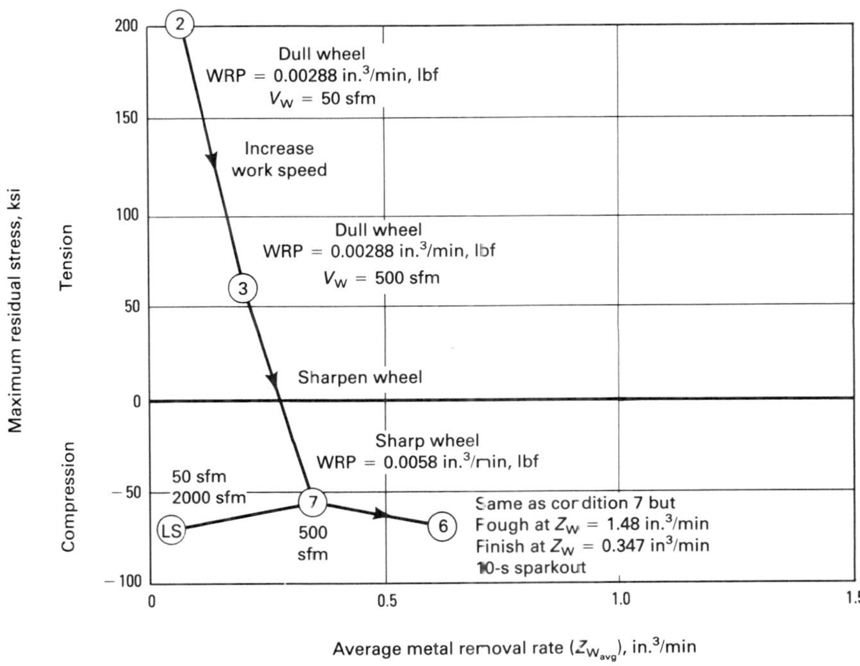


**Fig. 17** Changing tensile stress to compressive stress with sharp wheels and high work speed when grinding René 80 with an oil coolant. Source: Ref 5

Figure 18(c) shows how these three forces are resolved into $F_N$, a normal force. Drilling, boring, and milling operations can be similarly modeled.

Figure 19 shows an external grinding system. There are two forces: $F_N$ in the normal direction and $F_T$ in the tangential direction. As the slide is fed into the part at the rate $\overline{V}_F$, it induces $F_N$, which causes the workpiece to be ground at a radial rate $\overline{V}_W$ and the wheel to wear at a radial rate $\overline{V}_S$.

In Fig. 18 and 19, $Z$ is a volumetric removal rate. In cutting, $Z$ is the metal removal rate calculated as shown in Fig. 18(b). For grinding, $Z_W$ is the metal removal rate, and $Z_S$ is the wheel wear rate. In both cases, the tangential force, $F_C$ or $F_T$, multiplied by the work speed in cutting, $V_C$, or wheel speed in grinding, $V_S$, is the power used.

## System Graphs

**Turning.** Figure 20 shows turning data on 4130 steel tubing with four tools having different rake angles. Figure 20(a) plots metal removal rate against the normal force, and four linear relationships describe the slope. The tool with the 45° rake had a small threshold force and the steepest slope, while the 25° rake tool had the largest threshold force and the shallowest metal removal parameter (MRP) slope (metal removal rate per unit force). This means that for any $Z$ value (any horizontal line from the vertical axis) the 45° tool would always use the least force and that the force would

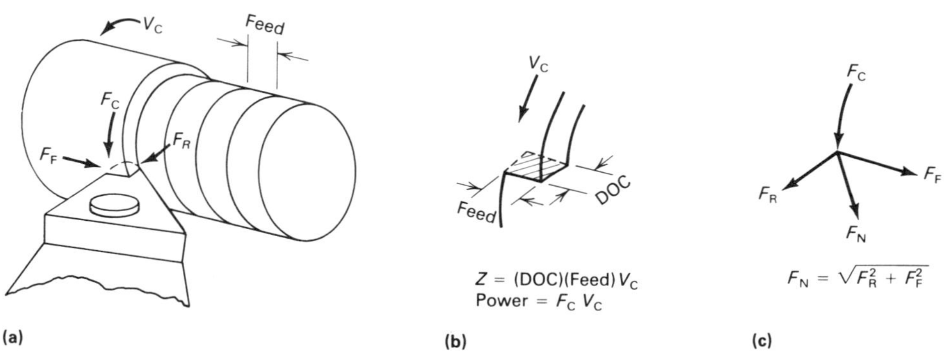


**Fig. 18** Nomenclature for turning operations on a lathe. (a) Schematic of workpiece and tool setup. (b) Relation of depth of cut to axial feed per revolution and workpiece speed, $V_C$. (c) Vector diagrams of the cutting ($F_C$), radial ($F_R$), and feed-force ($F_F$) components that form the resultant normal force ($F_N$)

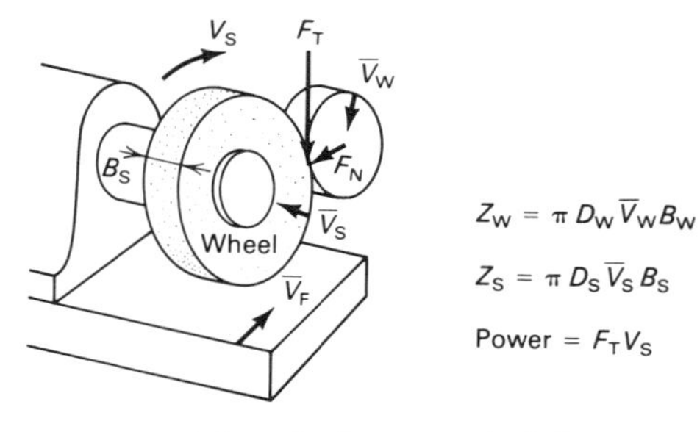


**Fig. 19** Schematic of an external grinding system. The terms in the schematic are used to calculate the metal removal rate ($Z_W$), the wheel wear rate ($Z_S$), and the power consumption.

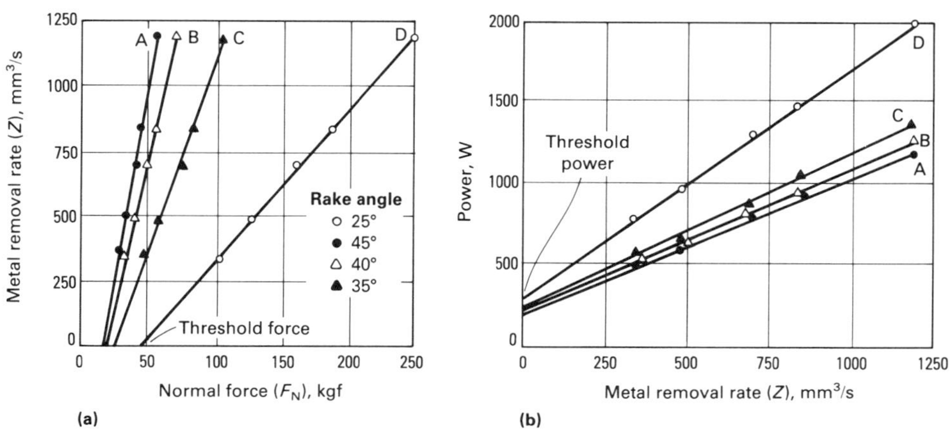


**Fig. 20** Turning data for 4130 steel tubing using four tool bits having different rake angles at $V_C$ of 0.45 m/s (90 sfm). (a) Metal removal rate plotted against normal force yields MRP slope (mm³/s, kgf): A, 36.3; B, 22.0; C, 13.7; D, 5.73. (b) Power plotted against metal removal rate yields SP slope (W, s/mm³): A, 0.846; B, 0.98; C, 1.037; D, 1.47

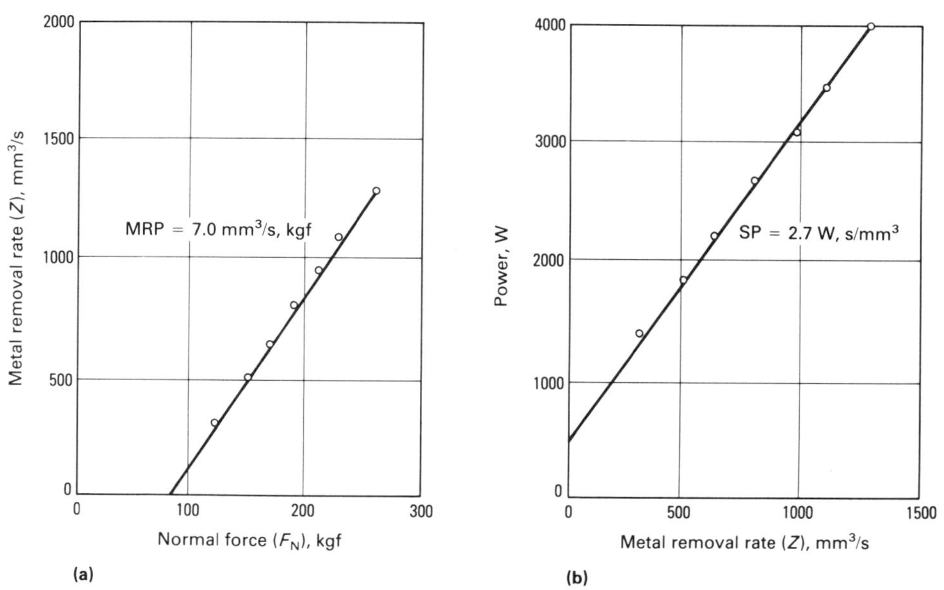


**Fig. 21** Turning data for 58 HRC high-speed tool steel cut with a CBN tool having a −6° rake. Depth of cut was 2 mm (0.08 in.), and $V_C$ was 1.33 m/s (260 sfm). (a) Metal removal rate plotted against normal force yields the metal removal parameter of 7.0 mm³/s, kgf. (b) Power consumption plotted against metal removal rate yields the specific power of 2.7 W, s/mm³.

increase as the rake angle is reduced to 40, 35, and 25°, which would always use the highest force.

Figure 20(b) plots power against metal removal rate. Again, there are four linear relationships, with the slopes designated specific power, which is the power required to cut at some removal rate. The worst tool for cutting efficiency, the 25° rake, used the most power at any $Z$ value; the best cutting tool, the 45° rake, used the least power.

This method provides two values, the metal removal parameter and the specific power, which describe the cutting and power requirements of each system. Usually, for cutting, the normal forces are ignored (but sometimes measured); each power is divided by its own metal removal rate, and different values of power/metal removal rate are obtained from five individual tests. These values are usually plotted against metal removal rate, showing that specific energy decreases with higher removal rates. In fact, Fig. 20(b) would simply show that a threshold power exists. If there were no threshold power, all five power/metal removal rate values would be nearly identical.

Figure 21 shows graphs of the hard turning of 58 HRC high-speed tool steel with a cubic boron nitride tool. Again, there are two numbers describing this system: the MRP slope (Fig. 21a) and the SP slope (Fig. 21b).

**External Grinding.** Figure 22 shows external grinding results on soft steel using CBN-electroplated wheels (one layer of abrasive on a steel hub). The 36-grit abrasive wheel cut freer (higher MRP slope) and used less power (shallower SP slope) than the 80-grit abrasive wheel.

These and other tests, including turning with negative-rake tools to simulate grinding, are shown in Fig. 23, which is a plot of the SP slope versus the MRP slope. The best direction on Fig. 23 is down (low power) and to the right (low normal force). All cutting results are in the best direction in the lower-right section, but the negative-rake cutting (simulating grinding) is in the upper-left or worst position. The upper-right section, which offers higher power but low normal force, contains the grinding results, although there was a 6:1

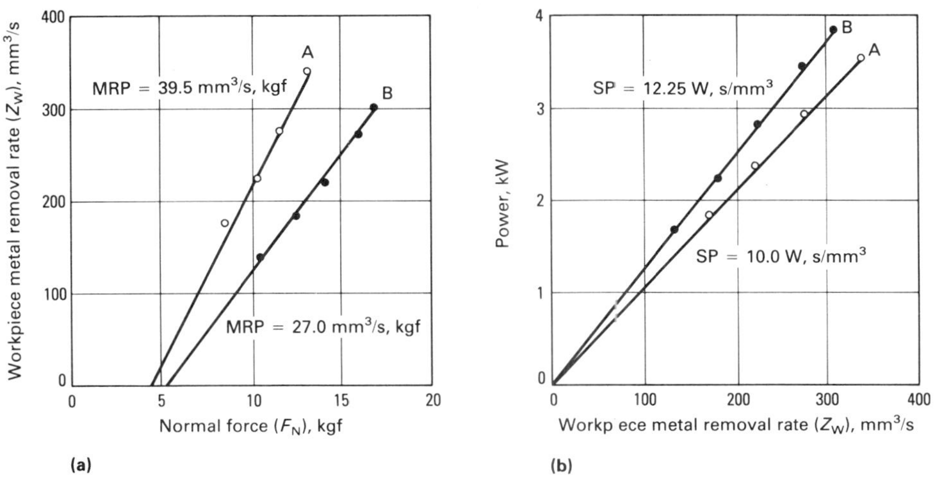


**Fig. 22** External cylindrical grinding of 4150 steel at 23 HRC using CBN-electroplated wheels. An oil coolant was used with $V_s$ of 57 m/s (11 200 sfm). Wheel grit sizes: A, 36 grit; B, 80 grit. (a) Workpiece metal removal rate plotted against normal force to obtain metal removal parameter. (b) Power consumption plotted against workpiece metal removal rate to obtain specific power

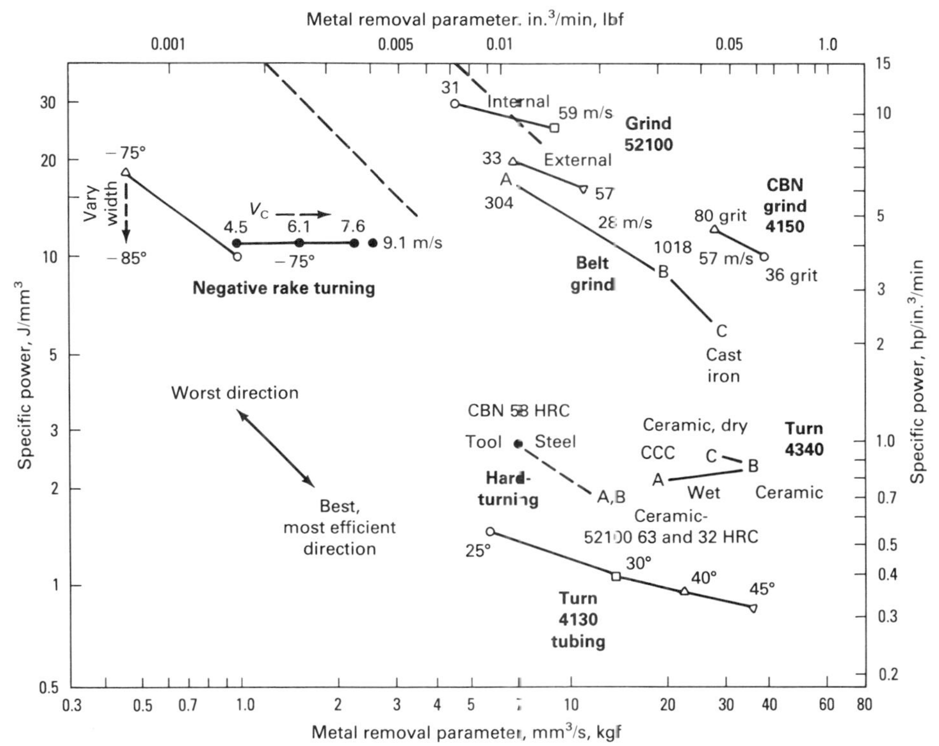


**Fig. 23** Specific power plotted against metal removal parameter to forecast optimum operating conditions for turning and grinding. CCC, ceramic-coated carbide

difference between the worst and the best grinding results. This means that ordinary cutting uses power much more efficiently than grinding, except where negative-rake tools are used. Therefore, the low-power aspects of cutting influence the ability to send a chip sliding up an inclined tool plane (positive-rake tools). However, cutting uses normal forces as high as those in grinding because both sets of data are generally in the same area on the right of the graph. This means that the normal or deflection forces in cutting can be as large as those in grinding. There was an 8:1 change in normal force in both the cutting and grinding data.

## REFERENCES

1. R.P. Lindsay, The Effects of Grinding Fluids on the Performance of CBN and a New High Alumina Abrasive, *Japan Society of Precision Engineers*, 1987
2. R.S. Hahn and R.P. Lindsay, Principles of Grinding, *Mach. Mag.*, July-Nov 1971
3. R. Snoeys and J. Peters, The Significance of Chip Thickness in Grinding, *Ann. CIRP*, Vol 23 (No. 2), 1974, p 227-237
4. G. Pahlitzsch and R. Schmitt, Abrichten Von Schleifscheiben Mit Diamantstucken Rollen, *Ann. CIRP*, Vol 17 (No. 2), 1969
5. R.S. Hahn and R.P. Lindsay, "The Production of Fine Surface Finishes While Maintaining Good Surface Integrity by Grinding," Paper presented at the International Grinding Conference, Carnegie-Mellon University, 1972
6. R. Snoeys, M. Maris, and J. Peters, Thermally Induced Damage in Grinding, *Ann. CIRP*, Vol 27 (No. 1), 1978, p 571-581; C.P. Bhateja and R.P. Lindsay, Ed., *Grinding: Theory, Techniques and Troubleshooting*, Society of Manufacturing Engineers, 1982

## SELECTED REFERENCES

- J.R. Besse, "Practical Creep Feed Grinding," Paper MR87-820 SME, Society of Manufacturing Engineers, 1987
- *Injury in Ground Surfaces*, Norton Company, 1973
- R. Komanduri, Some Aspects of Machining With Negative Rake Tools Simulating Grinding, *Int. J. Mach. Tool Des. Res.*, Vol 11, 1971, p 223-233
- R. Komanduri, W. Konig, and H. Tonshoff, Machining of Hard Materials, *Ann. CIRP*, 1984
- R.P. Lindsay, "On the Metal Removal and Wheel Removal Parameters Surface Finish, Geometry and Thermal Damage in Precision Grinding," PhD. thesis, Worcester Polytechnic Institute, 1971
- R.P. Lindsay, "On the Surface Finish—Metal Removal Relationship in Precision Grinding," ASME Paper 72-WA/Prod-13, American Society of Mechanical Engineers, 1972
- R.P. Lindsay, "Sparkout Behavior in Precision Grinding," Paper 72-205, Society of Manufacturing Engineers, 1972
- R.P. Lindsay, The Effect of Wheelwear Rate on the Grinding Performance of Three Wheel Grades, *Ann. CIRP*, Vol 32 (No.1), 1983, p 247-249
- R.P. Lindsay, "The Effect of Contact Time on Forces, Power and Metal Removal Rate in Precision Grinding," Paper presented at the International Grinding Conference, Lake Geneva, WI, Society of Manufacturing Engineers, 1984
- R.P. Lindsay, System Parameters for Cutting and Grinding, in *Proceedings of the Ninth Annual Conference on Composites and Advanced Ceramic Materials*, American Ceramic Society, 1985
- S. Malkin and N. Joseph, Minimum Energy in Abrasive Processes, *Wear Mag.*, Vol 32, 1975, p 15-23
- R.F. Pugh and R.F. Pohl, "High-Speed Metal Removal," Special Publication ARLOD-SP-82004, U.S. Army Armament Research and Development Command, 1982

# Grinding Equipment and Processes

William N. Ault, Norton Company

METAL IS REMOVED from the workpiece by the mechanical action of irregularly shaped abrasive grains in all grinding operations. This article will discuss grinding wheels and disks, coated abrasives, and grinding machines and processes. These processes include:

- Rough grinding
- Precision grinding
- Surface grinding
- Cylindrical grinding
- Centerless grinding
- Internal grinding
- Tool grinding

The quantitative relationships among important grinding process variables are described in the preceding article in this Section. In addition, the grinding fluids described briefly in this article are discussed in greater detail in the article "Metal Cutting and Grinding Fluids" in this Volume.

## Grinding Wheels and Disks

In their simplest form, grinding wheels can be thought of as multitooth cutters. They consist of three primary components:

- Abrasive (the cutting tool)
- Bond (the toolholder)
- Porosity or air for chip clearance and/or the introduction of coolant

Each of these components has a profound effect on the grinding process.

**Standard Marking Systems for Grinding Wheels.** Abrasives can be classified as conventional abrasives or superabrasives (see the article "Superabrasives" in this Volume). Figure 1 shows the standard marking system for conventional abrasive products (aluminum oxide or silicon carbide abrasives). Figure 2 shows the standard marking system for superabrasive products (diamond or cubic boron nitride, CBN, abrasives). Although standard marking systems are available, many parts of the markings have no standard of measurement.

**Abrasive Type and Grit Size.** In both marking systems, the first two components of the marking deal with the abrasive type. The second component defines the chemistry of the abrasive, while the first defines the specific type of that abrasive. The second two components deal with the size of the abrasive particle. The third position defines the sieve spacing to which the particle corresponds. In other words, the grit size is roughly equal to the linear holes per inch of a sieve that the particle would just pass through. A 60-grit particle, for example, would pass through a 56-mesh screen but would be caught on a 64-mesh screen. Table 1 lists the mean particle sizes for various grit sizes. The grit size varies indirectly with the particle size. The fourth position further describes the particle size distribution by defining the combination of grit sizes that has been used to manufacture the grinding wheel. There is no industry standard for grit size combination.

**Bond Designation and Grain Spacing.** The latter part of the marking deals mainly with the bond and the spacing of the grain in the bond. For conventional abrasive wheels, the fifth, sixth, and seventh positions are the bond hardness (or the amount of bond), the porosity or grain spacing, and the bond type. For superabrasive wheels, those positions are again the bond hardness, the concentration or the amount of grain in the abrasive section (and actually, by inference, the grain spacing), and the bond type.

*The bond hardness* or grade of a wheel is defined by an alpha character: the letter A being soft or having very little bond and the letter Z being hard and holding the abrasive tightly into the grinding wheel. The definition of what constitutes a given grade letter for hardness or harshness of the grinding action varies by bond and by manufacturer because there are no industry standards.

*Porosity*. There are no standards for structure or porosity in conventional abrasive grinding wheels or for concentration in superabrasive wheels other than the relative scales. In conventional abrasive wheels, low structure numbers are dense and have little porosity (consider that the number is the distance between the abrasive grains), while high numbers denote porous products. In superabrasive wheels, higher concentrations have more superabrasive particles.

*Bond Types*. Bond markings are generally standardized throughout the abrasives industry. The eighth position in conventional and superabrasive marking systems further defines the bond type and is vendor specific. In superabrasive product markings, the ninth position denotes abrasive layer thickness. Superabrasive wheels often contain thin rims of expensive abrasive in a bond matrix or a nonabrasive preform or holder.

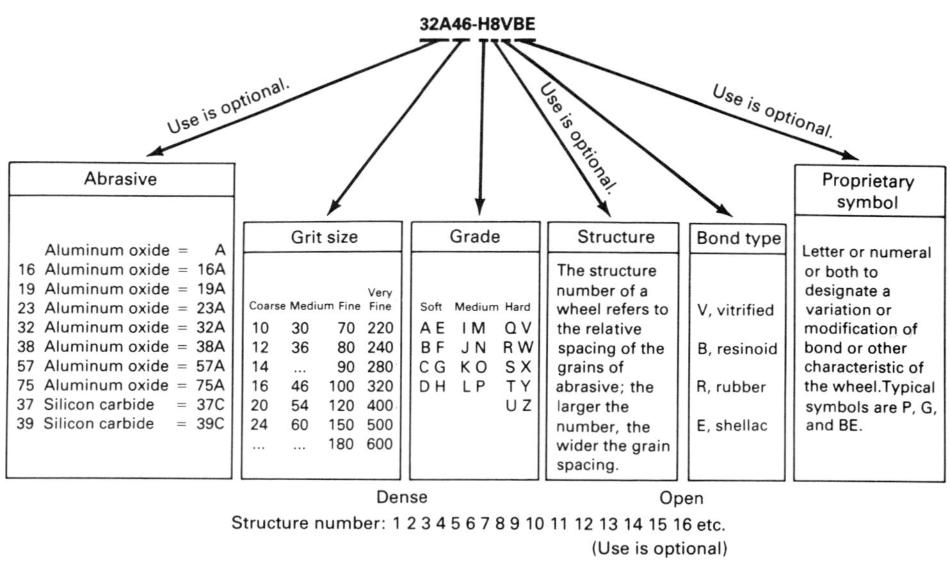


**Fig. 1** Standard marking system for conventional aluminum oxide and silicon carbide abrasive grinding wheels

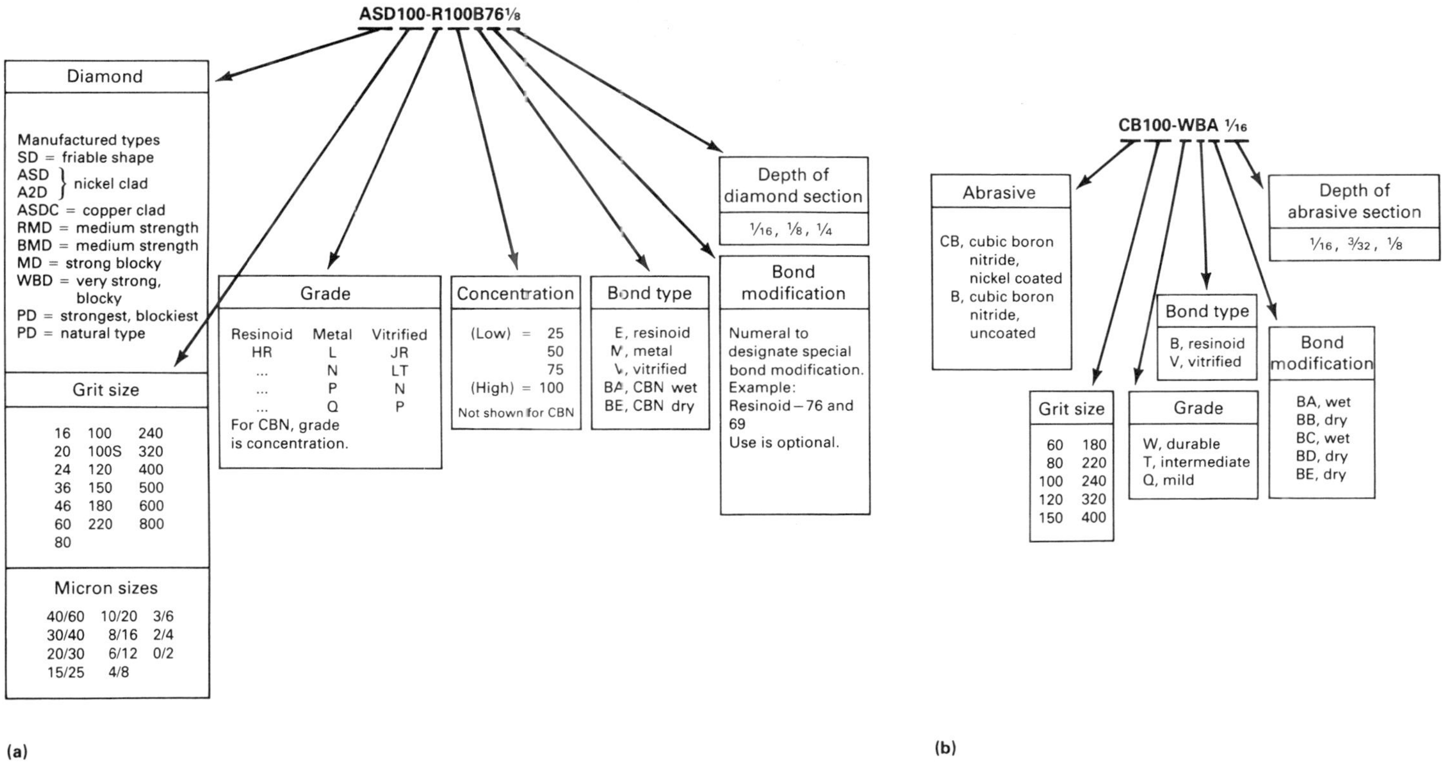


**Fig. 2** Standard marking system for diamond (a) and cubic boron nitride (b) superabrasive grinding wheels

**Table 1 Mean particle sizes for grits used in conventional abrasive grinding wheels**

| Grit size | Particle size (mean) µm | in. |
|---|---|---|
| 4 | 6848 | 0.2577 |
| 6 | 5630 | 0.2117 |
| 8 | 4620 | 0.1817 |
| 10 | 3460 | 0.1366 |
| 12 | 2550 | 0.1003 |
| 14 | 2100 | 0.0830 |
| 16 | 1660 | 0.0655 |
| 20 | 1340 | 0.0528 |
| 24 | 1035 | 0.0408 |
| 30 | 930 | 0.0365 |
| 36 | 710 | 0.0280 |
| 46 | 508 | 0.0200 |
| 54 | 430 | 0.0170 |
| 60 | 406 | 0.0160 |
| 70 | 328 | 0.0131 |
| 80 | 266 | 0.0105 |
| 90 | 216 | 0.0085 |
| 100 | 173 | 0.0068 |
| 120 | 142 | 0.0056 |
| 150 | 122 | 0.0048 |
| 180 | 86 | 0.0034 |
| 220 | 66 | 0.0026 |
| 240 | 63 | 0.0024 |
| 280 | 44 | 0.0017 |
| 320 | 32 | 0.0012 |
| 400 | 23 | 0.0009 |
| 500 | 16 | 0.0006 |
| 600 | 8 | 0.0003 |
| 900 | 6 | 0.0002 |
| Levigated alumina | 3 | 0.0001 |

Note: Grit size varies indirectly with particle size.

## Abrasives

In addition to particle or grit size, abrasives have a number of properties that determine their efficacy in the grinding process. These properties include chemistry, crystal structure, hardness, durability, friability, and sharpness.

### Properties of Abrasives

**The chemistry** of an abrasive can affect its ability to cut at grinding interface temperatures. Diamond and silicon carbide are harder than aluminum oxide, but when steel is ground under high pressures, a chemical reaction occurs that degrades these abrasives compared to the relatively chemically inert aluminum oxide. In a different vein, the chemical purity of an abrasive is often an indicator of crystal structure.

**Crystal Structure.** There are basically three types of abrasive particle: monocrystalline, multicrystalline, and microcrystalline.

*Monocrystalline* grains contain a single crystal. They tend to be relatively durable, and they wear along crystal planes.

*Multicrystalline* grains, usually made up of two to ten crystals, vary greatly with regard to durability and sharpness. These abrasives fracture along crystal boundaries (often catastrophically) and along crystal planes.

*Microcrystalline* abrasives (which may contain crystals smaller than 1 µm) tend to retain their sharpness during grinding, and they fracture along crystal boundaries. In some applications, microcrystalline abrasives may have more usable volume than mono- or multicrystalline grains before being shed by the bond.

**Hardness.** Figure 3 shows the relative hardness of various abrasives on the Knoop hardness scale. Hardness is an advantage; it is inefficient to abrade a material with an abrasive that is not significantly harder than the material. The ability of an abrasive wheel to grind a material is normally measured by the *G* ratio (see the article "Principles of Grinding" in this Volume), usually defined as the volume of metal removed per volume of wheel used. Under optimum conditions, superabrasive wheels, primarily because of their hardness, will yield *G* ratios hundreds of times larger than those of conventional abrasive products.

**Durability.** Durable grains tend to withstand heavy grinding pressures without catastrophic wear. Under light grinding pressures, they tend to dull, drawing higher power and giving better surface finishes unless there is vibration due to lack of material penetration.

**Friability.** Friable grains fracture to expose new sharp cutting points and may do so a number of times before the bond sheds the grain (Fig. 4). Friable grains tend to remain sharp and tend to draw lower power, often giving rougher finishes because they do not dull as readily. Very friable grains may not be efficient at high power levels and heavy grinding pressures because of

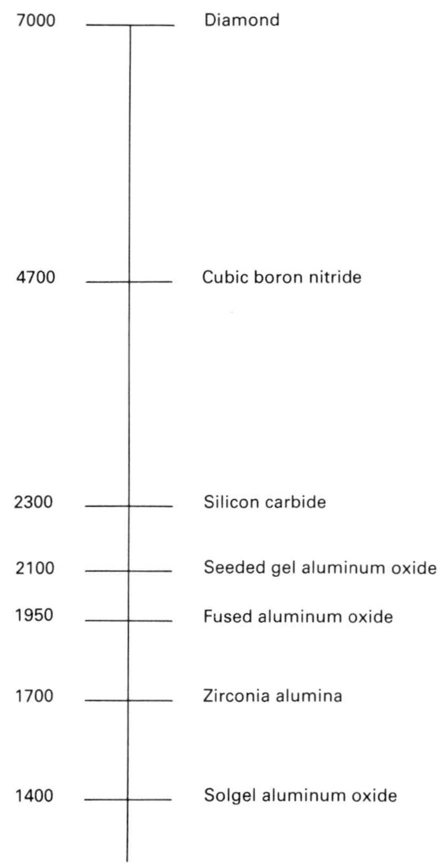


**Fig. 3** Knoop hardness ratings of various abrasives

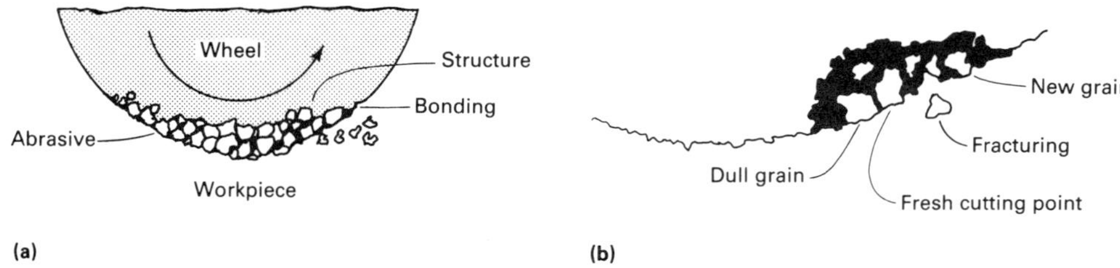


**Fig. 4** Abrasive wear in a grinding wheel due to attrition and fracture. (a) Attrition deteriorates abrasive grains by the loss of fine particles. This flattens and dulls the edges. (b) Fracture deteriorates abrasive grains by the breaking away of relatively large pieces of abrasive crystals from the wheel surface.

premature wear. With regard to conventional abrasives, friable grains are normally used on heat-sensitive and hardened steels to ensure material penetration at relatively low power and low frictional heat levels in precision grinding.

**Grain sharpness** is a function of crystal structure and the inherent ability of a given particle to cut prior to wear. Grain sharpness can affect abrasives selection, particularly in such extremely low pressure applications as honing or superfinishing, in which there may not be sufficient force to fracture the abrasive particle.

## Types of Conventional Abrasives

The three prevalent types of conventional abrasives are aluminum oxide, silicon carbide, and zirconia alumina.

**Aluminum Oxides.** A number of aluminum oxides have been developed for grinding applications. Aluminum oxides are divided into two initial subgroups: fused and unfused.

*Fused aluminum oxides* are available in a variety of types. These abrasives are categorized as dark, white, monocrystalline, and specialty abrasives.

Dark aluminum oxides tend to be less pure, more durable, less friable, and multicrystalline. They can also be mulled to a dull but even more durable shape. They are normally used in medium- to high-pressure, relatively heat insensitive operations on medium-to-soft materials.

White aluminum oxides are the most pure of the fused aluminas. They are multicrystalline, very friable, not durable, and relatively sharp. They are used in heat-sensitive operations on hard, ferrous materials.

Monocrystalline fused aluminum oxides tend to combine a level of durability and friability. This makes them efficient in medium-pressure heat-sensitive operations on a variety of ferrous materials.

Specialty abrasives are variations of the above types. They can be blends of dark and white aluminum oxides, ceramic-coated dark alumina (which tends to increase its impact resistance), or chemically treated white aluminas (which tend to be less friable and more durable).

*The unfused aluminas* (also known as ceramic aluminum oxides) are microcrystalline abrasives that are sintered after green crushing. Therefore, they can be harder than fused aluminas that have been crushed after furnacing.

A significant type of unfused aluminum oxide abrasive is seeded gel alumina abrasive. Seeded gel is the purest of the aluminum oxides and the hardest ($\sim$2150 HK). It is also durable, friable, and inherently sharp.

Seeded gel is made by a ceramic process in which submicron particles are sintered to form microcrystalline abrasive grit particles. A 60-grit particle of seeded gel contains billions of individual crystals. Seeded gel is purer, harder (because it is not crushed after sintering), and maintains its sharpness longer than fused aluminum oxide.

Seeded gel vitrified grinding wheels were introduced in 1987 and were commercialized in 1988. They have demonstrated the greatest utility for difficult-to-grind materials in which tight tolerances and no metallurgical damage are specified. Typical optimized results are 3 to 5 times the life and 1.5 to 2 times the cut rate of fused aluminum oxides. For the precision grinding of difficult-to-grind steels and alloys, seeded gel will grind at higher pressures and infeeds than fused aluminum oxides, without metallurgical damage and at significantly higher *G* ratios.

**Silicon carbide** is manufactured in two purities: black or green.

*Black silicon carbide* is less pure, more durable, and less friable than green silicon carbide. Both silicon carbides are harder and sharper than aluminum oxide, but are less impact resistant. Black silicon carbide tends to be efficient for grinding soft nonferrous materials; in such applications, its inherent sharpness inhibits heat generation.

*Green silicon carbide* is used for the precision grinding of materials that are hard, such as hard chromium (70 HRC), and/or materials that tend to fracture rather than form chips in the grinding process, such as ferrous carbides and other relatively soft ceramics.

**Zirconia aluminas** are distinguished by their extremely high impact resistance. These alloyed abrasives are of two chemistries: a eutectic ($\sim$44% $Zr_2O$) alloy and a 25% $Zr_2O$ alloy. The eutectic alloy has some friability and is the sharper of the two zirconia aluminas. The 25% Zr alloy is the more durable and has a slightly higher impact resistance. The zirconia aluminas are normally used in high-pressure, high stock removal operations on ferrous materials and are normally not associated with precision grinding.

## Types of Superabrasives

As previously noted, superabrasives consist of diamond and CBN. Diamond has a relatively low impact resistance and is therefore used for grinding such materials as carbides, glass, and ceramics that fracture rather than chip when ground. Cubic boron nitride has a high impact resistance and exhibits optimum results in the grinding of hard ferrous materials (>50 HRC). Superabrasives and their applications are discussed in the article "Superabrasives" in this Section.

# Abrasive Wheel Bonds

The properties of bonds that are important to the grinding process include the way in which the bond holds the abrasive, the rigidity and/or flexibility of the bond, the wear mechanism of the bond, and the inherent bursting strength or rotational strength of the bond under stress. (Grinding wheels should never be used at speeds higher than

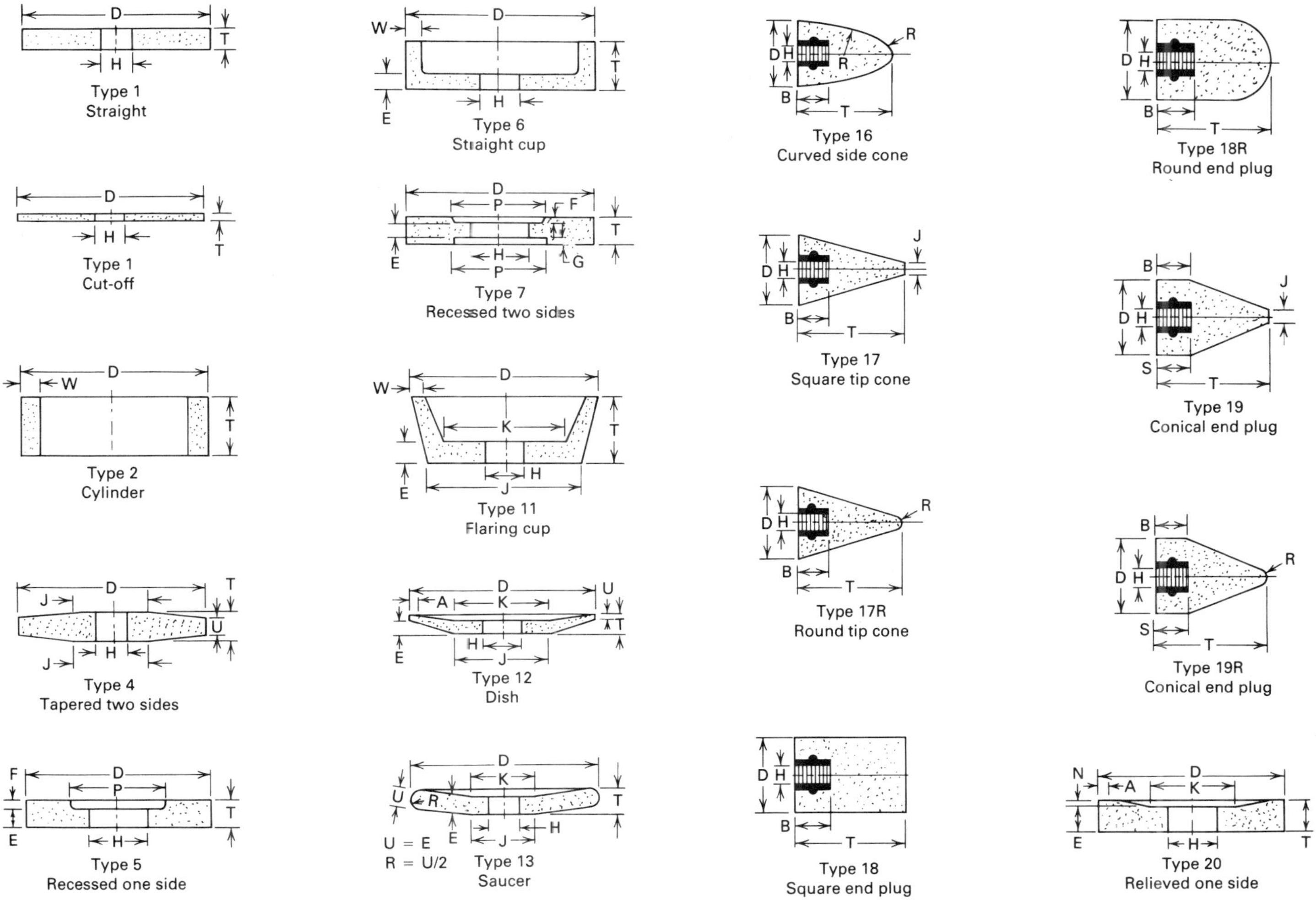


**Fig. 5** Standard wheel configurations for conventional abrasive grinding wheels

the rated speed shown on the grinding wheel.)

Conventional abrasive grinding wheels are held together by two types of bond: vitrified and organic. Organic bonds are further divided into resin bond (and its subgroup plastic bond), rubber, and shellac. In addition to vitrified and resin bonds, superabrasive wheels are available in metal and electroplated bonds.

## Conventional Abrasive Wheel Bonds

**Vitrified bonds** are chemical bonds that alloy with the abrasive at the abrasive/bond interface. This rigid, glassy bond family is relatively inflexible and wears as a result of pressure fracturing the bond. As the abrasive dulls and the power consumption needed to remove a chip increases, the bond fractures under the higher pressure. Because of the rigidity of vitrified bonds, the abrasive tends to act sharper than in flexible bonds, causing easier chip formation but rougher surface finishes. Vitrified bonds do not have high impact resistance and should not be used in high-impact, heavy-pressure operations such as foundry snagging or steel conditioning. Vitrified wheels are typically used at 25 to 33 m/s (5000 to 6500 sfm) and can be run at more than 43 m/s (8500 sfm) only on specially designed machines and applications.

**Organic bonds** are physical bonds that surround the grain to hold it in the wheel. They tend to break down as the abrasives dull and frictional heat increases.

*Resin bonds* vary in rigidity from some of the more brittle phenolic or bakelite bonds to the more flexible epoxy or other plastic bonds. Resin bonds can be reinforced with fiberglass layers to increase the rigidity of the wheels. Some resin wheels can be operated at up to 81 m/s (16 000 sfm). Resin bonds are used in most rough grinding applications, including floorstand, portable snagging, weld grinding, and most cutoff applications. In precision grinding, resin bonds are used in three situations: where system flexibility would break down a vitrified wheel (such as steel mill roll grinding and most centerless-bar grinding applications), where wheel geometry dictates some flexibility (such as the spiral fluting of drills), and for generating fine and cosmetic finishes (for example, hypodermic needle grinding).

*Rubber bonds* are the most flexible of the organic bonds. The use of rubber bonds is decreasing because of the development of resin bond technology, particularly plastic bonds. Rubber bonds have historically been used for polishing applications, wet cutoff operations, and centerless regulating or feed wheels.

*Shellac bonds* are the most heat sensitive of the organic bonds. As such, they have historically been used for toolroom cutoff applications in which burn-free cuts were essential, for some heat-sensitive roll grinding applications, and for very fine grit polishing operations in which geometry is not critical and stock removal is minimal.

## Superabrasive Wheel Bonds

Although conventional abrasive grinding wheels are limited to vitrified and organic bonds, superabrasive grinding wheels can utilize metal and electroplated bonds in addition to vitrified and organic bonds.

**Metal bonds** are used in superabrasive wheels primarily in ceramic and glass grinding operations, in which their tendency to wear by abrasion is an advantage. Metal bonds range from relatively soft bronze

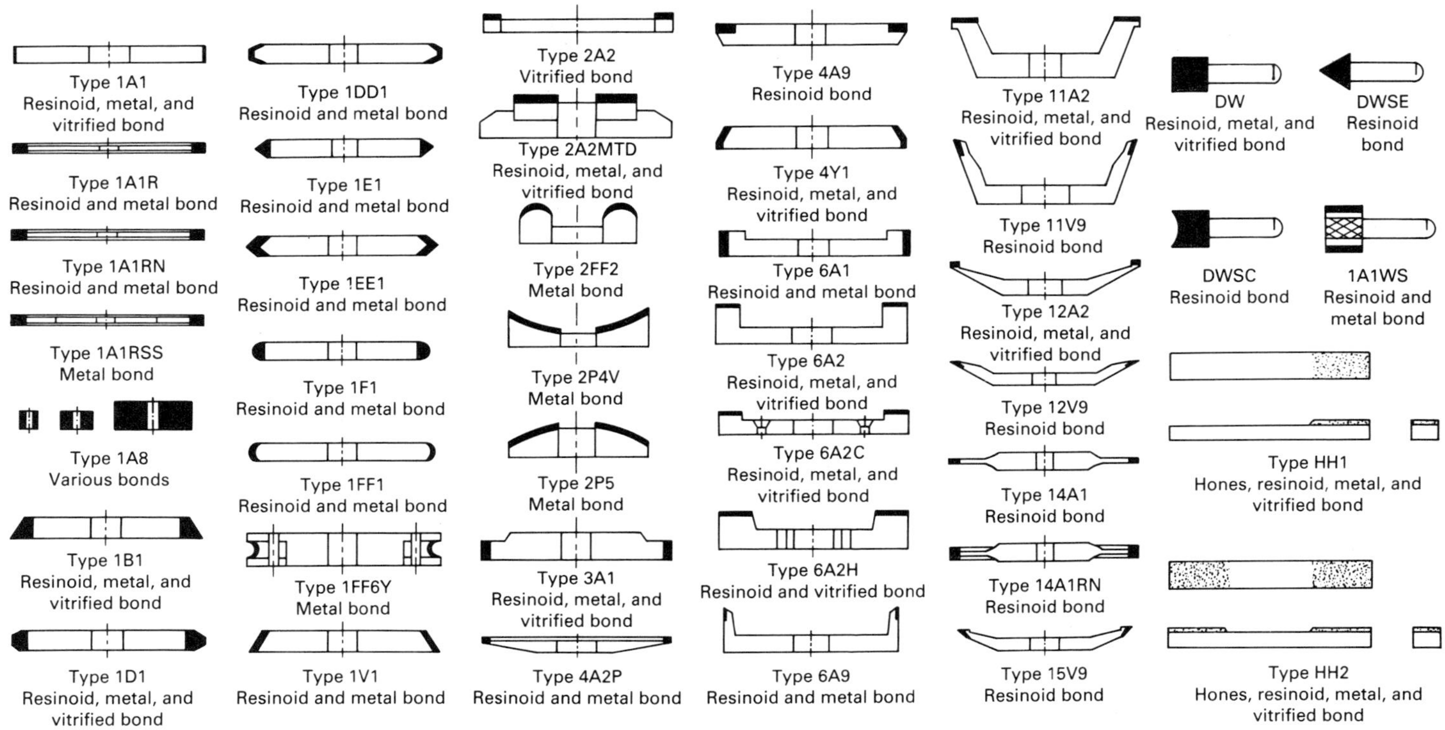


**Fig. 6** Typical wheel configurations for superabrasive grinding wheels

bonds to carbide bonds. In particular, they are used on materials that do not generate long chips in the grinding process.

**Electroplated** superabrasive wheels consist of a single layer of abrasive plated to a metal preform. These wheels are used for cutting exotic materials (such as fiberglass), for form grinding in which machinery limitations do not allow regeneration of the form in the wheel, and for various applications in which the inherent sharpness of electroplated wheels is the key to rapid stock removal.

## Abrasive Wheel Porosity

Porosity or the grain spacing in grinding wheels serves two major functions: coolant transfer and chip clearance. Heat transfer, chip removal, and lubricity are provided to varying degrees by coolants (including air) in a grinding operation (coolants are discussed in the section "Grinding Fluids" in this article). The grinding wheel is a balance between having enough cutting points for the work required and providing enough porosity to ensure freeness of cut.

**In conventional abrasive wheels,** low structure products (that is, those with tight grain spacings) tend to be used in high-pressure applications such as rough grinding and cutoff. Medium structure products (typically 5 to 9) are used for the precision grinding of small areas of contact, such as centerless and cylindrical grinding. High structure numbers are used in operations in which the large area of contact dictates the need for chip clearance. High structure numbers also aid in carrying coolant into the cutting zone in such machining operations as vertical-spindle rotary-table surface grinding or creep-feed grinding.

**In superabrasive wheels,** grain spacing is less influential in wheel performance because the abrasive, when properly exposed in the bond and when grinding the hard (or brittle in the case of glass) materials normally ground with superabrasives, tends to cut a very small chip. The amount of superabrasive grain (related directly to the grain spacing) usually determines the product life of the wheel, and higher concentrations tend to give better surface finishes.

## Abrasive Wheel Configurations

Figure 5 shows the standard wheel configurations for conventional abrasive wheels. Figure 6 shows some of the wheel configurations for superabrasive wheels.

It is essential when using abrasive wheels to apply the grinding forces in the direction of support. In other words, a type 1 conventional abrasive wheel should be used on the periphery, while a type 6 wheel would be used on the top of the rim. Side grinding on straight wheels or grinding on the periphery of straight cup wheels is an unsafe practice and can lead to wheel breakage.

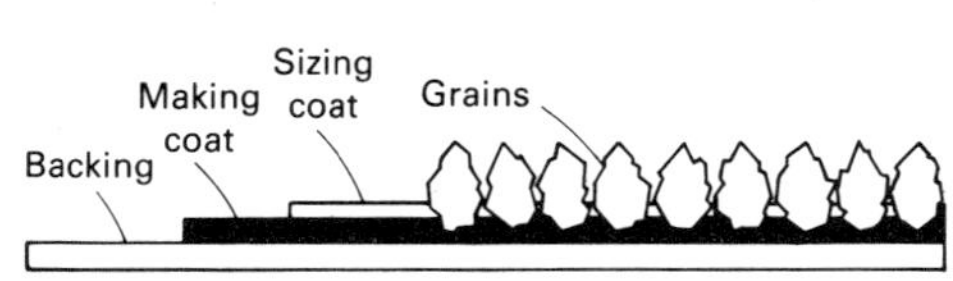


**Fig. 7** Typical construction of a coated abrasive product

## Coated Abrasives

Coated abrasive products, historically thought of as sandpaper, consist of an abrasive whose grain spacing is tightly controlled, a bond, and a backing. Although a grinding wheel can be thought of as regenerating itself to expose new cutting points during grinding, the coated abrasive product (like the electroplated superabrasive grinding wheel) has one layer of abrasive to do the work. Figure 7 shows a typical coated abrasive product construction.

### Coated Abrasive Composition

**Abrasives.** Conventional abrasives are almost exclusively used for coated abrasive applications. With few exceptions, superabrasives are not used in coated abrasives. In addition to aluminum oxide, silicon carbide, and zirconia alumina, coated abrasives are available in natural abrasives such as garnet, emery, flint, and crocus (iron oxide). However, natural abrasives are not normally used for metalworking and will not be discussed in this article.

**Bonds.** As shown in Fig. 7, the bond in a coated abrasive product consists of two coats: the making coat and the sizing coat. The maker coat ensures the adherence of the grain to the backing in the proper orientation, while the size coat controls the ex-

**Fig. 8** Various grinding wheels and coated abrasive products used in industrial applications

**Fig. 9** Coated abrasive belts with cloth, film, or paper backings

posure of the grain to the grinding operation. A low size coat exposes more of the grain, thus making the product more aggressive, while a high size coat tends to inhibit the formation of large chips and generates better surface finishes on the ground part.

Resin bonds predominate in metalworking applications, but some hand operations and light finishing are done with products having a resin maker coat and a glue size coat. Glue bond is very heat sensitive and tends to give fine finishes while not generating heat in finishing. Other bonds (such as vinyl acetate and varnish) are used on some specialty products, but these are atypical.

**Grain spacing** in coated abrasive products tends to be designated by either open coat (abrasives uniformly distributed over 50 to 70% of the surface) or closed coat (abrasives completely covering the surface). An open coat product has significantly less abrasive and has widely spaced grains. Open coat products are primarily used in woodworking applications either where heat generation while sanding relatively soft woods or where pine pitch, or other similar substances that tend to load the belts before the abrasive wears out are a problem. Coated abrasive products for metalworking are almost exclusively closed coat products.

**The backing** can be considered the toolholder in the coated abrasive product. The major properties that determine the selection of the backing are as follows:

- Backing strength versus backing cost
- Backing rigidity or flexibility
- Aggressiveness of cut versus ability to polish
- Wet or dry grinding
- The friction characteristics of the backing

Backings are available in various weights, expressed as pounds per papermaker's ream. In general, the heavier the weight of the backing, the stronger and more expensive it is. In addition, the heavier the weight of the backing, the more rigid and less flexible it is. Heavy, rigid backings that can withstand high grinding pressures are used in coarse-grit rough grinding operations. Light, flexible backings that may be able to conform to intricate shapes are used in finishing, polishing, and blending applications.

The backing material and any surface treatment applied to the backing determine whether it can be used in water-soluble coolants. The roughness of the back of a backing and its material will determine whether it has enough friction to grind under heavy pressures without slippage or whether it generates too much friction and causes heat problems during grinding. There are four major types of backings: paper, cloth, fiber, and film.

*Paper backings* are the lightest and have the least strength of all backing types. Paper backings are available in five weights: A (the lightest), C, D, E, and F (the heaviest). Paper is normally used in light-pressure applications. Unless specially treated, paper backings are not used with water-soluble coolants.

*Cloth backings* range from J weights (which are light, flexible, and conformable) to the standard X weight, the heavier Y weight, and the superheavy H weight. Cloth backing materials range from cotton to synthetics (such as rayon) to polyesters. Resin bond polyester belts can be used in water-soluble coolants, while cotton and rayon belts are normally not used in such applications.

Cloth backings have two architectures: woven and stitched. Woven cloth is manufactured by interlacing yarns at 90° angles to each other in the traditional weaving process. Stitched cloth is made by a relatively

**Table 2 Relative rating of the four types of grinding fluids on the basis of their properties**

| | Grinding fluids | | | | |
|---|---|---|---|---|---|
| Fluid property | Petroleum-base and mineral-base cutting oils | Water-soluble oils | Synthetic fluids | Semisynthetic fluids | Water plus additives |
| Cooling | D | B-C | A | B | A |
| Lubricity | A | B-C | B-C | B-C | D |
| Rust protection | A | B-C | B-C | B-C | A |
| Cleanliness | D | C-D | A | B | A |
| Stability | A | C-D | A | B-C | C |
| Tolerance to contamination | A-B | C-D | A-B | B-C | C |
| System life | A | B-C | A | B-C | C |
| Health and safety | C-D | C | A-B | B | B |
| Disposal | B-C | C | A-C | B-C | C |
| Fire hazard | D | A-B | A | A | A |

Note: A, excellent; B, very good; C, good; D, poor. Source: Van Straaten Corporation

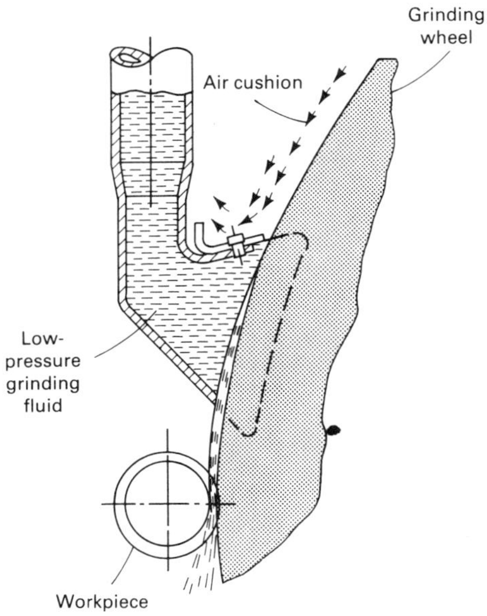


**Fig. 10** Typical air deflectors or chamber-type nozzle used to maximize fluid access to the grinding zone

modern process in which the yarns are not interwoven but laid on each other at 90° angles and then stitched together so that the yarns remain flat and uncrimped (and are therefore stronger). A unique variation of cloth backings is a mesh or screen cloth, which is used primarily in wet low-pressure finishing operations.

*Fiber backings* are made from multiple layers of impregnated paper and are very rigid and strong. They are normally used for dry, heavy-pressure applications.

*Film Backings.* Film products backed with plastic film are normally used wet and are flexible and brittle; therefore, the abrasive tends to cut freely without much heat. Film backings are normally used for polishing and blending operations under light pressures.

**Fig. 11** Dry cutoff of a pipe using a reinforced resin bond wheel

## Flexing

Once produced, coated abrasive products are flexed before being converted into finished form. Flexing is a controlled breaking of the adhesive bond holding the abrasive to the backing. Light flexing is used to make a belt act aggressively but relatively unconformable. Multiple flexing increases the flexibility and conformability of the product, usually at the expense of product life.

## Coated Abrasive Applications

Coated abrasive products are available in several forms: belts, disks, rolls, sheets, and specialty products. Various abrasive products are shown in Fig. 8.

**Fig. 12** Use of a portable cutoff machine to dry cut a metal pipe

**Abrasive belts** can be made from cloth, film, or E- or F-weight paper (Fig. 9). A length and a width of coated abrasive are joined together either by lapping the ends over and gluing them or by butting the ends and applying a patch to the nonabrasive side of the backing. Lap joints can be skived or ground on the abrasive side to ensure that the joint is no thicker than the rest of the belt. Skiving can weaken the belt at the joint and is normally done in finish-sensitive operations in which unskived joints leave marks on the ground part. Butt joints can be straight (usually at a 55° angle with the side of the belt), or they can be scalloped to offer resistance to hinging or fatigue, particularly in form grinding applications. Some appli-

**Fig. 13** Use of a type 27 reinforced resin bond abrasive wheel to grind welds

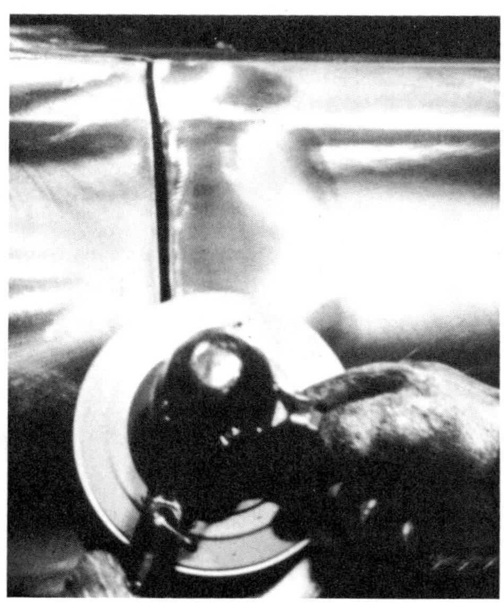

**Fig. 14** Use of a zirconia-alumina cloth abrasive disk to blend stainless steel seams with a portable grinding machine

**Fig. 15** Floorstand rough grinding of a casting using a zirconia-alumina resin bond wheel. Note the pressure bar used to increase the grinding rate.

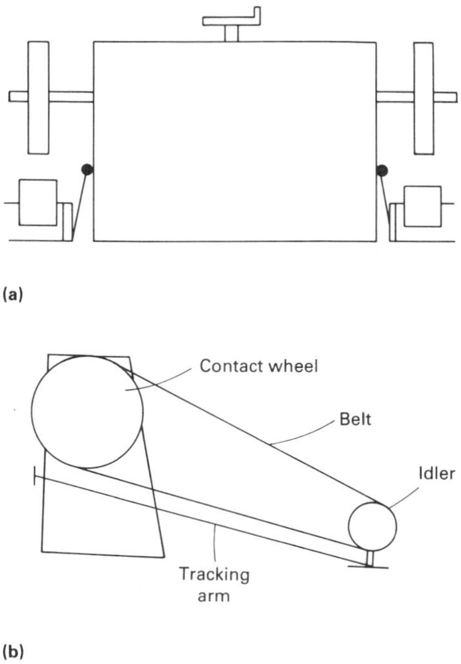


**Fig. 16** Front view (a) and side view (b) of a backstand grinder having coated abrasive belts for use in offhand rough grinding operations

cations demand special joint manufacture in order to eliminate marking of parts by the joints or to inhibit joint failure caused by heat or the severity of the form to which the belt must conform or flex. Belt sizes are specified by width and length.

**Abrasive disks** can be made from any of the backings. Fiber or heavy-cloth backed disks are usually mechanically held to the grinder backup pad by either a flanging arrangement or a built-in metal twist-on locking device. Cloth, paper, and film disks are normally glued to the backup pad, often by pressure-sensitive adhesive on the back of the disk. Disks are specified by diameter and hole size (or blank if no hole) and for special applications may be slotted or may have a special contour (for example, an orthogonal shape).

**Roll and sheet products** for metalworking are often used in hand finishing and deburring operations and can be made from any backing. Roll goods, specified as width (usually in inches or millimeters) by length (usually in yards or meters) are torn or cut to the desired length and are used by hand or in holders. Sheet goods, specified by width and length, are used similarly.

**Specialty products,** such as flap wheels and cartridge rolls, are used in a number of applications, most commonly on portable or bench-stand grinders. They are made from cloth products.

## Grinding Fluids

Grinding fluids are introduced into grinding operations to:

- Reduce and transfer heat during grinding
- Lubricate during chip formation
- Wash the grinding wheel or belt of loose chips and swarf
- Chemically aid the grinding action or machine maintenance

Grinding fluids or coolants are used almost exclusively in precision grinding applications in which metallurgical damage in ground parts must be minimized.

**The heat transfer characteristics** of water are well documented. The lubricity provided by straight oil is practically dictated in high-conformity operations. Water-soluble coolants provide varying degrees of compromise between these extremes.

**Lubricity** is more important in high-conformity operations with materials that generate relatively large chips, such as Inconel and the bearing steels. Lubricity is less important in low-conformity operations or with materials such as the carbides that tend almost to powder when ground.

**Chip Removal.** Chip size also dictates the nature of the swarf transfer characteristics of a coolant. Large chips may indicate that vigorous pressure is necessary to clean the wheel and that heavy coolant flow is necessary to carry the chips away. Small chips may mean that the coolant may need to have the ability to agglomerate chips to ensure that they can be removed from the system by filtering.

**Chemical Grinding Aids.** Some coolants are chemically defined to aid in the grinding process. For example, specific coolants for grinding titanium combine chemically with the titanium chips to inhibit the exothermic oxidation of the chips during grinding, resulting in a minimum of metallurgical damage.

### Types of Grinding Fluids

Grinding fluids can be classified into five categories:

- Petroleum-base and mineral-base cutting oils
- Water-soluble oils
- Synthetic fluids
- Semisynthetic fluids
- Water plus additives, including rust inhibitors, water treatments (for example, softeners), and defoaming agents

The advantages and disadvantages of each type of grinding fluid must be considered for each material machined, the specific grinding method used, and the speed and feed of the abrasive grinding wheel.

**Petroleum-base or mineral-base cutting oils** can be combined with polar additives and/or extreme-pressure (EP) additives, such as chlorine, sulfur, and phosphorus. These oils are used undiluted for very severe or specialized applications (for example, form grinding, gear grinding, and flute grinding). Danger from explosion or fires, poor cleanliness, and health con-

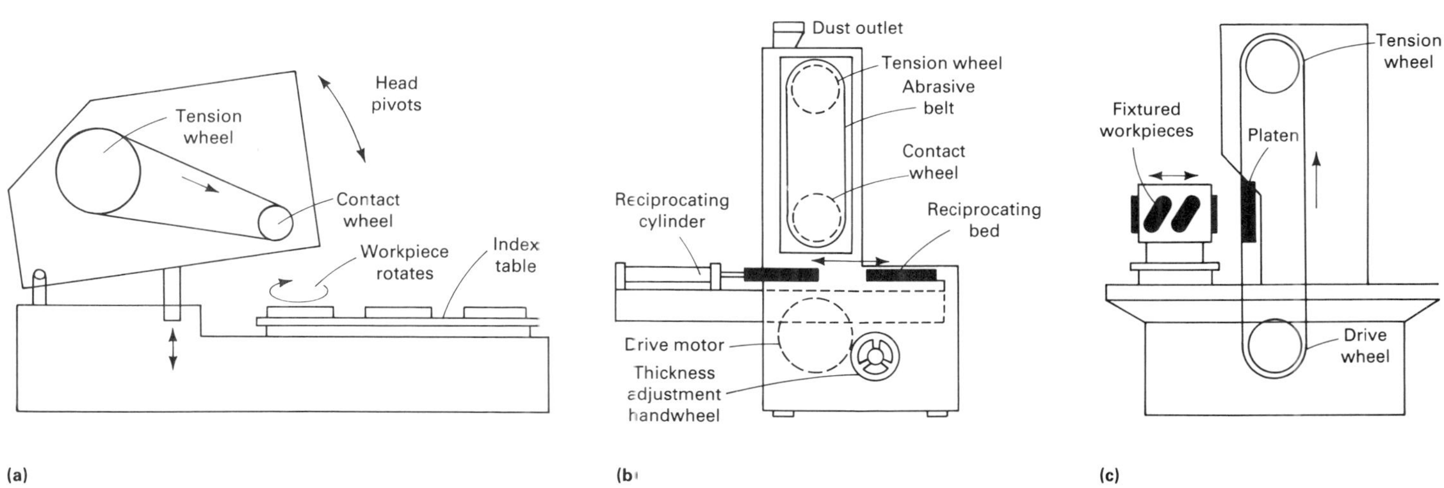


**Fig. 17** Schematics illustrating the primary components of rotary-table type (a), reciprocating (bed) table type (b), and vertical-platen type (c) (with hydraulic feed table) coated abrasive precision grinding machines

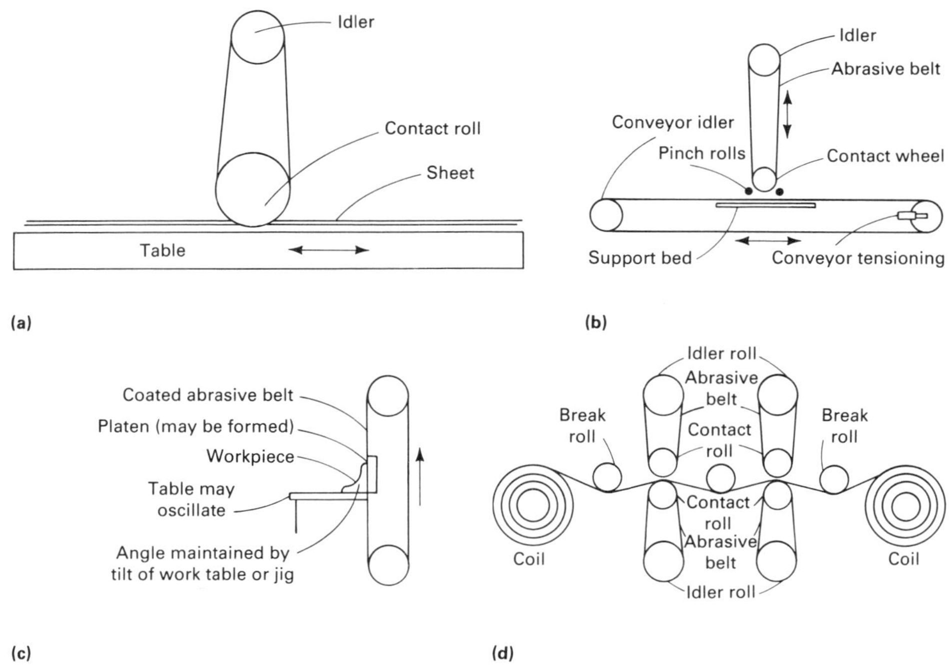


**Fig. 18** Schematics illustrating the primary components of several types of coated abrasive precision grinding operations. (a) Sheet dimensioning. (b) Conveyor grinding. (c) Vertical-platen sanding. (d) Coil grinding

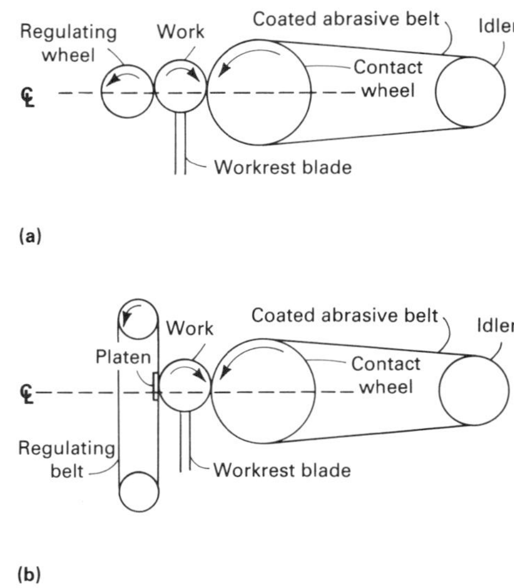


**Fig. 19** Schematic of two different coated abrasive centerless grinding setups. (a) Abrasive belt centerless grinder with regulating wheel. (b) Abrasive belt centerless grinder with regulating belt

cerns generally limit applications to those operations in which water-dilutable fluids do not provide enough physical (fluid film or boundary) lubricity.

**Water-soluble oils** consist of a suspension of oil droplets in water. The oil-base stock can also contain chlorine, sulfur, phosphorus, or fatty additives. These coolants incorporate additives that emulsify or disperse the product when the concentrate is added to water. These emulsions have very wide applications in all types of metal grinding.

The addition of polar additives and/or EP additives can produce emulsions of greater lubricating value. These solutions usually have a cloudy or hazy appearance. They provide a good combination of the cooling and lubricating properties required for higher-speed, higher heat producing grinding situations. Other advantages of water-soluble coolants over oils include:

- Higher heat removal and better cooling of the workpiece
- Improved operator acceptance due to improved cleanliness
- Reduced fire and health hazards

Water-soluble coolants can be used in most light-, moderate-, and heavy-duty grinding operations. Water-soluble coolants with EP additives are replacing many oils in severe grinding applications.

**Synthetic fluids** are water-dilutable coolants that contain no oil. These products are also called chemical solutions or ionic coolants. They usually contain synthetic lubricity and rust-preventive components and very effective preservative packages. The older synthetic technology was based on nitrited components, but most of these products have been removed from the market because of their potential for forming the possibly carcinogenic nitrosamines. Because they do not contain these materials, the newer synthetics offer a much safer alternative to oil-containing products. Other advantages of synthetic fluids include:

- Easy mixing because all product components are water soluble
- Light or imperceptible residues that are easy to clean using the coolant itself or a mild alkaline cleaner
- Easy coolant maintenance and concentration control, with very low additive requirements
- Excellent rancidity resistance and long system life

**Semisynthetic fluids** are basically mixtures of synthetic and soluble-oil product components. These products, sometimes referred to as microemulsions, include synthetic dispersions and some oil-accepting synthetics.

Semisynthetic coolants usually contain a low percentage of oil (generally 5 to 30%). They are normally used in grinding applications in which high heat removal and moderate lubricity are needed. Such operations range from general-purpose to heavy-duty classifications. Semisynthetic fluids are gaining acceptance because they incorporate the positive qualities of synthetic and soluble-oil coolants. The advantages of these products include:

- Better heat dissipation than the soluble oils
- Very good rust protection
- Very good rancidity resistance
- Very good acceptance by both the machine operators and maintenance personnel

Table 2 presents a generalized comparison of the characteristics of the grinding fluids and coolants.

**Fluid Cooling and Lubricity Properties of Oils and Synthetics.** Petroleum-base and mineral-base cutting oils and water-soluble oils provide excellent boundary and fluid-film lubrication, and with added EP components, they can provide good chemical lubricity. The chemical lubricant additives help to reduce the amount of heat generated in the grinding process. In contrast to the oil products, synthetic and semisynthetic fluids provide low boundary and fluid-film lubrication as well as very good chemical lubricity.

Because of the recent advances in machine technology and synthetic coolant technology, many grinding operations are being converted to accept water-base synthetic and semisynthetic coolants. In most conversions from one type of fluid to another, it is recommended that the grinding wheels be reevaluated for the new lubricant. Finer and/or softer grinding wheels may be optimal when switching from oil or water-soluble oil coolants to compensate for the more aggressive grinding action of synthetic and semisynthetic fluids. These water-base fluids act more aggressively because they do not have a high degree of fluid-film and boundary lubrication; therefore, they allow less slipping and plowing of the metal. Other considerations for grinding with a water-base fluid include the fluid-handling system and filter design as well as possible waste treatment or disposal requirements. Additional information on grinding fluids can be obtained in the article "Metal Cutting and Grinding Fluids" in this Volume.

## Disadvantages of Grinding Fluids

The disadvantages of using grinding fluids include:

- Loss of visibility
- Rust
- Flash point
- Foaming
- Bacterial growth
- Chemical disposal
- Effect of the coolant on the surrounding environment

Loss of visibility has become less important as machinery has become more automated and is often justified by removing from the environment the dust particles prevalent in dry grinding. Rust, foaming, and bacterial growth—problems normally associated with water-soluble coolants—can be controlled through additives and proper fluid maintenance.

Flash point and the effect of the coolant on the surrounding environment—problems normally associated with oil coolants—have been addressed by manufacturers. Synthetic coolants have been developed that approach the lubricity of oil with higher flash points and less tendency to pollute the local atmosphere. Machine manufacturers have also further enclosed grinding machines to address these problems.

The chemical disposal of used coolant has become a significant problem. Coolant manufacturers have designed and are implementing coolant systems for which the need to dump coolant is minimized.

Fluid maintenance is essential in any grinding fluid system. The removal of grinding swarf by filtering and other methods to ensure clean coolant at the grinding interface is necessary to reduce random scratching of parts and premature wheel wear or the need to dress the face of the wheel to clean it. Controlling the pH of water-soluble coolants (normally between 7.5 and 9.0) helps control dermatological problems by inhibiting bacterial growth, lessens rust problems, and, in the case of organic grinding wheels, may enhance wheel life because they are adversely affected by high-pH coolants. Controlling the concentration of water-soluble coolants to ensure lubricity and proper viscosity is essential to process control.

## Optimum Grinding Fluid Parameters

It is also important that coolant supply be at the grinding interface and at the proper volume, pressure, and temperature. Underwater grinding ensures maximum cooling, but is seldom feasible. Coolant nozzle design and placement is important to ensure maximum cooling and cleaning of the wheel. Pressurized nozzles are used to ensure wheel cleaning and may be used to introduce the coolants to the wheel face at a velocity similar to the wheel surface speed to aid coolant transfer to the grinding interface (Fig. 10). Refrigeration or other means of temperature control of the coolant as a process control ensures constant viscosity and heat transfer characteristics as well as temperature stability of the grinding machine and the accompanying assembly containing the fixture and workpiece.

**Fig. 20** Rotary diamond tools. Pictured are cups, thin rolls that traverse along the wheel face, and rolls that conform to the grinding wheel face.

## Grinding Machines and Processes

The grinding of metals can be divided, with some overlap, into rough and precision grinding. Precision grinding is defined as grinding to exacting tolerances and finishes. Rough grinding is defined as dimensioning or removing excess metal.

Some grinding operations are specific to the type of abrasive product used. Two examples are grinding wheels for cutoff and wide coated abrasive belts for sheet polishing. Other operations, such as centerless and weld grinding, may not be as clear cut as to what abrasive option should be chosen.

In rough grinding, in which high stock removal is important and parts tend not to be as sensitive to metallurgical damage, higher wheel speeds in the range of 45 to 63 m/s (9000 to 12 500 sfm) are often used. In precision grinding, the power required to run at higher speeds tends to make burn-free grinding more difficult, and minimizing metallurgical damage is more important; therefore, speeds of 25 to 43 m/s (5000 to 8500 sfm) are more prevalent.

Most decisions on abrasive usage are to some extent a process of elimination. For rough grinding, coated abrasive belts and disks and conventional abrasive organic wheels are used with few exceptions. Coated abrasive products are used only on materials softer than 50 HRC. In heavy-pressure applications with high horsepower requirements, grinding wheels are generally used. In light-pressure operations on soft materials, coated abrasive belts or disks are much more prevalent. For precision grinding, coated abrasive products tend to be used on cosmetic rather than tight-tolerance applications. Superabrasive grinding wheels are used almost exclusively on very hard (≥50 HRC) materials. In rough grinding, dry grinding is the accepted practice, with

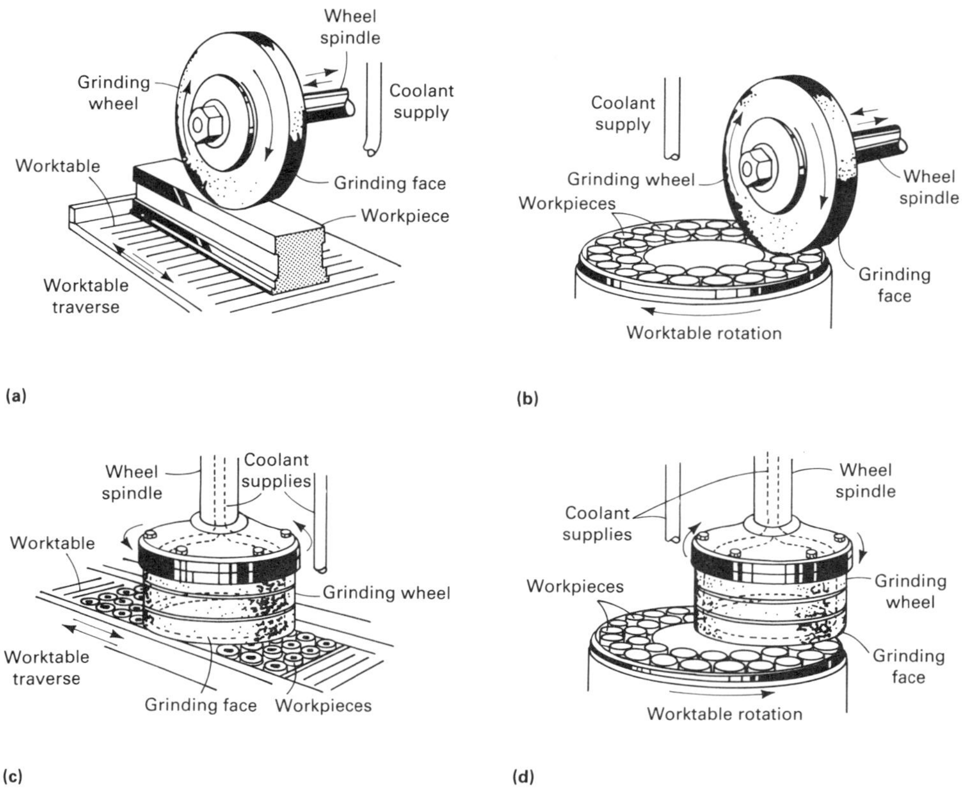


**Fig. 21** Surface grinding operations that utilize either the periphery or face of the grinding wheel to grind flat surfaces. Horizontal-spindle surface grinders, which use the periphery of the abrasive wheel, can be either reciprocating (a) or rotary (b). Vertical-spindle surface grinders, which use the face (side) of the abrasive wheel, can also be either reciprocating (c) or rotary (d).

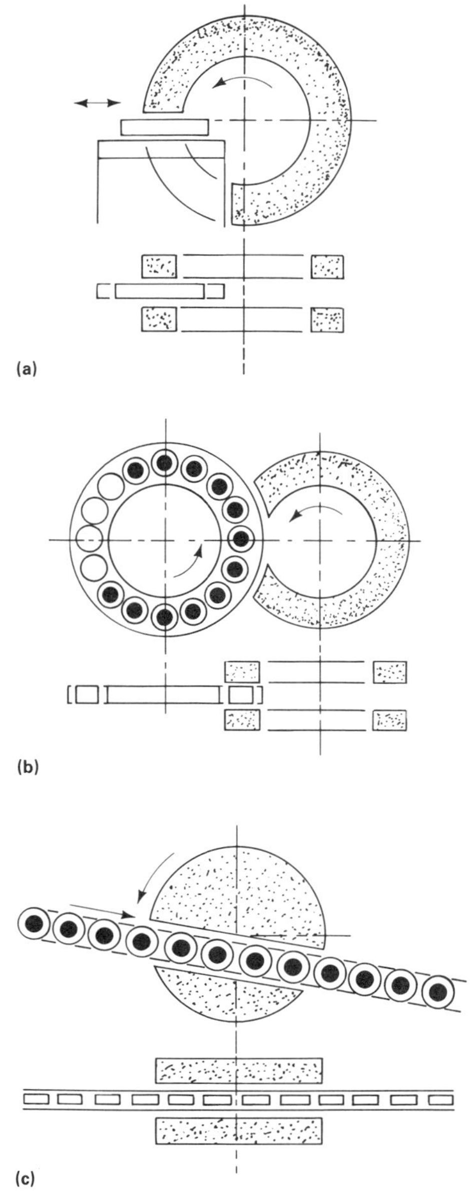


**Fig. 22** Use of the wheel face in double-disk grinders to machine the surface while the workpieces, sandwiched between two abrasive disks (top and side view of each setup is shown), traverse along a straight line or in an arcuate path. (a) Parts to be machined are pushed in and retracted by the drawer like movement of a feeding slide. (b) Parts to be ground move in an arcuate path while being transported in the nests of a rotating feed wheel. (c) Parts to be machined move diagonally while advancing along a rail.

abrasive product caused by the pressure and heat involved in the grinding process. Power consumption is excessively high for removing workpiece material, with little regard for dimensional tolerances or slight metallurgical damage.

## Rough Grinding

Rough grinding can be classified into four categories: cutoff, portable, floorstand/swingframe, and steel conditioning. Although cutoff is exclusively a grinding wheel operation, coated abrasives are often used in portable grinding, floorstand (typically referred to as offhand) and swingframe grinding, and steel conditioning (that is, the preparation of steel billets by removing surface impurities prior to rolling or forming).

**Cutoff.** Most cutoff operations are done on ferrous materials using aluminum oxide or zirconia-alumina resin bond reinforced or nonreinforced type 1 wheels. Reinforced wheels are used to withstand the incidental side pressures caused by part movement during the grind. Typical grinding wheel specifications would be 20 to 36 grit in grades P through U. Figure 11 shows a typical cutoff application, while Fig. 12 shows a portable cutoff machine.

**Portable grinding** operations (listed from heavy to light stock removal) include snagging, weld grinding (Fig. 13), shaping, blending (Fig. 14), and polishing. Although snagging tends to be done with zirconia-alumina or aluminum oxide resin bond grinding wheels and polishing with aluminum oxide or zirconia-alumina-coated abrasive disks and specialties, there is a good deal of overlap based on material configuration and composition, total stock removal, and the need for cosmetic finishes.

**Floorstand** (Fig. 15) or offhand (Fig. 16) and swingframe operations tend to split upon the subsequent use of the ground part. If the part is subsequently polished (such as a hand tool, plumbing fixture, or turbine blade), it is usually offhand ground with a coated abrasive belt. If the part is subsequently machined, ground, and so on, it is usually ground on a floorstand machine with a grinding wheel.

**Steel conditioning** is done on high-speed, high-horsepower grinders typically using zirconia-alumina reinforced resin bond wheels (8 to 12 grit, W to Z grade) or on specialized coated abrasive belt machines.

## Precision Grinding

Precision grinding can be divided between coated abrasive and grinding wheel operations, although there are areas of overlap. Coated abrasive precision grinding is generally on softer (50 HRC and softer and more commonly 45 HRC and softer) materials at light grinding pressures and where part tolerances are measured only to the closest 0.05 mm (0.002 in.) or looser.

**Coated Abrasive Precision Grinding.** A number of coated abrasive precision grinding operations are shown in Fig. 17 to 19. Abrasive machining, coil grinding, conveyor grinding, platen grinding, and sheet and plate dimensioning and polishing can be considered surface grinding operations (as discussed below), while centerless, cylindrical, and roll grinding have grinding wheel equivalents. Coated abrasive belts are very

**Table 3 Specifications for several typical surface grinders**

| Type (table-spindle) | Size mm | Size in. | Main motor kW | Main motor hp | Weight Mg | Weight lb |
|---|---|---|---|---|---|---|
| Reciprocating horizontal | 205 × 635 | 8 × 25 | 4.1 | 5.5 | 2.2 | 4 840 |
| Reciprocating horizontal | 610 × 3050 | 24 × 120 | 150 | 200 | 18.6 | 41 000 |
| Fixed, horizontal | 150 × 915 | 6 × 36 | 4.3 | 5.8 | ... | ... |
| Rotary horizontal | 610 | 24 | 15 | 20 | 5.5 | 12 200 |
| Rotary vertical | 915 | 36 | 26 | 35 | 6.4 | 14 000 |

Source: Ref 1

sharp compared to grinding wheels; they maintain their cut rate through being replaced when dull, and as a result they work well in applications where light grinding pressures are dictated by part configuration (such as the centerless grinding of thin-wall tubing or sheet polishing). Coated abrasive products are also commonly used on soft nonferrous materials, which tend to load and clog grinding wheels.

**Precision Grinding With Grinding Wheels.** This precision grinding system consists of the grinding machine (and its inherent variables of rigidity, horsepower, ability to deliver and process coolant, throughput rates, ability to true and dress, and so on), the grinding wheel, the coolant, and the truing tool. This section will discuss the machines; because grinding wheels and coolants have been covered previously in this article, a discussion of truing and dressing is appropriate.

In rough grinding, heat and pressure are used to maintain a somewhat steady-state grinding condition, but the tolerances and need for metallurgical integrity in precision grinding dictates a tighter control on the profile and sharpness of the wheel face. A dull wheel face generates a smoother finish, but can introduce unacceptable vibration and heat into the system. A sharp wheel face may remove material more quickly, but will not maintain tight finish and dimensional requirements in the finished part. Precision grinding wheels are commonly trued and dressed to regenerate or correct grinding performance.

Truing is defined as regenerating the dimensional integrity of the wheel, both in terms of roundness and profile. Dressing is defined as the regeneration of the desired cutting characteristics of the wheel. Truing and dressing are commonly combined into one operation for conventional abrasive grinding wheels, but are often two separate operations for superabrasive wheels.

*Truing* a wheel is important for minimizing system vibration caused by lack of concentricity and for ensuring that part geometries are maintained. Superabrasive grinding wheels trued to within 0.0013 mm (0.000050 in.) concentricity when first mounted can outlast (by a factor of five) similar wheels not carefully trued.

*Dressing* a wheel prepares the wheel face for the grind. Vigorous dressing, which causes the abrasive particles to be exposed more in the bond matrix, tends to create sharp, free-cutting wheels that may wear more quickly. Light, slow dressing that is relatively dull will give better surface finishes and longer wheel life.

*Regeneration of Wheel Geometry.* With conventional abrasive wheels, regenerating the wheel geometry is relatively easy with most truing tools; therefore, the wheels tend to be trued under conditions that generate the desired dressed face. A sharp, dressed face is called an open face; a dulled face is called a closed face. When truing superabrasive wheels (which is typically

**Fig. 23** Thread rolling dies (foreground) produced by surface grinding methods using the abrasive grinding wheel shown in background

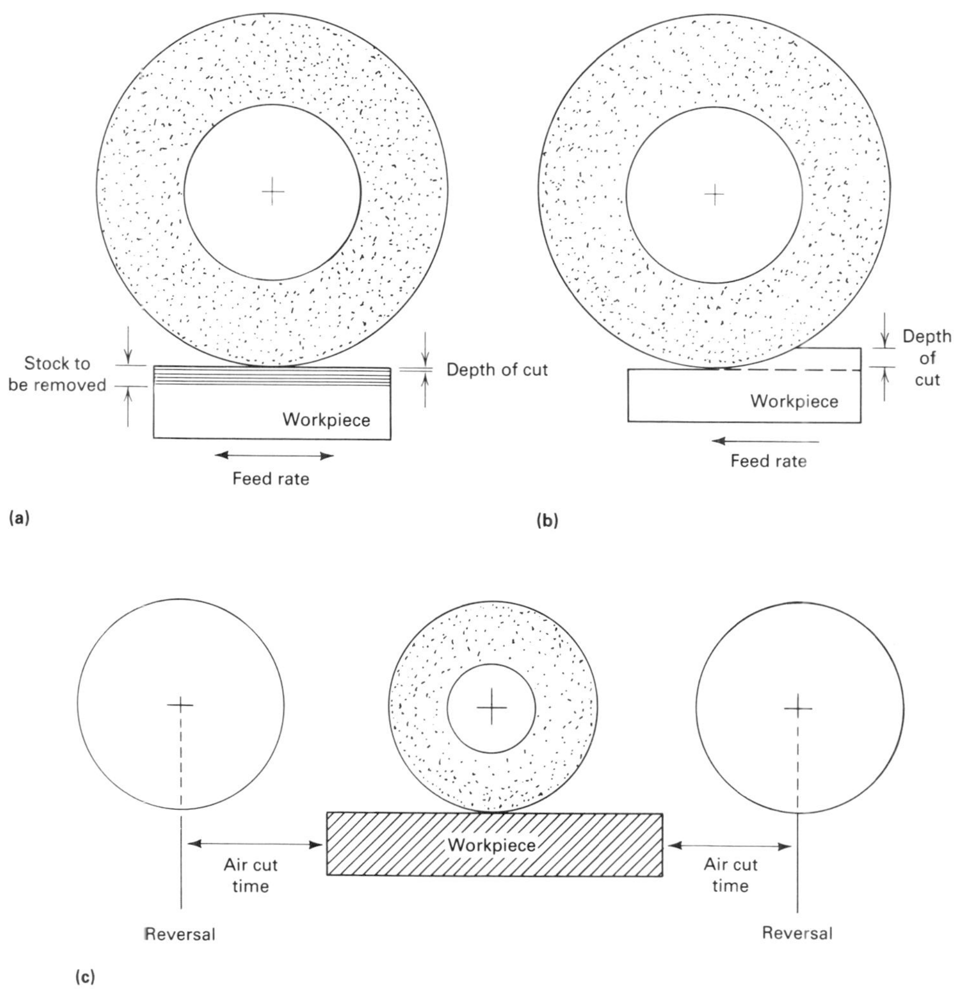



**Fig. 24** Pendulum surface grinding (a) compared to creep-feed surface grinding (b), resulting in decreased air-cutting time (c). Air-cutting time in horizontal-spindle reciprocating-table grinding, with its many light passes over the workpiece, is eliminated in creep-feed grinding because the full depth of cut is accomplished in a single pass.

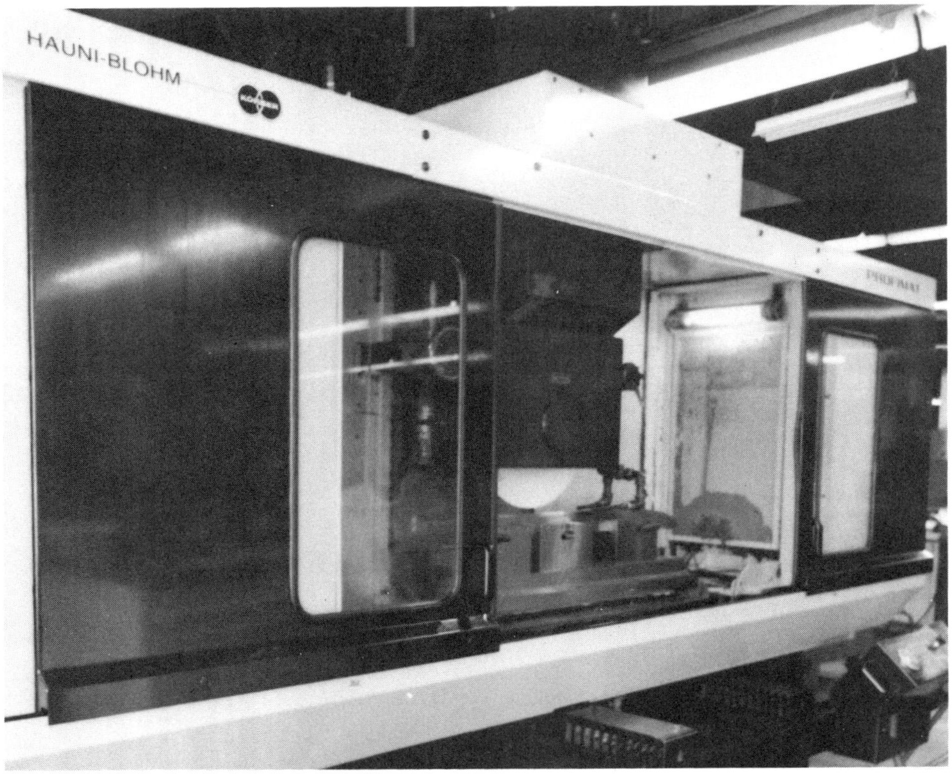


**Fig. 25** Creep-feed grinders with ball-screw drives to ensure rigidity

done in production grinding with diamond tools), the truing conditions tend to be set to minimize the damage to either the truing tool or the wheel because they are similar in hardness. The resulting light-pressure/infeed conditions close up the faces of superabrasive wheels, often to the point of optical reflectivity.

Dressing is required for creating chip clearance and porosity in the trued wheel. The dressing of superabrasive wheels is commonly done with soft conventional abrasive vitrified sticks, which relieve the bond without disturbing the superabrasive particles.

There are four major types of truing tools:

- Steel cutters
- Conventional abrasive vitrified or boron carbide sticks or wheels
- Steel or carbide crush rolls
- Diamond tools

Steel cutters are used to roughly true coarse-grit conventional abrasive wheels to ensure freeness of cut after truing. Vitrified and boron carbide sticks are used for the offhand truing of conventional abrasive wheels. Vitrified truing wheels, such as those used on a brake-controlled truing device, are normally used for truing resin bond superabrasive wheels. Crush rolls are used to crush true intricate forms into vitrified wheels, leaving a sharp, open-face wheel.

Diamond tools are available as single-point tools, cluster or bar tools, multipoint nibs, dressing blocks, and rotary tools. Superabrasive wheels for production grinding are usually trued with rotary diamond tools. Diamond tools can be natural, synthetic, or polycrystalline diamond in a metal matrix.

Single-point diamond tools consist of a single diamond in a holder. These are the oldest of the diamond tools and have been replaced by more efficient tools in many applications. Single-point tools are still used where imparting a truing pattern (such as a spiral on a lapping wheel) is beneficial or where constantly changing profiles make diamond rolls too costly.

Cluster or bar tools are used on straight or mildly shaped wheel faces, particularly on wide wheels. Multipoint nibs are used for truing straight faces into wheels. Diamond dressing blocks are used almost exclusively for horizontal-spindle reciprocating-table surface grinders to generate specific forms.

Rotary diamond tools are available as cups, thin rolls that traverse along the wheel face, and rolls that conform to the face of the grinding wheel (Fig. 20). There are three types of manufacturing methods for rotary diamond tools. In the first method, sintered random set tools have the diamond set randomly on the preform and then held in by a sintered powder metal bond. In the second method, the diamond pattern is meticulously set by hand and similarly affixed. Lastly, reverse-plated tools are made by generating a negative of the preform, setting the diamonds on the negative, then plating the diamonds onto the positive preform. Hand set rolls are used for mild forms, random set rolls for more intricate forms and where accuracy is of prime importance, and reverse-plated rolls

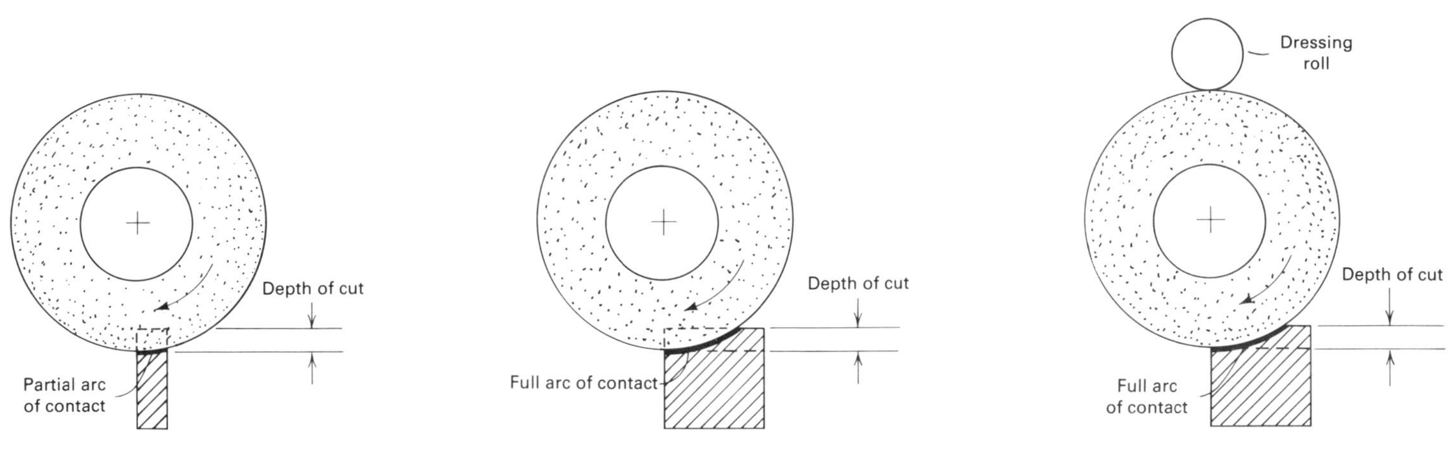


**Fig. 26** Schematics illustrating three types of creep-feed grinding. (a) Pseudo creep feed. (b) True creep feed. (c) Continuous-dress creep feed. See text for discussion.

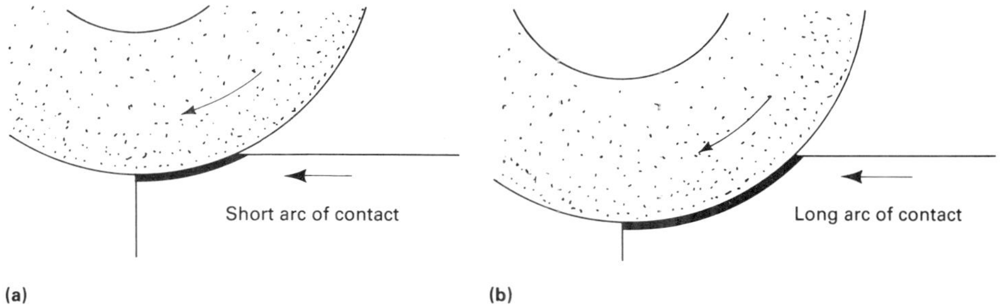


**Fig. 27** Comparison of contact arc lengths in conventional surface grinding (a) and creep-feed surface grinding (b)

for high-production, intricate form truing operations (for example, continuous-dress creep-feed grinding).

With the exception of dressing blocks and most diamond rolls, diamond tools are traversed along the wheel face to generate the wheel profile (see the article "Thread Grinding" in this Volume). The distance the diamond tool travels per revolution of the grinding wheel is known as the dress lead. The effect of dress lead is further discussed in the article "Principles of Grinding" in this Volume.

## Surface Grinding

The subsegments of surface grinding are defined by wheel type and fixture or table type. Common conventional abrasive wheel types are straight wheels, straight cup wheels, cylinder wheels, segments, and disk and lapping wheels. The parts can be held on rotary tables, reciprocating tables, conveyors (Fig. 21), or, in the case of disk and lapping wheels, between opposing wheels (Fig. 22). Table 3 lists typical specifications for several types of surface grinders. Surface grinding is by definition a high-conformity grinding operation requiring relatively free cutting, open wheels to remove material.

**Peripheral Surface Grinding.** Horizontal-spindle reciprocating- or rotary-table machines typically use straight or recessed wheels. These machines can be used to grind either flat surfaces or intricate forms (Fig. 23).

*Creep-Feed Grinding.* A significant subset of horizontal-spindle reciprocating-table grinding is creep-feed grinding. In conventional reciprocating-table grinding, many light passes are taken at rapid traverse rates (typically 0.025 to 0.05 mm, or 0.001 to 0.002 in., deep at 1270 to 2540 mm/min, or 50 to 100 in./min); in creep-feed grinding, full depth of cut is ground at slow traverse rates (typically 203 to 381 mm/min, or 8 to 15 in./min). By reducing the time spent not grinding and by eliminating the multiple impacts of the wheel on the ends of the part, productivity is increased while wheel geometry is maintained longer (Fig. 24). Creep-feed machines are specially designed for rigidity, typically using ball screws rather than hydraulic table drives to ensure control of slow, heavy cuts (Fig. 25).

There are three subtypes of creep-feed grinding (Fig. 26):

- Pseudo creep feed, in which the full depth of cut is taken from a part that is thin in the traverse direction, thus lowering the length of the contact arc
- Conventional creep feed with a very long arc of contact (Fig. 27) and intermittent truing and dressing (between parts)
- Continuous-dress creep-feed grinding, in which the diamond roll truing tool is in constant contact with the grinding wheel during the grind, infeeding typically 0.00051 to 0.0013 mm/rev (0.000020 to 0.000050 in./rev) of the wheel

Continuous-dress creep-feed grinding is the ultimate in grinding from a dressed face because the steady state is dictated by the truing tool. In continuous-dress grinding, constant machine compensation through numerical control ensures control of the grinding process (Fig. 28).

A British aerospace manufacturer increased productivity in the manufacture of turbine blades by replacing an automated seven-cell creep-feed grinding line with two creep-feed grinding centers. With a maximum of six cells in use at any one time, one cell is left free in case of a breakdown. The cells were arranged in pairs, one on each side of a central conveyor, and each cell contained two creep-feed grinding machines. Parts were transferred between the conveyor and the grinding machines by a robot in each cell. To facilitate handling and location, each turbine blade was encapsulated by casting a zinc alloy around its airfoil section. Automatic cleaning and inspection are also included. Previously, blades were moved in batches from one grinding machine to another for successive operations. With this setup, an individual blade progressed through the grinding line from one operation to another. Typically, 12 creep-feed grinding operations were performed on a single turbine blade. With the seven-cell line, each of these grinding operations was performed on a separate grinding machine.

A more recent development is that of the creep-feed grinding center, a single machine, designed by a German machine tool manufacturer, on which a number of these operations can be performed. Only two of these grinding centers were required in order to perform all 12 operations, the breakdown being seven on one machine, and five on the other. Understandably, the grinding center is much more complex than a conventional creep-feed grinder, as can be seen from the basic construction shown in Fig. 29. The machine has five grinding wheels, three on the upper spindle and two on the lower, with automatic wheel changing. There are three diamond rollers to dress these wheels, and there is the capability for continuous dressing. The power of each of the two spindle-drive motors is 35 kW (47 hp). There are ten roller-guided numerical control axes with resolutions of 1 μm (40 μin.) and 0.001°.

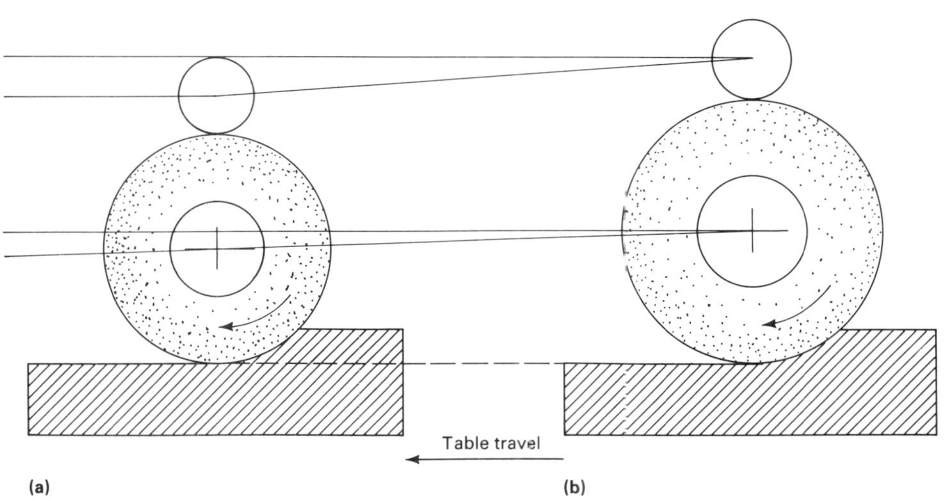


**Fig. 28** Continuous-dress creep-feed grinding system compensation. Compensation for grinding wheel wear through loss of diameter is accomplished by both grinding wheel and dressing wheel feeding toward workpiece (a) from initial positions (b) to maintain parallel cut on workpiece.

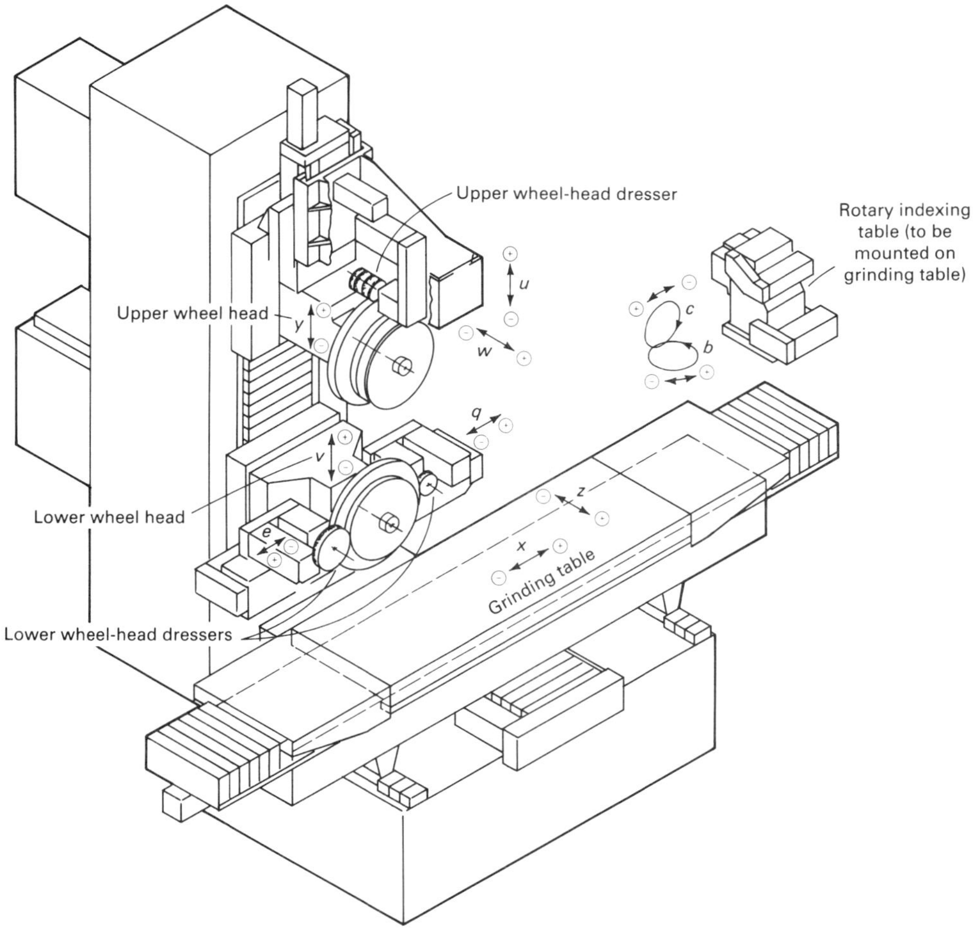


**Fig. 29** Schematic showing the primary components (and their range of motion) of a creep-feed grinding machine center, two of which in tandem can perform all 12 grinding operations previously requiring an automated seven-cell transfer line

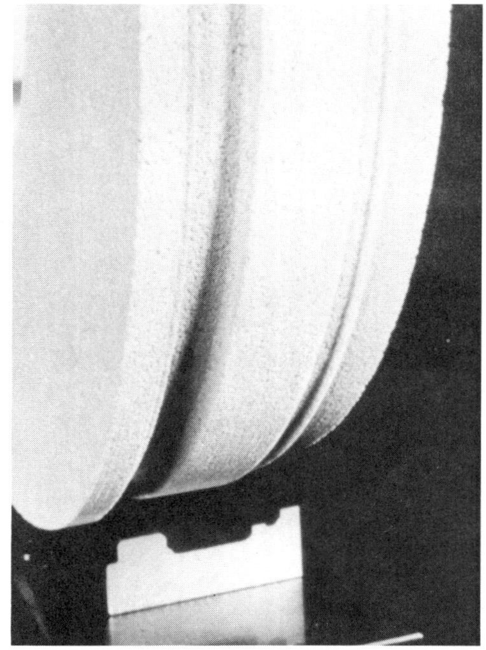

**Fig. 30** Applying creep-feed grinding methods to a form grinding operation requiring large stock removal

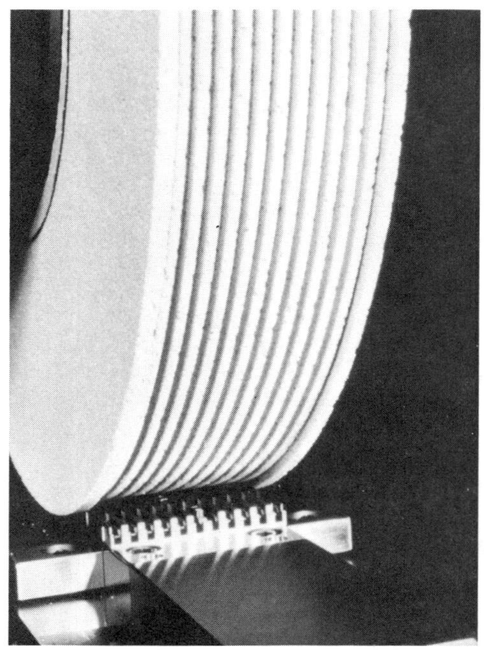

**Fig. 31** Creep-feed grinding methods used to form grind a workpiece

Creep-feed grinding is most effective in large stock removal form grinding operations (Fig. 30 to 32), in which the impact of the form on the ends of the part during reciprocation degrade the form holding capability of the wheel. Often, by grinding a form to full depth, greater part production rates are achieved by creep-feed grinding.

Conventional abrasive creep-feed grinding wheels are typically very high structure or porosity vitrified wheels because of the long arc of contact and the need for coolant transfer and chip clearance. Superabrasive creep-feed wheels vary in composition, but electroplated CBN wheels are widely used for the form grinding of aerospace alloys because they are inherently sharp, open, and require no truing.

**Wheel Face Grinding.** Vertical-spindle rotary-table, reciprocating-table, or conveyor machines use cup and cylinder wheels and segments to grind flat surfaces. Because of the virtually flat-to-flat contact of wheel to work, these machines require free-cutting, open wheels. Segments are used to ensure additional chip clearance and coolant transfer.

*Disk grinding wheels* and their fine-grit counterparts, lapping wheels, are defined as plate-mounted or inserted-nut wheels that are bolted onto the machine back plate (Fig. 22). Opposed-wheel disk grinders and conveyor machines are the exceptions to grinding as a batch process. Because of the even larger flat-to-flat contact that may exist in disk grinding compared to normal wheel face grinding, perforations or slots can be manufactured into the wheel for coolant transfer or chip clearance. For fine-grit lapping wheels, which are typically used in high-lubricity coolants, chip clearance is often dressed into a conventional abrasive (often silicon carbide for sharpness) vitrified wheel with a large single-point diamond tool, which dresses in a spiral pattern in the wheel.

## Cylindrical Grinding

Cylindrical grinding (Fig. 33) is defined as the grinding of round parts between centers. Cylindrical grinding is performed either by traversing the surface to be ground or by plunge grinding. Figure 34 shows eight workholding methods and devices utilized in cylindrical grinding.

Roll grinding is a specific case of cylindrical grinding in which machines with swing diameters in excess of 508 mm (20 in.) are used to grind large workpieces (up to 1.8 m, or 6 ft, diameter and 9 m, or 30 ft, long). Other subsets of cylindrical grinding include crankshaft grinding, thread and slot grinding (see the article "Thread Grinding" in this Volume), camshaft grinding (which, because of the nature of the cam shape, emulates both cylindrical and surface grinding at different parts of the lobe), and angular-approach grinding.

**Crankshaft or crank-pin grinders** resemble cylindrical center-type grinders, but

**Fig. 32** Pseudo creep-feed grinding application. In true creep-feed operations, properly applied coolant would obscure the entire grinding interface.

**Fig. 33** Cylindrical grinding of a bar between two lathe centers. Coolant flow was reduced to aid photo focusing.

are implemented to grind the offset pins in the throws of crankshafts (Fig. 35).

**Cam and camshaft grinders** are essentially modifications of center-type cylindrical grinders to finish various forms of round cams, camshafts, and pistons. The headstock and footstock may be on a cradle and rock to and from the grinding wheel in response to a master cam that rotates in unison with the workpiece. Newer machines use numerical control to generate cam shapes.

A **universal cylindrical center-type grinder** has all the units and movements of a plain grinder (which incorporates dead centers and rotates the workpiece using a dog or driver on the faceplate), but in addition:

- Its headstock spindle can be used alive or dead (rotated or not) so that work can be held and revolved by a chuck as well as between centers
- Its headstock can be swiveled in a horizontal plane so that any angle, even a flat plane, can be ground on a workpiece chucked on the headstock spindle
- Its wheel head and slide can be swiveled and traversed at any angle in the manner indicated in Fig. 36 so that any taper can be ground on work between centers

Also, most universal grinders can be arranged for internal grinding by the addition of an auxiliary wheel head to revolve small wheels at high speeds.

Universal grinders can grind surfaces such as steep tapers and holes not accommodated on plain grinders, but they sacrifice rigidity, power, and rapid output because of their flexibility. They are found in toolrooms and jobbing shops and sometimes on production jobs for shapes that are difficult to grind on plane grinders. Table 4 lists several types of cylindrical grinding machines, their uses, and machine specifications.

*Angular-approach grinding* can be carried out on universal-type cylindrical grinding machines, which permit swiveling the wheel head and operating the infeed in the direction of that swivel setting (Fig. 37). In low-volume production on universal-type grinding machines, angular-approach grinding can also be combined with hand-actuated table traverse, thus increasing the length of the cylindrical work section ground in the operation without being limited by the width of the grinding wheel.

For production-type grinding, special angular, plain grinding machines are available, commonly built with wheel-head angles at 30°. These machines are designed specifically for plunge grinding, and the table movement by hand wheel only serves the purpose of machine setup. By offering the use of relatively wide wheels, the production-type angular grinding machines permit the machining of rather long cylindrical

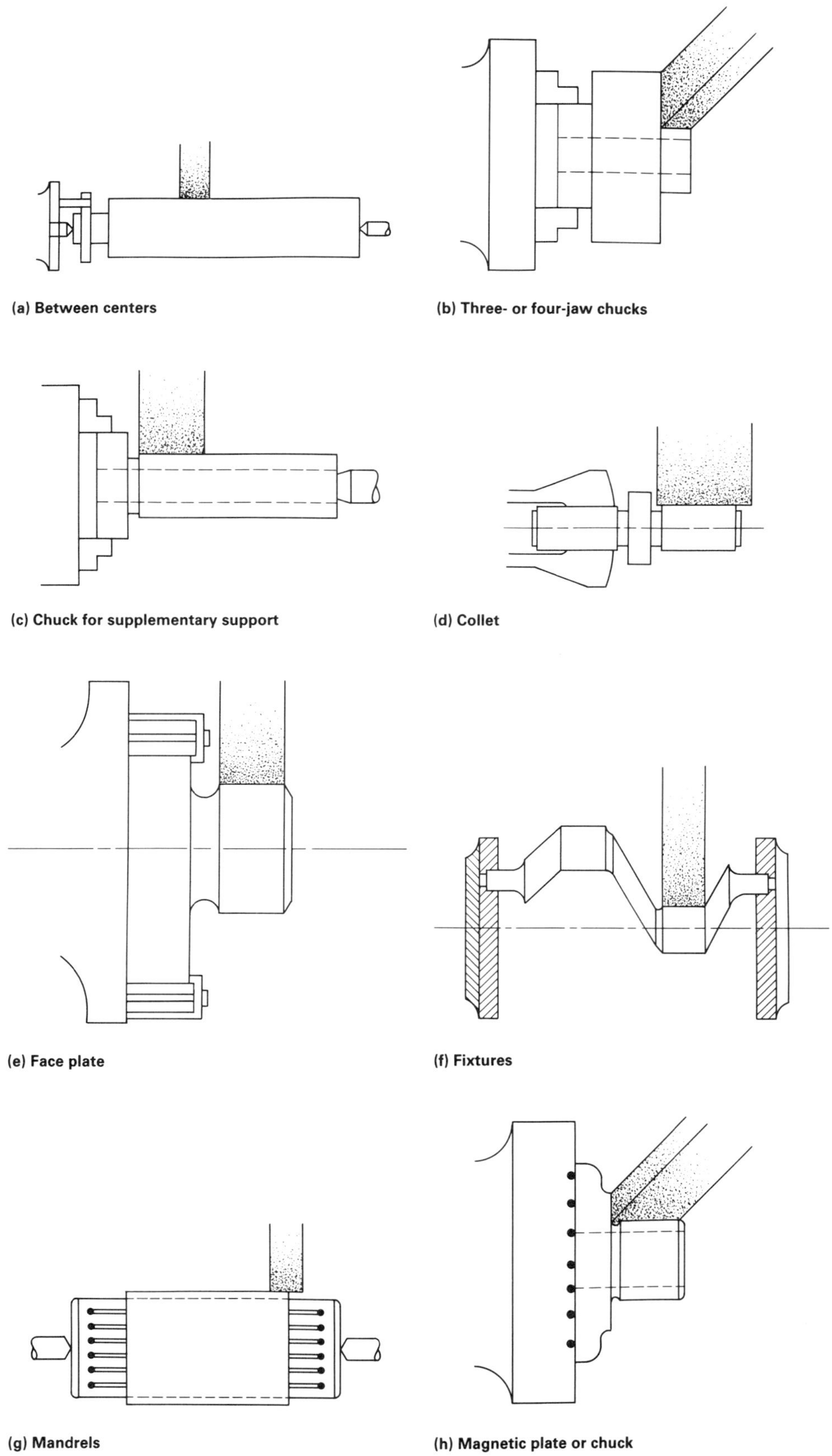


**Fig. 34** Workholding methods and devices utilized in cylindrical grinding. Chuck for supplementary support (c) includes tailstock center such as bull nose type for bores, back centers for parts susceptible to deflection, and center rest to leave the other end free for face or external grinding. Fixtures (f) are for parts of unusual configuration or those having nonidentical axes for the part or the section being ground, such as crankpin sections of crankshafts. Mandrels (g) are for parts with through bores. Magnetic plates or chucks (h) are usually used for short-length parts.

sections together with the adjoining flanges and, when needed, the fillet areas.

Aside from the process advantages resulting from the geometric conditions associated with an angular-wheel approach, certain beneficial effects on the quality of the grinding should also be considered.

First, the mutual perpendicularity of the ground surfaces is inherent in the shape of the trued wheel faces. This perpendicularity is not dependent on the positional variables to which operations requiring more than a single loading of the work are exposed.

Second, the wheel contact on the flat shoulder surface is limited to a narrow area, a condition often referred to as line contact. A narrow wheel contact on a flat work surface generally permits a more effective stock removal, without damaging the work surface, than does grinding with the face of the wheel.

Lastly, the contact of the wheel rotating around an axis inclined to both work surfaces produces an arcuate surface pattern combined with some wiping action. This produces a better surface finish value and a surface lay that can have a beneficial effect on lubricant retention in the service of the ground part.

Cylindrical grinding parameters include relative work-to-wheel-speed ratios, truing tool and method, infeed and/or traverse rate, horsepower, and the output variables (for example, taper, finish, and so on) required. The arc of contact in cylindrical grinding depends on the relative diameter of the wheel to the work, but is typically very small, almost a line contact. As a result, unit grinding pressures tend to be higher, requiring more durable wheels. Conventional abrasive cylindrical wheels are typically three to four grades harder than similarly sized surface grinding wheels because the bond needs to hold the grain more securely. Similarly, higher-concentration superabrasive wheels can be used because they are more durable.

## Centerless Grinding

Centerless grinding (Fig. 38) is the grinding of the outside diameters of round parts but not between centers. In centerless grinding, the part is supported by the work support blade, rotated by means of a regulating or feed wheel, and ground by the grinding wheel.

The advantages of centerless grinding over cylindrical include the grinding of long, thin parts without having to use steady rests for support, lessened taper problems, and higher productivity (Fig. 39). The disadvantages of centerless grinding are that it does not grind concentric with centers, short larger-diameter parts are difficult to control in the process, and it may not improve part perpendicularity.

The types of centerless grinding include throughfeed or traverse grinding, infeed

**Fig. 35** Cylindrical grinding of a crankshaft. Coolant flow was turned off to improve photo clarity and focus.

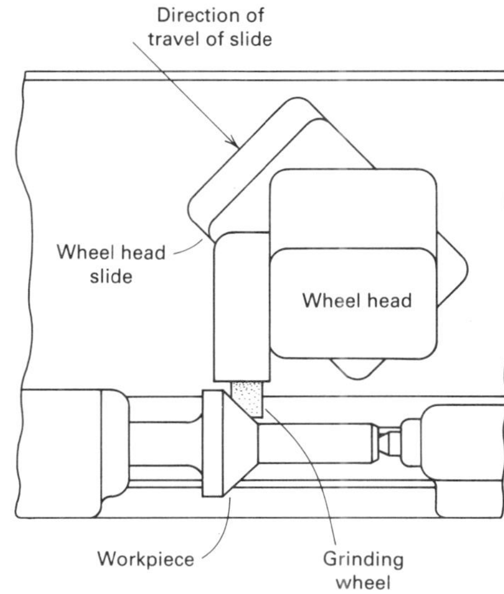


**Fig. 36** Universal center-type grinder equipped for grinding a steep taper

grinding (which is traverse grinding to a stop and then ejecting, endfeed grinding (used only on workpieces of tapered shape), or a combination of infeed and throughfeed grinding (Fig. 40). Plunge cuts can often be made over the entire length of the work-

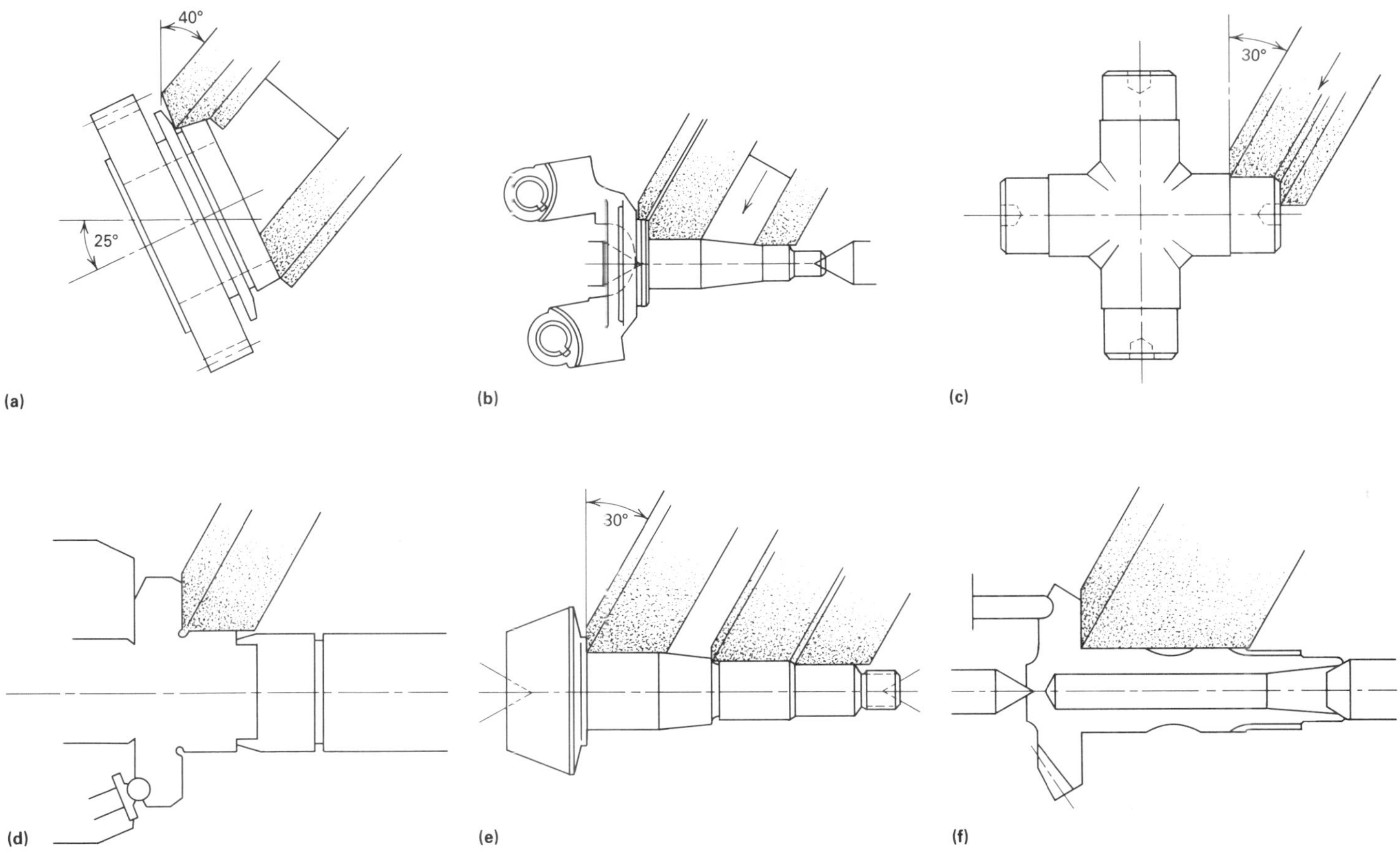


**Fig. 37** Application of angular-approach cylindrical grinding in automotive component production. (a) Gear held in a headstock swiveled at 25° to grind a taper section and the gear face. (b) Steering knuckle held between centers to grind three cylindrical sections, a shoulder, and a connecting fillet. (c) Universal joint spider ground on the shoulder, a cylindrical section, chamfer, and face. (d) Differential bevel gear secured in a quick-acting expanding collet to grind two adjacent and mutually perpendicular surfaces. (e) Gear shaft ground on two cylindrical sections, two shoulders, a chamfer, and a fillet. (f) Gear with integral shank ground on two cylindrical sections and the shoulder

**Table 4 Categories, applications, and specifications of cylindrical grinding machines**

| Designation | Applications | Typical capacity data and characteristic dimensions: Work swing, mm | Work swing, in. | Distance between centers, mm | Distance between centers, in. | Grinding wheel diameter × width, mm | Grinding wheel diameter × width, in. | Wheel drive, kW | Wheel drive, hp | Net weight, kg | Net weight, lb |
|---|---|---|---|---|---|---|---|---|---|---|---|
| Instrument-type cylindrical grinding machine | Very small cylindrical or tapered parts for instruments, and so on, requiring very close tolerances; flexibility of adaptation for diverse configurations | 100 | 4 | 125 | 5 | 125 × 13 | 5 × ½ | ⅛ | ⅙ | 540 | 1200 |
| Precision-type production grinding machine | Small and medium-size parts ground with high accuracy to be manufactured in large or intermediate lots | 100–125 | 4–5 | 255–455 | 10–18 | 255 × 25 | 10 × 1 | 0.75 | 1 | 725 | 1600 |
| Universal tool grinding machine | Tools and fixture elements often utilizing adjustment flexibility and optional accessories (for example, internal grinding spindles) | 255–355 | 10–14 | 510–1220 | 20–48 | 255–355<br>13–50 | 10–14<br>½–2 | 1.5–3.7 | 1.1–5 | 2300–4500 | 5000–10 000 |
| Plain cylindrical grinding machine | Mostly cylindrical or slenderly tapered regular features on parts with defined axes of rotation | 255–455 | 10–18 | 455–1830 | 18–72 | 610–915<br>100–205 | 24–36<br>4–8 | 5.6–19 | 7½–25 | 3200–8200 | 7000–18 000 |
| Heavy-duty plain cylindrical grinding machine | Work of large dimensions and heavy weight, often ground with very high stock removal rates | 455–610 | 18–24 | 1220–6100 | 48–240 | 760<br>205–305 | 30<br>8–12 | 19–30 | 25–40 | 13 500–27 000 | 30 000–60 000 |
| Plunge-type cylindrical production grinders | For grinding cylindrical or profiled round sections on generally short parts, also those having several coaxial features in different planes and diameters | 255–760 | 10–30 | 610–2030 | 24–80 | 355–610<br>50–150 | 14–24<br>2–6 | 7–22 | 10–30 | 6800–27 000 | 15 000–60 000 |
| Roll grinding machines | For large-size rolls such as are used in rolling mills; also for calender rolls in the paper and cloth industries | 915–1525 | 36–60 | 3050–6600 | 120–260 | 915<br>100 | 36<br>4 | 30–60 | 40–80 | 68 000–91 000 | 150 000–200 000 |
| Limited-purpose cylindrical grinding machines | Typical examples are crankshaft (pin) grinders and camshaft grinders; characteristic dimensions are based on these types. | 205–610 | 8–24 | 610–1830 | 24–72 | 510–915<br>50–100 | 20–36<br>2–4 | 19–30 | 25–40 | 11 500–18 000 | 25 000–40 000 |

Source: Ref 2

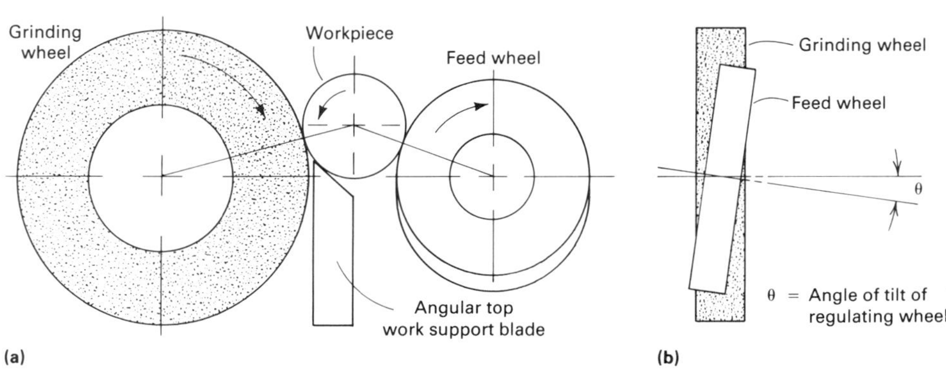


**Fig. 38** Basic elements of the centerless grinding operation showing the workpiece located between the grinding wheel and the feed wheel (a) and the inclination angle (θ) of the regulating or feed wheel (b)

piece. Grinding wheel selection for centerless grinding is similar to that for cylindrical grinding except that wide wheels may be a grade softer in conventional abrasive wheels and may be of lower concentrations in superabrasive wheels.

Regulating or feed wheels drive the part rotation and provide a braking action so that the grinding wheel does not drive the part (at its typically 60 times faster surface speed). Feed wheels need to conform to part geometry and have traditionally been aluminum oxide rubber bond wheels for most applications. Table 5 lists specifications for several types of centerless grinding machines grouped according to work size capacity.

## Internal Grinding

Internal grinding (Fig. 41) is grinding the inside diameter or bore of a part. Typically, conventional abrasive vitrified wheels or superabrasive wheels in all bonds are used to traverse or plunge grind the bore diameters.

Internal grinding machines are of two types: chucking and centerless.

**Chucking-type internal grinders** hold the workpiece on a faceplate or in a chuck or fixture and rotate it around the axis of the

**Fig. 39** High-production plunge centerless grinding operation

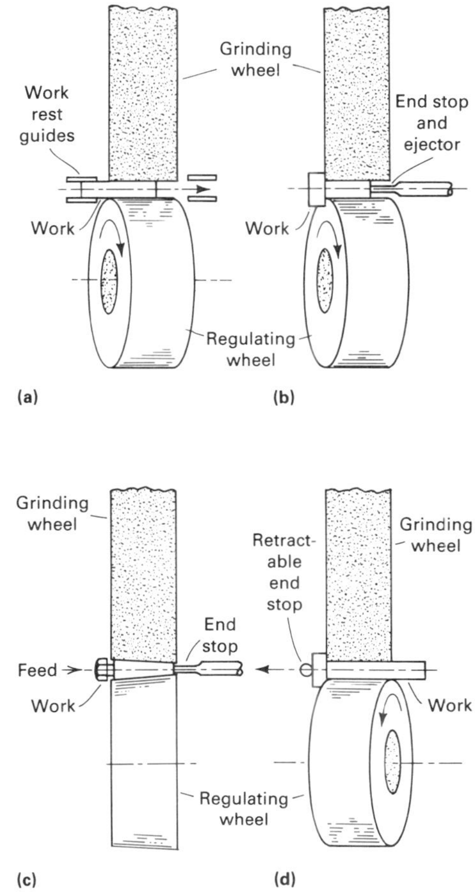


**Fig. 40** Four basic types of centerless grinding operations. (a) Throughfeed. (b) Infeed. (c) Endfeed. (d) Combination infeed/throughfeed. An inclined regulating wheel (always used for throughfeed as well as endfeed grinding and in some cases for infeed grinding) imparts a light axial force to the workpiece in all four cases.

hole ground, as shown in Fig. 42(a). The revolving grinding wheel is reciprocated lengthwise through the hole and is fed crosswise on a slide to engage the workpiece. The work head, and sometimes the wheel head, can be swiveled to adjust for straight and tapered holes and faces.

*A planetary internal grinder* is designed for parts that are too large or unwieldy to be rotated conveniently. The workpiece is not revolved. Instead, the grinding wheel is orbited around the axis of the hole being ground (Fig. 42b).

**Centerless internal grinders** grind the bore of a round workpiece concentric with the outside surface. Straight, tapered, continuous, interrupted, open-end, and blind holes and grooves are ground on centerless internal grinders. The wall thickness of a finished piece is quite uniform. The manner of holding the work lends itself well to automatic unloading and loading of pieces.

*Roll-Type Internal Centerless Grinders.* The workpiece is held and revolved amid three rolls in a roll-type centerless internal grinder, as shown in Fig. 43(a). The large roll is the driver; the rolls provide rigid support for the workpiece.

*Shoe-Type Internal Centerless Grinders.* On the shoe-type centerless grinder, the workpiece is revolved against two fixed and hard shoes by the action of the grinding wheel and a rotating end backing plate (Fig. 43b). The shoes contact more area than rolls and bridge irregularities and produce better average concentricity between the hole and the outside.

**Special Internal Grinding Operations.** Subsets of internal grinding include internal bearing-race grinding, internal thread grinding, and jig grinding (most commonly of dies). Internal grinding with conventional wheels can be done with wheel types 1 and 5 (typically for grinding blind holes) or with type 54 mounted wheels.

Conventional abrasive internal wheels tend to be slightly softer than cylindrical wheels because of the higher conformity of internal grinding. In superabrasive products, high concentrations are used to ensure maximum wheel life.

As in cylindrical grinding, the machine parameters for internal grinding include horsepower, truing tool/method, wheel-to-work rotational speed, infeed and/or traverse rates, and the output variables. Table 6 lists specifications and applications for the basic types of internal grinding machines.

## Tool Grinding

Tool grinding is divided into two subgroups: production tool manufacturing and tool resharpening.

**Production tool manufacturing** operations include fluting, thread grinding, clearing, pointing, gashing, squaring, point splitting, insert grinding, and formed wheel cutoff. These operations tend to be specific to tool manufacturers and will not be discussed here.

**Fig. 41** Internal grinding of a workpiece

**Table 5 Typical dimensions of major categories of centerless grinding machines**

Grouped according to worksize capacity. The listed values express orders of magnitude only and do not represent the dimensional data of particular models.

| Machine category | Nominal infeed work capacity, mm | Nominal infeed work capacity, in. | Through-feed capacity for solid bars, mm | Through-feed capacity for solid bars, in. | Grinding wheel diameter × width (max), mm | Grinding wheel diameter × width (max), in. | Regulating wheel diameter × width (max), mm | Regulating wheel diameter × width (max), in. | Speed range of regulating wheel, rev/min | Grinding wheel drive power, kw | Grinding wheel drive power, hp | Approximate machine weight, kg | Approximate machine weight, lb |
|---|---|---|---|---|---|---|---|---|---|---|---|---|---|
| Light, for small workpieces | 38 | 1½ | 25 | 1 | 305 × 75 | 12 × 3 | 150 × 75 | 6 × 3 | 30–480 | 3.7 | 5 | 1000 | 2200 |
| General purpose | 75 | 3 | 38 | 1½ | 510 × 205 | 20 × 8 | 305 × 205 | 12 × 8 | 15–300 | 11 | 15 | 3500 | 7700 |
| Production type | 150 | 6 | 100 | 4 | 510 × 305 | 20 × 12 | 355 × 305 | 14 × 12 | 10–300 | 19 | 25 | 7000 | 15 400 |
| Heavy duty | 305 | 12 | 205 | 8 | 610 × 305 | 24 × 12 | 405 × 305 | 16 × 12 | 0–120 | 30–75 | 40–100 | 9100–12 000 | 20 000–26 000 |
| Limited-purpose wide wheel | 255 | 10 | ... | ... | 510 × 610 | 20 × 24 | 355 × 610 | 14 × 24 | 0–150 | 55 | 75 | 10 000 | 22 000 |

Source: Ref 2

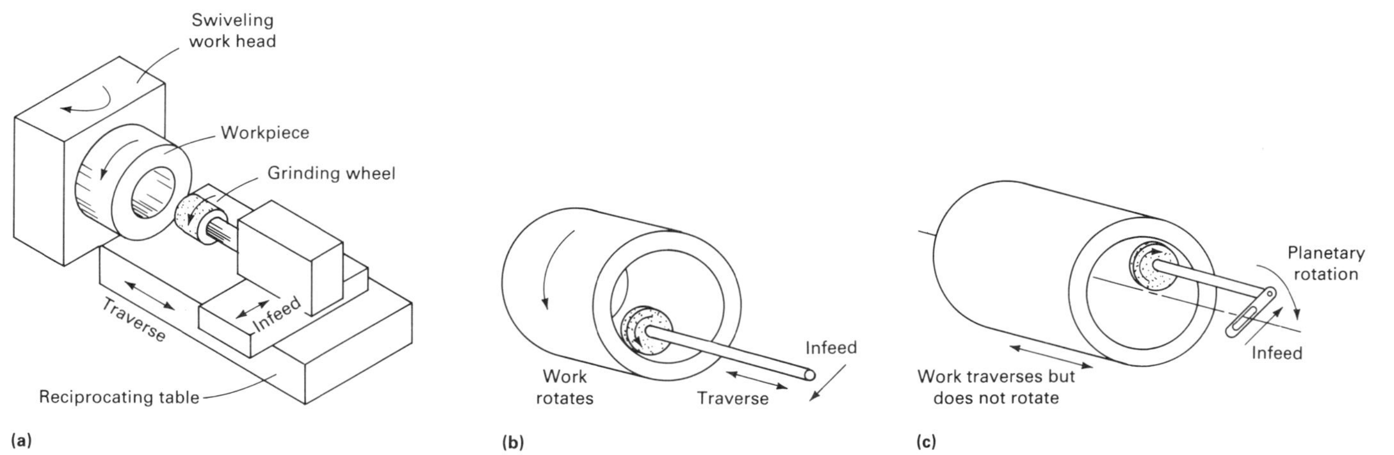


**Fig. 42** Schematic illustrating key components and their movement in chucking-type internal grinding machines (a). Two basic categories of chucking-type internal grinders are plain internal grinders (b) and planetary-type internal grinders (c).

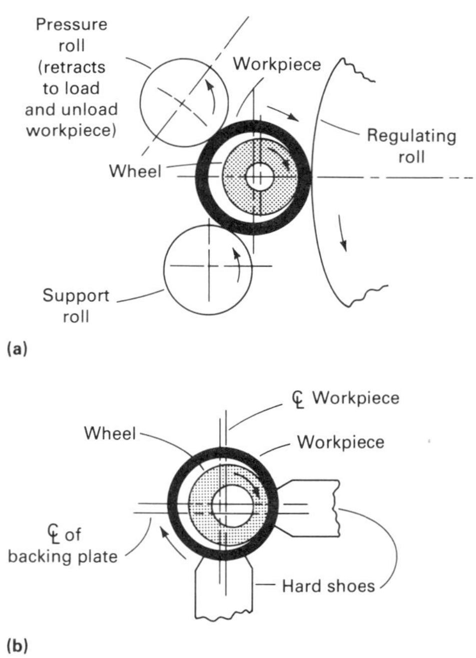


**Fig. 43** Two types of internal centerless grinding machines. (a) Roll-type centerless. (b) Shoe-type centerless

**Tool resharpening** includes hob grinding, drill grinding, tap grinding, saw sharpening (Fig. 44), grinding other multitooth rotary tools (for example, end mills and reamers), and grinding single-point tools. Hob, tap, and drill grinding are typically done on specialized equipment. High-speed tool steel parts are ground with CBN resin or electroplated bond products or aluminum oxide vitrified wheels. Carbide parts are typically ground with diamond resin or electroplated wheels. When grinding other multitooth rotary cutters (which is typically done on universal tool-and-cutter grinders), CBN resin or aluminum oxide vitrified wheels are used on high-speed tool steel, and diamond resin or silicon carbide vitrified wheels are used on carbide.

**Fig. 44** Application of tool grinding techniques to sharpen a rotary saw

Single-point tool grinding, generally done offhand, utilizes diamond vitrified or electroplated wheels or silicon carbide vitrified wheels on carbide or CBN electroplated or aluminum oxide vitrified wheels on high-speed tool steel. Among the special grinding machines used in tool applications are profile grinders and monoset cutter and tool grinders.

*Profile (contour) grinders* can reproduce a template form on a flat or round cutter. Some can grind metal surfaces to conform to outlines drawn on paper. An optical profile grinder is one on which a view of the zone of contact between wheel and workpiece is highly magnified (10, 20, or 50 times, for example) and cast on a screen. The form produced by the wheel can be seen as it is ground and can be compared to a large-scale drawing on the screen of the form desired. Machine movements can be simultaneously controlled by the operator.

*Monoset Grinders*. A type of tool grinder that is adaptable to a large variety of tools, but particularly to spiral tools, is the monoset cutter and tool grinder. The work head spindle can be indexed and synchronized with the table movement to grind

**Table 6 Basic types, applications, and specifications of internal grinding machines**

| | | Range of capacities of different models within the category | | | | | | | |
|---|---|---|---|---|---|---|---|---|---|
| | | Maximum work diameter | | Maximum hole diameter | | Maximum hole length | | Machine weight | |
| Category | Applications | mm | in. | mm | in. | mm | in. | Mg | lb |
| Miniature part internal grinder | Specially designed and equipped internal grinders for very small and miniature parts, with hole diameters starting at about 1 mm (0.04 in.). Some models are built as universal machines with versatile adaptations. Miniature part internal grinders are also built as centerless grinding machines, specially developed for very small ball bearing rings. | 100 | 4 | 15.2 | 0.600 | 25 | 1 | 0.55 | 1 200 |
| Small work-type internal grinder | Designed for grinding primarily limited size lots of small parts within the capacity range, with emphasis on easy setup; operated either entirely manually or in semiautomatic cycle with hand loading of the work | 150–255 | 6–10 | 50–100 | 2–4 | 50–100 | 2–4 | 1.8–2.7 | 4 000–6 000 |
| Medium-size internal grinder | Medium-size general-purpose internal grinding machines are available either in the plain or the more versatile toolroom type with different bed lengths; some are also designed to grind long bores. | 405–610 | 16–24 | 255–355 | 10–14 | 255–355 | 10–14 | 3.6–4.1 | 8 000–9 000 |
| Universal internal grinding machine | Offer the capacity and versatility needed in general toolroom work and in fixture and machine building. At the same time, these machines must satisfy severe accuracy requirements and are occasionally used for grinding external round and flat surfaces. | 610–760 | 24–30 | 455–610 | 18–24 | 305–455 | 12–18 | 2.7–4.1 | 6 000–9 000 |
| Production-type internal grinding machine | Fully automatic operation is one of the essential criteria of machines classified as production types; that requirement also acts as a limiting factor regarding maximum workpiece size and configuration. Many of the models in this category operate as centerless internal grinders with roll or shoe support. | 100–305 | 4–12 | 50–75 | 2–3 | 50–205 | 2–8 | 4.5–8.2 | 10 000–18 000 |
| Large internal grinding machine | Large internal grinding machines have the common characteristic of accepting heavy workpieces and are capable of grinding holes comparable to the outside diameter dimension of that work. A wide range of differences exist, however, between various models with respect to the maximum work length, resulting from the overall design length of the machines. | 815–915 | 32–36 | 510–610 | 20–24 | 510–1270 | 20–50 | 9.1–18 | 20 000–40 000 |
| Extralarge internal grinding machine | Machines referred to in this category are often custom designed for accommodating specific groups of very large and heavy workpieces. Such components usually have very deep bores requiring extended bridge-type machine construction and extralong table travels. | 2030 | 80 | 610 | 24 | 1525–1780 | 60–70 | 54 | 120 000 |
| Planetary-type internal grinding machine | Planetary internal grinders are used for parts that are too large to be rotated and/or the hole(s) to be ground are far out-of-center. For the acceptable workpiece size, the mounting surface of the worktable, the center height of the wheel spindle, and the cross-slide adjustment are the main controlling factors. | Not a characteristic dimension | | 75–355 | 3–14 | 380–760 | 15–30 | 4.5–14 | 10 000–30 000 |

Source: Ref 2

helices. The major units are fully adjustable to enable much work to be done in one setting that would otherwise require several setups.

## REFERENCES

1. L.E. Doyle, C.A. Keyser, J.L. Leach, G.F. Schrader, and M.B. Singer, *Manufacturing Processes and Materials for Engineers*, Prentice-Hall, 1985, p 722
2. F.T. Farago, *Abrasive Methods Engineering*, Vol 1, Industrial Press, 1976

## SELECTED REFERENCES

- C. Bhateja and R. Lindsay, Ed., *Grinding: Theory, Techniques and Troubleshooting*, Society of Manufacturing Engineers, 1982
- *Coated Abrasives—Modern Tool of Industry*, 1st ed., Coated Abrasive Manufacturer's Institute, 1965
- F.T. Farago, *Abrasive Methods Engineering*, Industrial Press, 1976
- R.I. King and R.S. Hahn, *Handbook of Modern Grinding Technology*, Chapman & Hall, 1986
- S.F. Kvar and J.W. Oswald, *Grinding Technology*, Delmar Publishers, 1974
- R.K. Springborn, Ed., *Cutting and Grinding Fluids: Selection and Application*, American Society of Tool and Manufacturing Engineers, 1967

# Superabrasives

K. Subramanian, Norton Company

SUPERABRASIVES collectively refer to the diamond and cubic boron nitride abrasives used in grinding applications. Diamond is the hardest material known, and cubic boron nitride is the second hardest. Because of their high hardness, abrasion resistance, and other unique properties, these materials find extensive use in a wide variety of abrasive (see the articles "Lapping" and "Honing" in this Section) or cutting applications ranging from the rough cutting of stone and concrete to the ultraprecision machining of electronic ceramics. Additional applications for diamond and cubic boron nitride are discussed in the article "Ultrahard Tool Materials" in this Volume.

Figure 1 shows the hardness of abrasives typically used in grinding operations along with a few work materials. Aluminum oxide and silicon carbide abrasives are about one-half the hardness of cubic boron nitride and about one-third the hardness of diamond. The hardness values shown in Fig. 1 are typical and do not represent the full range of hardnesses obtainable for each material. For example, hardness values as high as 11 000 HK have been reported for natural diamond. Table 1 compares selected properties of superabrasives with aluminum oxide and silicon carbide abrasives.

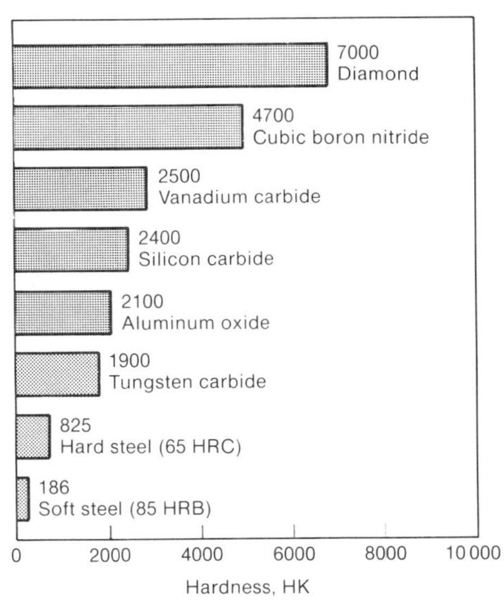


**Fig. 1** Typical hardness of abrasive materials and some work materials

**Table 1 Typical properties of abrasives**

| | Abrasive | | | |
|---|---|---|---|---|
| | Superabrasive | | Conventional | |
| | Diamond | Cubic boron nitride | Aluminum oxide | Silicon carbide |
| Chemical composition | Carbon | BN | $Al_2O_3$ | SiC |
| Density, g/cm$^3$ | 3.52 | 3.48 | 3.92 | 3.21 |
| Knoop hardness, HK (GPa) | 60–110 | 40–70 | 21 | 24 |
| Relative thermal conductivity | 100–350 | 35–120 | 1 | 10 |
| Coefficient of thermal expansion, $\times 10^{-6}$ mm/mm/°C | 4.8 | 5.6 | 8.6 | 4.5 |
| Threshold temperature for degradation (ambient conditions), °C (°F) | 800 (1470) | 1400 (2550) | 1750 (3180) | 1500 (2730) |

**Diamond** is an allotropic form of carbon having a cubic crystal structure (Fig. 2). Diamond has been used for engraving tools and for polishing long before the 20th century. An early industrial use of natural diamond was in truing tools for shaping conventional abrasives such as aluminum oxide. The synthesis of diamond on a laboratory scale and on a reproducible basis was demonstrated by the General Electric Company in the 1950s. The invention of the synthetic diamond caused the rapid growth

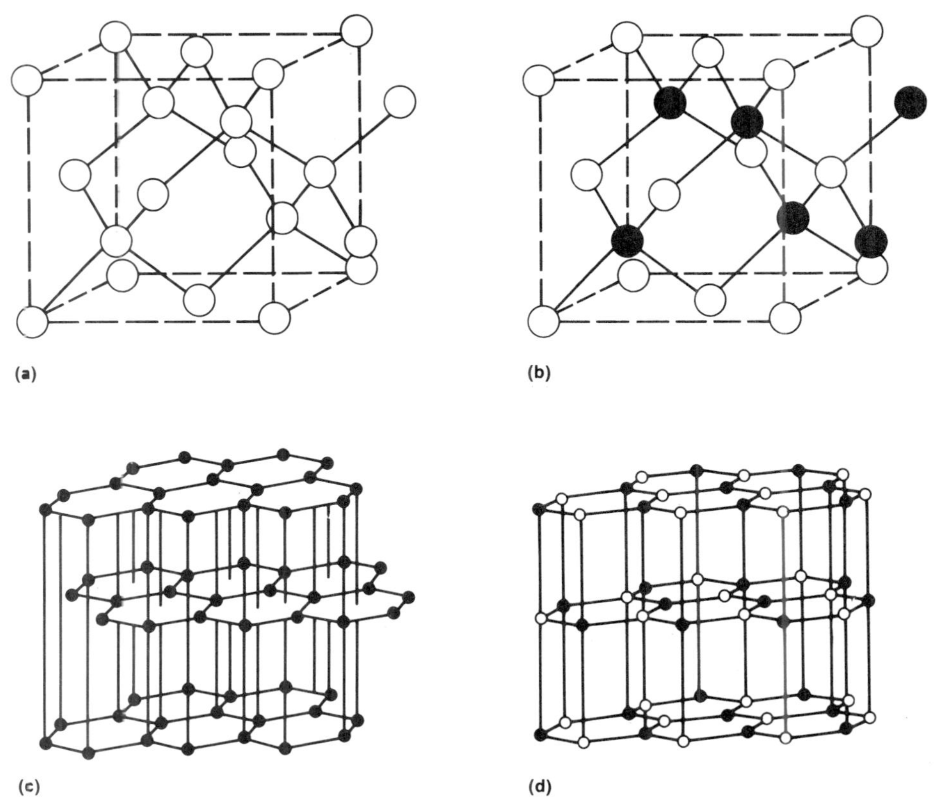


**Fig. 2** Crystal structure of diamond (a) and CBN (b) and their allotropes, graphite (c) and hexagonal boron nitride (d)

(a) (b) (c)

**Fig. 3** Typical shapes of diamond abrasive grains. (a) Strong and blocky. (b) Intermediate strength. (c) Weak and friable

of diamond in a wide variety of abrasive applications. In addition to its high hardness, diamond is an excellent heat conductor and has an extremely low coefficient of friction. However, at temperatures above 800 °C (1475 °F), diamond tends to graphitize (Fig. 2c), that is, convert back to disordered carbon. Thus, it loses its value as a wear-resistant abrasive. Diamond suffers rapid wear and chemical dissolution (erosion) when abraded against iron. Therefore, it is not used as an abrasive against ferrous material except in special machining operations such as honing.

The shape, size, and distribution of the diamond particles or abrasive grains are strictly controlled to achieve desirable performance characteristics. Figures 3(a), (b), and (c) show strong and blocky diamond grains, intermediate-strength grains, and weak and friable grains, respectively. The impact of the abrasive grain also changes with the particle size or mesh size, as shown in Fig. 4. Very small quantities (parts per million) of such impurities such as nitrogen, iron, and boron in diamond contribute to the color, strength, and other property variations in diamond abrasives.

**Cubic boron nitride (CBN)** is the second member of the superabrasives family. It is also cubic in structure (Fig. 2b). In its hexagonal close-packed structure, hexagonal boron nitride is similar to graphite and is used as a solid lubricant (Fig. 2d). Borazon, the General Electric trade name for CBN, was first synthesized in 1959. Unlike diamond, CBN is not very reactive with iron and therefore is highly wear resistant in grinding ferrous materials. Cubic boron nitride is effectively used in the precision grinding of a wide range of ferrous alloys and high-nickel alloys. The application principles of CBN are in most cases subtle modifications of the application principles of such conventional abrasives as aluminum oxide. The industry eventually began to appreciate these differences, and the effective use of CBN abrasives has been growing steadily in the late 1970s and throughout the 1980s.

Cubic boron nitride is not a naturally occurring substance; it is formed by synthetic processes. As with diamond grains, the shape, size, and characteristics of the CBN grains are modified to achieve variations in strength and performance (Fig. 5). Cubic boron nitride is much more stable than diamond at elevated temperatures and in oxidizing atmospheres. Figure 6 shows typical variations in CBN grit morphology and their effects on fracture resistance or friability.

## Superabrasive Applications

Table 2 lists the wide range of applications for diamond and CBN abrasives. The applications for diamond are much more extensive than those for CBN.

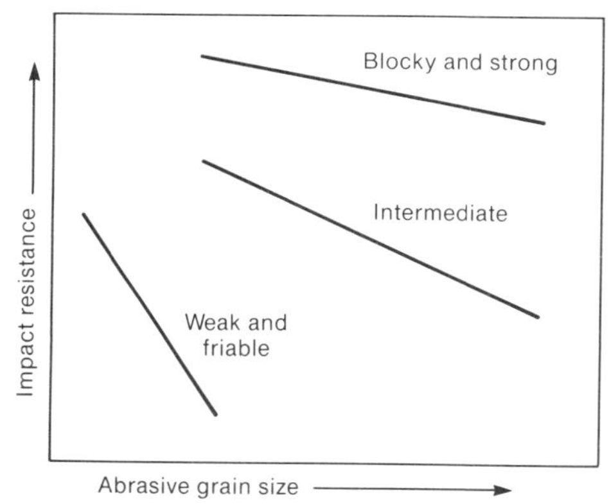


**Fig. 4** Plot of impact resistance against abrasive grain size showing relative toughness of synthetic diamond grains

**Diamond abrasives** are used in a wide variety of construction applications. These applications include the cutting and dimensioning of stones, quarry operations, polishing and shaping of stone and marbles, and cutting of concrete to repair buildings, highways, and airport runways. Diamond abrasives are also used as mining bits in rock drilling and oil-well drilling operations.

The use of diamond abrasives in cutoff operations is the basis for a variety of

(a) (b) (c)

**Fig. 5** Typical shapes of CBN abrasive grains. (a) Microcrystalline and strong. (b) Intermediate. (c) Weak and friable

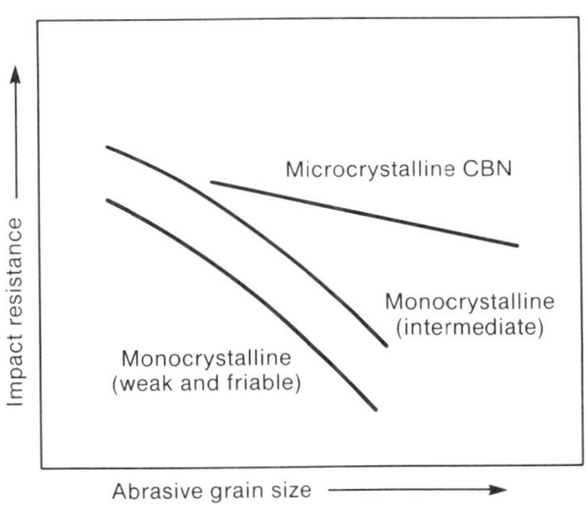


**Fig. 6** Plot of impact resistance against abrasive grain size showing relative toughness of CBN grains

**Table 2 Typical applications of superabrasives**

| Superabrasive | Work material | End use/operation |
|---|---|---|
| **Diamond** | | |
| Construction products | Stone | Quarrying, cutting, polishing |
| | Concrete (for construction and repair) | Buildings, airports, highways |
| Mining products | Stone, shale, rock | Drilling |
| Precision grinding wheels | Carbide | Tool production and resharpening; wear parts and others |
| | Glass | Optical, furniture, automotive, crystal glass |
| | Industrial ceramics | Electrical parts and insulators, magnets, kiln furniture |
| | Plastics and composites | Fiber-reinforced plastics, friction materials, optical lenses |
| | Electronic ceramics | Integrated circuit chip fabrication, magnetic heads, sensors, optical fibers |
| | Engineering ceramics | Internal combustion engine parts, wear-resistant parts and bearings, bioceramics |
| Precision grinding hones | Ferrous alloys | Automotive parts |
| Powders, slurries, and compounds | Structural ceramics, ferrous alloys, electronic ceramics | Wear-resistant parts, metallurgical samples, magnetic heads |
| **Cubic boron nitride** | | |
| Precision production grinding wheels | Alloy steels | Automotive parts, gears, pump parts, appliance parts |
| | Bearing steels | Bearing components |
| | High-nickel alloys | Aerospace parts |
| | Tool steels | Cutting tool production, die grinding, resharpening |
| Precision production honing operations | Alloy steels | Hydraulic cylinders, automotive parts |

precision grinding applications. Carbide tools, parts, and drill bits are ground and shaped with diamond wheels. Diamond grinding wheels are used to grind the glass used in optics, the flat glass used in furniture and automotive applications, and crystal glass having intricate designs. A variety of low- and high-density ceramics used in kiln furniture, magnet, capacitor, spark plug, and similar applications are ground with diamond grinding wheels. Electronic ceramics such as silicon wafers, magnetic heads, optical fibers, and sensors are machined to tight tolerances and fine surface finishes with diamond grinding wheels. The potential use of engineering, technical, or fine ceramics for a wide variety of thermal and mechanical applications will call for more extensive use of diamond grinding wheels in the future. Typical products ma-

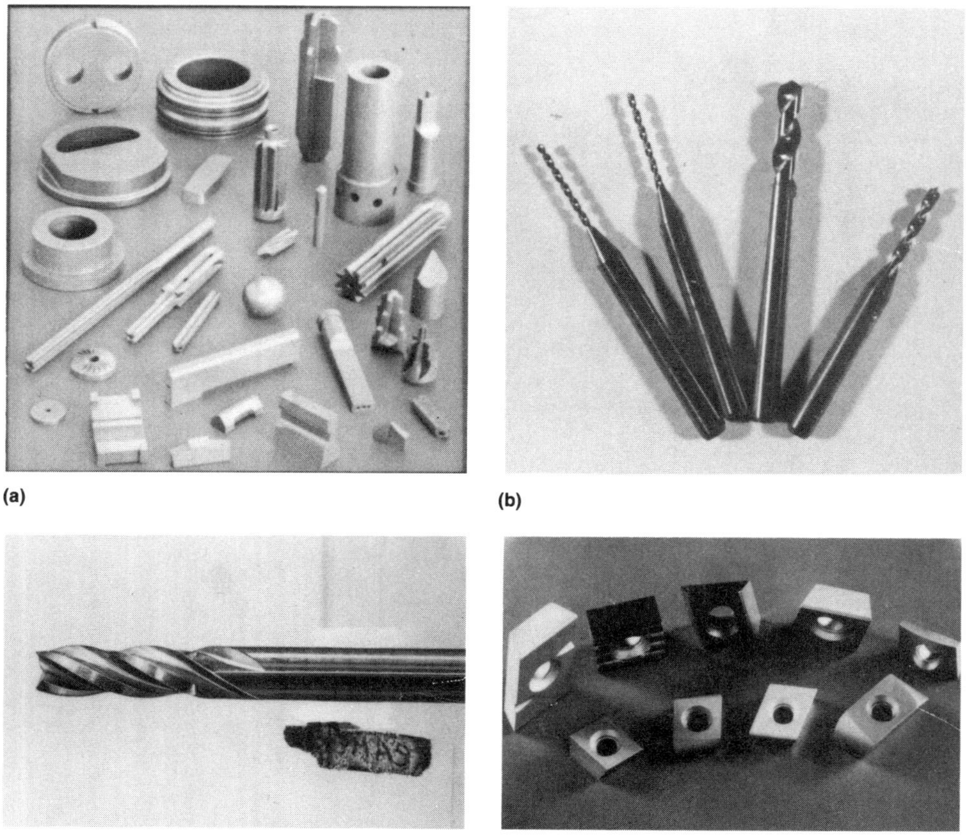

(a) (b) (c) (d)

**Fig. 7** Tungsten carbide components having surfaces machined with diamond abrasives. (a) Variety of parts having machined surfaces. (b) Variety of solid carbide twist drills. (c) Carbide reamer having ground helical grooves. (d) Indexable carbide inserts having ground surfaces

**Fig. 8** Solid carbide saw surfaces ground with diamond wheels

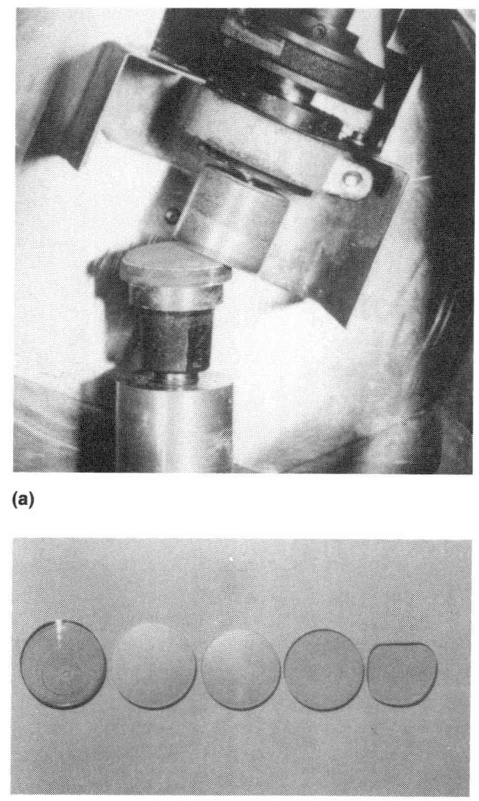

(a) (b)

**Fig. 9** Diamond abrasive machining of optical lenses. (a) Lens being ground on an apparatus using a metal bond diamond wheel. (b) Sequence of operations to produce a photochromic lens from the lens blank (left), after generating, after lapping, after polishing, and following shaping and beveling (right)

chined with diamond abrasives are shown in Fig. 7 to 9. Processes that utilize diamond abrasives are illustrated in Fig. 10. Fine surface-finishing processes such as lapping, honing, and polishing also use fine-grit diamond abrasives in loose abrasive form, such as powders, compounds, and slurries (see the articles "Lapping" and Honing" in this Volume).

**Cubic Boron Nitride Abrasives.** As stated earlier, CBN abrasives are used in the precision production grinding of a wide variety of ferrous and high-nickel alloy materials. Because of their better abrasive wear resistance and thermal shock resistance, CBN grains retain their cutting geometry and efficiency longer than aluminum oxide abrasive. Higher hardness and better thermal conductivity are reported as reasons for the improved residual properties (such as better fatigue life) achieved with CBN abrasive grinding. The advantages of using CBN in precision production grinding operations are listed in Table 3. Because of this wide range of benefits, CBN is often referred to as the ideal precision grinding abrasive for production grinding. Therefore, abrasives are used in automotive, aerospace, cutting tool production, tool maintenance, and a wide variety of other industrial applications.

## Superabrasive Wheels

Superabrasive wheels are available in a variety of sizes and shapes and are classified according to construction, concentration, and bond system.

**Size and Shape.** The superabrasive wheels used in the precision grinding applications shown in Fig. 11 vary greatly in size and shape (see the article "Grinding Equipment and Processes" in this Volume). Precision wheels ranging in thickness from 0.025 mm (0.001 in.) or less to 255 mm (10 in.) or more are used, depending on the application. Similarly, wheels ranging from 1 to 760 mm (0.040 to 30 in.) in diameter are used. Because of their lower wear rate and longer life, smaller superabrasive wheels can do the work of larger conventional abrasive wheels, provided the machining system can accommodate smaller wheels and lower spindle speeds. The typical shape designations of superabrasive wheels are shown in Fig. 12.

**Construction.** Because of their long wheel life and the higher cost of the superabrasives, these wheels are generally used in a rim-type construction. Exceptions are extremely small inside diameter or very thin grinding wheels.

The annular region of the wheel containing the superabrasives, called the rim, is integrally bonded to the core or structural part of the superabrasive grinding wheel (Fig. 13). The core is generally made of composites, aluminum, bronze, steel, or ceramic, depending on such performance requirements as strength, stiffness, and dimensional stability.

**Concentration.** The rim, or grinding face, consists of a bond, or matrix, that contains the superabrasive grains. The volume frac-

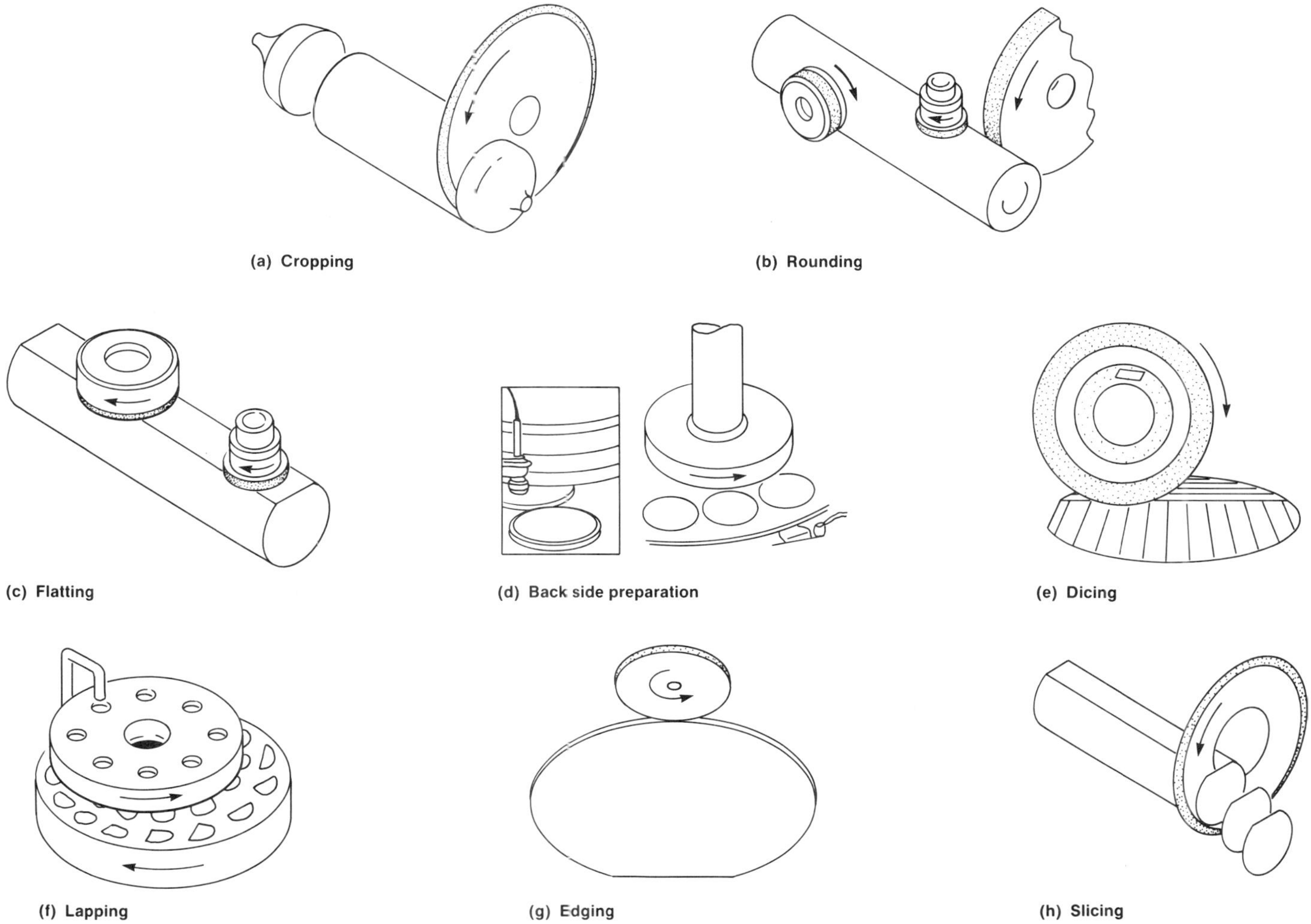

(a) Cropping (b) Rounding (c) Flatting (d) Back side preparation (e) Dicing (f) Lapping (g) Edging (h) Slicing

**Fig. 10** Diamond abrasive machining operations used to fabricate semiconductor components

tion of the abrasive grains in the rim is known as the concentration. This often determines the performance or behavior of superabrasive wheels.

**Bond Systems.** Four bond systems are typically used in superabrasive wheels:

- Resin
- Vitrified
- Metal
- Layered product

The details of the constituents of the bond and the processing techniques result in specific bond types or bond designation and determine the response of the bond to the applications. The manufacturer should be consulted for optimum bond selection. Each bond system has its own unique properties, as follows (Table 4).

*Resin* bond wheels provide good resilience and vibration-absorbing characteristics, which reduce chatter at the grinding zone. Wheels with resin bonds are easy to true and dress and are commonly selected for a wide range of applications.

*Vitrified* bond wheels offer controlled porosity, which facilitates chip removal and coolant flow to the grinding zone. They generally last longer than resin bond wheels and are suitable for producing accurate and complex forms.

*Metal* bond wheels, although difficult to true and dress, offer long life, good form-

**Table 3 Advantages of CBN abrasives in precision production grinding**

| Properties | Technology (research and development) | End results (factory) | Economic results (corporate) |
|---|---|---|---|
| High hardness | Wear-resistant abrasive for long wheel life | Fewer wheel changes (30 min versus days); (1 shift versus 3 months) | Lower total cost/part |
| Thermal stability | High grinding efficiency resulting in: | Consistent part quality | Better asset utilization |
| Chemical inertness | Lower grinding power | Better geometry | More value-added products |
| High thermal conductivity | Lower grinding forces | Lower inspection | Factory automation |
| Produces compressive residual stress | Better fatigue life | Lower rejections | |
| | Close tolerance/geometry | Lower in-process inventory | |
| | Better component performance | Shorter grinding cycle | |
| | Processing of difficult-to-grind parts, difficult materials, and difficult component geometries | New products | |
| | High unit-width metal removal rate | Less labor intensive | |
| | New processes | More automation | |

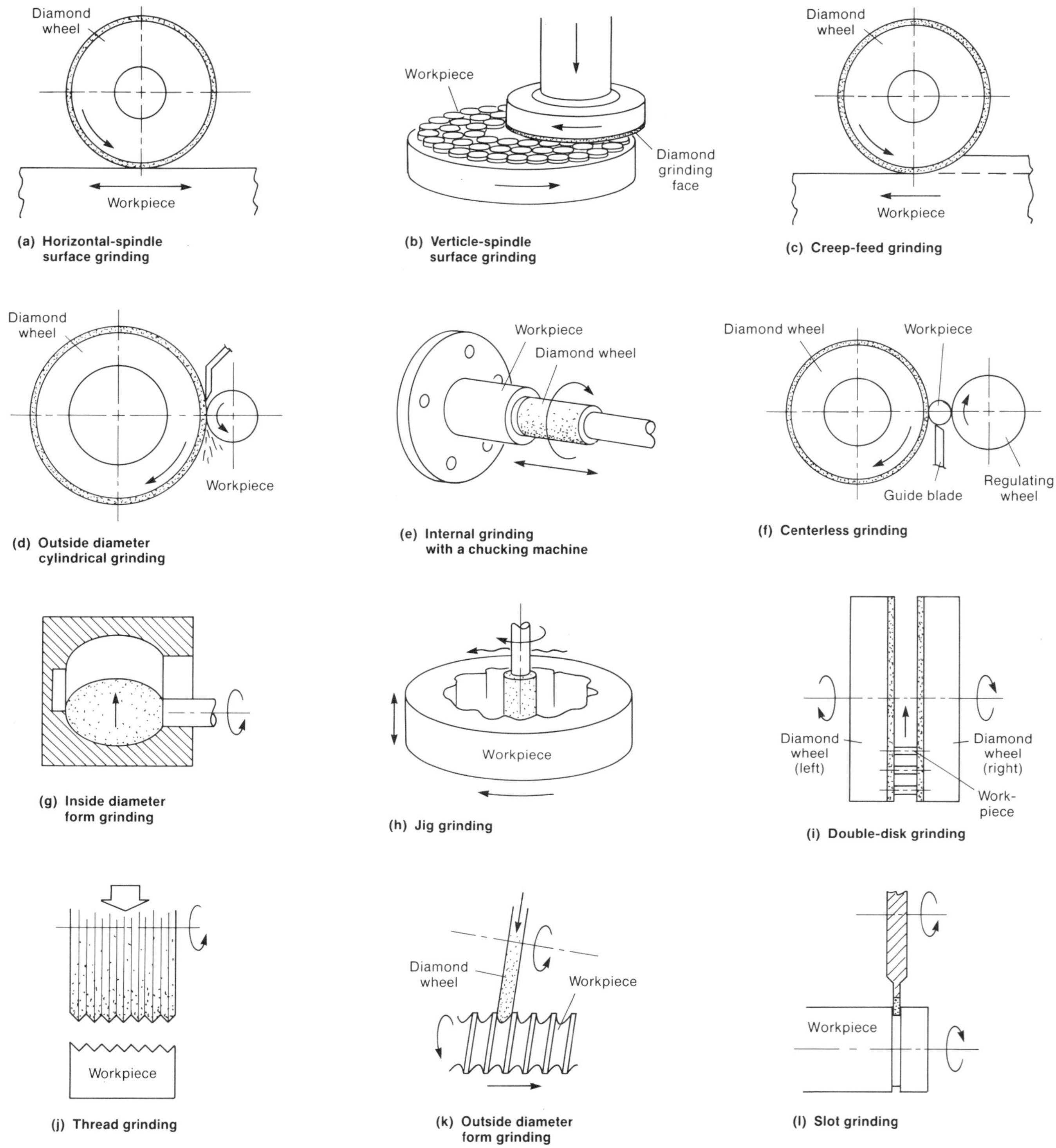


**Fig. 11** Production grinding applications of diamond grinding wheels

holding characteristics, and good thermal conductivity. They are excellent for simple form grinding, but usually require greater grinding forces and more power than resin or vitrified bond wheels.

*Layered product* wheels utilize a single abrasive layer plated or brazed to a premachined preform. They usually produce a poorer surface finish than bonded abrasive wheels. Layered product wheels are used for small production runs or where tolerance and surface finish are not very critical.

## Superabrasive Wheel Applications

The proper use of superabrasive wheels often requires careful evaluation of all factors of the grinding system, such as:

- Machine tool
- Work material
- Wheel selection
- Operational factors

Table 5 lists some of the key variables that affect each of these factors.

Figure 14 shows a schematic of the grinding process. In a superabrasive wheel, the abrasive grit serves as a wear-resistant cut-

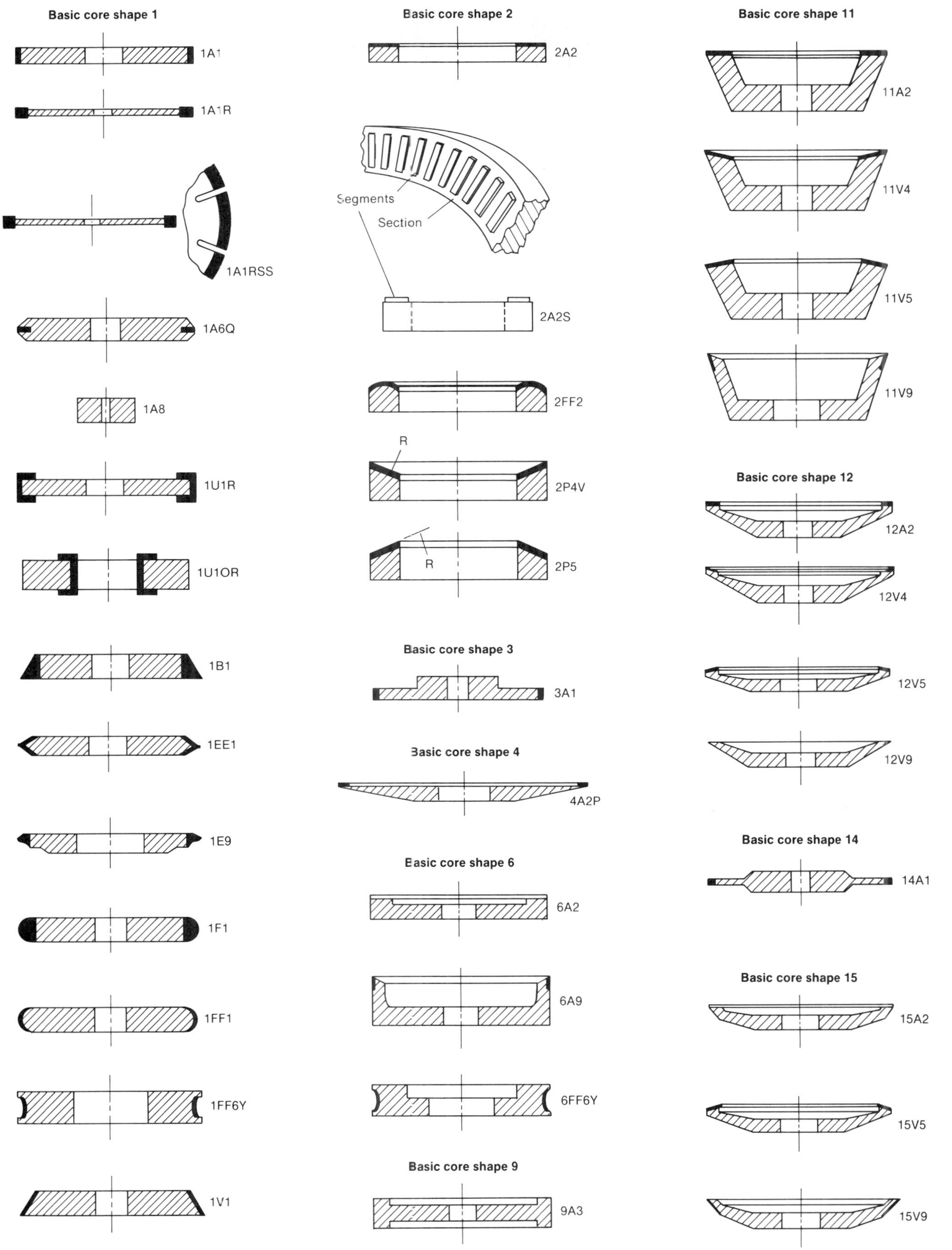


**Fig. 12** Superabrasive wheel configurations and their designations

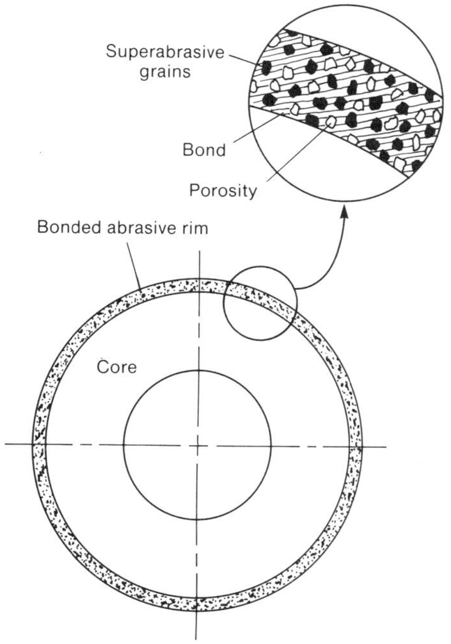


**Fig. 13** Construction of a typical superabrasive wheel

ting edge for long periods of time, unless it is pulled out of the bond prematurely. Therefore, it is imperative to maximize the abrasive/work interactions leading to chip generation and grinding efficiency and to minimize the rubbing or interaction at the bond/work, chip/bond, or chip/work interfaces.

## Machine Tool Variables

Machine tool developments in the past 20 years have contributed to innovative superabrasive applications. Precision spindles and slides, rigid machine frames, accurate positioning methods, multiaxis computer numerical control (CNC) movement to achieve complex geometries with a high degree of accuracy, high-speed spindles, and high-pressure flow coolant systems are some of the features incorporated into grinding machines using superabrasive wheels.

High-speed tool steel end mills are produced in the conventional method by milling the flutes on a cylindrical rod and then heat treating (Fig. 15a). During heat treating, the flutes become distorted and require finish grinding to restore flute geometry and to generate other features of the cutting geometry. It is also expensive to maintain an in-process inventory of the premachined blanks. Capital and labor costs for this process are high. The conventional process has been significantly improved upon by heat treating the rod and then grinding in the flutes with a CBN wheel (Fig. 15b), thus eliminating the milling operation used in the conventional process. The improved process utilizes the high material removal rate capability of CBN superabrasive wheels while maintaining the form or geometry of the wheel face for a relatively long time. However, the improved process requires the following:

- Multiaxial CNC machines of high rigidity that have the flexibility to be programmed for a range of part geometries
- Oil coolant systems able to withstand high pressures and high fluid flows
- Suitable enclosure to ensure operator safety
- High-precision truing and dressing equipment that lends itself to automation should production quantities warrant such an investment

Figure 16 shows the setup used to grind in flutes in an end mill and illustrates the array of coolant lines required for the machining operation.

**Table 4 Advantages of diamond abrasive bond types**

| |
|---|
| **Resin bond** |
| Readily available |
| Easy to true and dress |
| Moderate freeness of cut |
| Applicable for a range of operations |
| First selection for learning the use of diamond wheels |
| **Vitrified bond** |
| Free cutting |
| Easy to true |
| Does not need dressing (if selected and trued properly) |
| Controlled porosity to enable coolant flow to the grinding zone and chip removal |
| Intricate forms can be crush formed on the wheels |
| Suitable for creep-feed or deep grinding, inside diameter grinding, or high-conformity grinding |
| Potential for longer wheel life than resin bond |
| Excellent under oil as coolant |
| **Metal bond** |
| Very durable |
| Excellent for thin slot, groove, cutoff, simple form, or slot grinding |
| High stiffness |
| Good form holding |
| Good thermal conductivity |
| Potential for high-speed operation |
| Generally requires high grinding forces and power |
| Difficult to true and dress |
| **Layered products** |
| Single abrasive layer plated on a premachined steel preform |
| Extremely free cutting |
| High unit-width metal removal rates |
| Form wheels, easily produced |
| Form accuracy dependent on preform and plating accuracy |
| High abrasive density |
| Generally not truable |
| Generally poorer surface finish than bonded abrasive wheels |

**Table 5 Variables influencing grinding operations with superabrasives**

| Machine tool | Operational factors |
|---|---|
| Design<br>  Rigidity<br>  Precision<br>  Dynamic stability<br>Features<br>  Controls<br>  Power, speed, and so on<br>  Slide movements<br>  Truing and dressing equipment<br>Coolant<br>  Type<br>  Pressure<br>  Flow<br>  Filtration systems | Fixtures<br>Wheel balancing<br>Truing, dressing, and conditioning techniques and devices<br>Grinding cycle optimization<br>Inspection methods |
| **Wheel selection** | **Work material** |
| Core material<br>Wheel design<br>Abrasive<br>  Type, properties<br>  Particle size<br>  Size distribution<br>  Content<br>Bond<br>  Type<br>  Hardness/grade<br>  Porosity<br>  Thermal conduction | Properties<br>  Mechanical<br>  Thermal<br>  Microstructural<br>Geometry<br>  Wheel/part conformity<br>  Access to coolant<br>  Shape/profile required<br>Part quality<br>  Geometry<br>  Consistency |

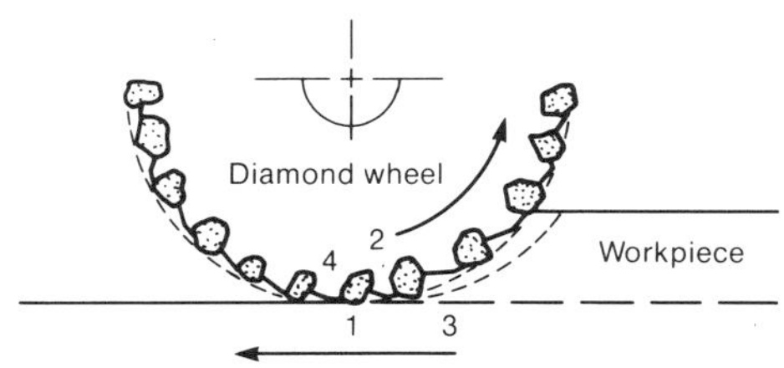


**Fig. 14** Schematic illustrating interactions in the grinding zone of a grinding wheel/workpiece interface. 1, abrasive/work interface; 2, chip/bond interface; 3, chip/work interface; 4, bond/work interface

## Wheel Selection

The shape, size, configuration, and features of superabrasive bond types are described in the section "Bond Systems" in this article. A typical example of superabrasive wheel designation is shown in Fig. 17. Wheel manufacturers should be consulted on the details of each specification and their influence on grinding results. Additional information on wheel selection is available in the article "Grinding Equipment and Processes" in this Volume. The principles of superabrasive wheel applications are discussed in the sections "Diamond Grinding Wheels" and "Cubic Boron Nitride Grinding Wheels" in this article.

**Diamond Grinding Wheels.** Diamond grinding wheels are used for a wide variety of work materials, such as carbide, glass, industrial ceramics, plastics, electronic ceramics, and composites and high-density/structural ceramics. Specification data for the materials are extensive and readily available.

*Grinding of Carbide Materials.* Figure 18 shows the effect of diamond abrasive particle size, concentration, and type on surface finish, metal removal rate, and *G* ratio (volume of work removal/volume of wheel

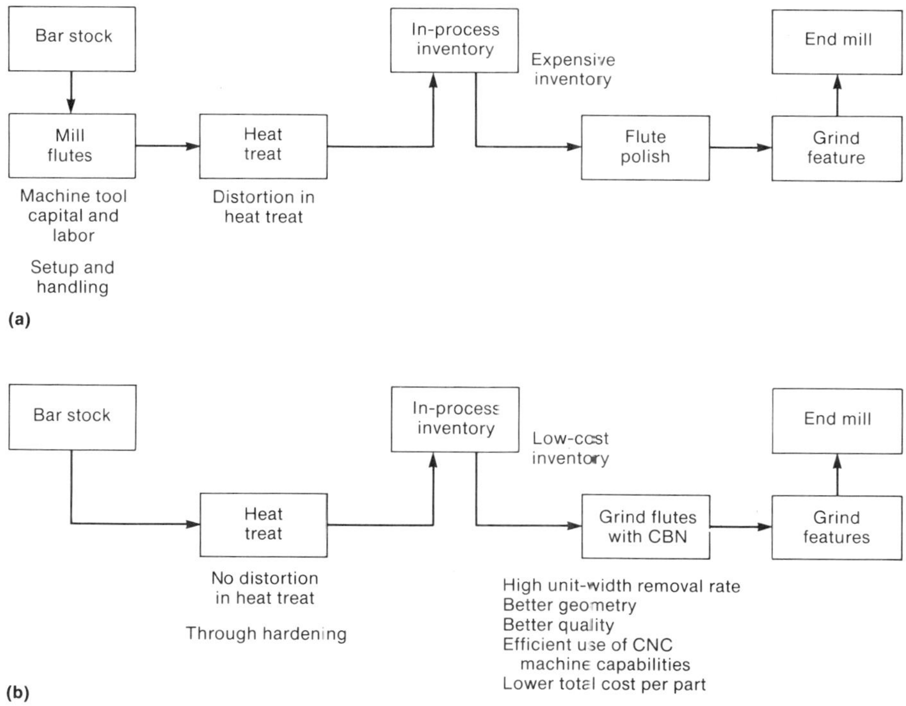


**Fig. 15** Methods of producing high-speed tool steel end mills. (a) Conventional process in which flutes are milled in prior to hardening. (b) Improved process in which flutes are ground in with a CBN wheel after hardening. The new process proved to be more cost effective and produced an end mill with more exact tool geometry.

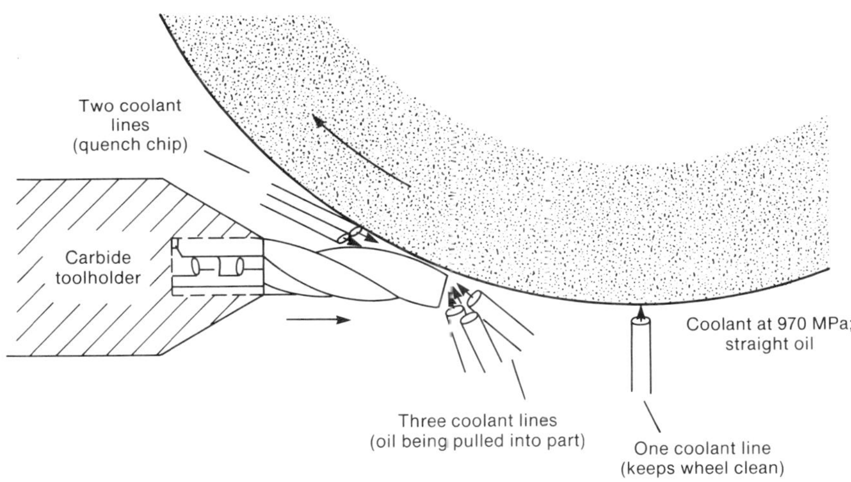


**Fig. 16** Schematic of tool setup for grinding in flutes on an end mill (the improved process) showing positions of coolant lines

worn). These qualitative curves are based on data compiled from grinding carbide with diamond abrasives.

Figure 19 shows the effect of work material structure (toughness), chip size produced, and abrasion resistance of the bond used on surface finish, metal removal rate, and $G$ ratio. Figure 20 shows the effect of wheel speed, $V_s$, and grinding pressure on surface finish, metal removal rate, and $G$ ratio. Grinding pressure is a control variable commonly used in toolroom or insert grinding operations.

*Grinding of Ceramics*. The advent of high-strength structural ceramics and their possible use in a variety of high-performance applications offer the potential for even wider use of diamond abrasive wheels. Some recent results are discussed in this section.

Figure 21 shows the influence of grit size, bond type, and material removal rate in the grinding of hot-pressed silicon nitride (HPSN). Figure 21(a) shows the grinding forces measured normal to the workpiece surface ($F_N$), and Fig. 21(b) shows the tangential grinding forces measured parallel to the work surface or in the direction of table traverse ($F_T$). Figures 21(a) and (b) show the forces in a normalized scale; therefore, a direct comparison is possible between the bonds used and the grit sizes. The normal

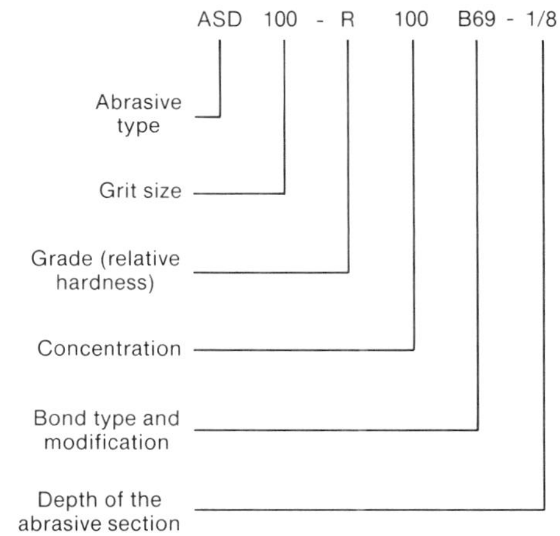


**Fig. 17** Typical specifications used for superabrasive wheels

and tangential forces are generally higher for the finer grit, within the experimental conditions of 50 to 255 mm/min (2 to 10 in./min) of table speed, 2.5 mm (0.100 in.) downfeed, and the corresponding normalized (unit-width) metal removal rate.

Higher forces were generally observed at finer grit sizes for the three bond systems evaluated. This conclusion may be more suitable for the grinding of structural components (such as cutting tools), than for electronic applications, in which the normalized metal removal rates used are generally well below the values tested in this example.

Among the bond systems evaluated, the vitrified bond generated the lowest normal forces, while the resin bond generated the lowest tangential forces (or lowest grinding power). The metal bonds evaluated required higher normal and tangential forces. The modification to the metal bond (M2) to improve the free cutting action of the wheel appeared to be significantly beneficial at the higher normalized metal removal rate.

Figure 22 shows the wheel wear measured as $G$ ratio for the test conditions shown in Fig. 21. In general, finer-grit diamond that required higher forces exhibited lower $G$ ratios (within the experimental conditions evaluated). However, the metal bond wheels that required higher forces also exhibited higher $G$ ratios. High tangential forces and lower $G$ ratios (despite the lower normal forces) are measured for the vitrified bond wheel. This highlights the significance of coolant application in the grinding of ceramic materials, particularly with vitrified bond diamond wheels. When the test was conducted in a machine setup with better coolant application conditions, the vitrified bond wheel showed relatively high $G$ ratios at the higher unit-width metal removal rate (Fig. 22).

The normal forces measured are shown in Fig. 23. Hot-pressed silicon nitride requires higher grinding forces than any of the other ceramics evaluated, including tungsten car-

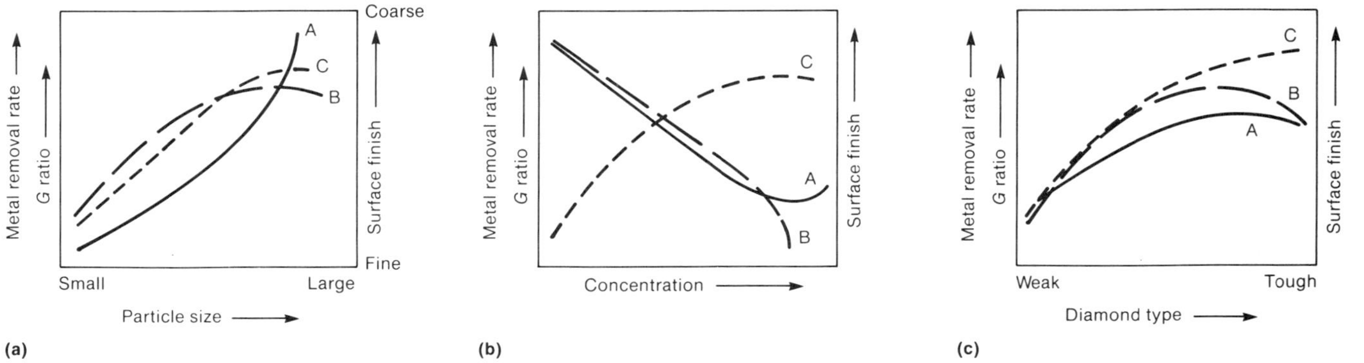


**Fig. 18** Plots of surface finish (curve A), metal removal rate (curve B), and *G* ratio (curve C) against particle size (a), concentration (b), and diamond type (c) to show the relative properties of diamond abrasives in the grinding of carbides

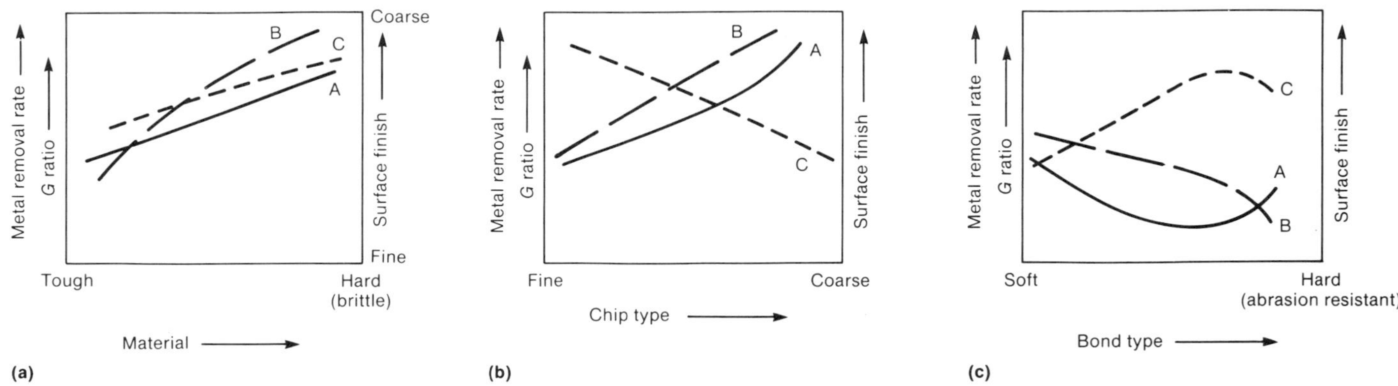


**Fig. 19** Plots of surface finish (curve A), metal removal rate (curve B), and *G* ratio (curve C) against material (a), chip type (b), and bond type (c), to show the relative properties of diamond abrasives in the grinding of carbides

bide. Ferrite, $Al_2O_3$-TiC, and zirconia require relatively lower forces. The relative grinding power required for the ceramics evaluated is shown in Fig. 23(b). The wheel wear measurements indicate that ferrite has the highest *G* ratio, followed by zirconia, $Al_2O_3$-TiC, hot-pressed silicon nitride, and tungsten carbide (in decreasing order).

**Cubic boron nitride grinding wheels** can be evaluated with regard to wheel life, concentration, equivalent diameter, surface finish, grit size, and coolant application.

*Wheel Life*. The wear of a grinding wheel is measured in terms of the *G* ratio, which is the ratio of the volume of work material removed divided by the wheel wear volume. This *G* ratio represents a measure of the real life of the abrasive as well as the useful life of the abrasive wheel itself.

The wheel lives of CBN wheels are often 100 to 1000 times longer than those of conventional abrasive wheels. Figure 24 compares the properties of a conventional aluminum oxide wheel with those of a CBN wheel for grinding 52100 bearing steel, M7 high-speed tool steel, and M50 and Inconel 718. The long life of the CBN wheel offers the advantages of lower wheel change frequency, lower machine downtime, consistent part geometry, and ease of automation. Because of this long life, the frequency of dressing superabrasive wheels is substantially lower than that of conventional abrasive wheels. Proper dressing procedures must be followed to obtain the maximum useful life of CBN wheels and to achieve maximum economic benefits.

*Concentration* denotes the amount of superabrasive present in the wheel and often represents the volume fraction of the abrasive. Figure 25 shows the effect of concentration in CBN wheels. In general, the higher the concentration, the longer the wheel life and the better the surface finish. In electronic ceramics, higher-concentration diamond wheels are used to achieve tighter tolerances in part geometry and to minimize chipping of the parts.

Equivalent diameter, $D_E$, determines the conformity between the wheel and the workpiece. The closer the conformity, the higher the equivalent diameter. Internal grinding generally creates higher conformity than external cylindrical grinding. Figure 26 shows that external grinding with lower conformity grinds more efficiently (lower power consumption for given material removal rate) but produces a poor surface finish.

*The surface finish* produced by CBN grinding wheels is determined by the dressing or exposure of the abrasive grains achieved in resin and metal bond wheels. In vitrified bond CBN wheels, the truing pro-

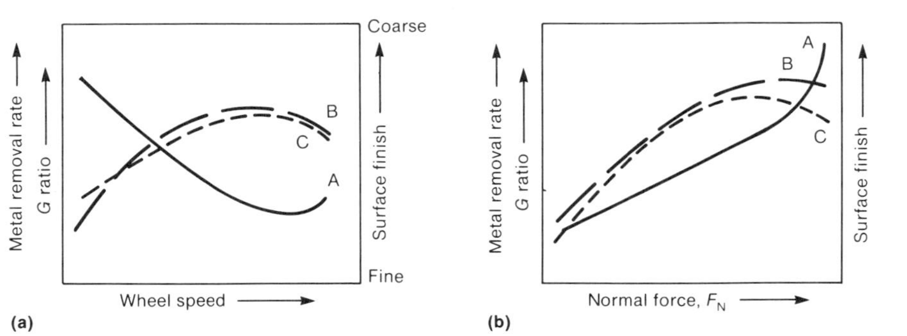


**Fig. 20** Plots of surface finish (curve A), metal removal rate (curve B), and *G* ratio (curve C) against wheel speed (a), and normal force (b), to illustrate the relative effect of operating conditions on carbide grinding

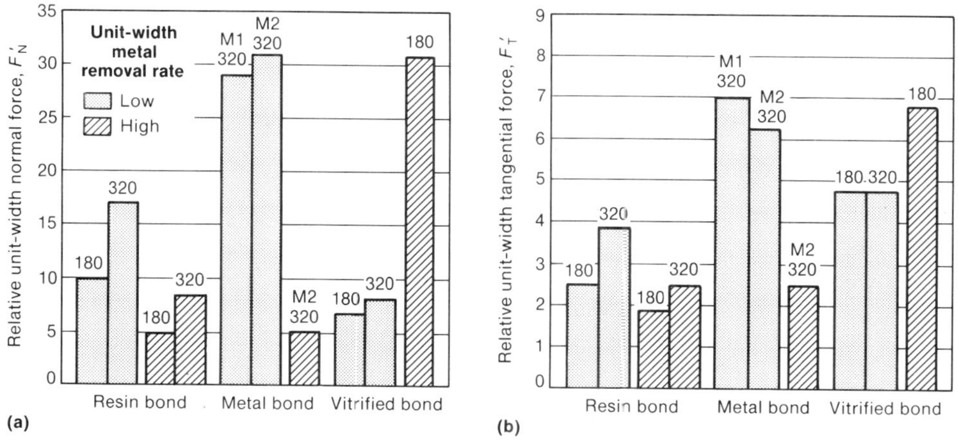


**Fig. 21** Effect of bond type and grit size on normal (a) and tangential (b) forces in the grinding of hot-pressed silicon nitride. Wheel speed was 28 m/s (5500 sfm) at both low (2 $mm^3/s$, mm; or 0.2 $in.^3/min$, in.) and high (10 $mm^3/s$, mm; or 1.0 $in.^3/min$, in.) unit-width metal removal rates. M2 indicates a modification of the original metal bond (M1). Grit sizes are 180 and 320.

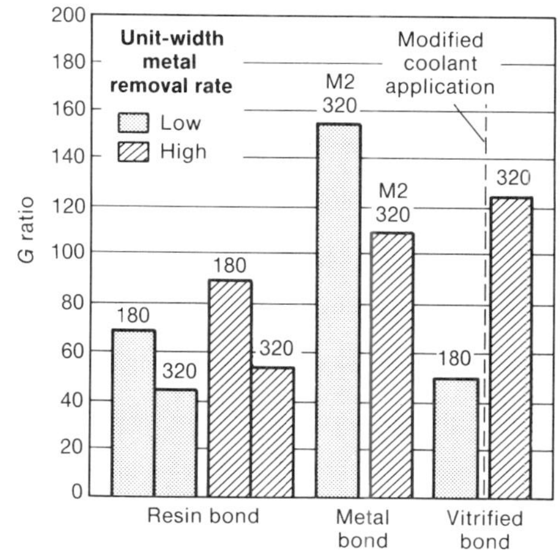


**Fig. 22** Effect of bond type and grit size on *G* ratio in the grinding of hot-pressed silicon nitride for conditions shown in Fig. 21. Grit sizes are 180 and 320. Wheel speed was 28 m/s (5500 sfm) at both low (2 $mm^3/s$, mm; or 0.2 $in.^3/min$, in.) and high (10 $mm^3/s$, mm; or 1.0 $in.^3/min$, in.) unit-width metal removal rates. M2 indicates a modification of the original metal bond.

cedure and wheel specification dictate the surface finish achieved.

Unlike conventional abrasive wheels, CBN wheels are generally not dressed within a grinding cycle to achieve a dull wheel face and therefore improve the work surface finish. In most cases, the work surface finish is governed by the highest material removal rate that the CBN wheel is subjected to, which generally occurs during the rough grinding cycle. Therefore, setting the cycle to achieve a short cycle time while maintaining the lowest normalized metal removal rate during rough grinding is a desirable strategy when using CBN wheels for production grinding.

*Grit Size.* In any application in which a conventional abrasive such as aluminum oxide is performing successfully, two conditions can be envisioned:

- The abrasive undergoes self-sharpening due to its inherently low thermal fatigue resistance and consequent microfracturing
- The bond system can be designed to self-sharpen and thus shed the low-cost abrasives (when they become dull and require high grinding forces without serious economic consequences

When applying the CBN superabrasive, the conditions discussed below occur.

The CBN abrasive has high thermal fatigue resistance and high wear resistance. Any flats (flank face wear) generated in the abrasive grain during the truing process or during grinding cannot be easily removed and will cause high grinding forces. This is similar to a cutting tool with large flank wear. Therefore, selection of the proper grit size is critical for avoiding excessive forces and consequent poor grinding results (such as chatter, burn, and hard grinding action). The bond system cannot be designed to be excessively forgiving (soft), because it is necessary to minimize the loss of expensive CBN superabrasive. This information would suggest that the selection of CBN superabrasive grit size should be the minimum size possible that is consistent with the metal removal rate required. Typically, this grit size is in the range of 150 to 320 for replacing aluminum oxide abrasives used at 60 to 120 grit. In addition, the truing parameters should be optimized to avoid the generation of excessive flats or large flank face wear (Fig. 27). For example, if the influence of the abrasive grit size and truing procedures (leading to flank face rubbing) is not recognized, hard grinding action will be mistakenly attributed to the entire line of CBN abrasives.

*Effect of Coolants.* The importance of pressure, flow, and direction of coolant application for superabrasive wheels is covered in the section "Machine Tool Variables" in this article. In addition, a 100% oil coolant applied at high pressure and flow rate and properly directed at the grinding zone generally improves CBN wheel performance significantly (Fig. 28). However, heat removal from the grinding zone may be the predominant requirement of the coolant in order to maintain part geometry and to avoid burn marks on thin sections or in heat-sensitive materials. In such cases, water-soluble oil may be preferable as a coolant. The application of high-pressure coolants and the use of 100% oil as a coolant necessitate proper guarding and ventilation of the system to avoid contamination of the grinding room atmosphere with oil mist.

*Effect of Wheel Speed.* The optimum wheel speed for CBN grinding is approximately 45 m/s (9000 sfm), but the speeds typically used range from 25 to 35 m/s (5000 to 7000 sfm). Higher grinding wheel speeds produce lower grinding forces, better wheel life, and better surface finish (Fig. 29). These benefits are generally utilized in higher-accuracy grinding to achieve better-toleranced parts or in higher-productivity grind-

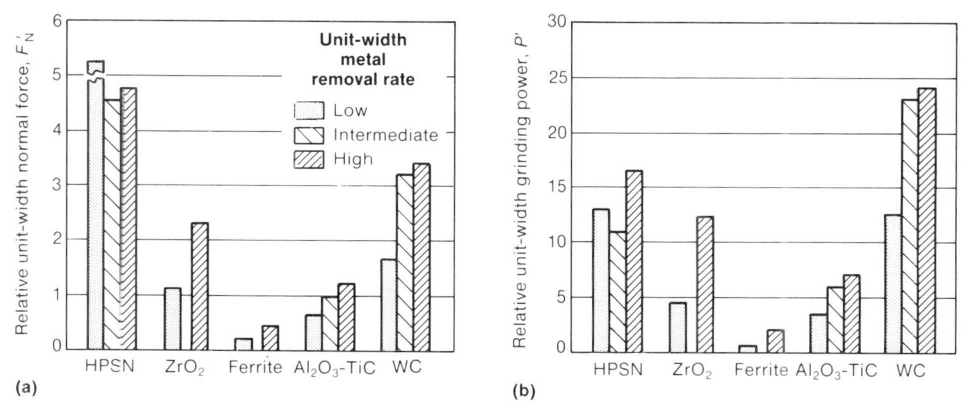


**Fig. 23** Relative unit-width normal force (a) and relative unit-width grinding power (b) required to machine various structural and electronic ceramics. Unit-width metal removal rates classified as low (2 $mm^3/s$, mm; or 0.2 $in.^3/min$, in.), medium (5 $mm^3/s$, mm; or 0.5 $in.^3/min$, in.), and high (10 $mm^3/s$, mm; or 1.0 $in.^3/min$, in.)

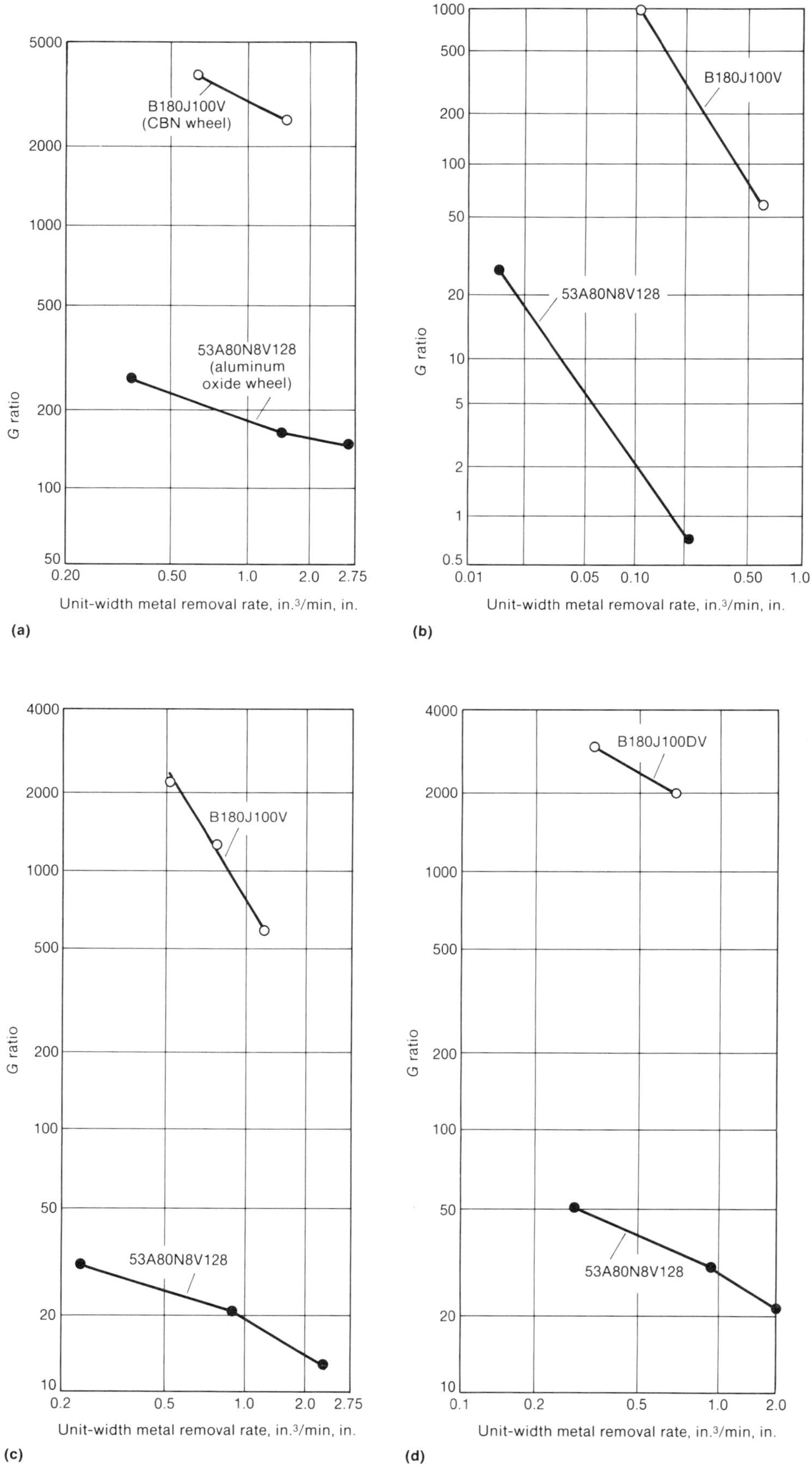


**Fig. 24** Comparison of CBN and conventional abrasive wheel performance in the grinding of 52100 bearing steel (a), M50 high-speed tool steel (b), M7 high-speed tool steel (c), and Inconel 718 (d). With the exception of the M50 material, which used a water-soluble oil and was inside diameter ground, all of the materials were outside diameter ground using a 100% oil coolant.

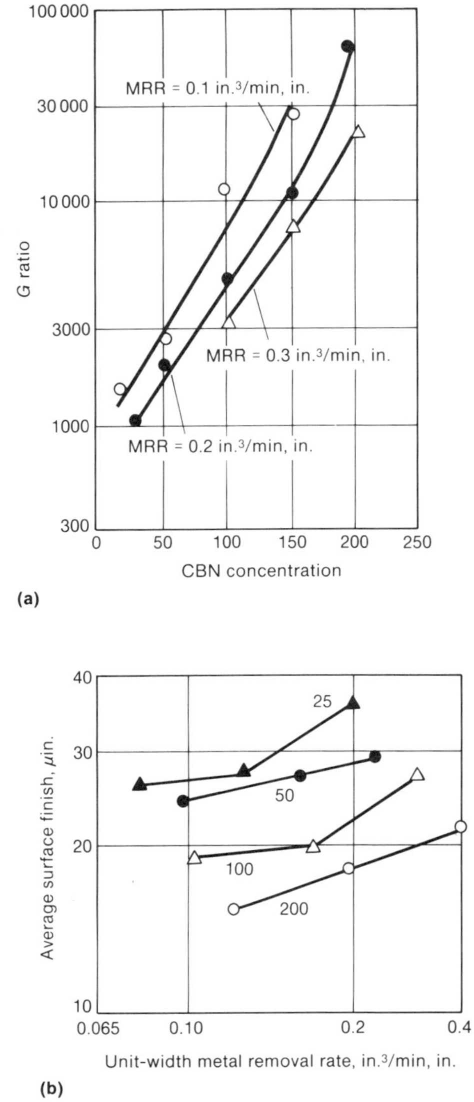


**Fig. 25** Effect of CBN concentration on wheel and workpiece properties. The data are for a vitrified bond CBN wheel used to grind 52100 bearing steel. (a) Normalized metal removal rate values for plot of *G* ratio versus CBN concentration. (b) CBN concentration for plot of surface finish against unit-width metal removal rate

ing to obtain more parts per hour without sacrificing tolerance requirements. However, high-speed grinding is always associated with higher grinding power (for a given unit-width metal removal rate) and therefore requires better coolant systems to prevent workpiece burn or damage to the part. High-speed grinding is also often associated with special machine designs having a high-accuracy spindle and low vibration levels which lead to improved wheel design for safe operation at the higher speeds.

*The fatigue life* of parts ground with CBN wheels is reported to be higher than that of unground case-hardened components such as gears. The reason for this improvement is not well understood. However, the use of CBN for gear grinding is being pursued to obtain increased gear life, smaller gear size

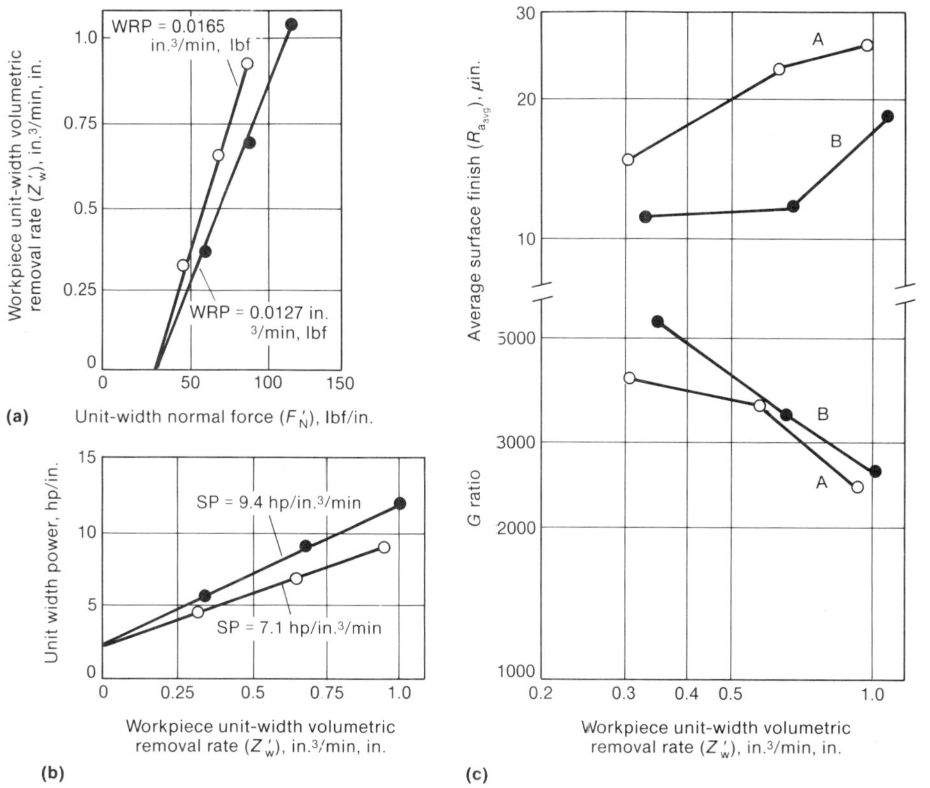


**Fig. 26** Graphs demonstrating that external grinding cuts faster than internal grinding but produces a poorer finish. Material is 52100 bearing steel at 60 HRC that is being ground by a B240J150V CBN wheel dressed with a rotary diamond wheel. Curve A (which is external mode), $D_E$ is 30.5 mm (1.2 in.) and $V_S$ is 55 m/s (11 000 sfm); curve B, $D_E$ is 178 mm (7.0 in.) and $V_S$ is 60 m/s (12 000 sfm). (a) Workpiece unit-width volumetric removal rate plotted against unit-width normal force. (b) Unit-width power plotted against workpiece unit-width volumetric removal rate to obtain slope, which equals specific power. (c) *G* ratio and average surface finish plotted against workpiece unit-width volumetric removal rate

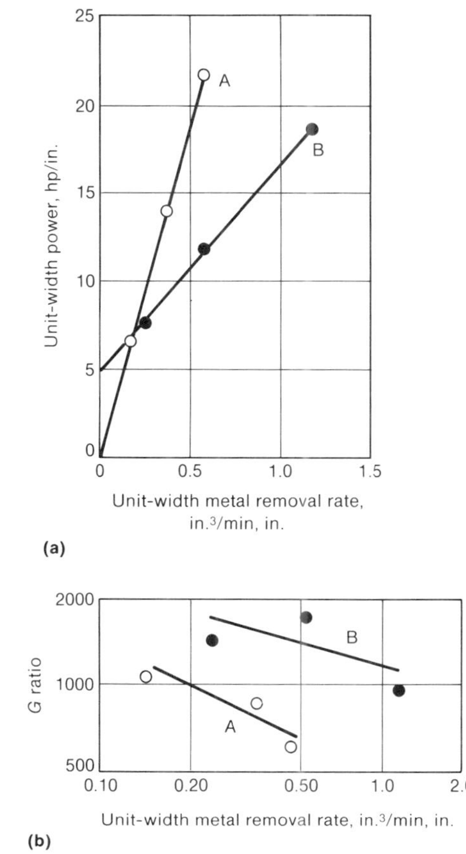


**Fig. 28** Effect of coolant on grinding performance with CBN wheels. The operation is the inside diameter grinding of M7 high-speed tool steel using a B180J100V wheel. A, 5% water-soluble oil; B, 100% oil coolant. (a) Unit-width power plotted against unit-width metal removal rate. (b) *G* ratio plotted against unit-width metal removal rate

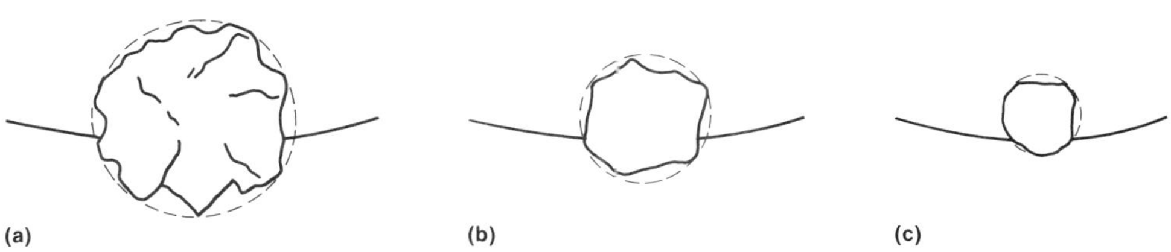


**Fig. 27** Selection of superabrasive grit size. 60-grit aluminum oxide (a) yields a bond system comparable to 100-grit CBN having wear flats (b) or 150-grit CBN without wear flats (c).

at equal fatigue life or higher efficiency, and lower noise levels for transmission applications.

## Operational Factors

Balancing, truing, dressing, and conditioning are necessary for the effective use of superabrasive wheels. Each of these operational factors will be discussed in turn.

**Balancing** of the superabrasive wheels prior to mounting and subsequent on-machine balancing are critical to achieving high wheel performance, micro-smooth surface finishes, and high-precision geometry of the surfaces produced with minimal chatter or chipping. Once wheel selection, machine tool settings, adjustments, and maintenance are understood, the successful use of superabrasives depends to a large extent on the truing, dressing, and conditioning of the wheel. It is generally thought that superabrasive wheels are hard, that they cannot be trued and dressed, or that special machines are required. When properly selected, superabrasive wheels can be trued, dressed, and made to operate in current production operations. However, a systematic understanding of the difference between conventional wheel truing and superabrasive wheel truing is required.

**Truing.** For the successful use of superabrasive wheels, the wheel surface must be concentric, free of round lobes, and straight across the thickness of the wheel, and it must have the correct profile for form wheels.

The process used to generate geometrically correct wheel faces is called truing. A properly trued wheel will grind with minimum or no chatter and will generate a straight cylinder, flat surface, or accurate profile on the workpiece, provided it is also dressed properly (Fig. 30).

*Dressing.* After truing, the wheel face is generally very smooth without much exposure of the superabrasive grits. For efficient grinding, the bond adjacent to the superabrasive grits will have to be eroded away, exposing the superabrasive grits. A properly exposed superabrasive wheel surface contains abrasive grits supported in the trailing side with tails. In the leading side, the bond is removed for about 30% of the grit size for free cutting action. The grits are connected by shallow grooves for chip clearance and cool flow. Exposing the superabrasive grits on the wheel surface for efficient grinding action is called dressing (Fig. 31).

A properly trued and dressed wheel produces a workpiece with the required geometry, tolerance, and surface finish; draws minimum grinding power; and produces a workpiece without burn, surface damage, or chatter marks. Truing and dressing are required during the first setting up of the

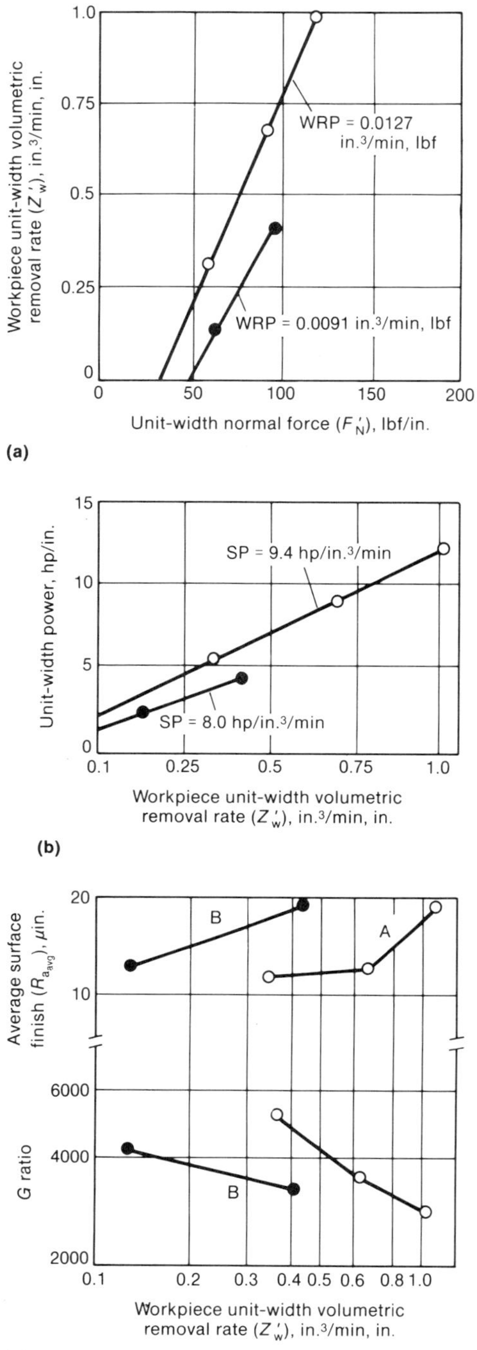


**Fig. 29** High wheel speed cuts faster and improves surface finish. Material is 52100 bearing steel at 60 HRC that is being internally ground by a B240J150V CBN wheel dressed with a rotary diamond wheel. Equivalent diameter is 127 mm (5.0 in.). Curve A, $V_S$ is 60 m/s (12 000 sfm); curve B, $V_S$ is 30 m/s (6000 sfm). (a) Workpiece unit-width volumetric removal rate plotted against unit-width normal force. (b) Unit-width power plotted against workpiece unit-width volumetric removal rate to obtain slope, which equals specific power. (c) G ratio and average surface finish plotted against workpiece unit-width volumetric removal rate

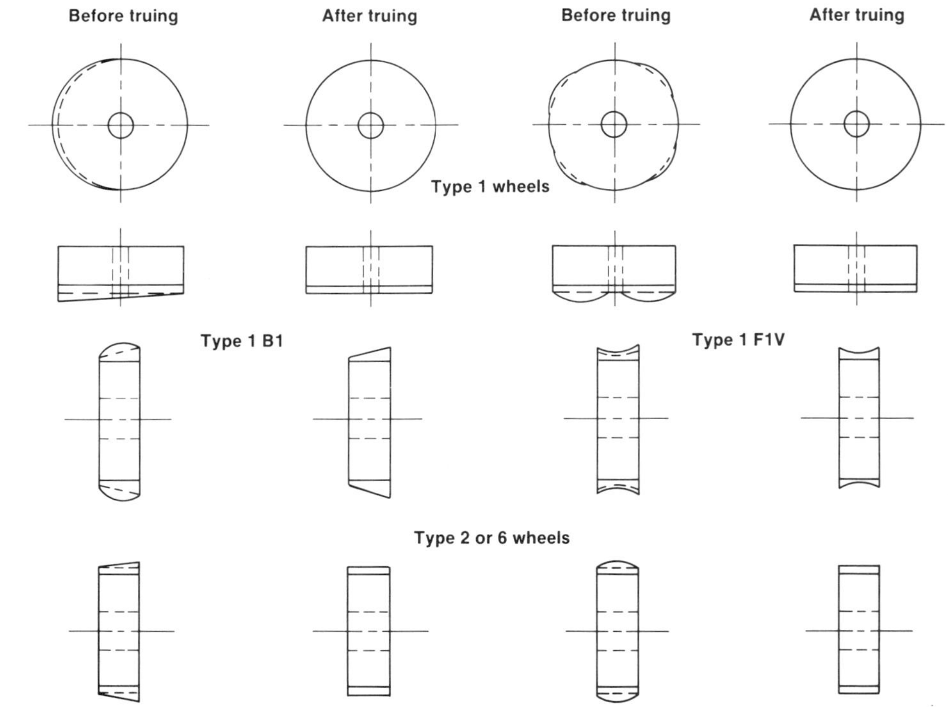


**Fig. 30** Typical examples of conditions that require truing

wheel. Resin and metal bond wheels must be trued and dressed. Vitrified bond wheels generally require truing only and do not require dressing. Electroplated superabrasive wheels do not require either truing or dressing, except in special situations.

**Conditioning.** When parts of acceptable quality are being ground, the superabrasive wheel may require periodic conditioning to restore the geometry of the part, the surface finish, or both. The conditioning may involve minor truing and/or dressing (Fig. 32).

Conventional abrasive wheels for precision production grinding are generally vitrified bond wheels. They do not require dressing. In this case, the terms truing and dressing are used synonymously. Truing and dressing are not synonymous terms for superabrasive wheels.

## Truing Methods for Production Grinding

A variety of methods are available for superabrasive wheel truing in production grinding operations. These methods can be broadly divided into three categories:

- Stationary tool truing
- Powered truing
- Form truing

The truing of diamond wheels in production grinding is still evolving. Although most of the comments in this section are based on results with CBN wheels, it is anticipated that similar results would be obtained with diamond wheels.

**Stationary tool truing** involves passing a nonrotating truing tool across the face of a rotating wheel. The tool may consist of a single diamond point, a cluster tool containing three to five points, or a nib containing many diamond particles.

*Single-point truing* can be used for very small vitrified wheels or for small scale infrequent production. However, the diamond point of the tool wears rapidly, requires frequent turning, and demands considerable operator attention. As a result, the single-point truing of CBN wheels is generally not recommended.

*Cluster tool truing* generates high truing forces. This method is not recommended.

*The nib* (Fig. 33a), containing many small diamond particles, is the preferred tool for

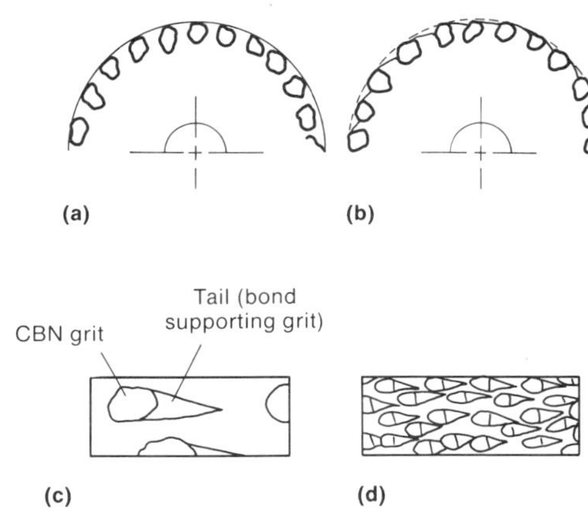


**Fig. 31** Schematic of a wheel that has been trued and dressed. (a) After truing. Wheel face is smooth and closed. (b) After dressing. Wheel face is open with grits exposed and ready for efficient grinding. (c) After dressing. Bond supports the grit. (d) After dressing. Note path connecting the tails for efficient coolant and chip flow.

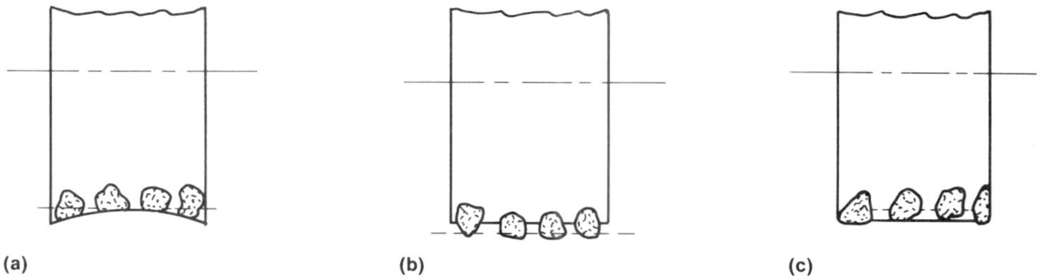


**Fig. 32** Typical conditions that require periodic correction to the wheel face (conditioning). (a) Out-of-form wheel requires precision truing to restore geometry. (b) Poor surface finish caused by bond erosion can be corrected by precision truing of the wheel face. (c) Freeness of cut is restored by precision dressing to reduce worn out CBN grit and exposed grit.

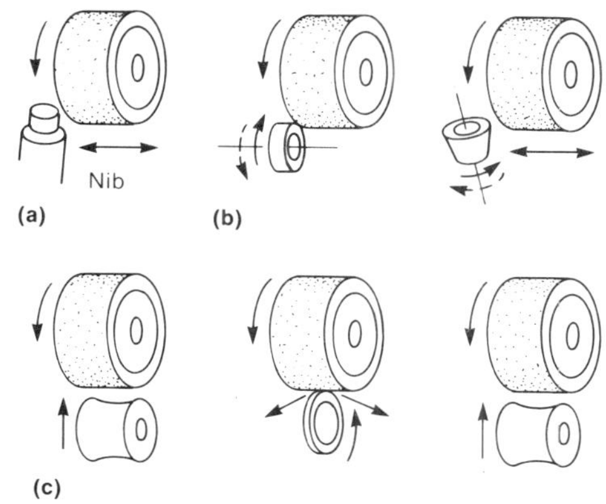


**Fig. 33** Truing methods used for CBN wheels. (a) Stationary tool truing with a nib. (b) Powered truing using a rotary cutter (left) or a rotary cup (right). (c) Form truing using a diamond roll (left), tracing with rotary cutters (center), and a crush roll (right)

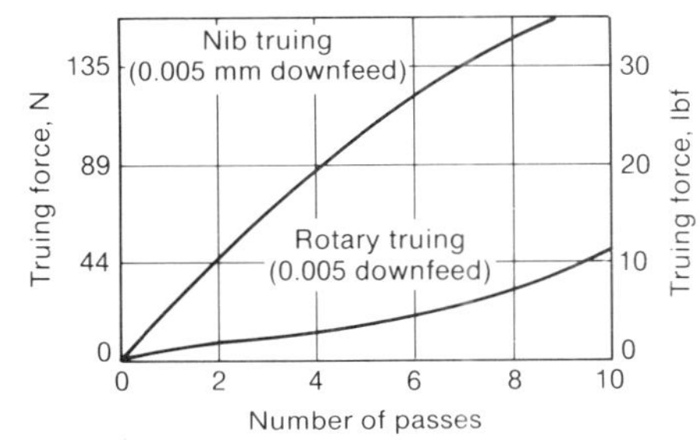


**Fig. 34** Comparison of truing forces in the nib truing and rotary truing of CB100WBA wheel

stationary truing. The truing forces generated are acceptable.

**Powered Truing Methods.** A systematic study of the truing process indicates that powered truing methods utilizing a rotating tool offer the best results (Fig. 33b). Figure 34 shows the truing forces generated by a diamond nib and a rotary cutter under identical conditions on a resin bond CBN wheel. The truing forces are considerably lower with a rotary tool than with the stationary tool. In addition, the rate of increase in truing force with a rotary tool is considerably smaller.

Commercially available rotary powered truing devices are versatile to use. They also generate low truing forces, offer consistent results, and are easily automated. Wheels with a straight face or simple forms can be trued with a traversing diamond rotary cutter or a rotary cup; plunge truing with a diamond roll can also be used.

**Form Truing.** Complex form wheels can be trued by plunge truing with diamond rolls or crush truing methods (Fig. 33c). Another method of truing form wheels is the use of a thin rotary cutter (considerably thinner than the wheel width). This rotary cutter can be traversed across the wheel profile with a mechanical arrangement that generates the wheel profile, for example, using a form bar or a CNC-controlled form generator. Cubic boron nitride form wheels have been successfully trued with this approach in order to generate simple radius or complex profiles.

## Wheel Truing Objectives for Superabrasives

The essential feature of all CBN wheel truing methods is to provide accurate and precise relative motion between the wheel and the truing tool in order to generate the straight cylinders, tapers, flat surfaces, or forms required on the wheel. Successful wheel truing depends on the five factors discussed below.

**Minimizing Truing Tool Wear.** Wear of the truing tool causes loss of form, crowning, or a taper on the wheel face. Excessive wear is generally associated with poor tool selection, large truing forces, very small traverse speed, high infeed rates, abusive conditions of wheel truing, or excessive wheel runout that could have been reduced during mounting of the wheel.

**Ensuring Consistent Wheel Surface Quality.** Unlike conventional abrasive wheels, a CBN wheel is not trued after every part or grinding cycle. If properly operated, it should be conditioned after typically 20 to 100 grinding cycles and as often as every 3 months. Once the conditioning frequency is established, the wheel should be conditioned at regular intervals. If this is not done, the production schedule will be disrupted, and increased operator attention will be required.

The truing mechanism should be reliable, accurate, and rigid to ensure consistent wheel surface quality. The machine should have the capability for precision movement and skip dressing.

**Roughing/Finishing Using the Same Wheel Face Without Conditioning.** Cubic boron nitride wheels can grind at high unit-width metal removal rates and can produce good surface finishes at low unit-width metal removal rates. This is possible because the CBN grit resists wear. The open condition of the wheel allows for high unit-width metal removal rates, coolant flow, and chip removal without burning. The flats generated on the CBN grits during the truing operation determine the surface finish at finish grinding rates. Excessive flats on coarse grits will generate a fine surface finish, but burning or smearing can occur.

**Machine Tool and Truing Device Requirements.** The truing of CBN wheels can be satisfactorily accomplished on moderately rigid (18 kN/mm, or 100 000 lbf/in.) machines. Excessive rigidity is useful but not necessary. However, a machine with worn ways, inconsistent feed rates, stick slip, and lack of fine infeed control (for example, 0.005 mm, or 0.0002 in. per step or lower) will degrade any truing effort. Similarly, high precision, low runout, and moderate rigidity are required in the truing devices. Good-quality coolant flow, proper diamond selection, and smaller-grit diamonds generally improve the truing.

**Goals for Successful Truing.** In a properly managed truing effort, it is necessary to:

- Minimize truing forces
- Minimize truing time
- Control the exposure and extent of flatness on CBN grits to obtain the desired surface finish
- Minimize loss of superabrasive grain
- Automate the truing process to achieve consistent wheel surface condition

## Truing Methods for Batch Production

For operations in which the wheel is used for short production runs, with each run containing different parts (for example, toolroom grinding), powered rotary truing methods are preferred and should be used whenever possible. However, for batch production, the truing operations can be carried out using the methods discussed below.

**Truing With Abrasive Wheels.** Brake-controlled and powered rotary truing devices use conventional abrasive wheels. The precision with which a trued wheel surface is generated by these methods is lower than in the methods described earlier. In general, a relatively harder wheel with silicon carbide abrasive will true the superabrasive wheel faster at lower levels of abrasive consumption. However, an open wheel surface is obtained with a softer wheel contain-

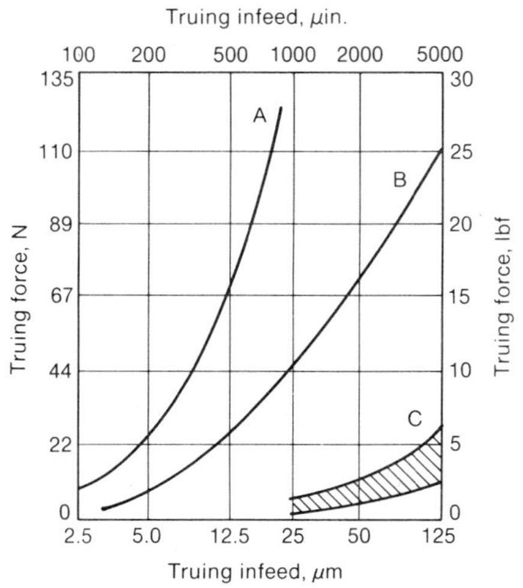


**Fig. 35** Effect of truing infeed on truing force. A, resin and metal bond CBN wheel; B, vitrified CBN wheel; C, conventional vitrified wheel

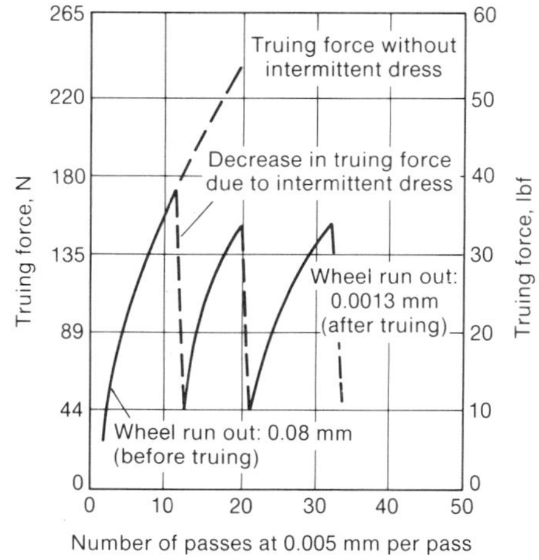


**Fig. 36** Effect of intermittent dressing of a CBN wheel on truing forces

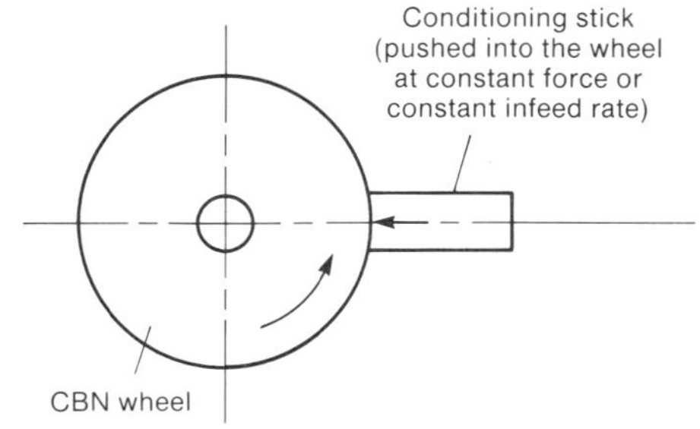


**Fig. 37** Schematic illustrating abrasive stick dressing. Low coolant flow is required during the stick dressing process.

ing aluminum oxide abrasive. Occasionally, the superabrasive wheels can be finished to the required form or geometry by using a grinder with an optical attachment.

**Truing With Hard Ceramics.** Resin bond CBN wheels can be trued by grinding with hard ceramics such as tungsten carbide or boron carbide. In this method, 0.005 mm (0.0002 in.) downfeed per pass is used in a surface grinder. Slightly more than half the wheel thickness should be crossfed, and table speed should be moderate. The wheel speed is the same as the grinding speed. The grinding power will gradually increase as the wheel becomes dull while being trued. When the power exceeds the normal power drawn during workpiece grinding, truing should be halted. The wheel face should be stick dressed with an abrasive stick. Truing is then continued. This cycle is repeated until the wheel is completely trued. Diamond nibs and, occasionally, single-point tools can also be used for truing CBN wheels in batch operations.

## Quantitative Understanding of Truing Parameters

The following variables apply to any truing operation:

- *Truing method used*: Stationary or powered rotary truing
- *Type of truing tool*: Stationary (single-point or nib) or rotary (disk, cup, diamond roll, bonded or plated cutter, or reverse-plated cutter)
- *Specification of truing tool*: Diamond type, grit size, concentration, and geometry
- *CBN wheels to be used*: Bond type, grit size, concentration, and wheel size
- *Truing process variables*: Speed, infeed increments, speed ratio, overlap ratio, relative motion, crossfeed, coolant flow, and machine rigidity
- *Output variables*: Truing forces, power, chatter or lobing, total indicator reading, accuracy of form, wheel surface quality and truing time, and grind quality after truing

Some of the key variables are discussed below.

**Infeed** is an incremental advance of the truing tool or cutter relative to the wheel face. The infeed used for superabrasive wheel truing with a diamond tool or cutter should be about one-tenth the values used for conventional abrasive wheels. This generally implies an infeed of 0.0025 to 0.005 mm (0.0001 to 0.0002 in.) per increment in traverse truing and 0.050 to 0.10 μm (2 to 4 μin.) per revolution in plunge truing.

The effect of infeed on truing forces is shown in Fig. 35. Control of infeed is a key aspect of achieving outstanding performance of superabrasive wheels without chatter or hard grinding action.

**Intermittent Dressing.** Figure 36 shows the gradual increase in truing forces as a function of time or number of traversals of the cutter across the wheel face. However, if the truing forces exceed the normal grinding forces, deflection of the spindle, wheel lobing, and chatter will occur. Intermittent or periodic dressing of the wheel minimizes these problems.

**Parameters for Rotary Truing.** The ratio of cutter speed to wheel speed is called the speed ratio. Selection of the speed ratio and its relative direction determines truing forces and such grinding results as work surface roughness and wheel wear.

*Grit Size.* The diamond cutter dressing tool for superabrasive wheels generally requires smaller grit sizes than those used for conventional abrasive wheels. Large diamond grit in the cutter generates higher truing forces and poor grinding results.

*Overlap ratio* is the ratio of the width or thickness of the cutter to the lead. The smaller the overlap ratio, the more open the wheel face, the lower the grinding power, and generally the shorter the wheel life. However, high overlap ratios can lead to a dull grinding wheel that results in poor wheel performance.

## Practical Dressing Methods

Several methods are available for dressing CBN wheels. These include:

- Abrasive stick dressing using mechanical devices
- Abrasive-jet dressing
- Slurry dressing
- High-pressure waterjet dressing

**Abrasive stick dressing using mechanical devices** is the most frequently recommended practice and is suitable for a wide variety of applications. This method consists of pushing an abrasive stick (called a conditioning stick) into the wheel face with a constant force or a constant infeed rate (Fig. 37). Several variations of the stick dressing arrangement are used in industry.

Dressing devices are available that can be adapted to grinding machines. In this method, the abrasive stick can be pushed into the wheel face at a constant infeed or a constant force level. As the stick is pushed into an as-trued CBN wheel face at a constant infeed rate, the forces increase rapidly and then reach a lower steady-state level (Fig. 38). For a given wheel and stick specification, the steady-state force will reach a constant value that is determined only by the infeed rate and is independent of the abrasive volume consumed after a minimum quantity. In general, if the operator determines that the wheel is consuming the stick excessively but is not grinding satisfactorily, he should consider a coarser-grit stick or a higher infeed rate.

Figure 39 shows the results of constant-force dressing tests. The CBN wheel was trued under identical conditions. The wheel was then dressed using abrasive sticks of 600, 320, and 220 grit; the same grade; and two force levels: 65 and 135 N (15 and 30 lbf). In each case, the wheel was dressed using 25, 50, 75, and 102 mm (1, 2, 3, and 4 in.) of stick length. The resultant wheel face was used to grind M2 high-speed tool steel

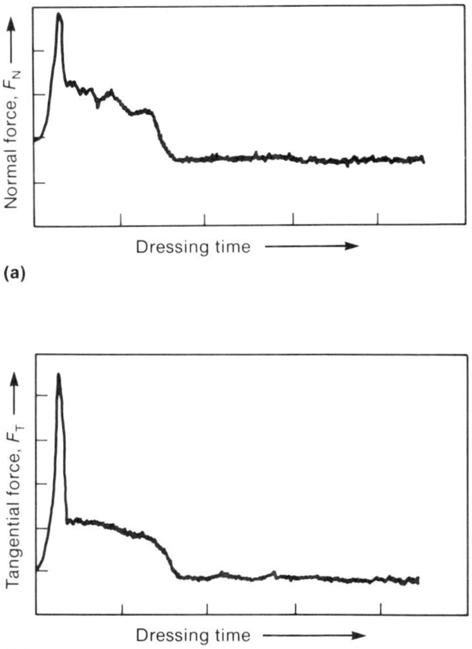


**Fig. 38** Variation of relative dressing force with time at a constant infeed rate of the dressing stick. (a) Normal force. (b) Tangential force

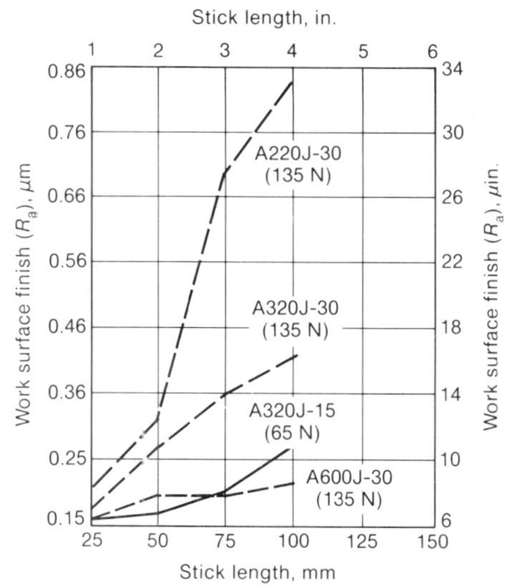


**Fig. 39** Effect of dressing stick grit size and dressing stick consumption on wheel roughness (as measured by work surface finish) during constant-force dressing. Work material ground was M2 high-speed tool steel using a CBN wheel (CB 220 WBA) measuring 255 × 13 × 3.2 × 32 mm (10 × ½ × ⅛ × 1¼ in.) and a wheel speed of 46 m/s (9000 sfm). The dressing sticks measured 13 × 13 mm (½ × ½ in.).

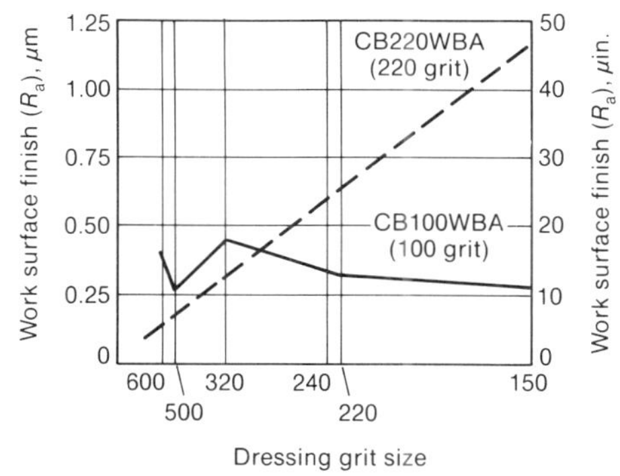


**Fig. 40** Effect of dressing stick parameters on CBN wheels for constant infeed rate dressing

under identical conditions. The work surface finish was measured to reflect indirectly the wheel face roughness. The wheel surface becomes rougher as the abrasive grit in the dressing stick becomes coarser. For the same abrasive stick (A32OJ), the increase in dressing force also increases wheel surface roughness.

The exposure of CBN is not the only factor influencing the work surface finish; the CBN grit size also has a strong influence. Figure 40 shows the grinding performance of two CBN wheels of the same concentration but different grit size that were dressed under the same conditions (constant force, constant volume of abrasive used, and constant grinding conditions). The finer-grit CBN wheel produces workpiece surface roughness that is dependent on the abrasive grit size used for dressing.

Such is not the case with the coarser-grit CBN wheel. The coarser-grit CBN produces finer work surface roughness than CBN of smaller grit size. The quality of work surface produced by 100-grit CBN is poorer because of smearing or a burnished appearance. Therefore, in grinding with CBN, it is preferable to use finer grit sizes without sacrificing wheel life for a given metal removal rate.

**Abrasive-jet dressing** consists of impacting the CBN wheel face with a jet of fine-grit abrasive propelled by high-pressure air. This method is also suitable for dressing a contour or form on the wheel face. The system is expected to be portable from machine to machine within a plant.

**Slurry Dressing.** Some machines that are specifically designed for CBN use an abrasive slurry (abrasive entrained in water) pumped across a gap between the wheel and a steel roll.

**High-pressure waterjet dressing** consists of a high-pressure waterjet directed at the wheel face to erode the bond in a controlled manner. Preliminary test results show that these methods generally required an abrasive waterjet rather than an ordinary waterjet.

The cost of such units for slurry dressing and waterjet dressing and the specialized nature of the equipment involved do not justify these methods for a wide variety of applications. However, the methods may become useful under special machine and production conditions.

## Dressing Methods for Batch Production

**Abrasive Stick Dressing.** Manual stick dressing is used in a number of batch production operations. However, from the standpoints of safety and consistency, manual stick dressing should be discouraged. The stick dressing can be mechanized whenever possible by mounting the stick on the machine table and rapidly feeding it into the wheel with machine controls. Proper guarding is essential in such cases. The conditioning sticks used for dressing CBN wheels should be distinguished from the soft rubber sticks used for cleaning the wheel face.

The dressing block can be used as a work material in surface grinding, and the wheel can be dressed by grinding such blocks. Care should be taken to prevent excessive wear of the CBN wheel or abrasive conditions that can lead to uneven wear.

**Dressing While Grinding.** Cubic boron nitride wheels, particularly resin or vitrified bond wheels, can be dressed by erosion of the bond during grinding. This invariably results in a loss of work material during dressing due to burn or improper work quality. In addition, depending on the grinding conditions, the wheel may not be adequately dressed at any time.

## Conditioning

Conditioning is a periodic correction to the wheel face to restore its geometry or to expose the superabrasive grains. When the superabrasive wheel is trued and dressed properly and the application engineering is satisfactory, parts of good quality (that is, those having the required geometry, surface finish, and surface quality) are produced with a minimum of grinding power. In a properly specified CBN wheel, this situation will continue for a number of grinding cycles—typically, 50 to 200 cycles in inside diameter grinding, approximately one shift in cylindrical grinding, and as long as 3 months in disk grinding. Therefore, the performance of the superabrasive wheel with consistent part quality will depend on the operation, but in every case it will involve a significantly longer duration than the conventional abrasive wheels. As a result, the conditioning interval for superabrasive wheels is much longer than that for conventional abrasive (which may be as frequent as once every cycle to a few times within one cycle). This is an important factor because frequent conditioning of the wheels may render the entire superabrasive wheel application uneconomical.

Figure 41 illustrates the conditioning scheme. A properly trued and dressed CBN wheel grinds with a minimum of grinding power (point A), which remains constant as good-quality parts are produced from the beginning. In practice, however, $A_1$ or $A_2$ may be the starting point, and the wheel quickly reaches an equilibrium grinding

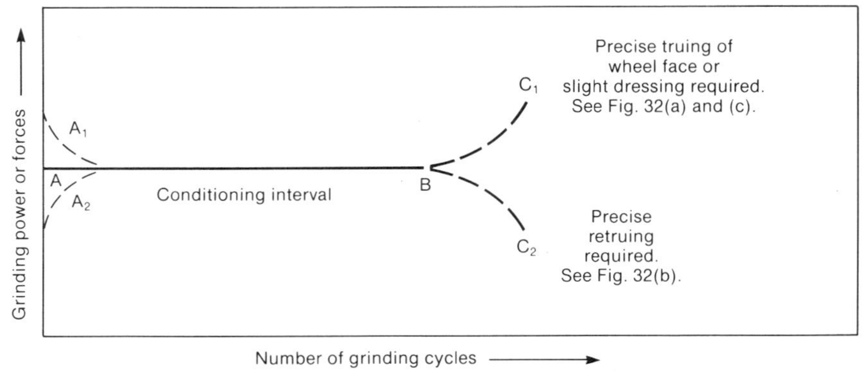


**Fig. 41** Diagram showing that minimum grinding power is required during conditioning interval (line A–B). An increase in grinding power or force (line B–$C_1$) can be corrected at point B by precise truing of the wheel face (see Fig. 32a) or slight dressing (see Fig. 32c). A gradual decrease in grinding power caused by excessive bond wear (line B–$C_2$) can be corrected by precision retruing (see Fig. 32b) of the wheel at point B.

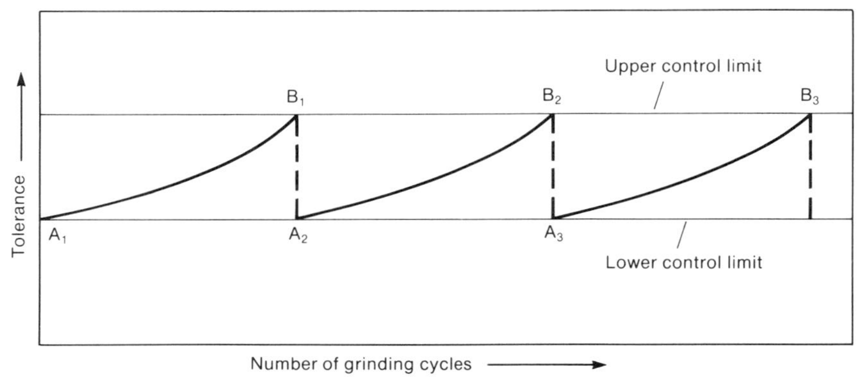


**Fig. 42** Schematic of conditioning needs for maintaining part tolerance

condition. The wheel wears gradually until the point B surface finish changes or until burn or a change in geometry occurs. Any one of these conditions could lead to an increase in grinding power or grinding forces, as shown by line B–$C_1$. Therefore, it is necessary to correct the wheel face at point B. This may involve precise truing of the wheel face or slight dressing. The precision involved and the extent of truing and/or dressing depend greatly on the application, that is, part size and tolerances required and wheel specification. For example, in one case, a truing infeed of 1.00 μm (40 μin.) was sufficient to restore the geometry in an inside diameter grinding application using a vitrified CBN wheel. In a resin bond CBN wheel for cam grinding, a standard dressing cycle (constant dressing time and volume of dressing abrasive used) was required once every shift.

If the bond wear is excessive, then at point B the wheel could produce parts of poor surface finish. This will lead to a gradual decrease in grinding power, as shown by line B–$C_2$. In this case, precision retruing of the wheel will be necessary.

Therefore, the conditioning interval for the operation will be the time duration or the number of grinding cycles between points A and B. Figure 42 illustrates this result with reference to part tolerances. The repetitive and predictable nature of the conditioning interval is a key aspect of superabrasive applications in production grinding. This strongly depends on wheel selection, the initial truing and dressing scheme, and the grinding process parameters chosen.

From the above discussion, it is observed that conditioning of the CBN wheel is a high-precision operation. The increments of infeed for retruing or dressing are much smaller in the conditioning operation than in the initial truing and dressing of the superabrasive wheel. It is imperative that grinding machines have such precision capabilities (consistent infeed of 0.0025 mm, or 0.0001 in., or less for retruing) and mechanical means for dressing. It is also necessary to set the machine to the skip dressing mode so that the superabrasive wheel is conditioned only at the preset interval and not after every cycle.

## SELECTED REFERENCES


- "American National Standard Specifications for Shapes and Sizes of Diamond and CBN Abrasive Products," ANSI B74.3-1986, Grinding Wheel Institute, 1986
- "CBN Wheel Truing and Dressing Devices," Catalog 118, Norton Company, 1986
- S. Chandrasekar *et al.*, Morphology of Ground and Lapped Surfaces of Ferrite and Metal, in *Proceedings of the ASME Winter Annual Meeting*, Vol 17, American Society of Mechanical Engineers, 1985
- B.A. Cooley and H.O. Juchem, Vitrified Bonds—No Longer a Synonym for Conventional Abrasive Tools, in *Superabrasives '85*, Publication MR 85-268, Society of Manufacturing Engineering, 1985, p 2-14 to 2-56
- I. Inasaki, Grinding of Hard and Brittle Materials, *Ann. CIRP*, Vol 36 (No. 2), 1987
- *Proceedings of Intersociety Symposium on Machining of Advanced Ceramic Materials and Components*, The American Ceramic Society, 1987
- J. Jablonowski, What is New in Grinders, Special Report 783, *Am. Mach. Auto. Mfg.*, Feb 1986, p 110-124
- H.O. Juchem and H. Wapler, Machining of Ceramics, Ferrites, Carbon and Graphite With Diamond Tools, *Ind. Diamond Rev.*, Feb 1979, p 43-50
- R. Komanduri and D. Maas, Ed., *Proceedings of the Milton C. Shaw Grinding Symposium*, Vol 16, American Society of Mechanical Engineers, 1985
- R.P. Lindsay, "On the Surface Finish-Metal Removal Relationship in Precision Grinding," Paper 72-WA/Prod-13, American Society of Mechanical Engineers, 1972
- R.P. Lindsay, Characteristics of Precision Production Grinding With CBN Wheels, in *Proceedings of North American Metalworking Research Conference*, NAMRC, 1974
- R.P. Lindsay, "Simulated Production Grinding Using Plated CBN Wheels," Paper MR75-105, Society of Manufacturing Engineers, 1975
- R.P. Lindsay, Laboratory Investigations in Support of CBN Production Grinding, *ASME Proc.*, Vol 16, 1985
- Machining of Ceramic Materials and Components, *ASME Proc.*, Vol 17, 1985
- R.L. Mahar, Progress in CBN Production Grinding, in *Superabrasives '85*, Publication MR85-268, Society of Manufacturing Engineering, 1985
- S. Malkin, Current Trends in CBN Grinding Technology, *Ann. CIRP*, Vol 34 (No. 2), 1985
- E.L. Manson and C.M. Sung, Design and Properties of Superabrasives, *Abrasive Eng. Soc. Mag.*, Nov-Dec 1981, p 8-14
- R.E. McEachron, Grinding of Structural Ceramics With Diamond Abrasives, in *Superabrasives '85*, Publication MR85-268, Society of Manufacturing Engineering, 1985
- *Proceedings of the 6th International Conference on Production Engineering* (Osa-

ka), The Japan Society of Precision Engineering, 1987

- L.G. Pukaite and K. Subramanian, Creepfeed Grinding of Silicon Nitride Tool Material, in *Proceedings of The Society of Carbide and Tool Materials Annual Meeting*, The Society of Carbide and Tool Materials, 1987
- E. Salje *et al.*, Internal Grinding of High Strength Ceramic Work Materials with Diamond Grinding Wheels, *Ann. CIRP*, Vol 34 (No. 2), 1985
- K. Subramanian, Make the Best Use of CBN Wheels by Proper Truing, Dressing, and Conditioning, in *Superabrasives '85*, Publication MR85-268, Society of Manufacturing Engineering, 1985
- K. Subramanian and P.P. Keat, Parametric Study of Grindability of Structural and Electronic Ceramics, in *Proceedings of the ASME Winter Annual Meeting*, Vol 17, American Society of Mechanical Engineers, 1985
- K. Subramanian, Superabrasives for Precision Production Grinding—A Case for Interdisciplinary Effort, in *Proceedings of the Symposium on Interdisciplinary Issues in Materials Processing and Manufacturing*, Vol 2, American Society of Mechanical Engineers, 1987, p 665-676
- K. Subramanian, Grinding Technology Smooths Way to Volume Production, *Ceram. Ind. J.*, May 1988
- K. Subramanian, Precision Finishing of Ceramic Components With Diamond Abrasives, *Ceram. Bull.*, Vol 67 (No. 8), 1988, p 1026-1029
- K. Subramanian, Strategic Machine Tool Developments for Precision Production Grinding, in *Proceedings of the 4th Biennial International Machine Tool Technology Conference*, National Machine Tool Builders' Association, 1988
- H. Wapler and H.O. Juchem, Glass and Ceramics Grindings—The Optimum Diamond Grits in the DeBeers Product Range, *Ceram. Ind. J.*, Aug 1979, p 36-40

# Honing

HONING is a controlled, low-speed sizing and surface-finishing process in which stock is removed by the shearing action to the bonded abrasive grains of a honing stone, or stick. Honing machines simultaneously apply several sticks (although one can be used) mounted on the periphery of a cylindrical body to the work surface. The usual purpose of honing is to produce uniform high accuracy and fine finish, most often on inside cylindrical surfaces, and in the most common applications only a few thousandths of an inch of stock is removed although heavy stock removal up to 6.35 mm (0.250 in.) can also be cost effective.

In honing, a simultaneous rotating and reciprocating action of the stone (Fig. 1a) results in a characteristic crosshatch lay pattern (Fig. 1b and c). This combination of motions gives the stones a figure eight travel path. For some applications, such as cylinder bores, angles between the crosshatched lines are important and may be specified within a few degrees. Because honing is a low-speed operation, metal is removed without the increase in workpiece temperature that accompanies grinding, and thus surface damage caused by heat is avoided.

In addition to removing stock, honing serves the important purpose of generating specified functional characteristics for surfaces and involves the correction of errors resulting from previous machining operations. Functional characteristics generated by honing include geometric accuracy (diametric roundness and straightness, and axial straightness), dimensional accuracy, and surface character (roughness, lay pattern, and integrity). Ten common bore errors caused by machining, heat treating, or workholding are illustrated in Fig. 2. Honing can correct all of these conditions with the least possible amount of material removal.

The most frequent application of honing is for finishing inside cylindrical surfaces, but numerous outside surfaces also are honed. Gear teeth, valve components, and races for ball bearings and roller bearings are typical applications of external honing. Microhoning, also known as superfinishing, microfinishing, superfinish honing, short-stroke honing, and microstoning, is closely related to honing. Microhoning uses finer grit (or grain) sizes (320 to 1200) than honing (60 to 600 grit). The primary distinction between honing and microhoning is that the tool rotates in honing, while in microhoning it is always the workpiece that rotates (see Fig. 3). Microhoning is a special honing operation in which the bonded abrasive stone is subjected to very light pressure and a short, high-frequency stroke.

Honing is generally used to produce surface finishes in the range of 0.80 to 0.20 μm (32 to 8 μin.) $R_a$, (roughness average) while microhoning can be cost effective in producing surface finishes of the order of 0.10 to 0.050 μm (4 to 2 μin.) $R_a$. Each method can produce even finer finishes, but productivity, as a result, suffers.

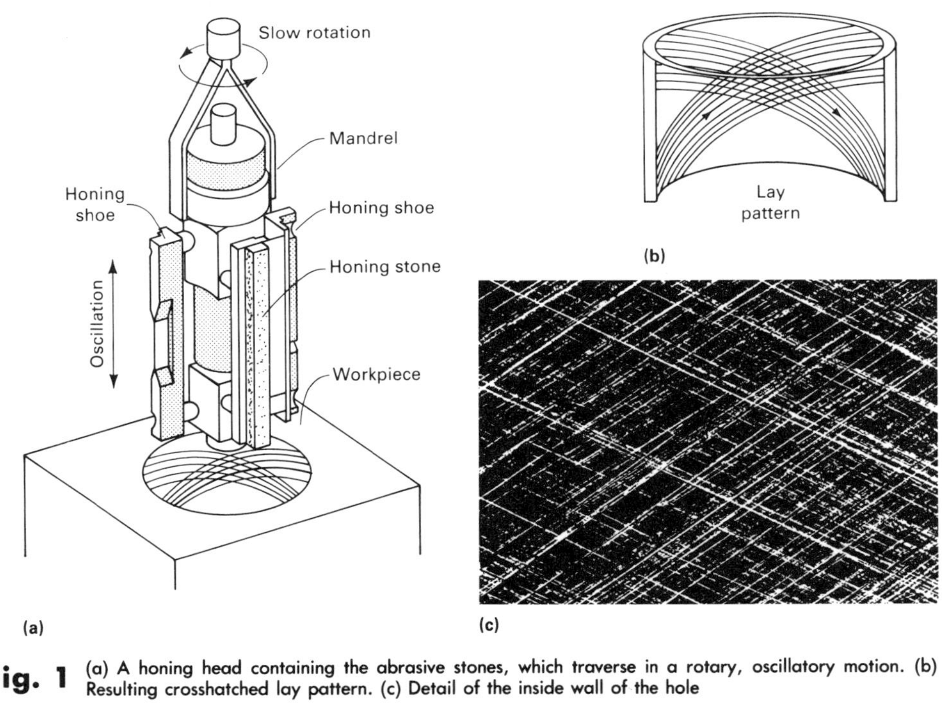


**Fig. 1** (a) A honing head containing the abrasive stones, which traverse in a rotary, oscillatory motion. (b) Resulting crosshatched lay pattern. (c) Detail of the inside wall of the hole

## Honing

Although cast iron and steel are the materials most commonly honed, the process has been used for finishing materials ranging from the softer metals, like aluminum alloys, to extremely hard materials, like nitrided cases or tungsten carbide. Honing has also been used for finishing ceramics and plastics.

## Process Capabilities

**Bore Size.** Bores as small as 1.6 mm (1/16 in.) in diameter can be honed. The maximum diameter of bore that can be honed is governed mainly by the ability of the machine to drive the honing tool and accommodate the workpiece. Machines powered by motors of up to 37 kW (50 hp) are available that can hone bores up to about 1270 mm (50 in.) in diameter. Honing bores up to 760 mm (30 in.) in diameter is common practice.

Bores of almost any length-to-diameter ratio can be honed. In oil-well applications, holes 32 mm (1¼ in.) in diameter and 9.8 m (32 ft) long (a length-to-diameter ratio of 307:1) are honed. At the opposite extreme, the process has been used for 38 mm (1½ in.) diam arbor holes as short as 0.4 mm (1/64 in.) (a length-to-diameter ratio of 1:96).

**Bore Shape.** Although most internal honing is done on simple, straight-through holes, blind holes and tapered holes also can be honed. It is not feasible to hone the sides of a blind hole flush with the bottom; because of the two-motion action that characterizes honing, some relief must be provided at the end of the hole. In holes like engine cylinder bores, a minimum relief of 3.2 mm (⅛ in.) is preferred, although in some operations it has

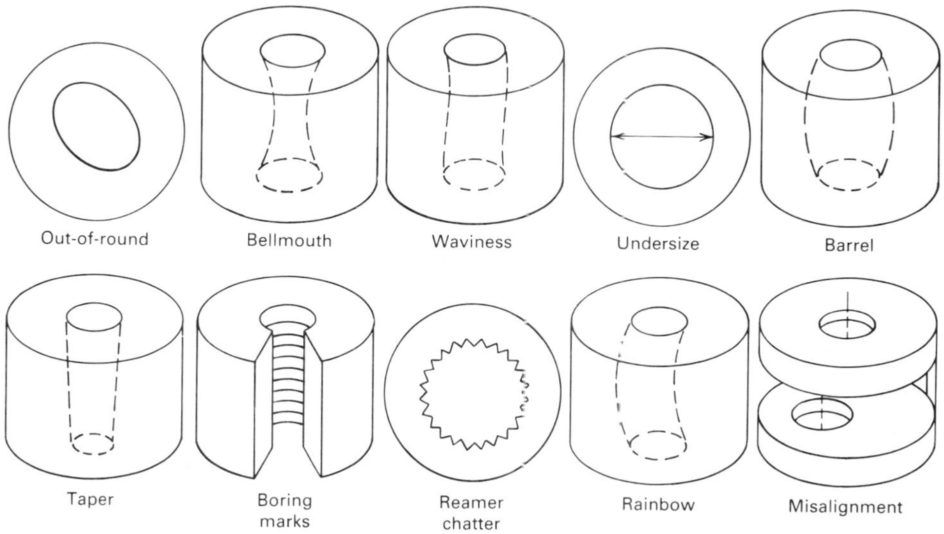


**Fig. 2** Ten common bore errors that can be corrected by honing

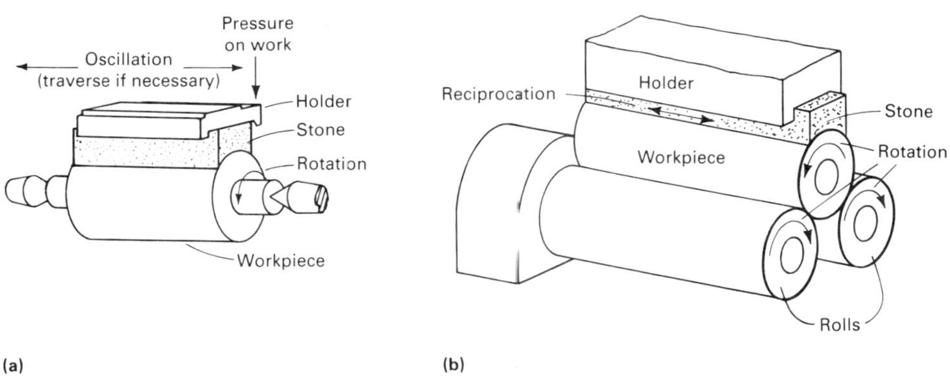


**Fig. 3** Principal components and their motion in (a) cylindrical and (b) centerless microhoning

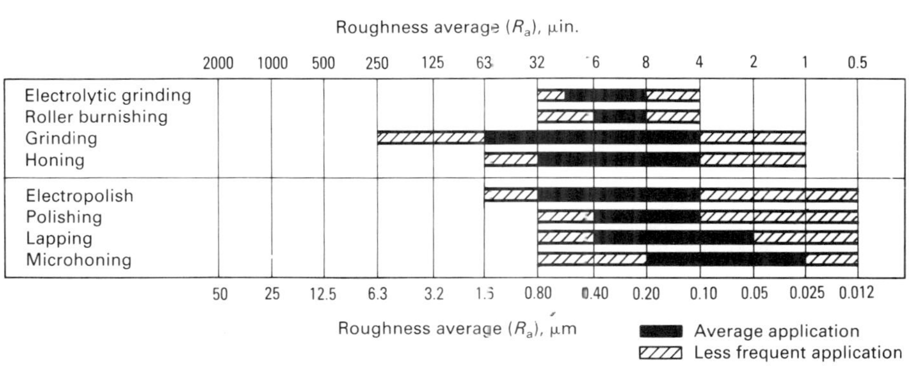


**Fig. 4** Typical ranges of surface finish from common production processes. Higher or lower values may be obtained under various conditions.

been possible to hone within 0.38 mm (0.015 in.) of the blind end.

Bores having keyways can be honed, and so can male or female splines.

**Stock Removal.** In honing, a general rule is to remove twice as much stock as the existing error in the workpiece. For instance, if a cylinder is 0.050 mm (0.002 in.) out-of-round or tapered, the removal of about 0.10 mm (0.004 in.) will be required for complete cleanup.

Because honing is seldom economical for removing large amounts of stock, preceding operations in high-production work are usually planned so that the amount of stock removed in honing is minimized. On the other hand, stock removal of up to 6.4 mm (¼ in.) may be practical in some applications. For instance, as much as 2.54 mm (0.100 in.) is honed from the inside diameter of hydraulic cylinders, because stock removal by honing is more practical and economical than attaining close preliminary dimensions by grinding or boring. In finishing the bores of long tubes, even larger amounts (as much as 6.35 mm, or 0.250 in.) may be removed by honing, because it is the only practical method. Such tubes are finished by honing immediately after drawing. In these operations as much as 32 000 $mm^3$/min (2 $in.^3$/min) can be removed from soft steel surfaces. For steel hardened to about 60 HRC, the rate of removal decreases to about 16 000 $mm^3$/min (1 $in.^3$/min).

Rough honing usually is employed before finish honing when large amounts of stock are to be removed and specific finishes are required. Stones containing abrasive as coarse as 80 grit, or even coarser, are used for rough honing to obtain maximum rate of stock removal. Finish honing would use abrasives of 180 to 320 grit, or finer.

## Honing Versus Other Processes

For some shapes, honing is the only practical means of attaining the required accuracy. A typical example is holes whose length greatly exceeds the diameter. For finishing these holes, grinding is precluded because of the long spindle overhang, and lapping is excessively tedious and expensive.

When inside diameter approaches or exceeds bore length, acceptable results can often be produced by either grinding or honing. A choice between the two methods then depends on cost, on availability of equipment, and sometimes on established policies within a plant. Figure 4 compares typical ranges of surface finishes obtained by honing and microhoning to other common production finishing processes.

## Machines

When only a few parts are to be finished and operators are skilled, parts can be honed successfully on a drill press or an engine lathe, on which arrangements can be made for simultaneous rotating and reciprocating motions. Vertical drill presses have often been used to drive honing tools. The tool is mounted in the chuck, and the workpiece is fixtured in such a way that it can float. The spindle provides rotary motion, and the reciprocating motion (stroking) can be done manually or by power, depending on the equipment. Feed-out of the honing stones is necessarily a manual operation. Various types of lathes and boring mills also have been tooled for honing. For some applications, portable machines like electric drills have been used to rotate the honing tools, while stroking is done manually. Portable machines are used for honing bores in structures (such as mill or refinery equipment) that are too large to be transported to a stationary machine. The bores honed need not be large; holes ranging from 65 to 510 mm (2½ to 20 in.) in diameter have been honed using portable machines.

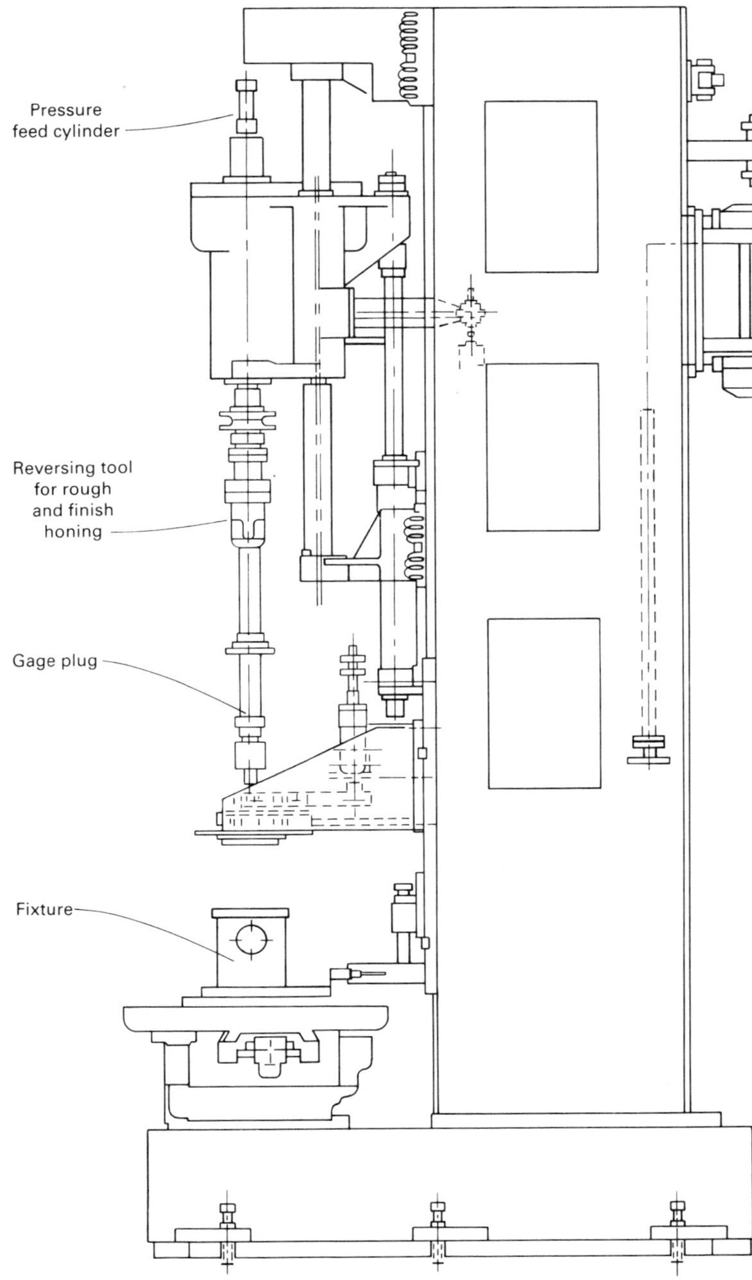


**Fig. 5** Vertical single-spindle hydraulic machine for honing long, large-diameter bores

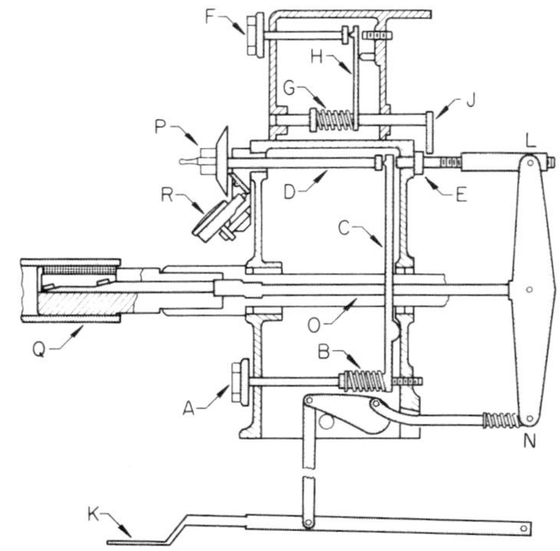


**Fig. 6** Typical manual-stroke honing machine employing automatic size control. Functions of components listed are described in the text. A, heavy cutting pressure control; B, spring; C, lever; D, feed screw; E, collar; F, light cutting pressure control; G, spring; H, lever; J, rod-and-fork assembly; K, foot pedal; L, M, and N, feed arm; O, honing stone expansion link; P, feed-out dial; Q, workpiece; R, honing dial

Most production honing, however, is done with machines built for the purpose. These machines are available in a wide range of sizes and designs, in both vertical and horizontal types.

A vertical spindle honing machine resembles an upright drill press or boring machine (Fig. 5). The machine normally incorporates a mechanism not only to rotate but also to reciprocate the honing tool or tools and expand the stones until size is reached. It may have a single spindle or a number of spindles for production, similar to a vertical boring machine. An example of eight-spindle honing machine capability is the removal of 0.010 mm (0.004 in.) of stock in each of eight cylinders in 30 s to a tolerance of less than 0.01 mm (0.0005 in.).

Typical for larger work, the tool is rotated and reciprocated by the head on the end of the bed of the horizontal hydraulic honing machine. Such a single-stroke horizontal honing machine with a 3.7 m (12 ft) spindle travel would be used to finish 105 mm (4.13 in.) gun barrels. Machines like this have been made with strokes up to approximately 25 m (80 ft) to hone holes over 990 mm (39 in.) in diameter.

Special machines that drive the honing tools at an angle are frequently used for honing bores in V-engine blocks. For high-production honing of similar parts, installations may even include automatic gaging.

Some honing machines require manual stroking of the workpiece; others have power-stroking mechanisms.

**In manual stroking,** the machine rotates the tool at a preestablished speed while the workpiece has free lateral movement and is stroked back and forth over the tool. These machines may be equipped with devices for controlled feed-out of the honing stones.

Figure 6 illustrates and identifies important components of a typical manual-stroke honing machine. In this machine, feed-out and subsequent size control are accomplished by spring loading. Simpler machines of this general type are available that use manual feed-out. Functions of the machine components shown in Fig. 6 are described below.

Setting heavy cutting pressure control A preloads spring B to the desired honing pressure in the heavy range (for roughing and stock removal on larger bores). The spring force, through lever C, forces feed screw D to the left until stopped by collar E.

When relatively light cutting pressure is required (as in small-bore honing and finishing operations), heavy cutting pressure control A is turned all the way counterclockwise, thus entirely unloading spring B. Then, setting the light cutting pressure control F preloads spring G, through lever H, to the desired honing pressure in the light range. This spring pressure forces rod and fork assembly J to the left, which in turn forces feed screw D to the left by acting on collar E. The resultant action on feed screw D is the same as that produced by heavy cutting pressure control A, except that forces are much lighter.

Depressing foot pedal K causes feed arm L-M-N to pivot around point L and advance the honing stone expansion link O by a fixed amount. Clockwise rotation of feed-out dial P also advances the honing stone by moving point L to the left with N as a pivot, advancing link O until the honing stone contacts workpiece Q.

Additional clockwise rotation of feed-out dial P cannot expand the honing stone further, and points L, M, and N remain fixed. However, this additional rotation moves feed screw D to the right and "compresses" either spring B through level C or spring G through rod-and-fork assembly J (depending on which cutting-pressure control is being used). This endwise movement of feed screw D to the right also actuates honing dial R. The reading on this dial now shows the amount of stock to be removed, which was preselected by advancing feed-out dial P.

During the honing operation, either spring B or spring G (depending on which cutting-pressure control is being used) acts on feed screw D through its respective linkage and causes points L and M, as well as link O, to move to the left, thus forcing the honing stone outward under the preset honing pressure. As the diameter of the honed hole gets larger, feed screw D moves to the left until honing dial R registers zero, at which time the honing is stopped by releasing foot pedal K. The work is now ready for gaging.

The operating cycle is repeated from part to part. The honing tool expanding mechanism is set for size on the first part and, except for a slight advance of the stone to compensate for wear, additional parts require no resetting of controls.

**In power stroking,** the workpiece usually is held stationary in a rigid fixture, while the honing tool (which is generally powered hydraulically) is rotated and reciprocated.

Length of workpiece often dictates whether a vertical or a horizontal machine can be used. Vertical machines are often preferred because they permit easier fixturing of workpieces. Also, it is easier to obtain straight holes when vertical machines are used. Vertical machines have been built that can hone workpieces 1.8 m (6 ft) long, but these machines are 7.0 m (23 ft) high—a practical limit. Workpieces longer than 1.8 m (6 ft) (even this length is unusual for vertical stroking) are stroked on a horizontal machine. As on vertical machines, the workpiece usually remains stationary while the tool rotates and reciprocates, although there are numerous instances in which it is advantageous to rotate the workpiece. For example, in honing bores in long tubes that are held stationary during the operation, occasionally the entire finished bore is slightly off center. The error is caused by the weight of the honing head and driving shank; it is corrected by rotating the tube during honing.

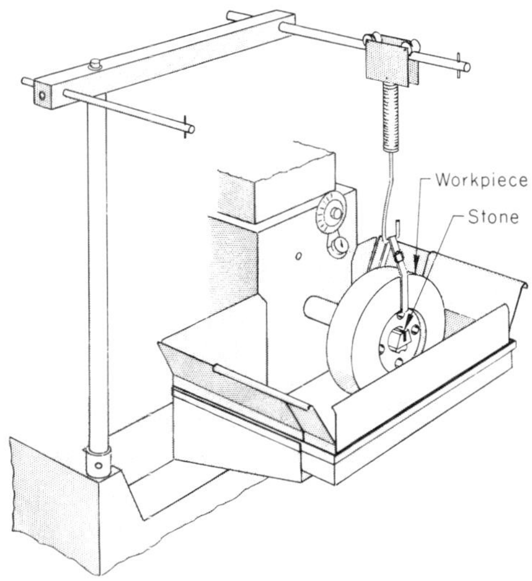


**Fig. 7** Typical device used for supporting heavy or cumbersome workpieces during manual-stroke honing

Feed-out of the honing stones can be controlled manually, but automatic control is more common in production honing. Either hydraulic or pneumatic force can be used to keep a predetermined constant pressure on the honing stones. Some machines employ an electromechanical means, which feeds out the stones at a constant rate instead of under constant pressure.

## Selection of Machine

Size and shape of the workpieces are usually the major factors that determine whether manual or power stroking is more appropriate. However, the quantity of similar pieces to be honed, tolerance requirements, availability of equipment, availability of skilled operators, and specific plant policies also affect the choice of honing method and subsequently the choice of machine.

**Manual stroking** is widely used and often preferred for parts that an operator is able and permitted to hold. Most bores honed by manual stroking are 25 mm (1 in.) in diameter or less, although bores up to about 127 mm (5 in.) in diameter (and even much larger, in rare instances) have been successfully finished. Bores up to about 460 mm (18 in.) long have been satisfactorily honed by manual stroking.

Supports may be used for workpieces that are too heavy or cumbersome for an operator to hold (Fig. 7). However, supports are practical only for honing a few parts of a kind, as in the reconditioning of tools by honing.

Manual stroking may be used in preference to power stroking for large quantities when tolerances are extremely close and size and shape of the workpiece permit. One advantage of manual stroking is that workpieces need not be fixtured, which reduces tooling investment and permits immediate changeover from one job to another.

Because manual honing is controlled largely by the operator, who can instantly gage the part he holds, closer tolerances often can be obtained. Techniques used to obtain the accuracy required include end-for-end reversal of the workpiece, change of stroke length, and quick hone-and-try. Also, when various areas of a given part require differing degrees of correction (because of bellmouth, taper, or other irregularities), the operator can favor those areas as stroking proceeds.

Manual honing is sometimes used as an adjunct to power honing in high production. Most of the honing is done by power, after which the final touch for tolerance of finish is achieved manually by skilled operators.

**Power stroking** can be used for honing virtually all types and sizes of workpieces. In high production of small parts, power stroking may prove more economical than manual stroking. Valve lifters are typical of small parts that are power stroked when production quantities are high (in excess of 50 000). Power stroking and fixtured honing are required for workpieces that exceed the size or weight that can be handled manually.

**Table 1 Grit sizes and corresponding particle sizes used in honing**

| Grit size | Average particle size, μm | in. |
|---|---|---|
| 36 | 710 | 0.0280 |
| 46 | 508 | 0.0200 |
| 54 | 430 | 0.0170 |
| 60 | 406 | 0.0160 |
| 70 | 328 | 0.0131 |
| 80 | 266 | 0.0105 |
| 90 | 216 | 0.0085 |
| 100 | 173 | 0.0068 |
| 120 | 142 | 0.0056 |
| 150 | 122 | 0.0048 |
| 180 | 86 | 0.0034 |
| 220 | 66 | 0.0026 |
| 240 | 63 | 0.00248 |
| 280 | 44 | 0.00175 |
| 320 | 32 | 0.00128 |
| 400 | 23 | 0.00090 |
| 500 | 16 | 0.00065 |
| 600 | 8 | 0.00033 |

Source: Metcut Research Associates Inc.

## Honing Stones

Honing stones (known also as honing sticks) consist of particles of aluminum oxide, silicon carbide, or diamond bonded together with vitrified clay, resinoid, cork, carbon, or metal. The abrasive particles, or grits, which provide the cutting action, must be able to withstand the pressure required for removing metal. The bond must be strong enough to hold the grit, but it must not be hard enough to rub the bore and thus retard cutting. The hardness and type of bond are indicated by code letters in the identification of the stone. The porosity of

**Table 2 Stone selection for honing internal diameters on a variety of materials**

| Material | Burr and rough hone | | | Stock removal | | | Fine finish | | |
|---|---|---|---|---|---|---|---|---|---|
| | Stone type(a) | Grit size | Stone hardness(b) | Stone type(a) | Grit size | Stone hardness(b) | Stone type(a) | Grit size | Stone hardness(b) |
| Agate | A | 150 | Extremely hard | D | 150 | Medium hard | D | 600 | Medium hard |
| Alnico | A | 150 | Extremely hard | B | 100 | Medium soft | C | 500 | Medium soft |
| Aluminum | A | 150 | Extremely hard | C | 220 | Medium soft | C | 500 | Medium soft |
| Aluminum, hard coat | A | 150 | Extremely hard | C | 220 | Medium soft | C | 500 | Medium soft |
| Boron carbide | A | 150 | Extremely hard | C | 150 | Medium hard | D | 600 | Medium hard |
| Brass (yellow) | A | 150 | Extremely hard | C | 280 | Medium soft | C | 400 | Soft |
| Bronze (soft) | A | 150 | Extremely hard | C | 280 | Medium hard | C | 400 | Medium soft |
| Carbon | A | 150 | Extremely hard | C | 280 | Soft | C | 500 | Soft |
| Cast iron | A | 150 | Extremely hard | C | 220 | Medium hard | C | 500 | Medium soft |
| Cemented carbides | A | 150 | Extremely hard | D | 220 | Medium hard | D | 600 | Medium hard |
| Ceramics | A | 150 | Extremely hard | D | 150 | Medium hard | D | 600 | Medium hard |
| Chilled iron | A | 150 | Extremely hard | C | 280 | Soft | C | 500 | Soft |
| Hard chromium plate | A | 150 | Extremely hard | D | 150 | Medium hard | C | 400 | Soft |
| Porous chromium plate | A | 150 | Extremely hard | A | 150 | Soft | C | 400 | Soft |
| Chrome carbide | A | 150 | Extremely hard | D | 150 | Medium hard | D | 600 | Medium hard |
| Colmonoy | A | 150 | Extremely hard | D | 150 | Medium hard | D | 600 | Medium hard |
| Copper | A | 150 | Extremely hard | C | 280 | Soft | C | 500 | Soft |
| Glass | A | 150 | Extremely hard | D | 220 | Medium hard | D | 600 | Medium hard |
| Gold | A | 150 | Extremely hard | C | 280 | Soft | C | 400 | Soft |
| Inconel | A | 150 | Extremely hard | B | 100 | Medium soft | C | 500 | Medium soft |
| Lucite | A | 150 | Extremely hard | C | 280 | Soft | C | 500 | Medium soft |
| Magnesium | A | 150 | Medium hard | C | 220 | Medium soft | C | 500 | Medium soft |
| Malleable iron | A | 150 | Extremely hard | C | 220 | Medium hard | C | 500 | Medium soft |
| Molybdenum | A | 150 | Extremely hard | A | 150 | Extremely hard | C | 500 | Extremely hard |
| Monel metal | A | 150 | Extremely hard | A | 220 | Medium hard | C | 500 | Medium soft |
| Niobium | A | 150 | Extremely hard | A | 220 | Extremely hard | C | 500 | Medium hard |
| Nitralloy | | | | | | | | | |
| Before nitriding | A | 150 | Extremely hard | A | 220 | Medium hard | C | 500 | Medium soft |
| After nitriding | A | 150 | Medium soft | B | 100 | Medium soft | C | 500 | Soft |
| Porcelain | D | 150 | Medium hard | D | 150 | Medium hard | D | 600 | Medium hard |
| Rexalloy and similar cemented and cast cutting tool materials | D | 150 | Medium hard | D | 150 | Medium hard | D | 600 | Medium hard |
| Rubber, hard | C | 150 | Medium hard | C | 280 | Medium soft | C | 400 | Medium soft |
| Silver | C | 150 | Medium soft | C | 280 | Soft | C | 500 | Soft |
| Steel, hard | A | 150 | Extremely hard | B | 100 | Medium soft | C | 500 | Soft |
| Steel, soft | A | 150 | Extremely hard | A | 220 | Medium hard | C | 500 | Medium soft |
| Steel, high speed | A | 150 | Extremely hard | A | 280 | Soft | C | 500 | Soft |
| | | | | C | 280 | Soft | B | 400 | Medium soft |
| | | | | B | 100 | Medium soft | | | |
| Steel, soft stainless | A | 150 | Extremely hard | A | 200 | Medium hard | C | 500 | Medium soft |
| Steel, hard (over 60 HRC) | A | 150 | Extremely hard | B | 100 | Medium soft | C | 500 | Soft |
| Stellite | A | 150 | Extremely hard | B | 100 | Medium soft | C | 500 | Soft |
| Titanium | A | 150 | Extremely hard | A | 220 | Medium hard | C | 500 | Medium soft |
| Tantalum carbide | D | 150 | Medium hard | D | 150 | Medium hard | D | 600 | Medium hard |
| Titanium carbide | D | 150 | Medium hard | D | 150 | Medium hard | D | 600 | Medium hard |
| Tungsten carbide | D | 150 | Medium hard | D | 150 | Medium hard | D | 600 | Medium hard |
| Tungsten, pure | A | 150 | Extremely hard | A | 220 | Extremely hard | C | 500 | Medium hard |
| Vanadium, chrome steel alloys | A | 150 | Extremely hard | C | 400 | Soft | C | 500 | Soft |
| Wrought iron | A | 150 | Extremely hard | A | 220 | Medium hard | C | 500 | Medium soft |

(a) Abrasive type: A, aluminum oxide; B, cubic boron nitride (metal bonded); C, silicon carbide; D, (metal bonded). (b) The term extremely hard denotes stones approximately in grades Q to S; medium hard denotes stones approximately in grade P; medium soft denotes stones approximately in grade L; soft denotes stones approximately in grade H. Stones with different surface areas should be of different grades to have the same effective hardness. Some honing stone manufacturers have taken size differences into account, making it unnecessary to select different grades to compensate for surface area. Consulting suppliers is recommended. Source: Metcut Research Associates Inc.

the honing stone, which facilitates chip clearance and minimizes heat generation, is controlled during molding.

**Grit size** may range from 36 to 600, but 120 to 320 is the range most widely used (see Table 1). Suitable sizes are available for every type and size of bore honed.

Selection of grit size depends mainly on the desired rate of metal removal and the required finish. Coarse grit removes metal faster, but results in a rougher finish. For this reason, rough honing followed by finish honing is often economical. In some plants, stones of different grit sizes are alternated to obtain a compromise. The value of this practice is controversial, but it has been found that by alternating 150-grit and 180-grit stones on the same tool, cutting action approached that of 150 grit and surface finish approached that normally obtained with 180 grit.

**Selection of abrasive** depends mainly on the composition and hardness of the metal being honed, the finish required, and cost factors. Typical abrasives for honing various metals are listed in Table 2 according to surface finish; parameters needed to hone various metals and alloys according to hardness are listed in Table 3; parameters needed to hone nonmetals are shown in Table 4.

Table 5 lists guidelines for selecting aluminum oxide, silicon carbide, cubic boron nitride (CBN), and diamond and shows typical cutting speeds and stone pressures for these abrasives. Abrasive selection for honing in terms of workpiece material alone is:

- Aluminum oxide is widely used on steels
- Silicon carbide is generally used on cast iron and nonferrous materials
- Cubic boron nitride is used on all steels (both soft and hard), nickel- and cobalt-base superalloys, stainless steels, other alloys such as beryllium copper, and metals such as zirconium
- Diamonds are used on chromium plate, carbides, ceramics, glass, cast irons, brass, bronze, and surfaces nitrided to depths greater than 0.03 mm (0.001 in.)

Table 6 provides detailed information on the honing of specific materials with cubic boron nitride.

**Table 3 Operating parameters for honing metals and alloys using power or manual stroke**

| Hardness, HB | Honing stone material: Type(a) | Honing stone material: Grade(b) ANSI and ISO | Grit size, for surface roughness ($R_a$), μm (μin.): 0.025–0.125 (1–5) | 0.15–0.25 (6–10) | 0.30–0.50 (11–20) | 0.53–0.75 (21–30) | >0.75 (>30) | Spindle motion: Rotation speed m/min | Rotation speed sfm | Reciprocation speed m/min | Reciprocation speed sfm | Working pressure kPa | Working pressure psi | Cutting fluid(c) |
|---|---|---|---|---|---|---|---|---|---|---|---|---|---|---|
| **Carbon and alloy steels, wrought or cast** | | | | | | | | | | | | | | |
| 100–300 | A or B | Q | 600 | 500 | 320 | 220 | 180 | 30–91 | 100–300 | 5.5–33 | 18–110 | 276 | 40 | A |
| 300–400 | A or B | Q | 500 | 500 | 320 | 220 | 180 | 27–84 | 90–275 | 4.9–30 | 16–100 | 276 | 40 | A |
| 45–52 HRC | A or B | N | 500 | 400 | 280 | 220 | 120 | 24–76 | 80–250 | 4.3–27 | 14–90 | 414 | 60 | A |
| 52–58 HRC | A or B | K | 500 | 320 | 280 | 120 | 120 | 23–69 | 75–225 | 3.9–25 | 13–82 | 517 | 75 | A |
| 58–60 HRC | B or A | I | 400 | 280 | 220 | 120 | 120 | 20–69 | 65–205 | 3.4–23 | 11–75 | 517 | 75 | A |
| >60 HRC | B or A | I | 320 | 220 | 120 | 120 | 120 | 18–56 | 60–185 | 3.0–21 | 10–68 | 690 | 100 | A |
| **Tool steels, wrought** | | | | | | | | | | | | | | |
| 150–300 | A or B | Q | 600 | 500 | 320 | 220 | 180 | 23–64 | 75–210 | 4.0–23 | 13–75 | 345 | 50 | A |
| 300–400 | A or B | N | 500 | 500 | 320 | 220 | 180 | 21–61 | 70–200 | 3.7–22 | 12–73 | 345 | 50 | A |
| 45–52 HRC | A or B | N | 400 | 320 | 220 | 150 | 120 | 20–55 | 65–180 | 3.4–20 | 11–65 | 517 | 75 | A |
| 52–58 HRC | A or B | K | 400 | 280 | 220 | 150 | 120 | 15–46 | 50–150 | 2.7–16.5 | 9–55 | 690 | 100 | A |
| >60 HRC | B or A | I | 320 | 220 | 180 | 120 | 120 | 15–46 | 50–150 | 2.7–16.5 | 9–55 | 830 | 120 | A |
| **Stainless steels, wrought and cast** | | | | | | | | | | | | | | |
| 135–315 | A or B | O | 600 | 500 | 320 | 280 | 180 | 30–76 | 100–250 | 5.5–27 | 18–90 | 276 | 40 | A |
| 315–420 | A or B | N | 500 | 500 | 320 | 220 | 150 | 26–72 | 85–235 | 4.6–26 | 15–85 | 276 | 40 | A |
| 45–52 HRC | A or B | N | 500 | 400 | 280 | 220 | 150 | 21–67 | 70–220 | 3.7–24 | 12–80 | 414 | 60 | A |
| 52–58 HRC | A or B | I | 500 | 320 | 220 | 150 | 120 | 15–61 | 50–200 | 2.7–23 | 9–75 | 517 | 75 | A |
| **Gray, ductile, and malleable cast iron** | | | | | | | | | | | | | | |
| 110–315 | C or D | Q | 500 | 500 | 320 | 220 | 180 | 36.5–100 | 120–330 | 6.4–36.5 | 21–120 | 276 | 40 | B |
| 315–420 | C or D | Q | 500 | 500 | 320 | 220 | 180 | 26–91 | 85–300 | 4.6–33.5 | 15–110 | 276 | 40 | B |
| 45–52 HRC | C or D | N | 500 | 400 | 320 | 220 | 120 | 20–67 | 65–220 | 3.4–25 | 11–82 | 414 | 60 | B |
| **Austenitic (Ni-Resist), gray, and malleable cast irons** | | | | | | | | | | | | | | |
| 100–270 | C or D | J to S | 500 | 400 | 280 | 220 | 120 | 38–68.5 | 125–225 | 12.2–24 | 40–80 | 345 | 50 | A |
| **Aluminum alloys, wrought and cast** | | | | | | | | | | | | | | |
| 30–450 (500 kg, or 1100 lb) | C or D | L | 600 | 500 | 400 | 280 | 220 | 15–64 | 50–210 | 2.7–23 | 9–75 | 276 | 40 | A |
| **Titanium alloys, wrought** | | | | | | | | | | | | | | |
| 110–315 | C or D | N | 600 | 500 | 400 | 280 | 180 | 15–61 | 50–200 | 2.7–23 | 9–75 | 276 | 40 | A |
| 315–440 | C or D | N | 500 | 500 | 400 | 280 | 180 | 9–46 | 30–150 | 1.5–16.5 | 5–55 | 414 | 60 | A |
| **Copper alloys, wrought and cast** | | | | | | | | | | | | | | |
| 40–200 (500 kg, or 1100 lb) | C or D | I | 600 | 500 | 500 | 320 | 180 | 15–76 | 50–250 | 2.7–27 | 9–90 | 276 | 40 | A |
| **Nickel alloys, wrought and cast** | | | | | | | | | | | | | | |
| 80–315 | A or B | J | 600 | 500 | 400 | 280 | 180 | 30–76 | 100–250 | 5.5–27 | 18–90 | 276 | 40 | A |
| **Nickel alloys, wrought and cast** | | | | | | | | | | | | | | |
| 315–420 | A or B | J | 500 | 500 | 400 | 220 | 180 | 27–69 | 90–225 | 4.9–24 | 16–80 | 276 | 40 | A |
| 45–52 HRC | A or B | J | 500 | 400 | 320 | 220 | 180 | 24–61 | 80–200 | 4.3–23 | 14–75 | 517 | 75 | A |
| **High-temperature alloys, wrought and cast** | | | | | | | | | | | | | | |
| 200–315 | C or D | J | 500 | 500 | 320 | 280 | 180 | 24–61 | 80–200 | 4.3–23 | 14–75 | 414 | 60 | A |
| 315–475 | C or D | J | 500 | 400 | 320 | 280 | 180 | 15–49 | 50–160 | 2.7–18 | 9–60 | 414 | 60 | A |
| **Refractory alloys** | | | | | | | | | | | | | | |
| 180–320 | A or B | Q | 500 | 400 | 320 | 280 | 180 | 15–61 | 50–200 | 2.7–23 | 9–75 | 414 | 60 | A |
| **Chromium plate** | | | | | | | | | | | | | | |
| Hard | A or D | L | 320 | 280 | 220 | 180 | 120 | 27–91 | 90–300 | 4.9–33.5 | 16–110 | 690 | 100 | A |

(a) Abrasive type: A, aluminum oxide; B, cubic boron nitride (metal bonded); C, silicon carbide; D, diamond (metal bonded). (b) Abrasive grade: soft (F, G, H, and I); medium (J, K, L, M, and N); and hard (O, P, Q, R, and S). (c) Honing fluid: A, 70% kerosene-30% sulfurized or chlorinated oil; B, kerosene. Source: Metcut Research Associates Inc.

Aluminum oxide and silicon carbide stones are comparable in initial cost. However, one may be more economical than the other because of increased service life.

Diamond stones initially cost up to 20 times as much as aluminum oxide or silicon carbide stones, depending on diamond concentration and thickness of abrasive layer. Diamond stones are almost mandatory, however, for honing extremely hard, wear-resistant materials such as tungsten carbide or ceramics. For honing nitrided cases or chromium plate, diamond stones have often been more economical on the basis of metal removed per unit of time. Table 7 lists parameters associated with single-stroke honing when using diamond-plated tools.

**Designations.** An abrasive marking system that applies to both grinding wheels and honing stones has been established by the abrasives manufacturers. Details of this system are given in the article "Grinding Equipment and Processes" in this Volume. The four most significant characteristics are type of grit, grit size, hardness grade, and type of bond. These characteristics are identified by letters and numbers in the above order. For example, an aluminum oxide stone having a grit size of 180, a grade of R, and a vitrified bond is identified as A-180-R-V. This method of stone designation is used in this article.

**Table 4 Selection of honing stone grit sizes for nonmetals using power or manual-stroke operations**

| Material | Honing stone material Type(a) | Grade(b) ANSI and ISO | Grit size for surface roughness ($R_a$), µm (µin.) 0.025–0.125 (1–5) | 0.15–0.25 (6–10) | 0.30–0.50 (11–20) | 0.53–0.75 (21–30) | >0.75 (>30) | Working pressures |
|---|---|---|---|---|---|---|---|---|
| Glass | A | S | 600 | 400 | 280 | 220 | 150 | Heavy, roughing |
| | D | P | 400 | 280 | 220 | 150 | ... | Normal |
| | D | L | 400 | 220 | 150 | ... | ... | Light, feathering out |
| Ceramic | A | S | ... | ... | 600 | 400 | 280 | Heavy, roughing |
| | D | P | ... | 600 | 400 | 280 | 220 | Normal |
| | D | L | 600 | 400 | 280 | 220 | 150 | Light, feathering out |
| Carbide | A | S | 600 | 400 | 280 | 220 | 150 | Heavy, roughing |
| | D | P | 600 | 400 | 280 | 220 | 150 | Normal |
| | D | L | 400 | 280 | 220 | 150 | ... | Light, feathering out |

(a) Abrasive type: A, aluminum oxide; D, diamond (bonded metal). (b) Abrasive grade: medium (L); hard (P and S). Source: Metcut Research Associates Inc.

## Tools

A phenomenon of honing that makes it possible to develop a round, straight bore is the relationship of the cutting faces of the stones to the surface being honed, which is completely independent of the machine. The fact that either the tool or the workpiece floats (see Fig. 8) enables the tool to exert equal pressure on all sides of the bore, regardless of vibration in the machine or its environment.

Except in special cases, such as with the square-axis requirement described next, honing will not change the axial location cf a hole. The centerline of the tool follows the neutral axis of the bore as established by previous operations; the tool or workpiece must float so that the bore and the tool may align themselves. If practical, the workpiece should be mounted in a fixture that permits it to float. If the part is too large or unbalanced to float, however, the tool should be designed with universal ball joints. The fixtures must be designed to locate the workpiece approximately in line with the machine spindle and take the torque and axial thrust of the tool without distorting the workpiece.

For bores that must be held square with an end surface, the part location method has been developed. In this technique, both the tool and fixturing are rigid. The workpiece is slipped over the tool and automatically aligned with it before clamping.

Honing tools differ in design depending on whether they are to be used with manual or power stroking.

**Manual Stroking.** In manually stroked honing, the tool is rotating around a fixed axis, but the work has free lateral motion and is effectively guided by the honing mandrel. A tool for manual stroking consists of one or more abrasive stones, a mandrel and a wedge, and guide shoes. The simplest type is illustrated in Fig. 9(a) and (b). The purpose of the unequal angular spacing of the stones with respect to the shoes (Fig. 9a) is to facilitate removal of stock from the high spots until roundness is attained. The wedge, which controls feed-out of the stone, can be actuated manually or by an automatic mechanism such as that shown in Fig. 6. The two parallel shoes (Fig. 9b) stabilize and guide the workpiece. Because the shoes wear very slowly, alignment with the bore is maintained. Shoes are made of materials that are wear resistant with respect to the workpiece material. Bronze, cast iron, steel, or even plastics have been used. Sintered shoes made by powder metallurgy can retain oil.

Honing tools of the type illustrated in Fig. 9 are available for honing bores with diameters from 1.6 to 100 mm (1/16 to 4 in.). Tools for manual-stroke honing of long bores (up to 460 mm, or 18 in.) have two or more short stones mounted in line, as shown in Fig. 9(c). Tools and stones should be long enough to permit bridging of common irregularities (Fig. 10).

Tools incorporating modifications of the design shown in Fig. 9 are available for honing holes having surface irregularities such as keyways (Fig. 11a) and for honing blind holes with and without relief (Fig. 11b and c). When honing blind holes, stone and shoe should be of equal length and shorter than the depth of the hole. In Fig. 11(b) and (c), stone and shoes have been equally shortened to ensure uniform wear. Note the shoe clearance at the blind end.

**Power Stroking.** Heavier workpieces, which, because of their bulk and weight, are not adapted to free floating, are honed with power-stroked heads whose shafts, which provide the connection with the rotating and reciprocating machine spindle, are equipped with a set of universal joints. Tools for power stroking (fixtured honing) usually have stones spaced at equal distances around the circumference and may or may not include guide elements (Fig. 12). Expanding cones (Fig. 13) control feed-out. Although the cones may be actuated manually, they are usually actuated automatically by constant-pressure or constant-feed methods. As in manual stroking, the tool may incorporate numerous modifications, depending on the application.

Figure 12 shows a typical tool used in fixtured honing. Tools of this type are used for multiple-spindle honing of parts such as motor blocks. For this application, tools must be self-aligning because it is impossible to align each cylinder bore with each spindle centerline. Therefore, each tool incorporates a double universal joint. The nonabrading guides that are placed between

**Table 5 Guidelines for selecting honing abrasives**

| Abrasive type | Applications | Cutting speed(a) m/min | sfm | Honing stone pressure(b) kPa | psi |
|---|---|---|---|---|---|
| Aluminum oxide ($Al_2O_3$) | Low cost, lowest hardness. Used for both soft or hard but rough, stringy materials; ordinary steels; and some nonferrous alloys. Wear ratio is 5:10. Good for deburring, interrupted cuts, poorly qualified bores, and scale and weld distortions | 5–30 | 16–100 | 100–400 | 15–60 |
| Silicon carbide (SiC) | Low cost. Used for either soft or hard but brittle materials or low shear-resistant materials. Results in more uniform surface finishes. Frequently used for finishing. Widely used on cast iron | 5–30 | 16–100 | 100–400 | 15–60 |
| Cubic boron nitride | High cost. Rapid cutting and long-life stones. Wear ratio is usually 800:1200. Low noise levels. Used for hard steels, tool steels, stellites, nickel- and cobalt-base alloys, and superalloys. Minimum speed is 75m/min (250 sfm) when honing mild steels (AISI 1015-1026) | 35–90 | 110–180 | 400–2000(c) | 60–300(c) |
| Diamond | High cost. Used for tungsten carbide, ceramics, glass, and cast iron. Very long life | 40–80 | 130–260 | 600–1200 | 85–170 |

(a) Cutting speed is the combination (vector sum) of axial and peripheral speed. The crosshatch angle generally runs between 45° and 80°—some evidence indicates the best removal rates come at 60° to 75°. A convenient way to arrive at a good rev/min for large-diameter bores (≥64 mm, or 2½ in.) is: optimum rev/min = 30 000/diam in mm or 1200/diam in in. These values represent approximately 90 m/min (300 sfm). For small diameters, use safer lower speeds shown in Table 3. (b) These values represent conservative starting values. A convenient starting point for any abrasive is the low end of the pressure range. To improve productivity, trials at successively higher pressures should be made until the stone wears too rapidly. (c) When substituting CBN abrasives for conventional abrasives, a convenient starting point is at ¾ of the pressure used with the former abrasive, and then increase the pressure until the stone wears too rapidly. Source: Metcut Research Associates Inc.

**Table 6 Operating parameters when honing with cubic boron nitride**

| Material | Hardness, HRC | Stock removal rate, mm/min | Stock removal rate, in./min | Stone wear rate per 0.025 mm (0.001 in.) of stock removal, mm | Stone wear rate per 0.025 mm (0.001 in.) of stock removal, in. | Diameter, mm | Diameter, in. | Length, mm | Length, in. | Honing oil | Pressure on stone surface, kPa | Pressure on stone surface, lb | Spindle speed, rev/min | Surface roughness ($R_a$), μm | Surface roughness ($R_a$), μin. |
|---|---|---|---|---|---|---|---|---|---|---|---|---|---|---|---|
| 1855 soderfors 16 | Hard | 0.18 | 0.007 | 0.0050 | 0.0002 | 12 | 0.470 | 54 | 2⅛ | MB-30 | 830 | 120 | 1600 | 1.3 | 50 |
| M4 | Hard | 0.06 | 0.0025 | 0.0008 | 0.00003 | 30 | 1.190 | 44 | 1¾ | MB-30 | 860 | 125 | 640 | 0.63 | 25 |
| M4 | Hard | 0.14 | 0.0054 | 0.0030 | 0.00012 | 19 | 0.750 | 44 | 1¾ | MB-30 | 1300 | 190 | 800 | 0.75 | 30 |
| M4 | Hard | 0.04 | 0.0016 | 0.0015 | 0.00006 | 18 | 0.700 | 63 | 2½ | MB-30 | 1720 | 250 | 1270 | 0.38 | 15 |
| D3 | 61–63 | 0.15 | 0.006 | 0.0013 | 0.00005 | 11 | 0.437 | 23 | 29/32 | MAN-852 | 2070 | 300 | 1600 | . . . | . . . |
| D7 | Hard | 0.11 | 0.0043 | 0.0036 | 0.00014 | 20 | 0.780 | 38 | 1½ | MB-30 | 2000 | 290 | 800 | 0.5 | 20 |
| 440C stainless | 59–62 | 0.05 | 0.002 | 0.0018 | 0.00007 | 57 | 2.247 | 50 | 2 | MB-30 | 860 | 125 | 400 | 0.18 | 7 |
| M2 | 60+ | 0.06 | 0.0025 | 0.0018 | 0.00007 | 31 | 1.250 | 75 | 3 | MB-30 | 1380 | 200 | 310 | . . . | . . . |
| M2 | Hard | 0.08 | 0.003 | 0.0036 | 0.00014 | 31 | 1.250 | 100 | 4 | MB-30 | 1380 | 200 | 230 | . . . | . . . |
| Inconel 718 | Soft | 0.04 | 0.0016 | 0.0050 | 0.0002 | 7.5 | 0.300 | 125 | 5 | MB-30 | 2480 | 360 | 1600 | 0.38 | 15 |
| H13 | 46 | 0.006 | 0.00025 | 0.033 | 0.0013 | 63 | 2.500 | 455 | 18 | MB-30 | Light | Light | 320 | 0.58 | 23 |

Source: Metcut Research Associates Inc.

**Table 7 Single-stroke honing with diamond-plated tools**

| Workpiece diameter, mm | Workpiece diameter, in. | Spindle speed, rev/min | Stroking feed per revolution at stroke rate, mm/min (in./min)(a): 1500 (60), mm/min | 1500 (60), in./min | 2300 (90), mm/min | 2300 (90), in./min | 3400 (134), mm/min | 3400 (134), in./min | 5000 (200), mm/min | 5000 (200), in./min | Results |
|---|---|---|---|---|---|---|---|---|---|---|---|
| ≤10 | ≤⅜ | 3000 | 0.51 | 0.020 | 0.76 | 0.030 | 1.14 | 0.045 | 1.70 | 0.067 | Slow production |
| 10–13 | ⅜–½ | 2400 | 0.64 | 0.025 | 0.91 | 0.036 | 1.42 | 0.056 | 2.11 | 0.083 | Best operation |
| 13–16 | ½–⅝ | 1900 | 0.81 | 0.032 | 1.19 | 0.047 | 1.80 | 0.071 | 2.67 | 0.105 | Best operation |
| 16–19 | ⅝–¾ | 1500 | 1.02 | 0.040 | 1.52 | 0.060 | 2.26 | 0.089 | 3.38 | 0.133 | Fast production, |
| 19–25 | ¾–1 | 1200 | 1.27 | 0.050 | 1.91 | 0.075 | 2.84 | 0.112 | 4.24 | 0.167 | but caution is advised |
| 25–32 | 1–1¼ | 950 | 1.60 | 0.063 | 2.41 | 0.095 | 3.58 | 0.141 | . . . | . . . | Do not use. |
| 32–41 | 1¼–1⅝ | 750 | 2.01 | 0.079 | 3.02 | 0.119 | 4.50 | 0.177 | . . . | . . . | Do not use. |
| 41–51 | 1⅝–2 | 600 | 2.54 | 0.100 | 3.81 | 0.150 | . . . | . . . | . . . | . . . | Do not use. |

(a) One stroke into and out of a bore can achieve respectable removal rates (0.025 to 0.038 mm, or 0.001 to 0.0015 in.) on the diameter, close tolerances (±0.0025 mm, or 0.0001 in.) and reasonable finish (0.050 to 0.63 μm, or 20 to 25 μin., $R_a$) when a crowned, adjustable, 220-grit, plated-diamond hone is used on cast iron.

the stones (Fig. 12) are kept within 0.50 to 0.75 mm (0.020 to 0.030 in.) of the cylinder wall to prevent excessive eccentric wear of the stone, which will cause out-of-roundness or other deviations in bore shape.

Figure 13 shows a typical tool assembly for honing large-diameter bores. In this assembly, which was used for honing a bore 1040 mm (41 in.) in diameter, the stones are mounted in a holder that will accept as many as four stones in line. Each stone is 200 mm (8 in.) long, making the total effective length 810 mm (32 in.). Depending on bore size, up to 24 stone holders can be spaced around a tool of this type (although 12 holders are indicated in Fig. 12). Because this tool is used on a single-spindle machine, no universal joints are needed. Guide elements are seldom used with this type of tool, but in horizontal honing of long bores, outside support of the spindle (drive shaft) may be required.

In power stroking, the honing stones must be encased to prevent wear on the tool body and to permit quick changing of worn stones. The material used for mounting the stones must not damage the bore surface by scratching or spalling. Two widely used types of holders are shown in Fig. 14: (a) a stone mounted in a plastic holder and (b) a stone cemented to a metal die casting.

The body of the honing tool must have open areas that permit a copious flow of honing fluid around the abrasive stones during honing. These open spaces, both axial and radial, are evident in the tool shown in Fig. 12.

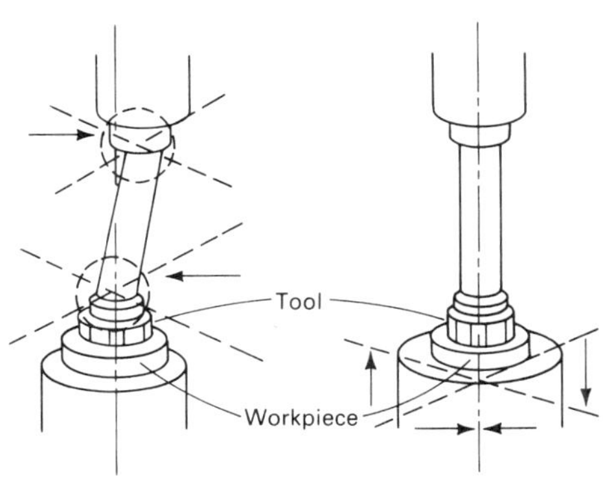


**Fig. 8** (a) Tool or (b) the workholding fixture floats to permit the bore and the tool to align.

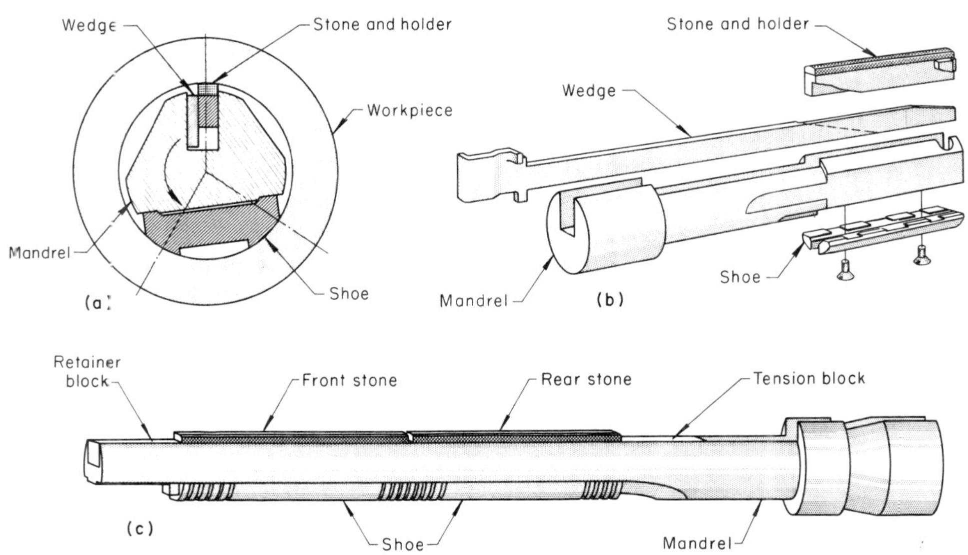


**Fig. 9** Tools for manual-stroke honing. (a) End view showing how self-alignment is provided by three-point contact of shoes and single honing stone with bore. (b) Exploded view showing how axial displacement of wedge controls the feed of the honing stone. (c) Manual-stroke honing tool with two stones; on tools for honing longer bores (up to 460 mm, or 18 in.), as many as five stones can be mounted in line

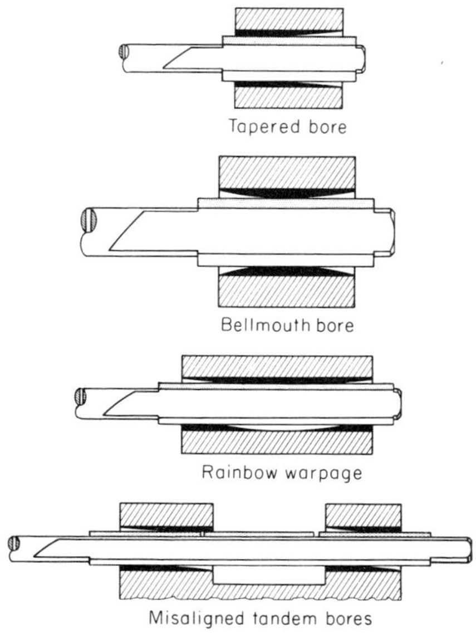


**Fig. 10** Common irregularities that dictate minimum length of honing tool and stone

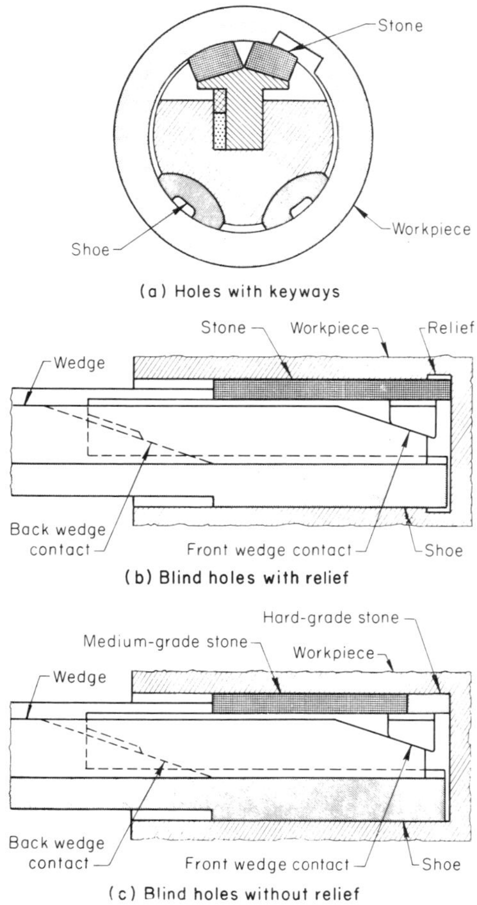


**Fig. 11** Tools for manual-stroke honing of holes having keyways and blind holes with and without relief

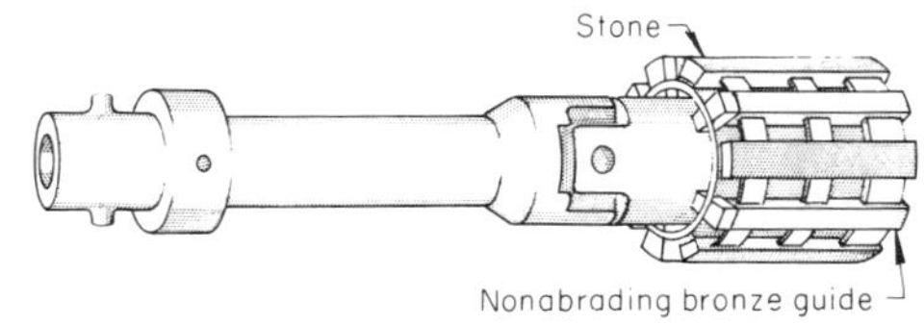


**Fig. 12** Tool used for power-stroke (fixtured) honing of automotive cylinder bores

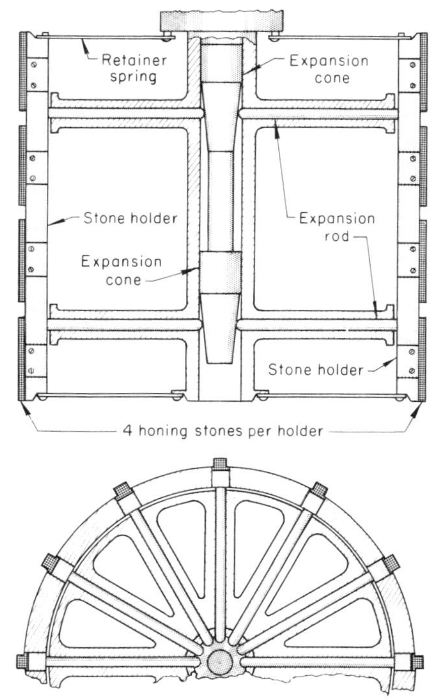


**Fig. 13** Tool used for power-stroke (fixtured) honing of large bores. Expansion cones are used for feed-out of honing stones. See text for discussion.

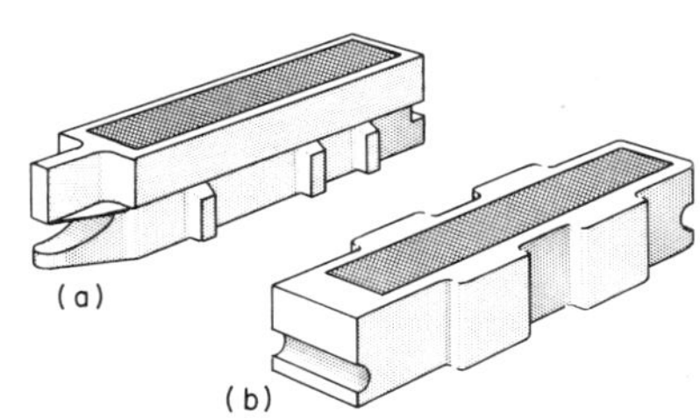


**Fig. 14** Mounting for abrasive stones used in fixtured honing. (a) Stone encased in a plastic holder. (b) Stone cemented to a metal die casting

**Part Location Method.** The principles of a proprietary system of work location for ensuring dependable alignment between the fixed honing tool axis and the hole, in a firmly clamped workpiece, are shown and described in the diagrams in Fig. 15. This system avoids reliance on a floating motion, which is accompanied by some friction; the workpiece is located by aligning its hole with the honing head and is clamped only after the necessary aligning adjustments have taken place. Connecting rods and a variety of other short bore parts are successfully honed individually in a two-position air-float fixture, as illustrated. The part is inserted and approximately located in the fixture at the load position.

The part location method allows the workpiece to be positioned concentric with the tool before it is clamped. The part is manually indexed into hone position on a thin cushion of air. The tool is then moved into the part bore, and the abrasives are expanded, moving the part into final hone position. The tool is stationary at this time, and the part is then clamped.

This type of location, when used with a rigid tool having no universal action, creates a better geometry. The clamp is brought down on the part with a confirming surface to compensate for any out-of-parallel condition.

This type of clamping eliminates torque influences. Therefore, there is no need for outside torque arms, and part distortion is eliminated. Another advantage in clamping the part in this fashion is that a support is put into the part allowing an increase in stone pressure and surface area without misalignment of the part and tool.

## Gages

Automatic size control in power stroking may be accomplished by adaptations of at least five types of gages—air, ring, expanding, plug, and bar gages. Use of these gages is described below, and methods are illustrated in Fig. 16 and 17.

**An air gage** is an integral part of the honing tool. With each stroke of the tool, the gage portion enters the bore with sufficient clearance to permit pressurized air emitted from gage orifices to escape to the atmosphere. As the bore size increases, back pressure in the air system decreases. When the bore is to size, a previously calibrated pressure switch cuts off the cycle. This type of gage is particularly well adapted to controlling multispindle machines in honing cylinder bores. By adjusting the calibrated control, various diameters can be honed without interrupting the cycle. Air gaging can hold diametral tolerances of 0.075 mm (0.0003 in.) and can be adjusted within a range of 0.10 mm (0.004 in.). The device shown in Fig. 16, an air-to-electronic gaging system, converts the back pressure drop into an electronic signal that is fed to a programmable controller that automatically terminates the honing cycle when the bore diameter reaches its programmable size.

**A ring gage** is mounted just above the workpiece (Fig. 17a). The ring, with an inside diameter equal to the specified bore diameter, is positioned so that only the plastic or metal tabs placed on the upper ends of the abrasive stones enter the ring at the top of each stroke. Because the abrasives and the tabs wear at the same rate, the diameter of the expanded tool is equal to the bore diameter. When the tool reaches gage ring size, friction from the tabs causes the ring to swing through a small arc and initiates the end of the cycle. This method of control is best adapted to bore diameters of 3.05 to 100 mm (0.120 to 4.0 in.). It is capable of obtaining diametral accuracy of

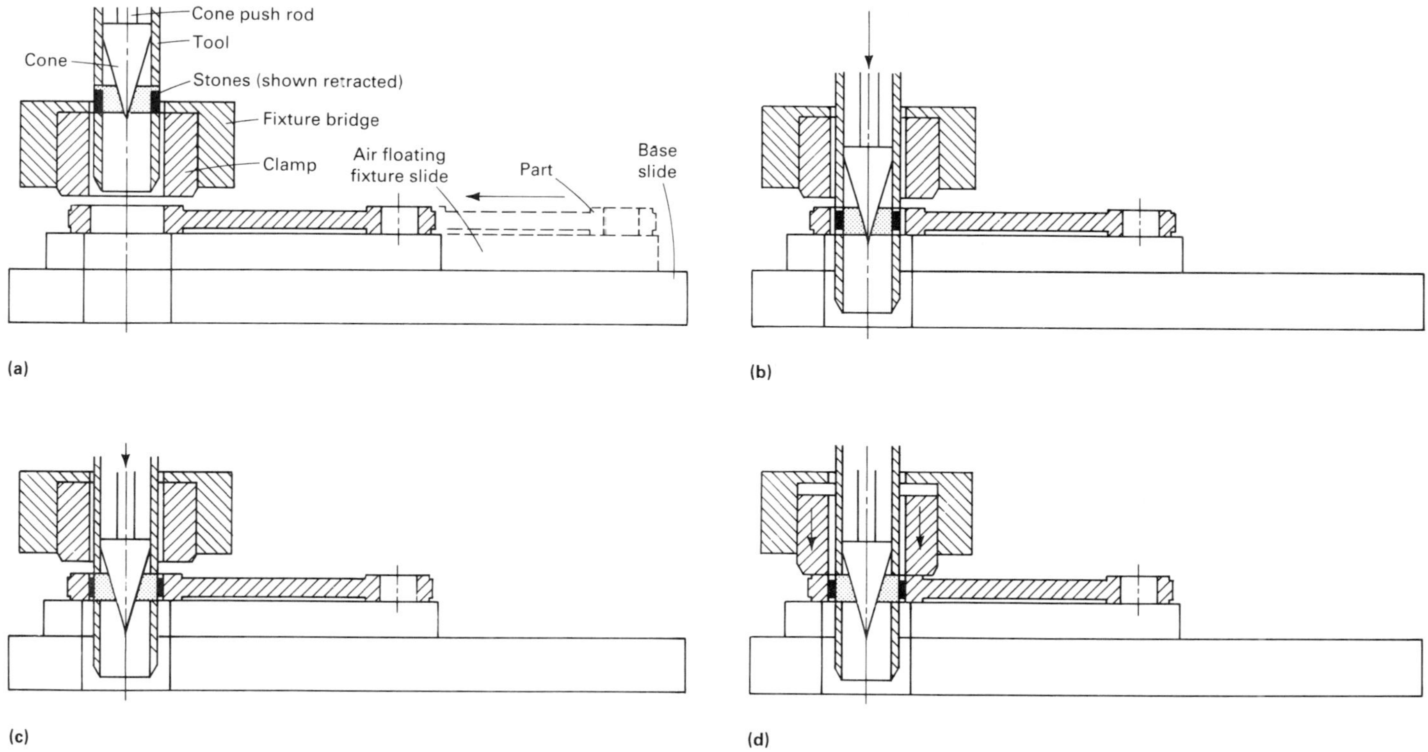


**Fig. 15** Automatic four-step process for aligning the workpiece and the honing tool to hone connecting rods. (a) Part is loaded in fixture, and fixture is moved into machining position on a cushion of air. (b) The tool, which is not rotating, is lowered into the bore with the abrasives retracted. (c) Stones are expanded outward against the bore wall, while the abrasives center the bore and the tool. (d) With the abrasives still expanded, the clamp is lowered onto the part to ready the tool for cycling.

0.0076 mm (0.0003 in.) or better, and adjustments within 0.013 mm (0.0005 in.) can be made to allow for wear or to vary the desired size.

**An expanding gage,** which consists mainly of a split sleeve held together with a ring (Fig. 17b), reciprocates with the tool but is not attached to it. The gaging member, whose diameter is smaller than the bore diameter, enters the bore with each downstroke. At the bottom of the stroke, a lever on the side of the sleeve contacts a post, causing the split sleeve to expand until the gaging member touches the bore surface. Controls are preset so that two electrical contacts on the lever of the split sleeve meet and end the cycle when the diameter of the gaging member equals the desired bore diameter. This type of gage has been used successfully for size control of bores larger than 19 mm (0.750 in.) in diameter (there is no maximum size). A calibrated dial allows fine adjustment of the size within 0.25 mm (0.010 in.). A tolerance of 0.0075 mm (0.0003 in.) can be maintained at any dial setting.

**A plug gage** is independent of the tool and approaches the bore from the end opposite that at which the honing tool is introduced (Fig. 17c). The plug, whose diameter equals the desired bore size, attempts to enter the bore with each stroke. When entry is made, controls are activated and the cycle is terminated. This method can be used to control accuracy within 0.005 mm (0.0002 in.).

**A bar gage** consists of two bars that float in the body of the tool (Fig. 17d). The bars are fastened to a split ring and held against the bore surface by spring pressure. Bore diameter is measured by the distance between the contact faces of the bars. When the tool enters a bore, the bars are pressed inward, and two low-potential electrical contacts on the split ring are held open. As the bore becomes larger, the contacts move closer together; when they meet (as preadjusted), the cycle is stopped. The use of this type of gage is usually restricted to size control of bores larger than 50 mm (2 in.) in diameter. However, this gage accepts adjustments within 1 mm (0.040 in.). Also, control of honed diameters within 0.0075 mm (0.0003 in.) is feasible.

## Rotation Speed

Spindle speed depends mainly on the diameter of the bore being honed, because surface speed is usually the basic consideration in honing. The choice of an optimum surface speed is influenced by:

- *Material being honed*: Higher speed can be used for metals that shear easily, such as cast iron and some of the softer nonferrous metals
- *Hardness*: Harder workpiece surfaces require lower speed
- *Surface roughness*: Rougher surfaces that mechanically dress the abrasive stone permit higher speed
- *Number and width of stones in the tool*: Speed should be decreased as the area of abrasive per unit area of bore increases
- *Finish requirements*: Higher speed usually results in finer surface finish

Because of the variables listed above, rotation speed cannot be standardized. However, the speeds given in Table 3 serve as a starting

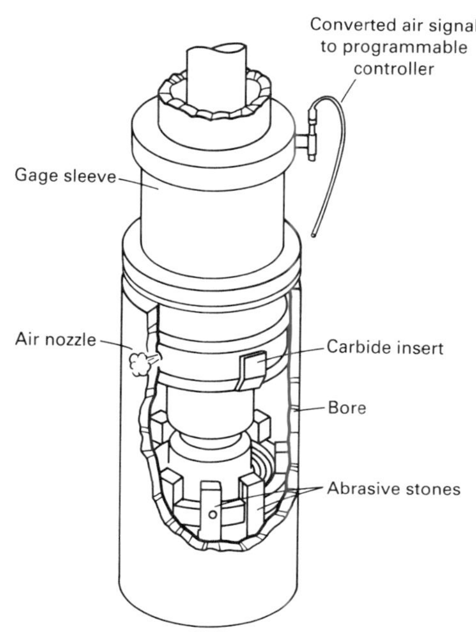


**Fig. 16** Air-to-electronic-gage sizing for automatic size control in honing

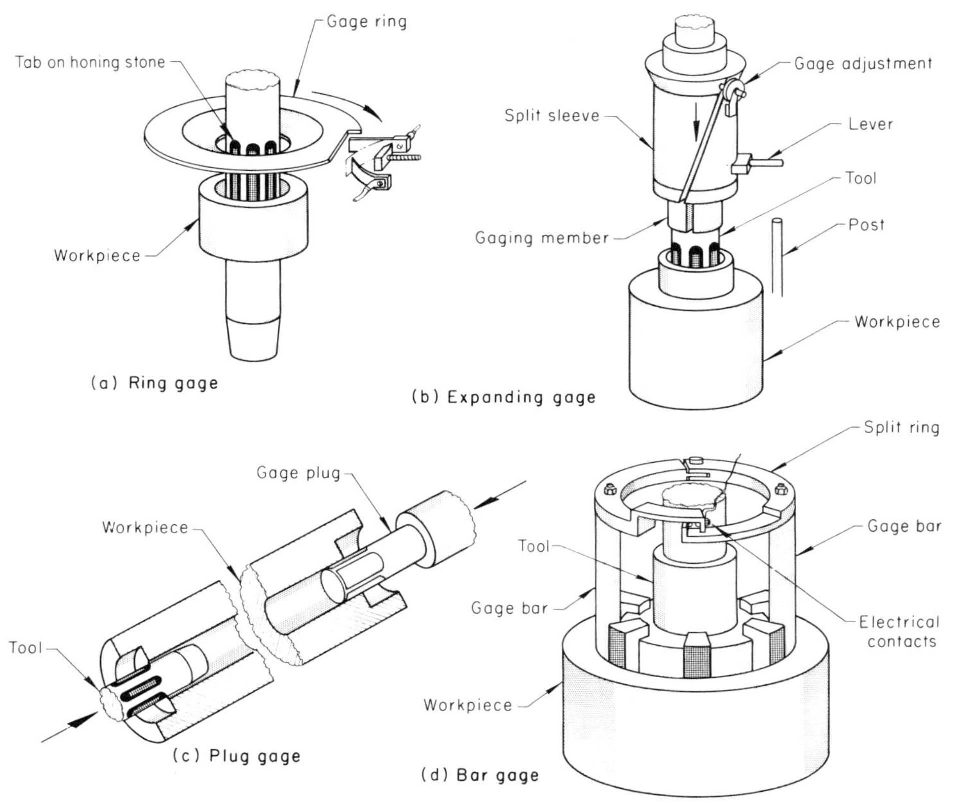


**Fig. 17** Gages for automatic size control of fixtured workpieces honed in power-stroking machines. (a) Ring gage. (b) Expanding gage. (c) Plug gage. (d) Bar gage. See text for discussion.

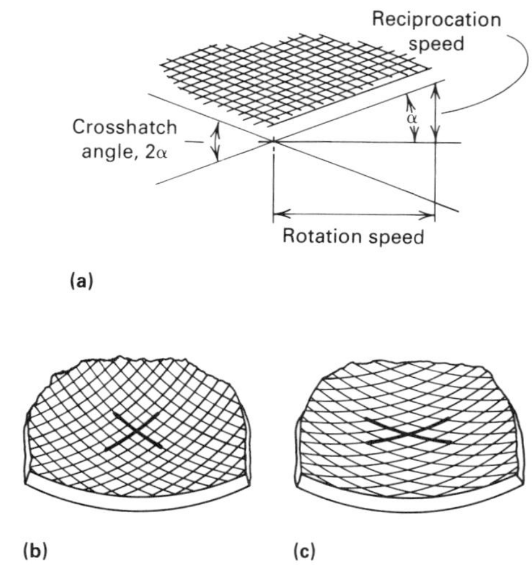


**Fig. 18** Relation of crosshatch angle (a) to rotation and reciprocation speeds. (b) Rotation approximately equal to reciprocation. (c) Rotation faster than reciprocation

point, and, in general, are close to those given in the examples in this article.

Experience with a particular application may indicate advantages for higher or lower speed. Rotation speeds as high as 183 m/min (600 sfm) have been used successfully. Conversely, a reduction in spindle speed, and thus in surface speed, can reduce the number of rejects.

Excessive speeds contribute to decreased dimensional accuracy, overheating of the workpiece, and glazing (dulling) of the abrasive. Overheating causes breakdown of honing fluid and distortion of the workpiece; the latter frequently affects final dimensions.

## Reciprocation Speed

Speed of reciprocation, which depends largely on the length of the honing tool and the depth of the bore, is most usefully expressed in meters per minute (or surface feet per minute), the product of the number of stroke cycles per minute and twice the stroke length. Reciprocation speeds commonly used with a variety of work metals and alloys are listed in Table 3.

Because reciprocation speed, rotation speed, and crosshatch angle are related functions (see the next section), crosshatch angle can be controlled by varying the reciprocation speed when rotation is constant. Reciprocation speed also has some effect on the action of the abrasive; higher speed increases dressing action and thus usually produces a rougher finish.

## Control of Crosshatch Angle

The crosshatch angle (Fig. 18a) obtained on a honed surface depends on the ratio of surface speed of reciprocation (stroking) to surface speed of rotation. When the rotation and reciprocation speeds are equal, the crosshatch angle is 90° (Fig. 18b). When rotation speed exceeds reciprocation speed, the crosshatch angle is less than 90° (Fig. 18c).

The following formula can be used to determine the approximate angle that will result from given speeds (Fig. 18a):

$$\text{Tan } \alpha = \frac{V_r}{V_p} \quad \text{(Eq 1)}$$

where $V_r$ is the speed of reciprocation, $V_p$ is the speed of rotation, and $2\alpha$ is the crosshatch angle.

Although the above formula may be useful as a guide when determining the speeds necessary for obtaining the desired angle for a new job, it is often more practical to resort to trial and error. In power stroking, a common practice is to establish rotation speed and then to vary reciprocation speed to get the desired crosshatch angle.

Reciprocation speed is less critical than rotation speed. However, an increase in the rate of reciprocation will sometimes improve the self-dressing characteristic of some abrasives and thus increase the stock removal rate; therefore, increased stock removal rates and fine finishes do not generally go together. If several thousandths of an inch of stock must be removed at a high production rate and less than a 0.50 μm (20 μin.) finish is required, two honing operations should be used.

**Table 8 Conditions for producing a 30° crosshatch angle in manual-stroke honing**

| Bore diameter, mm | Bore diameter, in. | Rotation, rev/min | Bore length, mm | Bore length, in. | Stone length, mm | Stone length, in. | Stroke length(a), mm | Stroke length(a), in. | Stroke cycles/min(a) |
|---|---|---|---|---|---|---|---|---|---|
| 6.4 | ¼ | 1600 | 38 | 1½ | 32 | 1¼ | 27.4 | 1.08 | 155 |
| | | | 75 | 3 | 32 | 1¼ | 65.5 | 2.58 | 65 |
| | | | 75 | 3 | 57 | 2¼ | 57.1 | 2.25 | 75 |
| 19 | ¾ | 640 | 75 | 3 | 64 | 2½ | 55.1 | 2.17 | 94 |
| | | | 150 | 6 | 64 | 2½ | 131.3 | 5.17 | 39 |
| | | | 150 | 6 | 114 | 4½ | 114.3 | 4.50 | 45 |
| 38 | 1½ | 320 | 100 | 4 | 83 | 3¼ | 74.2 | 2.92 | 70 |
| | | | 200 | 8 | 83 | 3¼ | 175.8 | 6.92 | 29 |
| | | | 200 | 8 | 165 | 6½ | 148.1 | 5.83 | 35 |
| 64 | 2½ | 200 | 100 | 4 | 83 | 3¼ | 74.2 | 2.92 | 72 |
| | | | 200 | 8 | 83 | 3¼ | 175.8 | 6.92 | 30 |
| | | | 200 | 8 | 165 | 6½ | 148.1 | 5.83 | 36 |

(a) In computing reciprocating surface speed: The distance traveled during one stroke cycle comprises a forward and a return stroke.

**Table 9 Relation of stone contact area and pressure with 220 N (50 lbf)**

| Bore diameter | | Bore length | | Stone length | | Stone width | | Stone area | | Pressure(a) | |
|---|---|---|---|---|---|---|---|---|---|---|---|
| mm | in. | mm | in. | mm | in. | mm | in. | $mm^2$ | $in.^2$ | MPa | psi |
| 1.6 | 1/16 | 16 | 5/8 | 11 | 7/16 | 0.71 | 0.028 | 7.75 | 0.012 | 28.72 | 4166 |
| 3.2 | 1/8 | 19 | 3/4 | 14 | 9/16 | 0.89 | 0.035 | 12.9 | 0.020 | 17.24 | 2500 |
| 4.8 | 3/16 | 25 | 1 | 19 | 3/4 | 1.27 | 0.050 | 24.5 | 0.038 | 9.074 | 1316 |
| 6.4 | 1/4 | 38 | 1½ | 32 | 1¼ | 1.78 | 0.070 | 56.8 | 0.088 | 3.915 | 568 |
| 13 | 1/2 | 75 | 3 | 57 | 2¼ | 3.68 | 0.145 | 210 | 0.326 | 1.062 | 154 |
| 16 | 5/8 | 90 | 3½ | 64 | 2½ | 5.59 | 0.220 | 354 | 0.550 | 0.627 | 91 |
| 16 | 5/8 | 230 | 9 | 190 | 7½ | 5.59 | 0.220 | 1064 | 1.650 | 0.207 | 30 |
| 38 | 1½ | 115 | 4½ | 83 | 3¼ | 6.22 | 0.245 | 513 | 0.796 | 0.434 | 63 |
| 38 | 1½ | 510 | 20 | 413 | 16¼ | 6.22 | 0.245 | 2568 | 3.981 | 0.089 | 13 |

Note: For manual honing with a single stone. (a) Pressure under 220 N (50 lbf)

A coarse abrasive (60 to 180 grit) is usually used first to generate geometry and size. Then a second, finer abrasive (240 to 600 grit) is used to refine the surface finish. The finish achieved with specific stones, however, depends on which material is honed.

For some applications (engine cylinder bores are a notable example), crosshatch angle is important, and is noted in specifications. In the majority of applications, however, although an angle of 30° is commonly sought, any angle within the range of 20° to 45° usually is suitable. Angles within this range are practical with either manual or power stroking.

In manual honing, an experienced operator can instantly alter the practice to suit conditions. For instance, the workpiece can first be stroked at a rate that has proved most efficient for stock removal. The technique can then be changed so that the desired crosshatch pattern is produced in the few final strokes. Conditions that will produce a crosshatch angle of approximately 30° in manual-stroke honing are given in Table 8 for a range of workpiece diameters and lengths.

## Honing Pressure

A relatively wide range of pressures will yield acceptable efficiency and results. For example, in equipment using hydraulic force for feed-out, gage pressures have varied from 1040 to 3100 kPa (150 to 450 psi). However, honing is more often controlled by rate of feed-out than by gage pressure. In general, high feed-out rates are used for large diameters, and low feed-out rates are used for small diameters. As an example, a feed-out rate of 0.23 mm/min (0.009 in./min) could be used for roughing and a 0.18 mm/min (0.007 in./min) rate for finishing 215 mm (8½ in.) diam cylinder bores in gray iron blocks for V-8 engines; for a 99.31 mm (3.910 in.) diam bore having a 150 mm (6 in.) length and made of the same material, the feed rate would be 0.075 mm/min (0.003 in./min) to 0.150 mm/min (0.006 in./min).

Insufficient pressure will result in a subnormal rate of metal removal. When pressure is excessive, rougher finishes are obtained, because the abrasive is broken down too fast. This will result in increased stone cost, as well as decreased productivity caused by downtime required for replacing stones.

Trial and error is the usual method of determining optimum pressure for a new application in production honing. A common procedure is to start with low pressure and then gradually build up, using workpiece finish as an indicator, until best conditions are found. Pressure must be kept up by automatic feed-out to compensate for stone breakdown and hole growth.

The data in Table 9, based on manual honing with a single stone, show why machines for honing various bores must apply a wide range of force. If a force of 220 N (50 lbf) is applied to a stone with a contact area of 7.74 $mm^2$ (0.012 $in.^2$), the pressure will be 29 MPa (4.2 ksi). Such a pressure would shatter the stone. But the same force applied to a stone with a contact area of 355 $mm^2$ (0.550 $in.^2$) creates a pressure of only 630 kPa (91 psi)—insufficient for cutting most materials.

## Honing Fluids

Lubrication is more critical in honing than in most other metal-removing operations. No single honing fluid possesses a maximum of all properties needed for honing. Therefore, some compromise must be made, and mixtures of two or more liquids are commonly used. The oils used in honing serve two main purposes:

- They promote cutting action by flushing workpiece metal and particles of abrasives from the honing stones, thus preventing the stones from loading and glazing
- They maintain an almost constant workpiece temperature, and thus minimize dimensional variation due to expansion and contraction

Honing fluid characteristics can directly influence quality or economy, or both. Viscous, gummy fluids, or fluids containing suspended solids, can cause a soft-bonded abrasive to lose efficiency and generate excessive heat from friction.

Water-base solutions are superior as coolants, but they are poor lubricants and have insufficient viscosity to prevent chatter. Water also causes rust. Because of the latter two characteristics, water-base solutions are seldom used for honing fluids.

Mineral seal oil is effective and is widely used for production honing. Mineral seal oil is a water-white product having a higher viscosity than kerosine (about 40 Saybolt Universal seconds, or SUS, compared to 31 SUS for kerosine). Its flash point also is higher than that of kerosine, and it is less likely to cause skin irritation. Mineral oils similar to those used for other machining operations have also proved satisfactory when one part of oil is diluted with four parts of kerosine.

Buffers are often added to honing fluids, to minimize or prevent chatter of the honing stones and thus prevent their premature disintegration. Buffer materials absorb shock and recoil from fluctuations in force. Animal oils, including tallow, lanolin, and lard oils, are usually the most economical buffers, despite their relatively high cost, because of their long lasting qualities. These oils generally flow under pressure and cling to metal surfaces better than mineral oils. Prepared proprietary oils that contain buffers are widely used. Before use, these oils (which may also contain rust inhibitors and deodorants) are diluted up to 95% with kerosine.

The use of too much buffer may detract from its beneficial effects. An excessive amount:

- Reduces the cutting action of the abrasive
- Produces smoother finishes
- Requires higher pressures or lower rotational speeds
- Lowers the ability of the fluid to dissipate heat
- Impairs fluid distribution
- Increases requirements for refrigeration and filtering

Regardless of the type of fluid used, it should be delivered to the honing stones in a constant and generous supply. The fluid also should be filtered through a system that removes particles coarser than 15 μm (600 μin.). The system should be kept free of water and stray oil (such as from the hydraulic system), which adversely affect the properties of honing fluids.

In many plants, 17 to 20 °C (62 to 68 °F) is the preferred temperature range for honing

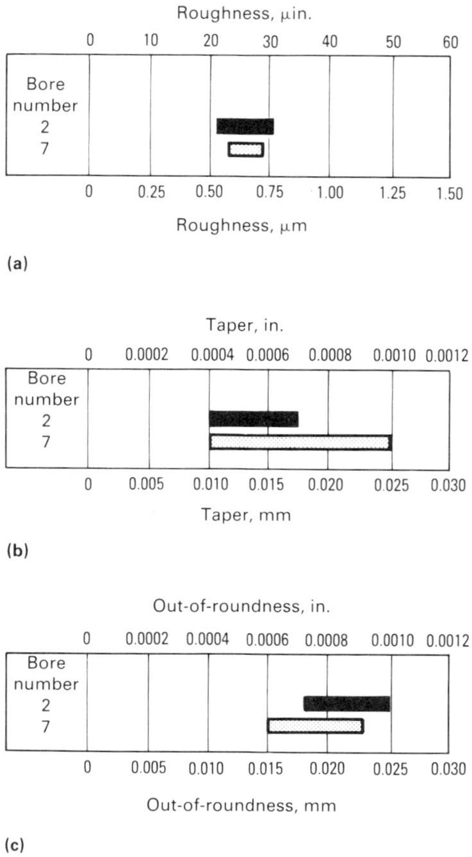


**Fig. 19** (a) Surface finish, (b) taper, and (c) out-of-roundness variations obtained in honing. Data represent measurements on cylinder bores 2 and 7 in 11 gray iron blocks for V-8 engines, selected from a run of 900. Measurements were made on blocks 1, 50, 100, 200, 300, 400, 500, 600, 700, 800, and 900.

fluids. Controlling the temperature becomes more important as tolerances become closer. If temperature is allowed to rise, dimensions may become inaccurate and the fluid may break down, causing excessive stone wear and changes in cutting characteristics. In production installations, heat exchangers are often used to maintain close control of honing fluid temperature.

## Dimensional Accuracy

Internal honing to tolerances of 0.025 to 0.0025 mm (0.001 to 0.0001 in.) is common. For some high-precision parts, tolerances as close as a few millionths of an inch are specified and achieved.

Close tolerances can be produced and repeated in fixtured honing (power stroking) if sources of variation, such as machine and honing fluid condition, are closely controlled, as described in the example below.

**Example 1: Variations in Dimensions and Finish for 900 Cylinder Blocks.** The data plotted in Fig. 19 represent results of a quality control check made on 99.31 mm (3.910 in.) diam cylinder bores in gray iron blocks for V-8 engines. Bores 2 and 7 were

**Table 10 Processing details for honing cylinder bores in V-8 engine blocks**

| Processing details(a) | |
|---|---|
| Machine production rate | 70 blocks/h |
| Spindle speed | 204 rev/min |
| Spindle reciprocation | 78 strokes/min |
| Stock removal: | |
| Amount | 0.051–0.102 mm (0.002–0.004 in.) |
| Time | ~39 s |
| Honing fluid | Mineral seal oil(b) |
| Stone life per set(c) | 450 blocks |
| Size control | Spindle-mounted plug gage |
| Dimensional tolerance | Max out-of-roundness and taper, 0.025 mm (0.001 in.) |
| Finish | 0.50–0.89 μm (20–35 μin.) |
| Crosshatch angle | 22½° |

(a) Bores are classified in five sizes differing 0.013 mm (0.0005 in.) in diameter, for selective fitting of pistons. (b) At 20 °C (68 °F), heat exchanger is required for maintaining this temperature, and honing fluid must be free of water and tramp hydraulic oil. (c) Silicon carbide stones

measured in 11 blocks from a production run of 900. The conditions employed in honing these bores are presented in Table 10.

The honing fluid was maintained at 20 °C (68 °F) by the use of a heat exchanger, and was constantly filtered. Less than 2% of the 7200 bores honed required a repair operation because either taper or out-of-roundness exceeded the specified 0.025 mm (0.001 in.).

## Surface Finish

Surface finish of 0.25 to 0.38 μm (10 to 15 μin.) can be obtained easily in production honing, and finish of less than 0.050 μm (2 μin.) can be achieved and reproduced. A range of roughness is sometimes specified. In other applications, a maximum surface roughness is specified. Under carefully controlled conditions, surface roughness can be maintained within a close range, as indicated in Fig. 19.

Size of grit in the honing stones is the main factor controlling surface finish. When grit is fine, the finish will be fine (other factors being equal); but as grit size is decreased, rate of stock removal is also decreased, as described in the following example.

**Example 2: Honing Gray Iron to a Finish of 0.25 to 0.38 μm (10 to 15 μin.).** In honing gray iron (hardness, 170 to 195 HB), a finish of 0.25 to 0.38 μm (10 to 15 μin.) was desired. Silicon carbide stones with a grit size of 180 produced a roughness of 0.63 to 0.75 μm (25 to 30 μin.). The required finish could be obtained with 320-grit stones, but the time required for honing made the use of this grit size impractical. The problem was solved by first rough honing with 180-grit stones and then finish honing, in another setup, with 320-grit stones.

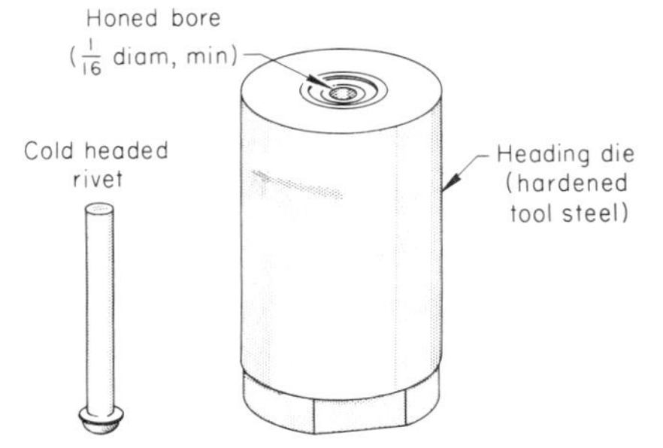


**Fig. 20** Bore of die for cold heading the rivet shown at the left is typical of small bores finished by manual-stroke honing. Dimensions given in inches

Rough finishes are sometimes improved by using a dwell time at the end of the honing cycle—that is, by continuing the rotation and reciprocation action for a few strokes after feed-out ceases and pressure drops off. In manual honing of a particular bore, use of this technique reduced surface roughness from the normal 0.50 to ~0.25 μm (20 to ~10 μin.).

## Honing Practice for Internal Diameters

Honing is widely used for finishing bores in engine cylinders, cylinder liners, and bearing bores. Procedures for honing similar parts may vary from one plant to another, depending on quantity, available equipment, and established plant practice.

As a rule, honing stones and techniques used for honing cast iron are different from those used for aluminum alloys. However, there are exceptions as in the case of an assembly of cast iron and aluminum in which the two metals were honed simultaneously, with the same abrasive, because it was the simplest way to achieve a proper fit.

**Small Bores.** Conventional manual-stroking honing tools (Fig. 9) are available for use in bores as small as 1.6 mm (1/16 in.) in diameter in parts such as fuel nozzles, miniature bearings, and heading dies, as in the following example.

**Example 3: Honing Very Small Bores.** Dies for cold heading tiny rivets and screw blanks had bores as small as 1.6 mm (1/16 in.) in diameter. Bore length varied, but was usually 25 to 50 mm (1 to 2 in.). Figure 20 shows one of the heading dies, which was made of tool steel, and a typical product of the die. Holes were drilled and reamed about 0.075 to 0.13 mm (0.003 to 0.005 in.) undersize before heat treatment. After hardening, they were honed, using manual stroking, to an accuracy of 0.0025 mm (0.0001 in.) for both roundness and straightness.

**Large Bores.** The maximum diameter and length of bore that can be honed is limited mainly by the size of the equipment required for the workpiece and by the power

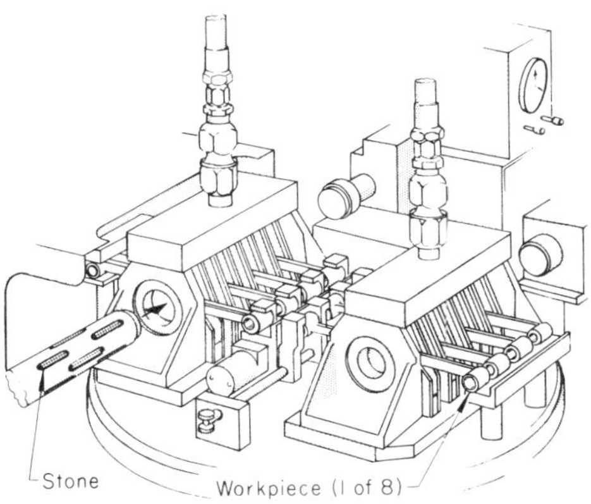


**Fig. 21** High-production honing of automotive parts. Fixture designed to hone crankpin bores on eight automobile connecting rods simultaneously, using a single honing tool. Rotating fixture permitted loading and unloading on one side while parts on the opposite side were honed.

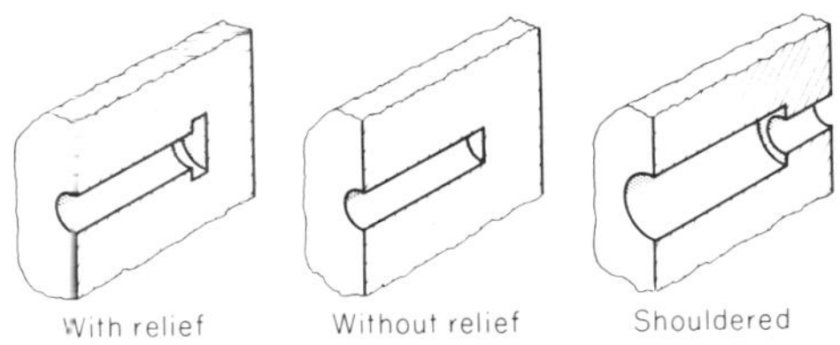


**Fig. 22** Three types of blind holes

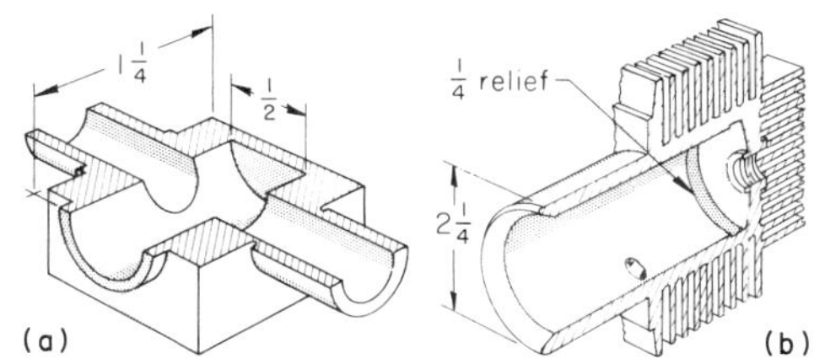


**Fig. 23** Blind holes honed by different methods. (a) Unrelieved blind hole that required a special tool (see Fig. 11c) for honing. (b) Relief that permitted use of a conventional honing tool in the bore of a cylinder head for a lawn mower engine. Dimensions given in inches

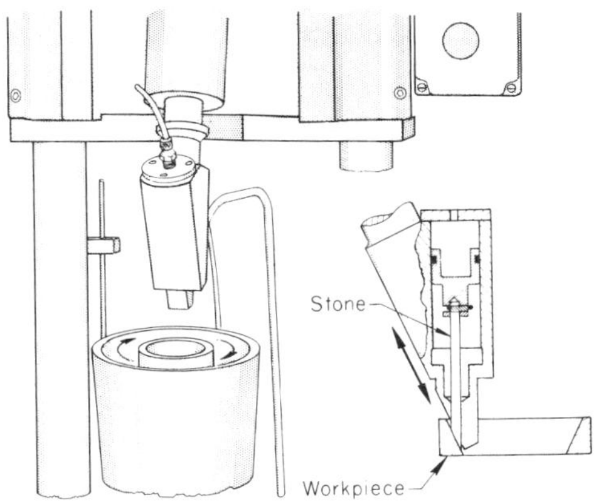


**Fig. 24** Machine and tooling for honing short, tapered bores

required for the tools. Equipment with drive motors of up to 37 kW (50 hp) is available for honing steel shells of 1040 mm (41 in.) inside diameter (ID) and 19 m (63 ft) length.

Cylinder shells for hydraulic hoists on regulating gates for dams are examples of large bores that are honed. In one honing operation, 0.75 mm (0.030 in.) of stock is removed from a 6.4 Mg (14 000 lb) shell of 760 mm (30 in.) ID and 7.9 m (26 ft) length to obtain a total envelope tolerance of 0.050 mm (0.002 in.). Before honing, the average out-of-roundness is 0.41 mm (0.016 in.).

**Short Bores.** Several different techniques are used for honing short bores. These are particularly applicable when bore diameter exceeds length. In the simplest method, several pieces are stacked with bores aligned, clamped tightly by any suitable means, and honed as a unit. For example, 13 mm (½ in.) long rings 38 mm (1½ in.) in inside diameter can be honed in stacks of eight. In effect, this would be the same as honing a single piece 100 mm (4 in.) long. Stacked parts may be either manually or power stroked. However, for successful results from this technique, the parts must have parallel sides to permit building a straight stack that can be clamped tightly and provide a straight bore. Another technique that has proved successful for honing short bores is shown in the following example involving automotive-engine connecting rods (compare with newer method shown in Fig. 15).

**Example 4: Short-Bore Honing Technique for Connecting Rods.** A power-stroking horizontal machine was used in high production for honing 61.54 mm (2.423 in.) ID crankpin bores simultaneously in eight connecting rods. Figure 21 shows the fixture and the honing tool. Eight rods were stacked between 4.8 mm (3/16 in.) wide parallel separator plates, resulting in an effective bore length of 260 mm (10¼ in.). The tool had three banks of four honing stones. Each stone was 9.5 mm (⅜ in.) square and 57 mm (2¼ in.) long. A two-station, rotary index table allowed the operator to unload eight completed rods and to load eight unfinished rods while eight other rods were being honed. A precheck plug probed the rods in the loading station to determine whether they had been bored to proper rough size. The operation completed one bank of rods in 45 s, floor-to-floor time, and a production rate of about 600 rods/h was obtained. In honing, 0.075 mm (0.003 in.) of stock was removed, a finish of 0.75 to 1.14 μm (30 to 45 μin.) was produced, and inside diameter was controlled within 0.13 mm (0.0005 in.).

**Blind holes** are bores that have a bottom, shoulder, or other obstruction that prevents a tool from passing completely through. The three most common types of blind holes are shown in Fig. 22. Most unrelieved blind holes can be honed satisfactorily, but there will always be some unfinished area at the bottom. The amount depends on length of bore, type of material, tolerance required, and amount of stock removed. Under the best conditions, dead-blind holes can be honed to within about 0.38 mm (0.015 in.) of the end. Any relief will improve results; as much relief as possible is preferred. Sometimes an unrelieved blind hole is in effect provided with a relief because specified tolerance and finish need not be met at the bottom of the hole.

Special tools may be required, depending on whether or not relief (or on how much relief) is provided. For example, the unrelieved 13 mm (½ in.) diam bore shown in Fig. 23(a) was manual-stroke honed to within about 0.38 mm (0.015 in.) of the end with a special tool having a hard-tipped honing stone (Fig. 11c). If adequate relief is provided, conventional tools are satisfactory. For example, cylinder heads in lawn mower engines (Fig. 23b) can be manual-stroke honed in high production with conventional tools, because of the generous relief (about 6.4 mm, or ¼ in., wide) at the blind end. Although both parts shown in Fig. 23 were manual-stroke honed, similar parts are frequently honed by power stroking.

Delivering enough honing fluid to the work area is often a problem in honing blind holes. When a hole has a bottom opening (Fig. 23b), fluid can be pumped through a plastic tube inserted in the opening. When a hole has no bottom opening, the flow of fluid should be directed parallel to the mandrel, into the mouth of the bore.

In manual honing, blind holes are more difficult to keep straight than open holes. A truing sleeve (dummy workpiece) is frequently used to keep the shoes and stones straight and parallel; also, the stone and shoe are made shorter than the blind hole. Experienced operators have found that using a series of short strokes with an occasional stroke all the way out of the mouth is the best practice, until the hole is close to final diameter. This keeps the bottom slightly larger than the mouth. Straight strokes are then used for finish honing.

**Tapered Bores.** Part size, angle of taper, and length-to-diameter ratio determine the method used in taper honing. Short tapers are honed using a machine and tool such as that shown in Fig. 24. The machine has a head that can be positioned for any desired degree of taper, and the reciprocating tool holds a single stone. The workpiece is rigidly clamped in a fixture that rotates. This method is most commonly used for producing tapers on parts for which the length of honed area is less than the diameter. As the length of the taper increases in proportion to the diameter, however, the practicality of the method decreases, because the longer and more slender tools lack adequate rigidity.

Applications of this method of taper honing include special bearing rings and parts that use end tapers for sealing, and bores in gears that must fit tapered shafts. For ex-

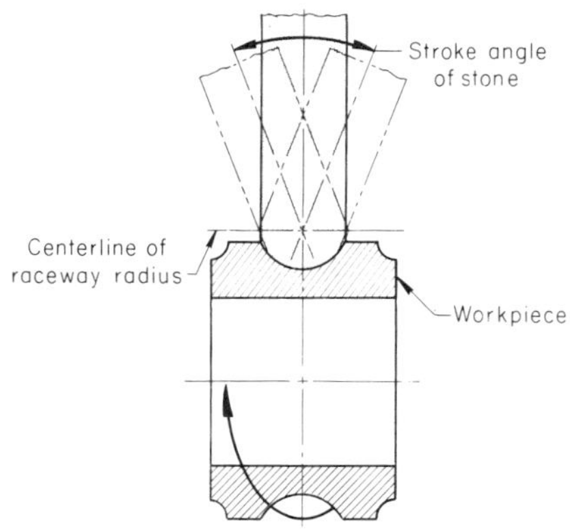


**Fig. 25** Fixtured honing of grooves on external surface of bearing rings with simultaneous oscillation of honing stone and rotation of workpiece

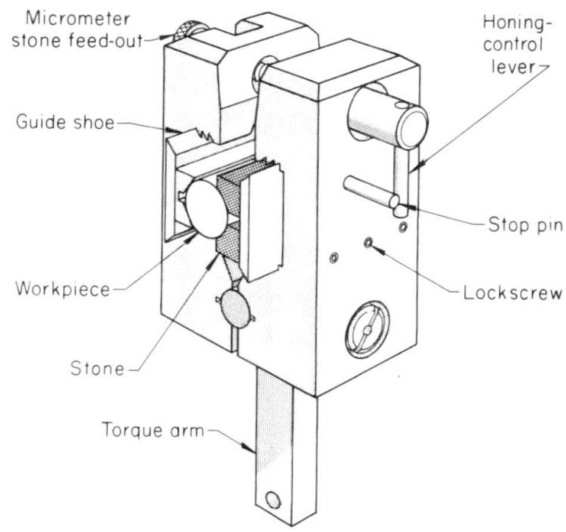


**Fig. 26** Assembly used for manual-stroke honing of outside diameters. See text for discussion.

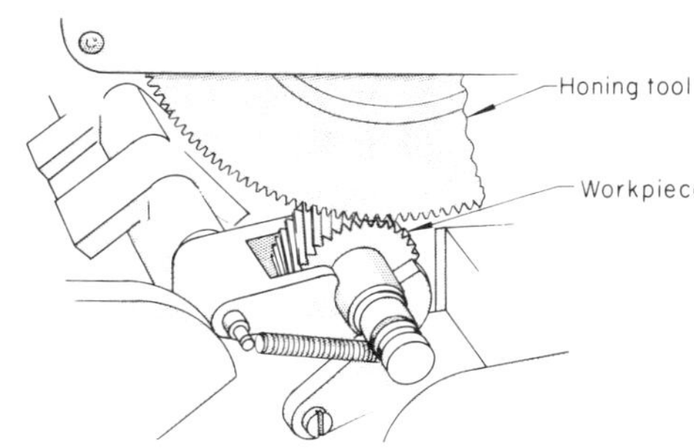


**Fig. 27** Honing teeth of helical gears

ample, drum-to-barrel seals in a 20 mm gun must have a taper of 0.050 mm/mm (0.050 in./in.) of length at each end and roughness less than 0.25 μm (10 μin.). To meet these requirements, 0.01 to 0.05 mm (0.0005 to 0.002 in.) of stock must be removed from the critical surfaces.

Taper honing long bores in large parts is far more complex than honing short tapers. A major portion of the stock is removed by step honing. In this operation, a straight stroke is used, its length being progressively reduced to form a rough taper consisting of a series of small steps. The taper is then finished in a second operation in which a sine bar regulates the increase and decrease of the diameter on the return and forward stroke of the honing cone.

**Special Shapes.** Machines and tools have been developed for honing various special shapes. For female splines, honing stones must be narrower than the spline width (preferably no wider than half the spline width) to allow for oscillation. Machines and tools for honing splines are designed to produce simultaneous reciprocation and oscillation, rather than reciprocation and rotation. Relief bores are commonly honed by contour boring.

## Special Applications of Honing

A few special uses of honing should be enumerated as a means of indicating the potential of the honing method beyond the field of its basic and most extensively accepted applications. These are related to the honing of internal cylindrical surfaces by using regular abrasives for obtaining specific dimensional conditions of the work surface. Among these related processes are:

- External honing
- Gear tooth honing
- Plateau honing
- Flat honing
- Electrochemical honing
- Hone forming

**External Honing.** Honing has been used to only a limited extent for finishing outside diameters, largely because required dimensions and finish can be produced at less expense by other processes, such as centerless grinding. However, advances in metrology and improved honing techniques have resulted in an increase in the number and scope of applications of external honing.

Special machines and special adaptations of conventional machines (such as lathes) have been tooled to hone outside surfaces of metal parts. With these machines, either power or manual stroking may be employed.

*Fixtured external honing* (power stroking) is widely used for pieces that are not adaptable to competitive methods. A notable example is the finishing of grooves in bearing races. Special machines that simultaneously rotate the workpiece and oscillate the stones (Fig. 25) produce the crosshatch lay pattern characteristic of a honed surface.

*Manual external honing* is applicable to the removal of small amounts of stock from external diameters of a wide variety of sizes and shapes. The honing of lengths up to 3 m (10 ft) is common practice. Conventional honing machines are generally used for rotating workpieces up to 610 mm (24 in.) long. Lathes or drill presses are preferred for longer workpieces.

Tools such as that illustrated in Fig. 26 are available for honing parts ranging in outside diameter from about 3.05 to 69.85 mm (0.120 to 2.750 in.). With this setup, the sides of the tool are gripped and stroked over the rotating workpiece. Feed-out and cutting rate are controlled by applying pressure to the honing-control lever, which will move through a preset distance. Size is controlled automatically by setting the micrometer stone feed-out so that the honing-control lever will be against the stop pin when the correct size is attained. The only adjustment needed during the honing operation, even in production runs, is a slight additional stone feed-out to compensate for stone wear. A turn of the honing-control lever will instantly disengage the stone from the work for quick gaging or unloading, but will not change the setting on the micrometer stone feed-out.

With the setup shown in Fig. 26, a line of stones with opposing guide shoes, or opposing stones, can be used. For honing long parts (up to 610 mm, or 24 in.), multiple holders that contain as many as three stones (or shoes) in line may be used for correcting waviness. The torque arm can be used to offset the tendency of the tool to turn. A guide bar mounted on the machine acts as a stop for the torque arm. This type of tool can produce dimensional accuracy to 0.0025 mm (0.0001 in.) or better and surface roughness as low as 0.050 μm (2 μin.).

Manual-stroke external honing has replaced lapping in some applications, because:

- Honing is usually faster
- Soft metals can be honed without being impregnated with abrasive
- The use, in honing, of multiple-length stones and shoes allows better control of bow and waviness

Long anodized aluminum tubes for inflight refueling are honed externally in a lathe, the honing tool being moved by hand, and the nozzle for the honing fluid moving with the tool. Crankpins of some crankshaft are honed the same way at overhaul.

**Gear-tooth honing** is an abrasive process designed to improve geometric accuracy and surface conditions of a hardened gear. The teeth of hardened gears are honed to remove nicks and burrs, to improve finish, and to make minor corrections in tooth shape. Gear teeth are honed on high-speed machines specially designed for the process (Fig. 27). The honing tool is like a gear driving the workpiece at high speed (up to 30 m/min, or 100 sfm) while oscillating so that the teeth slide axially against the workpiece.

Spur gears and internal or external helical gears ranging in diametral pitch from 24 to 2.5, in outside diameter from 19 to 673 mm (3/4 to 26 1/2 in.), and up to 75 mm (3 in.) in face width have been honed on these machines. Finishes of 0.75 μm (30 μin.) are easily achieved, and finishes of 0.075 to 0.10

μm (3 to 4 μin.) are possible. Both taper and crown honing can be done.

*Tools* used in honing gear teeth are of two types, a helical gear shape tool made of abrasive impregnated plastic, and a metal helical gear with a bonded abrasive coating that is renewable. The plastic tool, which is discarded at the end of its useful life, is widely used. The metal tool is used mainly for applications in which plastic tools would be likely to break; also, it is used primarily for fine-pitch gears.

Plastic tools are supplied with abrasives of 60-grit to 500-grit size. Size of abrasive, gear pitch, and desired finish are usually related as:

| Grit size | Gear pitch | Finish μm | Finish μin. |
|---|---|---|---|
| 60 | ≤16 | 0.75–0.89 | 30–35 |
| 100 | 16–20 | 0.63–0.75 | 25–30 |
| 180 | >20 | 0.38–0.50 | 15–20 |
| 280 | >20 | 0.25–0.30 | 10–12 |
| 500 | >20 | 0.075–0.10 | 3–4 |

Honing tools do not load up, and a plastic honing tool can wear until its teeth break. Stock removal of 0.025 to 0.050 mm (0.001 to 0.002 in.) measured over pins is the recommended maximum.

*Methods*. The two methods used to hone gear teeth are the zero-backlash method and the constant-pressure method. In the zero-backlash method, which is used for gears made to commercial tolerances, the tool head is locked so that the distance between the center of the work gear and the center of the honing tool is fixed throughout the honing cycle. In the constant-pressure method, which is used for gears produced to dimensions outside commercial tolerance ranges, the tool and the work gear are kept in pressure-controlled tight mesh.

*Applicability*. The use of honing for removing nicks and burrs from hardened gears can result in a considerable cost saving in comparison to the usual method. In the usual method, the gears are tested against master specimens on sound test machines. Nicks indicated are searched for and removed using a hand grinder. The gear is then retested to make certain the nick has been removed. When honing is used, all of these various tests and procedures can be eliminated.

Some shape correction can be achieved in the removal of 0.050 mm (0.002 in.) of stock by honing. A helical gear 127 mm (5 in.) in diameter may show lead correction of 0.010 mm (0.0004 in.), involute profile correction of 0.0075 mm (0.0003 in.), and eccentricity correction of 0.010 mm (0.0004 in.).

The advisability of using honing for salvaging hardened gears hinges on cost considerations. As the error in tooth shape increases, honing time increases and tool life decreases. On the other hand, if the gears represent a large investment in production time and material, honing may be the most economical method.

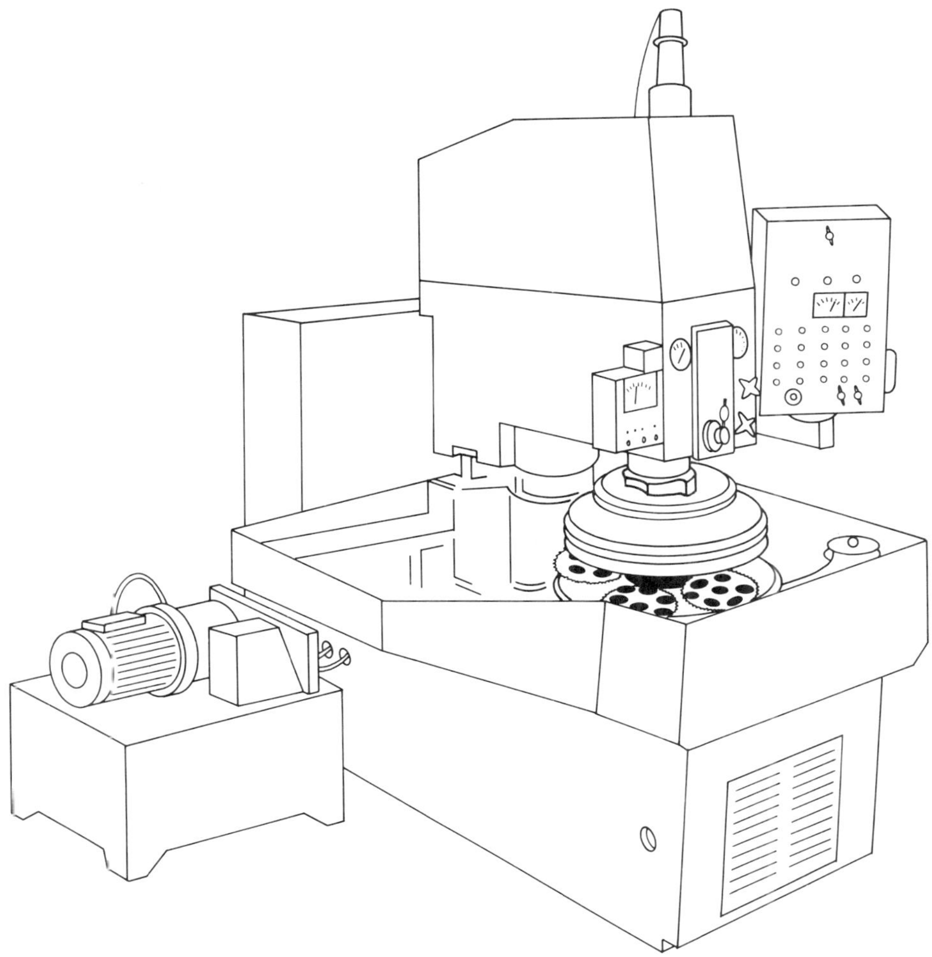

**Fig. 28** Two-wheel flat honing machine

Because honing is not designed for heavy stock removal or tooth correction, it cannot be substituted for grinding or shaving of gears. Rotary shaving usually leaves gear teeth smooth within 0.25 to 1.00 μm (10 to 40 μin.).

**Plateau honing** produces a special plateau finish, which removes the surface peaks but retains the deep valleys. Such a finish has been found desirable in engine performance because the valleys act as oil reservoirs for improved lubrication, especially during engine break-in.

A plateau finish is produced by first rough honing to final size. Then the surface is finished with a finer-grit stone for about 45 s, depending upon the amount of plateauing desired. The plateauing operation, with a 600-grit stone, removes so little stock that the bore diameter is not measurably increased.

**Flat honing** is a term designating a method and the equipment by which the flat surfaces of component parts produced by other methods are improved with regard to both flatness and parallelism of opposite surfaces. One of these surfaces may be that on which the part is located during the honing of the opposite face, or both faces may be honed simultaneously on machines operating with two honing disks.

The equipment used is similar in appearance to rotary face-grinding machines, but it is adapted to honing, a method which differs from grinding particularly in the low cutting speed of the abrasive disk, the applied speed being comparable to that used in conventional honing. The bonded abrasive disks used in flat honing are generally not intended for substantial rates of stock removal and thus can have very fine grains, promoting the development of a high-grade finish, even of the order of 0.025 μm (1.0 μin.) $R_a$ when needed. The spindle of the honing disk used on flat honing machines can be raised and lowered by an air or hydraulic cylinder.

Single- or double-surface flat honing machines are designed for high-production uses, finishing typically 1200 to 1800 parts per hour in a fully automated operation controlled by a timer.

On two-wheel machines (see Fig. 28), the top wheel, lower wheel, and workholder each have separate drives and controls. Machines are available with automatic con-

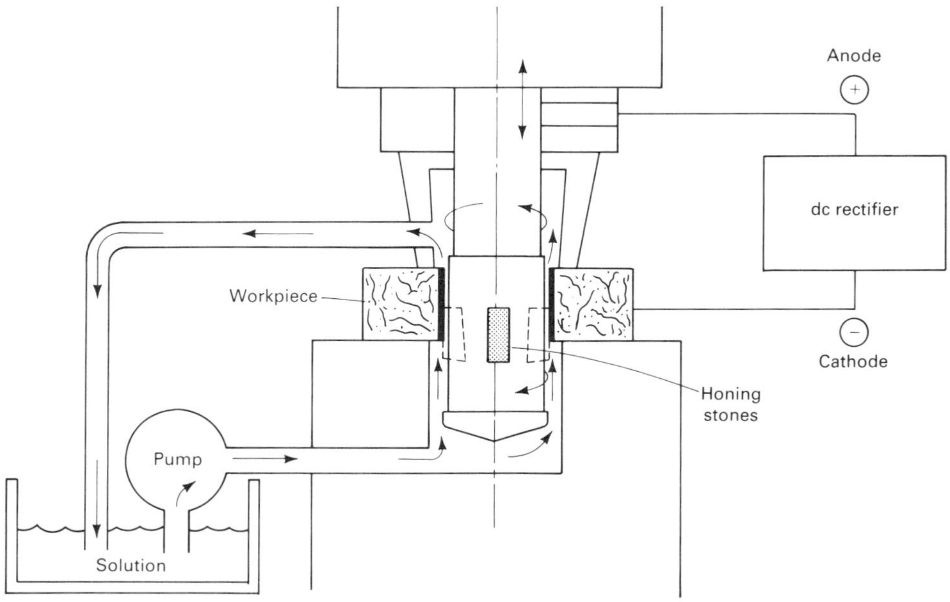


**Fig. 29** Principal components of a typical hone forming machine

trols to gradually increase pressure on the top wheel during the honing cycle. Automatic size control is also available. The workpiece carrier is part of an epicyclic sprocket holder.

**Electrochemical Honing.** In this process, metal is removed by introducing an electrolyte into a gap between a cathodic honing tool body and an anodic workpiece. Direct current from the power source is conducted to the tool through a brush assembly that acts as the cathode, while the workpiece becomes the anode.

The tool mandrel is made of metal and has a series of small holes which provide channels for the electrolyte circulating under controlled pressure. The electrolyte has the additional role of being a coolant and flushes away the chips that have been sheared off by the honing stones.

Abrasive honing stones are nonconductive and are limited to removing the electrochemically loosened metal particles, thereby controlling the geometric form and the size and texture of the produced surface. Thus, the electrochemically honed surface has the same characteristic crosshatch pattern and is essentially stress free, as is the surface produced in a regular honing process by purely mechanical action.

The abrasive stones continuously remove from the work surface the oxides developed by the electrochemical action, leaving the surface clean. This makes use of an electrolyte much less corrosive than that needed in the electrochemical machining process. Nor is it necessary for the electrolyte to be dispensed at high pressure, a process characteristic that calls for great structural strength of the equipment in the mechanically unassisted electrochemical machining.

The electrochemical honing process is carried out in a manner quite similar to regular honing, combining the rotation and reciprocation of the tool at controlled speeds. However, the tool mandrel must have good conductivity and all elements of the equipment that are exposed to the corrosive environment must be made of corrosion-resistant materials.

**Hone forming (HF)** is a recent development that constitutes a marriage of two different processes, honing and electrodeposition. The process is used to simultaneously abrade the work surface and deposit metal. It produces work surfaces that combine the benefits of both processes, that is, a geometrically and dimensionally controlled surface with a functionally favorable texture developed on a cladding that is concurrently electrodeposited on the base metal. In some of its basic principles, the method is a reversal of electrochemical honing.

The gap between the anodic honing tool body and cathodic workpiece is kept small. Current densities used in HF operations are many times those employed in other electrodepositing methods. Rate of deposition varies from 0.018 to 0.05 mm/min (0.0007 to 0.002 in./min) and up to 0.64 mm (0.025 in.) of metal has been deposited effectively in the laboratory.

In the following description of the equipment and process, a cylindrical internal surface or bore is considered; however, the method is applicable to flat or external round surfaces as well.

Hone forming is carried out on special equipment, which includes the machine for supporting the fixture and for actuating the rotating and reciprocating tool with its controlled feed motion. Additional elements of the equipment are the rectifier to supply the direct current, the solution tank, and the circulating system. The insoluble anode is part of the honing tool and is connected to the positive side of the rectifier. The workpiece functions as the cathode and is connected to the negative side of the rectifier. The gap between the anode (tool) and the cathode (workpiece) is kept small, while through that gap at controlled velocity passes the solution, which, together with the workpiece, is contained in a sealed circuit. Figure 29 shows the components of a typical hone forming machine.

After the workpiece has been cleaned of oil and other potential contaminants, it is mounted in a special honing machine and, as the first phase of the process, the bore is honed in a conventional manner. This part of the operation produces a clean base metal surface with accurate cylindrical form.

Subsequently, the metal-depositing process is started, keeping the anode at a controlled distance from the work surface, which acts as the cathode. The electric power supply is turned on, and the electrolyte is circulated between the anode and the cathode.

During this process the honing action continues, but with reduced stone pressure, thus ensuring that the deposited metallic layer will develop into a work surface of the desired form (roundness), size (diameter), and surface texture (crosshatched). The process is terminated when the desired size of the bore has been reached, as determined either indirectly, by the elapsed time, or directly, by means of a gage.

The entire process is carried out within a very short time, of the order of about one minute, because the cladding of the surface is many times faster than in conventional bath plating and has the added benefit of obtaining a precisely honed work surface with controlled characteristics. Other advantages of the new process are the elimination of most of the time-consuming preparatory and postprocess operations, such as masking, washing, neutralizing, and so forth, which are necessary for attaining the same results with the application of conventional procedures.

In hone forming applications, the workpiece must be conductive. To date, most materials used in hone forming are copper, bronze, tin, nickel, cobalt, or chromium plated and are hone formed on workpiece materials such as iron, steel, stainless steel, and bronze. Theoretically, any surface that can be honed can be processed with hone forming.

Although the immediate uses of this new method are mostly in the area of work salvaging out-of-tolerance parts and reconditioning worn surfaces, its potential is much wider and includes production line operations.

## Microhoning

The term microhoning is considered a descriptive designation for a method that uses hones (bonded abrasives) for tools and produces an accurately controlled surface whose parameters are measured in micro (very small) units. Essentially, the method is applied to round surfaces, both external and internal, of cylindrical, tapered, or spherical general form, and less frequently, to flat surfaces with limited areas. All of these contain surface elements that are either straight lines or circular arcs.

The method operates with tools made of bonded abrasives, with the active surfaces having forms complying with the contacted area of the work surface. The abrasive tools, while being held against the surface of the rotating work with a controlled, light force, generally effect a rapid, short-stroke, reciprocating (oscillating) motion in a direction parallel with the surface elements of the processed work surface.

Microhoning can be done on centers or with centerless through-feed machines. With centerless microhoning, 60 to 70° of the total workpiece circumference is in contact with one or more fine-grit stones.

## Comparison of Microhoning with Regular Honing

Characteristics common to regular honing and microhoning methods are:

- The use of bonded abrasives for tools
- Operation with low cutting speeds and producing light cutting force
- The combination of a rotational and a linear motion between the tool and the work; both methods produce a cross-hatch pattern, but in microhoning it is less distinct
- The objectives of the process, that is, improving the form, size, and surface texture of the work surface without causing metallurgical damage

Particular differences between regular honing and microhoning are described below.

**The Stroking Length.** Honing is essentially a long stroking method, capable of covering work surface lengths several times that of the stones. During the process the stones travel a substantial distance (usually expressed as a fraction of the stone length) over the ends of the processed work surface.

In microhoning, the stone reciprocation is over a very short length, typically of the order of about 1.0 to 4.1 mm (0.040 to 0.160 in.). For that reason the stone motion is often referred to as an oscillation.

**Frequency of the Reciprocation.** In honing, the number of strokes per minute varies from very few, in the case of very long strokes, to about 100 or 200 in general operations. The stroking, although usually hydraulically actuated, in the case of low-production or medium-production machines may also be manual.

Microhoning operates with a very rapid reciprocating motion, its frequency varying, according to the applied system and stroke length, in the range of 300 to 2500 cycles per minute. The reciprocating motion has mechanical or pneumatic drive.

**Rotating Member.** In honing, it is the tool that rotates (with the rare exception of the external honing of very long parts; in microhoning, it is always the workpiece that rotates in front of the tool whose reciprocation is along a fixed path.

**Contact Area.** In honing, usually several stones, at essentially uniform circumferential distances, make simultaneous contact with the work surface, each stone taking part in the stock removal.

In microhoning, a single stone, except for the rare multistone applications on large diameters, covers only one segment of the work surface, although the width of the stone produces a wraparound effect.

**Work Surface Configuration.** Stroking in honing is always along a straight-line path, which limits the general form of the surface adapted to honing. The long stroking of honing also excludes surfaces that do not permit free stroking, although the honing of surfaces with one free end, for example, blind holes, is possible. Honing is used nearly exclusively for bores.

Microhoning, which operates with very short strokes, can be applied also to surfaces with circular cross-sectional contours, such as spherical or toroidal shapes. Equally adaptable to external and internal surfaces, the short stroking also permits the microhoning of surfaces bounded by shoulders on both ends, requiring a very narrow undercut, comparable to that provided in cylindrical or internal grinding.

**Amount of Stock Removal.** Honing can be used for efficiently removing substantial amounts of stock. In rough honing, of the order of 0.25 to 0.38 mm (0.010 to 0.015 in.) can be removed, occasionally, considerably more. In finish honing, about 0.05 to 0.15 mm (0.002 to 0.006 in.) can be removed from the part diameter. This stock-removing capacity of the honing method permits the correction of various types of bore form irregularities.

Stock removal in microhoning is usually of the order of 0.0025 to 0.0075 mm (0.0001 to 0.0003 in.), consequently, the general form irregularities, for example, ovality, taper, and so on, can be corrected only within a very limited range.

**Characteristics of the Produced Work Surface.** Honing is generally used to produce surface finishes in the range of 0.8 to 0.2 μm (32 to 8 μin.) $R_a$. Such surface texture values are adequate, or even desirable, for many types of bores finished by honing. Often even rougher surfaces are required, or, in rare cases, a finish of the order of 0.10 μm (4 μin.) $R_a$ is specified. It is feasible to produce even finer finish values by honing, but generally at the price of rapidly decreasing productivity.

Microhoning can produce, in efficient operation, surface finishes of the order of 0.10 to 0.05 μm (4 to 2 μin.) $R_a$, and, upon occasion, even finer finishes. This single parameter, the average surface roughness, does not convey, however, all the functionally important surface characteristics that can be developed in regular production by microhoning. These will be discussed in greater detail under a separate heading.

## Basics of the Microhoning Process

The operating principles of the microhoning process were illustrated earlier in Fig. 3. The diagram illustrates that an abrasive stone, whose operating face complies with the general form of the work surface, is forced against the rotating workpiece to exert a specific pressure while carrying out a short-stroke reciprocating motion.

In operation, the abrasive stone will first make contact and act upon the protruding elements of the work surface. Protuberances can result from surface form irregularities such as lobbing, chatter marks, waviness, and so on, and can also constitute the peaks and crests of a rough surface. Generally, irregularities from both sources are present, with the roughness superimposed on the form irregularities.

By abrading the protruding elements of the surface, the action of the microhoning stone will gradually equalize the work surface, correcting some of the form irregularities and reducing the roughness. The major protuberances of the surface will be abraded first by the concentrated action of the total stone pressure; consequently, the rate of surface improvement will be rapid in the starting phase of the operation. As the contacted work surface area continuously extends, the rate of stone penetration will become slower, partly because the specific pressure on the contacted areas decreases, and partly because the stone surface starts to glaze, due to the reduced dressing action of the work surface. The latter condition, although detrimental to the abrading ability of the stone, is beneficial to the development of a better finish on the microhoned surface, which is the prime objective of the process. The optimum balance between the effective abrading and the dependable smoothening ability of the stones can be controlled by the proper selection of the process variables, such as reciprocating and rotating speeds; applied pressure; and form, dimensions, and composition of the stones.

The described progress from fast abrading to the slower smoothening action could be extended to produce very substantial improvements of the work surface by microhoning. However, it would take an excessive amount of time for a stone that is capable of rapid abrading action to lose its cutting ability to the degree that it could produce a very smooth surface. For that as well as for other reasons, that is, to retain the unimpeded abrading capacity of the stone, the microhoning process is usually operated in the range of its highest efficiency. The approximate extent of that range can be expressed conveniently by the ratio of surface roughness improvement, which usually varies from 4:1 to 8:1, although these values are not absolute limits. The lower ratios commonly apply to the effective range for coarse surfaces, such as improving the finish from 6.3 to 1.5 μm (250 to 60 μin.) $R_a$, while the roughness of a fine-ground surface can be reduced efficiently from 0.40 to 0.050 μm (16 to 2 μin.) $R_a$.

**Surface texture** is seldom measured, although it can affect certain functional characteristics, such as the sealing capacity and lubricant retainment of the surface. At a relatively higher roughness level, a crosshatch pattern can be observed on the microhoned surface; that pattern tends to change, as the process progresses, into a nondirectional pattern, or to disappear entirely, resulting in a reflective surface.

These changes of the surface texture, brought about by microhoning, when applied to the distinctly directional pattern produced by grinding are generally beneficial to the functional properties and service life of the work surface.

A rough surface will make contact with a mating part through its protruding crests, and the extensive lower-lying areas will have no functional role. Reducing the roughness of the surface without changing its alternating crests and valleys will only partially remedy the unfavorable conditions of high wear and friction in the contact area.

The optimum contact condition is created by cutting down the crests and transforming them into wide plateaus, a process brought about by microhoning. The effect of microhoning on an originally rough surface is visualized by the schematic cross-sectional diagrams in Fig. 30, which illustrates in five steps (corresponding to the indicated processing times) the manner in which the rugged surface resulting from a preceding machining is improved by changing it into a surface containing wide plateaus.

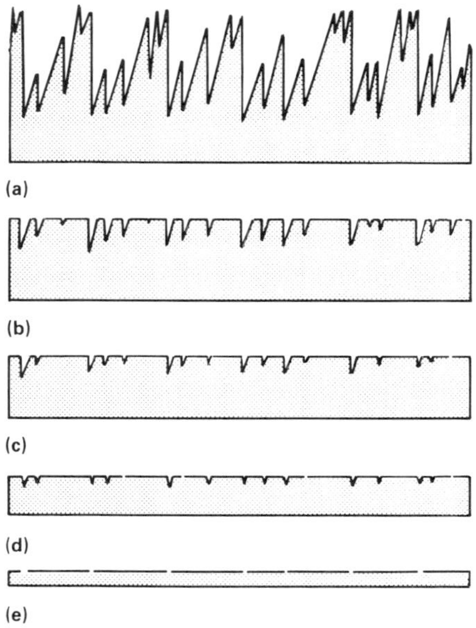


**Fig. 30** A rough surface, (a), being gradually improved by microhoning, that is, by cutting down the protruding crests and finally developing a surface consisting of wide plateaus. The elapsed time and corresponding surface roughness obtained at each interval for the five cross sections are: (a) 0 s, 1.25 μm (50 μin.), (b) 10 s, 0.38 μm (15 μin.), (c) 20 s, 0.28 μm (11 μin.), (d) 30 s, 0.20 μm (8 μin.), (e) 0.050 μm (2 μin.). The diagrams exaggerate the cross-sectional contours of the work surface for visualizing the effect of microhoning on the work surface texture.

## Process Parameters in Microhoning

**Type and Grit of Abrasive Stones.** Both types of abrasive materials are used for microhoning stones:

- Aluminum oxide abrasives (which fracture less easily), for carbon and alloyed steel
- The more friable silicon carbide, for very soft or very tough types of steels, as well as for cast iron and most types of nonferrous metals

The hardness of the stones, controlled by the percentage of the bond, varies from J (very soft), to P (very hard). The former is used for extremely hard alloys, chromium plates, and so on, while the hardest bonds are needed for cast iron and nonferrous metals.

The dimensions of the active stone face are determined by the size of the work surface. For external cylindrical surfaces, the width of the stone is about 60 to 80% of the part diameter, but generally not more than about 25 mm (1 in.). For work diameters larger than about 150 mm (6 in.), microhoning heads, with several stones arranged along an arc in compliance with the work surface, are frequently used. The length of the stone is usually somewhat less than the length of the work surface, but not more than about three times the width of the stone. For the microhoning of longer work surfaces, an additional traverse movement is needed. The thickness of the stones is controlled by the mounting dimensions, the access length (which is particularly essential for honing inside shoulders), and an adequate wear depth; this latter component is substantially more, in some applications, than the stone width.

The grit size of the stones is generally selected from a wide range to suit the condition of the machined surface and the objectives of the process. The type of microhoning work required can vary from roughing (applied either as a preparatory operation to a subsequent finishing or as a single process for general-purpose work) to fine, or even very fine finishing. For coarse microhoning applied to improve the finish and the functional properties of rough-machined surface, grit in the range of 60 to 220 is used. For general-purpose microhoning work to produce a finish of the order of 0.40 to 0.20 μm (16 to 8 μin.) $R_a$, grit sizes 320 to 500 are needed. Fine finishing is carried out with grit sizes in the 600 to 800 range, while for very fine finishing, stones with 1000 to 1200 grit may be selected.

**Stone Pressure.** One of the characteristics of the microhoning method is the use of a relatively low stone pressure to avoid a deep penetration into the work surface, which could leave furrows and generate heat. The shearing effect of the grains is limited to the protruding elements of the

**Table 11 Typical microhoning production rates for automotive components**

| Automotive component | Finish after grinding μm | Finish after grinding μin. | Number of spindles on microhoning machine | Surface configuration | Finish after microhoning μm | Finish after microhoning μin. | Production rate, parts/h |
|---|---|---|---|---|---|---|---|
| Tappet head | 0.76–1.00 | 30–40 | 1 | Spherical or flat | 0.125–0.20 | 5–8 | 900 |
| Crankshaft | 0.76–1.00 | 30–40 | 10 | Cylindrical | 0.125–0.20 | 5–8 | 80 |
| Stem pinion bearing | 0.38–0.63 | 15–25 | 2 | Cylindrical | 0.050–0.10 | 2–4 | 120 |
| Distributor shaft | 0.76–1.00 | 30–40 | 12 | Cylindrical | 0.075–0.125 | 3–5 | 720 |
| Pressure plate(a) | 2.50–5.0 | 100–200 | 2 | Flat | 0.18–0.30 | 7–12 | 100 |
| Brake drum(a) | 5.0–6.3 | 200–250 | 1 | Internal cylindrical | 0.38–0.63 | 15–25 | 150 |
| Tappet body | 0.25–0.50 | 10–20 | 12 | Cylindrical | 0.050–0.10 | 2–4 | 800 |
| Camshaft main bearing | 0.38–0.63 | 15–25 | 10 | Cylindrical | 0.050–0.10 | 2–4 | 80 |
| Gear thrust face | 0.25–0.50 | 10–20 | 9 | Flat | 0.050–0.10 | 2–4 | 500 |
| Tapered bearing race | 1.00–1.25 | 40–50 | 2 | Cylindrical | 0.050–0.10 | 2–4 | 450 |

(a) Starting finishes for pressure plates and brake drums are for turned surfaces rather than ground. In some cases, pressure plates are ground to a finish of 0.50 to 0.63 μm (20 to 25 μin.).

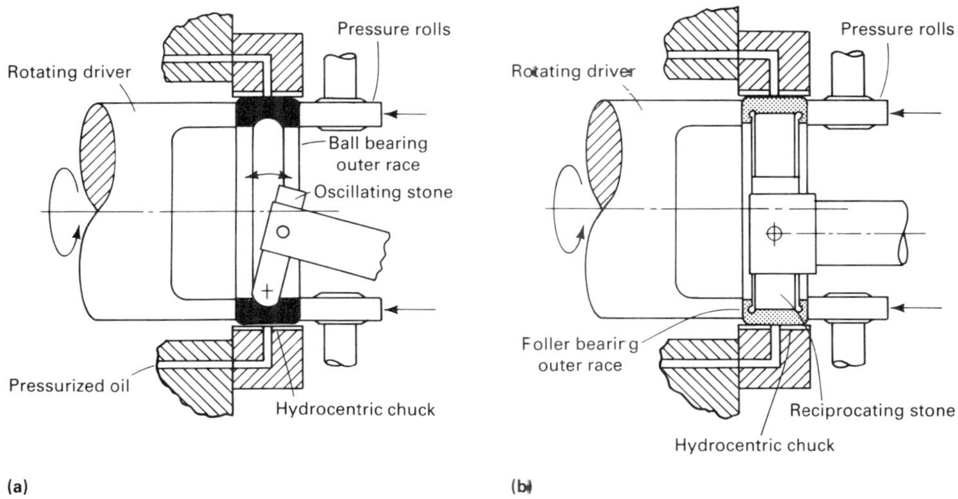


**Fig. 31** Setups for microhoning (a) ball bearing races and (b) roller bearing races

surface, and that action can be achieved by causing the stones to bear against the workpiece with a light force.

For average work, stone pressure in the range of 140 to 275 kPa (20 to 40 psi) is generally used, raising the pressure for roughing to about double these values. For very fine finishing, particularly for soft material, stone pressure as low as 14 to 34 kPa (2 to 5 psi) may be applied.

**Work Speed.** Most commonly, microhoning is applied to the surface of rotating workpieces that are of cylindrical and, less frequently, of tapered shape, referred to as OD. However, flat surfaces, either uninterrupted, such as the end faces of bearing rollers, or of annular shape, such as flange areas, are also adapted to microhoning. Longitudinal flat surfaces, which require linear traverse movement, can also be microhoned, although that method is seldom applied because flat surfaces, unless they are located in a recessed position, are generally adaptable to lapping (see the article "Lapping" in this Volume).

The work speed, therefore, is generally specified as the surface speed of the rotating workpiece, with its most commonly applied values falling within the following ranges:

- For roughing, 12 to 15 m/min (40 to 50 sfm)
- For finishing, 30 to 60 m/min (100 to 200 sfm)

To produce a very fine finish and also to transverse workpieces at a high feed rate, substantially higher surface speeds, up to about 120 m/min (400 sfm) are also applied.

At the lower work speed, the microhoning process generally develops a very fine but still distinguishable crosshatch pattern, which may be the desirable surface condition in many applications, although it reduces the reflectivity or shine of the work surface. At higher work speeds, that pattern disappears, and a brighter surface is developed.

**Stroke Length and Speed of the Stone Reciprocation.** The fast reciprocation of the stones in a short stroke is one of the essential characteristics that sets microhoning apart from regular honing. The length of the stroke differs somewhat in various systems, some of which employ a single stroke length, for example, 4.76 mm (0.1875 in.), while others are designed to provide variable stroke length, adjustable over a range of about 2.00 to 5.10 mm (0.080 to 0.200 in.).

The actual linear speed of the motion of the stone is the function of the stroke length and the rate of reciprocation. These two factors are also referred to as the amplitude and the frequency of stone reciprocation. Stone speed can be adjusted to the requirements of the operation by varying either of these factors.

The resulting linear stone speeds differ over a rather wide range according to the system used and the condition of adjustment. Typical extreme values are 3.18 to 20.3 m/min (125 and 800 in./min).

## Microhoning Applications

Microhoning lends itself to cylindrical (internal and external), flat, tapered, toroidal, spherical, and barrel-shaped surfaces on automotive products such as ball-and-roller bearings, transmission shafts, crankshafts, shock absorber piston rods, valve shaft diameters, and universal-joint spiders. Figure 31 shows two typical setups used to machine two different types of bearing races. Table 11 lists typical production rates and finishes obtained with microhoning on automotive components.

Microhoning is very effective for finishing ball-shape components to a unique degree of sphericity. Such shapes, executed to an excellent finish, are required in engineering production for valve balls, pump pistons, and so on. An uncommon, but important application of spherical microhoning is in the manufacture of prostheses, specifically, artificial hip joints. The accurate sphericity and very high finish of the ball and socket surfaces, which is accomplished by microhoning these surgically installed elements, ensures their dependable functioning and capability to withstand load of 1.18 kN (265 lbf) or more.

# Lapping

Revised by Pel Lynah, P.R. Hoffman Machine Products

LAPPING is a low-speed low-pressure abrading operation that accomplishes one or more of the following results:

- Extreme dimensional accuracy of lapped surface (flat or spherical)
- Close parallelism of double-lapped faces
- Refinement of surface finish
- Extremely close fit between mating surfaces
- Removal of damaged surface and subsurface layers that degrade the electrical or optical properties

In general, the quality that can be obtained by lapping is not easily or economically obtained by other processes.

Loose abrasive, carried in an appropriate vehicle, is used on cast iron laps in 99% of the lapping applications, but there are some isolated fixed-abrasive applications that are classified as lapping. It is difficult to make a clear distinction between lapping and honing. Lapping is the lower-pressure, lower-speed, and lower-power application of the use of fixed abrasives. Furthermore, the fixed abrasives of honing are usually limited to the conventional resinoid or vitrified face wheels, while the fixed-abrasive lapping operation often uses unconventional media, such as urethane-impregnated pads, polyvinyl alcohol and abrasive mixed, foamed, and cured to a hard, cellular block, plated or surface-bonded diamond, or thin abrasive-filled vinyl films. The usual definition of lapping is the random rubbing of a part against a lap, usually of cast iron composition, using an abrasive mixture in order to improve fit and finish. Lapping operations usually fall into one of two categories: individual-piece lapping and matched-piece lapping.

**In individual-piece lapping,** abrasive is rubbed against the workpiece with a special tool called a lap (usually of material softer than the workpiece), rather than with a mating workpiece surface. When loose abrasive is used, the lap is usually made of soft cast iron (typically close-grain cast iron or meehanite metal) or a soft nonferrous metal. Laps made of bonded abrasive also can be used, as discussed above.

Individual-piece lapping is most effective on hard metals or other hard materials. It is used to produce optically flat surfaces, to produce accurate planes from which other planes can be located (as for gage blocks), and to finish parallel faces.

The machine lapping of individual pieces, either one per cycle or in multiple-piece loads, represents the bulk of the production lapping currently done in industry. Single-side flat lapping machines, double-side planetary machines, cup lapping machines for spherical surfaces, and specialized single- or double-plate machines (such as ball, roller, or pin laps) constitute the vast majority of lapping installations.

**In matched-piece lapping,** sometimes called equalizing, two workpiece surfaces separated only by a layer of abrasive mixed with a vehicle are rubbed against each other. Each workpiece drives the abrasive so that the grit particles act on the opposing surfaces. Irregularities that prevent the surfaces from fitting together precisely are thus eliminated, and the surfaces are mated. In many cases, a part is first lapped individually and is then mated with another part by this method, before the two are stocked as a pair of lapped-together parts.

Matched-piece lapping enables mating parts (such as the heads and blocks of internal combustion engines) to form liquid-tight or gas-tight seals without the need for gaskets. It also eliminates the need for piston rings in fitting some plungers to cylinders. Other common uses of matched-piece lapping include fitting tapered valve components (Fig. 1) and mating two or more gears in a set.

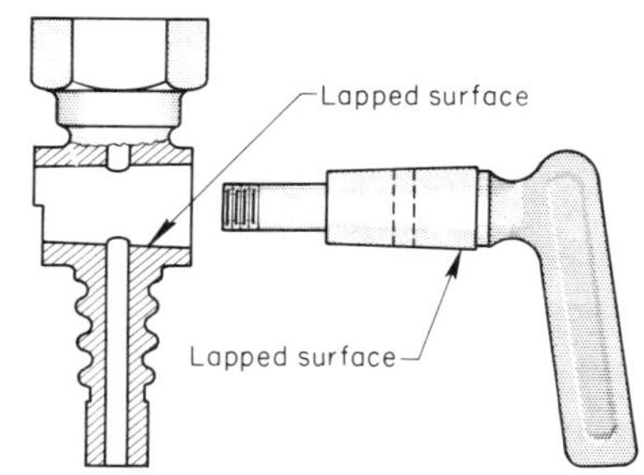


**Fig. 1** Tapered valve components finished by matched-piece lapping for precise fit of mating surfaces

## Process Capabilities

Parts that are processed by lapping are constructed of a variety of materials, ranging from metal parts for tooling, gaging, or sealing to electronic crystals such as quartz piezoelectric frequency devices and silicon semiconductor material for integrated circuit manufacture. Tungsten carbide, ceramic, and glass components; aluminum computer disks; tool steel slitter blades; saw blanks; and jade decorative tiles are among the applications that demonstrate the diversity of the lapping process.

**The size or weight of the workpieces** that can be lapped is limited only by the available equipment. Parts finished by lapping range in weight from a fraction of an ounce to hundreds of pounds.

**Workpiece Shape.** Tools and methods have been devised for lapping virtually every shape of workpiece on which a lapped surface is desired. Lapping is most widely used for finishing flat surfaces or outside and inside cylindrical surfaces. The process can also be applied to balls, rollers, cones, double-curved surfaces, assembled bearings, and shapes such as gear teeth.

**Material Removal.** Lapping is intended as a final finishing process that would be, in general, an impractical or uneconomical means of removing stock. In most applications, less than 0.13 mm (0.005 in.) of material is removed from a surface by lapping. However, occasionally (usually in flat lapping), 0.38 mm (0.015 in.) or even more may be removed. In a few cases, it has proved more economical to remove stock by lapping than to add a preliminary grinding operation.

## Selection of Abrasive

Silicon carbide and fused alumina are the abrasives most widely used for lapping. Silicon carbide is extremely hard (2500 HV). Its grit is sharp and brittle, making it nearly ideal as an abrasive for many lapping applications because it continually breaks down to expose new cutting edges. Silicon carbide is used for lapping hardened steel or cast iron, particularly when an appreciable amount of stock is to be removed.

Fused alumina (2000 HV) is also sharp, but it is tougher than silicon carbide and

**Table 1 Relative cost (as of 1988) of abrasives used in lapping**

| Type of abrasive | Average size µm | Average size µin. | Grade | Cost $/lb 5 lb | Cost $/lb 10 lb | Cost $/lb 25 lb |
|---|---|---|---|---|---|---|
| Aluminum oxide | 5.0 | 200 | 1950 | 3.66 | 2.99 | 2.80 |
| | 17.5 | 700 | 1600 | 3.09 | 2.44 | 2.27 |
| | 64.0 | 2520 | 1220 | 2.11 | 1.49 | 1.34 |
| Silicon carbide | 5.0 | 200 | 2950 | 11.89 | 10.99 | 10.57 |
| | 22.5 | 900 | 2400 | 3.85 | 3.18 | 3.01 |
| | 55 | 2200 | 2240 | 3.07 | 2.46 | 2.26 |
| Boron carbide | 10 | 400 | 3800 | 47.91 | 45.94 | 44.45 |
| | 20 | 800 | 3400 | 42.23 | 40.43 | 39.10 |
| | 40 | 1600 | 3280 | 36.63 | 35.00 | 33.84 |

Source: Lapmaster International

breaks down less readily. Fused alumina is generally more suitable for lapping soft steels or nonferrous metals than silicon carbide.

Boron carbide (2800 HV) is next to diamond in hardness and is an excellent abrasive for lapping. However, because it costs 10 to 25 times as much as silicon carbide or fused alumina, boron carbide is usually used only for lapping dies and gages, which is often done by hand and in small quantities using little abrasive. An example is synthetic sapphire for electronic applications. The raw material cost is expensive, justifying a high abrasive-processing cost. Relative costs for various quantities and grit sizes of silicon carbide, fused alumina, and boron carbide are compared in Table 1.

Diamond (6500 HV), the hardest of all materials is also used as an abrasive for lapping metals. It is available as a paste or a slurry.

Softer abrasives, such as emery, garnet, unfused alumina, and chromium oxide, also are used for lapping, but to a far lesser extent. As indicated in Table 2, these softer abrasives are used for lapping soft metals or for the final lapping of parts on which a highly reflective surface is required. In lapping to produce a reflective surface, no significant amount of stock is removed, and the finish is not necessarily finer than a matte finish.

**The grit sizes** most commonly employed in lapping range from 100 to 1000 (Table 2). However, abrasives are usually available in grit sizes from about 50 to 3800, and even finer.

For lapping hardened steel to remove about 0.0051 mm (0.0002 in.) of stock and to produce a finish of less than 0.050 µm (2 µin.), a grit size of 280 is appropriate. If finishing requirements are less stringent, 180-grit abrasive will be more economical because it removes metal faster than finer grit does.

As the amount of stock to be removed increases, coarser grits are required. For the removal of considerable amounts of stock, it is more economical to employ a roughing operation, followed by a finishing operation.

When a substantial amount of stock is being removed, a fine finish can be produced without the use of a small grit size, because the originally coarse grit breaks down as lapping proceeds and progressively produces a finer finish. If this technique is used, there must be enough stock allowance for lapping so that the deeper scratches formed initially by the coarse grit will be removed by the time final dimensions are reached.

**Grading.** When an abrasive of a specified grit size is purchased, some of it will be finer and some coarser than the stated size. The degree of grading is an important consideration in the selection of any abrasive. Abrasives increase in cost as the grading becomes closer. However, the use of a low-cost, loosely graded abrasive is not always economical, as demonstrated in the following example.

**Example 1: Closely Graded Abrasive for Greater Economy.** A low-cost grade of silicon carbide that ranged in grit size from 100 to 800 was used for lapping piston rings. A change to an abrasive that was closely graded to a grit size of 600 reduced the overall cost of abrasive by 50%, even though the initial cost of the 600-grit abrasive was twice that of the low-cost grade. The savings was made possible because the 600-grit abrasive contained more of the grit size that is most efficient for lapping; consequently, only one-fourth as much of it was required for removing the same amount of stock. The 600-grit abrasive also gave a smoother finish with less smudging.

**Table 2 Types and grit sizes of abrasives for various applications of lapping**

| Abrasive | Relative hardness | Grit size | Typical applications |
|---|---|---|---|
| **All-purpose compounds** | | | |
| Silicon carbide | Hard and sharp | 100, 220, 320, 400 | Tool-room lapping |
| Corundum | Medium soft | 220, 240, 280 | Tool-room lapping |
| **Compounds for roughing, finishing, or polishing** | | | |
| Corundum | Medium soft | 400, 500, 600 | Roughing softer steels |
| | | 700, 800 | Finishing softer steels |
| Alumina | Hard | 500, 600, 900 | Roughing harder steels, stainless, chromium plate |
| | | 2–10 µm (80–400 µin.) | Finishing hard steels |
| Alumina | Medium hard | 900 | Polishing hard steels |
| | | 5, 10, 15 µm (200, 400, 600 µin.) | Polishing hard steels |
| | | 1–3 µm (40–120 µin.) | Polishing stainless, chromium plate |
| Alumina | Soft | 1, 2 µm (40–80 µin.) | Polishing |
| Silicon carbide | Hard and sharp | 600, 800, 1000 | Roughing hardened steels; cast iron |
| Garnet | Medium soft | 600, 800 | Finishing brass, bronze |
| | | 10 µm (400 µin.) | Polishing brass, bronze |
| Emery | Medium soft | 800 | Polishing softer steels |
| Chromium oxide | Medium soft | 1 µm (40 µin.) | Polishing stainless |
| Ferric oxide | Soft | 1 µm (40 µin.) | Polishing soft metals |
| Cerium oxide | Medium hard | 1, 2 µm (40, 80 µin.) | Polishing |

## Selection of Vehicle

Vehicles, or binders, for loose abrasives include a wide variety of compounds. Some shops prepare their own formulations or modify standard compositions. However, more consistent results can be obtained with standardized, commercial compounds.

Two major factors in the selection of a vehicle are the material being lapped and the lapping method to be employed (inside or outside diameter, flat or spherical). Any vehicle should:

- Retain abrasives in uniform suspension and deagglomeration
- Serve as a cushion between surfaces being lapped (to minimize lap-to-part contact and yet avoid rolling action on abrasive particles)
- Adhere to laps and therefore minimize waste of compound
- Be noncorrosive to the material being lapped
- Be nontoxic to operators
- Be easily removable by cleaning
- Respond to temperature variations with the viscosity characteristics (stability or flexibility) desired in a given application. Although rapid changes in viscosity are usually undesirable, in some applications it is important that the vehicle be able to change quickly from a grease to an oil when under slight heat and pressure and then revert quickly to greaselike consistency when pressure is released

Most vehicles have an oil or grease base, although some are made of water-soluble compounds. The consistency of oil-base vehicles varies from that of mineral seal oil (a water-white product having a viscosity slightly higher than kerosene) to that of heavy grease. Common spindle oil is often used as a vehicle. Commercial compounds contain mixtures of animal fat, vegetable oils, and mineral oils.

Vehicles with an oil or grease base are usually used for lapping ferrous metals. For specific applications in which grease or oil would be objectionable (such as copper-base alloys and other nonferrous metals), water-soluble vehicles are available. These vehicles, which are readily removed with water, are low-viscosity compositions of starches, bentonite, and soluble oils with rust inhibitors.

Contamination is a potential problem in many nonmetal applications, especially in electronic components; therefore, cleanability is the primary consideration. Most commercial and proprietary vehicles for these materials are glycerine-base formulations; this provides a good-quality suspension, good film-forming properties (and therefore good lubricity), and a water-soluble mixture that is readily cleanable. Clay or mica is occasionally mixed in to fill the voids between the abrasive particles, thus enhancing the suspension. Ionic, charged, and submicron particles are also used as suspension agents. These particles affix themselves to the abrasive grain and produce electric repulsion forces to disperse and suspend the abrasive.

Pure water is used in the lapping of glass and ceramic. The abrasive slurry is generously applied and recirculated. The resulting high fluid flow is used to keep the abrasive stirred and dispersed.

## Lapping Outer Cylindrical Surfaces

Outer cylindrical surfaces are usually lapped by one of the following methods:

- Ring lapping (a manual operation)
- Machine lapping between plates
- Centerless roll lapping with loose abrasive
- Centerless lapping with bonded abrasives

In addition to these techniques, special methods are used for specific applications, such as the lapping of piston rings and crankshafts. Choice of method depends on part configuration, size of the production lot to be lapped, and cost. Many outer cylindrical surfaces can be lapped with equal success by two or more methods.

### Ring Lapping

Ring lapping is the simplest method of lapping outer surfaces. Designs of ring laps vary, but the assembly illustrated in Fig. 2 is typical. The lapping ring (or ring lap), usually made of cast iron, is manually stroked back and forth over the workpiece, which is chucked in a lathe or polishing head and rotated. Lapping compound (usually of paste consistency) is often applied to the surface of the workpiece. A manually adjusted screw is tightened, as required, to maintain a slight drag on the lap.

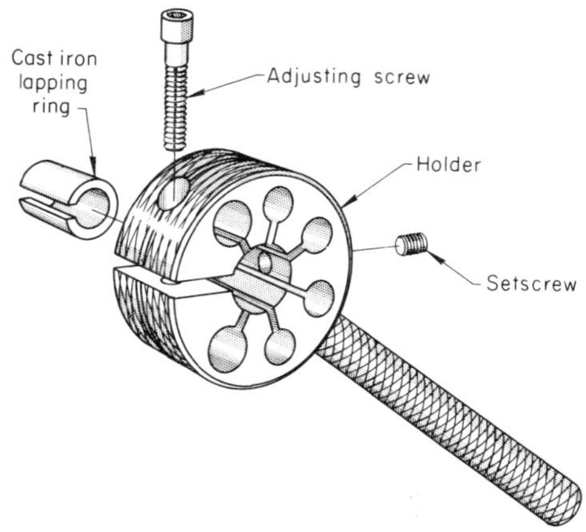


**Fig. 2** Typical ring lapping assembly. Drilled holes and slots permit uniform adjustment.

The ring lap should always be shorter than the workpiece, and if size permits, it should have adjustable slots. Finishing the bore of a ring lap is critical. It should be drilled, reamed, and honed (or lapped) to a size very close to the starting diameter of the workpiece; the screw adjustment should be used only to compensate for the slight decrease in workpiece diameter as lapping proceeds.

**Applicability.** Ring lapping, when performed by a skilled operator, offers at least two advantages over machine methods:

- Parts can be produced to extremely close tolerances
- Out-of-roundness can be corrected to a degree not feasible by machine lapping

Aside from requiring operator skill, however, ring lapping is tedious and expensive, and it should be considered only when one or more of the following conditions prevail:

- Equipment for any other method is not available
- Workpiece is out-of-round
- Weight of workpiece is unbalanced
- Workpiece has two or more different diameters that must be lapped
- Workpiece has flats, keyways, or other interruptions on its cylindrical surface
- Only a few pieces are to be finished

The following example describes a specific application in which some of the above conditions existed and in which ring lapping was therefore the most suitable method.

**Example 2: Preproduction Versus Production Lapping.** In many cases, ring lapping is used for parts being developed, and a more economical method is used when the parts are in production. This procedure was followed for the valve needle shown in Fig. 3, which was ring lapped in small quantities during development but was machine lapped between plates (see Example 3) in production lots. These needles, in diameters of 6.4 to 9.5 mm (¼ to ⅜ in.), were made of alloy tool steel and hardened to 60 to 65 HRC.

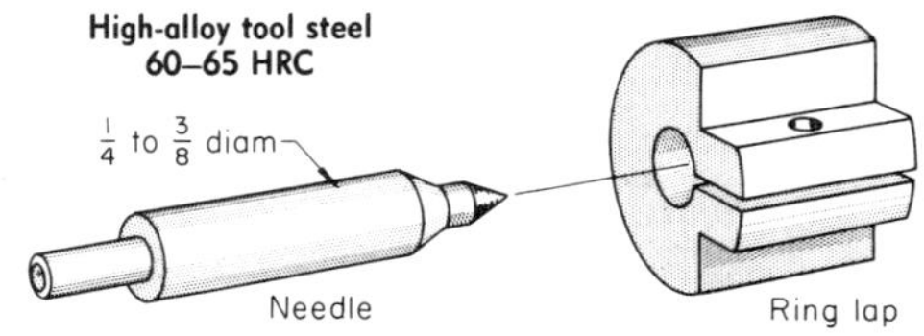


**Fig. 3** Valve needle and ring lap for finishing small, preproduction quantities. Dimensions given in inches

For ring lapping, each needle was chucked by its stem and rotated in a lathe at 650 rev/min. The lap (Fig. 3), which was made of cast iron, was stroked back and forth over the needle until grinding marks were eliminated. The lapping medium with which the needle was coated consisted of chromium oxide mixed with spindle oil. Lapping produced a finish of 0.050 μm (2 μin.) and maintained tolerances of 0.0013 mm (0.000050 in.) for straightness and 0.00064 mm (0.000025 in.) for roundness.

To ensure straightness, the laps used had to be at least three-fourths as long as the area to be lapped. The laps also had to be inspected frequently and had to be reconditioned by being lapped with internal laps of similar material.

### Machine Lapping Between Plates

In the machine lapping of outer cylindrical surfaces between plates, the laps are two opposed cast iron or bonded-abrasive circular plates that are held on vertical spindles of the machine (Fig. 4). The plates are usually 200 to 710 mm (8 to 28 in.) in diameter, although larger sizes are available. For the most part, plain-face laps are used, and for the greatest accuracy, the width of the lap face should not exceed the length of the surface being lapped. The workpieces are retained between these laps in slotted plates and are caused to rotate and slide. They are given an eccentric, or in-and-out, motion to break the pattern of motion and to ensure that they move over the inside and outside edges of the lap. This prevents grooving of the lap. For short runs, an eccentric motion is unnecessary if the laps are kept flat by reconditioning.

**Cast Iron Laps.** When cast iron laps are used, the lower lap is usually rotated and drives the workpieces. The upper lap is held stationary, but it is free floating so that it can adjust to the variations in workpiece size. The

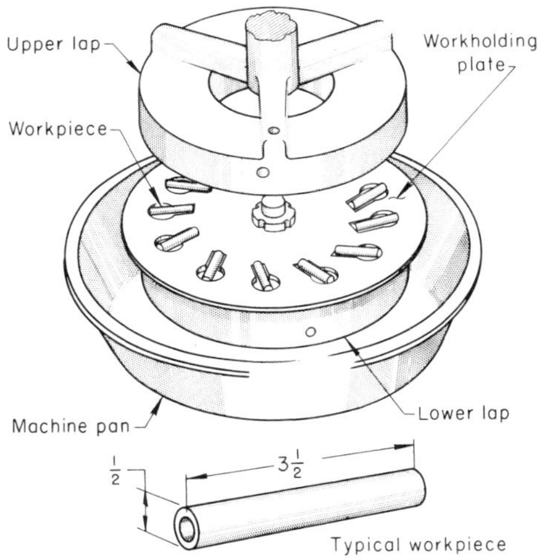


**Fig. 4** Typical vertical lapping machine for finishing cylindrical surfaces in production quantities. Dimensions given in inches

lower lap regulates the speed of rotation because the workholder is not driven.

The abrasive is used with a paste-type vehicle and is swabbed on the laps before the cycle is started. Oil or kerosene is then added during the cycle to prevent drying of the vehicle, which may result in scratching.

Because the upper lap floats, several parts must be lapped simultaneously. A quantity of three parts will support the upper lap, but when only three parts are lapped, the machine will not produce straightness or a common size. Therefore, it is advisable to lap a minimum of five parts; if this quantity is not available, the machine should be loaded with dummy parts. The best practice is to put as many parts as possible in a load. This reduces the pressure on each part and slows the operation. Thus, the operator has more control and can secure desired tolerances more easily.

Finishes as fine as 0.025 μm (1 μin.), with stock removal of 0.0025 to 0.010 mm (0.0001 to 0.0004 in.), are feasible when cast iron laps are used. Diametral tolerances as low as 0.00050 mm (0.000020 in.), roundness within 0.00013 mm (0.000005 in.), and taper of less than 0.00025 mm (0.000010 in.) have been achieved. However, such accuracy depends greatly on the accuracy achieved in prior machining operations.

**Bonded-Abrasive Laps.** When bonded-abrasive laps are used, both laps are rotated, with kerosene or a similar lubricant used as a coolant and to wash away chips or loose abrasive. Because both laps are driven at higher speeds than those used for cast iron laps, the lapping action is more severe. Consequently, the machine will not produce the extreme accuracy possible with machines using cast iron laps. In addition, because bonded-abrasive laps must be dressed with diamond tools, it is not possible to make them as flat as cast iron laps, on which the machines regenerate flatness.

The quantity of parts being lapped is less critical for machines using bonded-abrasive laps than for machines using cast iron laps, because both bonded-abrasive laps are rigidly supported on spindles and separately driven. As few as three parts can be successfully processed in this type of machine.

**Applicability.** Machine lapping between plates is an economical method of finishing outside cylindrical surfaces, provided its use is warranted by production quantities and is permitted by part configuration. The process can be used for lapping parts a few hundredths of a millimeter to 75 or 100 mm (3 or 4 in.) in diameter and 6 to 230 mm (¼ to 9 in.) long. Parts commonly lapped by this method include plug gages, piston pins, hypodermic plungers, ceramic pins, small valve pistons, cylindrical valves, small engine pistons, roller bearings, diesel injector valves, plungers, small rolls, and miscellaneous cylindrical pins.

Either hard or soft materials can be lapped, provided they are rigid enough to accept the pressure of the laps. Hard materials respond well to lapping and achieve luster. Hard materials are also easier to control for tolerance because the hardness slows the operation. Soft materials lap more rapidly and (especially when bonded-abrasive laps are used) often have a scratchy or dull appearance. This can be prevented by using a polishing abrasive, such as levigated alumina, which reduces the cutting ability of the bonded abrasive.

**Limitations.** A part with a diameter greater than its length is difficult or impossible to machine lap between plates. For parts of this type, other methods of outer cylindrical-surface lapping are more practical.

Parts with shoulders require special workholders that permit the shoulder section to be placed on the inside or outside of the lap face. Parts with keyways, flats, or interrupted surfaces are difficult to lap by machine, because the variations in pressure that occur are likely to cause out-of-roundness. If the relief extends over the entire length of the piece, this method of lapping cannot be used.

Parts with raised hubs in the middle require special laps that are cut in such a way that they clear the hub. Clearance is necessary between the hub and the work surface to allow for oscillation of the workpiece.

Thin-wall tubing can be lapped, but if the walls are so thin that deflection is significant, it will be difficult to maintain roundness. Parts that are hollow on one end but solid on the opposite end present problems in obtaining roundness and straightness, because the hollow end will deflect more under the weight of the upper lap. Plugging the hollow end of the part will sometimes solve these problems.

Because it is impractical to keep more than one working surface on the face of cast iron laps flat, workpieces with work surfaces that have different diameters require a separate operation for each surface. It is usually impractical to machine lap workpieces with diameters that are greater than the diameter to be lapped.

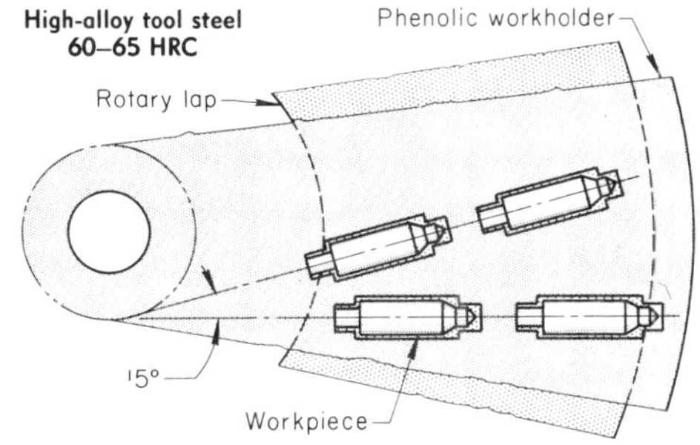


**Fig. 5** Setup for lapping production quantities of the valve needle shown in Fig. 3

The outside edges of the plates lap at a faster rate than the inside edges; therefore, care must be taken to prevent the workpieces from becoming tapered. One method of overcoming the problem consists of using short lapping cycles and, at the end of each cycle, turning the workpieces end for end in the slots in the workholder. In addition, the workpieces should be removed from the slots after each short cycle, mixed, and then replaced at random in different slots. This prevents the inadvertent placement of all larger pieces at one side of the workholder and smaller pieces at the opposite side. Because the upper lap floats, this placement of the workpieces would make it difficult to produce accurate parts. Taper can be minimized by positioning the workholder so that the parts in the slots are at a 15° angle to a radius, as illustrated in Fig. 5.

Because machine lapping between plates uses diametrically opposed laps, it cannot correct the out-of-roundness produced by centerless grinding. However, out-of-roundness of the type produced by grinding on centers can be corrected. The following example describes procedures for machine lapping between plates.

**Example 3: Lapping Valve Needles to Close Tolerance.** The valve needles described in Example 2 and illustrated in Fig. 3, although ring lapped in preproduction, were machine lapped between cast iron plates in production. Before being machine lapped, the parts were carefully ground for roundness and then (because the lap would ride on those parts that were largest in diameter) segregated into groups of 0.0025 to 0.005 mm (0.0001 to 0.0002 in.) diametral variation. Both upper and lower laps were grooved to prevent the breakdown of sharp edges during lapping. A laminated phenolic workholder designed to hold a maximum load of parts (Fig. 5) was eccentric to the laps to provide an oscillating motion.

In this operation, the cycle was stopped so that the parts could be measured with an

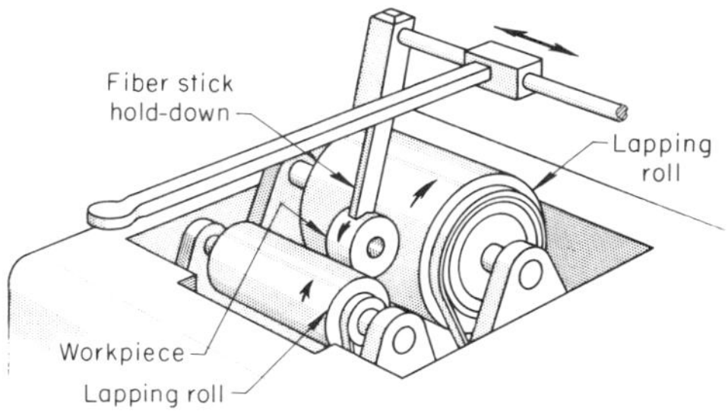


**Fig. 6** Typical centerless roll lapping

electrolimit gage, a visual shadow gage, or an air gage. If the desired size had not been attained, more finish lapping compound was added and lapping was continued. Lapping produced a finish of 0.050 μm (2 μin.), roundness within 0.00064 mm (0.000025 in.), and straightness of 0.0013 mm (0.000050 in.).

To recondition the laps, finish lapping compound was applied to the bottom lap. The laps were then brought together and rotated until the edges of the grooves were sharp. The laps required occasional regrinding to maintain a minimum groove depth of 1.6 mm (1/16 in.) and width of 0.76 mm (0.030 in.).

## Centerless Roll Lapping

In centerless roll lapping, only a single piece is processed at a time. Therefore, this method is best suited to the lapping of small quantities of parts (usually, fewer than ten).

A typical machine (Fig. 6) consists essentially of two 150 mm (6 in.) long cast iron rolls (one 150 mm, or 6 in., and one 75 mm, or 3 in., in diameter) and a reciprocating device for holding down the workpiece and controlling size. The end of the fiber stick that holds down the work is provided with a 120° V-groove. In operation, abrasive compound is applied to the rolls, and both rolls are rotated in the same direction—away from the operator and counter to the direction of workpiece rotation. The larger roll rotates at about 180 rev/min, the smaller roll at about 90 rev/min. The workpiece feeds across the rolls at about 50 mm/min (2 in./min) as the hold-down device is stroked back and forth to within 13 mm (½ in.) of each end of the workpiece.

The rate at which the workpiece feeds depends on the diameter of the piece. For example, if a 13 mm (½ in.) diam workpiece feeds at 50 mm/min (2 in./min), a 25 mm (1 in.) diam piece in the same setup will feed at approximately 25 mm/min (1 in./min). Slow stroking is necessary to obtain the best surface finish and control of size. Stock removal in centerless roll lapping is usually 0.005 to 0.0075 mm (0.0002 to 0.0003 in.), depending on the finish obtained in the previous operation.

The main advantage offered by centerless roll lapping is quick setup. Therefore, the process is readily adaptable to frequent size changes in short production runs. Limitations on the shape of parts for centerless roll lapping are similar to, but more stringent than, those that apply to machine lapping between plates. When both processes are equally suitable for a given application, the quantity of parts to be lapped determines which will be used.

## Centerless Lapping With Bonded Abrasives

Centerless lapping is a variation of centerless grinding (see the Section "Grinding, Honing, and Lapping" in this Volume). The machines for the two processes are similar in appearance, but the lapping machine is constructed to produce finishes of 0.050 μm (2 μin.) or better, diametral accuracy of 0.0013 mm (0.000050 in.), and roundness within 0.00064 mm (0.000025 in.). The lapping and regulating wheels are 560 mm (22 in.) wide, which is much wider than those ordinarily used for centerless grinding. Therefore, the work remains in contact with the lapping wheel longer and receives a finer finish.

The regulating and the lapping wheels (both are bonded abrasive) can be angled so that their axes are not parallel. Ordinarily, the regulating wheel is adjusted to a positive angle of 1 to 3° (depending on the production and finish requirements), and the lapping wheel is adjusted to a negative angle of about −4°. When trued, both wheels assume a slight hourglass shape, which then allows them to wrap around the workpiece as it passes between them. They also contact the workpiece at an angle to its axis, which is different from the axial-line contact of a grinding wheel. This eliminates lapping marks.

The finest finish obtainable in a centerless lapping machine requires at least three operations, each with a progressively finer lapping wheel, and a full flow of clean fluid (such as kerosene) as a coolant. During the first operation, the workpiece is supported on a blade faced with hard steel or carbide. For correcting out-of-roundness, the center of the workpiece should be slightly above the center of the wheels. In the first operation, a maximum of 0.013 mm (0.0005 in.) of stock is removed, and a finish of 0.10 to 0.15 μm (4 to 6 μin.) is obtained. For the second and third operations, the workpiece is supported on a rubber blade and is centered on the wheels so that scratches are minimized. During the second operation, a maximum of 0.0025 mm (0.0001 in.) of stock is removed, and a finish of 0.050 to 0.075 μm (2 to 3 μin.) is obtained. During the third operation, practically no stock is removed, and a finish of about 0.050 μm (2 μin.) is obtained.

**Applicability.** Centerless lapping is a high-production operation that is particularly suited to centerless ground parts that can be continuously fed, either manually or automatically. Parts 6 to 150 mm (¼ to 6 in.) in diameter by 380 mm (15 in.) long can be centerless lapped, and when a long bar feed is used, it is possible to lap parts 13 to 75 mm (½ to 3 in.) in diameter and 4.6 m (15 ft) long. Typical parts finished by centerless lapping are pistons, piston pins, shafts, and bearing races.

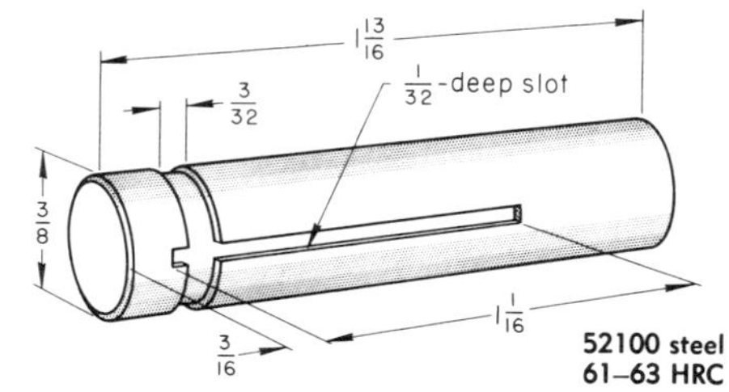


**Fig. 7** Part lapped by three different methods. Dimensions given in inches

Because little stock is removed in this process, only a small amount of correction can be made. Therefore, parts must be previously ground to the required straightness and roundness.

Parts with shapes that have no irregularities are ideally suited to centerless lapping, but irregularities such as those on the part shown in Fig. 7 can be tolerated. Such parts may, however, present problems in holding tolerances because of the undercut and the keyway. Cross holes also add to the difficulty of holding extremely close dimensions in centerless lapping. The production rates attained in centerless lapping and in two other processes for the part illustrated in Fig. 7 are compared in the following example.

**Example 4: Production Rates for Centerless Versus Centerless Roll Versus Two-Plate Machine Lapping.** The part shown in Fig. 7 (52100 steel hardened to 61 to 63 HRC) was lapped by three methods for the removal of 0.005 mm (0.0002 in.) of stock to produce a finish of 0.025 μm (1 μin.). Productivity was as follows:

| Method of lapping | Pieces per hour |
|---|---|
| Centerless | 700 |
| Centerless roll | 10 |
| Two-plate machine | 100 |

Centerless lapping was done with bonded-abrasive wheels (355 mm, or 14 in., in diameter and 560 mm, or 22 in. long; grit size: 500) at a rotation speed of 52 rev/min. A blade fixture was used.

Centerless roll lapping was carried out in a machine with two 150 mm (6 in.) long cast iron rolls—one 150 mm (6 in.) in diameter and one 75 mm (3 in.). The 1000-grit abrasive used was contained in a paste vehicle. Rotation of the smaller roll was at 100 rev/min.

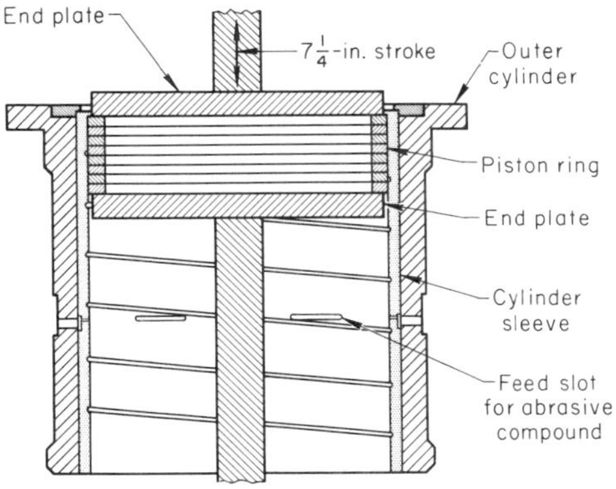


**Fig. 8** Setup and fixture for lapping eight piston rings simultaneously

In machine lapping between plates, the two laps, which were of cast iron, were 400 mm (16 in.) in diameter and 75 mm (3 in.) thick. The spider fixture was used. Rotation speed was 100 rev/min, and an 800-grit abrasive was used in a water vehicle.

## Lapping of Outer Surfaces of Piston Rings

Special procedures are required for lapping the outer cylindrical surfaces of parts that are considerably greater in diameter than in axial length. Piston rings are typical of such parts. Lapping is especially necessary for a chromium-plated piston ring because the ring will quickly ruin a cylinder unless the minute chromium nodules are removed.

Because of slight variations in machining, rings may have areas that exert low pressure. During lapping, material will be removed faster from the high-pressure areas, thus causing a more even distribution of pressure around the ring, removal of chromium nodules, and smoothing of the surface. A specific procedure employed in the lapping of piston rings is described in the following example.

**Example 5: Lapping Eight Piston Rings Simultaneously.** The stacking setup shown in Fig. 8 was used for simultaneously lapping eight chromium-plated steel piston rings (hardness: 775 HV). The lap consisted of an outer cylinder that was a solid casting and an inner sleeve that could be replaced when worn out. Replacing only the inner sleeve was more economical than replacing the entire cylinder. The piston rings were reciprocated in the sleeve at 150 cycles/min by a special machine. During each reciprocation, the stack of piston rings was rotated 45°. The abrasive, which was fed in through slots near the center of the sleeve, contained 10% of 600-grit aluminum oxide mixed with 90% (by weight) of a commercial lapping oil.

This practice was used for rings 51 to 216 mm (2 to 8½ in.) in diameter and 1.6 to 6.4 mm (1⁄16 to ¼ in.) thick. Productivity was 2 to 48 rings/min, depending on ring size. Metal removal ranged from about 0.025 to 0.038 mm (0.0010 to 0.0015 in.). Size, which was controlled by the number of cycles, was checked by measuring the gap between the ends of the ring when installed in a gage.

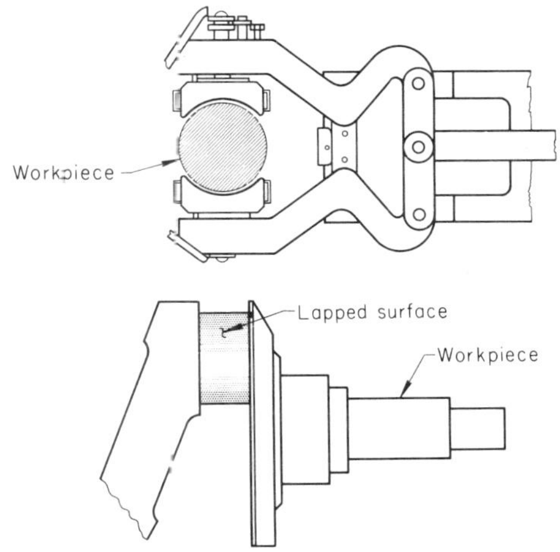


**Fig. 9** Setup for lapping a cylindrical surface on a crankshaft

## Lapping of Crankshafts

Crankshaft journals and pins and a variety of similar cylindrical surfaces are often lapped when they require a finish better than that ordinarily produced by production grinding. The pin and journal surfaces on crankshafts, for example, are ground to finishes of 0.63 to 1.40 μm (25 to 55 μin.), but a finish of 0.10 to 0.20 μm (4 to 8 μin.) is required for some applications. This finer finish has been inexpensively achieved by the method described in the following example, in which all surfaces are lapped in one setup.

**Example 6: Machine Lapping of Crankshafts.** Crankshafts were lapped in a machine with a 355 mm (14 in.) swing and 810 mm (32 in.) between centers. Work shoes automatically clamped coated-abrasive paper or cloth (fused alumina, 240 to 320 grit) around the crankshaft surfaces to be lapped (Fig. 9). Mineral seal oil was constantly applied to cool the surface being lapped. At a rotation speed of 125 rev/min and a reciprocation of 80 cycles/min, about 0.005 mm (0.0002 in.) of metal was removed to produce a finish of less than 0.25 μm (10 μin.). Production rate was 70 crankshafts per hour.

## Lapping of Inner Cylindrical Surfaces

Holes or bores are lapped by using either solid or adjustable laps, usually made of cast iron. The laps can be rotated by any one of a variety of machines, including honing machines, but lathes or polishing heads are most commonly used. The lap, which carries the abrasive, is rotated while the workpiece is manually stroked over it, a procedure similar to manual honing. Machines that reciprocate either the workpiece or the lap (usually the lap) in addition to rotating the lap also are used. These machines resemble those used for power stroking in honing (see the article "Honing" in this Volume).

In internal-surface lapping, virtually no stock is removed because stock can be removed at lower cost by honing. When lapping follows honing, it is usually just a touch-up operation.

A variation of internal-surface lapping is employed for matching or mating male and female cylindrical components. For example, fuel injection plungers are often lapped into cylinders to produce matched pairs. In these applications, the male part, which rotates, becomes the lap. It is swabbed with abrasive, and the mating part is then manually stroked over it. Skilled operators are required for these operations.

**Adjustable laps** (Fig. 10) are available in almost any diameter larger than 1.6 mm (1⁄16 in.). These laps are expanded manually, as required, during operation. Although various means can be used for expanding the lap, adjusting screws (Fig. 10) are most often used. Ring gages are examples of parts that are lapped to extremely accurate dimensions with adjustable laps. Tolerances specified for various diameters and classes of gages are given in Table 3. To obtain the accuracy required by the specifications in Table 3, it is necessary to adhere strictly to the following sequence of operations in lapping:

- Determine the initial condition of the workpiece (amount of stock to be removed, and taper and roundness)
- Insert the shank of the lap into the polishing-head chuck
- Coat the lap with abrasive compound
- Loosen the screw in the lap
- Place the workpiece on the lap
- Tighten the screw in the lap until the workpiece turns with slight resistance
- Start the machine and reciprocate the workpiece evenly by hand over the length of the rotating lap, maintaining uniform resistance, until the piece runs smoothly from end to end
- Stop machine, remove the workpiece and allow it to cool
- Measure the workpiece to check for size, taper, barrel or hourglass shape, out-of-roundness, and bellmouth
- Repeat the process, except for the method of reciprocation, until the desired results are obtained. At this point, the method of reciprocation depends on the conditions found in measuring the ring. Over-size and out-of-roundness require slow, even reciprocation. Bellmouth requires reversal before the point of low-

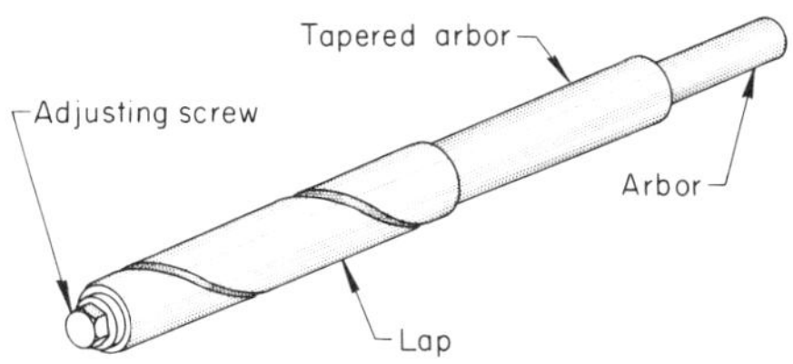


**Fig. 10** Components of an adjustable lap

ered resistance is reached. Barrel or hourglass shape and taper require faster reciprocation, localized at the high areas. Extremes of the above (0.005 mm, or 0.0002 in.) must be corrected by grinding

**Solid laps** are low-cost round bars that are accurately finished to size. They are usually made of cast iron, but for extremely small workpieces, they can be of copper or other nonferrous metal. Ordinarily, solid laps are used only when one or two odd-size pieces require finishing. However, solid laps must be used for finishing inside diameters of less than 1.6 mm (1/16 in.) (because this is the minimum diameter of adjustable laps), and they can also be used by skilled operators to correct bow or snake. The chief disadvantage of solid laps is that they are useless once they become only slightly worn, unless they are refinished for lapping smaller bores.

**Blind Holes.** Lapping blind or partly blind holes presents problems because of uneven distribution of abrasive. The difficulty can sometimes be overcome by an improvement in lap design, as in the following example.

**Example 7: Lapping Blind Holes.** The part shown in section in the lower portion of Fig. 11 was made of 52100 steel and hardened to 60 to 62 HRC. Parts of this type had main bores 6 to 9 mm (1/4 to 3/8 in.) in diameter and 25 to 38 mm (1 to 1½ in.) in length. The bores were originally lapped with expansion laps made of cast iron and slotted in a straight line (Fig. 11). This type of lap did not allow enough lapping compound to reach the bottom of the bore; consequently, lapping action at the outer end of the hole was more rapid, and excessive taper was produced.

The problem was solved by redesigning the lap, providing it with a right-hand spiral that formed a carrier for the lapping compound (Fig. 11). The redesigned lap was held in a speed lathe (horizontal chucking head) and rotated clockwise at about 650 rev/min; the part was rotated counterclockwise. A coarse abrasive (grit size: 600) was used. Lapping removed 0.005 mm (0.0002 in.) of stock at a rate of 40 pieces per hour. The bores were held to a diametral tolerance of 0.0013 mm (0.000050 in.) and to straightness within 0.00064 mm (0.000025 in.). Size and straightness were controlled by the use of an air gage and two master setting rings.

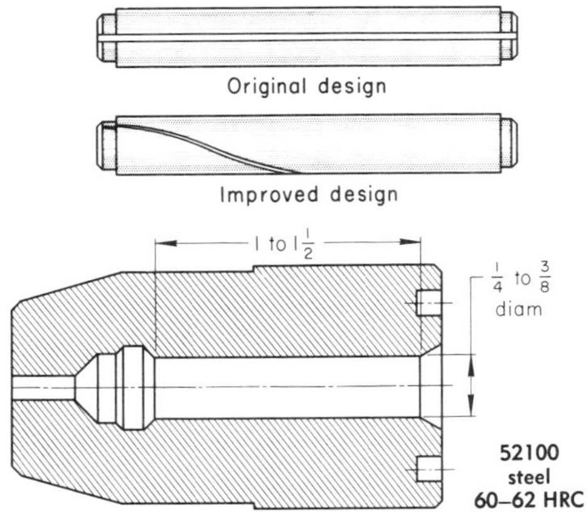


**Fig. 11** Substitution of spiral for straight slot in lap for better distribution of abrasive in lapping a blind hole. Dimensions given in inches

## Lapping Flat Surfaces

Flat surfaces can be lapped by either manual or mechanical methods. In general, manual methods are used only when small quantities of parts are to be lapped or when special requirements must be met. Most flat lapping on a production basis is done with rotating two-plate machines similar to those used for lapping cylindrical surfaces.

### Manual Methods

The hand rubbing of flat workpieces on a block or plate lap charged with loose abrasive is the simplest method of flat lapping. The lap, usually made of cast iron, has regularly spaced grooves about 1.6 mm (1/16 in.) deep to retain the lapping medium (usually an abrasive-containing paste). The flat workpiece is rubbed on the lap in a figure eight or similar motion that covers almost the entire lap surface, so that the lap will remain flat for a considerable amount of work. This method of lapping is slow and tedious and requires a high degree of skill for optimum results. It is used only when a few parts must be lapped or when more efficient equipment is not available.

Another and somewhat faster method of flat lapping single pieces makes use of a single-spindle vertical drill press. The lap, which is stationary, is mounted on the stand of the drill press. The workpiece is held by the spindle, which rotates it against the lap. Light pressures are applied by hand. This method is slow, and the lap is likely to wear unevenly. However, it may be preferable to other flat lapping methods for some applications. For example, certain round, flat sealing parts require a lapped surface with the concentric-line pattern produced by this method because random scratch patterns are not leakproof. To achieve the concentric-line pattern, the rotating part is held against a rotating or nonrotating lap of the same size so that there is no motion except the rotation between the workpiece and the lap.

When concentricity of scratch pattern is not important, the method described above is sometimes modified for faster lapping by oscillating the rotating workpiece back and forth over the lap. The workpiece should be mounted on the spindle of the drill press so that it is flexible and can float as it turns. A layer of rubber can be placed behind the workpiece to provide a suitably flexible mounting.

Care should be taken when lapping for concentric patterns because this process violates one of the basic rules of lapping, namely, the random and nonrepeating motion between the lap and the work. Without this random motion, a small scratch passing over the same spot soon becomes a deep groove, with the removed material picked up and welded to the lap.

### Mechanical Methods

The two general types of machines for flat lapping are:

- Single-face lapping machines having a single horizontal rotating lap
- Dual-face or four-way planetary lapping machines having two laps, one above or on top of the work and one below

**Single-Face Lapping Machines.** The simplest single-face lapping machine is the horizontal rotating annular lap, with (usually) three or four conditioning rings held in place on the lap face but free to rotate (Fig. 12). The dynamic action of this system drives the rings in the same rotational direction as the lap (Fig. 13). When abrasive is applied to the lap, the rings and the lap wear into intimate contact and are of ball-and-socket shape. This is the fundamental oper-

**Table 3 Lapping tolerances for various sizes of ring gages**

| Diameter | | Tolerance for gages of class: | | | | | |
|---|---|---|---|---|---|---|---|
| | | X | | Y | | Z | |
| mm | in. | mm | in. | mm | in. | mm | in. |
| 0.737–20.96 | 0.029–0.825 | 0.0010 | 0.000040 | 0.0018 | 0.000070 | 0.00254 | 0.000100 |
| 20.96–38.35 | 0.825–1.510 | 0.0015 | 0.000060 | 0.0023 | 0.000090 | 0.00305 | 0.000120 |
| 38.35–63.75 | 1.510–2.510 | 0.0020 | 0.000080 | 0.0030 | 0.000120 | 0.00406 | 0.000160 |
| 63.75–114.55 | 2.510–4.510 | 0.0025 | 0.000100 | 0.0038 | 0.000150 | 0.00508 | 0.000200 |
| 114.55–165.35 | 4.510–6.510 | 0.0033 | 0.000130 | 0.0048 | 0.000190 | 0.00635 | 0.000250 |
| 165.35–228.85 | 6.510–9.010 | 0.0040 | 0.000160 | 0.0061 | 0.000240 | 0.00813 | 0.000320 |
| 228.85–305.05 | 9.010–12.010 | 0.0050 | 0.000200 | 0.0076 | 0.000300 | 0.01020 | 0.000400 |

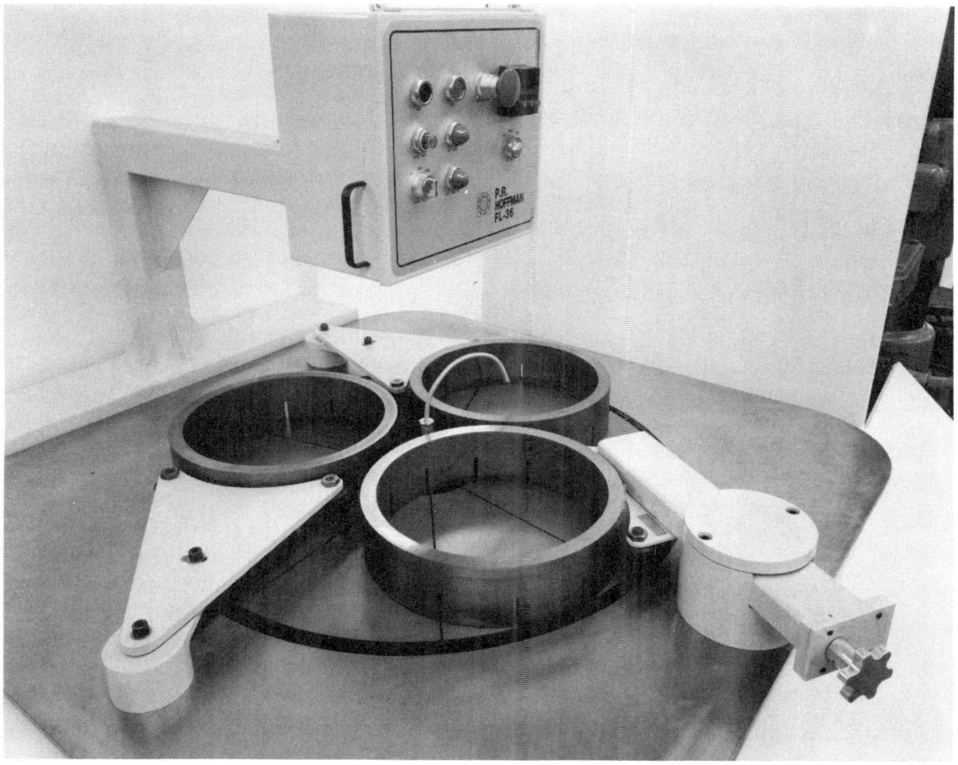


**Fig. 12** Typical horizontal single-face flat lap lapping machine with three conditioning rings. Ring position flattening is achieved by using gravity pressure. Courtesy of P.R. Hoffman Machine Products

ation of the single-face lap. If the rings are manipulated so that the lap is alternately the ball and the socket, then the lap flatness, and consequently the work flatness, can be held to within the required flatness tolerance.

The manipulation of the rings usually follows one of two methods. If the ring is positioned with the center of the ring on the center of the plate track, the result will be a lap socket and ring ball. This is because the relative velocity between plate and ring, with both turning in the same direction, is higher at the lap inside diameter than at the outside diameter, thus wearing the center faster and creating a socket or bowl shape.

If a tooth form is generated on the outside diameter of the conditioning ring and a corresponding driver is mounted at the lap center in mesh with the teeth, then the conditioning ring will be forced to rotate counter to the lap, the relative velocity ratio will be reversed, and the lap will wear in a ball shape.

An alternative method for reversing the wear pattern is to shift the ring position toward the lap outside diameter. As the ring overhangs the plate, higher unit pressure is exerted on the lap outside diameter than on the inside diameter. This pressure increase causes increased wear, which exceeds the velocity wear and causes the lap to become ball shaped.

With either type of machine, the rings act as tooling. The parts are placed in the ring or into carriers, which are then placed in the rings. For parts that are too large to fit in the rings, all but one can be removed, and fixtures can be made to hold the parts on the lap. The flatness is then controlled with both the ring and the parts, using the pressure flatting concept previously discussed.

Various options are available to enhance the capability of this machine:

- Pressure heads to apply down pressure to the parts through a pressure platen, which loosely fits the inside diameter of the rings
- Abrasive systems that automatically apply a predetermined rate of slurry flow
- Variable-speed drive to the lap
- Variable-speed driven rings
- Plate cooling
- Sizing devices

A variety of mechanical and electronic gaging systems permit the in-process monitoring of stock removal rates and workpiece thickness. These systems are frequency linked to the lapping machine so that a specific depth of material is removed and the machine stops when the target thickness is reached. These gaging systems include timers, hardened stops, electronic devices, and piezoelectric devices.

*Timers* are extensively used when the process is rigidly controlled and repetitive parts are produced. Because the removal rate is very constant, final size can be held closely, especially if the raw part sizes are in a narrow band.

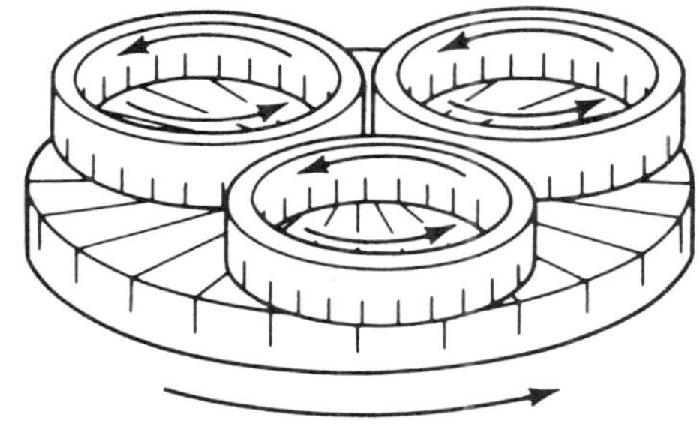

**Fig. 13** Dynamics of the lapping machine shown in Fig. 12. Work to be lapped is placed within the conditioning rings, which are held in place but are free to rotate. The work tends to abrade the lap plate, but the rotating action of the conditioning rings causes the lap plate to wear evenly, maintaining a flat surface. Standard machines handle parts from 3.2 to 810 mm (1/8 to 32 in.) in cross section. Steel, tool steel, bronze, cast iron, stainless steel, aluminum, magnesium, brass, quartz, ceramics, plastics, and glass can be lapped on the same lap plate.

*Hardened stops* are used in conjunction with the timer to space the pressure plates, thus stopping the lapping action.

*Electronic devices* are used to measure the drop of the pressure plate. This is an indication of the material removed.

*Piezoelectric Devices*. The most unique sizing method is used on thin material and involves placing a piezoelectric wafer in with the parts. The wafer should be of the same thickness as the parts being lapped. As the wafer is lapped together with the parts being machined, the size is monitored by reading the frequency of the piezoelectric wafer. As an electrical signal is passed through quartz or similar piezoelectric materials, the oscillation frequency of the crystal changes as the lapped wafer decreases in thickness. This phenomenon is exploited by using in-process gaging to stop the lapping process at a fixed frequency/thickness set point. In a production environment, size control of ±0.0013 mm (±0.000050 in.) has been achieved using this method.

**Dual-Face Lapping Machines.** Most of the dual-face lapping machines being produced are of the planetary type, with the workholders (carriers) nested between a center drive and a ring drive (Fig. 14). These drives can be either gear- or pin-type configurations, but must have positive engagement.

The parts being worked are propelled by the carrier in a serpentine path between two flat lap plates, on which abrasive has been placed (charged) or is continuously fed in the form of a slurry. In the simplest form, the bottom lap is fixed, and the top lap is restrained from rotating but is allowed to float so that it always bears on the largest pieces and laps all the pieces to the same size. In a machine of this type, the part is dragged between the plates by the carrier, and all the lapping power is directed into the system through the center (sun) and ring drives.

**Fig. 14** A 355 mm (14 in.) planetary fixed-plate dual-face lapping machine. The top plate has been removed to expose the center drive gear, ring drive, and parts carriers. Courtesy of P.R. Hoffman Machine Products

*The planetary fixed-plate machine* has a decided disadvantage in that no successful method has been developed to flat the plates in the machine while processing parts. An additional disadvantage is that the carrier forces are high, thus causing teeth failure of thin carriers and edge chipping on fragile parts.

*The top plate drive machine* was developed to eliminate the need to remove the plates to grind them flat (Fig. 15). In a machine of this type, the top plate is driven twice as fast in the same direction as the carriers translate the bottom plate. As a result, the removal from the top and bottom of the work is equal, but the top and bottom plates wear as mirror images of each other. If the bottom plate is concave, the top plate is convex, and vice versa. Using variable speed on the ring and center drive, the operator can work the flatness of the bottom lap plate in the same manner as the single-face lap and keep both top and bottom laps flat within a required tolerance.

The machine shown in Fig. 15 is a 1220 mm (48 in.) double-face lap that has been loaded with sample parts appropriate for this type of processing. It should be noted that this machine is loaded in this manner for demonstration purposes only.

## Workpiece Size and Shape

Parts 3.2 mm (⅛ in.) in diameter or smaller and as large as 810 mm (32 in.) in diameter can be flat lapped. Table 4 indicates the number of parts of given diameters that can be accommodated by single-face flat lapping machines with laps 305 to 2135 mm (12 to 84 in.) in diameter, in conjunction with various numbers and sizes of carrier rings. In addition, machines having plate diameters of 2440, 2690, 3050, and 3660 mm (96, 106, 120, and 144 in.) are available. Large quantities of small parts can be lapped on large machines. Large parts can be lapped only on large machines.

With the correct workholders and operating conditions, parts thinner than 0.05 mm (0.002 in.) have been lapped. Workpiece height limitation is determined by machine clearances and by the ratio of height to lapped surface area. This ratio must be kept low enough to prevent tipping or rocking of the part during lapping, unless a workholder or fixture is provided. The restraint or support required for typical shapes that can be flat lapped is indicated in Fig. 16.

## Flatness

Flat lapping can produce flatness within one light band (0.29 μm, or 11.6 μin.) and smoothness better than 0.050 μm (2 μin.). In some optical applications, flatness is better than one-tenth light band. To obtain these results, the workpieces must be stable during the operation so that the possibility of stress relief that will result in distortion is eliminated. In addition, flatness of the lap or laps must be kept within the flatness tolerance for the workpiece.

## Size Tolerance and Parallelism

Parts having parallel faces, such as disks and seal rings, can be held to tolerances varying from ±0.0025 mm (±0.0001 in.) for small parts to ±0.025 mm (±0.001 in.) for large parts. The difficulty in maintaining accuracy increases for parts of uneven configuration; such parts may require fixtures, which then determine the accuracy attainable. The closer the tolerance requirements, the more difficult the operation is likely to be.

Surfaces of disks, seal rings, and similar parts having parallel surfaces can be lapped on either the double-lap machines, which lap both sides in one operation, or the single-lap machines, which require two operations (one for each side). The single-face lap can be used to produce parallel parts, but is very skill sensitive and requires attention to such details as cleanliness, lap and pressure plate flatness, and the use of multiple transpositions. Parallelism requirements of less than 0.0002 mm/mm (0.0002 in./in.) of diameter or length in production lots dictate the use of a dual-face machine. These machines, with adherence to a process specification, can produce parts with opposing faces parallel within 0.00002 mm/mm (0.000020 in./in.).

The operation becomes simpler as the number of parts to be lapped at one time increases and as the distribution of thick and thin parts improves. When only a few workpieces are lapped, or if the thickness of the part varies more than about 0.075 mm (0.003 in.), it may be necessary to distribute the thicker parts selectively or to remove and redistribute the parts before lapping the second side. Allowance for stock removal in this type of operation should be 1½ to 2 times the amount that the parts are out-of-parallel plus the amount of the variation in part sizes.

## Procedure

Flat lapping is typically employed on piston rings when side flatness is extremely critical and on other ringlike parts. Although most ring-shaped parts can be lapped without elaborate procedures, special problems often arise.

There are two main problems in lapping piston ring sides:

- The rings are flexible and can be easily distorted under load
- Some piston rings have a very smooth, polished finish on the circumference that must not be marred during lapping

Some load must be applied during lapping because the rings are too light to lap without added load, but the applied load must not be great enough to distort the ring. An aluminum pressure plate is used because it is light

**Fig. 15** A 1220 mm (48 in.) planetary top plate drive dual-face lapping machine loaded with a variety of components having various configurations and thicknesses to demonstrate the processing capabilities of the equipment. Shown clockwise, starting with the bottom-most carriers, are the hardened high-carbon steel shear blades; 102 by 102 mm (4 by 4 in.) photomask blanks used to project microcircuits onto electronic substrates; stainless steel sender bars, which will be etched into optical position indicators; valve plates machined from cutoff hot-rolled bar stock for use in hydraulic motors; and a hardened alloy steel rotary slitter blade.

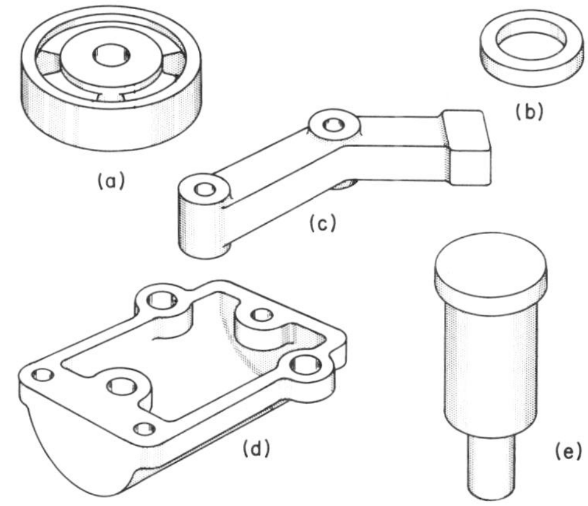


**Fig. 16** Parts requiring different amounts of restraint or support in flat lapping. Symmetrical components similar to (a) and (b) do not require workholders for flat lapping. Asymmetrical components similar to (c) and (d) require workholders to keep them separated from each other. Parts similar to (e) require workholders to keep them from tipping.

but also rigid enough to stay flat. In the modern dual-face lap, parameters such as the applied pressure due to the weight of the top plate are computer controlled, and the machines can be easily set up to lap parts requiring specialized processing. In the two examples that follow, procedures successfully employed for the flat lapping of a ringlike part (Example 8) and a more complex configuration (Example 9) are described in detail.

**Example 8: Lapping of Bearing Races.** The setup shown in Fig. 17 was used for lapping the flat sides of inner and outer bearing races made of carburized or through-hardened steel in diameters from 30 to 215 mm (1³⁄₁₆ to 8½ in.). The machine had two bonded-abrasive laps (400-grit silicon carbide) that rotated in opposite directions

## Table 4 Load capacities of planetary flat lapping machines

| Workpiece diameter(a) | | Production capacity (quantity of parts): 305 mm (12 in.) diam laps; three 108 mm (4¼ in.) ID conditioning rings | | 610 mm (24 in.) diam laps; three 248 mm (9¾ in.) ID conditioning rings | | 610 mm (24 in.) diam laps; four 210 mm (8¼ in.) ID conditioning rings | | 915 mm (36 in.) diam laps; three 368 mm (14½ in.) ID conditioning rings | | 910 mm (36 in.) diam laps; four 321 mm (12⅝ in.) ID conditioning rings | | 1220 mm (48 in.) diam laps; four 432 mm (17 in.) ID conditioning rings | | 1830 mm (72 in.) diam laps; four 692 mm (27¼ in.) ID conditioning rings | | 2135 mm (84 in.) diam laps; four 813 mm (32 in.) ID conditioning rings | |
|---|---|---|---|---|---|---|---|---|---|---|---|---|---|---|---|---|---|
| mm | in. | Each ring | Full load | Each ring | Full load | Each ring | Full load | Each ring | Full load | Each ring | Full load | Each ring | Full load | Each ring | Full load | Each ring | Full load |
| 6.4 | ¼ | 235 | 705 | 1300 | 3900 | 920 | 3680 | ... | ... | ... | ... | ... | ... | ... | ... | ... | ... |
| 13 | ½ | 55 | 165 | 310 | 930 | 220 | 880 | 700 | 2100 | 560 | 2240 | ... | ... | ... | ... | ... | ... |
| 19 | ¾ | 23 | 69 | 130 | 390 | 90 | 360 | 300 | 900 | 245 | 980 | ... | ... | ... | ... | ... | ... |
| 25 | 1 | 13 | 39 | 70 | 210 | 55 | 220 | 165 | 495 | 130 | 520 | 235 | 940 | 620 | 2480 | ... | ... |
| 32 | 1¼ | 8 | 24 | 45 | 135 | 32 | 128 | 105 | 315 | 80 | 320 | 145 | 580 | 390 | 1560 | ... | ... |
| 38 | 1½ | 5 | 15 | 31 | 93 | 22 | 88 | 70 | 210 | 55 | 220 | 100 | 400 | 270 | 1080 | 375 | 1500 |
| 50 | 2 | 2 | 6 | 17 | 51 | 12 | 48 | 39 | 117 | 31 | 124 | 55 | 220 | 145 | 580 | 205 | 820 |
| 75 | 3 | 1 | 3 | 7 | 21 | 5 | 20 | 16 | 48 | 13 | 52 | 23 | 92 | 60 | 240 | 85 | 340 |
| 100 | 4 | 1 | 3 | 4 | 12 | 2 | 8 | 9 | 27 | 7 | 28 | 13 | 52 | 34 | 136 | 48 | 192 |
| 150 | 6 | ... | ... | 1 | 3 | 1 | 4 | 4 | 12 | 3 | 12 | 5 | 20 | 14 | 56 | 21 | 84 |
| 200 | 8 | ... | ... | 1 | 3 | 1 | 4 | 1 | 3 | 1 | 4 | 2 | 8 | 8 | 32 | 11 | 44 |
| 250 | 10 | ... | ... | ... | ... | ... | ... | 1 | 3 | 1 | 4 | 1 | 4 | 5 | 20 | 7 | 28 |
| 360 | 14 | ... | ... | ... | ... | ... | ... | ... | ... | ... | ... | ... | ... | ... | ... | 3 | 12 |
| 380 | 15 | ... | ... | ... | ... | ... | ... | ... | ... | ... | ... | ... | ... | ... | ... | 2 | 8 |
| 400 | 16 | ... | ... | ... | ... | ... | ... | ... | ... | ... | ... | ... | ... | ... | ... | 2 | 8 |
| 360–660 | 14–26 | ... | ... | ... | ... | ... | ... | ... | ... | ... | ... | ... | ... | 1 | 4 | ... | ... |
| 430–790 | 17–31 | ... | ... | ... | ... | ... | ... | ... | ... | ... | ... | ... | ... | ... | ... | 1 | 4 |

(a) When using this table for determining capacity of machine with work separator, use the value for diameter next larger than that of the workpiece to be processed. Conditioning rings are carrier rings.

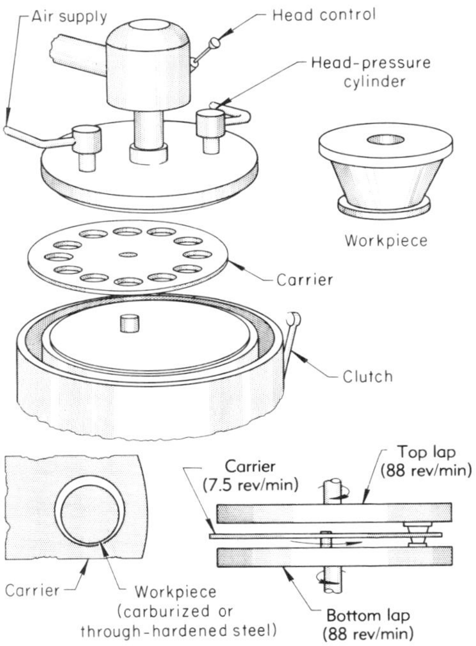


**Fig. 17** Setup for lapping inner or outer bearing races

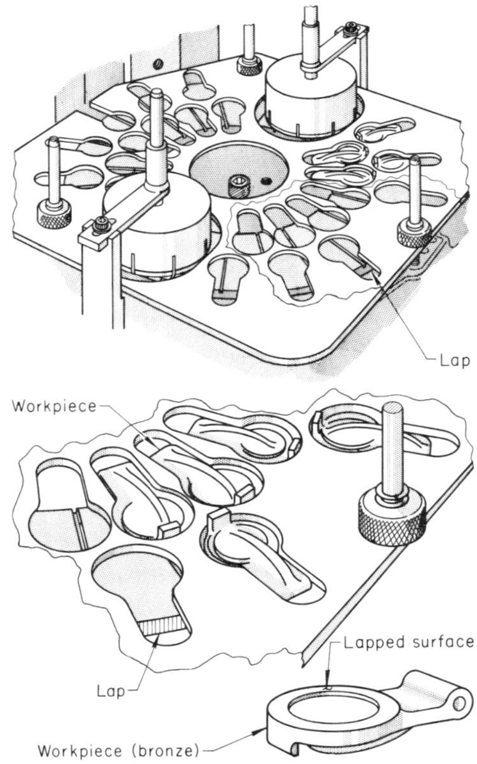


**Fig. 18** Setup for lapping seats of swing check valves

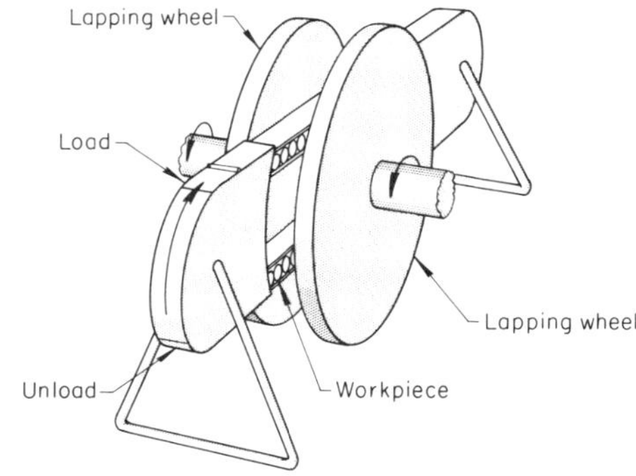


**Fig. 19** Setup for end lapping large quantities of hardened steel rollers

at 88 rev/min. The head of the machine, which was like that of a vertical drill press, was air actuated. It raised and lowered the top lap and could apply adjustable downward pressure (up to 414 kPa, or 60 psi) to the top lap. The bearing races were hand loaded in a horizontal circular fiber carrier that was eccentrically mounted over the bottom lap. The carrier rotated at 7½ rev/min, and was adjusted so that eccentricity was up to 25 mm (1 in.) and part clearance was about 6.4 mm (¼ in.). A viscous cutting oil was fed to the laps during the operation, and the laps were dressed two or three times during an 8-h shift. A maximum of 0.013 mm (0.0005 in.) of stock was removed from the races, leaving a finish of 0.075 μm (3 μin.). When checked with an indicator (0.0025 mm, or 0.0001 in., per division), no deviation from parallelism was detected.

**Example 9: Flat Lapping of Swing Check Valve With Special Laps and Fixtures.** The seats of swing check valves were lapped to the required flatness using the setup and relatively simple fixture shown in Fig. 18. This type of setup can be used for lapping similar offset flat surfaces. Cast iron laps were used, along with loose abrasive. The portion of the workpiece that protruded beyond the lapped surface extended over the inside or outside diameter of the lap and was loosely held in position by a cutout nesting fixture. To apply additional weight on the uneven back surfaces of the workpieces, rubber buttons on the underside of a pressure plate were placed so that they rested on the high points of the workpieces when lowered into lapping position. The pressure plate (not shown in Fig. 18) was indexed so that the buttons always rested in the correct position on the workpieces.

## Lapping End Surfaces

End lapping and conventional flat lapping are similar in principle, and the same machines are often used for both processes. As in flat lapping, fixtures are often necessary to prevent workpieces from tilting. In many cases, a workpiece whose height is several times the width of its lapped area is supported in a special fixture and processed in standard flat lapping machines. Engine connecting rods are notable examples. Faces that hold the shims for the connecting rod caps in assembly are supported by means of specially designed fixtures, and they are forced, by weight or air pressure, against a rotating lap. Their mating caps are often lapped at the same time. For straight cylindrical parts (not tapered), fixtures for end lapping are simple, and various methods can be used, as in the following example.

**Example 10: End Lapping of Straight Cylindrical Rollers.** A special machine was used for the high-production end lapping of rollers. Figure 19 shows the important components of this machine, which included an enclosed track that allowed the rollers to travel in a loop. The rollers were thus exposed to the bonded-abrasive wheels twice per trip around the track. The number of cycles required varied with the initial condition of the rollers. The laps, which were commercial vitrified wheels, rotated in the same direction at 1350 rev/min and were cooled with a low-viscosity honing oil.

## Problems in Flat and End Lapping

The problems encountered in flat and end lapping (and often in other methods of lapping, as well) are usually related to surface roughness or scratches, drop-off at edges, low rate of stock removal, improper size control, and failure to attain desired flatness and parallelism. These problems can be solved through proper control of the following conditions:

- Preparation of the workpiece
- Selection and application of the abrasive and the vehicle
- Sweep path of the workpiece over the lap
- Condition of the lap surface
- Interchange or transposition of workpieces during the lapping cycle
- Duration of the lapping cycle
- Cleanness of the environment

**Preparation of the Workpiece.** Parts to be lapped must have smooth, regular edges because irregularities at the edge cause the abrasive grains in the lapping compound to burst free and form the foxtail pattern shown in Fig. 20(a). (This effect may result also from etching on the lapped surface.) To provide the smooth edges, it may be necessary to barrel finish workpieces in fine abrasive, or to buff them, prior to lapping. Glass parts are frequently beveled on automatic machines, and the edges of electronic materials are rounded off by grinding prior to lapping. This not only eliminates foxtailing but also presents a much stronger edge to the driving carrier, thus minimizing edge chips and scratches.

Workpieces that have drop-off at edges, from previous operations, are subject to hydraulic cut in lapping. In hydraulic cut (Fig. 20b), the lapping abrasive builds up in the area of drop-off and perpetuates the condition. If drop-off cannot be eliminated, hydraulic cut can be minimized by the use of lesser amounts of abrasive in lapping.

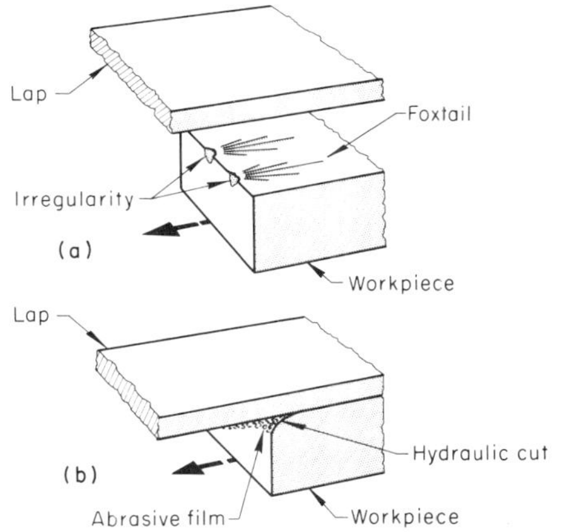


**Fig. 20** Two undesirable consequences of improper preparation of workpieces for lapping. See text for discussion.

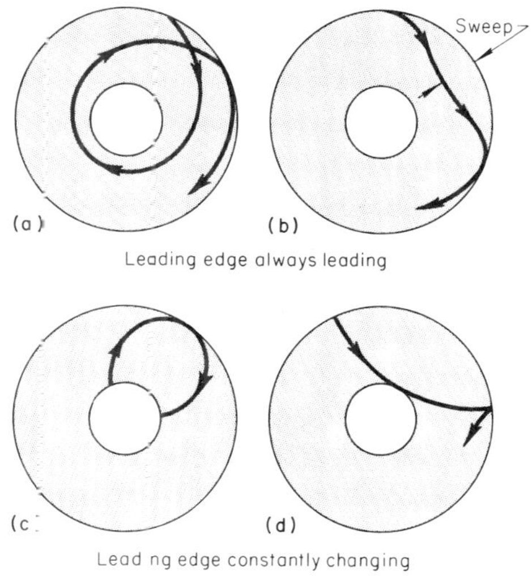


**Fig. 21** Typical sweep paths of workpiece over lap face in flat lapping. See text for discussion.

This remedy is usually uneconomical, however, because the stock removal rate decreases when a lesser amount of abrasive is used, and the likelihood of glazing is increased. Glazing can be prevented by the use of a diamond-base abrasive that contains aluminum oxide as the main cutting agent. The diamond abrasive should be of smaller grit size than the aluminum oxide; ¼ μm (10 μin.) diamond has been successfully used with 1200-grit aluminum oxide.

Metal particles must be removed from the surfaces of workpieces to be lapped so that the laps do not become charged with these particles and scratch the workpiece. Workpieces that have been ground should be freed of metal particles by being barrel finished in fine abrasive before being lapped. Previously lapped workpieces also may have particles adhering to the surface. To remove these particles, the workpieces must be rubbed on a dead lapping block (a lightly charged, noncutting block).

**Abrasives and Vehicles.** Coarse finish, scratches, loss of flatness or parallelism, and inadequate or excess stock removal rate often result from the use of improper abrasive or vehicle or from improper method of application of the lapping medium. In the sections "Selection of Abrasive" and "Selection of Vehicle" in this article, the basic considerations in the choice of these media for a given application are discussed in detail. Excessive amounts of lapping abrasive cause dull finishes, loss of size control, edge drop-off, and low rates of stock removal. Abrasive films that are too thin react similarly, except that they produce bright finishes and may result in glazing.

**Selection of Sweep Path.** The sweep path illustrated in Fig. 21(a) is advantageous for rectangular pieces with height-to-area ratios greater than 2 to 1. One of the longer edges of the lapped area should be the leading edge. To keep the lap flat, the carrier must be positioned so that part of a workpiece sweeps over the outside and the inside of the lap face. The eccentric throw of the carrier must be reduced when the workpiece is large in relation to the width of the lap face; thus, practically straight-line lapping occurs. As a result, it is difficult to keep the lap flat, to control the size of the workpiece, and to obtain parallelism. The best lapping action is obtained when every new grain cut line crosses every previous grain cut line at an angle.

The path shown in Fig. 21(b) is advantageous for small rectangular pieces over 1.0 mm (0.040 in.) thick and with a height-to-area ratio less than 2 to 1. The amount of sweep is determined by the throw of the carrier. Excessive throw of workpieces of small area is detrimental because of the high rate of side motion at reversal points. Insufficient throw of workpieces of large area is detrimental because the lapping pattern approaches a straight line.

The sweep paths illustrated in Fig. 21(c) and (d) are derived from planetary action and are advantageous for cylindrical and square workpieces, as well as rectangular workpieces with height-to-area ratios of less than 2 to 1. These sweep paths help in reducing hydraulic cut because the leading edge constantly changes, especially on shorter sweep paths. This provides a nondirectional pattern. As workpiece size is increased in relation to carrier size, the center of the workpiece tends to travel in a circular path and may increase the difficulty of maintaining flatness and parallelism.

**Condition of Lap Surface.** A dull gray surface on the lap is best for stock removal but not for producing highly reflective finishes. Conversely, bright lap surfaces are excellent for producing highly reflective finishes, but poor for stock removal. When using laps having bright surfaces, the abrasive film must be thin. If not, scratching will result. However, a film that is too thin may cause glazing and loss of control of size and flatness. Sand and other impurities in the lap surface sometimes break free and cause scratches, especially when heavy, viscous films of lapping compound are used.

A scleroscope hardness range of 27 to 32 (89 to 99 HRB) has proved optimum for cast iron lap surfaces. Harder laps often cause glazing and scratching. Softer lap surfaces cause loss of flatness and parallelism and also produce grayer finishes.

**Interchanging or transposing workpieces** during the cycle is advantageous for producing consistent results. An increase in handling time and cost is the only disadvantage of this practice. Either of two methods can be used:

- Workpieces in a load are interchanged between runs, and each load is processed independently until finished size is attained. Drop-off can be minimized by turning the workpieces end for end (reversing their leading edges) during the interchange
- Workpieces in a series of loads are intermixed at random outside the machine after each load is lapped for a given time. The process is repeated until size is reached. This method is usually the more economical

**Duration of Lapping Cycle.** Required lapping time is usually determined experimentally for a new job. Extremely short cycles (in the range of 1 to 3 min) are not economical, because too great a portion of the total time is used for loading and unloading. However, short cycles are preferred for size control. When abrasive is not continually fed to the laps, there is an optimum time for any job that must be determined by trial.

**Cleanliness of the environment** is important to the success of any lapping operation. Scratches, edge drop-off, and related difficulties can often be attributed to the dropping of air-borne particles into open machines, the use of contaminated abrasives, or inadequate cleaning of workpieces.

## Lapping Spherical Surfaces

Any of several methods can be used for lapping spherical surfaces. Size, quantity, and required accuracy of the workpieces are the main factors that determine choice of method.

**In single-piece lapping,** concave or convex laps are individually contoured to the workpiece. The laps should be made of fine-grain cast iron, which is suitable for lapping virtually any metal if the process is continued for enough time. Any of several types of machines that have one or two rotating spindles, such as drill presses or milling machines, can be used.

When machines with one spindle are used, rotating the lap against the workpiece

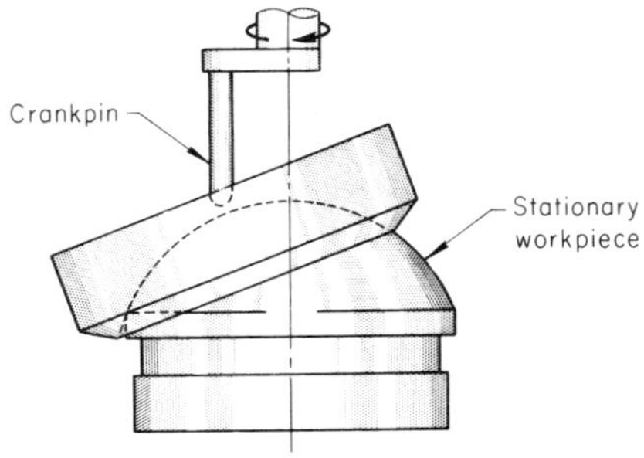


**Fig. 22** Setup for single-piece spherical lapping with a one-spindle machine using a rotating lap. See text for discussion.

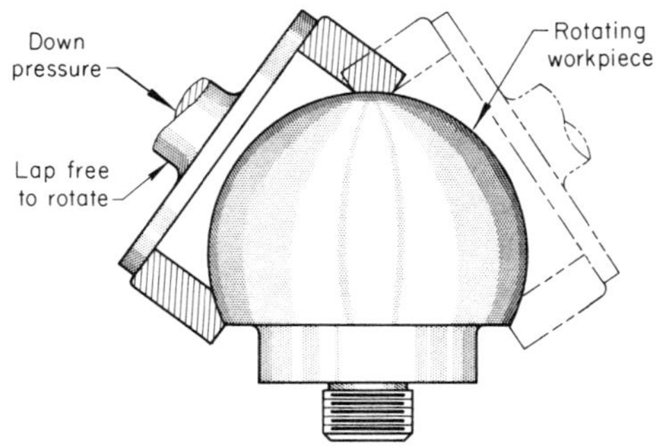


**Fig. 23** Setup for single-piece spherical lapping with a two-spindle machine

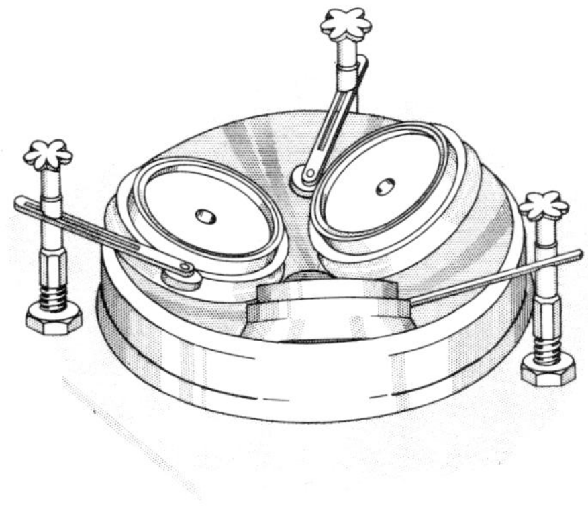

**Fig. 24** Setup for multiple-piece spherical lapping

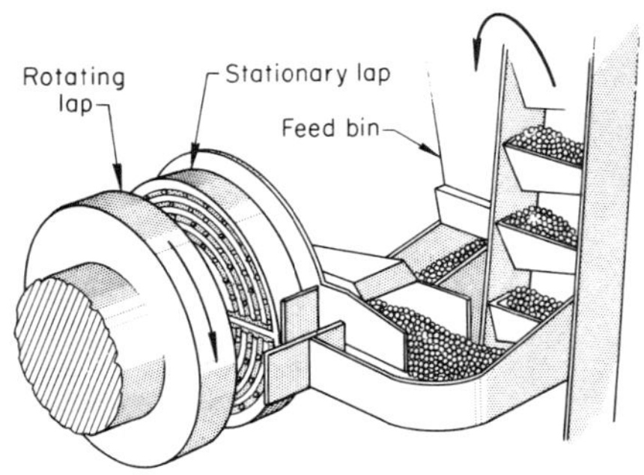


**Fig. 25** Multigroove lapper used for lapping bearing balls

is preferred because preparing a lap with either a straight or tapered shank to be held in a chuck is simpler than improvising a method of holding a workpiece for rotation. A magnetic chuck can hold workpieces made of ferrous metals. Clamping devices must be used for parts made of nonferrous materials. A crank, which is held by the chuck of the turning machine, is constructed with a ball-end crankpin that fits a drilled hole in the back of the lap (Fig. 22) and causes the lap to rotate over the surface of the workpiece. The part is in line with the spindle of the turning machine; the crankpin is offset from center as required for the diameter of the workpiece. For best results, the lap should be heavy enough to provide the required lapping pressure (about 70 kPa, or 10 psi) on the workpiece. If necessary, added pressure can be applied to the lap by a hand feed lever, such as that on a drill press, but it should be applied carefully because excessive pressure can throw the lap off the work.

When machines with two spindles are used, one spindle holds and rotates the workpiece while the other holds the lap in a floating position and oscillates it through an angle large enough to lap the required portion of the surface. A typical setup is illustrated in Fig. 23. One of the spindles must be designed so that it slides as the lap and the workpiece wear and thus keeps a constant pressure on the workpiece.

**Multiple-Piece Lapping.** Quantities of spherical parts are lapped on a concave or convex lap (Fig. 24). The lap can replace the lap on a conventional planetary lapping machine, or it can be fastened into the counterbore of the existing flat lap with holding brackets. Contoured laps should be made of fine-grain cast iron and should be machined to the required radii.

For continuous lapping operations, the conditioning (carrier) rings used should be made of the same material as the lap so that the rings and the lap wear at the same rate. When more than one part is lapped within each conditioning ring, workholders made of fiber, wood, or similar materials are required. Workholders are machined to the same curvature as the lap plate and cut out so that the workpieces can fit into them.

Heavy workpieces can be lapped without additional weight applied, but pressure plates can be placed atop lighter pieces to provide sufficient weight for lapping. These plates can be of various weights to suit requirements, but they must be shaped to conform to the final workpiece shape.

Planetary lapping machines converted to use for spherical workpieces are used to lap parts having radii of 1525 to 3050 mm (60 to 120 in.), the applications for a specific machine being limited by the dimension of its conditioning rings. The capacity of the machine will be the same as for flat lapping (Table 4).

In plain planetary lapping machines that are prepared for contour lapping but not equipped with conditioning rings, workpieces are held by hand or by improvised fixtures. In this case, the workpieces should be moved over the entire area of the lap with reasonable consistency so that the lap is worn evenly and its life is prolonged.

## Lapping of Balls

Spherical parts such as balls for ball bearings can be economically lapped to close dimensional tolerances and smooth finishes. Two machines used in production are the multigroove lapper and the single-groove lapper.

A multigroove lapper is shown in Fig. 25. The machine consists of two cast iron laps (130 to 150 HB); one is stationary and one rotates. Each lap has a series of concentric grooves that correspond with the grooves in the other lap. The radii of the grooves are the same as the radii of the balls being lapped. The stationary lap is cut out for the entrance and exit of the balls, which are fed in at spaced intervals.

In the multigroove machine, the speed of the rotating lap is normally between 50 and 65 rev/min, and the load is usually 33 to 44 kN (7500 to 10 000 lbf). Stock of 0.010 to 0.015 mm (0.0004 to 0.0006 in.) is removed from the diameters of the balls; the time required is 8 to 24 h, depending on the dimensions and finish specified. Balls up to about 44.5 mm (1¾ in.) in diameter can be lapped in this type of machine to an accuracy better than 0.00064 mm (0.000025 in.) and a finish better than 0.050 μm (2 μin.).

The single-groove machine is used for balls larger than about 44.5 mm (1¾ in.) in diameter and for balls of any size that require greater accuracy and better surface finish than are obtainable in the multigroove machine. In the single-groove machine, which also employs two opposing cast iron laps, the shaft may be either horizontal or vertical. There is no cutout in either lap.

Balls to be lapped are alternated with spacer balls of slightly smaller diameter until the groove is filled. The speed of the rotating lap may vary from 30 to 150 rev/min, depending on ball size and material and on the accuracy and surface finish desired. The load may be as low as 135 N (30 lbf) when extremes of accuracy and finish are sought, and the lapping time may extend to three days. Usually, 0.005 to 0.0075 mm (0.0002 to 0.0003 in.) is removed from the ball diameters. Accuracy to less than 0.125 μm (5 μin.) and surface finish of 0.01 μm (0.5 μin.) are possible.

Assuming that the pitch diameter of the groove is 38 mm (15 in.) and that spacer balls are used, 188 balls 3.2 mm (⅛ in.) in diameter or 9 balls 67 mm (2⅝ in.) in diameter could be finished in a single-groove machine in one run. However, less than the groove complement (even a single ball) can be lapped by using dummy balls of

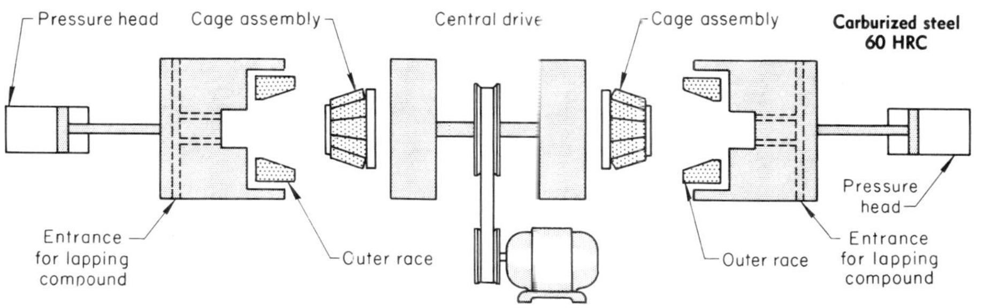


**Fig. 26** Schematic view of setup for matched-piece lapping roller-bearing cage assemblies into mating outer races

another material between the spacer balls to fill the groove. The material used to make the dummy balls must be different in some way (for example, color, density, magnetic properties, or reflectivity) from the material in the workpieces for easy identification.

In both types of machines, there is a constant flow of fluid lapping compound, which usually consists of fused alumina of grit size 5 to 10 μm (200 to 400 μin.) suspended in 10% soluble oil and 90% water. Mineral oil can be used instead of soluble oil and water. Although costs are greater for single-groove lapping than for multigroove lapping, in either process costs are proportional to the time required for obtaining desired accuracy and finish.

Most hard materials can be lapped to greater accuracy and better surface finish than soft materials. Rate of stock removal can be increased by using harder abrasives such as boron carbide or diamond. However, the use of harder or coarser abrasives results in some loss of surface smoothness.

## Lapping to Accelerate Wear-In

Internal lapping is often employed as an accelerated wearing-in process for matching and aligning components of bearing assemblies. In most applications of this type, virtually no stock is removed, and sometimes the desired surface correction is so slight that no abrasive is needed. Typical tooling and techniques are described in the following example.

**Example 11: Lapping of Bearing Assemblies.** The setup illustrated in Fig. 26 was used for the simultaneous lapping of two roller-bearing cage assemblies into mating outer races. All components of these assemblies except cages were made of steel carburized and hardened to 60 HRC. As shown in Fig. 26, the cage assemblies were mounted at each end of a horizontal driving spindle. The outer races were positioned by horizontally actuated pressure heads through which the lapping compound was supplied.

Lapping was performed by rotating the cage assemblies against the races at 1063 rev/min under pressure of 70 to 100 kPa (10 to 15 psi) for 5 min. The lapping compound was prepared by mixing 0.68 kg (24 oz) of medium-hard, 800-grit silica with 38 mL (10 gal.) of paraffin oil. The compound was recirculated and was changed every ten shifts (80 h). This machine could process cage assemblies of various outside diameters to a maximum of 190 mm (7½ in.).

## Lapping of Springlike Parts

Special equipment and techniques are sometimes required for lapping unusual parts. Springlike parts such as those used in numerous precision mechanisms are examples of parts requiring specially designed equipment and techniques.

## Lapping of Gears

Gear lapping corrects the minute errors in involute profile, helix angle, tooth spacing, and concentricity created in the forming or cutting or in the heat treatment of the gears. The lapping can be done by running a set of gears in mesh or by running one gear with a gear-shaped master lapping tool.

Gear lapping is most often applied to sets of hardened gears that are required to run silently in service. Gear lapping is strictly a mating process and is not intended for stock removal. Two gears that have been matched by lapping should be operated as a set, and they should be replaced as a set, rather than singly.

Gears are lapped in special machines, which can be arranged for manual, semiautomatic, or automatic operation. In semiautomatic operation, loading and unloading are manual; in automatic operation, loading and unloading are done automatically in accordance with a programmed cycle.

Angular, spur, and helical gears can be lapped, but the process is mainly applied to spiral bevel gears and hypoid gears. A typical setup used for lapping hypoid gears is shown in Fig. 27. Gears of up to 915 mm (36 in.) pitch diameter can be lapped in semiautomatic or automatic operation. Manual lapping, in special equipment, is employed for gears of approximately 2540 mm (100 in.) pitch diameter down to the smallest gear that can be manufactured. Production lapping machines can be adjusted to lap gears with shaft angles of 0 to 180°.

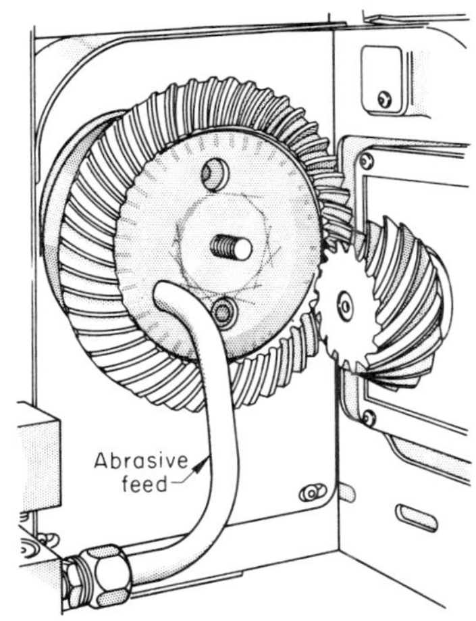


**Fig. 27** Setup for the lapping of hypoid gears

**Lapping Media.** Optimum grit size varies with different types and sizes of gears. A 280-grit abrasive is used for spiral bevel gears; a finer abrasive (about 400-grit) is more suitable for hypoid gears because the sliding action is greater. As a rule, coarser grit is used for gears having a coarse pitch, and finer grits are used for gears having a fine pitch. When compound is brushed on, as in manual operation, a paste-type vehicle is used. However, in semiautomatic or automatic lapping, the abrasive should be mixed with a thin oil (such as mineral seal oil) so that it can be pumped to the workpieces (Fig. 27).

**Processing Techniques.** It is important to roll all mating gears together before lapping to detect nicks and burrs, which can be removed by a small portable hand grinder before the gears are lapped. This preliminary rolling also inspects tooth contact, which should be in the same location for each set of gears and is especially important in automatic lapping.

During the lapping operation, the pinion (smaller gear) is used as the driver, and the larger gear is the driven member. The driven spindle is also used for applying the necessary tooth-contact load by adjusting to a slight drag. Running cycles as short as 15 s at about 76 m/min (250 sfm) can frequently produce desired results. However, longer time cycles may be necessary, depending on initial gear-tooth finish and service requirements.

Low noise level is the criterion of successful gear lapping. Because gear lapping is strictly a mating process and no stock removal is intended, measurements are not made as in most other lapping processes. Minor corrections in tooth bearing shape and position can be obtained, however. Lapping does improve the finish of gear teeth, but improved finish is seldom the purpose of gear lapping.

# Nontraditional Machining Processes

# Introduction

Gary F. Benedict, Allied-Signal Aerospace Company, Garrett Engine Division

NONTRADITIONAL MACHINING PROCESSES have evolved out of the increasingly complex needs of modern society. Inventions have been created to meet these needs, and new tools have been devised to enable the creation of increasingly sophisticated inventions.

One invention that has created new manufacturing challenges is the airplane. In the effort to increase performance, slow-flying aircraft have evolved into transatmospheric vehicles, and this has required continual improvements in materials to meet the demands for improved engine and aircraft-skin operating temperatures. Over the last 60 years, engine operating temperatures have been increased by more than 1100% and aircraft-skin operating temperature requirements have been increased by 2500% to withstand the increased friction resulting from extremely high airspeeds (Ref 1).

Technological developments of this type have prompted the creation of new, difficult-to-machine materials, such as metal-matrix composites, monolithic and composite ceramics, aluminides, and high-performance polymers. The difficulty in machining these and other new materials results from their high hardness and brittleness, high refractoriness, poor thermal properties, chemical reactivity with the cutting tool, and inhomogeneous microstructures (Ref 2). In many cases, the only effective way to machine such materials is by nontraditional methods.

A concise definition of nontraditional machining processes is difficult to establish because of the widely differing processes that fall into this category. Generally, nontraditional processes are considered to be manufacturing processes adopted in the last 50 years that use common energy forms in new ways or that apply forms of energy never used before. Nontraditional processes are subdivided according to the form of energy being harnessed. The generally accepted categories are mechanical, electrical, thermal, and chemical. Table 1 lists the processes that fall within each of these areas.

**Mechanical Methods.** Mechanical nontraditional processes harness direct mechanical abrasive action to remove material. Mechanical processes are usually applied to workpiece materials that are difficult to machine by traditional techniques because of material hardness, toughness, or brittleness. Ceramics, composites, and organic materials are particularly good candidates for mechanical machining because most of them are not electrically conductive (a requirement for processing by electrical methods) and because they are damaged by burning, charring, or cracking when thermal processes are applied.

**Electrical Methods.** The electrical nontraditional processes are limited in application to electrically conductive workpiece materials. Workpiece materials that are difficult to machine by conventional means constitute a large percentage of the applications in this category; however, numerous applications are selected because of the ability of the electrical processes to produce complex shapes in a single pass of the tool and (with the exception of the electrical grinding variations) to process parts without tool wear.

**Thermal Methods.** Primarily because of rapid increases in the sales of wire electrical discharge machining and laser equipment, thermal processes have become the fastest-growing segment of the nontraditional market. Given the diversity of energy sources used in this category (electrons, photons, electrical sparks, and so forth), generalizations are difficult to make concerning the application of these processes. Thermal processes are generally unaffected by the physical properties of the materials being processed and are therefore often applied to extremely hard or low-machinability workpiece materials. Because the mechanism for material removal is thermal, workpieces that will be used for critical applications may require the removal of thermally affected zones.

**Chemical Methods.** High-volume, high-production manufacturing is often performed by chemical nontraditional processes. Although sometimes applied to low-quantity or even one-of-a-kind parts because of low initial tooling costs, chemical machining has gained wide acceptance for the economical manufacture of high-volume products such as springs, electric motor laminations, and television picture tube masks. Because material is removed by means of chemical action, there are no forces acting on the workpiece. This enables parts to be machined without concern for distortion or damage. In addition, because the machining action occurs on all surfaces of the workpiece simultaneously, effective throughput can be extremely high, even compared to high production rate processes such as punching and stamping.

**The future** of nontraditional machining processes will no doubt be characterized by steady growth. Although nontraditional processes will probably never replace the conventional tools currently used by industry, the newer methods are ensured an increasingly important role because of their steadily improving capabilities and because of the beneficial effects of computer con-

**Table 1 Categories of nontraditional machining processes**

| Mechanical | Electrical | Thermal | Chemical |
|---|---|---|---|
| Abrasive jet machining (AJM)<br>Abrasive flow machining (AFM)<br>Waterjet machining (WJM)<br>Abrasive waterjet machining (AWJM)<br>Ultrasonic machining (USM) | Electrochemical machining (ECM)<br>Electrochemical grinding (ECG)<br>Electrochemical discharge grinding (ECDG)<br>Electrostream drilling (ES)<br>Capillary drilling (CD)<br>Shaped tube electrolytic machining (STEM) | Electrical discharge machining (EDM)<br>Electrical discharge wire cutting (EDWC)<br>Electrical discharge grinding (EDG)<br>Electron beam machining (EBM)<br>Laser beam machining (LBM)<br>Thermal energy method (TEM) | Chemical milling (CM)<br>Photochemical machining (PCM) |

trol, adaptive control, and education (Ref 3).

Compared to conventional processes, nontraditional processes possess almost unlimited capabilities, with one exception—volumetric material removal rates. Currently, conventional processes excel in rapid bulk material removal rates. However, many improvements in nontraditional process removal rates have been made in recent years, and there is every reason to believe that this trend will continue. This will enhance the competitiveness of nontraditional processes and will increase the range of applications.

Most nontraditional machining processes are computer controlled with respect to process parameters. Computer control simplifies processes that might otherwise intimidate potential users and, in the long term, accelerates acceptance of the process. In addition, the computer control of process parameters ensures process reliability and repeatability, and this further accelerates acceptance and implementation.

Unlike many conventional processes, most nontraditional processes can easily be adaptively controlled through the use of a broad range of in-process inspection sensors. For example, if a sensor detects that the holes being produced in a product are decreasing in diameter, the hole size can be modified without changing the hard tools by simply changing the process parameters or the offset values in the computer. In contrast, hard tools such as drills or punch and die inserts must be changed when conventional techniques are used. This ability to monitor, respond to, and correct undesirable situations automatically ensures the increasing importance of nontraditional processes in unattended machining cells and automated factories.

Every year more attention is directed toward nontraditional processes, as evidenced by the increasing number of technical papers, conferences, college courses, books, and technical symposia on the subject. These activities not only educate the manufacturing and product design communities about the unique capabilities of these processes, but also ensure a successful future for nontraditional machining processes.

## REFERENCES

1. M.A. Steinberg, Materials for Aerospace, *Sci. Am.*, Vol 255 (No. 4), 1986, p 67-72
2. R. Komanduri, On the Machining of Difficult-to-Machine Materials, in *Proceedings of the 2nd International Machine Tool Engineers Conference* (Kobe, Japan), IMEC, 1986
3. G.F. Benedict, *Nontraditional Manufacturing Processes*, Marcel Dekker, 1987, p 3

# Abrasive Jet Machining

Glenn E. Storck, S.S. White Industrial Products, Division of Pennwalt Corporation

ABRASIVE JET MACHINING (AJM) is a process that removes material from a workpiece through the use of abrasive particles entrained in a high-velocity gas stream. The AJM process removes material by the impingement of abrasive particles on the work surface. The process differs from conventional sandblasting in that abrasive jet machining has smaller-diameter abrasives (from 10 to 50 μm) and a more finely controlled delivery system. Abrasive jet machining is typically used to cut, clean, peen, deburr, deflash, or etch glass, ceramics, or hard metals. Applications include the following:

- Drilling and cutting small sections of glass, ceramics, or hardened metals
- Etching part numbers onto metal and plastic components (Fig. 1)
- Deburring and deflashing metal and plastic parts
- Frosting glass
- Cutting intricate patterns in hard, brittle materials
- Cleaning oxides from metal surfaces
- Cleaning metallic smears from ceramics
- Trimming, beveling, and cleaning electronic components
- Removing smudges and films from documents and museum artifacts

**Fig. 1** Identification of hard plastic embedded voltage regulator using AJM equipment and rubber mask

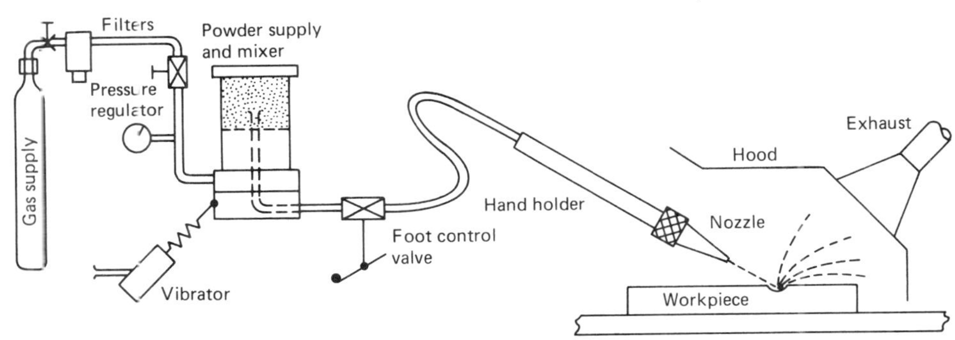


**Fig. 2** Components of an abrasive jet machining system

## System Components

Figure 2 illustrates the principal components of an AJM system. A gas supply of at least 690 to 860 kPa (100 to 125 psi) is necessary to pressurize the system to a regulated pressure of 520 to 830 kPa (75 to 120 psi). The gas supply, which can be compressed air, nitrogen, carbon dioxide, or a suitable industrial inert gas, must be free of foreign particles such as oil and water. Therefore, a gas filter and a water separator should be used in the gas supply.

The abrasive powder is held in a hopper that feeds into a vibrating chamber, where the powder is metered and mixed uniformly with the gas. The jet stream of abrasive particles is then created as the pressurized mixture emerges from a hand-held nozzle at a velocity of 150 to 300 m/s (500 to 1000 ft/s). The flow rate of the jet stream is controlled by a pneumatic pinch valve, which can be controlled by a foot valve for pulsed operation or a manual valve for continuous operation. (Good grounding within the system bleeds off the static electric charge created by the high-velocity stream.)

A dust hood or collection cabinet used in conjunction with a vacuum dust collector is needed to limit operator exposure to dust. If workpieces with toxic substances are involved, then dust collection requires special disposal considerations in accordance with local and federal regulations.

**Nozzle tips** for abrasive jet machining are normally made of tungsten carbide, although sapphire nozzles are available. Sapphire nozzles last longer than tungsten carbide nozzles, but are more expensive and are only available with round holes. Tungsten carbide nozzles can have either round or rectangular holes. Round nozzles are

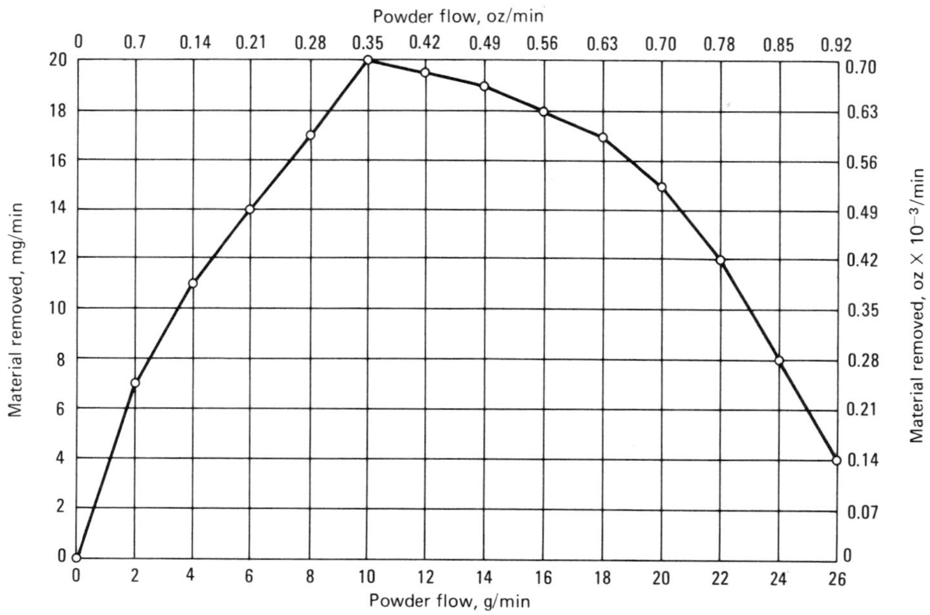


**Fig. 3** Effect of powder flow rate on material removal. Workpiece material: plate glass

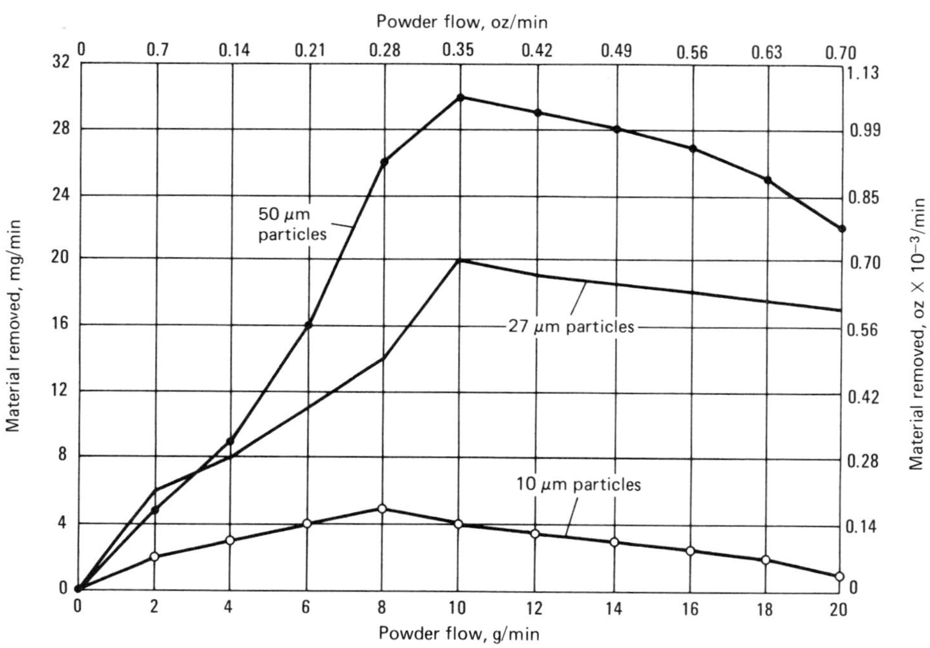


**Fig. 4** Effect of particle size on material removal. Workpiece material: plate glass

available in diameters from 0.13 to 1.25 mm (0.005 to 0.050 in.). Rectangular nozzles range from 0.08 × 0.50 to 0.18 × 3.8 mm (0.003 × 0.020 to 0.007 × 0.150 in.). Rectangular nozzles may reduce stray cutting.

Nozzle life depends on the type of abrasive used and the operating pressure. The service life of a tungsten carbide nozzle is about 8 to 15 h with a silicon carbide abrasive and about 20 to 35 h with aluminum oxide. With glass beads or sodium bicarbonate, nozzle life is indefinite. Sapphire nozzles have an average service life of 300 h.

## Process Characteristics

The AJM process is primarily a finishing and cleaning operation with low material removal rates. The typical material removal rate for glass is 16 mm$^3$/min (0.001 in.$^3$/min). Cutting rates for metals vary from 1.6 to 4.1 mm$^3$/min (0.0001 to 0.00025 in.$^3$/min). For the harder ceramics, cutting rates are about 50% higher than those for glass.

**Advantages and Disadvantages.** The AJM process has several advantages despite its low material removal rate. Some of the general advantages of the process are:

- The gas stream dissipates generated heat when cutting heat-sensitive materials
- The small loads transmitted to the workpiece allow the cutting of fragile pieces
- The chipping action of the process is effective on hard materials
- The nozzles can be directed toward small, difficult-to-reach areas

The disadvantages of the AJM process, in addition to its low material removal rate, are stray cutting, and the possibility of abrasive particles becoming embedded in the workpiece. Excessive taper on deep cuts may also be a disadvantage, although the amount of taper can be reduced by tilting the nozzle.

**Material Removal.** Several factors determine the characteristics of the AJM process. The major variables include:

- Type of material being worked
- Flow rate of the jet stream
- Type and size of abrasive particles
- The distance between the workpiece and nozzle

Generally, the material removal rate is higher when the material being removed is harder or when the abrasive particles are larger and have a higher velocity at impact.

*Flow Rates.* To obtain high material removal rates, abrasive flows of 10 to 20 g/min (0.4 to 0.7 oz/min) are normally used. For fine work, flow rates of 3 to 5 g/min (0.1 to 0.2 oz/min) have been successfully used.

There is also an optimum mass flow rate for maximizing material removal (Fig. 3). Flow rates greater than the optimum decrease the material removal rate. A number of variables determine the optimum, but for a given system and abrasive powder, the point is readily determined during start-up.

*Abrasive Powders.* The application dictates the size and type of abrasive powder used. Aluminum oxide and silicon carbide powders are used for heavy cleaning, cutting, and deburring. Magnesium carbonate is recommended for use in light cleaning and etching, while sodium bicarbonate is used for fine cleaning and the cutting of soft materials. For polishing and fine deflashing, glass beads or crushed glass is recommended.

The abrasive particles are available in sizes ranging from 10 to 50 μm. The size of the abrasive particle influences material removal rates (Fig. 4). The larger sizes have higher material removal rates and are best for cutting and peening. Smaller sizes are used for polishing and cleaning.

Commercial grade powders are not suitable for the AJM process, because their sizes are not well classified. Commercial powders may contain silica dust, which can be a health hazard. Powders should not be

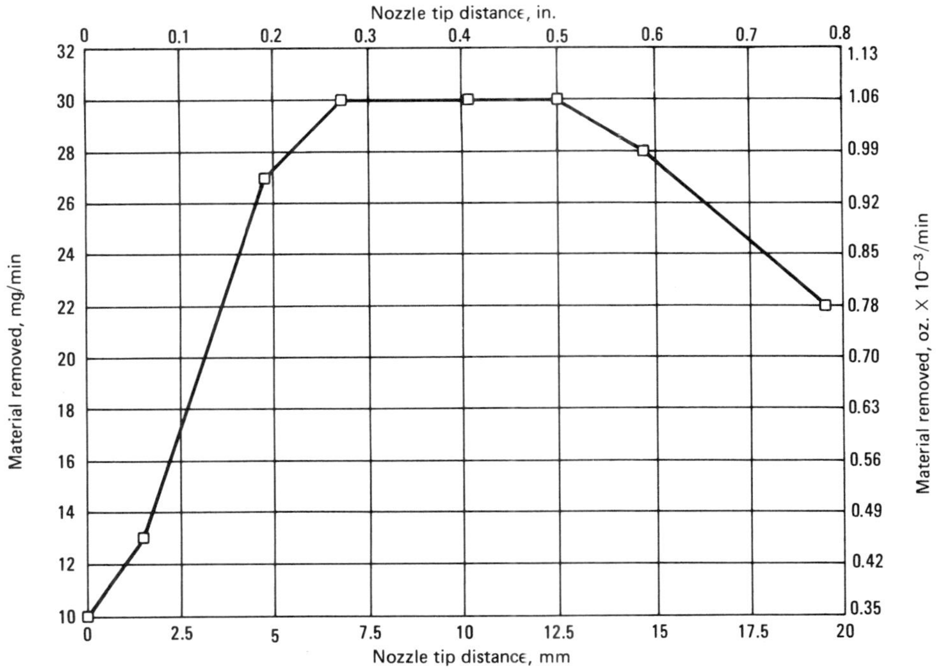


**Fig. 5** Effect of nozzle tip distance on material removal rate. Workpiece material: plate glass

reused, because the contaminated powder may clog the nozzle. Used powders should be disposed of in accordance with federal, state, and local regulations.

*Nozzle tip distance*, the distance between the nozzle tip and the workpiece, also influences the AJM process. The jet stream remains cylindrical only for a short distance of 1.6 mm (0.062 in.) after exiting the nozzle. At greater distances, the jet stream begins to diverge into a cone-shaped pattern with an included angle of about 7°. The amount of divergence probably depends on the pressure of the system.

The nozzle tip distance is varied according to the application. A nozzle tip distance of 5 to 13 mm (3/16 to 1/2 in.) lends itself to cleaning and peening, while 13 to 75 mm (0.5 to 3 in.) is suitable for light operations. For work requiring accuracy and the elimination of taper, the nozzle is positioned about 0.8 mm (0.032 in.) from the workpiece.

The nozzle tip distance also affects material removal rates. For a given abrasive, flow rate, and base material, the most effective cutting occurs with a nozzle tip distance between 7 and 13 mm (0.27 and 0.50 in.). Figure 5 illustrates this relationship between nozzle tip distance and material removal rate.

**Tolerance and Finish.** The AJM process generates surfaces with a granular, matte texture. Surface finishes range from 0.15 to 1.5 μm (6 to 60 μin.), depending on grit size. Table 1 lists some surface finishes attained with the AJM process. Tolerances of the AJM process can range from ±0.13 to 0.05 mm (±0.005 to 0.002 in.).

**Table 1 Surface finishes for annealed 316 stainless steel with different AJM abrasives**

| Abrasive type | Grit size μm | Grit size μin. | Surface roughness(a) μm | Surface roughness(a) μin. |
|---|---|---|---|---|
| Aluminum oxide ..... | 10 | 400 | 0.20–0.50 | 8–20 |
| | 25 | 1000 | 0.25–0.53 | 10–21 |
| | 50 | 2000 | 0.38–0.96 | 15–38 |
| Silicon carbide ...... | 20 | 800 | 0.30–0.50 | 12–20 |
| | 50 | 2000 | 0.43–0.86 | 17–34 |
| Glass bead.......... | 50 | 2000 | 0.30–0.96 | 12–38 |

(a) Initial surface was ground to 0.47 μm (18.5 μin.). Source: Metcut Research Associates

# Abrasive Flow Machining

Lawrence J. Rhoades and Hilary A. Clouser, Extrude Hone Corporation

ABRASIVE FLOW MACHINING (AFM) finishes surfaces and edges by extruding viscous abrasive media through or across the workpiece. Abrasion occurs only where the flow of the media is restricted; other areas remain unaffected. Abrasive flow machining can process many inaccessible passages on a workpiece simultaneously, and it can accommodate several dozen parts in one fixture. Tooling can also be designed so that tooling changes can be performed in minutes in production applications.

Abrasive flow machining is used to deburr, polish, or radius surfaces and edges. A variety of finishing results can be achieved by altering the process parameters. The process embraces a wide range of applications—from critical aerospace and medical components to high-production volumes of parts. The process can yield production rates of up to hundreds, or even thousands, of parts per hour.

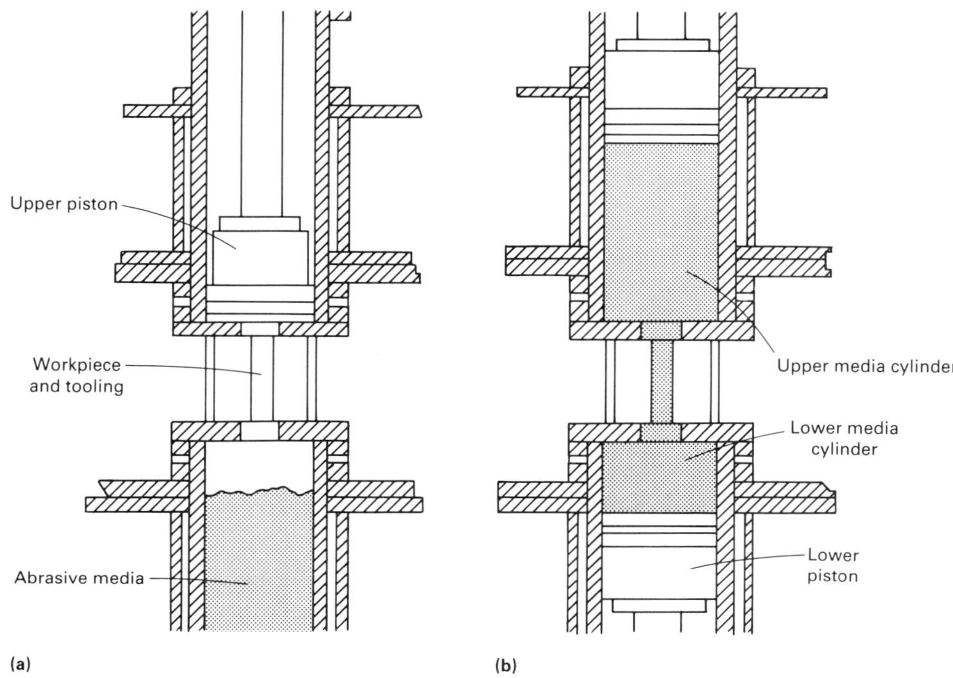


**Fig. 1** Schematic of the AFM process. (a) Abrasive media in the lower cylinder. (b) Abrasive media being extruded through the workpiece and tooling

## Process Characteristics

The AFM process uses two opposed cylinders to extrude semisolid abrasive media back and forth through the passages formed by the workpiece and the tooling (Fig. 1). By repeatedly extruding the media from one cylinder to the other, an abrasive action is produced as the media enter a restrictive passage and travel through or across the workpiece. The machining action is similar to a grinding or lapping operation as the abrasive media gently polish the surfaces or edges.

The process is abrasive only in the extrusion area, where the flow is restricted. When forced into a restrictive passage, the polymer carrier in the media temporarily increases in viscosity; this holds the abrasive grains rigidly in place. They abrade the passages only when the matrix is in this thicker viscous state. The viscosity returns to normal when the thickened portion of media exits the restrictive passage.

The viscosity and the flow rate of the media affect the uniformity of stock removal and the edge radius size. If the objective of abrasive flow machining involves the uniform polishing of walls within the restricted passages, as in die polishing, for example, the media chosen should maintain a uniform flow rate as it travels through the passage (Fig. 2). Low flow rates are best for uniform material removal. For deburring or

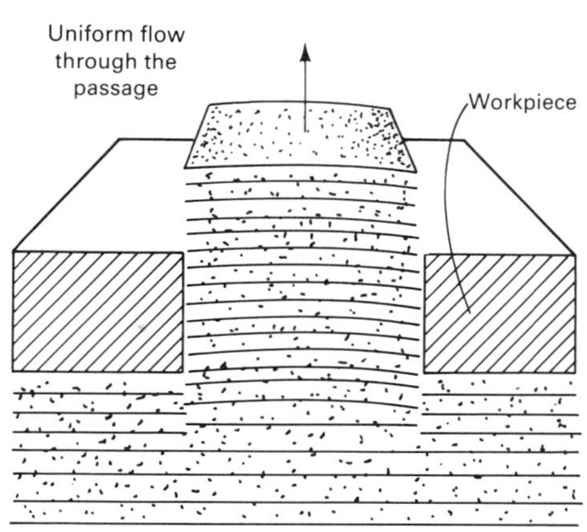


**Fig. 2** Schematic showing uniform flow through the passage. This removes an even amount of stock from the passage walls.

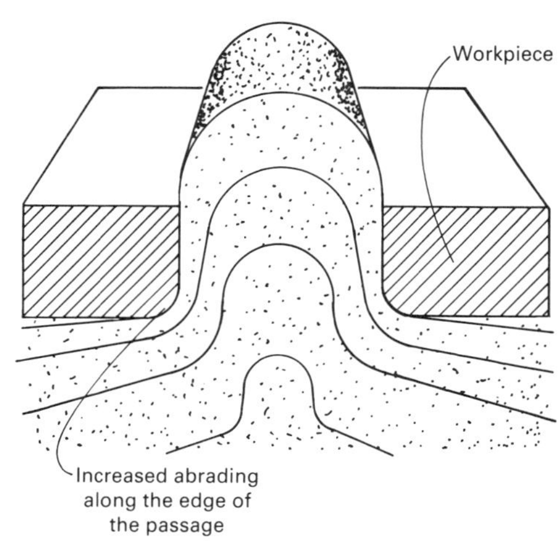


**Fig. 3** Schematic showing the flow pattern of media entering a passage, which generates the machining action used for deburring and radiusing. When radiusing or deburring passage corners, a lower media viscosity is used, providing faster flow through the center of the passage than along its periphery. Because one process cycle normally consists of flow in both directions, both ends of the passage would be abraded.

**Fig. 4** Tooling for restricting the flow of the media. To process external edges, the part is contained within a fixture that directs and restricts media flow in the appropriate areas.

**Fig. 5** Vertical column of AFM abrasive media after extrusion

**Fig. 6** Multiple-port workpiece processed by abrasive flow machining

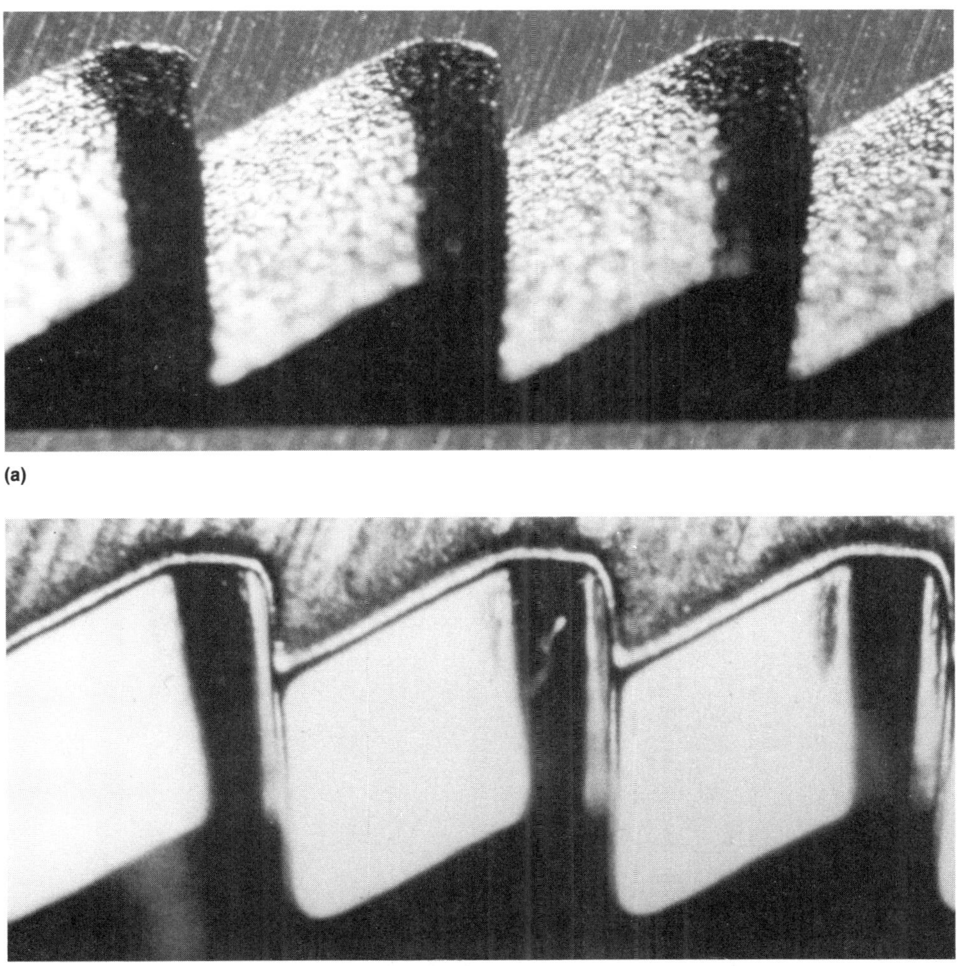

(a)

(b)

**Fig. 7** Teeth on the extrusion die shown in Fig. 6. (a) Before AFM. (b) After AFM

radiusing the edges of a passage, the higher flow rate of lower viscosity media within the passage causes the edges to be abraded more than the passage walls (Fig. 3). The flow rate depends on the machine settings, the formulation of the media, and the workpiece and tooling configuration.

## Equipment

The major elements of an AFM system include the machine, the tooling, and the abrasive media. Each is discussed below.

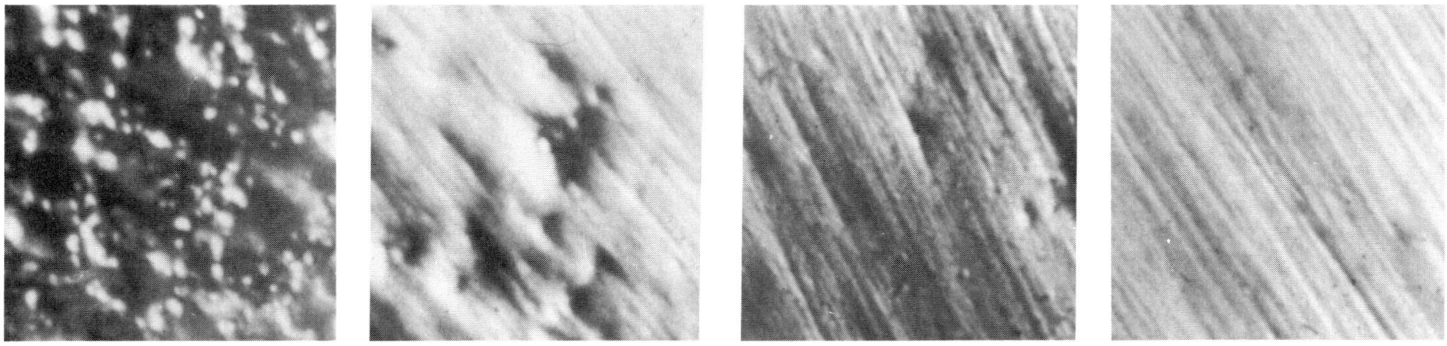

**Fig. 8** Progressive stages in AFM polishing from a 1.9 μm (75 μin.) finish (left) to a 0.2 μm (8 μin.) finish

**Machines** for abrasive flow machining are available with extrusion pressures ranging from 700 to 22 000 kPa (100 to 3200 psi), with flow rates up to 380 L/min (100 gal./min). (Typical flow rates range from 1 to 50 L/min, or 0.3 to 13 gal./min.) The flow rate of the media depends on the extrusion pressure, the viscosity of the media, and the workpiece and tooling configuration. Extrusion pressure and flow volume (the displacement of each cylinder of media multiplied by the total number of stroke cycles) are both preset at the machine.

Control systems can be added to monitor and control additional process parameters, such as the temperature, viscosity, wear, and flow speed of the media. Abrasive flow machining systems designed for production applications often include part-cleaning and unload/reload stations as well as media maintenance and cooling units. Automated systems can process thousands of parts per day, with processing times typically ranging between 1 and 3 min for each pallet loaded with workpieces.

**Tooling** holds the workpiece in position and directs the abrasive media to the appropriate areas. Many AFM applications require only simple fixturing. Dies, for example, typically need no special tooling, because the die passage itself provides the restriction for the flow path.

For external edges or surfaces, tooling is used to restrict the flow between the outside of the part and the inside of the fixture (Fig. 4). The tooling restricts flow at areas where abrasion is desired. The tooling can also block flow from areas that are to remain unaffected.

High-production fixtures are designed to facilitate part loading, unloading, and cleaning. Often mounted to indexing tables, these fixtures can hold multiple parts for processing in one operation.

**The media** consist of a pliable polymer carrier and a concentration of abrasive grains. The viscosity of the carrier and the type, concentration, and size of the abrasive grains can be varied to achieve specific results. Higher-viscosity media (Fig. 5) are

**Table 1 AFM medium selector for passageways with a 2:1 length-to-width ratio**

| | Media type/viscosity | | | | | | | | | |
|---|---|---|---|---|---|---|---|---|---|---|
| | LV/low | | LMV/low-medium | | MV/medium | | HMV/high-medium | | HV/high | |
| Passage size(a) | mm | in. | mm | in. | mm | in. | mm | in. | mm | in. |
| Minimum | 0.4 | 1/64 | 0.8 | 1/32 | 1.6 | 1/16 | 3.2 | 1/8 | 6.4 | 1/4 |
| Maximum | 3.2 | 1/8 | 6.4 | 1/4 | 13 | 1/2 | 25 | 1 | 50 | 2 |

(a) Indicated passage sizes are widths or diameters and assume that passage length is two times the width.

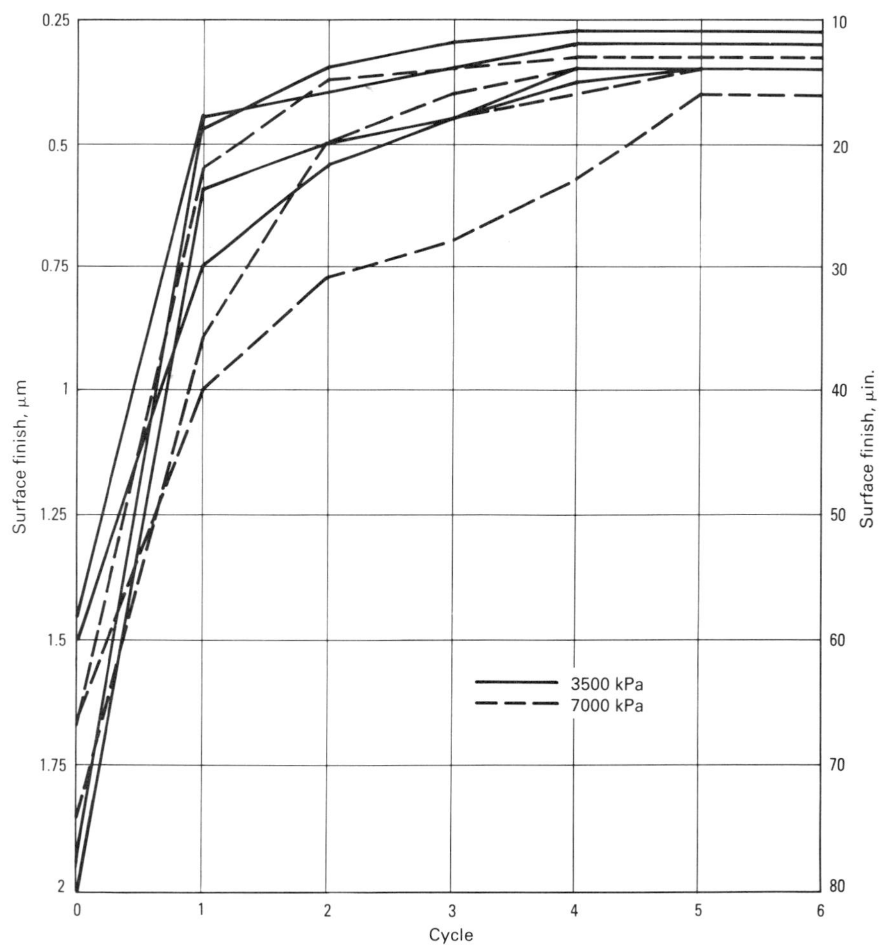


**Fig. 9** Comparison of surface improvement and the number of AFM cycles for various die polishing media formulations in an electric discharge machined tool steel passage with a cross-sectional area of 160 mm² (0.25 in.²)

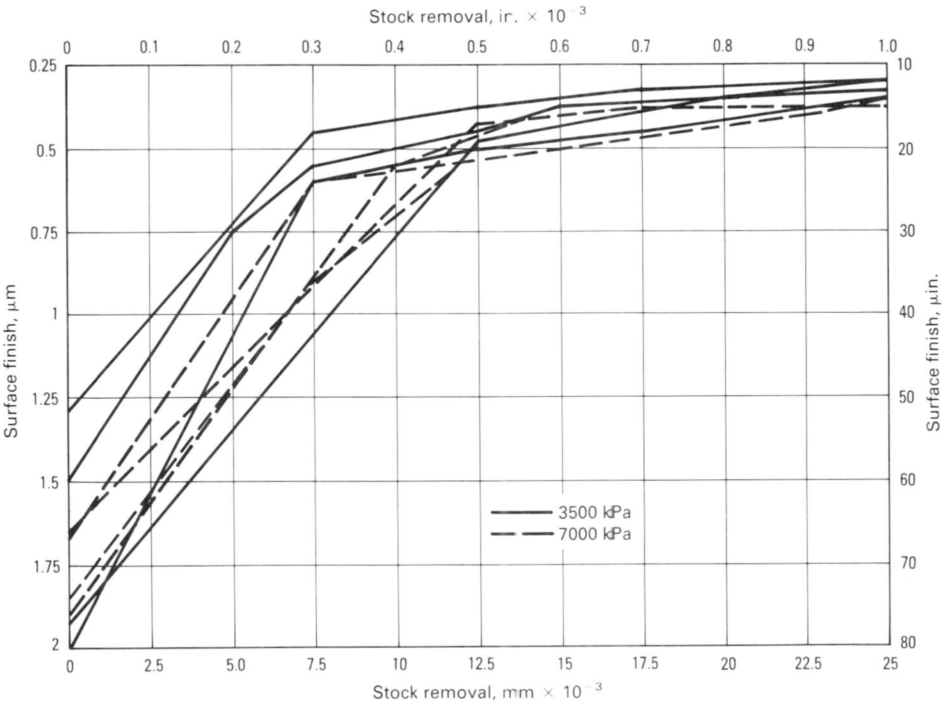


**Fig. 10** Correlation between surface finish and stock removal on an electrical discharge machined surface with an original $R_a$ of 2 μm (80 μin.)

nearly solid and are used for uniform polishing or for abrading the walls of large passages. Lower-viscosity media are generally appropriate for radiusing edges and for processing small passages.

The carrier is a mixture of a rubberlike polymer and a lubricating fluid. By varying the ratio of the polymer and the lubricating fluid, the carrier can have viscosities ranging continuously from very high to very low.

Table 1 lists some general guidelines for selection of the viscosity of the media for passageways with a length-to-width ratio of 2:1. If the passage length is substantially smaller than two times the passage width, higher-viscosity media or lower extrusion pressure should be used. Conversely, if the passage length is substantially greater than two times the passage width, lower-viscosity media or higher extrusion pressure is required.

The passage size range of a given media formulation can also be extended by increasing the temperature of the media during processing. An increase in temperature decreases the viscosity of the media, and decreased viscosity allows extrusion through a smaller passage of a given length.

**The abrasive grains** used in abrasive flow machining are most commonly made of silicon carbide, although boron carbide, aluminum oxide, and diamond can also be used. Particle sizes range from 0.005 to 1.5 mm (0.0002 to 0.060 in.). The better the starting finish, the smaller the grit size used for processing. The larger abrasives cut at a faster rate, while the smaller sizes provide finer finishes and accessibility to small holes. The depth of cut made by the abrasive grains at the surface depends on the sharpness and size of the abrasive grains, the extrusion pressure applied, and the stiffness of the media.

In the AFM process, the abrasive cutting particles break and become dull, and the abraded material becomes part of the abrasive media. The effective life of the media depends on a number of factors, including the initial quantity of media, the abrasive size and type, the flow speed, and the part configuration. Typically, a machine load of media can be used for weeks to process thousands of parts before replacement. Air or vacuum can be used to remove the media from accessed areas. Final traces can be extracted in a solvent wash.

## Process Capabilities

The AFM process is particularly useful in the polishing, radiusing, and edge finishing of inaccessible internal passages. Surface finishes of 0.05 μm (2 μin.) can be obtained, and materials from soft aluminum to tough ceramics, carbides, and nickel alloys can be successfully micromachined with this process.

**Example: Finishing an Aluminum Die.** The multiple-port aluminum extrusion die

**Fig. 11** Aircraft valve bodies and spools deburred by abrasive flow machining

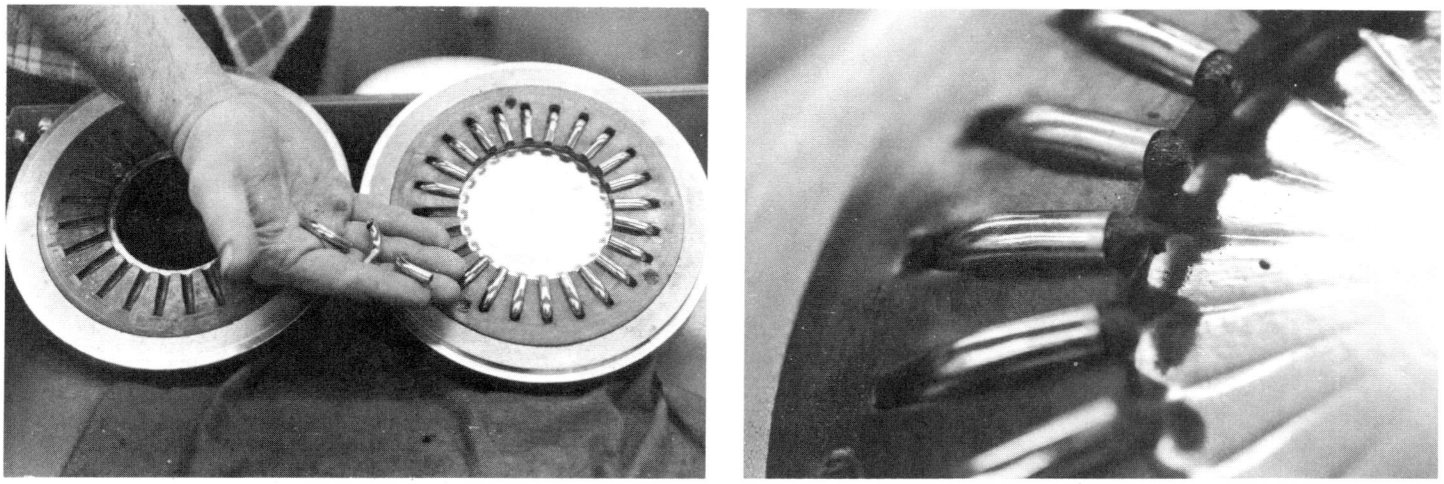

(a) (b)

**Fig. 12** Cannula tubes, for medical applications, finished by abrasive flow machining. (a) Twenty four parts are processed in one fixture. (b) Detail of (a)

shown in Fig. 6 used AFM for improving a finish of 1.9 μm (75 μin.) to 0.18 μm (7 μin.) in a 5-min cycle. Figure 7 shows the teeth before and after processing.

Figure 8 shows the progressive polishing action of the bearing surface on the extrusion die shown in Fig. 6; the processing time between each successive stage is 80 s. Hand polishing smears the surface peaks, causing edge rolling and an inconsistent surface, but the AFM process grinds away the high spots of the area, yielding a more uniform unidirectional surface and unidirectional lay. In many cases, this produces a more reliable and longer lasting part.

**Surface Finish and Stock Removal.** Figure 9 compares the number of AFM cycles with the resultant surface finish on a passage having an area of 160 $mm^2$ (0.25 $in.^2$) and an original electrical discharge machined finish with an average roughness ($R_a$) of 2 μm (80 μin.) that was processed with media formulations of varying abrasive grain size and viscosity. Despite the varying parameters for the media, the surface improvement after only one cycle with $2 \times 10^6$ $mm^3$ (120 $in.^3$) of medium ranges from 60 to 75%.

Figure 10 illustrates the relationship between surface finish and stock removal. Given beginning finishes ranging from 1.3 to 2 μm (52 to 80 μin.), surface improvement ranged from 50 to 70% after just 0.0075 mm (0.0003 in.) of stock removal. Removing 0.0125 mm (0.0005 in.) of material per side resulted in a 70 to 80% improvement in surface finish; removing 0.0125 mm (0.0005 in.) of additional stock yielded another 5 to 10% improvement. Therefore, despite varying viscosities and abrasive grit sizes in the medium, a stock removal amount of approximately 0.025 mm (0.001 in.) resulted in an 85 to 90% improvement over the starting electrical discharge machined surface finish. The uniformity of stock removal typically falls within 20% of the stock removed.

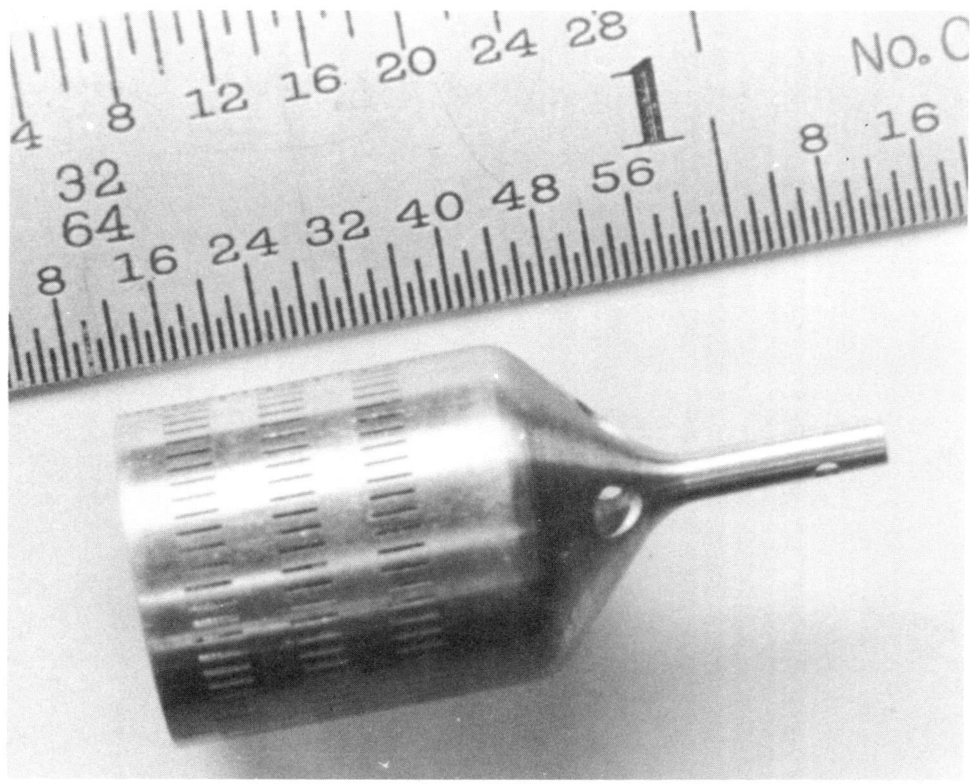

**Fig. 13** Finishing of an accelerometer pendulum by abrasive flow machining. Variable extrusion pressures permit the finishing of this part with a wall thickness of 0.25 mm (0.010 in.). The recast left by electrical discharge machining was removed from 180 slots, 0.15 mm (0.006 in.) wide and 1.5 mm (0.060 in.) long.

## Process Applications

Abrasive flow machining is a flexible method for achieving precise and uniform surfaces in a variety of finishing operations. The AFM process is used in the deburring, radiusing, sizing, and polishing of parts, and it can be applied to a wide range of part and passage sizes—from gears as small as 1.5 mm (0.060 in.) in diameter or orifices as small as 0.15 mm (0.006 in.) to splined die passages 50 mm (2 in.) across or turbine disks nearly 1.2 m (4 ft) in diameter.

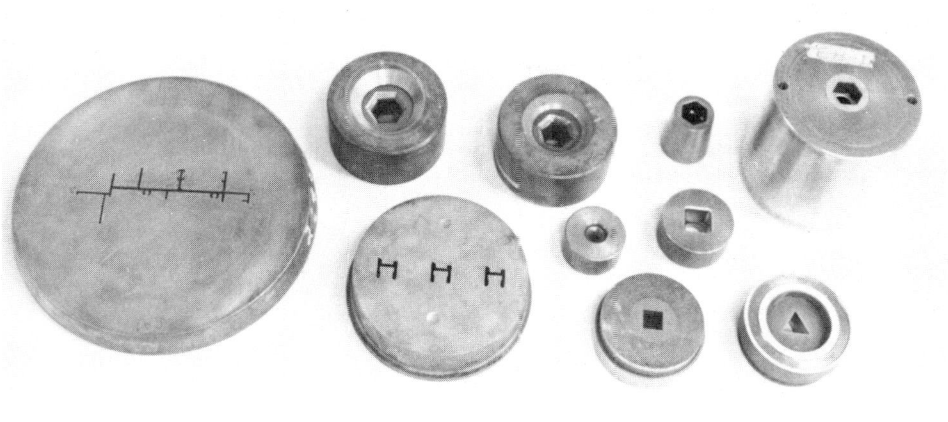

**Fig. 14** Various types of dies made of ceramic, carbide, and tool steel processed by AFM for improved surface profile and increased die life

**Fig. 15** Milled surfaces of an aircraft engine IBR finished by AFM. This 15 to 30 min operation improves the surface finish while retaining the shape within the profile envelope and generating leading and trailing edge shapes, eliminating hours of hand finishing.

Abrasive flow machining was initially developed for the critical deburring of aircraft valve bodies and spools (Fig. 11). The method provides burr-free internal edges that routinely pass 20× microscopic inspection, and it creates precisely controlled edge radii. Other applications include:

- Deburring, radiusing, and polishing the pocket of bearing retainers in a single operation
- Deburring and polishing fuel injector bodies at a rate of 30 000 parts per day with an automated AFM system
- Finishing cannula tubes for surgical implantation (Fig. 12)
- Removing recast layers from fragile workpieces (Fig. 13)
- Deburring fuel nozzles
- Finishing dies (Fig. 14) for extruding, drawing, forging, cold heading, and compacting
- Finishing impellers, integrally bladed rotor (IBR) (Fig. 15), compressor wheels, turbine disks, and gears

## SELECTED REFERENCES


- G.F. Benedict, *Nontraditional Manufacturing Processes*, Vol 19, *Manufacturing Engineering and Materials Processing Series*, Marcel Dekker, 1987
- T. Kohut, Surface Finishing With Abrasive Flow Machining, in *Proceedings of the 4th International Aluminum Extrusion Technology Seminar*, Vol 2, The Aluminum Association, April 1988
- L.J. Rhoades, Ed., *Cost Guide for Automatic Finishing Processes*, Society of Manufacturing Engineers, 1981
- L.J. Rhoades, "Automation of the Nontraditional Processes," SME Technical Paper MR85-475, Society of Manufacturing Engineers, 1985
- L.J. Rhoades, Abrasive Flow Machining With Not-So-Silly Putty, *Met. Finish.*, July 1987, p 27-29
- L.J. Rhoades, Abrasive Flow Machining, *Final Finish Technol.*, Winter 1988, p 8-19
- E.J. Weller, ed., *Nontraditional Machining Processes*, 2nd ed., Society of Manufacturing Engineers, 1984
- C. Wick and R.F. Veilleux, Ed., *Materials, Finishing and Coating*, Vol 3, 4th ed., *Tool and Manufacturing Engineers Handbook*, Society of Manufacturing Engineers, 1985

# Waterjet/Abrasive Waterjet Machining

C.E. Johnston, Flow Systems, Inc.

WATERJET MACHINING (WJM), also called hydrodynamic machining, uses a high-velocity stream of water as a cutting tool. This process is limited to the cutting of nonmetallic materials when the jet stream consists solely of water. However, when fine abrasive particles are injected into the water stream, the process can be used to cut harder and denser materials. Abrasive waterjet machining has expanded the range of fluid jet machining to include the cutting of metals, glass, and ceramics.

## Equipment

Figure 1 illustrates the major components of a WJM system. The major components are:

- Hydraulic unit
- Intensifier
- Accumulator
- Filters
- Water transmission lines
- On/off valve
- Waterjet nozzles
- Abrasive waterjet nozzle
- Waterjet catchers
- Fluid additives

**The hydraulic unit** consists of an electrically driven, variable-displacement, pressure-compensated hydraulic pump. Typical hydraulic pressures are adjustable to about 20 MPa (3 ksi).

**The intensifier** (Fig. 2) is used to increase the water pressure up to 380 MPa (55 ksi). The hydraulic pressure is applied to the low-pressure cylinder of the intensifier, and the water pressure is developed in the high-pressure cylinder. The pressure increase is determined by the ratio of the working areas of the two cylinders.

**The accumulator,** or shock attenuator, is plumbed in parallel with the high-pressure output of the intensifier. The accumulator is added to smooth the pressure spikes that occur at the reversal of the reciprocating stroke of the intensifier. These pressure spikes result from the compression of water. With an accumulator, the pressure variations from water compression can be reduced to ±2.5%.

**Filters** protect the nozzle orifice from possible damage by foreign material. For most applications, the inlet water is mechanically filtered to 0.45 μm. Other processing beyond this point must be determined on a site-by-site basis. Figure 3 shows the condition of the nozzle orifice after operation under various water conditions.

**Water transmission lines** consist of flexible hose, hard tubing, swivels, and flex joints. Flexible hose capable of operating at 380 MPa (55 ksi) is available and greatly simplifies plumbing. If the hosing cannot operate at the desired water pressure, then hard tubing, swivels, and flex joints must be used. The hard tubing used is typically stainless steel with an inside-to-outside-diameter ratio of 1:3. Flex joints (tubing in springlike coils) are more reliable than swivels, but do not have the range of movement of swivels. Flex joints can be used for rotations of 360° or less. Swivels are used for continuous or multiple 360° rotations, or to reduce forces exerted on the manipulator if coils are used.

**The on/off valve** is a two-way, two-position, rapid-acting valve used to turn the jet stream on or off. The tightness of the valve is important in applications that do not allow fluid on the workpiece. The response time of the valve is also important.

**Waterjet Nozzles.** Figure 4 illustrates the basic components of a typical waterjet nozzle. The collimating chamber helps to reduce the divergence of the waterjet once it exits the orifice. The quality of the orifice edge is also crucial in producing a coherent waterjet. Consequently, the orifice must be protected from foreign material in the water.

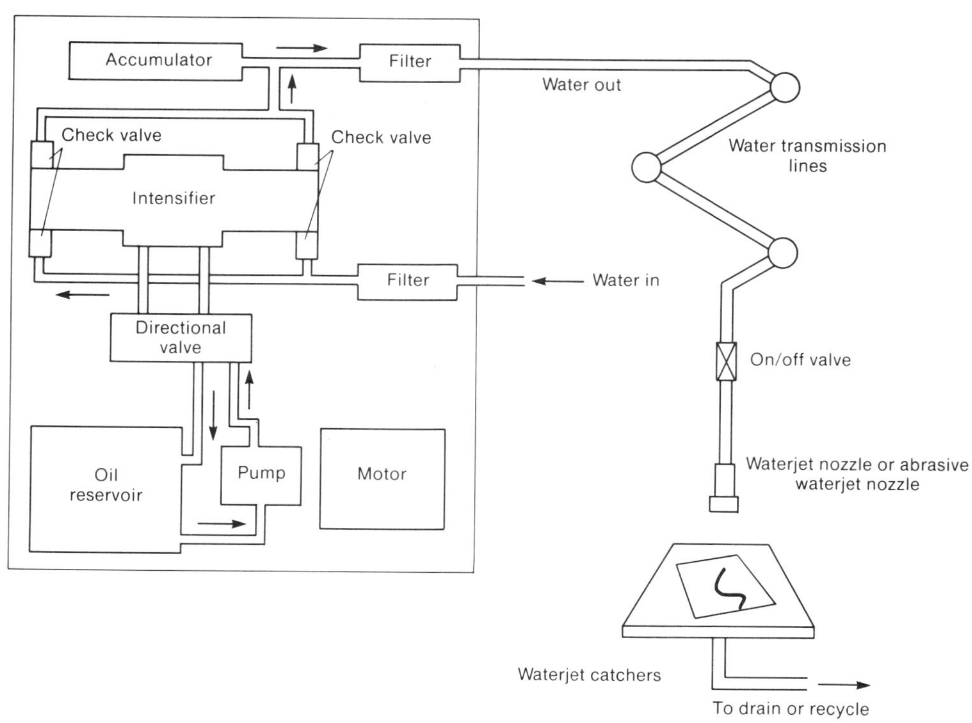


**Fig. 1** Schematic of a WJM system

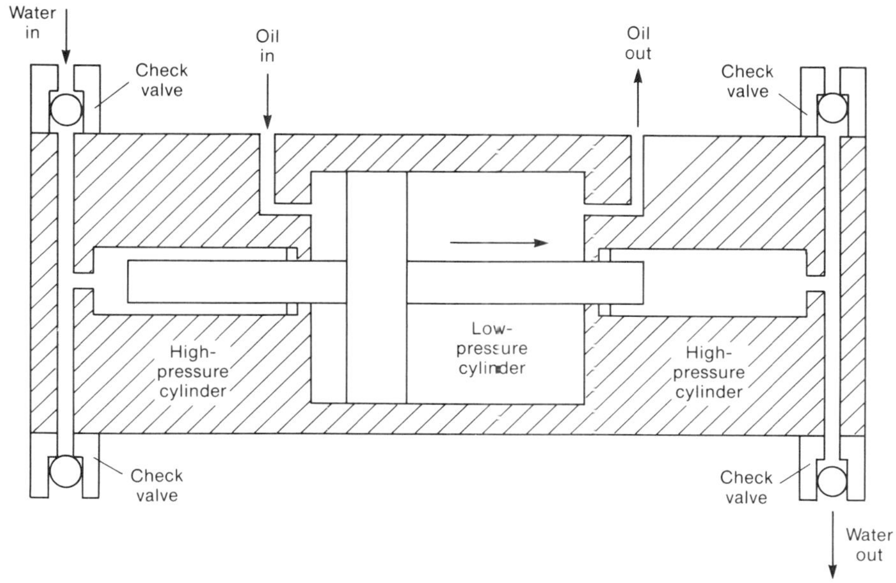


**Fig. 2** Schematic of a WJM intensifier

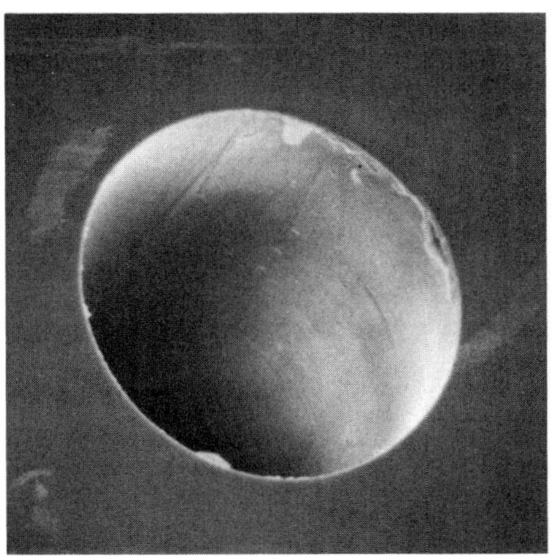

(a)

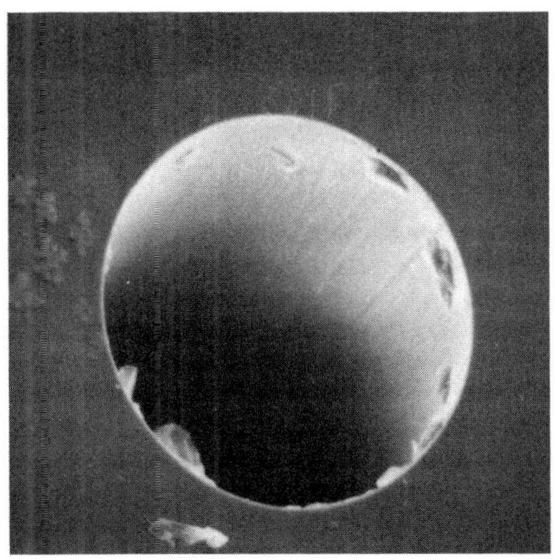

(b)

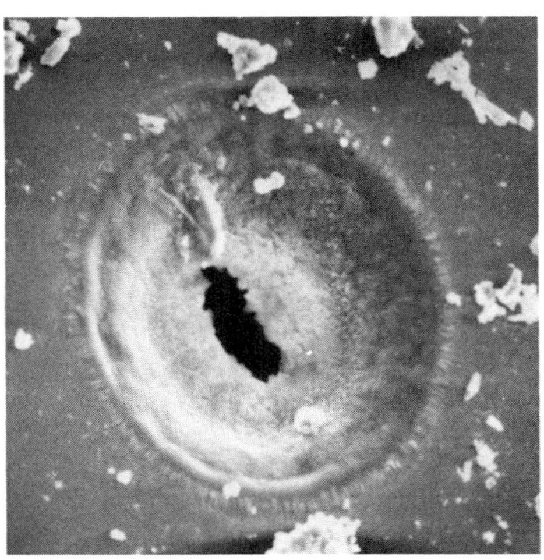

(c)

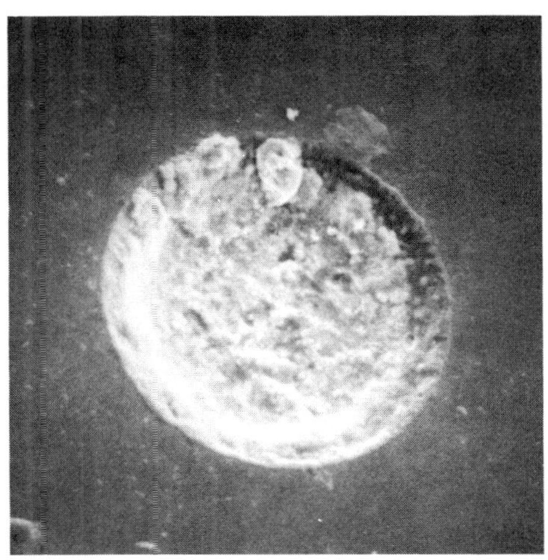

(d)

**Fig. 3** Orifice after testing in four types of water. (a) Deionized water, after 200 h. (b) Water treated by reverse osmosis, after 200 h. (c) Softened water, after 104 h. (d) Tap water, after 48 h. The operating lives of the orifices averaged 34 h with tap water, 78 h with softened water, 190 h with water treated by reverse osmosis, and 200 h or more with deionized water.

The orifice itself is most often produced from man-made sapphire. Orifices as small as 0.075 mm (0.003 in.) are available. More recently, sapphire orifices are being replaced with diamond orifices. A diamond orifice will normally operate ten times as long as the sapphire orifice before replacement, thus reducing the number of shutdowns for service. In addition, a diamond orifice can be cleaned several times and will continue to produce a coherent high-pressure jet. However, the cost of a diamond orifice is seven to ten times that of a sapphire orifice.

**Abrasive Waterjet Nozzle.** The only differences between an abrasive WJM system and a standard WJM system are the addition of an abrasive jet nozzle and an abrasive feed mechanism. Figure 5 illustrates a nozzle for abrasive WJM operations. The abrasive is added to the high-pressure waterjet in a mixing chamber. When the high-velocity waterjet passes through the mixing chamber, the high flow creates an area of lower pressure, which draws the abrasive from a delivery line. Abrasive flow is controlled with a metering mechanism. At the mixing chamber, the abrasive is randomly mixed with the high-pressure waterjet, which is then refocused through a secondary nozzle. Orifice sizes of the secondary nozzle typically range from 0.75 to 2.5 mm (0.030 to 0.100 in.) in diameter. The power of the waterjet ranges from 7 to 45 kW (10 to 60 hp) at the nozzle. The abrasives typically used in this process are garnet, olivine sand, and in some cases (especially in cleaning applications) silica sand.

**Waterjet Catchers.** The waterjet (or the abrasive waterjet) must be trapped and dissipated as it exits the material being cut. Containers normally referred to as catchers are used for this purpose. Catchers also reduce the noise associated with the breakup of the jet after it passes through the target material. Mist or splashback can be a problem if proper catcher systems are not used.

Figure 6 shows a waterjet catcher. The catcher must be deep enough so that the waterjet breaks up before reaching the bot-

**Table 1 Flow rates with a 0.7 orifice coefficient**

| Water pressure, MPa | Water pressure, ksi | Orifice diameter, mm | Orifice diameter, in. | Flow rate, L/min | Flow rate, gal./min |
|---|---|---|---|---|---|
| 380 | 55 | 0.15 | 0.006 | 0.64 | 0.17 |
| | | 0.20 | 0.008 | 0.79 | 0.31 |
| | | 0.30 | 0.012 | 2.6 | 0.70 |
| | | 0.35 | 0.014 | 3.6 | 0.95 |
| 275 | 40 | 0.15 | 0.006 | 0.57 | 0.15 |
| | | 0.20 | 0.008 | 0.98 | 0.26 |
| | | 0.30 | 0.012 | 2.2 | 0.60 |
| | | 0.35 | 0.014 | 3.1 | 0.81 |
| 200 | 30 | 0.15 | 0.006 | 0.49 | 0.13 |
| | | 0.20 | 0.008 | 0.87 | 0.23 |
| | | 0.30 | 0.012 | 2.0 | 0.52 |
| | | 0.35 | 0.014 | 2.6 | 0.70 |

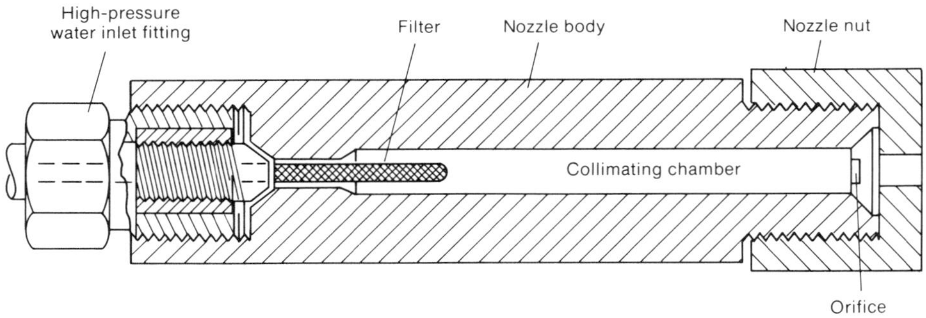


**Fig. 4** Schematic of a waterjet nozzle assembly

tom. This normally requires a depth of 300 to 600 mm (12 to 24 in.), but shorter tanks with steel balls or replaceable hard inserts at the bottom can be used where space is limited.

**Fluid additives** are sometimes used to aid coherency of the waterjet. Long-chain polymers are used as additives. However less than 1% of the installations worldwide use these additives.

## Process Characteristics

Several variables influence the WJM process, and empirical testing is the best way to determine the effects of waterjet-machining and abrasive waterjet machining on a given workpiece. The important parameters of the process are:

- Pressure, flow, and nozzle diameter
- Standoff distance
- Traverse rate
- Type and size of abrasive (if used)

**Pressure, flow, and nozzle diameter** are the main parameters in WJM fluid flow. Based on Bernoulli's equation for incompressible fluid flow (Eq 2), the volumetric flow rate, $Q$, in a WJM system is:

$$Q = 24C_D D^2 \sqrt{0.22P/\rho} \qquad \text{(Eq 1)}$$

where $C_D$ is the orifice coefficient, $D$ is the orifice diameter, $P$ is the fluid pressure, and $\rho$ is the average fluid density. Figure 7 illustrates the pressure and flow relationships for various nozzle diameters with an orifice coefficient of 1.0. The typical orifice coefficient for sapphire or diamond orifices is 0.7. Table 1 lists typical pressures and flow rates with an orifice coefficient of 0.7. The pressure-flow values of Table 1 can be determined by multiplying the flow rates given in Fig. 7 by 0.7. Figure 7 also provides the necessary pump size for a given flow rate and pressure.

*Effects of Cutting*. The flow rate and the pressure of the waterjet have different effects on cutting. The flow rate affects the rate of material removal, while the pressure affects not only the rate of material removal but also the cutting mechanism, which shears material away from the workpiece. This is apparent because high flow rates at low pressure will not cut. However, if the

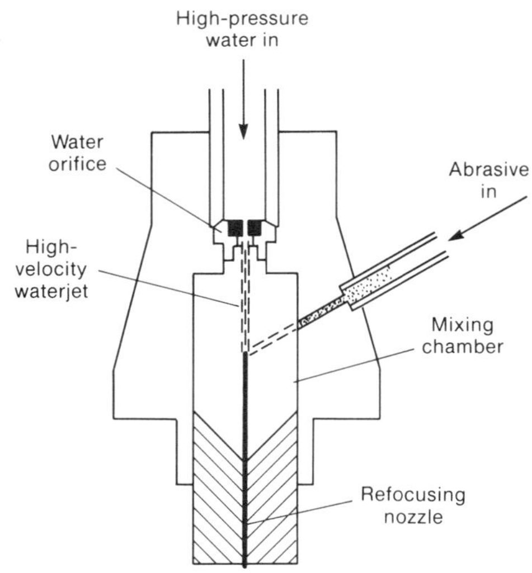


**Fig. 5** Schematic of an abrasive waterjet nozzle

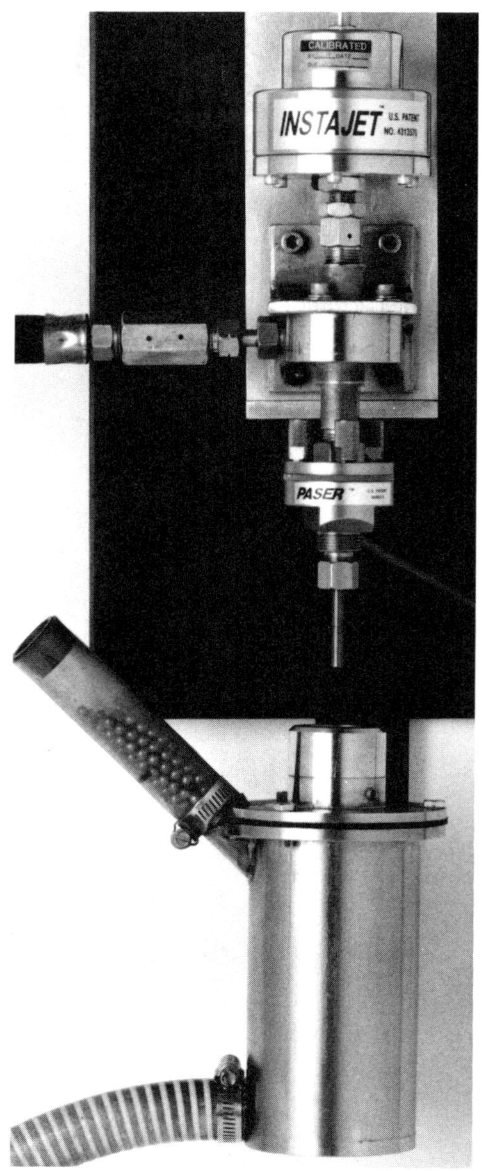


**Fig. 6** Waterjet assembly with waterjet catcher

**Table 2 Water jet cutting rates (without abrasives)**

| Material | Thickness mm | Thickness in. | Cutting rate mm/s | Cutting rate in./s | Nozzle diameter mm | Nozzle diameter in. | Jet pressure MPa | Jet pressure ksi |
|---|---|---|---|---|---|---|---|---|
| Corrugated board | 7 | 0.28 | 3300 | 130 | 0.25 | 0.001 | 385 | 56 |
| Sand paper (120 grit) | 1 layer | | 2030 | 80 | 0.13 | 0.005 | 275 | 40 |
| Rubber-backed carpet | 18 | 0.71 | 120 | 4.7 | 0.15 | 0.006 | 380 | 55 |
| Finished leather | 1 | 0.04 | 1270 | 50 | 0.13 | 0.005 | 275 | 40 |
| Asbestos board | 18 | 0.71 | 1520 | 60 | 0.20 | 0.008 | 190 | 28 |
| Insulation board | 50 | 2.0 | 1650 | 65 | 0.20 | 0.008 | 190 | 28 |
| Rubber tile | 3 | 0.12 | 150 | 6 | 0.13 | 0.005 | 380 | 55 |
| Urethane | 2 | 0.08 | 100 | 4 | 0.10 | 0.004 | 380 | 55 |
| Glass insulation | 300 | 12.0 | 420 | 16.5 | 0.15 | 0.006 | 350 | 50 |
| Polycarbonate | 2 | 0.08 | 20 | 0.8 | 0.15 | 0.006 | 380 | 55 |
| Polypropylene | 2 | 0.08 | 60 | 2.4 | 0.10 | 0.004 | 380 | 55 |
| Polyester | 12 | 0.47 | 600 | 24 | 0.15 | 0.006 | 380 | 55 |
| Polyvinyl chloride | 0.75 | 0.03 | 300 | 12 | 0.10 | 0.004 | 380 | 55 |
| Printed circuit boards (FR-4) | 1.5 | 0.06 | 50 | 2 | 0.15 | 0.006 | 380 | 55 |
| Sheet molding compound | 3 | 0.12 | 1000 | 40 | 0.15 | 0.006 | 350 | 50 |
| Graphite composite | 1.6 | 0.06 | 10 | 0.4 | 0.15 | 0.006 | 380 | 55 |
| Glass/epoxy laminate | 1.7 | 0.07 | 40 | 1.6 | 0.20 | 0.008 | 385 | 56 |
| Graphite/epoxy laminate | 3 | 0.12 | 30 | 1.2 | 0.20 | 0.008 | 385 | 56 |
| Peanut bar | 10 | 0.39 | 100 | 4 | 0.13 | 0.005 | 275 | 40 |
| Granola | 12 | 0.47 | 100 | 4 | 0.10 | 0.004 | 275 | 40 |
| Kidney filter | 100 | 4.0 | 100 | 4 | 0.30 | 0.012 | 380 | 55 |
| Formica | 1 | 0.04 | 600 | 24 | 0.13 | 0.005 | 350 | 50 |
| Foam board | 50 | 2.0 | 100 | 4 | 0.13 | 0.005 | 380 | 55 |

Source: Ref 1

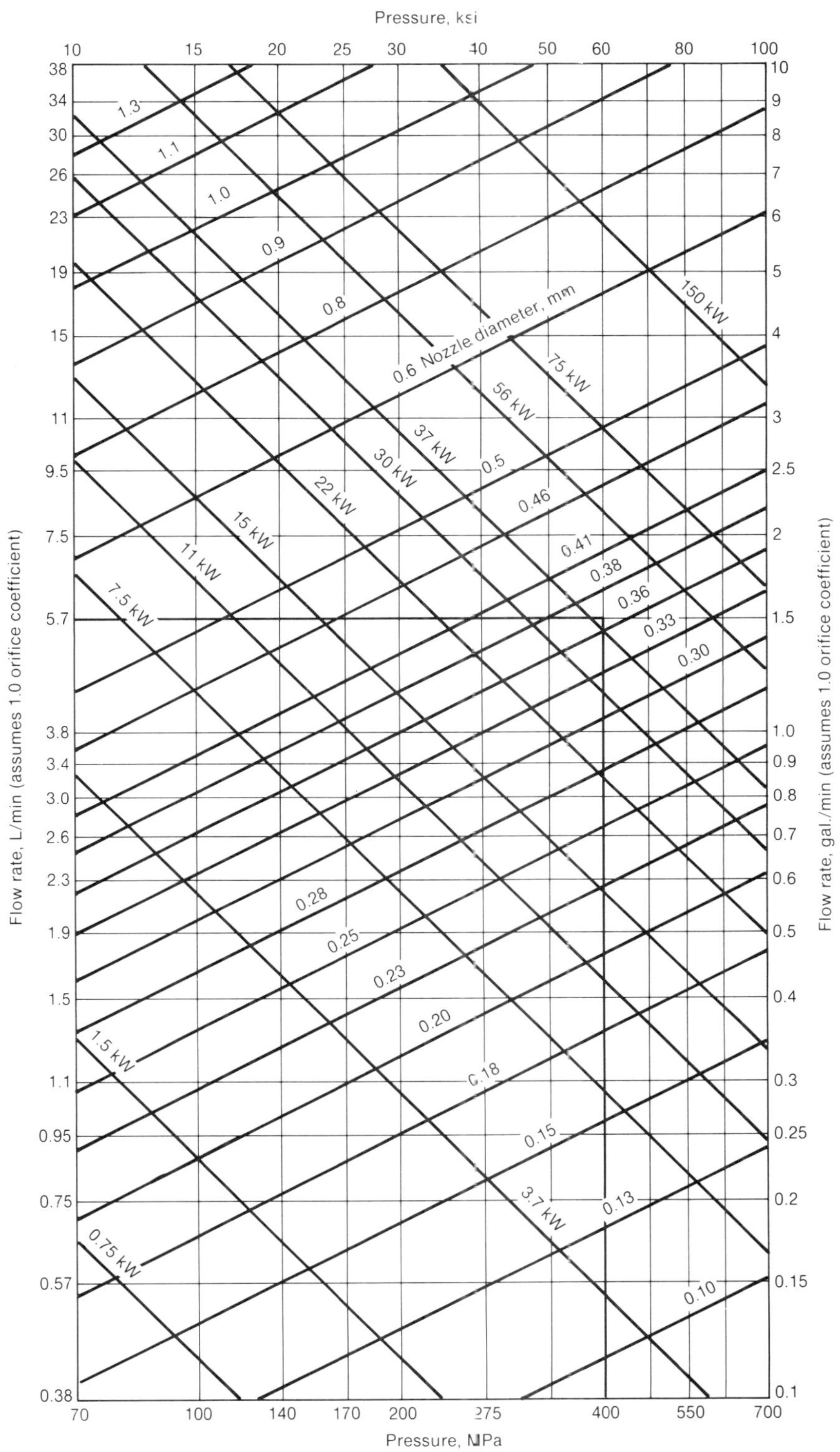


**Fig. 7** Pressure versus flow rate with various WJM nozzle sizes

pressure is increased while the orifice is decreased to keep the flow rate constant, the amount of cutting will increase. This suggests that a relationship exists between the pressure and the cutting mechanism.

The role of waterjet pressure in the WJM process can be demonstrated by considering Bernoulli's equation for incompressible fluid flow:

$$V^2 = \frac{2P}{\rho} \quad \text{(Eq 2)}$$

where $V$ is the velocity of the waterjet. The pressure ($P$) is proportional to the square of the velocity ($V^2$), which is also proportional to the kinetic energy ($\frac{1}{2}mV^2$) of a particle with mass $m$. Consequently, the pressure of the waterjet determines the average kinetic energy of a particle in the waterjet. At higher pressures, the average kinetic energy of the water molecules is higher and is more likely to overcome the molecular binding forces of the workpiece material. Therefore, the waterjet pressure required for cutting must increase as the strength of the workpiece increases.

If the waterjet pressure is adequate for cutting a given material, and is held constant, the rate of material removal is determined by the flow rate. Table 2 lists typical waterjet cutting rates for various materials. Increases in either the pressure or the orifice diameter produce a larger flow rate, which increases the cutting rate (Fig. 8) or the ability to cut thicker workpieces by increasing the total energy available at the nozzle. Larger pressures also increase the energy density of the waterjet.

*Pressure Drop*. In most waterjet cutting installations, the pressure drop has little effect on system operation. However, when materials such as printed circuit boards are being cut, the pressure drop becomes critical because the pressure at the nozzle must remain between 360 and 380 MPa (52 and 55 ksi). Furthermore, in abrasive waterjet machining, the nozzles used are generally larger, and a significant pressure drop can occur with higher flow rates. Pressure drops reduce available energy at the nozzle, thus reducing cutting efficiency.

**Standoff Distance.** The distance between the nozzle and the workpiece (the standoff) is generally 2.5 to 6.35 mm (0.10 to 0.25 in.). Because there is little change in the shape or diameter of the jet within 25 mm (1.0 in.), the standoff can be increased to 25 mm (1.0 in.) without significantly affecting material removal. Standoffs should be increased (up to about 50 mm, or 2 in.) for materials susceptible to shattering or delamination (such as printed circuit boards) or for applications requiring a smoother cut.

**Traverse Rate.** Thicker and denser materials can be cut as the traverse rate decreases. Materials that are too thick to be cut in a single pass can be cut with multiple passes if the first pass can produce a well-defined slot.

**Abrasives,** when used, also affect the cutting capabilities of the WJM process. Tables 3 and 4 list the cutting rates of abrasive waterjet machining for various materials and operating parameters.

## Advantages and Disadvantages

The advantages of waterjet and abrasive waterjet cutting include the following:

- Omnidirectional cutting with a zero radius for outside corners and a radius equal to the radius of the jet for inside corners
- Reduced tooling costs because of the minimal force (5 to 130 N, or 1 to 30 lbf) imparted to the workpiece

**Table 3 Effect of material thickness on abrasive waterjet cutting speed for various metals and a ceramic material**

Abrasive used is garnet.

| Parameters | | | | | | Maximum cutting rate, mm/min (in./min) | | | | | | | | |
|---|---|---|---|---|---|---|---|---|---|---|---|---|---|---|
| Diameters, mm (in.) | | Abrasive | | | | Metals | | | | | | | | Ceramic |
| Orifice | Nozzle | Feed rate kg/min (lb/min) | Mesh size | Nozzle pressure MPa (ksi) | Material thickness mm (in.) | Aluminum and aluminum alloys | Brass | Carbon steel | Copper | Alloy 718 | Stainless steel | Titanium | Tool steel 38 HRC | 99.6% Aluminum oxide |
| 0.23 (0.009) | 0.79 (0.031) | 0.23 (0.5) | 100 | 310 (45) | 0.8 (0.031) | 4570 (180) | 1270 (50) | 1520 (60) | 1270 (50) | 1520 (60) | 1140 (45) | 2030 (80) | 890 (35) | 127 (5) |
| | | | | | 1.6 (0.063) | 2030 (80) | 762 (30) | 1270 (50) | 1020 (40) | 1140 (45) | 762 (30) | 1520 (60) | 762 (30) | 61 (2.4) |
| | | | | | 3.2 (0.125) | 1270 (50) | 457 (18) | 762 (30) | 559 (22) | 559 (22) | 610 (24) | 1140 (45) | 635 (25) | 38 (1.5) |
| 0.33 (0.013) | 1.19 (0.047) | 0.68 (1.5) | 80 | 240 (35) | 6.4 (0.250) | 762 (30) | 254 (10) | 508 (20) | 305 (12) | 305 (12) | 483 (16) | 762 (30) | 432 (17) | 23 (0.9) |
| | | | | | 12.7 (0.500) | 457 (18) | 102 (4) | 305 (12) | 152 (6) | 152 (6) | 254 (10) | 457 (18) | 330 (13) | 15 (0.6) |
| 0.46 (0.018) | 1.19 (0.047) | 0.91 (2.0) | 80 | 240 (35) | 19.0 (0.750) | 305 (12) | 25 (1) | 203 (8) | 75 (3) | 75 (3) | 152 (6) | 305 (12) | 254 (10) | 8 (0.3) |
| | | | | | 25.4 (1.00) | 203 (8.0) | 13 (0.5) | 152 (6) | 38 (1.5) | 38 (1.5) | 102 (4) | 152 (6) | 191 (7.5) | ... |
| 0.56 (0.022) | 1.57 (0.062) | 1.46 (3.2) | 60 | 240 (35) | 50.8 (2.00) | 152 (6.0) | 8 (0.3) | 75 (3) | 15 (0.6) | 5 (0.2) | 57 (2.25) | 75 (3) | 127 (5) | ... |
| | | | | | 76.2 (3.00) | 127 (5.0) | 5 (0.2) | 50 (2) | 8 (0.3) | 3 (0.1) | 38 (1.5) | 50 (2) | 50 (2) | ... |
| | | | | | 102 (4.00) | 102 (4.0) | 3 (0.1) | 25 (1) | 3 (0.1) | ... | 25 (1) | 25 (1) | 25 (1) | ... |

Source: Flow Systems, Inc.

**Table 4 Effect of material thickness on abrasive waterjet cutting speed for nonmetallic materials**

Abrasive used is garnet.

| Parameters | | | | | | Maximum cutting rate, mm/min (in./min) | | | | | | | | | |
|---|---|---|---|---|---|---|---|---|---|---|---|---|---|---|---|
| Diameters, mm (in.) | | Abrasive | | | | Plastics and composites | | | | | | | Glass | | |
| Orifice | Nozzle | Feed rate kg/min (lb/min) | Mesh size | Nozzle pressure MPa (ksi) | Material thickness mm (in.) | Acetal | Acrylic | Carbon/carbon composite | Epoxy/glass composite | Graphite/epoxy composite | Aramid fiber composite | Poly-propylene | Laminate | Plate | Stained |
| 0.23 (0.009) | 0.79 (0.031) | 0.23 (0.5) | 100 | 310 (45) | 0.8 (0.031) | 3180 (125) | 3050 (120) | 2540 (100) | 6350 (250) | 4450 (175) | 2540 (100) | 2540 (100) | ... | 7620 (300) | 7620 (300) |
| | | | | | 1.6 (0.063) | 2290 (90) | 2030 (80) | 1910 (75) | 5720 (225) | 3810 (150) | 1520 (60) | 1910 (75) | ... | 6350 (250) | 6350 (250) |
| | | | | | 3.2 (0.125) | 1780 (70) | 1400 (55) | 1400 (55) | 4570 (180) | 3180 (125) | 1020 (40) | 1220 (48) | 660 (26) | 5080 (200) | 5080 (200) |
| | | | | | 6.4 (0.250) | 1270 (50) | 915 (36) | 1020 (40) | 2540 (100) | 2540 (100) | 510 (20) | 915 (36) | 559 (22) | 3810 (150) | 3810 (150) |
| | | | | | 12.7 (0.500) | 890 (35) | 457 (18) | 508 (20) | 1020 (40) | 1270 (50) | 279 (11) | 610 (24) | 457 (18) | 2540 (100) | 2540 (100) |
| 0.33 (0.013) | 1.19 (0.047) | 0.68 (1.5) | 80 | 240 (35) | 19.0 (0.750) | 610 (24) | 305 (12) | 254 (10) | 711 (28) | 635 (25) | 152 (6) | 381 (15) | 305 (12) | 1270 (50) | 1270 (50) |
| | | | | | 25.4 (1.00) | 381 (15) | 254 (10) | 127 (5) | 559 (22) | 508 (20) | 75 (3) | 203 (8) | 203 (8) | 635 (25) | ... |
| 0.46 (0.018) | 1.19 (0.047) | 0.91 (2.0) | 80 | 240 (35) | 50.8 (2.00) | 152 (6) | 102 (4) | 20 (0.8) | 305 (12) | 254 (10) | 25 (1.0) | 102 (4) | 127 (5) | 381 (15) | ... |
| | | | | | 76.2 (3.00) | 64 (2.5) | 25 (1.0) | 8 (0.3) | 127 (5) | 127 (5) | 13 (0.5) | 50 (2) | 25 (1) | 127 (5) | ... |
| | | | | | 102 (4.00) | 25 (1) | 8 (0.3) | 3 (0.1) | 25 (1) | 50 (2) | 3 (0.1) | 38 (1.5) | 20 (0.8) | 50 (2) | ... |

Source: Flow Systems, Inc.

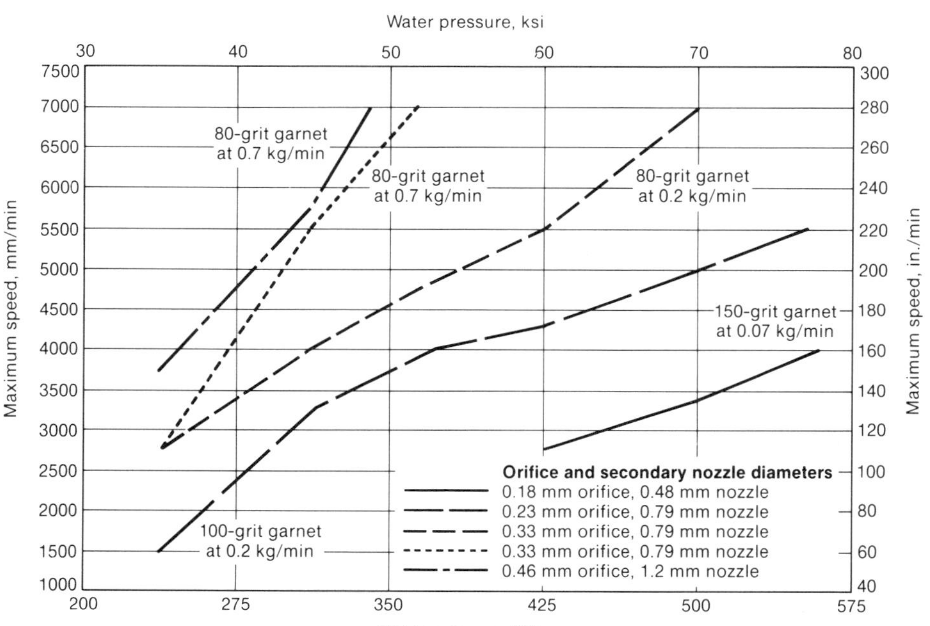


**Fig. 8** Maximum abrasive waterjet cutting speed versus pressure. The workpiece is 1.6 mm (0.63 in.) thick aluminum.

- Predrilling is not required when shape cutting
- No work hardening or thermal effects
- Reduction or elimination of airborne dust due to cutting
- Capable of cutting a variety of materials without a substantial change in system components
- Easily integrated with automated systems
- Multilayer cutting while maintaining quality of cut for all layers
- Small kerf width

The primary disadvantage is that mainly nonmetallic materials are effectively cut with the WJM process. This disadvantage is partly addressed by abrasive waterjet machining, which can cut harder and denser materials.

## Waterjet Applications

Waterjet machining has a wider variety of applications when lower strength materials are involved. Applications include the following:

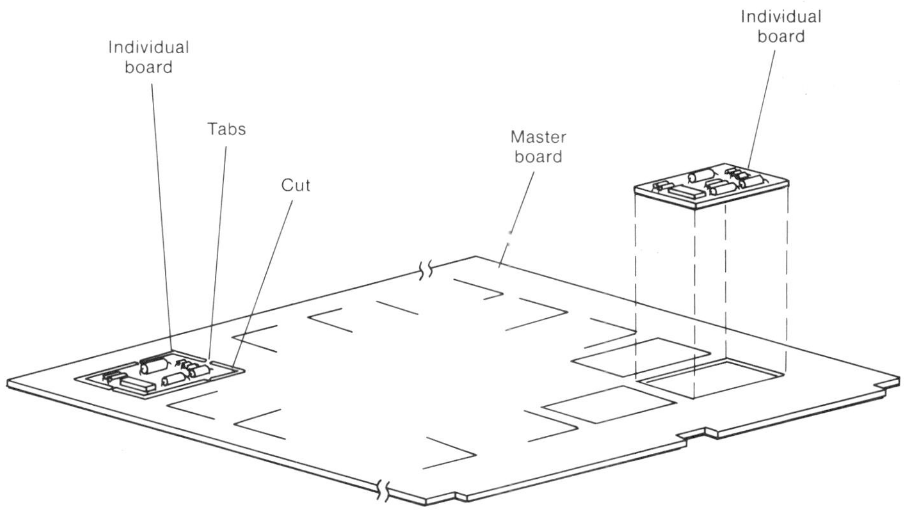


**Fig. 9** Circuit board cuts by the tab-cut method (left) and the full-cut method (right)

- Cutting of printed circuit boards
- Cutting of board materials
- Cutting of lightweight fiber-reinforced plastics
- Integration into robotic systems
- Wire stripping
- Crosscutting
- Cutting foods
- Cutting web materials

**The cutting of printed circuit boards** demonstrates many of the advantages of waterjet machining. The process typically involves the removal of several individual boards from one master board, either by a tab-cut method (Fig. 9, left) or a full-cut method (Fig. 9, right). These methods have been devised primarily to increase the ability of circuit board manufacturers to automate production.

Waterjet machining systems provide an easily automated process that can cut populated printed circuit boards of various sizes and shapes. The small diameter of the waterjet enables it to pass adjacent to any components mounted near the edge of the board. In addition, waterjet machining introduces little or no longitudinal or lateral cutting forces with respect to the jet when printed circuit boards are being cut. This enables the manufacturer to suspend the master board from its perimeter and use a small number of interior supports to prevent vibration during the cutting process. In this application, cutting speeds range from 1 to 8 m/min (3 to 26 sfm), with the material characteristics of the board determining the maximum cutting speed. Waterjets mounted on X-Y tables have contouring speeds of 25 m/min (82 sfm) and an accuracy of ±0.13 mm (±0.005 in.).

One concern about using a waterjet for cutting circuit boards is the introduction of water between the laminates. An extensive study performed by an independent testing service concluded that waterjets have no detrimental effect on printed circuit boards (Ref 2).

Another concern is noise. Catchers have been produced that eliminate the objectionable noise levels. In some cases, the entire machine has been enclosed when an open catcher capable of the required noise reduction could not be produced.

**Cutting of Board Materials.** Waterjets apply low forces to board materials (or any type of web material) during the cut. This can produce a higher-quality edge. Figure 10 shows the difference in edge quality in cutting double-wall corrugated board with circular knives and with high-pressure waterjets. The cutting rate can often be doubled while maintaining a clean, uncrushed edge in the finished product. Waterjets can also eliminate shutdown due to jam-ups caused by slitting knives.

**Robotics.** Waterjet machining can be easily integrated into robotic systems. A robotic application in the automotive industry is shown in Fig. 11. A minimal amount of tooling is necessary when trimming and cutting the lightweight fiber-reinforced plastic shown. If standard machining processes were used (for example, a router), the edge would need to be supported very rigidly up to the cut line to avoid cracking the part and breaking the cutting tool. These problems are eliminated with the waterjet because of the low forces exerted on the material being cut. The cut part is illustrated in Fig. 12.

**Wire Stripping.** The ability of the waterjet to cut insulating materials but not metals makes it useful in wire stripping. The waterjet cuts extremely fast, and there is less scrap due to wire damage. With waterjet machining, the 40 man-hours necessary for stripping a certain lot of wires is reduced to 8 man-hours. The waterjet will also cut the wire insulation without marring the tinning on the copper wire.

**The crosscutting** of continuous-web material that cannot be stopped during the cutting procedure is easily accomplished with a waterjet. Two types of crosscutting systems are currently used. The first process, often referred to in traditional mechanical cutting systems as flying cutoff, involves a waterjet nozzle support mechanism that is clamped to

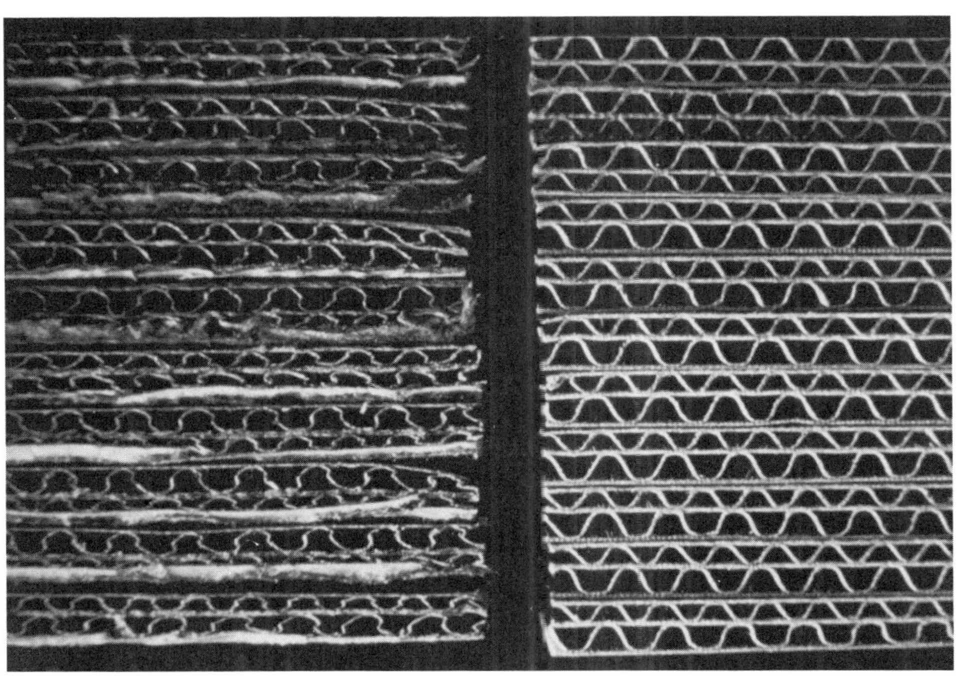

**Fig. 10** Corrugated board cut with circular knives (left) and waterjets (right)

**Fig. 11** Waterjet system used with robotic instrumentation

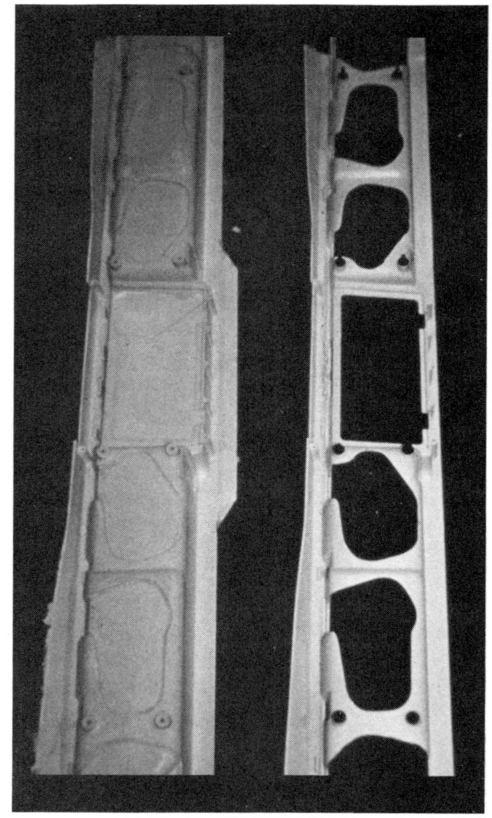

**Fig. 12** Cutouts performed with waterjets

a material being moved on a conveyor. While clamped and moving with the material, the jet is driven across the web. After crossing the web, the mechanism releases its clamp on the material and returns to its rest position. The mechanism remains at the rest position until it receives the signal to clamp the moving material and make another cut. A typical flying crosscutter mechanism is shown in Fig. 13.

The second type of crosscutting mechanism is a slant-bar crosscutter. In this mechanism, the path of the waterjet is not perpendicular to the movement of the web; it is set at an angle, or slant (Fig. 14). This slanted motion of the waterjet compensates for the motion of the web material and results in a perpendicular cut. The systems currently in operation monitor the web speed and vary the crosscutting speed to yield a perpendicular cut. Both types of crosscutters greatly reduce the material jam-ups and breaks that occur when using mechanical knives, shears, and saws.

**Cutting of Foods.** Conventional mechanical cutting methods tend to crack or break the food being cut. The low forces applied to the material by a waterjet can reduce breakage from 40 to 60% to less than 2%.

**Cutting of Web Material.** Baby diaper manufacturers are currently the largest users of waterjet systems as a means for the contour-cutting of web material. A continuously moving web is fed into the system at high speeds. The waterjet nozzle moves laterally in and out of the web area, cutting the desired shape in the web. The waterjet nozzle is normally controlled with a cam linkage, which is driven by the conveyor.

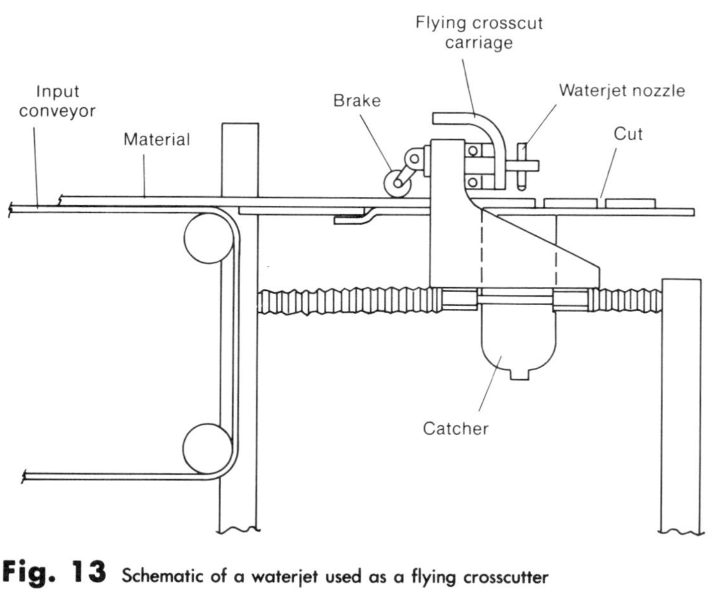


**Fig. 13** Schematic of a waterjet used as a flying crosscutter

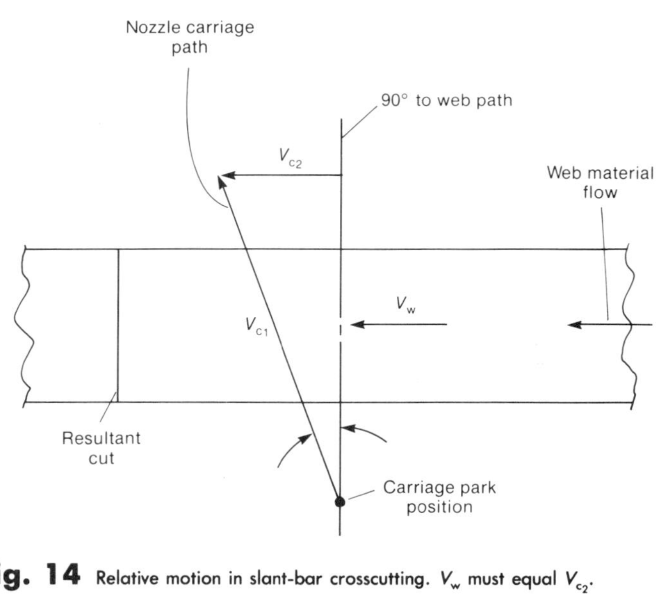


**Fig. 14** Relative motion in slant-bar crosscutting. $V_w$ must equal $V_{c2}$.

(a)

(b)

**Fig. 15** Abrasive waterjet machining of an Inconel disk (c) to produce turbine wheel blades (b)

## Abrasive Waterjet Applications

Abrasive waterjet cutting not only possesses the versatility of waterjet cutting but also extends the applications to harder and denser workpieces. The addition of abrasives allows the cutting of metals, glass, and ceramics.

**Cutting of Tool Steel.** Abrasive waterjet cutting compared favorably with traditional machining and electrical discharge machining in the production of a 100 mm (4 in.) diam hole in a 50 mm (2 in.) thick tool steel. Both electrical discharge machining and abrasive waterjet machining could process this particular material in a heat-treated state. Both methods also produced a useful slug of material as a by-product.

**Cutting of Inconel.** A 760 mm (30 in.) diam turbine wheel was machined with an abrasive waterjet from a solid, 45 mm (1.75 in.) thick disk of Inconel (Fig. 15). The objective was to remove the material from between the turbine blades. Final shaping was performed by electrochemical machining. The average cutting time was 48 h.

**Fatigue test coupons** made of titanium, Cres 321, and Inconel were cut at a rate of 250 mm/min (10 in./min). Size tolerance for the lot of 45 pieces was ±0.025 mm (±0.001 in.).

**Carbon fiber,** because of its strength and high abrasive content, is extremely difficult to cut with traditional machine tools. However, with the abrasive waterjet, the cut is consistent throughout the operation. Abrasive waterjet machining also eliminates the sanding, filing, and other operations that follow the traditional machining of this type of material. In one application, the abrasive WJM process significantly reduced tooling costs in the machining of irregularly shaped wing ribs made of carbon fiber.

### REFERENCES

1. T.J. Labus and R. Pilarski, Fluid Jet Technology for Industrial Applications, in *Machining Source Book*, ASM INTERNATIONAL, 1988, p 315
2. "Independent Evaluation Comparison of Through Systems Used to Profile Printed Wiring Boards—Waterjet Cut, Routed, and Sheared," Report 3512, Trace Laboratories, 1983

### SELECTED REFERENCES

- G.F. Benedict, *Nontraditional Manufacturing Processes*, Marcel Dekker, 1987
- *Proceedings of the Seventh International Symposium on Jet Cutting Technology*, British Hydrodynamic Research Association, 1984

# Ultrasonic Machining

William R. Tyrrell, Branson Ultrasonics Corporation

ULTRASONIC MACHINING (USM) is a process that utilizes the ultrasonic (~20 kHz) vibration of a tool in the machining of hard, brittle, nonmetallic materials. The USM process consists of two methods:

- *Ultrasonic impact grinding*: Involves an abrasive slurry and the ultrasonic vibration of a nonrotating tool
- *Rotary ultrasonic machining*: Involves the ultrasonic vibration of a rotating diamond core drill or milling tool

In ultrasonic impact grinding, an abrasive slurry flows through a gap between the workpiece and the vibrating tool (Fig. 1). Material removal occurs when the abrasive particles, suspended in the slurry, are struck on the downstroke of the vibrating tool. The velocity imparted to the abrasive particles causes microchipping and erosion as the particles impinge on the workpiece. The amplitude of vibration in ultrasonic impact grinding varies from 0.025 to 0.09 mm (0.001 to 0.0035 in.). The process forms a cavity in the shape of the tool.

Rotary ultrasonic machining is similar to the conventional drilling of glass and ceramic with diamond core drills, except that the rotating core drill is vibrated at an ultrasonic frequency of 20 kHz. Rotary ultrasonic machining does not involve the flow of an abrasive slurry through a gap between the workpiece and the tool. Instead, the tool contacts and cuts the workpiece, and a liquid coolant, usually water, is forced through the bore of the tube to cool and flush away the removed material. The amplitude of vibration in rotary ultrasonic machining is usually about 0.025 to 0.05 mm (0.001 to 0.002 in.), and the longitudinal extension and contraction of the drill tip improves drilling performance by:

- Reducing the friction between the tool and the material at the point of cutting
- Eliminating seizure of the core piece by allowing greater coolant flow in the bore of the tool
- Preventing the diamond tool from loading with removed material
- Increasing drilling speed with less pressure on the tool and reduced tool wear

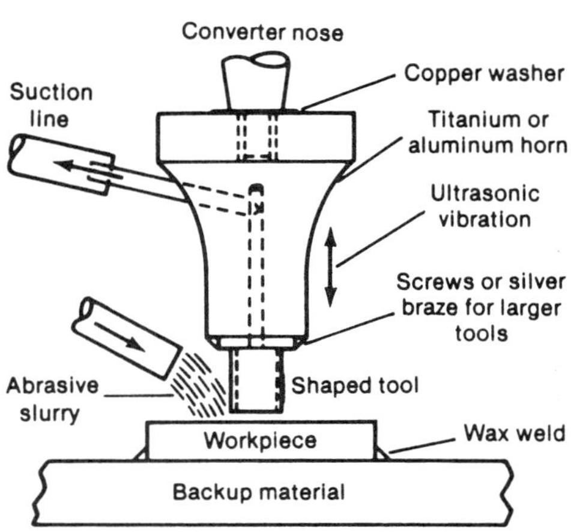


**Fig. 1** Principal elements of ultrasonic impact grinding

## Equipment

Rotary ultrasonic machining and ultrasonic impact grinding both use similar ultrasonic components (Fig. 2). The major components include:

- A power supply to provide electrical output at an ultrasonic frequency of 20 kHz. A 450-W power supply is used for rotary machining. Impact machining will use power supplies from 150 to 2500 W as tools with large surface areas require higher power than those with small surface areas
- A converter to change the electrical energy into mechanical vibrations. Two types of converters are available for USM systems. One type uses a magnetostrictive device, and the other a quartz or a lead zirconate titanate piezoelectric transducer
- A horn to focus the vibratory energy on the tool. Horns as large as 130 mm (5 in.) in diameter have been used, and the horn is very accurately machined to be axially resonant at 20 kHz in order to develop the maximum amplitude of vibration at the tool
- A booster can be used between the converter and the horn to increase the amplitude of the ultrasonic vibration; boosters are available with different amplitude gain factors

**A rotary ultrasonic machine** is similar to a standard milling-drilling machine with an ultrasonic spindle mounted above an *X-Y* table (Fig. 3). The power supply, rotary machining head, and cutting tool constitute the basic system, and this system can be used on a conventional milling machine or can be attached to special machinery. If required, the rotary ultrasonic spindle can be used remotely on special or automatic equipment.

The ultrasonic spindle consists of a motor-driven bearing-mounted spindle with an ultrasonic converter (Fig. 4). The machine also has a pneumatic system for raising and lowering the spindle and a motor speed control with associated meters and indicators. The heart of the spindle is two lead zirconate titanate piezoelectric disks, which convert the electrical energy from the power supply into a mechanical vibration. These disks expand and contract in sympathy with the polarity of the voltage applied to them and are attached to a titanium alloy horn. In practice, the assembly is tuned to a frequency slightly higher than 20 kHz to enable it to accommodate the diamond tool attached to the tip.

Standard diamond core drills and mandrels are used, but to achieve the desired acoustic coupling, they are brazed to a

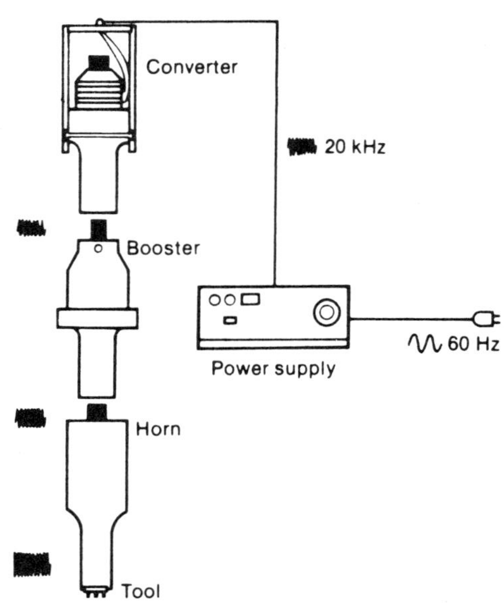


**Fig. 2** Schematic of components used in ultrasonic machining

**Fig. 3** Rotary ultrasonic machine tool

threaded adaptor that screws firmly into the end of the spindle. Mechanical coupling devices such as chucks and collets are acoustically inefficient and limit energy transfer. Because the horn and tool assembly must be resonant at 20 kHz to operate effectively, the size of the tools that can be used is limited. The maximum diameter is about 38 to 50 mm (1.5 to 2 in.), depending on wall thickness and length. Grinding and thread-cutting wheels up to 32 mm (1¼ in.) in diameter have also been successfully used. Typical core drills and wheels are shown in Fig. 5.

A water jacket is positioned at the bottom of the spindle and immediately above the tool. Water from a recirculating pump system or the main water supply is fed through the water jacket, into the center of the rotating spindle, and down through the bore of the core drill. When solid tools are used, water is applied directly at the drill/material interface.

**Ultrasonic impact grinding machines** are available in sizes that vary from a floor-standing machine for large parts to more compact units for light-duty, small-part applications (Fig. 6). The machine also includes a recirculating pump system for the abrasive slurry and vertical-axis feed rate controls. More sophisticated machines incorporate such features as:

- A programmable controller for the *X-Y* table
- Automatic compensation for tool wear to ensure repeatable depth of cut
- Complete computer numerical control (CNC) of all axes, and camera vision (Fig. 7)
- An ultrasonic frequency display to indicate any deviation from 20 kHz due to excessive tool wear or breakage
- Automatic control of the complete machining cycle

*The cutting tool*, custom shaped to the hole or cavity required, is most commonly made of type 304 stainless steel for durability. For low-volume production, brass can be used. Brass is easier to shape than stainless steel, but is less wear resistant. Tools are attached to the horn by one of two methods. If the tool is small, it can be machined from, or soldered to, a round threaded tip, which will be screwed into the end of the horn. Larger tools are silver soldered directly to the face of the horn. Ultrasonic impact grinding tools cost less than an equivalent diamond drill and can be purchased from suppliers of such machines or can be made by the user (given some understanding of the weight limitations and the attachment methods). Figure 8 shows selection of cutting tools.

*Horns* that accept threaded tool tips are usually made of titanium. Larger horns are made of aluminum. Both aluminum and titanium have high tensile strength.

*Abrasive Slurry*. Three types of abrasives are commonly used in the slurry: aluminum oxide, silicon carbide, and boron carbide. They are available in the following grit sizes:

| Number | Grit size mm | Grit size in. |
|---|---|---|
| 240 | 0.05 | 0.002 |
| 320 | 0.04 | 0.0015 |
| 400 | 0.03 | 0.0012 |
| 600 | 0.014 | 0.00057 |
| 800 | 0.009 | 0.00035 |

The abrasive and the grit size used will relate to the application; the larger the grit size, the faster the cutting but the coarser the surface finish. Small grit sizes cut slower but produce a better surface finish. A surface finish of 0.38 to 0.25 μm (15 to 10 μin.) can be expected from a 240-grit and an 800-grit abrasive, respectively, but such factors as the surface of the tool, tool amplitude, and the material being machined also affect surface finish. The abrasive material used should be harder than the workpiece. Aluminum oxide, for example, would be suitable for glass and the more brittle materials, while harder materials, such as sapphire, will require boron carbide, the hardest of the abrasive materials.

The slurry consists of 50 vol% abrasive particles and 50 vol% water. The slurry is pumped through a nozzle close to the tool/

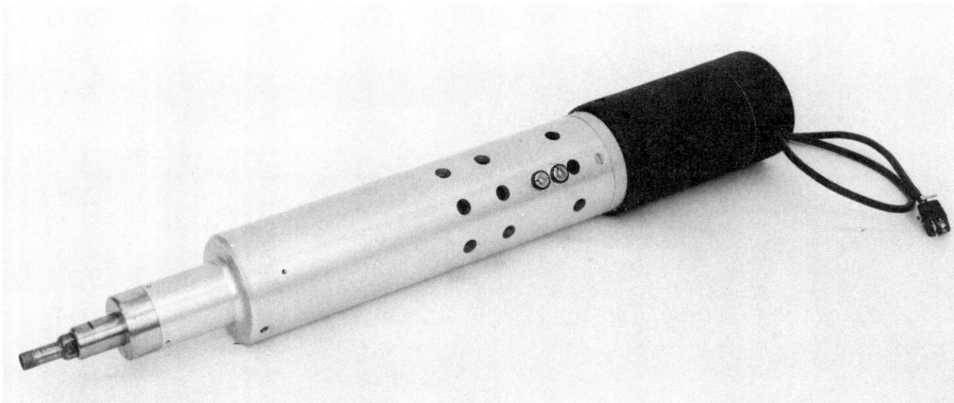

**Fig. 4** Rotary ultrasonic spindle

**Fig. 5** Diamond tools used in rotary ultrasonic machining

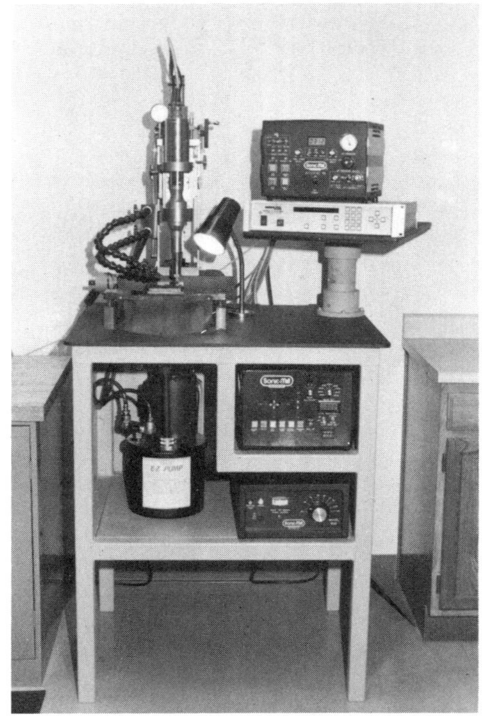

**Fig. 6** Ultrasonic impact grinding machine. Courtesy of Sonic Mill

workpiece interface (Fig. 9) at a rate of about 25 L/min (7 gal./min). In addition to its cutting action, the slurry provides cooling for the tool and the workpiece. The slurry will eventually become less effective as the particles wear and break down. Expected life is between 150 and 200 h of ultrasonic exposure.

## Rotary Ultrasonic Machining

Rotary ultrasonic machining is applied only to nonmetallic material, such as glass, alumina, ceramic, ferrite, quartz, zirconium oxide, ruby, sapphire, beryllium oxide, and some composite materials. Softer ductile materials clog the diamond tool to the point at which it will cease cutting.

The advantages of rotary ultrasonic machining are faster material removal, lower tool pressure for delicate parts, improved deep-hole drilling, less breakout on through holes, and no core seizing when core drilling. Drilled, milled, and threaded parts are shown in Fig. 10.

**Drilling** is the primary application of the ultrasonic rotary machine because it drills faster than a conventional drilling machine for a given tool pressure and revolution rate. For example, with a 120-grit, 13 mm (0.5 in.) diam core drill rotating at 3000 rev/min with a constant pressure of 83 kPa (12 psi), an ultrasonic unit will drill through 13 mm (½ in.) thick aluminum oxide (99.5%) in an average time of 27 s. In the same drilling operation under identical conditions without ultrasonics, the drill reached a depth of 6 mm (¼ in.) in 80 s and then stopped cutting. When the ultrasonics were activated, the tool again began to cut, and it rapidly reached the previous ultrasonic cutting rate. The accelerated cutting of ultrasonic-aided drilling is greater at low tool pressures than at high tool pressures.

Rotary ultrasonic machining also allows drilling at lower pressures; this is particularly advantageous when drilling or machining delicate components. Low tool pressure enables the drilling of holes without breakage and, if required, with only a few hundredths of a millimeter of separation. Holes can also be drilled close to an edge or corner without breakage.

The use of ultrasonics reduces the problem of edge cracking at the point of breakout when drilling through holes in brittle materials. The lighter pressure on the tool enables the cutting action to be maintained to a greater depth before the pushout occurs. Rigid clamping or backing of the workpiece will further minimize or eliminate breakout.

The drilling of straight small-diameter deep holes with conventional drilling methods can be difficult when excessive pressure causes the drill to wander from its vertical path. An ultrasonically vibrated drill can significantly improve the degree of accuracy for such an operation. For example, two 1.0 mm (0.040 in.) diam holes can be drilled into glass 65 mm (2½ in.) deep, with the holes separated by only 0.5 mm (0.020 in.). Another example of the deep-hole drilling capabilities of the rotary ultrasonic technique was the drilling of a 1.0 mm (0.040 in.) diam hole through a 90 mm (3½ in.) length of beryllium oxide. Upon completion of the drilling operation, a 0.43 mm (0.017 in.) diam core piece measuring 90 mm (3½ in.) in length was extracted from the bore of the hole.

Rotary ultrasonic machining also allows the uninterrupted drilling of small-diameter holes, while drilling without ultrasonics often necessitates frequent retraction of the tool to flush away debris and to loosen the core, which may jam in the bore of the tool. This frequent retraction increases drilling time.

Another application involved the drilling of four 2.4 mm (0.095 in.) diam holes through the sides of a 190 mm (7½ in.) square block of quartz. Accurate intersection of the holes at a depth of 165 mm (6½ in.) was required and was achieved within 0.1 mm (0.005 in.). In tests with a 3.2 mm

**Fig. 7** Ultrasonic impact grinding machine with computer numerical control and camera vision. Courtesy of Bullen Ultrasonics

**Fig. 8** Ultrasonic impact grinding tools. Courtesy of Sonic Mill

(1/8 in.) diam core drill through a 115 mm (4½ in.) square block of similar material, a misalignment of only 0.02 mm (0.0008 in.) resulted at a depth of 90 mm (3½ in.).

**Milling.** When ultrasonics are applied to a diamond mandrel or a wheel for milling, the operation benefits in the same way as when drilling. The pressure required for cutting is much less, progress through the material is more rapid, and the operation becomes smoother. When end milling, it is possible to mill parallel grooves and to maintain a dividing wall as thin as 0.75 mm (0.030 in.).

**Surface grinding** can be accomplished with small-diameter (32 mm, or 1¼ in.) diamond-plated wheels when the bottom of the tool is the cutting surface. The same tool can be used for slotting when the edge of the diamond wheel would be the cutting surface. Because of the stresses induced in the material by ultrasonic vibration, wheels thinner than 2.0 mm (0.080 in.) are impractical. As a result, the rotary ultrasonic machine cannot be used for slicing operations.

## Ultrasonic Impact Grinding

The capabilities of ultrasonic impact grinding are different from those of rotary

**Fig. 9** Abrasive slurry being flowed beneath the tool. Courtesy of Sonic Mill

ultrasonic machining. Because the cutting tool has no rotary motion, ultrasonic impact grinding can cut irregular shapes (Fig. 11) or drill multiple small holes. Although single holes can be drilled faster with rotary ultrasonic machining, ultrasonic impact grinding can drill smaller holes (as small as 0.1 mm, or 0.005 in., in diameter). Ultrasonic impact grinding can also drill multiple close-pitch, small-diameter holes. A tool for such an application is shown in Fig. 12.

**Machining Characteristics.** Cutting is accomplished by very small abrasive particles suspended in the liquid medium that flows between the vibrating tool and the workpiece. The rate of cutting depends on the amplitude of the vibratory motion of the tool. The amplitude for a given horn is determined by the booster selected and by an adjustable control on the ultrasonic power supply. It will vary from 0.025 to 0.09 mm (0.001 to 0.0035 in.) for most applications. Optimum results are obtained when the amplitude exceeds the grit size.

In operation, the tool does not contact the workpiece, but is held 0.025 to 0.075 mm (0.001 to 0.003 in.) above the surface. The feed rate ($z$-axis down speed) is related to the rate of cutting and may vary from 0.0025 to 0.1 mm/s (0.0001 to 0.005 in./s). Some approximate penetration rates with materials that can be machined by ultrasonic impact grinding are as follows:

| Material | Penetration rate, mm/min | Penetration rate, in./min |
|---|---|---|
| Glass | 3.8 | 0.150 |
| Ferrite | 3.2 | 0.125 |
| Carbon-graphite | 2.0 | 0.080 |
| Ceramic | 1.5 | 0.060 |
| Quartz | 1.7 | 0.065 |
| Tungsten carbide | 0.25 | 0.010 |
| Boron carbide | 0.20 | 0.008 |

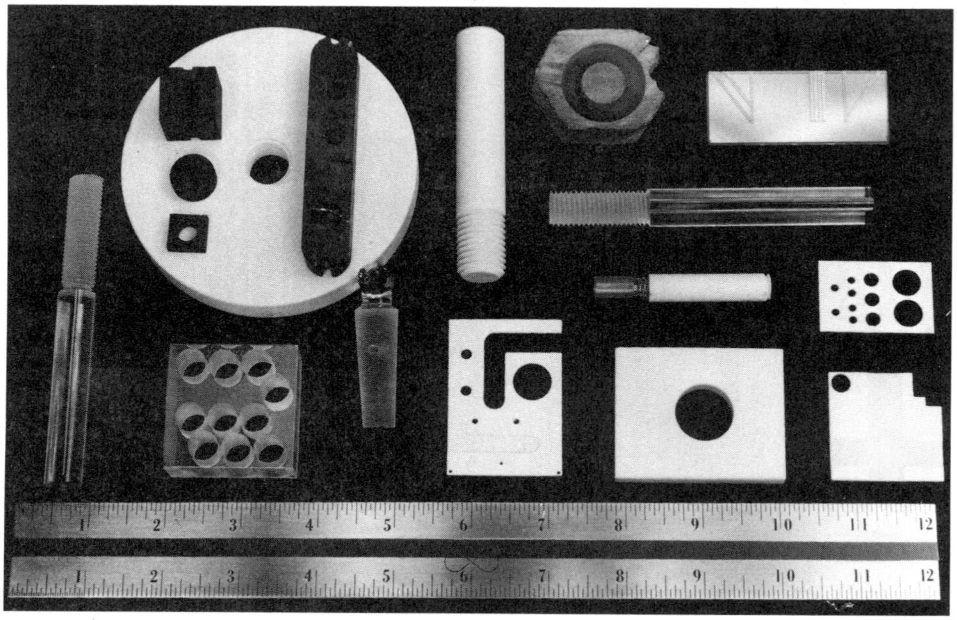

**Fig. 10** Parts machined by rotary ultrasonic machining

**Fig. 11** Decorative objects machined by ultrasonic impact grinding. Courtesy of Sonic Mill

**Fig. 12** Multiple-pin tool and horn. Courtesy of Sonic Mill

Other machinable materials include alumina, granite, ruby, sapphire, silicon, laminated fiber composites, and some plastics.

The cutting or penetration rate will vary according to the size and depth of the cavity. Smaller holes require more time, and the rate of cut declines with depth of penetration because of the difficulty in maintaining a continuous supply of new slurry at the tool face.

Precision will also decline with hole depth. As slurry flows down the side of the tool to its cutting face, some cutting will also occur on the sides of the bore, resulting in tapering with an increase in diameter at the point of entry. A depth-to-diameter ratio of 2.5 is considered the workable limit. Tapering can be minimized with a two-stage operation. In this technique, an undersize tool is used for the first cut, and the final cut is made with the required-size tool. Because it has less material to remove, the required-size tool will cut faster, thus minimizing the tapering.

# Electrochemical Machining

Revised by Terry L. Lievestro, Lehr Precision, Inc.

ELECTROCHEMICAL MACHINING (ECM) is the controlled removal of metal by anodic dissolution in an electrolytic cell in which the workpiece is the anode and the tool is the cathode. The electrolyte is pumped through the cutting gap between the tool and the workpiece, while direct current is passed through the cell at a low voltage, to dissolve metal from the workpiece.

Electrochemical machining can be used for a variety of tasks, including some that would be difficult, impossible, or time consuming by mechanical machining. The tasks include machining hard materials (such as hardened steel and heat-resistant alloys) and oddly shaped, small, deep holes. Electrochemical machining is used for operations as widely different as milling, drilling, deburring, etching, and marking.

Although electrochemical machining is sometimes applicable to small-lot production, the process is best suited to larger-lot production applications because of high tooling and setup costs and high capitol equipment costs. A 15 000 A machine tool for removing metal at about $25 \times 10^3$ mm$^3$/min (1.5 in.$^3$/min) may cost \$400 000 to \$700 000 (not including the cost of peripheral equipment and facilities). The most frequent application of the ECM process is in the production of gas turbine engine parts and for other aerospace applications. Electrochemical machining is also used selectively in automotive and general manufacturing applications.

## System Description

Figure 1 shows a typical setup for electrochemical machining. The electric current varies from 50 to 20 000 A with a current density of 0.2 to 3 A/mm$^2$ (100 to 2000 A/in.$^2$) and 30 Vdc applied across a gap of 0.025 to 1.3 mm (0.001 to 0.050 in.) between the tool (cathode) and the workpiece (anode). The electrolyte flows through this cutting gap at a velocity of 30 to 60 m/s (100 to 200 ft/s), forced by a pressure of 70 to 2800 kPa (10 to 400 psi). Typical temperature of the electrolyte varies from 24 to 65 °C (75 to 150 °F). Suspended solids are removed by settling, centrifuging, or filtering (or a combination of these), and the clarified electrolyte is circulated for reuse.

Electrolytes are usually aqueous solutions of inorganic salts such as sodium chloride, potassium chloride, sodium nitrate, or sodium chlorate. They may also contain proprietary additives. Other electrolytes, such as strong acids or sodium hydroxide solutions, can be used for special ECM applications or for select metals.

The tool can be fed into the workpiece at the rate of 0.25 to 20 mm/min (0.010 to 0.750 in./min). In some applications, machining continues while the tool is held stationary for a specified dwell interval after completion of the feed travel. The tool can be made of brass, bronze, copper, stainless steel, Monel, titanium, a sintered copper-tungsten alloy, or aluminum. The workpiece must be electrically conductive.

**Equipment.** The ECM machine must be rigid enough to withstand the forces caused by the high pressures of the electrolyte, which tend to separate the tooling from the workpiece. These forces are caused by the hydraulic pressure of the electrolyte as it flows through the gap between the tool and the workpiece. The electrolyte system includes pumps, filters, tanks, and a heat exchanger.

A single-axis machine is the most common configuration, although five-axis machines are in use. The simpler machines have manual voltage and feed rate controls,

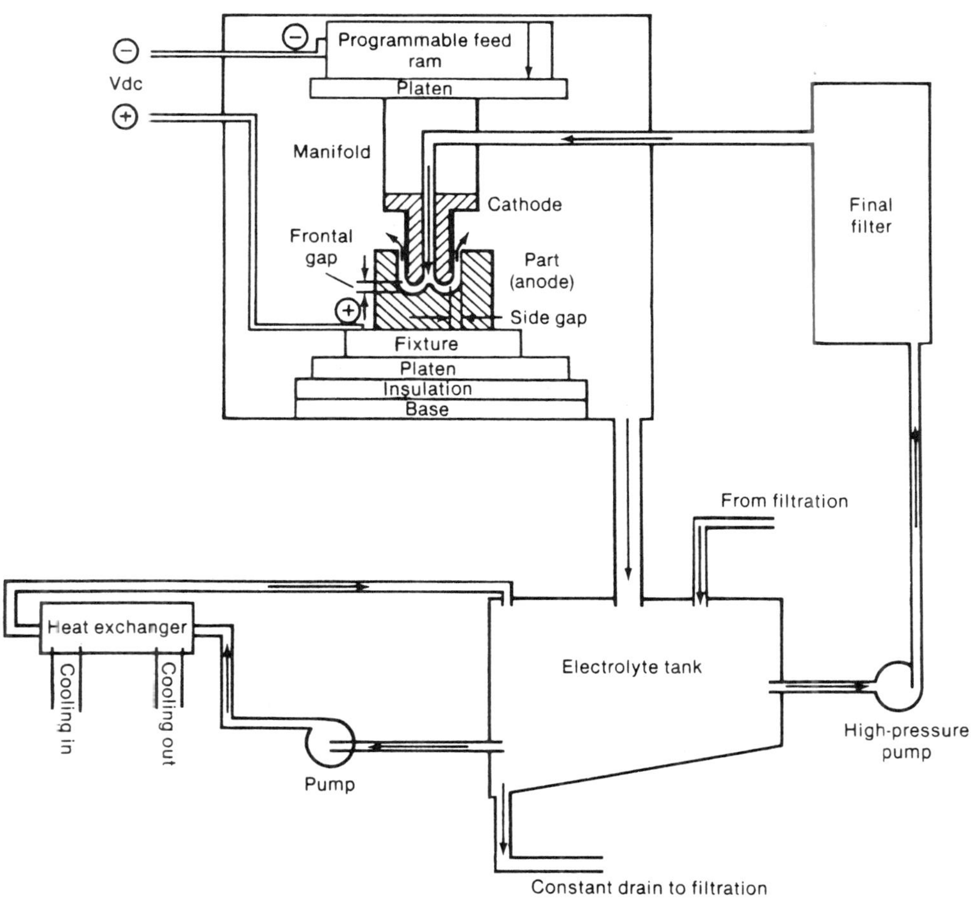


**Fig. 1** Schematic of the ECM system

while more advanced machines have computer numerical control (CNC) of the machine axes and process parameters. Most machines control feed rate and voltage.

**Power supplies** are equipped with devices for monitoring amperage, which is a product of feed rate and voltage. By monitoring changes in amperage, a problem can be detected as or before a short circuit occurs between the cathode and the workpiece. If a short circuit is detected, the power supply cuts the power to the cathode and dissipates the power built up in the transformers into another device. This action minimizes damage to the workpiece and cathode.

## Process Control

The rate of material removal in electrochemical machining is governed by Faraday's law and is a function of current density. The primary variables that affect the current density and the material removal rate are:

- Voltage
- Feed rate
- Electrolyte conductivity
- Electrolyte composition
- Electrolyte flow
- Workpiece material

**The voltage** across the cutting gap influences the current and the material removal rate and is controlled in most ECM operations. However, for a constant voltage, the current also depends on the electrical resistance of the cutting gap. Resistance is much more difficult to control because it depends on the conductivity of the electrolyte and the distance across the gap.

**The feed rate,** or penetration rate, is also controlled in most ECM operations. At a constant voltage, both the frontal gap and side gap (Fig. 1) are inversely proportional to the feed rate; the effect on the side gap is about 60% of that on the frontal gap. The distance across the frontal gap is a function of feed rate because, as the cathode is fed into the workpiece at a higher rate, the gap closes, causing resistance to drop. As resistance drops, amperage increases; therefore, machining rate also increases until an equilibrium is reached. At slower feed rates, the material removal rate decreases as the gap increases because the cathode is not keeping up with the workpiece surface. As the gap increases, the resistance rises and amperage drops. Side gap is also a function of feed rate. Frontal gaps are usually between 0.1 to 0.8 mm (0.005 to 0.030 in.), and side gaps are about 0.5 to 1.3 mm (0.020 to 0.050 in.).

The feed rate also varies directly with the current. For example, a hole machined at 2.5 mm/min (0.100 in./min) at 10 V and 1000 A would require 2000 A if the feed were increased to 5.0 mm/min (0.200 in./min). This would also require a potential of about 20 V and would increase power consumption ($V \cdot I$) from 10 to 40 kW.

The feed rate also depends on the application. Typical feed rates for different ECM operations on Inconel 718 are:

| Operation | Feed rate mm/min | in./min |
|---|---|---|
| Round holes (blind) | 2.2 | 0.085 |
| Simple cavities | 2.2 | 0.085 |
| $19 \times 10^3$ mm$^2$ (30 in.$^2$) faces | 1.3 | 0.050 |
| Broaching (angled ECM) | 7.6 | 0.300 |

Accurate estimates of feed rates usually require pilot testing in the desired ECM configuration.

**Electrolyte conductivity** will also affect resistance across the cutting gap. Increasing the concentration of an electrolyte will cause conductivity to rise, which causes a decrease in resistance. Temperature increases of the electrolyte also increase conductivity. Therefore, electrolyte concentration and temperature must be controlled.

**Electrolyte composition** directly influences conductivity, material removal rates, and surface characteristics. The parameters used for a given application may not yield the same ECM results if a different type of electrolyte is used. The normal development of an operation usually begins with the selection of the correct electrolyte. The other parameters and the cathode are then adjusted to obtain the desired result.

**Electrolyte flow rate** is also a factor in ECM process control. The temperature increase of the electrolyte passing through the gap is dependent on the flow rate. In addition, the rate at which hydrogen bubbles are carried away is thought to influence conductivity. Pressure control is the method of controlling flow rate (especially when a centrifugal pump is used). The flow rate also affects the level of turbulence of the electrolyte as it passes through the gap, and this influences the surface finish. The flow rate must also be great enough to remove machining by-products (sludge).

**The workpiece material** also affects the material removal rates and the current density. Theoretical removal rates for various metals are listed in Table 1. These removal rates are derived from Faraday's law, which is described in the article "Electrochemical Grinding" in this Volume.

## Electrolyte

The electrolyte has three main functions in electrochemical machining. It carries the current between the tool and the workpiece, it removes the products of the reaction from the cutting region, and it removes the heat produced in the operation.

Electrolytes must have high electrical conductivity, low toxicity and corrosivity, chemical and electrochemical stability, and a controllable passivating effect. The workpiece should dissolve at a consistent valence. The electrolytes perform better when the crystal structure of the workpiece is fine grained and when its constituents have a

### Table 1 Theoretical removal rates in electrochemical machining

| Metal | Valence | Density g/cm$^3$ | Density lb/in.$^3$ | Removal rate (1000 A current, 100% efficiency(a)) Mass kg/h | Mass lb/h | Volume mm$^3$ × 10$^3$/min | Volume in.$^3$/min |
|---|---|---|---|---|---|---|---|
| Aluminum | 3 | 2.7 | 0.098 | 0.34 | 0.74 | 2.1 | 0.13 |
| Beryllium | 2 | 1.9 | 0.067 | 0.17 | 0.37 | 1.5 | 0.09 |
| Copper | 1 | 9.0 | 0.324 | 2.37 | 5.22 | 4.4 | 0.27 |
| | 2 | 9.0 | 0.324 | 1.18 | 2.61 | 2.1 | 0.13 |
| Iron | 2 | 7.9 | 0.284 | 1.04 | 2.30 | 2.3 | 0.14 |
| | 3 | 7.9 | 0.284 | 0.69 | 1.53 | 1.5 | 0.09 |
| Magnesium | 2 | 1.7 | 0.063 | 0.45 | 1.00 | 4.4 | 0.27 |
| Molybdenum | 3 | 10.2 | 0.369 | 1.19 | 2.63 | 2.0 | 0.12 |
| | 4 | 10.2 | 0.369 | 0.89 | 1.97 | 1.5 | 0.09 |
| | 6 | 10.2 | 0.369 | 0.60 | 1.32 | 1.0 | 0.06 |
| Nickel | 2 | 8.9 | 0.322 | 1.09 | 2.41 | 2.1 | 0.13 |
| | 3 | 8.9 | 0.322 | 0.73 | 1.61 | 1.3 | 0.08 |
| Niobium | 3 | 8.6 | 0.310 | 1.16 | 2.55 | 2.3 | 0.14 |
| | 4 | 8.6 | 0.310 | 0.87 | 1.92 | 1.6 | 0.10 |
| | 5 | 8.6 | 0.310 | 0.69 | 1.53 | 1.3 | 0.08 |
| Tantalum | 5 | 16.6 | 0.600 | 1.35 | 2.98 | 1.3 | 0.08 |
| Titanium | 3 | 4.5 | 0.163 | 0.59 | 1.31 | 2.1 | 0.13 |
| | 4 | 4.5 | 0.163 | 0.45 | 0.99 | 1.6 | 0.10 |
| Tungsten | 6 | 19.3 | 0.697 | 1.14 | 2.52 | 1.0 | 0.06 |
| | 8 | 19.3 | 0.697 | 0.86 | 1.89 | 0.8 | 0.05 |
| **Commercial alloys** | | | | | | | |
| 4340 | ... | ... | ... | ... | ... | 2.18 | 0.133 |
| 17-4 PH | ... | ... | ... | ... | ... | 2.02 | 0.123 |
| A-286 | ... | ... | ... | ... | ... | 1.92 | 0.117 |
| M-252 | ... | ... | ... | ... | ... | 1.80 | 0.110 |
| René 41 | ... | ... | ... | ... | ... | 1.77 | 0.108 |
| U-500 | ... | ... | ... | ... | ... | 1.80 | 0.110 |
| U-700 | ... | ... | ... | ... | ... | 1.77 | 0.108 |
| L-605 | ... | ... | ... | ... | ... | 1.75 | 0.107 |

(a) It is not always possible to predict the valence at which some metals will dissolve nor how much current will flow through the gap. Also, practical factors, such as the shape of the electrode, can limit current flow.

**Table 2 Electrolytes for the electrochemical machining of various metals**

| Work metal | Electrolyte — Major constituent | Electrolyte — Concentration (max), kg/L (lb/gal.) of $H_2O$ | Removal rate, $mm^3 \times 10^3$/min ($in.^3$/min) per 1000 A |
|---|---|---|---|
| Steel; iron-, nickel-, and cobalt-base alloys | NaCl or KCl | 0.30 (2½) | 2.1 (0.13) |
| | $NaNO_3$ | 0.60 (5) | 2.1 (0.13) |
| Steel; hardened tool steel | $NaClO_3$ | 0.78 (6½) | 2.0 (0.12) |
| Gray iron | NaCl | 0.30 (2½) | 2.0 (0.12)(a)(b) |
| | $NaNO_3$ | 0.60 (5) | 2.0 (0.12)(a)(b) |
| White cast iron | $NaNO_3$ | 0.60 (5) | 1.6 (0.10)(c) |
| Aluminum and aluminum alloys(d) | $NaNO_3$ | 0.60 (5) | 2.1 (0.13) |
| | NaCl or KCl | 0.30 (2½) | 2.1 (0.13) |
| Titanium alloys | NaCl or KCl(e) | 0.12 (1) | 1.6 (0.10) |
| Tungsten | NaOH(f) | 0.18 (1½)(g) | 1.0 (0.06) |
| Molybdenum | NaOH(h) | 0.18 (1½) | 1.0 (0.06) |
| | NaCl or KCl | 0.30 (2½) | 1.0 (0.06) |
| Copper and copper alloys(d) | NaCl or KCl | 0.30 (2½) | 4.4 (0.27) |
| | $NaNO_3$ | 0.60 (5) | 3.3 (0.20) |
| Zirconium | NaCl or KCl | 0.30 (2½) | 2.1 (0.13) |

(a) Feed rates limited by graphite particle size. (b) Maximum; can vary widely. (c) Rough surface finish. (d) $NaNO_3$ electrolyte provides better surface finish. (e) Voltage must be greater than 11. (f) NaOH used up in process and must be replenished. (g) Minimum of 0.09 kg/L (¾ lb/gal.) (h) pH of electrolyte decreases with use; maintain pH by adding NaOH or KOH.

sufficiently uniform metal removal rate. It is sometimes advantageous to electrochemically machine the alloy in a fully solutioned condition.

General compositions of electrolytes for the ECM processing of different types of metals are given in Table 2. Compositions that have been used in specific applications are listed in the examples at the end of this article.

**Sludging electrolytes** are solutions of typical salts, such as sodium chloride. These salts are used primarily because they provide substantial conductivity to the solutioning water. The salts are generally not depleted in the ECM process. Instead, the water is depleted, yielding hydrogen gas ($H_2$) and the important hydroxide ions. The hydroxide ions combine with the metal ions that are being removed by ECM, thus forming insoluble reaction products, or sludge.

Sodium chloride solution is the most widely used electrolyte. It can be kept at constant strength by adding water to keep the concentration constant, and its electrical conductivity is stable over a wide pH range of 4 to 13. In ECM operations, the chemical and electrochemical reactions and the presence of metallic hydroxides have a buffering effect, holding pH in the neutral range of 4 to 10 without special control procedures.

As an electrolyte, sodium chloride also has some disadvantages. It is fairly corrosive and produces large amounts of sludge. Some metals, such as tungsten and pure titanium, cannot be machined in a sodium chloride electrolyte. Rough finishes are produced on aluminum alloys that contain silicon when this electrolyte is used.

Sodium nitrate, another salt, is used singly or in combination with sodium chloride in some applications (Table 2). Sodium nitrate gives smoother finishes on aluminum or copper alloys and is less corrosive than sodium chloride, but it is more expensive and is more likely to cause passivity of some work surfaces. Its electrochemical action is also less efficient than that of sodium chloride.

Sodium chlorate salt solution has some unusual properties as an ECM electrolyte, and it can be used in special applications. Additional precautions must be taken with sodium chlorate because of its flammability.

**Nonsludging Electrolytes.** Strong alkali solutions (for example, NaOH) are used in the electrochemical machining of heavy metals (such as tungsten and molybdenum) and their alloys. The salt in these electrolytes also provides the substantial conductivity required. However, the salt is depleted because the sodium ions of the salt join with the metals being removed by the ECM process. Therefore, new compounds, such as sodium tungstate ($Na_2WO_4{\cdot}2H_2$) and sodium molybdate ($Na_2MoO_4{\cdot}2H_2$), form during the process, and makeup of both the alkali salts and the water are required for process control. This type of ECM operation is referred to as a depletion mode because the alkali salt is depleted during the process. The new compounds in this process are quite soluble in water and heavy precipitate volumes do not occur. However, there is a tendency for the heavy metals to plate out from the solution and onto the cathode. Therefore, cathodes must be periodically cleaned because this coating is objectionable in a well-controlled, precision ECM process.

Strong acid electrolytes (for example, hydrochloric, nitric, and sulfuric) are also used in a depletion mode. The metals being removed by electrochemical machining also form new compounds (such as nickel chloride, nickel nitrate, or nickel sulfate) with the electrolyte. Hydrogen gas is also produced. These new compounds in acid electrolytes must be reduced by conditioning procedures or by periodic replacement with unused acid. The attendant plating-out conditions on the cathode occur early, adherently, and continuously. It is often necessary to introduce a periodic voltage polarity reversal to keep the cathode surfaces clean. When using virtual electrodes (glass impingement nozzles), the plate out is essentially negated. Nevertheless, the contaminant level in solution keeps increasing and requires conditioning of electrolyte.

**Sludge,** consisting of insoluble hydroxides or hydrated oxides of the work metal, is produced in conventional salt electrolytes at the rate of about 6 to 8 kg of sludge per kilogram of metal after filter pressing. A sodium chloride solution of 0.2 kg/L (2 lb/gal.) of water becomes more viscous soon after its sludge content exceeds 2.3 wt%. A thinner (less viscous) electrolyte is necessary when smaller cutting gaps or higher electrolyte flow rates are used. The electrolyte preferably should contain no more than 2 wt% sludge.

An electrolyte with up to 2 wt% sludge is nearly as efficient as a clean solution. There is little difference in the conductivity, rate of penetration, surface finish, or accuracy of machining. A dirty solution may deposit more precipitate on the sides of the workpiece, depending mainly on the flow rate of the electrolyte. There is some danger that dried precipitate may become lodged in the cutting gap or flow passages, causing a short circuit or a flow disturbance. For this reason in particular, a clean electrolyte is preferred.

**Filtering,** centrifuging, and settling are generally done to remove sludge. It is usually more practical to clarify even the cheapest electrolyte (sodium chloride) than to discard and replace it.

**The concentration level** of the electrolyte solution for electrochemical machining is usually a compromise. A concentrated solution has the advantages of lower voltage and power requirements because of better conductivity. Faster rate of penetration and greater precision are also possible because the conductivity of a concentrated solution varies less with the changes in temperature and concentration encountered in the ECM process. On the other hand, diluted, low-concentration solutions cost less, dissolve more readily, and give a smoother surface on some work.

Low-concentration solutions also make crystallization less likely. An overconcentrated solution may become saturated and allow the formation of crystals that can damage pumps, valves, instruments, and pipes. However, a very weak solution causes local or intermittent passivity of the work, which makes machining difficult. A sodium chloride solution of 0.12 to 0.30 kg/L (1 to 2.5 lb/gal.) of water performs well as an electrolyte.

**The conductivity** of the electrolyte depends primarily on its concentration and temperature. It must be controlled because it directly affects power requirements and

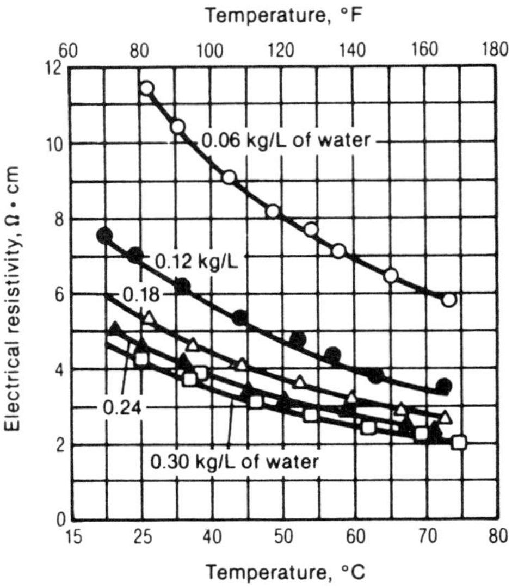


**Fig. 2** Effect of temperature on the resistivity of five sodium chloride solutions

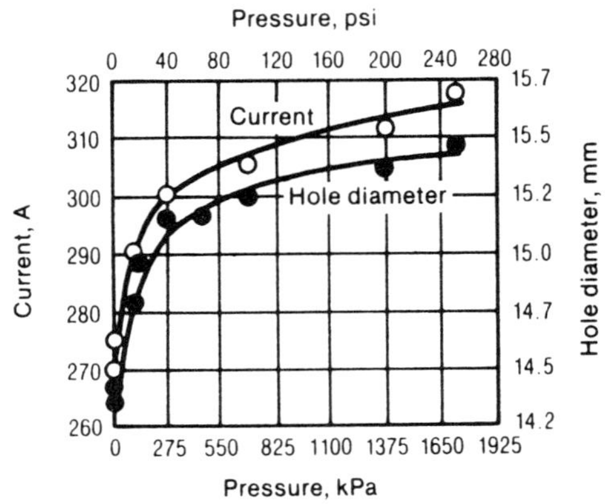


**Fig. 3** Effect of electrolyte back pressure on hole size and current at constant voltage

rate of penetration. For higher conductivity, the penetration rates are faster.

The conductivity of an electrolyte changes substantially with temperature. For example, solutions of sodium chloride are 100% more conductive at 70 °C (160 °F) than at 24 °C (75 °F). Figure 2 shows the effect of temperature on the resistivity of five solutions of sodium chloride. Any change in electrolyte conductivity will affect ECM geometry because the dimensions of the cutting gap and the side gap depend on the conductivity. This is why electrolyte temperature is closely controlled. Selection of the operating temperature usually involves a compromise. Maintaining a lower temperature on a high-amperage operation requires additional cooling requirements. A low-amperage operation may require heating to maintain a higher temperature. To some extent, the conductivity of the electrolyte limits the current that can be used.

**The flow rate** of the electrolyte is important because the electrolyte must adequately remove the heat and the products of the chemical reaction. Flow can be related to the amount of current used; the larger the ratio of flow to current, the better the removal of heat and reaction products. However, the cost of pumping increases as the flow increases, and excessive flow rates can cause local erosion on the workpiece or tool. A flow rate of 0.95 L/min (0.25 gal./min) per 100 A of current is a guiding minimum for machining steel with an electrolyte of sodium chloride.

The flow rate also influences surface finish and machining accuracy. High electrolyte flow rates often improve the uniformity of metal removal with a given electrolyte without reducing the metal removal rate. Low flow rates may cause precipitates to stick to the workpiece. This may cause changes in the pattern of flow that could affect the work. Local chemical attack may also vary with flow rate, causing changes in the surface finish. The flow rate sometimes affects the location and the thickness of deposits on the cathode; high rates minimize the formation of such deposits. In addition, flow rate has a small effect on the voltage required.

Tool design also influences the rate, pattern, and uniformity of electrolyte flow. To produce a smooth, uniform surface on the workpiece, the tool design must provide a uniform flow rate over the entire machining area. However, a uniform flow rate at all points is difficult to achieve given the typical flow velocities (30 to 60 m/s, or 100 to 200 ft/s) in the gap. Moreover, the effects of changes in tool design on flow rates and patterns are difficult to predict, although tool design can improve flow characteristics (see the section "Tool Design" in this article).

**Machining with electrolyte back pressure** is a useful form of electrochemical machining in some applications. In this technique, the work is enclosed, restricting the outflow of the electrolyte. Therefore, the average pressure in the gap is increased. This increase in pressure results in increased conductivity, which results in the tool cutting a proportionately larger hole (Fig. 3). Even a small increase in pressure makes the tool cut a larger hole, but at pressures exceeding 550 kPa (80 psi), the gap increase becomes slight. Studies suggest that the current rises because the pressure makes the hydrogen bubbles smaller in the cutting gap, causing a greater effective conductivity of the electrolyte. Another effect of using back pressure in electrochemical machining is the reduction or the elimination of flow lines on the work. Disadvantages of back pressure include an increase in hydraulic forces and an increase in tooling costs.

**Safety Precautions.** Sodium chloride and potassium chloride, used as electrolytes, present no unusual hazards. Dry oxidizing salts such as sodium nitrate and particularly sodium chlorate are hazard prone. They should not be allowed to impregnate and dry on organic or combustible materials. The alkalies (wet or dry) are dangerous to bodily tissue. Mists, vapors, and dusts of alkalies must be avoided. Ventilation must be provided on equipment and for handling chemicals. Protective gloves and face shields or masks may be required for handling most chemicals. Hydrogen, freed at the ECM cathode, must be vented from the high point of the work enclosure to prevent hydrogen explosions.

Electrolytes used to machine alloys containing chromium or other heavy metals may become hazardous to the environment. Proper storage, handling, and disposal make it essential to recycle these electrolytes. The metals can be reclaimed or properly converted to eliminate environmental disposal problems.

Proper design of equipment and tools, together with sound operating procedures, minimizes the likelihood of major short circuits, which can cause melting of the tool or workpiece. Acids and chloride salts can be extremely corrosive; therefore, their use necessitates suitable construction materials and safe operating practice.

## Tools (Cathodes)

Critical aspects of electrochemical machining are the material design and construction of tools. The method of determining the proper tool for a particular application is largely empirical.

**Tool Materials.** The metal used for a cathode or associated tooling should have the necessary stiffness and machinability in addition to electrical and thermal conductivity and chemical resistance to the electrolyte. Copper, brass, bronze, stainless steel, and titanium are the materials most frequently used for ECM tools. Bronze or brass is usually the optimum choice, except where greater stiffness is needed. Titanium is especially useful when machining with an acid electrolyte that anodizes it (sulfuric acid, for example). The current can then be reversed periodically to remove plated deposits without excessive electrochemical machining of the cathode.

Protective devices on ECM equipment limit the amount of current transferred by a short circuit. Copper and brass are less likely than other metals to be seriously damaged by such a current-limited short circuit because these highly conductive metals dissipate heat rapidly. The same type of short circuit, when using stainless steel or titanium cathodes because of low thermal conductivity, could concentrate much more heat at the affected point and cause severe local damage. Copper tungsten is very resistant to damage from short circuits because of its high melting point. Damaged tools and work usually would

**Table 3 Relative properties of four metals used as tools (cathodes) in electrochemical machining**

| Property | Tool (electrode) material: Copper | Brass | Stainless | Titanium |
|---|---|---|---|---|
| Electrical resistivity | 1 | 4 | 53 | 48 |
| Stiffness | 1.1 | 1 | 1.9 | 1.1 |
| Machinability | 6.0 | 8.0 | 2.5 | 1.0 |
| Thermal conductivity | 25 | 7.5 | 1.0 | 2.6 |

have to be removed for repair or replacement. Table 3 compares the properties of four metals used for ECM tools, and Table 4 lists materials for ECM tools.

**Tool Design.** The tool or cathode is usually shaped like a mirror image of the machined area of the completed part. Its function is to provide a defined path for the ECM current. The designer must determine the nature and the extent of the required deviation or gap allowances from the mirror image configuration, while providing for a uniform and sufficiently high flow rate of electrolyte in the gap to allow a practical rate of metal removal.

Tool dimensions must be slightly different from the nominal mirror dimensions of the completed part to allow for ECM overcut, which can range from 0.025 to 1.3 mm (0.001 and 0.050 in.), depending on electrolyte flow and the required dimensional accuracy. Both the part and the cathode must have adequate current-carrying capacity. The ECM cell must have strength and rigidity to avoid flutter and arcing. When possible, the cathode and workpiece surfaces should approximately conform at the start of cutting; stock is sometimes added to, or removed from, the workpiece to accomplish this.

**Table 4 Materials for ECM tools (cathodes)**

Copper alloys
- 110, electrolytic tough pitch copper
- 145, free-cutting copper (½% Te)
- 187, free-cutting copper (1% Pb)
- 220, commercial bronze, 90%
- 260, cartridge brass
- 360, free-cutting brass
- 464, naval brass
- 639, aluminum-silicon bronze
- 706, copper nickel, 10%
- Copper-manganese alloy

316 stainless steel
Titanium (99%)
Tungsten 80%, copper 20% (sintered)

*Tool Types.* General types of tool design for various ECM applications are illustrated in Fig. 4. The most common tool design is the open-flow type (Fig. 4a), in which electrolyte enters the gap through a channel in the center of the tool and exits around the outside of the tool to the atmosphere. Taper-free cavities can be machined with the insulated construction shown. The dimensions of the lip on the leading edge of the tool and the location of the insulation govern the overcut and the extent of side cutting. For optimum surface finish, side cutting can be removed by making the lip thinner.

Tools for external machining may be of the cross-flow type (Fig. 4b), in which electrolyte enters the gap at one side of the workpiece and exits at the opposite side. This may necessitate special fixtures to confine the electrolyte flow to the space between the workpiece and the tool. With reverse-flow tools, the electrolyte is pumped through the gap and out through the tool by using special fixtures to confine the liquid. This type of tool is sometimes needed to provide more uniform flow for long or deep cuts or to meet close-tolerance or low-micron finish requirements.

Figures 4(c) to (h) show other tools for a variety of ECM applications. Because of the interaction of working-tip shape and dimensions, location of insulation, current density, and feed rate, the design of tools for machining complex shapes requires an understanding of fluid flow, electrical, and

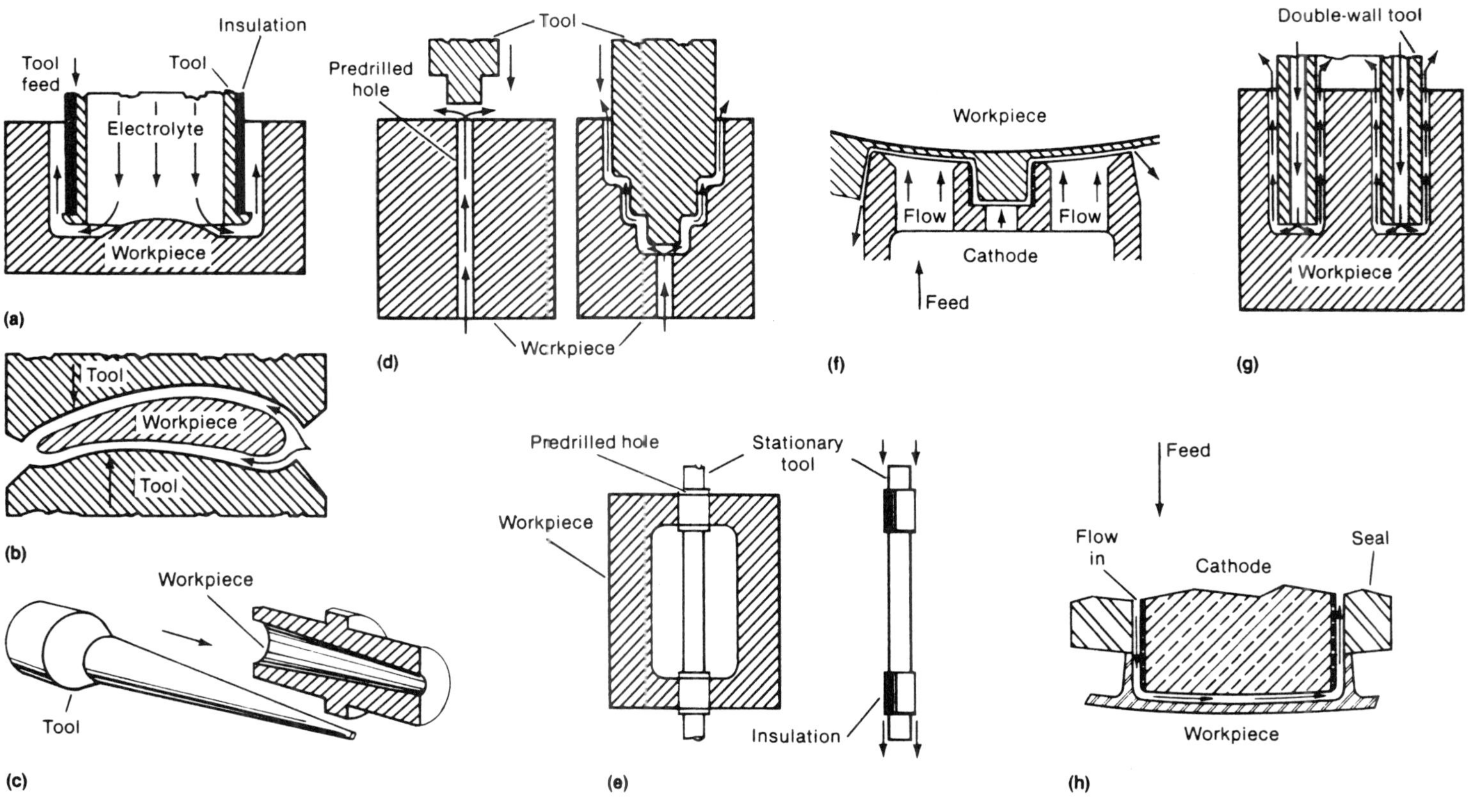


**Fig. 4** Tools for electrochemical machining. (a) Hole-sinking tool of the open-flow type, with insulated sidewall. Holes of uniform curvature can be cut with curved tool. (b) Dual external-cutting tool, cross-flow type. Special fixtures are required to confine electrolyte. (c) Tool for tapering a predrilled hole. (d) Tool for sinking a stepped through hole. Electrolyte enters gap through predrilled hole. (e) Tool for enlarging interior section of a hole. A stub tool of similar design can be used to enlarge the interior of a blind hole. (f) Open-flow cathode used to generate the outside diameter wall and leave an embossment. (g) Double-wall trepanning tool; entire wall is insulated. A single-wall tool can be used, with electrolyte entering on one side and leaving on the other side, but flow rates must be high and control is less precise. Special types are available for slug cutoff. Oddly-shaped or tapered holes can be made. (h) Cross-flow tool used to generate ribs on a surface without leaving flow lines on the part. Flow is fed down one side, across the face, and up the second side.

electrochemical principles, as well as experience and ingenuity. Although tool design may be difficult and time consuming and may require tool trials for complex shapes or close-tolerance applications, the cost of additional or replacement ECM tooling is usually much less than that for conventional machining.

*Flow and Tool Design.* The tool design must permit electrolyte flow to and from all machining areas at a rate sufficient to keep a film of fresh fluid in the machining area and to dissipate the generated heat. The tool design should also allow a uniform flow rate in the gap.

A common peculiarity of electrochemical machining is the occurrence of striations, ridges, or protuberances on the workpiece. These are found opposite or near the electrolyte flow channels in the tool or at other points where sudden changes in flow direction cause stagnation. The flow pattern can be improved and the surface irregularities reduced in number and size by using a larger number of smaller holes or slots in the tool, but it is sometimes still necessary to conduct an additional operation to remove these irregularities.

Tool surfaces should be smoothly finished. Any slight defect in the tool may affect the ECM machining, leaving marks on the machined surface. Nicks, scratches, notches, burrs, and lines can all cause matching defects in the work. Any flaw in the tool that disturbs the flow of electrolyte can cause a flow line or flow pattern to be machined into the work.

*Computer-Aided Design and Manufacture (CAD/CAM).* With computer-integrated manufacturing, cathodes can be produced at lower costs and with greater accuracy. Computer-aided systems are used first to design a cathode with the proper offsets from the workpiece. The computer-aided designs are programmed for the CNC system of the precision milling machines (or lathes), which produce the cathodes. Programs for the computer numerical control of multiple-axis ECM machines can also be produced. Next, inspection data and programs for a coordinate-measuring machine are produced and down loaded. The coordinate-measuring machine inspects the workpiece or cathode and sends data back to the CAD/CAM unit for analysis of the results. Educated iterations of the tool are then possible so that optimum geometries can be produced.

In the past, the development of ECM tools and parameters was essentially an art. Cathodes were produced by an experienced and skilled toolmaker using much hand finishing. Today, with the advent of computer systems, cathodes can be machined completely by computer numerical control. This introduces new levels of accuracy and substantially reduces time and cost. In addition, with computer numerical control of the ECM machine drives and peripheral systems, much more flexibility, producibility, and precision are possible. Many studies have been conducted on parameters and their effects during the ECM process, and the result has been a much better understanding of the process and its optimization.

**Insulation** is important in the control of the electrical current. The tool can be insulated in a number of ways, depending on its shape, as discussed in the section "Tool Design" in this article. Spraying or dipping is generally the simplest method of applying insulation. However, a more durable insulation is often needed. The following qualities are essential:

- Adhesion to the tool. Preformed insulation can be held to the tool by shrinkfitting, adhesives, or fasteners
- Sealing without pores or leaks that could cause stray machining by current leakage
- Adequate thickness (0.05 mm, or 0.002 in., is minimum)
- Smoothness to avoid disturbing the flow of electrolyte
- Resistance to heat for continuous service at 200 °C (400 °F) without breakdown
- Curing temperature of 230 °C (450 °F) or less to prevent damage to the tool
- Durability to resist wear in guides and fixtures
- Chemical resistance to the electrolyte
- High electrical resistivity
- Uniform application to minimize disturbance of the flow of electrolyte and to prevent interference
- Low water absorption

Teflon, urethane, phenolic, epoxy, powder coating, and other materials are commonly used for insulation. At low flow rates and low current density, most of these materials are satisfactory insulators and provide adequate life when the tool has a lip to protect the edge of the insulation from the flow force of the electrolyte.

Sprayed or dipped coatings of epoxy resins are among the most effective insulating materials. For optimum effectiveness, these coatings should be used with a protective lip on the tool. Nylon, acetal, and fiberglass-reinforced epoxy are among the better self-supporting preformed materials for insulation. However, such insulation should have a 0.5 mm (0.020 in.) minimum thickness to obtain long life in heavy-duty ECM service and should be protected at the edge against the force of electrolyte flow.

**Table 5 Effects of double tempering treatments on metal removal rates for 4340 steel in electrochemical machining**

| Double tempering treatment (after quenching to 49 °C, or 120 °F) | Hardness, HRC | Removal rate, $mm^3 \times 10^3$/min ($in.^3$/min) per 1000 A | Appearance of machined surface |
|---|---|---|---|
| 218 °C (425 °F), 2 h; 230 °C (450 °F), 2 h | 54 | 1.72 (0.105) | Dark, dull |
| 425 °C (800 °F), 2 h; 425 °C (800 °F), 2 h | 44 | 1.88 (0.115) | Dark, dull |
| 650 °C (1200 °F), 2 h; 650 °C (1200 °F), 2 h | 32 | 2.02 (0.123) | Bright |

## Workpiece

Workpieces of almost any material that conducts electricity can be electrochemically machined. Table 2 provides the selection of electrolytes and rates of machining for various materials.

**Microstructure.** Little information is available on the effects of different work-alloy microstructures on ECM results. However, experience in similar processes such as electropickling and electropolishing suggests that the following effects may occur:

- Large grain size may cause a rougher finish than fine grain size
- Insoluble inclusions (graphite in cast iron, for example) may cause roughness and process problems
- Variations in composition, as in case-hardened steel, may cause differences in machining
- Precipitation of intermetallic compounds at the grain boundaries, as in the heat sensitizing of some stainless steels, may cause serious intergranular attack

**Hardness.** Electrochemical machining can be used to cut hardened metals. Although metal removal rate is not directly related to work metal hardness, it may be affected by heat treatment and other operations that alter the microstructure of the material. Because of this dependence of machining rate on microstructure, constant voltage and feed rate cannot ensure close tolerances and a fine finish; a consistent workpiece material is also important. Surface finish and tolerances can also vary with changes in material hardness. Table 5 provides data on the rate of metal removal (volume of metal removed per minute per 1000 A) for 4340 steel after three double tempering treatments. All machining was done with a sodium chloride solution concentration testing 22° Baumé at 27 °C (80 °F). Table 5 indicates that metal removal was faster at lower hardness; other alloys and heat treatments might exhibit the reverse relation or no change in removal rate relative to hardness.

**Removal rates** depend on the valence and atomic weight of the workpiece. As indicated in Table 1, theoretical maximum rates range from 650 to 4400 $mm^3$/min (0.04 to 0.27 $in.^3$/min) per 1000 A. With suitable selection of electrolyte and operating conditions, workpieces can be electrochemically machined close to the theoretical removal rates. If metal dissolution proceeds at a level of efficiency substantially lower than the theoretical efficiency, the evolution of oxygen or chlorine or another undesirable product at the workpiece is indicated, and such conditions should be avoided. As a first approximation, a value of 1600 $mm^3$/min (0.1 $in.^3$/min) per 1000 A (Table 1) can be assumed for most metals in preliminary planning; this assumption disregards the effects of valence, work metal density, and other factors.

**Cleaning** of the workpiece is nearly always necessary after electrochemical machining. Cleaning should be done before any residue has hardened on the work. A rinse in water may be all that is needed for metals that resist corrosion. Steel and cast iron are usually treated in alkaline cleaning solutions (with or without a water rinse) or in a water-displacing compound that may leave a protective coating on the workpiece. A mild hydrochloric acid solution is commonly used before a clean water rinse. Glass bead peening or vapor blasting is also used in numerous applications.

**Hydrogen embrittlement** of the work surface does not normally occur during electrochemical machining, because hydrogen is given off at the cathode and not at the workpiece. However, the entry of hydrogen may be possible when a workpiece is in an acid solution with no electrical current. A hydrogen-sensitive workpiece should not be immersed in an acidic electrolyte unless the current is flowing. Immediate, thorough rinsing should follow electrochemical machining. Neutral electrolytes, such as salt solutions, are unlikely to cause hydrogen embrittlement. However, laboratory checks are sometimes made.

## Surface Finish

Surface finish of electrochemically machined parts is usually 0.30 to 1.9 μm (12 to 75 μin.) for the frontal cutting-gap area and as rough as 5 μm (200 μin.) or more for the side-gap area. Important variables that affect surface finish are current density, feed rate, gap dimensions, electrolyte composition, viscosity, temperature, flow, and workpiece microstructure.

Some metals can be polished in simple salt solutions if the current density is great enough. The polishing action is more pronounced at higher current densities and is influenced by such factors as the flow and temperature of the electrolyte. The bottom of a hole made by electrochemical machining is often polished, while the sides have a dull finish. This is partly explained by local variations in current density. Most of the current flows through the cutting face of the tool. Some current strays to the sides, behind the leading edge. The sides of the hole are machined somewhat by the stray current at a lower current density, producing a dull finish. The greater current density at the cutting face yields a finer finish.

Microscopic surface defects, such as intergranular attack, may be caused by selective ECM attack on certain constituents in an alloy. Such defects are usually associated with low current densities and intermetallic precipitates at grain boundaries. Intergranular attack can shorten the fatigue life of metal and cannot be tolerated in parts subjected to alternating high stress. Intergranular attack can usually be avoided with proper selection of electrolyte and ECM parameters.

Surfaces that have been electrochemically machined have little or no residual stress. When electrochemical machining replaces mechanical machining, the effect on fatigue life should be considered because a compressive stress may be required on the surface of the workpiece for adequate fatigue life. Many applications benefit from shot peening or glass bead operations to induce compressive stress at the surface after electrochemical machining. This will increase the fatigue life of an electrochemically machined workpiece.

## Accuracy

The usual dimensional tolerances for electrochemical machining are ±0.13 mm (±0.005 in.) for the frontal gap and ±0.25 mm (±0.010 in.) for the side gap. However, it is possible to control the process to ±0.025 mm (±0.001 in.). It is difficult to machine internal radii smaller than 0.8 mm (0.030 in.); external radii can be 0.5 mm (0.02 in.) or larger. Overcut, taper, and corner radii depend on the configuration of the cathode. Common results are 0.001 mm/mm (0.001 in./in.) of taper, 0.5 mm (0.020 in.) overcut, and 2.5 mm (0.10 in.) corner radius. The contour of the internal radii can be approximated by adding the amount of side cutting gap to the radius on the cathode.

**Table 6 Comparison of time and cost for ECM versus previous machining method in five production applications**

| Item | Two slots in Waspaloy | | Pocket in A-286 | | Through hole in M-252 | | Square hole in tool steel (52 HRC) | | Parallel holes in U-700 Application of electrostream drilling |
|---|---|---|---|---|---|---|---|---|---|
| | ECM | End milling | ECM | End milling(a) | ECM | End milling(b) | ECM(c) | EDM | |
| **Time per piece, min** | | | | | | | | | |
| Machining | 6 | 126 | 20 | 84 | 10 | 81 | 6 | 31 | ... |
| Loading and unloading | 4 | 3 | 10 | 5 | 5 | 2 | 2 | 2 | ... |
| Changing tools | ... | 6 | ... | 12 | ... | 12 | ... | ... | ... |
| **Total** | **10** | **135** | **30** | **101** | **15** | **95** | **8** | **33** | |
| **Ratio of cost per piece (costs include labor, overhead, and tool use)** | | | | | | | | | |
| ECM cost/cost of other process | 1/9 | ... | 1/4 | ... | 1/5 | ... | 1/4 | ... | ... |

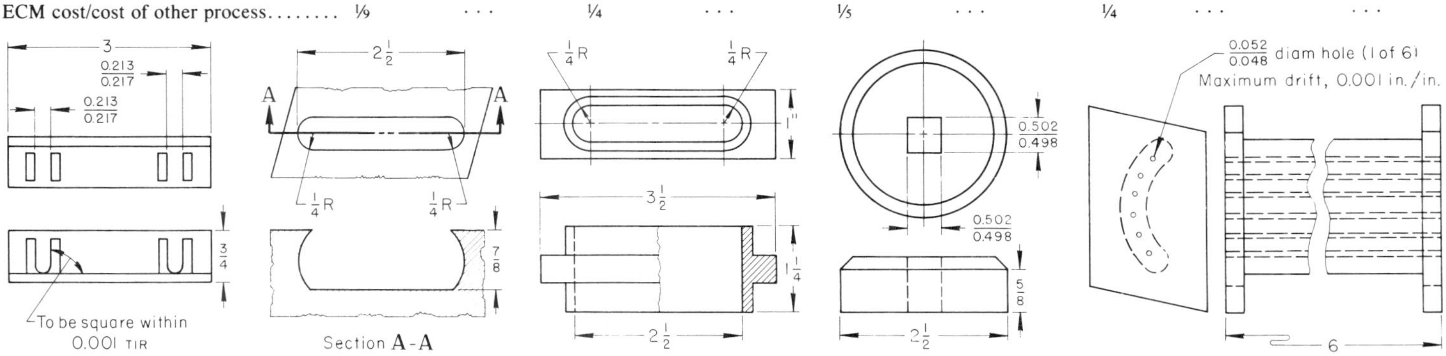


Note: Dimensions in figure given in inches. (a) Tool breakage averaged four cutters per pocket. (b) Cutter required three regrinds per hole. (c) Final sizing (not shown) was done on electrochemically machined parts in one third the time required for electrical discharge machined parts. Source: SAE technical paper S350

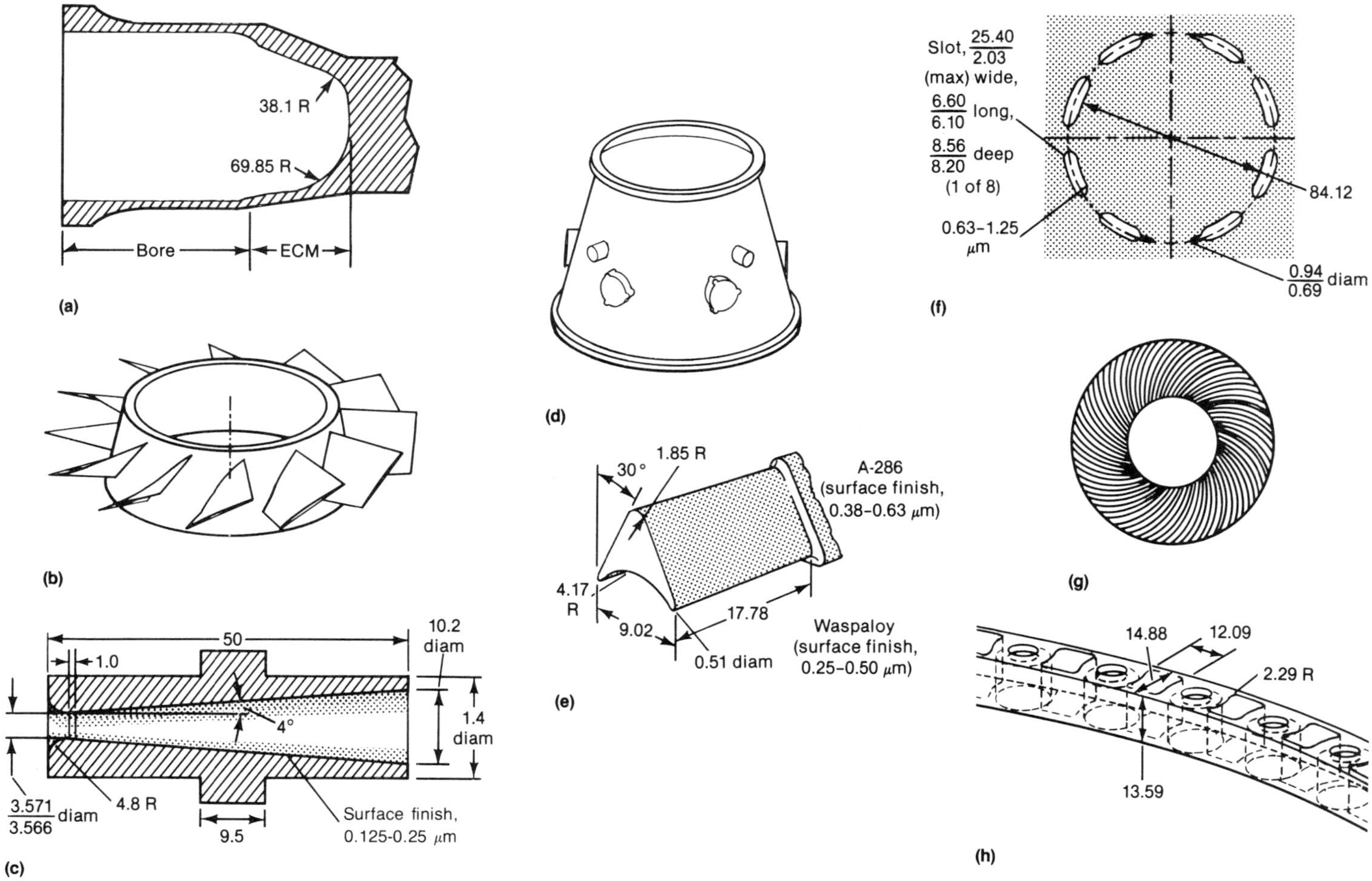


**Fig. 5** Eight applications of electrochemical machining. (a) Bottom contour of a deep hole. (b) Airfoils machined directly on a compressor disk. (c) Finishing a conical hole in a nozzle. (d) Machining a thin-wall casing with embossments. (e) Contouring a turbine blade surface. (f) Cutting slots in a valve plate. (g) Cutting spiral grooves in a friction plate. (h) Cutting multiple small cavities in Inconel 718. All dimensions given in millimeters

## Applications

Table 6 compares the processing time for machining four different parts by end milling, electrochemical machining, and electrical discharge machining. In each case, electrochemical machining reduces the required processing time. The cost advantage of the ECM process in these four examples also ranges from 4 to 1 to 9 to 1.

The fifth part illustrated in Table 6 applies more to electrostream drilling, which is a modified ECM process. A description of this process is provided in the article "Electrostream and Capillary Drilling" in this Volume.

A major application of electrochemical machining is deburring in automated, high-production applications, particularly on points not practically accessible by conventional machining methods. The ECM process is also used to rough machine massive forgings and other large, complex shapes. The variety of work done by electrochemical machining is further illustrated by the following examples.

**Example 1: Electrochemically Machined Bottom Contour of a Deep Hole.** Figure 5(a) shows a deep hole with a bottom contour. The part material, Ti-6Al-4V, was rough bored approximately 250 mm (10 in.) deep. A 1 to 20 Vdc machine with a maximum amperage of 15 000 A was used to rough and finish the bottom contour. Brass cathodes were used, with electrolyte flowing into flow slots in the face and out through the center.

The roughing operation started at 17 Vdc and a 2 mm/min (0.080 in./min) feed rate and ramped down to a feed rate of 0.8 mm/min (0.030 in./min) so that the amperage did not exceed 13 000 A. The finishing started at 4.8 mm/min (0.190 in./min) at 14 Vdc and ramped down to 0.71 mm/min (0.028 in./min), also limiting amperage to 13 000 A. The current was limited by the cooling capacity of the part because of the large mass and its electrical resistance.

**Example 2: Electrochemically Machined Airfoils.** Airfoils (Fig. 5b) were machined directly on a compressor disk to eliminate the mechanical joining of a disk and the airfoils. Electrochemical machining was used to rough and finish the airfoils out of a turned hub. A multiaxis ECM machine with computer numerical control was used. The specifics of the process and the machine are proprietary. Electrochemical machining yielded a consistent surface finish and satisfactory part tolerances; therefore, no further processing (machining or polishing) of the airfoils was required.

**Example 3: Electrochemical Machining of a Stainless Steel Nozzle.** Figure 5(c) shows the cross section of a nozzle of type 316 stainless steel, made from a tube in a machine that could provide 500 A at 24 Vdc or less. A copper tool was fed into one end of the tube to machine the hole to a conical shape, with a current beginning at 20 A and ending at 310 A and 17 V.

The electrolyte was sodium chloride (0.12 kg/L, or 1 lb/gal., of water) at a temperature of 27 °C (80 °F). A pressure of 550 kPa (80 psi) caused the electrolyte to flow at 7.6 L/min (2 gal./min) with 50 μm filtration. Tool penetration was 44.2 mm (1.74 in.), with a feed rate of 5.8 mm/min (0.230 in./min). A cutting time of 7.6 min produced internal smoothness of 0.125 to 0.25 μm (5 to 10 μin.).

After this first operation, the part was inverted, and another copper tool was used to machine the radius at the other end of the hole. The current began at 20 A and ended at 220 A. The electrolyte was the same as before, with the same circulation. Tool penetration was 4.95 mm (0.195 in.), with a feed of 2.5 mm/min (0.100 in./min). Cutting time was 2 min, producing an internal finish of 0.125 to 0.25 μm (5 to 10 μin.).

**Example 4: Electrochemical Machining of a Casing.** Figure 5(d) shows a thin-wall casing with embossments on the outside. The material used was Inconel 718. The part was finish machined on the inside and had stock on both flanges. Multiple cathodes were required for generating the different embossments and the walls between them. The part was divided in such a way that each ECM cut covered a certain angle around the circumference. The direction of feed was normal to the outside surface.

Typically, an open-flow cut is used for this type of part. In an open-flow cut, the electrolyte is fed through the center of the cathode and out to atmosphere. The position at which the cathode is stopped determines the wall thickness of the part.

Sodium chloride at a concentration of 0.138 kg/L (1.15 lb/gal.) of water was pumped through the gap at 230 kPa (330 psi). Voltages of 12 to 18 Vdc are most common. Feed rates are normally 0.8 to 1.8 mm/min (0.030 to 0.070 in./min). The maximum amperage of the machine or the pump size is sometimes the limiting factor in maximum material removal rates.

Surface finishes of 1.6 μm (63 μin.) on the wall and 3.8 μm (150 μin.) on the sides of the embossments are typical. Tolerances on the wall thickness are usually ±0.25 mm (±0.010 in.) and are ±0.38 mm (±0.015 in.) on boss sizes and position for a production part. If required, tighter tolerances can be achieved, but costs will increase.

**Example 5: Electrochemical Machining of a Turbine Blade.** The constant-section portion of the turbine blade shown in Fig. 5(e) was electrochemically machined from either of two heat-resistant alloys (A-286 or Waspaloy) in a machine that could provide 5000 A at 2 to 20 Vdc, using a tool made of a copper-tungsten alloy.

To make the blade of A-286, the current began at 100 A, rose to 170 A, and ended at 150 A, at 11 volts. The electrolyte was a sodium nitrate solution (0.26 kg/L, or 2.2 lb/gal. of water) at a temperature of 43 °C (110 °F). Pressure of 900 to 1400 kPa (130 to 200 psi) caused electrolyte flow of 7.6 L/min (2 gal./min), with 50 μm filtration. Penetration of the 1.8 mm (0.7 in.) length at the rate of 7.6 mm/min (0.300 in./min) took 2.33 min, machining the surface to a final roughness of 0.38 to 0.63 μm (15 to 25 μin.).

To make the blade of Waspaloy, the current began at 100 A, rose to 160 A, and ended at 150 A, at 12 V. The electrolyte was a sodium chloride solution (0.20 kg/L, or 1.7 lb/gal. of water) at a temperature of 32 °C (90 °F). A pressure of 900 to 1400 kPa (130 to 200 psi) caused electrolyte flow of 9.5 L/min (2.5 gal./min), with 50 μm filtration. Penetration of the 1.8 mm (0.7 in.) length at the rate of 8.25 mm/min (0.325 in./min) took 2.15 min; surface finish was 0.25 to 0.50 μm (10 to 20 μin.).

**Example 6: Electrochemical Machining of Slots in 52100 Steel Plates.** Eight slots 8.4 ± 0.2 mm (0.330 ± 0.007 in.) deep were simultaneously machined in a valve plate (Fig. 5f) made of 52100 steel hardened to 65 HRC. The holes were made in a setup that could provide 5000 A at 2 to 20 Vdc. A brass tool was used, with a current beginning at 1500 A and ending at 2200 A, at 13 V. The electrolyte was a sodium chloride solution (0.20 kg/L, or 1.7 lb/gal. of water) at a temperature of 32 °C (90 °F). A pressure of 690 to 1000 kPa (100 to 150 psi) caused the electrolyte to flow at 68 to 76 L/min (18 to 20 gal./min), with 50 μm filtration. Tool penetration was 8.4 mm (0.330 in.), with a feed rate of 3.8 mm/min (0.150 in./min). Cutting time was 2.2 min; finish was 0.63 to 1.25 μm (25 to 50 μin.).

**Example 7: Electrochemical Machining of Spiral Grooves.** Seventy two equally spaced spiral grooves (Fig. 5g), 0.38 mm (0.015 in.) deep by 1.0 mm (0.040 in.) wide, were machined in a friction disk (1020 steel, 60 to 75 HRB) in an ECM machine that could provide 12 800 A at 14 to 24 Vdc. A brass tool was used, with current beginning at 650 A and ending at 600 A, at 20 V. The electrolyte was a solution of sodium chloride (0.15 kg/L, or 1.25 lb/gal., of water) at a temperature between 30 and 32 °C (85 and 90 °F). A pressure of 690 kPa (100 psi) caused the electrolyte to flow at 40 L/min (10 gal./min) through a clarifier. A tool penetration of 0.4 ± 0.1 mm (0.015 ± 0.005 in.) was performed with a feed rate of 1.3 mm/min (0.050 in./min) and with a 5 s final dwell. Cutting time was 0.3 min. The 72 grooves were machined at one time, and the operation was repeated on the other side.

**Example 8: Electrochemical Machining of Multiple Small Cavities in Inconel 718.** Milling small cavities into Inconel 718 can be very costly because small cutters must be used. With electrochemical machining, however, the cavities can be machined in groups, and perishable cutters are not required. Figure 5(h) shows the basic cavity characteristics of a part for which about 7 h of conventional milling is needed to machine 60 cavities. This part can be electrochemically machined in approximately 1.5 h.

A vertical-ram ECM machine with a 10 000 A, 2 to 20 Vdc power supply was used to machine ten cavities per cut. The ten brass cathodes were produced on precision CNC milling machines, thus making replacement cathodes more economical and accurate. Lot sizes were held to offset the 8-h setup over as many parts as possible.

The electrolyte was a sodium chloride solution (0.138 kg/L, or 1.15 lb/gal., of water) at 38 °C (100 °F). The feed rate was 2.2 mm/min (0.085 in./min) for a depth of 13.5 mm (0.530 in.). A voltage of 14 Vdc and a maximum current of 2600 A were used. Surface finishes of 1.60 μm (63 μin.) on the bottom and 3.8 μm (150 μin.) on the sidewalls were typical.

# Electrochemical Grinding

Robert E. Phillips, Everite Machine Products Company

ELECTROCHEMICAL GRINDING (ECG), also called electrolytic grinding, is similar to electrochemical machining (ECM), except that the cathode is an electrically conductive abrasive grinding wheel instead of a tool shaped like the contour to be machined. Electrochemical grinding is used primarily to machine difficult-to-machine alloys (such as stainless steel, Hastelloy, Inconel, Monel, Waspalloy, and tungsten carbide), heat-treated workpieces (60 to 65 HRC and harder), fragile or thermosensitive parts, or parts for which stress-free and burr-free results are required. The process was introduced in the early 1950s, evolving from developments in the U.S.S.R. on electrical discharge machining. Both processes, electrochemical machining and electrical discharge machining (EDM), make use of the direct application of the electrical energy in metal removal. However, ECM and ECG should not be confused with EDM because ECM and ECG involve electrochemical processes, not a discharge process.

## Process Characteristics

Electrochemical grinding removes metal by a combination of electrochemical and grinding actions. The electrochemical action is the reverse plating or oxidation process that occurs when a metallic workpiece passes an electrical current in the presence of an electrolyte. This electrochemical action is responsible for most (90%) of the material removal, but the grinding action of the cathode wheel removes the buildup of oxide film on the surface of the workpiece.

In electrochemical grinding (Fig. 1), the workpiece is a positive electrode (the anode), and the wheel is a negative electrode (the cathode). The two electrodes are connected to a direct current source and are immersed in a conductive ionic solution (the electrolyte) containing positively and negatively charged ions. Under the influence of electric potential, the positive ions migrate to the cathode, and negative ions migrate to the anode. At the cathode, the positive ions of the electrolyte take on electrons (reduction) and are discharged as neutral atoms or molecules. At the anode, the negative ions release electrons (oxidation) and are discharged as neutral atoms or molecules. As electrolysis occurs, an oxide film is formed on the surface of the anodic workpiece. The oxide film is an electrically insulating dielectric and, if allowed to remain, would slow or stop the process, depending on the porosity of the film. The abrasive material in the cathode wheel wipes the oxide away, exposing metal for continued oxidation. The mechanics of the process are shown in Fig. 1.

Less power is needed for electrochemical grinding than for electrochemical machining because the machining area is smaller and the abrasive in the wheel is removing the oxide. Current ranges from 5 to 1000 A are most common, with a voltage of 3 to 15 V over an electrolyte gap of approximately 0.025 mm (0.001 in.) or less. The process is usually operated at room temperature, at a wheel speed of 1100 to 1800 m/min (3500 to 6000 sfm). New high-speed slotting applications using metal bond diamond wheels are sometimes run at speeds of 2500 to 3000 m/min (8000 to 10 000 sfm) or more. Operating voltages are generally adjusted to give a maximum metal removal rate consistent with surface finish requirements and acceptable wheel wear. This adjustment is often made by increasing the voltage setting until audible sparking is observed and then reducing the voltage by ½ to 1 V. Lower voltages are used if the objective is close tolerances, reduced overcut, sharp edges, bright finishes, or other special requirements. Amperage, rate of chemical change, and rate of stock removal increase linearly with increases in voltage, until severe arcing and sparking occur. This reflects a change from electrochemical grinding to electrical discharge machining, which can result in severe pitting of both electrodes. Depending on the type of cathode wheel used, this change takes place at a point somewhere between 10 and 15 V. This condition can be quickly corrected by voltage adjustment.

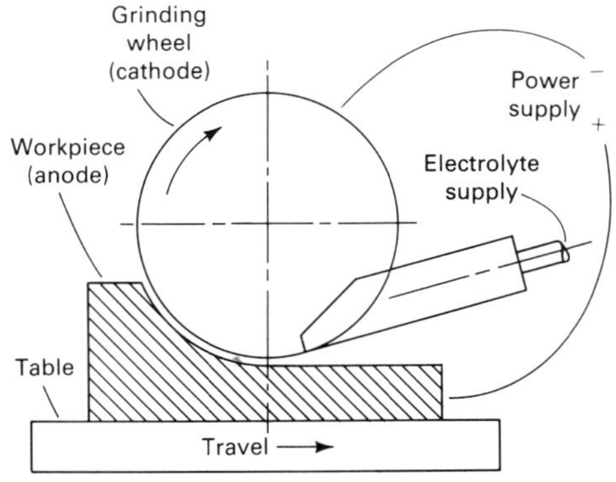


**Fig. 1** Schematic of electrochemical grinding

Wheel-workpiece force is generally considered to be very low. Depending on the area of contact between the wheel and workpiece, it is frequently in the range of 140 kPa (20 psi) or less. On larger cuts, tests have shown it can be as high as 1400 kPa (200 psi), but this is not normal.

**Removal Rates.** Although 99% of the material removed can be accomplished by electrochemical action, usually 5 to 10% of the metal removed is by abrasive action. For most metals, a removal rate of 1600 mm$^3$/min (0.1 in.$^3$/min) per 1000 A of current can be assumed for preliminary purposes.

Theoretically, ECG removal rates are limited by Faraday's Second Law, which states that 1 Faraday (96 494 coulombs or ampere seconds) will liberate 1 g equivalent weight of a substance, or its atomic weight divided by the valence. For example, the gram equivalent weight of iron is the atomic weight divided by the valence of the dissolved iron, or 56 divided by 2 equals 28. Thus, under laboratory conditions, 28 g of iron will dissolve during the passage of each Faraday of electricity. Table 1 lists the theoretical metal removal rates at 1000 A of current flow.

Removal rates for a given current are, of course, less than the theoretical limit, and variations in temperature, metallurgy, and electrochemical reactivity of the electrolyte affect the metal removal rate. The most accurate method of determining removal rates is by empirical testing.

The electrical current in a particular application is determined by the current density and the area of the gap between the anode (wheel) and the cathode (workpiece). Current density most often depends on the material being processed, although it is also affected by gap distance and voltage. Table 2 lists current densities for various work-

**Table 1 Theoretical limits of electrochemical grinding**

| Metal | Valence | Density g/cm³ | Density lb/in.³ | Metal removal rate at 1000 A cm³/min | Metal removal rate at 1000 A in.³/min |
|---|---|---|---|---|---|
| Aluminum | 3 | 2.67 | 0.098 | 2.06 | 0.126 |
| Beryllium | 2 | 1.85 | 0.067 | 1.50 | 0.092 |
| Chromium | 2 | 7.19 | 0.260 | 2.25 | 0.137 |
| | 3 | 7.19 | 0.260 | 1.51 | 0.092 |
| | 6 | 7.19 | 0.260 | 0.75 | 0.046 |
| Cobalt | 2 | 8.85 | 0.322 | 2.05 | 0.125 |
| | 3 | 8.85 | 0.322 | 1.38 | 0.084 |
| Niobium | 3 | 8.57 | 0.310 | 2.16 | 0.132 |
| | 4 | 8.57 | 0.310 | 1.69 | 0.103 |
| | 5 | 8.57 | 0.310 | 1.34 | 0.082 |
| Copper | 1 | 8.96 | 0.324 | 4.39 | 0.268 |
| | 2 | 8.96 | 0.324 | 2.20 | 0.134 |
| Iron | 2 | 7.86 | 0.284 | 2.21 | 0.135 |
| | 3 | 7.86 | 0.284 | 1.47 | 0.090 |
| Magnesium | 2 | 1.74 | 0.063 | 4.34 | 0.265 |
| Manganese | 2 | 7.43 | 0.270 | 2.28 | 0.139 |
| | 4 | 7.43 | 0.270 | 1.15 | 0.070 |
| | 6 | 7.43 | 0.270 | 0.77 | 0.047 |
| | 7 | 7.43 | 0.270 | 0.66 | 0.040 |
| Molybdenum | 3 | 10.22 | 0.369 | 1.95 | 0.119 |
| | 4 | 10.22 | 0.369 | 1.47 | 0.090 |
| | 6 | 10.22 | 0.369 | 0.98 | 0.060 |
| Nickel | 2 | 8.90 | 0.322 | 2.11 | 0.129 |
| | 3 | 8.90 | 0.322 | 1.36 | 0.083 |
| Silicon | 4 | 2.33 | 0.084 | 1.87 | 0.114 |
| Silver | 1 | 10.49 | 0.379 | 6.39 | 0.390 |
| Tin | 2 | 7.30 | 0.264 | 5.05 | 0.308 |
| | 4 | 7.30 | 0.264 | 2.52 | 0.154 |
| Titanium | 3 | 4.51 | 0.163 | 2.19 | 0.134 |
| | 4 | 4.51 | 0.163 | 1.65 | 0.101 |
| Tungsten | 6 | 19.30 | 0.697 | 0.98 | 0.060 |
| | 8 | 19.30 | 0.697 | 0.74 | 0.045 |
| Uranium | 4 | 19.10 | 0.689 | 1.92 | 0.117 |
| | 6 | 19.10 | 0.689 | 1.29 | 0.078 |
| Vanadium | 3 | 6.10 | 0.220 | 1.74 | 0.106 |
| | 5 | 6.10 | 0.220 | 1.05 | 0.064 |
| Zinc | 2 | 7.13 | 0.258 | 2.85 | 0.174 |

**Table 2 Current densities for various materials**

| Material | Current density A/cm³ | Current density A/in.³ |
|---|---|---|
| Cast iron cutting grade of carbide | 78–124 | 500–800 |
| Steel cutting grade of carbide | 124–155 | 800–1000 |
| Low-carbon steels | 465–620 | 3000–4000 |
| High-carbon steels | 310–465 | 2000–3000 |
| Stainless steels | 465–620 | 3000–4000 |
| Stellite | 310–465 | 2000–3000 |
| Udimet 500 | 465–620 | 3000–4000 |

piece materials. To obtain maximum removal rates, the grinding area should be as large as possible so that it will draw greater current.

**Feed rates** vary with different parameters, depending on the grinding method. In plunge, or face, grinding (as described in the section "Methods" in this article), for example, the entire surface being ground is exposed to the face of the wheel. Consequently, the proper feed rate can be estimated from the material removal rate and the current density of the workpiece. For example, with a material removal rate of 1600 mm³/min (0.1 in.³/min) per 1000 A and a current density of 1.2 A/mm² (800 A/in.²), the proper feed rate is about 1.9 mm/min (0.08 in./min).

In surface grinding (as described in the section "Methods" in this article) the area of the electrode gap is determined by the wheel diameter and the depth of cut. Consequently, feed rates in electrochemical surface grinding depend on current density, material removal rate, depth of cut, and wheel diameter. Figure 2 shows some typical feed rates in electrochemical surface grinding.

Generally, if the feed rate is too slow for a particular application, an overcut is produced, resulting in poor surface finishes and tolerances. If the feed rate is too fast, excessive wheel wear results.

**Surface Finish.** Electrochemical grinding involves a chemical reaction with minimal mechanical abrasion. Therefore, the surface finish generated by electrochemical grinding depends on the metallurgical crystalline structure of the workpiece. The finer the crystalline structure, the finer the finish. Instead of generating a surface that has tool marks, the process generates a surface that has the topography of the crystalline structure. To the naked eye it looks like a gray/black matte finish. The color that is generated is the color of the oxide finish on all but a few alloys that produce a bright surface.

Surface roughness generally runs from 10 to 40 μm (400 to 1600 μin.) depending on the material being machined. If a better finish is required, a finish pass is incorporated in the procedure. The finish pass is performed at a low voltage of 3 to 5 V and a relatively high feed rate of 250 to 500 mm/min (10 to 20 in./min) or more. The surface metal is burnished, with the crests of the hills being pushed into the valleys of the crystalline structure, thereby generating a smoother finish. The use of electric current while performing this sparkout pass, even though it is at a low rate, is very effective in minimizing stress and pressure on the workpiece. The selection of the proper ECG wheel is important. Because the abrasive action of the wheel is the major portion of the process during a sparkout operation, the selection of the grit determines the end finish quality. Fine-grit wheels of 220 to 320 grit (or grain) size have been successful, for example, for grinding surgical instruments and industrial knives.

**Accuracy of Electrochemical Grinding.** The ability to hold close tolerances depends upon the current, electrolyte flow, feed rate, and metallurgy of the workpiece itself. The more reactive the workpiece is to the ECG decomposition, the more difficult it is to hold close tolerances. As a result, metals that are very reactive, such as alloys with high chromium content, require mild electrolytes that will not produce great amounts of overcut. In extreme conditions, the electrolyte has to be monitored for temperature, specific gravity, and conductivity on a regular, automatic cycle. With normal ECG operations, tolerance expectations are generally ±0.025 mm (±0.001 in.), and in well-controlled operations, they are ±0.01 mm (±0.0005 in.). When extreme monitoring has been implemented, as described above, with the temperature of the electrolyte maintained within ±0.6 °C (±1 °F) along with control of conductivity and specific gravity, companies have achieved tolerance-holding capabilities of ±0.0025 mm (±0.0001 in.). Such a case is definitely the exception and should not be expected with normal ECG operations. By using a sparkout pass, tolerance-holding capabilities are greatly enhanced.

The selection of grit size, current flow, electrolyte, and electrolyte flow determine minimum corner radii. An inside corner of 0.25 to 0.40 mm (0.010 to 0.015 in.) is normally accepted as the standard. This has been accomplished with 100-grit wheels. Finer-grit wheels produce sharper corners, but require more frequent wheel dressing because of more rapid wheel wear and form breakdown. As always, the material being machined has a major effect on any accuracies achieved. External corners normally have a radius of approximately 0.1 to 0.2 mm (0.005 to 0.008 in.). The radius can be reduced by taking the finish pass at a lower voltage.

**Workpieces With Dissimilar Materials.** In electrochemical grinding of dissimilar materials such as carbides brazed to tool steel or parts that have been pressed togeth-

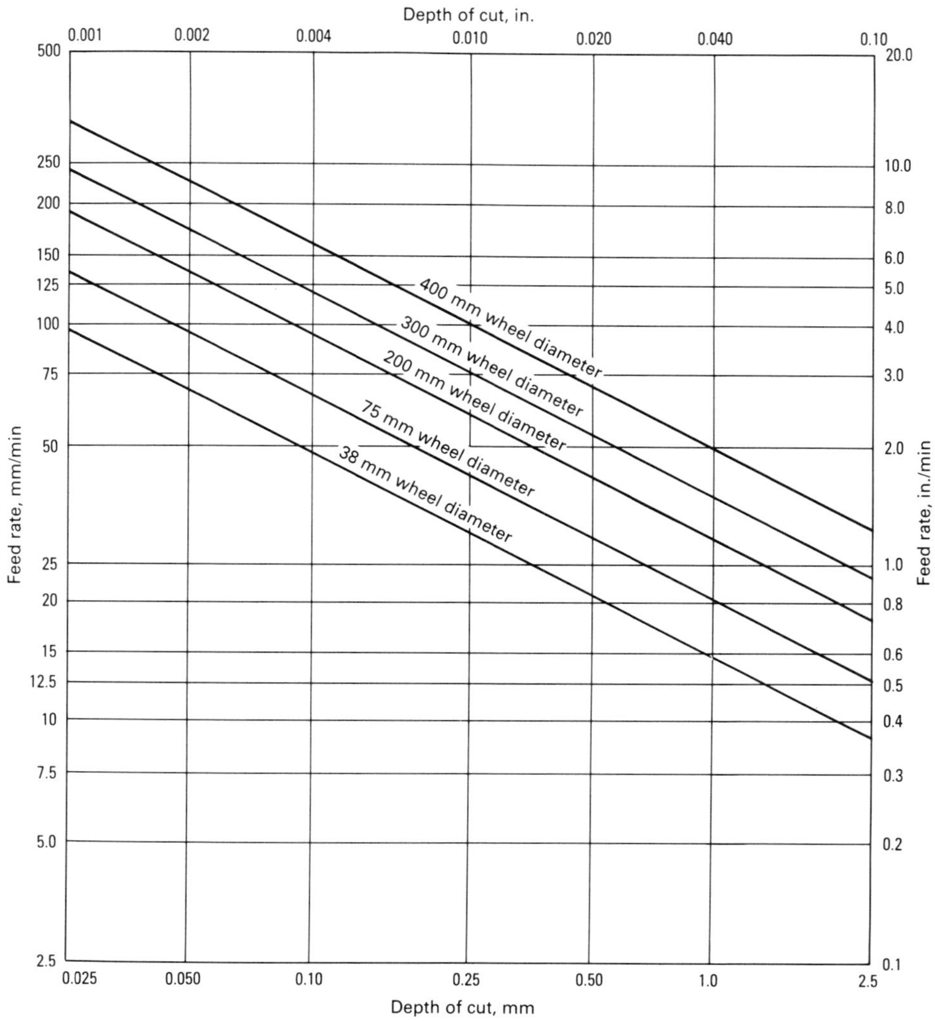


**Fig. 2** Typical feed rates for surface electrochemical grinding, determined by a current density of 1.5 A/mm$^2$ (1000 A/in.$^2$) and wheel speeds of 1200 to 2100 m/min (4000 to 7000 sfm)

er, the selection of the proper voltage and the resultant current density are based on the ratio of the areas of the two materials. In surface grinding of carbide-tipped tools in which the wheel first grinds carbide and then passes to steel, the power needed is calculated for the current density of steel, while the feed rate is based on the current density for the carbide. It is necessary to prevent overcut of the more reactive material at the point at which the two materials are in contact. This is especially noticeable in a material such as cast iron combined with carbides. Lowering the voltage sometimes successfully corrects this problem.

**Oxygen and Hydrogen Generation.** The chemical or oxidation process that takes place at the workpiece generates not only an oxide or hydroxide film that must be removed, but also minute amounts of oxygen. The reduction process at the wheel also releases minute amounts of hydrogen. The amount of oxygen and hydrogen released in this process is negligible, and extensive testing has shown that because the hydrogen is generated at the (cathode) wheel, there is no hydrogen embrittlement of the workpiece (anode). When using sodium nitrate base electrolytes, another chemical (reduction reaction) change that may occur at the cathode is the formation of ammonia gas.

## Equipment

Figure 3 illustrates a typical ECG system. The major components of the system are:

- Electrolyte delivery and circulating system
- The electrolyte
- The dc power supply
- The grinding wheel
- The ECG machine
- The workpiece

**Supply of Electrolyte.** In electrochemical grinding, the electrodes are not totally immersed, yet there must be an ample supply of electrolyte. Therefore, the system is designed to supply high volumes of electrolytes. Even more critical is the application of electrolytes to the wheel and workpiece interface.

*Nozzles* are used to ensure proper wetting action of the wheel. These nozzles wrap around the wheel and are designed to break the air layer that is normally found around a grinding wheel. By rubbing against the wheel, the nozzle creates a partial vacuum and causes the electrolyte to be sucked up, filling the cavities around the grit. The rotation of the wheel then carries the electrolyte into the area of contact between the workpiece and the wheel, ensuring that sufficient electrolyte is present to permit the ECG process to proceed. The point at which the electrolyte is applied to the wheel should be as close as practicable to the area of machining so that the maximum possible amount of electrolyte is available at the point of contact between the wheel and the workpiece.

*Wheel speed* is also critical. Excessive revolutions of the wheel create centrifugal forces that will throw the electrolyte from the wheel and starve the process. Flow control valves are required to ensure that proper amounts of electrolyte are available

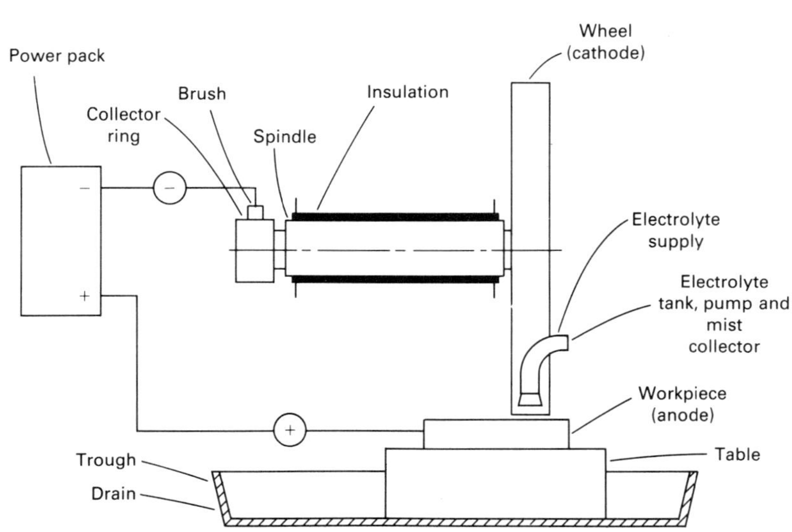


**Fig. 3** A typical electrochemical grinding system

for the process. In the past it was always assumed that the maximum amount of electrolyte was always best, but recent developments have shown that the objective of the ECG process is not always maximum metal removal. There are times when minimum amounts of electrolyte are applied for high-quality high-tolerance work, for example, for sharp edges or for thermosensitive or fragile workpieces.

**Electrolytes** developed for the ECG process resemble those used for the ECM process in that they are electrically conductive aqueous solutions containing chiefly inorganic salts. Many of the chemicals used in ECM electrolytes are also present in ECG electrolytes. However, the formulations for ECG are distinctly different. They are designed to enable faster formation of oxide films on the positively charged workpiece, whereas in ECM many electrolytes formed on the workpiece must dissolve at once in the electrolyte. Because the abrading action of the ECG wheel is continuously exposing fresh metal, faster-forming oxide deposits in ECG protect surrounding metal from being decomposed.

Electrolytes must be chosen for their intended applications and preferably should be formulated with additives to provide corrosion inhibition. When used in the machining of ferrous, nickel, and cobalt alloys, it is also desirable that the formulation efficiently chelate these ions. Formulas can be developed for each application. However, with costs as an additional item of consideration, standardized electrolytes have been developed.

A desirable ECG electrolyte should provide high conductivity, high stock removal efficiency, passivation to limit stray currents, good surface finishes, and corrosion inhibition. However, no single electrolyte provides all these properties for all metals. Sodium chloride, for example, provides high stock removal efficiency, but it does not provide the passivation protection of oxidizing electrolytes such as sodium nitrate or sodium chlorate. Yet sodium nitrate and sodium chloride cannot efficiently machine titanium. Neither can sodium chloride, sodium chlorate, nor sodium nitrate machine tungsten. For tungsten, one must use alkaline electrolytes such as sodium carbonate or sodium hydroxide. Sodium chloride is also an extremely corrosive electrolyte and thus can be very damaging to the machinery or workpiece.

The electrolyte salts used most often for the ECG process are sodium nitrate and potassium salts. Major components such as nitrites or alkaline sodium or potassium compounds (borate, carbonate, or phosphate) reduce corrosive attack on the equipment. The selection of electrolytes also varies with the materials being machined. Specification of salts varies among suppliers. There is more than one source for electrolytes; each supplier has its own reasons for specifying particular salts. In most cases, a single salt is sufficient for the majority of metals being machined. It is only by exception that special electrolytes would be required.

**ECG Wheels.** The abrasive grains on the ECG wheel serve three major purposes. First, they act to wipe the oxide from the workpiece, exposing new metal and allowing the process to continue. Second, the grit serves as a spacer to keep the conductive media in the wheel from making direct contact with the workpiece and generating a short circuit. Third, the cavities between the grit are filled with electrolyte, and the grit acts as a carrier bringing the electrolyte to the work area between the workpiece and the wheel, making the ECG process possible. For this reason, as previously stated, it is very important that the nozzle position relative to the wheel and the workpiece be correct. The nozzle must ride against the wheel to break the air layer that normally surrounds a grinding wheel, thereby allowing the electrolyte to be absorbed into the cavities that exist around the grit. After the wheel is dressed, it must be conditioned to remove loose particles of metal and resin from around the grit and expose the cavities. The conditioning is done by passing a conductive cable across the face of the wheel, allowing an ECG reaction between the cable and the wheel.

ECG wheels are made of an abrasive material, a bonding agent, and a conductive medium. Most ECG wheels have aluminum oxide as the abrasive and contain copper-impregnated resins for conductivity. The particle size of the aluminum oxide ranges from 60 to 320 grit. Other abrasive materials used for ECG include silicon carbide and diamond, and recently some attempts have been made to use borazon.

Wheels must be matched to the unique objective in each application. The diamond wheel, either metal bond or resin bond, is used for the grinding of carbides, heat-treated steels with a hardness of 60 to 65 HRC or higher, and thin slots. Slots 0.35 mm (0.014 in.) thick and 2.5 mm (0.100 in.) deep have been successfully machined. Diamond grit used in electrochemical grinding should be blocky rather than sharp, because the purpose is to burnish the oxide residue from the workpiece, not to physically abrade the material. When grinding materials such as Stellite, it has been found that a 60-grit wheel outperforms the 100-grit wheel by approximately 2:1. What actually happens is that more abrasive metal removal is introduced into the process.

The most common wheel is a resin bond, copper-impregnated, aluminum oxide grit wheel. Other wheels that can be found in service are metal bond diamond, resin bond diamond, diamond-plated, and silver-impregnated wheels. The copper-impregnated wheels used for electrochemical grinding are easily formed, are homogeneous in construction, and have superior conductivity. They are ideal for form grinding because they can readily be dressed with single-point diamond dressers, diamond dressing blocks, or roll dressers. In most applications, because the physical abrasion of metal is minimal, the form-holding capability of the wheel is superior, with dressing in many applications required at 4 h or even 8 h intervals. Form dressing is satisfactorily accomplished with sharp, chisel point diamond forms at 0.025 mm (0.001 in.) per pass at a cross feed from 300 to 380 mm/min (12 to 15 in./min).

New wheel formulations have allowed changes in the overall concept of ECG. It is now possible to increase feed rates and thereby increase the percentage of metal removed by abrasive action to generate higher production rates. Coarser or harder grits are used with the current applied as in traditional ECG grinding. The results are higher production with the retained benefits of burr-free and stress-free workpieces at a cost of greater wheel wear.

**The power supplies** for ECG should be designed with the following characteristics:

- The voltage should be continuously variable in the range of 0 to 15 V
- Once a specific voltage is set, the amperage variation should automatically vary (the amount of variation depending on the contact area between the wheel and the work) in such a manner that the current density remains constant
- The current should have a minimum of ripple. Recent studies have shown that the elimination of ripple is definitely advantageous

To determine the required current of an ECG operation, the average current density of the workpiece is multiplied by the area of the gap between the electrodes. ECG power supplies up to 3000 A are available, but power supplies greater than the 1000 A are seldom economical for electrochemical grinding.

Voltage control up to 15 V is sufficient because sparkover occurs with most materials when this voltage is exceeded. Variable voltage up to 20 to 25 V can be designed into power sources, but there is no application in the ECG process for which this would be required. Over the years of development and application it has been found that the constant current density controls are more efficient than other mechanisms. Typically the dc power cannot be started until the spindle and the electrolyte pump are running. By the same token, the feed rate motor control system cannot operate without power so the machine shuts down safely in the event of an overload. The wheel spindle must also be electrically insulated from the rest of the machine.

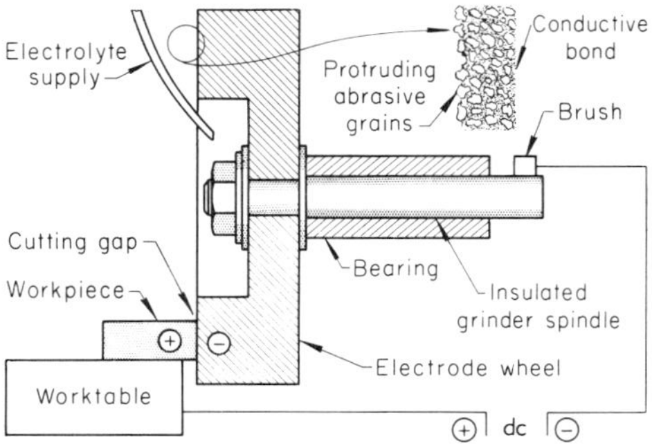


**Fig. 4** Schematic diagram of a setup for electrochemical plunge grinding. Nonconductive abrasive grains separate the conductive body of the wheel from the workpiece and determine the thickness of the cutting gap, which is flooded with electrolyte during grinding.

**ECG machines** are available with either vertical or horizontal spindles having outer- and inner-diameter grinders. Motorized rotary fixtures allow the surface grinding of small cylindrical parts. Larger cylindrical parts should be ground with inner- or outer-diameter grinders. Table feeds should be designed so that there is no slip-stick effect on the table or slides when motion is slow.

## Methods

Five different grinding methods can be performed with ECG equipment. The methods are face grinding, surface grinding, internal grinding, form grinding, and cylindrical grinding.

**Face, or plunge, grinding** (Fig. 4) is a method in which the entire surface to be ground is fed against the face of the wheel. To prevent uneven wheel wear from excessive truing of the wheel, either the work is oscillated across the face of the wheel, or the wheel is moved in an oscillating pattern. Oscillating the work is usually preferred.

This method produces a better surface finish; the other methods produce a spiral pattern on the workpiece because of the reciprocating action of the wheel. Face grinding is also the fastest method. By exposing the entire surface to the wheel, the maximum current and metal removal rate can be obtained. When plunge grinding a cylindrical surface with a diameter larger than the wheel, it is generally more economical to plunge grind in steps with a final sparkout to attain the desired diameter.

**Surface grinding** is performed by reciprocating the workpiece under the wheel (Fig. 1). The primary difference between conventional and electrochemical surface grinding is that the depth of cut for each wheel pass is much deeper with electrochemical grinding. The maximum depth of cut for a single ECG wheel pass is usually considered to be about 2.5 mm (0.100 in.). In surface grinding, the area of the electrode gap varies with wheel diameter and depth of cut (Fig. 5). Therefore, maximum current and material removal is achieved at the maximum depth of cut. Typical feed rates for surface electrochemical grinding are shown in Fig. 2.

**Internal grinding** also can be performed with the ECG process. It is similar to conventional internal grinding, although the largest-diameter wheel possible is selected for maximum current and material removal rates.

**Form grinding** is performed with soft wheels that facilitate the production of a form. Because formed wheels are used, this is the most costly method. Once formed, however, subsequent dressings are performed much less frequently than with conventional form grinding.

**Cylindrical grinding** is the slowest method because of the limited contact area.

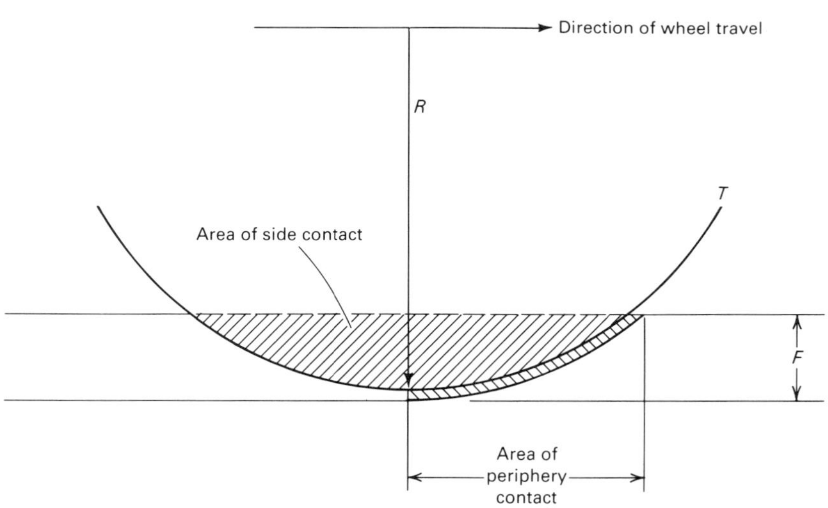


Total area of contact = Area of periphery contact + Area of side contact (Arccos is in degrees, not radians.)

$$\text{Area} = \frac{T\pi R}{180}\arccos\left(\frac{R-F}{R}\right) + 2\left[\frac{\pi R^2}{180}\arccos\left(\frac{R-F}{R}\right) - (R-F)\sqrt{2RF-F^2}\right]$$

**Fig. 5** Calculation of contact area of wheel and workpiece for the surface electrochemical grinding of slots. $R$, radius; $T$, wheel thickness; $F$, down feed. Source: Ref 1

## Advantages and Disadvantages

Table 3 compares electrochemical grinding with milling and conventional grinding for various workpiece materials. In general, ECG superiority increases with hardness or toughness of the material. Other advantages of ECG include:

- Absence of work hardening
- Elimination of grinding burn
- Burr-free surfaces
- Absence of distortion of thin, fragile, or thermosensitive parts
- Less wheel truing
- Surface finishes of 0.125 to 1.0 μm (5 to 40 μin.) depending on material, wheel, and other variables
- Tolerances of ±0.025 mm (±0.001 in.) under normal conditions and ±0.0025 mm (±0.0001 in.) under special conditions
- Longer wheel life

Disadvantages of electrochemical grinding are:

- Higher capital cost than for conventional machines
- Use limited to electrically conductive workpieces
- Corrosive nature of electrolytes
- Required disposal and filtering of electrolytes
- Noncompetitive removal rates compared to conventional methods for readily machinable metals

## Applications

The largest single use for electrochemical grinding today is the manufacturing and remanufacturing of turbine blades and vanes for aircraft turbine engines. The process is fast and very effective on these difficult-to-machine alloys, particularly for the machining of Stellite materials used in the remanufacturing of turbine components. The minimal wheel wear is a definite asset in form grinding. Other well-accepted applications in the turbine engine industry include the machining of honeycomb seals, for which burr-free and stress-free results are critical.

Machines have been designed for burr-free and stress-free electrolytic cutting of tubing. These machines have found extensive application in the specialty mill supply houses that provide stainless steel and more exotic materials cut to specific customer requirements.

**Table 3 Comparison of electrochemical grinding and precision conventional milling and grinding**

| Material | Stock removal (milling) | Stock removal (grinding) | Tooling costs and replacement | Size control | Production of fragile parts | Potential of heat damage | Quality of surface finish | Problems with burrs |
|---|---|---|---|---|---|---|---|---|
| Machinery steel | − | = | − | − | + | = | = | + |
| Tool steel, soft | − | = | − | − | + | + | = | + |
| Tool steel, hard | + | + | − | − | + | + | = | + |
| Cast iron | − | − | − | − | + | = | − | = |
| Copper | − | + | − | = | + | = | + | + |
| Brass | − | + | − | = | + | = | = | + |
| Aluminum | − | − | − | − | + | = | − | + |
| Tungsten | − | + | − | = | + | + | + | + |
| Tungsten carbide | + | + | + | = | + | + | + | = |
| Beryllium | − | = | − | − | = | = | − | + |
| 300 stainless | − | + | + | = | + | + | + | + |
| 400 stainless | − | + | + | = | + | + | + | + |
| Titanium | − | = | = | = | + | + | = | + |
| A-236 | − | + | + | = | + | + | = | + |
| Waspaloy | + | + | + | = | + | + | + | + |
| Inconel 718 | + | + | + | = | + | + | + | + |
| Inconel X | + | + | + | = | + | + | + | + |
| René 41 | + | + | + | = | + | + | + | + |
| HS-21 | + | + | + | + | + | + | = | = |
| Hastelloy alloy X | + | + | + | = | + | + | = | + |
| PWA 1004 | + | + | + | = | + | + | = | + |
| PWA 689 | + | + | + | = | + | + | = | + |
| AMS 5668 | = | + | + | = | + | + | = | + |
| Udimet 500 | + | + | + | = | + | + | + | + |
| Udimet 700 | + | + | + | = | + | + | + | + |
| Greek ascoloy | − | = | + | = | + | + | = | + |
| PWA 90 | + | + | + | = | + | + | + | + |

Note: =, processes about equal; −, ECG process inferior to conventional machining; +, ECG process superior to conventional machining. Source: Ref 2

The medical-instrument industry has been the newest area of growth. The ECG process is now being employed in the manufacture of surgical instruments, hypodermic needles, and numerous other types of medical apparatus. Because of the use of finer-grit wheels and new techniques using low-voltage finish passes, sharp edges can be generated when demanded. This has also opened up applications in the industrial knife market.

## REFERENCES


1. J.R. Thompson, Operating Parameters in Electrochemical Grinding With Aluminum Oxide Wheels, in *New Developments in Grinding*, Carnegie Press, 1972, p 810
2. R.R. Brandi, Basics of Electrochemical Grinding, *Am. Mach.*, Vol 118, 1974, p 48


## SELECTED REFERENCE


- "Metal Removal by Electrochemical Methods and Its Effect on Mechanical Properties of Metals," Report 213, Defense Metals Information Center, Battelle Memorial Institute

# Electrochemical Discharge Grinding

ELECTROCHEMICAL DISCHARGE GRINDING (ECDG), sometimes referred to as ECDM grinding, is a combination of electrochemical grinding (ECG) and electrical discharge grinding (EDG), with some modification of each. The process resembles ECG in the electrochemical formation of oxides on a positively charged workpiece (anode); however, it employs alternating current or pulsing direct current and does not use an abrasive-coated wheel. It resembles EDG in the use of a graphite wheel that does no mechanical grinding and in the use of intermittent spark discharges to remove material from the workpiece surface, but differs from EDG in using a highly conductive electrolyte instead of a dielectric fluid and in using low-voltage, pulsating direct current. Like ECG, electrochemical discharge grinding is most useful for grinding carbide tools, hardened tool steel, nickel-base alloys, and parts that are fragile or sensitive to heat.

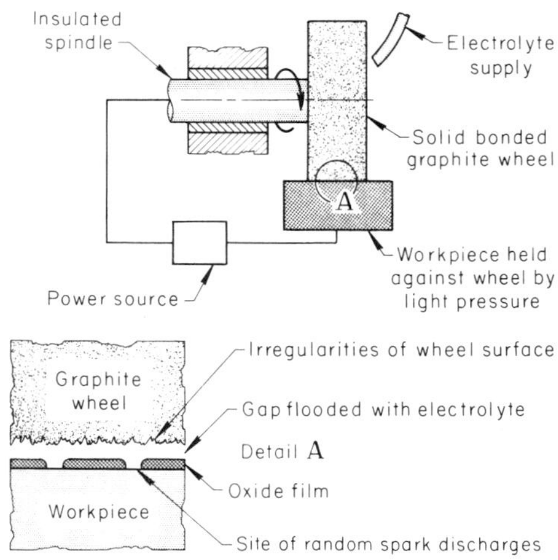


**Fig. 1** Schematic of setup for ECDG. The power source provides either an ac output at 60 Hz or a pulsing dc output at 120 Hz (see text). The workpiece is positively charged (anode) when direct current is used.

## Equipment

Figure 1 shows a setup for ECDG using a solid bonded graphite wheel. An enlarged sectional view of the grinding interface is also shown. The wheel is made of graphite and a bonding agent; most commercial grades with particle size of 300 mesh or finer are satisfactory. Like an EDG wheel, it contains no abrasive; dressing gives it a finish of about 6.4 μm (25 μin.) or less, usually producing the final contour in a single pass.

The spindle is electrically insulated to isolate the electrical circuit. Versatile table positioning and feed mechanisms are provided. In sharpening single-point tools, the workpiece (such as a disposable carbide insert) is held against a flat surface of the graphite wheel. In profile grinding, the workpiece passes over the periphery of a preformed graphite wheel.

The equipment includes an electrolyte mist collector as well as a supply tank in which the electrolyte is stored and to which used electrolyte is returned. The electrolyte is pumped through a filter before flowing again onto the work and the wheel. Accessories may include equipment for forming, trueing, and dressing the wheel without removing it from the machine. Formed plastic or graphite scrapers are used during profile grinding to provide a uniform distribution of electrolyte.

**Single-point grinders** for sharpening carbide tools have a self-contained 200 A, 12 V, 60 Hz ac power source. Commercial ECDG units for profile grinding and surface grinding are equipped with dual power sources to provide either an ac output like that of the single-point grinder or a pulsing dc output of 200 A or more at 4 to 12 V and 120 Hz. At a given rate of power consumption, metal removal on steel and most other metals is greater for pulsing direct current than for alternating current. However, alternating current must be used in grinding carbide.

**Electrolytes.** As with ECG, the electrolytes used for ECDG are conductive liquids (dilute aqueous solutions of inorganic salts). A solution containing 110 g/L (14 oz/gal.) of $KNO_3$ plus 55 g/L (7 oz/gal.) of $Na_2CO_3$ (anhydrous) is a general-purpose electrolyte suitable for the grinding of carbide and most metals; the $Na_2CO_3$ can be omitted except for carbide. The concentration of the solution is usually maintained at about 0.18 to 0.24 kg/L (1½ to 2 lb/gal.) (specific gravity of 1.11 to 1.15).

## Process Characteristics

The mechanism of ECDG can be understood in relation to the ECM, ECG, and EDG processes. In the EDG process, metal removal takes place entirely by high-voltage spark discharges through a dielectric fluid, while anodic oxidation in a conductive electrolyte is the primary mechanism for metal removal in ECM, ECG, and ECDG. In these three electrochemical processes, work metal can continue to dissolve anodically at a practical rate only if any continuous and adherent oxide films that are formed electrochemically on the surface of the work are continuously removed at a sufficiently rapid rate.

Both ECM and ECG make use of low-voltage, high-amperage direct current to dissolve metal from the workpiece. In ECM, electrolyte composition and current density are regulated so that no interfering adherent oxide films are produced. Any oxide films that form dissolve chemically in the electrolyte as rapidly as they are formed. In ECG, the mechanical abrading action of the protruding surface grains of abrasive on the wheel removes interfering films and nonreacting particles, thus exposing fresh metal for the electrochemical reaction.

In ECDG, metal is removed by first transforming it into a continuous, adherent, poorly conductive oxide film about 0.25 to 1.25 μm (10 to 50 μin.) thick. When alternating current is used, the oxide film builds up during a positive current pulse (workpiece positively charged), and it is removed at discharge sites when the breakdown voltage of the film is exceeded during the following negative pulse. The cycle is then repeated. When pulsing direct current is used, the oxide film builds up during the low-voltage portion of a pulse, and it is removed at discharge sites when the breakdown voltage of the film is exceeded. Over a period of a few seconds, discharge sites are randomly distributed in the cutting area.

Depending on its energy level, an individual spark can remove a small amount of metal (as well as the oxide film) from the work, leaving a small crater in the surface. However, cratering is not observed on the workpiece under normal operating conditions. Cratering occurs at spark sites on the graphite wheel and is a source of wheel wear.

**Current density** is usually controlled by a voltage adjustment on the current source. Current density can also be regulated by adjusting the gap distance. On a tool grinder, this is done by changing the spring loading that holds the workpiece against the wheel; on a profile grinder, the feed rate is changed. The relationship between pressure

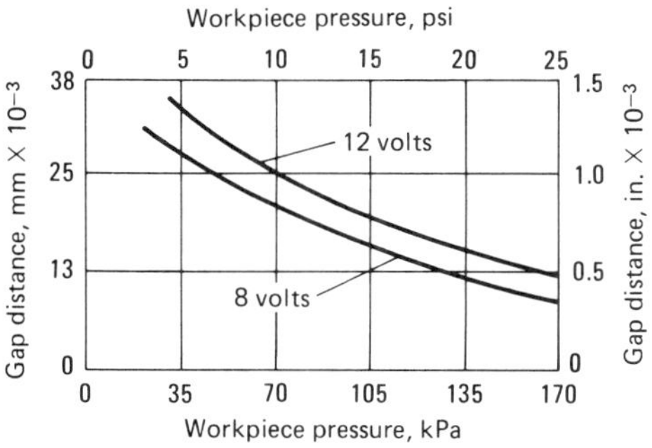


**Fig. 2** Effect of workpiece pressure on gap distance in ECDG

and gap at two gap voltages is shown in Fig. 2. A greater pressure (for a given wheel speed and electrolyte flow rate) is required at higher voltage to maintain the gap because of increased gassing. Wheel speed and electrolyte flow also affect gassing. Workpiece pressure is held between 35 to 140 kPa (5 and 20 psi) to maintain gap distance between 0.0125 and 0.030 mm (0.0005 and 0.0012 in.).

Total current in typical ECDG applications is 200 to 2000 A at a potential of 4 to 12 V. Metal could be removed more rapidly by increasing the current, but current density when using alternating current is restricted to about 0.9 A/mm$^2$ (600 A/in.$^2$) for carbides and to about 1.2 A/mm$^2$ (800 A/in.$^2$) for other metals. When a pulsing dc source is employed, a current density of about 1.9 A/mm$^2$ (1200 A/in.$^2$) can be used in grinding steel. Intermediate current densities are used for other metals. Heat damage on the work and wheel wear are greater at high current density.

**Wheel speed** for most applications is between 1200 to 1800 m/min (4000 and 6000 sfm). Lower speed does not permit the flushing of fresh electrolyte through the grinding gap at a rate sufficient to maintain the electrolyte composition in the gap. Lower speed also increases the electrical resistance across the gap by permitting the formation of larger gas bubbles in the gap. The maximum operational current density decreases steeply for wheel speeds below about 600 m/min (2000 sfm).

**Wear ratio** is the volume of metal removed from the workpiece divided by the volume of graphite worn from the wheel, and it is affected by gap voltage, as shown in Fig. 3. Operation at a gap voltage of about 8 V gives maximum wheel life or wear ratio. Below about 4 V, wheel wear by mechanical abrasion increases, and above about 12 V, cratering of the wheel surface results from high-energy spark discharges. In either situation, wheel wear is excessive. When using alternating current on carbides or on tool steel, the wear ratio is about 7:1; when using direct current on other ferrous workpieces, the wear ratio is about 40:1.

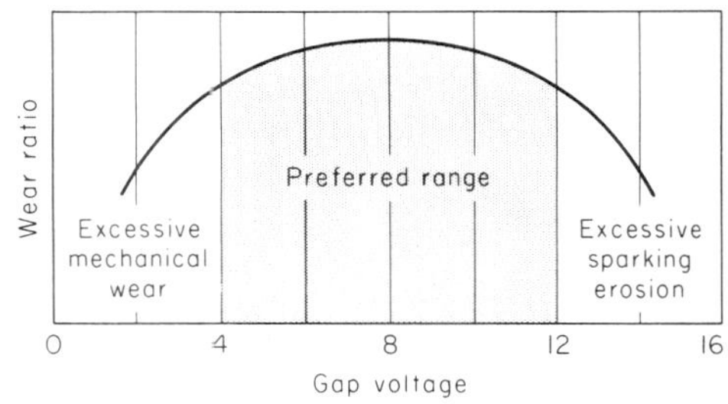


**Fig. 3** Effect of gap voltage on wear ratio in ECDG

**Table feed** rates are tabulated below for typical operations at a current density of 0.8 to 0.9 A/mm$^2$ (500 to 600 A/in.$^2$):

| Work metal | Plunge grinding mm/min | Plunge grinding in./min | Surface grinding(a) mm/min | Surface grinding(a) in./min |
|---|---|---|---|---|
| Carbide | 0.5 | 0.020 | 3.8 | 0.15 |
| Steel | 1.5 | 0.060 | 12.7 | 0.50 |

(a) For 2.5 mm (0.10 in.) depth of cut; 200 mm (8 in.) diam wheel

Typical metal removal rates are 100 mm$^3$/min (0.006 in.$^3$/min) for carbide and 250 mm$^3$/min (0.015 in.$^3$/min) for steel, at a machining current of 200 A. Typical depth of cut in a single pass is about 6.4 mm (¼ in.).

**Accuracy and Finish.** Dimensional accuracy of ±0.0125 mm (±0.0005 in.) can be obtained under carefully controlled conditions in plunge or surface grinding. Similar tolerances can be met in profile grinding by using auxiliary plastic or graphite scrapers to control the distribution of the electrolyte on the wheel surface. The accuracy obtained in ordinary production operations is ±0.025 mm (±0.001 in.). Typical surface roughness values are 0.125 to 0.375 μm (5 to 15 μin.) for carbide and 0.375 to 0.75 μm (15 to 30 μin.) for steel.

**Wheel Maintenance.** Flat surfaces of a graphite wheel can be dressed with a single-point carbide or high-speed steel tool on the grinder. Frequent dressing of the wheel is unnecessary; good results can be obtained even with a wheel surface that shows scratches or gouges. Wheel life depends mainly on the grade of graphite, operating voltage, and type of power supply.

For profile or form grinding, the wheel can be formed in place on the grinder by plunge cutting with a high-speed steel master. While in use, the wheel can be dressed in the same manner when necessary.

**Profile Grinding.** The accuracy with which ECDG profile grinding reproduces the dimensions of the master is primarily limited by the uniformity of electrolyte film thickness on the formed wheel at the grinding interface. Figure 4 illustrates the nonuniform pattern of electrolyte distribution obtained at the grinding interface of a rotating formed wheel unless auxiliary devices such as scrapers are provided to distribute

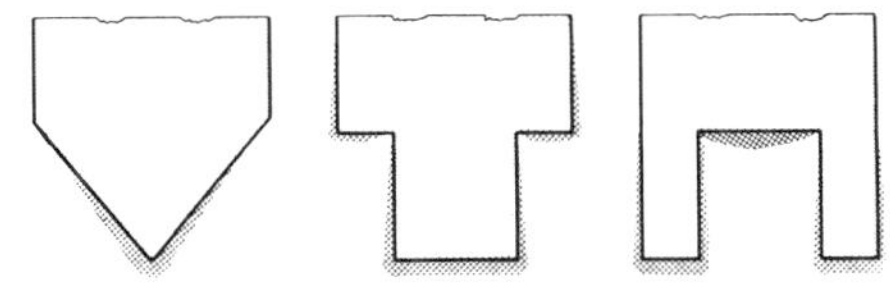

**Fig. 4** Nonuniform electrolyte film thickness on the surface of a rotating graphite ECDG wheel. The shaded areas show the variation of film thickness for various wheel shapes.

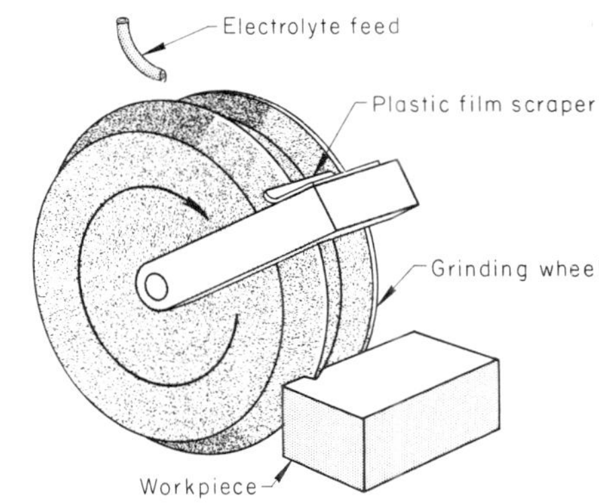


**Fig. 5** Use of a plastic film scraper to equalize the distribution of electrolyte on the graphite wheel in profile grinding by ECDG

the liquid more uniformly over the contours of the wheel.

For the profile grinding of single-point tools that do not have flat vertical or horizontal surfaces, a flat piece of 0.15 mm (0.006 in.) thick fluorocarbon plastic sheet is held against the wheel peak, as shown in Fig. 5. This plastic sheet scrapes off excessive accumulation of electrolyte at the wheel peak.

For grinding multiple-contour profiles that do not have flat vertical or horizontal surfaces (such as chaser tools), conforming graphite scrapers about 9.5 mm (3/8 in.) thick are held against the wheel at a pressure of about 70 kPa (10 psi). The graphite scrapers are formed on the profile grinding machine in a normal ECDG operation, using pulsing direct current (0.3 to 0.7 A/mm$^2$, or 200 to 450 A/in.$^2$) and a workpiece pressure of 0.03 to 0.05 MPa (4 to 7 psi) (gap spacing of 0.0125 to 0.0375 mm, or 0.0005 to 0.0015 in.). At the pressures and current densities used, wear on the wheel is negligible because the gassing and the electrolyte fluid pressure almost completely eliminate direct physical contact between the workpiece and the wheel.

For grinding complex profiles that have both curved contours and flat vertical or horizontal surfaces, compressed air is injected through a hollow conforming graphite scraper to control the electrolyte distribution, particularly on the flat vertical or horizontal surfaces. Optimum machining accuracy is obtained by using a scraper pressure of about 100 to 200 kPa (15 to 30 psi) while feeding air at a pressure of about 300 to 400 kPa (40 to 60 psi) through a 2 mm

(0.08 in.) wide slot extending across the scraper. Under these conditions, the distance between the scraper and the wheel is about 0.05 to 0.2 mm (0.002 to 0.007 in.).

In a well-controlled operation that makes use of scrapers, dimensional deviation from the wheel-forming master averages less than ±0.0125 mm (±0.0005 in.).

**Comparison with EDG and ECG.** The ECDG process has the same general range of applications as EDG and ECG and shows the same advantages in grinding materials and shapes that present problems in conventional grinding with abrasive wheels. Electrochemical discharge grinding requires 10 to 15 times as much current as EDG to remove metal at the same rate, but ECDG has much higher metal removal capability and produces smooth surfaces at much faster metal removal rates, as shown below:

| Work metal | Finish μm (μin.) | Removal, $mm^3 \times 10^3$/h ($in.^3$/h) EDG | ECDG (a) |
|---|---|---|---|
| Tool steel | 0.75 (30) | 0.3 (0.02) | 6 (0.4) |
| Carbide | 0.25 (10) | 0.07 (0.004) | 15 (0.9) |

(a) Current of 200 A and a depth of cut of 13 mm (0.5 in.)

However, dimensional accuracy is lower for ECDG than for EDG:

| | EDG | ECDG |
|---|---|---|
| Close control, mm (in.) | ±0.0025 (±0.0001) | ±0.0125 (±0.0005) |
| Routine production, mm (in.) | ±0.0125 (±0.0005) | ±0.025 (±0.0010) |

Electrochemical discharge grinding has the advantage over ECG of using much less expensive wheels than the specially manufactured abrasive-coated wheels required for ECG. However, electrochemical grinding does have a much higher metal removal rate than ECDG. Electrochemical grinding also leaves a metallurgically clean surface, whereas electrochemical discharge grinding leaves a recast layer.

The ECDG graphite wheels are readily formed to intricate shapes and dressed while in place on the grinder, and they have long life. A single cup-type ECDG wheel can be used to grind more than 35 000 disposable carbide inserts to a depth of 0.25 mm (0.010 in.). Furthermore, it is necessary to stock wheels of only one material, instead of the several different types needed for a range of applications in ECG. Because ECDG uses a lower pressure of workpiece against wheel, it can do more delicate operations in grinding stress-free, burr-free surfaces on honeycomb and other fragile structures that are difficult to grind by other methods.

**Applications.** Single-point carbide disposable inserts have been ground to a depth of 0.25 mm (0.010 in.) with a surface finish of 0.375 to 0.635 μm (15 to 25 μin.) at a rate of one insert per minute. In a typical profile grinding operation, three C-5 carbide chaser tools were machined at the same time, removing 2.5 mm (0.1 in.) of metal in a single pass in 15 min (plus 8 min for forming a graphite scraper). Mean thread height deviation from master dimensions was ±0.0075 mm (±0.0003 in.), and pitch deviation was half this amount. Using air injection, brazed C-6 carbide dovetail form tools were ground to an average depth of 8 mm (0.3 in.) over a 20 mm (0.8 in.) length in 37 min per piece (plus 25 min for forming the scraper).

# Electrostream and Capillary Drilling

M. Anthony Newton, NItech, Inc.

ELECTROSTREAM DRILLING (ES) AND CAPILLARY DRILLING (CD) are electrochemical machining processes that were developed by General Electric and Rolls Royce, respectively, for drilling holes that are too deep to be drilled by electrical discharge machining and too small to be drilled by shaped tube electrolytic machining. Electrostream and capillary drilling both cover the same general size range and are similar in that they both use glass tube cathodes and acid electrolytes, but there are distinct differences in the process and in the equipment. Both processes differ from shaped tube electrolytic machining in that they use glass tubes for cathodes instead of coated titanium tubes. The ES and CD processes also do not require periodic voltage reversals.

Applications for ES and CD include:

- Drilling small rows of cooling holes in gas turbine blades and vanes
- Drilling wire electrical discharge machining (EDM) start holes less than 0.5 mm (0.020 in.) in diameter
- Any application where higher-strength materials have to be substituted to improve quality, performance, reliability, and weight and machinability is a problem

## Process Capabilities

**Feed rates** for ES and CD typically range from 0.75 to 2.5 mm/min (0.030 to 0.100 in./min). Feed rates depend on the material being drilled, but not on the number of holes being drilled simultaneously.

**Normal hole sizes and depths** are:

- 0.15 mm (0.006 in.) in diameter to a depth of 6 mm (0.250 in.) to 0.5 mm (0.020 in.) in diameter to a depth of 25 mm (1.000 in.) for ES
- 0.25 mm (0.010 in.) in diameter to a depth of 12 mm (0.500 in.) to 0.5 mm (0.020 in.) in diameter to a depth of 25 mm (1.000 in.) for CD

**Normal hole size tolerances** for ES and CD are:

- 0.25 mm (0.010 in.) in diameter, ±0.025 mm (±0.001 in.)
- 0.5 mm (0.020 in.) in diameter, ±0.05 mm (±0.002 in.)

These tolerances can be halved by special process control and with very pure metals. The normal hole depth tolerance, where applicable, is ±0.05 mm (±0.002 in.).

**Normal hole wander** is negligible unless the cathode hits an impurity in the material being drilled; this can result in either cathode breakage or cathode deflection. Hole wander of 0.002 mm/mm (0.002 in./in.) is possible.

**Competitive Processes.** Electrical discharge machining is usually much slower than the electrolytic processes and generally is not capable of drilling to such depths. The EDM process can be used to drill deep holes, but the electrode wear rate makes the electrode cost prohibitive.

In addition, electrical discharge machining is less effective if there is not sufficient space behind the exit of the hole to enable the electrodes to proceed past the exit until the unworn electrode makes a parallel-sided hole all the way through. The EDM process also produces a recast layer, which may not be acceptable.

Conventional drilling can produce much smaller holes, but the depth-to-diameter ratio obtainable is much less than that possible with the electrolytic processes. In addition, machinability becomes a major factor. Hole diameter tolerance is usually better with conventionally machined holes.

**Limitations and Advantages.** Electrostream and capillary drilling can be used only on corrosion-resistant metals (stainless steels and cobalt- and nickel-base turbine engine alloys) and electrically conductive workpieces. Generally, the CD and ES processes cannot drill commercially pure titanium and refractory metals. Other limitations include the following:

- The process is slow when drilling single holes
- Problems can result when the hole breaks through the bottom of the part
- The handling of acid requires special workplace and environmental precautions
- A hazardous waste is generated
- Oblique entry is difficult

Additional limitations relating to ES are the following. High voltage requires additional precautions and a more complex electrolyte system, and the tube geometry limits the minimum spacing between holes. For CD, cathode arrays are more complex.

The advantages of CD and ES drilling include the following:

- The hardness (machinability) of the metal is not a consideration
- Depth-to-diameter ratios are very high
- Large numbers of holes can be drilled simultaneously
- Blind and intersecting holes can be drilled
- No recast layer or metallurgical defects are produced
- Powder metallurgy alloys can be machined
- A parallel-sided hole can be drilled without the need for clearance beyond the hole exit (as in EDM)
- No burrs are produced

An additional advantage of ES is that smaller holes can be drilled than when using CD. However, CD is a simpler machine system and process; there are also no hole spacing limitations using CD.

## Equipment and Tooling

**Electrostream drilling** is a high-voltage process in which a voltage is applied between the workpiece and an acid electrolyte that flows through a manifold and nozzle-shaped glass tubes (Fig. 1). The voltage, which must range from 600 to 900 V, drives a current through an electrolyte column measuring 50 to 100 mm (2 to 4 in.).

Electrostream drilling also requires an extra tank in the electrolyte system. The acid is pumped from one tank and, after traveling through the machine, overflows into a separate tank, which is electrically insulated from the first. Thus, current is prevented from making complete circuits around the system. The disadvantage is that periodically one tank must be emptied and the other refilled.

The machine used with ES is also necessarily much more elaborately grounded and guarded to prevent 600 V mishaps.

**Capillary drilling** uses an array of shorter glass tubes with straight sides (instead of the nozzle shape of the ES tubes). In each one of

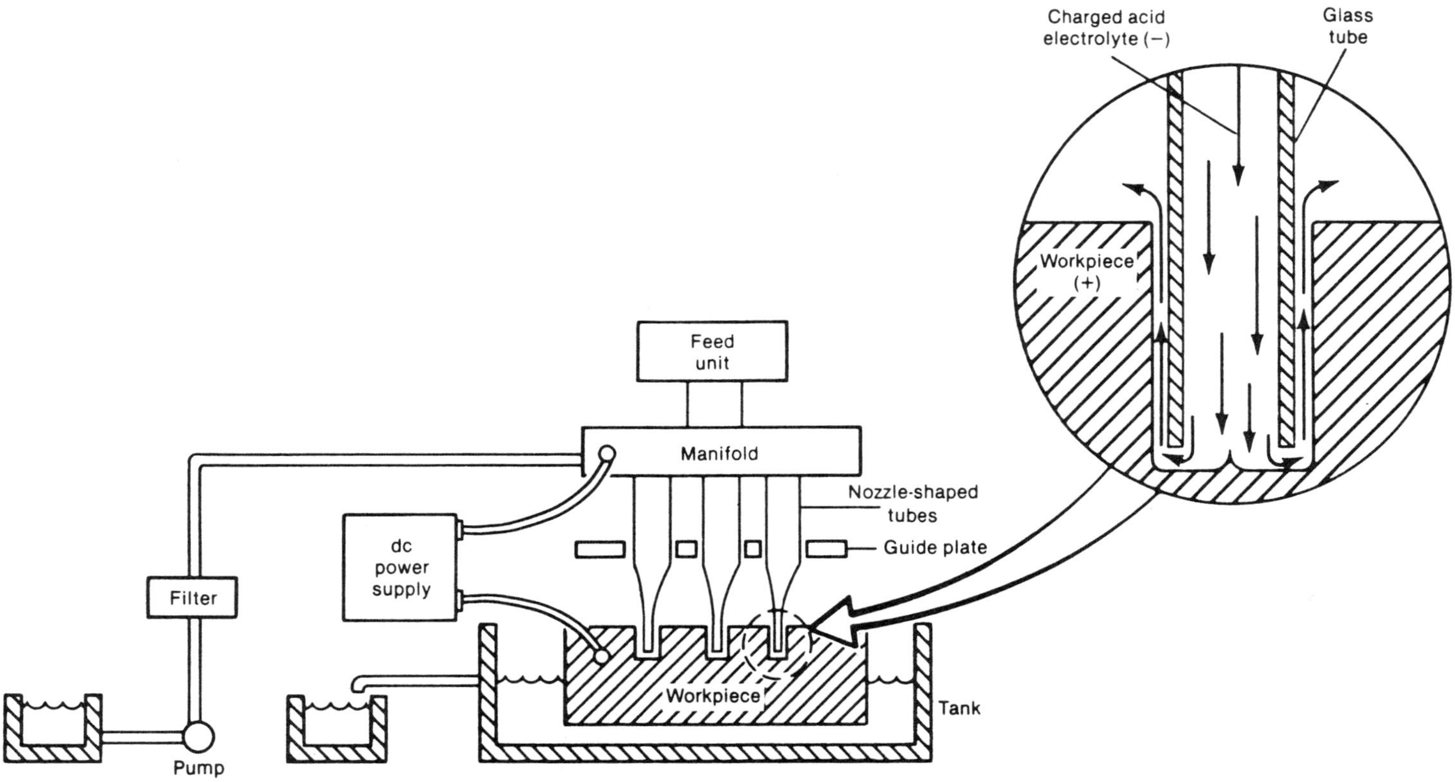


**Fig. 1** Schematic of material removal by electrostream drilling

the tubes is a small platinum wire, which is inserted down to approximately 1 mm (0.040 in.) from the cutting tip (all the wires in an array must be the same distance from the tips). This requires a lower voltage (70 to 150 V), and some of the safeguards needed with electrostream drilling are not necessary.

**Electrostream and capillary drilling machines** have one feed axis capable of producing constant feed rates between 25 and 0.125 mm/min (1 and 0.005 in./min) as well as a jogging movement. Multiaxis machines are also available; these units allow rotation of the part or allow an array of tubes to be indexed across the part. The use of computer numerical control enables feed rates and voltages to be programmed. The control can also monitor process variables. Electrostream and capillary drilling machines must have an interlocked enclosure and a vent to remove the hydrogen produced by the electrolytic process and must be well grounded to prevent accidents.

**Power Supplies.** The power supply for ES is typically a 900 V, 25 A unit (full-wave rectified). It can be designed to have a multichannel output if required, each channel feeding a separate manifold. For CD, a 150 V, 25 A unit is required; multichannel output may also be required.

**The Electrolyte System.** For ES, a double-electrolyte system is used (see the discussion of electrostream drilling above). The pumping is often achieved by air pressure in a pressure vessel. This avoids the need for a pump and metal parts, which could be adversely affected by the high operating voltage.

For a CD system, it is usually sufficient to have two tanks separated by a few feet of high-pressure piping. The acid is pumped from one tank to the machine, and it drains into the other. This acid then overflows back to the supply tank.

The electrolyte system must be environmentally secure, and local and federal regulations must be considered and complied with. A line break in the electrolyte system pipework can cause acid to be jetted out under pressure. The system, which should be grounded and guarded against such accidents, must include:

- A pump
- Filters (final filter: 0.5 μm)
- Heat exchanger
- Temperature controller: ±1 °C (±2 °F)
- Acid-mixing facility
- Waste acid storage facility

In addition, electrolyte concentration and metal contamination monitoring devices are desirable but not essential. However, it is necessary to have the means at hand to measure acid concentration. Metal contamination should be determined based on the volume of metal removed by drilling so that acid can be discarded at a predetermined point. The maximum limit of metal contamination is about 3 g/L (0.4 oz/gal.), depending on the metal being drilled.

**Part holding fixtures** should be made of commercially pure titanium. They can have carbide locating points, but these will have to be replaced regularly. It must be remembered that the tool has to transfer current into the part and that any restrictions relative to this must be complied with.

**Cathode holders** are built so that the cathodes are parallel to the feed axis of the machine (unless a multiaxis machine tool is being used). Many stations are often placed in one machine; each station will be used to drill all or some of the holes, depending on the hole spacing and the minimum possible tube spacing. If the glass tubes cannot be held close enough together to drill all the holes simultaneously in one piece, the workpieces will be passed through each station until all the holes are drilled.

**Cathode Tubes.** The ES cathodes consist of nozzle-shaped glass tubes 3 or 6 mm (⅛ or ¼ in.) in diameter, with the front drawn down to the necessary cutting tube size of 0.5 to 0.15 mm (0.020 to 0.006 in.). The length of this small-diameter section of the tube is that needed to drill the hole depth. The tubes in one setup should all be the same overall length, 75 to 100 mm (3 to 4 in.). Because the tubes are held on the larger diameter, the concentricity of the two diameters is critical. For CD, short, straight glass tubes of the correct drilling size are also used.

The importance of accurate tubes cannot be overemphasized. The tips must all be lapped flat, with no chips or cracks. The diameters and lengths must be identical, and the concentricity of the outside and inside diameters is critical. The inside diameter must be kept free of obstructions at all times.

## Process Parameters

**Electrolytes** for electrostream and capillary drilling are solutions of sulfuric, nitric, or hydrochloric acid with a concentration of 15 to 20 wt% water. The electrolyte temperature is 40 °C (100 °F) for sulfuric acid and is 20 °C (70 °F) for the others. The electrolyte pressure is 275 to 400 kPa (40 to 60 psig). The maximum limit of metal contamination is about 3 g/L (0.4 oz/gal.), depending on the metal being drilled. A minimum metal contamination of about 0.5 g/L (0.07 oz/gal.) is also desired because some contamination enhances the process.

Additional key process parameters for ES and CD include:

- *Voltage*: 600 to 900 V for ES; 70 to 150 V for CD
- *Electrolyte pressure*: 275 to 400 kPa (40 to 60 psig)
- *Feed rate*: 0.75 to 3.0 mm/min (0.030 to 0.100 in./min)

# Shaped Tube Electrolytic Machining

M. Anthony Newton, NItech, Inc.

SHAPED TUBE ELECTROLYTIC MACHINING (STEM) was originally developed and refined by General Electric Aircraft Engine Group for drilling holes with large depth-to-diameter ratios, which could not be drilled conventionally. Initially, such holes had been attempted by electrochemical machining (ECM), but the normal ECM process produces insoluble precipitates that inhibit the process by obstructing the flow path of the electrolyte. Essentially, the STEM process is a modified ECM process that uses an acid electrolyte so that the removed metal goes into solution instead of forming a precipitate.

Because the process uses an acid electrolyte, its use is generally limited to drilling holes in stainless steel or other corrosion-resistant metals. In parts designed for jet engines and gas turbines, some specific applications of the STEM process include the machining of:

- Turbine blade cooling holes
- Turbine vane cooling holes
- Turbine disk cooling passages
- Oil passages
- Fuel nozzles
- Any holes where electrical discharge machining (EDM) recast is not allowable

Other applications of the STEM process include:

- Starting holes for wire EDM cuts, especially where the length of cut exceeds 100 mm (4 in.)
- Drilling regular arrays of holes in corrosion-resistant metals of low machinability (for example, strainers and dies)
- Drilling oil passages in bearings in which EDM will cause cracks
- Many applications in general industry where higher-strength materials are used or where machinability is a problem

## Process Capabilities

**Feed Rates.** The STEM process has a feed rate of 0.75 to 3 mm/min (0.030 to 0.120 in./min), depending on the material being drilled but regardless of the number of holes being drilled at once.

**Normal hole sizes and depths** are 0.50 mm (0.020 in.) in diameter up to a depth of 75 mm (3 in.), and 6 mm (0.250 in.) in diameter up to a depth of 900 mm (36 in.).

**Normal hole size tolerances** are:

- 0.5 mm (0.020 in.) in diameter, ±0.05 mm (±0.002 in.)
- 1.5 mm (0.060 in.) in diameter, ±0.075 mm (±0.003 in.)
- 6 mm (0.250 in.) in diameter, ±0.10 mm (±0.004 in.)

These tolerances can be halved by special process control and with very pure metals.

The normal hole depth tolerance, where applicable, is ±0.05 mm (±0.002 in.), regardless of hole size (although larger holes have a cusp in the bottom of the hole).

**Normal hole wander** is 0.001 mm/mm (0.001 in./in.) of depth for holes 0.5 mm (0.020 in.) in diameter, and 0.003 mm/mm (0.003 in./in.) of depth for holes 6 mm (0.250 in.) in diameter. Again, these values can be halved through the use of special process control and very pure metals.

**Competitive Processes.** Shaped tube electrolytic machining competes with the following processes:

- *Gun drilling*: Limited to 1.9 mm (0.076 in.) minimum diameter in easier-to-machine metals. It is possible to drill nonstainless metals with closer tolerances, but usually only a small number of parallel holes can be drilled at once
- *Conventional drilling*: Limited depth-to-diameter ratio
- *Electrical discharge machining*: Limited depth-to-diameter ratio, blind holes cannot be drilled easily, and a recast layer is formed

Shaped tube electrolytic machining can often compete with gun drilling because, although essentially a slow process, it is possible to drill many (up to 200) holes at once. These holes need not be parallel and can have axial directions of up to 20° from the machine motion axis.

**Limitations and Advantages.** The limitations of shaped tube electrolytic machining include the following:

- The process can be used only on corrosion-resistant metals and generally cannot drill commercially pure titanium and refractory metals
- A complex machining and tooling system is required
- The process is slow when single holes are being drilled
- Problems can result when the hole breaks through the bottom of the part
- The handling of acid requires special workplace and environmental precautions
- A hazardous waste is generated
- Oblique entry is difficult

The advantages of the STEM process are as follows:

- The hardness (machinability) of the metal is not a consideration
- Depth-to-diameter ratios are very high (up to 300)
- Large numbers of holes can be drilled simultaneously
- Nonparallel holes can be drilled simultaneously
- Blind holes can be drilled
- No recast layer or metallurgical defects are produced
- The system can be operated unattended
- Powder metallurgy alloys can be machined
- Shaped holes (limited) can be drilled
- Curved holes (limited) can be drilled
- Slots (limited) can be drilled

## Equipment

Figure 1 illustrates the basic components of a STEM system. Shaped tube electrolytic machining is similar to ECM systems in that a low-voltage dc power supply creates electrolytic action between the anode (the workpiece) and cathode (the tool); however, the STEM system also has some differences:

- It must be acid resistant
- Less rigidity is required
- The power supply must periodically reverse polarity because the acid electrolyte increases the tendency for the cathodes to become plated with workpiece material

The basic STEM machine is like an ECM machine and has one feed axis capable of constant feed rates from 25 to 0.125 mm/min (1 to 0.005 in./min) as well as a jogging movement. However, multiaxis ECM/

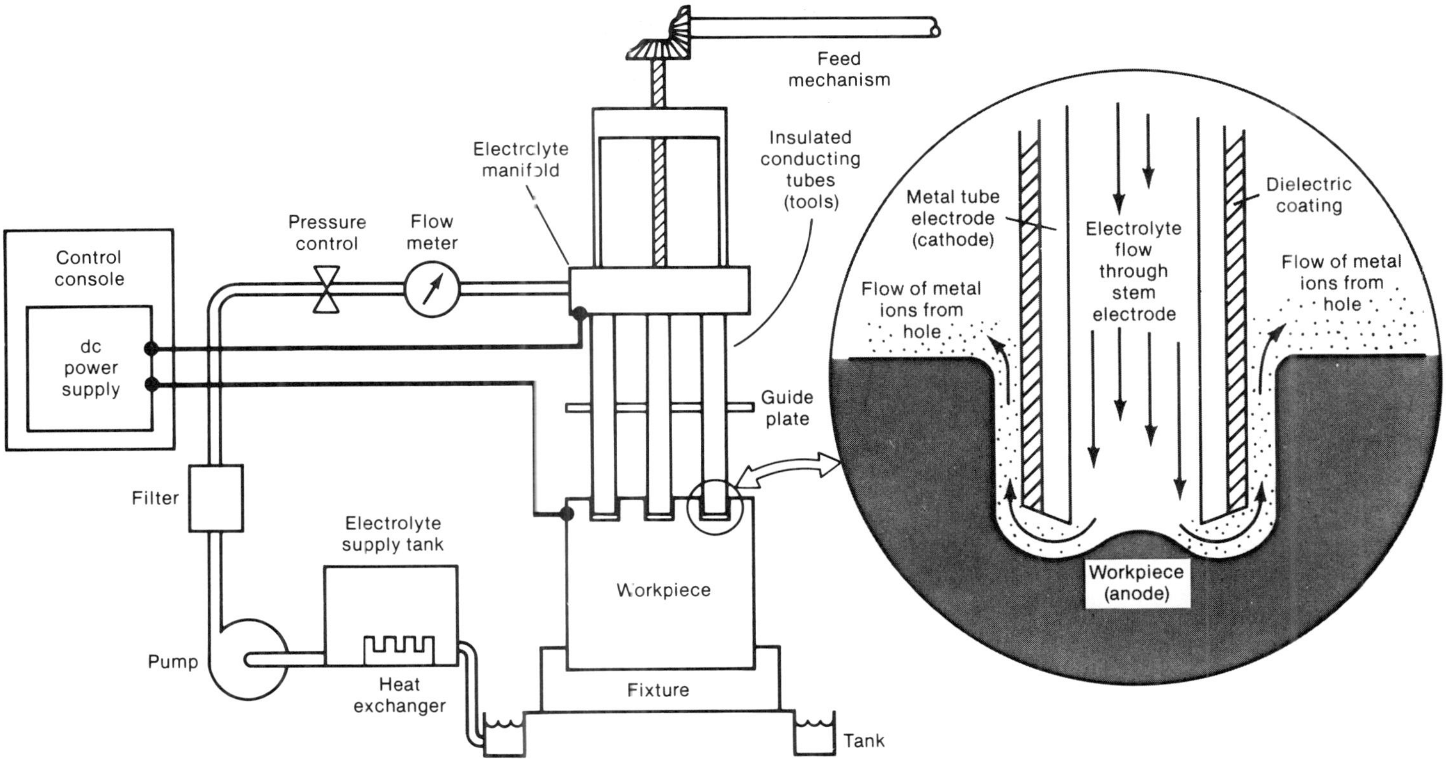


**Fig. 1** Schematic of the shaped tube electrolytic machining process

STEM machines are also used for specific applications. The use of computer numerical control enables variable feed rates to be programmed, and the computer control can also regulate the process parameters. Process feed rates are typically 0.75 to 3.0 mm/min (0.03 to 0.12 in./min).

Although it is an unusual practice, it is sometimes recommended to electrically insulate the worktable, where the workpieces are mounted, from the rest of the machine frame. This prevents corrosion damage to any metallic parts in contact with the electrolyte during the application of voltage.

**The power supply** for the STEM system has a voltage output of up to 20 V, and the polarity reverses every 5 to 10 s. The duration of the polarity reversal is controlled to vary from 75 to 250 ms. The reverse voltage can vary from 0.1 to 1 times the forward voltage, which is typically 8 to 14 V.

Power supplies for shaped tube electrolytic machining are capable of carrying a current of up to 600 A at forward polarity and 50 A at reversed polarity. Power supplies can be air cooled or water cooled, and normal three-phase full-wave rectification is sufficient.

**Electrolyte systems** for shaped tube electrolytic machining include the following:

- A pump capable of producing the required flow at pressures up to 550 kPa (80 psi) (the amount of flow depends on the application)
- Filters (final filter: 1 μm)
- Heat exchanger
- Temperature control: ±1 °C (±2 °F)
- Acid mixing facility
- Waste acid storage facility

The STEM machine must also have an enclosure that will prevent splashing of the electrolyte and a vent system able to remove the hydrogen produced by the electrolytic process. In addition, electrolyte concentration and metal contamination monitoring devices are desirable but not essential. However, it is necessary to have the means at hand to measure acid concentration. Metal contamination can be determined based on the volume of metal removed so that the acid can be discarded at a predetermined point.

In multiple-hole drilling of holes with different diameters, it may be necessary to prevent acid starvation in the smaller-diameter holes. This can be achieved by using multiple electrolyte manifolds and closed-loop control of electrolytic pressure and flow.

**Tooling** is the most critical part of the STEM process, and it must be correctly designed to obtain optimum results. Tooling for the STEM system consists of four parts:

- *Part holding fixture*: Same as a conventional tool
- *Guide for cathodes*: Analogous to a drill bushing
- *Cathodes*: Analogous to a drill
- *Cathode holder/manifold*: Similar to the collet in a gun drill

In general, the part holding fixtures and the cathode guides should be made entirely of commercially pure titanium, which forms an anodic layer that protects it from the acid. The cathodes are insulated commercially pure titanium tubes, and the cathode holders and manifold are made of 20Cb-3 stainless steel and Lexan or acrylic plastic, respectively.

*Part holding fixtures* should be designed in the same manner as a normal machining fixture. The actual locating points can be made of carbide, but they will have to be replaced regularly. The fixtures can be built in multiples, and they should be aligned so that the nonparallel holes are angled approximately equally from the motion axis of the machine.

*Cathode guides* should be attached to the part holding fixture or should be lined up with it, and there should be a small amount of clearance between the bottom of the guide and the top of the part. The guided length should be approximately 50 times the diameter of the largest cathode being guided (up to 100 mm, or 4 in., for very large cathodes).

*Cathodes* are commercially pure titanium tubes coated with a special resin for electrical insulation, with the front (drilling face) lapped flat. The straightness of the cathodes is critical. Restraightening and tip relapping must be done periodically.

*The cathode holder and manifold* must be designed in such a way as to drive the cathodes through the cathode guides. The cathode holder and manifold must also conduct electricity into the cathodes and feed acid under pressure into the cathodes with a minimum of leakage.

## Process Parameters

**The electrolytes** for shaped tube electrolytic machining are sulfuric, nitric, or hydrochloric acid in concentrations of 10 to 25 wt% in water. The electrolyte temperature is 38 °C (100 °F) for sulfuric acid and 21 °C (70 °F) for the others.

The maximum metal contamination of STEM electrolytes is about 3 g/L (0.4 oz/gal.), depending on the metal being drilled. A minimum metal concentration of 0.5 g/L (0.07 oz/gal.) is also desirable because the process is usually more effective if the acid is precontaminated.

Other than electrolyte composition and concentration, the primary STEM process parameters are:

- *Forward voltage*: 8 to 14 V
- *Forward time*: 5 to 7 s
- *Reverse voltage*: 0.1 to 1 times forward voltage
- *Reverse time*: 75 to 250 ms
- *Electrolyte pressure*: 275 to 500 kPa (40 to 70 psig)
- *Feed rate*: 0.75 to 3.0 mm/min (0.030 to 0.120 in./min)

# Electrical Discharge Machining

John E. Fuller, Rockwell International

ELECTRICAL DISCHARGE MACHINING (EDM) is a process that removes metal with sparks. A shaped electrode (usually graphite or copper) is used to make a cavity that is the mirror image of the electrode. There is no direct contact between electrode and workpiece. The sparks travel through a dielectric fluid (typically a light oil) at a controlled distance. Both electrode and work must be electrically conductive.

The advantages of electrical discharge machining are:

- Cavities with thin walls and fine features are possible to machine because there is no contact between electrode and work
- Difficult geometry in general is possible to machine
- Although the removal rate is related to the melting point of the metal being machined, the use of electrical discharge machining is not affected by the hardness of the work. Materials with poor machinability, such as cemented tungsten carbide and hardened tool steel, can be cut by the EDM process
- Electrical discharge machining is burr free

Electrical discharge machining is typically applied to molds and dies in many forms. Plastic injection molds, extrusion dies, forging dies, and die casting dies are just a few examples that show the versatility of this method, which is also used directly in production processes.

In addition to discussing the use of a shaped electrode for EDM, this article will discuss traveling-wire EDM. This process differs from conventional EDM in that a thin-diameter brass wire acts as the electrode.

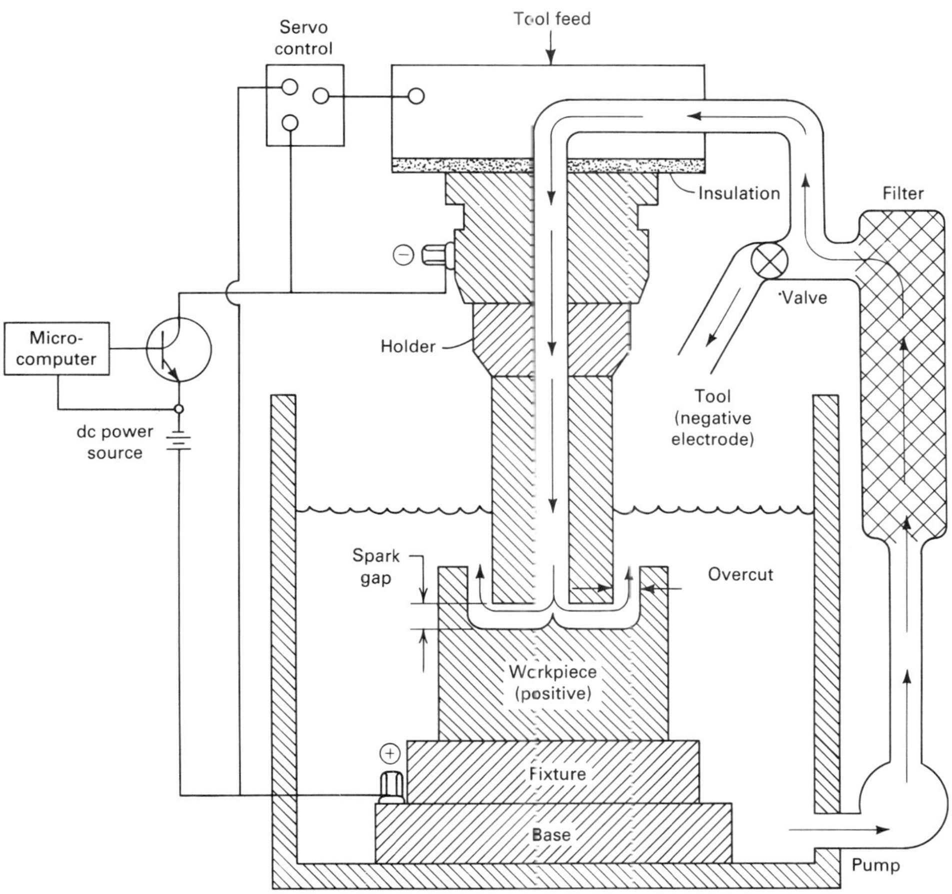


**Fig. 1** Typical setup for electrical discharge machining. Negative (standard) polarity is shown. Positive (reverse) polarity is also extensively used.

## Method of Operation

Although some EDM machines are capable of movement in several axes, the EDM process is best illustrated by a machine with the vertical axis only under servo control (Fig. 1). Negative polarity indicates that the electrode is negative. If the ram were allowed to move forward unchecked, there would be direct contact between electrode and workpiece, causing an electrical short circuit. This is prevented by a servo mechanism in which the potential is monitored and compared with a reference. If the potential is greater than the reference, the ram advances; if it is less, the ram retracts. The movement may be accomplished by a hydraulic cylinder or a direct-drive servomotor. As the work is machined by spark erosion, the distance between electrode and workpiece increases. The potential goes up, and the ram advances until the potential matches the reference. Thus, the servo mechanism maintains a constant gap. Erosion continues until a preset depth is reached. At this point, the electrode is retracted from the workpiece.

**Mechanism of Metal Removal.** Direct-current power is supplied to the circuit in pulses that can be approximated by a square wave. Ideally, each pulse creates a spark. Each spark occurs where the resistance is smallest, usually near the last spark. The erosion occurs over the surface of the electrode and workpiece interface. The square wave, plotting current against time, is the basis for common terminology of the EDM power supply. The variables are on-time, off-time, and peak current (Fig. 2).

Although Fig. 2 is a convenient reference, the process is more complex (Ref 1). When the electrode is separated from the work, the potential is the open circuit voltage,

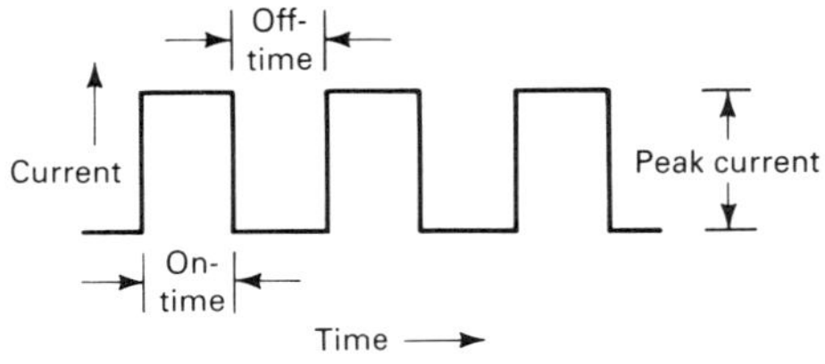


**Fig. 2** Typical variables of the EDM pulse train. On-time and peak current determine the spark energy.

usually about 100 V. As the electrode approaches the work, the dielectric fluid begins to ionize at the point of closest approach. The potential drops to a lower level (about 35 V) as current begins to flow. This is the spark. It occurs at a spark-gap distance of 0.01 to 0.4 mm (0.0004 to 0.016 in.). With each spark, material is removed by vaporization and melting, and small craters appear on both electrode and work. Figure 3 gives a more detailed picture, plotting voltage and current as a function of time. The on-time can be separated into ionization time, spark time, and deionization time. The interruption of current (off-time) allows the particles ejected from the crater to be flushed away by the flowing dielectric fluid. The ionized fluid is dissipated and replaced by fresh fluid. The off-time must be greater than the deionization time to prevent continued sparking at one point. This condition is known as a direct current (dc) arc.

**Power supplies** for electrical discharge machining have progressed from resistance-capacitance (RC) circuits and vacuum tube types to the present solid-state power supplies. RC circuits are still used for microhole machining. Currently, there is a trend to metal oxide field effect transistor (MOSFET) devices. These are useful for EDM power supplies because of their rapid switching capabilities at higher power. A modern power supply allows independent setting of on-times and off-times. A typical range is from 2 to 1000 μs. The total energy of the individual spark is proportional to the volume of the rectangular block graphing time, current, and voltage on three orthogonal axes (Ref 2). Only the effective on-time, that is, the time following ionization, is effective in removing metal (Fig. 4). As a first approximation, the crater diameter is proportional to the applied current, and the depth is proportional to the on-time.

Removal rates range from almost nil at finish settings to 410 $mm^3$/min (1.5 $in.^3$/hr) for a 25 A power supply. A 400 A power supply can remove up to 4350 $mm^3$/min (16 $in.^3$/h) (Ref 3). It should be noted that the increase is not linear. The pulse train from a standard power supply makes only one spark at a time. When using many small electrodes at the same time, this is inefficient. A solution is the so-called multilead power supply (Ref 4). This actually combines several low current supplies into one unit and enables several simultaneous sparks (but only one at a time for each electrode).

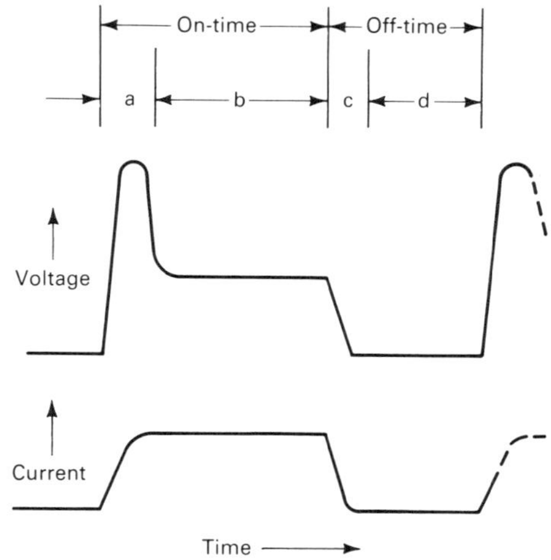


**Fig. 3** Profile of a single EDM pulse. a, ionization time; b, discharge time; c, deionization time; d, rest time. Most of the electrode wear occurs during the ionization time.

With the arrival of solid-state power supplies, it has become more common to use positive polarity on the electrode. Polarities for various combinations of electrode and workpiece are shown in Table 1. Some power supplies insert an opposite polarity "swing pulse" at fixed intervals to try to prevent arcing. A typical ratio is 1 swing pulse for every 15 standard pulses. Power supplies are rated in amperes and range from 10 to 1000 A.

## The EDM Surface

**Topography.** Electric discharge machining has a distinctive surface signature. The surface resembles a cratered surface with all craters the same size (Fig. 5). There is no "lay" or directionality to the surface as in conventional machining. Because the crater size depends on spark energy, and spark energy varies widely, the EDM surface finish surface can range from 0.2 to 12.5 μm $R_a$ (8 to 500 μin. $R_a$). The average factor to convert from the root mean square ($R_q$) to the arithmetic average ($R_a$) varies with machining method and is 1.26 for EDM (Ref 5). A small fraction of the total surface is near the peak elevation. Therefore, when polishing, the surface finish improves rapidly at first, then more slowly.

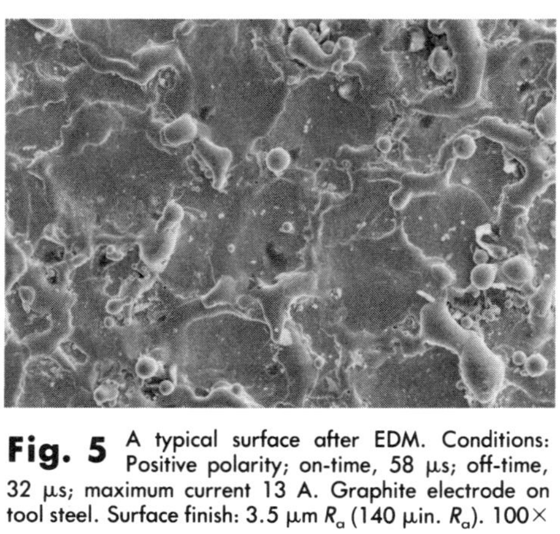

**Fig. 5** A typical surface after EDM. Conditions: Positive polarity; on-time, 58 μs; off-time, 32 μs; maximum current 13 A. Graphite electrode on tool steel. Surface finish: 3.5 μm $R_a$ (140 μin. $R_a$). 100×

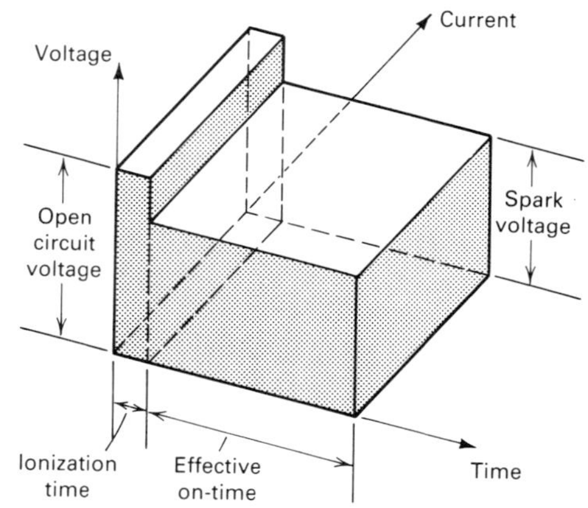


**Fig. 4** Three-dimensional schematic of the EDM spark. The spark energy available for material removal is proportional to the product of effective on-time, current, and spark voltage.

**Metallurgical and Chemical Effects.** Because of the rapid quenching by the dielectric and the heat sink effect of the workpiece, the surface layer affected by the EDM process is quite thin—less than 0.13 mm (0.005 in.) for roughing settings and 0.01 mm (0.0004 in.) for finish settings (Ref 3). Because different criteria are used to judge the thickness of the layer (hardness alteration, residual stress, metallographic evidence), the method should be specified. The layer can be separated into a so-called recast layer and a heat-affected zone. The recast layer is characterized by a rapidly quenched structure, while the heat-affected zone has an annealed or tempered structure. The effects are highly dependent on the alloy being machined. Because a large part of electrical discharge machining is done on high-carbon tool steel used for dies and molds, much work has been done on the characterization of EDM layers on steel. The recast layer on tool steel is harder than the bulk material because of martensite precipitation, while the heat-affected zone is softer, because of overtempered martensite. A single-phase alloy or precipitation-

**Table 1 Polarity for various electrode-work combinations**

| Electrode materials | Work materials: Steel | Tungsten carbide | Copper | Aluminum | Ni-base alloys |
|---|---|---|---|---|---|
| Graphite | +, − | − | − | + | +, − |
| Copper | + | +, − | − | + | + |
| Cu-W | + | +, − | − | + | + |
| Steel | +, − | + | − | − | − |
| Brass | − | ... | ... | + | − |

hardening alloy would have a different behavior.

Small amounts of electrode material as well as carbon from the dielectric can be deposited on the EDM surface. In the case of steel, extra carbon in the surface layer can cause more martensite, with increased susceptibility to cracking.

**Surface integrity** has always been of concern in electrical discharge machining. Both the recast layer and the heat affected zone are in tension (Ref 6). Microcracks occur in the recast layer and can act as initiation points for failure. The questions to be answered are: Can the bulk material propagate a surface crack? If it is necessary to remove the surface layer, how can this be done? Shot peening removes a small part of the recast layer and improves the fatigue properties. To obtain fatigue properties representative of the bulk material, however, both the recast layer and the heat-affected zone must be removed. The most common methods of removal are by abrasive (Ref 7) and electrochemical methods. In one reported method (Ref 8), the dielectric is replaced with an electrolyte, and the EDM electrode becomes a static electrochemical electrode. This has the advantage of high current density, resulting in little grain boundary attack. In some cases, the EDM surface is beneficial. The higher hardness and ability to retain lubrication can result in increased die life.

## Electrodes and Electrode Manufacturing Processes

Electrode costs are often the most expensive part of an EDM operation. Material, fabrication, wear, and redressing costs must be carefully weighed to determine the best electrode material and EDM machine settings. The ideal electrode material has high electrical conductivity, is easy to fabricate, has a high melting point, and is strong enough to stand up under conventional machining without deformation.

**Wear.** Melting point is most important in determining wear (Ref 9). Electrode wear ratios are expressed as end wear, side wear, corner wear, and volume wear. These terms are defined in Fig. 6. The term no-wear EDM is generally applied to cases in which the electrode-to-workpiece wear ratio is 1% or less. Wear first appears on corners. Thus, corner wear is most likely to determine when an electrode must be redressed.

**Materials.** In the following discussion of electrode materials, cost comparisons are made by assigning tough pitch copper a value of 1. Electrode material costs are usually less than fabrication costs and EDM machine time. Thus, the cheapest material does not necessarily result in the lowest overall cost.

*Graphite* is the most commonly used electrode material because of good machinability and EDM wear characteristics. Small, flush holes are easy to drill. It is available in a great variety of size and form. Commercially available EDM grades range in grain size from 0.001 to 0.1 mm (0.00004 to 0.004 in.), in flexural strength by a factor of five, and in hardness by a factor of two. The price increases with decreasing grain size, increasing flexural strength, and increasing hardness. At the time of publication, the cost factor, compared to copper, was 1.3 to 24, with most grades falling between 2.6 and 10. A drawback of graphite is that it is dirty to machine; vacuum systems are needed for all but the smallest jobs.

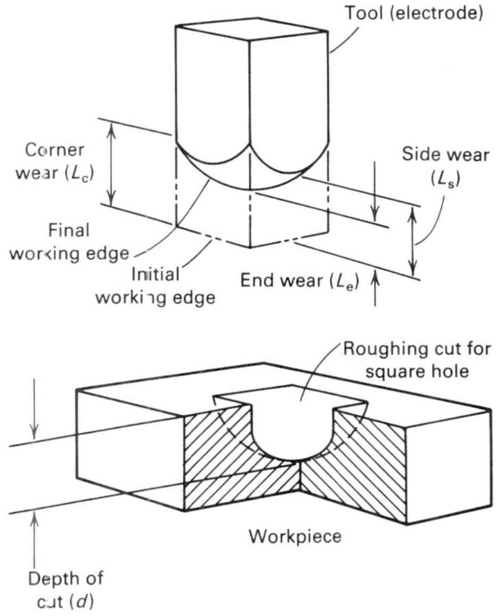


End wear ratio = $d/L_e$
Side wear ratio = $d/L_s$
Corner wear ratio = $d/L_c$

$$\text{Volume wear ratio} = \frac{\text{Volume of workpiece removed}}{\text{Volume of electrode consumed}}$$

**Fig. 6** Terms used in expressing wear of EDM electrodes

*Copper* has good EDM wear, good conductivity, and is economical. It does not machine as well as brass or graphite. Nonetheless, it is used almost as much as graphite, and is especially good for machining tungsten carbide. Copper is preferred for finishes better than 0.5 μm $R_a$ (20 μin. $R_a$). As the standard of cost comparison for this article, it has a cost factor of 1. Electrodes can be made by a variety of methods. An alternative is the Cu-0.5Te alloy. It is only slightly more expensive than copper (cost factor, 1.2), and it is nearly as machinable as brass.

*Copper tungsten and silver tungsten* are expensive materials with cost factors of 18 and 100, respectively. Their main use is in making deep slots under poor flushing conditions, and they are often used on tungsten carbide. They are not true alloys. The tungsten is a pressed and sintered powder that is infiltrated with copper or silver. The machinability is fair, but because the material is brittle, it cannot be formed after sintering.

*Copper graphite* is a graphite that has been infiltrated with copper. It is 1.5 to 2 times more expensive than the same graphite without copper, thus making it from 5 to 20 times more expensive than copper. The flexural strength is higher than the comparable grade of graphite, making it good for thin cross-section electrodes. Electrical conductivity is greatly improved, but corner wear is not as good as it is for the same grade of pure graphite. This material works well on tungsten carbide.

*Brass* is inexpensive and easy to machine, but does not wear well. It is often used for tubular electrodes in specialized small-hole EDM drilling machines where high wear is acceptable.

*Steel* is not a satisfactory electrode material for normal use. Its main use is to match the parting planes of molds in which half of the mold is used as the electrode and the other half is used as the workpiece.

*Tungsten* is used to make small holes (<0.2 mm, or 0.008 in.) for which electrodes with flush holes are not available.

**Electrode Manufacture.** The most common method of electrode making is on lathes, milling machines, and grinders. One reason graphite has become dominant as an electrode material is that it is easy to machine. The design of EDM tooling systems is based on the assumption that the electrode is machined while attached to the holder.

*Traveling-wire EDM* is an excellent method for making vertical EDM electrodes. All metallic electrode materials cut well. The cutting speed on graphite is about one-third as fast as on copper, but recently introduced ultrafine grained graphite cuts more easily. Extremely thin walls and intricate detail are possible. The method can be combined with other methods of electrode manufacture.

*Abraded-graphite electrodes* require a master in the shape of the cavity. The master is used as a tool to make a mirror image of its shape, much like the EDM process. When the abrasive is in the master as a mixture with epoxy, the master itself does the cutting. This is accomplished by an orbiting action between master and electrode known as orbital abrading. In ultrasonic abrading (Ref 10), the abrasive in a slurry is injected between the master and the electrode, and the master is vibrated at about 20 000 cycles per second. Orbital abrading is used for high removal rates, while ultrasonic abrasion is used when finer detail is required. Graphite electrodes can be redressed many times by both methods. Because of the cost of making the master, the method is most economical for repetitive work. Tooling systems to maintain location between EDM machine and abrading machine are used.

*Molded copper tungsten* (Ref 11) is produced from an original master that is made from any material. The master is used to make epoxy duplicates, which in turn are used to make molds for the electrodes. Identical molded copper tungsten electrodes are made. The method is used for cases in which many complicated electrodes are needed because the difficult machining must be done only once.

*Electroforming* is a process in which a form is made, coated with a conducting paint, and electroplated. After electroplating, the original form is removed. The form is often made of wax, which can be melted out; thus, the process is like investment casting in reverse. Aluminum, etched out with a caustic solution, is also used, with copper as the plated metal. Recent developments in no-wear EDM electrical discharge machining have made the use of thin-shell electrodes more feasible.

*Extruded, coined, or pressed copper* use the good forming properties of copper to advantage. Copper extrusions of intricate cross section and good dimensional quality can be produced. High-detail cavities are often made in hardened tool steel or tungsten carbide by the EDM process. These cavities are in turn used as pressing or coining dies for multiple copper electrodes.

*Die casting* is a high-production process capable of reproducing intricate geometry. It is limited as an electrode-making method because good die casting alloys (usually Sn-Zn) are poor electrode materials. The low melting points that help the die-casting process result in poor electrode wear.

**Tooling Systems.** The purpose of an EDM tooling system is to allow electrode fabrication and redressing on standard shanks and mounting plates. The datum surfaces on the shanks and plates that locate the electrodes during fabrication are the same as those used to locate the electrodes in the EDM machine. The EDM machine location should be so reliable that once dimensional inspection is done outside the EDM machine, the location is ensured. Auxiliary hardware is usually included to verify this. The result is reduced EDM machine setup time and increased time for effective machine use. The electrode-making methods normally used are turning, milling, grinding, and wire EDM. One reason effective tooling systems are valuable is that a severe penalty is paid for misalignment of rough and finish electrodes. The finish electrode uses low removal rates with high wear. When extra material resulting from misalignment is encountered, longer finish times result. The problem is compounded by loss of effective flushing.

## Dielectric Fluids

The dielectric fluid performs several functions: It is a spark conductor that must ionize under an applied voltage. It is a coolant for work and electrode. It is a flushing medium that carries away the swarf (EDM chips) resulting from the process. Desirable properties are low viscosity, high flash point, and low cost. Because of small working gaps at finish spark settings, a low-viscosity dielectric is especially important. Low viscosity also helps in settling of the swarf, thus keeping the dielectric fluid clean. The most common dielectric fluid is petroleum-base oil. Also used are kerosene, silicone oils, and water-base dielectrics. Deionized water has some desirable characteristics (Ref 12), these being fire safety, low cost, low viscosity, and absence of carbon to react with the work. On balance, however, the petroleum-base oils give better control of the spark gap and are the most-used dielectric.

**Flushing** is the single most important component of successful EDM. In multilead power supplies, the multiple electrodes are a single mechanical unit, even though they are electrically separate. Flushing problems at one electrode cause the entire assembly to retract. Thus, good flushing is even more important when using the multilead method. Changing from very poor to very good flushing conditions can improve efficiency and thus reduce machining time by a factor of six (Ref 13). Some common-sense rules apply:

- Through-the-tool flushing is preferred to side flushing
- Many small flush holes are preferable to a few large ones. Besides giving better fluid distribution, smaller and more-easily-removed spikes (the column of metal left from a flush hole) result
- A steady flow of dielectric fluid over the entire electrode-workpiece interface is desirable
- Dead spots created by pressure flushing from opposite sides of a piece should be avoided
- A vent hole should be provided for any upwardly concave part of the electrode to prevent accumulation of explosive gases
- A flush box can be useful if there is a hole in the cavity

## Recent Advances in EDM

**No-wear EDM** is usually defined as machining that results in a work to electrode wear ratio of 100 or more. It is based on the fact that at a given current, the wear per pulse is nearly the same, regardless of the length of on-time. Thus, the traits of no-wear EDM are positive electrode polarity, long on-times, and low peak current. It is not a separate form of electrical discharge machining. Spark settings near no-wear settings give wear ratios near 100. Graphite, copper-impregnated graphite, copper, copper tungsten, and silver tungsten can all be used as electrode materials. Work materials can be steels, Stellite alloys (cobalt-chromium base), high-nickel alloys, or aluminum. Tungsten carbide cannot be used in this manner. Traditionally, long on-times have been thought to result in a high-energy pulse that is unsuitable for finishing. However, power supplies have progressed to the point of lowering no-wear settings to 20 μs on-times and 3 A peak currents, thus enabling the machining of fine surfaces. No-wear electrical discharge machining can also be used for roughing cuts. However, no-wear settings do not give the highest removal rates. These usually occur at negative polarity. Figure 7 shows maximum removal rate and best wear ratio as a function of on-time (Ref 14). The peak current is held constant. Figure 8 shows a case in which high removal rate at negative polarity, combined with efficient redressing of electrodes, results in an overall time savings (Ref 11). Thus, the choice of the most economical process depends on the cost of electrode material, and redressing, labor, and machine time.

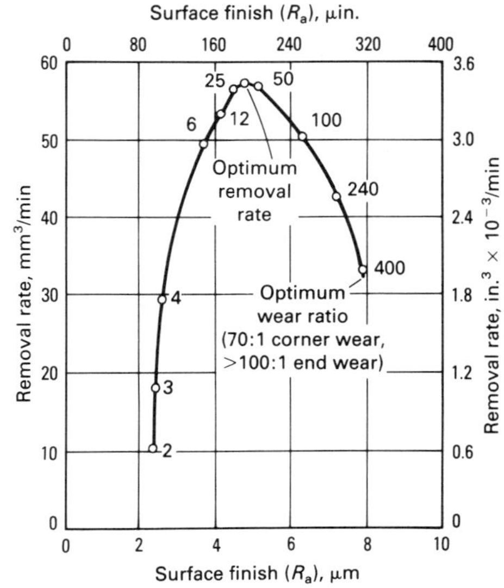


**Fig. 7** Increasing on-times at positive polarity results in a removal rate maximum. No such maximum occurs at negative polarity. Best wear is at maximum on-time with a positive polarity. Workpiece: tool steel. Current: 12 A. Electrode material: graphite. The numbers on the curve refer to on-times in microseconds. Adapted from: Ref 13

**CNC Vertical EDM.** Computer numerically controlled (CNC) machines provide additional capability in servocontrolled motion and erosion. Nearly all machines have at least three controlled axes; machines with as many as six axes are available. They have the ability to produce contoured surfaces with ball-nosed electrodes in a manner similar to that of a CNC mill. However, it is more common to use shaped electrodes in the same manner as on standard EDM machines. Computer numerically controlled EDM machines provide advantages in other capabilities:

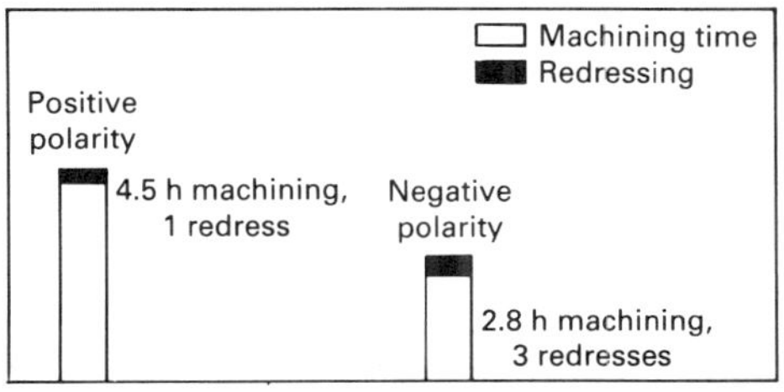


**Fig. 8** An example of net time savings with negative polarity and high electrode wear. The key is an efficient electrode redressing method, in this case, orbital abrading of graphite (Ref 10).

- Location of electrodes with respect to some reference location such as a tooling ball, a reference hole in the work, or reference surfaces
- Multiple cavities with the same spark parameters
- Electrode wear compensation using touchoffs on reference surfaces
- On-machine inspection with probes held in the electrode holder
- Fast retrieval of spark settings for a given work-electrode combination. Selection can be based on removal rate or surface finish
- Storage and retrieval of electrode offset data when the electrode axis is not coincident with the shank on which it is mounted. Angular orientation errors can also be corrected
- Multiple coordinate systems for workpieces at different locations and orientations
- Centerline offsets used with milling cutter-type electrodes. Large offsets are used for roughing cuts, and smaller offsets are used for finish cuts

This partial list demonstrates the wealth of possibilities that are available using CNC electrical discharge machining. The ultimate goal is to produce multiple cavities unattended. It is important to note that without a tool changer, CNC EDM loses many of its advantages. For unattended operation, aggressive spark settings (characterized by short off-times) should not be used because arcing can ruin the part before the problem is discovered.

**Orbiting** is relative movement between electrode and workpiece in a set pattern. Round and square orbits in the $x$-$y$ plane are the most common type. There is no limit, however, to the complexity of the orbit; even curvilinear orbits are possible (Ref 15). Orbiting can be accomplished with movement of the electrode on a stationary workpiece using an orbiting device on the quill. This gives a non-CNC vertical EDM machine orbiting capability. On CNC machines, orbiting is done almost universally with movement of the table. Such a system places no limits on the size of the orbit. Orbiting on CNC machines can also be done in the $x$-$z$ and $y$-$z$ planes. Orbiting has important advantages:

**Table 2 Power settings for a typical orbiting sequence**

As settings become finer, wear ratios increase. The polarity is positive.

| On-time, μs | Peak current, A | Surface finish ($R_q$) μm | Surface finish ($R_q$) μin. | Volume removal, %(a) | Total electrode wear, %(a) | Orbit radius mm | Orbit radius in. |
|---|---|---|---|---|---|---|---|
| 110 | 55 | 7.0 | 275 | 97.8 | 79.4 | 0 | 0 |
| 90 | 27 | 5.2 | 205 | 1.1 | 7.9 | 0.079 | 0.0031 |
| 60 | 19 | 4.6 | 181 | 0.6 | 4.0 | 0.122 | 0.0048 |
| 45 | 15 | 4.3 | 169 | 0.3 | 3.2 | 0.147 | 0.0058 |
| 30 | 12 | 3.8 | 150 | 0.2 | 5.5 | 0.167 | 0.0066 |

(a) In the first power setting

- Flushing is improved by alternately widening and narrowing the gap between the electrode and the work
- A given electrode, often in a stock size, can be used for machining a wide range of cavity sizes, eliminating the need for custom machining the electrode
- The same electrode can be used for rough and finish machining. As orbit size increases, spark energy is decreased. The effect is similar to using progressively finer grades of sandpaper

A sequence of spark settings and corresponding orbit radii and surface finishes is given in Table 2.

**EDM Hole Drilling.** An important branch of EDM cavity sinking is drilling. There are several specialized machines built specifically for drilling. One is a single-hole machine using tubular electrodes. Without through-the-electrode flushing, hole drilling beyond ten diameters is very difficult. These machines use brass or copper electrodes with negative polarity. Some machines use a guide bushing that is ultrasonically vibrated to aid flushing. Because of its low viscosity, water is the dielectric. The drilling speeds can be as high as 25 mm/min (1 in./min). A wide variety of brass and copper electrodes are available, including those with a spiral metal core that removes the spike that normally occurs at the point of a flush hole. This spike is troublesome in deep-hole drilling. Also worth mention are the multiple-hole production machines with automatic electrode refeed mechanisms. They are used to make cooling holes in gas turbine blades.

## Applications

Although the application of electrical discharge machining is limited to the machining of electrically conductive workpiece materials, the process has the capability of cutting these materials regardless of their hardness or toughness. Nonconductors such as glass, ceramics, or plastics cannot be machined using EDM techniques, but the machining of hardened steel using EDM eliminates the need for subsequent heat treatment with possible distortion. Complex shapes can be cut in hardened steel or carbide without costly sectional construction being necessary.

The EDM process is most widely used by the mold-making and tool and die industries, but it is increasingly being applied to make prototype and production parts, especially in the aerospace and electronics industries, in which production requirements are relatively low. The production of stamping dies is a major application of this process because of the favorable economics afforded by using the EDM process to match one portion of the die made conventionally. Extruding, heading, drawing, forging, and die casting dies, as well as molds for plastics, can also be done with EDM techniques.

Electrical discharge machining is particularly well suited for parts that are made from materials that are difficult to machine and/or contain small or odd-shaped holes, a large number of holes, or holes having shallow entrance angles, intricate cavities, or intricate contours. Miniature parts and parts made from material too thin or fragile to withstand conventional, mechanical cutting forces are also good applications. Round or irregular-shape holes as small as 0.05 mm (0.002 in.) in diameter can be produced with length-to-diameter ratios of about 20:1. Narrow slots as small as 0.05 to 0.30 mm (0.002 to 0.012 in.) wide are cut routinely.

Burr-free cutting is characteristic of the EDM process. Many features of modern EDM equipment, such as multiple electrodes, automated dressing, and numeric control, contribute to broad process applicability. The scope of EDM applications ranges from tool and die and cavity work to automated transfer line operations.

## Traveling-Wire EDM

Traveling-wire EDM is a specialized form of EDM in which the electrode is a continuously spooling conducting wire. The wire moves with respect to the work by a numerically controlled table. The result is a precision jigsaw. Either the inside (a punch, for example) or outside (a die) piece may be the part of interest. A diagram of a traveling wire EDM machine is shown in Fig. 9. Machines have at least two controlled axes ($x$ and $y$). Most machines can tilt the wire to produce tapered work using two auxiliary horizontal axes

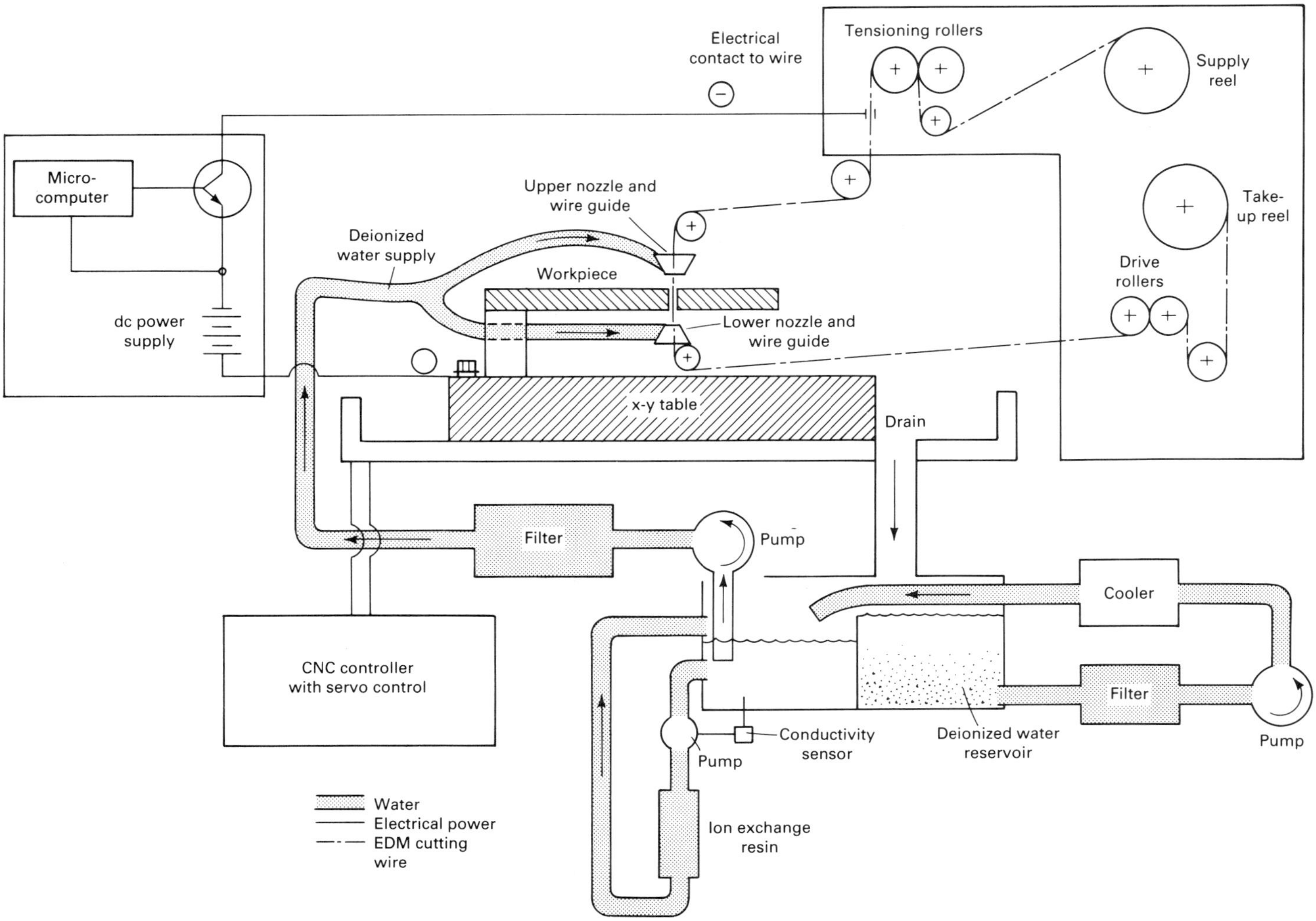


**Fig. 9** Diagram of a typical traveling wire EDM machine

(called *u* and *v*). Controlled vertical and rotary axes are also available.

Wire (normally brass) is used once. Servo control of the machine is similar to that of vertical CNC machines. Flushing problems or workpiece distortion sometimes makes it necessary for the wire and work to separate. The controller must ensure that the separation takes place along the path previously cut. Work thickness capacity of 150 mm (6 in.) is average with some machines capable of up to 420 mm (16.5 in.). Flushing is provided by nozzles at the upper and lower wire guides, with the stream of dielectric, usually deionized water, being coaxial with the wire. Swarf is removed by filtration and settling.

Normal accuracy is about ±0.013 mm (±0.0005 in.). Special measures such as multiple passes and precise temperature control are used for a higher accuracy of ±0.005 mm (±0.0002 in.). Power settings may be decreased with each pass for improved surface, as with vertical EDM.

Constant wire tension is maintained. The tension must be high enough to keep the wire straight in the cutting zone, but not high enough to break the wire. The right tension is a function of the wire material and diameter. Wire guides are made of a hard material such as sapphire or diamond to minimize wear. The guides are required to keep the wire vertical, which is necessary to maintain straight sides on the workpiece. Hydraulic servo mechanisms are not used in wire EDM. The servo motion is done with direct-drive ac servo motors, dc servo motors, or stepper motors. The monitored quantity is the voltage between the wire and the workpiece.

A typical job of cutting flat stock with wire EDM is shown in Fig. 10. It is good practice to use a start hole. If entry is made from outside the stock, the effect (near the end of the cut) is similar to a picture frame with one corner not joined. Internal stresses in the frame cause distortion during the cut, with resulting loss of accuracy. The effect can be lessened if the frame is massive compared to the slug. A so-called glue strap, or piece of metal with spots of glue on frame and slug, is often used to bridge the slug and frame. The purpose is to maintain electrical continuity and keep the slug from falling out at the end of the cut.

**Wire EDM Applications.** Electrical discharge wire cutting is used to make stamp-

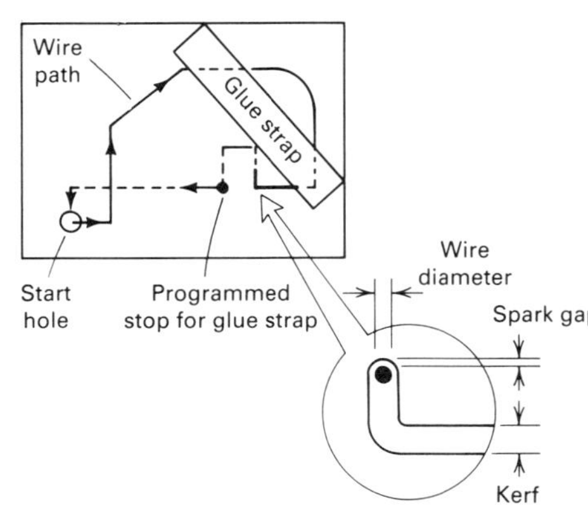


**Fig. 10** Traveling wire EDM job. The inside piece is the part to be kept (a punch, for example). If the outside is to be kept, as in a die, the start hole is in the slug. The glue strap keeps the slug from falling and maintains electrical continuity to the finish.

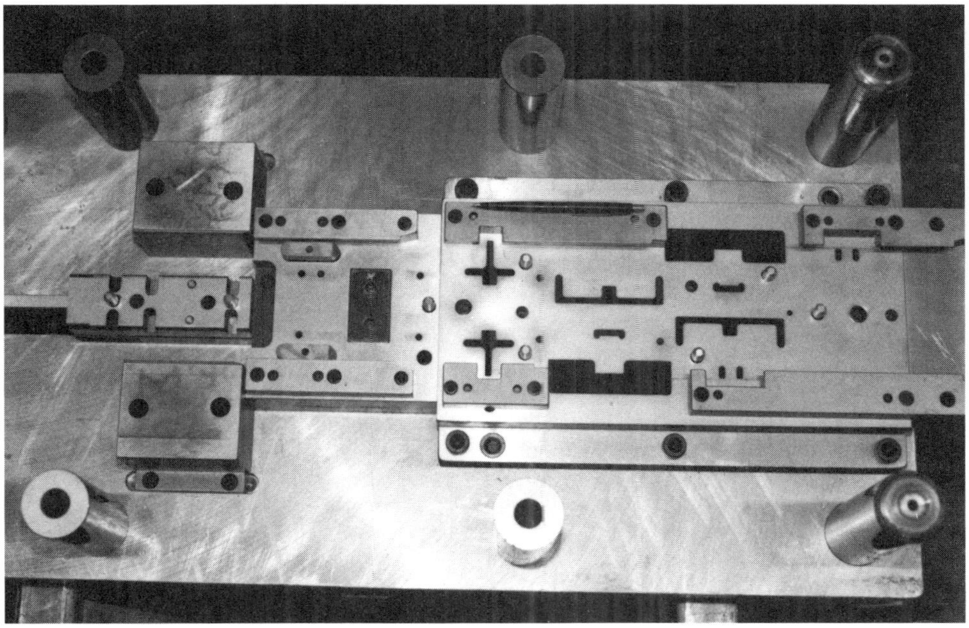

**Fig. 11** Progressive die cut by traveling-wire EDM. Courtesy of Progressive Tool Company

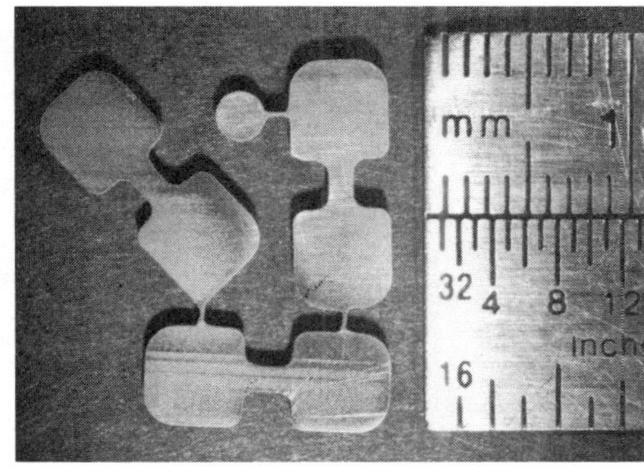

**Fig. 12** Set of microtensile specimens cut by wire EDM for determination of anisotropy in rolled sheet

ing dies, tools for lathes, templates for use in tracer lathes, electrodes for vertical EDM, broaches, and extrusion dies (Ref 16). It can also be used for prototype production of parts to be made later by die stamping or CNC milling.

The progressive die shown in Fig. 11 is one of the most important applications of electrical discharge wire cutting. A strip of sheet metal is advanced through the die, one station at a time. A separate operation is done at each station. The finished part is removed from the strip at the last operation.

Wire EDM also has many applications in metallurgy, such as the removal of core samples from castings to determine variation in chemistry; the sectioning of welds for metallography, and the making of mechanical property specimens (Fig. 12).

**Power Supplies.** As in vertical EDM, peak current and on-time are the major variables controlling spark energy. In modern pulse generating power supplies, these variables, along with off-time, can be set independently. The wire has limited current carrying capacity, and the current rating of the power supplies rarely exceeds 30 A. The potential difference between wire and work is usually between 55 and 60 V. Because electrode wear is of little importance, negative polarity on the wire is always used, and better cutting speed is the result. Larger-diameter wire can handle higher energy sparks and can therefore cut faster.

**Surfaces.** The character of the wire-cut EDM surface is the same as that of vertical EDM. Lower spark energies generally give finer surfaces. Surface finishes range from 1.25 to 0.2 μm $R_a$ (50 to 8 μin. $R_a$). Multiple skim cuts with progressively lower spark energies are used for the best finishes. Workpiece material and thickness must be considered when determining spark settings. Small amounts of the wire material are deposited on the workpiece surface. Experiments by the author have shown that even graphite-coated molybdenum wire deposits carbon and molybdenum on the work surface. Copper and zinc deposits from brass wire can be detrimental to welding, but can be removed with nitric acid. Wire EDM applications to cemented tungsten carbide have had difficulty with preferential attack of the cobalt binder and with cracking. Measures have been developed to deal with these problems (Ref 17).

**Flushing and Dielectrics.** Good flushing is as important in wire EDM as in vertical EDM. Nozzles should be as close as possible to the work. Workpieces with large variations in thickness prevent this and are especially troublesome. Edge entry is also difficult, and the two problems combine when cutting round bar stock. Tricks, such as making an artificial hole with tape or a split soda straw, are often used. The usual result of poor flushing is a wire break. To avoid wire breaks, the off-time is usually increased. The penalty is slower cutting. The reason for using off-time for this purpose is that the spark energy is not affected, thus maintaining the same cutting gap and surface finish. A constant cutting gap is essential for accuracy.

Deionized water is used almost exclusively as a dielectric. The low viscosity is ideal for the difficult flushing conditions found in wire EDM. Conductivity is maintained between set limits with a circuit containing an ion exchange resin. Although additives are not normally needed, they are sometimes used as antirust compounds or (ethylene-glycol base compounds) to make the dielectric more "slippery." Light oils are sometimes used as dielectrics.

Good filtration is also important because contaminants affect the gap distance and, thus, accuracy. Cartridge-type paper and diatomaceous earth filters are common. Because significant heat is generated by electric discharge wire cutting, dielectric cooling and temperature-controlled environments are used.

**Wire.** The most widely used wire is brass wire. It has most of the qualities needed for wire EDM, that is, high tensile strength, high electrical conductivity, and good wire-drawing ability to close tolerances. Research in wire is active. The trend is toward layered wires, with different materials for different functions. One example is a wire with a steel core for tensile strength, a copper layer for conductivity, and a graphite outer layer for speed. The outer layer in particular can be selected to suit the application. Although more expensive, such wires cut faster than brass, and a case can be made for their net economy (Ref 18). Zinc-coated brass and molybdenum core wires are also available. High tensile strength wires are especially good when a small-diameter wire must be used to make small-corner radii. Wire diameters range from 0.05 to 0.3 mm (0.002 to 0.012 in.).

**Cutting Speed.** An informal industry standard of cutting speed is the linear rate at which 25 mm (1 in.) D2 tool steel can be cut. The standard is expressed in terms of $mm^2$/h (in.$^2$/h) of surface cut. Early machines cut at 1300 $mm^2$/h (2 in.$^2$/h). That speed has increased by a factor of ten or more for present machines. Most of the progress has resulted from improvements in power supplies. Ideal flushing conditions are needed for high speed.

**Programming.** The geometry of the piece to be cut is defined in terms of points, lines, and circles. The cutting path is then specified. When a line intersects a circle, a clockwise or counterclockwise path must be chosen. The tool path is then converted to codes understood by the machine tool. Programming is done on centerline, with offset introduced at the machine tool. The wire can be likened to a small milling cutter. Mathematically defined curves can be generated by a sequence of points, and interpolation can be linear or circular. Points from figures not defined mathematically (artwork, for example) can be obtained from a digitizing tablet. The top and bottom surfaces can be independently programmed. The *u* and *v* auxiliary horizontal axes are used to make the necessary continuous transitions. This allows transition from a circular to a square cross section, for example. The method finds particular application in extrusion dies.

## Health and Safety

Electrical discharge machining is one of the safest machining methods. The inherent dangers of chucks rotating at high speeds, sharp chips, and large milling cutters are not present. The low voltage used makes the electrical hazard small. The EDM process is not without danger, however. The largest risk is fire, the most common cause being a drop in the dielectric level below the level where sparking occurs. If this occurs, the dielectric becomes heated above its flashpoint, and the EDM sparks provide a ready-made source of ignition. Large electrodes increase the problem. As a safety measure, the machine should turn off automatically when the fluid level drops below a preset level—the top of the work plus a safety factor. Also, automatic fire extinguishers should be mounted on any unattended machine.

Skin irritation and smoke are the other potential EDM health problems. Hand creams specifically formulated for EDM users are available. If the work is large, ventilation should be provided. Self-contained smoke-consuming units that remove the need for vented hoods are available.

### REFERENCES

1. Technological Aspects of Spark Erosion, *EDM Dig.*, Vol 6 (No. 4), 1984, p 16-36
2. L. Houman, "What's New in EDM Power Supplies and Machines," MR73-816, Society of Manufacturing Engineers, 1973
3. *Machining Data Handbook*, 3rd ed., Machinability Data Center, Metcut Research Associates, Inc., 1980, p 12-22
4. D. Kisner, Multi-Lead: To Successful EDM, *EDM Dig.*, Vol 3 (No. 3), 1981
5. J. Peters *et al.*, Assessment of Surface Typology Analysis Techniques, *Ann. CIRP*, Vol 28 (No. 2), 1978
6. W.R. Koster and J.B. Kohls, "Relation of Surface Integrity to Cost and Reliability of Structural Components," IQ72-207, Society of Manufacturing Engineers, 1972
7. L.J. Rhoades, "Removing the EDM Recast," MR87-643, Society of Manufacturing Engineers, 1987
8. J.A. McGeough, EDM-ECM Finishing Hybrid Technology, *EDM Dig.*, Vol 10 (No. 1), 1988, p 8-12
9. F.J. Demaine and B. Schneider, "A Microscopic Model of the EDM Process," TR 01.1178, Endicott Laboratories, International Business Machines Corporation, 1970, p 19
10. H. Clauser, Ultrasonically Forming Intricate Multiple Graphite Electrodes, *EDM Dig.*, Vol 9 (No. 5), 1987, p 12-15
11. Molded EDM Electrodes: Shortcut to Quality Cavities, *Manuf. Eng.*, Vol 18 (No. 1), July 1978
12. S. Tariq and P.C. Pandey, Experimental Investigations in the Performance of Water as Dielectric in EDM, *Int. J. Mach. Tool Des. Res.*, Vol 24 (No. 1), 1984, p 31-44
13. L. Houman, How to Estimate EDM Time Requirements, *Met. Stamp.*, Vol 11, June 1977, p 15
14. *EDM Technical Manual*, 4th ed., Poco Graphite, Inc., 1977
15. J.E. Fuller, "EDM Cutting and Orbital Paths Using Parametric Equations," MR87-642, Society of Manufacturing Engineers, 1987
16. T.J. Drozda and C. Wick, Ed., *Tool and Manufacturing Engineers Handbook*, Vol 1, Machining, 4th ed., Society of Manufacturing Engineers, 1983, p 14-59, 60
17. G.P. McCleary, Cutting Tungsten Carbide With Wire EDM—From the Carbide Manufacturer's Point of View, *Carbide Tool J.*, Nov-Dec 1984
18. D.S. Tomalin, Fundamentals of the Past, Present, and Future of Wire Technology for EDM Applications, Part 2, the Present and the Future, *EDM Dig.*, Vol 10 (No. 5), 1988, p 8-11

### SELECTED REFERENCES

- G. Bellows and D.N. Tishler, "Impact of Surface Integrity on Material Properties," TM70-975, General Electric Company, 1970
- E.C. Jameson, Ed., *Thermal Machining Processes*, Society of Manufacturing Engineers, 1979
- E.C. Jameson, Ed., *Electric Discharge Machining*, Society of Manufacturing Engineers, 1983
- C.H. Kahng and K.P. Radukar, "Fundamental Theories of the Parameters of EDM Process," MR77-285, Society of Manufacturing Engineers, 1977
- "Surface Integrity (Sinking)," Agietron Corporation, 1988
- E.J. Weller, Ed., *Nontraditional Machining Processes*, 2nd ed., Society of Manufacturing Engineers, 1984
- Y.S. Wong, *et al.*, An Investigation of the Power Output and Gap Voltage During Electrical Discharge Machining of Metals Under Geometrically Defined Conditions, *Int. J. Mach. Tools Manuf.*, Vol 27 (No. 2), 1987, p 191-214

# Electrical Discharge Grinding

ELECTRICAL DISCHARGE GRINDING (EDG) is much like electrical discharge machining except that the electrode (tool) is a rotating graphite wheel. The work is fed to the wheel by a servo-controlled worktable. The workpiece is cut by the action of a stream of electric sparks between a negatively charged wheel and a positively charged workpiece immersed in a dielectric fluid. Each spark discharge melts or vaporizes a small amount of metal from the workpiece surface, producing a small crater at the discharge site, as in electrical discharge machining.

## Equipment and Operation

The stream of sparks is produced by high-frequency pulses of direct current in an arrangement such as that shown in Fig. 1. The power supply and the dielectric fluid are similar to those used in electrical discharge machining, but lower amperage is used in most EDG applications because the method usually involves a smaller cutting area and is primarily used to achieve higher accuracy and a smoother finish.

The power supply can be rated at 30 to 400 V, 0.5 to 100 A, with a pulse frequency of 0.2 to 260 kHz. Machining current for most applications is 0.5 to 15 A at a potential of 40 to 80 V.

When work loads in grinding do not justify an EDG machine, it is possible to use a numerically controlled EDM machine in which the EDG wheel is mounted and rotated on the vertical spindle. The table can then be fed automatically into the rotating wheel by means of servo-controlled feed (instead of tool feed).

**Wheels.** Electrical discharge grinding wheels are typically constructed from porous, low-grade graphite. Brass wheels are sometimes used when the sections to be cut are thinner than 0.375 mm (0.015 in.) or when dielectric circulation is difficult. Because of their composition, EDG wheels can be easily formed and dressed with inexpensive high-speed steel tools.

The graphite wheel rotates at 30 to 180 m/min (100 to 600 sfm), and the spindle is insulated from the wheel for electrical isolation. High wheel speed causes oil splash; excessively low speed may result in nonuniform wheel wear and out-of-roundness.

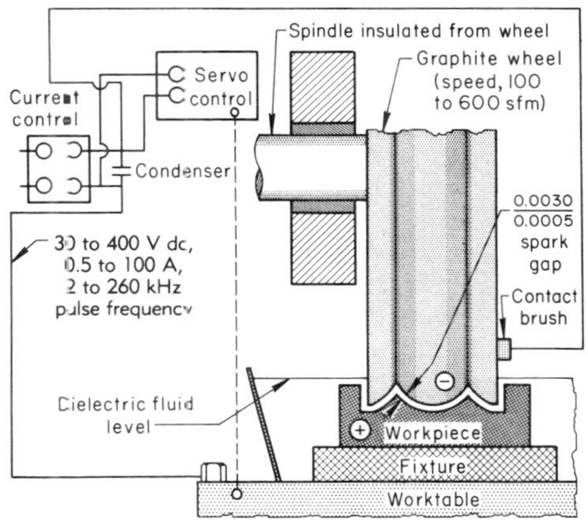


**Fig. 1** Setup for electrical discharge grinding. Spark gap given in inches

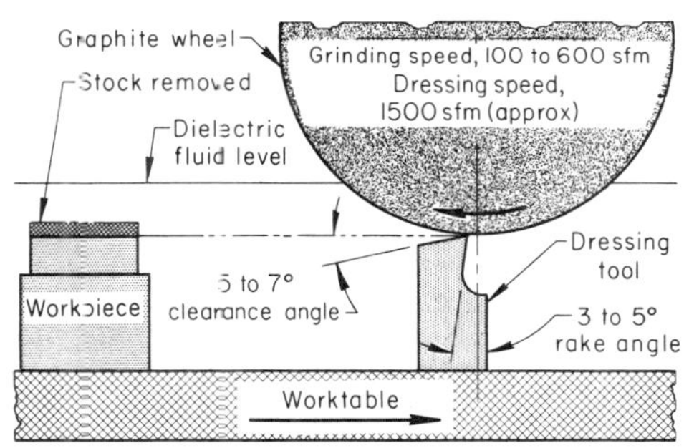


**Fig. 2** Setup for the in-process dressing of an electrical discharge grinding wheel

The grinding wheel, rotating at approximately 450 m/min (1500 sfm), can be dressed by a tool arranged on the machine as shown in Fig. 2. The grinding wheel can be rough dressed at 1.25 mm (0.050 in.) per pass of the tool of hardened steel, followed by a finishing pass of 0.125 mm (0.005 in.).

For precision grinding, extreme rigidity is needed for minimum deflection during grinding and in-process wheel dressing. Spindle runout can be held to a maximum total indicator reading (TIR) of 0.001 mm (0.00004 in.). Deflection in wheel dressing can be held to 0.004 mm (0.00015 in.) for traversing 200 mm (8 in.) on hardened and ground V-ways.

**Control.** The spark gap is normally held at 0.0125 to 0.075 mm (0.0005 to 0.0030 in.) by a servomechanism that controls the motion of the worktable. The machining gap is immersed in a dielectric fluid similar to those used in electrical discharge machining—usually oil having a viscosity of 40 SUS (Saybolt universal seconds) at 38 °C (100 °F).

Higher voltage permits larger spark gaps; this is of value in plunge grinding and for ease of flushing with dielectric fluid. An average spark gap is about 0.025 mm (0.001 in.). Higher electric current cuts faster but leaves a rougher, more damaged surface; crater size is proportional to spark energy. Less current is used to produce smoother, less damaged surfaces, especially in the grinding of carbide. Higher pulse frequency makes a smoother surface. Greater capacitance is generally used for faster cutting.

Once the machining cycle has been started, operator attention is not required on automatic EDG machines. The servo drive for the worktable is controlled by the voltage between the grinding wheel and the workpiece. When the voltage is high (~80 V), the workpiece is rapidly traversed toward the grinding wheel until the first spark occurs. At this point, the voltage drops to the normal machining range (40 to 60 V), and the servo device maintains a constant spacing between the grinding wheel and the workpiece (normally in the range of 0.0125 to 0.075 mm, or 0.0005 to 0.0030 in., depending on the power supply settings). The servo system automatically backs the work away from the grinding wheel if the voltage drops for any reason.

## Process Characteristics

Electrical discharge grinding is used for work that demands great accuracy, such as making carbide form tools. Tolerances of ±0.005 mm (±0.0002 in.) are common; ±0.001 mm (±0.00005 in.) can be obtained in some applications. It is a slow process, removing only 160 to 2500 $mm^3$ (0.01 to 0.15 $in.^3$) of stock per hour. Higher metal removal rates result in a surface finish that is

### Table 1 Typical applications and conditions for the electrical discharge grinding of stainless steels using 300 mm (12 in.) diam wheels

Dimensions in figure given in inches

| Work | | | | | Operating conditions | | | | | Results | | | | |
|---|---|---|---|---|---|---|---|---|---|---|---|---|---|---|
| Part No. | Type of stainless steel | Description of workpiece | Shape produced | Dimensions of shape produced | Speed, rev/min | Volts | Amp | Capacitance, µF | Pulse frequency, kHz | Grinding rate, $mm^3 \times 10^3$/h ($in.^3$/h) | Cutting time, min | Volume wear ratio | Overcut, mm (in.) per side | Finish, µm (µin.) |
| 1 | 303 | Rod, 13.5 mm (0.531 in.) diam × 31.0 mm (1 7/32 in.) long | 3 grooves | 4.8 mm (3/16 in.) wide, 5.54 mm (0.218 in.) deep; full-radius bottom | 20 | 70 | 10 | 6 | 32 | 7.7 (0.470) | 6 | 3:1 | 0.043 (0.0017) | 5.0 (200) |
| 2 | 304 | Strip, 2.64 × 101.6 × 165.1 mm (0.104 × 4.000 × 6.500 in.) | Rib | 15.2 mm (0.600 in.) wide, 0.25 mm (0.010 in.) deep; central web 0.25 mm (0.010 in.) wide | 20 | 70 | 0.5 | 0.5 | 32 | 1.0 (0.06) | 44 | 3:1 | 0.025 (0.001) | 1.5 (60) |
| 3 | 304 | Strip, 3.2 × 75 × 500 mm (1/8 × 3 × 20 in.) | 2 grooves | 71.1 mm (2.800 in.) wide, 0.51 mm (0.020 in.) deep | 20 | 70 | 2 | 2 | 130 | 2.56 (0.156) | 400 | 6:1 | 0.025 (0.001) | 1.5 (60) |
| 4 | 304 | Hexagonal tubing, 57 mm (2 1/4 in.) × 1.25 mm (0.050 in.) wall | Face, step-cut end | Step, 4.75 mm (0.187 in.) deep | 34 | 80 | 8 | 2 | 32 | ... | 30 | ... | 0.025 (0.001) | 3.2 (125) |

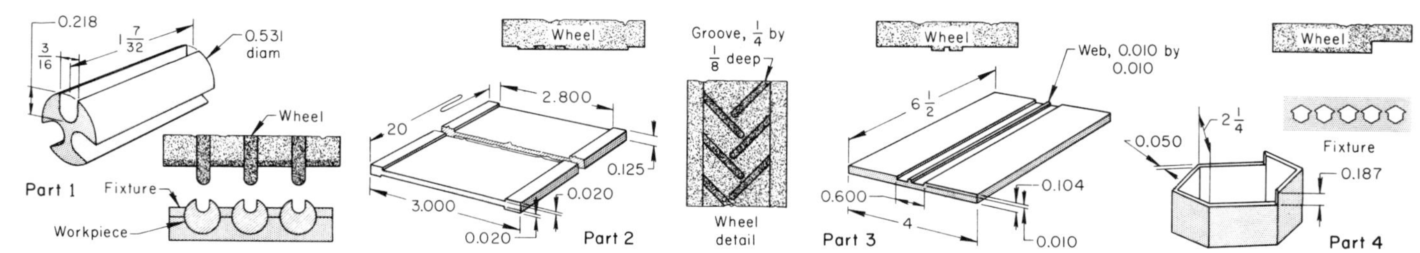


usually acceptable only for certain roughing applications.

**Applications.** In addition to making carbide form tools, electrical discharge grinding is used to grind carbide in shapes such as lamination dies, crushing rolls, and tools of other kinds. It is also used for grinding other hard materials such as gear racks, for making closely spaced thin slots in hard materials, for grinding brittle or fragile parts, and for the production grinding of complex shapes.

The grinding of stainless steel is illustrated in Table 1, which shows workpiece and wheel configurations for four parts and gives processing details. The wheel used in making Part 3 has a herringbone pattern cut in the working surface to improve the flushing action. Cast iron is not ordinarily processed by electrical discharge grinding, because it is readily ground by conventional methods and because nonconductive impurities (sand and slag) are not removed by the spark discharge and can damage the wheel.

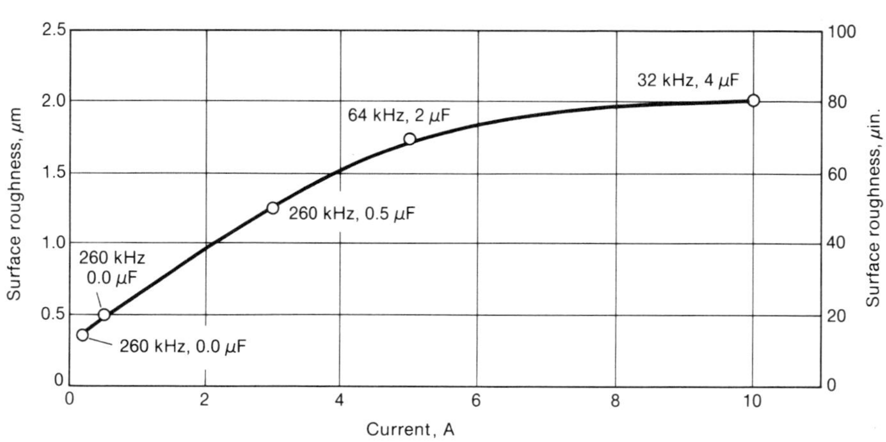


**Fig. 3** Effect of operating conditions on surface roughness in the electrical discharge grinding of tungsten carbide. Pulse frequency and capacitance are shown for each data point. Voltage was 60 V, except 50 V at 0.2 A. The dielectric fluid was a hydrocarbon (petroleum) oil with a viscosity of about 40 SUS at 38 °C (100 °F). The wheel was graphite.

**Surface Finish.** In electrical discharge grinding, because of the wiping action of the wheel past the workpiece, a surface finish of 0.25 µm (10 µin.) is practical. Surface roughness depends primarily on the metal removal rate. In grinding carbide, for example, a metal removal rate of 200 $mm^3$/h (0.012 $in.^3$/h) results in a surface finish of 0.38 µm (15 µin.), while a metal removal rate of 2500 $mm^3$/h (0.150 $in.^3$/h) results in a finish of 3.2 µm (125 µin.).

The effect of operating conditions on the grinding of tungsten carbide is illustrated in Fig. 3. Further decreases in frequency and increases in capacitance beyond the values shown in Fig. 3, while holding the current at 10 A, will increase surface roughness substantially. For example, roughness was 2.8 µm (110 µin.) at 16 kHz and 10 µF, and 3.8 µm (150 µin.) at 8 kHz and 14 µF. Results vary for other metals and grinding conditions.

**Corner radius** depends on overcut. The minimum corner radius in typical EDG applications is 0.0125 to 0.125 mm (0.0005 to 0.005 in.).

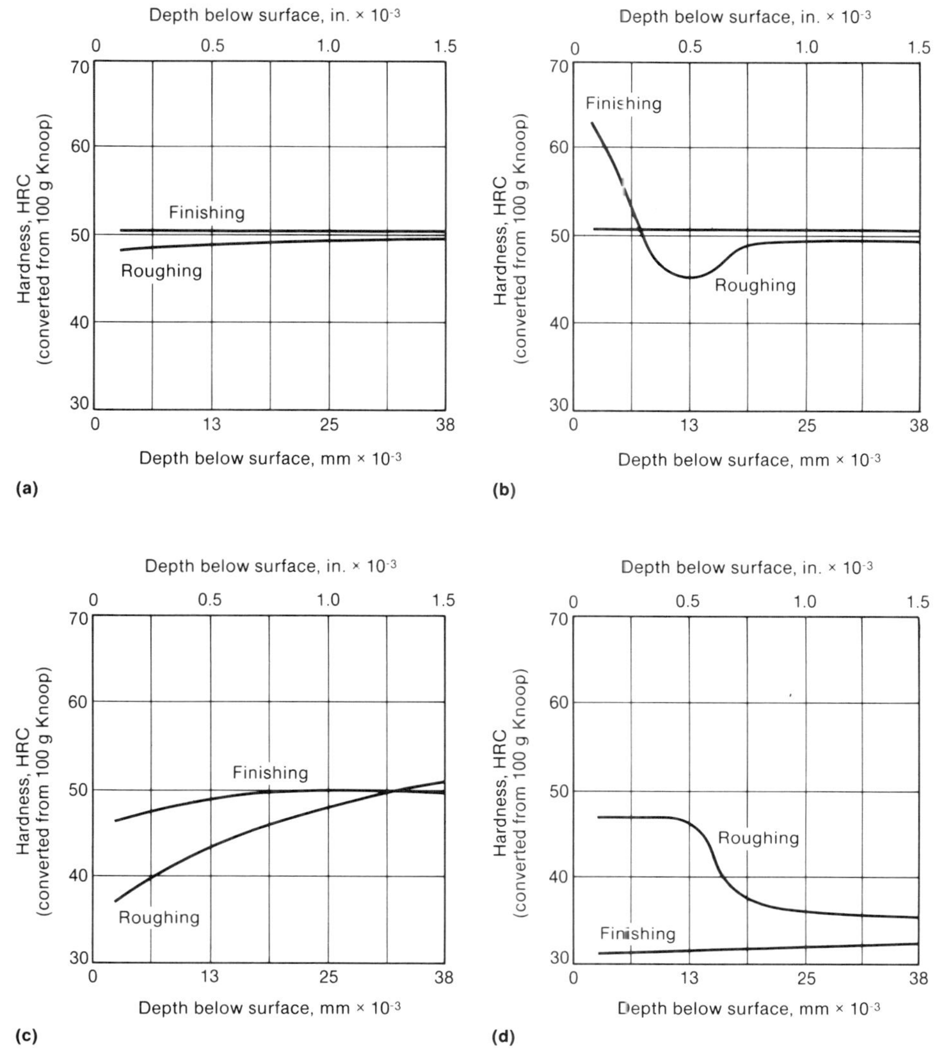


**Fig. 4** Effect of heat in electrical discharge grinding on the surface hardness of various work metals. (a) 4340 steel, 50 HRC. (b) D-6ac steel, 50 HRC. (c) Grade 250 maraging steel, 50 HRC. (d) Ti-8Al-1Mo-1V, 50 HRC

**Wheel Wear.** The wear ratio in electrical discharge grinding depends on current density, tool material, work material, and dielectric fluid. Volume-wear ratio (ratio of work volume removed to wheel volume removed) can range from 100:1 to 0.1:1. The average volume-wear ratio is about 3:1.

Wheel wear, however, is spread over the entire periphery of the wheel. Therefore, assuming a uniform distribution of wear, the linear-wear ratio (ratio of depth of cut to depth of wheel wear) is much higher. For a 25 mm (1 in.) length of cut, the average linear-wear ratio would be 75:1 for a 200 mm (8 in.) wheel or 110:1 for a 300 mm (12 in.) wheel. In grinding irregular contours, wheel wear is concentrated on the high points and sharp edges of the profile, and more frequent redressing is required than for uniform wear.

**Effect on the Workpiece.** As in electrical discharge machining, the high temperatures attained in melting and vaporizing metal from the workpiece affect a thin layer (0.0025 to 0.035 mm, or 0.0001 to 0.0015 in.) of the surfaces ground by electrical discharge grinding. Figure 4 shows typical changes caused by electrical discharge grinding in the hardness of four materials in roughing and in finishing cuts. Carbide and other materials are also affected. For this reason, the EDG process is more widely used in forming carbide tools than in sharpening them. Although some carbides are damaged by electrical discharge grinding and although transverse strength may be reduced, form tools shaped by this method are satisfactory. For high-stress applications, the surface layer affected by electrical discharge grinding may have to be removed.

# Electron Beam Machining

Ronald W. Schneider, MG Industries, Systems Division

ELECTRON BEAM MACHINING (EBM) uses a focused beam of high-velocity electrons to remove material. In this process, a stream of electrons strikes an object and causes rapid melting and vaporization of the material. For drilling applications, a backing material is used on the rear side of the workpiece. When the beam penetrates through the workpiece and contacts the backing material, the high vapor pressure produced by the vaporized backing material expels the molten workpiece material, leaving a hole in the workpiece.

The EBM process can be used to machine a variety of materials, and material properties such as hardness, ductility, electrical conductivity, thermal conductivity, and melting points are usually not limiting factors. The greatest industrial use of EBM is the precision drilling of small holes (0.1 to 1.0 mm, or 0.004 to 0.04 in.) in metals. Although some work has been done to apply the EBM process to nonmetallic and curvilinear cutting applications, this industrial application is very limited. Therefore, unless otherwise noted, the descriptions in this article apply only to metallic hole-drilling applications.

## Equipment Description

The high-velocity electron beam is generated in an electron beam gun (Fig.1) that accelerates free electrons to speeds in excess of 60% of the speed of light. The free electrons are created when the tungsten filament of the cathode is heated by passing an electrical current through it. A high voltage potential (typically 120 kV) between the filament (cathode) and an anode accelerates the free electrons toward the anode. Because the anode contains a hole, the accelerated electrons pass through the anode and continue their motion toward the workpiece. A bias cup (Wehnelt electrode) located between the cathode and anode acts as a grid that controls the number of electrons being accelerated (beam current). The bias cup also acts as a switch for the pulsing of the beam current. Beam current typically is adjustable from 1.0 to 80 mA, and pulse powers up to 12 kW per pulse are possible. For drilling purposes, the beam is pulsed once per hole. Drilling requires very close control of the high voltage, the beam current, and the pulse length.

The electron beam gun also focuses the beam to a very high power density (at least $10^6$ W/mm$^2$ or $6 \times 10^8$ W/in.$^2$). Magnetic coils are used as lenses to focus the beam (Fig. 1). Between these two lenses is a diaphragm, which removes electrons from the fringe area of the beam. This further reduces the spot diameter and helps create the proper power density distribution. Finally, a stigmator coil is used to correct minor beam aberrations and ensure a round beam at the workpiece.

There also must be some means of deflecting the beam. Again, electromagnetic coils are used to accomplish this function. First, a coil is used to bend the beam around a corner so that the electron source is somewhat isolated and protected from line-of-sight exposure to vapors and expelled material. A deflection system is also used to deflect the beam so that it moves in sequence with the part, thus allowing drilling while the part is moving (called on-the-fly drilling).

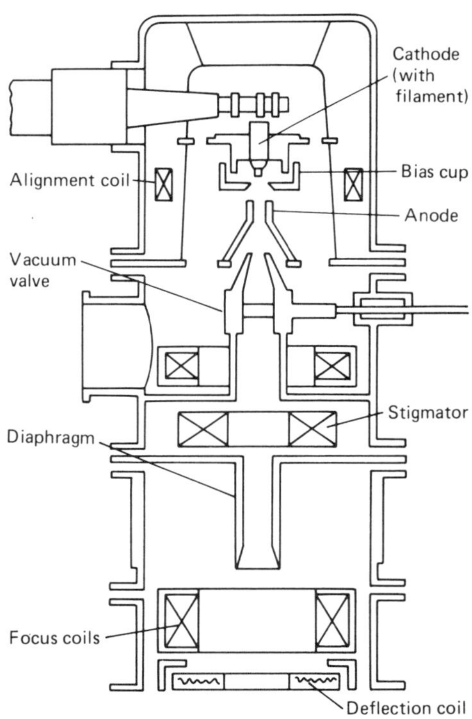


**Fig. 1** Electron beam gun

**Vacuum Chamber.** EBM is performed with the workpiece in a vacuum chamber. Drilling is usually done at 1 Pa ($10^{-2}$ mbar). Drilling is done in a vacuum to avoid dispersion of the beam because electrons have mass and are deflected by air molecules. Approximately 3 min/m$^3$ (5 s/ft$^3$) of chamber volume is needed to pump down to 1 Pa ($10^{-2}$ mbar). Vacuum chambers usually should have a minimum volume of at least 1 m$^3$ (35 ft$^3$) to minimize the amount of debris from the drilling process that sticks to the walls. The debris that is deflected upward during the drilling process is caught by counterrotating, overlapping disks on the ceiling of the chamber. The debris, which adheres to these disks, is scraped off as the disks rotate over scraper pans. The disks also minimize the entry of debris into the electron beam gun itself.

**Workpiece Manipulator.** The next major piece of hardware is the workpiece manipulator. The most elementary manipulator (called a tensioning drum) includes a rotary axis with translational-motion capability (Fig. 2). In this case, the workpiece is formed from sheet into a cylinder and tacked to maintain its shape. Backing material is placed inside the cylindrical workpiece, and the workpiece is placed over a tensioning drum. Inflatable bladders hold the backing material in place and center the workpiece. Typically a "flying" drilling operation is used with this manipulator. In other words, the drum and workpiece rotate at a constant speed; as each beam pulse starts hole drilling, the beam is deflected while the drum rotates. When the hole is completed, the pulse is turned off, and the beam instantaneously goes back to its original position so the next hole can be started. This on-the-fly drilling process is continuously repeated. Also, the translational axis of the manipulator is moved in synchronization with the beam to advance the drum and the workpiece under the beam. In this design, the chamber door is usually part of the workpiece manipulator.

A more sophisticated multiaxis manipulator can be used for complicated shapes or for drilling holes in different directions (Fig. 3). Up to five translation and rotary axes

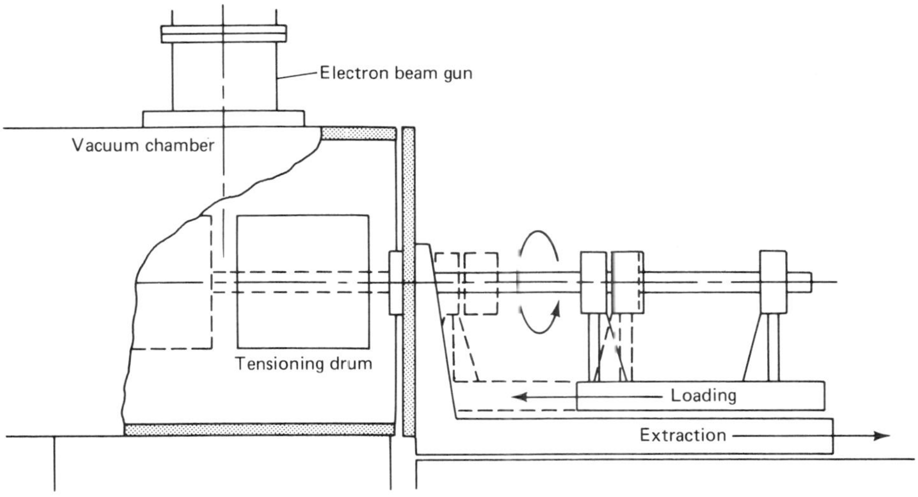


**Fig. 2** Typical electron beam drilling unit with a tensioning drum

can be provided. The workpiece is usually held in a chuck or between a chuck and a center. The motion of the axes is computer numerically controlled. It is worth noting that bearings must be carefully sealed to protect against damage by the drilling debris. Also, the door, usually part of the chamber itself, is a powered sliding door. Finally, special fixturing and tooling can be provided for special applications.

**System Control.** Generally, a computerized numerical control system is required for precise control and integration of all the process parameters. The fast drilling speeds and accuracies now being achieved are possible only through the use of a fast and efficient computer control. Specialized hardware and software is usually required.

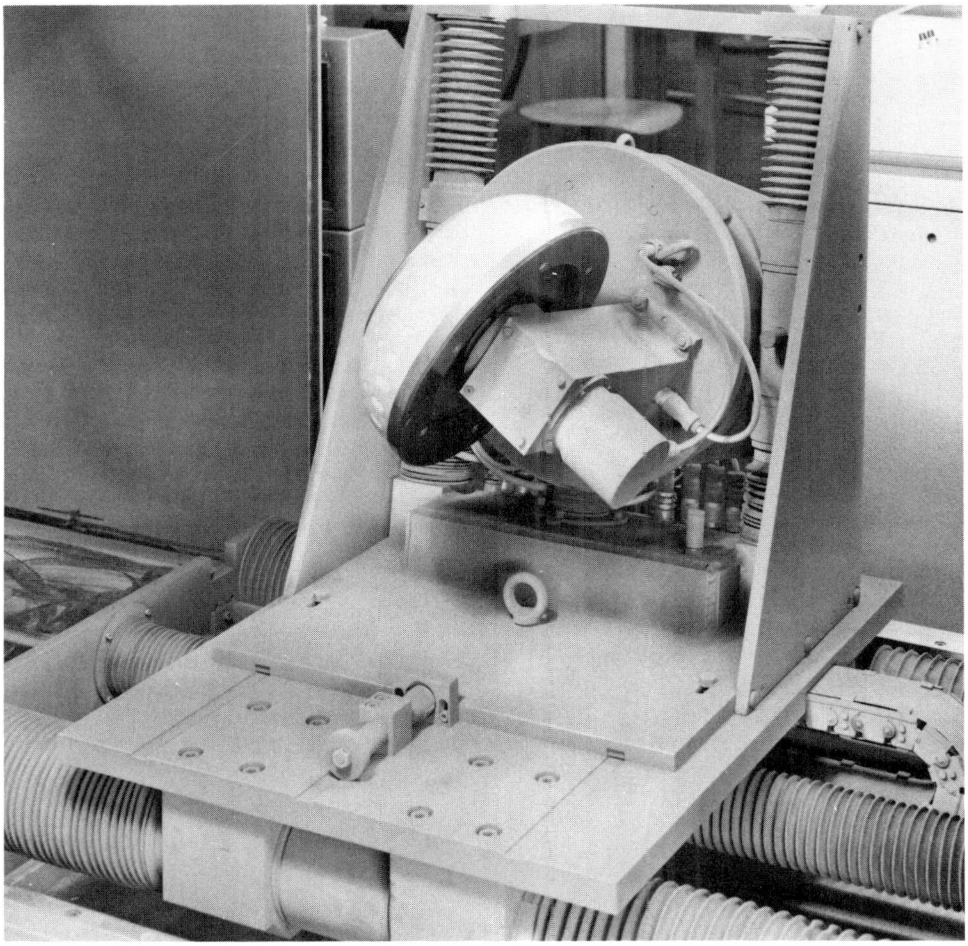

**Fig. 3** Multiaxis manipulator with clamped workpiece. Courtesy of MG Industries/Steigerwald

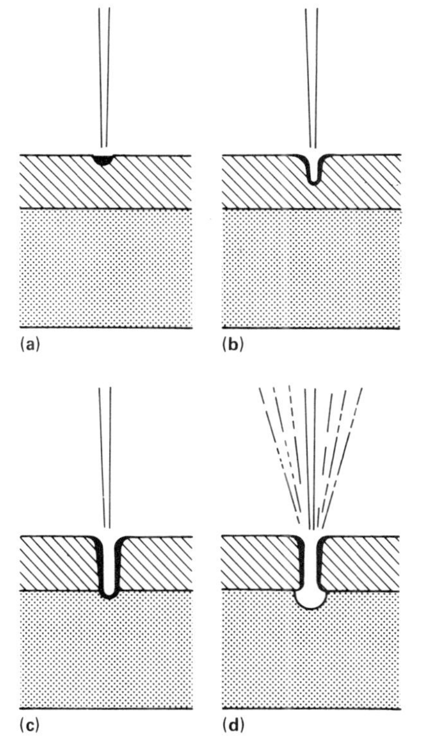


**Fig. 4** Schematic of the electron beam drilling process. (a) The focused electron beam causes local heating and fusion at the surface of the workpiece. (b) Some vapor is produced by the high power density of the beam, and the escaping vapors form a small capillary channel encircled by molten material. (c) The electron beam penetrates the workpiece and begins to vaporize the backing material. (d) Vapor pressure from the backing material builds up until the molten material of the workpiece is expelled.

## Process Characteristics

The physical process of EBM drilling can be examined by dividing it into steps (Fig. 4):

- The sharply focused beam causes local heating and melting, but only a small amount of vaporization occurs (Fig. 4a)
- The pressure of escaping vapor forms and maintains a small capillary channel and pushes the molten material back against the walls of the capillary channel (Fig. 4b)
- The beam penetrates the material by means of this process of vaporization and melting, and a high vapor pressure develops when the beam begins to penetrate the backing material (Fig. 4c)
- The high vapor pressure expels the molten material on the walls, leaving a hole in the workpiece and a small cavern in the backing material (Fig. 4d)

A typical hole cross section is shown in Fig. 5. A small lip remains on the entrance side of the hole after drilling, but this lip can

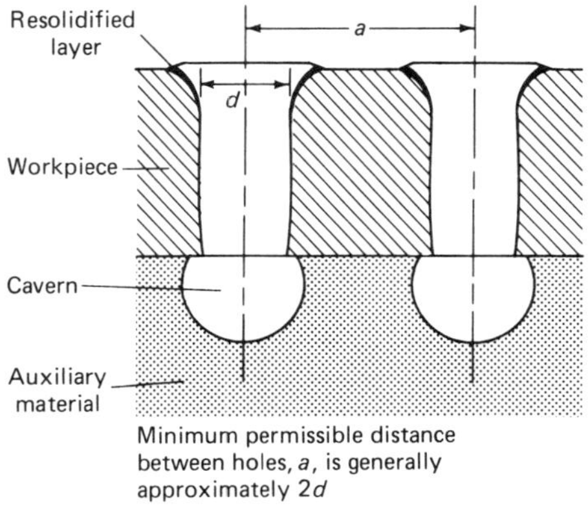


**Fig. 5** Normal configuration of a hole drilled with an electron beam

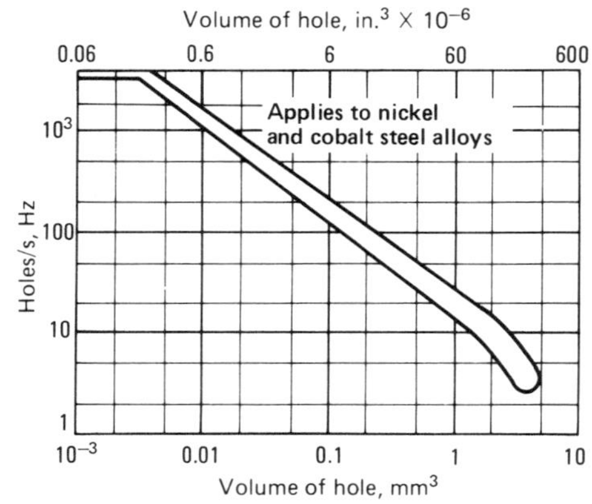


**Fig. 6** Maximum electron beam drilling frequencies for nickel and cobalt steel alloys

**Table 1 Typical results of electron beam drilling**

| Work material | Thickness mm | Thickness in. | Hole diameter mm | Hole diameter in. | Total number of holes | Cycle time, min |
|---|---|---|---|---|---|---|
| Cobalt alloy steel (mixer disk) | 1.6 | 0.063 | 0.9 | 0.035 | 2 340 | 30 |
| CrNiCoMoW steel (combustion chamber housing) | 1.1 | 0.043 | 0.9 | 0.035 | 3 748 | 60 |
| Cobalt alloy steel (spinning head) | 4.3–6.3 | 0.17–0.25 | 0.81 | 0.032 | 11 766 | 40 |
| Inconel 100 (test piece) | 1.5–2.5 | 0.06–0.1 | 0.46 | 0.018 | 162 | 3.5 |
| Copper sheet | 5 | 0.2 | 0.2 | 0.008 | 254 364 | 720 |
| Stainless steel wire cloth/foil composite(a) | 1.1 | 0.043 | 13.2 | 0.52 | 1 | 0.03 |

(a) Beam used in trepanning mode

be removed using standard finishing procedures if a sharp edge is required. There are also some small thermal effects. Electron beam drilling produces a recast layer and a small heat-affected zone of the order of 0.025 mm (0.001 in.).

**Backing material** is required in electron beam drilling to provide high vapor pressure to expel the molten material and leave a hole. Backing material also protects interior features from being affected by the penetrating beam. The material used for backing depends on the workpiece material and the geometry. A typical material is called HM/S and contains three parts brass powder to one part silicone rubber. The silicone rubber has a 5% T-hardener. This material has very little shrinkage, but is not reusable. Other backing materials include an epoxy brass that is cast in place and a reusable wax that is often used when drilling surfaces enclose blind cavities. All backing materials can be easily made by the user.

**Process parameters** of electron beam drilling include:

- Voltage (usually kept constant at 120 kV)
- Beam current
- Pulse duration
- Lens or focus current
- Beam deflection
- Position of the rotation and translation axes
- Speed of the rotation and translation axes

Generally, these parameters are optimized through trial and error for a particular application. Pulse duration and pulse current (beam current) affect the depth and diameter of a hole. Pulse duration ranges from 0.05 to 100 ms. The lens or focus current influences the hole taper, because the focal point position in the vertical direction determines the actual hole shape (tapered, straight, or other).

Deflection coils are used to enable on-the-fly drilling. In this operation, the beam is actually deflected during the period of pulsing so that the beam and workpiece move together during the hole-drilling operation. When the pulse has ended, the beam jumps back to its original position for drilling the next hole.

**Drilling speed** in EBM is determined by hole volume, geometry of the workpiece, and duty cycle limitations. Figure 6 shows the maximum drilling frequency in terms of hole volume. The hole-drilling speeds shown in Fig. 6 are based on an optimum hole spacing of two diameters. Speeds for wider spacings can be calculated assuming that the beam can be moved at the rate of 100 mm/s (4 in./s). Beam power also determines drilling speeds. Relative power requirements to remove equal volumes of various metals in equal time, based on aluminum being 1.0, are:

| Material | Relative power |
|---|---|
| Aluminum | 1.0 |
| Titanium | 1.5 |
| Iron/steel | 1.8 |
| Molybdenum | 2.2 |
| Tungsten | 2.9 |

**Workpieces Drilled in EBM.** Currently, material thickness up to 8 mm (0.3 in.) in steel can be drilled for general industrial application. Material thickness capability ranges from 0.05 to 8 mm (0.002 to 0.3 in.).

Table 1 shows some results of drilling various materials. Electron beam drilling can be used on a variety of materials, but most of the work to date has been with metals, such as stainless steel, nickel and cobalt steel alloys, copper, aluminum, and titanium. There has been some experience with ceramics and artificial leather, and theoretically, drilling plastics with EBM is also possible. The hardness, ductility, electrical conductivity, thermal conductivity, and melting point of the material are not

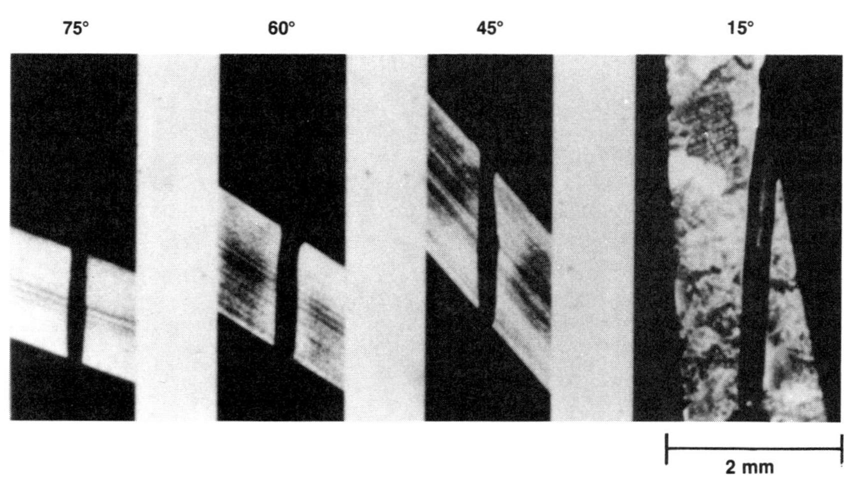


**Fig. 7** Holes drilled at various angles with an electron beam. Courtesy of MG Industries/Steigerwald

**Fig. 8** Jet engine combustion chamber housing with electron beam machined holes. Courtesy of MG Industries/Steigerwald

limiting factors when drilling with the EBM process.

**Hole Specifications.** The holes that can presently be obtained range from 0.1 to 1.0 mm (0.004 to 0.04 in.) in diameter, and work is being done to produce holes down to 0.05 mm (0.002 in.) in diameter. Various shapes of holes can be generated. In the longitudinal direction, cylindrical, tapered, bell-shape, and center-bowed holes can be produced. In the transverse section, round holes are produced. Electron beam drilling can achieve hole depth to diameter ratios of up to 10:1 or even 15:1 in special cases. Another unique capability of electron beam drilling is that the angle to the surface can be quite acute, as small as 20° (Fig. 7).

Hole tolerances in electron beam drilling are quite good. The hole can be located to within ±0.1 mm (±0.004 in.), and the diameter of the hole at the narrowest point can be held to within ±5%, but not less than ±0.03 mm (±0.001 in.). Typically, hole centers must be spaced at least two diameters apart.

## Applications

EBM applications are almost completely confined to drilling. Electron beam drilling is useful when many small holes must be drilled, or when holes are difficult to drill because of hole geometry or material hardness. The fluid and chemical industries use electron beam drilling as a perforating process that produces a multitude of holes for filters and screens.

One aerospace manufacturer uses the electron beam process to cut a 13 mm (0.5 in.) diam circle in a multilayered, stainless steel wire cloth/foil composite using high linear speeds. However, this use represents a very limited application. Experimentation with electron beam cutting of 0.5 mm (0.020 in.) motor laminations has also been conducted, but not carried into production.

**Example 1: Application of Electron Beam Hole Drilling.** A jet engine combustor dome is made of chrome nickel cobalt molybdenum tungsten steel (Fig. 8). It has a 1.1 mm (0.043 in.) wall thickness and requires hole diameters of 0.9 mm, ±0.05 mm (0.035 in., ±0.002 in.) for cooling purposes. The electron beam drilling machine is used to provide almost 3800 holes in 60 min.

**Example 2: Drilling Holes for a Spinning Head.** In the insulation industry, spinning heads are required to manufacture fiberglass. These spinning heads are typically made of cobalt alloy steel and have multiple holes. Figure 9 shows a spinning head that varies in thickness from 4.3 to 6.3 mm (0.17 to 0.25 in.) and requires 0.81 mm, ±0.03 mm (0.032 in., ±0.001 in.) diam holes. In 40 min, 11 766 holes are drilled in this material. This is 100 times faster than electrical discharge machining and 20 to 100 times faster than laser drilling.

## Advantages and Disadvantages

The advantages of electron beam drilling are

- Drilling rates up to 4000 holes/s
- Drills in many different configurations
- No limitations imposed by the hardness, thermal capacity, ductility, electrical properties or surface properties (reflectivity) of the workpiece
- No difficulty encountered with acute angles
- No mechanical distortion because no force is imparted to the part during the drilling process
- Limited thermal effects because only one pulse is required to generate each hole and pulse durations are short
- Drilling parameters that are easily changed during the drilling of a single workpiece, even from row to row of holes
- Capability to maintain high accuracy and repeatability ±0.1 mm (±0.004 in.) for position of the hole and ±5% for the hole diameter
- No tool wear
- The best obtainable finish, compared to other processes (some finishing may be required to remove the slight rounding on the entrance side)

**Fig. 9** Spinning head, used to produce fiberglass. Courtesy of MG Industries/Steigerwald

- Relatively low operating costs compared to other processes used to obtain very small holes

The disadvantages of electron beam drilling are:

- High capital equipment cost
- Unavailability at the present time in the United States from an independent job shop, but Steigerwald Strahltechnik in West Germany can provide job shop capability
- Some nonproductive pumpdown time in the drilling cycle. However, the pumpdown time is generally small compared to the total drilling time
- The presence of a thin recast layer because electron beam machining is a thermal process, which may be a consideration in some applications
- Requirement for operator skill to run EBM machines, although the use of a computer numerically controlled system does ease this task considerably
- Necessity for auxiliary backing material

## SELECTED REFERENCES


- G. Bellows and J.B. Kohls, Drilling Without Drills, *Am. Mach.*, March 1982
- G.F. Benedict, *Nontraditional Manufacturing Processes*, Marcel Dekker, 1987
- "EBOPULS CNC Electron Beam Drilling Machines," Publication 47.2010e, Steigerwald Strahltechnik, 1982
- T. Lyman, in *Machining*, Vol 3, 8th ed., *Metals Handbook*, American Society for Metals, 1967
- D. Von Dobeneck, Drilling With Electron Beams, *Ingenieur Dig.*, Vol 16 (No. 38), 1977

# Laser Beam Machining

Dana Elza and Gene White, Coherent General

LASER BEAM MACHINING (LBM) removes, melts, or thermally modifies a material by focusing a coherent beam of monochromatic light on the workpiece. The LBM process does not involve mass material removal, but does provide rapid material removal with an easily controlled, noncontact, nonwearing tool. The LBM process also places minimal demands on workpiece fixturing in cutting operations and performs a variety of other metal processing functions such as drilling, welding, marking, and heat treating. The effectiveness of LBM in a particular application depends on the pulsing and focusing of the beam and on the reflectivity, absorption coefficient, thermal conductivity, specific heat, and heat of vaporization of the workpiece.

The greatest use of lasers is in metal cutting, primarily two-axis profiling of sheet goods that might otherwise have been blanked out by punch press or fabricated by hand after laborious layout of the pattern. Currently, most metal cutting falls within 9.5 mm (0.375 in.) and thinner, although $CO_2$ lasers are now competitive with plasma arc cutting for metal thicknesses of 13 mm (0.500 in.) and greater. The principal factor in the use of laser metal cutting is manufacturing methodology. Laser cutting is ideal for batch processes, just-in-time, or low-to-medium volume production. Most laser-cutting work is performed on generic or multipurpose material handling systems, as opposed to dedicated automation controlled systems.

Generally, there is a clear-cut reason for using one type of laser over another. Carbon dioxide lasers are capable of delivering much higher average power than Nd:YAG lasers. As such, $CO_2$ lasers can cut faster and produce deeper weld penetration than Nd:YAG lasers. Pulsed Nd:YAG lasers develop a high pulse energy that allows percussion drilling and the cutting of metals at angles and thicknesses not possible with $CO_2$ lasers. There are some applications—spot welding and hole cutting, for example—where either laser type can provide acceptable results at comparable speeds. Capabilities and operating ranges for $CO_2$ and Nd:YAG lasers are shown in Tables 1 and 2.

Laser welding is becoming increasingly used in both low-volume specialty production and high-volume assembly operations. Laser welding operations are typically more dependent on lengthy process development than are cutting applications. Often parts are specifically designed to be laser welded. Laser welding generally requires more precise part fit-up than arc welding processes. While most laser welding is autogenous, that is, without added filler metal, it is becoming apparent that the addition of filler metal to bridge gaps or to act as a buffer when joining metals has great potential.

## Equipment

The principal lasers used in metal processing are the neodymium-glass (Nd: glass), the carbon dioxide ($CO_2$), and the neodymium-doped yttrium aluminum garnet (Nd:YAG) lasers. In some instances, the same laser can perform cutting, welding, marking, and heat-treating functions by varying the energy density, focus (spot size), and pulse duration. Other lasers with wavelengths outside of the infrared spectrum have also been used in material processing. Excimer lasers, for example, produce energy in the ultraviolet spectrum and are being successfully used in the micromachining of thin metal sections, as well as metal marking applications.

**Nd:YAG lasers** have wavelengths of 1.06 μm (4.17 μin.) and can operate in either a pulsed or continuous-wave (CW) mode, but not both. Pulse frequency is limited by the maximum power rating, and ranges from 1 to 10 000 pulses per second. The pulse energy for Nd:YAG lasers ranges from 5 to 80 J per pulse, although pulse energies as high as 100 J per pulse can be obtained with an oscillator-amplifier in the resonator. Most industrial types of Nd:YAG lasers have an average output power of 100 to 400 W, but larger lasers with 500 W average output power are also available.

**$CO_2$ lasers** have wavelengths of 10.6 μm (41.7 μin.) and are capable of delivering a much higher average power than are the Nd:YAG lasers. The typical average output power of a $CO_2$ laser in the continuous-wave (CW) mode is 250 to 5000 W. In pulsed operations, average output power is reduced. The average output power when cutting metals with a $CO_2$ laser in the pulsed mode ranges from 100 to 2000 W.

Carbon dioxide lasers are available in two designs, the axial-flow laser and the transverse-flow, or gas transport, laser. Axial-flow $CO_2$ lasers may be either fast-flow or slow-flow types. Fast axial-flow lasers can achieve 600 W/m (180 W/ft) of tube length and have an average power up to 5000 W. The slow-flow axial-flow $CO_2$ laser is limited to 100 W/m (30 W/ft) of

**Table 1 Process capabilities of $CO_2$ lasers by power range**

| | 200–300 W | 300–500 W | 500–800 W | 800–1500 W | 1500–3000 W | >3000 W |
|---|---|---|---|---|---|---|
| Seam welds, spot welds, maximum penetration, mm (in.) | 0.75 (0.030) | 1.3 (0.050) | 2.0 (0.080) | 3.2 (0.125) | 6.4 (0.250) | 19.0 (0.750) |
| Cutting, maximum thickness, mm (in.) | 1.5 (0.060) | 5.0 (0.20) | 9.5 (0.375) | 12.7 (0.500) | 19.0 (0.750) | <25.0 (<1.00) |
| Heat treating case depth, mm (in.) | ... | ... | 0.75 (0.030) | 1.3 (0.050) | 1.3 (0.050) | 1.3 (0.050) |
| Cladding, surfacing, glazing, annealing | ... | ... | Yes | Yes | Yes | Yes |

**Table 2 Process capabilities of Nd:YAG lasers by power range**

| | <100 W | 150–200 W | 200–400 W |
|---|---|---|---|
| Microwelding, soldering, marking | Yes | ... | ... |
| Seam welds, spot welds, maximum penetration, mm (in.) | ... | 1.3 (0.050) | 2.0 (0.080) |
| Cutting, maximum thickness, mm (in.) | ... | 5.0 (0.2) | 38 (1.5) |
| Drilling, maximum thickness, mm (in.) | ... | 5.0 (0.2) | 38 (1.5) |

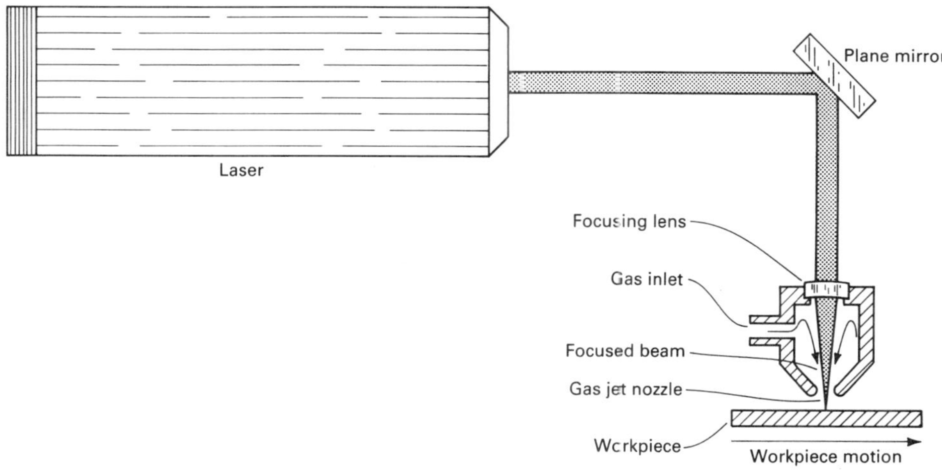


**Fig. 1** Schematic of laser beam cutting with a gas jet

tube length and to less than 1500 W of average output power. Most axial-flow $CO_2$ lasers can be operated in either the CW or the pulsed mode. Pulse frequencies range from 1 to 10 000 pulses per second.

Transverse-flow $CO_2$ lasers operate only in the CW mode, although the beam can be modulated with a beam chopper. Transverse-flow $CO_2$ lasers are used when compact medium-power lasers are needed or when high powers are required. Average powers range from 2500 to 15 000 W.

**Gas-assisted laser beam machining** (Fig. 1) may be used in cutting, welding, and heat-treating operations. The gas type can be oxygen, inert gas, or air, depending on material type and quality requirements. The gas device is designed to produce either a columnar or diffuse flow of gas, depending on the process. In cutting operations, a coaxial, columnar flow of gas is used to purge molten metal from the cutting area. In welding operations, a diffuse flow of gas is used.

**A lens** is used to focus the laser beam on the workpiece surface. Short focal length lenses provide small focus spot sizes and high power density. Long focal length lenses provide narrow beam convergence angles, which are necessary to penetrate thick metal sections. Initial beam diameter is an important consideration in lens selection. Power densities and depth of focus can change with different beam diameters.

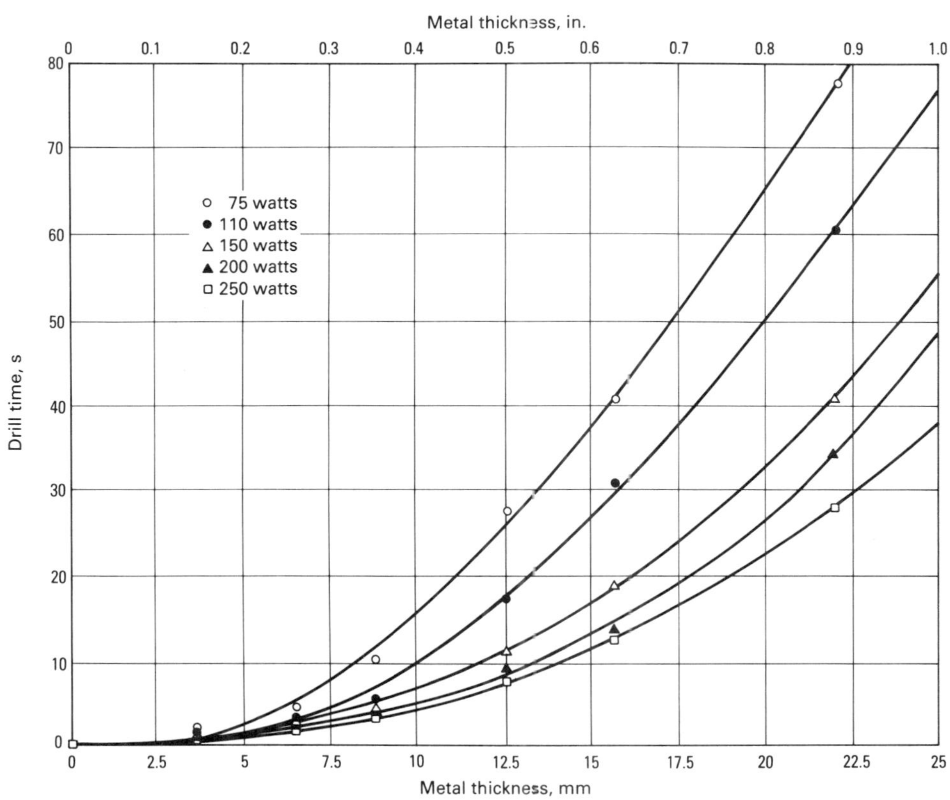


**Fig. 2** Drill time with a Nd:YAG laser. Workpiece: Inconel 718; pulse energy, 40 J

## Drilling

Laser beam drilling is done either by percussion drilling or trepanning. Percussion drilling is used for producing relatively small holes (<1.3 mm, or 0.050 in., in diameter) through metal sections up to 25 mm (1 in.) thick. While holes less than 0.25 mm (0.010 in.) in diameter can be produced by percussion drilling, the maximum metal thickness for small holes is limited. Holes greater than 1 mm (0.04 in.) are usually made by trepanning.

**Percussion drilling** is most often performed with Nd:YAG lasers because of their higher energy per pulse. The laser is operated in a pulsed mode with each pulse removing a certain volume of metal. As metal thickness increases, the range of hole diameter decreases. Energy is transmitted to the bottom of the hole by the laser beam being reflected off the inside walls of the hole. The operating parameters of concern in percussion drilling are discussed below.

*Power.* Percussion drilling is accomplished using average power ranging from less than 100 to 250 W. Drill times at 40 J per pulse for various power levels are illustrated in Fig. 2.

*Pulse length* is selected to optimize the quality of the hole. Shorter pulse lengths may limit the maximum energy achievable in a single pulse. Typical drilling pulse lengths range from 0.5 to 2 ms.

*Pulse frequency* is determined by optimizing throughput and quality requirements. Pulse frequency used for percussion drilling ranges from 5 to 20 Hz with Nd:YAG lasers and up to 100 Hz with $CO_2$ lasers.

*Pulse energy* required is fundamentally determined by material thickness, composition, and hole diameter. Higher pulse energy also provides faster drilling rates, but can be detrimental to hole quality.

*The focusing lens* determines the spot size, which corresponds to the desired hole diameter. The spot size is equal to the hole size in thinner metals (<6 mm, or 0.25 in., thick). As metal thickness increases, the range of hole diameters that can be percussion drilled decreases. Thicker metals and larger holes dictate the use of longer focal length lenses. Focal lengths usually range from 100 to 250 mm (4.0 to 10.0 in.).

*Focal position* can be optimized above, below, or on the surface of a workpiece, depending on the desired results. Most often, the focus lies below the surface at a depth that is 5 to 15% of the metal thickness. The best focus for a desired result is most often determined empirically after evaluation of hole quality. Hole quality is determined by roundness, taper, recast, and microcracks.

*Gas Jets*. Percussion drilling can be accomplished with or without a gas jet. If no gas jet is used, an alternative lens protection method must be employed. Oxygen can be used to produce an exothermic reaction with many metals. Exothermic percussion drilling provides more efficient metal removal. Air or inert gas can also be used in percussion drilling. Nozzle orifice diameters and standoffs are larger for percussion drilling than those used for cutting (or for trepanning, which is essentially a cutting operation). Nozzle designs vary, but typical orifice diameters for percussion drilling range from 2.5 to 6 mm (0.10 to 0.25 in.). Nozzle standoffs range from 4 to 40 mm (0.15 to 1.5 in.).

**Trepanning** is commonly performed with both $CO_2$ lasers and Nd:YAG lasers. Trepanning requires a percussion-drilled pilot hole and is performed by translating either the workpiece or the focusing optic (typically with circular interpolation of two or more axes of motion). Metal thickness range is the same as percussion drilling. However, as metal thickness increases, process speeds decrease. Trepanning is essentially cutting and can be performed by operating the laser in the CW or pulsed mode. The operating parameters of concern in trepanning are given below.

*Power* is established at the level needed to maximize throughput rates. Average power is determined by laser design and the maximum rate at which the part and/or beam can be translated without loss of specified tolerances.

*Pulse length* is selected by the quality requirements, material composition, material thickness, and laser type. Pulse lengths are increased as material thickness increases. Pulse lengths used with both $CO_2$ and Nd:YAG lasers are generally less than 2.0 ms.

*Pulse frequency* defines the process rate at which a material can be trepanned. Lower pulse frequencies are typically used as metal thickness increases.

*Lens focal length* is determined by material thickness. Longer focal length lenses are used with thicker material or when power density drops below the level required to overcome surface reflectivity. Lens focal lengths are similar to those used for percussion drilling, although $CO_2$ lasers require a focal length of 125 mm (5 in.) or shorter for trepanning.

*A coaxial gas jet* is used to facilitate trepanning of holes. Gas type and pressures vary, depending on material. Oxygen gas is used to promote an exothermic reaction, which improves process efficiency. Inert gas is used when an oxide free layer is desired. Air can also be used if it produces acceptable quality holes. Gas pressure mainly affects the quality of the trepanned holes. Oxygen pressures range from 100 to 350 kPa (15 to 50 psi). Inert gas and air pressures range from 200 to 620 kPa (30 psi to 90 psi). Nozzles are used to constrict and direct the gas flow. Nozzles vary in design, but typical orifice openings range from 0.75 to 2.5 mm (0.030 to 0.100 in.) for trepanning. Nozzle standoff varies from 0 to 15 mm (0 to 0.60 in.) depending on laser type, nozzle design, and gas flow rates. Nozzle stand-off with Nd:YAG lasers is 5 mm (0.20 in.).

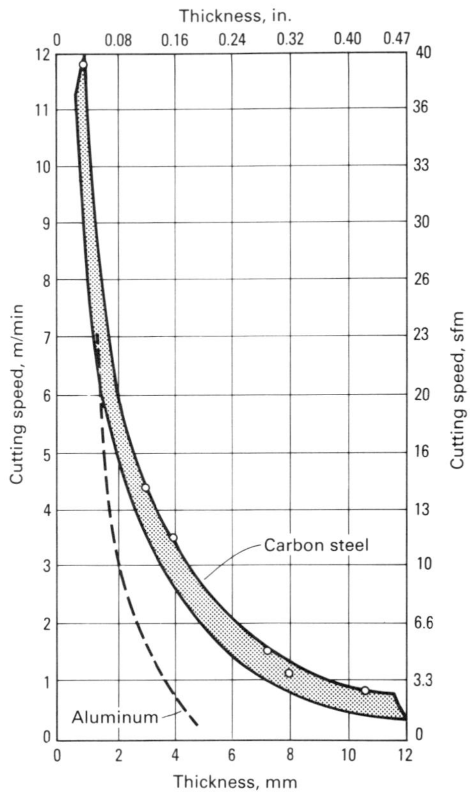


**Fig. 3** Cutting rates for a 1250 W $CO_2$ laser

## Cutting

Cutting can be accomplished by a laser operating in either a CW or pulsed mode. With $CO_2$ lasers, the CW mode is used for thicker metal sections, while the pulsed mode is used for thinner metal sections. Pulsed high-power Nd:YAG lasers are used to cut thick sections of superalloy metals. Lasers are also used to cut hard workpieces with low electrical conductivity, which cannot be cut by electrical discharge machining (EDM). Solid cubic boron nitride (see the article "Ultrahard Tool Materials" in this Volume), for example, is cut with lasers instead of by EDM.

Cutting with lasers is accomplished with $CO_2$ and Nd:YAG lasers. More often, cutting is done with $CO_2$ lasers because of the faster cutting rates. Typical cutting rates for a 1250 W $CO_2$ laser are illustrated in Fig. 3.

**The process variables** in laser cutting are described below.

*Average power* for cutting with a CW $CO_2$ laser is typically 250 to 5000 W. In the pulsed mode, lower average powers can cut metal because of higher peak instantaneous powers. The power ranges for cutting metal in the pulsed mode range from less than 100 to 2000 W for $CO_2$ lasers. Cutting with Nd:YAG lasers is accomplished with a power range of less than 100 to more than 400 W.

*Pulse length* is selected to optimize the quality of the cut surface. Shorter pulse lengths (<0.75 ms) are used for the intricate cutting of thin metals. Shorter pulse lengths may limit the maximum energy achievable in a single pulse. Longer pulse lengths (up to 2 ms) provide greater pulse energies, allowing thicker metals to be cut. Similar pulse lengths would be used for both $CO_2$ and Nd:YAG lasers.

*Pulse frequency* is adjusted to give the maximum cutting speed for the quality required. In general, higher frequencies are used to cut thinner metals. For $CO_2$ lasers, higher frequencies range from 200 to 500 Hz. In Nd:YAG lasers, higher frequencies range from 30 to 100 Hz. Lower frequencies are used to cut thicker metal sections.

*Pulse energy* needed in cutting is related to material thickness. As metal thickness increases, greater pulse energy is required. In practice, $CO_2$ lasers deliver maximum pulse energies up to 2 J at longer pulse lengths and lower pulse frequencies. Nd:YAG lasers can produce much higher pulse energies up to 80 J with pulse frequency limited by the maximum power rating of the laser.

*Lens choice* is based on metal thickness, composition, and quality requirements and on beam diameter. Wider kerf widths are obtained by using longer focal length lenses, and some materials, such as aluminum, require a larger kerf width for good results. The following guidelines are for beam diameters from 13 to 25 mm (0.5 to 1 in.).

For $CO_2$ lasers, a general rule is to use a 65 mm (2.5 in.) focal length for cutting metals up to 6 mm (0.25 in.) thick. A 125 mm (5 in.) focal-length lens can be used for metals from 5 to 15.8 mm (0.2 to 0.625 in.) thick. A 190 or 250 mm (7.5 or 10 in.) lens can be used for materials thicker than 13 mm (0.5 in.).

Typical lenses for Nd:YAG lasers have focal lengths of 100 mm (4 in.) for metals thinner than 3 mm (0.125 in.). For cutting metals up to 25 mm (1 in.), the focal lengths range from 150 to 250 mm (6 to 10 in.). Thicker metals require even longer focal lengths.

**Gas jets** provide a coaxial, columnar flow of gas through the cut slot to remove molten metal. The gas type can be oxygen, inert gas, or air depending on material type and

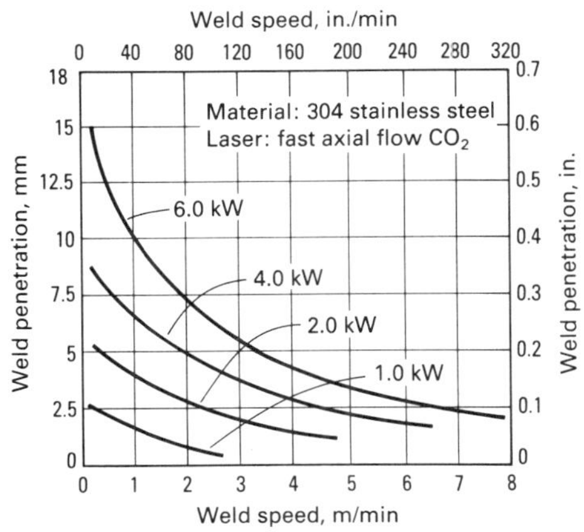


**Fig. 4** Weld penetration for CW $CO_2$ lasers. Source: Ref 1

quality requirements. Oxygen is most commonly used for cutting steels. When an oxide-free surface is desired, an inert gas, such as helium, is used.

Gas pressures for oxygen range from 100 to 350 kPa (15 to 50 psi), while pressures for inert gases range from 200 to 620 kPa (30 to 90 psi). Typical nozzle orifices range from 0.75 to 2.5 mm (0.030 to 0.100 in.). Nozzle standoff, depending on laser type, nozzle design, and gas flow rates, varies from 0 to 1.5 mm (0.0 to 0.06 in.) for $CO_2$ lasers and up to 5 mm (0.200 in.) for Nd:YAG lasers. Nozzle standoff and gas pressure to a large degree determine the quality of the cut. Using a $CO_2$ laser, a clean cut (oxide free and dross free) can be achieved in thinner sections of metals such as stainless steel, aluminum, or titanium when using inert coaxial gas at 620 kPa (90 psi) with minimum nozzle standoff.

## Laser Welding

The high energy density achievable with lasers can produce a keyhole vapor cavity that allows for deep-penetration, high aspect ratio weld nuggets. Lasers can also be used in more conventional power density ranges that yield conduction welds of lower aspect ratio.

**Process Parameters.** Laser welds can be produced in either the CW or pulsed mode, with the selection being determined by material type, thickness, and joint design. The welding can be accomplished with either $CO_2$ or Nd:YAG lasers in either the CW or pulsed mode. Weld penetrations for various power levels and feed rates are shown in Fig. 4. The operating parameters of concern in laser welding are given below.

*Power* is the determining factor for maximum penetration and feed rate obtainable. $CO_2$ lasers produce higher average power than Nd:YAG and therefore can achieve deeper penetration welds at greater rates. Welding is performed by CW or pulsing depending on metal composition, weld type, and process speed.

*Pulse length* is selected to achieve melting and to control the cooling rate. Welding pulse lengths for Nd:YAG lasers vary from 2 to 10 ms. Welding pulse lengths for $CO_2$ lasers vary from <0.5 to 5 ms. Longer pulse lengths are selected to reduce the solidification stresses by slowing down the cooling rate in more crack-sensitive alloys.

*Pulse frequency* in welding determines process speed, percent of overlap, and cooling rate. Overlaps vary from 40 to up to 70% in hermetic sealing. Cooling rate requirements are a function of material composition and mass.

*Pulse energy* for welding is lower than for laser metal removal processes. Laser energies for $CO_2$ usually range in fractions of joules, while the Nd:YAG laser range may exceed several joules. Higher energy levels disturb the weld pool.

*Lens* considerations include power density required to achieve the needed penetration, joint type, and fit-up, as well as shielding and lens protection techniques. Joint design with poor fit-up or critical alignment can be accommodated by the use of a longer focal length lens. When the welding of metals creates excessive weld spatter, longer focal length lens are recommended to increase lens service life. For $CO_2$ lasers at subkilowatt power levels, a lens with a focal length of 125 mm (5 in.) is primarily used. Multikilowatt $CO_2$ lasers primarily use 125 mm (5 in.) or greater focal lengths. Nd:YAG lasers generally use focal lengths of 63 to 200 mm (2.5 to 8 in.) for laser welding.

*Beam oscillations*, either circular or linear, are used to overcome poor fit-ups and produce low aspect ratio welds. Filler metal additions, in the form of wire, powder, or preforms, are used to bridge gaps and to alter weld composition for improved weldability.

**A gas jet** (or lens shroud) is used to protect the focusing lens from spatter and to direct a diffuse gas flow to protect weld pools from the atmosphere. Some welds require a trailing shield of inert gas or gas mixture to protect the weldment from contamination. The orifice of gas jet nozzles in welding operations is commonly 2.5 to 6.4 mm (0.100 to 0.250 in.) in diameter. Nozzle standoff distances are usually from 3.2 to 9.5 mm (0.125 to 0.375 in.). Cross-jets are used for lens protection when a coaxial gas jet causes interference with the reflective focusing optics or the part. Coaxial gas jet pressure is low (35 to 100 kPa, or 5 to 15 psi) to prevent weld pool disruption. Backside shielding techniques are similar to those employed in arc welding.

## Surface Treatment

Surface treatment is the modification of metals to produce structures suitable for the intended component function. The categories of surface treatment discussed are heat treating, cladding, surfacing, glazing, and marking.

Heat treating consists of two processes: transformation hardening and annealing. Transformation hardening involves the modification of the surface physical properties of metal by raising the metal temperature to a point below the liquidus, followed by rapid cooling. This produces a hard microstructure. Annealing involves the modification of the surface physical properties of metal by raising the metal temperature to a point below the liquidus, followed by controlled cooling. This produces a softer microstructure.

The other types of surface treatment are:

- *Cladding*: Adding material of a composition different from the base metal to improve the life of a component
- *Surfacing*: Adding metal of the same or similar composition to alter or restore the physical dimensions of a component
- *Glazing*: Altering a metal surface by raising the metal temperature to a point above the liquidus with conduction and radiation of laser energy, thereby removing surface defects and modifying the microstructure
- *Marking*: Altering a metal surface either by metal removal or superficial alteration to produce a legible, contrasting image

**Process Variables.** A wide range of process variables are used in surface modification. Many ancillary devices are needed to produce the desired surfaces, including special optics, powder, or wire feed mechanisms, and preheating and/or postheating elements. Some of the process variables are described below.

*Power* needed for various surface modifications ranges from below 20 W to well above 10 kW. Heat treating, surfacing, cladding, and glazing are accomplished with power levels of 500 W or greater. Transformation hardening or annealing with higher power levels does not increase treatment depth, but does increase the surface area that can be modified in a single pass. The marking of metals is usually accomplished with Nd:YAG lasers, with average power levels ranging from 20 to 75 W. Marking can be achieved in the CW or pulsed mode.

*Pulse parameters*, such as pulse length, frequency, and energy, vary greatly in surface modifications. Heat treating, surfacing, cladding, and glazing are predominately performed with CW $CO_2$ lasers. Metal marking pulse frequencies range from 500 to 10 000 Hz, while pulse lengths are approximately 50 ns in width.

*Focusing lenses and gas jets* also differ vastly, depending on the desired surface treatment. Beam profile intensity is often changed optically to yield an even power distribution when heat treating. A galvo-optic device is a common method of directing and focusing the laser beam for metal marking to produce the specified characters. Inert-gas chambers are used in many cladding and surfacing applications, while arc lamps are used to preheat crack-sensitive alloys to reduce the cooling rate.

Gas jets may or may not be used in surface treatment. Galvo-optic devices used in laser marking prohibit the use of any nozzle, although a side shield of oxygen can be used to enhance the contrast of the mark. Inert gas side shields prevent oxidation, thereby providing bright, shiny characters. Heat treating, cladding, and surfacing may use water-cooled nozzles for prolonged life, especially in a preheating and/or post-heating environment. Nozzle design may incorporate a powder delivery mechanism in cladding and surfacing operations.

## Future Developments

Developments in laser technology today are being directed toward meeting requirements identified by users of earlier lasers. Beam quality (analogous to tool sharpness) has improved, particularly with higher power $CO_2$ and Nd:YAG lasers. Better beam quality at higher power is providing the ability to cut thick metals with finishes once thought not possible. Advanced design Nd:YAG lasers have cut high nickel alloys up to 50 mm (2.0 in.) thick at speeds greater than electrical discharge machining. Better beam quality at higher power produces deep penetration welds with very low aspect ratios.

Laser machining systems are also changing. Fiber optic cables are being used with Nd:YAG lasers to simplify laser beam delivery. Robotic devices are being designed specifically for integration with lasers and will soon be operating in automated assembly lines. Vision systems and seam tracking devices are beginning to show promise as means to process parts with unpredictable geometry.

# Appendix: Laser-Enhanced Etching

Laser-enhanced etching is a photochemical process that generates micron-sized or submicron-sized patterns on the surfaces of metals, ceramics, and semiconductors. In this process, a workpiece is enveloped in either a gaseous or liquid etchant, and the surface of the workpiece is irradiated with a focused laser beam. The etching action occurs when the focused laser beam produces either photolysis or a thermal effect in the etchant near the surface of the workpiece. This photoetching process, although confined primarily to research investigations, represents a potential technique for generating patterns without contact masks or, in some cases, without external biasing of electrolyte solutions.

**Methods and Mechanisms.** The laser initiates or enhances the reaction between the workpiece and the etchant by photolysis, a thermal effect, or a combination of the two. In those cases in which gas is used as an etchant, photolysis is the mechanism of laser-enhanced etching. The laser radiation excites or dissociates the gas molecules into chemically active constituents, which then diffuse to and react with the surface of the workpiece.

Thermal effects are the predominant mechanism when the workpiece is the anode in an electrolytic cell. This etching process, which is similar to the additive process of laser-enhanced electroplating (Fig. 5), has been described as a thermal effect in which the enhancement is attributed to the temperature increases and thermal gradients produced by the absorption of laser light by the workpiece. These pyrolytic etching processes are inherently isotropic in action and are resolved over dimensions no smaller than a thermal diffusion length.

There are also methods of laser-enhanced etching that use liquid etchants but no external bias or electromotive driving sources (Ref 3-5). These methods involve photocatalyzed electrochemical dissolution of the workpiece, although some of the methods also induce melting (Ref 4, 5). Each of these methods utilizes different etchant solutions.

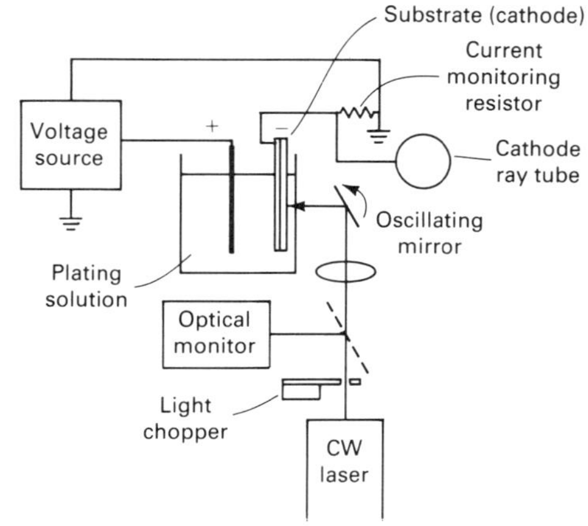


**Fig. 5** Schematic for laser plating enhancements in the region of optical absorption on the cathode. Source: Ref 2

**Bromine and iodine etchants** are used in one method of laser-enhanced etching (Ref 3). Photocatalyzed etching can occur without an external bias, and the process is expected to achieve submicron resolution.

**Potassium hydroxide etchants** have been used in the laser-enhanced etching of (111) silicon and an alumina-titanium carbide ceramic (Ref 4). This method differs from other etching schemes in that laser irradiation causes local melting and some vaporization while the workpiece is submerged in the liquid etchant.

**Neutral salt solutions** also have been considered an etchant in laser-enhanced etching (Ref 5). This technique demonstrates the ability to perform laser-enhanced etching without the use of corrosive acidic or alkaline solutions. Experiments with a sodium nitrate neutral salt solution indicate that melt or near melt temperatures are required to obtain etching.

### REFERENCES

1. D. Belforte, *Industrial Laser Handbook*
2. R.J. von Gutfeld *et al.*, *Appl. Phys. Lett.*, Vol 35, 1979, p 651
3. R.W. Haynes *et al.*, *Appl. Phys. Lett.*, Vol 37, 1980, p 344
4. R.J. von Gutfeld and R.T. Hodgson, *Appl. Phys. Lett.*, Vol 40, 1982, p 352
5. M. Datta *et al.*, *Appl. Phys. Lett.*, Vol 51, 1987, p 2040

# Thermal Energy Method

Thomas Fisher, Surftran, Division of Robert Bosch Corporation

THE THERMAL ENERGY METHOD (TEM) of deburring was introduced in 1970 as a method of deburring and deflashing parts by the application of intense heat. The thermal energy method can remove even the largest burrs in less than 20 ms, and it removes all burrs simultaneously. This method produces repeatable results and can be automated.

**Principles of Operation.** Parts to be processed are loaded in bulk or in fixtures (Fig. 1), and the basket or fixture containing the parts is indexed into position in line with the chamber (Fig. 2). The lower closure on which parts are located is lifted by a ram and toggle, pressing it with 2200 kN (250 tonf) of force to the upper chamber and sealing the chamber. A fuel mixture of natural gas and oxygen is then injected into the chamber under pressure. The initial fire or heat is supplied by an ignition system in the form of a spark from an ignitor. The spark causes the combustible gas mixture to ignite, and all the fuel gas is consumed in approximately 2 ms to form a 3300 °C (6000 °F) heat wave.

The heat wave hits everything within the chamber. It hits the water-cooled walls of the chamber, the parent metal of the part or parts, and the burrs within blind holes, external edges, and intersecting holes (small or large) that cannot be reached by hand. The main body of a part becomes warm (usually under 150 °C, or 300 °F), while the flash or burrs (having less mass per surface area) heat up instantly and burst into flames. Heat created by the burning of fuel gas thus triggers the start of a second fire of the burr material itself. Because there is an abundance of oxygen in the initial fuel mixture, burrs will continue to burn until the heat is dispersed throughout the main body of the part.

In each case, the metal combines with oxygen to form the oxide of the metal that is present. This metal oxide settles on the chamber walls, the fixturing or baskets, and the surface of the part. The oxides can be removed from the parts with an aqueous solution. In the case of steels that will later be heat treated, cleaning can be omitted. For zinc (which is dichromated), aluminum (which is anodized), or brass (which is brite dipped), additional cleaning is not neces-

(a) (b)

**Fig. 1** Parts loaded in baskets or fixtures for TEM. (a) A typical bulk run. (b) A nest-type fixture for nick-free parts or a reduction in distortion

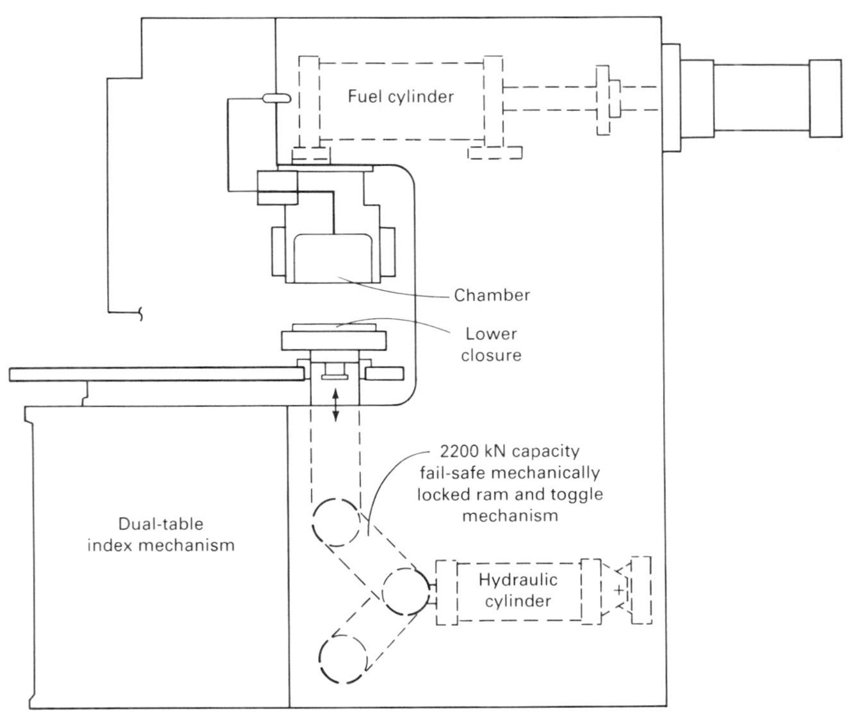


**Fig. 2** Schematic of a thermal energy method system

**Fig. 3** Typical bearing race before TEM (left) and after TEM (right)

**Fig. 4** Cast iron housing before TEM (left) and after TEM (right)

sary. In the case of inexpensive components, the oxides are sometimes simply left on the parts.

**Capabilities.** The thermal energy method is effective on many noncombustible materials, such as ferrous and nonferrous metals, plastics, and rubber. The process requires more energy to deburr corrosion-resistant metals than unalloyed metals. The second fire extinguishes because of heat flowing into the component; therefore, materials that are better heat conductors cause the fire to extinguish sooner than poor conductors. As a result, it is easier to obtain an edge break or a radius on a poor conductor, such as steel or cast iron, than on a relatively good heat conductor, such as brass.

The thermal energy method can instantly remove fingernail-size burrs as thick as 0.5 mm (0.020 in.) under ideal conditions. Ideal means that the component has no sections thinner than 15 times the thickness of the largest burr and that it is capable of sustaining the detonation without damage.

Figure 3 illustrates the capabilities of TEM on a bearing race. This method is equally effective on large and small parts. Figure 4 shows a cast iron housing before and after TEM.

**Limitations.** Many deburring processes remove the flash formed in a casting or die casting process. The thermal energy method is also used to remove flash, although it will not remove parting line flash entirely because the second fire extinguishes as the flash widens at its root. Unlike a burr, flash is wider at its root than at its tip. The tip of the flash ignites in TEM, but the heat flow into the body of the part is different in flash from that in burrs. Therefore, TEM might leave a rounded lump where the root of the flash was located. Also, the thermal energy method does not buff or polish surfaces.

Another limitation of TEM is the shock effect of the heat waves. The heat wave can cause distortion in thin or fragile parts, but in some cases the problem can be solved with fixturing. Figure 1(b) shows a typical nest-type fixture that can be used to solve a distortion problem or to keep the parts from bouncing and hitting one another.

# Chemical Milling

Eugene M. Langworthy, Aerochem, Inc.

CHEMICAL MILLING (CM), also termed chem milling, chemical machining, chemical contouring, and chemietching, is a method of processing structural metal parts by controlled chemical etching. The process either thins the parts in specific areas or removes metal from all surfaces. Areas from which metal is not to be removed are protected by special coatings termed maskants or masks, and parts are etched by immersion in a tank of etchant. The process is rarely electrically assisted. Most of the process applications are for producing parts with high strength-to-weight ratios.

Chemical milling differs from chemical blanking, which involves the etching of very thin metal parts in spray chambers. Chemical blanking etchants are mild compared to those used most often in the chemical milling process. The photosensitive resists most frequently used for masking in chemical blanking facilitate the etching of very intricate and close-tolerance patterns. Parts are normally completely perforated in etched areas, similar to mechanically blanked parts, rather than etched to a given final thickness, as in chemical milling. The article "Photochemical Machining" in this Volume gives a more detailed description of chemical blanking.

Chemical milling services are offered on a subcontract basis by a small group of specialized companies. Most of the large aerospace manufacturers also have CM facilities, although many of them perform chemical milling of aluminum only. The most complex etching processes and the most sophisticated scribing technology are not widely available. The birth of this technology as a machining process is credited to Manuel C. Sanz, who, as an engineer with North American Aviation in 1953, resolved an urgent weight problem on a missile casing by chemical etching. A more detailed account of the history and technology of the process is given in Ref 1. Today chemical milling is an accepted and widely applied process within the aerospace industry, with major applications in military and commercial airframes, launch vehicles, turbine and rocket engines, helicopters, and missiles. There are very few nonaerospace, commercial applications.

## Equipment and Process Characteristics

The two key materials used in the chemical milling process are the etchant and the maskant. Etchants are acid or alkaline solutions maintained within controlled ranges of chemical composition and temperature. Maskants are specially designed elastomeric products that are hand strippable and chemically resistant to the harsh etchants.

Virtually all metals used in the aerospace and commercial metalworking industries are chemically milled. Aluminum alloys are the simplest to mill chemically, in terms of both facility requirements and process controls. Magnesium and titanium alloys are next hardest, followed by the low-alloy and stainless steels. The most difficult to process are the superalloys.

The steps in the chemical milling process for a detailed part (one requiring masking and scribing, for example) are:

- A metal workpiece is cleaned and coated with a liquid maskant
- The maskant is dried and cured
- The pattern outline is scribed through the mask
- The mask is peeled away in the areas to be etched
- The desired reduction in thickness is made by immersion in the etchant for a controlled period
- The preceding two steps are repeated if more than one removal step is required
- The remainder of the mask is removed
- The part is mechanically benched as required

A simplified flow chart of the principal steps is also shown in Fig. 1. Required rinsing, drying, and certain other steps specific to given metals are not shown in this chart.

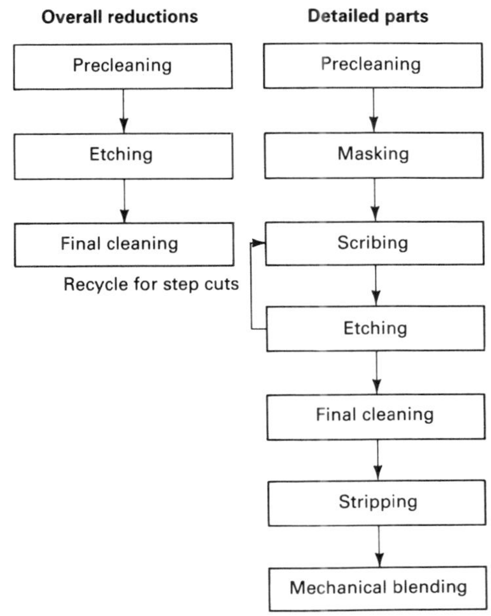


**Fig. 1** Flow chart of principal process steps for chemical milling

**Precleaning.** If the part surface is not clean prior to maskant application, the adhesion properties of the maskant may be seriously impaired. The most likely problems are extremely high adhesion, making the maskant very difficult to strip, or very low adhesion, resulting in poor line definition between etched and nonetched areas. If oily contaminants on the part surface are the only concern, vapor degreasing or solvent wiping afford sufficient cleaning. However, if surface scale and chemical contamination are present from previous handling and thermal treatments, they must be removed by more extensive chemical treatment. Many standard and proprietary cleaners are available for the various metals.

Grit blasting is frequently used for cleaning titanium. If coarse grit must be used to remove heavy scale, the resulting rough surface finish may require special low-adhesion maskants that are available. Grit blasting imparts surface tensions (up to 0.04 mm, or 0.0015 in. deep) that are relieved in chemically milled areas. This can produce warping in light-gage parts.

**Masking.** Maskant is most commonly applied by immersing the parts into dip tanks or covering the parts with hot airless spray equipment. Conventional air spraying, flow coating, brushing, and roller coating are less frequently used. Maskants must provide impermeable, chemically resistant coatings that can be easily hand stripped before and after etching, yet that strongly adhere at cut edges during etching. Special so-called top coats are often sprayed (or dipped) on top of the normal maskant to give it improved chemical and abrasion resistance.

Cure of the maskant varies, depending on the product and its application. Solvent evaporation normally takes from 4 to 12 h at ambient temperatures above 20 °C (70 °F). Oven curing above 50 °C (120 °F) can be used to hasten solvent removal. Curing maskants above 90 °C (200 °F) usually increases their adhesion and chemical resistance.

The thickness of maskant coatings usually varies from 0.2 to 0.4 mm (0.008 to 0.015 in.). Top coat applications, when required, range from 0.025 to 0.075 mm (0.001 to 0.003 in.). Applied to properly cleaned metal surfaces, the cured maskant film yields peel adhesion values ranging from 7 to 36 g/mm of width (6 to 32 oz./in. of width), and a tensile strength of at least 4000 kPa (600 psi). Higher maskant adhesions require increased coating thickness and make manual stripping more difficult; lower adhesion values increase the likelihood of pattern edge failure either from handling or during the etching process.

Carbon adsorption equipment for solvent recapture is often installed to meet air quality requirements. The solvent is frequently returned to the maskant supplier for reuse, resulting in a significant cost savings.

New chemical milling maskants are continually being developed, adding to the excellent products now available. Product bulletins issued by the manufacturers detail the properties, application, and function of the maskants. Manufacturers also provide field support to customers, as well as state-of-the-art information about the chemical milling industry.

**Scribing.** The pattern is normally scribed through the mask by guiding a scribe knife along the edges of a template located precisely on the part. The mask is then peeled off the areas to be etched. When etching to more than one thickness is required (step cuts), the mask is initially peeled from only the first cut, that is, the thinnest final area. All lines are normally scribed at this time, but a line-sealer is brushed over the scribe lines that will be peeled with later cuts. Although this special coating keeps the etchant from penetrating the scribe cuts, it still allows the maskant to be easily peeled. Potential flaws in the coating are detected visually and, through the use of special electrical flaw detectors, repaired with patching compounds or tapes.

A heated knife is sometimes used for scribing soft metals such as alclad, because the use of a normal scribe knife could result in deep gouging of the metal. The heated knife blade melts the maskant, greatly reducing the amount of hand pressure needed to cut through it, and minimizing the danger of making cuts too deep.

Computer-guided lasers are also used for maskant cutting. This scribing method is justified with relatively simple-shape parts if the part quantity is great, or if the scribing pattern is complex. Laser scribing is often used when producing the novel I-beam land cross section, as described in the following example.

**Fig. 2** Integrally stiffened engine fan and exhaust ducts

**Example 1: Integrally Stiffened Titanium Engine Ducts.** The turbine engine fan ducts shown in Fig. 2 were chemically milled from 0.250 gage 6Al-4V titanium alloy to produce integral stiffening lands and attachment bosses. This design used the concept of an I-beam land cross section (Ref 2). Previous designs for parts of this type were either a skin stringer or a honeycomb sandwich structure.

The structurally efficient I-beam cross section is produced by the two-step chemical milling procedure depicted in Fig. 3. The first milling step, of about one-half the total depth, produces an initial radius (Fig. 3b); the second step undercuts the first to create the I-beam section (Fig. 3d). Computer-directed lasers are often used to locate the scribe cuts precisely, because exceptional accuracy is required to maintain the finished dimensions. Of equal importance is the control of the cleaning, masking, and etchant variables that influence undercut.

**Etching.** The etching procedure typically consists of immersing stationary parts in a tank of etchant. Parts are held by clamps, springs, or wires, in a basket or other handling fixture that is transported in and out of the tank. Normally the work is exposed to the etchant for only a portion of the estimated removal time, remeasured to calculate an actual rate, and then returned for a recalculated final period of etching. One or more extra trips in and out of the tank for remeasuring are frequently employed to control the removal step more closely.

Parts with different chemically milled thicknesses are easily produced by etching in a series of step cuts, requiring that a new area of the part be exposed before each of a series of etching steps. The thickest areas

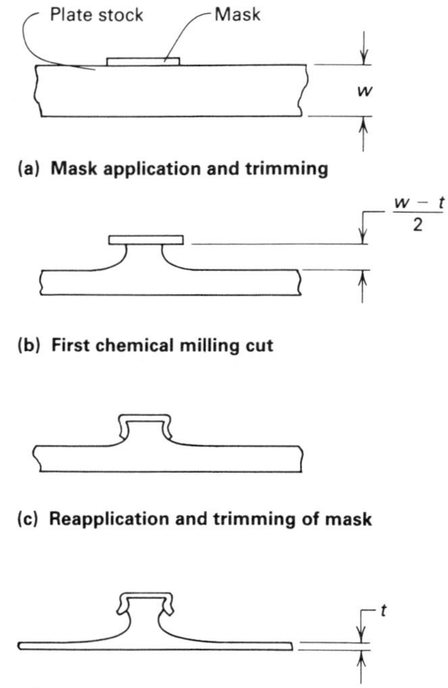


**Fig. 3** Steps in two-stage chemical milling of I-beam ribs

are usually adjacent to a structure that is attached by riveting or welding. The thicker areas are often termed lands; they border the thinner areas, or pockets. In some cases, the thickest areas are relatively narrow ribs, which impart integral stiffness to an otherwise thin skin. It is often a simple matter to incorporate tapered areas within the part to meet special design requirements. Tapering is done by either immersing or withdrawing the workpiece from the etchant at a controlled, slow rate.

The etching reaction creates a turbulent evolution of gases from part surfaces. In conventional immersion etching, the parts must be positioned to prevent entrapment of these gases in areas to be etched. Rapid liquid flow caused by the rising gases can produce uneven etching patterns at the pocket edges. These process characteristics can be minimized by means of careful etchant control and by changing the position of the part intermittently during the etching process. Better control can be obtained by continuously changing the part position while it is being etched. Part rotation is a common practice for the chemical milling of cylindrical and conical parts made from metals other than aluminum.

Other unconventional methods of etching have been employed for special applications. Examples are:

- Spray etching the inside of a rotating tube
- Spraying or flooding the outside surface of a rotating tube (Ref 1)
- Flowing etchant through tubes
- Spinning floating tubes or spheres on the surface of the etchant

The following example illustrates an application of spray etching.

**Example 2: Spray Etching a Rotating Tube.** Cruise missile launch tubes 530 mm (21 in.) in diameter, and 6 m (20 ft) long have been chemically milled. In this application, the removal of material from the inner surface is needed for functional purposes rather than weight removal, although some 111 kg (245 lb) of material were removed from each tube. The initial 5.5 mm (0.215 in.) wall thickness was reduced to 4.8 mm (0.190 in.) over an 810 mm (32 in.) length and to 4.1 mm (0.160 in.) over a 4200 mm (165 in.) length. The work-hardenable 304 alloy and the relatively inaccessible inside removal made mechanical milling a more costly approach.

The size and shape of the part made normal immersion etching impractical. The enclosed part configuration would have resulted in relatively poor uniformity of metal removal and created sizing requirements that would be extremely difficult to accomplish. The normal mixed-acid immersion etchant generates oxides of nitrogen that in this case would have required a special fume scrubber. Thus, a spray etching approach was used. The closed-off tube itself formed the spray chamber, an oscillating spray tube was placed at the tube axis, and the part was rotated during the spray cycle. The etchant used was ferric chloride, commonly used in the chemical blanking process. The advantages of this approach were improved uniformity of removal, elimination of oxides of nitrogen, and the ability to regenerate or sell the spent etchant.

**Etchants.** The etchants used for chemical milling are usually mixtures of standard industrial chemicals and special additives, such as wetting agents. Although the etchant composition for each alloy and heat treat condition may vary, the principles of selection and control are the same for each. The main technical goals in etchant development, followed by practical considerations, are:

- Good surface finish
- Good uniformity of metal removal
- Control of selective and intergranular attack
- For titanium alloys, control of hydrogen absorption
- Personnel safety
- Cost and reliability of materials of construction required for process tanks
- Air quality issues
- Low chemical cost per unit weight dissolved
- The ability to regenerate the etchant solution and/or readily neutralize and dispose of its waste products

**Final Cleaning and Stripping.** After each etching step, the part is thoroughly rinsed. In the case of aluminum, the surface is brightened by immersion in a nitric acid or deoxidizer solution. Stainless steels are normally passivated. Other treatments are performed, depending on the specific metal and the customer's requirements. After etching, the remaining maskant is usually peeled off by hand, but occasionally is removed by vapor degreaser or other solvent treatment.

**Mechanical Finishing.** The final manufacturing step is called a hand benching operation. Here step mismatches and sharp edges are removed by the use of sanding and deburring tools. In addition, dimensional requirements beyond the normal process capabilities, such as special fillet radii and extremely close-tolerance step locations, are accomplished in this operation.

## Specifications and Controls

Important characteristics of a chemically milled part, such as its surface texture and the amount of hydrogen it absorbs in the process, depend on the properties of both the metal and the etching solution. The optimum etchant for different parts made from the same alloy can therefore vary substantially because of factors that influence grain structure: heat treat condition; degree of strain hardening; and the original form of the alloy, that is, sheet, plate, forging, extrusion, or casting. The exact etchant control ranges for the more difficult applications are often considered trade secrets, as are certain of the processing techniques.

Most large aerospace manufacturers maintain chemical milling specifications to control in-house and supplier processing. A military specification, MIL-C-81769, "Chemical Milling of Metals," is available. It covers all types of metals to the specifications maintained by aerospace companies.

**Definitions of CM Specifications.** Some features of chemically milled parts are illustrated in Fig. 4. Etching proceeds at approximately the same rate under the mask as in all other directions. The result is a fillet radius approximately equal to the cut depth, $d$. The undercut, $u$, is also approximately equal to the cut depth. The etch factor is defined as $u/d$, the undercut divided by the depth of cut, the exact opposite of the definition used in connection with chemical blanking. Unlike conventional machining, the fillet is seldom an exact radius and should not be defined and measured as such. The terms radius as etched and modified radius are two blueprint notes frequently used.

Chemical milling specifications normally define another process characteristic, a transition zone that extends beyond the base of the fillet, illustrated in Fig. 4(c). The part thickness may exceed blueprint tolerance in this area. MIL-C-81769 allows the length of this zone to be four times the depth of cut for aluminum, and ten times the depth of cut for other metals. For applications in which a transition zone would cause interference for mating parts, it can often be controlled by special procedures.

The sharp corner edge usually produced at the top of the fillet radius is normally rounded or chamfered mechanically in a subsequent operation. In parts with step cuts, the etching action often rounds the edges of the earlier steps, as depicted in Fig. 4(c).

**Lateral Tolerance of Steps.** The location of the finished step depends on the accuracy of scribing and the control of the undercut. The initial scribe location depends on the tooling or layout accuracy (which is often influenced by variables in part contour) and careful scribing. The variables that affect undercut are:

- Control of the cleaning and masking steps
- The amount of dissolved metal in the etchant
- The etchant range
- The depth of cut
- The grain direction
- Positioning and movement of the part during the etching step

Normal blueprint tolerances for step locations range from ±0.4 to ±0.8 mm (±0.015 to ±0.030 in.) for steps under 3.2 mm (0.125 in.), to a range of ±0.8 to ±1.5 mm (±0.030 to ±0.060 in.) for cuts over 3.2 mm (0.125 in). Special methods can produce closer tolerances when required. Specifications normally allow waviness in the chemically milled line, as long as sharp notches are not present. The undercut allowances for tooling or layout purposes can normally be estimated with sufficient accuracy from data accumulated from previous tests or parts of similar nature. The undercut allowances for step cuts, shown as $u'$ in Fig. 4(b), are more difficult to predict and are often determined by processing test panels.

**Depth of Etch.** There are no basic process limitations on the etching depth. Any limits are based on cost trade-offs with alternative methods of manufacture and the value to the customer of the weight removed. Etchant costs can be very significant when deep cuts are made on large parts. The deeper the cut, the more variation in thickness produced in the etching step, and chemical sizing labor is often required to meet the final thickness tolerances.

**Thickness Tolerance.** In the case of chemically milled aluminum sheet metal parts, final blueprint tolerances should allow about ±0.05 to ±0.08 mm (±0.002 to ±0.003 in.) more than the thickness variation expected in the starting blank. For depths of cut over 5 mm (0.200 in.), an allowance for chemical milling variation should be made, the amount depending on the particular application. As a general rule, etching removal is much more uniform with aluminum than with the other metals. Al-

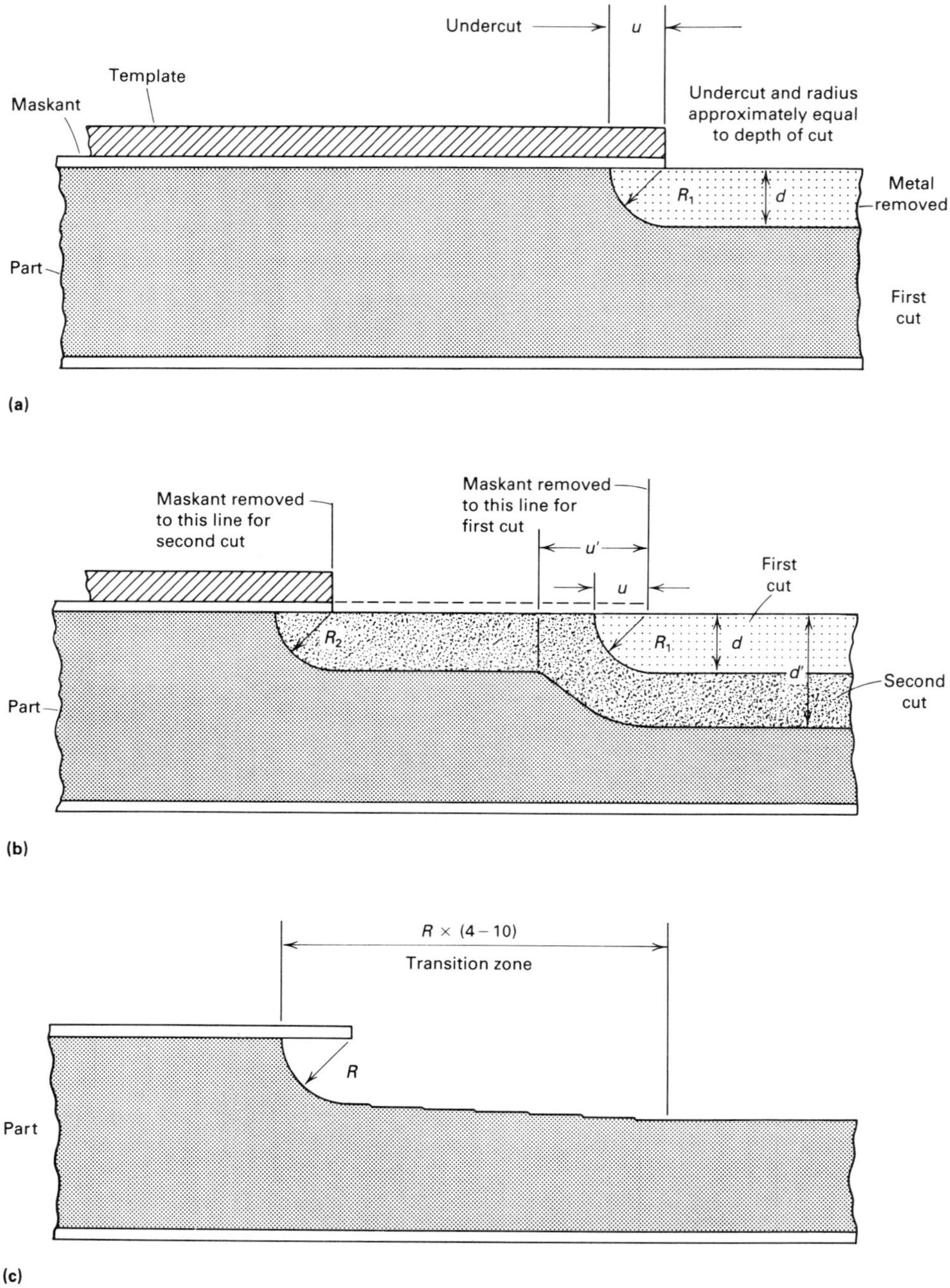


**Fig. 4** Design features for chemical milling. (a) Single chemical milling step. (b) Second chemical milling step. (c) Transition zone of greater thickness adjacent to step

though parts made from the denser alloys should allow wider finished tolerances from a producibility standpoint, weight considerations often dictate even closer thickness tolerances than for aluminum. Chemical sizing is thus a frequent requirement for steel and superalloy parts.

**Taper Cuts.** Continuous tapers as great as 0.060 mm/mm (0.060 in./in.) for aluminum and 0.010 mm/mm (0.010 in./in.) for steel alloys have been produced on a production basis. Step tapers, a series of adjacent steps of diminishing thickness, are often used as a substitute for continuous tapers. This approach can result in a weight penalty, but it also has more design flexibility and is often simpler to produce.

**Sheet and Plate Specifications.** Special attention should be given to the specifications for sheet or plate stock to be used for chemical milling applications. Unlike machining, in which completely new surfaces are created, the thickness variation and surface quality of the incoming material strongly contribute to the thickness variation and surface quality of the chemically milled part. Another extremely important variable to control is residual stress in the procured stock. Minimum residual stress (MRS) and chem-mill-quality are terms used by the aluminum industry and are frequently specified for heat-treated stock intended to be used for chemically milled parts. If the material is to be stretch formed before chemical milling, this condition is not required.

**Process Defects.** The following list of process defects (illustrated in Fig. 5) are essentially as presented in MIL-C-81769:

- *Channeling*: The formation of a channel or a groove at the base of the fillet radius. Acceptable if corrected mechanically with all dimensions within blueprint
- *Gas channeling*: Vertical grooves or channels in the etched surface, resulting from gas bubbles rising to the surface. Acceptable if corrected mechanically, maintaining all dimensions within blueprint
- *Dishing*: Gradual pattern of thinning from all edges of an etched pocket to the center area. Acceptable if all dimensions are within blueprint (the allowance of a transition zone adjacent to the fillet area often brings this condition within blueprint)
- *Pitting*: Localized depressions in the etched surface is a condition that is not acceptable unless pits are mechanically blended, with all thicknesses remaining within blueprint tolerances
- *Overhanging*: The presence of overhanging metal at the top edge of the fillet radius. Generally, a 0.05 mm (0.002 in.) radius on the top edge is required for fighter aircraft; larger radii are required for commercial aircraft
- *Islands*: Local raised areas that result from maskant or other material left on the surface, preventing etching from taking place. Must be mechanically blended flush with surrounding area, holding blueprint tolerances
- *Ridging*: The formation of a ridge at the base of the fillet, which must be removed mechanically, and all blueprint dimensions maintained
- *Fillet notching*: Notches at the base of the fillet radius, a result of scribing too deeply into the metal. Acceptable if the etching has made the depression smooth, with width greater than twice the depth, and if thickness is within blueprint tolerances

**Surface Finish.** In chemical milling, the exposed metal is dissolved by an alkaline or acid etchant. At a microscopic level, the chemical attack takes place both at the individual grain surfaces and at the grain boundaries, usually with different reaction rates. The etching mechanism is discussed in Ref 1 (see Fig. 6).

Fine grain size and homogeneous structure are necessary if fine surface finishes and uniform appearance are to be attained on chemically milled parts. Titanium and magnesium, chemically milled in the standard etchants that produce good uniformity

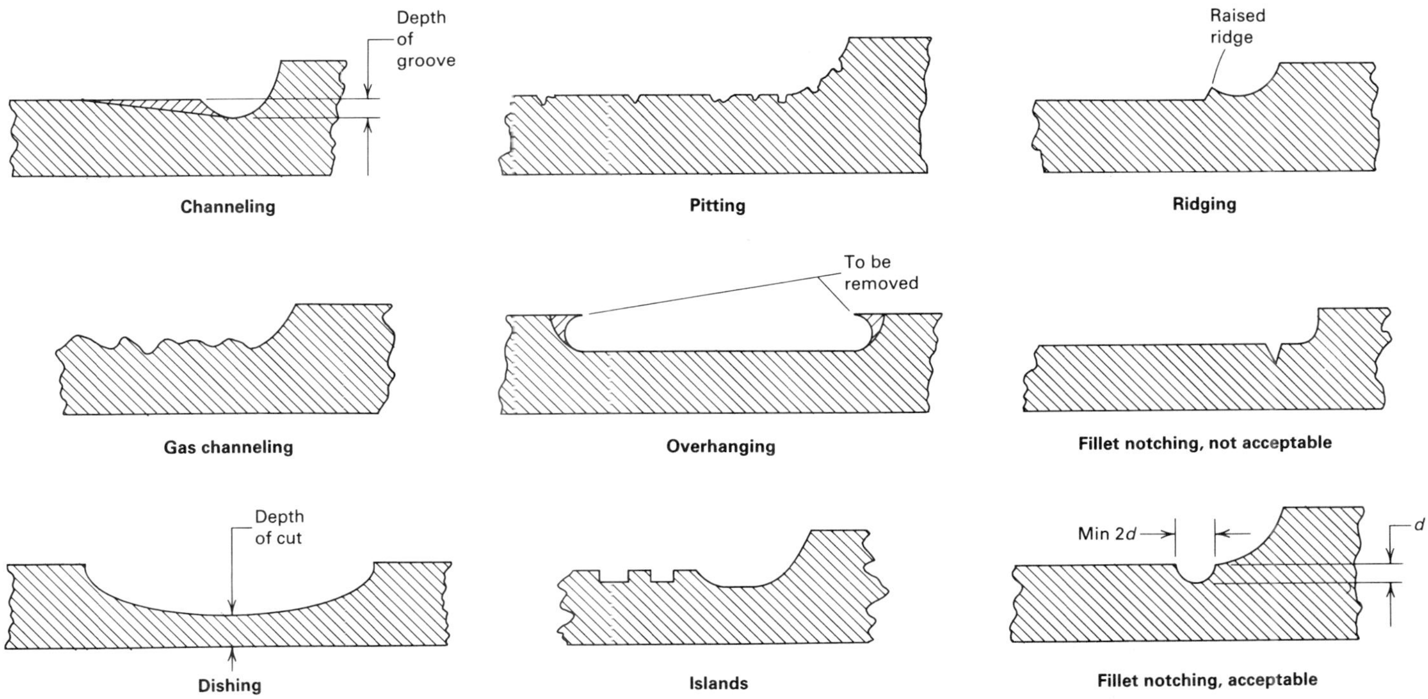


**Fig. 5** Typical process defects in chemically milled parts

of removal, produce exceptionally smooth surfaces. Most castings and 2000-series aluminum extrusions are examples of metals with abnormally large and uneven grain structure, and good surface finishes are very difficult to attain with these materials. Typical values for surface finishes produced on sheet, forgings, or castings of various metals are given in Table 1.

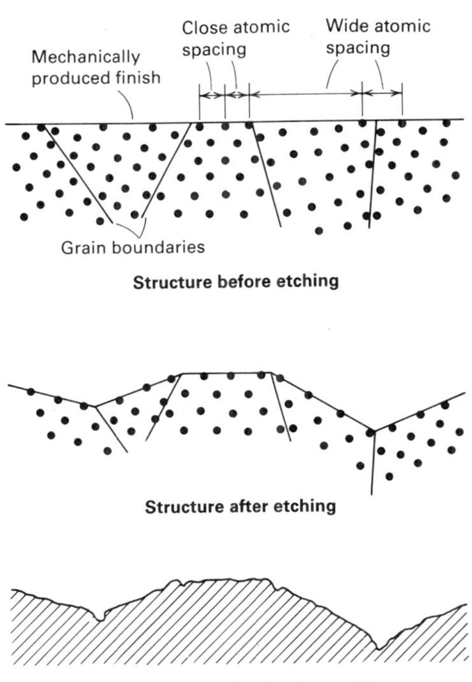


**Fig. 6** The microeffect of etching action at the surface of a material

Special etchants can often be developed to produce extremely smooth chemically milled surfaces for special applications. However, this smoothing characteristic is often gained at the expense of the uniformity of metal removal.

The addition of triethanolamine (TEA) to the normal caustic aluminum etchant produces a marked improvement in surface finish for most aluminum alloys. This discovery was disclosed in a patent in 1959 (Ref 3). Graphs exhibiting the dramatic effect of the TEA component on aluminum surface finish are presented in Fig. 7 and 8. Figure 7 shows the improvement in surface finish of 2024-T3 and 7075-T6 samples that had been previously chemically milled in a standard etchant. Roughness of 3.0 μm (120 μin.) $R_a$ of the 2024-T3 specimen was reduced to about 1.25 μm (50 μin.) $R_a$ with the removal of 0.5 mm (0.020 in.). The initial 3.3 μm (130 μin.) $R_a$ of the 7075-T6 specimen was reduced to about 1.25 μm (50 μin.) $R_a$ with the removal of 0.75 mm (0.030 in.). Figure 8 shows the results of chemically milling 1.0 mm (0.040 in.) from as-rolled stock samples of 7075-T6 in the standard and the TEA etchants. Micrographs of the fillet area of the same 7075-T6 material, chemically milled in the two etchants, are shown in Fig. 9. The TEA etchant not only improves surface finish, but also significantly reduces the extent of grain boundary attack.

TEA-base etchants have been commercially available since the 1970s. A sudden increase in the use of TEA-base etchants occurred about 1975, when a group of 747 cargo area skins that had previously been metal bonded were redesigned for chemical milling. Since that time, there has been a tremendous growth in the number of commercial airframe applications that specify the use of the TEA etchant.

**Intergranular and selective attack** are inherent problems in the etching of metals, but are minimized through proper etchant control. Most chemical milling specifications allow minor conditions to exist. Mil-C-81769, for example, allows 0.008 mm (0.0003 in.) at grain boundaries for titanium and ferrous alloys, and 0.013 mm (0.0005 in.) at the grain ends (fillet areas). On nickel and cobalt alloys, the acceptable limit is 0.013 mm (0.0005 in.) at the grain boundaries and 0.025 mm (0.001 in.) at the grain ends. The acceptable limits for aluminum are significantly higher. Monitoring is accomplished by microscopic examination of test coupons attached to or etched concurrently with the parts.

## Mechanical Properties of Chemically Milled Parts

With the exception of possible changes in fatigue strength due to changes in surface texture, and the effect of hydrogen absorption in certain alloys, the mechanical properties of metals are not affected by chemical milling. Stresses are not induced by the process, as neither mechanical deformation nor exposure to high temperatures is involved.

When surface layers have different mechanical properties from those of the base metal, their removal results in a change in the average mechanical properties of the finished part. Chemical milling is widely used in such instances and frequently

**Table 1 Surface finishes normally obtained by chemical milling**

| Alloy | Form | Etchant type | Surface finish, $R_a$, μm | Surface finish, $R_a$, μin. |
|---|---|---|---|---|
| **Aluminum** | | | | |
| 2024-T3 | Sheet, plate, forging | $NaOH + Na_2S$ | 2.5–4.0 | 100–160 |
| 2024-T3 | Sheet, plate, forging | NaOH + TEA(a) | 1.0–1.5 | 40–60 |
| 2024-T3511 | Extrusion | NaOH + TEA(a) | 2.0–4.0 | 80–160 |
| 7075-T6 | Sheet, plate, forging | $NaOH + Na_2S$ | 2.5–4.0 | 100–160 |
| 7075-T6 | Sheet, plate, forging | NaOH + TEA(a) | 1.0–1.5 | 40–60 |
| 2219-T37 | Sheet, plate | HCl | 1.5–2.5 | 60–100 |
| 2219-T87 | Sheet, plate | $NaOH + Na_2S$ | 3.2–5.0 | 125–200 |
| 2219-T87 | Sheet, plate | NaOH + TEA(a) | 3.2–5.0 | 125–200 |
| A-356 | Castings | NaOH + additives | 6.3–12.5 | 250–500 |
| **Mild steels** | | | | |
| 1010, 1020 | Sheet, plate | $HNO_3$ | 3.2–5.0 | 125–200 |
| **Low-alloy steels** | | | | |
| 4130, 4340 | Sheet | $HNO_3$ | 1.0–2.0 | 40–80 |
| 4130, 4340 | Forgings | $HNO_3$ | 1.0–2.0 | 40–80 |
| **High-alloy steels** | | | | |
| H-11 | Sheet | $HNO_3$, HCl, HF | 1.5–3.2 | 60–125 |
| **Stainless steels** | | | | |
| 301, 304, 316, 321 | Sheet | $HNO_3$ | 0.75–1.5 | 30–60 |
| 347 | Sheet | $HNO_3$ | 0.75–1.5 | 30–60 |
| 410 | Sheet | $HNO_3$ | 1.0–1.5 | 40–60 |
| **PH steels** | | | | |
| 13-8, 15-7, 17-7 | Sheet | $HNO_3$, HCl, HF | 0.75–1.5 | 30–60 |
| AM 350, AM 355 | Sheet | $HNO_3$, HCl, HF | 0.75–1.5 | 30–60 |
| **Superalloys** | | | | |
| N-155 | Sheet | $HNO_3$, HCl, HF | 2.5–3.6 | 100–140 |
| L-605 | Sheet | $HNO_3$, HCl, HF | 2.5–3.6 | 100–140 |
| H-188 | Sheet | $HNO_3$, HCl, HF | 1.5–2.3 | 60–90 |
| Hastelloy B | Sheet | $HNO_3$, HCl, HF | 1.5–2.3 | 60–90 |
| Hastelloy C | Sheet | $HNO_3$, HCl, HF | 1.5–2.3 | 60–90 |
| Hastelloy X | Sheet | $HNO_3$, HCl, HF | 1.5–2.3 | 60–90 |
| Inconel 600 | Sheet | $HNO_3$, HCl, HF | 1.5–2.3 | 60–90 |
| Inconel 625 | Sheet | $HNO_3$, HCl, HF | 1.5–2.3 | 60–90 |
| Incoloy 903 | Sheet | $HNO_3$, HCl, HF | 1.5–2.3 | 60–90 |
| Inconel 718 | Sheet | $HNO_3$, HCl, HF | 1.0–2.0 | 40–80 |
| René 41 | Sheet | $HNO_3$, HCl, HF | 1.0–2.0 | 40–80 |
| Waspaloy | Sheet | $HNO_3$, HCl, HF | 1.0–2.0 | 40–80 |
| **Titanium alloys** | Sheet, plate, forging | HF, $HNO_3$ | 0.25–1.0 | 10–40 |
| | Casting | HF, $HNO_3$ | 0.75–1.5 | 30–60 |
| **Magnesium alloys** | Sheet | $H_2SO_4$ | 1.0–1.5 | 40–60 |
| **Niobium alloys** | Sheet | $HNO_3$, HCl, HF | 1.0–1.5 | 40–60 |

(a) The NaOH + TEA etchants normally contain $Na_2S$ to precipitate out Zn and Cu components in the alloys. If not precipitated out, these metals will plate to the surface being chemically milled, thereby increasing surface roughness.

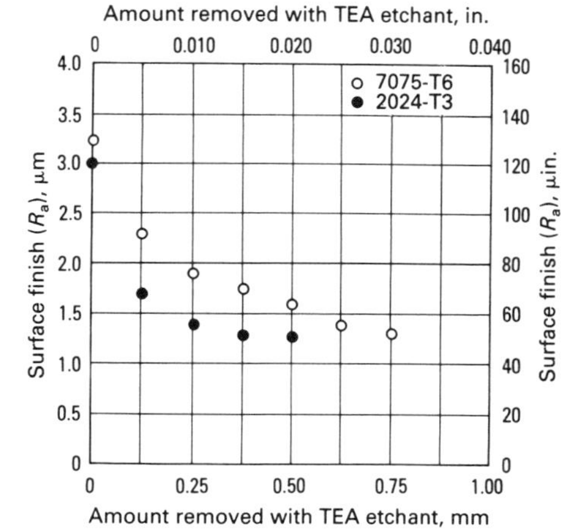


**Fig. 7** Surface finish improvement produced by TEA etchant. The initial surface was produced with a standard etchant.

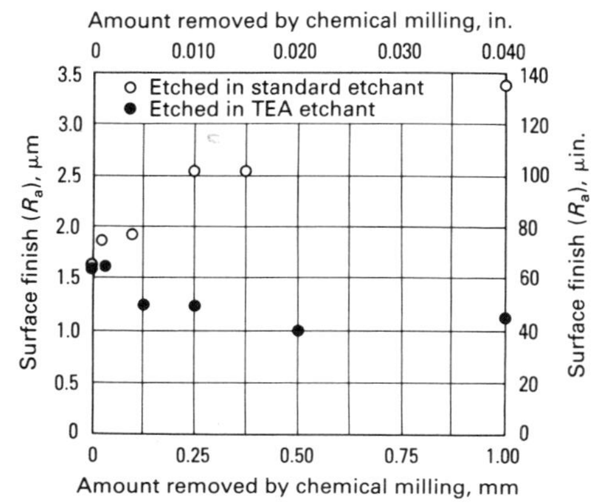


**Fig. 8** Comparison of surface finishes produced by standard aluminum etchant and by TEA etchant. Material: 7075-T6 as received from mill

changes the mechanical properties of a base material by removing material at the surface. Detrimental surface conditions such as alpha case, decarburized layer, and recast structure are easily removed by chemical milling, resulting in an improvement in properties of the finished part.

**Fatigue Properties.** Extensive fatigue studies of TEA-etched aluminum were conducted by the Boeing Company about 1975, but their data have not been released. Previous studies of chemically milled aluminum, generally indicating some loss of fatigue properties (Ref 1), were based on non-TEA etchants.

Impaired fatigue strength can be improved by imparting a uniform compressive stress to the chemically milled surface by peening, wet blasting, or other suitable techniques. Additionally, adverse surface conditions are sometimes removed by a final mechanical polishing operation.

**Hydrogen Absorption.** Hydrofluoric acid (HF) is the active chemical used to dissolve titanium in chemical milling etchants and pickling solutions, and this reaction produces hydrogen at the metal surface. Straight HF is often used as an etchant for alpha alloys and some forms of alpha-beta alloys. However, because hydrogen tends to be absorbed into the metal structure if beta phase is present, nitric acid is usually added to the HF to counter this tendency. Figure 10 demonstrates this effect of nitric acid for a pickling solution (Ref 1). As the amount of beta phase present in the titanium increases, control is maintained by increasing the concentration of nitric acid in the etchant. For the all-beta alloys and alloys that are nearly all-beta, the special MCAIR process is available. This process was developed, patented, and is licensed by McDonnell Douglas Corporation.

Identical parts chemically milled from titanium of the same heat lot, gage, and thermal processing are generally found to absorb the same amount of hydrogen. The rolling mill variables and the thermal cycles used in hot forming, superplastic forming, and beta stabilizing can affect beta phase, and thus can greatly affect hydrogen absorption. Etchant temperature and etch rate also affect the rate at which hydrogen is absorbed.

Chemical milling process specifications normally specify either a maximum allowable final hydrogen content, or a maximum allowed hydrogen pickup. The latter control requires a before-and-after measurement and assumes that the initial and the final hydrogen contents are evenly dispersed thoughout the sample thickness. However, the absorption mechanism is considered to be a surface effect with a depth of 0.25 to 0.50 mm (0.010 to 0.020 in.). As a result, etchant control can be much more critical for thin parts, especially for thin parts that must be etched from both sides.

## Applications

Chemical milling is used extensively to etch preformed aerospace parts to obtain maximum strength-to-weight ratios. Integrally stiffened aluminum wing and fuselage sections are chemically milled to produce an optimum cross section and minimum skin thickness. Spars, stringers, floor beams, and frames are frequent applications as well. Chemical milling also can be readily applied to most superalloys, and thus many applications can be found in the sizing and thinning of sheet metal components for the hot section of the jet engine. Spherical or

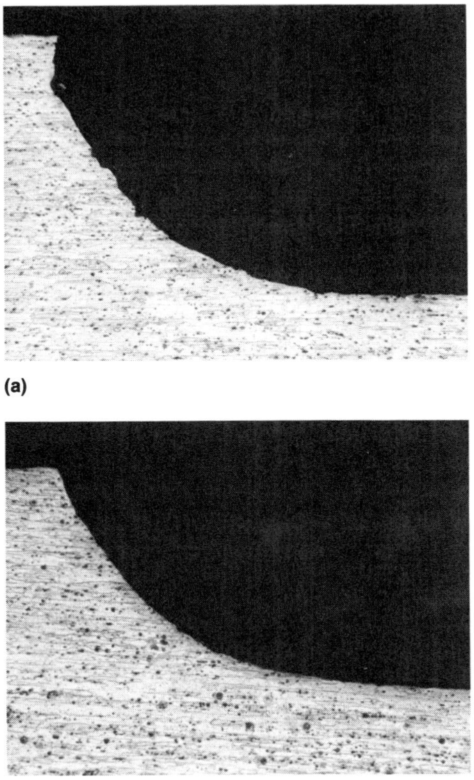

(a)

(b)

**Fig. 9** Fillet area of 0.5 mm (0.020 in.) step 7075-T6 sheet chemically milled in (a) standard etchant and (b) TEA etchant. 55×

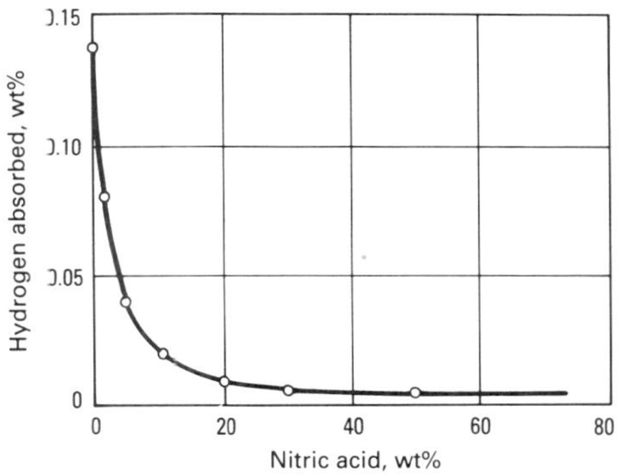


**Fig. 10** Hydrogen absorption of Ti-8Mn as a function of the nitric acid content of the pickling solution. The pickling solution contained 2% hydrofluoric acid and was held at 70 °C (158 °F). An etch of 0.05 mm (0.002 in.) was made on 0.8 mm (0.030 in.) sheet.

**Fig. 11** Delta booster tank bulkhead

**Fig. 12** Engine cowl inlet duct skins

ellipsoidal bulk heads for large launch fuel tanks, as described in the following example, are also chemically milled from stretch-formed segments or from complete domes. Initial part thickness of 13 mm (0.500 in.) and final thicknesses of 0.25 to 0.50 mm (0.010 to 0.020 in.) in chemically milled areas are not uncommon. Removal rates per surface vary from 0.005 to 0.02 mm/min (0.0002 to 0.0008 in./min) for steel alloys to as high as 0.08 mm/min (0.003 in./min) for aluminum alloys.

**Example 3: Thinning and Sizing of a Delta Booster Tank Bulkhead.** The 2.4 m (8 ft) diam dome shown in Fig. 11 is an end closure for the McDonnell Douglas Delta booster fuel tank. The part has been formed in one piece from 0.250-gage 2014 aluminum alloy, aged to the T6 condition prior to chemical milling. The radial lands are 7.6 mm (0.300 in.) wide and remain at the 6.3 mm (0.250 in.) stock thickness to help stiffen the structure. Other areas remain at 6.3 mm (0.250 in.) to facilitate welding when the connecting fuel lines or the cylindrical tank section are welded to the dome.

The thinnest sections of the bulkhead are 1.9 mm (0.075 in.), with a transition area 2.8 mm (0.110 in.) thick adjacent to both circumferential and the apex stock areas. The wall thickness tolerance is ±0.25 mm (±0.010 in.), and the land width tolerance is ±2.5 mm (±0.10 in.). The tolerances are usually generous.

**Gage Reduction or Chemical Sizing.** The term chemical sizing has been given to various techniques for etching raw stock or unfinished parts to a given thickness range, with or without the use of selective masking. One of the simplest applications is the etching of aluminum sheet stock into a thinner gage by overall reduction.

A much more complex procedure is often applied to detailed parts to remove thickness variation either present in the furnished material or produced during the process. Parts are "mapped" with contour lines of constant thickness, for example, in 0.05 mm (0.002 in.) intervals. The part is remasked and then etched in 0.05 mm (0.002 in.) steps starting with the area of greatest thickness. Constant-thickness lines are located either by sliding an ultrasonic transducer over the part surface or by interpolating from a close grid of thickness measurements marked on the part.

Although chemical sizing can add significantly to the normal chemical milling costs, it is frequently justified by the resulting weight savings.

**Example 4: Chemical Sizing of Engine Cowl Inlet Duct Skins.** Engine inlet cowl skins, shown in Fig. 12, are produced from aluminum alloy 2219 and are chemically sized at two separate stages. The initial sizing step reduces flat, circular 3.2 mm (0.125 in.) gage material to 2.3 mm (0.090 in.) prior to forming. The forming process produces thickness variations of about 0.5 mm (0.020 in.), requiring a second sizing operation to bring the part within a tolerance of ±0.08 mm (±0.003 in.) for weight control. This second operation includes a detailed mapping of thickness variation over the entire surface and requires from three to six sizing steps, depending on the thickness variation to be removed.

The forming process also produces an extremely varied grain size over the part surface. Such problems are frequently encountered when forming 2219 alloy. Surface finishes range from smooth to extremely rough over any given part surface. The fine-grained smooth areas also etch at a faster rate, making the sizing operation more difficult. These parts are sanded and polished to a fine surface finish after chemical milling.

**Example 5: Undercut on Clad Aluminum.** The part shown in Fig. 13 is a nacelle access door used for both the Boeing 747 and the 767. The door is made by stretch forming 6.4 mm (0.250 in.) Alclad 7075 and then chemical milling to final thicknesses of 1.0, 2.0, and 3.2 ± 0.1 mm (0.040, 0.079, and 0.125 ± 0.005 in.).

Accurate control of undercut on clad aluminum is difficult. Therefore this extremely deep cut requires the removal of the cladding from the inside surface in order to maintain land width tolerances of ±0.8 mm (±0.030 in.). Another special requirement for this part is a polished finish on the outside surface.

**Thinning Machined Parts, Castings, Forgings, and Extrusions.** Chemical milling is frequently used to produce thin sections that would be uneconomical or impossible to make by mechanical means alone. Thousands of aluminum forgings for airline seat frames have been produced by this

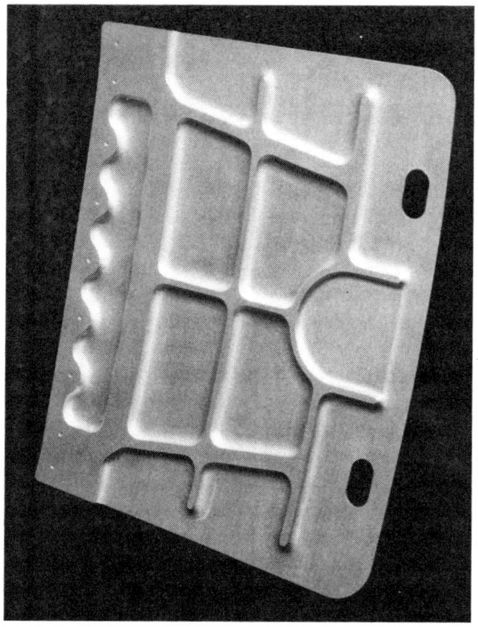

**Fig. 13** Engine cowl access door

process. The thousands of stringers used in the manufacture of the C-5A and C-5B were made from long aluminum extrusions designed oversized for chemical milling. Many turbine engine components, made from titanium and superalloy castings, are also processed with chemical milling, as illustrated in the following example.

**Example 6: Removing the Alpha Case From a Titanium Casting.** The highly complex casting pictured in Fig. 14 is a critical component of a turbine engine, made from titanium alloy 6Al-2Sn-4Zr-2Mo. The chemical milling task is to remove the alpha case that forms on the surface in the casting process. Because of the deeply recessed areas of the casting, careful positioning and good etchant control are required to guarantee the required metal removal range of 0.9 to 1.0 mm (0.035 to 0.040 in.). Small masked areas (witness pads) are located over the part surface to monitor the etched depth. This part has a diameter of approximately 660 mm (26 in.) and a height of 460 mm (18 in.).

**Improving Surfaces.** Material that is structurally weak or objectionable for some other reason can be easily removed by etching. Some examples are:

- Elimination of alpha case from titanium forgings and superplastic-formed parts
- Elimination of decarburized layer from low-alloy steel forgings
- Elimination of recast layers from EDM

Deep gouges, pits, and scratches are also often removed by the chemical milling process. The defect is filled with maskant, the part is etched to a depth greater than the defect, and the resulting island is mechanically removed. Significant defects should not be blended out prior to chemical milling because the resulting dished low areas would be much more difficult to locate and more costly to repair.

**Fig. 14** Titanium casting for combustor case

*Prepenetrant Etching.* Removal of a thin surface layer from forgings, castings, or machined parts is often advantageous prior to penetrant inspection. Latent defects hidden by smeared metal just below the surface can be detected, and false indications of defects can be screened through this procedure.

*Chemical Deburring.* Etching to remove from 0.03 to 0.13 mm (0.001 in. to 0.005 in.) per surface is often a very cost-effective way to remove sharp burrs from machined or pierced metal parts of complex shape.

**Chemical Leaching.** Etchants can often be used to dissolve one metallic component within a structure without affecting the other components. The structure is simply immersed until the selected metal is completely etched away. One example of this application is the fabrication of carbon fiber reinforced spars for man-powered aircraft. The layup is made on a temporary form of aluminum tubing, and the aluminum is subsequently etched away in a normal caustic etchant. Hollow-core titanium parts have also been produced by using nitric acid to leach low-alloy steel inserts from diffusion-bonded structures.

## Advantages and Disadvantages

The advantages of chemical milling include:

- Metal removal is completely stress free
- Complex shapes and deeply recessed areas can be uniformly chemically milled
- Chemical milling after forming often allows the use of the simplest forming method
- Extremely thin sections can be chemically milled
- Metal hardness or brittleness is not a factor
- Part size is limited only by tank dimensions
- Tapered sections can be chemically milled
- Most alloys and forms can be chemically milled
- Fine surface finishes are produced on many alloys
- Extremely close thickness tolerances are achievable, regardless of initial thickness variation, through chemical sizing methods
- Tooling and tool maintenance costs are low
- The process adapts well to prototype quantities because a pattern can be laid out directly on masked parts, avoiding lead time for tooling
- Cutouts and the periphery of difficult-to-machine parts can be rough trimmed by etching through the metal (chemical blanking), at minimal added cost

The disadvantages of chemical milling include:

- A fillet radius is approximately equal to depth of cut
- Extremely deep cuts are usually not cost effective
- A homogeneous metal structure is normally required for good results
- The surface texture of a chemically milled part is very dependent on the grain structure of the metal
- Welds and castings often produce pitted surfaces when chemically milled
- Process costs depend on the quality of the original workpiece (thickness variation, presence of surface scratches, and corrosion)
- The surface is affected by chemical milling, and sometimes mechanical polishing is required
- Lateral tolerances closer than ±0.8 mm (±0.030 in.) are often difficult to achieve
- It is impractical to make grooves of width less than twice the depth
- Alpha-beta titanium alloys often require special etchants to limit hydrogen pickup. Beta alloys are extremely difficult to control
- Process control coupons are often required, adding to cost
- Hazardous chemicals used in the process present difficult safety, waste disposal, and air pollution problems

### REFERENCES


1. W.T. Harris, *Chemical Milling*, Oxford University Press, 1974, p 109-139 and 160-161
2. D.J. Brimm, "Unistructure—A New Concept for Light Weight Integrally Stiffened Skin Structures," SAE Technical Paper 801231, Society of Automotive Engineers, 1980
3. E. Holman, "Method of Etching Aluminum and Aluminum Alloys," U.S. Patent 2,869,267, 1959

# Photochemical Machining

Howard Friedman, Fotofabrication Corporation

PHOTOCHEMICAL MACHINING (PCM), also known as chemical blanking, is a metal-etching process that uses a photoresist (photosensitive masking) to define the locations where the metal will be etched. The PCM process is used to produce intricate and close-tolerance patterns in a variety of flat, metal parts with a thickness range of 0.01 to 1.6 mm (0.0005 to 0.062 in.).

The PCM process, which has been in commercial use since the mid-1950s, is used in the production of a variety of precision parts, such as color television shadow masks; integrated circuit lead frames; magnetic recording head laminations; prototype motor laminations; heat sinks; encoder disks; grids, sieves, and screens; washers, shims, and gaskets; jewelry; and decorative plaques and nameplates. Most of the commonly used metals and alloys available in sheet form can be etched with the PCM process.

## Process Description

A flow chart showing the principal steps in the PCM process is given in Fig. 1. The major steps are:

- Preparation of phototool
- Selection of metal
- Preparation of workpiece
- Masking with photoresists
- Etching
- Stripping and inspection

**Selection of Metal.** Most of the commonly used industrial metals can be etched with the PCM process. More exotic alloys and metals can also be etched, but these require more sophisticated chemistry and operator knowledge. Except for a few special PCM installations, the process is used to etch rectangular metal sheets or panels. In choosing a standard-size sheet, some questions to be considered are:

- How wide a coil is available?
- What are the limitations of photographic processing equipment?
- How will the dimensional stability of the master affect quality?

The panel size must be matched with the capabilities of the etcher used. A reasonable choice is 300 × 450 mm (12 × 18 in.), but larger production capability is readily available.

Some of the qualities of a metal that produce good parts with the PCM process are:

- A grain size that is as fine as possible, because the smoothness of the edges of the parts decreases with increased grain size
- A metal that is soluble in the etchant chosen
- A material that is flat and of uniform thickness
- A surface finish that is uniform and free from scratches, embedded particles, or inclusions

Table 1 gives some of the effects obtained from poor-quality metal.

The etchability of a metal is dependent on its chemical composition. Table 2 lists etchability ratings of some metals and alloys used in photochemical machining.

**Preparation of Workpiece.** In order to promote good adhesion between metal and photoresist, the metal surface must be free of contaminants. Most photoresists have been developed for use in the printed circuit industry and have good adhesion to copper and copper-base materials. Other metals, especially stainless steels, require surface treatment to promote good photoresist adhesion. Pre-etching and passivation help. In all cases, a mechanical abrasion provided by machines with horizontal, cylindrical brushes helps to prepare the surface. The use of abrasives should be avoided because they must be completely removed to avoid surface contamination. If flatness is a problem, the material should be tension leveled at the factory.

A good test for efficient cleaning is spraying water on the metal surface in question and determining whether the individual droplets spread out to form a uniform, continuous, thin film of water. Formation of such a film indicates adequate cleaning,

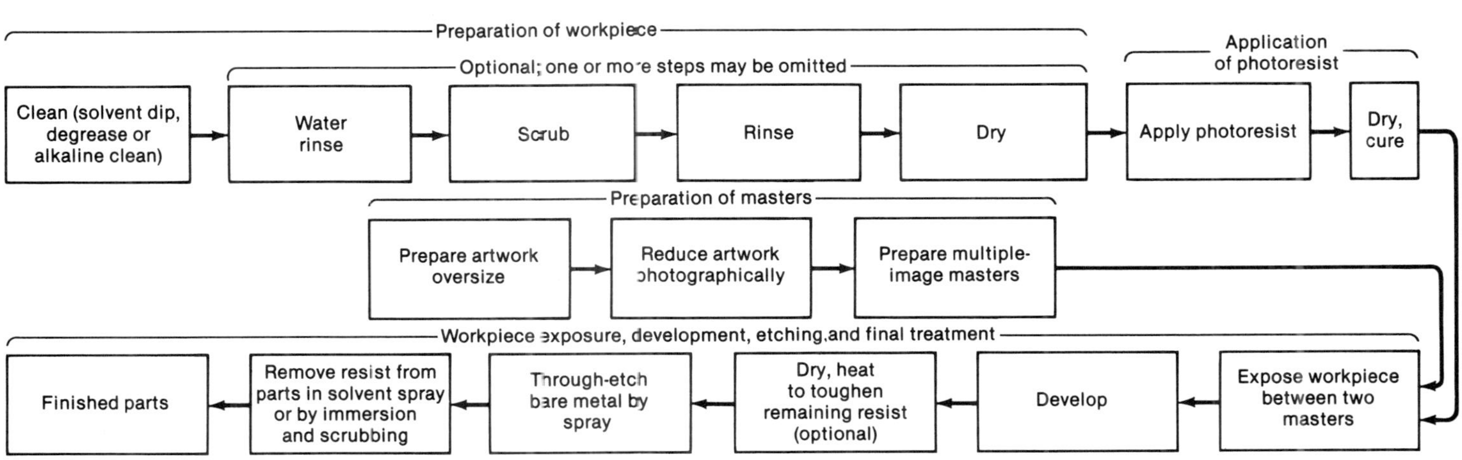


**Fig. 1** Flow chart of principal process steps for photochemical machining

**Table 1 Effects of poor-quality metal on PCM**

| Metal defect | Effect on PCM process |
|---|---|
| Coil set (metal curved along its length) and dents | Difficult to coat with photoresist and to contact print against the phototool, resulting in loss of registration and detail |
| Gage variation (for example, crown, where the thickness increases from the edges to the center of a strip of metal) | Difficult to determine optimum etching time as this is dependent on metal thickness |
| Too large a grain size | Loss of resolution in etched features |
| Surface scratches | Difficult to coat with photoresist, with the result that etchant flows into the scratches and produces cosmetic defects on the surfaces of the product |
| Embedded particles and inclusions such as oxides, sulfides, and silicates | Can produce defects on etching (for example, pits, pimples and loss of resolution in product) because they etch at different rates compared with the base material |

Source: Ref 1

whereas inadequate cleaning produces uneven wetting of the surface. Selecting a pretreatment cycle involves the same principles as selecting a preparation cycle for electroplating or painting. A rigorous pretreatment cycle used on nickel parts is described in Example 2 of this article.

**Preparation of masters,** which serve as the tools in photochemical machining, begins with the generation of oversize artwork on paper, polyester drafting film, or polyester-base or glass-base scribing film (listed in order of increasing dimensional stability and permanence). The most common method used is to cut the pattern into a material called cut-and-peel rubylith, which consists of a red, strippable layer coated onto a stable, clear polyester base. The part is drawn with the various etching factors taken into account.

The artwork is usually generated at some magnification factor to minimize dimensional error. The original artwork may range from 2 to 200 times actual size, depending on equipment, part size, and accuracy required. The various methods of artwork generation are:

- Manual drafting
- Precision coordinatograph plotting
- Computer aided drafting with a plotting table, light beam plotter, laser plotter, or integrated circuit pattern generator

The artwork is photographed with a precision reduction camera that reduces the artwork to 1:1. In high-volume production of parts, multiple-image masters are developed because it is often economical and convenient to expose, develop, and etch a number of parts from a single sheet of metal. In preparing multiple-image masters, the images are stepped and repeated to fill the master with an optimum number of parts. The stepping and replication of images can be done on the reduction camera or on a step-and-repeat camera. When a photoplotter or laser plotter is used, the array may be stepped at 1:1, eliminating the camera operation entirely. Any of these methods results in a master film that contains the pattern to be etched. From the master, working sets of films are made to facilitate simultaneous exposure of a panel from both sides. The highest-quality results are obtained by using matched sets of high-resolution glass plates to image the sensitized metal. In actual practice, most work is done using dimensionally stable photographic film, which is punched and registered to form a workable master. Film is also bonded to blank registered glass for precise registration.

During the preparation of masters, provisions are made for the reclaiming of parts after etching. Each part must have an etch line surrounding it so the part can eventually be separated from the metal sheet. The usual width chosen is twice the metal thickness.

In many cases tabs are added to the artwork to produce connections that hold the parts together after etching; these tabs must be removed later. If a protrusion from the broken-off tab is objectionable, a recessed-tab design may be used.

Small holes are usually etched completely away. In order to conserve etchant, larger openings can be made, surrounded by an etch line, which will allow a portion to drop out. When this is done, a corrosion-resistant screen is used to transport the panels through the etcher.

**Masking with photoresists** provides a photosensitive surface that resists the action of the etchant by providing a tightly adhering protective coating. After imaging and developing, the photoresist defines the shape of the part and therefore must not break down or lift off during etching.

*The photoresist* can be either a positive or a negative masking. In the positive acting system, the area that is exposed to light washes away during development. In the more commonly used negative system, the exposed areas are made insoluble in the photoresist developing solution. In either case, the final result is a panel in which the areas to be etched are bare metal, and the area representing the part is covered with a tightly adhering, acid-resistant coating.

Photoresists may either be applied in thin liquid form or laminated as a thicker, solid, dry film. Liquid resists have better photographic resolution; the best resolution obtainable is with positive-acting photoresists. However, the combination of ease of application and short exposure time has made

**Table 2 Etchability ratings of some selected metals and alloys used in the PCM process**

| Good | Good to fair | Fair to poor | Poor |
|---|---|---|---|
| Copper (rolled) | AISI 215 stainless steel | Molybdenum | Tungsten |
| Copper (electrolytic) | AISI 301 stainless steel | Nichrome (Ni, 20% Fe, 15% Cr) | Hastelloy C (Ni, 15% Mo, 14% Cr, 5% Fe, 3% W, 2.5% Co, 0.08% C) |
| Beryllium copper | AISI 302 stainless steel | Udimet alloys (for example, Ni, 42% Fe, 12.5% Cr, 2.7% Ti) | Titanium |
| Brass (Cu, Zn) | AISI 304 stainless steel | Vanadium | René 41 (Ni, 19% Cr, 11% Co, 10% Mo, 3% Ti, 1.5% Al) |
| Oxygen-free high-conductivity copper | AISI 305 stainless steel | Chromium | Niobium (Colombium) |
| Phosphor bronze (Cu, 10% Sn, ≤0.5% P) | AISI 316 stainless steel | Gold | |
| 90–10 copper-nickel | AISI 321 stainless steel | Lead | Tantalum |
| Zinc | AISI 347 stainless steel | Manganese | |
| Carbon steel | PH 15-7 stainless steel | Rhenium | |
| Kovar (54% Fe, 29% Ni, 17% Co) | PH 17-7 stainless steel | Zirconium | |
| Nickel | AISI 410 stainless steel | | |
| Monel (for example, Ni, 31.5% Cu, 1.3% Fe) | AISI 420 stainless steel | | |
| Nickel silver (Cu, 25% Zn, 10% Ni) | AISI 430 stainless steel | | |
| Magnesium | Inconel alloys (Ni, 15% Cr, 7% Fe) | | |
| Aluminum | Hastelloy B (Ni, 28% Mo, 5% Fe, 2.5% Co, 1% Cr, 0.5% V, 0.05% C) | | |
| Aluminum (anodized) | | | |

Source: Photo Chemical Machining Institute

the dry-film resists a popular choice in many job shops. As a general rule, the recommendations of the photoresist manufacturer should be closely followed.

*Application* of photoresist to the workpiece is best done immediately after cleaning and pretreatment, although work can be covered and stored in a controlled environment from several hours to several days before applying the photoresist. The work must not be handled before the resist is applied.

Photoresists are applied by dipping, whirl-coating, or spraying. Dipping is the simplest method, but whirl-coating gives the most uniform thickness, and spraying is the most versatile method.

Resists are dried at room temperature and then baked for about 15 min at a maximum of 120 °C (250 °F) to remove residual solvent. Baking temperature is not critical, and longer baking times at lower temperatures can be used.

*Exposure* of conventional photoresists to ultraviolet light partly polymerizes the exposed areas of light-sensitive resin, increasing the resistance of these areas to the chemicals used as developers. Exposure equipment used in the PCM process consists of a vacuum contact printing frame and a source of high-intensity ultraviolet light. In some cases, a highly collimated light source will resolve and reproduce finer detail.

Each side of the workpiece panel can be exposed individually, or the two sides can be exposed at the same time between a pair of mirror image masters in precise register. The vacuum printing frame is used with the emulsion side of the master pattern toward the work panel. A vacuum of about 500 mm (20 in.) of mercury will ensure good contact. Also, the printing frame must be padded appropriately, or glass masters will break in the evacuated frame.

Ideally, the wavelength of the light source should correspond to the maxima of the photoresist absorption spectra. The most commonly used light sources include high-pressure mercury lamps, metal halide lamps, mercury xenon lamps, pulsed xenon lamps, and actinic fluorescent tubes. Commercial exposure units invariably contain an upper and lower light source so that the two sides of a coated material can be exposed simultaneously.

*Development* of photoresists is usually accomplished in conveyor spray equipment in which the liquid developer removes the unpolymerized photoresist that covers the areas of the metal to be etched. Development by immersion with agitation in a suitable spray developer is another method.

The developers are often proprietary solutions formulated for specific photoresists. Developer temperature, spray pressure, spray pattern oscillation, and rinse water temperature are all important developing

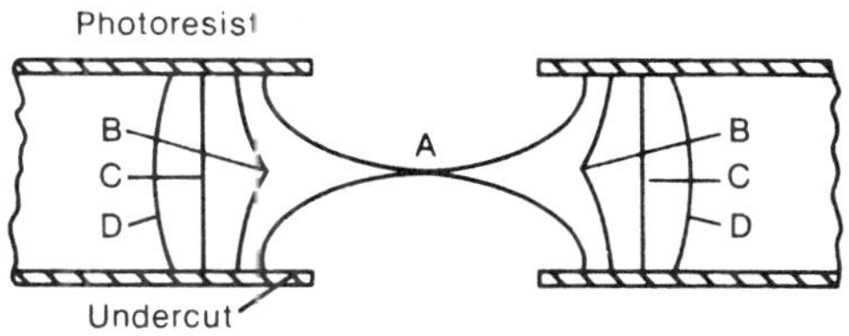


**Fig. 2** Development of two-sided etch profile: A, Breakthrough point; B, Biconvex; C, Straight; D, Biconcave

parameters. Developing conditions and duration are determined by the nature and thickness of the photoresist.

After the panel is removed from the developer, it should be rinsed in running water or spray rinsed, to remove the developer and dissolved polymer from the nonimage areas without leaving residual scum. If the nonimage areas are still not clean (not fully developed), development can be repeated without damaging the image.

In some cases it is desirable to bake the panels after development. Convection or infrared ovens are used. Postbaking helps to evaporate the remaining solvents in the photoresist and secures a tougher molecular bond to the substrate. This provides more acid resistance and is particularly important when using solvent-base photoresists in which swelling occurs during development. Care must be taken not to exceed the recommended baking temperature and times; otherwise the photoresist can become brittle and difficult to remove after etching.

**Etching.** The developed and postbaked panel of metal is exposed to a heated acid during the etching process. A chemical reaction takes place in which the acid reacts with the exposed metal and oxidizes it to form a soluble reaction product. Etching in the PCM process can be done either by immersion in a mechanically agitated or air agitated chemical bath or by spraying with heated acid. The force of the spray washes away the reaction products so that a fresh metal surface is always in contact with fresh etchant. When etching has proceeded to a point at which penetration from each surface has gone halfway through the sheet, the breakthrough occurs (Fig. 2). After this happens, the etchant can run through the opening, causing the edges to smooth out and produce a vertical sidewall. It should be noted that while the etchant is penetrating the metal, the lateral width of the etch line is increasing and undercutting the photoresist. The thicker the metal, the greater the amount of undercut. A properly designed phototool compensates for this undercut so that the resulting parts are in tolerance. It is this portion of the process that requires good statistical process control.

In most production operations, the composition and concentration of the etchant, and other process conditions, are regulated for a removal rate of about 0.01 to 0.05 mm/min (0.0005 to 0.002 in./min). Faster rates can be obtained, particularly in spray etching, but usually only at the cost of reduced accuracy, more frequent failure of the resist by penetration or by bond failure at pattern edges, and accelerated corrosion of equipment.

*Etchants* for photochemical machining of various metals are listed in Table 3, together with operating characteristics of each. Etch factor for the PCM process equals depth of cut (from one side) divided by undercut. It is important to note that in chemical machining, the same term, etch factor, means just the opposite (undercut divided by depth of cut).

Ferric chloride solutions can be used for the photochemical machining of a wide variety of metals, as indicated in Table 3. Ferric chloride will attack a wide range of metals and has become the most widely used etchant in the PCM industry. The range of etchants is usually restricted to the less dangerous ones, namely ferric chloride (often modified with additives), cupric chloride, diluted mineral acids, and some alkaline etchants based on sodium hydroxide or ammonium salts. Sodium hydroxide is used extensively on aluminum and aluminum alloys.

Most acid etchants can be maintained by the addition of fresh acid until their chemical activity is altered excessively by the dissolved metal content, at which point they are discarded. Sodium hydroxide etchants can be maintained by additions and regenerated when aluminum content becomes excessive. The effect of aging on etching rate and uniformity varies with the etchant and the work metal.

Etchant compositions can be adjusted to meet the requirements of specific applications, and proprietary additives can be included to control foaming or wetting characteristics, increase or decrease etching rate, or make etching more uniform. Many formulations resemble those used in bright dipping, chemical polishing, or electropolishing. A comparison of compositions of solutions used in the etching of printed circuit cards is given in Table 4.

*Etching machines* are made from materials (such as polyvinyl chlorides and titanium) that can withstand corrosion from ferric chloride and other etchants. Etchant temperature must be maintained below 55 °C (130 °F) to avoid distortion of the plastics used in machine construction.

Most etching machines have similar general construction. At the base of the machine there is a sump that contains the etchant. A titanium heating element and cooling coil help maintain a constant etchant temperature, which is necessary for accurate etching. The etchant is pumped through a manifold system to spray nozzles, which direct the etchant to both surfaces of the panel being etched. Etcher designs

**Table 3 Etchants for sheet metals and foils**

| Metal (composition) | Etchant formulation(a) | Temperature, °C (°F) |
|---|---|---|
| Alfenol (Fe, 16% Al) | 42 °Bé $FeCl_3$ | 49 (120) |
| | Conc $HNO_3$:conc HCl:$H_2O$ (1:1:2) | 38–49 (100–120) |
| Aluminum | Conc HCl:$H_2O$ (1:4) | 22–65 (72–150) |
| | 20% NaOH solution | 60–90 (140–190) |
| | 12–20 °Bé $FeCl_3$ | 49 (120) |
| | Alkaline potassium ferricyanide solution, for example, $K_3Fe(CN)_6$ (329 g/L, or 44 oz/gal.): NaOH (16 g/L, or 2 oz/gal.): $Na_3PO_4{\cdot}12H_2O$ (30 g/L, or 4 oz/gal.) | 55 (131) |
| | Conc HCl:conc $HNO_3$:$H_2O$ (10:1:9) | 49 (120) |
| | 40 °Bé $FeCl_3$: conc HCl (100:9) | 43 (110) |
| Anodized aluminum | 20% NaOH solution | 60–90 (140–190) |
| | 12–20 °Bé $FeCl_3$ | 49 (120) |
| Beryllium | Ammonium bifluoride ($NH_4HF_2$) (90–180 g/L, or 12–24 oz/gal.) | 27–32 (80–90) |
| Beryllium copper | See copper alloys | ... |
| Chromium | 42 °Bé $FeCl_3$:conc HCl (2:1) | 32 (90) |
| Constantan (55% Cu, 45% Ni) | 35–42 °Bé $FeCl_3$ | 49 (120) |
| Copper and copper alloys | 30–42 °Bé $FeCl_3$ | 43–49 (110–120) |
| | 33 °Bé $CuCl_2$ (acidic) | 54 (129) |
| | Modified chromic acid (for example, Hunt Multi-Circuit Etch) | 27–38 (80–100) |
| | Ammonium persulfate (220 g/L, or 29 oz/gal.) | 32–49 (90–120) |
| | Cupric ammonium chloride $[Cu(NH_3)_4]Cl_2$(2M) | |
| | Hydrogen peroxide/sulfuric acid (for example, Shipley Hydro-etch 536) | 45–50 (113–122) |
| Germanium | HF + $HNO_3$ | ... |
| Gold | Aqua regia, that is, conc HCl:conc $HNO_3$ (3:1) | 20–32 (68–90) |
| | Potassium iodide (saturated) + iodine (20 g/L, or 3 oz/gal.) | 50 (122) |
| HyMu 80, 800 (80% Ni, 4% Mo, rem Fe | 42 °Bé $FeCl_3$: conc HCl (9:1) | 43–49 (110–120) |
| Inconels (Ni, Cr, Fe) | 42 °Bé $FeCl_3$ | 54 (129) |
| Kovar (Fe, 29% Ni, 17% Co) | 42 °Bé $FeCl_3$ | 49 (120) |
| Lead | 33 °Bé $CuCl_2$ (acidic) | 54 (129) |
| Magnesium | 10–20% conc $HNO_3$ (plus additive such as Hunt X-flex to improve etch factor if required) | 35 (95) |
| Molybdenum | Alkaline potassium ferricyanide solution, for example, $K_3Fe(CN)_6$ (200 g/L, or 27 oz/gal.):NaOH (20 g/L, or 3 oz/gal.): sodium oxalate (5 g/L, or 0.7 oz/gal.) | 55 (131) |
| | 40 °Bé Fe $(NO_3)_3$ | 40–55 (104–131) |
| | Conc $HNO_3$:HF:$H_2O$ | |
| | Conc $HNO_3$:conc $H_2SO_4$:$H_2O$ (1:1:3) | 50–54 (122–129) |
| | Conc $HNO_3$:conc HCl:$H_2O$ (3:3:4) | |
| Moly Permalloy (Ni, 13% Fe, 5% Cu, 4% Mo) | 42 °Bé $FeCl_3$:conc HCl (9:1) | 54 (129) |
| Monel (67% Ni, 33% Cu) | 42 °Bé $FeCl_3$ | 49–54 (120–129) |
| Mumetal (Ni, 16% Fe, 5% Cu, 1½% Cr) | 42 °Bé $FeCl_3$ | 49–54 (120–129) |
| Nickel | 38–42 °Bé $FeCl_3$ | 49 (120) |
| | Conc $HNO_3$:conc HCl:$H_2O$ (1:1:3) | |
| Nickel-iron alloys (for example, Invar, Alloy 42) | 42 °Bé $FeCl_3$ (with $HNO_3$ additions if necessary) | 49 (120) |
| Nimonics (~80% Ni, 20% Cr) | $FeCl_3$:$HNO_3$:HCl | |
| Niobium | Conc $HNO_3$:HF:$H_2O$ (7:1:2) | |
| Phosphor bronze (Cu, 5% Sn, ½% P) | 42 °Bé $FeCl_3$ | 25–49 (77–120) |
| Platinum | Aqua regia, that is, conc HCl:conc $HNO_3$ 3:1) | 20–25 (68–77) |
| Silver | 36 °Bé $Fe(NO_3)_3$ | 43–54 (110–129) |
| | 50–90% $HNO_3$ | 38–49 (100–120) |
| Stainless steels | 35–48 °Bé $FeCl_3$ | 35–55 (95–131) |
| Stainless steels (containing molybdenum) | 36–42 °Bé $FeCl_3$ with $HNO_3$ additions | 35–55 (95–131) |
| Steels (mild, spring, silicon, and tool) | 36–42 °Bé $FeCl_3$ | 52 (126) |
| Tin | 42 °Bé $FeCl_3$ | 54 (129) |
| Titanium | 10–50% HF (optionally with additions of $HNO_3$) | 30–49 (86–120) |
| | Ammonium bifluoride ($NH_4HF_2$):conc HCl:$H_2O$ | 30–49 (86–120) |
| Tungsten | Alkaline potassium ferricyanide solution, for example, $K_3Fe(CN)_6$ (200 g/L, or 27 oz/gal.):NaOH (20 g/L, or 3 oz/gal.):sodium oxalate (5 g/L, or 0.7 oz/gal.) | 55 (131) |
| Vanadium | Conc $HNO_3$:$H_2O$ (1:1) | 25 (77) |
| Zinc | 20–25% conc $HNO_3$ (plus additives such as Hunt Rocket Etch to improve etch factor if required) | 36 (97) |
| Zirconium | Conc $H_2SO_4$:HF | 36 (97) |
| | Conc $HNO_3$:HF:$H_2O$ | 36 (97) |

(a) Volume ratios if known, except where stated. °Bé, Baumé scale; conc, concentrated; rem, remainder. Source: Ref 1

vary. In some machines the spray bars oscillate, and in others, they are fixed. Etchers with conveyor systems (Fig. 3) allow continuous processing and are the most widely used. Batch-type machines, in which one sheet at a time is processed, are used for short runs and high-precision parts.

**Stripping and Inspection.** After etching, the panels are still coated with photoresist. In most cases, removal (stripping) is necessary. The art of good stripping is to be able to remove the photoresist completely without staining or corroding the metal surface. Some formulas only soften and lift the coatings, which then have to be gently brushed off the surface. Methylene chloride base strippers will completely dissolve the photoresists, but require proper ventilation and care in handling.

Handling of parts during stripping is easier when the parts remain attached to the metal sheet. Generally, one or more small connecting tabs are located in such a manner that they do not interfere with the function of the part. It should be noted that sheets of tabbed parts facilitate electroplating. The parts may be detached by cutting or by manually twisting. When the requirements of the part preclude the use of a tab, the parts may be made to drop out during etching. This requires the use of a carrying screen and can cause tolerance problems.

Once the thickness of the part is checked, the rest of the inspection is two dimensional. The best methods of inspection involve the use of plug gages and horizontal optical comparators. The use of statistical process controls, especially during etching, can reduce final inspection requirements.

## Design Considerations

The mechanical properties of the work metal have little effect on etching rate, nor does workpiece shape (unless it prevents uniform access of fresh etchant to the surface). The maximum thickness of sheet from which parts are photochemically machined on a production basis is 1.6 mm (1⁄16 in.). However, cold-rolled steel parts 3.2 mm (1⁄8 in.) thick have been etched to a tolerance of 0.4 mm (±0.015 in.) under special conditions.

Figure 4 shows progressive stages of metal removal and the effect of undercut on design. Because photochemical machining is concerned primarily with the etched surfaces that become the edges of the etched part, undercut refers only to attack just below the resist edge on one side of the cut.

**Table 4 Effect of etchant composition on production of 250 μm (10 mil) lines in 60 g (2 oz) copper foil bonded to a plastic base**

Undercut and etching taper are influenced by etchant composition, operating conditions, and foil thickness; undercut and taper would be substantially smaller for 30 g (1 oz) (0.036 mm, or 0.0014 in. thick) copper foil; controlled overetching will reduce taper.

| Six etchant compositions: Constituent | g/L (oz/gal.) | Etching time, s(a) | Undercut, $R - T$(a) mm (in.) | Etching taper, $B - T$(b) mm (in.) |
|---|---|---|---|---|
| $(NH_4)_2S_2O_8$ | 217 (29) | 88 | 0.091 (0.0036) | 0.066 (0.0026) |
| $+HgCl_2$ | 0.0064 (0.00086) | | | |
| $+H_2SO_4$ | 17.6 (2.26)(c) | | | |
| +Copper | 4.9–8.1 (0.65–1.08) | | | |
| $(NH_4)_2S_2O_8$ | 217 (29) | 90 | 0.089 (0.0035) | 0.089 (0.0035) |
| $+NH_4Cl$ | 16.2 (2.16) | | | |
| $+HgCl_2$ | 0.0064 (0.00086) | | | |
| +Copper | 4.9–13.3 (0.65–1.78) | | | |
| $(NH_4)_2S_2O_8$ | 217 (29) | 134 | 0.074 (0.0029) | 0.11 (0.0045) |
| $+NH_4Cl$ | 16.2 (2.16) | | | |
| +Copper | 4.9–13.3 (0.65–1.78) | | | |
| $(NH_4)_2S_2O_8$ | 217 (29) | 182 | 0.053 (0.0021) | 0.10 (0.0040) |
| +NaCl | 140 (19) | | | |
| +Copper | 6.5–9.9 (0.87–1.32) | | | |
| $CuCl_2$ | 270 (36) | 143 | 0.074 (0.0029) | 0.046 (0.0018) |
| $+NH_4Cl$ | 64.6 (8.64) | | | |
| +Copper | 4.9–7.6 (0.65–1.02) | | | |
| $FeCl_3$ | 380 (51) | 76 | 0.066 (0.0026) | 0.046 (0.0018) |

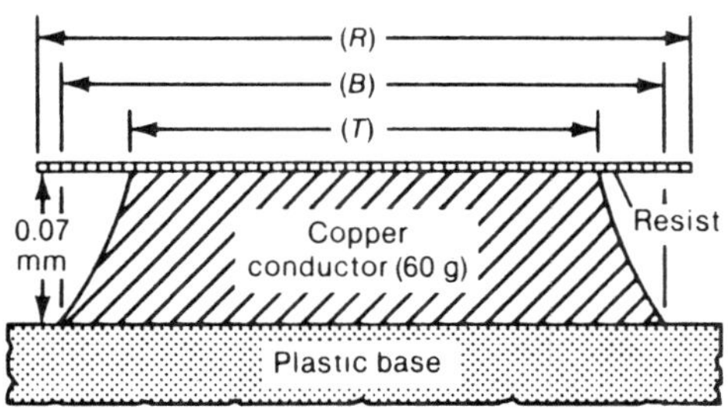


(a) At 37 to 38 °C (99 to 100 °F). (b) See illustration; etch factor as calculated for other PCM operations is not applicable to printed circuit etching; etching taper ($B - T$) is used instead. (c) mL/L (fluid oz/gal.)

The etch factor is defined as depth of cut divided by undercut; higher values are preferred to lower values. In Fig. 4(d), stock 150 μm (6 mils) thick is shown blanked with an undercut of 12 μm (1 mil) and an etch factor of 3.0. Straightness of the side walls of the cut can be improved substantially by overetching, as shown in Fig. 4(f). Allowance is made for the undercut and the etch factor in the layout of artwork, on the basis of trial processing or experience.

**Dimensional Limitations.** Figure 5 shows the typical relationships between metal thickness ($T$) and various dimensional limitations. As a general rule, the smallest hole diameter or slot width that can be produced by the PCM process cannot be less than 1.2 times the metal thickness. Exceptions to this occur as the thickness decreases. The minimum practical hole diameter or slot width is 0.10 mm (0.004 in.). The minimum slot width or hole diameter may also depend on the workpiece. The approximate limits for some commonly used metals are:

| Metal | Minimum slot width or hole diameter |
|---|---|
| Steels, carbon and alloy | 1.0 $T$ |
| Stainless steels | 1.4 $T$ |
| Aluminum alloys | 1.4 $T$ |
| Copper alloys | 0.7 $T$ |

As a general rule, the minimum spacing between holes and slots is at least 0.9 times the metal thickness. With metals less than 0.1 mm (0.005 in.) thick, the remaining metal must be at least equal to the metal thickness. Center-to-center accuracy is determined photographically and will vary only in accordance with the limitations of the photographic medium and the artwork.

**Tolerances.** It is not possible to establish hard-and-fast rules regarding etching tolerances because too many factors are involved, such as the type of metal, size of the sheet, type of exposure equipment, type of etching equipment, condition of the etchant, and of course, the skill of the operator. However, some typical PCM tolerances of various metals are listed in Table 5 for thicknesses of 0.05 to 1.5 mm (0.002 to 0.060 in.). The values shown range from ±10% of stock thickness for the photochemical processing of steel, nickel, copper and copper alloys with rigid control to ±100% of stock thickness for routine photochemical processing of 0.05 mm (0.002 in.) thick aluminum alloy foil. Under special conditions, accuracy of ±5% of thickness and part reproducibility of ±2% to ±3% of thickness can be maintained.

**Edge Quality.** Although burrs are not produced in photochemical machining, the edges obtained at the instant of complete breakthrough may be irregular and rough to varying degrees. Edge uniformity depends on the uniformity and fineness of the grain structure of the workpiece, local stresses, edge orientation, uniformity of resist edge, and adherence of resist, as well as on local variation in etchant concentration, impingement velocity, and direction of flow against the surface being cut. Slower etching rates usually produce more uniform edges. Edge quality can be improved by controlled overetching (see Fig. 4) or by subsequent bright dipping or chemical or electrolytic polishing.

## Advantages and Disadvantages

The PCM process is an alternative to conventional stamping (or blanking) when intricate patterns or low production volumes are involved. The advantage exists because tooling changes in the PCM process involve artwork changes, while stamping requires the fabrication of precision dies of very tough materials such as tungsten carbide. The PCM process also has the capability to produce intricate patterns that cannot be duplicated by other processes.

Some of the chief advantages of the PCM process are:

- Magnetically soft materials can be fabricated and retain optimum permeability
- Setup and tooling costs are extremely low
- Initial quantities of newly designed parts can be produced on short notice—some-

**Table 5 Tolerances in photochemical machining of various metals**

| Stock thickness mm | Stock thickness in. | Tolerance (±): Aluminum alloys mm | in. | Copper and alloys mm | in. | Nickel mm | in. | Stainless steel mm | in. | Low-alloy steel mm | in. |
|---|---|---|---|---|---|---|---|---|---|---|---|
| **For routine production** | | | | | | | | | | | |
| 0.05 | 0.002 | 0.05 | 0.002 | 0.025 | 0.001 | 0.025 | 0.001 | 0.025 | 0.001 | 0.025 | 0.001 |
| 0.13 | 0.005 | 0.08 | 0.003 | 0.05 | 0.002 | 0.08 | 0.003 | 0.05 | 0.002 | 0.05 | 0.002 |
| 0.25 | 0.010 | 0.10 | 0.004 | 0.08 | 0.003 | 0.13 | 0.005 | 0.08 | 0.003 | 0.10 | 0.004 |
| 0.50 | 0.020 | 0.15 | 0.006 | 0.13 | 0.005 | 0.25 | 0.010 | 0.13 | 0.005 | 0.15 | 0.006 |
| 1.0 | 0.040 | 0.20 | 0.008 | 0.15 | 0.006 | | ... | 0.25 | 0.010 | 0.25 | 0.010 |
| 1.5 | 0.060 | 0.30 | 0.012 | 0.18 | 0.007 | | ... | 0.36 | 0.014 | 0.30 | 0.012 |
| **For production with rigid control** | | | | | | | | | | | |
| 0.05 | 0.002 | 0.008 | 0.0003 | 0.005 | 0.0002 | 0.005 | 0.0002 | 0.013 | 0.0005 | 0.005 | 0.0002 |
| 0.13 | 0.005 | 0.018 | 0.0007 | 0.013 | 0.0005 | 0.013 | 0.0005 | 0.025 | 0.001 | 0.013 | 0.0005 |
| 0.25 | 0.010 | 0.04 | 0.0015 | 0.025 | 0.001 | 0.025 | 0.001 | 0.05 | 0.002 | 0.025 | 0.001 |
| 0.50 | 0.020 | 0.08 | 0.003 | 0.05 | 0.002 | 0.05 | 0.002 | 0.10 | 0.004 | 0.05 | 0.002 |
| 1.0 | 0.040 | 0.13 | 0.005 | 0.10 | 0.004 | 0.13 | 0.005 | 0.20 | 0.008 | 0.13 | 0.005 |
| 1.5 | 0.060 | 0.18 | 0.007 | 0.15 | 0.006 | 0.20 | 0.008 | 0.30 | 0.012 | 0.18 | 0.007 |

Note: Tolerances are applicable to a dimension of 50 mm (2 in.) or less; proportionate value applies for a larger dimension. Closer tolerances are possible.

**Fig. 3** Two-stage chemical clean system with dryer. Courtesy of Chemcut Corporation

times within a few hours after the design has been conceived
- Design change costs are low, because only artwork is altered
- The process does not produce burrs
- Extremely thin metal can easily be blanked without distortion. Accuracy actually increases as metal thickness decreases, because the extent of undercutting is decreased
- Metal hardness has no significant effect on ease of blanking
- The temper, stress, and other physical properties of the metal are not changed
- Brittle metals (such as the silicon electrical steels, which often fracture during conventional blanking) are blanked without difficulty
- The process allows great design flexibility

Disadvantages associated with photochemical machining are:

- A relatively high level of operator skill is required
- Special safety precautions are needed in handling of chemicals; the etchant vapors also are very corrosive. Etching equipment must usually be isolated from other plant equipment
- Suitable photographic facilities are not always available
- Maximum metal thickness that can be blanked is about 1.6 mm (1/16 in.)
- Sharp radii cannot be produced

## Applications

One of the major applications of the PCM process is the manufacture of burr-free, intricate, thin "stampings." The PCM process often is more economical than punch-press or press-brake blanking for quantities up to 5000, depending on the size of the part.

Typical blanked parts processed with PCM include laminations for electric motors or magnetic recording heads, rotors, slotted spring disks and gaskets, meter parts, vanes for high-speed cameras, strain-gage loops or grids, electrical contacts and terminals, fine screens, helicopter vent screens, templates, electronic-conductor matrices, and printed circuit cards. Shadow masks for color television also are made by photochemical machining; about 320 000 conical holes, each with a minimum diameter of 0.28 mm (0.011 in.), are blanked in 0.15 mm (0.006 in.) thick, full-hard 1010 steel foil, for a 530 mm (21 in.) television picture tube. Corrosion-resistant filter screens for chemical applications have been made with 0.15 mm (0.006 in.) diam holes on 0.28 mm (0.011 in.) centers in type 316 stainless steel foil 0.1 mm (0.005 in.) thick.

Photochemical machining is sometimes used to eliminate deburring or to reduce assembly time. Electronic chassis are made in small-to-medium quantities by using punch presses for routine holes and chemical blanking for odd-shape holes and slots.

Thin intricate parts that are impractical to blank mechanically because of hardness or brittleness of the work material can often be made with ease by photochemical machining. The following two examples illustrate this.

**Example 1: Molybdenum Alloy Parts That Delaminated in Punch-Press Blanking.** The conventional blanking of small,

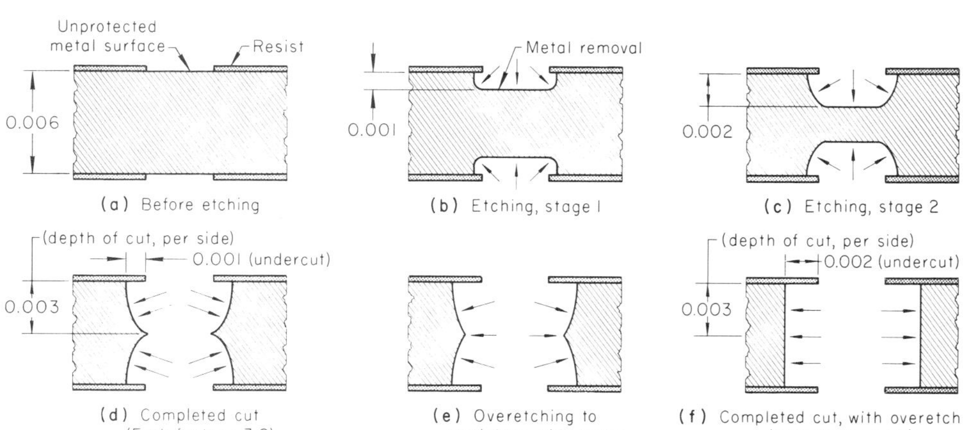


**Fig. 4** Progressive stages of metal removal in photochemical machining. Metal is removed by etching on all work metal surfaces not covered by resist. Workpieces can be taken from etchant as soon as blanking is completed (d), for an undercut of 0.025 mm (0.001 in.); or etching can be continued to improve straightness of side walls, for an undercut of 0.05 mm (0.002 in.) (f). Workpiece dimensions depend on resist dimensions and control of undercut. Dimensions given in inches

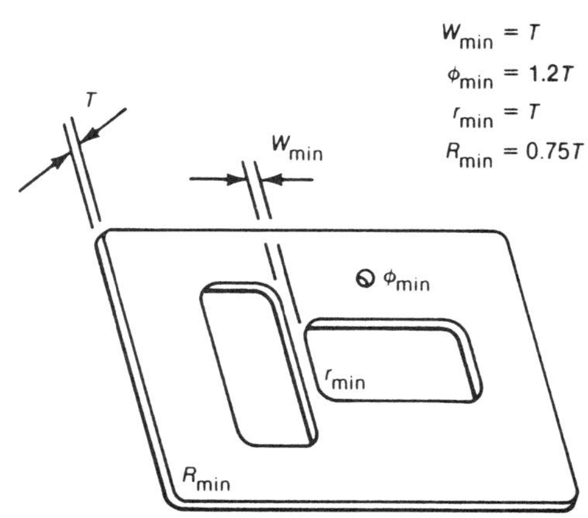


**Fig. 5** Etched dimensional limitations

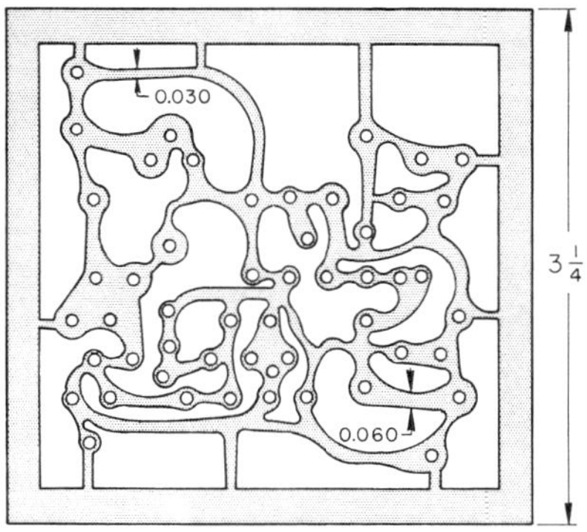


**Fig. 6** Electronic-connector network photochemically blanked from predrilled 1.1 mm (0.042 in.) thick nickel sheet. Dimensions given in inches

complex, close-tolerance aerospace parts from sheet molybdenum alloys was excessively slow, and rejection rates were high because of edge delamination. Production rate and yield were increased by blanking with the PCM process and a nitric-hydrofluoric acid etchant.

Parts were spaced as close together as possible on a panel (minimum spacing 2.5 mm, or 0.100 in.), and the area to be etched away was kept as small as possible to conserve etchant and to simplify control of the etching process. The multiple-part panel was then etched through on the unmasked areas at a penetration rate of 0.025 mm/min (0.001 in./min) by immersion in the nitric-hydrofluoric acid bath. The masking material was removed from the completed parts by a combination of solvent spray and brushing. A dimensional accuracy of ±0.25 mm (±0.010 in.) was maintained on material up to 1.5 mm (0.060 in.) thick. A controlled period of overetching was used on some parts for deburring and radiusing.

**Example 2: Nickel Electronic-Connector Network.** Chemical blanking was the only practical way to manufacture the connector network shown in Fig. 6 from predrilled 1.1 mm (0.042 in.) thick nickel. This matrix, used in high-density packaging of electronic circuits, was made of nickel for strength, and had electrical conductivity, corrosion resistance, and suitability for spot welding. The processing sequence and the etchant composition were:

- Remove mill markings from precut sheet in organic solvent
- Immerse for 5 min in nonsilicated alkaline cleaner 30 to 45 g/L (4 to 6 oz/gal.) at 77 to 88 °C (170 to 190 °F)
- Rinse thoroughly
- Treat anodically in hydrochloric acid (25% by volume) for 5 s at current density of 1100 to 2200 A/m$^2$ (100 to 200 A/ft$^2$)
- Rinse thoroughly
- Oven dry at 90 to 120 °C (200 to 250 °F)
- Coat with diluted photosensitive resist (proprietary material)
- Bake at 120 °C (250 °F) for 10 min (typical)
- Mount artwork
- Expose to ultraviolet light
- Develop image for 2 to 3 min by immersion in resist developer (organic solvent)
- Postbake 20 min at 120 °C (250 °F) (typical)
- Preactivate by anodic treatment as the fourth item above
- Rinse thoroughly without delay
- Without delay, etch by immersion for 20 to 30 min in the following solution at 43 to 49 °C (110 to 120 °F), using work agitation:

| Concentrate | Etchant, vol% |
|---|---|
| Nitric acid, 70% | 48 |
| Sulfuric acid, 96% | 5.5 |
| Phosphoric acid, 85% | 11 |
| Acetic acid, 99% | 5.5 |
| Water | 30 |
| **Total** | **100** |

- Remove from bath as soon as blanking is complete
- Rinse thoroughly
- Remove resist with organic solvent plus scrubbing

The completed matrix was deburred and smoothed, when necessary, in a bright dip or chemical polishing bath. Large areas of bare nickel adjacent to the masked area were avoided to conserve etchant and to minimize gas evolution and local heating, which could affect adhesion of the resist or cause local variation in removal rate. A chemically resistant screen at the bottom of the tank was used to remove pieces of nickel that became detached from the work and would unnecessarily shorten bath life if left to dissolve.

Support legs at the periphery of the matrix were cut away after processing. Minimum line width was 0.53 mm (0.021 in.). Time required for complete blanking, 20 to 30 min, was reduced to about 4 min in test runs by spray etching preactivated surfaces. Monel matrices were produced experimentally by chemical blanking in a similar but more dilute etchant.

**Printed Circuit Etching.** A special use of chemical blanking is in the manufacture of printed circuit cards, to etch away the copper cladding from defined areas of plastic laminate, leaving a pattern of resist-protected conductor circuits on the plastic panel. The techniques are essentially the same as those for other PCM applications, but usually are applied to 30 or 60 g (1 or 2 oz) copper foil bonded to plastic panels about 0.25 to 6.4 mm (0.010 to 0.250 in.) thick. In some applications, the masking material is used only as an intermediate coating to define the areas on which a metallic etch resist (such as solder, nickel, tin-nickel alloy, palladium, or gold) is then electroplated over the copper foil. Etching is done in specially designed spray etchers, using horizontal conveyor-type machines for volume production.

Etchants used on printed circuit cards include solutions of chromic acid and either sulfuric acid or sodium sulfate, ammonium persulfate, ferric chloride, and cupric chloride. Selection of an etchant depends on type of product; nature of resist or resists; subsequent soldering or welding requirements; removal rate; undercutting behavior; compatibility with plastic substrate (chemical attack, absorption, deposition of insoluble films); material cost; and ease and cost of control, regeneration, and waste disposal. Table 4 compares the performance of six different etching solutions in producing 250 μm (10 mil) lines in 60 g (2 oz) copper foil bonded to a plastic base. Space between lines was 125 to 760 μm (5 to 30 mils). The data in Table 4 show that etching rate, undercut, and etching taper can be changed by altering the etchant composition; these variables can be changed also by adjusting the operating temperature or the spray velocity.

Under rigidly controlled conditions, patterns with line widths and spacings of 125 μm (5 mils) can be etched on 30 or 60 g (1 or 2 oz) copper foil at constant etching time, with ammonium persulfate, chromic acid and sulfuric acid, or ferric chloride etchants. Etching of patterns with lines and spacings narrower than 125 μm (5 mils) presents serious problems, chiefly because of variation in foil thickness. Patterns with line widths and spacings of 500 to 760 μm (20 to 30 mils) can be etched on a production basis without the critical control of etchant composition, temperature, uniformity of spray distribution, etching time, foil thickness, and surface condition required for fine-line etching.

## REFERENCE

1. D.M. Allen, *The Principles and Practice of Photochemical Machining and Photoetching*, Adam Hilger, 1986

## SELECTED REFERENCES

- H. Friedman, Manufacturing Metal Parts by the Photo Chemical Machining Process, in *Nontraditional Machining, Proceedings of 1985 Conference*, American Society for Metals, Dec 1986, p 141-144
- "Photo Chemical Machining—The Answer to Precision Flat Metal Parts," Fotofabrication Corporation, 1986

# High-Productivity Machining

# High-Speed Machining

HIGH-SPEED MACHINING is a relative term from a materials viewpoint because of the vastly different speeds at which different materials can be machined with acceptable tool life (Ref 1). For example, it is easier to machine aluminum at approximately 1800 m/min (6000 sfm) than titanium at 180 m/min (600 sfm). Because of this difference and because speed determines to a significant degree whether a material will form continuous chips or segmented shear-localized chips, one way of defining high-speed machining is to relate it to the chip formation process (see the section "Mechanics of Chip Formation" in this article). Localized shear occurs when the negative effect on strength of increasing temperature due to intense plastic deformation is equal to or greater than the positive effect of strain hardening. In this context, high-speed machining for a given material can be defined as that speed above which shear localization develops completely in the primary shear zone.

Although appealing from a technical standpoint, the foregoing is not very useful as a practical definition. For this reason, it is generally preferable to define machining speeds quantitatively in terms of specific ranges. One suggestion is that 600 to 1800 m/min (2000 to 6000 sfm) should be termed high-speed machining, 1800 to 18 000 m/min (6000 to 60 000 sfm) very high-speed machining, and greater than 18 000 m/min (60 000 sfm) ultrahigh-speed machining (Ref 2). In the case of very-difficult-to-machine alloys such as titanium, it is preferable to use the term high-throughput machining rather than high-speed machining in order to maintain a proper focus on realistic machining conditions.

The use of high-speed machining technology in industry has become more prevalent in recent years. The development of tougher, more refractory tool materials and of high-speed machining spindles has contributed to this growth in acceptance. High-speed machining can be used to machine parts that require the removal of significant amounts of material and to machine long, thin webs. The need throughout industry and the defense establishment to reduce costs and increase productivity has created new interest in high-speed machining.

The rationale for continuing research and development in high-speed machining technology is evident when one considers the funds expended in metal removal annually in the United States. Of the more than $100 billion spent for metal removal, 75% can be attributed to the four conventional processes of turning, milling, drilling, and grinding (Ref 3). Machining costs can be significantly reduced only by determining ways to increase metal removal rates.

## Historical Background (Ref 4)

The concept of high-speed machining was conceived by Dr. Carl J. Salomon during a series of experiments from 1924 to 1931 (German patent 523594, 1931). The patent was based on a series of curves of cutting speeds plotted against generated cutting temperatures. These experiments were performed on nonferrous metals such as aluminum, copper, and bronze. Salomon obtained speeds up to 16 500 m/min (54 200 sfm) using helical milling cutters on aluminum. His contention was that the cutting temperature reached a peak at a given cutting speed; however, as the cutting speed was further increased, the temperature decreased. Figure 1 shows a simplified presentation of this concept.

As the cutting speed is increased from 0 in the normal mode, $V_1$, the temperature will increase in a direct relationship until a peak value, $T_{cr}$, is achieved. The cutting speed at $T_{cr}$ is commonly called the critical cutting speed, $V_{cr}$. If the cutting speed is further increased, it was predicted that the cutting temperature would decline. Near $V_{cr}$, Salomon suggested that there was an unworkable regime in which cutters were not able to withstand the severe process temperatures and forces. The shape of the curve was thought to be dependent on the exact nature of the base material being cut. When the cutting speed was sufficiently increased, the resulting temperatures, at $V_2$, were reduced to those of the normal cutting temperatures, at $V_1$, and the materials and cutters would once again permit practical cutting procedures. The same cutting temperature, $T_a$, found in the normal speed range, $V_1$, could possibly be reproduced in the high-speed range, $V_2$.

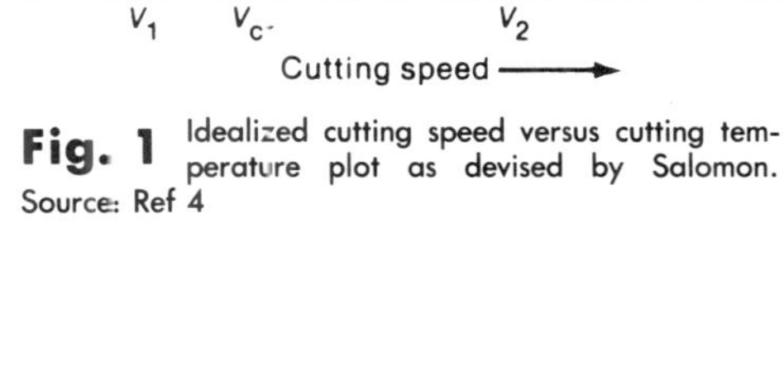


**Fig. 1** Idealized cutting speed versus cutting temperature plot as devised by Salomon. Source: Ref 4

The studies conducted by Salomon are now mainly of historical interest since current research is developing more definitive data using more sophisticated techniques. In addition, interpretations of Salomon's theory have been responsible for confusion and false expectations concerning high-speed machining.

The first systematic investigation of high-speed machining in the United States was undertaken by R.L. Vaughn at Lockheed Aircraft Corporation (Ref 5-7). In 1958, Vaughn studied a series of variables involved in traditional machining that became very important in high-speed machining. According to Vaughn, the rate at which metal can be machined is affected by:

- Size and type of machine
- Power available
- Cutting tool used
- Material to be cut
- Speed, feed, and depth of cut

These five general variables can be broken down further into:

- Rigidity of machine, cutter, and workpiece
- Variations in speed from the slowest to the fastest, depending on machine used
- Variations in feed and depth of cut from light to heavy and whether cut dry or with the aid of lubricant and/or coolant
- Type and material of cutting tool
- Variations in cutter shape and geometry
- Type and physical characteristics of work material
- Specific requirements of desired cutting speed, tool life, surface finish, horsepower required, residual stress, and heat effects

Recent advances in the development of computer control systems have provided the capability of accurately manipulating high-performance, automatic production machines. Progress in the field of bearing design, alternative spindle power sources, automatic tool changing, tool retention devices, and cutter materials has also made contributions toward proving Vaughn's experiments.

A second series of studies was initiated in the 1970s. These studies were contracted by the U.S. Navy with Lockheed Missiles & Space Company (Ref 8-10). The objective of these studies was to determine the feasibility of using high-speed machining in a production mode, initially with aluminum alloys and later with nickel-aluminum-bronze. A team of researchers demonstrated that it was economically feasible to introduce high-speed machining procedures into the production environment to realize major improvements in productivity. This resulted in a significant increase in overall interest and a very active period of both experimental and applied research, as can be noted in a review of the literature. For example, important data concerning the effect of cutting speed and cutter geometry on cutting temperature when turning 2014-T652 aluminum were provided by F.J. McGee (Ref 11). These data indicated that cutting temperature curves tend to peak near the melting point of the aluminum alloys.

In the late 1970s and early 1980s, a third series of contracts was awarded by the U.S. Air Force to the General Electric Company to provide a data base for the machining of aluminum alloys, titanium alloys, nickel-base superalloys, and steels. These studies were spearheaded by D.G. Flom (Ref 1, 12-19) and R. Komanduri (Ref 20-27), and the results will be the focus of this article.

## Mechanics of Chip Formation (Ref 26)

Two types of chip formation have been observed in high-speed machining, depending on the type of work material to be machined and its metallurgical conditions. They are the continuous chip and the shear-localized (segmental) chip, a term arising from the intense deformation (or shear localization) between the segments. Continuous chips are likely to occur in the high-speed machining of a metal or alloy of body-centered cubic/face-centered cubic crystal structure, high thermal diffusivity, and low hardness, such as aluminum alloys or soft low-carbon steel. The shear-localized chip formation process has been studied and understood in some detail only recently. Shear localization occurs with such materials as titanium alloys, nickel-base superalloys, and hardened alloy steels, which are characterized by low thermal diffusivity, hexagonal close-packed crystal structures, and high hardness.

Figure 2 shows the various surfaces involved in the formation of shear-localized chips. The process is divided into two stages. The first involves plastic instability and strain localization in a narrow band in the primary shear zone, leading to catastrophic shear failure along a shear surface. The surface originates from the tool tip almost parallel to the cutting velocity vector and gradually curves concavely upward until it meets the free surface. In the second stage, a gradual buildup of the segment (a low-deformation process) is caused by the flattening of the wedge-shaped work material ahead of the advancing tool. Initial contact between the segment being formed and the tool face takes place at the apex of the tool and is of extremely short duration. The contact increases as the flattening progresses. There is almost no relative motion between the bottom surface of the chip segment being formed and the tool face until near the end of the flattening stage. The chip segment being formed pushes gradually against the previously formed chip segment. The contact between the segment being formed and the one before it shifts gradually, beginning close to the work surface and gradually shifting toward the tool face as flattening progresses. As the upsetting of the segment being formed progresses, intense concentrated shear bands (white etched bands) appear between the segments in a longitudinal midsection of a chip (the regions are marked by arrows in Fig. 3b). These bands are caused by the buildup of stresses in the primary zone and are actually formed during this upsetting stage.

Once shear-localized chips are formed above a certain speed, they persist with increases in speed. No further transition into different chip forms occurs, at least up to ~30 000 m/min (~100 000 sfm).

Because they are easier to dispose of, shear-localized chips are preferable to continuous chips, especially at higher speeds where individual segments of a chip are completely isolated. Formation of the shear-localized chip, however, has not been accompanied by any rapid reduction of tool wear at high speeds. Whether or not formation of this chip reduces the forces and stresses on the tool has yet to be proved.

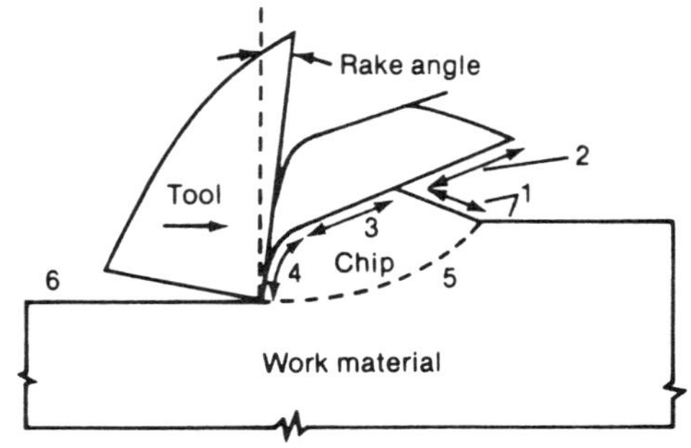


**Fig. 2** Schematic of the shear-localized chip formation process that occurs in the high-speed machining of certain materials. 1, undeformed surfaces; 2, part of the catastrophically shear-failed surface separated from the following segment due to intense shear; 3, intense shear band formed due to catastrophic shear during the upsetting stage of the segment being formed; 4, intensely sheared surface of a segment in contact with the tool and subsequently slid along the tool face; 5, intense localized deformation in the primary shear zone; 6, machined surface. Source: Ref 26

With several metals and alloys, the degree of segmentation depends directly on cutting speed (Ref 1). An example is AISI 4340 steel, for which continuous chips are formed at 120 m/min (400 sfm), as shown in Fig. 4(a). At 975 m/min (3200 sfm), however, completely segmented and detached chips are formed (Fig. 4b). Similarly, Inconel 718, a nickel-base superalloy, forms relatively continuous chips below 60 m/min (200 sfm), but within the range of 60 to 120 m/min (200 to 400 sfm), segmentation begins. At higher speeds, severe detachment occurs (Fig. 5). Titanium alloys such as Ti-6Al-4V are unique in that they form segmented chips at all speeds regardless of their heat treatment conditions.

**Analytical modeling** of the chip formation process has been conducted in high-speed machining within the framework of continuum mechanics (Ref 28-33). The theoretical and metallurgical features of chip

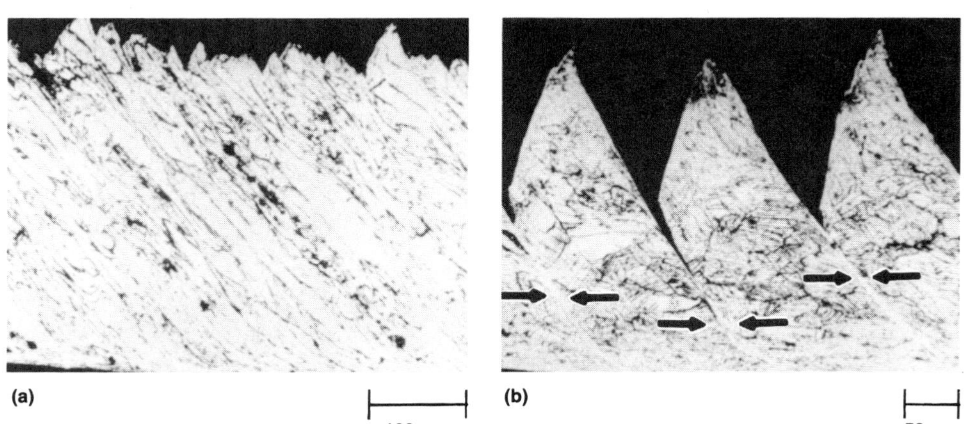


**Fig. 3** Examples of continuous (a) and segmented (b) chip formation. Arrows indicate areas of shear localization. Source: Ref 1

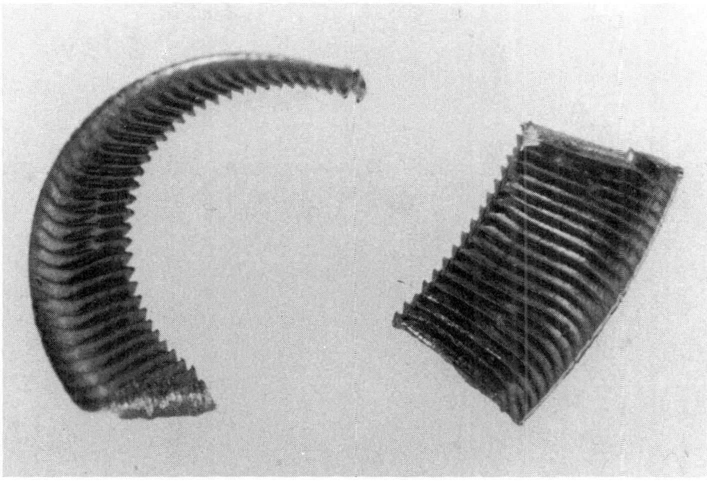

(a)

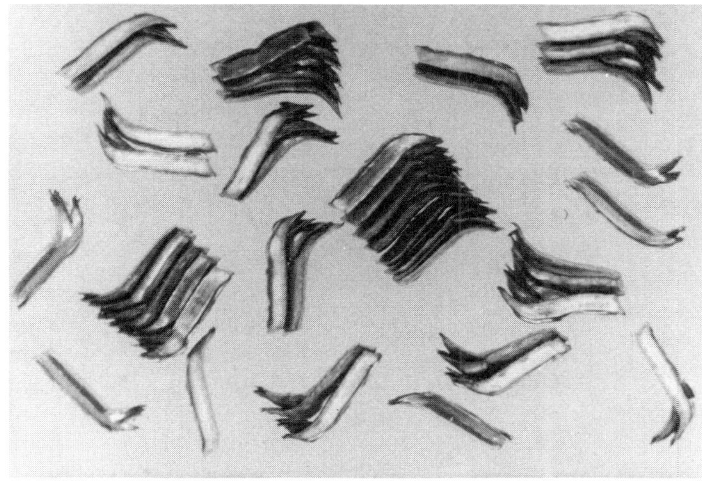

(b)

**Fig. 4** Effect of cutting speed on chip formation of AISI 4340 steel. (a) Cutting speed of 120 m/min (400 sfm). (b) Cutting speed of 975 m/min (3200 sfm). Source: Ref 1

formation have been organized into a detailed engineering analysis, the main features of which are:

- Chip geometry and morphology
- Kinematics
- Deformation zones
- Cutting forces

The theory applies both to continuous and segmental types of chips and is based on constitutive laws for the machined materials. The continuous, ribbonlike chip can be described in terms of shear lamellae joined by intense, thin shear bands. The thickness of a lamella depends on how far a crack can propagate before being arrested by the plastic state of the material at the crack tip. This distance is a material property and does not depend on cutting speed. Initial confirmation of this prediction is obtained in machining aluminum. Chips in the speed range of 150 to 4500 m/min (500 to 15 000 sfm) show virtually no change in lamellar thickness with speed. The metallurgy associated with shear fracture is important. The theoretical work strongly suggests that a miniature shear crack in front of the tool edge is essential for the observed chip formation.

Analysis of the transition from the continuous-type chip to the segmental-type chip as a function of cutting speed has led to the formulation of a stability principle, based on the properties of the stress-strain-strain rate surface under adiabatic deformation conditions. Such surfaces exhibit a locus of instability expressed by:

$$\frac{\delta\tau}{\delta\gamma} = 0$$

where $\tau$ is stress and $\gamma$ is strain for various strain rates (Fig. 6). Because the strain and velocity conditions in metal cutting are a coupled set of variables, a speed can be identified at which the chip formation mechanism becomes unstable and chip segmentation begins. Within the range of practical cutting speeds, the onset of chip segmentation depends primarily on speed through influence of the latter on strain rate. In addition, thermal and mechanical properties, for example, hardness, play a significant role. The onset of segmentation changes the character of the cutting force and the stress distribution on the tool face.

In addition to the modeling just described, a two-dimensional finite-difference computer program has been investigated for modeling high-speed machining processes (Ref 28). The code solves the basic equations of continuum mechanics throughout a fixed mesh to obtain the deviatoric stress components, hydrostatic pressure, mass density, material velocity components, and internal energy as functions of space and time. Three calculations using the code have been performed—one at 60 000 m/min (200 000 sfm) and two at 15 000 m/min (50 000 sfm). In the latter calculations, one workpiece was twice as hard as the other.

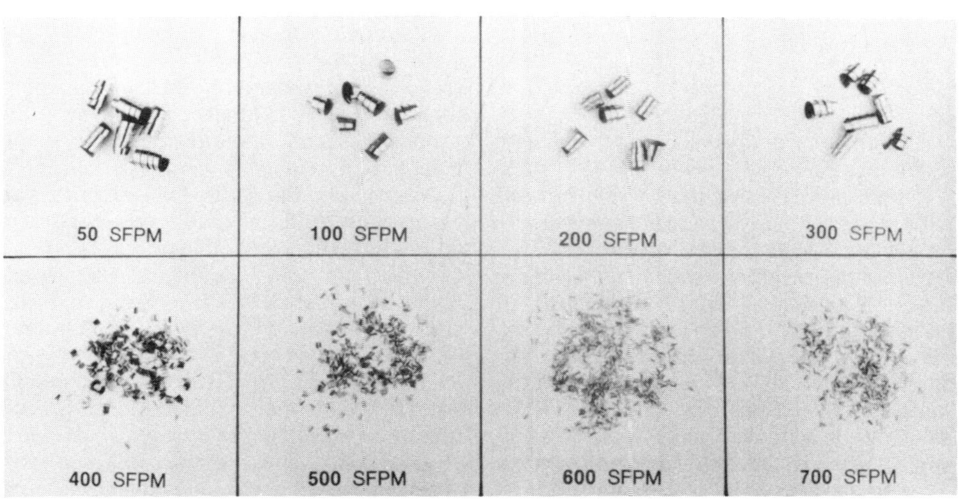


**Fig. 5** Inconel 718 chips formed at increasingly higher cutting speed. Units are given in surface feet per minute. Source: Ref 1

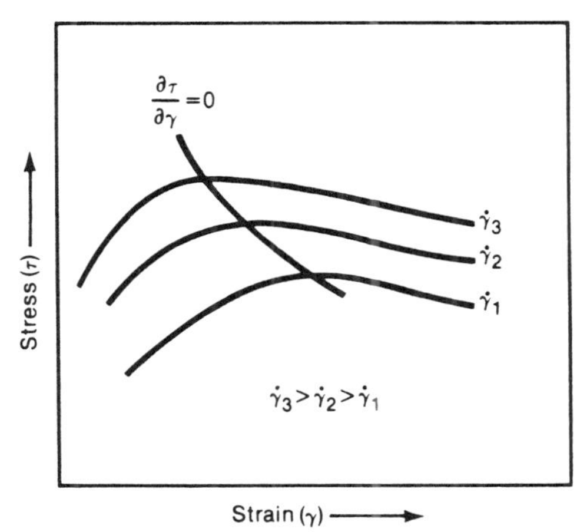


**Fig. 6** Variation of shear stress, $\tau$, versus shear strain, $\gamma$, at different strain rates $\dot{\gamma}$. The focus of shear instability is given by $\partial\tau/\partial\gamma = 0$. Source: Ref 1

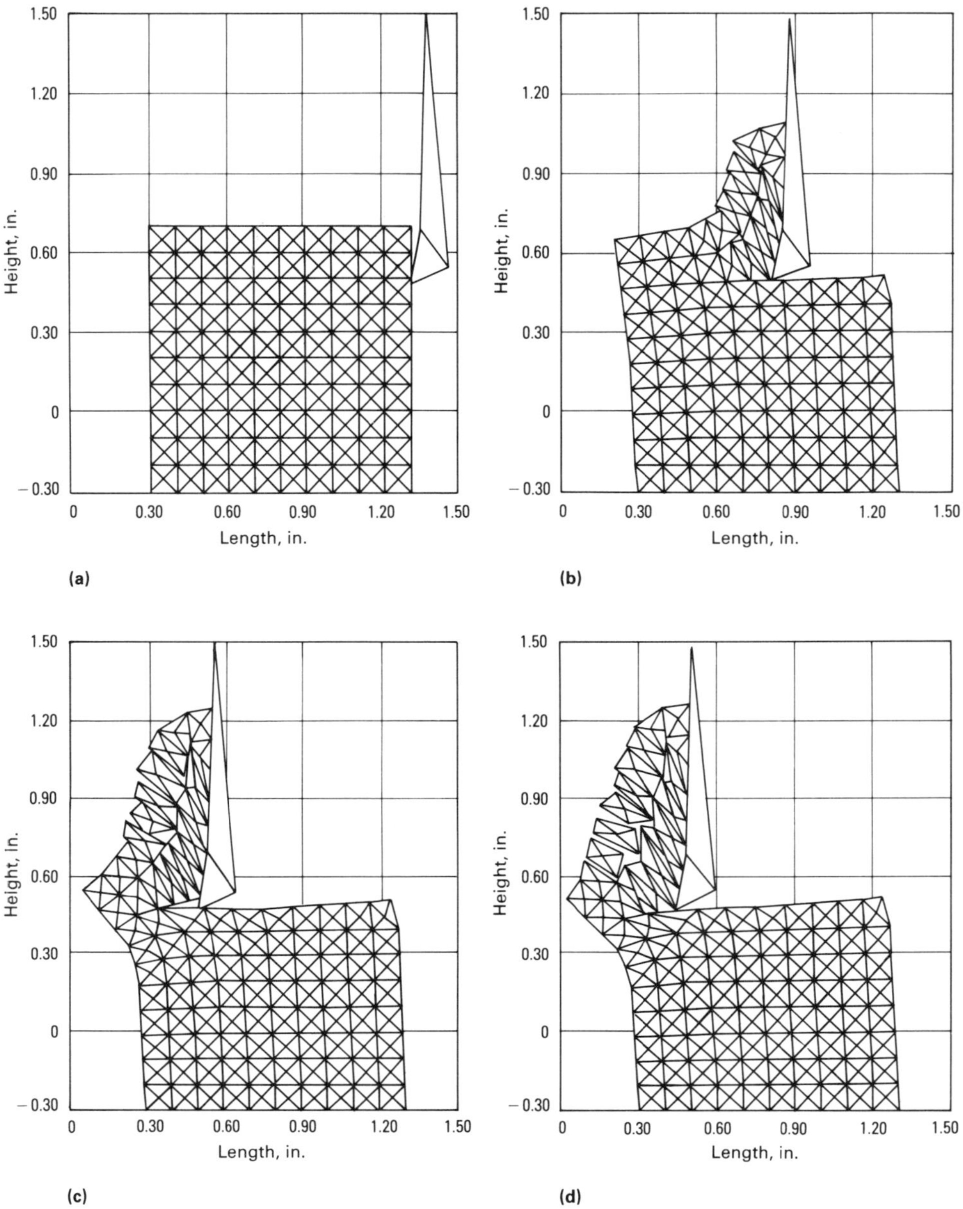


**Fig. 7** Results of a finite-element analysis used to simulate chip segmentation during high-speed machining. The results correspond to a cutting speed of 1800 m/min (6000 sfm) and a rake angle of 5°. (a) Initial geometry, time = 0.0 s. (b) Geometry at 0.005 s. (c) Geometry at 0.008 s. (d) Geometry at 0.0085 s. Source: Ref 34

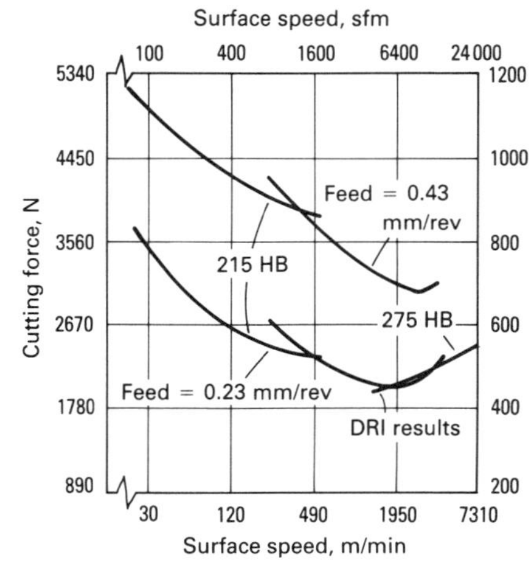


**Fig. 8** Variation of cutting force with speed for AISI 4340 steel. Source: Ref 1

At the highest speed used, the workpiece behaves like a fluid; material strength effects become more important as cutting velocity is reduced. In addition, the material yields to a smaller depth beneath the cutting tool as velocity is decreased. The depth of yielding is further reduced by doubling the yield strength of the workpiece (simulating strain hardening). It is significant that the results of computer calculations are consistent with trends observed in practice. In addition to use of the finite-difference method, a partially successful finite-element analysis has been made of a simplified, continuous-chip, orthogonal machining process. Figure 7 shows the results of a finite-element analysis used to simulate segmented chipping.

Simple constitutive equations can be used to describe the plastic-flow behavior of materials over wide ranges of strain, strain rate, and temperature (Ref 28). Of the three materials studied—6061-T6 aluminum, Ti-6Al-4V, and AISI 4340 steel—the steel exhibits the greatest strain-hardening rate. Orthogonal machining tests conducted on the same three materials at very low speeds have been used to determine strain distributions in and adjacent to the deforming chips. A modified thermal-mechanical criterion of instability predicts that the critical shear strains for aluminum and steel are very much higher than those for the titanium alloy at moderate cutting speed; the differences in these strains increase markedly with increasing speed. The strain rate for transition from a simple mechanical to a thermally induced shear failure process cannot yet be determined from first principles.

## High-Speed Machining Parameters

**Cutting Force Versus Speed.** It has been confirmed that cutting force decreases with increasing speed until a minimum is reached at a speed characteristic of the given workpiece material. Beyond this characteristic speed, the force tends to slowly increase. For example, force for AISI 4340 steel continues to decrease with increasing speed until about 1500 m/min (5000 sfm), at which point the force begins to increase with speed (Fig. 8). Most of these data were generated on two lathes—a high-speed 500 mm (20 in.), 110 kW (150 hp) engine lathe and an 800 mm (32 in.), heavy-duty, 95 kW (125 hp) engine lathe (Ref 35). The two curves at each feed in Fig. 8 represent data from each lathe. Additional confidence in these results is provided by force measurements for AISI 4340 steel machined at the Denver Research Institute (DRI) during ballistic experiments.

As shown in Fig. 8, the DRI data lie on an extrapolation of the curve generated at a feed of 0.23 mm/rev (0.009 in./rev). Similar to AISI 4340 steel, aluminum 6061-T6 exhibits an initial decrease in force with increasing speed up to about 3000 m/min (10 000 sfm), beyond which the force increases slightly. In contrast to the results for AISI 4340 steel and aluminum, the cutting force for titanium is relatively unaffected by speed. The decrease in force with speed observed for several materials does not mean a lowering of horsepower requirements; however, one can take advantage of

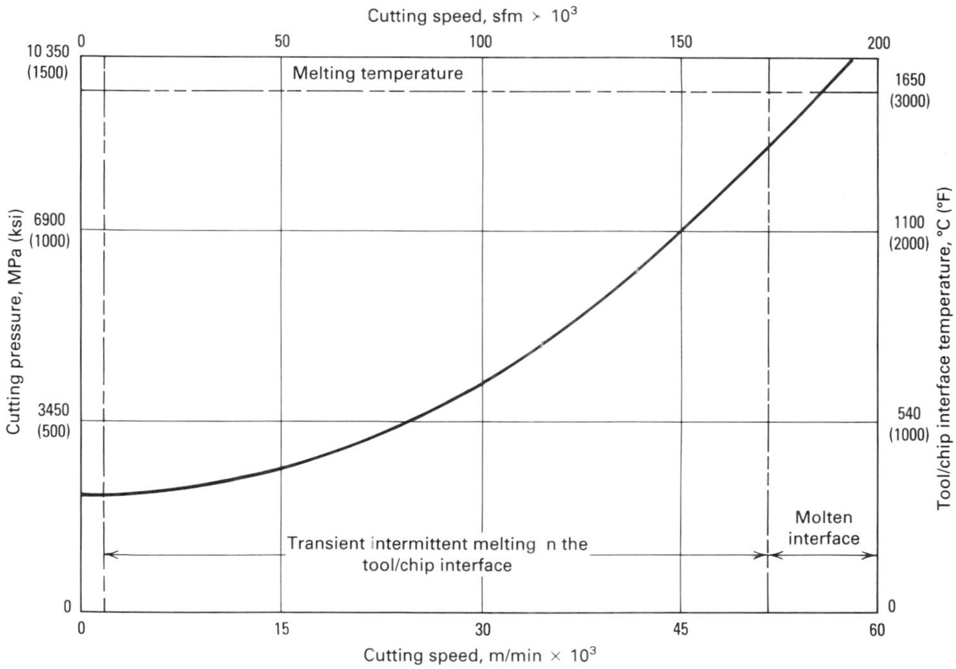


**Fig. 9** Variation of chip/tool interface temperature with speed for AISI 4340 steel. Adapted from Ref 36

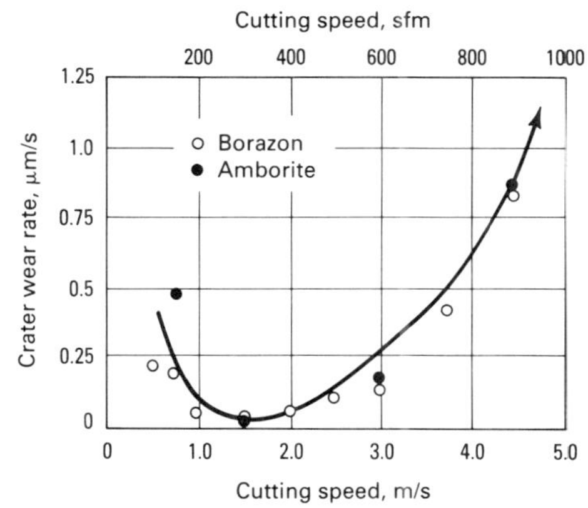


**Fig. 10** Effect of cutting speed on the wear rate of cubic boron nitride tooling. Workpiece: AISI 4340 steel (35 HRC). Source: Ref 37

the lower forces at high speed (especially with lighter chip loads) to machine accurately both thin webs and unsupported sections of a part. The practical top cutting speed for aluminum does not appear to be limited by cutting tool wear; spindle speed and power are the controlling factors. It should be obvious, however, that high spindle speeds alone do not ensure high metal removal rates. High feed rates and adequate depths of cut are also needed. Depending on the types of cut being made (straight or contoured), the speed of response of the machine tool and its control may be critical.

**Cutting Temperature Versus Speed.** The evidence indicates that the chip/tool interface temperature increases with speed, approaching the melting point of the work material (Fig. 9), rather than falling off at very high speeds, as had been claimed by Salomon (Ref 36). Because the melting temperature of aluminum alloys is low (~540 °C, or 1000 °F) and well below the temperature limitations of carbide and coated carbide tools, the top cutting speed for aluminum alloys appears unlimited from a cutting tool point of view. Table 1 lists the softening points of tool materials, along with the melting points of some common workpiece materials.

**Surface Finish Versus Speed.** There are indications that surface finish tends to improve with increasing speed, but these results are not conclusive. Some of the reasons could be due to:

- Burnishing action due to tool wear
- Smaller chip load used
- Dynamic response of the cutter

## Cutting Tool Materials (Ref 37)

Contrary to predictions based on early work in high-speed machining (see the section "Historical Background" in this article), there does not appear to be an upper speed range at which cutting temperatures decrease with increasing speed. The general observation is that as speed is increased, the temperature approaches the melting point of the workpiece material as a maximum. Because tool wear is usually strongly dependent on temperature, tool wear is the major factor limiting cutting speed, the main exception to this being aluminum machining.

**Wear Mechanisms.** A 10-year research program on tool materials for high-speed machining concluded that the two major wear mechanisms associated with high-speed machining are high-speed chemical dissolution wear and high-speed diffusion-limited wear (Ref 37). In the range of cutting speeds used in high-speed machining, the chemical dissolution of the tool material into the workpiece is the most important contributor to wear. In essence, the tool material dissolves into the flowing chip. The tool material that is the most resistant to dissolution exhibits the least wear.

The second consideration is that of diffusion-limited wear. As the cutting speed is increased, the cutting temperature rises to a level at which seizure of the chip material occurs everywhere on the tool face. This layer of adherent material becomes saturated with tool constituents and serves as a diffusion-boundary layer, reducing the rate of transport of tool material into the chip and consequently the wear rate. The wear phenomenon becomes increasingly diffusion-limited, and the observed dramatic decrease in wear occurs with increasing speed. Because the diffusivity increases exponentially with temperature, a further increase in cutting speed beyond the speed for minimum wear produces a rapid increase in wear rate, as shown in Fig. 10.

**Cutting Tool Selection.** Consideration of the wear mechanisms of tool materials during high-speed machining suggests three possibilities for tool development:

- Pick a tool material that is so chemically stable with respect to the workpiece that chemical dissolution of the tool does not occur to a significant extent, even at the melting point of the workpiece
- Promote the transition to the diffusion-limited wear regime
- Isolate the tool from the workpiece rather than a modification of the tool material

**Table 1 Softening points of tool materials and melting points of workpiece materials**

| Tool material | Softening point °C | Softening point °F | Workpiece material | Melting point °C | Melting point °F |
|---|---|---|---|---|---|
| High-speed steel | 600 | 1110 | Aluminum | 600–660 | 1110–1220 |
| Tungsten carbide | 1100 | 2010 | Superalloys | 1300–1400 | 2370–2550 |
| Aluminum oxide | 1400 | 2550 | Steel | 1450–1500 | 2640–2730 |
| Cubic boron nitride | 1500 | 2730 | Titanium | 1600–1650 | 2910–3000 |
| Diamond | 1500 | 2730 | Zirconium | 1800–1850 | 3270–3360 |

Source: Ref 37

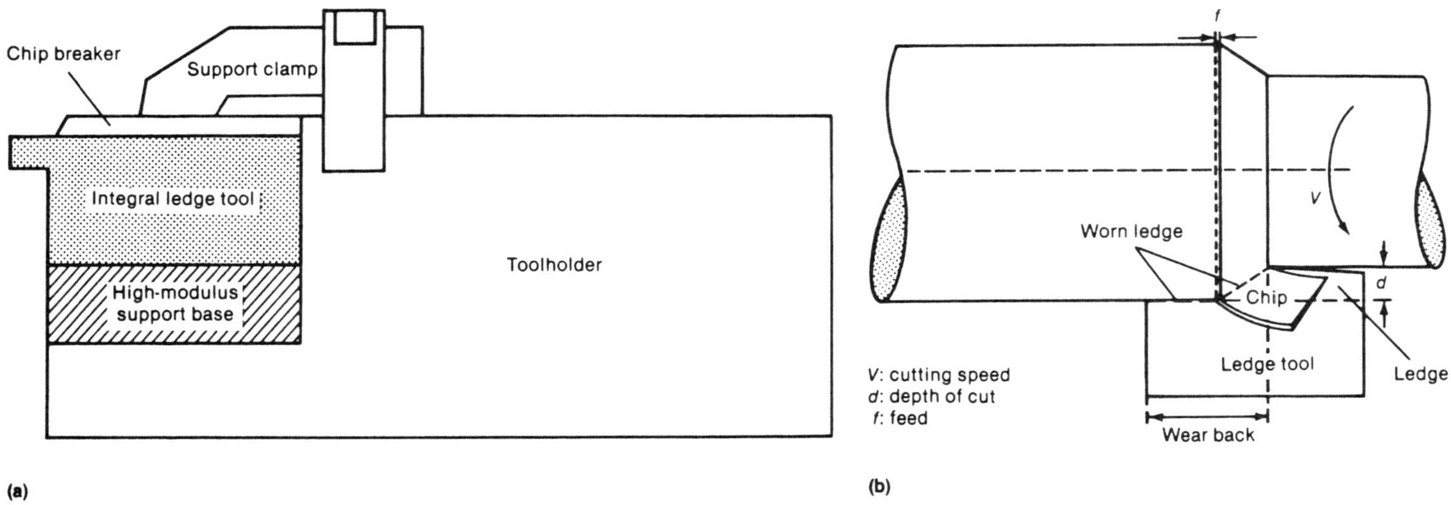


**Fig. 11** Schematic of ledge tool, which is designed to increase productivity during the high-throughput machining of titanium. (a) Ledge tool mounted on a conventional toolholder. (b) A turning operation using a ledge tool. Source: Ref 27

itself. If a protective layer could be introduced between the tool and the chip, transport of the tool constituents into the chip could be prevented. The use of viscous lubricants in the profile milling of titanium has proved successful (Ref 38)

**Tool Systems for Aluminum Alloys.** In aluminum alloys, the cutting temperature is limited by the low melting point and high thermal conductivity of aluminum. Chemical effects are minimal, and wear is primarily a result of the abrasion of the tool material by hard second-phase particles. Abrasion resistance increases with the hardness of the tool material; high-speed steel tooling and cemented carbide tooling are suitable for machining most structural alloys, and polycrystalline diamond is preferred for the highly abrasive cast aluminum-silicon (10 to 20% Si) alloys (see the article "Machining of Aluminum and Aluminum Alloys" in this Volume for a discussion of machining high-silicon aluminum alloys). Existing tool materials are adequate for machining aluminum alloys at any conceivable speed, with spindle speed and horsepower design limitations setting the upper limit on cutting speed.

**Tool Systems for Steel.** The oxides are the only potential tool materials that are not limited by their chemical stability. Therefore, the most promising area for tool material development is in improving the toughness and flow strength of the oxides. A second possibility involves the development of tool materials with improved hot strength. The development of tool materials other than cubic boron nitride (CBN) with high hot strength in the range of 1300 to 1400 °C (2400 to 2600 °F) might allow a transition from dissolution-limited to diffusion-limited wear, with a corresponding increase in tool life. Finally, the wear of CBN in the machining of steels of moderate hardness (35 to 50 HRC) should be investigated to determine whether a transition to a low-wear regime (similar to that which has been observed in hard steels) occurs at sufficiently high cutting speeds. If so, machining in the range of 1200 m/min (4000 sfm) and above may be feasible.

**Tool Systems for Superalloys.** The recommendations are similar to those for steel. The oxides are quite chemically stable with respect to nickel and cobalt, making the development of tough oxide materials a priority. In addition, the transition to diffusion-limited wear is known to occur at high speeds; therefore, any new high hot strength compositions will likely find application in the machining of superalloys. A possible example is the class of tool materials based on silicon nitride ($Si_3N_4$) and alloys of $Si_3N_4$ and aluminum oxide ($Al_2O_3$), referred to as SiAlON (see the article "Ceramics" in this Volume). These tool materials are very effective in machining nickel-base alloys at high speeds. In light of the relatively poor chemical stability of these materials, it is suspected that they represent the second example (in addition to CBN) of a tool material that has sufficient hot strength to enable the very high speed wear transition.

**Tool Systems for Titanium Alloys.** Titanium has low thermal conductivity, low specific heat, and a high melting point. These properties ensure that cutting temperatures will be high at even moderate cutting speeds. In addition, titanium is highly chemically reactive with all known tool materials, causing rapid wear. It is quite likely that the most wear resistant materials—tungsten carbide and diamond—have already been identified. Therefore, the most promising area for investigation is the development of effective lubrication techniques to reduce the interaction between the tool and the chip. In addition, new tool geometries, such as the ledge tool described below, have also increased the productivity of machining titanium alloys.

## Alternative Cutting Tool Geometries (Ref 27, 28)

Rapid tool wear remains a problem in the machining of titanium and other difficult-to-machine alloys, even though cutting speeds for titanium have been recently increased three- to fivefold through judicious choice of cutter grades and geometries, fluids, and machining parameters. Partial solutions to the tool life problem lie in new tool geometries and the use of rotating cutters.

**The Ledge Tool.** The concept underlying the ledge tool (Fig. 11) is very simple: The tool contains a ledge that is allowed to wear away at a controlled rate with only minimal increase in force. Thus, tool wear has not been reduced, but tool life has been greatly increased. The size of the ledge (overhang) equals the depth of cut desired, and its thickness equals the ultimate flank wear width to be tolerated. In turning, the square tool with the ledged side is brought against the workpiece so that clearance is available between the edge of the ledge and the finished surface of the workpiece (that is, the end cutting edge angle, $\leqq 1°$), as shown in Fig. 11. Because the depth of cut is the same or less than the width of the ledge, only the ledge portion of the tool does the cutting and wearing. The ledge wear back is due to a combination of flank wear and microchipping wear. With this tool, cutting speeds for titanium alloys can be increased five times over conventional speeds, with long tool life (~30 min) and good finish (Fig. 12). The sparking that accompanies tool wear at high cutting speeds has been eliminated by surrounding the tool/workpiece interface with inert gas ($N_2$) or by submerging the workpiece in a cutting fluid. A combination of a flood lubricant and an inert gas also serves to eliminate sparks and to keep the workpiece cool.

Ledge tools have also been evaluated in the face milling of forged Ti-6Al-2Sn-4Zr-

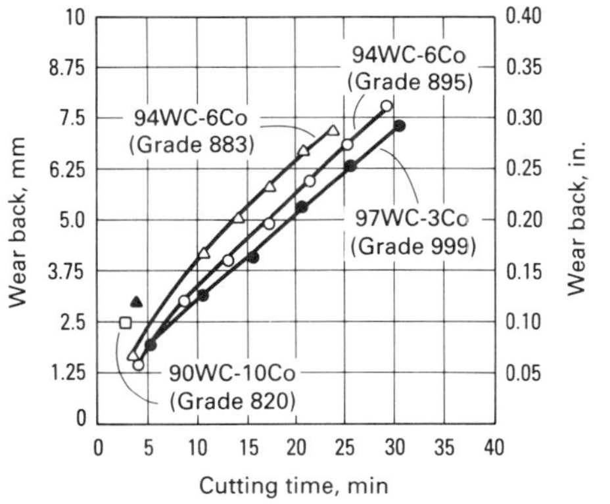


**Fig. 12** Variation of ledge wear back with cutting time for different carbide grades when turning Ti-6Al-4V at 180 m/min (600 sfm) with a depth of cut of 0.75 mm (0.030 in.) and a feed of 0.023 mm/rev (0.009 in./rev). Source: Ref 27

2Mo. The results have been comparable to those in turning except the rate of ledge wear back was about three times faster in milling; on a microscopic level, the mode of tool wear is the same. Figure 13 shows the variation of ledge wear with cutting time and volume of material removed.

Another incremental approach for increasing tool life when machining titanium alloys is a new cutting geometry comprised of a high clearance angle (10 to 15°) together with a high negative-rake angle (−10 to −15°). This geometry will also allow the use of a conventional insert on a modified toolholder (Ref 39).

**Rotary tool machining** is another technique that holds promise for extending tool life. In this method, a tool with a circular cutting edge is allowed to rotate about its own axis, either self-propelled by the cutting process or driven externally at the desired speed. As cutting proceeds, new portions of the cutting edge are continuously brought into contact with the workpiece at the cutting zone. Thus, increased tool life can be anticipated because of lower temperatures and reduced chemical reactions at the moving chip/tool interface, and reduced cutting forces can be anticipated because of increased options for modifying the chip formation.

## Applications of High-Speed Machining (Ref 26)

High-speed machining is used in the defense and airframe industries to manufacture aircraft engine propulsion components and in the automobile industry. When high-speed machining is accompanied by higher feed rates and spindle power, the higher spindle speeds allow higher removal rates. Increased productivity, however, necessitates high-speed, high-power, compact spindle designs; low-inertia feed tables; fast feed drives; quick-response numerical control; and a totally integrated machining system.

A typical integrated machining system consists of multiple machining cells that will fabricate large machine parts under hierarchical computer control in an effective, cost-efficient manner (Ref 40). Benefits of the integrated machining system include:

- Reduced labor requirements
- Improved throughput with the application of high-speed and high-throughput machining
- Enhanced production flexibility and reduced work-in-process through establishment of a serial production environment
- Improved product quality and enhanced industrial base

Figure 14 shows an isometric view of an integrated machining system. Such a system can take up to more than 9000 $m^2$ (100 000 $ft^2$) of floor space.

**Airframe and Defense.** Most airframe manufacturers have implemented high-speed machining. Its primary application is in end milling with small-size cutters. Aluminum alloys are the common work materials used, so tool wear is not a limitation, especially with carbide cutters. The ideal candidates for high-speed machining are parts whose machining time is a significant fraction of the floor-to-floor time. The ideal cuts are long, straight ones that enable the use of high feed rates and consequently high removal rates. Although some parts may fall under this category, there are others that are suitable for high-speed machining but require extensive pocketing and complex contouring. Both pocketing and contouring can involve frequent accelerations and decelerations in feed rates.

Tool-changing time can be reduced by using as few cutters as possible, eventually using only one, smallest-diameter cutter capable of generating all the radii on the part. Modifying the design of the part may enable the use of high-speed machining and may be desirable if the modification can be achieved without compromising the part performance specifications. Further, although high-speed steel tools may be satisfactory for some applications, carbide cutters will not only extend tool life but also provide about three times the stiffness, which is essential for machining long, thin webs with slender end mills.

The integrated machining system shown in Fig. 14 is used to machine both aluminum and titanium. For the high-speed machining of aluminum parts, metal removal rates of 3300 $cm^3$/min (200 $in.^3$/min) are possible using machines with up to 55 kW (75 hp) spindle drives and spindle speeds approaching 20 000 rev/min. For the high-throughput machining of titanium alloys, metal removal rates of the order of 165 $cm^3$/min (10 $in.^3$/min) using 95 kW (125 hp) spindle drives and up to 833 rev/min spindle speeds are possible. Many high-speed machining applications, however, do not require such machine capabilities. For example, a 15 kW (20 hp), 20 000 rev/min spindle is used to machine A7 wing spars made of 7057-T6 aluminum (Ref 26). The feed rate is 15 000 mm/min (600 in./min) on long, external tapered flanges and 7500 mm/min (300 in./min) in pocket areas; the metal removal rate is 1300 $cm^3$/min (80 $in.^3$/min).

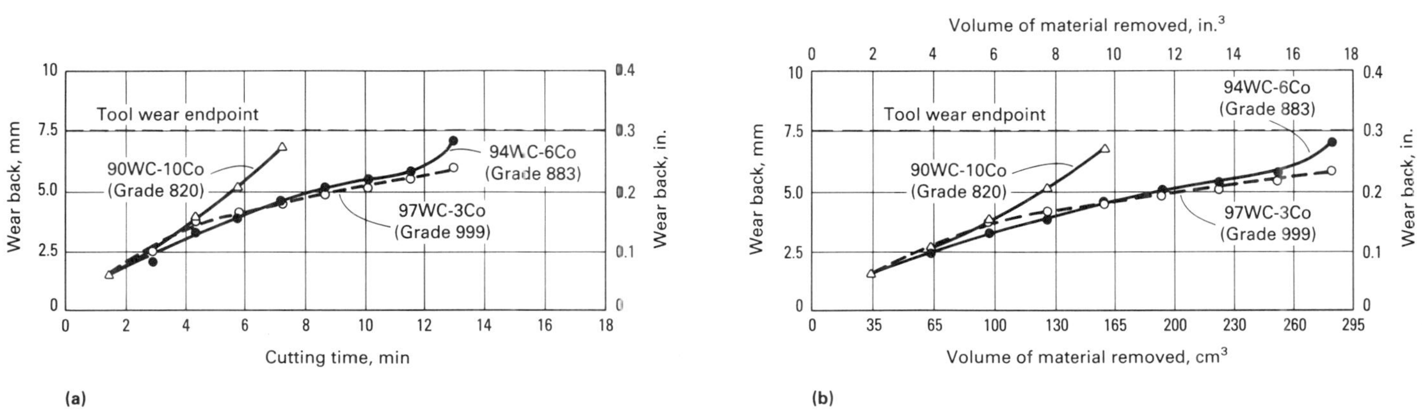


**Fig. 13** Variation of ledge wear back with cutting time (a) and with the volume of material removed (b) for different carbide grades when face milling a Ti-6Al-2Sn-4Zr-2Mo alloy (36 HRC). Tool: 0.75 mm × 1.0 mm (0.030 in. × 0.040 in.). Cutting speed: 155 m/min (515 sfm). Chip load per tooth: 0.23 mm (0.009 in.). Axial depth: 0.75 mm (0.03 in.). Source: Ref 28

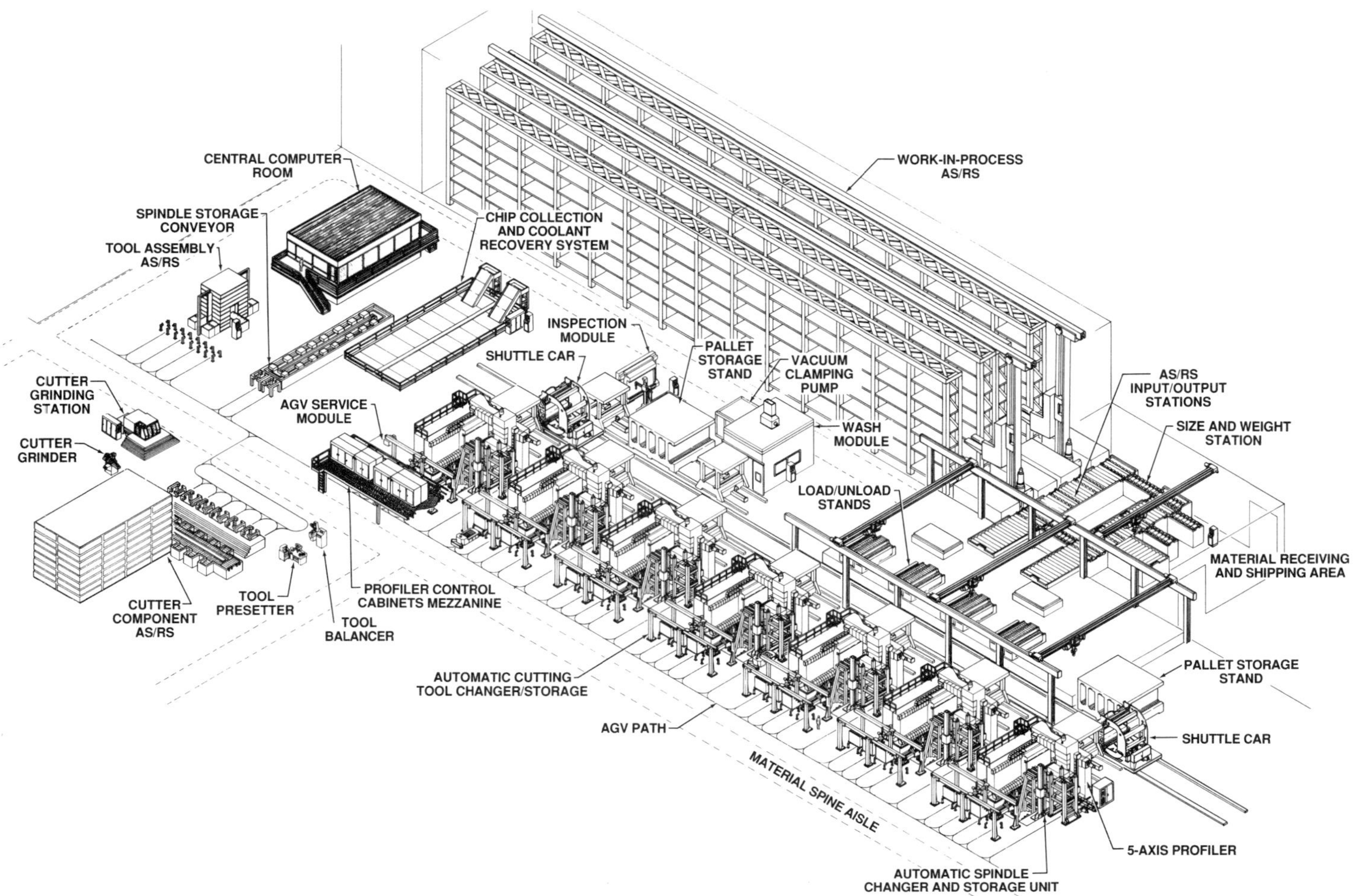


**Fig. 14** Integrated machining center for the high-speed and high-throughput machining of aluminum and titanium, respectively. AGV, automated guide vehicle; AS, automated storage; RS, retrieval system. Source: Ref 40

Retrofitted machining centers with 20 kW (26 hp), 18 000 rev/min high-speed spindles are used for the high-speed machining of 7075-T6 aluminum parts for the Trident missile (Ref 26). The worktable has a feed capacity of 5000 mm/min (200 in./min). The high-speed system incorporates such safety features as a double lock for the toolholder and sensors that can shut down the machine quickly if tool breakage creates an imbalance.

A major commercial airline manufacturer uses six-axis machining centers for the contour milling of aluminum honeycomb for engine nacelles (Ref 26). The system has a 3.3 m (11 ft) diam rotary table capable of a full 360° rotation. The spindle is mounted on the gantry and can tilt 240° and rotate 360°. A unique feature of this system is the use of a graphite-fiber reinforced plastic ram. This ram provides the required stiffness and reduced weight necessary for the high feed rates of low-inertia parts. The high-speed machining systems utilize high-speed (24 000 rev/min) 11 kW (15 hp) spindles.

**Aircraft Engine Propulsion.** Nickel-base superalloys and titanium alloys are the work materials most often used in aircraft engine propulsion components. These cause rapid tool wear at high speeds, which constitute a major limitation in the high-speed machining of propulsion parts. Until recently, the cutting speeds possible with nickel-base superalloys were about 30 m/min (100 sfm) with carbide cutters and about 9 m/min (30 sfm) with high-speed steel tools. The development of CBN and ceramics such as hot-pressed $Al_2O_3$ plus TiC, SiAlON, and silicon carbide (SiC) whisker-reinforced $Al_2O_3$ has made possible an increase in cutting speeds from 30 to 180 m/min (100 to 600 sfm) for machining nickel-base superalloys. These ceramic tool materials are much tougher and more consistent in performance than the ceramics introduced in the 1950s. SiAlON and SiC whisker-reinforced $Al_2O_3$ are recommended for roughing, and hot-pressed $Al_2O_3$-TiC for finishing applications. Although new, tougher ceramic tool materials, perhaps based on ceramic composites, will undoubtedly be developed in the future, it is unlikely that cutting speeds will exceed 600 m/min (2000 sfm), because tool wear will continue to be a limitation. As described earlier in this article, innovative tool designs and geometries, such as the ledge tool, have improved productivity in titanium machining.

**Automobile Industry.** High-speed machining in the automobile industry is performed on gray cast iron and aluminum alloys, especially the high-silicon type. Silicon nitride ceramic and polycrystalline diamond are the important tool materials that permit higher speed with longer tool life. Gray cast iron can be machined at a speed of about 1500 m/min (5000 sfm) with $Si_3N_4$ tools, and aluminum alloys with high silicon content (10 to 20%) can be machined at about 750 m/min (2500 sfm) with polycrystalline diamond tools.

In working with these materials, the current high-speed machining systems need to provide more power and stiffness and improved chip-handling means, controls, and safety features. High-speed machining with these materials may require spindle power from 150 to 375 kW (200 to 500 hp). Chip removal rates can reach 16 000 $cm^3$/min (1000 $in.^3$/in.), necessitating efficient chip disposal systems. Because products are mass produced in this industry, the nonmachining time should be minimal.

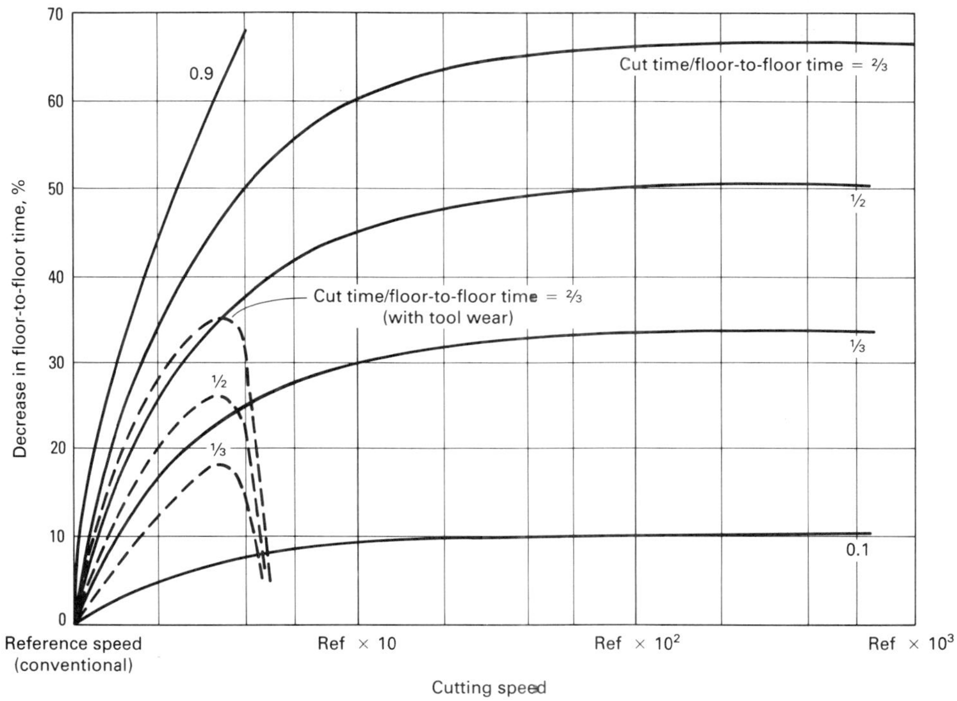


**Fig. 15** Variation of percent decrease in floor-to-floor time with cutting speed. See text for details. Source: Ref 26

## Implementing High-Speed Machining (Ref 26)

There are several factors to be considered by companies planning to employ high-speed machining systems. Whether high-speed machining is economically appropriate to the application, the relative value of various systems, and how well the system can be integrated into existing operations must all be evaluated. Other areas to consider are the research and development support required, system reliability and maintenance, overall safety aspects, general acceptance on the manufacturing shop floor, capital and other investment costs, the skills required, corporate goals, and the financial health of the company. The overall goal should be to improve productivity, reduce costs, and produce parts of a given size, shape, finish, and accuracy at competitive cost.

Productivity and overall costs depend on cutting time, noncutting time, labor, and overhead. High-speed machining can decrease cutting time by increasing cutting speed. Noncutting time can be decreased by the automatic loading and unloading of parts, automatic tool changing, in-process inspection, in-process sensing, and adaptive control. Labor costs can be decreased because with high-speed machining fewer operators are needed to work on fewer, more efficient machine tool systems. Similarly, overhead costs can be reduced by operating fewer, more efficient machine tools on two or more shifts and during holidays and at night.

In order to evaluate the influence of cutting speed on productivity, floor-to-floor time can be regarded as the sum of the cutting time and the noncutting time. Figure 15 shows the percentage decrease in floor-to-floor time with an increase in cutting speed for different ratios of cutting time to floor-to-floor time, using conventional, currently used speeds of whatever the value may be for a given material and process as a base. When the cutting time is a significant fraction of the floor-to-floor time and when tool wear at high speed is not significant (solid lines, Fig. 15), the cutting speed can be increased considerably to effect a significant reduction in floor-to-floor time. If, however, this ratio is low (bottom curve, Fig. 15), an increase in cutting speed by even an order of magnitude or more will result in only a marginal decrease in floor-to-floor time. In this case, unless the noncutting time could be decreased significantly, high-speed machining would not be advantageous.

When an increase in cutting speed does not contribute to a significant reduction in floor-to-floor time and when tool wear is significant at high speeds, as in the machining of titanium alloys (dashed line, Fig. 15), high-throughput machining can be adopted to reduce noncutting time and to increase productivity. Similarly, where heavier cuts can be made on the part without affecting either the part finish and accuracy requirements or the cutting tool performance, greater depths of cut (as in high removal rate machining) are recommended instead of higher speed (see the article "High Removal Rate Machining" in this Volume).

**Future Needs (Ref 1).** High-speed machining can be cost effective only if other aspects of machining, including reduction of noncutting times and labor costs, can also be improved. For this reason, automated machining, such as the integrated machining system discussed earlier, is receiving much attention. Dynamic in-process inspection is an integral part of automated machining. Sensors and diagnostics have been identified as critical to effective in-process inspection. Recently, proximity sensors for measuring tool wear, vibration sensors for measuring tool-touch (between cuts), and lasers for measuring workpiece dimensions have been studied. The integrated machining system shown in Fig. 14 utilizes a coordinate-measuring machine as its inspection module.

Some of the specific needs for the increased use of high-speed machining are:

- Integrated sensing techniques for monitoring machining processes
- Sensors and diagnostics for detecting tool breakage (including incipient tool breakage)
- Linkage of signal analysis to tool wear mechanisms and machining variables
- Self-teaching adaptive control systems
- Faster response in machine tool controls when machining aluminum
- Better cutting tools (composition and geometry) for machining titanium
- Water-base cutting fluids for machining titanium
- More reliability and predictability in cutting tools
- Greater machine tool rigidity for increased productivity
- More notch-resistant cutting tools for superalloys

In addition to the above-mentioned technical needs, significant reductions in noncutting time can be achieved by procedures that to a large extent involve improved management. These procedures include set-up, part load/unload, queing, chip cleanup, tool change, maintenance, scheduling, and general operator efficiency. Until improvements in these procedures are achieved, cutting times will remain only a small fraction of the overall manufacturing sequence, and the full advantages of high-speed machining will not be realized.

## ACKNOWLEDGMENT

ASM INTERNATIONAL is grateful to D.G. Flom (retired: General Electric Corporate Research & Development) and R. Komanduri (National Science Foundation) for their valuable contributions to this article. Much of this article is based on the work of Flom and Komanduri conducted from 1980 to 1985.

## REFERENCES


1. D.G. Flom, High-Speed Machining, in *Innovations in Materials Processing*, G. Bruggeman and V. Weiss, Ed., Plenum Press, 1985, p 417-439
2. B.F. von Turkovich, Influence of Very High Cutting Speed on Chip Formation Mechanics, in *Proceedings of NAMRC-VII*, 1979, p 241-247
3. The 12th American Machinist Inventory of Metalworking Equipment 1976-78, *Am. Mach.*, Dec 1978, p 133-148
4. R.I. King and R.L. Vaughn, A Synoptic Review of High-Speed Machining from Salomon to the Present, in *High-Speed Machining*, American Society of Mechanical Engineers, 1984, p 1-13
5. R.L. Vaughn, Ultra-High Speed Machining, *Am. Mach.*, Vol 107 (No. 4), 1960, p 111-126
6. R.L. Vaughn, "Recent Developments in Ultra-High Speed Machining," Technical paper 255, Vol 60, Book 1, Society of Manufacturing Engineers, 1960
7. R.L. Vaughn, "Ultra-High Speed Machining—Feasibility Study," Final Report, Contract AF 33 (600) 36232, Production Engineering Department, Lockheed Aircraft Corporation, June 1960
8. R.L. Vaughn, L.J. Quackenbush, and L.V. Colwell, "Shock Waves and Vibration in High-Speed Milling," Technical Paper 62-WA-282, American Society of Mechanical Engineers, Nov 1962
9. R.L. Vaughn and L.J. Quackenbush, "The High-Speed Milling of Titanium Alloys," Technical Paper MR 66-151, Society of Manufacturing Engineers, April 1966
10. R.I. King and J. McDonald, Product Design Implications of New High-Speed Milling Techniques, *Trans. ASME*, Nov 1976
11. F.J. McGee, "Final Technical Report for Manufacturing Methods for High-Speed Machining of Aluminum," Technical Registry No. 6089, Manufacturing Methods and Technology Branch (DRDMI-EAT), U.S. Army Missile Research and Development Command, Feb 1978
12. D.G. Flom, Ed., "Advanced Machining Research Program (AMRP)," Semiannual Technical Report, Air Force Contract No. F33615-79-C-5119, GE Report No. SRD-80-018, Feb 1980
13. D.G. Flom, Ed., "Advanced Machining Research Program (AMRP)," Annual Technical Report, Air Force Contract No. F33615-79-C-5119, GE Report No. SRD-80-118, Aug 1980
14. D.G. Flom, Ed., "Advanced Machining Research Program (AMRP)," Semiannual Technical Report, Air Force Contract No. F33615-79-C-5119, GE Report No. SRD-81-018, Feb 1981
15. D.G. Flom, Ed., "Advanced Machining Research Program (AMRP)," Semiannual Technical Report, Air Force Contract No. F33615-79-C-5119, GE Report No. SRD-81-062, Aug 1981
16. D.G. Flom, Ed., "Advanced Machining Research Program (AMRP)," Semiannual Technical Report, Air Force Contract No. F33615-79-C-5119, GE Report No. SRD-82-027, Feb 1982
17. D.G. Flom, Ed., "Advanced Machining Research Program (AMRP)," Annual Technical Report, Air Force Control No. F33615-79-C-5119, GE Report No. SRD-82-070, Aug 1982
18. D.G. Flom, Ed., "Advanced Machining Research Program (AMRP)," Semiannual Technical Report, Air Force Contract No. F33615-79-C-5119, GE Report No. 83-SRD-012, Feb 1983
19. D.G. Flom, R. Komanduri, and M. Lee, "Review of Past Work in High-Speed Machining," Paper presented at the TMS-AIME Meeting (Louisville, KY), The Metallurgical Society, Oct 1981
20. R. Komanduri and J. Hazra, A Metallurgical Investigation of Chip Morphology in Machining an AISI 1045 Steel at Various Speeds up to 10,100 SFPM, in *Proceedings of NAMRC-IX*, Society of Manufacturing Engineers, 1981
21. R. Komanduri and B.F. von Turkovich, New Observations on the Mechanism of Chip Formation When Machining Titanium Alloys, *Wear*, Vol 69, 1981, p 179-188
22. R. Komanduri, "Titanium—A Model Material for Studying the Mechanism of Chip Formation in High-Speed Machining," Paper presented at the TMS-AIME Meeting (Louisville, KY), The Metallurgical Society, Oct 1981
23. R. Komanduri, Some Clarifications on the Mechanics of Chip Formation When Machining Titanium Alloys, *Wear*, Vol 76, 1982, p 15-34
24. R. Komanduri and R.H. Brown, The Mechanics of Chip Segmentation in Machining, *J. Eng. Ind. (Trans. ASME)*, Vol 103, Feb 1981, p 33-51
25. R. Komanduri, T. Schroeder, J. Hazra, B.F. von Turkovich, and D.G. Flom, On the Catastrophic Shear Instability in High-Speed Machining of an AISI 4340 Steel, *J. Eng. Ind. (Trans. ASME)*, Vol 104, May 1982, p 121-131
26. R. Komanduri, High-Speed Machining, *Mech. Eng.*, Dec 1985, p 65-76
27. R. Komanduri, D.G. Flom, and M. Lee, Highlights of the DARPA Advanced Machining Research Program, *J. Eng. Ind. (Trans. ASME)*, Vol 107, Nov 1985, p 325-335
28. D.G. Flom, "Advanced Machining Research Program (AMRP) Final Technical Report," Air Force Contract No. F33615-79-C-5119, GE Report No. 83-SRD-040, Oct 1983
29. D.R.C. Durham, "Physical Metallurgy of Deformation Localization," Topical report prepared for DARPA, UV-81-DD1, Aug 1981
30. B.F. von Turkovich, Influence of Very High Cutting Speed on Chip Formation Mechanics, in *Proceedings of NAMRC-VII*, 1979, p 291-297
31. D.R.C. Durham and B.F. von Turkovich, Material Deformation Characteristics at Moderate Strains and High Strain Rates, from Metal Cutting Data, in *Proceedings of NAMRC-X*, 1982, p 324-331
32. B.F. von Turkovich and D.R.C. Durham, Machining of Titanium and Its Alloys, in *Proceedings of the Symposium on Advanced Processing Methods for Titanium* (Louisville, KY), The Metallurgical Society, 1981, p 241-256
33. C.A. Brown, A Practical Method for Estimating Machining Forces from Tool-Chip Contact Length, *Ann. CIRP*, Vol 32 (No. 2), 1983, p 91-96
34. G. Meir, J. Hashemi, and P.C. Chou, "Finite-Element Simulation of Segmented Chipping in High-Speed Machining," Report MR88-120, Society of Manufacturing Engineers, 1988
35. T.A. Schroeder and J. Hazra, High Speed Machining Analysis of Difficult-to-Machine Materials, in *Proceedings of NAMRC-IX*, Society of Manufacturing Engineers, 1981
36. J.P. Kottenstette and R.F. Recht, Ultra-High-Speed Machining Experiments, in *Proceedings of NAMRC-X*, 1982, p 263-270
37. B.M. Kramer, On Tool Materials for High-Speed Machining, in *High-Speed Machining*, American Society of Mechanical Engineers, 1984, p 127-140
38. J. Jensen, "High-Speed Milling of Titanium," M.S. thesis, Massachusetts Institute of Technology, 1983
39. R. Komanduri and W. Reed, Jr., Evaluation of Carbide Grades and a New Cutting Geometry for Machining Titanium Alloys, *Wear*, Vol 92, 1983, p 113-123
40. *Integrated Machining System—An Overview*, LTV Aircrafts Products Group, Military Aircraft Division, 1988

# High Removal Rate Machining

HIGH REMOVAL RATE (HRR) MACHINING involves the use of extremely rigid, high-power, high-precision machines, such as roll turning lathes, to achieve material removal rates far beyond the capacity of conventional machine tools. Material removal rates as high as 6000 $cm^3$/min (370 $in.^3$/min) have been reached using multiple ceramic cutters on high-carbon (0.8% C) heat treated cast steel rolls. This article will review the machine requirements, cutting parameters, and applications associated with HRR machining. Additional information can be found in Ref 1 to 5.

## Machine Requirements

The key to success in HRR machining is extreme rigidity in machine, workpiece, and cutting tool setups. Figure 1 shows a "Super-Lathe" capable of making cuts as deep as 25 mm (1 in.) at feeds up to 1.3 mm/rev (0.050 in./rev) on hardened steel and chilled cast iron rolls. Metal removal rates as high as 4500 kg/h (10 000 lb/h) can be achieved with such machines, which have power ranges from 100 to 450 kW (150 to 600 hp) and can produce up to 400 kN (95 000 lbf) of output torque at the spindle. Workpieces of softer materials can also be turned, producing large, segmented chips, as shown in Fig. 2.

**The lathe bed,** which is the main structural member of the "Super-Lathe," is designed to withstand applied loads with minimal deflection. The structure is an elongated steel box fabricated from hot-rolled ASTM A 36 steel that has a high elastic modulus. The six exterior plates of the box become the principally stressed members. When fabricated correctly, the unit needs no vibrational damping, and when subjected to maximum rated loads, maximum unit stresses of only approximately 3450 kPa (500 psi) will develop in the critically stressed members.

**The tailstock** of the machine features a rectangular quill that can withstand radial loads two or three times as great, for a given amount of deflection, as an equivalent circular quill with an identical bore and wall thickness. The tailstock is also fabricated from A 36 structural steel plate.

The method of clamping the tailstock to the bed again provides for maximum rigidity. There are four clamp bars at the bottom edge of the tailstock; when pulled up with 100 J (75 ft · lbf) of torque, the force required to separate the tailstock from the bed is approximately 5500 kN (625 tonf).

**The headstock** is also made of heat-treated wrought steel. In the power shafting, no shaft between gears is smaller in diameter than the length between the gears. The main spindle bearing, the bull gear, and the ring gear (when used) are all the same nominal diameter as the maximum size of the roll the lathe will turn. Additional information on superlathe gearing and electrical drive systems can be found in Ref 2 and 3. Workpieces are gripped by an eight-jaw chuck.

**The carriage,** which carries the cutter, must be capable of withstanding the maximum torque force and feed forces with only a fraction of a thousandth of an inch of deflection. The weight and mass of the cross slide and saddle are important considerations because inertia can substantially reduce the changes in deflection of the carriage as the feed and cut forces vary at high frequency. For example, cutting forces suddenly drop from a maximum to almost zero when each segment of the chip ruptures from the workpiece. Antifriction ways (with linear ball and roller bearings) to the cross slide and bed are a major factor in steadying the carriage.

**Fig. 1** Superlathe used for HRR machining. This 300 kW (400 hp) machine, which is computer numerically controlled, can perform rough cuts on 1290 mm (50.75 in.) diam rolls and finishing cuts on workpieces up to 1200 mm (47.25 in.) in diameter at maximum roll lengths of 7.8 m (25 ft, 7 in.). Courtesy of J. Binns, Sr., Binns Machinery Products

**Fig. 2** A large AISI 1055 steel (180 HB) chip cut on an HRR machining lathe. Courtesy of J. Binns, Sr., Binns Machinery Products

**Table 1 Transverse rupture strengths of common cutting tool materials**

| Tool material | Transverse rupture strength MPa | ksi |
|---|---|---|
| $Al_2O_3$ | 690 | 100 |
| $Al_2O_3$ + TiC | 860 | 125 |
| SiAlON | 780 | 113 |
| Tungsten carbide | 1380–2760 | 200–400 |
| Coated tungsten carbide | 1030–2070 | 150–300 |
| High-speed steel | 2760 (min) | 400 (min) |

**Tooling** for HRR machining is also of considerable importance. Because the cutter shank is the link between the insert cutting edge and the machine, the same considerations in terms of rigidity must be given it as the bed, headstock, and carriage. A number of criteria must be considered in shank design. The first is that the larger and more massive the shank, the better it will cut. A large monobloc shank is both more rigid and a better heat sink than either a small shank or one comprised of a small interchangeable toolholder placed in a large shank. In addition, a properly matched shank socket and insert will reduce insert breakage. Because aluminum oxide ($Al_2O_3$) and $Al_2O_3$ + TiC inserts are low in transverse rupture strength as compared to alternative cutting tool materials (Table 1), it is essential that the internal stresses from the cutting forces and thermal distortion be confined as much as possible to the compressive state.

Inserts should be mechanically clamped in place with only enough force to secure them while the cut is starting. After the cut has begun, the cut force will hold the inserts. When initially clamping the inserts, it is important to have them solidly against the stops in the socket.

**Evaluating Machine Rigidity.** The manner in which chips flow from the work indicates the variation in deflections of the cutting edge with respect to the workpiece and therefore provides a good evaluation of rigidity. Adequate rigidity has been achieved when the chip peels off as a smooth, continuous ribbon from the work (Fig. 3). After a short period of smooth flow, the ribbon will fall apart into short lengths from its own weight or movement, then drop into the chip chute. If the chips fly off like snowflakes, there is insufficient rigidity, and the insert will soon break.

## Cutting Tools and Cutting Parameters

Although cemented carbide and cubic boron nitride (CBN) tools can be used for HRR machining, $Al_2O_3$ + TiC tooling is normally used, particularly for finishing cuts on ferrous alloys (surface finishes of 0.25 to 0.75 μm, or 10 to 30 μin., can be produced). Alumina-base inserts with a density of 4.30 g/cm$^3$ and a hardness of ~95 HRA are commonly used. Carbide tools are used for some roughing cuts. Both round and rectangular insert configurations are used, with the latter being the most common (Fig. 4). For rectangular inserts, negative-rake angles of 5 to 25° with 5 to 10° clearance are used; side cutting edge angles (lead angles) range from at least 30° up to 90° for lap cutting at fast speeds. Depths of cut to 75 mm (3 in.) are possible. Speeds range from 6 to 240 m/min (20 to 800 sfm), with feeds reaching as high as 6.4 mm/rev (0.250 in./rev) at low speeds. The effects of cutting parameters on tool life are reviewed in Ref 3 and 4. Detailed information on cemented carbide, CBN, and ceramics can be found in the Section "Cutting Tool Materials" in this Volume.

As stated earlier in this article, metal removal rates for ferrous materials can range as high as 6000 cm$^3$/min (370 in.$^3$/min) with multiple inserts. A 375 kW (500 hp) lathe can make cuts at more than 105 m/min (350 sfm) at a feed of 2 mm/rev (0.08 in./rev) and depths of cut of 75 mm (3 in.) in as-forged 52100 bearing steel rolls. This results in about 7700 kg (17 000 lb) of chips produced per hour, which equals an energy rate of metal removal of 44 cm$^3$/min/kW (2 in.$^3$/min/hp).

**Fig. 4** Ceramic insert (bottom) used for HRR machining, along with the resulting chip (top). The insert cutting edge is that adjacent to the scale. Courtesy of J. Binns, Sr., Binns Machinery Products

Turning tests on Inconel 718 and Ti-6Al-4V have produced metal removal rates of more than 800 cm$^3$/min (50 in.$^3$/min). The workpiece finish was between 0.75 and 2.5

**Fig. 3** Continuous 115 mm (4½ in.) wide ribbon of machined material peeling off an HRR cutter. Courtesy of J. Binns, Sr., Binns Machinery Products

μm (30 and 100 μin.) With aluminum, it has been shown that removal rates of 3300 $cm^3$/min (200 $in.^3$/min) are attainable on each insert of a gang tool (Ref 5).

## ACKNOWLEDGMENT

The editors would like to thank Jack Binns, Sr., inventor of the "Super-Lathe," for his valuable contributions to this article.

## REFERENCES

1. Staff Report, *Mech. Eng.*, March 1984, p 55-59
2. J. Binns, Sr., "Rough Turning and Hogging With Ceramic Cutters," Papers presented at the Creative Manufacturing Seminars, American Society of Tool and Manufacturing Engineers, 1963-1964
3. J. Binns, Sr., "Ceramic Cutter Performance on Rough Turning and Hogging Cuts," Technical Paper 633, American Society of Tool and Manufacturing Engineers, 1964
4. J. Binns, Sr., Super-Lathe for Roll Turning, *Iron Steel Eng.*, Oct 1961
5. D.G. Flom, Ed., "Advanced Machining Research Program," Annual Technical Report, Air Force Contract No. F33615-79-C-5119, General Electric Co. Report No. SRD-82-070, Aug 1982

# Machine Controls and Computer Applications in Machining

# Numerical Control

Yoram Koren, The University of Michigan

THE METAL CUTTING INDUSTRY changed drastically during World War II. The ambitious aircraft and missile projects of the U.S. Air Force, which required the manufacturing of complicated and accurate parts, made it clear that conventional machine tools could not fill future needs. Under contract to the U.S. Air Force, the Parsons Corporation in Traverse City, MI, undertook the development of a flexible manufacturing system designed to maximize productivity and to achieve desired accuracies for small and medium-sized production runs. The Parsons Corporation subcontracted the development of the control system to the Servomechanism Laboratory of the Massachusetts Institute of Technology (MIT). The challenge was met successfully by MIT, and in 1952 a three-axis Cincinnati Hydrotel milling machine controlled with digital technology was developed. This digital technology was termed numerical control (NC). Thus, the primary motivations for the development of NC systems for machine tools were the demand for high accuracy in the manufacture of complicated parts and the desire to shorten production time.

## Evolution of Numerical Control

The first NC controllers in the 1950s used vacuum tubes and were extremely large. Controllers in the early 1960s used transistors in their logic circuits and digital control loops. These units were more reliable. The third generation of NC controllers used integrated circuit chips and consequently became less expensive, more reliable, and smaller. Many of these third-generation controllers are still in operation. In all of these systems, the numerical data required for producing a part is maintained on punched tapes and inserted to the controller through a built-in tape reader.

An important advance in the philosophy of machine tool numerical control that took place in the early 1970s was the shift toward the use of computers instead of controller units in NC systems. This produced both computer numerical control (CNC) and direct numerical control (DNC). Computer numerical control is a self-contained NC system for a single machine tool including a dedicated computer controlled by stored instructions to perform some or all of the basic NC functions. With DNC, several machine tools are directly controlled by a central computer. Of the two types of computer control, CNC has become much more widely used for manufacturing systems (for example, machine tools, welders, and laser beam cutters) mainly because of its flexibility and the lower investment required. The preference of CNC over DNC is increasing as a result of the availability and declining costs of minicomputers and microcomputers.

One of the objectives of CNC systems is to replace as much of the conventional NC hardware as possible with software and to simplify the remaining hardware. There are many ways in which functions can be shared between software and hardware in such systems, but all involve some hardware in the controller dedicated to the individual machine (Ref 1). This hardware must contain at least the servoamplifiers, the transducer circuits, and the interface components, as shown in Fig. 1.

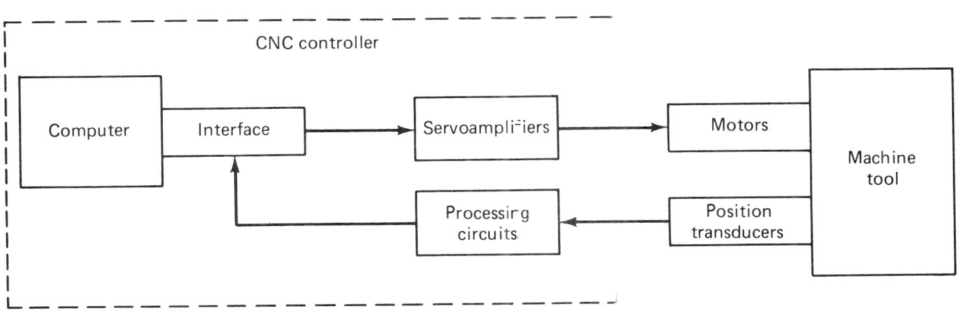


**Fig. 1** Block diagram of a CNC machine

The software of a CNC system consists of at least three major programs:

- A part program
- A service program
- A control program

The part program contains a description of the geometry of the part being produced and the cutting conditions, such as spindle speed and feed rate. The service program is used to check, edit, and correct the part program and to run system diagnostics. The control program accepts the part program as input data and produces signals to drive the axes of motion of the machine.

The CNC controllers of the 1980s are more powerful and more user friendly than earlier units. They incorporate troubleshooting features such as on-board diagnostics, which allow self-testing of the controller, and simulation mode, which is used to test part programs without generating axes motions. Many controllers offer high-level programming facilities, three-dimensional tool path animation with graphics, tool data base, and preselection of cutting parameters. The modern machine controller is a workstation in a Local Area Network (Ref 2). It is capable of communicating with other controllers and of being integrated into a flexible manufacturing system in the future, thus permitting the gradual construction of a full flexible manufacturing system.

Numerical control was introduced and developed in the metalworking industry, and the largest concentration of NC equipment remains in metalworking shops. Numerical control has been successfully implemented for turning, milling, drilling, grinding, boring, punching, and electrical discharge machines. It is interesting to note that numerical control has made possible the development of machines with basic capabilities that far surpass those of conventional machines. For example, sophisticated NC milling machines maintain control over five axes of motion and can literally sculpt complex surfaces (Ref 3). A new breed of NC machine tool is the machining center and the turning center, which incor-

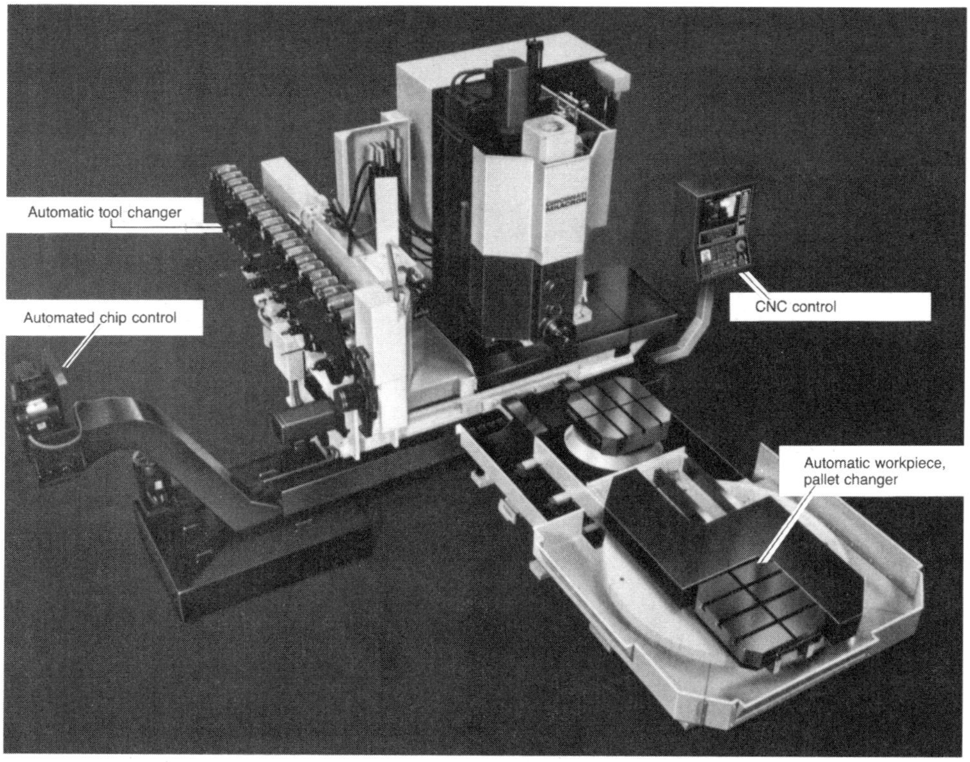


**Fig. 2** CNC machining center. Courtesy of Cincinnati Milicron

porate the functions of many machines into a single device. A machining center can access multiple tools to perform such operations as milling, drilling, boring, and tapping (Fig. 2). A turning center is a powerful lathe equipped with an automatic tool changer. Other types of NC machines include welding machines, drafting machines, tube benders, inspection machines, and wiring machines in the electronics industry.

## Fundamentals of Numerical Control

Numerical control equipment has been defined by the Electronic Industries Association (EIA) as a system in which actions are controlled by the direct insertion of numerical data at some point. The system must automatically interpret at least some portion of these data.

In a typical NC or CNC system, the numerical data required for producing a part are maintained on a disk or on a tape and called the part program. The part program is arranged in the form of blocks of information. Each block contains the numerical data required to produce one segment of the workpiece profile. The block contains, in coded form, all the information needed for processing a segment of the workpiece: the segment length, its cutting speed, feed rate, and so on. Dimensional information (length, width, and radii of circles) and the contour form (linear, circular, or other) are taken from an engineering drawing. Dimensions are given separately for each axis of motion ($X$, $Y$, and so on). Cutting speed, feed rate, and auxiliary functions (such as coolant on and off, spindle direction, clamp, and gear changes) are programmed according to surface finish, tolerance, and machining requirements (Ref 1-6).

Compared to a conventional machine tool, the NC system replaces the manual actions of the operator. In conventional machining, a part is produced by moving a cutting tool along a workpiece by means of powered slides that are engaged and disengaged by an operator. Contour cuttings are performed by an expert operator by sight. On the other hand, the operators of NC machine tools need not be skilled machinists. They only have to monitor the operations of the machine, operate the tape reader, and load and remove the workpiece. Most intellectual operations that were formerly done by the operator are now included in the part program. However, because the operator works with a sophisticated and expensive system, intelligence, clear thinking, and good judgment are essential qualifications of a good NC operator.

Preparing the part program of an NC machine tool requires a part programmer. The part programmer should possess knowledge and experience in mechanical engineering fields. Knowledge of tools, cutting fluids, fixture design techniques, machinability data, and process engineering are all of considerable importance. The part programmer must be familiar with the function of NC machine tools and machining processes and must decide on the optimum sequence of operations. The part program can be written manually, or a computer-assisted language, such as the automatically programmed tool language, can be used.

In NC machines, the part dimensions are presented in part programs by integers. In CNC machines, the dimensions in part programs are sometimes expressed as numbers with a decimal point, but are always stored in the computer as integers. Each unit of these integers corresponds to the position resolution of the axes of motion and is referred to as the basic length unit (BLU). The BLU is also known as the increment size or bit weight, and in practice it corresponds approximately to the accuracy of the NC system.

In NC and CNC machine tools, each axis of motion is equipped with a separate driving device, which replaces the handwheel of the conventional machine. The driving device may be a dc motor, a hydraulic actuator, or a stepping motor. The type selected is determined by the power requirements and the machine.

An axis of motion in numerical control means an axis in which the cutting tool moves relative to the workpiece. This movement is achieved by the motion of the machine tool slides. The main three axes of motion are referred to as the $X$-, $Y$-, and $Z$-axes. The $Z$-axis is perpendicular to both $X$ and $Y$ in order to create a right-hand coordinate system. For example, in a vertical drilling machine (as one faces the machine), a $+X$ command moves the worktable from left to right, a $+Y$ command moves it from front to back, and a $+Z$ command moves the drill up, away from the workpiece. The $X$-, $Y$-, and $Z$-axes are always assigned to create a right-hand cartesian coordinate system.

Each axis of motion also has a separate control loop. The control loops of NC or CNC systems use two types of feedback devices (Fig. 1): tachometers to monitor velocity and encoders or other position transducers (for example, resolvers) to measure position. The controller compares the actual position with the required one and generates an error. The control loop is designed in such a way as to reduce the error, that is, the loop is a negative-feedback type. A common requirement of continuous-path NC and CNC systems is the generation of coordinated movement of the separately driven axes in order to achieve the desired path of the tool. This coordination is accomplished by interpolators. Numerical control systems contain hardware interpolators, but in CNC systems interpolators are implemented by software.

## Advantages of NC Systems

It has been clearly shown that NC manufacturing reduces the number of direct employees because fewer multipurpose NC machines are required and, in some cases, one operator can operate more than one machine. However, the ratio of indirect to direct employees might increase, and in turn, overall employment in the industry might rise despite increased automation and mechanization (Ref 6). Output would of course also increase.

Realistically, then, the advantage is not in lowering labor costs but in increasing the output per man-hour. Handling costs with NC technology have decreased—in some cases remarkably. Setup times have been substantially reduced, and actual productive time has been substantially increased. A furthur savings of time is achieved while passing from one operation to another during the machining of the workpiece. With a conventional machine tool, the work must be stopped at such points because the operator must go to the next step. The operator must stop the cutting process frequently and measure the part dimensions to ensure that the material is not overcut. It has been proved that the time wasted on measurements is frequently 70 to 80% of the total working time (Ref 1). The rate of production is also decreased because of operator fatigue. In NC systems, these problems do not exist. Because the accuracy is repeatable with numerical control, inspection time is also reduced.

Numerical control produces higher-quality parts and makes possible the accurate manufacture of more complex designs without the usual loss in accuracy encountered in conventional manufacturing. Producing a part that must be cut with an accuracy of 0.01 mm (0.0004 in.) or better may take a considerable amount of time using conventional methods. In numerical control using single-axis motion, obtaining such accuracies is the state-of-the-art, and they are maintained throughout the entire range of cutting speeds and feed rates.

Another intangible advantage of numerical control is the production of complex parts that are not feasible in conventional manufacturing. Complex-contour cutting in three dimensions cannot be performed by manual operation. Even when it is possible, the operator must manipulate the two handwheels of the table simultaneously while maintaining the required accuracy; thus, it becomes possible only when the part is simple and requires relatively low accuracies. It is obvious that in such work the NC machines save a considerable amount of time.

Compared to conventional machining methods, the NC machine tool has the following advantages:

- Complete flexibility; a part program is needed only for producing a new part
- Accuracy is maintained through the full range of speeds and feeds
- The possibility of manufacturing a part of complicated contour
- A shorter production time
- Higher productivity achieved by saving indirect time, such as setting up and adjusting the machine and using one operator to monitor several machining operations, or by using completely automatic operation in unmanned production

## Programming

Most NC and CNC machine tools use off-line programming methods, which can be either manual or computer assisted, such as programming with the aid of the automatically programmed tool (APT) language. During off-line programming, the machine remains in operation while a new part program is being written. Typically, when a part program is ready, it is stored on a punched tape or a floppy disk. The tape or disk is taken to the machine shop and loaded into the machine tool controller, and the part is subsequently produced.

Manual part programs are written by programmers. First, the programmers must determine the machining parameters and the optimum sequence of operations to be performed. Based on this sequence, they calculate the tool path and write a manuscript. Each line of the manuscript, which is referred to as a block, contains the required data for transferring the cutting tool from one point to the next, including all machining instructions that should be executed either at the point or along the path between the points.

The EIA standard RS-273-A ("Interchangeable Perforated Tape Variable Block Format for Positioning and Straight Cut Numerical Controlled Machines") provides a line format for point-to point (PTP) and straight-cut NC machines. A typical line according to this standard is as follows:

N102 G01 X-52000 Y9100 F315 S717
T65432 M03 (EB)

The letter and the number that follows it are referred to as a word. For example, X-52000 and M03 are words. The first letter of the word is the word address. Word addresses are denoted as follows: *N*, sequential number, G, preparatory function; X and Y, dimensional words; F, feed rate code; S, spindle speed code; T, tool code; M, miscellaneous function; and EB, end-of-block character. The EB character is not printed but only punched or coded, and it is usually the carriage-return code, thus permitting a new line to begin immediately afterward. The EB character indicates to the NC controller that the current reading is completed and that the axes of motion must start up. When this motion is accomplished, the next block is read.

According to EIA standards RS-273-A and RS-274-D ("Interchangeable Variable Block Data Contouring Format for Positioning and Contouring\Positioning for Numerically Controlled Machines"), two block formats are available: the word address format and the tab sequential format. In the word address format, each word must be headed by the word address. The controller uses the address letter to identify the word that follows it. In this type of format, words need not be arranged in any specific order within the block because the letter identifies the corresponding word. However, because the standards prescribe a definite sequencing of words, the format recommended above is used for PTP programming.

In the tab sequential format, a tab character is inserted between each two words in the block, and the address letter is omitted. In this format, words must be arranged in a specific order. When a word is not needed in a particular block, it can be omitted, but the corresponding tab character must be punched.

Manual part programming can be applied to PTP systems, but it is too complex for contouring systems. Therefore, a great number of computer software systems have been developed to assist in NC part programming. The APT system is the most widespread and the most comprehensive one. It is available on many computers and is widely used by many manufacturers of NC equipment.

The first prototype of the APT software system was developed in 1956 by the Electronic System Laboratory of MIT. The program was further developed by the cooperative efforts of 21 industrial companies sponsored by the Aerospace Industries Association with assistance from MIT. As a result of these efforts, a system called APT II was produced in 1958, and a more effective system, the APT III, was distributed in 1961. The Illinois Institute of Technology Research Institute was elected to direct the future expansion of the program, and the capabilities of APT are continually being expanded. The current APT language has a vocabulary of more than 600 words.

The APT program is a series of instructions that specifies the path the tool must follow to produce a part. To communicate the tool path to the computer, one must provide the computer with geometric descriptions of the part surfaces. The APT language enables the programmer to accomplish this and then to specify the manner in which the tool should move along these surfaces. The geometric expressions describing the part and the motion statements represent about 70% of the average program.

A geometric expression defines a geometric shape or form. For each geometric form, there are many different methods of definition. Definitions of at least 16 different geometric forms are contained in APT: the most useful ones are POINT, LINE,

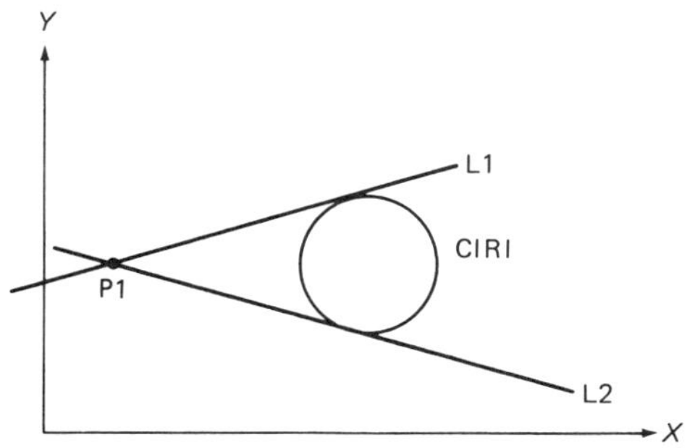


**Fig. 3** Line definition by a point and a tangent circle

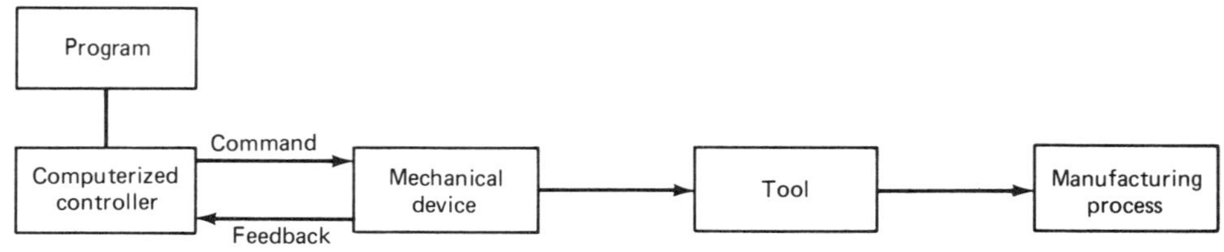


**Fig. 4** Schematic of an NC or industrial robot system

PLANE, CIRCLE, CYLINDER, ELLIPS, HYPERB, CONE, and SPHERE. Several examples of definitions are presented below.

Most APT statements are divided into two sections, major and minor, which are separated by a slash. The major section appears to the left of the slash and is generally one word containing one to six letters. The minor section, if required, appears to the right of the slash and contains modifiers to the major portion. For example, a point can be defined by:

POINT/*X*-coordinate, *Y*-coordinate, *Z*-coordinate

An example of a corresponding geometric statement is:

PT2 = POINT/3,4

where PT2 is the symbolic designation of a point whose *X*-coordinate is 3 and whose *Y*-coordinate is 4.

A line can be defined by a point and a tangent circle (Fig. 3):

LINE/symbol for a point, LEFT/RIGHT, TANTO, symbol for a circle

The modifiers LEFT or RIGHT are applied looking from the point toward the circle. Examples are:

L1 = LINE/P1, LEFT, TANTO, CIRI

L2 = LINE/P1, RIGHT, TANTO, CIRI

Once the required part has been defined with the geometric expressions, tool movements are specified using the motion statements. Each motion statement will move the tool either to a new location or along a surface specified by the statement. Examples of motion statements are:

GOTO/HOLE2

GOLFT/L1, PAST, L2

It should be noted that L1, L2, and HOLE2 are identifying names. Identifying names are given to geometric expressions and cannot be APT words. In addition to the geometric and motion statements, there are other kinds of statements and features. One of the most useful statements is the CLPRNT. The CLPRNT is an instruction to the APT system to produce a printed list of all the cutter location coordinates that have been computed. The computation results are those of the APT program before postprocessing.

The output of the APT program is sent as input to another program called the postprocessor. The latter is a program written specifically for each CNC machine tool system, and it includes information about the particular machine (size, accelerations, coolant, and so on). The output of the postprocessor is the NC part program that is loaded onto the machine controller to produce the required part.

## Numerical Control

Numerically controlled machine tool systems are in some ways related to industrial robot systems. In both, the axes of a mechanical device are controlled to guide a tool that performs a manufacturing process task (Fig. 4). In both systems, each axis of motion is equipped with a separate control loop and a separate driving device. The difference is in the process and the mechanical device. The process in numerical control may be drilling, milling, grinding, or welding, and in robotics, painting, welding, assembly, or handling. In numerical control, the mechanical handling device is the machine tool, and in robotics, the manipulator. Machine tools, however, are more rigid and usually have perpendicular axes of motion, which result in simpler control strategies and better position accuracy than those of robot arms.

A common feature in numerical control and robotics is that the required path of the tool is generated by the combined motion of the individual axes. In numerical control, the tool is the cutting tool, such as a milling cutter or a drill; in robotics, the tool is the instrument at the far end of the manipulator, which might be a gripper, a welding gun, or a paint-spraying gun. Robot systems, however, are more complex than machine tool CNC systems for the reason discussed below.

Machine tools require control of the position of the tool cutting edge in space. In many cases, the control of three axes is adequate. Robots require the control of both the position of the tool center point and the orientation of the tool; this is achieved by controlling six axes of motion (or degrees of freedom).

Some robot systems use more than six axes, and some CNC machines use more than three axes of motion. A typical example is the addition of a rotary table to a three-axis milling machine. On the other hand, many robot systems use fewer than six axes, and in such cases the wrist section contains fewer than three degrees of freedom (Ref 7). Similarly, there are CNC machines with only two numerically controlled axes of motion. For example, only two numerically controlled axes are required for turning parts on a lathe because the parts are symmetrical.

## System Structure

Computer numerical control systems can be divided into PTP and continuous-path or contouring systems (Ref 1-6). A typical PTP system is encountered in a CNC drilling machine. In a drilling operation, the machine table moves until the point to be drilled is exactly under the tool, and then the hole is drilled. The table then moves to a new point, and another hole is drilled. This process is repeated until all the required holes on the part are drilled. The machine table is then brought to the starting point, and the system is ready for the next part.

In more general terms, the PTP operation is described as follows. The machine tool moves to a numerically defined position, and then the motion is stopped. The tool performs the required task with the machine table stationary. Upon completion of the task, the machine tool moves to the next point, and the cycle is repeated.

In a PTP system, the path and the velocity while traveling from one point to the next are insignificant. Therefore, as shown in Fig. 5, a basic PTP system would require only position counters for controlling the final position of the machine table in order to bring it to the target point (Ref 8). The coordinate values for each desired position are loaded into the counters with a resolution that depends on the basic length unit of the system. During the motion of the table, the encoder at each axis transmits pulses, each representing the tool travel of 1 BLU along the axis. Each axis of motion is equipped with a counter to which the corresponding encoder pulses are transmitted. At the beginning of a motion from a point, each axial counter is loaded by the corresponding required axial incremental distance (in BLUs) to the next point. During

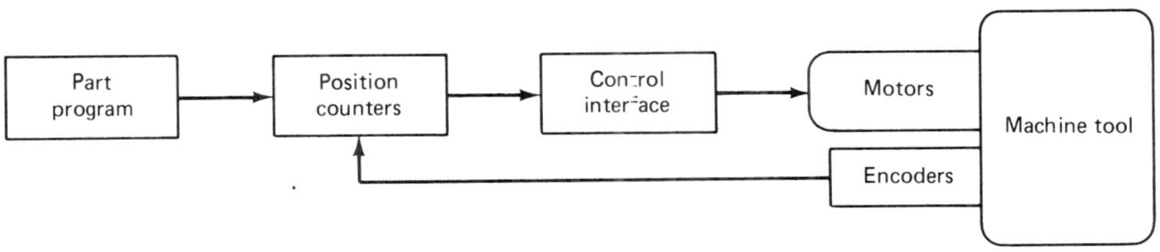


**Fig. 5** Block diagram of a point-to-point system

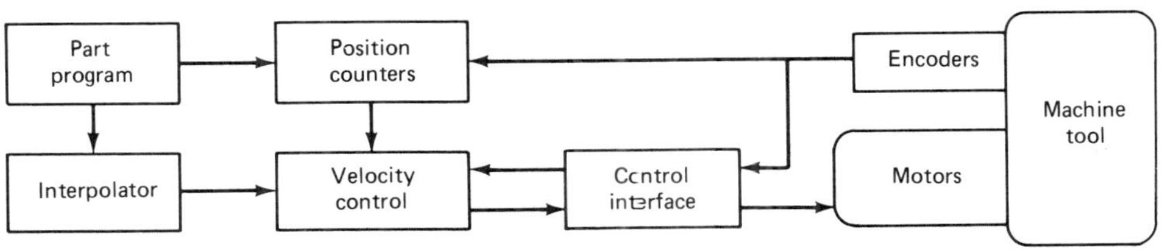


**Fig. 6** Block diagram of a contouring system

the motion of the table, the contents of each counter are gradually decremented by the pulses arriving from the corresponding encoder. When all counters are at zero, the machine table is in its new desired position.

In continuous-path CNC machines, the tool performs the task while the axes of motion are moving, as in milling machines. The task of the controller in milling is to guide the tool along the programmed path. In continuous-path systems, all axes of motion may move simultaneously, each at a different velocity. These velocities, however, are coordinated under computer control in order to trace the required path or trajectory. In a continuous-path operation, the position of the tool at the end of each segment (motion step), together with the ratio of axes velocities, determines the generated contour, and at the same time the resultant velocity also affects the quality of the surface.

Continuous-path systems require interpolators for determining the path between given end points (Fig. 6). Part programs supply only the end points of the segments along the contour, and the system computer interpolates the path between the points and generates in real time the commands to the individual axes of motion. Typically, NC and CNC machines are capable of linear and circular interpolation.

The principle of operation for linear interpolators of machine tools is relatively simple (Ref 9). The axes of motion in machine tools are perpendicular to each other, thus creating a cartesian coordinate system. A motion along a straight path with a length $L$ employs the following relationships:

$$V_x = \left(\frac{x}{L}\right)V \quad \text{(Eq 1)}$$

$$V_y = \left(\frac{y}{L}\right)V \quad \text{(Eq 2)}$$

and

$$V_z = \left(\frac{z}{L}\right)V \quad \text{(Eq 3)}$$

where

$$L = x^2 + y^2 + z^2$$

The distances $x$, $y$, and $z$ are the components of $L$ in the $X$, $Y$, and $Z$ directions, respectively, and $V$ is the required velocity along the path (referred to as feed rate in machining). Equations 1 to 3 provide the basis of linear interpolator algorithms for machine tools. Circular interpolator algorithms are more complicated and can be found in the literature (Ref 10-12).

Acceleration and deceleration are achieved by varying $V$ in Eq 1 to 3 according to a predetermined formula. Machining feed rates (the tangential velocity of the tool along the contour) in machine tools are small; and therefore, paths that include acceleration and deceleration periods are not often encountered.

## Adaptive Systems

Computer numerical control machines in production are programmable systems that can repeat a sequence of programmed operations as long as necessary. However, these systems are unable to sense and respond to any changes in their working environments. For example, assume that a CNC lathe is used to turn a batch of cylindrical workpieces to a 200 mm (8 in.) radius from raw material having radii from 204 to 207 mm (8.04 to 8.16 in.). The machining feed rate in the part program is calculated as the maximum allowable feed rate needed to remove a maximum load of 4 mm (0.16 in.) without cutter breakage, although with the smaller load the feed rate can be increased without risking the cutter. This and similar situations can be remedied only if the systems are able to adapt to the changing conditions in their environments; such changes would be sensed with suitable sensors. This is basically the motivation for introducing adaptive control systems in machining. The objectives of these adaptive controls are to increase productivity in rough machining and to improve part accuracy in fine cutting (Ref 13). The article that immediately follows in this Section provides a more thorough discussion of adaptive control for machine tools.

## REFERENCES

1. Y. Koren, *Computer Control of Manufacturing Systems*, McGraw-Hill, 1983
2. P. Ranky, *Computer Integrated Manufacturing*, Prentice-Hall, 1986
3. R.S. Pressman and J.E. Williams, *Numerical Control and Computer-Aided Manufacturing*, John Wiley & Sons, 1977
4. N.O. Olesten, *Numerical Control*, Wiley-Interscience, 1970
5. M.P. Groover, *Automation, Production Systems, and Computer-Aided Manufacturing*, Prentice-Hall, 1980
6. J. Pusztaai and M. Sava, *Computer Numerical Control*, Reston Publishing, 1983
7. Y. Koren, *Robotics for Engineers*, McGraw-Hill, 1985
8. J.T. Beckett and H.W. Mergler, Analysis of an Incremental Digital Positioning Servosystem With Digital Rate Feedback, *J. Dyn. Sys. Meas. Contr. (Trans. ASME)*, Vol 87, March 1965
9. G. Ertell, *Numerical Control*, Wiley-Interscience, 1969
10. B.W. Jordan, W.J. Lenon, and B.D. Holm, An Improved Algorithm for Generation of Nonparametric Curves, *IEEE Trans. Comp.*, Conf. C-22, No. 12, Dec 1973, p 1052-1060
11. O. Masory and Y. Koren, Reference-Word Circular Interpolators for CNC Systems, *J. Eng. Ind. (Trans. ASME)*, Vol 104, 1982
12. A.N. Poo and J.G. Bollinger, Dynamic Errors in Type I Contouring Systems, *IEEE Trans. Ind. Appl.*, Vol 1A-8 (No. 4), July 1972, p 477-484
13. G.A. Ulsoy, Y. Koren, and F. Rasmussen, Principal Developments in Adaptive Control of Machine Tools, *J. Dyn. Sys. Meas. Contr. (Trans. ASME)*, Vol 105 (No. 2), June 1983, p 107-112

# Adaptive Control

Yoram Koren and A. Galip Ulsoy, University of Michigan

ADAPTIVE CONTROL (AC) systems for metal cutting processes are a logical extension of computer numerical control (CNC), which was described in the preceding article in this Volume. The term adaptive control in metal cutting is in fact a misnomer because these are actually process control systems and may or may not be adaptive in the sense commonly used in the control theory literature (Ref 1, 2). In this article, AC systems for metal cutting are described first, and then classified according to the AC strategies used. Finally, the economic benefits of AC systems are enumerated.

In CNC systems, the relative position between the tool and the workpiece is controlled; however, the part programmer must specify the cutting speed and feed rates. The determination of these cutting parameters requires experience and knowledge regarding workpiece and tool materials, machine characteristics, coolant effects, and so on. Their selection directly affects such important economic factors as product dimensional accuracy, surface finish, metal removal rates, tool wear rates, and tool breakage. The main focus of adaptive control is the improvement of these production and product quality related factors in the actual machining process. This is accomplished using the measurement and control of certain process variables in real time as well as several alternative strategies for the AC algorithm. A schematic of a typical AC configuration for a machine tool is shown in Fig. 1. It is clear that the adaptive control represents a process control system that operates in addition to the CNC position, or servo, control system. In this discussion, the different types of AC systems that have been proposed are consid-

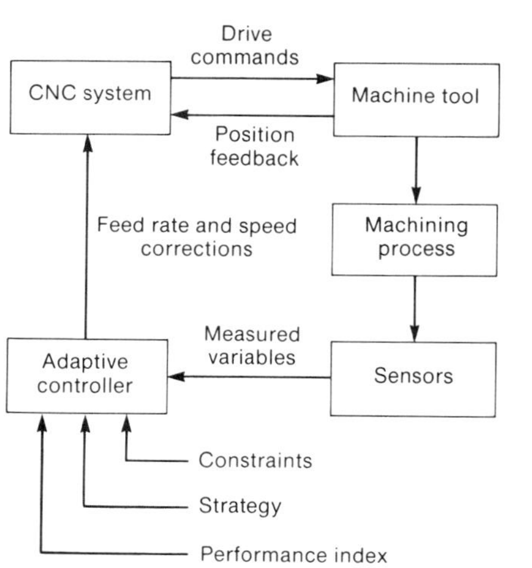


**Fig. 1** Schematic of an adaptive control system incorporated into a CNC machine tool system. Source: Ref 3

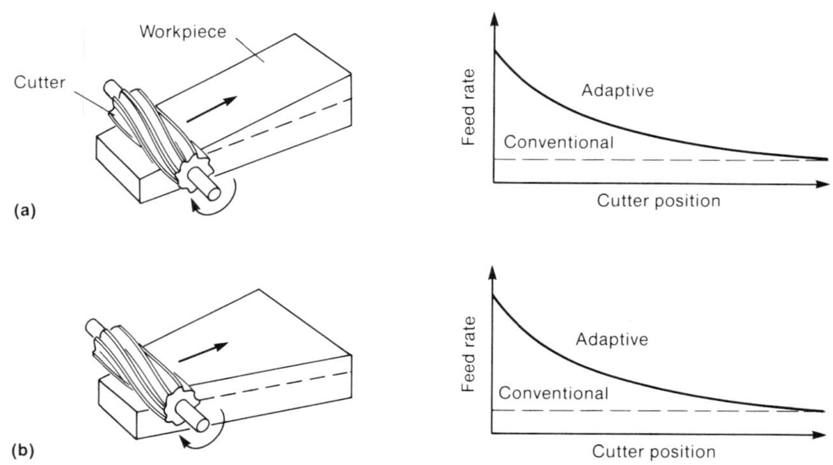


**Fig. 2** Comparison of feed rates in adaptive and conventional (nonadaptive) milling when cut varies. (a) Variable depth. (b) Variable width. Source: Ref 3

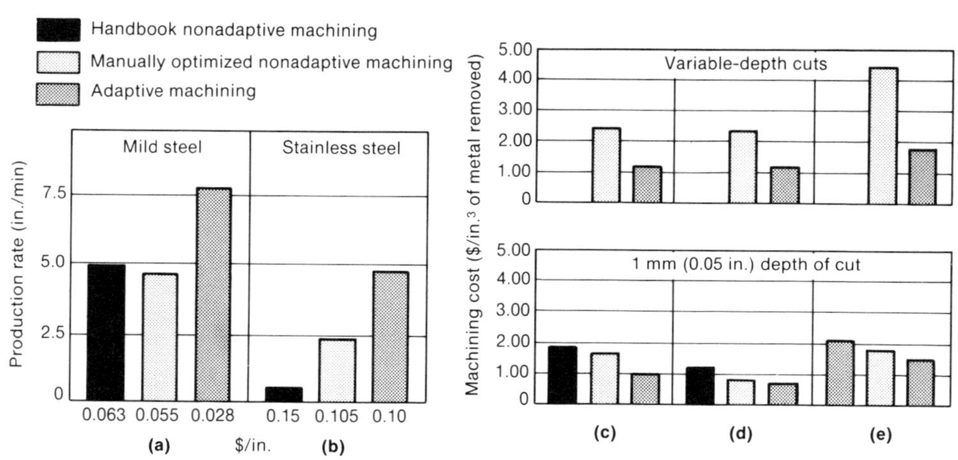


**Fig. 3** Machining cost comparison for adaptive and conventional (nonadaptive) machining. Bar graphs (a) and (b) compare a Cincinnati Milacron system with nonadaptive methods under the following conditions: (a) the machining of mild steel with a 0.05 mm (0.002 in.) tolerance and a cut 0.75 mm (0.003 in.) deep and 25 mm (1 in.) wide, and (b) the machining of stainless steel with a 0.25 mm (0.01 in.) tolerance and a cut 2.5 mm (0.1 in.) deep and 25 mm (1 in.) wide. Bar graphs (c), (d), and (e) compare a Bendix system with nonadaptive methods when variable-depth cuts (top graph) and a 1 mm (0.05 in.) depth of cut (bottom graph) are used. Machining conditions are as follows: (c) machining of 4140 steel with a high-speed steel cutter, (d) machining of 4140 steel with a carbide cutter, and (e) machining of stainless steel with a high-speed steel cutter. Source: Ref 3

ered first and then the economic benefits of AC systems.

Adaptive control systems use a variety of sensors and control strategies. Depending on these factors, as well as the particular process being controlled, the AC systems can conveniently be classified as (Ref 3, 4):

- Adaptive control with optimization (ACO), in which an economic index of performance is used to optimize the cutting process using on-line measurements
- Adaptive control with constraints (ACC), in which the process is controlled using on-line measurements to maintain a particular process constraint (for example, a desired force or power level)
- Geometric adaptive control (GAC), in which the process is controlled using on-line measurements to maintain desired product geometry (for example, dimensional accuracy or surface roughness)

These three types of AC systems are discussed in some detail in the sections that follow. It is worth mentioning here that the ACO system is the most general. However, because it is difficult to implement an ACO system, the suboptimal ACC and GAC strategies are often used. These can typically provide much of the benefit to be expected from ACO systems and are easier to implement. The ACC-type systems are well suited for rough cutting, and the GAC-type systems are typically used in finishing operations.

The economic benefits of AC systems can be significant, particularly under varying cutting conditions (Ref 5). However, the main objective is improvement in productivity, for example, by increasing the metal removal rate (MRR). This is illustrated in Fig. 2 for a milling operation with a variable depth and/or width of cut. With adaptive control, the feed rate can be increased when the depth and/or width of cut is small, and reduced if either becomes larger. By contrast, in conventional milling, the smallest feed rate would be selected based on the worst-case conditions for that particular part.

Some typical results demonstrating the economic benefits of adaptive control are given in Fig. 3. These compare machining costs in adaptive and conventional (nonadaptive) machining and show that improvements in productivity of 20% to 80% can be expected with the use of AC systems. The actual improvement depends on the material being machined and the complexity of the part being produced. Improvements or consistency in product quality can also be an objective of AC systems (usually GAC systems). Another benefit is the reduced part programming time because feeds and speeds are adjusted on-line. The economic impact of adaptive control also depends on the percentage of total production time allocated to actual cutting, and currently this is often only 5 to 20%. As automation reduces the percentage of production time required for part setup and tool changing, the percentage of time the machine spends in actual cutting increases, and the advantages of adaptive control become even more significant.

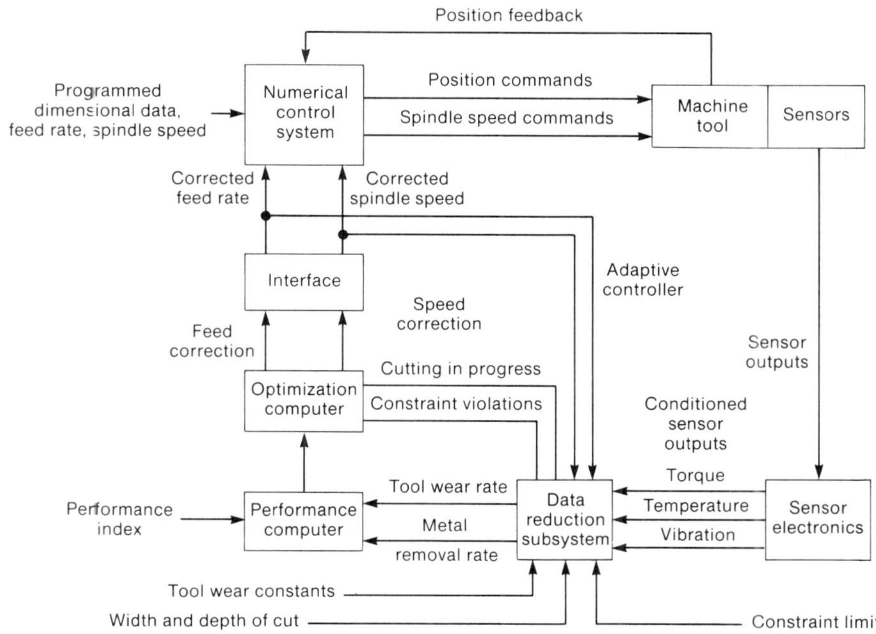


**Fig. 4** Schematic of the Bendix ACO system for milling. Source: Ref 3

## Adaptive Control With Optimization

This section describes two systems, a milling and a grinding system, based on an ACO strategy. While ACO systems represent the most general class of AC systems, they have not found application in industry. The two systems described here indicate the ACO strategy and also help to illustrate the limitations that have prevented their use in industry. The first system described is a laboratory system developed in the mid-1960s for milling and used industrially in a few installations. The ACO system described for grinding is a more recent devel-

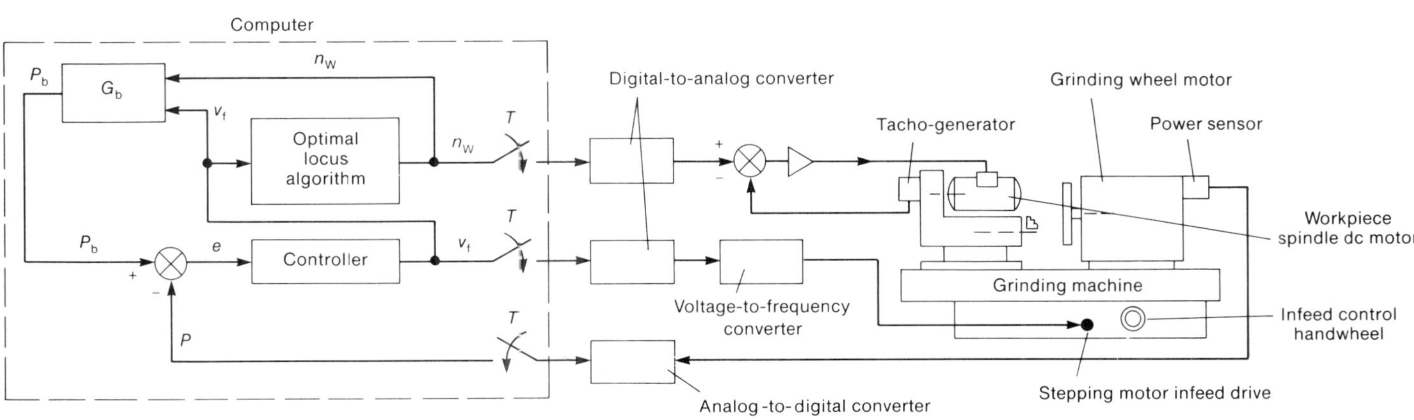


**Fig. 5** Schematic of an ACO system for grinding. Source: Ref 7

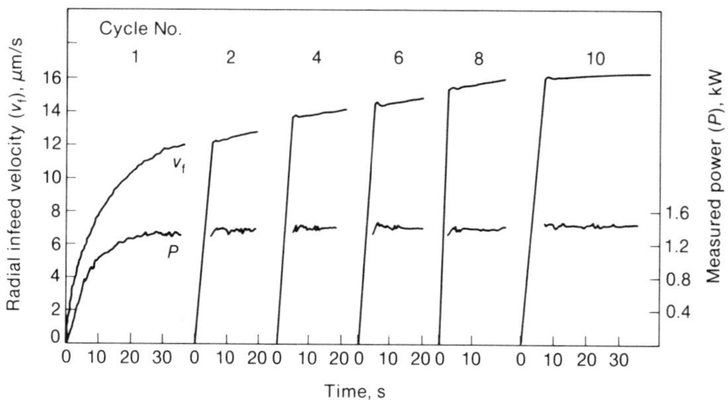


**Fig. 6** Grinding of repetitive cycles with power limit constraint wheel, 32A46K8VBE; workpiece, SAE 4340 steel, 240 HB; wheel diameter, 440 mm (17.3 in.); workpiece diameter, 165 mm (6.50 in.); grinding width, 32 mm (1.26 in.); material removed from workpiece radius per cycle, 0.3 mm (0.012 in.); single-point dressing only before cycle 1 with a dressing depth of 10 μm (400 μin.), and a dressing lead of 5 μm (200 μin.). Source: Ref 7

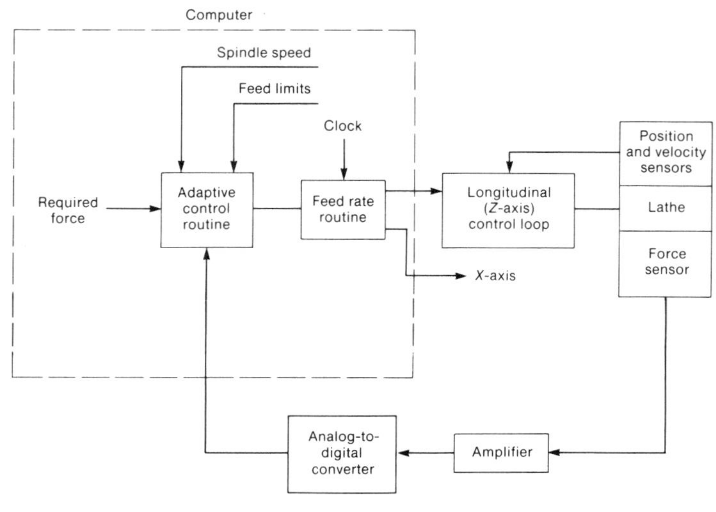


**Fig. 7** An ACC system for turning. Source: Ref 9

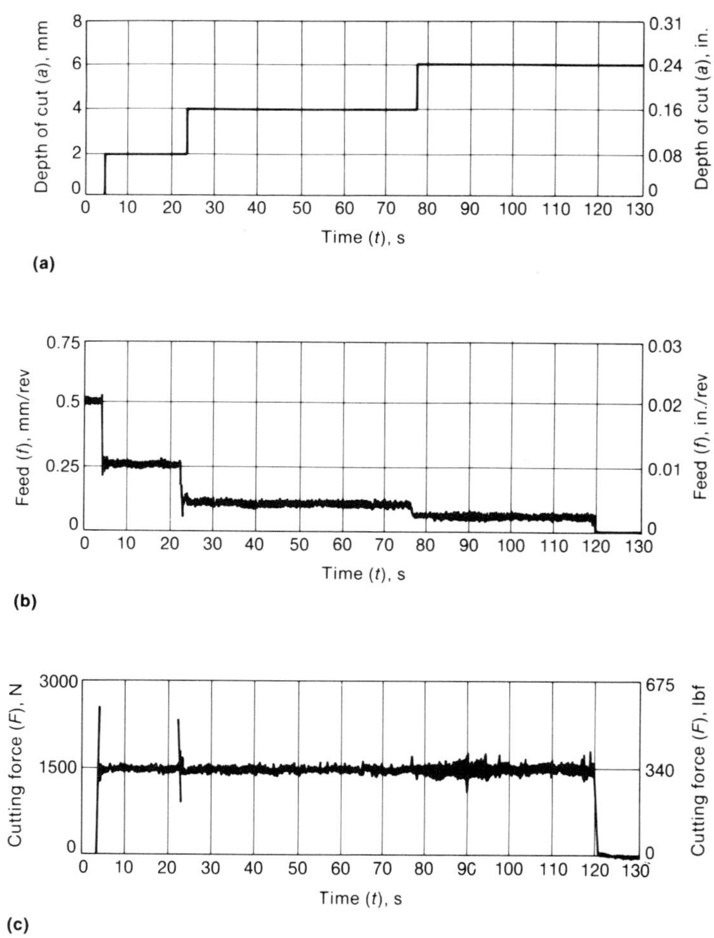


**Fig. 8** Turning experiments with an ACC system with a medium controller gain and step changes in depth of cut. The plots show the effect of changes in (a) depth of cut on (b) feed and (c) cutting force with increasing time. Source: Ref 9

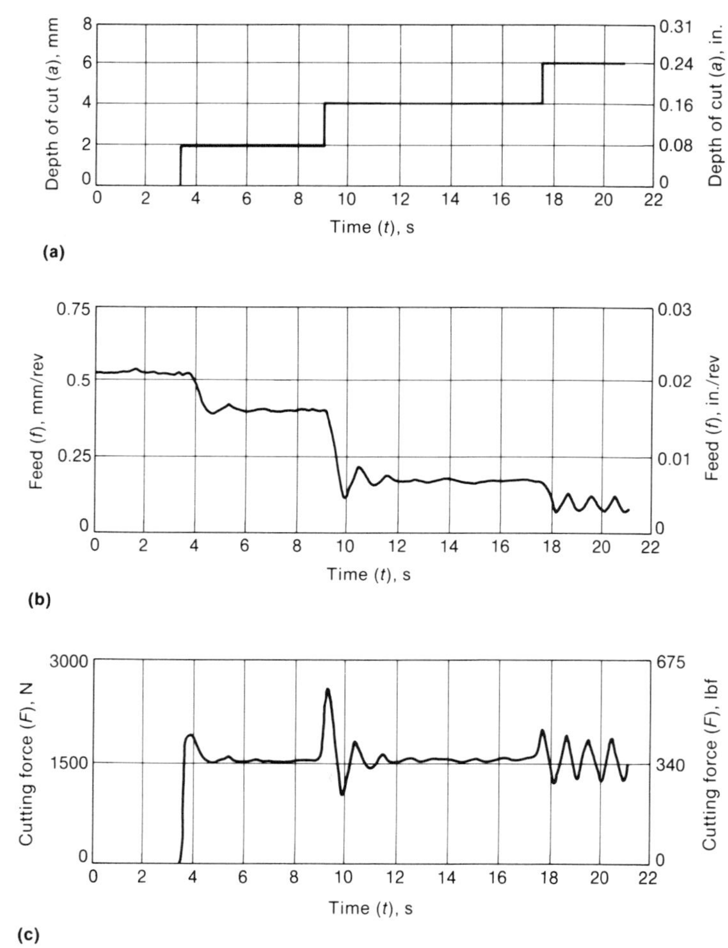


**Fig. 9** Turning experiments with an ACC system with a high controller gain and step changes in depth of cut. Plots show the effect of changes in (a) depth of cut on (b) feed and (c) cutting force with increasing time. Source: Ref 9

opment that has found off-line application in industry. However, its on-line implementation has so far been limited to the laboratory.

**ACO System for Milling.** The best-known ACO system is probably the system developed by The Bendix Corporation for the U.S. Air Force in the early 1960s (Ref 5, 6). The structure of this ACO system for milling is illustrated in Fig. 4. This system will not be described in detail here, but requires some mention because of its historical significance and because it illustrates many aspects of ACO systems. It uses several measurements of process variables (cutting torque, tool temperature, and machine vibration) and an economic index of performance for on-line adjustment of the

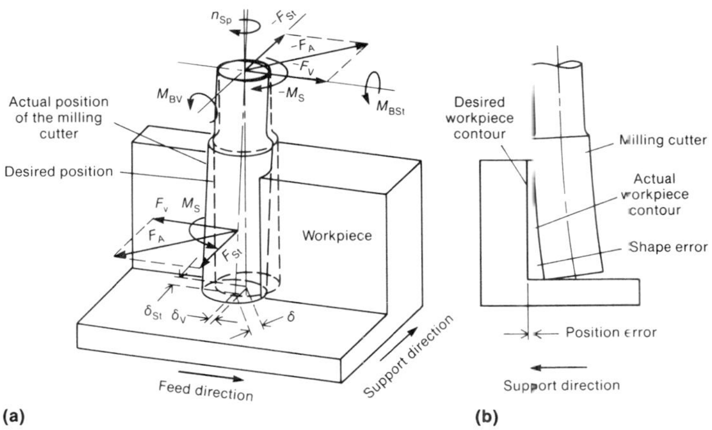


**Fig. 10** Two views of the deflection of an end mill during machining showing the various components of force, *F*; moment, *M*; and deflection, δ. (a) Front view. (b) Side view. Source: Ref 12

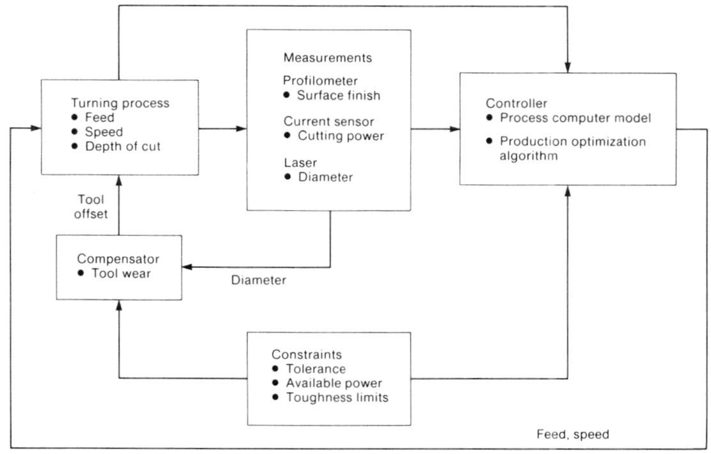


**Fig. 11** A geometric adaptive control system for turning cylindrical parts. Source: Ref 14

cutting feed and speed. The performance index ($J$) can be expressed as:

$$J(f,N) = \frac{MRR}{[C_1 + (C_1t_1 + C_2\beta)(TWR)/W_0]} \quad \text{(Eq 1)}$$

where $MRR = waV$ is the metal removal rate (in.$^3$/min), $w$ is the milling width (in.), $a$ is the depth-of-cut (in.), $V$ is the milling feed rate (in./min), $TWR = K_1(MRR) + K_2\Theta + K_3(dT/dt)$ is the tool wear rate (in./min), $\Theta$ is the tool temperature (°F), $T$ is the cutting torque (lbf · ft), $W_0$ is the allowable width of flank wear (in.), $f$ is the feed (in./rev), $N$ is the spindle speed (rev/min), $C_1$ is the cost of machine and operator per unit time ($/min), $C_2$ is the cost of tool and regrind per tool change ($/change), and $t_1$ is the tool changing time (min).

The constants $K_1$, $K_2$, and $K_3$ depend on tool and workpiece materials, and $\beta$ is an adjustable parameter in the range of $0 \leq \beta \leq 1$ that determines the type of performance index, $J$. When $\beta = 1$, the $J$ represents the reciprocal of the cost per unit produced; when $\beta = 0$, the $J$ represents the production rate; and for intermediate values of $\beta$, the $J$ represents a weighted combination of these two objectives, which can be adjusted to represent the profit (that is, maximum production with minimum cost).

The objective of the ACO system, then, is to maximize this economic index of performance, $J$, subject to constraints on maximum spindle speed, minimum spindle speed, maximum torque, maximum feed, maximum temperature, and maximum vibration amplitude. This constrained optimization problem is solved on-line using gradient methods, with increments in feed of 0.003 mm/rev (0.0001 in./rev) and increments in spindle speed of 10 rev/min. Although this ACO system was successfully demonstrated in the laboratory, it was not widely accepted by industry. The main problem was the need for a tool wear sensor (for tool wear rate, or $TWR$, in Eq 1) that can operate in an industrial environment. The calculation of the tool wear rate from the temperature and torque measurements as used in the Bendix system was not satisfactory for the entire range of feeds and speeds, and reliable temperature measurements were difficult to obtain in a production environment. Accurate on-line tool wear estimation remains an important topic of current research, as discussed in the section "Trends in AC Systems" in this article.

**ACO System for Grinding.** A successful ACO system for grinding is based on the measurement of the power at the grinding wheel, as illustrated in Fig. 5. This measured power, $P$, is compared to the maxi-

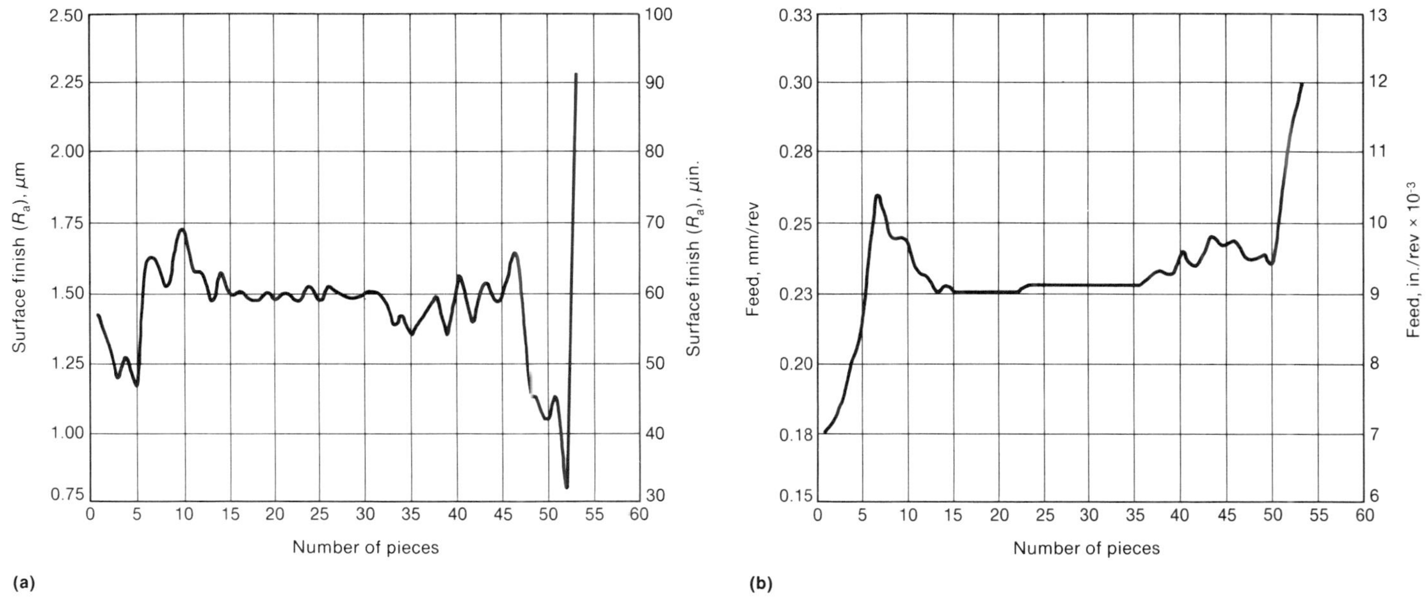


**Fig. 12** (a) Surface finish in a GAC system as determined by (b) feed rate. Source: Ref 14

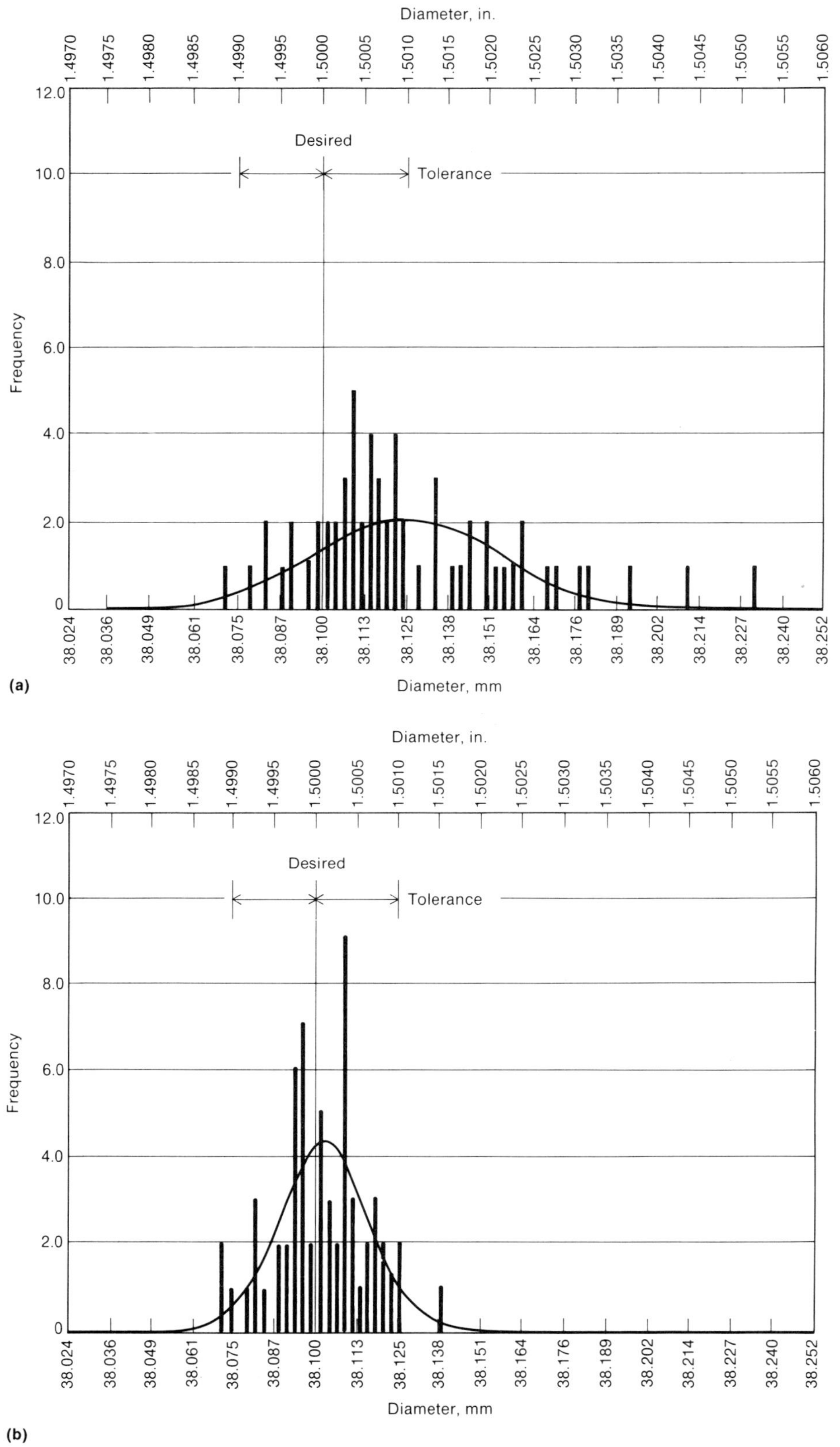


**Fig. 13** Histogram plots of diameter dimensional accuracy with (a) conventional (nonadaptive) control and (b) geometric adaptive control. Source: Ref 14

mum allowable power for burning, $P_b$, and the power error generated, $e$, is used to adjust the workpiece spindle speed, $n_w$, and the radial infeed velocity, $v_f$. The adjustment strategy uses a function $G_b$ to calculate the allowable power for burning and the optimal locus algorithm, based on a model of the grinding process mechanics, to maximize $v_f$ and calculate the corresponding value of $n_w$ such that $P = P_b$. Thus, the system is designed to operate at (or near) the power constraint for burning. The sampling period is denoted by $T$, and is 0.5 s. A surface roughness constraint can also be included. Experimental evaluation of the system was performed in the laboratory, and excellent results were obtained, as shown in Fig. 6. These results illustrate the successful operation of the system at the specified power level through ten cycles, despite the progression of grinding wheel wear; this was achieved by increasing the radial infeed velocity, $v_f$. This optimal locus approach may also be applicable to the development of ACO systems for metal removal operations other than grinding, and as such is a topic of current research (Ref 8).

## Adaptive Control With Constraints

In general, commercial AC systems used in production today for rough milling and turning are of the ACC type. This is because ACO systems are rather complex and will require further research before they can be fully implemented. The ACC systems can provide much of the benefit of ACO systems and are relatively easy to implement. In this section, a general discussion of ACC systems is followed by a description of a particular ACC system for a turning operation.

**Fundamentals of ACC Systems.** Typically, ACC systems are based on the measurement of a single process variable, such as force, torque, or motor current, and try to maintain that variable at some predetermined reference value. If this reference value is determined to ensure a relatively high production rate without excessive wear rates or breakage, it provides good, although suboptimal, performance. Most ACC systems attempt to maximize the metal removal rate, *MRR*, by maximizing one of the machining variables, for example, operating at a maximum feed rate compatible with maintaining a constant load on the cutter, as was illustrated in Fig. 2 for a slab milling operation. In this case the average feed rate with the ACC system is larger than the programmed feed rate, particularly when there are significant variations in the depth and/or width of cut in the particular milling operation. The most commonly used constraints in ACC systems are the cutting force, cutting torque, and the machining power. The machining variables commonly manipulated to meet these constraints are the feed rate ($V$) and the spindle speed ($N$). The machining feed is defined by the ratio:

$$f = \frac{V}{(pN)} \qquad \text{(Eq 2)}$$

where $p = 1$ in turning and drilling and is the number of teeth on the cutter in milling. The

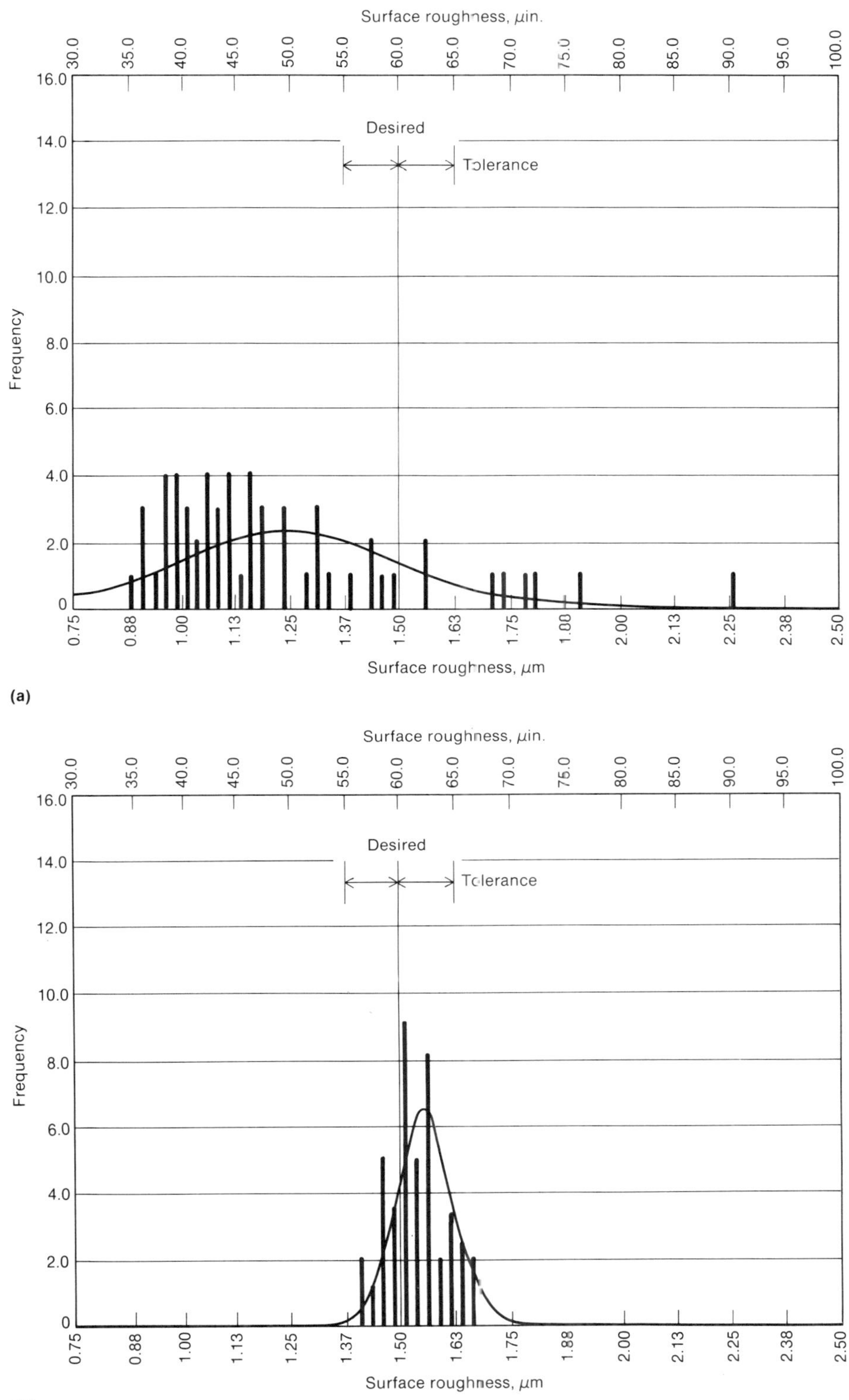


**Fig. 14** Histogram plots of surface roughness with (a) conventional (nonadaptive) control and (b) geometric adaptive control. Source: Ref 14

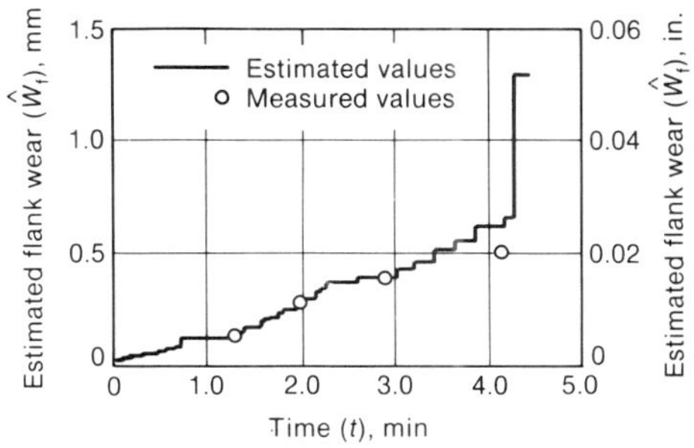


**Fig. 15** Estimated versus measured flank wear in turning. Estimated values were obtained from force measurement. Source: Ref 16

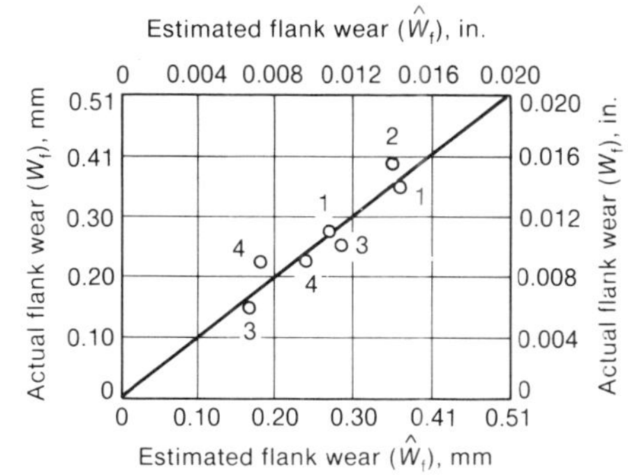


**Fig. 16** Actual flank wear versus estimated flank wear in four separate tests with varying depths-of-cut. Straight line indicates the expected results for perfect estimation. Source: Ref 17

main cutting force component is typically modeled as being proportional to the depth of cut and the feed:

$$F = K_S a f^u \quad \text{(Eq 3)}$$

where $K_S$ is the specific cutting force coefficient, and the exponent, $u$, is typically in the range $0.6 < u < 1.0$. Both $K_S$ and $u$ depend on workpiece and tool material properties and must be empirically determined by means of machining tests. The cutting torque $T$ is proportional to the force times the workpiece radius in turning and to the force times the tool radius in milling and drilling. The machining power $P$ is proportional to the force times the spindle speed.

**An ACC system for turning** is shown in Fig. 7 and described in Ref 9. The main cutting force component is measured and sampled (typically every 0.1 s). The sampled force value $F$ is then compared to a desired, or reference, force value $F_r$. The force error is provided as input to the ACC controller, which in turn produces a feed rate command signal. A positive error increases the feed rate and consequently increases the actual force $F$ (see Eq 3). Some results of laboratory experiments in which the depth of cut varies in a stepwise fashion are shown in Fig. 8. To completely eliminate the force error, an integral control strategy has been used; that is, the feed rate command signal is proportional to the integral of the force error.

However, the ACC system can still have poor performance problems, and even tool breakage or controller instability (Ref 10, 11). The experimental results in Fig. 9 illustrate the instability that occurs because of changes in the cutting process parameters (for example, depth of cut, spindle speed, and even feed rate). The cutting process, as shown in Fig. 7, is part of the control loop, and unless the controller gain is varied to compensate for the process parameter variations, consistent closed-loop performance

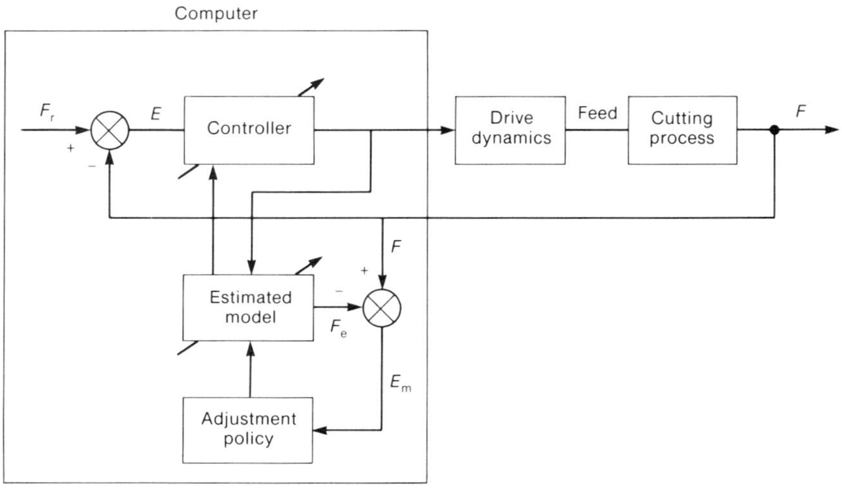


**Fig. 17** A variable-gain ACC system for turning. The estimated force, $F_e$, is compared to the measured force, $F$, to obtain the model error, $E_m$. This is used to adjust the model parameters. The error, $E$, between the reference force, $F_r$, and the measured force, $F$, is used for the control. The controller gains are dependent on the model parameters. Source: Ref 19

cannot be obtained. Thus, the process parameter variations require the use of techniques from adaptive control theory (Ref 1). In other words, the controller gains must be adapted on-line to the changing process parameters. Such variable-gain ACC systems have been developed and are described in the section "Trends in AC Systems" in this article.

## Geometric Adaptive Control

Geometric adaptive control (GAC) systems combine inspection of the finished product with on-line adjustment of feeds and speeds. In this section, some basic characteristics of GAC systems are described and then a particular GAC system for the turning of cylindrical parts is presented.

**Fundamentals of GAC Systems.** Another AC strategy is to base the selection of feed rates and speeds on the dimensional accuracy or surface roughness of the part being machined. Such an approach puts the emphasis on product quality rather than metal removal rate. Thus, such AC systems are most suited for finishing operations. These GAC systems require a direct or indirect measurement of the product simultaneously as it is being machined or possibly after it has been machined. For example, as illustrated in Fig. 10, indirect measurement can be obtained by means of the sensing of spindle deflections in a milling operation (Ref 12, 13). Direct measurements typically require laser gaging systems or other optical methods (Ref 14), and are usually performed immediately after machining.

**GAC System for Turning.** An example of a GAC system is illustrated in Fig. 11. This system, developed at the research laboratories of a major automobile manufacturer, measures the dimensional accuracy and surface roughness of the part after machining to manipulate the feed rate in a turning operation. Near real time measurement and control are used in the turning of cylindrical parts. For a given cutting speed, the principal factors affecting surface roughness are the feed and the tool wear. Surface finish and dimensional accuracy are measured for each part at the completion of the cutting operation, and corrections then affect subsequent parts. The goal is to adjust the feed rate to reduce variability in the surface finish that arises because of tool wear. The results obtained from some laboratory experiments are shown in Fig. 12 to 14. Figure 12 shows the feed being adjusted to maintain the desired surface finish. Roughly speaking, there are four distinguishable regions:

- Break-in region, where the initial feed is adjusted (pieces 1 to 10)
- Steady-state region, where the desired feed is maintained (pieces 11 to 32)
- Wear out region, where roughness increases because of wear (pieces 33 to 46)
- Failure region, where the tool must be replaced (pieces 47 to 53)

A comparison of the diameter and surface finish histograms in Fig. 13 and 14 shows the significant improvement in dimensional accuracy and surface finish achieved with the GAC system.

## Trends in AC Systems

In this section, current research issues and future trends in AC systems are briefly summarized. First, the tool wear and breakage problem is discussed. This is followed by a discussion of variable-gain AC systems, which can adapt controller gains to changing process parameters. Finally, some brief comments are made regarding the role of AC systems in computer-integrated manufacturing (CIM).

**ACO Systems and Tool Wear/Breakage.** As discussed in the previous sections, the most general AC strategy is adaptive control with optimization. However, ACO systems are typically very difficult to implement, the major difficulty being the need for a practical on-line tool wear sensor. Current research on the use of force sensing together with model-based estimation techniques for on-line tool wear sensing appears promising (Ref 15). These methods have already been shown to work successfully when flank wear, $W_f$, is dominant (see Fig. 15) and have also been applied to flank wear estimation in turning with varying depths of cut and feed rates (see Fig. 16). Acoustic emissions from the metal cutting process have also been extensively investigated for tool wear and breakage detection, as well as for chip entanglement detection (Ref 18). Acoustic emission signals, together with frequency domain signal processing and pattern recognition techniques, will need further development before they can be used effectively for tool wear sensing in an industrial environment.

**Variable-Gain ACC Systems.** Another area of recent research is related to variable-gain ACC systems (Ref 2). These are AC systems for metal cutting that are in fact adaptive in the sense commonly used in the control literature. In these process controllers, the controller gains are varied to maintain consistent closed-loop performance despite process parameter variations (Fig. 17). A process model is used to estimate on-line the changing process parameters, and the controller gain is then varied according to the adjustment policy. The turning system described previously in the section "Adaptive Control With Constraints," which incorporates a variable gain controller, has been shown to yield excellent results (Fig. 18). There has been considerable interest in designing variable-gain ACC systems for turning (Ref 19, 20) and for milling (Ref 11, 21).

**Integration of Adaptive Control Into CAD/CAM/CIM Systems.** An important issue in the future development of AC systems is their complete integration into CNC systems, as well as their role in a CIM system hierarchy. Such issues are extremely important for unmanned manufacturing and will require additional research to extend our current understanding of AC systems. Important issues include the interface between computer-aided design (CAD) and computer-aided manufacturing (CAM), and the application of expert systems and other methods, from artificial intelligence to AC systems, as well as process monitoring and diagnostics. Detailed information on interfacing adaptive control techniques with CAD/CAM systems is available in the arti-

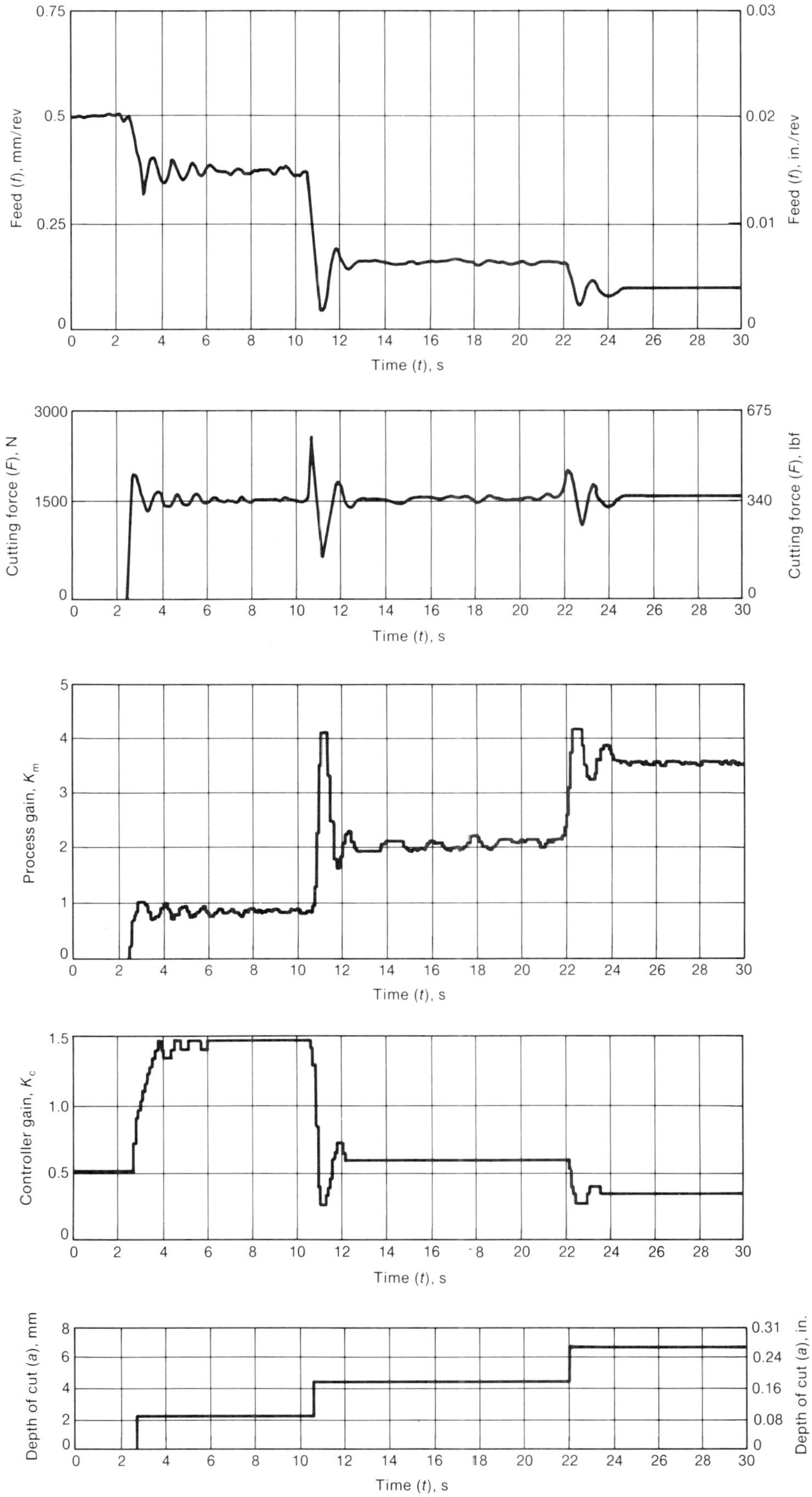


**Fig. 18** A variable-gain ACC system for turning with estimation of the process gain, $K_m$, and adjustment of the controller gain, $K_c$, for step changes in the depth of cut, $a$, with increasing time. Also shown are the variations of cutting force, $F$, and feed rate, $f$. Source: Ref 19

cle "CAD/CAM Applications in Machining" in this Volume.

## REFERENCES


1. G.C. Goodwin and K. Sin, *Adaptive Filtering, Prediction, and Control*, Prentice-Hall, 1984
2. A.G. Ulsoy, Y. Koren, and F. Rasmussen, Principal Developments in the Adaptive Control of Machine Tools, *J. Dyn. Syst., Meas., Cont. (Trans. ASME)*, Vol 105 (No. 2), 1983, p 107-112
3. Y. Koren, *Computer Control of Manufacturing Systems*, McGraw-Hill, 1983
4. R.S. Pressman and J.E. Williams, *Numerical Control and Computer-Aided Manufacturing*, John Wiley & Sons, 1977
5. J. Huber and R. Centner, "Test Results With an Adaptive Controlled Milling Machine," Paper MS68-638, American Society of Tool and Manufacturing Engineers, 1968
6. R. Centner, Final Report on the Development of an Adaptive Control Technique for a Numerically Controlled Milling Machine, "Technical Documentary Report ML-TDR-64-279," U.S. Air Force, Aug 1964
7. G. Amitay, S. Malkin, and Y. Koren, Adaptive Control Optimization of Grinding, *J. Eng. Ind. (Trans. ASME)*, Vol 103 (No. 1), 1981, p 102-111
8. Y. Koren, The Optimal Locus Method for Control of Manufacturing Processes, *J. Dyn. Syst., Meas., Cont. (Trans. ASME)*, 1988
9. O. Masory and Y. Koren, Adaptive Control System for Turning, *Ann. CIRP*, Vol 29 (No. 1), 1980
10. O. Masory and Y. Koren, Stability Analysis of a Constant Force AC System for Turning, *J. Eng. Ind. (Trans. ASME)*, Vol 107 (No. 4), 1985, p 295-300
11. L.K. Lauderbaugh and A.G. Ulsoy, Model Reference Adaptive Force Control in Milling, *J. Eng. Ind. (Trans. ASME)*, 1988
12. G. Stute, Adaptive Control, in *Proceedings of the Machine Tool Task Force Conference*, Vol 4, Sect 7.14, Wright Aeronautical Laboratories, U.S. Air Force, 1980
13. T. Watanabe and S. Iwai, A Control System to Improve the Accuracy of Finished Surfaces in Milling, *J. Dyn. Syst., Meas., Cont. (Trans. ASME)*, Vol 105, 1983, p 192-199
14. C.L. Wu, R.K. Haboush, D.R. Lymburner, and G.H. Smith, Closed Loop Machining Control for Cylindrical Turning, *Modeling, Sensing and Control of Manufacturing Systems*, American Society of Mechanical Engineers, Nov 1986, p 189-204
15. Y. Koren, Flank Wear Model of Cutting Tools Using Control Theory, *J.*

*Eng. Ind. (Trans. ASME)*, Vol 100 (No. 1), Feb 1978

16. K. Danai and A.G. Ulsoy, An Adaptive Observer for On-Line Tool Wear Estimation in Turning, Part I, Theory; Part II, Results, *Mech. Syst. Signal Proc.*, Vol 1 (No. 2), 1987, p 211-240
17. Y. Koren, A.G. Ulsoy, and T.R. Ko, Estimation of Tool Wear Under Varying Cutting Conditions, in *Proceedings of the 14th NSF Conference on Production Research and Technology*, National Science Foundation, 1987, p 73-77
18. E. Kannatey-Asibu, Jr. and D. Dornfeld, Quantitative Relationships for Acoustic Emission From Orthogonal Metal Cutting, *J. Eng. Ind. (Trans. ASME)*, Vol 103 (No. 3), 1981, p 330-340
19. O. Masory and Y. Koren, Variable-Gain Adaptive Control Systems for Machine Tools, *J. Manuf. Syst.*, Vol 2 (No. 2), 1983, p 165-174
20. L.K. Daneshmend and H.A. Pak, Model Reference Adaptive Control of Feed Force in Turning, *J. Dyn. Syst., Meas., Cont. (Trans. ASME)*, Vol 108 (No. 3), 1986, p 215-222
21. M. Tomizuka, J. Oh, and D. Dornfeld, Model Reference Adaptive Control of the Milling Process, *Control of Manufacturing Processes and Robotic Systems*, American Society of Mechanical Engineers, p 55-64, Nov 1983

# CAD/CAM Applications in Machining

Emory W. Zimmers, Jr., Lehigh University

CAD/CAM is a term that refers to computer-aided design and computer-aided manufacturing. This technology is moving in the direction of greater integration of design and manufacturing, which have traditionally been treated as distinct and separate functions in a production firm. When the machining function is included as part of a plantwide CAD/CAM system, many benefits accrue. For example, more accurate information is provided to and received from the machining operation, and the total cycle time from concept to finished machined part is reduced.

**Computer-aided design** (CAD) can be defined as the use of computer systems to assist in the creation, modification, analysis, or optimization of a design. The computer systems consist of the hardware and software used to perform the specialized design functions required by a particular customer. The CAD hardware typically includes the computer, one or more graphics display terminals, keyboards, and other peripheral equipment. The CAD software consists of the computer programs necessary to implement computer graphics on the system and application programs to facilitate the engineering functions of the customer.

Examples of these application programs include stress-strain analysis of components, dynamic response of mechanisms, heat-transfer calculations, and numerical control part programming. Many of these techniques, which were historically available only to the component designer, are becoming available to personnel responsible for planning and performing machining operations. This method replaces intuitive decision making with a more scientific approach leading to higher product quality and improved production rates.

**Computer-aided manufacturing** (CAM) can be defined as the use of computer systems to plan, manage, and control the operations of a manufacturing plant through either direct or indirect computer interface with the production resources of the plant. As indicated by the definition, the applications of CAM fall into two broad categories, computer monitoring and control and manufacturing support.

Computer process monitoring involves a direct computer interface with the manufacturing process for the purpose of observing the process and associated equipment, for example, collecting data from the machining process. The computer is not used to control the operation directly. The control of the process remains in the hands of machinists, who may be guided by the information compiled by the computer.

Computer process control goes one step further than monitoring by not only observing the process but also controlling it, based on the observations. With computer monitoring, the flow of data between the machining process and the computer is in one direction only, from the process to the computer. In process control, the computer interface allows a two-way flow of data. Signals are transmitted from the machining process to the computer, just as in the case of computer monitoring, but, in addition, the computer issues command signals directly to the machining process, based on control algorithms contained in its software. An example of this is adaptive control machining (see the article "Adaptive Control" in this Volume).

In addition to the applications involving a direct computer process interface for the purpose of process monitoring and control, CAM also includes indirect applications in which the computer serves a support role in the manufacturing operations of the plant. In these applications, the computer is not directly linked to the manufacturing process. Instead, the computer is used off-line to provide plans, schedules, forecasts, instructions, and information by which production resources can be managed more effectively. An example of CAM for manufacturing support is numerical control (NC) part programming by computers in which control programs are prepared for automated machine tools (see the article "Numerical Control" in this Volume).

In the presentation to follow, elements of a CAD/CAM system relevant to machining are presented. In addition, the hardware, interfaces, and benefits of integrating machining with the CAD/CAM system of the manufacturing plant are discussed.

## Functions of a CAD/CAM System

The basic functions that should be included when considering the relationship of CAD/CAM and machining are:

- Design
- Analysis
- Drafting
- Process planning
- Part programming
- Program verification
- Part machining
- Inspection

When electronically linked, these provide the foundation for efficient machining operations. Each of these functional areas will be examined to gain a better understanding of the activity involved in each and the interrelationships among the various activities.

**Design.** Starting from a rough sketch or possibly just an idea, the designer, manufacturing engineer, and other qualified personnel use the graphic display as if it were a drafting board to construct the geometry of a part, tool, or fixture design. Several parts and/or fixtures can be simultaneously displayed so that the designer can ensure that they will fit together properly and that there will be no interference upon assembly.

Computer-aided design systems are at their best when designing families of parts or machining fixtures (parametric design). A component can be designed so that the variable parameters are input data. The component can then be called up as a library function, and the values of the required parameters can be specified. In this way, the design of the entire family of parts, and ultimately the machining instruction set, can be completed in little more time than is required to produce the design for one part or fixture.

Most CAD/CAM systems provide some form of layering. A layer is simply one way of describing a part or machining fixture. This can be a particular geometric view, for example, a rough cast shape or a finished shape, or a nonpictorial way of describing the part, such as a bill of material or Automatically Programmed Tools (APT) part program. Layering is used heavily through-

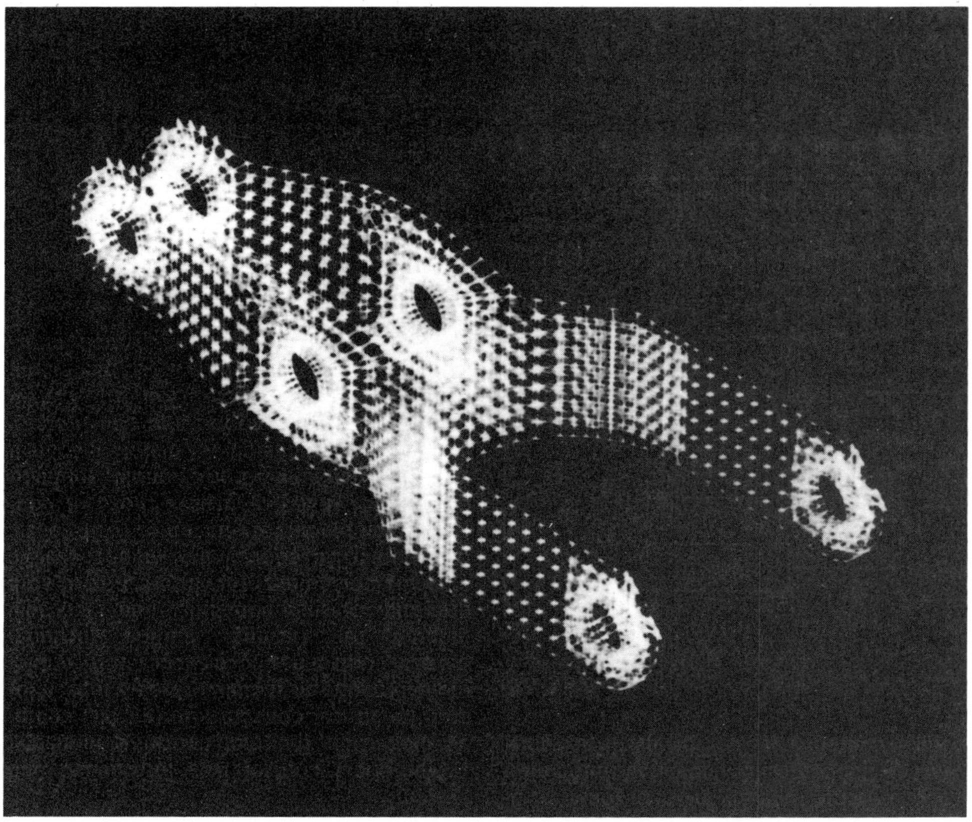

**Fig. 1** A finite-element model mesh used for stress and thermal analysis of machined parts. Courtesy of L. Niggemann, Schlumberger Technologies, CAD/CAM Division

out the design process, in drafting, in planning for machining, and in NC part programming.

**Analysis Capabilities.** Although traditionally thought of as being applicable only in the initial stages of the manufacturing cycle, analysis capabilities are being more fully utilized in support of the machining process.

*Mass Properties.* The analysis capability of a CAD/CAM system that has perhaps the widest application is that of mass properties. This feature automatically gives the perimeter, area, center of gravity, and moments of inertia of any cross section. The properties of the cross section when rotated about any axis to form a solid are also given. These include exposed surface area, volume, weight, center of gravity, and moments of inertia. Applications range from moments of inertia (for nonsymmetric parts to be machined) to contour roll die design. Many problems in machining relating to chatter or poor surface finish can be analyzed and solved using these capabilities.

*Stackups.* The ability to display an assembly of various components is used to good advantage in axial stack-up calculations. In this application, the engineer or machinist has all the components of interest displayed on the screen. In the time it would normally take to describe his requirements to a draftsman, he is able to identify the areas for which he wants a nominal clearance and to obtain their values.

*Finite-Element Analysis.* Many parts or sections of parts are subjected to severe heat transfer and stress conditions during the machining process, for example, in pocket milling or machining thin-wall sections. Typically, few calculations are made because of the time or skill level involved. One mathematical tool used to calculate the effects of these on a given part is finite-element model analysis. In this technique, the three-dimensional model is divided into a number of small cube- or wedge-shape elements, which are analyzed separately (Fig. 1). The total effect on the part is the sum total of all of the subdivisions.

The finite-element grid is generated automatically through software developed for finite elements. The user has the ability to change any portion of this grid pattern if so desired. The elements are numbered, and the precise coordinates of the nodal points are obtained automatically by the graphics system. This information is output from the system in a format that requires no human intervention before being input to a computer for analysis. Should the analysis indicate that the geometry requires modification, this is done rapidly at the cathode ray tube (CRT) terminal, and the entire process is repeated. With the availability of more local computing capability and with improvements in networking, use of these techniques has increased at the factory level. When this occurs, more precise calculations are substituted for intuition and trial and error approximations by the machinist, setup person, or NC part programmer.

**Drafting.** To perform this function, a draftsman accesses the data base and retrieves the part design. A series of steps is followed:

- The detail work is performed
- Lines and fillets are trimmed
- Layout lines are removed
- Areas are crosshatched
- Drill holes are identified
- Dimensions are added (generally available in English or metric units, or both)
- Tolerances are noted
- All necessary text is placed on the drawing

Company standards for drawings can be easily enforced by programming the standard into the system. Aids to the draftsman include transformations to assist in developing oblique, isometric, and perspective views, and the capability to replicate details that appear at various places on the design. Information critical to the machining process is prepared at this stage.

Common applications include surface contour and styling drawings, layouts for preliminary design, exploded views, perspective drawings, dimensioned drawings, isometric drawings, nondimensioned master drawings for production design, fixture layouts, interference problem layouts, assembly drawings, and subassembly drawings.

**Process Planning.** The engineering drawing of the workpart must be interpreted in terms of the manufacturing processes to be used. This step, referred to as process planning, is concerned with the preparation of a route sheet. The route sheet is a listing of the sequence of operations that must be performed on the workpart. It is called a route sheet because it also lists the machine tools through which the part must be routed in order to accomplish the sequence of operations.

The foundation of the most commonly used retrieval-type computer-aided process planning (CAPP) systems is parts classification and coding and group technology. In this approach, the parts produced and their associated machining fixtures are grouped into families according to their manufacturing characteristics. For each part family, a standard process plan is established. The standard process plan is stored in computer files and retrieved for new workparts that belong to that family. Some form of parts classification and coding system is required to organize the computer files and to permit efficient retrieval of the appropriate process plan for a new workpart. Figure 2 illustrates the flow of information in a CAPP system.

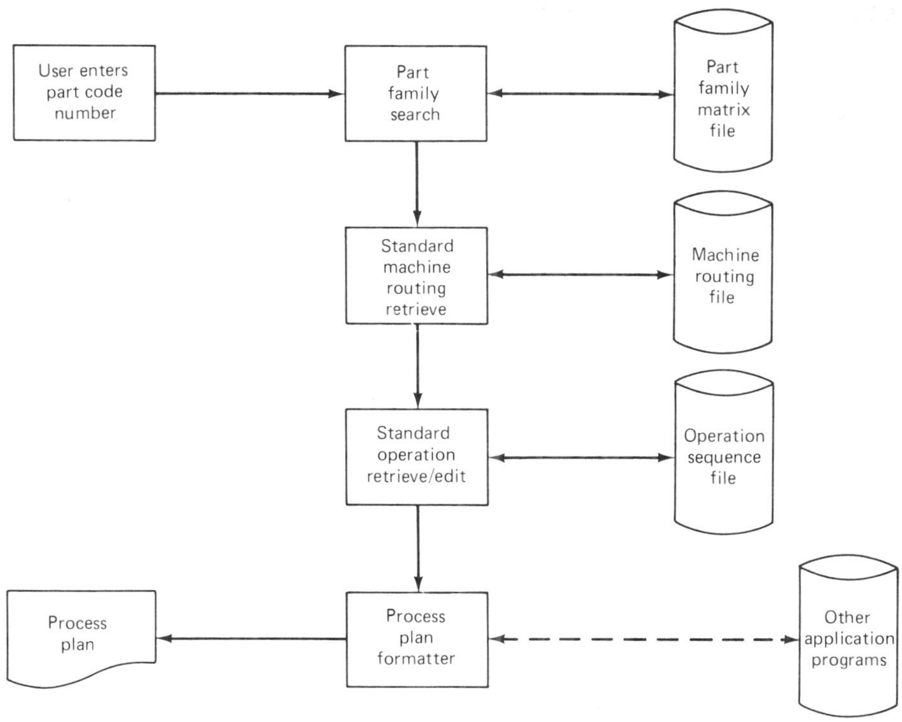


**Fig. 2** Information flow in a retrieval-type computer-aided process planning system

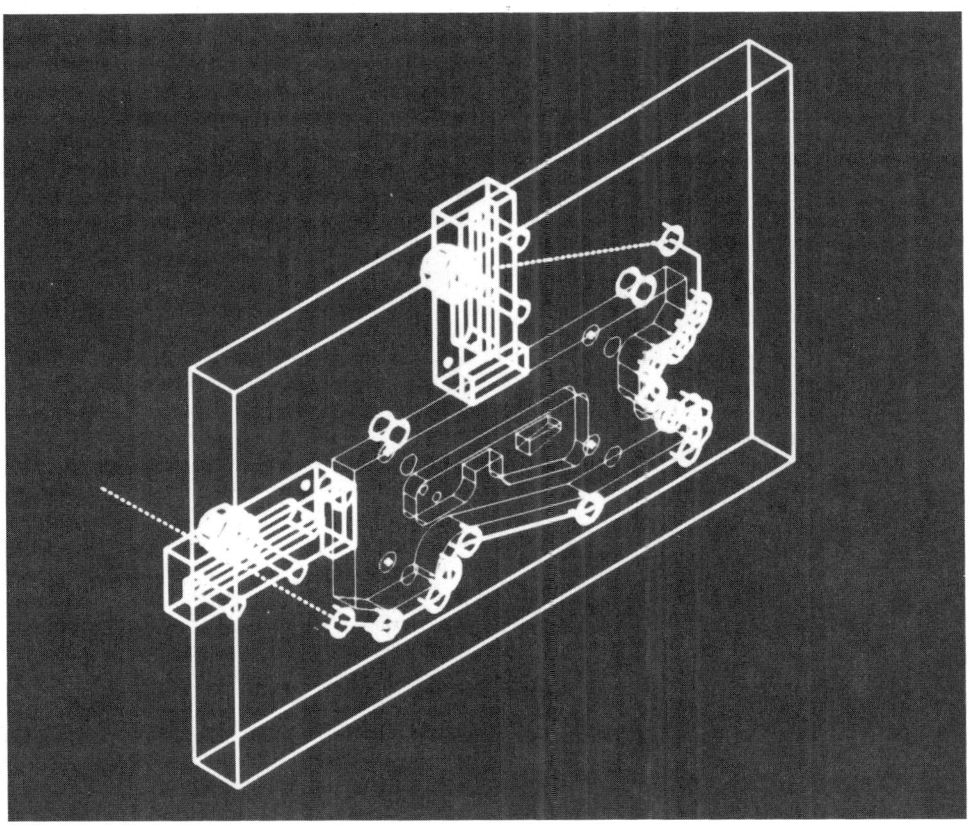

**Fig. 3** Graphics display of a tooling fixture and component part with the associated tool path clearly defined to generate the necessary CNC machine code output. Courtesy of [illegible]. Niggemann, Schlumberger Technologies, CAD/CAM Division

For some new parts, editing the existing process plan may be required. This is done when the manufacturing requirements of the new part are slightly different from the standard. The machine routing may be the same for the new part, but the specific operations required at each machine may be different. The complete process plan must document the operations as well as the sequence of machines through which the part must be routed. Because of the alterations that are made in the retrieved process plan, a CAPP system such as this is sometimes also called a variant system.

It is important that accurate, updated machining information be included as part of the overall process. This requires feedback and updating as the inevitable machining improvements are made on the factory floor. In many CAD/CAM systems, this occurs as electronic data transfer from a shop floor terminal. If this is not done, it is likely that the system will produce suboptimal machine information on such things as feeds and speeds.

**Part Programming.** In producing an NC part program, the accurate geometric definition of the part that is present in the data base is used by the part programmer. This data base geometric definition eliminates the step of defining the part geometry from drawings, which is the first step in traditional (APT-like) NC programming. Auxiliary statements concerning cutting speed, feed rate, coolants, and tool changes must be entered. The definition of the cutter path is then normally specified interactively, rather than by the traditional APT process of mentally envisioning the cutter path. The part program can often be done through a graphic presentation of the machining operation (if the CAD system is of the refresh type, this is often accomplished through animation). Figure 3 shows such a graphic presentation. The resulting NC part program is usually in APT/CL (cutter location) file format.

Typically, NC programming can be done for two, two and one-half, three, four, and five axis machining and for machines including milling machines, lathes, punch presses, drilling machines, and most of the other digitally controlled machining operations.

**Program Verification.** With many CAD/CAM systems, verification using computer simulation software is available at varying levels of sophistication. The basic objective is to reveal what can be anticipated during the actual machining process, primarily by visual verification of the images created during the simulation. The image may appear on a graphics display terminal, a plotter, or other hard copy device. Often the hard copy output can be included in the operator setup and operations sheets used by the machinist. The advantages include making the verification process more productive and predictable, providing an efficient mechanism for allowing NC data to be optimized, reducing scrap, reducing inspection of first-run parts, and freeing production machines for production work.

It should be noted that the level of acceptance of the simulation model is dependent upon the accuracy and completeness with

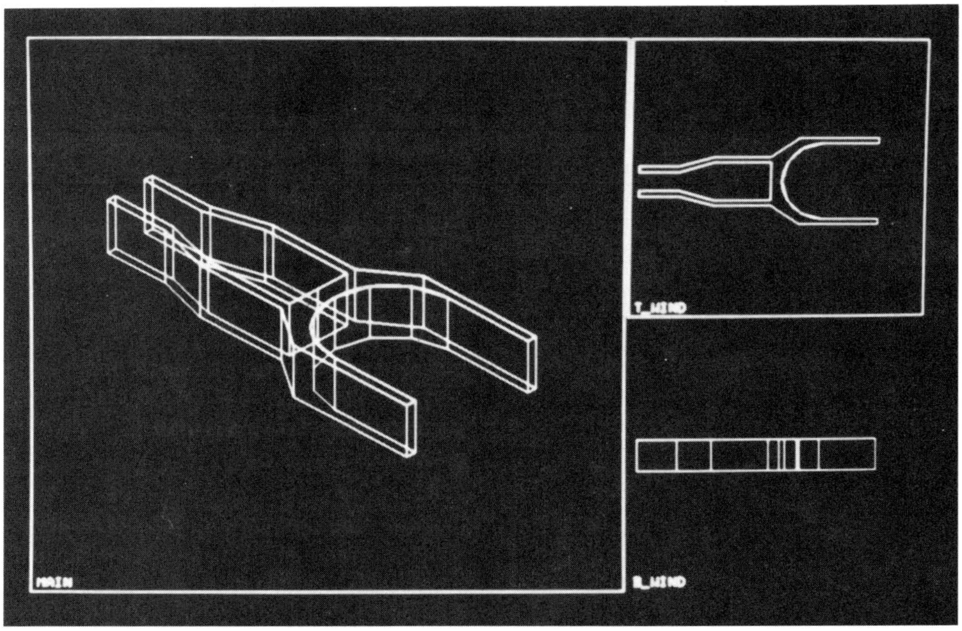

**Fig. 4** Preliminary wire frame geometry of a connecting bracket shown in three different window orientations to facilitate visualization and enhance the design process. Courtesy of L. Niggemann, Schlumberger Technologies, CAD/CAM Division

which it depicts the real world. The graphic representation may be a wire frame model, a solid model, or a shaded image. There may be the capability of displaying the raw stock from which the part is to be cut. The software may or may not be able to display the full geometry of cutting tools, as well as toolholders, and/or the clamps and fixtures for locating and holding the workpiece. Desirable features for manipulating the image include the capability to orient the part for optimum viewing advantage, to display more than one view on the screen, and to modify a view by sectioning, reorienting, zooming, or other operations. Color can be effectively used to enhance comprehension, for example, by indicating rapid moves (in red), a gouge cut in a fixture, or the path cut by each tool (using different colors). Of great use is the ability to interrogate the model to identify the command in the tool path that caused a detected error.

CAD/CAM and NC programming systems usually offer wire frame models (Fig. 4) for tool path simulation. There are usually several options for tool path display. These include the display of the centerline of the cutter (which may include the tool axis vector), the display of a two-dimensional representation of the cutter (a disk for mills and the insert, and perhaps a toolholder outline for lathes), a display of a three-dimensional representation of the cutter, and a display of textual information about the cut, for example, identification of point-to-point cycles (drill, bore, and so on), programmed feed rates, and cutting sequence number identification.

The display of the tool path is usually fairly rapid; therefore, the ability to slow down the display, step through each cutter location position in individual steps, or zoom in on specific details may be required to discover whether the workpiece is being cut correctly (or whether the fixture is inadvertently being cut incorrectly). Also, actions or cycles that are initiated by Machine Control Data (MCD) codes, such as special machining cycles, touch probe or robotic functions, or repetitive programming codes for loops or macros that compress the length of MCD to fit Machine Control Unit (MCU) memory, are usually not understood by the NC programming software. Therefore, the actions or movements resulting from those commands would not be available to the tool path simulation display. Any hidden errors could only be discovered during a tryout on the machine tool. In many cases, however, the NC program is uncomplicated enough to instill a high level of confidence by means of the wire frame simulation model alone. A machine tryout is still required on complex parts, however, to reduce the potential for catastrophic error.

**Part Machining.** In the machining process itself, the CAD/CAM system provides information (often to the machine tool site) in the form of instructions for the operator and/or instructions to the machine tool. To fully utilize the CAD/CAM technology, this linkage should be electronic and timely. Information provided can include setup and operation instructions, along with detailed machining settings such as feed and speeds. On advanced systems, some of this information is provided in a graphic format.

When a direct interface to a numerically controlled machine tool is installed, an additional benefit is often derived. Information can be provided to the CAD/CAM system regarding operational status, which is valuable for optimizing overall plant performance. This type of information often includes uptime status of the machine tool and the causes of delay, such as a broken tool or a waiting interval for tooling.

**Inspection.** The CAD/CAM system can facilitate the inspection of machined parts with complex shapes. The use of various cut planes allows the part to be measured by an inspector, after which the CAD/CAM system generates a view of the part. In some systems, provision is made for rapid electronic feedback to specific machining operations by means of the CAD/CAM system. When this is done, scrap is reduced and "drift" resulting from tool wear or other causes can be detected in time to take corrective action.

## Benefits of Using CAD/CAM for Machining

**Tool Design.** The ability to obtain a view of a part from any desired angle (provided by CAD/CAM) is a great aid in designing tooling. For example, in the design of holding fixtures for machining compound angles, information that takes days to calculate by traditional methods can be easily obtained by automatic measurement from the CAD/CAM system.

**Analysis.** The analysis routines available in a CAD/CAM system help to consolidate the analysis process into a more logical work pattern. Rather than having a back and forth exchange between manufacturing and analysis groups, a single person can perform the analysis while remaining at an engineering work station. Generally, the engineer can retain his train of thought more easily. There is a time savings to be derived from analysis routines, both in engineer time and in absolute time. This savings results both from the rapid performance of the analysis and from the time no longer lost when the design is routed by traditional means from the shop to the analyst and back again.

**Understandable Drawings.** Interactive graphics are just as adept at creating and maintaining isometric and oblique drawings as they are at producing the simpler orthographics. All drawings can be generated and updated with equal ease. Thus, an up-to-date version of any drawing type is always available.

In general, understanding a drawing relates directly to the projection used. Orthographic views are less comprehensible than isometrics. Both are less understandable than a perspective view. Most conventional construction drawings are line drawings. The addition of shading increases comprehension (Fig. 5). Different colors further enhance understanding. Thus, the machinist or machine tool setup person can refer-

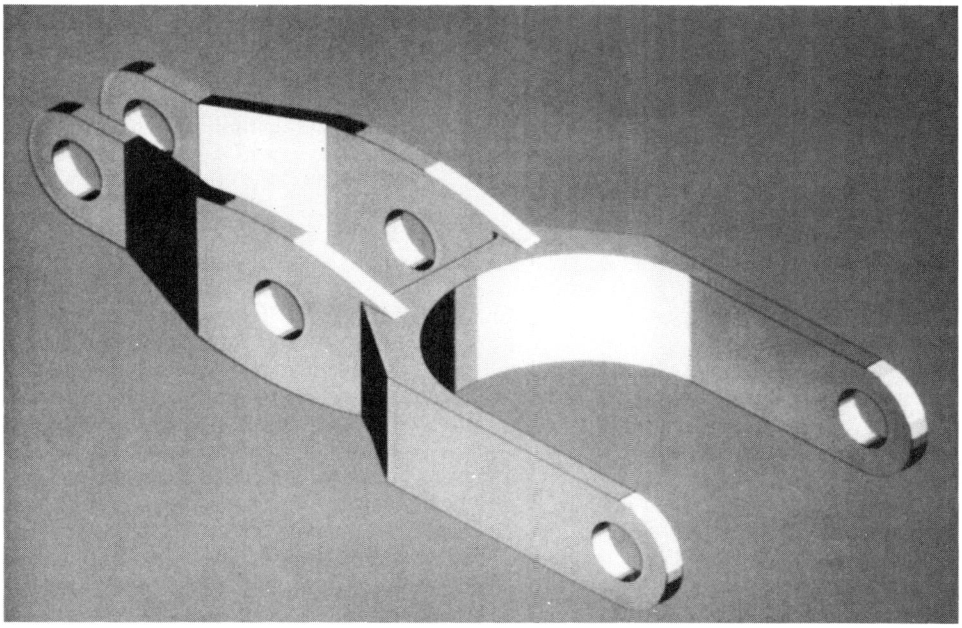

**Fig. 5** Computer-generated shaded model of a connecting bracket. Courtesy of L. Niggeman, Schlumberger Technologies, CAD/CAM division

ence an up-to-date part description in the format most useful for the machining task to be performed.

**Numerical Control Part Programming.** One typical NC part programming language is Automatically Programmed Tools, described in the article "Numerical Control" in this Volume. A powerful and versatile language, APT has three potential disadvantages. The first is that the user must learn a language with its own syntax and grammar. Unless he already has computer programming experience, he is exposed to some concepts that are entirely alien to him. A second disadvantage is that the part programmer must interpret the engineering drawings (with the possibility of error) and define the geometry of the part for APT. The third disadvantage is that the programmer must mentally visualize the tool path as he is programming. Experience has shown that these disadvantages either prevent capable machine shop personnel from becoming part programmers or else make the learning time unacceptably long. On the other hand, CAD/CAM uses language easily understood by machine shop personnel. This eases the problem of computer shock and eliminates the need to learn a completely new language. Because the geometry of the part is defined in the CAD/CAM data base, there is no need to go through the process of extracting the part geometry from the drawings. The geometry is already given, precisely as the designer specified it. The graphic display and interactive nature of the system eliminate the need to envision the cutter path, because the user is provided with visual verification of every step in the process.

These advantages have a significant impact on the time necessary to produce an NC tape and to train an NC part programmer. The experience that the General Electric Company had with the design and manufacture of the T-700 engine indicated that NC tapes were produced in half the time taken by conventional APT programming and that the time required to train a methods person to make NC tapes was reduced by a factor of six. Additional information on NC part programming can be found later in this article.

**Improved Accuracy.** When CAD/CAM is used, there is a high level of dimensional control, far beyond the levels of accuracy attainable manually. Mathematical accuracy is often to 14 significant decimal places. The accuracy delivered by interactive CAD/CAM systems for machining three-dimensional curved-space designs is much greater than that offered by manual methods.

Computer-based accuracy is beneficial in many ways. Parts are labeled by the same, recognizable nomenclature and number throughout all drawings. In some CAD/CAM systems, a change entered on a single item can appear throughout the entire documentation package, effecting the change on all drawings that use that part.

Such accuracy also shows up in the form of fewer engineering change orders, more accurate material and cost estimates, and tighter procurement scheduling. These last two points are especially important, for example, for long lead time material purchases.

## CAD/CAM Hardware Components

As described below, the hardware for a CAD/CAM system includes a graphics terminal(s), digitizer (optional), alpha-numeric terminal, hard copy unit, output plotter(s), secondary storage device, machine control unit, and auxiliary hardware, all of which interface with the central processing unit. Figure 6 illustrates the general hardware setup for a CAD/CAM system.

**The Graphics Terminal.** The most visible part of a CAD/CAM system from the point of view of the user is the graphics terminal. Virtually all state-of-the-art CAD/CAM systems are provided with some form of CRT similar to those used in televisions, oscilloscopes, and radars. In addition, CRTs are available in many sizes and configurations and with various capabilities. The majority of CAD/CAM systems provide either a "storage" CRT or a "refresh" CRT.

**Cursor Control.** The cursor is generally a bright spot on the screen that indicates where lettering or drawing will be placed. There is normally a method by which the position of the cursor can be read by the computer. Therefore, controlling the cursor position enables one to enter locational information into the computer. This information can be used in a geometric sense, such

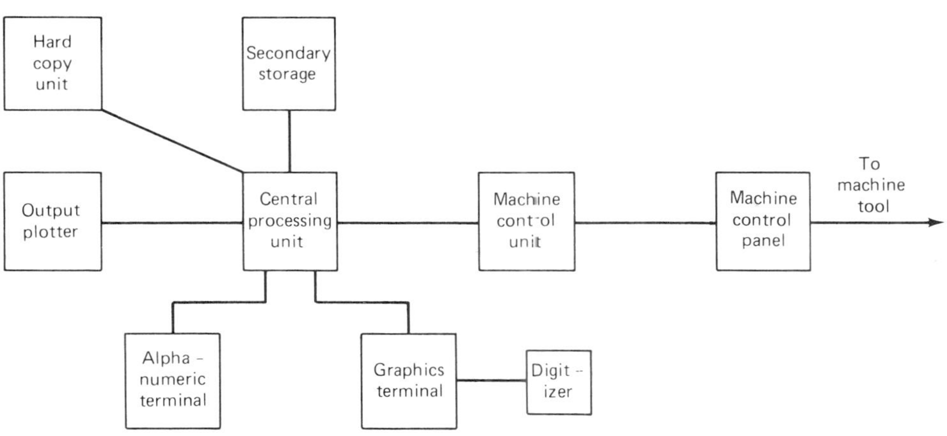


**Fig. 6** General hardware configuration of a CAD/CAM system

as to define a point. A more sophisticated use interprets the cursor position as the selection of an item from a menu (the screen is divided into sections, one of which corresponds to each menu section).

There are three basic divisions in cursor control devices. The first type includes those that control the cursor without a direct physical relationship to the screen (from the viewpoint of a user). For example, there are two thumbwheels used to control the cursor; one controls the horizontal position of the cursor, and the other controls the vertical position. The thumbwheels are turned to change the position of the cursor.

Another type of device is the light pen. The cursor follows the light pen, with the computer seeking the position of the light by trying to place the bright cursor under the pen.

The third type of device is a digitizing tablet and pen. This type of tablet detects the position of the pen and moves the cursor to a corresponding position on the screen.

**Digitizers.** A manual digitizer has a cursor that can be moved about a large, smooth table. It is oriented by means of a standard *x-y* coordinate system through which the computer can detect its location. The cursor generally has a cross hair target.

Drafters find the digitizer not unlike their drafting boards. It tilts, raises, and lowers. Some digitizers are available with backlighting for easy tracing or with free-floating cursors. Automatic raster scanners can be used to digitize continuous-tone photographs and line drawings. Automatic line followers reduce a drawing to a series of lines rather than a stream of points. By recording only the information needed to reproduce the lines, the output is condensed. However, a line follower operator must intervene in the entry process to select the direction to be followed at line intersections.

**Alpha-Numeric Terminal.** Nearly all CAD/CAM systems employ an alpha-numeric terminal along with the graphics terminal. Often terminals are accessible at the machine tool itself. Such a terminal is used to enter supplemental information, commands (if there is no cursor-type menu), alpha-numeric input, and system commands by means of a keyboard. The terminal receives messages from the system, design analysis output, program listings, and error messages. The advantage of a separate terminal is that these messages can be displayed and entered without erasing or overwriting the image on the graphics screen.

The terminal is either a CRT or a hard copy terminal. Each type has its advantages. The CRT provides much faster output and does not produce masses of scrap paper. The hard copy terminal provides a permanent listing, often desirable in debugging or for permanent records.

**Hard Copy Unit.** There is normally a hard copy unit provided to make copies of the screen in a few seconds. These copies can be used as permanent records of intermediate steps or on occasions when a rough hard copy of the screen is desired. These units are not suitable for final drawings because they are not as accurate as plotters.

**Output Plotters.** There are many types of plotters available for use on CAD/CAM systems, including high-speed drum plotters, flatbed plotters, and computer output microfilm plotters.

**The Central Processing Unit.** Functioning as the heart of a CAD/CAM system, the central processing unit (CPU) has traditionally been a mainframe for large installations or a minicomputer for smaller turnkey systems. The CPU does the necessary mathematical calculations and directs all activity within the system. It acts as the master controller and manager of workstation activity such as storage and plotting.

It is important to note that the traditional view of CAD/CAM systems as a stand-alone entity is evolving. The CAD/CAM functions and support hardware/software should be considered in the context of an integrated factorywide machining system. As this occurs, computational activity will be distributed in a cost-effective manner, and cross-departmental communication of technical/operational information will be more efficient.

The host CPU in most configurations directs plotters in precisely what to draw, copies disk data onto magnetic tapes for semiactive storage, reads magnetic tapes of drawings and/or documentation for revision or other reuse, and transmits data to and from other computers in the CAD/CAM system.

**Secondary Storage.** The CAD/CAM data base is normally maintained on magnetic disks along with much of the CAD/CAM system software. Disks or similar direct-access devices provide rapid access to the information stored on them because of their random access configuration. CAD/CAM systems generally have the capability of handling multiple disk drives, totaling many hundreds of megabytes of storage.

Magnetic tape drives are normally available. They are used as disk backup and for permanent archival file storage. Information access is not nearly as rapid from a magnetic tape as it is from a disk because the tape is a sequential storage device. Therefore, to access any piece of information, all the information preceding it must first be read. For archival files and backup, however, the infrequent use makes the slow access time perfectly acceptable. Information transfer rates from magnetic tape are typically half as fast as for disks. However, this is still extremely rapid; magnetic tape is an extremely compact and inexpensive method of storing large amounts of data.

**Machine Control Unit.** This hardware device is located where the machining operation occurs. It consists of the electronics and hardware that read and interpret the program of instructions and convert it into mechanical actions of the machine tool. The typical elements of a conventional NC controller unit include the interface to the CAD/CAM factory system or stand-alone reader, a data buffer, signal output channels to the machine tool, feedback channels from the machine tool, and the sequence controls that coordinate the overall operation of the foregoing elements. It should be noted that nearly all modern NC systems have a microcomputer as the controller unit.

A reader is an electromechanical device for reading the tape or other magnetic medium containing the program of instructions. The data contained on the tape are read into the data buffer. The signal output channels are connected to the servomotors and to other controls in the machine tool. Through these channels, the instructions are sent to the machine tool from the controller unit. To make certain that the instructions have been properly executed by the machine, feedback data are sent to the controller by means of the feedback channels. The most important function of this return loop is to ensure that the table and the workpart have been properly located with respect to the tool.

**Machine Control Panel.** The control panel, or control console, contains the dials and switches by which the machine operator runs the NC system. It may also contain data displays to provide information to the operator. Although the NC system is an automatic system, an operator is still needed to turn the machine on and off, to change tools (some NC systems have automatic tool changers), to load and unload the machine, and to perform various other duties. To be able to discharge these duties, the operator must be able to control the system, and this is done through the control panel.

**Additional Hardware.** The hardware described in the above sections is mainly based on mainframe use. In addition to this, an interface module is required to transfer the data from the mainframe to the machine tool. Two types of interface modules are widely used. A direct-interface module is used when a postprocessing program is available on the mainframe. A postprocessor converts the code generated by the NC program into machine-readable instructions. A direct interface, for example, the NC machine controller, transfers the machine-readable instructions from the mainframe to the machine. An indirect-interface module is used when a postprocessing program is not available on the mainframe. Examples of this type of controller are modified personal computers (PCs) configured as part of the machine tool controller and advanced NC machine controllers. In this case, code generated by the NC program on the mainframe is transferred to the PC or the machine controller. A postprocessing program on a PC or controller converts the code into machine-readable instructions and is then transferred to the machine.

As personal computers have become more powerful and less expensive, they are becoming extremely popular in NC machining. Many of the functions (design, analysis, NC programming, and so on) are now available on PCs. Minimum hardware requirements to perform these functions on PCs are an 80286 or equivalent processor, a 640 KB (kilobyte) random access memory, a 10 MB (megabyte) hard drive, an RS232 port, a math coprocessor, and a high-quality plotter for hard copy.

## CAD/CAM Data Base

All of the functions of a CAD/CAM system revolve around its data base. Depending on the overall structure of the machining equipment, the physical hardware location or locations may vary. The discussion below will cover logical, not physical, organization. The use of a common data base allows the simplification of many tasks. Once an item is entered into the data base, it can be accessed and used by any subsequent operation. For example, in creating an NC part program prior to machining, there is no need to reenter geometry of the part because it can be recalled from the data base.

Essentially, everything contained on the disks of the system is part of the data base. Thus, the data base includes much of the CAD/CAM system software (such as system commands, multiple menus, macros, plotter output routines, NC postprocessors, and compilers).

The data base also contains programming routines, part models, designs, drawings, and assemblies, as well as the alpha-numeric information associated with the machining task, such as the results of optimized metal cutting routines and text for machinist or technician reference.

The data base can be set up in various ways, depending on the requirements of the vendor. Much of this activity centers around the part drawing. Some vendors store a drawing as a copy of the design geometry, while others store it as a pointer to the appropriate model along with the appropriate transformation information. This compacts the data base and simplifies the process of updating all the drawings of a part when the design is changed, because the only change necessary would be one to the model.

The value of these transformation techniques can easily be seen. If one wished to rotate an entire drawing by a certain amount and also make it larger, one could simply multiply each line in the drawing by the proper transformation matrix. Three-dimensional work is much more complex (and involves about three times as much calculation), but it is easy to imagine the extension of the matrix transformation concept to points located in three-dimensional space. Use of the proper transformation matrix can produce combinations of shearing, local scaling, rotation, reflection, translation, perspective, and overall scaling.

The matrix transformation technique allows the construction of views from any desired viewpoint. Stored in the data base is a full, three-dimensional, geometric description of the part. To obtain a view of the part, one multiplies each component by appropriate transformation matrices. Thus, it is possible to store a drawing of a part essentially as a transformation matrix and a pointer to the specific mathematical model of the part. For example, many systems are configured to store the mathematical description of a part and generate drawings or views based on matrix mathematics transformations. The pointer references the specific part description, and the transformation matrix generates the required view. This allows a tremendous space savings if more than one drawing of a part is stored.

Of course, the entire process is invisible to the machinist or other user: The vendor-provided CAD/CAM software determines the transformation matrix, and the computer performs the numerical calculations. A general understanding of this concept is useful, however, in that it helps to remove the mystique from a CAD/CAM system and reinforces the fact that no individual item in a CAD/CAM system is incomprehensible when viewed in isolation without the large number of components surrounding and camouflaging it.

## CAD to CAM Interfaces

Experience has shown that the machining benefits obtained from an integrated CAD/CAM system are far greater than those realized by applying CAD and CAM as separate technologies. Hence, bridging the gap between CAD and CAM is of great importance. Ideas start out as concepts of the three-dimensional (3D) part that must be machined. Unfortunately, traditional documentation usually consists of transferring those ideas into two-dimensional (2D) representations. To automate production, the 2D drawing must be reinterpreted and, in a certain sense, recreated as a 3D part for the purposes of creating an NC part program prior to machining.

With an integrated CAD/CAM system, the common data base allows the use of the 3D part design already created while the model was being built during the design and drafting process. All dimensions are stored to an accuracy typically far greater than that of the most precise machining tool; hence, the preparation for NC machining essentially consists of generating the tool path and creating the necessary files.

Nearly all CAD/CAM systems that are capable of producing NC tapes do so with output in the form of a cutter location (CL) file. This is suitable for input to a postprocessor. Typically, postprocessing is accomplished on another computer, but many vendors also provide postprocessors that can be run on the same computer as the CAD/CAM system.

There are many advantages associated with using the CAD/CAM data as a basis for NC programming. Among the advantages are:

- Tooling is standardized
- Most simple, calculational errors are eliminated
- Documentation is kept up to date in the common data base
- Programmers are freed from many tedious operations, leaving more time for truly creative work and eliminating many errors caused by boredom
- Personnel do not have to be sophisticated programmers
- Frequently a user-defined library of macros can be set up to handle difficult tasks, allowing semitechnical personnel to operate the system
- Productivity and task turnaround times improve immensely. Figures typically quoted range from a 2:1 to a 10:1 increase in productivity
- Visual verification of tool paths prior to machining provides for less tooling (tool collisions are essentially eliminated), less wasted motion and discarded material, and fewer dry runs

In systems that take full advantage of the CAD/CAM data base, a tool path is often generated from a geometric boundary. This is in contrast to the conventional method of describing a path by means of a series of commands, such as "go right," "go forward," or "tangent to." The surface to be followed can normally be specified with the cursor control by pointing the cursor to the proper surface.

For example, APT postprocessor statements are inserted at appropriate points in the specification of the cutter path. Once a tool path has been created, it contains all the necessary information to produce the complete machine motion. Tool paths range from a two and one-half axis motion to full 3D contouring. Automatic generation of pocketing around open and closed boundaries and point-to-point operation proceed directly from an initial setup specification from the user. Full 3D contouring includes profiling and pocketing to a surface and/or surface intersection. Three and five axis motions may also be programmed.

The tool paths may be displayed as they are created as an interactive means for establishing optimum machine tool flow or in a determined sequence showing intermediate tool points, in other words, in the actual machining sequence. If a refresh-type terminal is used, animation of the cutter path may be available. This means that the operator has all the necessary capabilities to create full simulation of the NC process before actually attempting a machining operation; he can "see" the part being run on the graphics display with geometric verification of output.

Typically, it is possible for a library of user-created routines to be set up that allow the performance of complicated tasks by merely inserting a few parameters. An example is preset tool paths for a family of parts that are recognizable as such by associated descriptive text and other parameters. Routines for automatic calculation of the number and depth of cuts (rough and final finish cuts) based on tables contained in common storage can also be set up. Another example is separate user routines that are called (invoked) by standard NC routines. These special routines are essentially nontechnical in nature and contain extensive prompting. This opens up the possibility of having fewer technically qualified people operate the system by providing closely tailored programming using simple, conversational questions in a menu-type format.

## NC Part Programming

Part programming involves the planning and specification of the sequence of processing steps to be performed on the NC machine. It also involves, although less directly, the preparation of the input medium by which the processing instructions are communicated to the machine tool or cell.

With computer-assisted part programming, the machining instructions are written in Englishlike statements in the NC programming language, which are processed by a computer either to prepare the physical medium or for electronic filing.

When using one of the NC programming languages, part programming essentially consists of two tasks:

- Defining the geometry of the workpart
- Specifying the tool path and/or operation sequence

No matter how complicated the workpart may appear, it is composed of basic geometric elements. Nearly any component that can be conceived of by a designer can be described by points, straight lines, planes, circles, cylinders, and other mathematically defined surfaces. It is the part programmer's task to enumerate the component elements out of which the workpart can be formed. Each geometric element must be identified, and the dimensions and location of the element must be explicitly defined.

After defining the workpart geometry, the programmer must construct the path that the cutter will follow to machine the part. This tool path specification consists of a detailed step-by-step sequence of cutter moves. The moves are made along the geometry elements that have previously been defined. By using NC programming language statements, the tool can be directed to machine along the workpart surfaces, proceed to point locations, drill holes at those locations, and so on.

The job of the computer in computer-assisted part programming consists of:

- Input translation
- Arithmetic calculations
- Cutter offset computation
- Postprocessing

The part programmer submits his NC part program to the computer in the NC programming language he is using. The input translation component converts the coded instructions contained in the program into computer-usable form in preparation for further processing.

The arithmetic calculations unit of the system consists of a comprehensive set of subroutines for solving the mathematics required to generate the part surface. These subroutines are called (invoked) by the various part programming language statements. The arithmetic calculations unit is in fact the fundamental element in the part programming package. This unit frees the programmer from time-consuming geometry and trigonometry calculations so he can concentrate on the workpart processing.

The purpose of the cutter offset computation is to offset the tool path from the desired part surface by the radius of the cutter. This allows the part programmer to define the exact part outline in his geometry statements.

Numerically controlled machine tools have different features and capabilities. They use different NC tape formats. Therefore, the final task of computer-assisted part programming is to take the general instructions and make them specific to a particular machine tool system. The unit that performs this task is called a postprocessor.

The postprocessor is really a separate computer program that has been written to prepare the punched tape for a specific machine tool. The input to the postprocessor is the output from the other three components. The output is the NC tape written in the correct format for the machine on which it is to be used.

## Direct Numerical Control

There are a number of problems inherent in conventional NC that have motivated machine tool builders to seek improvements in the basic NC system. Among the difficulties encountered in using conventional NC systems are problems with the tape or the tape reader and the fact that the controller unit is hard wired; that is, its control features cannot be easily altered to incorporate improvements.

It was with these problems in mind that the machine tool builders developed the concept of using the general-purpose computer to control NC machines. Their concept has come to be called direct numerical control (DNC).

Direct numerical control can be defined as a manufacturing system in which a number of machines are controlled by a computer through a direct connection and in real time. The tape reader, which was the least reliable component of the NC system, is omitted in DNC. Instead of using the tape reader, the part program is transmitted to the machine tool directly from the computer memory. The system consists of four components:

- A central computer
- Bulk memory, which stores the NC part programs
- Communication links
- Machine tools

The computer calls the part program instructions from bulk storage and sends them to the individual machines as the need arises. It also receives feedback data from the machines. This bidirectional information flow occurs in real time, which means that requests from each machine for instructions must be satisfied almost instantaneously. A diagram of a typical DNC system is shown in Fig. 7.

## Computer Numerical Control

Over the years, the cost and physical size of the digital computer have been greatly reduced. At the same time, its computational capabilities have increased. The result of these improvements has been the maturing of computer numerical control (CNC).

Computer numerical control is an NC system that uses a dedicated, stored-program computer to perform the basic NC functions. Because a digital computer is used in both CNC and DNC, there is often confusion surrounding the two system types. The principal differences between the two are:

- Direct numerical control computers distribute instructional data to, and collect data from, a number of machines or machining cells. With CNC, only one machine or a small number of machines in a cell can be controlled
- Direct numerical control computers occupy a location that is typically remote from the machines under their control. Computer numerical control computers are located close to their machine tools

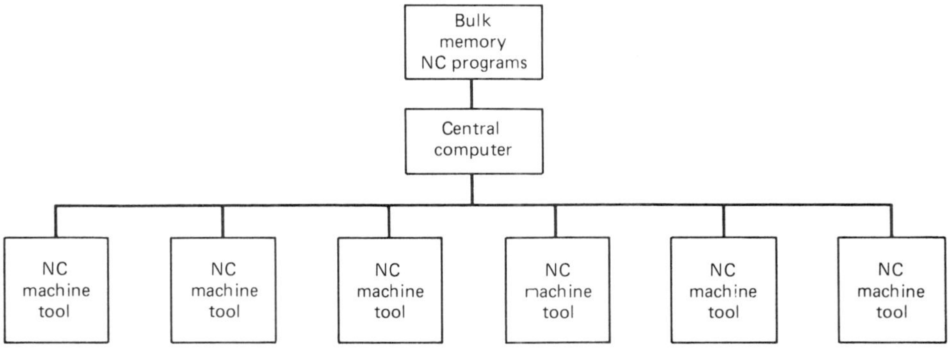


**Fig. 7** Diagram of a direct numerical control system

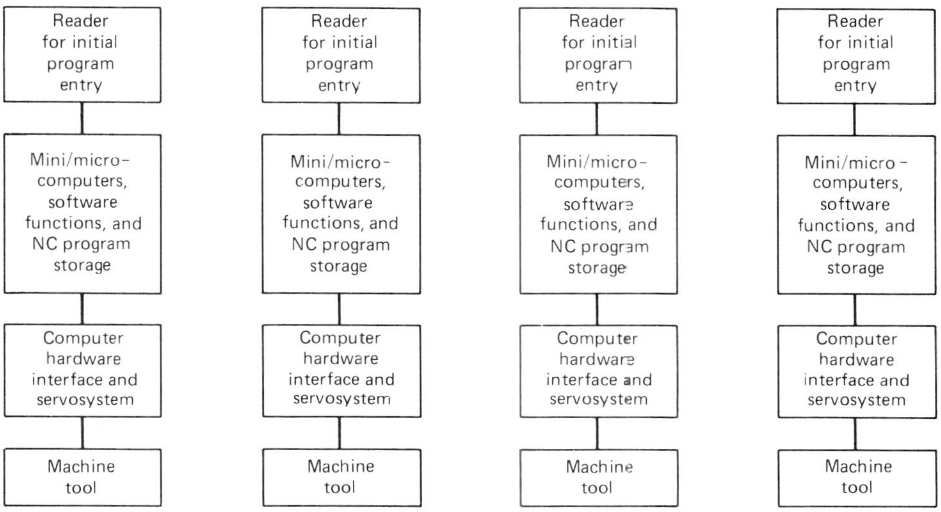


**Fig. 8** General configuration of a CNC system

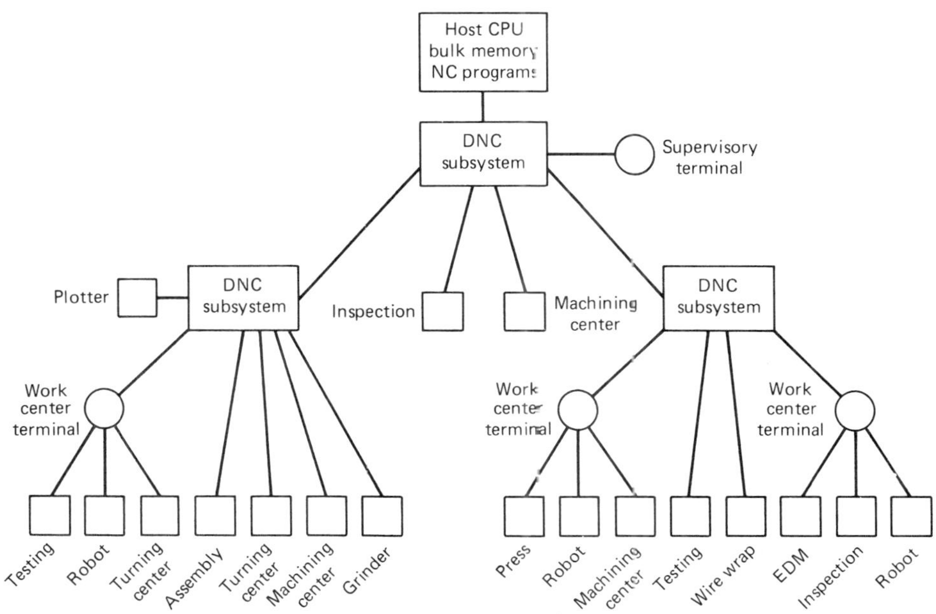


**Fig. 9** Diagram of a distributed numerical control system. EDM, electrical discharge machining

- Direct numerical control software is developed not only to control individual pieces of production equipment but also to serve as part of a management information system in the manufacturing sector of a firm. Computer numerical control software is developed to augment the capabilities of a particular machine tool

Almost all CNC machine tools provide tape editing on the shop floor. This means that the NC tape can be optimized during tape tryout at the site of the machine tool. Frequently, CNC machines have local CAD capabilities, allowing the operator to access the workpiece or other graphic information needed on the shop floor. The general configuration of a CNC system is shown in Fig. 8.

## Distributed Numerical Control

A logical extension of direct numerical control is referred to as distributed numerical control. This is complementary to and consistent with the concept of an integrated CAD/CAM system for machining. The new distributed numerical control combines the centralized data base feature of DNC with the distributed computer power available in CNC machine tools. This creates a communications network for the shop floor that eliminates paper tape and provides an automated part program library, status reporting on machine tool operation, and the capability to run part programs of practically infinite length.

The configuration of a DNC system is such that the CL file is generated by compiling the APT part program or by generating a tool path from geometry created with a CAD software package. This CL file is sent to the machine-specific postprocessor, which generates the instruction set to operate the NC machine tool. The postprocessor is a software package running on either the CAD/CAM computer system or a satellite computer. A statement in the CL file specifies which postprocessor is to be used to generate the instruction set, which is then sent to the proper NC machine tool over a communications link. This same link is used to gather production data from the NC machine tools. A block diagram of a typical distributed numerical control system is shown in Fig. 9.

Some of the reasons for acceptance of DNC as an effective approach to manufacturing include:

- The availability of low-cost microprocessor adapters that can interface to any vendor's machine tool
- The price reduction of microcomputers and magnetic disk drives, which makes these computers cost-effective management tools for the distribution of part programs
- The development of data communications facilities such as fiber optics that can operate reliably in the often hostile environment of the shop floor
- The growth in the population of CNC machine tools, which makes their integration into a shop network a logical productivity improvement

- The establishment of standards for part programs and data communications
- The familiarity with computers that shop owners and managers have acquired
- The success of CAD/CAM and Materials Requirement Planning (MRP), which is prompting managers to look for additional ways of increasing productivity through computers
- The realization by management that the success of the factory of the future is dependent upon the integration of the production process with the engineering and planning operations

This integration process is the basis of distributed numerical control. The integration of production tools and CAD systems will provide the means and flexibility whereby the real benefits of flexible manufacturing can be realized. In order to optimize the benefits of DNC, certain elements of a state-of-the-art system should be recognized.

Editing, printing, and plotting part programs are all necessary functions of a distributed numerical control system. Editing provides a means of changing part programs on the shop floor or even manually writing a part program without tying up an expensive machine tool for this purpose. Distributed numerical control terminals should be provided for editing and printing part programs. Plotting or other graphic verification of part programs should also be a by-product of a distributed numerical control system.

File management provides an automatic library for part programs. It should include automatic storage, purging, deleting, copying, scheduling, and renaming of the revision of a part program. The distributed numerical control file manager must provide a technique for tracking, storing, and updating revisions of part programs.

The DNC system should also offer rugged terminals that can exist in the manufacturing environment. The terminals should have a flexible security structure that allows management to assign responsibility and functionality where it is needed, not where the distributed numerical control system supplies it.

Many operators on a factory or shop floor do not have computer training or experience. Most CNC operators have keyboard experience and a sense for computer operation, but they are neither computer operators nor NC programmers. A DNC system should provide a menu technique for accessing the functions of the system. A well-designed system will also allow NC programmers to access additional functions where needed.

A distributed numerical control system must be designed to connect economically to any machine. It is an acknowledged industry fact that NC controllers have not been standardized. Different shops have machine tools of varying NC vintage and capability. A distributed numerical control system must interface with all NC machine tools as well as robots and other numerically driven machines.

A state-of-the-art DNC system will provide interfaces and integrations with a CAD/CAM or part programming system. Computer networks are also a future productivity tool. Many computer manufacturers have advanced computer networks, and many offer a standard network. It is important that the distributed numerical control system be easily integrated into a widely accepted computer network by standard, well-maintained products. The integration into a standard computer network will simplify future extension of the distributed numerical control system of the plant to many of the shared vendor data bases that major manufacturers will be establishing. An industry-accepted operating system will permit the integration of other manufacturing tasks to the distributed numerical control system of the plant.

## SELECTED REFERENCES


- *Autofact '88 Conference Proceedings*, Society of Manufacturing Engineers, 1988
- K. Blache, *Success Factors for Implementing Change: A Manufacturing Viewpoint*, Society of Manufacturing Engineers, 1988
- *CAD/CAM*, 2nd ed., Society of Manufacturing Engineers, 1985
- *CAD/CAM*, Videotape, Society of Manufacturing Engineers
- *CAD/CAM Networking*, Videotape, Society of Manufacturing Engineers
- N. Chiantella, Ed., *Management Guide for CIM*, CASA/SME, 1986
- S. Cousins, Ed., *Integrating the Automated Factory*, Society of Manufacturing Engineers, 1988
- D. Goetsch, *Fundamentals of CIM Technology*, Delmar Publications, 1988
- G. Graham, Ed., *Automation Encyclopedia: A to Z in Advanced Manufacturing*, Society of Manufacturing Engineers, 1988
- M.P. Groover, *Automation, Production Systems, and Computer Integrated Manufacturing*, Prentice-Hall, 1987
- T.G. Gunn, *Computer Applications in Manufacturing*, Industrial Press, 1981
- J. Harrington, *Computer Integrated Manufacturing*, Industrial Press, 1973
- L.G. Lamit and V. Paige, *Computer-Aided Design and Drafting (CADD)*, Merrill Publishing, 1987
- Y.C. Pao, *Elements of Computer-Aided Design and Manufacturing*, John Wiley & Sons, 1984
- P.G. Ranky, *Computer Integrated Manufacturing*, Prentice-Hall, 1986
- U. Rembold, C. Blume, and R. Dillman, *Computer Integrated Manufacturing Technology and Systems*, Marcel Dekker, 1985
- K. Taraman, Ed., *CAD/CAM Integration and Innovation*, Society of Manufacturing Engineers, 1984
- E. Teicholz, Ed., *CAD/CAM Handbook*, Society of Manufacturing Engineers, 1985
- E. Teicholz and J.N. Orr, *Computer-Integrated Manufacturing Handbook*, McGraw-Hill, 1987
- R.W. Yeomans, A. Choudry, and P.J.W. Ten Hagen, *Design Rules for A CIM System*, North-Holland, 1985

# Machining of Specific Metals and Alloys

# Machinability Test Methods

C. Zimmerman, S.P. Boppana, and K. Katbi, GTE Valenite

MEASURING THE PERFORMANCE of cutting tool materials and geometries over a broad range of materials and conditions is crucial to design improvement. Modern machining methods require that cutting tools be more versatile to handle a wide range of operations in an effort to reduce inventories and tool change time. A wide variety of machine tests are regularly conducted to meet this goal and to understand the phenomena associated with machining. These tests encompass workpiece materials, cutting tools, and the cutting operation and its characterization (Ref 1-8). This article will discuss the considerations used in selecting and evaluating machining tests for the purpose of evaluating cutting tool performance and workpiece machinability. Additional information on machining tests for steels can be found in the article "Machining of Carbon and Alloy Steels" in this Volume (see the section "Appendix on Machinability Testing").

## Cutting Tool Materials

The evolution of cutting tools has proceeded from high-speed steel (HSS) tools, first introduced at the beginning of the century, to modern cemented carbides, coated carbides, cermets, ceramics, and finally polycrystalline diamond (PCD) and cubic boron nitride (CBN) materials. All these materials, which are reviewed below, are used in current machining practice. The application and speed ranges of various cutting tool materials are shown in Fig. 1. More detailed information and data can be found in the Section "Cutting Tool Materials" in this Volume.

**High-speed steels** are characterized by moderate wear resistance and high transverse rupture strengths, giving them wide applicability on machining operations. Their primary limitation is speed/metal removal rate, typically suffering plastic deformation at relatively low cutting speeds (30 to 60 m/min, or 100 to 200 sfm). As a result, their primary applications are in form cutters, reamers, taps, drills, and small-diameter end mills. High-speed steels are also used in the milling of high-temperature materials and on older machines with less rigidity and limited speed capability.

**Uncoated and Coated Carbides.** Carbides dominate the metal removal market. They possess high wear resistance and compressive strengths that enable them to cut a wide variety of materials at favorably high material removal rates. Uncoated carbides are typically found in the machining of cast irons, steels, stainless steels, and nonferrous and high-temperature materials at speeds up to 150 m/min (500 sfm).

Coatings such as titanium nitride (TiN), titanium carbide (TiC), and aluminum oxide ($Al_2O_3$) were added to enable still higher metal removal rates to be achieved. These coatings enhance the wear and crater resistance of cemented carbides with a modest loss in strength. As a result, a major portion of the market in cast iron, steel, and stainless steel machining is served by these materials, with machining speeds up to 275 m/min (900 sfm).

**Cermets,** like carbides, are composite materials, the primary difference being the use of nickel instead of cobalt as a binder material. The high TiC content results in a material with very high hardness and thermal deformation resistance. However, they are not as tough as conventional carbides and so are used primarily in semifinishing and finishing operations at speeds up to 365 m/min (1200 sfm).

**Ceramics.** Although ceramics have been used for machining since 1905, only the recent advancements in materials and processing methods have opened up a larger market share by improving the toughness and reliability of the various compositions. Modern ceramics are available in a variety of compositions tailored to market needs. These include a more or less pure $Al_2O_3$ ceramic for finishing applications, a composite $Al_2O_3$-TiC ceramic for semifinishing applications, and a whisker-reinforced alumina-silicon carbide ($Al_2O_3$-$SiC_w$) material for the roughing of nickel-base high-temperature materials. Additionally, silicon nitride ($Si_3N_4$) base materials are used for the high-speed machining of cast irons and the roughing of nickel-base high-temperature alloys. The high hardness and thermal deformation resistance of these materials allow metal removal at speeds as high as 1220 m/min (4000 sfm).

**Cubic boron nitride** has an extremely high hardness and compares favorably in toughness to conventional ceramic materials. Although very costly to manufacture, it can provide an economical advantage when machining steels and cast irons where the hardness exceeds 35 HRC, especially in interrupted cutting. This material is also occasionally used in the machining of aerospace alloys.

**Polycrystalline Diamond.** The hardest material known to man, PCD is used for machining nonferrous and composite materials at very high speeds. Limited by its toughness and chemical resistance, this material is commonly used in the machining of 390 aluminum-silicon alloy and other highly abrasive materials. Polycrystalline diamond is not recommended for the machining of ferrous alloys.

## Cutting Tool Material Test Methods

During the course of cutting tool material development and the qualification of production processes, a variety of machining tests are used to determine market position, cutting ranges, and overall quality of products. Because the goal is to produce a cutting tool with the widest possible application range, a matrix test approach is used, with testing being conducted over a wide range of workpiece materials and cutting conditions. Ultimately this allows the cutting tool material supplier to target the tool in the marketplace and to provide useful application information to the end user at product introduction. The following matrices represent the range of tests performed on candidate cutting tool materials.

### Workpiece Matrix

Cutting tool compositions vary widely because of the variety of workpiece materials they are designed to machine. A grade or geometry targeted for a particular market niche might be tested on only one or two materials, while a general-purpose grade or geometry may be tested on three or more materials. These workpiece materials are generally from one or more of the following groups:

- Plain carbon and alloy steels
- Cast irons

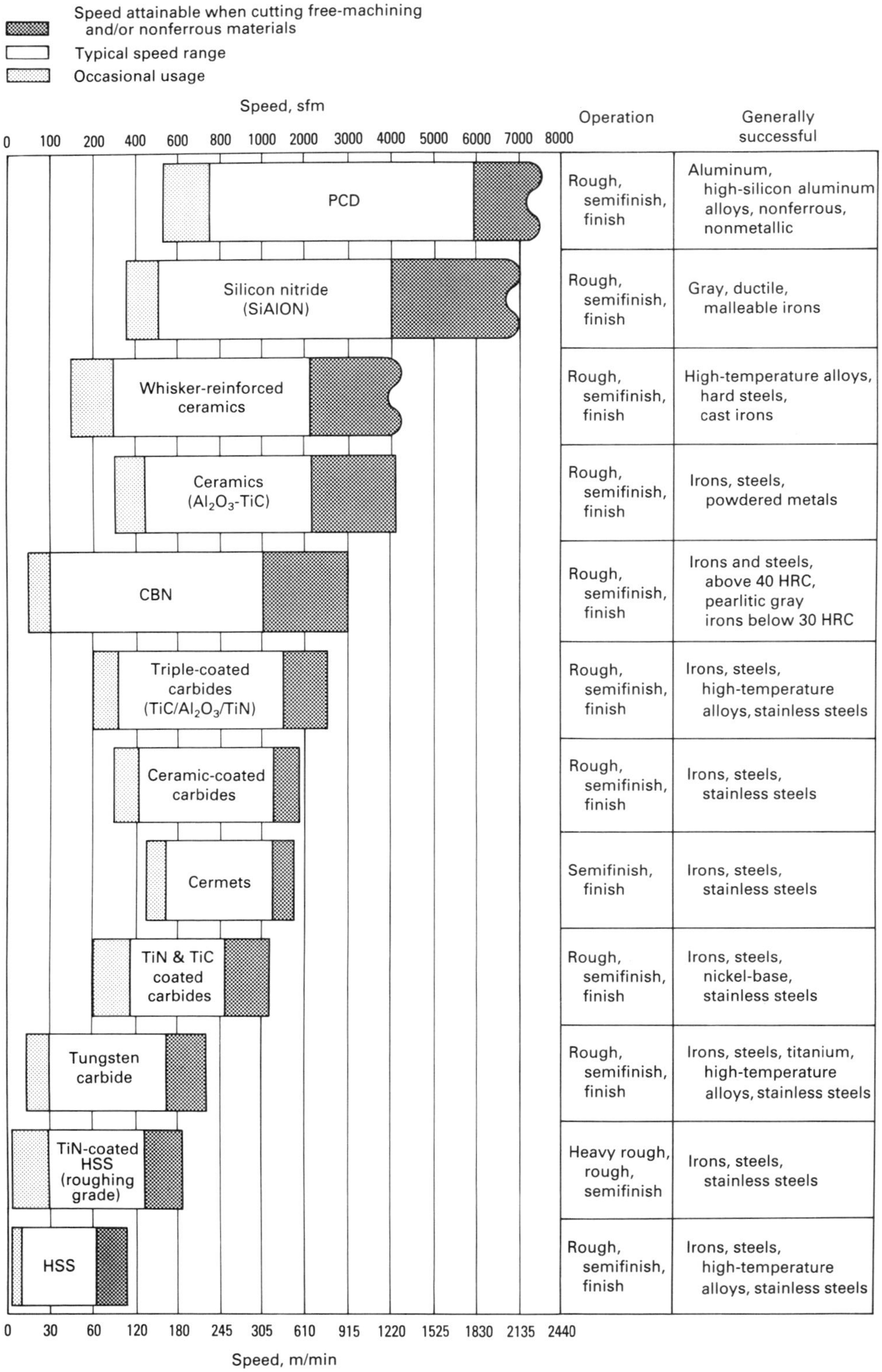


**Fig. 1** Approximate speed ranges and applications of various cutting tool materials. Source: GTE Valenite Corporation

- Stainless steels
- Tool steels
- High-temperature (aerospace) alloys
- Nonferrous materials
- Composites

Utilizing single-point turning or milling test conditions appropriate to the workpiece material and the tool being tested, an evaluation of the tool applicability and projected tool life can be made. Typical workpiece-tool testing combinations include the following:

- *Uncoated carbide*: Any of the materials listed above
- *Coated carbide*: AISI 1045, 4140, 4150, 4340, and 8620, gray and ductile cast irons, and stainless steels
- *Cermets*: Steels at finishing conditions
- *$Al_2O_3$-TiC ceramics*: Tool and hardened steels
- *$Al_2O_3$ ceramics*: Cast irons and steels at finishing conditions
- *Whisker-reinforced ceramics*: Inconel 718 and most nickel-base high-temperature alloys
- *$Si_3N_4$ ceramics*: Cast irons at high speeds and nickel-base high-temperature alloys
- *CBN*: Chilled cast iron (>40 HRC), tool steels, Ni-Hard cast iron, and hardened steels
- *PCD*: Aluminum-silicon alloys (380, 390, and so on), composites, and other abrasive materials

## Property Matrix

In evaluating a cutting tool material in a machining operation, the applicability is dependent on having the correct combination of physical properties. Because maximizing one property typically means lowering some other property (for example, extremely wear-resistant materials generally have poor toughness), the workpiece properties and cutting operations strongly influence the selection of a cutting tool material. Similarly, developing an understanding of the tool properties and their relationship to cutting performance is the key to understanding the potential application range.

Properties of primary concern in cutting tool design are fracture resistance (toughness), plastic or thermal deformation resistance, and wear resistance. Also of concern is the resistance of the material to cracking, notching, cratering, and, in the case of coated inserts, spalling (poor coating adherence).

Machine tests are designed to measure the resistance of cutting tools to these failure mechanisms. This is accomplished by matching the workpiece and tool materials, adjusting machining parameters over the likely application range, and then carefully evaluating the failure modes and their relationship to basic physical properties of materials. Common failure mechanisms are shown in Fig. 2. The cutting tool properties of most concern in evaluating these failures are described below.

**Toughness** is commonly defined by cutting tool users as the resistance of the cutting edge to breakage or fracture under unfavorable conditions, which typically might include one or more of the following:

- High feed rates
- Moderate-to-severe interruptions
- Inconsistent or difficult workpiece material
- Lack of rigidity

Because milling incorporates many of these factors, a flycut milling test is commonly

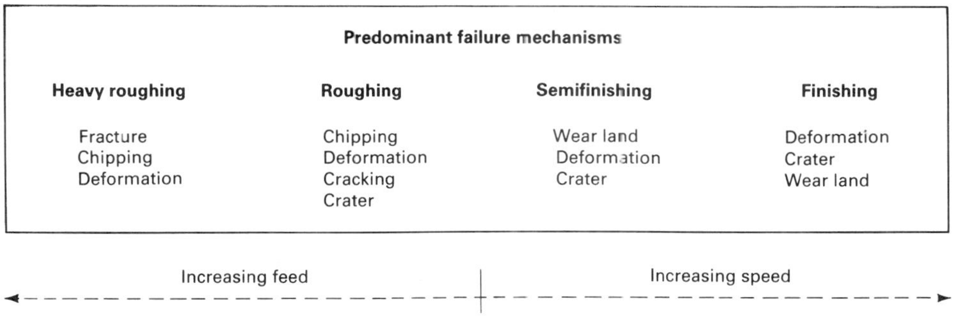


**Fig. 2** Predominant failure mechanisms of coated carbides when machining steels

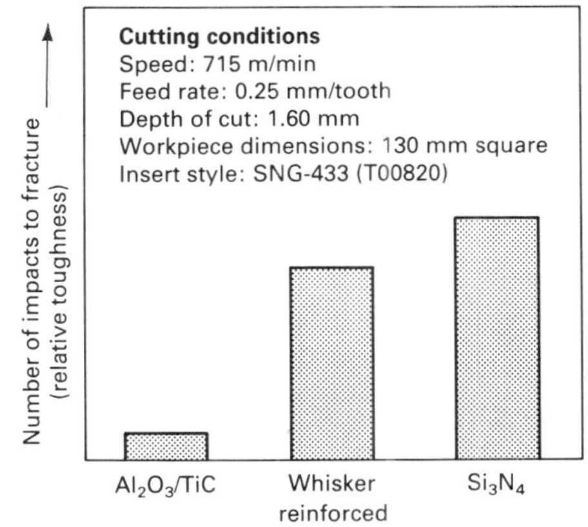


**Fig. 3** Relative toughness of three types of ceramic tools in milling gray cast iron

used to evaluate toughness. Figure 3 shows the relative toughness of three different types of ceramics by comparing the number of impacts required before failure occurs by breakage, chipping, or cracking.

**Flank Wear Resistance.** The resistance of the cutting tool to flank wear is a primary concern in the applicability of the material to a cutting operation. Because the tool must have adequate resistance to both abrasive and chemical wear, the chemical inertness of the material as well as its hardness must be considered. Figure 4(a) shows a typical wear land generated during a turning or facing operation. It is the balancing of the abrasive and chemical wear resistance, along with toughness, that gives each tool material family its own application range, as shown in Fig. 1. Generally, as the cutting speed increases, chemical wear resistance becomes more important; therefore, inert materials, such as oxides and nitrides, perform better at higher cutting speeds.

**Deformation.** A secondary effect of high cutting speeds is the occurrence of plastic deformation due to increased cutting temperatures. Under these conditions, the binder phase of the cutting tool material may soften and deform causing a bulging effect, as shown in Fig. 4(b). Such an effect can cause breakage, a higher flank wear rate, or spalling of the coatings. This effect is accentuated under high chip loads and is a typical failure mode for rough cutting in which the toughness requirements prevent the use of ceramic or other binderless cutting materials.

**Crater Resistance.** A secondary result of abrasive and chemical wear is the formation of a crater on the rake surface of the insert. In this case, the chips formed during the operation rub and weld on the rake face, causing a dish-shaped depression to be formed, as shown in Fig. 4(c). As a result of this formation, cutting forces can increase and, coupled with the weakening of the insert, can result in catastrophic failure due to chipping or breakage. Crater formation can be inhibited, however, with the proper chip-breaker geometry.

**Cracking** occurs under heavy mechanical load and/or as a result of rapid and repeated temperature changes during the cutting operation. Cracks typically initiate at high-stress areas and extend parallel or perpendicular to the cutting edge, as shown in Fig. 4(d). Commonly seen in milling operations, these cracks eventually lead to catastrophic failure due to chipping or breakage.

**Notching** occurs most commonly in the machining of high-temperature materials such as Inconel 718. It arises in a localized area at the depth of cut and is a result of the high stresses in that area. Under these conditions, minor chipping, coupled with an accelerated wear rate, results in a failure of the type shown in Fig. 4(e). As this notch grows, the likelihood of crack formation and subsequent breakage is increased. The resistance of whisker-reinforced ceramics to this crack formation is one of the reasons why it has been successful in cutting this type of material.

**Chipping** is a less severe and more common form of breakage. This is the result of small fragments of the cutting edge being broken away during the cutting operation, as shown in Fig. 4(f). This is a result of the strength of the cutting tool material being exceeded in localized areas due to chatter, variations in the workpiece, or buildup on the cutting edge. Edge preparation plays an important role in minimizing the occurrence of chipping.

Ultimately, the intent in targeting application ranges is to avoid fracture of the insert, as shown in Fig. 4(g). When this occurs, the insert (and often the workpiece) is rendered unusable.

## Operation Matrix

In order to evaluate the performance range of a comparable group of cutting tools in single-point turning or in milling tests, cutting conditions are selected to cover the expected application range of the group (particularly cutting speed and feed rate). The choice of workpiece materials is likewise based on the expected application range of the group of cutting tools. Tool life criteria are defined and, along with cutting conditions and workpiece materials, are organized into a table called a machine test matrix (Table 1).

Although such testing is time consuming, it needs to be conducted only once for any group of materials. The testing is conducted to:

- Determine any performance differences in wear, toughness, deformation resistance, and other failure mechanisms
- Establish curves showing the relationship between cutting speed and tool life, as shown in Fig. 5. Tool life curves allow an accurate comparison between grades and provide a tool life exponent, which is widely used in tool life formulas and machining economics
- Determine the type of failure mechanism(s) encountered in such conditions and to study the influence of each mechanism on tool performance. This allows the definition of cutting conditions in terms of failure mechanisms, such as the example shown in Fig. 2
- Study the effect of different materials and conditions on substrates and/or coatings—for example, the effect of speed on one substrate with two different types of coating

The same procedures are applied in turning and milling. It should be noted that the tool life criteria shown in Table 1 can differ, depending on tool material, workpiece material, test type, and machining conditions.

## Cutting Conditions

A wide variety of machine tests are used to evaluate cutting tools. In addition to a variety of workpiece types, these tests employ variations in speed, feed rate, and depth of cut. These variables can be manipulated to generate the desired failure modes and to determine the overall application range of the cutting tool material. Similarly, the type of machining operation itself determines the primary properties required for success. Milling grades typically require higher strengths than their turning counterparts.

Determining the application range of a cutting tool, then, requires the variation of speed, feed rate, and depth of cut, as well as

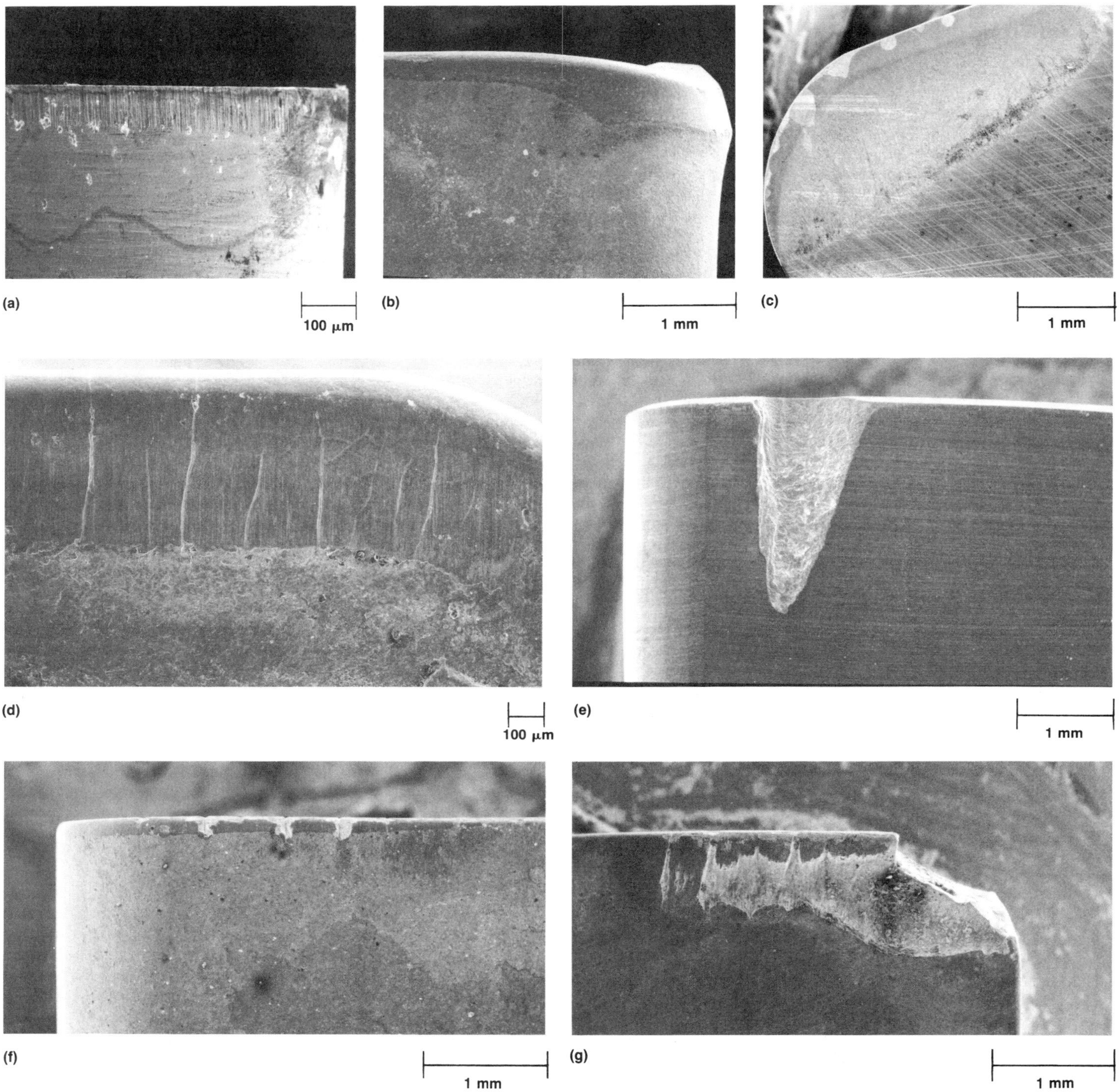


**Fig. 4** Failure mechanisms of cutting tools. (a) Typical flank wear on a carbide insert. (b) Typical edge deformation on a carbide insert. (c) Typical crater wear on a carbide insert. (d) Typical perpendicular cracks on a carbide insert. (e) Typical notching at depth of cut on a whisker-reinforced ceramic insert. (f) Typical chipping on a carbide insert. (g) Typical fracture on a carbide insert

workpiece material. The following tests are typical of those used in this determination.

**The impact test** is a flycut milling test that uses a 150 mm (6.0 in.) diam cutter body loaded with a single insert. This test is conducted at a constant speed and depth of cut, with the feed rate being varied to determine the overall toughness of the cutting tool. Typical conditions for this test in carbide testing are as follows:

- *Material*: AISI 4150 steel (resulfurized), 265 HB, 75 × 40 mm (3.0 × 1.5 in.) cross section
- *Speed*: 230 m/min (750 sfm)
- *Depth of cut*: 3.2 mm (0.125 in.)

The centerline of the cutter is aligned with the entry of bar, and the insert is run one pass per corner at successively higher feed rates until chipping or breakage occurs.

**The turning test** is generally used to determine wear resistance and high-speed deformation resistance. In this test, speed, feed, and depth of cut are constant, and tool life is measured as the time required to form a 0.38 mm (0.015 in.) wear land or a 0.13 mm (0.005 in.) crater depth. A wide variety of conditions are used, depending on the material being tested, but typical conditions might be:

**Table 1 An example of a machine test matrix**

| Test | Work material | Speed m/min | Speed sfm | Feed mm/rev | Feed in./rev | Depth of cut |
|---|---|---|---|---|---|---|
| 1 | A | 150 | 500 | 0.25 | 0.010 | Constant |
| 2 | | | | 0.50 | 0.020 | |
| 3 | | | | 0.75 | 0.030 | |
| 4 | A | 210 | 700 | 0.25 | 0.010 | Constant |
| 5 | | | | 0.50 | 0.020 | |
| 6 | | | | 0.75 | 0.030 | |
| 7 | A | 275 | 900 | 0.25 | 0.010 | Constant |
| 8 | | | | 0.50 | 0.020 | |
| 9 | | | | 0.75 | 0.030 | |
| 1 | B | 90 | 300 | 0.25 | 0.010 | Constant |
| 2 | | | | 0.38 | 0.015 | |
| 3 | | | | 0.50 | 0.020 | |
| 4 | B | 150 | 500 | 0.25 | 0.010 | Constant |
| 5 | | | | 0.38 | 0.015 | |
| 6 | | | | 0.50 | 0.020 | |
| 7 | B | 210 | 700 | 0.25 | 0.010 | Constant |
| 8 | | | | 0.38 | 0.015 | |
| 9 | | | | 0.50 | 0.020 | |

Note: Tool life criteria for carbide tools: 0.380 mm (0.015 in.) flank/nose wear, 0.13 mm (0.005 in.) crater, chipping, or breaking of the cutting edge, whichever occurs first

- *Material*: AISI 1045 steel, 180 HB
- *Speed*: 210 m/min (700 sfm)
- *Feed*: 0.50 mm/rev (0.020 in./rev)
- *Depth of cut*: 2.55 mm (0.100 in.)

During the test, inserts are checked and wear lands measured at regular intervals with the tool life to achieve a 0.38 mm (0.015 in.) wear land or a 0.13 mm (0.005 in.) crater depth expressed in minutes.

**A facing test** is used to provide a rapid means of evaluating the wear and crater resistance of cutting tools (particularly coated materials). In this test, speed, feed, and depth of cut are held constant while the insert is run through multiple facing passes. The total number of passes required to achieve a 0.38 mm (0.015 in.) wear land or a 0.13 mm (0.005 in.) crater depth is considered the tool life. Typical conditions for testing a coated carbide material are as follows:

- *Material*: AISI 4150 steel, 220 HB, facing 175 to 50 mm (7 to 2 in.) diameter
- *Speed*: 275 m/min (900 sfm)
- *Feed*: 0.40 mm/rev (0.015 in./rev)
- *Depth of cut*: 1.6 mm (0.0625 in.)

**Nodular Iron Impact Test.** A specialized test used to determine relative coating adherence is conducted using a gummy nodular (ductile) iron. This is a flycut test utilizing a 150 mm (6.0 in.) diam cutter loaded with a single insert. This insert is exposed to a single pass at a specified speed, feed, and depth of cut. Adhered iron is then removed in an acid bath, and the insert is inspected to determine the amount of coating that has spalled away from the substrate. This is expressed as a percentage of the total chip contact area. Typical conditions for this test are as follows:

- *Material*: 80-55-06 ductile iron, 235 HB, 50 mm (2 in.) diam cross section
- *Speed*: 275 m/min (900 sfm)
- *Feed*: 0.23 mm/tooth (0.0092 in./tooth)
- *Depth of cut*: 3.2 mm (0.125 in.)

## Machinability Ratings

Machinability is defined as the relative ease or difficulty of removing metal in transforming a raw material into a finished product. The concept of a machinability rating, that is, a number assigned to indicate the ease or difficulty of machining for materials, was introduced early in this century when high-speed steel tools dominated the marketplace. The tool life obtained in turning B1112 steel with a high-speed steel tool at 55 m/min (180 sfm) was assigned a rating of 100% (Ref 9). Machinability ratings were assigned to other materials based on the relative tool life achieved when machining them under the same conditions with a high-speed steel tool.

The development of new cutting tool and workpiece materials, along with the wide variety of chip-breaker geometries, has complicated this somewhat. In measuring the machinability of a material, the following factors must be taken into consideration:

- Cutting tool material
- Machining operation (continuous or interrupted)
- Geometry of cutting tool
- Edge preparation
- Machining conditions
- Cutting fluid
- Rigidity of the machine

The machinability rating is also influenced by material hardness, tensile strength, chemical composition, microstructure, strain hardenability, and manufacturing process.

When machining steels, hardness is often used as a relative measure of machinability; the harder the material, the less machinable. However, steels with low carbon levels are more gummy and tend to develop a buildup on the edge of the cutting tool, resulting in a lower-than-expected machinability rating. Small additions of sulfur or phosphorus can significantly improve machinability. Therefore, chemical composition is equally important in estimating the machinability of steels.

Microstructure is also influential in determining the machinability of materials. Figure 6 shows the variation that can occur in a gray cast iron from 260 to 200 mm (10.3 to 8.0 in.) in diameter. This resulted in a considerable difference in the machinability of the material, with a $Si_3N_4$ test tool failing catastrophically at diameters greater than 215 mm (8.5 in.).

The influence of the manufacturing process is illustrated in an A390 aluminum-silicon alloy. In this case, a material manufactured by high-pressure conventional die casting is more machinable than the same alloy manufactured by sand casting (Ref 10). The die cast process produces very fine, evenly distributed silicon particles, while the sand cast process has larger particles and poorer distribution. The result is a large difference in machinability.

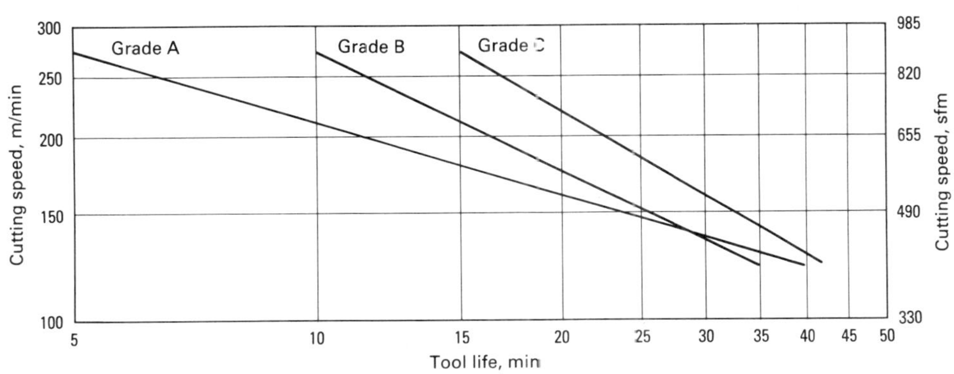


**Fig. 5** Typical tool life lines of carbide tools

## Development of Machinability Ratings

Machinability ratings for different materials can be developed by using one type of cutting tool at a constant set of cutting conditions. These ratings, however, may

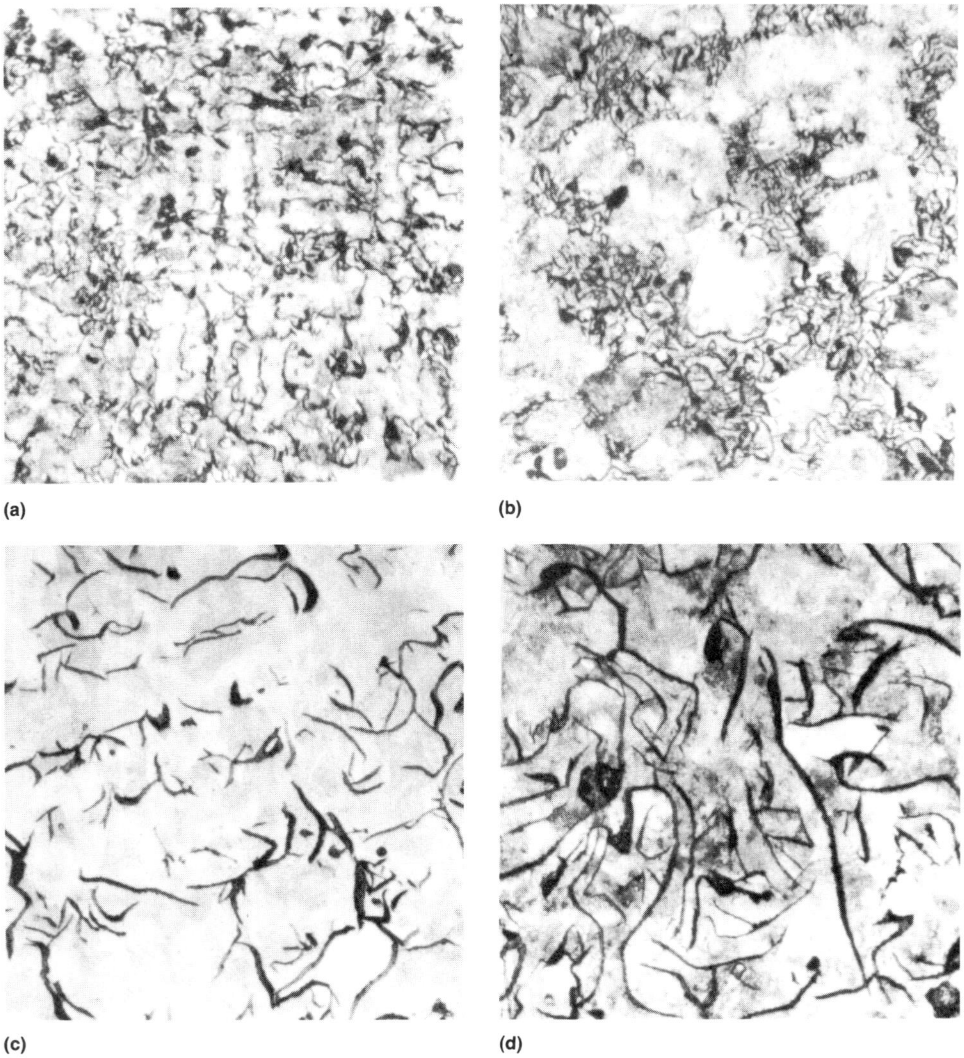

**Fig. 6** Microstructural variations that influence machinability in class 40 gray iron 260 mm (10.3 in.) diam bar samples. Machinability increases from (a) low machinability to (d) highest machinability. (a) This gray iron has type D and E distribution of graphite flakes approximately 260 to 250 mm (10.3 to 9.8 in.) in diameter. The graphite flakes are surrounded by pearlite and isolated pools of ferrite. (b) This gray iron has type D distribution of graphite flakes approximately 250 to 235 mm (9.8 to 9.3 in.) in diameter. The graphite flakes are surrounded by pearlite and small isolated pools of ferrite. (c) This gray iron has type D (coarse) distribution of graphite flakes surrounded by pearlite approximately 235 to 205 mm (9.3 to 8.1 in.) in diameter. (d) This gray iron has type A distribution of graphite flakes surrounded by pearlite below 205 mm (8.1 in.) in diameter. All at 150×

change as conditions or cutting tool materials change. Machinability tests are conducted in different forms, but generally include the following:

- Tool life or wear tests
- Surface finish tests
- Cutting force tests
- Power consumption tests
- Cutting temperature tests

Tool life is generally considered to be the amount of time the tool produces parts with acceptable finish and/or tolerance while not showing sufficient wear to be in danger of catastrophic failure.

**Tool Life or Wear Tests.** In determining a machinability rating based on tool wear, tests are conducted at one speed on a variety of workpiece materials, utilizing one type of cutting tool material and one geometry. The time or work required to generate a predefined amount of wear is found, and machinability ratings are assigned based on the relative tool life for the respective materials. For example, materials A, B, and C are to be evaluated for machinability rating using a carbide tool. The wear generated as a function of time for all the materials is shown in Fig. 7. The tool lives for materials A, B, and C to generate 0.38 mm (0.015 in.) of wear are 50, 32, and 14 min, respectively. The relative assigned machinability ratings are given in the table accompanying Fig. 7.

In Taylor's tool life tests, a series of operations is run to determine the relationship between tool life and cutting speed (Ref 11). The machinability of a material is then based on the cutting speed that produces a predetermined tool life of 60 min. The relationship between tool life and cutting speed for three workpiece materials, designated A, B, and C, that are to be machined using carbide tools is established in Fig. 8 using the Taylor methodology.

The speeds that produced a tool life of 60 min for materials A, B, and C are 189, 146, and 122 m/min (620, 480, and 400 sfm), respectively. The assigned relative machinability ratings are given in the table accompanying Fig. 8.

**Surface Finish, Cutting Force, and Power Consumption Tests.** Surface finish tests consider the finish produced on the machined part as the criterion for assessing the machinability of materials. The better the finish produced under a given set of conditions, the more machinable the material. Similarly, cutting force tests measure the feed or longitudinal force required to machine a material under a given set of conditions (Ref 12). The less force required, the more machinable the material. In a similar way, power consumption tests measure the horsepower required for machining materials.

**Cutting temperature tests** measure the temperature at the tip of the cutting tool during machining operations (Ref 13). Harder materials and/or higher cutting speeds generally produce higher cutting tip temperatures, with a drop in tool life. These tests, although difficult and time consuming to conduct, have established some correlation to the results produced by tool life and tool wear test methods and probably represent the best means by which to determine tool and workpiece performance.

## Variables in Machinability Ratings

The variation in machinability ratings between materials and operations is indicated in Table 2. The machinability ratings of four workpiece materials are listed along with the variation that occurs as a result of machining with high-speed steel or carbide tools in turning or milling operations. For example, the machinability of maraging steel with a high-speed steel tool can vary from 54 to 92, depending on the operation. Similarly, the machinability rating of these four materials in milling with carbide tools can vary from 20 to 88.

Much of this variability can be attributed to the inverse relationship between toughness and wear resistance in cutting tools. Tools that have high abrasion resistance, such as ceramics, tend to demonstrate lower toughness. Such tools, when used in machinability studies, will produce a higher rating in turning and a lower rating in milling for most of the commonly used workpiece materials.

Most machinability ratings are developed under a fixed set of cutting conditions, even though a wide variety of machining condi-

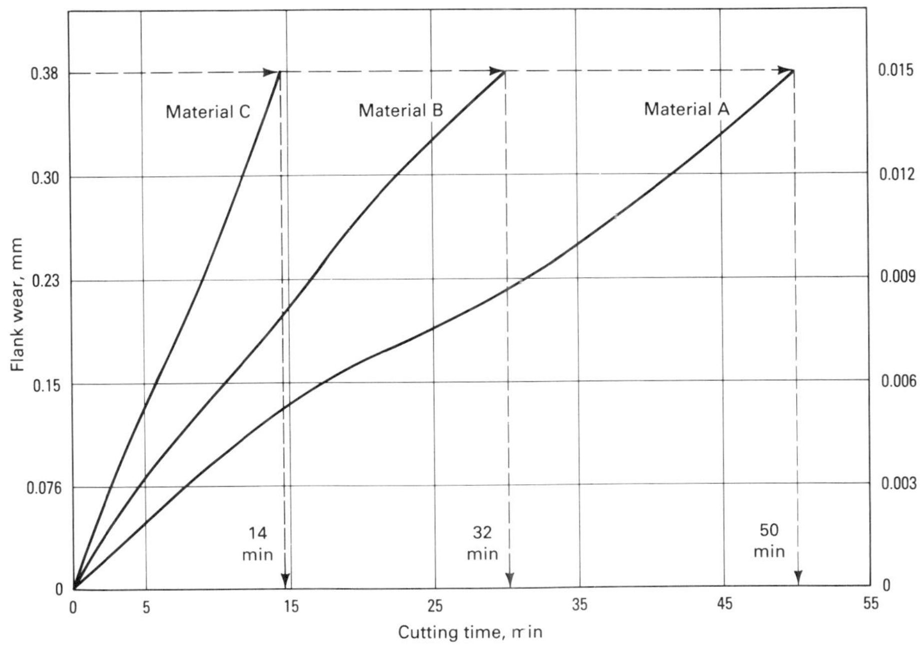


| Material | Tool life, min | Machinability rating |
|---|---|---|
| A | 50 | 100 |
| B | 32 | 64 (32/50) × 100 |
| C | 14 | 28 (14/50) × 100 |

**Fig. 7** Relationship between flank wear and cutting speed and the resulting machinability rating for three different materials

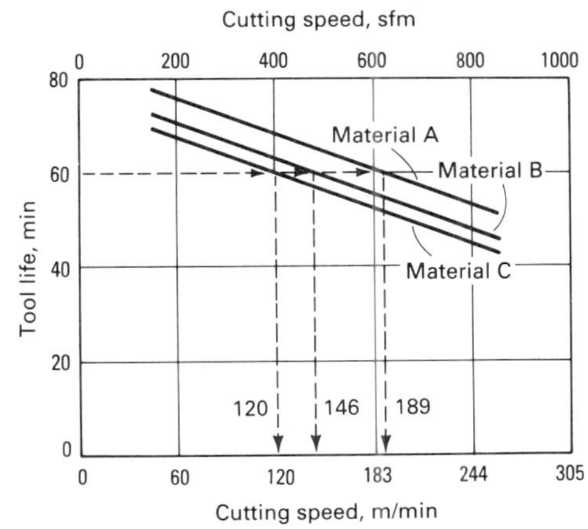


| Material | Cutting speed for 60 min tool life m/min | sfm | Machinability rating |
|---|---|---|---|
| A | 189 | 620 | 100 |
| B | 146 | 480 | 77 (146/189) × 100 |
| | | | (480/620) × 100 |
| C | 122 | 400 | 65 (122/189) × 100 |
| | | | (400/620) × 100 |

**Fig. 8** Relationship between tool life and cutting speed and the resulting machinability rating for three different materials

tions are encountered in practice. This of course serves to simplify a difficult problem by eliminating the variables most easily controlled. However, the cutting conditions can change.

**Variations in Tool Material.** The results from a machinability study conducted on microalloyed dual-phase sheet steel are shown in Fig. 9. In this study, a constant feed rate and depth of cut were maintained over different cutting speeds. Machining was done with coated and uncoated carbide tools. In machining with uncoated carbide, the tool life fell from 50 min at 183 m/min (600 sfm) to 10 min at 305 m/min (1000 sfm). Coated carbide tools, which begin with four times the tool life of uncoated carbides at 183 m/min (600 sfm), demonstrated a similar drop.

**Variations in feed and depth of cut** can create similar effects in the machinability of materials, depending on the affinity of the cutting tool and workpiece materials. The reason is the additional heat generated by the cutting operation at higher speeds, feeds, or depths of cut. As mentioned earlier, at these higher cutting temperatures, chemical wear takes over as the predominant component of overall tool wear. The result is a rapid acceleration of the tool wear rate unless extremely inert tool or workpiece materials are being used.

**Cutting tool geometries** can strongly influence machinability ratings. These geometries include positive versus negative rake, flat top versus chip groove geometry, radius versus chamfer edge preparation, and ground versus polished rake surfaces. Negative rake geometries generate higher cutting forces than positive rake geometries. When machining gummy materials such as aluminum, titanium, and some stainless steels, negative rake geometries can result in built-up edge and a lowering of tool life and the machinability rating. Negative rake inserts with a chip-breaker groove present a higher shear angle to the workpiece than a flat top insert, thus lowering forces and extending tool life. Larger hones (edge radius) will accelerate wear in a turning operation while improving tool life in milling by preventing chipping. One study on the machining of aluminum found that a tool with a polished rake face reduces built-up edge, providing better tool life than a conventionally ground rake face (Ref 15).

It is obviously a difficult and complex task to develop universal and meaningful machinability ratings for workpiece materials, given the variables in the machining process. However, machinability ratings can be developed, provided the cutting tool and process conditions are held constant. Experience in machinability testing suggests that the end user should become familiar with the capabilities of the different cutting tools and the properties of the workpiece materials employed in order to use machinability ratings wisely and productively.

## Application/Grade Selection

A knowledge of machinability and cutting tools is the starting point for selecting a grade/geometry combination for any machining operation. The task of the tooling engineer is to select, from the vast numbers of possibilities, the grade that will give the best performance, predictability, reliability, and economic justification. To accomplish this, operation characteristics, insert and chip groove selection criteria, and machining economics need to be evaluated (Ref 16).

### Operation Characteristics

**The operation category** involves determining whether heavy roughing, roughing, semifinishing, or finishing is needed. For example, the high-speed finishing of steel

**Table 2 Machinability rating inconsistencies**

| Operation | Work material: 4340 steel at 340 HB | Ti-3Al-13V-11Cr at 285 HB | 17-7 PH steel at 170 HB | 250 maraging steel at 330 HB |
|---|---|---|---|---|
| **Turning** | | | | |
| High-speed steel | 25 | . . . | 32 | 54 |
| Carbide | 31 | 16 | 18 | 59 |
| **Milling** | | | | |
| High-speed steel | 25 | 16 | 37 | 74 |
| Carbide | 88 | 20 | 53 | 83 |
| **Drilling** | | | | |
| High-speed steel | 42 | 17 | 42 | 92 |

Source: Metcut Research Associates Inc.

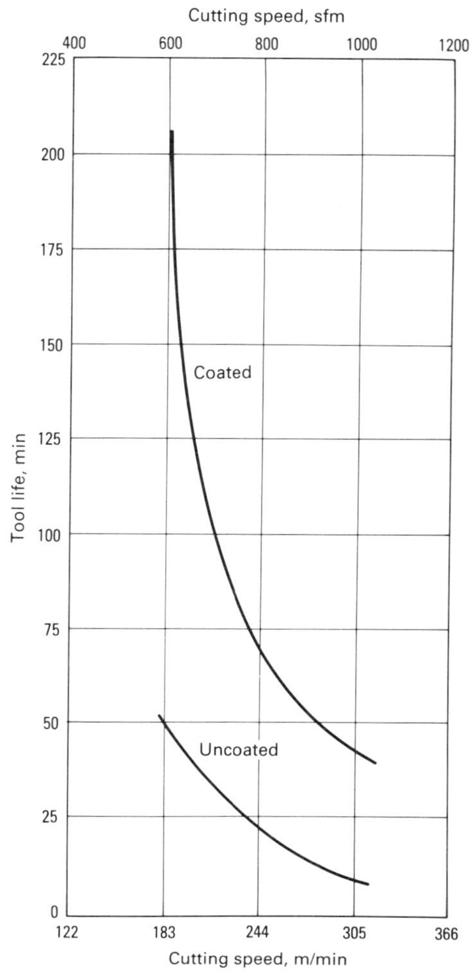


| | |
|---|---|
| Tool | C-5 uncoated and coated carbide |
| Feed, mm/rev (in./rev) | 0.23 (0.009) |
| Insert style | TNEX-333-2M |
| Depth of cut, mm (in.) | 0.38 (0.015) |
| Cutting fluid | Dry |
| Wear, mm (in.) | 0.25 (0.010) |
| Notching, mm (in.) | 0.38 (0.015) |

**Fig. 9** Effect of cutting speed on tool life in turning dual-phase sheet steels. Source: Ref 14

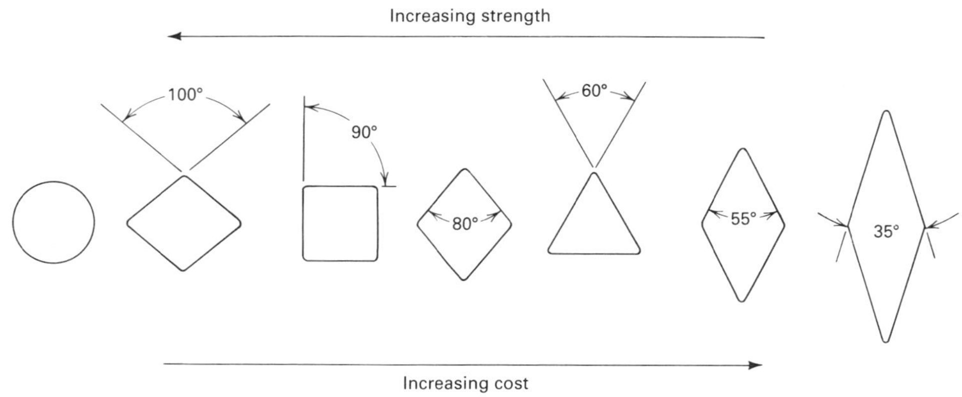


**Fig. 10** Relationship among insert shape, strength, and cost

requires high wear and deformation resistance. Coated grades, cermets, and ceramics are candidates for this type of operation.

**Workpiece Material Type.** Whether a material is hard, soft and gummy, or difficult to machine must also be determined. For example, the machining of soft and gummy low-carbon steel at moderate speeds requires a TiN-coated insert or a cermet to resist buildup at the cutting edge and to provide a good surface finish.

**The production volume** (large or small scale, periodic or long-range production) is a third important consideration. For example, the machining of difficult-to-cut Inconel 718 in large-scale production using whisker-reinforced ceramics will increase productivity and lower cost. In small-scale production, conventional $Al_2O_3$-base ceramics may prove to be more economically justified.

**Surface finish, part geometry, and part tolerance** should also be considered. For example, coated carbide inserts generally produce a better surface finish than uncoated carbides.

**Machine capabilities and limitations** represent the final operation characteristic that must be determined. For example, older machines with poor rigidity and low horsepower cannot make effective use of advanced ceramic materials.

## Insert and Chip Groove Selection

A variety of insert shapes, sizes, nose radii, and chip grooves are available for any machining operation. Proper selection is essential to optimize the operation for productivity and cost. The differences, as well as the advantages and disadvantages of various geometry options, provide flexibility in determining the best choice.

The shape of an insert determines its relative strength and cost, as shown in Fig. 10. In general, the stronger insert has more available cutting edges and therefore can be more economical on a cost per index basis. However, the part geometry may limit the use of the desired insert shape. The general rule is to select the strongest insert capable of producing the required part configuration.

**Sizes and Nose Radii.** The size of an insert is determined by the inscribed circle. The depth of cut should not exceed one-half the inscribed circle size, while the insert thickness should be at least four times the operating feed rate. Finally, the largest possible nose radius should be selected. Larger radii produce better surface finishes, handle heavier feed rates, and strengthen the insert.

**Chip groove selection** is almost as crucial as selecting the proper grade. The performance of the cutting tool is determined not only by the grade properties and coating type but also by a chip groove style that will allow better utilization of the machine tool through lower cutting forces, better chip handling and control, extended tool life, and lower machining costs. Modern numerically controlled machining operations put a high demand on reliable chip flow. Long chips cause interruptions of the machining cycle, a loss of productivity, and damage to the cutting tool and workpiece. Long chips also represent a personal hazard to the operator.

Commercially available inserts with chip grooves are designed to produce acceptable chips throughout the widest possible range of feed rates and depths of cut while maintaining high edge strength. General recommendations are listed below for proper chip groove selection on carbide tools:

- For general-purpose applications, select a chip groove that has its nominal feed range as near as possible to the intended operating feed rate. This will ensure acceptable chips, predictable and reliable tool life, and a lower cost per index
- For heavy feeds/high material removal rates, select single-sided inserts. The improved rigidity and resultant higher effective edge strength permit higher speeds and feeds while generating lower forces than double-sided inserts. In most cases, this will improve productivity
- For finishing or for low feeds and depths of cut applications, select a chip groove that produces a good combination of chip control, low cost per index, and reliable tool life

**Edge preparation** is a final consideration. This is the addition of a radius or chamfer to the cutting edge of the insert. Although the general rule for hones is that less is better, the toughness of the cutting tool material and the integrity of the insert cutting edge should be considered. A larger hone or chamfer gives additional protection against catastrophic failure by fracture, which results in the loss of all cutting edges.

Ceramic edge preparation is even more crucial than that for carbide. The brittle nature of the material requires that special edge preparation or T-lands be used to increase their effective edge strength. As

**Table 3 Different types of ceramic edge preparation with recommended application**

| Operation | Rake angle | Chamfer |
|---|---|---|
| General purpose | Negative | 0.20 mm × 20° (0.008 in. × 20°) |
| Finishing | Negative or positive | 0.075 mm × 25° (0.003 in. × 25°) |
| General purpose and milling | Negative | 0.15 mm × 30° (0.006 in. × 30°) |
| Heavy roughing | Negative | 0.38 mm × 25° (0.015 in. × 25°) |
| Special | Negative or positive | Special |

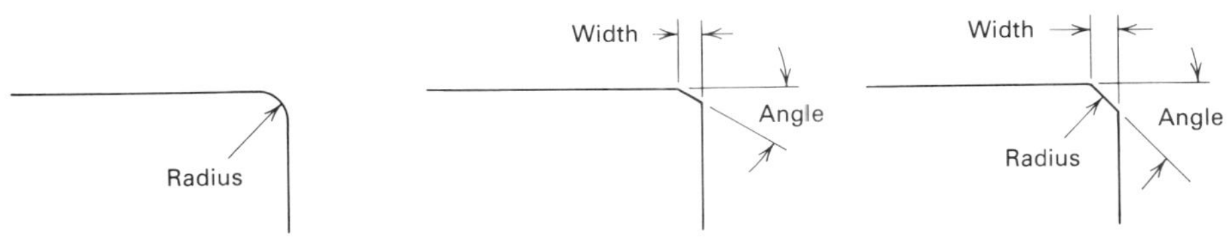


indicated in Table 3, different edge preparations are necessary, depending on the insert shape, cutting speed, feed rate, depth of cut, and required surface finish.

## Machining Economics

The primary variables affecting machining economics vary from operation to operation. However, there are some factors that are important in optimizing the economics of any operation (Ref 17).

The cutting speed is the most important factor in economical machining. Too slow a cutting speed results in poor throughput and an increase in overall production costs. Too fast a cutting speed can result in lower tool life with higher tool costs and more downtime. Somewhere between these extremes lies the optimum.

Figure 11 shows production costs plotted against cutting speed. The optimum cutting range lies between the point of minimum cost and maximum production. The size of this range depends on the operation, the cutting tool material, and the material being machined. Generally, cutting tool manufacturers supply the optimum cutting speed range for all grades in their grade handbooks.

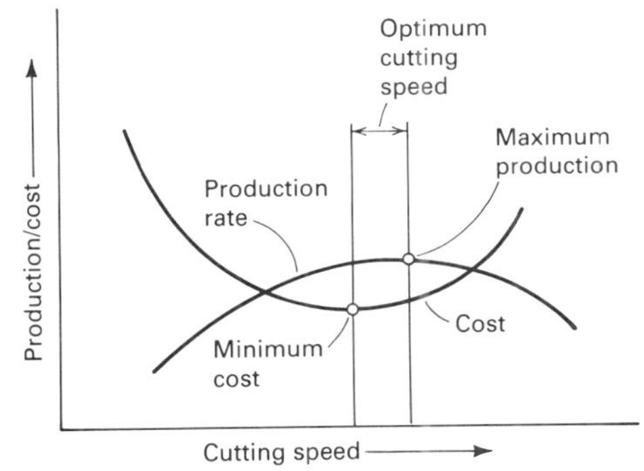


**Fig. 11** Determination of minimum cost, maximum production, and optimum cutting speed. Source: Ref 17

## REFERENCES


1. *Machinability Testing and Utilization of Machining Data*, Proceedings of the International Conference on Machinability Testing and Utilization of Machining Data, American Society for Metals, Sept 1978
2. "Evaluating Machining Performance of Ferrous Metals Using an Automatic Screw/Bar Machine," E 618, *Annual Book of ASTM Standards*, American Society for Testing and Materials
3. *Influence of Metallurgy on Machinability*, Proceedings of the International Symposium Sponsored by the Machinability Activity of the Mechanical Working and Forming Division, American Society for Metals, Oct 1975
4. "Tool Life Testing With Single Point Turning Tools," ISO 3685-1977, International Organization for Standardization, 1977
5. W.B. Heginbotham and P.C. Pandey, Taper Turning Tests Produce Reliable Wear Equations, *Adv. Mach. Tool Des. Res.*, 1966
6. J.L. Thomas and B.K. Lambert, "Reliability of Accelerated Tool Life Testing Method," Technical Paper MR74-703, Society of Manufacturing Engineers, 1974
7. T.S. Kiang and G. Barrow, Determination of Tool-Life Equations by Step Turning Test, in *International Machine Tool Design Research Conference Proceedings*, Sept 1971, p 379-385
8. B. Mills and A.H. Redford, *Machinability of Engineering Materials*, Applied Science, 1983
9. *Machining Data Handbook*, Vol 2, 3rd ed., Metcut Research Associates, 1980, p 19-15
10. "*Machining of 390 Alloy*," Internal Report, Reynolds Metals Company, 1966
11. F.W. Taylor, On the Art of Cutting Metals, *Trans. ASME*, Vol 28, 1907
12. D.W. Murphy and P.T. Aylward, *Machinability of Steel*, Homer Research Laboratories, Bethlehem Steel Corporation
13. G.E. Kane and M.P. Groover, "The Use of Cutting Temperature as a Measure of the Machinability of Steels," Technical Paper MR67-199, American Society of Tool and Manufacturing Engineers, 1967
14. W.J. Riffe and S.P. Boppana, in *Machinability of Sheet Steels*, Proceedings of the SAE International Congress & Exposition (Detroit, MI), Society of Automotive Engineers, 1985, p 7
15. D. Stashko, in *Aluminum Machining—A Practical Approach*, Proceedings of the SAE International Congress & Exposition (Detroit, MI), Society of Automotive Engineers, 1980, p 8
16. H. Pastor, Present Status and Development of Tool Materials: Part 1—Cutting Tools, *Refract. Hard Met.*, Dec 1987, p 196-209
17. C. Donaldson, G.H. Lecain, and V.C. Goold, *Tool Design*, 3rd ed., McGraw-Hill, 1973, p 270-271


## SELECTED REFERENCES


- G. Brandt, "Flank and Crater Wear Mechanisms of Alumina-Based Cutting Tools When Machining Steel," *Wear*, Vol 112, 1986, p 39-56
- T.J. Drozda and C. Wick, Ed., *Tool and Manufacturing Engineer's Handbook*, Vol 1, 4th ed., Society of Manufacturing Engineers, 1983
- R.L. Hatscheck, Take a New Look at Ceramics/Cermets, *Am. Mach.*, May 1981, p 165-176
- T. Hoshi, High-Productivity Machining Research in Japan, *J. Appl. Metalwork.*, Vol 4 (No. 3), July 1986, p 226-237
- R. Komanduri and J.D. Desai, Tool Materials for Machining, Part One, *Carbide Tool J.*, Sept-Oct 1983, p 3-10
- R. Komanduri and J.D. Desai, Tool Materials for Machining, Part Two, *Carbide Tool J.*, Nov-Dec 1983, p 2-10
- R. Komanduri and J.D. Desai, Tool Materials for Machining, Part Three, *Carbide Tool J.*, Jan-Feb 1984, p 3-11
- R. Komanduri, Highlights of the DARPA Advanced Machining Research Program, *J. Eng. Ind.*, Vol 107, Nov 1985, p 325-335
- B.M. Kramer, On Tool Materials for High Speed Machining, *J. Eng. Ind.*, Vol 109, May 1987, p 87-91
- K. Ogawa and Y. Hara, Cutting Performance of Carbide Ceramics in Milling Operations, *Carbide J.*, Nov-Dec 1974, p 4-8
- J.B. Pond, New Tools Change Cutting Economics, *Iron Age*, 7 Feb 1986, p 35-38
- *Turning Handbook of High-Efficiency Metal Cutting*, General Electric Corporation, 1980
- P.N. Vaidyanathan and E.D. Whitney, Materials Technology Meets High Speed Cutting Demands, *Mach. Tool Bluebook*, April 1985, p 64-68

# Machining of Cast Irons*

CAST IRONS vary greatly in machinability. For example, white irons, which are harder than glass, are extremely difficult to machine, while ferritic cast irons are among the easiest metals to machine. Between these two extremes lie the various commercial grades of gray, ductile, and malleable cast irons.

The factors that influence machining, the selection of cutting fluid and cutting tool material, and comparisons of the machinability of different types of cast iron are included in the following sections. Additional information on the machining of cast irons can be found throughout this Volume, particularly in the Sections "Cutting Tool Materials" and "Traditional Machining Processes" and the preceding article "Machinability Test Methods."

## Effect of Microstructure on Machinability

The machinability of iron relates specifically to its microstructure. The shape of the graphite in the iron determines the type of iron, that is, gray, ductile, malleable, or compacted graphite. The presence of graphite provides the free-machining characteristic of iron, and the shape and amount of graphite establish the potential surface finish obtainable with a cutting process and the necessary cutting force. The microstructure of the metal around the graphite determines tool life and establishes the most advantageous cutting speeds and feeds.

An understanding of the influence of microstructure on machinability can provide an insight into more efficient machining and the correct solution to problems without the influence of traditional opinions. Small variations in microstructure can make a difference in machining that would be very significant in high production. Brinell hardness is an indicator of machinability because hardness is primarily dependent on microstructure, but hardness alone is not consistent as a measure of machinability. Gray, ductile, and malleable cast irons are typical of materials for which hardness testing is used to assess machinability. A preferred method is to combine hardness testing with a microstructure evaluation because some microconstituents that adversely affect machinability have only a minor influence on hardness.

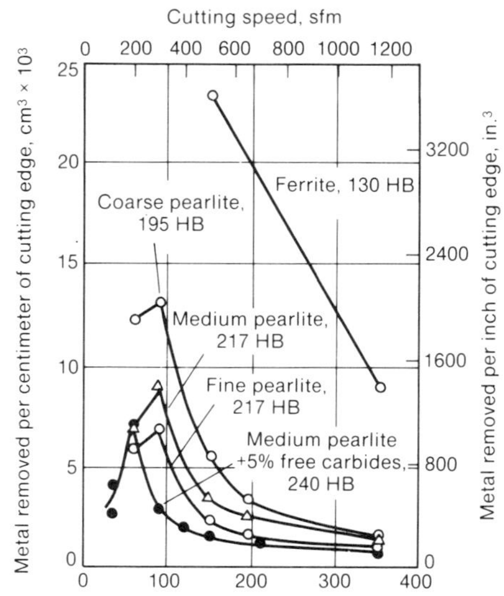


**Fig. 1** Relationship between the matrix microstructure of cast iron and tool life. Cutting operation: milling

In the machining of cast irons, tool life and cutting speeds are directly influenced by microstructure. The graphite in iron is important to its free-machining characteristics, but tool life is dependent on the microstructure of the matrix metal around the graphite. This is demonstrated by the test data shown in Fig. 1. The influence of the various matrix constituents on tool life is discussed below.

**Ferrite** is the essentially carbon-free constituent that constitutes the entire matrix of fully annealed irons. With the exception of graphite, ferrite has the lowest hardness of any constituent in iron (Table 1). However, it is not as soft as the ferrite of low-carbon steel because the ferrite in iron contains silicon. The moderate hardening effect of the dissolved silicon gives ferrite a clean-cutting property. Silicon contents in the usual range of 1.0 to 3.0% have very little influence on tool life, but special irons with higher silicon contents have reduced machinability. High-silicon irons with 14% Si must be abrasively machined.

Figure 2 illustrates the microstructures of typical gray cast irons and a white cast iron.

**Table 1 Hardness of cast iron microconstituents**

| Microconstituent | Knoop hardness, 100 g load |
|---|---|
| Graphite | 15–40 |
| Ferrite | 215–270 |
| Pearlite | 300–390 |
| Steadite | 600–1200 |
| Carbide | 1000–2300 |

As the percentage of ferrite decreases and the pearlite increases, the tool life obtained in such machining operations as turning, face milling, and drilling decreases.

Figures 3 and 4 illustrate typical microstructures of ductile and malleable irons. The annealed structures of both types have a matrix of practically all ferrite and can be machined at relatively high speeds comparable to those used for the machining of gray iron.

Table 2 lists the cutting speeds that result in a 30-min tool life in turning for various microstructures of gray, ductile, and malleable cast irons. It is interesting to note that even though the 100% ferrite structure of ductile iron is 170 HB—considerably harder than the ferritic gray iron at 100 HB—the cutting speeds for a 30-min tool life are comparable.

**Pearlite** is the common constituent in irons of medium strength and hardness. It is composed of alternate plates of soft ferrite and hard iron carbide. The thickness of the alternate plates can vary from coarse to fine. The finer structure is stronger and harder and machines at lower speeds. The pearlite in iron provides the best combination of machinability and wear resistance.

**Martensite** is the full-hard constituent that is formed in heat treating by quenching. It is unmachinable, but when tempered to a lower hardness, it will machine very well —how well depends on the amount of tempering. It is somewhat more machinable than pearlite of the same hardness. When martensite is tempered to a relatively low hardness, it forms a structure of spheroidized carbides in ferrite, which is next to ferrite in machinability.

**Acicular** structures, including bainite, usually occur in alloyed irons or are obtained by a hot-quenching heat treatment. They are generally of intermediate hardness and are machinable with difficulty, a little

*Portions of this article were adapted with permission from *Iron Castings Handbook*, Iron Castings Society, 1981.

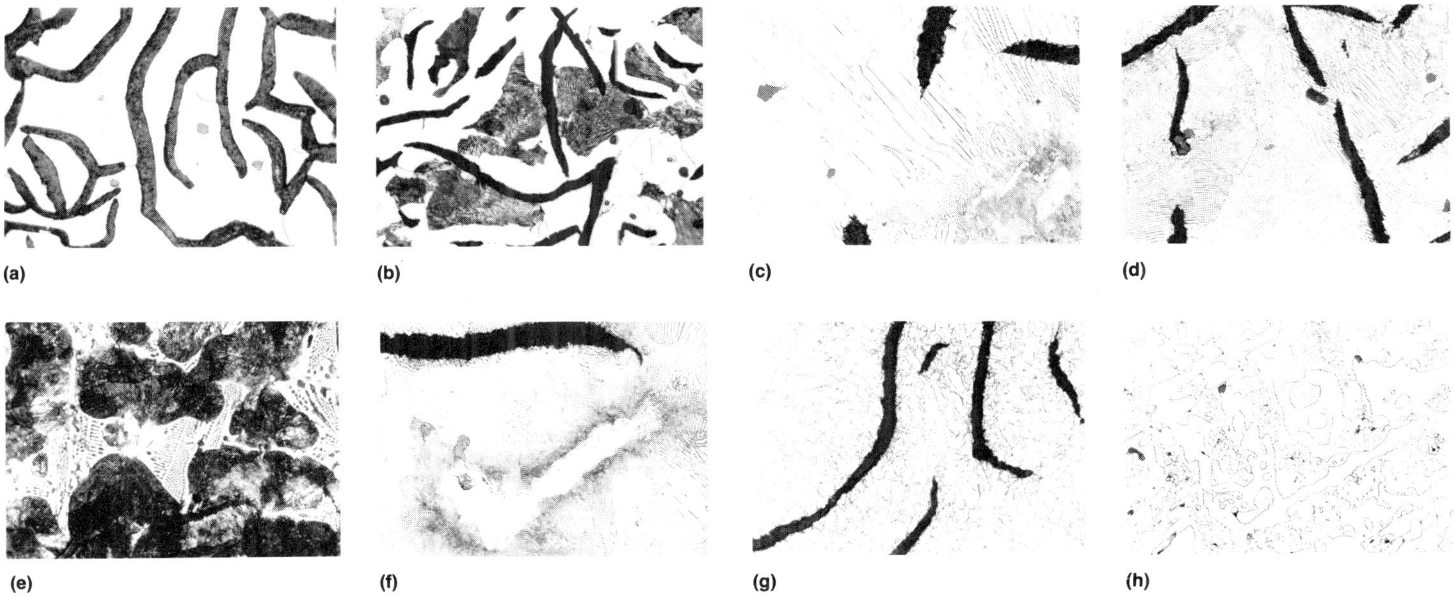

**Fig. 2** Microstructures of gray irons. (a) 100% ferrite; 120 HB. (b) 50% pearlite, 50% ferrite; 150 HB. (c) Coarse pearlite; 195 HB. (d) Fine pearlite; 215 HB. (e) Pearlite plus steadite; 200 HB. (f) Pearlite plus carbide; 240 HB. (g) Acicular; 263 HB. (h) White iron, pearlite plus carbide; 550 HB. All except white iron contain flake graphite. Etched with nital. 100×

less machinable than tempered martensite of the same hardness.

**Austenite** is the principal constituent in the high-nickel gray and ductile irons that are nonmagnetic. Austenite is a relatively soft constituent and is comparable to ferrite in machinability. However, some types of the austenitic irons contain sufficient chromium to produce chromium carbides in the microstructure. This reduces the machinability more than the increased hardness would imply.

**Carbides** are extremely hard constituents whether they are plain iron carbides or contain alloying elements. When in the thin plates in pearlite, carbide can be readily sheared, but in larger particles in iron as a separate constituent, carbides are very detrimental to tool life. As little as 5% free carbides in the matrix structure has a significant detrimental effect (Fig. 5). Carbides can be present at the edge of a casting because of very rapid solidification of the iron at this point.

**Steadite** is a hard constituent that is formed by phosphorus in the iron. With a phosphorus content of less than 0.20%, there is no significant effect on tool life. At 0.40% P, a detrimental effect on machinability can be evident in critical operations such as tapping. Higher phosphorus contents significantly reduce tool life.

**Mixed-matrix structures** result in a tool life that is intermediate to those of the individual components, but the combined effect is likely to be somewhat less than a prorated amount. For example, a matrix structure with 50% ferrite and pearlite will probably produce a tool life that is somewhat less than the average to be expected with an entirely ferritic or pearlitic structure and an entirely pearlitic matrix of equivalent fineness.

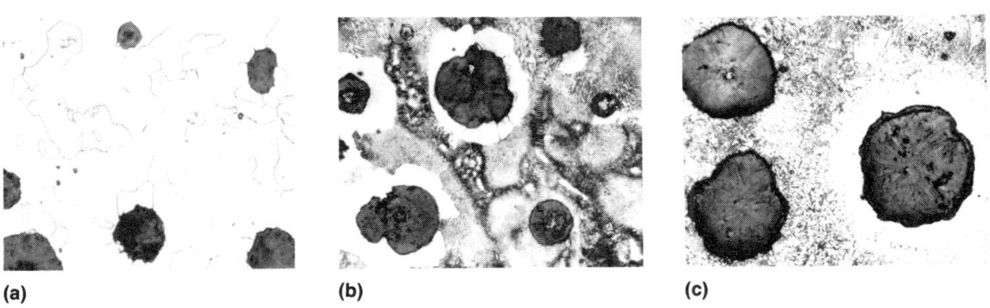

**Fig. 3** Microstructures of ductile irons. (a) 100% ferrite; 170 HB. (b) 50% ferrite, 50% pearlite; 207 HB. (c) Spheroidite; 265 HB. All contain spheroidal graphite. Etched with nital. 500×

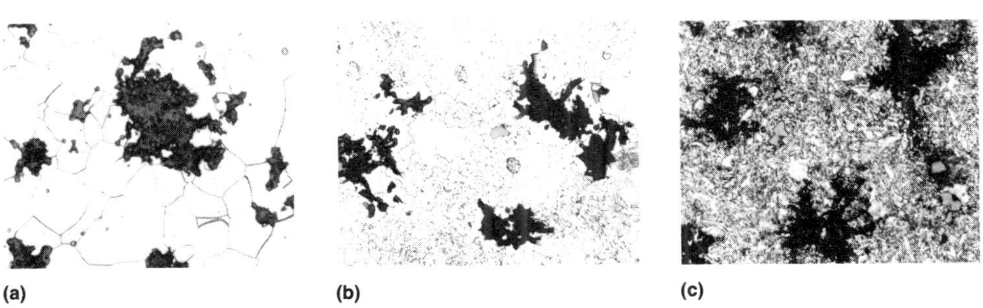

**Fig. 4** Microstructures of malleable irons. (a) Ferritic malleable ASTM 32510; ferritic structure. (b) Pearlitic malleable ASTM 48004; spheroidized structure. (c) Pearlite malleable ASTM 80002; spheroidized structure. All contain ductile graphite or temper carbon. Etched with nital. 500×

## Machining Characteristics of Gray Iron

The machinability of most gray cast irons is superior to that of most other cast irons of equivalent hardness and to that of virtually all steel. The flake graphite introduces discontinuities in the metal matrix, which act as chip breakers. The graphite itself serves as a lubricant for the cutting tool. However, economical cutting depends on more than inherent machinability. Difficulties in machining gray iron can often be traced to one or more of the following factors:

- The presence of chill at corners and in light sections
- The presence of adhering sand on the surface of the casting

**Table 2 Effect of matrix microstructure on tool life**

| Material | Matrix microstructure | Type of graphite | Hardness, HB | Cutting speed(a) m/min | Cutting speed(a) sfm | Ultimate tensile strength MPa | Ultimate tensile strength ksi |
|---|---|---|---|---|---|---|---|
| Gray iron | 100% ferrite | Flake | 100 | 270 | 880 | 108 | 15.7 |
| | Coarse pearlite | Flake | 195 | 110 | 360 | 241 | 35 |
| | Fine pearlite | Flake | 225 | 105 | 340 | 310 | 45 |
| | Acicular | Flake | 263 | 60 | 200 | 407 | 59 |
| Ductile iron | 100% ferrite | Spheroidal | 170 | 250 | 810 | 483 | 70 |
| | 97% ferrite, 3% pearlite | Spheroidal | 183 | 175 | 570 | 531 | 77 |
| | 60% ferrite, 40% pearlite | Spheroidal | 207 | 130 | 430 | 584 | 84.7 |
| | 60% ferrite, 40% pearlite | Spheroidal | 215 | 110 | 360 | 641 | 93 |
| | 20% ferrite, 80% pearlite | Spheroidal | 265 | 75 | 240 | 671 | 97.25 |
| Ferritic malleable | | | | | | | |
| ASTM 32510 | 100% ferrite | Temper carbon | 109 | 290 | 950 | 345 | 50 |
| Pearlitic malleable | | | | | | | |
| ASTM 48004 | Spheroidite | Temper carbon | 179 | 140 | 450 | 483 | 70 |
| ASTM 60003 | Spheroidite | Temper carbon | 230 | 85 | 280 | 552 | 80 |
| ASTM 80002 | Spheroidite | Temper carbon | 250 | 80 | 260 | 690 | 100 |

(a) Cutting speed for 30 min tool life

- Swells, usually the result of soft molds
- Shifted castings
- Shrinks

**Chill** at corners and in light sections is more likely to be encountered with small castings, with higher-strength irons, and with designs that have light sections in the cope or top of the mold. Most foundries control iron with a chill test that gives an indication of the tendency of the iron to form white or mottled iron in light sections. The iron can be treated with a small amount (0.5 to 2.5 kg/Mg, or 1 to 5 lb/ton) of a graphitizing alloy such as calcium-bearing ferrosilicon or other proprietary inoculant, which effectively decreases chill. Inoculation to achieve control of the tendency to chill usually does not result in significant changes in the composition or the physical properties of the iron, although it does produce changes in the mechanical properties. Light sections (5 mm, or 3/16 in.) usually cannot be cast in gray iron of higher than class 25. Class 30 iron can be cast in 6 mm (1/4 in.) sections. These values are different for different designs, depending on how the casting is made and gated. It is important to note that cooling in the mold at the time of freezing controls whether the iron will be gray, white, or mottled. If the thin section is in the drag or near the gate, the flow of hot metal heats the mold, thus decreasing the rate of cooling and enhancing the formation of the gray iron.

If chill is encountered, it is generally best to correct the trouble at its origin. It is usually uneconomical to anneal castings to remove chill because, in addition to heat treating for 2 h at 900 °C (1650 °F) for unalloyed irons, recleaning may be required for removing scale. Distortion beyond tolerance often occurs, and there are sacrifices in hardness and strength.

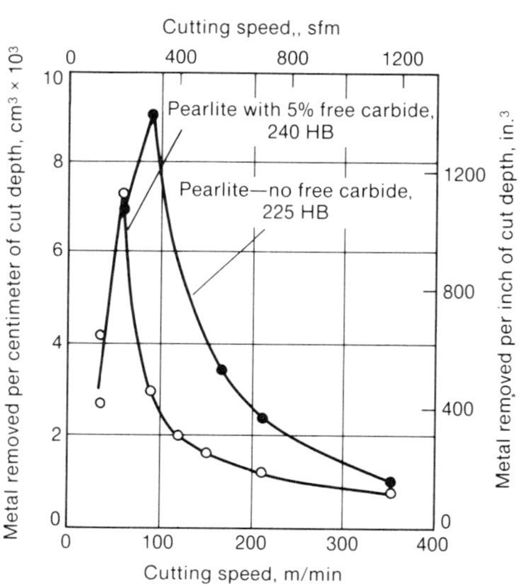


**Fig. 5** Effect of carbides on the machining characteristics of cast iron. Although carbides cause only a small increase in hardness, tool life is substantially reduced.

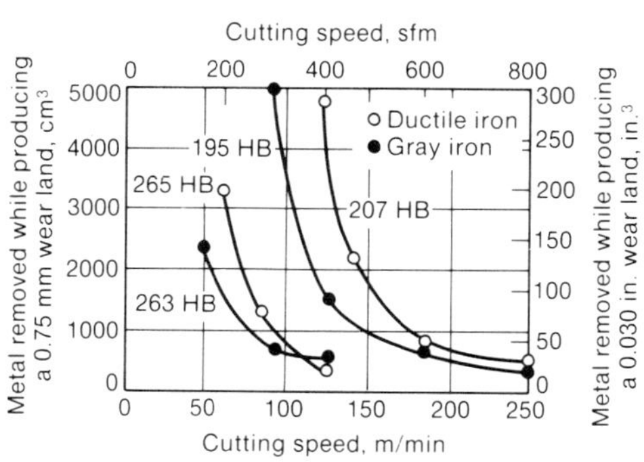


| Hardness and type of iron | Tensile strength MPa | Tensile strength ksi | Relative speed(a) m/min | Relative speed(a) sfm |
|---|---|---|---|---|
| 263 HB, gray | 407 | 59 | 45 | 150 |
| 265 HB, ductile | 669 | 97 | 60 | 200 |
| 195 HB, gray | 241 | 35 | 100 | 325 |
| 207 HB, ductile | 586 | 85 | 130 | 420 |

(a) Cutting speed for removing 3280 $cm^3$ (200 $in.^3$) of metal while producing a wear land of 0.75 mm (0.030 in.)

**Fig. 6** Machinability of gray and ductile irons

**Adhering sand** can usually be removed by effective cleaning, but sand present as the result of penetration of the iron into the mold wall is extremely difficult to blast clean. This is a foundry defect that is best corrected at the source. Slowing the speed of machining and increasing the rate of feed is the best approach to salvaging castings of this type. Carbide tools are better than high-speed steel tools for resisting the extreme abrasion.

Swells are most troublesome in operations such as broaching and in other setups tooled for high production. The additional metal often places an excessive load on the tool, which may chip or dull but not actually fail until some time after the troublesome parts have been machined.

**Shifted castings** are similar to swells in their action on cutting tools. Shifts or swells may also cause excessive tool loading if the locating points are affected. It is important to consider the positions of such locating points when designing the castings and to avoid indiscriminate grinding of locating points in the foundry cleaning room.

**Shrinks** are rarely present, but can be troublesome when encountered in such operations as drilling. For example, the drill may tend to drift from its intended path to follow the shrink, which offers less resistance to the drill. Sometimes, a drill may break because it encounters a region of higher hardness, which is often associated with an area of shrink. Cast iron is the easiest metal to cast without internal shrink. Eutectic freezing is accompanied by expansion due to the precipitation of low-density graphite, which aids in obtaining internal soundness.

**Machining Data.** Because machinability is complex, it is difficult to evaluate it in tables. Criteria such as power per unit volume in unit time are of primary importance in selecting a machine tool and the size of its motor. Machinability ratings based on tool life under standard test conditions are help-

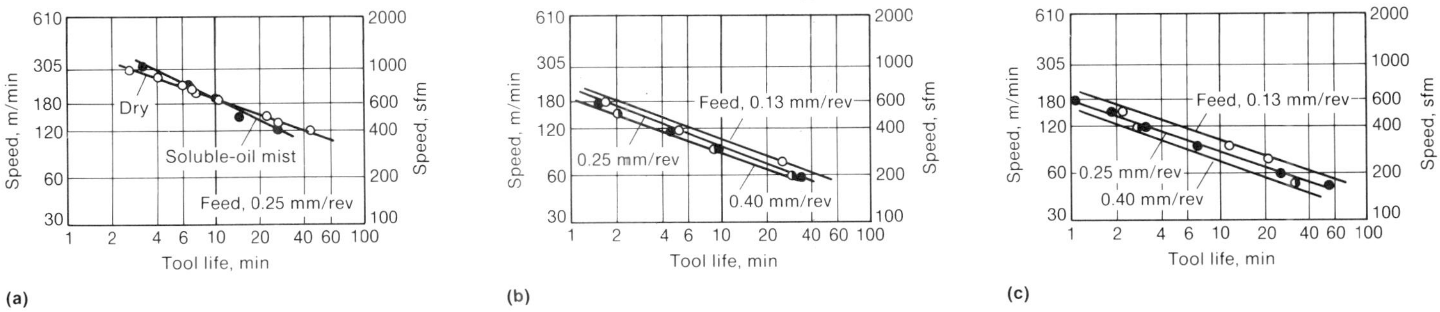


| Material | Hardness, HB | Graphitic carbon(d) | Combined carbon(d) | Total carbon | Si | Mn | P | S | Cu | Ni |
|---|---|---|---|---|---|---|---|---|---|---|
| | | Composition % | | | | | | | | |
| Ferritic 60-45-10(a) | 159–170 | 3.61 | 0.18 | 3.79 | 2.22 | 0.35 | 0.024 | 0.010 | 0.061 | 0.62 |
| Pearlitic 80-60-03(b) | 204–232 | 3.06 | 0.46 | 3.52 | 2.23 | 0.40 | 0.059 | 0.022 | 0.08 | 0.42 |
| Pearlitic 100-70-03(c) | 229–248 | 2.66 | 0.70 | 3.36 | 2.35 | 0.84 | 0.058 | 0.018 | 0.44 | 0.05 |

(a) Heat treatment: held at 885 °C (1625 °F) for 8 h, cooled to 690 °C (1275 °F), held for 5 h, cooled to 425 °C (800 °F) at 55 °C/h (100 °F/h), doors opened. (b) Heat treatment: held at 885 °C (1625 °F) for 2 h, air cooled to 540 °C (1000 °F) and held for 4 h, furnace cooled. (c) Heat treatment: held at 540 °C (1000 °F) for 2 h, furnace cooled. (d) Based on analysis of material as received

**Fig. 7** Effect of variables on relation between cutting speed and tool life for three ductile irons. (a) Ferritic ductile iron (60-45-10). (b) Pearlitic ductile iron (80-60-03). (c) Pearlitic ductile iron (100-70-03). For all three irons the results plotted were obtained in turning test specimens (165 mm, or 6½ in., outside diameter and 480 mm, or 19 in., long) in a 7.5 kw (10 hp) engine lathe equipped with a variable-speed drive motor, using C-2 carbide tools. Depth of cut was 2.5 mm (0.100 in.); tool life limit, 0.40 mm (0.015 in.) flank wear.

ful, but are not readily interpreted into the economics of machining.

One method of demonstrating the effect on machinability of changing from one grade of iron to another is indicated in Table 3, which is based on an experiment in which metal was removed from four types of gray iron using single-point carbide tools. For each type of iron, the cutting speed was varied until the removal of 3280 $cm^3$ (200 $in.^3$) of iron resulted in a 0.75 mm (0.030 in.) wear land on the tool. The values of cutting speed thus obtained serve as qualitative evaluations of machinability, but not as quantitative indexes. Optimum cutting speeds, tool materials, cutting fluids, feeds, and finish requirements must be studied for any given machine tool setup. Tool life is an important factor because the machine must be stopped to change tools and the tools must be resharpened. The use of solid-carbide inserts or bits is one method of decreasing both machine downtime for tool changes and resharpening cost.

The greatly improved machinability obtainable in gray iron by annealing has been advantageous to the automotive and other industries for many years. Annealing is usually of the subcritical type, such as 1 h at 730 to 760 °C (1350 to 1400 °F). A cycle of heating to 785 to 815 °C (1450 to 1500 °F) and cooling at 22 °C/h (40 °F/h) to about 600 °C (1100 °F) is sometimes used. These treatments graphitize the carbide in the pearlite and result in a ferritic matrix. Finding it uneconomical to graphitize primary carbide, most users try to avoid obtaining it. Annealing for improved machinability is most economical when the casting is small and the amount of machining is large.

The annealing treatments described result in sacrifices in hardness and strength. A typical class 35 iron will be downgraded to about class 20 in strength by this treatment. In applications in which wear resistance is important (for example, cylinder blocks), gray iron is not annealed, because of the unsatisfactory performance obtained with a ferritic matrix.

**Table 3 Machinability of gray iron**

| Microstructure | ASTM class | Tensile strength MPa | Tensile strength ksi | Hardness, HB | Cutting speed(a) m/min | Cutting speed(a) ft/min |
|---|---|---|---|---|---|---|
| Acicular iron | 50 | 407 | 59 | 263 | 46 | 150 |
| Fine pearlite, alloy | 40 | 310 | 45 | 225 | 95 | 310 |
| Ferrite (annealed) | ··· | 108 | 15.7 | 100 | 293 | 960 |
| Coarse pearlite, no alloy | 35 | 241 | 35 | 195 | 99 | 325 |

(a) Cutting speed at which removal of 3280 $cm^3$ (200 $in.^3$) produced 0.75 mm (0.030 in.) wear land on single-point carbide tools

**Table 4 Identification codes for cutting fluid types**

| Code No. | Type |
|---|---|
| 0 | Dry |
| 1 | Oils, light duty (general purpose) |
| 2 | Oils, medium duty |
| 3 | Oils, heavy duty |
| 4 | Emulsifiable oils, light duty (general purpose) |
| 5 | Emulsifiable oils, heavy duty |
| 6 | Chemicals and synthetics, light duty (general purpose) |
| 7 | Chemicals and synthetics, heavy duty |
| 8 | Specials, light duty<br>80 honing oils<br>81 chemical grinding fluids<br>82 cutting fluids for aluminum<br>83 cutting fluids for titanium |
| 9 | Specials, heavy duty<br>90 honing oils<br>91 cutting fluids for titanium<br>92 biodegradable soluble oils |

**Table 5 Cutting fluid recommendations for cast irons**

See Table 4 for identification of cutting fluid code numbers.

| Operation | Tool material | Cutting fluid code No. |
|---|---|---|
| Turning(a) | High-speed steel | 4, 6 |
| | Carbide | 0, 4, 6 |
| Face milling | High-speed steel | 4, 6 |
| | Carbide | 0, 4, 6 |
| End milling (peripheral) | High-speed steel | 4, 6 |
| | Carbide | 0, 4, 6 |
| Drilling | High-speed steel | 4, 6 |
| | Carbide | 0, 4, 6 |
| Reaming | High-speed steel | 4, 6 |
| | Carbide | 4, 6 |
| Tapping | High-speed steel | 4, 6 |
| Surface grinding(b) | $Al_2O_3$, SiC, CBN | 4, 6(c) |

(a) Single-point and box tools. (b) Horizontal spindle, reciprocating table. (c) For cast irons over 400 HB

## Machining Characteristics of Ductile Iron

In ductile (nodular) iron, the graphite is present as tiny balls or spherulites (not as flakes, as in gray iron), which serve the same useful purposes of chip breaking and lubrication in machining as for gray iron. Although ductile iron has considerably

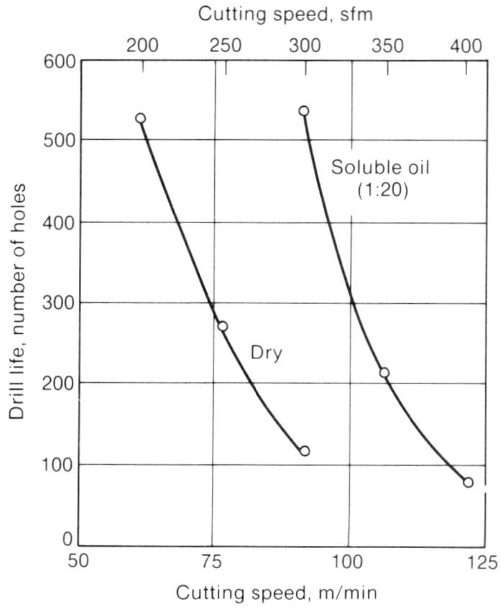


**Fig. 8** Effect of cutting fluid on tool life when drilling grade 32510 malleable iron with high-speed steels. Feed: 0.13 mm/rev (0.005 in./rev). Depth of hole: 25 mm (1 in.). The drill life end point was chosen to be 0.40 mm (0.015 in.) wear land or drill breakdown.

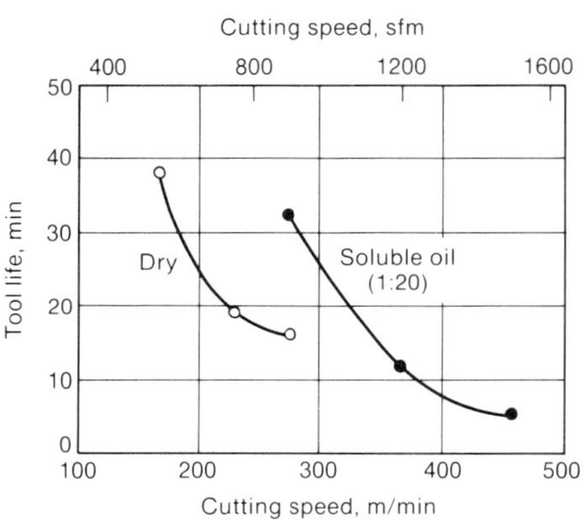


**Fig. 9** Effect of cutting speed and cutting fluid on tool life when turning ferritic malleable iron. Feed, 0.25 mm/rev (0.01 in./rev). Depth of cut, 1.57 mm (0.062 in.). The tool life end point was chosen to be 0.04 mm (0.015 in.) of uniform wear or 0.75 mm (0.03 in.) of localized wear.

higher tensile strength than gray iron at the same hardness, the machinability of ductile iron is superior in some applications. The data in Fig. 6, obtained in turning tests using carbide tools, show the differences in machinability at two different hardness levels. In the same series of turning tests, gray iron annealed to a fully ferritic condition (100 HB; 108 MPa, or 15.7 ksi) had the same

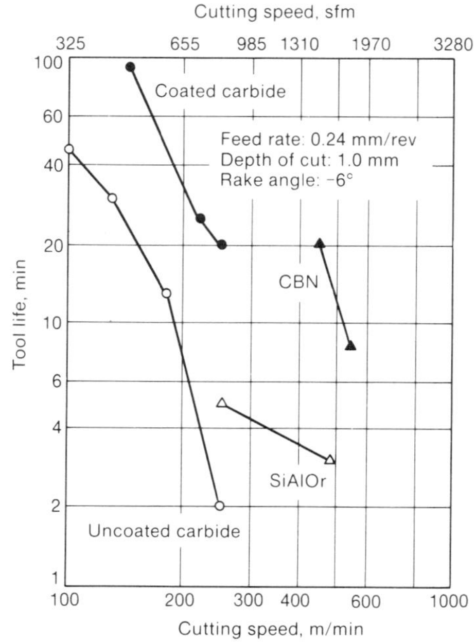


**Fig. 10** Tool life diagram for turning gray iron

cutting speed for 3280 cm³ (200 in.³) tool life (290 m/min, or 960 sfm) as fully annealed (ferritized) ductile iron (170 HB). This comparison is not necessarily valid for machining operations in general, because contrary results have been observed in drilling gray iron and pearlitic ductile iron at the same hardness.

Machinability data are given in the following example. These data compare cutting speed and tool life for a ferritic ductile iron (grade 60) with two pearlitic ductile irons (grades 80 and 100) and show the effect of a spray cutting fluid on the machining characteristics of the ferritic iron.

**Example 1: Cutting Speed and Tool Life for Ferritic Versus Pearlitic Ductile Irons.** The relation between cutting speed and life of grade C-2 carbide tools in turning one ferritic and two pearlitic ductile irons is shown in Fig. 7. Figure 7(a) shows the relation between cutting speed and tool life for ferritic iron (grade 60-45-10), affected by the use and nonuse of a cutting fluid (a 2% mixture of soluble oil in water applied as a spray mist). Improved tool life was obtained with the use of cutting fluid at speeds above about 215 m/min (700 sfm) because of the ability of the fluid to reduce tool-flank buildup. Figures 7(b) and (c) show this relation for pearlitic irons (grades 80-60-03 and 100-70-03) as a function of feed rate. An increase in feed caused a decrease in tool life. These two pearlitic irons were machined dry.

## Machining Characteristics of Malleable Iron

A frequently quoted general rule is that the machinability of malleable iron is 25% better than that of free-cutting steel. Although this rule applies to most applications of turning (the usual basis for machinability ratings), it does not necessarily apply to broaching or some other machining operation. Data on the machining of malleable iron are few, probably because machining is seldom a problem, except when some difficulty arises because of surface variations. Pearlitic or decarburized structures at or just below the surface of ferritic malleable castings reduce machinability.

## Effects of Alloying Elements on Machining Characteristics

With two principal exceptions, alloying elements affect the machining characteristics of cast iron largely through their effects on the relative stability of iron carbide and graphite in the microstructure. The exceptions are phosphorus and tin, both of which affect the matrix.

**Phosphorus** can have an adverse effect on the machinability of cast iron. When present in quantities above about 0.15%, it occurs as the iron-iron phosphide eutectic, steadite. Steadite may occur in the microstructure independently in gray or ductile cast iron or in association with the carbide in white or mottled cast irons. The extreme hardness of steadite (Table 1) and its abrasion resistance markedly reduce tool life in machining pearlitic iron.

Although the presence of more than about 0.2% P causes a significant decrease in machinability, the decrease is structure sensitive and can be largely offset by converting the matrix in which steadite particles are embedded from its normal (pearlitic) condition to a relatively soft (ferritic) structure. In the latter, which can be produced by using inoculants or by annealing, the soft matrix lacks the rigidity to force steadite particles into abrasive contact with

**Table 6 Comparison of metal removal rates of various cutting tool materials for gray cast iron machining**

| Cutting tool material | Speed | | Feed | | Depth of cut | | Metal removal rate | | | |
|---|---|---|---|---|---|---|---|---|---|---|
| | m/s | sfm | cm/rev | in./rev | cm | in. | m³/s | in.³/min | kg/s | lb/min |
| $Al_2O_3$-coated cemented carbide(a) | 6 | 1200 | 0.025 | 0.010 | 0.318 | 0.125 | 0.00005 | 18 | 0.03 | 4.6 |
| $Al_2O_3$(a) | 15 | 3000 | 0.013 | 0.005 | 0.152 | 0.060 | 0.00003 | 10.8 | 0.02 | 2.8 |
| $Al_2O_3$ + TiC(a) | 15 | 3000 | 0.025 | 0.010 | 0.152 | 0.060 | 0.00006 | 21.6 | 0.04 | 5.5 |
| $Si_3N_4$ based(b) | 25 | 5000 | 0.075 | 0.030 | 0.478 | 0.188 | 0.00091 | 337.5 | 0.65 | 86.3 |

(a) Catalog recommended conditions. (b) Field data

## Table 7 Nominal speeds and feeds for the turning of cast iron with single-point and box tools

| Material | Hardness, HB | Condition | Depth of cut, mm (in.) | High-speed steel tool: Speed, m/min (sfm) | High-speed steel tool: Feed, mm/rev (in./rev) | High-speed steel tool: Tool material, ISO (AISI) | Carbide tool, Uncoated: Speed, m/min (sfm), Brazed | Carbide tool, Uncoated: Speed, m/min (sfm), Indexable | Carbide tool, Uncoated: Feed, mm/rev (in./rev) | Carbide tool, Uncoated: Tool material grade, ISO (C) | Carbide tool, Coated: Speed, m/min (sfm) | Carbide tool, Coated: Feed, mm/rev (in./rev) | Carbide tool, Coated: Tool material grade, ISO (C) |
|---|---|---|---|---|---|---|---|---|---|---|---|---|---|
| **Ferritic gray irons** | | | | | | | | | | | | | |
| ASTM A48: class 20; SAE J431c: grade G1800 | 120–150 | Annealed | 1 (0.040) | 56 (185) | 0.18 (0.007) | S4, S5 (M2, M3) | 200 (650) | 220 (725) | 0.25 (0.010) | K20, M20 (C-3) | 290 (950) | 0.18 (0.007) | CK20, CM20 (CC-3) |
| | | | 4 (0.150) | 44 (145) | 0.40 (0.015) | S4, S5 (M2, M3) | 150 (500) | 170 (550) | 0.50 (0.020) | K20,M20 (C-2) | 185 (600) | 0.40 (0.015) | CK20, CM20 (CC-2) |
| | | | 8 (0.300) | 37 (120) | 0.50 (0.020) | S4, S5 (M2, M3) | 120 (400) | 135 (450) | 0.75 (0.030) | K20, M20 (C-2) | 175 (575) | 0.50 (0.020) | CK20, CM20 (CC-2) |
| | | | 16 (0.625) | 30 (100) | 0.75 (0.030) | S4, S5 (M2, M3) | 100 (325) | 110 (360) | 1.0 (0.040) | K20, M20 (C-2) | ... | ... | ... |
| **Pearlitic gray irons** | | | | | | | | | | | | | |
| ASTM A48: classes 30, 35, 40; SAE J431c: grade G3000 | 190–220 | As-cast | 1 (0.040) | 37 (120) | 0.18 (0.007) | S9, S11(a) (T15, M42)(a) | 115 (370) | 125 (410) | 0.18 (0.007) | K20, M20 (C-3) | 160 (525) | 0.18 (0.007) | CK20, CM20 (CC-3) |
| | | | 4 (0.150) | 24 (80) | 0.40 (0.015) | S9, S11(a) (T15, M42)(a) | 90 (300) | 105 (340) | 0.40 (0.015) | K20, M20 (C-2) | 135 (450) | 0.40 (0.015) | CK20, CM20 (CC-2) |
| | | | 8 (0.300) | 20 (65) | 0.50 (0.020) | S9, S11(a) (T15, M42)(a) | 76 (250) | 84 (275) | 0.75 (0.030) | K20, M20 (C-2) | 105 (350) | 0.50 (0.020) | CK20, CM20 (CC-2) |
| | | | 16 (0.625) | 17 (55) | 0.75 (0.030) | S9, S11(a) (T15, M42)(a) | 60 (200) | 69 (225) | 1.0 (0.040) | K20, M20 (C-2) | ... | ... | ... |
| **Ferritic ductile irons** | | | | | | | | | | | | | |
| ASTM A536: grades 60-40-18, 65-45-12; SAE J434c: grades D4018, D4512 | 140–190 | Annealed | 1 (0.040) | 60 (200) | 0.18 (0.007) | S4, S5 (M2, M3) | 215 (700) | 235 (775) | 0.25 (0.010) | P10, M10 (C-7) | 290 (950) | 0.25 (0.010) | CP10, CM10 (CC-7) |
| | | | 4 (0.150) | 46 (150) | 0.40 (0.015) | S4, S5 (M2, M3) | 170 (550) | 185 (600) | 0.50 (0.020) | P10, M10 (C-7) | 235 (775) | 0.50 (0.020) | CP10, CM10 (CC-7) |
| | | | 8 (0.300) | 38 (125) | 0.50 (0.020) | S4, S5 (M2, M3) | 135 (450) | 150 (500) | 0.75 (0.030) | P20, M20 (C-6) | 200 (650) | 0.75 (0.030) | CP20, CM20 (CC-6) |
| | | | 16 (0.625) | 30 (100) | 0.75 (0.030) | S4, S5 (M2, M3) | 110 (360) | 120 (400) | 1.0 (0.040) | P30, M30 (C-6) | ... | ... | ... |
| **Ferritic-pearlitic ductile irons** | | | | | | | | | | | | | |
| ASTM A536: grade 80-55-06; SAE J434c: grade D5506 | 190–225 | As-cast | 1 (0.040) | 43 (140) | 0.18 (0.007) | S4, S5 (M2, M3) | 145 (480) | 165 (540) | 0.25 (0.010) | P10, M10 (C-7) | 215 (700) | 0.25 (0.010) | CP10, CM10 (CC-7) |
| | | | 4 (0.150) | 34 (110) | 0.40 (0.015) | S4, S5 (M2, M3) | 115 (375) | 130 (425) | 0.50 (0.020) | P10, M10 (C-7) | 170 (550) | 0.50 (0.020) | CP10, CM10 (CC-7) |
| | | | 8 (0.300) | 26 (85) | 0.50 (0.020) | S4, S5 (M2, M3) | 95 (310) | 105 (350) | 0.75 (0.030) | P20, M20 (C-6) | 135 (450) | 0.75 (0.030) | CP20, CM20 (CC-6) |
| | | | 16 (0.625) | 21 (70) | 0.75 (0.030) | S4, S5 (M2, M3) | 76 (250) | 84 (275) | 1.0 (0.040) | P30, M30 (C-6) | ... | ... | ... |
| **Ferritic malleable irons** | | | | | | | | | | | | | |
| ASTM A47: grades 32510, 35018; ASTM A602: grade M3210; SAE J158: grade M3210 | 110–160 | Malleablized | 1 (0.040) | 58 (190) | 0.18 (0.007) | S4, S5 (M2, M3) | 230 (750) | 245 (800) | 0.25 (0.010) | P10, M10 (C-7) | 305 (1000) | 0.18 (0.007) | CP10, CM10 (CC-7) |
| | | | 4 (0.150) | 44 (145) | 0.40 (0.015) | S4, S5 (M2, M3) | 175 (575) | 190 (625) | 0.50 (0.020) | P10, M10 (C-7) | 245 (800) | 0.40 (0.015) | CP10, CM10 (CC-7) |
| | | | 8 (0.300) | 37 (120) | 0.50 (0.020) | S4, S5 (M2, M3) | 145 (475) | 150 (500) | 0.75 (0.030) | P20, M20 (C-6) | 200 (650) | 0.50 (0.020) | CP20, CM20 (CC-6) |
| | | | 16 (0.625) | 29 (95) | 0.75 (0.030) | S4, S5 (M2, M3) | 115 (375) | 120 (400) | 1.0 (0.040) | P20, M20 (C-6) | ... | ... | ... |
| **Pearlitic malleable irons** | | | | | | | | | | | | | |
| ASTM A220: grades 40010, 45006, 45008, 50005; ASTM A602: grade M4504, M5003; SAE J158: grades M4504, M5003 | 160–200 | Malleablized and heat treated | 1 (0.040) | 40 (130) | 0.18 (0.007) | S4, S5 (M2, M3) | 160 (520) | 185 (600) | 0.25 (0.010) | P10, M10 (C-7) | 235 (775) | 0.25 (0.010) | CP10, CM10 (CC-7) |
| | | | 4 (0.150) | 30 (100) | 0.40 (0.015) | S4, S5 (M2, M3) | 120 (400) | 135 (450) | 0.50 (0.020) | P10, M10 (C-7) | 185 (600) | 0.40 (0.015) | CP10, CM10 (CC-7) |
| | | | 8 (0.300) | 24 (80) | 0.50 (0.020) | S4, S5 (M2, M3) | 100 (325) | 115 (375) | 0.75 (0.030) | P20, M20 (C-6) | 145 (475) | 0.50 (0.020) | CP20, CM20 (CC-6) |
| | | | 16 (0.625) | 20 (65) | 0.75 (0.030) | S4, S5 (M2, M3) | 79 (260) | 90 (300) | 1.0 (0.040) | P20, M20 (C-6) | ... | ... | ... |
| | 200–240 | Malleablized and heat treated | 1 (0.040) | 30 (100) | 0.18 (0.007) | S4, S5 (M2, M3) | 120 (400) | 135 (450) | 0.25 (0.010) | P10, M10 (C-7) | 185 (600) | 0.25 (0.010) | CP10, CM10 (CC-7) |
| | | | 4 (0.150) | 23 (75) | 0.40 (0.015) | S4, S5 (M2, M3) | 90 (300) | 105 (350) | 0.50 (0.020) | P10, M10 (C-7) | 135 (450) | 0.40 (0.015) | CP10, CM10 (CC-7) |
| | | | 8 (0.300) | 18 (60) | 0.50 (0.020) | S4, S5 (M2, M3) | 76 (250) | 85 (280) | 0.75 (0.030) | P20, M20 (C-6) | 105 (350) | 0.50 (0.020) | CP20, CM20 (CC-6) |
| | | | 16 (0.625) | 15 (50) | 0.75 (0.030) | S4, S5 (M2, M3) | 60 (200) | 70 (230) | 1.0 (0.040) | P20, M20 (C-6) | ... | ... | ... |

(a) Any premium high-speed steel can be used. Source: Metcut Research Associates Inc.

**Table 8 Nominal speeds and feeds for the turning of cast irons with ceramic tools**

| Material | Hardness, HB | Condition | Depth of cut mm | Depth of cut (in.) | Speed m/min | Speed (sfm) | Feed mm/rev | Feed (in./rev) | Type of ceramic(a) |
|---|---|---|---|---|---|---|---|---|---|
| **Ferritic gray irons** | | | | | | | | | |
| ASTM A48: class 20 | 120–150 | Annealed | 1 | (0.040) | 610 | (2000) | 0.25 | (0.010) | CPA |
| SAE J431c: grade G1800 | | | 4 | (0.150) | 460 | (1500) | 0.40 | (0.015) | CPA |
| | | | 8 | (0.300) | 365 | (1200) | 0.65 | (0.025) | CPA |
| **Pearlitic gray irons** | | | | | | | | | |
| ASTM A48: classes 30, 35, 40 | 190–220 | As-cast | 1 | (0.040) | 460 | (1500) | 0.25 | (0.010) | HPC |
| SAE J431c: grade G3000 | | | 4 | (0.150) | 305 | (1000) | 0.40 | (0.015) | HPC |
| | | | 8 | (0.300) | 215 | (700) | 0.50 | (0.020) | HPC |
| **Ferritic ductile irons** | | | | | | | | | |
| ASTM A536: grades 60-40-18, 65-45-12 | 140–190 | Annealed | 1 | (0.040) | 365 | (1200) | 0.25 | (0.010) | HPC |
| SAE J434c: grades D4018, D4512 | | | 4 | (0.150) | 305 | (1000) | 0.40 | (0.015) | HPC |
| | | | 8 | (0.300) | 230 | (750) | 0.65 | (0.025) | HPC |
| **Ferritic-pearlitic ductile irons** | | | | | | | | | |
| ASTM A536: grade 80-55-06 | 190–225 | As-cast | 1 | (0.040) | 335 | (1100) | 0.25 | (0.010) | HPC |
| SAE J434c: grade D5506 | | | 4 | (0.150) | 275 | (900) | 0.40 | (0.015) | HPC |
| | | | 8 | (0.300) | 200 | (650) | 0.50 | (0.020) | HPC |
| | 225–260 | As-cast | 1 | (0.040) | 275 | (900) | 0.13 | (0.005) | HPC |
| | | | 4 | (0.150) | 215 | (700) | 0.25 | (0.010) | HPC |
| | | | 8 | (0.300) | 170 | (550) | 0.40 | (0.015) | HPC |
| **Ferritic malleable irons** | | | | | | | | | |
| ASTM A47: grades 32510, 35018 | 110–160 | Malleablized | 1 | (0.040) | 460 | (1500) | 0.25 | (0.010) | HPC |
| ASTM A602: grade M3210 | | | 4 | (0.150) | 365 | (1200) | 0.40 | (0.015) | HPC |
| SAE J158: grade M3210 | | | 8 | (0.300) | 275 | (900) | 0.50 | (0.020) | HPC |
| **Pearlitic malleable irons** | | | | | | | | | |
| ASTM A220: grades 40010, 45006, 45008, 50005 | 160–200 | Malleablized | 1 | (0.040) | 365 | (1200) | 0.25 | (0.010) | HPC |
| ASTM A602: grade M4504, M5003 | | and heat | 4 | (0.150) | 290 | (950) | 0.40 | (0.015) | HPC |
| SAE J158: grades M4504, M5003 | | treated | 8 | (0.300) | 230 | (750) | 0.50 | (0.020) | HPC |

(a) CPA, cold-pressed alumina; HPC, hot-pressed $Al_2O_3$-TiC ceramic. Source: Metcut Research Associates Inc.

the cutting tools, and the adverse effect on machinability is minimized.

**Tin.** In contrast to the effect of phosphorus, the addition of tin in small amounts (0.05 to 0.15%) to otherwise unalloyed gray iron increases machinability and tool life by reducing the range of hardness values throughout the castings and by eliminating hard spots and chilling at corners.

## Cutting Fluids

Iron castings can be satisfactorily machined without a cutting fluid, but the use of fluids for the machining of iron castings is increasing. Cutting fluids are in common use with modern production machine tools that provide high metal removal rates. Fluids are always used in production abrasive machining to achieve high metal removal rates.

Three common functions of a cutting fluid are cooling, lubrication, and chip removal. A fluid is not necessary in the conventional machining of iron, except for some operations such as drilling and tapping that are very confined and in which a fluid can help dispel a concentration of heat or chips. The improvement in drilling of malleable iron with high-speed steel and a coolant is shown in Fig. 8. A similar effect would be expected with ductile iron, but the use of a coolant is not as effective with gray iron. Because of their graphite content, iron castings are free machining, have a relatively low power requirement, and form discontinuous chips without a breaker. However, with high rates of metal removal, both heat and chips can accumulate very rapidly so that the use of a fluid is advantageous in prolonging tool life, maintaining accuracy, and removing chips. The effectiveness of a coolant at very high cutting speeds with carbide tools is demonstrated by the comparative turning tests of malleable iron summarized in Fig. 9. In this test, a soluble-oil coolant doubled the tool life at a cutting speed of 275 m/min (900 sfm). Graphite dust that may be a nuisance in dry machining is also eliminated.

Iron may have two reactions with fluids that do not occur in the machining of steel. The flake graphite in gray iron has an affinity for oil and can actually extract oil from an emulsion. Therefore, the concentration of oil-emulsion cutting fluids should be monitored to maintain proper levels. Some gray irons and all ductile irons contain a small amount of alkaline metals such as calcium, magnesium, and cerium. The carbides of these elements can react with water to form acetylene, and the odor of this is sometimes evident when iron is machined. These elements are present in very small quantities (a total of less than 0.1%), and only the alkaline carbides that are exposed on a fresh surface can react. The amount of gas formed is infinitesimal, but it may be detected because of our extreme sensitivity to this odor. In dry machining, the humidity in the air may be sufficient to cause the faint odor of acetylene to be evident.

Cutting fluids are divided into ten general types with subdivisions (Table 4). Specific recommendations for the various machining and grinding operations on iron are given in Table 5. Additional information can be found in the article "Metal Cutting and Grinding Fluids" in this Volume.

## Tool Materials

With the exception of polycrystalline diamond, all commercially produced cutting tool materials are used to machine cast irons. These cutting tool materials include high-speed steels, cast cobalt alloys, cemented carbides, cermets, ceramics, and cubic boron nitride. Each will be discussed below. More detailed information on these materials is available in the Section "Cutting Tool Materials" in this Volume.

**High-Speed Tool Steels.** Molybdenum high-speed steels are frequently used to machine cast irons. For cast irons less than 200 HB in hardness, molybdenum high-speed steels that do not contain cobalt can be used. For hardness levels exceeding 200 HB, cobalt-containing grades should be used. Straight-tungsten high-speed steels are not commonly used on cast irons. High-

**Table 9 Nominal speeds and feeds for the boring of cast iron with high-speed steel and carbide tools**

| Material | Hardness, HB | Condition | Depth of cut, mm (in.) | High-speed steel tool: Speed, m/min (sfm) | High-speed steel tool: Feed, mm/rev (in./rev) | High-speed steel tool: Tool material, ISO (AISI) | Carbide tool, Uncoated: Speed, m/min (sfm), Brazed | Carbide tool, Uncoated: Speed, m/min (sfm), Indexable | Carbide tool, Uncoated: Feed, mm/rev (in./rev) | Carbide tool, Uncoated: Tool material grade, ISO (C) | Carbide tool, Coated: Speed, m/min (sfm) | Carbide tool, Coated: Feed, mm/rev (in./rev) | Carbide tool, Coated: Tool material grade, ISO (C) |
|---|---|---|---|---|---|---|---|---|---|---|---|---|---|
| **Ferritic gray irons** | | | | | | | | | | | | | |
| ASTM A48: class 20; SAE J431c: grade G1800 | 120–150 | Annealed | 0.25 (0.010) | 58 (190) | 0.075 (0.003) | S4, S5 (M2, M3) | 160 (530) | 190 (625) | 0.15 (0.006) | K10, M10 (C-3) | 255 (840) | 0.075 (0.003) | CK10, CM10 (CC-3) |
| | | | 1.25 (0.050) | 46 (150) | 0.13 (0.005) | S4, S5 (M2, M3) | 130 (425) | 150 (500) | 0.20 (0.008) | K20, M20 (C-2) | 205 (675) | 0.13 (0.005) | CK20, CM20 (CC-2) |
| | | | 2.5 (0.100) | 35 (115) | 0.30 (0.012) | S4, S5 (M2, M3) | 98 (320) | 115 (375) | 0.40 (0.015) | K20, M20 (C-2) | 130 (425) | 0.30 (0.012) | CK20, CM20 (CC-2) |
| **Pearlitic gray irons** | | | | | | | | | | | | | |
| ASTM A48: classes 30, 35, 40; SAE J431c: grade G3000 | 190–220 | As-cast | 0.25 (0.010) | 37 (120) | 0.075 (0.003) | S9, S11(a) (T15, M42)(a) | 95 (305) | 110 (360) | 0.075 (0.003) | K10, M10 (C-3) | 145 (470) | 0.075 (0.003) | CK10, CM10 (CC-3) |
| | | | 1.25 (0.050) | 29 (95) | 0.13 (0.005) | S9, S11(a) (T15, M42)(a) | 75 (245) | 88 (290) | 0.13 (0.005) | K20, M20 (C-2) | 115 (375) | 0.13 (0.005) | CK20, CM20 (CC-2) |
| | | | 2.5 (0.100) | 20 (65) | 0.30 (0.012) | S9, S11(a) (T15, M42)(a) | 62 (205) | 73 (240) | 0.30 (0.012) | K20, M20 (C-2) | 95 (315) | 0.30 (0.012) | CK20, CM20 (CC-2) |
| **Ferritic ductile irons** | | | | | | | | | | | | | |
| ASTM A536: grades 60-40-18, 65-45-12; SAE J434c: grades D4018, D4512 | 140–190 | Annealed | 0.25 (0.010) | 60 (200) | 0.075 (0.003) | S4, S5 (M2, M3) | 178 (585) | 210 (690) | 0.15 (0.006) | P01, M10 (C-8) | 260 (850) | 0.15 (0.006) | CP01, CM10 (CC-8) |
| | | | 1.25 (0.050) | 49 (160) | 0.13 (0.005) | S4, S5 (M2, M3) | 145 (475) | 170 (550) | 0.20 (0.008) | P10, M10 (C-7) | 205 (675) | 0.20 (0.008) | CP10, CM10 (CC-7) |
| | | | 2.5 (0.100) | 37 (120) | 0.30 (0.012) | S4, S5 (M2, M3) | 110 (360) | 130 (425) | 0.40 (0.015) | P20, M20 (C-6) | 170 (550) | 0.40 (0.015) | CP20, CM20 (CC-6) |
| **Ferritic-pearlitic ductile irons** | | | | | | | | | | | | | |
| ASTM A536: grade 80-55-06; SAE J434c: grade D5506 | 190–225 | As-cast | 0.25 (0.010) | 43 (140) | 0.075 (0.003) | S4, S5 (M2, M3) | 120 (400) | 145 (470) | 0.15 (0.006) | P10, M10 (C-7) | 190 (625) | 0.15 (0.006) | CP10, CM10 (CC-7) |
| | | | 1.25 (0.050) | 34 (110) | 0.13 (0.005) | S4, S5 (M2, M3) | 98 (320) | 115 (375) | 0.20 (0.008) | P20, M20 (C-6) | 150 (500) | 0.20 (0.008) | CP20, CM20 (CC-6) |
| | | | 0.25 (0.100) | 27 (90) | 0.30 (0.012) | S4, S5 (M2, M3) | 78 (255) | 90 (300) | 0.40 (0.015) | P30, M30 (C-6) | 115 (375) | 0.40 (0.015) | CP30, CM30 (CC-6) |
| | 225–260 | As-cast | 0.25 (0.010) | 30 (100) | 0.075 (0.003) | S9, S11(a) (T15, M42)(a) | 84 (275) | 95 (315) | 0.15 (0.006) | P10, M10 (C-7) | 120 (400) | 0.15 (0.006) | CP10, CM10 (CC-7) |
| | | | 1.25 (0.050) | 24 (80) | 0.13 (0.005) | S9, S11(a) (T15, M42)(a) | 64 (210) | 76 (250) | 0.20 (0.008) | P20, M20 (C-6) | 100 (325) | 0.20 (0.008) | CP20, CM20 (CC-6) |
| | | | 2.5 (0.100) | 18 (60) | 0.30 (0.012) | S9, S11(a) (T15, M42)(a) | 50 (165) | 59 (195) | 0.40 (0.015) | P30, M30 (C-6) | 76 (250) | 0.40 (0.015) | CP30, CM30 (CC-6) |
| **Ferritic malleable irons** | | | | | | | | | | | | | |
| ASTM A47: grades 32510, 35018; ASTM A602: grade M3210; SAE J158: grade M3210 | 110–160 | Malleablized | 0.25 (0.010) | 58 (190) | 0.075 (0.003) | S4, S5 (M2, M3) | 185 (600) | 215 (700) | 0.15 (0.006) | P10, M10 (C-7) | 265 (875) | 0.075 (0.003) | CP10, CM10 (CC-7) |
| | | | 1.25 (0.050) | 46 (150) | 0.13 (0.005) | S4, S5 (M2, M3) | 145 (475) | 170 (560) | 0.20 (0.008) | P20, M20 (C-6) | 215 (700) | 0.13 (0.005) | CP20, CM20 (CC-6) |
| | | | 2.5 (0.100) | 35 (115) | 0.30 (0.012) | S4, S5 (M2, M3) | 115 (370) | 135 (435) | 0.40 (0.015) | P30, M30 (C-6) | 168 (550) | 0.30 (0.012) | CP30, CM30 (CC-6) |
| **Pearlitic malleable irons** | | | | | | | | | | | | | |
| ASTM A220: grades 40010, 45006, 45008, 50005; ASTM A602: grade M4504, M5003; SAE J158: grades M4504, M5003 | 160–200 | Malleablized and heat treated | 0.25 (0.010) | 40 (130) | 0.075 (0.003) | S4, S5 (M2, M3) | 135 (450) | 160 (530) | 0.15 (0.006) | P01, M10 (C-8) | 205 (675) | 0.15 (0.006) | CP01, CM10 (CC-8) |
| | | | 1.25 (0.050) | 32 (105) | 0.13 (0.005) | S4, S5 (M2, M3) | 110 (360) | 130 (425) | 0.20 (0.008) | P10, M10 (C-7) | 170 (550) | 0.20 (0.008) | CP10, CM10 (CC-7) |
| | | | 2.5 (0.100) | 24 (80) | 0.30 (0.012) | S4, S5 (M2, M3) | 82 (270) | 95 (315) | 0.40 (0.015) | P20, M20 (C-6) | 130 (425) | 0.30 (0.012) | CP20, CM20 (CC-6) |
| | 200–240 | Malleablized and heat treated | 0.25 (0.010) | 30 (100) | 0.075 (0.003) | S4, S5 (M2, M3) | 100 (330) | 120 (390) | 0.15 (0.006) | P10, M10 (C-7) | 160 (525) | 0.15 (0.006) | CP10, CM10 (CC-7) |
| | | | 1.25 (0.050) | 24 (80) | 0.13 (0.005) | S4, S5 (M2, M3) | 82 (270) | 95 (315) | 0.20 (0.008) | P20, M20 (C-6) | 130 (425) | 0.20 (0.008) | CP20, CM20 (CC-6) |
| | | | 2.5 (0.100) | 18 (60) | 0.30 (0.012) | S4, S5 (M2, M3) | 64 (210) | 75 (245) | 0.40 (0.015) | P30, M30 (C-6) | 95 (315) | 0.30 (0.012) | CP30, CM30 (CC-6) |

(a) Any premium high-speed steel can be used. Source: Metcut Research Associates Inc.

speed steel tools are generally used in low cutting speed machining operations and for tools of a complex shape, and they are usually used in machining iron castings for end milling, drilling, reaming, tapping, broaching, and gear cutting.

**Cast alloy cutting tools** are not widely used. These tools are usually made of a cobalt-chromium-tungsten alloy, and because they are cast to shape, they offer some advantage in specially shaped form tools. They are more abrasion resistant than high-speed steel tools and are more resistant to chipping than carbides.

**Carbides.** Expendable carbide insert tooling is the type most commonly used for cast iron. It is most efficiently used in automatic machining systems. Sintered car-

**Table 10 Nominal speeds and feeds (chip loads) for the broaching of cast irons with high-speed steel and carbide tools**

| Material | Hardness, HB | Condition | Speed | | Chip load | | Tool material grade | |
|---|---|---|---|---|---|---|---|---|
| | | | m/min | sfm | mm/tooth | in./tooth | ISO | AISI or C |
| **Ferritic gray irons** | | | | | | | | |
| ASTM A48: class 20 | 120–150 | Annealed | 9 | 30 | 0.13 | 0.005 | S4, S2 | M2, M7 |
| SAE J431c: grade G1800 | | | 46 | 150 | 0.075–0.13 | 0.003–0.005 | K20 | C-2 |
| **Pearlitic gray irons** | | | | | | | | |
| ASTM A48: classes 30, 35, 40 | 190–220 | As-cast | 8 | 25 | 0.075 | 0.003 | S4, S2 | M2, M7 |
| SAE J431c: grade G3000 | | | 46 | 150 | 0.075–0.13 | 0.003–0.005 | K20 | C-2 |
| **Ferritic ductile irons** | | | | | | | | |
| ASTM A536: grades 60-40-18, 65-45-12<br>SAE J434c: grades D4018, D4512 | 140–190 | Annealed | 8 | 25 | 0.13 | 0.005 | S4, S2 | M2, M7 |
| **Ferritic-pearlitic ductile irons** | | | | | | | | |
| ASTM A536: grade 80-55-06 | 190–225 | As-cast | 8 | 25 | 0.10 | 0.004 | S4, S2 | M2, M7 |
| SAE J434c: grade D5506 | 225–260 | As-cast | 6 | 20 | 0.10 | 0.004 | S4, S2 | M2, M7 |
| **Ferritic malleable irons** | | | | | | | | |
| ASTM A47: grades 32510, 35018<br>ASTM A602: grade M3210<br>SAE J158: grade M3210 | 110–160 | Malleablized | 9 | 30 | 0.10 | 0.004 | S4, S2 | M2, M7 |
| **Pearlitic malleable irons** | | | | | | | | |
| ASTM A220: grades 40010, 45006, 45008, 50005 | 160–200 | Malleablized and heat treated | 6 | 20 | 0.075 | 0.003 | S4, S2 | M2, M7 |
| ASTM A602: grade M4504, M5003<br>SAE J158: grades M4504, M5003 | 200–240 | Malleablized and heat treated | 5 | 15 | 0.075 | 0.003 | S4, S2 | M2, M7 |

Source: Metcut Research Associates Inc.

bides are also commonly braze welded to special toolholders.

Eight general grades of carbides, from C-1 to C-8, have been established to identify carbide types of increasing hardness and decreasing chip resistance. The generally preferred C-grade for each machining operation and type of cast iron is given in the tables of speeds and feeds for specific machining operations that can be found at the end of this article. However, there are numerous carbide manufacturers, and each produces a variety of grades. Some producers have several grades within each general C-grade classification, and each grade has individual characteristics that are suited to specific applications. The International Organization for Standardization (ISO) carbide classification is more detailed and has three general categories, P, M, and K. Each category has a number of individual grades. The article "Cemented Carbides" in this Volume provides a detailed classification.

Micrograin carbides have recently been developed and possess outstanding strength and impact resistance. These tool materials provide an overlap between high-speed and conventional carbides. They can be used effectively in form tools and for interrupted cuts.

The development of coated carbide inserts has also improved cast iron machining. Very thin surface coatings of titanium carbide, titanium nitride, or aluminum oxide are applied to a carbide base. Coated carbides have been especially effective in the milling of iron castings. The coating reduces flank wear and cratering so that the tool life or metal removal rate can be appreciably increased.

**Cermets** are titanium carbide and titanium carbonitride powder metallurgy materials. As described in the article "Cermets" in this Volume, these powder metallurgy products are used to turn and mill cast irons. Detailed information on the processing, composition, properties, and applications of these materials can be found in Volume 7 of the 9th Edition of *Metals Handbook*.

**Ceramics** are increasingly used to machine cast irons, particularly by industries dominated by high-volume manufacture (such as the automotive and earth-moving equipment industries) in which high-speed (high-productivity) machining is critical. Speeds as high as 1500 m/min (5000 sfm) have been achieved with ceramic cutting tools. Both alumina-base ($Al_2O_3$-TiC) and silicon nitride base ($Si_3N_4$ plus sintering aids such as $Al_2O_3$, $Y_2O_3$, and MgO) cutting tools are used to machine cast irons.

At this point, silicon nitride base cutting tools have captured applications in cast iron machining that have previously been exclusively served by cemented carbide and alumina-base tools. In high-speed finishing and semiroughing applications, these tools, with their superior wear resistance, outperformed both $Al_2O_3$-base tools and $Al_2O_3$-coated cemented carbides. The capabilities of $Si_3N_4$ cutting tools allow their use on old and new machine tools. By applying $Si_3N_4$ tools, reported maximum metal removal rates in production exceed by far the productivity obtained with coated carbide or alumina tools (Table 6).

**Cubic Boron Nitride (CBN).** Next to diamond, CBN is the hardest material known. With the use of CBN indexable inserts, the turning of white irons and high-nickel irons is possible; previously, these materials could only be abrasively machined. Gray irons are also machined with CBN inserts. Machining parameters and applications related to the use of CBN for cast irons can be found in the article "Ultrahard Tool Materials" in this Volume. A tool life curve comparing CBN, tungsten carbide, an $Al_2O_3$-coated carbide, and SiAlON (an $Si_3N_4$-base ceramic) for the turning of gray iron is shown in Fig. 10.

## Turning

All of the common types of tool material are used in turning operations on cast irons. High-speed steel and cast alloy tools are generally purchased as solid bits; carbides, cermets, and ceramics are purchased as inserts.

When using inserts, the size of the end cutting edge angle depends on the shape of the insert and the size of the lead angle. A square insert is generally stronger because it has a 90° included angle nose; a triangular insert has a 60° nose. A square insert with a 15° lead angle also has a 15° end cutting edge angle. In triangular inserts, a 15° lead angle has a 45° end cutting edge angle, while a 0° lead angle has an end cutting edge angle of 30°.

**Table 11 Nominal speeds and feeds for the planing of cast iron with high-speed steel and carbide tools**

| | | | | High-speed steel tool | | | Carbide tool | | |
|---|---|---|---|---|---|---|---|---|---|
| Material | Hardness, HB | Condition | Depth of cut, mm (in.) | Speed, m/min (sfm) | Feed, mm/stroke (in./stroke) | Tool material, ISO (AISI) | Speed, m/min (sfm) | Feed, mm/stroke (in./stroke) | Tool material grade, ISO (C) |
| **Ferritic gray irons** | | | | | | | | | |
| ASTM A48: class 20; SAE J431c: grade G1800 | 120–150 | Annealed | 0.1 (0.005) | 21 (70) | (a) | S4, S5 (M2, M3) | 72 (235) | (a) | K10, M10 (C-2) |
| | | | 2.5 (0.100) | 24 (80) | 3.20 (0.125) | S4, S5 (M2, M3) | 90 (300) | 3.20 (0.125) | K20, M20 (C-2) |
| | | | 12 (0.500) | 18 (60) | 2.30 (0.090) | S4, S5 (M2, M3) | 90 (300) | 2.30 (0.090) | K30, M30 (C-1) |
| **Pearlitic gray irons** | | | | | | | | | |
| ASTM A48: classes 30, 35, 40; SAE J431c: grade G3000 | 190–220 | As-cast | 0.1 (0.005) | 11 (35) | (a) | S4, S5 (M2, M3) | 46 (150) | (a) | K10, M10 (C-2) |
| | | | 2.5 (0.100) | 12 (40) | 2.55 (0.100) | S4, S5 (M2, M3) | 90 (300) | 3.20 (0.125) | M20, P20 (C-6) |
| | | | 12 (0.500) | 9 (30) | 2.30 (0.090) | S4, S5 (M2, M3) | 69 (225) | 2.30 (0.090) | M30, P30 (C-6) |
| **Ferritic ductile irons** | | | | | | | | | |
| ASTM A536: grades 60-40-18, 65-45-12; SAE J434c: grades D4018, D4512 | 140–190 | Annealed | 0.1 (0.005) | 18 (60) | (a) | S4, S5 (M2, M3) | 69 (225) | (a) | P10, M10 (C-6) |
| | | | 2.5 (0.100) | 21 (70) | 1.25 (0.050) | S4, S5 (M2, M3) | 76 (250) | 2.05 (0.080) | P20, M20 (C-6) |
| | | | 12 (0.500) | 15 (50) | 1.50 (0.060) | S4, S5 (M2, M3) | 60 (200) | 1.50 (0.060) | P30, M30 (C-5) |
| **Ferritic-pearlitic ductile irons** | | | | | | | | | |
| ASTM A536: grade 80-55-06; SAE J434c: grade D5506 | 190–225 | As-cast | 0.1 (0.005) | 14 (45) | (a) | S4, S5 (M2, M3) | 58 (190) | (a) | P10, M10 (C-6) |
| | | | 2.5 (0.100) | 17 (55) | 1.25 (0.050) | S4, S5 (M2, M3) | 73 (240) | 2.05 (0.080) | P20, M20 (C-6) |
| | | | 12 (0.500) | 12 (40) | 1.50 (0.060) | S4, S5 (M2, M3) | 55 (180) | 1.50 (0.060) | P30, M30 (C-5) |
| | 225–260 | As-cast | 0.1 (0.005) | 8 (25) | (a) | S4, S5 (M2, M3) | 30 (100) | (a) | P10, M10 (C-6) |
| | | | 2.5 (0.100) | 11 (35) | 1.25 (0.050) | S4, S5 (M2, M3) | 38 (125) | 2.05 (0.080) | P20, M20 (C-6) |
| | | | 12 (0.500) | 8 (25) | 1.50 (0.060) | S4, S5 (M2, M3) | 30 (100) | 1.50 (0.060) | P30, M30 (C-5) |
| **Ferritic malleable irons** | | | | | | | | | |
| ASTM A47: grades 32510, 35018; ASTM A602: grade M3210; SAE J158: grade M3210 | 110–160 | Malleablized | 0.1 (0.005) | 18 (60) | (a) | S4, S5 (M2, M3) | 72 (235) | (a) | P10, M10 (C-6) |
| | | | 2.5 (0.100) | 23 (75) | 1.25 (0.050) | S4, S5 )M2, M3) | 90 (300) | 1.50 (0.060) | P20, M20 (C-6) |
| | | | 12 (0.500) | 15 (50) | 1.50 (0.060) | S4, S5 (M2, M3) | 90 (300) | 1.25 (0.050) | P30, M30 (C-5) |
| **Pearlitic malleable irons** | | | | | | | | | |
| ASTM A220: grades 40010, 45006, 45008, 50005; ASTM A602: grade M4504, M5003; SAE J158: grades M4504, M5003 | 160–200 | Malleablized and heat treated | 0.1 (0.005) | 15 (50) | (a) | S4, S5 (M2, M3) | 60 (200) | (a) | P10, M10 (C-6) |
| | | | 2.5 (0.100) | 20 (65) | 1.25 (0.050) | S4, S5 (M2, M3) | 73 (240) | 1.50 (0.060) | P20, M20 (C-6) |
| | | | 12 (0.500) | 14 (45) | 1.50 (0.060) | S4, S5 (M2, M3) | 58 (190) | 1.25 (0.050) | P30, M30 (C-5) |
| | 200–240 | Malleablized anc heat treated | 0.1 (0.005) | 11 (35) | (a) | S4, S5 (M2, M3) | 27 (90) | (a) | P10, M10 (C-6) |
| | | | 2.5 (0.100) | 12 (40) | 1.25 (0.050) | S4, S5 (M2, M3) | 38 (125) | 1.50 (0.060) | P20, M20 (C-6) |
| | | | 12 (0.500) | 8 (25) | 1.50 (0.060) | S4, S5 (M2, M3) | 30 (100) | 1.25 (0.050) | P30, M30 (C-5) |

(a) Feed equals three-fourths the width of a square-nose finishing tool. Source: Metcut Research Associates Inc.

For the best surface finishes, the use of a 1° end cutting edge angle combined with a 1.5 mm (1⁄16 in.) nose radius is recommended. This combination is intended for shallow finishing cuts (0.4 to 0.5 mm, or 0.015 to 0.020 in., deep).

**The nose radius** should be as large as possible to obtain optimum surface finish, but not so large as to cause work deflection or chatter. Normal insert radii range from 0.4 to 3 mm (1⁄64 to 1⁄8 in.). The finished part configuration often dictates the required shoulder radius and consequently the nose radius to be used. When round inserts are used, the radius can vary from 5 to 16 mm (3⁄16 to 5⁄8 in.).

**Rake Angles.** High-speed steel tools are provided with positive rake angles to obtain better cutting action at normal cutting speeds. Because these tool materials are much stronger than many of the carbides, they accommodate positive rake angles more easily. Most standard brazed carbide tools are ground with a 4 to 10° side or back

### Table 12 Nominal speeds and feeds for the drilling of cast iron with high-speed steel and carbide drills

| Material | Hardness, HB | Condition | Speed(a), m/min (sfm) | Feed(a), mm/rev (in./rev), at a nominal hole diameter of: 1.5 mm (1/16 in.) | 3 mm (1/8 in.) | 6 mm (1/4 in.) | 12 mm (1/2 in.) | 18 mm (3/4 in.) | 25 mm (1 in.) | 35 mm (1½ in.) | 50 mm (2 in.) | Tool material grade, ISO (AISI or C) |
|---|---|---|---|---|---|---|---|---|---|---|---|---|
| **Ferritic gray irons** | | | | | | | | | | | | |
| ASTM A48, class 20<br>SAE J431c: grade G1800 | 120–150 | Annealed | 27<br>(90) | 0.025<br>(0.001) | . . . | . . . | . . . | . . . | . . . | . . . | . . . | S2, S3<br>(M10, M7, M1) |
| | | | 49<br>(160) | . . . | 0.075<br>(0.003) | 0.15<br>(0.006) | 0.30<br>(0.012) | 0.45<br>(0.018) | 0.55<br>(0.022) | 0.65<br>(0.025) | 0.75<br>(0.030) | S2, S3<br>(M10, M7, M1) |
| | | | 105<br>(350) | . . . | 0.075<br>(0.003) | 0.15<br>(0.006) | 0.30<br>(0.012) | 0.45<br>(0.018) | 0.55<br>(0.022) | 0.65<br>(0.025) | 0.75<br>(0.030) | K10<br>(C-2) |
| **Pearlitic gray irons** | | | | | | | | | | | | |
| ASTM A48: classes 30, 35, 40<br>SAE J431c: grade G3000 | 190–220 | As-cast | 23<br>(75) | 0.025<br>(0.001) | . . . | . . . | . . . | . . . | . . . | . . . | . . . | S2, S3<br>(M10, M7, M1) |
| | | | 29<br>(95) | . . . | 0.075<br>(0.003) | 0.13<br>(0.005) | 0.30<br>(0.012) | 0.45<br>(0.018) | 0.55<br>(0.022) | 0.65<br>(0.025) | 0.75<br>(0.030) | S2, S3<br>(M10, M7, M1) |
| | | | 69<br>(225) | . . . | 0.075<br>(0.003) | 0.13<br>(0.005) | 0.30<br>(0.012) | 0.45<br>(0.018) | 0.55<br>(0.022) | 0.65<br>(0.025) | 0.75<br>(0.030) | K10<br>(C-2) |
| **Ferritic ductile irons** | | | | | | | | | | | | |
| ASTM A536: grades 60-40-18, 65-45-12<br>SAE J434c: grades D4018, D4512 | 140–190 | Annealed | 26<br>(85) | 0.025<br>(0.001) | . . . | . . . | . . . | . . . | . . . | . . . | . . . | S2, S3<br>(M10, M7, M1) |
| | | | 35<br>(115) | . . . | 0.075<br>(0.003) | 0.15<br>(0.006) | 0.25<br>(0.010) | 0.33<br>(0.013) | 0.40<br>(0.016) | 0.53<br>(0.021) | 0.65<br>(0.025) | S2, S3<br>(M10, M7, M1) |
| **Ferritic-pearlitic ductile irons** | | | | | | | | | | | | |
| ASTM A536: grade 80-55-06<br>SAE J434c: grade D5506 | 190–225 | As-cast | 21<br>(70) | 0.025<br>(0.001) | 0.075<br>(0.003) | 0.15<br>(0.006) | 0.25<br>(0.010) | 0.33<br>(0.013) | 0.40<br>(0.016) | 0.53<br>(0.021) | 0.65<br>(0.025) | S2, S3<br>(M10, M7, M1) |
| | 225–260 | As-cast | 15<br>(50) | 0.025<br>(0.001) | 0.050<br>(0.002) | 0.10<br>(0.004) | 0.18<br>(0.007) | 0.25<br>(0.010) | 0.30<br>(0.012) | 0.40<br>(0.015) | 0.45<br>(0.017) | S9, S11(b)<br>(T15, M42)(b) |
| **Ferritic malleable irons** | | | | | | | | | | | | |
| ASTM A47: grades 32510, 35018<br>ASTM A602: grade M3210<br>SAE J158: grade M3210 | 110–160 | Malleablized | 40<br>(130) | 0.025<br>(0.001) | 0.075<br>(0.003) | 0.15<br>(0.006) | 0.25<br>(0.010) | 0.33<br>(0.013) | 0.40<br>(0.016) | 0.53<br>(0.021) | 0.65<br>(0.025) | S2, S3<br>(M10, M7, M1) |
| **Pearlitic malleable irons** | | | | | | | | | | | | |
| ASTM A220: grades 40010, 45006, 45008, 50005<br>ASTM A602: grade M4504, M5003 | 160–200 | Malleablized and heat treated | 30<br>(100) | 0.025<br>(0.001) | 0.075<br>(0.003) | 0.15<br>(0.006) | 0.25<br>(0.010) | 0.33<br>(0.013) | 0.40<br>(0.016) | 0.53<br>(0.021) | 0.65<br>(0.025) | S2, S3<br>(M10, M7, M1) |
| SAE J158: grades M4504, M5003 | 200–240 | Malleablized and heat treated | 26<br>(85) | . . . | 0.050<br>(0.002) | 0.10<br>(0.004) | 0.18<br>(0.007) | 0.25<br>(0.010) | 0.30<br>(0.012) | 0.40<br>(0.015) | 0.45<br>(0.017) | S2, S3<br>(M10, M7, M1) |

(a) For drilling deep holes with twist drills, reduce speed and feed as follows. If hole depth is three drill diameters, reduce speed and feed both by 10%; four diameters, speed by 20%, feed by 10%; five diameters, speed by 30% feed by 20%. (b) Any premium high-speed steel can be used. Source: Metcut Research Associates Inc.

rake, depending on the direction of cutting and the style of tool used. As the hardness of the iron or the degree of surface roughness increases, rake angle is decreased.

With expendable inserts, it is possible to select a positive rake, a negative rake, or a combination negative rake with a positive-rake groove. Tool costs vary, depending on which of these designs is selected.

In addition to their cost advantage over positive-rake inserts, negative-rake inserts are much stronger and yield better results on workpieces that are of higher hardness or exhibit a scaly or uneven surface. Negative rakes are particularly effective in cutting the segmental, discontinuous type of chip that is characteristic of cast irons. In general, turning operations should begin with negative-rake tooling, except for workpieces of certain shapes, such as those with thin walls, that require the use of positive rakes.

**Relief (clearance) angles** for normal turning and facing tools of the 5° negative-rake type are 5°; the normal side clearance on 6° positive-rake insert tooling is 5°, with 11° end clearance, and on brazed tools it is 5 to 7°. If a positive-rake insert is placed in a neutral-rake holder (0° back and side rake), an 11° clearance angle is obtained. This is because the insert has an 11° angle. This combination is sometimes used in finishing operations. Under these conditions, the wear land does not develop as rapidly as on a 5° clearance insert because the wear is spread over a larger area and therefore cutting or feeding pressure remains lower.

**Honing of Cutting Edges.** To strengthen the cutting edges of carbide and ceramic tools, the entire cutting edge is sometimes lightly honed or rounded. This minimizes chipping or breaking of the tool edge.

**Speed and Feed.** Among the major variables that affect the selection of turning speeds and feeds are the microstructure and hardness of the casting and the type of tool material used. Nominal speeds and feeds for turning the frequently used grades of gray iron, ductile iron, and malleable iron with several types of high-speed steel and carbide tools are listed in Table 7. Table 8 lists speeds and feeds for turning with ceramic tools. As described in the article "High-Speed Machining" in this Volume, speeds ranging from 900 to 1200 m/min (3000 to 4000 sfm) are common using ceramic tools.

## Boring

General aspects of the machining methods of boring and discussions of the equipment, tools, and procedures used, together with examples of boring cast iron parts, are given in the article "Boring" in this Volume. Nominal speeds and feeds for boring the frequently used grades of gray iron, ductile iron, and malleable iron with high-speed steel and carbide tools are listed in Table 9.

## Broaching

When high-speed steel broaches are used, cutting speeds up to 9 m/min (30 sfm) are feasible for certain grades of cast iron; cutting speeds do not exceed 6 m/min (20 sfm) for most grades. With carbide broaches, much higher speeds (up to about 45 m/min, or 150 sfm) are attainable.

**Table 13 Nominal speeds and feeds for the reaming of cast iron with high-speed steel and carbide tools**

| Material | Hardness, HB | Condition | Roughing: Speed, m/min (sfm) | Roughing: Feed(a), mm/rev (in./rev), at a reamer diameter of: 3 mm (1/8 in.) | 6 mm (1/4 in.) | 12 mm (1/2 in.) | 25 mm (1 in.) | 35 mm (1.5 in.) | 50 mm (2 in.) | Roughing: Tool material grade, ISO (AISI or C) | Finishing: Speed, m/min (sfm) | Finishing: Feed(a), mm/rev (in./rev), at a reamer diameter of: 3 mm (1/8 in.) | 6 mm (1/4 in.) | 12 mm (1/2 in.) | 25 mm (1 in.) | 35 mm (1.5 in.) | 50 mm (2 in.) | Finishing: Tool material grade, ISO (AISI or C) |
|---|---|---|---|---|---|---|---|---|---|---|---|---|---|---|---|---|---|---|
| **Ferritic gray irons** | | | | | | | | | | | | | | | | | | |
| ASTM A48: | 120–150 | Annealed | 37 | 0.15 | 0.20 | 0.30 | 0.50 | 0.65 | 0.75 | S3, S4, S2 | 20 | 0.15 | 0.25 | 0.40 | 0.65 | 0.75 | 0.90 | S3, S4, S2 |
| class 20 | | | (120) | (0.006) | (0.008) | (0.012) | (0.020) | (0.025) | (0.030) | (M1, M2, M7) | (65) | (0.006) | (0.010) | (0.015) | (0.025) | (0.030) | (0.035) | (M1, M2, M7) |
| SAE J431c: | | | 60 | 0.15 | 0.20 | 0.30 | 0.50 | 0.65 | 0.75 | K20 | 27 | 0.15 | 0.20 | 0.25 | 0.50 | 0.65 | 0.75 | K20 |
| grade G1800 | | | (200) | (0.006) | (0.008) | (0.012) | (0.020) | (0.025) | (0.030) | (C-2) | (90) | (0.006) | (0.008) | (0.010) | (0.020) | (0.025) | (0.030) | (C-2) |
| **Pearlitic gray irons** | | | | | | | | | | | | | | | | | | |
| ASTM A48: | 190–220 | As-cast | 27 | 0.13 | 0.15 | 0.25 | 0.40 | 0.50 | 0.65 | S3, S4, S2 | 12 | 0.13 | 0.20 | 0.30 | 0.50 | 0.65 | 0.75 | S3, S4, S2 |
| classes 30, | | | (90) | (0.005) | (0.006) | (0.010) | (0.015) | (0.020) | (0.025) | (M1, M2, M7) | (40) | (0.005) | (0.008) | (0.012) | (0.020) | (0.025) | (0.030) | (M1, M2, M7) |
| 35, 40 | | | 37 | 0.13 | 0.15 | 0.25 | 0.40 | 0.50 | 0.65 | K20 | 20 | 0.13 | 0.20 | 0.30 | 0.50 | 0.65 | 0.75 | K20 |
| SAE J431c: | | | (120) | (0.005) | (0.006) | (0.010) | (0.015) | (0.020) | (0.025) | (C-2) | (65) | (0.005) | (0.008) | (0.012) | (0.020) | (0.025) | (0.030) | (C-2) |
| grade G3000 | | | | | | | | | | | | | | | | | | |
| **Ferritic ductile irons** | | | | | | | | | | | | | | | | | | |
| ASTM A536: | 140–190 | Annealed | 21 | 0.10 | 0.15 | 0.25 | 0.40 | 0.50 | 0.65 | S3, S4, S2 | 14 | 0.13 | 0.20 | 0.30 | 0.50 | 0.65 | 0.75 | S3, S4, S2 |
| grades | | | (70) | (0.004) | (0.006) | (0.010) | (0.015) | (0.020) | (0.025) | (M1, M2, M7) | (45) | (0.005) | (0.008) | (0.012) | (0.020) | (0.025) | (0.030) | (M1, M2, M7) |
| 60-40-18, | | | 37 | 0.10 | 0.15 | 0.25 | 0.40 | 0.50 | 0.65 | K20 | 18 | 0.13 | 0.20 | 0.30 | 0.50 | 0.65 | 0.75 | K20 |
| 65-45-12 | | | (120) | (0.004) | (0.006) | (0.010) | (0.015) | (0.020) | (0.025) | (C-2) | (60) | (0.005) | (0.008) | (0.012) | (0.020) | (0.025) | (0.030) | (C-2) |
| SAE J434c: | | | | | | | | | | | | | | | | | | |
| grades | | | | | | | | | | | | | | | | | | |
| D4018, | | | | | | | | | | | | | | | | | | |
| D4512 | | | | | | | | | | | | | | | | | | |
| **Ferritic-pearlitic ductile irons** | | | | | | | | | | | | | | | | | | |
| ASTM A536: | 190–225 | As-cast | 20 | 0.10 | 0.15 | 0.25 | 0.40 | 0.50 | 0.65 | S3, S4, S2 | 12 | 0.13 | 0.20 | 0.30 | 0.50 | 0.65 | 0.75 | S3, S4, S2 |
| grade | | | (65) | (0.004) | (0.006) | (0.010) | (0.015) | (0.020) | (0.025) | (M1, M2, M7) | (40) | (0.005) | (0.008) | (0.012) | (0.020) | (0.025) | (0.030) | (M1, M2, M7) |
| 80-55-06 | | | 34 | 0.10 | 0.15 | 0.25 | 0.40 | 0.50 | 0.65 | K20 | 15 | 0.102 | 0.15 | 0.25 | 0.40 | 0.50 | 0.65 | K20 |
| SAE J434c: | | | (110) | (0.004) | (0.006) | (0.010) | (0.015) | (0.020) | (0.025) | (C-2) | (50) | (0.004) | (0.006) | (0.010) | (0.015) | (0.020) | (0.025) | (C-2) |
| grade D5506 | 225–260 | As-cast | 18 | 0.10 | 0.15 | 0.20 | 0.30 | 0.40 | 0.50 | S3, S4, S2 | 11 | 0.102 | 0.18 | 0.25 | 0.40 | 0.50 | 0.65 | S3, S4, S2 |
| | | | (60) | (0.004) | (0.006) | (0.008) | (0.012) | (0.016) | (0.020) | (M1, M2, M7) | (35) | (0.004) | (0.007) | (0.010) | (0.016) | (0.020) | (0.025) | (M1, M2, M7) |
| | | | 30 | 0.10 | 0.15 | 0.20 | 0.30 | 0.40 | 0.50 | K20 | 14 | 0.102 | 0.15 | 0.20 | 0.30 | 0.40 | 0.50 | K20 |
| | | | (100) | (0.004) | (0.006) | (0.008) | (0.012) | (0.016) | (0.020) | (C-2) | (45) | (0.004) | (0.006) | (0.008) | (0.012) | (0.016) | (0.020) | (C-2) |
| **Malleable cast irons** | | | | | | | | | | | | | | | | | | |
| ASTM A47: | 110–160 | Malleablized | 26 | 0.15 | 0.20 | 0.30 | 0.50 | 0.65 | 0.75 | S3, S4, S2 | 17 | 0.15 | 0.25 | 0.40 | 0.65 | 0.75 | 0.90 | S3, S4, S2 |
| grades | | | (85) | (0.006) | (0.008) | (0.012) | (0.020) | (0.025) | (0.030) | (M1, M2, M7) | (55) | (0.006) | (0.010) | (0.015) | (0.025) | (0.030) | (0.035) | (M1, M2, M7) |
| 32510, 35018 | | | 43 | 0.15 | 0.20 | 0.30 | 0.50 | 0.65 | 0.75 | K20 | 24 | 0.13 | 0.20 | 0.30 | 0.50 | 0.65 | 0.75 | K20 |
| ASTM A602: | | | (140) | (0.006) | (0.008) | (0.012) | (0.020) | (0.025) | (0.030) | (C-2) | (80) | (0.005) | (0.008) | (0.012) | (0.020) | (0.025) | (0.030) | (C-2) |
| grade M3210 | | | | | | | | | | | | | | | | | | |
| SAE J158: | | | | | | | | | | | | | | | | | | |
| grade M3210 | | | | | | | | | | | | | | | | | | |
| **Pearlitic malleable irons** | | | | | | | | | | | | | | | | | | |
| ASTM A220: | 160–200 | Malleablized | 24 | 0.10 | 0.15 | 0.25 | 0.40 | 0.50 | 0.65 | S3, S4, S2 | 14 | 0.13 | 0.20 | 0.30 | 0.50 | 0.65 | 0.75 | S3, S4, S2 |
| grades | | and heat | (80) | (0.004) | (0.006) | (0.010) | (0.015) | (0.020) | (0.025) | (M1, M2, M7) | (45) | (0.005) | (0.008) | (0.012) | (0.020) | (0.025) | (0.030) | (M1, M2, M7) |
| 40010, 45006, | | treated | 41 | 0.10 | 0.15 | 0.25 | 0.40 | 0.50 | 0.65 | K20 | 21 | 0.13 | 0.20 | 0.30 | 0.50 | 0.65 | 0.75 | K20 |
| 45008, 50005 | | | (135) | (0.004) | (0.006) | (0.010) | (0.015) | (0.020) | (0.025) | (C-2) | (70) | (0.005) | (0.008) | (0.012) | (0.020) | (0.025) | (0.030) | (C-2) |
| SAE J158: | 200–240 | Malleablized | 18 | 0.10 | 0.15 | 0.20 | 0.30 | 0.40 | 0.50 | S3, S4, S2 | 11 | 0.102 | 0.18 | 0.25 | 0.40 | 0.50 | 0.65 | S3, S4, S2 |
| grades | | and heat | (60) | (0.004) | (0.006) | (0.008) | (0.012) | (0.016) | (0.020) | (M1, M2, M7) | (35) | (0.004) | (0.007) | (0.010) | (0.015) | (0.020) | (0.025) | (M1, M2, M7) |
| M4504, | | treated | 30 | 0.10 | 0.15 | 0.20 | 0.30 | 0.40 | 0.50 | K20 | 18 | 0.102 | 0.15 | 0.20 | 0.30 | 0.40 | 0.50 | K20 |
| M5003 | | | (100) | (0.004) | (0.006) | (0.008) | (0.012) | (0.016) | (0.020) | (C-2) | (60) | (0.004) | (0.006) | (0.008) | (0.012) | (0.016) | (0.020) | (C-2) |
| ASTM A602: | | | | | | | | | | | | | | | | | | |
| grade | | | | | | | | | | | | | | | | | | |
| M4504, | | | | | | | | | | | | | | | | | | |
| M5003 | | | | | | | | | | | | | | | | | | |

(a) Based on four flutes for 3 and 6 mm (1/8 and 1/4 in.) reamers, six flutes for 12 mm (1/2 in.) reamers, and eight flutes for 25 mm (1 in.) and larger reamers. Source: Metcut Research Associates Inc.

Table 10 lists nominal speeds and feeds for broaching cast iron with high-speed steel and carbide tools. Face angles of about 8 to 12° are recommended for the cutting teeth of broaches used on cast irons. Roughing teeth are usually given clearance of 2 to 3°, semifinishing teeth about 1°, and finishing teeth about ½°.

## Planing and Shaping

Planing and shaping are principally used for the production of flat and angular surfaces. The machines employed in these operations are designed to make straight-line cuts, and the cutting tools used in both operations are identical except for size.

**Tool Design.** The principles that govern the shape of turning tools also apply to tools for planing (see the section "Turning" in this article). The amount of back rake and side rake provided depends on the hardness of the material, and the resultant rake and inclination should be negative. The end relief should be about 4 to 8° for planer tools because a planer tool is held about square with the platen, while a lathe tool, whose height and inclination can be varied, may not always be clamped in the same position. Illustrations showing the details of planing and shaping tools suitable for use on cast iron are provided in the articles "Planing" and "Shaping and Slotting" in this Volume.

**Speeds and feeds** for planing vary widely, depending on the composition and hardness of the work metal and on the tool material used. Feed is also governed by the depth of cut, the nature of cut (whether roughing or

**Table 14 Nominal speeds and feeds for the counterboring and spotfacing of cast iron with high-speed steel and carbide tools**

| Material | Hardness, HB | Condition | Speed, m/min (sfm) | Feed(a), mm/rev (in./rev), at a nominal hole diameter of: 6 mm (¼ in.) | 12 mm (½ in.) | 18 mm (¾ in.) | 25 mm (1 in.) | 35 mm (1.5 in.) | 50 mm (2 in.) | 62 mm (2.5 in.) | 75 mm (3 in.) | Tool material grade, ISO (AISI or C) |
|---|---|---|---|---|---|---|---|---|---|---|---|---|
| **Ferritic gray irons** | | | | | | | | | | | | |
| ASTM A48: class 20; SAE J431c: grade G1800 | 120–150 | Annealed | 37 (120) | 0.15 (0.006) | 0.20 (0.008) | 0.20 (0.008) | 0.20 (0.008) | 0.25 (0.010) | 0.25 (0.010) | 0.30 (0.012) | 0.30 (0.012) | S4, S5 (M2, M3) |
| | | | 64 (210) | 0.15 (0.006) | 0.20 (0.008) | 0.20 (0.008) | 0.20 (0.008) | 0.25 (0.010) | 0.25 (0.010) | 0.30 (0.012) | 0.30 (0.012) | K10 (C-2) |
| **Pearlitic gray irons** | | | | | | | | | | | | |
| ASTM A48: classes 30, 35, 40; SAE J431c: grade G3000 | 190–220 | As-cast | 18 (60) | 0.075 (0.003) | 0.10 (0.004) | 0.15 (0.006) | 0.15 (0.006) | 0.15 (0.006) | 0.20 (0.008) | 0.20 (0.008) | 0.25 (0.010) | S4, S5 (M2, M3) |
| | | | 38 (125) | 0.075 (0.003) | 0.10 (0.004) | 0.15 (0.006) | 0.15 (0.006) | 0.15 (0.006) | 0.20 (0.008) | 0.20 (0.008) | 0.25 (0.010) | K10 (C-2) |
| **Ferritic ductile irons** | | | | | | | | | | | | |
| ASTM A536: grades 60-40-18, 65-45-12; SAE J434c: grades D4018, D4512 | 140–190 | Annealed | 35 (115) | 0.10 (0.004) | 0.15 (0.006) | 0.15 (0.006) | 0.15 (0.006) | 0.20 (0.008) | 0.20 (0.008) | 0.20 (0.008) | 0.25 (0.010) | S4, S5 (M2, M3) |
| | | | 60 (200) | 0.10 (0.004) | 0.15 (0.006) | 0.15 (0.006) | 0.15 (0.006) | 0.20 (0.008) | 0.20 (0.008) | 0.20 (0.008) | 0.25 (0.010) | K10 (C-2) |
| **Ferritic-pearlitic ductile irons** | | | | | | | | | | | | |
| ASTM A536: grade 80-55-06; SAE J434c: grade D5506 | 190–225 | As-cast | 26 (85) | 0.075 (0.003) | 0.10 (0.004) | 0.10 (0.004) | 0.10 (0.004) | 0.15 (0.006) | 0.15 (0.006) | 0.20 (0.008) | 0.20 (0.008) | S4, S5 (M2, M3) |
| | | | 44 (145) | 0.075 (0.003) | 0.10 (0.004) | 0.10 (0.004) | 0.10 (0.004) | 0.15 (0.006) | 0.15 (0.006) | 0.20 (0.008) | 0.20 (0.008) | K10 (C-2) |
| | 225–260 | As-cast | 20 (65) | 0.075 (0.003) | 0.10 (0.004) | 0.10 (0.004) | 0.10 (0.004) | 0.15 (0.006) | 0.15 (0.006) | 0.15 (0.006) | 0.20 (0.008) | S4, S5 (M2, M3) |
| | | | 38 (125) | 0.075 (0.003) | 0.10 (0.004) | 0.10 (0.004) | 0.10 (0.004) | 0.15 (0.006) | 0.15 (0.006) | 0.15 (0.006) | 0.20 (0.008) | K10 (C-2) |
| **Ferritic malleable irons** | | | | | | | | | | | | |
| ASTM A47: grades 32510, 35018; ASTM A602: grade M3210; SAE J158: grade M3210 | 110–160 | Malleablized | 50 (165) | 0.15 (0.006) | 0.20 (0.008) | 0.20 (0.008) | 0.20 (0.008) | 0.25 (0.010) | 0.25 (0.010) | 0.30 (0.012) | 0.30 (0.012) | S4, S5 (M2, M3) |
| | | | 84 (275) | 0.15 (0.006) | 0.20 (0.008) | 0.20 (0.008) | 0.20 (0.008) | 0.25 (0.010) | 0.25 (0.010) | 0.30 (0.012) | 0.30 (0.012) | K10 (C-2) |
| **Pearlitic malleable irons** | | | | | | | | | | | | |
| ASTM A220: grades 40010, 45006, 45008, 50005; ASTM A602: grade M4504, M5003 | 160–200 | Malleablized and heat treated | 23 (75) | 0.075 (0.003) | 0.10 (0.004) | 0.10 (0.004) | 0.102 (0.004) | 0.15 (0.006) | 0.15 (0.006) | 0.15 (0.006) | 0.20 (0.008) | S4, S5 (M2, M3) |
| | | | 41 (135) | 0.075 (0.003) | 0.10 (0.004) | 0.10 (0.004) | 0.102 (0.004) | 0.15 (0.006) | 0.15 (0.006) | 0.15 (0.006) | 0.20 (0.008) | K10 (C-2) |
| SAE J158: grades M4504, M5003 | 200–240 | Malleablized and heat treated | 18 (60) | 0.075 (0.003) | 0.10 (0.004) | 0.10 (0.004) | 0.102 (0.004) | 0.15 (0.006) | 0.15 (0.006) | 0.15 (0.006) | 0.20 (0.008) | S4, S5 (M2, M3) |
| | | | 35 (115) | 0.075 (0.003) | 0.10 (0.004) | 0.10 (0.004) | 0.102 (0.004) | 0.15 (0.006) | 0.15 (0.006) | 0.15 (0.006) | 0.20 (0.008) | K10 (C-2) |

(a) Feeds are based on three flutes for 6 mm (¼ in.), four flutes for 12 to 62 mm (0.5 to 2.5 in.), and six flutes for 75 mm (3 in.) diameters. Source: Metcut Research Associates Inc.

finishing), and the rigidity of the workpiece when clamped in position for machining.

Table 11 lists nominal speeds and feeds for planing the frequently used grades of gray iron, ductile iron, and malleable iron with high-speed steel and carbide tools. Speeds for carbide tools are about four to six times those for comparable operations using high-speed steel tools; recommended feeds are the same in most cases. In shaping, rapid metal removal is of less importance than in planing. Speeds and feeds are often restricted to values substantially below those used for planing because of limitations on power and rigidity, the generally smaller size of tools, and the higher frequency of interruptions in cutting.

## Drilling

High-speed steel drills are widely used for drilling cast irons. Except for drilling out cored holes and gun drilling, carbide drills are seldom used because they cost much more than high-speed steel drills and require more attention in handling and sharpening. Standard drills have a point angle of 118° and a helix angle of 32°. Table 12 lists nominal speeds and feeds for the drilling of cast irons.

## Reaming

Both high-speed steel and carbide-tip reamers are used for reaming cast irons; the latter are most often used in high production or for reaming the more-difficult-to-machine alloys. When carbide-tip reamers are used, chatter must be completely eliminated to avoid tool breakage. Table 13 lists nominal speeds and feeds for reaming the cast irons.

## Counterboring and Spotfacing

Procedures and equipment for these two closely related machining methods are described in the article "Countersinking, Counterboring, and Spotfacing" in this Volume. Table 14 lists nominal speeds and feeds for counterboring and spotfacing the cast irons with high-speed steel and carbide tools.

## Tapping

High-speed steel taps are recommended for tapping cast irons. These tools usually employ a rake angle of 0°, although a rake angle of 3 to 5° is often preferred for tapping annealed material. Taps should be operated at the highest speed permitted by equipment, lubrication, and type of work material. High speed not only increases production and thread accuracy but also results in longer tap life. Most tap breakage occurs because of insufficient speed. Table 15 lists nominal speeds for tapping cast irons with high-speed steel taps.

## Milling

The basic principles of the three general types of milling—face, peripheral (slab), and end—are discussed in the article "Milling" in this Volume. Equipment, tool materials, tool design, and operating conditions are also considered in that article in relation to the composition and hardness of the work metal.

**Table 15 Nominal speeds for the tapping of cast iron with high-speed steel taps**

| Material | Hardness, HB | Condition | Speed, m/min (sfm), at a pitch (threads/in.) of: >3 mm | (≤7) | 1.5–3 mm | (8–15) | 1–1.5 mm | (16–24) | ≤1 mm | (>24) | High-speed steel tool material ISO | (AISI) |
|---|---|---|---|---|---|---|---|---|---|---|---|---|
| **Ferritic gray cast irons** | | | | | | | | | | | | |
| ASTM A48: class 20<br>SAE J431c: grade G1800 | 120–150 | Annealed | 11 | (35) | 18 | (60) | 23 | (75) | 24 | (80) | S2, S3 | (M10, M7, M1) |
| **Pearlitic gray irons** | | | | | | | | | | | | |
| ASTM A48: classes 30, 35, 40<br>SAE J431c: grade G3000 | 190–220 | As-cast | 5 | (15) | 9 | (30) | 14 | (45) | 15 | (50) | S2, S3 | (M10, M7, M1) |
| **Ferritic ductile irons** | | | | | | | | | | | | |
| ASTM A536: grades 60-40-18, 65-45-12<br>SAE J434c: grades D4018, D4512 | 140–190 | Annealed | 6 | (20) | 12 | (40) | 17 | (55) | 18 | (60) | S2, S3 | (M10, M7, M1) |
| **Ferritic-pearlitic ductile irons** | | | | | | | | | | | | |
| ASTM A536: grade 80-55-06 | 190–225 | As-cast | 5 | (15) | 9 | (30) | 12 | (40) | 14 | (45) | S2, S3 | (M10, M7, M1) |
| SAE J434c: grade D5506 | 225–260 | As-cast | 3 | (10) | 6 | (20) | 8 | (25) | 9 | (30) | S2, S3 | (M10, M7, M1) |
| **Ferritic malleable irons** | | | | | | | | | | | | |
| ASTM A47: grades 32510, 35018<br>ASTM A602: grade M3210<br>SAE J158: grade M3210 | 110–160 | Malleablized | 8 | (25) | 12 | (40) | 17 | (55) | 18 | (60) | S2, S3 | (M10, M7, M1) |
| **Pearlitic malleable irons** | | | | | | | | | | | | |
| ASTM A220: grades 40010, 45006, 45008, 50005 | 160–200 | Malleablized and heat treated | 6 | (20) | 9 | (30) | 14 | (45) | 15 | (50) | S2, S3 | (M10, M7, M1) |
| ASTM A602: grade M4504, M5003<br>SAE J158: grades M4504, M5003 | 200–240 | Malleablized and heat treated | 5 | (15) | 8 | (25) | 11 | (35) | 12 | (40) | S2, S3 | (M10, M7, M1) |

Source: Metcut Research Associates Inc.

In the face and end milling of cast iron, either high-speed steel or carbide cutters are used, depending primarily on the nature and hardness of the work metal. Most peripheral milling of cast iron is done with high-speed steel cutters. Tables 16 and 17 list nominal speeds and feeds for milling cast iron.

## Grinding

Grinding is a commonly used finishing operation for iron castings that is characterized by minimal stock removal, fine surface finish, and close dimensional tolerances. With the advent of high-horsepower grinding machines, the term abrasive machining has come into general use to describe grinding operations used for heavy stock removal as well as fine finishing.

In all grinding operations, small chips of metal are removed from the workpiece by the mechanical action of randomly spaced grains of abrasive. These grains of abrasive can be bonded into wheels or blocks, bonded onto endless belts, or used loose. Two constituents in grinding wheels control their grinding action. These are the abrasive and the bond that holds the abrasive grains together. These two components can be selected to provide a large number of operating characteristics for optimizing the grinding operation.

The abrasives most frequently recommended for use on iron castings are silicon carbide and aluminum oxide, although cubic boron nitride is used in some applications. The most suitable specific grade of abrasive wheel will depend on the hardness and microstructure of the casting. The most suitable process or type of machine is determined by the casting configuration, its finish and tolerance requirements, the amount of metal to be removed, and production quantity.

As mentioned earlier in this article, some cast irons are extremely difficult to machine. For example, high-silicon irons (14.5% Si) can only be abrasively machined. Various grinding operations carried out on cast irons are discussed below. Detailed information on grinding wheels, abrasives, and types of grinding equipment and operations can be found in the articles "Grinding Equipment and Processes" and "Superabrasives" in this Volume.

**The horizontal-spindle surface grinding** machine is primarily used for finishing flat surfaces. These machines can have either a reciprocating table or a rotary table. Frequently, they have fixtures or magnetic chucks to hold the work. On these machines, metal is removed by the periphery of the grinding wheel. The reciprocating-table machine is the most common type of horizontal-spindle grinder. Surfaces ground on these machines have a scratch pattern parallel to the motion of the reciprocating table. Rotary-table machines produce a scratch pattern of concentric rings. Some machines are equipped with a tilting table so that they can produce either concave or convex surfaces with a concentric scratch pattern. Disks and wedges for large valves are often produced on this type of grinder, with the concentric scratch pattern serving to aid in seating of the valve. Table 18 provides data on the surface grinding of cast irons.

**Vertical-spindle surface grinding** machines commonly have rotary tables. These machines are used for high-production heavy grinding. Stock removal of 3 to 6 mm (⅛ to ¼ in.) is common on parts with flat, unencumbered surfaces. Wherever possible, parts should be designed with flat, parallel surfaces to accommodate this type of grinding. Dimensional tolerances of 0.05 mm (0.002 in.) and surface finishes of 0.8 μm (32 μin.) can be readily achieved in production applications. On this type of machine, metal is removed by contact with the circular face of the wheel, which can be either a single piece or a series of segments mechanically held in a chuck to form the wheel.

The work is most commonly held by the magnetic table. Very large parts, which have only a small proportion of their total area in contact with the table, and very small parts may require supplementary blocking. Special fixtures are used for the production grinding of castings that cannot be held directly on the table. This method of grinding can usually produce finished surfaces more economically than milling. The possibility of grind finishing should be considered when cast components are being designed. Bosses or pads having elevations different from the main surface should be avoided if possible in order to take advantage of this low-cost machining process.

**Cylindrical grinding** is used to abrasively machine the outside diameters of round parts. In this type of grinding, stepped or multidiameter shafts are slowly rotated on centers while the grinding wheel traverses the part, with the periphery of the grinding wheel removing metal from the workpiece.

**Table 16 Nominal speeds and feeds for the face milling of cast iron with high-speed steel and carbide cutters**

| | | | | High-speed steel tool | | | Carbide tool: Uncoated | | | | Carbide tool: Coated | | |
|---|---|---|---|---|---|---|---|---|---|---|---|---|---|
| Material | Hardness, HB | Condition | Depth of cut(a), mm (in.) | Speed, m/min (sfm) | Feed, mm/tooth (in./tooth) | Tool material, ISO (AISI) | Speed, m/min (sfm), Brazed | Speed, m/min (sfm), Indexable | Feed, mm/tooth (in./tooth) | Tool material grade, ISO (C) | Speed, m/min (sfm) | Feed, mm/tooth (in./tooth) | Tool material grade, ISO (C) |
| **Ferritic gray irons** | | | | | | | | | | | | | |
| ASTM A48: class 20; SAE J431c: grade G1800 | 120–150 | Annealed | 1 (0.040) | 72 (235) | 0.25 (0.010) | S4, S2 (M2, M7) | 190 (630) | 210 (695) | 0.25 (0.010) | K10, M20 (C-2) | 305 (1000) | 0.25 (0.010) | CK10, CM20 (CC-2) |
| | | | 4 (0.150) | 56 (185) | 0.36 (0.014) | S4, S2 (M2, M7) | 145 (475) | 160 (525) | 0.40 (0.015) | K20, M30 (C-2) | 205 (675) | 0.36 (0.014) | CK20, CM30 (CC-2) |
| | | | 8 (0.300) | 44 (145) | 0.45 (0.018) | S4, S2 (M2, M7) | 100 (335) | 125 (410) | 0.50 (0.020) | K30, M40 (C-2) | 160 (525) | 0.45 (0.018) | CK30, CM40 (CC-2) |
| **Pearlitic gray irons** | | | | | | | | | | | | | |
| ASTM A48: classes 30, 35, 40; SAE J431c: grade G3000 | 190–220 | As-cast | 1 (0.040) | 37 (120) | 0.20 (0.008) | S4, S2 (M2, M7) | 140 (465) | 155 (510) | 0.20 (0.008) | K10, M20 (C-2) | 235 (775) | 0.18 (0.007) | CK10, CM20 (CC-2) |
| | | | 4 (0.150) | 27 (90) | 0.30 (0.012) | S4, S2 (M2, M7) | 105 (350) | 117 (385) | 0.30 (0.012) | K20, M30 (C-2) | 150 (500) | 0.25 (0.010) | CK20, CM30 (CC-2) |
| | | | 8 (0.300) | 21 (70) | 0.40 (0.016) | S4, S2 (M2, M7) | 75 (245) | 90 (300) | 0.40 (0.016) | K30, M40 (C-2) | 120 (400) | 0.36 (0.014) | CK30, CM40 (CC-2) |
| **Ferritic ductile irons** | | | | | | | | | | | | | |
| ASTM A536: grades 60-40-18, 65-45-12; SAE J434c: grades D4018, D4512 | 140–190 | Annealed | 1 (0.040) | 59 (195) | 0.25 (0.010) | S4, S2 (M2, M7) | 205 (665) | 222 (730) | 0.25 (0.010) | M20, P20 (C-6) | 335 (1100) | 0.20 (0.008) | CM20, CP20 (CC-6) |
| | | | 4 (0.150) | 46 (150) | 0.36 (0.014) | S4, S2 (M2, M7) | 150 (500) | 170 (550) | 0.40 (0.015) | M30, P30 (C-6) | 220 (715) | 0.30 (0.012) | CM30, CP30 (CC-6) |
| | | | 8 (0.300) | 35 (115) | 0.45 (0.018) | S4, S2 (M2, M7) | 105 (350) | 130 (430) | 0.50 (0.020) | M40, P40 (C-6) | 170 (560) | 0.40 (0.016) | CM40, CP40 (CC-6) |
| **Ferritic-pearlitic ductile irons** | | | | | | | | | | | | | |
| ASTM A536: grade 80-55-06; SAE J434c: grade D5506 | 190–225 | As-cast | 1 (0.040) | 44 (145) | 0.20 (0.008) | S4, S2 (M2, M7) | 140 (465) | 155 (510) | 0.20 (0.008) | M20, P20 (C-6) | 235 (765) | 0.20 (0.008) | CM20, CP20 (CC-6) |
| | | | 4 (0.150) | 34 (110) | 0.30 (0.012) | S4, S2 (M2, M7) | 105 (350) | 115 (385) | 0.30 (0.012) | M30, P30 (C-6) | 150 (500) | 0.30 (0.012) | CM30, CP30 (CC-6) |
| | | | 8 (0.300) | 26 (85) | 0.40 (0.016) | S4, S2 (M2, M7) | 75 (245) | 90 (300) | 0.40 (0.016) | M40, P40 (C-6) | 120 (400) | 0.40 (0.016) | CM40, CP40 (CC-6) |
| | 225–260 | As-cast | 1 (0.040) | 35 (115) | 0.20 (0.008) | S4, S2 (M2, M7) | 120 (400) | 135 (440) | 0.18 (0.007) | M20, P20 (C-6) | 200 (650) | 0.18 (0.007) | CM20, CP20 (CC-6) |
| | | | 4 (0.150) | 27 (90) | 0.30 (0.012) | S4, S2 (M2, M7) | 95 (310) | 100 (330) | 0.25 (0.010) | M30, P30 (C-6) | 130 (425) | 0.25 (0.010) | CM30, CP30 (CC-6) |
| | | | 8 (0.300) | 21 (70) | 0.40 (0.016) | S4, S2 (M2, M7) | 64 (210) | 78 (255) | 0.36 (0.014) | M40, P40 (C-6) | 100 (325) | 0.36 (0.014) | CM40, CP40 (CC-6) |
| **Ferritic malleable irons** | | | | | | | | | | | | | |
| ASTM A47: grades 32510, 35018; ASTM A602: grade M3210; SAE J158: grade M3210 | 110–160 | Malleablized | 1 (0.040) | 100 (330) | 0.25 (0.010) | S4, S2 (M2, M7) | 245 (800) | 270 (880) | 0.25 (0.010) | M20, P20 (C-6) | 395 (1300) | 0.20 (0.008) | CM20, CP20 (CC-6) |
| | | | 4 (0.150) | 76 (250) | 0.36 (0.014) | S4, S2 (M2, M7) | 185 (600) | 200 (660) | 0.40 (0.015) | M30, P30 (C-6) | 260 (855) | 0.30 (0.012) | CM30, CP30 (CC-6) |
| | | | 8 (0.300) | 60 (200) | 0.40 (0.016) | S4, S2 (M2, M7) | 130 (425) | 160 (520) | 0.50 (0.020) | M40, P40 (C-6) | 205 (675) | 0.40 (0.016) | CM40, CP40 (CC-6) |
| **Pearlitic malleable irons** | | | | | | | | | | | | | |
| ASTM A220: grades 40010, 45006, 45008, 50005; ASTM A602: grade M4504, M50003; SAE J158: grades M4504, M5003 | 160–200 | Malleablized and heat treated | 1 (0.040) | 53 (175) | 0.20 (0.008) | S4, S2 (M2, M7) | 150 (500) | 170 (550) | 0.20 (0.008) | M20, P20 (C-6) | 250 (825) | 0.20 (0.008) | CM20, CP20 (CC-6) |
| | | | 4 (0.150) | 41 (135) | 0.30 (0.012) | S4, S2 (M2, M7) | 115 (375) | 125 (415) | 0.30 (0.012) | M30, P30 (C-6) | 165 (540) | 0.30 (0.012) | CM30, CP30 (CC-6) |
| | | | 8 (0.300) | 32 (105) | 0.40 (0.016) | S4, S2 (M2, M7) | 81 (265) | 100 (325) | 0.40 (0.016) | M40, P40 (C-6) | 130 (425) | 0.40 (0.016) | CM40, CP40 (CC-6) |
| | 200–240 | Malleablized and heat treated | 1 (0.040) | 37 (120) | 0.18 (0.007) | S4, S2 (M2, M7) | 130 (430) | 145 (475) | 0.18 (0.007) | M20, P20 (C-6) | 215 (700) | 0.18 (0.007) | CM20, CP20 (CC-6) |
| | | | 4 (0.150) | 27 (90) | 0.28 (0.011) | S4, S2 (M2, M7) | 105 (350) | 120 (400) | 0.25 (0.010) | M30, P30 (C-6) | 160 (525) | 0.25 (0.010) | CM30, CP30 (CC-6) |
| | | | 8 (0.300) | 21 (70) | 0.40 (0.016) | S4, S2 (M2, M7) | 75 (245) | 90 (300) | 0.36 (0.014) | M40, P40 (C-6) | 120 (400) | 0.36 (0.014) | CM40, CP40 (CC-6) |

(a) Horsepower requirements should be checked on heavier depths of cut. Source: Metcut Research Associates Inc.

Cylindrical grinding can produce very accurate round parts. Dimensional tolerances of 0.01 mm (0.0004 in.) and surface finishes of 0.4 μm (16 μin.) can be achieved.

When designing parts that will be cylindrically ground, it should be remembered that sharp shoulders are difficult to achieve on stepped shafts. Large fillets should be considered with loose tolerances on the radii. These will provide maximum economy in cylindrical grinding operations. Table 19 provides data on the cylindrical grinding of cast irons.

**Centerless grinding** is a method of grinding the outside diameters of round parts without mounting the part between centers. Instead, the workpiece is supported on a work-rest blade between a nonabrasive regulating wheel and the abrasive wheel. Metal removal is accomplished with the periphery of the abrasive wheel. If the workpiece is straight and has only one diameter, the

**Table 17 Nominal speeds and feeds for the end milling (peripheral) of cast iron with high-speed steel and carbide cutters**

| | | | | High-speed steel tool | | | | | | Carbide tool | | | | | |
|---|---|---|---|---|---|---|---|---|---|---|---|---|---|---|---|
| | | | | | Feed, mm/tooth (in./tooth) at a cutter diameter of: | | | | | | Feed, mm/tooth (in./tooth) at a cutter diameter of: | | | | |
| Material | Hardness, HB | Condition | Radial depth of cut(a), mm (in.) | Speed, m/min (sfm) | 10 mm (3/8 in.) | 12 mm (1/2 in.) | 18 mm (3/4 in.) | 25–50 mm (1–2 in.) | Tool material, ISO (AISI) | Speed, m/min (sfm) | 10 mm (3/8 in.) | 12 mm (1/2 in.) | 18 mm (3/4 in.) | 25–50 mm (1–2 in.) | Tool material grade, ISO (C) |
| **Ferritic gray irons** | | | | | | | | | | | | | | | |
| ASTM A48: class 20; SAE J431c: grade G1800 | 120–150 | Annealed | 0.5 (0.020) | 55 (180) | 0.025 (0.001) | 0.050 (0.002) | 0.13 (0.005) | 0.18 (0.007) | S4, S5, S2 (M2, M3, M7) | 190 (625) | 0.025 (0.001) | 0.075 (0.003) | 0.18 (0.007) | 0.23 (0.009) | K20, M20 (C-2) |
| | | | 1.5 (0.060) | 44 (145) | 0.050 (0.002) | 0.102 (0.004) | 0.15 (0.006) | 0.20 (0.008) | | 145 (480) | 0.050 (0.002) | 0.13 (0.005) | 0.23 (0.009) | 0.25 (0.010) | |
| | | | diam/4 (diam/4) | 37 (120) | 0.038 (0.0015) | 0.075 (0.003) | 0.13 (0.005) | 0.15 (0.006) | | 120 (400) | 0.038 (0.0015) | 0.075 (0.003) | 0.15 (0.006) | 0.18 (0.007) | |
| | | | diam/2 (diam/2) | 30 (100) | 0.025 (0.001) | 0.050 (0.002) | 0.10 (0.004) | 0.13 (0.005) | | 115 (375) | 0.025 (0.001) | 0.050 (0.002) | 0.13 (0.005) | 0.15 (0.006) | |
| **Pearlitic gray irons** | | | | | | | | | | | | | | | |
| ASTM A48: classes 30, 35, 40; SAE J431c: grade G3000 | 190–220 | As-cast | 0.5 (0.020) | 38 (125) | 0.025 (0.001) | 0.050 (0.002) | 0.075 (0.003) | 0.10 (0.004) | S4, S5, S2 (M2, M3, M7) | 120 (390) | 0.025 (0.001) | 0.075 (0.003) | 0.13 (0.005) | 0.18 (0.007) | K20, M20 (C-2) |
| | | | 1.5 (0.060) | 29 (95) | 0.050 (0.002) | 0.075 (0.003) | 0.10 (0.004) | 0.13 (0.005) | | 90 (300) | 0.050 (0.002) | 0.10 (0.004) | 0.15 (0.006) | 0.20 (0.008) | |
| | | | diam/4 (diam/4) | 26 (85) | 0.038 (0.0015) | 0.050 (0.002) | 0.075 (0.003) | 0.10 (0.004) | | 78 (255) | 0.038 (0.0015) | 0.075 (0.003) | 0.13 (0.005) | 0.15 (0.006) | |
| | | | diam/2 (diam/2) | 23 (75) | 0.025 (0.001) | 0.038 (0.0015) | 0.050 (0.002) | 0.075 (0.003) | | 72 (235) | 0.025 (0.001) | 0.050 (0.002) | 0.10 (0.004) | 0.13 (0.005) | |
| **Ferritic ductile irons** | | | | | | | | | | | | | | | |
| ASTM A536: grades 60-40-18, 65-45-12; SAE J434c: grades D4018, D4512 | 140–190 | Annealed | 0.5 (0.020) | 38 (125) | 0.025 (0.001) | 0.050 (0.002) | 0.13 (0.005) | 0.18 (0.007) | S4, S5, S2 (M2, M3, M7) | 135 (450) | 0.025 (0.001) | 0.075 (0.003) | 0.18 (0.007) | 0.23 (0.009) | P20, M20 (C-5) |
| | | | 1.5 (0.060) | 32 (105) | 0.050 (0.002) | 0.075 (0.003) | 0.15 (0.006) | 0.20 (0.008) | | 105 (345) | 0.050 (0.002) | 0.13 (0.005) | 0.23 (0.009) | 0.25 (0.010) | |
| | | | diam/4 (diam/4) | 26 (85) | 0.025 (0.001) | 0.050 (0.002) | 0.13 (0.005) | 0.15 (0.006) | | 90 (295) | 0.038 (0.0015) | 0.075 (0.003) | 0.15 (0.006) | 0.20 (0.008) | |
| | | | diam/2 (diam/2) | 23 (75) | 0.018 (0.0007) | 0.025 (0.001) | 0.10 (0.004) | 0.13 (0.005) | | 84 (275) | 0.025 (0.001) | 0.050 (0.002) | 0.13 (0.005) | 0.15 (0.006) | |
| **Ferritic-pearlitic ductile irons** | | | | | | | | | | | | | | | |
| ASTM A536: grade 80-55-06; SAE J434c: grade D5506 | 190–225 | As-cast | 0.5 (0.020) | 32 (105) | 0.025 (0.001) | 0.050 (0.002) | 0.10 (0.004) | 0.15 (0.006) | S4, S5, S2 (M2, M3, M7) | 115 (375) | 0.025 (0.001) | 0.050 (0.002) | 0.15 (0.006) | 0.18 (0.007) | P20, M20 (C-5) |
| | | | 1.5 (0.060) | 27 (90) | 0.050 (0.002) | 0.075 (0.003) | 0.13 (0.005) | 0.18 (0.007) | | 85 (280) | 0.050 (0.002) | 0.10 (0.004) | 0.18 (0.007) | 0.20 (0.008) | |
| | | | diam/4 (diam/4) | 21 (70) | 0.025 (0.001) | 0.050 (0.002) | 0.10 (0.004) | 0.13 (0.005) | | 75 (245) | 0.038 (0.0015) | 0.075 (0.003) | 0.13 (0.005) | 0.15 (0.006) | |
| | | | diam/2 (diam/2) | 18 (60) | 0.018 (0.0007) | 0.025 (0.001) | 0.075 (0.003) | 0.10 (0.004) | | 69 (225) | 0.025 (0.001) | 0.050 (0.002) | 0.10 (0.004) | 0.13 (0.005) | |
| | 225–260 | As-cast | 0.5 (0.020) | 27 (90) | 0.013 (0.0005) | 0.025 (0.001) | 0.050 (0.002) | 0.10 (0.004) | S4, S5, S2 (M2, M3, M7) | 88 (290) | 0.025 (0.001) | 0.050 (0.002) | 0.13 (0.005) | 0.15 (0.006) | P20, M20 (C-5) |
| | | | 1.5 (0.060) | 21 (70) | 0.025 (0.001) | 0.050 (0.002) | 0.075 (0.003) | 0.13 (0.005) | | 69 (225) | 0.050 (0.002) | 0.10 (0.004) | 0.15 (0.006) | 0.18 (0.007) | |
| | | | diam/4 (diam/4) | 18 (60) | 0.018 (0.0007) | 0.025 (0.001) | 0.050 (0.002) | 0.10 (0.004) | | 59 (195) | 0.038 (0.0015) | 0.075 (0.003) | 0.10 (0.004) | 0.13 (0.005) | |
| | | | diam/2 (diam/2) | 15 (50) | 0.013 (0.0005) | 0.018 (0.0007) | 0.038 (0.0015) | 0.075 (0.003) | | 53 (175) | 0.025 (0.001) | 0.050 (0.002) | 0.075 (0.003) | 0.10 (0.004) | |
| **Ferritic malleable irons** | | | | | | | | | | | | | | | |
| ASTM A47: grades 32510, 35018; ASTM A602: grade M3210; SAE J158: grade M3210 | 110–160 | Malleablized | 0.5 (0.020) | 55 (180) | 0.025 (0.001) | 0.050 (0.002) | 0.13 (0.005) | 0.18 (0.007) | S4, S5, S2 (M2, M3, M7) | 187 (615) | 0.025 (0.001) | 0.075 (0.003) | 0.18 (0.007) | 0.23 (0.009) | P20, M20 (C-5) |
| | | | 1.5 (0.060) | 44 (145) | 0.050 (0.002) | 0.102 (0.004) | 0.15 (0.006) | 0.20 (0.008) | | 145 (470) | 0.050 (0.002) | 0.13 (0.005) | 0.23 (0.009) | 0.25 (0.010) | |
| | | | diam/4 (diam/4) | 37 (120) | 0.038 (0.0015) | 0.075 (0.003) | 0.10 (0.004) | 0.15 (0.006) | | 120 (400) | 0.038 (0.0015) | 0.075 (0.003) | 0.15 (0.006) | 0.20 (0.008) | |
| | | | diam/2 (diam/2) | 30 (100) | 0.025 (0.001) | 0.038 (0.0015) | 0.075 (0.003) | 0.13 (0.005) | | 115 (375) | 0.025 (0.001) | 0.050 (0.002) | 0.13 (0.005) | 0.15 (0.006) | |

(continued)

(a) Depth of cut is measured parallel to the axis of the cutter. Source: Metcut Research Associates Inc.

workpiece traverses the abrasive wheel. This is known as through-feed grinding. When the workpiece has multiple diameters, the part is placed on the blade rest, and the grinding wheel is fed into the part. This is known as infeed grinding. Centerless grinding can produce a 0.5 μm (20 μin.) finish. Table 20 provides data on the centerless grinding of cast irons.

**Internal grinding** is used to grind the inside surfaces of holes. When the hole is round, the workpiece is held in a chuck and rotated while the arbor-mounted abrasive wheel removes metal by peripheral contact. A surface finish of 0.2 to 0.3 μm (8 to 12 μin.) can be attained with this process. Table 21 provides data on the internal grinding of cast irons.

**Abrasive Belt Grinding.** Coated abrasive belts are used to remove metal at a high rate. Iron castings are especially suitable for finishing with abrasive belts. Because the abrasive grains can be preferentially oriented and ideally spaced on the belt without influencing its strength, metal can be removed more efficiently with less heat by belt grinding than with an abrasive wheel.

### Table 17 (continued)

| Material | Hardness, HB | Condition | Radial depth of cut(a), mm (in.) | High-speed steel tool: Speed, m/min (sfm) | High-speed steel tool: Feed, mm/tooth (in./tooth) at a cutter diameter of: 10 mm (3/8 in.) | 12 mm (1/2 in.) | 18 mm (3/4 in.) | 25–50 mm (1–2 in.) | Tool material, ISO (AISI) | Carbide tool: Speed, m/min (sfm) | Carbide tool: Feed, mm/tooth (in./tooth) at a cutter diameter of: 10 mm (3/8 in.) | 12 mm (1/2 in.) | 18 mm (3/4 in.) | 25–50 mm (1–2 in.) | Tool material grade, ISO (C) |
|---|---|---|---|---|---|---|---|---|---|---|---|---|---|---|---|
| **Pearlitic malleable irons** | | | | | | | | | | | | | | | |
| ASTM A220: grades 40010, 45006, 45008, 50005; ASTM A602: grade M4504, M5003; SAE J158: grades M4504, M5003 | 160–200 | Malleablized andheat treated | 0.5 (0.020) | 43 (140) | 0.025 (0.001) | 0.050 (0.002) | 0.10 (0.004) | 0.15 (0.006) | S4,S5,S2 (M2,M3,M7) | 130 (425) | 0.025 (0.001) | 0.075 (0.003) | 0.13 (0.005) | 0.18 (0.007) | P20, M20 (C-5) |
| | | | 1.5 (0.060) | 34 (110) | 0.050 (0.002) | 0.075 (0.003) | 0.13 (0.005) | 0.18 (0.007) | | 100 (325) | 0.050 (0.002) | 0.10 (0.004) | 0.18 (0.007) | 0.20 (0.008) | |
| | | | diam/4 (diam/4) | 27 (90) | 0.025 (0.001) | 0.050 (0.002) | 0.10 (0.004) | 0.13 (0.005) | | 85 (280) | 0.038 (0.0015) | 0.075 (0.003) | 0.10 (0.004) | 0.15 (0.006) | |
| | | | diam/2 (diam/2) | 24 (80) | 0.018 (0.0007) | 0.025 (0.001) | 0.075 (0.003) | 0.10 (0.004) | | 79 (260) | 0.025 (0.001) | 0.050 (0.002) | 0.075 (0.003) | 0.13 (0.005) | |
| | 200–240 | Malleablized andheat treated | 0.5 (0.020) | 34 (110) | 0.025 (0.001) | 0.050 (0.002) | 0.075 (0.003) | 0.10 (0.004) | S4,S5,S2 (M2,M3,M7) | 115 (375) | 0.025 (0.001) | 0.050 (0.002) | 0.10 (0.004) | 0.15 (0.006) | P20, M20 (C-5) |
| | | | 1.5 (0.060) | 27 (90) | 0.050 (0.002) | 0.075 (0.003) | 0.10 (0.004) | 0.13 (0.005) | | 85 (280) | 0.050 (0.002) | 0.075 (0.003) | 0.13 (0.005) | 0.18 (0.007) | |
| | | | diam/4 (diam/4) | 21 (70) | 0.025 (0.001) | 0.050 (0.002) | 0.075 (0.003) | 0.10 (0.004) | | 75 (245) | 0.038 (0.0015) | 0.063 (0.0025) | 0.10 (0.004) | 0.13 (0.005) | |
| | | | diam/2 (diam/2) | 18 (60) | 0.018 (0.0007) | 0.025 (0.001) | 0.050 (0.002) | 0.075 (0.003) | | 69 (225) | 0.025 (0.001) | 0.050 (0.002) | 0.075 (0.003) | 0.10 (0.004) | |

(a) Depth of cut is measured parallel to the axis of the cutter. Source: Metcut Research Associates Inc.

### Table 18 Conditions for the surface grinding of cast iron—horizontal spindle, reciprocating table

| Material | Hardness, HRC | Condition(a) | Wheel speed: m/s | Wheel speed: sfm | Table speed: m/min | Table speed: sfm | Down feed(b): mm/pass | Down feed(b): in./pass | Cross feed: mm/pass | Cross feed: in./pass | Wheel identification(c), ANSI |
|---|---|---|---|---|---|---|---|---|---|---|---|
| Gray, ductile, and malleable iron | 45 max | A-C, A, Q&T, N&T, M | 28–33 | 5500–6500 | 15–30 | 50–100 | R: 0.075<br>F: 0.025 max | R: 0.003<br>F: 0.001 max | 1.25–12.5 | 0.050–0.500 | C36JV or A46JV |
| | | | | | | | | | (Max: 1/3 of wheel width) | | |
| | 45–52 | A-C, A, Q&T, N&T, M | 28–33 | 5500–6500 | 15–30 | 50–100 | R: 0.050<br>F: 0.013 max | R: 0.002<br>F: 0.0005 max | 1.25–12.5 | 0.050–0.500 | C36IV or A46IV |
| | | | | | | | | | (Max 1/5 of wheel width) | | |
| | 48–60 | F or IH | 28–33 | 5500–6500 | 15–30 | 50–100 | R: 0.050<br>F: 0.013 max | R: 0.002<br>F: 0.0005 max | 0.65–6.5 | 0.025–0.250 | C46HV or A46HV |
| | | | | | | | | | (Max: 1/10 of wheel width) | | |

(a) A, annealed; A-C, as-cast; Q&T, quenched and tempered; N&T: normalized and tempered; M, malleablized; F, flame hardened; IH, induction hardened. (b) R, rough; F, finish. (c) Wheel recommendations are for wet grinding. For dry grinding, use a softer-grade wheel. See the article "Grinding Equipment and Processes" in this Volume for wheel marking designations. Source: Metcut Research Associates Inc.

### Table 19 Conditions for the cylindrical grinding of cast iron

| Material | Hardness, HRC | Condition(a) | Wheel speed: m/s | Wheel speed: sfm | Work speed: m/min | Work speed: sfm | Infeed on diameter(b): mm/pass | Infeed on diameter(b): in./pass | Traverse, wheel width per revolution of work | Wheel identification(c), ANSI |
|---|---|---|---|---|---|---|---|---|---|---|
| Gray, ductile, and malleable iron | 45 max | A-C, A, Q&T, N&T, M | 28–33 | 5500–6500 | 21–30 | 70–100 | R: 0.050<br>F: 0.025 max | R: 0.002<br>F: 0.001 max | 1/2<br>1/6 | C46KV |
| | 45–52 | A-C, A, Q&T, N&T, M | 28–33 | 5500–6500 | 21–30 | 70–100 | R: 0.050<br>F: 0.013 max | R: 0.002<br>F: 0.0005 max | 1/3<br>1/6 | C54JV |
| | 48–60 | F or IH | 28–33 | 5500–6500 | 21–30 | 70–100 | R: 0.050<br>F: 0.013 max | R: 0.002<br>F: 0.0005 max | 1/4<br>1/8 | C60IV |

(a) A, annealed; A-C, as-cast; Q&T, quenched and tempered; N&T: normalized and tempered; M, malleablized; F, flame hardened; IH, induction hardened. (b) R, rough; F, finish. (c) Wheel recommendations are for the wet grinding of 50–100 mm (2–4 in.) diam work. For dry grinding, use a softer-grade wheel. For larger-diameter work, use a softer and/or coarser-grit wheel. For smaller-diameter work, use a harder-grade wheel. Wheel recommendations also apply to plunge grinding applications. See the article "Grinding Equipment and Processes" in this Volume for wheel marking designations. Source: Metcut Research Associates Inc.

The endless belts are run between a drive wheel and a contact wheel. The belt is held against the work by the contact wheel, and its hardness is selected to provide the desired cutting action. With a very rigid machine and adequate power, metal removal rates can be as high as 13 $cm^3$ per minute per millimeter of belt width (20 $in.^3$ per minute per inch of belt width). Surface finishes under 1 μm (40 μin.) and tolerances of 0.05 mm (0.002 in.) in flatness are not difficult to produce.

## Honing and Lapping

**Honing** is the usual method of finishing the cylinder bores of automotive engines. The process can be applied to the finishing of parts made from cast iron in any hardness range, although it is most widely used for the various grades of gray iron. The outside diameters of cast iron parts can also be finished by honing, but grinding is usually satisfactory and is less expensive than honing. The machines, tools, and techniques applicable to the honing of cast iron are described in the article "Honing" in this Volume.

Although there are exceptions, silicon carbide is almost always used as the abrasive in honing stones, and vitrified bond is the most common. The abrasive in stones for the honing of cast iron may vary from 150 to 500 grit size, depending on finish requirements. A finish of 0.075 to 0.125 μm (3 to 5 μin.) can be produced on as-cast and hardened cast irons by stones having a grit

**Table 20 Conditions for the centerless grinding of cast iron**

| Material | Hardness, HRC | Condition(a) | Wheel speed | | Throughfeed of work(b) | | In feed on diameter(c) | | Wheel identification(d), ANSI |
|---|---|---|---|---|---|---|---|---|---|
| | | | m/s | sfm | m/min | in./min | mm/pass | in./pass | |
| Gray, ductile, and malleable iron | 45 max | A-C, A, Q&T, N&T, M | 28–33 | 5500–6500 | 1.3–3.7 | 50–100 | R: 0.13<br>F: 0.050 max | R: 0.005<br>F: 0.002 max | C46LV |
| | 45–52 | A-C, A, Q&T, N&T, M | 28–33 | 5500–6500 | 1.3–3.7 | 50–100 | R: 0.13<br>F: 0.038 max | R: 0.005<br>F: 0.0015 max | C54KV |
| | 48–60 | F or IH | 28–33 | 5500–6500 | 1.3–3.7 | 50–100 | R: 0.10<br>F: 0.025 max | R: 0.004<br>F: 0.001 max | C60JV |

(a) A, annealed; A-C, as-cast; Q&T, quenched and tempered; N&T: normalized and tempered; M, malleablized; F, flame hardened; IH, induction hardened. (b) As recommended starting conditions for work traverse rates, use a regulating wheel angle with a positive inclination of 3° and a regulating wheel speed of 25 to 40 rev/min. (c) R, rough; F, finish. (d) Wheel recommendations are for wet grinding 20–50 mm (0.8–2.0 in.) diam work. For larger-diameter work, use a softer and/or coarser-grit wheel. For smaller-diameter work, use a harder-grade wheel. Source: Metcut Research Associates Inc.

**Table 21 Conditions for the internal grinding of cast iron**

| Material | Hardness, HRC | Condition(a) | Wheel speed | | Work speed | | Infeed on diameter(b) | | Traverse, wheel width per revolution of work | Wheel identification(c), ANSI |
|---|---|---|---|---|---|---|---|---|---|---|
| | | | m/s | sfm | m/min | sfm | mm/pass | in./pass | | |
| Gray, ductile, and malleable iron | 45 | A-C, A, Q&T, N&T, M | 25–33 | 5000–6500 | 23–60 | 75–200 | R: 0.050<br>F: 0.005 max | R: 0.002<br>F: 0.0002 max | ⅓<br>⅙ | C46JV or A46KV |
| | 45–52 | A-C, A, Q&T, N&T, M | 25–33 | 5000–6500 | 23–60 | 75–200 | R: 0.050<br>F: 0.005 max | R: 0.002<br>F: 0.0002 max | ⅓<br>⅙ | C54JV or A60JV |
| | 48–60 | F or IH | 25–33 | 5000–6500 | 23–60 | 75–200 | R: 0.025<br>F: 0.005 max | R: 0.001<br>F: 0.0002 max | ⅓<br>⅙ | C60IV or A60IV |

(a) A, annealed; A-C, as-cast; Q&T, quenched and tempered; N&T: normalized and tempered; M, malleablized; F, flame hardened; IH, induction hardened. (b) R, rough; F, finish. (c) Wheel recommendations are for wet grinding 20–50 mm (0.8–2.0 in.) diam holes. For larger holes, use the same or softer-grade wheel. For smaller holes, use a harder-grade wheel. Source: Metcut Research Associates Inc.

size of 400 to 500. For a finish of 0.15 to 0.225 μm (6 to 9 μin.), stones having a grit size of 280 to 400 are used. A grit size of 280 will usually produce a surface finish of 0.375 to 0.63 μm (15 to 25 μin.).

In most applications involving the honing of cast iron, a range of surface finish is specified (as is usually true for automobile engine bores) rather than a maximum value. The most common ranges of specified finish can usually be accommodated by abrasive grit sizes of 150 to 300. Detailed information on the surface finish of cast irons due to various machining processes, including honing, can be found in the article "Surface Finish and Surface Integrity" in this Volume.

Because most applications for honing cast iron require a specific lay pattern, honing speed is critical. The speed of rotation for the honing of cast irons with a hardness of 160 to 210 HB is usually 50 to 65 m/min (160 to 210 sfm), which is decreased to 25 to 40 m/min (80 to 130 sfm) for work metal hardnesses of 50 to 65 HRC. A speed of reciprocation of 25 m/min (80 sfm) is common, although this may vary considerably, depending on the specified angle of crosshatch. The article "Honing" in this Volume provides a formula for determining the crosshatch angle that will result from different combinations of reciprocation and rotation speeds. However, in practice, the desired angle is more often obtained by trial and error. The most common procedure is to establish a rotation speed and then to vary the speed of reciprocation until the desired angle of crosshatch is attained.

Acceptable results in honing cast iron depend greatly on a copious supply of lubricant at the areas being honed. Mineral seal oil is widely used, although higher-viscosity mineral oils diluted with kerosene and other lubricants are also used.

**Lapping** with loose abrasive is not extensively used for finishing cast iron parts, mainly because the workpieces become charged with abrasive, which is difficult to remove. Matched-piece lapping is sometimes used for the mating of cast iron parts such as valve components.

# Machining of Carbon and Alloy Steels

M.E. Finn, Steltech, a subsidiary of Stelco, Inc.

THE COMPETITIVE COST PERFORMANCE of machining steel is dependent on many factors, such as tool material, tool geometry, cutting velocity, cutting fluid, and tool/work support systems, as well as the properties of the steel workpiece. The workpiece mechanical properties influencing machinability are hardness, yield strength, and ductility. These properties are controlled by the chemical composition and microstructure. Recent developments in primary and secondary steelmaking processes, such as powder injection, wire feeding, and argon stirring technologies, coupled with uniform temperature control using ladle reheat furnaces, permit the consistent production of standard and engineered steel grades. Modern ingot bottom pouring and continuous-strand casting methods also ensure consistent as-cast microstructures with minimal segregation of the chemical elements. Primary rolling mills and secondary cold finishing and/or heat treatment provide the final control of the microstructure and mechanical properties of the steel workpiece.

The machining performance or machinability of carbon and alloy steel is a systems property. Nevertheless, an attempt is made in this article to isolate the influence of steel chemical composition and microstructure on the machining processes. The first section of this article describes various microstructural phases of standard carbon and alloy steels which influence machinability. The subsequent sections describe the expected response of several traditional machining operations, such as turning, drilling, milling, shaping, thread cutting, and grinding, to the microstructure of standard steel grades. The final section discusses emerging technologies in nontraditional machining processes, such as abrasive waterjet cutting, electrical chemical grinding, and laser drilling.

## Compositions and Microstructures of Steels

Plain carbon and alloy steels are classified by their chemical composition. Each classification requires unique machining conditions recommended in the *Machining Data Handbook* (Ref 1). Matrix and second-phase variations of the microstructure will often require machining parameters different from those recommended.

Plain carbon steels can be divided into three groups, depending on their carbon content:

- Low-carbon steel with less than 0.3% C (UNS grades G10050 to G10290) (Ref 2)
- Medium-carbon steel with 0.3 to 0.6% C (UNS grades G10300 to G10600)
- High-carbon steel with more than 0.6% C (UNS grades G10640 to G10950)

The low-carbon steels consist principally of ferrite, a solid-solution phase of carbon dissolved in α-iron, a body-centered cubic crystal. Ferrite (Fig. 1) is the softest phase of steel, ranging in hardness from 150 to 200 HK, or 136 to 185 HV (Ref 3). As the carbon content increases, there is an increasing amount of pearlite in the microstructure of slowly cooled steel. Pearlite (Fig. 2) is a microconstituent consisting of alternate layers of ferrite and iron carbide (cementite). The hardness of pearlite is dependent on the lamellar spacing; coarse pearlite has a hardness of approximately 250 HK (237 HV), while fine pearlite has a hardness of 350 HK (345 HV) (Ref 3). As-rolled medium-carbon steels consist of ferrite and pearlite (Fig. 3), with pearlite dominating in steels containing more than 0.4% C (UNS grades G10400 to G10590).

As-rolled high-carbon steels with more than 0.8% C have a pearlitic matrix with a cementite network (Fig. 4). Cementite is the

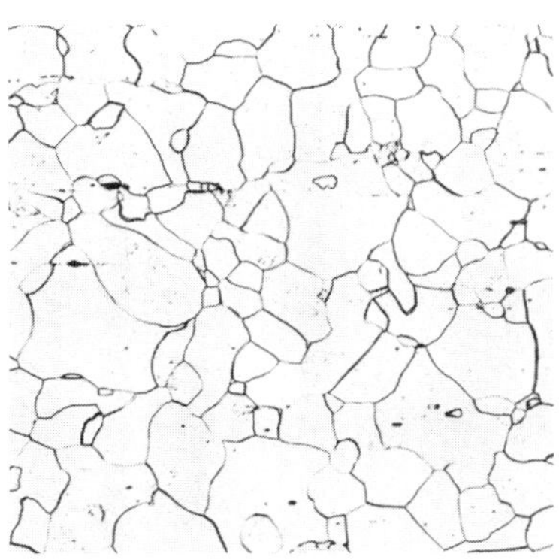

**Fig. 1** Ferrite microstructure of low-carbon (0.012% C) steel. Ferrite (white) grains are surrounded by grain boundaries (black). Etched with 2% nital. 100×

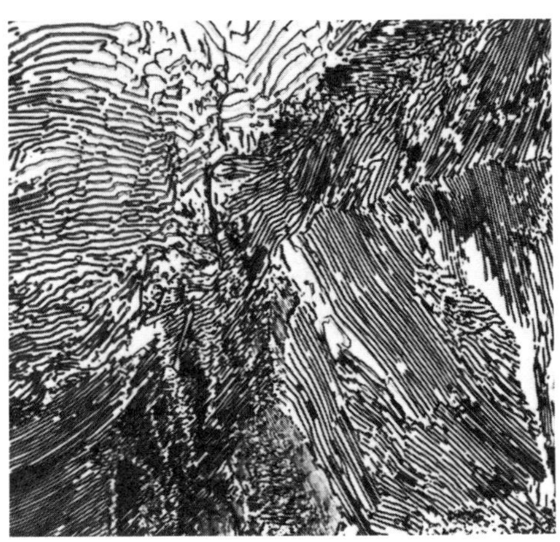

**Fig. 2** Pearlite microstructure of high-carbon (0.95% C) steel with ferrite (white) and cementite (black) platelets. Etched with 4% picral. 500×

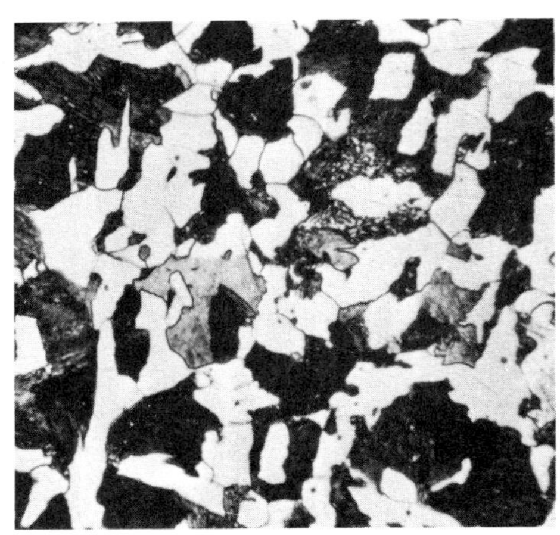

**Fig. 3** Ferrite (white) and pearlite (black) microstructure of medium-carbon (0.45% C) steel. Etched with 2% nital. 500×

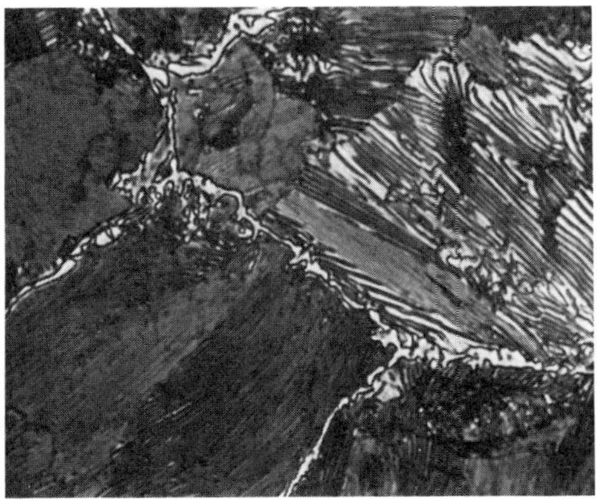

**Fig. 4** Pearlite colonies surrounded by cementite network in high-carbon (1.0% C) steel. Etched with equal parts of 4% picral + 4% nital. 1000×

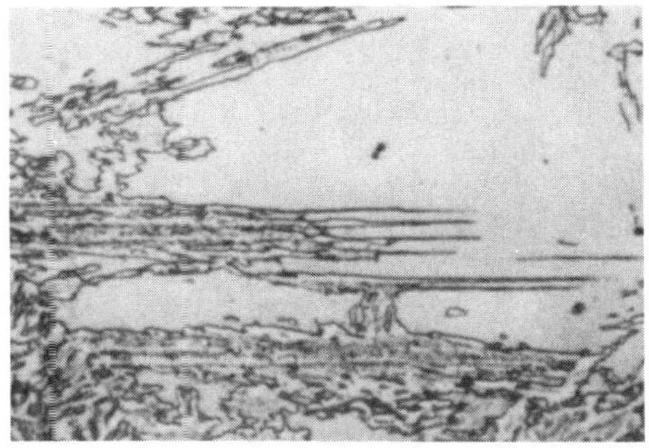

**Fig. 5** Upper bainite in UNS 43600 steel. Etched with picral. 500×

**Fig. 6** Lower bainite in UNS 43600 steel. Matrix is untempered martensite. 500×

hardest phase of steel (1500 HK) (Ref 3). Air-cooled or as-rolled high-carbon steels have limited applications because of the hard and brittle cementite network.

Heat treating can be used to transform steel microstructures into bainite and/or tempered martensite in order to achieve different combinations of mechanical properties, especially in medium- and high-carbon steels. Bainite consists of ferrite and cementite in a feathery or acicular pattern in which the ferrite and cementite cannot be resolved with an optical microscope. The hardness of bainite will range from 45 to 60 HRC (450 to 700 HV) and is dependent on carbon content and its morphology. Upper bainite (Fig. 5) is softer than lower bainite (Fig. 6).

Martensite (Fig. 7) is another phase of steel that forms at lower temperatures by rapid cooling (oil or water quenching). Rapid cooling traps carbon in a body-centered tetragonal structure that is very hard (up to 65 HRC, or 840 HV) and brittle. Tempering generally reduces brittleness at the expense of hardness; the higher the tempering temperature, the higher the toughness and the lower the hardness. Other solid-state heat treatments such as annealing, normalizing, and spheroidizing modify the microstructural phases and the mechanical properties.

Most alloying elements dissolve in ferrite and/or combine with carbon as carbides to increase the hardness and strength of steels. Elements such as silicon, phosphorus, nitrogen, nickel, and copper dissolve in ferrite to harden and strengthen steel. Other elements, such as manganese, chromium, tungsten, molybdenum, vanadium, and titanium, dissolve in ferrite and also combine with carbon to form simple and complex carbides, which increase the hardness and wear resistance of as-rolled and annealed steels. Alloying elements also increase the hardenability (depth of hardness) of quenched-and-tempered steels. In general, at a given carbon content (Table 1), alloy steel has a greater influence on cutting tool wear than its plain carbon counterpart. As alloying content increases, the disparity in machining performance between carbon and alloy steel continues to increase.

Microalloyed (high-strength low-alloy) steel is strengthened by the precipitation of the microalloying element as carbides, nitrides, or carbonitrides, which affect the grain (crystal) size and hardness of the microstructural phases. In general, microalloyed steels require greater cutting forces than plain carbon steel with similar thermal processing. Cost performance improvements, however, can be achieved by substituting a ferrite-pearlite microalloyed steel for a tempered martensite alloy or carbon steel.

Smaller second phases or particles such as oxide and sulfide inclusions (Fig. 8) do not generally influence the type or distribution of the larger microstructural phases; however, under certain conditions, high melting temperature particles will induce crystal nucleation and will affect

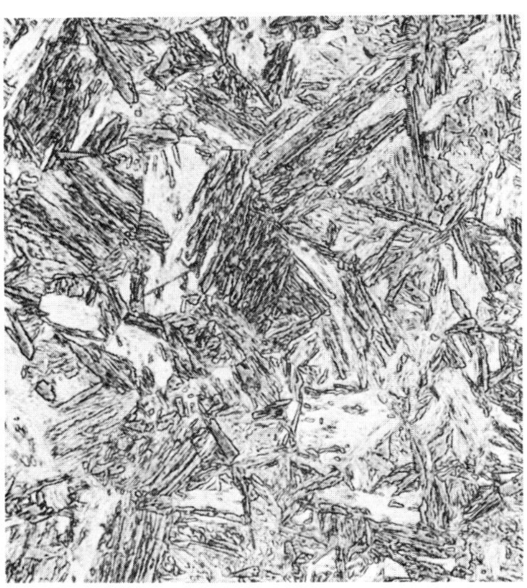

**Fig. 7** Lath martensite in low-carbon (0.2% C) steel. Etched with 8% $Na_2S_2O_5$. 500×

**Table 1 Standard alloy steels**

| UNS grade series | Principal alloying elements with approximate percentages |
|---|---|
| G15*xx*0, G13*xx*0 and H13*xxx* | Mn 1.3 and 1.7 |
| G40*xx*0 and H40*xx*0 | Mo 0.20 and 0.25 |
| G41*xx*0 and H41*xx*0 | Cr 0.5, 0.80, and 0.95, Mo 0.12, 0.20, and 0.30 |
| G43*xx*0 and H43*xx*0 | Cr 0.50 and 0.80, Mo 0.25, and Ni 1.83 |
| G44*xx*0 | Mo 0.53 |
| G46*xx*0 and H46*xx*0 | Mo 0.25 and 0.25 Ni 0.85 and 1.83 |
| G48*xx*0 and H48*xx*0 | Mo 0.25 and Ni 3.50 |
| G51*xx*0 and H51*xx*0 | Cr 0.80, 0.88, and 0.93 0.95 and 1.00 |
| G61*xx*0 and H61*xx*0 | Cr 0.60 and 0.95 V 0.13 and 0.15 |
| G87*xx*0 and H87*xx*0 | Cr 0.50, Mo 0.25, and Ni 0.55 |
| G88*xx*0 and H88*xx*0 | Cr 0.50, Mo 0.35, and Ni 0.55 |
| G92*xx*0 and H92*xx*0 | Si 2.00 |
| GF93100 | Cr 1.20, Mo 0.12, and Ni 3.25 |
| G94*xx*0 | Cr 0.40, Mo 0.12, and Ni 0.45 |

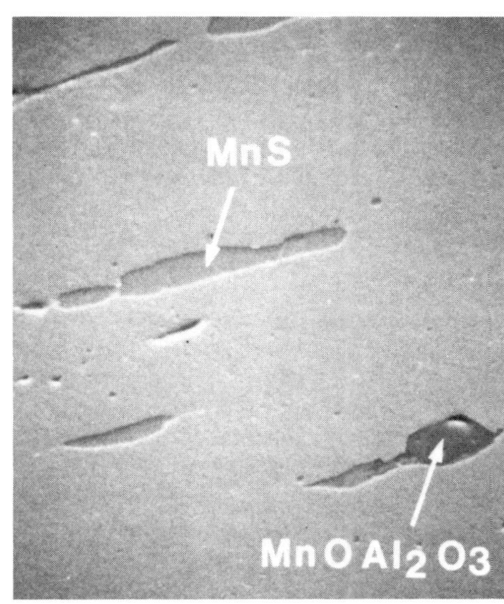


**Fig. 8** Deformable manganese sulfide and nondeformable oxide ($MnOAl_2O_3$) inclusions in as-rolled and cold-drawn bars of UNS 12L140 steel. Unetched. 900×

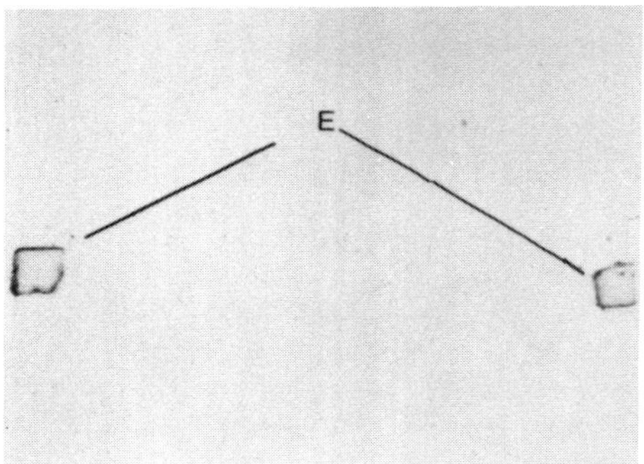


**Fig. 9** Titanium nitrides (E) in UNS G10600 steel microalloyed with titanium and vanadium. Unetched. 270×

grain size. Hard particles, such as alumina, silica, and titanium nitride (Fig. 9), are particularly harmful on cutting tool flank wear by abrasion. Soft particles, such as manganese sulfide (MnS) and free-machining additives (for example, lead and bismuth), promote ductile fracture during machining. Under certain conditions, soft particles of lead or MnS particles encapsulated with lead will lubricate the interface between the metal chip and the cutting tool.

## Machinability of Carbon and Alloy Steels

The machining performance, or machinability, of carbon and alloy steel is influenced by its carbon and alloy content, microstructure, amount of soft or free-machining particles, oxide inclusions, hardness, and work-hardening rate. A reasonable index for comparing the machining performance of carbon and alloy steels was published several years ago for turning operations (Tables 2 and 3). The reference steel, UNS G12120, has a 1-h cutting tool life at 50 m/min (165 sfm) and was assigned an index of 100. All other steels had a cutting speed below or above 50 m/min (165 sfm) under the same conditions, including the same high-speed steel cutting tools. It is doubtful that these ratings could be duplicated exactly today because of the recent advances in steelmaking and processing technology. The machinability index does, however, facilitate steel selection by ranking the relative machinability of a steel grade.

The following sections and corresponding references in this article will assist engineers and operators in selecting steel, treatments, and conditions for maximizing the machinability of carbon and alloy steel workpieces. Emphasis is placed on the metallurgical factors (microstructures, alloying elements, and free-machining additives) that influence steel machinability. Information on machining parameters, such as speeds, feeds, and depth of cut, can be found in Ref 1 and in the Sections "Traditional Machining Processes", "Grinding, Honing, and Lapping," and "Nontraditional Machining Processes" in this Volume. In addition, the effects of high cutting speeds (>600 m/min, or 2000 sfm) on the chip formation characteristics of steel are discussed in the article "High-Speed Machining" in this Volume. Data pertaining to the selection and properties of tool materials and cutting fluids can be found in the Sections "Cutting Tool Materials" and "Cutting Fluids," respectively, in this Volume.

## Turning Operations

**Turning of Low-Carbon Steels.** The turning and boring of steel cylindrical surfaces are steady-state processes in which there is continuous engagement of the cutting tool at a constant speed (*V*, Fig. 10) (Ref 4). After engagement, temperatures reach equilibrium, and there is very limited access for cooling or lubrication of the cutting tool/metal chip interface. The depth of cut (*d*, Fig. 10) corresponds to the chip width, and the undeformed chip thickness is slightly less than the linear feed rate (*f*, Fig. 10). Facing, parting-off, and form turning of steel rods and bars are similar to turning and boring except that the surface speed diminishes as the tool moves toward the center of the workpiece. The temperature gradient of the cutting tool and the chip will also diminish slightly as the surface speed decreases. The chip (undeformed) thickness is proportional to the feed rate.

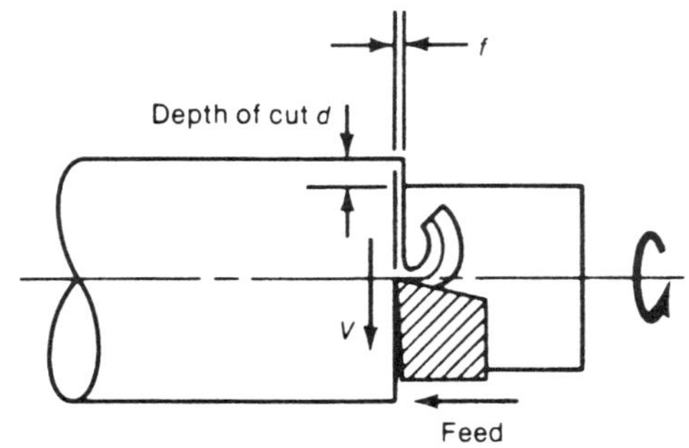


**Fig. 10** Schematic of turning operation. Source: Ref 5

The strain rates in turning low-carbon steels are extremely high, with very complex stresses in the deformation zones (Fig. 11). These rates and frictional forces raise the temperature near the nose of the cutting tool (Fig. 12). The maximum temperature is reached at the rake face away from the nose before the chip lifts off (Fig. 12). Friction conditions and the thermal properties of the steel workpiece and cutting tool affect the partitioning of heat among the workpiece/cutting tool/chip interfaces (Ref 4). A cut-

**Table 2 Machinability rating of carbon steels in turning relative to G12120 at 50 m/min (160 sfm)**

| UNS grade(a) | Hardness, HB | Machinability rating, % | UNS grade(a) | Hardness, HB | Machinability rating, % |
|---|---|---|---|---|---|
| G10050 | ... | 45 | G10490 | 197 | 45 |
| G10060 | 95 | 50 | G10490(c) | 187 | 55 |
| G10080 | 95 | 55 | G10500 | 197 | 55 |
| G10100 | 105 | 55 | G10500(c) | 189 | 55 |
| G10110 | ... | 53 | G10530 | ... | 55 |
| G10120 | 105 | 55 | G10550(c) | 197 | 55 |
| G10130 | ... | 53 | G10590 | ... | 52 |
| G10150 | 111 | 60 | G10600(d) | 183 | 60 |
| G10160 | 121 | 70 | G10640(d) | 183 | 60 |
| G10170 | 116 | 65 | G10650(d) | 187 | 60 |
| G10180 | 126 | 70 | G10660 | ... | 48 |
| G10190 | 131 | 70 | G10690 | ... | 48 |
| G10200 | 121 | 65 | G10700(d) | 192 | 55 |
| G10210 | 131 | 70 | G10740(d) | 192 | 55 |
| G10220 | 137 | 70 | G10750 | 192 | 48 |
| G10230 | 121 | 65 | G10780(d) | 192 | 55 |
| G10250 | 126 | 65 | G10800(d) | 192 | 45 |
| G10260 | 143 | 75 | G10840(d) | 192 | 45 |
| G10290 | ... | 68 | G10850(d) | 192 | 45 |
| G10300 | 149 | 70 | G10860(d) | 192 | 45 |
| G10350 | 163 | 65 | G10900(d) | 197 | 45 |
| G10370 | 167 | 65 | G10950(d) | 197 | 45 |
| G10380 | 163 | 65 | G11170 | 137 | 90 |
| G10390 | 179 | 60 | G11180 | 143 | 85 |
| G10400 | 170 | 60 | G11370 | 197 | 70 |
| G10420 | 179 | 60 | G11400 | 170 | 70 |
| G10420(b) | 179 | 70 | G11410 | 212 | 70 |
| G10430 | 179 | 60 | G11440 | 217 | 80 |
| G10430(b) | 179 | 70 | G11460 | 187 | 75 |
| G10450 | 179 | 55 | G11510 | 207 | 65 |
| G10450(c) | 170 | 65 | G12120 | ... | 100 |
| G10460 | 187 | 55 | G12130 | ... | 136 |
| G10460(c) | 179 | 65 | G12150 | ... | 136 |
| | | | G12L140 | 163 | 160 |

(a) Unless otherwise indicated, all values are for as-rolled cold-drawn steels. (b) Normalized and cold drawn. (c) Annealed and cold drawn. (d) Spheroidized and cold drawn

**Table 3 Machinability rating of alloy steels in turning relative to G12120 at 50 m/min (160 sfm)**

| UNS grade(a) | Hardness, HB | Machinability rating, % | UNS grade | Hardness, HB | Machinability rating, % |
|---|---|---|---|---|---|
| G13300(b) | 179–235 | 55 | G50B600(d) | 170–212 | 55 |
| G13350(b) | 179–235 | 55 | G51150(c) | 163–201 | 65 |
| G13400(b) | 183–241 | 50 | G51200(c) | 163–201 | 70 |
| G13450(c) | 183–241 | 45 | G51300(b) | 174–212 | 70 |
| G40120(c) | 149–196 | 70 | G51320(b) | 174–212 | 70 |
| G40230(c) | 156–207 | 70 | G51350(b) | 179–217 | 70 |
| G40240(c) | 156–207 | 75 | G51400(b) | 179–217 | 65 |
| G40270(c) | 167–212 | 70 | G51450(b) | 179–229 | 65 |
| G40280(b) | 167–212 | 75 | G51470(b) | 179–229 | 65 |
| G40320(b) | 174–217 | 70 | G51500(b) | 183–235 | 60 |
| G40370(b) | 174–217 | 70 | G51550(b) | 183–235 | 55 |
| G40420(b) | 179–229 | 65 | G51600(d) | 179–217 | 55 |
| G40470(b) | 179–229 | 65 | G51B600(d) | 179–217 | 55 |
| G41180(c) | 170–207 | 60 | G501000(d) | 183–241 | 40 |
| G41300(b) | 187–229 | 70 | G511000(d) | 183–241 | 40 |
| G41350(b) | 187–229 | 70 | G521000(d) | 183–241 | 40 |
| G41370(b) | 187–229 | 70 | G61180(c) | 179–217 | 60 |
| G41400(b) | 187–229 | 65 | G61500(d) | 183–241 | 55 |
| G41L40(b) | . . . | 85 | G81150(c) | 163–202 | 65 |
| G41420(b) | 187–229 | 65 | G81B450(b) | 179–223 | 65 |
| G41450(b) | 187–229 | 60 | G86150(c) | 179–235 | 70 |
| G41470(b) | 187–235 | 60 | G86170(c) | 179–235 | 70 |
| G41500(b) | 187–241 | 55 | G86200(c) | 179–235 | 65 |
| G41610(d) | 187–241 | 50 | G86L200(f) | . . . | 85 |
| G43200(e) | 187–229 | 60 | G86220(c) | 179–235 | 65 |
| G43400(d) | 187–241 | 50 | G86250(b) | 179–223 | 60 |
| G43400(d) | 187–241 | 50 | G86270(b) | 170–223 | 60 |
| G44190(b) | 170–212 | 65 | G86300(b) | 179–229 | 70 |
| G44220(c) | 170–212 | 65 | G86370(b) | 179–229 | 65 |
| G44270(b) | 170–212 | 65 | G86400(b) | 184–229 | 65 |
| G46150(c) | 174–223 | 65 | G86420(b) | 184–229 | 65 |
| G46170(c) | 174–223 | 65 | G86450(b) | 184–235 | 65 |
| G46200(c) | 183–229 | 65 | G86B450(b) | 184–235 | 65 |
| G46210(c) | 183–229 | 60 | G86500(b) | 187–248 | 60 |
| G46260(c) | 170–212 | 70 | G86550(b) | 187–248 | 55 |
| G47180(c) | 187–229 | 60 | G86600(d) | 179–217 | 55 |
| G47200(c) | 187–229 | 65 | G87200(c) | 179–235 | 65 |
| G48150(e) | 187–229 | 50 | G87400(b) | 184–235 | 65 |
| G48170(e) | 187–229 | 50 | G88220(d) | 179–223 | 55 |
| G48200(e) | 187–229 | 50 | G92540(d) | 187–241 | 45 |
| G50150(c) | 156–196 | 65 | G92550(d) | 179–229 | 40 |
| G50B400(b) | 174–223 | 65 | G92600(d) | 184–235 | 40 |
| G50B440(b) | 174–223 | 65 | G93100(e) | 184–229 | 50 |
| G50460(b) | 174–223 | 60 | G94B150(c) | 163–202 | 70 |
| G50B460(b) | 174–223 | 60 | G94B170(c) | 163–202 | 70 |
| G50B500(b) | 183–235 | 55 | G94B300(b) | 170–223 | 70 |
| G50600(d) | 170–212 | 55 | | | |

(a) All ratings apply to as-rolled and cold-finished bars. (b) Microstructure composed primarily of lamellar pearlite and ferrite. (c) Microstructure composed primarily of blocky pearlite and bainite, as found in hot-rolled steels. (d) Microstructure composed primarily of spheroidized cementite. (e) Microstructure composed primarily of coarse pearlite resulting from subcritical anneal. (f) Microstructure not specified

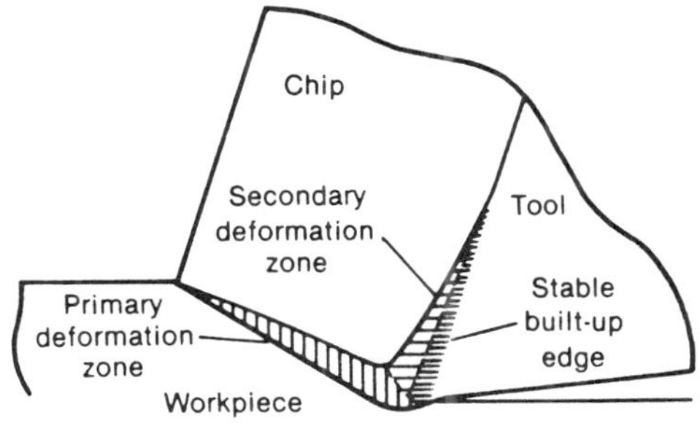


**Fig. 11** Deformation zones and stable built-up edge. Source: Ref 6

ting tool of low heat conductivity will localize heat in the interfaces or secondary shear zone. Generally, the greatest portion of heat is carried away by the chip.

Metal removal during the turning of clean (desulfurized, low oxygen) low-carbon steel is by plastic shear and ductile fracture along a plane inclined to the direction of cutting. As the steel is progressively deformed, microvoids begin to form at the ferrite grain boundaries and at any inclusions present. Microvoid initiation occurs because of the inherent deformation incompatibilities that exist between grain boundaries (also particles, when present) and the ferrite grains. The increasing plastic strain promotes microvoid growth. With continuing deformation, crystal lattice damage accumulates to a critical value at which microstructural instability occurs, resulting in microvoid coalescence, microcrack formation, and finally ductile fracture. The fracture process in chip formation can be defined in terms of damage accumulation to ductile fracture (Ref 7).

The shear, however, does not take place in an infinitely narrow plane (Fig. 13 and 14). At low cutting speeds (lower than those recommended in Table 1.1 in Ref 1), chip formation is discontinuous, and seizure or sticking friction may occur, which causes a dead-metal zone and the nonsymmetrical shape on the cutting tool tip known as built-up edge (Ref 4), as shown in Fig. 11 and 15. The built-up edge acts as an extension of the tool, causing changes in part dimensions and rough-machined surfaces. The built-up edge is seldom stable (it frequently breaks off and re-forms); therefore, tolerance allowances on rough cuts do not overcome undersize parts. At higher-than-recommended cutting speeds, the heat generated is sufficient to localize shearing in the ferrite matrix, and the chip becomes continuous with an increase in cutting force and tool wear.

Cold working increases the hardness of the ferrite and reduces the strain-hardening rate during metal turning. The chip length is shorter and the surface finish generally smoother because of smaller amounts of built-up edge when turning cold-worked or drawn low-carbon steel. Rephosphorized and renitrogenized low-carbon steels have lower strain-hardening rates than standard steels in the as-rolled condition, but in some turning operations, tool wear may be slightly greater because of the strengthened ferrite phase.

Higher sulfur levels (>0.020% S) also improve the machine turning characteristics of low-carbon steel. The sulfur combines with manganese to form deformable inclusions (MnS) that have low strength and low interfacial energy between the sulfide particle and the ferrite matrix. Therefore, lower strain rates are required to initiate microvoids at the MnS particle/matrix interface.

Rapid reheating of low-carbon steel to the austenitizing region with a short soaking time and quenching to a dual-phase structure containing both martensite and/or bainite in the soft ferrite matrix may also improve machinability (Ref 9). The harder second phases and ferrite matrix reduce cutting forces and tool wear (Fig. 16) by initiating microvoids at the interface of the second phase and the ferrite matrix. Increasing plastic strain during machining promotes microvoid growth, coalescence, and cracking in the shear zone.

The overall machinability of as-rolled or annealed low-carbon steel improves with increasing pearlite content and with smaller ferrite and prior-austenite grain size. During turning of the pearlite-ferrite low-carbon steels, microvoids form at the interface of the pearlite and ferrite because of the inherent deformation incompatibilities that exist between the pearlite and the surrounding matrix. Machinability is enhanced by the initiation, growth, and coalescence of mi-

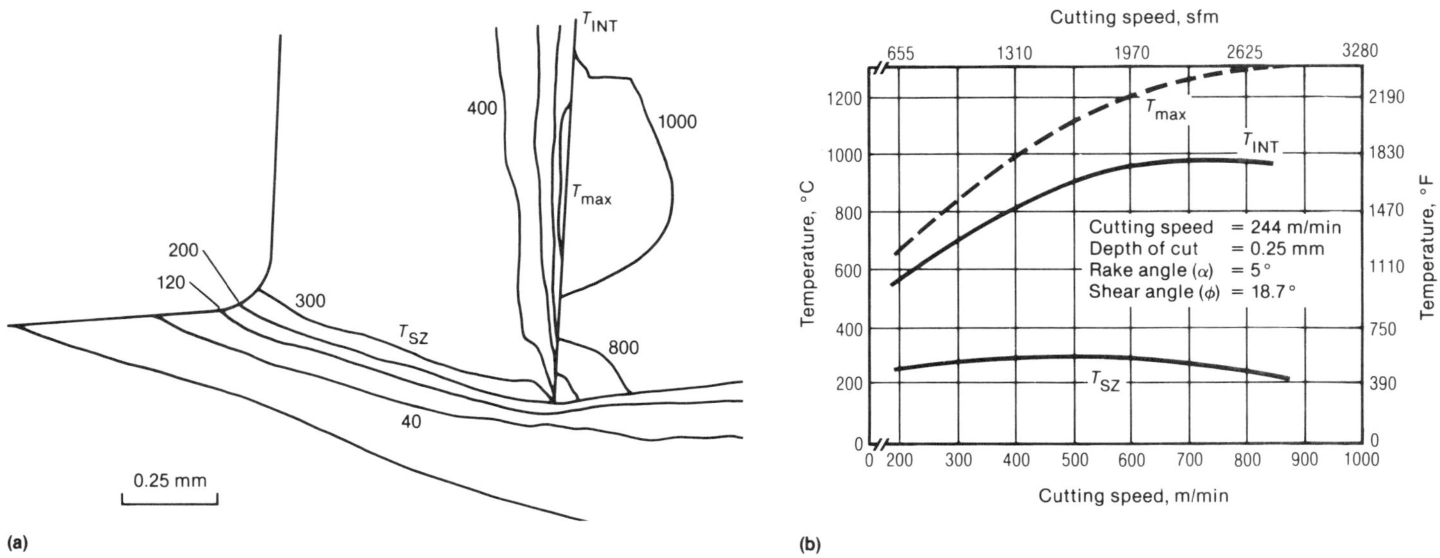


**Fig. 12** Calculated temperature (°C) distribution in chip and cutting tool (a) and variation of temperature with cutting speed (b) using carbide cutting tool on UNS G10160 steel. $T_{max}$: maximum temperature; $T_{INT}$: tool chip interface temperature; $T_{SZ}$: shear zone temperature. Source: Ref 4

crovoids. Maximum machinability of low-carbon steels is achieved at 0.15 to 0.25% C in the as-rolled or annealed condition.

Oxide inclusions that are hard and nondeformable cause excessive tool wear in machine turning. Even deformable sulfides containing oxides are abrasive to the cutting tool. Low-carbon steels decarburized in the steelmaking furnace with longer refining periods are prone to heavier oxide inclusions. Even though oxides are harder than the matrix, microvoid nucleation will begin after the oxide particles crack during the high strain rate deformation of machine turning.

**Turning of Medium-Carbon Steels.** Metal removal during the turning of medium-carbon steel occurs by both plastic shear and microcracking. These plain carbon steels contain 40 to 75% pearlite. The cementite phase, which is hard and nondeformable, causes microvoids in the shear zone that enhance metal removal. Surface finish is smoother on turned medium-carbon than on low-carbon steel parts. Cutting forces and tool wear, however, increase as the carbon content of medium-carbon steels is increased. Cutting speeds should therefore be reduced with increasing carbon (Table 1.1 in Ref 1). The increased amount of pearlite and the cementite platelets are abrasive to the cutting tool. Annealing or normalizing medium-carbon steel will improve machinability by coarsening the pearlite. Coarse-grain structures are also preferred for machining medium-carbon steel.

Higher sulfur levels up to the maximum specified for medium-carbon steels (0.050% S) reduce cutting forces and tool wear. The sulfur combines with manganese to form MnS inclusions, which interrupt matrix plastic flow by creating microvoids and microcracks in the weaker planes containing MnS (Ref 11). These microdefects propagate to enhance chip formation. Some steel producers may charge a premium for higher sulfur contents because yield during processing is reduced due to increased surface defects.

**Turning of High-Carbon Steels.** As-rolled and air-cooled high-carbon steels

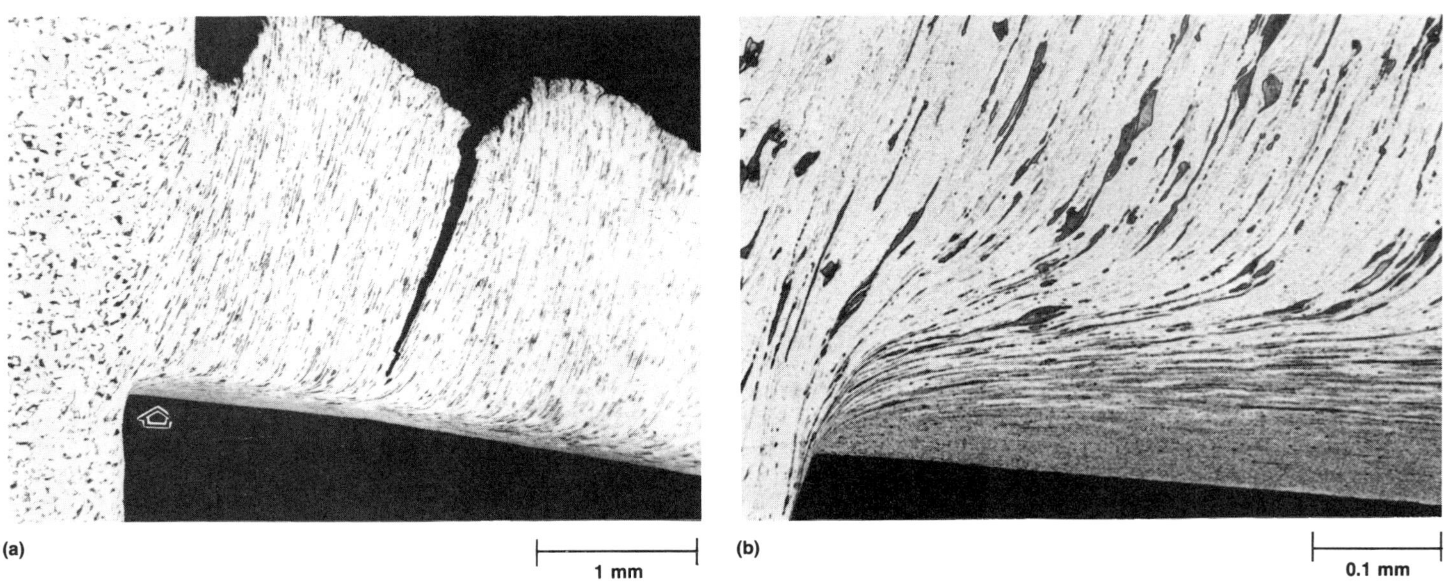


**Fig. 13** Section through quick-stop showing flow zone in 0.1% C carbon steel after cutting at high speed. (a) Low-magnification view. (b) Higher-magnification view of the lower left region of (a) indicated by arrow. Source: Ref 8

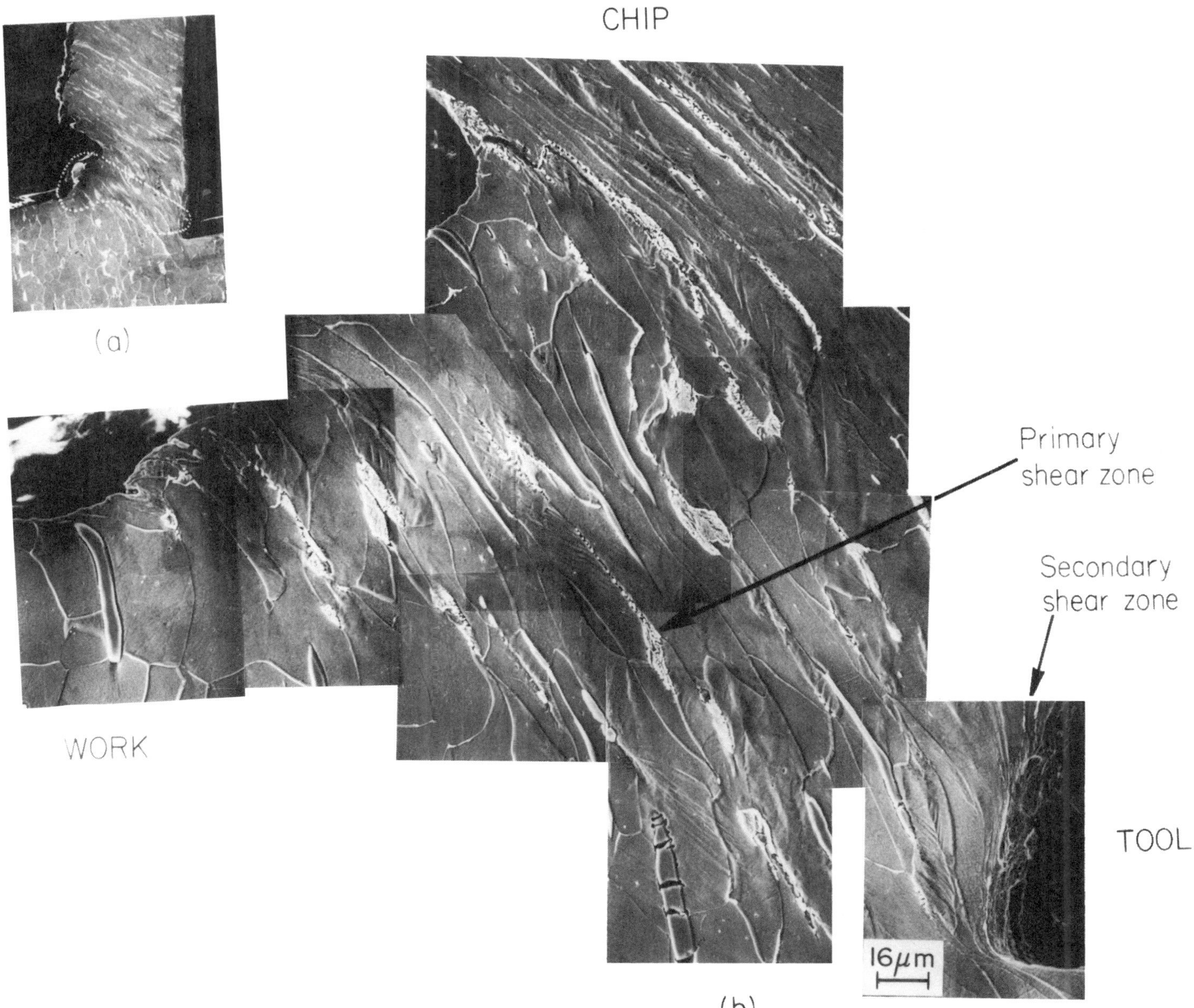


**Fig. 14** Shear formation in steel. (a) Section through low-carbon resulfurized steel chip after quick-stop. (b) Scanning electron microscope montage corresponding to the region marked on the inset in (a) showing primary and secondary shear zones. Source: Ref 7

contain 75 to 100% pearlite, with excess cementite in steels containing more than 0.8% C. Cutting forces and tool wear are higher than those for medium-carbon steels because of the greater amounts of cementite in high-carbon steel. Therefore, lower feeds and speeds (Table 1.1 in Ref 1) are necessary to minimize tool wear. Metal removal occurs mainly by the microvoid-to-microcrack sequence around the hard cementite platelets and grain-boundary network. Machinability can be enhanced by heat treating. Isothermal annealing or spheroidizing will agglomerate the cementite into larger platelets or spheroids, which reduce the number of fragmented, abrasive particles. Annealing isothermally to coarse pearlite (Fig. 17) reduces the amount of fine pearlite compared to full annealing (Fig. 18).

Hardened high-carbon steels contain martensite and/or bainite, which are very hard. Tempering reduces hardness but increases toughness, and larger cutting forces are required to remove the steel during turning. Hardening treatments should be performed after the part is rough machined.

Higher amounts of MnS in high-carbon steel will enhance machinability. However, most high-carbon steel applications require some toughness, which may be adversely influenced by increased sulfur levels. Careful consideration should be given to the component service loading before increasing the sulfur up to the maximum of 0.050% specified for high-carbon steels.

**Turning of Alloy Steels.** In general, the addition of alloying elements to steel tends to raise the as-rolled hardness and shorten turning tool life. Lower feeds and speeds (Table 1.1 in Ref 1) are required to offset these conditions. Alloy elements strengthen and harden the phases of steel by increasing the hardness of ferrite and/or increasing the amount of carbide particles. Alloying elements also change the solid-phase transformation characteristics of the steel. Bainite, for example, may form upon slow cooling with sufficient alloying elements. A bainite matrix is of course harder and tougher to machine turn than ferrite and pearlite mi-

**Fig. 15** Section through quick-stop showing built-up edge after cutting 0.15% C steel at low speed in air. Source: Ref 8

crostructures. The machine turning of alloy steels can be improved by increasing the amount of sulfur (the amount of MnS) and/or by full annealing or spheroidizing.

Lean alloy steels, in the as-rolled or annealed condition, often behave more like their plain carbon counterparts than do the more highly alloyed types (Ref 13). For example, UNS G40230 may have the same effect on tool life as UNS G10220 and G10260, while the more highly alloyed G86200 has a greater effect on tool wear (similar to the effect of G10400 on tool wear). Algorithms for carbon equivalence are very useful for predicting the effect of alloy steel on tool wear when replacing plain carbon steel workpieces (Ref 12).

Highly alloyed steels, like high-carbon steels, should be machined to finished dimensions before hardening to martensite (Fig. 19).

**Turning of Free-Machining Steels.** Free-machining additives such as lead and sulfur (as MnS) form soft particles that are large enough to enhance turning by forming interfaces in the steel matrix. During the high strain of turning, microvoids form at the interfaces and continue to grow and coalesce into microcracks. Cutting forces are drastically reduced (Ref 7). Under certain cutting conditions, the temperature in the secondary deformation/shear zone may be high enough to melt the free-machining additives to form a viscous film between the cutting tool and the steel, thus reducing friction. Therefore, faster

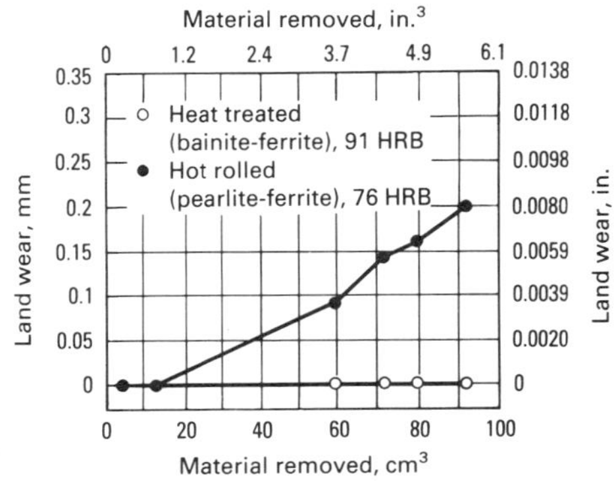


**Fig. 16** High-speed steel wear while turning UNS G10080 steel at 200 m/min (660 sfm). Depth of cut was 1.9 mm (0.74 in.), and feed rate was 0.406 mm/rev (0.0160 in.). Source: Ref 10

feeds and speeds (Table 1.1 in Ref 1) are possible when leaded carbon and alloy steels are used instead of the standard grades. Surface finish is also generally better for leaded free-machining steels than for the standard steels.

Resulfurized steels (UNS grades G11*xx*0), which contain up to 0.35% S, have lower ductility and impact strength than plain carbon steels with the same carbon content (Ref 15). Nevertheless, these steels are used to make such components as automotive transmission gears, furniture wheel assemblies, and high-pressure hydraulic hose fittings.

Rephosphorized low-carbon steels have better machinability than standard steels because the strain-hardening rate of the ferrite phase is reduced by phosphorus additions. Generally, built-up edge is lower with rephosphorized steels, and the finished surface is smoother after turning. Under some machining conditions, however, the higher steel hardness may reduce tool life.

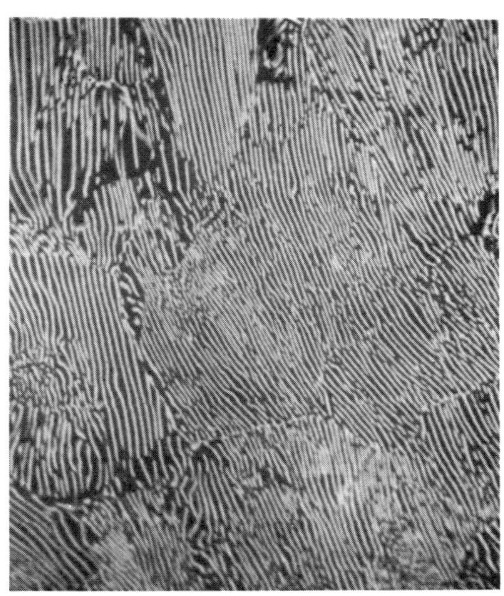

**Fig. 18** Fine pearlite of fully annealed UNS G10700 steel. Etched with nital. 2000×. Source: Ref 12

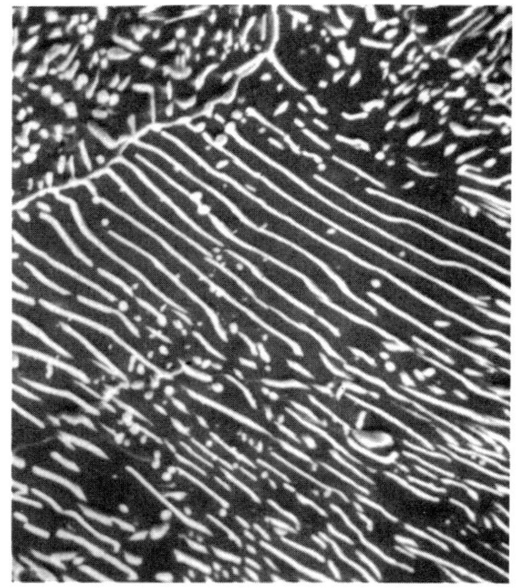

**Fig. 17** Coarse pearlite of isothermally annealed UNS G10700 steel. Etched with nital. 2000×. Source: Ref 12

Leaded free-machining steel (UNS G12L140) combines the effects of three free-machining additives—lead, MnS, and phosphorus—in a low-carbon steel (<0.15% C). Traditionally, ingot-teemed steel with high oxygen content (200 to 300 ppm) promoted globular MnS + Pb inclusions and produced the best machining of UNS G12L140 grade. However, recent advances in continuously bloom cast steel technology have produced steel with equivalent machinability (Ref 15, 16). Recommended

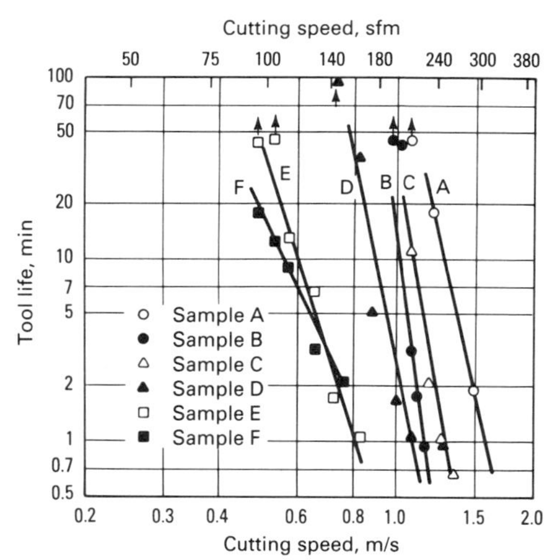


| Sample | Hardness, HB | Heat treatment |
|---|---|---|
| A.............. | 146 | Spheroidized |
| B............. | 153 | Spheroidized |
| C............. | 150 | Annealed |
| D............. | 187 | Normalized and tempered |
| E............. | 238 | Hardened and tempered |
| F............. | 308 | Hardened and tempered |

**Fig. 19** Effect of heat treatment on tool life while turning UNS 41350 steel. Feed: 0.2 mm/rev (0.008 in./rev). Depth of cut: 2.0 mm (0.08 in.). Source: Ref 14

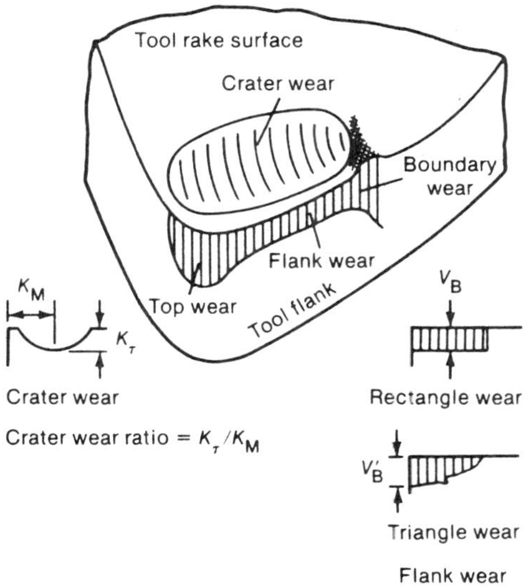


**Fig. 20** Schematic illustrating typical cutting tool wear. Source: Ref 21

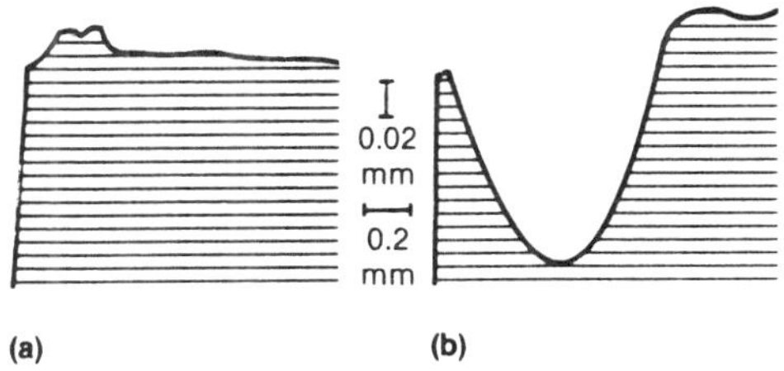


**Fig. 21** Wear of the tool rake surface after cutting different DIN CK-45 (UNS G10450) steels. (a) CaSi-deoxidized steel. Cutting time, 100 min; crater wear ratio, 0. (b) FeSi-deoxidized steel. Cutting time, 20 min; crater wear ratio, 0.28. Source: Ref 21

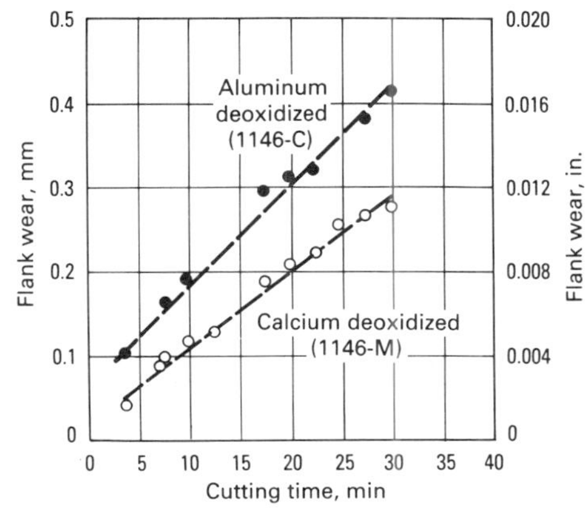


**Fig. 22** Average high-speed steel tool wear while machining UNS G11460 steel bars at 40 m/min (130 sfm). Depth of cut was 2 mm (0.08 in.), feed rate was 0.1 mm/rev (0.004 in./rev), and rake angle (α) was 20°. Source: Ref 22

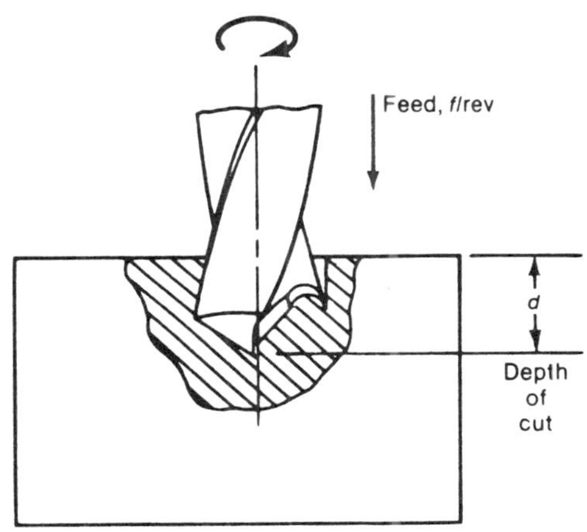


**Fig. 23** Schematic of drilling operation. Source: Ref 5

speeds and feeds (Table 1.1 in Ref 1) should be used for UNS G12L140 grade steel because higher cutting speeds generate too much heat and destroy the lead-base lubricating film. Cold-drawing UNS G12L140 grade will reduce tool life slightly, but will improve the surface finish. Cold-drawn or cold-finished bars are also straighter, more dimensionally accurate, and smoother (for example, scale free) than hot-rolled bars (Ref 15). Smoother machined surfaces are also obtained on renitrogenized UNS G12L140 grade with a slight reduction in tool life.

Bismuth, when added to steel, also forms insoluble particles in the solid state. Bismuth can be used as an additive to UNS G12L140 grade and as a nontoxic substitute for lead in the same grade. According to the Bismuth Institute, bismuth is nontoxic and is used in various salt compounds as a suspension taken orally for gastrointestinal disorders (Ref 17). As an additive, bismuth combines with lead and provides more effective free-machining particles than lead alone. As a substitute, bismuth reacts in the same manner as lead and provides similar performance in turning (Ref 15, 18, 19).

Selenium and tellurium are also effective in enhancing the machine turning of resulfurized steels. They promote the retention of globular manganese sulfides, which are more effective in initiating microcracks in the shear zone than elongated sulfides. Selenium or tellurium can be used to reduce the amount of MnS when higher mechanical properties are required. These additives combine with manganese to form inclusions analogous to MnS that act as free-machining second-phase particles.

As single or combined additives, lead, selenium, and tellurium are effective in reducing cutting forces during the turning of both plain and alloy medium-carbon steels. However, some mechanical property values may be reduced with these additions.

**Turning of Calcium-Treated Steels.** Steels, especially high-carbon and alloy steels, deoxidized with calcium are generally less abrasive to the turning tool than steels deoxidized with aluminum or silicon. During deoxidation of the steel, calcium combines with these hard inclusions to form particles that are softer than alumina and silica. Calcium treatment modifies silicon/manganese deoxized steels to form soft anorthitic inclusions ($CaO \cdot Al_2O_3 \cdot 2SiO_2$), while calcium modifies aluminum-killed steel to form soft gehlenitic ($2CaO \cdot Al_2O_3 \cdot SiO_2$) and duplex calcium aluminate/calcium sulfide inclusions (Ref 20, 21). In the latter case, the abrasive alumina inclusions are modified to calcium aluminates on which soft calcium sulfide can precipitate. The softer calcium-containing particles form interfaces with the matrix that are sites for microvoids during machine turning.

At high cutting speeds (160 to 250 m/min, or 525 to 820 sfm), the calcium-containing inclusions form a film over the carbide cutting tool that reduces crater wear by preventing direct contact of the cutting chips and the machined surface with the rake and flank of the tool (Fig. 20 and 21). Longer tool life is obtained when calcium-treated silicon-killed steel is machined with titanium-containing alloyed carbide tools coated with TiN/TiC.

Calcium is also effective in improving tool life in the turning of resulfurized free-machining steel. The tool life, as measured by flank wear, was longer while turning UNS G11460 grade steel modified with a calcium deoxidation plus niobium grain control practice than the same grade of steel with an aluminum deoxidation practice (Fig. 22).

## Drilling Operations

**Drilling of Low-Carbon Steel.** Drilling (Fig. 23) with a twist or spade drill is a highly complex deformation and shearing process. Conditions along the cutting edges are similar to those in turning except that shear stresses and temperatures reach a maximum at about three-quarters distance to the outside diameter (Ref 4). The center point of the drill penetrates the workpiece by indentations and causes high thrust forces before the cutting edges engage. Friction is high at the point and along the side surfaces of the flute (Fig. 24). The heat produced from this rubbing can be reduced with a recommended coolant. The chip (undeformed) thickness is about one-half the linear feed rate, $f$ (Fig. 23). The undeformed chip width is equivalent to the length of the drill lip, which depends on the point angle as well as the drill size.

When starting a drilled hole in low-carbon steel, a sharp three-cornered punch rather than a prick punch can be used to minimize work hardening of the steel at the mark (see the article "Machining of Stainless Steel" in this Volume). Once the drill is engaged, low-carbon steel under the chisel edge is first extruded by the high-thrust force, then it shears under the action of the negative-rake cutting angle. Built-up edge occurs along the chisel edge and can be reduced by lowering the frictional heat. The extruded steel under the center of the drill is forced out into the outer portions of the chisel edge

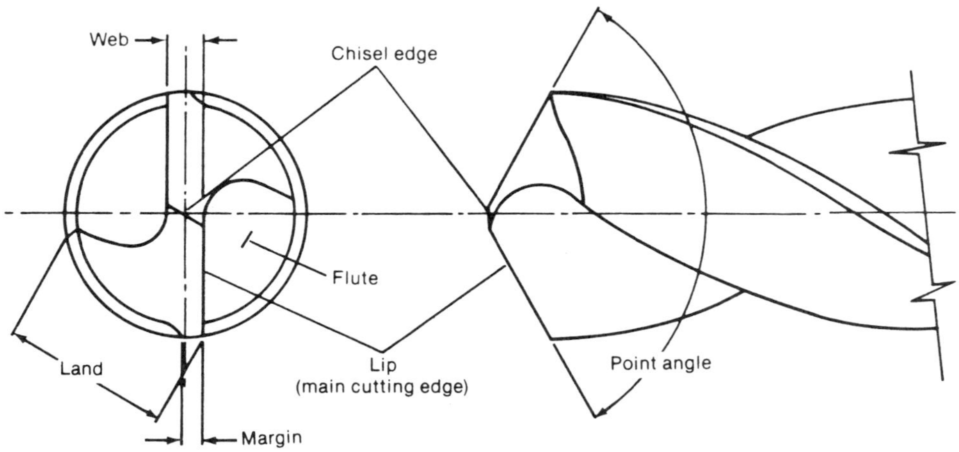


**Fig. 24** Elements of the cutting end of a twist drill. Source: Ref 23

**Table 5 Effect of feed rate on the drillability of a microalloyed and an alloyed steel**

| Grade (UNS) | Feed mm/rev | Feed in./rev | Drill force N | Drill force lbf | Drill torque N · m | Drill torque lbf · in. | Resultant force N | Resultant force lbf |
|---|---|---|---|---|---|---|---|---|
| G10450 fV(a) | 5.161 | 0.2032 | 1425 | 320 | 6.2 | 55 | 1475 | 332 |
| | 8.387 | 0.3302 | 2355 | 529 | 9.0 | 80 | 2225 | 500 |
| G41400(b) | 5.161 | 0.2032 | 1890 | 425 | 7.7 | 68 | 1880 | 423 |
| | 8.387 | 0.3302 | 2705 | 608 | 11.6 | 103 | 2790 | 627 |

(a) Air-cooled. (b) Quenched and tempered. Source: Ref 22

while steel near the outer edge is sheared. The chip from the center is more highly work hardened, and the abrupt change in direction of the cutting edge between the lip and chisel edge may cause separation of the chip. The sliding of the chips up the drill flutes also increases thrust forces. Smooth, polished flutes often help reduce forces by decreasing the resistance to sliding of the chip.

Low-carbon steel, especially when clean, work hardens rapidly. Frequent sharpening of twist drills, constant feed rates (Table 3.1 in Ref 1), and adequate coolants will reduce work hardening. Coolant hole twist drills have been used to reduce frictional heat and built-up edge while drilling low-carbon ASTM A36 plate (UNS K02600 grade) (Ref 24).

An ejector drill can also be used to reduce thrust forces because the chips will slide across the cutting edge (Ref 25). The ejector drill consists of a multiple-insert, self-centering, high-penetration drill made up of several cemented carbide tips.

Cold-worked low-carbon steel, because of its decreased deformation (strain hardening), will form shorter chips in drilling than as-rolled steel. Higher thrust forces, however, will be noticed.

Steels with higher phosphorus and/or nitrogen, which strengthen the ferrite phase, drill more readily because of improved chip segmentation. Steels with larger amounts of MnS have a larger number of soft particles available for microcracking, and this enhances drilling. Abrasive oxide inclusions are detrimental in drilling low-carbon steel.

A general-purpose twist drill, which usually has a point angle of 118°, a helix angle of 30°, a relief angle of 12°, and a chisel edge angle of 135°, may be suitable for the short-hole drilling of clean low-carbon steel (Table 4). However, cold-worked, rephosphorized, and/or renitrogenized low-carbon steel will require a lower chisel angle. Harder medium-carbon, high-carbon, and alloy steels require a higher point angle, a lower chisel edge angle, and less lip relief (Table 4).

Gun drills, which consist of a single cutting tip and coolant holes, are effective in the deep-hole drilling of low-carbon steels. Long, continuous chips are more readily removed. Coolant hole and coolant tube drills are also effective in forcing coolant under pressure to the tool point; this reduces heat and friction. Chips will also slide more easily up a higher helix angle in twist drills, which are better suited to deep-hole drilling.

**Drilling of Medium- and High-Carbon Steels.** Medium- and high-carbon steels require higher thrust forces and lower feeds and speeds (Table 3.1 in Ref 1) than low-carbon steel. Drilling can be improved by heat treating the steel workpiece, that is, annealing, normalizing, and/or spheroidizing (especially high-carbon steel). Heat treating will also reduce centerline segregation or coring of steel rod and bar stock. Increasing the sulfur-to-manganese ratio will also improve drillability. Mechanical properties, however, may be reduced with higher sulfur contents.

**Table 4 Recommended twist drill geometries for steel**

| Steel class | Angle, degrees Point | Chisel edge | Helix | Lip relief |
|---|---|---|---|---|
| Low carbon | 118 | 130 | 30 | 12 |
| Medium carbon | 130 | 120 | 30 | 6 |
| High carbon and alloy | 150 | 120 | 25 | 5 |

Medium- and high-carbon steels require a higher point angle, a lower chisel edge angle, and less lip relief for a stronger cutting edge on twist drills than low-carbon steels (Table 4). A high helix angle will make chips slide more easily at the expense of a weaker cutting edge, so there is a practical upper limit to the size of the helix angle for medium- and high-carbon steels.

**Drilling of Alloy Steels.** Alloy steels, like medium- and high-carbon steels, should be annealed or spheroidized before drilling. The optimum hardness for a spheroidized chromium-alloyed steel (UNS 51200 grade) is 170 to 190 HV (Ref 26). Cold working will reduce thrust force but will increase drill wear (Ref 26) at the recommended speeds and feeds (Table 3.1 in Ref 1). Alloy steels with added sulfur or tellurium require lower thrust force, which reduces drill wear (Ref 26).

For short-hole drilling with twist drills, alloy steels generally require a strong cutting edge as a result of a high point angle, low chisel edge angle, and low lip relief angle (Table 4).

Microalloyed steels that are air cooled from hot-shaping operations (hot forging) drill more readily than alloy steels that are quenched and tempered after hot shaping (Ref 27-29). Cutting forces are generally lower for air-cooled microalloyed steel than for a quenched-and-tempered alloy steel at equivalent hardness (see, for example, the 22 to 23 HRC workpieces in Table 5).

The deep-hole drillability of microalloyed steel is also superior to that of quenched-and-tempered alloy steel, mainly because the built-up edge in air-cooled microalloyed steel breaks off more frequently and lowers thrust force intermittently (Ref 27). Manganese sulfide and lead additions will also improve the drilling of microalloyed steels. Microalloyed steels should contain less than 0.03% Ti to prevent abrasive wear of titanium nitride particles (Ref 27).

A split point and a wide chip disposal channel can be used on twist drills to improve the deep-hole drilling of air-cooled microalloyed steel. Up to three times the drill life may be achieved (Ref 29).

**Drilling of Free-Machining Steels.** Additions of bismuth, lead, and tellurium reduce the drilling forces of low-carbon steel (Ref 30). These additions, in association with larger amounts of MnS, deform plastically during chip formation to form planes

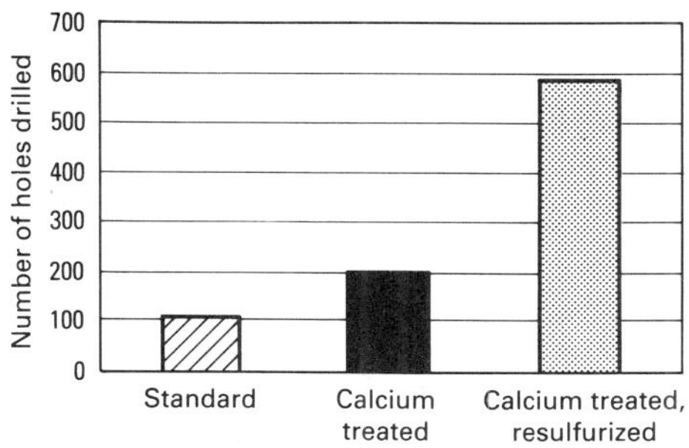


**Drilling conditions**

| | |
|---|---|
| Drill material | M2 high-speed steel |
| Drill diameter, mm (in.) | 5 (0.2) |
| Point angle, degrees | 118 |
| Blind hole depth, mm (in.) | 20 (0.8) |
| Cutting speed, rev/min | 1400 |
| Feed rate, mm/rev (in./rev) | 0.25 (0.01) |
| Coolant | Soluble oil |

**Fig. 25** Machinability benefits of inclusion-modified UNS G41400 steels. Source: Ref 32

of low strength, along which the energy for microvoid and microcrack formation is low. These microdefects quickly propagate in the shear area, thus reducing cutting forces. Tellurium and bismuth appear to be more effective than lead in reducing drilling forces (Ref 30). Cutting speeds can therefore be increased by more than 25% when drilling free-machining low-carbon steel (Table 3.1 in Ref 1). Nitrogen, which is added to improve the surface finish after turning, increases thrust and torque forces in drilling.

Calcium-treated medium-carbon, high-carbon, and alloy steels contain soft inclusions that are less abrasive on the drill and form a protective layer on the cutting edge of the drill, especially at cutting speeds higher than those recommended for nontreated steels (Ref 31). When calcium treatment is combined with resulfurization in alloy steels, drill lives can be increased by up to six times (Ref 32), as shown in Fig. 25.

As mentioned previously, MnS inclusions cause microdefects, which increase the shear angle and chip curl and reduce chip thickness, tool/chip contact length, frictional forces, and cutting temperatures. The manganese sulfides and oxysulfides are also extended into the tool/chip interface as a lubricating film.

## Milling Operations

**Milling of Low-Carbon Steel.** Slab and end milling with a peripheral cut differ in several aspects from turning and drilling. In milling, a formed surface is generated by the action of a rotary cutter, each tooth of the cutter removing a small amount of metal with each revolution of the spindle (Fig. 26). The time of tooth-workpiece engagement can vary, but there will always be a period of rest for each tooth during the cycle. The length of engagement for each tooth is brief. The undeformed chip thickness changes from zero to some maximum value by the chosen value of feed per tooth (Ref 4). In climb (or down) milling (Fig. 27), the chip thickness is greatest at engagement and tapers out to nothing. In conventional (or up) milling (Fig. 27), the cut begins at zero chip thickness.

The life of a milling cutter compares favorably with that of a single-point tool when machining low-carbon steel at equal or recommended cutting speeds (Section 2 of Ref 1). Entry and exit shocks, possible vibration, and variation in chip thickness during the cut are factors that tend to reduce the life of the mill to less than that of a turning tool (Ref 23). On the other hand, the point of engagement keeps moving around the circumference of the milling tool; this allows it to cool and enables more efficient use of coolants to prolong tool life.

Built-up edge on the cutting teeth is less of a problem in milling low-carbon steels than in turning and drilling operations.

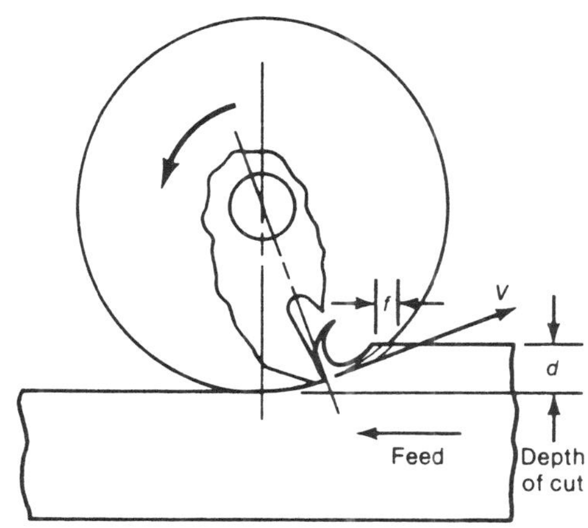


**Fig. 26** Schematic of milling operation. Source: Ref 5

Built-up edge is influenced by friction and temperature, which are intermittent in milling. The tooth engagement cycle also breaks off small areas of built-up edge and prevents accumulation. The shearing action of the chip from low-carbon steel in milling is similar to turning. Low-carbon steel work hardens during milling; work hardening can be reduced by cold working the steel prior to milling. Low-carbon steel with added nitrogen or phosphorus to strengthen the ferrite will have improved chip formation but reduced tool life due to reduced work hardening. Steels with a higher sulfur-to-manganese ratio yield longer cutter life in milling. The improved tool life may, however, occur at the expense of lower mechanical properties. Higher sulfur levels will also allow the use of higher speeds and feeds (Section 2 of Ref 1). Oxide inclusions are abrasive and are detrimental to tool life.

**Milling of Medium- and High-Carbon Steels.** The annealing of medium- and high-carbon steels improves the tool life of milling centers. For medium-carbon steels, a coarse pearlitic microstructure is generally best for combining good tool life and surface finish. High-carbon grades (such as

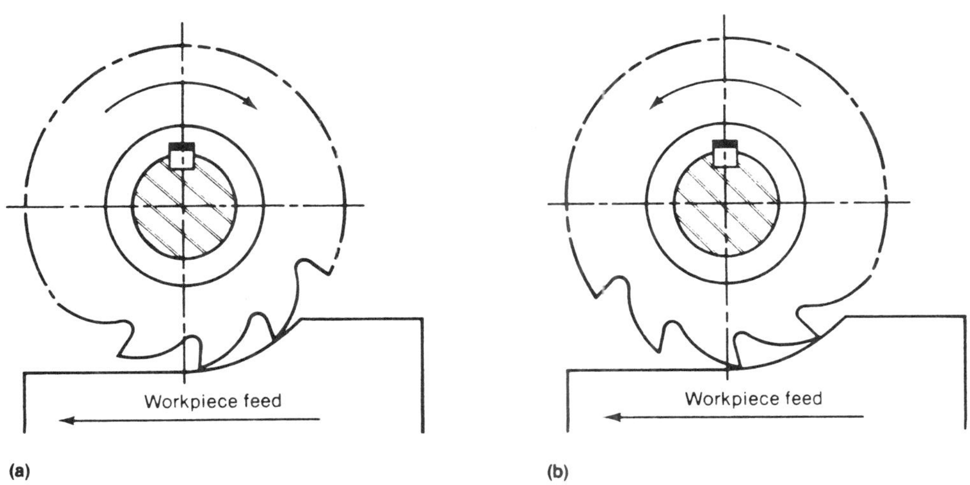


**Fig. 27** Schematics of climb (down) milling (a) and conventional (up) milling (b). Source: Ref 23

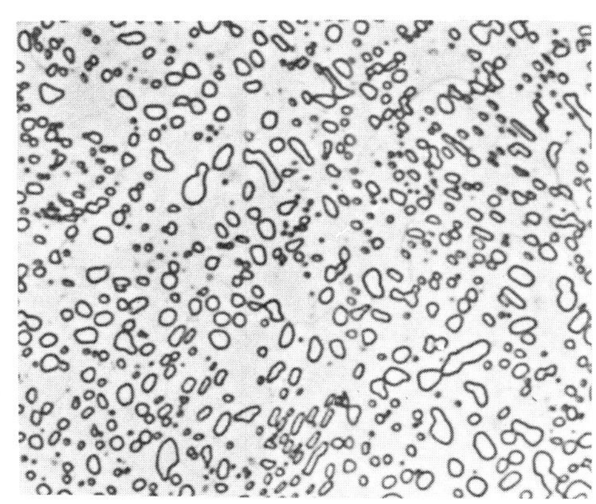

**Fig. 28** Dispersion of spheroidal carbides in a matrix of ferrite (UNS G521000 steel) obtained by spheroidizing. Etched with 4% picral + 0.05% HCl. 1000×

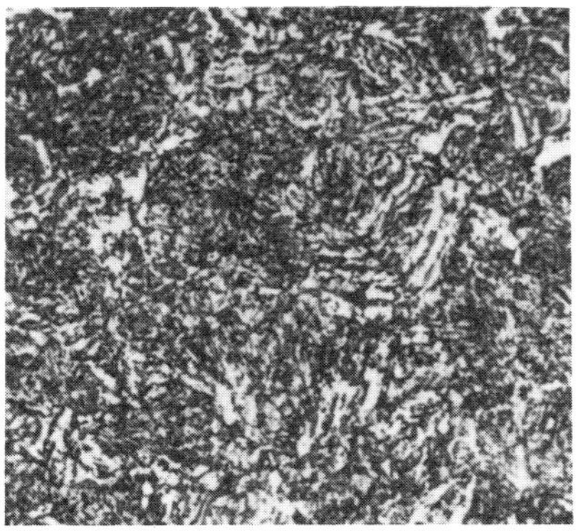

**Fig. 29** Tempered martensite microstructure (UNS G41400 steel) obtained by quenching and tempering. Etched with 2% nital. 750×

UNS 52100) should be spheroidized (Fig. 28) by annealing for optimum milling cutter performance (Ref 25). Hardening by quenching and tempering (Fig. 29) should be avoided before milling because the speed must be reduced by at least 75% (Section 2 of Ref 1) to avoid rapid wear or tooth breakage. Medium- and high-carbon steels will mill more readily with maximum amounts of sulfur. As mentioned previously in this article, MnS inclusions deform plastically, forming weak shear planes and interfaces with the matrix, which form microdefects that initiate crack propagation in the shear zone; oxide inclusions do not deform and are abrasive to the milling cutter.

**Milling of Alloy Steels.** The annealing or spheroidizing of alloy steels will improve the performance of milling. The higher hardness of alloy steels requires lower speed and feed rates than carbon steel (Section 2 of Ref 1). Hardening heat treatments should be performed after milling. Alloy steels with sulfur levels approaching the specified maximum will mill more readily, again with a slightly detrimental effect on mechanical properties, particularly ductility and toughness.

At a hardness of about 250 HB (263 HV), air-cooled, vanadium-microalloyed medium-carbon steel mills more readily than its equivalent quenched-and-tempered steel (UNS grade G10450) at the same hardness. At a higher hardness of 300 HB (318 HV), both steels have similar millability (Ref 29).

**Milling of Free-Machining Steels.** Low-carbon free-machining steels can be milled at speeds up to 53% faster than plain carbon steels (Section 2 of Ref 1). As explained previously in this article, sulfide and/or lead, bismuth, and tellurium in free-machining steels deform plastically with weak planes in the shear zone under the milling tooth. The weak planes and interfaces with the matrix form microdefects, which propagate to enhance metal chip parting. Rephosphorized and renitrogenized low-carbon free-machining steels have a smoother finish after milling, although possibly at the expense of a slight decrease in milling cutter life compared to lower-phosphorus and lower-nitrogen free-machining steels.

## Shaping and Thread Cutting Operations

**Shaping of Carbon and Alloy Steels.** Shaping, with a linearly moving tool (Fig. 30), and planing, with a linearly moving workpiece, are similar to the chip-parting mechanics described in the section "Turning Operations" in this article, except that tool engagement is typically interrupted more frequently. The engagement travel may be short enough, especially in shaping, to prevent the development of maximum temperatures. Better coolant or lubricant access to the metal/tool interface and the tool disengagement also maintain lower shear zone temperatures than those found in turning. If a built-up edge develops in the shaping or planing of plain and free-machining low-carbon steels, the built-up edge may break off between engagement cycles (Ref 4). Shaping and planing speeds and feeds (Table 5.1 in Ref 1) are influenced by the same steel characteristics as those described in the section "Turning Operations" in this article.

**Thread Cutting of Carbon and Alloy Steels.** External and internal thread cutting operations are similar to form turning except that the cutting edges during the tapping of an internal thread are simultaneously engaged and the undeformed chip thickness is influenced by tap design (Ref 4). Threading parameters (Table 1.5 in Ref 1) are influenced by the steel characteristics described in the section "Turning Operations" in this article.

Thread rolling of an external surface and cold form tapping shape threads by plastic indentation. Low- and medium-carbon steels that have been hardened in the ferrite phase by cold working or solid-solution strengthening may require higher driving forces, but will work harden less in the rolled or form-tapped thread root thus reducing concentrated stresses in service. Dimensional tolerances may also be easier to maintain.

## Grinding Operations

The grinding process involves the removal of excess stock from the steel workpiece through the cutting action of abrasive, hard, angular particles (grains or grits). Aluminum oxide and silicon carbide are the most widely used abrasives for the wheel grinding of steel. The basic differences relative to other cutting operations such as turning or milling are that individual grits are more or less randomly oriented to the cutting direction and that the undeformed chip thickness is generally small and short with a variable mean chip cross-sectional dimension along the contact path.

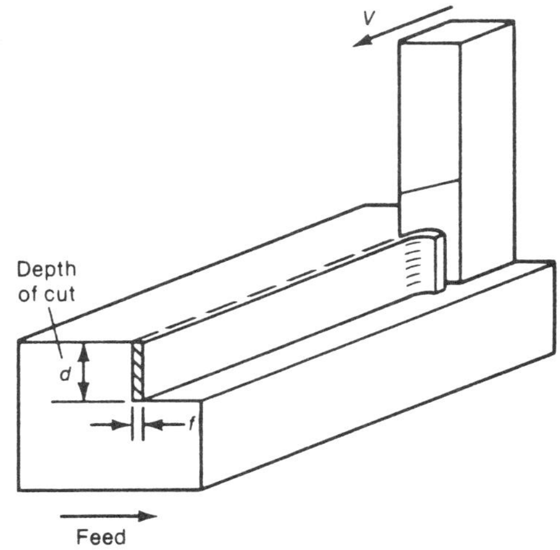


**Fig. 30** Schematic of shaping operation. Source: Ref 5

Grains or grits, which are manufactured to be irregular in shape, present sharp cutting edges that remove steel by plastic shear. The undeformed chip thickness is variable and small. Worn grits will fracture or break away from the bond to expose new cutting edges. Rake angles, therefore, range from positive to generally negative. The particles can displace or plow the steel workpiece sideways. Deformation may become purely elastic with high friction and heat without metal removal. Plowing of the steel surface creates a frontal bulge, also known as prow, ahead of the grit. Rake angles can be highly negative, and steel flow may be directed into both the chip and the workpiece. Work (energy) is lost in the thick deformation zone, penetrating up to ten times the depth of indentation. Work is also expended to overcome friction that may become a substantial part of the total work. The clearance angle can be zero or even negative, which increases friction and heat (Ref 4).

**Grinding of Low-Carbon Steels.** Clean low-carbon steels should be ground with slow wheel speeds, infeed, and workpiece feed to minimize prow (Section 8 of Ref 1). The formation of prow changes the effective rake angle and leads to a poorer workpiece finish. Grinding fluids will reduce prow by cooling and lubricating the contact surface, flushing away the prow debris, and reducing wheel clogging. Cold-worked low-carbon steel will grind with less prow. Rephosphorized and renitrogenized low-carbon steels will also grind with a better finish. Silicon carbide grits may be more effective in grinding low-carbon steel than wheels with aluminum oxide grits (Ref 23).

**Grinding of Medium- and High-Carbon Steels.** Medium- and high-carbon steels can be ground at higher speeds (Section 8 of Ref 1). For as-rolled or annealed steels, grind-

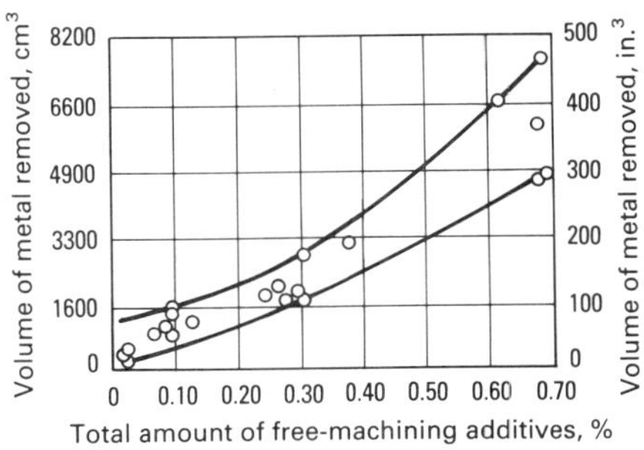


**Fig. 31** Effect of the total amount of free-machining additives present in the steel on the volume of metal removed in a grindability test. Source: Ref 33

ability improves as the amount of pearlite is increased up to about 70% (the amount found in UNS G10550) (Ref 33).

**Grinding of Free-Machining Steels.** Free-machining additives, such as manganese sulfide, lead, bismuth, and selenium, will improve the grindability of carbon and alloy steel thirtyfold (Ref 33). These soft particles weaken the plastic shear plane and probably melt because of the high heat generated in grinding to lubricate the contact surface. In one case, the overall effect of the total amount of free-machining additives in a study of several steel grades was present as a linear relationship to grindability (Fig. 31), which clearly illustrates the improvement in grindability with increasing amounts of soft particles (Ref 33).

## Nontraditional Machining Processes

Nontraditional machining processes such as abrasive waterjet and laser cutting may be superior in cost and performance to traditional metal removal processes in special applications. See the Section "Nontraditional Machining Processes" in this Volume for detailed information on these methods.

**Abrasive waterjet cutting** has been used to cut through 50 mm (2 in.) thick steel plate, leaving a smooth, clean edge that requires no further processing, such as grinding (Ref 34). The abrasive waterjet cuts straight and contour edges without heat; this eliminates thermal distortion and localized microstructural change.

**Electrical chemical grinding** is a process that removes metal by chemical action resulting from the passage of electric current in the presence of an electrolytic fluid. This process may be cost effective for finishing small, expensive, hardened steel parts. Electrical chemical grinding removes iron at rates up to 2.21 $cm^3$/min (0.135 $in.^3$/min) from medium- and high-carbon steels, leaving a burr-free and stress-free surface (Ref 35).

**Lasers** are used to drill holes having the following benefits compared to traditional drill bit holemaking:

- Lower aspect ratios, as small as 0.10
- Less deformation of the hole walls
- Higher accuracy and repeatability
- Less taper
- Higher production rates (one hole per second) (Ref 36)

Laser drilling with carbon dioxide lasers usually involves pulsing a focused laser beam over the workpiece and vaporizing the steel, leaving a hole.

### ACKNOWLEDGMENTS

The comments and guidance of C.A. Apple, Bethlehem Steel Corporation; J.E. Hood, Steltech; R. Millsop, Steltech; H. Yaguchi, Inland Steel Company; and S.V. Subramanian, McMaster University, were very helpful in the preparation of this article. The contributions of B.A. Buck, Steltech; E.M. Trent, University of Birmingham, Great Britain; and the ASM Handbook editorial staff are also greatly appreciated.

# Appendix: Machinability Testing of Carbon and Alloy Steels

Hiroshi Yaguchi, Inland Steel Company

A test method for a particular machining application is selected based on steel grade, type of machining operation to be used, amount of steel available, and the purpose of the test. Multiple tests are often carried out on the same sample. This is necessary because machinability is a system property that depends on machining conditions; therefore, different results may be obtained under quite different conditions. As a result, it is desirable to use several tests to characterize machinability under many possible conditions. This section will summarize six machinability tests commonly carried out on carbon and alloy steel rod and bar products at the author's company. It should be noted that numerous other machinability tests are applicable to steels. Information on these tests can be found in the article "Machinability Testing" in this Volume and in Ref 37 and 38.

## Automatic Screw Machine Test

Many types of grades, especially free-machining steels, are machined into parts in automatic screw machines. Because it is desirable for machinability test conditions to simulate actual conditions as much as possible, this test is regarded as the most reliable test available for free-machining steels. However, a disadvantage is that this test requires a large amount of steel and manpower.

The machine used for the screw machine test is a 32 mm (1¼ in.) six-spindle automatic screw machine, and the material used is 25 mm (1 in.) round cold-drawn bars. The test methodology used is a modified form of ASTM E 618. Figure 32 shows the part produced in the test and the relative positions of the rough form and finish form tools. The form tools are made of M2 high-speed tool steel, and the drills are made of M7 high-speed tool steel. Straight mineral oil without additives is used as the cutting fluid. The rough- and finish-formed part growths and surface roughnesses are measured at 1-h intervals on parts produced from each of the six spindles. In this test, the rough form tool part growth is considered the major criterion, and a part growth of more than 0.13 mm (0.005 in.) or catastrophic tool failure is considered the end point for the test. The finish-formed surface roughness is also measured by a surface profilometer. The drilled holes are monitored for drill wear and eccentricity. The maximum machining speed at which the material could be machined to obtain a rough form tool life of 8 h and the tool life obtained at a common speed for all the steels tested are considered the measure of machinability. Figure 33 illustrates typical curves showing rough-formed part growth as a function of machine time. Additional information on automatic screw machine tests can be found in Ref 39 to 41.

## Plunge Test

The plunge machinability test has been developed to simulate the forming operation in the automatic screw machine test. The advantages of this test are the small amount of steel necessary and the ease of control over machining conditions. Therefore, this test has been primarily used for the purpose of basic understanding in developing superior free-machining steels (Ref 42).

A 14.7 kW (19.7 hp) computer numerical control (CNC) lathe with variable spindle speed capability is instrumented with a three-component piezoelectric dynamometer that has good rigidity and small cross sensitivity among the force components. In

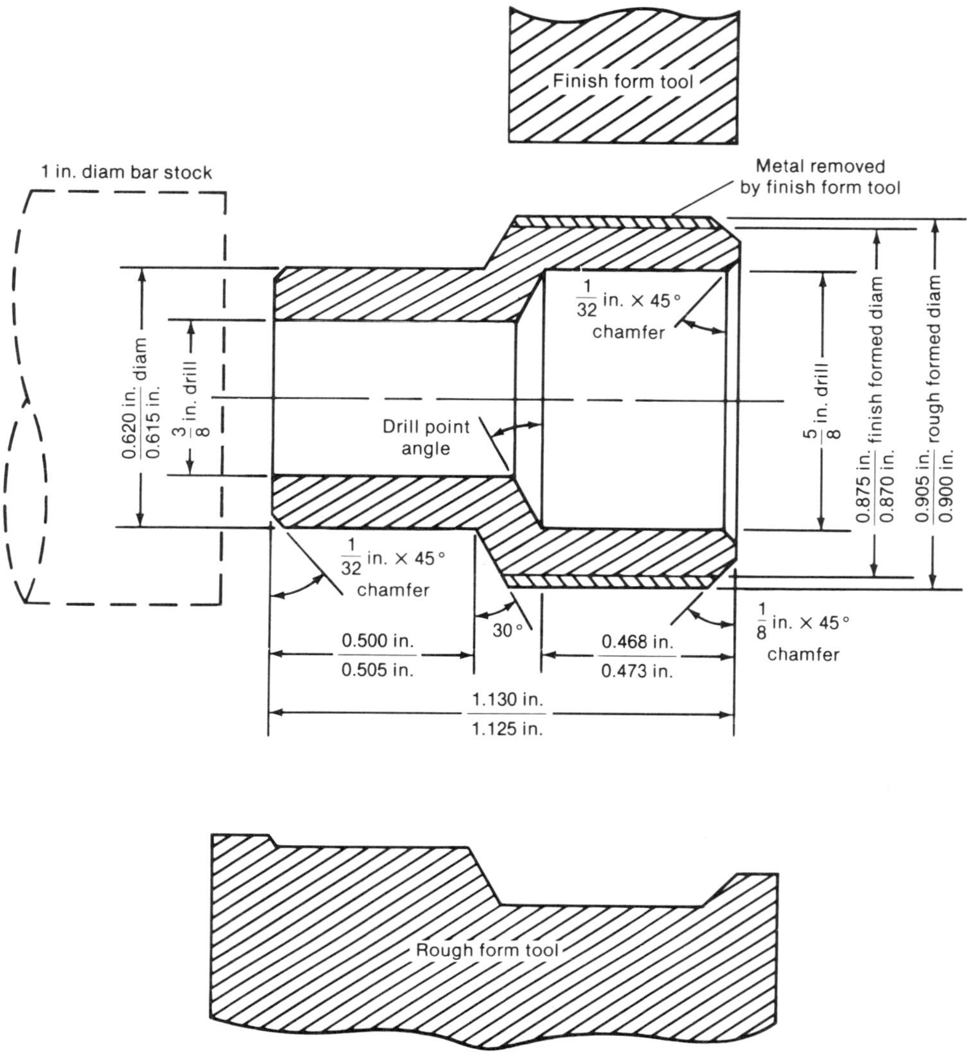


**Fig. 32** Details of the ASTM E 618 machinability test specimen and the relative positions of the form tools

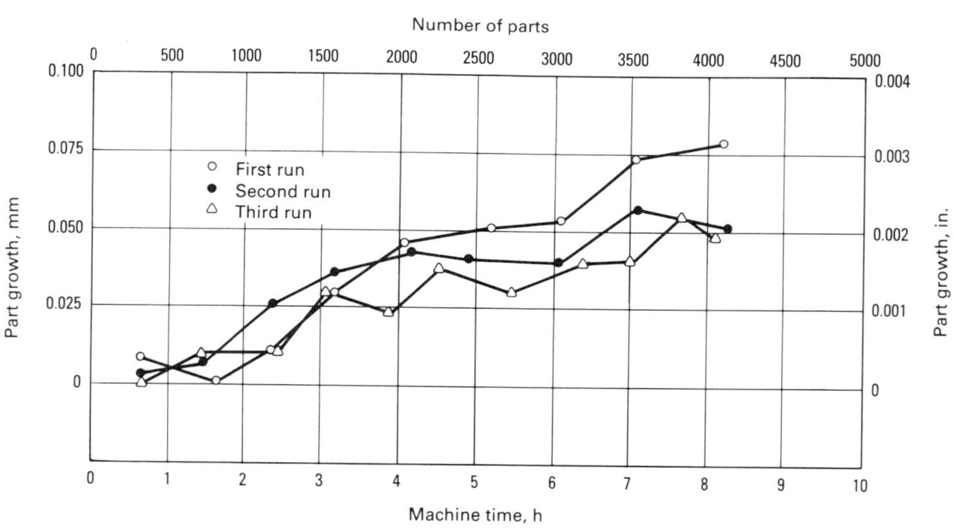


**Fig. 33** Part growth of rough-formed surface during three tests on the same material

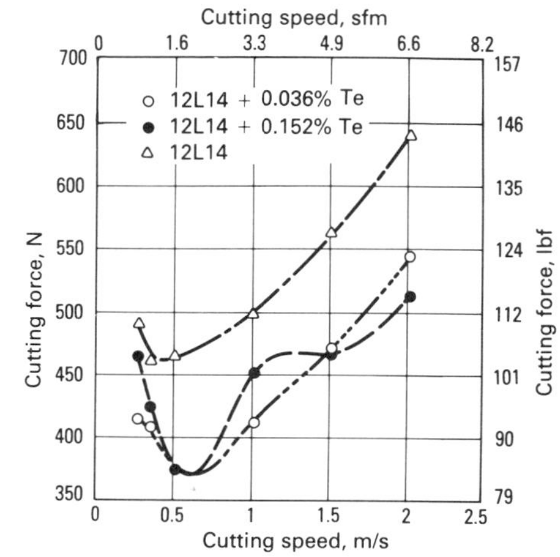


**Fig. 34** Cutting force as a function of cutting speed

this test, only two force components—cutting force and thrust force—are measured. The force data are collected and analyzed with a personal computer. Cutoff tools of M2 high-speed tool steel, 3.2 mm (⅛ in.) wide, are used to make plunges under orthogonal cutting. The rake angle is set at 0° and the clearance angle at 15°. Generally, no cutting fluid is used.

A maximum of 20 plunges is made with a feed rate of 0.064 mm/rev (0.0025 in./rev) at each of several predetermined speeds. Surface cutting speed is kept constant within a plunge by adjusting spindle speed for the diameter of the bar at the position of machining. Each test at a particular speed is conducted with a freshly ground tool. The total depth of cut is 6.35 mm (0.250 in.). When the cutting edge is completely worn and catastrophic tool failure has occurred, the testing is terminated before 20 plunges. Occasionally, representative chips are collected from each test and observed. The tools are also observed for built-up edge and tool wear. The results of force measurement and tool life expressed in terms of the number of plunges until catastrophic tool failure are used as the major machinability criterion. A typical example of cutting force measured as a function of cutting speed is shown in Fig. 34.

## Drill Force Test

Occasionally, only a very small amount of steel is available for a machinability test, especially when testing experimental heats or confirming customer complaints. This test has been developed for situations in which insufficient steel is available for other tests or when tests should be carried out with the minimum manpower required (Ref 43). Therefore, this test is not expected to characterize machinability fully, but to present only a relative rating of machinability.

A six-spindle indexing-turret drilling machine with electric power feed is instrumented with a two-component piezoelectric dynamometer, which measures torque and thrust force. The force data, analyzed in a personal computer, are used as the major machinability criteria. Drills 9.5 mm (⅜ in.)

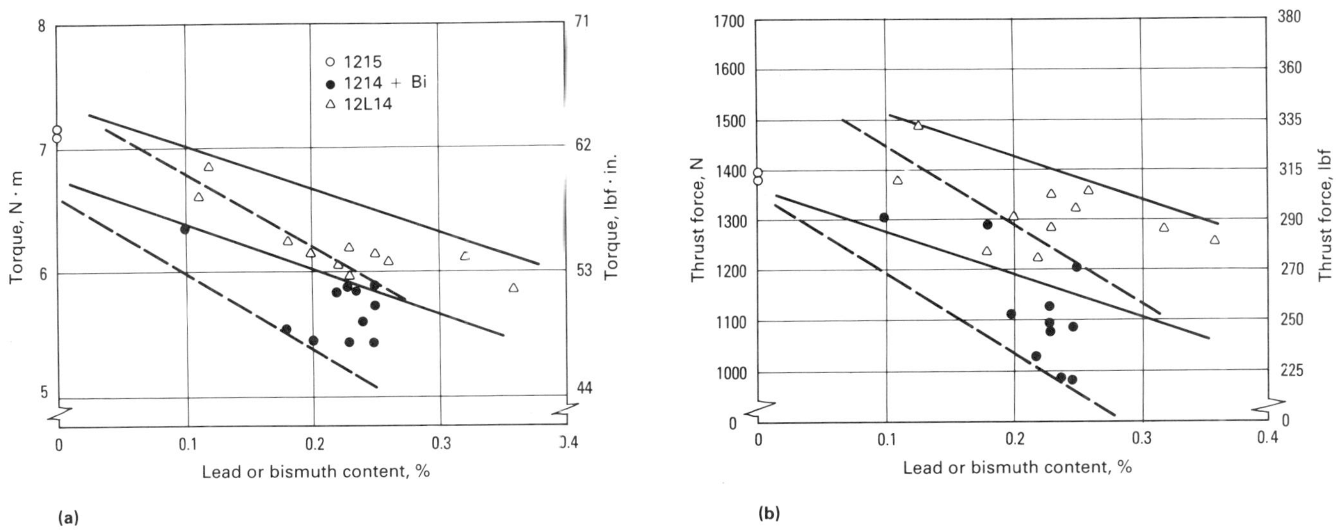


**Fig. 35** Effect of lead and bismuth content on torque (a) and thrust force (b)

in diameter of M7 high-speed tool steel are used. A large number of drills are purchased at one time from the same lot to minimize possible compositional or microstructural variations. Further, web thickness is measured, and only drills having web thickness within the predetermined range are used. A special faceted drill point instead of the factory-made chisel edge is employed. Finally, four drills are used for one sample to reduce variability further. A total of 24 blind holes 25 mm (1 in.) deep are made in one sample, and average torque and thrust force are compared. Figure 35 shows the effect of lead or bismuth content on the torque and thrust force.

## Single-Point Turning Test for Carbide Tools

Carbide tools are often used to machine parts made from high-strength steel grades. The single-point turning test can be used for testing such conditions. The test methodology used is a modified form of ISO 3685-1977 (E) (see discussion below).

The turning operation is carried out with either TiN-coated or noncoated carbide tools with a 5° negative rake angle, 0.38 mm/rev (0.015 in./rev) feed rate, and 1.9 mm (0.075 in.) depth of cut on a CNC lathe. Various cutting speeds are used, depending on steel grade and other conditions. The tool is periodically disassembled, and flank wear is measured in a toolmakers' microscope. Machinability is compared based on either wear rate at a common cutting speed among all the samples tested or the higher cutting speed used to reach a certain length of tool life, which is determined by the length of flank wear. The articles "Tool Wear and Tool Life" and "Machinability Testing" in this Volume provide more detailed information on the testing of carbides.

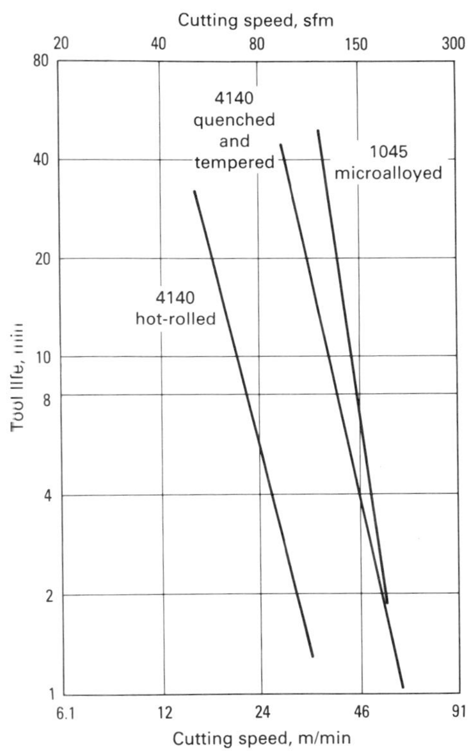


**Fig. 36** Taylor tool life versus cutting speed. Curves were obtained in a single-point turning test.

## Single-Point Turning Test for High-Speed Steel Tools

The ISO 3685-1977 (E) machinability test, in which Taylor tool life curves are generated (Fig. 36), is perhaps the most widely known and used. This turning test is carried out with M42 high-speed steel tools with a 5° positive rake angle, 0.13 mm (0.005 in./rev) feed rate, and 1.3 mm (0.05 in.) depth of cut. Various cutting speeds are used, and testing at a certain cutting speed is continued until catastrophic tool failure occurs. Machinability is defined by the highest cutting speed used to reach a predetermined length of tool life.

## Drill Eccentricity Test

This test has been developed to investigate the propensity of bar samples for off-center or eccentric drill hole formation. The samples are 25 mm (1 in.) diam bars of 40 mm (1.5 in.) length. A 6.4 mm (¼ in.) drill is used to drill through at the center of the samples along the bar axis. Complete drilling is achieved in six penetrations of 6.4 mm (¼ in.) depth. The feed rate is 0.15 mm/rev (0.006 in./rev), and the drill point is a standard chisel edge commercial 118° point. The speed used is 1300 rev/min, which is approximately 26 m/min (85 sfm). A mist spray of a synthetic water-base lubricant is used. After drilling, the tendency of the test material to produce axially drilled holes that are off-center is monitored by measuring the variation in circumferential wall thickness, reported as the total indicated runout.

### REFERENCES


1. *Machining Data Handbook*, Vol 1 and 2, 3rd ed., Metcut Research Associates, 1980
2. *Metals and Alloys in the Unified Numbering System*, 4th ed., Society of Automotive Engineers, 1986
3. "United States Air Force Machinability Report," Curtis-Wright Corporation, 1950
4. J.A. Schey, *Tribology in Metalworking: Friction, Lubrication and Wear*, American Society for Metals, 1984
5. F.W. Wilson, *Fundamentals of Tool Design*, American Society of Tool and Manufacturing Engineers, Prentice-Hall, 1962

6. B. Mills and A.H. Redford, *Machinability of Engineering Materials*, University of Salford, Applied Science, 1983
7. S.V. Subramanian, D.A.R. Kay, N. Chandrasekanan, and R. Sowerby, Design of Clean Steel for High Strain Rate Machining in *Strategies for Automation of Machining: Materials and Processes*, Proceedings of an International Conference, ASM INTERNATIONAL, 1987
8. E.M. Trent, *Metal Cutting*, University of Birmingham, 1983
9. W.J. Riffe and S.P. Boppana, "Machining of Sheet Steels," Paper 850410, presented at the International Congress and Exposition (Detroit), Society of Automotive Engineers, 1985
10. M.E. Finn, "Product Process Integration—Machined Parts," Discussion Meeting, Canadian Steel Industries Research Association, April 1988
11. L.H.S. Long, Influence of Microcracks in the Machinability of Metals, *Met. Technol.*, Nov 1980, p 465-470
12. M.E. Finn, G.A. Beaudoin, R.M. Nishizaki, S.V. Subramanian, and D.A.R. Kay, Influence of Steel Matrix on Machinability of High Carbon Steel in Automatic Machining, in *Strategies for Automation of Machining: Materials and Processes*, Proceedings of an International Conference, ASM INTERNATIONAL, 1987, p 43
13. D.W. Murray, Machining Performance of Steels, *Iron Steel Eng.*, April 1967, p 123-128
14. T. Araki, H. Fukunaga, T. Sata, and N. Nakntaki, Some Results of Cooperative Research of the Effect of Heat Treated Structure on the Machinability of Low Alloy Steel, in *Influence of Metallurgy on Machinability*, Proceedings of an International Conference, ASM INTERNATIONAL, 1975, p 381
15. J.A. Vaccari, Free-Machining Steels, *Am. Mach.*, Oct 1986, p 116-130
16. D.M. Keane and P.H. Wannell, Machinability of Continuous Cast Steel, in *High Productivity Machining: Materials and Processes*, Proceedings of an International Conference, ASM INTERNATIONAL, 1985, p 151
17. P. Lechat and R. Kisch, *Conditions of Use in Relation to the Neurotoxicity of Ingested Bismuth Salts*, French Academie National de Pharmacie, March 1986
18. D. Bhattacharya, "New Free-Machining Steel With Bismuth, Part 1—Developments," Paper presented at the 22nd Mechanical Working and Processing Conference (Toronto, Ontario), American Institute of Mining, Metallurgical, and Petroleum Engineers, 1980
19. L.B. Schmitt, "New Free-Machining Steel With Bismuth, Part II—Evaluation," Paper presented at the 22nd Mechanical Working and Processing Conference (Toronto, Ontario), American Institute of Mining, Metallurgical, and Petroleum Engineers, 1980
20. H. Opitz, M. Gappisch, W. Konig, R. Pape, and W. Wicher, Einfluss Oxidischer Einschliise auf die Bearbirtbarkiit von Stake CK45 mit Hartmetall-Drekwerkzengen, *Arch. Eisenhüttenwes.*, Vol 33 (No. 12), 1962, p 841
21. T. Ototani, *Calcium Clean Steel*, Springer-Verlag, 1986, p 132
22. S.V. Subramanian and D.A.R. Kay, "Development of Nonleaded Free-Machining Steels—Phase II," DSS Contract No. USU80—0011, McMaster University, March 1983
23. *Machining Principles*, Metals Engineering Institute, ASM INTERNATIONAL
24. J.E. Repars and R.W. Brockman, How Coolants Affect Drill Wear, *Am. Mach.*, Feb 1971
25. J. Wolfe, An Engineer's Guide to Deep Hole Drilling Cutting Fluids, *Cutting Tool Eng.*, Nov-Dec 1980
26. S. Abeyama, A. Kimura, and S. Nakamara, The Influence of Heat Treatment and Cold Forging on Machinability of Low Alloyed Steels, *J. Appl. Metalwork.*, Jan 1983, p 243-248
27. T. Shirago, S. Ishizakai, and H. Osuzu, Deep Hole Drillability of Microalloyed Steels in Hot Forged Components, in *Strategies for Automation of Machining: Materials and Processes*, Proceedings of an International Conference, ASM INTERNATIONAL, 1987, p 51
28. D. Bhattacharya, Machinability of a Medium Carbon Microalloyed Bar Steel, in *Fundamentals of Microalloyed Forging Steels*, Proceedings of an International Conference (Golden, CO), Ferrous Metallurgy Committee, The Metallurgical Society, July 1986
29. V. Ollilainen, I. Lahti, H. Pontinen, and H. Heiskala, Machinability Comparison When Substituting Microalloyed Forging Steel for Quenched and Tempered Steel, in *Fundamentals of Microalloyed Forging Steels*, Proceedings of an International Conference (Golden, CO), Ferrous Metallurgy Committee, The Metallurgical Society, July 1986
30. H. Yaguchi, "Development of Drill Machinability Test for Initial and Limited Screening Purposes," Paper presented at the 28th Mechanical Working and Steel Processing Conference (Pittsburgh, PA), Iron and Steel Society, Oct 1986
31. H. Pontinen, Calcium Treated Steel Shows Improved and Uniform Machinability, in *Strategies for Automation of Machining: Materials and Processes*, Proceedings of an International Conference, ASM INTERNATIONAL, 1987.
32. M.L. Pickett, M. Cristinacce, and D.J. Naylor, Development and Application of Improved Machinable Engineering Steels, in *High Productivity Machining: Materials and Processes*, Proceedings of an International Conference, ASM INTERNATIONAL, 1985, p 253
33. H.J. Tata, "Effect of Composition and Additives on the Grinding Characteristics of Carbon and Low-Alloy Steel Bars," Paper presented at the International Grinding Conference (Fontana, WI), Society of Manufacturing Engineers, Aug 1984
34. M. Hashish, Application of Abrasive Waterjets to Metal Cutting, in *Conference Proceedings of Nontraditional Machining* (Cincinnati, OH), Society of Carbide and Tool Engineers, Dec 1985
35. R.E. Phillips, What is Electrochemical Grinding and How Does It Work?, in *Conference Proceedings of Nontraditional Machining* (Cincinnati, OH), Society of Carbide and Tool Engineers, Dec 1985
36. M. Pellicchia, Lasers: The Manufacturer's Outlook, in *Conference Proceedings of Nontraditional Machining* (Cincinnati, OH), Society of Carbide and Tool Engineers, Dec 1985
37. B. Mills and A.H. Redford, *Machinability of Engineering Materials*, Applied Science, 1983
38. *Machinability Testing and Utilization of Machining Data*, Materials/Metalworking Technology Series, American Society for Metals, 1979
39. S. Ohtani, S. Katayama, T. Akasawa, D. Bhattacharya, and H. Yaguchi, Qualities of Strand Cast Resulfurized Free-Machining Steels, *J. Appl. Metalwork.*, Vol 4, 1986, p 245
40. H. Yaguchi and D. Bhattacharya, An Investigation of Factors Controlling Part Growth and Surface Finish of Form Tools in the Automatic Screw Machine Test, *J. Mater. Shaping Technol.*, Vol 5, 1987
41. H. Yaguchi and N. Onodera, Effect of Cold Work on the Machinability of AISI 12L14 Steel, in *Strategies for Automation of Machining: Materials and Processes*, Proceedings of an International Conference, ASM INTERNATIONAL, 1987, p 15
42. H. Yaguchi, Effect of MnS Inclusion Size on Machinability of Low-Carbon, Leaded, Resulfurized Free-Machining Steel, *J. Appl. Metalwork.*, Vol 4, 1986, p 214
43. H. Yaguchi, Development of Drill Machinability Test for Initial and Limited Screening Purposes, in *Proceedings of the 28th Mechanical Working and Steel Processing Conference*, Iron and Steel Society, 1987, p 223

# Machining of Stainless Steels

Ted Kosa and Ronald P. Ney, Sr., Carpenter Technology Corporation

THE STAINLESS STEELS that are available, how and why they differ in machining performance, and the machining methods used are the topics addressed in this article. With regard to machining methods, emphasis is given to conventional machining techniques, such as turning, drilling, and sawing, because most stainless steels are currently machined using such techniques. However, special or nontraditional machining techniques are overviewed in the closing portion of the article.

## General Material and Machining Characteristics

The common aspect of stainless steels is the presence of a minimum of about 11% Cr to provide the excellent corrosion and oxidation resistance that is the chief characteristic of the materials. However, stainless steels do not constitute a single, well-defined material, but instead comprise several families of alloys, each generally having its own characteristic microstructure, alloying elements, and range of properties. Compositional differences within each family can also produce a variety of alloys suited to a wide range of applications.

Because of the wide variety of stainless steels available, a simple characterization of their machinability can be somewhat misleading. As shown in later sections of this article, the machinability of stainless steels varies from low to very high, depending on the final choice of alloy. In general, however, stainless steels are considered more difficult to machine than other metals, such as aluminum or low-carbon steels. Stainless steels have been characterized as gummy during cutting, showing a tendency to produce long, stringy chips, which seize or form a built-up edge on the tool. This may result in reduced tool life and degraded surface finish. These general characteristics are due to the following properties, which are possessed by stainless steels to different extents (Ref 1, 2; see also the article "Forming of Stainless Steel" in Volume 14 of the 9th Edition of *Metals Handbook*):

- High tensile strength (Fig. 1)
- Large spread between yield strength (YS) and ultimate tensile strength (UTS) (Fig. 1)
- High ductility and toughness
- High work-hardening rate (Fig. 1)
- Low thermal conductivity (Fig. 2)

Despite these properties, stainless steels can be machined under the appropriate conditions. In general, more power is required to machine stainless steels than carbon steels, cutting speeds must often be lower, a positive feed must be maintained, tooling and fixtures must be rigid, chip breakers or curlers may be needed on the tools, and care must be taken to ensure good lubrication and cooling during cutting (Ref 4). These and other practices are discussed in more detail in the sections on individual conventional machining techniques in this article.

## Classification of Stainless Steels

Stainless steels can be divided into five families. Four are based on the characteristic microstructure of the alloys in the family: ferritic, martensitic, austenitic, or duplex (austenitic plus ferritic). The fifth family, the precipitation-hardenable alloys, is based on the type of heat treatment used, rather than microstructure. In addition, stainless steels can be divided into the nonfree-machining alloys and the free-machining alloys. Free-machining alloys form a limited group that includes some of the

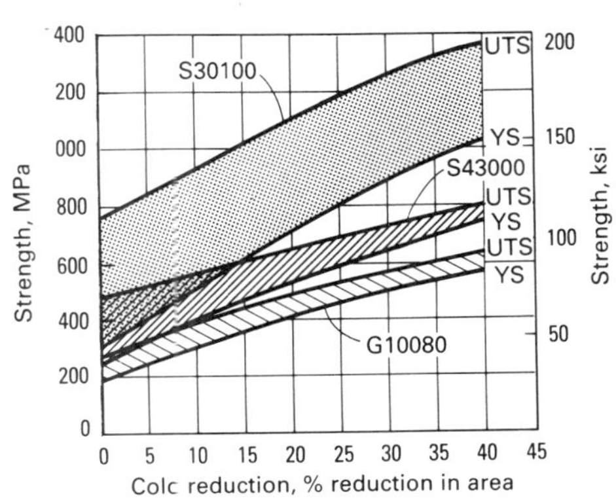


**Fig. 1** Comparison of tensile strengths, spread between yield and ultimate tensile strengths, and work-hardening rate for S30100 austenitic stainless steel, S43000 ferritic stainless steel, and G10080 carbon steel

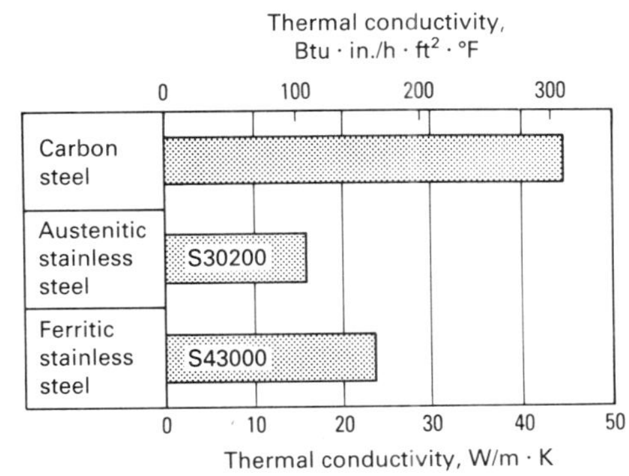


**Fig. 2** Comparison of thermal conductivities for carbon steel, S30200 austenitic stainless steel, and S43000 ferritic stainless steel. Source: Ref 3

alloys of the five basic families. Nonfree-machining and free-machining alloys may be available in versions having enhanced machining properties.

The designations for stainless steels are complicated by a proliferation of competing systems in both the United States and abroad. Perhaps the most widely used system is the type designations of the American Iron and Steel Institute (AISI). More recently, the Unified Numbering System (UNS) for identifying metallic materials, including stainless steels, has been introduced. Whenever possible, the UNS numbers incorporate the numerical portion of the AISI type designations to simplify recognition. Table 1 provides compositional limits for selected stainless steels listed by UNS and AISI type designations and separated into the basic families. Where AISI type designations are not available, Table 1 uses three alternatives: American Society for Testing and Materials (ASTM) XM designations, type designations used in the industry that resemble AISI types (but are not specifically used by AISI), or AISI numbers used for the alloys in their role as high-temperature high-strength alloys.

The identification of stainless steels is further complicated by the various trade names used by steel producers. In most cases, such alloys have their own UNS designation or fall within the compositional limits of a standard alloy. For this reason, trade names have been avoided.

## Table 1 Selected standard and special stainless steels

| UNS designation | AISI type(a) | Composition, wt%(b) C | Mn | Si | P | S | Cr | Ni | Mo | N | Others |
|---|---|---|---|---|---|---|---|---|---|---|---|
| **Ferritic alloys** | | | | | | | | | | | |
| S40500...... | 405 | 0.08 | 1.00 | 1.00 | 0.040 | 0.030 | 11.50–14.50 | . . . | . . . | . . . | 0.10–0.30 Al |
| S40900...... | 409 | 0.08 | 1.00 | 1.00 | 0.045 | 0.045 | 10.50–11.75 | 0.50 | . . . | . . . | 6 × C-0.75 Ti |
| S43000...... | 430 | 0.12 | 1.00 | 1.00 | 0.040 | 0.030 | 16.00–18.00 | . . . | . . . | . . . | . . . |
| S43020...... | 430F | 0.12 | 1.25 | 1.00 | 0.060 | 0.15 min | 16.00–18.00 | . . . | 0.60 | . . . | . . . |
| S43023...... | 430FSe | 0.12 | 1.25 | 1.00 | 0.060 | 0.060 | 16.00–18.00 | . . . | . . . | . . . | 0.15 min Se |
| S43400...... | 434 | 0.12 | 1.00 | 1.00 | 0.040 | 0.030 | 16.00–18.00 | . . . | 0.75–1.25 | . . . | . . . |
| S44200...... | 442 | 0.20 | 1.00 | 1.00 | 0.040 | 0.030 | 18.00–23.00 | . . . | . . . | . . . | . . . |
| S44300...... | (443) | 0.20 | 1.00 | 1.00 | 0.040 | 0.030 | 18.00–23.00 | 0.50 | . . . | . . . | 0.90–1.25 Cu |
| S44400...... | (444) | 0.025 | 1.00 | 1.00 | 0.040 | 0.030 | 17.5–19.5 | 1.00 | 1.75–2.50 | 0.025 | [0.20+4(C+N)]–0.80 Ti+Nb |
| S44600...... | 446 | 0.20 | 1.50 | 1.00 | 0.040 | 0.030 | 23.00–27.00 | . . . | . . . | 0.25 | . . . |
| S18200...... | [XM-34] | 0.08 | 1.25–2.50 | 1.00 | 0.04 | 0.15 min | 17.50–19.50 | . . . | 1.50–2.50 | . . . | . . . |
| S18235...... | . . . | 0.025 | 0.50 | 1.00 | 0.030 | 0.15–0.35 | 17.50–18.50 | 1.00 | 2.00–2.50 | 0.025 | 0.30–1.00 Ti, 0.035 C+N |
| **Martensitic alloys** | | | | | | | | | | | |
| S40300...... | 403 | 0.15 | 1.00 | 0.50 | 0.040 | 0.030 | 11.50–13.00 | . . . | . . . | . . . | . . . |
| S41000...... | 410 | 0.15 | 1.00 | 1.00 | 0.040 | 0.030 | 11.50–13.00 | . . . | . . . | . . . | . . . |
| S41400...... | 414 | 0.15 | 1.00 | 1.00 | 0.040 | 0.030 | 11.50–13.50 | 1.25–2.50 | . . . | . . . | . . . |
| S41600...... | 416 | 0.15 | 1.25 | 1.00 | 0.060 | 0.15 min | 12.00–14.00 | . . . | 0.60 | . . . | . . . |
| S41610...... | [XM-6] | 0.15 | 1.50–2.50 | 1.00 | 0.06 | 0.15 min | 12.00–14.00 | . . . | 0.60 | . . . | . . . |
| S41623...... | 416Se | 0.15 | 1.25 | 1.00 | 0.060 | 0.060 | 12.00–14.00 | . . . | . . . | . . . | 0.15 min Se |
| S42000...... | 420 | 0.15 min | 1.00 | 1.00 | 0.040 | 0.030 | 12.00–14.00 | . . . | . . . | . . . | . . . |
| S42010...... | . . . | 0.15–0.30 | 1.00 | 1.00 | 0.040 | 0.030 | 13.5–15.0 | 0.25–1.00 | 0.40–1.00 | . . . | . . . |
| S42020...... | 420F | 0.15 min | 1.25 | 1.00 | 0.060 | 0.15 min | 12.00–14.00 | . . . | 0.60 | . . . | . . . |
| S42023...... | (420FSe) | 0.30–0.40 | 1.25 | 1.00 | 0.06 | 0.06 | 12.00–14.00 | . . . | 0.60 | . . . | 0.15 min Se; 0.60 Zr or Cu |
| S43100...... | 431 | 0.20 | 1.00 | 1.00 | 0.040 | 0.030 | 15.00–17.00 | 1.25–2.50 | . . . | . . . | . . . |
| S44002...... | 440A | 0.60–0.75 | 1.00 | 1.00 | 0.040 | 0.030 | 16.00–18.00 | . . . | 0.75 | . . . | . . . |
| S44003...... | 440B | 0.75–0.95 | 1.00 | 1.00 | 0.040 | 0.030 | 16.00–18.00 | . . . | 0.75 | . . . | . . . |
| S44004...... | 440C | 0.95–1.20 | 1.00 | 1.00 | 0.040 | 0.030 | 16.00–18.00 | . . . | 0.75 | . . . | . . . |
| S44020...... | (440F) | 0.95–1.20 | 1.25 | 1.00 | 0.040 | 0.10–0.35 | 16.00–18.00 | 0.75 | 0.40–0.60 | 0.08 | . . . |
| S44023...... | (440FSe) | 0.95–1.20 | 1.25 | 1.00 | 0.040 | 0.030 | 16.00–18.00 | 0.75 | 0.60 | 0.08 | 0.15 min Se |
| **Austenitic alloys** | | | | | | | | | | | |
| S20100...... | 201 | 0.15 | 5.50–7.50 | 1.00 | 0.060 | 0.030 | 16.00–18.00 | 3.50–5.50 | . . . | 0.25 | . . . |
| S20300...... | [XM-1] | 0.08 | 5.00–6.50 | 1.00 | 0.040 | 0.18–0.35 | 16.00–18.00 | 5.00–6.50 | 0.50 | . . . | 1.75–2.25 Cu |
| S20910...... | [XM-19] | 0.06 | 4.00–6.00 | 1.00 | 0.040 | 0.030 | 20.50–23.50 | 11.50–13.50 | 1.50–3.00 | 0.20–0.40 | 0.10–0.30 Nb; 0.10–0.30 V |
| S21904...... | [XM-11] | 0.04 | 8.00–10.00 | 1.00 | 0.060 | 0.030 | 19.00–21.50 | 5.50–7.50 | . . . | 0.15–0.40 | . . . |
| S24100...... | [XM-28] | 0.15 | 11.00–14.00 | 1.00 | 0.060 | 0.030 | 16.50–19.50 | 0.50–2.50 | . . . | 0.20–0.45 | . . . |
| S28200...... | . . . | 0.15 | 17.00–19.00 | 1.00 | 0.045 | 0.030 | 17.00–19.00 | . . . | 0.50–1.50 | 0.40–0.60 | 0.50–1.50 Cu |
| S30100...... | 301 | 0.15 | 2.00 | 1.00 | 0.045 | 0.030 | 16.00–18.00 | 6.00–8.00 | . . . | . . . | . . . |
| S30200...... | 302 | 0.15 | 2.00 | 1.00 | 0.045 | 0.030 | 17.00–19.00 | 8.00–10.00 | . . . | . . . | . . . |
| S30300...... | 303 | 0.15 | 2.00 | 1.00 | 0.20 | 0.15 min | 17.00–19.00 | 8.00–10.00 | 0.60 | . . . | . . . |
| S30310...... | [XM-5] | 0.15 | 2.50–4.50 | 1.00 | 0.20 | 0.25 min | 17.00–19.00 | 7.00–10.00 | 0.75 | . . . | . . . |
| S30323...... | 303Se | 0.15 | 2.00 | 1.00 | 0.20 | 0.060 | 17.00–19.00 | 8.00–10.00 | . . . | . . . | 0.15 min Se |
| S30330...... | (303Cu) | 0.15 | 2.00 | 1.00 | 0.15 | 0.10 min | 17.00–19.00 | 6.00–10.00 | . . . | . . . | 2.50–4.00 Cu; 0.10 Se |
| S30345...... | [XM-2] | 0.15 | 2.00 | 1.00 | 0.05 | 0.11–0.16 | 17.00–19.00 | 8.00–10.00 | 0.40–0.60 | . . . | 0.60–1.00 Al |
| S30360...... | [XM-3] | 0.15 | 2.00 | 1.00 | 0.040 | 0.12–0.30 | 17.00–19.00 | 8.00–10.00 | 0.75 | . . . | 0.12–0.30 Pb |
| S30400...... | 304 | 0.08 | 2.00 | 1.00 | 0.045 | 0.030 | 18.00–20.00 | 8.00–10.50 | . . . | . . . | . . . |
| S30403...... | 304L | 0.03 | 2.00 | 1.00 | 0.045 | 0.030 | 18.00–20.00 | 8.00–12.00 | . . . | . . . | . . . |
| S30430...... | [XM-7] | 0.10 | 2.00 | 1.00 | 0.045 | 0.030 | 17.00–19.00 | 8.00–10.00 | . . . | . . . | 3.00–4.00 Cu |
| S30431...... | . . . | 0.06 | 2.00 | 1.00 | 0.040 | 0.14 | 16.00–19.00 | 9.00–11.00 | . . . | . . . | 1.30–2.40 Cu |
| S30452...... | [XM-21] | 0.08 | 2.00 | 1.00 | 0.045 | 0.030 | 18.00–20.00 | 8.00–10.50 | . . . | 0.16–0.30 | . . . |
| S30500...... | 305 | 0.12 | 2.00 | 1.00 | 0.045 | 0.030 | 17.00–19.00 | 10.00–13.00 | . . . | . . . | . . . |
| S30900...... | 309 | 0.20 | 2.00 | 1.00 | 0.045 | 0.030 | 22.00–24.00 | 12.00–15.00 | . . . | . . . | . . . |
| S30908...... | 309S | 0.08 | 2.00 | 1.00 | 0.045 | 0.030 | 22.00–24.00 | 12.00–15.00 | . . . | . . . | . . . |
| S31000...... | 310 | 0.25 | 2.00 | 1.50 | 0.045 | 0.030 | 24.00–26.00 | 19.00–22.00 | . . . | . . . | . . . |
| S31008...... | 310S | 0.08 | 2.00 | 1.50 | 0.045 | 0.030 | 24.00–26.00 | 19.00–22.00 | . . . | . . . | . . . |
| S31600...... | 316 | 0.08 | 2.00 | 1.00 | 0.045 | 0.030 | 16.00–18.00 | 10.00–14.00 | 2.00–3.00 | . . . | . . . |
| S31603...... | 316L | 0.030 | 2.00 | 1.00 | 0.045 | 0.030 | 16.00–18.00 | 10.00–14.00 | 2.00–3.00 | . . . | . . . |
| S31620...... | 316F | 0.08 | 2.00 | 1.00 | 0.20 | 0.10 min | 17.00–19.00 | 12.00–14.00 | 1.75–2.50 | . . . | . . . |
| S31700...... | 317 | 0.08 | 2.00 | 1.00 | 0.045 | 0.030 | 18.00–20.00 | 11.00–15.00 | 3.00–4.00 | . . . | . . . |
| S31703...... | 317L | 0.030 | 2.00 | 1.00 | 0.045 | 0.030 | 18.00–20.00 | 11.00–15.00 | 3.00–4.00 | . . . | . . . |
| S32100...... | 321 | 0.08 | 2.00 | 1.00 | 0.045 | 0.030 | 17.00–19.00 | 9.00–12.00 | . . . | . . . | 5 ×C min Ti |
| S34700...... | 347 | 0.08 | 2.00 | 1.00 | 0.045 | 0.030 | 17.00–19.00 | 9.00–13.00 | . . . | . . . | 10 ×C min Nb |
| S34720...... | (347F) | 0.08 | 2.00 | 1.00 | 0.040 | 0.18–0.35 | 17.00–19.00 | 9.00–12.00 | . . . | . . . | 10 ×C–1.10 Nb |
| S34723...... | (347FSe) | 0.08 | 2.00 | 1.00 | 0.11–0.17 | 0.030 | 17.00–19.00 | 9.00–12.00 | . . . | . . . | 10 ×C–1.10 Nb; 0.15–0.35 Se |
| S38400...... | 384 | 0.08 | 2.00 | 1.00 | 0.045 | 0.030 | 15.00–17.00 | 17.00–19.00 | . . . | . . . | . . . |
| N08020 ..... | . . . | 0.07 | 2.00 | 1.00 | 0.045 | 0.035 | 19.00–21.00 | 32.00–38.00 | 2.00–3.00 | . . . | 8 ×C–1.00 Nb; 3.00–4.00 Cu |
| **Duplex alloys** | | | | | | | | | | | |
| S31803...... | . . . | 0.030 | 2.00 | 1.00 | 0.030 | 0.020 | 21.0–23.0 | 4.50–6.50 | 2.50–3.50 | 0.08–0.20 | . . . |
| S32550...... | . . . | 0.04 | 1.50 | 1.00 | 0.04 | 0.03 | 24.0–27.0 | 4.50–6.50 | 2.00–4.00 | 0.10–0.25 | 1.50–2.50 Cu |
| S32900...... | 329 | 0.20 | 1.00 | 0.75 | 0.040 | 0.030 | 23.00–28.00 | 2.50–5.00 | 1.00–2.00 | . . . | . . . |
| S32950...... | . . . | 0.03 | 2.00 | 0.60 | 0.035 | 0.010 | 26.0–29.0 | 3.50–5.20 | 1.00–2.50 | 0.15–0.35 | . . . |

(continued)

(a) AISI types are listed except as follows: ( ), designations resembling AISI types but not used in that system; [ ], designations used by ASTM; ⟨ ⟩, AISI numbers used for high-strength high-temperature alloys. (b) Except as noted, single elements indicate maximum values. All compositions include iron as balance. (c) Martensitic. (d) Semiaustenitic. (e) Austenitic

**Table 1 (continued)**

| UNS designation | AISI type(a) | Composition, wt%(b) | | | | | | | | |
|---|---|---|---|---|---|---|---|---|---|---|
| | | C | Mn | Si | P | S | Cr | Ni | Mo | N | Others |
| **Precipitation-hardenable alloys** | | | | | | | | | | | |
| S13800(c) ..... | [XM-13] | 0.05 | 0.20 | 0.10 | 0.01 | 0.008 | 12.25–13.25 | 7.50–8.50 | 2.00–2.50 | 0.01 | 0.90–1.35 Al |
| S15500(c) ..... | [XM-12] | 0.07 | 1.00 | 1.00 | 0.040 | 0.030 | 14.00–15.50 | 3.50–5.50 | ... | ... | 0.15–0.45 Nb; 2.50–4.50 Cu |
| S17400(c) ..... | ⟨630⟩ | 0.07 | 1.00 | 1.00 | 0.040 | 0.030 | 15.50–17.50 | 3.00–5.00 | ... | ... | 0.15–0.45 Nb; 3.00–5.00 Cu |
| S17700(d) ..... | ⟨631⟩ | 0.09 | 1.00 | 1.00 | 0.040 | 0.040 | 16.00–18.00 | 6.50–7.75 | ... | ... | 0.75–1.50 Al |
| S35000(d) ..... | ⟨633⟩ | 0.07–0.11 | 0.50–1.25 | 0.50 | 0.040 | 0.030 | 16.00–17.00 | 4.00–5.00 | 2.50–3.25 | 0.07–0.13 | ... |
| S35500(d) ..... | ⟨634⟩ | 0.10–0.15 | 0.50–1.25 | 0.50 | 0.040 | 0.030 | 15.00–16.00 | 4.00–5.00 | 2.50–3.25 | 0.07–0.13 | ... |
| S45000(c) ..... | [XM-25] | 0.05 | 1.00 | 1.00 | 0.030 | 0.030 | 14.00–16.00 | 5.00–7.00 | 0.50–1.00 | ... | 8 × C min Nb; 1.25–1.75 Cu |
| S45500(c) ..... | [XM-16] | 0.05 | 0.50 | 0.50 | 0.040 | 0.030 | 11.00–12.50 | 7.50–9.50 | 0.50 | ... | 0.10–0.50 Nb; 1.50–2.50 Cu 0.80–1.40 Ti |
| S66286(e) ..... | ⟨660⟩ | 0.08 | 2.00 | 1.00 | 0.040 | 0.030 | 13.50–16.00 | 24.0–27.0 | 1.00–1.50 | ... | 0.35 Al; 0.0010–0.010 B 1.90–2.35 Ti; 0.10–0.50 V |

(a) AISI types are listed except as follows: ( ), designations resembling AISI types but not used in that system; [ ], designations used by ASTM; ⟨ ⟩, AISI numbers used for high-strength high-temperature alloys. (b) Except as noted, single elements indicate maximum values. All compositions include iron as balance. (c) Martensitic. (d) Semiaustenitic. (e) Austenitic

Because of the variety of stainless steels available, it is usually possible to obtain an alloy possessing the desired set of attributes, although this same wealth of alloys can create problems during the selection process. Although aids have been published to assist in the selection process, it is advisable to consult the producer to determine the alloy that best meets the requirements of the application (Ref 5, 6).

**Ferritic stainless steels** have a body-centered cubic structure and are magnetic. In the annealed condition, they have a tensile yield strength of about 275 to 350 MPa (40 to 50 ksi). They are generally hardenable only by cold working, but not to the same extent as the austenitic stainless steels (Fig. 1). The alloys have relatively good ductility in the annealed condition, but are not used where toughness is of concern. They have a broad range of corrosion resistance, depending on alloy content. However, as a class, they are considered less corrosion resistant than the austenitic alloys.

The most well-known alloy of this family is S43000, which is an iron-base alloy containing 16 to 18% Cr. Other alloys, such as S40500 or S40900, contain lower chromium. Higher levels of chromium are used in alloys such as S44200 or S44600 for improved corrosion or oxidation resistance. Molybdenum is added to certain alloys, such as S43400, to improve corrosion resistance, particularly in chloride-containing solutions. Titanium or niobium is used to stabilize carbon and nitrogen to improve the as-welded properties of alloys such as S40900. Extralow-interstitial (ELI) alloys with carbon and nitrogen reduced by special melting techniques have been more recently introduced. The ELI ferritic alloys usually contain higher chromium, along with molybdenum and sometimes nickel. The more highly alloyed ones are often limited to thin sections because of toughness restrictions.

**Martensitic stainless steels** have a body-centered cubic/tetragonal structure and are magnetic. In the annealed condition, they have a tensile yield strength of about 275 MPa (40 ksi) and can be moderately hardened by cold working, similarly to ferritic alloys. However, martensitic alloys are normally heat treated by hardening plus tempering to yield strength levels up to 1900 MPa (280 ksi), depending primarily on carbon level. The alloys have good ductility and toughness, which decrease as strength increases.

The most commonly used alloy of this family is S41000, which contains about 12% Cr and 0.1% C to provide strengthening. Carbon level and consequently strength capability increase in the series S42000, S44002, S44003, and S44004 alloys. Chromium is increased, particularly in the latter three alloys, to maintain corrosion resistance because increasing carbon removes chromium from solution by forming chromium carbides. Molybdenum can be added to improve mechanical properties or corrosion resistance, as in S42010. Nickel can be added for the same reasons, as in S41400 and S43100. Nickel also serves to maintain the desired microstructure, preventing excessive free ferrite, when higher chromium levels are used to improve corrosion resistance. The limitations on alloy content required to maintain the desired fully martensitic structure restrict to moderate levels the corrosion resistance obtainable with martensitic alloys.

**Austenitic stainless steels** have a face-centered cubic structure and are nonmagnetic in the annealed condition. The alloys can be subdivided into two categories: the standard alloys, such as S30200, containing nickel to maintain an austenitic structure at room temperature and those alloys containing a substantial quantity of manganese, usually with higher levels of nitrogen and in many cases nickel. Examples of the latter are S20100, S20910, S21904, S24100, and S28200. Nitrogen can also be used to provide strengthening in the chromium-nickel grades, as in S30452. The standard chromium-nickel alloys with lower nitrogen levels have tensile yield strengths of 200 to 275 MPa (30 to 40 ksi) in the annealed condition, while alloys containing higher nitrogen have yield strengths up to about 500 MPa (70 ksi).

Austenitic stainless steels possess good ductility and toughness, even at cryogenic temperatures, and can be substantially hardened by cold working (Fig. 1). The degree of work hardening depends on alloy content. Austenitic stainless steels with a lower alloy content may become magnetic because of the transformation of austenite to martensite if heavily deformed during cold working or even machining. A corrective anneal or the selection of an alloy with a lower work-hardening rate may be necessary if a low magnetic permeability is required for the intended application. The corrosion resistance of austenitic alloys varies from good to excellent, again depending on alloy content.

The most common austenitic stainless steel is S30400, which contains approximately 18% Cr and 8% Ni. In addition to the alloying variations noted above, higher chromium, nickel, molybdenum, or copper can be added to improve particular aspects of corrosion or oxidation resistance. Examples are S31600, S31700, S30900, S31000, and N08020. Many of the more corrosion-resistant alloys, such as N08020, have nickel levels high enough to rate classification as nickel-base alloys. To prevent intergranular corrosion after elevated-temperature exposure, titanium or niobium is added to stabilize carbon in alloys such as S32100 or S34700. Conversely, carbon levels are reduced to low levels during melting to produce the AISI L or S alloys, such as S30403 or S30908.

**Duplex stainless steels** contain a mixture of ferrite and austenite and are magnetic. They have tensile yield strengths of about 550 MPa (80 ksi) in the annealed condition, or about twice that of the standard austenitic alloys. Strength can be increased by cold working. The alloys have good ductility and toughness, along with excellent corrosion resistance.

The original alloy in this classification was S32900, which contains chromium, molybdenum, and sufficient nickel to provide the desired balance of ferrite and austenite. More recent alloys, such as S31803 and S32950, also contain nitrogen.

**Precipitation-hardenable stainless steels** are categorized by their ability to be age hardened to various strength levels. The alloys can be subdivided into the austenitic (for example, S66286), martensitic (for example, S17400), or semiaustenitic (for example, S17700) classifications. The latter alloys may have an austenitic structure for formability, but can be subsequently transformed to martensite and then aged to the desired strength level. Depending on the type of alloy, precipitation-hardenable stainless steels can reach tensile yield strength levels of up to 1700 MPa (250 ksi). Cold working prior to aging can result in even higher strengths. The alloys generally have good ductility and toughness with moderate-to-good corrosion resistance. A better combination of strength and corrosion resistance is achieved than with the martensitic alloys.

The most well-known precipitation-hardenable stainless steel is S17400. It contains chromium and nickel, as do all precipitation-hardenable stainless steels, with copper for age hardening and niobium for stabilizing the carbon. The age-hardening agents used in other alloys include titanium (S45500), aluminum (S13800), and niobium (S45000). Molybdenum can be added to improve mechanical properties or corrosion resistance. Both molybdenum and copper are added for corrosion resistance in S45000. Carbon is normally restricted, except in semiaustenitic alloys such as S35500, which require it to provide the desired phase transformations.

**Free-machining alloys** contain a free-machining additive such as sulfur to form inclusions that significantly improve overall machining characteristics. In some cases, other compositional changes can be made either within or outside the broad compositional ranges of the corresponding nonfree-machining alloy. Such additional changes in composition may improve machining characteristics beyond those obtained by the simple addition of the free-machining additives.

The benefit of improved machining characteristics by the addition of sulfur or other free-machining additives is not obtained without changes in other properties. In particular, the following properties may be degraded by the addition of a free-machining agent:

- Corrosion resistance
- Transverse ductility and toughness
- Hot workability
- Cold formability
- Weldability

In some cases, variants of the basic free-machining alloy are available to provide an optimum combination of machinability with another property. However, the trade-off among the various properties must still be considered when selecting an alloy; that is, the ease of machining must be balanced against the possible reduction in other important properties, such as corrosion resistance.

**Table 2 Correspondence of nonfree-machining and free-machining stainless steels**

| Nonfree-machining alloys | Related free-machining alloys Selenium-bearing | Sulfur-bearing |
|---|---|---|
| **Ferritic** | | |
| S43000 | S43023 | S43020 |
| S44400 | ... | S18200(a) |
| | | S18235 |
| **Martensitic** | | |
| S41000 | S41623 | S41600 |
| | | S41610(b) |
| S42000 | S42023 | S42020 |
| S44004 | S44023 | S44020 |
| **Austenitic** | | |
| ... | ... | S20300 |
| S30200/S30400 | S30323 | S30300 |
| | | S30310(b) |
| | | S30345(c) |
| | | S30360(d) |
| S30430 | ... | S30330 |
| | | S30431(e) |
| S31600 | ... | S31620 |
| S34700 | S34723 | S34720 |

(a) Does not contain titanium. (b) Contains high manganese. (c) Contains aluminum. (d) Contains lead. (e) Contains lower copper

Table 2 lists some nonfree-machining and free-machining alloys within the ferritic, martensitic, and austenitic families. Free-machining alloys are currently unavailable in the duplex or precipitation-hardenable families. Because duplex alloys are noted for excellent corrosion resistance but have somewhat limited hot workability, the addition of a free-machining agent, which would likely degrade both properties, would be undesirable. Similarly, precipitation-hardenable alloys are noted for good toughness at high strength levels, making it undesirable to add a free-machining agent, which would degrade toughness.

The best-known alloys in the three families represented, S43000 (ferritic), S41000 (martensitic), and S30400 (austenitic), have corresponding free-machining alloys (Table 2). In addition, the more corrosion-resistant molybdenum-bearing alloys S44400 and S31600 have free-machining versions in the ferritic and austenitic families, respectively, and the higher-carbon, higher-strength alloys S42000 and S44004 have free-machining versions in the martensitic family. Consequently, a variety of basic free-machining alloys are available to satisfy the two most important selection criteria for stainless steels, namely, corrosion resistance and mechanical properties (strength/hardness).

A variety of additional distinctions can be made among the other alloys listed in Table 2. Free-machining versions are available for S34700, a niobium-stabilized austenitic alloy, and for S30430, a copper-bearing alloy noted for a low work-hardening rate and excellent cold formability for an austenitic alloy. The free-machining versions of S30430 (S30330 and S30431) are intended to offer a good combination of cold formability and machinability. Another alloy that can offer this combination of properties is S30345. The selenium-bearing free-machining alloys are also noted for better cold formability than the sulfur-bearing alloys and can be used where machined surface finish is more important than tool life.

Alloy S20300, which lacks a corresponding nonfree-machining version, is a high-manganese, high-copper alloy with excellent machinability for an austenitic alloy. It can be substituted for S30300, specifications permitting. Higher manganese is also present in S41610 and S30310; aluminum in S30345; and lead in S30360. These compositional modifications are intended to improve machining performance.

Versions of S30300, S41600, and S43020 are also available to provide combinations of properties not obtainable with the standard alloys. The compositions of such versions still fall within the broad ranges of the standard alloy. For example, S30300 and S41600 are available in forging-quality versions, which are intended to provide a good combination of hot workability and machinability. Alloy S41600 is also available in a modified version intended to provide a higher quenched hardness level after bright hardening. Alloy S43020 is available in a modified version for optimum soft-magnetic properties.

**Enhanced-Machining Alloys.** The compositions and processing of alloys can also be modified within the broad limits to provide an optimum level of machining performance. This approach has been taken with both nonfree-machining and free-machining alloys, resulting in enhanced-machining alloys that meet the same specifications as the standard alloys. It should be noted that the enhanced-machining versions of the nonfree-machining alloys provide machining performance superior to that of the corresponding standard alloys, but still do not provide the machinability of comparable free-machining alloys. However, other properties, such as corrosion resistance, ductility, toughness, weldability, and cold formability, will be superior to those of the corresponding free-machining alloy. Thus, the enhanced-machining versions of the nonfree-machining alloys provide a means of obtaining improved machining performance without significant degradation of other properties.

There are many possibilities for tailoring enhanced-machining alloys because the latitude in the composition and processing of stainless steel alloys is great enough to

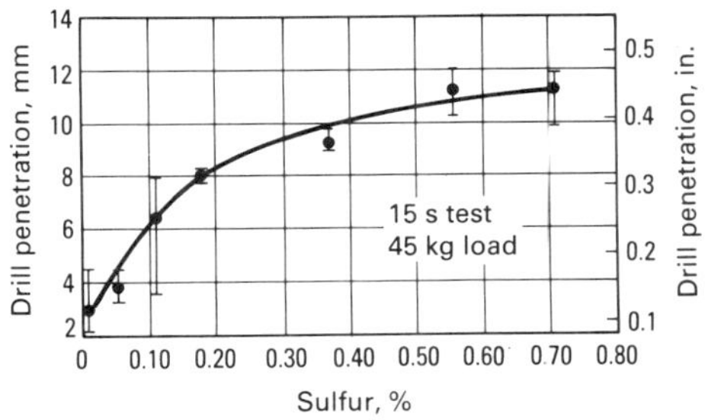


**Fig. 3** Effect of sulfur content on machinability in a drill penetration test for an 18Cr-9Ni austenitic stainless steel

provide a variety of levels of machinability. As a result, some alloys are available in more than one enhanced-machining version. For example, S41600 is available in an enhanced-machining version for meeting certain minimum hardness requirements and in a version providing even higher machining performance but having limited hardness capabilities. Obviously, the level of machinability necessary and the compromises to be made with other properties depend on the needs of the user. Before specifying or purchasing an alloy, the producer should be consulted to determine the proper alloy or, more important, the proper version of the alloy.

## Machinability

Machinability refers to several specific aspects of machining a material, and the definition of machinability must be precisely specified in a particular application. Some of the specific criteria for defining machinability include:

- Tool life or tool wear
- Machined surface finish
- Chip removal
- Cutting rate
- Productivity

The relative importance of these criteria varies, and the criteria may be interdependent in various ways.

Machinability also depends on many variables in the machining process. Therefore, because of the different criteria of machinability and because of the many variables in the machining process, the ranking or extrapolation of machinability must be viewed with caution. Some of the machining variables that may affect the evaluation of machinability are:

- Rigidity of tooling or fixtures
- Tool material and tool geometry
- Type of cutting fluid
- Type of machining operation

Generally, the harder the material, the more difficult it is to machine. However, machinability is more directly influenced by microstructure than by hardness. The machinability of many classes of alloys can be improved if the microstructure is a two-phase structure consisting of either a brittle or easily sheared second phase dispersed throughout a moderately ductile matrix. Stainless steels have a high alloy content that reduces machinability, but free-machining stainless steels are available that compare favorably with some free-machining carbon steels.

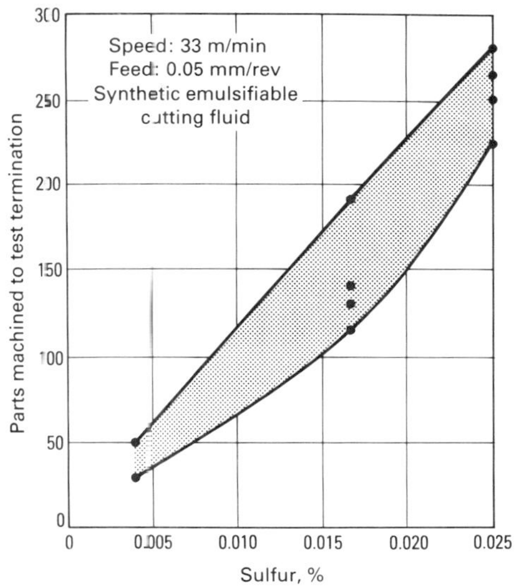


**Fig. 4** Effect of low levels of sulfur on machinability in a screw machine test for S30400 austenitic stainless steel. Termination is defined as a 0.075 mm (0.003 in.) increase in the diameter of the part being cut.

### Machinability Additives

The most important machinability additives are those that form inclusions in the metal. Such additives include sulfur, selenium, tellurium, lead, bismuth, and certain oxides. The role of these inclusions in improving machining performance has been the basis of much study, and theories include lubrication, chip embrittlement, and stress-concentration effects as the mechanisms for improving machinability. Whatever the reason, free-machining inclusions increase tool life, allow higher cutting rates, and may also affect chip breakage/disposability and the machined surface finish.

Elements that modify the matrix of the metal or the size, distribution, and mechanical properties of the free-machining inclusions can also improve machinability beyond that obtained with the free-machining additive alone, although these additional additives are of secondary importance. Compositional changes that modify the inclusions are discussed in the sections below that cover the appropriate inclusion. Modifications of the matrix are discussed in the section "Machinability of Alloy Families" in this article. An exception is phosphorus, which has traditionally been discussed with the inclusion-forming machinability additives.

**Sulfur.** The use of sulfur to improve the overall machining characteristics of stainless steels dates back to the early 1930s, when alloys similar to the current S41600 and S30300 were first developed (Ref 7, 8). Since then, sulfur has been the primary element used to provide free-machining characteristics in stainless steels. The amount of sulfur that can be added to a stainless steel is limited by the allowable degradation of other properties, particularly hot and cold workability in the producer's mill.

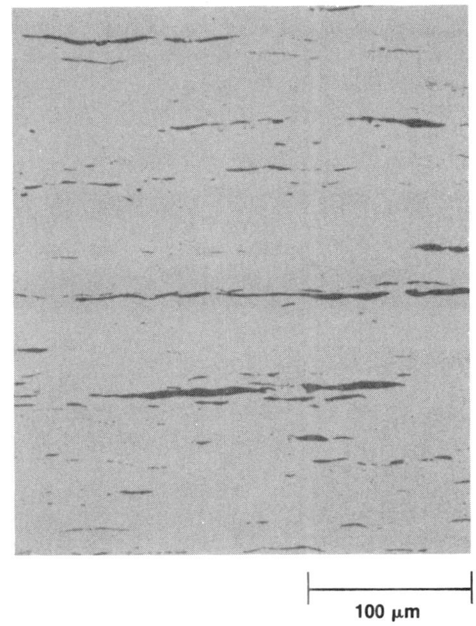


**Fig. 5** Sulfide inclusions in S30300 austenitic stainless steel

Increases in sulfur continuously increase the ease of cutting, although generally at a decreasing rate (Fig. 3). Increasing sulfur also produces relatively large increases in tool life, even within the typical limit of sulfur content (0.030% maximum) allowed for nonfree-machining alloys (Fig. 4). Close control of sulfur to the upper end of this range is at least part of the explanation for the enhanced-machining versions of the standard nonfree-machining alloys (Ref 9, 10).

Sulfur is present in stainless steels as discrete sulfide inclusions (Fig. 5). Sulfides that are larger and more globular provide a greater benefit to machining than smaller and elongated ones (Ref 9-11), as shown in Fig. 6 and 7. Although larger sulfides benefit tool life or ease of cutting, they can be detrimental to machined surface finish (Fig. 8). This is a case in which the definition of machinability is important.

Except at low levels of manganese, the sulfides in free-machining stainless steels are basically manganese sulfides containing chromium and iron. As the manganese level or manganese-to-sulfur ratio increases in a stainless steel, the composition of the sulfide will change (Ref 12-14), as shown in

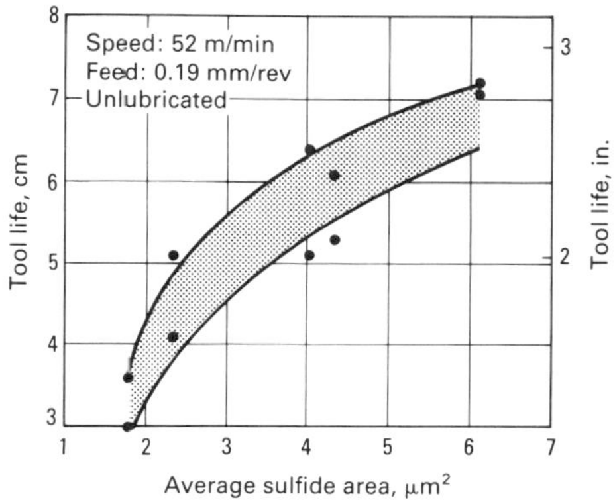


**Fig. 6** Effect of sulfide size on machinability in a tool life test for S30300 austenitic stainless steel. Tool life is measured as the distance traveled along a 23.2 mm (0.915 in.) diam bar until tool failure.

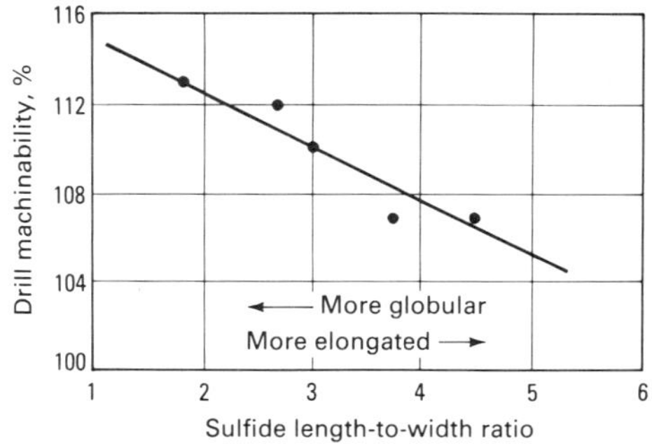


**Fig. 7** Effect of sulfide shape on machinability in a drill test for S30300 austenitic stainless steel. Source: Ref 9

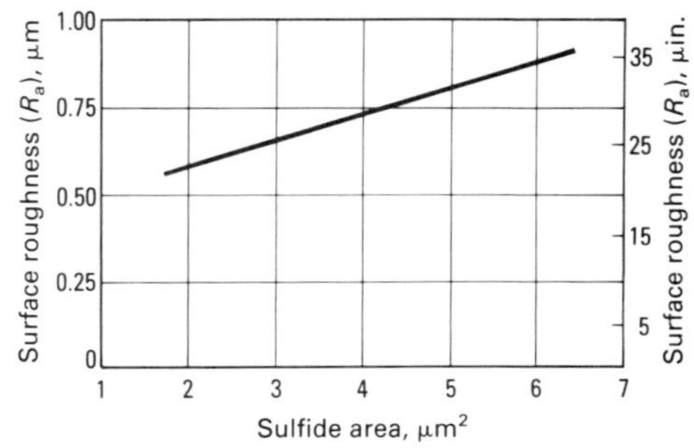


**Fig. 8** Effect of sulfide size on machined surface finish for a free-machining austenitic stainless steel. Source: Ref 11

Fig. 9. Higher manganese or a higher manganese-to-sulfur ratio will further improve the machinability of free-machining ferritic, martensitic, and austenitic stainless steels (Ref 15-19), as shown in Fig. 10. The use of higher manganese to increase machinability also has an effect on corrosion resistance. Stainless steels containing manganese-rich sulfides can be less corrosion resistant than those containing sulfides with a higher chromium content (Ref 13-15). Whether any loss in corrosion resistance is significant depends on the specific environment. Free-machining austenitic alloys are not affected to the same extent as free-machining ferritic and martensitic alloys (Fig. 11).

Stainless steels specifically containing higher manganese levels for improved machinability include S20300, S30310, and S41610. Other free-machining alloys, such as S30300, can be melted with manganese on the high end of the allowable range to improve machinability. On the other hand, manganese can also be intentionally limited in S30300 or S41600 to give these free-machining alloys improved corrosion resistance.

Instead of limiting manganese to improve the corrosion resistance of a free-machining stainless steel, another approach has been to produce alternative sulfides, such as titanium sulfides (Ref 20-22). An example of this approach is S18235.

**Selenium** is the next most commonly used free-machining agent in stainless steels, forming inclusions generally analogous to sulfides. Typically, selenium is less effective than an equivalent weight percent of sulfur in improving the overall machining characteristics of stainless steels (Ref 11, 17, 23-26), as shown in Fig. 12. However, there are reports to the contrary (Ref 2). Selenium-bearing alloys can provide a better machined surface finish than sulfur-bearing alloys (Fig. 13). In addition to the possibility of a better machined surface finish, selenium-bearing stainless steels may offer improved cold formability and somewhat improved corrosion resistance compared to the corresponding sulfur-bearing alloys (Ref 11, 23-27). Selenium additions have also been used in sulfur-bearing alloys to promote sulfides that are larger and more globular and therefore more beneficial to machinability (Ref 28, 29).

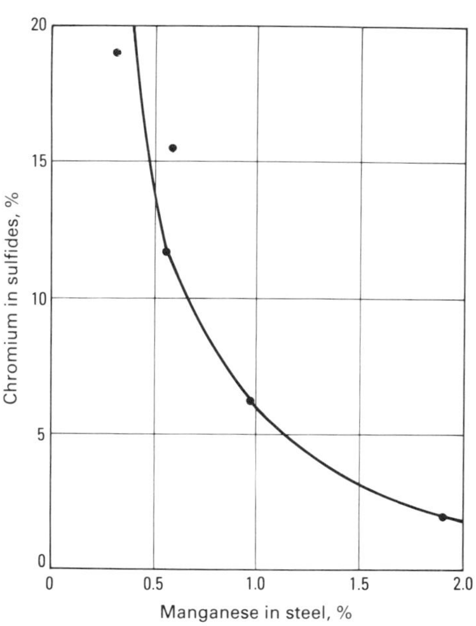


**Fig. 9** Effect of manganese content in steel on chromium content in sulfides for a free-machining 13% Cr martensitic stainless steel. Source: Ref 14

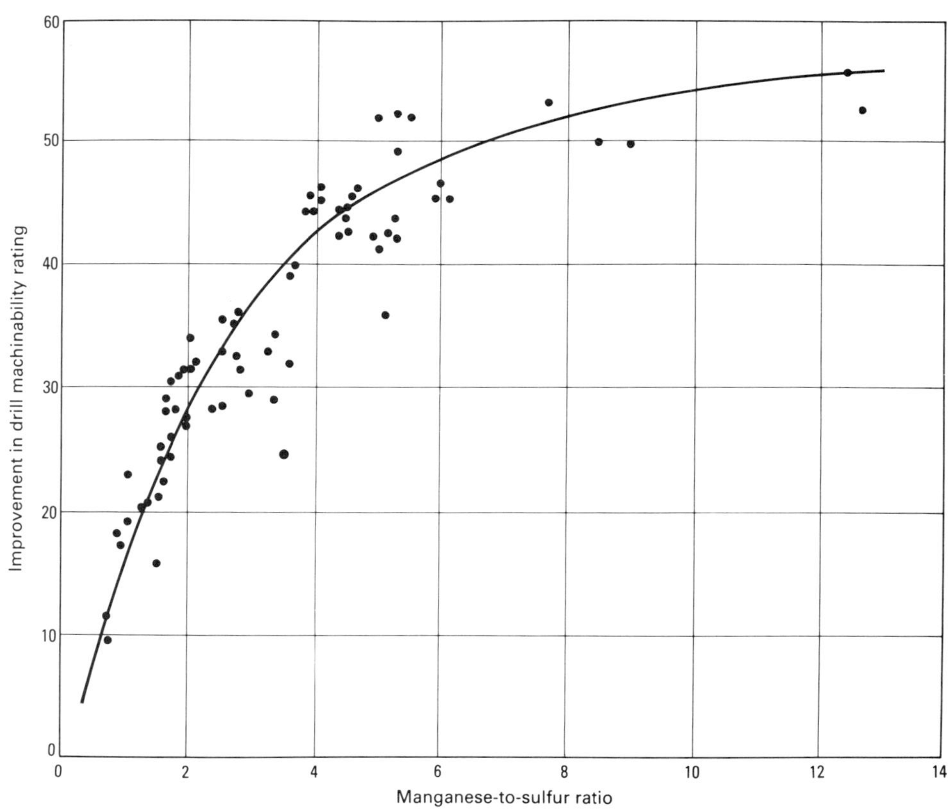


**Fig. 10** Effect of manganese-to-sulfur ratio on machinability in a drill test for a free-machining martensitic stainless steel. Source: Ref 16

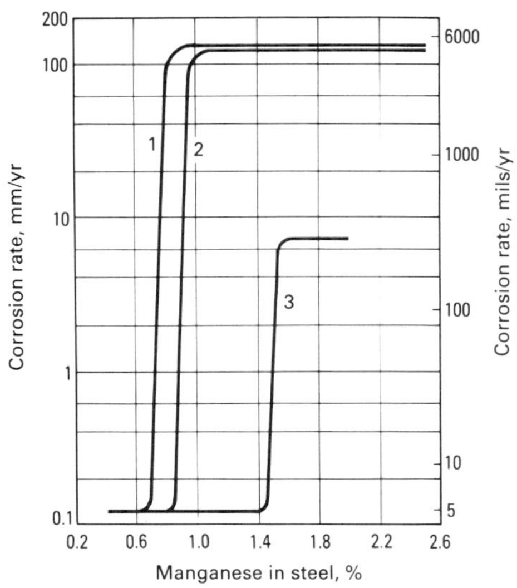


**Fig. 11** Effect of manganese content on the corrosion of free-machining stainless steels in a 10% nitric acid solution at 24 °C (75 °F). 1, 13% Cr martensitic alloy; 2, 17% Cr ferritic alloy; 3, 18Cr-9Ni austenitic alloy. Source: Ref 14

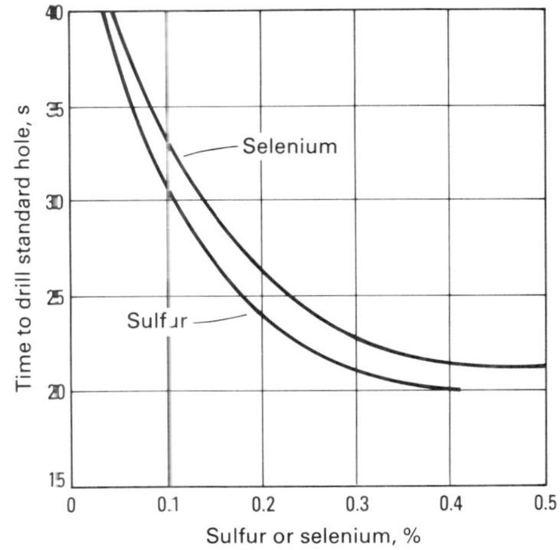


**Fig. 12** Comparison of the effect of selenium and sulfur on machinability in a drill test for a 13.5% Cr martensitic stainless steel at 19 to 21 HRC. Source: Ref 24

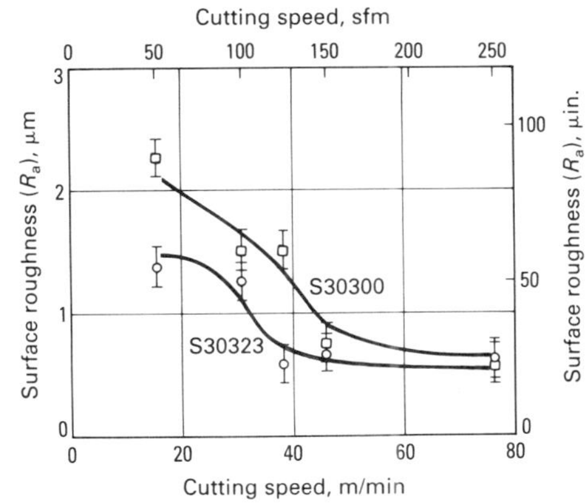


**Fig. 13** Comparison of machined surface finish obtained for sulfur-bearing (S30300) and selenium-bearing (S30323) austenitic stainless steels in a plunge machining test. Source: Ref 2

**Tellurium,** like selenium, forms inclusions similar to sulfides. Tellurium has been shown to be more effective than sulfur in improving the machinability of an austenitic stainless steel (Fig. 14). Like selenium, tellurium can be used to promote sulfides that tend to retain their globular shape (Ref 28, 29), as shown in Fig. 15. The utility of tellurium is limited, primarily because of its adverse effect on hot workability. This is particularly true for austenitic alloys (Ref 11, 17, 26). However, tellurium has recently been used in a ferritic alloy in conjunction with sulfur and lead (Ref 25).

**Lead and bismuth** have low solubility in stainless steels, forming metallic inclusions that benefit machinability, particularly in austenitic stainless steels (Ref 17, 23-25, 30-34). Lead is more beneficial to the machinability of an austenitic stainless steel than other common free-machining additives (Fig. 14). The use of lead, with or without limited sulfur, also reportedly results in better machined surface finish, corrosion resistance, and cold formability than the use of higher sulfur alone (Ref 25, 30, 32). Somewhat similar benefits have been attributed to the use of bismuth (Ref 34).

Despite the benefits associated with the use of lead or bismuth, alloys containing these elements are commercially available only to a limited extent. Problems with the use of lead include toxicity, reduction in hot workability, and erratic machinability associated with the difficulty in obtaining a uniform dispersion in the steel (Ref 17, 23, 34). Boron additions have been used to

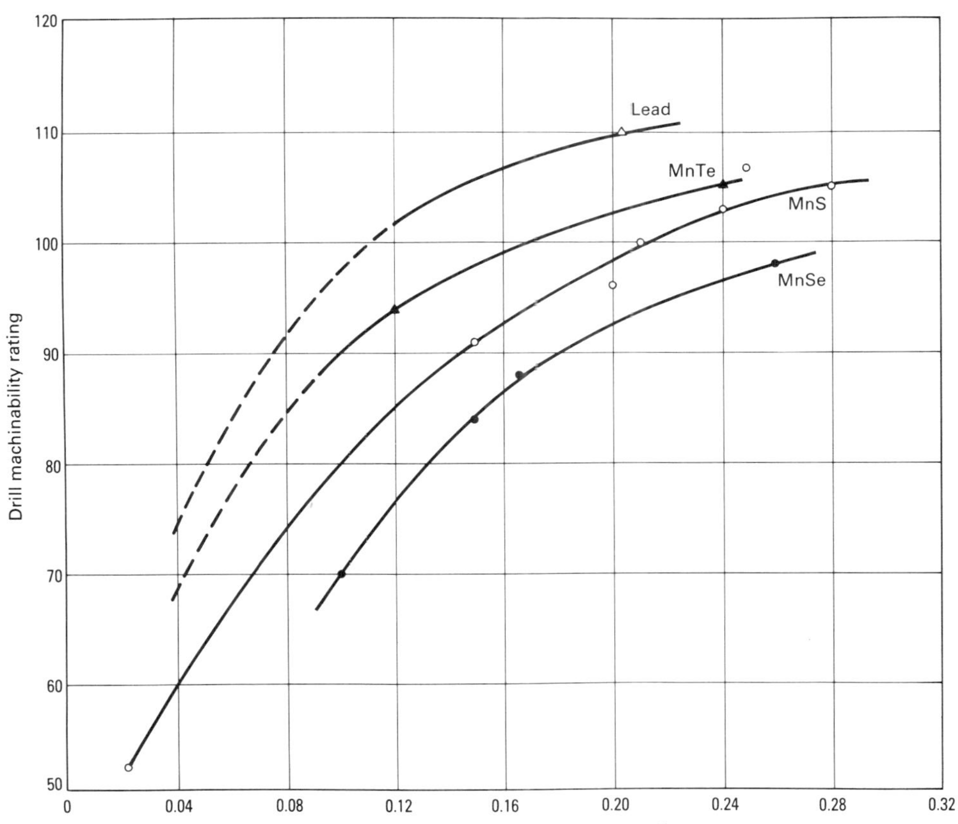


**Fig. 14** Comparison of the effect of lead or tellurium (MnTe) with sulfur (MnS) and selenium (MnSe) on machinability in a drill test for an 18Cr-9Ni austenitic stainless steel. Source: Ref 17

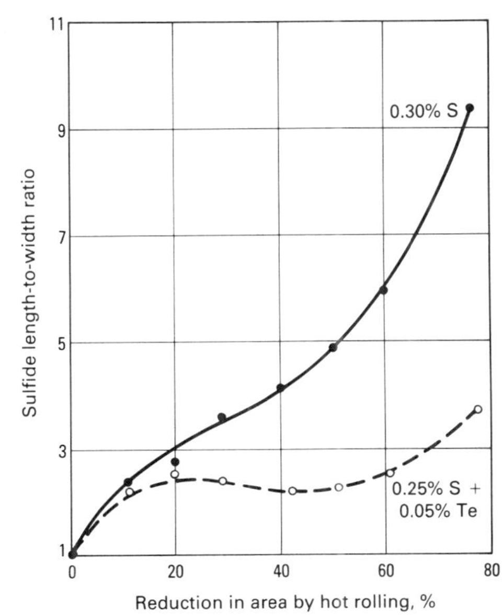


**Fig. 15** Effect of tellurium on sulfide shape in a 13% Cr martensitic stainless steel hot rolled various percent reductions in area at 1177 °C (2150 °F)

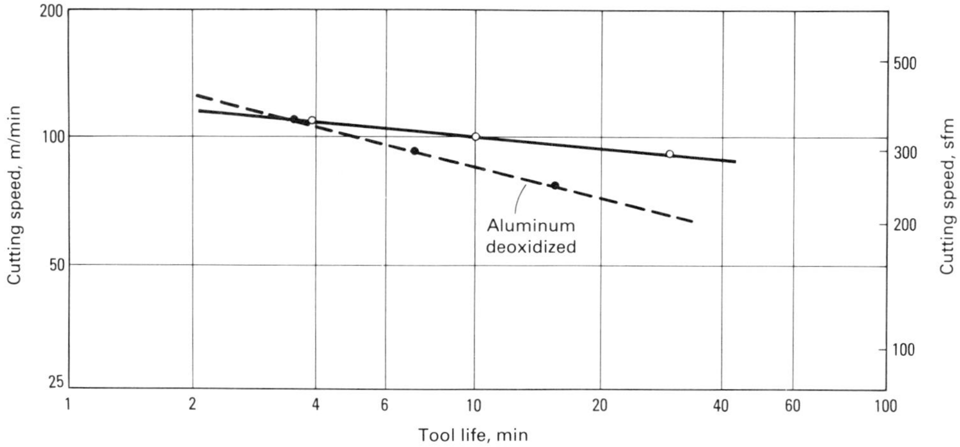


**Fig. 16** Effect of aluminum oxide inclusions on the tool life of S43020. Source: Ref 15

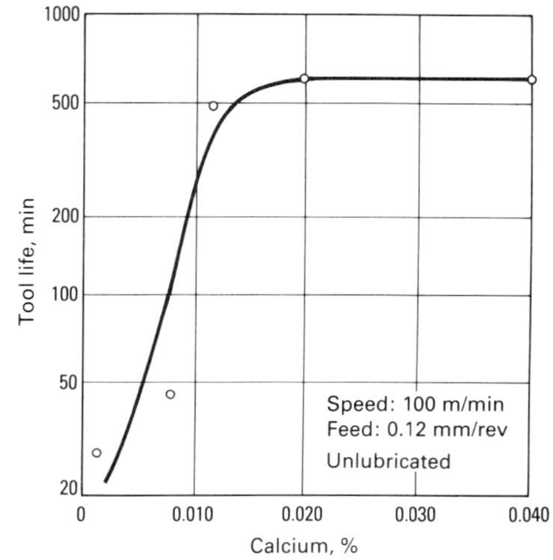


**Fig. 17** Effect of calcium content on carbide tool life for an austenitic stainless steel. Source: Ref 37

alleviate the hot-workability problems (Ref 32). The use of bismuth involves similar problems, except for toxicity (Ref 33, 34).

**Controlled Oxides.** Stainless steels may contain oxides as a result of deoxidation or reoxidation. In some cases, such oxides may be deleterious to machinability. The deoxidation of standard air-melted stainless steels is generally accomplished with silicon. The deoxidation of machining grades with aluminum should be avoided because aluminum oxides are highly abrasive and degrade tool life (Ref 11, 15, 28, 29, 35), as shown in Fig. 16.

However, calcium deoxidation can form inclusions that are actually beneficial to machinability under certain conditions (Ref 35-37). This involves the formation of calcium-(aluminum)-silicates in the steel. Such inclusions will soften during high-speed machining with titanium carbide tools, forming a protective layer on the tool. The increase in carbide tool life with increasing calcium content is shown in Fig. 17. The composition of the inclusions, that is, the proportions of CaO, $SiO_2$, and $Al_2O_3$, must be tightly controlled to obtain oxides that will properly soften in the cutting speed range used (Ref 36).

**Phosphorus** is added in conjunction with sulfur or selenium to enhance machinability, not by forming inclusions but by modifying the matrix properties of the alloy. The original purpose in adding phosphorus was to embrittle tough austenitic alloys (Ref 8). Phosphorus has only a small beneficial effect in an austenitic alloy and is detrimental above a certain level (Fig. 18).

## Machinability of Alloy Families

Significant differences in machinability exist between different alloy systems and alloy families, including the various free-machining alloys. This section will discuss machinability both within and among the five basic families of stainless steels.

**Ferritic and Martensitic Alloys.** Free-machining ferritic alloys (such as S43020) and annealed, low-carbon, free-machining martensitic alloys (such as S41600) are the easiest to machine of the stainless steels (Ref 2, 27, 38-40). In fact, their machinability ratings approach and in some cases are comparable to those of certain free-machining carbon steels (Ref 27, 38, 39). The nonfree-machining lower-chromium ferritic alloys (S40500, S43000) and annealed, low-carbon, straight-chromium martensitic alloys (S40300, S41000) are also generally easier to machine than most other nonfree-machining alloys (Ref 27, 38-40). The higher-chromium ferritic alloys, such as S44600, are considered by some to be somewhat more difficult to machine than the lower-chromium alloys because of gumminess and stringy chips (Ref 41, 42).

Other than the presence or lack of a free-machining additive, the machining characteristics of martensitic stainless steels are influenced by the following variables:

- Hardness level
- Carbon content
- Nickel content
- Phase balance, that is, the percentage of free or δ-ferrite in the martensitic matrix

Increasing the hardness level for a particular alloy results in a decrease in machinabil-

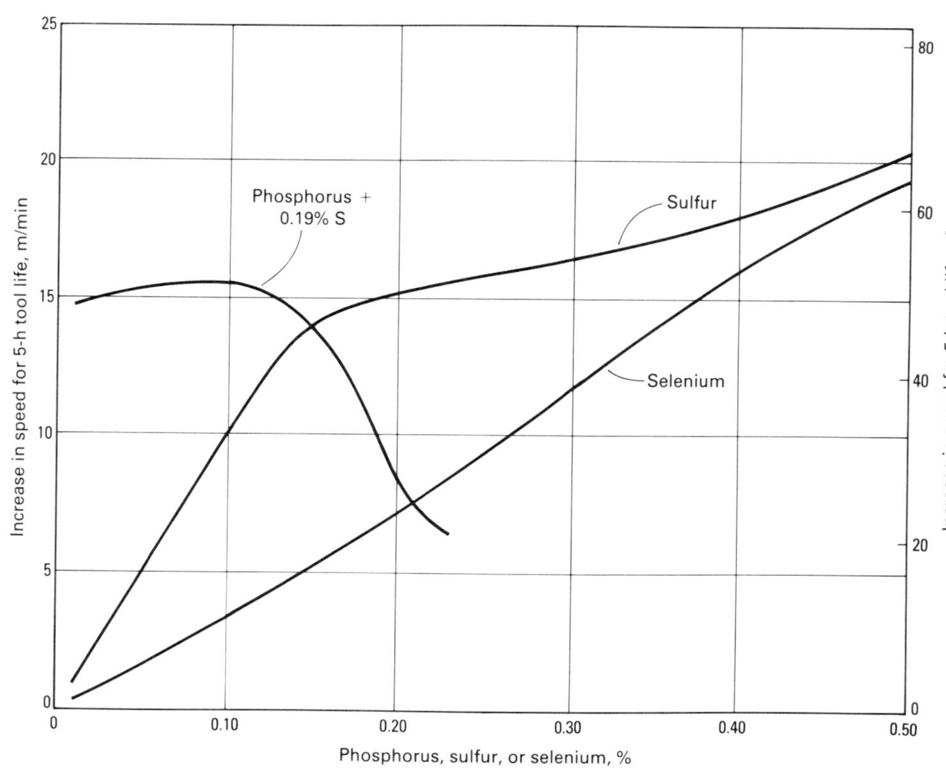


**Fig. 18** Effect of phosphorus on cutting speed (tool life) for a 0.19S-18Cr-8Ni austenitic stainless steel compared to a similar steel containing only sulfur or selenium. Source: Ref 23

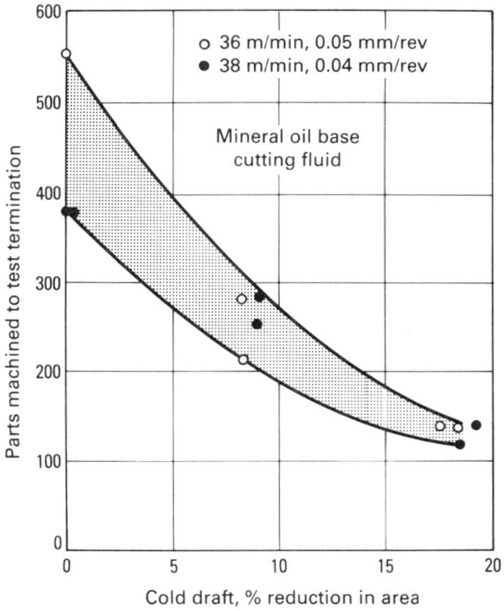


**Fig. 19** Effect of percent cold draft on machinability in a screw machine test for an enhanced-machining version of S30400. Termination is defined as a 0.075 mm (0.003 in.) increase in the diameter of the part being cut.

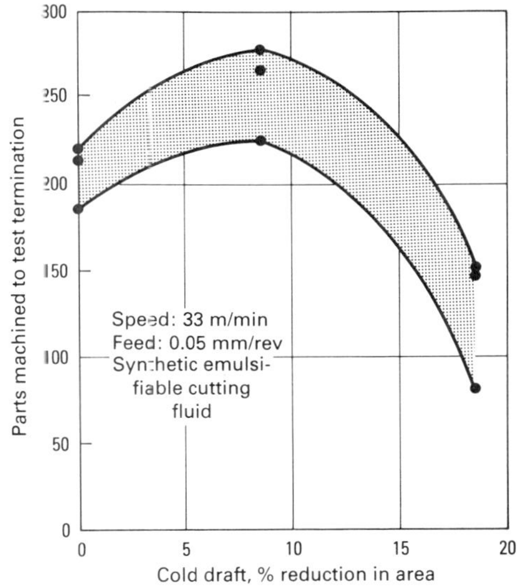


**Fig. 20** Effect of percent cold draft on machinability in a screw machine test for an enhanced-machining version of S30400. Termination is defined as a 0.075 mm (0.003 in.) increase in the diameter of the part being cut.

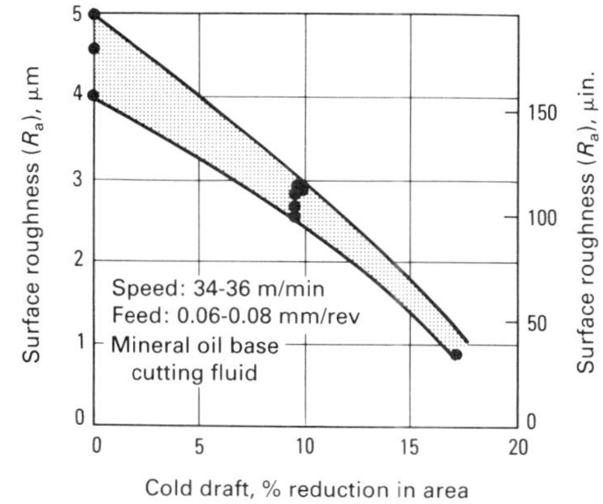


**Fig. 21** Effect of percent cold draft on machined surface finish in a screw machine test for an enhanced-machining version of S31600

ity as measured by various criteria (tool life, drillability, and so on) (Ref 2, 24, 40, 43). Within certain limits, however, surface finish can be improved by machining harder material (Ref 24, 40).

In the martensitic grades, machinability decreases as the carbon content increases from S41000 to S42000 to S44004 or from S41600/S41623 to S42020/S42023 to S44020/S44023. With higher carbon levels, there also tends to be a smaller difference in machinability between the corresponding free-machining and nonfree-machining versions. These effects are primarily due to the increasing quantities of abrasive chromium carbides present as carbon level increases in this series of alloys. As a further detriment to machinability, annealed hardness level increases with increasing carbon level (Ref 24, 40-42).

Nickel content also influences machinability by increasing annealed hardness levels. Consequently, alloys such as S41400 and S43100 will be more difficult to machine than S41000 in the annealed condition (Ref 38, 39).

Changing phase balance has been used to improve the machining characteristics of S41600. It has generally been found that increasing free or δ-ferrite content results in improved machinability, including tool life and surface finish (Ref 9, 24, 40, 43-45). The introduction of a higher ferrite content also results in decreasing hardness capability.

**Austenitic Alloys.** The difficulties in machining attributed to stainless steels in general are more specifically attributable to the austenitic stainless steels (Ref 27, 39-42, 45). Compared to ferritic and martensitic alloys, typical austenitic alloys have a higher work-hardening rate, a wider spread between yield and ultimate tensile strengths, and higher toughness and ductility. When machining austenitic stainless steels, particularly the nonfree-machining alloys, several factors become more pronounced:

- Tools will run hotter with more tendency to form a large built-up edge
- Chips will be stringier with a tendency to tangle, making their removal difficult
- Chatter will be more likely if tool rigidity is inadequate or marginal
- Cut surfaces will be work hardened and more difficult to machine if cutting is interrupted or if the feed rate is too low

Because of these factors, the general precautions for machining stainless steels are particularly important for austenitic alloys.

Although there have been differing opinions (Ref 23), a moderate amount of cold work has been regarded as beneficial to the overall machining characteristics of austenitic stainless steels (Ref 40, 42). The cold working will reduce the ductility of the material, which results in cutting with a cleaner chip and less tendency for a built-up edge. This produces a better machined surface finish but with some loss of tool life due to the higher hardness level (Ref 40).

Automatic screw machine testing has shown that the effects of cold working and hardness are variable and may or may not be seen, depending on the type of alloy and machining conditions. In such testing, tool life has been lowered by an increasing level of cold work for both free-machining (S30300) and nonfree-machining (S30400, S31600) austenitic stainless steels. This effect is shown in Fig. 19 for S30400. On the other hand, there have also been indications under different cutting conditions of an optimum level of tool life at an intermediate level of cold work (Fig. 20).

Machined surface finish can be improved by an increasing level of cold work for nonfree-machining alloys (S30400, S31600). Figure 21 shows this effect for S31600. A decreasing tendency for tool chatter with increasing cold work has also been seen for these alloys. On the other hand, the use of cold-drawn bar does not consistently benefit the machined surface finish of a free-machining alloy (S30300).

Additions of manganese or copper can increase the machinability (Fig. 22) and decrease the high work-hardening rate of the lower-alloy austenitic stainless steels (Ref 19, 45-48). Austenitic free-machining alloys that have additions of manganese and/or copper include S20300, S30310, S30330, and S30431. Although higher alloy content generally reduces work-hardening rate, it may not necessarily benefit machinability. Highly alloyed austenitic stainless steels, such as S30900, S31000, and N08020, tend to be more difficult to machine (Ref 38, 39).

Carbon and nitrogen can affect work-hardening rate and will increase the strength and hardness of austenitic stainless steels. Higher levels of either or both elements will decrease machinability (Fig. 23). Consequently, the high-nitrogen austenitic alloys, such as S20910 and S28200, are more difficult to machine than the standard lower-nitrogen austenitic alloys (Ref 38).

Strong carbide/nitride-forming elements, including titanium and niobium, are used in stainless steels such as S32100 and S34700 to prevent grain-boundary carbide, which can reduce intergranular corrosion resistance. However, the carbide/nitride inclusions are abrasive and will increase tool wear (Fig. 24).

**Duplex Alloys.** The machinability of duplex stainless steels is limited by their high annealed strength level. Figures 25 and 26 compare the machinability of a duplex alloy, S32950, with that of a high-nitrogen aus-

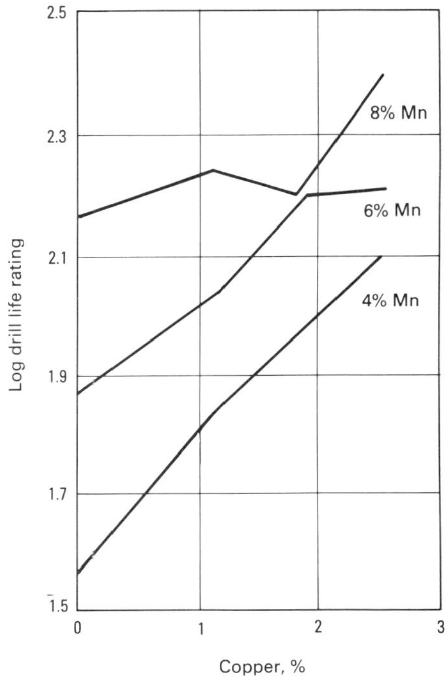


**Fig. 22** Effect of copper and manganese contents on machinability in a drill test for a free-machining chromium-manganese-nickel austenitic stainless steel. Source: Ref 19

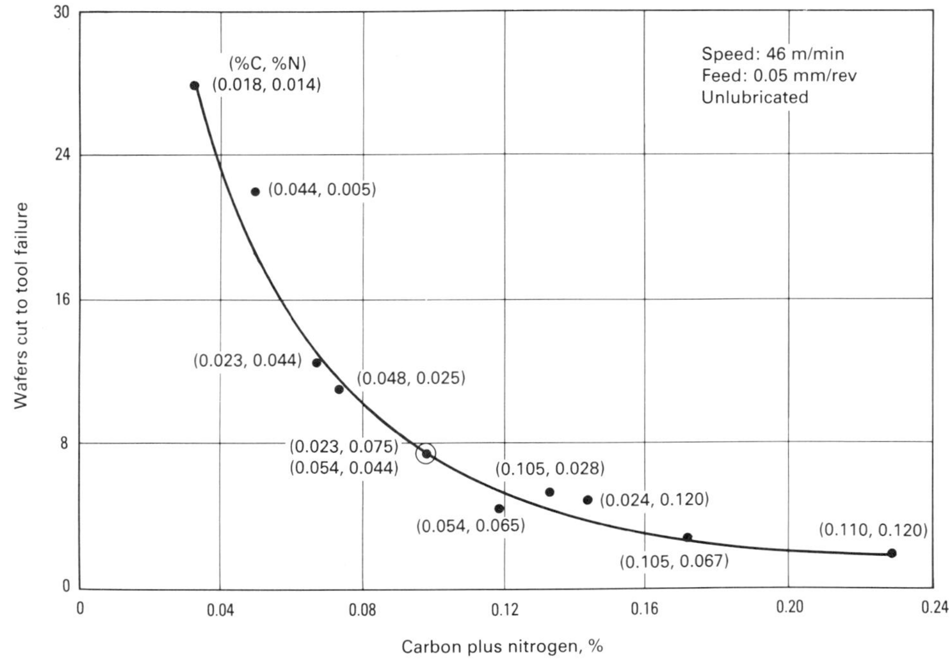


**Fig. 23** Effect of carbon and nitrogen contents on machinability in a tool life test for a free-machining 18Cr-9Ni-3Mn austenitic stainless steel. Source: Ref 49

tenitic alloy, S20910, and a conventional austenitic alloy, S31600, in standard (0.004% S) and enhanced-machining (0.027% S) versions. The duplex alloy (S32950) has a hardness level comparable to that of the high-nitrogen austenitic alloy (S20910), but provides better machinability. However, it does not machine as well as either the standard or the enhanced-machining S31600 alloy.

Other nitrogen-bearing duplex alloys are expected to machine similarly to S32950. No enhanced-machining versions of duplex alloys are available.

**Precipitation-Hardenable Alloys.** The machinability of precipitation-hardenable stainless steels depends on the type of alloy and its hardness level. Martensitic precipitation-hardenable stainless steels are often machined in the solution-treated condition; therefore, only a single aging treatment is required afterward to reach the desired strength level. In this condition, the relatively high hardness limits machinability. Most of these alloys machine comparably to, or somewhat worse than, a standard austenitic alloy such as S30400. Alloy S17400 is available in enhanced-machining versions that allow machining at higher speeds with a significantly reduced tendency toward chatter.

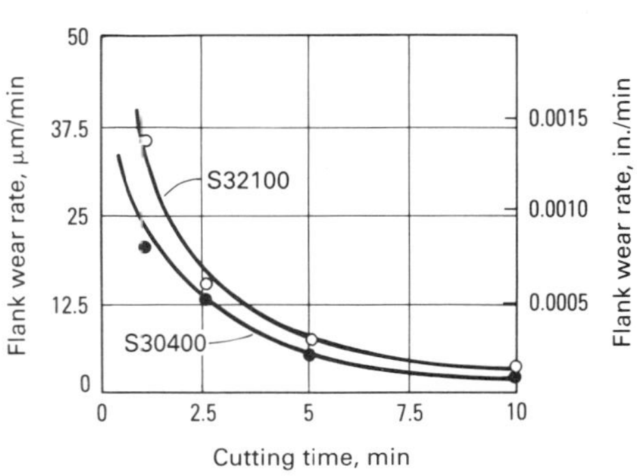


**Fig. 24** Comparison of tool wear for austenitic stainless steels with (S32100) and without (S30400) titanium carbide inclusions. Source: Ref 50

Martensitic precipitation-hardenable stainless steels can also be machined in an aged condition so that heat treating can be avoided and closer tolerances maintained. The ease of cutting generally varies with the hardness or heat-treated condition (Table 3).

In the annealed, austenitic condition, semiaustenitic alloys can be expected to machine with difficulty, somewhat worse than an alloy such as S30200, which has a high work-hardening rate. Alloys S35000 and S35500 can be supplied in an equalized and overtempered condition, which will provide the best machinability. As with the martensitic precipitation-hardenable alloys, machining difficulties increase with aged hardness level.

Austenitic precipitation-hardenable alloys, such as S66286, machine quite poorly, requiring slower cutting rates than even the highly alloyed austenitic stainless steels (Ref 38). Machining in an aged condition will require even slower speeds.

## General Guidelines

The characteristics of stainless steels that have a large influence on machinability include:

- Relatively high tensile strength
- High work-hardening rate, particularly for the austenitic alloys
- High ductility

These factors explain the tendency of the material to form a built-up edge on the tool during traditional machining operations. The chips removed in machining exert high pressures on the nose of the tool; these pressures, when combined with the high temperature at the chip/tool interface, cause pressure welding of portions of the chip to the tool. In addition, the low thermal conductivity of stainless steels contributes to a continuing heat buildup.

The difficulties involved in the traditional machining of stainless steels can be minimized by observing the following points:

- Because more power is generally required to machine stainless steels, equipment should be used only up to about 75% of the rating for carbon steels
- To avoid chatter, tooling and fixtures must be as rigid as possible. Overhang or protrusion of either the workpiece or the tool must be minimized. This applies to turning tools, drills, reamers, and so on
- To avoid glazed, work-hardened surfaces, particularly with austenitic alloys, a positive feed must be maintained. In some cases, increasing the feed and reducing the speed may be necessary. Dwelling, interrupted cuts, or a succession of thin cuts should be avoided
- Lower cutting speeds may be necessary, particularly for nonfree-machining austenitic alloys, precipitation-hardenable stainless steels, or higher-hardness mar-

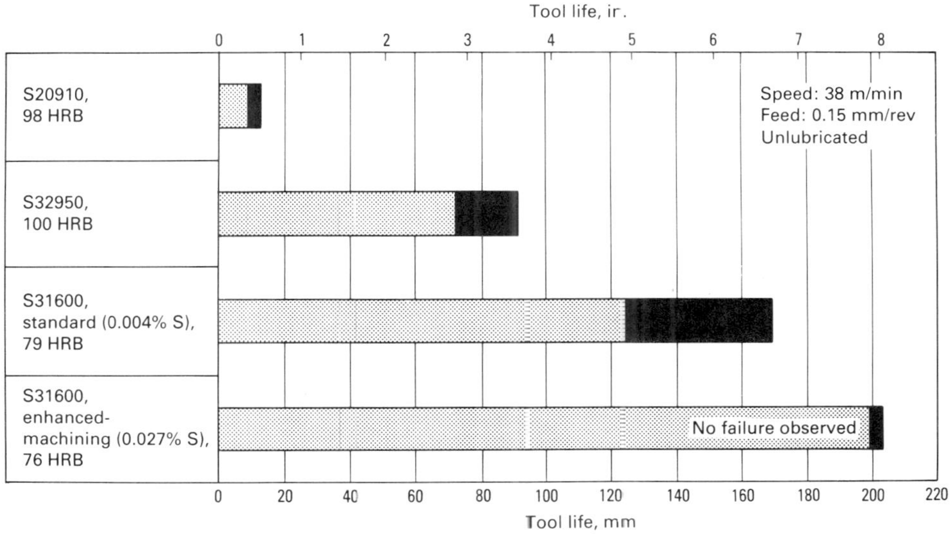


**Fig. 25** Comparison of tool life for a duplex stainless steel (S32950), a high-nitrogen austenitic stainless steel (S20910), and a lower-nitrogen austenitic stainless steel (S31600). Tool life is measured as the distance traveled along a 25 mm (1.0 in.) diam bar until tool failure. Shaded areas represent distance to failure.

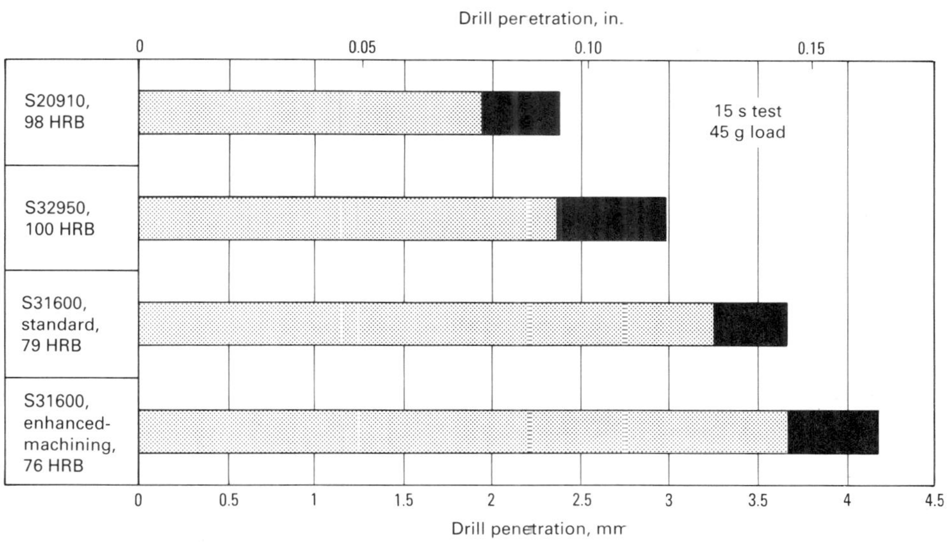


**Fig. 26** Comparison of machinability in a drill penetration test for a duplex stainless steel (S32950), a high-nitrogen austenitic stainless steel (S20910), and a lower-nitrogen austenitic stainless steel (S31600)

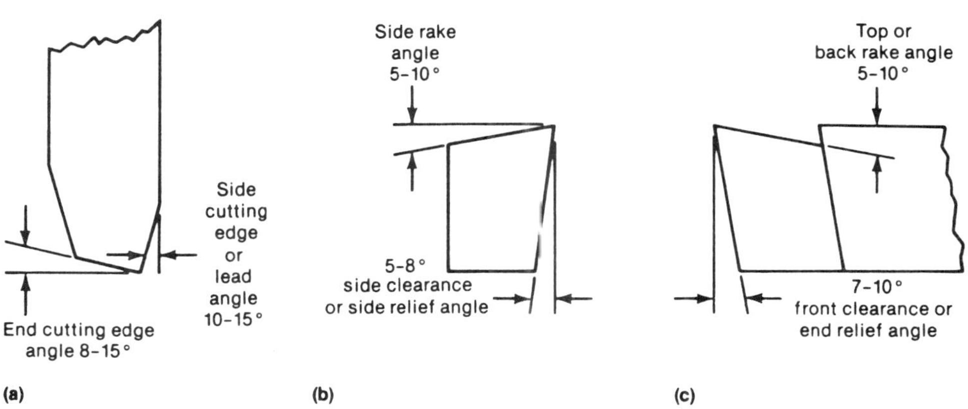


**Fig. 27** Suggested geometries for single-point turning tools used on stainless steels. (a) Plan view. (b) End view. (c) Side view

**Table 3 Relative machinability of S17400 in various heat-treated conditions**

| Condition | Typical hardness, HRC |
|---|---|
| **Improved machinability (higher cutting speed)** | |
| H1150M | 27 |
| H1150 | 33 |
| H1075 | 36 |
| A (solution treated) | 34 |
| H1025 | 38 |
| H900 | 44 |
| **Improved surface finish** | |

Source: Ref 51

tensitic alloys. Excessive cutting speeds result in tool wear or tool failure and shutdown for tool regrinding or replacement. Slower speeds with longer tool life are often the answer to higher output and lower costs

- Tools, both high-speed steel and carbides, must be kept sharp, with a fine finish to minimize friction with the chip. A sharp cutting edge produces the best surface finish and provides the longest tool life. To produce the best cutting edge on high-speed steel tools, 60-grit roughing should be followed by 120- and 150-grit finishing. Honing produces an even finer finish
- Cutting fluids must be selected or modified to provide proper lubrication and heat removal. Fluids must be carefully directed to the cutting area at a sufficient flow rate to prevent overheating

## Cutting Fluid

Machining stainless steel without a cutting fluid is usually restricted to low production quantities or to specific operating conditions in which use of cutting fluid is impractical, such as the machining of large workpieces or complex shapes. Complex shapes are sometimes machined dry to avoid entrapment of fluid that would impair subsequent processing or service.

The use of a cutting fluid is more desirable for machining stainless steels than for machining carbon or alloy steels for two reasons. First, stainless steels are generally less machinable than carbon or alloy steels, and second, the lower heat conductivity of stainless steels increases the need for cooling.

Water-soluble oils (usually in proportions of 1 part oil to 12 to 20 parts water) and sulfochlorinated cutting oils (having a viscosity of 300 SUS maximum at 38 °C, or 100 °F) are the cutting fluids most widely used in machining stainless steels.

Water-soluble oils are used for most single-point tool operations such as turning and boring. Soluble oils cost less than sulfochlorinated oils, are easy to supply to the cutting area in copious amounts, and func-

## Table 4 Nominal turning parameters for wrought stainless steels—single point and box tools

Depending on actual machining conditions, optimum parameters may be higher or lower than those given.

| | | | High-speed steel tools(c) | | | | | Carbide tools | | | | | | | | |
|---|---|---|---|---|---|---|---|---|---|---|---|---|---|---|---|---|
| | Depth of cut(b) | | Speed | | Feed | | Tool | Speed, brazed | | Speed, indexable | | Speed, coated | | Feed | | Tool |
| UNS designation(a) | mm | in. | m/min | sfm | mm/rev | in./rev | material | m/min | sfm | m/min | sfm | m/min | sfm | mm/rev | in./rev | material |
| **Free-machining alloys** | | | | | | | | | | | | | | | | |
| **Ferritic alloys** | | | | | | | | | | | | | | | | |
| S18235, S43020, S43023 | 3.8 | 0.150 | 50 | 165 | 0.38 | 0.015 | M2/M3 | 155 | 510 | 175 | 575 | 229 | 750 | 0.38 | 0.015 | C6,CC6/ |
| | 0.75 | 0.030 | 56 | 185 | 0.18 | 0.007 | | 175 | 575 | 198 | 650 | 259 | 850 | 0.18 | 0.007 | C7,CC7 |
| S18200 | 3.8 | 0.150 | 53 | 175 | 0.38 | 0.015 | M2/M3 | 160 | 525 | 183 | 600 | 236 | 775 | 0.38 | 0.015 | C6,CC6/ |
| | 0.75 | 0.030 | 61 | 200 | 0.18 | 0.007 | | 183 | 600 | 206 | 675 | 267 | 875 | 0.18 | 0.007 | C7,CC7 |
| **Martensitic alloys** | | | | | | | | | | | | | | | | |
| S41600, S41610, S41623 | 3.8 | 0.150 | 53 | 175 | 0.38 | 0.015 | M2/M3 | 160 | 525 | 183 | 600 | 236 | 775 | 0.38 | 0.015 | C6,CC6/ |
| | 0.75 | 0.030 | 61 | 200 | 0.18 | 0.007 | | 183 | 600 | 206 | 675 | 267 | 875 | 0.18 | 0.007 | C7,CC7 |
| S42020, S42023 | 3.8 | 0.150 | 32 | 105 | 0.38 | 0.015 | T15/M42 | 130 | 425 | 152 | 500 | 183 | 600 | 0.38 | 0.015 | C6,CC6/ |
| | 0.75 | 0.030 | 40 | 130 | 0.18 | 0.007 | | 152 | 500 | 175 | 575 | 206 | 675 | 0.18 | 0.007 | C7,CC7 |
| S44020, S44023 | 3.8 | 0.150 | 26 | 85 | 0.38 | 0.015 | T15/M42 | 114 | 375 | 130 | 425 | 160 | 525 | 0.38 | 0.015 | C6,CC6/ |
| | 0.75 | 0.030 | 30 | 100 | 0.18 | 0.007 | | 130 | 425 | 145 | 475 | 175 | 575 | 0.18 | 0.007 | C7,CC7 |
| **Austenitic alloys** | | | | | | | | | | | | | | | | |
| S20300, S30300, S30310, S30360 | 3.8 | 0.150 | 34 | 110 | 0.38 | 0.015 | M2/M3 | 130 | 425 | 145 | 475 | 175 | 575 | 0.38 | 0.015 | C2,CC2/ |
| | 0.75 | 0.030 | 40 | 130 | 0.18 | 0.007 | | 152 | 500 | 168 | 550 | 213 | 700 | 0.18 | 0.007 | C3,CC3 |
| S30323, S30330, S30345, S30431, S31620, S34720, S34723 | 3.8 | 0.150 | 27 | 90 | 0.38 | 0.015 | T15/M42 | 107 | 350 | 122 | 400 | 160 | 525 | 0.38 | 0.015 | C2,CC2/ |
| | 0.75 | 0.030 | 32 | 105 | 0.18 | 0.007 | | 122 | 400 | 137 | 450 | 183 | 600 | 0.18 | 0.007 | C3,CC3 |
| **Nonfree-machining alloys** | | | | | | | | | | | | | | | | |
| **Ferritic alloys** | | | | | | | | | | | | | | | | |
| S40500, S40900, S43000, S43400 | 3.8 | 0.150 | 30 | 100 | 0.38 | 0.015 | M2/M3 | 122 | 400 | 137 | 450 | 183 | 600 | 0.38 | 0.015 | C6,CC6/ |
| | 0.75 | 0.030 | 38 | 125 | 0.18 | 0.007 | | 152 | 500 | 168 | 550 | 229 | 750 | 0.18 | 0.007 | C7,CC7 |
| S44200, S44300, S44400 S44600 | 3.8 | 0.150 | 27 | 90 | 0.38 | 0.015 | T15/M42 | 107 | 350 | 122 | 400 | 168 | 550 | 0.38 | 0.015 | C2,CC2/ |
| | 0.75 | 0.030 | 34 | 110 | 0.18 | 0.007 | | 137 | 450 | 152 | 500 | 213 | 700 | 0.18 | 0.007 | C3,CC3 |
| **Martensitic alloys** | | | | | | | | | | | | | | | | |
| S40300, S41000 | 3.8 | 0.150 | 30 | 100 | 0.38 | 0.015 | M2/M3 | 122 | 400 | 137 | 450 | 183 | 600 | 0.38 | 0.015 | C6,CC6/ |
| | 0.75 | 0.030 | 38 | 125 | 0.18 | 0.007 | | 152 | 500 | 168 | 550 | 229 | 750 | 0.18 | 0.007 | C7,CC7 |
| S41400, S42000, S42010, S43100 | 3.8 | 0.150 | 26 | 85 | 0.38 | 0.015 | T15/M42 | 107 | 350 | 114 | 375 | 152 | 500 | 0.38 | 0.015 | C6,CC6/ |
| | 0.75 | 0.030 | 30 | 100 | 0.18 | 0.007 | | 114 | 375 | 137 | 450 | 183 | 600 | 0.18 | 0.007 | C7,CC7 |
| S44002, S44003, S44004 | 3.8 | 0.150 | 20 | 65 | 0.38 | 0.015 | T15/M42 | 84 | 275 | 91 | 300 | 114 | 375 | 0.38 | 0.015 | C6,CC6/ |
| | 0.75 | 0.030 | 24 | 80 | 0.18 | 0.007 | | 107 | 350 | 114 | 375 | 152 | 500 | 0.18 | 0.007 | C7,CC7 |
| **Austenitic alloys** | | | | | | | | | | | | | | | | |
| S20100, S30100, S30200, S30400, S30403, S30430, S30500, S31600, S31603, S32100, S34700, S38400 | 3.8 | 0.150 | 26 | 85 | 0.38 | 0.015 | T15/M42 | 99 | 325 | 107 | 350 | 137 | 450 | 0.38 | 0.015 | C2,CC2/ |
| | 0.75 | 0.030 | 30 | 100 | 0.18 | 0.007 | | 107 | 350 | 122 | 400 | 160 | 525 | 0.18 | 0.007 | C3,CC3 |
| S30900, S30908, S31000, S31008, S31700, S31703 | 3.8 | 0.150 | 23 | 75 | 0.38 | 0.015 | T15/M42 | 91 | 300 | 99 | 325 | 130 | 425 | 0.38 | 0.015 | C2,CC2/ |
| | 0.75 | 0.030 | 27 | 90 | 0.18 | 0.007 | | 99 | 325 | 114 | 375 | 152 | 500 | 0.18 | 0.007 | C3,CC3 |
| S20910, S21904, S24100, S28200, S30452, N08020 | 3.8 | 0.150 | 17 | 55 | 0.38 | 0.015 | T15/M42 | 53 | 175 | 61 | 200 | 76 | 250 | 0.38 | 0.015 | C2,CC2/ |
| | 0.75 | 0.030 | 21 | 70 | 0.18 | 0.007 | | 61 | 200 | 76 | 250 | 91 | 300 | 0.18 | 0.007 | C3,CC3 |
| **Duplex alloys** | | | | | | | | | | | | | | | | |
| S31803, S32550, S32900, S32950 | 3.8 | 0.150 | 23 | 75 | 0.38 | 0.015 | T15/M42 | 91 | 300 | 99 | 325 | 130 | 425 | 0.38 | 0.015 | C2,CC2/ |
| | 0.75 | 0.030 | 27 | 90 | 0.18 | 0.007 | | 99 | 325 | 114 | 375 | 152 | 500 | 0.18 | 0.007 | C3,CC3 |
| **Precipitation-hardenable alloys(a)** | | | | | | | | | | | | | | | | |
| S15500, S17400, S17700, S45000 | 3.8 | 0.150 | 24 | 80 | 0.38 | 0.015 | T15/M42 | 91 | 300 | 107 | 350 | 137 | 450 | 0.38 | 0.015 | C6,CC6/ |
| | 0.75 | 0.030 | 29 | 95 | 0.18 | 0.007 | | 107 | 350 | 122 | 400 | 160 | 525 | 0.18 | 0.007 | C7,CC7 |
| S13800, S35000, S35500, S45500 | 3.8 | 0.150 | 23 | 75 | 0.38 | 0.015 | T15/M42 | 84 | 275 | 99 | 325 | 130 | 425 | 0.38 | 0.015 | C6,CC6/ |
| | 0.75 | 0.030 | 27 | 90 | 0.18 | 0.007 | | 99 | 325 | 114 | 375 | 152 | 500 | 0.18 | 0.007 | C7,CC7 |
| S66286 | 3.8 | 0.150 | 15 | 50 | 0.38 | 0.015 | T15/M42 | 76 | 250 | 91 | 300 | 107 | 350 | 0.38 | 0.015 | C6,CC6/ |
| | 0.75 | 0.030 | 18 | 60 | 0.18 | 0.007 | | 91 | 300 | 107 | 350 | 122 | 400 | 0.18 | 0.007 | C7,CC7 |

(a) All machining parameters given are for material in the annealed condition (140–270 HB) except the precipitation-hardenable grades, which are solution treated (150–325 HB) or equalized and overtempered for S35000 and S35500. For material heat treated or cold drawn to hardnesses higher than their normal annealed (or softest) level, best results are obtained by reducing speeds by approximately 10% to as much as 60% at the higher hardnesses. Premium high-speed steel or carbide tooling may also be required to cut high-hardness material. (b) The 0.75 mm (0.030 in.) depth of cut represents conditions for a finishing cut, while the 3.8 mm (0.150 in.) depth of cut represents conditions for a roughing cut. (c) Any premium high-speed steel (T15, M33, M41–47) can be used where the designated tool material is T15, M42. Source: Ref 38, 52

tion as an acceptable means of cooling and flushing away chips. Soluble oils are also successfully used for milling and for machining in turret lathes and bar and chucking machines when several operations take place consecutively or simultaneously.

Sulfochlorinated or other additive-type cutting oils are infrequently used for the turning, boring, or milling of stainless steels, although there are operating conditions for which sulfochlorinated oils have proved to be the best choice. Examples of

### Table 5 Nominal turning parameters for wrought stainless steels—cutoff and form tools

Depending on actual machining conditions, the optimum parameters may be higher or lower than those given.

| UNS designation(a) | Speed, m/min (sfm) | Feed, mm/rev (in./rev), for tool width of: 1.6 mm (1/16 in.) | 3.2 mm (1/8 in.) | 6.4 mm (1/4 in.) | 13 mm (1/2 in.) | 25 mm (1 in.) | 38 mm (1½ in.) | 50 mm (2 in.) | Tool material(a)(b)(c) |
|---|---|---|---|---|---|---|---|---|---|
| **Free-machining alloys** | | | | | | | | | |
| **Ferritic alloys** | | | | | | | | | |
| S18235, S43020, S43023 | 46 (150) | 0.038 (0.0015) | 0.05 (0.002) | 0.064 (0.0025) | 0.064 (0.0025) | 0.05 (0.002) | 0.038 (0.0015) | 0.025 (0.001) | M2, M3 |
| | 122 (400) | 0.038 (0.0015) | 0.05 (0.002) | 0.064 (0.0025) | 0.064 (0.0025) | 0.05 (0.002) | 0.038 (0.0015) | 0.025 (0.001) | C6 |
| S18200 | 50 (165) | 0.038 (0.0015) | 0.05 (0.002) | 0.064 (0.0025) | 0.064 (0.0025) | 0.05 (0.002) | 0.038 (0.0015) | 0.025 (0.001) | M2,M3 |
| | 130 (425) | 0.038 (0.0015) | 0.05 (0.002) | 0.064 (0.0025) | 0.064 (0.0025) | 0.05 (0.002) | 0.038 (0.0015) | 0.025 (0.001) | C6 |
| **Martensitic alloys** | | | | | | | | | |
| S41600, S41610, S41623 | 46 (150) | 0.05 (0.002) | 0.064 (0.0025) | 0.075 (0.003) | 0.064 (0.0025) | 0.05 (0.002) | 0.038 (0.0015) | 0.025 (0.001) | M2,M3 |
| | 122 (400) | 0.05 (0.002) | 0.064 (0.0025) | 0.075 (0.003) | 0.064 (0.0025) | 0.05 (0.002) | 0.038 (0.0015) | 0.025 (0.001) | C6 |
| S42020, S42023 | 30 (100) | 0.038 (0.0015) | 0.05 (0.002) | 0.064 (0.0025) | 0.05 (0.002) | 0.038 (0.0015) | 0.025 (0.001) | 0.025 (0.001) | M2,M3 |
| | 99 (325) | 0.038 (0.0015) | 0.05 (0.002) | 0.064 (0.0025) | 0.05 (0.002) | 0.038 (0.0015) | 0.025 (0.001) | 0.025 (0.001) | C6 |
| S44020, S44023 | 23 (75) | 0.025 (0.001) | 0.038 (0.0015) | 0.051 (0.002) | 0.05 (0.002) | 0.038 (0.0015) | 0.025 (0.001) | 0.025 (0.001) | M2,M3 |
| | 76 (250) | 0.025 (0.001) | 0.038 (0.0015) | 0.051 (0.002) | 0.05 (0.002) | 0.038 (0.0015) | 0.025 (0.001) | 0.025 (0.001) | C6 |
| **Austenitic alloys** | | | | | | | | | |
| S20300, S30300, S30310, S30360 | 30 (100) | 0.038 (0.0015) | 0.05 (0.002) | 0.064 (0.0025) | 0.05 (0.002) | 0.038 (0.0015) | 0.025 (0.001) | 0.025 (0.001) | M2,M3 |
| | 99 (325) | 0.038 (0.0015) | 0.05 (0.002) | 0.064 (0.0025) | 0.05 (0.002) | 0.038 (0.0015) | 0.025 (0.001) | 0.025 (0.001) | C6 |
| S30323, S30330, S30345, S30431, S31620, S34720, S34723 | 24 (80) | 0.038 (0.0015) | 0.05 (0.002) | 0.05 (0.002) | 0.05 (0.002) | 0.038 (0.0015) | 0.025 (0.001) | 0.025 (0.001) | M2,M3 |
| | 91 (300) | 0.038 (0.0015) | 0.05 (0.002) | 0.05 (0.002) | 0.05 (0.002) | 0.038 (0.0015) | 0.025 (0.001) | 0.025 (0.001) | C2 |
| **Nonfree-machining alloys** | | | | | | | | | |
| **Ferritic alloys** | | | | | | | | | |
| S40500, S40900, S43000, S43400 | 27 (90) | 0.025 (0.001) | 0.038 (0.0015) | 0.05 (0.002) | 0.05 (0.002) | 0.038 (0.0015) | 0.038 (0.0015) | 0.025 (0.001) | M2,M3 |
| | 99 (325) | 0.025 (0.001) | 0.038 (0.0015) | 0.05 (0.002) | 0.05 (0.002) | 0.038 (0.0015) | 0.038 (0.0015) | 0.025 (0.001) | C6 |
| S44200, S44300, S44400, S44600 | 24 (80) | 0.025 (0.001) | 0.038 (0.0015) | 0.05 (0.002) | 0.05 (0.002) | 0.038 (0.0015) | 0.038 (0.0015) | 0.025 (0.001) | M2,M3 |
| | 91 (300) | 0.025 (0.001) | 0.038 (0.0015) | 0.05 (0.002) | 0.05 (0.002) | 0.038 (0.0015) | 0.038 (0.0015) | 0.025 (0.001) | C6 |
| **Martensitic alloys** | | | | | | | | | |
| S40300, S41000 | 27 (90) | 0.038 (0.0015) | 0.05 (0.002) | 0.064 (0.0025) | 0.05 (0.002) | 0.05 (0.002) | 0.038 (0.0015) | 0.025 (0.001) | M2,M3 |
| | 99 (325) | 0.038 (0.0015) | 0.05 (0.002) | 0.064 (0.0025) | 0.05 (0.002) | 0.05 (0.002) | 0.038 (0.0015) | 0.025 (0.001) | C6 |
| S41400, S42000, S42010, S43100 | 23 (75) | 0.025 (0.001) | 0.038 (0.0015) | 0.05 (0.002) | 0.05 (0.002) | 0.038 (0.0015) | 0.038 (0.0015) | 0.025 (0.001) | M2,M3 |
| | 84 (275) | 0.025 (0.001) | 0.038 (0.0015) | 0.05 (0.002) | 0.05 (0.002) | 0.038 (0.0015) | 0.038 (0.0015) | 0.025 (0.001) | C6 |
| S44002, S44003, S44004 | 15 (50) | 0.025 (0.001) | 0.038 (0.0015) | 0.05 (0.002) | 0.05 (0.002) | 0.038 (0.0015) | 0.038 (0.0015) | 0.025 (0.001) | M2,M3 |
| | 61 (200) | 0.025 (0.001) | 0.038 (0.0015) | 0.05 (0.002) | 0.05 (0.002) | 0.038 (0.0015) | 0.038 (0.0015) | 0.025 (0.001) | C6 |
| **Austenitic alloys** | | | | | | | | | |
| S20100, S30100, S30200, S30400, S30403, S30430, S30500, S31600, S31603, S32100, S34700, S38400 | 23 (75) | 0.038 (0.0015) | 0.038 (0.0015) | 0.05 (0.002) | 0.05 (0.002) | 0.038 (0.0015) | 0.025 (0.001) | 0.025 (0.001) | M2,M3 |
| | 84 (275) | 0.038 (0.0015) | 0.038 (0.0015) | 0.05 (0.002) | 0.05 (0.002) | 0.038 (0.0015) | 0.025 (0.001) | 0.025 (0.001) | C2 |
| S30900, S30908, S31000, S31008, S31700, S31703 | 21 (70) | 0.038 (0.0015) | 0.038 (0.0015) | 0.05 (0.002) | 0.05 (0.002) | 0.038 (0.0015) | 0.038 (0.0015) | 0.025 (0.001) | M2,M3 |
| | 76 (250) | 0.038 (0.0015) | 0.038 (0.0015) | 0.05 (0.002) | 0.05 (0.002) | 0.038 (0.0015) | 0.038 (0.0015) | 0.025 (0.001) | C2 |
| S20910, S21904, S24100, S28200, S30452, N08020 | 12 (40) | 0.025 (0.001) | 0.038 (0.0015) | 0.038 (0.0015) | 0.038 (0.0015) | 0.038 (0.0015) | 0.025 (0.001) | 0.025 (0.001) | M42,T15 |
| | 46 (150) | 0.025 (0.001) | 0.038 (0.0015) | 0.038 (0.0015) | 0.038 (0.0015) | 0.038 (0.0015) | 0.025 (0.001) | 0.025 (0.001) | C2 |
| **Duplex alloys** | | | | | | | | | |
| S31803, S32550, S32900, S32950 | 21 (70) | 0.038 (0.0015) | 0.038 (0.0015) | 0.05 (0.002) | 0.05 (0.002) | 0.038 (0.0015) | 0.038 (0.0015) | 0.025 (0.001) | M2,M3 |
| | 76 (250) | 0.038 (0.0015) | 0.038 (0.0015) | 0.05 (0.002) | 0.05 (0.002) | 0.038 (0.0015) | 0.038 (0.0015) | 0.025 (0.001) | C2 |
| **Precipitation-hardenable alloys(a)** | | | | | | | | | |
| S15500, S17400, S17700, S45000 | 20 (65) | 0.025 (0.001) | 0.038 (0.0015) | 0.05 (0.002) | 0.038 (0.0015) | 0.025 (0.001) | 0.025 (0.001) | 0.013 (0.0005) | M42,T15 |
| | 61 (200) | 0.025 (0.001) | 0.038 (0.0015) | 0.05 (0.002) | 0.038 (0.0015) | 0.025 (0.001) | 0.025 (0.001) | 0.013 (0.0005) | C2 |
| S13800, S35000, S35500, S45500 | 15 (50) | 0.025 (0.001) | 0.038 (0.0015) | 0.05 (0.002) | 0.038 (0.0015) | 0.025 (0.001) | 0.025 (0.001) | 0.013 (0.0005) | M42,T15 |
| | 53 (175) | 0.025 (0.001) | 0.038 (0.0015) | 0.05 (0.002) | 0.038 (0.0015) | 0.025 (0.001) | 0.025 (0.001) | 0.013 (0.0005) | C2 |
| S66286 | 14 (45) | 0.025 (0.001) | 0.038 (0.0015) | 0.05 (0.002) | 0.038 (0.0015) | 0.025 (0.001) | 0.025 (0.001) | 0.013 (0.0005) | M42,T15 |
| | 49 (160) | 0.025 (0.001) | 0.038 (0.0015) | 0.05 (0.002) | 0.038 (0.0015) | 0.025 (0.001) | 0.025 (0.001) | 0.013 (0.0005) | C2 |

(a) All machining parameters given are for material in the annealed condition (140–270 HB) except the precipitation-hardenable grades, which are solution treated (150–325 HB) or equalized and overtempered for S35000 and S35500. For material heat treated or cold drawn to hardnesses higher than their normal annealed or softest level, best results are obtained by reducing speeds by approximately 10% to as much as 60% at the higher hardnesses. Premium high-speed steel or carbide tooling may also be required to cut high-hardness material. (b) Any premium high-speed steel (T15, M33, M41–47) can be used where the designated tool material is T15, M42. (c) Tool widths of 1.6–13 mm (1/16–1/2 in.) are for cutoff tools and 25–50 mm (1–2 in.) for form tools. Source: Ref 38, 52

such conditions are excessively hard work metal, the use of high-speed steel tools (because of interrupted cutting), and critical surface finish requirements.

Stainless steels are drilled with twist drills using soluble oil, although as the depth of hole increases, the frequency of using sulfochlorinated or other additive oils increases. Soluble oil is typically used for the reaming of stainless steels. The gun drilling or other deep-hole drilling of stainless steels is generally done with sulfochlorinated or other additive-type oils. When additive oils are used, viscosity should be no higher than 300 SUS at 38 °C (100 °F) in order to permit the necessary flow.

For the broaching of stainless steels, best results are obtained with additive-type cut-

## Table 6 Nominal drilling parameters for wrought stainless steels

Depending on actual machining conditions, the optimum parameters may be higher or lower than those given.

| UNS designation | Speed(a), m/min (sfm) | Feed, mm/rev (in./rev), at a nominal hole diameter of: 1.6 mm (1/16 in.) | 3.2 mm (1/8 in.) | 6.4 mm (1/4 in.) | 13 mm (1/2 in.) | 25 mm (1 in.) | 38 mm (1 1/2 in.) | 50 mm (2 in.) | Tool material(a) |
|---|---|---|---|---|---|---|---|---|---|
| **Free-machining alloys** | | | | | | | | | |
| **Ferritic alloys** | | | | | | | | | |
| S18235, S43020, S43023 | 46 (150) | 0.025 (0.001) | 0.075 (0.003) | 0.15 (0.006) | 0.25 (0.010) | 0.43 (0.017) | 0.53 (0.021) | 0.635 (0.025) | M1,M7,M10 |
| S18200 | 49 (160) | 0.025 (0.001) | 0.075 (0.003) | 0.15 (0.006) | 0.25 (0.010) | 0.43 (0.017) | 0.53 (0.021) | 0.635 (0.025) | M1,M7,M10 |
| **Martensitic alloys** | | | | | | | | | |
| S41600, S41610, S41623 | 40 (130) | 0.025 (0.001) | 0.075 (0.003) | 0.15 (0.006) | 0.25 (0.010) | 0.43 (0.017) | 0.53 (0.021) | 0.635 (0.025) | M1,M7,M10 |
| S42020, S42023 | 26 (85) | 0.025 (0.001) | 0.075 (0.003) | 0.15 (0.006) | 0.25 (0.010) | 0.43 (0.017) | 0.53 (0.021) | 0.635 (0.025) | M1,M7,M10 |
| S44020, S44023 | 18 (60) | 0.025 (0.001) | 0.075 (0.003) | 0.15 (0.006) | 0.25 (0.010) | 0.43 (0.017) | 0.53 (0.021) | 0.635 (0.025) | M1,M7,M10 |
| **Austenitic alloys** | | | | | | | | | |
| S20300, S30300, S30310 S30360 | 30 (100) | 0.025 (0.001) | 0.075 (0.003) | 0.15 (0.006) | 0.25 (0.010) | 0.43 (0.017) | 0.53 (0.021) | 0.635 (0.025) | M1,M7,M10 |
| S30323, S30330, S30345, S30431, S31620, 34720, S34723 | 23 (75) | 0.025 (0.001) | 0.05 (0.002) | 0.10 (0.004) | 0.175 (0.007) | 0.30 (0.012) | 0.38 (0.015) | 0.46 (0.018) | M1,M7,M10 |
| **Nonfree-machining alloys** | | | | | | | | | |
| **Ferritic alloys** | | | | | | | | | |
| S40500, S40900, S43000 S43400 | 18–21 (60–70) | 0.025 (0.001) | 0.05 (0.002) | 0.10 (0.004) | 0.175 (0.007) | 0.30 (0.012) | 0.38 (0.015) | 0.46 (0.018) | M1,M7,M10 |
| S44200, S44300, S44400, S44600 | 15–20 (50–65) | 0.025 (0.001) | 0.05 (0.002) | 0.10 (0.004) | 0.175 (0.007) | 0.30 (0.012) | 0.38 (0.015) | 0.46 (0.018) | M1,M7,M10 |
| **Martensitic alloys** | | | | | | | | | |
| S40300, S41000 | 18–21 (60–70) | 0.025 (0.001) | 0.075 (0.003) | 0.15 (0.006) | 0.25 (0.010) | 0.41 (0.016) | 0.53 (0.021) | 0.635 (0.025) | M1,M7,M10 |
| S41400, S42000, S42010, S43100 | 15–18 (50–60) | 0.025 (0.001) | 0.075 (0.003) | 0.15 (0.006) | 0.25 (0.010) | 0.41 (0.016) | 0.53 (0.021) | 0.635 (0.025) | M1,M7,M10 |
| S44002, S44003, S44004 | 12–15 (40–50) | 0.025 (0.001) | 0.075 (0.003) | 0.15 (0.006) | 0.25 (0.010) | 0.41 (0.016) | 0.53 (0.021) | 0.635 (0.025) | M1,M7,M10 |
| **Austenitic alloys** | | | | | | | | | |
| S20100, S30100, S30200, S30400, S30403, S30430, S30500, S31600, S31603, S32100, S34700, S38400 | 15–18 (50–60) | 0.025 (0.001) | 0.05 (0.002) | 0.10 (0.004) | 0.175 (0.007) | 0.30 (0.012) | 0.38 (0.015) | 0.46 (0.018) | M1,M7,M10 |
| S30900, S30908, S31000, S31008, S31700, S31703 | 12–15 (40–50) | 0.025 (0.001) | 0.05 (0.002) | 0.10 (0.004) | 0.175 (0.007) | 0.30 (0.012) | 0.38 (0.015) | 0.46 (0.018) | M1,M7,M10 |
| S20910, S21904, S24100, S28200, S30452, N08020 | 9–14 (30–45) | 0.025 (0.001) | 0.05 (0.002) | 0.10 (0.004) | 0.175 (0.007) | 0.30 (0.012) | 0.38 (0.015) | 0.46 (0.018) | M42,T15(b) |
| **Duplex alloys** | | | | | | | | | |
| S31803, S32550, S32900, S32950 | 12–15 (40–50) | 0.025 (0.001) | 0.05 (0.002) | 0.10 (0.004) | 0.175 (0.007) | 0.30 (0.012) | 0.38 (0.015) | 0.46 (0.018) | M1,M7,M10 |
| **Precipitation-hardenable alloys** | | | | | | | | | |
| S15500, S17400, S17700, S45000 | 15–18 (50–60) | 0.025 (0.001) | 0.05 (0.002) | 0.10 (0.004) | 0.175 (0.007) | 0.25 (0.010) | 0.30 (0.012) | 0.38 (0.015) | M42,T15(b) |
| S13800, S35000, S35500, S45500 | 12–15 (40–50) | 0.025 (0.001) | 0.05 (0.002) | 0.10 (0.004) | 0.175 (0.007) | 0.25 (0.010) | 0.30 (0.012) | 0.38 (0.015) | M42,T15(b) |
| S66286 | 8–12 (25–40) | 0.025 (0.001) | 0.05 (0.002) | 0.10 (0.004) | 0.175 (0.007) | 0.25 (0.010) | 0.30 (0.012) | 0.38 (0.015) | M42,T15(b) |

(a) All speeds given are for material in the annealed condition (140–270 HB) except for the precipitation-hardenable grades, which are solution treated (150–325 HB) or equalized and overtempered for S35000 and S35500. For material heat treated or cold drawn to hardnesses higher than their normal annealed or softest level, best results are obtained by reducing speeds by approximately 25% to as much as 75% depending on hardness. Premium high-speed steel or carbide drills may also be necessary at the higher hardnesses. (b) Any premium high-speed steel (T15, M33, M41–47) can be used where the designated tool material is T15, M42. Source: Ref 38, 52

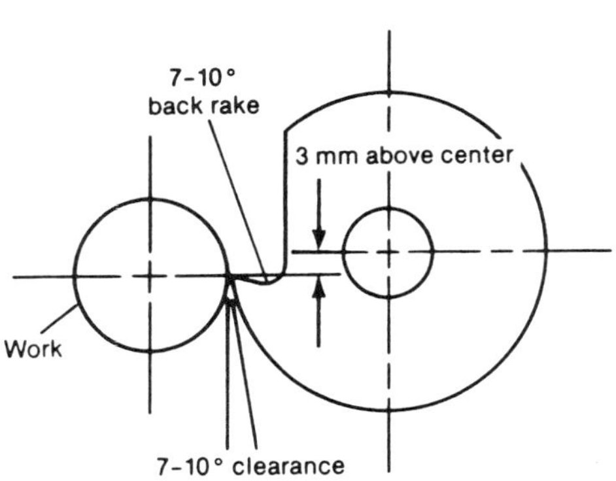


**Fig. 28** Suggested geometries for circular cutoff tools used on stainless steels

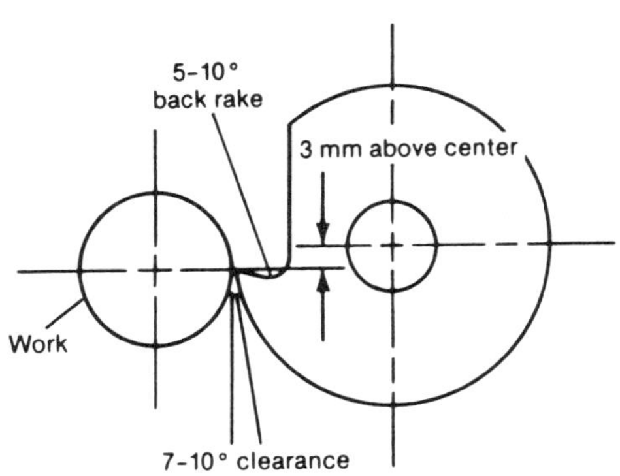


**Fig. 29** Suggested geometries for circular form tools used on stainless steels

ting oils. The same applies to tapping and to die threading. Soluble oil is used for sawing stainless steels.

**Sulfochlorinated Oils.** In the machining of stainless steels, two types of sulfurized oil are recommended:

- A sulfochlorinated petroleum oil containing active sulfur and approximately 8 to 10% fatty oil. Viscosity is approximately 200 SUS at 38 °C (100 °F)
- A sulfochlorinated petroleum oil containing active sulfur without a fatty-base oil. Viscosity is approximately 130 SUS at 38 °C (100 °F)

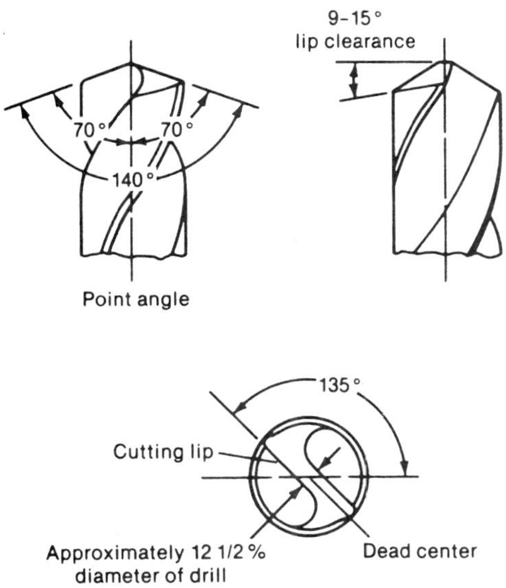


**Fig. 30** Suggested geometries for drills used on stainless steels

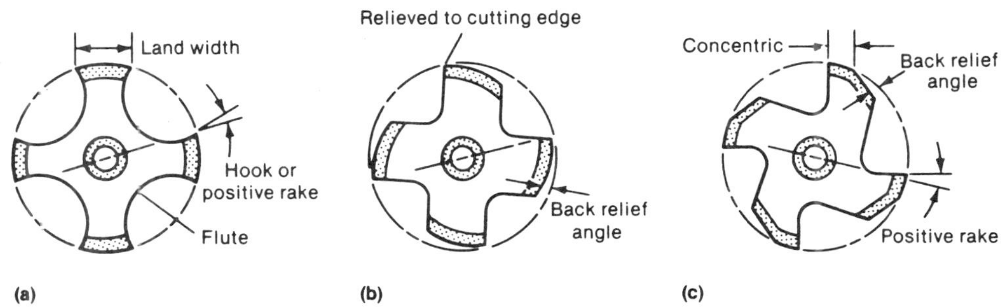


**Fig. 31** Relief of lands on taps. (a) Lands concentric with hole. (b) Lands fully eccentric with hole, with full relief. (c) Lands partly concentric with hole, but with final relief

The first oil is generally used for the non-free-machining types of stainless steels, while the second oil is usually preferred for the free-machining alloys.

Either of these oils, depending on the work being done, can be used straight or thinned with a blending oil, the best of which is paraffin oil. Normally, a 1:1 mixture of sulfurized oil with paraffin oil is used initially. If excessive tool wear occurs, more paraffin blending oil should be added. If the chips weld to the tool or if the tool burns, there may be too much paraffin blending oil in the mixture. In that case, an addition of more sulfurized oil is indicated.

The following are suggested starting points for mixtures of sulfurized oils and paraffin oil for various types of machining operations:

- When using high-speed, light-feed automatic screw machines, principally for turning, forming, and cutoff operations with the free-machining alloys, one part sulfurized oil to five parts paraffin oil is suggested
- When using screw machines at normal or average speed or performing operations that include threading, tapping, drilling, and milling of the free-machining alloys, one part sulfurized oil to three parts paraffin oil is suggested
- For coarse bolt threading, nut tapping, pipe threading, and broaching, one part sulfurized oil to two parts paraffin oil is suggested
- In general, when machining the austenitic stainless steels (except the free-machining type), a mixture of one to one of the first oil type (8 to 10% fatty oil) and paraffin oil is suggested. Care must be exercised to ensure that the viscosity of the cutting fluid is sufficiently low to allow it to reach the point of cutting. Usually 120 SUS at 20 °C (70 °F) is satisfactory
- A good criterion for starting new jobs is to remember that the more difficult the steel is to machine, the more sulfurized oil is needed. For the free-machining steels, the mixture can contain a larger percentage of paraffin-base oil

**Emulsifiable Fluids.** Water-emulsifiable (water-soluble) cutting fluids are applicable to machining stainless steels, particularly in situations that require greater cooling capability. However, the water-base fluids are unsuitable in those cases in which the cutting fluid

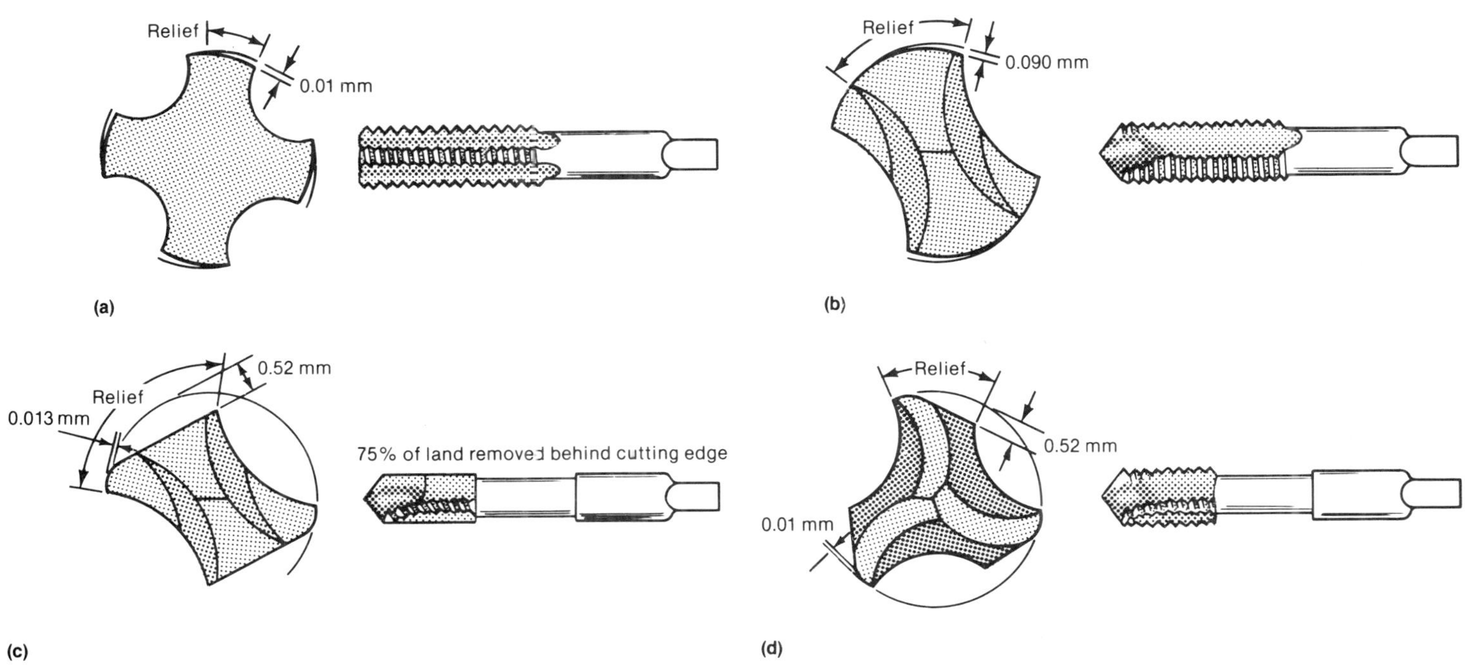


**Fig. 32** Progressive modification of tap design for tapping ¼-32 threads in stainless steel valve bonnets. (a) Original design: four-flute plug tap with conventional relief. Life: 2 holes. (b) Changed to two-flute tap with spiral point and 70% land relief. Life: 8 holes. (c) Two-flute tap improved by full relief and removal of all but 17 threads. Life: 264 holes. (d) Final design: three-flute tap with full relief. Life: 3000 holes. Tap shown in (a) underwent five revisions before final design (d), which produced a 1500-fold increase in tool life in tapping 6 mm (¼ in.) diam holes in valve bonnets of soft austenitic stainless steels (S30400, S31600, and N08330). Two intermediate redesigns are shown in (b) and (c). Taps of all designs were of M2 high-speed steel and had standard lead and hook angles.

### Table 7 Nominal tapping parameters for wrought stainless steels

Depending on actual machining conditions, optimum parameters may be higher or lower than those given.

| UNS designation | Speed(a), m/min (sfm), at a pitch (threads per 25 mm, or 1 in.) of: ≤7 | 8–15 | 16–24 | >24 | Tool material(b) |
|---|---|---|---|---|---|
| **Free-machining alloys** | | | | | |
| **Ferritic alloys** | | | | | |
| S43020, S18235, S43023 | 6 (20) | 9 (30) | 14 (45) | 15 (50) | M1,M7,M10 |
| S18200 | 6 (20) | 9 (30) | 14 (45) | 15 (50) | M1,M7,M10 |
| **Martensitic alloys** | | | | | |
| S41600, S41610, S41623 | 6 (20) | 9 (30) | 14 (45) | 15 (50) | M1,M7,M10 |
| S42020, S42023 | 5 (15) | 8 (25) | 11 (35) | 12 (40) | M1,M7,M10 |
| S44020, S44023 | 4 (12) | 5.5 (18) | 6.7 (22) | 8 (25) | M1,M7,M10 |
| **Austenitic alloys** | | | | | |
| S20300, S30300, S30310, S30360 | 5 (15) | 8 (25) | 11 (35) | 12 (40) | M1,M7,M10 |
| S30323, S30330, S30345, S30431, S31620, S34720, S34723 | 4 (12) | 6 (20) | 9 (30) | 11 (35) | M1,M7,M10 |
| **Nonfree-machining alloys** | | | | | |
| **Ferritic alloys** | | | | | |
| S40500, S40900, S43000, S43400 | 5 (15) | 8 (25) | 11 (35) | 12 (40) | M1,M7,M10 |
| S44200, S44300, S44400, S44600 | 4 (12) | 6 (20) | 9 (30) | 11 (35) | M1,M7,M10 |
| **Martensitic alloys** | | | | | |
| S40300, S41000 | 5 (15) | 8 (25) | 11 (35) | 12 (40) | M1,M7,M10 |
| S41400, S42000, S42010, S43100 | 4 (12) | 5.5 (18) | 8 (25) | 9 (30) | M1,M7,M10 |
| S44002, S44003, S44004 | 3 (10) | 4 (12) | 5 (16) | 6 (20) | Nitrided M1,M7,M10 |
| **Austenitic alloys** | | | | | |
| S20100, S30100, S30200, S30400, S30403, S30430, S30500, S31600, S31603, S32100, S34700, S38400 | 4 (12) | 5.5 (18) | 6.7 (22) | 8 (25) | M1,M7,M10 |
| S30900, S30908, S31000, S31008, S31700, S31703 | 4 (12) | 4.5 (15) | 6 (20) | 7 (23) | M1,M7,M10 |
| S20910, S21904, S24100, S28200, S30452, N08020 | 3 (10) | 4 (12) | 5 (16) | 6 (20) | Nitrided M1,M7,M10 |
| **Duplex alloys** | | | | | |
| S31803, S32550, S32900, S32950 | 4 (12) | 4.5 (15) | 6 (20) | 7 (23) | M1,M7,M10 |
| **Precipitation-hardenable alloys** | | | | | |
| S15500, S17400, S17700, S45000 | 4 (12) | 5.5 (18) | 6.7 (22) | 8 (25) | M1,M7,M10 |
| S13800, S35000, S35500, S45500 | 3 (10) | 4.5 (15) | 6 (20) | 6.7 (22) | M1,M7,M10 |
| S66286 | 2 (8) | 4 (12) | 5.5 (18) | 6 (20) | Nitrided M1,M7,M10 |

(a) All speeds given are for tapping 65–75% threads in shallow through holes. Speeds should be reduced when tapping blind holes, deep holes, or higher percentage of thread. All speeds are for material in the annealed condition (140–270 HB) except the precipitation-hardenable grades, which are solution treated (150–325 HB) or equalized and overtempered for S35000 and S35500. For material heat treated or cold drawn to hardnesses higher than the annealed or softest level, best results are obtained by reducing speeds by approximately 25% to as much as 60% depending on hardness. (b) Nitrided taps may be necessary for some grades at the higher hardness levels. Source: Ref 38, 52

and machine lubrication oils have the potential of mixing. Many of the water-emulsifiable fluids will not withstand some of the more severe cutting operations. Nevertheless, water-base fluids are successfully used in many machining operations, as discussed earlier.

The cutting fluid should contain polar and extreme-pressure (EP) additives, and favorable results have been obtained with some of the synthetic emulsifying systems. Compared to the cutting oils, the water-emulsifiable fluids can result in better machined surface finishes, less heat-affected zone in the tool, and therefore less regrinding and increased productivity for some grades of stainless steel.

## Turning

Turning operations on automatic screw machines and turret lathes involve so many variables that it is impossible to make specific recommendations that would apply to all conditions. Suggested tool angles, cutting speeds, and feeds are primarily starting points for each specific job. Table 4 lists reasonable speeds and feeds for single-point turning; Table 5, for cutoff and forming operations.

**Single-Point Turning Tools.** Grinding tools properly is particularly important in machining stainless steels. Figure 27 shows suggested starting geometries for high-speed steel single-point turning tools. Tools with a 5 to 10° positive top rake angle will generate less heat and cut more freely with a cleaner surface. It is also beneficial to select as large a tool as possible to provide a greater heat sink as well as a more rigid setup. To ensure adequate support for the cutting edge, the front clearance angle should be kept to a minimum, that is, 7 to 10°, as shown in Fig. 27. Austenitic stainless steels, because of their toughness and work-hardening characteristics, require tools ground with top rake angles on the high side of the 5 to 10° range to control the chips and may require increased side clearance angles to prevent rubbing and localized work hardening.

The nonfree-machining stainless steels tend to produce long, stringy chips that can be very troublesome. This difficulty can be alleviated by using chip curlers or chip breakers that, in addition to controlling long chips, reduce friction on the cutting edge of the tool. Chip breakers or curlers for the free-machining stainless steels do not need to be as deep as those for the nonfree-machining alloys. Otherwise, the depth of cut and feed rate usually govern the width and depth of the curler or breaker. Heavier chips require deeper curlers or breakers, but they must be ground without weakening the cutting edge. If a chip curler or breaker cannot be ground into the tool, it is advisable to have a steep top rake angle.

Carbide tools can be used in single-point turning operations and will allow higher speeds than high-speed steel tools. However, carbide tooling requires even greater attention to the rigidity of the tooling and the workpiece, and interrupted cuts should be avoided.

**Cutoff Tools.** Either blade-type or circular cutoff tools are used for stainless steel applications. Blade-type tools usually have sufficient bevel for side clearance, that is, 3° minimum, but may need greater clearance for deep cuts. In addition, they should be ground to provide for top rake and front clearance. The front clearance angle is 7 to 10°. A similar angle is used for top rake, or a radius or shallow concavity can be ground instead. The end cutting edge angle can range from less than 5 to 15°, with the angle decreasing for larger-diameter material.

Angles for circular cutoff tools are similar to those for blade-type tools, including a top rake angle of 7 to 10° (Fig. 28). Circular cutoff tools are more rigid than the blade-type and can withstand more shock. Therefore, they may be preferred for automatic screw machine operations in which they are fed into drilled or threaded holes. Because of their size, they also dissipate heat better.

Carbide-tip cutoff tools can be used. However, shock loading from interrupted cuts must be taken into consideration when selecting the carbide.

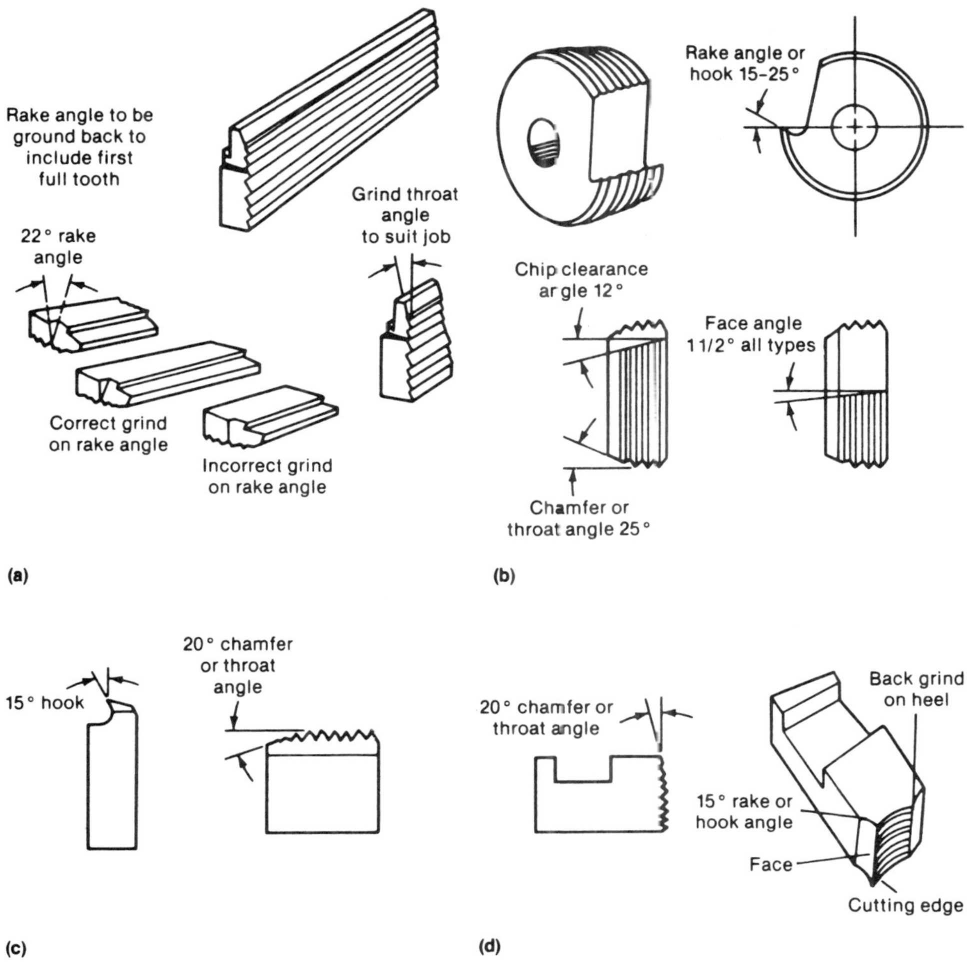


**Fig. 33** Suggested geometries of thread chasers used for stainless steels. (a) Tangent-type chaser. (b) Circular-type chaser. (c) Insert-type chaser. (d) Radial-type chaser

**Form tools** are usually dovetail or circular. The speeds and feeds for form tools are influenced by the width of the tool in relation to the diameter of the bar, the amount of overhang, and the contour or shape of the tool. Generally, the width of the form tool should not exceed 1½ times the diameter of the workpiece; otherwise, chatter may be a problem.

Dovetail form tools should be designed with a front clearance angle of 7 to 10° and should be ground with a top rake angle of 5 to 10°. Angles for circular form tools are similar (Fig. 29). Higher rake angles within the 5 to 10° range can be used for roughing operations and lower rake angles for finishing. The design of the tool should incorporate sufficient side clearance or relief angles, typically 1 to 5° depending on depth of cut, to prevent rubbing and localized heat buildup, particularly during rough forming. It may also be necessary to round corners. A finish form or shave tool may be necessary to obtain the final shape, especially for deep or intricate cuts.

Carbide-tip tooling can be used for forming operations. However, as with cutoff tools, shock loading from interrupted cuts must be taken into consideration.

**A shaving tool** can be used to obtain optimum machined surface finish or close tolerances on formed parts. Shaving tools remove metal with a tangential rather than a radial cut, with the workpiece supported by integral rollers. Usually a thin layer of metal (approximately 0.13 to 0.25 mm, or 0.005 to 0.010 in.) is removed at relatively high speeds. The tool must have a very smooth finish on the cutting edge because the finish of the tool will influence the finish of the part. A clearance angle of about 10° is normally needed.

## Drilling

In any drilling operation, the following factors are important:

- Work must be kept clean and chips removed frequently because dirt and chips act as an abrasive to dull the drill
- Drills must be correctly ground
- Drills must be properly aligned and the work firmly supported
- A stream of cutting fluid must be properly directed at the hole
- Drills should be chucked for shortest drilling length to avoid whipping or flexing, which may break drills or cause inaccurate work
- Heavier feeds and lower speeds may be necessary to reduce or eliminate work hardening

When working with stainless steels, particularly the austenitic alloys, it is advisable to use a sharp three-cornered punch rather than a prick punch to avoid work hardening the material at the mark. Drilling templates or guides may also be helpful in locating the mark.

To relieve chip packing and congestion, drills must occasionally be backed out. The general rule is to drill to a depth of three to four times the diameter of the drill for the first bite, one or two diameters for the second bite, and around one diameter for each of the subsequent bites. A groove ground parallel to the cutting edge in the flute for chip clearance will allow deeper holes to be drilled per bite, particularly with larger-size drills. The groove breaks up the chip for easier removal.

Drills should not be allowed to dwell during cutting, particularly with austenitic stainless steels. Allowing the drill to dwell or ride causes glazing on the bottom of the hole, which makes restarting difficult. Therefore, when relieving chip congestion, drills must be backed out quickly and reinserted at full speed to avoid glazing.

**Drilling Parameters.** Drill feed is an important factor in determining the rate of production. Because proper feed increases drill life and production between grinds, it should be carefully selected for each job. Table 6 lists feeds and speeds for drills of various sizes.

**Small-Diameter Drills.** Procedures used with large or normal-size drills will not always prove successful for small-diameter drills, that is, 1.8 mm (0.070 in.) and under. These very small drills are subject to deflection, both torsional and longitudinal, because of their length in relation to their diameter. In addition, the web on small drills is proportionally heavier than on large drills. This thicker web adds the strength needed for the required work pressure, but decreases chip clearance. Therefore, the depth of each bite may have to be reduced. All small-diameter drilling is actually deep-hole drilling; therefore, careful resharpening, frequent and adequate chip removal, and feeds and speeds that are properly adjusted to the strength and load-carrying capacity of the drill are very important. Good small-hole drilling is dependent on feeds that produce chips instead of powder.

**Grinding of Drills.** It is especially important to grind drills correctly. Figure 30

**Table 8 Nominal die-threading parameters for wrought stainless steels**

Depending on actual machining conditions, optimum parameters may be higher or lower than those given.

| UNS designation | Speed(a), m/min (sfm), at a pitch (threads per 25 mm, or 1 in.) of: ≤7 | 8–15 | 16–24 | >24 | Tool material |
|---|---|---|---|---|---|
| **Free-machining alloys** | | | | | |
| **Ferritic alloys** | | | | | |
| S18235, S43020, S43023 | 5–8 (15–25) | 8–11 (25–35) | 11–14 (35–45) | 12–15 (40–50) | M1,M2,M7,M10 |
| S18200 | 5–8 (15–25) | 8–11 (25–35) | 11–14 (35–45) | 12–15 (40–50) | M1,M2,M7,M10 |
| **Martensitic alloys** | | | | | |
| S41600, S41610, S41623 | 5–8 (15–25) | 8–11 (25–35) | 11–14 (35–45) | 12–15 (40–50) | M1,M2,M7,M10 |
| S42020, S42023 | 3–5 (10–15) | 5–8 (15–25) | 8–11 (25–35) | 11–12 (35–40) | M1,M2,M7,M10 |
| S44020, S44023 | 2–5 (8–15) | 3–6 (10–20) | 5–8 (15–25) | 8–9 (25–30) | M1,M2,M7,M10 |
| **Austenitic alloys** | | | | | |
| S20300, S30300, S30310, S30360 | 3–5 (10–15) | 5–8 (15–25) | 8–11 (25–35) | 11–12 (35–40) | M1,M2,M7,M10 |
| S30323, S30330, S30345, S30431, S31620, S34720, S34723 | 2–5 (8–15) | 4–7 (12–22) | 6–9 (20–30) | 9–11 (30–35) | M1,M2,M7,M10 |
| **Nonfree-machining alloys** | | | | | |
| **Ferritic alloys** | | | | | |
| S40500, S40900, S43000, S43400 | 1.5–5 (5–15) | 3–8 (10–25) | 6–11 (20–35) | 8–12 (25–40) | M1,M2,M7,M10 |
| S44200, S44300, S44400, S44600 | 1.5–4 (5–12) | 2–5 (8–15) | 3–6 (10–20) | 5–8 (15–25) | M1,M2,M7,M10 |
| **Martensitic alloys** | | | | | |
| S40300, S41000 | 1.5–5 (5–15) | 3–8 (10–25) | 6–11 (20–35) | 8–12 (25–40) | M1,M2,M7,M10 |
| S41400, S42000, S42010, S43100 | 1.5–5 (5–15) | 3–6 (10–20) | 6–9 (20–30) | 8–11 (25–35) | M1,M2,M7,M10 |
| S44002, S44003, S44004 | 1.5–4 (5–12) | 2–5 (8–15) | 3–6 (10–20) | 5–8 (15–25) | M1,M2,M7,M10 |
| **Austenitic alloys** | | | | | |
| S20100, S30100, S30200, S30400, S30403, S30430, S30500, S31600, S31603, S32100, S34700, S38400 | 2–5 (8–15) | 3–6 (10–20) | 5–8 (15–25) | 8–9 (25–30) | M1,M2,M7,M10 |
| S30900, S30908, S31000, S31008, S31700, S31703 | 1.5–3 (5–10) | 2–4 (8–13) | 3–5 (10–15) | 5–6 (15–20) | M1,M2,M7,M10 |
| S20910, S21904, S24100, S28200, S30452, N08020 | 1–2 (4–8) | 2–3 (6–10) | 2–4 (8–12) | 3–5 (10–15) | M1,M2,M7,M10 |
| **Duplex alloys** | | | | | |
| S31803, S32550, S32900, S32950 | 1.5–3 (5–10) | 2–4 (8–13) | 3–5 (10–15) | 5–6 (15–20) | M1,M2,M7,M10 |
| **Precipitation-hardenable alloys** | | | | | |
| S15500, S17400, S17700, S45000 | 1.5–4 (5–12) | 2–5 (8–15) | 3–6 (10–20) | 5–8 (15–25) | M1,M2,M7,M10 |
| S13800, S35000, S35500, S45500 | 1.5–3 (5–10) | 2–4 (8–13) | 3–5 (10–15) | 5–6 (15–20) | M1,M2,M7,M10 |
| S66286 | 1–2 (4–8) | 2–3 (6–10) | 2–4 (8–12) | 3–5 (10–15) | M1,M2,M7,M10 |

(a) All speeds are for material heat treated or cold drawn to low-to-intermediate hardness levels. All speeds should be reduced 20–30% for taper threads. Source: Ref 38, 52

shows suggested geometries for high-speed steel drills to be used with stainless steels. The point angle should be 140°, although a smaller angle can be used for easier-to-machine alloys. A larger angle produces a more easily removed chip when drilling hard or tough alloys. The lip clearance should be between 9 and 15°, and the two cutting edges must be equal in length and angle. The web thickness at the point should be about 12.5% of the drill diameter or less. A thinner web reduces feed pressure, heat generation, and glazing or work hardening of the bottom of the hole. Grinding fixtures should be used when regrinding drills. For best results in grinding high-speed steel drills, medium-grain, soft-grade dry wheels should be used. Blueing or burning is to be avoided, as is quenching. Quenching will often check or crack the drill if it has been overheated.

**Special Drills.** Drills should be chucked for the shortest drilling length. Some jobs, however, require exceptionally deep drilled holes where the depth of the hole is eight to ten times the diameter. In such cases, short chucking is impossible, and special drills known as crankshaft hole drills may be useful. These drills were originally designed for drilling oil holes in forged crankshafts and connecting rods, but have found widespread use in drilling deep holes. They are made with a very heavy web and a higher spiral or helix angle to aid in chip removal. They usually have a notched point type of web thinning, which is done on a sharp-cornered hard grinding wheel.

A cotter pin drill should be used to drill small cross holes in the heads of bolts, screws, pins, and so on. Like a crankshaft hole drill, it is a more heavily constructed drill that withstands abnormal strains and has a faster or higher helix angle to aid in chip removal.

## Tapping

Whenever possible, it is desirable to use free-machining grades for tapping, especially when blind holes are specified, to minimize the difficulty of removing and disposing of chips. The gummy, stringy chips resulting from the nonfree-machining grades of stainless steel are a source of difficulty in tapping.

In tapping all grades of stainless steel and especially the nonfree-machining grades, the following recommendations are made:

- Use taps that are as large as permissible in diameter (high side of the tolerance)
- Keep the thread pitch as fine as possible because the finer the pitch, the less metal removed per tooth
- Keep the percentage of full thread as low as permissible (60 to 75% is a preferred range, but where the depth of the tapped hole exceeds twice the diameter of the tap, it is economical to use only 50% thread)
- Use a flood of cutting oil (under pressure of at least 35 kPa, or 5 psi, when tapping blind holes)

The difficulties encountered in tapping stainless steel are often a result of work-hardened surfaces caused by previous drilling or reaming operations. Thus, when tapping problems arise, an investigation of the drilling and reaming conditions is recommended. Chips must be cut in either drilling or reaming; consequently, when drills or reamers are permitted to burnish rather than cut, surfaces become work hardened. The degree of work hardening varies among the different grades of stainless steel; the austenitic grades are the most susceptible.

**Tap Material and Design.** High-speed steel taps are nearly always used for tapping stainless steel. Chip removal is important when tapping tight-fitting, close-tolerance (class 3) threads, and the flutes should not be too shallow nor the lands too wide.

Consideration must also be given to the number of flutes on the tap. With small holes, a tap with four flutes is more likely to produce chip congestion than one with fewer flutes. Therefore, general practice is to use a tap with fewer flutes as the size of the hole decreases. For cutting class 3 threads, taps should have two flutes for holes up to size No. 6 (5.2 mm, or 0.204 in., in diameter). Three-flute taps are satisfactory for holes between 5.2 and 13 mm (0.204 and ½ in.) in diameter. Three-flute taps reduce the possibility of cutting oversize but provide less chip clearance than two-flute taps. Holes larger than 13 mm (½ in.) in diameter are tapped with four-flute taps.

*Types*. Heavy-duty spiral-point and heavy-duty spiral-flute taps are generally used for the tapping of stainless steel alloys (see the article "Tapping" in this Volume for a description of spiral-point and spiral-flute taps). Spiral-point taps push the chips ahead of the tap and should not be used for tapping blind holes unless there is sufficient untapped depth to accommodate the chips at the bottom of the hole. Spiral-flute taps pull the chips from the tapped hole and can be used for tapping either blind or through holes. Straight-flute taps are used for holes larger than 13 mm (½ in.) in diameter.

*Tap Geometry*. For austenitic alloys, a hook grind is best. The hook angle is commonly 15 to 20° for austenitic alloys with hardnesses from 135 to 275 HB. Occasionally, because of a combination of factors on certain work, the hook grind will not produce satisfactory results. In such cases, an interrupted-thread tap with an uneven number of flutes may be desirable because it requires 40 to 50% less power than regular taps (this may be important when the tapping machine lacks power).

For ferritic, martensitic, and precipitation-hardenable alloys, the common rake angles are 8 to 12° for alloys with a hardness of 135 to 275 HB and 0 to 5° for alloys with a hardness of 275 to 325 HB. Above 325 HB, 0° is the common rake angle for the martensitic and precipitation-hardenable alloys.

The problem of roughness on the back face of the thread can sometimes be overcome by grinding the heel of the lands so that the bearing area is reduced. This back relief prevents tearing of the threads when backing the tap out. Roughness can also be caused by insufficient rake or hook angle.

Back relief (or relieving the thread) may either have the lands fully eccentric with the hole or may have the lands partly concentric with the hole (Fig. 31). Suggested back relief angles (Fig. 31) are 10° for austenitic alloys with hardnesses of 135 to 275 HB and 8° for ferritic, martensitic, and precipitation-hardenable alloys with hardnesses of 135 to 325 HB. Back relief is approximately 5° for deep holes and 6 to 8° for the martensitic, and precipitation-hardenable alloys with hardnesses of 325 to 425 HB.

**Table 9 Nominal milling parameters for wrought stainless steels**

Depending on actual machining conditions, optimum parameters may be higher or lower than those given. See Ref 52 for speeds and feeds in face, side, and hollow milling.

| UNS designation | Speed(a)(b), m/min (sfm) | Feed, mm/tooth (in./tooth), at a cutter diameter of: 6 mm (¼ in.) | 13 mm (½ in.) | 19 mm (¾ in.) | 25–50 mm (1–2 in.) | Tool material(a) |
|---|---|---|---|---|---|---|
| **Free-machining alloys** | | | | | | |
| **Ferritic alloys** | | | | | | |
| S18235, S43020, S43023 | 43 (140) | 0.025 (0.001) | 0.05 (0.002) | 0.075 (0.003) | 0.10 (0.004) | M2,M7 |
| | 122 (400) | 0.025 (0.001) | 0.05 (0.002) | 0.13 (0.005) | 0.175 (0.007) | C6 |
| S18200 | 44 (145) | 0.05 (0.002) | 0.075 (0.003) | 0.13 (0.005) | 0.15 (0.006) | M2,M7 |
| | 125 (410) | 0.05 (0.002) | 0.075 (0.003) | 0.13 (0.005) | 0.175 (0.007) | C6 |
| **Martensitic alloys** | | | | | | |
| S41600, S41610, S41623 | 43 (140) | 0.05 (0.002) | 0.075 (0.003) | 0.13 (0.005) | 0.15 (0.006) | M2,M7 |
| | 122 (400) | 0.05 (0.002) | 0.075 (0.003) | 0.13 (0.005) | 0.175 (0.007) | C6 |
| S42020, S42023 | 30 (100) | 0.05 (0.002) | 0.075 (0.003) | 0.13 (0.005) | 0.15 (0.006) | M2,M7 |
| | 91 (300) | 0.025 (0.001) | 0.075 (0.003) | 0.13 (0.005) | 0.175 (0.007) | C6 |
| S44020, S44023 | 26 (85) | 0.025 (0.001) | 0.05 (0.002) | 0.075 (0.003) | 0.10 (0.004) | M2,M7 |
| | 76 (250) | 0.025 (0.001) | 0.075 (0.003) | 0.13 (0.005) | 0.175 (0.007) | C6 |
| **Austenitic alloys** | | | | | | |
| S20300, S30300, S30310, S30360 | 35 (115) | 0.05 (0.002) | 0.075 (0.003) | 0.13 (0.005) | 0.15 (0.006) | M2,M7 |
| | 107 (350) | 0.05 (0.002) | 0.075 (0.003) | 0.13 (0.005) | 0.15 (0.006) | C2 |
| S30323, S30330, S30345, S30431, S31620, S34720, S34723 | 30 (100) | 0.025 (0.001) | 0.05 (0.002) | 0.10 (0.004) | 0.13 (0.005) | M2,M7 |
| | 91 (300) | 0.025 (0.001) | 0.05 (0.002) | 0.10 (0.004) | 0.13 (0.005) | C2 |
| **Nonfree-machining alloys** | | | | | | |
| **Ferritic alloys** | | | | | | |
| S40500, S40900, S43000, S43400 | 34 (110) | 0.05 (0.002) | 0.075 (0.003) | 0.13 (0.005) | 0.15 (0.006) | M2,M7 |
| | 107 (350) | 0.025 (0.001) | 0.05 (0.002) | 0.10 (0.004) | 0.15 (0.006) | C6 |
| S44200, S44300, S44400, S44500 | 30 (100) | 0.05 (0.002) | 0.075 (0.003) | 0.13 (0.005) | 0.15 (0.006) | M2,M7 |
| | 91 (300) | 0.025 (0.001) | 0.05 (0.002) | 0.10 (0.004) | 0.15 (0.006) | C6 |
| **Martensitic alloys** | | | | | | |
| S40300, S41000 | 34 (110) | 0.05 (0.002) | 0.075 (0.003) | 0.13 (0.005) | 0.15 (0.006) | M2,M7 |
| | 107 (350) | 0.025 (0.001) | 0.05 (0.002) | 0.10 (0.004) | 0.15 (0.006) | C6 |
| S41400, S42000, S42010, S43100 | 27 (90) | 0.025 (0.001) | 0.05 (0.002) | 0.075 (0.003) | 0.10 (0.004) | M2,M7 |
| | 82 (270) | 0.025 (0.001) | 0.05 (0.002) | 0.10 (0.004) | 0.15 (0.006) | C6 |
| S44002, S44003, S44004 | 23 (75) | 0.025 (0.001) | 0.05 (0.002) | 0.075 (0.003) | 0.10 (0.004) | M2,M7 |
| | 69 (225) | 0.025 (0.001) | 0.05 (0.002) | 0.10 (0.004) | 0.15 (0.006) | C6 |
| **Austenitic alloys** | | | | | | |
| S20100, S30100, S30200, S30400, S30403, S30430, S30500, S31600, S31603, S32100, S34700, S38400 | 24 (80) | 0.05 (0.002) | 0.075 (0.003) | 0.13 (0.005) | 0.15 (0.006) | M2,M7 |
| | 82 (270) | 0.025 (0.001) | 0.05 (0.002) | 0.075 (0.003) | 0.13 (0.005) | C2 |
| S30900, S30908, S31000, S31008, S31700, S31703 | 23 (75) | 0.05 (0.002) | 0.075 (0.003) | 0.13 (0.005) | 0.15 (0.006) | M2,M7 |
| | 79 (260) | 0.025 (0.001) | 0.05 (0.002) | 0.075 (0.003) | 0.13 (0.005) | C2 |
| S20910, S21904, S24100, S28200, S30452, N08020 | 15 (50) | 0.025 (0.001) | 0.05 (0.002) | 0.05 (0.002) | 0.075 (0.003) | M2,M7 |
| | 49 (160) | 0.025 (0.001) | 0.05 (0.002) | 0.075 (0.003) | 0.10 (0.004) | C2 |
| **Duplex alloys** | | | | | | |
| S31803, S32550, S32900, S32950 | 23 (75) | 0.05 (0.002) | 0.075 (0.003) | 0.13 (0.005) | 0.15 (0.006) | M2,M7 |
| | 79 (260) | 0.025 (0.001) | 0.05 (0.002) | 0.075 (0.003) | 0.13 (0.005) | C2 |
| **Precipitation-hardenable alloys** | | | | | | |
| S15500, S17400, S17700, S45000 | 26 (85) | 0.025 (0.001) | 0.05 (0.002) | 0.075 (0.003) | 0.10 (0.004) | M2,M7 |
| | 84 (275) | 0.025 (0.001) | 0.05 (0.002) | 0.10 (0.004) | 0.15 (0.006) | C2 |
| S13800, S35000, S35500, S45500 | 23 (75) | 0.025 (0.001) | 0.05 (0.002) | 0.075 (0.003) | 0.10 (0.004) | M2,M7 |
| | 82 (270) | 0.025 (0.001) | 0.05 (0.002) | 0.10 (0.004) | 0.15 (0.006) | C2 |
| S66286 | 5 (50) | 0.025 (0.001) | 0.05 (0.002) | 0.05 (0.002) | 0.075 (0.003) | M2,M7 |
| | 53 (175) | 0.025 (0.001) | 0.05 (0.002) | 0.075 (0.003) | 0.10 (0.004) | C2 |

(a) All speeds given are for material in the annealed condition (140–270 HB) except for the precipitation-hardenable grades, which are solution treated (150–325 HB) or equalized and overtempered for S35000 and S35500. For material heat treated or cold drawn to hardnesses higher than their normal annealed or softest level, best results are obtained by reducing speeds approximately 15% to as much as 75% depending on hardness. Premium high-speed steel or carbide cutters may also be necessary at the higher hardnesses. (b) Depth of cut for all speeds and feeds given is 1.3 mm (0.050 in.). Source: Ref 38, 52

**Example: Progressive Modification of Tap Design for Soft Austenitic Stainless.** As the result of a series of modifications in tap design, the tool life of taps used to thread S30400, S31600, and N08330 stainless steels on a single-spindle automatic machine was extended from 2 to 3000 holes per tap. Before the final (and most satisfactory) tap design was developed, tap breakage had been caused by the jamming of chips in the lands of the tap and by the tapping of holes previously work hardened in a drilling operation. In addition, the generation of excessive heat during tapping had caused breakdown of the tap teeth at the cutting tips.

The three austenitic stainless grades tapped were all in the annealed condition, with a hardness of 208 HB or less. The application consisted of drilling and tapping 6 mm (¼ in.) diam holes in a valve bonnet.

**Table 10 Nominal broaching parameters for wrought stainless steels**

Depending on actual machining conditions, optimum parameters may be higher or lower than those given.

| UNS designation | Speed(a) m/min | Speed(a) sfm | Chip load mm/tooth | Chip load in./tooth | Tool material(a)(b) |
|---|---|---|---|---|---|
| **Free-machining alloys** | | | | | |
| **Ferritic alloys** | | | | | |
| S18235, S43020, S43023 | 9 | 30 | 0.10 | 0.004 | M2,M7 |
| S18200 | 11 | 35 | 0.10 | 0.004 | M2,M7 |
| **Martensitic alloys** | | | | | |
| S41600, S41610, S41623 | 9 | 30 | 0.10 | 0.004 | M2,M7 |
| S42020, S42023 | 8 | 25 | 0.075 | 0.003 | M2,M7 |
| S44020, 44023 | 6 | 20 | 0.075 | 0.003 | M2,M7 |
| **Austenitic alloys** | | | | | |
| S20300, S30300, S30310, S30360 | 8 | 25 | 0.10 | 0.004 | M2,M7 |
| S30323, S30330, S30345, 30431, S31620, S34720, S34723 | 6 | 20 | 0.10 | 0.004 | M2,M7 |
| **Nonfree-machining alloys** | | | | | |
| **Ferritic alloys** | | | | | |
| S40500, S40900, S43000, 43400 | 6 | 20 | 0.075 | 0.003 | M2,M7 |
| S44200, S44300, S44400, S44600 | 6 | 20 | 0.075 | 0.003 | M2,M7 |
| **Martensitic alloys** | | | | | |
| S40300, S41000 | 8 | 25 | 0.10 | 0.004 | M2,M7 |
| S41400, S42000, S42010, S43100 | 5 | 15 | 0.075 | 0.003 | T15,M42 |
| S44002, S44003, S44004 | 5 | 15 | 0.05 | 0.002 | T15,M42 |
| **Austenitic alloys** | | | | | |
| S20100, S30100, S30200, S30400, S30403, S30430, S30500, S31600, S31603, S32100, S34700, S38400 | 6 | 20 | 0.075 | 0.003 | M2,M7 |
| S30900, S30908, S31000, S31008, S31700, S31703 | 5 | 15 | 0.075 | 0.003 | T15,M42 |
| S20910, S21904, S24100, S28200, S30452, N08020 | 3 | 10 | 0.075 | 0.003 | T15,M42 |
| **Duplex alloys** | | | | | |
| S31803, S32550, S32900, S32950 | 5 | 15 | 0.075 | 0.003 | T15,M42 |
| **Precipitation-hardenable alloys** | | | | | |
| S15500, S17400, S17700, S45000 | 5 | 15 | 0.05 | 0.002 | T15,M42 |
| S13800, S35000, S35500, S45500 | 3 | 10 | 0.05 | 0.002 | T15,M42 |
| S66286 | 2 | 8 | 0.05 | 0.002 | T15,M42 |

(a) All speeds are for material in the annealed condition (140–270 HB) except the precipitation-hardenable grades, which are solution treated (150–325 HB) or equalized and overtempered for S35000 and S35500. For material heat treated or cold drawn to hardnesses of 275–375 HB (325–375 for precipitation-hardenable grades) speed reductions of approximately 15–60% may be necessary for best results. Premium high-speed steel broaches may be necessary for higher-hardness material. (b) Any premium high-speed steel (T15, M33, M41–47) can be used where the designated tool material is T15,M42. Source: Ref 38, 52

Tap size was ¼-32. All parts were required to have a class 3 fit on 80% of the threads, and the threads had to be smooth and free of tear marks.

The final tap design was achieved in six steps. Taps of all six designs were made of grade M2 high-speed steel and received no special surface treatment. The original tap and three of the redesigns—including the final one—are illustrated in Fig. 32.

In the initial setup, a standard four-flute, conventionally relieved, ground plug tap (Fig. 32a) was used. With two machine speeds available, tapping was performed at 275 rev/min (~5.5 m/min, or 18 sfm), and the tap was withdrawn at a speed of 610 rev/min (~12 m/min, or 40 sfm). Thus, the ratio of withdrawal speed to threading speed was 2.2:1. For lubrication, a mineral-base oil containing fatty oils and active sulfur was used; its viscosity was similar to that of SAE 20 oil. Under these conditions, not more than two holes could be tapped before heavy chips became packed in the flutes of the tap, which resulted in tap breakage.

Because greater chip relief was obviously required, a two-flute tap (Fig. 32b) was tried. This tap, which provided a land that was larger and stronger in cross section, was ground to a spiral point that afforded a path for chip disposal. Tapping was performed with this two-flute tap at the same speeds and feeds and with the same cutting oil that had been used with the original four-flute taps. This tap resulted in more rubbing; galling occurred during withdrawal of the tap at high speed; and chips jammed the flutes, restricting flow of the cutting fluid.

The two-flute tap was modified to full relief, which allowed a greater flow of cutting oil around the cutting tips. This modification helped preserve cutting edges, reduced galling, and increased tap life to 22 holes.

Next, to increase the flow of cutting fluid to the tips, all threads to the rear of the 17th tooth (17 being the minimum number of teeth required for threading the part) were removed. Under those conditions, thread quality was improved, and tool life was extended to 85 holes per tap.

The two-flute tap was modified again by removing approximately 75% of the land behind the cutting edge to the depth of the root diameter at the trailing edge of the land (Fig. 32c). This decrease in cutting tool area permitted an increase in the flow of cutting fluid to the cutting edges and reduced the area of the thread drag that occurred when the tap was withdrawn from the hole at high speeds.

The three-flute tap, representing the final improvement in design, incorporated an increase in the number of cutting edges (from two to three), thus providing increased distribution of cutting fluid around each row of cutting edges. Except for its having three flutes (Fig. 32d), it retained the same design as that shown for the two-flute tap.

To reduce the amount of work-hardened metal developed as a result of drilling, drill size was changed from a No. 5 to a No. 6, and a 5.3 mm (0.2087 in.) drill was used to finish the hole before tapping. After producing 3000 holes, the new tap showed little sign of wear. Galling and metal flow at the tips of the stainless steel threads were virtually eliminated. The threads were smooth and were gaged satisfactorily with a class 3 thread gage.

**Tapping Parameters and Cutting Fluid.** Typical cutting speeds for tapping various stainless steels are listed in Table 7. Proper lubrication also plays an important role in tapping operations. Although the sulfurized oils have proved most successful, specific compositions or ratios of sulfurized oils to paraffin-base oils are required on certain jobs. After the best mixture has been determined, it is equally important that the flow of the cutting fluid to the taps and the work be considered. It is often desirable to use two streams of cutting fluid—one on each side of the tap. The flow should be started before the tap begins to cut and should not be shut off until after the cut is finished.

## Die Threading

Thread chasers for self-opening die heads are made of high-speed steel. The standard commercial chasers generally have a satis-

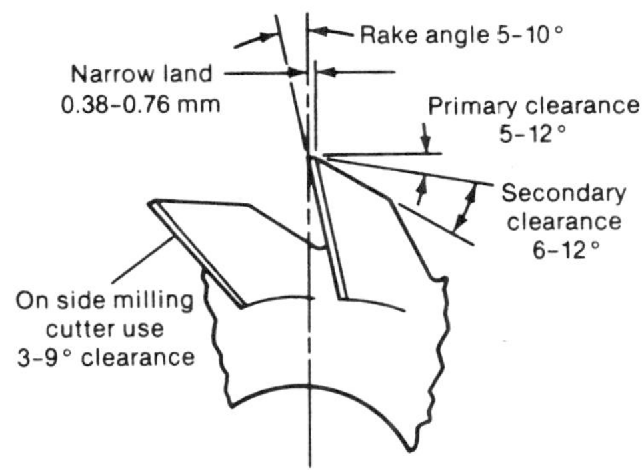


**Fig. 34** Suggested geometries of milling cutters used for stainless steels

factory life when they are kept sharp and correctly ground for the materials cut.

Figure 33 shows suggested geometries for the four main types of thread cutting tools. The most frequently used chaser for close-tolerance threads is the tangent type (Fig. 33a). It is particularly adaptable for heavy-duty jobs, such as producing long, coarse threads. The tangent-type chaser maintains thread size better on heavy-duty work and provides good life between grinds. Whenever possible, a 20° throat angle should be used. However, where the threads do not run into a shoulder, a 15° throat is better.

The circular type (Fig. 33b) is considered to be the universal thread chaser because it is adaptable to all types of threads and will work equally well on tubing. This type of chaser should generally have a 25° throat angle.

The insert-type chaser (Fig. 33c) is widely used. It produces good threads on nonfree-machining alloys at a low cost. A 20° throat angle is usually suitable for this type of chaser.

The radial-type chaser (Fig. 33d) will produce very smooth threads because it is ground to follow the shape or contour of the threaded piece. This type of chaser has been successfully used where extremely fine threads are required when form plunge tools are used with screw machines. Radial chasers typically have a 20° throat angle.

The throat angle or chamfer will vary slightly from the typical geometries given for each of the chasers based on the type of thread being cut and the grade of stainless steel machined. In general, it is advisable to use a 1½- to 3-thread chamfer or lead on the throat. This will usually produce a smooth thread with a fine finish and will increase chaser life between grinds. The advantage of using a long throat angle is that each tooth makes a smaller cut and consequently produces cleaner threads. For example, in one case, a 45° throat angle produced a chip approximately 0.46 mm (0.018 in.) thick, while a 15° throat angle produced a chip only 0.17 mm (0.0065 in.) thick.

When threading close to a shoulder where a long throat angle cannot be used, it may be necessary to grind only a ½- or 1-thread chamfer on the chaser. If this short throat angle produces a rough thread, a smooth finish can be obtained by running the chaser over the workpiece a second time, or the thread can be cut in two operations by first taking a rough cut and then finishing with a second cut.

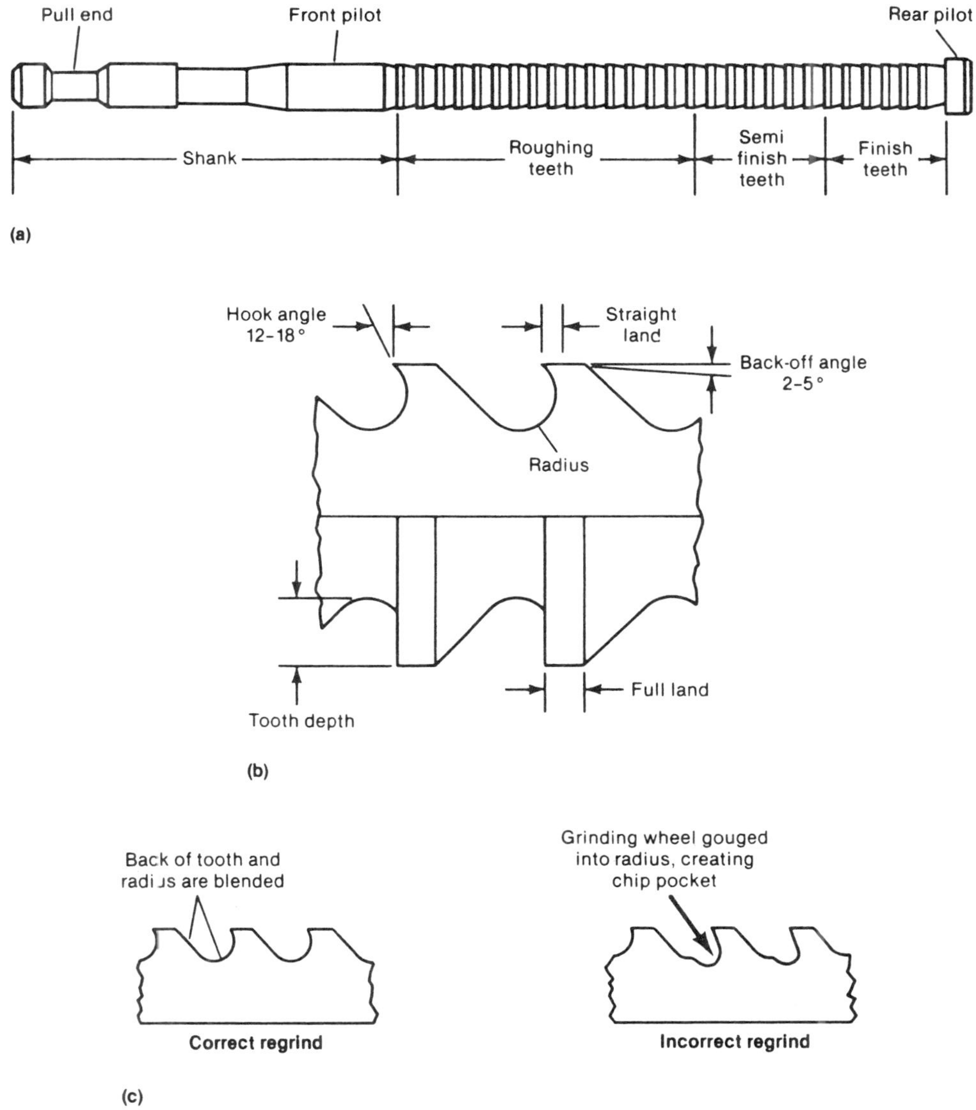


**Fig. 35** Typical one-pass internal broaching tool (a), along with suggested geometries (b) and grinding techniques (c)

**Die-Threading Parameters and Cutting Fluid.** Suggestions for cutting speeds are found in Table 8. Regardless of the type of chaser being used, speeds will vary somewhat with the type of thread being cut. Acme threads are usually cut at somewhat slower speeds. Where extremely fine threads are required, it might be desirable to decrease speeds to 1.5 to 4.5 m/min (5 to 15 sfm). In addition, when cutting fine threads, an advantage will be found in diluting heavy sulfurized oil with paraffin oil, generally one part sulfurized oil to five parts paraffin. The oil must be kept clean and free of chips, which could damage the threads.

**Percentage of Thread.** As with tapping, the percentage of thread being cut influences production rate and cost. Cutting threads with the maximum major diameter and/or minimum pitch diameter allowed for the class of thread involved means that more metal must be removed. This decreases tool life and does not necessarily produce a stronger thread. The blank should be turned to leave the minimum allowable stock.

## Thread Rolling

Thread rolling can be done on automatic screw machines and turret lathes. However, the equipment must have sufficient power and rigidity for stainless steels. As discussed previously, stainless steels have high strength and work-hardening rates; therefore, substantial pressure is required to form the threads. These characteristics of stainless steels may also limit the amount of thread that can be formed, and care must be

## Table 11 Nominal reaming parameters for wrought stainless steels

Depending on actual machining conditions, the optimum parameters may be higher or lower than those given.

| UNS designation | Speed(a) m/min | Speed(a) sfm | Feed, mm/rev (in./rev), at a reamer diameter of: 6 mm (¼ in.) | 13 mm (½ in.) | 25 mm (1 in.) | 50 mm (2 in.) | Tool material(a)(b) | |
|---|---|---|---|---|---|---|---|---|
| **Free-machining alloys** | | | | | | | | |
| **Ferritic alloys** | | | | | | | | |
| S18235, S43020, S43023 | 40 | 130 | 0.13 (0.005) | 0.20 (0.008) | 0.25 (0.010) | 0.38 (0.015) | M7 | Roughing |
| | 46 | 150 | 0.20 (0.008) | 0.30 (0.012) | 0.41 (0.016) | 0.61 (0.024) | C2 | |
| | 18 | 60 | 0.10 (0.004) | 0.15 (0.006) | 0.20 (0.008) | 0.30 (0.012) | M7 | Finishing |
| | 21 | 70 | 0.15 (0.006) | 0.20 (0.008) | 0.25 (0.010) | 0.30 (0.012) | C2 | |
| S18200 | 40 | 130 | 0.18 (0.005) | 0.20 (0.008) | 0.25 (0.010) | 0.38 (0.015) | M7 | Roughing |
| | 46 | 150 | 0.20 (0.008) | 0.30 (0.012) | 0.41 (0.016) | 0.61 (0.024) | C2 | |
| | 18 | 60 | 0.10 (0.004) | 0.15 (0.006) | 0.20 (0.008) | 0.30 (0.012) | M7 | Finishing |
| | 21 | 70 | 0.15 (0.006) | 0.20 (0.008) | 0.25 (0.010) | 0.30 (0.012) | C2 | |
| **Martensitic alloys** | | | | | | | | |
| S41600, S41610, S41623 | 40 | 130 | 0.13 (0.005) | 0.20 (0.008) | 0.25 (0.010) | 0.38 (0.015) | M7 | Roughing |
| | 46 | 150 | 0.20 (0.008) | 0.30 (0.012) | 0.41 (0.016) | 0.61 (0.024) | C2 | |
| | 18 | 60 | 0.10 (0.004) | 0.15 (0.006) | 0.20 (0.008) | 0.30 (0.012) | M7 | Finishing |
| | 21 | 70 | 0.15 (0.006) | 0.20 (0.008) | 0.25 (0.010) | 0.30 (0.012) | C2 | |
| S42020, S42023 | 30 | 100 | 0.13 (0.005) | 0.20 (0.008) | 0.25 (0.010) | 0.38 (0.015) | M7 | Roughing |
| | 37 | 120 | 0.20 (0.008) | 0.30 (0.012) | 0.41 (0.016) | 0.61 (0.024) | C2 | |
| | 15 | 50 | 0.10 (0.004) | 0.15 (0.006) | 0.20 (0.008) | 0.30 (0.012) | M7 | Finishing |
| | 18 | 60 | 0.15 (0.006) | 0.20 (0.008) | 0.25 (0.010) | 0.30 (0.012) | C2 | |
| S44020, S44023 | 23 | 75 | 0.13 (0.005) | 0.20 (0.008) | 0.25 (0.010) | 0.38 (0.015) | M7 | Roughing |
| | 27 | 90 | 0.20 (0.008) | 0.30 (0.012) | 0.41 (0.016) | 0.61 (0.024) | C2 | |
| | 12 | 40 | 0.10 (0.004) | 0.15 (0.006) | 0.20 (0.008) | 0.30 (0.012) | M7 | Finishing |
| | 15 | 50 | 0.15 (0.006) | 0.20 (0.008) | 0.25 (0.010) | 0.30 (0.012) | C2 | |
| **Austenitic alloys** | | | | | | | | |
| S20300, S30300, S30310, S30360 | 27 | 90 | 0.15 (0.006) | 0.23 (0.009) | 0.28 (0.011) | 0.46 (0.018) | M7 | Roughing |
| | 34 | 110 | 0.20 (0.008) | 0.30 (0.012) | 0.41 (0.016) | 0.61 (0.024) | C2 | |
| | 15 | 50 | 0.10 (0.004) | 0.15 (0.006) | 0.20 (0.008) | 0.25 (0.010) | M7 | Finishing |
| | 20 | 65 | 0.15 (0.006) | 0.20 (0.008) | 0.25 (0.010) | 0.30 (0.012) | C2 | |
| S30323, S30330, S30445, S30431, S31620, S34720, S34723 | 23 | 75 | 0.15 (0.006) | 0.23 (0.009) | 0.28 (0.011) | 0.41 (0.018) | M7 | Roughing |
| | 30 | 100 | 0.20 (0.008) | 0.30 (0.012) | 0.41 (0.016) | 0.61 (0.024) | C2 | |
| | 12 | 40 | 0.10 (0.004) | 0.15 (0.006) | 0.20 (0.008) | 0.25 (0.010) | M7 | Finishing |
| | 17 | 55 | 0.15 (0.006) | 0.20 (0.008) | 0.25 (0.010) | 0.30 (0.012) | C2 | |
| **Nonfree-machining alloys** | | | | | | | | |
| **Ferritic alloys** | | | | | | | | |
| S40500, S40900, S43000, S43400 | 24 | 80 | 0.10 (0.004) | 0.13 (0.005) | 0.20 (0.008) | 0.30 (0.012) | M7 | Roughing |
| | 29 | 95 | 0.20 (0.008) | 0.30 (0.012) | 0.41 (0.016) | 0.61 (0.024) | C2 | |
| | 12 | 40 | 0.10 (0.004) | 0.15 (0.006) | 0.20 (0.008) | 0.25 (0.010) | M7 | Finishing |
| | 18 | 60 | 0.15 (0.006) | 0.20 (0.008) | 0.25 (0.010) | 0.30 (0.012) | C2 | |
| S44200, S44300, S44400, S44600 | 23 | 75 | 0.10 (0.004) | 0.13 (0.005) | 0.20 (0.008) | 0.30 (0.012) | M7 | Roughing |
| | 27 | 90 | 0.20 (0.008) | 0.30 (0.012) | 0.41 (0.016) | 0.61 (0.024) | C2 | |
| | 11 | 35 | 0.10 (0.004) | 0.15 (0.006) | 0.20 (0.008) | 0.25 (0.010) | M7 | Finishing |
| | 15 | 50 | 0.15 (0.006) | 0.20 (0.008) | 0.25 (0.010) | 0.30 (0.012) | C2 | |
| **Martensitic alloys** | | | | | | | | |
| S40300, S41000 | 24 | 80 | 0.10 (0.004) | 0.13 (0.005) | 0.20 (0.008) | 0.30 (0.012) | M7 | Roughing |
| | 29 | 95 | 0.20 (0.008) | 0.30 (0.012) | 0.41 (0.016) | 0.61 (0.024) | C2 | |
| | 12 | 40 | 0.10 (0.004) | 0.15 (0.006) | 0.20 (0.008) | 0.25 (0.010) | M7 | Finishing |
| | 18 | 60 | 0.15 (0.006) | 0.20 (0.008) | 0.25 (0.010) | 0.30 (0.012) | C2 | |
| S41400, S42000, S42010, S43100 | 20 | 65 | 0.10 (0.004) | 0.13 (0.005) | 0.20 (0.008) | 0.30 (0.012) | M7 | Roughing |
| | 24 | 80 | 0.20 (0.008) | 0.30 (0.012) | 0.41 (0.016) | 0.61 (0.024) | C2 | |
| | 11 | 35 | 0.10 (0.004) | 0.15 (0.006) | 0.20 (0.008) | 0.25 (0.010) | M7 | Finishing |
| | 15 | 50 | 0.15 (0.006) | 0.20 (0.008) | 0.25 (0.010) | 0.30 (0.012) | C2 | |
| S44002, S44003, S44004 | 17 | 55 | 0.075 (0.003) | 0.13 (0.005) | 0.20 (0.008) | 0.30 (0.012) | T15,M42 | Roughing |
| | 21 | 70 | 0.20 (0.008) | 0.30 (0.012) | 0.41 (0.016) | 0.61 (0.024) | C2 | |
| | 11 | 35 | 0.075 (0.003) | 0.10 (0.004) | 0.15 (0.006) | 0.20 (0.008) | T15,M42 | Finishing |
| | 15 | 50 | 0.10 (0.004) | 0.15 (0.006) | 0.20 (0.008) | 0.25 (0.010) | C2 | |

(continued)

(a) All speeds given are for material in the annealed condition (140–270 HB) except for the precipitation-hardenable grades, which are solution treated (150–325 HB) or equalized and overtempered for S35000 and S35500. For material heat treated or cold drawn to hardnesses higher than their normal annealed or softest level, best results are obtained by reducing speeds by approximately 25% to as much as 75% depending on hardness. Premium high-speed steel or carbide tooling may also be necessary at the higher hardnesses. (b) Any premium high-speed steel (T15, M33, M41–47) can be used where the designated tool material is T15, M42. Source: Ref 38, 52

**Table 11 (continued)**

| UNS designation | Speed(a) m/min | Speed(a) sfm | Feed, mm/rev (in./rev), at a reamer diameter of: 5 mm (¼ in.) | 13 mm (½ in.) | 25 mm (1 in.) | 50 mm (2 in.) | Tool material(a)(b) | |
|---|---|---|---|---|---|---|---|---|
| **Austenitic alloys** | | | | | | | | |
| S20100, S30100, S30200, S30400, S30403, S30430, S30500, S31600, S31603, S32100, S34700, S38400 | 21 | 70 | 0.13 (0.005) | 0.20 (0.008) | 0.25 (0.010) | 0.38 (0.015) | M7 | Roughing |
| | 26 | 85 | 0.20 (0.008) | 0.30 (0.012) | 0.41 (0.016) | 0.61 (0.024) | C2 | |
| | 11 | 35 | 0.075 (0.003) | 0.10 (0.004) | 0.15 (0.006) | 0.20 (0.008) | M7 | Finishing |
| | 15 | 50 | 0.10 (0.004) | 0.15 (0.006) | 0.20 (0.008) | 0.25 (0.010) | C2 | |
| S30900, S30908, S31000, S31008, S31700, S31703 | 21 | 70 | 0.13 (0.005) | 0.20 (0.008) | 0.25 (0.010) | 0.38 (0.015) | T15,M42 | Roughing |
| | 26 | 85 | 0.20 (0.008) | 0.30 (0.012) | 0.41 (0.016) | 0.61 (0.024) | C2 | |
| | 11 | 35 | 0.075 (0.003) | 0.10 (0.004) | 0.15 (0.006) | 0.20 (0.008) | T15,M42 | Finishing |
| | 15 | 50 | 0.10 (0.004) | 0.15 (0.006) | 0.20 (0.008) | 0.25 (0.010) | C2 | |
| S20910, S21904, S24100, S28200, S30452, N08020 | 18 | 60 | 0.13 (0.005) | 0.20 (0.008) | 0.25 (0.010) | 0.38 (0.015) | T15,M42 | Roughing |
| | 23 | 75 | 0.20 (0.008) | 0.30 (0.012) | 0.41 (0.016) | 0.61 (0.024) | C2 | |
| | 9 | 30 | 0.075 (0.003) | 0.10 (0.004) | 0.15 (0.006) | 0.20 (0.008) | T15,M42 | Finishing |
| | 12 | 40 | 0.10 (0.004) | 0.15 (0.006) | 0.20 (0.008) | 0.25 (0.010) | C2 | |
| **Duplex alloys** | | | | | | | | |
| S31803, S32550, S32900, S32950 | 21 | 70 | 0.13 (0.005) | 0.20 (0.008) | 0.25 (0.010) | 0.38 (0.015) | T15,M42 | Roughing |
| | 26 | 85 | 0.20 (0.008) | 0.30 (0.012) | 0.41 (0.016) | 0.61 (0.024) | C2 | |
| | 11 | 35 | 0.075 (0.003) | 0.10 (0.004) | 0.15 (0.006) | 0.20 (0.008) | T15,M42 | Finishing |
| | 15 | 50 | 0.10 (0.004) | 0.15 (0.006) | 0.20 (0.008) | 0.25 (0.010) | C2 | |
| **Precipitation-hardenable alloys** | | | | | | | | |
| S15500, S17400, S17700, S45000 | 18 | 60 | 0.13 (0.005) | 0.20 (0.008) | 0.25 (0.010) | 0.38 (0.015) | T15,M42 | Roughing |
| | 23 | 75 | 0.20 (0.008) | 0.30 (0.012) | 0.41 (0.016) | 0.61 (0.024) | C2 | |
| | 11 | 35 | 0.10 (0.004) | 0.15 (0.006) | 0.20 (0.008) | 0.25 (0.010) | T15,M42 | Finishing |
| | 15 | 50 | 0.15 (0.006) | 0.20 (0.008) | 0.25 (0.010) | 0.30 (0.012) | C2 | |
| S13800, S35000, S35500, S45500 | 17 | 55 | 0.13 (0.005) | 0.20 (0.008) | 0.25 (0.010) | 0.38 (0.015) | T15,M42 | Roughing |
| | 21 | 70 | 0.20 (0.008) | 0.30 (0.012) | 0.41 (0.016) | 0.61 (0.024) | C2 | |
| | 9 | 30 | 0.10 (0.004) | 0.15 (0.006) | 0.20 (0.008) | 0.25 (0.010) | T15,M42 | Finishing |
| | 14 | 45 | 0.15 (0.006) | 0.20 (0.008) | 0.25 (0.010) | 0.30 (0.012) | C2 | |
| S66286 | 15 | 50 | 0.13 (0.005) | 0.20 (0.008) | 0.25 (0.010) | 0.38 (0.015) | T15,M42 | Roughing |
| | 20 | 65 | 0.20 (0.008) | 0.30 (0.012) | 0.41 (0.016) | 0.61 (0.024) | C2 | |
| | 8 | 25 | 0.10 (0.004) | 0.15 (0.006) | 0.20 (0.008) | 0.25 (0.010) | T15,M42 | Finishing |
| | 12 | 40 | 0.15 (0.006) | 0.20 (0.008) | 0.25 (0.010) | 0.30 (0.012) | C2 | |

(a) All speeds given are for material in the annealed condition (140–270 HB) except for the precipitation-hardenable grades, which are solution treated (150–325 HB) or equalized and overtempered for S35000 and S35500. For material heat treated or cold drawn to hardnesses higher than their normal annealed or softest level, best results are obtained by reducing speeds by approximately 25% to as much as 75% depending on hardness. Premium high-speed steel or carbide tooling may also be necessary at the higher hardnesses. (b) Any premium high-speed steel (T15, M33, M41–47) can be used where the designated tool material is T15, M42. Source: Ref 38, 52

taken to avoid work hardening the surface prior to thread rolling. Despite these cautions, however, thread rolling offers certain advantages. The threads produced are stronger and tougher than cut threads and can be more accurate in size. Generally, the nonfree-machining alloys will produce smoother and cleaner threads than the free-machining alloys.

## Milling

High-speed steel cutters are used in milling stainless steels, although tooling with carbide inserts can also be used, particularly for alloys that are more difficult to machine. Generally, the smoothest finishes are obtained with helical or spiral cutters running at high speed, particularly for cuts over 19 mm (0.75 in.) wide. Helical cutters cut with a shearing action and thus cut more freely and with less chatter than straight-tooth cutters. Coarse-tooth or heavy-duty cutters work under less stress and permit higher speeds than fine-tooth or light-duty cutters. They also have more space between the teeth to aid in chip disposal.

For heavy, plain milling work, a heavy-duty cutter with a faster, 45° left-hand spiral is preferred. The higher angle allows more teeth to contact the work at the same time. This applies steady pressure on the arbor and spindle, thus reducing chatter. In wide, slab milling, such cutters are particularly necessary to produce smooth finishes and to avoid chatter.

Milling deep slots in stainless steel sometimes presents the problems of chatter, binding, and jamming of wide chips. These difficulties can be eliminated by using a staggered-tooth cutter. Its alternating teeth cut only one-half of the slot, thus taking a smaller bite and producing a shorter chip. For the end milling of stainless steels, the solid-shank end mill is preferred because of its high strength.

**Grinding of Milling Cutters.** Figure 34 shows the rake angle, land width, and primary and secondary clearance for high-speed steel cutters. The geometries shown provide sufficient strength and clearance. On cutters up to 100 mm (4 in.) in diameter, the maximum clearance shown in Fig. 34 should be used, remembering that small cutters require a greater clearance angle than large cutters. Sufficient clearance behind the cutting edge of every tooth is necessary to avoid a rubbing or burnishing action. Excessive vibration may indicate that the cutter has insufficient clearance (rigidity of the tooling and fixtures should also be considered). Hogging-in generally indicates too much rake (or possibly too high a cutting speed).

**Milling Parameters and Cutting Fluid.** Table 9 lists speeds and feeds for end (peripheral) milling stainless steels using either high-speed steel or carbide tooling. It may be necessary to vary feeds from the nominal values. If the feed is too light, the tool will burnish the work; if too heavy, tool life will be shortened.

Speeds and feeds also vary for heavy roughing cuts and lighter finishing cuts. For

**Table 12 Nominal power hacksawing parameters for wrought stainless steels**

Depending on actual machining conditions, optimum parameters may be higher or lower than those given.

| UNS designation(a) | Pitch (teeth per 25 mm (1 in.) at a material thickness of: <6 mm (¼ in.) | 6–19 mm (¼–¾ in.) | 19–50 mm (¾–2 in.) | >50 mm (2 in.) | Speed, strokes/min | Feed mm/stroke | Feed in./stroke |
|---|---|---|---|---|---|---|---|
| **Free-machining alloys** | | | | | | | |
| **Ferritic alloys** | | | | | | | |
| S18235, S43020, S43023 | 10 | 6 | 4 | 4 | 130 | 0.15 | 0.006 |
| S18200 | 10 | 6 | 4 | 4 | 130 | 0.15 | 0.006 |
| **Martensitic alloys** | | | | | | | |
| S41600, S41610, S41623 | 10 | 6 | 4 | 4 | 130 | 0.15 | 0.006 |
| S42020, S42023 | 10 | 6 | 4 | 4 | 120 | 0.15 | 0.006 |
| S44020, S44023 | 10 | 6 | 4 | 4 | 110 | 0.15 | 0.006 |
| **Austenitic alloys** | | | | | | | |
| S20300, S30300, S30310, 30360 | 10 | 6 | 6 | 4 | 110 | 0.15 | 0.006 |
| S30323, S30330, S30345, S30431, S31620, S34720, S34723 | 10 | 6 | 6 | 4 | 95 | 0.15 | 0.006 |
| **Nonfree-machining alloys** | | | | | | | |
| **Ferritic alloys** | | | | | | | |
| S40500, S40900, S43000, 43400 | 10 | 6 | 4 | 4 | 120 | 0.15 | 0.006 |
| S44200, S44300, S44400, S44600 | 10 | 6 | 4 | 4 | 110 | 0.15 | 0.006 |
| **Martensitic alloys** | | | | | | | |
| S40300, S41000 | 10 | 6 | 4 | 4 | 110 | 0.15 | 0.006 |
| S41400, S42000, S42010 S43100 | 10 | 10 | 6 | 4 | 85 | 0.10 | 0.004 |
| S44002, S44003, S44004 | 10 | 10 | 6 | 4 | 50 | 0.10 | 0.004 |
| **Austenitic alloys** | | | | | | | |
| S20100, S30100, S30200, S30400, S30403, S30430, S30500, S31600, S31603, S32100, S34700, S38400 | 10 | 10 | 6 | 4 | 90 | 0.10 | 0.004 |
| S30900, S30908, S31000, 31008, S31700, S31703 | 10 | 10 | 6 | 4 | 65 | 0.10 | 0.004 |
| S20910, S21904, S24100, S28200, S30452, N08020 | 10 | 10 | 6 | 6 | 45 | 0.075 | 0.003 |
| **Duplex alloys** | | | | | | | |
| S31803, S32550, S32900, S32950 | 10 | 10 | 6 | 4 | 65 | 0.10 | 0.004 |
| **Precipitation-hardenable alloys(a)** | | | | | | | |
| S15500, S17400, S17700, S45000 | 10 | 10 | 6 | 4 | 60 | 0.10 | 0.004 |
| S13800, S35000, S35500, S45500 | 10 | 10 | 6 | 4 | 60 | 0.10 | 0.004 |
| S66286 | 10 | 10 | 6 | 4 | 45 | 0.075 | 0.003 |

(a) All machining parameters given are for material in the annealed condition (140–270 HB) except the precipitation-hardenable grades, which are solution treated (150–325 HB) or equalized and overtempered for S35000 or S35500. For material heat treated or cold drawn to hardnesses higher than the normal annealed or softest condition, speed reductions of approximately 10% to as much as 50% at the higher hardnesses are necessary for best results. Light-to-medium pressures are generally used (medium pressure at highest material thickness); heavy pressure can be used on the martensitic free-machining grades in the annealed condition at higher material thicknesses. Source: Ref 38, 52

example, a roughing cut would be run with heavier feeds and slower speeds than those used for finishing cuts. Regardless of the type of cut, both work and tool should be flooded with a good sulfurized oil properly diluted with paraffin oil, or preferably a good emulsifiable fluid. Milling generates considerable heat, which must be carried off by the cutting fluid; otherwise, the work will distort and the tool edges will dull or chip rapidly.

Once a milling cut has been started, it should not be stopped unless absolutely necessary because the tool will undercut when starting again. When it is necessary to back out and start again, the tool should be placed two or three turns behind the work before starting again. This eliminates the danger of backlash and guards against undercutting.

## Broaching

Broaches for stainless alloys are made of high-speed steel. A broach is a simple tool to handle because the broach manufacturer builds into it the necessary feed and depth of cut by steps from one tooth to another. Basically, a broach can provide a roughing cut, a semifinished cut, a final precision cut, or any combination of these operations (Fig. 35). Some broaches are made with burnishing buttons when a burnished finish is required. Because the form or shape of a broach tooth is unlimited, there is no limitation to the shape or contour of broached surfaces.

**Table 13 Cutting rates for stainless steels in abrasive water jet machining**

| Alloy | Thickness mm | Thickness in. | Cutting rate mm/min | Cutting rate in./min |
|---|---|---|---|---|
| S15500(a) | 3 | 0.13 | 230–380 | 9–15 |
| S15500(a) | 64 | 2.50 | 13–25 | 0.5–1 |
| S31600(a) | 75 (diam) | 3 (diam) | 13–50 | 0.5–2 |
| S17400(b) | 25 | 1 | 50 | 2 |

(a) 275–345 MPa (40–50 ksi) pressure with 60-mesh garnet. (b) 200 MPa (30 ksi) pressure with 60-mesh garnet. Source: Ref 53, 54

In designing the radius for the broach, the manufacturer provides maximum tooth strength and a pocket for chips. Depending on the job, the broach manufacturer can also incorporate into the broach such items as side relief (flat broaches), undercut and clearance (spline broaches), and chip breakers (to handle wide chips).

When a broach becomes dull, it should be resharpened only on a broach grinder or returned to the manufacturer for regrinding. For internal broaches, the back-off angle should be held to a minimum, preferably 2° and not more than 5° (Fig. 35). Too much back-off angle will shorten broach life because of size reductions from resharpening.

Any nicks on the cutting edges of the broach will score the surface of the work. Therefore, careful handling is very important.

**Broaching Parameters and Cutting Fluid.** Table 10 lists nominal feeds and speeds for broaching various stainless steels. Sulfochlorinated oils diluted with paraffin oil, rather than water-soluble oils, are suggested.

When broaching hard (35 HRC) stainless steel, it is necessary to reduce the speed and to use a lubricant with very little thinning. Broaches have been used to cut stock harder than 35 HRC, but the cost is usually prohibitive because the broach dulls rapidly. Where higher hardnesses are required, pieces should first be broached and then heat treated. To remove any light scale and to correct possible distortion resulting from heat treating, a special broach known as a hard gear broach can be used on machines having sufficient power.

## Reaming

The difficulties involved in reaming stainless steel are most often caused by previous operations, particularly with the nonfree-machining austenitic alloys. For example, if the feed in drilling is too light, the hole wall can be severely work hardened and can resist cutting by the reamer. It is also important that ample material be left from the previous operation to allow the reamer to get below any work-hardened layer.

**Table 14 Parameters used to drill holes with electron beam machining**

| Alloy | Workpiece thickness, mm | Workpiece thickness, in. | Hole diameter, mm | Hole diameter, in. | Drilling time, s | Accelerating voltage, kV | Average beam current, μA | Pulse width, μs | Pulse frequency, Hz |
|---|---|---|---|---|---|---|---|---|---|
| Ferritic/martensitic | 0.25 | 0.010 | 0.013 | 0.0005 | <1 | 130 | 60 | 4 | 3000 |
| Stainless steel | 1.0 | 0.040 | 0.13 | 0.005 | <1 | 140 | 100 | 80 | 50 |
| Stainless steel | 2.0 | 0.080 | 0.13 | 0.005 | 10 | 140 | 100 | 80 | 50 |
| Stainless steel | 2.5 | 0.100 | 0.13 | 0.005 | 10 | 140 | 100 | 80 | 50 |
| Stainless steel | 6.4 | 0.250 | 0.5–1.0 | 0.020–0.040 | 180 | 145 | 4000 | 2100 | 12.5 |

Source: Ref 55

**Reamer Design.** Reamers of straight or spiral-flute design are used for both straight and tapered holes. Spiral-flute reamers are usually preferred because they are less susceptible to chatter, can better dispose of chips from deep holes, and are capable of producing a better finish.

Recommended design for high-speed steel reamers for all grades of stainless steel at 135 to 425 HB is:

| | |
|---|---|
| Margin (or land) width, mm (in.) | 0.13–0.38 (0.005–0.015) |
| Chamfer angle, degrees | 30–40 |
| Chamfer length, mm (in.) | 1.5 (1/16) |
| Chamfer relief angle, degrees | 4–5 |
| Radial rake angle, degrees | 3–8 |
| Helix angle, degrees | 0–10 |
| Primary relief angle, degrees | 4–8 |

For carbide-tip reamers, margin (land) width should be 0.05 to 0.13 mm (0.002 to 0.005 in.) for reaming all free-machining stainless steels (135 to 425 HB) and nonfree-machining martensitic grades up to 52 HRC. For reaming the nonfree-machining ferritic and austenitic grades and the precipitation-hardening grades, the margin should be increased to 0.13 to 0.25 mm (0.005 to 0.010 in.). For reaming all grades, carbide-tip reamers should have the following geometry:

| | |
|---|---|
| Chamfer (lead) angle, degrees | 2 |
| Chamfer length, mm (in.) | 4.8 (3/16) |
| Radial rake angle, degrees | 7–10 |
| Helix angle, degrees | 5–8 |
| Primary relief angle, degrees | 7–15 |

**Reaming Parameters and Cutting Fluid.** Feeds and speeds for reaming stainless steels are listed in Table 11 for high-speed steel or carbide tooling and for roughing or finishing operations. When finish of the hole is not critical, the parameters for roughing can be used. Smooth finishes require significantly lower speeds. When both size and finish are important, a two-step operation should be used with both roughing and finishing cuts.

The cutting fluid must be given careful consideration to avoid overheating. In addition to providing good lubrication, the cutting fluid must be a coolant to carry away the heat that would otherwise burn the cutting edges of the reamer. The cutting fluid must also be kept clean. Reaming produces slivers and very fine chips, which can float in the cutting fluid and damage the surface finish of the work, especially if the machine is equipped with a recirculating system.

## Sawing

High-speed steel blades are used for cutting wrought and cast stainless steels by band sawing and power hacksawing. Pitches for band saw blades are eight to ten teeth per 25 mm (1 in.) for cutting material up to 6 mm (1/4 in.) thick, six to eight teeth per 25 mm (1 in.) for thicknesses of 6 to 32 mm (1/4 to 1 1/4 in.), and three to six teeth per 25 mm (1 in.) for materials thicker than 38 mm (1 1/2 in.).

Power hacksaw blades of ten teeth per 25 mm (1 in.) are recommended for material up to 20 mm (3/4 in.) thick. Coarser blades of six teeth per 25 mm (1 in.) are used for material from 20 to 50 mm (3/4 to 2 in.) in thickness; blades of four teeth per 25 mm (1 in.) are used for material 50 mm (2 in.) or more in thickness.

**Feeds and speeds** for sawing stainless steel are primarily influenced by the hardness of the material being cut. Table 12 lists some nominal feeds and speeds for the hacksawing of stainless steels. Table 9 in the article "Sawing" in this Volume lists speeds for the contour band sawing of stainless steels.

## Grinding

Aluminum oxide wheels are most commonly used for stainless steels. Silicon carbide wheels can also be used, but at a reduced wheel life; therefore, their use is limited to special applications. Medium-density wheels of hardness grades H to L are generally selected for stainless steels, although harder wheels are used for thread grinding. Grit sizes commonly used are 46, 54, or 60; finer grits can be used to produce a finer finish. Vitrified bond wheels are normally used, although the stronger resinoid bond wheels are preferred for equipment operated at higher speeds. Grinding wheels used previously to grind another metallic material should not be used to grind a stainless steel, because particles of the other material may be embedded in the stainless steel, affecting its corrosion resistance.

**Table 15 Plasma arc cutting parameters for stainless steels**

| Thickness, mm | Thickness, in. | Cutting speed, mm/min | Cutting speed, in./min | Power selection, A |
|---|---|---|---|---|
| 6 | 0.25 | 1800 | 70 | 105 |
| | | 2500 | 100 | 140 |
| | | 3800 | 150 | 210 |
| 13 | 0.50 | 500 | 20 | 135 |
| | | 1000 | 40 | 190 |
| | | 1800 | 70 | 250 |
| | | 2500 | 100 | 270 |
| | | 3800 | 150 | 700 |
| | | 5300 | 210 | 1000 |
| 25 | 1 | 250 | 10 | 175 |
| | | 500 | 20 | 210 |
| | | 760 | 30 | 270 |
| | | 1000 | 40 | 350 |
| | | 2000 | 80 | 540 |
| | | 2800 | 110 | 1000 |
| 38 | 1.5 | 250 | 10 | 280 |
| | | 500 | 20 | 420 |
| | | 1000 | 40 | 620 |
| | | 1800 | 70 | 1000 |
| 50 | 2 | 130 | 5 | 320 |
| | | 250 | 10 | 420 |
| | | 500 | 20 | 610 |
| | | 1000 | 40 | 950 |
| 65 | 2.5 | 130 | 5 | 410 |
| | | 250 | 10 | 550 |
| | | 500 | 20 | 820 |
| 75 | 3 | 130 | 5 | 510 |
| | | 250 | 10 | 675 |
| | | 500 | 20 | 1020 |
| 90 | 3.5 | 130 | 5 | 550 |
| | | 250 | 10 | 730 |
| | | 500 | 20 | 1110 |
| 100 | 4 | 130 | 5 | 675 |
| | | 250 | 10 | 900 |
| 115 | 4.5 | 130 | 5 | 900 |
| 130 | 5 | 75 | 3 | 1100 |
| 140 | 5.5 | 75 | 3 | 1100 |

Source: Ref 55

**Grinding Parameters and Cutting Fluid.** For many grinding operations, typical wheel speeds are 1500 to 2000 m/min (5000 to 6500 sfm). For surface grinding, table speeds are 15 to 30 m/min (50 to 100 sfm), with a down feed of up to 0.050 mm/pass (0.002 in./pass) for roughing and 0.013 mm/pass (0.0005 in./pass) for finishing and a cross feed of 1.3 to 13 mm/pass (0.050 to 0.500 in./pass). Thread grinding is done at higher speeds with harder wheels, as mentioned previously.

Because of the lower thermal conductivity of stainless steels, a good coolant is necessary when grinding them. Conventional water-soluble fluids generally provide lower grinding wheel life than heavy-duty sulfochlorinated soluble oils at a concentration of about 10%.

## Nontraditional Machining Operations

Although most stainless steels are machined with conventional techniques, nontraditional techniques are used when machining alloys with extreme toughness or hardness or when machining intricate shapes. This section will briefly describe some of the nontraditional machining techniques that have been successfully applied to stainless steels. More detailed information on the processes and basic operating parameters can be found elsewhere in this Volume.

**Abrasive jet machining** is not intended for bulk material removal. This process is used for cleaning and deburring operations.

**Abrasive waterjet machining** provides omnidirectional cutting and is highly suitable for automation and the cutting of complex shapes from stainless steel. It is slower than plasma arc machining, but does not introduce residual stress or a heat-affected zone. Table 13 lists some abrasive waterjet cutting rates for stainless steel.

**Electrochemical machining** is used for machining hard materials into complex shapes. Electrochemical machining is faster than electrical discharge machining and does not introduce metallurgical changes or distortion at the surface.

**Electrostream and shaped tube electrolytic machining** are modified electrochemical machining processes that are used for drilling small holes in corrosion-resistant metals. The articles "Electrostream and Capillary Drilling" and "Shaped Tube Electrolytic Machining" in this Volume describe their capabilities in more detail.

**Electron beam machining** is used in the high-precision drilling of fine holes or narrow slots in thin materials. Table 14 lists some electron beam drilling times of stainless steel.

**Laser beam machining** is also used to drill small-diameter holes. Unlike electron beam machining, laser beam machining does not require a vacuum.

**Plasma arc machining** is used to perform straight cuts of tough materials at high speeds. Table 15 provides some typical cutting rates with plasma arc machining.

## REFERENCES


1. Machining and Abrasive Wheel Grinding of Carpenter Stainless Steels, in *Carpenter Stainless Steels, Selection, Alloy Data, Fabrication*, Carpenter Technology Corporation, 1987, p 240-241
2. V.A. Tipnis, Machining of Stainless Steels, *Wire*, Aug 1971, p 153-161
3. Metallurgy of Welding Stainless Steels, in *Stainless Steels*, American Society for Metals, 1978, p 11-1 to 11-22
4. R.A. Lula, Fabrication of Stainless Steels—Machining, in *Stainless Steels*, American Society for Metals, 1986, p 112-114
5. "Simplifying Stainless Steel Selection With Carpenter's Selectaloy Method," Carpenter Technology Corporation, 1969
6. Corrosion Control, in *Carpenter Stainless Steels, Selection, Alloy Data, Fabrication*, Carpenter Technology Corporation, 1987, p 211-222
7. F.R. Palmer, Antifriction Corrosion-Resisting Steel, U.S. Patent 1,835,960, 1931
8. F.R. Palmer, Ferrous Alloy, U.S. Patent 1,961,777, 1934
9. F.M. Richmond, A Decade of Progress in Machinability, Finishing and Forming, *Met. Prog.*, Aug 1967, p 85-86
10. C.W. Kovach, Sulfide Inclusions and the Machinability of Steel, in *Sulfide Inclusions in Steel*, American Society for Metals, 1975, p 459-479
11. C. Sparre, Stainless Free-Cutting Steel, *Wire*, April 1972, p 56-60
12. C.W. Kovach *et al.*, The Sulfide Phases in Iron-Chromium Steels, *Trans. ASM*, Vol 61, 1968, p 575-581
13. A.J. Sedriks, Role of Sulfide Inclusions in Pitting and Crevice Corrosion of Stainless Steels, *Int. Met. Rev.*, Vol 28 (No. 5), 1983, p 295-307
14. M. Henthorne, Corrosion of Resulfurized Free-Machining Stainless Steels, *Corrosion*, Vol 26 (No. 12), Dec 1970, p 511-528
15. H.W. Garvin and R.M. Larrimore, Jr., Metallurgical Factors Affecting the Machining of a Free Machining Stainless Steel, in *Mechanical Working of Steel II*, American Institute of Mining, Metallurgical, and Petroleum Engineers, 1965, p 133-150
16. C.W. Kovach and A. Moskowitz, Effects of Manganese and Sulfur on the Machinability of Martensitic Stainless Steels, *Trans. Met. Soc. AIME*, Vol 245, Oct 1969, p 2157-2164
17. C.W. Kovach and J.J. Eckenrod, Free Machining Austenitic Stainless Steels, in *Proceedings of the 13th Mechanical Working and Steel Processing Conference*, American Institute of Mining, Metallurgical, and Petroleum Engineers, 1971, p 300-325
18. C.W. Kovach *et al.*, Improved Free-Machining Ferritic Stainless Steel, British Patent 1,137,295, 1968
19. J.A. Ferree, Jr., Free Machining Austenitic Stainless Steel, U.S. Patent 3,888,659, 1975
20. R. Kiessling, Sulfide Inclusions: A Summary and Study of Sulfides in Resulfurized Ferritic ELI-Steels of the 18:2 Type, in *Sulfide Inclusions in Steel*, American Society for Metals, 1975, p 104-122
21. R. Kiessling, Influence of Sulfide Composition on the Machinability and Corrosion Properties of a Resulfurized Ferritic 18:2 Steel, in *Influence of Metallurgy on Machinability of Steels*, Proceedings of an International Symposium, American Society for Metals, 1977, p 253-261
22. H. Kiesheyer and H. Brandis, Titanium Alloyed Free-Machining Stainless Steels, *Bull. Cercle d'Études Metaux*, Vol 15 (No. 11), 1986
23. W.C. Clarke, Which Free-Machining Stainless?, *Metalwork. Prod.*, 27 May 1964, p 43-45
24. W.C. Clarke, Which Free-Machining Chromium Stainless?, *Metalwork. Prod.*, 9 Sept 1964, p 68-71
25. T. Fujiwara *et al.*, Effects of Free-Machining Additives on Machinability of the 18Cr-2Mo Stainless Steel, in *Influence of Metallurgy on Machinability of Steel*, Proceedings of an International Symposium, American Society for Metals, 1977, p 231-240
26. E. Herzog and J. Bellot, Influence of Tellurium and Selenium on the Properties of Some Steels, in *Selenium and Tellurium in Iron and Steel*, Swedish Institute for Metal Research, 1969
27. "Free-Machining Stainless Steels," American Iron and Steel Institute, 1975
28. V.A. Tipnis, Stainless Steel Having Improved Machinability, U.S. Patent 3,846,186, 1974
29. V.A. Tipnis, Method of Making Stainless Steel Having Improved Machinability, U.S. Patent 3,933,480, 1976
30. N.L. McClymonds, Machinability of Leaded Stainless, *Met. Prog.*, Aug 1964, p 166, 168
31. N.L. McClymonds, Measuring Machinability, *Met. Prog.*, Aug 1967, p 183-184
32. K. Ono *et al.*, Development of Leaded Free-Machining Austenitic Stainless Steel, *Denki Seiko (Electro. Furn. Steel)*, Vol 54 (No. 4), Nov 1983, p 265-274
33. H. Pray *et al.*, Addition of Bismuth for Producing Free-Machining Stainless Steels, *Proc. ASTM*, Vol 41, 1941, p 646-658
34. A. Kimura *et al.*, Super STARCUT Stainless Steel 304BF With Bismuth, *Bull. Bismuth Inst.*, 1986, p 1-5
35. B. Heritier *et al.*, Machinability of Stainless Steels, Evolution and Perspectives, *Bull. Cercle d'Études Metaux*, Vol 15 (No. 11), 1986
36. T. Kato *et al.*, The Effect of Ca Oxide Inclusions on the Machinability of Heavy Duty Steels, in *The Machinability of Engineering Materials*, American Society for Metals, 1983, p 323-337
37. T. Fujiwara and T. Ito, Calcium-Bearing Free-Machining Steels, *Bull. Jpn. Inst. Met.*, Vol 15 (No. 10), 1976, p 613-620 (BISI 17016)
38. *Guide to Machining Stainless Steels and Other Specialty Metals*, Carpenter Technology Corporation, 1985
39. D.M. Blott, Machining Wrought and Cast Stainless Steels, in *Handbook of Stainless Steels*, McGraw-Hill, 1977, p 24-2 to 24-30
40. C.A. Divine, Jr., What to Consider in Choosing an Alloy, *Met. Prog.*, Feb 1968, p 19-23

41. L. Colombier and J. Hochmann, Manufacturing, Forming and Finishing Techniques—Machining, in *Stainless and Heat Resisting Steels*, St. Martin's Press, 1968, p 508-514
42. Machining Operations, in *Stainless Steel Fabrication*, Allegheny Ludlum Steel Corporation, 1959, p 223-259
43. A. Moskowitz *et al.*, Free-Machining Stainless Steels, U.S. Patent 3,401,035, 1968
44. C.W. Kovach and A. Moskowitz, How to Upgrade Free-Machining Properties, *Met. Prog.*, Aug 1967, p 173-180
45. J.R. Blank *et al.*, Improved and More Consistent Steels for Machining, in *Influence of Metallurgy on Machinability of Steel*, Proceedings of an International Symposium, American Society for Metals, 1977, p 397-419
46. W.C. Clarke, Jr., Free-Machining Stainless Steel and Method, U.S. Patent 2,697,035, 1954
47. J.J. Eckenrod and C.W. Kovach, Effects of Manganese on Austenitic Stainless Steels, *Met. Eng. Quart.*, Feb 1972, p 5-10
48. R.P. Ney, Sr., Free Machining, Cold Formable Austenitic Stainless Steel, U.S. Patent 4,444,588, 1984
49. J.J. Eckenrod *et al.*, Low Carbon Plus Nitrogen, Free-Machining Austenitic Stainless Steel, U.S. Patent 4,613,367, 1986
50. P.K. Wright and A. Bagchi, Wear Mechanisms That Dominate Tool-Life in Machining, *J. Appl. Metalwork.*, Vol 1 (No. 4), 1981, p 15-23
51. "Carpenter Custom 630 (17Cr-4Ni)," Carpenter Technology Corporation, 1971
52. *Machining Data Handbook*, Vol 1, Metcut Research Associates Inc., 1980
53. B.L. Schwartz, Principles and Applications of Water and Abrasive Jet Cutting, in *High Productivity Machining: Materials and Processes*, American Society for Metals, 1985, p 291-298
54. A.L. Hitchcox, Vote of Confidence for Abrasive Waterjet Cutting, *Met. Prog.*, July 1986, p 33-42
55. *Machining Data Handbook*, Vol 2, Metcut Research Associates, Inc., 1980


## SELECTED REFERENCES


- D.M. Blott, How to Cut the Standard Grades, *Met. Prog.*, Feb 1968, p 337-343
- D. Daniels, AWJ Cutting, A New Tool for Metal Fabricators, *Met. Stamp.*, Sept 1986, p 3-6
- C.A. Divine, Jr., How to Machine the Stainless Steels, Carpenter Steel Company; originally published in *Mod. Mach. Shop*, 1968
- D. Elza and S. Burns, Lasers Take Their Place in Metalworking, *Mach. Tool Blue Book*, July 1985, p 34-38
- M. Field *et al.*, The Surface Effects Produced in Nonconventional Metal Removal—Comparison With Conventional Machining Techniques, *Met. Eng. Quart.*, Aug 1966, p 32-45
- T.E. Gittens, Plasma Arc Cutting—Process Fundamentals, *Trans. AFS*, Vol 92, 1984, p 29-35
- S. Holden, The Plasma Cutting of Stainless Steel, *Stainless Steel Ind.*, Vol 13 (No. 74), July 1985, p 12-25
- B. Kellock, Have a Ball With a Cinderella Process!, *Mach. Prod. Eng.*, May 19, 1982, p 40-46
- "Machining the Austenitic Chromium-Nickel Stainless Steels," International Nickel Company, 1972
- N. McClymonds *et al.*, Shop Practice, in *Machining, Estimating Manual for Stainless Steels*, Joslyn Stainless Steels, 1963, p 79-122
- R.K. Mosavi, Comparing Laser and Waterjet Cutting, *Lasers Optronics*, Vol 6 (No. 7), July 1987, p 65-68
- *Nontraditional Machining*, Conference Proceedings, American Society for Metals, 1986
- V.K. Sarin, Ed., *High Productivity Machining: Materials and Processes*, Conference Proceedings, American Society for Metals, 1985, p 291-298
- R.J. Saunders, Laser Metalworking, *Met. Prog.*, July 1984, p 45-51
- W.F. Tutte, The Application of Laser Cutting to Stainless Steel, *Stainless Steel Ind.*, Vol 15 (No. 86), July 1987, p 9, 11

# Machining of Tool Steels

THE HIGH CARBON and alloy contents that make tool steels serviceable as tools also make them more difficult to machine than the lower-carbon and the lower-alloy constructional steels. Table 1 gives nominal compositions of the alloy tool steels covered in this article (additional information can be found in the articles "High-Speed Tool Steels" and "P/M High-Speed Tool Steels" in this Volume). Several of the alloying elements used in tool steel, especially chromium, tungsten, molybdenum, and vanadium, readily form carbides that have adverse effects in machining. These effects are markedly influenced by the size, shape, and distribution of the carbide particles in the matrix of the steel.

Most tool steels (notably the higher-carbon tool steels) are easiest to machine when they have been annealed to a microstructure that consists of small spheroidal carbides uniformly distributed in a matrix of ferrite.

In tool steels containing less than about 0.75% C, the spheroids are more likely to become large and widely dispersed in relatively large areas of ferrite. Carbides dispersed in this manner cause poor finish and low tool life. If the alloy content of the steel is low, the ferrite is characteristically gummy; if the alloy content is high, the ferrite is tough. Thus, the properties of the ferrite matrix, as well as the size and distribution of carbides, can be a major factor in the machinability of tool steel. The preferred structure usually is a mixture of spheroidite and lamellar pearlite obtained by controlled annealing. As the carbon content is increased to approximately 1% in unalloyed tool steel, the spheroids become finer, more numerous, and more closely spaced.

To provide a basis for comparing the relative machinability of different types of tool steel, carbon tool steel containing 1% C is rated 100; other tool steels are rated as a percentage of 100, as shown in Table 2.

The micrographs in Fig. 1 indicate the differences in microstructure characteristic of three different tool steels (W1 containing 1% C, A2, and M2) in the annealed condition. As shown in Fig. 1(a), the microstructure of the W1 steel is a uniform distribution of small spheroids. In the A2 steel (Fig. 1b), massive alloy carbides appear, and in the M2 steel (Fig. 1c), the carbides are larger and more numerous than in A2. As annealed, the hardness ranges are 156 to 196 HB for W1, 202 to 217 HB for A2, and 217 to 235 HB for M2. However, the increase in hardness is only partly indicative of the relative decrease in machinability encountered in a comparison of these steels. Compared to the iron carbide in the unalloyed steel, the more abrasive alloy carbides have a greater adverse effect on machining characteristics than the slightly higher hardness of the annealed steels containing them would indicate.

**Hardness.** Low-carbon tool steels are relatively soft in the annealed condition and have machining characteristics similar to those of their counterparts in tonnage carbon and alloy steels. For instance, P1 mold steel has machining characteristics similar to those of 1010, and P20 closely resembles 4130 in machinability, under conditions of similar annealing practice.

W1 steel can be annealed to a hardness of less than 200 HB, whereas the typical hardness of annealed T15 and M15 high-speed steels is about 255 to 275 HB. Tool steels containing high percentages of cobalt, which strengthens the ferrite, also have lower machinability because of generally higher hardness, as annealed. T5 and T6 cobalt-containing high-speed steels can be annealed no softer than about 250 to 275 HB.

As machinability decreases because of hardness or composition, or both, greater power and rigidity are required in the machine tool to perform a given machining operation.

**Tool Steel Selection.** An aid to the selection of tool steels is the graphs given in Fig. 2, in which the major selective factors (wear resistance, toughness, and hot hardness) are plotted against each other.

Tool steels vary widely in wear resistance, and the numbers 1 to 9 used to rate them are not intended to indicate the quantitative differences, which are considerably greater than 9 to 1. Hardness too can significantly influence this property, while toughness of tool steels varies widely with carbon content, heat treatment, surface condition of tools, and size of section.

This method of selection is recommended. The user should learn the technique of analyzing an application for tool steel on the basis of important required metallurgical properties and select the group or class of tool steel that most closely matches required properties. On many applications, two or more classes need to be considered, depending on the interrelation of the major selective factors (wear resistance, toughness, hot hardness) and the minor selective factors in addition to those affecting manufacturing and cost.

Table 3 is a quick reference guide for selecting tool steels on the basis of specific product applications.

## Turning

Tools made from high-speed tool steel, carbide, and cast Co-Cr-W alloy are all used for turning tool steels. For continuous cutting with sufficient rigidity to prevent chatter, carbide tools usually provide the greatest productivity at the lowest tool cost per piece. However, for shock loads (as in interrupted cutting), for the turning of forged drill blanks, or for setups lacking rigidity, high-speed tool steel tools are more suitable than carbide. Cast alloy tools are especially suited for turning applications in which rugged cutting is necessary or in which the use of cutting fluids is impractical.

Indexable coated carbide, cubic boron nitride (CBN), and ceramic tooling inserts have also been used for turning because of their higher resistance to wear and hence longer tool life.

**Indexable carbide inserts** have now replaced ground carbide tools in most applications, with the exception of form tools, which consist of electrical discharge machined (EDM) micrograin carbide.

Standard ANSI inserts and toolholders to match the shape of the parts being machined should be used. Negative rake is the most cost effective for straight carbide and coated carbide inserts.

**Ceramic inserts** are currently being used in machining operations involving the turning and boring of tool steels. Aluminum oxide-base ceramics are used almost exclusively to machine tool steels; $Si_3N_4$-base ceramics are unsuitable because of chemical interactions that occur at the cutting interface. Thus, while $Si_3N_4$-base ceramics are now commonly used to machine aerospace materials such as high nickel-base alloys and automotive components made of cast irons, $Al_2O_3$-base ceramics have found widespread use in the machining of tool steel materials.

Both cold-pressed and hot-pressed $Al_2O_3$-base ceramics are used to machine

## Table 1 Composition limits of principal types of tool steels

| Designations | | | Composition, %(a) | | | | | | | | |
|---|---|---|---|---|---|---|---|---|---|---|---|
| AISI | SAE | UNS | C | Mn | Si | Cr | Ni | Mo | W | V | Co |
| **Molybdenum high-speed steels** | | | | | | | | | | | |
| M1 | M1 | T11301 | 0.78–0.88 | 0.15–0.40 | 0.20–0.50 | 3.50–4.00 | 0.30 max | 8.20–9.20 | 1.40–2.10 | 1.00–1.35 | . . . |
| M2 | M2 | T11302 | 0.78–0.88; 0.95–1.05 | 0.15–0.40 | 0.20–0.45 | 3.75–4.50 | 0.30 max | 4.50–5.50 | 5.50–6.75 | 1.75–2.20 | . . . |
| M3, class 1 | M3 | T11313 | 1.00–1.10 | 0.15–0.40 | 0.20–0.45 | 3.75–4.50 | 0.30 max | 4.75–6.50 | 5.00–6.75 | 2.25–2.75 | . . . |
| M3, class 2 | M3 | T11323 | 1.15–1.25 | 0.15–0.40 | 0.20–0.45 | 3.75–4.50 | 0.30 max | 4.75–6.50 | 5.00–6.75 | 2.75–3.75 | . . . |
| M4 | M4 | T11304 | 1.25–1.40 | 0.15–0.40 | 0.20–0.45 | 3.75–4.75 | 0.30 max | 4.25–5.50 | 5.25–6.50 | 3.75–4.50 | . . . |
| M6 | . . . | T11306 | 0.75–0.85 | 0.15–0.40 | 0.20–0.45 | 3.75–4.50 | 0.30 max | 4.50–5.50 | 3.75–4.75 | 1.30–1.70 | 11.00–13.00 |
| M7 | . . . | T11307 | 0.97–1.05 | 0.15–0.40 | 0.20–0.55 | 3.50–4.00 | 0.30 max | 8.20–9.20 | 1.40–2.10 | 1.75–2.25 | . . . |
| M10 | . . . | T11310 | 0.84–0.94; 0.95–1.05 | 0.10–0.40 | 0.20–0.45 | 3.75–4.50 | 0.30 max | 7.75–8.50 | . . . | 1.80–2.20 | . . . |
| M30 | . . . | T11330 | 0.75–0.85 | 0.15–0.40 | 0.20–0.45 | 3.50–4.25 | 0.30 max | 7.75–9.00 | 1.30–2.30 | 1.00–1.40 | 4.50–5.50 |
| M33 | . . . | T11333 | 0.85–0.92 | 0.15–0.40 | 0.15–0.50 | 3.50–4.00 | 0.30 max | 9.00–10.00 | 1.30–2.10 | 1.00–1.35 | 7.75–8.75 |
| M34 | . . . | T11334 | 0.85–0.92 | 0.15–0.40 | 0.20–0.45 | 3.50–4.00 | 0.30 max | 7.75–9.20 | 1.40–2.10 | 1.90–2.30 | 7.75–8.75 |
| M36 | . . . | T11336 | 0.80–0.90 | 0.15–0.40 | 0.20–0.45 | 3.75–4.50 | 0.30 max | 4.50–5.50 | 5.50–6.50 | 1.75–2.25 | 7.75–8.75 |
| M41 | . . . | T11341 | 1.05–1.15 | 0.20–0.50 | 0.15–0.50 | 3.75–4.50 | 0.30 max | 3.25–4.25 | 6.25–7.00 | 1.75–2.25 | 4.75–5.75 |
| M42 | . . . | T11342 | 1.05–1.15 | 0.15–0.40 | 0.15–0.65 | 3.50–4.25 | 0.30 max | 9.00–10.00 | 1.15–1.85 | 0.95–1.35 | 7.75–8.75 |
| M43 | . . . | T11343 | 1.15–1.25 | 0.20–0.40 | 0.15–0.65 | 3.50–4.25 | 0.30 max | 7.50–8.50 | 2.25–3.00 | 1.50–1.75 | 7.75–8.75 |
| M44 | . . . | T11344 | 1.10–1.20 | 0.20–0.40 | 0.30–0.55 | 4.00–4.75 | 0.30 max | 6.00–7.00 | 5.00–5.75 | 1.85–2.20 | 11.00–12.25 |
| M46 | . . . | T11346 | 1.22–1.30 | 0.20–0.40 | 0.40–0.65 | 3.70–4.20 | 0.30 max | 8.00–8.50 | 1.90–2.20 | 3.00–3.30 | 7.80–8.80 |
| M47 | . . . | T11347 | 1.05–1.15 | 0.15–0.40 | 0.20–0.45 | 3.50–4.00 | 0.30 max | 9.25–10.00 | 1.30–1.80 | 1.15–1.35 | 4.75–5.25 |
| **Tungsten high-speed steels** | | | | | | | | | | | |
| T1 | T1 | T12001 | 0.65–0.80 | 0.10–0.40 | 0.20–0.40 | 3.75–4.00 | 0.30 max | . . . | 17.25–18.75 | 0.90–1.30 | . . . |
| T2 | T2 | T12002 | 0.80–0.90 | 0.20–0.40 | 0.20–0.40 | 3.75–4.50 | 0.30 max | 1.00 max | 17.50–19.00 | 1.80–2.40 | . . . |
| T4 | T4 | T12004 | 0.70–0.80 | 0.10–0.40 | 0.20–0.40 | 3.75–4.50 | 0.30 max | 0.40–1.00 | 17.50–19.00 | 0.80–1.20 | 4.25–5.75 |
| T5 | T5 | T12005 | 0.75–0.85 | 0.20–0.40 | 0.20–0.40 | 3.75–5.00 | 0.30 max | 0.50–1.25 | 17.50–19.00 | 1.80–2.40 | 7.00–9.50 |
| T6 | . . . | T12006 | 0.75–0.85 | 0.20–0.40 | 0.20–0.40 | 4.00–4.75 | 0.30 max | 0.40–1.00 | 18.50–21.00 | 1.50–2.10 | 11.00–13.00 |
| T8 | T8 | T12008 | 0.75–0.85 | 0.20–0.40 | 0.20–0.40 | 3.75–4.50 | 0.30 max | 0.40–1.00 | 13.25–14.75 | 1.80–2.40 | 4.25–5.75 |
| T15 | . . . | T12015 | 1.50–1.60 | 0.15–0.40 | 0.15–0.40 | 3.75–5.00 | 0.30 max | 1.00 max | 11.75–13.00 | 4.50–5.25 | 4.75–5.25 |
| **Chromium hot-work steels** | | | | | | | | | | | |
| H10 | . . . | T20810 | 0.35–0.45 | 0.25–0.70 | 0.80–1.20 | 3.00–3.75 | 0.30 max | 2.00–3.00 | . . . | 0.25–0.75 | . . . |
| H11 | H11 | T20811 | 0.33–0.43 | 0.20–0.50 | 0.80–1.20 | 4.75–5.50 | 0.30 max | 1.10–1.60 | . . . | 0.30–0.60 | . . . |
| H12 | H12 | T20812 | 0.30–0.40 | 0.20–0.50 | 0.80–1.20 | 4.75–5.50 | 0.30 max | 1.25–1.75 | 1.00–1.70 | 0.50 max | . . . |
| H13 | H13 | T20813 | 0.32–0.45 | 0.20–0.50 | 0.80–1.20 | 4.75–5.50 | 0.30 max | 1.10–1.75 | . . . | 0.80–1.20 | . . . |
| H14 | . . . | T20814 | 0.35–0.45 | 0.20–0.50 | 0.80–1.20 | 4.75–5.50 | 0.30 max | . . . | 4.00–5.25 | . . . | . . . |
| H19 | . . . | T20819 | 0.32–0.45 | 0.20–0.50 | 0.20–0.50 | 4.00–4.75 | 0.30 max | 0.30–0.55 | 3.75–4.50 | 1.75–2.20 | 4.00–4.50 |
| **Tungsten hot-work steels** | | | | | | | | | | | |
| H21 | H21 | T20821 | 0.26–0.36 | 0.15–0.40 | 0.15–0.50 | 3.00–3.75 | 0.30 max | . . . | 8.50–10.00 | 0.30–0.60 | . . . |
| H22 | . . . | T20822 | 0.30–0.40 | 0.15–0.40 | 0.15–0.40 | 1.75–3.75 | 0.30 max | . . . | 10.00–11.75 | 0.25–0.50 | . . . |
| H23 | . . . | T20823 | 0.25–0.35 | 0.15–0.40 | 0.15–0.60 | 11.00–12.75 | 0.30 max | . . . | 11.00–12.75 | 0.75–1.25 | . . . |
| H24 | . . . | T20824 | 0.42–0.53 | 0.15–0.40 | 0.15–0.40 | 2.50–3.50 | 0.30 max | . . . | 14.00–16.00 | 0.40–0.60 | . . . |
| H25 | . . . | T20825 | 0.22–0.32 | 0.15–0.40 | 0.15–0.40 | 3.75–4.50 | 0.30 max | . . . | 14.00–16.00 | 0.40–0.60 | . . . |
| H26 | . . . | T20826 | 0.45–0.55(b) | 0.15–0.40 | 0.15–0.40 | 3.75–4.50 | 0.30 max | . . . | 17.25–19.00 | 0.75–1.25 | . . . |
| **Molybdenum hot-work steels** | | | | | | | | | | | |
| H42 | . . . | T20842 | 0.55–0.70(b) | 0.15–0.40 | . . . | 3.75–4.50 | 0.30 max | 4.50–5.50 | 5.50–6.75 | 1.75–2.20 | . . . |
| **Air-hardening, medium-alloy, cold-work steels** | | | | | | | | | | | |
| A2 | A2 | T30102 | 0.95–1.05 | 1.00 max | 0.50 max | 4.75–5.50 | 0.30 max | 0.90–1.40 | . . . | 0.15–0.50 | . . . |
| A3 | . . . | T30103 | 1.20–1.30 | 0.40–0.60 | 0.50 max | 4.75–5.50 | 0.30 max | 0.90–1.40 | . . . | 0.80–1.40 | . . . |
| A4 | . . . | T30104 | 0.95–1.05 | 1.80–2.20 | 0.50 max | 0.90–2.20 | 0.30 max | 0.90–1.40 | . . . | . . . | . . . |
| A6 | . . . | T30106 | 0.65–0.75 | 1.80–2.50 | 0.50 max | 0.90–1.20 | 0.30 max | 0.90–1.40 | . . . | . . . | . . . |
| A7 | . . . | T30107 | 2.00–2.85 | 0.80 max | 0.50 max | 5.00–5.75 | 0.30 max | 0.90–1.40 | 0.50–1.50 | 3.90–5.15 | . . . |
| A8 | . . . | T30108 | 0.50–0.60 | 0.50 max | 0.75–1.10 | 4.75–5.50 | 0.30 max | 1.15–1.65 | 1.00–1.50 | . . . | . . . |
| A9 | . . . | T30109 | 0.45–0.55 | 0.50 max | 0.95–1.15 | 4.75–5.50 | 1.25–1.75 | 1.30–1.80 | . . . | 0.80–1.40 | . . . |
| A10 | . . . | T30110 | 1.25–1.50(c) | 1.60–2.10 | 1.00–1.50 | . . . | 1.55–2.05 | 1.25–1.75 | . . . | . . . | . . . |
| **High-carbon, high-chromium, cold-work steels** | | | | | | | | | | | |
| D2 | D2 | T30402 | 1.40–1.60 | 0.60 max | 0.60 max | 11.00–13.00 | 0.30 max | 0.70–1.20 | . . . | 1.10 max | 1.00 max |
| D3 | D3 | T30403 | 2.00–2.35 | 0.60 max | 0.60 max | 11.00–13.50 | 0.30 max | . . . | 1.00 max | 1.00 max | . . . |
| D4 | . . . | T30404 | 2.05–2.40 | 0.60 max | 0.60 max | 11.00–13.00 | 0.30 max | 0.70–1.20 | . . . | 1.00 max | . . . |
| D5 | D5 | T30405 | 1.40–1.60 | 0.60 max | 0.60 max | 11.00–13.00 | 0.30 max | 0.70–1.20 | . . . | 1.00 max | 2.50–3.50 |
| D7 | D7 | T30407 | 2.15–2.50 | 0.60 max | 0.60 max | 11.50–13.50 | 0.30 max | 0.70–1.20 | . . . | 3.80–4.40 | . . . |
| **Oil-hardening cold-work steels** | | | | | | | | | | | |
| O1 | O1 | T31501 | 0.85–1.00 | 1.00–1.40 | 0.50 max | 0.40–0.60 | 0.30 max | . . . | 0.40–0.60 | 0.30 max | . . . |
| O2 | O2 | T31502 | 0.85–0.95 | 1.40–1.80 | 0.50 max | 0.35 max | 0.30 max | 0.30 max | . . . | 0.30 max | . . . |
| O6 | O6 | T31506 | 1.25–1.55(c) | 0.30–1.10 | 0.55–1.50 | 0.30 max | 0.30 max | 0.20–0.30 | . . . | . . . | . . . |
| O7 | . . . | T31507 | 1.10–1.30 | 1.00 max | 0.60 max | 0.35–0.85 | 0.30 max | 0.30 max | 1.00–2.00 | 0.40 max | . . . |
| **Shock-resisting steels** | | | | | | | | | | | |
| S1 | S1 | T41901 | 0.40–0.55 | 0.10–0.40 | 0.15–1.20 | 1.00–1.80 | 0.30 max | 0.50 max | 1.50–3.00 | 0.15–0.30 | . . . |
| S2 | S2 | T41902 | 0.40–0.55 | 0.30–0.50 | 0.90–1.20 | . . . | 0.30 max | 0.30–0.60 | . . . | 0.50 max | . . . |
| S5 | S5 | T41905 | 0.50–0.65 | 0.60–1.00 | 1.75–2.25 | 0.35 max | . . . | 0.20–1.35 | . . . | 0.35 max | . . . |
| S6 | . . . | T41906 | 0.40–0.50 | 1.20–1.50 | 2.00–2.50 | 1.20–1.50 | . . . | 0.30–0.50 | . . . | 0.20–0.40 | . . . |
| S7 | . . . | T41907 | 0.45–0.55 | 0.20–0.80 | 0.20–1.00 | 3.00–3.50 | . . . | 1.30–1.80 | . . . | 0.20–0.30(d) | . . . |

(a) All steels except group W contain 0.25 max Cu, 0.03 max P, and 0.03 max S; group W steels contain 0.20 max Cu, 0.025 max P, and 0.025 max S. Where specified, sulfur may be increased to 0.06 to 0.15% to improve machinability of groups H, M, and T. (b) Available in several carbon ranges. (c) Contains free graphite in the microstructure. (d) Optional. (e) Also contains 1.05–1.25 Al. (f) Specified carbon ranges are designated by suffix numbers.

(continued)

**Table 1 (continued)**

| Designations AISI | SAE | UNS | Composition, %(a) C | Mn | Si | Cr | Ni | Mo | W | V | Co |
|---|---|---|---|---|---|---|---|---|---|---|---|
| **Low-alloy special-purpose tool steels** | | | | | | | | | | | |
| L2 | ... | T61202 | 0.45–1.00(b) | 0.10–0.90 | 0.50 max | 0.70–1.20 | ... | 0.25 max | ... | 0.10–0.30 | ... |
| L6 | L6 | T61206 | 0.65–0.75 | 0.25–0.80 | 0.50 max | 0.60–1.20 | 1.25–2.00 | 0.50 max | ... | 0.20–0.30(d) | ... |
| **Low-carbon mold steels** | | | | | | | | | | | |
| P2 | ... | T51602 | 0.10 max | 0.10–0.40 | 0.10–0.40 | 0.75–1.25 | 0.10–0.50 | 0.15–0.40 | ... | ... | ... |
| P3 | ... | T51603 | 0.10 max | 0.20–0.60 | 0.40 max | 0.40–0.75 | 1.00–1.50 | ... | ... | ... | ... |
| P4 | ... | T51604 | 0.12 max | 0.20–0.60 | 0.10–0.40 | 4.00–5.25 | ... | 0.40–1.00 | ... | ... | ... |
| P5 | ... | T51605 | 0.10 max | 0.20–0.60 | 0.40 max | 2.00–2.50 | 0.35 max | ... | ... | ... | ... |
| P6 | ... | T51606 | 0.05–0.15 | 0.35–0.70 | 0.10–0.40 | 1.25–1.75 | 3.25–3.75 | ... | ... | ... | ... |
| P20 | ... | T51620 | 0.28–0.40 | 0.60–1.00 | 0.20–0.80 | 1.40–2.00 | ... | 0.30–0.55 | ... | ... | ... |
| P21(e) | ... | T51621 | 0.18–0.22 | 0.20–0.40 | 0.20–0.40 | 0.20–0.30 | 3.90–4.25 | ... | ... | 0.15–0.25 | ... |
| **Water-hardening tool steels** | | | | | | | | | | | |
| W1 | W108, W109 W110, W112 | T72301 | 0.70–1.50(f) | 0.10–0.40 | 0.10–0.40 | 0.15 max | 0.20 max | 0.10 max | 0.15 max | 0.10 max | ... |
| W2 | W209, W210 | T72302 | 0.85–1.50(f) | 0.10–0.40 | 0.10–0.40 | 0.15 max | 0.20 max | 0.10 max | 0.15 max | 0.15–0.35 | ... |
| W5 | ... | T72305 | 1.05–1.15 | 0.10–0.40 | 0.10–0.40 | 0.40–0.60 | 0.20 max | 0.10 max | 0.15 max | 0.10 max | ... |

(a) All steels except group W contain 0.25 max Cu, 0.03 max P, and 0.03 max S; group W steels contain 0.20 max Cu, 0.025 max P, and 0.025 max S. Where specified, sulfur may be increased to 0.06 to 0.15% to improve machinability of groups H, M, and T. (b) Available in several carbon ranges. (c) Contains free graphite in the microstructure. (d) Optional. (e) Also contains 1.05–1.25 Al. (f) Specified carbon ranges are designated by suffix numbers.

tool steels. These $Al_2O_3$-base ceramics can be classified as:

- *Group A-1*: $Al_2O_3$ with up to about 10% of other oxides or carbides, primarily those of titanium, magnesium, molybdenum, chromium, nickel, or cobalt. The mixtures are cold pressed and sintered
- *Group A-2*: Essentially pure $Al_2O_3$, hot pressed
- *Group A-3*: $Al_2O_3$ and 25 to 30% of a refractory carbide such as titanium carbide, hot pressed

More detailed information on $Si_3N_4$-base and $Al_2O_3$-base ceramics can be found in the article "Ceramics" in this Volume.

**Cubic Boron Nitride Tools.** It is now possible to turn and face mill almost any type of tool steel in the fully hardened condition, typically 63 to 65 HRC with CBN. Typical CBN cutting parameters are:

- Speed range, 45 to 105 m/min (150 to 350 sfm)
- Feed, up to 0.10 to 0.13 mm/rev (0.004 to 0.005 in./rev) or 0.025 to 0.075 mm/tooth (0.001 to 0.003 in./tooth) (milling) for negative rake inserts
- Surface finish, 0.25 to 0.38 μm (10 to 15 μin.) $R_a$ (depending on feed and tool nose radius)
- The metal removal rate can be as much as ten times greater than the grinding rate

Additional information on CBN can be found in the article "Ultrahard Tool Materials" in this Volume.

**Tool Angles.** Many form tools are difficult to grind because of their shape or because they must maintain high accuracy. It is economical to use slower speeds and lighter feeds to prolong the life of such tools. Conversely, when the maximum rate of metal removal is the primary objective, as in rough turning, tool grinding is usually simpler. For these conditions it is more economical to remove metal at the maximum rate and grind tools more frequently.

## Drilling

Most tool steels, particularly those with a carbon content of 1% or greater, respond best to drilling when the microstructure consists of ferrite and small spheroidized carbides. However, the smallest carbides should be large enough to be resolved clearly at a magnification of 1000×.

Carbon and lower-alloy tool steels, such as those in the W, O, and L groups, are drilled with standard drills of high-speed steel at speeds of 12 to 24 m/min (40 to 80 sfm).

For drilling more highly alloyed tool steels, such as high-speed steels (and particularly the high-carbon high-vanadium types), special drills are used. These highly alloyed tool steels are not only harder (often up to 280 HB) but also more abrasive. Heavy-duty drills made of cobalt-bearing high-speed steel (such as T5, T15, M6, or M36) are preferred. Nitriding the drills will prolong drill life. Lower speeds (often as low as 6 m/min, or 20 sfm) are usually best for drilling the more highly alloyed tool steels. More recently, the titanium nitride (TiN) coated drills and indexable drills have been applied successfully in drilling some of these tool steels.

**Drill points** with the conventional included angle of 118° and normal lip relief angle (about 12° for a 13 mm, or ½ in., diam drill) are suitable for drilling tool steels of the W, O, and L groups. However, drill points with a 130° included angle and lip relief about two-thirds the normal angle are more suitable for drilling steels of the D and A groups and the high-speed steels, particularly when hardness of the work material is 260 HB or higher. A 135° crankshaft point is used on high-speed tool steels of high hardness, such as M42 and T15.

**Cutting Fluid.** Water-soluble or sulfurized oil should be used as a cutting fluid when drilling tool steels. The cutting fluid should be directed onto the cut, and the rate of flow must be sufficient to cool the drill point. In the vertical drilling of deep holes, an intermittent feed is best; that is, the drill should be withdrawn at regular intervals to remove chips, cool the point, and allow more cutting fluid to reach the cut. This is termed peck feed. In horizontal drilling, the cutting fluid is directed along flutes of the drill to the point. Intermittent withdrawal of the drill flushes away chips and allows the drill to cool. For deep-hole drilling, regardless of workpiece composition, oil-hole drills should be used to permit a continuous supply of cutting fluid at the critical area.

**Table 2 Approximate machinability ratings for annealed tool steels**

Based on the average cutting speeds for turning, boring, drilling, slab milling, and end milling given in Tables 5, 6, 7, 9, 12, and 13.

| Steel or group(a) | Machinability rating(b) |
|---|---|
| W | 100 |
| S | 60–70 |
| O | 45–60 |
| A (except A7) | 45–60 |
| D, A7 | 30–40 |
| H10–H19(c) | 60–70 |
| H20–H43(c) | 45–55 |
| M2, T1 | 40–50 |
| M3 (class 1), T4 | 35–40 |
| M15, T15 | 25–30 |
| L | 65–75 |
| F | 55–60 |
| P1–P6 | 75–90 |
| P20, P21 | 65–80 |

(a)For specific steels included in the H, M, and T groups, see the tables mentioned in the table subtitle above. (b) As a percentage of the rating of 100 for W steels. (c) Applies only to hardness range of 150 to 200 HB

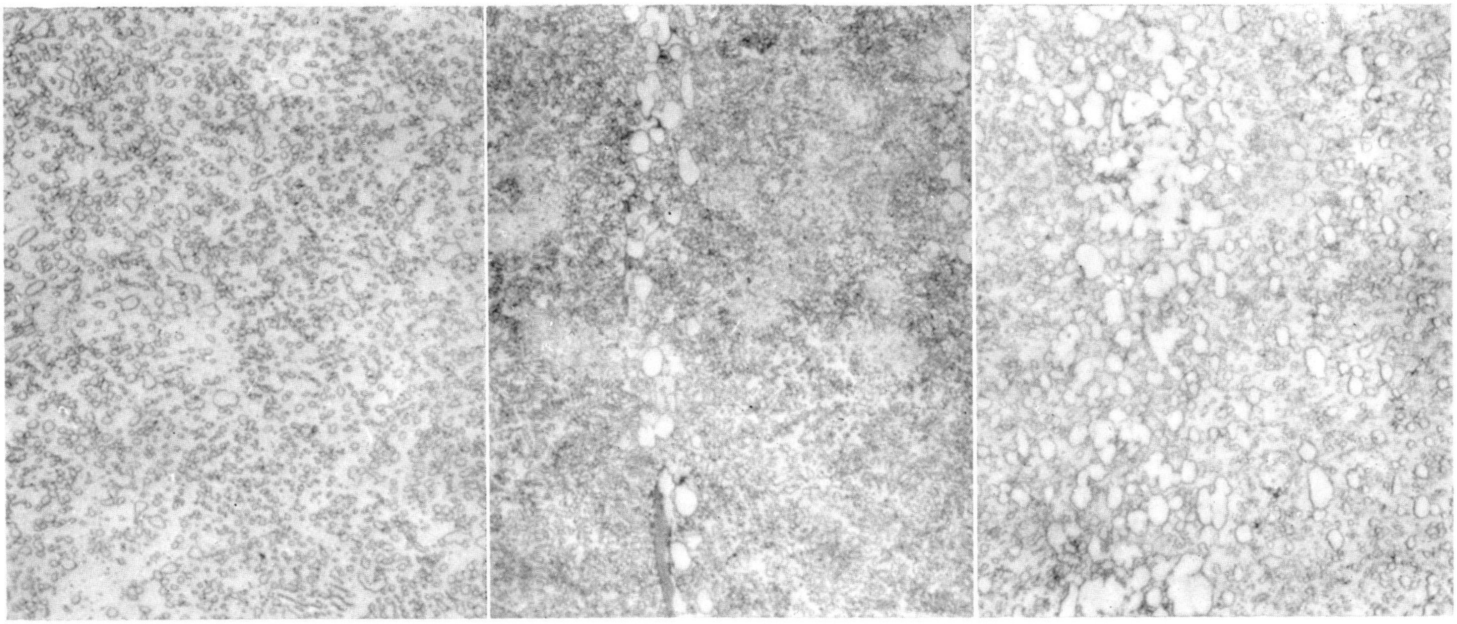

(a) (b) (c)

**Fig. 1** Microstructure of three annealed tool steels, indicating differences resulting from chemical composition that affect machining characteristics. (a) W1 (1.00% C), 156 to 196 HB, etched in 2% nital. (b) A2, 202 to 217 HB, etched in 2% nital. (c) M2, 217 to 235 HB, etched in 5% nital. 750×

## Speed and Feed

Nominal speeds and feeds for machining the various tool steels are given in Tables 4 to 16. These tables include turning, boring, broaching, drilling, reaming, tapping, milling, and sawing. Some of the tables also consider variables such as depth of cut, width of tool (in form turning) and diameter of the hole to be drilled or reamed.

**Speed.** In Tables 4 to 16 the cutting speed is based on the use of rigid setups and effective cutting fluids. Speed depends largely on the composition and hardness of the steel being machined and on the tool material. Speeds given in some of the processing examples in this article do not agree with those given in Tables 4 to 16. These variations are generally caused by specific conditions such as lack of rigidity, workpiece configuration, machine capacity, or other factors that are not considered in the tabulated nominal speeds.

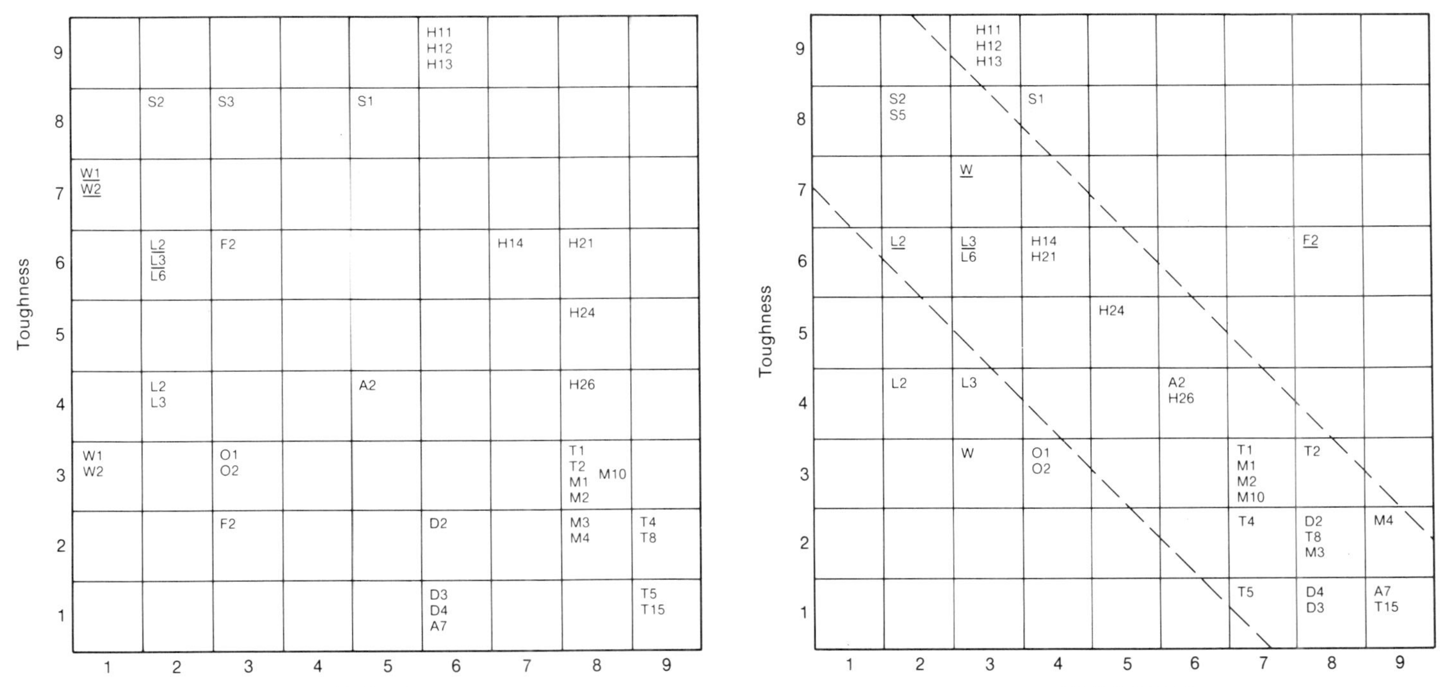


(a) (b)

**Fig. 2** Plots of (a) toughness against hot hardness and (b) wear resistance for tool steels. Types underlined indicate shallow-hardened tool steels. The area between the dashed lines in (b) represents average values.

**Table 3 Reference guide for tool steel selection**

| Application areas | Tool steel categories, AISI letter symbols, and typical applications: High-speed tool steels, M and T | Hot-work tool steels, H | Cold-work tool steels, D, A, and O | Shock-resisting tool steels, S | Mold steels, P | Special-purpose tool steels, L and F | Water-hardening tool steels, W |
|---|---|---|---|---|---|---|---|
| **Cutting tools** .......... Single-point types (lathe, planer, boring) Milling cutters Drills Reamers Taps Threading dies Form cutters | General-purpose production tools: M2, T1 For increased abrasion resistance: M3, M4, M10 Heavy-duty work calling for high hot hardness: T5, T15 Heavy-duty work calling for high abrasion resistance: M42, M44 | . . . | Tools with sharp edges (knives, razors) Tools for operations in which no high speed is involved, yet stability in heat treatment and substantial abrasion resistance are needed | Pipe cutter wheels | . . . | . . . | Uses that do not require hot hardness or high abrasion resistance Examples with carbon content of applicable group: Taps (1.05/1.10% C) Reamers (1.10/1.15% C) Twist drills (1.20/1.25% C) Files (1.35/1.40% C) |
| **Hot forging tools and dies** .......... Dies and inserts Forging machine plungers and piercers | For combining hot hardness with high abrasion resistance: M2, T1 | Dies for presses and hammers: H20, H21 For severe conditions over extended service periods: H22–H26, H43 | Hot trimming dies: D2 | Hot trimming dies Blacksmith tools Hot swaging dies | . . . | . . . | Smith tools (0.65/0.70% C) Hot chisels (0.70/0.75% C) Drop forging dies (0.90/1.00% C) Applications limited to short-run production |
| **Hot extrusion tools and dies** ...... Extrusion dies and mandrels Dummy blocks Valve extrusion tools | Brass extrusion dies: T1 | Extrusion dies and dummy blocks: H20–H26 For tools that are exposed to less heat: H10–H19 | . . . | Compression molding: S1 | . . . | . . . | . . . |
| **Cold forming dies** ... Bending, forming, drawing, and deep drawing dies and punches | Burnishing tools: M1, T1 | Cold heading die casings: H13 | Drawing dies: O1 Coining tools: O1, D2 Forming and bending dies: A2 Thread rolling dies: D2 | Hobbing and short-run applications: S1, S7 Rivet sets and rivet busters | . . . | Blanking, forming, and trimmer dies when toughness has precedence over abrasion resistance: L6 | Cold heading dies: W1 or W2 (C ≈ 1.00%) Bending dies: W1 (C ≈ 1.00%) |
| **Shearing tools** ....... Dies for piercing, punching, and trimming Shear blades | Special dies for cold and hot work: T1 For work requiring high abrasion resistance: M2, M3 | For shearing knives: H11, H12 For severe hot-shearing applications: M21, M25 | Dies for medium runs: A2, A6, O1, O4 Dies for long runs: D2, D3 Trimming dies (also for hot trimming): A2 | Cold and hot shear blades Hot punching and piercing tools Boilermaker tools | . . . | Knives for work requiring high toughness: L6 | Trimming dies (0.90/0.95% C) Cold blanking and punching dies (1.00% C) |
| **Die casting and molding dies**....... | . . . | For aluminum and lead: H11, H13 For brass: H21 | A2, A6, O1 | . . . | Plastic molds: P2–P4, P20 | . . . | . . . |
| **Structural parts for severe service conditions**......... | Roller bearings for high temperature environment: T1 Lathe centers: M2, T1 | For aircraft components (landing gears, arrester hooks, rocket cases): H11 | Lathe centers: D2, D3 Arbors: O1 Bushings: A4 Gages: D2 | Pawls Clutch parts | . . . | Spindles, clutch parts (if high toughness is needed): L6 | Spring steel (1.10/1.15% C) |
| **Battering tools for hand and power**.... | . . . | . . . | . . . | Pneumatic chisels for cold work: S5 For higher performances: S7 | . . . | . . . | For intermittent use: W1 (0.80% C) |

Source: E. Oberg, F. Jones, and H. Horton, *Machinery's Handbook*, 23rd ed., H. Ryffel, Ed., Industrial Press Inc., 1988, p 483–484

**Feed.** Nominal feeds given in Tables 4 to 16 are also affected by work metal composition and hardness, although generally to a lesser degree than are speeds. Depth of cut has a marked influence on optimum feed in operations such as turning, boring, and milling.

## Water-Hardening Tool Steels

The machining characteristics of water-hardening tool steels are covered here in three examples that describe the milling and drilling of these steels in production applications.

**Example 1: Face Milling W2.** After being cold sawed to desired lengths, various sizes of hot-rolled sections of W2 tool steel (202 HB maximum) were face milled on a semiproduction basis using the setup shown in Fig. 3. To eliminate the clamping problem normally encountered in milling a mixture of steel sizes, the machine table was equipped with a 305 × 915 mm (12 × 36 in.) magnetic chuck. However, because the rough surface of the hot-rolled sections did not always make full contact with the ground surface of the magnetic chuck, the feed rate had to be varied according to the operator's judgment, and backup blocks sometimes had to be placed around the workpiece to prevent movement during milling (Fig. 3).

The sections were milled in a 23 kW (30 hp) vertical knee-type machine with a 230 mm (9 in.) diam cutter that had 14 blades

**Table 4 Nominal speeds and feeds for the turning of tool steels with high-speed steel tools**

High-speed steels M2 and T5 for groups W, S, O, A, D, H2O, L, F, and P and for group H11 to 200 HB; M2, T5, and T15 for groups T1 and T4 and for group H11 at 200 to 375 HB; T15, M41, M42, M43, and M44 for group T15 and for group H11 at hardness above 375 HB

| Steel | Hardness, HB(a) | Single-point and box tool speed: Roughing 3.8 mm (0.150 in.) cut; feed, 0.38 mm/rev (0.015 in./rev), m/min | Roughing sfm | Finishing 1.0 mm (0.040 in.) cut; feed, 0.18 mm/rev (0.007 in./rev), m/min | Finishing sfm | Form tools: Speed, m/min (sfm) | Form tools feed, mm/rev (in./rev), for tool width of 13 mm (½ in.) | 25 mm (1 in.) | 50 mm (2 in.) | Cutoff tools: Speed, m/min (sfm) | Cutoff tools feed, mm/rev (in./rev), for tool width of 1.6 mm (1⁄16 in.) | 3.2 mm (⅛ in.) | 6.4 mm (¼ in.) |
|---|---|---|---|---|---|---|---|---|---|---|---|---|---|
| W group | 150–200 | 30 | 100 | 38 | 125 | 34 (110) | 0.043 (0.0017) | 0.036 (0.0014) | 0.028 (0.0011) | 34 (110) | 0.036 (0.0014) | 0.041 (0.0016) | 0.048 (0.0019) |
| S group | 175–225 | 27 | 90 | 35 | 115 | 23 (75) | 0.038 (0.0015) | 0.030 (0.0012) | 0.023 (0.0009) | 23 (75) | 0.030 (0.0012) | 0.036 (0.0014) | 0.043 (0.0017) |
| O group | 200–250 | 23 | 75 | 29 | 95 | 18 (60) | 0.033 (0.0013) | 0.025 (0.0010) | 0.023 (0.0009) | 18 (60) | 0.025 (0.0010) | 0.030 (0.0012) | 0.038 (0.0015) |
| A group(b) | 200–250 | 23 | 75 | 29 | 95 | 18 (60) | 0.033 (0.0013) | 0.025 (0.0010) | 0.018 (0.0007) | 18 (60) | 0.025 (0.0010) | 0.030 (0.0012) | 0.038 (0.0015) |
| D group and A7 | 200–250 | 14 | 45 (0.25 mm/rev, or 0.010 in./rev) | 18 | 60 (0.13 mm/rev, or 0.005 in./rev) | 14 (45) | 0.033 (0.0013) | 0.025 (0.0010) | 0.018 (0.0007) | 14 (45) | 0.025 (0.0010) | 0.030 (0.0012) | 0.038 (0.0015) |
| H11(c) | 150–200 | 27 | 90 | 34 | 110 | 24 (80) | 0.036 (0.0014) | 0.028 (0.0011) | 0.020 (0.0008) | 24 (80) | 0.028 (0.0011) | 0.033 (0.0013) | 0.041 (0.0016) |
| | 200–250 | 23 | 75 | 29 | 95 | 18 (60) | 0.033 (0.0013) | 0.025 (0.0010) | 0.018 (0.0007) | 18 (60) | 0.025 (0.0010) | 0.030 (0.0012) | 0.038 (0.0015) |
| | 325–375 | 17 | 55 | 21 | 70 | 11 (35) | 0.030 (0.0012) | 0.023 (0.0009) | 0.015 (0.0006) | 11 (35) | 0.020 (0.0008 | 0.028 (0.0011) | 0.036 (0.0014) |
| | 48–50 HRC | 8 | 25 (0.25 mm/rev, or 0.010 in./rev) | 9 | 30 (0.13 mm/rev, or 0.005 in./rev) | 6 (20) | 0.028 (0.0011) | 0.020 (0.0008) | 0.013 (0.0005) | 6 (20) | 0.018 (0.0007) | 0.025 (0.0010) | 0.033 (0.0013) |
| | 50–52 HRC | 6 | 20 (0.18 mm/rev, or 0.007 in./rev) | 8 | 25 (0.08 mm/rev, or 0.003 in./rev) | ... | ... | ... | ... | ... | ... | ... | ... |
| H20(d) | 150–200 | 20 | 65 | 24 | 80 | 20 (65) | 0.036 (0.0014) | 0.028 (0.0011) | 0.020 (0.0008) | 20 (65) | 0.028 (0.0011) | 0.033 (0.0013) | 0.041 (0.0016) |
| | 200–250 | 18 | 60 | 23 | 75 | 18 (60) | 0.033 (0.0013) | 0.025 (0.0010) | 0.018 (0.0007) | 18 (60) | 0.025 (0.0010) | 0.030 (0.0012) | 0.038 (0.0015) |
| M2 and T1(e) | 200–250 | 20 | 65 | 24 | 80 | 18 (60) | 0.041 (0.0016) | 0.033 (0.0013) | 0.023 (0.0009) | 18 (60) | 0.033 (0.0013) | 0.041 (0.0016) | 0.051 (0.0020) |
| M3-1 and T4(f) | 225–275 | 18 | 60 | 23 | 75 | 17 (55) | 0.038 (0.0015) | 0.030 (0.0012) | 0.020 (0.0008) | 17 (55) | 0.030 (0.0012) | 0.038 (0.0015) | 0.048 (0.0019) |
| M15 and T15(g) | 225–275 | 12 | 40 (0.25 mm/rev, or 0.010 in./rev) | 15 | 50 (0.13 mm/rev, or 0.005 in./rev) | 15 (50) | 0.038 (0.0015) | 0.030 (0.0012) | 0.020 (0.0008) | 15 (50) | 0.030 (0.0012) | 0.038 (0.0015) | 0.048 (0.0019) |
| L group | 150–200 | 29 | 95 | 37 | 120 | 23 (75) | 0.043 (0.0017) | 0.036 (0.0014) | 0.028 (0.0011) | 23 (75) | 0.036 (0.0014) | 0.041 (0.0016) | 0.048 (0.0019) |
| F group | 200–250 | 24 | 80 | 32 | 105 | 20 (65) | 0.038 (0.0015) | 0.030 (0.0012) | 0.023 (0.0009) | 20 (65) | 0.030 (0.0012) | 0.036 (0.0014) | 0.043 (0.0017) |
| P1–P6 | 100–150 | 37 | 120 | 46 | 150 | 27 (90) | 0.043 (0.0017) | 0.036 (0.0014) | 0.028 (0.0011) | 27 (90) | 0.036 (0.0014) | 0.041 (0.0016) | 0.048 (0.0019) |
| P20–P21 | 150–200 | 32 | 105 | 41 | 135 | 24 (80) | 0.043 (0.0017) | 0.036 (0.0014) | 0.028 (0.0011) | 24 (80) | 0.036 (0.0014) | 0.041 (0.0016) | 0.048 (0.0019) |

(a) Unless otherwise noted. (b) Except A7. (c) Includes also H10, H12, H13, H14, H16, and H19. (d) Includes also H21, H22, H23, H24, H25, H41, H42, and H43. (e) Includes also M1, M6, M10, T2, T6, and T7. (f) Includes also M4, M7, M30, M33, M34, M35, M36, M41, M42, M43, M44, T5, and T8. (g) Includes also M3 (class 2) and T9. Source: Metcut Research Associates Inc.

tipped with C-6 carbide. Axial rake of the blades was 7°; radial rake was 3°. The cutter was mounted on the face of the spindle. Additional processing conditions are listed in the table accompanying Fig. 3.

**Example 2: Straddle Milling W2.** Bars of W2 tool steel hot rolled to the T-section shown in Fig. 4 were machined simultaneously on four faces by straddle milling, using the setup shown in Fig. 4. Bars of four different lengths (645 mm, or 25⅜ in.; 679 mm, or 26¾ in.; 781 mm, or 30¾ in.; and 949 mm, or 37⅜ in.) were milled in batches of 50. After milling, each length was sawed in half, and the halves were drilled; ultimately, these parts served as die guides.

Straddle milling was done in a 23 kW (30 hp) horizontal knee-type machine using two 180 mm (7 in.) diam staggered-tooth cutters, each with 14 insert-type high-speed steel blades. The blades had 15° axial rake and 11° radial rake. The cutters were mounted on an arbor to permit simultaneous milling of the width of the T and both adjacent flanges. Other processing conditions are tabulated with Fig. 4.

**Milling.** In the production milling of flutes in carbon tool steel drills, accuracy is of the utmost importance. Rigidity of the milling machine and setup is essential. However, even with required rigidity, excessive variations in drill web and land dimensions may be experienced if the steel is not properly annealed. Specifically, the presence of lamellar pearlite in the annealed microstructure is objectionable because it may contribute to large variations in web and land dimensions and tearing of the surface metal. Also, despite the higher hardness of the pearlitic material, it has a lower resistance to machining as a result of the lower shear strength caused by the lamellae. Web and land dimensions, in addition to varying widely, are often undersize. Because it is extremely difficult to produce lamellar pearlite in suitably small and controlled amounts during annealing, it is necessary that the annealed structure be completely spheroidized. This is demonstrated by the application described in the following example.

**Example 3: Drilling W2.** Eight 16.5 mm (21⁄32 in.) diam by 38 mm (1.5 in.) through holes were drilled in W2 die guide plates using the two-operation setup shown in Fig. 5. These plates ranged from 320 to 470 mm (12½ to 18½ in.) in length and from 16.0 to 23.8 kg (35.3 to 52.4 lb) in weight. They were drilled in the annealed condition at 202 HB max. The 3.7 kW (5 hp) box column, vertical drill press had a geared drill head with four adjustable spindles and a guide rod fixture base to accommodate two guide plates.

In the first operation (top view in Fig. 5), parts were placed in the fixture with one end located firmly against a removable gage block and were clamped in this position. The drill press was cycled, and two holes were drilled in each workpiece.

For the second operation (middle view in Fig. 5), the removable gage block was withdrawn, and the parts were moved 31.75 mm

### Table 5 Nominal speeds and feeds for the turning of tool steels with carbide tools

For carbide grades C-6 for roughing and C-7 for finishing, except that grade C-8 is more suitable for turning steels at hardness higher than approximately 48 HRC

| Steel | Hardness, HB(a) | Single-point and box tool speed — Brazed — Roughing 3.8 mm (0.150 in.) cut; feed, 0.38 mm/rev (0.015 in./rev), m/min | sfm | Brazed — Finishing 1.0 mm (0.040 in.) cut; feed, 0.18 mm/rev (0.007 in./rev), m/min | sfm | Indexable — Roughing 3.8 mm (0.150 in.) cut; feed, 0.38 mm/rev (0.015 in./rev), m/min | sfm | Indexable — Finishing 1.0 mm (0.040 in.) cut; feed, 0.18 mm/rev (0.007 in./rev), m/min | sfm |
|---|---|---|---|---|---|---|---|---|---|
| W group | 150–200 | 107 | 350 | 137 | 450 | 130 | 425 | 167 | 550 |
| S group | 175–225 | 99 | 325 | 126 | 415 | 122 | 400 | 160 | 525 |
| O group | 200–250 | 84 | 275 | 106 | 350 | 102 | 335 | 134 | 440 |
| A group(b) | 200–250 | 84 | 275 | 106 | 350 | 102 | 335 | 134 | 440 |
| D group and A7 | 200–250 | 52 | 170 (0.25 mm/rev, or 0.010 in./rev) | 64 | 210 (0.13 mm/rev, or 0.005 in./rev) | 64 | 210 (0.25 mm/rev or 0.010 in./rev) | 79 | 260 (0.13 mm/rev, or 0.005 in./rev) |
| H11(c) | 150–200 | 84 | 275 | 114 | 375 | 101 | 330 | 130 | 425 |
| | 200–250 | 84 | 275 | 106 | 350 | 98 | 320 | 122 | 400 |
| | 325–375 | 43 | 140 (0.25 mm/rev, or 0.010 in./rev) | 53 | 175 (0.13 mm/rev, or 0.005 in./rev) | 49 | 160 (0.25 mm/rev, or 0.010 in./rev) | 61 | 200 (0.13 mm/rev, or 0.005 in./rev) |
| | 48–50 HRC | 32 | 105 (0.25 mm/rev, or 0.010 in./rev) | 40 | 130 (0.13 mm/rev, or 0.005 in./rev) | 35 | 115 (0.25 mm/rev, or 0.010 in./rev) | 44 | 145 (0.13 mm/rev, or 0.005 in./rev) |
| | 50–52 HRC | 24 | 80 (0.18 mm/rev, or 0.007 in./rev) | 32 | 105 (0.13 mm/rev, or 0.005 in./rev) | 27 | 90 (0.18 mm/rev, or 0.007 in./rev) | 35 | 115 (0.13 mm/rev, or 0.005 in./rev) |
| | 52–54 HRC | 23 | 75 (0.18 mm/rev, or 0.007 in./rev) | 29 | 95 (0.13 mm/rev, or 0.005 in./rev) | 24 | 80 (0.18 mm/rev, or 0.007 in./rev) | 32 | 105 (0.13 mm/rev, or 0.005 in./rev) |
| | 54–56 HRC | 21 | 70 (0.18 mm/rev, or 0.007 in./rev) | 26 | 85 (0.13 mm/rev, or 0.005 in./rev) | 23 | 75 (0.18 mm/rev, or 0.007 in./rev) | 29 | 95 (0.13 mm/rev, or 0.005 in./rev) |
| H20(d) | 150–200 | 84 | 275 | 106 | 350 | 99 | 325 | 126 | 415 |
| | 200–250 | 76 | 250 | 98 | 320 | 94 | 310 | 122 | 400 |
| M2 and T1(e) | 200–250 | 84 | 275 | 106 | 350 | 99 | 325 | 126 | 415 |
| M3-1 and T4(f) | 225–275 | 76 | 250 | 98 | 320 | 88 | 290 | 113 | 370 |
| M15 and T15(g) | 225–275 | 61 | 200 | 76 | 250 | 69 | 225 | 85 | 280 |
| L group | 150–200 | 99 | 325 | 126 | 415 | 122 | 400 | 160 | 525 |
| F group | 200–250 | 99 | 325 | 126 | 415 | 117 | 385 | 152 | 500 |
| P1–P6 | 100–150 | 121 | 400 (0.50 mm/rev, or 0.020 in./rev) | 152 | 500 | 145 | 475 (0.50 mm/rev, or 0.020 in./rev) | 183 | 600 |
| P20–P21 | 150–200 | 107 | 350 | 137 | 450 | 137 | 450 | 175 | 575 |

| Steel | Hardness, HB(a) | Form tools — Speed, m/min | sfm | Feed for tool width of 13 mm (½ in.), mm/rev | in./rev | 25 mm (1 in.), mm/rev | in./rev | 50 mm (2 in.), mm/rev | in./rev |
|---|---|---|---|---|---|---|---|---|---|
| W group | 150–200 | 114 | 375 | 0.043 | 0.0017 | 0.036 | 0.0014 | 0.028 | 0.0011 |
| S group | 175–225 | 78 | 255 | 0.038 | 0.0015 | 0.030 | 0.0012 | 0.023 | 0.0009 |
| O group | 200–250 | 62 | 205 | 0.033 | 0.0013 | 0.025 | 0.0010 | 0.023 | 0.0009 |
| A group(b) | 200–250 | 62 | 205 | 0.033 | 0.0013 | 0.025 | 0.0010 | 0.018 | 0.0007 |
| D group and A7 | 200–250 | 46 | 150 | 0.033 | 0.0013 | 0.025 | 0.0010 | 0.018 | 0.0007 |
| H11(c) | 150–200 | 78 | 255 | 0.036 | 0.0014 | 0.028 | 0.0011 | 0.020 | 0.0008 |
| | 200–250 | 58 | 190 | 0.033 | 0.0013 | 0.025 | 0.0010 | 0.018 | 0.0007 |
| | 325–375 | 35 | 115 | 0.030 | 0.0012 | 0.023 | 0.0009 | 0.015 | 0.0006 |
| | 48–50 HRC | 20 | 65 | 0.028 | 0.0011 | 0.020 | 0.0008 | 0.013 | 0.0005 |
| H20(d) | 150–200 | 62 | 205 | 0.036 | 0.0014 | 0.028 | 0.0011 | 0.020 | 0.0008 |
| | 200–250 | 58 | 190 | 0.033 | 0.0013 | 0.025 | 0.0010 | 0.018 | 0.0007 |
| M2 and T1(e) | 200–250 | 58 | 190 | 0.041 | 0.0016 | 0.033 | 0.0013 | 0.023 | 0.0009 |
| M3-1 and T4(f) | 225–275 | 53 | 175 | 0.038 | 0.0015 | 0.030 | 0.0012 | 0.020 | 0.0008 |
| M15 and T15(g) | 225–275 | 46 | 150 | 0.038 | 0.0015 | 0.030 | 0.0012 | 0.020 | 0.0008 |
| L group | 150–200 | 76 | 250 | 0.043 | 0.0017 | 0.036 | 0.0014 | 0.028 | 0.0011 |
| F group | 200–250 | 62 | 205 | 0.038 | 0.0015 | 0.030 | 0.0012 | 0.023 | 0.0009 |
| P1–P6 | 100–150 | 88 | 290 | 0.043 | 0.0017 | 0.036 | 0.0014 | 0.028 | 0.0011 |
| P20–P21 | 150–200 | 78 | 255 | 0.043 | 0.0017 | 0.036 | 0.0014 | 0.028 | 0.0011 |

| Steel | Hardness, HB(a) | Cutoff tools — Speed, m/min | sfm | Feed for tool width of 1.6 mm (1/16 in.), mm/rev | in./rev | 3.2 mm (⅛ in.), mm/rev | in./rev | 6.4 mm (¼ in.), mm/rev | in./rev |
|---|---|---|---|---|---|---|---|---|---|
| W group | 150–200 | 114 | 375 | 0.036 | 0.0014 | 0.041 | 0.0016 | 0.048 | 0.0019 |
| S group | 175–225 | 78 | 255 | 0.030 | 0.0012 | 0.036 | 0.0014 | 0.043 | 0.0017 |
| O group | 200–250 | 62 | 205 | 0.025 | 0.0010 | 0.030 | 0.0012 | 0.038 | 0.0015 |
| A group(b) | 200–250 | 62 | 205 | 0.025 | 0.0010 | 0.030 | 0.0012 | 0.038 | 0.0015 |
| D group and A7 | 200–250 | 46 | 150 | 0.025 | 0.0010 | 0.030 | 0.0012 | 0.038 | 0.0015 |
| H11(c) | 150–200 | 78 | 255 | 0.028 | 0.0011 | 0.033 | 0.0013 | 0.041 | 0.0016 |
| | 200–250 | 58 | 190 | 0.025 | 0.0010 | 0.030 | 0.0012 | 0.038 | 0.0015 |
| | 325–375 | 35 | 115 | 0.020 | 0.0008 | 0.028 | 0.0011 | 0.036 | 0.0014 |
| | 48–50 HRC | 20 | 65 | 0.018 | 0.0007 | 0.025 | 0.0010 | 0.033 | 0.0013 |
| H20(d) | 150–200 | 62 | 205 | 0.028 | 0.0011 | 0.033 | 0.0013 | 0.041 | 0.0016 |
| | 200–250 | 58 | 190 | 0.025 | 0.0010 | 0.030 | 0.0012 | 0.038 | 0.0015 |
| M2 and T1(e) | 200–250 | 58 | 190 | 0.033 | 0.0013 | 0.041 | 0.0016 | 0.051 | 0.0020 |
| M3-1 and T4(f) | 225–275 | 53 | 175 | 0.030 | 0.0012 | 0.038 | 0.0015 | 0.048 | 0.0019 |
| M15 and T15(g) | 225–275 | 46 | 150 | 0.030 | 0.0012 | 0.038 | 0.0015 | 0.048 | 0.0019 |
| L group | 150–200 | 76 | 250 | 0.036 | 0.0014 | 0.041 | 0.0016 | 0.048 | 0.0019 |
| F group | 200–250 | 62 | 205 | 0.030 | 0.0012 | 0.036 | 0.0014 | 0.043 | 0.0017 |
| P1–P6 | 100–150 | 88 | 290 | 0.036 | 0.0014 | 0.041 | 0.0016 | 0.048 | 0.0019 |
| P20–P21 | 150–200 | 78 | 255 | 0.036 | 0.0014 | 0.041 | 0.0016 | 0.048 | 0.0019 |

(a) Unless otherwise noted. (b) Except A7. (c) Includes also H10, H12, H13, H14, H16, and H19. (d) Includes also H21, H22, H23, H24, H25, H41, H42, and H43. (e) Includes also M1, M6, M10, T2, T6, and T7. (f) Includes also M4, M7, M30, M33, M34, M35, M36, M41, M42, M43, M44, T5, and T8. (g) Includes also M3 (class 2) and T9. Source: Metcut Research Associates Inc.

**Table 6 Nominal speeds and feeds for the turning of tool steels with coated-carbide and ceramic tools**

| | | Single-point and box tool speed | | | | | | | |
|---|---|---|---|---|---|---|---|---|---|
| | | Coated carbide | | | | Ceramic | | | |
| | | Roughing 3.3 mm (0.150 in.) cut; feed, 0.38 mm/rev (0.015 in./rev) | | Finishing 1.0 mm (0.040 in.) cut; feed, 0.18 mm/rev (0.007 in./rev) | | Roughing 3.8 mm (0.150 in.) cut; feed, 0.25 mm/rev (0.010 in./rev) | | Finishing 1.0 mm (0.040 in.) cut; feed, 0.13 mm/rev (0.005 in./rev) | |
| Steel | Hardness, HB(a) | m/min | sfm | m/min | sfm | m/min | sfm | m/min | sfm |
| W group | 150–200 | 167 | 550 | 213 | 700 | 305 | 1000 | 488 | 1600 |
| S group | 175–225 | 160 | 525 | 206 | 675 | 267 | 875 | 450 | 1475 |
| O group | 200–250 | 137 | 450 | 175 | 575 | 259 | 850 | 442 | 1450 |
| A group(b) | 200–250 | 137 | 450 | 175 | 575 | 259 | 850 | 442 | 1450 |
| D group and A7 | 200–250 | 84 | 275 (0.25 mm/rev, or 0.010 in./rev) | 107 | 350 (0.13 mm/rev, or 0.005 in./rev) | 229 | 750 | 396 | 1300 |
| H11(c) | 150–200 | 130 | 425 | 168 | 550 | 320 | 1050 | 488 | 1600 |
| | 200–250 | 122 | 400 | 160 | 525 | 259 | 850 | 442 | 1450 |
| | 325–375 | 61 | 200 | 76 | 250 | 213 | 700 | 366 | 1200 |
| | 48–50 HRC | ... | ... | ... | ... | 122 | 400 | 213 | 700 |
| | 50–52 HRC | ... | ... | ... | ... | 107 | 350 | 183 | 600 |
| | 52–54 HRC | ... | ... | ... | ... | 91 | 300 | 152 | 500 |
| | 54–56 HRC | ... | ... | ... | ... | 76 | 250 | 122 | 400 |
| H20(d) | 150–200 | 130 | 425 | 168 | 550 | 312 | 1025 | 480 | 1575 |
| | 200–250 | 122 | 400 | 160 | 525 | 251 | 825 | 434 | 1425 |
| M2 and T1(e) | 200–250 | 130 | 425 | 160 | 525 | 274 | 900 | 457 | 1500 |
| M3-1 and T4(f) | 225–275 | 122 | 400 | 152 | 500 | 259 | 850 | 427 | 1400 |
| M15 and T15(g) | 225–275 | 91 | 300 | 99 | 325 | 251 | 825 | 419 | 1375 |
| L group | 150–200 | 160 | 525 | 213 | 700 | 259 | 850 | 442 | 1450 |
| F group | 200–250 | 152 | 500 | 198 | 650 | 251 | 825 | 434 | 1425 |
| P1–P6 | 100–150 | 190 | 625 | 236 | 775 | 335 | 1100 | 503 | 1650 |
| P20–P21 | 150–200 | 183 | 600 | 229 | 750 | 350 | 1150 | 518 | 1700 |

(a) Unless otherwise noted. (b) Except A7. (c) Includes also H10, H12, H13, H14, H16, and H19. (d) Includes also H21, H22, H23, H24, H25, H41, H42, and H43. (e) Includes also M1, M6, M10, T2, T6, and T7. (f) Includes also M4, M7, M30, M33, M34, M35, M36, M41, M42, M43, M44, T5, and T8. (g) Includes also M3 (class 2) and T9. Source: Metcut Research Associates Inc.

(1.25 in.) to locate against a stationary gage block. Then the drill press was cycled, and two more holes were drilled in each workpiece. Additional processing details are given in the table that accompanies Fig. 5.

**Sawing.** For removing stock from flat surfaces, sawing is often more economical than shaping, provided surface finish is not of primary importance and that only straight cuts, or simple or compound angles, are involved. Hydraulically fed band saws can perform such operations with a minimum of setup time.

### Shock-Resisting Tool Steels (S Steels)

The hardness of annealed S steels is usually between those of the W and O steels (Tables 4 to 15). Most machining of the S steels is done at speeds and feeds nearer those used for the O group than the W group. Accordingly, the data in Examples 1 to 3 of the previous section can be used as a starting point in the machining of S steels, with a moderate downward adjustment of speed and feed. The speed of the W tool steels is typically twice that of the O tool steels; the feed of the W tool steels is typically 0 to 50% greater than that of the O tool steels. Likewise, the data in Examples 4 and 5 and in the section that follows can be used in the same way with a slight upward adjustment of speed and feed.

## Type O Cold-Work Tool Steels

The machinability of tool steels in group O is lower than that of the water-hardening tool steels discussed earlier (45 to 60, compared to 100). Hence, the data in Examples 1 to 3 can be applied to the O steels, provided speeds and feeds are reduced accordingly. Similarly, the two examples that follow, for O steels, are applicable to similar operations and setups for W (as well as S) steels with proportionately increased speeds and feeds.

**Example 4: Turning O1 Tap Blanks.** Before being heat treated and ground, tap blanks 36.5 mm (1 7/16 in.) in diameter by 95 mm (3 3/4 in.) long, made from O1 bar stock (hardness, 196 HB), were turned on a 15 kW (20 hp) tracer lathe. These blanks weighed 0.77 kg (1.7 lb) each before turning and 0.68 kg (1.5 lb) after. The turning tool (indexable carbide tips) had 0 to 5° back rake, 6° side rake, and 1.6 mm (1/16 in.) nose radius.

Turning speed was 1225 rev/min (1400 m/min, or 460 sfm); feed was 0.41 mm/min (0.016 in./min); depth of cut was 8.2 mm (0.323 in.); length of cut was 50 mm (2 in.); time per cut was 6 s. A 200:1 mixture of water and chemical solution was the cutting fluid. Tool life for the indexable tips was 45 pieces per corner.

**Example 5: Milling Flutes in O1 Taps.** A No. 2 plain mill was used for cutting the flutes in 19 mm (3/4 in.) diam four-flute hand taps made of O1 tool steel. Blanks for these taps were milled at a hardness of 196 HB. Each blank weighed 0.11 kg (0.25 lb) before machining and 0.077 kg (0.17 lb) after.

The milling cutter, of M10 high-speed tool steel, was 100 mm (4 in.) in diameter by 6.60 mm (0.260 in.) wide and had 32 teeth (10° clearance angle, 0 to 2° positive radial rake, 4.83 mm, or 0.190 in. tooth radius). An average of 600 flutes could be cut between grinds.

Speed was 95 rev/min (30 m/min, or 100 sfm), at a feed of 90 mm/min (3.5 in./min). A sulfur-base oil was used as the cutting fluid. Length of cut was 60 mm (2 3/8 in.), and depth of cut was 4.95 mm (0.195 in.). Time for milling each flute was 0.67 min.

**Sawing.** The sawing characteristics of W2 and O6 tool steels are similar; cutting rates for both steels are almost identical.

## Types A and D Cold-Work Tool Steels

The high carbon and alloy content of tool steels in groups A and D promotes the formation of large alloy carbides and makes it difficult to obtain a low hardness in the annealed steels; both conditions adversely affect machining characteristics. Listed in order of decreasing machinability, the steels in the A and D groups may be rated as follows: A2, A6, A4, A5, D1, D2, D5, D6, D3, D4, A7, D7.

Because of their high content of carbon, chromium, and vanadium, A7 and D7 have extremely poor machinability and are highly susceptible to grinding damage in the hardened condition. In general, the D steels are more susceptible to grinding damage than are the A steels, because they have more massive carbides, but this susceptibility can be reduced somewhat by tempering at higher temperatures. The following example indicates the relative machinability of one of these tool steels.

**Example 6: Face Milling D2.** A 23 kW (30 hp) vertical-spindle knee-type milling

**Table 7 Nominal speeds and feeds for the boring of tool steels with high-speed steel, carbide, coated-carbide, and ceramic tools**

| Steel | Hardness, HB(a) | Roughing 2.54 mm (0.100 in.) cut; feed, 0.30 mm/rev (0.012 in./rev) | | | | | | | | Ceramic 3.18 mm (0.125 in.) cut; feed, 0.25 mm/rev (0.010 in./rev) | |
|---|---|---|---|---|---|---|---|---|---|---|---|
| | | High-speed steel(b) | | Carbide(c) | | | | | | | |
| | | | | Brazed | | Indexable | | Coated carbide | | | |
| | | m/min | sfm | m/min | sfm | m/min | sfm | m/min | sfm | m/min | sfm |
| W group | 150–200 | 24 | 80 | 78 | 255 | 91 | 300 | 117 | 385 | 244 | 800 |
| S group | 175–225 | 21 | 70 | 73 | 240 | 85 | 280 | 111 | 365 | 221 | 725 |
| O group | 200–250 | 18 | 60 | 61 | 200 | 72 | 235 | 96 | 315 | 213 | 700 |
| A group(d) | 200–250 | 18 | 60 | 61 | 200 | 72 | 235 | 96 | 315 | 213 | 700 |
| D group and A7 | 200–250 | 11 | 35 (0.20 mm/rev, or 0.008 in./rev) | 40 | 130 (0.20 mm/rev, or 0.008 in./rev) | 46 | 150 (0.20 mm/rev, or 0.008 in./rev) | 58 | 190 (0.20 mm/rev, or 0.008 in./rev) | 183 | 600 |
| H11(e) | 150–200 | 21 | 70 | 59 | 195 | 70 | 230 | 91 | 300 | 259 | 850 |
| | 200–250 | 18 | 60 | 58 | 190 | 69 | 225 | 85 | 280 | 213 | 700 |
| | 325–375 | 14 | 45 | 38 | 125 (0.20 mm/rev, or 0.008 in./rev) | 44 | 145 (0.20 mm/rev, or 0.008 in./rev) | 64 | 210 | 168 | 550 (0.20 mm/rev, or 0.008 in./rev) |
| | 48–50 HRC | 8 | 25 (0.20 mm/rev, or 0.008 in./rev) | 29 | 95 (0.20 mm/rev, or 0.008 in./rev) | 34 | 110 (0.20 mm/rev, or 0.008 in./rev) | ... | ... | 91 | 300 (0.15 mm/rev, or 0.006 in./rev) |
| | 50–52 HRC | 6 | 20 (0.15 mm/rev, or 0.006 in./rev) | 26 | 85 (0.15 mm/rev, or 0.006 in./rev) | 30 | 100 (0.15 mm/rev, or 0.006 in./rev) | ... | ... | 76 | 250 (0.15 mm/rev, or 0.006 in./rev) |
| | 52–54 HRC | ... | ... | 17 | 55 (0.15 mm/rev, or 0.006 in./rev) | 20 | 65 (0.15 mm/rev, or 0.006 in./rev) | ... | ... | 61 | 200 (0.15 mm/rev, or 0.006 in./rev) |
| | 54–56 HRC | ... | ... | 14 | 45 (0.10 mm/rev, 0.004 in./rev) | 17 | 55 (0.10 mm/rev, or 0.004 in./rev) | ... | ... | 46 | 150 (0.15 mm/rev, or 0.006 in./rev) |
| H20(f) | 150–200 | 15 | 50 | 58 | 190 | 69 | 225 | 91 | 300 | 251 | 825 |
| | 200–250 | 15 | 50 | 55 | 180 | 66 | 215 | 85 | 280 | 206 | 675 |
| M2 and T1(g) | 200–250 | 15 | 50 | 58 | 190 | 69 | 225 | 91 | 300 | 229 | 750 |
| M3-1 and T4(h) | 225–275 | 15 | 50 | 53 | 175 | 62 | 205 | 85 | 280 | 213 | 700 |
| M15 and T15(i) | 225–275 | 9 | 30 (0.20 mm/rev, or 0.008 in./rev) | 40 | 130 | 47 | 155 | 64 | 210 | 206 | 675 |
| L group | 150–200 | 23 | 75 | 73 | 240 | 85 | 280 | 111 | 365 | 213 | 700 |
| F group | 200–250 | 20 | 65 | 70 | 230 | 82 | 270 | 107 | 350 | 206 | 675 |
| P1–P6 | 100–150 | 29 | 95 | 85 | 280 | 101 | 330 | 134 | 440 | 274 | 900 |
| P20–P21 | 150–200 | 26 | 85 | 81 | 265 | 96 | 315 | 128 | 420 | 290 | 950 |

| Steel | Hardness, HB(a) | Finishing 0.25 mm (0.010 in.) cut; feed, 0.08 mm/rev (0.003 in./rev) | | | | | | | | Ceramic 0.50 mm (0.020 in.) cut; feed, 0.13 mm/rev (0.005 in./rev) | |
|---|---|---|---|---|---|---|---|---|---|---|---|
| | | High-speed steel(b) | | Carbide(c) | | | | | | | |
| | | | | Brazed | | Indexable | | Coated carbide | | | |
| | | m/min | sfm | m/min | sfm | m/min | sfm | m/min | sfm | m/min | sfm |
| W group | 150–200 | 38 | 125 | 125 | 410 | 146 | 480 | 189 | 620 | 427 | 1400 |
| S group | 175–225 | 34 | 110 | 119 | 390 | 139 | 455 | 181 | 595 | 390 | 1275 |
| O group | 200–250 | 29 | 95 | 101 | 330 | 119 | 390 | 152 | 500 | 381 | 1250 |
| A group(d) | 200–250 | 29 | 95 | 101 | 330 | 119 | 390 | 152 | 500 | 381 | 1250 |
| D group and A7 | 200–250 | 18 | 60 (0.05 mm/rev, or 0.002 in./rev) | 59 | 195 (0.05 mm/rev, or 0.002 in./rev) | 70 | 230 (0.05 mm/rev, or 0.002 in./rev) | 93 | 305 (0.05 mm/rev, or 0.002 in./rev) | 335 | 1100 |
| H11(e) | 150–200 | 34 | 110 | 98 | 320 | 114 | 375 | 146 | 480 | 427 | 1400 |
| | 200–250 | 29 | 95 | 90 | 295 | 107 | 350 | 139 | 455 | 381 | 1250 |
| | 325–375 | 21 | 70 | 61 | 200 (0.05 mm/rev, or 0.002 in./rev) | 72 | 235 (0.05 mm/rev, or 0.002 in./rev) | 107 | 350 | 305 | 1000 (0.10 mm/rev, or 0.004 in./rev) |
| | 48–50 HRC | 12 | 40 (0.05 mm/rev, or 0.002 in./rev) | 46 | 150 (0.05 mm/rev, or 0.002 in./rev) | 53 | 175 (0.05 mm/rev, or 0.002 in./rev) | ... | ... | 168 | 550 (0.08 mm/rev, or 0.003 in./rev) |
| | 50–52 HRC | 11 | 35 (0.05 mm/rev, or 0.002 in./rev) | 40 | 130 (0.05 mm/rev, or 0.002 in./rev) | 47 | 155 (0.05 mm/rev, or 0.002 in./rev) | ... | ... | 137 | 450 (0.08 mm/rev, or 0.003 in./rev) |
| | 52–54 HRC | ... | ... | 26 | 85 (0.05 mm/rev, or 0.002 in./rev) | 30 | 100 (0.05 mm/rev, or 0.002 in./rev) | ... | ... | 122 | 400 (0.08 mm/rev, or 0.003 in./rev) |
| | 54–56 HRC | ... | ... | 21 | 70 (0.05 mm/rev, or 0.002 in./rev) | 26 | 85 (0.05 mm/rev, or 0.002 in./rev) | ... | ... | 91 | 300 (0.08 mm/rev, or 0.003 in./rev) |
| H20(f) | 150–200 | 24 | 80 | 93 | 305 | 110 | 360 | 146 | 480 | 419 | 1375 |
| | 200–250 | 23 | 75 | 91 | 300 | 107 | 350 | 139 | 455 | 373 | 1225 |
| M2 and T1(g) | 200–250 | 24 | 80 | 93 | 305 | 110 | 360 | 139 | 455 | 396 | 1300 |
| M3-1 and T4(h) | 225–275 | 23 | 75 | 84 | 275 | 99 | 325 | 133 | 435 | 366 | 1200 |
| M15 and T15(i) | 225–275 | 15 | 50 (0.05 mm/rev, or 0.002 in./rev) | 62 | 205 | 73 | 240 | 99 | 325 | 358 | 1175 |
| L group | 150–200 | 37 | 120 | 117 | 385 | 139 | 455 | 186 | 610 | 381 | 1250 |
| F group | 200–250 | 32 | 105 | 113 | 370 | 133 | 435 | 174 | 570 | 373 | 1225 |
| P1–P6 | 100–150 | 46 | 150 | 136 | 445 | 160 | 525 | 206 | 675 | 442 | 1450 |
| P20–P21 | 150–200 | 43 | 140 | 130 | 425 | 152 | 500 | 200 | 655 | 457 | 1500 |

(a) Unless otherwise noted. (b) High-speed steels M2 and T5 for groups W, S, O, A, D, H20, L, F, and P, and for group H11 to 200 HB; M2, T5 and T15 for groups T1 and T4, and for group H11 at 200 to 375 HB; T15, M41, M42, M43, and M44 for groups T15, and for group H11 at hardnesses above 375 HB. (c) Carbide grade C-7, except that grade C-8 is more suitable for boring steels at hardness above about 48 HRC. (d) Includes all of group A except A7. (e) Includes also H10, H12, H13, H14, H16, and H19. (f) Includes also H21, H22, H23, H24, H25, H41, H42, and H43. (g) Includes also M1, M6, M10, T2, T6, and T7. (h) Includes also M4, M7, M30, M33, M34, M35, M36, M41, M42, M43, M44, T5, and T8. (i) Includes also M3 (class 2) and T9. Source: Metcut Research Associates Inc.

machine was used for face milling flat die sections of various lengths cut from hot-rolled D2 bar stock at 255 HB. Several of these sections were machined simultaneously. The 200 mm (8 in.) diam, 16-tooth milling cutter was mounted on the face of the spindle. It contained carbide-tip inserted blades with a 7° axial rake and a 3° radial rake.

Depth of cut was 0.75 to 4.6 mm (0.030 to 0.180 in.), width of cut was 50 to 150 mm (2 to 6 in.), and length of cut was 150 to 760 mm (6 to 30 in.). Milling speed was 60 m/min (200 sfm), feed was 230 mm/min (9 in./min) or 0.15 mm/tooth (0.006 in./tooth). No cutting fluid was used. Metal removal rate was 8.8 to 160 $cm^3$/min (0.54 to 9.7 $in.^3$/min).

**Table 8 Nominal speeds and feeds for the broaching of wrought tool steels with high-speed tool steels**

| Steel | Hardness, HB | Condition | Speed, m/min | Speed, sfm | Chip load, mm/tooth | Chip load, in./tooth | Tool material grade, ISO | Tool material grade, AISI or C |
|---|---|---|---|---|---|---|---|---|
| **High speed** M1 M6 T1 T6 M2 M10 T2 | 200–250 | Annealed | 3 | 10 | 0.050 | 0.002 | S9, S11(a) | T15, M42(a) |
| **High speed** M3-1 M33 M42 M47 M4 M34 M43 T4 M7 M36 M44 T5 M30 M41 M46 T8 | 225–275 | Annealed | 1.5 | 5 | 0.050 | 0.002 | S9, S11(a) | T15, M42(a) |
| **High speed** M3-2 T15 | 225–275 | Annealed | 1.5 | 5 | 0.050 | 0.002 | S9, S11(a) | T15, M42(a) |
| **Hot work** H10 H13 H11 H14 H12 H19 | 150–200 | Annealed | 6 | 20 | 0.075 | 0.003 | S4, S2 | M2, M7 |
| | 325–375 | Quenched and tempered | 3 | 10 | 0.050 | 0.002 | S9, S11(a) | T15, M42(a) |
| **Hot work** H21 H24 H42 H22 H25 H23 H26 | 150–200 | Annealed | 5 | 15 | 0.050 | 0.002 | S9, S11(a) | T15, M42(a) |
| | 200–250 | Annealed | 3 | 10 | 0.050 | 0.002 | S9, S11(a) | T15, M42(a) |
| **Cold work** A7 D4 D2 D5 D3 D7 | 200–250 | Annealed | 3 | 10 | 0.050 | 0.002 | S9, S11(a) | T15, M42(a) |
| **Cold work** A2 A6 A10 O6 A3 A8 O1 O7 A4 A9 O2 | 200–250 | Annealed | 5 | 15 | 0.075 | 0.003 | S9, S11(a) | T15, M42(a) |
| **Shock resisting** S1 S6 S2 S7 S5 | 175–225 | Annealed | 5 | 15 | 0.075 | 0.003 | S9, S11(a) | T15, M42(a) |
| **Mold** P2 P5 P4 P6 | 100–150 | Annealed | 6 | 20 | 0.075 | 0.003 | S4, S2 | M2, M7 |
| **Mold** P20 P21 | 150–200 | Annealed | 6 | 20 | 0.075 | 0.003 | S4, S2 | M2, M7 |
| **Special purpose** L2 L6 L7 | 150–200 | Annealed | 6 | 20 | 0.075 | 0.003 | S4, S2 | M2, M7 |
| **Special purpose** F1 F2 | 200–250 | Annealed | 5 | 15 | 0.075 | 0.003 | S9, S11(a) | T15, M42(a) |
| **Water hardening** W1, W2, W5 SAE J438b: Types W108, W109, W110, W112, W209, W210, W310 | 150–200 | Annealed | 6 | 20 | 0.075 | 0.003 | S4, S2 | M2, M7 |

(a) Any premium high-speed steel (T15, M33, M41–M47 or S9, S10, S11, S12). Source: Metcut Research Associates Inc.

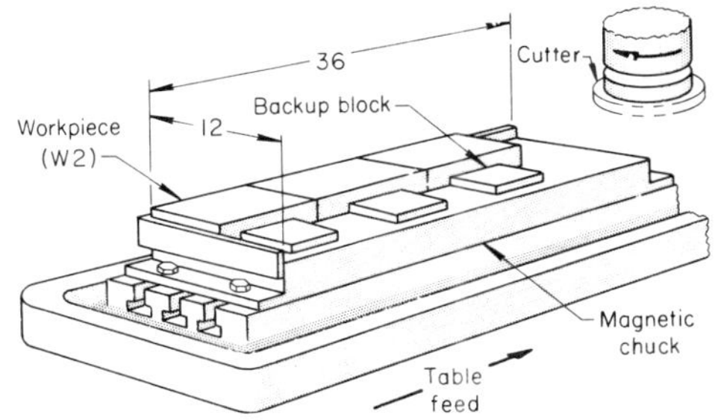


| | |
|---|---|
| Speed, m/min (sfm) | 106 (350) |
| Feed at 330–685 mm/min (13–27 in./min), mm/tooth (in./tooth) | 0.15–0.33 (0.006–0.013) |
| Depth of cut, mm (in.) | 0.76–4.6 (0.030–0.180) |
| Width of cut, mm (in.) | 50–150 (2–6) |
| Length of cut, mm (in.) | 150–760 (6–30) |
| Cutting rate, $cm^3/min$ ($in.^3/min$) | 13–479 (0.8–29.2) |
| Cutting fluid | None |
| Tool life per grind, h | 4–8 |

**Fig. 3** Face milling sections of W2 tool steel using a magnetic chuck and backup blocks. Dimensions in figure given in inches

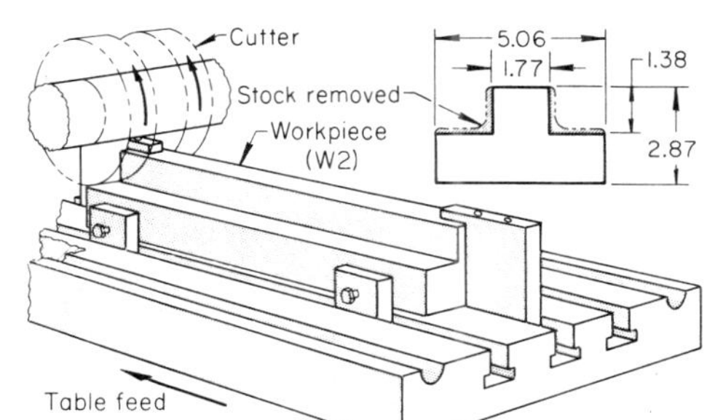


| | |
|---|---|
| Speed, m/min (sfm) | 16 (54) |
| Feed at 38 mm/min (1.5 in./min), mm/tooth (in./tooth) | 0.19 (0.0073) |
| Cross-sectional area of cut, $cm^2$ ($in.^2$) | 5.20 (0.806) |
| Length of cut, mm (in.) | 645 (25⅜), 679 (26¾), 781 (30¾), 949 (37⅜) |
| Cutting rate, $cm^3/min$ ($in.^3/min$) | 19.8 (1.21) |
| Cutting fluid, water: low-sulfur water-soluble synthetic | 40:1 |
| Tool life per grind: Stock removed, $cm^3$ ($in.^3$) | 18 000 (1100) |
| Tool life per grind: Length of cut, m (in.) | 34.3 (1350) |

| Length of blank, mm (in.) | Original weight, kg (lb) | Weight after milling, kg (lb) |
|---|---|---|
| 645 (25⅜) | 35 (77½) | 32.5 (71¾) |
| 679 (26¾) | 37 (81¾) | 34 (75¾) |
| 781 (30¾) | 43 (94) | 39.5 (87) |
| 949 (37⅜) | 52 (114) | 48 (105½) |

**Fig. 4** Straddle milling of four different lengths of T-section W2 tool steel, using two staggered-tooth cutters. Dimensions in figure given in inches

## Hot-Work Tool Steels (H Steels)

When used for tools, the H steels are machined in the annealed condition, with machinability somewhat less than the A steels and better than the D steels. Type H11, usually containing 0.40% C instead of 0.35% C, is also used for structural parts in aerospace applications. Such parts are generally machined in the quenched-and-tempered condition at higher hardness. The example that follows describes applications in which sulfurized H13 permitted a reduction in machining time from that required for the regular nonsulfurized grade of H13 tool steel.

**Example 7: Facing, Turning, and Milling H13 Versus Sulfurized H13.** Porthole extrusion dies (Fig. 6) were machined from H13 and sulfurized H13 forgings 305 mm (12 in.) in diameter and 130 mm (5⅛ in.) thick, weighing 84.4 kg (186 lb). Machining consisted of facing and turning, and of milling the four ports.

Facing and turning, in which 39 kg (86 lb) of metal were removed, required 9 h for the H13, compared to 6⅔ h for the sulfurized H13. Milling required 10 h for the H13 and 7 h for the sulfurized H13.

## High-Speed Steels

General-purpose types of high-speed steel, such as T1, ordinarily machine at 50% of the rate for the W group (the most easily machined). Annealed T1 steel, however, has been machined still faster, at 40% of the rate for 1112 steel. For instance, with a 1.5 mm (0.060 in.) depth of cut and at a feed of 0.25 mm/rev (0.010 in./rev), annealed T1 has been turned with a high-speed steel tool at 35 m/min (120 sfm), whereas for similar conditions 1112 steel is turned at 90 m/min (300 sfm).

Machinability of the more highly alloyed high-speed steels decreases as hardness and

**Table 9 Nominal speeds and feeds for the drilling of wrought tool steels with high-speed tool steel and carbide tools**

| Steel | Hardness, HB(a) | Condition | Speed, m/min (sfm)(b) | Feed, mm/rev (in./rev)(b), with nominal hole diameter of 1.5 mm (1/16 in.) | 3 mm (1/8 in.) | 6 mm (1/4 in.) | 12 mm (1/2 in.) | 18 mm (3/4 in.) | 25 mm (1 in.) | 35 mm (1 1/2 in.) | 50 mm (2 in.) | Tool material grade, ISO (AISI or C) |
|---|---|---|---|---|---|---|---|---|---|---|---|---|
| **High speed**<br>M1 M10 T6<br>M2 T1<br>M6 T2 | 200–250 | Annealed | 14 (45) | 0.025 (0.001) | 0.050 (0.002) | 0.075 (0.003) | 0.13 (0.005) | 0.18 (0.007) | 0.23 (0.009) | 0.28 (0.011) | 0.33 (0.013) | S2, S3 (M10, M7, M1) |
| **High speed**<br>M3-1 M36 M47<br>M4 M41 T4<br>M7 M42 T5<br>M30 M43 T8<br>M33 M44<br>M34 M46 | 225–275 | Annealed | 11 (35) | 0.025 (0.001) | 0.050 (0.002) | 0.075 (0.003) | 0.13 (0.005) | 0.18 (0.007) | 0.20 (0.008) | 0.28 (0.011) | 0.33 (0.013) | S2, S3 (M10, M7, M1) |
| **High speed**<br>M3-2<br>T15 | 225–275 | Annealed | 8 (25) | 0.025 (0.001) | 0.025 (0.001) | 0.050 (0.002) | 0.10 (0.004) | 0.13 (0.005) | 0.15 (0.006) | 0.18 (0.007) | 0.20 (0.008) | S2, S3 (M10, M7, M1) |
| **Hot work**<br>H10 H13<br>H11 H14<br>H12 H19 | 150–200 | Annealed | 18 (60) | 0.025 (0.001) | 0.050 (0.002) | 0.075 (0.003) | 0.15 (0.006) | 0.23 (0.009) | 0.28 (0.011) | 0.36 (0.014) | 0.40 (0.016) | S2, S3 (M10, M7, M1) |
| | 325–375 | Quenched and tempered | 12 (40) | ... | 0.050 (0.002) | 0.075 (0.003) | 0.13 (0.005) | 0.18 (0.007) | 0.20 (0.008) | 0.25 (0.010) | 0.28 (0.011) | S2, S3 (M10, M7, M1) |
| | 48–50 HRC | Quenched and tempered | 6 (20) | ... | 0.025 (0.001) | 0.050 (0.002) | 0.075 (0.003) | 0.075 (0.003) | 0.10 (0.004) | 0.10 (0.004) | 0.10 (0.004) | S9, S11(c) (T15, M42)(c) |
| | 54–56 HRC | Quenched and tempered | 18 (60) | ... | ... | 0.025 (0.001) | 0.025 (0.001) | 0.038 (0.0015) | ... | ... | ... | K10 (C-2) |
| **Hot work**<br>H21 H25 | 150–200 | Annealed | 14 (45) | 0.025 (0.001) | 0.050 (0.002) | 0.075 (0.003) | 0.13 (0.005) | 0.18 (0.007) | 0.23 (0.009) | 0.28 (0.011) | 0.33 (0.013) | S2, S3 (M10, M7, M1) |
| H22 H26<br>H23 H42<br>H24 | 200–250 | Annealed | 11 (35) | 0.025 (0.001) | 0.050 (0.002) | 0.075 (0.003) | 0.13 (0.005) | 0.18 (0.007) | 0.23 (0.009) | 0.28 (0.011) | 0.33 (0.013) | S2, S3 (M10, M7, M1) |
| **Cold work**<br>A7 D4<br>D2 D5<br>D3 D7 | 200–250 | Annealed | 9 (30) | 0.025 (0.001) | 0.025 (0.001) | 0.075 (0.003) | 0.13 (0.005) | 0.18 (0.007) | 0.20 (0.008) | 0.30 (0.012) | 0.40 (0.015) | S2, S3 (M10, M7, M1) |
| **Cold work**<br>A2 A8 O2<br>A3 A9 O6<br>A4 A10 O7<br>A6 O1 | 200–250 | Annealed | 14 (45) | 0.025 (0.001) | 0.025 (0.001) | 0.075 (0.003) | 0.13 (0.005) | 0.18 (0.007) | 0.20 (0.008) | 0.30 (0.012) | 0.40 (0.015) | S2, S3 (M10, M7, M1) |
| **Shock resisting**<br>S1 S6<br>S2 S7<br>S5 | 175–225 | Annealed | 17 (55) | 0.025 (0.001) | 0.050 (0.002) | 0.075 (0.003) | 0.18 (0.007) | 0.23 (0.009) | 0.28 (0.011) | 0.40 (0.015) | 0.45 (0.018) | S2, S3 (M10, M7, M1) |
| **Mold**<br>P2 P5<br>P4 P6 | 100–150 | Annealed | 23 (75) | 0.025 (0.001) | 0.050 (0.002) | 0.10 (0.004) | 0.18 (0.007) | 0.25 (0.010) | 0.30 (0.012) | 0.45 (0.018) | 0.50 (0.020) | S2, S3 (M10, M7, M1) |
| **Mold**<br>P20<br>P21 | 150–200 | Annealed | 18 (60) | 0.025 (0.001) | 0.050 (0.002) | 0.075 (0.003) | 0.18 (0.007) | 0.23 (0.009) | 0.28 (0.011) | 0.40 (0.015) | 0.45 (0.018) | S2, S3 (M10, M7, M1) |
| **Special purpose**<br>L2<br>L6<br>L7 | 150–200 | Annealed | 18 (60) | 0.025 (0.001) | 0.075 (0.003) | 0.10 (0.004) | 0.18 (0.007) | 0.25 (0.010) | 0.30 (0.012 | 0.45 (0.018) | 0.50 (0.020) | S2, S3 (M10, M7, M1) |
| **Special purpose**<br>F1<br>F2 | 200–250 | Annealed | 15 (50) | 0.025 (0.001) | 0.050 (0.002) | 0.075 (0.003) | 0.13 (0.005) | 0.18 (0.007) | 0.20 (0.008) | 0.28 (0.011) | 0.40 (0.015) | S2, S3 (M10, M7, M1) |
| **Water hardening**<br>W1, W2, W5<br>SAE J438b: Types<br>W108, W109,<br>W110, W112,<br>W209, W210, W310 | 150–200 | Annealed | 29 (95) | 0.025 (0.001) | 0.050 (0.002) | 0.10 (0.004) | 0.18 (0.007) | 0.25 (0.010) | 0.30 (0.012) | 0.40 (0.015) | 0.45 (0.018) | S2, S3 (M10, M7, M1) |

(a) Unless otherwise noted. (b) For holes more than two diameters deep, reduce speed and feed. (c) Any premium HSS (T15, M33, M41–M47) or (S9, S10, S11, S12). Source: Metcut Research Associates Inc.

abrasiveness increase. This is especially true of the high-carbon high-vanadium types, such as M4, M15, T9, and T15. These steels are considerably more difficult to machine than T1, requiring lower cutting speeds or causing increased tool wear, or both. These more highly alloyed grades of high-speed steel are comparable to the A7 and D7 cold-work die steels with respect to machinability.

The examples that follow provide details for two machining operations on tungsten and molybdenum high-speed steel workpieces.

**Example 8: Turning M10 Versus Sulfurized M10 Reamer Blanks.** In turning 50 mm (2 in.) diam reamer blanks to remove 6.35 mm (0.250 in.) of stock from the diameter, cutting speed was increased from 37 to 52 m/min (120 to 170 sfm) when a sulfurized M10 steel was substituted for the regular grade. Both steels were machined in the annealed condition (228 HB). Processing conditions are given in Table 17.

**Example 9: Milling Flutes in T15 End Mills.** In milling helical flutes in 19 mm (3/4 in.) diam four-flute end mills made of T15 high-speed steel, the machining characteristics of cutters made of M2 high-speed steel were compared with those of carbide-tip cutters. Because T15 steel is much more

## Table 10 Nominal speeds and feeds for the reaming of wrought tool steels with high-speed tool steels and carbide tools

| Steel | Hard-ness, HB(a) | Condi-tion | Roughing: Speed, m/min (sfm) | Roughing: Feed, mm/rev (in./rev)(b), with reamer diameter of 3 mm (1/8 in.) | 6 mm (1/4 in.) | 12 mm (1/2 in.) | 25 mm (1 in.) | 35 mm (1.5 in.) | 50 mm (2 in.) | Roughing: Tool material grade, ISO (AISI or C) | Finishing: Speed, m/min (sfm) | Finishing: Feed, mm/rev (in./rev)(b), with reamer diameter of 3 mm (1/8 in.) | 6 mm (1/4 in.) | 12 mm (1/2 in.) | 25 mm (1 in.) | 35 mm (1.5 in.) | 50 mm (2 in.) | Finishing: Tool material grade, ISO (AISI or C) |
|---|---|---|---|---|---|---|---|---|---|---|---|---|---|---|---|---|---|---|
| **High speed** | 200–250 | Ann. | 14 | 0.075 | 0.15 | 0.20 | 0.30 | 0.40 | 0.50 | S3, S4, S2 | 11 | 0.10 | 0.18 | 0.25 | 0.40 | 0.50 | 0.65 | S3, S4, S2 |
| M1 M10 | | | (45) | (0.003) | (0.006) | (0.008) | (0.012) | (0.015) | (0.020) | (M1, M2, M7) | (35) | (0.004) | (0.007) | (0.010) | (0.015) | (0.020) | (0.025) | (M1, M2, M7) |
| M2 T1 | | | 18 | 0.10 | 0.15 | 0.20 | 0.30 | 0.40 | 0.50 | K20 | 14 | 0.10 | 0.18 | 0.25 | 0.40 | 0.50 | 0.65 | K20 |
| M6 T6 | | | (60) | (0.004) | (0.006) | (0.008) | (0.012) | (0.015) | (0.020) | (C-2) | (45) | (0.004) | (0.007) | (0.010) | (0.015) | (0.020) | (0.025) | (C-2) |
| **High speed** | 225–275 | Ann. | 11 | 0.075 | 0.15 | 0.20 | 0.30 | 0.40 | 0.50 | S3, S4, S2 | 9 | 0.10 | 0.18 | 0.25 | 0.40 | 0.50 | 0.65 | S3, S4, S2 |
| M3-1 M43 | | | (35) | (0.003) | (0.006) | (0.008) | (0.012) | (0.015) | (0.020) | (M1, M2, M7) | (30) | (0.004) | (0.007) | (0.010) | (0.015) | (0.020) | (0.025) | (M1, M2, M7) |
| M4 M44 | | | 15 | 0.10 | 0.15 | 0.20 | 0.30 | 0.40 | 0.50 | K20 | 12 | 0.10 | 0.18 | 0.25 | 0.40 | 0.50 | 0.65 | K20 |
| M7 M46 | | | (50) | (0.004) | (0.006) | (0.008) | (0.012) | (0.015) | (0.020) | (C-2) | (40) | (0.004) | (0.007) | (0.010) | (0.015) | (0.020) | (0.025) | (C-2) |
| M30 T4 | | | | | | | | | | | | | | | | | | |
| M34 T5 | | | | | | | | | | | | | | | | | | |
| M36 T8 | | | | | | | | | | | | | | | | | | |
| M41 | | | | | | | | | | | | | | | | | | |
| **High speed** | 225–275 | Ann. | 9 | 0.050 | 0.10 | 0.15 | 0.20 | 0.25 | 0.30 | S3, S4, S2 | 8 | 0.075 | 0.13 | 0.15 | 0.25 | 0.30 | 0.40 | S3, S4, S2 |
| M3-2 | | | (30) | (0.002) | (0.004) | (0.006) | (0.008) | (0.010) | (0.012) | (M1, M2, M7) | (25) | (0.003) | (0.005) | (0.006) | (0.010) | (0.012) | (0.015) | (M1, M2, M7) |
| T15 | | | 14 | 0.10 | 0.15 | 0.20 | 0.25 | 0.30 | 0.40 | K20 | 11 | 0.10 | 0.15 | 0.20 | 0.25 | 0.30 | 0.40 | K20 |
| | | | (45) | (0.004) | (0.006) | (0.008) | (0.010) | (0.012) | (0.015) | (C-2) | (35) | (0.004) | (0.006) | (0.008) | (0.010) | (0.012) | (0.015) | (C-2) |
| **Hot work** | 150–200 | Ann. | 20 | 0.075 | 0.15 | 0.20 | 0.30 | 0.40 | 0.50 | S3, S4, S2 | 11 | 0.13 | 0.20 | 0.30 | 0.50 | 0.65 | 0.75 | S3, S4, S2 |
| H10 H13 | | | (65) | (0.003) | (0.006) | (0.008) | (0.012) | (0.015) | (0.020) | (M1, M2, M7) | (35) | (0.005) | (0.008) | (0.012) | (0.020) | (0.025) | (0.030) | (M1, M2, M7) |
| H11 H14 | | | 24 | 0.10 | 0.15 | 0.20 | 0.30 | 0.40 | 0.50 | K20 | 14 | 0.13 | 0.20 | 0.30 | 0.50 | 0.65 | 0.75 | K20 |
| H12 H19 | | | (80) | (0.004) | (0.006) | (0.008) | (0.012) | (0.015) | (0.020) | (C-2) | (45) | (0.005) | (0.008) | (0.012) | (0.020) | (0.025) | (0.030) | (C-2) |
| | 325–375 | Q&T | 11 | 0.050 | 0.10 | 0.15 | 0.20 | 0.025 | 0.30 | S3, S4, S2 | 8 | 0.075 | 0.13 | 0.15 | 0.25 | 0.30 | 0.40 | S3, S4, S2 |
| | | | (35) | (0.002) | (0.004) | (0.006) | (0.008) | (0.010) | (0.012) | (M1, M2, M7) | (25) | (0.003) | (0.005) | (0.006) | (0.010) | (0.012) | (0.015) | (M1, M2, M7) |
| | | | 15 | 0.10 | 0.15 | 0.20 | 0.25 | 0.30 | 0.40 | K20 | 11 | 0.10 | 0.13 | 0.20 | 0.25 | 0.30 | 0.40 | K20 |
| | | | (50) | (0.004) | (0.006) | (0.008) | (0.010) | (0.012) | (0.015) | (C-2) | (35) | (0.004) | (0.005) | (0.008) | (0.010) | (0.012) | (0.015) | (C-2) |
| | 48–50 | Q&T | 5 | 0.025 | 0.038 | 0.050 | 0.050 | 0.050 | 0.050 | S9, S11(c) | 5 | 0.025 | 0.038 | 0.038 | 0.075 | 0.075 | 0.075 | S9, S11(c) |
| | HRC | | (15) | (0.001) | (0.0015) | (0.002) | (0.002) | (0.002) | (0.002) | (T15, M42)(c) | (15) | (0.001) | (0.0015) | (0.002) | (0.003) | (0.003) | (0.003) | (T15, M42)(c) |
| | | | 9 | 0.050 | 0.075 | 0.10 | 0.13 | 0.15 | 0.15 | K20 | 8 | 0.050 | 0.075 | 0.10 | 0.13 | 0.15 | 0.15 | K20 |
| | | | (30) | (0.002) | (0.003) | (0.004) | (0.005) | (0.006) | (0.006) | (C-2) | (25) | (0.002) | (0.003) | (0.004) | (0.005) | (0.006) | (0.006) | (C-2) |
| | 54–56 | Q&T | 6 | 0.050 | 0.050 | 0.075 | 0.10 | 0.10 | 0.10 | K20 | 5 | 0.050 | 0.050 | 0.075 | 0.10 | 0.10 | 0.10 | K20 |
| | HRC | | (20) | (0.002) | (0.002) | (0.003) | (0.004) | (0.004) | (0.004) | (C-2) | (15) | (0.002) | (0.002) | (0.003) | (0.004) | (0.004) | (0.004) | (C-2) |
| **Hot work** | 150–200 | Ann. | 14 | 0.075 | 0.15 | 0.20 | 0.30 | 0.40 | 0.50 | S3, S4, S2 | 11 | 0.13 | 0.20 | 0.30 | 0.50 | 0.65 | 0.75 | S3, S4, S2 |
| H21 H25 | | | (45) | (0.003) | (0.006) | (0.008) | (0.012) | (0.015) | (0.020) | (M1, M2, M7) | (35) | (0.005) | (0.008) | (0.012) | (0.020) | (0.025) | (0.030) | (M1, M2, M7) |
| H22 H26 | | | 18 | 0.10 | 0.15 | 0.20 | 0.30 | 0.40 | 0.50 | K20 | 14 | 0.13 | 0.20 | 0.30 | 0.50 | 0.65 | 0.75 | K20 |
| H23 H42 | | | (60) | (0.004) | (0.006) | (0.008) | (0.012) | (0.015) | (0.020) | (C-2) | (45) | (0.005) | (0.008) | (0.012) | (0.020) | (0.025) | (0.030) | (C-2) |
| H24 | 200–250 | Ann. | 11 | 0.075 | 0.15 | 0.20 | 0.30 | 0.40 | 0.50 | S3, S4, S2 | 9 | 0.13 | 0.20 | 0.30 | 0.50 | 0.65 | 0.75 | S3, S4, S2 |
| | | | (35) | (0.003) | (0.006) | (0.008) | (0.012) | (0.015) | (0.020) | (M1, M2, M7) | (30) | (0.005) | (0.008) | (0.012) | (0.020) | (0.025) | (0.030) | (M1, M2, M7) |
| | | | 15 | 0.10 | 0.15 | 0.20 | 0.30 | 0.40 | 0.50 | K20 | 12 | 0.13 | 0.20 | 0.30 | 0.50 | 0.65 | 0.75 | K20 |
| | | | (50) | (0.004) | (0.006) | (0.008) | (0.012) | (0.015) | (0.020) | (C-2) | (40) | (0.005) | (0.008) | (0.012) | (0.020) | (0.025) | (0.030) | (C-2) |
| **Cold work** | 200–250 | Ann. | 8 | 0.075 | 0.15 | 0.20 | 0.30 | 0.40 | 0.50 | S3, S4, S2 | 6 | 0.10 | 0.18 | 0.25 | 0.40 | 0.50 | 0.65 | S3, S4, S2 |
| A7 D4 | | | (25) | (0.003) | (0.006) | (0.008) | (0.012) | (0.015) | (0.020) | (M1, M2, M7) | (20) | (0.004) | (0.007) | (0.010) | (0.015) | (0.020) | (0.025) | (M1, M2, M7) |
| D2 D5 | | | 12 | 0.10 | 0.15 | 0.20 | 0.30 | 0.40 | 0.50 | K20 | 9 | 0.10 | 0.18 | 0.25 | 0.40 | 0.50 | 0.65 | K20 |
| D3 D7 | | | (40) | (0.004) | (0.006) | (0.008) | (0.012) | (0.015) | (0.020) | (C-2) | (30) | (0.004) | (0.007) | (0.010) | (0.015) | (0.020) | (0.025) | (C-2) |
| **Cold work** | 200–250 | Ann. | 14 | 0.075 | 0.15 | 0.20 | 0.30 | 0.40 | 0.50 | S3, S4, S2 | 9 | 0.10 | 0.18 | 0.25 | 0.40 | 0.50 | 0.65 | S3, S4, S2 |
| A2 A10 | | | (45) | (0.003) | (0.006) | (0.008) | (0.012) | (0.015) | (0.020) | (M1, M2, M7) | (30) | (0.004) | (0.007) | (0.010) | (0.015) | (0.020) | (0.025) | (M1, M2, M7) |
| A3 O1 | | | 18 | 0.10 | 0.15 | 0.20 | 0.30 | 0.40 | 0.50 | K20 | 12 | 0.10 | 0.18 | 0.25 | 0.40 | 0.50 | 0.65 | K20 |
| A6 O2 | | | (60) | (0.004) | (0.006) | (0.008) | (0.012) | (0.015) | (0.020) | (C-2) | (40) | (0.004) | (0.007) | (0.010) | (0.015) | (0.020) | (0.025) | (C-2) |
| A8 | | | | | | | | | | | | | | | | | | |
| **Shock resisting** | 175–225 | Ann. | 18 | 0.10 | 0.15 | 0.25 | 0.40 | 0.50 | 0.65 | S3, S4, S2 | 11 | 0.13 | 0.20 | 0.30 | 0.50 | 0.65 | 0.75 | S3, S4, S2 |
| S1 S6 | | | (60) | (0.004) | (0.006) | (0.010) | (0.015) | (0.020) | (0.025) | (M1, M2, M7) | (35) | (0.005) | (0.008) | (0.012) | (0.020) | (0.025) | (0.030) | (M1, M2, M7) |
| S2 S7 | | | 23 | 0.10 | 0.15 | 0.25 | 0.40 | 0.50 | 0.65 | K20 | 14 | 0.13 | 0.20 | 0.30 | 0.50 | 0.65 | 0.75 | K20 |
| S5 | | | (75) | (0.004) | (0.006) | (0.010) | (0.015) | (0.020) | (0.025) | (C-2) | (45) | (0.005) | (0.008) | (0.012) | (0.020) | (0.025) | (0.030) | (C-2) |
| **Mold** | 100–150 | Ann. | 23 | 0.10 | 0.18 | 0.30 | 0.50 | 0.65 | 0.75 | S3, S4, S2 | 14 | 0.15 | 0.25 | 0.40 | 0.65 | 0.75 | 0.90 | S3, S4, S2 |
| P2 P5 | | | (75) | (0.004) | (0.007) | (0.012) | (0.020) | (0.025) | (0.030) | (M1, M2, M7) | (45) | (0.006) | (0.010) | (0.015) | (0.025) | (0.030) | (0.035) | (M1, M2, M7) |
| P4 P6 | | | 27 | 0.10 | 0.18 | 0.30 | 0.50 | 0.65 | 0.75 | K20 | 15 | 0.15 | 0.25 | 0.40 | 0.65 | 0.75 | 0.90 | K20 |
| | | | (90) | (0.004) | (0.007) | (0.012) | (0.020) | (0.025) | (0.030) | (C-2) | (50) | (0.006) | (0.010) | (0.015) | (0.025) | (0.030) | (0.035) | (C-2) |
| **Mold** | 150–200 | Ann. | 20 | 0.10 | 0.18 | 0.30 | 0.50 | 0.65 | 0.75 | S3, S4, S2 | 11 | 0.13 | 0.20 | 0.30 | 0.50 | 0.65 | 0.75 | S3, S4, S2 |
| P20 | | | (65) | (0.004) | (0.007) | (0.012) | (0.020) | (0.025) | (0.030) | (M1, M2, M7) | (35) | (0.005) | (0.008) | (0.012) | (0.020) | (0.025) | (0.030) | (M1, M2, M7) |
| P21 | | | 24 | 0.10 | 0.18 | 0.30 | 0.50 | 0.65 | 0.75 | K20 | 14 | 0.13 | 0.20 | 0.30 | 0.50 | 0.65 | 0.75 | K20 |
| | | | (80) | (0.004) | (0.007) | (0.012) | (0.020) | (0.025) | (0.030) | (C-2) | (45) | (0.005) | (0.008) | (0.012) | (0.020) | (0.025) | (0.030) | (C-2) |
| **Special purpose** | 150–200 | Ann. | 20 | 0.10 | 0.18 | 0.30 | 0.50 | 0.65 | 0.75 | S3, S4, S2 | 11 | 0.13 | 0.20 | 0.30 | 0.50 | 0.65 | 0.75 | S3, S4, S2 |
| L2 | | | (65) | (0.004) | (0.007) | (0.012) | (0.020) | (0.025) | (0.030) | (M1, M2, M7) | (35) | (0.005) | (0.008) | (0.012) | (0.020) | (0.025) | (0.030) | (M1, M2, M7) |
| L6 | | | 24 | 0.10 | 0.18 | 0.30 | 0.50 | 0.65 | 0.75 | K20 | 14 | 0.13 | 0.20 | 0.30 | 0.50 | 0.65 | 0.75 | K20 |
| L7 | | | (80) | (0.004) | (0.007) | (0.012) | (0.020) | (0.025) | (0.030) | (C-2) | (45) | (0.005) | (0.008) | (0.012) | (0.020) | (0.025) | (0.030) | (C-2) |
| **Special purpose** | 200–250 | Ann. | 14 | 0.075 | 0.15 | 0.20 | 0.30 | 0.40 | 0.50 | S3, S4, S2 | 9 | 0.10 | 0.18 | 0.25 | 0.40 | 0.50 | 0.65 | S3, S4, S2 |
| F1 | | | (45) | (0.003) | (0.006) | (0.008) | (0.012) | (0.015) | (0.020) | (M1, M2, M7) | (30) | (0.004) | (0.007) | (0.010) | (0.015) | (0.020) | (0.025) | (M1, M2, M7) |
| F2 | | | 18 | 0.10 | 0.15 | 0.20 | 0.30 | 0.40 | 0.50 | K20 | 12 | 0.10 | 0.18 | 0.25 | 0.40 | 0.50 | 0.65 | K20 |
| | | | (60) | (0.004) | (0.006) | (0.008) | (0.012) | (0.015) | (0.020) | (C-2) | (40) | (0.004) | (0.007) | (0.010) | (0.015) | (0.020) | (0.025) | (C-2) |
| **Water hardening** | 150–200 | Ann. | 30 | 0.10 | 0.18 | 0.30 | 0.50 | 0.65 | 0.75 | S3, S4, S2 | 14 | 0.15 | 0.25 | 0.40 | 0.65 | 0.75 | 0.90 | S3, S4, S2 |
| W1, W2, | | | (100) | (0.004) | (0.007) | (0.012) | (0.020) | (0.025) | (0.030) | (M1, M2, M7) | (45) | (0.006) | (0.010) | (0.015) | (0.025) | (0.030) | (0.035) | (M1, M2, M7) |
| W5(d) | | | 34 | 0.10 | 0.18 | 0.30 | 0.50 | 0.65 | 0.75 | K20 | 17 | 0.15 | 0.25 | 0.40 | 0.65 | 0.75 | 0.90 | K20 |
| | | | (110) | (0.004) | (0.007) | (0.012) | (0.020) | (0.025) | (0.030) | (C-2) | (55) | (0.006) | (0.010) | (0.015) | (0.025) | (0.030) | (0.035) | (C-2) |

Note: Ann., annealed; Q&T, quenched and tempered. (a) Unless otherwise noted. (b) Based on four flutes for 3 and 6 mm (1/8 and 1/4 in.) reamers, six flutes for 13 mm (1/2 in.) reamers, and eight flutes for 25 mm (1 in.) and larger reamers. (c) Any premium high-speed steel (T15, M33, M41–M47 or S9, S10, S11, S12). (d) Also includes SAE J438b: Types W108, W109, W110, W112, W209, W210, and W310.
Source: Metcut Research Associates Inc.

## Table 11 Nominal speeds for the tapping of wrought tool steels with high-speed tool steels

| | | | Speed, m/min (sfm)(b) | | | | | | | | HSS tool material | |
|---|---|---|---|---|---|---|---|---|---|---|---|---|
| Steel | Hardness, HB(a) | Condition | Pitch, mm >3 | Threads/in. ≤7 | Pitch, mm 1.5–3 | Threads/in. 8–15 | Pitch, mm 1–1.5 | Threads/in. 16–24 | Pitch, mm ≤1 | Threads/in. >24 | AISI | ISO |
| **High speed**<br>M1 M10 T6<br>M2 T1<br>M6 T2 | 200–250 | Annealed | 2.1 | (7) | 5 | (15) | 6 | (20) | 8 | (25) | S2, S3 | M10, M7, M1 |
| **High speed**<br>M3-1 M33 M42 M47<br>M4 M34 M43 T4<br>M7 M36 M44 T5<br>M30 M41 M46 T8 | 225–275 | Annealed | 2.1 | (7) | 5 | (15) | 5.5 | (18) | 6 | (20) | S2, S3 | M10, M7, M1 |
| **High speed**<br>M3-2<br>T15 | 225–275 | Annealed | 1.5 | (5) | 3 | (10) | 4.3 | (14) | 5 | (15) | S2, S3 | M10, M7, M1 |
| **Hot work**<br>H10 H13 | 150–200 | Annealed | 2.4 | (8) | 5.5 | (18) | 8 | (25) | 9 | (30) | S2, S3 | M10, M7, M1 |
| H11 H14<br>H12 H19 | 325–375 | Quenched and tempered | 2.1 | (7) | 5 | (15) | 5.5 | (18) | 6 | (20) | S2, S3 | M10, M7, M1 |
| | 48–50 HRC | Quenched and tempered | 0.6 | (2) | 1.5 | (5) | 1.8 | (6) | 2.1 | (7) | Nitrided S2, S3 | Nitrided M10, M7, M1 |
| | 50–52 HRC | Quenched and tempered | 0.6 | (2) | 0.9 | (3) | 1.2 | (4) | 1.5 | (5) | Nitrided S2, S3 | Nitrided M10, M7, M1 |
| **Hot work**<br>H21 H25 | 150–200 | Annealed | 2.1 | (7) | 4.6 | (15) | 6 | (20) | 8 | (25) | S2, S3 | M10, M7, M1 |
| H22 H26<br>H23 H42<br>H24 | 200–250 | Annealed | 2.1 | (7) | 5 | (15) | 5.5 | (18) | 6 | (20) | S2, S3 | M10, M7, M1 |
| **Cold work**<br>A7 D4<br>D2 D5<br>D3 D7 | 200–250 | Annealed | 1.5 | (5) | 3 | (10) | 4.3 | (14) | 5 | (15) | S2, S3 | M10, M7, M1 |
| **Cold work**<br>A2 A8 O2<br>A3 A9 O6<br>A4 A10 O7<br>A6 O1 | 200–250 | Annealed | 2.1 | (7) | 5 | (15) | 6 | (20) | 8 | (25) | S2, S3 | M10, M7, M1 |
| **Shock resisting**<br>S1 S6<br>S2 S7<br>S5 | 175–225 | Annealed | 2.4 | (8) | 5.5 | (18) | 8 | (25) | 9 | (30) | S2, S3 | M10, M7, M1 |
| **Mold**<br>P2<br>P4<br>P5<br>P6 | 100–150 | Annealed | 5 | (15) | 9 | (30) | 12 | (40) | 14 | (45) | S2, S3 | M10, M7, M1 |
| **Mold**<br>P20<br>P21 | 150–200 | Annealed | 3 | (10) | 8 | (25) | 9 | (30) | 11 | (35) | S2, S3 | M10, M7, M1 |
| **Special purpose**<br>L2<br>L6<br>L7 | 150–200 | Annealed | 3 | (10) | 8 | (25) | 9 | (30) | 11 | (35) | S2, S3 | M10, M7, M1 |
| **Special purpose**<br>F1<br>F2 | 200–250 | Annealed | 2.1 | (7) | 5 | (15) | 6 | (20) | 8 | (25) | S2, S3 | M10, M7, M1 |
| **Water hardening**<br>W1, W2, W5<br>SAE J438b: Types<br>W108, W109, W110,<br>W112, W209, W210,<br>W310 | 150–200 | Annealed | 5 | (15) | 9 | (30) | 14 | (45) | 15 | (50) | S2, S3 | M10, M7, M1 |

(a) Unless otherwise noted. (b) These speeds are for tapping 65 to 75% threads in shallow through holes. Reduce the speed when tapping deep or blind holes or with a higher percentage of thread. Source: Metcut Research Associates Inc.

difficult to machine than are M1, M2, or M10, surface speed and feed were reduced to about 50% of the rates ordinarily used for these M steels, in the interest of tool life. Nevertheless, as shown in Table 18, which presents the results of the comparison, carbide tooling doubled the machining rate and nearly tripled the tool life provided by M2 tooling.

On the basis of these data, the use of carbide tooling for machining difficult high-speed steels such as T15, M15, and M4 was recommended, provided machine tools are in good repair and the setup is rigid.

**Tangential thread chasers** are ordinarily made of M1 high-speed steel, but M3 may be substituted when longer tool life is desired or a more difficult material is to be threaded. The example below compares the tool life obtained in milling step-and-angle dovetails in thread chasers made of these two steels.

**Example 10: Milling M1 Versus M3 Thread Chasers.** The dovetail shown in Fig. 7 was produced in tangential thread chasers made of M1 and M3 high-speed steel, each in one pass by a single milling cutter. (In service, this dovetail locks the chaser into the holder.) Blanks for these chasers weighed 0.099 kg (3½ oz) before machining and 0.071 kg (2½ oz) after. At a speed of 95 rev/min and a table feed of 79 mm/min (3⅛ in./min), tool life in milling the M1 chasers was 2400 pieces per grind, but only 1200 to 1440 for M3 chasers.

**Table 12 Nominal speeds and feeds for peripheral (slab) milling of tool steels with high-speed steel tools**

High-speed steels M2 and M7, except M33 and T15 for milling of the H11 group at hardnesses above 48 HRC

| Steel | Hardness, HB | Rough milling, depth of cut, 3.8 mm (0.150 in.) | | | | Finish milling, depth of cut, 1.0 mm (0.040 in.) | | | |
|---|---|---|---|---|---|---|---|---|---|
| | | Speed | | Feed | | Speed | | Feed | |
| | | m/min | sfm | mm/tooth | in./tooth | m/min | sfm | mm/tooth | in./tooth |
| W group | 150–200 | 37 | 120 | 0.20 | 0.008 | 47 | 155 | 0.15 | 0.006 |
| S group | 175–225 | 34 | 110 | 0.20 | 0.008 | 41 | 135 | 0.15 | 0.006 |
| O and A groups(a) | 200–250 | 27 | 90 | 0.18 | 0.007 | 35 | 115 | 0.13 | 0.005 |
| D group and A7 | 200–250 | 15 | 50 | 0.18 | 0.007 | 18 | 60 | 0.13 | 0.005 |
| H11(b) | 150–200 | 29 | 95 | 0.20 | 0.008 | 38 | 125 | 0.15 | 0.006 |
| | 200–250 | 24 | 80 | 0.18 | 0.007 | 30 | 100 | 0.13 | 0.005 |
| H20(c) | 150–200 | 23 | 75 | 0.20 | 0.008 | 29 | 95 | 0.15 | 0.006 |
| | 200–250 | 21 | 70 | 0.18 | 0.007 | 26 | 85 | 0.13 | 0.005 |
| M2 and T1(d) | 200–250 | 20 | 65 | 0.18 | 0.007 | 26 | 85 | 0.13 | 0.005 |
| M3-1 and T4(e) | 225–275 | 17 | 55 | 0.18 | 0.007 | 21 | 70 | 0.13 | 0.005 |
| M15 and T15(f) | 225–275 | 14 | 45 | 0.18 | 0.007 | 18 | 60 | 0.11 | 0.0045 |
| L group | 150–200 | 35 | 115 | 0.20 | 0.008 | 41 | 135 | 0.15 | 0.006 |
| P group | 200–250 | 27 | 90 | 0.18 | 0.007 | 37 | 120 | 0.13 | 0.005 |
| P1–P6 | 100–150 | 40 | 130 | 0.20 | 0.008 | 49 | 160 | 0.15 | 0.006 |
| P20–P21 | 150–200 | 37 | 120 | 0.20 | 0.008 | 44 | 145 | 0.15 | 0.006 |

(a) Includes all of group A except A7. (b) Includes also H10, H12, H13, H14, H16, and H19. (c) Includes also H21, H22, H23, H24, H25, H41, H42, and H43. (d) Includes also M1, M6, M10, T2, T6, and T7. (e) Includes also M4, M7, M30, M33, M34, M35, M36, M41, M42, M43, M44, T5, and T8. (f) Includes also M3 (class 2) and T9. Source: Metcut Research Associates Inc.

## Low-Alloy Special-Purpose Tool Steels

In the spheroidize-annealed condition (179 to 207 HB), steels L1, L2, L3, and L7 are 65 to 75% as machinable as a W1 steel containing 1% C. Steel L6, which contains 1.50% Ni, is slightly more difficult to machine.

Completely or nearly completely spheroidized stock is preferred for most machining operations. Occasionally, however, microstructures that contain some lamellar pearlite are preferred. Such applications usually involve those steels of lower-carbon content, which can be specified for L2.

For the L steels with 1% carbon, microstructures of ferrite and spheroidized carbides are almost invariably preferred (and specified) for all machining operations. For these steels, microstructures containing lamellar pearlite usually result in a rough surface finish and shorter tool life. The following example describes an application in which the sequence of operations was changed to avoid machining a microstructure that contained pearlite.

**Example 11: Deferred Forging for Improved Machinability.** The original and improved methods used for producing wood drill shanks from L1 tool steel are shown in Fig. 8. In the original method, the blank was first forged to obtain the square tapered end, after which the recess was form turned in a 3.7 kW (5 hp) lathe. Even though the blanks were spheroidized, at least half of the recessed area had been heated to the forging temperature, and this resulted in the mixed microstructure denoted as poor machinability in Fig. 8. Severe wear of the M34 form tools, especially at the corner nearest the forged end, occurred in form turning of workpieces made by the original method. Form tools required grinding after 5 to 20 pieces.

In the improved method, the first and second steps were reversed, so that the blanks were formed before the microstructure (denoted in Fig. 8 as good machinability) was altered by heating for forging. Form tool life was increased to 150 pieces per grind. Machining conditions for the forming operations are given in the table in Fig. 8.

## Carbon-Tungsten Special-Purpose Tool Steels

As-annealed (183 to 207 HB) F1 carbon-tungsten steel is approximately 75% as machinable as 1% carbon W1 steel of comparable structure. Steels F2 and F3, with normally higher hardnesses of 207 to 235 HB in the annealed condition, machine approximately 60% as well as the annealed 1% carbon W1.

**Milling.** Recommended tool design for milling the F steels with high-speed steel cutters is given in Table 19.

## Low-Carbon Mold Steels

Many dies made from low-carbon mold steel are formed by a hubbing operation, in which a hard master form is pressed into the die to produce the die impression. Many of these steels have very low hardness in the annealed condition. This is especially true for P1, which is frequently supplied at 80 to 100 HB. As the alloy content increases, hardness of the annealed steel also increases, and suitability for hubbing decreases. Thus, P4, P6, P20, and P21 are generally considered unsuitable for hubbing, and molds of these steels must be produced by machining.

The low-carbon steels of the P group (P1 to P6) have poor machinability because of their softness; chips are gummy and readily build up on cutting edges and drill margins. Because of the low hardness of types P1 to P6, the surface of the mold impression must be carburized and hardened to resist wear, and the entire die must be heat treated to resist deformation of the impression. Considerable distortion occurs in quenching, particularly in the water quenching of a relatively small die. Thus, size control is difficult, and dies must be finished by grinding after hardening. When dies are ground, care must be taken not to penetrate the carburized case, because core hardness is insufficient for resistance to abrasion or deformation.

Because the group P steels are used almost exclusively in the manufacture of molds, few, if any, turning operations are encountered. However, steels P1 to P6, depending on their composition, have machining characteristics like those of production carburizing steels of similar composition. The P20 steel resembles 4140.

**Example 12: Drilling, Fly Cutting, and Milling P20.** In the production of a P20 steel mold for forming plastic housings for televisions, the prehardened block (285 HB) was first ground square in a surface grinder to facilitate layout and provide reference planes. The sequence of operations (see Fig. 9) used to machine the block (760 × 810 × 255 mm, or 30 × 32 × 10 in.) was:

- After the block had been clamped into a large boring mill, an 89 mm (3½ in.) diam hole was drilled in the center of one of the 760 × 810 mm (30 × 32 in.) faces, to a depth of 150 mm (6 in.)
- With the block still in position in the boring mill, the drilled hole was enlarged to a 660 mm (26 in.) diameter by a series of fly cutters
- The block was then placed in a tracer-controlled milling machine, which squared the corners of the impression and produced the final symmetrical contour
- The block was returned to the boring mill to drill holes for the circulation of cooling water. Four 17.5 mm (11/16 in.) diam through holes (later plugged at one end) were drilled through under the impression and along the sides

**Table 13 Nominal speeds and feeds for the end milling of tool steels with high-speed steel and carbide cutters**

| Steel | Hardness, HB(a) | Rough milling, depth of cut, 1.5 mm (0.060 in.): Speed m/min | sfm | Feed for cutter diameter of 9.5 mm (3/8 in.): mm/tooth | in./tooth | 13 mm (1/2 in.): mm/tooth | in./tooth | 19 mm (3/4 in.): mm/tooth | in./tooth | 25–50 mm (1–2 in.): mm/tooth | in./tooth | Finish milling, depth of cut, 0.5 mm (0.020 in.): Speed m/min | sfm | Feed for cutter diameter of 9.5 mm (3/8 in.): mm/tooth | in./tooth | 13 mm (1/2 in.): mm/tooth | in./tooth | 19 mm (3/4 in.): mm/tooth | in./tooth | 25–50 mm (1–2 in.): mm/tooth | in./tooth |
|---|---|---|---|---|---|---|---|---|---|---|---|---|---|---|---|---|---|---|---|---|---|
| **High-speed steel cutters(b)** | | | | | | | | | | | | | | | | | | | | | |
| W group | 150–200 | 32 | 105 | 0.05 | 0.002 | 0.08 | 0.003 | 0.13 | 0.005 | 0.15 | 0.006 | 43 | 140 | 0.025 | 0.001 | 0.05 | 0.002 | 0.10 | 0.004 | 0.13 | 0.005 |
| S group | 175–225 | 26 | 85 | 0.05 | 0.002 | 0.08 | 0.003 | 0.10 | 0.004 | 0.13 | 0.005 | 34 | 110 | 0.025 | 0.001 | 0.05 | 0.002 | 0.08 | 0.003 | 0.10 | 0.004 |
| O group | 200–250 | 21 | 70 | 0.025 | 0.001 | 0.05 | 0.002 | 0.08 | 0.003 | 0.10 | 0.004 | 27 | 90 | 0.018 | 0.0007 | 0.025 | 0.001 | 0.05 | 0.002 | 0.08 | 0.003 |
| A group(c) | 200–250 | 21 | 70 | 0.025 | 0.001 | 0.05 | 0.002 | 0.08 | 0.003 | 0.10 | 0.004 | 27 | 90 | 0.018 | 0.0007 | 0.025 | 0.001 | 0.05 | 0.002 | 0.08 | 0.003 |
| D group and A7 | 200–250 | 15 | 50 | 0.025 | 0.001 | 0.05 | 0.002 | 0.08 | 0.003 | 0.10 | 0.004 | 20 | 65 | 0.018 | 0.0007 | 0.025 | 0.001 | 0.05 | 0.002 | 0.08 | 0.003 |
| H11(d) | 150–200 | 23 | 75 | 0.025 | 0.001 | 0.05 | 0.002 | 0.08 | 0.003 | 0.10 | 0.004 | 30 | 100 | 0.018 | 0.0007 | 0.025 | 0.001 | 0.05 | 0.002 | 0.08 | 0.003 |
| | 200–250 | 21 | 70 | 0.025 | 0.001 | 0.05 | 0.002 | 0.08 | 0.003 | 0.10 | 0.004 | 29 | 95 | 0.018 | 0.0007 | 0.025 | 0.001 | 0.05 | 0.002 | 0.08 | 0.003 |
| | 325–375 | 15 | 50 | 0.025 | 0.001 | 0.025 | 0.001 | 0.05 | 0.002 | 0.08 | 0.003 | 20 | 65 | 0.013 | 0.0005 | 0.025 | 0.001 | 0.025 | 0.001 | 0.05 | 0.002 |
| | 48–50 HRC | 9 | 30 | ... | ... | 0.013 | 0.0005 | 0.025 | 0.001 | 0.038 | 0.0015 | 11 | 35 | 0.013 | 0.0005 | 0.013 | 0.0005 | 0.018 | 0.0007 | 0.025 | 0.001 |
| H20(e) | 150–200 | 20 | 65 | 0.025 | 0.001 | 0.038 | 0.0015 | 0.05 | 0.002 | 0.08 | 0.003 | 27 | 90 | 0.013 | 0.0005 | 0.025 | 0.001 | 0.038 | 0.0015 | 0.05 | 0.002 |
| | 200–250 | 18 | 60 | 0.025 | 0.001 | 0.038 | 0.0015 | 0.05 | 0.002 | 0.08 | 0.003 | 24 | 80 | 0.013 | 0.0005 | 0.025 | 0.001 | 0.038 | 0.0015 | 0.05 | 0.002 |
| M2 and T1(f) | 200–250 | 20 | 65 | 0.025 | 0.001 | 0.05 | 0.002 | 0.08 | 0.003 | 0.10 | 0.004 | 26 | 85 | 0.018 | 0.0007 | 0.025 | 0.001 | 0.05 | 0.002 | 0.08 | 0.003 |
| M3-1 and T4(g) | 225–275 | 17 | 55 | 0.025 | 0.001 | 0.05 | 0.002 | 0.08 | 0.003 | 0.10 | 0.004 | 23 | 75 | 0.018 | 0.0007 | 0.025 | 0.001 | 0.05 | 0.002 | 0.08 | 0.003 |
| | 225–275 | 14 | 45 | 0.018 | 0.0007 | 0.025 | 0.001 | 0.038 | 0.0015 | 0.064 | 0.0025 | 18 | 60 | 0.013 | 0.0005 | 0.025 | 0.001 | 0.038 | 0.0015 | 0.05 | 0.002 |
| L group | 150–200 | 24 | 80 | 0.05 | 0.002 | 0.08 | 0.003 | 0.10 | 0.004 | 0.13 | 0.005 | 34 | 110 | 0.025 | 0.001 | 0.05 | 0.002 | 0.08 | 0.003 | 0.10 | 0.004 |
| F group | 200–250 | 23 | 75 | 0.025 | 0.001 | 0.05 | 0.002 | 0.08 | 0.003 | 0.10 | 0.004 | 30 | 100 | 0.018 | 0.0007 | 0.025 | 0.001 | 0.05 | 0.002 | 0.08 | 0.003 |
| P1–P6 | 100–150 | 30 | 100 | 0.05 | 0.002 | 0.08 | 0.003 | 0.13 | 0.005 | 0.15 | 0.006 | 41 | 135 | 0.025 | 0.001 | 0.05 | 0.002 | 0.10 | 0.004 | 0.13 | 0.005 |
| P2–P21 | 150–200 | 27 | 90 | 0.05 | 0.002 | 0.08 | 0.003 | 0.10 | 0.004 | 0.13 | 0.005 | 37 | 120 | 0.025 | 0.001 | 0.05 | 0.002 | 0.08 | 0.003 | 0.10 | 0.004 |
| **Carbide cutters(i)** | | | | | | | | | | | | | | | | | | | | | |
| W group | 150–200 | 152 | 500 | 0.064 | 0.0025 | 0.10 | 0.004 | 0.18 | 0.007 | 0.20 | 0.008 | 198 | 650 | 0.038 | 0.0015 | 0.089 | 0.0035 | 0.15 | 0.006 | 0.18 | 0.007 |
| S group | 175–225 | 113 | 370 | 0.05 | 0.002 | 0.08 | 0.003 | 0.13 | 0.005 | 0.18 | 0.007 | 145 | 475 | 0.025 | 0.001 | 0.05 | 0.002 | 0.10 | 0.004 | 0.13 | 0.005 |
| O group | 200–250 | 91 | 300 | 0.038 | 0.0015 | 0.064 | 0.0025 | 0.10 | 0.004 | 0.13 | 0.005 | 119 | 390 | 0.025 | 0.001 | 0.038 | 0.0015 | 0.08 | 0.003 | 0.10 | 0.004 |
| A group(c) | 200–250 | 91 | 300 | 0.038 | 0.0015 | 0.064 | 0.0025 | 0.10 | 0.004 | 0.13 | 0.005 | 119 | 390 | 0.025 | 0.001 | 0.038 | 0.0015 | 0.08 | 0.003 | 0.10 | 0.004 |
| D group and A7 | 200–250 | 61 | 200 | 0.038 | 0.0015 | 0.064 | 0.0025 | 0.10 | 0.004 | 0.13 | 0.005 | 79 | 260 | 0.025 | 0.001 | 0.038 | 0.0015 | 0.08 | 0.003 | 0.10 | 0.004 |
| H11(d) | 150–200 | 91 | 300 | 0.038 | 0.0015 | 0.064 | 0.0025 | 0.10 | 0.004 | 0.13 | 0.005 | 119 | 390 | 0.025 | 0.001 | 0.038 | 0.0015 | 0.064 | 0.0025 | 0.10 | 0.004 |
| | 200–250 | 82 | 270 | 0.038 | 0.0015 | 0.064 | 0.0025 | 0.10 | 0.004 | 0.13 | 0.005 | 110 | 360 | 0.025 | 0.001 | 0.038 | 0.0015 | 0.064 | 0.0025 | 0.10 | 0.004 |
| | 325–375 | 53 | 175 | 0.025 | 0.001 | 0.05 | 0.002 | 0.08 | 0.003 | 0.10 | 0.004 | 70 | 230 | 0.025 | 0.001 | 0.025 | 0.001 | 0.05 | 0.002 | 0.08 | 0.003 |
| | 48–50 HRC | 27 | 90 | 0.013 | 0.0005 | 0.025 | 0.001 | 0.05 | 0.002 | 0.13 | 0.005 | 37 | 120 | 0.013 | 0.0005 | 0.013 | 0.0005 | 0.038 | 0.0015 | 0.05 | 0.002 |
| | 50–52 HRC | 17 | 55 | ... | ... | 0.025 | 0.001 | 0.05 | 0.002 | 0.08 | 0.003 | 23 | 75 | 0.013 | 0.0005 | 0.013 | 0.0005 | 0.038 | 0.0015 | 0.05 | 0.002 |
| | 52–54 HRC | 14 | 45 | ... | ... | 0.025 | 0.001 | 0.05 | 0.002 | 0.08 | 0.003 | 18 | 60 | 0.013 | 0.0005 | 0.013 | 0.0005 | 0.038 | 0.0015 | 0.05 | 0.002 |
| | 54–56 HRC | 11 | 35 | ... | ... | 0.025 | 0.001 | 0.05 | 0.002 | 0.08 | 0.003 | 14 | 45 | 0.013 | 0.0005 | 0.013 | 0.0005 | 0.038 | 0.0015 | 0.05 | 0.002 |
| H20(e) | 150–200 | 70 | 230 | 0.025 | 0.001 | 0.05 | 0.002 | 0.08 | 0.003 | 0.10 | 0.004 | 91 | 300 | 0.025 | 0.001 | 0.038 | 0.0015 | 0.05 | 0.002 | 0.08 | 0.003 |
| | 200–250 | 69 | 225 | 0.025 | 0.001 | 0.038 | 0.0015 | 0.064 | 0.0025 | 0.089 | 0.0035 | 88 | 290 | 0.025 | 0.001 | 0.038 | 0.0015 | 0.05 | 0.002 | 0.064 | 0.0025 |
| M2 and T1(f) | 200–250 | 73 | 240 | 0.038 | 0.0015 | 0.064 | 0.0025 | 0.10 | 0.004 | 0.13 | 0.005 | 98 | 320 | 0.025 | 0.001 | 0.038 | 0.0015 | 0.064 | 0.0025 | 0.10 | 0.004 |
| M3-1 and T4(g) | 225–275 | 66 | 215 | 0.038 | 0.0015 | 0.064 | 0.0025 | 0.10 | 0.004 | 0.13 | 0.005 | 85 | 280 | 0.025 | 0.001 | 0.038 | 0.0015 | 0.064 | 0.0025 | 0.10 | 0.004 |
| M15 and T15(h) | 225–275 | 53 | 175 | 0.038 | 0.0015 | 0.05 | 0.002 | 0.08 | 0.003 | 0.10 | 0.004 | 69 | 225 | 0.025 | 0.001 | 0.038 | 0.0015 | 0.05 | 0.002 | 0.08 | 0.003 |
| L group | 150–200 | 113 | 370 | 0.05 | 0.002 | 0.08 | 0.003 | 0.13 | 0.005 | 0.18 | 0.007 | 148 | 485 | 0.025 | 0.001 | 0.05 | 0.002 | 0.10 | 0.004 | 0.13 | 0.005 |
| F group | 200–250 | 99 | 325 | 0.038 | 0.0015 | 0.064 | 0.0025 | 0.10 | 0.004 | 0.13 | 0.005 | 128 | 420 | 0.025 | 0.001 | 0.038 | 0.0015 | 0.064 | 0.0025 | 0.10 | 0.004 |
| P1–P6 | 100–150 | 140 | 460 | 0.064 | 0.0025 | 0.10 | 0.004 | 0.15 | 0.006 | 0.20 | 0.008 | 184 | 605 | 0.038 | 0.0015 | 0.089 | 0.0035 | 0.13 | 0.005 | 0.18 | 0.007 |
| P20–P21 | 150–200 | 122 | 400 | 0.05 | 0.002 | 0.08 | 0.003 | 0.13 | 0.005 | 0.18 | 0.007 | 158 | 520 | 0.025 | 0.001 | 0.05 | 0.002 | 0.10 | 0.004 | 0.13 | 0.005 |

(a) Unless otherwise noted. (b) High-speed steels M2 and M7 except T15, M33, M41, M42, M43, and M44 for the milling of M3 (class 1) and T4 groups, M15 and T15 groups, and group H11 at hardnesses of 48 to 50 HRC. (c) Includes all of group A except A7. (d) Includes also H10, H12, H13, H14, H16, and H19. (e) Includes also H21, H22, H23, H24, H25, H41, H42, and H43. (f) Includes also M1, M6, M10, T2, T6, and T7. (g) Includes also M4, M7, M30, M33, M34, M35, M36, M41, M42, M43, M44, T5 and T8. (h) Also M3 (class 2). (i) Carbide grade C-2. Source: Metcut Research Associates Inc.

**Variation in Practice for P20.** Practice described in Example 12 is typical for making plastics molds. Because P20 steel can be processed in several ways, it is used for many different mold types and sizes. The molds are machined at three or more different hardnesses.

The most economical way to make a plastic mold is to machine prehardened steel (usually 275 to 325 HB) for service without further heat treatment. Machinability for most operations is reasonably good at this hardness, and the cost of further heat treatment is eliminated.

At a hardness of 275 to 325 HB a high-speed steel cutter can be operated at about 18 m/min (60 sfm) and at a feed rate up to 0.13 mm/tooth (0.005 in./tooth) (somewhat less if the cutter shank lacks rigidity). When carbide cutters are used, the feed will be about the same, but the speed can be increased to about 73 m/min (240 sfm). Cooling ducts can be drilled with carbide-tip gun drills at 76 m/min (250 sfm) using a feed of 0.010 to 0.015 mm/rev (0.0004 to 0.0006 in./rev) for holes no larger than 13 mm (½ in.) in diameter.

For molds used in molding plastics that contain abrasive fillers, mold hardness must be higher to attain acceptable mold life. Under these conditions one of three procedures is followed.

The simplest procedure is to machine the steel in the fully annealed condition (175 to 225 HB) to finished or nearly finished size. At this hardness, metal can be removed with high-speed steel cutters at 27 m/min (90 sfm) or at 107 m/min (350 sfm) with carbide cutters. Feed is commonly 0.13 to 0.18 mm/tooth (0.005 to 0.007 in./tooth), although lesser feeds are necessary if the cutter lacks rigidity. Cooling ducts can be drilled with carbide-tip gun drills at 90 m/min (300 sfm) and feeds of 0.010 to 0.015 mm/rev (0.0004 to 0.0006 in./rev) for holes up to 13 mm (½ in.) in diameter. After machining, the mold is carburized and hardened so that the surface is approximately 60 HRC and the core is approximately 50 HRC. Some grinding or polishing, or both, are needed to finish the mold.

**Table 14 Nominal speeds for the power band sawing of wrought tool steels with high-speed tool steels**

| Steel | Hardness, HB | Condition | Material thickness, mm | Material thickness, in. | Tooth form(a) | Pitch, mm | Pitch, Teeth/in. | Band speed, m/min | Band speed, sfm |
|---|---|---|---|---|---|---|---|---|---|
| **High speed** | 200–250 | Annealed | <13 | <½ | P | 2.5–1.8 | 10–14 | 43 | 140 |
| M1 T1 | | | 13–25 | ½–1 | P | 3–2.5 | 8–10 | 38 | 125 |
| M2 T2 | | | 25–75 | 1–3 | P | 4–3 | 6–8 | 32 | 105 |
| M6 T6 | | | >75 | >3 | C,P | 8.5–6.3 | 3–4 | 26 | 85 |
| M10 | | | | | | | | | |
| **High speed** | 225–275 | Annealed | <13 | <½ | P | 2.5–1.8 | 10–14 | 40 | 130 |
| M3-1 M36 M47 | | | 13–25 | ½–1 | P | 3–2.5 | 8–10 | 35 | 115 |
| M4 M41 T4 | | | 25–75 | 1–3 | P | 4–3 | 6–8 | 29 | 95 |
| M7 M42 T5 | | | >75 | >3 | C,P | 8.5–6.3 | 3–4 | 24 | 80 |
| M30 M43 T8 | | | | | | | | | |
| M33 M44 | | | | | | | | | |
| M34 M46 | | | | | | | | | |
| **High speed** | 225–275 | Annealed | <13 | <½ | P | 2.5–1.8 | 10–14 | 27 | 90 |
| M3-2 | | | 13–25 | ½–1 | P | 3–2.5 | 8–10 | 24 | 80 |
| T15 | | | 25–75 | 1–3 | P | 4–3 | 6–8 | 20 | 65 |
| | | | >75 | >3 | C,P | 8.5–6.3 | 3–4 | 17 | 55 |
| **Hot work** | 150–200 | Annealed | <13 | <½ | P | 2.5–1.8 | 10–14 | 60 | 200 |
| H10 | | | 13–25 | ½–1 | P | 3–2.5 | 8–10 | 55 | 180 |
| H11 | | | 25–75 | 1–3 | P | 4–3 | 6–8 | 46 | 150 |
| H12 | | | >75 | >3 | C,P | 8.5–6.3 | 3–4 | 37 | 120 |
| H13 | 325–375 | Quenched | <13 | <½ | P | 2.5–1.8 | 10–14 | 38 | 125 |
| H14 | | and | 13–25 | ½–1 | P | 3–2.5 | 8–10 | 34 | 110 |
| H19 | | tempered | 25–75 | 1–3 | P | 4–3 | 6–8 | 27 | 90 |
| | | | >75 | >3 | C,P | 8.5–6.3 | 3–4 | 23 | 75 |
| **Hot work** | 150–200 | Annealed | <13 | <½ | P | 2.5–1.8 | 10–14 | 46 | 150 |
| H21 H25 | | | 13–25 | ½–1 | P | 3–2.5 | 8–10 | 41 | 135 |
| H22 H26 | | | 25–75 | 1–3 | P | 4–3 | 6–8 | 34 | 110 |
| H23 H42 | | | >75 | >3 | C,P | 8.5–6.3 | 3–4 | 27 | 90 |
| H24 | 200–250 | Annealed | <13 | <½ | P | 2.5–1.8 | 10–14 | 43 | 140 |
| | | | 13–25 | ½–1 | P | 3–2.5 | 8–10 | 38 | 125 |
| | | | 25–75 | 1–3 | P | 4–3 | 6–8 | 32 | 105 |
| | | | >75 | >3 | C,P | 8.5–6.3 | 3–4 | 26 | 85 |
| **Cold work** | 200–250 | Annealed | <13 | <½ | P | 2.5–1.8 | 10–14 | 38 | 125 |
| A7 | | | 13–25 | ½–1 | P | 3–2.5 | 8–10 | 34 | 110 |
| D2 | | | 25–75 | 1–3 | P | 4–3 | 6–8 | 27 | 90 |
| D3 | | | >75 | >3 | C,P | 8.5–6.3 | 3–4 | 23 | 75 |
| D4 | | | | | | | | | |
| D5 | | | | | | | | | |
| D7 | | | | | | | | | |
| **Cold work** | 200–250 | Annealed | <13 | <½ | P | 2.5–1.8 | 10–14 | 52 | 170 |
| A2 A10 | | | 13–25 | ½–1 | P | 3–2.5 | 8–10 | 46 | 150 |
| A3 O1 | | | 25–75 | 1–3 | P | 4–3 | 6–8 | 38 | 125 |
| A4 O2 | | | >75 | >3 | C,P | 8.5–6.3 | 3–4 | 32 | 105 |
| A6 O6 | | | | | | | | | |
| A8 O7 | | | | | | | | | |
| A9 | | | | | | | | | |
| **Shock resisting** | 175–225 | Annealed | <13 | <½ | P | 2.5–1.8 | 10–14 | 49 | 160 |
| S1 | | | 13–25 | ½–1 | P | 3–2.5 | 8–10 | 44 | 145 |
| S2 | | | 25–75 | 1–3 | P | 4–3 | 6–8 | 37 | 120 |
| S5 | | | >75 | >3 | C,P | 8.5–6.3 | 3–4 | 30 | 100 |
| S6 | | | | | | | | | |
| S7 | | | | | | | | | |
| **Mold** | 100–150 | Annealed | <13 | <½ | P | 2.5–1.8 | 10–14 | 73 | 240 |
| P2 | | | 13–25 | ½–1 | P | 3–2.5 | 8–10 | 66 | 215 |
| P4 | | | 25–75 | 1–3 | P | 4–3 | 6–8 | 53 | 175 |
| P5 | | | >75 | >3 | C,P | 8.5–6.3 | 3–4 | 44 | 145 |
| P6 | | | | | | | | | |
| **Mold** | 150–200 | Annealed | <13 | <½ | P | 2.5–1.8 | 10–14 | 64 | 210 |
| P20 | | | 13–25 | ½–1 | P | 3–2.5 | 8–10 | 56 | 185 |
| P21 | | | 25–75 | 1–3 | P | 4–3 | 6–8 | 47 | 155 |
| | | | >75 | >3 | C,P | 8.5–6.3 | 3–4 | 38 | 125 |
| **Special purpose** | 150–200 | Annealed | <13 | <½ | P | 2.5–1.8 | 10–14 | 56 | 185 |
| L2 | | | 13–25 | ½–1 | P | 3–2.5 | 8–10 | 50 | 165 |
| L6 | | | 25–75 | 1–3 | P | 4–3 | 6–8 | 41 | 135 |
| L7 | | | >75 | >3 | C P | 8.5–6.3 | 3–4 | 34 | 110 |
| **Special purpose** | 200–250 | Annealed | <13 | <½ | P | 2.5–1.8 | 10–14 | 52 | 170 |
| F1 | | | 13–25 | ½–1 | P | 3–2.5 | 8–10 | 46 | 150 |
| F2 | | | 25–75 | 1–3 | P | 4–3 | 6–8 | 38 | 125 |
| | | | >75 | >3 | C P | 8.5–6.3 | 3–4 | 32 | 105 |
| **Water hardening** | 150–200 | Annealed | <13 | <½ | P | 2.5–1.8 | 10–14 | 60 | 200 |
| W1, W2, W5 | | | 13–25 | ½–1 | P | 3–2.5 | 8–10 | 55 | 180 |
| SAE J438b: Types | | | 25–75 | 1–3 | P | 4–3 | 6–8 | 46 | 150 |
| W108, W109, W110, | | | >75 | >3 | C P | 8.5–6.3 | 3–4 | 37 | 120 |
| W112, W209, W210, | | | | | | | | | |
| W310 | | | | | | | | | |

(a) P, precision; C, claw; B, buttress. Source: Metcut Research Associates Inc.

When molds are too intricate to withstand the carburizing treatment without excessive distortion, a second procedure is sometimes used. The prehardened steel is completely machined and then gas nitrided for 30 h.

The third machining procedure is also used for plastic molds that are too intricate to carburize in the finish machined condition. The fully annealed stock is rough machined to within about 1.6 mm (1/16 in.) of finish surface, then hardened to 45 to 50 HRC. After heat treatment, the mold is finish machined. Carbide cutters are usually used. A speed of 26 m/min (85 sfm) with a feed of about 0.05 mm/tooth (0.002 in./tooth) with grade C-2 carbide is common practice.

## Machining of Tool Steel Gears

Gears that function in a hot or abrasive environment are sometimes made of a hot-work tool steel such as H11, H12, or H13. Milling is the gear-production process most often used, because it is practical for the small quantities that are usually needed.

One of three procedures is usually followed, the choice of procedure depending largely on whether the primary requirement is abrasion resistance or heat resistance in service.

One procedure is to mill the gear teeth in the fully annealed condition (200 to 250 HB) followed by quenching and tempering to approximately 50 HRC. If this procedure causes excessive distortion, a final grinding operation can be included. When this procedure is used, a speed of about 30 m/min (100 sfm) may be used for rough milling and 40 m/min (130 sfm) for finishing with high-speed steel cutters. For gears having diametral pitch of 1 to 3, a feed of 0.15 mm/tooth (0.006 in./tooth) is usually optimum. As pitch becomes finer, the feed should be decreased gradually to 0.08 mm/tooth (0.003 in./tooth) for gears having diametral pitch of 25 to 32.

When the gears must resist abrasion, common practice is to quench and temper the blanks to 325 to 375 HB, mill the teeth, and then nitride the machined gear. At this higher hardness, speed must be reduced to about 20 m/min (65 sfm) for rough cuts and 30 m/min (100 sfm) for finishing with high-speed steel cutters. Feed is also reduced to about 0.10 mm/tooth (0.004 in./tooth) for gears having diametral pitch of 1 to 3 or about 0.08 mm/tooth (0.003 in./tooth) for gears having diametral pitch of 25 to 32.

In some instances it is desirable to mill the teeth at higher hardness and employ no further finishing operations. Milling H11 at up to 55 HRC is common practice.

## Grinding of Tool Steels

The wide differences in composition among tool steels give rise to wide varia-

**Table 15 Nominal speeds and feeds for the power hack sawing of wrought tool steels with high-speed tool steels**

| Steel | Hardness, HB | Condition | Material thickness, mm | Material thickness, in. | Pitch, mm(a) | Teeth/in. | Speed, strokes/min | Feed: Positive, mm/stroke | Feed: Positive, in./stroke | Feed: Pressure(b) |
|---|---|---|---|---|---|---|---|---|---|---|
| **High speed** | 200–250 | Annealed | <6 | <¼ | 2.5 | 10 | 60 | 0.075 | 0.003 | L |
| M1 T1 | | | 6–19 | ¼–¾ | 2.5 | 10 | 60 | 0.075 | 0.003 | L |
| M2 T2 | | | 19–50 | ¾–2 | 4 | 6 | 55 | 0.15 | 0.006 | M |
| M6 T6 | | | >50 | >2 | 6.3 | 4 | 55 | 0.15 | 0.006 | M |
| M10 | | | | | | | | | | |
| **High speed** | 225–275 | Annealed | <6 | <¼ | 2.5 | 10 | 55 | 0.075 | 0.003 | L |
| M3-1 M41 T4 | | | 6–19 | ¼–¾ | 2.5 | 10 | 55 | 0.075 | 0.003 | L |
| M4 M42 T5 | | | 19–50 | ¾–2 | 4 | 6 | 50 | 0.15 | 0.006 | M |
| M7 M43 T8 | | | >50 | >2 | 6.3 | 4 | 50 | 0.15 | 0.006 | M |
| M30 M44 | | | | | | | | | | |
| M33 M46 | | | | | | | | | | |
| M34 M47 | | | | | | | | | | |
| M36 | | | | | | | | | | |
| **High speed** | 225–275 | Annealed | <6 | <¼ | 2.5 | 10 | 45 | 0.075 | 0.003 | L |
| M3-2 | | | 6–19 | ¼–¾ | 2.5 | 10 | 45 | 0.075 | 0.003 | L |
| T15 | | | 19–50 | ¾–2 | 4 | 6 | 40 | 0.075 | 0.003 | L |
| | | | >50 | >2 | 4 | 6 | 40 | 0.15 | 0.006 | M |
| **Hot work** | 150–200 | Annealed | <6 | <¼ | 2.5 | 10 | 100 | 0.075 | 0.003 | L |
| H10 | | | 6–19 | ¼–¾ | 2.5 | 10 | 100 | 0.15 | 0.006 | M |
| H11 | | | 19–50 | ¾–2 | 4 | 6 | 90 | 0.15 | 0.006 | M |
| H12 | | | >50 | >2 | 6.3 | 4 | 90 | 0.15 | 0.006 | M |
| H13 | 325–375 | Quenched | <6 | <¼ | 2.5 | 10 | 55 | 0.075 | 0.003 | L |
| H14 | | and | 6–19 | ¼–¾ | 2.5 | 10 | 55 | 0.075 | 0.003 | L |
| H19 | | tempered | 19–50 | ¾–2 | 4 | 6 | 50 | 0.15 | 0.006 | M |
| | | | >50 | >2 | 6.3 | 4 | 50 | 0.15 | 0.006 | M |
| **Hot work** | 150–200 | Annealed | <6 | <¼ | 2.5 | 10 | 60 | 0.075 | 0.003 | L |
| H21 H25 | | | 6–19 | ¼–¾ | 2.5 | 10 | 60 | 0.075 | 0.003 | L |
| H22 H26 | | | 19–50 | ¾–2 | 4 | 6 | 55 | 0.15 | 0.006 | M |
| H23 H42 | | | >50 | >2 | 6.3 | 4 | 55 | 0.15 | 0.006 | M |
| H24 | 200–250 | Annealed | <6 | <¼ | 2.5 | 10 | 55 | 0.075 | 0.003 | L |
| | | | 6–19 | ¼–¾ | 2.5 | 10 | 55 | 0.075 | 0.003 | L |
| | | | 19–50 | ¾–2 | 4 | 6 | 50 | 0.15 | 0.006 | M |
| | | | >50 | >2 | 6.3 | 4 | 50 | 0.15 | 0.006 | M |
| **Cold work** | 200–250 | Annealed | <6 | <¼ | 2.5 | 10 | 45 | 0.075 | 0.003 | L |
| A7 | | | 6–19 | ¼–¾ | 2.5 | 10 | 45 | 0.075 | 0.003 | L |
| D2 | | | 19–50 | ¾–2 | 4 | 6 | 40 | 0.075 | 0.003 | L |
| D4 | | | >50 | >2 | 4 | 6 | 40 | 0.15 | 0.006 | M |
| D5 | | | | | | | | | | |
| D7 | | | | | | | | | | |
| **Cold work** | 200–250 | Annealed | <6 | <¼ | 2.5 | 10 | 70 | 0.075 | 0.003 | L |
| A2 A10 | | | 6–19 | ¼–¾ | 2.5 | 10 | 70 | 0.075 | 0.003 | L |
| A3 O1 | | | 19–50 | ¾–2 | 4 | 6 | 60 | 0.15 | 0.006 | M |
| A4 O2 | | | >50 | >2 | 6.3 | 4 | 60 | 0.15 | 0.006 | M |
| A6 O6 | | | | | | | | | | |
| A8 O7 | | | | | | | | | | |
| A9 | | | | | | | | | | |
| **Shock resisting** | 175–225 | Annealed | <6 | <¼ | 2.5 | 10 | 100 | 0.075 | 0.003 | L |
| S1 | | | 6–19 | ¼–¾ | 2.5 | 10 | 100 | 0.15 | 0.006 | M |
| S2 | | | 19–50 | ¾–2 | 4 | 6 | 90 | 0.15 | 0.006 | M |
| S5 | | | >50 | >2 | 6.3 | 4 | 90 | 0.15 | 0.006 | M |
| S6 | | | | | | | | | | |
| S7 | | | | | | | | | | |
| **Mold** | 100–150 | Annealed | <6 | <¼ | 2.5 | 10 | 120 | 0.075 | 0.003 | L |
| P2 | | | 6–19 | ¼–¾ | 4 | 6 | 120 | 0.15 | 0.006 | M |
| P4 | | | 19–50 | ¾–2 | 4 | 6 | 110 | 0.15 | 0.006 | M |
| P5 | | | >50 | >2 | 6.3 | 4 | 110 | 0.15 | 0.006 | M |
| P6 | | | | | | | | | | |
| **Mold** | 150–200 | Annealed | <6 | <¼ | 2.5 | 10 | 115 | 0.075 | 0.003 | L |
| P20 | | | 6–19 | ¼–¾ | 4 | 6 | 115 | 0.15 | 0.006 | M |
| P21 | | | 19–50 | ¾–2 | 4 | 6 | 105 | 0.15 | 0.006 | M |
| | | | >50 | >2 | 6.3 | 4 | 105 | 0.15 | 0.006 | M |
| **Special purpose** | 150–200 | Annealed | <6 | <¼ | 2.5 | 10 | 110 | 0.075 | 0.003 | L |
| L2 | | | 6–19 | ¼–¾ | 4 | 6 | 110 | 0.15 | 0.006 | M |
| L6 | | | 19–50 | ¾–2 | 4 | 6 | 100 | 0.15 | 0.006 | M |
| L7 | | | >50 | >2 | 6.3 | 4 | 100 | 0.15 | 0.006 | M |
| **Special purpose** | 200–250 | Annealed | <6 | <¼ | 2.5 | 10 | 90 | 0.075 | 0.003 | L |
| F1 | | | 6–19 | ¼–¾ | 2.5 | 10 | 90 | 0.075 | 0.003 | L |
| F2 | | | 19–50 | ¾–2 | 4 | 6 | 80 | 0.15 | 0.006 | M |
| | | | >50 | >2 | 6.3 | 4 | 80 | 0.15 | 0.006 | M |
| **Water hardening** | 150–200 | Annealed | <6 | <¼ | 2.5 | 10 | 110 | 0.075 | 0.003 | L |
| W1, W2, W5 | | | 6–19 | ¼–¾ | 4 | 6 | 110 | 0.15 | 0.006 | M |
| SAE J438b: Types | | | 19–50 | ¾–2 | 4 | 6 | 100 | 0.15 | 0.006 | M |
| W108, W109, | | | >50 | >2 | 6.3 | 4 | 100 | 0.15 | 0.006 | M |
| W110, W112, | | | | | | | | | | |
| W209, W210, | | | | | | | | | | |
| W310 | | | | | | | | | | |

(a) Conversions per ISO 2336. (b) L, light; M, medium; H, heavy. Source: Metcut Research Associates Inc.

tions in grinding characteristics. The grindability index of a material is a measure of the ease of removing stock by grinding, expressed in terms of wheel wear. Numerically it is equal to the grinding ratio obtained under a specified set of grinding conditions, the grinding ratio being the volume of work material removed per volume of wheel wear. The higher this index, the easier the material is to grind. This concept of grindability does not include grinding sensitivity, which is the susceptibility of the work material to cracking during or after grinding. Nor does this concept involve the ease of obtaining a good surface finish. The grindability index is about 100 times as great for the easiest tool steel to grind as for the most difficult. This relation is evident in Fig. 10.

In general, the grindability index has direct significance only below some limiting value that decreases as the proportion of wheel wear ascribable to dressing increases. Indirectly, however, the grindability index can be helpful in rating a group of work materials, even when most of the wheel wear is from dressing. Thus, the indexes for a number of similar materials can sometimes be related to the number of pieces per dressing for a particular operation. Likewise, as size of the wheel is increased relative to the area of the ground surface, the highest value of the grindability index that is directly significant becomes less, but the index may have indirect significance even though it exceeds the limiting value.

The grindability index is less of a factor with easy-to-grind materials. It is much more important when difficult-to-grind tool steels are considered. However, the development in CBN wheels with a resinoid or vitrified matrix and in metal-plated monolayer CBN wheels has advanced tremendously. This superabrasive has a very low wear rate and has increased both the metal removal rates and the wheel life significantly in many applications. Specifically, form wheels with CBN abrasives have gained wide acceptance in numerous form tools for grinding (drills, end mills, and so on). More detailed information on the use of CBN for grinding can be found in the article "Superabrasives" in this Volume.

Because the grindability index is an inverse measure of wheel wear, its ranking of work materials can be used to predict the relative ease of maintaining dimensional accuracy and also the relative production rate, provided surface roughness or grinding damage are not controlling factors. These may be affected by entirely different properties of the work material than those that affect wheel performance. For example, the susceptibility of hardened steel to the formation of grinding cracks, known as surface damage (see the article "Surface Finish and Surface Integrity" in this Volume) can be

## Table 16 Nominal speeds and feeds for the circular sawing of wrought tool steels with high-speed tool steels

| Steel | Hardness, HB(a) | Condition | Solid stock diameter or thickness | | Pitch | | Cutting speed | | Feed | | High-speed steel tool material | |
|---|---|---|---|---|---|---|---|---|---|---|---|---|
| | | | mm | in. | mm/tooth | in./tooth | m/min | sfm | mm/tooth | in./tooth | ISO | AISI |
| **High speed** | 200–250 | Annealed | 6–80 | ¼–3 | 5–20 | 0.20–0.75 | 15 | 50 | 0.13 | 0.005 | S4, S2 | M2, M7 |
| M1 T1 | | | 80–160 | 3–6 | 12–30 | 0.50–1.10 | 12 | 40 | 0.13 | 0.005 | | |
| M2 T2 | | | 160–250 | 6–9 | 20–35 | 0.75–1.30 | 9 | 30 | 0.18 | 0.007 | | |
| M6 T6 | | | 250–400 | 9–15 | 25–45 | 0.90–1.75 | 8 | 25 | 0.18 | 0.007 | | |
| M10 | | | | | | | | | | | | |
| **High speed** | 225–275 | Annealed | 6–80 | ¼–3 | 5–20 | 0.20–0.75 | 12 | 40 | 0.10 | 0.004 | S4, S2 | M2, M7 |
| M3-1 M36 M47 | | | 80–160 | 3–6 | 12–30 | 0.50–1.10 | 11 | 35 | 0.10 | 0.004 | | |
| M4 M41 T4 | | | 160–250 | 6–9 | 20–35 | 0.75–1.30 | 8 | 25 | 0.15 | 0.006 | | |
| M7 M42 T5 | | | 250–400 | 9–15 | 25–45 | 0.90–1.75 | 6 | 20 | 0.15 | 0.006 | | |
| M30 M43 T8 | | | | | | | | | | | | |
| M33 M44 | | | | | | | | | | | | |
| M34 M46 | | | | | | | | | | | | |
| **High speed** | 225–275 | Annealed | 6–80 | ¼–3 | 5–20 | 0.20–0.75 | 11 | 35 | 0.10 | 0.004 | S4, S2 | M2, M7 |
| M3-2 | | | 80–160 | 3–6 | 12–30 | 0.50–1.10 | 9 | 30 | 0.10 | 0.004 | | |
| T15 | | | 160–250 | 6–9 | 20–35 | 0.75–1.30 | 6.7 | 22 | 0.15 | 0.006 | | |
| | | | 250–400 | 9–15 | 25–45 | 0.90–1.75 | 5.2 | 17 | 0.15 | 0.006 | | |
| **Hot work** | 150–200 | Annealed | 6–80 | ¼–3 | 5–20 | 0.20–0.75 | 18 | 60 | 0.15 | 0.006 | S4, S2 | M2, M7 |
| H10 | | | 80–160 | 3–6 | 12–30 | 0.50–1.10 | 15 | 50 | 0.15 | 0.006 | | |
| H11 | | | 160–250 | 6–9 | 20–35 | 0.75–1.30 | 12 | 40 | 0.20 | 0.008 | | |
| H12 | | | 250–400 | 9–15 | 25–45 | 0.90–1.75 | 9 | 30 | 0.20 | 0.008 | | |
| H13 | 325–375 | Quenched and tempered | 6–80 | ¼–3 | 5–20 | 0.20–0.75 | 6 | 20 | 0.10 | 0.004 | S4, S2 | M2, M7 |
| H14 | | | 80–160 | 3–6 | 12–30 | 0.50–1.10 | 5 | 15 | 0.10 | 0.004 | | |
| H19 | | | 160–250 | 6–9 | 20–35 | 0.75–1.30 | 3.6 | 12 | 0.15 | 0.006 | | |
| | | | 250–400 | 9–15 | 25–45 | 0.90–1.75 | 3 | 10 | 0.15 | 0.006 | | |
| | 48–50 HRC | Quenched and Tempered | 6–80 | ¼–3 | 4–15 | 0.15–0.60 | 3 | 10 | 0.050 | 0.002 | S4, S2 | M2, M7 |
| | | | 80–160 | 3–6 | 10–25 | 0.40–0.90 | 2.5 | 8 | 0.050 | 0.002 | | |
| | | | 160–250 | 6–9 | 18–30 | 0.70–1.10 | 1.5 | 5 | 0.075 | 0.003 | | |
| | | | 250–400 | 9–15 | 20–35 | 0.80–1.35 | 1.5 | 5 | 0.075 | 0.003 | | |
| **Hot work** | 150–200 | Annealed | 6–80 | ¼–3 | 5–20 | 0.20–0.75 | 18 | 60 | 0.15 | 0.006 | S4, S2 | M2, M7 |
| H21 H25 | | | 80–160 | 3–6 | 12–30 | 0.50–1.10 | 15 | 50 | 0.15 | 0.006 | | |
| H22 H26 | | | 160–250 | 6–9 | 20–35 | 0.75–1.30 | 12 | 40 | 0.20 | 0.008 | | |
| H23 H42 | | | 250–400 | 9–15 | 25–45 | 0.90–1.75 | 9 | 30 | 0.20 | 0.008 | | |
| H24 | 200–250 | Annealed | 6–80 | ¼–3 | 5–20 | 0.20–0.75 | 15 | 50 | 0.13 | 0.005 | S4, S2 | M2, M7 |
| | | | 80–160 | 3–6 | 12–30 | 0.50–1.10 | 12 | 40 | 0.13 | 0.005 | | |
| | | | 160–250 | 6–9 | 20–35 | 0.75–1.30 | 9 | 30 | 0.18 | 0.007 | | |
| | | | 250–400 | 9–15 | 25–45 | 0.90–1.75 | 8 | 25 | 0.18 | 0.007 | | |
| **Cold work** | 200–250 | Annealed | 6–80 | ¼–3 | 5–20 | 0.20–0.75 | 12 | 40 | 0.13 | 0.005 | S4, S2 | M2, M7 |
| A7 | | | 80–160 | 3–6 | 12–30 | 0.50–1.10 | 11 | 35 | 0.13 | 0.005 | | |
| D2 | | | 160–250 | 6–9 | 20–35 | 0.75–1.30 | 8 | 25 | 0.18 | 0.007 | | |
| D3 | | | 250–400 | 9–15 | 25–45 | 0.90–1.75 | 6 | 20 | 0.18 | 0.007 | | |
| D4 | | | | | | | | | | | | |
| D5 | | | | | | | | | | | | |
| D7 | | | | | | | | | | | | |
| **Cold work** | 200–250 | Annealed | 6–80 | ¼–3 | 5–20 | 0.20–0.75 | 21 | 70 | 0.13 | 0.005 | S4, S2 | M2, M7 |
| A2 A10 | | | 80–160 | 3–6 | 12–30 | 0.50–1.10 | 17 | 55 | 0.13 | 0.005 | | |
| A3 O1 | | | 160–250 | 6–9 | 20–35 | 0.75–1.30 | 14 | 45 | 0.18 | 0.007 | | |
| A4 O2 | | | 250–400 | 9–15 | 25–45 | 0.90–1.75 | 11 | 35 | 0.18 | 0.007 | | |
| A6 O6 | | | | | | | | | | | | |
| A8 O7 | | | | | | | | | | | | |
| A9 | | | | | | | | | | | | |
| **Shock resisting** | 175–225 | Annealed | 6–80 | ¼–3 | 5–20 | 0.20–0.75 | 24 | 80 | 0.13 | 0.005 | S4, S2 | M2, M7 |
| S1 | | | 80–160 | 3–6 | 12–30 | 0.50–1.10 | 20 | 65 | 0.13 | 0.005 | | |
| S2 | | | 160–250 | 6–9 | 20–35 | 0.75–1.30 | 15 | 50 | 0.18 | 0.007 | | |
| S5 | | | 250–400 | 9–15 | 25–45 | 0.90–1.75 | 12 | 40 | 0.18 | 0.007 | | |
| S6 | | | | | | | | | | | | |
| S7 | | | | | | | | | | | | |
| **Mold** | 100–150 | Annealed | 6–80 | ¼–3 | 5–20 | 0.20–0.75 | 27 | 90 | 0.15 | 0.006 | S4, S2 | M2, M7 |
| P2 | | | 80–160 | 3–6 | 12–30 | 0.50–1.10 | 23 | 75 | 0.15 | 0.006 | | |
| P4 | | | 160–250 | 6–9 | 20–35 | 0.75–1.30 | 17 | 55 | 0.20 | 0.008 | | |
| P5 | | | 250–400 | 9–15 | 25–45 | 0.90–1.75 | 14 | 45 | 0.20 | 0.008 | | |
| P6 | | | | | | | | | | | | |
| **Mold** | 150–200 | Annealed | 6–80 | ¼–3 | 5–20 | 0.20–0.75 | 30 | 100 | 0.15 | 0.006 | S4, S2 | M2, M7 |
| P20 | | | 80–160 | 3–6 | 12–30 | 0.50–1.10 | 24 | 80 | 0.15 | 0.006 | | |
| P21 | | | 160–250 | 6–9 | 20–35 | 0.75–1.30 | 20 | 65 | 0.20 | 0.008 | | |
| | | | 250–400 | 9–15 | 25–45 | 0.90–1.75 | 15 | 50 | 0.20 | 0.008 | | |
| **Special purpose** | 150–200 | Annealed | 6–80 | ¼–3 | 5–20 | 0.20–0.75 | 34 | 110 | 0.15 | 0.006 | S4, S2 | M2, M7 |
| L2 | | | 80–160 | 3–6 | 12–30 | 0.50–1.10 | 27 | 90 | 0.15 | 0.006 | | |
| L6 | | | 160–250 | 6–9 | 20–35 | 0.75–1.30 | 21 | 70 | 0.20 | 0.008 | | |
| L7 | | | 250–400 | 9–15 | 25–45 | 0.90–1.75 | 17 | 55 | 0.20 | 0.008 | | |
| **Special purpose** | 200–250 | Annealed | 6–80 | ¼–3 | 5–20 | 0.20–0.75 | 24 | 80 | 0.13 | 0.005 | S4, S2 | M2, M7 |
| F1 | | | 80–160 | 3–6 | 12–30 | 0.50–1.10 | 20 | 65 | 0.13 | 0.005 | | |
| F2 | | | 160–250 | 6–9 | 20–35 | 0.75–1.30 | 15 | 50 | 0.18 | 0.007 | | |
| | | | 250–400 | 9–15 | 25–45 | 0.90–1.75 | 12 | 40 | 0.18 | 0.007 | | |
| **Water hardening** | 150–200 | Annealed | 6–80 | ¼–3 | 5–20 | 0.20–0.75 | 27 | 90 | 0.15 | 0.006 | S4, S2 | M2, M7 |
| W1, W2, W5 | | | 80–160 | 3–6 | 12–30 | 0.50–1.10 | 23 | 75 | 0.15 | 0.006 | | |
| SAE J438b: Types | | | 160–250 | 6–9 | 20–35 | 0.75–1.30 | 17 | 55 | 0.20 | 0.008 | | |
| W108, W109, W110, | | | 250–400 | 9–15 | 25–45 | 0.90–1.75 | 14 | 45 | 0.20 | 0.008 | | |
| W112, W209, W210, | | | | | | | | | | | | |
| W310 | | | | | | | | | | | | |

(a) Unless otherwise noted. Source: Metcut Research Associates Inc.

**Table 17 Turning M10 versus sulfurized M10 reamer blanks**

| Item | M10 | Sulfurized M10 |
|---|---|---|
| Speed, rev/min | 230 | 325 |
| Speed, m/min (sfm) | 37 (120) | 52 (170) |
| Feed, mm/rev (in./rev) | 0.64 (0.025) | 0.64 (0.025) |
| Depth of cut, mm (in.) | 3.18 (0.125) | 3.18 (0.125) |
| Length of cut, mm (in.) | 95 (3¾) | 95 (3¾) |
| Power input, kW (hp) | 3.4 (4.5) | 3.7 (5) |
| Force on tool bit, kN (lbf) | 4.67 (1050) | 4.00 (900) |
| Cutting rate, cm$^3$/min (in.$^3$/min) | 69.6 (4.25) | 98.3 (6) |
| Cutting rate, kg/min (lb/min) | 0.567 (1.25) | 0.82 (1.8) |
| Tool life per grind, pieces | 18 | 25 |

Note: Reamer blanks 50 mm (2 in.) in diameter were turned for removal of 6.35 mm (0.250 in.) of stock on the diameter. A 1370 mm (54 in.) lathe with 3.7 kW (5 hp) capacity and 19 mm (0.75 in.) square tools or M2 high-speed steel were used for turning. Cutting fluid was a 1:25 mixture of soluble oil and water.

markedly affected by changes in the processing (especially heat treating) to which the steel is subjected before grinding. At the same time, these processing changes may have no effect on the grindability index. The relative ease of producing acceptable parts from different steels may depend on either the grindability index or the grinding sensitivity. A similar situation may arise with respect to the ease of producing a desired finish.

In operations such as cylindrical or thread grinding, in which the wheel normally wears only to a negligible extent before it has to be dressed, the grindability index has less significance.

## Steel Classification for Grindability

Classification of steels as low, medium, or high in grindability is usually precise enough for practical purposes. The classification of that kind shown in Table 20 correlates satisfactorily with the grindability index relationships of Fig. 10. All the steels shown in Table 20 as having low grindability are generally comparable to T15 (Fig. 10) in terms of grindability index. Similarly, the grades listed as medium are similar to D3, and those indicated as having high grindability usually show an index equal to, or slightly higher than, that of O2 (Fig. 10). It is worth noting that all the steels classified as having low grindability contain 3% or more vanadium (see also Fig. 11). Some tool steels that contain about 2% V and cobalt additions, such as M43 and M44, are also difficult to grind and are placed in the category of low grindability (see "Vanadium" in the section "Effect of Steel Composition on Grindability" below).

In general, as the wear resistance of steel is increased, grindability is sacrificed. The abrasion-resistant, high-vanadium, high-speed and die steels, when properly applied, offer considerably longer tool life than the lower-alloy easier-to-grind materials, and this helps to compensate for the extra cost of fabricating them into tools. However, the use of a difficult-to-grind steel for an intricate tool such as a tap, small twist drill, compacting die, or spline broach may increase the cost of fabrication enough to defeat the application.

**Table 18 M2 versus carbide-tip cutters for milling flutes in T15 end mills**

| Item | Cutter: M2 | Cutter: Carbide |
|---|---|---|
| Speed, rev/min | 125 | 322 |
| Speed, m/min (sfm) | 17 (57) | 45 (146) |
| Feed, mm/min (in./min) | 44 (1¾) | 89 (3½) |
| Machining time per cut, min | 1.43 | 0.71 |
| Tool life, number of cuts | 70 | 200 |

Note: Milling four helical flutes in 19 mm (¾ in.) diam end mills in a 2.2 kW (3 hp) universal mill. Depth of cut was 3.2 to 1.6 mm (⅛ to 1⁄16 in.), width was 9.5 mm (⅜ in.), and length was 57 mm (2¼ in.). A sulfur-base oil was used as the cutting fluid. Both M2 and carbide-tip cutters were 44 mm (1¾ in.) in diameter and 9.5 mm (⅜ in.) wide.

## Effect of Steel Composition on Grindability

The ease of grinding after heat treatment is usually inversely proportional to the base hardness, the volume of carbide phase present, and the hardness of individual carbide particles. Steels containing high percentages of carbon and carbide-forming alloying elements are difficult to grind. Tool steels relatively high in carbide-forming elements are more difficult to grind than the carbon or low-alloy types, even though hardness and carbon content may be higher for the low-alloy steels.

**Vanadium** has the greatest effect of all alloying elements on grindability, and all steels containing more than about 1% V are difficult to grind. Vanadium up to about 1% is soluble in austenite in most tool steel compositions, and little if any undissolved vanadium carbide remains in the normal heat-treated structure. Beyond 1% V, however, excess undissolved vanadium carbide begins to appear in the microstructure, and the carbide particles increase rapidly in size and number at higher vanadium concentrations. T1 high-speed steel contains only about 0.5% undissolved vanadium-rich carbide after conventional heat treatment, whereas T15 contains approximately 8%. Because of this relationship, increasing the vanadium content of a high-speed steel from 1 to 4 wt% reduces the grindability index by

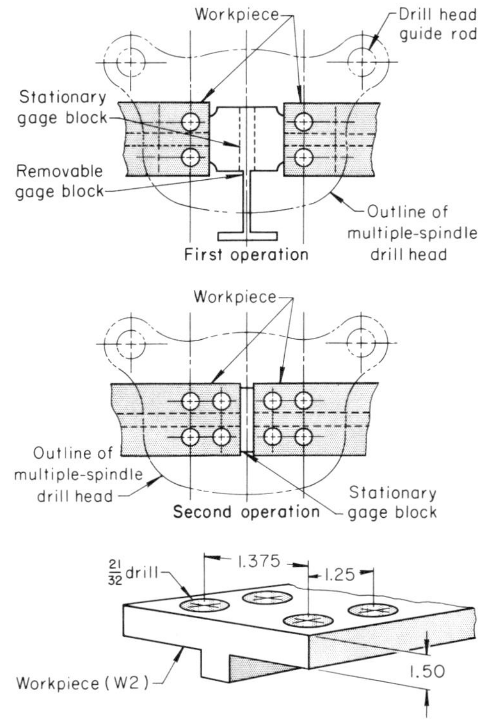


| Drill details | |
|---|---|
| Material | M7 |
| Diameter, mm (in.) | 16.5 (21⁄32) |
| Length, mm (in.) | 130 (5⅛) |
| Helix angle, degrees | 30 |
| Point angle (included), degrees | 118 |
| Lip relief angle, degrees | 8–10 |
| **Machining conditions** | |
| Speed at 220 rev/min, m/min (sfm)(a) | 12 (38) |
| Feed at 0.13 mm/rev (0.005 in./rev), mm/min (in./min)(a) | 29.2 (1.15) |
| Depth of hole, mm (in.) | 38.1 (1.50) |
| Cutting fluid, water: nonsulfurized water-soluble synthetic | 40:1 |
| Drilling time per cycle of four holes, min | 1.5 |
| Tool life per drill, holes | 300–500 |

(a) Speed and feed were slightly lower than normal for drilling W2 to conform to the power (3.7 kW, or 5 hp) and rigidity of the machine and drill head.

**Fig. 5** Setup and conditions for the multiple-spindle drilling of holes in W2 die guide plates. Dimensions in figure given in inches

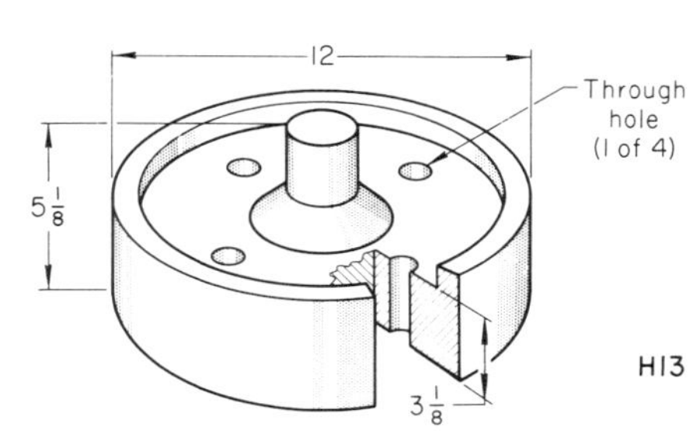


**Fig. 6** Porthole extrusion die that was faced, turned, and milled in 28% less time when made of sulfurized H13 than when made of H13. Dimensions given in inches

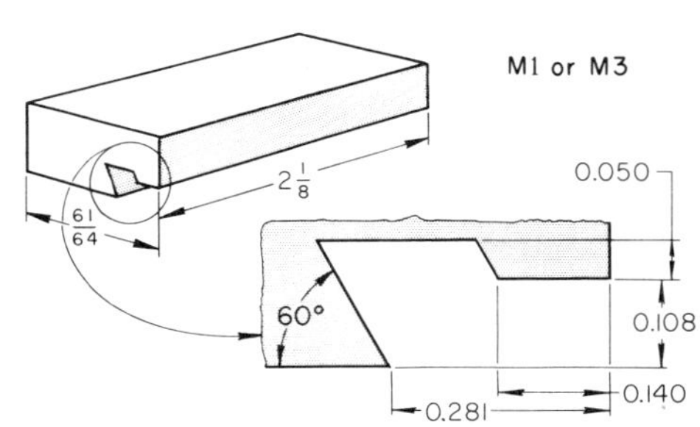


**Fig. 7** Tangential thread chaser in which dovetail was milled. Dimensions given in inches

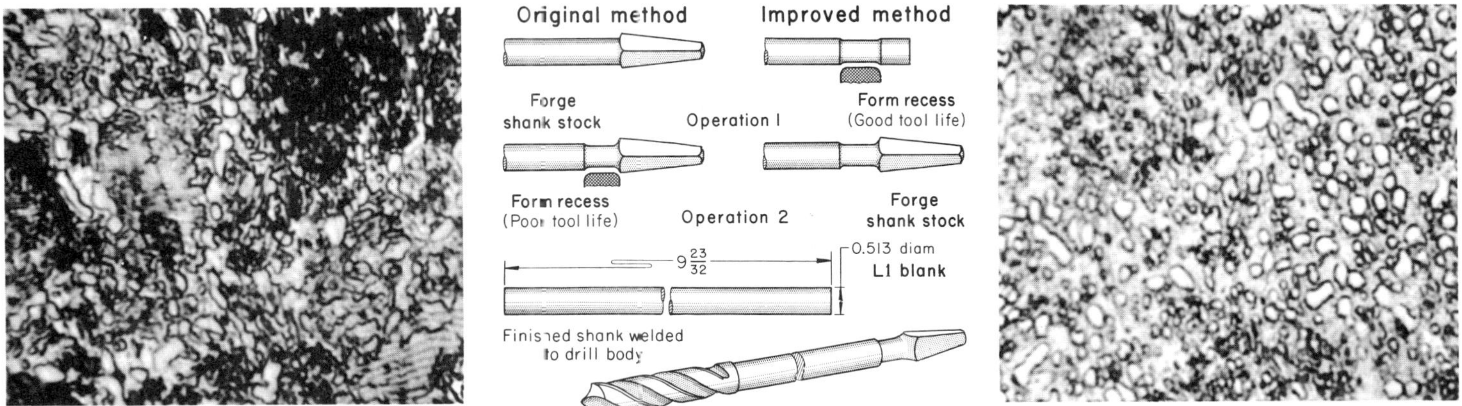


**Poor machinability** **Good machinability**

Weight of blank, kg (lb) .............. 0.257 (0.566)
Hardness, HB ........................ 215
Speed at 830 rev/min, m/min (sfm) ..... 34 (110)
Feed mm/rev (in./rev)............... 0.016 (0.00063)
Depth of cut, mm (in.)............... 2.20 (0.087)
Cutting fluid, soluble oil:water........ 1:70
Metal removal per blank, kg (lb)....... 0.0073 (0.016)
Rate of metal removal, kg/min (lb/min) ........................... 0.44 (0.096)

**Fig. 8** Change of method that improved machinability and tool life in forming recess in wood-drill shanks made of hot-rolled and annealed L1 tool steel. Dimensions in figure given in inches. Specimens for micrographs were etched in 2% nital. 2000×

a factor of close to 30, if there is enough carbon in the steel to convert the vanadium into vanadium carbide.

Vanadium carbides are relatively pure carbides formed during solidification of the steel; they contain very little iron or other metallic elements. These carbides have a hardness of 2300 to 2700 HK, averaging about 2500 HK. Aluminum oxide grinding wheel abrasive has a hardness of 1900 to 2900 HK, averaging 2440 HK. The near equality in hardness of these two substances partly explains the difficulty in grinding high-vanadium steels.

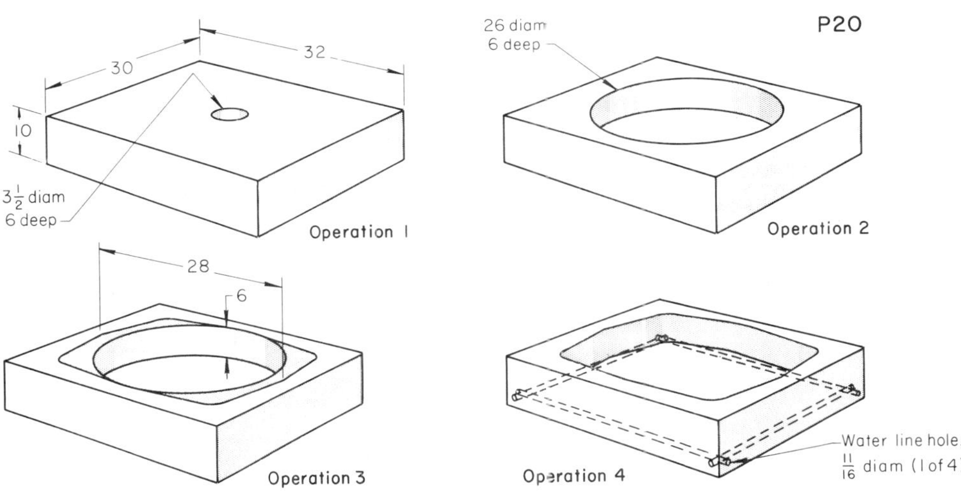


**Operation 1 Drilling**

Tool material(a) .............M2 high-speed steel
Hole diameter, mm (in.)......90 (3½)
Hole depth, mm (in.).........150 (6)

**Operation 2 Boring**

Tool material(b) .............T5 or M34 high-speed steel
Speed, rev/min ..............32-16(c)
Feed, mm/rev (in./rev) .......0.05–0.13 (0.002–0.005)
Final diameter, mm (in.)......660 (26)

**Operation 3 Die-sink milling**

Tool material(d) ............ M2 high-speed steel
Speed, rev/min ............. 80
Feed, mm/min (in./min) ...... 20–35 (0.8–1.4)
Depth of cut, mm (in.) ....... 50 (2)

**Operation 4 Drilling waterline holes**

Tool material(e) ............ M10 high-speed steel
Speed, rev/min ............. 80
Feed, mm/rev (in./min)....... 0.05 (0.002)

(a) A 90 mm (3½ in.) diam twist drill of standard design was used; speed and feed were not reported. (b) Fly-cutting tool bits used had 15° top rake angle, 10° end rake angle, 5° end cutting-edge angle, and 0° side cutting-edge angle. (c) Speed was decreased as diameter of hole was progressively increased by use of larger-diameter fly cutters. (d) Taper die-sinking cutter was 50 mm (2 in.) in diameter by 152 mm (6 in.) in length. (e) A 17.5 mm (11/16 in.) diam crankshaft drill was used.

**Fig. 9** Four-operation machining of prehardened P20 steel mold for forming plastic television cabinet housings. Blank weighed 1230 kg (2720 lb); finished mold weighed 650 kg (1430 lb). Dimensions in figure given in inches

Figure 11 shows the relative length of time for grinding high-speed steels of different vanadium contents. The same relative time to grind T1 and M2 steels is indicated, even though M2 has a higher vanadium content. However, wheel wear is greater when grinding M2 than when grinding T1, and this factor has not been taken into consideration in the presentation of data on grinding time. The adverse effect of vanadium on grindability is reflected also in the grindability index comparison plotted in Fig. 12.

**Chromium.** Although less potent than vanadium, chromium also has a marked effect on the grindability index. In the A and D series of cold-work die steels, the same effects of vanadium on grindability have been observed as in high-speed steel. A7 and D7, however, are difficult to grind not only because of the high volume of vanadium carbide present, but also because of the presence of a large number of complex iron-chromium carbides. The effect of chromium on the grindability index is indicated in Fig. 13.

**Sulfur.** The effect of sulfide inclusions in tool steel appears to be predominantly chemical in nature; sulfide inclusions greatly reduce wheel wear without marked degradation of the mechanical properties of the steel. In the hardened condition, sulfurized tool steels have better grindability than the same grades without sulfur. Data are summarized in Tables 21 and 22.

## Effect of Hardness of Steel on Grindability

Grindability of tool steel decreases as hardness increases. However, the magnitude of the difference contributed by hardness varies considerably with different compositions.

**Table 19 Tool design for milling carbon-tungsten special-purpose tool steels with high-speed steel cutters**

| | |
|---|---|
| Rake angle, degrees | 10–15 |
| Clearance angle, degrees | 10–15 |
| End relief angle, degrees | 3–5 |
| Tooth land radius, mm (in.) | 0.8 (1/32) |

For tool steels that contain few if any particles of alloy carbide, such as the W and O types, hardness has a relatively small effect on the grindability index, as indicated in Fig. 10. For instance, O2 steel has a grindability of 60 at a hardness of 20 HRC (annealed), but as hardness is increased to 60 HRC, the grindability decreases to about 30.

The decrease in the grindability index with increasing hardness is more pronounced (on a logarithmic scale) for the more highly alloyed tool steels such as D3 and T15 (Fig. 10). For instance, the grindability index for T15 is only about 10 in the annealed condition (23 HRC) and decreases rapidly to about 0.6 at 60 HRC. Thus, it is always more economical to do as much of the grinding as possible prior to heat treatment. However, from the relationship of the grindability index and hardness shown in Fig. 10, it is evident that for highly alloyed steels such as T15 the critical factor is whether stock is removed in the soft or hard condition.

## Effect of Hardness on Surface Finish

Hardness influences the surface finish that results from grinding tool steel when other factors remain the same. However, grinding parameters are the major influence on surface finish and can easily overcome the effects of material hardness. Factors that increase grindability generally cause a rougher finish. Figure 14 shows the relationship between hardness and surface roughness for one of the easier-to-grind steels (O2) and for one of the more highly alloyed and more difficult-to-grind steels (D3). For both steels, surface roughness increases as hardness decreases. However, this increase is more pronounced for O2 than it is for D3. To maintain a constant surface roughness, it is necessary to change to a harder wheel for the softer steels, which is in keeping with general grinding practice.

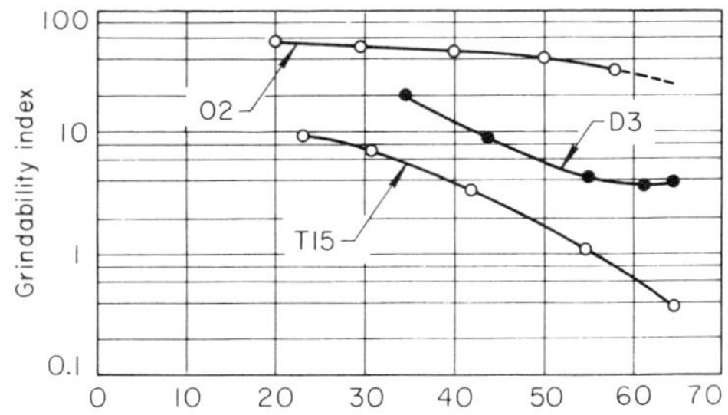


**Fig. 10** The effect of hardness on the grindability of selected tool steels

## Effect of Steel Composition on Grinding Cost

In the production of intricate tools and dies involving considerable finish grinding after heat treatment, not only is the cost of finish grinding an important consideration, it is usually of greater significance than the comparative machinability or workability of the steel in the annealed state. Finishing operations that involve form grinding, thread grinding, or the internal grinding of small holes separate the difficult-to-grind tool steels from the easy-to-grind grades more than does surface grinding of flats. However, in surface grinding of the more abrasion-resistant and longer-wearing grades of high-speed steel and die steel, the rate of wheel wear often equals or exceeds the rate of metal removal (grinding ratio, *G*, is less than one). The consequent wheel dressing accounts in part for the higher costs.

## General Practice

Equipment for grinding tool steel is seldom different from that used for similar operations on other metals. Equipment is discussed in detail in the article "Grinding Equipment and Processes" in this Volume.

**Selection of Wheel.** Composition and hardness of the tool steel being ground, the workpiece shape, and tolerance and finish requirements are significant factors in selecting a grinding wheel for a given application. The type of grinding operation may also influence the characteristics of the wheel used (Table 23). Details of the ASA system of grinding wheel identification and of a modified system used in these articles and, in addition, a more complete discussion of grinding wheels are given in the

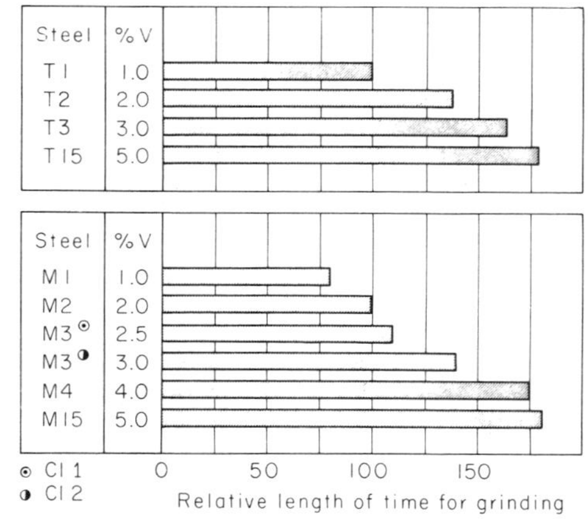


**Fig. 11** The effect of vanadium content of high-speed steel on the grinding time required for the removal of equal volumes of metal with the same type of wheel

article "Grinding Equipment and Processes" in this Volume (structure number has been omitted from some wheel designations in this article).

In most grinding applications on tool steel, aluminum oxide wheels are used. Silicon carbide wheels are sometimes used for grinding tool steels that have an extremely low grindability index. For the finish grinding of cutting tools, CBN wheels are preferred; aluminum oxide wheels are the most cost effective in roughing operations prior to finishing with CBN wheels. CBN wheels wear longer and produce less grinding damage to the part. Table 24 lists data for CBN wheels used in three different grinding applications.

The selection of abrasive-grain size depends mainly on the surface finish required. As grain size decreases, finish improves, but the rate of metal removal usually decreases. Wheels having grain size as coarse as 36 are used for rough grinding. Grain sizes of 60 to 120 are most common for finish grinding. Much finer grain sizes are available for special applications, such as roll and thread grinding.

**Table 21 Effect of sulfurization of tool steel on grindability**

Data are for surface grinding.

| Steel | Grindability index: Regular grade | Sulfurized(a) | Increase, % |
|---|---|---|---|
| **Dry grinding** | | | |
| M2 | 17 | 34 | 100 |
| M3 (class 1) | 5.5 | 15.4 | 180 |
| M4 | 2.8 | 9.5 | 240 |
| **Wet grinding** | | | |
| M2 | 6.4 | 9.6 | 50 |
| M3 (class 1) | 2.7 | 4.2 | 60 |
| M4 | 0.9 | 2.1 | 130 |
| D7 | 0.9 | 3.1 | 240 |

(a) Free machining

**Table 20 Relative grindability of tool steels**

| Grindability | Steels |
|---|---|
| Low(a) | A7, D7, M3 (class 2), M4, M15, M43, M44, T3, T9, T15 |
| Medium(b) | A2–A6, A8–A10, D1–D5, H steels, M1, M2, M7, M10, T1, T4, F steels, P4 |
| High(c) | O steels, W steels, S steels, L steels, P steels except P4 |

(a) Grindability index is generally comparable to that for T15 in Fig. 10. (b) Grindability index is generally comparable to that for D3 in Fig. 10. (c) Grindability index is equivalent to or greater than that for O2 in Fig. 10.

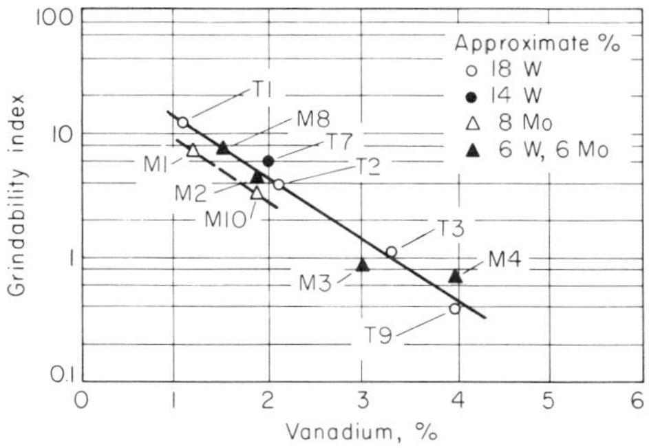


**Fig. 12** The effect of vanadium content on the grindability of tool steels

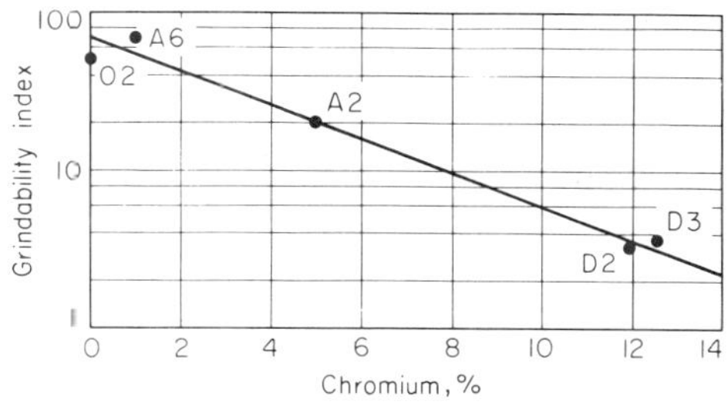


**Fig. 13** The effect of chromium content on the grindability of tool steels

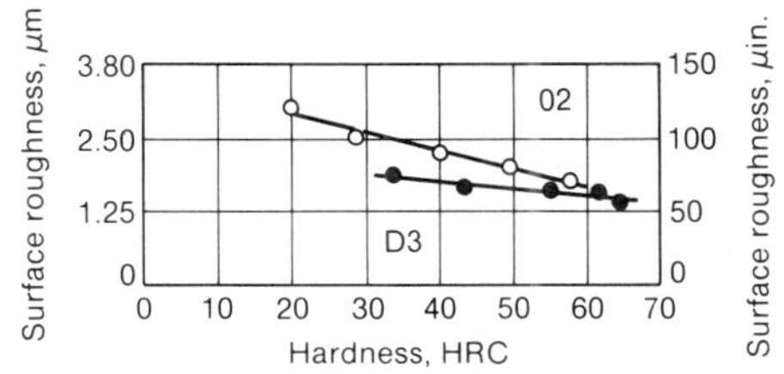


**Fig. 14** The effect of hardness on ground finish of tool steels

As in the grinding of production steels, softer wheels are normally used for the harder materials, and vice versa. Most tool steels are ground in the hardened condition, for which soft-to-medium-grade wheels are selected because wheel breakdown should be rapid, to expose new cutting edges of the abrasive. Hard wheels are likely to glaze and burn the surface of hardened steel. Wheels softer than E in a porous structure or H in a normal structure are seldom used on tool steel.

The selection of wheels for thread grinding does not necessarily follow the pattern for other types of tool steel grinding, because it is important to minimize the rate of wheel breakdown to keep the wheel accurate in thread form. Therefore, wheels of R to Z hardness are commonly used for thread grinding.

Structure (density) numbers from 5 to 9 are common, and the majority of wheels fall within this range. Structure numbers, specified by each manufacturer, are related to other wheel characteristics and to the specific operation. Structure numbers are not necessarily related to the hardness of the steel being ground.

*Vitrified bond* is the most widely used for grinding tool steel. Such a bond is strong, has a high yield in terms of metal removal rate, is not affected by any grinding fluid, and resists all temperature changes normally encountered in grinding operations. With some exceptions, a vitrified-bond wheel should be operated at speeds not exceeding 2000 m/min (6500 sfm).

*Resinoid Bond.* Wheels bonded with resinoid are generally less efficient for grinding tool steel than those having a vitrified bond. However, the resinoid bond is stronger than the vitrified bond, and thus wheels with resinoid bond can be safely operated at higher speeds. They are often used for centerless and thread grinding of hardened tool steel and also for cutting off, slotting, and flute grinding.

*Rubber bond* wheels are often used in specialized applications such as flute grinding. Rubber bonds have the disadvantage of being affected by oil.

### Table 22 The effect of sulfurization and hardening temperature of M3 high-speed steel on grindability

| Temperature °C | °F | Dry grinding: Regular grade | Dry grinding: Sulfurized(a) | Wet grinding: Regular grade | Wet grinding: Sulfurized(a) |
|---|---|---|---|---|---|
| **Grindability index** | | | | | |
| 1210 | 2210 | 5.7 | 18.7 | 2.9 | 4.4 |
| 1225 | 2240 | 5.5 | 15.4 | 2.7 | 4.2 |
| 1245 | 2270 | 5.5 | 14.6 | 2.5 | 3.2 |
| **Net power input(b)** | | | | | |
| 1210 | 2210 | 8.7 | 8.3 | 10.8 | 10.3 |
| 1225 | 2240 | 9.2 | 8.9 | 10.6 | 10.2 |
| 1245 | 2270 | 10.2 | 9.2 | 10.7 | 9.4 |

(a) Free machining. (b) Expressed in arbitrary units

### Table 23 Typical conditions for the grinding of tool steel

| Operating condition | Work material condition and hardness: Annealed (150–275 HB) | Work material condition and hardness: Quenched and tempered (56–65 HRC) |
|---|---|---|
| **Surface grinding(a)** | | |
| Wheel classification | A46IV | A46IV |
| Wheel speed, m/min (sfm) | 1700–2000 (5500–6500) | 1700–2000 (5500–6500) |
| Table speed, m/min (sfm) | 15–30 (50–100) | 15–30 (50–100) |
| Down feed, mm/pass (in./pass) | | |
| Roughing | 0.05 (0.002) | 0.05 (0.002) |
| Finishing | 0.013 (0.0005) max | 0.013 (0.0005) max |
| Cross feed, fraction of wheel width per pass | 1/5 | 1/10 |
| **Cylindrical grinding(b)** | | |
| Wheel classification | A46KV | A46LV |
| Wheel speed, m/min (sfm) | 1700–2000 (5500–6500) | 1700–2000 (5500–6500) |
| Work speed, m/min (sfm) | 18 (60) | 15 (50) |
| Infeed, mm/pass (in./pass) | | |
| Roughing | 0.05 (0.002) | 0.05 (0.002) |
| Finishing | 0.013 (0.0005) max | 0.013 (0.0005) max |
| Traverse per work rotation, fraction of wheel width | | |
| Roughing | 1/2 | 1/4 |
| Finishing | 1/6 | 1/8 |
| **Centerless grinding(c)** | | |
| Wheel classification | A60MV | A80LV |
| Wheel speed, m/min (sfm) | 1700–2000 (5500–6500) | 1700–2000 (5500–6500) |
| Work feed, mm/min (in./min) | 15 (50) | 15 (50) |
| Infeed, mm/pass (in./pass) | | |
| Roughing | 0.13 (0.005) | 0.13 (0.005) |
| Finishing | 0.038 (0.0015) max | 0.038 (0.0015) max |
| Regulating wheel | | |
| Angle, degrees | 3 | 3 |
| Speed, rev/min | 30 | 30 |
| **Internal grinding(d)** | | |
| Wheel classification | A80LV | A80LV |
| Wheel speed, m/min (sfm) | 1500–2000 (5000–6500) | 1500–2000 (5000–6500) |
| Work speed, m/min (sfm) | 23–46 (75–150) | 23–46 (75–150) |
| Infeed, mm/pass (in./pass) | | |
| Roughing | 0.013 (0.0005) | 0.013 (0.0005) |
| Finishing | 0.005 (0.0002) max | 0.005 (0.0002) max |
| Traverse per work rotation, fraction of wheel width | | |
| Roughing | 1/3 | 1/3 |
| Finishing | 1/6 | 1/6 |

(a) Applicable to all tool steels except A7 and the D, T, M, and F groups when they are quenched and tempered. For these groups an A60IV wheel is recommended. (b) Applicable to all tool steels except A7, D7, T9, T15, M3, M4, M15, and the F group when they are quenched and tempered. For these steels either an A46JV or a C46JV should be used, and traverse for rough grinding should be no more than one-third the wheel width per work rotation. (c) Applicable to all tool steels except A7 and the D, T, M, and F groups when they are quenched and tempered. For these steels infeed per pass should be no more than 0.08 mm (0.003 in.) per pass for roughing and 0.025 mm (0.001 in.) per pass for finishing. For grinding A7, D7, T9, T15, M3, M4, M15, and the F group in the quenched-and-tempered condition, an A80KV wheel should be used. (d) Applicable to all tool steels in all conditions. Source: Metcut Research Associates Inc.

**Table 24 Typical grinding practice for wrought tool steels in excess of 50 HRC hardness using cubic boron nitride (CBN) wheels**

| Grinding operation | Wheel identification | Wheel speed, m/min (sfm) | Table speed, m/min (sfm) | Work speed, m/min (sfm) | Down feed, mm/pass (in./pass) | Cross feed, mm/pass (in./pass) | Infeed on diameter, m (in.) | Traverse wheel width/ work rev |
|---|---|---|---|---|---|---|---|---|
| Surface grinding, horizontal spindle, reciprocating table | B100T75B | 1700–2400 (5500–8000) | 15–38 (50–125) | . . . | 0.013–0.038 (0.0005–0.0015)(a) | 3.0–25 (0.12–1.00) half of wheel width, max | . . . | . . . |
| Cylindrical grinding | B100T100B | 1500–2300 (5000–7500) | . . . | 15–30 (50–100) | 0.013 (0.0005) max(b) | . . . | 0.025 (0.001)(a) | 1/4 |
| | | | | | | | 0.005 (0.0002) max(b) | 1/8 |
| Internal grinding | B120TB | 1500–3000 (5000–10 000) | . . . | 24–46 (80–150) | . . . | . . . | 0.013 (0.0005)(a) | 1/3 |
| | B180TV | | | | | | 0.005 (0.0002) max(b) | 1/8 |

(a) Rough. (b) Finish

*Resin bond* wheels are useful for finish grinding and for producing burrfree edges on cutting tools using CBN abrasives.

**Speed.** Most tools and other workpieces made of tool steel are ground at speeds of 1200 to 2000 m/min (4000 to 6500 sfm) although low-stress grinding (LSG) requires speeds below 1200 m/min (4000 sfm) while finish grinding with CBN can exceed 3000 m/min (10 000 sfm). However, the specific grinding operation, as indicated in the section "Types of Grinding" in this article, is a major factor in the selection of wheel speed (Table 23).

**Grinding Fluid.** All computer numerical control (CNC) tool and cutter grinding (sharpening) is done wet. Only manual machines are run dry, which usually results in loss of surface integrity on the finished cutters. Fluids used are:

- Soluble-oil emulsion
- Synthetic solution
- Grinding oil

Emulsions of about one part soluble oil to 25 to 40 parts water are the most widely used fluids for grinding tool steel. For grinding threads or for operations involving heavy cuts and hard wheels, specially compounded oils give greater accuracy and smoother finish. Where possible, a sulfurized oil should be used to maximize the grinding ratio and minimize surface integrity damage.

In applications in which the conventional use of a flow of grinding fluid is not practical, oil or a water-base fluid sprayed in a mist over the work has proved effective. Compressed air has also proved effective for applications in which liquids are impractical.

## Types of Grinding

Grinding of tool steel includes surface, cylindrical, centerless, internal, thread, and flute grinding, and sharpening or clearing of tools. Flute grinding is mainly used in the manufacture of end mills, taps, drills, and reamers. Sharpening or clearing is used for finishing or reconditioning various tools and usually involves minimal stock removal.

**Surface grinding** of tool steel is commonly done in both of the principal types of surface grinders: horizontal spindle and vertical spindle (a description of the two types can be found in the article "Grinding Equipment and Processes" in this Volume). The first method is more flexible in accommodating various workpiece shapes and sizes, and the second is especially well adapted to the grinding of many identical pieces in one load, as in grinding the sides of milling cutter blanks. Lower speeds (often as low as 1200 m/min, or 4000 sfm) are used in reciprocating-table operations, in which it is difficult to keep grinding fluid trained on the working surfaces or in which workpiece sections are thin, or both. The surface grinding of knives made of carbon tool steel is an application in which extreme care (including low speed) is necessary. One advantage of rotary-table grinding is the ease of supplying a flood of grinding fluid to the critical areas.

**Cylindrical grinding** of workpieces between centers is extensively used for tool steel because of the large number of round tools that require grinding of two or more diameters. This type of grinding is least likely to damage workpieces, because of the small area of contact between wheel and workpiece. Minimum contact makes possible the most efficient use of the grinding fluid and, therefore, the use of speeds as high as allowable for the particular wheel. Speed of workpiece rotation in grinding tool steel is essentially the same as in cylindrical grinding of any other metal (23 to 46 m/min, or 75 or 150 sfm, is a common range). Stock removal per pass may be as high as 0.13 mm (0.005 in.) on the diameter in the rough grinding of hardened tool steel, and from 0.013 to 0.025 mm (0.0005 to 0.001 in.) in finishing.

**Centerless grinding** is used in preference to cylindrical grinding on centers when the workpiece shape permits. It is not applicable if more than one diameter is to be ground or if a shoulder or protuberance interferes.

**Internal grinding** involves greater contact of wheel and workpiece than does external grinding. Thus, there is a greater possibility of surface damage. Although this danger exists for internal grinding of all metals, it is more critical for highly hardened tool steel. Common practice is to use a wheel diameter no larger than 90% of the diameter of the bore to be ground when the bore diameter is smaller than 25 mm (1 in.). For grinding bores larger than 25 mm (1 in.), the wheel diameter should be no larger than 70% of the bore diameter.

Stock removal per pass must be light (seldom more than 0.013 mm, or 0.0005 in., on the diameter) or burning may result. When it is impractical to use grinding fluid, as when fluid would impair the operator's vision into small holes, prevention of damage from excessive heat may be difficult. Under these conditions, air blast cooling is sometimes used.

**Thread Grinding.** Normal practice for thread grinding tool steel is not necessarily different from that for grinding of threads in production steels of similar hardness; see the article "Thread Grinding" in this Volume. Thread grinding tool steel is often critical because of the end use. For instance, in grinding the cutting teeth in taps from the hardened solid, extreme care must be used to avoid burning the crests of the teeth.

A relatively fine grit is ordinarily used for grinding threads in hardened tool steel. The type of bond depends largely on the accuracy required. Vitrified wheels give greater accuracy because they deflect less than resinoid bond wheels. The latter are used when a high production rate is more important than accuracy.

A constant flood of grinding oil over the area being ground is mandatory for successful thread grinding. Water-base cutting fluids are unsuitable for thread grinding.

**Flute Grinding.** Spiral or straight flutes are often ground in tool blanks after heat

**Table 25 Process parameters and machine condition guidelines for low-stress grinding results**

Parameters are listed in descending order of significance for low stress generation in the workpiece surface.

| Process parameters and machine conditions | Guidelines |
|---|---|
| **Process parameters** | |
| Grinding wheel dressing technique | Frequent and coarse to maintain sharpness<br>Maintain sharpness of dressing tool<br>Avoid dwell in using crush, roll, or single-point dressing tools |
| Wheel speed | Low (<1100 m/min, or 3500 sfm) |
| Down feed (or infeed) rate | 0.005–0.013 mm/pass (0.0002–0.0005 in./pass) with programmed reduction from conventional rates |
| Grinding fluid | Oil-base fluid is preferred |
| Wheel classification | Soft grade (G, H, or I)(a)<br>Open structure (≥6)<br>Grain size (60 or coarser) |
| Table (workpiece) speed | High (≥15 m/min, or 50 sfm)<br>Cross feed is preferred to plunge motion |
| Grinding fluid flow control | Adequate to high fluid flow<br>Placement of fluid between wheel and workpiece ensured<br>Flow-controlling nozzle design<br>Reduce air film on wheel |
| **Machine conditions** | |
| Machine type and quality | Solid, vibration free, and well maintained<br>Clean tanks, filters, and lines |
| Rigidity of setup | Good fixture design<br>Cleanliness of fixture maintained |

(a) Cylindrical grinding frequently requires the use of harder wheels (with J grade prevalent). However, the other parameters must be selected to compensate for the extra hardness. Source: *Low Stress Grinding for Quality Production*, 2nd ed., Metcut Research Associates Inc., 1983

treatment, rather than being cut in a prior milling operation. Grinding is preferred for producing accurate flutes. However, because of the relatively long contact between the wheel and the work and because the wheel is buried in the work, this method calls for extreme caution.

In most applications, relatively hard aluminum oxide resinoid bond wheels are used. Larded grinding oil (or some sulfurized-type oil specially compounded for the purpose) is recommended over water-oil emulsions, which are used for most other grinding operations.

**Low-Stress Grinding (LSG).** Tool steels are especially prone to alterations such as microcracks, burns, and phase changes when subjected to excessively intense grinding conditions due to the high rate of energy input into the material surface. Use of low-stress grinding, which ranks just below lapping in ascending order of energy density (rough grinding and abusive grinding being the most intense conditions), can minimize these surface alterations and maximize surface integrity. Low-stress grinding typically yields a surface roughness of 0.10 to 1.60 μm (4 to 63 μin.) $R_a$ compared to 0.80 to 3.8 μm (32 to 150 μin.) $R_a$ for high-stress grinding. Low-stress grinding uses special combinations of grinding parameters that reduce the heat shock and plastic deformation inherent in all grinding operations. Control of these parameters produces a workpiece surface with excellent dimensional tolerances, fine surface finish, a minimum of residual stress, and low distortion, in sum, high surface integrity. When good dimensional stability of a component is needed, or for surfaces subjected to high fatigue stresses or possible stress corrosion, the enhanced integrity from LSG is a positive contribution to product integrity. Table 25 lists typical low-stress grinding parameters.

The advantages of low-stress grinding must be weighed against the process disadvantages. The principal drawback to low-stress grinding is the generally slower rates of material removal. Lower infeed values, softer wheels with lower grinding ratios, and more frequent wheel dressing all contribute to slower material removal and extra cycle time. The use of higher workpiece speeds, however, can offset this lower productivity to some extent. The down feed or infeed rates for finishing must be kept at very low levels for good LSG results. Feed rates can be scheduled to decrease progressively from several thousandths of an inch per pass to finishing values of a few ten-thousandths of an inch per pass. If the work material has not cracked in roughing, the finishing process levels need not be applied until 0.13 to 0.25 mm (0.005 to 0.010 in.) of the finish size is reached. The other grinding parameters have little, if any, effect on production rates, although they can affect the operating cost.

**Sharpening or clearing** is needed in the final finishing of new tools, as well as in reconditioning almost all types of cutting tools and many types of dies. Machines for this use are special in most instances; tools such as end mills and cutters need one type of machine, whereas drills need a different type.

The hand grinding of tools is not recommended, because it is impossible to obtain accuracy as high as is possible in machine grinding, and damage to cutting edges is more likely in hand grinding.

## Grinding of W, O, S, and L Steels

Although there is considerable variation in grindability among steels in the W, O, S, and L groups, they are all within the general classification of high-grindability steels (Table 20). Depending on hardness, a grindability index approximating that shown for O2 in Fig. 10 can be expected.

All tool steels are to some degree susceptible to tempering from overheating during grinding. However, the W, O, S, and L steels are more susceptible than the more highly alloyed steels. Surface hardness begins to decrease from the maximum value when these steels are heated higher than 120 to 150 °C (250 to 300 °F).

## Grinding of A and D Steels

With the exception of A7 and D7, the cold-work tool steels of the A and D groups are classified as medium in grindability (Table 20). Although there is considerable variation within the group, a grindability index approximating that for D3 (Fig. 10) can be expected. The variation in chromium content among the different types has some effect on grindability, as indicated in Fig. 13. The amount of vanadium in the A and D steels (other than A7 and D7) is not sufficient to affect grindability significantly.

Because they contain 4% or more vanadium, A7 and D7 are classified as low in grindability (Table 20), and a grindability index approximating that shown for T15 in Fig. 1 can be expected. Steels such as A7 and D7 are used if resistance to abrasion is necessary and thus they would be expected to be extremely low in grindability.

## Grinding of Hot-Work Steels (H Types)

The chromium, tungsten, and molybdenum hot-work steels are generally classified as medium in grindability (Table 20). A grindability index approximating that of D3 (shown in Fig. 10) would be expected for the higher-alloy steels in the H group, and a somewhat higher index would be expected for the lower-alloy steels, such as H10.

There are some differences in grinding a low-alloy steel like H10 (total alloy content, about 9.5%) and in grinding the highest-alloy steel of the group (H26, with an alloy content of 23%). However, none of these steels contains more than about 2% V, which is a major factor affecting grindability. In addition, all hot-work steels are relatively low in carbon content (0.65% is the highest, and most of them have less than 0.55%). Therefore, massive carbides are not

**Table 26 Grinding ratios for high-speed tool steels**

| Steel | Hardness, HRC | Ratio for wheel type A46H8V | A60H8V | A80H8V |
|---|---|---|---|---|
| T15 | 65.7 | 0.49 | 0.62 | 0.51 |
| M44 | 67.7 | 0.97 | 0.99 | 0.88 |
| M41 | 68.7 | 1.2 | 1.6 | 1.4 |
| M43 | 67.5 | 1.4 | 2.2 | 1.7 |
| M42 | 68.8 | 4.8 | 6.5 | 3.8 |
| M2 | 64.9 | 6.1 | 7.2 | 6.7 |
| M1 | 64.9 | 7.8 | 8.0 | 11.9 |

Note: For grinding hardened high-speed steel in a horizontal-spindle, reciprocating surface grinder

likely to occur. These steels are seldom harder than 55 HRC when they are ground. Thus, their grindability is generally better than that of tool steel hardened to 60 HRC or higher.

Loading the grinding wheel is a greater problem in grinding hot-work steels than in grinding high-carbon tool steels, and more frequent dressing of wheels is usually required. Selection of soft (K or softer) aluminum oxide wheels that have a structure in density of 5 or lower is usually helpful in minimizing wheel loading.

## Grinding of High-Speed Steels

The T and M types of high-speed steel comprise a group of about 30 members. The lower-alloy general-purpose types are usually classified as medium in grindability (Table 20). However, this depends to some extent on their carbon content. The grades M2, M7, and M10, with 1.0% carbon, have a grindability approximately the same as M3 (class 2) (Table 20). The more highly alloyed types, particularly those containing 3% or more vanadium, are low in grindability and have a grindability index approximately the same as that shown for T15 in Fig. 10. The low grindability and consequent high grinding cost of the highly alloyed steels are major factors in deterring their more widespread use.

A more recent development in high-speed steels is the M40 group. These grades can be heat treated to as high as 70 HRC in small sections, but are less difficult to grind than the high-vanadium types such as T15, which are two or three points lower in hardness. The reason for the appreciably better grindability of the M40 steels is that they do not contain the large amounts of the extremely hard and abrasion-resistant particles of vanadium carbide that are present in T15.

The M40 steels, all heat treated to approximately the same hardness, differ considerably among themselves in grindability, presumably because of differences in the amount and hardness of the carbide particles present.

As shown in Table 26, the grindability of the M40 steels is distributed over most of the range between that of T15 on the low side and M1 on the high side. Measurements of grinding power showed the 46-grit wheel to be somewhat harder than the 60-grit wheel, although both wheels were well within the normal hardness range for H-grade wheels. The consistently lower grinding ratios obtained with the coarser grit size indicate that 60 grit was superior to 46 grit from the standpoint of stock removal for all those steels. That is, more material was removed at the same power consumption.

This is equivalent to removing the same amount of material at a lower power level, which is the same as saying that the 60-grit abrasive is cooler cutting than 46 under these grinding conditions. A cooler-cutting wheel is less likely to damage the work surface, and because a finer grit size normally results in a better finish, 60 grit is superior to 46 with respect to both grinding damage and surface finish. In many applications, however, a cooler-cutting wheel produces a poorer finish, and the relative quality of two wheels then depends on whether it is more important to minimize grinding heat at the same production rate or to obtain a better finish.

Similar power comparisons have shown that the 80-grit wheel was a little softer than the 60-grit wheel, which accounts for the somewhat lower grinding ratios obtained with the finer wheel for all the steels except M1. When the slight grade difference between the two wheels was taken into account, the 60-grit abrasive was found to provide slightly cooler cutting than the 80-grit material for the T15, M44, and M42 steels, while the reverse was true for M2, M41, and M43 steels. The finer grit was markedly superior for the M1 steel, as shown in Table 26.

These results illustrate two conditions. First, in grinding two high-speed steels of the same type having essentially the same grindability rating under a specified set of grinding conditions, one wheel may perform better on one steel, and another one may have better results on the other steel. Second, the grindability ratings of two similar steels may differ by a small amount for one wheel and considerably for another, as can be seen by comparing the grinding ratios for the M1 and M2 steels and the three grit sizes.

Present knowledge on the effects of alloying elements on the grindability of high-speed steel can be summarized:

- Vanadium carbide has a pronounced effect in reducing grindability
- Iron-tungsten carbide, in which molybdenum may substitute for part or all of the tungsten, has very little effect in comparison with vanadium carbide
- Reducing the hardness of the steel by tempering increases the grindability. The effect is small for a change of a few points on the HRC scale, but can be quite large if the hardness is reduced considerably below the range in which these steels are used in practice
- Sulfurizing the steel to make it free machining can increase the grindability appreciably. The degree of improvement appears to be affected by the grindability of the steel in the unsulfurized condition, but it is unclear whether high-speed steels of low or high grindability benefit more from sulfur
- Cobalt moderately reduces the grindability of high-speed steel
- In molybdenum high-speed steel, raising the tungsten content and reducing the molybdenum appears to decrease the grindability

## Grinding of F Steels

The carbon-tungsten special-purpose tool steels (F types) are classified as medium in grindability, despite their relatively low alloy content. They are all high in carbon content and are nearly always used in the fully hardened condition (65 to 67 HRC) and thus are placed lower in grindability than the W or L types.

Grinding practice for the general-purpose types of high-speed steel is also applicable for grinding the F types.

## Grinding of P Steels

The low-carbon mold steels are seldom ground in the soft condition. Steels P1 to P6 are commonly carburized, hardened, and tempered, after which a grinding operation is required for correcting dimensions. Steels P20 and P21 also are hardened before being ground. Provided reasonable care is exercised to avoid grinding burn, overheating, and heavy grinding of sharp internal angles, these steels, with the exception of P4, are rated high in grindability (Table 20). In general, the grinding characteristics of the carburized surfaces are comparable to those of 1% C water-hardening tool steel or to low-alloy tool steel. Because of its higher chromium content, however, carburized P4 is more susceptible to grinding burn and cracking, and hence is usually classified as medium in grindability (Table 20).

### ACKNOWLEDGMENT

ASM INTERNATIONAL would like to express its thanks to Chi-Hung Shen, General Motors Technical Center; Nick Cerwin, A. Finkl & Sons, Inc.; and John D. Christopher, Metcut Research Associates, Inc., for their assistance in updating this article.

# Machining of P/M Tool Steels

Kenneth E. Pinnow and Carl J. Dorsch, Crucible Materials Corporation

DURING MANUFACTURING, as much as 50% of a tool blank can be removed by sawing, turning, drilling, milling, grinding, and other machining operations. Consequently, the machinability and the final surface finish obtainable with tool steels are important factors that affect material selection and the cost of a finished tool. Tool steels are generally more difficult to machine and grind than plain carbon and low-alloy construction steels because of their high carbide contents and their high hardnesses in both the annealed and hardened conditions. Further, the large size and the uneven distribution of the carbides (caused by ingot segregation in conventionally produced tool steels) are often unfavorable for optimum machinability.

Rapid solidification of the atomized powders used in the production of wrought powder metallurgy (P/M) tool steels eliminates the segregation present in conventional tool steels and produces a very fine microstructure with a very uniform distribution of small carbides and nonmetallic inclusions. As a result, wrought P/M high-speed tool steels exhibit better machinability, dimensional control and safety in heat treatment, grindability, and edge toughness during cutting than conventional high-speed tool steels of the same composition (Ref 1).

A variety of Anti-Segregation Process (ASP) and Crucible Particle Metallurgy (CPM) wrought P/M tool steels are available (Tables 1 and 2). For the most part, wrought P/M processing has been applied to the production of standard AISI-type high-speed tool steels (M-2, M-3, M-35, T-15, and so on) with or without sulfur additions for enhanced machinability. However, several P/M-processed, highly alloyed super-high-speed tool steels (CPM Rex 20, CPM Rex 45/ASP 30, CPM Rex 76, and ASP 60) have been developed for more demanding applications. Further, wrought P/M processing has been applied to the production of high-performance cold/warm work P/M tool steels (Table 2), resulting in substantially improved wear resistance and toughness (Ref 2), and to P/M warm/hot work tool steels (Table 2), resulting in improved thermal fatigue properties, wear resistance, and toughness (Ref 3). Detailed information on the production and properties of wrought P/M tool steels is given in the articles "Tool Steels" and "P/M Tool Steels" in Volumes 3 and 7, respectively, of the 9th Edition of *Metals Handbook*.

As with conventional tool steels, P/M tool steels are generally machined in two stages: rough machining of the workpiece with the steel in the annealed condition, followed by

**Table 1 Commercial ASP tool steel compositions**

| ASP grade | Composition, % C | Cr | Mo | W | V | Co | Typical hardness, HRC |
|---|---|---|---|---|---|---|---|
| 23 | 1.28 | 4.2 | 5.0 | 6.4 | 3.1 | ... | 65/67 |
| 30 | 1.28 | 4.2 | 5.0 | 6.4 | 3.1 | 8.5 | 66/68 |
| 60 | 2.30 | 4.0 | 7.0 | 6.5 | 6.5 | 10.5 | 67/69 |

Note: The ASP P/M process is described in the article "P/M High-Speed Tool Steels" in this Volume.

**Table 2 Commercial CPM tool steel compositions**

| Material | AISI | Composition, % C | Cr | W | Mo | V | Co | S | Typical hardness, HRC |
|---|---|---|---|---|---|---|---|---|---|
| **High-speed steels** | | | | | | | | | |
| CPM Rex M2 HCHS | M2 | 1.00 | 4.15 | 6.40 | 5.00 | 2.00 | ... | 0.27 | 64–66 |
| CPM Rex M3 HCHS | M3 | 1.30 | 4.00 | 6.25 | 5.00 | 3.00 | ... | 0.27 | 65–67 |
| CPM Rex M4 | M4 | 1.35 | 4.25 | 5.75 | 4.50 | 4.00 | ... | 0.06 | 64–66 |
| CPM Rex M4 HS | M4 | 1.35 | 4.25 | 5.75 | 4.50 | 4.00 | ... | 0.22 | 64–66 |
| CPM Rex M35 HCHS | M35 | 1.00 | 4.15 | 6.00 | 5.00 | 2.00 | 5.00 | 0.27 | 65–67 |
| CPM Rex M42 | M42 | 1.10 | 3.75 | 1.50 | 9.50 | 1.15 | 8.00 | ... | 66–68 |
| CPM Rex 45 | ... | 1.30 | 4.00 | 6.25 | 5.00 | 3.00 | 8.25 | 0.03 | 66–68 |
| CPM Rex 45 HS | ... | 1.30 | 4.00 | 6.25 | 5.00 | 3.00 | 8.25 | 0.22 | 66–68 |
| CPM Rex 20 | M62 | 1.30 | 3.75 | 6.25 | 10.50 | 2.00 | ... | 0.06 | 66–68 |
| CPM Rex T15 | T15 | 1.55 | 4.00 | 12.25 | ... | 5.00 | 5.00 | 0.06 | 65–67 |
| CPM Rex T15 HS | T15 | 1.55 | 4.00 | 12.25 | ... | 5.00 | 5.00 | 0.22 | 65–67 |
| CPM Rex 76 | M48 | 1.50 | 3.75 | 10.00 | 5.25 | 3.10 | 9.00 | 0.06 | 67–69 |
| CPM Rex 76 HS | M48 | 1.50 | 3.75 | 10.00 | 5.25 | 3.10 | 9.00 | 0.22 | 67–69 |
| **Cold/warm work tool steels** | | | | | | | | | |
| CPM 9V | ... | 1.75 | 5.25 | ... | 1.30 | 9.00 | ... | ... | 53–55 |
| CPM 10V | A11 | 2.45 | 5.25 | ... | 1.30 | 9.75 | ... | 0.07 | 60–62 |
| CPM 440V | ... | 2.15 | 17.50 | ... | 0.50 | 5.75 | ... | ... | 57–59 |
| **Warm/hot work tool steels** | | | | | | | | | |
| CPM H-19V | ... | 0.80 | 4.25 | 4.25 | 0.40 | 4.00 | 4.25 | ... | 54–56 |

Note: The CPM process is described in the article "P/M High-Speed Tool Steels" in this Volume.

**Table 3 Typical machining conditions for P/M and conventional grades of AISI high-speed tool steels**

| Operation | Tool width or depth of cut, mm | in. | High-speed tooling: Speed, m/min | sfm | Feed, mm/rev | in./rev | Carbide tooling: Speed, m/min | sfm | Feed, mm/rev | in./rev |
|---|---|---|---|---|---|---|---|---|---|---|
| Single-point turning | 3.8 | 0.150 | 18 | 60 | 0.38 | 0.015 | 91 | 300 | 0.38 | 0.015 |
| | 0.64 | 0.025 | 23 | 75 | 0.18 | 0.007 | 111 | 365 | 0.18 | 0.007 |
| Drilling | 6.4 | ¼ | 12 | 40 | 0.08 | 0.003 | ... | ... | ... | ... |
| | 13 | ½ | 12 | 40 | 0.13 | 0.005 | ... | ... | ... | ... |
| | 25 | 1 | 12 | 40 | 0.23 | 0.009 | ... | ... | ... | ... |
| | 50 | 2 | 12 | 40 | 0.33 | 0.013 | ... | ... | ... | ... |
| Broaching | ... | ... | 3 | 10 | 0.05 | 0.002 | ... | ... | ... | ... |
| Face milling | 3.2 | 0.125 | 20 | 65 | 0.20 | 0.008 | 78 | 255 | 0.30 | 0.012 |
| | 0.64 | 0.025 | 26 | 85 | 0.15 | 0.006 | 101 | 330 | 0.25 | 0.010 |
| Cutoff | 1.6 | 0.062 | 14 | 45 | 0.03 | 0.001 | 53 | 175 | 0.05 | 0.002 |
| | 3.2 | 0.125 | 14 | 45 | 0.03 | 0.001 | 53 | 175 | 0.08 | 0.003 |
| | 6.4 | 0.250 | 14 | 45 | 0.04 | 0.0015 | 53 | 175 | 0.11 | 0.0045 |
| Cutting fluid | ... | ... | Sulfurized oil, light duty | | | | Water-soluble oil | | | |

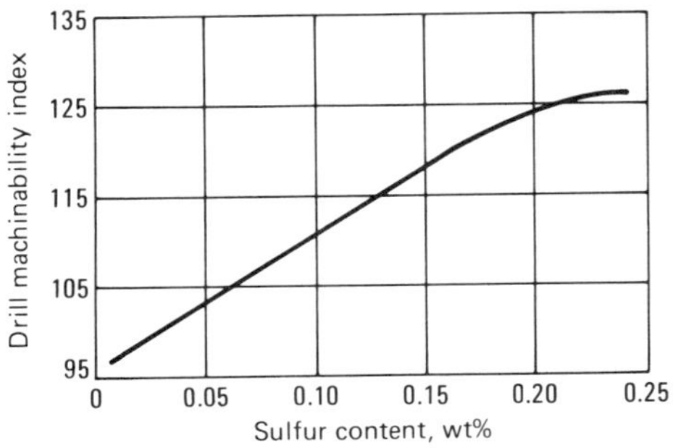


**Fig. 1** Effect of sulfur content on the drill machinability of CPM M-4 high-speed steel. Source: Crucible Materials Corporation

finish machining (typically grinding) after heat treatment when the steel is in the hardened-and-tempered condition. These operations are discussed separately in the following sections of this article.

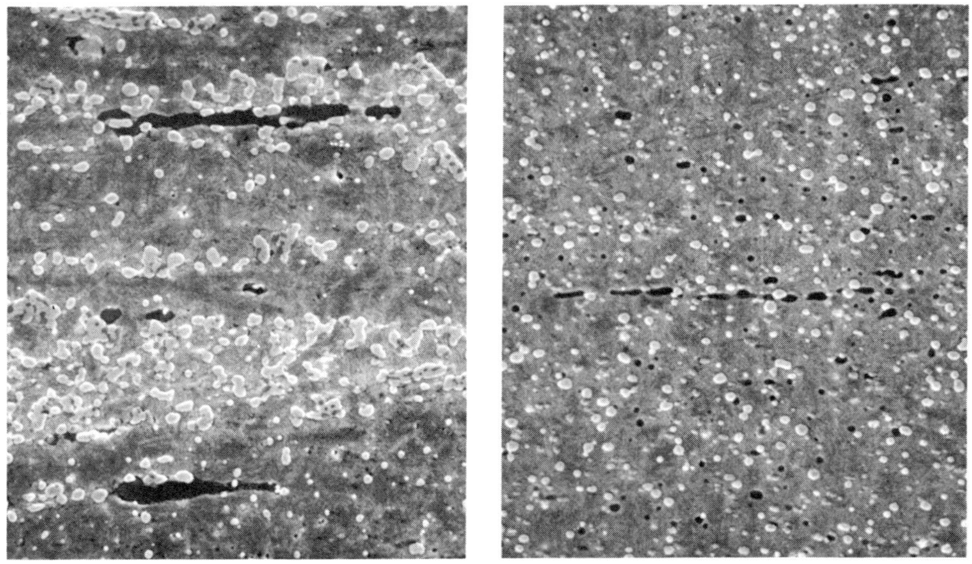

(a) (b)

**Fig. 2** Scanning electron micrographs showing the distribution and size of carbides and sulfides in (a) conventional M-35 (0.10% S) and (b) P/M M-3 (0.28% S) high-speed tool steels. Both 500×. Courtesy of Crucible Materials Corporation

## Machining

In general, machining operations are performed on annealed tool steels to bring the size and the shape of the workpiece close to those of the finished part, while leaving enough extra stock on the part to allow for distortion and removal of the surface deterioration (decarburization, oxidation, and so on) that may occur during heat treatment. Less extra stock is generally needed with P/M tool steels than with conventional tool steels; this is because the distortion and size changes that occur during heat treatment are more consistent and predictable as a result of the more uniform microstructure of P/M tool steels. In general, the machinability of the P/M tool steels in the annealed condition is comparable to, and in many cases superior to, that of conventional AISI tool steels of similar composition. There are no conventional counterparts of the highly resulfurized P/M tool steels and the more highly alloyed high-performance P/M tool steels such as CPM 10V. Therefore, the machinability of these steels must be considered separately. Table 3 lists the typical cutting conditions for P/M and conventional AISI high-speed steels of similar composition.

An important advantage of the P/M process relates to the fact that the machinability and grindability of P/M tool steels can be improved by increasing their sulfur content to much-higher-than-conventional levels without sacrificing toughness or cutting performance. Therefore, as indicated in Table 2, several P/M tool steels are available that contain as much as 0.30% S. Depending on specific conditions, the highly resulfurized P/M tool steels offer as much as a 30% improvement in machinability over lower-sulfur-containing P/M and conventional tool steels of similar base composition. A specific example is shown in Fig. 1, which illustrates the beneficial effect of sulfur on the drill machinability of CPM M-4 high-speed steel.

The machinability advantage of resulfurized P/M tool steels largely relates to the uniform distribution of the carbides and sulfides in their microstructures. As shown in Fig. 2 for M-3 high-speed tool steel, the carbides and the sulfides in the conventional steel are typically segregated in bands and are larger than those of the P/M steel. Because of the small size and uniform distribution of the sulfides, more sulfur (with a corresponding greater improvement in machinability) can be used in P/M tool steels than in conventional tool steels before hot workability or mechanical properties are degraded.

Another important class of P/M tool steels consists of the cold/warm work grades such as CPM 10V. These grades contain much greater-than-normal amounts of carbon and vanadium for increased wear resistance. Consequently, these steels are somewhat more difficult to machine than high-speed steels or cold-work steels with lower vanadium contents. In general, CPM 10V requires lower speeds and feeds than those listed in Table 3. Further, nitrided or black oxide coated drills are recommended because of the high wear resistance of CPM 10V.

## Grinding

Grinding is generally one of the last operations performed in the production of a hardened-and-tempered tool steel part. In addition to bringing the workpiece to the required final dimensions, the grinding process forms the sharp working edges on cutting tools, punches, and dies. Grinding is similar to machining in that the tool cutting

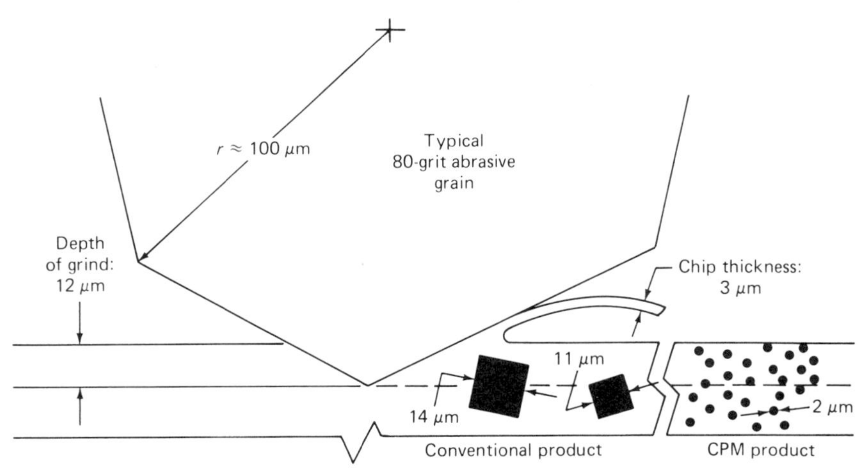


**Fig. 3** Schematic showing the relative sizes of carbides and chip thickness in the grinding of conventional and CPM tool steels. Source: Crucible Materials Corporation

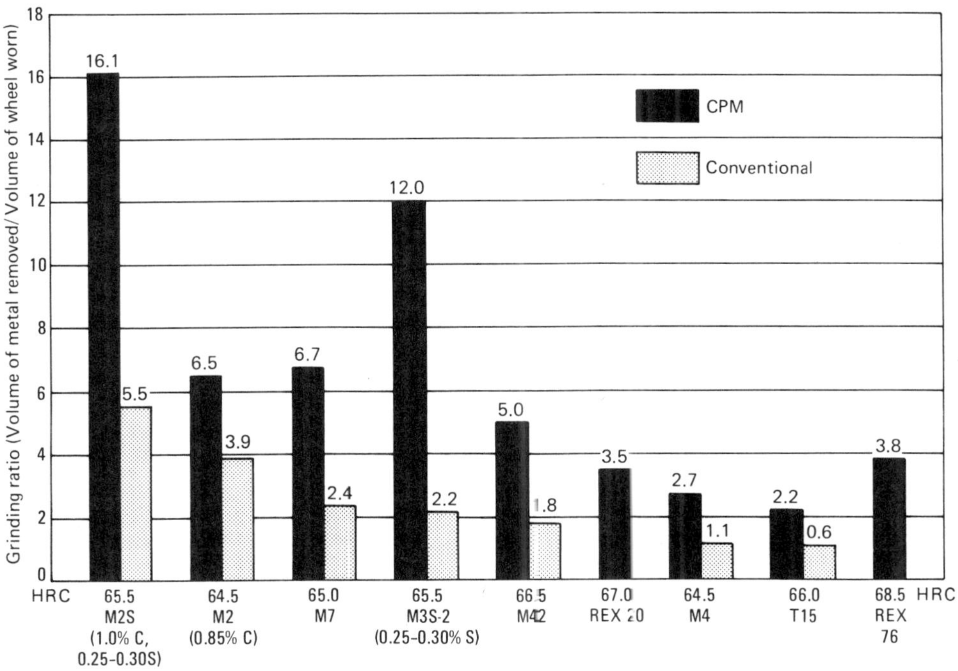


**Fig. 4** Comparagraph showing the relative grindability of CPM and conventional high-speed tool steels. Source: Crucible Materials Corporation

edge—in this case, an abrasive particle in the grinding wheel—cuts material from the workpiece, and the material is removed in the form of chips. Grinding appears to be different from machining because in grinding the workpiece is typically in a hardened-and-tempered condition and because the chips are many times smaller than those produced in machining.

As shown in Fig. 3, the carbides in P/M tool steels are about the same size as the thickness of a typical chip produced by a grinding wheel abrasive particle. In contrast, the carbides in conventional tool steels are larger and many times the thickness of the grinding chip. Therefore, during the grinding operation, fewer carbides must be fractured or worn down in the P/M steel because most of the carbides can be completely removed within a grinding chip. This results in significantly less grinding wheel wear and a better surface finish.

The distribution of the carbides in the tool steel matrix also affects grindability. In conventional tool steel, the carbides typically lie in bands within the steel, as discussed previously. The grinding wheel encounters alternating regions of high carbide density (and typically larger carbides) and low carbide density as it traverses the workpiece. This can result in nonuniform grinding wheel wear, a nonuniform surface finish, and a tool working edge that is not uniformly sharp. Because the carbides in P/M tool steel are uniformly distributed throughout the steel matrix, grinding proceeds at a consistent rate, and the wheel wear, workpiece surface finish, and edge sharpness are all uniform.

The relative grindability of several conventional and P/M high-speed tool steels is illustrated in Fig. 4. The grinding ratio (volume of metal removal to the volume of wheel worn, as explained in the article "Principles of Grinding" in this Volume) is clearly superior for the P/M tool steels. As expected, the grinding ratios generally decrease for both the conventional and the P/M tool steels as their alloy and carbon contents increase. The grinding conditions suggested for the CPM tool steels are similar to those recommended for conventional tool steels in the article "Machining of Tool Steels" in this Volume. Some specific conditions for grinding CPM 10V are given in Table 4.

## REFERENCES


1. R. Riedl *et al.*, Developments in High Speed Tool Steels, *Steel Res.*, Vol 58 (No. 8), 1987, p 339-352
2. R. Dixon, "Advances in the Development of Wear-Resistant High-Vanadium Tool Steels for Both Tooling and Non-Tooling Applications," Paper 8201-085, presented at the ASM Metals Congress (St. Louis, MO), American Society for Metals, 1982
3. W. Stasko *et al.*, Tool and Die Materials From Rapidly Solidified Powders, in *Rapidly Solidified Materials: Properties and Processing*, Proceedings of the 2nd International Conference on Rapidly Solidified Materials (San Diego, CA), ASM INTERNATIONAL, March 1988


**Table 4 Grinding recommendations for CPM 10V cold-work tool steel**

**Toolroom grinding (sharpening)**(a)

Abrasive: very sharp 38A or 32A
Grit sizes: 60 to 120 depending on removal and finish requirements
Grade: Grade I most effective, but grades as soft as G can work
Bond: Vitrified
Wheel example: Norton 32A60-I8VBE

**Wet surface grinding**(a)

Grit sizes: 100–150
Wheel example: CBN (Borazon) CB 120TBA

**Internal grinding**(a)

Grit sizes: 100–150
Wheel example: CBN (Borazon) CB 150WBA

**Field reports concerning abrasives**

Cubic boron nitride (Borazon) grinders must be rigid, in good condition, and able to mount wheels with very good accuracy.
Crystolon (silicon carbide) such as 39C60-I8VK are recommended.
Use very sharp, very friable alundum abrasive that remains sharp during grinding, such as 38A.
Wet grinding recommended

(a) Based on in-house laboratory testing. Note: Grinding wheel symbols and nomenclature are defined in the article "Grinding Equipment and Processes" in this Volume.

## SELECTED REFERENCES


- B.-A. Cehlin, "Improving Productivity With High Strength P/M High Speed Steel Cutting Tools," SME Technical Paper MR82-948, presented at the Increasing Productivity With Advanced Machining Concepts Clinic (Los Angeles, CA), Society of Manufacturing Engineers, 1982
- CPM Rex M3 HCHS (High Speed Steel: AISI Type M-3 Except Carbon, Manganese, Sulfur, and Silicon), *Alloy Dig.*, April 1987
- E.J. Dulis and T.A. Neumeyer, Particle-Metallurgy of High-Speed Tool Steel, in *Materials for Metal Cutting*, Publication 126, The Iron and Steel Institute, 1970, p 112-118
- P. Hellman *et al.*, The ASEA-STORA Process, in *Modern Developments in Powder Metallurgy*, Vol 4, Plenum Press, 1970, p 573-582
- P. Hellman and H. Wisell, "Effect of Structure on Toughness and Grindability of High Speed Steels," Paper presented at the Colloquium on High Speed Steels (Saint-Étienne, France), 1975
- G. Lutz, Grindability of Powder-Metallurgy High Speed Steels and Ledeburitic Chromium Tool Steels, *Werstatt. Betr.*, Vol 120 (No. 10), 1987, p 859-863
- M.A. Pailleux, "Study of the Workability of Powder Metallurgy High Speed Steels," Paper presented at the Centre Technique des Industries Mécaniques (Seulis), April 1983

# Machining of Heat-Resistant Alloys

HEAT-RESISTANT ALLOYS are usually classified as iron-base, nickel-base, and cobalt-base. Most of the cobalt-base alloys are more accurately referred to as cobalt-chromium-nickel-base alloys, and one of the iron-base alloys (N-155) contains 60% Co + Cr + Ni. The three basic categories of alloys include wrought and cast products with a wide range of compositions, some of which are listed in Tables 1 to 4. Heat-resistant alloys also include mechanically alloyed products, which are prepared by powder metallurgy techniques. The mechanically alloyed products discussed in this article are the oxide dispersion strengthened products listed in Table 5.

Machining is used in the manufacture of heat-resistant parts, although advances in near-net shaping with precision casting, precision forging, and powder metallurgy processing provide important alternatives

**Table 1 Nominal compositions of wrought heat-resistant alloys covered in this article**

| Alloy | UNS number | Composition, % | | | | | | | | | | |
|---|---|---|---|---|---|---|---|---|---|---|---|---|
| | | Cr | Ni | Co | Mo | W | Nb | Ti | Al | Fe | C | Other |
| **Iron-base alloys** | | | | | | | | | | | | |
| 16-25-6 | ... | 16.0 | 25.0 | ... | 6.00 | ... | ... | ... | ... | 50.7 | 0.06 | 1.35 Mn, 0.70 Si, 0.15 N |
| 17-14CuMo | ... | 16.0 | 14.0 | ... | 2.50 | ... | 0.4 | 0.3 | ... | 62.4 | 0.12 | 0.75 Mn, 0.50 Si, 3.0 Cu |
| 19-9DL | K63198 | 19.0 | 9.0 | ... | 1.25 | 1.25 | 0.4 | 0.3 | ... | 66.8 | 0.30 | 1.10 Mn, 0.60 Si |
| A-286 | K66286 | 15.0 | 26.0 | ... | 1.25 | ... | ... | 2.0 | 0.2 | 55.2 | 0.04 | 0.005 B, 0.3 V |
| Discaloy | K66220 | 14.0 | 26.0 | ... | 3.0 | ... | ... | 1.7 | 0.25 | 55.0 | 0.06 | ... |
| Incoloy 800 | N08800 | 21.0 | 32.5 | ... | ... | ... | ... | 0.38 | 0.38 | 45.7 | 0.05 | ... |
| Incoloy 801 | N08801 | 20.5 | 32.0 | ... | ... | ... | ... | 1.13 | ... | 46.3 | 0.05 | ... |
| Incoloy 802 | ... | 21.0 | 32.5 | ... | ... | ... | ... | 0.75 | 0.58 | 44.8 | 0.35 | ... |
| N-155 | R30155 | 21.0 | 20.0 | 20.0 | 3.00 | 2.5 | 1.0 | ... | ... | 32.2 | 0.15 | 0.15 N, 0.02 La, 0.02 Zr |
| V-57 | ... | 14.8 | 27.0 | ... | 1.25 | ... | ... | 3.0 | 0.25 | 48.6 | 0.08 max | 0.01 B, 0.5 max V |
| W-545 | K66545 | 13.5 | 26.0 | ... | 1.5 | ... | ... | 2.85 | 0.2 | 55.8 | 0.08 | 0.05 B |
| **Cobalt-base alloys** | | | | | | | | | | | | |
| AR 213 | ... | 19.0 | 0.5 max | 65.0 | ... | 4.5 | ... | ... | 3.5 | 0.5 max | 0.17 | 6.5 Ta, 0.15 Zr, 0.1 Y |
| Haynes 25 (L-605) | R30605 | 20.0 | 10.0 | 50.0 | ... | 15.0 | ... | ... | ... | 3.0 | 0.10 | 1.5 Mn |
| Haynes 188 | R30188 | 22.0 | 22.0 | 37.0 | ... | 14.5 | ... | ... | ... | 3.0 max | 0.10 | 0.90 La |
| S-816 | R30816 | 20.0 | 20.0 | 42.0 | 4.0 | 4.0 | 4.0 | ... | ... | 4.0 | 0.38 | ... |
| J1570 | ... | 20.0 | 28.0 | 38.8 | ... | 7.00 | ... | 4.00 | ... | 2.0 | 0.20 | ... |
| **Nickel-base alloys** | | | | | | | | | | | | |
| Astroloy | ... | 15.0 | 56.5 | 15.0 | 5.25 | ... | ... | 3.5 | 4.4 | <0.3 | 0.06 | 0.03 B, 0.06 Zr |
| Hastelloy B | N10001 | 1.0 max | 63.0 | 2.5 max | 28.0 | ... | ... | ... | ... | 5.0 | 0.05 max | 0.03 V |
| Hastelloy B-2 | N10665 | 1.0 max | 69.0 | 1.0 max | 28.0 | ... | ... | ... | ... | 2.0 max | 0.02 max | ... |
| Hastelloy C | N10002 | 16.5 | 56.0 | ... | 17.0 | 4.5 | ... | ... | ... | 6.0 | 0.15 max | ... |
| Hastelloy C-276 | N10276 | 15.5 | 59.0 | ... | 16.0 | 3.7 | ... | ... | ... | 5.0 | 0.02 max | ... |
| Hastelloy S | ... | 15.5 | 67.0 | ... | 15.5 | ... | ... | ... | 0.2 | 1.0 | 0.02 max | 0.02 La |
| Hastelloy X | N06002 | 22.0 | 49.0 | 1.5 max | 9.0 | 0.6 | ... | ... | 2.0 | 15.8 | 0.15 | ... |
| Inconel 600 | N06600 | 15.5 | 76.0 | ... | ... | ... | ... | ... | ... | 8.0 | 0.08 | 0.25 max Cu |
| Inconel 601 | N06601 | 23.0 | 60.5 | ... | ... | ... | ... | ... | 1.35 | 14.1 | 0.05 | 0.5 max Cu |
| Inconel 617 | ... | 22.0 | 55.0 | 12.5 | 9.0 | ... | ... | ... | 1.0 | ... | 0.07 | ... |
| Inconel 625 | N06625 | 21.5 | 61.0 | ... | 9.0 | ... | 3.6 | 0.2 | 0.2 | 2.5 | 0.05 | ... |
| Inconel 700 | ... | 15.0 | 45.0 | 30.0 | 3.0 | ... | ... | 22 | 3.2 | 1.0 | 0.13 | ... |
| IN 102 | N06102 | 15.0 | 67.0 | ... | 2.9 | 3.0 | 2.9 | 0.5 | 0.5 | 7.0 | 0.06 | 0.005 B, 0.02 Mg, 0.03 Zr |
| Incoloy 901 | N09901 | 12.5 | 42.5 | ... | 6.0 | ... | ... | 2.7 | ... | 36.2 | 0.10 max | ... |
| Inconel 706 | N09706 | 16.0 | 41.5 | ... | ... | ... | ... | 1.75 | 0.2 | 37.5 | 0.03 | 2.9 (Nb + Ta), 0.15 max Cu |
| Inconel 718 | N07718 | 19.0 | 52.5 | ... | 3.0 | ... | 5.1 | 0.9 | 0.5 | 18.5 | 0.08 max | 0.15 max Cu |
| Inconel 751 | ... | 15.5 | 72.5 | ... | ... | ... | 1.0 | 2.3 | 1.2 | 7.0 | 0.05 | 0.25 max Cu |
| Inconel X750 | N07750 | 15.5 | 73.0 | ... | ... | ... | 1.0 | 2.5 | 0.7 | 7.0 | 0.04 | 0.25 max Cu |
| M252 | N07252 | 19.0 | 56.5 | 10.0 | 10.0 | ... | ... | 2.6 | 1.0 | <0.75 | 0.15 | 0.005 B |
| Nimonic 80A | N07080 | 19.5 | 73.0 | 1.0 | ... | ... | ... | 2.25 | 1.4 | 1.5 | 0.05 | 0.10 max Cu |
| Nimonic 90 | N07090 | 19.5 | 55.5 | 18.0 | ... | ... | ... | 2.4 | 1.4 | 1.5 | 0.06 | ... |
| Nimonic 95 | ... | 19.5 | 53.5 | 18.0 | ... | ... | ... | 2.9 | 2.0 | 5.0 max | 0.15 max | +B, +Zr |
| Nimonic 75 | ... | 19.5 | 75.0 | ... | ... | ... | ... | 0.4 | 0.15 | 2.5 | 0.12 | 0.25 max Cu |
| Refractaloy 26 | ... | 18.0 | 38.0 | 20.0 | 3.2 | ... | ... | 2.6 | 0.2 | 16.0 | 0.03 | 0.015 B |
| René 41 | N07041 | 19.0 | 55.0 | 11.0 | 10.0 | ... | ... | 3.1 | 1.5 | <0.3 | 0.09 | 0.01 B |
| René 95 | ... | 14.0 | 61.0 | 8.0 | 3.5 | 3.5 | 3.5 | 2.5 | 3.5 | <0.3 | 0.16 | 0.01 B, 0.05 Zr |
| TD Nickel | ... | ... | 98.0 | ... | ... | ... | ... | ... | ... | ... | ... | 2.0 $ThO_2$ |
| Udimet 500 | N07500 | 19.0 | 48.0 | 19.0 | 4.0 | ... | ... | 3.0 | 3.0 | 4.0 max | 0.08 | 0.005 B |
| Udimet 630 | ... | 17.0 | 50.0 | ... | 3.0 | 3.0 | 6.5 | 1.0 | 0.7 | 18.0 | 0.04 | 0.004 B |
| Udimet 700 | ... | 15.0 | 53.0 | 18.5 | 5.0 | ... | ... | 3.4 | 4.3 | <1.0 | 0.07 | 0.03 B |
| Udimet 710 | ... | 18.0 | 55.0 | 14.8 | 3.0 | 1.5 | ... | 5.0 | 2.5 | ... | 0.07 | 0.01 B |
| Waspaloy | N07001 | 19.5 | 57.0 | 13.5 | 4.3 | ... | ... | 3.0 | 1.4 | 2.0 max | 0.07 | 0.006 B, 0.09 Zr |

**Table 2 Nominal compositions of nickel-base heat-resistant casting alloys covered in this article**

| Alloy designation | Nominal composition, % | | | | | | | | | | |
|---|---|---|---|---|---|---|---|---|---|---|---|
| | C | Ni | Cr | Co | Mo | Fe | Al | B | Ti | W | Zr | Others |
| B-1900 | 0.1 | 64 | 8 | 10 | 6 | . . . | 6 | 0.015 | 1 | . . . | 0.10 | 4 Ta(a) |
| Hastelloy B | 0.10 | . . . | 0.6 | 2.5 max | 28 | 5 | . . . | . . . | . . . | . . . | . . . | . . . |
| Hastelloy C | 0.07 | . . . | 16 | 2.5 max | 17 | 5 | . . . | . . . | . . . | 4 | . . . | . . . |
| IN-100 | 0.18 | 60.5 | 10 | 15 | 3 | . . . | 5.5 | 0.01 | 5 | . . . | 0.06 | 1 V |
| IN-738X | 0.17 | 61.5 | 16 | 8.5 | 1.75 | . . . | 3.4 | 0.01 | 3.4 | 2.6 | 0.1 | 1.75 Ta, 0.9 Nb |
| IN-792 | 0.2 | 60 | 13 | 9 | 2.0 | . . . | 3.2 | 0.02 | 4.2 | 4 | 0.1 | 4 Ta |
| Inconel 713C | 0.12 | 74 | 12.5 | . . . | 4.2 | . . . | 6 | 0.012 | 0.8 | . . . | 0.1 | 2 Nb |
| Inconel 718 | 0.04 | 53 | 19 | . . . | 3 | 18 | 0.5 | . . . | 0.9 | . . . | . . . | 0.1 Cu, 5 Nb |
| M-252 | 0.15 | 56 | 20 | 10 | 10 | . . . | 1 | 0.005 | 2.6 | . . . | . . . | . . . |
| MAR-M 200 | 0.15 | 59 | 9 | 10 | . . . | 1 | 5 | 0.015 | 2 | 12.5 | 0.05 | 1 Nb(b) |
| MAR-M 246 | 0.15 | 60 | 9 | 10 | 2.5 | . . . | 5.5 | 0.015 | 1.5 | 10 | 0.05 | 1.5 Ta |
| René 80 | 0.17 | 60 | 14 | 9.5 | 4 | . . . | 3 | 0.015 | 5 | 4 | 0.03 | . . . |
| René 100 | 0.18 | 61 | 9.5 | 15 | 3 | . . . | 5.5 | 0.015 | 4.2 | . . . | 0.06 | 1 V |
| TRW-NASA VIA | 0.13 | 61 | 6 | 7.5 | 2 | . . . | 5.5 | 0.02 | 1 | 6 | 0.13 | 0.4 Hf, 0.5 Nb, 0.5 Re, 9 Ta |
| Udimet 500 | 0.1 | 53 | 18 | 17 | 4 | 2 | 3 | . . . | 3 | . . . | . . . | . . . |
| Udimet 700 | 0.1 | 53.5 | 15 | 18.5 | 5.25 | . . . | 4.25 | 0.03 | 3.5 | . . . | . . . | . . . |
| Grades HW50(c), HW(d) | 0.55 | 60 | 12 | . . . | . . . | 25 | . . . | . . . | . . . | . . . | . . . | 2.5 max Si |
| Grades HX50(c), HX(d) | 0.55 | 66 | 17 | . . . | . . . | 14 | . . . | . . . | . . . | . . . | . . . | 2.5 max Si |

(a) B-1900 + Hf also contains 1.5% Hf. (b) MAR-M 200 + Hf also contains 1.5% Hf. (c) Alloy Casting Institute (ACI) designation per American Society for Testing and Materials (ASTM) specification A608. (d) ACI designation per ASTM specification A297

**Table 3 Compositions of cobalt-base heat-resistant casting alloys covered in this article**

| Alloy designation | Nominal composition, % | | | | | | | | | |
|---|---|---|---|---|---|---|---|---|---|---|
| | C | Co | Cr | Ni | Al | B | Fe | Ta | W | Zr | Others |
| AiResist 13 | 0.45 | 62 | 21 | . . . | 3.4 | . . . | . . . | 2 | 11 | . . . | 0.1 Y |
| AiResist 215 | 0.35 | 63 | 19 | 0.5 | 4.3 | . . . | 0.5 | 7.5 | 4.5 | 0.1 | 0.1 Y |
| Haynes 21 | 0.25 | 64 | 27 | 3 | . . . | . . . | 1 | . . . | . . . | . . . | 5 Mo |
| MAR-M 302 | 0.85 | 58 | 21.5 | . . . | . . . | 0.005 | 0.5 | 9 | 10 | 0.2 | . . . |
| MAR-M 322 | 1.0 | 60.5 | 21.5 | . . . | . . . | . . . | 0.5 | 4.5 | 9 | 2 | 0.75 Ti |
| MAR-M 509 | 0.6 | 54.5 | 23.5 | 10 | . . . | . . . | . . . | 3.5 | 7 | 0.5 | 0.2 Ti |
| NASA Co-W-Re | 0.40 | 67.5 | 3 | . . . | . . . | . . . | . . . | . . . | 25 | 1 | 2 Re, 1 Ti |
| WI-52 | 0.45 | 63.5 | 21 | . . . | . . . | . . . | 2 | . . . | 11 | . . . | 2 Nb + Ta |
| X-40 | 0.50 | 57.5 | 22 | 10 | . . . | . . . | 1.5 | . . . | 7.5 | . . . | 0.5 Mn, 0.5 Si |

**Table 4 Compositions of iron-base heat-resistant casting alloys covered in this article**

| ACI designation | UNS number | ASTM specifications(a) | Composition, %(b) | | | |
|---|---|---|---|---|---|---|
| | | | C | Cr | Ni | Si (max) |
| HC | J92605 | A297, A608 | 0.50 max | 26–30 | 4 max | 2.00 |
| HD | J93005 | A297, A608 | 0.50 max | 26–30 | 4–7 | 2.00 |
| HE | J93403 | A297, A608 | 0.20–0.50 | 26–30 | 8–11 | 2.00 |
| HF | J92603 | A297, A608 | 0.20–0.40 | 19–23 | 9–12 | 2.00 |
| HH | J93503 | A297, A608 | 0.20–0.50 | 24–28 | 11–14 | 2.00 |
| HI | J94003 | A297, A567, A608 | 0.20–0.50 | 26–30 | 14–18 | 2.00 |
| HK | J94224 | A297, A351, A567, A608 | 0.20–0.60 | 24–28 | 18–22 | 2.00 |
| HL | J94604 | A297, A608 | 0.20–0.60 | 28–32 | 18–22 | 2.00 |
| HN | J94213 | A297, A608 | 0.20–0.50 | 19–23 | 23–27 | 2.00 |
| HP | . . . | A297 | 0.35–0.75 | 24–28 | 33–37 | 2.00 |
| HT | J94605 | A297, A351, A567, A608 | 0.35–0.75 | 13–17 | 33–37 | 2.50 |
| HU | . . . | A297, A608 | 0.35–0.75 | 17–21 | 37–41 | 2.50 |

(a) ASTM designations are the same as ACI designations. (b) Rem Fe in all compositions. Manganese content: 0.35 to 0.65% for HA, 1% for HC, 1.5% for HD, and 2% for the other alloys. Phosphorus and sulfur contents: 0.04% max for all but HP-50WZ. Molybdenum is intentionally added only to HA, which has 0.90 to 1.20% Mo; maximum for other alloys is set at 0.5% Mo. HH also contains 0.2% max N.

because of the difficulty and expense involved in machining the heat-resistant alloys.

In discussing the machining of heat-resistant alloys, this article classifies the alloys into groups (Table 6) that are used in presenting the nominal feeds and speeds for traditional machining processes. The section "Nontraditional Machining Processes" in this Volume discusses several other machining methods that are useful with difficult-to-machine materials such as heat-resistant alloys.

## Machining Comparisons

Heat-resistant alloys are difficult to machine because of the very characteristics that make them good high-temperature materials. The iron-base alloys, some of which can be properly designated as stainless steels (Table 7), usually machine more easily than the nickel-base and cobalt-base alloys under similar conditions of heat treatment. However, the iron-base alloys do present chip-breaking and disposal problems, which often require special tool geometries.

The nickel-base and cobalt-base alloys have several characteristics in common that contribute to difficulty in machining cost. These characteristics are:

- High shear strength, which causes high forces at the cutting edges of tools
- High capacity for work hardening
- Presence of hard, abrasive intermetallic compounds in the microstructure
- Low thermal conductivity, resulting in heat concentration in the cutting area

The mechanically alloyed products are included in the best-fit generic groups of Table 6. The $Y_2O_3$ dispersion has no significant effect on machinability, and the mechanically alloyed products can be machined by practices appropriate for wrought alloys of similar composition and hardness. Alloy MA 754 (277 HB) appears to have better machining characteristics than Inconel 718 and A-286, but not as good as the 410 stainless steels. Alloy MA 956 (270 HB) appears to have better machining characteristics than alloy MA 754, comparable to 410 stainless steel, but not as good as conventional ferritic stainless steels. The machining characteristics of alloy MA 6000 (450 HB) appear similar to those of Udimet 700, but the alloy is listed in the Ni-Wrt 3 group for preliminary consideration of speeds and feeds.

**Table 5 Composition (wt%) of mechanically alloyed products**

| Alloy | Composition, wt % Ni | Fe | Cr | Al | Ti | Ta | W | Mo | Zr | B | $Y_2O_3$ |
|---|---|---|---|---|---|---|---|---|---|---|---|
| Inconel MA 754 | rem | ... | 20 | 0.3 | 0.5 | ... | ... | ... | ... | ... | 0.6 |
| Inconel MA 6000 | rem | ... | 15 | 4.5 | 2.5 | 2.0 | 4.0 | 2.0 | 0.15 | 0.01 | 1.1 |
| Incoloy MA 956 | ... | rem | 20 | 4.5 | 0.5 | ... | ... | ... | ... | ... | 0.5 |

**Table 6 Group designations used in tables of nominal feeds and speeds**

| Group(a) | Alloys |
|---|---|
| Ni Wrt 1 | Incoloy 901, Incoloy 903, Inconel 617, Inconel 625, Inconel 706, Inconel 718, Inconel X-750, Inconel 751, M252, Nimonic 75, Nimonic 80A, Waspaloy, Inconel MA 754(b) |
| Ni Wrt 2 | Astroloy IN-102, Inconel 700, Nimonic 90, Nimonic 95, René 41, René 63, Udimet 500, Udimet 700, Udimet 710 |
| Ni Wrt 3 | René 95, René 77, Inconel MA 6000(b) |
| Ni Wrt 4 | Hastelloy B, Hastelloy B-2, Hastelloy C, Hastelloy C-276, Hastelloy S, Hastelloy X, Inconel 600, Inconel 601, Refractoloy 26, Udimet 630 |
| Ni Wrt 5 | TD-nickel |
| Ni Cast 1 | Hastelloy B, Hastelloy C, ASTM A297 grades HW and HX, ASTM A608 grades HW50, HX50 |
| Ni Cast 2 | B-1900, IN 100 (René 100), IN 738, IN 792, Inconel 713C, Inconel 718, M252, MAR-M200, MAR-M246, René 80, TRW VI A, Udimet 500, Udimet 700 |
| Co Wrt | AiResist 213, Haynes 25 (L 605), Haynes 188, J-1570, S-816 |
| Co Cast | AiResist 13, AiResist 215, Haynes 21, MAR-M302, MAR-M322, MAR-M509, NASA Co-W-Re, WI-52 |
| Fe Wrt | A-286, Discaloy, Incoloy 800, Incoloy 801, Incoloy 802, N-155, V-57, W-545, 16-25-6, 19-9DL, Incoloy MA 956(b) |
| Fe Cast 1 | ASTM A297 grade HC |
| Fe Cast 2 | ASTM 351 grades HK-30, HK-40, HT-30 |
| Fe Cast 3 | ASTM A297 grades HD, HE, HF, HH, HI, HK, HL, HN, HP, HT, HU; ASTM A608 grades HD50, HE35, HF30, HH30, HH33, HI35, HK30, HK40, HL30, HL40, HN40, HT50, HU50 |

(a) Ni Wrt, nickel-base wrought alloy; Ni Cast, nickel-base casting alloy; Co Wrt, cobalt-base wrought alloy; Co Cast, cobalt-base casting alloy; Fe Wrt, iron-base wrought alloy; Fe Cast, iron-base casting alloy. (b) In the tables on speeds and feeds, the mechanically alloyed products are sometimes listed separately. Otherwise the grouping and hardness of the mechanically alloyed products provides nominal feeds and speeds.

**Table 7 Wrought heat-resistant stainless steels**

| Designation | UNS number |
|---|---|
| Ferritic stainless steels | |
| 405 | S40500 |
| 406 | ... |
| 409 | S40900 |
| 430 | S43000 |
| 434 | S43400 |
| 439 | S43027 |
| 18 SR | ... |
| 18Cr-2Mo | ... |
| 446 | S44600 |
| E-Brite 26-1 | S44627 |
| 26-1Ti | ... |
| 29Cr-4Mo | ... |
| Quenched-and-tempered martensitic stainless steels | |
| 403 | S40300 |
| 410 | S41000 |
| 416 | S41600 |
| 422 | S42200 |
| H-46 | ... |
| Moly Ascoloy | ... |
| Greek Ascoloy | ... |
| Jethete M-152 | ... |
| Almar 363 | ... |
| 431 | S43100 |
| Precipitation-hardening martensitic stainless steels | |
| Custom 450 | ... |
| Custom 455 | ... |
| 15-5 PH | S15500 |
| 17-4 PH | S17400 |
| PH 13-8 Mo | S13800 |
| Precipitation-hardening semiaustenitic stainless steels | |
| AM-350 | S35000 |
| AM-355 | S35500 |
| 17-7 PH | S17700 |
| PH 15-7 Mo | S15700 |
| Austenitic stainless steels | |
| 304 | S30400 |
| 304L | S30403 |
| 304N | S30451 |
| 309 | S30900 |
| 310 | S31000 |
| 316 | S31600 |
| 316L | S31603 |
| 316N | S31651 |
| 317 | S31700 |
| 321 | S32100 |
| 347 | S34700 |
| 19-9 DX | K63199 |
| 202 | S20200 |
| 216 | S21600 |
| 21-6-9 | S21900 |
| Nitronic 32 | ... |
| Nitronic 33 | ... |
| Nitronic 50 | ... |
| Nitronic 60 | ... |
| Carpenter 18-18 Plus | ... |

**The relative machinability** of several heat-resistant alloys, stainless steels, refractory metals, and alloy steels is compared in Fig. 1 in terms of cutting speed in face milling. For the alloy steels and most of the other metals, cutters with carbide inserts were employed. For the nickel alloys, the speeds shown are based on cutters with high-speed steel inserts. The slowest speeds (~6 m/min, or 20 sfm) were required for the three nickel-base alloys, even though the hardness of two of these was only slightly higher (and of one alloy, lower) than that of the 4340 steel (340 HB, in this comparison), which was milled at the highest speed shown, 160 m/min (525 sfm).

In addition to the large variations in machinability among different alloys, the same alloy will differ in its response to different machining operations. This is illustrated by Tables 8 and 9.

In Table 8, the machinability of PH 15-7 Mo and A-286, both precipitation-hardening

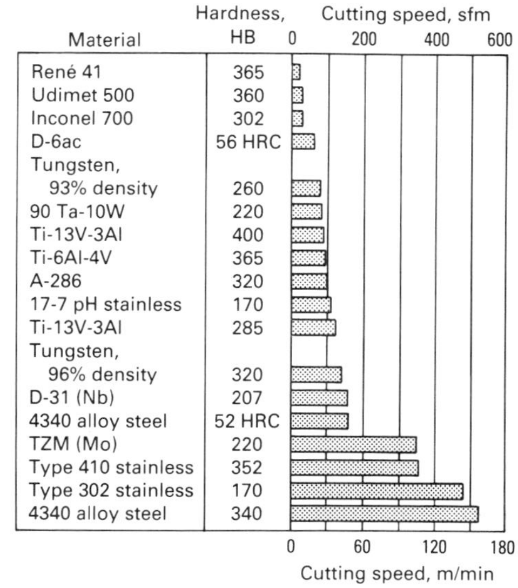


**Fig. 1** Typical speeds in face milling 12 heat-resistant alloys and 6 steels

**Table 8 Comparison of machining characteristics of PH 15-7 Mo (43 HRC) and A-286 (35 HRC) (relative to 4130 steel at 15 HRC as 100%)**

| Operation | Rate of metal removal relative to 4130, % PH 15-7 Mo | A-286 |
|---|---|---|
| Face milling | 10.5 | 8.5 |
| End milling | 9.5 | 25.0 |
| Straddle milling | 13.9 | 11.5 |
| Turning | 11.5 | 15.6 |
| Threading, 32–300 mm (1¼–12 in.) thread | 37.7 | 47.0 |
| Band sawing | 26.7 | 28.6 |
| Drilling, 6 mm (¼ in.) diam | 14.2 | 3.6 |
| Drilling, 13 mm (½ in.) diam | 7.0 | 7.1 |
| Reaming, 6 mm (¼ in.) diam | 38.1 | 20.9 |
| Reaming, 13 mm (½ in.) diam | 17.4 | 22.4 |
| Tapping, 6–710 mm (¼–28 in.) thread | 9.8 | 15.2 |
| Tapping, 13–500 mm (½–20 in.) thread | 22.7 | 14.2 |
| **Average, all operations** | **18.25** | **18.30** |

**Table 9 Comparison of machining characteristics of Inconel X-750 (35 HRC), Haynes 25 (24 HRC), and A-286 (35 HRC) (relative to 4130 steel at 15 HRC as 100%)**

| Operation | Inconel rate of metal removal relative to 4130, % X-750 | HS-25 | A-286 |
|---|---|---|---|
| Face milling | 4.5 | 2.6 | 8.5 |
| End milling | 11.4 | 10.2 | 25.0 |
| Straddle milling | 11.6 | 9.6 | 11.5 |
| Turning | 15.2 | 23.2 | 15.6 |
| Threading, 32–300 mm (1¼–12 in.) thread | 83.0 | 96.0 | 47.0 |
| Band sawing | 22.8 | 19.0 | 28.6 |
| Drilling, 6 mm (¼ in.) diam | 10.6 | 12.0 | 3.6 |
| Drilling, 13 mm (½ in.) diam | 9.3 | 9.4 | 7.1 |
| Reaming, 6 mm (¼ in.) diam | 7.2 | 15.7 | 20.9 |
| Reaming, 13 mm (½ in.) diam | 10.0 | 16.7 | 22.4 |
| Tapping, 6–710 mm (¼–28 in.) thread | 18.9 | 7.4 | 15.2 |
| Tapping, 13–500 mm (½–20 in.) thread | 13.9 | 14.1 | 14.2 |
| **Average, all operations** | **18.2** | **19.7** | **18.3** |

**Table 10 Manufacturing time required to produce the same part from Hastelloy X, René 41, and HS-25, and for total production of each part (see Table 11 for manufacturing ratings)**

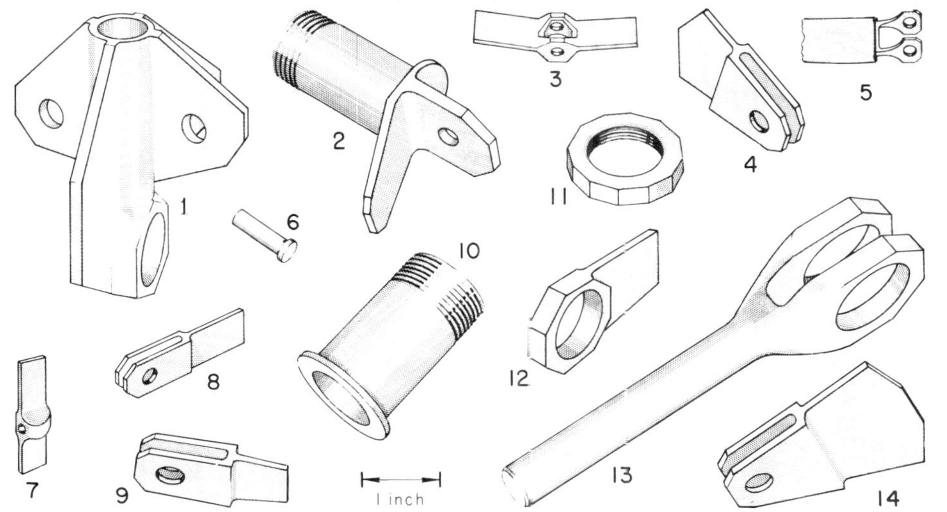


| Part number | Production time per piece, h: Hastelloy X | René 41 | HS-25 | Quantity machined of each alloy | Total production time, h |
|---|---|---|---|---|---|
| 1 | 71.8 | 116.5 | 126.0 | 4 | 1257.2 |
| 2 | 27.0 | 39.3 | 25.0 | 1 | 91.3 |
| 3 | 15.8 | 9.5 | 6.0 | 1 | 31.3 |
| 4 | 11.2 | 10.0 | 15.8 | 4 | 148.0 |
| 5 | 13.0 | 10.5 | 13.5 | 1 | 37.0 |
| 6 | 7.0 | 13.0 | 11.5 | 1 | 31.5 |
| 7 | 10.5 | 14.0 | 21.3 | 1 | 45.8 |
| 8 | 4.6 | 7.3 | 5.3 | 5 | 86.0 |
| 9 | 5.4 | 6.3 | 4.6 | 4 | 65.2 |
| 10 | 4.1 | 7.3 | 6.1 | 4 | 70.0 |
| 11 | 1.6 | 1.8 | 2.0 | 5 | 27.0 |
| 12 | 20.5 | 20.0 | 51.0 | 1 | 91.5 |
| 13 | 18.0 | 25.0 | 19.5 | 1 | 62.5 |
| 14 | 18.5 | 36.0 | 20.5 | 1 | 75.0 |
| **Total** | **229.0** | **316.5** | **328.1** | **34** | **2119.3** |

iron-base alloys, is compared with that of 4130 steel, arbitrarily rated at 100% for each of the operations considered. The 4130 steel was heat treated to 15 HRC (tensile strength: 700 MPa, or 100 ksi). The iron-base heat-resistant alloys were solution treated and aged to 43 HRC (tensile strength: 1400 MPa, or 200 ksi) for PH 15-7 Mo and to 35 HRC (tensile strength: 1120 MPa, or 163 ksi) for A-286.

As indicated in Table 8, the average rates of metal removal for the 2 heat-resistant alloys for the 12 machining operations are almost identical—18.25% for PH 15-7 Mo and 18.30% for A-286. Nevertheless, in comparing the ratings for specific machining operations, several marked differences appear. For example, although the machinability ratings for both alloys are similar in face milling and straddle milling, PH 15-7 Mo is approximately three times as difficult to machine as A-286 in end milling. Other marked differences are recorded for turning, for drilling and reaming the smaller-diameter holes, and for tapping.

Table 9 compares the machining characteristics of three widely used heat-resistant alloys representing the three alloy classes—iron-base (A-286), nickel-base (Inconel X-750), and cobalt-base (Haynes 25). Although the average machinability ratings of the three alloys are closely similar, marked differences occur for several processes—for example, the threading, reaming, end milling, and drilling of 6 mm (¼ in.) diam holes.

**Manufacturing Comparisons.** Machinability ratings based on cutting speed or metal removal rate have limited utility. Ratings that permit estimations of machining costs and shop load for production scheduling are more useful. Such manufacturing ratings reflect the consideration of noncutting time as well as cutting time. Therefore, allowance is made for such time-consuming factors as tool changes and setup adjustments (to obtain the greater rigidity needed in machining a more difficult material).

In a development program for an aerospace vehicle, simulated service tests were conducted on identical structural parts made from two nickel-base alloys (Hastelloy X and René 41) and a cobalt-base alloy (Haynes 25). Although the production of acceptable parts was the primary concern, the machining program afforded an opportunity to compile comparative information on the three alloys. Milling predominated in the machining schedule, but enough drilling, single-point cutting, shaping, and abrasive sawing operations were involved to provide a fair knowledge of the production capabilities of the three alloys in these operations.

**Table 11 Manufacturing ratings for the alloys for which production time is given in Table 10**

| Alloy | Development program: Time, h | Rating, % | Subsequent manufacturing rating, % |
|---|---|---|---|
| Hastelloy X | 531.3 | 100 | 100 |
| René 41 | 773.2 | 69 | 72 |
| HS-25 | 814.8 | 65 | 61 |

Fittings for assembling one body joint, one wing joint, and four wing-to-body joints were produced in 14 different shapes of various degrees of complexity (see illustrations in Table 10). In all, 102 pieces were machined, 34 of each of the three alloys, in approximately 1 man-year (2119 h) of manufacturing time. Table 10 analyzes production time per alloy for each of the 14 parts. Table 11 rates the three alloys on the basis of the development program and again on the basis of subsequent manufacturing experience with the same alloys.

## Turning

Because the heat-resistant alloys retain most of their strength at cutting temperatures, more heat is generated in the shear zone, and greater tool wear occurs for a given cutting speed, than with most other metals. Moreover, because the cutting of heat-resistant alloys requires a larger force (about twice the force for cutting medium-carbon alloy steel in turning operations), tool geometry, tool strength, and/or rigidity of the toolholder are also important concerns.

**Carbide tools** are usually used in turning heat-resistant alloys, although ceramic, coated carbide, cubic boron nitride (CBN), and high-speed steel (HSS) tools are also used. A C-2 grade containing about 94% WC and 6.0% Co is frequently selected for roughing. A C-3 grade is used in finishing.

Standard carbide inserts with positive or negative rakes are suitable for the roughing and finishing of heat-resistant alloys. Negative rake tools, which depend on generated heat to reduce the shear strength of the material, require a larger cutting force than positive rake tools. The cutting force with a standard negative rake insert is about 10% larger than that of a standard positive rake insert.

For rough cutting, it is advantageous to use a "K" land. Such a tool will not cut as cleanly as a sharper-angled tool, but it is stronger at the cutting edge and will better withstand the knocks encountered in removing any rough surfaces.

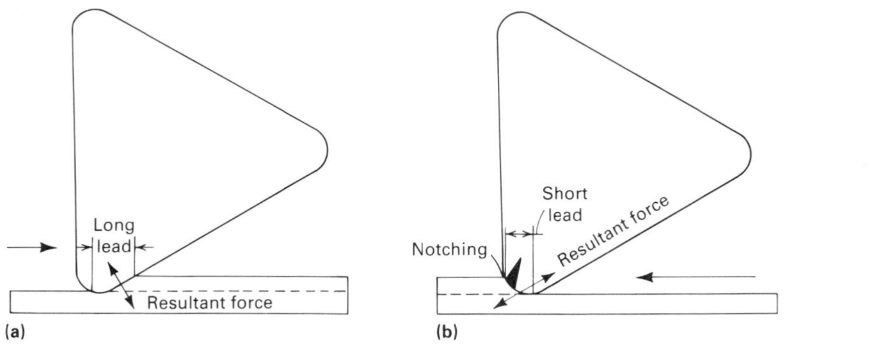


**Fig. 2** Increases in the lead by changing the direction of the feed. (a) A longer lead occurs in the approach on the left, but the push-away effect of this approach may not be feasible on thin sections. (b) A shorter lead results in a larger force and more notching.

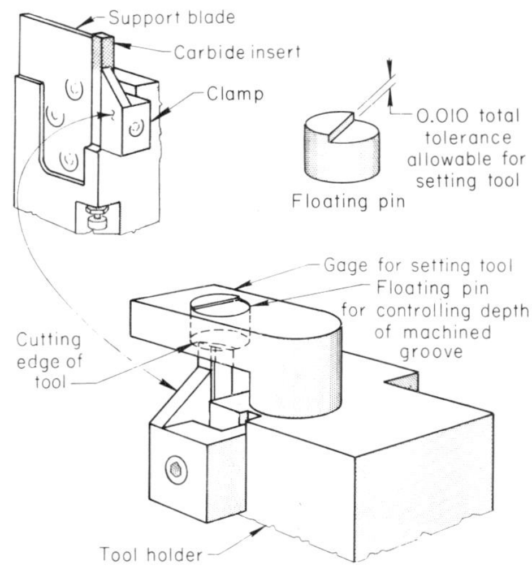


**Fig. 3** Cutoff tool, holder, and tool-setting gage used in plunge grooving. Dimension given in inches

**High-speed steel tools** are seldom used in turning heat-resistant alloys, except for interrupted cuts. In such applications, high-speed steel tools are more practical than carbide tools because of their greater shock resistance. The general-purpose, highly alloyed grades such as T15, M36, or M44 are usually selected despite higher cost. Tools of these grades have longer life than general-purpose grades such as M2 or T1. The geometry of high-speed steel tools in the single-point turning of heat-resistant alloys typically consists of a 0° back rake angle, a 10° side rake angle, and 5° relief angles (end and side clearance angles).

**Ceramic Tools.** The careful application of ceramic tools can improve productivity by allowing higher cutting speeds in the turning of heat-resistant alloys. Speeds generally range from 150 to 200 m/min (500 to 700 sfm) with feeds of about 80% those of carbide.

Depth-of-cut notching is more pronounced when cutting heat-resistant alloys with ceramic tools because of the relatively low fracture toughness of ceramics and because of the high-temperature strength of the heat-resistant alloys. The tougher ceramics (SiAlON and SiC whisker reinforced $Al_2O_3$) exhibit less depth-of-cut notching than $Al_2O_3$-TiC ceramics. However, these tougher ceramics are as chemically inert as the $Al_2O_3$-TiC ceramics (see the section "Properties" in the article "Ceramics" in this Volume), and they are more apt to react chemically with steel or cobalt-base workpieces. Consequently, SiAlON and SiC whisker reinforced $Al_2O_3$ are more effective when cutting wrought nickel-base alloys. The cast nickel-base alloys, because of their grain structure, chip even the tougher ceramics.

Depth-of-cut notching may also preclude the use of delicate geometries (such as triangles and diamond shapes). Round inserts should be used whenever possible. Round inserts are inherently the strongest, although other shapes can be used when the depth of cut does not exceed about two-thirds the nose radius. Some of the tougher ceramics are applicable in square styles and perform better in heavy cuts (where the notch is in a thicker cross section).

When notching is a severe problem (primarily in rough cuts on forgings with surface scale), the lead angle and tool path are also important considerations. Longer lead angles reduce tool pressure, reduce work hardening, and improve surface finish. Figure 2 shows how feeding the tool in the opposite direction changes the lead angle. The cutter path should also be carefully controlled so that chips wedged in the notch area do not put the insert in tension. The use of cutting fluids is also desirable whenever practical.

Roughing cuts of wrought nickel-base alloys when scale and runout are present are best made with the tougher SiAlON ceramics at approximately 150 m/min (500 sfm). Early experience with SiC whisker reinforced $Al_2O_3$ ceramics indicates an increase in toughness but a more severe notching problem. Both of these ceramics also have a higher thermal shock resistance than $Al_2O_3$-TiC ceramics, which thus promotes the use of cutting fluids. Semifinish and finish turning operations can tolerate less insert toughness and can be done with $Al_2O_3$-TiC ceramics.

**Coated carbides** yield small increases in the metal removal rate when cutting the iron-base alloys. In general, metal removal rates can be increased only about 25%, and higher tool costs limit their application.

Coated carbide tools have not yet been proved effective in significantly increasing the metal removal rates of fully aged nickel-base alloys. The limitation is the ability of the substrates to resist deformation at substantially higher cutting temperatures, regardless of the coating materials used.

**Cubic boron nitride tools** are used when turning the harder nickel-base (wrought and cast) and cobalt-base cast alloys. The article "Ultrahard Tool Materials" in this Volume discusses CBN tools in more detail.

**Tool-holding devices** must be given consideration equal to that of tool design when heat-resistant alloys are being turned. A fivefold difference in tool life could result from variations in tool positioning. Holders designed for specific operations are described in the following two examples.

**Example 1: Mechanical Toolholder and Tool-Setting Gage for Plunge Cutting Tool.** To reduce the time required for accurately positioning a brazed carbide tool used for the close-tolerance (0.25 mm, or 0.010 in., total) plunge grooving of a 6.4 mm (0.250 in.) wide A-286 weldment flange, the mechanical holder and tool illustrated in Fig. 3 (ordinarily used for cutting off) were employed. A tool-setting gage of the flush-pin type helped position the tool accurately. The ease of inserting the tools, together with the positive positioning afforded by the gage, resulted in a decrease of tool-setting time from 30 min per tool to less than 5 min per tool. Tool life also increased by 20% because of the mechanical toolholder.

**Example 2: Tool for Turning a Radius on Waspaloy Turbine Blades.** A special tool and holder (Fig. 4a) were designed for turning a radius on turbine blades made of Waspaloy (32 to 38 HRC). The geometry of the tool, which was made of M3 or T15 high-speed steel, is given in Fig. 4(a). A special tool post functioned with a cam-actuated tool slide. To turn the radius, the blade was clamped at the serrated butt with a conforming-serration holder. After turning, the surface finish was 2.0 to 2.5 μm (80 to 100 μin.).

**Cutoff Tool.** In cutting off Inconel 600 tubing having a 75 mm (3 in.) diameter and a 9.5 mm (⅜ in.) wall thickness, the cast cobalt-base alloy tool shown in Fig. 4(b) provided excellent surface finish and had a minimum life of 40 pieces per grind. The principal features of this tool are the 4° side relief on each side and the curved nose. The tool was used at a speed of 9 m/min (30 sfm) and a feed of 0.1 mm/rev (0.004 in./rev), along with a 1 to 20 mixture of soluble oil and water as the cutting fluid.

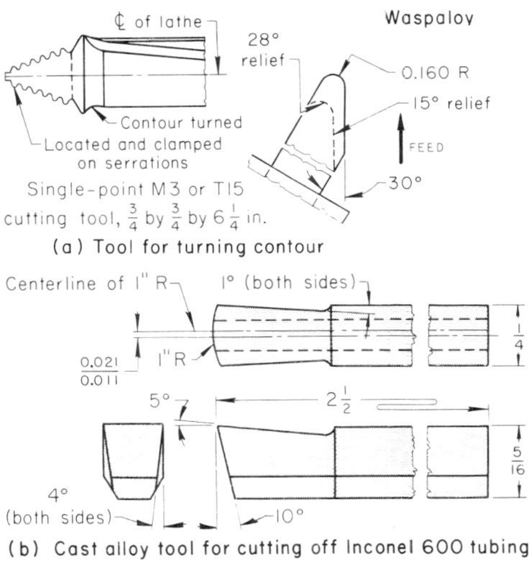


**Fig. 4** Tool and holder (a) to turn radius on blade. (b) Cutoff tool. Dimensions given in inches

**Speeds and feeds** for turning heat-resistant alloys with high-speed steel and carbide box or single-point tools are given in Tables 12 and 13. Speeds and feeds for typical applications of CBN and coated carbide tools are given in Table 14. Tool life usually increases as feed decreases, but both feed and depth of cut must be great enough to avoid glazing.

**Cutting Fluid.** Water-soluble oils in mixtures of 1 part oil with 20 to 40 parts water are most frequently used in turning heat-resistant alloys. Water-base chemical emulsions have also proved acceptable. Supplying a constant flood of cutting fluid to the cutting area is frequently more important than the composition of the fluid.

Sulfurized or chlorinated cutting oils, applied straight or diluted 1 to 1 with low-viscosity mineral oil, are used in some applications. Diluting the cutting oil with mineral oil permits better mobility (and cooling) without seriously impairing the properties of these chemically active oils. Active cutting oils are preferred to soluble oils when surface finishes are critical and when high-speed steel cutting tools are being used.

If sulfurized or chlorinated oil is used as a cutting fluid, the workpieces must be thoroughly cleaned before heat treatment or high-temperature service. Serious damage to workpieces during heating cycles may result if any residue remains.

**Effect of Alloy Composition on Conditions for Turning.** In general, when an alloy is more difficult to machine, the microstructure has a greater impact on machinability. This is especially true for the nickel-base materials.

## Boring and Trepanning

Heat-resistant alloys are bored by methods similar to those used for turning, al-

**Table 12 Nominal feeds and speeds for turning heat-resistant alloys with high-speed steel (HSS) and carbide box or single-point tools**

| Alloy group from Table 6 | Condition | Hardness, HB | Roughing speeds, m/min (sfm)(a) HSS(c) | Roughing: Carbide, brazed(d) | Roughing: Carbide, indexable(d) | Finishing speeds, m/min (sfm)(b) HSS(e) | Finishing: Carbide, brazed(f) | Finishing: Carbide, indexable(f) |
|---|---|---|---|---|---|---|---|---|
| Ni Wrt 1 ...... | Annealed or solution treated | 200–300 | 6 (20) | 15(g) (50)(g) | 18(g) (60)(g) | 8 (25) | 27 (90) | 30 (100) |
| Ni Wrt 1 ..... | Solution treated and aged | 300–400 | 5 (15) | 12(g) (40)(g) | 15(g) (50)(g) | 8 (25) | 24 (80) | 29 (95) |
| Ni Wrt 2 ..... | Solution treated | 225–300 | 3.6 (12) | 12 (40) | 17 (55) | 5 (15) | 21 (70) | 24 (80) |
| Ni Wrt 2 ..... | Solution treated and aged | 300–400 | 3 (10) | 11(g) (35)(g) | 15(g) (50)(g) | 3.6 (12) | 20 (65) | 23 (75) |
| Ni Wrt 3 ..... | Solution treated | 275–390 | 3.6 (12) | 8 (25) | 11 (35) | 5 (15) | 14 (45) | 15 (50) |
| Ni Wrt 3 ..... | Solution treated and aged | 400–475 | 3 (10) | 6 (20) | 9 (30) | 3.6 (12) | 12 (40) | 14 (45) |
| Ni Wrt 4 ..... | Annealed or solution treated | 140–220 | 6 (20) | 21(g) (70)(g) | 24(g) (80)(g) | 8 (25) | 30 (100) | 35 (115) |
| Ni Wrt 4 ..... | Cold drawn or aged | 240–310 | 5 (15) | 12(g) (40)(g) | 15(g) (50)(g) | 6 (20) | 24 (80) | 27 (90) |
| Ni Wrt 5 ..... | As-rolled | 180–200 | 18(h) (60)(h) | 53(g) (175)(g) | 60(g) (200)(g) | 30 (100) | 84 (275) | 90 (300) |
| Ni Cast 1..... | As-cast or cast and aged | 200–375 | 3.6 (12) | 11 (35) | 14 (45) | 5 (15) | 23 (75) | 26 (85) |
| Ni Cast 2..... | As-cast or cast and aged | 250–320 | 3.6(i) (12)(i) | 9 (30) | 11 (35) | 5 (15) | 14 (45) | 18 (60) |
| Ni Cast 2..... | As-cast or cast and aged | 320–425 | 3.0(i) (10)(i) | 8 (25) | 9 (30) | 3.6 (12) | 12 (40) | 15 (50) |
| Co Wrt....... | Solution treated | 150–230 | 6 (20) | 14 (45) | 17 (55) | 8 (25) | 24 (80) | 27 (90) |
| Co Wrt....... | Solution treated and aged | 270–320 | 5 (15) | 12 (40) | 15 (50) | 6 (20) | 21 (70) | 24 (80) |
| Co Cast ...... | As-cast or cast and aged | 220–290 | 3.6 (12) | 12(j) (40)(j) | 15(j) (50)(j) | 5 (15) | 15 (50) | 18 (60) |
| Co Cast ...... | As-cast or cast and aged | 290–425 | 3.0(i) (10)(i) | 11(k) (35)(k) | 12(k) (40)(k) | 3.6 (12) | 14 (45) | 17 (55) |
| Fe Wrt....... | Solution treated | 180–230 | 11 (35) | 30 (100) | 37 (120) | 14 (45) | 49 (160) | 58 (190) |
| Fe Wrt....... | Solution treated and aged | 250–320 | 9 (30) | 26 (85) | 30 (100) | 12 (40) | 44 (145) | 52 (170) |
| Fe Cast 1..... | Annealed | 135–185 | 23(l)(m) (75)(l)(m) | 69(n) (225)(n) | 76(n) (250)(n) | 37(o)(m) (120)(o)(m) | 105(o)(p) (305)(o)(p) | 115(o)(p) (375)(o)(p) |
| Fe Cast 2..... | Annealed or normalized | 135–185 | 15(l)(m) (50)(l)(m) | 53(l) (175)(l) | 60(l) (200)(l) | 24(o)(m) (80)(o)(m) | 84(o) (275)(o) | 90(o) (300)(o) |
| Fe Cast 3..... | As-cast | 160–210 | 12(l)(m) (40)(l)(m) | 46(l) (150)(l) | 53(l) (175)(l) | 21(o)(m) (70)(o)(m) | 69(o) (225)(o) | 76(o) (250)(o) |

(a) Depth of cut of 5 mm (0.2 in.) and a feed rate of 0.25 mm/rev (0.01 in./rev) except for HSS tools (see footnote c) and some carbide applications. (b) Depth of cut of 0.8 mm (0.03 in.) and a feed rate of 0.13 mm/rev (0.005 in./rev) except as specified in footnotes. (c) Any premium high-speed steel (T15, M33, M41-M47, or S9, S10, S11, S12) except as specified in footnote (m); roughing speeds with HSS tools are with a depth of cut of 2.5 mm (0.1 in.) and a feed rate of 0.18 mm/rev (0.007 in./rev) except as specified in footnotes (h), (i), and (l). (d) Grade C-2 (K30 or M30) except as specified in footnotes. (e) Any premium high-speed steel (T15, M33, M41-M47, or S9, S10, S11, S12) except as specified in footnotes. (f) Grade C-3 (K10 or M10) except as specified in footnotes. (g) With a feed rate of 0.4 mm/rev (0.015 in./rev). (h) With a depth of cut of 5 mm (0.2 in.) and a feed rate of 0.4 mm/rev (0.015 in./rev). (i) With a feed rate of 0.13 mm/rev (0.005 in./rev). (j) With a depth of cut of 2.5 mm (0.1 in.) and a feed rate of 0.18 mm/rev (0.007 in./rev). (k) With a depth of cut of 2.5 mm (0.1 in.) and a feed rate of 0.13 mm/rev (0.005 in./rev). (l) With a depth of cut of 8 mm (0.3 in.) and a feed rate of 0.5 mm/rev (0.02 in./rev). (m) M2 or M3 (S4 or S5) high-speed steel tool. (n) C-6 (P20, K20) carbide grade with a depth of cut of 8 mm (0.3 in.) and a feed rate of 0.75 mm/rev (0.03 in./rev). (o) With a depth of cut of 1 mm (0.04 in.) and a feed rate of 0.18 mm/rev (0.007 in./rev). (p) C-7 (P10, K10) carbide grade. Source: Metcut Research Associates Inc.

**Table 13 Nominal speeds and feeds for single-point turning of mechanically alloyed heat-resistant products**

| Workpiece material and hardness, HB | Depth of cut, mm (in.) | HSS tools: Speed, m/min (sfm) | HSS tools: Feed, mm/rev (in./rev) | HSS tools: Tool material | Carbide tools: Speed, m/min (sfm) Brazed | Carbide tools: Speed, m/min (sfm) Indexable | Carbide tools: Feed, mm/rev (in./rev) | Carbide tools: Tool material grade |
|---|---|---|---|---|---|---|---|---|
| Alloy MA 754, 277 HB | 6.5 (0.250) | 3.5–5.5 (12–18) | 0.25 (0.010) | T-15 | 9–12 (30–40) | 12–18 (40–60) | 0.25 (0.010) | C-2 |
| | 1.3 (0.050) | 4.5–6 (15–20) | 0.20 (0.008) | M-36 | 12–15 (40–50) | 15–30 (50–100) | 0.20 (0.008) | C-2 |
| Alloy MA 956, 270 HB | 6.5 (0.250) | 12–15 (40–50) | 0.75 (0.030) | T-15 | 53–69 (175–225) | 61–76 (200–250) | 0.5 (0.020) | C-6 |
| | 1.3 (0.050) | 18–21 (60–70) | 0.25 (0.010) | M-36 | 69–84 (225–275) | 76–91 (250–300) | 0.2 (0.008) | C-8 |
| Alloy MA 6000, 450 HB | 1.25–2.5 (0.05–0.1) | 3–4.5 (10–15) | 0.1–0.3 (0.004–0.012) | T-15 | 8–15 (25–50) | 9–18 (30–60) | 0.1–0.3 (0.004–0.012) | C-2 |
| | 0.125–0.25 (0.005–0.01) | 3–4.5 (10–15) | 0.08–0.15 (0.003–0.006) | M-36 | 8–15 (25–50) | 9–18 (30–60) | 0.08–0.15 (0.003–0.006) | C-2 |

**Table 14 Nominal speeds and feeds for single-point turning of heat-resistant alloys with CBN and coated carbide tools**

| | | | Roughing speeds | | | | Finishing speeds | | | |
|---|---|---|---|---|---|---|---|---|---|---|
| | | | CBN tool(a) | | Coated carbide(b) | | CBN tool(c) | | Coated carbide(d) | |
| Alloy group from Table 6 | Alloy condition | Alloy hardness, HB | m/min | sfm | m/min | sfm | m/min | sfm | m/min | sfm |
| Ni Wrt 1 | Solution treated and aged | 300–400 | 135 | 450 | ... | ... | 185 | 600 | ... | ... |
| Ni Wrt 2 | Solution treated and aged | 300–400 | 120 | 400 | ... | ... | 185 | 600 | ... | ... |
| Ni Wrt 3 | Solution treated and aged | 400–475 | 120 | 400 | ... | ... | 185 | 600 | ... | ... |
| Ni Cast 2 | As-cast or cast and aged | 320–425 | 120 | 400 | ... | ... | 185 | 600 | ... | ... |
| Co Cast | As-cast or cast and aged | 290–425 | 120 | 400 | ... | ... | 185 | 600 | ... | ... |
| Fe Cast 1 | Annealed | 135–185 | ... | ... | 100(e) | 325(e) | ... | ... | 145(f) | 475(f) |
| Fe Cast 2 | Annealed or normalized | 135–185 | ... | ... | 76 | 250 | ... | ... | 115 | 375 |
| Fe Cast 3 | As-cast | 160–210 | ... | ... | 69 | 225 | ... | ... | 100 | 325 |

(a) With a depth of cut of 5 mm (0.2 in.) and a feed rate of 0.13 mm/rev (0.005 in./rev). (b) CC-2 grade with a depth of cut of 8 mm (0.3 in.) and a feed rate of 0.5 mm/rev (0.02 in./rev). (c) With a depth of cut 0.8 mm (0.03 in.) and a feed rate of 0.075 mm/rev (0.003 in./rev). (d) CC-3 grade with a depth of cut of 1 mm (0.04 in.) and a feed rate of 0.18 mm/rev (0.007 in./rev). (e) CC-6 grade. (f) CC-7 grade

**Table 15 Nominal feeds and speeds for boring heat-resistant alloys with HSS and carbide tools**

| | | | Rough boring(a) | | | Finish boring(b) | | |
|---|---|---|---|---|---|---|---|---|
| Alloy group from Table 6 | Condition | Alloy hardness, HB | HSS, m/min (sfm)(c) | Carbide, brazed, m/min (sfm)(d) | Carbide, indexable, m/min (sfm)(d) | HSS, m/min (sfm)(c) | Carbide, brazed, m/min (sfm)(e) | Carbide, indexable, m/min (sfm)(e) |
| Ni Wrt 1 | Annealed or solution treated | 200–300 | 3 (10) | 12 (40) | 14 (45) | 8 (25) | 21 (70) | 26 (85) |
| Ni Wrt 1 | Solution treated and aged | 300–400 | 3 (10) | 11 (35) | 12 (40) | 8 (25) | 20 (65) | 24 (80) |
| Ni Wrt 2 | Solution treated | 225–300 | 2 (7) | 9 (30) | 11 (35) | 5 (15) | 18 (60) | 21 (70) |
| Ni Wrt 2 | Solution treated and aged | 300–400 | 1.5 (5) | 9 (30) | 11 (35) | 5 (15) | 18 (60) | 23 (75) |
| Ni Wrt 3 | Solution treated | 275–390 | 1.8 (6) | 6 (20) | 8 (25) | 3.6 (12) | 12 (40) | 14 (45) |
| Ni Wrt 3 | Solution treated and aged | 400–475 | 1.5 (5) | 5 (15) | 6 (20) | 3 (10) | 11 (35) | 12 (40) |
| Ni Wrt 4 | Annealed or solution treated | 140–220 | 3 (10) | 12 (40) | 15 (50) | 8 (25) | 26 (85) | 30 (100) |
| Ni Wrt 4 | Cold drawn or aged | 240–310 | 2.4 (8) | 9 (30) | 11 (35) | 6 (20) | 18 (60) | 21 (70) |
| Ni Wrt 5 | As-rolled | 180–200 | 15 (50) | 50 (165) | 59 (195) | 29 (95) | 67 (220) | 79 (260) |
| Ni Cast 1 | As-cast or cast and aged | 200–375 | 2 (7) | 9 (30) | 11 (35) | 5 (15) | 15 (50) | 18 (60) |
| Ni Cast 2 | As-cast or cast and aged | 250–320 | 2 (7) | 5 (15) | 6 (20) | 5 (15) | 12 (40) | 15 (50) |
| Ni Cast 2 | As-cast or cast and aged | 320–425 | 1.5 (5) | 3 (10) | 5 (15) | 3.6 (12) | 12 (40) | 14 (45) |
| Co Wrt | Solution treated | 180–230 | 3 (10) | 9 (30) | 11 (35) | 8 (25) | 18 (60) | 21 (70) |
| Co Wrt | Solution treated and aged | 270–320 | 2 (7) | 9 (30) | 11 (35) | 6 (20) | 15 (50) | 18 (60) |
| Co Cast | As-cast or cast and aged | 220–290 | 1.8 (6) | 5 (15) | 6 (20) | 5 (15) | 12 (40) | 15 (50) |
| Co Cast | As-cast or cast and aged | 290–425 | 1.5(f) (5)(f) | 5(f) (15)(f) | 6(f) (20)(f) | 3.6 (12) | 9 (30) | 11 (35) |
| Fe Wrt | Solution treated | 180–230 | 5 (15) | 21 (70) | 26 (85) | 14 (45) | 40 (130) | 47 (155) |
| Fe Wrt | Solution treated and aged | 250–320 | 3 (10) | 18 (60) | 21 (70) | 12 (40) | 38 (125) | 44 (145) |
| Fe Cast 1 | Annealed | 135–185 | 24(g)(h) (80)(g)(h) | 58(h)(i) (190)(h)(i) | 69(h)(i) (225)(h)(i) | 37(g)(j) (120)(j)(g) | 84(j)(k) (275)(j)(k) | 100(j)(k) (325)(j)(k) |
| Fe Cast 2 | Annealed or normalized | 135–185 | 15(g)(h) (50)(g)(h) | 44(h) (145)(h) | 53(h) (175)(h) | 24(g)(j) (80)(j)(g) | 67(j) (220)(j) | 79(j) (260)(j) |
| Fe Cast 3 | As-cast | 160–210 | 14(g)(h) (45)(g)(h) | 38(h) (125)(h) | 46(h) (150)(h) | 21(g)(j) (70)(j)(g) | 58(j) (190)(j) | 69(j) (225)(j) |

(a) Depth of cut of 2.5 mm (0.1 in.) and a feed rate of 0.15 mm/rev (0.006 in./rev) except as specified in footnotes. (b) Depth of cut of 0.25 mm (0.01 in.) and a feed rate of 0.05 mm/rev (0.002 in./rev) except as specified in footnotes. (c) Any premium high-speed steel T15, M33, M41–M47, or S9, S10, S11, S12. (d) Grade C-2. (e) Grade C-3. (f) With a feed rate of 0.10 mm/rev (0.004 in./rev). (g) M2, M3 high-speed steel. (h) With a feed rate of 0.3 mm/rev (0.012 in./rev). (i) Grade C-6 carbide tool. (j) With a feed rate of 0.075 mm/rev (0.003 in./rev). (k) Grade C-8 carbide tool. Source: Metcut Research Associates Inc.

**Table 16 Nominal speeds and feeds for typical applications of CBN and coated carbide tools in the boring of heat-resistant alloys**

| | | | Rough boring speeds(a) | | | | Finish boring speeds(b) | | | |
|---|---|---|---|---|---|---|---|---|---|---|
| | | | CBN(c) | | Coated carbide(d) | | CBN | | Coated carbide | |
| Alloy group from Table 6 | Condition | Hardness, HB | m/min | sfm | m/min | sfm | m/min | sfm | m/min | sfm |
| Ni Wrt 1 | Solution treated and aged | 300–400 | 120 | 400 | ... | ... | 185 | 600 | ... | ... |
| Ni Wrt 2 | Solution treated and aged | 300–400 | 120 | 400 | ... | ... | 185 | 600 | ... | ... |
| Ni Wrt 3 | Solution treated and aged | 400–475 | 120 | 400 | ... | ... | 185 | 600 | ... | ... |
| Ni Cast 2 | As-cast or cast and aged | 320–425 | 120 | 400 | ... | ... | 185 | 600 | ... | ... |
| Co Cast | As-cast or cast and aged | 290–425 | 120 | 400 | ... | ... | 185 | 600 | ... | ... |
| Fe Cast 1 | Annealed | 135–185 | ... | ... | 90(e) | 300(e) | ... | ... | 120(f) | 400(f) |
| Fe Cast 2 | Annealed or normalized | 135–185 | ... | ... | 64(g) | 210(g) | ... | ... | 100(h) | 325(h) |
| Fe Cast 3 | As-cast | 160–210 | ... | ... | 58(g) | 190(g) | ... | ... | 85(h) | 280(h) |

(a) Depth of cut of 2.5 mm (0.10 in.). (b) Depth of cut of 0.25 mm (0.010 in.) and a feed rate of 0.075 mm/rev (0.003 in./rev). (c) With a feed rate of 0.2 mm/rev (0.008 in./rev). (d) With a feed rate of 0.3 mm/rev (0.012 in./rev). (e) Carbide grade CC-6 (CP20 or CK20). (f) Carbide grade CC-8 (CP01 or CK01). (g) Carbide grade CC-2 (CK30 or CM30). (h) Carbide grade CC-3 (CK10 or CM10). Source: Metcut Research Associates Inc.

though the speeds and feeds (Tables 15 and 16) must be reduced in most cases because the same cooling and lubricating efficiency cannot be obtained as in cutting. In addition, in boring operations, the cutting tool cannot be held as rigidly as in turning.

The selection and application of tool materials are similar to turning, but tool geometry varies. The end relief angle of boring tools must be varied inversely with the diameter being bored.

Trepanning has not been extensively used as a method of machining heat-resistant alloys, although iron-base casting alloys have been trepanned in some applications. Limited experience indicates that the speeds, feeds, and tool materials suitable for boring are satisfactory for trepanning under similar conditions. Several cast alloys have been successfully trepanned in the as-cast condition (160 to 210 HB) with cutting tools made of M2 and T5 high-speed steels; speeds of 12 to 15 m/min (40 to 50 sfm) and feeds of 0.13 mm/rev (0.005 in./rev) were used.

## Planing and Shaping

Planing has been done on some large heat-resistant alloy castings, but is seldom done on wrought heat-resistant products. Nominal speeds for the rough and finish planing of the casting alloys are given in

**Table 17 Nominal speeds for planing heat-resisting alloys**

| Alloy group from Table 6 | Hardness, HB | Speed: HSS tools(a) m/min | sfm | Carbide tools(b) m/min | sfm |
|---|---|---|---|---|---|
| **Rough planing(c)** | | | | | |
| Fe Cast 1 | 160–210 | 9 | 30 | 58 | 190 |
| Fe Cast 2, Fe Cast 3 | 160–210 | 6 | 20 | 43 | 140 |
| **Finish planing(d)** | | | | | |
| Fe Cast 1 | 160–210 | 14 | 45 | 59 | 195 |
| Fe Cast 2, Fe Cast 3 | 160–210 | 8 | 25 | 46 | 150 |

(a) M2 or T5. (b) Grade C-6. (c) Depth of cut, 13 mm (0.500 in.); feed, 1.5 mm/stroke (0.060 in./stroke). (d) Depth of cut, 0.13 mm (0.005 in.); feed, ¾ the width of square-nose tool

**Table 18 Nominal speeds and rise per tooth for broaching heat-resistant alloys with HSS tools**

| Alloy group from Table 6 | Condition | Hardness, HB | Speed(a) m/min | sfm |
|---|---|---|---|---|
| Ni Wrt 1 | Annealed or solution treated | 200–300 | 2.4 | 8 |
| Ni Wrt 1 | Solution treated and aged | 300–400 | 1.8 | 6 |
| Ni Wrt 2 | Solution treated | 225–300 | 2.4 | 8 |
| Ni Wrt 2 | Solution treated and aged | 300–400 | 1.8 | 6 |
| Ni Wrt 3 | Solution treated | 275–390 | 1.8 | 6 |
| Ni Wrt 3 | Solution treated and aged | 400–475 | ... | ... |
| Ni Wrt 4 | Annealed or solution treated | 140–220 | 3 | 10 |
| Ni Wrt 4 | Cold drawn or aged | 240–310 | 2.4 | 8 |
| Ni Wrt 5 | As-rolled | 180–200 | 6(b) | 20(b) |
| Ni Cast 1 | As-cast or cast and aged | 200–375 | 2.4 | 8 |
| Ni Cast 2 | As-cast or cast and aged | 250–320 | 2.4 | 8 |
| Ni Cast 2 | As-cast or cast and aged | 320–425 | 1.8 | 6 |
| Co Wrt | Solution treated | 180–230 | 2.4 | 8 |
| Co Wrt | Solution treated and aged | 270–320 | 1.8 | 6 |
| Co Cast | As-cast or cast and aged | 220–290 | 2.4 | 8 |
| Co Cast | As-cast or cast and aged | 290–425 | 1.8 | 6 |
| Fe Wrt | Solution treated | 180–230 | 3.6 | 12 |
| Fe Wrt | Solution treated and aged | 250–320 | 3 | 10 |
| Fe Cast 1 | Annealed | 135–185 | 6(c) | 20(c) |
| Fe Cast 2 | Annealed or normalized | 135–185 | 5(b) | 15(b) |
| Fe Cast 3 | As-cast | 160–210 | 5(b) | 15(b) |

(a) With any premium HSS (T15, M33, M41–M47, or S9, S10, S11, S12) broach and a chip load (rise per tooth) of 0.05 mm/tooth (0.002 in./tooth). (b) With M2 or M7 (S4 or S2) HSS broach and a chip load of 0.075 mm/tooth (0.003 in./tooth). (c) With M2 or M7 (S4 or S2) HSS broach and a chip load of 0.005 mm/tooth (0.002 in./tooth). Source: Metcut Research Associates Inc.

Table 17. Workpieces are usually planed without a cutting fluid, but synthetic emulsions are sometimes used.

Shaping tools with +8° side rake, 0 to 3° back rake, 4 to 6° relief angle, and 1.1 to 1.5 mm (0.045 to 0.060 in.) nose radius are suitable for heat-resistant alloys. Ram speeds must be slow; 2 to 4 m/min (7 to 13 sfm) is optimum, using a feed of 0.5 to 0.75 mm/stroke (0.020 to 0.030 in./stroke) for roughing and 0.25 to 0.40 mm/stroke (0.010 to 0.015 in./stroke) for finishing. Depths of cut range from 1.3 to 2.5 mm (0.050 to 0.100 in.) for roughing and 0.4 to 0.75 mm (0.015 to 0.030 in.) for finishing. Sulfur-free chlorinated oil applied with a brush is recommended for use as a cutting fluid in shaping.

## Broaching

Although broaching is one of the more difficult machining operations, it is extensively used on heat-resistant alloys because it is often the only practical method of machining the complex contours of blades, wheels, and related components of gas turbines. The successful broaching of heat-resistant alloys requires careful consideration of broach design, broach material, and technique. The following points should be emphasized:

- Broach design that provides ample clearance for swarf
- Good rigidity of the machine and work combined with adequate power
- Avoidance of the tool edge rubbing against the workpiece
- Careful selection of cutting oil

**Broach Design.** Face (hook) angle, back angle, and gullet shape are important because of the behavior of heat-resistant alloys in shearing and chip formation. The use of short, replaceable broach inserts can provide cost savings as well as better control of surface finish and accuracy.

The pitch of the teeth should be approximately 25% more than that for broaching plain-carbon and low-alloy steels in order to provide the necessary greater chip clearance. The large pitch also decreases the total load by reducing the number of teeth in engagement. For the same desired chip thickness, it is therefore necessary to use longer broaches or more broaches to a set. The front cutting angle (rake) can be increased by a maximum of 15°, promoting a freer chip flow and minimizing the work-hardening tendency. Rubbing contact should be avoided by providing as large a relief angle as possible, consistent with tool strength and support for the cutting edge.

In producing gas-turbine components, a change in work metal is sometimes necessary. Such a change may require revisions in broach design. Examples 7 and 8 in this article describe operations in which broach design was altered because of a change in work metal.

**Broach Material.** High-speed steel is usually used for broaching heat-resistant alloys. The selection of solid broaches versus those with inserted cutting edges depends on the size and design of the broach and on cost. Cost is usually the deciding factor. In many applications (particularly when large broaches are being used), cost can be decreased by using high-speed steel inserts in an alloy steel body. Assuming that the other factors are constant, whether broaches are solid or have inserts will not influence broach performance.

The more highly alloyed grades of high-speed steel, such as T4, T5, and T6, are generally superior in terms of broach wear and life. Although acceptable results can be obtained with an M2 broach for some applications involving iron-base heat-resistant alloys (such as A-286), M3 (class 2) broaches are near the minimum in alloy content (only slightly higher than general-purpose grades) usually considered suitable for broaching heat-resistant alloys.

**Speed and Rise Per Tooth.** Nominal speeds for broaching different groups of heat-resistant alloys are given in Table 18. These values are based on the use of T15 broaches (with four exceptions noted in the table) and sulfurized or sulfochlorinated oil as the cutting fluid. The speeds shown in Table 18 are conservative, particularly for the iron-base alloys. In many applications, speeds of 3 to 5 m/min (10 to 18 sfm) will provide acceptable results when broaching the iron-base alloys. Common practice is to use a speed near the high side of the range for roughing and then to finish at a lower speed.

To obtain acceptable tool life when broaching the nickel-base and cobalt-base alloys, lower speeds are necessary; 1.5 to 3 m/min (5 to 10 sfm) has proved satisfactory for broaching most of these alloys. For example, René 41 (one of the most difficult-to-broach alloys) is successfully broached at 1.8 to 2.4 m/min (6 to 8 sfm); common practice is to rough and finish broach at the same low speed.

The nominal rise per tooth (chip load) for broaching heat-resistant alloys is included in Table 18; the values tabulated are conservative. For rough broaching, values of chip load up to 0.13 mm/tooth (0.005 in./tooth) are often used. For the finishing operation, 0.01 to 0.025 mm/tooth (0.0005 to 0.001 in./tooth) is not uncommon. A rise of less than 0.01 mm/tooth (0.0005 in./tooth) is not recommended, because the broach is likely to burnish rather than cut. Burnishing causes variations in the work metal finish and excessive broach wear.

**Cutting Fluid.** The main difficulty in broaching superalloys usually lies in a fusion buildup on the tip of the tooth and consequent rapid wearing of the edges. A

**Table 19 Details and results of changes in design of broach for spline broaching turbine wheels made of 16-25-6 alloy**

Length of cut was 30.2 to 46.9 mm (1.189 to 1.846 in.).

| Broach change No. | Item | Original design | Improved design |
|---|---|---|---|
| 1(a) | Face angle, degrees | 15 | 18 |
| 2(b) | Tooth pitch, mm (in.) | 13–16 (½–⅝) | 14, 15, 12 (9/16, 19/32, 31/64)(c) |
| 3(d) | Length of teeth, mm (in.) | 3.2 (0.125) | 4.0 (0.156) |
| 4(e) | Gullet radius, mm (in.) | 4.0 (5/32) | 4.0 (5/32) |
| 5(e) | Tooth depth, mm (in.) | 6.3 (0.250) | 6.3 (0.250) |
| 6(f) | Backoff clearance, outside diameter, degrees | 2–½ | 2°; 2°; 1°(c) |
| 7(g) | Number of roughing teeth | 33 | 43 (2° backoff) |
| 8(g) | Number of semifinishing teeth | 32 | 42 (2° backoff) |
| 9(e) | Number of finishing teeth | 6 | 6 (1° backoff) |
| 10(h) | Length of broach, mm (in.) | 1500 (59.0) | 1650 (65.0) |
| 11(j) | Sides of teeth relieved up to 0.8 mm (1/32 in.) of top of teeth, degrees | . . . | 1 ± ¼ backoff(k) |

(a) Enabled bar to cut freely and provided efficient chip curl. (b) Eliminated harmonic rings. (c) Roughing, semifinishing, and finishing, respectively. (d) Increased life of teeth. (e) No change. (f) Reduced drag and gall on broach. (g) Increased number of teeth; reduced rise per tooth. (h) Used full stroke of machine and improved broach design. (j) Eliminated all galling and tears. (k) Full form on sides of all spline teeth up sharp (no land)

**Table 20 Improved tool design for broaching a root form in turbine vanes made of X-40**

| Broach detail | Original | Improved |
|---|---|---|
| Face angle, degrees | 18 | 12 |
| Backoff clearance, degrees | 2 | 4 |
| Pitch, mm (in.) | 7.1 (9/32) | 8.7 (11/32) |
| Tooth depth, mm (in.) | 6 (¼) | 4.8 (3/16) |
| Shear angle, degrees | 5 | 45 |
| Tool life, number of pieces per setup | 300 | 13 000 |

good cutting fluid can help considerably. By rapidly wetting the surfaces in contact, the cutting fluid inhibits welding between the chip and the tool edge.

A flood of sulfochlorinated oil over the area being broached is preferred and in most applications is mandatory for acceptable results. In some applications, cutting fluids similar to thread cutting oil have been used successfully, but the use of such fluids is usually a compromise, especially when broaching nickel-base or cobalt-base alloys. In the application described in Example 3, a change to sulfochlorinated oil improved results in broaching alloy 16-25-6 (ordinarily one of the easier-to-broach heat-resistant alloys).

Oils containing about 1% active sulfur, along with chlorine and synthetic additions, are often used. A plentiful supply of cutting fluid in the area being broached is of equal, if not greater, importance than fluid composition. Preferably, cutting fluid is supplied under pressure up to about 35 kPa (5 psi). To obtain the viscosity required for use in pressure systems, the cutting oil can be diluted with plain mineral oil. A mixture of one part concentrated cutting oil and one part mineral oil has lower viscosity than concentrated cutting oil and is adequate for most applications. Cutting fluids with viscosity higher than 300 SUS are not recommended for broaching. Thorough cleaning of workpieces broached with chemically active oils (sulfurized or chlorinated) is extremely important before heat treating or high-temperature service in order to prevent damage to the workpiece.

**Tool modifications** that improved results in broaching specific alloys and contours are described in the following six examples. Several of these examples describe the broaching of fir-tree slots in turbine blades. However, it is worth noting that fir-tree slots in MA6000 turbine blades have been satisfactorily cut by wire electrodischarge machining followed by abrasive grinding.

**Example 3: Original and Improved Tool Designs for Broaching Splines in 16-25-6 Turbine Wheels.** In broaching splines in turbine wheels of 16-25-6 alloy (32 to 36 HRC), a change of broach design, broach material, and type of cutting fluid resulted in a savings of 67% in broach costs and 576 man-hours per year. In addition, galling, tearing, and undesirable tapering of splines were completely eliminated. Table 19 lists the changes made in broach design, and the footnotes indicate the improvement achieved by each change.

As broach material, T15 (66 to 68 HRC) was substituted for M3 because T15 performed more satisfactorily over a wide range of tool hardness. Because the heavy-duty, high-sulfur thread cutting oil that had been originally used resulted in excessive burning, galling, and buildup and welding of chips, a cutting fluid composed of mineral oil with 1% active sulfur, chlorine, and synthetic additions was substituted. The defects were eliminated with this cutting fluid.

With these changes and the changes in broach design listed in Table 19, a greater number of pieces could be broached per tool sharpening. Thus, the number of setups, as well as setup certifications, was reduced, saving a total of 576 man-hours per year.

**Example 4: Redesign of Tool for Broaching Incoloy 901 Turbine Wheels.** Poor surface finish, severe galling in critical areas, and excessive tool wear were encountered in broaching Incoloy 901 turbine wheels (255 HB min) with a conventional broach made of T5 (64 to 66 HRC). Broaching was required to form 52 two-branch fir-tree contours in a wheel 578 mm (22.750 in.) in outside diameter by 41 mm (1.600 in.) thick. The angle of broaching was 19° 30′.

On the original broach, the teeth had a 13 mm (0.500 in.) pitch, a 15° face angle, and a 2° backoff clearance angle. Finishing teeth were full form and truncated.

An improved broach tooth more compatible with the cutting characteristics of the alloy solved the problem. In the new design, the face angle was increased to 18° and the backoff angle to 4°. Improved chip control enabled the full-form cutting to be postponed to the semifinishing stage, and final sizing could be done by inserted finishing teeth. The finishing teeth had a chip load (rise per tooth) of 0.018 to 0.0075 mm/tooth (0.0007 to 0.0003 in./tooth) and removed the final 0.13 mm (0.005 in.) of metal. Grade T5 high-speed steel was used, and the cutting speed was 3.4 m/min (11 sfm).

**Example 5: Redesign of Tool for Broaching X-40 Turbine Vanes.** Improvement in both tool life and surface finish was obtained in broaching the root form in X-40 turbine vanes at low hardness (10 to 15 HRC) when tools were altered to a large (45°) shear angle and a sharper positive face angle. For this soft cast alloy, chip formation was a problem when a small (5°) shear angle was tried because of the probability of chipping the work at the exit of the cut. Details of the original and improved tools are given in Table 20. The changes in face angle and backoff clearance, and especially the increase in shear angle, almost completely eliminated chipping. The productive life of the high-speed steel broach of the original design averaged 300 pieces per setup. However, after the tooth profile was redesigned, the broach averaged 13 000 pieces per setup. The operation utilized rigid fixturing, pressurized cutting fluid, and a horizontal broaching machine.

**Example 6: Redesigned Broach for N-155 Nozzle-Diaphragm Blades.** Nozzle-diaphragm blades were fabricated from flat-rolled N-155 at 30 to 40 HRC. A form die was used for folding over the sheet to form a hollow blade. Following this operation, the open ends were welded, using an A-286 welding rod. The blade face and weld area were broached to a smooth contour. The blade is illustrated in Fig. 5.

Both the convex and concave sides of the blade were broached. Because the problems encountered with both sides were similar, only the broach used to cut the concave side will be considered.

The broach originally used consisted of two sections, each 330 mm (13 in.) long and containing 48 teeth. All teeth were made

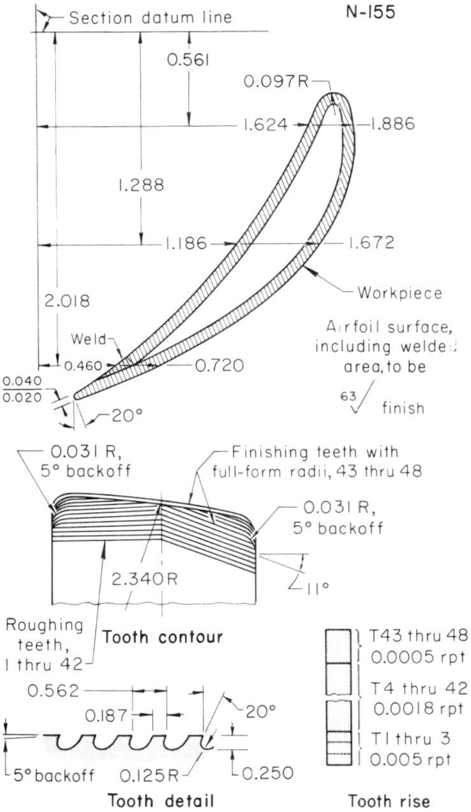


**Fig. 5** Tooth details for redesigned broach used to finish both sides of the hollow blade shown. Dimensions given in inches

with full-form radii and 2° backoff angles. This broach was unsuccessful because the full-form radii did not permit the chips to curl, a condition that resulted in excessive weld buildup on the teeth, tooth breakage, and damaged work. Resharpening was required after each 100 to 125 pieces, and the broach was scrapped after broaching only 1800 pieces.

In a redesigned broach, the roughing teeth were changed from full-form radii to straight teeth cutting on both sides of center on all but the last five teeth. Thus, all but the last five teeth were used primarily for stock removal; the split-angle arrangement provided excellent chip control. The backoff clearance on all teeth was changed to 5°. Tooth details and tooth rise for the redesigned broach are shown in Fig. 5.

The new design provided free cutting action, with a minimum of weld buildup and tooth damage. From 350 to 600 blades could be broached before resharpening was required. Total life of the broach was extended to between 6000 and 9000 pieces.

The broach material was T15 steel (66 to 68 HRC). The machine was a 915 mm (36 in.) vertical broaching machine (11 kW, or 15 hp, hydraulic ram), and the speed was 3.7 m/min (12 sfm) for a cutting length of 100 mm (4 in.). A cutting fluid consisting of mineral oil with 1% active sulfur and chlorine and synthetic additions was used.

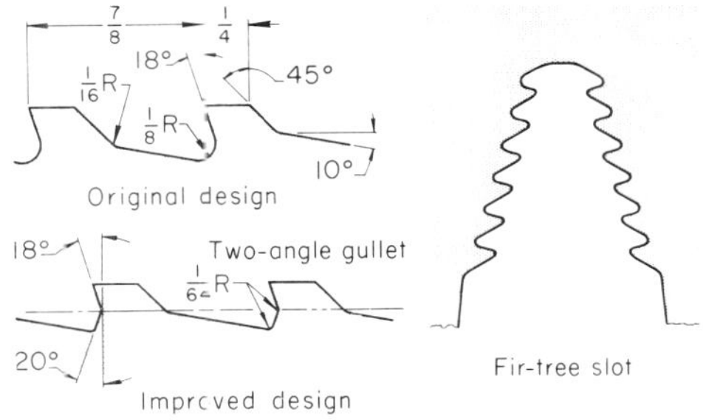


| Detail No. | Length of insert, mm (in.) | Pitch, mm (in.) | Land width, mm (in.) | Rise per tooth, mm (in.) |
|---|---|---|---|---|
| **Roughing(a)** | | | | |
| 1 | 422.2 | 12.7 | 3.96 | 0.1 |
| | (16.625) | (0.500) | (0.156) | (0.004) |
| 2 | 476.2 | 12.4 | 3.96 | 0.1 |
| | (18.750) | (0.490) | (0.156) | (0.004) |
| 3 | 476.2 | 12.4 | 3.96 | 0.1 |
| | (18.750) | (0.490) | (0.156) | (0.004) |
| 4 | 428.6 | 12.4 | 3.96 | 0.1 |
| | (16.875) | (0.490) | (0.156) | (0.004) |
| 5 | 422.2 | 12.7 | 3.96 | 0.1 |
| | (16.625) | (0.500) | (0.156) | (0.004) |
| 6 | 476.2 | 12.4 | 3.96 | 0.1 |
| | (18.750) | (0.490) | (0.156) | (0.004) |
| 7 | 476.2 | 12.4 | 3.96 | 0.1 |
| | (18.750) | (0.490) | (0.156) | (0.004) |
| 8 | 428.6 | 19.5 | 6.35 | 0.17 |
| | (16.875) | (0.767)(b) | (0.250)(b) | (0.0067) |
| **Semifinishing** | | | | |
| Side cutting only, using detail No. 8 | | | | |
| **Finishing(c)** | | | | |
| 1 | 40.00 | 12.7 | 4.76 | 0.046 |
| | (15.750) | (0.500)(d) | (0.1875)(d) | (0.0018)(d) |
| | | 18.0 | 12.7 | 0.015 |
| | | (0.710)(e) | (0.500)(e) | (0.0006)(e) |
| 2 | 40.00 | 22.2 | 6.4 | 0.041 |
| | (15.750) | (0.875)(e) | (0.250)(e) | (0.0016) |
| 3 | 40.00 | 22.2 | 6.4 | 0.046 |
| | (15.750) | (0.875)(e) | (0.250)(e) | (0.0018) |
| 4 | 40.00 | 22.2 | 6.4 | 0.05 |
| | (15.750) | (0.875)(e) | (0.250) | (0.002)(d) |

(a) Rough cutting with eight two-pass broach details; face angle was 10° plus 20°, except for detail 8 (side cutting only), where it was 15°; backoff clearance was 2°. (b) Alternate. (c) Detail 1: top cut, 13 teeth; side cut, 12 teeth. Details 2 and 3: side cut only. Detail 4: side cut, 13 teeth; straight cut, 6 teeth; top cut, 6 teeth; 0.005-in. step on top radius. Face angle was 18° plus 20°, and backoff clearance, 2°. (d) Top cut. (e) Side cut

**Fig. 6** Original broach design for broaching a fir-tree slot in 16-25-6 and redesign for broaching the same slot in A-286. Fir-tree slot is shown at right. Table lists broach details for the improved design. Dimensions in figure given in inches

**Example 7: Redesign of Broach Because of Change in Work Metal From 16-25-6 to A-286.** In finish broaching fir-tree slots (internal broaching) in 50 mm (2 in.) thick, 453.5 mm (17.856 in.) diam turbine wheels made of 16-25-6 (235 to 293 HB), tool life was 4050 slots per regrind. However, tool life was reduced to only 38 slots when A-286 alloy (143 to 202 HB) was substituted as the turbine-wheel material, even though the same tooling setup was used as for broaching the 16-25-6 wheels.

The short tool life was caused by the greater tendency of the chips of A-286 to weld and abrade the tooth face and cutting edge. This welding prevented succeeding

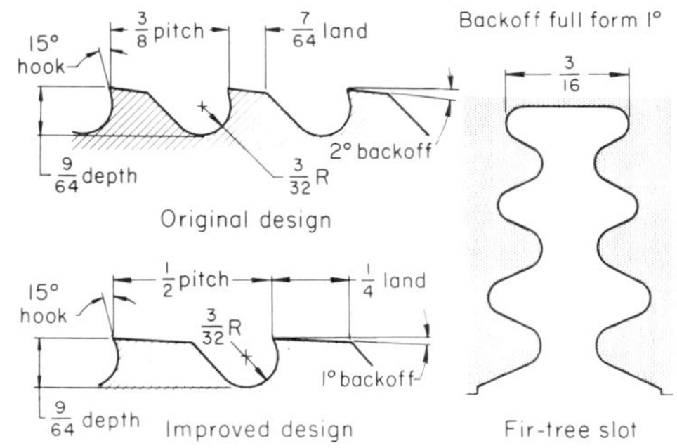


**Fig. 7** Original broach design for broaching fir-tree slot in turbine wheel made from A-286 and redesign for broaching the same slot in René 41. Dimensions given in inches

chips from sliding along the full face of the tooth and thus increased friction and heat. Teeth broke down at 0.9 to 1.3 mm (0.035 to 0.050 in.) from the cutting edges.

To extend tool life, the broach material was changed from M3 to T15. Only one wheel (54 slots) was broached with T15 inserts, and the finish was poor because the chips welded to the cutting edges, again causing greater friction and heat, which in turn increased welding.

By changing from a full-radius gullet to a two-angle gullet (Fig. 6) while retaining T15 as the tool material, parts could be finish broached satisfactorily. The 18° face angle that had blended to a radius at the root of the gullet was changed to a combination of two angles, one of 18° and one of 20°. Redesigning the angles and radii of the T15 broach inserts permitted chips to flow into the gullet with minimum welding on the tooth face and edges, and broach life was extended to 378 slots before resharpening was required. In addition, the surface finish of the turbine wheels was greatly improved because chips no longer interfered with the cutting edges but flowed away, curling neatly to clear the workpiece. Broach breakdown was appreciably reduced.

The two-angle design was also used to advantage in roughing in which a 10°/20° combination was used for all but the last roughing detail and the semifinish detail, which did not present a chip problem. Broach details are listed in the table accompanying Fig. 6.

Speed was 5.5 m/min (18 sfm) for roughing and semifinishing and for finishing detail 1. For finishing details 2, 3, and 4, speed was 3.7 m/min (12 sfm). Cutting fluid was mineral oil with 1% active sulfur, chlorine, and synthetic additions.

**Example 8: Redesign of Broach Because of Change in Work Metal From A-286 to René 41.** Redesigning a broach by increasing pitch length and land width enabled the broaching of 7½ times as many fir-tree slots (internal broaching) in René 41 turbine wheels as with the conventional design. The wheels were first stage and second stage, requiring 119 and 109 slots,

**Table 21 Typical range of hole sizes when drilling heat-resistant alloys with gun, twist, or oil-hole drills**

For holes larger than 50 mm (2 in.) in diameter, spade drills or trepanning tools are needed.

| Alloy group from Table 6 | Condition | Hardness, HB | Range of nominal hole size: Gun drills(a), mm | Gun drills(a), in. | Twist drills(b), mm | Twist drills(b), in. | Oil-hole drills(c), mm | Oil-hole drills(c), in. |
|---|---|---|---|---|---|---|---|---|
| Ni Wrt 1 | Annealed or solution treated | 200–300 | 1.5–50 | 1/16–2 | 3–19 | 1/8–3/4 | 3–38 | 1/8–1 1/2 |
| Ni Wrt 1 | Solution treated and aged | 300–400 | 1.5–50 | 1/16–2 | 3–19 | 1/8–3/4 | 3–38 | 1/8–1 1/2 |
| Ni Wrt 2 | Solution treated | 225–300 | 1.5–50 | 1/16–2 | 3–19 | 1/8–3/4 | 3–38 | 1/8–1 1/2 |
| Ni Wrt 2 | Solution treated and aged | 300–400 | 1.5–50 | 1/16–2 | 3–19 | 1/8–3/4 | 3–38 | 1/8–1 1/2 |
| Ni Wrt 3 | Solution treated | 275–390 | 1.5–50 | 1/16–2 | 3–19(d) | 1/8–3/4(d) | 3–38 | 1/8–1 1/2 |
| Ni Wrt 3 | Solution treated and aged | 400–475 | ... | ... | 3–19(d) | 1/8–3/4(d) | 3–38 | 1/8–1 1/2 |
| Ni Wrt 4 | Annealed or solution treated | 140–220 | 1.5–50 | 1/16–2 | 3–25 | 1/8–1 | 3–38 | 1/8–1 1/2 |
| Ni Wrt 4 | Cold drawn or aged | 240–310 | 1.5–50 | 1/16–2 | 3–25 | 1/8–1 | 3–38 | 1/8–1 1/2 |
| Ni Wrt 5 | As-rolled | 180–200 | 1.5–50 | 1/16–2 | 3–50 | 1/8–2 | 3–38 | 1/8–1 1/2 |
| Ni Cast 1 | As-cast or cast and aged | 200–375 | 1.5–50 | 1/16–2 | 3–19 | 1/8–3/4 | 3–38 | 1/8–1 1/2 |
| Ni Cast 2 | As-cast or cast and aged | 250–320 | 1.5–50 | 1/16–2 | 3–19 | 1/8–3/4 | 3–38 | 1/8–1 1/2 |
| Ni Cast 2 | As-cast or cast and aged | 320–425 | ... | ... | 3–19(d) | 1/8–3/4(d) | 6–25(e) | 1/4–1(e) |
| Co Wrt | Solution treated | 180–230 | 1.5–50 | 1/16–2 | 3–19 | 1/8–3/4 | 3–38 | 1/8–1 1/2 |
| Co Wrt | Solution treated and aged | 270–320 | 1.5–50 | 1/16–2 | 3–19 | 1/8–3/4 | 3–38 | 1/8–1 1/2 |
| Co Cast | As-cast or cast and aged | 220–290 | 1.5–50 | 1/16–2 | 3–25 | 1/8–1 | 3–38 | 1/8–1 1/2 |
| Co Cast | As-cast or cast and aged | 290–425 | ... | ... | 3–19(d) | 1/8–3/4(d) | 6–25(e) | 1/4–1(e) |
| Fe Wrt | Solution treated | 180–230 | 1.5–50 | 1/16–2 | 3–25 | 1/8–1 | 3–38 | 1/8–1 1/2 |
| Fe Wrt | Solution treated and aged | 250–320 | 1.5–50 | 1/16–2 | 3–25 | 1/8–1 | 3–38 | 1/8–1 1/2 |
| Fe Cast 1 | Annealed | 135–185 | 1.5–50 | 1/16–2 | 3–50 | 1/8–2 | 3–50 | 1/8–2 |
| Fe Cast 2 | Annealed or normalized | 135–185 | 1.5–50 | 1/16–2 | 3–50 | 1/8–2 | 3–50 | 1/8–2 |
| Fe Cast 3 | As-cast | 160–210 | 1.5–50 | 1/16–2 | 3–50 | 1/8–2 | 3–50 | 1/8–2 |
| MA 754 | Mechanically alloyed | 277 | 1.5–50 | 1/16–2 | 1.5–17 | 1/16–11/16 | 1.5–38 | 1/16–1 1/2 |
| MA 956 | Mechanically alloyed | 270 | 1.5–50 | 1/16–2 | 1.5–17 | 1/16–11/16 | 1.5–38 | 1/16–1 1/2 |
| MA 6000 | Mechanically alloyed | 450 | 1.5–50 | 1/16–2 | 3–17(d) | 1/8–11/16(d) | 3–38 | 1/8–1 1/2 |

(a) Single-flute carbide gun drill. (b) HSS drills except as specified in footnotes. (c) HSS or carbide drills except as specified in footnotes. (d) Carbide twist drill. (e) Carbide drill only. Source: Metcut Research Associates Inc.

respectively, at an angle of 10° 52′, and they were 768 mm (30.250 in.) in diameter and 25 mm (1 in.) thick.

The broach of the original conventional design had been used with reasonable success on similar parts made of A-286. However, when it was tried on the René 41 wheels using the same operating conditions as for A-286, only eight slots could be obtained per broach resharpening. Not only were tools being expended by wear and excessive grinding, but tooth breakout occurred after several grinds.

The revised design, with stronger teeth, enabled the broaching of 60 or more slots per broach sharpening. Use of a backoff angle of only 1° for the full-form slot shown in Fig. 7 extended the life of a broach to 12 or more sharpenings. Broaches were sharpened as soon as chips fused to the cutting edge and could not be brushed off freely. Broaches had smooth surfaces (0.25 μm, or 10 μin.), and teeth were ground to a sharp edge (no flat land).

A broaching speed of 1.8 m/min (6 sfm) yielded the best results, and the machine used had sufficient capacity to provide smooth cutting at this low speed. Grade M2 high-speed steel was an acceptable tool material if nitrided and oxide coated by steam, but best results were obtained using T15 (65 to 67 HRC). All tools made from T15 were tempered three times and were oxide coated after grinding.

## Drilling

The heat-resistant alloys can be drilled by conventional drilling methods, but several nontraditional machining methods provide important alternatives because of the forces required in the conventional

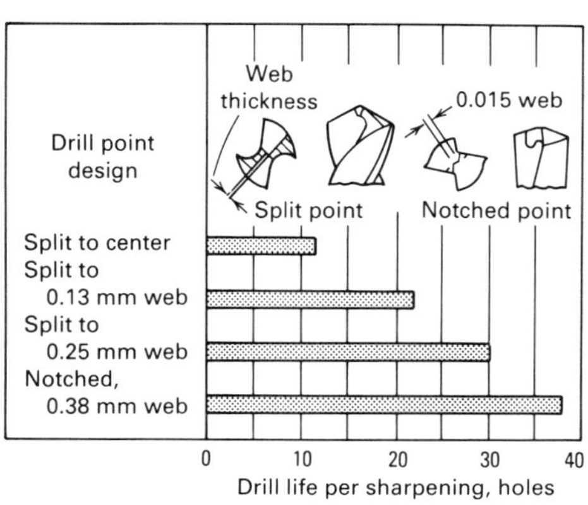


**Fig. 8** Effect of drill-point design on life of T15 drills when drilling 7.87 mm (0.310 in.) deep holes in 20 mm (0.800 in.) thick Astroloy

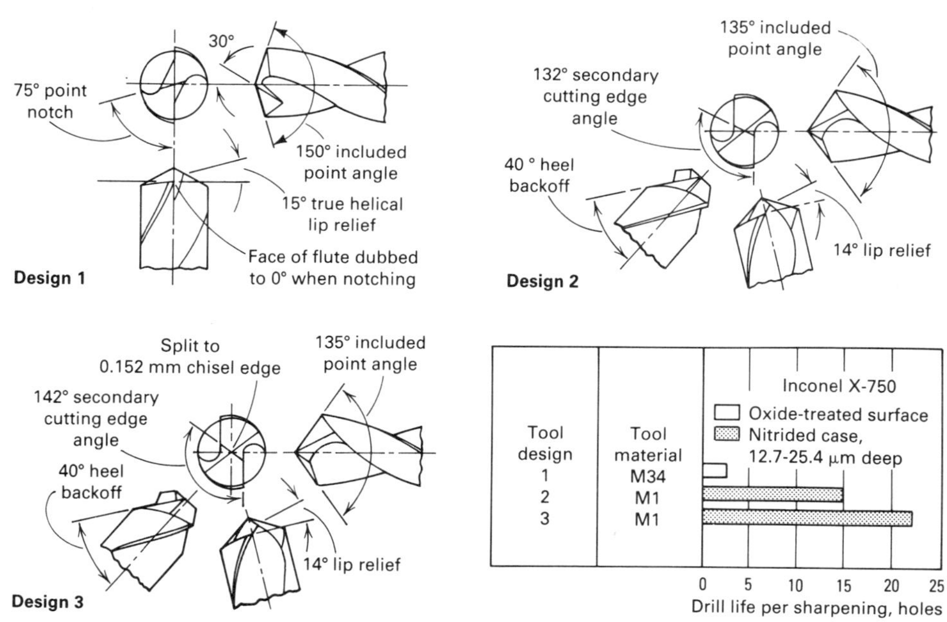


**Fig. 9** Changes in drill design to improve tool life in hand drilling Inconel X-750. Graph shows effects on tool life. All drills were of screw-machine length, with a diameter of 2.5 mm (0.098 in.) and a flute helix angle of 29°. Designs 1 and 2 had a web thickness of 0.5 mm (0.020 in.) and a 132° secondary cutting edge angle; design 3 had a web thickness of 0.61 mm (0.024 in.) and a secondary cutting edge angle of 142°.

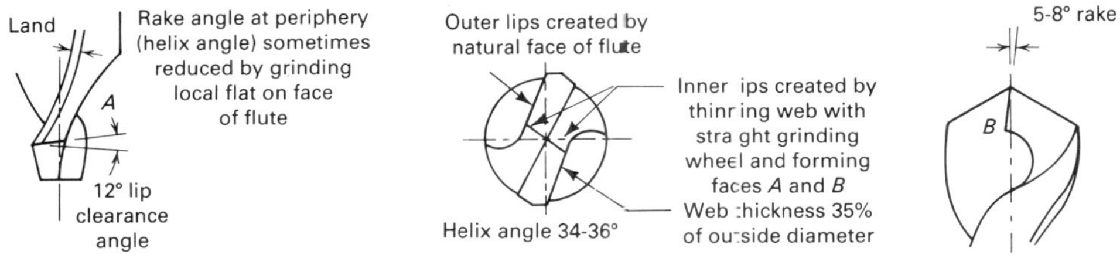


**Fig. 10** Suggested grind of carbide twist drills when drilling heat-resistant alloys

drilling of these alloys. Nontraditional drilling methods are particularly attractive when deep, small-diameter holes must be drilled in heat-resistant alloys. The articles "Electrostream and Capillary Drilling" and "Shaped Tube Electrolytic Machining" in this Volume describe two nontraditional methods that are capable of drilling deep, small holes in difficult-to-machine metals.

When conventional drilling is employed, the high forces produced necessitate maximum rigidity of the machine, tools, and setup. In terms of tool design, the most important single requirement is that the drills be as short and rigid as possible within the limiting requirements of the workpiece and setup.

The conventional drilling of heat-resistant alloys is performed with gun drills, twist drills, or oil-hole drills. Table 21 lists the typical applications of these drills in terms of hole size and the various heat-resistant alloys. Gun drills have the largest range of hole size applications, while twist drills have the smallest range. Oil-hole drills are helpful in deep-hole drilling and can sometimes achieve hole depths up to 30 diameters.

## Twist Drills

Stub-length screw-machine drills, type-C aircraft drills with accurately ground split points, rail drills, and extraheavy-web drills are recommended. These heavy-duty drills yield much better results than standard jobbers-length drills because of their greater rigidity.

**Drill Design.** The crankshaft or split points, which are standard for type-C aircraft and heavy-web drills, are preferred for drilling all iron-base heat-resistant alloys harder than 400 HB and other heat-resistant alloys harder than 350 HB. Drills with standard chisel-edge points can be used for softer alloys.

Drill wear and life can be controlled to some extent by modifying the drill point. Chipping of drill corners can be reduced by decreasing the point angle, and severe wear at the point can be reduced by increasing the point angle to 135°. Excessive margin wear can be eliminated by using a dual-angle (118°/90°) lip. All drills should be machine ground to very close accuracy. A slight amount of runout or point eccentricity will greatly reduce drill life. The following two examples demonstrate the importance of drill design in drill life.

**Example 9: Effect of Drill-Point Design on Drill Life.** Drills with split points were tested against drills with notched points to determine the effect of point design on drill life in drilling solution-treated and aged Astroloy (42 HRC). The web thickness of the split point was varied from none to 0.25 mm (0.010 in.). The service lives of three split points and of one notched point are plotted in Fig. 8. Of the split points, the least satisfactory results were obtained with a point split to center (no web thickness). Split-point designs with web thicknesses of 0.1 to 0.25 mm (0.005 to 0.010 in.) exhibited longer service lives. However, a notched point, with a 0.4 mm (0.015 in.) web, yielded the best results. All drills had 135° point angles. Additional tests confirmed the finding that a 135° point angle is superior to 118°. Clearance angles of 9 to 10° proved superior in supplementary tests to clearance angles of 7 to 8° and 11 to 12°. All drills were made of T15 high-speed steel and were operated at 3.41 m/min (11.2 sfm) with a feed of 0.05 mm/rev (0.002 in./rev). Cutting fluid was a sulfochlorinated concentrate of mineral and fatty oil.

**Example 10: Effect of Drill Design on Cost of Drilling Inconel X-750.** For lower drilling costs in hand drilling parts made of heat-treated Inconel X-750 (36 to 38 HRC), changes in tool material and drill design were required. The workpieces ranged in thickness from 0.5 to 3.2 mm (0.020 to 0.125 in.), and a speed of 3.90 m/min (12.8 sfm) was used with a 2.5 mm (0.098 in.) diam drill. No cutting fluid was used.

The first drills tested were made of M34 high-speed steel and had been developed

**Table 22 Nominal speeds and the range of feeds for drilling heat-resistant alloys with twist drills**

| Alloy group from Table 6 | Condition | Hardness, HB | Speed m/min | Speed sfm | Range of nominal hole size diameter mm | Range of nominal hole size diameter in. | Range of feed(a) mm/rev | Range of feed(a) in./rev | Tool material grade |
|---|---|---|---|---|---|---|---|---|---|
| Ni Wrt 1 | Annealed or solution treated | 200–300 | 6 | 20 | 3–19 | ⅛–¾ | 0.05–0.1 | 0.002–0.004 | (b) |
| Ni Wrt 1 | Solution treated and aged | 300–400 | 4.5 | 15 | 3–19 | ⅛–¾ | 0.05–0.1 | 0.002–0.004 | (b) |
| Ni Wrt 2 | Solution treated | 225–300 | 4.5 | 15 | 3–19 | ⅛–¾ | 0.05–0.1 | 0.002–0.004 | (b) |
| Ni Wrt 2 | Solution treated and aged | 300–400 | 3.7 | 12 | 3–19 | ⅛–¾ | 0.05–0.1 | 0.002–0.004 | (b) |
| Ni Wrt 3 | Solution treated | 275–390 | 6 | 20 | 3–19 | ⅛–¾ | 0.025–0.1 | 0.001–0.004 | C-2 (K10) carbide |
| Ni Wrt 3 | Solution treated and aged | 400–475 | 4.5 | 15 | 3–19 | ⅛–¾ | 0.025–0.1 | 0.001–0.004 | C-2 (K10) carbide |
| Ni Wrt 4 | Annealed or solution treated | 140–220 | 6 | 20 | 3–25 | ⅛–1 | 0.05–0.1 | 0.002–0.004 | M10, M7, M1 (S2, S3) |
| Ni Wrt 4 | Cold drawn or aged | 240–310 | 4.5 | 15 | 3–25 | ⅛–1 | 0.05–0.1 | 0.002–0.004 | M10, M7, M1 (S2, S3) |
| Ni Wrt 5 | As-rolled | 180–200 | 17 | 55 | 3–50 | ⅛–2 | 0.05–0.45 | 0.002–0.018 | M10, M7, M1 (S2, S3) |
| Ni Cast 1 | As-cast or cast and aged | 200–375 | 1.5 | 5 | 3–19 | ⅛–¾ | 0.05–0.1 | 0.002–0.004 | (b) |
| Ni Cast 2 | As-cast or cast and aged | 250–320 | 2 | 7 | 3–19 | ⅛–¾ | 0.05–0.1 | 0.002–0.004 | (b) |
| Ni Cast 2 | As-cast or cast and aged | 320–425 | 4.5 | 15 | 3–19 | ⅛–¾ | 0.025–0.1 | 0.001–0.004 | C-2 (K10) carbide |
| Co Wrt | Solution treated | 180–230 | 6 | 20 | 3–19 | ⅛–¾ | 0.05–0.1 | 0.002–0.004 | (b) |
| Co Wrt | Solution treated and aged | 270–320 | 4.5 | 15 | 3–19 | ⅛–¾ | 0.05–0.1 | 0.002–0.004 | (b) |
| Co Cast | As-cast or cast and aged | 220–290 | 2 | 7 | 3–25 | ⅛–1 | 0.05–0.15 | 0.002–0.006 | (b) |
| Co Cast | As-cast or cast and aged | 290–425 | 4.5 | 15 | 3–19 | ⅛–¾ | 0.025–0.1 | 0.001–0.004 | C-2 (K10) carbide |
| Fe Wrt | Solution treated | 180–230 | 8 | 25 | 3–25 | ⅛–1 | 0.05–0.25 | 0.002–0.010 | (b) |
| Fe Wrt | Solution treated and aged | 250–320 | 6 | 20 | 3–25 | ⅛–1 | 0.05–0.25 | 0.002–0.008 | (b) |
| Fe Cast 1 | Annealed | 135–185 | 18 | 60 | 3–50 | ⅛–2 | 0.05–0.45 | 0.002–0.018 | M10, M7, M1 (S2, S3) |
| Fe Cast 2 | Annealed or normalized | 135–185 | 15 | 50 | 3–50 | ⅛–2 | 0.05–0.45 | 0.002–0.018 | M10, M7, M1 (S2, S3) |
| Fe Cast 3 | As-cast | 160–210 | 14 | 45 | 3–50 | ⅛–2 | 0.05–0.33 | 0.002–0.013 | M10, M7, M1 (S2, S3) |
| MA 754, MA956 | Mechanically alloyed | 277, 270 | 6–23 | 20–75 | 1.5–8 | 1⁄16–5⁄16 | 0.025–0.13 | 0.001–0.005 | (b) |
| MA 754, MA956 | Mechanically alloyed | 277, 270 | 2.5–3.5 | 8–12 | 9.5–17.5 | ⅜–11⁄16 | 0.10–0.25 | 0.004–0.01 | (b) |
| MA 6000 | Mechanically alloyed | 450 | 2.5–3.5 | 8–12 | 9.5–17.5 | ⅜–11⁄16 | 0.10–0.25 | 0.004–0.01 | C-2 (K10) carbide |

(a) Increases within the range occur as the nominal hole size is increased. (b) Any premium high-speed steel (T15, M33, M41–M47, or S9, S10, S11, S12). Source: Metcut Research Associates Inc.

**Table 23 Reduction of speed and feed for drilling deep holes with twist drills**

| Hole depth-to-diameter ratio | Reduce speed by, % | Reduce feed by, % |
|---|---|---|
| 3:1 | 10 | 10 |
| 4:1 | 20 | 10 |
| 5:1 | 30 | 20 |
| 6:1 | 35 | 20 |
| 8:1 | 40 | 20 |

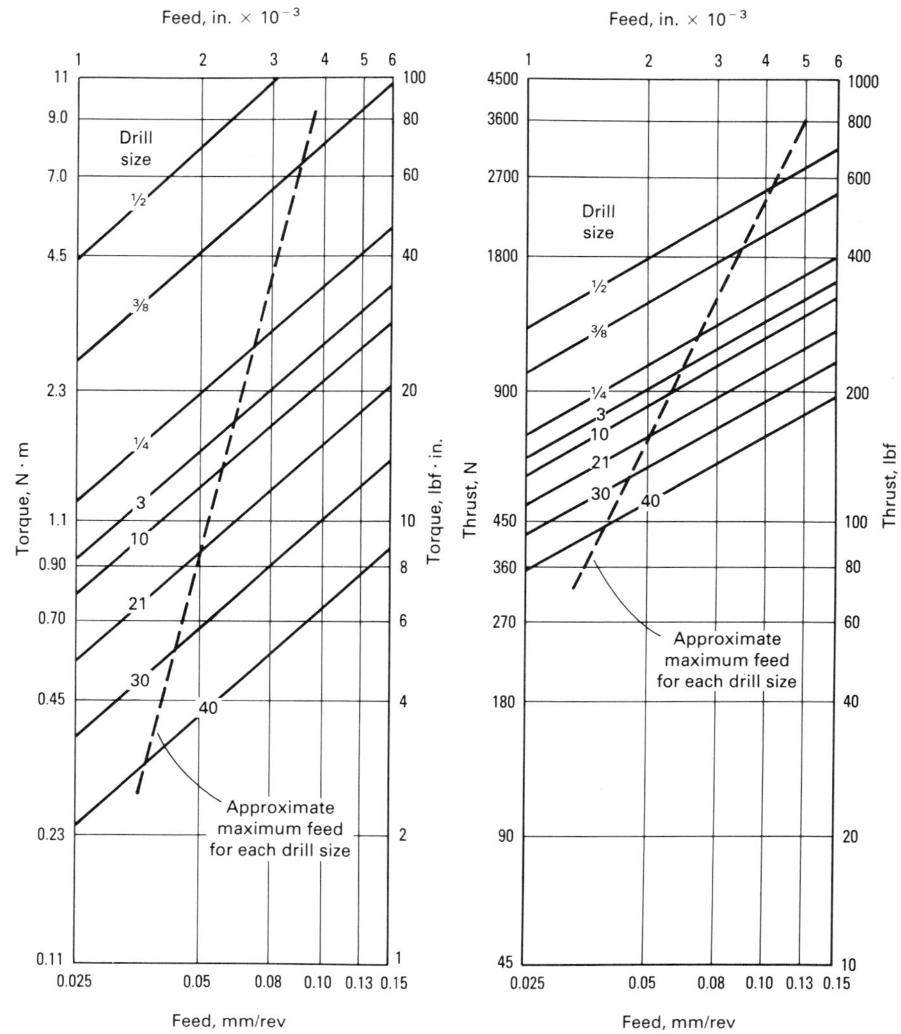


**Fig. 11** Relations between torque and feed (a) and between thrust and feed (b) in drilling through holes in heat-treated, 3 mm (⅛ in.) thick René 41 (41 to 42 HRC) with split-point rail drills. Drill size given in inches

under ideal conditions of rigid setup and positive feed. They performed satisfactorily under these conditions. These drills, however, proved impractical when used for hand drilling operations, using light, limited-production tooling. Brittleness of the tool material caused rapid deterioration of the drill, which, when coupled with the premium price of the special point (design 1, Fig. 9), resulted in tool costs that were prohibitive.

A change to a type-C aircraft drill (design 2, Fig. 9) made of M1 steel liquid nitrided to a depth of 0.01 to 0.025 mm (0.0005 to 0.001 in.) resulted in increased tool life and reduced costs. In operating the new drills, if the drill point was split past center, as permitted by the specification, tool life would usually be greatly shortened, and the tool would chip just before it failed. Chipping would occur at the point of intersection of the secondary cutting edge and the flute face.

These problems were overcome with the drill designated design 3 in Fig. 9. In this drill, the possibility of splitting past center was eliminated by the drill dimensions; the strength of the area that had chipped in the previous design was increased by enlarging the 132° secondary cutting edge angle to 142°, and the web thickness was increased to improve the overall stability of the tool. The change in web thickness helped considerably to extend tool life.

**High-speed steel twist drills** are used when drilling heat-resistant alloys, although carbide drills are mostly used in the aircraft engine industries. In many applications, twist drills made of the general-purpose grades have proved satisfactory, as judged by the number of holes drilled and by the initial cost of the drills and the cost of resharpening. The premium grades of high-speed steel, such as T15, M33, or M36, are preferred for drilling many of the heat-resistant alloys, and their use is often mandatory for obtaining acceptable drill life. The higher cost of the drills made from the more highly alloyed high-speed steels (commonly about four times the cost of their general-purpose counterparts) and the higher cost of resharpening are often warranted by increased tool life. However, a new application can usually be started with drills made of a general-purpose high-speed steel, and a premium grade is then used only when the need for it has been established. Various surface treatments such as nitriding can also be applied to high-speed steel drills to improve drill life.

In applications involving small-diameter (6 mm, or ¼ in.) twist drills, a T15 drill material has shown erratic performance. Grade T15 is capable of developing very high hardness, 67 HRC or higher, and small drills of T15 usually fail by microscopic chipping of the cutting edges. This is probably the result of the inherent lack of rigidity of small drills and of the slightly coarser grain structure of T15 high-speed steel.

**Carbide-tip twist drills** are more suitable for drilling the alloys in the Ni Wrt 3 group (Table 6) and the mechanically alloyed product MA 6000. These alloys present high resistance to the conventional type of high-speed steel drill. A carbide drill is also preferred when the alloys in the Ni Cast 2 and Co Cast groups (Table 6) are at higher hardnesses (320 to 425 HB and 290 to 425 HB, respectively).

A suitable grind of a carbide drill is shown in Fig. 10. Whether or not carbide drills can be used for drilling heat-resistant alloys depends largely on the rigidity of the setup.

**Speeds, Feeds, and Drill Life.** Nominal feeds and speeds (and the preferred drill material) for drilling heat-resistant alloys with twist drills are given in Table 22. When drilling deep holes with twist drills, the speeds and feeds should be reduced as the holes become deeper (Table 23).

Speeds of 4.5 to 9 m/min (15 to 30 sfm) are extensively used for drilling with high-speed steel drills, although in drilling the more difficult-to-machine alloys, such as René 41, speeds as low as 3 m/min (10 sfm) may be required. Speeds lower than 3 m/min (10 sfm) are seldom feasible, because shear action is poor and the drills are likely to fail by chipping. When carbide drills are used, speeds are usually two or three times as fast.

**Table 24 Nominal speeds for drilling heat-resistant alloys with carbide (C-2 or K20) gun drills**

| Alloy group from Table 6 | Condition | Hardness, HB | Speed(a) m/min | Speed(a) sfm |
|---|---|---|---|---|
| Ni Wrt 1 | Annealed or solution treated | 200–300 | 24 | 80 |
| Ni Wrt 1 | Solution treated and aged | 300–400 | 18 | 60 |
| Ni Wrt 2 | Solution treated | 225–300 | 24 | 80 |
| Ni Wrt 2 | Solution treated and aged | 300–400 | 18 | 60 |
| Ni Wrt 3 | Solution treated | 275–390 | 18 | 60 |
| Ni Wrt 3 | Solution treated and aged | 400–475 | . . . | . . . |
| Ni Wrt 4 | Annealed or solution treated | 140–220 | 24 | 80 |
| Ni Wrt 4 | Cold drawn or aged | 240–310 | 18 | 60 |
| Ni Wrt 5 | As rolled | 180–200 | 24 | 80 |
| Ni Cast 1 | As-cast or cast and aged | 200–375 | 18 | 60 |
| Ni Cast 2 | As-cast or cast and aged | 250–320 | 18 | 60 |
| Ni Cast 2 | As-cast or cast and aged | 320–425 | . . . | . . . |
| Co Wrt | Solution treated | 180–230 | 20 | 65 |
| Co Wrt | Solution treated and aged | 270–320 | 15 | 50 |
| Co Cast | As-cast or cast and aged | 220–290 | 15 | 50 |
| Co Cast | As-cast or cast and aged | 290–425 | . . . | . . . |
| Fe Wrt | Solution treated | 180–230 | 46 | 150 |
| Fe Wrt | Solution treated and aged | 250–320 | 30 | 100 |
| Fe Cast 1 | Annealed | 135–185 | 64 | 210 |
| Fe Cast 2 | Annealed or normalized | 135–185 | 50 | 165 |
| Fe Cast 3 | As-cast | 160–210 | 46 | 150 |
| MA 754 | Mechanically alloyed | 277 | 18–30 | 60–100 |
| MA 956 | Mechanically alloyed | 270 | 45–76 | 150–250 |
| MA 6000 | Mechanically alloyed | 450 | 15–23 | 50–75 |

(a) Speeds apply for the feeds defined in the text. Source: Metcut Research Associates Inc.

The proper feed for a job must be determined from chip-breaking characteristics. A steady rate of feed is also important, especially when drilling alloys that work harden readily. Hand feeding, although sometimes used, is not recommended for drilling heat-resistant alloys. The relationship among torque, thrust, and feed when drilling through holes in 3 mm (⅛ in.) René 41 with T15 drills is presented in Fig. 11.

Drill life varies with speed, feed, the workpiece, and the length-to-diameter ratio. Figure 12 relates drill life in three different workpieces at various speeds and feeds. Drill life in the N-155 workpiece exhibits an optimum value at a cutting speed of about 12 m/min (40 sfm), at which drill life decreases with an increase or decrease in cutting speed. No such optimum point is observed in the data of the other two workpiece materials.

## Gun Drills

Gun drills are recommended for the deep-hole drilling (depths greater than 3 diameters) of larger diameters (up to 50 mm, or 2 in.) in the more difficult-to-machine alloys, such as the NiWrt 3, NiCast 2, and Co Cast groups (Table 6). This type of drill tends to steady itself, and it avoids the work hardening that occurs at the extremes of point with a standard twist drill (where the cutting speed varies from a maximum at the hole periphery to zero at the center). The true top rake should not exceed 10° (positive). Nominal feed rates depend on the diameter of the hole and are:

| Hole diameter mm | Hole diameter in. | Feed rate mm/rev | Feed rate in./rev |
|---|---|---|---|
| 2–4 | 0.78–0.15 | 0.004–0.006 | 0.00015–0.00025 |
| 4–6 | 0.15–0.25 | 0.008–0.013 | 0.0003–0.0005 |
| 6–13 | 0.25–0.50 | 0.013–0.018 | 0.0005–0.0007 |
| 13–19 | 0.50–0.75 | 0.018–0.023 | 0.0007–0.0009 |
| 19–25 | 0.75–1 | 0.020–0.025 | 0.0008–0.001 |
| 25–50 | 1–2 | 0.025–0.030 | 0.001–0.0012 |

These feed rates are for carbide (C-2, or K20) single-flute gun drills, and they apply for all the heat-resistant alloys. The nominal speeds with these feed rates are given in Table 24.

## Oil-Hole Drills

Oil-hole drills are used to advantage for the deep drilling of small holes, in which frequent retraction for lubrication and chip clearance might otherwise be required. Because these drills have polished flutes and chip breakers ground into the cutting edge, they provide forced lubrication at the cutting area and avoid coiled chips that might clog flutes. Table 25 lists nominal feeds and speeds. For holes more than five diameters deep, the feed rate should be reduced as the holes become deeper (Table 26).

## Fixturing

Because the successful drilling of most heat-resistant alloys depends on the rigid-

**Table 25 Nominal speeds and feeds for drilling heat-resistant alloys with oil-hole drills**

| Alloy group from Table 6 | Condition | Hardness, HB | Speed(a) m/min | Speed(a) sfm | Feed(b), mm/rev (in./rev), for a hole diameter of: 3 mm (⅛ in.) | 6–13 mm (¼–½ in.) | 19–25 mm (¾–1 in.) | 38 mm (1½ in.) |
|---|---|---|---|---|---|---|---|---|
| Ni Wrt 1 | Annealed or solution treated | 200–300 | 8 | 25 | 0.025 (0.001) | 0.05–0.1 (0.002–0.004) | 0.15–0.2 (0.006–0.008) | 0.25 (0.01) |
| Ni Wrt 1 | Solution treated and aged | 300–400 | 6 | 20 | 0.025 (0.001) | 0.05–0.1 (0.002–0.004) | 0.15–0.2 (0.006–0.008) | 0.25 (0.01) |
| Ni Wrt 2 | Solution treated | 225–300 | 6 | 20 | 0.025 (0.001) | 0.05–0.075 (0.002–0.003) | 0.15–0.18 (0.006–0.007) | 0.20 (0.008) |
| Ni Wrt 2 | Solution treated and aged | 300–400 | 4.5 | 15 | 0.025 (0.001) | 0.05 (0.002) | 0.10–0.15 (0.004–0.006) | 0.20 (0.008) |
| Ni Wrt 3 | Solution treated | 275–390 | 4.5 | 15 | 0.025 (0.001) | 0.05–0.075 (0.002–0.003) | 0.15–0.20 (0.006–0.008) | 0.25 (0.01) |
| Ni Wrt 3 | Solution treated and aged | 400–475 | 3 | 10 | 0.025 (0.001) | 0.05–0.075 (0.002–0.003) | 0.15–0.20 (0.006–0.008) | 0.25 (0.01) |
| Ni Wrt 4 | Annealed or solution treated | 140–220 | 11 | 35 | 0.025 (0.001) | 0.05–0.10 (0.002–0.004) | 0.15–0.20 (0.006–0.008) | 0.25 (0.01) |
| Ni Wrt 4 | Cold drawn or aged | 240–310 | 8 | 25 | 0.025 (0.001) | 0.05–0.10 (0.002–0.004) | 0.15–0.20 (0.006–0.008) | 0.25 (0.01) |
| Ni Wrt 5 | As-rolled | 180–200 | 20 | 65 | 0.025 (0.001) | 0.10–0.15 (0.004–0.006) | 0.20–0.25 (0.008–0.01) | 0.30 (0.012) |
| Ni Cast 1 | As-cast or cast and aged | 200–375 | 6 | 20 | 0.025 (0.001) | 0.05–0.10 (0.002–0.004) | 0.15–0.20 (0.006–0.008) | 0.25 (0.01) |
| Ni Cast 2 | As-cast or cast and aged | 250–320 | 4.5 | 15 | 0.025 (0.001) | 0.05–0.075 (0.002–0.003) | 0.15–0.18 (0.006–0.007) | 0.20 (0.008) |
| Ni Cast 2 | As-cast or cast and aged | 320–425 | 8(c) | 25(c) | . . . | 0.025–0.05(c) (0.001–0.002)(c) | 0.075–0.10(c) (0.003–0.004)(c) | . . . |
| Co Wrt | Solution treated | 180–230 | 8 | 25 | 0.025 (0.001) | 0.05–0.10 (0.002–0.004) | 0.15–0.20 (0.006–0.008) | 0.25 (0.01) |
| Co Wrt | Solution treated and aged | 270–320 | 6 | 20 | 0.025 (0.001) | 0.05–0.075 (0.002–0.003) | 0.15–0.18 (0.006–0.007) | 0.20 (0.008) |
| Co Cast | As-cast or cast and aged | 220–290 | 3 | 10 | 0.025 (0.001) | 0.025–0.05 (0.001–0.002) | 0.1–0.15 (0.004–0.006) | 0.15 (0.006) |
| Co Cast | As-cast or cast and aged | 290–425 | 8(c) | 25(c) | . . . | 0.025–0.05(c) (0.001–0.002)(c) | 0.075–0.10 (0.003–0.004)(c) | . . . |
| Fe Wrt | Solution treated | 180–230 | 9 | 30 | 0.025 (0.001) | 0.05–0.075 (0.002–0.003) | 0.15–0.18 (0.006–0.007) | 0.20 (0.008) |
| Fe Wrt | Solution treated and aged | 250–320 | 8 | 25 | 0.025 (0.001) | 0.05–0.075 (0.002–0.003) | 0.15–0.18 (0.006–0.007) | 0.20 (0.008) |
| Fe Cast 1 | Annealed | 135–185 | 23(d) | 75(d) | 0.05(d) (0.002)(d) | 0.13–0.25(d) (0.005–0.01)(d) | 0.30–0.36(d) (0.012–0.014)(d) | 0.40(d) (0.016)(d) |
| Fe Cast 2 | Annealed or normalized | 135–185 | 18(d) | 60(d) | 0.025(d) (0.001)(d) | 0.13–0.20(d) (0.005–0.008)(d) | 0.25–0.30(d) (0.01–0.012)(d) | 0.36(d) (0.014)(d) |
| Fe Cast 3 | As-cast | 160–210 | 17(d) | 55(d) | 0.025(d) (0.001)(d) | 0.10–0.20(d) (0.004–0.008)(d) | 0.25–0.30(d) (0.01–0.012)(d) | 0.36(d) (0.014)(d) |

(a) Recommended speeds with any premium (T15, M33, M41–M47, or S9, S10, S11, S12) HSS drill except as specified in footnotes. Cutting speeds are 100 to 150% higher with carbide drills. (b) Feed rates with HSS drills except as specified in footnotes. Feeds with carbide drills are about 0 to 0.05 mm/rev (0 to 0.002 in./rev) less than those with T15 or M42 drills and about 0.10 to 0.15 mm/rev (0.004 to 0.006 in./rev) less than those with M10, M7, M1 (S2, S3) drills. (c) Speeds and feeds with carbide (C-2 or K10) drill. (d) Speeds and feeds with M10, M7, M1 (S2, S3) drills. Source: Metcut Research Associates Inc.

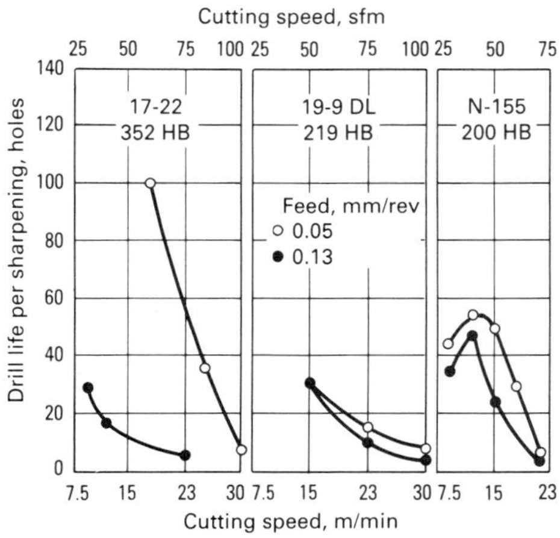


| | |
|---|---|
| Depth of holes, mm (in.) | 13 (0.500) (through) |
| Drill wear end point, mm (in.) | 0.38 (0.015) wear land |
| Cutting fluid | Sulfurized and mineral oils (1:1) |

**Drill details**

| | |
|---|---|
| Material | M2 high-speed steel |
| Diameter, mm (in.) | 6 (0.250) |
| Length, mm (in.) | 100 (4) |
| Point angle, degrees | 118 |
| Helix angle, degrees | 29 |
| Primary clearance, degrees | 7 |
| Point grind | Standard |

**Fig. 12** Effect of speed and feed on drill life in drilling 17-22 steel, 19-9 DL, and N-155

ity of the workpiece, fixturing is important. When the workpiece is too thin or too weak to withstand clamping forces, special techniques must be employed, such as filling portions of the workpiece with a low-melting alloy to give it added support and rigidity. An application of this technique is described in the following example.

**Example 11: Use of Low-Melting Alloy Filler for Improved Workpiece Rigidity.**

**Table 26 Reduction of feed for deep-hole drilling with oil-hole drills**

| Hole depth-to-diameter ratio | Reduce feed by, % |
|---|---|
| 5–7:1 | 10 |
| 8–10:1 | 20 |
| >10:1 | 30 |

For greater rigidity and better drill life in drilling 2.5 to 2.6 mm (0.098 to 0.102 in.) diam holes in a Haynes 25 casting, the central portion of the piece, between the two webs, was filled with a low-melting alloy (Fig. 13). Before it was made rigid, the part could not properly support the pressure of the drill, and there was enough deflection to cause the alloy to work harden (initial hardness was 96 HRB). Drill life was less than one hole per drill before the use of the low-melting alloy filler. With the rigidity of the improved setup, drill life increased as follows:

| Speed, m/min (sfm) | Feed, mm/rev (in./rev) | Holes per grind |
|---|---|---|
| 1.5 (5) | 0.04 (0.0015) | 50 |
| 3 (10) | 0.04 (0.0015) | 50 |
| 3 (10) | 0.075 (0.003) | 10 |

## Cutting Fluid

Active cutting oils—sulfurized, chlorinated, or sulfochlorinated—usually provide better drill life and productivity than soluble-oil emulsions or other water-base cutting fluids. The superiority of active oils becomes more significant as the hardness of the work metal increases and as the depth of the drilled hole increases. When sulfochlorinated or other active oils are used as

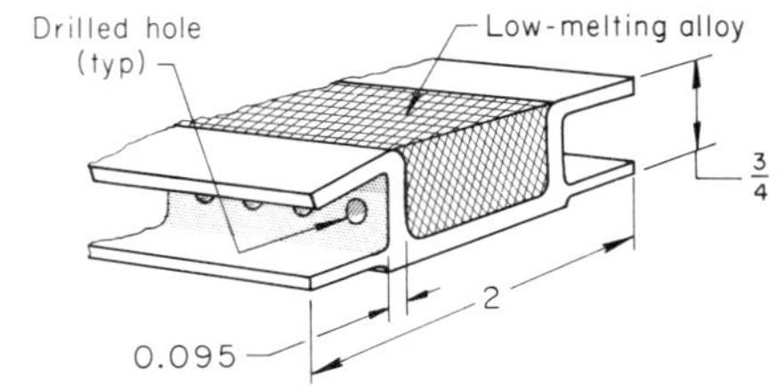


**Fig. 13** Use of low-melting alloy filler for improved rigidity in drilling through webs of castings made of HS-25 (L-605) alloy. Dimensions given in inches

cutting fluids, all traces of oil must be removed from the workpiece before it is heat treated or placed in service at high temperature. Blind holes require special attention.

## Reaming

Rigidity is of primary importance in reaming heat-resistant alloys, and unsupported lengths of tool holders and reamers should be kept at a minimum. It is generally preferable to use carbide-tip reamers, although high-speed steel reamers are also used on the wrought alloys and the iron-base casting alloys. Carbide-tip reamers should be used when reaming the mechanically alloyed MA 6000 product.

Although it is generally desirable to use reamers having six or eight flutes, in reaming small holes (for example, less than 6 mm, or 1/4 in., in diameter), four-flute reamers are often used. Optimum tool angles may vary somewhat, depending on the alloy being reamed, size of hole, and number of flutes in the reamer. However, the variation in angles is seldom great. For example, in one plant, eight-flute reamers with 2° radial

**Table 27 Nominal speeds for reaming heat-resistant alloys**

| Alloy group from Table 6 | Condition | Hardness, HB | Roughing speeds(a): HSS tools(b) m/min | sfm | Carbide tools(c) m/min | sfm | Finishing speeds(a): HSS tools(b) m/min | sfm | Carbide tools(c) m/min | sfm |
|---|---|---|---|---|---|---|---|---|---|---|
| Ni Wrt 1 | Annealed or solution treated | 200–300 | 6 | 20 | 12 | 40 | 4.5 | 15 | 8 | 25 |
| Ni Wrt 1 | Solution treated and aged | 300–400 | 4.5 | 15 | 9 | 30 | 3.6 | 12 | 4.5 | 15 |
| Ni Wrt 2 | Solution treated | 225–300 | 4.5 | 15 | 9 | 30 | 3.6 | 12 | 4.5 | 15 |
| Ni Wrt 2 | Solution treated and aged | 300–400 | 3.6 | 12 | 8 | 25 | 3 | 10 | 4.5 | 15 |
| Ni Wrt 3 | Solution treated | 275–390 | 3.6 | 12 | 6 | 20 | 2 | 7 | 3.6 | 12 |
| Ni Wrt 3 | Solution treated and aged | 400–475 | 3 | 10 | 4.5 | 15 | 1.5 | 5 | 3 | 10 |
| Ni Wrt 4 | Annealed or solution treated | 140–220 | 6 | 20 | 14 | 45 | 4.5 | 15 | 8 | 25 |
| Ni Wrt 4 | Cold drawn or aged | 240–310 | 4.5 | 15 | 11 | 35 | 3.6 | 12 | 4.5 | 15 |
| Ni Wrt 5 | As-rolled | 180–200 | 17 | 55 | 34 | 110 | 9 | 30 | 14 | 45 |
| Ni Cast 1 | As-cast or cast and aged | 200–375 | ... | ... | 9 | 30 | ... | ... | 3.6 | 12 |
| Ni Cast 2 | As-cast or cast and aged | 250–320 | ... | ... | 9 | 30 | ... | ... | 3 | 10 |
| Ni Cast 2 | As-cast or cast and aged | 320–425 | ... | ... | 3 | 10 | ... | ... | 1.5 | 5 |
| Co Wrt | Solution treated | 180–230 | 4.5 | 15 | 9 | 30 | 3.6 | 12 | 6 | 20 |
| Co Wrt | Solution treated and aged | 270–320 | 3.6 | 12 | 8 | 25 | 3 | 10 | 4.5 | 15 |
| Co Cast | As-cast or cast and aged | 220–290 | ... | ... | 9 | 30 | ... | ... | 3 | 10 |
| Co Cast | As-cast or cast and aged | 290–425 | ... | ... | 3 | 10 | ... | ... | 1.5 | 5 |
| Fe Wrt | Solution treated | 180–230 | 9 | 30 | 21 | 70 | 8 | 25 | 12 | 40 |
| Fe Wrt | Solution treated and aged | 250–320 | 8 | 25 | 20 | 65 | 6 | 20 | 9 | 30 |
| Fe Cast 1 | Annealed | 135–185 | 21 | 70 | 26 | 85 | 11 | 35 | 14 | 45 |
| Fe Cast 2 | Annealed or normalized | 135–185 | 17 | 55 | 21 | 70 | 9 | 30 | 12 | 40 |
| Fe Cast 3 | As-cast | 160–210 | 15 | 50 | 20 | 65 | 8 | 25 | 11 | 35 |

(a) See Table 28 for feed rates and tool material grades. A chemical emulsion is recommended as the cutting fluid. (b) High-speed steel tool design: 0.18 to 0.23 mm (0.007 to 0.009 in.) margin width, 1.5 mm (1/16 in.) by 45°, chamfer angle, 5 to 6° chamfer relief angle, 2 to 3° radial rake angle, 0 to 10° helix angle, 12 to 17° primary relief angle. (c) Carbide tool design: 0.13 to 0.25 mm (0.005 to 0.010 in.) margin width, 5 mm (3/16 in.) by 2° lead angle, 7 to 10° radial rake angle, 5 to 8° helix angle, 7 to 15° primary relief angle. Source: Metcut Research Associates Inc.

**Table 28 Nominal feeds for reaming heat-resistant alloys**

| Alloy group from Table 6 | Hardness, HB | Condition | Roughing; feed, mm/rev (in./rev)(a), for a reamer diameter of: 3 mm (1/8 in.) | 6 mm (1/4 in.) | 13 mm (1/2 in.) | 25 mm (1 in.) | 50 mm (2 in.) | Finishing; feed, mm/rev (in./rev)(a), for a reamer diameter of: 3 mm (1/8 in.) | 6 mm (1/4 in.) | 13 mm (1/2 in.) | 25 mm (1 in.) | 50 mm (2 in.) | Tool material trade |
|---|---|---|---|---|---|---|---|---|---|---|---|---|---|
| Ni Wrt 1 | 200–300 | Annealed or solution treated | 0.05 (0.002) | 0.15 (0.006) | 0.20 (0.008) | 0.25 (0.010) | 0.40 (0.015) | 0.075 (0.003) | 0.10 (0.004) | 0.15 (0.006) | 0.20 (0.008) | 0.30 (0.012) | (b) or (c) |
| | 300–400 | Solution treated and aged | 0.05 (0.002) | 0.15 (0.006) | 0.20 (0.008) | 0.25 (0.010) | 0.40 (0.015) | 0.075 (0.003) | 0.10 (0.004) | 0.15 (0.006) | 0.20 (0.008) | 0.30 (0.012) | (b) or (c) |
| Ni Wrt 2 | 225–300 | Solution treated | 0.05 (0.002) | 0.15 (0.006) | 0.20 (0.008) | 0.25 (0.010) | 0.40 (0.015) | 0.075 (0.003) | 0.10 (0.004) | 0.15 (0.006) | 0.20 (0.008) | 0.30 (0.012) | (b) or (c) |
| | 300–400 | Solution treated and aged | 0.05 (0.002) | 0.15 (0.006) | 0.20 (0.008) | 0.25 (0.010) | 0.40 (0.015) | 0.075 (0.003) | 0.10 (0.004) | 0.15 (0.006) | 0.20 (0.008) | 0.30 (0.012) | (b) or (c) |
| Ni Wrt 3 | 275–390 | Solution treated | 0.05 (0.002) | 0.10 (0.004) | 0.15 (0.006) | 0.20 (0.008) | 0.30 (0.012) | 0.075 (0.003) | 0.10 (0.004) | 0.13 (0.005) | 0.15 (0.006) | 0.25 (0.010) | (b) or (c) |
| | 400–475 | Solution treated and aged | 0.05 (0.002) | 0.10 (0.004) | 0.15 (0.006) | 0.20 (0.008) | 0.30 (0.012) | 0.075 (0.003) | 0.10 (0.004) | 0.13 (0.005) | 0.15 (0.006) | 0.25 (0.010) | (b) or (c) |
| Ni Wrt 4 | 140–220 | Annealed or solution treated | 0.05 (0.002) | 0.15 (0.006) | 0.20 (0.008) | 0.25 (0.010) | 0.40 (0.015) | 0.075 (0.003) | 0.10 (0.004) | 0.15 (0.006) | 0.20 (0.008) | 0.30 (0.012) | (b) or (c) |
| | 240–310 | Cold drawn or aged | 0.05 (0.002) | 0.15 (0.006) | 0.20 (0.008) | 0.25 (0.010) | 0.40 (0.015) | 0.075 (0.003) | 0.10 (0.004) | 0.15 (0.006) | 0.20 (0.008) | 0.30 (0.012) | (b) or (c) |
| Ni Wrt 5 | 180–200 | As-rolled | 0.075 (0.003) | 0.15 (0.006) | 0.25 (0.010) | 0.30 (0.012) | 0.40 (0.015) | 0.15 (0.006) | 0.20 (0.008) | 0.25 (0.010) | 0.40 (0.015) | 0.50 (0.020) | (c) or (d) |
| Ni Cast 1 | 200–375 | As-cast or cast and aged | 0.05 (0.002) | 0.10 (0.004) | 0.15 (0.006) | 0.20 (0.008) | 0.30 (0.012) | 0.075 (0.003) | 0.10 (0.004) | 0.15 (0.006) | 0.20 (0.008) | 0.30 (0.012) | (c) |
| Ni Cast 2 | 250–320 | As-cast or cast and aged | 0.05 (0.002) | 0.10 (0.004) | 0.15 (0.006) | 0.20 (0.008) | 0.30 (0.012) | 0.05 (0.002) | 0.10 (0.004) | 0.15 (0.006) | 0.20 (0.008) | 0.30 (0.012) | (c) |
| | 320–425 | As-cast or cast and aged | 0.05 (0.002) | 0.10 (0.004) | 0.15 (0.006) | 0.20 (0.008) | 0.30 (0.012) | 0.05 (0.002) | 0.10 (0.004) | 0.15 (0.006) | 0.20 (0.008) | 0.30 (0.012) | (c) |
| Co Wrt | 180–230 | Solution treated | 0.05 (0.002) | 0.15 (0.006) | 0.20 (0.008) | 0.25 (0.010) | 0.40 (0.015) | 0.075 (0.003) | 0.10 (0.004) | 0.15 (0.006) | 0.20 (0.008) | 0.30 (0.012) | (b) or (c) |
| | 270–320 | Solution treated and aged | 0.05 (0.002) | 0.15 (0.006) | 0.20 (0.008) | 0.25 (0.010) | 0.40 (0.015) | 0.075 (0.003) | 0.10 (0.004) | 0.15 (0.006) | 0.20 (0.008) | 0.30 (0.012) | (b) or (c) |
| Co Cast | 220–290 | As-cast or cast and aged | 0.05 (0.002) | 0.10 (0.004) | 0.15 (0.006) | 0.20 (0.008) | 0.30 (0.012) | 0.075 (0.003) | 0.10 (0.004) | 0.15 (0.006) | 0.20 (0.008) | 0.30 (0.012) | (c) |
| | 290–425 | As-cast or cast and aged | 0.05 (0.002) | 0.10 (0.004) | 0.15 (0.006) | 0.20 (0.008) | 0.30 (0.012) | 0.05 (0.002) | 0.10 (0.004) | 0.15 (0.006) | 0.20 (0.008) | 0.30 (0.012) | (c) |
| Fe Wrt | 180–230 | Solution treated | 0.05(e) (0.002)(e) | 0.15 (0.006) | 0.25 (0.010) | 0.30 (0.012) | 0.40 (0.015) | 0.075 (0.003) | 0.10 (0.004) | 0.15 (0.006) | 0.20 (0.008) | 0.30 (0.012) | (b) or (c) |
| | 250–320 | Solution treated and aged | 0.05(e) (0.002)(e) | 0.15 (0.006) | 0.25 (0.010) | 0.30 (0.012) | 0.40 (0.015) | 0.075 (0.003) | 0.10 (0.004) | 0.15 (0.006) | 0.20 (0.008) | 0.30 (0.012) | (b) or (c) |
| Fe Cast 1 | 135–185 | Annealed | 0.10 (0.004) | 0.20 (0.008) | 0.30 (0.012) | 0.40 (0.016) | 0.60 (0.024) | 0.10 (0.004) | 0.15 (0.006) | 0.20 (0.008) | 0.25 (0.010) | 0.30 (0.012) | (c) |
| | 135–185 | Annealed | 0.075 (0.003) | 0.10 (0.004) | 0.13 (0.005) | 0.20 (0.008) | 0.30 (0.012) | 0.10 (0.004) | 0.10 (0.004) | 0.15 (0.006) | 0.20 (0.008) | 0.25 (0.010) | (d) |
| Fe Cast 2 | 135–185 | Annealed or normalized | 0.10 (0.004) | 0.20 (0.008) | 0.30 (0.012) | 0.40 (0.016) | 0.60 (0.024) | 0.075 (0.003) | 0.10 (0.004) | 0.15 (0.006) | 0.20 (0.008) | 0.25 (0.010) | (c) |
| | 135–185 | Annealed or normalized | 0.075 (0.003) | 0.13 (0.005) | 0.20 (0.008) | 0.25 (0.010) | 0.40 (0.015) | 0.075 (0.003) | 0.075 (0.003) | 0.10 (0.004) | 0.15 (0.006) | 0.20 (0.008) | (d) |
| Fe Cast 3 | 160–210 | As-cast | 0.10 (0.004) | 0.20 (0.008) | 0.30 (0.012) | 0.40 (0.016) | 0.60 (0.024) | 0.075 (0.003) | 0.10 (0.004) | 0.15 (0.006) | 0.20 (0.008) | 0.25 (0.010) | (c) |
| | 160–210 | As-cast | 0.075 (0.003) | 0.13 (0.005) | 0.20 (0.008) | 0.25 (0.01) | 0.40 (0.015) | 0.075 (0.003) | 0.075 (0.003) | 0.10 (0.004) | 0.15 (0.006) | (0.008) | (d) |

(a) Based on four flutes for 3–6 mm (1/8–1/4 in.) diam reamers, six flutes for 13 mm (1/2 in. diam reamers, and eight flutes for larger reamers. (b) Any premium high-speed steel (T15, M33, M41–M47; or S9, S10, S11, S12). (c) C-2 (K20) carbide. (d) M1, M2, M7 (S3, S4, S2) high-speed steel. (e) Feed rate with high-speed steel reamer is 0.075 mm/rev (0.003 in./rev). Source: Metcut Research Associates Inc.

rake, 12° radial relief, 42° chamfer, and 7° chamfer relief are successfully used for reaming 13 mm (½ in.) holes in A-286 and other iron-base alloys. The same angles are used for reaming 13 mm (½ in.) holes in nickel-base alloys, with the exception of the chamfer relief angle, which is reduced to 5°. For reaming 6 mm (¼ in.) holes in the same metals with four-flute reamers, the radial rake is increased to 3° and the radial relief to 17°. Otherwise, angles are the same as those used for reaming 13 mm (½ in.) holes. Narrow margins of 0.18 to 0.23 mm (0.007 to 0.009 in.) are used for reaming nickel-base alloys, but margin widths are increased to 0.25 to 0.43 mm (0.010 to 0.017 in.) for reaming iron-base alloys.

**Nominal speeds and feeds** for reaming heat-resistant alloys are presented in Tables 27 and 28. Speeds of less than 3 m/min (10 sfm) are seldom practical, because cutting edges are likely to chip. The feed must be great enough to maintain cutting action with a practical size of chip. Size of hole is the major factor influencing feed, as indicated in Table 28. Relief angles, rake, and clearances are given in Table 27.

The amount of stock left for reaming is especially critical with heat-resistant alloys. Removing an excessive amount of stock overloads the tools, while insufficient stock will induce burnishing, which causes the alloy to work harden. The optimum amount varies with hole size, but 0.13 mm (0.005 in.) on the radius should be the minimum, even for the smallest holes that can be reamed.

## Counterboring and Spotfacing

Tools used for counterboring and spotfacing heat-resistant alloys are not necessarily different from those used for similar operations on other metals. As in other machining operations, great emphasis should be placed on rigidity of setup in these two processes.

**Nominal speeds and feeds** for counterboring and spotfacing heat-resistant alloys are given in Table 29. Feeds are the same for high-speed steel and carbide tools. The size of hole to be counterbored or spotfaced influences the feed that should be used. The speeds and feeds given in Table 29 are based on the use of sulfurized, fat-containing mineral oil for cutting fluid; however, in practice, spotfacing is often done without a cutting fluid because the process involves the removal of only a small amount of metal.

## Tapping and Thread Milling

Machining internal threads in heat-resistant alloy workpieces is especially difficult, mainly because the surface to be threaded

**Table 29 Nominal speeds and feeds for counterboring and spotfacing heat-resistant alloys**

| Alloy group from Table 6 | Hardness, HB | Condition | Speed, m/min (sfm)(a) | Feed, mm/rev (in./rev)(b), with a nominal hole diameter of: 6 mm (¼ in.) | 13 mm (½ in.) | 19 mm (¾ in.) | 25–65 mm (1–2.5 in.) |
|---|---|---|---|---|---|---|---|
| Ni Wrt 1 | 200–300 | Annealed or solution treated | 4.5 (15) | 0.05 (0.002) | 0.075 (0.003) | 0.1 (0.004) | . . . |
| | 300–400 | Solution treated and aged | 3 (10) | 0.05 (0.002) | 0.075 (0.003) | 0.1 (0.004) | . . . |
| Ni Wrt 2 | 225–300 | Solution treated | 4.5 (15) | 0.05 (0.002) | 0.075 (0.003) | 0.1 (0.004) | . . . |
| | 300–400 | Solution treated and aged | 3 (10) | 0.05 (0.002) | 0.075 (0.003) | 0.1 (0.004) | . . . |
| Ni Wrt 3 | 275–390 | Solution treated | 3 (10) | 0.05 (0.002) | 0.075 (0.003) | 0.075 (0.003) | . . . |
| | 400–475 | Solution treated and aged | 3 (10) | 0.05 (0.002) | 0.075 (0.003) | 0.075 (0.003) | . . . |
| Ni Wrt 4 | 140–220 | Annealed or solution treated | 6 (20) | 0.05 (0.002) | 0.075 (0.003) | 0.075 (0.003) | . . . |
| | 240–310 | Cold drawn or aged | 4.5 (15) | 0.05 (0.002) | 0.075 (0.003) | 0.075 (0.003) | . . . |
| Ni Wrt 5 | 180–200 | As-rolled | 20 (65) | 0.05 (0.002) | 0.075 (0.003) | 0.075 (0.003) | . . . |
| Ni Cast 1 | 200–375 | As-cast or cast and aged | 3 (10) | 0.05 (0.002) | 0.075 (0.003) | 0.075 (0.003) | . . . |
| Ni Cast 2 | 250–320 | As-cast or cast and aged | 3 (10) | 0.05 (0.002) | 0.075 (0.003) | 0.075 (0.003) | . . . |
| | 320–425 | As-cast or cast and aged | 8 (25)(c) | 0.05 (0.002) | 0.075 (0.003) | 0.075 (0.003) | . . . |
| Co Wrt | 180–230 | Solution treated | 4.5 (15) | 0.05 (0.002) | 0.075 (0.003) | 0.075 (0.003) | . . . |
| | 270–320 | Solution treated and aged | 3 (10) | 0.05 (0.002) | 0.075 (0.003) | 0.075 (0.003) | . . . |
| Co Cast | 220–290 | As-cast or cast and aged | 3 (10) | 0.05 (0.002) | 0.075 (0.003) | 0.075 (0.003) | . . . |
| | 290–425 | As-cast or cast and aged | 8 (25)(c) | 0.05 (0.002) | 0.075 (0.003) | 0.075 (0.003) | . . . |
| Fe Wrt | 180–230 | Solution treated | 11 (35) | 0.05 (0.002) | 0.075 (0.003) | 0.075 (0.003) | . . . |
| | 250–320 | Solution treated and aged | 8 (25) | 0.05 (0.002) | 0.075 (0.003) | 0.075 (0.003) | . . . |
| Fe Cast 1 | 135–185 | Annealed | 29 (95)(d) | 0.075 (0.003) | 0.10 (0.004) | 0.1 (0.004) | 0.1–0.15 (0.004–0.006) |
| Fe Cast 2 | 135–185 | Annealed or normalized | 24 (80)(d) | 0.075 (0.003) | 0.1 (0.004) | 0.1 (0.004) | 0.1–0.15 (0.004–0.006) |
| Fe Cast 3 | 160–210 | As-cast | 17 (55)(d) | 0.075 (0.003) | 0.075 (0.003) | 0.075 (0.003) | 0.1–0.1 (0.004–0.004) |

(a) Speeds are with any premium high-speed steel (T15, M33, M41–M47, or S9, S10, S11, S12) except as specified in footnotes. Speeds with carbide tools (C-2 or K10) are three to four times higher except for the iron-base alloys (wrought and cast) and the nickel-base alloys in the Ni Wrt 4 and Ni Wrt 5 groups. For these alloys, speeds are about two times higher with carbide tools. (b) Feeds are the same for both HSS and carbide tools. Feeds are based on three flutes for a 6 mm (¼ in.) diam hole and four flutes for a 13–62 mm (0.5–2.5 in.) diam hole. (c) Speed with C-2 (K10) carbide. (d) M2 or M3 (S4 or S5) high-speed steel. Source: Metcut Research Associates Inc.

work hardens during the operation (drilling, reaming, or boring) that prepares the hole for threading. Therefore, the preliminary operations should be planned so that the tools continuously cut chips of substantial thickness; this is done to prevent burnishing of the workpiece. Because reamed holes are the most likely to cause difficulty in tapping (particularly in nickel-base and cobalt-base alloys), chip thickness in reaming should be no less than 0.13 mm (0.005 in.) and preferably 0.25 mm (0.010 in.) or more.

Most production problems in tapping heat-resistant alloys are more readily solved by some alteration in the method of preparing the holes than by changes in the tapping operation. Subject to the limitation on chip thickness mentioned in the preceding paragraph, holes should be made to maximum rather than minimum size to reduce the amount of metal to be removed, thus prolonging tap life.

**Table 30 Nominal feeds and speeds for thread milling heat-resistant alloys with HSS tools**

| Alloy group from Table 6 | Hardness, HB | Speed(a) m/min | sfm | Feed mm/rev | in./rev |
|---|---|---|---|---|---|
| Ni Wrt 1 | 200–300 | 6 | 20 | 0.038 | 0.0015 |
| Ni Wrt 1 | 300–400 | 6 | 20 | 0.025 | 0.001 |
| Ni Wrt 2 | 225–300 | 4.5 | 15 | 0.038 | 0.0015 |
| Ni Wrt 2 | 300–400 | 3 | 10 | 0.025 | 0.001 |
| Ni Wrt 3 | 275–390 | 3 | 10 | 0.025 | 0.001 |
| Ni Wrt 3 | 400–475 | 2 | 7 | 0.025 | 0.001 |
| Ni Wrt 4 | 140–220 | 6 | 20 | 0.038 | 0.0015 |
| Ni Wrt 4 | 240–310 | 6 | 20 | 0.025 | 0.001 |
| Ni Wrt 5 | 180–200 | 21 | 70 | 0.05 | 0.002 |
| Ni Cast 1 | 200–375 | 3 | 10 | 0.025 | 0.001 |
| Ni Cast 2 | 250–320 | 3.6 | 12 | 0.025 | 0.001 |
| Ni Cast 2 | 320–425 | 3 | 10 | 0.025 | 0.001 |
| Co Wrt | 180–230 | 6 | 20 | 0.025 | 0.001 |
| Co Wrt | 270–320 | 4.5 | 15 | 0.025 | 0.001 |
| Co Cast | 220–290 | 3.6 | 12 | 0.025 | 0.001 |
| Co Cast | 290–425 | 3 | 10 | 0.025 | 0.001 |
| Fe Wrt | 180–230 | 11 | 35 | 0.038 | 0.0015 |
| Fe Wrt | 250–320 | 9 | 30 | 0.025 | 0.001 |
| Fe Cast 1 | 135–185 | 27(b) | 90(b) | 0.05(b) | 0.002(b) |
| Fe Cast 2 | 135–185 | 18(b) | 60(b) | 0.05(b) | 0.002(b) |
| Fe Cast 3 | 160–210 | 17(b) | 55(b) | 0.05(b) | 0.002(b) |

(a) With any premium high-speed steel (T15, M33, M41–M47 or S9, S10, S11, S12) except as otherwise specified. (b) M2, M7 (S4, S2) high-speed steel. Source: Metcut Research Associates Inc.

**General Tapping Practice.** Drill presses are ordinarily used for the tapping of heat-resistant alloys because production lots are usually small. For large production lots, the cost of tapping can be decreased by using turret lathes or automatic chucking machines. Regardless of the machine used, it should be equipped with a mechanical feed, such as lead control, and whenever possible, torque-limiting tapping heads with axial float should be used in conjunction with automatic feed. (Devices for controlling feed in tapping are discussed in the article "Tapping" in this Volume.) Electrodischarge machining has also been used to produce internal threads on heat-resistant alloys.

Selection of the proper drill size is also an important factor. Decreasing the thread engagement decreases the torque necessary to drive the tap. A thread engagement of 55 to 60% is recommended.

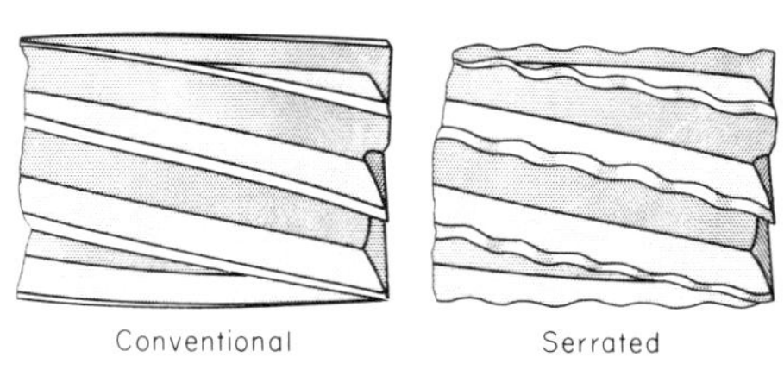


**Fig. 14** Two types of high-speed steel end mills used in the profile milling of Waspaloy rings

**Tap Design.** Spiral-point plug (or gun) taps are preferred for tapping heat-resistant alloys. Taps with three or four 7° spiral flutes are also used. A four-flute tap is preferred to a three-flute tap because it has opposing cutting edges.

The hook angle of the tap should never exceed 15° and should be less for higher work metal hardness. For example, a hook angle of 4 to 6° is suitable for the tapping of work metal having a hardness of about 40 HRC. Taps with lands of minimum width are preferred in order to minimize work hardening of the metal being tapped.

The chamfer should be short; the typical recommendation is three to five threads. A chamfer longer than five threads does not permit cutting deep enough to penetrate the work-hardened layer, and chamfers shorter than three threads cause excessive tooth loads, resulting in chipped or broken taps.

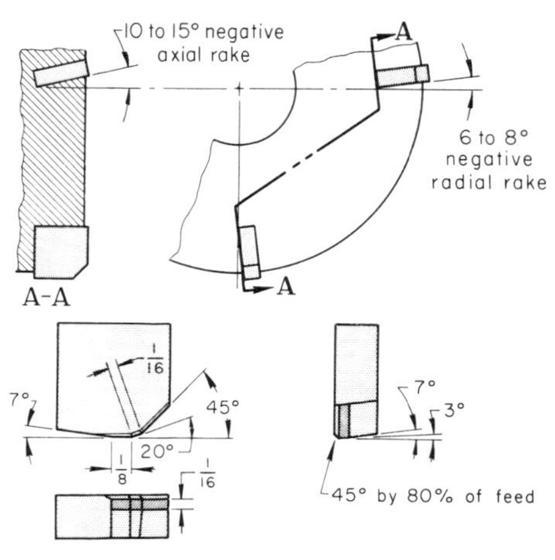


**Fig. 15** Details of face milling cutter used to climb cut age-hardened Inconel X-750

**Table 31 Nominal speeds and feeds for the face milling of heat-resistant alloys with T15 or M42 (S9 or S11) HSS cutters**

| Alloy group from Table 6 | Hardness, HB | Depth of cut, 1 mm (0.04 in.) Speed m/min | sfm | Feed mm/tooth | in./tooth | Depth of cut, 4 mm (0.15 in.) Speed m/min | sfm | Feed mm/tooth | in./tooth | Depth of cut, 8 mm (0.30 in.) Speed m/min | sfm | Feed mm/tooth | in./tooth |
|---|---|---|---|---|---|---|---|---|---|---|---|---|---|
| Ni Wrt 1 | 200–300 | 8 | 25 | 0.10 | 0.004 | 6 | 20 | 0.15 | 0.006 | ... | ... | ... | ... |
| Ni Wrt 1 | 300–400 | 6 | 20 | 0.075 | 0.003 | 4.5 | 15 | 0.13 | 0.005 | ... | ... | ... | ... |
| Ni Wrt 2 | 225–300 | 6 | 20 | 0.10 | 0.004 | 4.5 | 15 | 0.15 | 0.006 | ... | ... | ... | ... |
| Ni Wrt 2 | 300–400 | 4.5 | 15 | 0.075 | 0.003 | 3 | 10 | 0.13 | 0.005 | ... | ... | ... | ... |
| Ni Wrt 3 | 275–390 | 4.5 | 15 | 0.05 | 0.002 | 3 | 10 | 0.075 | 0.003 | ... | ... | ... | ... |
| Ni Wrt 3 | 400–475 | 3.6 | 12 | 0.05 | 0.002 | 2.6 | 8 | 0.075 | 0.003 | ... | ... | ... | ... |
| Ni Wrt 4 | 140–220 | 9 | 30 | 0.075 | 0.003 | 8 | 25 | 0.13 | 0.005 | 6 | 20 | 0.18 | 0.007 |
| Ni Wrt 4 | 240–310 | 8 | 25 | 0.05 | 0.002 | 6 | 20 | 0.10 | 0.004 | 4.5 | 15 | 0.15 | 0.006 |
| Ni Wrt 5 | 180–200 | 35 | 115 | 0.10 | 0.004 | 27 | 90 | 0.15 | 0.006 | 21 | 70 | 0.20 | 0.008 |
| Ni Cast 1 | 200–375 | 8 | 25 | 0.10 | 0.004 | 6 | 20 | 0.15 | 0.006 | 4.5 | 15 | 0.20 | 0.008 |
| Ni Cast 2 | 250–320 | 6 | 20 | 0.05 | 0.002 | 3.6 | 12 | 0.05 | 0.002 | 2.8 | 9 | 0.075 | 0.003 |
| Ni Cast 2 | 320–425 | 4.5 | 15 | 0.05 | 0.002 | 3 | 10 | 0.05 | 0.002 | 2 | 7 | 0.075 | 0.003 |
| Co Wrt | 180–230 | 9 | 30 | 0.05 | 0.002 | 8 | 25 | 0.075 | 0.003 | 6 | 20 | 0.1 | 0.004 |
| Co Wrt | 270–320 | 4.5 | 15 | 0.05 | 0.002 | 3 | 10 | 0.075 | 0.003 | 2 | 7 | 0.1 | 0.004 |
| Co Cast | 220–290 | 4.5 | 15 | 0.05 | 0.002 | 3 | 10 | 0.075 | 0.003 | 2 | 7 | 0.1 | 0.004 |
| Co Cast | 290–425 | 3.6 | 12 | 0.05 | 0.002 | 2.6 | 8 | 0.05 | 0.002 | 1.5 | 5 | 0.075 | 0.003 |
| Fe Wrt | 180–230 | 18 | 60 | 0.13 | 0.005 | 9 | 30 | 0.20 | 0.008 | 4.5 | 15 | 0.25 | 0.010 |
| Fe Wrt | 250–320 | 12 | 40 | 0.10 | 0.004 | 6 | 20 | 0.15 | 0.006 | 4.5 | 15 | 0.20 | 0.008 |
| Fe Cast 1 | 135–185 | 43(a) | 140(a) | 0.20(a) | 0.008(a) | 34(a) | 110(a) | 0.30(a) | 0.012(a) | 24(a) | 80(a) | 0.40(a) | 0.016(a) |
| Fe Cast 2 | 135–185 | 30(a) | 100(a) | 0.20(a) | 0.008(a) | 24(a) | 80(a) | 0.30(a) | 0.012(a) | 18(a) | 60(a) | 0.40(a) | 0.016(a) |
| Fe Cast 3 | 160–210 | 27(a) | 90(a) | 0.20(a) | 0.008(a) | 21(a) | 70(a) | 0.30(a) | 0.012(a) | 15(a) | 50(a) | 0.40(a) | 0.016(a) |

(a) M2 or M7 (S4 or S2) high-speed steel cutters. Source: Metcut Research Associates Inc.

**Table 32 Nominal speeds and feeds for the face milling of heat-resistant alloys with indexable carbide cutters**

| Alloy group from Table 6 | Hardness, HB | Depth of cut, 1 mm (0.04 in.)(a) Speed m/min | sfm | Feed mm/tooth | in./tooth | Depth of cut, 4 mm (0.15 in.)(b) Speed m/min | sfm | Feed mm/tooth | in./tooth | Depth of cut, 8 mm (0.30 in.) Speed m/min | sfm | Feed mm/tooth | in./tooth |
|---|---|---|---|---|---|---|---|---|---|---|---|---|---|
| Ni Wrt 1 | 200–400 | ... | ... | ... | ... | ... | ... | ... | ... | ... | ... | ... | ... |
| Ni Wrt 2 | 225–400 | ... | ... | ... | ... | ... | ... | ... | ... | ... | ... | ... | ... |
| Ni Wrt 3 | 275–475 | ... | ... | ... | ... | ... | ... | ... | ... | ... | ... | ... | ... |
| Ni Wrt 4 | 140–220 | 24 | 80 | 0.13 | 0.005 | 23 | 75 | 0.15 | 0.006 | ... | ... | ... | ... |
| Ni Wrt 4 | 240–310 | 20 | 65 | 0.13 | 0.005 | 18 | 60 | 0.15 | 0.006 | ... | ... | ... | ... |
| Ni Wrt 5 | 180–200 | 84 | 275 | 0.15 | 0.006 | 67 | 220 | 0.20 | 0.008 | 52 | 170 | 0.25 | 0.010 |
| Ni Cast 1 | 200–375 | 23 | 75 | 0.13 | 0.005 | 21 | 70 | 0.15 | 0.006 | ... | ... | ... | ... |
| Ni Cast 2 | 250–425 | ... | ... | ... | ... | ... | ... | ... | ... | ... | ... | ... | ... |
| Co Wrt | 180–230 | 21 | 70 | 0.13 | 0.005 | 20 | 65 | 0.13 | 0.005 | ... | ... | ... | ... |
| Co Wrt | 270–320 | 18 | 60 | 0.13 | 0.005 | 17 | 55 | 0.15 | 0.006 | ... | ... | ... | ... |
| Co Cast | 220–290 | 15 | 50 | 0.13 | 0.005 | 14 | 45 | 0.15 | 0.006 | ... | ... | ... | ... |
| Co Cast | 290–425 | 9 | 30 | 0.13 | 0.005 | 6 | 20 | 0.15 | 0.006 | ... | ... | ... | ... |
| Fe Wrt | 180–230 | 30 | 100 | 0.15 | 0.006 | 24 | 80 | 0.20 | 0.008 | 18 | 60 | 0.25 | 0.010 |
| Fe Wrt | 250–320 | 21 | 70 | 0.15 | 0.006 | 20 | 65 | 0.20 | 0.008 | 12 | 40 | 0.25 | 0.010 |
| Fe Cast 1 | 135–185 | 120 | 400 | 0.20 | 0.008 | 105 | 350 | 0.30 | 0.012 | 84 | 275 | 0.40 | 0.016 |
| Fe Cast 2 | 135–185 | 100 | 325 | 0.20 | 0.008 | 84 | 275 | 0.30 | 0.012 | 66 | 215 | 0.40 | 0.016 |
| Fe Cast 3 | 160–210 | 84 | 275 | 0.20 | 0.008 | 70 | 230 | 0.30 | 0.012 | 55 | 180 | 0.40 | 0.016 |

(a) Carbide grade C-2 or K10, M20. (b) Carbide grade C-2 or K20, M30. (c) Carbide grade C-2 or K30, M40. Source: Metcut Research Associates Inc.

When it is necessary to rough and finish tap heat-resistant alloys, the roughing tap should be undersize on the pitch diameter, and its sharp crests should be backed off. A plug tap should not be followed by a tap of the same pitch diameter, because tap and work are likely to bind and oversize holes may be produced.

**Tap Material.** For small production quantities, taps made of a general-purpose grade of high-speed steel (such as M1) will produce satisfactory results in heat-resistant alloys, but surface treatment of the taps by liquid nitriding is recommended. When larger quantities are tapped, the higher cost of taps made of one of the more highly alloyed high-speed steels (such as M4, M36, or T15) is usually warranted.

**Speeds** of 1.5 to 4.5 m/min (5 to 15 sfm) are satisfactory for tapping most heat-resistant alloys. For tapping iron-base heat-resistant alloys, speeds near the high end of the range result in satisfactory tool life, and speeds higher than 4.5 m/min (15 sfm) have occasionally been used. In tapping nickel-base and cobalt-base heat-resistant alloys, speeds near the low end of the range are necessary.

**Cutting Fluid.** Sulfochlorinated oils should be used for tapping all heat-resistant alloys, and the oil should be supplied in plentiful amounts during the tapping operation Recommended practice is to force the cutting fluid under pressure of about 35 kPa (5 psi) through a nozzle directly into the hole being tapped. If the sulfochlorinated oil is too viscous for the pressure system, it can be diluted with a thinner mineral oil without seriously impairing its characteristics. Chemically active cutting fluids must be removed from the tapped holes to prevent damage to the work metal during subsequent heat treatment or high-temperature service.

**Thread Milling.** Nominal speeds and feeds for thread milling heat-resistant alloys are given in Table 30. Thread milling is most frequently used for cutting external threads, but if tapping is impractical because of alloy characteristics, high hardness, or work-hardened surfaces, thread milling is sometimes used for producing internal threads. The principal limitation of thread milling for internal threads is the size (diameter and length) of the hole to be threaded. It is unlikely that a milling cutter could be operated in holes much smaller than 25 mm (1 in.) in diameter. The length of the hole is also a limiting factor because rigidity is reduced when the spindle is too long for its diameter.

**Example 12: Milling 1⅛-12 Threads in X-40.** Difficulty was encountered in tapping 1⅛-12 3B threads approximately 13 mm (0.500 in.) long (through holes) in X-40 castings. When holes were tapped in one

**Table 33 Nominal feeds and speeds for end milling (peripheral) heat-resistant alloys with high-speed steel (T15, M33, or M41–M47) tool**

| | | | Depth of cut, cutter diameter/2 | | | Depth of cut, 0.5 mm (0.020 in.) | | |
|---|---|---|---|---|---|---|---|---|
| | | | | Feed, mm/tooth (in./tooth), for a cutter diameter of: | | | Feed, mm/tooth (in./tooth), for a cutter diameter of: | |
| Alloy group from Table 6 | Condition | Hardness, HB | Speed, m/min (sfm) | 13 mm (½ in.) | 25–50 mm (1–2 in.) | Speed, m/min (sfm) | 13 mm (½ in.) | 25–50 mm (1–2 in.) |
| Ni Wrt 1 | Annealed or solution treated | 200–300 | 3 (10) | 0.025 (0.001) | 0.050 (0.002) | 6 (20) | 0.025 (0.001) | 0.05 (0.002) |
| Ni Wrt 1 | Solution treated and aged | 300–400 | 2.4 (8) | 0.018 (0.0007) | 0.038 (0.0015) | 4.5 (15) | 0.025 (0.001) | 0.05 (0.002) |
| Ni Wrt 2 | Solution treated | 225–300 | 3 (10) | 0.025 (0.001) | 0.050 (0.002) | 6 (20) | 0.025 (0.001) | 0.05 (0.002) |
| Ni Wrt 2 | Solution treated and aged | 300–400 | 2.4 (8) | 0.018 (0.0007) | 0.038 (0.0015) | 4.5 (15) | 0.025 (0.001) | 0.05 (0.002) |
| Ni Wrt 3 | Solution treated | 275–390 | 2.4 (8) | 0.018 (0.0007) | 0.038 (0.0015) | 4.5 (15) | 0.025 (0.001) | 0.05 (0.002) |
| Ni Wrt 3 | Solution treated and aged | 400–475 | 2 (7) | 0.018 (0.0007) | 0.038 (0.0015) | 3.6 (12) | 0.025 (0.001) | 0.05 (0.002) |
| Ni Wrt 4 | Annealed or solution treated | 140–220 | 4.5 (15) | 0.025 (0.001) | 0.063 (0.0025) | 9 (30) | 0.025 (0.001) | 0.05 (0.002) |
| Ni Wrt 4 | Cold drawn or aged | 240–310 | 3.6 (12) | 0.025 (0.001) | 0.05 (0.002) | 8 (25) | 0.025 (0.001) | 0.05 (0.002) |
| Ni Wrt 5 | As-rolled | 180–200 | 15 (50) | 0.025 (0.001) | 0.05 (0.002) | 27 (90) | 0.025 (0.001) | 0.05 (0.002) |
| Ni Cast 1 | As-cast or cast and aged | 200–375 | 3.6 (12) | 0.025 (0.001) | 0.063 (0.0025) | 8 (25) | 0.025 (0.001) | 0.05 (0.002) |
| Ni Cast 2 | As-cast or cast and aged | 250–320 | 3 (10) | 0.018 (0.0007) | 0.05 (0.002) | 6 (20) | 0.025 (0.001) | 0.05 (0.002) |
| Ni Cast 2 | As-cast or cast and aged | 320–425 | 1.5 (5) | 0.018 (0.0007) | 0.05 (0.002) | 4.5 (15) | 0.025 (0.001) | 0.05 (0.002) |
| Co Wrt | Solution treated | 180–230 | 2.4 (8) | 0.018 (0.0007) | 0.05 (0.002) | 4.5 (15) | 0.025 (0.001) | 0.05 (0.002) |
| Co Wrt | Solution treated and aged | 270–320 | 1.5 (5) | 0.018 (0.0007) | 0.038 (0.0015) | 3.6 (12) | 0.025 (0.001) | 0.05 (0.002) |
| Co Cast | As-cast or cast and aged | 220–290 | 1.5 (5) | 0.018 (0.0007) | 0.038 (0.0015) | 3.6 (12) | 0.025 (0.001) | 0.05 (0.002) |
| Co Cast | As-cast or cast and aged | 290–425 | 1.5 (5) | 0.018 (0.0007) | 0.038 (0.0015) | 3 (10) | 0.025 (0.001) | 0.05 (0.002) |
| Fe Wrt | Solution treated | 180–230 | 6 (20) | 0.025 (0.001) | 0.063 (0.0025) | 12 (40) | 0.025 (0.001) | 0.05 (0.002) |
| Fe Wrt | Solution treated and aged | 250–320 | 4.5 (15) | 0.025 (0.001) | 0.063 (0.0025) | 9 (30) | 0.025 (0.001) | 0.05 (0.002) |
| Fe Cast 1 | Annealed | 135–185 | 26(a) (85)(a) | 0.038(a) (0.0015)(a) | 0.10(a) (0.004)(a) | 44(a) (145)(a) | 0.05(a) (0.002)(a) | 0.13(a) (0.005)(a) |
| Fe Cast 2 | Annealed or normalized | 135–185 | 14(a) (45)(a) | 0.038(a) (0.0015)(a) | 0.10(a) (0.004)(a) | 24(a) (80)(a) | 0.05(a) (0.002)(a) | 0.13(a) (0.005)(a) |
| Fe Cast 3 | As-cast | 160–210 | 12(a) (40)(a) | 0.038(a) (0.0015)(a) | 0.10(a) (0.004)(a) | 21(a) (70)(a) | 0.05(a) (0.002)(a) | 0.13(a) (0.005)(a) |

(a) M2, M3, M7 high-speed steel. Source: Metcut Research Associates Inc.

**Table 34 Nominal feeds and speeds for end milling (peripheral) of heat-resistant alloys with carbide (C-2) tools**

| | | | Depth of cut, cutter diameter/2 | | | Depth of cut, 0.5 mm (0.020 in.) | | |
|---|---|---|---|---|---|---|---|---|
| | | | | Feed, mm/tooth (in./tooth), for a cutter diameter of: | | | Feed, mm/tooth (in./tooth), for a cutter diameter of: | |
| Alloy group from Table 6 | Condition | Hardness, HB | Speed, m/min (sfm) | 19 mm(a) (¾ in.)(a) | 25–50 mm (1–2 in.) | Speed, m/min (sfm) | 13 mm (½ in.) | 25–50 mm (1–2 in.) |
| Ni Wrt 1 | Annealed or solution treated | 200–300 | 14 (45) | 0.038 (0.0015) | 0.05 (0.002) | 24 (80) | 0.025 (0.001) | 0.05 (0.002) |
| Ni Wrt 1 | Solution treated and aged | 300–400 | 11 (35) | 0.025 (0.001) | 0.038 (0.0015) | 18 (60) | 0.025 (0.001) | 0.05 (0.002) |
| Ni Wrt 2 | Solution treated | 225–300 | 12 (40) | 0.038 (0.0015) | 0.05 (0.002) | 21 (70) | 0.025 (0.001) | 0.05 (0.002) |
| Ni Wrt 2 | Solution treated and aged | 300–400 | 11 (35) | 0.025 (0.001) | 0.038 (0.0015) | 18 (60) | 0.025 (0.001) | 0.05 (0.002) |
| Ni Wrt 3 | Solution treated | 275–390 | 8 (25) | 0.025 (0.001) | 0.025 (0.001) | 14 (45) | 0.025 (0.001) | 0.05 (0.002) |
| Ni Wrt 3 | Solution treated and aged | 400–475 | 6 (20) | 0.025 (0.001) | 0.025 (0.001) | 12 (40) | 0.025 (0.001) | 0.05 (0.002) |
| Ni Wrt 4 | Annealed or solution treated | 140–220 | 18 (60) | 0.025 (0.001) | 0.038 (0.0015) | 30 (100) | 0.025 (0.001) | 0.05 (0.002) |
| Ni Wrt 4 | Cold drawn or aged | 240–310 | 17 (55) | 0.025 (0.001) | 0.038 (0.0015) | 27 (90) | 0.025 (0.001) | 0.05 (0.002) |
| Ni Wrt 5 | As-rolled | 180–200 | 52 (170) | 0.038 (0.0015) | 0.05 (0.002) | 84 (275) | 0.025 (0.001) | 0.075 (0.003) |
| Ni Cast 1 | As-cast or cast and aged | 200–375 | 14 (45) | 0.025 (0.001) | 0.038 (0.0015) | 23 (75) | 0.025 (0.001) | 0.05 (0.002) |
| Ni Cast 2 | As-cast or cast and aged | 250–320 | 14 (45) | 0.025 (0.001) | 0.038 (0.0015) | 23 (75) | 0.025 (0.001) | 0.05 (0.002) |
| Ni Cast 2 | As-cast or cast and aged | 320–425 | 8 (25) | 0.025 (0.001) | 0.025 (0.001) | 14 (45) | 0.025 (0.001) | 0.05 (0.002) |
| Co Wrt | Solution treated | 180–230 | 12 (40) | 0.025 (0.001) | 0.038 (0.0015) | 21 (70) | 0.025 (0.001) | 0.05 (0.002) |
| Co Wrt | Solution treated and aged | 270–320 | 11 (35) | 0.025 (0.001) | 0.025 (0.001) | 18 (60) | 0.025 (0.001) | 0.05 (0.002) |
| Co Cast | As-cast or cast and aged | 220–290 | 11 (35) | 0.025 (0.001) | 0.038 (0.0015) | 18 (60) | 0.025 (0.001) | 0.05 (0.002) |
| Co Cast | As-cast or cast and aged | 290–425 | 6 (20) | 0.025 (0.001) | 0.025 (0.001) | 12 (40) | 0.025 (0.001) | 0.05 (0.002) |
| Fe Wrt | Solution treated | 180–230 | 27 (90) | 0.038 (0.0015) | 0.05 (0.002) | 46 (150) | 0.025 (0.001) | 0.05 (0.002) |
| Fe Wrt | Solution treated and aged | 250–320 | 18 (60) | 0.038 (0.0015) | 0.05 (0.002) | 30 (100) | 0.025 (0.001) | 0.05 (0.002) |
| Fe Cast 1 | Annealed | 135–185 | 84(b) (275)(b) | 0.025(b)(c) (0.001)(b)(c) | 0.075(b) (0.003)(b) | 135(b) (450)(b) | 0.025(b) (0.001)(b) | 0.13(b) (0.005)(b) |
| Fe Cast 2 | Annealed or normalized | 135–185 | 60 (200) | 0.025(c) (0.001)(c) | 0.075 (0.003) | 100 (325) | 0.025 (0.001) | 0.10 (0.004) |
| Fe Cast 3 | As-cast | 160–210 | 55 (180) | 0.038(c) (0.0015)(c) | 0.075 (0.003) | 90 (300) | 0.05 (0.002) | 0.10 (0.004) |

(a) The nickel-base, cobalt-base, and wrought iron-base alloys should be milled with a 19 mm (¾ in.) diam cutter (minimum) when the depth of cut is diam/2. (b) With carbide C-5 (P20, M20) grade. (c) Feed rate with a 13 mm (½ in.) diam cutter. Source: Metcut Research Associates Inc.

pass, either the casting cracked or the tap seized or broke off. Progressive tapping was tried with some success, but tapping became more difficult with each succeeding pass, and picking up the previous thread became a problem. The work metal resisted any appreciable amount of stock removal, making it difficult to cut the work-hardened surface.

Ultimately, the thread milling process was tried and it proved successful, partly because it decreased chip load per tooth. A tungsten carbide cutter and sulfochlorinated cutting fluid were used. Tool life was extended to 28 parts per grind. Threading required 13 min per piece. Approximately 3000 parts were successfully threaded by milling.

## Milling

Climb milling is generally preferred to conventional milling if suitable equipment is available. Climb milling requires the ultimate in rigidity and a machine equipped with a backlash eliminator. However, cuts deeper than 1.5 mm (0.060 in.) are seldom attempted with the climb milling of heat-resistant alloys, because it is virtually impossible to obtain the required rigidity.

**Cutter Design.** For milling heat-resistant alloys, two principles of cutter design must be given special consideration. First, tooth strength must be greater than that required for milling steel or cast iron, and second, relief angles must be large enough to prevent rubbing action and consequent work hardening of the alloy being cut.

Regardless of the cutter material, inserted blades are used on nearly all but the smallest cutters, because even under the most favorable machining conditions, the life of the cutting edges is short. Mechanical methods of securing the blades in the cutter body are preferred because replacement of chipped or broken blades is easier. Cutter life can sometimes be greatly increased by small design changes, as

**Table 35 Nominal speeds for power band sawing of heat-resistant alloys with high-speed steel blades**

Table 36 gives tooth form and nominal pitches of the blades.

| Alloy group from Table 6 | Condition | Hardness, HB | Speed, m/min (sfm), with a material thickness of: <13 mm (<½ in.) | 13–25 mm (½–1 in.) | 25–75 mm (1–3 in.) | >75 mm (>3 in.) |
|---|---|---|---|---|---|---|
| Ni Wrt 1 | Annealed or solution treated | 200–300 | 15 (50) | 14 (45) | 11 (35) | 9 (30) |
| Ni Wrt 1 | Solution treated and aged | 300–400 | 12 (40) | 11 (35) | 9 (30) | 8 (25) |
| Ni Wrt 2 | Solution treated | 225–300 | 15 (50) | 14 (45) | 11 (35) | 9 (30) |
| Ni Wrt 2 | Solution treated and aged | 300–400 | 9 (30) | 8 (25) | 6 (20) | 4.5 (15) |
| Ni Wrt 3 | Solution treated | 275–390 | 9 (30) | 8 (25) | 6 (20) | 4.5 (15) |
| Ni Wrt 3 | Solution treated and aged | 400–475 | ... | ... | ... | ... |
| Ni Wrt 4 | Annealed or solution treated | 140–220 | 18 (60) | 17 (55) | 14 (45) | 11 (35) |
| Ni Wrt 4 | Cold drawn or aged | 240–310 | 15 (50) | 14 (45) | 11 (35) | 9 (30) |
| Ni Wrt 5 | As-rolled | 180–200 | 27 (90) | 24 (80) | 20 (65) | 17 (55) |
| Ni Cast 1 | As-cast or cast and aged | 200–375 | 15 (50) | 14 (45) | 11 (35) | 9 (30) |
| Ni Cast 2 | As-cast or cast and aged | 250–320 | 12 (40) | 11 (35) | 9 (30) | 8 (25) |
| Ni Cast 2 | As-cast or cast and aged | 320–425 | ... | ... | ... | ... |
| Co Wrt | Solution treated | 180–230 | 12 (40) | 11 (35) | 9 (30) | 8 (25) |
| Co Wrt | Solution treated and aged | 270–320 | 11 (35) | 9 (30) | 8 (25) | 6 (20) |
| Co Cast | As-cast or cast and aged | 220–290 | 9 (30) | 8 (25) | 6 (20) | 4.5 (15) |
| Co Cast | As-cast or cast and aged | 290–425 | ... | ... | ... | ... |
| Fe Wrt | Solution treated | 180–230 | 27 (90) | 24 (80) | 20 (65) | 17 (55) |
| Fe Wrt | Solution treated and aged | 250–320 | 23 (75) | 20 (65) | 17 (55) | 14 (45) |
| Fe Cast 1 | Annealed | 135–185 | 41 (135) | 37 (120) | 30 (100) | 24 (80) |
| Fe Cast 2 | Annealed or normalized | 135–185 | 24 (80) | 21 (70) | 18 (60) | 15 (50) |
| Fe Cast 3 | As-cast | 160–210 | 23 (75) | 20 (65) | 17 (55) | 14 (45) |
| MA 754 and MA 956 | Mechanically alloyed | 270–277 | 15 (50) | 14 (45) | 11 (35) | 9 (30) |
| MA 6000 | Mechanically alloyed | 450 | ... | ... | ... | ... |

Source: Metcut Research Associates Inc.

demonstrated in the following two examples.

**Example 13: Serrated Versus Straight End Mills for the Profile Milling of Waspaloy.** Waspaloy rings 1170 mm (46 in.) in diameter were profile milled on the outside with a conventional cutter made of M3, class 2, high-speed steel and with a serrated cutter made of the same material. Both tools are shown in Fig. 14.

Using the conventional cutter, four roughing cuts and one finishing cut were necessary to remove the required amount of metal. Total machining time was 12½ h. Using the same speed (4.5 m/min, or 15 sfm) and feed (0.10 mm/tooth, or 0.004 in./tooth), the serrated cutter removed the same amount of metal in 7½ h by eliminating two of the roughing passes. Given the same feed, the serrated cutter produced a thicker chip and penetrated the work-hardened layer better.

Both types of cutters required sharpening after 145 min of cutting, which produced a wear land of approximately 0.30 mm (0.012 in.). The serrated cutter was used only for roughing because the finish it produced was not acceptable as a final finish. However, this was not a disadvantage, because a freshly sharpened cutter had always been used for finish milling; therefore, no extra tool change was involved.

**Example 14: Change of Carbide Cutter for Increased Tool Life.** Chamfering the front cutting edge of the carbide tip of a 150 mm (6 in.) diam face milling cutter resulted in a 25% increase in tool life and a reduction in costs in climb milling aged Inconel X-750. The tip was of grade C-5 carbide. The improved cutter is shown in Fig. 15.

The use of a cutter that had been designed especially for climb milling, and of a soluble-oil cutting fluid (1 part oil to 15 parts water), minimized cutter cratering and chipping. The major contribution to tool life, however, resulted from the chamfer (45° by 80% of feed; or 1.8 mm, or 0.072 in.). The cutter was operated at a speed of 18 m/min (60 sfm), a feed of 0.23 mm/tooth (0.009 in./tooth), and a depth of cut of 1.5 mm (0.060 in.). Width of cut was 63 mm (2½ in.). The cutting fluid was brushed on.

**Cutter Material.** Because of the interrupted cutting action, high-speed steel is used for cutters in most applications for milling heat-resistant alloys. However, carbide is frequently more economical than high-speed steel when milling the more difficult-to-machine alloys, such as René 41 and MA 6000. Small solid-carbide end mills have been successfully used in a few applications. The more highly alloyed grades of high-speed steel usually outperform the general-purpose grades, but there is less difference in performance between the two grades in milling cutters than in some other tools used for machining heat-resistant alloys.

**Speed, Feed, and Depth of Cut.** Nominal speeds and feeds for the face and end (peripheral) milling of heat-resistant alloys are given in Tables 31 to 34. The speeds listed in these tables assume that the operations are conducted under conditions of best practice, that is, adequate rigidity in the setup, optimum tool angles, and a plentiful supply of cutting fluid.

When one or more factors are less favorable in a specific application (such as excessive work metal hardness or inadequate rigidity), speed must be reduced. In some applications, it is necessary to mill nickel-base and cobalt-base alloys at speeds as low as 1.5 m/min (5 sfm) for acceptable tool life.

Optimum feed depends primarily on the type of milling operation and the depth of cut. The rate of feed is usually decreased for finishing cuts. The feeds used for slab milling do not differ greatly from those used for face milling (Table 31).

**Table 36 Nominal pitches for power band sawing of heat-resistant alloys**

| Alloy group from Table 6 | Material thickness, mm | in. | Tooth form(a) | Pitch, teeth/25 mm (1 in.) |
|---|---|---|---|---|
| Fe Cast 1, Fe Cast 2, Fe Cast 3 | <13 | <½ | P | 10–14 |
| | 13–25 | ½–1 | P | 8–10 |
| | 25–75 | 1–3 | P | 6–8 |
| | >75 | >3 | C, P | 3–4 |
| Remaining alloy groups listed in Table 35(b) | <13 | <½ | P | 10–14 |
| | 13–25 | ½–1 | P | 6 |
| | 25–75 | 1–3 | P | 4 |
| | >75 | >3 | C | 3 |

(a) P, precision; C, claw. (b) The following alloy groups from Table 35 are not band sawed: MA 6000, Ni Wrt 3 (solution treated and aged, 400–475 HB), Ni Cast 2 (320–425 HB), Co Cast (290–425 HB). Source: Metcut Research Associates Inc.

**Table 37 Nominal speeds and feeds for the circular sawing of heat-resistant alloys with high-speed steel (M2, M7, or S4, S2) blades**

See Table 38 for saw pitch.

| Alloy group from Table 6 | Condition | Hardness, HB | Speed, m/min (sfm), at a solid stock diameter thickness of: 6–75 mm (¼–3 in.) | 75–150 mm (3–6 in.) | 150–225 mm (6–9 in.) | 225–400 mm (9–15 in.) |
|---|---|---|---|---|---|---|
| Ni Wrt 1 | Annealed or solution treated | 200–300 | 4.5 (15) | 3.6 (12) | 3 (10) | 2.5 (8) |
| Ni Wrt 1 | Solution treated and aged | 300–400 | 4.5 (15) | 3.6 (12) | 3 (10) | 2.5 (8) |
| Ni Wrt 2 | Solution treated | 225–300 | 4.5 (15) | 3.6 (12) | 3 (10) | 2.5 (8) |
| Ni Wrt 2 | Solution treated and aged | 300–400 | 4.5 (15) | 3.6 (12) | 3 (10) | 2.5 (8) |
| Ni Wrt 3 | Solution treated | 275–390 | 4.5 (15) | 3.6 (12) | 3 (10) | 2.5 (8) |
| Ni Wrt 3 | Solution treated and aged | 400–475 | 3 (10) | 2.5 (8) | 1.5 (5) | 1.5 (5) |
| Ni Wrt 4 | Annealed or solution treated | 140–220 | 9 (30) | 8 (25) | 6 (20) | 4.5 (15) |
| Ni Wrt 4 | Cold drawn or aged | 240–310 | 8 (25) | 6 (20) | 4.5 (15) | 3.6 (12) |
| Ni Wrt 5 | As-rolled | 180–200 | 18 (60) | 15 (50) | 12 (40) | 9 (30) |
| Ni Cast 1 | As-cast or cast and aged | 200–375 | 4.5 (15) | 3.6 (12) | 3 (10) | 2.5 (8) |
| Ni Cast 2 | As-cast or cast and aged | 250–320 | 6 (20) | 4.5 (15) | 3.6 (12) | 3 (10) |
| Ni Cast 2 | As-cast or cast and aged | 320–425 | 3 (10) | 2.5 (8) | 1.5 (5) | 1.5 (5) |
| Co Wrt | Solution treated | 180–230 | 6 (20) | 4.5 (15) | 3.6 (12) | 3 (10) |
| Co Wrt | Solution treated and aged | 270–320 | 4.5 (15) | 3.6 (12) | 3 (10) | 2.5 (8) |
| Co Cast | As-cast or cast and aged | 220–290 | 4.5 (15) | 3.6 (12) | 3 (10) | 2.5 (8) |
| Co Cast | As-cast or cast and aged | 290–425 | 3 (10) | 2.5 (8) | 1.5 (5) | 1.5 (5) |
| Fe Wrt | Solution treated | 180–230 | 8 (25) | 6 (20) | 4.5 (15) | 3.6 (12) |
| Fe Wrt | Solution treated and aged | 250–320 | 8 (25) | 6 (20) | 4.5 (15) | 3.6 (12) |
| Fe Cast 1 | Annealed | 135–185 | 30 (100) | 24 (80) | 20 (65) | 15 (50) |
| Fe Cast 2 | Annealed or normalized | 135–185 | 21 (70) | 17 (55) | 14 (45) | 11 (35) |
| Fe Cast 3 | As-cast | 160–210 | 15 (50) | 12 (40) | 9 (30) | 8 (25) |

| Alloy group from Table 6 | Condition | Hardness, HB | Feed, mm/tooth (in./tooth), at a solid stock diameter thickness of: 6–75 mm (¼–3 in.) | 75–150 mm (3–6 in.) | 150–225 mm (6–9 in.) | 225–400 mm (9–15 in.) |
|---|---|---|---|---|---|---|
| Ni Wrt 1 | Annealed or solution treated | 200–300 | 0.075 (0.003) | 0.075 (0.003) | 0.10 (0.004) | 0.10 (0.004) |
| Ni Wrt 1 | Solution treated and aged | 300–400 | 0.075 (0.003) | 0.075 (0.003) | 0.10 (0.004) | 0.10 (0.004) |
| Ni Wrt 2 | Solution treated | 225–300 | 0.075 (0.003) | 0.075 (0.003) | 0.10 (0.004) | 0.10 (0.004) |
| Ni Wrt 2 | Solution treated and aged | 300–400 | 0.075 (0.003) | 0.075 (0.003) | 0.10 (0.004) | 0.10 (0.004) |
| Ni Wrt 3 | Solution treated | 275–390 | 0.075 (0.003) | 0.075 (0.003) | 0.10 (0.004) | 0.10 (0.004) |
| Ni Wrt 3 | Solution treated and aged | 400–475 | 0.050 (0.002) | 0.050 (0.002) | 0.075 (0.003) | 0.075 (0.003) |
| Ni Wrt 4 | Annealed or solution treated | 140–220 | 0.075 (0.003) | 0.075 (0.003) | 0.10 (0.004) | 0.10 (0.004) |
| Ni Wrt 4 | Cold drawn or aged | 240–310 | 0.075 (0.003) | 0.075 (0.003) | 0.10 (0.004) | 0.10 (0.004) |
| Ni Wrt 5 | As-rolled | 180–200 | 0.10 (0.004) | 0.10 (0.004) | 0.15 (0.006) | 0.15 (0.006) |
| Ni Cast 1 | As-cast or cast and aged | 200–375 | 0.075 (0.003) | 0.075 (0.003) | 0.10 (0.004) | 0.10 (0.004) |
| Ni Cast 2 | As-cast or cast and aged | 250–320 | 0.075 (0.003) | 0.075 (0.003) | 0.10 (0.004) | 0.10 (0.004) |
| Ni Cast 2 | As-cast or cast and aged | 320–425 | 0.050 (0.002) | 0.050 (0.002) | 0.075 (0.003) | 0.075 (0.003) |
| Co Wrt | Solution treated | 180–230 | 0.075 (0.003) | 0.075 (0.003) | 0.10 (0.004) | 0.10 (0.004) |
| Co Wrt | Solution treated and aged | 270–320 | 0.075 (0.003) | 0.075 (0.003) | 0.10 (0.004) | 0.10 (0.004) |
| Co Cast | As-cast or cast and aged | 220–290 | 0.075 (0.003) | 0.075 (0.003) | 0.10 (0.004) | 0.10 (0.004) |
| Co Cast | As-cast or cast and aged | 290–425 | 0.050 (0.002) | 0.050 (0.002) | 0.075 (0.003) | 0.075 (0.003) |
| Fe Wrt | Solution treated | 180–230 | 0.075 (0.003) | 0.075 (0.003) | 0.10 (0.004) | 0.10 (0.004) |
| Fe Wrt | Solution treated and aged | 250–320 | 0.075 (0.003) | 0.075 (0.003) | 0.10 (0.004) | 0.10 (0.004) |
| Fe Cast 1 | Annealed | 135–185 | 0.15 (0.006) | 0.15 (0.006) | 0.20 (0.008) | 0.20 (0.008) |
| Fe Cast 2 | Annealed or normalized | 135–185 | 0.15 (0.006) | 0.15 (0.006) | 0.20 (0.008) | 0.20 (0.008) |
| Fe Cast 3 | As-cast | 160–210 | 0.15 (0.006) | 0.15 (0.006) | 0.20 (0.008) | 0.20 (0.008) |

Source: Metcut Research Associates Inc.

**Table 38 Nominal pitches for the circular sawing of heat-resistant alloys with high-speed steel saws**

| Alloy group from Table 37 | Diameter or thickness of solid stock, mm | in. | Pitch, mm/tooth | in./tooth |
|---|---|---|---|---|
| Ni Wrt 1 (200–300 HB), Ni Wrt 2 (225–300 HB), Ni Wrt 4 (140–220 HB), Ni Wrt 5 (180–200 HB), Co Wrt (180–230 HB), Fe Wrt (180–230 HB), Fe Cast 1, Fe Cast 2, Fe Cast 3 | 6–75 | ¼–3 | 5–19 | 0.20–0.75 |
| | 75–150 | 3–6 | 13–28 | 0.50–1.10 |
| | 150–225 | 6–9 | 20–33 | 0.75–1.30 |
| | 225–400 | 9–15 | 23–44 | 0.90–1.75 |
| Ni Wrt 1 (300–400 HB), Ni Wrt 2 (300–400 HB), Ni Wrt 3 (275–390 HB), Ni Wrt 4 (240–310 HB), Ni Cast 1, Ni Cast 2 (250–320 HB), Co Wrt (270–320 HB), Co Cast (220–290 HB), Fe Wrt (250–320 HB) | 6–75 | ¼–3 | 5–18 | 0.20–0.7 |
| | 75–150 | 3–6 | 13–25 | 0.50–1.00 |
| | 150–225 | 6–9 | 20–30 | 0.75–1.20 |
| | 225–400 | 9–15 | 23–40 | 0.90–1.50 |
| Ni Wrt 3 (400–475 HB), Ni Cast 2 (320–425 HB), and Co Cast (290–425 HB) | 6–75 | ¼–3 | 4–15 | 0.15–0.60 |
| | 75–150 | 3–6 | 10–23 | 0.40–0.90 |
| | 150–225 | 6–9 | 19–28 | 0.75–1.10 |
| | 225–400 | 9–15 | 20–35 | 0.80–1.35 |

Source: Metcut Research Associates Inc.

The feeds used for end milling are necessarily less than those used for face milling because of the inherent lack of rigidity in end milling. In addition to depth of cut, the size of the cutter is a significant variable influencing optimum feed in end milling. Nominal feeds for end milling, covering a range of cutter sizes, are given in Tables 33 and 34.

Depths of cut for rough face and slab milling are usually in the range of 1.5 to 2.5 mm (0.060 to 0.100 in.). The availability of sufficient power is frequently a factor in limiting the depth of cut. Finish face or peripheral milling is usually done with a depth of cut of 0.6 mm (0.025 in.) or less, depending to some extent on the required surface finish. To minimize transverse streaks and to obtain the best surface finish, depth of cut often ranges from 0.1 to 0.4 mm (0.005 to 0.015 in.). Depth of cut less than 0.1 mm (0.005 in.) is not recommended because it is likely to result in excessive work hardening of the alloy being milled.

**Cutting Fluid.** Sulfochlorinated oil introduced in copious amounts at the exhaust side of the cutter is the preferred condition for milling heat-resistant alloys. Soluble-oil emulsions are often used, and they provide better cooling for the tools and workpieces than straight oils. However, some sacrifice in surface finish and tool life attends the use of soluble-oil emulsions compared to sulfochlorinated oils. The latter are often diluted with mineral oil (up to 50%) to obtain fluidity with no large sacrifice in ability to promote cutting action and good surface finish. Workpieces milled with sulfochlorinated or other chemically active oils must be thoroughly cleaned before being placed in service at elevated temperature.

## Sawing

Because of their applications, heat-resistant alloys are less frequently machined by sawing than by most other processes.

**Band Sawing.** Nominal pitches (teeth per 25 mm, or 1 in.) and speeds for the power band sawing of several groups of heat-resistant alloys are given in Tables 35 and 36. Power band sawing is not applicable for heat-resistant alloys with hardnesses higher than 400 HB.

The effect of band speed on tool life is indicated in the following tabulation for the cutoff band sawing of bars of A-286 (35 HRC) 38 mm (1½ in.) thick with a high-speed steel band:

| Band speed, m/min | sfm | Band life, $cm^2$ cut | $in.^2$ cut |
|---|---|---|---|
| 21 | 70 | 1080 | 168 |
| 27 | 90 | 390 | 60 |

The sawing comparison was made at a hydraulically controlled feed rate of 25 mm/min (1.0 in./min) for a cutting rate of 960 $mm^2$/min (1.5 $in.^2$/min), with flow application of a 1 to 1 mixture of sulfurized and mineral oils.

**Circular Sawing and Slitting.** Nominal pitches, speeds, and feeds for circular saw-

**Table 39 Nominal speeds and feeds for the power hacksawing of heat-resistant alloys with a high-speed steel blade**

See Table 40 for blade pitch.

| Alloy group from Table 6 | Condition | Hardness, HB | Speed, strokes/min, with a light feed pressure(a) and a material thickness of: <6 mm (<¼ in.) | 6–19 mm (¼–¾ in.) | 19–50 mm (¾–2 in.) | >50 mm (>2 in.) |
|---|---|---|---|---|---|---|
| Ni Wrt 1 | Annealed or solution treated | 200–300 | ... | ... | ... | ... |
| Ni Wrt 1 | Solution treated and aged | 300–400 | ... | ... | ... | ... |
| Ni Wrt 2 | Solution treated | 225–300 | 35 | 35 | 35 | 35 |
| Ni Wrt 2 | Solution treated and aged | 300–400 | ... | ... | ... | ... |
| Ni Wrt 3 | Solution treated | 275–390 | 35 | 35 | 35 | 35 |
| Ni Wrt 3 | Solution treated and aged | 400–475 | ... | ... | ... | ... |
| Ni Wrt 4 | Annealed or solution treated | 140–220 | 35 | 35 | 35 | 35 |
| Ni Wrt 4 | Cold drawn or aged | 240–310 | 30 | 30 | 30 | 30 |
| Ni Wrt 5 | As-rolled | 180–200 | 70 | 70 | 60(b) | 60(b) |
| Ni Cast 1 | As-cast or cast and aged | 200–375 | ... | ... | ... | ... |
| Ni Cast 2 | As-cast or cast and aged | 250–320 | ... | ... | ... | ... |
| Ni Cast 2 | As-cast or cast and aged | 320–425 | ... | ... | ... | ... |
| Co Wrt | Solution treated | 180–230 | 35 | 35 | 35 | 35 |
| Co Wrt | Solution treated and aged | 270–320 | ... | ... | ... | ... |
| Co Cast | As-cast or cast and aged | 220–290 | 30 | 30 | 30 | 30 |
| Co Cast | As-cast or cast and aged | 290–425 | ... | ... | ... | ... |
| Fe Wrt | Solution treated | 180–230 | 45 | 45 | 40 | 40(b) |
| Fe Wrt | Solution treated and aged | 250–320 | 40 | 40 | 35 | 35 |
| Fe Cast 1 | Annealed | 135–185 | 120 | 120(b) | 110(b) | 110(b) |
| Fe Cast 2 | Annealed or normalized | 135–185 | 60 | 60 | 55(b) | 55(b) |
| Fe Cast 3 | As-cast | 160–210 | 50 | 50 | 45 | 45(b) |

(a) A light feed pressure with a feed rate of 0.075 mm/stroke (0.003 in./stroke) for all applicable alloys except as otherwise specified. (b) Medium feed pressure with a feed rate of 0.15 mm/stroke (0.006 in./stroke). Source: Metcut Research Associates Inc.

**Table 40 Nominal blade pitch for the power hacksawing of heat-resistant alloys**

| Alloy group from Table 39 | Material thickness, mm | in. | Pitch, teeth per 25 mm (1 in.) |
|---|---|---|---|
| Ni Wrt 5 | <6 | <¼ | 10 |
| | 6–19 | ¼–¾ | 10 |
| | 19–50 | ¾–2 | 6 |
| | >50 | >2 | 4 |
| Ni Wrt 2 (225–300 HB), Ni Wrt 3 (275–390 HB), Ni Wrt 4, Co Wrt (180–230 HB), Co Cast (220–290 HB), Fe Wrt | <6 | <¼ | 10 |
| | 6–19 | ¼–¾ | 10 |
| | 19–50 | ¾–2 | 6 |
| | >50 | >2 | 6 |
| Fe Cast 1 | <6 | <¼ | 10 |
| | 6–19 | ¼–¾ | 6 |
| | 19–50 | ¾–2 | 6 |
| | >50 | >2 | 4 |
| Fe Cast 2, Fe Cast 3 | <6 | <¼ | 10 |
| | 6–19 | ¼–¾ | 10 |
| | 19–50 | ¾–2 | 6 |
| | >50 | >2 | 4 |

Source: Metcut Research Associates Inc.

ing are presented in Tables 37 and 38. The diameter or thickness of the work metal is the major variable influencing pitch, although work metal composition and hardness also have some effect.

**Hacksawing.** Nominal speeds and feeds for power hacksawing are given in Table 39. As stated in the footnotes, a feed rate of 0.075 mm/stroke (0.003 in./stroke) is applicable for most of the nickel-base, cobalt-base, and wrought iron-base alloys with thicknesses from 6 to 50 mm (¼ to 2 in.). The iron-base casting alloys, the iron-base wrought alloys with hardnesses of 180 to 230 HB, and the alloys in the Ni Wrt 5 group allow slightly higher feeds for thicker materials. Pitch is varied according to workpiece thickness, as indicated in Table 40.

**Cutting Fluids.** Moderate-duty to heavy-duty chemical emulsions are widely used and are generally recommended as cutting fluids for sawing heat-resistant alloys. Plain soluble-oil emulsions have also provided satisfactory results. Regardless of the type of cutting fluid used, a plentiful supply at the area being sawed is important.

## Grinding

Even when operating conditions are favorable, heat-resistant alloys are more difficult and costly to grind than low-alloy steels. Because high-temperature nickel and cobalt alloys are sensitive to the level of energy used during processing, metallurgical alterations and microcracking may occur at the surface. The altered material zones or layers can attain a substantial proportion of the thickness of thin components, resulting in a deleterious effect on the requirements for the surface being produced. These requirements include low distortion, absence of cracks, fine finish, or high fatigue strength. Therefore, parameters and conditions for the grinding of heat-resistant alloys should be controlled to planned values in order to obtain high-integrity surfaces. Detailed information on the effects of processing on surface properties can be found in the article "Surface Finish and Surface Integrity" in this Volume.

**Table 41 Alloy groups referred to in Tables 42 to 46**

Compositions for these alloys can be found in Tables 1 to 3.

| Group | Condition | Typical alloys |
|---|---|---|
| Fe-1 | Wrought, solution treated, or solution treated and aged | A-286, Discaloy, Incoloy Alloy 800, Incoloy Alloy 800H, Incoloy Alloy 801, Incoloy Alloy 802, N-155, 16-25-6, 19-9DL |
| Ni-1 | Wrought, annealed, solution treated, or solution treated and aged | Astroloy, Inconel Alloy 700, Inconel Alloy 718, Inconel Alloy X-750, Nimonic 75, Nimonic 90, René 41, René 95, Udimet 500, Udimet 700, Waspaloy |
| Ni-2 | Wrought, annealed, solution treated, cold drawn, or aged | Hastelloy Alloy B, Hastelloy Alloy C, Hastelloy Alloy C-276, Hastelloy Alloy G, Incoloy Alloy 804, Incoloy Alloy 825, Inconel Alloy 600, Udimet 630 |
| Ni-3 | Cast, as-cast, or cast and aged | B-1900, Hastelloy Alloy D, IN-100, IN-738, Inconel Alloy 718, M252, MAR-M200, MAR-M432, René 80, René 125, TRW VI A, Udimet 700 |
| Co-1 | Wrought, solution treated, or solution treated and aged | AiResist 213, Haynes Alloy 25, Haynes Alloy 188, J-1570, MAR-M905, S-816 |
| Co-2 | Cast, as-cast, or cast and aged | AiResist 13, AiResist 215, FSX-414, HS-6, HS-21, MAR-M302, MAR-M509, WI-52, X-45 |

**Table 42 Nominal processing conditions for the surface grinding of heat-resistant alloys with straight wheels**

Conditions given are for a unit downfeed of 0.025 mm (0.001 in.) per series of passes in roughing and 0.013 mm (0.0005 in.) in finishing. Table speed is 15–30 m/min (50–100 sfm). Grinding fluid may be a water-base soluble-oil emulsion or a synthetic solution, or sulfurized oil.

| Alloy group(a) | Hardness, HB | Wheel: Classification | Wheel: Speed, m/s (sfm) | Cross feed per pass, fraction of wheel width |
|---|---|---|---|---|
| Fe-1 | 180–320 | A46HV | 15–20 (3000–4000) | 1/10 |
| Ni-1 | 200–475 | A46HV | 15–18 (3000–3500) | 1/12 |
| Ni-2 | 140–310 | A46HV | 15–20 (3000–4000) | 1/10 |
| Ni-3 | 200–425 | A46HV | 15–18 (3000–3500) | 1/12 |
| Co-1 | 180–320 | A46HV | 15–18 (3000–3500) | 1/12 |
| Co-2 | 220–425 | A46HV | 15–18 (3000–3500) | 1/12 |

(a) See Table 41 for designations of alloy groups and typical alloys in each group. Source: Metcut Research Associates Inc.

**Table 43 Nominal processing conditions for the cylindrical grinding of heat-resistant alloys**

Data given are for a traverse rate of one-fifth the wheel width per revolution of work in roughing and one-tenth the wheel width in finishing. Work speed is 15–30 m/min (50–100 sfm). Grinding fluid may be a water-base soluble-oil emulsion or a chemical solution, or sulfurized oil.

| | | Wheel | | Infeed, mm/pass (in./pass) | |
|---|---|---|---|---|---|
| Alloy group(a) | Hardness, HB | Classification | Speed, m/s (sfm) | Roughing | Finishing (max) |
| Fe-1 | 180–320 | A60JV | 15–20 (3000–4000) | 0.025 (0.001) | 0.005 (0.0002) |
| Ni-1 | 200–475 | A60JV | 15–18 (3000–3500) | 0.025 (0.001) | 0.005 (0.0002) |
| Ni-2 | 140–310 | A60JV | 15–18 (3000–3500) | 0.025 (0.001) | 0.005 (0.0002) |
| Ni-3 | 200–425 | A60JV | 15–18 (3000–3500) | 0.025 (0.001) | 0.005 (0.0002) |
| Co-1 | 180–320 | A60JV | 15–18 (3000–3500) | 0.025 (0.001) | 0.005 (0.0002) |
| Co-2 | 220–425 | A60JV | 15–18 (3000–3500) | 0.025 (0.001) | 0.005 (0.0002) |

(a) See Table 41 for identity of alloy groups and typical alloys. Source: Metcut Research Associates Inc.

**Table 44 Nominal processing conditions for the through-feed centerless grinding of heat-resistant alloys**

| | |
|---|---|
| Wheel classification | A60KV |
| Wheel speed, m/s (sfm), for alloy group(a) | |
| Fe-1 (180–320 HB) | 15–20 (3000–4000) |
| Ni-1 (200–475 HB) | 15–20 (3000–4000) |
| Ni-2 (140–310 HB) | 15–20 (3000–4000) |
| Ni-3 (200–425 HB) | 15–20 (3000–4000) |
| Co-1 (180–320 HB) | 15–20 (3000–4000) |
| Co-2 (220–425 HB) | 15–20 (3000–4000) |
| Infeed, mm/pass (in./pass) | |
| Roughing | 0.075 (0.003) |
| Finishing | 0.025 (0.001) |
| Work feed, m/min (in./min) | 1.3–3.8 (50–150) |
| Regulating wheel angle, degrees | 3 |
| Regulating wheel speed, rev/min | 25–40 |
| Grinding fluid | Water-base(b) or sulfurized oil |

(a) See Table 41 for designations of alloy groups and typical alloys in each group. (b) Either soluble-oil emulsion or synthetic solution. Source: Metcut Research Associates Inc.

**Table 46 Nominal processing conditions for surface grinding of heat-resistant alloys with CBN straight wheels**

Conditions given are for a unit downfeed of 0.025 mm (0.001 in.) per series of passes in roughing and 0.013 mm (0.0005 in.) in finishing. Table speed is 15–38 m/min (50–125 sfm). Grinding fluid may be heavy-duty oils or emulsifiable oils.

| | | Wheel | | |
|---|---|---|---|---|
| Alloy group(a) | Hardness, HB | Classification | Speed, m/s (sfm) | Crossfeed per pass, fraction of wheel width |
| Ni-1 | 200–475 | B100T75B | 28–41 (5500–8000) | ½ max |
| Ni-2 | 140–310 | B100T75B | 28–41 (5500–8000) | ½ max |
| Ni-3 | 200–425 | B100T75B | 28–41 (5500–8000) | ½ max |
| Co-1 | 180–320 | B100T75B | 28–41 (5500–8000) | ½ max |
| Co-2 | 220–425 | B100T75B | 28–41 (5500–8000) | ½ max |

(a) See Table 41 for designations of alloy groups and typical alloys in each group. Source: Metcut Research Associates Inc.

**Table 45 Nominal processing conditions for the internal grinding of heat-resistant alloys**

All data apply to holes of 2½ diameters maximum length and to grinding wheels of 1½ diameters maximum width.

| | |
|---|---|
| Wheel classification(a) | A60JV |
| Wheel speed, m/s (sfm), for alloy group(b) | |
| Fe-1 | 15–20 (3000–4000) |
| Ni-1 | 15–20 (3000–4000) |
| Ni-2 | 15–20 (3000–4000) |
| Ni-3 | 15–20 (3000–4000) |
| Co-1 | 15–20 (3000–4000) |
| Co-2 | 15–20 (3000–4000) |
| Infeed, mm/pass (in./pass) | |
| Roughing | 0.013 (0.0005) |
| Finishing | 0.005 (0.0002) |
| Traverse, wheel width per revolution of work | |
| Roughing | ⅓ |
| Finishing | ⅙ |
| Grinding fluid | Water-base(c) or sulfurized oil |

(a) Wheel recommendations are for wet grinding 20–50 mm (0.8–2.0 in.) diam holes. For larger holes, use the same or softer-grade wheel. For smaller holes, use a harder-grade wheel. (b) See Table 41 for designations of alloy groups and typical alloys in each group. (c) Soluble-oil emulsion or synthetic solution. Source: Metcut Research Associates Inc.

## Wheel Selection

Aluminum oxide ($Al_2O_3$) wheels are used for most grinding of heat-resistant alloys, although CBN is also used for some precision grinding applications. Table 41 classifies five groups of heat-resistant alloys for which grinding data are presented in this section. Tables 42 to 46 identify wheels suitable for grinding these five alloy groups by four different methods. Markings for identifying grinding wheels and bonded abrasives can be found in the articles "Grinding Equipment and Processes" ($Al_2O_3$ and SiC abrasives) and "Superabrasives" (CBN and diamond abrasives) in this Volume.

**The grade (hardness)** of the wheel is designated by letter, ranging from A (very soft) to Z (very hard). Most heat-resistant alloys are ground with medium-hard wheels (G to L).

**Structure.** Wheels with structure number from 7 to 10 are commonly used for grinding heat-resistant alloys. The definition of structure number varies among wheel manufacturers; however, all agree that as structure number increases, the distance between grains increases. Structure numbers are omitted from some wheel designations in this article.

**Bond.** Vitrified bond (designated by letter V) is most commonly used for grinding heat-resistant alloys. Resinoid bond wheels can be used for higher-speed work ($\geq$1400 m/min, or 4700 sfm).

## Grinding Fluid

Grinding fluids can be classified into four principal groups, as shown in Table 47. Because heat-resistant alloys have low thermal conductivity, grinding fluid must be applied at the grinding area in plentiful amounts to prevent heat checking of the work surface.

For fast removal of heat, highly sulfurized water-base soluble-oil emulsions are the best fluid for any wrought heat-resistant alloy. Sulfurized oils are appropriate grinding fluids for all heat-resistant alloys, but they remove heat less rapidly than the water-base soluble-oil emulsions.

Chlorinated oil, generally about 1% chlorine, is particularly useful for the wet dressing of form-grinding wheels to a tolerance of 0.005 mm (0.0002 in.) or less. Chlorinated water-base soluble-oil emulsions can be used in dressing form-grinding wheels within tolerances wider than 0.005 mm (0.0002 in.). Synthetic solutions and water-base soluble oil emulsions do not have this capability. The chlorinated straight oils and chlorinated soluble-oil emulsions are also applicable to other methods of grinding.

A disadvantage of chlorinated fluid is that any residual or entrapped fluid will react with the alloy during high-temperature service of the workpiece. Entrapment of fluid is especially likely in parts with small cavities such as blind holes and other difficult-to-clean recesses. For this reason, some users of heat-resistant alloy parts do not permit the use of chlorinated grinding fluids.

In many applications, selection of the fluid is based on cost. Emulsions of soluble oil and water are the least expensive and grinding oils the most expensive grinding fluids.

**Effect on Grindability.** Table 48 shows the effects of 15 water-base fluids on the grinding ratio (*G*-ratio) of selected heat-resistant alloys. The *G*-ratio is the volume of metal removed per volume of wheel wear. The higher this index, the easier the metal is to grind. It should be noted that the concept of grindability does not involve grinding sensitivity, which is the susceptibility of the metal to cracking during or after grinding, nor does it involve the ease of obtaining a good surface. The *G*-ratio is explained more fully in the article "Principles of Grinding" in this Volume.

In Table 48, a few data are included for the use of plain water or air. All the grinding fluids were at 10% concentration, which is

**Table 47 Identification and classification of grinding fluids**

| Fluid | Remarks |
|---|---|
| **Soluble-oil emulsions (regular)** | |
| S1 | Contains soap |
| S2 | Contains soaps and fatty materials |
| S3 | Emulsified kerosene |
| S4 | Contains soap; a brand different from that in S1 |
| **Soluble-oil emulsions (heavy-duty)** | |
| H1 | Contains sulfur and chlorine |
| H2 | Contains sulfurized fats; designed for stainless steel |
| H3 | Contains sulfur and extreme-pressure additives, high percentage of fats |
| H4 | Contains fatty materials, synthetic soaps; designed for stainless steel |
| H5 | Contains lead additive but no sulfur or chlorine |
| **Chemical (synthetic) solutions** | |
| C1 | Contains 35% potassium nitrite ($KNO_2$) before dilution for use |
| C2 | Contains 40% sodium nitrite ($NaNO_2$) before dilution for use; no organic compounds |
| C3 | Contains a moderate percentage of sodium nitrite |
| C4 | Based on synthetic wax |
| C5 | Synthetic lubricant and sulfurized fatty acid |
| C6 | Same as C5 but without sulfur |
| C7 | Contains fatty acid |
| C8 | Same as C7, with ionic additive |
| **Grinding oils** | |
| G1 | Transparent sulfochlorinated grinding oil containing fats, 4% sulfur and 2% chlorine (both active); viscosity, 230 sus at 40 °C (100 °F) |
| G2 | Dark sulfochlorinated grinding oil containing fats, 3% sulfur and 0.5% chlorine (both active); 190 SUS at 40 °C (100 °F) |
| G3 | Inactive grinding oil containing fats, no added sulfur or chlorine; 300 SUS at 40 °C (100 °F) |

higher than normal. The magnitude of improvement in *G*-ratio caused by grinding fluid increases with the concentration.

Grinding dry proved less satisfactory than grinding with most of the water-base fluids, and grinding with plain water resulted in a very low *G*-ratio. The highest ratios (greatest ease of grinding) were obtained with synthetic fluids containing nitrite ions in considerable quantity. Detailed information on grinding fluids can be found in the article "Metal Cutting and Grinding Fluids" in this Volume.

## Surface Grinding

Table 42 lists conditions for the surface grinding of heat-resistant alloys with straight wheels. The major differences among alloy groups are in wheel speed and cross feed.

Usually, an increase in wheel hardness causes an increase in the *G*-ratio and the grindability index. Although wheel speed can exert an important influence on the *G*-ratio in grinding certain alloys, the effects of changes in down feed, cross feed, and traverse speed are variable. Data on the effects of speed and feed on the *G*-ratio of various heat-resistant alloys in surface grinding are shown in Fig. 16 and 17.

**Table 48 Effect of fluid on the grinding ratio of four heat-resistant alloys**

| Grinding fluid(a) | Grinding ratio obtained for: U-500 | J-1570 | J-1570 | HS-31 |
|---|---|---|---|---|
| | Grinding wheel: A-60-H8-V | A-60-H8-V | A-60-J8-V | A-60-J8-V |
| C1 | 3.5 | 2.8 | ... | ... |
| C2 | ... | ... | 4.8 | 16 |
| C3 | ... | ... | 3.9 | 15 |
| H1 | 3.3 | ... | ... | ... |
| H2 | 1.5 | ... | ... | ... |
| C4 | ... | 1.7 | ... | ... |
| C5 | ... | 1.4 | ... | ... |
| H3 | ... | 1.3 | ... | ... |
| C7 | ... | 1.3 | ... | ... |
| S1 | ... | 1.1 | 2.6 | 14 |
| S2 | ... | ... | 2.5 | 12 |
| S3 | ... | 1.1 | ... | ... |
| C8 | ... | 1.1 | ... | ... |
| H4 | ... | ... | 2.4 | 12 |
| S4 | ... | ... | 2.3 | 13 |
| Air(b) | ... | 0.9 | 2.1 | 12 |
| Water | ... | ... | 1.7 | 7 |

(a) 10% concentration; see Table 47 for identification of fluids. (b) Dry grinding

**Wheel Speed.** As Fig. 16(a) shows, the *G*-ratio decreased only moderately (or not at all) for all alloys shown, except HS-21, when wheel speed was reduced from 1800 to 1200 m/min (6000 to 4000 sfm). Reductions in speed below 1200 m/min (4000 sfm), however, resulted in a more precipitous drop in the *G*-ratio. The only exception, alloy HS-21, responded like carbon or low-alloy steel in that the *G*-ratio decreased continuously with decreases in wheel speed from 1800 to 600 m/min (6000 to 2000 sfm).

Decreasing the wheel speed is one way to reduce grinding heat and the probability of workpiece cracking. Thus, it is possible to minimize cracking in these alloys by grinding them at 1200 m/min (4000 sfm) without significantly increasing wheel wear for a reasonable grinding rate.

**Down Feed.** The effects of changes in unit down feed on the *G*-ratio (Fig. 16b and c) indicate that no consistent influence is exerted by this variable. In some tests, the *G*-ratio decreased, rapidly or slowly, with larger down feed increments; in others, it remained nearly constant or increased slowly. The curve for HS-31 went through a maximum at a unit down feed of 0.05 mm/pass (0.002 in./pass), and it may well be that HS-21 would have shown the same behavior had it been tested at down feeds of less than 0.05 mm/pass (0.002 in./pass). The *G*-ratio for the U-500 alloy was reduced by more than half—from 13.5 to 6—as the down feed increased from 0.025 to 0.125 mm/pass (0.001 to 0.005 in./pass).

**Cross Feed.** The effect of cross feed on the *G*-ratio was not significant for HS-31 and J-1570 alloys, as shown in Fig. 17(a). The scatter of test results obtained with HS-31 makes it difficult to determine whether the slight downward trend to the right shown for both alloys is significant. A

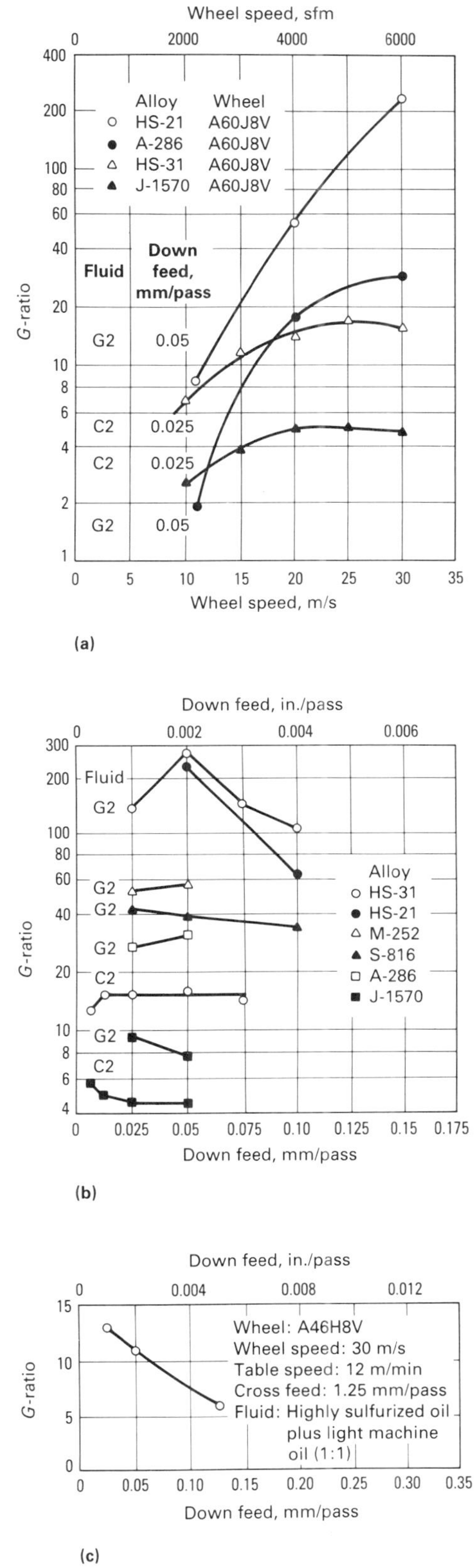


**Fig. 16** Effects of wheel speed and down feed on surface grinding ratio: (a) HS-21, A-286, HS-31, and J-1570 alloys. (b) HS-31, HS-21, M-252, S-816, A-286, and J-1570 alloys. (c) Udimet 500 alloy (solution treated and aged)

more pronounced effect of cross feed on the *G*-ratio for Udimet 500 is shown in Fig. 17(b).

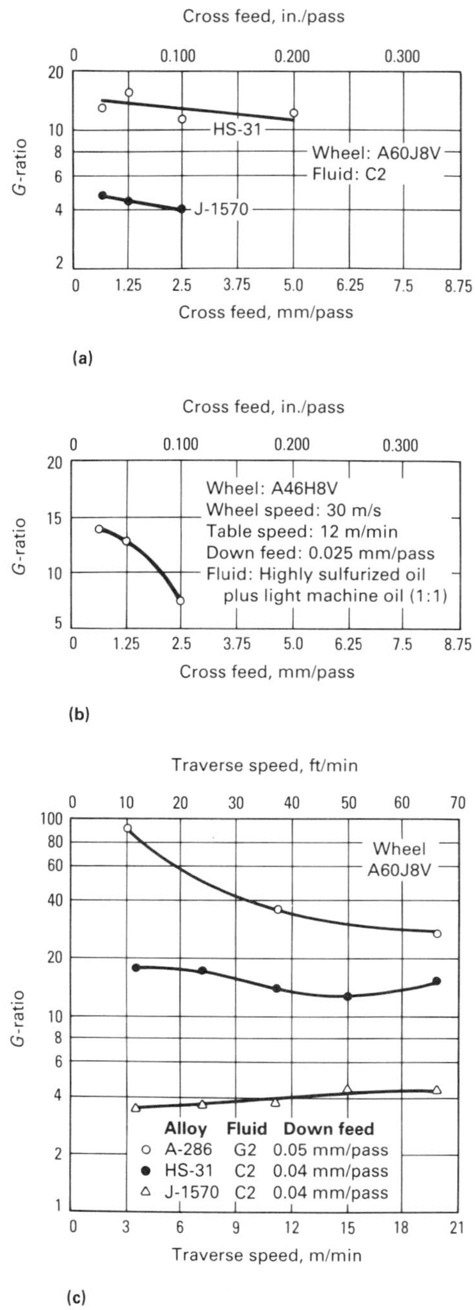


**Fig. 17** Effects of wheel speed and down feed on surface grinding ratio. (a) HS-31 and J-1570 alloys. (b) Udimet 500 alloy. (c) A-286, HS-31, and J-1570 alloys. See Table 47 for grinding fluid description.

**Traverse speed,** as shown in Fig. 17(c), had relatively little effect on *G*-ratio when a water-base grinding fluid was used; however, there was a considerable decrease in the *G*-ratio with increasing traverse speed when A-286 was ground with an oil. It is questionable whether this difference in behavior should be attributed to the alloy or the grinding fluid. However, reducing the traverse speed raises grinding temperature and introduces the possibility of damage to the work surface.

## Other Methods of Grinding

Conditions for grinding heat-resistant alloys by cylindrical, centerless, and internal methods are given in Tables 43 to 45. These three methods are discussed in the article "Grinding Equipment and Processes" in this Volume.

**Abrasive belt grinding** operations for heat-resistant alloys use either resinoid or serrated rubber bond $Al_2O_3$ or $Al_2O_3ZrO_2$ wheels with a grain size of 40 to 80 for roughing and 100 to 240 for finishing. The hardness (durometer scale) is 70 to 95 for roughing and 20 to 60 for finishing. Grinding is usually done wet, using a soluble-oil emulsion, synthetic solution, or sulfurized oil. The use of this method on vanes for the first stage of a turbine nozzle is described in the following example.

**Example 15: Removal of Coating From Turbine Vanes.** Investment cast turbine nozzle vanes made of HS-31 alloy were ground to remove the antioxidation diffusion coating from the tip-skirt and root-skirt sections (Fig. 18). Grinding was done in a turbine blade airfoil grinder equipped with a cloth-backed abrasive belt operating over a 25 mm (1 in.) diam contact wheel. The vanes were ground singly in a mechanical fixture. Each vane weighed 21.65 g (0.764 oz) before grinding and 20.25 g (0.714 oz) after. Processing data are given in Fig. 18.

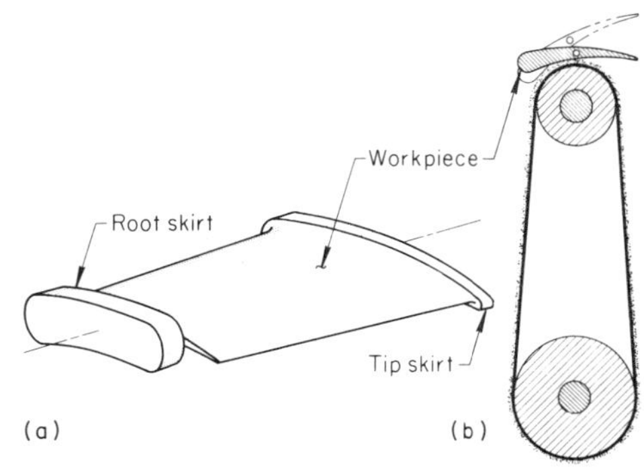


| | |
|---|---|
| Belt type | 60-grit $Al_2O_3$ resinoid bond(a) |
| Belt size, mm (in.) | 15 × 2285 (⅝ × 90) |
| **Operating conditions** | |
| Belt speed, m/s (sfm) | 30 (6000) |
| Metal removed, mm/side (in./side) | |
| Root skirt | 0.40 (0.015)(b) |
| Tip skirt | 0.30 (0.012)(c) |
| Grinding fluid | None |
| Production rate, s/piece | 93(d) |
| Belt life | 15 root skirts or 10 tip skirts |

(a) Low-flexibility, heavy-duty belt. (b) 0.65 g of metal removed. (c) 0.75 g of metal removed. (d) Each piece required 33 s for the root end and 60 s for the tip end.

**Fig. 18** Belt grinding of root and tip skirts of HS-31 turbine nozzle vanes

## Grinding Mechanically Alloyed Products

Mechanically alloyed products can be economically machine ground and present no problems with regard to stock removal. For example, in terms of grindability, Inconel Alloy MA 6000 is considered similar to Inconel Alloy 718 but less abrasive than IN-738. When grinding is to be the finishing operation and the material to be removed is not too great, all the work can be done on a grinding machine, using roughing and finishing steps. If an extremely accurate ground finish is required, particularly if the metal is of hard temper, it is advisable to remove the workpiece from the lathe or grinder after the next-to-last cut or grind and to set it aside for a day or two. This rest period allows sufficient time for the workpiece to adjust to any residual stresses present, possibly resulting in slight distortion, so that the piece can then be finish ground without danger of further distortion.

Wheels should be sufficiently soft to yield a moderate rate of metal removal. Hard wheels are likely to embed abrasive grains in the metal being ground. Aluminum oxide wheels with a grain size between 36 and 60 are commonly used. Grinding conditions should be severe enough to yield a slight breakdown action—for example, roughing cut of 0.01 mm (0.0004 in.) and finishing of 0.005 mm (0.0002 in.). A stream of soluble grinding lubricant should flow onto the metal. The oil produces a better finish, keeps the wheel from loading, reduces the frequency of dressing the wheels, and allows grinding to closer limits. Speeds and feeds are similar to those for wrought nickel-base alloys.

# Machining of Aluminum and Aluminum Alloys

ALUMINUM ALLOYS can be machined rapidly and economically. Because of their complex metallurgical structure, their machining characteristics are superior to those of pure aluminum.

The microconstituents present in aluminum alloys have important effects on machining characteristics. Nonabrasive constituents have a beneficial effect, and insoluble abrasive constituents exert a detrimental effect on tool life and surface quality. Constituents that are insoluble but soft and nonabrasive are beneficial because they assist in chip breakage; such constituents are purposely added in formulating high-strength free-cutting alloys for processing in high-speed automatic bar and chucking machines.

In general, the softer alloys—and, to a lesser extent, some of the harder alloys—are likely to form a built-up edge on the cutting lip of the tool. This edge consists of aluminum particles that have become welded to the tool edge because they were melted by the heat generated in cutting. Edge buildup can be minimized by using effective cutting fluids and by employing tools with surfaces that are free of grinding marks and scratches.

Alloys containing more than 10% Si are the most difficult to machine because hard particles of free silicon cause rapid tool wear. Alloys containing more than 5% Si will not finish to the bright machined surfaces of other high-strength aluminum alloys, but will have slightly gray surfaces with little luster. Chips are torn rather than sheared from the work, and special precautions (such as the use of lubricant-containing cutting fluids) must be taken to avoid the buildup of burrs on cutting edges.

A

B

C

D

E

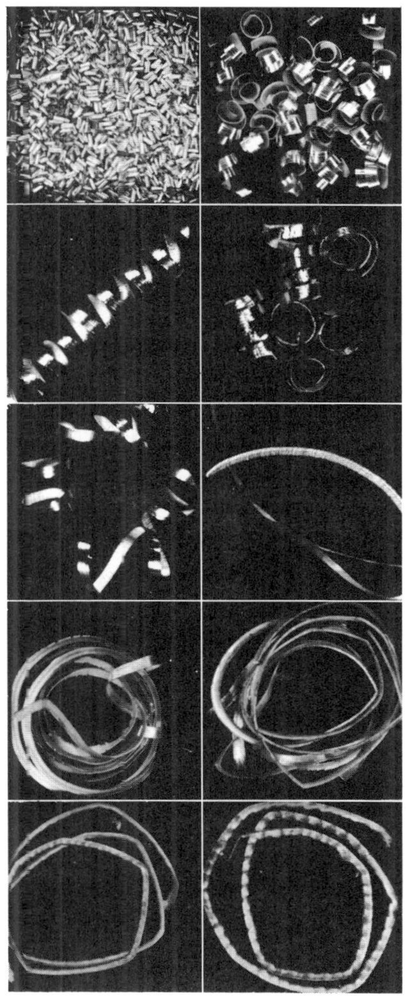

| Machinability rating | Alloy | Speed, m/min (sfm) | Feed, mm/rev (in./rev) Left photo | Right photo |
|---|---|---|---|---|
| A .......... | 2011-T3 | 120 (400) | 0.066 (0.0026) | 0.152 (0.0060) |
| B .......... | 2024-T4 | 30 (100) | 0.152 (0.0060) | 0.264 (0.0104) |
| C .......... | 6061-T6 | 120 (400) | 0.152 (0.0060) | 0.264 (0.0104) |
| D .......... | 3004-H32 | 120 (400) | 0.152 (0.0060) | 0.264 (0.0104) |
| E .......... | 1100-H12 | 120 (400) | 0.152 (0.0060) | 0.264 (0.0104) |

**Fig. 1** Typical chips for machinability ratings A to E (Table 1) for aluminum alloys. All chips were made with 20° rake tool and 2.54 mm (0.100 in.) depth of cut.

## Classification of Aluminum Alloys

Cast, wrought, strain hardenable, and heat treatable are the four major classifications of aluminum alloys. Machinability groupings are also used.

**Cast alloys** containing copper, magnesium, or zinc as the principal alloying elements impose few machining problems. Tools with small rake angles can normally be used with little danger of burring the part or of developing buildup on the cutting edges of tools. Alloys having silicon as the major alloying element require tools with larger rake angles, and they are more economically machined at lower speeds and feeds.

**Wrought Alloys.** Most wrought aluminum alloys have excellent machining characteristics; several are well suited to multiple-operation machining. A thorough understanding of tool designs and machining practices is essential for full utilization of the free-machining qualities of aluminum alloys.

**Strain-hardenable alloys** (including commercially pure aluminum) contain no alloying elements that would render them hardenable by solution heat treatment and precipitation, but they can be strengthened to some extent by cold work. In machining, a continuous chip is formed that must be directed away from the workpiece by tools with generous side and back rake angles, thus preventing scratching of the finished surface with the work-hardened chips. These alloys machine easily, although tool pressures are high as a result of high friction. To obtain good surface finish, sharp tools are mandatory because the alloys are gummy. Machinability is improved by cold working; alloys in the full-hard temper are easier to machine to a good finish than those in the annealed condition.

**Heat-Treatable Alloys.** Most of the alloys of this group contain fairly high percentages of alloying elements such as copper, silicon, magnesium, and zinc. They can be machined to a good finish with or without cutting fluid, but a cutting fluid is recommended for most operations. Turnings usually occur as long, continuous curls, except for the free-machining alloys, which contain chip-breaking constituents. Heat-treatable alloys are more machinable in the heat-treated tempers than in the softer as-fabricated or annealed solution.

**Machinability groupings** for aluminum alloys are useful in specifying tool forms. For this purpose, alloys are classified into five groups: A, B, C, D, and E, in increasing order of chip length and in decreasing order

## Table 1(a) Machinability ratings of wrought aluminum alloys

| Alloy designation | Temper | Product form | Hardness, HB (500 kg load, 10 mm ball) | Machinability rating(a) |
|---|---|---|---|---|
| 1060 | O | Extruded rod, bar, | 19 | E |
| | H12 | extruded and drawn | 23 | E |
| | H14 | tube, pipe | 26 | D |
| | H16 | | 30 | D |
| | H18 | | 35 | D |
| 1100 | O | Sheet, plate; rolled and | 23 | E |
| | H12 | extruded rod, bar; | 28 | E |
| | H14 | extruded and drawn | 32 | D |
| | H16 | tube, pipe; other | 38 | D |
| | H18 | | 44 | D |
| 2011 | T3 | Rod, bar, tube, pipe | 95 | A |
| | T8 | | 100 | A |
| 2014 | O | Plate, rod, bar, tube, pipe; | 45 | D |
| | T4 | other | 105 | B |
| | T6 | | 135 | B |
| 2017 | O | Rolled rod, bar; other | 45 | C |
| | T4 | | 105 | B |
| 2018 | T61 | Forging stock | 120 | B |
| 2024 | O | Sheet, plate, rod, bar, tube, | 47 | D |
| | T3 | pipe; other | 120 | B |
| | T4 | | 120 | B |
| | T61 | | 130 | B |
| 2025 | T6 | Forging stock | 110 | B |
| 2117 | T4 | Rivet wire, rod | 70 | C |
| 2218 | T72 | Forging stock | 95 | B |
| 2219 | O | Sheet, plate; extruded rod, | ... | ... |
| | T42 | bar; extruded and drawn | ... | ... |
| | T351 | tube, pipe; forging stock | 100 | B |
| | T37 | | 117 | B |
| | T62 | | 115 | B |
| | T851 | | 130 | B |
| | T87 | | 130 | B |
| 2618 | T61 | Forgings | 115 | B |
| 3002 | O | Sheet | 25 | ... |
| | H25 | | ... | ... |
| 3003 | O | Sheet, plate; rolled and | 28 | E |
| | H12 | extruded rod, bar; | 35 | E |
| | H14 | extruded and drawn | 40 | D |
| | H16 | tube, pipe; other | 47 | D |
| | H18 | | 55 | D |
| 3004 | O | Sheet, plate; drawn tube, | 45 | D |
| | H32 | pipe | 52 | D |
| | H34 | | 63 | C |
| | H36 | | 70 | C |
| | H38 | | 77 | C |
| 4032 | T6 | Forging stock | 120 | B |
| 5005 | O | Sheet, plate; rolled rod and | 28 | E |
| | H12 | bar; other | 36 | E |
| | H14 | | 41 | D |
| | H16 | | 46 | D |
| | H18 | | 51 | D |
| | H32 | | 36 | E |
| | H34 | | 41 | D |
| | H36 | | 46 | D |
| | H38 | | 51 | D |
| 5050 | O | Sheet, plate; drawn tube, | 36 | E |
| | H32 | pipe | 46 | D |
| | H34 | | 53 | D |
| | H36 | | 58 | C |
| | H38 | | 63 | C |
| 5052 | O | Sheet, plate; rolled rod, | 47 | D |
| | H32 | bar; drawn tube, pipe; | 60 | D |
| | H34 | other | 68 | C |
| | H36 | | 73 | C |
| | H38 | | 77 | C |
| 5056 | O | Rivet rod, wire | 65 | D |
| | H18 | | 105 | C |
| | H38 | | 100 | C |
| 5083 | O | Sheet, plate, rod, bar, tube, | 67 | D |
| | H321 | pipe, forgings | 82 | D |
| 5086 | O | Sheet, plate; extruded rod, | 60 | D |
| | H32 | bar; extruded and drawn | 72 | D |
| | H34 | tube, pipe | 82 | C |
| | H112 | | 64 | D |
| 5154 | O | Sheet, plate; welding wire | 58 | D |
| | H32 | and rod | 67 | D |
| | H34 | | 73 | C |
| | H36 | | 78 | C |
| | H38 | | 80 | C |
| | H112 | | 63 | D |
| 5252 | H25 | Sheet | 68 | C |
| | H38 | | 75 | C |
| 5254 | O | Sheet, plate | 58 | D |
| | H32 | | 67 | D |
| | H34 | | 73 | C |
| | H36 | | 78 | C |
| | H38 | | 80 | C |
| | H112 | | 63 | D |
| 5257 | H25 | Sheet | 32 | C |
| | H28 | | 43 | C |
| 5357 | O | Sheet, plate | 32 | D |
| | H25 | | 50 | C |
| | H28 | | 55 | C |
| 5454 | O | Sheet, plate; other | 62 | D |
| | H32 | | 73 | D |
| | H34 | | 81 | C |
| | H111 | | 70 | D |
| | H112 | | 62 | D |
| 5456 | O | Sheet, plate; extruded rod | 70 | D |
| | H111 | bar; extruded tube, pipe; | 75 | D |
| | H112 | forgings | 70 | D |
| | H116 | | 90 | D |
| 5457 | O | Sheet | 32 | E |
| | H25 | | 48 | C |
| | H28 | | 55 | C |
| 5557 | O | Sheet | 27 | E |
| | H25 | | 46 | D |
| | H28 | | 55 | D |
| 5652 | O | Sheet, plate | 47 | D |
| | H32 | | 60 | D |
| | H34 | | 68 | C |
| | H36 | | 73 | C |
| | H38 | | 77 | C |
| 5657 | H25 | Sheet | 40 | D |
| | H28 | | 50 | D |
| 6005 | T5 | Extruded rod, bar | 95 | C |
| 6061 | O | Sheet, plate, rod, bar, tube, | 30 | D |
| | T4 | pipe; forging; other | 65 | C |
| | T6 | | 95 | C |
| 6063 | O | Extruded rod, bar; | 25 | D |
| | T1 | extruded and drawn | 42 | D |
| | T4 | tube, pipe | 60 | D |
| | T5 | | 60 | C |
| | T6 | | 73 | C |
| | T83 | | 82 | C |
| | T831 | | 70 | C |
| | T832 | | 95 | C |
| 6066 | O | Extruded rod, bar; forging | 43 | D |
| | T4 | stock | 90 | C |
| | T6 | | 120 | B |
| 6070 | T6 | Rod, bar, tube, pipe | 120 | C |
| 6151 | T6 | Forging stock | 100 | ... |
| 6262 | T9 | Rod, bar, tube, pipe | 120 | B |
| 6463 | T1 | Extruded rod, bar; | 42 | D |
| | T5 | extruded and drawn | 60 | C |
| | T6 | tube, pipe | 74 | C |
| 6951 | O | Sheet | 28 | ... |
| | T6 | | 82 | ... |
| 7001 | O | Extruded rod, bar | 60 | B |
| | T6 | | 160 | B |
| 7005 | T53 | Rod, bar, tube, pipe | ... | ... |
| 7075 | O | Sheet, plate, rod, bar, tube, | 60 | D |
| | T6 | pipe; forging stock | 150 | B |
| 7079 | O | Sheet, plate, rod, bar, tube, | ... | ... |
| | T6 | pipe; forging stock | 145 | B |
| 7178 | O | Sheet, plate, rod, bar, tube, | 60 | ... |
| | T6 | pipe | 160 | B |
| | T76 | | ... | ... |
| 8280 | O | Sheet, plate | ... | B |
| | H12 | | ... | A |

(a) A, B, C, D, and E are relative ratings in increasing order of chip length (see Fig. 1) and decreasing order of quality of finish. A, free cutting, very small broken chips and excellent finish; B, curled or easily broken chips and good-to-excellent finish; C, continuous chips and good finish; D, continuous chips and satisfactory finish; E, optimum tool design and machine settings required to obtain satisfactory control of chip and finish. Source: Ref 1

## Table 1(b) Machinability ratings of cast aluminum alloys

| Alloy designation | Temper | Casting type | Hardness, HB (500 kg load, 10 mm ball) | Machinability rating(a) | Alloy designation | Temper | Casting type | Hardness, HB (500 kg load, 10 mm ball) | Machinability rating(a) |
|---|---|---|---|---|---|---|---|---|---|
| 208 | F | Sand | 55 | B | A356 | T6 | Permanent mold | 80 | ... |
| 213 | F | Permanent mold | 85 | ... | | T61 | Permanent mold | 80 | B |
| 222 | T52 | Permanent mold | 100 | ... | 357 | F | Permanent mold | ... | ... |
| | T551 | Permanent mold | 115 | ... | | T51 | Permanent mold | ... | ... |
| | T65 | Permanent mold | 140 | ... | | T6 | Sand | 90 | B |
| 238 | F | Permanent mold | 100 | B | | T6 | Permanent mold | 85 | B |
| A240 | F | Sand | 90 | A | | T7 | Sand | 60 | ... |
| 242 | F | Sand | ... | ... | | T7 | Permanent mold | 70 | ... |
| | T21 | Sand | 70 | B | A357 | T6 | Sand | 85 | ... |
| | T571 | Sand | 85 | B | | T6 | Permanent mold | 85 | ... |
| | T571 | Permanent mold | 105 | B | B358 | T6 | Permanent mold | 90 | B |
| | T61 | Permanent mold | 110 | B | | T62 | Permanent mold | ... | ... |
| | T77 | Sand | 75 | B | 359 | T61 | Permanent mold | 90 | ... |
| A242 | T77 | Sand | 70 | ... | | T62 | Permanent mold | 100 | B |
| 295 | T4 | Sand | 60 | B | 360 | F | Die | 75 | C |
| | T6 | Sand | 75 | B | A360 | F | Die | 75 | C |
| | T62 | Sand | 90 | B | 364 | F | Die | ... | C |
| B295 | T4 | Permanent mold | 75 | B | 380 | F | Die | 80 | B |
| | T6 | Permanent mold | 90 | B | A380 | F | Die | 80 | B |
| | T7 | Permanent mold | 80 | B | 384 | F | Die | ... | C |
| 308 | F | Permanent mold | 70 | B | 390 | F | Die | 120 | ... |
| 319 | F | Sand, permanent mold | 70 | C | A390 | F | Sand | 100 | ... |
| | T5 | Sand | 80 | B | | F | Permanent mold | 110 | ... |
| | T6 | Sand | 80 | B | | T5 | Sand | 100 | ... |
| | T6 | Permanent mold | 95 | B | | T5 | Permanent mold | 110 | ... |
| A332 | T551 | Permanent mold | 105 | C | | T6 | Sand | 140 | ... |
| | T65 | Permanent mold | 125 | C | | T6 | Permanent mold | 145 | ... |
| F332 | T5 | Permanent mold | 105 | C | | T7 | Sand | 115 | ... |
| 333 | F | Permanent mold | 90 | C | | T7 | Permanent mold | 120 | ... |
| | T5 | Permanent mold | 100 | B | 413 | F | Die | 80 | E |
| | T6 | Permanent mold | 105 | B | A413 | F | Die | 80 | ... |
| | T7 | Permanent mold | 90 | B | 443 | F | Sand | 40 | E |
| 354 | T61 | Permanent mold | 100 | B | | F | Permanent mold | 45 | E |
| | T62 | Permanent mold | 110 | B | | F | Die | 50 | E |
| 355 | F | Sand | ... | ... | A444 | F | Sand | ... | ... |
| | T51 | Sand | 65 | B | | F | Permanent mold | 44 | ... |
| | T51 | Permanent mold | 75 | B | | T4 | Sand | ... | ... |
| | T6 | Sand | 80 | B | | T4 | Permanent mold | 45 | ... |
| | T6 | Permanent mold | 90 | B | 514 | F | Sand | 50 | B |
| | T61 | Sand | 90 | B | A514 | F | Permanent mold | 60 | B |
| | T62 | Permanent mold | 105 | B | | F | Die | ... | B |
| | T7 | Sand | 85 | B | B514 | F | Sand | 50 | ... |
| | T7 | Permanent mold | 85 | B | F514 | F | Sand | 50 | B |
| | T71 | Sand | 75 | B | L514 | F | Die | ... | ... |
| | T71 | Permanent mold | 85 | B | 518 | F | Die | 80 | B |
| C355 | T6 | Sand | 85 | ... | 520 | T4 | Sand | 75 | B |
| | T6 | Permanent mold | 90 | ... | 535 | F | Sand | 70 | ... |
| | T61 | Permanent mold | 100 | B | A535 | F | Sand | 65 | B |
| 356 | F | Sand | ... | ... | B535 | F | Sand | 65 | A |
| | F | Permanent mold | ... | ... | 705 | F | Sand | 65 | B |
| | T51 | Sand | 60 | C | 707 | F | Sand | 85 | B |
| | T51 | Permanent mold | ... | C | A712 | F | Sand | 75 | B |
| | T6 | Sand | 70 | C | C712 | F | Permanent mold | 70 | B |
| | T6 | Permanent mold | 90 | C | D712 | F | Sand | 75 | B |
| | T7 | Sand | 75 | C | 713 | F | Sand | 75 | B |
| | T7 | Permanent mold | 70 | C | 850 | T5 | Sand | 45 | A |
| | T71 | Sand | 60 | C | | T5 | Permanent mold | 45 | A |
| A356 | F | Sand | ... | ... | A850 | T5 | Sand, permanent mold | 45 | A |
| | T51 | Sand | ... | ... | B850 | T5 | Sand | 65 | A |
| | T6 | Sand | 75 | ... | | T5 | Permanent mold | 70 | A |

(a) A, B, C, D, and E are relative ratings in increasing order of chip length (see Fig. 1) and decreasing order of quality of finish. A, free cutting, very small broken chips and excellent finish; B, curled or easily broken chips and good-to-excellent finish; C, continuous chips and good finish; D, continuous chips and satisfactory finish; E, optimum tool design and machine settings required to obtain satisfactory control of chip and finish. Source: Ref 1

of finish quality, as defined in the footnotes of Tables 1(a) and 1(b). Ratings for most commercial aluminum alloys are given in Tables 1(a) and 1(b), and typical chips for each rating are illustrated in Fig. 1.

## Cutting Force and Power

The cutting force, and therefore the power, required to machine aluminum is less than might be expected on the basis of its mechanical properties. Although the cutting force required to machine similar metals is often in direct proportion to tensile strength, this proportion is not necessarily valid with dissimilar metals. For example, the common mechanical properties of 2017-T4 aluminum alloy and of hot-rolled low-carbon steel are quite similar (Table 2), but as Fig. 2 shows, the cutting force required in turning aluminum is only about 35% of that required in turning low-carbon steel. Consequently, as shown in Fig. 3, the number of cubic millimeters of metal that can be removed per minute per unit kilowatt expended is approximately three times as great for aluminum alloy 2017-T4 as for hot-rolled low-carbon steel of closely similar tensile strength.

**Table 2 Comparison of common mechanical properties of 2017-T4 aluminum alloy and hot-rolled low-carbon steel**

| Mechanical property | Aluminum alloy 2017-T4 | Hot-rolled low-carbon steel |
|---|---|---|
| Tensile strength, MPa (ksi) | 440 (63.8) | 449 (65.2) |
| 0.2% yield strength, MPa (ksi) | 278 (40.3) | 277 (40.2)(a) |
| Elongation in four diameters, % | 26.0 | 32.7 |
| Shear strength, MPa (ksi) | 279 (40.5) | 324 (47.0) |
| Hardness, HB | | |
| 500 kg, 10 mm ball | 113 | 110 |
| 3000 kg, 10 mm ball | 132 | 128 |

(a) Yield point is 309 MPa (44.8 ksi).

## Selection of Alloy and Temper

An application often dictates the use of a specific alloy or temper or both. Under these conditions, composition cannot be changed for the sake of improving machinability. However, there is often a marked difference in machinability among different tempers of the same alloy. Therefore, it may be feasible to do some or all of the machining operations with the alloy in the most favorable condition for machining and then to convert the alloy to the temper specified for the end use.

For some applications, two or more alloys are equally acceptable. Under these conditions, machinability can be a major consideration in making the final selection. For example, high-strength, free-cutting alloy 2011 can be machined to an excellent surface finish at high speed and feed, with a low rate of tool wear. The chips formed are finely broken. Alloy 2011 is therefore recommended for all general and high-production machining where a free-cutting alloy is desired. Alloy 2011 is especially desirable for multiple-operation machining, mainly because it machines with a broken chip. Stock for multiple-operation machining is also available in alloys 2017, 2024, 6061, and 6262 in several heat-treated tempers.

Alloys 2024-T4, 2017-T4, and 7075-T6 produce continuous chips that must be broken by a chip breaker in the tool. Alloys 6061-T6 and 5056-H38 are slightly more difficult to machine, and they produce chips that are difficult to control. The softer alloys 5052, 3003, and 1100 are likely to produce gummy chips. Wrought alloy 4032 and cast alloys 220, 13, and A132 are quite abrasive, and high rates of tool wear result. The following example describes an application in which a change to an alloy and temper having better machinability improved results.

**Example 1: Change of Alloy 6061-T6 to 6262-T9 to Improve Machinability of Hexagonal Nut.** When alloy 6061-T6 was used for a hexagonal nut (Fig. 4), a lead of three threads chamfer on the tap form was required to produce a thread without tearing. A length of 4.8 mm (3/16 in.) had to be cut off after tapping. After changing to the freer-cutting 6262-T9, lead was reduced to 1½ threads, so that only 3.2 mm (1/8 in.) had to be cut off. Material savings was 8.5%. Because 6262-T9 produced fewer burrs on the tapped threads, frequency of tool grinding was reduced.

With 6061-T6, the workpiece before tapping weighed 0.0254 kg (0.0560 lb), and with 6262-T9, it weighed 0.0232 kg (0.0512 lb). Production rate was 890 pieces per hour with both alloys.

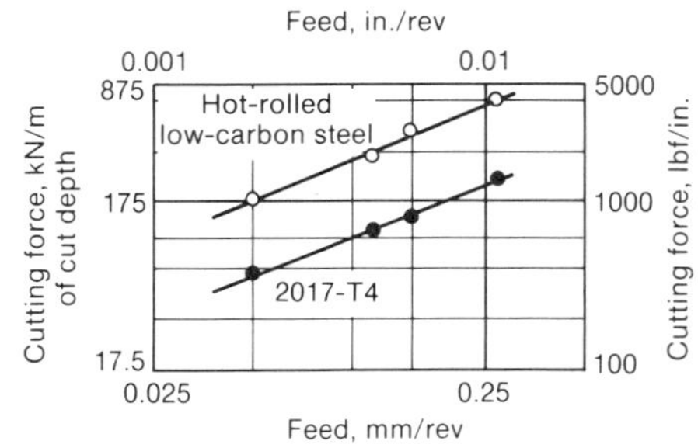


**Fig. 2** The effect of feed on cutting force for low-carbon steel and aluminum alloy 2017-T4 with comparable mechanical properties

## Bar Versus Tubing

Many parts can be machined equally well from either bar stock or tubing. When either of these product forms can be used, cost per piece machined is the determining factor. Initial cost of bar versus tube, cost of additional machining to make the part from bar stock, and value of the additional scrap that results from machining bar stock must be considered to determine cost. Because of the high cost of producing small tubing, total cost per piece machined may be greater when tubing of small sizes (less than about 32 mm, or 1¼ in., in diameter) is used. For parts requiring a diameter greater than 32 mm (1¼ in.), tubing is usually less expensive.

## General Machining Conditions

Power requirements for machining are proportional to speed and cutting force, and the power lost in the bearings and gears of the machine increases with speed. Power requirements for machining aluminum decrease somewhat as the rake angle of the cutting tool is increased. Table 3 lists typical power requirements for several wrought and cast alloys, as measured at the cutter with single-point tools having 0 and 20° rake angles.

**The cutting force** for aluminum alloys can vary widely at low speeds, such as 30 to 60 m/min (100 to 200 sfm), rising momentarily to peak values several times higher than normal. At higher speeds, the cutting

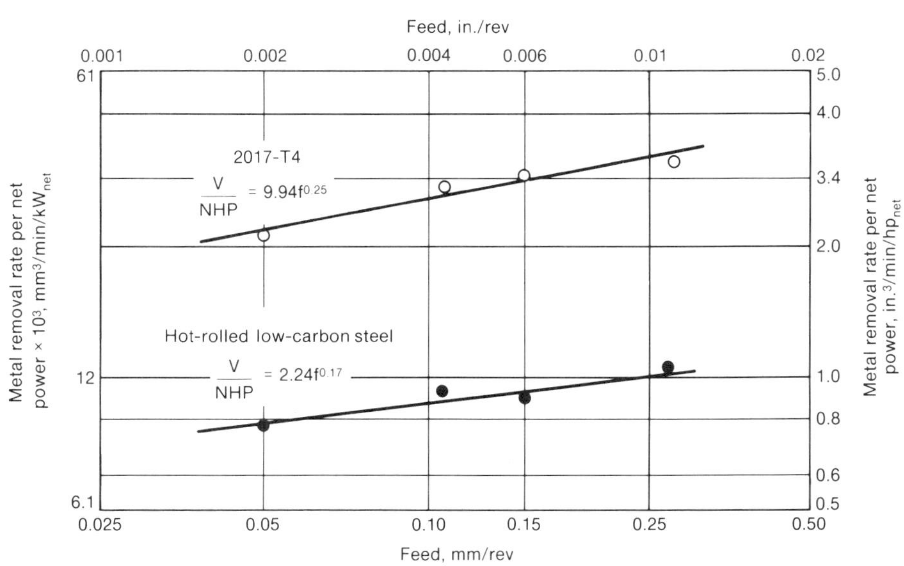


**Fig. 3** The effect of feed on metal removal rate per net power for low-carbon steel and aluminum alloy 2017-T4 with comparable mechanical properties. V/NHP, volume per net horsepower

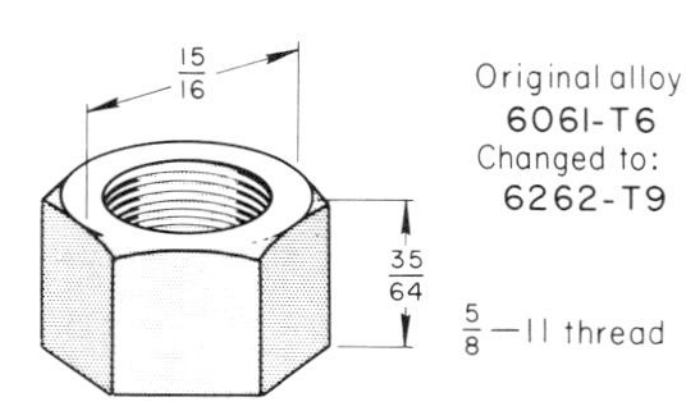


**Fig. 4** Hexagonal nut for which a change from alloy 6061-T6 to 6262-T9 resulted in a savings of material and less burring of the tapped thread. Dimensions given in inches

**Table 3 Power requirements for machining aluminum alloys**

| Alloy | Power, 0° rake, (kW/mm³/min) × 10⁻⁴ | Power, 0° rake, hp/in.³/min | Power, 20° rake, (kW/mm³/min) × 10⁻⁴ | Power, 20° rake, hp/in.³/min |
|---|---|---|---|---|
| F132-T5 | 0.16–0.32 | 0.20–0.40 | 0.16–0.41 | 0.20–0.50 |
| 356-T51 | 0.20–0.36 | 0.25–0.45 | 0.24–0.81 | 0.30–1.00 |
| 2011-T3 | 0.12–0.20 | 0.15–0.25 | 0.16–0.24 | 0.20–0.30 |
| 2024-T351 | 0.16–0.32 | 0.20–0.40 | 0.24–0.41 | 0.30–0.50 |
| 6061-T651 | 0.20–0.28 | 0.25–0.35 | 0.24–0.41 | 0.30–0.50 |

**Table 4 Design of single-point tools for machining aluminum alloys of A and B machinability ratings**

| Tool details | High-speed steel, Roughing | High-speed steel, Finishing | Carbide, Roughing | Carbide, Finishing |
|---|---|---|---|---|
| Back rake, degrees | 20 | 20 | 20 | 20 |
| Side rake, degrees | 20 | 20 | 20 | 20 |
| End relief, degrees | 10 | 10 | 7 | 7 |
| Side relief, degrees | 10 | 10 | 7 | 7 |
| End cutting edge, degrees | 5 | 5 | 5 | 5 |
| Side cutting edge, degrees | 10 | 10 | 10 | 10 |
| Nose radius, mm (in.) | 1.6 (0.063) | 5.1 (0.20) | 1.6 (0.063) | 5.1 (0.20) |

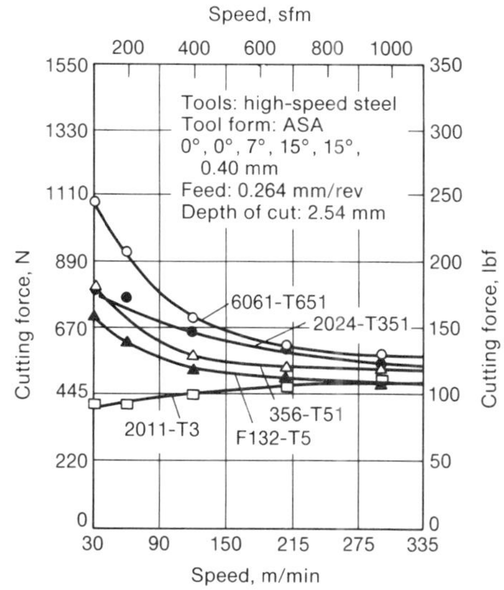


**Fig. 5** The effect of speed on cutting force for five aluminum alloys

force for machining 2011-T3 alloy rises slightly with increasing speed, but for most alloys it decreases. The overall effect of speed on cutting force is small. As speed increases up to about 300 m/min (1000 sfm), cutting force changes slightly; above 300 m/min (1000 sfm), the effect of speed is negligible. Increasing the speed does not produce much more heat, but it does shorten the time available for removing the heat from the tool. The effect of speed on cutting force for several aluminum alloys is plotted in Fig. 5. Heating of tool surfaces is not sufficient to have a harmful effect on a high-speed steel tool until the speed exceeds about 215 m/min (700 sfm). High-speed steel can be used for speeds well beyond this limit, but carbide tools are recommended for long tool life.

**The cutting speed** for aluminum alloys is determined by the limits of the machine tool and by the workpiece. Speeds as high as 4600 m/min (15 000 sfm) have been used in aerospace applications (see the article "High-Speed Machining" in this Volume). Even higher speeds have been achieved with experimental equipment. However, in most practice, mainly because of the limitations imposed by available spindle speed, available horsepower, and dynamic balance of the part, machining speeds are seldom higher than 900 m/min (3000 sfm) and they are more commonly less than 300 m/min (1000 sfm), as indicated by the examples in this article.

Cutting speed should be as high as is practical in order to save time and to minimize temperature rise in the part, as described in the section "Thermal Expansion" in this article. As cutting speed is increased above 30 to 60 m/min (100 to 200 sfm), the probability of forming a built-up edge on the cutter is reduced, chips break more readily, and finish is improved.

**Depth of cut** should be as great as possible within the limits of part strength, chucking equipment, power of the machine tool, and amount of stock to be removed in order to minimize the number of cuts required. As depth of cut is increased, cutting force increases. Depth of cut must be limited to a value that will not distort the workpiece or cause it to slip, nor overload the machine. Depth of cut in roughing may be as high as 6.35 mm (0.250 in.) for small work or up to 38.10 mm (1.500 in.) for medium or large work. At the opposite extreme, depth of cut in finishing is often less than 0.635 mm (0.025 in.).

**Feed** will depend on the finish desired and on the strength and rigidity of the workpiece and of the machine. Finishing cuts require a light feed of 0.05 to 0.15 mm/rev (0.002 to 0.006 in./rev); rough cuts may use a feed of 0.15 to 2.03 mm/rev (0.006 to 0.080 in./rev). Alloys with machinability ratings of D and E are best machined with a feed in the lower end of the range.

**Table 5 Current machining practice for 390 aluminum alloy**

| Operation | Tool | Speed, mm/min (sfm) | Feed | Depth, mm (in.) | Coolant |
|---|---|---|---|---|---|
| Turning | Tungsten carbide, with J polish (SPG 422 insert) | 30–150 (100–500) | 0.13–0.51 mm/rev (0.005–0.02 in./rev) | 0.13–5.1 (0.005–0.20) | Soluble oil (20:1) |
| | Diamond (SPG 422 insert) | 310–910 (1000–3000) | 0.08–0.38 mm/rev (0.003–0.015 in./rev) | 0.13–3.8 (0.005–0.15) | Soluble oil (20:1) |
| Milling | Tungsten carbide, C-2 or C-3, 5–10° rake, 7–15° shear | 99–175 (325–575) | 0.10–0.30 mm/tooth (0.004–0.012 in./tooth) | 0.13–5.1 (0.005–0.20) | Soluble oil (15:1) or dry |
| | Diamond, 5–10° rake, 7–15° shear | 400–790 (1300–2600) | 0.10–0.20 mm/tooth (0.004–0.008 in./tooth) | 0.20–1.0 (0.008–0.040) | Soluble oil (15:1) or dry |
| Drilling | High-speed steel, high helix | 20 (70) | 381 mm/min (15 in./min) | 6 × diam | Soluble oil (20:1), 19 L/min (5 gal./min) at 170 kPa (25 psi) |
| | Tungsten carbide tipped, coolant fed | 90 (300) | 508 mm/min (20 in./min) | 6 × diam | Soluble oil (20:1), 34 L/min (9 gal./min) at 3.4 MPa (500 psi) |
| Tapping | Form taps, with four lobes | 20 (70) | By pitch | 6 × diam | Soluble oil (20:1), 19 L/min (5 gal./min) at 170 kPa (25 psi) |
| Broaching | Tungsten carbide, 0° shear, 5° rake | 45 (150) | 229 mm/min (9 in./min) | 0.13 (0.005) | Soluble oil (20:1) or dry |

Source: Ref 2

**Table 6 Cutting fluids for aluminum**

| Type of lubricant | Principal ingredients | Viscosity range | Application; maintenance | Relative effectiveness | Necessary precautions |
|---|---|---|---|---|---|
| Mineral oils (fatty-additive type preferred) | Mineral oil, lard, or neats-foot oil; oleic acid or butyl stearate | 40 SUS at 40 °C (100 °F) for high-speed machining to 300 SUS for low speeds | Generous flow at all cutting edges; keep recirculating fluid clean and cool. | Good lubricity and chip flushing; fair cooling; excellent finish as built-up edge is minimized | Control air above oil where mist application endangers shop air; remove oil from finished parts (also from chips to reclaim and reduce fire hazard). |
| Soluble oils (emulsions) (3–5% soluble oil in water) | Soluble oil, petroleum sulfonate emulsifying agents, water (oil is added); rust inhibitor, germicide, stain inhibitor | Generally low | Generous flow at all cutting edges; keep recirculating fluid clean; cool when necessary. | Good chip flushing; lubrication adjustable by varying concentration; excellent cooling; good finish | For high-speed machining, cooling is more important than lubrication; where emulsion is applied as mist, keep oil content as low as possible to reduce air and shop contamination. |
| Aqueous chemical solutions | Water; soluble synthetics (usually clear); sometimes, fatty materials; rust inhibitors; germicides | Generally low | Generous flow at all cutting edges; keep recirculating fluid clean; cool as required. | Good chip flushing; excellent visibility of cut; excellent cooling; adjustable lubrication; good finish | Keep oil content low; control mist; consider cost (significantly higher than soluble-oil emulsions). |
| Stick lubricants | Mineral compounds, animal fats, waxes, synthetics | Various hardnesses | Applied as required to blades, wheels, disks, or files or to workpiece | Prevents the loading of abrasive surfaces or of teeth of saws and files | Application is intermittent, but should be monitored and made as required throughout run. |

Source: Ref 1

## Tool Design

Tools intended for machining aluminum and its softer alloys should be ground to allow considerably more side rake and back rake than are customary when machining steel. Therefore, they approach the contours of tools designed for cutting hardwood. The larger rake angles are recommended for finishing tools and for the machining of alloys that are not free cutting, especially the softer alloys, which require exceptionally acute and keen cutting edges. Smaller rake angles can be used for the free-cutting alloys and for roughing cuts, which require a sturdy tool for the heavier cuts and feeds employed. Suggested rake angles, as related to alloy machinability rating (Table 1), are:

| Rating | Rake angle, degrees |
|---|---|
| A | 0–20 |
| B | 20 |
| C | 20–30 |
| D | 40 |
| E | 40 |

Tool forms for machining aluminum alloys with machinability ratings of A and B with single-point tools are given in Table 4. Variations of these may be desirable, depending on machining conditions or shape of the workpiece.

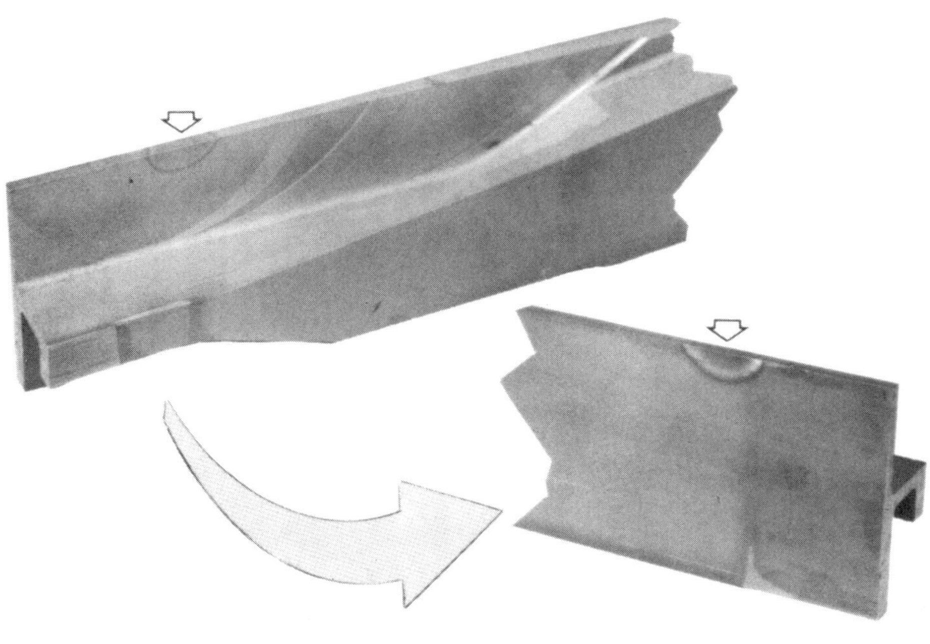

**Fig. 6** Typical spar mill defect at a section change. Courtesy of McDonnell Douglas Canada, Ltd.

**Clearance angle** is important to proper functioning of the tool. Too small a clearance will permit the side or heel of the tool to rub the work and generate heat, while too large an angle will cause the tool to dig into the work and chatter. This angle must be carried around the side of the tool that advances into the work. For most applications, clearance angles of 6 to 10° are suitable. The side rake angle imparts to the tool a slicing action that assists materially in shearing the chip from the stock. This angle is important and should be held within the ranges recommended in subsequent sections of this article.

**Cutting Edge Finish.** For maximum performance, it is essential that tool cutting edges be keen, smooth, and free of grinding-wheel scratches, burrs, or wire edges. Keen edges can be obtained by finish grinding on a fine abrasive wheel and then lapping, or hand stoning with a fine oilstone. Neither the angles nor the contour of the cutting edge should be appreciably modified during tool finishing.

## Tool Material

Water-hardening tool steel, such as W1, heat treated to 65 to 68 HRC, is an adequate cutting tool material when production runs are short and speeds are low. However, tool steels of this type soften rapidly if the temperature of the cutting edge exceeds 150 °C (300 °F). In addition, tools made from water-hardening tool steel have low resistance to edge wear.

**High-speed steel tools** are generally satisfactory for machining all but the high-silicon alloys, which are quite abrasive and should be machined with carbide or diamond tools, unless runs are short. For short runs, high-speed steel is usually satisfactory. Complexly shaped cutters such as twist drills, reamers and counterbores, taps and other thread-cutting tools, end mills and many other types of milling cutters, and form tools are widely used in turning operations. For general machining and for relatively short-run production, single-point tools of high-speed steel are also commonly used.

Grade M2 high-speed tool steel is commonly used in machining aluminum, although M7 is sometimes preferred because it is more suitable for fine-edge tools and is more abrasion resistant. Grade T2 is also used sometimes, as are high-speed steels containing cobalt.

**Carbide Tools.** Because of the brittleness of the tool tip, the lip angle for carbide tools is usually greater than those recommended for high-speed steel tools in order to provide maximum support to the edge. This is indicated in Table 4, in which smaller relief angles are given for carbide tools. The rake angles can also be decreased to zero or negative to increase the lip angle, but negative rake angles are generally not recommended. When very light finishing cuts are made, it is sometimes feasible to reverse this practice and use higher rake angles and smaller lip angles.

Carbide tools retain sharp edges over a longer period between regrinds than carbon or high-speed steel tools, provided they are not used for heavy, intermittent cuts. A better finish is obtained because of the hardness of the tool tip compared to that of the stock. Carbide tools are particularly useful for machining high-silicon alloys, many of which cannot otherwise be machined satisfactorily under production conditions.

For longer production runs, cemented tungsten carbide tooling is generally the choice for machining low-silicon alloys and is often used for high-silicon alloys. The primary advantage offered by carbides in the machining of easily machined aluminum alloys is the greatly increased cutting speed possible compared to high-speed steel tooling—double or more the metal removal rates possible with the latter.

Both brazed-tip and indexable inserts are suitable. Straight-tungsten carbides are recommended in the C-2 to C-3 application categories. Some carbide specialists recommend fine-grain carbides for aluminum because a keener edge can be produced and because any edge chipping is small and therefore has a smaller effect on sharpness. Edge sharpness is desirable for cutting aluminum.

Coatings, which have proved to be highly advantageous in many steel machining operations, do not appear to provide any benefits in the machining of aluminum alloys. Indeed, vacuum-deposited coatings have a detrimental effect on the smoothness of the rake face of an insert, and all aluminum machining specialists stress the need for highly polished cutters in any material.

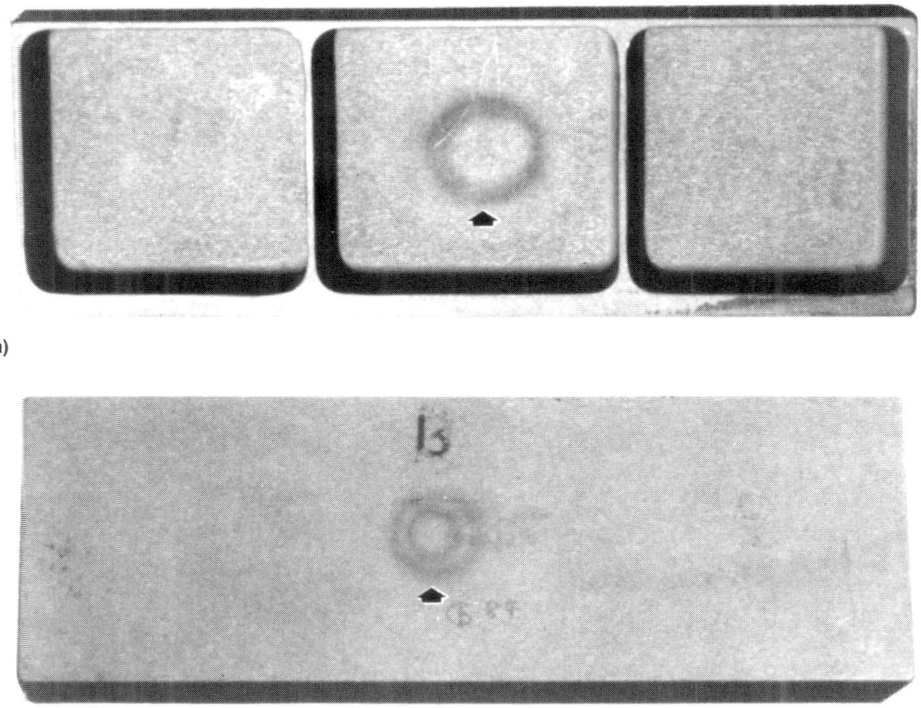

(a) (b)

**Fig. 7** Typical end mill defect associated with vertical plunge cut pocketing operations. (a) Front view. (b) Back view. Courtesy of McDonnell Douglas Canada, Ltd.

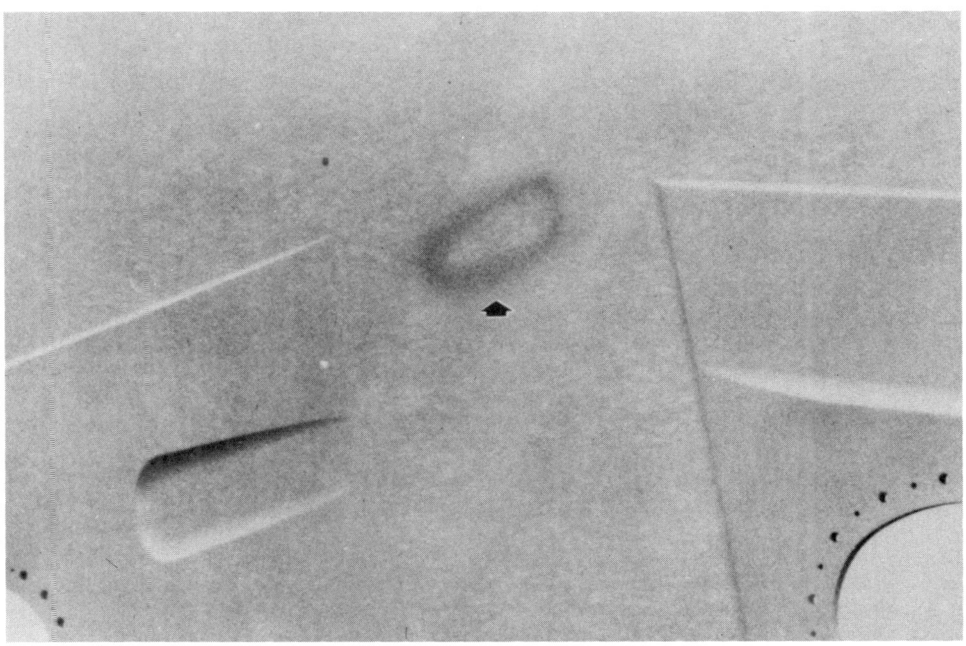

**Fig. 8** Milling defect associated with the NC machining of 7075-T7651 aluminum plate with a 75 mm (3 in.) diam milling cutter. Courtesy of McDonnell Douglas Canada, Ltd.

**Diamond tools** are used only in operations requiring an exceptionally high finish, particularly on high-silicon alloys, in which particles of free silicon will in time slightly dull the cutting edge of even carbide tools. Finishing cuts with diamond tools seldom exceed a few thousandths of an inch.

Single-crystal natural diamond cutting tools have long been used for producing very smooth finishes on turned aluminum workpieces—finishes of the order of 0.125 μm (5 μin.) roughness and better. For example, the lens barrels of even relatively inexpensive cameras are often turned with

**Fig. 9** Attachment tee made of 7075-T6 aluminum with the soft spot area having a hardness of 73 to 83 HRB compared to 90 HRB for the rest of the structure. Courtesy of McDonnell Douglas Canada, Ltd.

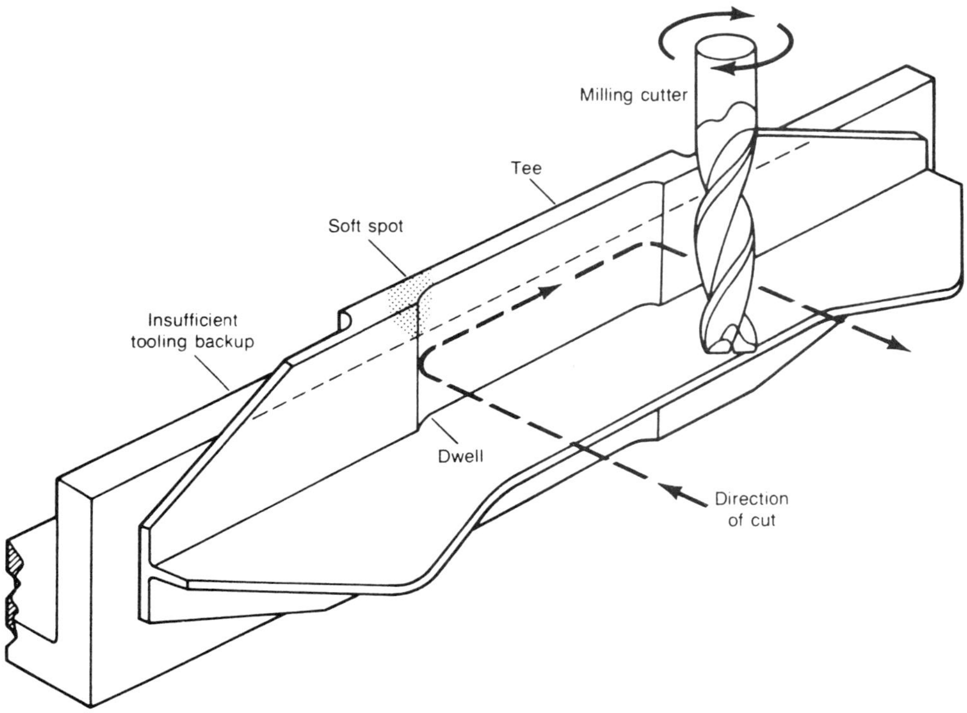


**Fig. 10** Schematic of the milling cutter path in machining the attachment tee shown in Fig. 9

single-crystal diamond tools, and no further finishing operations are performed.

The principal advantages of single-crystal diamond for machining aluminum are the ability to take an extremely fine edge, the highest known hardness and wear resistance, and an extremely low coefficient of friction with no tendency to adhere. In addition, there is no chemical reaction between the carbon of diamond and aluminum.

Diamond tools are usually made with either circular or faceted cutting edges, the latter being the more common. With the faceted cutter, there may be as many as five facets on one cutting edge, each varying in size from 0.5 to 1.5 mm (0.02 to 0.06 in.). Cutting angles of 74 to 90° and top rake angles of 6 to 10° are used. Rake should not be less than −6°. The tool should be set on, or slightly above, the centerline of the work.

**Polycrystalline Diamond Tooling (Ref 1).** A development of recent years that is having a major effect in many aluminum machining operations has been the introduction of compacted polycrystalline diamond cutting tools. Polycrystalline diamond is produced by a technique that includes pressing gritlike diamond particles in a mold under pressures and temperatures approaching those necessary to convert carbon into synthetic diamond. Natural or synthetic diamond particles can be used as the raw material, and it is possible to produce a compacted polycrystalline diamond surface on a tungsten carbide substrate.

Polycrystalline diamond, being true diamond, possesses the basic physical and chemical properties that make single-crystal diamond a valuable tool material for machining aluminum, but it does differ in several respects. Because it is polycrystalline, it cannot produce as fine an edge as single-crystal diamond. Therefore, it is not suitable for producing the finest finishes; about 0.125 to 0.25 μm (5 to 10 μin.) is the practical limit, which is acceptable in most applications.

Because the crystal orientation is random, impact resistance is high, and the properties are essentially omnidirectional. In contrast, single-crystal stones require careful orientation of the crystal for optimum utilization of properties. Another advantage over natural stones is that tool tips can be produced virtually to final shape, saving considerable effort for the toolmaker, and the use of a carbide substrate permits easy brazing of the polycrystalline tip to a tool steel shank or standard-size insert.

Initially introduced in the United States in 1972, polycrystalline diamond tooling has gained an important position in the machining of the highly abrasive silicon-aluminum alloys. Typical production operations include turning, boring, and milling. Reports of polycrystalline diamond tools outlasting tungsten carbide by factors as great as 100 in the machining of high-silicon aluminum are not uncommon. More detailed information on polycrystalline diamond tooling can be found in the article "Ultrahard Tool Materials" in this Volume. Application data for aluminum-silicon alloys follow.

## Machining of High-Silicon Aluminum Alloy 390 (Ref 2)

**Microstructural Considerations.** Because of the presence of primary silicon in the microstructure, the machining of aluminum-silicon alloys such as 390 must be approached somewhat differently from the machining of conventional aluminum casting alloys. The major alloying ingredients are 17% Si, 4.5% Cu, and 0.5% Mg. The eutectic composition is approximately 12% Si; the remaining 5 to 6% Si occurs as a primary phase. This primary-phase silicon is desirable in wear parts, but it affects cutting tool life drastically when machining alloy 390. The 390 silicon-aluminum alloy machines differently from other cast aluminum alloys, such as 319 (5.5 to 6.5% Si), and 380 (7.5 to 9.5% Si), in that the primary silicon causes rapid tool wear. Alloy 390 was originally developed for use in engines in which the cylinder bore is finished in such a manner that primary silicon stands in relief and prevents piston rings, and so on, from contacting the aluminum matrix.

Most aluminum casting alloys contain silicon as a major alloying element. Silicon, in either eutectic or primary form, is much harder than any other phase of the microstructure. The hardness of silicon crystals generally ranges from 1000 to 1300 HK, while the microhardness of an aluminum casting alloy matrix seldom exceeds 180 HK. Silicon, like the heavy-element intermetallic impurity phases of iron, manganese, and chromium, is a very abrasive material in an otherwise soft matrix and is

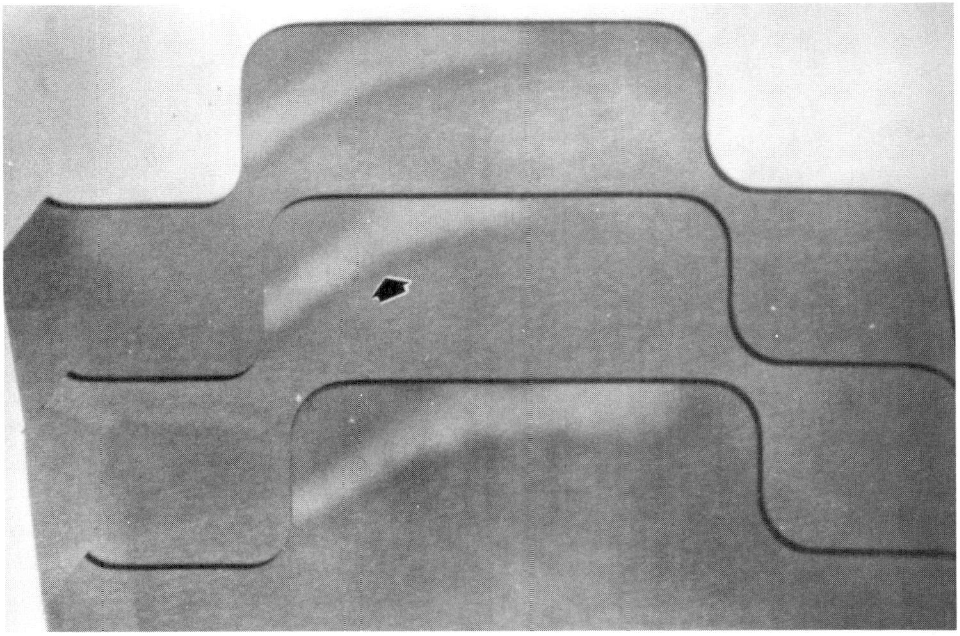

**Fig. 11** Occurrence of soft spots in the same location on each of three 7075-T651 aluminum plate components. Of the 19 parts produced, 13 had a soft zone hardness of 68 to 87 HRB and conductivity of 36 to 39%. Unaffected regions of the plate had a hardness of 90 to 91 HRB and conductivity of 31 to 32%. Courtesy of McDonnell Douglas Canada, Ltd.

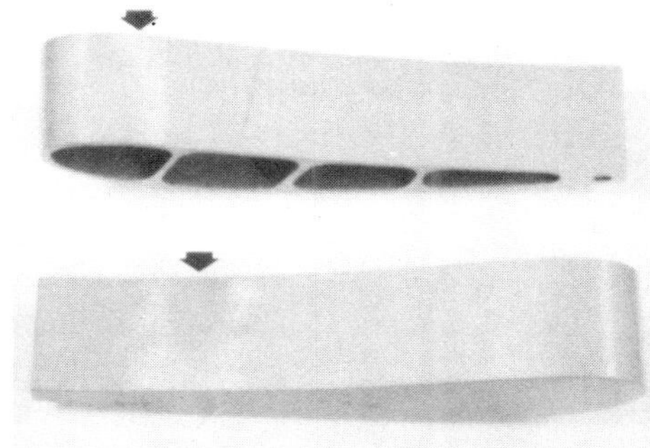

**Fig. 12** Two views of a milling defect on an unsupported vertical flange in a 7075-T651 aluminum plate. Courtesy of McDonnell Douglas Canada, Ltd.

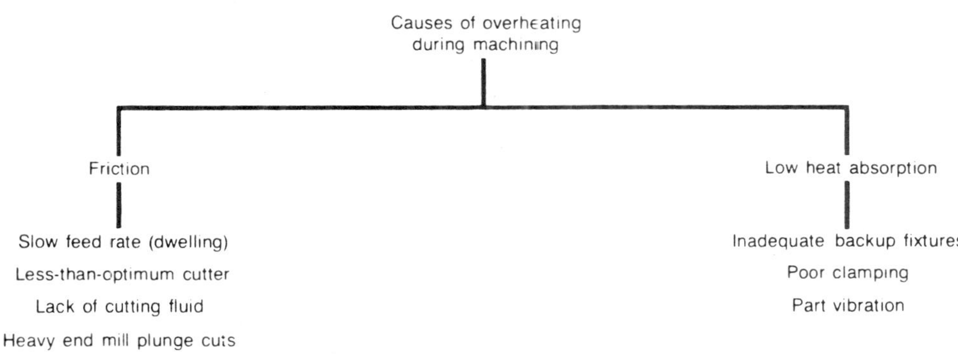


**Fig. 13** Causes of overheating during machining process attributable to friction and low heat absorption problems

the element that singularly has the greatest tendency to decrease cutting tool life.

A fine, well-modified eutectic silicon structure is far less detrimental to tool life than heavy-element intermetallic impurity phases. However, the rate of wear on cutting tools increases as silicon particle size increases. If the eutectic silicon structure is coarse, tool life suffers. Primary silicon crystals, even if well refined and distributed, are more detrimental than eutectic silicon, and large unrefined primary crystals can seriously degrade tool life.

In permanent mold or sand casting, primary silicon is controlled by a refinement treatment, adding a small amount of phosphorus to the molten alloy. Unrefined primary silicon is eight to ten times the size of refined silicon crystal, and it drastically affects machinability.

In conventional die casting, a refinement treatment is not necessary. Primary silicon size in conventional die castings is very small even without phosphorus refinement, and the silicon size and distribution are controlled by such process parameters as melt temperature, die temperature, and die fill rate.

Regardless of the casting method employed, primary silicon acts as a chip breaker. Although primary silicon crystals are very hard, they are also quite friable. When the cutting tool passes through the matrix of the alloy, it fractures primary silicon particles. This fracturing, along with the natural hardness of the alloy and work hardening of the chips, causes the close, tightly curled chips to break into short ringlets similar in appearance to gray iron chips.

**Machining Parameters.** Table 5 lists information regarding tools and cutting conditions for machining alloy 390. The turning of alloy 390 can be performed using C-3 tungsten carbide or polycrystalline diamond cutting tools. Turning with carbide should be limited to relatively slow cutting speeds (30 to 150 m/min, or 100 to 500 sfm) to ensure adequate tool life. Tool life can be greatly improved and cutting speeds substantially increased through the use of polycrystalline diamond tools. The productivity increase (300 to 900 m/min, or 1000 to 3000 sfm) can more than offset the higher cost of the diamond.

Similar results have been experienced for milling alloy 390. Milling with carbide should be done at slow speeds (99 to 175 m/min, or 325 to 575 sfm) with or without coolant. Milling with polycrystalline diamond can be done at speeds of 400 to 790 m/min (1300 to 2600 sfm) (with or without coolant) and with much longer tool life.

The drilling of alloy 390 can be accomplished using high-speed steel, high-helix drills with parabolic flutes at a speed of 20 m/min (70 sfm) for holes up to six times diameter in depth. Much better productivity can be achieved by using tungsten-carbide-tip, coolant-fed drills at 90 m/min (300 sfm). In addition, diamond-tip, coolant-fed gun drills have achieved even higher productivity gains with much better tool life in holes up to eight times diameter.

Thread forming taps have the longest tool life and can be run at speeds to 20 m/min (70 sfm). They also produce stronger threads than thread cutting taps.

The use of generous quantities of coolant is imperative when drilling and tapping alloy 390. The coolant must be in sufficient volume and pressure to reach the bottom of the hole and flush out the chips. These chips are very abrasive to the cutting tool.

The broaching of alloy 390 can be accomplished with the same speeds, tool geometries, and materials used to broach gray iron. Tool life will be much better, and the surface finish will be comparable to, or better than, the finish typical of gray iron.

The machining of alloy 390 in recent years has been much improved. This has been accomplished through the application of new technology in cutting tool materials and in tool designs, optimization of cutting

### Table 7 Suggested practices for turning aluminum on general-purpose equipment

The suggested practices are commonly used, but higher speeds, feeds, and depths of cut can be employed in many cases (depending on the nature of the part, machine tool, tool design, lubrication, and other cutting conditions) to increase production rates.

| Variable and condition | Suggested practice |
|---|---|
| Back rake angle, degrees | 5–15 |
| Side rake angle, degrees | |
| Machinability rating A | 0–20 |
| Machinability rating B | 20 |
| Machinability rating C | 20–30 |
| Machinability rating D, E | 40 |
| End relief angle clearance, degrees | |
| High-speed steel tool | 8–15 |
| Carbide tool | 6–8 |
| Side relief angle, degrees | |
| High-speed steel tool | 8–15 |
| Carbide tool | 6–8 |
| End cutting edge angle, degrees | 5 |
| Side cutting edge angle, degrees | 5–15 |
| Nose radius, mm (in.) | |
| Rough cutting | 0.8–3.2 (1/32–1/8) |
| Finish cutting | 0.25–0.76 (0.010–0.030) |
| Depth of cut, mm (in.) | |
| Small work | 0.4–6.4 (1/64–1/4) |
| Large work | 13–38 (1/2–1 1/2) |
| Feed, mm/rev (in./rev) | |
| Rough cutting | 0.15–2.0 (0.006–0.080) |
| Finish cutting | 0.05–0.15 (0.002–0.006) |
| Speed, m/min (sfm) | |
| High-speed steel tool | ≤300 (≤1000) |
| Carbide tool | ≤4600 (≤15 000) |

Source: Ref 4

### Table 8 Nominal speeds for turning aluminum alloys

| Operation | Speed, m/min (sfm): Nonheat-treated cast alloys | All other alloys and tempers |
|---|---|---|
| **High-speed steel tools (M2 or T5)** | | |
| Single-point roughing | 230 (750) | 180 (600) |
| Single-point finishing | 300 (1000) | 240 (800) |
| Forming and cutoff | 170 (550) | 140 (450) |
| **Carbide tools (C-2)** | | |
| Single-point roughing(a) | | |
| Brazed tips | 490 (1600) | 340 (1100) |
| Disposable tips | 610 (2000) | 430 (1400) |
| Forming and cutoff | 360 (1200) | 250 (825) |

Note: Speeds for single-point turning are based on a 3.81 mm (0.150 in.) depth of cut and a feed of 0.38 mm/rev (0.015 in./rev) for roughing and a 0.64 mm (0.025 in.) depth of cut and a feed of 0.18 mm/rev (0.007 in./rev) for finishing. Speeds for form turning are based on feeds of 0.089, 0.08, and 0.05 mm/rev (0.0035, 0.003, and 0.002 in./rev) for tool widths of 13, 25, and 50 mm (1/2, 1, and 2 in.), respectively. Speeds for cutoff are based on a feed rate of 0.05 mm/rev (0.002 in./rev). (a) For finish turning, use the maximum speed of the machine and C-3 carbide.

conditions, and the replacement of outdated transfer lines with machine tools designed for aluminum alloys.

## Cutting Fluids

A cutting fluid for aluminum can be a soluble-oil emulsion, a mineral oil, or an aqueous chemical solution. Cutting oils that contain compounds of sulfur or chlorine or both are seldom used and are not usually required for machining aluminum. In addition, many of them will stain the work. Table 6 lists various cutting fluids and their applications.

**Soluble oil** mixed with water in ratios of 1 part oil to 20 to 30 parts water is the cutting fluid most widely used for machining aluminum alloys. Soluble-oil emulsions are inexpensive, highly efficient for cooling and removing chips, and are usually adequate for preventing built-up edges.

**Mineral oil** used as a cutting fluid may contain a fatty additive, such as lard oil, neat's-foot oil, oleic acid, or butyl stearate. However, mineral oil that contains no additive and that has a viscosity of 40 to 300 Saybolt universal seconds (SUS) at 40 °C (100 °F) is most often used. As cutting speed is increased, the viscosity of the oil should be decreased to provide easier flow and therefore greater cooling. Straight mineral seal oil (~40 SUS at 40 °C, or 100 °F) has been effective in many applications. Kerosene, which is slightly less viscous than mineral seal oil, is also used.

**Chemical solutions** are effective as cutting fluids for machining aluminum and are especially desirable when a transparent fluid is needed to permit viewing the work during machining. These solutions vary in composition, but most contain amines, nitrides, phosphates, borates, soaps, wetting agents, glycols, and germicides. Some of these solutions stain aluminum alloys.

**Stick grease** is sometimes used in band sawing, circular sawing, and abrasive belt, abrasive disk, or abrasive wheel polishing and grinding when requirements are not too severe and a flood of lubricant is not required.

**Continuous filtering** of cutting fluids for the removal of chips, slivers, grindings, and other foreign material is especially desirable because aluminum alloys are relatively soft and are easily damaged by a contaminated cutting fluid.

**The mist application of cutting fluids** is sometimes preferred for machining aluminum, especially on such machine tools as small, vertical mills and high-speed routers. Compressed air, at pressures ranging from 70 to 550 kPa (10 to 80 psi) is used to atomize the cutting fluid and propel it into the cutting zone. The technique can be used with straight oils (especially of lower viscosities), emulsions, or chemical solutions as long as there are no solid additives contained in the fluid. Some producers of spray-mist equipment, however, recommend the use of water-base fluids over oils because the latter may tend to clog and may present a hazard.

Although the use of compressed air adds a new element of cost, the overall cost can often be reduced because of the smaller volume of fluid used and lower losses of fluid carried out by chips. The cooling effect

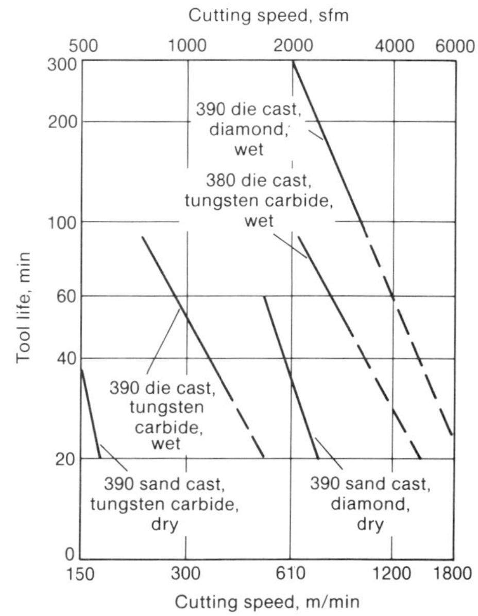


**Fig. 14** Tool life curves for the dry turning of sand cast alloy 390 using carbide and diamond, the wet turning of die cast alloy 390 using carbide and diamond, and the wet turning of die cast alloy 380 using carbide. Source: Ref 2

of a mist, while not equal to that of a copious flood, is considerable because the rapid evaporation of the finely divided droplets adds the latent heat of vaporization to the cooling capability of the liquid.

Proponents of the compressed air method also claim that the high velocity of the droplets tends to get the fluid into the cutting zone effectively and that the evaporation of the water increases the lubricity at the cut. Visibility into the cutting area is improved, and splash shielding does not pose as much of a problem as flooding does. Most experts recommend the use of automatic shutoff devices for mist systems and venting to prevent potential air contamination.

## Distortion and Dimensional Variation

Because aluminum alloys have a low modulus of elasticity (~70 GPa, or $10 \times 10^6$ psi), they will distort more than most metals for a given clamping or chucking force. Moderate clamping forces should be used to avoid dimensional variations due to distortion. High clamping pressures are not required, because cutting forces are low.

**Thermal Expansion.** The coefficient of thermal expansion of aluminum alloys (18 to 25 μm/m·K, or 10 to 14 μin./in./°F) is higher than that of most metals commonly machined. Therefore, the dimensional accuracy of finished parts requires that the part be kept cool during machining. When turning between centers, it is important to avoid expansion, which will put excessive pres-

**Table 9 Nominal speeds and feeds for the boring of aluminum alloys with high-speed tool steels and carbide tools**

| Material | Hardness, HB (500 kg) | Condition | Depth of cut, mm | Depth of cut, in. | High-speed steel tool: Speed, m/min | Speed, sfm | Feed, mm/rev | Feed, in./rev | Tool material grade, ISO | Tool material grade, AISI | Uncoated carbide tool: Speed, Brazed, m/min | Brazed, sfm | Indexable, m/min | Indexable, sfm | Feed, mm/rev | Feed, in./rev | Tool material grade, ISO | Tool material grade, C |
|---|---|---|---|---|---|---|---|---|---|---|---|---|---|---|---|---|---|---|
| **Wrought** | | | | | | | | | | | | | | | | | | |
| EC, 3005, 6066, 1060, 4032, 6070, 1100, 5005, 6101, 1145, 5050, 6151, 1175, 5052, 6253, 2011, 5083, 6463, 1235, 5056, 6262, 2014, 5086, 6951, 2017, 5154, 7001, 2018, 5252, 7004, 2021, 5254, 7005, 2024, 5454, 7039, 2025, 5456, 7049, 2117, 5457, 7050, 2218, 5652, 7075, 2219, 5657, 7079, 2618, 6053, 7175, 3003, 6061, 7178, 3004, 6063 | 30–80 | Cold drawn | 0.25 | 0.010 | 305 | 1000 | 0.075 | 0.003 | S4, S5 | M2, M3 | 610 | 2000 | Max | Max | 0.15 | 0.006 | K01, M10 | C-3 |
| | | | 1.25 | 0.050 | 245 | 800 | 0.13 | 0.005 | S4, S5 | M2, M3 | 550 | 1800 | Max | Max | 0.20 | 0.008 | K10, M10 | C-2 |
| | | | 2.5 | 0.100 | 215 | 700 | 0.30 | 0.012 | S4, S5 | M2, M3 | 365 | 1200 | Max | Max | 0.40 | 0.015 | K20, M20 | C-2 |
| | 75–150 | Solution treated and aged | 0.25 | 0.010 | 305 | 1000 | 0.075 | 0.003 | S4, S5 | M2, M3 | 610 | 2000 | Max | Max | 0.15 | 0.006 | K01, M10 | C-3 |
| | | | 1.25 | 0.050 | 245 | 800 | 0.13 | 0.005 | S4, S5 | M2, M3 | 550 | 1800 | Max | Max | 0.20 | 0.008 | K10, M10 | C-2 |
| | | | 2.5 | 0.100 | 215 | 700 | 0.30 | 0.012 | S4, S5 | M2, M3 | 365 | 1200 | Max | Max | 0.40 | 0.015 | K20, M20 | C-2 |
| **Cast** | | | | | | | | | | | | | | | | | | |
| **Sand and permanent mold** | | | | | | | | | | | | | | | | | | |
| A140, F332.0, 520.0, 201.0, 333.0, 535.0, 208.0, 354.0, 705.0, 213.0, 355.0, 707.0, 222.0, C355.0, A712.0, 224.0, 356.0, D712.0, 242.0, A356.0, 713.0, 295.0, 357.0, 771.0, B295.0, 359.0, 850.0, 308.0, B443.0, A850.0, 319.0, 514.0, B850.0, 328.0, A514.0, A332.0, B514.0, Hiduminium RR-350 | 40–100 | As-cast | 0.25 | 0.010 | 305 | 1000 | 0.075 | 0.003 | S4, S5 | M2, M3 | 550 | 1800 | Max | Max | 0.15 | 0.006 | K01, M10 | C-3 |
| | | | 1.25 | 0.050 | 245 | 800 | 0.13 | 0.005 | S4, S5 | M2, M3 | 365 | 1200 | Max | Max | 0.20 | 0.008 | K10, M10 | C-2 |
| | | | 2.5 | 0.100 | 215 | 700 | 0.30 | 0.012 | S4, S5 | M2, M3 | 305 | 1000 | Max | Max | 0.40 | 0.015 | K20, M20 | C-2 |
| | 70–125 | Solution treated and aged | 0.25 | 0.010 | 230 | 750 | 0.075 | 0.003 | S4, S5 | M2, M3 | 550 | 1800 | Max | Max | 0.15 | 0.006 | K01, M10 | C-3 |
| | | | 1.25 | 0.050 | 185 | 600 | 0.13 | 0.005 | S4, S5 | M2, M3 | 365 | 1200 | Max | Max | 0.20 | 0.008 | K10, M10 | C-2 |
| | | | 2.5 | 0.100 | 170 | 550 | 0.30 | 0.012 | S4, S5 | M2, M3 | 305 | 1000 | Max | Max | 0.40 | 0.015 | K20, M20 | C-2 |
| **Die castings** | | | | | | | | | | | | | | | | | | |
| 360.0, A380.0, A360.0, C443.0, 380.0, 518.0 | 40–100 | As-cast | 0.25 | 0.010 | 305 | 1000 | 0.075 | 0.003 | S4, S5 | M2, M3 | 550 | 1800 | Max | Max | 0.15 | 0.006 | K01, M10 | C-3 |
| | | | 1.25 | 0.050 | 245 | 800 | 0.13 | 0.005 | S4, S5 | M2, M3 | 365 | 1200 | Max | Max | 0.20 | 0.008 | K10, M10 | C-2 |
| | | | 2.5 | 0.100 | 215 | 700 | 0.30 | 0.012 | S4, S5 | M2, M3 | 305 | 1000 | Max | Max | 0.40 | 0.015 | K20, M20 | C-2 |
| | 70–125 | Solution treated and aged | 0.25 | 0.010 | 230 | 750 | 0.075 | 0.003 | S4, S5 | M2, M3 | 490 | 1600 | Max | Max | 0.15 | 0.006 | K01, M10 | C-3 |
| | | | 1.25 | 0.050 | 185 | 600 | 0.13 | 0.005 | S4, S5 | M2, M3 | 425 | 1400 | Max | Max | 0.20 | 0.008 | K10, M10 | C-2 |
| | | | 2.5 | 0.100 | 145 | 550 | 0.30 | 0.012 | S4, S5 | M2, M3 | 245 | 800 | Max | Max | 0.40 | 0.015 | K20, M20 | C-2 |
| 383.0, 413.0, A384.0, A413.0 | 40–100 | As-cast | 0.25 | 0.010 | 230 | 750 | 0.075 | 0.003 | S4, S5 | M2, M3 | 490 | 1600 | Max | Max | 0.15 | 0.006 | K01, M10 | C-3 |
| | | | 1.25 | 0.050 | 185 | 600 | 0.13 | 0.005 | S4, S5 | M2, M3 | 425 | 1400 | Max | Max | 0.20 | 0.008 | K10, M10 | C-2 |
| | | | 2.5 | 0.100 | 170 | 550 | 0.30 | 0.012 | S4, S5 | M2, M3 | 245 | 800 | Max | Max | 0.40 | 0.015 | K20, M20 | C-2 |
| | 70–125 | Solution treated and aged | 0.25 | 0.010 | 205 | 675 | 0.075 | 0.003 | S4, S5 | M2, M3 | 410 | 1350 | Max | Max | 0.15 | 0.006 | K01, M10 | C-3 |
| | | | 1.25 | 0.050 | 170 | 550 | 0.13 | 0.005 | S4, S5 | M2, M3 | 335 | 1100 | Max | Max | 0.20 | 0.008 | K10, M10 | C-2 |
| | | | 2.5 | 0.100 | 145 | 480 | 0.30 | 0.012 | S4, S5 | M2, M3 | 260 | 850 | Max | Max | 0.40 | 0.015 | K20, M20 | C-2 |
| 390.0, 392.0 | 40–100 | As-cast | 0.25 | 0.010 | 46 | 150 | 0.075 | 0.003 | S4, S5 | M2, M3 | 125 | 410 | 145 | 480 | 0.15 | 0.006 | K10, M10 | C-2 |
| | | | 1.25 | 0.050 | 37 | 120 | 0.13 | 0.005 | S4, S5 | M2, M3 | 100 | 325 | 115 | 385 | 0.20 | 0.008 | K10, M10 | C-2 |
| | | | 2.5 | 0.100 | 34 | 110 | 0.30 | 0.012 | S4, S5 | M2, M3 | 90 | 300 | 105 | 350 | 0.30 | 0.012 | K20, M20 | C-2 |
| | 70–125 | Solution treated and aged | 0.25 | 0.010 | 43 | 140 | 0.075 | 0.003 | S4, S5 | M2, M3 | 105 | 350 | 125 | 410 | 0.15 | 0.006 | K10, M10 | C-2 |
| | | | 1.25 | 0.050 | 34 | 110 | 0.13 | 0.005 | S4, S5 | M2, M3 | 85 | 280 | 100 | 330 | 0.20 | 0.008 | K10, M10 | C-2 |
| | | | 2.5 | 0.100 | 30 | 100 | 0.30 | 0.012 | S4, S5 | M2, M3 | 70 | 230 | 84 | 275 | 0.30 | 0.012 | K20, M20 | C-2 |

Source: Ref 5

sure on the centers. High cutting speed helps keep the part cool because most of the heat introduced into the part during a given rotation is removed with the chip during the next rotation, and the time for diffusion of heat is short. A cutting fluid is effective in removing heat that is not removed with the chips. Live centers are recommended to minimize frictional heating at the center. Dull tools cause a heat rise in the workpiece; therefore, cutting tools should be kept sharp.

**Residual stress** can be induced by a dull or improperly designed cutter that cold works the surface, by excessive chucking or clamping force, or by faulty clamping. Distortion from residual stress is most noticeable in slender parts.

The distortion resulting from machining stresses can be minimized or eliminated either by employing a series of light cuts as the part approaches finished size or by stress relieving the part between rough and finish machining. For heat-treatable alloys, it is preferable to do all rough machining on material in the solution-treated and aged condition, rather than in the annealed con-

**Table 10 Nominal speeds and feeds for the boring of aluminum alloys with diamond tools**

| Material | Hardness, HB (500 kg) | Condition | Depth of cut, mm (in.) | Speed, m/min (sfm) | Feed, mm/rev (in./rev) |
|---|---|---|---|---|---|
| **Wrought** | | | | | |
| EC 2218 5252 6253 | 30–150 | All | 0.13–0.40 | 460 | 0.075–0.15 |
| 1060 2219 5254 6262 | | | (0.005–0.015) | (1500) | (0.003–0.006) |
| 1100 2618 5454 6463 | | | 0.40–1.25 | 305 | 0.15–0.30 |
| 1145 3003 5456 6951 | | | (0.015–0.050) | (1000) | (0.006–0.012) |
| 1175 3004 5457 7001 | | | 1.25–3.2 | 150 | 0.30–0.50 |
| 1235 3005 5652 7004 | | | (0.050–0.125) | (500) | (0.012–0.020) |
| 2011 4032 5657 7005 | | | | | |
| 2014 5005 6053 7039 | | | | | |
| 2017 5050 6061 7049 | | | | | |
| 2018 5052 6063 7050 | | | | | |
| 2021 5056 6066 7075 | | | | | |
| 2024 5083 6070 7079 | | | | | |
| 2025 5086 6101 7175 | | | | | |
| 2117 5154 6151 7178 | | | | | |
| **Cast** | | | | | |
| **Sand and permanent mold** | | | | | |
| A140 319.0 357.0 A712.0 | 40–100 | As-cast | 0.13–0.40 | 760 | 0.075–0.15 |
| 201.0 328.0 359.0 D712.0 | | | (0.005–0.015) | (2500) | (0.003–0.006) |
| 208.0 A332.0 B443.0 713.0 | | | 0.40–1.25 | 610 | 0.15–0.30 |
| 213.0 F332.0 514.0 771.0 | | | (0.015–0.050) | (2000) | (0.006–0.012) |
| 222.0 333.0 A514.0 850.0 | | | 1.25–3.2 | 305 | 0.30–0.50 |
| 224.0 354.0 B514.0 A850.0 | | | (0.050–0.125) | (1000) | (0.012–0.020) |
| 242.0 355.0 520.0 B850.0 | 70–125 | Solution treated | 0.13–0.40 | 670 | 0.075–0.15 |
| 295.0 C355.0 535.0 | | and aged | (0.005–0.015) | (2200) | (0.003–0.006) |
| B295.0 356.0 705.0 | | | 0.40–1.25 | 520 | 0.15–0.30 |
| 308.0 A356.0 707.0 | | | (0.015–0.050) | (1700) | (0.006–0.012) |
| Hiduminium RR-350 | | | 1.25–3.2 | 215 | 0.30–0.50 |
| | | | (0.050–0.125) | (700) | (0.012–0.020) |
| **Die castings** | | | | | |
| 360.0 A380.0 | 40–100 | As-cast | 0.13–0.40 | 915 | 0.075–0.15 |
| A360.0 C443.0 | | | (0.005–0.015) | (3000) | (0.003–0.006) |
| 380.0 518.0 | | | 0.40–1.25 | 760 | 0.15–0.30 |
| 383.0 | | | (0.015–0.050) | (2500) | (0.006–0.012) |
| | | | 1.25–3.2 | 610 | 0.30–0.50 |
| | | | (0.050–0.125) | (2000) | (0.012–0.020) |
| A384.0 | 40–100 | As-cast | 0.13–0.40 | 760 | 0.075–0.15 |
| 413.0 | | | (0.005–0.015) | (2500) | (0.003–0.006) |
| A413.0 | | | 0.40–1.25 | 610 | 0.15–0.30 |
| 390.0 | | | (0.015–0.050) | (2000) | (0.006–0.012) |
| | | | 1.25–3.2 | 460 | 0.30–0.50 |
| | | | (0.050–0.125) | (1500) | (0.012–0.020) |
| 392.0 | 40–100 | As-cast | 0.13–0.40 | 610 | 0.075–0.15 |
| | | | (0.005–0.015) | (2000) | (0.003–0.006) |
| | | | 0.40–1.25 | 460 | 0.15–0.30 |
| | | | (0.015–0.050) | (1500) | (0.006–0.012) |
| | | | 1.25–3.2 | 305 | 0.30–0.50 |
| | | | (0.050–0.125) | (1000) | (0.012–0.020) |

Source: Ref 5

dition, because the less ductile structure is more machinable.

The distortion resulting from machining stress can often be minimized by purchasing the alloy in a stress-relieved condition, normally designated by T451, T651, or T851 if the metal has been stress relieved by stretching. The code T*x*52 denotes stress relief by compression; T*x*53, stress relief by heat treating.

A major source of dimensional variation arises from the presence of movement, or play, in the feed mechanism of the machine. When the tool or the machining conditions cause forces that take up the play completely, no variations in dimensions are encountered. However, when low cutting forces (typical of those required for machining aluminum) are combined with a small feed and light cut, the total force may be insufficient to overcome tool-slide friction. Then, some movement may still occur, and the tool may float.

## Soft Spots in Machined Aluminum Parts (Ref 3)

Machining practices that are less than optimum may cause localized overheating in aluminum alloys. In alloys being ma-

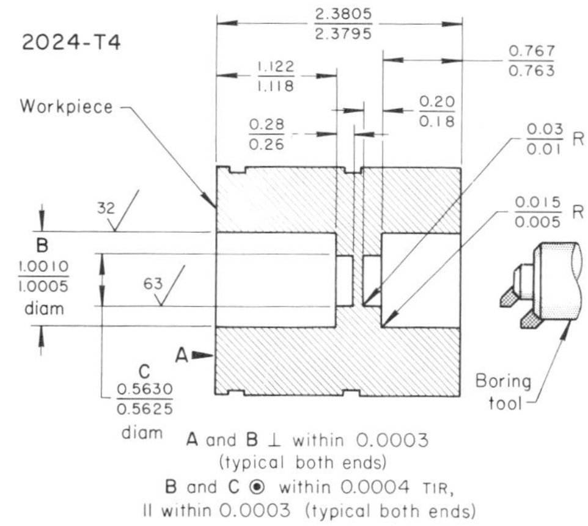


| | |
|---|---|
| Speed, at 1200 rev/min, m/min (sfm) | 97 (318) |
| Feed, mm/min (in./min) | 35.6 (1.4) |
| Cutting fluid | Kerosene plus proprietary oil |
| Production rate, pieces/h | 9 |
| Tool life/grind, lineal m (lineal in.) | 20 (800) |

**Fig. 15** Rough and finish boring of a heat-treated aluminum alloy forging. Dimensions in figure given in inches

chined in the postsolution heat treatment tempers, the resultant effect is overaging, softening, and therefore loss of strength.

Such defects, referred to as soft spots, are usually invisible and may remain undetected at this stage except where hardness and/or conductivity testing is performed in the overheated area. However, anodizing has the capability to reveal these defects. Locally overheated and softened areas are indicated by a color contrast against the background of the normal anodic film. The area is always very local on the parts, and its shape is usually related to the machine cutting tool that profiled the particular area. Frequently, the soft areas are at or near changes of section thickness. In addition, there is often exact repeatability of the soft spot location in all other parts of the batch.

Figure 6 illustrates a half-moon defect associated with spar milling deficiencies in an area adjacent to a change in section thickness. Figure 7 shows a circular indication that is associated with vertical plunge cut pocketing operations using an end mill. The 25 mm (1 in.) diameter of the soft spot is a function of the end mill diameter. Figure 8 shows a similar example of a milling problem that occurred in a numerically controlled (NC) machining application using a 75 mm (3 in.) diam cutter.

A different type of problem is illustrated in Fig. 9. Two silver-white defective areas are clearly visible on the tee section part machined from an extrusion. All 36 parts in the batch were similarly affected. The defects are attributed to a lack of full-support tooling backup that did not provide an adequate heat sink. The milling cutter was allowed to dwell at the start of the rebate cut, thus introducing frictional heating. The

**Table 11 Nominal speeds and feeds for the planing of aluminum alloys with high-speed tool steels and carbide tools**

| Material | Hardness, HB (500 kg) | Condition | Depth of cut, mm (in.) | High-speed steel tool: Speed, m/min (sfm) | High-speed steel tool: Feed, mm/stroke (in./stroke) | High-speed steel tool: Tool material grade, ISO (AISI) | Carbide tool: Speed, m/min (sfm) | Carbide tool: Feed, mm/stroke (in./stroke) | Carbide tool: Tool material grade, ISO (C grade) |
|---|---|---|---|---|---|---|---|---|---|
| **Wrought** | | | | | | | | | |
| EC 2218 5252 6253<br>1060 2219 5254 6262<br>1100 2618 5454 6463<br>1145 3003 5456 6951<br>1175 3004 5457 7001<br>1235 3005 5652 7004<br>2011 4032 5657 7005<br>2014 5005 6053 7039<br>2017 5050 6061 7049<br>2018 5052 6063 7050<br>2021 5056 6066 7075<br>2024 5083 6070 7079<br>2025 5086 6101 7175<br>2117 5154 6151 7178 | 30–80 | Cold drawn | 0.25 (0.010) | 90 (300) | (a) (a) | S4, S5 (M2, M3) | 90 (300) | (a) (a) | K10, M10 (C-2) |
| | | | 2.5 (0.100) | 90 (300) | 3.20 (0.125) | S4, S5 (M2, M3) | 90 (300) | 3.20 (0.125) | K20, M20 (C-2) |
| | | | 12 (0.500) | 90 (300) | 2.30 (0.090) | S4, S5 (M2, M3) | 90 (300) | 2.30 (0.090) | K30, M30 (C-2) |
| | 75–150 | Solution treated and aged | 0.25 (0.010) | 90 (300) | (a) (a) | S4, S5 (M2, M3) | 90 (300) | (a) (a) | K10, M10 (C-2) |
| | | | 2.5 (0.100) | 90 (300) | 3.20 (0.125) | S4, S5 (M2, M3) | 90 (300) | 3.20 (0.125) | K20, M20 (C-2) |
| | | | 12 (0.500) | 90 (300) | 2.30 (0.090) | S4, S5 (M2, M3) | 90 (300) | 2.30 (0.090) | K30, M30 (C-2) |
| **Cast** | | | | | | | | | |
| Sand and permanent mold<br>A140 319.0 357.0 A712.0<br>201.0 328.0 359.0 D712.0<br>208.0 A332.0 B443.0 713.0<br>213.0 F332.0 514.0 771.0<br>222.0 333.0 A514.0 850.0<br>224.0 354.0 B514.0 A850.0<br>242.0 355.0 520.0 B850.0<br>295.0 C355.0 535.0<br>B295.0 356.0 705.0<br>308.0 A356.0 707.0<br>Hiduminium RR-350 | 40–100 | As-cast | 0.25 (0.010) | 90 (300) | (a) (a) | S4, S5 (M2, M3) | 90 (300) | (a) (a) | K10, M10 (C-2) |
| | | | 2.5 (0.100) | 90 (300) | 3.20 (0.125) | S4, S5 (M2, M3) | 90 (300) | 3.20 (0.125) | K20, M20 (C-2) |
| | | | 12 (0.500) | 90 (300) | 2.30 (0.040) | S4, S5 (M2, M3) | 90 (300) | 2.30 (0.090) | K30, M30 (C-2) |
| | 70–125 | Solution treated and aged | 0.25 (0.010) | 90 (300) | (a) (a) | S4, S5 (M2, M3) | 90 (300) | (a) (a) | K10, M10 (C-2) |
| | | | 2.5 (0.100) | 90 (300) | 3.20 (0.125) | S4, S5 (M2, M3) | 90 (300) | 3.20 (0.125) | K20, M20 (C-2) |
| | | | 12 (0.500) | 90 (300) | 2.30 (0.090) | S4, S5 (M2, M3) | 90 (300) | 2.30 (0.090) | K30, M30 (C-2) |

(a) Feed = ¾ the width of square nose finishing tool. Source: Ref 5

part was then turned over, and the operation was repeated, thus producing an opposite soft spot. The milling sequence is outlined in Fig. 10. This illustrates clearly the phenomenon of soft spot location repeatability.

Another example of soft spot indications is shown in Fig. 11. Of the 19 parts produced in this batch, 13 were affected. The defects are attributed to vibration or distortion at an inadequately clamped edge of the parts during milling on a vacuum chuck. As a result, heat dissipation was restricted.

All soft spots have a lower hardness and a higher electrical conductivity than the adjacent normal material. The strength reduction can be significant.

## Sources of Machining Problems

Some of the more common machining problems are outlined in this section. It should be kept in mind that the basic cause of the defects under discussion is always excessive heat. This may arise directly by frictional heating and/or indirectly by insufficient cooling. Because the cause(s) of some problems may be obscure, on-site analysis of the hardware, machining techniques, tooling, and so on, is always desirable.

**Cutters.** All tools should be kept sharp and in good condition at all times. Tool geometry must be maintained within the established requirements for aluminum alloys.

Cutters must have the optimum number of flutes and the optimum spiral configuration for each application. For most applications, the minimum number of flutes and a coarse spiral are more desirable. Clogging results in rubbing and frictional heating; it can be minimized by maintaining a good surface finish in the flutes.

**Feeds and speeds** must be correct for the operation being performed. In particular, a slow feed rate (dwelling) combined with high spindle speeds is especially troublesome. Great care must be taken not to reduce feeds any more than is absolutely necessary when the cutter is approaching or is actually machining areas requiring section changes. On multispindle equipment, it is vital that sufficient net horsepower be available to each spindle during heavy roughing cuts. Where practical, when using end mills, deep plunge cuts should ramp in or start from a predrilled hole and should progress on a multipass basis.

**Cutting Fluids.** An adequate and continuous flow of cutting fluid (flood or mist) directed at the cutting edges is essential. The flow of cutting fluid should begin before cutting and must continue until the cutter has been removed from the part. Fluid circulation must be continuous because even a momentary reduction in flow, such as could be caused by an air lock, may result in local overheating.

**Support Fixtures and Clamping Methods.** Machining backup fixtures must be rigid and must have sufficient contact area and mass to provide an adequate heat absorption capacity. Gaps in modular fixtures should not be excessive. This is especially true in areas adjacent to section changes. In addition, it is vital that clamping methods maintain continuous contact between part and fixtures because even a minute air gap due to part distortion or vibration, together with the resultant deeper cutting action, may produce overheated areas. On flat parts that are being machined on a vacuum chuck, it is important that the edge seal be located as close to the periphery as practical in order to eliminate vibration and consequent gapping around the outer edges. In general, all thin edges are susceptible to heat buildup. Similarly, thin-section vertical

**Table 12 Nominal speeds and feeds per tooth (chip load) for the broaching of aluminum alloys with high-speed tool steels**

| Material | | | | Hardness, HB (500 kg) | Condition | Speed, m/min (sfm) | Chip load, mm/tooth (in./tooth) | Tool material grade, ISO (AISI or C) |
|---|---|---|---|---|---|---|---|---|
| **Wrought** | | | | | | | | |
| EC | 2218 | 5252 | 6253 | 30–80 | Cold drawn | 12 (40) | 0.15 (0.006) | S4, S2 (M2, M7) |
| 1060 | 2219 | 5254 | 6262 | 75–150 | Solution treated and aged | 12 (40) | 0.15 (0.006) | S4, S2 (M2, M7) |
| 1100 | 2618 | 5454 | 6463 | | | | | |
| 1145 | 3003 | 5456 | 6951 | | | | | |
| 1175 | 3004 | 5457 | 7001 | | | | | |
| 1235 | 3005 | 5652 | 7004 | | | | | |
| 2011 | 4032 | 5657 | 7005 | | | | | |
| 2014 | 5005 | 6053 | 7039 | | | | | |
| 2017 | 5050 | 6061 | 7049 | | | | | |
| 2018 | 5052 | 6063 | 7050 | | | | | |
| 2021 | 5056 | 6066 | 7075 | | | | | |
| 2024 | 5083 | 6070 | 7079 | | | | | |
| 2025 | 5086 | 6101 | 7175 | | | | | |
| 2117 | 5154 | 6151 | 7178 | | | | | |
| **Cast** | | | | | | | | |
| **Sand and permanent mold** | | | | | | | | |
| A140 | 319.0 | 357.0 | A712.0 | 40–100 | As-cast | 12 (40) | 0.15 (0.006) | S4, S2 (M2, M7) |
| 201.0 | 328.0 | 359.0 | D712.0 | 70–125 | Solution treated and aged | 12 (40) | 0.15 (0.006) | S4, S2 (M2, M7) |
| 208.0 | A332.0 | B443.0 | 713.0 | | | | | |
| 213.0 | F332.0 | 514.0 | 771.0 | | | | | |
| 222.0 | 333.0 | A514.0 | 850.0 | | | | | |
| 224.0 | 354.0 | B514.0 | A850.0 | | | | | |
| 242.0 | 355.0 | 520.0 | B850.0 | | | | | |
| 295.0 | C355.0 | 535.0 | | | | | | |
| B295.0 | 356.0 | 705.0 | | | | | | |
| 308.0 | A356.0 | 707.0 | | | | | | |
| Hiduminium RR-350 | | | | | | | | |

Source: Ref 5

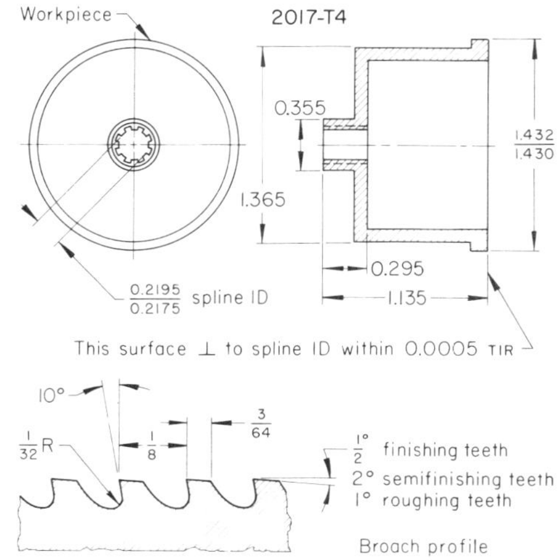


| | |
|---|---|
| Speed, m/min (sfm) | 9.1 (30) |
| Feed, mm/tooth (in./tooth) | 0.05 (0.002) |
| Length of stroke, mm (in.) | 152 (6) |
| Cutting fluid | Soluble oil:water (1:25) |
| Setup time, min | 30 |
| Production rate, pieces/h | 180 |
| Tool life/grind | 2000 pieces |
| Total tool life | 40 000 pieces |

**Fig. 16** Clutch housing and details of broach used for producing the internal splines. Dimensions in figure given in inches

webs of pocketed or stiffened parts, for which no backup support is practical, are also prone to these problems. Figure 12 illustrates this latter problem on a small pocketed end fitting in which the effect is noticeable on the vertical flanges.

**Numerically Controlled Machining Operations.** All of the foregoing considerations regarding sources of machining problems also apply to NC operations. In addition, however, there is generally less operator involvement or awareness of the actual machining operation in NC processes. For example, in some plants, one operator may attend several machines. Futhermore, the programmer may be less aware of possible pitfalls, and the optimum mathematical cutter path may create local potential problems. Dwelling at corners or between subroutines must be strictly avoided.

**Quality Control.** The causes of overheating during machining are well recognized (Fig. 13). However, the safety margin between acceptable and unacceptable practices is not wide; therefore, the utmost care and attention to detail is warranted at all times.

The maximum impact of these problems obviously occurs over the short term. The tendency of these defects to be repetitive in one batch of parts can raise havoc with a tight production schedule if none of the parts is acceptable. Wholesale rejections can be avoided by close monitoring of material conductivity values in critical areas during machining, particularly on new parts. As a result, the impact of downstream economic losses and potential major operations difficulties (cost of replacement parts plus the associated disruptive factors) can be minimized.

## Turning

As indicated in Fig. 2 and 3, the cutting force, and therefore the power, required to turn aluminum is considerably less than that needed to turn low-carbon steel of approximately the same hardness and tensile properties.

**Tool Design.** The recommended angles for a single-point lathe tool are given in Table 7. Cutting angles should be on the high end of the ranges for alloys that are not heat treatable or that are more ductile than the free-cutting alloys.

Low modulus of elasticity, high ductility, and the tendency to stick to tool materials are the principal factors dictating differences in the cutting tools used on aluminum. Because of its low modulus of elasticity, aluminum has a greater tendency to yield, or spring, under the pressure of a cutting tool. It deforms more than steel before the cutter can start to bite. For this reason, a fine cutting edge is desirable, and higher clearance angles are generally recommended—8 to 15° for high-speed steel tools, 6 to 10° for carbides, and somewhat less for diamonds.

Traditional practice in the machining of aluminum has specified high, positive rake angles, both back and side, for high-speed steel and carbide tools, especially for the gummier alloys rated D and E in machinability. Although these recommendations are specifically for single-point turning operations, they are also generally applicable to other types of tools. In addition, as the alloy in question increases in Brinell hardness, in alloying complexity, and in machinability rating, lower rake angles can be used.

One of the major reasons for using high-rake tooling is to prevent, or at least delay, the formation of a built-up edge on the cutter. For the same reason and to facilitate chip flow, it is universally recommended that all tool surfaces that contact either the workpiece or the chip be given a high polish. Any grinding marks on the rake face or flank of the tool should be parallel to the direction of the chip-flow or the cut, whichever is applicable. Grinding marks should be removed by stoning or polishing.

Because carbide tools are more brittle than high-speed steel tools, they will chip or break when the cutting angles are too large. In turning free-cutting alloys, increasing the side relief or side rake angle, or both, will reduce the power required. Because the power needed for turning aluminum alloys

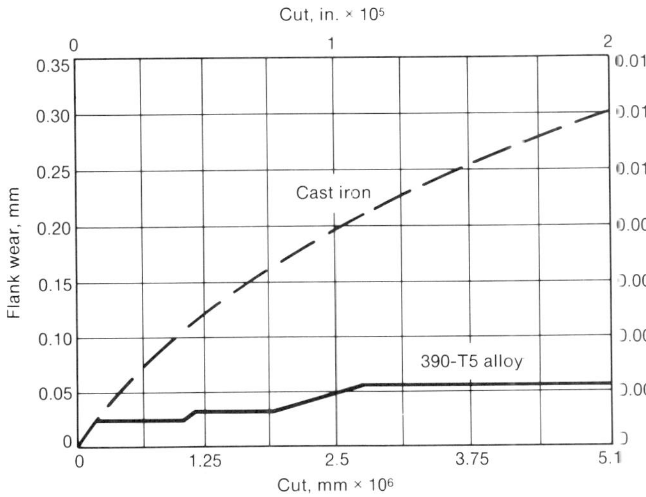


**Fig. 17** Typical tool wear when broaching with a carbide cutting tool at 45 m/min (150 sfm). Source: Ref 2

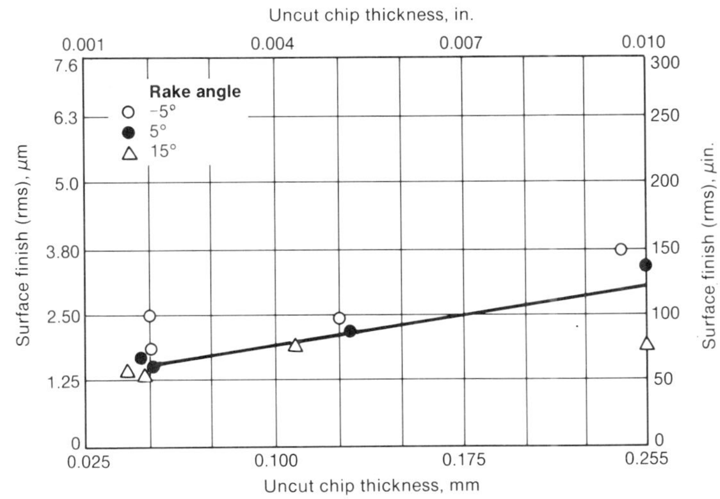


**Fig. 18** Surface finish at various depths of cut when broaching 390 aluminum alloy of 100 HB at 45 m/min (150 sfm) with a carbide tool. Although the rake angle varied, the shear angle of the tool was 0° in all three cases. Source: Ref 2

is small (about one-third that needed for machining soft low-carbon steel), a turning tool ground with angles that are too large is likely to hog the work metal. Because the tool floats in operation, any backlash or play in the spindle or machine ways causes difficulty in holding tolerances. Tool life and surface finish also suffer. Surface finish can be improved by grinding a large nose radius on the turning tool. However, the maximum nose radius depends partly on the allowable fillet on the workpiece when turning to a shoulder. As the nose radius is made larger, cutting force and horsepower requirements increase. On small-diameter stock, increased cutting force may cause the work metal to bend away from the tool; therefore, stock supports or rests are required. A nose radius that is too large also causes chatter, which results in poor surface finish and possibly tool breakage.

Another way to improve surface finish is to grind the end cutting edge of the tool parallel to the work for a width equal to 1½ to 2 times the feed rate in millimeters per revolution. This flat edge will cut behind the nose of the tool to smooth out the ridges caused by the feed. Too wide a flat will cause chatter, and a negative angle will leave a taper equal in length to the width of the flat, on the work at the end of the turn.

**Tool Honing.** Honing a carbide tool with a diamond-impregnated stone will improve the surface finish on the workpiece and will extend the life of the tool. Disposable carbide inserts have a honed edge when purchased, and no additional honing is necessary.

**Tool Material.** Carbide tools, either brazed-insert or disposable-insert, last far longer than high-speed steel tools for turning any aluminum alloy. For turning the high-silicon alloys, the use of carbide tools or polycrystalline diamond is mandatory for optimum results (see the section "Machining of High-Silicon Aluminum Alloy 390" in this article).

**The speed** used in turning aluminum alloys depends to some extent on the alloy and condition, but far more on tool material and type of tool (single-point, form, or cutoff). The effect of alloy composition and condition is small in selecting a turning speed, except for the nonheat-treated cast alloys.

Nominal speeds for turning aluminum with high-speed steel and carbide tools are given in Table 8. These speeds are based on feeds and depths of cut that are typical for turning aluminum alloys and on the assumptions that the setup is rigid, that a cutting fluid is used, and that the workpiece can be rotated to attain these surface speeds without causing excessive vibration. When conditions are below normal, speeds must be scaled down from those given in Table 8. Similarly, under nearly ideal conditions, higher speeds are often used successfully.

**Feed.** For rough turning with single-point tools, regardless of tool material, a feed of 0.38 mm/rev (0.015 in./rev) is common for all aluminum alloys. For finish turning, a feed of 0.18 mm/rev (0.007 in./rev) is recommended; this feed will usually result in a surface finish of 1.60 to 3.20 μm (63 to 125 μin.). Lighter feeds are sometimes used to provide a better finish.

In form turning, the width of the form tool is the major variable that affects the rate of feed. For all alloys and tool materials, a feed rate of 0.089 mm/rev (0.0035 in./rev) is generally satisfactory for form tools no wider than 12.7 mm (0.500 in.). Rate of feed should be decreased as the width of the form tool increases. Feed rates of 0.08 and 0.05 mm/rev (0.003 and 0.002 in./rev) are recommended for form tool widths of 25 and 50 mm (1 and 2 in.), respectively.

For cutoff tools, a feed rate of 0.05 mm/rev (0.002 in./rev) is usually satisfactory, regardless of the alloy being turned, the tool material, or the width of the cutoff tool.

**Depth of cut** in turning aluminum often depends on the power available. The speeds given in Table 7 are based on a 3.81 mm (0.150 in.) depth of cut for roughing and 0.64 mm (0.025 in.) for finishing. When power is available, roughing cuts of 6.35 mm (0.250 in.) are common. Finishing cuts of less than 0.64 mm (0.025 in.) often result in better surface finish.

**Cutting fluid** is recommended for turning all aluminum alloys (see the section "Cutting Fluids" in this article).

**The procedures and equipment** for turning that are the same for all metals are discussed in the article "Turning" in this Volume (special considerations for high-silicon content alloys are given below). Because aluminum alloys are far less sensitive to abrupt changes in speed, feed, and depth

**Table 13 Tool angles for drilling aluminum alloys with high-speed steel drills in drill presses**

| | |
|---|---|
| **Point angle (θ), degrees** | |
| Thin stock | 118–150 |
| Thinned point | (a) |
| General work | 118–140 |
| Aluminum-silicon alloys | ≥90 |
| **Helix angle, degrees** | |
| Thin stock | 24 |
| Medium depth(b) | 28 |
| Deep holes(c) | 40–48 |
| **Lip clearance, degrees** | |
| Soft alloys | 17 |
| Strong alloys | 15 |
| Aluminum-silicon alloys | 12 |

(a) Diameter ÷ 1.8 stock thickness = tan θ/2. (b) Up to six times drill diameter. (c) Over six times drill diameter

## Table 14 Nominal speeds and feeds for the drilling of aluminum alloys with high-speed tool steel tool

| Material | Hardness, HB (500 kg) | Condition | Speed, m/min (sfm) | Feed, mm/rev (in./rev)(a) at a nominal hole diameter of: 1.5 mm (1/16 in.) | 3 mm (1/8 in.) | 6 mm (1/4 in.) | 12 mm (1/2 in.) | 18 mm (3/4 in.) | 25 mm (1 in.) | 35 mm (1 1/2 in.) | 50 mm (2 in.) | Tool material grade, ISO (AISI or C) |
|---|---|---|---|---|---|---|---|---|---|---|---|---|
| **Wrought** | | | | | | | | | | | | |
| EC 3005 6066 | 30–80 | Cold drawn | 43 | 0.025 | . . . | . . . | . . . | . . . | . . . | . . . | . . . | S2, S3 |
| 1060 4032 6070 | | | (140) | (0.001) | . . . | . . . | . . . | . . . | . . . | . . . | . . . | (M10, M7, M1) |
| 1100 5005 6101 | | | 105 | . . . | 0.075 | 0.18 | 0.30 | 0.40 | 0.50 | 0.65 | 0.75 | S2, S3 |
| 1145 5050 6151 | | | (350) | . . . | (0.003) | (0.007) | (0.012) | (0.016) | (0.019) | (0.025) | (0.030) | (M10, M7, M1) |
| 1175 5052 6253 | 75–150 | Solution treated | 43 | 0.025 | . . . | . . . | . . . | . . . | . . . | . . . | . . . | S2, S3 |
| 1235 5056 6262 | | and aged | (140) | (0.001) | . . . | . . . | . . . | . . . | . . . | . . . | . . . | (M10, M7, M1) |
| 2011 5083 6463 | | | 84 | . . . | 0.075 | 0.18 | 0.30 | 0.40 | 0.50 | 0.65 | 0.75 | S2, S3 |
| 2014 5086 6951 | | | (275) | . . . | (0.003) | (0.007) | (0.012) | (0.016) | (0.019) | (0.025) | (0.030) | (M10, M7, M1) |
| 2017 5154 7001 | | | | | | | | | | | | |
| 2018 5252 7004 | | | | | | | | | | | | |
| 2021 5254 7005 | | | | | | | | | | | | |
| 2024 5454 7039 | | | | | | | | | | | | |
| 2025 5456 7049 | | | | | | | | | | | | |
| 2117 5457 7050 | | | | | | | | | | | | |
| 2218 5652 7075 | | | | | | | | | | | | |
| 2219 5657 7079 | | | | | | | | | | | | |
| 2618 6053 7175 | | | | | | | | | | | | |
| 3003 6061 7178 | | | | | | | | | | | | |
| 3004 6063 | | | | | | | | | | | | |
| **Cast** | | | | | | | | | | | | |
| **Sand and permanent mold** | | | | | | | | | | | | |
| A410 F332.0 520.0 | 40–100 | As-cast | 43 | 0.025 | . . . | . . . | . . . | . . . | . . . | . . . | . . . | S2, S3 |
| 201.0 333.0 535.0 | | | (140) | (0.001) | . . . | . . . | . . . | . . . | . . . | . . . | . . . | (M10, M7, M1) |
| 208.0 354.0 705.0 | | | 105 | . . . | 0.075 | 0.18 | 0.30 | 0.40 | 0.50 | 0.65 | 0.75 | S2, S3 |
| 213.0 355.0 707.0 | | | (350) | . . . | (0.003) | (0.007) | (0.012) | (0.016) | (0.019) | (0.025) | (0.030) | (M10, M7, M1) |
| 222.0 C355.0 A712.0 | 70–125 | Solution treated | 43 | 0.025 | . . . | . . . | . . . | . . . | . . . | . . . | . . . | S2, S3 |
| 224.0 356.0 D712.0 | | and aged | (140) | (0.001) | . . . | . . . | . . . | . . . | . . . | . . . | . . . | (M10, M7, M1) |
| 242.0 A356.0 713.0 | | | 84 | . . . | 0.075 | 0.18 | 0.30 | 0.40 | 0.50 | 0.65 | 0.75 | S2, S3 |
| 295.0 357.0 771.0 | | | (275) | . . . | (0.003) | (0.007) | (0.012) | (0.016) | (0.019) | (0.025) | (0.030) | (M10, M7, M1) |
| B295.0 359.0 850.0 | | | | | | | | | | | | |
| 308.0 B443.0 A850.0 | | | | | | | | | | | | |
| 319.0 514.0 B850.0 | | | | | | | | | | | | |
| 328.0 A514.0 | | | | | | | | | | | | |
| A332.0 B514.0 | | | | | | | | | | | | |
| Hiduminium RR-350 | | | | | | | | | | | | |
| **Die castings** | | | | | | | | | | | | |
| 360.0 A380.0 | 40–100 | As-cast | 46 | 0.025 | . . . | . . . | . . . | . . . | . . . | . . . | . . . | S2, S3 |
| A360.0 C443.0 | | | (150) | (0.001) | . . . | . . . | . . . | . . . | . . . | . . . | . . . | (M10, M7, M1) |
| 380.0 518.0 | | | 115 | . . . | 0.075 | 0.18 | 0.30 | 0.40 | 0.50 | 0.65 | 0.75 | S2, S3 |
| | | | (375) | . . . | (0.003) | (0.007) | (0.012) | (0.016) | (0.019) | (0.025) | (0.030) | (M10, M7, M1) |
| | 70–125 | Solution treated | 46 | 0.025 | . . . | . . . | . . . | . . . | . . . | . . . | . . . | S2, S3 |
| | | and aged | (150) | (0.001) | . . . | . . . | . . . | . . . | . . . | . . . | . . . | (M10, M7, M1) |
| | | | 90 | . . . | 0.075 | 0.18 | 0.30 | 0.40 | 0.50 | 0.65 | 0.75 | S2, S3 |
| | | | (300) | . . . | (0.003) | (0.007) | (0.012) | (0.016) | (0.019) | (0.025) | (0.030) | (M10, M7, M1) |
| 383.0 413.0 | 40–100 | As-cast | 30 | 0.025 | . . . | . . . | . . . | . . . | . . . | . . . | . . . | S2, S3 |
| A384.0 A413.0 | | | (100) | (0.001) | . . . | . . . | . . . | . . . | . . . | . . . | . . . | (M10, M7, M1) |
| | | | 43 | . . . | 0.075 | 0.18 | 0.30 | 0.40 | 0.50 | 0.65 | 0.75 | S2, S3 |
| | | | (140) | . . . | (0.003) | (0.007) | (0.012) | (0.016) | (0.019) | (0.025) | (0.030) | (M10, M7, M1) |
| | 70–125 | Solution treated | 27 | 0.025 | . . . | . . . | . . . | . . . | . . . | . . . | . . . | S2, S3 |
| | | and aged | (90) | (0.001) | . . . | . . . | . . . | . . . | . . . | . . . | . . . | (M10, M7, M1) |
| | | | 37 | . . . | 0.075 | 0.18 | 0.30 | 0.40 | 0.50 | 0.65 | 0.75 | S2, S3 |
| | | | (120) | . . . | (0.003) | (0.007) | (0.012) | (0.016) | (0.019) | (0.025) | (0.030) | (M10, M7, M1) |
| 390.0 | 40–100 | As-cast | 8 | 0.025 | . . . | . . . | . . . | . . . | . . . | . . . | . . . | S2, S3 |
| 392.0 | | | (25) | (0.001) | . . . | . . . | . . . | . . . | . . . | . . . | . . . | (M10, M7, M1) |
| | | | 15 | . . . | 0.075 | 0.18 | 0.30 | 0.40 | 0.50 | 0.65 | 0.75 | S2, S3 |
| | | | (50) | . . . | (0.003) | (0.007) | (0.012) | (0.016) | (0.019) | (0.025) | (0.030) | (M10, M7, M1) |
| | 70–125 | Solution treated | 8 | 0.025 | . . . | . . . | . . . | . . . | . . . | . . . | . . . | S2, S3 |
| | | and aged | (25) | (0.001) | . . . | . . . | . . . | . . . | . . . | . . . | . . . | (M10, M7, M1) |
| | | | 14 | . . . | 0.075 | 0.18 | 0.30 | 0.40 | 0.50 | 0.65 | 0.75 | S2, S3 |
| | | | (45) | . . . | (0.003) | (0.007) | (0.012) | (0.016) | (0.019) | (0.025) | (0.030) | (M10, M7, M1) |

(a) For holes more than two diameters deep, reduce speed and feed. Source: Ref 5

of cut than many other metals, they are especially well adapted to turning in automatic equipment.

**The turning of high-silicon aluminum alloys** (Ref 2) such as alloy 390 is best accomplished with polycrystalline diamond tools at speeds up to 900 m/min (3000 sfm). Carbide tools are satisfactory for speeds up to 150 m/min (500 sfm). Carbide tools should be polished for best results. Either tool material should have at least a slight positive rake. The use of a coolant will always improve tool life; however, alloy 390 can be turned dry, if required, but with somewhat reduced tool life. Dry turning speeds should be limited to 300 m/min (1000 sfm) for diamond and 30 m/min (100 sfm) for carbide tools.

**Table 15 Nominal speeds and feeds for the spade drilling of aluminum alloys with high-speed tool steels and carbide tools**

| Material | Hardness, HB (500 kg) | Condition | Speed, m/min (sfm) | Feed, mm/rev (in./rev), at a nominal hole diameter of: 18–25 mm (3/4–1 in.) | 35 mm (1 1/4 in.) | 50 mm (2 in.) | 75 mm (3 in.) | 100 mm (4 in.) | 150 mm (6 in.) | 200 mm (8 in.) | 200 mm and up (8 in. and up) | Tool material grade, ISO (AISI or C) |
|---|---|---|---|---|---|---|---|---|---|---|---|---|
| **Wrought** | | | | | | | | | | | | |
| EC 3005 6066 | 30–80 | Cold drawn | 90 | 0.40 | 0.45 | 0.65 | 0.90 | 1.10 | 1.65 | 1.90 | 2.30 | S3, S4, S5, S2 |
| 1060 4032 6070 | | | (300) | (0.015) | (0.018) | (0.026) | (0.035) | (0.042) | (0.065) | (0.075) | (0.090) | (M1, M2, M3, M10) |
| 1100 5005 6101 | | | 185 | 0.13 | 0.15 | 0.20 | 0.30 | 0.36 | 0.55 | 0.65 | 0.75 | K20 |
| 1145 5050 6151 | | | (600) | (0.005) | (0.006) | (0.008) | (0.012) | (0.014) | (0.022) | (0.025) | (0.030) | (C-2) |
| 1175 5052 6253 | 75–150 | Solution treated | 84 | 0.40 | 0.45 | 0.65 | 0.90 | 1.10 | 1.65 | 1.90 | 2.30 | S3, S4, S5, S2 |
| 1235 5056 6262 | | and aged | (275) | (0.015) | (0.018) | (0.026) | (0.035) | (0.042) | (0.065) | (0.075) | (0.090) | (M1, M2, M3, M10) |
| 2011 5083 6463 | | | 185 | 0.13 | 0.15 | 0.20 | 0.30 | 0.36 | 0.55 | 0.65 | 0.75 | K20 |
| 2014 5086 6951 | | | (600) | (0.005) | (0.006) | (0.008) | (0.012) | (0.014) | (0.022) | (0.025) | (0.030) | (C-2) |
| 2017 5154 7001 | | | | | | | | | | | | |
| 2018 5252 7004 | | | | | | | | | | | | |
| 2021 5254 7005 | | | | | | | | | | | | |
| 2024 5454 7039 | | | | | | | | | | | | |
| 2025 5456 7049 | | | | | | | | | | | | |
| 2117 5457 7050 | | | | | | | | | | | | |
| 2218 5652 7075 | | | | | | | | | | | | |
| 2219 5657 7079 | | | | | | | | | | | | |
| 2618 6053 7175 | | | | | | | | | | | | |
| 3003 6061 7178 | | | | | | | | | | | | |
| 3004 6063 | | | | | | | | | | | | |
| **Cast** | | | | | | | | | | | | |
| **Sand and permanent mold** | | | | | | | | | | | | |
| A140 F332.0 520.0 | 40–100 | As-cast | 90 | 0.40 | 0.45 | 0.65 | 0.90 | 1.10 | 1.65 | 1.90 | 2.30 | S3, S4, S5, S2 |
| 201.0 333.0 535.0 | | | (300) | (0.015) | (0.018) | (0.026) | (0.035) | (0.042) | (0.065) | (0.075) | (0.090) | (M1, M2, M3, M10) |
| 208.0 354.0 705.0 | | | 185 | 0.13 | 0.15 | 0.20 | 0.30 | 0.36 | 0.55 | 0.65 | 0.75 | K20 |
| 213.0 355.0 707.0 | | | (600) | (0.005) | (0.006) | (0.008) | (0.012) | (0.014) | (0.022) | (0.025) | (0.030) | (C-2) |
| 222.0 C355.0 A712.0 | 70–125 | Solution treated | 84 | 0.40 | 0.45 | 0.65 | 0.90 | 1.10 | 1.65 | 1.90 | 2.30 | S3, S4, S5, S2 |
| 224.0 356.0 D712.0 | | and aged | (275) | (0.015) | (0.018) | (0.026) | (0.035) | (0.042) | (0.065) | (0.075) | (0.090) | (M1, M2, M3, M10) |
| 242.0 A356.0 713.0 | | | 185 | 0.13 | 0.15 | 0.20 | 0.30 | 0.36 | 0.55 | 0.65 | 0.75 | K20 |
| 295.0 357.0 771.0 | | | (600) | (0.005) | (0.006) | (0.008) | (0.012) | (0.014) | (0.022) | (0.025) | (0.030) | (C-2) |
| B295.0 359.0 850.0 | | | | | | | | | | | | |
| 308.0 B443.0 A850.0 | | | | | | | | | | | | |
| 319.0 514.0 B850.0 | | | | | | | | | | | | |
| 328.0 A514.0 | | | | | | | | | | | | |
| A332.0 B514.0 | | | | | | | | | | | | |
| Hiduminium RR-350 | | | | | | | | | | | | |

Source: Ref 5

High-speed steel tools yield consistently poor results. Tool wear is rapid, and workpiece buildup on the cutting tool is rapid and inconsistent, resulting in poor surface finish and difficult-to-control dimensions.

Ceramic tools exhibit rapid tool edge buildup at high speeds with consequent loss of finish and geometry. At lower speeds, these conditions improve, but excessive heat transmitted to the ceramic material results in premature failure due to fracture of the tool insert.

The best overall carbide grade has been found to be C-3, with C-2 as next best. Both are typical grades used for machining gray iron. It is desirable to have a good polish on the top rake face of the tool to minimize tool edge buildup. Any good natural or synthetic soluble-oil emulsion is recommended in order to prolong tool life. Generally, positive rake and relief angles and a sharp nose radius promote good surface finish and minimize tool edge buildup for both carbide and polycrystalline diamond tools. Tool grades and geometry for turning 390 aluminum alloy are:

| | |
|---|---|
| Tools | Tungsten carbide, C-3, SPG422 insert with J polish |
| | Polycrystalline diamond, SPG422 insert |
| Geometry | 0° Back rake angle |
| | 5° Side rake angle |
| | 5° Side rake angle |
| | 5° End clearance angle (relief) |
| | 5° Side clearance angle (relief) |
| | 15° Minor cutting edge angle (end) |
| | 15° Lead angle (side) |
| | 0.76 mm (0.030 in.) corner radius |

Polycrystalline diamond is destined to become the most cost-effective tool material for machining harder and more abrasive casting alloys such as 390. It can ensure a potential improvement in transfer line productivity.

Tool lives for the machining of alloys 380 and 390 are shown in Fig. 14. The data represent the cutting times required to achieve a 0.38 mm (0.015 in.) uniform wear land on the flank face of the cutting tool (except for the die cast alloy 390 machined with diamond; in this case, cutting was stopped at a uniform wear land of 0.25 mm, or 0.010 in.; and the 0.38 mm, or 0.015 in., tool life would be even greater, to the right in Fig. 14). All of the turning tests, except for sand cast alloy 390, were conducted using soluble-oil emulsion coolant at the ratio of 20:1. The dashed lines in Fig. 14 represent extrapolations of actual data (solid lines).

## Boring

The boring of aluminum alloys, particularly the alloys with high silicon content, requires the use of tools with acute rake and clearance angles. In general, rake angles are increased as silicon content is decreased. The same tool geometry and cutting tools used to turn high-silicon aluminum alloys such as 390 apply to boring operations of this same alloy (see the section "Turning of High-Silicon Aluminum Alloys" in this article).

**Tool Material.** Although high-speed steel tools are used to some extent for boring aluminum, carbide tools permit higher surface speeds (up to 300 m/min, or 1000 sfm, or more), with markedly longer tool life.

**Table 16 Maximum depths per drill entry for drilling in automatic bar and chucking machines**

| Drill type | Depth, in drill diameters — First drill: First entry | Second entry | Third entry | Second drill: First entry | Second entry | Third entry | Subsequent entries |
|---|---|---|---|---|---|---|---|
| Standard twist drill | 4 | 1½ | ¾ | ... | ... | ... | ... |
| High or low helix, or straight flute | 5 | 2 | 1 | 1½ | ¾ | ½ | ½ |

**Table 17 Feeds for drilling aluminum alloys in automatic bar and chucking machines**

| Drill diameter, mm | Drill diameter, in. | Tolerance, mm | Tolerance, in. | Feed, 2011-T3, mm | Feed, 2011-T3, in. | Feed, Other alloys, mm | Feed, Other alloys, in. |
|---|---|---|---|---|---|---|---|
| 1.59 | 0.0625 | ±0.038 | ±0.0015 | 0.102 | 0.0040 | 0.10 | 0.004 |
| 3.18 | 0.125 | ±0.05 | ±0.002 | 0.305 | 0.0120 | 0.25 | 0.010 |
| 4.75 | 0.187 | ±0.05 | ±0.002 | 0.366 | 0.0144 | 0.30 | 0.012 |
| 6.35 | 0.250 | ±0.05 | ±0.002 | 0.426 | 0.0168 | 0.36 | 0.014 |
| 9.53 | 0.375 | ±0.064 | ±0.0025 | 0.518 | 0.0204 | 0.43 | 0.017 |
| 12.70 | 0.500 | ±0.064 | ±0.0025 | 0.518 | 0.0204 | 0.43 | 0.017 |
| 19.05 | 0.750 | ±0.08 | ±0.003 | 0.518 | 0.0204 | 0.43 | 0.017 |

**Table 18 Relation of drill diameter to approximate maximum depth of hole in gun drilling**

| Drill diameter, mm | Drill diameter, in. | Depth of hole, number of drill diameters |
|---|---|---|
| ≤13 | ≤½ | 12 |
| 13–25 | ½–1 | 7 |
| 25–38 | 1–1½ | 6 |
| 38–50 | 1½–2 | 5 |
| 50–64 | 2–2½ | 4 |
| 64–75 | 2½–3 | 3 |

Carbide cutters readily yield surface finishes of 0.25 to 0.50 μm (10 to 20 μin.).

When used in conjuction with precision boring machines, diamond tools can produce surface finishes of 0.025 μm (1 μin.). They are also capable of holding size over a long production run. If the cut is continuous and the work metal contains no hard spots, diamond tools are most effective in boring the abrasive high-silicon alloys.

**Tool Sharpening.** For optimum tool life, cutting edges and adjacent surfaces must be free of burrs and scratches. The hand stoning of cutting edges with an oil stone is recommended. When a carbide boring tool is sharpened with a 400- or 500-grit diamond wheel, surface finishes of 0.075 to 0.10 μm (3 to 4 μin.) can be obtained.

**Speed and Feed.** The optimum speed for boring aluminum alloys depends somewhat on alloy and temper, but largely on tool material and whether the operation is roughing or finishing. Selection of feed depends largely on tool material and whether the operation is roughing or finishing.

Nominal speeds and feeds for boring with high-speed steel and carbide tools are given in Table 9; speeds and feeds for boring aluminum alloys with polycrystalline diamond tools are given in Table 10. Speed is increased for a shallower depth of cut and lighter feed, regardless of the alloy or tool material.

**Depth of Cut.** The speeds and feeds given in Tables 9 and 10 are based on a 2.5 mm (0.10 in.) depth of cut for roughing and 0.25 mm (0.010 in.) for finishing. Depth of cut is sometimes greater than 2.5 mm (0.10 in.) for rough boring if power is available and the setup can be made sufficiently rigid. Finishing cuts significantly less than 0.25 mm (0.010 in.) in depth are seldom used.

**Cutting fluid** is recommended, but boring has been done dry. There is usually some sacrifice of productivity, dimensional accuracy, or surface finish when aluminum is bored without a cutting fluid. A mixture of one part soluble oil to 20 to 30 parts water is most commonly used. In addition, mineral oil, or mineral oil mixed with up to 50% lard oil, is often used, especially when the best possible surface finish is desired. Mixtures of kerosenes and lubricating oil are sometimes used.

**Procedures and equipment** common to the boring of all metals are covered in detail in the article "Boring" in this Volume. Specific procedures for boring an aluminum alloy are described in the example that follows.

**Example 2: The Boring of Concentric Holes.** Two small and two large holes were rough and finish bored from opposite sides of a heat-treated aluminum forging, as shown in Fig. 15. It was required that the large and small holes be concentric within 0.010 mm (0.0004 in.) total indicator reading (TIR), parallel within 0.008 mm (0.0003 in.), and square with the surface within 0.008 mm (0.0003 in.). The operation was done in a double-end precision boring mill, and carbide tools were used. Details are given with Fig. 15.

## Planing and Shaping

The techniques and equipment used for the planing and shaping of other metals are generally applicable to aluminum alloys. Planing methods are discussed in the article "Planing" in this Volume. Similar information for shaping appears in the article "Shaping and Slotting" in this Volume.

**Planing.** Either high-speed or carbide tools for planing aluminum usually have a back rake angle of 30° or more (sometimes as much as 60°) and a 0° side rake. A speed of 90 m/min (300 sfm) is generally recommended as maximum; however, this speed is higher than can be obtained with most planers.

Roughing feeds are about 2.3 mm (0.090 in.) per stroke when depth of cut is near the recommended maximum of 13 mm (½ in.). When shallow cuts are necessary (for lack of power or other reasons), a heavier feed can be used. For a 2.5 mm (0.10 in.) depth of cut, feed can be increased to about 3.18 mm (0.125 in.) per stroke. The finish planing of aluminum alloys is usually done with a cut no deeper than 0.25 to 0.38 mm (0.010 to 0.015 in.) and a feed equal to three-fourths the width of the broad-nose finishing tool.

Aluminum alloys are often planed without a cutting fluid. A cutting fluid is helpful in producing a better surface finish, but flooding with cutting fluid is seldom feasible. When surface finish is a primary objective, application of a mixture of kerosene and lubricating oil, or of kerosene and lard oil, by means of a swab is common practice. Sprayed cutting fluid is effective. Soluble-oil mixtures are ordinarily used in cutting fluid spray systems. Table 11 lists nominal speeds and feeds for the planing of wrought and cast aluminum.

**Shaping.** Tool forms, tool materials, and cutting fluids for shaping aluminum are the same as those for planing. Maximum speed in shaping is usually the maximum ram speed of the machine. Feed and depth of cut also depend to some extent on machine capabilities. A feed of 0.51 to 0.76 mm/stroke (0.020 to 0.030 in./stroke) for roughing and about 0.25 mm/stroke (0.010 in./stroke) for finishing are common. Depth of cut is often as much as 0.25 mm (0.10 in.) for roughing and 0.51 mm (0.020 in.) or less for finishing.

## Broaching

All aluminum alloys can be broached successfully with standard broaching equipment and with the same general procedures as for other metals. However, better surface finish and dimensional accuracy are obtained with heat-treated alloys. Details of equipment and procedures that are common to the broaching of all metals are covered in the article "Broaching" in this Volume. The principal limitation in broaching aluminum is the difficulty in maintaining an accurate relation between the broached hole and

**Table 19 Nominal speeds and feeds for the gun drilling of aluminum alloys with carbide tools**

| Material | Hardness, HB (500 kg) | Condition | Speed, m/min (sfm) | Feed, mm/rev (in./rev)(a), at a nominal hole diameter of: 2–4 mm (0.078–0.156 in.) | 4–6 mm (0.156–0.250 in.) | 6–12 mm (¼–½ in.) | 12–18 mm (½–¾ in.) | 18–25 mm (¾–1 in.) | 25–50 mm (1–2 in.) | Carbide tool material grade, ISO (C) |
|---|---|---|---|---|---|---|---|---|---|---|
| **Wrought** | | | | | | | | | | |
| EC, 3005, 6066, 1060, 4032, 6070, 1100, 5005, 6101, 1145, 5050, 6151, 1175, 5052, 6253, 1235, 5056, 6262, 2011, 5083, 6463, 2014, 5086, 6951, 2017, 5154, 7001, 2018, 5252, 7004, 2021, 5254, 7005, 2024, 5454, 7039, 2025, 5456, 7049, 2117, 5457, 7050, 2218, 5652, 7075, 2219, 5657, 7079, 2618, 6053, 7175, 3003, 6061, 7178, 3004, 6063 | 30–80 | Cold drawn | 215 | 0.004–0.006 | 0.008–0.013 | 0.025–0.063 | 0.038–0.075 | 0.050–0.102 | 0.075–0.13 | K20 |
| | | | (700) | (0.00015–0.00025) | (0.0003–0.0005) | (0.001–0.0025) | (0.0015–0.003) | (0.002–0.004) | (0.003–0.005) | (C-2) |
| | 75–150 | Solution treated and aged | 200 | 0.004–0.006 | 0.008–0.013 | 0.025–0.063 | 0.038–0.075 | 0.050–0.102 | 0.075–0.13 | K20 |
| | | | (650) | (0.00015–0.00025) | (0.0003–0.0005) | (0.001–0.0025) | (0.0015–0.003) | (0.002–0.004) | (0.003–0.005) | (C-2) |
| **Cast** | | | | | | | | | | |
| *Sand and permanent mold* | | | | | | | | | | |
| A140, F332.0, 520.0, 201.0, 333.0, 535.0, 208.0, 354.0, 705.0, 213.0, 355.0, 707.0, 222.0, C355.0, A712.0, 224.0, 356.0, D712.0, 242.0, A356.0, 713.0, 295.0, 357.0, 771.0, B295.0, 359.0, 850.0, 308.0, B443.0, A850.0, 319.0, 514.0, B850.0, 328.0, A514.0, A332.0, B514.0, Hiduminium RR-350 | 40–100 | As-cast | 200 | 0.004–0.006 | 0.008–0.013 | 0.025–0.063 | 0.038–0.075 | 0.050–0.102 | 0.075–0.13 | K20 |
| | | | (650) | (0.00015–0.00025) | (0.0003–0.0005) | (0.001–0.0025) | (0.0015–0.003) | (0.002–0.004) | (0.003–0.005) | (C-2) |
| | 70–125 | Solution treated and aged | 200 | 0.004–0.006 | 0.008–0.013 | 0.025–0.063 | 0.038–0.075 | 0.050–0.102 | 0.075–0.13 | K20 |
| | | | (650) | (0.00015–0.00025) | (0.0003–0.0005) | (0.001–0.0025) | (0.0015–0.003) | (0.002–0.004) | (0.003–0.005) | (C-2) |
| *Die castings* | | | | | | | | | | |
| 360.0, A380.0, A360.0, C443.0, 380.0, 518.0 | 40–100 | As-cast | 200 | 0.004–0.006 | 0.008–0.013 | 0.025–0.063 | 0.038–0.075 | 0.050–0.102 | 0.075–0.13 | K20 |
| | | | (650) | (0.00015–0.00025) | (0.0003–0.0005) | (0.001–0.0025) | (0.0015–0.003) | (0.002–0.004) | (0.003–0.005) | (C-2) |
| | 70–125 | Solution treated and aged | 200 | 0.004–0.006 | 0.008–0.013 | 0.025–0.063 | 0.038–0.075 | 0.050–0.102 | 0.075–0.13 | K20 |
| | | | (650) | (0.00015–0.00025) | (0.0003–0.0005) | (0.001–0.0025) | (0.0015–0.003) | (0.002–0.004) | (0.003–0.005) | (C-2) |
| 383.0, A384.0, 413.0, A413.0 | 40–100 | As-cast | 200 | 0.004–0.006 | 0.008–0.013 | 0.025–0.063 | 0.038–0.075 | 0.050–0.102 | 0.075–0.13 | K20 |
| | | | (650) | (0.00015–0.00025) | (0.0003–0.0005) | (0.001–0.0025) | (0.0015–0.003) | (0.002–0.004) | (0.003–0.005) | (C-2) |
| | 70–125 | Solution treated and aged | 200 | 0.004–0.006 | 0.008–0.013 | 0.025–0.063 | 0.038–0.075 | 0.050–0.102 | 0.075–0.13 | K20 |
| | | | (650) | (0.00015–0.00025) | (0.0003–0.0005) | (0.001–0.0025) | (0.0015–0.003) | (0.002–0.004) | (0.003–0.005) | (C-2) |
| 390.0, 392.0 | 40–100 | As-cast | 150 | 0.004–0.006 | 0.008–0.013 | 0.020–0.050 | 0.025–0.063 | 0.038–0.075 | 0.050–0.102 | K20 |
| | | | (500) | (0.00015–0.00025) | (0.0003–0.0005) | (0.0008–0.002) | (0.001–0.0025) | (0.0015–0.003) | (0.002–0.004) | (C-2) |
| | 70–125 | Solution treated and aged | 135 | 0.004–0.006 | 0.008–0.013 | 0.020–0.050 | 0.025–0.063 | 0.038–0.075 | 0.050–0.102 | K20 |
| | | | (450) | (0.00015–0.00025) | (0.0003–0.0005) | (0.0008–0.002) | (0.001–0.0025) | (0.0015–0.003) | (0.002–0.004) | (C-2) |

(a) Feeds are specified for single-flute gun drills. Source: Ref 5

other surfaces of the workpiece, even when the starting hole is accurately located.

**Tool Material.** The general-purpose high-speed steels, such as M2, are used for most tools for broaching aluminum. In some high-production operations, especially when broaching high-silicon alloys, broaches made of the more highly alloyed high-speed steels or of carbide have proved economical. Surface treatments such as chromium plating or oxidizing help to prolong the life of high-speed steel broaches.

A fine finish on the tool can be important. In one application, the life of a high-speed steel broach was increased from 2000 to 7400 pieces when the cutting edges were wet blasted with a superfine abrasive.

**Broach Design.** In rough broaching, a coarse tooth pitch is desirable, with only two or three teeth in contact and cutting at any one time. For internal finish broaching, the best results are obtained if only two teeth are cutting; in external finish broaching, it is often best to have only one tooth engaging at a time.

Broaches used for aluminum should have a face angle of 10°. For external surface broaching, the clearance angle should be 3½°; for internal broaching, 2°. Although a large clearance angle provides better cutting action, it markedly reduces broach life. Therefore, clearance angles should be kept to a minimum to reduce loss of size when the broach is sharpened.

**Speed and Feed.** A speed range of 9.1 to 15 m/min (30 to 50 sfm) is generally recommended for broaching aluminum alloys. When rigidity, supply of cutting fluid, and hardness of the work metal are nearly ideal, the upper portion of the range can be used. When one or more of these conditions is less than ideal, the speed can be reduced.

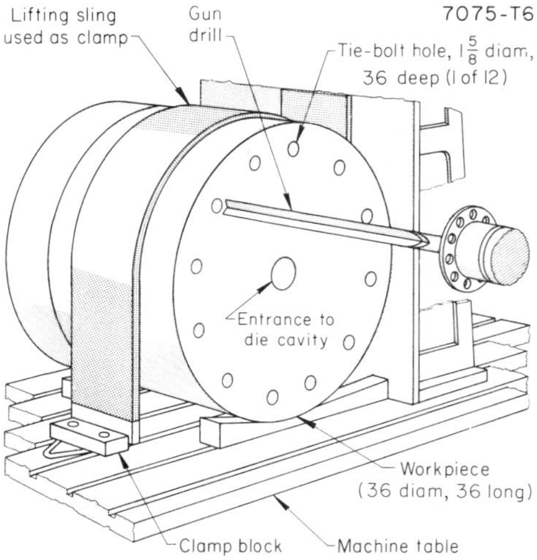


**Fig. 19** Setup for deep-hole drilling an explosive forming die in a boring mill. Dimensions given in inches

Greater feed per tooth (chip load) is recommended for broaching aluminum alloys than for steel. A feed of 0.15 mm/tooth (0.006 in./tooth) is usually optimum for spline broaching and about 0.08 mm/tooth (0.003 in./tooth) for broaching round holes. Table 12 lists nominal speeds and feeds for machining both wrought and cast aluminum alloys.

**Cutting fluid** is required for best results. A copious supply of soluble oil mixed with water is satisfactory for most applications. However, mineral oil, or mineral oil mixed with lard oil, will usually improve surface finish (see the section "Cutting Fluids" in this article).

**Production Practice.** The following example provides details of typical practice in broaching aluminum. For the workpiece described in this example, broaching was the most feasible and least costly method of obtaining the required results.

**Example 3: The Broaching of Internal Splines.** Internal splines were broached in a 2017-T4 aluminum clutch housing (Fig. 16). The splined hole contained eight teeth with 32 diametral pitch, 20° pressure angle, and partial dedendum; the internal diameter was 5.550 mm (0.2185 in.). The length of the broaching cut was 7.49 mm (0.295 in.). The pull broach (details in Fig. 16) was made of high-speed steel, and broaching was done in a 9 kN (1 tonf) vertical machine. Processing data are given in the table accompanying Fig. 18.

**Broaching of 390 High-Silicon Aluminum Alloy (Ref 2).** Horizontal broaching has long been the preferred method for rough and finish machining for such complex cast iron castings as cylinder heads and engine blocks. Broaching permits several surfaces on such parts to be machined in exact relationship to each other because this relationship can be build into the broach tooling. Broaching also permits very high stock removal rates, and machine downtime is typically low.

Because of their low hardness, modulus, and strength, aluminum casting alloys have traditionally been considered unlikely candidates for this type of broaching. The cutting speed capability typical of a broach designed to machine cast iron is slow (7.6 to 46 m/min, or 25 to 150 sfm). Most aluminum alloys are gummy at such speeds; because of relatively low hardness, chips clog the cutting teeth of the broach, and the surface of the resulting broached casting is torn and rough. The unit tool forces, characteristic of such a broach, result in breakage of thinner casting sections and edge breakout on other sections (chipping of the trailing edge of the workpiece as the tool exits). The low modulus of most aluminum castings, when compared to that of iron, also results in additional fixturing and clamping requirements.

One study concluded that alloy 390 could be broached at cutting speeds and tool geometries typical of existing iron cutting broaches with carbide tooling. Tool life was much better than that achieved for cast iron (Fig. 17). It also becomes evident that surface finishes better than 2.50 mm (100 μin.) could be obtained as long as the uncut chip thickness of each cutting edge was kept below 0.10 μm (0.004 in.) (Fig. 18). Such chip loads are typical of finishing cuts when broaching cast iron, and the surface finishes obtained are comparable to or better than those obtained with cast iron. The use of coolant had little effect on tool life or surface finish; however, coolant would tend to aid in chip disposal.

Tests have shown that varying the rake angles, shear angles, and chip loads can be effective in reducing cutting forces during broaching. This approach to low-force tooling may not be needed for alloy 390, because the cutting forces are less than one-half the forces for cast iron in comparable roughing cuts using negative-rake tooling. However, these concepts have proved successful in reducing breakage while broaching lighter-weight, thin-wall gray iron parts and will also perform successfully when broaching alloy 390.

## Drilling in Drill Presses

Although the standard twist drills and drilling equipment used for steel can be employed in drilling aluminum alloys, optimum results require drills of special design as well as higher rotational speeds and heavier feeds. Drills for aluminum are usually made with deep, well-polished flutes, narrow margins, and large helix angles. Proper drill design and drilling practice will frequently permit the removal of three to four times as much aluminum as steel per unit of power. The article "Drilling" in this Volume discusses general equipment and practice.

**Table 20 Reamer design and operating conditions for reaming aluminum alloys**

| | |
|---|---|
| **Reamer design** | |
| Reamer size | |
| Hand reaming, number times drill diameter | 1.01 |
| Machine reaming, number times drill diameter | 1.02 |
| Flute type | Straight to 10° spiral |
| Tooth spacing | |
| Straight flute | Uneven(a) |
| Spiral flute | Even |
| Tooth style | |
| Roughing | Solid or nicked |
| Finishing | Solid |
| Top rake, degrees | 5–8 |
| Clearance angles, degrees | |
| Primary | 4–7 |
| Secondary | 15–20 |
| Cutting angle, degrees | 85–91 |
| Land width, mm (in.) | 0.51–1.52 (0.020–0.060)(b) |
| **Operating conditions** | |
| Speed, roughing, m/min (sfm) | |
| Hard alloys | ≤60 (200) |
| Soft alloys | 18–30 (60–100) |
| Speed, finishing, m/min (sfm) | |
| Straight reamers | ≤120 (400) |
| Taper reamers | ≤90 (300) |
| Feed, mm/rev (in./rev) | |
| Roughing | 0.33–0.89 (0.013–0.035) |
| Finishing | 0.08–0.25 (0.003–0.010) |

(a) To avoid chatter. (b) Land width is approximately 1/25 of reamer diameter.

For drilling in drill presses, the helix angle of the drill should be increased with the depth of the hole to be drilled, ranging from a low-helix 24° angle for very shallow holes in thin stock to a high-helix (40 to 48°) angle for deep holes, for which freer cutting is important (Table 13). The high-helix drill has a more acute cutting angle, resulting in more rapid penetration and freer and cleaner cutting. Lands and margin are narrower than on the low-helix drill, resulting in reduced friction and increased chip space in the flutes. The slightly greater resistance to chip movement can be overcome by polishing the flutes and supplying ample cutting fluid. This type of drill is recommended for deep holes, but is unsuitable for drilling thin stock because of its tendency to hog in.

A twist drill with a 28° helix is suitable for holes of medium depth up to about six drill diameters. A drill with a 24° (or less) helix is recommended for thin stock because it has less tendency to overfeed.

The point angle supplied on standard twist drills is 116 to 118°. The angle should be about 130 to 140° for drilling most aluminum alloys to facilitate chip removal and to minimize burring. However, drills for high-silicon alloys should have a less obtuse point, down to about 90°, for ease of penetration. For drilling thin sheet, the point angle should be very obtuse to permit the drill to cut to its full diameter before the point breaks through. With this type of drill,

**Table 21 Nominal speeds and feeds for the reaming of aluminum alloys with high-speed tool steel and carbide tools**

| | | | Roughing | | | | | | | | Finishing | | | | | | | |
|---|---|---|---|---|---|---|---|---|---|---|---|---|---|---|---|---|---|---|
| | | | Speed, m/min (sfm) | Feed, mm/rev (in./rev)(a), at a reamer diameter, mm (in.), of: | | | | | | Tool material grade, ISO (AISI or C) | Speed, m/min (sfm) | Feed, mm/rev (in./rev)(a), at a reamer diameter, mm (in.), of: | | | | | | Tool material grade, ISO (AISI or C) |
| Material | Hardness, HB (500 kg) | Condition | | 3 (1/8) | 6 (1/4) | 12 (1/2) | 25 (1) | 35 (1.5) | 50 (2) | | | 3 (1/8) | 6 (1/4) | 12 (1/2) | 25 (1) | 35 (1.5) | 50 (2) | |
| **Wrought** | | | | | | | | | | | | | | | | | | |
| EC 3005 6066 | 30–80 | Cold drawn | 150 | 0.18 | 0.25 | 0.40 | 0.50 | 0.65 | 0.75 | S3, S4, S2 | 55 | 0.102 | 0.15 | 0.25 | 0.40 | 0.45 | 0.50 | S3, S4, S2 |
| 1060 4032 6070 | | | (500) | (0.007) | (0.010) | (0.015) | (0.020) | (0.025) | (0.030) | (M1, M2, M7) | (180) | (0.004) | (0.006) | (0.010) | (0.015) | (0.018) | (0.020) | (M1, M2, M7) |
| 1100 5005 6101 | | | 305 | 0.13 | 0.13 | 0.25 | 0.40 | 0.50 | 0.65 | K20 | 76 | 0.102 | 0.15 | 0.25 | 0.40 | 0.45 | 0.50 | K20 |
| 1145 5050 6151 | | | (1000) | (0.005) | (0.007) | (0.010) | (0.015) | (0.020) | (0.025) | (C-2) | (250) | (0.004) | (0.006) | (0.010) | (0.015) | (0.018) | (0.020) | (C-2) |
| 1175 5052 6253 | 75–150 | Solution treated and aged | 150 | 0.18 | 0.25 | 0.40 | 0.50 | 0.65 | 0.75 | S3, S4, S2 | 55 | 0.102 | 0.15 | 0.25 | 0.40 | 0.45 | 0.50 | S3, S4, S2 |
| 1235 5056 6262 | | | (500) | (0.007) | (0.010) | (0.015) | (0.020) | (0.025) | (0.030) | (M1, M2, M7) | (180) | (0.004) | (0.006) | (0.010) | (0.015) | (0.018) | (0.020) | (M1, M2, M7) |
| 2011 5083 6463 | | | 305 | 0.13 | 0.13 | 0.25 | 0.40 | 0.50 | 0.65 | K20 | 76 | 0.102 | 0.15 | 0.25 | 0.40 | 0.45 | 0.50 | K20 |
| 2014 5086 6951 | | | (1000) | (0.005) | (0.007) | (0.010) | (0.015) | (0.020) | (0.025) | (C-2) | (250) | (0.004) | (0.006) | (0.010) | (0.015) | (0.018) | (0.020) | (C-2) |
| 2017 5154 7001 | | | | | | | | | | | | | | | | | | |
| 2018 5252 7004 | | | | | | | | | | | | | | | | | | |
| 2021 5254 7005 | | | | | | | | | | | | | | | | | | |
| 2024 5454 7039 | | | | | | | | | | | | | | | | | | |
| 2025 5456 7049 | | | | | | | | | | | | | | | | | | |
| 2117 5457 7050 | | | | | | | | | | | | | | | | | | |
| 2218 5652 7075 | | | | | | | | | | | | | | | | | | |
| 2219 5657 7079 | | | | | | | | | | | | | | | | | | |
| 2618 6053 7175 | | | | | | | | | | | | | | | | | | |
| 3003 6061 7178 | | | | | | | | | | | | | | | | | | |
| 3004 6063 | | | | | | | | | | | | | | | | | | |
| **Cast** | | | | | | | | | | | | | | | | | | |
| **Sand and permanent mold** | | | | | | | | | | | | | | | | | | |
| A140 F332.0 520.0 | 40–100 | As-cast | 150 | 0.18 | 0.25 | 0.40 | 0.50 | 0.65 | 0.75 | S3, S4, S2 | 49 | 0.102 | 0.15 | 0.25 | 0.40 | 0.45 | 0.50 | S3, S4, S2 |
| 201.0 333.0 535.0 | | | (500) | (0.007) | (0.010) | (0.015) | (0.020) | (0.025) | (0.030) | (M1, M2, M7) | (160) | (0.004) | (0.006) | (0.010) | (0.015) | (0.018) | (0.020) | (M1, M2, M7) |
| 208.0 354.0 705.0 | | | 305 | 0.13 | 0.13 | 0.25 | 0.40 | 0.50 | 0.65 | K20 | 76 | 0.102 | 0.15 | 0.25 | 0.40 | 0.45 | 0.50 | K20 |
| 213.0 355.0 707.0 | | | (1000) | (0.005) | (0.007) | (0.010) | (0.015) | (0.020) | (0.025) | (C-2) | (250) | (0.004) | (0.006) | (0.010) | (0.015) | (0.018) | (0.020) | (C-2) |
| 222.0 C335.0 A712.0 | 70–125 | Solution treated and aged | 120 | 0.18 | 0.25 | 0.40 | 0.50 | 0.65 | 0.75 | S3, S4, S2 | 49 | 0.102 | 0.15 | 0.25 | 0.40 | 0.45 | 0.50 | S3, S4, S2 |
| 224.0 356.0 D712.0 | | | (400) | (0.007) | (0.010) | (0.015) | (0.020) | (0.025) | (0.030) | (M1, M2, M7) | (160) | (0.004) | (0.006) | (0.010) | (0.015) | (0.018) | (0.020) | (M1, M2, M7) |
| 242.0 A356.0 713.0 | | | 260 | 0.13 | 0.13 | 0.25 | 0.40 | 0.50 | 0.65 | K20 | 76 | 0.102 | 0.15 | 0.25 | 0.40 | 0.45 | 0.50 | K20 |
| 295.0 357.0 771.0 | | | (850) | (0.005) | (0.007) | (0.010) | (0.015) | (0.020) | (0.025) | (C-2) | (250) | (0.004) | (0.006) | (0.010) | (0.015) | (0.018) | (0.020) | (C-2) |
| B295.0 359.0 850.0 | | | | | | | | | | | | | | | | | | |
| 308.0 B443.0 A850.0 | | | | | | | | | | | | | | | | | | |
| 319.0 514.0 B850.0 | | | | | | | | | | | | | | | | | | |
| 328.0 A514.0 | | | | | | | | | | | | | | | | | | |
| A332.0 B514.0 | | | | | | | | | | | | | | | | | | |
| Hiduminium RR-350 | | | | | | | | | | | | | | | | | | |
| **Die castings** | | | | | | | | | | | | | | | | | | |
| 360.0 A380.0 | 40–100 | As-cast | 150 | 0.18 | 0.25 | 0.40 | 0.50 | 0.65 | 0.75 | S3, S4, S2 | 37 | 0.102 | 0.15 | 0.25 | 0.40 | 0.45 | 0.50 | S3, S4, S2 |
| A360.0 C443.0 | | | (500) | (0.007) | (0.010) | (0.015) | (0.020) | (0.025) | (0.030) | (M1, M2, M7) | (120) | (0.004) | (0.006) | (0.010) | (0.015) | (0.018) | (0.020) | (M1, M2, M7) |
| 380.0 518.0 | | | 305 | 0.13 | 0.13 | 0.25 | 0.40 | 0.50 | 0.65 | K20 | 60 | 0.102 | 0.15 | 0.25 | 0.40 | 0.45 | 0.50 | K20 |
| | | | (1000) | (0.005) | (0.007) | (0.010) | (0.015) | (0.020) | (0.025) | (C-2) | (200) | (0.004) | (0.006) | (0.010) | (0.015) | (0.018) | (0.020) | (C-2) |
| | 70–125 | Solution treated and aged | 120 | 0.18 | 0.25 | 0.40 | 0.50 | 0.65 | 0.75 | S3, S4, S2 | 37 | 0.102 | 0.15 | 0.25 | 0.40 | 0.45 | 0.50 | S3, S4, S2 |
| | | | (400) | (0.007) | (0.010) | (0.015) | (0.020) | (0.025) | (0.030) | (M1, M2, M7) | (120) | (0.004) | (0.006) | (0.010) | (0.015) | (0.018) | (0.020) | (M1, M2, M7) |
| | | | 260 | 0.13 | 0.13 | 0.25 | 0.40 | 0.50 | 0.65 | K20 | 60 | 0.102 | 0.15 | 0.25 | 0.40 | 0.45 | 0.50 | K20 |
| | | | (850) | (0.005) | (0.007) | (0.010) | (0.015) | (0.020) | (0.025) | (C-2) | (200) | (0.004) | (0.006) | (0.010) | (0.015) | (0.018) | (0.020) | (C-2) |
| 383.0 | 40–100 | As-cast | 105 | 0.18 | 0.25 | 0.40 | 0.50 | 0.65 | 0.75 | S3, S4, S2 | 30 | 0.102 | 0.15 | 0.25 | 0.40 | 0.45 | 0.50 | S3, S4, S2 |
| A384.0 | | | (350) | (0.007) | (0.010) | (0.015) | (0.020) | (0.025) | (0.030) | (M1, M2, M7) | (100) | (0.004) | (0.006) | (0.010) | (0.015) | (0.018) | (0.020) | (M1, M2, M7) |
| 413.0 | | | 215 | 0.13 | 0.13 | 0.25 | 0.40 | 0.50 | 0.65 | K20 | 53 | 0.102 | 0.15 | 0.25 | 0.40 | 0.45 | 0.50 | K20 |
| A413.0 | | | (700) | (0.005) | (0.007) | (0.010) | (0.015) | (0.020) | (0.025) | (C-2) | (175) | (0.004) | (0.006) | (0.010) | (0.015) | (0.018) | (0.020) | (C-2) |
| | 70–125 | Solution treated and aged | 105 | 0.18 | 0.25 | 0.40 | 0.50 | 0.65 | 0.75 | S3, S4, S2 | 30 | 0.102 | 0.15 | 0.25 | 0.40 | 0.45 | 0.50 | S3, S4, S2 |
| | | | (350) | (0.007) | (0.010) | (0.015) | (0.020) | (0.025) | (0.030) | (M1, M2, M7) | (100) | (0.004) | (0.006) | (0.010) | (0.015) | (0.018) | (0.020) | (M1, M2, M7) |
| | | | 215 | 0.13 | 0.13 | 0.25 | 0.40 | 0.50 | 0.65 | K20 | 53 | 0.102 | 0.15 | 0.25 | 0.40 | 0.45 | 0.50 | K20 |
| | | | (700) | (0.005) | (0.007) | (0.010) | (0.015) | (0.020) | (0.025) | (C-2) | (175) | (0.004) | (0.006) | (0.010) | (0.015) | (0.018) | (0.020) | (C-2) |
| 390.0 | 40–100 | As-cast | 30 | 0.18 | 0.25 | 0.40 | 0.50 | 0.65 | 0.75 | S3, S4, S2 | 24 | 0.13 | 0.20 | 0.30 | 0.45 | 0.50 | 0.65 | S3, S4, S2 |
| 392.0 | | | (100) | (0.007) | (0.010) | (0.015) | (0.020) | (0.025) | (0.030) | (M1, M2, M7) | (80) | (0.005) | (0.008) | (0.012) | (0.018) | (0.020) | (0.025) | (M1, M2, M7) |
| | | | 60 | 0.18 | 0.25 | 0.40 | 0.50 | 0.65 | 0.75 | K20 | 30 | 0.13 | 0.20 | 0.30 | 0.45 | 0.50 | 0.65 | K20 |
| | | | (200) | (0.007) | (0.010) | (0.015) | (0.020) | (0.025) | (0.030) | (C-2) | (100) | (0.005) | (0.008) | (0.012) | (0.018) | (0.020) | (0.025) | (C-2) |
| | 70–125 | Solution treated and aged | 30 | 0.18 | 0.25 | 0.40 | 0.50 | 0.65 | 0.75 | S3, S4, S2 | 24 | 0.13 | 0.20 | 0.30 | 0.45 | 0.50 | 0.65 | S3, S4, S2 |
| | | | (100) | (0.007) | (0.010) | (0.015) | (0.020) | (0.025) | (0.030) | (M1, M2, M7) | (80) | (0.005) | (0.008) | (0.012) | (0.018) | (0.020) | (0.025) | (M1, M2, M7) |
| | | | 60 | 0.18 | 0.25 | 0.40 | 0.50 | 0.65 | 0.75 | K20 | 30 | 0.13 | 0.20 | 0.30 | 0.45 | 0.50 | 0.65 | K20 |
| | | | (200) | (0.007) | (0.010) | (0.015) | (0.020) | (0.025) | (0.030) | (C-2) | (100) | (0.005) | (0.008) | (0.012) | (0.018) | (0.020) | (0.025) | (C-2) |

(a) Based on four flutes for 3 and 6 mm (1/8 and 1/4 in.) reamers, six flutes for 12 mm (1/2 in.) reamers, and eight flutes for 25 mm (1 in.) and larger reamers. Source: Ref 5

a spur point may be necessary to assist in centering.

The standard lip clearance of 12 to 13° should be increased to about 17° for heavy feeds and for softer alloys. Insufficient lip clearance will cause excessive drill breakage. The drill cutting lips must be keen and smooth, and all surfaces over which the chip passes must be polished to minimize friction and chip buildup. Recommended point angles, helix angles, and lip clearances are summarized in Table 13.

**Drill Material.** Most drills for aluminum are made of high-speed tool steels; M1, M7, and M10 are the most common grades. Only rarely can the additional cost for drills made of a more highly alloyed grade of high-speed tool steel or of carbide be justified for conventional drilling in drill presses.

**Speed and Feed.** With most drill presses, the peripheral speed of small-diameter drills is relatively low; therefore, such drills can be operated at the maximum efficient rotational speed of the machine. In general, high-speed tool steel drills can be operated at a maximum of about 180 m/min (600 sfm). When variable speed is available, drill life can be increased in drilling deeper holes by bringing the drill up to speed gradually.

Because of the ease of penetrating most aluminum alloys, feeds up to twice those used for drilling steel can be employed. Feed varies with drill diameter; the larger-diameter drills permit heavier feeds, as indicated below:

| Feed, mm/rev | Feed, in./rev | Drill diameter, mm | Drill diameter, in. |
|---|---|---|---|
| 0.025 | 0.001 | 1.6 | 1/16 |
| 0.08 | 0.003 | 3.2 | 1/8 |
| 0.18 | 0.007 | 6.4 | 1/4 |
| 0.30 | 0.012 | 13 | 1/2 |
| 0.41 | 0.016 | 19 | 3/4 |
| 0.48 | 0.019 | 25 | 1 |
| 0.64 | 0.025 | 38 | 1 1/2 |
| 0.76 | 0.030 | 50 | 2 |

Nominal speeds and feeds for wrought and cast aluminum alloys are listed in Table 14. Table 15 provides speed and feed data using spade drills.

**Cutting Fluid.** Drilling of thin sections does not require a cutting fluid, but it is essential to both drill life and hole quality that a copious supply of cutting fluid be provided for all deep-hole drilling. Soluble-oil emulsions or kerosene and lard oil mixtures are satisfactory for general drilling. In drilling to a depth greater than six times the drill diameter, the workpiece should be kept cool by spraying. In addition, the drill should be withdrawn several times during drilling to ensure that the cutting fluid floods the hole completely.

## Drilling Operations in Automatic Bar and Chucking Machines

When drilling is done in multiple-operation machines, it is especially important that the tool be correctly ground and set (see the section "Automatic Lathes" in the article "Multiple-Operation Machining" in this Volume). Standard twist drills are generally used for drilling holes up to six diameters deep. For drilling deeper holes, drills that pass the chips up the flutes more readily are recommended. When machining 2017-T4, 2024-T4, and 2011-T3, straight-flute drills are generally used for drilling holes deeper than six diameters. Either high-helix or low-helix drills can be used for drilling deep holes. The web of the high-helix drill is uniform in thickness along the entire length of the body, providing large flutes for the chips. Wide flutes also characterize the low-helix drill. Both types give good results, particularly in the drilling of soft or gummy alloys. For holes with a high ratio of depth to diameter, half-round or gun drills are sometimes used.

When standard twist drills are used for drilling deeper than three diameters, it may be necessary to enlarge the flutes. This can be done with a thin grinding wheel that has been dressed to a radius. The flute is held at an angle to the wheel to ensure the proper curvature in the flute. All flutes on a drill should be ground alike.

**Drill Margin.** Margins along the edges of flutes support the drill in the hole and keep the drill cutting the correct diameter. When drilling aluminum alloys, standard drill margins can often be reduced in width without loss of necessary support. The narrower margin reduces friction between the drill and the hole, thus reducing the amount of heat generated. In many production jobs, a narrower margin greatly increases drill life.

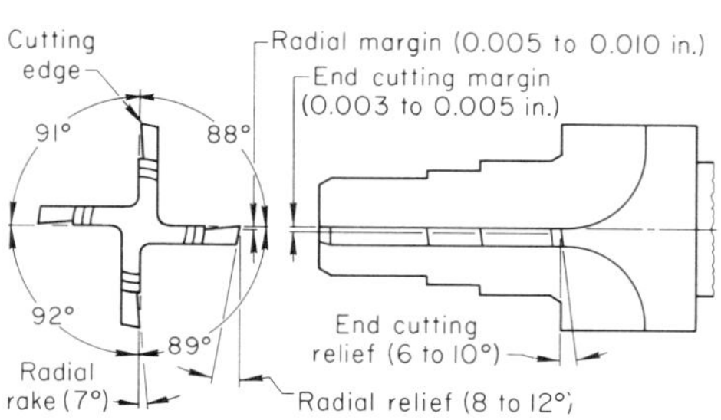


**Fig. 20** Design of a finishing reamer used for reaming aluminum in automatic bar and chucking machines. Dimensions given in inches

**Table 22 Nominal speeds for the tapping of aluminum alloys**

| Alloy | Speed, m/min | Speed, sfm |
|---|---|---|
| Nonheat-treated cast alloys | 35 | 115 |
| Heat-treated cast alloys, solution treated and aged | 27 | 90 |
| Cold-drawn wrought alloys | 38 | 125 |
| Heat-treated wrought alloys, solution treated and aged | 30 | 100 |

**Web Thickness.** For drilling aluminum alloys, web thickness at the point should be reduced as the drill is ground back. This will reduce the end pressure on the drill because the chisel point does not cut but compresses the metal ahead of it. Generally, the web can be somewhat thinner at the point without making the drill susceptible to breakage

**Table 23 Nominal speeds for the tapping of aluminum alloys with high-speed tool steels**

| Material | Hardness, HB (500 kg) | Condition | Speed, m/min (sfm) at a pitch, mm (threads/in.), of: >3 (≤7) | 1.5–3 (8–15) | 1–1.5 (16–24) | ≤1 (>24) | High-speed steel tool material grade, ISO (AISI) |
|---|---|---|---|---|---|---|---|
| **Wrought** | | | | | | | |
| EC 2218 5252 6253 | 30–80 | Cold drawn | 17 | 29 | 37 | 38 | S2, S3 |
| 1060 2219 5254 6262 | | | (55) | (95) | (120) | (125) | (M10, M7, M1) |
| 1100 2618 5454 6463 | 75–150 | Solution treated | 14 | 23 | 29 | 30 | S2, S3 |
| 1145 3003 5456 6951 | | and aged | (45) | (75) | (95) | (100) | (M10, M7, M1) |
| 1175 3004 5457 7001 | | | | | | | |
| 1235 3005 5652 7004 | | | | | | | |
| 2011 4032 5657 7005 | | | | | | | |
| 2014 5005 6053 7039 | | | | | | | |
| 2017 5050 6061 7049 | | | | | | | |
| 2018 5052 6063 7050 | | | | | | | |
| 2021 5056 6066 7075 | | | | | | | |
| 2024 5083 6070 7079 | | | | | | | |
| 2025 5086 6101 7175 | | | | | | | |
| 2117 5154 6151 7178 | | | | | | | |
| **Cast** | | | | | | | |
| *Sand and permanent mold* | | | | | | | |
| A140 319.0 357.0 A712.0 | 40–100 | As-cast | 15 | 26 | 34 | 35 | S2, S3 |
| 201.0 328.0 359.0 D712.0 | | | (50) | (85) | (110) | (115) | (M10, M7, M1) |
| 208.0 A332.0 B443.0 713.0 | 70–125 | Solution treated | 12 | 20 | 26 | 27 | S2, S3 |
| 213.0 F332.0 514.0 771.0 | | and aged | (40) | (65) | (85) | (90) | (M10, M7, M1) |
| 222.0 333.0 A514.0 850.0 | | | | | | | |
| 224.0 354.0 B514.0 A850.0 | | | | | | | |
| 242.0 355.0 520.0 B850.0 | | | | | | | |
| 295.0 C355.0 535.0 | | | | | | | |
| B295.0 356.0 705.0 | | | | | | | |
| 308.0 A356.0 707.0 | | | | | | | |
| Hiduminium RR-350 | | | | | | | |
| *Die castings* | | | | | | | |
| 360.0 A380.0 | 40–100 | As-cast | 20 | 34 | 43 | 46 | S2, S3 |
| A360.0 C443.0 | | | (65) | (110) | (140) | (150) | (M10, M7, M1) |
| 380.0 518.0 | 70–125 | Solution treated | 14 | 23 | 29 | 30 | S2, S3 |
| | | and aged | (45) | (75) | (95) | (100) | (M10, M7, M1) |
| 383.0 | 40–100 | As-cast | 20 | 34 | 43 | 46 | S2, S3 |
| A384.0 | | | (65) | (110) | (140) | (150) | (M10, M7, M1) |
| 413.0 | 70–125 | Solution treated | 14 | 23 | 29 | 30 | S2, S3 |
| A413.0 | | and aged | (45) | (75) | (95) | (100) | (M10, M7, M1) |
| 390.0 | 40–100 | As-cast | 17 | 30 | 34 | 37 | S2, S3 |
| 392.0 | | | (55) | (100) | (110) | (120) | (M10, M7, M1) |
| | 70–125 | Solution treated | 12 | 21 | 27 | 29 | S2, S3 |
| | | and aged | (40) | (70) | (90) | (95) | (M10, M7, M1) |

(a) These speeds are for tapping 65 to 75% threads in shallow through holes. Reduce the speed when tapping deep holes, blind holes, or a higher percentage of thread. Source: Ref 5

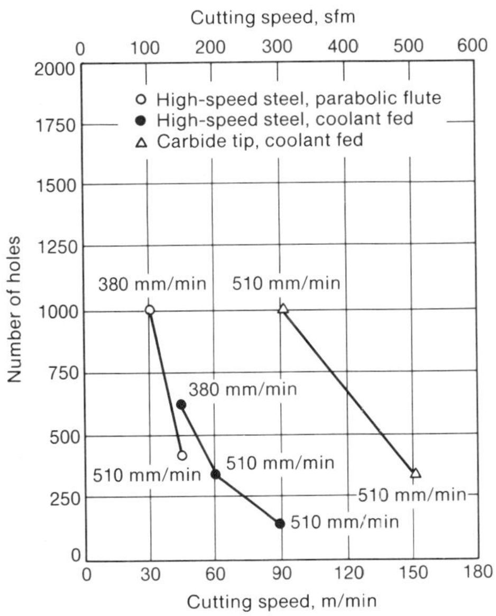


**Fig. 21** Drill life at various penetration rates for several types of drills when drilling permanent mold 390 aluminum alloy. Source: Ref 2

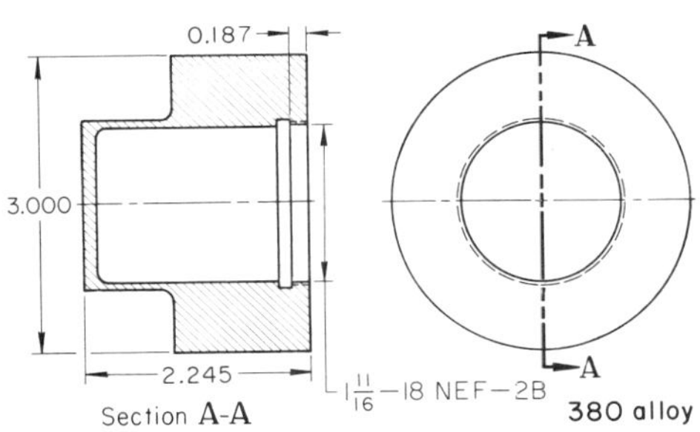


| Details of single-point threading tool | |
|---|---|
| Material | Carbide |
| Cutting edge angle, degrees | 60 |
| Top rake angle, degrees | 20 |
| Clearance angle, degrees | 7 |
| **Operating conditions** | |
| Speed, m/min (sfm) | 145 (475) |
| Diameter of bore, mm (in.) | 43 ($1^{11}/_{16}$) |
| Length of threads, mm (in.) | 4.75 (0.187) |
| Cutting fluid | Soluble oil:water (1:25) |
| Setup time, h | 1 |
| Cycle time, s | 40 |
| Production rate, pieces/h | 75 |

**Fig. 22** Die casting threaded in a single-spindle chucking machine with a threading attachment. Dimensions in figure given in inches

when drilling aluminum alloys than when drilling steel.

For small-diameter drills, notched-point thinning is common. On larger drills, this type of point may produce a poor chip; therefore, the entire flute is ground at the point. The notched point is obtained by using the sharp corner of an abrasive wheel with the side of the wheel following the angle of the chisel point. The drill should be held at an angle to the wheel to form a slight rake for the new cutting edge, which is ground to the center of the point of the drill.

When the entire flute at the point is thinned by grinding, a thin wheel that has been dressed to a radius is used. The drill is held so that the flute is at an angle to the wheel, as when widening the flute. Most of the metal should be ground off the back of the land, and care must be taken not to grind the rake formed by the helix angle from the cutting edge or to destroy the shape of the cutting edge. Thinning of the web must be done uniformly in each flute to ensure that the cut remains balanced on both sides of the centerline of the drill.

**Cutting-Lip Angle.** In general drilling practice, if holes are not too deep, a standard included cutting-lip angle of 118° will give a satisfactory performance. However, if deeper holes are drilled with this cutting-lip angle, the chip produced does not come out easily. For drilling deeper holes, larger included cutting-lip angles are used, forming a narrower chip that readily passes up the flutes.

Occasionally, smaller cutting-lip angles can be used because the transverse forces at the cutting edges are greater and the tool can drill without runout. However, a broader chip is produced. When several drills are used in the same hole, each succeeding tool should have a slightly more blunt cutting-lip angle than that of the preceding one, so that it can center at the outside of the cutting edges.

**The clearance angle** behind the cutting edges should be 12 to 20°. Larger clearances are used for straight-flute drills and for drills ground with large cutting-lip angles. Clearance should extend from periphery to center so that the chisel point is at an angle of 130 to 145° with the cutting edges.

**Rake angles** are set by the helix angle of the drill. For standard twist drills, this is usually 20 to 25°; for high-helix drills, 40 to 43°; for low-helix drills, 7 to 15°; and for straight-flute and half-round drills, 0°.

**Table 24 Geometry and cutting conditions for the drilling and tapping of alloy 390**

| Type | Speed m/min | Speed sfm | Feed | Coolant |
|---|---|---|---|---|
| **Drills** | | | | |
| Tungsten carbide tipped, coolant fed | 90 | 300 | 508 mm/min (20 in./min) | 34 L/min (9 gal./min) at 3.5 MPa (500 psi), soluble oil (20:1) |
| High-speed steel, high helix, parabolic flute, noncoolant fed | 20 | 70 | 381 mm/min (15 in./min) at 170 kPa (25 psi) | 19 L/min (5 gal./min), soluble oil (20:1) |
| High-speed steel, regular helix, coolant fed | 20 | 70 | 4.6 m/min (15 sfm) | 34 L/min (9 gal./min) at 3.5 MPa (500 psi), soluble oil (20:1) |
| **Taps** | | | | |
| Thread forming, high-speed steel, four lubricant grooves | 20 | 70 | Determined by pitch | 19 L/min (5 gal./min), soluble oil (20:1) |

Source: Ref 2

**Center Drills.** When a drill starts against a flat surface, it may skid sideways before cutting, particularly if it is small or protrudes considerably from the holder. This may cause the drill to break or to cut off-center. Therefore, except with large, rigidly held drills, a center drill is recommended to provide an accurate start.

For maximum rigidity, center drills should be of relatively large diameter and short length. Unless a countersink of a special angle is to be left in the workpiece, an included cutting-lip angle of 90° should be used. In general, for proper centering, the outside of the cutting lips of the drill that follows the center drill should strike the stock first, ensuring adequate support.

**Depth of Hole.** In multiple-operation machines, to prevent chips from jamming in the flutes and causing drill breakage when deep holes are being drilled, a limit must be set on the depth drilled with each entry of the tool. The maximum depths, in terms of drill diameter, that can be drilled per entry under normal production conditions are given in Table 16. Before reentry, the drill should be completely backed out of the hole so that the chips can be washed away.

The type of workpiece, the number of positions on the machine available for drilling, and the type of machine determine the practice for drilling deep holes. It is sometimes economical to use several drills for drilling deep holes.

For holes more than eight diameters deep, half-round or gun drills are often used. To ensure adequate support for the half-round drill when it starts cutting, another type of drill should be used to drill a starting hole three diameters or more deep. The half-round drill can drill four diameters deeper in the first entry of the started hole.

**Drill Speed.** Generally, small-diameter, high-speed tool steel drills can be operated at speeds up to 180 m/min (600 sfm). However, when larger drills are used to remove large quantities of metal at a rapid rate,

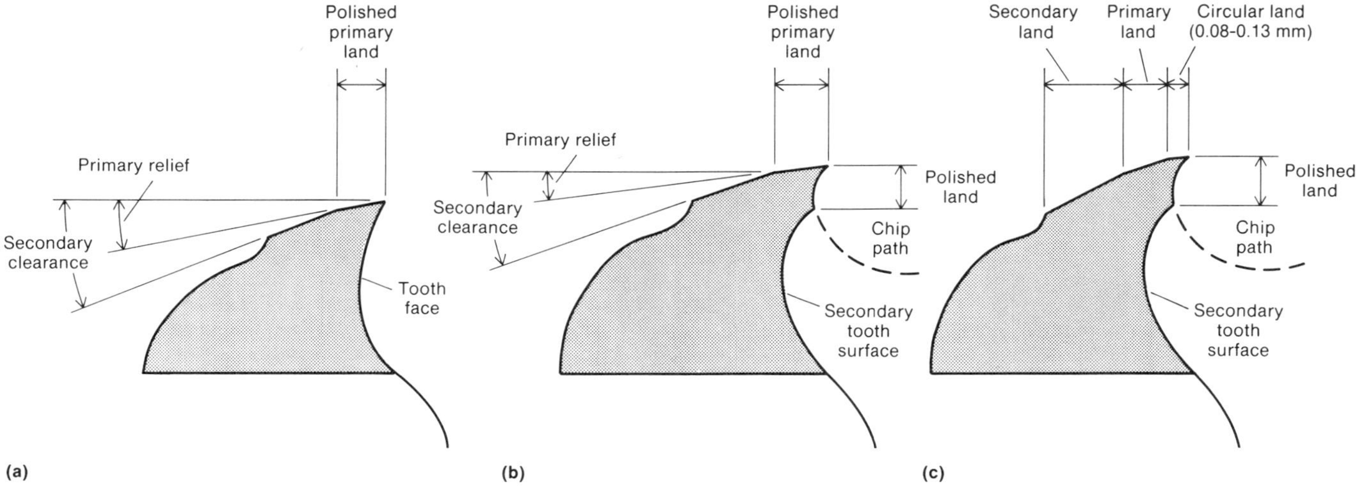


**Fig. 23** Three end mill geometries designed for the cutting of aluminum alloys. (a) Conventional milling. (b) Heavy metal removal. (c) Finishing operations. Source: Ref 1

lower speeds will give more economical drill life. A guide for choosing proper speeds for most conditions is:

| Drill diameter, mm | Drill diameter, in. | Speed, m/min | Speed, sfm |
|---|---|---|---|
| <25 | <1 | 180 | 600 |
| 25–38 | 1–1½ | 170 | 550 |
| >38 | >1½ | 140 | 450 |

**Feed** for drilling in multiple-operation machines depends on the size and strength of the drill; finish, tolerance, and concentricity desired; and power available. As feed is increased, torque on the drill increases, until the breaking point may be reached. On the other hand, lower feed produces thinner chips, which are more likely to clog the flutes than thicker chips. Clogging is likely to break the drill or mar the finish of the workpiece.

Recommended drill feeds are given in Table 17 for drill sizes up to 19.05 mm (0.750 in.) in diameter. Feeds for larger drills usually depend on the power available in the machine.

For greater accuracy and better finish, lower feed may be necessary. For deep holes, machined with several entries, feed should be decreased 15% for each successive entry. Lower feed should be used to drill thin-wall parts.

**Power Requirements.** When large quantities of metal are removed, lower feed may be necessary to keep the power within the limits available in the machine. When drilling with standard drills (ground to 118° included cutting-lip angle and 15° clearance) to four diameters deep in 2017-T4 and 2024-T4, the metal removal rate is 18 000 to 24 000 $mm^3$/min/kW (1.5 to 2 $in.^3$/min/hp). For 2011-T3, the rate is 31 000 to 37 000 $mm^3$/min/kW (2.5 to 3.0 $in.^3$/min/hp). These figures are based on a feed of 0.43 mm/rev (0.017 in./min). With lower feed, the rate of metal removal decreases.

**Drill Size Versus Hole Size.** Drills, when properly ground and set, will cut aluminum alloys to size or not more than 0.05 mm (0.002 in.) oversize. However, any condition that causes overheating is likely to decrease the size of the hole when it is measured after the workpiece has cooled. This is especially noticeable when drilling large holes. The drilling of deep holes at a high feed rate will decrease hole size in relation to drill size. Other tools working simultaneously will sometimes generate enough heat to cause the production of holes close to or even smaller than drill size, as measured after the workpiece has cooled to room temperature.

## Deep-Hole Drilling

Gun drills, tipped with either carbide or high-speed tool steel, have replaced twist drills for many deep-hole drilling applications, regardless of the metal being drilled. In gun drilling, the maximum depth of hole that can be drilled successfully is generally related to drill size (Table 18). The length-to-diameter relationships shown in Table 18 are generally standard, although gun drills having much greater length-to-diameter ratios are often used. Table 19 lists nominal speeds and feeds for the gun drilling of wrought and cast aluminum alloys.

Gun drills, with a single-flute cutting head and grooved shank, must have a sufficient flow of cutting fluid under high pressure at the point where the cutting edge contacts the work; this keeps the cutting edge cool and ensures that the chips will be forced out through the chip groove. Cutting fluid is normally applied at a pressure of 3.4 to 4.1 MPa (500 to 600 psi) for drills 9.5 mm (⅜ in.) in diameter or smaller and at 2.1 to 2.8 MPa (300 to 400 psi) for larger drills. A pump of 3.7 kW (5 hp) capacity is usually required for average applications. A paraffin-base oil with a viscosity of 100 to 125 SUS at 40 °C (100 °F) has been used satisfactorily in gun drilling aluminum with a 9.5 mm (⅜ in.) diam drill at a speed of 90 m/min (300 sfm) for high-speed tool steel or 180 m/min (600 sfm) for carbide and at feeds of 0.025 to 0.08 mm/rev (0.001 to 0.003 in./rev).

Although deep-drilling problems have been alleviated to some extent by the development of gun drills and special gun drilling equipment, this equipment is expensive and

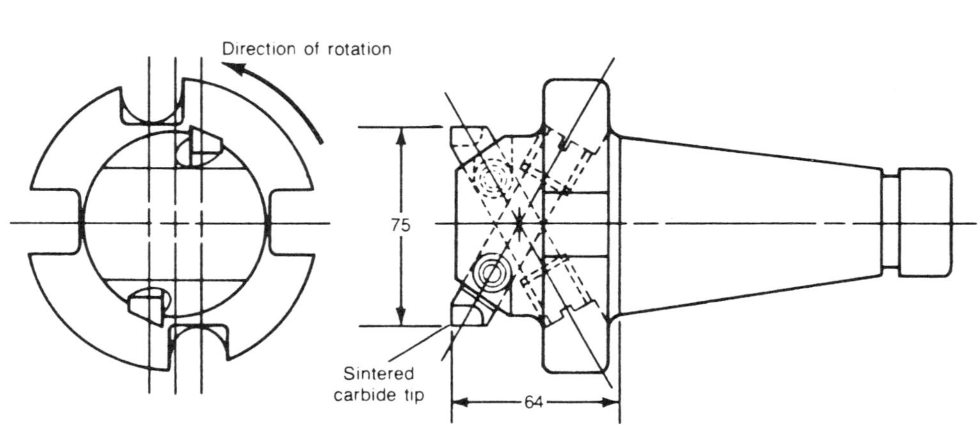


**Fig. 24** Special two-cut tooth end mill with sintered carbide cutters. Dimensions given in millimeters. Source: Ref 4

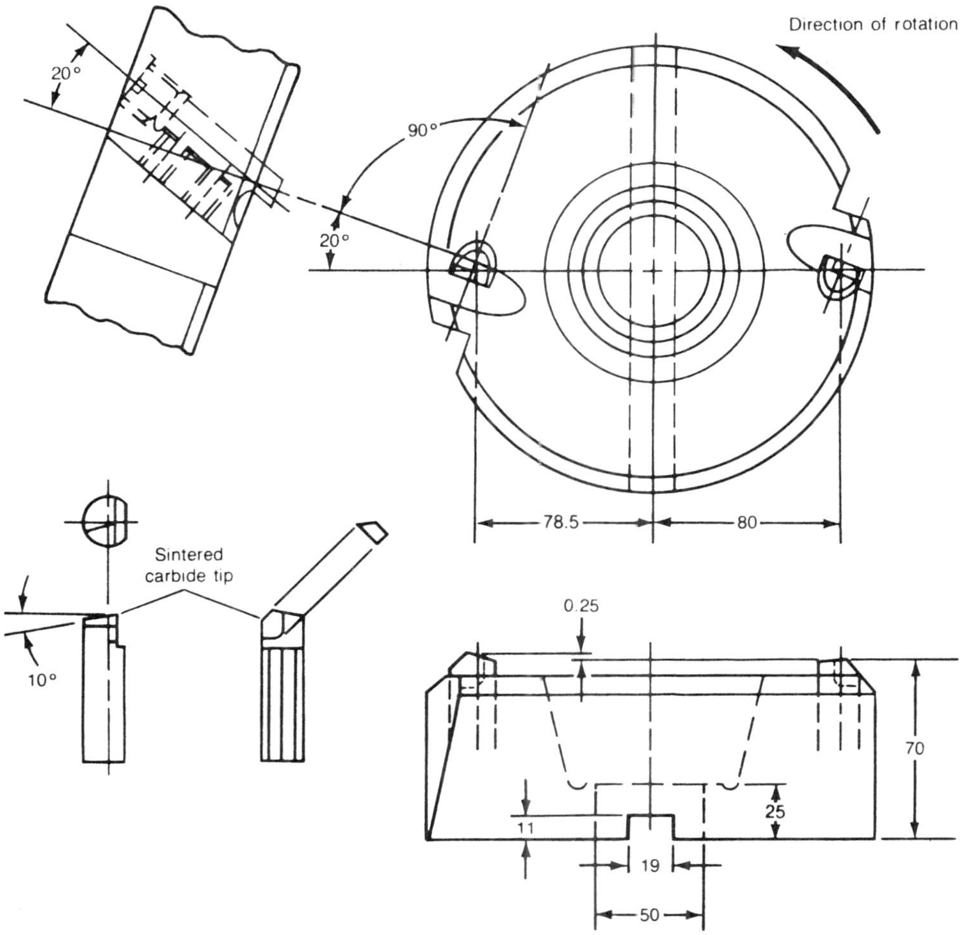


**Fig. 25** Tool geometry for special two-tooth sintered carbide-tip milling cutter used to cut aluminum alloys. Dimensions given in millimeters. Source: Ref 4

cannot always be justified. The following example describes a deep-hole drilling application that utilized available equipment (in this case, a boring mill).

**Example 4: Deep-Hole Drilling in a Boring Mill.** The 914 mm (36 in.) diam, 914 mm (36 in.) long explosive forming die illustrated in Fig. 19 was originally made up from nine 102 mm (4 in.) thick aluminum alloy plates. Because no plate stock obtainable was wide enough to make a 914 mm (36 in.) diam one-piece plate, each round plate was made by doweling together two 102 mm (4 in.) thick half-sections. As each 102 mm (4 in.) plate section was completed by machining the die cavity hole through its center and drilling 12 tie-bolt holes, it was mated to the next 102 mm (4 in.) thick plate until the die was completed. Each individual plate was placed on the preceding plate so that the half-section cut lines were 90° from those in the preceding plate. Then each section was doweled with two pins to the next, completing the die.

By the improved method of producing these dies, four 914 mm (36 in.) diam, 229 mm (9 in.) thick forgings were obtained. These forgings were machined (as were the 102 mm, or 4 in., thick plates) individually and doweled to one another to maintain alignment. The assembly was then set up in a boring mill as shown in Fig. 19. Each of the 41 mm (1⅝ in.) diam tie-bolt holes was drilled in one pass using a standard gun drill. Drilling time was reduced from 7 h by the original method to ½ h by the improved method.

The boring mill was modified by adding a hydraulic pump with a capacity of 154 L/min (40.6 gal./min) to supply cutting fluid to the drill point at 2.1 MPa (300 psi). A baffle plate and filter system were installed to handle the flow of oil.

## Reaming

The machines, basic designs of reamers, and techniques discussed in the article "Reaming" in this Volume are generally applicable to aluminum alloys.

**Reamer Design.** Because it is less likely to cause chatter, a spiral-flute reamer (solid, expandable, or adjustable) is generally preferable to a straight-flute reamer for finishing holes in aluminum alloys. In most applications, it is advantageous to use a reamer with a negative spiral (this is, one spiraled in the direction opposite to rotation) to prevent the reamer from feeding itself into the hole. Flutes should be large enough to pass the chips readily, and there should be enough flutes to provide adequate support to the tool. The margins of straight-flute reamers should be as narrow as possible to reduce friction between tool and work. Straight-flute reamers are designed with an even number of blades arranged opposite each other in pairs, but with flute spacing varied slightly to prevent chatter and marking.

Spiral flutes (right-hand cut, left-hand spiral) are frequently more effective than angular spacing in reducing chatter. However, a spiral-flute reamer must have sufficient spiral so that two or more flutes overlap in the length of the reamed hole. The spiral angle must be held to a minimum because the steeper the angle, the more end pressure is required to feed the reamer through the hole. Additional reamer design data are given in Table 20.

When the hole has close tolerances and rigid surface finish requirements, the reaming procedure must sometimes be altered. If the design of the workpiece permits, spiral-flute reamers with 7° hook will produce fine finishes, especially on angular surfaces. When reaming diameters of 19.05 mm (0.750 in.) or larger, a carbide-tip expandable reamer will produce good finish and provide extended tool life.

**Speed and Feed.** The nominal speed for reaming the nonheat-treated cast alloys with high-speed tool steel (M1, M2, or M7) reamers is 150 m/min (500 sfm). For all other cast and wrought alloys (excluding high-silicon alloys), speed should be about 120 m/min (400 sfm).

When reaming the nonheat-treated cast alloys with carbide tools, nominal speed is 300 m/min (1000 sfm). For other cast and wrought alloys (excluding high-silicon alloys), nominal reaming speed is 260 m/min (850 sfm).

Feed rate in reaming aluminum alloys is generally the same for all alloys and tool materials; hole size, however, does affect optimum feed, as indicated below:

| Feed, mm/rev | Feed, in./rev | Hole diameter, mm | Hole diameter, in. |
|---|---|---|---|
| 0.13 | 0.005 | 3.2 | ⅛ |
| 0.18 | 0.007 | 6.4 | ¼ |
| 0.30 | 0.012 | 13 | ½ |
| 0.38 | 0.015 | 25 | 1 |
| 0.51 | 0.020 | 38 | 1½ |
| 0.76 | 0.030 | 50 | 2 |

Table 21 lists nominal speeds and feeds for the rough and finish reaming of aluminum alloys.

**Cutting Fluid.** At high reaming speed, a cutting fluid is required for reducing the temperature in the workpiece, minimizing distortion, and preventing undersize reaming. For reaming with high-speed tool steel reamers, mixtures of lard oil and paraffin oil

**Table 25 Nominal speeds and feeds for the peripheral end milling of aluminum alloys with high-speed tool steel and carbide tools**

| | | | | High-speed steel tool | | | | | | Carbide tool | | | | | |
|---|---|---|---|---|---|---|---|---|---|---|---|---|---|---|---|
| | | | | | Feed, mm/tooth (in./tooth) with a cutter diameter of: | | | | | | Feed, mm/tooth (in./tooth) with a cutter diameter of: | | | | |
| Material | Hardness, HB (500 kg) | Condition | Radial depth of cut(a) mm (in.) | Speed, m/min (sfm) | 10 mm (3/8 in.) | 12 mm (1/2 in.) | 18 mm (3/4 in.) | 25–50 mm (1–2 in.) | Tool material grade, ISO (AISI) | Speed, m/min (sfm) | 10 mm (3/8 in.) | 12 mm (1/2 in.) | 18 mm (3/4 in.) | 25–50 mm (1–2 in.) | Tool material grade, ISO (C) |
| **Wrought** | | | | | | | | | | | | | | | |
| EC 3005 6066 | 30–80 | Cold drawn | 0.5 | 245 | 0.075 | 0.102 | 0.13 | 0.18 | S4, S5, S2 | 395 | 0.075 | 0.102 | 0.13 | 0.18 | K20, M20 |
| 1060 4032 6070 | | | (0.020) | (800) | (0.003) | (0.004) | (0.005) | (0.007) | (M2, M3, M7) | (1300) | (0.003) | (0.004) | (0.005) | (0.007) | (C-2) |
| 1100 5005 6101 | | | 1.5 | 185 | 0.102 | 0.15 | 0.20 | 0.25 | | 305 | 0.102 | 0.15 | 0.20 | 0.25 | |
| 1145 5050 6151 | | | (0.060) | (600) | (0.004) | (0.006) | (0.008) | (0.010) | | (1000) | (0.004) | (0.006) | (0.008) | (0.010) | |
| 1175 5052 6253 | | | diam/4 | 150 | 0.075 | 0.102 | 0.15 | 0.20 | | 275 | 0.075 | 0.13 | 0.15 | 0.20 | |
| 1235 5056 6262 | | | (diam/4) | (500) | (0.003) | (0.004) | (0.006) | (0.008) | | (900) | (0.003) | (0.005) | (0.006) | (0.008) | |
| 2011 5083 6463 | | | diam/2 | 120 | 0.050 | 0.075 | 0.13 | 0.15 | | 245 | 0.050 | 0.102 | 0.13 | 0.15 | |
| 2014 5086 6951 | | | (diam/2) | (400) | (0.002) | (0.003) | (0.005) | (0.006) | | (800) | (0.002) | (0.004) | (0.005) | (0.006) | |
| 2017 5154 7001 | 75–150 | Solution treated and aged | 0.5 | 245 | 0.075 | 0.102 | 0.13 | 0.18 | S4, S5, S2 | 395 | 0.075 | 0.102 | 0.13 | 0.18 | K20, M20 |
| 2018 5252 7004 | | | (0.020) | (800) | (0.003) | (0.004) | (0.005) | (0.007) | (M2, M3, M7) | (1300) | (0.003) | (0.004) | (0.005) | (0.007) | (C-2) |
| 2021 5254 7005 | | | 1.5 | 185 | 0.102 | 0.15 | 0.20 | 0.25 | | 305 | 0.102 | 0.15 | 0.20 | 0.25 | |
| 2024 5454 7039 | | | (0.060) | (600) | (0.004) | (0.006) | (0.008) | (0.010) | | (1000) | (0.004) | (0.006) | (0.008) | (0.010) | |
| 2025 5456 7049 | | | diam/4 | 150 | 0.075 | 0.102 | 0.15 | 0.20 | | 275 | 0.075 | 0.13 | 0.15 | 0.20 | |
| 2117 5457 7050 | | | (diam/4) | (500) | (0.003) | (0.004) | (0.006) | (0.008) | | (900) | (0.003) | (0.005) | (0.006) | (0.008) | |
| 2218 5652 7075 | | | diam/2 | 120 | 0.050 | 0.075 | 0.13 | 0.15 | | 245 | 0.050 | 0.102 | 0.13 | 0.15 | |
| 2219 5657 7079 | | | (diam/2) | (400) | (0.002) | (0.003) | (0.005) | (0.006) | | (800) | (0.002) | (0.004) | (0.005) | (0.006) | |
| 2618 6053 7175 | | | | | | | | | | | | | | | |
| 3003 6061 7178 | | | | | | | | | | | | | | | |
| 3004 6063 | | | | | | | | | | | | | | | |
| **Cast** | | | | | | | | | | | | | | | |
| **Sand and permanent mold** | | | | | | | | | | | | | | | |
| A140 F332.0 520.0 | 40–100 | As-cast | 0.5 | 305 | 0.075 | 0.102 | 0.13 | 0.18 | S4, S5, S2 | 395 | 0.075 | 0.102 | 0.15 | 0.18 | K20, M20 |
| 201.0 333.0 535.0 | | | (0.020) | (1000) | (0.003) | (0.004) | (0.005) | (0.007) | (M2, M3, M7) | (1300) | (0.003) | (0.004) | (0.005) | (0.007) | (C-2) |
| 208.0 354.0 705.0 | | | 1.5 | 245 | 0.102 | 0.15 | 0.20 | 0.25 | | 305 | 0.102 | 0.15 | 0.20 | 0.25 | |
| 213.0 355.0 707.0 | | | (0.060) | (800) | (0.004) | (0.006) | (0.008) | (0.010) | | (1000) | (0.004) | (0.006) | (0.008) | (0.010) | |
| 222.0 C355.0 A712.0 | | | diam/4 | 185 | 0.075 | 0.102 | 0.15 | 0.20 | | 275 | 0.075 | 0.13 | 0.15 | 0.20 | |
| 224.0 356.0 D712.0 | | | (diam/4) | (600) | (0.003) | (0.004) | (0.006) | (0.008) | | (900) | (0.003) | (0.005) | (0.006) | (0.008) | |
| 242.0 A356.0 713.0 | | | diam/2 | 150 | 0.050 | 0.075 | 0.13 | 0.15 | | 245 | 0.050 | 0.102 | 0.13 | 0.15 | |
| 295.0 357.0 771.0 | | | (diam/2) | (500) | (0.002) | (0.003) | (0.005) | (0.006) | | (800) | (0.002) | (0.004) | (0.005) | (0.006) | |
| B295.0 359.0 850.0 | 70–125 | Solution treated and aged | 0.5 | 245 | 0.075 | 0.102 | 0.13 | 0.18 | S4, S5, S2 | 395 | 0.075 | 0.102 | 0.13 | 0.18 | K20, M20 |
| 308.0 B443.0 A850.0 | | | (0.020) | (800) | (0.003) | (0.004) | (0.005) | (0.007) | (M2, M3, M7) | (1300) | (0.003) | (0.004) | (0.005) | (0.007) | (C-2) |
| 319.0 514.0 B850.0 | | | 1.5 | 185 | 0.102 | 0.15 | 0.20 | 0.25 | | 305 | 0.102 | 0.15 | 0.20 | 0.25 | |
| 328.0 A514.0 | | | (0.060) | (600) | (0.004) | (0.006) | (0.008) | (0.010) | | (1000) | (0.004) | (0.006) | (0.008) | (0.010) | |
| A332.0 B514.0 | | | diam/4 | 150 | 0.075 | 0.102 | 0.15 | 0.20 | | 275 | 0.075 | 0.13 | 0.15 | 0.20 | |
| Hiduminium RR-350 | | | (diam/4) | (500) | (0.003) | (0.004) | (0.006) | (0.008) | | (900) | (0.003) | (0.005) | (0.006) | (0.008) | |
| | | | diam/2 | 120 | 0.050 | 0.075 | 0.13 | 0.15 | | 245 | 0.050 | 0.102 | 0.13 | 0.15 | |
| | | | (diam/2) | (400) | (0.002) | (0.003) | (0.005) | (0.006) | | (800) | (0.002) | (0.004) | (0.005) | (0.006) | |
| **Die castings** | | | | | | | | | | | | | | | |
| 360.0 | 40–100 | As-cast | 0.5 | 305 | 0.075 | 0.102 | 0.13 | 0.18 | S4, S5, S2 | 395 | 0.075 | 0.102 | 0.13 | 0.18 | K20, M20 |
| A360.0 | | | (0.020) | (1000) | (0.003) | (0.004) | (0.005) | (0.007) | (M2, M3, M7) | (1300) | (0.003) | (0.004) | (0.005) | (0.007) | (C-2) |
| 380.0 | | | 1.5 | 245 | 0.102 | 0.15 | 0.20 | 0.25 | | 305 | 0.102 | 0.15 | 0.20 | 0.25 | |
| A380.0 | | | (0.060) | (800) | (0.004) | (0.006) | (0.008) | (0.010) | | (1000) | (0.004) | (0.006) | (0.008) | (0.010) | |
| C443.0 | | | diam/4 | 185 | 0.075 | 0.102 | 0.15 | 0.20 | | 275 | 0.075 | 0.13 | 0.15 | 0.20 | |
| 518.0 | | | (diam/4) | (600) | (0.003) | (0.004) | (0.006) | (0.008) | | (900) | (0.003) | (0.005) | (0.006) | (0.008) | |
| | | | diam/2 | 150 | 0.050 | 0.075 | 0.13 | 0.15 | | 245 | 0.050 | 0.102 | 0.13 | 0.15 | |
| | | | (diam/2) | (500) | (0.002) | (0.003) | (0.005) | (0.006) | | (800) | (0.002) | (0.004) | (0.005) | (0.006) | |
| | 70–125 | Solution treated and aged | 0.5 | 275 | 0.075 | 0.102 | 0.13 | 0.18 | S4, S5, S2 | 395 | 0.075 | 0.102 | 0.13 | 0.18 | K20, M20 |
| | | | (0.020) | (900) | (0.003) | (0.004) | (0.005) | (0.007) | (M2, M3, M7) | (1300) | (0.003) | (0.004) | (0.005) | (0.007) | (C-2) |
| | | | 1.5 | 215 | 0.102 | 0.15 | 0.20 | 0.25 | | 305 | 0.102 | 0.15 | 0.20 | 0.25 | |
| | | | (0.060) | (700) | (0.004) | (0.006) | (0.008) | (0.010) | | (1000) | (0.004) | (0.006) | (0.008) | (0.010) | |
| | | | diam/4 | 150 | 0.075 | 0.102 | 0.15 | 0.20 | | 275 | 0.075 | 0.13 | 0.15 | 0.20 | |
| | | | (diam/4) | (500) | (0.003) | (0.004) | (0.006) | (0.008) | | (900) | (0.003) | (0.005) | (0.006) | (0.008) | |
| | | | diam/2 | 120 | 0.050 | 0.075 | 0.13 | 0.15 | | 245 | 0.050 | 0.102 | 0.13 | 0.15 | |
| | | | (diam/2) | (400) | (0.002) | (0.003) | (0.005) | (0.006) | | (800) | (0.002) | (0.004) | (0.005) | (0.006) | |
| 383.0 | 40–100 | As-cast | 0.5 | 275 | 0.075 | 0.102 | 0.13 | 0.18 | S4, S5, S2 | 395 | 0.075 | 0.102 | 0.13 | 0.18 | K20, M20 |
| A384.0 | | | (0.020) | (900) | (0.003) | (0.004) | (0.005) | (0.007) | (M2, M3, M7) | (1300) | (0.003) | (0.004) | (0.005) | (0.007) | (C-2) |
| 413.0 | | | 1.5 | 215 | 0.102 | 0.15 | 0.20 | 0.25 | | 305 | 0.102 | 0.15 | 0.20 | 0.25 | |
| A413.0 | | | (0.060) | (700) | (0.004) | (0.006) | (0.008) | (0.010) | | (1000) | (0.004) | (0.006) | (0.008) | (0.010) | |
| | | | diam/4 | 150 | 0.075 | 0.102 | 0.15 | 0.20 | | 275 | 0.075 | 0.13 | 0.15 | 0.20 | |
| | | | (diam/4) | (500) | (0.003) | (0.004) | (0.006) | (0.008) | | (900) | (0.003) | (0.005) | (0.006) | (0.008) | |
| | | | diam/2 | 120 | 0.050 | 0.075 | 0.13 | 0.15 | | 245 | 0.050 | 0.102 | 0.13 | 0.15 | |
| | | | (diam/2) | (400) | (0.002) | (0.003) | (0.005) | (0.006) | | (800) | (0.002) | (0.004) | (0.005) | (0.006) | |
| | 70–125 | Solution treated and aged | 0.5 | 245 | 0.075 | 0.102 | 0.13 | 0.18 | S4, S5, S2 | 305 | 0.075 | 0.102 | 0.13 | 0.18 | K20, M20 |
| | | | (0.020) | (800) | (0.003) | (0.004) | (0.005) | (0.007) | (M2, M3, M7) | (1000) | (0.003) | (0.004) | (0.005) | (0.007) | (C-2) |
| | | | 1.5 | 185 | 0.102 | 0.15 | 0.20 | 0.25 | | 275 | 0.102 | 0.15 | 0.20 | 0.25 | |
| | | | (0.060) | (600) | (0.004) | (0.006) | (0.008) | (0.010) | | (900) | (0.004) | (0.006) | (0.008) | (0.010) | |
| | | | diam/4 | 120 | 0.075 | 0.102 | 0.15 | 0.20 | | 245 | 0.075 | 0.13 | 0.15 | 0.20 | |
| | | | (diam/4) | (400) | (0.003) | (0.004) | (0.006) | (0.008) | | (800) | (0.003) | (0.005) | (0.006) | (0.008) | |
| | | | diam/2 | 90 | 0.050 | 0.075 | 0.13 | 0.15 | | 215 | 0.050 | 0.102 | 0.13 | 0.20 | |
| | | | (diam/2) | (300) | (0.002) | (0.003) | (0.005) | (0.006) | | (700) | (0.002) | (0.004) | (0.005) | (0.008) | |

**Table 25 (continued)**

| | | | | High-speed steel tool | | | | | | Carbide tool | | | | | |
|---|---|---|---|---|---|---|---|---|---|---|---|---|---|---|---|
| Material | Hardness, HB (500 kg) | Condition | Radial depth of cut(a) mm (in.) | Speed, m/min (sfm) | Feed, mm/tooth (in./tooth) with a cutter diameter of: 10 mm (3/8 in.) | 12 mm (1/2 in.) | 18 mm (3/4 in.) | 25–50 mm (1–2 in.) | Tool material grade, ISO (AISI) | Speed, m/min (sfm) | Feed, mm/tooth (in./tooth) with a cutter diameter of: 10 mm (3/8 in.) | 12 mm (1/2 in.) | 18 mm (3/4 in.) | 25–50 mm (1–2 in.) | Tool material grade, ISO (C) |
| 390.0 | 40–100 | As-cast | 0.5 | 90 | 0.075 | 0.102 | 0.13 | 0.18 | S4, S5, S2 | 185 | 0.075 | 0.102 | 0.13 | 0.18 | K20, M20 |
| 392.0 | | | (0.020) | (300) | (0.003) | (0.004) | (0.005) | (0.007) | (M2, M3, M7) | (600) | (0.003) | (0.004) | (0.005) | (0.007) | (C-2) |
| | | | 1.5 | 60 | 0.102 | 0.15 | 0.20 | 0.25 | | 150 | 0.102 | 0.15 | 0.20 | 0.25 | |
| | | | (0.060) | (200) | (0.004) | (0.006) | (0.008) | (0.010) | | (500) | (0.004) | (0.006) | (0.008) | (0.010) | |
| | | | diam/4 | 53 | 0.075 | 0.102 | 0.13 | 0.18 | | 120 | 0.075 | 0.13 | 0.15 | 0.20 | |
| | | | (diam/4) | (175) | (0.003) | (0.004) | (0.005) | (0.007) | | (400) | (0.003) | (0.005) | (0.006) | (0.008) | |
| | | | diam/2 | 46 | 0.050 | 0.075 | 0.102 | 0.13 | | 90 | 0.050 | 0.102 | 0.13 | 0.15 | |
| | | | (diam/2) | (150) | (0.002) | (0.003) | (0.004) | (0.005) | | (300) | (0.002) | (0.004) | (0.005) | (0.006) | |
| | 70–125 | Solution treated | 0.5 | 76 | 0.075 | 0.102 | 0.13 | 0.18 | S4, S5, S2 | 150 | 0.075 | 0.102 | 0.13 | 0.18 | K20, M20 |
| | | and aged | (0.020) | (250) | (0.003) | (0.004) | (0.005) | (0.007) | (M2, M3, M7) | (500) | (0.003) | (0.004) | (0.005) | (0.007) | (C-2) |
| | | | 1.5 | 60 | 0.102 | 0.15 | 0.20 | 0.25 | | 120 | 0.102 | 0.15 | 0.20 | 0.25 | |
| | | | (0.060) | (200) | (0.004) | (0.006) | (0.008) | (0.010) | | (400) | (0.004) | (0.006) | (0.008) | (0.010) | |
| | | | diam/4 | 46 | 0.075 | 0.102 | 0.13 | 0.18 | | 90 | 0.075 | 0.13 | 0.15 | 0.20 | |
| | | | (diam/4) | (150) | (0.003) | (0.004) | (0.005) | (0.007) | | (300) | (0.003) | (0.005) | (0.006) | (0.008) | |
| | | | diam/2 | 38 | 0.050 | 0.075 | 0.102 | 0.13 | | 60 | 0.050 | 0.102 | 0.13 | 0.15 | |
| | | | (diam/2) | (125) | (0.002) | (0.003) | (0.004) | (0.005) | | (200) | (0.002) | (0.004) | (0.005) | (0.006) | |

(a) For standard-length end mills, maximum axial depth can be up to 1.5 times the cutter diameter. Source: Ref 5

or kerosene, or of petroleum and turpentine, are especially recommended. Sulfurized and chlorinated oils are often used, but they are likely to stain the work. With carbide reamers, emulsions of soluble oil and water are the most widely used cutting fluids.

**Reaming and Burnishing.** When difficulty is encountered in obtaining the desired tolerance and surface finish, it may be necessary to employ two-stage finishing—for example, to ream the hole 0.013 mm (0.0005 in.) under the specified dimension and then to burnish it to size and finish requirements (see the article "Roller Burnishing" in this Volume.)

**Reaming in Automatic Machines.** A finishing reamer for use in automatic bar and chucking machines is shown in Fig. 20. This reamer finishes three diameters and faces the part to length. A radial rake angle of approximately 7° helps improve surface finish, particularly in reaming angular surfaces. The varied angular spacing of the flutes is intended to minimize slip, deflection, and chatter.

## Burnishing

In the roller burnishing of aluminum, tools of hardened and polished steel are normally employed to finish a surface by compressing the surface while either the tool or the work is rotating. Holes in aluminum are also sometimes burnished in production by pressing a bearing-grade steel ball through the bore to improve the finish.

## Tapping

Taps for producing threads in aluminum are usually made of one of the general-purpose grades of high-speed tool steel such as M1, M7, or M10. Taps should have polished flutes and ample backoff behind the land to prevent the pickup of work metal when the tap is withdrawn. Pitch diameters should be one thread class higher than those normally used for steel to compensate for elastic deformation during the tapping process.

Taps with a diameter less than 9.5 mm (3/8 in.) should have no more than two flutes. Larger taps should have the maximum number of flutes that will give the relationships of land width to circumference shown below

| Tap circumference mm | in. | Optimum land, % |
|---|---|---|
| 13–25 | 1/2–1 | 33 |
| 25–44 | 1–1¾ | 25 |
| >44 | >1¾ | 20 |

The straight-flute tap is satisfactory for use with many aluminum alloys. The spiral-flute tap can be used for any of the alloys; this is better than the straight-flute type, especially for tapping soft material. A spiral-flute tap for cutting right-hand threads should have a right-hand spiral of about the same angle as that found on an ordinary twist drill; it should have a generous taper, and this taper should be backed off. The top rake on the front face should be approximately 25 to 45°. The rake on the back face enables the tap to cut rather than bind when it is reversed, and it provides a clean thread. The spiral-flute tap will bridge a keyway or slot in a hole and aid in the movement of chips from the hole.

Taps for through holes and the harder aluminum alloys should be provided with a hook angle of 10 to 20°, a spiral point, 3 to 4 threads chamfer, and a pitch diameter of GH2 (basic plus 0.013 to 0.025 mm, or 0.0005 to 0.001 in.) or GH3 (basic plus 0.025 to 0.038 mm, or 0.001 to 0.0015 in.). Taps for blind holes and soft alloys should have 10 to 15° hook and 40° right-hand spiral flutes.

For tapping blind holes less than 3.2 mm (1/8 in.) in diameter, spiral flutes should be avoided. For best size control, it is preferable not to bring the chip up the flute but to push it ahead of the tap with a spiral point. The efficiency of tapping may be greatly improved by a change in tooling or method—for example, a change from a solid tap to a collapsible tap.

**The tapping speed** recommended for aluminum is considerably higher than the speeds used for steel. Alloy composition and condition, thread pitch, and method of tapping are major factors affecting tapping speed.

Assuming that the thread pitch is fine (18 to 24) and that the operation can be closely controlled (as in leadscrew tapping), the nominal speeds given in Table 22 are generally suitable.

For coarser pitches, such as 8 to 12, any of the speeds given in Table 22 will be decreased because of the difficulty in controlling the machine. For short thread lengths, speeds for tapping a coarse thread should be about one-half the speeds given in Table 22. For long threads of the same pitch, the speed can be faster than for short threads because of the longer time between starting and stopping. The use of lead control devices permits higher tapping speeds than when no control is used. For example, a speed of 35 m/min (115 sfm) was used to tap 10-24 UNC threads in 2024-T4. The holes were blind and approximately 13 mm (1/2 in.) deep. However, leadscrew control was used. Under the same conditions but without leadscrew control, speed would have been 18 m/min (60 sfm) maximum. Table 23 is a more detailed listing of tapping speeds and tooling used.

**Cutting Fluid.** Lard oil diluted with kerosene or other low-viscosity mineral oil is usually preferred for tapping aluminum. Low-viscosity commercial cutting oils are also used successfully. Soluble-oil emul-

## Table 26 Nominal speeds and feeds for the face milling of aluminum alloys with high-speed tool steel and carbide tools

| Material | Hardness, HB (500 kg) | Condition | Depth of cut(a), mm (in.) | High-speed steel tool: Speed, m/min (sfm) | High-speed steel tool: Feed, mm/tooth (in./tooth) | High-speed steel tool: Tool material grade, ISO (AISI) | Carbide tool, uncoated: Speed: Brazed, m/min (sfm) | Carbide tool, uncoated: Speed: Indexable, m/min (sfm) | Carbide tool, uncoated: Feed, mm/tooth (in./tooth) | Carbide tool, uncoated: Tool material grade, ISO (C) |
|---|---|---|---|---|---|---|---|---|---|---|
| **Wrought** | | | | | | | | | | |
| EC 3005 6066 | 30–80 | Cold drawn | 1 | 365 | 0.25 | S4, S2 | 610 | Max | 0.25 | K10, M20 |
| 1060 4032 6070 | | | (0.040) | (1200) | (0.010) | (M2, M7) | (2000) | (Max) | (0.010) | (C-2) |
| 1100 5005 6101 | | | 4 | 245 | 0.40 | S4, S2 | 550 | Max | 0.50 | K10, M20 |
| 1145 5050 6151 | | | (0.150) | (800) | (0.015) | (M2, M7) | (1800) | (Max) | (0.020) | (C-2) |
| 1175 5052 6253 | | | 8 | 200 | 0.50 | S4, S2 | 365 | Max | 0.65 | K20, M30 |
| 1235 5056 6262 | | | (0.300) | (650) | (0.020) | (M2, M7) | (1200) | (Max) | (0.025) | (C-2) |
| 2011 5083 6463 | 75–150 | Solution treated | 1 | 365 | 0.25 | S4, S2 | 610 | Max | 0.25 | K10, M20 |
| 2014 5086 6951 | | and aged | (0.040) | (1200) | (0.010) | (M2, M7) | (2000) | (Max) | (0.010) | (C-2) |
| 2017 5154 7001 | | | 4 | 245 | 0.40 | S4, S2 | 550 | Max | 0.50 | K10, M20 |
| 2018 5252 7004 | | | (0.150) | (800) | (0.015) | (M2, M7) | (1800) | (Max) | (0.020) | (C-2) |
| 2021 5254 7005 | | | 8 | 200 | 0.50 | S4, S2 | 365 | Max | 0.65 | K20, M30 |
| 2024 5454 7039 | | | (0.300) | (650) | (0.020) | (M2, M7) | (1200) | (Max) | (0.025) | (C-2) |
| 2025 5456 7049 | | | | | | | | | | |
| 2117 5457 7050 | | | | | | | | | | |
| 2218 5652 7075 | | | | | | | | | | |
| 2219 5657 7079 | | | | | | | | | | |
| 2618 6053 7175 | | | | | | | | | | |
| 3003 6061 7178 | | | | | | | | | | |
| 3004 6063 | | | | | | | | | | |
| **Cast** | | | | | | | | | | |
| **Sand and permanent mold** | | | | | | | | | | |
| A140 F332.0 520.0 | 40–100 | As-cast | 1 | 365 | 0.20 | S4, S2 | 610 | 760 | 0.25 | K10, M20 |
| 201.0 333.0 535.0 | | | (0.040) | (1200) | (0.008) | (M2, M7) | (2000) | (2500) | (0.010) | (C-2) |
| 208.0 354.0 705.0 | | | 4 | 245 | 0.30 | S4, S2 | 550 | 610 | 0.40 | K20, M30 |
| 213.0 355.0 707.0 | | | (0.150) | (800) | (0.012) | (M2, M7) | (1800) | (2000) | (0.015) | (C-2) |
| 222.0 C355.0 A712.0 | | | 8 | 200 | 0.40 | S4, S2 | 365 | 460 | 0.50 | K30, M40 |
| 224.0 356.0 D712.0 | | | (0.300) | (650) | (0.016) | (M2, M7) | (1200) | (1500) | (0.020) | (C-2) |
| 242.0 A356.0 713.0 | 70–125 | Solution treated | 1 | 305 | 0.20 | S4, S2 | 550 | 760 | 0.25 | K10, M20 |
| 295.0 357.0 771.0 | | and aged | (0.040) | (1000) | (0.008) | (M2, M7) | (1800) | (2500) | (0.010) | (C-2) |
| B295.0 359.0 850.0 | | | 4 | 215 | 0.30 | S4, S2 | 425 | 550 | 0.40 | K20, M30 |
| 308.0 B443.0 A850.0 | | | (0.150) | (700) | (0.012) | (M2, M7) | (1400) | (1800) | (0.015) | (C-2) |
| 319.0 514.0 B850.0 | | | 8 | 170 | 0.40 | S4, S2 | 305 | 395 | 0.50 | K30, M40 |
| 328.0 A514.0 | | | (0.300) | (550) | (0.016) | (M2, M7) | (1000) | (1300) | (0.020) | (C-2) |
| A332.0 B514.0 | | | | | | | | | | |
| Hiduminium RR-350 | | | | | | | | | | |
| **Die castings** | | | | | | | | | | |
| 360.0 A380.0 | 40–100 | As-cast | 1 | 365 | 0.20 | S4, S2 | 425 | 550 | 0.13 | K10, M20 |
| A360.0 C443.0 | | | (0.040) | (1200) | (0.008) | (M2, M7) | (1400) | (1800) | (0.005) | (C-2) |
| 380.0 518.0 | | | 4 | 245 | 0.30 | S4, S2 | 350 | 460 | 0.25 | K10, M20 |
| | | | (0.150) | (800) | (0.012) | (M2, M7) | (1150) | (1500) | (0.010) | (C-2) |
| | 70–125 | Solution treated | 1 | 305 | 0.20 | S4, S2 | 380 | 490 | 0.13 | K10, M20 |
| | | and aged | (0.040) | (1000) | (0.008) | (M2, M7) | (1250) | (1600) | (0.005) | (C-2) |
| | | | 4 | 215 | 0.30 | S4, S2 | 305 | 395 | 0.25 | K10, M20 |
| | | | (0.150) | (700) | (0.012) | (M2, M7) | (1000) | (1300) | (0.010) | (C-2) |
| 383.0 | 40–100 | As-cast | 1 | 305 | 0.20 | S4, S2 | 350 | 460 | 0.13 | K10, M20 |
| A384.0 | | | (0.040) | (1000) | (0.008) | (M2, M7) | (1150) | (1500) | (0.005) | (C-2) |
| 413.0 | | | 4 | 215 | 0.30 | S4, S2 | 290 | 365 | 0.25 | K10, M20 |
| A413.0 | | | (0.150) | (700) | (0.012) | (M2, M7) | (950) | (1200) | (0.010) | (C-2) |
| | 70–125 | Solution treated | 1 | 215 | 0.15 | S4, S2 | 350 | 425 | 0.13 | K10, M20 |
| | | and aged | (0.040) | (700) | (0.006) | (M2, M7) | (1150) | (1400) | (0.005) | (C-2) |
| | | | 4 | 185 | 0.25 | S4, S2 | 275 | 335 | 0.25 | K10, M20 |
| | | | (0.150) | (600) | (0.010) | (M2, M7) | (900) | (1100) | (0.010) | (C-2) |
| 390.0 | 40–100 | As-cast | 1 | 50 | 0.15 | S4, S2 | 130 | 150 | 0.13 | K10, M20 |
| 392.0 | | | (0.040) | (165) | (0.006) | (M2, M7) | (425) | (500) | (0.005) | (C-2) |
| | | | 4 | 43 | 0.23 | S4, S2 | 115 | 135 | 0.25 | K10, M20 |
| | | | (0.150) | (140) | (0.009) | (M2, M7) | (370) | (450) | (0.010) | (C-2) |
| | 70–125 | Solution treated | 1 | 47 | 0.15 | S4, S2 | 115 | 145 | 0.13 | K10, M20 |
| | | and aged | (0.040) | (155) | (0.006) | (M2, M7) | (375) | (475) | (0.005) | (C-2) |
| | | | 4 | 40 | 0.23 | S4, S2 | 105 | 130 | 0.25 | K10, M20 |
| | | | (0.150) | (130) | (0.009) | (M2, M7) | (350) | (425) | (0.010) | (C-2) |

(a) Depth of cut is measured parallel to axis of cutter. Source: Ref 5

**Table 27 Nominal speeds and feeds for the face milling of aluminum alloys with diamond tools**

| Material | | | | Hardness, HB (500 kg) | Condition | Depth of cut(a) | | Speed | | Feed | |
|---|---|---|---|---|---|---|---|---|---|---|---|
| | | | | | | mm | in. | m/min | sfm | mm/tooth | in./tooth |
| **Wrought** | | | | | | | | | | | |
| EC | 2218 | 5252 | 6253 | 30–150 | All | 0.25–0.75 | 0.010–0.030 | 610 | 2000 | 0.13 | 0.005 |
| 1060 | 2219 | 5254 | 6262 | | | 0.75–1.25 | 0.030–0.050 | 425 | 1400 | 0.25 | 0.010 |
| 1100 | 2618 | 5454 | 6463 | | | 1.25–2.50 | 0.050–0.100 | 305 | 1000 | 0.40 | 0.015 |
| 1145 | 3003 | 5456 | 6951 | | | | | | | | |
| 1175 | 3004 | 5457 | 7001 | | | | | | | | |
| 1235 | 3005 | 5652 | 7004 | | | | | | | | |
| 2011 | 4032 | 5657 | 7005 | | | | | | | | |
| 2014 | 5005 | 6053 | 7039 | | | | | | | | |
| 2017 | 5050 | 6061 | 7049 | | | | | | | | |
| 2018 | 5052 | 6063 | 7050 | | | | | | | | |
| 2021 | 5056 | 6066 | 7075 | | | | | | | | |
| 2024 | 5083 | 6070 | 7079 | | | | | | | | |
| 2025 | 5086 | 6101 | 7175 | | | | | | | | |
| 2117 | 5154 | 6151 | 7178 | | | | | | | | |
| **Cast** | | | | | | | | | | | |
| **Sand and permanent mold** | | | | | | | | | | | |
| A140 | 319.0 | 357.0 | A712.0 | 40–100 | As-cast | 0.25–0.75 | 0.010–0.030 | 1005 | 3300 | 0.13 | 0.005 |
| 201.0 | 328.0 | 359.0 | D712.0 | | | 0.75–1.25 | 0.030–0.050 | 840 | 2750 | 0.25 | 0.010 |
| 208.0 | A332.0 | B443.0 | 713.0 | | | 1.25–2.50 | 0.050–0.100 | 550 | 1800 | 0.40 | 0.015 |
| 213.0 | F332.0 | 514.0 | 771.0 | 70–125 | Solution treated and aged | 0.25–0.75 | 0.010–0.030 | 915 | 3000 | 0.13 | 0.005 |
| 222.0 | 333.0 | A514.0 | 850.0 | | | 0.75–1.25 | 0.030–0.050 | 760 | 2500 | 0.25 | 0.010 |
| 224.0 | 354.0 | B514.0 | A850.0 | | | 1.25–2.50 | 0.050–0.100 | 550 | 1800 | 0.40 | 0.015 |
| 242.0 | 355.0 | 520.0 | B850.0 | | | | | | | | |
| 295.0 | C355.0 | 535.0 | | | | | | | | | |
| B295.0 | 356.0 | 705.0 | | | | | | | | | |
| 308.0 | A356.0 | 707.0 | | | | | | | | | |
| Hiduminium RR-350 | | | | | | | | | | | |
| **Die castings** | | | | | | | | | | | |
| 360.0 | A380.0 | | | 40–100 | As-cast | 0.25–0.75 | 0.010–0.030 | 1100 | 3600 | 0.13 | 0.005 |
| A360.0 | C443.0 | | | | | 0.75–1.25 | 0.030–0.050 | 975 | 3200 | 0.25 | 0.010 |
| 380.0 | 518.0 | | | | | 1.25–2.50 | 0.050–0.100 | 840 | 2750 | 0.40 | 0.015 |
| 383.0 | | | | 40–100 | As-cast | 0.25–0.75 | 0.010–0.030 | 1005 | 3300 | 0.13 | 0.005 |
| A384.0 | | | | | | 0.75–1.25 | 0.030–0.050 | 840 | 2750 | 0.25 | 0.010 |
| 413.0 | | | | | | 1.25–2.50 | 0.050–0.100 | 550 | 1800 | 0.40 | 0.015 |
| A413.0 | | | | | | | | | | | |
| 390.0 | | | | 40–100 | As-cast | 0.25–0.75 | 0.010–0.030 | 855 | 2800 | 0.13 | 0.005 |
| 392.0 | | | | | | 0.75–1.25 | 0.030–0.050 | 670 | 2200 | 0.25 | 0.010 |
| | | | | | | 1.25–2.50 | 0.050–0.100 | 470 | 1550 | 0.40 | 0.015 |

(a) Depth of cut is measured perpendicular to the axis of the cutter. Source: Ref 5

sions are sometimes used, but they will not provide as good a finish as the other cutting fluids mentioned. However, the following example describes an application in which a soluble oil was more effective than a mineral oil, probably because of the lower viscosity of the soluble-oil emulsion, which allowed it to penetrate more readily to the cutting edges of the tap and also allowed it to flow in greater volume for flushing away chips.

**Example 5: Mineral Oil Versus Soluble Oil for Tapping.** An automatic tapping machine was used for cutting 6-32 UNC-2B threads in through holes (6.4 mm, or ¼ in., deep) in alloy 2024-T4. Taps were straight-flute, spiral-point, with 15° chamfer angle and 18° hook, and they were operated at 4150 rev/min (37 m/min, or 122 sfm). Cutting fluid was mineral oil.

Chip congestion, together with buildup of work metal on the cutting edges of the tap, resulted in rough threads and tap breakage. Variations in tapping speed were tried but did not alleviate these problems. However, by substituting soluble oil (in a 1:20 mixture with water) for mineral oil as the cutting fluid, thread finish became acceptable, and productivity and tap life were increased:

| | Mineral oil | Soluble oil |
|---|---|---|
| Holes tapped/h | 1480 | 1520 |
| Total tap life, holes | 6250 | 8900 |

Optimum results are obtained when the cutting fluid is applied with pressure, especially when tapping blind holes.

**Form Tapping.** If chips are a serious problem and if the hole wall is thick enough to support the pressure of the tool, a form (chipless) tap can be used (see the article "Tapping" in this Volume). Although a 75% thread is generally recommended for cut threads, tap drilling for 55 to 65% thread is practical for threads produced with a form tap. All aluminum alloys except the high-silicon (12%) die casting alloys can be tapped with form taps.

In some applications, form taps have been successfully operated at speeds twice as fast as those used for their cutter-type counterparts. Limitations are the same for form tapping as for cutting tapping, and the same cutting fluids are used. The following example compares performance of cutting taps and form taps.

**Example 6: Tapping: Cutting Versus Forming.** The performance of No. 4-40 cutting taps was compared with that of form (chipless) taps in tapping a through hole in a 6.4 mm (¼ in.) section of alloy 380 die castings. Both taps were used in a leadscrew machine at 3000 rev/min. The cutting fluid was a high-grade cutting oil with low sulfur content. Although the production rate for both taps was 400 pieces per hour, the production rate per setup with the form tap (1200 pieces) was double the rate obtained when the cutting tap was used.

In tapping these die castings, a form tap was superior to a cutting tap in all sizes from No. 2 through No. 8, provided the hole wall was thick enough to support the pressure of the form tap.

**Table 28 Nominal speeds and feeds for the end milling and slotting of aluminum alloys with high-speed tool steel tools**

| Material | | | | Hardness, HB (500 kg) | Condition | Axial depth of cut, mm (in.) | Speed, m/min (sfm) | Feed, mm/tooth (in./tooth), at a width of slot of: 10 mm (3/8 in.) | 12 mm (1/2 in.) | 18 mm (3/4 in.) | 25–50 mm (1–2 in.) | High-speed tool steel material grade, ISO (AISI) |
|---|---|---|---|---|---|---|---|---|---|---|---|---|
| **Wrought** | | | | | | | | | | | | |
| EC | 2218 | 5254 | 6463 | 30–80 | Cold drawn | 0.75 | 150 | 0.075 | 0.13 | 0.15 | 0.25 | S4, S5, S2 |
| 1060 | 2219 | 5454 | 6951 | | | (0.030) | (500) | (0.003) | (0.005) | (0.006) | (0.010) | (M2, M3, M7) |
| 1100 | 2618 | 5456 | 7001 | | | 3 | 135 | 0.102 | 0.15 | 0.20 | 0.30 | |
| 1145 | 3003 | 5457 | 7004 | | | (0.125) | (450) | (0.004) | (0.006) | (0.008) | (0.012) | |
| 1175 | 3004 | 5652 | 7005 | | | diam/2 | 120 | 0.075 | 0.13 | 0.15 | 0.20 | |
| 1235 | 3005 | 5657 | 7039 | | | (diam/2) | (400) | (0.003) | (0.005) | (0.006) | (0.008) | |
| 2011 | 4032 | 6053 | 7049 | | | diam/1 | 105 | 0.050 | 0.075 | 0.13 | 0.15 | |
| 2014 | 5005 | 6061 | 7050 | | | (diam/1) | (350) | (0.002) | (0.003) | (0.005) | (0.006) | |
| 2014 | 5050 | 6063 | 7075 | 75–150 | Solution treated and aged | 0.75 | 145 | 0.075 | 0.13 | 0.15 | 0.25 | S4, S5, S2 |
| 2017 | 5052 | 6066 | 7079 | | | (0.030) | (475) | (0.003) | (0.005) | (0.006) | (0.010) | (M2, M3, M7) |
| 2018 | 5056 | 6070 | 7175 | | | 3 | 130 | 0.102 | 0.15 | 0.20 | 0.30 | |
| 2021 | 5083 | 6101 | 7178 | | | (0.125) | (425) | (0.004) | (0.006) | (0.008) | (0.012) | |
| 2024 | 5086 | 6151 | | | | diam/2 | 115 | 0.075 | 0.13 | 0.15 | 0.20 | |
| 2025 | 5154 | 6253 | | | | (diam/2) | (375) | (0.003) | (0.005) | (0.006) | (0.008) | |
| 2117 | 5252 | 6262 | | | | diam/1 | 100 | 0.050 | 0.075 | 0.13 | 0.15 | |
| | | | | | | (diam/1) | (325) | (0.002) | (0.003) | (0.005) | (0.006) | |
| **Cast** | | | | | | | | | | | | |
| **Sand and permanent mold** | | | | | | | | | | | | |
| A140 | 319.0 | 357.0 | A712.0 | 40–100 | As-cast | 0.75 | 120 | 0.075 | 0.13 | 0.15 | 0.25 | S4, S5, S2 |
| 201.0 | 328.0 | 359.0 | D712.0 | | | (0.030) | (400) | (0.003) | (0.005) | (0.006) | (0.010) | (M2, M3, M7) |
| 208.0 | A332.0 | B443.0 | 713.0 | | | 3 | 105 | 0.102 | 0.15 | 0.20 | 0.30 | |
| 213.0 | F332.0 | 514.0 | 771.0 | | | (0.125) | (350) | (0.004) | (0.006) | (0.008) | (0.012) | |
| 222.0 | 333.0 | A514.0 | 850.0 | | | diam/2 | 90 | 0.075 | 0.13 | 0.15 | 0.20 | |
| 224.0 | 354.0 | B514.0 | A850.0 | | | (diam/2) | (300) | (0.003) | (0.005) | (0.006) | (0.008) | |
| 242.0 | 355.0 | 520.0 | B850.0 | | | diam/1 | 76 | 0.050 | 0.075 | 0.13 | 0.15 | |
| 295.0 | C355.0 | 535.0 | | | | (diam/1) | (250) | (0.002) | (0.003) | (0.005) | (0.006) | |
| B295.0 | 356.0 | 705.0 | | 70–125 | Solution treated and aged | 0.75 | 115 | 0.075 | 0.13 | 0.15 | 0.25 | S4, S5, S2 |
| 308.0 | A356.0 | 707.0 | | | | (0.030) | (375) | (0.003) | (0.005) | (0.006) | (0.010) | (M2, M3, M7) |
| Hiduminium RR-350 | | | | | | 3 | 100 | 0.102 | 0.15 | 0.20 | 0.30 | |
| | | | | | | (0.125) | (325) | (0.004) | (0.006) | (0.008) | (0.012) | |
| | | | | | | diam/2 | 84 | 0.075 | 0.13 | 0.15 | 0.20 | |
| | | | | | | (diam/2) | (275) | (0.003) | (0.005) | (0.006) | (0.008) | |
| | | | | | | diam/1 | 69 | 0.050 | 0.075 | 0.13 | 0.15 | |
| | | | | | | (diam/1) | (225) | (0.002) | (0.003) | (0.005) | (0.006) | |
| **Die castings** | | | | | | | | | | | | |
| 360.0 | | | | 40–100 | As-cast | 0.75 | 150 | 0.075 | 0.13 | 0.15 | 0.25 | S4, S5, S2 |
| A360.0 | | | | | | (0.030) | (500) | (0.003) | (0.005) | (0.006) | (0.010) | (M2, M3, M7) |
| 380.0 | | | | | | 3 | 135 | 0.102 | 0.15 | 0.20 | 0.30 | |
| A380.0 | | | | | | (0.125) | (450) | (0.004) | (0.006) | (0.008) | (0.012) | |
| C443.0 | | | | 70–125 | Solution treated and aged | 0.75 | 145 | 0.075 | 0.13 | 0.15 | 0.25 | S4, S5, S2 |
| 518.0 | | | | | | (0.030) | (475) | (0.003) | (0.005) | (0.006) | (0.010) | (M2, M3, M7) |
| | | | | | | 3 | 130 | 0.102 | 0.15 | 0.20 | 0.30 | |
| | | | | | | (0.125) | (425) | (0.004) | (0.006) | (0.008) | (0.012) | |
| 383.0 | | | | 40–100 | As-cast | 0.75 | 135 | 0.075 | 0.13 | 0.15 | 0.25 | S4, S5, S2 |
| A384.0 | | | | | | (0.030) | (450) | (0.003) | (0.005) | (0.006) | (0.010) | (M2, M3, M7) |
| 413.0 | | | | | | 3 | 130 | 0.102 | 0.15 | 0.20 | 0.30 | |
| A413.0 | | | | | | (0.125) | (425) | (0.004) | (0.006) | (0.008) | (0.012) | |
| | | | | 70–125 | Solution treated and aged | 0.75 | 130 | 0.075 | 0.13 | 0.15 | 0.25 | S4, S5, S2 |
| | | | | | | (0.030) | (425) | (0.003) | (0.005) | (0.006) | (0.010) | (M2, M3, M7) |
| | | | | | | 3 | 120 | 0.102 | 0.15 | 0.20 | 0.30 | |
| | | | | | | (0.125) | (400) | (0.004) | (0.006) | (0.008) | (0.012) | |
| 390.0 | | | | 40–100 | As-cast | 0.75 | 53 | 0.075 | 0.13 | 0.15 | 0.25 | S4, S5, S2 |
| 392.0 | | | | | | (0.030) | (175) | (0.003) | (0.005) | (0.006) | (0.010) | (M2, M3, M7) |
| | | | | | | 3 | 46 | 0.102 | 0.15 | 0.20 | 0.30 | |
| | | | | | | (0.125) | (150) | (0.004) | (0.006) | (0.008) | (0.012) | |
| | | | | 70–125 | Solution treated and aged | 0.75 | 46 | 0.075 | 0.13 | 0.15 | 0.25 | S4, S5, S2 |
| | | | | | | (0.030) | (150) | (0.003) | (0.005) | (0.006) | (0.010) | (M2, M3, M7) |
| | | | | | | 3 | 38 | 0.102 | 0.15 | 0.20 | 0.30 | |
| | | | | | | (0.125) | (125) | (0.004) | (0.006) | (0.008) | (0.012) | |

Source: Ref 5

**Drilling and tapping of high-silicon alloy 390** (Ref 2) can be successfully accomplished with the same tool materials and geometries used for other aluminum casting alloys; however, cutting must be performed at lower speeds. Selection of the best tool material and geometry becomes more significant as the hole-depth-to-diameter ratio increases.

The use of natural or synthetic soluble-oil cutting fluids is strongly recommended. Coolant should be directed into the hole with a volume and pressure sufficient to reach the cutting lips of the tool and to flush the chips out of the hole. Inadequate coolant can result in chips packing in the flutes of a drill or the teeth of a tap, a condition that can cause rapid wear, poor hole finish,

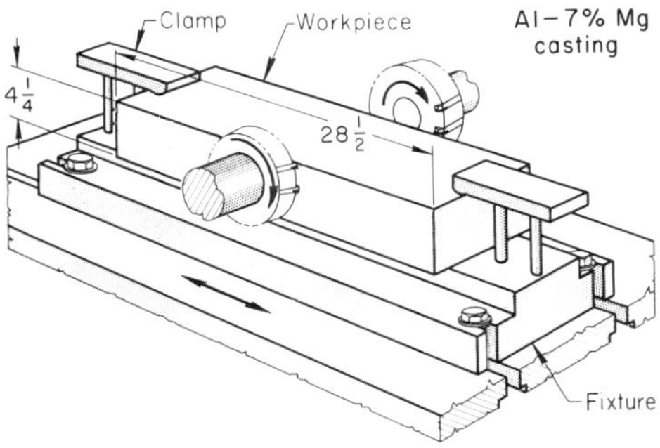


**Cutter details**

| | |
|---|---|
| Type | Face mill, inserted blade |
| Size, mm (in.) | 152 (6) diameter |
| Number of teeth | 12 |
| Material | Carbide-tip blades |

**Operating conditions**

| | |
|---|---|
| Speed, at 2345 rev/min, m/min (sfm) | 1120 (3680) |
| Feed, mm/tooth (in./tooth) | 0.14 (0.0055) |
| Depth of cut, mm (in.) | 3.2 (1/8) max |
| Cutting fluid | None |
| Setup time, h | 1.5 |
| Downtime for changing tools, min | 15 |
| Production rate, pieces/h | 31 |
| Cutter life, pieces/grind | ~100 |
| Tolerances, mm (in.) | |
| Flatness | Within 0.025 (0.001) |
| Width | ±0.08 (±0.003) |
| Parallelism error | 0.13 (0.005) max |
| Finish, μm (μin.) | 1.60 (63) |

**Fig. 26** Simultaneous face milling of parallel surfaces on opposite sides of a casting. Dimensions in figure given in inches

and blowout fracturing of the exit surface upon emergence of a through hole drill. Even more important with respect to alloy 390 is that the chips are exceptionally abrasive and can wear on the lip and margin of a drill if not quickly flushed from the hole.

In a recent study, several drill and tap materials and geometries were evaluated, and cutting conditions were optimized for the most promising combinations. Chip packing became a problem at a depth of approximately three diameters (13 mm, or ½ in., holes), with slow and regular helix drills (24 to 33°) causing breakthrough blowouts and rapid drill wear. Solid high-speed tool steel drills with a high helix angle (38°) and a parabolic flute shape showed less wear under all cutting conditions than high-helix (40°) drills that had a standard flute design.

A speed of 20 m/min (70 sfm) and a feed rate of 381 mm/min (15 in./min) worked well with a high-helix, parabolic flute shaped drill (Fig. 21). A tool life of over 1000 holes was achieved when drilling 13 mm (½ in.) holes to a depth of five to six diameters. Heavy feed rate (in terms of inches per revolution) produced a better tool life than light feed rate.

The best method of drilling alloy 390 is to move the chips quickly by flushing with coolant and by moving the tool through the work in as few revolutions as possible. This creates an easier-to-remove, substantial chip.

Coolant-fed drills were also evaluated, but with mixed results. Carbide-tip, coolant-fed drills produced 1000 holes at a speed of 90 m/min (300 sfm) and a feed rate of 508 mm/min (20 in./min) (Fig. 21). Not only did this accomplish a much higher production rate, but hole size control, roundness, straightness, and finish were much improved over noncoolant-fed drills. Drilling with coolant-fed, high-speed tool steel, on the other hand, resulted in no productivity or hole quality improvement over the best solid drills.

Far better tool life results have been reported using diamond-tip gun drills in similar-size holes (10.79 mm, or 0.425 in.) at a hole depth of eight diameters when boring through solid material. When used in the mass production of automobile engine cylinders, these diamond-tip gun drills give a service life 60 times that of customary twist drills.

Tapping tests were conducted with various types of thread cutting and thread forming tools. No thread cutting taps threaded more than 360 holes before they would no longer gage to size. The best performance, more than 1000 holes, was achieved with a standard form tap having four lobes. It has also been shown that form taps produce the strongest threads. The conditions that were most successful for both the drilling and tapping experiments are given in Table 24.

## Single-Point Threading

Single-point threading tools of conventional design are used to cut both internal and external threads on aluminum alloys. Speeds of 150 m/min (500 sfm) are common, although higher speeds have been used. Typical practice is represented by the following example.

**Example 7: Single-Point Threading of Die Castings.** A fine-pitch thread (pitch diameter: 42.113/41.946 mm, or 1.6580/1.6514 in.) was machined in an alloy 380 die casting (Fig. 22) with a single-point carbide tool. This alloy was susceptible to chipping and tearing if high cutting pressure was applied, particularly if the die casting was porous in the area being threaded.

## Die Threading

Circular chasers for die threading aluminum alloys should have a hook angle of 20° and a face angle of 2°. When threading with tangential chasers, a combination of nominal back rake angle of 20° and side rake angle of 0° is generally best.

To avoid damage to the first few threads, chasers should have a lead chamfer of 25 to 35° for 1½ threads. This chamfer will ensure a smooth, even start.

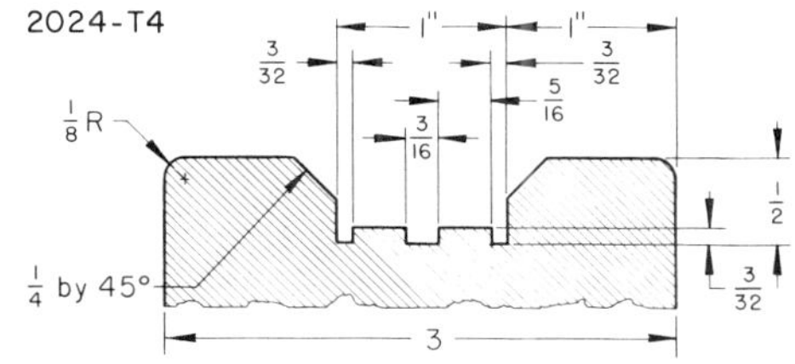


**Operating conditions(a)**

| | |
|---|---|
| Form milling (one pass), mm/min (in./min) | 75 (3) |
| Standard milling (five passes), mm/min (in./min) | |
| 1 Side mill slot | 203 (8) |
| 2 Side mill center slot | 203 (8) |
| 3 Straddle mill reliefs | 127 (5) |
| 4 Straddle mill two 6.4 mm (¼ in.) × 45° chamfers | 203 (8) |
| 5 Straddle mill two 3.2 mm (⅛ in.) radii | 203 (8) |

**Time analysis**

| | |
|---|---|
| Time/piece, min | |
| Form milling | 2.54 |
| Standard milling | 7.79 |
| Time saved by using form milling | 5.25 |

(a) For all operations, cutter speed was 23 m/min (75 sfm), and sulfurized oil was used as the cutting fluid.

**Fig. 27** Contour that could be milled in a single pass using a form cutter or in five operations using standard milling cutters. Dimensions in figure given in inches

Speeds up to 40 m/min (130 sfm) for the nonheat-treated cast alloys and about 30 m/min (100 sfm) for the other alloys can be used, if the length and pitch of the thread and the equipment used permit control at these speeds. Speeds no more than half of the above are more often used because at the higher speeds control is more difficult, especially when threading short lengths or close to a shoulder.

## Milling

The characteristics of chip formation sometimes cause difficulty in the milling of aluminum alloys. In some applications, chip ejection in the milling of deep slots or heavy cuts can be improved by changing from an alloy in the O or F temper to one in the T4 or T6 temper. In the T tempers, the chip is much less likely to clog the cutter. However, when aluminum is milled in the heat-treated condition, the production rate may be lower because a lighter feed rate may be necessary.

**Power Requirements.** Metal removal rates greater than 2600 $mm^3$/min/kW (3.0 $in.^3$/min/hp) can be readily attained in the production milling of aluminum alloys. Some rules applicable to milling aluminum are:

- If feed per tooth is doubled, the horsepower at the cutter must be increased in the ratio of 3:2, or 50%
- If the width of cut is doubled, the horsepower at the cutter must be doubled

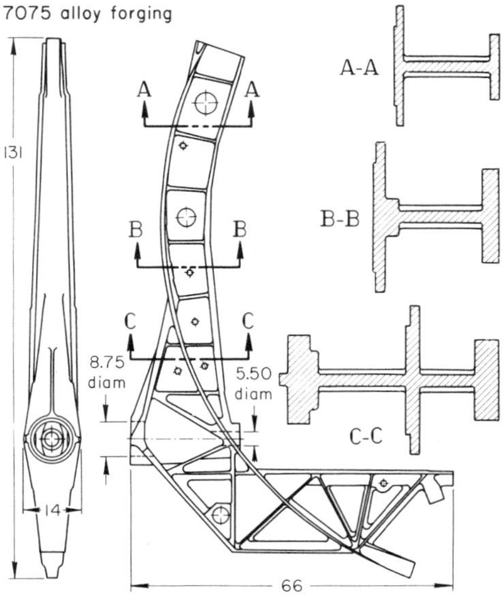


| | |
|---|---|
| Machine | Three-axis profile milling machine, numerically controlled, 4370 × 1320 × 457 mm (172 × 52 × 18 in.) |
| Cutter | |
| Material | Modified M2 high-speed tool steel |
| Number of cutters | 27 |
| Size of cutters, mm (in.) | 19.1 (0.75) diameter × 95 (3¾) long to 76.2 (3.00) diameter × 203 (8) long |
| Rake angles | 10° radial, 30° axial |
| **Operating conditions** | |
| Speed, m/min (sfm) | 60–135 (200–450) |
| Feed, mm/tooth (in./tooth) | 0.15–0.33 (0.006–0.013) |
| Depth of cut, axial dimension, mm (in.) | 3.0–178 (0.12–7.00) |
| Width of cut, radial dimension, mm (in.) | 9.40 (0.37) to cutter diameter |
| Machining time, each part, h | 91(a) |
| Cutting fluid | Soluble oil:water (1:25) |
| Tolerance on critical dimensions, mm (in.) | ±0.05 (±0.002) |
| Surface finish, μm (μin.) | 1.50–3.05 (60–120) |

(a) Roughing, 29 h; finishing, 62 h

**Fig. 28** Part that was machined from a large blocker forging by NC profile milling. Dimensions in figure given in inches

- If the depth of cut is doubled, the horsepower at the cutter must be increased in the ratio of 1.9:1, or 90%
- If the depth of cut is halved and horsepower at the cutter remains the same, the feed may be increased by a factor of 2.5
- If the speed of the cutter is doubled and the feed per tooth is halved, the horsepower at the cutter must be increased by about 30%

**Cutter Design.** For efficiency in milling aluminum, the cutter should have a radial rake angle of 10 to 20°, an axial rake angle of 15 to 45°, and end or peripheral clearance of 10 to 12°. Form-relieved cutters should have a relief of 10° on the profile.

Best results can be obtained with a cutter that has fewer teeth and larger positive-rake angles than cutters for milling steel. Fewer teeth permit more chip space, which is especially important in milling aluminum because speeds are usually much greater than those used in milling steel. To prevent chatter, however, the cutter should have enough teeth so that at least two teeth are engaged at all times.

Because of the high speeds used in milling aluminum, careful consideration must be given to the effect of centrifugal force on the cutter. This is more important with cutters of large diameter. Cutter bodies in which teeth are wedge locked and from which some of the teeth have been removed to provide greater chip space should not be operated at high speed. All cutters for high-speed milling that have inserted teeth (brazed or mechanically secured) should be dynamically balanced. Careful attention should be given to cutter adapters, cutter holders, arbors, spacers, and other components that rotate with the cutter.

End mills for aluminum should have sharp rake angles with sufficient clearance to prevent heeling. Neither rake nor clearance angles should be excessive, or digging-in will result. The cutter should contact the workpiece in such a manner that any torsional, or lateral, deflection will decrease the depth of cut.

Flute form and surface finish in the flute are important. Consequently, many of these cutters are made by grinding from the solid after heat treatment. The lips of an end mill (of either the plain two-flute or ball-nose type) that is to be used for making plunge cuts directly into solid aluminum should have slightly greater clearances than is normal for these cutters in order to prevent heeling. Preferably, the faces of the cutting edges should have a surface finish of 0.038 μm (1.5 μin.) and should never be rougher than 0.088 μm (3.5 μin.).

In making plunge cuts directly into aluminum, the problem of chip ejection becomes acute. For this reason, it is best to feed the cutter into the material in such a way that, for each axial advance of the cutter (equal to about half the cutter diameter), the cutter is fed laterally about one diameter in an oscillating pattern until the desired depth of cut is reached, rather than making a purely axial cut.

For milling slots, or pockets, where the end of the cutter is in contact with the workpiece, it is usually best to have the same direction for the cut and the helix. However, when profiling with the periphery of the cutter, where the end of the cutter is not in contact with the workpiece, a combination of right-hand cut, left-hand helix gives the best results.

Router bits are a special type of end mill that is modified to ensure chip ejection at the speeds normally used for these cutters.

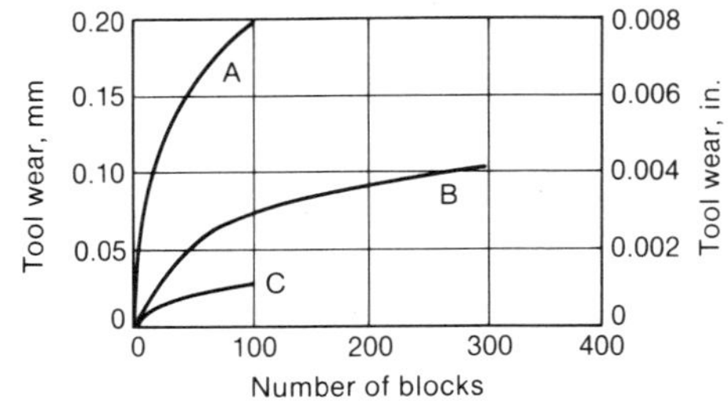


**Fig. 29** Tool wear curves for the single-tooth milling of alloy 390 engine blocks (wet) at 0.30 mm/rev (0.012 in./rev). A, carbide, 150 m/min (492 sfm); B, diamond, 1500 m/min (4920 sfm); C, diamond, 150 m/min (492 sfm). Source: Ref 2

Some cutters for general routing purposes may have two flutes with a helix angle of 25°. The best overall performance in routing accurate slots or grooves is with a two-flute cutter having a 45° helix angle, with the direction of helix and of cut the same.

Cutters for routing stacked sheet stock should be designed with a single flute and a helix angle of 25 to 45°. An integral pilot running in an outboard bearing is used on some of these cutters to help control deflection.

All cutters should have a 15° hook angle on the cutting lip. The flute should be of 2.3 mm (0.09 in.) uniform depth, with a smooth gullet extending 3.2 mm (⅛ in.) inward from the lip. The lips should be radially relieved on the periphery to 0.08 to 0.18 mm (0.003 to 0.007 in.) per 1.6 mm (1⁄16 in.) width of land behind the cutting edge.

**Special-Geometry End Mills for Aluminum (Ref 1, 4).** An interesting example of fluted cutters designed specifically for machining aluminum is provided by the end mills shown in Fig. 23. The first departure from standard end-mill design is an increase to 45° in the helix angle from the standard 30°—the equivalent of high, positive side rake. The second departure is in the form of the tooth, the face of which provides a curved, positive-rake, polished primary surface followed by a secondary undercut that blends into the web (Fig. 23). The polished primary face surface and the discontinuity provide smooth chip flow and minimize the adherence of chips that invariably leads to clogging.

Two versions of these aluminum cutting end mills are available—one for heavy metal removal applications (such as pocketing and rough profiling) and one for finishing operations in which the radial depth of cut approximates 5% of tool diameter. The latter type incorporates an unrelieved circular land of 0.08 to 0.13 mm (0.003 to 0.005 in.) width on the periphery to minimize any tendency toward chatter, vibration, or squealing.

Somewhat similar to this concept is the use of a molded chip-breaker style of carbide insert. This type of insert increases the positive rake (nominally 5° rake, with 5°

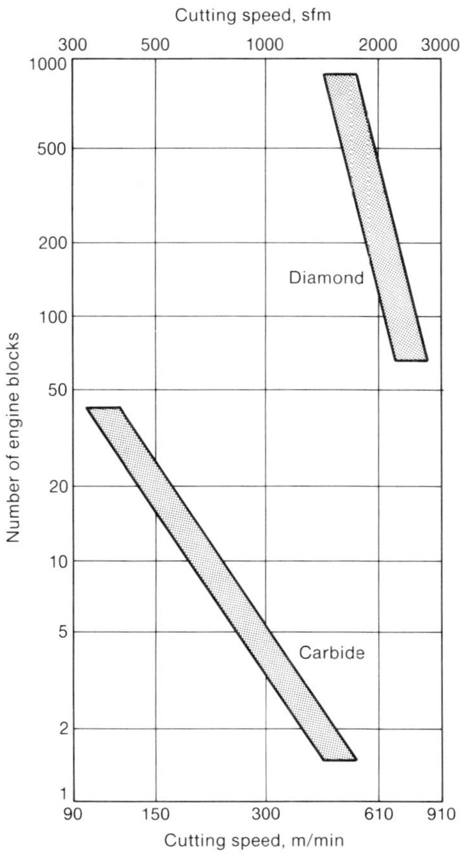


**Fig. 30** Tool life of carbide and diamond tools when milling alloy 390 engine blocks (dry). Source: Ref 2

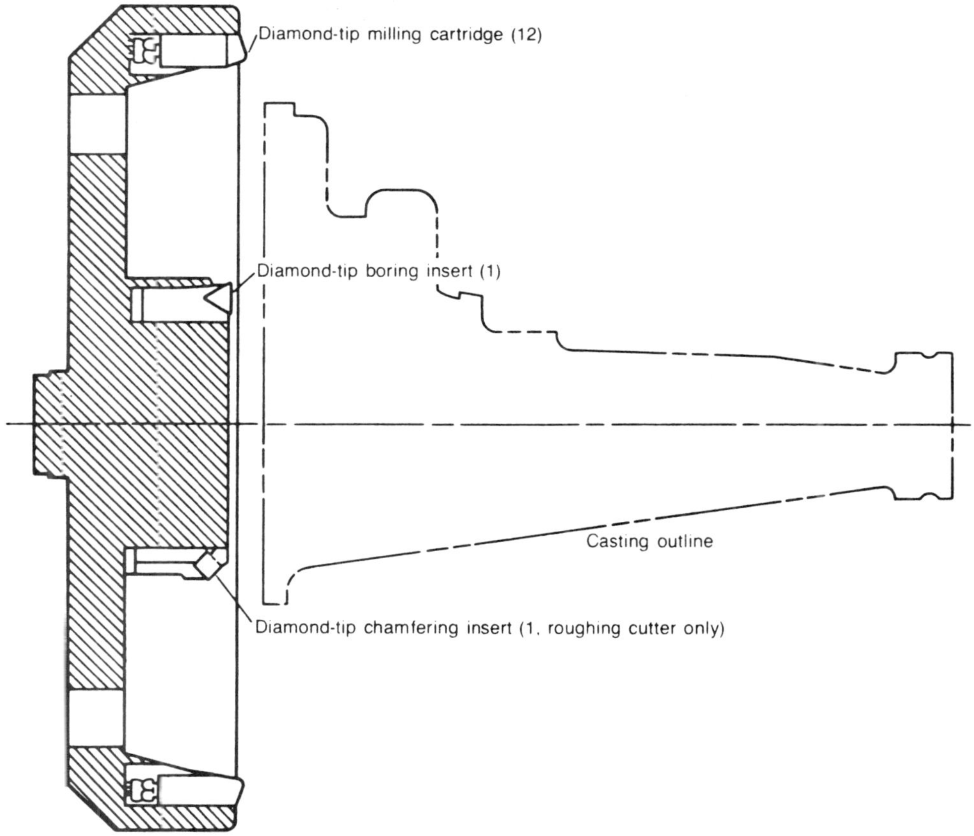


**Fig. 31** Schematic of a combination boring-milling cutter with diamond-tip cutters machining a 380 aluminum alloy transmission extension housing at 3840 m/min (12 600 sfm)

clearance in the matching toolholder) while maintaining stronger edge geometry.

Although the most common geometry for positive-rake inserts provides a relief of 10 to 11°, which is decreased by the top rake angle at which it is held, it is also possible to specify carbide inserts with higher relief angles. One supplier of these inserts recommends a 20° relief insert for turning applications and 26° for milling—and without any chip control groove, especially for the gummier alloys. This allows the top surface of the insert to be lapped or polished to the high finish desirable for improved chip flow.

Special end mills and face mills designed with two carbide inserts for high milling speeds can be used to reduce milling time. Cutters for operations at high speeds should be statically and dynamically balanced.

The carbide inserts in the cutter shown in Fig. 24 can be adjusted for the depth of cut and the most effective rake angle. The radial and axial rake angles of the end milling cutter shown in Fig. 25 result from the position of the insert in the body of the cutter. This cutter removed a large amount of aluminum and incorporates adequate chip space.

**The number of teeth** in a cutter is largely determined by its diameter, the depth of the cut, and the type of cut (rough or finishing cut). Various formulas have been developed by cutter manufacturers to determine the optimum number of teeth for all types and sizes of milling cutters for milling aluminum. For good cutter performance, as a general rule, not more than two of the teeth should engage the cut at the same time. The teeth should have a strong cross section but ample chip space.

**Speed.** Nominal speeds for peripheral, face (high-speed, carbide, and diamond), and end milling are given in Tables 25 to 28. Factors affecting speed are the alloy being milled, tool material, and depth of cut (roughing or finishing). A major variable that is not reflected in Tables 25 to 28 is chip disposal. If the chips can be adequately ejected from the cutting zone, speeds for carbide cutters are limited mainly by the capabilities of the machine. However, when the speed is too high for a particular setup, the cutter teeth may not have sufficient time to remove stock at the proper rate because of erratic feeding; the cutter teeth may dig and ride, or cut and skip. Optimum speed depends on the machine, work metal, depth of cut, and power.

Under certain conditions, milling speeds as high as 4600 m/min (15 000 sfm) have been used. However, speeds this high are rarely feasible. Speeds of 600 to 1200 m/min (2000 to 4000 sfm) are frequently used, as illustrated in the following example.

**Example 8: Face Milling Parallel Sides Simultaneously in a Duplex Machine.** Face milling two opposite sides was the first machining operation performed on the casting shown in Fig. 26. Maintaining required dimensions in subsequent operations depended greatly on the accuracy obtained in milling the sides. Parallelism was easily maintained by milling the two sides simultaneously because the workpiece did not have to be moved and reclamped.

A 7.5 kW (10 hp) duplex machine was used for this operation. This type of machine afforded two advantages in addition to good dimensional control. First, after the initial cut, the spindles were advanced slightly to take a skim (finish) cut as the table returned to the start position. Second, milling both sides at one time increased productivity, compared to milling each side separately. Details of the milling cutters and operation are tabulated below Fig. 26.

**Feed.** Nominal feeds for peripheral, face, and end milling are given in Tables 25 to 28. For wheel-type cutters (peripheral and face), the alloy being milled has little effect on feed. Depth of cut has some effect, although usually this is not large. Some reduction in feed is usually made when high-speed tool steel cutters are replaced by

**Table 29 Nominal speeds and feeds for circular sawing of aluminum alloys with high-speed tool steel blades**

| Material | Hardness, HB (500 kg) | Condition | Solid stock diameter or thickness, mm (in.) | Pitch, mm/tooth (in./tooth) | Cutting speed, m/min (sfm) | Feed, mm/tooth (in./tooth) | High-speed tool steel material grade, ISO (AISI) |
|---|---|---|---|---|---|---|---|
| **Wrought** | | | | | | | |
| EC, 2218, 5252, 6253, 1060, 2219, 5254, 6262, 1100, 2618, 5454, 6463, 1145, 3003, 5456, 6951, 1175, 3004, 5457, 7001, 1235, 3005, 5652, 7004, 2011, 4032, 5657, 7005, 2014, 5005, 6053, 7039, 2017, 5050, 6061, 7049, 2018, 5052, 6063, 7050, 2021, 5056, 6066, 7075, 2024, 5083, 6070, 7079, 2025, 5086, 6101, 7175, 2117, 5154, 6151, 7178 | 30–80 | Cold drawn | 6–80 (¼–3) | 5–30 (0.20–1.10) | 305 (1000) | 0.25 (0.010) | S4, S2 (M2, M7) |
| | | | 80–160 (3–6) | 18–40 (0.70–1.50) | 245 (800) | 0.25 (0.010) | |
| | | | 160–250 (6–9) | 25–45 (1.00–1.80) | 185 (600) | 0.30 (0.012) | |
| | | | 250–400 (9–15) | 30–60 (1.20–2.40) | 150 (500) | 0.30 (0.012) | |
| | 75–150 | Solution treated and aged | 6–80 (¼–3) | 5–30 (0.20–1.10) | 305 (1000) | 0.25 (0.010) | S4, S2 (M2, M7) |
| | | | 80–160 (3–6) | 18–40 (0.70–1.50) | 245 (800) | 0.25 (0.010) | |
| | | | 160–250 (6–9) | 25–45 (1.00–1.80) | 185 (600) | 0.30 (0.012) | |
| | | | 250–400 (9–15) | 30–60 (1.20–2.40) | 150 (500) | 0.30 (0.012) | |
| **Cast** | | | | | | | |
| **Sand and permanent mold** | | | | | | | |
| A140, 319.0, 357.0, A712.0, 201.0, 328.0, 359.0, D712.0, 208.0, A332.0, B443.0, 713.0, 213.0, F332.0, 514.0, 771.0, 222.0, 333.0, A514.0, 850.0, 224.0, 354.0, B514.0, A850.0, 242.0, 355.0, 520.0, B850.0, 295.0, C355.0, 535.0, B295.0, 356.0, 705.0, 308.0, A356.0, 707.0, Hiduminium RR-350 | 40–100 | As-cast | 6–80 (¼–3) | 5–30 (0.20–1.10) | 380 (1250) | 0.25 (0.010) | S4, S2 (M2, M7) |
| | | | 80–160 (3–6) | 18–40 (0.70–1.50) | 305 (1000) | 0.25 (0.010) | |
| | | | 160–250 (6–9) | 25–45 (1.00–1.80) | 230 (750) | 0.30 (0.012) | |
| | | | 250–400 (9–15) | 30–60 (1.20–2.40) | 185 (600) | 0.30 (0.012) | |
| | 70–125 | Solution treated and aged | 6–80 (¼–3) | 5–30 (0.20–1.10) | 305 (1000) | 0.25 (0.010) | S4, S2 (M2, M7) |
| | | | 80–160 (3–6) | 18–40 (0.70–1.50) | 245 (800) | 0.25 (0.010) | |
| | | | 160–250 (6–9) | 25–45 (1.00–1.80) | 185 (600) | 0.30 (0.012) | |
| | | | 250–400 (9–15) | 30–60 (1.20–2.40) | 150 (500) | 0.30 (0.012) | |
| **Die castings** | | | | | | | |
| 360.0, A380.0, A360.0, C443.0, 380.0, 518.0 | 40–100 | As-cast | 6–80 (¼–3) | 5–30 (0.20–1.10) | 380 (1250) | 0.25 (0.010) | S4, S2 (M2, M7) |
| | 70–125 | Solution treated and aged | 6–80 (¼–3) | 5–30 (0.20–1.10) | 305 (1000) | 0.25 (0.010) | S4, S2 (M2, M7) |
| 383.0, A384.0, 413.0, A413.0 | 40–100 | As-cast | 6–80 (¼–3) | 5–30 (0.20–1.10) | 380 (1250) | 0.25 (0.010) | S4, S2 (M2, M7) |
| | 70–125 | Solution treated and aged | 6–80 (¼–3) | 5–30 (0.20–1.10) | 305 (1000) | 0.25 (0.010) | S4, S2 (M2, M7) |
| 390.0, 392.0 | 40–100 | As-cast | 6–80 (¼–3) | 5–30 (0.20–1.10) | 150 (500) | 0.25 (0.010) | S4, S2 (M2, M7) |
| | 70–125 | Solution treated and aged | 6–80 (¼–3) | 5–30 (0.20–1.10) | 150 (500) | 0.25 (0.010) | S4, S2 (M2, M7) |

Source: Ref 5

**Table 30 Nominal speeds and feeds for the circular sawing of aluminum alloys with carbide-tip tools**

Quantity of available data was limited. Performance data are presented instead of the general recommendations given in other machining operations in this Volume.

| Hardness, HB (500 kg) | Condition | Solid stock diameter or thickness | | Pitch | | Feed | | Cutting speed | | Carbide tool material grade | |
|---|---|---|---|---|---|---|---|---|---|---|---|
| | | mm | in. | mm/tooth | in./tooth | mm/tooth | in./tooth | m/min | sfm | ISO | C |
| <60 | Cold drawn | 6–25 | 0.25–1 | 15–25 | 0.60–1.00 | 0.25 | 0.010 | 4575 | 15 000 | K30 | C-2 |
| | | 25–50 | 1–2 | 23–35 | 0.90–1.40 | 0.25 | 0.010 | 4575 | 15 000 | | |
| | | 50–100 | 2–4 | 25–50 | 1.00–2.00 | 0.13 | 0.005 | 3650 | 12 000 | | |
| | | 100–150 | 4–6 | 40–65 | 1.50–2.50 | 0.13 | 0.005 | 3650 | 12 000 | | |
| 60–150 | Solution treated and aged | 6–25 | 0.25–1 | 15–25 | 0.60–1.00 | 0.18 | 0.007 | 3650 | 12 000 | K30 | C-2 |
| | | 25–50 | 1–2 | 23–35 | 0.90–1.40 | 0.18 | 0.007 | 3650 | 12 000 | | |
| | | 50–100 | 2–4 | 25–50 | 1.00–2.00 | 0.075 | 0.003 | 3050 | 10 000 | | |
| | | 100–150 | 4–6 | 40–65 | 1.50–2.50 | 0.075 | 0.003 | 3050 | 10 000 | | |
| >150 | Solution treated and aged | 6–25 | 0.25–1 | 15–25 | 0.60–1.00 | 0.13 | 0.005 | 3050 | 10 000 | K30 | C-2 |
| | | 25–50 | 1–2 | 23–35 | 0.90–1.40 | 0.13 | 0.005 | 3050 | 10 000 | | |
| | | 50–100 | 2–4 | 25–50 | 1.00–2.00 | 0.050 | 0.002 | 2750 | 9 000 | | |
| | | 100–150 | 4–6 | 40–65 | 1.50–2.50 | 0.050 | 0.002 | 2750 | 9 000 | | |

Source: Ref 5

**Table 31 Nominal speeds for the power band sawing of aluminum alloys with high-speed tool steel blades**

| Material | Hardness, HB (500 kg) | Condition | Material thickness mm | Material thickness in. | Tooth form(a) | Thread designation Pitch, mm | Thread designation Teeth/in. | Band speed m/min | Band speed sfm |
|---|---|---|---|---|---|---|---|---|---|
| **Wrought** | | | | | | | | | |
| EC 2218 5252 6253 | 30–80 | Cold drawn | <12 | <½ | P | 1.8–1.4 | 14–18 | 305(b) | 1000(b) |
| 1060 2219 5254 6262 | | | 12–25 | ½–1 | P | 2.5–1.8 | 10–14 | 275(b) | 900(b) |
| 1100 2618 5454 6463 | | | 25–75 | 1–3 | P | 4–2.5 | 6–10 | 230(b) | 750(b) |
| 1145 3003 5456 6951 | | | >75 | >3 | C | 8.5 | 3 | 200(b) | 650(b) |
| 1175 3004 5457 7001 | 75–150 | Solution treated and aged | <12 | <½ | P | 1.8–1.4 | 14–18 | 275(b) | 900(b) |
| 1235 3005 5652 7004 | | | 12–25 | ½–1 | P | 2.5–1.8 | 10–14 | 245(b) | 800(b) |
| 2011 4032 5657 7005 | | | 25–75 | 1–3 | P | 4–2.5 | 6–10 | 200(b) | 650(b) |
| 2014 5005 6053 7039 | | | >75 | >3 | C | 8.5 | 3 | 170(b) | 550(b) |
| 2017 5050 6061 7049 | | | | | | | | | |
| 2018 5052 6063 7050 | | | | | | | | | |
| 2021 5056 6066 7075 | | | | | | | | | |
| 2024 5083 6070 7079 | | | | | | | | | |
| 2025 5086 6101 7175 | | | | | | | | | |
| 2117 5154 6151 7178 | | | | | | | | | |
| **Cast** | | | | | | | | | |
| **Sand and permanent mold** | | | | | | | | | |
| A140 319.0 357.0 A712.0 | 40–100 | As-cast | <12 | <½ | P | 1.8–1.4 | 14–18 | 305(b) | 1000(b) |
| 201.0 328.0 359.0 D712.0 | | | 12–25 | ½–1 | P | 2.5–1.8 | 10–14 | 275(b) | 900(b) |
| 208.0 A332.0 B443.0 713.0 | | | 25–75 | 1–3 | P | 4–2.5 | 6–10 | 230(b) | 750(b) |
| 213.0 F332.0 514.0 771.0 | | | >75 | >3 | C | 8.5 | 3 | 200(b) | 650(b) |
| 222.0 333.0 A514.0 850.0 | 70–125 | Solution treated and aged | <12 | <½ | P | 1.8–1.4 | 14–18 | 275(b) | 900(b) |
| 224.0 354.0 B514.0 A850.0 | | | 12–25 | ½–1 | P | 2.5–1.8 | 10–14 | 245(b) | 800(b) |
| 242.0 355.0 520.0 B850.0 | | | 25–75 | 1–3 | P | 4–2.5 | 6–10 | 200(b) | 650(b) |
| 295.0 C355.0 535.0 | | | >75 | >3 | C | 8.5 | 3 | 170(b) | 550(b) |
| B295.0 356.0 705.0 | | | | | | | | | |
| 308.0 A356.0 707.0 | | | | | | | | | |
| Hiduminium RR-350 | | | | | | | | | |
| **Die castings** | | | | | | | | | |
| 360.0 | 40–100 | As-cast | <12 | <½ | P | 1.8–1.4 | 14–18 | 200(b) | 650(b) |
| A360.0 | | | 12–25 | ½–1 | P | 2.5–1.8 | 10–14 | 185(b) | 600(b) |
| 380.0 | | | 25–75 | 1–3 | P | 4–2.5 | 6–10 | 150(b) | 500(b) |
| A380.0 | | | >75 | >3 | ... | ... | ... | ... | ... |
| C443.0 | 70–125 | Solution treated and aged | <12 | <½ | P | 1.8–1.4 | 14–18 | 170(b) | 550(b) |
| 518.0 | | | 12–25 | ½–1 | P | 2.5–1.8 | 10–14 | 150(b) | 500(b) |
| | | | 25–75 | 1–3 | P | 4–2.5 | 6–10 | 120(b) | 400(b) |
| | | | >75 | >3 | ... | ... | ... | ... | ... |
| 383.0 | 40–100 | As-cast | <12 | <½ | P | 1.8–1.4 | 14–18 | 170(b) | 550(b) |
| A384.0 | | | 12–25 | ½–1 | P | 2.5–1.8 | 10–14 | 150(b) | 500(b) |
| 413.0 | | | 25–75 | 1–3 | P | 4–2.5 | 6–10 | 120(b) | 400(b) |
| A413.0 | | | >75 | >3 | ... | ... | ... | ... | ... |
| | 70–125 | Solution treated and aged | <12 | <½ | P | 1.8–1.4 | 14–18 | 150(b) | 500(b) |
| | | | 12–25 | ½–1 | P | 2.5–1.8 | 10–14 | 135(b) | 450(b) |
| | | | 25–75 | 1–3 | P | 4–2.5 | 6–10 | 105(b) | 350(b) |
| | | | >75 | >3 | ... | ... | ... | ... | ... |
| 390.0 | 40–100 | As-cast | <12 | <½ | P | 1.8–1.4 | 14–18 | 76 | 250 |
| 392.0 | | | 12–25 | ½–1 | P | 2.5–1.8 | 10–14 | 67 | 220 |
| | | | 25–75 | 1–3 | P | 4–2.5 | 6–10 | 56 | 185 |
| | | | >75 | >3 | ... | ... | ... | ... | ... |
| | 70–125 | Solution treated and aged | <12 | <½ | P | 1.8–1.4 | 14–18 | 60 | 200 |
| | | | 12–25 | ½–1 | P | 2.5–1.8 | 10–14 | 55 | 180 |
| | | | 25–75 | 1–3 | P | 4–2.5 | 6–10 | 46 | 150 |
| | | | >75 | >3 | ... | ... | ... | ... | ... |

(a) P, precision; C, claw; B, buttress. (b) Recommended speed for high-carbon-steel blade. Source: Ref 5

carbide cutters, mainly because the speeds are usually greater for carbide cutters. Depth of cut (roughing or finishing), particularly in face milling, has some effect on the rate of feed selected. For end milling, the principal factor affecting rate of feed is cutter diameter (Table 28) because small end milling cutters lack rigidity.

**Depth of cut** is commonly about 6.35 mm (0.250 in.) for roughing and 0.64 (0.025 in.) or less for finishing. If power is available, depth of cut can be several times greater than the above when large amounts of metal are to be removed.

In machining aircraft components, skin milling, which involves deep cuts (sometimes 50 mm, or 2 in., or more), is common practice. Machines used for skin milling are basically planer mills that have evolved into elaborate tracer-controlled machines. The milling is ordinarily done by means of large peripheral-type (slabbing) cutters.

**Cutting fluid** should be supplied copiously and under pressure to the tool and workpiece. It is important, particularly with carbide cutters, that the cutting fluid be supplied uniformly and consistently to all parts of the cutter in order to prevent overheating and sudden chilling of the carbide tips. Rapid heating and cooling is extremely detrimental to the life of carbide cutters.

For high-speed milling, emulsions of soluble oil and water at a 1:15 ratio are recommended. For low-speed milling, a ratio of 1:30 is recommended.

Kerosene is effective with form cutters. When modified T-slot cutters were used to mill a blade-root slot in aluminum wheels for axial-flow compressors to a tolerance of less than 0.025 mm (0.001 in.), no cutting fluid except kerosene could produce the required accuracy and surface finish.

**Table 32 Nominal speeds and feeds for the power hacksawing of aluminum alloys with high-speed tool steel blades**

| Material | Hardness, HB (500 kg) | Condition | Material thickness, mm | Material thickness, in. | Thread designation: Pitch, mm | Thread designation: Teeth/in.(a) | Speed, strokes/min | Feed: mm/stroke | Feed: in./stroke | Feed: Pressure(b) |
|---|---|---|---|---|---|---|---|---|---|---|
| **Wrought** | | | | | | | | | | |
| EC 2218 5252 6253 | 30–80 | Cold drawn | <6 | <¼ | 4 | 6 | 160 | 0.15 | 0.006 | M |
| 1060 2219 5254 6262 | | | 6–18 | ¼–¾ | 4 | 6 | 160 | 0.23 | 0.009 | H |
| 1100 2618 5454 6463 | | | 18–50 | ¾–2 | 6.3 | 4 | 160 | 0.30 | 0.012 | H |
| 1145 3003 5456 6951 | | | >50 | >2 | 6.3 | 4 | 160 | 0.30 | 0.012 | H |
| 1175 3004 5457 7001 | 75–150 | Solution treated and aged | <6 | <¼ | 4 | 6 | 140 | 0.15 | 0.006 | M |
| 1235 3005 5652 7004 | | | 6–18 | ¼–¾ | 4 | 6 | 140 | 0.23 | 0.009 | H |
| 2011 4032 5657 7005 | | | 18–50 | ¾–2 | 6.3 | 4 | 140 | 0.30 | 0.012 | H |
| 2014 5005 6053 7039 | | | >50 | >2 | 6.3 | 4 | 140 | 0.30 | 0.012 | H |
| 2017 5050 6061 7049 | | | | | | | | | | |
| 2018 5052 6063 7050 | | | | | | | | | | |
| 2021 5056 6066 7075 | | | | | | | | | | |
| 2024 5083 6070 7079 | | | | | | | | | | |
| 2025 5086 6101 7175 | | | | | | | | | | |
| 2117 5154 6151 7178 | | | | | | | | | | |
| **Cast** | | | | | | | | | | |
| **Sand and permanent mold** | | | | | | | | | | |
| A140 319.0 357.0 A712.0 | 40–100 | As-cast | <6 | <¼ | 4 | 6 | 140 | 0.15 | 0.006 | M |
| 201.0 328.0 359.0 D712.0 | | | 6–18 | ¼–¾ | 4 | 6 | 140 | 0.23 | 0.009 | H |
| 208.0 A332.0 B443.0 713.0 | | | 18–50 | ¾–2 | 6.3 | 4 | 140 | 0.30 | 0.012 | H |
| 213.0 F332.0 514.0 771.0 | | | >50 | >2 | 6.3 | 4 | 140 | 0.30 | 0.012 | H |
| 222.0 333.0 A514.0 850.0 | 70–125 | Solution treated and aged | <6 | <¼ | 2.5 | 10 | 130 | 0.15 | 0.006 | M |
| 224.0 354.0 B514.0 A850.0 | | | 6–18 | ¼–¾ | 4 | 6 | 130 | 0.23 | 0.009 | H |
| 242.0 355.0 520.0 B850.0 | | | 18–50 | ¾–2 | 6.3 | 4 | 130 | 0.23 | 0.009 | H |
| 295.0 C355.0 535.0 | | | >50 | >2 | 6.3 | 4 | 130 | 0.30 | 0.012 | H |
| B295.0 356.0 705.0 | | | | | | | | | | |
| 308.0 A356.0 707.0 | | | | | | | | | | |
| Hiduminium RR-350 | | | | | | | | | | |
| **Die castings** | | | | | | | | | | |
| 360.0 A380.0 | 40–100 | As-cast | <6 | <¼ | 4 | 6 | 140 | 0.15 | 0.006 | M |
| A360.0 C443.0 | | | 6–18 | ¼–¾ | 4 | 6 | 140 | 0.23 | 0.009 | H |
| 380.0 518.0 | | | 18–50 | ¾–2 | 6.3 | 4 | 140 | 0.30 | 0.012 | H |
| | | | >50 | >2 | 6.3 | 4 | 140 | 0.30 | 0.012 | H |
| | 70–125 | Solution treated and aged | <6 | <¼ | 2.5 | 10 | 130 | 0.15 | 0.006 | M |
| | | | 6–18 | ¼–¾ | 4 | 6 | 130 | 0.23 | 0.009 | H |
| | | | 18–50 | ¾–2 | 6.3 | 4 | 130 | 0.23 | 0.009 | H |
| | | | >50 | >2 | 6.3 | 4 | 130 | 0.30 | 0.012 | H |
| 383.0 | 40–100 | As-cast | <6 | <¼ | 2.5 | 10 | 130 | 0.15 | 0.006 | M |
| A384.0 | | | 6–18 | ¼–¾ | 4 | 6 | 130 | 0.23 | 0.009 | H |
| 413.0 | | | 18–50 | ¾–2 | 6.3 | 4 | 130 | 0.23 | 0.009 | H |
| A413.0 | | | >50 | >2 | 6.3 | 4 | 130 | 0.30 | 0.012 | H |
| | 70–125 | Solution treated and aged | <6 | <¼ | 2.5 | 10 | 130 | 0.15 | 0.006 | M |
| | | | 6–18 | ¼–¾ | 4 | 6 | 130 | 0.23 | 0.009 | H |
| | | | 18–50 | ¾–2 | 6.3 | 4 | 130 | 0.23 | 0.009 | H |
| | | | >50 | >2 | 6.3 | 4 | 130 | 0.23 | 0.009 | H |
| 390.0 | 40–100 | As-cast | <6 | <¼ | 2.5 | 10 | 125 | 0.15 | 0.006 | M |
| 392.0 | | | 6–18 | ¼–¾ | 4 | 6 | 125 | 0.15 | 0.006 | M |
| | | | 18–50 | ¾–2 | 6.3 | 4 | 115 | 0.15 | 0.006 | M |
| | | | >50 | >2 | 6.3 | 4 | 115 | 0.15 | 0.006 | M |
| | 70–125 | Solution treated and aged | <6 | <¼ | 2.5 | 10 | 120 | 0.075 | 0.003 | L |
| | | | 6–18 | ¼–¾ | 4 | 6 | 120 | 0.15 | 0.006 | M |
| | | | 18–50 | ¾–2 | 4 | 6 | 110 | 0.15 | 0.006 | M |
| | | | >50 | >2 | 6.3 | 4 | 110 | 0.15 | 0.006 | M |

(a) Conversions per ISO 2336. (b) L, light; M, medium; H, heavy. Source: Ref 5

**Form milling** is often an economical method for machining complicated shapes in aluminum. The complexity of form that can be built into the design of a cutter is limited, as when the tangent of the curvature approaches 90° to the horizontal plane. As a result, secondary operations are sometimes required to complete a complex form.

Basic types of form cutters have cam-relieved teeth or shaped teeth. Cam-relieved teeth are sharpened by grinding the tooth faces without changing the form. If surface finish is a secondary consideration, straight-tooth cutters are used because they are easier to sharpen. However, if surface finish is important, helical cutters must be used. Although helical cutters can be operated at higher speeds and greater feeds than straight-tooth cutters, they require considerably greater care when being sharpened.

Shape form cutters are made either with integral teeth or blade inserts. These cutters are less expensive than integral-tooth, cam-relieved cutters, but they require more equipment and more skill in grinding.

The feasibility of form milling depends on several variables, including cutter cost, volume of production, design of workpiece, and type of milling machine available. Thus, for small production lots, the cost of a single form cutter must be compared with the cost of several cutters of less complex shape. For form milling, the workpiece must be relatively large, must have no thin sections, and must be capable of being well secured.

Finally, form milling requires high-power equipment that is rugged and in good operating condition. The following example describes an application of form milling and shows the production quantity required to justify a high-cost form cutter.

**Example 9: Cost of Standard Versus Form Milling.** The contour of the workpiece shown in Fig. 27 could be milled equally well by using five standard cutters for five separate operations or by using one specially designed form cutter that could mill the contour in a single operation. The operations performed by both methods are listed in the table accompanying Fig. 27. Although the cost of the cutters was the same for form and standard milling, the time required for form milling was only about one-third that for standard milling.

**Automatic Control.** The generally good machining characteristics of aluminum alloys make them especially well adapted to automatic control, by which extremely complex shapes can be milled efficiently. The following example gives details of an application in which automatic control was used.

**Example 10: Milling a 758 kg (1670 lb) Forging With Numerical Control.** The part shown in Fig. 28 was made from an alloy 7075 blocker forging that weighed 758 kg (1670 lb). Overall measurements of the forging were 3380 × 1730 × 406 mm (133 × 68 × 16 in.), and those of the finished part were 3330 × 1680 × 356 mm (131 × 66 × 14 in.). The weight of the finished part was 178 kg (393 lb), with 579 kg (1277 lb) of metal having been removed in machining.

With the part held in a special milling fixture, rough profile machining was performed in a tape-controlled three-axis profile milling machine. A special boring fixture was used for rough line boring the 140 and 222 mm (5.50 and 8.75 in.) diam trunnion holes. After roughing cuts were completed, approximately 3.05 mm (0.120 in.) of metal remained to be removed by finish machining after the forging was heat treated to the T6 temper. Special fixtures prevented warpage during heat treating and in finish milling. Machining was completed using the same NC machine. Details of the process, machine, and cutter are given with Fig. 28.

**Dimensional accuracy** is affected by residual stress in the workpiece, induced stress caused by milling, built-up edge on cutters, and dull cutters. Workpieces with complex shapes or variable section thickness, especially in the aged condition, are more difficult to mill to close tolerances. To achieve dimensional stability and to maintain close tolerances, residual stress must be avoided.

**Milling of high-silicon alloy 390** (Ref 2) is accomplished with carbide tools at speeds up to 175 m/min (575 sfm) or with polycrystalline diamond tools at speeds up to 790 m/min (2600 sfm). Moderate positive-rake and

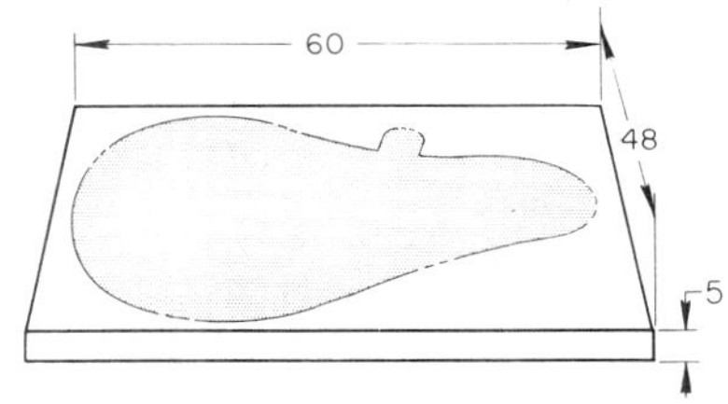


**Fig. 32** Shape that was contour band sawed from aluminum plate. Dimensions given in inches

shear angles should be used. Optimum tool life will be achieved if a coolant is used, but alloy 390 can be machined dry with good tool life and with little built-up edge problem.

Compared to other aluminum alloys, alloy 390 has traditionally been milled at rather slow speeds (90 to 175 m/min, or 300 to 575 sfm) using C-2 grade tungsten carbide inserts with a liberal application of flood coolant. Feed rates of 0.10 to 0.30 mm/tooth (0.004 to 0.012 in./tooth) and cutting depths ranging from 0.13 to 5.1 mm (0.005 to 0.20 in.) are common practice.

Cutter geometry is important for ensuring adequate surface finish of the machined casting and a long tool life for the cutting inserts. The entering angle should be 45° to minimize vibration and workpiece frittering. A large positive-rake angle, from 5 to 20°, extends tool life and reduces the tendency toward built-up edge. Clearance angles from 5 to 15° reduce buildup on the edge clearance flank. Pitch depends largely on the diameter of the cutter, depth of cut, and type of cut (roughing or finishing). As with turning, the productivity and tool life performance of the milling operation can be greatly improved through substitution of polycrystalline diamond for carbide.

The first large-volume alloy 390 engine manufactured in the United States was machined on a transfer line designed for gray iron. Instead of broaching, face milling was employed, with an average production tool life of 130 engine blocks per carbide tool.

When compared experimentally with polycrystalline diamond (Fig. 29), the performance of the milling operation was greatly enhanced. At ten times the speed and metal removal rate, tool life would be projected to be in excess of 1000 engine blocks. At the same speed and production rate used for carbide, the diamond tool life would be practically infinite.

This performance improvement was also experienced in a study in which alloy 390 engine blocks were face milled without coolant (Fig. 30). The best tool life achieved when milling with carbide from 100 to 175 m/min (328 to 574 sfm) was approximately 50 engine blocks at lower speed. The feed rate was 0.18 mm/tooth (0.007 in./tooth), and the depth of cut was 1.02 mm (0.040 in.) for roughing and 0.51 mm (0.020 in.) for finishing cuts. The best tool life was achieved using H10 grade cemented carbide. The tool life when dry milling with polycrystalline diamond at speeds ranging from 400 to 800 m/min (1312 to 2625 sfm) was approximately 1000 engine blocks at the lower speed. Surface finish was also greatly improved, indicating that polycrystalline diamond is much less sensitive than carbide to the effect of coolant when milling alloy 390. These results do not indicate that the application of coolant is not desirable to prolong tool life, but it does mean that milling with polycrystalline diamond without coolant, if desired, can be a commercial reality.

**Milling of Alloy 380 With Polycrystalline Cutting Tools (Ref 1).** An interesting application of diamond tooling is the machining of an aluminum automotive extension housing made of die cast 380 aluminum alloy, which contains approximately 8% Si. The component is about 330 mm (13 in.) long, tapering from about 203 × 254 mm ( 8 × 10 in.) at one end to a diameter of about 75 mm (3 in.) at the other. The overall shape of the casting is relatively complex, and the nominal wall thickness over most of its area is about 4.0 mm (5⁄32 in.). A 14-station pallet-type transfer machine performs a number of drilling, tapping, reaming, chamfering, boring, and facing operations on the transmission part.

The large end of the casting requires the machining of a shallow, interruped bore of about 140 mm (5½ in.) in diameter and 9.5 mm (3⁄8 in.) deep and the facing (milling) of the entire end. To maintain perpendicularity, the boring and the milling tools are combined into a single cutter body. Two of these are used—one for roughing (which also incorporates a tool for chamfering the bore inside diameter), and one for finishing.

The overall configuration of the cutter is essentially a hollow mill with a central boring bar. Tool and workpiece are both shown in Fig. 31. This is fed axially into the part to produce the bore and the chamfer, while the milling teeth rotate beyond the periphery of the part. The cutter is then withdrawn from the bore and fed radially across the surface to be face milled. The milling teeth are positioned farther out axially than the boring and chamfering teeth so that there is no interference.

All cutting edges are polycrystalline diamond. The boring tool is a special diamond-tip 9.5 mm (3⁄8 in.) inscribed circle triangular insert clamped in place with 2° radial rake and 5° axial rake. The chamfering tool is a 9.5 mm (3⁄8 in.) square insert with a brazed diamond cutting edge. This insert is presented to the cut with 0° radial and 5° axial rake. Although chamfering is done only at the roughing station, a dummy cartridge is used to maintain balance of the finishing cutter—a necessity at these speeds.

Twelve milling teeth are set on an effective diameter of 356 mm (14 in.). These are

## Table 33 Wheel recommendations for grinding aluminum alloys

| Material | Hardness, HB (500 kg) | Condition | Surface grinding: Horizontal spindle, reciprocating table: Wheel speed, m/s (sfm) | Table speed, m/min (sfm) | Down feed, mm/pass (in./pass) | Cross feed, mm/pass (in./pass) | Wheel identification, ISO (ANSI)(a) | Surface grinding: Vertical spindle, rotary table: Wheel speed, m/s (sfm) | Table (work) speed, m/min (sfm) | Down feed per revolution of table, mm (in.) | Operation | Wheel identification: Narrow work area, ISO (ANSI) | Wheel identification: Broad work area, ISO (ANSI) |
|---|---|---|---|---|---|---|---|---|---|---|---|---|---|
| **Wrought** | | | | | | | | | | | | | |
| EC, 1060, 1100, 1145, 1175, 1235, 2011, 2014, 2017, 2018, 2021, 2024, 2025, 2117, 2218, 2219, 2618, 3003, 3004, 3005, 4032, 5005, 5050, 5052, 5056, 5083, 5086, 5154, 5252, 5254, 5454, 5456, 5457, 5652, 5657, 6053, 6061, 6063, 6066, 6070, 6101, 6151, 6253, 6262, 6463, 6951, 7001, 7004, 7005, 7039, 7049, 7050, 7075, 7079, 7175, 7178 | 30–150 | Cold drawn or solution treated and aged | 20–25 (4000–5000) | 15–30 (50–100) | Rough: 0.075 (0.003)<br>Finish: 0.025 max (0.001 max) | 1.25–12.5 (0.050–0.500) (Max: ⅓ of wheel width) | C46JV(b) (C46JV)(b) | 18–30 (3500–6000) | 30–76 (100–250) | 0.025–0.100 (0.001–0.004) | Roughing<br>Finishing | CA46IB (CA46IB)<br>CA60IB (CA601B) | CA46GB (CA46GB)<br>CA60GB (CA60GB) |
| **Cast** | | | | | | | | | | | | | |
| **Sand and permanent mold**<br>A140, 201.0, 208.0, 213.0, 222.0, 224.0, 242.0, 295.0, B295.0, 308.0, 319.0, 355.0, C355.0, B443.0, 514.0, A514.0, B514.0, 520.0, 535.0, 705.0, 707.0, A712.0, D712.0, 713.0, 771.0, 850.0, A850.0, B850.0, Hiduminium RR-350<br>**Die castings**<br>C443.0, 518.0 | 40–125 | As-cast or solution treated and aged | 20–25 (4000–5000) | 15–30 (50–100) | Rough: 0.075 (0.003)<br>Finish: 0.025 max (0.001 max) | 1.25–12.5 (0.050–0.500) (Max: ⅓ of wheel width) | C46JV(b) (C46JV)(b) | 18–30 (3500–6000) | 30–76 (100–250) | 0.025–0.100 (0.001–0.004) | Roughing<br>Finishing | CA46IB (CA46IB)<br>CA60IB (CA601B) | CA46GB (CA46GB)<br>CA60GB (CA60GB) |
| **Sand and permanent mold**<br>328.0, A332.0, F332.0, 333.0, 354.0, 356.0, A356.0, 357.0, 359.0<br>**Die castings**<br>360.0, A360.0, 380.0, A380.0, 383.0, A384.0, 390.0, 392.0, 413.0, A413.0 | 40–125 | As-cast or solution treated and aged | 20–25 (4000–5000) | 15–30 (50–100) | Rough: 0.075 (0.003)<br>Finish: 0.025 max (0.001 max) | 1.25–12.5 (0.050–0.500) (Max: ⅓ of wheel width) | A46IV (A46IV) | 18–30 (3500–6000) | 30–76 (100–250) | 0.025–0.100 (0.001–0.004) | Roughing<br>Finishing | CA46IB (CA46IB)<br>CA60IB (CA601B) | CA46GB (CA46GB)<br>CA60GB (CA60GB) |

(continued)

(a) Wheel recommendations are for wet grinding. For dry grinding, use a softer-grade wheel. (b) Wax filled. (c) Wheel recommendations are for the wet grinding of 50–100 mm (2–4 in.) diam work. For dry grinding, use a softer-grade wheel. For larger-diameter work, use a softer-grade and/or coarser-grit wheel. For smaller-diameter work, use a harder-grade wheel. Wheel recommendations also apply to plunge grinding applications. (d) As recommended starting conditions, use a regulating wheel angle with a positive inclination of 3° and a regulating wheel speed of 25–40 rev/min. (e) Wheel recommendations are for wet grinding 20–50 mm (0.8–2 in.) diam work. For larger-diameter work, use a softer-grade and/or coarser-grit wheel. For smaller-diameter work, use a harder-grade wheel. (f) Maximum hole length is 2.5 times hole diameter. (g) Maximum wheel width is 1.5 times wheel diameter. (h) Wheel recommendations are for wet grinding 20–50 mm (0.8–2 in.) diam holes. For larger holes, use the same or softer-grade wheel. For smaller holes, use a harder-grade wheel. Source: Ref 5

## Table 33 (continued)

| Material | Hardness, HB (500 kg) | Condition | Cylindrical grinding: Wheel speed, m/s (sfm) | Cylindrical grinding: Work speed, m/min (sfm) | Cylindrical grinding: Infeed on diameter, mm/pass (in./pass) | Cylindrical grinding: Traverse wheel width per revolution of work | Cylindrical grinding: Wheel identification, ISO (ANSI)(c) | Centerless grinding: Wheel speed, m/s (sfm) | Centerless grinding: Through feed of work, m/min (in./min)(d) | Centerless grinding: Infeed on diameter, mm/pass (in./pass) | Centerless grinding: Wheel identification, ISO (ANSI)(e) | Internal grinding(f)(g): Wheel speed, m/s (sfm) | Internal grinding(f)(g): Work speed, m/min (sfm) | Internal grinding(f)(g): Infeed on diameter, mm/pass (in./pass) | Internal grinding(f)(g): Traverse wheel width per revolution of work | Internal grinding(f)(g): Wheel identification, ISO (ANSI)(h) |
|---|---|---|---|---|---|---|---|---|---|---|---|---|---|---|---|---|
| **Wrought** | | | | | | | | | | | | | | | | |
| EC 2218 5252 6253<br>1060 2219 5254 6262<br>1100 2618 5454 6463<br>1145 3003 5456 6951<br>1175 3004 5457 7001<br>1235 3005 5652 7004<br>2011 4032 5657 7005<br>2014 5005 6053 7039<br>2017 5050 6061 7049<br>2018 5052 6063 7050<br>2021 5056 6066 7075<br>2024 5083 6070 7079<br>2025 5086 6101 7175<br>2117 5154 6151 7178 | 30–150 | Cold drawn or solution treated and aged | 28–33 (5500–6500) | 15–46 (50–150) | Rough: 0.050 (0.002)<br>Finish: 0.013 max (0.0005 max) | ⅓<br>⅙ | C54JV(b) (C54JV)(b) | 28–33 (5500–6500) | 1.3–3.8 (50–150) | Rough: 0.13 (0.005)<br>Finish: 0.038 max (0.0015 max) | C54JV (C54JV) | 25–33 (5000–6500) | 23–60 (75–200) | Rough: 0.075 (0.003)<br>Finish: 0.005 max (0.0002 max) | ⅓<br>⅙ | C46JV (C46JV) |
| **Cast** | | | | | | | | | | | | | | | | |
| **Sand and permanent mold**<br>A140 295.0 514.0 A712.0<br>201.0 B295.0 A514.0 D712.0<br>208.0 308.0 B514.0 713.0<br>213.0 319.0 520.0 771.0<br>222.0 355.0 535.0 850.0<br>224.0 C355.0 705.0 A850.0<br>242.0 B443.0 707.0 B850.0<br>Hiduminium RR-350<br>**Die castings**<br>C443.0 518.0 | 40–125 | As-cast or solution treated and aged | 28–33 (5500–6500) | 15–46 (50–150) | Rough: 0.050 (0.002)<br>Finish: 0.013 max (0.0005 max) | ⅓<br>⅙ | C54JV(b) (C54JV)(b) | 28–33 (5500–6500) | 1.3–3.8 (50–150) | Rough: 0.13 (0.005)<br>Finish: 0.038 max (0.0015 max) | C54JV (C54JV) | 25–33 (5000–6500) | 23–60 (75–200) | Rough: 0.075 (0.003)<br>Finish: 0.005 max (0.0002 max) | ⅓<br>⅙ | C46JV (C46JV) |
| **Sand and permanent mold**<br>328.0 333.0 A356.0<br>A332.0 354.0 357.0<br>F332.0 356.0 359.0<br>**Die castings**<br>360.0 A360.0 390.0 A413.0<br>A360.0 383.0 392.0<br>380.0 A384.0 413.0 | 40–125 | As-cast or solution treated and aged | 28–33 (5500–6500) | 15–46 (50–150) | Rough: 0.050 (0.002)<br>Finish: 0.013 max (0.0005 max) | ⅓<br>⅙ | A54IV (A54IV) | 28–33 (5500–6500) | 1.3–3.8 (50–150) | Rough: 0.13 (0.005)<br>Finish: 0.038 max (0.0015 max) | A60IV (A60IV) | 25–33 (5000–6500) | 23–60 (75–200) | Rough: 0.075 (0.003)<br>Finish: 0.005 max (0.0002 max) | ⅓<br>⅙ | A60HV (A60HV) |

(a) Wheel recommendations are for wet grinding. For dry grinding, use a softer-grade wheel. (b) Wax filled. (c) Wheel recommendations are for the wet grinding of 50–100 mm (2–4 in.) diam work. For dry grinding, use a softer-grade wheel. For larger-diameter work, use a softer-grade and/or coarser-grit wheel. For smaller-diameter work, use a harder-grade wheel. Wheel recommendations also apply to plunge grinding applications. (d) As recommended starting conditions, use a regulating wheel angle with a positive inclination of 3° and a regulating wheel speed of 25–40 rev/min. (e) Wheel recommendations are for wet grinding 20–50 mm (0.8–2 in.) diam work. For larger-diameter work, use a softer-grade and/or coarser-grit wheel. For smaller-diameter work, use a harder-grade wheel. (f) Maximum hole length is 2.5 times hole diameter. (g) Maximum wheel width is 1.5 times wheel diameter. (h) Wheel recommendations are for wet grinding 20–50 mm (0.8–2 in.) diam holes. For larger holes, use the same or softer-grade wheel. For smaller holes, use a harder-grade wheel. Source: Ref 5

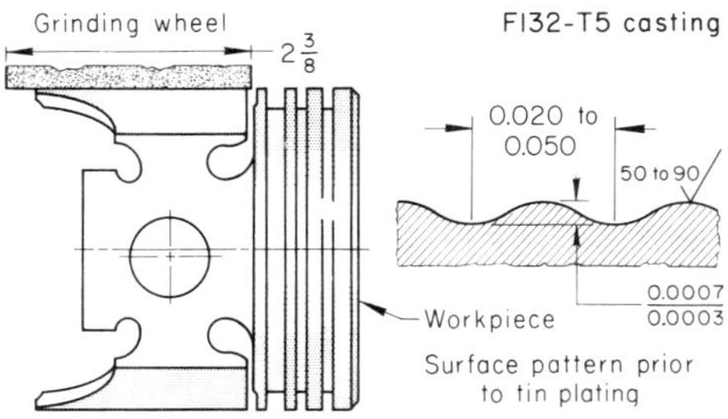


| | |
|---|---|
| Machine | Cylindrical grinder, 152 × 457 mm (6 × 18 in.), with automatic infeed electric-cycle control and semiautomatic profile dresser |
| Wheel | |
| Type | A-46-K5-V |
| Size, mm (in.) | 508 × 60 × 305 (20 × 2⅜ × 12) |
| Life per dressing | 100 pieces |
| **Operating conditions** | |
| Wheel speed, at 1300 rev/min, m/min (sfm) | 2100 (6800) |
| Work speed, rev/min | 96 |
| Feed/pass, mm (in.) | 0.41–0.46 (0.016–0.018) |
| Grinding fluid | Soluble oil:water (1:25) |
| Production rate, pieces/h | 106 |
| Time/pass, s(a) | 20 |
| Surface finish, μm (μin.) | 1.25–2.25 (50–90) |

(a) Infeed set with automatic timer

**Fig. 33** Permanent mold cast piston (90 to 120 HB) that was ground in a single pass, removing 0.41 to 0.46 mm (0.016 to 0.018 in.) of stock. Dimensions in figure given in inches

cartridge-type cutters with brazed diamond tips having the following geometry: 15° lead angle (the inner edge), 5° radial rake, 5° axial rake, 6° clearance, and 1.52 mm (0.060 in.) nose radius. On the finishing cutter, one of these is ground with a 0.76 mm (0.030 in.) wiper flat instead of the radius, which was found to improve significantly the surface finish produced on the part.

These combination cutters are run at 3450 rev/min for both roughing and finishing, yielding a cutting speed of 1514 m/min (4967 sfm) for the boring and chamfering operations and 3854 m/min (12 645 sfm) for the face milling operations. In roughing, material removal is approximately 2.03 mm (0.080 in.) in the bore and 1.65 mm (0.065 in.) on the face. In finishing, 0.38 mm (0.015 in.) is removed from both surfaces.

## Sawing

The contour cutting of aluminum alloys is usually done by band sawing; straight cutting is done on a circular saw, band saw, power hacksaw, or abrasive cutoff wheel. Circular saws or band saws are preferred for the rapid cutoff of rod and bar stock; either can be readily adapted for high-speed, automatic work handling.

Linear feed rates as high as 762 to 2030 mm/min (30 to 80 in./min) are sometimes used in the cutoff of stock 50 to 203 mm (2 to 8 in.) thick, for cutting rates of 0.032 to 0.16 $m^2$/min (50 to 250 in.$^2$/min). Recommended tooth angles and contours for the three types of saws are given in Tables 29 to 32.

**Circular Sawing.** Peripheral speeds are 45 to 4600 m/min (150 to 15 000 sfm) for high-speed steel and carbide-tip circular saws, depending on the saw material, the type of cut, and the ability of the machine to withstand high speeds (Tables 29 and 30). The limiting factor is usually the maximum safe operating speed of the machine and blade.

The feed rate for circular sawing ranges from 0.05 to 7.6 mm/tooth (0.002 to 0.30 in./tooth), depending mainly on the alloy being sawed. The width of the section being sawed and the speed of the saw have some bearing on the rate of feed.

**Band Sawing.** Hard-tooth, flexible-back carbon steel bands are used at 46 to 305 m/min (150 to 1000 sfm) on cast and wrought aluminum alloys (Table 31). The following example compares milling and contour band sawing for producing a shape from an aluminum alloy plate. In this application, band sawing was faster and less expensive.

**Example 11: Contour Sawing of a Large Plate.** An aluminum alloy plate weighing 653 kg (1440 lb) and measuring 1220 × 1525 × 127 mm (48 × 60 × 5 in.) was rough machined in a milling machine to obtain the contour shown in Fig. 32. Milling required about 4½ h. The same part was produced in 67 min in a contour band saw, cutting at a rate of 9700 $mm^2$/min (15 in.$^2$/min).

Sawing was done with a high-speed steel blade (hook tooth; raker set; 6 pitch; 25 mm, or 1 in., wide; 0.89 mm, or 0.035 in., thick), using a heavy-duty chemical solution as a cutting fluid. A dimensional tolerance of ±0.38 mm (±0.015 in.) was maintained.

**Power hacksawing** is done at 110 to 160 strokes per min with a feed of 0.15 to 0.30 mm/stroke (0.006 to 0.012 in./stroke) (Table 32). The higher speed is recommended for sawing nonheat-treated cast aluminum alloys, and the lower speed (110 strokes per min) for abrasive materials such as 390 and 392 solution-treated and aged die cast alloys.

**Cutting Fluid.** Although circular saws can be operated satisfactorily on aluminum alloys at moderate speeds and medium cuts

**Table 34 Abrasive cutoff wheels for cutting aluminum alloys**

| Material | | | | Hardness, HB (500 kg) | Condition | Wheel identification: Dry cutting ISO | Dry cutting ANSI | Wet cutting ISO | Wet cutting ANSI |
|---|---|---|---|---|---|---|---|---|---|
| **Wrought** | | | | | | | | | |
| EC | 2218 | 5252 | 6253 | 30–150 | Cold drawn or solution treated and aged | A36—RB | A36—RB | A46—PR | A46—PR |
| 1060 | 2219 | 5254 | 6262 | | | | | | |
| 1100 | 2618 | 5454 | 6463 | | | | | | |
| 1145 | 3003 | 5456 | 6951 | | | | | | |
| 1175 | 3004 | 5457 | 7001 | | | | | | |
| 1235 | 3005 | 5652 | 7004 | | | | | | |
| 2011 | 4032 | 5657 | 7005 | | | | | | |
| 2014 | 5005 | 6053 | 7039 | | | | | | |
| 2017 | 5050 | 6061 | 7049 | | | | | | |
| 2018 | 5052 | 6063 | 7050 | | | | | | |
| 2021 | 5056 | 6066 | 7075 | | | | | | |
| 2024 | 5083 | 6070 | 7079 | | | | | | |
| 2025 | 5086 | 6101 | 7175 | | | | | | |
| 2117 | 5154 | 6151 | 7178 | | | | | | |
| **Cast** | | | | | | | | | |
| **Sand and permanent mold** | | | | | | | | | |
| A140 | 295.0 | 514.0 | A712.0 | 40–125 | As-cast or solution treated and aged | A36—RB | A36—RB | A46—PR | A46—PR |
| 201.0 | B295.0 | A514.0 | D712.0 | | | | | | |
| 208.0 | 308.0 | B514.0 | 713.0 | | | | | | |
| 213.0 | 319.0 | 520.0 | 771.0 | | | | | | |
| 222.0 | 355.0 | 535.0 | 850.0 | | | | | | |
| 224.0 | C355.0 | 705.0 | A850.0 | | | | | | |
| 242.0 | B443.0 | 707.0 | B850.0 | | | | | | |
| Hiduminium RR-350 | | | | | | | | | |
| **Die castings** | | | | | | | | | |
| C443.0 | 518.0 | | | | | | | | |
| **Sand and permanent mold** | | | | | | | | | |
| 328.0 | 333.0 | A356.0 | | 40–125 | As-cast or solution treated and aged | A36—RB | A36—RB | A46—PR | A46—PR |
| A332.0 | 354.0 | 357.0 | | | | | | | |
| F332.0 | 356.0 | 359.0 | | | | | | | |
| **Die castings** | | | | | | | | | |
| 360.0 | 383.0 | 413.0 | | | | | | | |
| A360.0 | A384.0 | A413.0 | | | | | | | |
| 380.0 | 390.0 | | | | | | | | |
| A380.0 | 392.0 | | | | | | | | |

Source: Ref 5

**Table 35 Nominal speeds for the abrasive belt grinding of aluminum alloys**

| Material | Hardness, HB (500 kg) | Condition | Operation | Abrasive type | Grain size | Belt speed(a) m/s | Belt speed(a) sfm | Contact wheel Type(b) | Contact wheel Hardness, Durometer |
|---|---|---|---|---|---|---|---|---|---|
| Wrought and cast | 30–150 | Cold drawn, as-cast, or solution treated and aged | Roughing | $Al_2O_3$, SiC | 24–80 | 25–33 | 5000–6500 | SR, SFR | 70–95 |
| | | | Polishing | $Al_2O_3$, SiC | 100–240 | 25–40 | 5000–8000 | SR, SFR, B | 20–60 |

(a) Use lower values of belt speed when good surface integrity is required (b) SR, serrated rubber; SFR, smooth face rubber; B, buff type. Serrations are usually at a 45° angle, although some heavy stock removal operations use a 60° angle. Widths of lands and grooves vary from narrow lands and wide grooves for fast aggressive cuts to wide lands and narrow grooves for intermediate and finishing operations, depending on workpiece shape and operating conditions. Source: Ref 5

**Table 36 Honing recommendations for aluminum alloys**

| Material | Hardness, HB (500 kg) | Honing stone material: Type | Honing stone material: Grade, ANSI or ISO | Grain size at a surface roughness, $R_a$, µm (µin.), of: 0.025–0.125 (1–5) | 0.15–0.25 (6–10) | 0.30–0.50 (11–20) | 0.53–0.75 (21–30) | >0.75 (>30) | Spindle motion, m/min (sfm): Rotating speed | Reciprocating speed | Working pressure, kPa (psi) | Cutting fluid |
|---|---|---|---|---|---|---|---|---|---|---|---|---|
| Wrought and cast | 30–450 | Silicon carbide or diamond (metal bonded) | L(a) | 600 (600) | 500 (500) | 400 (400) | 280 (280) | 220 (220) | 15.2–64 (50–210) | 2.7–22.9 (9–75) | 276 (40) | 70–30 kerosene/oil (sulfurized or chlorinated) |

(a) L, a medium grade. Source: Ref 5

without a cutting fluid, it is advisable to supply copious amounts of soluble-oil:water emulsion (1:20) for all high-speed cutting. The cutting fluid should flood the blade and workpiece under slight pressure and it should be filtered or settled before recycling. In some applications, the addition of a little kerosene or lard oil to the emulsion has been beneficial. Soap solutions can be substituted for oil emulsions.

For band saws, some cutting fluid is essential for all but the lightest cuts. A wide selection of compounds is available, ranging from tallow or grease sticks to kerosene-thinned mineral-base lubricating oil or emulsions of soluble oil and water. It is often more convenient to use a fluid lubricant, supplied generously through a recycling system. For hacksawing, procedures are similar to those for band sawing.

## Grinding

The harder, free-cutting aluminum alloys are comparatively easy to grind. The non-free-cutting alloys, particularly in their softer tempers, are likely to clog grinding wheels, and they do not finish to as bright and smooth a surface as the harder alloys.

**Abrasive Wheels.** For grinding aluminum alloys, a silicon carbide abrasive in a flexible base is generally preferred. Aluminum oxide is seldom recommended, except for piston grinding and in cutoff wheels. Wheels of medium hardness, about 46-grit size, and with a synthetic resin bond work best for roughing. For finishing, wheels of finer grit size (to about 60) and with a vitrified bond are generally used. Recommendations for wheels for several different types of grinding operations are given in Table 33 (the ANSI system for identifying the characteristics of grinding wheels is explained in the articles "Grinding Equipment and Processes" and "Superabrasives" in this Volume). The following example describes specific grinding practice for pistons.

**Example 12: Grinding of Pistons.** After extensive testing, an A-46-K5-V wheel was selected for grinding the skirt of permanent mold cast F132-T5 pistons (Fig. 33). Requirements for the wheel were:

- Satisfactory performance without frequent redressing
- Ability to hold the required profile
- Production of a 1.25 to 2.25 µm (50 to 90 µin.) finish

The A-46-K5-V wheel was capable of grinding 100 parts between redressings and met the other requirements.

A special hydraulic dresser, equipped with a profile bar, was used to dress the wheel. The dressing tool was a ¾-carat diamond that was turned in its holder after eight dressings to maintain a sharp point. Details of this grinding operation are given in the table accompanying Fig. 33.

**Cutoff Wheels.** Abrasive cutoff wheels for aluminum are usually of aluminum oxide, as indicated in Table 34. For wet grinding, which is generally recommended, the bond is usually rubber. For dry grinding, either rubber or resinoid bond can be used. Typical specifications are A-20-U6-R (aluminum oxide, 20 grain size, U grade, 6 structure, rubber bond) and A-24-S7-B (aluminum oxide, 24 grain size, S grade, 7 structure, resinoid bond). The width of the wheels ranges from 0.08 to 3.96 mm (0.003 to 0.156 in.); diameter varies from 25 to 762 mm (1 to 30 in.). These wheels cut to an accuracy of a few thousandths of an inch.

**Reinforced (Flexible) Wheels.** In cutting off gates and sprues from castings of irregular shape, excessive wheel breakage can occur because of very high pressure on the sides of the wheel. To eliminate this problem and to provide for safe operation, special reinforced, flexible, abrasive wheels have been developed. These wheels are manufactured from laminated sheets of cotton fiber filled with abrasive grain and are used for operations ranging from heavy grinding to light sanding: cutting off, sharpening, deburring, and finishing. Such wheels used for aluminum are C-24-O14-B to C-24-U14-B.

**Speed.** Typical wheel speeds for several types of grinding are given in Table 33. Sometimes, by adjusting the speed, a wheel that was previously unsuited for a particular operation can be made to grind satisfactorily. For example, if a wheel is too soft, increasing the speed will give a harder action.

Using the recommended speed for a wheel is important, not only from the standpoint of grinding results but also to ensure safety. Stress from centrifugal force increases greatly as wheel velocity increases. The force tending to pull a wheel apart will be four times greater at 3600 rev/min than at 1800 rev/min.

**Grinding Fluid.** Neutral soluble-oil emulsions are satisfactory for grinding aluminum. The addition of a wetting agent (detergent) is sometimes helpful. Although an emulsion of 1 part soluble oil to 35 parts water is commonly used, a better cushioning effect, which also prevents clogging of the wheel, is obtained by using more oil and less water. One manufacturer, in grinding soft aluminum alloy castings, went progressively from a mixture of 1 part oil to 20 parts water to a mixture of 1 part oil to 6 parts water before obtaining satisfactory results.

In rough grinding, where stock removal is the primary objective, a generous application of wax or stick grease is often satisfactory. This type of lubricant is often used

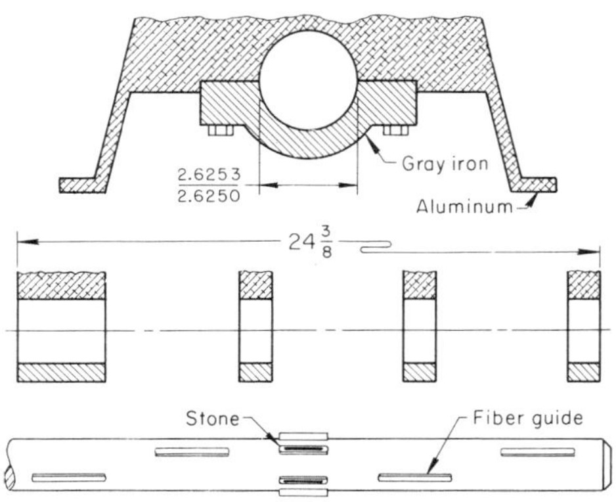


| | |
|---|---|
| Speed, at 900 rev/min, m/min (sfm) | 190 (620) |
| Spindle reciprocation rate, strokes/min | 60 |
| Hone speed, m/min (sfm) | 21 (70) |
| Stock removed, mm (in.) | 0.038–0.51 (0.0015–0.020) |
| Honing time, s | 25 |
| Cutting fluid | Mineral seal oil |
| Size control method | Air gage |
| Stone life, per set (average) | 300 assemblies |
| Production rate, pieces/h | 75 |
| Bore alignment before honing, mm (in.) | 0.05 (0.002) |
| Bore alignment after honing, mm (in.) | 0.018 (0.0007) |
| Specified tolerance, mm (in.) | ±0.008 (±0.0003) |
| Specified finish, μm (μin.) | 0.50–0.63 (20–25) |

**Fig. 34** Die cast engine block that was honed with gray iron bearing caps in place, and honing tool that was used (bottom). Dimensions in figure given in inches

when the application of liquids is not feasible.

**Belt Lubricant.** Grinding fluids or stick lubricants improve the finish produced by a coated abrasive belt and prevent glazing of the belt. Lubricants used range from water to stick waxes; standard cutting oils and soluble-oil mixtures are often used in high-production belt machines.

Greases cushion the penetrating action of the abrasive into the work and produce finer finishes than are obtained in dry grinding. Although the crest of each grain is free to cut, the grease prevents deep scratches.

Greases are used for the offhand grinding of aluminum die castings. When stock removal is the primary objective, a light grease should be used, preventing the belt from loading but permitting a maximum amount of abrasive penetration into the metal being ground. For a good finish, a heavy grease should be used.

Some operations use two types of grease, side by side, on the same abrasive belt. One side is used for a high rate of stock removal, and the other for a finer finish. If the belt becomes loaded despite the use of grease, an application of kerosene will free the belt of embedded particles.

Liquid lubricants are primarily used to prevent belt glazing. When water is used, a rust inhibitor should be added.

Soluble oil mixed with water is effective in grinding aluminum if stock removal is the primary objective. Table 35 lists data for abrasive belt grinding.

## Honing

Aluminum alloys are honed by methods similar to those used for other metals (see the article ''Honing'' in this Volume). Resin bond abrasives are preferred; sulfurized mineral-base oil or lard oil mixed with kerosene is used to flush the abrasive sticks clean and to carry away heat. Table 36 lists honing specifications for aluminum alloys.

Honing is used to finish anodized aluminum surfaces (primarily the bores of some small aluminum engine blocks) and most aircraft hydraulic cylinders. Long anodized aluminum tubes that are components of in-flight refueling apparatuses are finished by manual honing. The tubes are chucked in a lathe, and the honing tool is moved manually. The oil supply is attached to the honing tool in such a manner that the flow of oil is directed where most needed. This method of honing is also used for finishing connecting-rod journals or crankshafts in aircraft overhaul shops.

The following example describes the technique and conditions employed in the honing of main bearing bores in aluminum engine blocks. The aluminum blocks and the gray iron bearing caps were honed simultaneously with a silicon carbide abrasive, even though, if honed separately, gray iron and aluminum would be honed with different abrasives.

**Example 13: Simultaneous Honing of Aluminum and Gray Iron.** Main-bearing bores in six-cylinder, die cast aluminum engine blocks were honed with gray iron bearing caps bolted in place, as shown in Fig. 34. The blocks were fixtured for honing and were held in position by a light clamping force on the head faces. Each block was held securely so that it could resist torsional forces from the honing tool, but adjustment of clamping force was critical because excessive force would distort the block. Hydraulic cone expanders automatically fed out the single bank of six carbon-bonded silicon carbide stones (50 × 6.4 mm, or 2 × ¼ in., 120 grit). Fifteen fiber guides, each 75 × 6.4 mm (3 × ¼ in.), were incorporated in the honing tool. Processing details are given in the table with Fig. 34.

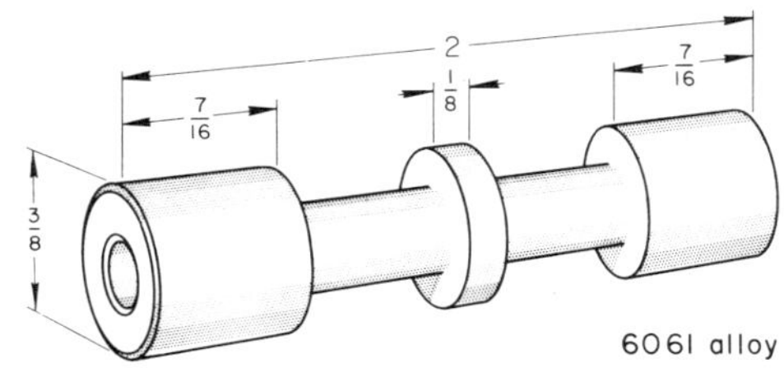


**Fig. 35** Part from which 0.005 mm (0.0002 in.) of metal was removed by lapping to smooth the surface to 0.025 μm (1 μin.). Dimensions given in inches

## Lapping

The methods used for lapping aluminum are the same as those used for other metals (see the article ''Lapping'' in this Volume). However, because similar finishes often can be produced by other methods at less cost, lapping is seldom used. In the following example, an extremely smooth finish of 0.025 μm (1 μin.) on an anodized part was produced by two methods of lapping.

**Example 14: Lapping Hard-Anodized Alloy 6061.** The hard-anodized aluminum part (surface hardness equivalent to 65 HRC) shown in Fig. 35 required removal of 0.005 mm (0.0002 in.) from the three lands to produce a finish of 0.025 μm (1 μin.). Diamond abrasive (8000 mesh) in a paste vehicle was used for both centerless roll lapping and lapping in a two-plate machine. The roll lapper, at a rotation speed of 100 rev/min (large roll) and a stroke speed of 50 mm/min (2 in./min), produced ten parts per hour. The two-plate machine, which had upper and lower cast iron laps 406 mm (16

**Table 37 Comparison of data and characteristics of systems for the chemical milling of titanium, steel, aluminum, and nickel- and cobalt-base alloys**

| Metal | Principal etchant | Etch rate, mm/min | Etch rate, in./min | Optimum etch depth, mm | Optimum etch depth, in. | Etchant temperature, °C | Etchant temperature, °F | Average surface roughness, $R_a$, μm | Average surface roughness, $R_a$, μin. |
|---|---|---|---|---|---|---|---|---|---|
| Titanium alloys | Hydrofluoric acid | 0.015–0.030 | 0.0006–0.0012 | 3.2 | 0.125 | 46±2.7 | 115±5 | 0.4–2.5 | 16–100 |
| Steels | Hydrochloric acid-nitric acid | 0.015–0.030 | 0.0006–0.0012 | 3.2 | 0.125 | 63±2.7 | 145±5 | 0.8–3.2 | 30–120 |
| Aluminum alloys | Sodium hydroxide | 0.020–0.030 | 0.0008–0.0012 | 3.2 | 0.125 | 90±2.7 | 195±5 | 2.0–3.2 | 80–120 |
| Nickel- and cobalt-base alloys | Nitric acid-hydrochloric acid-ferric chloride | 0.010–0.038 | 0.0004–0.0015 | 3.2 | 0.125 | 60±2.7 | 140±5 | 1.0–3.8 | 40–150 |

Source: Ref 5

**Table 38 Properties of maskant materials exposed to acids or alkalies that are widely used in the chemical milling of aluminum alloys**

| Property | Maskant material | | |
|---|---|---|---|
| | Butyl rubber | Acrylonitrile rubber | Neoprene rubber |
| Ease of manufacture | Fair | Fair | Good |
| Shelf life, months | 4–6 | 3–6 | 6–8 |
| Solids, % | 20–25 | 15–25 | 25–35 |
| Ease of application | | | |
| Dipping | Good | Good | Good |
| Flow coating | Good | Good | Good |
| Air spraying | Poor | Poor | Fair |
| Type of cure | Heat | Air or heat | Air or heat |
| Tensile strength, MPa (ksi) | | | |
| Air-dried, 24 h | ... | 7 (1.0) max | 14 (2.0) max |
| Heat cured | 6–11 (0.87–1.6) | 11–17.5 (1.6–2.5) | 11–21 (1.6–3.0) |
| Resistance to etchant | | | |
| Deterioration | Very good | Very good | Very good |
| Permeability | Excellent | Very good | Very good |
| Heat limit, °C (°F)(a) | 150 (300) | 120 (250) | 95 (200) |

(a) Maximum useful temperature for intermittent exposure and for curing temperature. Source: Ref 5

in.) in diameter and 75 mm (3 in.) thick, lapped 1000 parts per hour.

## Chemical Milling

Parts can be chemically milled over their entire surface, or the surface can be selectively machined by masking areas that do not require machining. The process is mainly used for parts having large surface areas requiring small amounts of metal removal (see the article "Chemical Milling" in this Volume).

Chemical milling is primarily used for parts having shallow cavities or pockets or requiring overall weight reduction. The main application has been in the aerospace industry to obtain maximum strength-to-weight ratios, but the process has also been receiving attention from other industries. Extremely large parts such as aircraft skin and fuselage sections and airframe extrusions are chemically milled.

Table 37 compares the chemical milling of aluminum to that of other metals typically machined by this process. In contrast to titanium alloys, steels, and nickel- and cobalt-base alloys, which use acid as an etchant, caustic soda is the usual etchant for aluminum alloys.

When selective machining is involved and the scribe-and-peel method is used, masking material is applied to the workpiece by spraying, dipping, or brushing. The dry maskant is scribed from a pattern and peeled from the workpiece to expose the areas that are to be chemically milled. Maskant can also be selectively applied by the silk-screening process or by photographic methods. Properties of maskants used to chemically mill aluminum alloys are listed in Table 38.

Machining takes place when the workpiece, contained in a work basket or held by a rack, is immersed in the chemical solution. Agitation of the work or circulation of the solution is necessary to ensure uniform rates of metal removal from all exposed surfaces. The maskant material is removed from the workpiece by hand or with the aid of a demasking solution.

When scribe-and-peel maskants are used, cuts as deep as 13 mm (0.5 in.) can be made. With the thinner and less chemical-resistant silk-screened masks, the depth of cut is limited to 1.5 mm (0.060 in.), but a more accurate and detailed cavity is possible. Photoresists provide still more detail and accuracy, but the cut is limited to a depth of 1.3 mm (0.050 in.). Sheets and plates are tapered in thickness by immersing and withdrawing the workpiece from the etching bath at a controlled rate. Steps are produced by repeated cycles of etching with different areas masked. The depth of cut is limited to about 13 mm (½ in.) in plate materials and to less than this with forgings, castings, and extrusions.

Table 39 lists data for the surface roughness of three forms of aluminum alloys that were subjected to the chemical milling process. The effect of depth of cut on the surface finish of 7075 aluminum sheet is shown in Fig. 36. Typical tolerances of various depths of cut are:

| Depth of cut | | Tolerance | |
|---|---|---|---|
| mm | in. | mm | in. |
| 0–1.25 | 0–0.050 | 0.025 | 0.001 |
| 1.25–2.50 | 0.051–0.100 | 0.038 | 0.0015 |
| 2.5–6.4 | 0.101–0.250 | 0.05 | 0.002 |
| 6.4–12.7 | 0.251–0.500 | 0.075 | 0.003 |

Additional information is available in the article "Chemical Milling" in this Volume. The following example illustrates the use of chemical milling to reduce the weight of a 7072 aluminum aircraft part.

**Example 15: Fuselage Section Made From Preformed 7072 Aluminum Alloy Plate.** The fuselage skin section shown in Fig. 37 was chemically milled after stretch forming. The web areas comprising about 80% of the surface area (one side) of the originally 6.4 mm (¼ in.) thick section were etched to a thickness of 0.64 mm (0.025 in.),

**Table 39 Surface roughness achieved by the chemical milling of aluminum alloys**

| Form | Surface roughness, $R_a$, after 0.25–0.40 mm (0.010–0.015 in.) removed | |
|---|---|---|
| | μm | μin. |
| Sheet | 2.00–3.8 | 80–150 |
| Casting | 3.8–7.6 | 150–300 |
| Forging | 2.50–6.3 | 100–250 |

Source: Ref 5

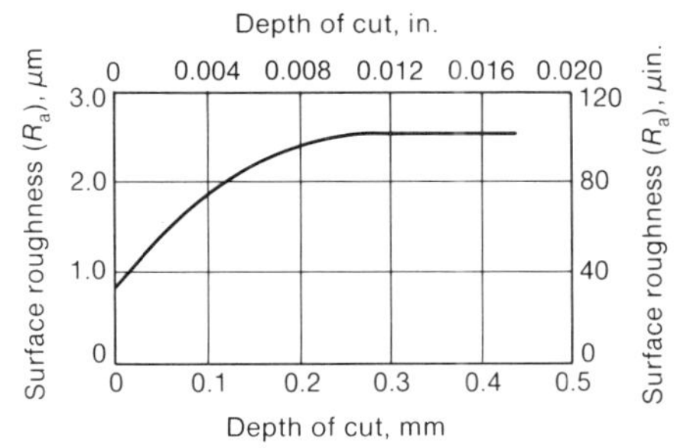


**Fig. 36** Relationship between surface finish and depth of cut for the chemical milling of 7075 aluminum sheet. Source: Ref 5

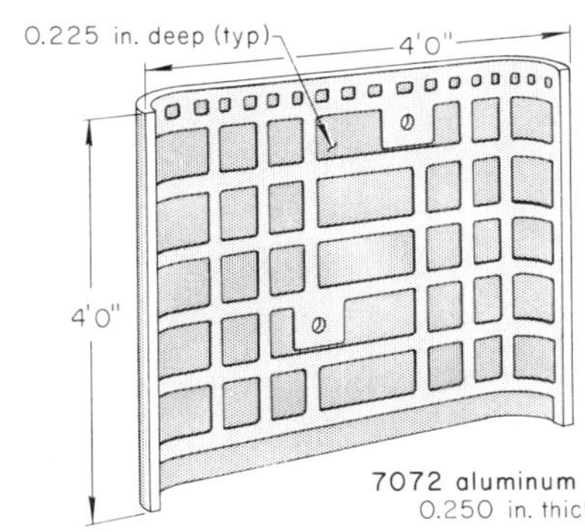


**Fig. 37** Preformed aircraft skin section chemically contoured to reduce weight. Dimensions given in inches

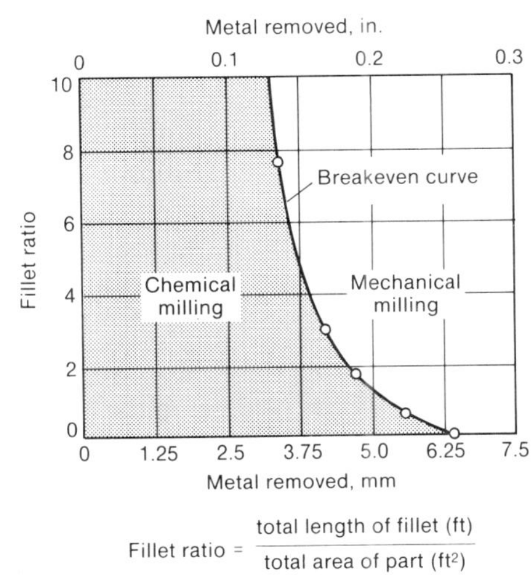


**Fig. 38** Breakeven between chemical and mechanical milling. Choice between them depends on fillet ratio and depth of cut.

reducing the weight of the part from 27 to 6.8 kg (60 to 15 lb).

Before the development of chemical milling, parts of this type were made by welding or riveting reinforcing sheet sections to a formed 0.64 mm (0.025 in.) thick skin; there was no practical method of forming such a panel after machining or of machining such a panel after forming. Chemical milling lowered manufacturing cost by greatly reducing the number of parts and eliminating assembly time.

**Chemical Versus Mechanical Milling.** Figure 38 presents a typical curve for choosing the most economical of these two methods for removing metal from flat parts on which large areas having complex or wavy peripheral outlines are to be reduced in thickness. Fillet ratio and thickness of metal to be removed are used as the basis for evaluation. For this specific application, metal thicknesses greater than 6.35 mm (0.250 in.) should be removed mechanically; thicknesses less than 3.18 mm (0.125 in.), chemically. Between these two values, the choice depends on fillet ratio, which governs the weight penalty.

## ACKNOWLEDGMENT

ASM INTERNATIONAL would like to thank Alan McMechan, McDonnell Douglas Canada; Eli Levy, The de Havilland Aircraft Company of Canada; and R.A. Morley, Reynolds Aluminum, for their assistance in updating this article. Mr. McMechan prepared the section "Soft Spots in Machined Aluminum Parts," Mr. Morley supplied data on the machining of high-silicon aluminum alloys, and Mr. Levy provided information on machining practices in the aerospace industry.

## REFERENCES

1. Machining Aluminum, Special Report 702, *Am. Mach.*, March 1978
2. J.C. Miller, "Machining High Silicon Aluminum," Paper G-T81-035, presented at the 11th International Die Casting Congress and Exposition, Cleveland, OH, Society of Die Casting Engineers, June 1981
3. A. McMechan, "Soft Spots in Machined Aluminum Parts," Paper presented at the Western Metal & Tool Exposition & Conference, Los Angeles, CA, McDonnell Douglas Corporation, March 1982
4. *Machining Alcoa Aluminum*, Aluminum Company of America, 1967
5. *Machining Data Handbook*, Vol 1 and 2, 3rd ed., Metcut Research Associates, Inc., 1980

# Machining of Copper and Copper Alloys

COPPER and copper alloys can be divided into three groups, with respect to machinability:

- Free-cutting alloys, to which an appreciable amount of lead, sulfur, or tellurium has been added to improve machining characteristics
- Moderately machinable alloys, chiefly the nonleaded brasses containing 60 to 85% Cu and the leaded nickel silvers
- Alloys that are difficult to machine, including the nonleaded coppers, low-zinc brasses and nickel silvers, the phosphor bronzes and aluminum bronzes, cupronickel, and some beryllium copper

## Machinability Ratings

Table 1 lists the nominal compositions and machinability ratings of the most widely used wrought coppers and copper alloys. Ratings are based on free-cutting brass (61.5 copper, 35.5 zinc, 3.0 lead) as 100.

In machining characteristics, most alloys of group 2 are closer to group 3 than to group 1. Leaded silicon bronze and nonleaded aluminum-silicon bronze (both rated 60) are exceptions.

**Type of Chip.** Group 1 alloys yield short, brittle chips. Because such chips readily free themselves from the tool or the machine, the alloys of group 1 are well suited to rapid machining.

Typical chips produced in machining group 2 alloys are a fairly open coil or a closely wound helix. These chips are relatively brittle, and a little more distortion will break them into separate fragments.

A typical chip from a group 3 alloy is long and continuous, and it is often tightly curled. Because it is strong and tough, such a chip will withstand considerable distortion without breaking, even to the extent of bending back flat.

**Casting Alloys.** Machinability ratings for copper casting alloys (Table 2) follow the same general pattern as for wrought alloys. Group 1 includes only those alloys containing lead, either as an alloying element or specifically for the purpose of improving machinability. As in wrought alloys, the lead additions facilitate chip breakage, thus making possible high machining speeds, good tool life, and good finish.

Alloys in group 2 contain secondary phases that are harder or more brittle than the matrix. In the machining of such alloys, short chips are produced. The silicon bronzes, several aluminum bronzes, and the high-tin alloys belong to this group.

The microstructure of the manganese bronzes would place them in group 2, but they produce a long spiral chip, smooth on both sides, that does not break. Some aluminum bronzes, on the other hand, produce a long spiral chip that is rough on the underside and breaks, thus acting like a short chip cut from an alloy in group 1.

Group 3 comprises the most difficult-to-machine alloys. It consists mainly of the high-strength manganese and aluminum bronzes that are high in iron or nickel content.

## Variations from Machinability Ratings

Machinability ratings do not precisely describe the actual degree of machinability in a particular application, and two alloys from the same group may differ considerably from the relative rankings implied by their nominal machinability ratings. Variations from the nominal machinability ratings may occur because of the nature of the machining process, the composition and microstructure of the alloy, or the precise criteria (tool life, cutting rate, and/or surface finish) of machinability.

Consider, for example, the effects of copper content (from about 58 to 62%) on the machinability of free-cutting (3.25% Pb) $\alpha$-$\beta$ alloys in drilling, sawing, and milling tests, as shown in Fig. 1. The upper curve in Fig. 1, plotted from ratings of two $\alpha$-$\beta$ alloys from Table 1, indicates qualitative agreement with the tests on the other four leaded alloys in the same range of copper content. However, the lack of precision in machinability ratings is shown by the variations among the three different machining processes represented in Fig. 1.

Another example is that the machinability of some alloys in group 3 may differ considerably in production machining, even though all the alloys in the group have the same nominal rating of 20. For example, tough pitch and other oxygen-bearing coppers contain cuprous oxide as a finely dispered phase that functions somewhat like the lead particles in group 1 alloys, although to a much lesser degree. Because of this microstructural characteristic, the tough pitch coppers are better in machinability, type of chip, and surface finish than oxygen-free coppers.

The machinability of the group 3 alloys can also be improved with chip curlers, or chip-breaking grooves, on the tool. By selecting the appropriate land width of the chip-breaking groove (as discussed in the section "Turning" of this article), the machining of group 3 alloys can produce short, broken chips similar to those of the free-machining alloys (group 1).

## Effect of Alloying Elements on Machinability

As indicated in the following sections, machinability is improved by lead, sulfur, tellurium, and zinc and deteriorates when tin and iron are used as alloying elements.

**Lead** functions in wrought copper alloys to improve machinability by increasing the brittleness of the alloy, causing chips to break off readily and in short curls rather than in long stringers. Lead does not dissolve in copper alloys, but is finely dispersed throughout the alloy matrix. Because it is soft, lead reduces the strain involved in chip formation and produces fine needles that break close to the shear plane so that the tool-chip content is small. It should also be noted that an increase in brittleness accompanies any lead addition, but chips are extremely short and brittle only when at least 2% Pb is present in the alloy.

Figure 2 shows machinability as a function of lead content for brasses containing 62 to 65% Cu. Machinability improves rapidly at first, but the rate of improvement drops off drastically to practically nothing at 3.25% Pb. An increase in lead content beyond 3.25% is seldom advantageous. The use of less than 1% Pb is common.

**Sulfur and Tellurium.** The effects of sulfur and tellurium on the drilling of copper are compared in Fig. 3. The basis of ap-

## Table 1 Compositions and machinability ratings of wrought copper and copper alloys

| Alloy No. | Alloy name | Nominal composition, % Cu | Ni | Zn | Pb | Sn | Machinability(a) |
|---|---|---|---|---|---|---|---|
| **Group 1: free-cutting alloys** | | | | | | | |
| C17300 | Beryllium copper | 97.50 | (0.2 min, Ni + Co) | (1.9 Be) | 0.4 | . . . | 50 |
| C18700 | Leaded copper | 99.00 | . . . | . . . | 1.00 | . . . | 80 |
| C14500 | Tellurium copper | 99.40 | | (0.50 Te) | | | 80 |
| C33000 | Low-leaded brass (tube) | 66.00 | . . . | 33.50 | 0.50 | . . . | 60 |
| C33500 | Low-leaded brass | 65.00 | . . . | 34.50 | 0.50 | . . . | 60 |
| C34900 | Brass | 62.00 | . . . | 37.50 | 0.30 | . . . | 50 |
| C34000 | Medium-leaded brass | 65.00 | . . . | 34.00 | 1.00 | . . . | 70 |
| C33200 | High-leaded brass (tube) | 66.00 | . . . | 32.40 | 1.60 | . . . | 80 |
| C34200 | High-leaded brass | 65.00 | . . . | 33.00 | 2.00 | . . . | 90 |
| C35600 | Extrahigh-leaded brass | 63.00 | . . . | 34.50 | 2.50 | . . . | 100 |
| C36000 | Free-cutting brass | 61.50 | . . . | 35.50 | 3.00 | . . . | 100 |
| C36500-C36800 | Leaded Muntz metal | 60.00 | . . . | 39.40 | 0.60 | . . . | 60 |
| C37000 | Free-cutting Muntz metal | 60.00 | . . . | 39.00 | 1.00 | . . . | 70 |
| C37700 | Forging brass | 59.00 | . . . | 39.00 | 2.00 | . . . | 80 |
| C48500 | Leaded naval brass | 60.00 | . . . | 37.50 | 1.75 | 0.75 | 80 |
| C48200 | Medium-leaded naval brass | 60.5 | . . . | 38 | 0.7 | 0.8 | 50 |
| C54400 | Free-cutting phosphor bronze | 88.00 | . . . | 4.00 | 4.00 | 4.00 | 90 |
| C38500 | Architectural bronze | 57.00 | . . . | 40 | 3.00 | . . . | 90 |
| C31400 | Leaded commercial bronze | 89 | . . . | 9.1 | 1.9 | . . . | 80 |
| C31600 | Leaded commercial bronze-nickel | 89 | 1 | 8.1 | 1.9 | . . . | 80 |
| C62300 | Aluminum bronze | 87 | (3 Fe) | (10 Al) | . . . | . . . | . . . |
| C62400 | Aluminum bronze | 86 | (3 Fe) | (11 Al) | . . . | . . . | 50 |
| C63800 | Coronze | 95 | (0.4 Co) | (2.8 Al) | (1.8 Si) | . . . | . . . |
| C64200 | Aluminum bronze | 90 | (0.4 Co) | (7 Al) | (1.8 Si) | . . . | . . . |
| C78200 | Leaded nickel silver | 65 | 8 | 25 | 2 | . . . | 60 |
| **Group 2: moderately machinable alloys** | | | | | | | |
| C19000 | High-copper alloy | 98.75 | 1.1 | . . . | 0.05 | (0.1 Fe) | . . . |
| C22600 | Jewelry bronze | 87.50 | . . . | 12.5 | . . . | . . . | 30 |
| C42500 | Tin brass | 88.50 | . . . | 9.50 | . . . | 2.00 | 30 |
| C23000 | Red brass, 85% | 85.00 | . . . | 15.00 | . . . | . . . | 30 |
| C24000 | Low brass, 80% | 80.00 | . . . | 20.00 | . . . | . . . | 30 |
| C26000 | Cartridge brass, 70% | 70.00 | . . . | 30.00 | . . . | . . . | 30 |
| C26800, C27000 | Yellow brass | 65.00 | . . . | 35.00 | . . . | . . . | 30 |
| C28000 | Muntz metal | 60.00 | . . . | 40.00 | . . . | . . . | 40 |
| C44300–C44500 | Inhibited admiralty | 71.00 | . . . | 28.00 | . . . | 1.00 | 30 |
| C46400–C46700 | Naval brass | 60.00 | . . . | 39.25 | . . . | 0.75 | 30 |
| C65100 | Low-silicon bronze (B) | 98.50 | (1.50 Si) | | . . . | . . . | 30 |
| C65500 | High-silicon bronze (A) | 97.00 | (3.00 Si) | | . . . | . . . | 30 |
| C67500 | Manganese bronze (A) | 58.50 | (0.10 Mn) | 39.00 | (1.40 Fe) | 1.00 | 30 |
| C61300 | Aluminum bronze | 90.00 | (7 Al) | . . . | (2.7 Fe) | 0.3 | . . . |
| C61800 | Aluminum bronze | 89.00 | (9.8 Al) | . . . | (1.0 Fe) | . . . | . . . |
| C63000 | Nickel-aluminum bronze | 82.00 | 5 | (10 Al) | (3.0 Fe) | . . . | 30 |
| C66100 | Leaded silicon bronze (D) | 95.60 | | (0.4 Pb, 1.5 Mn, 1.5 Zn, 3 Si) | | | 60 |
| C43500 | Tin brass | 81.00 | . . . | 18.1 | . . . | 0.9 | 30 |
| C68700 | Aluminum brass | 77.50 | . . . | 20.50 | (2.00 Al) | . . . | 30 |
| C69400 | Silicon red brass | 81.50 | . . . | 14.50 | (4.0 Si) | . . . | 30 |
| C77000 | Nickel silver | 55.00 | 18.00 | 27.00 | . . . | . . . | 60 |
| **Group 3: difficult-to-machine alloys** | | | | | | | |
| C10100–C10800 | Oxygen-free copper | 99.90+ | . . . | . . . | . . . | . . . | 20 |
| C11000 | Electrolytic tough pitch copper | 99.90+ | . . . | . . . | . . . | . . . | 20 |
| C11300–C11600 | Silver-bearing electrolytic tough pitch copper | 99.96 (including 0.4 Ag) | | | . . . | . . . | 20 |
| C12200 | Phosphorus-deoxidized copper | 99.90+ | (0.02 P) | . . . | . . . | . . . | 20 |
| C12500–C13000 | Fire-refined electrolytic tough pitch copper | 99.88+ | . . . | . . . | . . . | . . . | 20 |
| C16200 | Cadium copper | 98.78 | (1.2 Cd) | . . . | . . . | . . . | 20 |
| C16500 | . . . | 99.38 | . . . | . . . | . . . | 0.60 | 20 |
| C17000 | Beryllium copper | 98.00 | (0.3 Co) | (1.7 Be) | . . . | . . . | 20 |
| C17200 | Beryllium copper | 97.80 | . . . | (1.9 Be) | . . . | . . . | 20 |
| C17500 | Low-beryllium copper | 97.50 | (2.5 Co) | (0.5 Be) | . . . | . . . | . . . |
| C18200–C18500 | Chrome copper | 97.75+ | (0.4–1.2 Cr) | | . . . | . . . | 20 |
| C19500 | Strescon | 97.00 | (0.8 Co) | (1.5 Fe) | (0.1 P) | 0.6 | 20 |
| C41100 | Lubaloy | 95.00 | . . . | 3.00 | . . . | 2.00 | 20 |
| C21000 | Gilding, 95% | 95.00 | . . . | 5.00 | . . . | . . . | 20 |
| C22000 | Commercial bronze, 90% | 90.00 | . . . | 10.00 | . . . | . . . | 20 |
| C50500, C51000, C51100, C52100, C52400 | Phosphor bronze | 90.00+ | (0.03–0.35 P) | | . . . | 1.3–10 | 20 |
| C60800, C61000 | Aluminum bronze | 92–95 | (5–8 Al) | | . . . | . . . | 20 |
| C62500 | Ampco 21, Wearite 4–13 | 82.7 | (4.3 Fe) | (13 Al) | . . . | . . . | 20 |
| C61400 | Aluminum bronze (D) | 91.00 | (7.00 Al, 2.00 Fe) | | | . . . | 20 |
| C68800 | Alcoloy | 73.50 | (0.4 Co) | 22.70 | (3.4 Al) | . . . | . . . |
| C70600 | Copper nickel, 10% | 88.70 | 10.00 | (1.30 Fe) | . . . | . . . | 20 |
| C71500 | Copper nickel, 30% | 70.00 | 30.00 | . . . | . . . | . . . | 20 |
| C74500 | Nickel silver, 65–10 | 65.00 | 10.00 | 25.00 | . . . | . . . | 20 |
| C75200 | Nickel silver, 65–18 | 65.00 | 18.00 | 17.00 | . . . | . . . | 20 |
| C75400 | Nickel silver, 65–15 | 65.00 | 15.00 | 20.00 | . . . | . . . | 20 |
| C75700 | Nickel silver, 65–12 | 65.00 | 12.00 | 23.00 | . . . | . . . | 20 |
| C71000 | Copper nickel | 80.00 | 20.00 | . . . | . . . | . . . | 20 |
| C72500 | Copper nickel, tin-bearing | 88.20 | 9.5 | . . . | . . . | 2.3 | 20 |

(a) Approximate relative machinability rating (free-cutting brass = 100)

**Table 2 Compositions and machinability ratings of copper casting alloys**

| Alloy No. | Alloy name | Nominal composition, % Cu | Sn | Pb | Zn | Ni | Al | Others | Machinability(a) |
|---|---|---|---|---|---|---|---|---|---|
| **Group 1: free-cutting alloys** | | | | | | | | | |
| C83400 | Leaded red brass | 90 | 0.2 | 0.5 | 10 | ... | ... | ... | 90 |
| C83600 | Leaded red brass | 85 | 5.0 | 5.0 | 5 | ... | ... | ... | 90 |
| C83800 | Leaded red brass | 83 | 4.0 | 6.0 | 7 | ... | ... | ... | 90 |
| C84200 | Leaded semi-red brass | 80 | 5.0 | 2.5 | 13 | ... | ... | ... | 90 |
| C84400 | Leaded semi-red brass | 81 | 3.0 | 7.0 | 9 | ... | ... | ... | 90 |
| C84800 | Leaded semi-red brass | 76 | 3.0 | 6.0 | 15 | ... | ... | ... | 90 |
| C85200 | Leaded yellow brass | 71 | 1.0 | 3.0 | 25 | ... | ... | ... | 80 |
| C85400 | Leaded yellow brass | 67 | 1.0 | 3.0 | 29 | ... | ... | ... | 80 |
| C85500 | Leaded yellow brass | 61 | 0.2 | 0.2 | 38 | ... | ... | ... | 80 |
| C85700, C85800 | Leaded yellow brass | 63 | 1.0 | 1.0 | 35 | ... | ... | ... | 80 |
| C86400 | Manganese bronze | 59 | 1.0 | 1.0 | 38 | ... | 0.75 | 1.25 Fe, 0.5 Mn | 60(b) |
| C86700 | Manganese bronze | 57 | 1.5 | 1.0 | 34 | ... | 2.0 | 2.0 Fe, 2.25 Mn | ... |
| C87900 | Silicon brass | 63+ | 0.25 | 0.25 | 33 | 0.5 | ... | 1.0 Si | ... |
| C92800 | Leaded tin bronze | 80 | 16 | 5 | 0.8 | ... | ... | ... | ... |
| C93200 | High-leaded tin bronze | 83 | 7 | 7 | 3 | 0.5 max | ... | ... | 70 |
| C93400, C93500 | High-leaded tin bronze | 83–85 | 8–5 | 8–9 | 0.8–1 | ... | ... | ... | 70 |
| C93700–C93900 | High-leaded tin bronze | 78–80 | 10–6 | 10–15 | 0.7–1.5 max | ... | ... | ... | 80 |
| C95300 | Aluminum bronze | 89 | ... | ... | ... | ... | 10 | 1 Fe | 55(b) |
| C95400 | Aluminum bronze | 85 | ... | ... | ... | ... | 11 | 4 Fe | 60(b) |
| C95600 | Silicon aluminum bronze | 88 | ... | ... | ... | 0.25 | 7 | 2.5 Si | 50(b) |
| C97300 | Leaded nickel brass | 56 | 2 | 10 | 20 | 12 | ... | ... | 70 |
| C97400 | Leaded nickel brass | 60 | 3 | 5 | 16 | 16 | ... | ... | ... |
| C97600 | Leaded nickel bronze | 64 | 4 | 4 | 8 | 20 | ... | ... | 70 |
| C97800 | Leaded nickel bronze | 66.5 | 5 | 1.5 | 2 | 25 | ... | ... | 60(b) |
| **Group 2: moderately machinable alloys** | | | | | | | | | |
| C81700 | Beryllium copper | 94.2 min | ... | ... | ... | 0.9 | ... | 1.0 Ag, 0.4 Be | ... |
| C82100 | Beryllium copper | 95.5 min | ... | ... | ... | 0.9 | ... | 0.6 Be, 0.9 Be | ... |
| C83300 | Leaded red brass | 93 | 1.5 | 3 | 4 | ... | ... | ... | ... |
| C86100, C86200 | Manganese bronze | 64 | ... | ... | 24 | ... | 5 | 3 Fe, 4 Mn | 30 |
| C86500 | Manganese bronze | 57 | 1.0 | ... | 39 | ... | 1 | 1.3 Fe, 0.5 Mn | 26 |
| C86800 | Manganese bronze | 54 | 1.0 | ... | 34 | 3.25 | 2 | 1.75 Fe, 3.25 Mn | ... |
| C87200 | Silicon bronze | 89 min | 1.0 max | 0.5max | 5.0 max | ... | 1.5 max | 1.0–5.0 Si, 2.5 max Fe, 1.5 max Mn | 50 |
| C87400 | Silicon bronze | 79 min | ... | 1.0 | 14 | ... | ... | 2.5–4.0 Si | ... |
| C87500, C87600, C87800 | Silicon brass | 79–88 min | | 0.5–0.15 | 14 | ... | 0.5–0.15 | 4.0 Si | 50 |
| C90300 | Tin bronze | 88 | 8 | ... | 4 | ... | ... | ... | 30 |
| C90500 | Tin bronze | 88 | 10 | ... | 2 | ... | ... | ... | 30 |
| C92200 | Navy M bronze | 88 | 6 | 1.5 | 4.5 | ... | ... | ... | 42 |
| C92300 | Navy G bronze | 87 | 8 | 1.0 | 4.0 | ... | ... | ... | 42 |
| C92500 | Leaded tin bronze | 87 | 11 | 1.0 | ... | 1 | ... | ... | 30 |
| C92600 | Leaded tin bronze | 87 | 10 | 1.0 | 2.0 | ... | ... | ... | 40 |
| C92700 | Leaded tin bronze | 88 | 10 | 2.0 | ... | ... | ... | ... | 45 |
| C94700, C94800 | Nickel-tin bronze | 88–86 | 5 | ... | 2 | 5 | ... | ... | ... |
| C95200 | Aluminum bronze | 88 | ... | ... | ... | ... | 9 | 3 Fe | 20 |
| C95500 | Aluminum bronze | 81 | ... | ... | ... | 4 | 11 | 4 Fe | 50 |
| C95700 | Manganese aluminum bronze | 75 | ... | ... | ... | 2 | 8 | 3 Fe, 12 Mn | 50 |
| C95800 | Propeller bronze | 82 | ... | ... | ... | 4 | 9 | 4 Fe, 1 Mn | 50 |
| **Group 3: difficult-to-machine alloys** | | | | | | | | | |
| C80100–C81100 | Copper | 99+ | ... | ... | ... | ... | ... | ... | 10 |
| C81300, C81400 | Beryllium copper | 98.5 min | ... | ... | ... | ... | ... | 0.02–0.1 Be | 30(c) |
| C81500 | Chromium copper | 99 | ... | ... | ... | ... | ... | 1 Cr | 20 |
| C81800 | Beryllium copper | 97 | ... | ... | ... | ... | ... | 1.5 Co, 1 Ag, 0.4 Be | 30(c) |
| C82000 | Beryllium copper | 97 | ... | ... | ... | ... | ... | 2.5 Co, 0.5 Be | 30(c) |
| C82200 | Beryllium copper | 98 | ... | ... | ... | 1.5 | ... | 0.5 Be | 30(c) |
| C82400 | Beryllium copper | 98 | ... | ... | ... | ... | ... | 0.3 Co, 1.7 Be | 30(d) |
| C82500 | Beryllium copper | 97.2 | ... | ... | ... | ... | ... | 0.5 Co, 0.25 Si, 2 Be | 30(d) |
| C82600 | Beryllium copper | 97 | ... | ... | ... | ... | ... | 0.5 Co, 2.4 Be | 30(d) |
| C82700 | Beryllium copper | 96 | ... | ... | ... | 1.25 | ... | 0.15 Si, 2.45 Be | ... |
| C82800 | Beryllium copper | 96.6 | ... | ... | ... | ... | ... | 0.5 Co, 0.3 Si, 2.6 Be | 30(d) |
| C86300 | Manganese bronze | 64 | ... | ... | 26 | ... | 3 | 3 Fe, 4 Mn | 8 |
| C90200 | Tin bronze | 92 | 6.7 | 0.3 | 0.5 | 0.5 | ... | ... | ... |
| C90700, C90900 | Tin bronze | 87 | 13 | ... | ... | ... | ... | ... | 20 |
| C91000, C91100 | Tin bronze | 85 | 15 | ... | ... | ... | ... | ... | 20 |
| C91300 | Tin bronze | 81 | 19 | ... | ... | ... | ... | ... | ... |
| C91600, C91700 | Tin bronze | 87.5–85.5 | 11–13 | ... | ... | 1.5 | ... | ... | ... |
| C96200 | Copper nickel | 86 | ... | ... | ... | 10 | ... | 1.5 Fe, 1.5 Mn, 1 Nb | ... |
| C96400 | Copper nickel | 70 | ... | ... | ... | 30 | ... | ... | 20 |
| C96600 | Beryllium cupro-nickel | 69.5 | ... | ... | ... | 30 | ... | 0.5 Be | 40(e) |
| C99300 | Incrament 800 | 72 | | | | 15 | 11 | 1.5 Co, 0.5 Fe | ... |

(a) Machinability rating is expressed as a percentage of the machinability of C36000, free-cutting brass. The rating is based on relative speed for equivalent tool life. (b) Alloy has a moderate machinability rating, but is grouped with the speeds and feeds of the free-cutting alloys. (c) Machinability of 30 with TB00 temper or as-cast. Machinability rating of 40 with TF00 temper. (d) Machinability of 30 with TB00 temper or as-cast. Machinability rating of 10 to 20 with TF00 temper or cast and aged. (e) TF00 temper

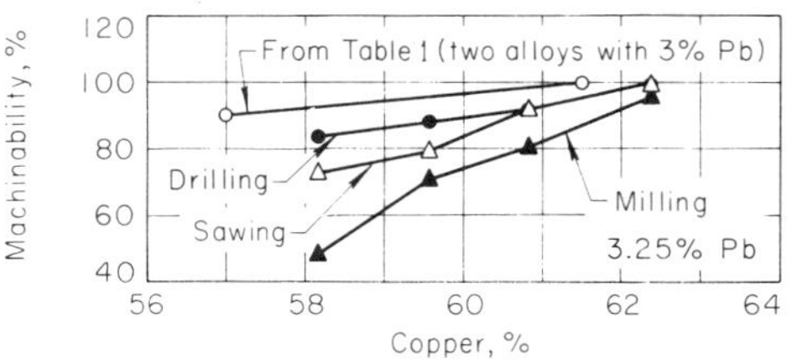


**Fig. 1** Effect of copper content on the machinability of free-cutting α-β alloys

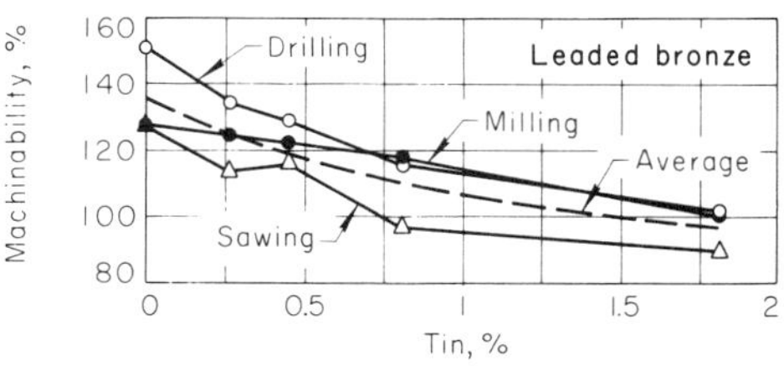


**Fig. 4** Effect of tin content on the machinability of leaded commercial bronze (machinability scale nonstandard)

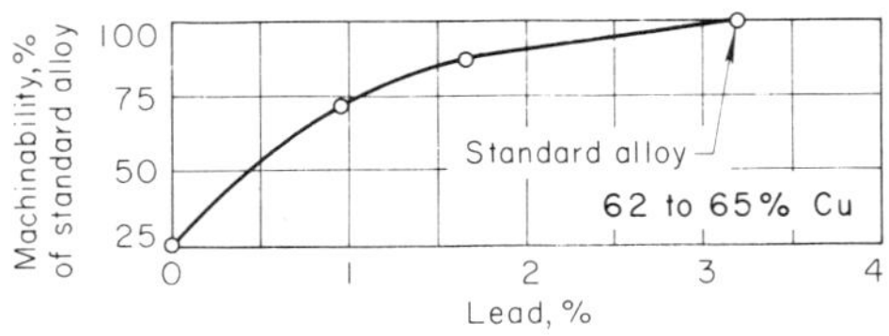


**Fig. 2** Effect of lead content on the machinability of yellow brass

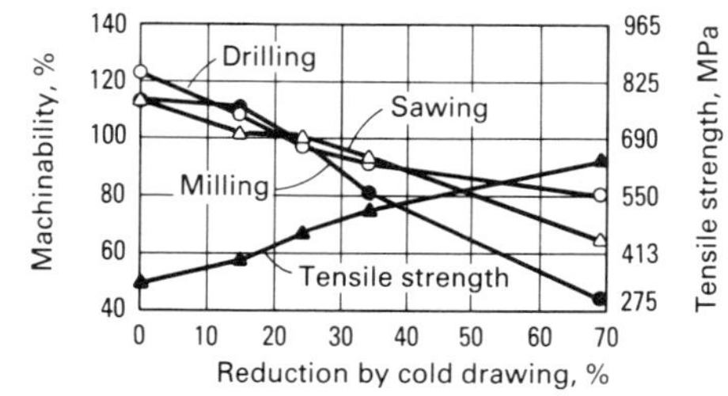


**Fig. 5** Effect of cold work on the machinability of free-cutting brass (machinability scale non-standard)

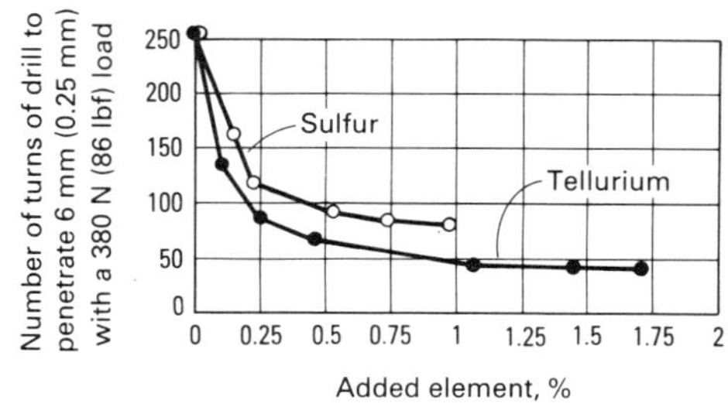


**Fig. 3** Effect of sulfur and tellurium on drillability

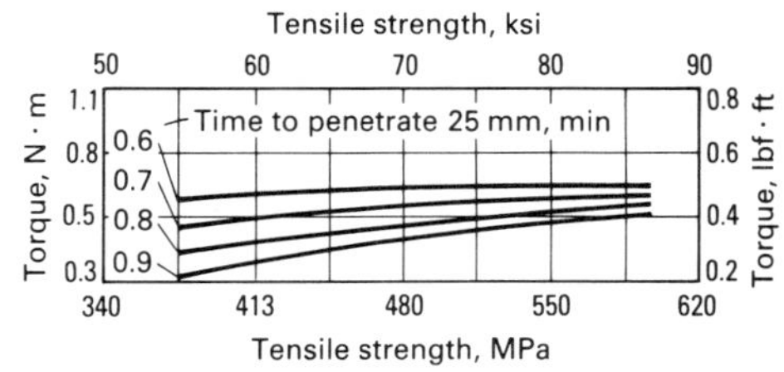


**Fig. 6** Effect of cold work (as measured by tensile strength) on the drillability (torque and time to penetrate) of free-cutting brass

praisal was the number of turns of a drill under constant load required to penetrate 6 mm (0.25 in.) into the alloy indicated. For both elements, the number of turns dropped abruptly with small additions and then more gradually as increased amounts were added. There is no significant advantage in adding more than 0.5% of either of these elements due to the accompanying loss of electrical conductivity.

**Zinc.** The hardness of α-brass increases with the addition of zinc up to the limit of α solubility, about 37% Zn. In lead-free brasses, increases in zinc content from 0 to 30% improve the machinability rating from 20 to 30.

**Tin.** Figure 4 shows the effect of tin content (up to 1.8%) in leaded commercial bronze (nominal composition: 89Cu-9.25Zn-1.75Pb) in drilling, sawing, and milling. The dashed curve shows the average effect of tin.

**Iron.** In drilling, sawing, and milling tests, 1.6% Fe in leaded commercial bronze was found to decrease machinability by about 10%.

## Effect of Cold Work

The adverse effect of cold work on the machinability of a typical free-cutting brass (Zn-62.3Cu-3.5Pb) is shown in Fig. 5. The 70% reduction by cold drawing indicated is well beyond the 25% maximum cold reduction normally given leaded brass bars of most sizes.

In drilling tests on a typical free-cutting brass rod to determine the effect of cold work on machinability, the torque developed by a 6 mm (¼ in.) drill in penetrating alloys of four different tempers was recorded and is plotted in Fig. 6. The alloys were drilled to a depth of 25 mm (1 in.) under a wide range of axial loads. The lower curve in Fig. 6 shows the torque used to drill the low-strength rod to be about 50% of that to drill the high-strength rod, while the upper curve shows only about 15% difference caused by cold work. In all tests, the torque developed increased with the tensile strength of the alloy being drilled. In evaluating these drilling tests, it was concluded that high tool pressures and high cutting speeds could offset the effects of cold work.

## Cutting Fluid

Some machining of copper alloys is done dry, but the use of a cutting fluid invariably increases the level of possible speed and feed, increases productivity, improves surface finish, enhances accuracy, and lengthens tool life. Unless otherwise noted, the speeds and feeds shown in tables throughout this article are based on the use of water-soluble oil as the cutting fluid. For most operations, water-soluble oil (20 parts water to 1 part oil) is adequate as an anti-weld agent, especially for the group 1 alloys. When machining group 2 and 3 alloys (more stringy chips) or when a very fine surface finish is required, results can be improved by using a light mineral oil, or mineral oil containing 5 to 20% lard oil. The higher content of lard oil is used for the alloys that produce extremely tough,

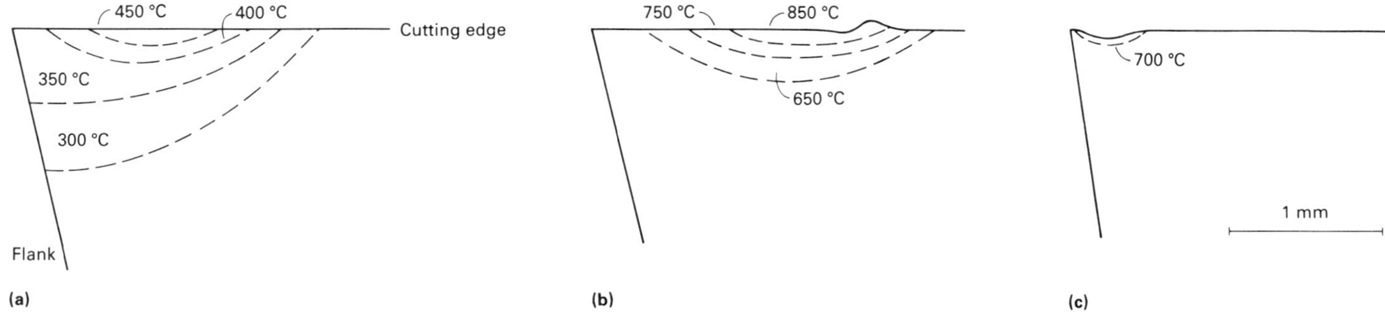


**Fig. 7** Temperature distribution in high-speed steel tools in cutting 60/40 brass (a), low-carbon steel (b), and CA 104 aluminum bronze (10% Al, 5% Fe, 5% Ni) (c). The proximity of the high-temperature region to the cutting edge is the reason aluminum bronzes are considered difficult to machine. This problem can be offset by using cemented carbide tools of the appropriate composition. Machining conditions of the temperature distributions are as follows. Depth of cut, 1.25 mm (0.05 in.); feed rate, 0.25 mm/rev (0.010 in./rev); cutting speeds: (a) 120 m/min (400 sfm). (b) 120 m/min (400 sfm). (c) 125 m/min (410 sfm)

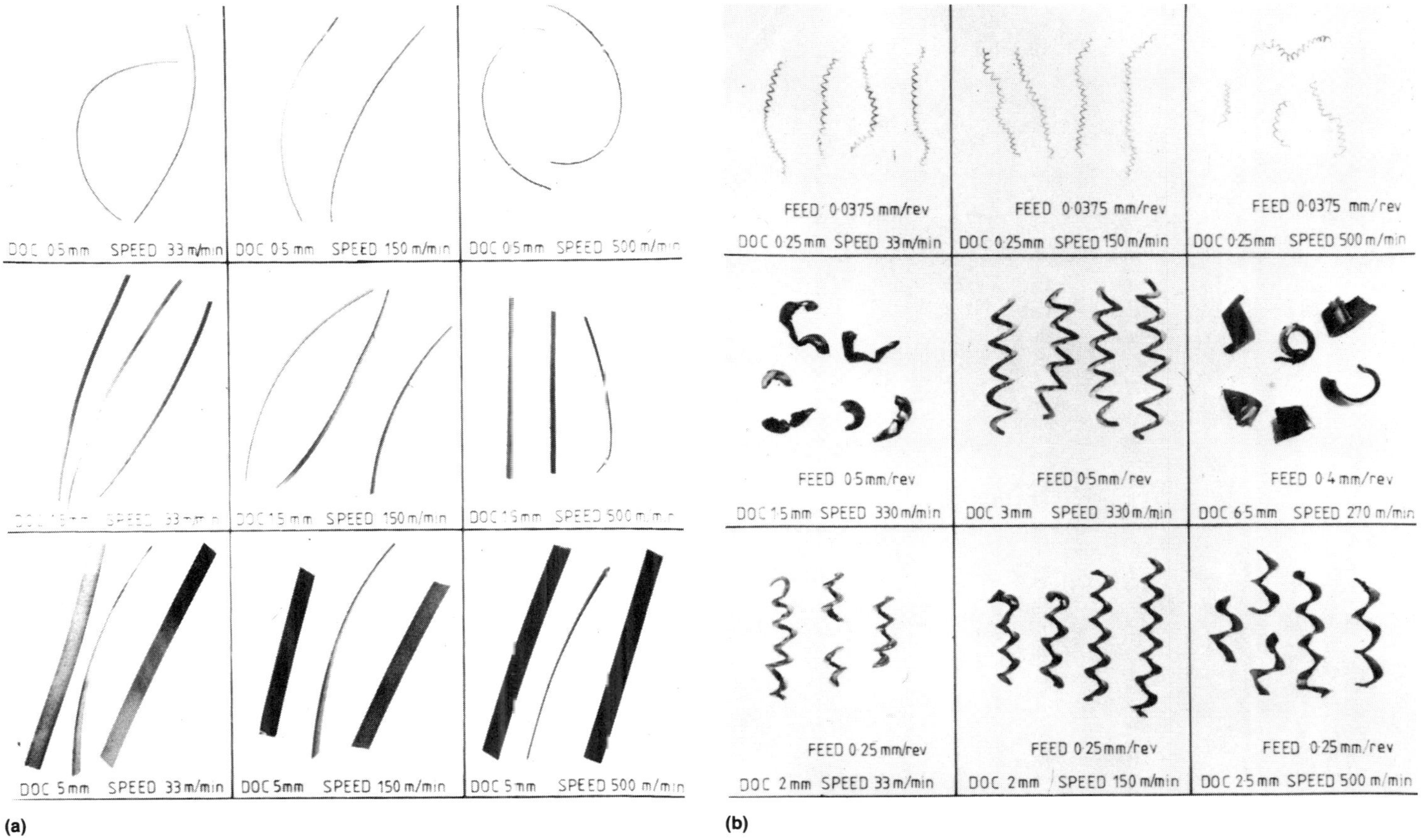


**Fig. 8** Various chip forms from the machining of oxygen-free high-conductivity copper. (a) Chips from a conventional open-faced tool (25° side rake, 20° back rake, and a nose radius of 1.25 mm, or 0.05 in.) and a feed rate of 0.2 mm/rev (0.008 in./rev). (b) Chips from a variety of tools with chip curlers. DOC, depth of cut

stringy chips or where finish is critical, as in cutting fine threads.

Sulfurized, chlorinated, or sulfochlorinated oil is sometimes used as a cutting fluid, particularly for machining copper alloys that contain substantial amounts of nickel. When sulfurized or chlorinated oil is used for machining any copper alloy, the workpieces should be cleaned immediately after machining to prevent staining. If the workpieces do become stained, the discoloration can be removed by immersing them for 20 min in a solution of 10% sodium cyanide in water.

## Turning

Because of the wide variation in machining characteristics among the various coppers and copper alloys, tool design, as well as speed and feed, must be varied for best results. Both high-speed steel and carbide are used as tool materials in the turning of copper alloys. However, with the aluminum and silicon bronzes, the heat produced during cutting is generated much closer to the cutting edge (Fig. 7), where the stress due to the cutting force is greatest. As a result, cemented carbide tools are preferred to high-speed steel tools when machining aluminum bronze. The most suitable cemented carbide tool material for machining aluminum bronze is one containing fine-grain (0.8 μm) tungsten carbide in a matrix of 10% Co (Ref 1).

Polycrystalline diamond is also used as a tool material in the turning of copper alloys. The article "Ultrahard Tool Materials" in this Volume describes diamond tools in more detail.

**Chip Breaking.** Group 3 alloys can exhibit the free-machining characteristic of short, broken chips when the following conditions are met:

- An exit angle (the angle between the chip and the bar axis) of 20 to 60°
- A suitable chip curler on the tool that prevents the chip from reverting to a flat, continuous strip (Fig. 8)

By selecting the correct land width of the chip curler (as discussed below), even oxygen-free high-conductivity (OFHC) copper can be machined without difficulty (Ref 2).

Successful chip breaking requires an exit angle of 20 to 60°, which is influenced by a variety of machining parameters, as shown in Fig. 9 for electrolytic tough pitch (ETP) and oxygen-free (OF) coppers. Assuming an exit angle of 20 to 60° and a tool with a 20° side rake and a 20° back rake, the appropriate land width of a chip curler for OFHC copper can be obtained from the diagram shown in Fig. 10. This type of chip curler is also successful in the chip breaking of ETP copper (Fig. 11).

Other materials in group 3 with chip-breaking difficulties can be successfully machined with chip curlers. Work on unleaded 60/40 brass has shown that the best chip-breaking behavior is given by cemented carbide inserts with the geometry shown in Fig. 12.

**Single-Point Turning.** Zero to moderate rake angles are recommended for group 1 alloys to reduce the probability that the tool will plow into the work. Rake angles are increased for turning alloys of groups 2 and 3 to provide a free flow of chips. Table 3 lists tool angles for turning copper alloys. The previous discussion on chip breaking also applies.

The speeds given in Table 4 are intended as starting points in establishing procedures. In many cases, speeds can be increased with experience.

**Form Turning.** Both circular and dovetail form tools are used for turning copper and copper alloys; selection depends on the same factors that influence the choice in the form turning of steel. Numerical control and computer numerical control systems also allow the machining of forms with single-point turning. Form tools for turning copper alloys are shown in Fig. 13(a); speeds and feeds, in Table 5.

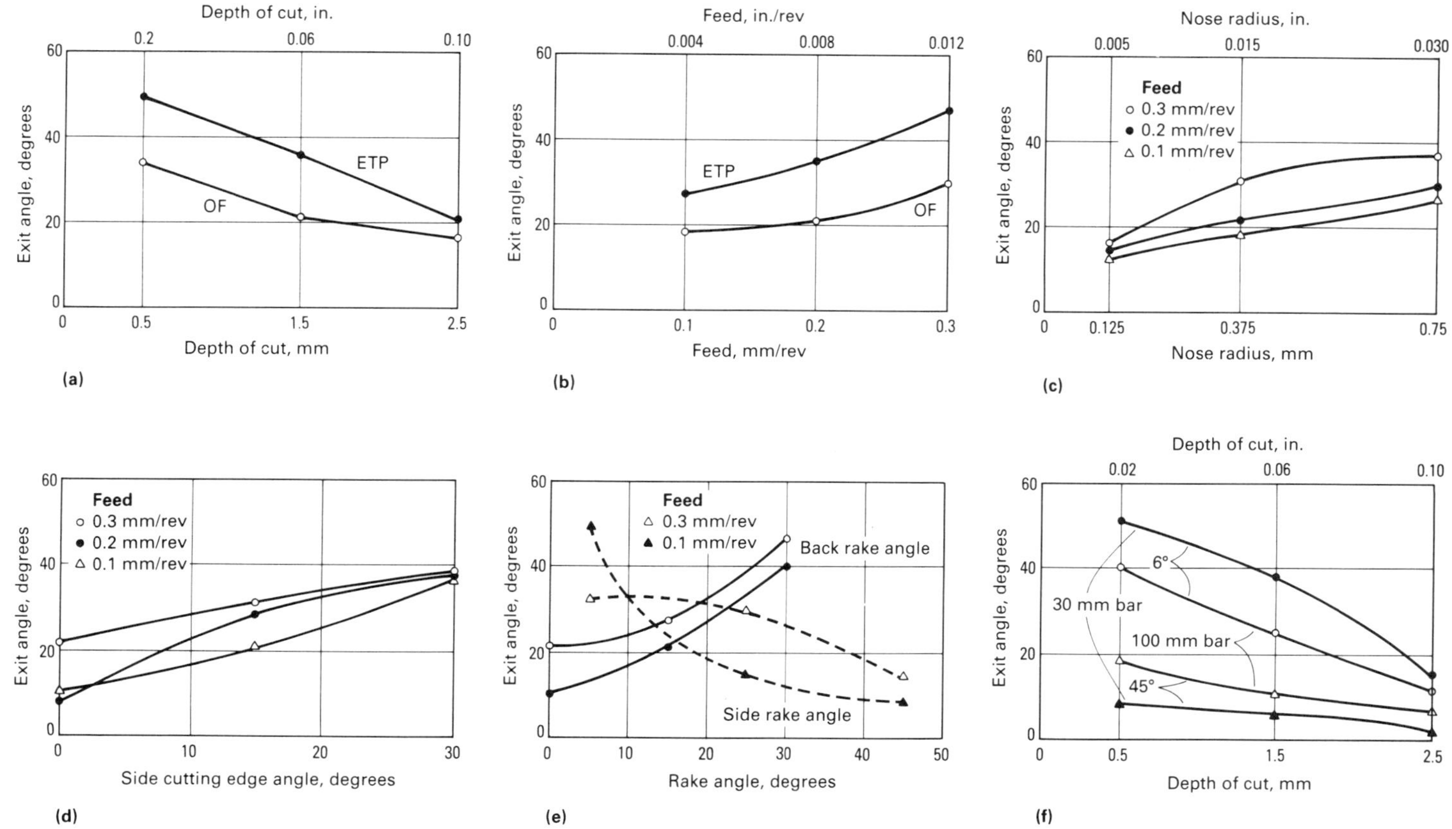


**Fig. 9** Parameters affecting the exit angle in the machining of ETP and OF coppers with high-speed steel tools. (a) Effect of depth of cut on exit angle for ETP and OF coppers. Side rake angle 6°; back rake angle, 0°; side cutting edge angle, 0°; nose radius, 0.125 mm (0.005 in.); feed, 0.2 mm/rev (0.008 in./rev); speed, 150 m/min (500 sfm). (b) Effect of feed on exit angle for ETP and OF coppers. Side rake angle, 6°; back rake angle, 0°; side cutting edge angle, 0°; nose radius, 0.125 mm (0.005 in.); depth of cut, 1.5 mm (0.06 in.); speed, 150 m/min (500 sfm). (c) Effect of nose radius and feed on exit angle for copper. Side rake angle, 6°; back rake angle 0°; side cutting edge angle 0°; depth of cut, 1.5 mm (0.06 in.); speed, 150 m/min (500 sfm). (d) Effect of side cutting edge angle on exit angle for OF copper. Side rake angle 6°; back rake angle, 0°; nose radius, 0.125 mm (0.005 in.); depth of cut, 1.5 mm (0.06 in.); speed, 50 m/min (165 sfm). (e) Effect of side rake and back rake angles on exit angle for 0.1 and 0.3 mm/rev (0.004 and 0.012 in./rev) feed for OF copper. Side rake angles, 6, 25, and 45° for 0° back rake angle; back rake angles, 0, 13, and 30° for 6° side rake angle; side cutting edge angle, 0°; nose radius, 0.125 mm (0.005 in.); depth of cut, 1.5 mm (0.06 in.); speed, 50 m/min (164 sfm). (f) Effect of depth of cut (bar diameters of 30 and 100 mm, or 1.2 and 4 in.) on the exit angle for OF copper. Side rake angles, 6 and 45°; back rake angle, 0°; side cutting edge angle, 0°; nose radius, 0.125 mm (0.005 in.); feed, 0.2 mm/rev (0.008 in./rev); speed, 50 m/min (165 sfm)

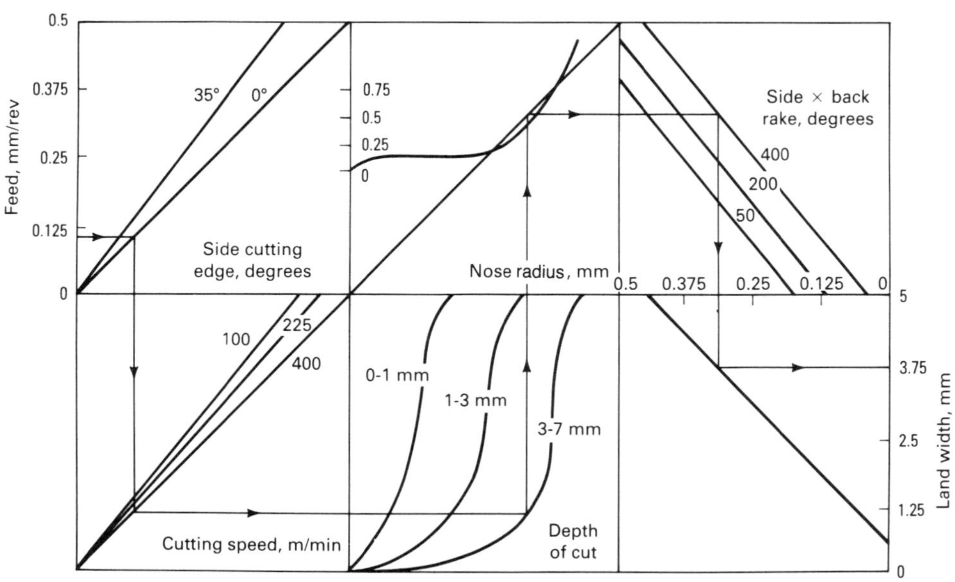


**Fig. 10** The appropriate land width of the chip breaker is determined as follows. Draw a line horizontally from the desired feed rate to intercept the value of side cutting edge angle, for example, 0°. Project down vertically to meet the cutting speed selected; horizontally to the depth of cut used, for example, 4 mm (0.16 in.); up to the nose radius value used, 0.5 mm (0.02 in.); across to the product of side × back rake, 400; then down to intercept the land width line. Lastly, project across to read off the value of land width, 3.75 mm (0.15 in.). This gives the best chip control for these cutting conditions, as shown in Fig. 11.

**Cutoff** tool designs for turning copper alloys are shown in Fig. 13(b); feeds and speeds, in Table 6.

## Boring

Techniques used for the boring of copper alloys do not differ greatly from those used for the boring of ferrous metals. Recommended angles for high-speed steel and carbide tools are:

| Angle | Degrees |
|---|---|
| Back rake | 0–10 |
| Side rake | 5–20 |
| End relief | 8–13 |
| Side relief | 5–8 |
| End cutting edge | 45 |
| Side cutting edge | 45 |

Table 7 lists speeds and feeds for the boring of copper alloys.

## Planing

Although planing is not used extensively on copper alloys, it may be preferred for

**Fig. 11** Overhead view of chip curler tool promoting chip breaking with ETP copper

some machining, such as on large workpieces or when quantity requirements are too small to justify the cost of a milling cutter. Techniques for planing other metals are generally applicable to planing copper alloys. Tool angles for high-speed steel and carbide planer tools are essentially the same as those for single-point turning (Table 3). Speeds for the rough and finish planing of copper alloys are given in Table 8. Recommendations of speeds and feeds for planing hard and soft bronzes at depths of cut different from those in Table 8 are given in Table 1 in the article "Planing" in this Volume.

## Broaching

Internal broaching has been used to cut narrow grooves in copper electrical fittings and to cut square and hexagonal shapes for valve and plumbing fittings. External broaching has been used in the manufacture of electric breaker arms. Sometimes broaching is one of a sequence of operations. Face and clearance angles for broaching copper alloys are given in Table 9; speeds, in Table 10.

## Drilling

For drilling copper alloys of groups 1 and 2, the included angle of the drill point is 118°, with 12 to 15° lip clearance (Fig. 14). For group 3 alloys, the included angle is 100 to 110° (for pure copper, smaller angles are sometimes used), and the lip clearance is 12 to 20°. Flat and straight-flute drills with no rake angle are sometimes used for the free-machining alloys in group 1, but a slow spiral drill with a helix angle of 10 to 22° is preferred for the deep-hole drilling of alloys in groups 1 and 2. For drilling group 3 alloys, drills with a greater helix angle (usually about 40°) are generally preferred. The greater angle is helpful in removing chips from deep holes. When drilling the group 3 alloys, it is sometimes helpful to notch the cutting edge, which splits the chips. Polished flutes are recommended for drilling copper and other tough group 3 alloys. One application for which drills with a high helix angle (fast spiral) were appropriate is described in the example that follows.

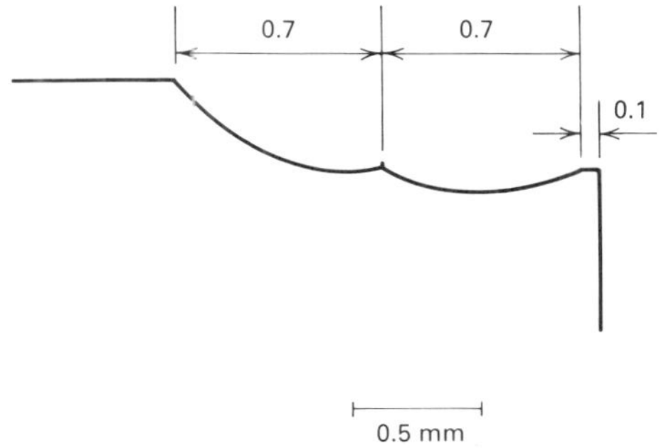


**Fig. 12** Profile of SNMG-61 chip-breaker groove found to be the best commercially available cemented carbide tool style for the chip breaking of unleaded 60/40 brass. Dimensions given in millimeters

**Example 1: Effect of Drill Geometry in Drilling Blind Holes in OFHC Cold-Drawn Copper.** High-speed steel drills (M2; 12 mm, or ½ in., in diameter) were used to produce a series of 10 holes, 50 mm (2 in.) deep, in blocks of cold-drawn OF copper at 650 rev/min and 0.2 mm/rev (0.008 in./rev) using an M2 cutting fluid (Ref 2). The drill geometry is given in Table 11.

The high-helix drill started by producing broken chips, which became long and continuous chips as cutting progressed, but this was remedied by reducing the feed. This chip-breaking effect with reduced feed was not seen with the other drills. The jobber drill, both types of low-helix drill, and the deep-hole drill produced long continuous chips, which could be broken only by external interference. The chips produced by the jobber and low-helix drills were thicker than those produced by the deep-hole and high-helix drills, the latter producing the thinnest chips.

Examination of the holes showed that during the early stages of cutting, the jobber drill produced a helical pattern in the hole surface that disappeared as cutting progressed, resulting in a clean hole apart from occasional areas of pickup and smearing. The helical pattern arose from the drilling of

**Table 3 Design of single-point tools for turning copper alloys**

| Angle | Alloys machined: Group 1 | Group 2 | Group 3 |
|---|---|---|---|
| **High-speed steel tools** | | | |
| Back rake, degrees | 0 | 5–10 | 10–20 |
| Side rake, degrees | 0–3 | 5–10 | 20–30 |
| End relief, degrees | 6 | 6–15 | 10–15 |
| Side relief, degrees | 0–5 | 5–10 | 10–20 |
| End cutting edge, degrees | 8–15 | 8–15 | 8–15 |
| Side cutting edge, degrees | 10–15 | 10–15 | 15 |
| **Carbide tools** | | | |
| Back rake, degrees | 0 | 0–5 | 4–8 |
| Side rake, degrees | 2–6 | 4–8 | 15–25 |
| End relief, degrees | 4–6 | 4–8 | 7–10 |
| Side relief, degrees | 4–6 | 4–8 | 7–10 |
| End cutting edge, degrees | 8–15 | 8–15 | 8–15 |
| Side cutting edge, degrees | 10–15 | 10–15 | 10–15 |

**Table 4 Nominal speeds for the turning of wrought and cast copper alloys with single-point and box tools**

| Alloy group | Condition | Hardness, HRB | Speed, m/min (sfm): High-speed steel tools(a) | Carbide (brazed)(b) | Carbide (indexable)(b) | Diamond |
|---|---|---|---|---|---|---|
| **Roughing(c)** | | | | | | |
| 1 (Table 1) | Annealed | 10–70 | 100 (335) | 230 (750) | 260 (860) | 230–460 (750–1500) |
| 2 | Annealed | 10–70 | 69 (225) | 155 (510) | 175 (580) | 150–305 (500–1000) |
| 3 | Annealed | 10–70 | 27 (90) | 64 (210) | 70 (230) | 120–245 (400–800) |
| 1 (Table 1) | Cold drawn | 60–100 | 120 (390) | 270 (880) | 305 (1000) | 305–550 (1000–1800) |
| 2 | Cold drawn | 60–100 | 76 (250) | 170 (560) | 190 (625) | 215–365 (700–1200) |
| 3 | Cold drawn | 60–100 | 30 (100) | 69 (225) | 76 (250) | 185–305 (600–1000) |
| 1 (Table 2) | As-cast | 40–150 HB (500 kg) | 100 (335) | 230 (750) | 260 (860) | 150–305 (500–1000) |
| 2 | As-cast | 40–175 HB (500 kg) | 69 (225) | 155 (510) | 175 (580) | 120–245 (400–800) |
| 3 | As-cast | 40–200 HB (500 kg) | 27 (90) | 64 (210) | 70 (230) | 90–150 (300–500) |
| **Finishing(d)** | | | | | | |
| 1 (Table 1) | Annealed | 10–70 | 145 (480) | 295 (960) | 335 (1100) | 760–1370 (2500–4500) |
| 2 | Annealed | 10–70 | 100 (325) | 200 (650) | 230 (750) | 610–1220 (2000–4000) |
| 3 | Annealed | 10–70 | 40 (130) | 84 (275) | 90 (300) | 460–915 (1500–3000) |
| 1 (Table 1) | Cold drawn | 60–100 | 175 (575) | 345 (1125) | 395 (1300) | 855–1460 (2800–4800) |
| 2 | Cold drawn | 60–100 | 110 (360) | 220 (725) | 250 (825) | 670–1280 (2200–4200) |
| 3 | Cold drawn | 60–100 | 46 (150) | 87 (285) | 100 (325) | 520–975 (1700–3200) |
| 1 (Table 2) | As-cast | 40–100 HB (500 kg) | 145 (480) | 295 (960) | 335 (1100) | 610–1220 (2000–4000) |
| 2 | As-cast | 40–100 HB (500 kg) | 100 (325) | 200 (650) | 230 (750) | 460–915 (1500–3000) |
| 3 | As-cast | 40–100 HB (500 kg) | 40 (130) | 84 (275) | 90 (300) | 305–760 (1000–2500) |

(a) Tool grade S4, S5 (M2, M3). (b) Carbide grade K20, M20, (C-2) for roughing operations and carbide grade K01, M10, (C-3) for finishing operations. (c) Depth of cut is 1.25–3.2 mm (0.05–0.125 in.) for diamond tools and 8 mm (0.3 in.) for high-speed steel and carbide tools. Roughing feed rates are 0.5 mm/rev (0.020 in./rev) for high-speed steel tools, 0.75 mm/rev (0.03 in./rev) for carbide tools, and 0.3–0.5 mm/rev (0.012–0.020 in./rev) for diamond tools. (d) Depth of cut is 0.13–0.4 mm (0.005–0.15 in.) for diamond and 1 mm (0.04 in.) for high-speed steel and carbide tools. Finishing feed rates are 0.075–0.15 mm/rev (0.003–0.006 in./rev) for diamond and 0.18 mm/rev (0.007 in./rev) for high-speed steel and carbide tools. Source: Metcut Research Associates Inc.

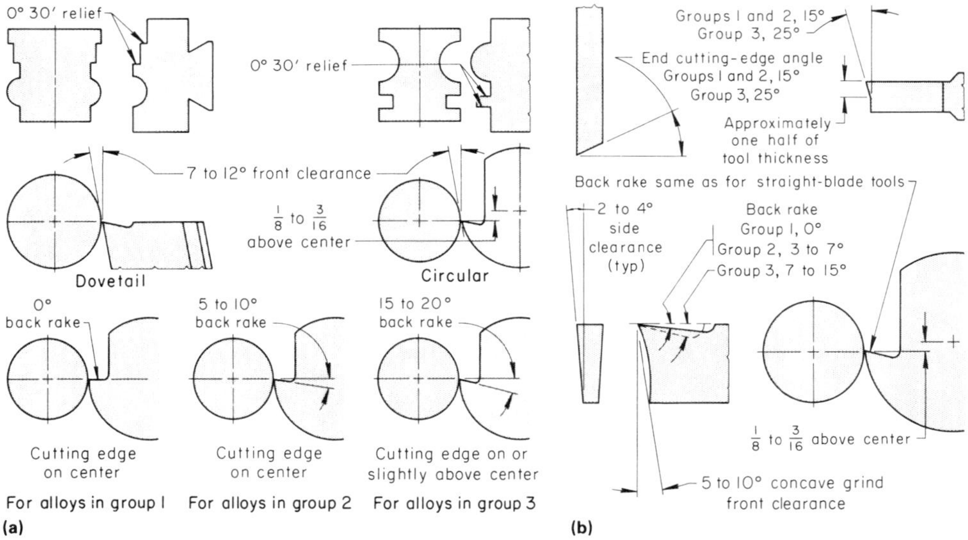


**Fig. 13** Design of form tools (a) and cutoff tools (b) for turning copper and copper alloys

### Table 5 Nominal speeds and feeds for the turning of copper alloys with form tools

| Alloy group | Condition | Hardness, HRB | Speed m/min | Speed sfm | Feed, mm/rev (in./rev), for tool width of: 13 mm (0.500 in.) | 25 mm (1.000 in.) | 50 mm (2.000 in.) |
|---|---|---|---|---|---|---|---|
| **High-speed steel tools** | | | | | | | |
| 1 | Annealed | 20–70 | 90 | 300 | 0.075 (0.003) | 0.05 (0.002) | 0.025 (0.001) |
| 2 | Annealed | 20–70 | 62 | 205 | 0.05 (0.002) | 0.05 (0.002) | 0.025 (0.001) |
| 3 | Annealed | 20–70 | 23 | 75 | 0.05 (0.002) | 0.05 (0.002) | 0.025 (0.001) |
| 1 | Cold drawn | 60–100 | 110 | 360 | 0.075 (0.003) | 0.05 (0.002) | 0.025 (0.001) |
| 2 | Cold drawn | 60–100 | 69 | 225 | 0.05 (0.002) | 0.05 (0.002) | 0.025 (0.001) |
| 3 | Cold drawn | 60–100 | 24 | 80 | 0.05 (0.002) | 0.05 (0.002) | 0.025 (0.001) |
| **Carbide tools** | | | | | | | |
| 1 | Annealed | 20–70 | 183 | 600 | 0.18 (0.007) | 0.13 (0.005) | 0.075 (0.003) |
| 2 | Annealed | 20–70 | 125 | 410 | 0.18 (0.007) | 0.13 (0.005) | 0.075 (0.003) |
| 3 | Annealed | 20–70 | 46 | 150 | 0.064 (0.0025) | 0.064 (0.0025) | 0.038 (0.0015) |
| 1 | Cold drawn | 60–100 | 213 | 700 | 0.18 (0.007) | 0.13 (0.005) | 0.075 (0.003) |
| 2 | Cold drawn | 60–100 | 137 | 450 | 0.18 (0.007) | 0.13 (0.005) | 0.075 (0.003) |
| 3 | Cold drawn | 60–100 | 52 | 170 | 0.064 (0.0025) | 0.064 (0.0025) | 0.038 (0.0015) |

Source: Metcut Research Associates Inc.

### Table 6 Nominal speeds and feeds for the turning of copper alloys with cutoff tools

| Alloy group | Condition | Hardness, HRB | Speed m/min | Speed sfm | Feed, mm/rev (in./rev), for tool width of: 1.6 mm (0.062 in.) | 3.2 mm (0.125 in.) | 6.4 mm (0.250 in.) |
|---|---|---|---|---|---|---|---|
| **High-speed steel tools** | | | | | | | |
| 1 | Annealed | 20–70 | 90 | 300 | 0.1 (0.004) | 0.13 (0.005) | 0.15 (0.006) |
| 2 | Annealed | 20–70 | 62 | 205 | 0.1 (0.004) | 0.13 (0.005) | 0.15 (0.006) |
| 3 | Annealed | 20–70 | 23 | 75 | 0.05 (0.002) | 0.05 (0.002) | 0.09 (0.0035) |
| 1 | Cold drawn | 60–100 | 110 | 360 | 0.1 (0.004) | 0.13 (0.005) | 0.15 (0.006) |
| 2 | Cold drawn | 60–100 | 69 | 225 | 0.1 (0.004) | 0.13 (0.005) | 0.15 (0.006) |
| 3 | Cold drawn | 60–100 | 24 | 80 | 0.05 (0.002) | 0.05 (0.002) | 0.09 (0.0035) |
| **Carbide tools** | | | | | | | |
| 1 | Annealed | 20–70 | 183 | 600 | 0.13 (0.005) | 0.15 (0.006) | 0.18 (0.007) |
| 2 | Annealed | 20–70 | 125 | 410 | 0.13 (0.005) | 0.15 (0.006) | 0.18 (0.007) |
| 3 | Annealed | 20–70 | 46 | 150 | 0.065 (0.0025) | 0.09 (0.0035) | 0.11 (0.0045) |
| 1 | Cold drawn | 60–100 | 213 | 700 | 0.13 (0.005) | 0.15 (0.006) | 0.18 (0.007) |
| 2 | Cold drawn | 60–100 | 137 | 450 | 0.13 (0.005) | 0.15 (0.006) | 0.18 (0.007) |
| 3 | Cold drawn | 60–100 | 52 | 170 | 0.65 (0.0025) | 0.09 (0.0035) | 0.11 (0.0045) |

Source: Metcut Research Associates Inc.

an oversize hole, with the oversize decreasing as the drill penetrated the hole. With the high-helix and low-helix (118° point angle) drills, the helical formation persisted over the entire length of the hole, which also had an extremely poor surface due to pickup and smearing. The cleanest hole was produced by the deep-hole drill because of the

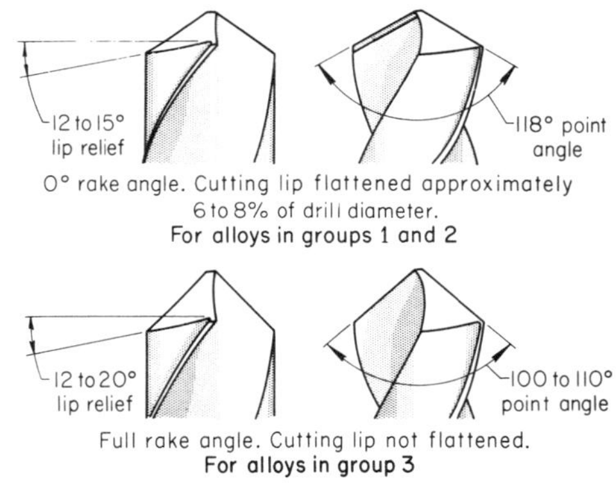


**Fig. 14** Drill point and clearance angles for twist drills

greater clearance of the deep-hole drill chisel edge corner and because reduction of the chisel edge length reduced the tendency of the swarf to snag the surface of the hole. Consequently, if chip disposal is important and surface finish is not, the high-helix drill should be used; conversely, the deep-hole drill should be used.

The least amount of oversizing occurred with the low-helix (90° point angle) drill, which produced a hole having a poor surface finish but without the helical pattern. However, hole accuracy is extremely dependent on the symmetry of the drill point and is altered by such factors as the lip height and asymmetry of the drill. The jobber, deep-hole, and low-helix (118°) drills had to penetrate a sufficient depth into the hole for the lands to support the drill, resulting in greater accuracy with increasing depth.

**Speeds and feeds** for drilling the three groups of copper alloys are given in Tables 12 and 13.

**Type of machine** or method of fixturing can contribute markedly to drilling efficiency. Mainly because of the shapes and sizes of workpieces commonly made from copper alloys, machines capable of performing several operations consecutively or simultaneously are often used.

## Reaming

Nominal feeds and speeds for the reaming of copper alloys are given in Table 14. Reamers for finishing holes in copper alloy parts should have a right-hand spiral with positive-rake flutes. Polished cutting edges produce optimum results, especially on alloys that machine with stringy chips. Reamers with inserted carbide blades are best for copper alloys, but they are more expensive than high-speed steel reamers, particularly in the smaller sizes, and therefore are usually used only in mass production. Reamer design is often altered to improve results, as in the following example.

**Example 2: Revised Reamer Design to Meet Close Tolerances.** Close tolerance (±0.01 mm, or ±0.0005 in.) was specified

**Table 7 Nominal speeds and feeds for the boring of wrought and cast copper alloys**

| Alloy group | Condition | Hardness, HRB | Speed, m/min (sfm): High-speed steel tools(a) | Carbide (brazed)(b) | Carbide (indexable)(b) | Diamond |
|---|---|---|---|---|---|---|
| **Roughing(c)** | | | | | | |
| 1 (Table 1) .. | Annealed | 10–70 | 100 (320) | 170 (550) | 200 (650) | 245 (800) |
| 2 ........... | Annealed | 10–70 | 67 (220) | 115 (380) | 137 (450) | 245 (800) |
| 3 ........... | Annealed | 10–70 | 27 (90) | 46 (150) | 53 (175) | 150 (500) |
| 1 (Table 1) .. | Cold drawn | 60–100 | 117 (385) | 200 (650) | 235 (775) | 365 (1200) |
| 2 ........... | Cold drawn | 60–100 | 73 (240) | 125 (405) | 145 (475) | 305 (1000) |
| 3 ........... | Cold drawn | 60–100 | 30 (100) | 49 (160) | 58 (190) | 245 (800) |
| 1 (Table 2) .. | As-cast | 40–150 HB (500 kg) | 100 (320) | 170 (550) | 200 (650) | 245 (800) |
| 2 ........... | As-cast | 40–175 HB (500 kg) | 67 (220) | 115 (375) | 134 (440) | 120 (400) |
| 3 ........... | As cast | 40–200 HB (500 kg) | 27 (90) | 46 (150) | 53 (175) | 120 (400) |
| **Finishing(d)** | | | | | | |
| 1 (Table 1) .. | Annealed | 10–70 | 145 (480) | 250 (825) | 295 (970) | 915 (3000) |
| 2 ........... | Annealed | 10–70 | 100 (325) | 170 (550) | 200 (650) | 760 (2500) |
| 3 ........... | Annealed | 10–70 | 40 (130) | 67 (220) | 79 (260) | 520 (1700) |
| 1 (Table 1) .. | Cold drawn | 60–100 | 175 (575) | 290 (950) | 345 (1125) | 1070 (3500) |
| 2 ........... | Cold drawn | 60–100 | 110 (360) | 185 (610) | 220 (720) | 915 (3000) |
| 3 ........... | Cold drawn | 60–100 | 46 (150) | 73 (240) | 85 (280) | 670 (2200) |
| 1 (Table 2) .. | As-cast | 40–150 HB (500 kg) | 145 (480) | 250 (825) | 295 (970) | 760 (2500) |
| 2 ........... | As-cast | 40–175 HB (500 kg) | 100 (325) | 170 (555) | 200 (655) | 365 (1200) |
| 3 ........... | As-cast | 40–200 HB (500 kg) | 40 (130) | 67 (220) | 79 (260) | 520 (1700) |

(a) Tool grade S4, S5 (M2, M3). (b) Carbide grade K20, M20, (C-2) for roughing operations and grade K01, M10, (C-3) for finishing operations. (c) Depth of cut is 1.25–3.2 mm (0.05–0.125 in.) with diamond tools and 2.5 mm (0.1 in.) with high-speed steel and carbide tools. Roughing feed rates are 0.3 mm/rev (0.012 in./rev) for high-speed steel tools, 0.4 mm/rev (0.015 in./rev) for carbide tools, and 0.3–0.5 mm/rev (0.012–0.02 in./rev) for diamond tools. (d) Depth of cut is 0.13–0.4 mm (0.005–0.015 in.) for diamond tools and 0.25 mm (0.01 in.) for high-speed steel and carbide tools. Finishing feed rates are 0.75–0.15 mm/rev (0.003–0.006 in./rev) for diamond tools and 0.075 mm/rev (0.003 in./rev) for high-speed steel and carbide tools. Source: Metcut Research Associates Inc.

**Table 8 Nominal speeds for the planing of copper alloys**

| Alloy group | Condition | Hardness, HRB | Speed, m/min (sfm): High-speed steel tools | Carbide tools |
|---|---|---|---|---|
| **Rough planing(a)** | | | | |
| 1 ........ | Annealed | 20–70 | 55 (180) | 90 (300) |
| 2 ........ | Annealed | 20–70 | 37 (120) | 90 (300) |
| 3 ........ | Annealed | 20–70 | 15 (50) | 76 (250) |
| 1 ........ | Cold drawn | 60–100 | 60 (200) | 90 (300) |
| 2 ........ | Cold drawn | 60–100 | 40 (130) | 90 (300) |
| 3 ........ | Cold drawn | 60–100 | 18 (60) | 76 (250) |
| **Finish planing(b)** | | | | |
| 1 ........ | Annealed | 20–70 | 67 (220) | 90 (300) |
| 2 ........ | Annealed | 20–70 | 49 (160) | 90 (300) |
| 3 ........ | Annealed | 20–70 | 21 (70) | 76 (250) |
| 1 ........ | Cold drawn | 60–100 | 82 (270) | 90 (300) |
| 2 ........ | Cold drawn | 60–100 | 52 (170) | 90 (300) |
| 3 ........ | Cold drawn | 60–100 | 23 (75) | 76 (250) |

(a) Depth of cut, 13 mm (0.500 in.); feed, 1.5 mm/stroke (0.060 in./stroke). (b) Depth of cut, 0.25 mm (0.010 in.); feed, ¾ the width of square-nose finishing tool. Source: Metcut Research Associates Inc.

**Table 9 Tool angles for the broaching of copper alloys**

| Alloy group | Face angle, degrees | Clearance angle, degrees |
|---|---|---|
| 1 .................. | −5–5 | 1–2 |
| 2 ................ | 0–10 | 1–2 |
| 3 ................ | 10–15 | 2–3 |

on the diameter of the flat-bottom blind hole reamed to 19.9/19.7 mm (0.786/0.776 in.) depth in the alloy C37700 (45 HRB) forging shown at the top of Fig. 15. Parts were rejected because the eight-flute reamer (lower left, Fig. 15) originally used was forced off center in bottoming out, causing out-of-roundness at the base of the hole.

Changing to a single-flute reamer (lower right, Fig. 15) and incorporating two wear strips to prevent the tool from being forced off center when bottoming out resulted in parts meeting the required tolerance. Both reamers had carbide cutting edges, and the same speed (37 m/min, or 120 sfm) and feed (0.13 mm/rev, or 0.005 in./rev).

**Table 10 Nominal speeds and chip loads for the broaching of copper alloys with high-speed steel tools**

| Alloy group | Speed: m/min | sfm | Chip load: mm/tooth | in./tooth |
|---|---|---|---|---|
| 1.................. | 9 | 30 | 0.13 | 0.005 |
| 2................ | 9 | 30 | 0.13 | 0.005 |
| 3................ | 7.6 | 25 | 0.10 | 0.004 |

**Table 11 Drill geometries used in drilling blind holes in OFHC copper**

| Drill style | Helix angle, degrees | Point angle, degrees | Lip clearance angle, degrees |
|---|---|---|---|
| Jobbers length............. | 34 | 118 | 13 ± 2 |
| High helix (jobbers length).. | 39 | 118 | 13 ± 2 |
| Low helix (jobbers length) .. | 15 | 118 | 13 ± 2 |
| | 15 | 90 | 13 ± 2 |
| Deep hole (long series length)................. | 38 | 118 | 13 ± 2 |

## Roller Burnishing

Bores in copper alloys are especially well suited to finishing by roller burnishing or simultaneous roller burnishing and peening, because copper alloys have low hardness and high ductility. The article "Roller Burnishing" in this Volume describes the finishing of bores by roller peen burnishing.

## Tapping and Threading

The equipment, tools, and techniques used for cutting internal and external threads in copper alloys are not greatly different from those used for ferrous metals. Straight, four-flute taps with a rake angle of 3 to 6°; chamfer relief angle of 10°; and chamfer angles of 5° on taper taps, 15° on plug taps, and 30° on bottom taps are recommended for tapping group 1 copper alloys. For tapping group 2 and 3 alloys, spiral-pointed taps with two or three flutes are preferred. Taps for group 2 should have rake angles of 9 to 12°, and for group 3 rake angles should be 15 to 18°. For tapping alloys of both groups 2 and 3, taps should have chamfer relief angles of 12°, with chamfer angles of 15° on plug taps and 30° on bottom taps.

General-purpose grades of high-speed steel such as M1, M7, and M10 are used for taps for copper alloys. The more highly alloyed grades of high-speed steel are rarely justified. However, the nitriding or chromium plating of taps is often advantageous, especially for tapping group 2 and 3 alloys. One application in which surface-treated taps made higher speeds possible is described in the following example.

**Example 3: Untreated Versus Nitrided and Chromium-Plated Taps.** In tapping the ⅝-11 UNC-2B threads in the alloy C61000 (aluminum bronze) connector sleeve shown in Fig. 16, the production rate was increased almost 40% when taps that had been nitrided and flash chromium plated were substituted for untreated taps. This increase was caused mainly by the higher speed at which the surface-treated taps could be operated (900 rev/min as compared to 575 rev/min for the untreated taps).

The sleeves were tapped in a vertical drill press, under the conditions given in the accompanying table in Fig. 16. The operator inserted the square shank end of the tap into the rotating tap driver and held the tap until it started threading the workpiece, which was held in an air-operated chuck. As the tap pulled itself through, enough pressure was applied to the handle to keep the tap driver engaged. The operator then retracted the tap driver, opened the chuck, and removed the workpiece over the shank end of the tap.

**Tapping Speed.** Conservative speeds used for tapping copper alloys are:

| Alloy group | Speed: m/min | sfm |
|---|---|---|
| 1........................... | 26–30 | 85–100 |
| 2......................... | 18–23 | 60–75 |
| 3......................... | 12–15 | 40–50 |

## Table 12 Nominal speeds and feeds for the drilling of copper alloys with high-speed steel tools

| | Speed, m/min (sfm) | | Feed, mm/rev (in./rev), for nominal diameter of: | | | | | | | |
|---|---|---|---|---|---|---|---|---|---|---|
| Alloy group | Annealed(a) | Cold drawn(b) | 1.6 mm (0.062 in.) | 3.2 mm (0.125 in.) | 6.4 mm (0.250 in.) | 13 mm (0.500 in.) | 19 mm (0.750 in.) | 25 mm (1.000 in.) | 38 mm (1.500 in.) | 50 mm (2.000 in.) |
| 1 | 43 (140) | 43 (140) | 0.025 (0.001) | 0.075 (0.003) | ... | ... | ... | ... | ... | ... |
| | 53 (175) | 60 (200) | ... | ... | 0.10 (0.004) | 0.20 (0.008) | 0.30 (0.012) | 0.46 (0.018) | 0.50 (0.020) | 0.56 (0.022) |
| 2 | 30 (100) | 37 (120) | 0.025 (0.001) | ... | ... | ... | ... | ... | ... | ... |
| | 37 (120) | 43 (140) | ... | 0.075 (0.003) | 0.15 (0.006) | 0.25 (0.010) | 0.25 (0.010) | 0.38 (0.015) | 0.38 (0.015) | 0.50 (0.020) |
| 3 | 15 (50) | 18 (60) | 0.025 (0.001) | ... | ... | ... | ... | ... | ... | ... |
| | 18 (60) | 23 (75) | ... | 0.075 (0.003) | 0.10 (0.004) | 0.20 (0.008) | 0.25 (0.010) | 0.30 (0.012) | 0.38 (0.015) | 0.50 (0.020) |

(a) 20–70 HRB. (b) 60–100 HRB. Source: Metcut Research Associates Inc.

## Table 13 Nominal speeds for the gun drilling of copper alloys with carbide tools

| | | Feed, mm/rev (in./rev), for nominal diameter of: | | | | |
|---|---|---|---|---|---|---|
| Alloy group | Speed, m/min (sfm) | 6.4 mm (0.25 in.) | 6.4–13 mm (0.25–0.50 in.) | 13–19 mm (0.50–0.75 in.) | 19–25 mm (0.75–1.00 in.) | 25–50 mm (1.00–2.00 in.) |
| 1 | 183 (600) | 0.025 (0.0010) | 0.075 (0.003) | 0.13 (0.005) | 0.20 (0.008) | 0.25 (0.010) |
| 2 | 152 (500) | 0.01 (0.0005) | 0.025 (0.001) | 0.075 (0.003) | 0.13 (0.005) | 0.20 (0.008) |
| 3 | 107 (350) | 0.01 (0.0005) | 0.025 (0.001) | 0.075 (0.003) | 0.13 (0.005) | 0.20 (0.008) |

## Table 14 Nominal speeds and feeds for the reaming of copper alloys with six-flute reamers

| | Speed, m/min (sfm) | | | | Feed, mm/rev (in./rev), for nominal diameter of: | | | | | |
|---|---|---|---|---|---|---|---|---|---|---|
| | Annealed(a) | | Cold drawn(b) | | | | | | | |
| Alloy group | High-speed steel | Carbide | High-speed steel | Carbide | 3.2 mm (0.125 in.) | 6.4 mm (0.250 in.) | 13 mm (0.500 in.) | 25 mm (1.000 in.) | 38 mm (1.500 in.) | 50 mm (2.000 in.) |
| 1 | 49 (160) | 98 (320) | 55 (180) | 108 (360) | 0.075 (0.003) | 0.15 (0.006) | 0.23 (0.009) | 0.30 (0.012) | 0.38 (0.015) | 0.50 (0.020) |
| 2 | 34 (110) | 76 (250) | 37 (120) | 84 (275) | 0.075 (0.003) | 0.15 (0.006) | 0.23 (0.009) | 0.30 (0.012) | 0.38 (0.015) | 0.50 (0.020) |
| 3 | 15 (50) | 55 (180) | 18 (60) | 61 (200) | 0.075 (0.003) | 0.13 (0.005) | 0.18 (0.007) | 0.25 (0.010) | 0.30 (0.012) | 0.38 (0.015) |

(a) 20–70 HRB. (b) 60–100 HRB. Source: Metcut Research Associates Inc.

However, higher speeds are often used successfully (as in Example 3). Pitch of thread, depth of hole, and type of cutting fluid all influence the selection of speed. Speed can usually be increased as thread pitch becomes finer.

The foregoing recommendations are based on the use of water-soluble oil as the cutting fluid. Mineral oil with lard oil additions will usually improve efficiency and thread quality in either tapping or die threading.

Difficulty in disposing of chips when blind holes are being tapped is not unique to copper alloys, but because copper alloys produce soft chips, they are likely to pack in blind holes. In some cases, chip disposal problems can be eliminated by converting blind holes into through holes.

**Cold form tapping** is often more successful for copper alloys than for ferrous metals because copper alloys are softer and more ductile. The following example describes an application for which the cold form method resulted in better efficiency in tapping blind holes.

**Example 4: Blind Hole Tapped by Cold Forming.** A fluteless tap in a single-spindle automatic lathe was used in the high-production cold forming of 5-40 UNC-2B threads in the blind hole in one end of the alloy C51000 (phosphor bronze) mounting stud shown in Fig. 17. In previous runs using regular taps, downtime for tool changes averaged 6 min per 100 pieces, and in a secondary process, cleaning of chips from the holes required 15 min per 100 pieces. With the conditions given in the table accompanying Fig. 17, use of the fluteless tap resulted in a saving of 35 h in a production run of 10 000 pieces.

**Threading With Dies.** When threading copper alloys of group 1 with tangential chasers, the rake angle should be 0 to 10° (positive). For threading alloys of group 2 the rake angle should be increased to 20 to 25°, and for threading commercially pure copper or other alloys of group 3 the rake angle should be 30 to 35° for tangential chasers. On circular chasers, face angles of 1½ to 2° are satisfactory for threading alloys of all three groups. The hole angle

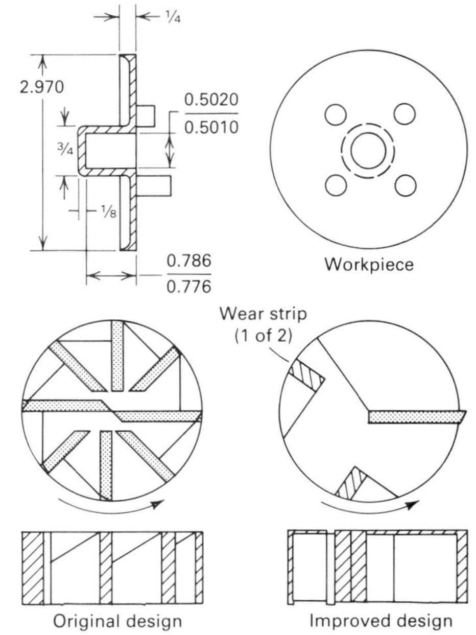


**Fig. 15** Change from eight-flute reamer (bottom left) to single-flute reamer with wear strips (bottom right) eliminated out-of-roundness in reaming the blind hole in the alloy C37700 forging shown at top. Dimensions given in inches

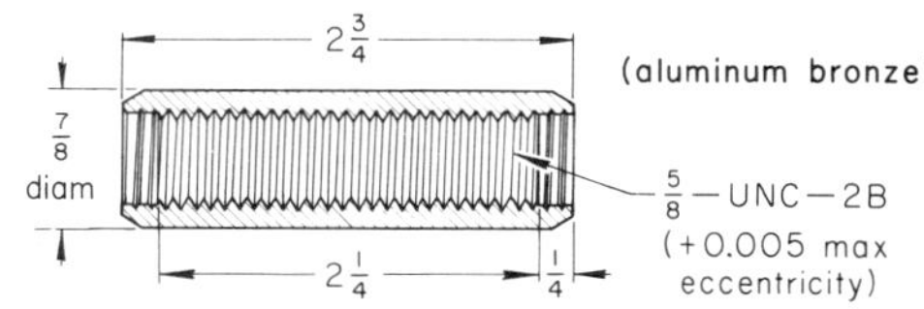


| Condition or result(a) | Untreated taps | Treated taps |
|---|---|---|
| Spindle speed, rev/min | 575 | 900 |
| Surface speed, m/min (sfm) | 29 (94) | 45 (147) |
| Production/h, pieces | 180 | 250 |
| Tool life/grind, pieces | ... | 1000 |
| **Tap defects** | | |
| Type | Straight-shank plug, modified(b) | |
| Size | ⅝-11 UNC-2B (GH6) | |
| Number of flutes | 4 | |
| Rake angle | Negative, 0–5° | |
| Overall length, mm (in.) | 213 (8⅜) | |
| Perfect-thread length, mm (in.) | 40 (1⅝) | |
| Material | High-speed steel | |

(a) For both taps, feed was established by tap lead, with hand pressure applied only to maintain engagement of the tap driver, and sulfochlorinated oil was used as the cutting fluid. (b) End was chamfered 45° from a 4.0 mm (5/32 in.) diameter (to ensure the concentricity of the first thread), beyond which point the tap had conventional plug chamfer.

**Fig. 16** Untreated versus nitrided and chromium-plated taps in tapping an alloy C61000 connector sleeve (95 HRB). Tapping was done in a vertical drill press. Dimensions in figure given in inches

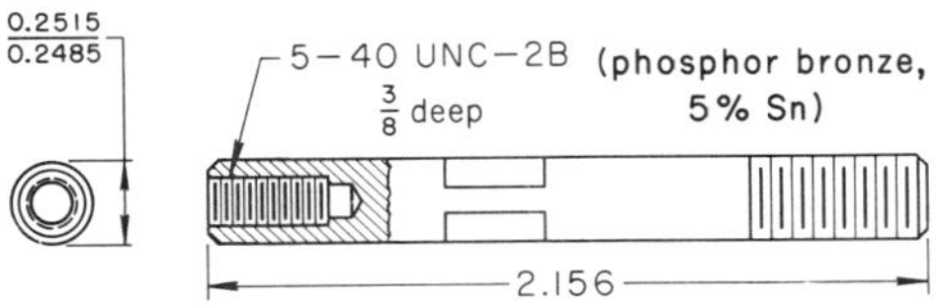


| | |
|---|---|
| Machine | Single-spindle automatic lathe |
| Tool | Fluteless tap, 3.18 mm (0.125 in.) in diameter |
| Speed, at 3770 rev/min, m/min (sfm) | 38 (123) |
| Cutting fluid | Light mineral oil |
| Setup time | 2 h |
| Cycle time | 6.5 s/piece |
| Production rate | 500 pieces/h |
| Tool life | (a) |
| Time saved | |
| Cleaning chips | 25 h/10 000 pieces |
| Tool change downtime | 10 h/10 000 pieces |

(a) Because data are for 10 000 pieces, tool life was not determined to point of failure or regrind. The tap used maintained thread size, quality, and concentricity with outside diameter within 0.075 mm (0.003 in.), as specified, throughout the entire production run.

**Fig. 17** Cold form tapping of a blind hole in an alloy C51000 mounting stud. Dimensions in figure given in inches

on circular chasers should be about zero for threading the alloys of group 1 (up to 5° negative for free-cutting brass) and 15 to 20° positive for threading alloys of groups 2 and 3.

Soluble oil is often used as the cutting fluid for die threading. However, low-viscosity mineral oil or mineral oil with an addition of about 10% lard oil will usually give better finish and tool life, especially when alloys of groups 2 and 3 are being threaded.

In external threading of group 1 alloys, a speed of 15 m/min (50 sfm) is recommended for cutting a 7-pitch thread, while speed may be about 55 m/min (180 sfm) for a 24-pitch thread. For the same pitches, recommended speeds are 9 and 30 m/min (30 and 100 sfm) for alloys of group 2 and less for those group 3 alloys that produce less

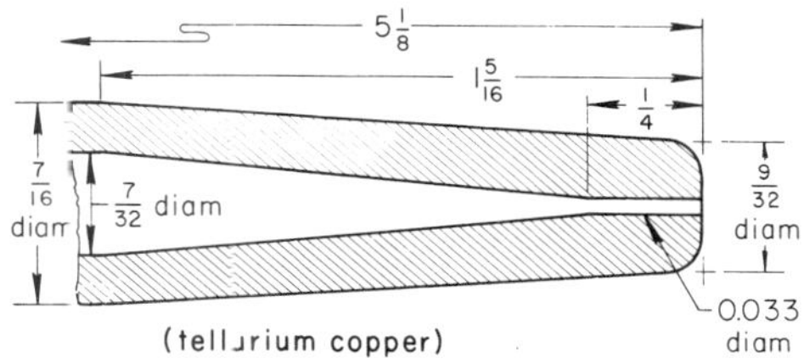


**Fig. 18** Swaged end of a spout produced from copper tubing (Alloy C14500). Production time was reduced by swaging the 6 mm (¼ in.) long hole over 0.84 mm (0.033 in.) diam music wire instead of producing the hole by drilling and reaming after swaging. Dimensions given in inches

stringy chips. For commercially pure copper, speeds of 5 and 12 m/min (15 and 40 sfm), respectively, for the 7 and 24 thread pitches are recommended. Speeds for threads of other pitches can be determined by interpolation.

## Multiple-Operation Machining

Copper alloys, particularly the leaded alloys, are especially well suited to machining in multiple-operation equipment such as turret lathes and automatic bar and chucking machines. Moreover, because the machining and forming characteristics of copper alloys are generally better than those of ferrous metals, it is often possible to use procedures and to incorporate operations in multiple-operation sequences for machining copper alloys that would not be practical for harder and less formable metals. The following two examples describe procedures in which the ductility of copper alloys was used to advantage in saving time or material in multiple-operation machining.

**Example 5: Swaging That Eliminated Drilling and Reaming.** Figure 18 shows the swaged end of a nozzle produced in a turret lathe from tellurium copper tubing. Originally, the 6 mm (¼ in.) long hole was produced after the end had been swaged, by drilling to 0.74 mm (0.029 in.) diameter and then reaming to size. Drilling required 17 s, reaming 15 s.

Drilling and reaming were eliminated by revising the processing so that the hole was formed to size during swaging by the use of a mandrel of 0.84 mm (0.033 in.) diam music wire. Insertion and removal of the wire required only 3 s, which resulted in a saving of 29 s in machining time per piece. The improved method also eliminated tool and downtime costs for the drills and reamers.

**Example 6: Spin Flaring That Saved Material.** Figure 19 illustrates the technique used in a small multiple-spindle automatic bar machine for providing an insert nut made from 5 mm (3⁄16 in.) bar stock of alloy 342 (high-leaded brass) with one end flared to 6 mm (¼ in.) in diameter. This technique (spinning, followed by trimming and facing with a form tool) enabled a 43.7% savings in material that would have been required had the part been more conventionally produced by turning 6 mm (¼ in.) diam bar.

Spinning was done with a two-lip, cam-ground, form-relieved tool at a speed of 23 m/min (75 sfm) and a feed of 0.064 mm/rev (0.0025 in./rev). Other machining opera-

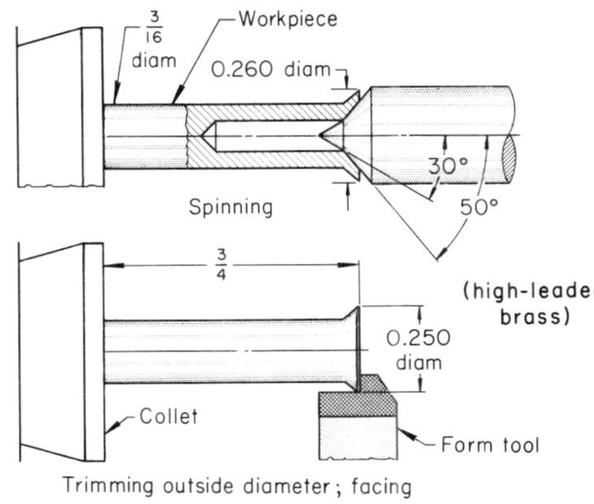


**Fig. 19** Spinning and trimming that eliminated necessity for oversize bar stock and material waste by machining. Dimensions given in inches

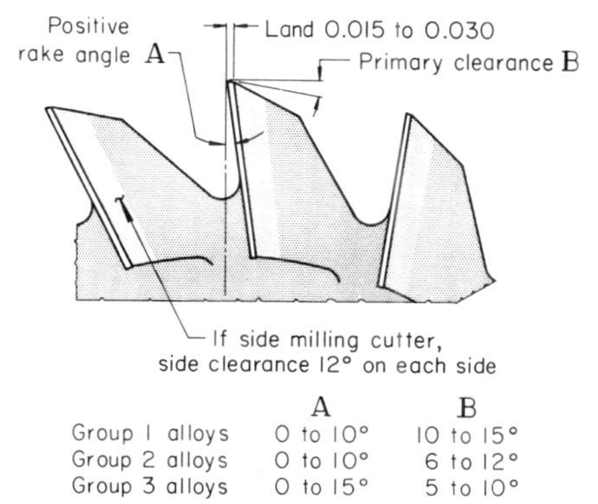


**Fig. 20** Recommended angles and land width for peripheral and side milling cutters. Dimensions given in inches

**Table 15 Nominal speeds and feeds for the peripheral milling of copper alloys with high-speed steel cutters**

| Alloy group | Condition | Hardness, HRB | Speed m/min | Speed sfm | Feed mm/rev | Feed in./rev |
|---|---|---|---|---|---|---|
| **Depth of cut, 3.2 mm (0.125 in.)** | | | | | | |
| 1 | Annealed | 20–70 | 130 | 430 | 0.46 | 0.018 |
| 2 | Annealed | 20–70 | 90 | 300 | 0.28 | 0.011 |
| 3 | Annealed | 20–70 | 37 | 120 | 0.18 | 0.007 |
| 1 | Cold drawn | 60–100 | 155 | 510 | 0.46 | 0.018 |
| 2 | Cold drawn | 60–100 | 107 | 350 | 0.28 | 0.011 |
| 3 | Cold drawn | 60–100 | 43 | 140 | 0.18 | 0.007 |
| **Depth of cut, 0.64 mm (0.025 in.)** | | | | | | |
| 1 | Annealed | 20–70 | 160 | 525 | 0.41 | 0.016 |
| 2 | Annealed | 20–70 | 114 | 375 | 0.23 | 0.009 |
| 3 | Annealed | 20–70 | 50 | 165 | 0.13 | 0.005 |
| 1 | Cold drawn | 60–100 | 190 | 625 | 0.41 | 0.016 |
| 2 | Cold drawn | 60–100 | 130 | 425 | 0.23 | 0.009 |
| 3 | Cold drawn | 60–100 | 60 | 200 | 0.13 | 0.005 |

Source: Metcut Research Associates Inc.

**Table 16 Nominal speeds and feeds for the face milling of copper alloys with high-speed steel cutters**

| Alloy group | Condition | Hardness, HRB | Speed | | Feed | |
|---|---|---|---|---|---|---|
| | | | m/min | sfm | mm/rev | in./rev |
| **Depth of cut, 3.2 mm (0.125 in.)** | | | | | | |
| 1 | Annealed | 20–70 | 137 | 450 | 0.5 | 0.020 |
| 2 | Annealed | 20–70 | 99 | 325 | 0.36 | 0.014 |
| 3 | Annealed | 20–70 | 30 | 100 | 0.25 | 0.010 |
| 1 | Cold drawn | 60–100 | 152 | 500 | 0.5 | 0.020 |
| 2 | Cold drawn | 60–100 | 107 | 350 | 0.36 | 0.014 |
| 3 | Cold drawn | 60–100 | 35 | 115 | 0.25 | 0.010 |
| **Depth of cut, 0.64 mm (0.025 in.)** | | | | | | |
| 1 | Annealed | 20–70 | 168 | 550 | 0.30 | 0.012 |
| 2 | Annealed | 20–70 | 130 | 425 | 0.25 | 0.010 |
| 3 | Annealed | 20–70 | 40 | 130 | 0.20 | 0.008 |
| 1 | Cold drawn | 60–100 | 183 | 600 | 0.30 | 0.012 |
| 2 | Cold drawn | 60–100 | 137 | 450 | 0.25 | 0.010 |
| 3 | Cold drawn | 60–100 | 46 | 150 | 0.20 | 0.008 |

Source: Metcut Research Associates Inc.

**Table 17 Nominal speeds and feeds for the face milling of copper alloys with carbide cutters**

| Alloy group | Condition | Hardness, HRB | Speed, m/min (sfm) | | Feed | |
|---|---|---|---|---|---|---|
| | | | Brazed | Disposable | mm/rev | in./rev |
| **Depth of cut, 3.2 mm (0.125 in.)** | | | | | | |
| 1 | Annealed | 20–70 | 274 (900) | 302 (990) | 0.46 | 0.018 |
| 2 | Annealed | 20–70 | 183 (600) | 201 (660) | 0.36 | 0.014 |
| 3 | Annealed | 20–70 | 76 (250) | 84 (275) | 0.25 | 0.010 |
| 1 | Cold drawn | 60–100 | 305 (1000) | 335 (1100) | 0.46 | 0.018 |
| 2 | Cold drawn | 60–100 | 198 (650) | 218 (715) | 0.36 | 0.014 |
| 3 | Cold drawn | 60–100 | 91 (300) | 100 (330) | 0.25 | 0.010 |
| **Depth of cut, 0.64 mm (0.025 in.)** | | | | | | |
| 1 | Annealed | 20–70 | 366 (1200) | 402 (1320) | 0.30 | 0.012 |
| 2 | Annealed | 20–70 | 244 (800) | 268 (880) | 0.25 | 0.010 |
| 3 | Annealed | 20–70 | 91 (300) | 100 (330) | 0.20 | 0.008 |
| 1 | Cold drawn | 60–100 | 396 (1300) | 436 (1430) | 0.30 | 0.012 |
| 2 | Cold drawn | 60–100 | 260 (850) | 285 (935) | 0.25 | 0.010 |
| 3 | Cold drawn | 60–100 | 107 (350) | 117 (385) | 0.20 | 0.008 |

Source: Metcut Research Associates Inc.

**Table 18 Nominal speeds and feeds for the end milling of copper alloys with high-speed steel cutters**

| Alloy group | Condition | Hardness, HRB | Speed, m/min (sfm) | Feed, mm/rev (in./rev), for cutter diameter of: | | |
|---|---|---|---|---|---|---|
| | | | | 6.4 mm (0.250 in.) | 19 mm (0.750 in.) | 25–50 mm (1.0–2.0 in.) |
| **Depth of cut, 1.3 mm (0.050 in.)** | | | | | | |
| 1 | Annealed | 20–70 | 114 (375) | 0.075 (0.003) | 0.20 (0.008) | 0.25 (0.010) |
| 2 | Annealed | 20–70 | 76 (250) | 0.05 (0.002) | 0.15 (0.006) | 0.20 (0.008) |
| 3 | Annealed | 20–70 | 38 (125) | 0.04 (0.0015) | 0.13 (0.005) | 0.18 (0.007) |
| 1 | Cold drawn | 60–100 | 122 (400) | 0.075 (0.003) | 0.20 (0.008) | 0.25 (0.010) |
| 2 | Cold drawn | 60–100 | 84 (275) | 0.05 (0.002) | 0.15 (0.006) | 0.20 (0.008) |
| 3 | Cold drawn | 60–100 | 38 (125) | 0.04 (0.0015) | 0.13 (0.005) | 0.18 (0.007) |
| **Depth of cut, 0.04 mm (0.015 in.)** | | | | | | |
| 1 | Annealed | 20–70 | 137 (450) | 0.05 (0.002) | 0.13 (0.005) | 0.18 (0.007) |
| 2 | Annealed | 20–70 | 99 (325) | 0.05 (0.002) | 0.13 (0.005) | 0.15 (0.006) |
| 3 | Annealed | 20–70 | 46 (150) | 0.025 (0.001) | 0.10 (0.004) | 0.13 (0.005) |
| 1 | Cold drawn | 60–100 | 152 (500) | 0.05 (0.002) | 0.13 (0.005) | 0.18 (0.007) |
| 2 | Cold drawn | 60–100 | 107 (350) | 0.05 (0.002) | 0.13 (0.005) | 0.15 (0.006) |
| 3 | Cold drawn | 60–100 | 53 (175) | 0.025 (0.001) | 0.10 (0.004) | 0.13 (0.005) |

Source: Metcut Research Associates Inc.

tions included drilling, tapping, knurling, recessing, and cutting off; production rate was 1200 pieces per hour.

**The dimensional accuracy** that can be maintained in machining copper alloys in automatic machines is similar to that obtained when machining similar shapes from other metals. The condition of the machine is usually the major controlling factor. Common practice for maintaining dimensions within specified tolerance is to measure the significant dimensions of a predetermined number of parts during a production run. From the distribution pattern of these measurements it is possible to judge machine condition.

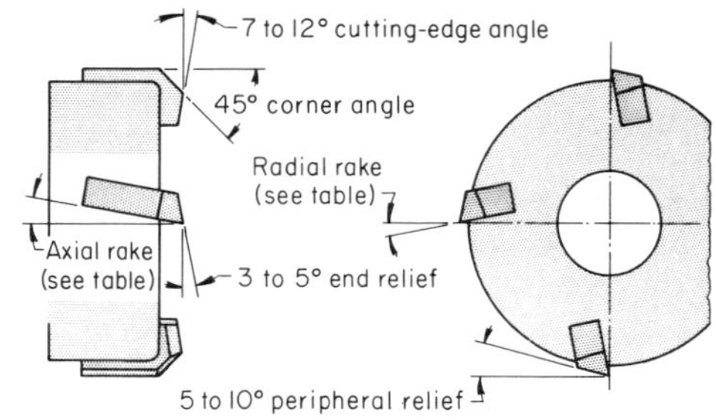


| Angle | High-speed steel, degrees | Disposable carbide, degrees | Brazed carbide, degrees |
|---|---|---|---|
| Axial rake | 12–25 | 5–7 | 3–10 |
| Radial rake | 10–12 | 0–5 | 3–10 |

**Fig. 21** Recommended angles for face milling cutters; all angles are positive.

## Milling

High-speed steel cutters, either solid or with inserted blades, are used for most applications in milling copper alloys, regardless of the type of milling (peripheral, face, or end). However, cutters with carbide inserts are more efficient than high-speed steel for some applications. The usual practice is to begin with high-speed steel cutters and then change to carbide only when the setup can be made rigid enough or when advantages are strongly indicated by experience.

**Peripheral Milling.** The angles and land width recommended for peripheral and side milling cutters for use in milling copper alloys are given in Fig. 20. Nominal speeds and feeds for the peripheral milling of copper alloys with high-speed steel cutters are given in Table 15. These values are conservative (as are other speeds and feeds tabulated in this section) and may often be increased under favorable conditions, such as maximum rigidity, a flood of cutting fluid, or a shallow cut.

**Face Milling.** Angles for cutters (high-speed steel and carbide) for the face milling of copper alloys are presented in Fig. 21. Nominal speeds and feeds for face milling with high-speed steel cutters are given in Table 16; speeds and feeds for carbide cutters are given in Table 17. Diamond tools are also used in the face milling of copper alloys.

**End Milling.** Recommended angles for end milling cutters used for copper alloys are shown in Fig. 22. Although these recommendations are for high-speed steel cutters, the same angles are applicable to carbide-tip cutters. Speeds for the end milling of copper alloys with high-speed steel and carbide cutters are given in Tables 18 and 19.

**Table 19 Nominal speeds and feeds for the end milling of copper alloys with carbide cutters**

| Alloy group | Condition | Hardness, HRB | Speed, m/min (sfm) | Feed, mm/rev (in./rev), for cutter diameter of: 6.4 mm (0.250 in.) | 19 mm (0.750 in.) | 25–50 mm (1.0–2.0 in.) |
|---|---|---|---|---|---|---|
| **Depth of cut, 1.3 mm (0.050 in.)** | | | | | | |
| 1 | Annealed | 20–70 | 229 (750) | 0.075 (0.003) | 0.20 (0.008) | 0.25 (0.010) |
| 2 | Annealed | 20–70 | 152 (500) | 0.075 (0.003) | 0.18 (0.007) | 0.23 (0.009) |
| 3 | Annealed | 20–70 | 91 (300) | 0.050 (0.002) | 0.15 (0.006) | 0.20 (0.008) |
| 1 | Cold drawn | 60–100 | 244 (800) | 0.075 (0.003) | 0.20 (0.008) | 0.25 (0.010) |
| 2 | Cold drawn | 60–100 | 160 (525) | 0.075 (0.003) | 0.18 (0.007) | 0.23 (0.009) |
| 3 | Cold drawn | 60–100 | 107 (350) | 0.05 (0.002) | 0.15 (0.006) | 0.20 (0.008) |
| **Depth of cut, 0.4 mm (0.015 in.)** | | | | | | |
| 1 | Annealed | 20–70 | 274 (900) | 0.05 (0.002) | 0.13 (0.005) | 0.18 (0.007) |
| 2 | Annealed | 20–70 | 213 (700) | 0.075 (0.003) | 0.15 (0.006) | 0.18 (0.007) |
| 3 | Annealed | 20–70 | 107 (350) | 0.04 (0.0015) | 0.13 (0.005) | 0.15 (0.006) |
| 1 | Cold drawn | 60–100 | 290 (950) | 0.05 (0.002) | 0.13 (0.005) | 0.18 (0.007) |
| 2 | Cold drawn | 60–100 | 229 (750) | 0.075 (0.003) | 0.15 (0.006) | 0.18 (0.007) |
| 3 | Cold drawn | 60–100 | 122 (400) | 0.04 (0.0015) | 0.13 (0.005) | 0.15 (0.006) |

Source: Metcut Research Associates Inc.

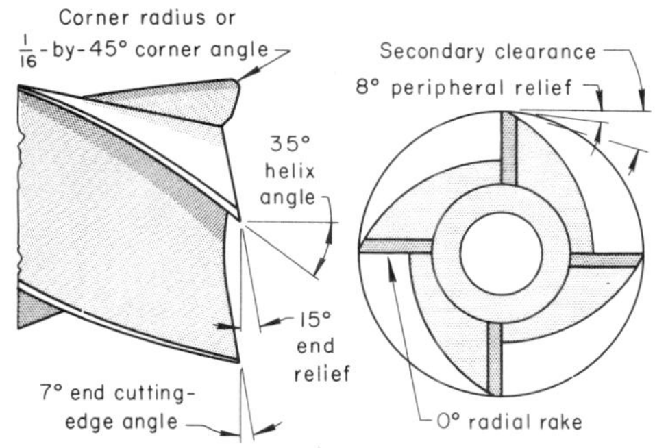


**Fig. 22** Recommended angles for end milling cutters. Dimensions given in inches

## Slitting and Circular Sawing

Copper alloys are sawed with relative ease, and because chips are removed rapidly, coarse-tooth saws are preferred. When sawing leaded alloys, saws with no hook angle are best. For sawing commercially pure copper or copper alloys that produce stringy chips, the hook angle should be as much as 18°. When surface finish on the cut surfaces is critical, the use of metal-slitting saws with side chip clearance will minimize rubbing and chip wash. Recommended shapes for solid and inserted saw teeth are shown in Fig. 23.

General-purpose high-speed steels are most often used for saws for copper alloys, although for some alloys, such as aluminum-silicon bronze, carbide-tip saws have proved more efficient. The free-cutting alloys such as high-leaded brass are often sawed without a cutting fluid. However, fluids are recommended for most circular sawing or slitting operations.

Nominal pitches, speeds, and feeds for the slitting and circular sawing of various compositions and thicknesses of copper alloys with high-speed steel and carbide-tip saws are given in Tables 20 and 21. Speeds several times as great as those shown in Table 20 are often used successfully in production applications.

**Circular Sawing Compared With Alternative Methods.** Cutoff operations are commonly performed in sequence with other operations, as in a multiple-spindle bar machine, although when cutoff is the principal operation (or the only one) a sawing machine is more efficient. A choice between circular sawing and abrasive cutting for copper alloys depends greatly on the

**Table 20 Nominal pitches, speeds, and feeds for the circular sawing and metal slitting of wrought and cast copper alloys with high-speed steel (M2, M7) saws**

Metal slitting includes cutters up to 200 mm (8 in.) in diameter; circular sawing normally includes cutters larger than 200 mm (8 in.) in diameter.

| Alloy group | Condition | Hardness, HRB | Saw pitch, tooth-to-tooth distance, mm (in.), for diameter or thickness of solid stock of: 6.4–75 mm (1/4–3 in.) | 75–150 mm (3–6 in.) | 150–230 mm (6–9 in.) | 230–380 mm (9–15 in.) |
|---|---|---|---|---|---|---|
| All | All | . . . | 5–20 (0.20–0.75) | 12–30 (0.50–1.10) | 20–35 (0.75–1.30) | 25–45 (0.90–1.75) |

| Alloy group | Condition | Hardness, HRB | Cutting speed, m/min (sfm), for diameter or thickness of solid stock of: 6.4–75 mm (1/4–3 in.) | 75–150 mm (3–6 in.) | 150–230 mm (6–9 in.) | 230–380 mm (9–15 in.) |
|---|---|---|---|---|---|---|
| 1 (Table 1) | Annealed | 10–70 | 120 (400) | 105 (350) | 76 (250) | 60 (200) |
| 2 | Annealed | 10–70 | 90 (300) | 76 (250) | 60 (200) | 46 (150) |
| 3 | Annealed | 10–70 | 30 (100) | 24 (80) | 20 (65) | 15 (50) |
| 1 (Table 1) | Cold drawn | 60–100 | 150 (500) | 120 (400) | 90 (300) | 76 (250) |
| 2 | Cold drawn | 60–100 | 120 (400) | 105 (350) | 76 (250) | 60 (200) |
| 3 | Cold drawn | 60–100 | 46 (150) | 37 (120) | 27 (90) | 23 (75) |
| 1 (Table 2) | As-cast | 40–150 HB (500 kg) | 120 (400) | 105 (350) | 76 (250) | 60 (200) |
| 2 | As-cast | 40–175 HB (500 kg) | 90 (300) | 76 (250) | 60 (200) | 46 (150) |
| 3 | As-cast | 40–200 HB (500 kg) | 30 (100) | 24 (80) | 20 (65) | 15 (50) |

| Alloy group | Condition | Hardness, HRB | Feed, mm/tooth (in./tooth), for diameter or thickness of solid stock of: 6.4–75 mm (1/4–3 in.) | 75–150 mm (3–6 in.) | 150–230 mm (6–9 in.) | 230–380 mm (9–15 in.) |
|---|---|---|---|---|---|---|
| 1 (Table 1) | Annealed | 10–70 | 0.2 (0.008) | 0.2 (0.008) | 0.25 (0.010) | 0.25 (0.010) |
| 2 | Annealed | 10–70 | 0.2 (0.008) | 0.2 (0.008) | 0.25 (0.010) | 0.25 (0.010) |
| 3 | Annealed | 10–70 | 0.18 (0.007) | 0.18 (0.007) | 0.23 (0.009) | 0.23 (0.009) |
| 1 (Table 1) | Cold drawn | 60–100 | 0.18 (0.007) | 0.18 (0.007) | 0.23 (0.009) | 0.23 (0.009) |
| 2 | Cold drawn | 60–100 | 0.18 (0.007) | 0.18 (0.007) | 0.23 (0.009) | 0.23 (0.009) |
| 3 | Cold drawn | 60–100 | 0.15 (0.006) | 0.15 (0.006) | 0.20 (0.008) | 0.20 (0.008) |
| 1 (Table 2) | As-cast | 40–150 HB (500 kg) | 0.18 (0.007) | 0.18 (0.007) | 0.23 (0.009) | 0.23 (0.009) |
| 2 | As-cast | 40–175 HB (500 kg) | 0.18 (0.007) | 0.18 (0.007) | 0.23 (0.009) | 0.23 (0.009) |
| 3 | As-cast | 40–200 HB (500 kg) | 0.15 (0.006) | 0.15 (0.006) | 0.20 (0.008) | 0.20 (0.008) |

Source: Metcut Research Associates Inc.

**Table 21 Nominal pitches, speeds, and feeds for the circular sawing of copper alloys with carbide-tip plate saws**

| | |
|---|---|
| **Saw pitch, mm (in.) (all alloys)(a)** | |
| 6.4–38 mm (1/4–1 1/2 in.) plate | 13–19 (0.500–0.750) |
| 38–90 mm (1 1/2–3 1/2 in.) plate | 19–25 (0.750–1.000) |
| 90–125 mm (3 1/2–5 in.) plate | 25–38 (1.000–1.500) |
| 125–150 mm (5–6 in.) plate | 35–50 (1.500–2.000) |
| **Speed, m/min (sfm), for 6.4–150 mm (1/4–6 in.) plate** | |
| Copper (20–60 HRB) | 460–910 (1500–3000) |
| Brasses (60–100 HRB) | 150 (500) |
| Bronzes (60–100 HRB) | 120 (400) |
| **Feed, mm/tooth (in./tooth), for 6.4–150 mm (1/4–6 in.) plate** | |
| Copper (20–60 HRB) | 0.17 (0.0065) |
| Brasses (60–100 HRB) | 0.15 (0.006) |
| Bronzes (60–100 HRB) | 0.15 (0.006) |

(a) Tooth-to-tooth distance, in inches. Source: Metcut Research Associates Inc.

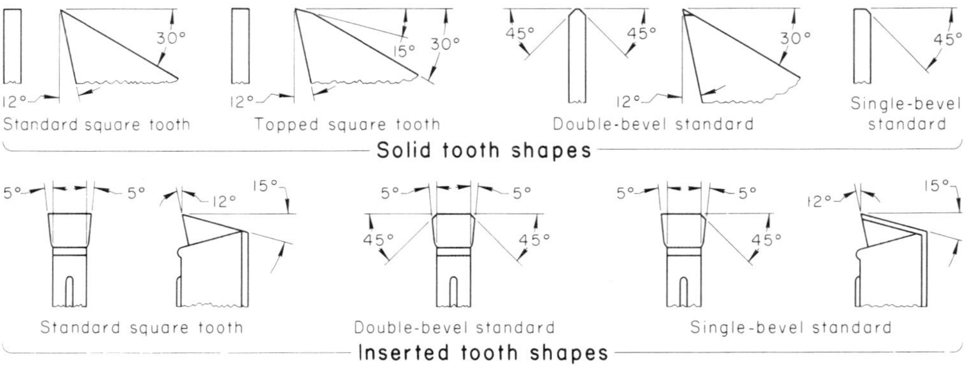


**Fig. 23** Solid and inserted tooth shapes for circular saws

### Table 22 Speeds for the power band sawing of copper alloys with high-speed steel bands

| Alloy group | Condition | Hardness, HRB | Speed, m/min (sfm), for a workpiece thickness of: <13 mm (½ in.) | 13–25 mm (½–1 in.) | 25–75 mm (1–3 in.) | >75 mm (3 in.) |
|---|---|---|---|---|---|---|
| 1 (Table 1) | Annealed | 10–70 | 115 (380) | 105 (340) | 85 (280) | 70 (230) |
| 2 | Annealed | 10–70 | 100 (320) | 87 (285) | 72 (235) | 59 (195) |
| 3 | Annealed | 10–70 | 60 (200) | 55 (180) | 46 (150) | 37 (120) |
| 1 (Table 1) | Cold drawn | 60–100 | 130 (425) | 115 (380) | 100 (320) | 79 (260) |
| 2 | Cold drawn | 60–100 | 110 (360) | 100 (320) | 82 (270) | 67 (220) |
| 3 | Cold drawn | 60–100 | 73 (240) | 66 (215) | 53 (175) | 44 (145) |
| 1 (Table 2) | As-cast | 40–150 HB (500 kg) | 115 (380) | 105 (340) | 85 (280) | 70 (230) |
| 2 | As-cast | 40–175 HB (500 kg) | 100 (320) | 87 (285) | 72 (235) | 59 (195) |
| 3 | As-cast | 40–200 HB (500 kg) | 60 (200) | 55 (180) | 46 (150) | 37 (120) |

### Table 24 Power hacksawing of copper alloys with high-speed steel blades

| Alloy group | Pitch, teeth/25 mm (1 in.), for thickness of: 6.4 mm (¼ in.) | 6.4–19 mm (¼–¾ in.) | 19–50 mm (¾–2 in.) | >50 mm (2 in.) | Speed, strokes/min | Feed mm/stroke | Feed in./stroke |
|---|---|---|---|---|---|---|---|
| **Annealed alloys (20–70 HRB)** | | | | | | | |
| 1 | 10 | 6 | 6 | 4 | 130 | 0.3 | 0.012 |
| 2 | 10 | 6 | 6 | 4 | 100 | 0.3 | 0.012 |
| 3 | 10 | 6 | 6 | 4 | 90 | 0.25 | 0.010 |
| **Cold-drawn alloys (60–100 HRB)** | | | | | | | |
| 1 | 10 | 10 | 6 | 4 | 150 | 0.3 | 0.012 |
| 2 | 10 | 10 | 6 | 4 | 120 | 0.3 | 0.012 |
| 3 | 10 | 10 | 6 | 4 | 100 | 0.25 | 0.010 |

Source: Metcut Research Associates Inc.

work metal composition. Some copper alloys, particularly the softest and most ductile, cause excessive loading of abrasive wheels. Under these conditions, circular sawing is more efficient than abrasive wheel cutting. On the other hand, for the most abrasive copper alloys, such as the aluminum bronzes or silicon bronzes, abrasive cutting is more efficient than sawing.

### Table 23 Pitch and tooth form for the power band sawing of copper alloys

| Workpiece thickness mm | in. | Tooth form | Pitch, teeth/25 mm (1 in.) |
|---|---|---|---|
| <13 | <½ | Precision | 10–14 |
| 13–25 | ½–1 | Precision | 8–10 |
| 27–75 | 1–3 | Precision | 6 |
| >75 | >3 | Claw, buttress | 3–4 |

## Power Band Sawing and Power Hacksawing

Nominal pitches and speeds that are successful for band sawing copper alloys are given in Tables 22 and 23. Similar data for power hacksawing are given in Table 24. The figures in these tables are for conditions that include the use of soluble oil as a cutting fluid. Speeds are based on an indefinitely long cut. If the length of cut is short (6 mm, or ¼ in.) and there is time between cuts, speed can be increased.

### Table 25 Conditions for the grinding of copper alloys

**Surface grinding**

Wheel classification
- Workpiece 20–70 HRB ........ C-46-K-V
- Workpiece 60–100 HRB ........ A-46-K-V

Wheel speed, m/min (sfm) ........ 1680–1980 (5500–6500)
Table speed, m/min (sfm) ........ 15–30 (50–100)
Downfeed
- Rough, mm/pass (in./pass) ........ 0.075 (0.003)
- Finish, mm/pass (in./pass) ........ 0.01 (0.0005) max

Crossfeed ........ ⅓ wheel width/pass

**Cylindrical grinding**

Wheel classification
- Workpiece 20–70 HRB ........ A-60-N-V
- Workpiece 60–100 HRB ........ A-46-L-V

Wheel speed, m/min (sfm) ........ 1680–1980 (5500–6500)
Work speed, m/min (sfm) ........ 30 (100)
Infeed
- Rough, mm/pass (in./pass) ........ 0.05 (0.002)
- Finish, mm/pass (in./pass) ........ 0.01 (0.0005) max

Traverse
- Rough ........ ⅓ wheel width/work rev
- Finish ........ ⅙ wheel width/work rev

**Centerless grinding**

Grinding wheel classification ........ A-60-L-V
Wheel speed, m/min (sfm) ........ 1680–1980 (5500–6500)
Work feed, mm/min (in./min) ........ 125 (50)
Infeed
- Rough, mm/pass (in./pass) ........ 0.13 (0.005)
- Finish, mm/pass (in./pass) ........ 0.04 (0.0015) max

Regulating wheel
- Angle ........ 3°
- Speed ........ 30 rev/min

**Internal grinding**

Wheel classification
- Workpiece 20–70 HRB ........ A-46-J-V
- Workpiece 60–100 HRB ........ A-60-L-V

Wheel speed, m/min (sfm) ........ 1520–1980 (5000–6500)
Work speed, m/min (sfm) ........ 30–60 (100–200)
Infeed
- Rough, mm/pass (in./pass) ........ 0.05 (0.002)
- Finish, mm/pass (in./pass) ........ 0.005 (0.0002)

Traverse
- Rough ........ ⅓ wheel width/work rev
- Finish ........ ⅙ wheel width/work rev

Source: Metcut Research Associates Inc.

## Abrasive Cutoff

The use of an abrasive cutoff wheel may be more economical than sawing, particularly for the most abrasive copper alloys. Aluminum oxide wheels are recommended—resinoid bond for dry cutting and rubber bond for wet cutting. Wheels should be relatively hard (grades N to R). Typical wheel recommendations are:

| Work metal hardness, HRB | Dry cutting | Wet cutting |
|---|---|---|
| 30–70 | A-30-R-B | A-46-N-R |
| 60–100 | A-46-R-B | A-46-N-R |

## Grinding

The grinding of copper alloys is not common, but in some applications grinding is the best means of obtaining accuracy and finish. Sometimes, when finish grinding must supplement machining, one grinding operation can be used to replace both operations.

Speeds, feeds, and wheels are given in Table 25. Aluminum oxide wheels are recommended for all types of grinding except for surface grinding softer alloys. For these conditions, silicon carbide wheels are preferred. In all cases, vitrified bond wheels of medium grade (J to N) are recommended. Emulsions of soluble oil and water are satisfactory grinding fluids.

## Honing

Honing is frequently used to finish bores to close dimensions, particularly in hard alloys such as aluminum bronze.

**Honing Stones (Sticks).** Resinoid bond silicon carbide stones of grade R should be used for honing all copper alloys. Selection of grain size depends on finish requirements as follows:

| Finish, μm | Finish, μin. | Grain size |
|---|---|---|
| 0.075–0.125 | 3–5 | 600 |
| 0.15–0.23 | 6–9 | 500 |
| 0.25–0.36 | 10–14 | 400 |
| 0.38–0.5 | 15–20 | 320 |
| 0.53–0.66 | 21–26 | 280 |
| 0.69–0.76 | 27–30 | 220 |

For a finish of over 0.76 μm (30 μin.), a grain size of 180 is recommended.

**Speed.** A reciprocating speed of 24 m/min (80 sfm) is recommended regardless of alloy hardness. Rotational speeds of 49 to 64 m/min (160 to 210 sfm) are recommended for alloys with hardness of 120 to 140 HB. For honing alloys of 180 to 200 HB, lower speeds (46 to 55 m/min, or 150 to 180 sfm) are better.

## ACKNOWLEDGMENT

ASM INTERNATIONAL would like to express its appreciation to M.L.H. Wise, M.A. Staley, and M. Samandi from the Department of Metallurgy and Materials, University of Birmingham (United Kingdom), for their significant contributions to this article. Their work was sponsored by the International Copper Research Association.

## REFERENCES

1. "Machinability of Copper-Based Alloys," Project Report 384, International Copper Research Association, 1985
2. "The Machining of High-Conductivity Copper," Project Report 343A, International Copper Research Association, 1984

# Machining of Magnesium and Magnesium Alloys*

MAGNESIUM, with a nominal alloy density of only 1.8 g/cm$^3$ (0.065 lb/in.$^3$), is considered to be the lightest structural metal. It is also one of the easiest metals to machine. Magnesium is often chosen because of its light weight, but its excellent machinability is a valuable advantage when a large amount of machining must be done.

Less power is required for removing a given volume of magnesium by machining than for any other commonly machined metal. The relative power required to machine some common metals is indicated below, with magnesium taken as unity:

| Metal | Relative power |
|---|---|
| Magnesium alloys | 1.0 |
| Aluminum alloys | 1.8 |
| Brass | 2.3 |
| Cast iron | 3.5 |
| Low-carbon steel | 6.3 |
| Nickel alloys | 10.0 |

Magnesium is machined in low-volume production on small, manually operated machine tools and on large, specially built, completely automated transfer machines operating at high production rates. The ease of machining magnesium results in significant benefits over more-difficult-to-machine metals. Heavy cuts can be taken at high speeds and feeds. Machining time is reduced; therefore, fewer machines, less capital investment, less floor space, and lower labor and overhead costs are required to do the same job. The low cutting pressures and the high thermal conductivity of the metal result in rapid dissipation of heat. Tool life is increased, and tool buildup is lower; this results in lower tool costs and less downtime for tool changing. The final surface finish is often produced in only one pass instead of two or more. Because magnesium is a free-cutting metal, the chips produced are well broken.

Initially, this article will focus on factors that affect the machining of magnesium:

- *Material*: Chip formation, distortion, and magnesium-matrix composites
- *Tools*: Materials, design, and sharpening
- *Cutting fluids*: Dry machining, coolants, mineral oil, and emulsions

Detailed examples are given on the various machining operations used on magnesium, including turning and boring, planing and shaping, broaching, drilling, reaming, counterboring, roller burnishing, tapping and die threading, multiple-operation machining, milling, sawing, and grinding.

The magnesium alloys mentioned include die casting alloy AZ91B (AZ91D); gravity casting alloys AZ91C, (AZ91E), and AZ92A; and wrought alloys AZ31B, AZ61A, AZ80A, ZK21A, and ZK60A. The machining parameters given generally apply to all magnesium alloys and forms. Safety is important to the machining of any material but is of particular importance to magnesium; the section "Safety" covers machining in general and grinding in particular, the handling of chips and fines, and fire extinguishment.

## Material Considerations

Certain physical characteristics of magnesium must be taken into account during machining. This section will discuss chip formation; distortion due to thermal expansion, cold work, and clamping; and magnesium-matrix composites.

**Chip Formation.** The type of chip formed during machining depends on the composition, form, and temper of the alloy and on the rate of feed. Rake angles and cutting speeds influence chip formation to a high degree when machining other metals, but have little or no effect on magnesium. The chips produced with single-point tools during turning, boring, shaping, and milling fall into three general types: short and well broken for heavy feeds, short and partially broken for medium feeds, and long and curled for light feeds. Cast alloys are more likely to yield broken or partially broken chips, depending on heat treatment. Forgings and extrusions typically yield partially broken or curled chips, depending on the feed rate used.

**Distortion.** Because of the high specific heat and good thermal conductivity of magnesium, frictional heat is rapidly diffused throughout the entire part. Consequently, the temperatures developed when cutting magnesium are usually low. With high speeds and feeds, however, considerable heating of the part can occur if a large amount of metal is being removed.

*Thermal Expansion.* When appreciable heat is generated under such conditions, the relatively high coefficient of thermal expansion of magnesium must be taken into account if close tolerances are required on the finished part. The mean coefficient of thermal expansion in the temperature range of 20 to 200 °C (68 to 392 °F) extends from 26.6 to 27.4 μm/m/°C (14.8 to 15.2 μin./in./°F), depending on the magnesium alloy. The coefficient of thermal expansion of magnesium is slightly higher than that of aluminum and appreciably higher than that of steel.

*Cold Working.* Distortion or warpage of magnesium parts due to cold working during machining seldom occurs. When it does occur, it can usually be traced to poor machining practices. Poor practices include the use of dull tools, slow feeds, tool dwell, or any operation that heats the part excessively. Cold working due to poor machining practices can build up extremely high stresses in the surface to a depth of as much as 0.5 mm (0.020 in.). Such stresses tend to distribute themselves throughout the part and can cause warpage. Warpage as a result of cold working during machining may appear shortly after the workpiece is removed from the clamps, or it might not appear until after several days or months of storage. It may also appear during the application of a heat-cured coating.

Rough cuts are more likely than fine finishing cuts to induce cold work into the magnesium surface. Rough machining can be carried down to within 0.5 mm (0.020 in.) of the final dimension desired, and the rest of the metal can be removed with one or two finishing cuts. Two finish cuts removing up to a total of 0.5 mm (0.020 in.) serve to remove the high-stress layer induced by the rough machining. The lower stresses that might be induced by the finish cuts are not likely to cause warpage.

*Revised through the cooperation of the International Magnesium Association Technical Committee

Stress relief should be used only as a last resort to relieve machining-induced stresses. Stress relief may be advisable after rough machining on complex parts or on parts in which extremely close tolerances are required. In these cases, finishing cuts should be taken after stress relieving. Parts should be cooled slowly and at a uniform rate from the stress-relief temperature. The times and temperatures for stress relieving wrought alloys after machining various tempers are given below:

| Alloy and temper | Time and temperature |
|---|---|
| AZ31B-F, -O | 1–4 h at 204 °C (400 °F) |
| AZ31B-H24 | 1–4 h at 135 °C (275 °F) |
| AZ61A-F | 1–4 h at 204 °C (400 °F) |
| AZ80A-T5 | 1–4 h at 204 °C (400 °F) |
| ZK21A-F | 1–4 h at 204 °C (400 °F) |
| ZK60A-F | 1–4 h at 204 °C (400 °F) |
| ZK60A-T5, -T6 | 4 h at 149 °C (300 °F) |

*Clamping* should always be done on the heavier sections of a magnesium part. In the case of die castings, the clamp locating pads should be on areas of the part that are formed by the same half of the die to minimize variation caused by the parting line. Care should be taken so that clamp pressures are not high enough to cause distortion. Shims should be used between the part and the tool bed when necessary. Special care should be taken with thin-section parts, which might be easily distorted by the pressures in a chuck or clamp or while making heavy cuts.

**Magnesium-Matrix Composites.** The parameters for machining magnesium-matrix composites have not been extensively studied, but some general guidelines have been developed. Standard magnesium tool designs are recommended when machining a magnesium-matrix composite material. At lower concentrations of reinforcing material (for example, <10% by weight), carbide-tip tooling can be used. At higher concentrations (>10% by weight), diamond-tip tooling should be used. In both cases, the tools must be sharpened frequently to avoid sparking at the point of contact between the tool and the composite material. Increasing the surface feed may help to keep the tool from becoming dull too rapidly.

## Tools

**Materials.** The selection of a tool material for machining magnesium depends on the amount of machining to be done. Ordinary carbon steel tools have exceptional service lives, but are not recommended for high-production jobs. Experience has shown that high-speed steel tools will last as long on magnesium as carbide-tip tools will last on other metals. For high-production jobs, carbide-tip tools are usually preferred because they last longer and are more accurate than high-speed steel tools. The higher cost of carbide-tip tools can also be justified for long-duration runs because the longer tool life will offset the higher initial cost. The hard, abrasion-resistant grades of any of the commercial brands of carbide material are satisfactory. Diamond-tip tools are usually not necessary when machining magnesium, although they have been effectively used when fine surface finishes of 0.075 to 0.125 μm (3 to 5 μin.) are required or when magnesium-matrix composites are being machined.

**Design.** Tools designed for machining steel or aluminum are normally suitable for machining magnesium. However, because of the low resistance of magnesium to cutting and its comparatively low heat capacity, tools should have smooth faces, large peripheral relief angles, large chip spaces, few blades (as on milling cutters), and small rake angles. Tool design specifications are provided throughout this article.

Special attention should be given to relief and clearance angles. These angles should be as large as possible to keep the areas of the tool behind the cutting edge from rubbing on the work and causing excessive heating and tool buildup. Larger clearance and relief angles than those usually permitted for other metals can be used on tools for magnesium because of the low cutting pressures.

Cutting pressure is greatly affected by cutting tool angles, especially rake angles. Decreasing the top rake angle of a turning tool from 25 to 15° can increase the cutting pressure by as much as 50%. An increase in the rake angle will naturally decrease cutting pressure, but at a sacrifice in tool life. For maximum tool life, back rake angles can range up to 20°. Rake angles on carbide-tip tools should be smaller than those on high-speed steel tools to reduce the possibility of tool chipping.

End and side cutting edge angles are usually not critical and are best determined by the requirements of each job. Extremely large side cutting edge angles may cause tool chatter. The nose radii of tools used for machining magnesium can be somewhat greater than those used on other metals, permitting greater feed rates for a given surface finish.

**Sharpening.** A very important rule is that tools must be kept as sharp and as smooth as possible when machining magnesium. They must also be free of scratches, burrs, and wire edges. Cutting tools that have been used on other metals should be resharpened and honed, even if the cutting angles of the tools have not changed. Tools should not be used when completely dull, but should be resharpened when only slightly dull to maintain good cutting edges and to conserve tool materials. Many shops engaged in production machining operate tools for a predetermined length of time and then replace them with freshly sharpened tools.

The rough grinding of cutting tools can be done on medium-grade wheels. To provide smooth tool faces and good cutting surfaces, finish grinding should be done on fine-grain wheels, followed, if necessary, by hand honing with fine or extrafine oilstone. Aluminum oxide 100-grit wheels are satisfactory for finishing high-speed steel, although finer grits produce smoother finishes. Silicon carbide 320-grit wheels or 200- to 300-grit diamond wheels are used for finishing carbide-tip tools.

To ensure long tool life and an excellent surface finish on parts, careful hand honing should be done to remove the burr and grinding wheel marks. It is imperative that surfaces on the clearance angles and on all areas over which the chips pass be well polished to minimize tool buildup. This is particularly important in high-production machining operations. Drills for deep-hole drilling should also have polished flutes.

A simple visual inspection often cannot detect whether a cutting edge is sharp enough to provide satisfactory service. Indications that tools need sharpening or replacement are:

- An inability to hold tolerances and to produce a good finish on the part
- The generation of excessive heat
- The formation of long chips with burnished surfaces
- Flashing or sparking at the tool edge

## Cutting Fluids

Magnesium is frequently machined without a cutting fluid. The two primary benefits of cutting fluid—cooling and lubrication—are often not required on magnesium. Magnesium dissipates heat rapidly, and this keeps the machined surface cool. In addition, the free-machining nature of magnesium and its resistance to galling against steel generally allow magnesium to be machined without a lubricant. Cutting fluids may be required for lubrication in deep boring or for cooling at very high feeds and speeds. Dry machining is usually cleaner, less costly, and therefore more desirable than machining with a cutting fluid.

Safe, efficient chip removal is important in the dry machining of magnesium. The cutting tool zones of single-unit and transfer machines must be as free of obstructions and pockets as possible to ensure a smooth flow of chips to the collection area. Chip shields, beveled pallets, slanted machine beds, and blowoff chambers are used to minimize chip accumulation on the machine. In transfer machines, screw conveyors under the length of the machine remove chips to a collector (see the section "Transfer Machines" in the article "Multiple-Operation Machining" in this Volume). Dry machining is preferred in turning and boring, drilling, milling, shaping and planing, and other operations involving machines of relatively simple design in which

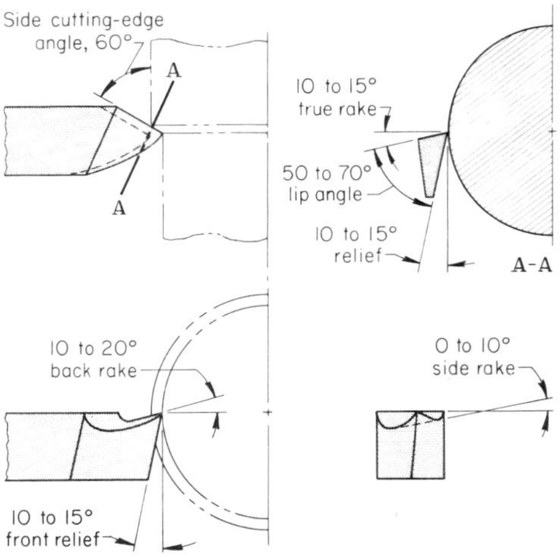


**Fig. 1** Lathe tool for the rough turning of magnesium

the operator can observe chip collection areas and extinguish a fire should one start.

**Coolants.** The improved surface finish and increased tool life that cutting fluids provide for many metals can be minor considerations in most machining operations on magnesium. Smooth surfaces are produced by machining at high or low speeds, with or without a cutting fluid. If required, such fluids are mainly used to cool the work, to minimize the possibility of distortion of the part, and to reduce the chances of ignition of chips (especially when the chips consist of fine particles that can easily be ignited unless submerged in a liquid). Consequently, cutting fluids are referred to as coolants when used in the machining of magnesium. In very high production operations, coolants can become a factor in increasing tool life.

Efficient machining practice demands that high speeds be used whenever possible. However, when high speeds are used, there is the danger of fire, particularly when fine chips are being produced. Less heat is generated in cutting magnesium than in cutting other metals, but the high cutting speeds used and the high thermal expansion and low heat capacity of magnesium make it necessary in some machining operations to provide for heat dissipation. In addition, iron or steel inserts and sand cast surfaces are likely to spark when hit with a cutting tool. Cutting feeds below 0.025 mm (0.001 in.) or tools rubbing on the work are more likely to generate enough heat to ignite chips. A stream of 15 to 19 L/min (4 to 5 gal./min) of cutting fluid per tool provides good cooling. If a particular job or machine tool precludes the use of a cutting fluid, cutting speed should be reduced to less than 150 m/min (500 sfm), and the recommendations regarding tools and feeds should be carefully followed.

When tapping and reaming, there is the possibility of chip jamming. Coolants can help flush away the chips. In the more complicated boring machines and transfer machines, which are designed for metals other than magnesium, it is more difficult to detect a fire visually, and it is often desirable to use a coolant to wash the chips out of the machining area and into a collection sump.

*Mineral Oil.* Oil-base coolants should always be mineral oils. Animal or vegetable oils are not suitable for use on magnesium. Various mineral oil cutting fluids are used on magnesium, and mineral seal oil and kerosene have been successfully used as coolants. For adequate cooling, the cutting oil should have a low viscosity. To prevent corrosion of the magnesium, the free acid content of the cutting fluid should be less than 0.2%. The use of low flash point oils, such as diesel oil, is strongly discouraged. Recommended properties of mineral oil coolants for the machining of magnesium are given below:

| Property | Value |
|---|---|
| Specific gravity | 0.79–0.86 |
| Viscosity at 40 °C (100 °F), SUS | 55 |
| Flash point, minimum (closed cup), °C (°F) | 135 (275) |
| Saponification number, maximum | 16 |
| Free acid, maximum, % | 0.2 |

*Emulsions.* In some applications, water-soluble oils or oil-water emulsions have been successfully used in machining magnesium; but the use of water-base coolants on magnesium can be dangerous, and extreme care must be observed. Water reacts with magnesium to form hydrogen, which is flammable and potentially explosive. Accumulations of small volumes of hydrogen represent a severe hazard during the shipment or storage of wet magnesium chips. In addition, water will intensify a magnesium fire should one start in wet chips. Wet magnesium chips are inefficiently reclaimed, may have no scrap value, and could present a disposal problem.

The use of an aqueous coolant has definite advantages in any machining operation. Off-the-shelf machining stations are often designed for the use of water-base coolants. Large coolant reservoirs can be used, and auxiliary cooling of the coolant is usually unnecessary. Machined parts and tools are cleaner. Chips are easily flushed to collection sites. When applied to magnesium, machining systems that use water-base coolants have been redesigned for the open-air collection of coolant and chips to allow for the dissipation of hydrogen. Hydrogen detectors have also been incorporated into the system. In addition, consideration must be given to the handling of wet magnesium chips, including drying, if the chips are to be recovered for resale or disposal. Even with these precautions, extreme attention to safety is required for minimizing the hazards of hydrogen formation and accumulation.

## Turning and Boring

Figure 1 shows a typical lathe tool for rough turning. The large relief angle (10 to 15°) and the small back rake (10 to 20°) are important. A larger back rake often causes feeding of the tool into the work. In some plants, 0° back rake is used to break chips and prevent curling. This practice is seldom recommended because it increases the roughness of the machined surface and increases power consumption. The side rake may vary from 0 to about 10°. The design of form tools for taking wide cuts or for turning specially shaped contours is similar to that of single-point finishing tools, except that the back rake should be reduced to about 3 to 8° to avoid chatter.

Figure 2 shows a typical parting tool. The large relief angles are important because they reduce friction and prevent tool breakage. The speeds, feeds, and depth of cut commonly used for the rough and finish turning or boring of magnesium are given in Table 1. The following example describes

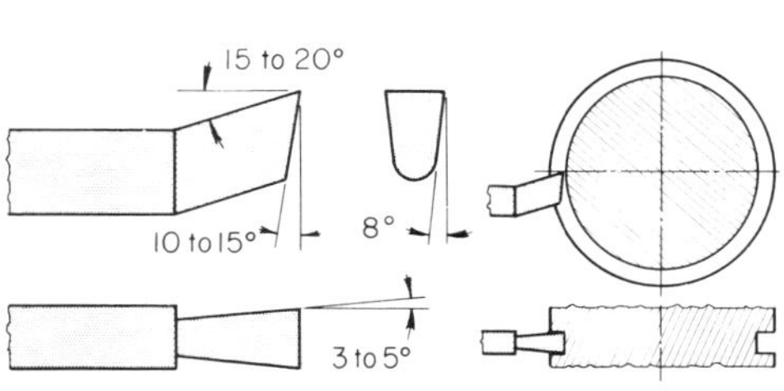


**Fig. 2** Lathe tool for the parting of magnesium

**Table 1 Speeds, feeds, and depths of cut for turning and boring magnesium alloys**

| Speed | | Feed | | Maximum depth of cut | |
|---|---|---|---|---|---|
| m/min | sfm | mm/rev | in./rev | mm | in. |
| **Roughing** | | | | | |
| 90–185 | 300–600 | 0.76–2.5 | 0.030–0.100 | 12.7 | 0.500 |
| 185–305 | 600–1000 | 0.51–2.0 | 0.020–0.080 | 10.2 | 0.400 |
| 305–460 | 1000–1500 | 0.25–1.5 | 0.010–0.060 | 7.62 | 0.300 |
| 460–610 | 1500–2000 | 0.25–1.0 | 0.010–0.040 | 5.08 | 0.200 |
| 610–1525 | 2000–5000 | 0.25–0.76 | 0.010–0.030 | 3.81 | 0.150 |
| **Finishing** | | | | | |
| 90–185 | 300–600 | 0.13–0.64 | 0.005–0.025 | 2.54 | 0.100 |
| 185–305 | 600–1000 | 0.13–0.51 | 0.005–0.020 | 2.03 | 0.080 |
| 305–1525 | 1000–5000 | 0.076–0.38 | 0.003–0.015 | 1.27 | 0.050 |

typical production practices for turning and boring magnesium.

**Example 1: Form Turning AZ91B Die Castings.** The die casting shown in Fig. 3 was turned to remove parting line flash and to form a sealing surface for a rubber gasket. A feed of 0.2 mm/rev (0.008 in./rev) was used to obtain the gasket surface and to produce a chip that could be easily handled. Because of the type of chip desired, the dwell normally used at the bottom of such a cut was omitted. No cutting fluid, as such, was used; however, chips were flushed away by a fire-resistant oil. Other processing details are given with Fig. 3.

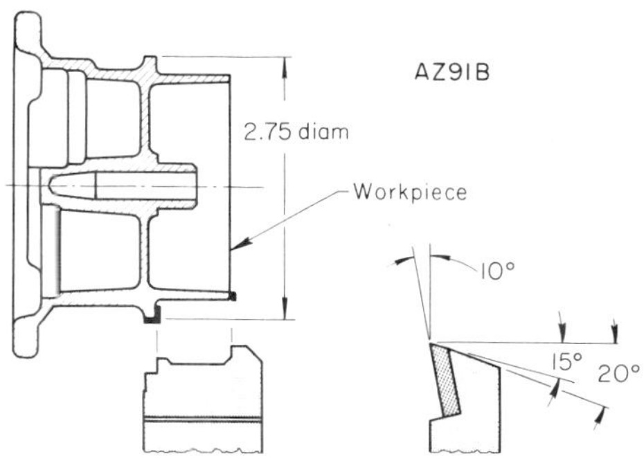


| | |
|---|---|
| Speed, at 1120 rev/min, m/min (sfm) | 245 (800) |
| Feed, mm/rev (in./rev) | 0.20 (0.008) |
| Metal removed, mm (in.) | ~0.38 (0.015) |
| Cutting fluid | None |
| Production rate, pieces/h | 700 |
| Tool life per grind, pieces | 100 000 |

**Fig. 3** Die casting that was machined by the carbide form tool shown. Dimensions in figure given in inches

## Planing and Shaping

Single-point tools for planing and shaping have essentially the same angles as the tools used for turning and boring (Fig. 1). Cutting fluids are seldom used for planing or shaping.

**Planing.** Because the maximum cutting speeds in planing are much lower than those for turning and boring, operating economy is obtained by heavy feeds and deep cuts. Depths of roughing cuts are commonly about 12.7 mm (0.50 in.) at a speed of 90 m/min (300 sfm) and a feed of 23 mm (0.90 in.) per stroke. For optimum surface finish, the depth of cut for finish planing should be 0.25 mm (0.010 in.). These heavy feeds and deep cuts are possible because of the low power needed for planing magnesium. The rigidity of the setup is the main limitation on the maximum size of the cut.

**Shaping.** Operating conditions for planing are also generally applicable to shaping. Speed is limited only by the capability of the shaper. Feed and depth of cut are more often limited by workpiece design or the rigidity of the setup.

## Broaching

Broaches for magnesium are usually made from M2 high-speed steel. Face (hook) angles of 10 to 15° are recommended. Relief angles are usually 1 to 3°. Broaching is done dry or with a broaching oil.

Speeds ranging from 6.1 to 15 m/min (20 to 50 sfm) are generally used; machine capability is the major factor limiting maximum speed. A rise per tooth (chip load) of 0.15 mm (0.006 in.) is usually near optimum. A chip load that is too light will cause burnishing, which will shorten broach life and distort the workpiece. The following example describes the practice for broaching splines in a magnesium alloy part.

**Example 2: Broaching Internal Splines in an AZ92A Casting.** A horizontal variable-speed broaching machine with a 1625 mm (64 in.) stroke was used to cut splines in a casting of magnesium alloy AZ92A. The hydraulically powered broaching machine was equipped with a 7.5 kW (10 hp) motor, and it had a pulling capacity of 180 kN (41 000 lbf) and a maximum cutting speed of 6.1 m/min (20 sfm).

Figure 4 shows the workpiece and broaching setup. The pull broach used, which was made of M3 high-speed steel, was 129 mm (5.08 in.) in diameter and 1725 mm (68 in.) long and had 131 rows of roughing teeth and 5 rows of finishing teeth. Each row of cutting teeth on the broach produced an internal form consisting of 19 teeth, with one space omitted, 30° involute spline, 4/8 diametral pitch, and 120.7 mm (4.750 in.) pitch diameter. Broach-teeth details and operating conditions are given with Fig. 4.

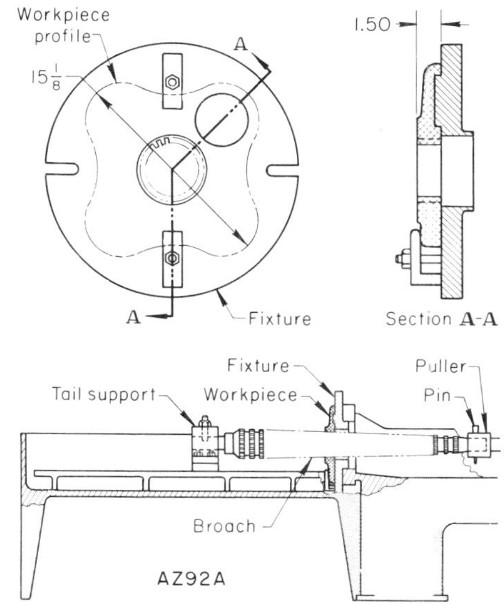


| Details of broach teeth | |
|---|---|
| Pitch, mm (in.) | 10.3 (13/32) |
| Land, mm (in.) | 3.2 (1/8) |
| Face (hook) angle | 15° |
| Backoff angle | 1° |
| Side relief angle | 1° |
| **Operating conditions** | |
| Speed, m/min (sfm)(a) | 6.1 (20) |
| Cutting fluid | Sulfurized oil |
| Tolerance on pitch diameter, mm (in.) | ±0.025 (±0.001) |
| Maximum permissible involute error, mm (in.) | 0.020 (0.0008) |
| Time per piece, min(b) | 3.75 |
| Production rate, pieces/h | 16 |
| Tool life per grind, pieces(c) | 500 |

(a) Maximum speed for this machine. (b) 16 s for the stroke; 5 s for the ram to return to the fixture; 10 s for unpinning the broach and retracting it; 14 s for pinning the broach; 3 min for loading, unloading, and cleaning. (c) Estimated; sharpening not required for production quantity

**Fig. 4** Broaching spline teeth in a cast magnesium part. Dimensions in figure given in inches

## Drilling

Magnesium alloys can be easily drilled with twist drills made of any high-speed steel. The speeds most commonly used are given in Table 2. Feed depends on hole size and usually ranges from 0.025 to 0.76 mm/rev (0.001 to 0.030 in./rev) for holes (1.6 to 51 mm (1/16 to 2 in.) in diameter, as shown in Table 2.

**Deep Holes.** Twist drills of the design shown at the left in Fig. 5 are suitable for drilling holes of depth-to-diameter ratio up to 20:1. A web of uniform thickness is preferred. The clearance (relief behind the

**Table 2 Nominal speeds and feeds for the drilling, gun drilling, reaming, and counterboring of magnesium alloys**

| Operation | Speed, m/min | Speed, sfm | Feed, mm/rev (in./rev), for hole diameter of: 1.6 mm (1/16 in.) | 3.2 mm (1/8 in.) | 6.4 mm (1/4 in.) | 13 mm (1/2 in.) | 19 mm (3/4 in.) | 25 mm (1 in.) | 38 mm (1½ in.) | 51 mm (2 in.) |
|---|---|---|---|---|---|---|---|---|---|---|
| Drilling | 43–100 | 140–330 | 0.025 (0.001) | 0.076 (0.003) | 0.18 (0.007) | 0.30 (0.012) | 0.41 (0.016) | 0.51 (0.020) | 0.64 (0.025) | 0.76 (0.030) |
| Gun drilling | 198 | 650 | 0.025 (0.001) | 0.025 (0.001) | 0.076 (0.003) | 0.13 (0.005) | 0.20 (0.008) | 0.25 (0.010) | 0.25 (0.010) | 0.25 (0.010) |
| Reaming | 120 | 400(b) | ... | 0.13 (0.005) | 0.20 (0.008) | 0.30 (0.012) | ... | 0.41 (0.016) | 0.51 (0.020) | 0.76 (0.030) |
| Counterboring | | | | | | | | | | |
| High-speed steel | 195 | 640(c) | ... | ... | 0.13 (0.005) | 0.15 (0.006) | 0.18 (0.007) | 0.22 (0.0085) | 0.28 (0.011) | 0.33 (0.013) |
| Carbide | 490 | 1600(c) | ... | ... | 0.15 (0.006) | 0.18 (0.007) | 0.20 (0.008) | 0.25 (0.010) | 0.30 (0.012) | 0.36 (0.014) |

(a) Limitations of machine, rigidity, or both may require lower speeds. (b) For high-speed steel; for carbide reamers, speed is 260 m/min (850 sfm). (c) For 76 mm (3 in.) diam holes, use 0.41 mm/rev (0.016 in./rev). Source: Metcut Research Associates Inc.

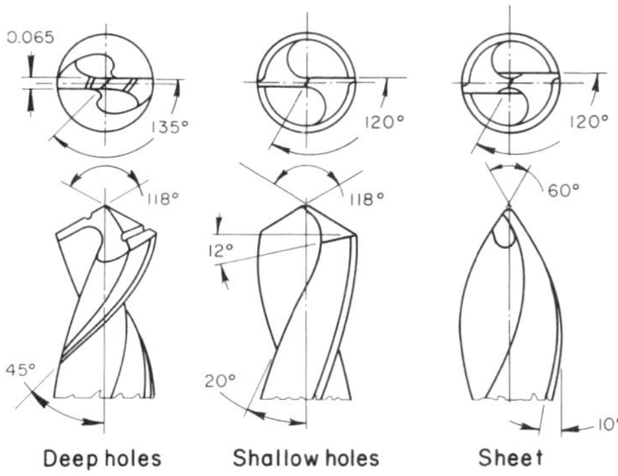


| | |
|---|---|
| **For deep holes** | |
| Point angle | 118° |
| Helix angle | 40–50° |
| Chisel-edge angle | 135–150° |
| Web | Constant thickness |
| Flutes | Polished |
| **For shallow holes** | |
| Point angle | 118° |
| Helix angle | 10–30° |
| Chisel-edge angle | 120–135° |
| Relief angle | 12° |
| Flutes | Polished |
| **For sheet** | |
| Point angle | 60° |
| Helix angle | 10° |
| Chisel-edge angle | 120–135° |
| Web | Thinned at point |
| Corners | Rounded |

**Fig. 5** Recommended designs of twist drills for magnesium. Margin dimension of drill in figure given in inches

land) should be twice as deep as that of standard drills, and the margin should be half the standard width. A point angle of 118° and a lip relief of 15° are recommended. A spur or pilot can be ground on the point to prevent drill marks and to keep the drill straight. Chisel-edge angles must be 135 to 150° to provide good surface finish and to minimize spiraling in the holes. When chisel-edge angles are smaller or larger, difficulties result due to improper relief at the cutting edge and improper centering of the drill.

A drill of this design cuts freely and will readily clear chips. Raising the drill to remove the chips is unnecessary unless the depth of the hole is more than about 20 diameters. However, if the drill bushing and the surface of the drilled part are too close, the chips may not flow out readily and can jam in the flutes. An approximate minimum distance of 1½ times the hole diameter is recommended. If a flat machined face is available, the drill bushing can be held tightly against the machined face during drilling. This allows chips to be brought back through the bushing.

**Shallow Holes.** Drilling holes in magnesium to a depth less than four times the diameter presents few difficulties, and standard twist drills (center, Fig. 5) can be used.

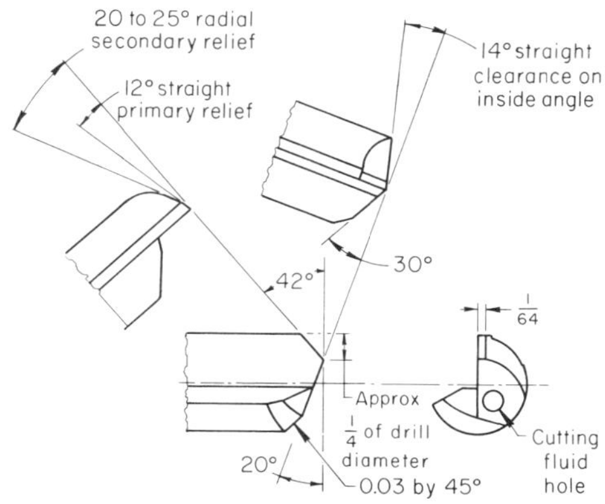


**Fig. 6** Gun drill for magnesium machining applications. Dimensions in figure given in inches

Polished flutes are recommended for aiding in clearing chips, and cutting edges must be kept sharp.

**Drilling of Sheet.** Magnesium sheet can be drilled with a sharp, standard twist drill (118° point angle), but a slightly modified drill (at right, Fig. 5) is recommended for the mass production of accurate holes with a good finish and minimum burr. The point angle of this drill is reduced to approximately 60° to prevent walking of the drill, to reduce thrust, and to prevent an abrupt change of thrust when breaking through. The chisel-edge angle is 120 to 135°. The web is thinned and the ends of the cutting edges rounded. A thin web at the point helps to center the drill and reduce thrust; the rounded corners provide a smooth finish and reduce burrs. A helix angle of about 10° prevents the work from climbing the drill on breakthrough.

**Gun Drilling.** A typical gun drill for magnesium (Fig. 6) consists of a cylindrical driver with a flat cutout, a crimped tubular shank through which cutting fluid can flow, and a carbide head or tip. Typical angles are shown in Fig. 6, although these vary in practice in different plants. For example, one plant uses cutting-edge angles of 10 and 10°, instead of 20 and 42° as shown in Fig. 6. The dimensional tolerance of the drill tip is from −0.005 to +0.000 mm (−0.0002 to +0.0000 in.). Gun drills are end cutting tools with an outer cutting angle and an inner angle and with no back or side rake and only one axial flute.

In gun drilling, the tip of the drill passes through an accurate starting bushing and into the workpiece. The inside diameter of the bushing, which should be replaced when it becomes 0.015 mm (0.0006 in.) oversize (depending on required accuracy), has a tolerance of −0.000 to +0.005 mm (−0.0000 to +0.0002 in.). After the tip enters the workpiece, the hole it has cut acts as a guide. The cutting fluid is delivered to the cutting point by a high-pressure system (2.8 MPa, or 0.4 ksi, minimum for 6.4 mm, or ¼ in., diam holes and higher pressure for

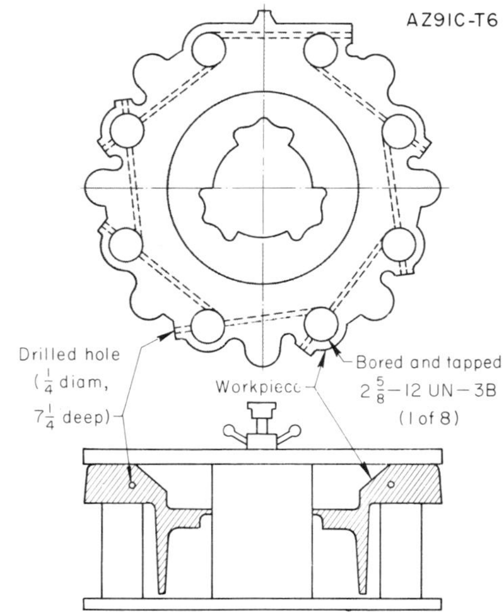


| | |
|---|---|
| **Drilling conditions** | |
| Speed, m/min (sfm) | 95 (313) |
| Feed, mm/rev (in./rev) | 0.038 (0.0015) |
| Cutting fluid | Mineral oil |
| Tolerance, mm (in.) | Straightness within 0.13 (0.005) |
| Specified finish, μm (μin.) | 0.75 (30) max |

**Fig. 7** Aircraft brake carrier drilled with carbide-tip gun drill. Dimensions in figure given in inches

smaller holes) that contains filters for removing metal chips and foreign particles. Cutting speeds for average machining conditions are usually 200 m/min (650 sfm) or less, depending on the capability of the machine. A speed of 200 m/min (650 sfm) is suggested in Table 2, along with feeds of 0.025 to 0.25 mm (0.001 to 0.010 in.), depending on hole size. Production application of gun drilling deep holes in magnesium is described in the following examples.

**Example 3: Gun Drilling Holes 29 Diameters Long in Brake Carriers.** Eight holes 184 mm (7¼ in.) deep by 6.4 mm (¼ in.) in diameter were gun drilled in sand cast AZ91C-T6 alloy aircraft brake carriers 394 mm (15½ in.) in diameter by 76 mm (3 in.) thick. Figure 7 shows the workpiece and the drilling setup. The work was mounted in a fixture that located it on a center bore. A 5.6 kW (7½ hp) deep-hole drilling machine and a carbide-tip gun drill with a top rake of 10°, a side clearance angle of 15°, and a chip breaker were used. The workpiece was indexed at each of eight locations to drill the required holes. The cutting fluid and the pressure of the gun drill cleared the chips. Further details are given with Fig. 7.

## Reaming

Reamers for magnesium should have fewer flutes than those used for most metals in order to provide more chip space. Reamers under 25 mm (1 in.) in diameter usually have four or six flutes. There should be an even number of flutes with opposing cutting edges spaced at 180° (Fig. 8).

Reamers for machining magnesium usually have 45° chamfer, 7° rake, a 0.15 to 0.30 mm (0.006 to 0.012 in.) margin with no relief, a primary relief angle of 5 to 8°, and a secondary relief angle of approximately 20°. Flutes can be straight or can have a negative helix of up to −10°. Reamers with either high-speed steel or carbide cutting edges are suitable for magnesium; the choice depends on the number of holes to be reamed.

Holes drilled for reaming should allow enough stock for the reamer to take a definite cut (a minimum of 0.25 mm, or 0.010 in., from the diameter) because, with less stock for reaming, burnishing is likely to occur. A maximum allowance of 0.38 mm (0.015 in.) on the diameter is preferred, although greater amounts are often removed. If too much stock is removed, chips will build up in the flutes and eventually jam the reamer.

Cutting speeds are generally 30 to 120 m/min (100 to 400 sfm) for high-speed steel reamers and up to 260 m/min (850 sfm) for reamers with carbide cutting edges. Table 2 lists speeds of 120 and 260 m/min (400 and 850 sfm) for high-speed steel and carbide, respectively. However, these speeds are maximum, and in many operations the machine used is not capable of such speeds. Sometimes lower speeds must be used because of the lack of rigidity in the setup. As a rule, the best surface finish and the most accurate holes are obtained by reaming at high speeds and medium feeds. Feeds are 0.13 to 0.76 mm/rev (0.005 to 0.030 in./rev), depending on hole size, as indicated in Table 2.

In addition to reaming, holes in magnesium alloy workpieces can be finished to close accuracy and low roughness by roller burnishing. Detailed information is available in the article "Roller Burnishing" in this Volume.

## Counterboring

A tool recommended for counterboring magnesium is illustrated in Fig. 9. A narrow margin of approximately 0.38 mm (0.015 in.) (not shown in Fig. 9), with adequate relief and clearance angles, allows adequate chip space and eliminates rubbing. Counterbores of many different designs are used. For example, the counterbore illustrated in Fig. 10 resembles a reamer. Counterbores with cutting edges of either high-speed steel or carbide can be used, depending on the number of holes to be counterbored.

Speeds can often be as high as 195 m/min (640 sfm) for counterboring with high-speed steel and 490 m/min (1600 sfm) with carbide (Table 2). However, many machines cannot develop these high speeds, and lower speeds must be used. Feed rates are 0.13 to 0.36 mm/rev (0.005 to 0.014 in./rev), depending on hole size and tool material. Recommended feeds are given in Table 2. Typical practice for counterboring magnesium is described in the following example.

**Example 4: Counterboring Cavities in AZ91B Die Castings.** Two cavities in the AZ91B die casting shown in Fig. 10 were counterbored to a diameter of 25.30 mm (0.996 in.) on a two-station nonindexing machine. The workpiece was placed by hand in the first station, and one cavity was counterbored while Dryseal threads were tapped in a cored hole. The work was then manually placed in the second station, and the second cavity was counterbored, after which the workpiece was ejected automatically. Both stations cycled simultaneously, producing one finished part per cycle.

A six-flute carbide counterbore (Fig. 10) with a cam-feed head was used. A 4° dwell on the highest point of the cam feed produced a smooth finish on the gasket surface at the bottom of the cavity, although an undesirable ribbon-type chip was produced. Fixtures were mounted at a 15° angle and had enough clearance to flush the chips. Additional operating details are given with Fig. 10.

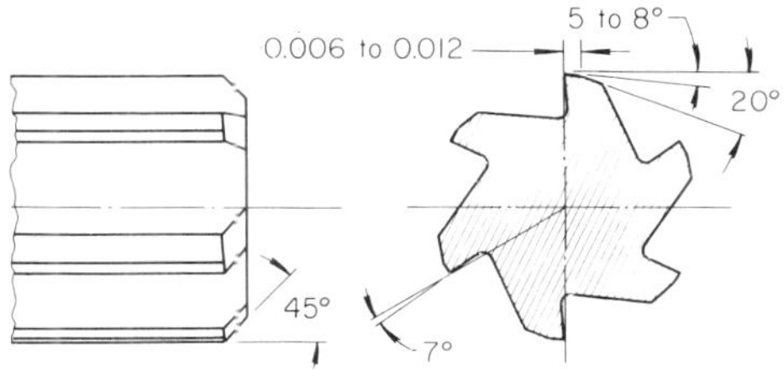


**Fig. 8** Reamer design recommended for machining magnesium. Dimensions in figure given in inches

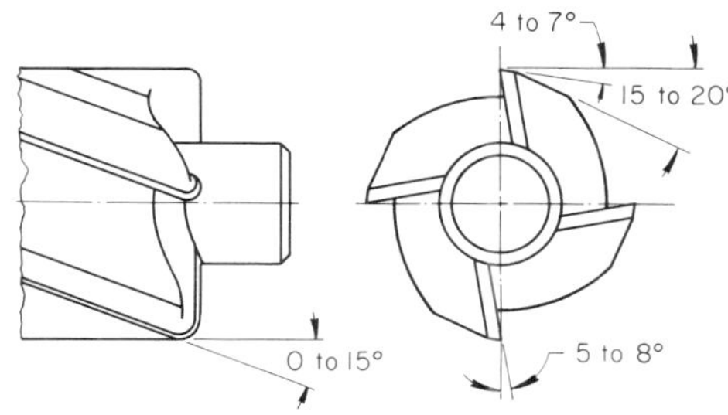


**Fig. 9** Recommended design of tool for counterboring magnesium

## Tapping and Die Threading

Standard taps are adequate if production quantities are small and if high quality tapping is not required. When a large number of parts or close tolerances are involved, standard taps must be modified, or taps designed especially for magnesium must be used.

Taps made of a general-purpose grade of high-speed steel such as M1, M7, or M10 are usually adequate. For long production runs, taps are sometimes made of T5 or T15. Carbide taps are rarely used for magnesium alloys.

Straight-flute or helical-flute taps are recommended. The number of flutes is determined by limiting the total land width to about 30% of the tap circumference. Generally, two flutes are used for taps up to 4.8 mm (3/16 in.) in diameter, three flutes for taps up to 19 mm (3/4 in.) in diameter, and four flutes for taps over 19 mm (3/4 in.) in diameter.

The tap design shown in Fig. 11(a) usually gives good finish and accuracy. However, if chips jam in the hole as the tap is being removed, a heel rake angle of 3 to 5° is recommended (Fig. 11b). This heel rake (with no relief) provides cutting action as

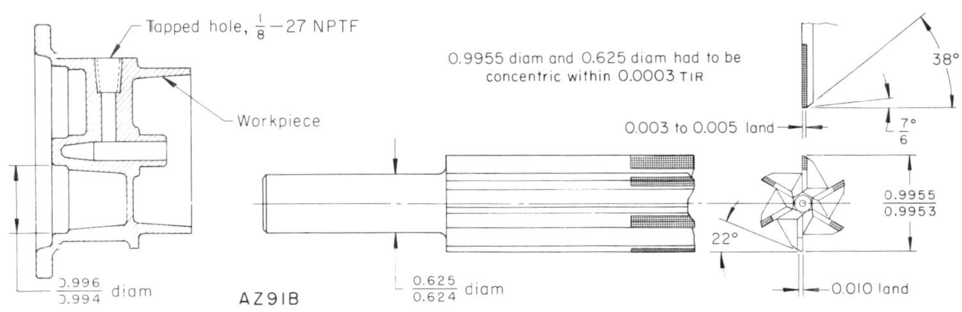


**Counterboring conditions**

| | | | |
|---|---|---|---|
| Speed at 1500 rev/min, m/min (sfm) | 120 (390) | Tolerance, mm (in.) | +0.000, −0.051 (+0.000, −0.002) |
| Feed, mm/rev (in./rev) | 0.25 (0.010) | Production rate, pieces/h | 414 |
| Metal removed, mm (in.) | 0.51 (0.020) from sides and bottom | Counterbore life, pieces/grind | 500 000 (1 million holes) |
| Cutting fluid | Mineral oil, with sperm oil added | | |

**Fig. 10** Counterbored die casting (left) and counterboring tool. The two cavities in the die casting were counterbored to 25.30 mm (0.996 in.) in diameter with the six-flute counterbore shown. The cored hole was tapped while the casting was in position for counterboring of the first cavity. Dimensions in figure given in inches

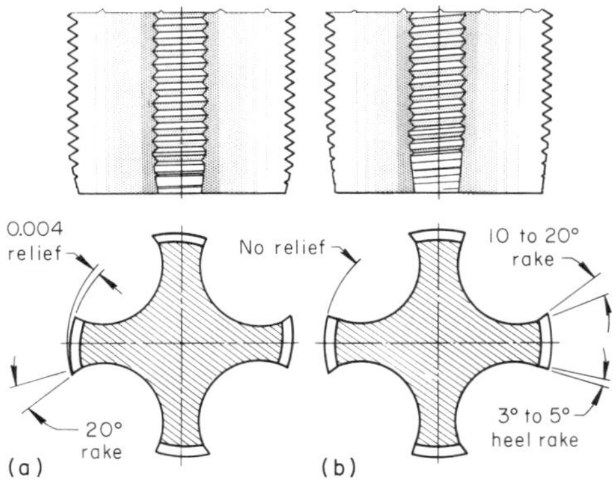


**Fig. 11** Recommended tap specifications for magnesium machining applications. (a) Eccentric-ground tap. (b) Heel-cutting tap, a concentric tap with no relief. Dimensions in figure given in inches

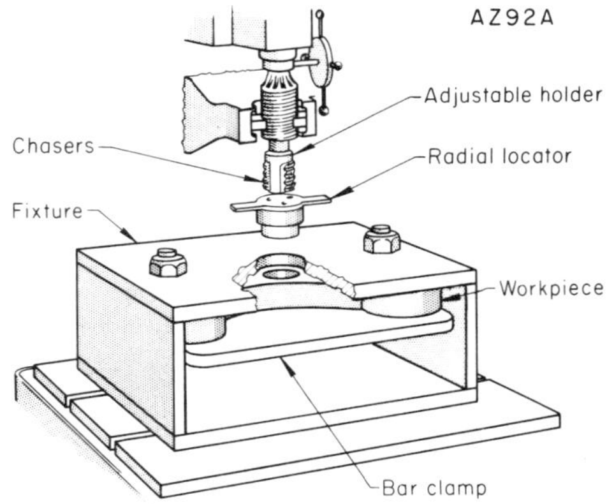


**Fig. 12** Tapping six cavities in a 76 mm (3 in.) thick cast brake housing

the tap is removed and results in a clean, accurately tapped hole.

If the tap cuts oversize, the rake angle should be decreased, and conversely, increasing the rake angle will make the tap cut larger. Sometimes, taps on the high side of pitch-diameter tolerance are used because tapped holes in magnesium are likely to close-in slightly after the tap is removed. This is especially true if high tapping speeds have been used.

Tapping speeds range from 23 to 53 m/min (75 to 175 sfm). The lower speeds are used when:

- Taps are small (because of their low capacity to conduct the heat of cutting)
- No cutting fluid is used
- The alloy is tough or abrasive

A cutting fluid is recommended for improving surface finish, for increasing tap life, and for producing accurate threads. The following example illustrates the production tapping of a cast brake housing.

**Example 5: Tapping 3⅛-16 UN-3B Holes in AZ92A.** A precision tapping-and-threading machine was used to tap six cavities in a 76 mm (3 in.) thick brake housing cast of AZ92A alloy (Fig. 12). This machine was equipped with a rapid-reversing drive motor and had a tapping capacity of 9.5 to 105 mm (⅜ to 4⅛ in.) diameter, a maximum spindle stroke of 140 mm (5½ in.), a maximum tapping stroke of 89 mm (3½ in.), and 16 forward and 8 reverse spindle speeds.

The holding fixture, a rectangular, boxlike unit enclosed on all sides except the front, was mounted on the machine table so that a precision-bored hole in its top plate was aligned with the machine spindle. The workpiece was loaded and unloaded through the front of the fixture. A bar clamp held the workpiece loosely under the top plate. A radial locator plug was inserted through the hole in the top plate into the first cavity of the workpiece to align the cavity with the machine spindle. The workpiece was then clamped securely in position, and the radial plug was removed before tapping was started. This procedure was repeated for each cavity.

Solid, adjustable holders capable of accommodating thread chasers ranging from 33 to 105 mm (1$\frac{5}{16}$ to 4⅛ in.) in major diameter were used. For this operation, a chaser with a 22° chamfer and a 33° hook was best for producing the 3⅛-16 UN-3B threads required for each cavity. Oil was used as cutting fluid.

Machining time (including loading and unloading) was 11.76 min per piece; production rate was 5.10 pieces per hour. Tool life was about 100 pieces per grind. Tool change and adjustment required 1 h.

**Cold form tapping** (see the article "Tapping" in this Volume) is not recommended for magnesium, unless a 45% thread will have enough holding power for the application. Attempting to form a fuller thread is likely to cause embrittlement and subsequent spalling of the metal.

Because of the excellent machinability of magnesium, the life of thread-cutting taps is usually greater than that for other metals. Therefore, the only advantage of using the cold forming methods is to eliminate the need for removing chips from the tapped hole.

**Die Threading.** Threading dies for magnesium should have approximately the same cutting angles as taps. A cutting edge relief of 0.10 mm (0.004 in.) is generally recommended, although there are applications that require extremely good surface finishes and high accuracy. Under these conditions, a 3° rake angle of the heel side of the land will clean up the thread when the die is removed. Self-opening dies help to provide smooth threads. Cutting angles of chasers for die threading should be about the same as those on single-point turning tools.

## Multiple-Operation Machining

When magnesium is being machined in multiple-operation bar or chucking machines, the maximum speed of the machine is usually permissible (see the section "Automatic Lathes" in the article "Multiple-Operation Machining" in this Volume). Heavy feeds are preferred because they produce well-broken chips; feeds 40% higher than those used in machining free-cutting brass are common. Feeds of 0.38 mm/rev (0.015 in./rev) produce short coils; below this rate, chips are likely to be long coils or ribbons.

Alloy AZ31B is preferred for multiple-operation bar machining, although any of the other extruded magnesium alloys can be cut satisfactorily. The use of a mineral oil type of cutting fluid is recommended. Tools used for cutting magnesium in multiple-operation machines require little if any modification from those used for brass or those used for single operations discussed in this article.

## Milling

Milling cutters for magnesium should have fewer teeth than conventional cutters used for steel or other metals—preferably only one-half to one-third as many. The larger chip space and the heavier cut resulting from the smaller number of teeth reduce the frictional heat and increase the chip clearance, and this leads to higher speeds, decreased distortion, lower power consumption, and smoother finish. Milling cutters for magnesium commonly have a 10° relief angle, 1.6 mm ($\frac{1}{16}$ in.) land width, and a 20° secondary clearance (Fig. 13a).

**Solid Cutters.** Figure 13 shows tool angles for five types of solid cutters for milling magnesium. Form cutters have a much smaller rake angle than plain cutters because the heavier cutting pressure of form cutters would cause chatter if a larger rake angle were used. Details of an inserted-blade cutter for face milling are shown in Fig. 14.

**Operating Conditions.** Under ordinary circumstances, magnesium alloys can be milled with standard equipment at maximum spindle speeds and with a feed rate and depth of cut as great as the capacity of the machine permits. As a rule, the only reason for using slower speeds is that higher speeds are likely to cause high residual stresses in the work metal.

Nominal speeds, feeds, and depths of cut used for the face, peripheral, and end milling of magnesium alloys are given in Tables 3 and 4. Equipment and operating conditions employed in the milling of magnesium alloy parts are described in the following example.

**Example 6: Peripheral Milling of an Irregularly Shaped ZK60-T5 Forging.** A 279 mm (11 in.) diam carbide-tip side-milling cutter was used on a vertical milling machine to produce a slot 38.89 mm (1.531 in.) wide in an irregularly shaped forging. The workpiece and setup are shown in Fig. 15. The milling machine, which had an 11

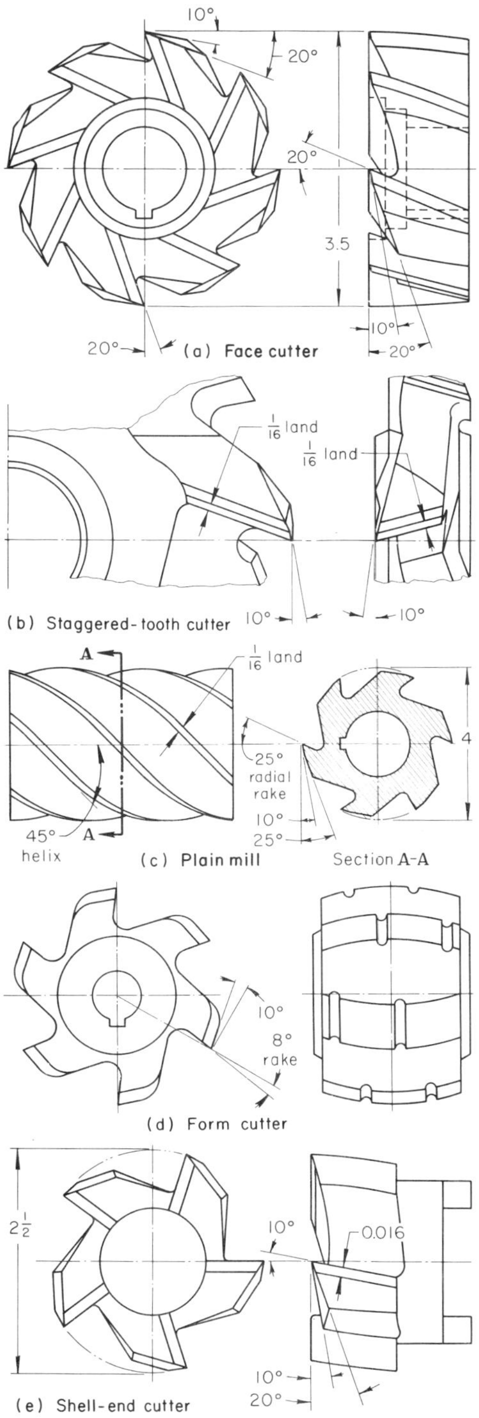


**Fig. 13** Solid cutters for milling magnesium. Dimensions in figure given in inches

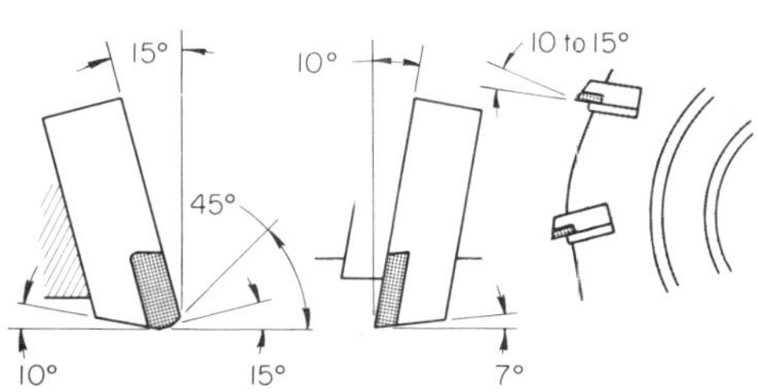


**Fig. 14** Inserted-blade cutter for the face milling of magnesium at high peripheral speeds

## Table 3 Nominal speeds and feeds for the face and peripheral (slab) milling of magnesium alloys

| Operation | Roughing (6.35 mm, or 0.250 in., depth) Speed m/min | Speed sfm | Feed mm/tooth | Feed in./tooth | Finishing (0.64 mm, or 0.025 in., depth) Speed m/min | Speed sfm | Feed mm/tooth | Feed in./tooth |
|---|---|---|---|---|---|---|---|---|
| Face milling, high-speed steel cutters | 275 | 900 | 0.56 | 0.022 | 460 | 1500 | 0.36 | 0.014 |
| Face milling, carbide cutters | max | max | 0.64 | 0.025 | max | max | 0.30 | 0.012 |
| Peripheral milling, high-speed steel cutters | 275 | 900 | 0.46 | 0.018 | 395 | 1300 | 0.41 | 0.016 |

## Table 4 Nominal speeds and feeds for the end milling of magnesium alloys

| Operation | Speed m/min | Speed sfm | Feed, mm/tooth (in./tooth), cutter diameter of: 6.4 mm (¼ in.) | 19 mm (¾ in.) | 25–50 mm (1–2 in.) |
|---|---|---|---|---|---|
| **High-speed steel cutters** | | | | | |
| Roughing(a) | 245 | 800 | 0.10 (0.004) | 0.23 (0.009) | 0.28 (0.011) |
| Finishing(b) | 305 | 1000 | 0.08 (0.003) | 0.13 (0.005) | 0.20 (0.008) |
| **Carbide cutters** | | | | | |
| Roughing(a) | max | max | 0.10 (0.004) | 0.25 (0.010) | 0.30 (0.012) |
| Finishing(b) | max | max | 0.08 (0.003) | 0.13 (0.005) | 0.23 (0.009) |

(a) 1.27 mm (0.050 in.) depth. (b) 0.38 mm (0.015 in.) depth

kW (15 hp) motor and 451 mm (17¾ in.) throat clearance, was capable of spindle speeds of 13 to 1300 rev/min and table feeds of 6.4 to 1520 mm/min (¼ to 60 in./min).

The workpiece was clamped in a fixture that was mounted on a power-driven 610 mm (24 in.) turntable on top of the machine table. The fixture registered on the machine face and the bore of the workpiece, as shown in Fig. 15. Angular rotation of the turntable was 116°.

The forging weighed 7.03 kg (15.5 lb) before machining and 5.48 kg (12.1 lb) after machining. Tools were changed when they were worn enough to produce surface roughness greater than 2.5 μm (100 μin.). Additional processing details are given in Fig. 15.

## Sawing

Magnesium can be cut readily with hand or power saws. Because only low cutting pressure is required, large cuts can be taken. This means that saws must have large chip spaces for a free cutting action. Too small a chip space rapidly loads the teeth and causes the saw to ride over the work. The set of the teeth on band saws and hacksaws must be relatively circular saw teeth. Relief angles on the teeth of circular saws must be adequate to minimize friction. Designs of circular saws, band saws, and power and hand hacksaws are summarized in Table 5. Blades of circular saws may be made of high-speed steel or may have carbide inserts.

For the general cutting of magnesium plate, the saw teeth should have square tops with the face either straight or tilted 5°, with an alternate bevel (Fig. 16). For slotting operations, a triple-chip blade with alternating chamfered roughing teeth and square finishing teeth is used to provide both cutting and clearing of the slot. Figure 17 shows the design of teeth for a 305 mm (12 in.) diam slotting saw with 72 teeth.

**Speed and Feed.** For circular saws, high-speed steel blades are limited to a peripheral speed of 610 m/min (2000 sfm); carbide-tip blades can be operated at speeds as high as 3000 m/min (10 000 sfm). Magnesium requires only one-tenth the power required for sawing soft steel. A 305 mm (12 in.) carbide-tip blade of proper design can cut 25 mm (1 in.) thick plate at 6.4 linear m (250 linear in.) per minute.

Power hacksaws are commonly used for sawing magnesium and can be operated at speeds as high as 160 strokes per minute, using a feed of 0.38 mm (0.015 in.) per stroke. Power band saws are commonly operated at 365 m/min (1200 sfm).

## Grinding

It is seldom necessary to grind magnesium, because machining produces good finishes. Surface roughness measurements of 0.075 to 0.125 μm (3 to 5 μin.) have been produced on magnesium machined with diamond-tip tools. Surface finishes of 0.25 to 0.75 μm (10 to 30 μin.) are common with standard tooling. When required, magnesium can be ground by any method.

Aluminum oxide grit is most widely used, although silicon carbide has also proved successful in many grinding applications. A relatively coarse grit is usually best; the surface finish is controlled mainly by the rate of feed, as in grinding most other metals. Wheels of hardness J or K are most commonly used. Most wheels for grinding magnesium have vitrified bond, although resinoid-bonded wheels are sometimes

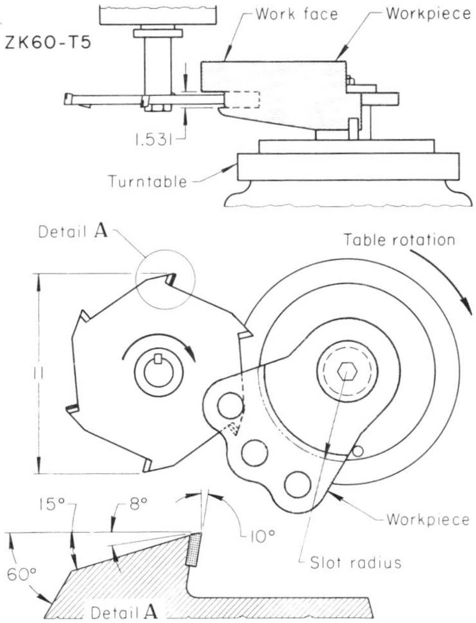


| | |
|---|---|
| **Cutter details** | |
| Peripheral cutting edge angle | 10° |
| Face relief angle | 3° |
| Radial rake angle | 10° |
| Face cutting edge angle | 5° |
| Peripheral relief angle | 8° |
| First peripheral relief angle | 15° |
| Second peripheral relief angle | 60° |
| **Operating conditions** | |
| Speed, at 1074 rev/min, m/min (sfm) | 945 (3100) |
| Feed, mm/tooth (in./tooth) | 0.13 (0.005) |
| Cutting fluid | None |
| Tolerance, side of slot to outer face, mm (in.) | ±0.13 (±0.005) |
| Surface finish, μm (μin.) | 2.5 (100) max |
| Production rate, pieces/h | 4.36 |
| Downtime for changing cutters, min | 25–36 |
| Cutter life, pieces/grind | 200–250 |

**Fig. 15** Milling of an irregularly shaped forging. Dimensions given in inches

used. Typical conditions for grinding magnesium are given in Table 6.

## Safety

The available information on safety in handling magnesium should be reviewed, including National Fire Protection Association (NFPA) standard NFPA 480 "Storage, Handling and Processing of Magnesium." Safety procedures should be fully understood by supervisors and operators before the machining of magnesium begins. Because the greatest danger in working with magnesium is complacency, proper machining practice and safety in handling magnesium should be reviewed frequently as part of an ongoing safety program. Continuous adherence to proper safety procedures is critical to the successful machining of magnesium. This section will discuss safety as it relates to machining (with emphasis on grinding), handling of chips and fines, and fire extinguishing.

**Table 5 Saws for cutting magnesium alloys**

| | |
|---|---|
| **Circular saws** | |
| Pitch, teeth/25 mm (1 in.) | ½–4 |
| Tooth set | None |
| End relief angle | 9–11° |
| Side relief (HSS saws) | 1–1°30′ |
| Clearance angle | 10–30° |
| Rake angle | 5–20° |
| Face bevel angle | 0–5° |
| Kerf, mm (in.) | 2.0–15 (0.080–0.600) |
| **Band saws** | |
| Pitch, teeth/25 mm (1 in.) | 4–6 |
| Tooth set, mm (in.) | 0.51–1.27 (0.02–0.05) |
| End relief angle | 10–12° |
| Clearance angle | 20–30° |
| **Power hacksaws** | |
| Pitch, teeth/25 mm (1 in.) | 2–6 |
| Tooth set, mm (in.) | 0.38–0.76 (0.015–0.030) |
| Clearance angle | 20–30° |
| **Hand hacksaws** | |
| Pitch, teeth/25 mm (1 in.) | 12–18 |
| Clearance angle | 20–30° |

### Machining

The magnesium chips and fines produced during machining operations can present a fire hazard. The likelihood of igniting magnesium chips and fines can be reduced by following the recommended machining practices given previously. Magnesium alloys must be heated to near the melting point before ignition can occur, although thin edges and fine particles are more easily ignited. In roughing operations, it is very difficult to attain such temperatures because the thermal conductivity of magnesium allows rapid dissipation of frictional heat. Rough cuts and medium finishing cuts produce chips of a size that cannot be readily ignited during machining. In finishing operations, in which light shavings or fines with high surface-to-volume ratios are produced, temperatures high enough for ignition can be reached.

Factors that will produce such temperatures include very fine feeds or cuts, excessive dwell, dull or chipped tools, tools with small relief angles and small chip space, and high cutting speeds in the absence of a cutting fluid. Stopping the feed and letting the tool dwell before disengagement or letting the tool or toolholder rub on the work can produce a combination of fine chips and excessive frictional heat. When cutting must be interrupted, the tool should be backed away from the work to prevent excessive heat buildup. The combination of high speeds and excessive drag with the production of small particles in a fine cut is particularly hazardous.

Sharp tools with ample clearance will reduce the possibility of heat buildup. Coarser feeds, deeper cuts, and lower speeds will have a similar effect. A mineral oil based cutting fluid is recommended in applications in which fine feeds and cuts and high speeds are essential, that is, speeds above 180 to 215 m/min (600 to 700 sfm) in combination with feeds of less than 0.25 mm/rev (0.010 in./rev). Under the most adverse conditions, which would be the use of dull tools and fine feeds, the chances of chip ignition are reduced at speeds below about 120 m/min (400 sfm).

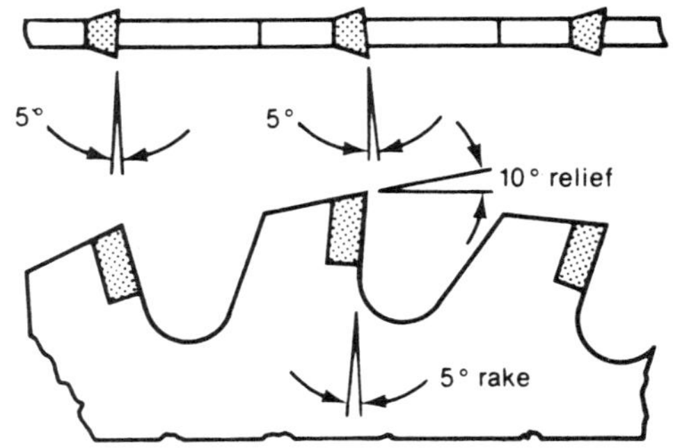


**Fig. 16** Design of a 305 mm (12 in.) diam, 48-tooth carbide (tipped) circular saw for cutoff operations at 1880 rev/min (1800 m/min, or 5900 sfm) on magnesium plate and extrusions

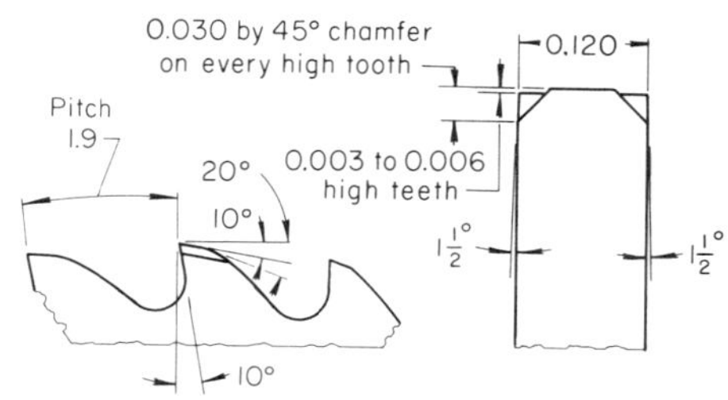


**Fig. 17** Design of a 72-tooth high-speed steel or carbide-tip circular saw for slotting magnesium. Dimensions given in inches

In machining cast magnesium parts with a cast-in-place insert of a dissimilar material such as a ferrous metal, sparking may occur when the tool strikes the insert. The hazard presented by this sparking is minimized by the use of a mineral oil based cutting fluid.

The most important precaution in preventing the spread of a magnesium fire is to avoid the accumulation of magnesium chips or turnings around, under, or on machines. The spreading of a fire will be minimized if the quantity of chips near the machine is small. Machines and work areas should be kept clean; dry magnesium chips should be stored in plainly labeled, covered noncombustible containers. Wet magnesium chips and fines are particularly hazardous (see the section "Handling of Chips and Fines" in this article). Machinists should not wear coarse-textured or fuzzy clothing, and chips, filings, and fine dust should not be allowed to accumulate in cuffs or pockets. Good magnesium machining practice includes the following:

**Table 6 Typical wheels and grinding conditions for magnesium alloys**

| | |
|---|---|
| **Surface grinding** | |
| Wheel classification | A-46-K-V |
| Wheel speed, m/min (sfm) | 1675–1980 (5500–6500) |
| Table speed, m/min (sfm) | 15–30 (50–100) |
| Downfeed per pass, mm (in.) | |
| Roughing | 0.076 (0.003) |
| Finishing | 0.025 (0.001) |
| Crossfeed per pass | ⅓ wheel width |
| **Cylindrical grinding** | |
| Wheel classification | C-46-J-V |
| Wheel speed, m/min (sfm) | 1675–1980 (5500–6500) |
| Work speed, m/min (sfm) | 46 (150) |
| Infeed per pass, mm (in.) | |
| Roughing | 0.051 (0.002) |
| Finishing | 0.013 (0.0005) |
| Traverse per work revolution | |
| Roughing | ⅓ wheel width |
| Finishing | ⅙ wheel width |
| **Centerless grinding** | |
| Wheel classification | C-60-K-V or A-60-J-V |
| Wheel speed, m/min (sfm) | 1675–1980 (5500–6500) |
| Infeed per pass, mm (in.) | |
| Roughing | 0.13 (0.005) |
| Finishing | 0.38 (0.0015) |
| Work feed, mm/min (in./min) | 1270 (50) |
| Regulating wheel | 30-rev/min rotation; 3° angle |
| **Internal grinding** | |
| Wheel classification | A-36-K-V |
| Wheel speed, m/min (sfm) | 1525–1980 (5000–6500) |
| Work speed, m/min (sfm) | 24–49 (80–160) |
| Infeed per pass, mm (in.) | |
| Roughing | 0.076 (0.003) |
| Finishing | 0.0051 (0.0002) |
| Traverse per work revolution | |
| Roughing | ⅓ wheel width |
| Finishing | ⅙ wheel width |

- Keep all cutting tools sharp and ground with adequate relief and clearance angles; avoid using dull, chipped, or broken tools
- Use heavy feeds to produce thick chips
- Avoid fine feeds, and do not permit the tool to dwell on the work
- When taking fine cuts, use a mineral oil coolant to reduce heat buildup
- Avoid striking the tool on iron or steel inserts in the magnesium part
- Practice good housekeeping
- Smoking or open flames should not be allowed in the machining area
- Keep an adequate, plainly labeled supply of recommended fire-extinguishing materials within reach of all machine operators, and be sure operators are trained in their use (see the section "Fire Extinguishing" in this article)

**Grinding.** Magnesium dust ignites readily and may explode if ignited while suspended in air. Every precaution should be taken to ensure the proper collection and disposal of magnesium grinding dust.

Belt or disk sanders for the dry grinding of magnesium should be designed so that no sparking will take place if a belt or disk breaks and strikes a bolt, nut, or any metal part that might cause a spark. The shield surrounding the grinding operation should be constructed of a nonsparking material and should be designed for easy opening to permit daily, or more frequent, removal of grinding dust.

The dust produced by the dry grinding of magnesium must be immediately removed from the work area with a properly designed wet-type dust collection system. Proper systems precipitate the magnesium dust using a heavy spray of water and should be designed so that dust or sludge cannot accumulate and dry out to a flammable state. The best dust collectors are small units serving one or two grinding stands. Short, straight ducts should connect the grinder to the collector. Self-opening vents prevent the collection of hydrogen during shutdown. The power supply to the grinders should be interlocked with the exhaust blower motor and the liquid level controller of the wet collector so that a malfunction of the dust collection system will shut down the grinding machine it serves. A time delay switch or equivalent device should be provided on the grinder to prevent start-up of the motor drive until the wet collector is in complete operation and several changes of air have swept out any residual hydrogen. Dust collectors should be kept clean and should be vented to the outdoors. The magnesium-water sludge should be frequently removed from the collector to prevent excessive accumulation. The sludge should be kept under water until disposal.

Grinding and polishing booths in which the work is done on a grating should follow the same principles of dust collection. The air in the hood flows past the work through the grating and through a water spray, which removes the dust. Booths should be designed to catch a maximum amount of dust, keeping it off the floor and worker's clothing. The power supply to the portable grinders should be interlocked with the booth dust collection system in the manner described above for grinding stands.

Good housekeeping is essential for the safe grinding of magnesium. Daily inspection and cleaning of the duct between the grinding wheel and the collector, as well as thorough cleaning of the entire dust collecting unit at least once a month, is mandatory. Magnesium dust should not be allowed to accumulate on benches, windows, pipes, or other horizontal surfaces. If there is any evidence of dust accumulation, the collection system is not functioning properly.

Connecting a number of dust-producing units to one exhaust system is not recommended. Central collection systems, which pass the dust through long, dry, or partly dry ducts and conventional dry-type filter collectors are not suitable for magnesium dust collection.

If magnesium is to be ground wet on a belt sander or disk grinder, sufficient cutting fluid should be used to collect all the dust and to convey it to a collection point. The dust should always be kept wet with a copious supply of fluid, and no dry or damp dust should be allowed to collect on the machine. Mineral oil can be used as the cutting fluid. The sludge formed by the mixture of magnesium dust and cutting fluid should be disposed of according to applicable regulations. The following precautions are recommended for magnesium grinding operations.

- Grinding equipment should be reserved for magnesium and should be plainly labeled "Magnesium Only." Before wheels are dressed, the dust collector should be thoroughly cleaned
- The grinding of chromium-pickled magnesium surfaces can cause sparking; therefore, dust or air-dust mixtures must not be allowed to accumulate near such sparks
- Operators of grinding equipment should wear smooth caps, smooth gloves, and smooth fire-retardant clothing without pockets or cuff. Aprons or protective clothing should be easily removable and should be kept free of dust
- Warning signs should be prominently displayed in order to prevent unnecessary fire hazards

## Handling of Chips and Fines

Dry chips should be placed in clean, tightly closed steel containers and stored away from possible contact with water. Clean, dry magnesium chips generally have resale value.

Wet chips or sludge should be stored in vented steel containers in a remote location. Venting must be sufficient to allow the escape of hydrogen. Tightly covered containers of wet chips and fines are particularly hazardous because of the possible accumulation of hydrogen in a small, confined space. Wet chips and sludge must be disposed of frequently and never stored for long periods or allowed to become partially dry. Containers should be kept full of water until disposal is arranged. Partially dry chips and sludge are extremely flammable and may ignite spontaneously.

Magnesium chips, fines, or sludge can be rendered noncombustible by reaction with a 5% solution of ferrous chloride ($FeCl_2 \cdot 2H_2O$). This process is outlined in standard NFPA 48-1974 and in the "NFPA Fire Protection Handbook." Reference should be made to these publications before using the treatment. (The ferrous chloride treatment method was apparently inadvertently omitted from the NFPA 48-1982 and NFPA 480-1987 editions of the standard.) Before implementing procedures and processes that use ferrous chloride solutions to chemically treat wet magnesium chips, fines, or sludge, all applicable federal, state, and local laws, regulations, and permit procedures must be considered.

The rate of the ferrous chloride reaction is such that, within 24 h, the magnesium is usually sufficiently converted to magnesium hydroxide and magnesium chloride that the residue will not burn. Because hydrogen is generated by the reaction, the process should be done in an open container placed outdoors so that natural air movement will prevent dangerous accumulations of hydrogen. Open flames and smoking should be prohibited near the process. The amount of ferrous chloride required to carry out the decomposition is approximately 0.6 kg/kg (0.6 lb/lb) of dry magnesium. The amount of water in the sludge should be considered in determining the amount of water needed to make up a 5% solution. The cycle can be repeated daily in the same container until it is necessary to clean out the brown, damp residue. The reactivity of the residue should be determined in accordance with applicable disposal regulations. Information on the safe handling of ferrous chloride should be obtained from the supplier of the material.

## Fire Extinguishing

Two powders, both Underwriters Laboratories listed, are available for extinguishing magnesium fires:

- G-1, an inhibited graphite-base powder
- Met-L-X, a sodium chloride-base powder

These powders smother the fire by excluding oxygen. Both powders are available in pressurized canister-type fire extinguishers. The G-1 powder is also available in bulk form suitable for scooping or shoveling onto a fire. A magnesium machine shop should have both types of powder on hand, with some of the G-1 powder in bulk. Full containers of the recommended powders should be located at each machine. Operators should be trained in the use of the fire-extinguishing materials, and they should not become alarmed at the brilliant white light produced when magnesium burns. In the unexpected absence of an approved extinguishing powder, clean, dry, rust-free cast iron chips can also be used. Cast iron chips cool the fire below its ignition point rather than smother it.

Under no circumstances should water or any standard fire extinguisher be used to combat a magnesium fire. Water or other liquids, carbon dioxide, foam, or sand will react with burning magnesium and will intensify rather than control the fire.

# Machining of Zinc Alloy Die Castings

Gary Adams, Cominco Metals

ZINC ALLOY DIE CASTING has the capability of producing complex parts to close tolerances. Often, little or no machining is required in finishing zinc die castings. However, flash formed at die parting surfaces usually has to be removed, and critical surfaces can require machining. Holes, undercuts, and threads are commonly and economically produced by machining when they are not readily cast in because of shape complexities and limitations in die casting machines and tooling.

## Alloys

Alloys 3 and 5 are the common hot-chamber zinc die casting alloys based on the zinc-aluminum eutectic at 4% aluminum. The third, less commonly used member of this Zn-4Al family is alloy 7. These alloys have been die cast and machined since the 1930s.

More recently, the higher-aluminum alloys ZA-8, ZA-12, and ZA-27 were developed for die casting as well as gravity and continuous casting. The ZA alloys have improved strength and lower density than the eutectic alloys and offer a broader range of properties to zinc die casters. The alloy ZA-8 can be hot-chamber die cast, while cold-chamber casting is generally required for the higher melting point alloys, ZA-12 and ZA-27.

Thus zinc die casting alloys are zinc-aluminum based with aluminum contents ranging from 4 to 27 wt%. Corresponding ranges in major properties are: melting point (liquidus) 387 to 484 °C (728 to 903 °F); density 6.6 to 5.0 g/cm$^3$ (0.24 to 0.18 lb/in.$^3$); hardness, 80 to 120 HB; and tensile strength 280 to 440 MPa (41 to 64 ksi). The machining behaviors and recommendations for machining are quite similar for die castings of all of these alloys.

## Machinability

Zinc alloy die castings give good machining performance over a wide range of machining conditions. High-quality surface finishes can be achieved at good machining rates. Tool wear and cutting forces are lower than for most other metals. Good (discontinuous) chip formation is common in machining zinc die castings.

**Cutting fluids** should be applied generously to the tool-workpiece contact area. Dry machining is not recommended, but is practiced successfully in some cases when small amounts of metal are removed. Emulsifiable oils mixed with water (ratio 1:20 to 1:10 for heavy cutting) are preferred to oil-base fluids because of their superior cooling characteristics. Cutting fluids are important for surface quality, accuracy, chip formation, tool life, and prevention of burr formation and built-up edge (or tool-workpiece welding).

**Built-up edges** lead to deterioration in finish and accuracy. In severe cases, flute loading and tool fracture can occur. Built-up edges can be prevented by reducing friction and heat generation at the rake and clearance surfaces adjacent to the cutting edge. Sharp tools, proper rake and clearance angles, polished tool surfaces, and properly balanced feeds and speeds help prevent the occurrence of tool-workpiece welding. Cutting fluids help both by lubricating the tool surfaces and by cooling the tool and workpiece.

**Cutting Tools.** In general, high-speed steel tools are suitable. Carbide tools are preferred by some for high-volume, heavy machining at high speeds if the higher tooling costs are offset by longer tool life and higher production rates. Cutting edges should be sharp, and rake and clearance faces should be positive (for example, 10°) and finely ground or polished to reduce friction and heat generation. Fluted tools (drills, taps, and so forth) should have large, polished flutes to minimize friction and assist in ejecting cutting wastes. Tool and workpiece holding systems should be rigid for good surface quality and accuracy.

**Hard Spots.** In the higher-aluminum alloys, ZA-27 and (to a lesser extent) ZA-12, severe tool wear can occur if hard iron-aluminum particles ($FeAl_3$ intermetallics) are allowed to form and enter the die cavity during melting and casting. This can be prevented by keeping iron impurity levels within specification and avoiding low melt holding temperatures. The intermetallic particles are buoyant in the molten alloys, and thus bath surface skimming is recommended prior to casting.

## Turning and Boring

**Turning Tool Angles.** For turning tools of high-speed steel or cast cobalt alloy, back and side rake angles of 5 to 15° (10° typical) are recommended. End relief angles should also be about 10°, large enough to prevent drag on the heel of the tool. When side cutting is involved, a side relief angle of 4° is recommended.

With carbide turning tools (in which the carbide tip must be rigidly supported), the end relief angle should be reduced to about 8°, and the back rake should be within the range from 5 to 10°. The side cutting-edge angle may be from 10 to 20°, but this is not critical. The end cutting-edge angle is commonly in the range of 5 to 10°. The nose of the tool is rounded to a maximum of 0.8 mm (1/32 in.) radius. Figure 1 shows the basic carbide-tipped tool for turning zinc alloys. For best accuracy, the tool should be set horizontal and on center, or slightly below center.

**Turning Speeds and Feeds.** For roughing cuts using high-speed steel or cast alloy tools, a speed of 60 m/min (200 sfm) is recommended. For finishing, this speed may be increased to 120 m/min (400 sfm) or higher. When using carbide tools, speeds of 120 to 180 m/min (400 to 600 sfm) are common, and range as high as 240 m/min (800 sfm) after experience is gained on a specific job.

Feeds of 0.13 to 0.25 mm/rev (0.005 to 0.010 in./rev) are most commonly used, mainly because cuts taken on die castings are usually light. When heavier cuts must be taken, higher feeds are feasible.

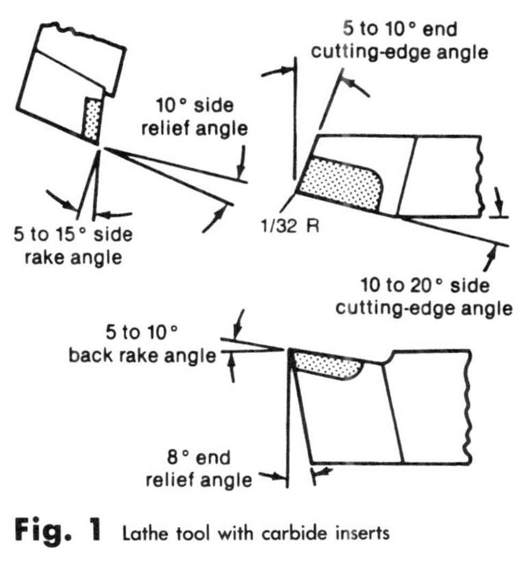


**Fig. 1** Lathe tool with carbide inserts

**Boring.** For boring tools, rake and relief angles are usually about the same as for tools used in turning. The same tool materials, speeds, and feeds can be employed. As in other machining operations on zinc alloys, light cuts and high speeds are recommended and commonly used to avoid seizure.

Diamond tools are infrequently used for boring zinc die castings, although they produce surfaces of exceptional smoothness, and when a precision machine is used, extreme accuracy can be held. Diamond tools are expensive and must be lapped, but when properly applied, they can bore many thousands of holes without relapping. Much higher speeds (for example, 300 m/min, or 1000 sfm) are feasible with diamond tools, but feed rates are lower (for example, 0.13 mm/rev, or 0.005 in./rev), and intermittent cuts should be avoided.

Tool and toolholders must be as rigid as possible in boring to prevent chatter and to maintain accuracy. Care must be taken in small-diameter bores to provide an adequate volume of cutting fluid to the cutting area and to ensure that clearance exists between the bore surface and all parts of the tooling away from the cutting edge.

## Drilling

Drilling is the most widely used operation in the machining of zinc die castings. Holes may be produced in their entirety by drilling, or cored holes may be deepened or enlarged. Cored holes, depending upon their depth and draft, may be drilled prior to tapping to remove drafts for more uniform threads. Fast-spiral or polished general-purpose jobber drills made of high-speed steel are generally recommended. Carbide-tipped drills can also be used. Cutting fluids should also be applied generously.

**Drill Design.** The standard 118° point angle is generally recommended (Fig. 2a); for special work, this angle may be varied from 90 to 136°. The flatter points are used to minimize runout when drilling heavy sections of castings in which porosity may be encountered. The sharper points are used to minimize burr formation after breaking through a section.

A lip clearance angle of 12° works satisfactorily in most drilling (Fig. 2b), but is often increased to 15°, and in at least one instance a lip clearance of 20° proved to be advantageous. In another application, a gradual increase of the lip clearance angle toward the center of the drill was the best configuration. The chisel edge angle should be 120 to 135° (Fig. 2c).

The fast-spiral drill geometry is designed for softer, nonferrous metals and is preferred for zinc die castings. Features of this drill include: a thin web, large polished flutes, and a large back rake angle. These all help prevent galling and improve chip ejection. Drill strength is sacrificed in thin-web

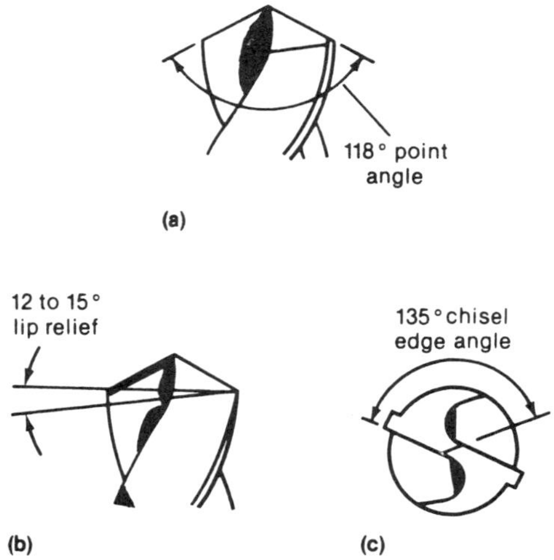


**Fig. 2** Drill angles for drilling zinc die castings

drills, and if drill breakage is a problem, the general-purpose jobber drill (polished) can be used. Web thinning at the drill point can be effective on heavier-web drills, but care must be taken in grinding to keep the proper tip geometry and to maintain polished surfaces.

One large producer of machined die castings recommends two-flute drills up to 6 mm (¼ in.) in diameter, three-flute drills in 6 to 9.5 mm (¼ to 3/8 in.) sizes, and four-flute drills for all work above the 9.5 mm (⅜ in.) diam. One manufacturer of drills recommends that two-flute drills be used for all drilling of new holes and that three or four flutes be used for enlarging holes. These two recommendations are not necessarily consistent, because cored holes are often less than 6 mm (¼ in.) in diameter. Drilling practice varies considerably among different shops.

The slow-spiral drill geometry, like the fast-spiral design, has large polished flutes and a thin web. The flute helix angle is much lower, resulting in a lower rake angle and different chip removal characteristics. The slow-spiral drill can provide good results in shallow holes or when grabbing is a problem with the higher-helix drill designs.

Carbide-tipped drills are sometimes used for long runs if the higher initial and regrinding costs can be offset by longer tool life and higher speeds. Details of such a drill are shown in Fig. 3.

**Speeds.** Speeds of 60 to 90 m/min (200 to 300 sfm) are generally satisfactory for high-speed steel drills. One plant reports a speed of 12 000 rev/min (180 m/min, or 580 sfm) in drilling deep holes 5 mm (3/16 in.) in diameter, but this rate is much higher than normal.

**Feeds** are:

| | | | | | |
|---|---|---|---|---|---|
| Diameter, mm (in.) | 3 (⅛) | 6 (¼) | 9.5 (⅜) | 13 (½) | 19 (¾) |
| Feed rate, mm/rev (in./rev) | 0.1 (0.004) | 0.2 (0.008) | 0.28 (0.011) | 0.33 (0.013) | 0.41 (0.016) |

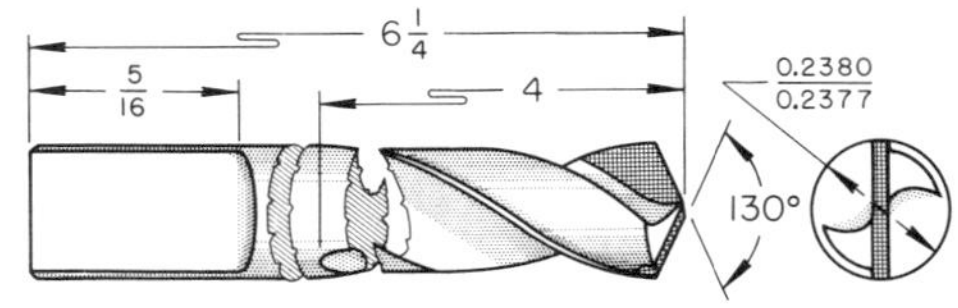


**Fig. 3** Carbide-tipped drill for high-production drilling of zinc castings. Dimensions given in inches

Dwells in feed should be avoided.

**Deep-Hole Drilling.** For deep holes (depths greater than 6 times the drill diameter), speeds and feeds should be reduced by up to 50%. Intermittent withdrawal is helpful to clear cutting wastes and to allow cutting fluids to reach the drill tip.

## Reaming

In zinc die castings, holes are often cored to reaming size, thus eliminating the need for drilling. However, reaming predrilled holes is also common. Reamers made of solid high-speed steel are commonly used. High-speed steel reamers have produced 30 000 reamed holes before being resharpened, although more frequent sharpening is usually required. Reamers with inserted blades of either high-speed steel or carbide are also used for reaming zinc, provided the size of the holes to be reamed permits insert construction. The generous application of cutting fluid is also important in reaming zinc die castings for dimensional accuracy and surface finish.

Some work is permitted to float during reaming to avoid drift of the hole. The uniform cutting action of the reamer holds the center fixed.

**Reamer Design.** The common six-fluted machine reamer design is generally recommended for zinc die castings. Both right-hand spiral-fluted and straight-fluted geometries are acceptable. Spiral-fluted reamers are required for holes having keyways or similar irregularities. Spiral-fluted reamers tend to assist in the removal of cutting wastes and prevention of flute loading. Rose reamers are not recommended.

Deep flutes are recommended to allow sufficient chip space. The polishing of flutes aids in preventing built-up edges. The chromium plating of reamer flutes after polishing has also helped prevent built-up edges.

Sharp cutting edges are necessary for successful reaming of zinc alloys. The faces of flutes are usually made radial (zero rake), and the clearance angle behind the land generally approximates 10°. A back taper from nose to shank of about 0.025 mm (0.001 in.) is generally used.

Some standard reamers have a land that is too wide for best results in zinc die castings. A 0.38 mm (0.015 in.) land width is common, but this is often ground down to 0.13 to 0.18 mm (0.005 to 0.007 in.) to aid in the prevention of built-up edges and to

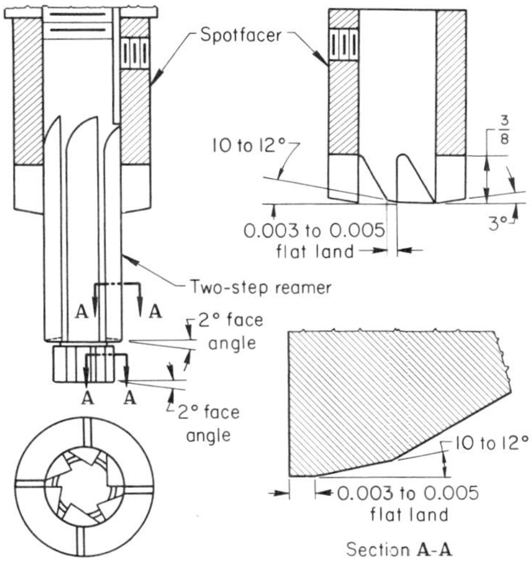


**Fig. 4** Combination tool for reaming a two-step hole and spotfacing a valve seat. Dimensions given in inches

reduce the amount of burnishing and generated heat. If heating is sufficient to cause the hole to expand, the hole may be too small after the reamer has been removed and the workpiece has cooled.

The hole diameter should be approximately 97% of the reamer diameter before reaming. If at least 0.8 mm (1/32 in.) of metal is to be removed by reaming, one procedure is to grind a stepped portion at the end of the reamer about 0.13 mm (0.005 in.) smaller than the body of the reamer. The step cuts away most of the metal, leaving the body portion to size the hole with a light cut. This practice, however, may cause excessive reamer wear.

Stepped reamers are used for simultaneously reaming a number of bores in stepped holes. Reamers are sometimes combined with spotfacing tools. Figure 4 shows a combination tool used for two-step reaming and the simultaneous spotfacing of a 3° conical valve seat.

In reaming, most of the cutting is done by the reamer nose, which is chamfered to provide a lead. This chamfer is usually 45°. Once the initial grinding of the flutes and lands is correct, little other grinding is required except at the lead end, although a very light cut may be taken off the radial face of the flutes. In grinding the lead chamfer, a clearance angle of about 10° is recommended.

For holes large enough to permit the use of inserted-blade reamers, this type is usually preferred, but blades should be firmly supported. The blade can be shaped to perform a simultaneous facing operation at the bottom or at the top of the hole.

**Speeds and Feeds.** Wide ranges of speed and feed have produced successful results in reaming zinc. Optimum speeds and feeds are usually determined by trial and error. Speeds of 90 to 120 m/min (300 to 400 sfm) have often been used for high-speed steel reamers, and up to twice these speeds have been employed for carbide-blade reamers. Lower speeds (reduced by up to 50%) are recommended for critical diameters. Feed rates recommended for drilling are also suitable for reaming in general.

## Tapping

It is nearly always cheaper to tap than to cast threads around a core that has to be unscrewed from the casting. Only very coarse threads and those of special shape are cored. Cutting fluids should be used in tapping.

**Cored Holes.** Holes for tapping are usually cored in zinc die castings except when coring would greatly increase die cost. Because core pins require some taper, cored holes are smaller at the bottom than at the top. Hence, unless allowance is made for taper or the hole is drilled or reamed before tapping, tap breakage or nonuniform threads will be produced, especially in deep holes. The hole diameter is usually designed to yield 75 to 80% of full thread after tapping.

**Tap Designs.** Commercially available tap geometries are suitable for tapping zinc die castings, and special grinding is not normally required. Spiral-point taps are preferred for through holes. The ends of the flutes on these taps are ground off at an angle, instead of being straight. This creates a side rake angle at the cutting edges at the end of the tap. The cutting wastes produced by the spiral-point tap are curled away from and ahead of the tap and do not normally enter the flutes. This results in clean cutting action as long as the collection of waste at the exit side of the hole can be accommodated. The spiral-point tap is thus not normally used for tapping blind holes where accumulations of chips at the bottom of the threaded holes are difficult to remove.

For blind holes, the standard straight-fluted general-purpose tap geometry is generally acceptable, but the "special aluminum" tap design is preferred. This tap has high hook (or rake) angles such as 12 to 15° (Fig. 5), and greater radial clearance (or eccentric relief) than the standard general-purpose tap geometry. These features, together with a slight back taper (Fig. 5), reduce the friction and galling when tapping zinc. Eccentric clearance also facilitates removal of the tap from the hole and reduces sticking. Sticking can also be reduced by grinding off or "breaking" the back corner of the lands at the side opposite the cutting edge. Right-hand spiral-fluted taps sometimes prove to be advantageous because they are effective in carrying the chips out of the hole. However, these taps produce high torques, and tap breakage can occur, especially in small-diameter holes.

Opinions differ as to whether the added cost of ground-thread taps is offset by the benefits of increased tap life and more precise threads. Ground-thread taps are often used because they produce smoother threads and minimize galling. They are recommended when close tolerances must be met. For class 3 fits, the use of a leadscrew control is recommended.

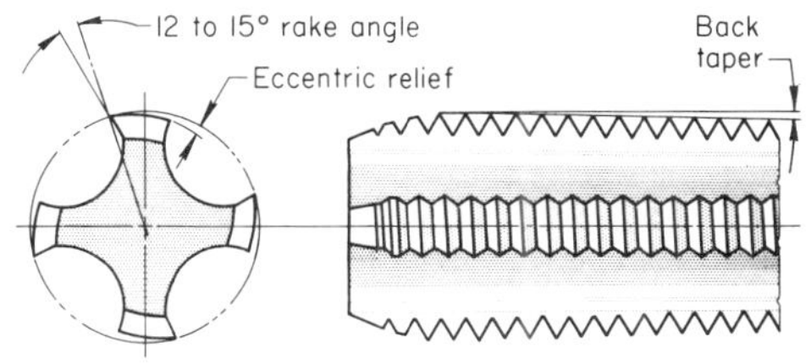


**Fig. 5** Tap design for zinc die castings

Generous chip clearance is mandatory for successful tapping. Two-flute taps for holes up to about 8 mm (5/16 in.) in diameter and other taps with one less flute than standard are commonly preferred. According to one large manufacturer, however, three-flute taps in sizes below 6 mm (1/4 in.) have longer life than two-flute taps because less tool surface bears on the threads cut in the casting.

For holes 30 mm (1 1/4 in.) in diameter or larger, adjustable or collapsible taps are often used to advantage. Collapsible taps make it easy to vary the size and fit of the thread and eliminate the need for reversing the tap.

In selecting taps for taper pipe threads, the general recommendations for bottoming taps should be followed. Some rake and some relief are desirable. One maker especially recommends interrupted-thread pipe taps.

**Form Tapping.** Form tapping is attractive for small-diameter holes, for which chip removal and tap breakage are potential problems. Form taps produce threads by plastic deformation instead of by cutting, and there are no cutting wastes. Form taps are fluteless and much stronger than cutting taps. Zinc die castings are readily form tapped over wide ranges of tapping conditions.

The disadvantages of form tapping are higher tool costs, poorer thread shape, and the need for more accurate starting-hole diameters. Form tapping is not commonly practiced for lower-ductility zinc alloys such as ZA-27. Also, very high power is required for large diameters.

**Speeds.** Tapping speeds up to 60 m/min (200 sfm) can be used for zinc die castings. In general, the speed should be reduced as the pitch of the thread decreases.

## Die Threading

Many die castings require external threading. The die threading operation may be only for breaking fins and cleaning up a cast thread at the die parting line. Die threading may also be used to produce full threads on studs or projections where the casting of threads is not feasible. Because zinc alloys are easily machined and chasing is likely to be needed anyway, it is often more economical to cast the stud or projection to threading size and then to chase the full thread rather than to cast rough threads that still require chasing to obtain acceptable results. Cutting fluids are recommended.

**Dies.** Threading can be done successfully with button or acorn dies in small sizes, but for sizes larger than 6 mm (¼ in.), chaser dies are recommended. Chasers made by milling are preferred for threading zinc, although those made by other methods are sometimes used.

Chasers may be of either the radial or the tangential type. For the radial type, one tool manufacturer recommends a 10° radial hook for straight threads and a 7° radial hook for tapered threads.

A maker of tangential chasers states that with zinc alloys the cutting edge of the chaser must be on or very near the center to avoid rapid wear of the chasers just beyond the cutting edge, whereas with other materials the cutting edge must be above the center. The type of grinding recommended for this type of chaser involves a 5° positive rake. Using this rake, about 30 000 threads (¾-24) have been cut between grinds.

For round dies, one manufacturer recommends a rake of 15 to 20° with narrow cutting sections and eccentric relief for an 11 mm (7/16 in.) diameter and larger. Self-opening dies or solid adjustable dies are recommended for cutting the larger-size threads.

**Speed.** Dies for threading zinc castings are commonly operated 10 to 15% slower than taps used for cutting similar internal threads. One manufacturer recommends that when radial chasers are to be used, speeds of 30 m/min (100 sfm) for 8 to 11 threads/25 mm (1 in.) and 60 m/min (200 sfm) for 12 to 32 threads/25 mm (1 in.) be used.

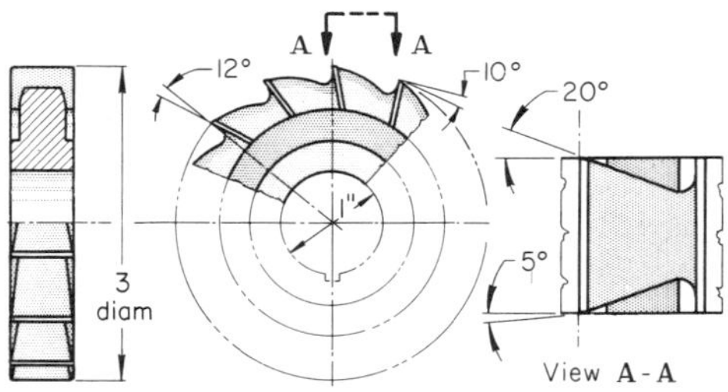


**Fig. 6** Typical cutter for milling zinc die castings

## Milling

Although not as common as some other operations in machining zinc die castings, milling is useful for producing slots and undercuts and accurate, flat surfaces. Both end milling and plain milling operations are used. Cutting fluids should be applied generously to the cutting area.

**Cutter Designs.** For end milling, the common right-hand spiral-fluted cutter and the "special aluminum" cutter designs give good results. Large polished flutes are preferred. Similarly, for plain milling, helical fluted cutter designs with large polished flutes are preferred. If possible, contact should not be lost between a cutter tooth and the workpiece until the next tooth engages the work. This reduces vibrations and contributes to improved surface finish and reduced cutter and machine wear.

Stock cutters can be used if the teeth have adequate primary clearance. About 10° is appropriate (Fig. 6), although larger angles are better for special form cutters. The radial rake angle of 10°, as found on most stock cutters, is satisfactory, but a 15° positive radial rake angle is often used.

Staggered-tooth cutters are sometimes used for deep slots, because they produce finer chips and a better surface finish in some cases than when chips are the full width of the slot. A similar effect can be obtained by nicking or chamfering the corners of alternate teeth. Land width should range from approximately 0.6 mm (0.025 in.) for small cutters (<75 mm, or 3 in. diam) to about 1.3 mm (0.050 in.) for larger cutters (>125 mm, or 5 in. diam).

**Milling Direction.** Climb milling is preferred for reduced burr formation. It also improves chip flow and disposal, carries cutting fluid into the cutting zone more readily, and reduces sensitivity of surface finish to cutter sharpness.

**Depth of Cut.** Shallow depths, such as 0.6 mm (0.025 in.) are preferred. Multiple passes should be considered when large amounts of metal are to be removed by milling.

**Speeds and Feeds.** For shallow cuts, 100 m/min (350 sfm) is a good cutting speed, although speeds in excess of 300 m/min (1000 sfm) have been used successfully. Cutter speeds should be reduced for deep milling cuts. Optimum feeds vary considerably, depending on cutter diameter, width and depth of cut, number of cutter teeth, cutter speed, machine and workpiece rigidity, and other factors. Optimum feeds usually are determined by trial and error, guided by the general behaviors noted above.

## Sawing

Sawing is not often used in machining zinc die castings, but is sometimes useful, for example in prototyping or short runs, or for rough trimming prior to a finishing operation such as milling or buffing.

**Blade Design.** High-speed steel blades of common design with raker set and standard tooth profile give good results in sawing zinc die castings. The coarsest blade possible should be used for high productivity and to prevent blade loading. Blade pitch should be selected so that at least two teeth are in contact with the workpiece at all times.

**Speeds and Feeds.** The recommended sawing speed range is 90 to 120 m/min (300 to 400 sfm). Much higher speeds have been used successfully in some applications. Feed rates should vary with workpiece shape, blade geometry, and surface finish requirements. Moderate feed pressure should be applied initially and adjusted as required.

# Machining of Nickel and Nickel Alloys

Revised by R.W. Breitzig, INCO Alloys International

NICKEL-BASE ALLOYS can be machined by techniques used for iron-base alloys. However, certain requirements are imposed by the high strength of the nickel alloys, their tendency to work harden, and their gumminess in some conditions.

## Machinability

Nickel alloys, like austenitic stainless steels, work harden rapidly. The high pressures developed during machining produce a hardening effect that retards further machining and may also cause distortion in parts that have small cross sections.

One method of reducing work hardening during machining is to work harden the material by cold deformation before machining. Cold-drawn stress-relieved material is always preferred for machining, particularly when the smoothest finish is desired. Hot-rolled material is next best, and annealed material is the least desirable in most applications. On age-hardenable alloys, optimum finish is produced by machining in the aged condition. The high strength and hardness of the aged material prevent heavy cuts; therefore, most roughing is done before age hardening. Solution annealing of age-hardenable alloys improves machinability by dissolving hard phases.

A second method of minimizing work hardening during machining is to employ careful machining practices. Sharp tools with positive rake angles, which cut the metal instead of pushing it, are required. Feed rate and depth of cut must be sufficient to prevent burnishing or glazing. Tools should not be allowed to rub the work, either because of improper clearance or by being allowed to dwell in the cut.

**Distortion.** Even under optimum machining conditions, some stresses are produced that may cause subsequent distortion of the work. For maximum dimensional stability, it is best to rough out the part almost to size, stress relieve it, and then finish it to size. Stress relieving has little effect on dimensions, but may affect mechanical properties.

**Microstructure.** Grain size has little direct effect on the machinability of nickel alloys. In general, microstructure affects machinability in two ways:

- The presence of graphite or sulfide phases greatly improves machinability
- Hard phases, such as carbides, nitrides, carbonitrides, oxides, silicates, and possibly the $\gamma'$ phase $Ni_3(Al,Ti)$, are abrasive and cause rapid tool wear

The nickel-chromium and nickel-chromium-iron alloys are less abrasive than the common grades of austenitic stainless steel because they have lower carbon content and therefore fewer carbide particles. Probably the hardest and most abrasive of all the phases is titanium carbide (TiC), which is present in most of the age-hardenable alloys; another hard phase is niobium carbide (NbC). These carbides are usually present in as-rolled or mill-annealed products. Solution annealing at a high temperature of 1100 °C (2000 °F) or above is required to dissolve them. Age hardening precipitates greater amounts of niobium and TiC, along with chromium carbide ($Cr_3C_2$) and the $\gamma'$ phase. The refractory nature of these phases, together with their strengthening effect, makes age-hardened material less machinable, although chip action is improved.

These refractory phases have little or no effect on the finishes obtainable. Softer and gummier alloys, such as those in the A, B, and C groups identified below, are likely to have torn surfaces after roughing operations, and greater care is necessary in the finishing operations. With proper technique, however, finishes of 0.10 to 0.20 μm (4 to 8 μin.) and finer can be obtained by grinding, honing, and lapping. It may not always be possible to produce these finishes on Monel R-405, because this alloy is likely to have sulfide and graphite inclusions.

## Classification of Alloys

Table 1 classifies 58 nickel alloys with respect to machining characteristics. Except as noted in subsequent discussion, all alloys in a given group require similar machining practices.

**Group A** consists of alloys containing 95% or more nickel. These alloys have moderate mechanical strength and high toughness. They are hardened only by cold work. The alloys are quite gummy in the annealed and the hot-worked condition, and cold-drawn material is recommended for best machinability and smoothest finish.

**Group B** consists of most of the nickel-copper alloys. The alloys in this group have higher strength and slightly lower toughness than those in group A. They are hardened only by cold work. Cold-drawn or cold-drawn and stress-relieved material gives the best machinability and smoothest finish.

**Group C** consists largely of the solid-solution nickel-chromium and nickel-chromium-iron alloys, which are similar to the austenitic stainless steels. They are hardened only by cold work and are machined most readily in the cold-drawn or cold-drawn and stress-relieved condition.

**Group D,** which consists primarily of the age-hardenable alloys, has two subgroups:

- Alloys in the unaged condition
- The alloys of group D-1 in the aged condition, plus several other alloys in both the aged and unaged conditions

The alloys in group D have high strength and hardness, particularly when aged. Material that has been solution annealed and quenched or rapidly air cooled is in the softest condition, and it machines most easily. Because of its comparative softness, the unaged condition is necessary for greater ease in drilling, tapping, and threading.

Age-hardenable alloys take heavy machining best when they are either solution annealed or hot worked and quenched. Although fully age-hardened material is usually too hard for tools with weak cutting edges, such as small drills and taps, and for rough machining, it can be finish machined to fine finishes and close tolerances.

The best way to machine the alloys of group D, therefore, is to machine almost to size in the unaged condition, age harden, and then finish to size. Because the age-

**Table 1 Classification of nickel alloys by machining characteristics**

| Alloy | Nominal composition, %(a) | | | | | | | | | | | | | | | | |
|---|---|---|---|---|---|---|---|---|---|---|---|---|---|---|---|---|---|
| | Ni | C | Mg | Mn | Fe | Al | Cr | Mo | Ce | Ti | Si | Nb | Cu | Co | $Y_2O_3$ | W | S |
| **Group A** | | | | | | | | | | | | | | | | | |
| Nickel 200 | 99.6 | 0.08 | ... | ... | ... | ... | ... | ... | ... | ... | ... | ... | ... | ... | ... | ... | ... |
| Nickel 201 | 99.6 | 0.01 | ... | ... | ... | ... | ... | ... | ... | ... | ... | ... | ... | ... | ... | ... | ... |
| Nickel 205 | 99.6 | 0.04 | 0.04 | ... | ... | ... | ... | ... | ... | ... | ... | ... | ... | ... | ... | ... | ... |
| Nickel 212 | 97.0 | 0.05 | ... | 2.0 | ... | ... | ... | ... | ... | ... | ... | ... | ... | ... | ... | ... | ... |
| Nickel 222 | 99.5 | ... | 0.075 | ... | ... | ... | ... | ... | ... | ... | ... | ... | ... | ... | ... | ... | ... |
| **Group B** | | | | | | | | | | | | | | | | | |
| Monel 400 | 66.5 | ... | ... | ... | ... | ... | ... | ... | ... | ... | ... | ... | 31.5 | ... | ... | ... | ... |
| Monel 401 | 42.5 | ... | ... | 1.6 | 0.3 | ... | ... | ... | ... | ... | ... | ... | 55.5 | ... | ... | ... | ... |
| Monel 450 | 30.0 | ... | ... | 0.7 | 0.7 | ... | ... | ... | ... | ... | ... | ... | 68.0 | ... | ... | ... | ... |
| Ferry alloy | 45.0 | ... | ... | ... | ... | ... | ... | ... | ... | ... | ... | ... | 55.0 | ... | ... | ... | ... |
| Nilo alloy 36 | 36.0 | ... | ... | ... | 64.0 | ... | ... | ... | ... | ... | ... | ... | ... | ... | ... | ... | ... |
| Nilo alloy 48 | 48.0 | ... | ... | ... | 52.0 | ... | ... | ... | ... | ... | ... | ... | ... | ... | ... | ... | ... |
| Nilo alloy K | 29.5 | ... | ... | ... | 53.0 | ... | ... | ... | ... | ... | ... | ... | ... | 17.0 | ... | ... | ... |
| Inco alloy MS 250 | 19.0 | ... | ... | ... | 76.0 | ... | ... | 3.0 | ... | 1.4 | ... | ... | ... | ... | ... | ... | ... |
| **Group C** | | | | | | | | | | | | | | | | | |
| Nickel 270 | 99.98 | 0.01 | ... | ... | ... | ... | ... | ... | ... | ... | ... | ... | ... | ... | ... | ... | ... |
| Monel K-500, unaged | 65.5 | ... | ... | ... | 1.0 | 2.7 | ... | ... | ... | 0.6 | ... | ... | 29.5 | ... | ... | ... | ... |
| Inconel 600 | 76.0 | ... | ... | ... | 8.0 | ... | 15.5 | ... | ... | ... | ... | ... | ... | ... | ... | ... | ... |
| Inconel 601 | 60.5 | ... | ... | ... | 14.0 | 1.4 | 23.0 | ... | ... | ... | ... | ... | ... | ... | ... | ... | ... |
| Inconel 690 | 61.0 | ... | ... | ... | 9.0 | ... | 29.0 | ... | ... | ... | ... | ... | ... | ... | ... | ... | ... |
| Nimonic 75 | 80.0 | ... | ... | ... | ... | ... | 19.5 | ... | ... | ... | ... | ... | ... | ... | ... | ... | ... |
| Nimonic 86 | 64.0 | ... | ... | ... | ... | ... | 25.0 | 10.0 | 0.03 | ... | ... | ... | ... | ... | ... | ... | ... |
| Incoloy 800 | 32.5 | 0.05 | ... | ... | 46.0 | ... | 21.0 | ... | ... | ... | ... | ... | ... | ... | ... | ... | ... |
| Incoloy 800HT | 32.5 | 0.08 | ... | ... | 46.0 | 1.0 Al + Ti | 21.0 | ... | ... | ... | ... | ... | ... | ... | ... | ... | ... |
| Incoloy 802 | 32.5 | 0.4 | ... | ... | 46.0 | ... | 21.0 | ... | ... | ... | ... | ... | ... | ... | ... | ... | ... |
| Incoloy 825 | 42.0 | ... | ... | ... | 30.0 | ... | 21.5 | 3.0 | ... | 1.0 | ... | ... | 2.2 | ... | ... | ... | ... |
| Incoloy DS | 37.0 | ... | ... | ... | 41.0 | ... | 18.0 | ... | ... | ... | 2.3 | ... | ... | ... | ... | ... | ... |
| Inco 330 | 35.5 | ... | ... | ... | 44.0 | ... | 18.5 | ... | ... | ... | 1.1 | ... | ... | ... | ... | ... | ... |
| Inco 020 | 35.0 | ... | ... | ... | 37.0 | ... | 20.0 | 2.5 | ... | ... | ... | 0.6 | 3.5 | ... | ... | ... | ... |
| **Group D-1** | | | | | | | | | | | | | | | | | |
| Duranickel 301, unaged | 94.0 | ... | ... | ... | ... | 4.4 | ... | ... | ... | 0.6 | ... | ... | ... | ... | ... | ... | ... |
| Incoloy 925 | 42.0 | ... | ... | ... | 32.0 | 0.3 | 21.0 | 3.0 | ... | 2.1 | ... | ... | 2.2 | ... | ... | ... | ... |
| Incoloy MA 956 | ... | ... | ... | ... | 74.0 | 4.5 | 20.0 | ... | ... | 0.5 | ... | ... | ... | ... | 0.5 | ... | ... |
| Ni-span-c 902, unaged | 42.5 | ... | ... | ... | 49.0 | 0.5 | 5.3 | ... | ... | 2.4 | ... | ... | ... | ... | ... | ... | ... |
| **Group D-2** | | | | | | | | | | | | | | | | | |
| Duranickel 301, aged | 94.0 | ... | ... | ... | ... | 4.4 | ... | ... | ... | 0.6 | ... | ... | ... | ... | ... | ... | ... |
| Monel K-500, aged | 65.5 | ... | ... | ... | 1.0 | 2.7 | ... | ... | ... | 0.6 | ... | ... | 29.5 | ... | ... | ... | ... |
| Inconel 617 | 52.0 | ... | ... | ... | 1.5 | 1.2 | 22.0 | 9.0 | ... | ... | ... | ... | ... | 12.5 | ... | ... | ... |
| Inconel 625 | 61.0 | ... | ... | ... | 2.5 | ... | 21.5 | 9.0 | ... | ... | ... | 3.6 | ... | ... | ... | ... | ... |
| Inconel 706 | 42.0 | ... | ... | ... | 36.5 | ... | 16.0 | ... | ... | 1.8 | ... | 3.1 | ... | ... | ... | ... | ... |
| Inconel 718 | 54.0 | ... | ... | ... | 18.5 | ... | 18.0 | 3.0 | ... | ... | ... | 5.1 | ... | ... | ... | ... | ... |
| Inconel X-750 | 73.0 | ... | ... | ... | 7.0 | 0.7 | 15.5 | ... | ... | 2.5 | ... | 1.0 | ... | ... | ... | ... | ... |
| Inconel 751 | 73.0 | ... | ... | ... | 7.0 | 1.1 | 15.5 | ... | ... | 2.5 | ... | 1.0 | ... | ... | ... | ... | ... |
| Inconel MA 754 | 77.5 | ... | ... | ... | 1.0 | 0.3 | 20.0 | ... | ... | 0.5 | ... | ... | ... | ... | 0.6 | ... | ... |
| Nimonic 80A | 76.0 | ... | ... | ... | ... | 1.4 | 19.5 | ... | ... | 2.4 | ... | ... | ... | ... | ... | ... | ... |
| Nimonic 81 | 67.0 | ... | ... | ... | ... | 0.9 | 30.0 | ... | ... | 1.8 | ... | ... | ... | ... | ... | ... | ... |
| Nimonic 90 | 60.0 | ... | ... | ... | ... | 1.5 | 19.0 | ... | ... | 2.5 | ... | ... | ... | 16.5 | ... | ... | ... |
| Nimonic 105 | 54.0 | ... | ... | ... | ... | 4.7 | 15.0 | 5.0 | ... | 1.3 | ... | ... | ... | 20.0 | ... | ... | ... |
| Nimonic 115 | 60.0 | ... | ... | ... | ... | 4.9 | 14.2 | 3.2 | ... | 3.8 | ... | ... | ... | 13.2 | ... | ... | ... |
| Nimonic 263 | 51.0 | ... | ... | ... | ... | 0.5 | 20.0 | 5.8 | ... | 2.2 | ... | ... | ... | 20.0 | ... | ... | ... |
| Nimonic 901 | 42.5 | ... | ... | ... | 36.0 | ... | 12.5 | 5.8 | ... | 2.9 | ... | ... | ... | ... | ... | ... | ... |
| Nimonic PE11 | 39.0 | ... | ... | ... | 34.0 | 0.8 | 18.0 | 5.2 | ... | 2.3 | ... | ... | ... | ... | ... | ... | ... |
| Nimonic PE16 | 43.5 | ... | ... | ... | 34.0 | 1.2 | 16.5 | 3.3 | ... | 1.2 | ... | ... | ... | ... | ... | ... | ... |
| Nimonic PK50 | 58.0 | ... | ... | ... | ... | 1.4 | 19.5 | 4.25 | ... | 3.0 | ... | ... | ... | 13.5 | ... | ... | ... |
| Incoloy 903 | 38.0 | ... | ... | ... | 41.5 | 0.9 | ... | ... | ... | 1.4 | ... | 3.0 | ... | 15.0 | ... | ... | ... |
| Incoloy 907 | 38.4 | ... | ... | ... | 42.0 | 0.03 | ... | ... | ... | 1.5 | 0.15 | 4.7 | ... | 13.0 | ... | ... | ... |
| Incoloy 909 | 38.4 | ... | ... | ... | 42.0 | 0.03 | ... | ... | ... | 1.5 | 0.4 | 4.7 | ... | 13.0 | ... | ... | ... |
| Ni-span-c 902, aged | 42.5 | ... | ... | ... | 49.0 | 0.5 | 5.3 | ... | ... | 2.4 | ... | ... | ... | ... | ... | ... | ... |
| Inco G-3 | 44.0 | ... | ... | ... | 19.5 | ... | 22.0 | 7.0 | ... | ... | ... | ... | 2.0 | 2.5 | ... | ... | ... |
| Inco C-276 | 57.0 | ... | ... | 0.5 | 5.5 | ... | 15.5 | 16.0 | ... | ... | ... | ... | ... | 1.2 | ... | 3.8 | ... |
| Inco HX | 47.5 | ... | ... | ... | 18.5 | ... | 21.8 | 9.0 | ... | ... | ... | ... | ... | 1.5 | ... | ... | ... |
| **Group E** | | | | | | | | | | | | | | | | | |
| Monel R-405 | 66.5 | ... | ... | 1.1 | 1.2 | ... | ... | ... | ... | ... | ... | ... | 31.5 | ... | ... | ... | 0.04 |

(a) Not for specifications

hardening treatment will relieve machining stresses, allowance must be made for warpage. A permanent contraction up to about 0.07% occurs during aging. Aged material has good dimensional stability.

**Group E** contains Monel R-405 only. This alloy is designed for high production rates in automatic bar and chucking machines and is discussed in the section "Multiple-Operation Machining" in this article.

## Cutting Fluids

Almost any cutting fluid, or none, can be used in machining nickel alloys. In many applications, nickel alloys respond well to ordinary sulfurized mineral oil; sulfur imparts improved lubricity and antiweld properties. If the temperature of the oil and workpiece becomes high enough during machining to cause brown sulfur staining of the work, the stain can be readily removed with a cleaning solution of the sodium cyanide or chromic-sulfuric acid type. This should be

**Table 2 Design of single-point tools for turning nickel alloys**

| Angle | Roughing | Finishing |
|---|---|---|
| Back rake | 0° | 8° |
| End relief | 6° | 8° |
| Side relief | 6° | 8° |
| End cutting edge | 6° | 60° |
| Side cutting edge | To 45°(a) | To 45°(a) |

| Hardness, HRB | Alloy group | Positive side rake angle, degrees(b) HSS | Cast(c) | Carbide |
|---|---|---|---|---|
| 45 | A | 45 | 22 | ... |
| 50 | A | 40 | 20 | ... |
| 65 | A, B | 30 | 15 | ... |
| 70 | A, B | 30 | 15 | 20 |
| 80 | A, B, C, D-1 | 25 | 12 | 20 |
| 95 | A, B, C, D-1 | 25 | 10 | 20 |
| 100 | B, C, D-1, D-2 | 20 | 10 | 10 |
| 25 HRC | C, D-1, D-2 | 15 | 10 | 10 |
| 40 HRC | D-2 | 12 | 10 | 10 |
| 45 HRC | D-2 | 10 | 10 | 5 |

| Depth of cut | | Nose radius | | Ratio of diameter to length | Nose radius | |
|---|---|---|---|---|---|---|
| mm | in. | mm | in. | | mm | in. |
| 0.8 | 1/32 | 0.33 | 0.013 | 1:30 | 0.25 | 0.010 |
| 1.6 | 1/16 | 0.5 | 0.020 | 1:25 | 0.30 | 0.012 |
| 2.4 | 3/32 | 0.79 | 0.031 | 1:20 | 0.38 | 0.015 |
| 3.2 | 1/8 | 0.89 | 0.035 | 1:15 | 0.46 | 0.018 |
| 4.8 | 3/16 | 1.0 | 0.040 | 1:10 | 0.58 | 0.023 |
| 6.4 | 1/4 | 1.6 | 0.062 | 1:5 | 0.74 | 0.029 |

(a) Depending on job requirements and chip load. (b) For finishing cuts and light-to-medium roughing cuts. Feed rates are up to 0.64 mm/rev (0.025 in./rev) and depth of cut to 6.35 mm (0.250 in.). (c) Cast cobalt-base alloy tools

done before any thermal treatment, including welding, because during further exposure to high temperature the staining may cause intergranular surface attack. To avoid intergranular corrosion, the parts should be immersed in cleaning solution only long enough to remove the stain. High-speed machining operations that create high temperatures might preclude the use of a sulfurized oil because of sulfur embrittlement of carbide tools. (Many sintered carbides have a nickel or cobalt matrix that is sensitive to sulfur attack at high temperature.) However, flooding the cutting area with cutting fluid generally cools the tool enough to avoid breakdown of the carbide bond.

Water-base fluids are preferred in high-speed turning, milling, and grinding because of their greater cooling effect. These may be soluble oils or chemical solutions. Except for grinding, which depends almost entirely on cooling and flushing, some chemical activity is always desired and is generally provided by chlorine, amines, or other chemicals.

For slower operations, such as drilling, boring, tapping, and broaching, heavy lubricants and very rich mixtures of chemical solutions are needed. Oils should be used when drilling Nickel 200 and Inconel X-750. In the drilling and tapping of small-diameter holes and in other operations in which lubricant flow and chip flushing are restricted, solvents will improve performance. These less viscous fluids can be used alone or can be used for diluting mineral and lard oils. A cutting fluid of the spray-mist type is adequate for simple turning operations on all alloys.

**Table 3 Typical depths and widths of chip breakers**

Dimensions may vary slightly with alloy.

| Feed rate mm/min | Feed rate in./min | Depth mm | Depth in. | Width mm | Width in. |
|---|---|---|---|---|---|
| 0.13 | 0.005 | 0.38 | 0.015 | 1.5 | 0.060 |
| 0.25 | 0.010 | 0.50 | 0.020 | 2.0 | 0.080 |
| 0.50 | 0.020 | 0.73 | 0.030 | 3.8 | 0.150 |

## Turning

Single-point turning tools used for cutting nickel alloys must have positive rake angles (Table 2) so that the metal is cut instead of pushed, as would occur if negative rake angles were used. A secondary function of the rake angle is to guide the chip away from the finished surface. The side cutting edge angle is second in importance only to the rake angle. It must be large enough to provide clearance, but small enough to give adequate support to the cutting edge.

The nose radius, which joins the end and side cutting edges, strengthens the tool nose and helps to dissipate the heat generated in cutting. Nose radii are given with other recommended tool angles in Table 2.

**Chip Control.** Nickel alloys present a minimum of chip disposal problems when cut with tools that have properly designed chip curlers or breakers. High-speed steel (HSS) tools require chip curlers, commonly referred to as lipped tools. The lip should include the proper rake angles for the alloy and should be wide and deep enough to cause the chip to curl and break but not to force it into a wad or tight knot.

Carbide tools should have chip breakers. With these devices, tool rake angles are plane surfaces that terminate at the chip breaker wall. The radius joining the wall of the chip breaker and the rake angle plane must be kept very small. The angle between the two surfaces must be 125 to 135°. A small radius and the proper angle will usually prevent the chip from welding in the chip breaker. Width and depth of the chip breaker depend on the feed rate used (Table 3).

**Tool Material.** Carbide tools permit the highest cutting rates and are recommended for most turning operations involving uninterrupted cuts. Cast alloy tools are recommended for turning group A alloys at optimum cutting rates. As with carbide tools, interrupted cutting is not included in this recommendation. High-speed steel tools should be used for interrupted cuts such as occur in the roughing of an uneven surface. They are also used for finishing to close tolerances, finishing to the smoothest surfaces, and cutting with the least work hardening.

**Feeds and speeds** for turning nickel alloys are given in Table 4. The centers of the ranges in Table 4 should be used as starting points in establishing the best conditions for specific jobs.

**Tool Life.** The effects of several machining variables on tool life in turning Monel R-405 are shown in Fig. 1 to 3 in comparison with data for other materials. All material was machined in the cold-drawn condition.

## Planing and Shaping

The tools used for planing and shaping are similar to lathe tools. For rough planing, the top rake angle of the tool is the most important; it must be positive and of large magnitude to achieve good cutting action. The optimum chip, resulting from a suitable combination of side cutting edge angle and rake angle, is a small curl that curves over ahead of the tool and breaks as it hits the work.

The gooseneck type of planer tool should be used for finishing. Its spring action makes smooth cuts. The cutting edge of a gooseneck tool should be behind the centerline of the clapper-box pin so that the tool will spring away from the cut and not dig in.

Cutting fluids are not essential for roughing, but sulfurized oil should be applied to the workpiece for smooth finishing cuts. Table 5 lists feeds, speeds, and depths of cut used for planing. Speeds are generally 80 to 85% of those used for turning.

Heavy sections can be parted in a planer with the aid of a gooseneck finishing tool. Only light cuts (0.13 to 0.25 mm/stroke, or 0.005 to 0.010 in./stroke) may be taken. Continuous soluble-oil lubrication should be provided. Practice for shaping operations is similar to that for planing.

## Broaching

Although group D alloys are broached more cleanly in the age-hardened condition, high pressures are required for these materials in either the aged or unaged condition. For example, when broaching fir-tree profiles in a jet engine turbine wheel of solution-annealed Inconel X-750 (93 HRB), forces of 70 to 55 kN (8 to 6 tonf) were required for roughing and 35 to 25 kN (4 to 3 tonf) for finishing. When broaching the same section in fully age-hardened Inconel X-750 (31 HRC), forces of 55 to 25 kN (6 to 3 tonf) were required. The decrease in power requirements for the harder material was attributed to cleaner cutting and less drag on the broach.

Sulfurized mineral oil is recommended as cutting fluid. Broaching speeds and face angles are as follows:

| Alloy group | Speed m/min | Speed sfm | Face angle, degrees |
|---|---|---|---|
| A and B | 3–5.5 | 10–18 | 12–18 |
| C and D-1 | 1.5–3.7 | 5–12 | 10–15 |
| D-2 | 1.8 | 6 | 8–10 |

**Table 4 Conditions for turning nickel alloys with single-point tools**

| Group (any condition) | Hardness, HRB | Depth of cut, mm (in.) | High-speed steel: Surface speed, m/min (sfm) | High-speed steel: Feed, mm/rev (in./rev) | High-speed steel: Material | High-speed steel: Cutting fluid(a) | Cast alloy: Surface speed, m/min (sfm) | Cast alloy: Feed, mm/rev (in./rev) | Cast alloy: Cutting fluid(a) | Carbide: Surface speed: Brazed tool, m/min (sfm) | Carbide: Surface speed: Throw-away, m/min (sfm) | Carbide: Feed, mm/rev (in./rev) | Carbide: Tool material | Carbide: Cutting fluid(a) |
|---|---|---|---|---|---|---|---|---|---|---|---|---|---|---|
| A | 45 | 6.35 (0.250) | 15–18 (50–60) | 0.75 (0.030) | T-5 | II | (b) | (b) | (b) | (b) | (b) | (b) | (b) | (b) |
| | 95 | 1.25 (0.050) | 52–61 (170–200) | 0.20 (0.008) | M-36 | II, III | 115–120 (370–400) | 0.20 (0.008) | II, III | (b) | (b) | (b) | (b) | (b) |
| B | 65 | 6.35 (0.250) | 18–21 (60–70) | 0.75 (0.030) | T-5 | II | (b) | (b) | (b) | 75–90 (250–300) | 85–100 (275–325) | 0.50 (0.020) | C-6 | II |
| | 100 | 1.25 (0.050) | 27–30 (90–100) | 0.25 (0.010) | M-36 | II | 45–60 (150–200) | 0.20 (0.008) | II | 90–110 (300–350) | 110–120 (350–400) | 0.20 (0.008) | C-7 | II |
| C | 75 | 6.35 (0.250) | 7.6–11 (25–35) | 0.75 (0.030) | T-5 | II | (b) | (b) | (b) | 45–60 (150–200) | 55–70 (175–225) | 0.50 (0.020) | C-6 | II |
| | 30 HRC | 1.25 (0.050) | 15–18 (50–60) | 0.25 (0.010) | M-36 | II | 30–45 (100–150) | 0.20 (0.008) | (b) | 100–115 (325–375) | 115–225 (375–425) | 0.20 (0.008) | C-2 | II |
| D-1 | 80 | 6.35 (0.250) | 12–15 (40–50) | 0.75 (0.030) | T-5 | II | (b) | (b) | (b) | 55–70 (175–225) | 60–75 (200–250) | 0.50 (0.020) | C-6 | II |
| | 35 HRC | 1.25 (0.050) | 18–24 (60–70) | 0.25 (0.010) | M-36 | II | 30–45 (100–150) | 0.20 (0.008) | II | 70–85 (225–275) | 75–90 (250–300) | 0.20 (0.008) | C-8 | II |
| D-2 | 85 | 6.35 (0.250) | 3.7–5.5 (12–18) | 0.25 (0.010) | T-5 | II | (b) | (b) | (b) | 10–12 (30–40) | 12–18 (40–60) | 0.25 (0.010) | C-2 | II |
| | 45 HRC | 1.25 (0.050) | 4.6–6.1 (15–20) | 0.20 (0.008) | M-36 | II | (b) | (b) | (b) | 12–15 (40–50) | 15–30 (50–100) | 0.20 (0.008) | C-2 | II |
| E | 65 | 6.35 (0.250) | 21–24 (70–80) | 0.75 (0.030) | T-5 | II | (b) | (b) | (b) | 75–90 (250–300) | 85–100 (275–325) | 0.50 (0.020) | C-6 | II |
| | 100 | 1.25 (0.050) | 37–40 (120–130) | 0.25 (0.010) | M-36 | II | 55–70 (175–225) | 0.20 (0.008) | II | 90–110 (300–350) | 110–120 (350–400) | 0.20 (0.008) | C-7 | II |

(a) II, water base—oil emulsion or chemical solution; III, sulfurized or chlorinated oil or mixed oil. (b) Not recommended

## Drilling

In drilling nickel alloys, steady feed rates should be used. If the drill is allowed to dwell, excessive work hardening of the metal at the bottom of the hole will make it difficult to resume cutting and may result in breaking of the drill when it does take hold. The setup should be as rigid as possible. Stub drills are recommended. Drill jigs should be used whenever possible.

Standard high-speed steel drills are satisfactory for general-purpose drilling of group A and B alloys. These drills have a 118° point angle, a helix of about 30°, a 12° lip relief angle, and a chisel-edge angle of 125 to 135°.

Heavy-duty high-speed steel drills with a heavy web are recommended for drilling group C and D alloys. Cobalt-bearing high-speed steel drills give longer tool life. Cutting pressures are reduced and a positive effective rake maintained if the web is thinned at the chisel point. Increasing the point angle to 135° is helpful.

**Crankshaft drills** are useful for producing deep holes. These drills have a heavy web and a helix angle slightly higher than normal; the web is thinned at the chisel point. Cutting action with drills larger than 19 mm (¾ in.) in diameter will be improved by grinding several small grooves through the lip, extending back along the lip clearance. The spacing of the grooves should be staggered between the two cutting edges. The effect of this serration will be to produce narrow chips with less tendency to foul in the helical flutes.

**Speeds and feeds** for drilling with twist drills are given in Table 6.

**Spade drills** are regularly used for deep-hole and heavy drilling—40 mm (1½ in.) in diameter and greater. The drill is secured in a steel head, which is attached to a rigid bar, with bearing support between the work and tailstock. Spade drills are made of high-speed steel; the cutting edges can be tipped with carbide. Lead holes should be made with a drill having a point smaller than that of the spade drill. Table 7 lists speeds and feeds for spade drilling.

**Gun drills** are primarily used for producing deep holes up to and including 50 mm (2 in.) in diameter, but they are occasionally used for holes as large as 60 mm (2½ in.) in diameter. Appropriate speeds and feeds are given in Table 7.

A highly sulfurized oil is recommended for gun drilling and other deep-hole drilling and finishing. Cutting fluid pressure should be about 5500 kPa (800 psi) for 5 mm (3/16 in.) holes, decreasing to about 1500 kPa (200 psi) for 50 mm (2 in.) holes.

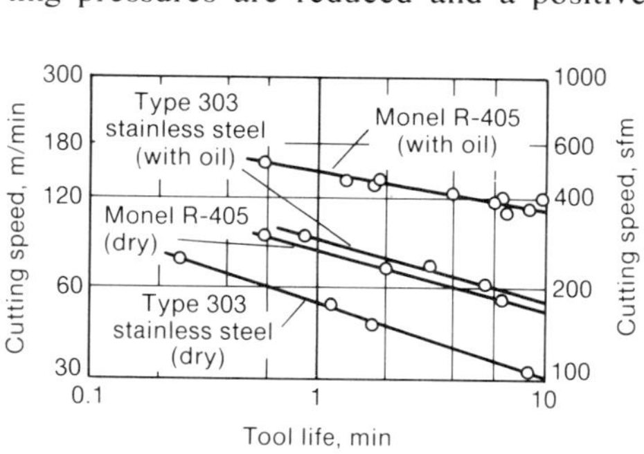


**Fig. 1** Effect of cutting oil on the relation of tool life to cutting speed of single-point turning tools. Depth of cut 2.5 mm (0.100 in.); feed, 0.032 mm/rev (0.00125 in./rev); T1 high-speed steel tools ground with 0° back rake angle and 0° side rake angle. Tool life was taken as the time required to develop a wear land of 1.3 mm (0.050 in.). Work was cold-drawn bars.

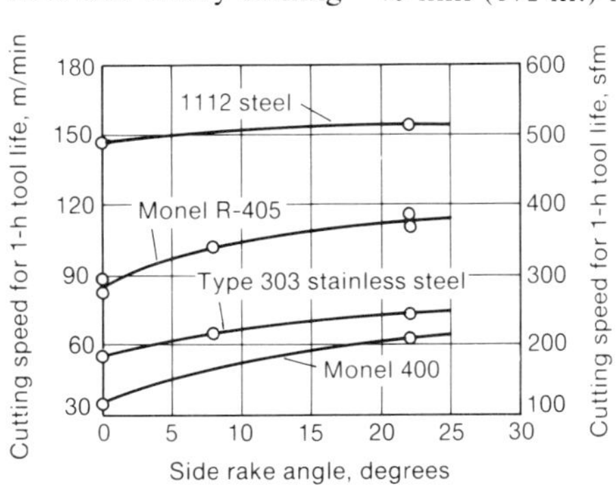


**Fig. 2** Effect of side rake angle on the speed for 1 h tool life of single-point turning tools. Depth of cut 2.5 mm (0.100 in.); feed, 0.032 mm/rev (0.00125 in./rev); T1 high-speed steel tools; sulfurized chlorinated cutting oil. Tool life was taken as the time required to develop a wear land of 1.3 mm (0.050 in.). Work was cold-drawn bars.

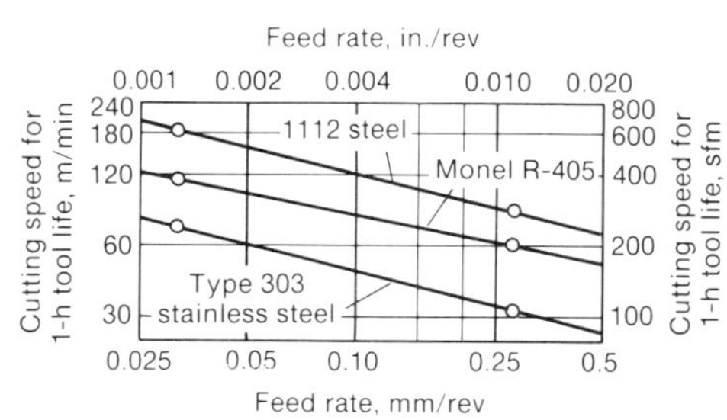


**Fig. 3** Effect of feed rate on the speed for 1 h tool life of single-point turning tools. Depth of cut 2.5 mm (0.100 in.); T1 high-speed steel tools ground with 8° back rake angle and 22° side rake angle; sulfurized chlorinated cutting oil. Tool life was taken as the time required to develop a wear land of 1.3 mm (0.050 in.). Work was cold-drawn bars.

**Table 5 Speeds and feeds for planing nickel alloys**

| Alloy group | Roughing: Depth of cut, mm | Roughing: Depth of cut, in. | Roughing: Feed, mm | Roughing: Feed, in. | Roughing: Table surface speed, m/min | Roughing: Table surface speed, sfm | Finishing: Depth of cut, mm | Finishing: Depth of cut, in. | Finishing: Feed, mm | Finishing: Feed, in. | Finishing: Table surface speed, m/min | Finishing: Table surface speed, sfm | Parting: Feed, mm | Parting: Feed, in. | Parting: Table surface speed, m/min | Parting: Table surface speed, sfm |
|---|---|---|---|---|---|---|---|---|---|---|---|---|---|---|---|---|
| A | 16 | 5/8 | 1.25 | 0.050 | 15–18 | 50–60 | 0.25 | 0.010 | 6.35 | 0.250 | 15 | 50 | 0.125–0.25 | 0.005–0.010 | 15 | 50 |
| B | 16 | 5/8 | 1.25 | 0.050 | 12–15 | 40–50 | 0.25 | 0.010 | 6.35 | 0.250 | 12 | 40 | 0.125–0.25 | 0.005–0.010 | 12 | 40 |
| C | 10 | 3/8 | 1.25 | 0.050 | 5–6 | 15–20 | 0.25 | 0.010 | 6.35 | 0.250 | 5 | 15 | 0.125–0.25 | 0.005–0.010 | 5 | 15 |
| D-1 | 10 | 3/8 | 1.25 | 0.050 | 6–9 | 20–30 | 0.25 | 0.010 | 6.35 | 0.250 | 6 | 20 | 0.125–0.25 | 0.005–0.010 | 6 | 20 |
| D-2, unaged | 10 | 3/8 | 1.0 | 0.040 | 1.5–3 | 5–10 | 0.25 | 0.010 | 6.35 | 0.250 | 1.5 | 5 | 0.125–0.25 | 0.005–0.010 | 1.5 | 5 |
| D-2, aged | 10 | 3/8 | 1.0 | 0.040 | 1.5–3 | 5–10 | 0.25 | 0.010 | 6.35 | 0.250 | 1.5 | 5 | 0.125–0.25 | 0.005–0.010 | 1.5 | 5 |

## Reaming

Fluted reamers for nickel alloys are produced as standard items and are characterized by:

- High-speed steel tool material
- Right-hand cut
- Right-hand helix (positive axial rake)
- Positive radial rake

The operating speed for reaming should be about two-thirds the speed for drilling the same material (Table 6), but not so high as to cause chatter. Other factors contributing to chatter are lack of rigidity in the setup, misalignment, and dull tools.

Reamer feed into the work should be 0.038 to 0.1 mm (0.0015 to 0.004 in.) per flute per revolution. Too low a feed rate will result in glazing of the work and excessive wear of the tool. An excessive feed rate reduces the accuracy of hole dimensions and the quality of the finish. In reaming nickel alloys, sufficient stock must be removed so that nonwork-hardened or nonglazed material is being cut. Good starting points for stock removal are 0.25 mm (0.010 in.) for a 6 mm (¼ in.) hole, 0.38 mm (0.015 in.) for a 13 mm (½ in.) hole, and up to 0.64 mm (0.025 in.) for a 40 mm (1½ in.) hole.

Reamers must be kept sharp. Honed reamers produce smoother surfaces and last longer between grinds.

Flat (solid) reamers and built-up (inserted-blade) reamers are used for holes 40 mm (1½ in.) in diameter or larger. Blind holes are reamed with flat reamers, and through or stepped holes with built-up reamers. A reaming allowance of 1.6 to 3.2 mm (1/16 to 1/8 in.) on the diameter should be provided for both types.

Conventional fluted reamers, flat solid reamers, and insert tools for built-up reamers are made of high-speed steel (usually M2 and M10). Composite tools having steel shanks tipped with carbide are also used for all types of reamers and are recommended for group D-2 alloys. Grades C-2 and C-6 carbide give good results. Sulfurized or chlorinated oil should be used as cutting fluid.

Because flat and built-up reamers constitute a specialized area of finishing inside diameters, cutting speeds and feeds must be developed for each job. Speeds of about two-thirds those listed for turning with similar tool material and feeds of 0.20 to 0.25 mm/rev (0.008 to 0.010 in./rev) should be used as a starting point.

## Tapping and Threading

The most important factor in tapping is the selection of the proper drill size. The standard tap drill selection tables, in use for many years, are based on 75% of full thread depth and were established by experience with low-strength materials such as brass. Modern high-strength materials, however, provide adequate holding strength with less thread depth. For most requirements, 55% is sufficient, and more than 60% is seldom required. This is particularly true for holes tapped to a length of 1½ times the bolt diameter. Tests of thread strength show that any increase in thread depth above 60% for the tapped member does not increase the static strength of a threaded joint. In general, the bolt will break first (before threads strip) in holes tapped to 55% of full thread.

Decreasing thread profile depth decreases the torque necessary to drive the tap and markedly decreases tap breakage. As a general rule, torque is doubled when thread depth is increased from 60 to 72% and tripled when thread depth is raised to 80%. Suggested percentages of thread depth for nickel alloys are given in Table 8.

For most applications, standard high-speed steel four-flute taps are recommended. These standard taps are readily obtained with a 7° hook angle. The alloys of group D-2 are best tapped with serial taps, which are standard taps modified in diameter so that each successive tap increases the thread diameter proportionately.

Tapping speeds are given in Table 8. The age-hardenable alloys should be tapped in the unaged condition whenever possible. Ample cutting fluid is essential for both hand and machine tapping, with liquid chlorinated wax preferred.

**Table 6 Speeds and feeds for drilling nickel alloys with twist drills**

| Drill diameter, mm | Drill diameter, in. | Feed(a), mm/rev | Feed(a), in./rev |
|---|---|---|---|
| <1.5 | <1/16 | 0.0125–0.025 | 0.0005–0.001 |
| 1.5–3 | 1/16–1/8 | 0.025–0.05 | 0.001–0.002 |
| 3–4.75 | 1/8–3/16 | 0.05–0.10 | 0.002–0.004 |
| 6–8 | 1/4–5/16 | 0.075–0.125 | 0.003–0.005 |
| 9.5–11 | 3/8–7/16 | 0.10–0.175 | 0.004–0.007 |
| 13–17 | 1/2–11/16 | 0.15–0.25 | 0.006–0.010 |
| 19–25 | 3/4–1 | 0.20–0.38 | 0.008–0.015 |

| Alloy group | Surface speed, m/min | Surface speed, sfm |
|---|---|---|
| A | 17–23 | 55–75 |
| B | 14–17 | 45–55 |
| C | 8–11 | 25–35 |
| D-1 | 6–9 | 20–30 |
| D-2, unaged | 3–4 | 10–12 |
| D-2, aged | 2–3 | 8–10 |
| E | 15–21 | 50–70 |

(a) Use the lower value for smaller drills in the range or for harder material.

**Table 7 Speeds and feeds for the spade and gun drilling of nickel alloys**

| Alloy group | Speed, m/min | Speed, sfm | Feed, mm/rev | Feed, in./rev |
|---|---|---|---|---|
| **Spade drilling with 25–50 mm (1–2 in.) diam tool of M2 HSS** | | | | |
| A | 17–23 | 55–75 | 0.125–0.175 | 0.005–0.007 |
| B | 14–17 | 45–55 | 0.125–0.175 | 0.005–0.007 |
| C | 8–11 | 25–35 | 0.125–0.175 | 0.005–0.007 |
| **Gun drilling with 1.6–50 mm (1/16–2 in.) drill** | | | | |
| A | 67 | 220 | 0.0025–0.05 | 0.0001–0.002 |
| B | 90 | 300 | 0.0038–0.10 | 0.00015–0.004 |
| C | 98 | 320 | 0.005–0.125 | 0.0002–0.005 |
| D-1 | 67 | 220 | 0.0025–0.075 | 0.0001–0.003 |
| D-2 unaged | 30 | 100 | 0.0025–0.075 | 0.0001–0.003 |
| D-2 aged | 18 | 60 | 0.0025–0.075 | 0.0001–0.003 |

| Alloy group | Insert material for gun drills |
|---|---|
| A | Cast alloy (40Co-30Cr-18.5W) |
| B | C-6 carbide |
| C, D-1, D-2 | C-2 carbide |

**Table 8 Speeds and thread-depth percentages for tapping nickel alloys**

| Alloy group | Surface speed m/min | Surface speed sfm | Recommended thread engagement, % |
|---|---|---|---|
| A | 5–8 | 15–25 | 60 |
| B | 5–8 | 15–25 | 60 |
| C | 3–5 | 10–15 | 55 |
| D-1 | 3–5 | 10–15 | 55 |
| D-2, unaged | 1.5–3 | 5–10 | 50 |
| D-2, aged | 1.5–3 | 5–10 | 50 |

**Lathe Threading.** Thread-cutting lathe tools are ground according to the principles described for turning tools, but the angles on threading tools are smaller than those on turning tools so that the small nose of the threading tool is well supported. Because of the weakness of the tool nose and the small volume of tool available to dissipate heat, single-point threading must be done at lower speeds and feeds than those used for turning. Conventional single-point lathe practices for threading steel are adequate for threading nickel alloys.

The tool nose should be flooded with sulfurized oil during threading. If the machine is not equipped to pump oil on the work, the workpiece should be brushed with a sulfurized oil during the cutting operation.

Threading speeds are listed in Table 9. The depth of cut will vary, decreasing as the work progresses and more of the tool cutting edge is engaged in removing metal.

**Die Threading.** Threading dies must be kept sharp and flooded with cutting fluid (sulfurized oil or a rich mixture of soluble oil or chemical solution) during use. A chamfer angle of 15 to 20° is recommended for producing V-threads in which no shoulder is involved. The rake angle is 15° for threading material of 30 HRC, and it is increased to 30° for threading material as soft as 65 HRB.

The workpiece diameter should be 1 to 1½% undersize to prevent binding in the die. The exact undersize required will vary with alloy and temper, but can be determined by one or two trials. The speeds given in Table 9 for single-point threading also apply to die threading.

**Thread Grinding.** External threads can be produced in group D-2 alloys (any condition) by form grinding. Aluminum oxide (150 to 320 grit) vitrified bond grinding wheels (medium hard, open structure) are used. The recommended grinding fluid is an oil of about 300 SUS viscosity at 20 °C (70 °F); it should be filtered.

## Multiple-Operation Machining

Monel R-405 was developed for good machinability and is recommended for machining in automatic bar and chucking machines. Other alloys in groups A, B, C, and D-1 can be machined in automatics, but the lower speeds required are seldom possible with this type of equipment.

**Table 9 Speeds for single-point or die-head threading**

| Alloy group | Surface speed m/min | Surface speed sfm |
|---|---|---|
| A, B, E | 8–9 | 25–30 |
| C, D-1 | 3.7–5.5 | 12–18 |
| D-2 | 0.9–1.1 | 3.0–3.5 |

Monel R-405 combines the toughness, strength, and corrosion resistance of Monel 400 with excellent machinability. This free-machining characteristic is obtained by careful adjustment of minor alloying constituents, including the controlled addition of a small amount of sulfur. The resulting nickel-copper sulfides in the alloy act as chip breakers. Because of these inclusions, the surface finish on a machined Monel R-405 part is inferior to that on a part made of Monel 400. Recommended cutting speeds and feeds for Monel R-405 are given in Table 10.

**Form tools,** which are extensively used for automatic machining, must be sturdy, and a rigid setup must be maintained to achieve efficient operation and adequate tool life. Recommended angles for form tools are shown in Fig. 4. The cutting edge of the form tool is set in line with the centerline of the work. Form tools with positive top rake angles are more efficient than those with no top rake. Sharp corners should be eliminated because they lead to early tool breakdown. Figure 4(c) shows a modified form tool design that will aid in minimizing the drag encountered on a conventional circular or dovetail form tool, because of its lack of side clearance. The excess stock left in the form by the modified tool can be removed by a typical cutoff or parting tool. The flat form tool shown in Fig. 4(d) is used for forms of simple design in which there is sufficient side clearance.

The recommended maximum width for form tools is 1½ to 2 times the minimum diameter that is to be machined. The use of wider form tools is likely to result in chatter and consequently in short tool life. Wider forms should be cut in two separate operations, using tools that have a combined width that is equal to the width of the form.

The plunge-cutting action of form tools is less desirable than longitudinal cutting, as with single-point tools. It might be possible first to rough with a longitudinal cut so that the least possible stock is removed in forming. If this cannot be done, separate tools for rough and finish forming will be more economical than a single form tool.

**Box tools** with V-shaped or roller work supports are used for both roughing and finishing cuts. Box tools are preferred to balance turning tools because they provide greater rigidity. Details of a suitable box tool cutter are shown in Fig. 5.

**Table 10 Recommended cutting speeds and feeds for Monel R-405**

| Operation | Surface speed m/min | Surface speed sfm | Feed mm/rev | Feed in./rev |
|---|---|---|---|---|
| Turning(a) | 43–49 | 140–160 | 0.075–0.125 | 0.003–0.005 |
| Forming(a) | 43–49 | 140–160 | 0.010–0.025 | 0.0004–0.001 |
| Drilling | 18–24 | 60–80 | 0.025–0.125 | 0.001–0.005 |
| Reaming | 9–14 | 30–45 | 0.075–0.30 | 0.003–0.012 |
| Tapping | 9–12 | 30–40 | ... | ... |
| Threading | 9–12 | 30–40 | ... | ... |
| Cutoff | 43–49 | 140–160 | 0.0125–0.025 | 0.0005–0.001 |

(a) For a single-spindle screw machine handling stock 25 mm (1 in.) in diameter and under, using high-speed steel tooling. For a multiple-spindle machine or a lathe handling stock exceeding 25 mm (1 in.) in diameter, use surface speeds of 27–38 m/min (90–125 sfm).

**Balance turning tools** are primarily used for roughing cuts. They offer no support for the work and should not be used on long, slender parts. The blades are held in the toolholder at a fixed angle, which must be considered when the tools are ground. Figure 6 shows cutting edge angles suitable for a balance turning tool, as well as the reduction of front relief, when a heavier cutting edge is desired.

**Cutoff Tools.** Circular cutoff tools are suitable for nickel alloys, but straight cutoff tools afford greater clearance and are preferred. The blade thickness should be greater than that used for cutting equivalent sections of carbon steel. A next-heavier-size blade than would be used for the same diameter of carbon steel is recommended.

## Milling

The essential requirements of milling are accuracy and smooth finish; therefore, it is imperative to have sharp tools and rigid machines and fixtures. High-speed steel cutters (M2 and M10) are the most suitable because milling usually involves an interrupted cutting action.

**Feeds and speeds** are listed in Table 11. Too light a feed, approximating rubbing, will cause an excessively work-hardened layer. Because rubbing at the beginning of the cut is avoided by climb milling, this technique is preferred to conventional (up) milling. In addition, the downward motion of the cut assists rigidity and diminishes chatter. The disadvantage of climb milling is the need for positive control of backlash in the table drive. Face milling is preferable to slab milling because it reduces the work hardening and chatter.

Chip problems in milling are the same as those in turning. Standard milling cutters provide adequate clearance for chips.

Heavy-duty milling cutters with 12° positive radial rake and 45° axial rake are preferred for rough milling all alloys except those of group D-2. Light-duty cutters with 12° positive radial rake and 18° axial rake (helical flutes) are best for the high-strength

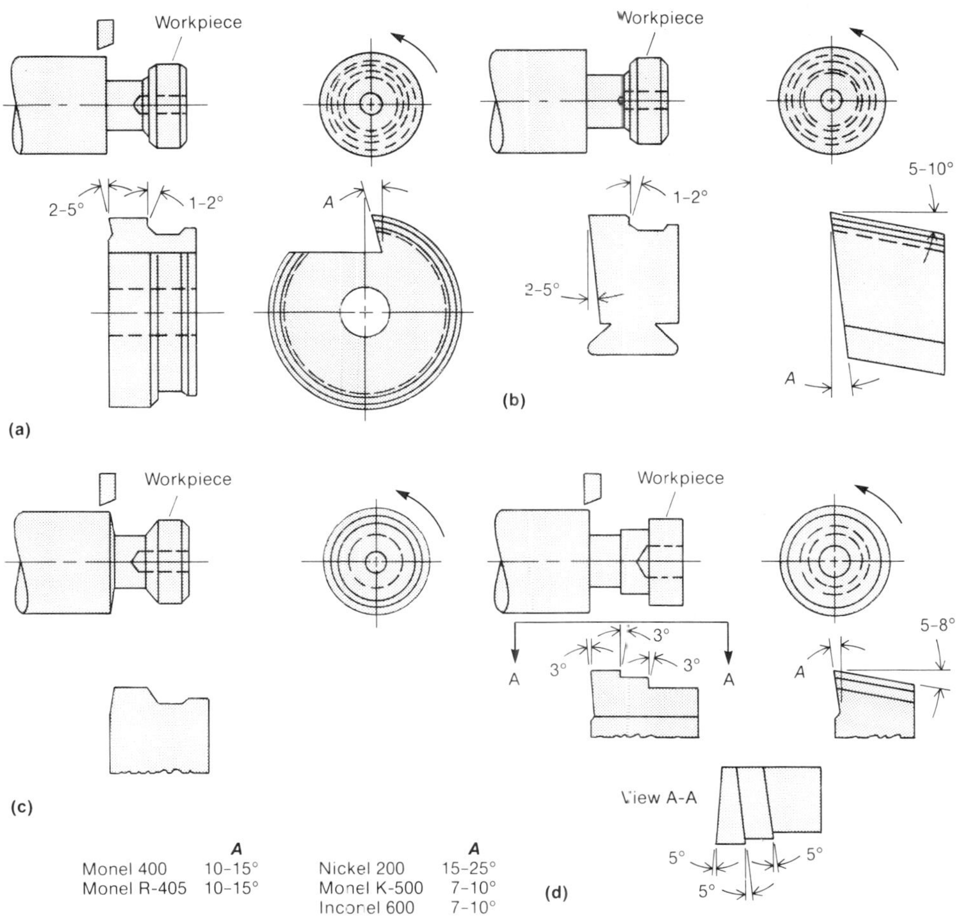


**Fig. 4** Form tools for turning nickel alloys in automatic bar and chucking machines. (a) Circular form tool. (b) Dovetail form tool. (c) Modified form tool for minimizing drag (d) Flat form tool

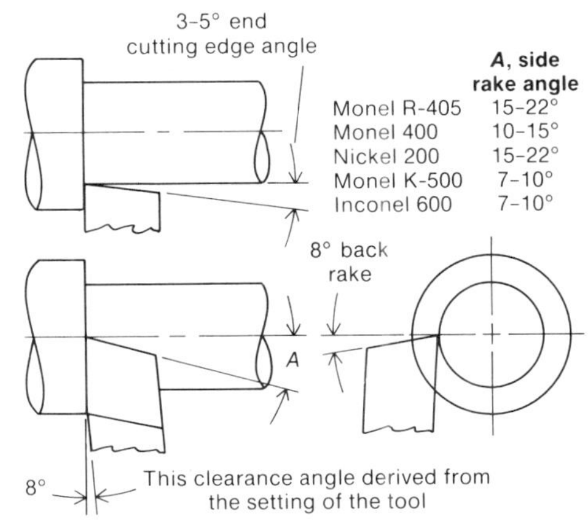


**Fig. 5** Box tool for turning nickel alloys in automatic bar and chucking machines

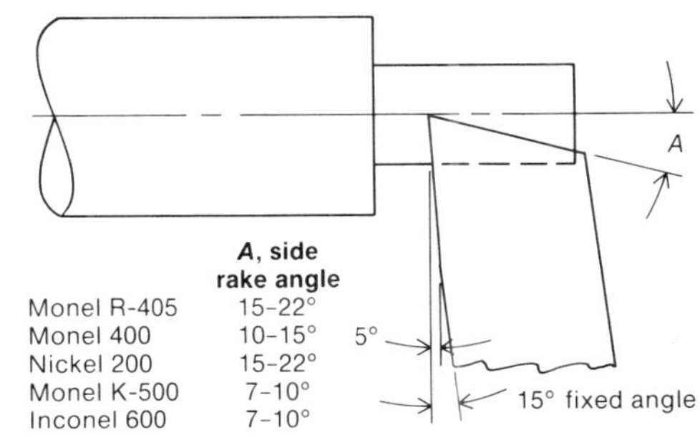


**Fig. 6** Balance tool for turning nickel alloys in automatic bar and chucking machines

alloys of this group. They require low cutting speeds (3 to 6 m/min, or 10 to 20 sfm) and light chip loads. The light-duty cutters have more teeth than the heavy-duty type; consequently, light-duty cutters operate at higher cutting rates for the cutting speeds allowed.

Finishing cutters for all alloys should be of the high-helix type with 15° positive radial rake and 52 to 65° helical flutes (positive axial rake). Staggered-tooth cutters, with alternate teeth of opposite helix, are best for milling grooves. High-speed steel slitting saws with side chip clearance are recommended for narrow slotting.

Face milling cutters with inserted teeth of high-speed steel should be designed so that the inserted teeth have positive rake and helix angles. A typical tool is made with cutters set into the head at a positive axial rake angle or helix of 7° and a positive radial rake angle of 15°. Primary relief angles should be 7 to 8°; secondary relief angles should be 12 to 14° on all except end mills and small-diameter cutters.

## Sawing and Cutting Off

Nickel alloys can be sawed or cut off by conventional methods.

**Hacksawing.** Hand and power hacksaws are suitable for cutoff operations involving the alloys of groups A, B, C, and D-1. Alloys of group D-2 are not readily cut by these tools.

Hand hacksaw blades should be made of high-speed steel. Blades with 14 to 18 teeth per 25 mm (1 in.), raker set, are used for general work. Blades with 24 to 32 teeth per 25 mm (1 in.), wavy set, are used for sawing thin-wall tube.

Power hacksaws can be operated at 90 strokes per minute for the alloys of groups A and B and about 60 strokes per minute for groups C and D-1. High-speed steel blades give satisfactory service. Heavy-duty power hacksaw blades with 6 to 10 teeth per 25 mm (1 in.), raker set, should be used for cutting bar stock. The same type of blade with 14 to 18 teeth per 25 mm (1 in.), raker or wavy set, is suitable for cutting off tube with a 1.6 mm (1⁄16 in.) or heavier wall. Tube with less than 1.6 mm (1⁄16 in.) wall thickness is not generally cut off on power hacksaws, but if necessary a blade with at least 18 teeth per 25 mm (1 in.), wavy set, should be used. The work should be kept flooded with water-soluble or sulfurized cutting oil.

**Circular Sawing.** The best method for cold cutting heavy sections (such as forgings, ingots, and blooms) of the alloys of groups A, B, C, and D-1 is by circular saws with insert teeth. A typical installation uses a 1100 mm (44 in.) diam blade with 56 inserted high-speed steel teeth. The teeth are ground and set into the blade to give a 15° rake angle. Square and round teeth alternate, with the round teeth projecting about 1.6 mm (1⁄16 in.) beyond the cutting edge of the square teeth. This saw cuts alloys of groups A and B at a speed of 15 m/min (50 sfm) with a feed of 8.5 mm/min (0.33 in./min). A water-soluble cutting oil is used. Alloys of group C and D-1 are cut at 7.5 m/min (25 sfm) with the same feed, but sulfurized oil is used as cutting fluid.

**Band sawing** can be used for cutting off all nickel alloys, although it is not recommended for group D-2 alloys of thick section. High-speed steel saws with flexible backs are recommended. Raker-set teeth are suggested for sawing all forms of material other than light-gage sheet and thin-wall tube. Saws with wavy-set teeth are best for sawing thin sections.

The speeds listed in Table 12 can be used as a guide. Medium feeding pressures should be used. The saw should constantly bite into the work; otherwise, the blade will work harden the material. The blade and workpiece should be flooded with a soluble oil. A sulfurized or sulfochlorinated oil can be brushed on the saw teeth to prevent chip welding.

**Table 11 Recommended speeds and feeds for milling nickel alloys**

| Alloy group | Operation | Surface cutting speed | | Feed | |
|---|---|---|---|---|---|
| | | m/min | sfm | mm/tooth | in./tooth |
| A | Helical milling | 24–30 | 80–100 | 0.075 | 0.003 |
| | Face milling | 24–30 | 80–100 | 0.10 | 0.004 |
| | Side milling | 24–30 | 80–100 | 0.05 | 0.002 |
| | End milling | 24–30 | 80–100 | 0.05 | 0.002 |
| | Slotting | 24–30 | 80–100 | 0.05 | 0.002 |
| | Sawing | 24–30 | 80–100 | 0.05 | 0.002 |
| B | Helical milling | 18–24 | 60–80 | 0.175 | 0.007 |
| | Face milling | 18–24 | 60–80 | 0.20 | 0.008 |
| | Side milling | 18–24 | 60–80 | 0.125 | 0.005 |
| | End milling | 18–24 | 60–80 | 0.10 | 0.004 |
| | Slotting | 18–24 | 60–80 | 0.125 | 0.005 |
| | Sawing | 18–24 | 60–80 | 0.05 | 0.002 |
| C | Helical milling | 9–12 | 30–40 | 0.125 | 0.005 |
| | Face milling | 9–12 | 30–40 | 0.15 | 0.006 |
| | Side milling | 9–12 | 30–40 | 0.10 | 0.004 |
| | End milling | 9–12 | 30–40 | 0.075 | 0.003 |
| | Slotting | 9–12 | 30–40 | 0.10 | 0.004 |
| | Sawing | 9–12 | 30–40 | 0.05 | 0.002 |
| D-1 | Helical milling | 8–11 | 25–35 | 0.125 | 0.005 |
| | Face milling | 8–11 | 25–35 | 0.15 | 0.006 |
| | Side milling | 8–11 | 25–35 | 0.10 | 0.004 |
| | End milling | 8–11 | 25–35 | 0.075 | 0.003 |
| | Slotting | 8–11 | 25–35 | 0.10 | 0.004 |
| | Sawing | 8–11 | 25–35 | 0.05 | 0.002 |
| D-2, unaged | Helical milling | 3–6 | 10–20 | 0.075 | 0.003 |
| | Face milling | 3–6 | 10–20 | 0.10 | 0.004 |
| | Side milling | 3–6 | 10–20 | 0.075 | 0.003 |
| | End milling | 3–6 | 10–20 | 0.05 | 0.002 |
| | Slotting | 3–6 | 10–20 | 0.075 | 0.003 |
| | Sawing | 3–6 | 10–20 | 0.025 | 0.001 |
| D-2, aged | Helical milling | 1.5–4.5 | 5–15 | 0.075 | 0.003 |
| | Face milling | 1.5–4.5 | 5–15 | 0.10 | 0.004 |
| | Side milling | 1.5–4.5 | 5–15 | 0.075 | 0.003 |
| | End milling | 1.5–4.5 | 5–15 | 0.05 | 0.002 |
| | Slotting | 1.5–4.5 | 5–15 | 0.075 | 0.003 |
| | Sawing | 1.5–4.5 | 5–15 | 0.025 | 0.001 |

**Table 12 Conditions for the band sawing of nickel alloys**

| Alloy group | Work thickness | | Teeth per 25 mm (1 in.) | Surface velocity | |
|---|---|---|---|---|---|
| | mm | in. | | m/min | sfm |
| A | 1.5 | 1/16 | 14 | 32 | 105 |
| | 6 | 1/4 | 10 | 23 | 75 |
| | 25 | 1 | 8 | 15 | 50 |
| | 75 | 3 | 6 | 15 | 50 |
| B | 1.5 | 1/16 | 18 | 38 | 125 |
| | 6 | 1/4 | 14 | 23 | 75 |
| | 25 | 1 | 10 | 15 | 50 |
| | 75 | 3 | 8 | 15 | 50 |
| C | 1.5 | 1/16 | 14 | 27 | 90 |
| | 6 | 1/4 | 12 | 23 | 75 |
| | 25 | 1 | 10 | 15 | 50 |
| | 75 | 3 | 8 | 15 | 50 |
| D-1 | 1.5 | 1/16 | 18 | 23 | 75 |
| | 6 | 1/4 | 12 | 12 | 40 |
| | 25 | 1 | 10 | 9 | 30 |
| | 75 | 3 | 8 | 9 | 30 |
| Inconel X-750 (cold-rolled, annealed) | 2.36 | 0.093 | 32 | 18 | 60 |

**Table 13 Conditions for friction sawing with a raker-set blade having 10 teeth per 25 mm (1 in.)**

| Alloy | Work thickness | | Blade width | | Blade surface speed | | Approximate lineal cutting rate | |
|---|---|---|---|---|---|---|---|---|
| | mm | in. | mm | in. | m/min | sfm | m/min | sfm |
| Monel 400 | 5 | 3/16 | 25 | 1 | 2700 | 9 000 | 13 | 43 |
| Monel K-500(a) | 5 | 3/16 | 25 | 1 | 3050 | 10 000 | 13 | 43 |
| Inconel 600(a) | 5 | 3/16 | 25 | 1 | 2700 | 9 000 | 22 | 72 |
| Inconel X-750(a) | 3 | 1/8 | 16 | 5/8 | 2900 | 9 500 | 9.5 | 31 |

(a) Annealed condition

**Friction Sawing.** The nickel alloys can be readily cut by friction sawing. Irregular and straight cuts can be made in material up to 25 mm (1 in.) thick. Material less than 13 mm (½ in.) thick can be fed by hand; hydraulic or power feeds should be used for material 13 to 25 mm (½ to 1 in.) thick. If it is necessary to use hand feed for thicker material, the workpiece should be rocked up and down slowly to expose a smaller surface to the saw.

Friction sawing depends on the heat developed between the saw and the workpiece; consequently, cutting fluids are not used for this type of cutting. A new saw blade is not required for friction sawing; the saw is efficient only after its teeth become blunt and create the frictional heat needed to soften the material.

Saw blades with raker-set teeth should be used. The saw blades can have 10, 14, or 18 teeth per 25 mm (1 in.). The saw with 10 teeth per 25 mm (1 in.) is used for the thicker materials, and the saw with 18 teeth per 25 mm (1 in.) is used for the thinnest stock. Generally, blades with 10 teeth per 25 mm (1 in.) are used for sawing material 16 mm (⅝ in.) thick and heavier. For straight cutting, a 25 mm (1 in.) wide blade is the most desirable for all thicknesses. However, for both straight and contour cutting, a blade at least 3.2 mm (⅛ in.) wider than the thickness of the material to be cut is preferred. Saw blades less than 13 mm (½ in.) wide should be used only to cut contours with radii less than 75 mm (3 in.).

Representative saw speeds are:

| Speed | | Material thickness | |
|---|---|---|---|
| m/min | sfm | mm | in. |
| 1500 | 5 000 | 0.8 | 1/32 |
| 2700 | 9 000 | 5 | 3/16 |
| 3350 | 11 000 | 13 | 1/2 |
| 4600 | 15 000 | 25 | 1 |

Table 13 gives data for friction sawing four nickel alloys.

**Lathe Cutoff.** Rounds larger than 22 mm (⅞ in.) can be cut off more rapidly on cutoff lathes than by hacksawing. These lathes usually have two high-speed steel blades 4.8 to 6.4 mm (3/16 to ¼ in.) thick, which should be set to give a positive rake. One tool should have a square nose, the other a rounded nose. Operating speeds are 15 to 18 m/min (50 to 60 sfm) for alloys of groups A and B, and 9 to 11 m/min (30 to 35 sfm) for those of groups C and D-1. Water-soluble oil is an adequate cutting fluid.

**Abrasive Cutoff.** All nickel alloys can be cut off with abrasive wheels. For dry cutting small sections (up to 25 mm, or 1 in.), aluminum oxide resinoid bond wheels, such as A-301-R6-B and A-602-Q8-B, are satisfactory.

Wet cutting is preferred for sections over 25 mm (1 in.) thick and for groups D-1 and D-2 alloys in all thicknesses. Aluminum oxide rubber bond wheels such as A-602-M-R are recommended. Water with a rust inhibitor is a satisfactory grinding fluid. Speed should be about 1500 to 1675 m/min (5000 to 5500 sfm); feed, the maximum permitted by machining capability.

## Grinding

Methods of grinding nickel alloys do not differ greatly from the practices used for steel. When only a small amount of metal must be removed, the finishing operation can be done on a grinding machine, using a rough and then a fine grind.

If an extremely accurate ground finish is required, particularly on material of hard temper, the work should be allowed to cool to room temperature after the final roughing cut or grind. This allows redistribution of internal stresses, and the resulting distortion, if any, can be corrected in the final grinding operation.

For best results, the alloys should be ground wet. A solution of 95 L (25 gal.) of water and 0.5 kg (1 lb) of sal soda, or a solution of 50 parts of water to 1 part soluble oil, is a suitable grinding lubricant for operations other than crush form and thread grinding. A good grinding oil is the best lubricant for crush form and thread grinding. Sodium chromate can be added to sal soda solutions to inhibit corrosion of the equipment.

In general, silicon carbide grinding wheels give best results on alloys of groups A, B, D-1, and E; aluminum oxide wheels are best for alloys of groups C and D-2. Grinding pressures should be great enough to cause slight wheel breakdown. Because of the many variables encountered in grinding, the wheel manufacturer should be consulted for information on specific applications.

**Surface Grinding.** Coarse-grit (46 to 60) aluminum oxide wheels produce the best finishes in surface grinding. To avoid warping during grinding, the workpiece should be in the stress-equalized condition. Low wheel contact and low pressure help prevent distortion during grinding, especially with annealed material. Reciprocating tables are preferred to rotary tables. Reciprocating tables reduce wheel contact, generate less heat, and cause less distortion of the work.

**Centerless grinding** should be done with a wheel having a face that will break down during operation and prevent the workpiece from becoming out of round. The breakdown of the wheel face depends on the diameter of the work, infeed per pass, and the angle and speed of the regulating wheel. Diamond-dressed wheels are more prone to cause ovality of the work than are wheels dressed on a Ross dressing device. This dresser produces a sharp wheel face similar to one that has been crushed-dressed. By taking light cuts on the material, finish grinding can be done without redressing the wheel after the roughing operation.

**Crush Form Grinding.** Vitrified-bond, medium-grade aluminum oxide wheels having medium-to-open structures produce good results in crush form grinding. A high-grade grinding oil is recommended and should be continuously filtered through all operations.

**Grinding With Abrasive Belts.** Abrasive belts (cloth belts coated with aluminum oxide) can be used for finishing group D alloys. One procedure used for the abrasive-belt finishing of precision-forged airfoil turbine blades consists of rough grinding with 80 grit, followed by semifinishing with 120 to 150 grit and final finishing with 180 to 220 grit. Rough grinding can be done dry or with a lubricant. A machine oil of high-flowing characteristic is suitable for rough grinding. Semifinish and finish grinding is done with a lubricant such as cottonseed oil. The addition of kerosene imparts a high-flowing characteristic to the oil.

**Honing** is done with aluminum oxide vitrified bond honing stones of medium-to-soft grade. The honing stone must have uniform breakdown characteristics. Ample coolant must be supplied; proprietary honing oils, either as-supplied or diluted by 2 to 3 parts of kerosene, are recommended. A mixture of 50% oleic acid, 35% kerosene, and 15% turpentine is also suitable.

Surface speeds for rotation of the hone are between 45 and 75 m/min (150 and 250 sfm). Reciprocation surface speeds are between 11 and 15 m/min (35 and 50 sfm). The lower speeds are used for roughing, and the higher speeds for finishing. Honing pressure should be about 3100 kPa (450 psi). The manufacturers of honing stones should be consulted for detailed recommendations on specific problems.

## Nontraditional Machining

Several methods of nontraditional machining are suitable for shaping high-temperature, high-strength alloys that are difficult to machine by conventional methods. A few of these techniques have potential application to nickel alloys, particularly to supplement conventional practice with the more difficult-to-machine alloys such as those of groups D-1 and D-2.

**Electrochemical machining** involves electrolytic deplating of the workpiece. For nickel alloys, electrolytes are salt solutions (sodium or potassium nitrate or chloride) or dilute acid (sulfuric).

The resistance of most nickel alloys to chemical attack does not seem to retard electrochemical solution. Some differences occur in processing the various alloys because of differences in their current-carrying capacities. For example, the nickel and nickel-copper alloys have fair-to-good electrical conductivity, while the chromium-bearing materials are essentially resistance alloys. Compensation must be made for greater power loss, voltage drop, and electrolyte heating with the latter alloys. Because mechanical properties do not affect the process, alloys can be shaped in the fully cold-worked and age-hardened conditions as easily as in the annealed condition. This is particularly advantageous with the group D-2 alloys and permits operations not feasible by conventional methods.

**Chemical and photochemical machining** involve selective metal removal by etching. The article "Photochemical Machining" in this Volume contains an example of etching a nickel electronic connector network.

Because nickel alloys are generally resistant to chemical attack, they require strong acids for metal removal at practical rates. These acids can cause intergranular attack or severe pitting if not properly used, especially on the chromium-bearing alloys. Quality control should include metallographic examination of a finished sample. Even under optimum conditions, chemical machining will lower fatigue life slightly.

One of the major problems to be overcome in the chemical machining of chromium-bearing alloys is the prevention of preferential attack at grain boundaries. The age-hardenable alloys should be solution treated or unaged to be chemically machined.

**Electron beam machining** and plasma-arc cutting are two similar techniques that remove metal by localized melting and vaporization with high-temperature, high-velocity beams of electrons or ionized gas. Plasma-arc torches are capable of cutting nickel alloys several inches thick at very high speed and are superior to all other methods in the rough cutting of plate and sheet.

**Laser beam machining** is capable of drilling holes through nickel- and cobalt-base superalloys in a relatively short time. For example, a 0.5 mm (0.020 in.) diam hole can be drilled through a 2.5 mm (0.1 in.) thick superalloy in about 1 to 3 s. Holes with diameters less than 0.25 mm (0.010 in.) can easily be laser beam drilled through nickel-base superalloys.

# Machining of Reactive Metals

Harry E. Chandler, ASM INTERNATIONAL

REACTIVE METALS refers collectively to the elements titanium, hafnium, and zirconium. Titanium is one of the more difficult metals to machine. The machining characteristics of zirconium are comparable to those of aluminum; hafnium is similar to zirconium in machinability.

All three of the reactive metals are nontoxic. Complicated or expensive equipment or procedures dictated by hazardous materials and environmental disposal considerations are not required. However, the reactive metals are all pyrophoric to varying degrees, presenting potential fire hazards that require special precautionary measures.

## Machinability of Reactive Metals

Because titanium, hafnium, and zirconium are all located in the same group in the Periodic Table, they pose similar problems when machined.

**Titanium.** Reasonable production rates and excellent surface finish can be obtained with conventional machining methods if the unique characteristics of this metal and its alloys are taken into account, such as, the reactivity of titanium with the cutting tools used.

Guidelines for coping with such problems are available. For instance, smearing, galling, and seizing are minimized if feeding is uninterrupted or continuous while the tool and work are in moving contact with each other. Only straight carbide and general-purpose high-speed tool steels can be used.

The correct use of cutting fluids will greatly extend cutting tool life. Both the tool and workpiece must be kept cool. In the past, it was believed that chlorinated cutting fluids should be used with great caution because of their potential to cause stress corrosion cracking. Evidence to the contrary has been found in various studies (Ref 1, 2).

**Zirconium and Hafnium.** Both of these materials can be readily machined with conventional methods. However, both have a tendency to gall and work harden.

These problems can be minimized by using sharp tools and by setting depth of cut at >0.08 mm (>0.003 in.) to penetrate the work-hardened area created by previous tool passes. Slow speeds and heavy feeds are recommended.

Satisfactory results are obtained with both carbide and high-speed tool steels. Carbide tools, however, provide better finishes and higher productivity. Tool life is comparable to that obtained when machining stainless steels.

It is recommended that zirconium and hafnium alloys be flooded with a water-soluble coolant when machined. Chips should not be allowed to accumulate in the machine bed and should be continually removed and stored, preferably under water, in isolated areas.

## Guidelines for Machining Titanium and Titanium Alloys

Success in machining titanium depends largely on overcoming several of the inherent properties of the metal:

- Titanium is chemically reactive and therefore has a tendency to weld to the tool during the machining process, thus leading to chipping and premature tool failure. Additionally, its low heat conductivity increases the temperature at the tool/workpiece interface, thereby also affecting tool life adversely
- The low elastic modulus of titanium permits greater deflections of workpieces, and proper backup may be required
- Susceptibility to surface damage during machining operations, particularly during grinding, is a disadvantage. Even properly processed grinding operations can result in surfaces that lower fatigue life appreciably

The following six guidelines contribute to the efficient machining of titanium:

- *Use low cutting speeds*: Tool tip temperature is strongly affected by cutting speed. A low cutting speed helps to minimize tool edge temperature and maximize tool life. Lower speeds are required for alloys such as Ti-6Al-4V than are necessary for unalloyed titanium
- *Maintain high feed rates*: Tool temperature is affected less by feed rate than by speed. The highest rate of feed consistent with good practice should be used. The depth of cut should be greater than the work-hardened layer resulting from the previous cut
- *Use a generous quantity of cutting fluid*: A coolant provides heat transfer in addition to washing away chips and reducing cutting forces, thus improving tool life
- *Maintain sharp tools*: Tool wear results in a buildup of metal on the cutting edges and causes poor surface finish, tearing, and deflection of the workpiece
- *Never stop feeding while tool and work are in moving contact*: Permitting a tool to dwell in moving contact with titanium causes work hardening and promotes smearing, galling, and seizing, which may lead to a total tool breakdown
- *Use rigid setups*: Rigidity of the machine tool and workpiece ensures a controlled depth of cut

**Tool Materials.** Cutting tools used to machine titanium require abrasion resistance and adequate hot hardness. Carbide tools (such as grades C-2 and C-3), if feasible, optimize production rates. General-purpose high-speed tool steels (such as grades M1, M2, M7, and M10) are often suitable. However, best results are generally obtained with more highly alloyed grades, such as T5, T15, M33, and the M40 series.

In recent years, new tool materials such as ceramics, coated carbides, and cubic boron nitride have increased the rate of metal removal of steels, cast irons, and heat-resistant alloys. Also, polycrystalline diamonds have made it possible to machine high-silicon aluminum alloys much more economically. None of these newer developments in cutting tool materials has found application in increasing the productivity of titanium machined parts. Some improvement, however, has been noted by the use of submicron grain size tungsten carbide-cobalt alloys (see the article "Cemented Carbides" in this Volume) and new tool geometries such as the ledge tool (see the article "High-Speed Machining" in this Volume).

**Tool Life.** Data have been developed experimentally on a wide variety of titanium

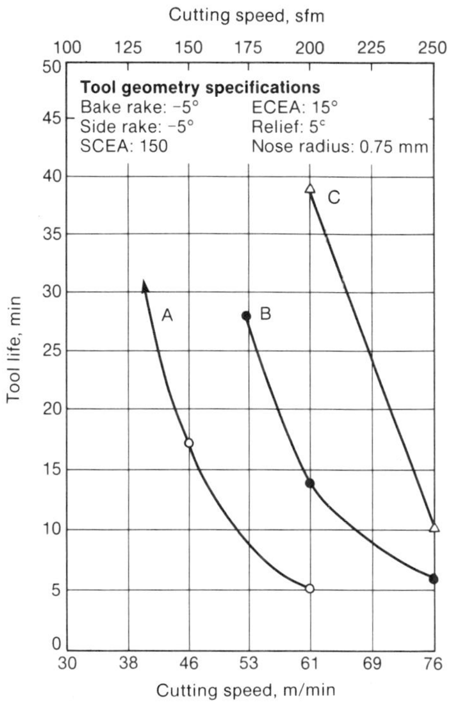


**Fig. 1** Effect of cutting speed and feed on tool life in turning annealed Ti-5Al-5Sn alloy with 321 HB hardness. Feed rates: A, 0.38 mm/rev (0.015 in./rev); B, 0.25 mm/rev (0.010 in./rev); and C, 0.18 mm/rev (0.007 in./rev). Tool material was C-2 (883) carbide throwaway inserts. The depth of cut was 1.6 mm (0.062 in.). Cutting fluid was a chemical emulsion (1:15). Tool life end point was 0.38 mm (0.015 in.) uniform wear and 0.75 mm (0.030 in.) localized wear.

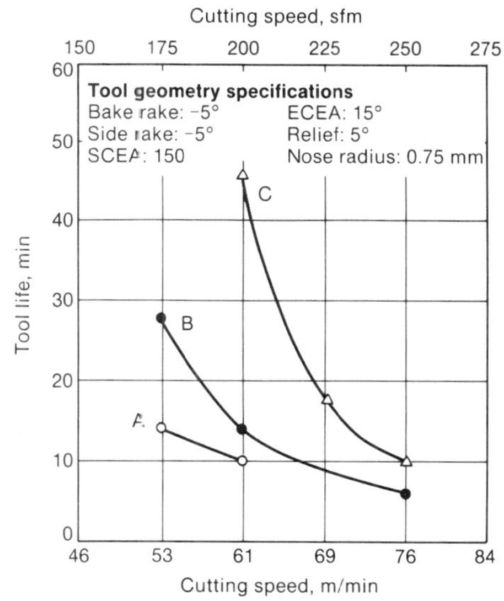


**Fig. 2** Effect of cutting speed and depth of cut on tool life in turning annealed Ti-5Al-2Sn having 321 HB hardness. Depths of cut: A, 3.18 mm (0.125 in.); B, 1.6 mm (0.062 in.); and C, 0.75 mm (0.030 in.). Tool material was C-2 (883) carbide throwaway inserts. Feed rate was 0.25 mm/rev (0.010 in./rev). A chemical emulsion cutting fluid (1:15) was used. Tool life end point was 0.38 mm (0.015 in.) uniform wear and 0.75 mm (0.030 in.) localized wear.

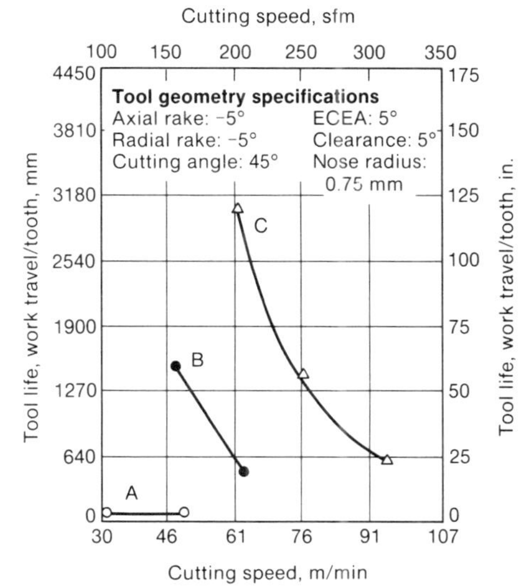


**Fig. 3** Effect of cutting speed and setup on tool life in face milling solution-treated and aged Ti-6Al-4V with 363 HB hardness. Type of setup: A, conventional milling; B, work on center; and C, climb milling. Cutter was a 100 mm (4 in.) diam single-tooth face mill with C-2 (883) carbide inserts. Cutting parameters: feed rate, 0.13 mm/tooth (0.005 in./tooth). Depth of cut, 1.5 mm (0.060 in.); width of cut, 50 mm (2 in.). The material was cut dry. Tool life end point was 0.038 mm (0.015 in.) uniform wear and 0.75 mm (0.030 in.) localized wear.

alloys. A typical way of presenting the information is shown in Fig. 1. Tool life is plotted against cutting speed for a given cutting tool material at a constant feed and depth. Tool life is extremely short at high cutting speeds. As speed is reduced, tool life improves dramatically.

*Speed.* When the tool life/cutting speed data are plotted on logarithmic coordinates, a straight line tends to be obtained, which makes it possible to obtain a simple mathematical relationship between tool life and cutting speed. Curve fitting of tool life to feed, speed, and other machining parameters using computer techniques is now common (Ref 3).

*Feed.* Another important variable affecting tool life is feed, which is measured in millimeters per revolution (inches per revolution) in turning, drilling, and reaming; or in millimeters per tooth (inches per tooth) in a milling operation. Often tool life is not dramatically changed with a change in feed. Titanium alloys, however, are very sensitive to changes in feed (Fig. 1). Whenever possible, it is desirable to operate at higher feeds. Higher feeds provide higher rates of metal removal and lead to a lower time to machine a given part.

*Depth of cut* must also be considered as indicated in Fig. 2. For example, increasing the depth of cut from 0.75 mm (0.030 in.) to 3 mm (0.125 in.), a four-fold increase, decreases tool life from 46 to 14 min at a cutting speed of 60 m/min (200 sfm).

Cutting tool performance in shops is influenced by many factors other than grade selection. Setup, processing methods, grinding techniques, material quality, and the condition of the machine tool and fixturing, all influence cutter performance. Figure 3 shows the importance of a setup using climb milling compared with conventional milling or having the work on center. With climb milling, there is less tendency for chipping because the chip leaving the tool is thinner than a chip produced in conventional milling.

**Cutting Fluids.** The correct use of coolants during machining operations greatly extends cutting tool life. Chemically active cutting fluids transfer heat efficiently and reduce cutting forces between tool and workpiece.

Large quantities of cutting fluid are needed to keep the workpiece and cutting tool cool during high-speed machining operations. Water-base fluids are more efficient than oils. A weak solution of rust inhibitor and/or water-soluble oil (5 to 10%) is the most practical fluid for high-speed cutting operations. Slow speed and complex operations may require chlorinated or sulfurized oils to minimize frictional forces and the galling and seizing tendency of titanium. The best tool life in intermediate-speed operations may be obtained by using a good coolant containing a chemically active additive.

If chlorinated cutting fluids are used on alloys that may be subject to stress corrosion cracking, carefully controlled postmachining cleaning operations must be followed.

*Chlorine-Containing Fluids.* The general prohibition against the use of chlorine-containing cutting fluids is not universally observed. A number of tests have been run in an attempt to develop specifics concerning the prohibition and when it should and should not be observed.

In one study (Ref 4), the U.S. Air Force Materials Laboratory arrived at certain conclusions such as:

- Sulfurized and chlorinated soluble-oil emulsions used in low-stress grinding and end milling/end cutting did not degrade the high-cycle fatigue strength of annealed Ti-6Al-4V (34 HRC) at 25 °C (75 °F) and 315 °C (600 °F) relative to results from a neutral soluble-oil emulsion
- Sulfurized and chlorinated soluble-oil emulsions used in abusive grinding did not degrade the 25 °C (75 °F) high-cycle fatigue strength of Ti-6Al-4V relative to results from a neutral soluble-oil emulsion
- Sulfurized and chlorinated oils and soluble-oil emulsions as crack tip environments did not accelerate 25 °C (75 °F) fatigue crack propagation rates in Ti-6Al-4V at 1 cpm and 1800 cpm relative to results in laboratory air environment
- A 100 h exposure under stress to sulfurized and chlorinated soluble oil emulsions did not affect 25 °C (75 °F) bend test results from low-stress ground and end

milled end cut Ti-6Al-4V relative to results from a neutral soluble oil emulsion

Another series of tests was run in West Germany. It was found through Auger analysis that when a cutting fluid containing chlorine was used, surface films were developed that had a thickness equal to or less than 150 nm (1500 Å) and a chlorine content of 3 at.% at the most (Ref 2). Similar films with 1.5 at.% and 100 to 150 nm (1000 to 1500 Å) thickness were obtained by machining titanium with demineralized water. The engineers concluded that the prohibition of machining titanium with lubricants containing chlorine additives can no longer be maintained.

**Fire Prevention.** Fine particles of titanium can ignite and burn. Use of water-base coolants or large volumes of oil-base coolants generally eliminates the danger of ignition during machining operations. However, an accumulation of titanium fines can pose a fire hazard. Chips, turnings, and other titanium fines should be collected regularly to prevent undue accumulation, and should always be removed from machines at the end of the day.

Salvageable material should be placed in covered, labeled, clean, dry, steel containers and stored, preferably in an outside yard area. Nonsalvageable fines should be properly disposed of. Titanium sludge should not be permitted to dry out before being removed to an isolated, outside location.

Dry powders developed for extinguishing combustible metal fines are recommended for the control of titanium fires. For maximum safety, such extinguishers should be readily available to each machinist working with titanium. Dry sand retards but does not extinguish titanium fires. Carbon dioxide and chlorinated hydrocarbons are not recommended. Water should never be applied directly to a titanium fire.

## Titanium Machining Methods

Conventional machining methods include turning; face milling, peripheral end milling, and climb cutting; drilling; tapping; reaming; wheel grinding, belt grinding, abrasive cutting, and hand abrasive grinding; hack sawing; and band sawing. Widely used nontraditional methods include electrochemical machining (ECM), chemical milling (CHM), and laser beam machining (LBM).

### Turning

Turning is the simplest of machining operations for titanium and its alloys. By means of proper machine parameters and the use of a coolant, surface finishes of 0.50 to 0.75 μm (20 to 30 μin.) rms are obtainable with ±0.025 mm (±0.001 in.) tolerances.

Carbide tools provide the highest production rates for continuous turning operations.

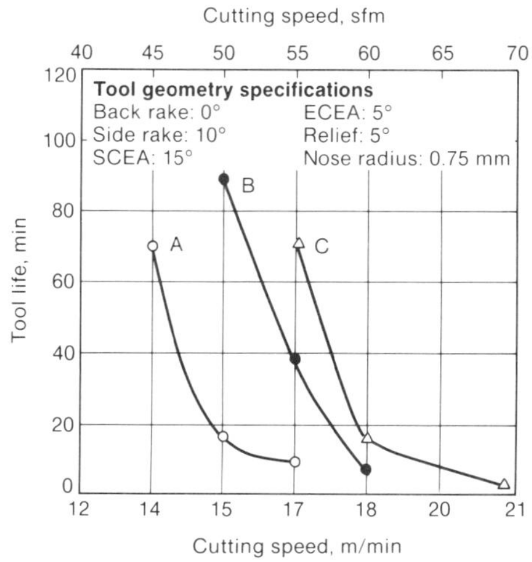


**Fig. 4** Effect of cutting speed and tool material in turning β-forged Ti-6Al-4V having 331 HB hardness. High-speed tool steel materials: A, M2; B, T15; and C, M42. Feed rate was 0.25 mm (0.010 in.). Depth of cut was 0.25 mm/rev (0.010 in./rev). Cutting fluid was phosphated oil. Tool life end point was 1.5 mm (0.060 in.).

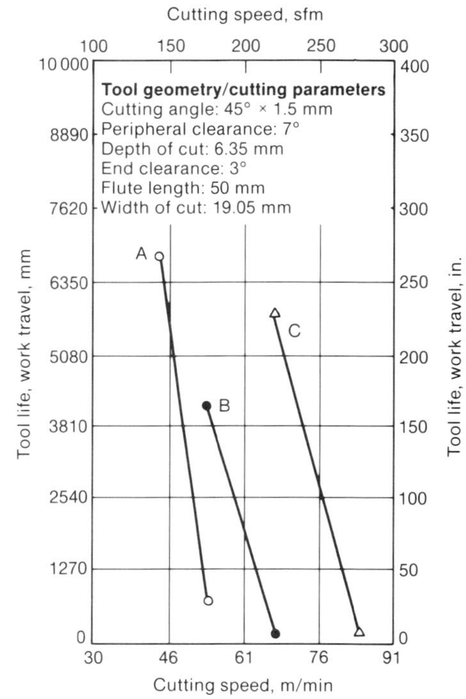


**Fig. 5** Effect of cutting speed and feed in the peripheral end milling of solution-treated and aged Ti-6Al-6Sn-4Zr-2Mo having 321 HB hardness. Feed rate: A, 0.08 mm/tooth (0.003 in./tooth); B, 0.05 mm/tooth (0.002 in./tooth); and C, 0.025 mm/tooth (0.001 in./tooth). Cutter was a four-flute, 25 mm (1 in.) diam tool made of M2 high-speed tool steel. Cutting fluid was chlorinated oil. Setup used climb milling. Tool life end point was 0.30 mm (0.012 in.) uniform wear and 0.51 mm (0.020 in.) localized wear.

Interrupted cuts, plunge cuts, and grooving are best performed by the softer but tougher high-speed tool steels (cast cobalt alloys can also be used). Tools must be resharpened or replaced before final tool failure occurs. A 0.38 mm (0.015 in.) wear land for carbide tools and 0.75 mm (0.030 in.) wear land for high-speed tool steel or cast alloy tools can be used as a guide for halting turning operations.

Tool geometry, particularly rake angle, is important. Negative rake angle is recommended for rough turning with carbide tools. Positive rakes are best for finish and semifinish turning and when high-speed tool steels or cast alloy tools are used.

Cutting speed and feed are two of the most important parameters in all types of machining operations. The effect of cutting speed and tool material in turning Ti-6Al-4V (β forged, 331 HB) is shown in Fig. 4.

In another instance, it was found that cutting speeds of 14 to 17 m/min (45 to 55 sfm) are reasonable for turning Ti-6Al-4V at a feed of 0.25 mm/rev (0.010 in./rev) and a depth of cut of 0.25 mm (0.010 in.). It was also learned that M42, high-cobalt, high-speed tool steel provides better tool life than M2 or T15 high-speed tool steel.

Large amounts of water-base soluble oils (5 to 10% solution) or chemically active coolants (5% sodium nitrate in water) are recommended. Sulfochlorinated oils may be used, if necessary, at low cutting speeds. Recommended machine parameters for the single-point turning of various titanium alloys are given in Table 1.

### Milling

Climb milling should be used as an alternative to conventional milling, when possible, to minimize tool chipping. Slow speeds and uniform, positive feeds help to minimize tool temperature and wear. Tools should not be allowed to dwell in the cut or rub across the workpiece.

High-speed tool steel cutters have proved to be satisfactory in milling. Carbide tools give the highest production rates, but are more susceptible to chipping. Regardless of the tool used, the smallest-diameter cutters with the largest number of teeth minimize deflection and chatter. Increased relief angles increase tool life by reducing pressure, deflection, and the tendency to load compared to standard cutter angles. Several specific guidelines must be considered in milling titanium and its alloys.

**Feed per tooth** must be given careful consideration in specifying machining conditions for Ti-6Al-2Sn-4Zr-2Mo when using peripheral end milling (Fig. 5). Reasonable cutting speeds range from 60 to 105 m/min (200 to 350 sfm). In peripheral end milling of Ti-6Al-4V (solution treated and aged, 363 HB), a depth of cut of 1.6 mm (0.062 in.) will not provide adequate tool life under the conditions specified in Fig. 6.

In face milling Ti-6Al-2Sn-4Zr-2Mo, feed per tooth is an extremely important parameter. Tool life is practically zero when feed is of the order of 0.25 mm (0.010 in.) (Fig.

**Table 1 Nominal speeds and feeds for turning titanium and titanium alloys with high-speed tool steel and carbide tools**

| Material | Hardness, HB | Condition | Depth of cut, mm (in.)(a) | High-speed tool steel: Speed, m/min (sfm) | High-speed tool steel: Feed, mm/rev (in./rev) | High-speed tool steel: Tool material grade, AISI | Carbide tool, uncoated: Speed, m/min (sfm), Brazed | Carbide tool, uncoated: Speed, m/min (sfm), Indexable | Carbide tool, uncoated: Feed, mm/rev (in./rev) | Carbide tool, uncoated: Tool material grade |
|---|---|---|---|---|---|---|---|---|---|---|
| Commercially pure Ti (99.0) | 110–170 | Annealed | 1.0 (0.040) | 76 (250) | 0.13 (0.005) | T15, M42(b) | 160 (525) | 172 (565) | 0.13 (0.005) | C-3 |
| | | | 4.0 (0.150) | 67 (220) | 0.25 (0.010) | T15, M42(b) | 137 (450) | 148 (485) | 0.25 (0.010) | C-2 |
| | | | 7.5 (0.300) | 53 (175) | 0.38 (0.015) | T15, M42(b) | 104 (340) | 110 (360) | 0.38 (0.015) | C-2 |
| | | | 16 (0.625) | ... | ... | ... | 52 (170) | 55 (180) | 0.50 (0.020) | C-2 |
| | 140–200 | Annealed | 1.0 (0.040) | 58 (190) | 0.13 (0.005) | T15, M42(b) | 137 (450) | 152 (500) | 0.13 (0.005) | C-3 |
| | | | 4.0 (0.150) | 52 (170) | 0.25 (0.010) | T15, M42(b) | 119 (390) | 130 (425) | 0.25 (0.010) | C-2 |
| | | | 7.5 (0.300) | 46 (150) | 0.38 (0.015) | T15, M42(b) | 88 (290) | 98 (320) | 0.38 (0.015) | C-2 |
| | | | 16 (0.625) | ... | ... | ... | 44 (145) | 49 (160) | 0.50 (0.020) | C-2 |
| | 200–275 | Annealed | 1.0 (0.040) | 35 (115) | 0.13 (0.005) | T15, M42(b) | 88 (290) | 113 (370) | 0.13 (0.005) | C-3 |
| | | | 4.0 (0.150) | 32 (105) | 0.25 (0.010) | T15, M42(b) | 76 (250) | 98 (320) | 0.20 (0.008) | C-2 |
| | | | 7.5 (0.300) | 29 (95) | 0.38 (0.015) | T15, M42(b) | 58 (190) | 73 (240) | 0.38 (0.015) | C-2 |
| | | | 16 (0.625) | ... | ... | ... | 29 (95) | 37 (120) | 0.50 (0.020) | C-2 |
| Alpha alloys | 300–340 | Annealed | 1.0 (0.040) | 24 (80) | 0.13 (0.005) | T15, M42(b) | 66 (215) | 76 (250) | 0.13 (0.005) | C-3 |
| (Ti-5Al-2.5Sn, Ti-5Al-2.5Sn-ELI, | | | 4.0 (0.150) | 21 (70) | 0.25 (0.010) | T15, M42(b) | 56 (185) | 66 (215) | 0.20 (0.008) | C-2 |
| Ti-6Al-2Nb-1Ta-0.80Mo | | | 7.5 (0.300) | 18 (60) | 0.38 (0.015) | T15, M42(b) | 43 (140) | 49 (160) | 0.25 (0.010) | C-2 |
| | | | 16 (0.625) | ... | ... | ... | 21 (70) | 24 (80) | 0.38 (0.015) | C-2 |
| Alpha-beta alloys | 310–350 | Annealed | 1.0 (0.040) | 21 (70) | 0.13 (0.005) | T15, M42(b) | 52 (170) | 69 (225) | 0.13 (0.005) | C-3 |
| (Ti-6Al-4V, Ti-6Al-4V-ELI, | | | 4.0 (0.150) | 18 (60) | 0.25 (0.010) | T15, M42(b) | 44 (145) | 59 (195) | 0.20 (0.008) | C-2 |
| Ti-6Al-2Sn-4Zr-2Mo, | | | 7.5 (0.300) | 15 (50) | 0.38 (0.015) | T15, M42(b) | 34 (110) | 44 (145) | 0.25 (0.010) | C-2 |
| Ti-6Al-2Sn-4Zr-2Mo-0.25Si, | | | 16 (0.625) | ... | ... | ... | 17 (55) | 21 (70) | 0.38 (0.015) | C-2 |
| Ti-6Al-2Sn-4Zr-6Mo | 320–380 | Solution treated and aged | 1.0 (0.040) | 20 (65) | 0.13 (0.005) | T15, M42(b) | 49 (160) | 58 (190) | 0.13 (0.005) | C-3 |
| | | | 4.0 (0.150) | 17 (55) | 0.25 (0.010) | T15, M42(b) | 41 (135) | 50 (165) | 0.20 (0.008) | C-2 |
| | | | 7.5 (0.300) | 14 (45) | 0.38 (0.015) | T15, M42(b) | 26 (85) | 37 (120) | 0.25 (0.010) | C-2 |
| | | | 16 (0.625) | ... | ... | ... | 15 (50) | 18 (60) | 0.38 (0.015) | C-2 |
| Beta alloys | 275–350 | Annealed or solution treated | 1.0 (0.040) | 12 (40) | 0.13 (0.005) | T15, M42(b) | 38 (125) | 49 (160) | 0.13 (0.005) | C-3 |
| Ti-3Al-8V-6Cr-4Mo-4Zr, | | | 4.0 (0.150) | 9 (30) | 0.25 (0.010) | T15, M42(b) | 32 (105) | 41 (135) | 0.20 (0.008) | C-2 |
| Ti-8Mo-8V-2Fe-3Al, Ti-11.5Mo-6Zr-4.5Sn, | | | 7.5 (0.300) | 7 (25) | 0.38 (0.015) | T15, M42(b) | 24 (80) | 26 (85) | 0.25 (0.010) | C-2 |
| Ti-10V-2Fe-3Al, Ti-13V-11Cr-3Al | | | 16 (0.625) | ... | ... | ... | 12 (40) | 15 (50) | 0.38 (0.015) | C-2 |
| | 350–440 | Solution treated and aged | 1.0 (0.040) | 11 (35) | 0.13 (0.005) | T15, M42(b) | 36 (110) | 38 (125) | 0.13 (0.005) | C-3 |
| | | | 4.0 (0.150) | 7 (25) | 0.25 (0.010) | T15, M42(b) | 27 (90) | 32 (105) | 0.20 (0.008) | C-2 |
| | | | 7.5 (0.300) | ... | ... | ... | 21 (70) | 24 (80) | 0.25 (0.010) | C-2 |
| | | | 16 (0.625) | ... | ... | ... | 11 (35) | 12 (40) | 0.38 (0.015) | C-2 |

Note: ELI, extra-low interstitial. (a) *Caution:* Check power requirements on heavier depths of cut. (b) Any premium high-speed tool steel can be used. Source: Metcut Research Associates Inc.

7). Tables 2 and 3 provide guidelines for end milling and face milling, respectively.

**Cutting Fluids.** Water-base coolants incorporating rust inhibitors or water-soluble oils are the best for most milling operations. Low-viscosity sulfochlorinated oil may be used when cutting speeds are low.

## Drilling

Sharp drills of proper geometry are needed to optimize cutting conditions. Heat removal by the use of large amounts of coolant and, if practical, heat sinks help maintain drill life. Dwell of the drill, which often occurs with hand drilling, should be avoided. The best results are obtained with positive-feed equipment.

Where practicable, carbide-tipped drills provide optimum life, particularly for deep holes. High-speed tool steel drills are suitable for many operations. Chromium plating or oxide coating may be useful in resisting galling of the drill margin. A machine-ground spiral-point configuration is preferable to conventional chisel points (Fig. 8).

Recommended parameters for drilling titanium are given in Table 4. When drilling holes over one diameter deep, the drill should be retracted frequently to clear the flutes and hole of chips. Chlorinated or sulfochlorinated oils and soluble-oil emulsions are satisfactory as cutting fluids. Oil-feeding drills may be required for deep holes. The performance of various cutting fluids in drilling Ti-6Al-4V is shown in Fig. 9.

## Tapping

Tapping of holes is relatively difficult. Holes to be tapped must be uniform and free of work hardening. Sharp, clean taps of proper design are essential. Replacement of taps at the first sign of wear is strongly recommended.

**Spiral-point interrupted-flute taps** with alternate teeth omitted have given good results at slow speeds. Modification of the tap by grinding away the trailing edge of the thread is beneficial (Fig. 10). High-speed tool steel taps are generally used. Because 75% threads are difficult to obtain with normal cutting speeds, 65% types are recommended when possible to maximize tap life. Surface treatments such as black-oxide coatings or nitriding can assist in reducing galling tendencies and improving tap life.

**Cutting Fluids and Pastes.** Paste-type cutting compounds (Lithopone paste, a mixture of zinc sulfide and barium sulfate) have given good results. Chlorinated or sulfochlorinated oils have also been used successfully in tapping titanium and its alloys. The superiority of chlorinated oil in tapping Ti-6Al-4V is shown in Fig. 11.

**Tapping speed** should be kept low to minimize heat buildup. Guidelines for tapping operations are given in Table 5.

## Reaming

Holes with a tolerance of +0.05 to −0.000 mm (+0.002 to −0.000 in.) are possible. Reamer margins tend to gall and seize, but proper tool design and operating conditions effectively eliminate the problem. Sufficient stock must be available to provide continuous cutting and prevent galling and work hardening.

High-speed tool steel reamers are generally effective when machining titanium. Carbide reamers allow higher surface speeds or longer tool life. Spiral-flute reamers generally provide longer life than do straight-flute reamers. Sulfochlorinated oils appear to be the best cutting fluid. However, water-base oil emulsions are also effective, particularly with the softer, unalloyed titanium grades. Recommended reaming parameters are given in Table 6.

## Grinding

The surfaces of titanium alloys can easily be damaged during machining and grinding operations (Ref 5-7). Damage appears in the form of microcracks, built-up edges, plastic deformation, heat-affected zones, and residual tensile stresses. In service, failures can occur as a result of fatigue and stress corrosion.

The three views in Fig. 12(a), (b), and (c) illustrate the effect of surface grinding on Ti-6Al-4V (32 to 34 HRC) with regard to

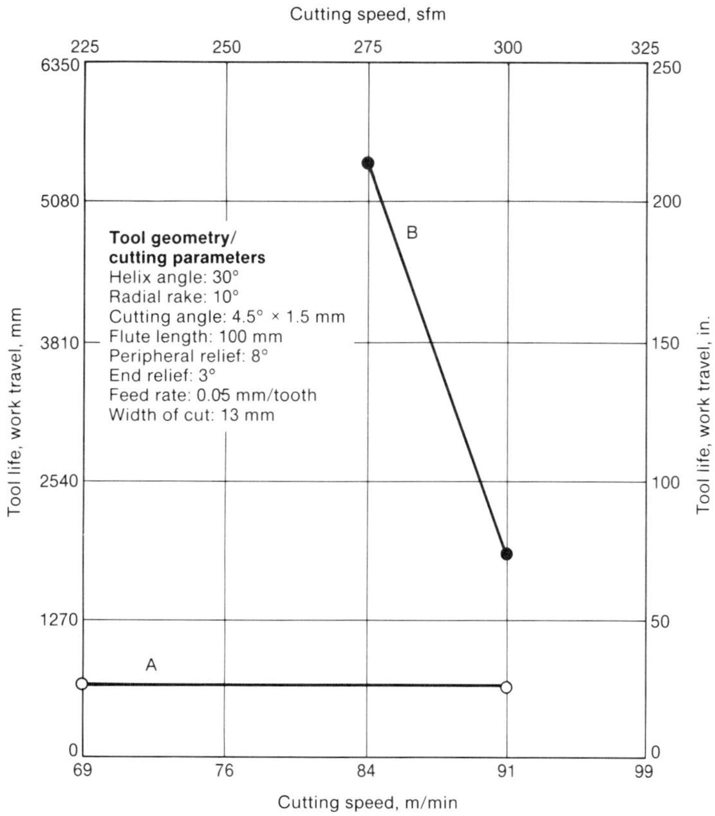


**Fig. 6** Effect of cutting speed and depth of cut in the peripheral end milling of solution-treated and aged Ti-6Al-4V having 363 HB hardness. Depth of cut: A, 1.6 mm (0.062 in.) and B, 0.8 mm (0.03 in.). Cutter was a four-flute 19 mm (¾ in.) diam tool made of M42 high-speed tool steel. Cutting fluid was a chemical emulsion (1:15). Setup used climb milling. Tool life end point was 0.30 mm (0.012 in.) uniform wear and 0.51 mm (0.020 in.) localized wear.

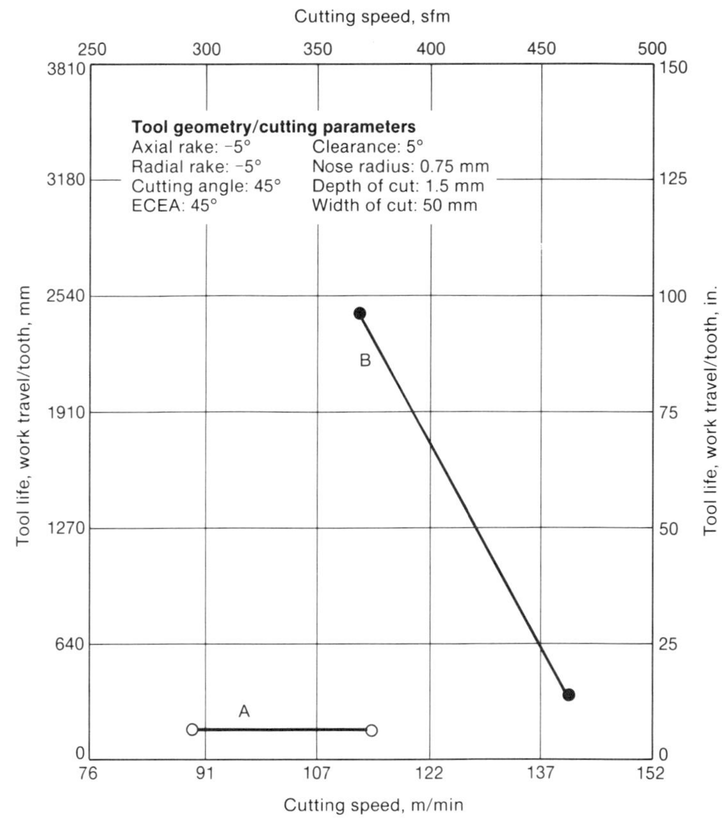


**Fig. 7** Effect of cutting speed and feed in the face milling of solution-treated and aged Ti-6Al-2Sn-4Zr-2Mo having 321 HB hardness. Feed rate: A, 0.25 mm/tooth (0.010 in./tooth) and B, 0.13 mm/tooth (0.005 in./tooth). The cutter was a 100 mm (4 in.) diam single-tooth face mill with a C-2 (883) carbide insert. Material cut dry. Setup was climb milling. Tool life end point was 0.38 mm (0.015 in.) uniform wear and 0.75 mm (0.030 in.) localized wear.

microstructure and hardness. In this study, gentle or low-stress grinding parameters showed no readily identifiable changes at the surface, but conventional and abusive practices altered the surface layer noticeably. There was an appreciable decrease in hardness in the gently ground specimen nonetheless; very respectable high-cycle fatigue values were reported.

It is noteworthy that gentle grinding produced a hardness loss of about ten points HRC at the surface, as indicated in Fig. 12(a). There were no changes in microstructure, however. Both conventional (Fig. 12b) and abusive grinding (Fig. 12c) resulted in a continuous layer of distorted metal at the surface, in addition to changes in microhardness. The abusively ground sample also showed evidence of plastic deformation at 0.05 to 0.08 mm (0.002 to 0.003 in.) beneath the surface. Abusively ground samples did not exhibit the same magnitude of hardness loss as was noted for gently and conventionally ground samples. Plastic deformation associated with abusive grinding may have offset the hardness loss produced by localized surface heating.

Figure 13 indicates a high-cycle fatigue endurance limit for Ti-6Al-4V (32 to 34 HRC) of 370 MPa (54 ksi) for the gentle grinding and values of 85 and 95 MPa (12 ksi and 14 ksi) for conventional and abusive conditions, respectively.

Table 7 gives recommended gentle (low-stress) grinding parameters. Except for grinding notches, soft hardness wheels are recommended. Additional suggestions include the use of low wheel speeds, low down feeds or infeeds per pass, chemically active cutting fluids, and coarse wheel dressing with diamond. It is evident that low-stress practices decrease production rates compared to conventional methods. The most important consideration, however, is to use methods that make undamaged parts.

Both abrasive wheel and belt grinding are used. Metal removal is slow compared to that of carbon steel. However, under proper conditions, abrasive wear is reasonably low, and a surface finish of 0.4 μm (15 μin.) is possible.

For more detailed information on the effects of cutting and grinding on properties of machined metals, including titanium, see the article "Surface Finish and Surface Integrity" in this Volume.

**Wheel Grinding.** Selection of wheel, wheel speed, and fluid is important. For hard wheel grinding, vitrified bonded wheels are the most effective. Aluminum oxide ($Al_2O_3$) wheels give good results when limited to grinding speeds of 10 m/s (2000 sfm) or less. Silicon carbide (SiC) wheels can be used at 20 to 30 m/s (4000 to 6000 sfm) if higher speeds are desirable. A feed of about 0.025 mm/pass (0.001 in./pass) is generally suitable for all wheels. Abrasive grit size of 60 to 80 and wheel hardness medium grades J to L are commonly used (see the article "Grinding Equipment and Processes" in this Volume for a description of grinding wheel nomenclature).

No appreciable sparking accompanies $Al_2O_3$ wheel grinding. The workpiece can

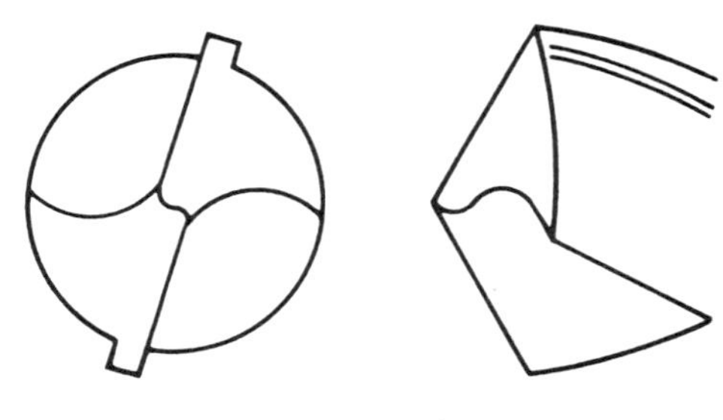


**Fig. 8** Front (a) and side view (b) of spiral-point drills used in drilling titanium. Grinding of spiral points produces a better drill for titanium than conventional chisel edge. Spiral points reduce the large negative rake angle of the chisel edge drill, provide a proper clearance angle along the entire surface of the cutting edge, and reduce thrust loading by 30%.

**Table 2 Nominal speeds and feeds for end milling (peripheral) titanium and titanium alloys with high-speed tool steel and carbide cutters**

| Material | Hardness, HB | Condition | Radial depth of cut, mm (in.)(a) | High-speed tool steel: Speed, m/min (sfm) | Feed, mm/tooth (in./tooth), Cutter diameter: 9 mm (3/8 in.) | 13 mm (1/2 in.) | 19 mm (3/4 in.) | 25–50 mm (1–2 in.) | Tool material grade, AISI |
|---|---|---|---|---|---|---|---|---|---|
| Commercially pure Ti (99.0) ......... | 110–170 | Annealed | 0.5 (0.020) | 53 (175) | 0.05 (0.002) | 0.08 (0.003) | 0.13 (0.005) | 0.15 (0.006) | M2, M3, M7 |
| | | | 1.5 (0.060) | 49 (160) | 0.08 (0.003) | 0.10 (0.004) | 0.15 (0.006) | 0.18 (0.007) | M2, M3, M7 |
| | | | diam/4 | 26 (85) | 0.038 (0.0015) | 0.05 (0.002) | 0.08 (0.003) | 0.10 (0.004) | M2, M3, M7 |
| | | | diam/2 | 18 (60) | 0.025 (0.001) | 0.038 (0.0015) | 0.05 (0.002) | 0.08 (0.003) | M2, M3, M7 |
| | 140–200 | Annealed | 0.5 (0.020) | 52 (170) | 0.038 (0.0015) | 0.08 (0.003) | 0.13 (0.005) | 0.15 (0.006) | M2, M3, M7 |
| | | | 1.5 (0.060) | 46 (150) | 0.05 (0.002) | 0.10 (0.004) | 0.15 (0.006) | 0.18 (0.007) | M2, M3, M7 |
| | | | diam/4 | 26 (85) | 0.038 (0.0015) | 0.05 (0.002) | 0.08 (0.003) | 0.10 (0.004) | M2, M3, M7 |
| | | | diam/2 | 18 (60) | 0.025 (0.001) | 0.038 (0.0015) | 0.05 (0.002) | 0.08 (0.003) | M2, M3, M7 |
| | 200–275 | Annealed | 0.5 (0.020) | 46 (150) | 0.025 (0.001) | 0.05 (0.002) | 0.10 (0.004) | 0.13 (0.005) | M2, M3, M7 |
| | | | 1.5 (0.060) | 40 (130) | 0.05 (0.002) | 0.08 (0.003) | 0.13 (0.005) | 0.15 (0.006) | M2, M3, M7 |
| | | | diam/4 | 23 (75) | 0.038 (0.0015) | 0.05 (0.002) | 0.08 (0.003) | 0.10 (0.004) | M2, M3, M7 |
| | | | diam/2 | 15 (50) | 0.025 (0.001) | 0.038 (0.0015) | 0.05 (0.002) | 0.08 (0.003) | M2, M3, M7 |
| Alpha alloys ...................... | 300–340 | Annealed | 0.5 (0.020) | 34 (110) | 0.025 (0.001) | 0.05 (0.002) | 0.10 (0.004) | 0.13 (0.005) | T15(b) |
| Ti-5Al-2.5Sn, Ti-5Al-2.5Sn-ELI, | | | 1.5 (0.060) | 30 (100) | 0.05 (0.002) | 0.08 (0.003) | 0.13 (0.005) | 0.15 (0.006) | T15 |
| Ti-6Al-2Nb-1Ta-0.80Mo | | | diam/4 | 17 (55) | 0.025 (0.001) | 0.038 (0.0015) | 0.05 (0.002) | 0.08 (0.003) | T15 |
| | | | diam/2 | 12 (40) | 0.025 (0.001) | 0.025 (0.001) | 0.038 (0.0015) | 0.05 (0.002) | T15 |
| Alpha-beta alloys ................. | 310–350 | Annealed | 0.5 (0.020) | 30 (100) | 0.025 (0.001) | 0.05 (0.002) | 0.10 (0.004) | 0.13 (0.005) | T15(b) |
| Ti-6Al-4V, Ti-6Al-4V-ELI, | | | 1.5 (0.060) | 27 (90) | 0.05 (0.002) | 0.08 (0.003) | 0.13 (0.005) | 0.15 (0.006) | T15 |
| Ti-6Al-2Sn-4Zr-2Mo, | | | diam/4 | 15 (50) | 0.025 (0.001) | 0.038 (0.0015) | 0.05 (0.002) | 0.08 (0.003) | T15 |
| Ti-6Al-2Sn-4Zr-2Mo-0.25Si, | | | diam/2 | 11 (35) | 0.025 (0.001) | 0.025 (0.001) | 0.038 (0.0015) | 0.05 (0.002) | T15 |
| Ti-6Al-2Sn-4Zr-6Mo | 320–380 | Solution treated and aged | 0.5 (0.020) | 26 (85) | 0.025 (0.001) | 0.05 (0.002) | 0.08 (0.003) | 0.13 (0.005) | T15(b) |
| | | | 1.5 (0.060) | 23 (75) | 0.05 (0.002) | 0.08 (0.003) | 0.10 (0.004) | 0.15 (0.006) | T15 |
| | | | diam/4 | 12 (40) | 0.025 (0.001) | 0.038 (0.0015) | 0.05 (0.002) | 0.08 (0.003) | T15 |
| | | | diam/2 | 9 (30) | 0.025 (0.001) | 0.025 (0.001) | 0.038 (0.0015) | 0.05 (0.002) | T15 |
| Beta alloys ....................... | 275–350 | Annealed or solution treated | 0.5 (0.020) | 15 (50) | 0.025 (0.001) | 0.05 (0.002) | 0.10 (0.004) | 0.13 (0.005) | T15(b) |
| Ti-3Al-8V-6Cr-4Mo-4Zr, | | | 1.5 (0.060) | 14 (45) | 0.038 (0.0015) | 0.08 (0.003) | 0.13 (0.005) | 0.15 (0.006) | T15 |
| Ti-8Mo-8V-2Fe-3Al, | | | diam/4 | 8 (25) | 0.025 (0.001) | 0.05 (0.002) | 0.08 (0.003) | 0.10 (0.004) | T15 |
| Ti-11.5Mo-6Zr-4.5Sn, | | | diam/2 | 6 (20) | 0.018 (0.0007) | 0.038 (0.0015) | 0.05 (0.002) | 0.08 (0.003) | T15 |
| Ti-10V-2Fe-3Al, | 350–440 | Solution treated and aged | 0.5 (0.020) | 12 (40) | 0.018 (0.0007) | 0.038 (0.0015) | 0.05 (0.002) | 0.10 (0.004) | T15(b) |
| Ti-13V-11Cr-3Al | | | 1.5 (0.060) | 11 (35) | 0.025 (0.001) | 0.05 (0.002) | 0.08 (0.003) | 0.13 (0.005) | T15 |
| | | | diam/4 | 6 (20) | 0.018 (0.0007) | 0.038 (0.0015) | 0.05 (0.002) | 0.08 (0.003) | T15 |
| | | | diam/2 | 5 (15) | 0.013 (0.0005) | 0.025 (0.001) | 0.038 (0.0015) | 0.05 (0.002) | T15 |

| Material | Hardness, HB | Condition | Radial depth of cut, mm (in.)(a) | Carbide tool: Speed, m/min (sfm) | Feed, mm/tooth (in./tooth), Cutter diameter: 9 mm (3/8 in.) | 13 mm (1/2 in.) | 19 mm (3/4 in.) | 25–50 mm (1–2 in.) | Tool material grade |
|---|---|---|---|---|---|---|---|---|---|
| Commercially pure Ti (99.0) ......... | 110–170 | Annealed | 0.5 (0.020) | 130 (425) | 0.025 (0.001) | 0.08 (0.003) | 0.13 (0.005) | 0.18 (0.007) | C-2 |
| | | | 1.5 (0.060) | 119 (390) | 0.05 (0.002) | 0.10 (0.004) | 0.15 (0.006) | 0.20 (0.008) | C-2 |
| | | | diam/4 | 73 (240) | 0.038 (0.0015) | 0.08 (0.003) | 0.13 (0.005) | 0.18 (0.007) | C-2 |
| | | | diam/2 | 55 (180) | 0.025 (0.001) | 0.05 (0.002) | 0.10 (0.004) | 0.15 (0.006) | C-2 |
| | 140–200 | Annealed | 0.5 (0.020) | 122 (400) | 0.025 (0.001) | 0.08 (0.003) | 0.13 (0.005) | 0.18 (0.007) | C-2 |
| | | | 1.5 (0.060) | 113 (370) | 0.05 (0.002) | 0.10 (0.004) | 0.15 (0.006) | 0.20 (0.008) | C-2 |
| | | | diam/4 | 70 (230) | 0.038 (0.0015) | 0.08 (0.003) | 0.13 (0.005) | 0.18 (0.007) | C-2 |
| | | | diam/2 | 53 (175) | 0.025 (0.001) | 0.05 (0.002) | 0.10 (0.004) | 0.15 (0.006) | C-2 |
| | 200–275 | Annealed | 0.5 (0.020) | 107 (350) | 0.025 (0.001) | 0.05 (0.002) | 0.13 (0.005) | 0.18 (0.007) | C-2 |
| | | | 1.5 (0.060) | 99 (325) | 0.05 (0.002) | 0.08 (0.003) | 0.15 (0.006) | 0.20 (0.008) | C-2 |
| | | | diam/4 | 61 (200) | 0.038 (0.0015) | 0.05 (0.002) | 0.13 (0.005) | 0.18 (0.007) | C-2 |
| | | | diam/2 | 46 (150) | 0.025 (0.001) | 0.025 (0.001) | 0.10 (0.004) | 0.15 (0.006) | C-2 |
| Alpha alloys ...................... | 300–340 | Annealed | 0.5 (0.020) | 91 (300) | 0.025 (0.001) | 0.05 (0.002) | 0.13 (0.005) | 0.18 (0.007) | C-2 |
| Ti-5Al-2.5Sn, Ti-5Al-2.5Sn-ELI, | | | 1.5 (0.060) | 84 (275) | 0.05 (0.002) | 0.08 (0.003) | 0.15 (0.006) | 0.20 (0.008) | C-2 |
| Ti-6Al-2Nb-1Ta-0.80Mo | | | diam/4 | 52 (170) | 0.038 (0.0015) | 0.05 (0.002) | 0.13 (0.005) | 0.15 (0.006) | C-2 |
| | | | diam/2 | 40 (130) | 0.025 (0.001) | 0.025 (0.001) | 0.10 (0.004) | 0.13 (0.005) | C-2 |
| Alpha-beta alloys ................. | 310–350 | Annealed | 0.5 (0.020) | 88 (290) | 0.025 (0.001) | 0.05 (0.002) | 0.13 (0.005) | 0.18 (0.007) | C-2 |
| Ti-6Al-4V, Ti-6Al-4V-ELI, | | | 1.5 (0.060) | 79 (260) | 0.05 (0.002) | 0.08 (0.003) | 0.15 (0.006) | 0.20 (0.008) | C-2 |
| Ti-6Al-2Sn-4Zr-2Mo, | | | diam/4 | 49 (160) | 0.038 (0.0015) | 0.05 (0.002) | 0.13 (0.005) | 0.15 (0.006) | C-2 |
| Ti-6Al-2Sn-4Zr-2Mo-0.25Si, | | | diam/2 | 38 (125) | 0.025 (0.001) | 0.025 (0.001) | 0.10 (0.004) | 0.13 (0.005) | C-2 |
| Ti-6Al-2Sn-4Zr-6Mo | 320–380 | Solution treated and aged | 0.5 (0.020) | 69 (225) | 0.025 (0.001) | 0.05 (0.002) | 0.10 (0.004) | 0.15 (0.006) | C-2 |
| | | | 1.5 (0.060) | 61 (200) | 0.05 (0.002) | 0.08 (0.003) | 0.13 (0.005) | 0.18 (0.007) | C-2 |
| | | | diam/4 | 38 (125) | 0.038 (0.0015) | 0.05 (0.002) | 0.13 (0.005) | 0.15 (0.006) | C-2 |
| | | | diam/2 | 30 (100) | 0.025 (0.001) | 0.025 (0.001) | 0.10 (0.004) | 0.13 (0.005) | C-2 |
| Beta alloys ....................... | 275–350 | Annealed or solution treated | 0.5 (0.020) | 46 (150) | 0.025 (0.001) | 0.05 (0.002) | 0.13 (0.005) | 0.18 (0.007) | C-2 |
| Ti-3Al-8V-6Cr-4Mo-4Zr, | | | 1.5 (0.060) | 40 (130) | 0.038 (0.0015) | 0.08 (0.003) | 0.15 (0.006) | 0.20 (0.008) | C-2 |
| Ti-8Mo-8V-2Fe-3Al, | | | diam/4 | 23 (75) | 0.025 (0.001) | 0.05 (0.002) | 0.13 (0.005) | 0.15 (0.006) | C-2 |
| Ti-11.5Mo-6Zr-4.5Sn, | | | diam/2 | 15 (50) | 0.018 (0.0007) | 0.025 (0.001) | 0.10 (0.004) | 0.13 (0.005) | C-2 |
| Ti-10V-2Fe-3Al, | 350–440 | Solution treated and aged | 0.5 (0.020) | 38 (125) | 0.018 (0.0007) | 0.038 (0.0015) | 0.08 (0.003) | 0.13 (0.005) | C-2 |
| Ti-13V-11Cr-3Al | | | 1.5 (0.060) | 34 (110) | 0.025 (0.001) | 0.05 (0.002) | 0.10 (0.004) | 0.15 (0.006) | C-2 |
| | | | diam/4 | 18 (60) | 0.018 (0.0007) | 0.038 (0.0015) | 0.08 (0.003) | 0.13 (0.005) | C-2 |
| | | | diam/2 | 14 (45) | 0.013 (0.0005) | 0.025 (0.001) | 0.05 (0.002) | 0.10 (0.004) | C-2 |

Note: ELI, extra-low interstitial. (a) For standard-length end mills, the maximum axial depth may be one and one-half times the cutter diameter. (b) Any premium high-speed tool steel can be used. Source: Metcut Research Associates Inc.

**Table 3 Nominal speeds and feeds for face milling titanium and titanium alloys with high-speed tool steel and carbide cutters**

| | | | | High-speed tool steel | | | Carbide tool, uncoated | | | |
|---|---|---|---|---|---|---|---|---|---|---|
| Material | Hardness, HB | Condition | Depth of cut, mm (in.)(a) | Speed, m/min (sfm) | Feed, mm/tooth (in./tooth) | Tool material grade, AISI | Speed, m/min (sfm) Brazed | Speed, m/min (sfm) Indexable | Feed, mm/tooth (in./tooth) | Tool material grade |
| Commercially pure Ti (99.0) | 110–170 | Annealed | 1.0 (0.040) | 53 (175) | 0.15 (0.006) | T15, M42(b) | 162 (530) | 178 (585) | 0.13 (0.005) | C-2 |
| | | | 4.0 (0.150) | 41 (135) | 0.23 (0.009) | T15, M42(b) | 122 (400) | 134 (440) | 0.25 (0.010) | C-2 |
| | | | 7.5 (0.300) | 32 (105) | 0.30 (0.012) | T15, M42(b) | 85 (280) | 105 (345) | 0.38 (0.015) | C-2 |
| | 140–200 | Annealed | 1.0 (0.040) | 44 (145) | 0.10 (0.004) | M2, M7 | 122 (400) | 134 (440) | 0.10 (0.004) | C-2 |
| | | | 4.0 (0.150) | 34 (110) | 0.15 (0.006) | M2, M7 | 91 (300) | 101 (330) | 0.15 (0.006) | C-2 |
| | | | 7.5 (0.300) | 26 (85) | 0.20 (0.008) | M2, M7 | 61 (200) | 76 (250) | 0.20 (0.008) | C-2 |
| | 200–275 | Annealed | 1.0 (0.040) | 32 (105) | 0.10 (0.004) | M2, M7 | 99 (325) | 107 (350) | 0.10 (0.004) | C-2 |
| | | | 4.0 (0.150) | 24 (80) | 0.15 (0.006) | M2, M7 | 84 (275) | 91 (300) | 0.15 (0.006) | C-2 |
| | | | 7.5 (0.300) | 18 (60) | 0.20 (0.008) | M2, M7 | 58 (190) | 72 (235) | 0.20 (0.008) | C-2 |
| Alpha alloys | 300–340 | Annealed | 1.0 (0.040) | 21 (70) | 0.10 (0.004) | T15, M42(b) | 79 (260) | 88 (290) | 0.10 (0.004) | C-2 |
| Ti-5Al-2.5Sn, Ti-5Al-2.5Sn-ELI, | | | 4.0 (0.150) | 17 (55) | 0.15 (0.006) | T15, M42(b) | 69 (225) | 73 (240) | 0.15 (0.006) | C-2 |
| Ti-6Al-2Nb-1Ta-0.80Mo | | | 7.5 (0.300) | 12 (40) | 0.20 (0.008) | T15, M42(b) | 46 (150) | 56 (185) | 0.20 (0.008) | C-2 |
| Alpha-beta alloys | 310–350 | Annealed | 1.0 (0.040) | 17 (55) | 0.10 (0.004) | T15, M42(b) | 52 (170) | 56 (185) | 0.10 (0.004) | C-2 |
| Ti-6Al-4V, Ti-6Al-4V-ELI, | | | 4.0 (0.150) | 14 (45) | 0.15 (0.006) | T15, M42(b) | 40 (130) | 44 (145) | 0.15 (0.006) | C-2 |
| Ti-6Al-2Sn-4Zr-2Mo, | | | 7.5 (0.300) | 11 (35) | 0.20 (0.008) | T15, M42(b) | 29 (95) | 35 (115) | 0.20 (0.008) | C-2 |
| Ti-6Al-2Sn-4Zr-2Mo-0.25Si, | 320–380 | Solution | 1.0 (0.040) | 17 (55) | 0.08 (0.003) | T15, M42(b) | 44 (145) | 49 (160) | 0.10 (0.004) | C-2 |
| Ti-6Al-2Sn-4Zr-6Mo | | treated | 4.0 (0.150) | 15 (50) | 0.13 (0.005) | T15, M42(b) | 34 (110) | 37 (120) | 0.15 (0.006) | C-2 |
| | | and aged | 7.5 (0.300) | 12 (40) | 0.18 (0.007) | T15, M42(b) | 24 (80) | 29 (95) | 0.20 (0.008) | C-2 |
| Beta alloys | 275–350 | Annealed or | 1.0 (0.040) | 12 (40) | 0.08 (0.003) | T15, M42(b) | 40 (130) | 44 (145) | 0.10 (0.004) | C-2 |
| Ti-3Al-8V-6Cr-4Mo-4Zr, | | solution | 4.0 (0.150) | 9 (30) | 0.13 (0.005) | T15, M42(b) | 30 (100) | 34 (110) | 0.15 (0.006) | C-2 |
| Ti-8Mo-8V-2Fe-3Al, | | treated | 7.5 (0.300) | 6 (20) | 0.18 (0.007) | T15, M42(b) | 21 (70) | 26 (85) | 0.20 (0.008) | C-2 |
| Ti-11.5Mo-6Zr-4.5Sn, | 350–440 | Solution | 1.0 (0.040) | 9 (30) | 0.05 (0.002) | T15, M42(b) | 24 (80) | 27 (90) | 0.10 (0.004) | C-2 |
| Ti-10V-2Fe-3Al, | | treated | 4.0 (0.150) | 8 (25) | 0.10 (0.004) | T15, M42(b) | 18 (60) | 20 (65) | 0.15 (0.006) | C-2 |
| Ti-13V-11Cr-3Al | | and aged | 7.5 (0.300) | 6 (20) | 0.15 (0.006) | T15, M42(b) | 12 (40) | 15 (50) | 0.20 (0.008) | C-2 |

Note: ELI, extra-low interstitial. (a) Depth of cut is measured parallel to the axis of the cutter. (b) Any premium high-speed tool steel (T15, M33, M41-M47). Source: Metcut Research Associates Inc.

**Table 4 Nominal speeds and feeds for drilling titanium and titanium alloys**

| | | | | Feed, mm/rev (in./rev)(a) — Nominal hole diameter | | | | | | | | |
|---|---|---|---|---|---|---|---|---|---|---|---|---|
| Material | Hardness, HB | Condition | Speed, m/min (sfm)(a) | 1.5 mm (1/16 in.) | 3 mm (1/8 in.) | 6 mm (1/4 in.) | 13 mm (1/2 in.) | 19 mm (3/4 in.) | 25 mm (1 in.) | 38 mm (1½ in.) | 50 mm (2 in.) | Tool material grade, AISI |
| Commercially pure Ti (99.0) | 110–170 | Annealed | 24 (80) | 0.013 (0.0005) | ... | ... | ... | ... | ... | ... | ... | M10, M7, M1 |
| | | | 34 (110) | ... | 0.05 (0.002) | 0.13 (0.005) | 0.20 (0.008) | 0.25 (0.010) | 0.30 (0.012) | 0.38 (0.015) | 0.43 (0.017) | M10, M7, M1 |
| | 140–200 | Annealed | 20 (65) | 0.013 (0.0005) | ... | ... | ... | ... | ... | ... | ... | M10, M7, M1 |
| | | | 27 (90) | ... | 0.05 (0.002) | 0.13 (0.005) | 0.20 (0.008) | 0.25 (0.010) | 0.30 (0.012) | 0.38 (0.015) | 0.43 (0.017) | M10, M7, M1 |
| | 200–275 | Annealed | 12 (40) | 0.025 (0.001) | ... | ... | ... | ... | ... | ... | ... | M10, M7, M1 |
| | | | 17 (55) | ... | 0.05 (0.002) | 0.13 (0.005) | 0.20 (0.008) | 0.25 (0.010) | 0.30 (0.012) | 0.38 (0.015) | 0.43 (0.017) | M10, M7, M1 |
| Alpha alloys Ti-5Al-2.5Sn, Ti-5Al-2.5Sn-ELI, Ti-6Al-2Nb-1Ta-0.80Mo | 300–340 | Annealed | 14 (45) | ... | 0.05 (0.002) | 0.13 (0.005) | 0.18 (0.007) | 0.20 (0.008) | 0.25 (0.010) | 0.30 (0.012) | 0.38 (0.015) | T15, M42(b) |
| Alpha-beta alloys Ti-6Al-4V, Ti-6Al-4V-ELI, | 310–350 | Annealed | 11 (35) | ... | 0.05 (0.002) | 0.10 (0.004) | 0.15 (0.006) | 0.18 (0.007) | 0.20 (0.008) | 0.25 (0.010) | 0.30 (0.012) | T15, M42(b) |
| Ti-6Al-2Sn-4Zr-2Mo, Ti-6Al-2Sn-4Zr-2Mo-0.25Si, Ti-6Al-2Sn-4Zr-6Mo | 320–380 | Solution treated and aged | 9 (30) | ... ... | 0.05 (0.002) | 0.08 (0.003) | 0.13 (0.005) | 0.15 (0.006) | 0.18 (0.007) | 0.23 (0.009) | 0.25 (0.010) | T15, M42(b) |
| Beta alloys Ti-3Al-8V-6Cr-4Mo-4Zr, Ti-8Mo-8V-2Fe-3Al, | 275–350 | Annealed or solution treated | 8 (25) | ... ... | 0.025 (0.001) | 0.08 (0.003) | 0.10 (0.004) | 0.13 (0.005) | 0.15 (0.006) | 0.18 (0.007) | 0.20 (0.008) | T15, M42(b) |
| Ti-11.5Mo-6Zr-4.5Sn, Ti-10V-2Fe-3Al, Ti-13V-11Cr-3Al | 350–440 | Solution treated and aged | 6 (20) | ... ... | 0.025 (0.001) | 0.05 (0.002) | 0.08 (0.003) | 0.10 (0.004) | 0.10 (0.004) | 0.13 (0.005) | 0.15 (0.006) | T15, M42(b) |

Note: ELI, extra-low interstitial. (a) For drilling deep holes with twist drills, reduce speed and feed as follows: If hole depth is three drill diameters, reduce speed by 10%, feed by 10%; four diameters, speed by 20%, feed by 10%; five diameters, speed by 30%, feed by 20%; six diameters, speed by 35%, feed by 20%; eight diameters, speed by 40%, feed by 20%. (b) Any premium high-speed steel can be used. Source: Metcut Research Associates Inc.

be flooded with standard grinding oils. Water-soluble nitrite amine solutions (rust inhibitors) also work well with aluminum oxide wheels. Silicon carbide wheels, however, operate best with sulfochlorinated grinding oils. Complete flooding of the workpiece minimizes the possibility of fire. A 10% solution of nitride rust inhibitor in water eliminates the risk of fire but is less effective than oil with SiC wheels. Water-soluble oils are also useful but are less effective. Some wheel specifications for various grinding operations and parameters are given in Table 8.

When it is necessary to grind by hand or if coolants cannot be used, care should be taken to provide protection (against hot sparks) for nearby personnel or equipment.

**Belt Grinding.** Coated-abrasive grinding requires proper selection of the belt, coolant, and operating parameters. Resin-bonded cloth belts with SiC abrasive generally provide the best performance. A 50-grit belt is typically used for coarse grinding, and a 120-

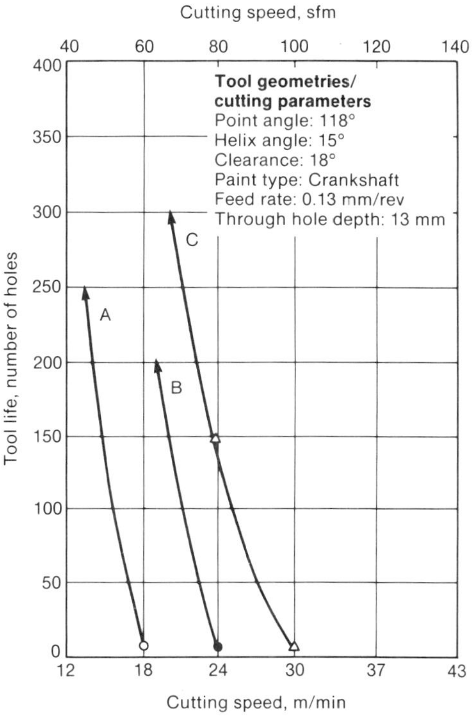


**Fig. 9** Effect of cutting speed and cutting fluids in drilling solution-treated and aged Ti-6Al-4V having 375 HB hardness. Cutting fluids, wear rates, and tool life: A, chemical emulsion (1:15), 0.15 mm (0.006 in.) wear at 13 m/min (43 sfm), 250 holes; B, heavy-duty soluble oil (1:15), 0.15 mm (0.006) wear at 18 m/min (60 sfm), 200 holes; and C, chlorinated oil, 0.13 mm (0.005) wear at 18 m/min (60 sfm), 300 holes. Drill used was a 6.4 mm (¼ in.) diam tool made of M7 high-speed tool steel, was of screw-machine length, and incorporated an oil hole. Tool life end point was 0.38 mm (0.015 in.) of wear.

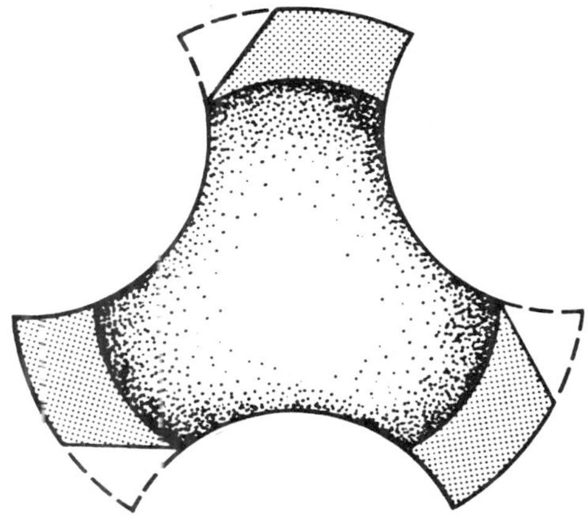

**Fig. 10** Stock tap modified by grinding away trailing edge of thread to optimize the threading of titanium. It is essential to use taps with interrupted threads and with alternate teeth removed. The greatest chip clearance is obtained by grinding a large chamfer on the trailing edge of the tap.

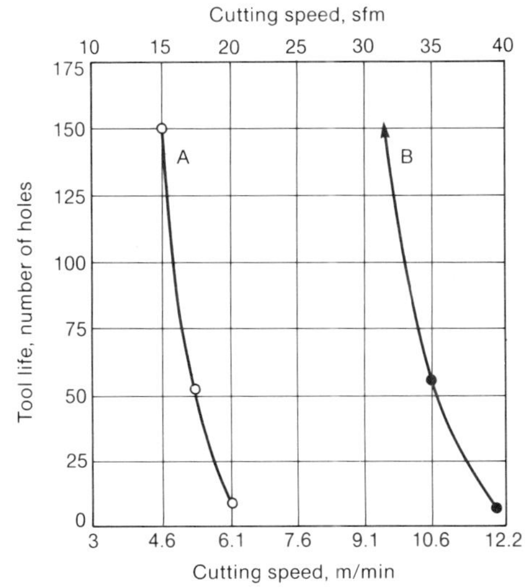


**Fig. 11** Effect of cutting speed and cutting fluid in tapping solution-treated and aged Ti-6Al-4V with 375 HB hardness. Cutting fluid; A, phosphated (nonchlorinated) oil, 150 holes, and B, chlorinated oil, 9.1 m/min (30 sfm). Tap was a chromium-plated, spiral-point, three-flute 8 mm (5⁄16 in.)-24 NF plug tap made of M1 high-speed tool steel. Percentage of thread was 75. Depth of through hole was 13 mm (½ in.). Tool life end point occurs with tap breakage or undersize thread.

grit or finer belt is used for finish grinding. A surface finish of 0.125 to 0.25 μm (5 to 10 μin.) is obtainable in commercial practice.

Fluids should always be used to protect the workpiece and eliminate sparks, which might cause fires. Spraying and flooding techniques are used. Water solutions of 15% tripotassium phosphate or 5% potassium or sodium nitrite have been effective. Detailed information on coated abrasives is available in the article "Grinding Equipment and Processes" in this Volume.

Belt grinding performance generally improves with an increase in load and a decrease in speed. Speeds of the order of 5 to 10 m/s (1000 to 2000 sfm) and pressures in the vicinity of 0.7 MPa (100 psi) provide optimum productivity and belt life.

**Abrasive Cutting.** Rubber-bonded 60-grit SiC cutoff wheels flooded with a water solution of 10% nitrite amine (rust inhibitor) have been used. Machines with oscillating cutting heads give the best results. If the workpiece diameter is greater than 75 mm (3 in.), rotation is recommended to minimize wheel breakage and/or heat checking. Operating guidelines for abrasive cutting are given in Table 9.

**Hand Abrasive Grinding.** A clean wheel used only on titanium is important. An open-type wheel containing large grains has been found to minimize clogging. Excessive buildup of heat should be avoided to minimize metal contamination. Ground surfaces should be filed or mechanically finished to remove abrasive particles and particularly any visible metal oxide (burns).

Sandpaper or steel wool should be avoided, wheel-type mechanical burrs (rotary files) should be operated at low speeds to avoid burning and to maximize tool life. Additionally, when titanium is ground, measures must be taken to protect adjacent titanium surfaces as well as the surroundings from the extremely hot sparks.

## Sawing

**Band Sawing.** Coarse-pitched (6 teeth/25 mm, or 1 in.) high-speed tool steel blades, 25 mm (1 in.) wide, employed at speeds of 24 to 27 m/min (80 to 90 sfm) have given good results. Cutting rates of the order of 650 mm²/min (1 in.²/min) are optimum. Water-soluble or sulfochlorinated cutting fluids are required. Rigid setups are necessary for precision work.

**Hacksawing.** Rigid setups and either water-soluble or sulfochlorinated cutting fluids are suggested. Low surface speeds and positive feed, combined with coarse-pitched (3, 4, and 6 teeth/25 mm, or 1 in.) high-speed tool steel blades have proved to be effective. Surface scale or contaminated surfaces cause rapid blade wear if not removed. Operating guidelines for hacksawing are given in Table 10.

**Table 5 Nominal speeds for tapping titanium and titanium alloys with high-speed tool steel taps**

| | | | Speed, m/min (sfm)(a) | | | | |
|---|---|---|---|---|---|---|---|
| | | | Pitch, mm (threads/in.) | | | | |
| Material | Hardness, HB | Condition | >3 (≤7) | 1.5–3 (8–15) | 1–1.5 (16–24) | ≤1 (>24) | Tool material grade, AISI |
| Commercially pure Ti (99.0) | 110–170 | Annealed | 6.1 (20) | 12.2 (40) | 16.8 (55) | 18.3 (60) | Nitrided M10, M7, M1 |
| | 140–200 | Annealed | 4.6 (15) | 9.1 (30) | 13.7 (45) | 15.2 (50) | Nitrided M10, M7, M1 |
| | 200–275 | Annealed | 3.7 (12) | 7.6 (25) | 10.7 (35) | 12.2 (40) | Nitrided M10, M7, M1 |
| Alpha alloys<br>Ti-5Al-2.5Sn, Ti-5Al-2.5Sn-ELI,<br>Ti-6Al-2Nb-1Ta-0.80Mo | 300–340 | Annealed | 3.0 (10) | 4.6 (15) | 6.1 (20) | 7.6 (25) | Nitrided M10, M7, M1 |
| Alpha-beta alloys<br>Ti-6Al-4V, Ti-6Al-4V-ELI, | 310–350 | Annealed | 2.1 (7) | 4.6 (15) | 5.5 (18) | 6.1 (20) | Nitrided M10, M7, M1 |
| Ti-6Al-2Sn-4Zr-2Mo,<br>Ti-6Al-2Sn-4Zr-2Mo-0.25Si,<br>Ti-6Al-2Sn-4Zr-6Mo | 320–380 | Solution treated and aged | 0.9 (3) | 2.1 (7) | 2.7 (9) | 3.0 (10) | Nitrided M10, M7, M1 |
| Beta alloys<br>Ti-3Al-8V-6Cr-4Mo-4Zr,<br>Ti-8Mo-8V-2Fe-3Al, | 275–350 | Annealed or solution treated | 1.5 (5) | 3.0 (10) | 4.3 (14) | 4.6 (15) | Nitrided M10, M7, M1 |
| Ti-11.5Mo-6Zr-4.5Sn,<br>Ti-10V-2Fe-3Al,<br>Ti-13V-11Cr-3Al | 350–440 | Solution treated and aged | 0.6 (2) | 0.9 (3) | 1.2 (4) | 1.5 (5) | Nitrided M10, M7, M1 |

Note ELI, extra-low interstitial. (a) These speeds are for tapping 65 to 75% threads in shallow through holes. Reduce speed when tapping deep holes, blind holes, or higher percentages of thread. Source: Metcut Research Associates Inc.

**Table 6 Nominal speeds and feeds for reaming titanium and titanium alloys with high-speed tool steel and carbide reamers**

| | | | Roughing | | | | | | | |
|---|---|---|---|---|---|---|---|---|---|---|
| | | | | Feed, mm/rev (in./rev)(a) | | | | | | |
| | | | | Reamer diameter | | | | | | |
| Material | Hardness, HB | Condition | Speed, m/min (sfm) | 3 mm (1/8 in.) | 6 mm (1/4 in.) | 13 mm (1/2 in.) | 25 mm (1 in.) | 38 mm (1 1/2 in.) | 50 mm (2 in.) | Tool material grade, AISI or C grade |
| Commercially pure Ti (99.0) | 110–170 | Annealed | 53 (175) | 0.10 (0.004) | 0.20 (0.008) | 0.30 (0.012) | 0.45 (0.018) | 0.55 (0.022) | 0.64 (0.025) | M1, M2, M7 |
| | | | 114 (375) | 0.10 (0.004) | 0.20 (0.008) | 0.30 (0.012) | 0.45 (0.018) | 0.55 (0.022) | 0.64 (0.025) | C-2 |
| | 140–200 | Annealed | 43 (140) | 0.10 (0.004) | 0.20 (0.008) | 0.30 (0.012) | 0.45 (0.018) | 0.55 (0.022) | 0.64 (0.025) | M1, M2, M7 |
| | | | 114 (375) | 0.10 (0.004) | 0.20 (0.008) | 0.30 (0.012) | 0.45 (0.018) | 0.55 (0.022) | 0.64 (0.025) | C-2 |
| | 200–275 | Annealed | 37 (120) | 0.10 (0.004) | 0.20 (0.008) | 0.30 (0.012) | 0.45 (0.018) | 0.55 (0.022) | 0.64 (0.025) | M1, M2, M7 |
| | | | 91 (300) | 0.10 (0.004) | 0.20 (0.008) | 0.30 (0.012) | 0.45 (0.018) | 0.55 (0.022) | 0.64 (0.025) | C-2 |
| Alpha alloys | 300–340 | Annealed | 21 (70) | 0.08 (0.003) | 0.18 (0.007) | 0.23 (0.009) | 0.30 (0.012) | 0.38 (0.015) | 0.43 (0.017) | T15, M42(b) |
| Ti-5Al-2.5Sn, | | | 76 (250) | 0.08 (0.003) | 0.20 (0.008) | 0.30 (0.012) | 0.40 (0.016) | 0.50 (0.020) | 0.58 (0.023) | C-2 |
| Ti-5Al-2.5Sn-ELI, | | | | | | | | | | |
| Ti-6Al-2Nb-1Ta-0.80Mo | | | | | | | | | | |
| Alpha-beta alloys | 310–350 | Annealed | 20 (65) | 0.08 (0.003) | 0.15 (0.006) | 0.25 (0.010) | 0.30 (0.012) | 0.35 (0.014) | 0.40 (0.016) | T15, M42(b) |
| Ti-6Al-4V, Ti-6Al-4V-ELI, | | | 61 (200) | 0.08 (0.003) | 0.15 (0.006) | 0.25 (0.010) | 0.30 (0.012) | 0.35 (0.014) | 0.40 (0.016) | C-2 |
| Ti-6Al-2Sn-4Zr-2Mo, | 320–380 | Solution | 15 (50) | 0.08 (0.003) | 0.18 (0.007) | 0.25 (0.010) | 0.30 (0.012) | 0.35 (0.014) | 0.40 (0.016) | T15, M42(b) |
| Ti-6Al-2Sn-4Zr-2Mo-0.25Si, | | treated | 49 (160) | 0.08 (0.003) | 0.18 (0.007) | 0.25 (0.010) | 0.30 (0.012) | 0.35 (0.014) | 0.40 (0.016) | C-2 |
| Ti-6Al-2Sn-4Zr-6Mo | | and aged | | | | | | | | |
| Beta alloys | 275–350 | Annealed or | 9 (30) | 0.05 (0.002) | 0.13 (0.005) | 0.18 (0.007) | 0.25 (0.010) | 0.30 (0.012) | 0.35 (0.014) | T15, M42(b) |
| Ti-3Al-8V-6Cr-4Mo-4Zr, | | solution | 23 (75) | 0.05 (0.002) | 0.13 (0.005) | 0.18 (0.007) | 0.25 (0.010) | 0.30 (0.012) | 0.35 (0.014) | C-2 |
| Ti-8Mo-8V-2Fe-3Al, | | treated | | | | | | | | |
| Ti-11.5Mo-6Zr-4.5Sn, | 350–440 | Solution | 6 (20) | 0.05 (0.002) | 0.10 (0.004) | 0.15 (0.006) | 0.20 (0.008) | 0.25 (0.010) | 0.30 (0.012) | T15, M42(b) |
| Ti-10V-2Fe-3Al, | | treated | 15 (50) | 0.05 (0.002) | 0.10 (0.004) | 0.15 (0.006) | 0.20 (0.008) | 0.25 (0.010) | 0.30 (0.012) | C-2 |
| Ti-13V-11Cr-3Al | | and aged | | | | | | | | |

| | | | Finishing | | | | | | | |
|---|---|---|---|---|---|---|---|---|---|---|
| | | | | Feed, mm/rev (in./rev)(a) | | | | | | |
| | | | | Reamer diameter | | | | | | |
| Material | Hardness, HB | Condition | Speed, m/min (sfm) | 3 mm (1/8 in.) | 6 mm (1/4 in.) | 13 mm (1/2 in.) | 25 mm (1 in.) | 38 mm (1 1/2 in.) | 50 mm (2 in.) | Tool material grade, AISI or C grade |
| Commercially pure Ti (99.0) | 110–170 | Annealed | 18 (60) | 0.10 (0.004) | 0.15 (0.006) | 0.25 (0.010) | 0.30 (0.012) | 0.38 (0.015) | 0.50 (0.020) | M1, M2, M7 |
| | | | 23 (75) | 0.10 (0.004) | 0.15 (0.006) | 0.25 (0.010) | 0.30 (0.012) | 0.38 (0.015) | 0.50 (0.020) | C-2 |
| | 140–200 | Annealed | 15 (50) | 0.10 (0.004) | 0.15 (0.006) | 0.25 (0.010) | 0.30 (0.012) | 0.35 (0.014) | 0.40 (0.016) | M1, M2, M7 |
| | | | 20 (65) | 0.10 (0.004) | 0.15 (0.006) | 0.25 (0.010) | 0.30 (0.012) | 0.35 (0.014) | 0.40 (0.016) | C-2 |
| | 200–275 | Annealed | 14 (45) | 0.10 (0.004) | 0.15 (0.006) | 0.25 (0.010) | 0.30 (0.012) | 0.35 (0.014) | 0.40 (0.016) | M1, M2, M7 |
| | | | 18 (60) | 0.10 (0.004) | 0.15 (0.006) | 0.25 (0.010) | 0.30 (0.012) | 0.35 (0.014) | 0.40 (0.016) | C-2 |
| Alpha alloys | 300–340 | Annealed | 6 (20) | 0.10 (0.004) | 0.15 (0.006) | 0.25 (0.010) | 0.30 (0.012) | 0.35 (0.014) | 0.40 (0.016) | T15, M42(b) |
| Ti-5Al-2.5Sn, | | | 11 (35) | 0.10 (0.004) | 0.15 (0.006) | 0.25 (0.010) | 0.30 (0.012) | 0.35 (0.014) | 0.40 (0.016) | C-2 |
| Ti-5Al-2.5Sn-ELI, | | | | | | | | | | |
| Ti-6Al-2Nb-1Ta-0.80Mo | | | | | | | | | | |
| Alpha-beta alloys | 310–350 | Annealed | 6 (20) | 0.10 (0.004) | 0.15 (0.006) | 0.25 (0.010) | 0.30 (0.012) | 0.35 (0.014) | 0.40 (0.016) | T15, M42(b) |
| Ti-6Al-4V, Ti-6Al-4V-ELI, | | | 11 (35) | 0.10 (0.004) | 0.15 (0.006) | 0.25 (0.010) | 0.30 (0.012) | 0.35 (0.014) | 0.40 (0.016) | C-2 |
| Ti-6Al-2Sn-4Zr-2Mo, | 320–380 | Solution | 6 (20) | 0.08 (0.003) | 0.13 (0.005) | 0.20 (0.008) | 0.25 (0.010) | 0.30 (0.012) | 0.35 (0.014) | T15, M42(b) |
| Ti-6Al-2Sn-4Zr-2Mo-0.25Si, | | treated | 11 (35) | 0.08 (0.003) | 0.13 (0.005) | 0.20 (0.008) | 0.25 (0.010) | 0.30 (0.012) | 0.35 (0.014) | C-2 |
| Ti-6Al-2Sn-4Zr-6Mo | | and aged | | | | | | | | |
| Beta alloys | 275–350 | Annealed or | 5 (15) | 0.08 (0.003) | 0.13 (0.005) | 0.20 (0.008) | 0.25 (0.010) | 0.30 (0.012) | 0.35 (0.014) | T15, M42(b) |
| Ti-3Al-8V-6Cr-4Mo-4Zr, | | solution | 9 (30) | 0.08 (0.003) | 0.13 (0.005) | 0.20 (0.008) | 0.25 (0.010) | 0.30 (0.012) | 0.35 (0.014) | C-2 |
| Ti-8Mo-8V-2Fe-3Al, | | treated | | | | | | | | |
| Ti-11.5Mo-6Zr-4.5Sn, | 350–440 | Solution | 3 (10) | 0.08 (0.003) | 0.13 (0.005) | 0.20 (0.008) | 0.25 (0.010) | 0.30 (0.012) | 0.35 (0.014) | T15, M42(b) |
| Ti-10V-2Fe-3Al, | | treated | 8 (25) | 0.08 (0.003) | 0.13 (0.005) | 0.20 (0.008) | 0.25 (0.010) | 0.30 (0.012) | 0.35 (0.014) | C-2 |
| Ti-13V-11Cr-3Al | | and aged | | | | | | | | |

Note: ELI, extra-low interstitial. (a) Based on four flutes for 3 and 6 mm (1/8 and 1/4 in.) reamers, six flutes for 13 mm (1/2 in.) reamers, and eight flutes for 25 mm (1 in.) and larger reamers. (b) Any premium high-speed tool steel can be used. Source: Metcut Research Associates Inc.

## Nontraditional Machining Methods

Probably the most widely used of these methods for titanium and titanium alloys are electrochemical machining (ECM), chemical milling (CHM), and laser beam machining (LBM). Detailed information on these processes can be found in the Section "Nontraditional Machining Processes" in this Volume.

**Electrochemical Machining.** An example of typical ECM operating conditions is the following. For Ti-6Al-6V-2Sn (annealed), the electrolyte is sodium chloride at a concentration of 120 g/L (16 oz/gal.), the maximum starting voltage is 3.2 V, and the metal removal rate is approximately 1.64 $cm^3$/min/1000 A (0.10 $in.^3$/min/1000 A). The electrolyte temperature is 40 °C (100 °F).

**Chemical Milling.** Typical operating conditions for titanium alloys are:

| | |
|---|---|
| Principal etchant | Hydrofluoric acid |
| Etch rate, mm/min (in./min) | 0.015–0.030 (0.0006–0.0012) |
| Optimum etch depth, mm (in.) | 3.18 (0.125) |
| Etchant temperature, °C (°F) | 46 ± 2.7 (115 ± 5) |
| Average surface roughness ($R_a$), μm (μin.) | 0.40–2.50 (16–100) |

Tolerances on depth of cut up to 12.7 mm (0.5 in.) for titanium alloys are:

| Depth of cut | | Tolerance | |
|---|---|---|---|
| mm | in. | mm | in. |
| 0–1.25 | 0–0.050 | 0.05 | 0.002 |
| 1.25–2.55 | 0.050–0.100 | 0.075 | 0.003 |
| 2.55–6.35 | 0.100–0.250 | 0.10 | 0.004 |
| 6.35–12.7 | 0.250–0.500 | 0.15 | 0.006 |

**Laser Beam Machining.** Cutting rates for titanium alloys are shown in Table 11. Cutting speed data are based on the use of a continuous-wave $CO_2$ laser with an oxygen assist.

## Guidelines for Machining Zirconium Alloys

These materials can be machined with conventional methods, including turning and boring, milling, drilling, wheel grinding, belt grinding, and sawing. A nontraditional method, electrical discharge machining (EDM), is also used.

The three basic parameters are:

- Slow speeds
- Heavy feeds

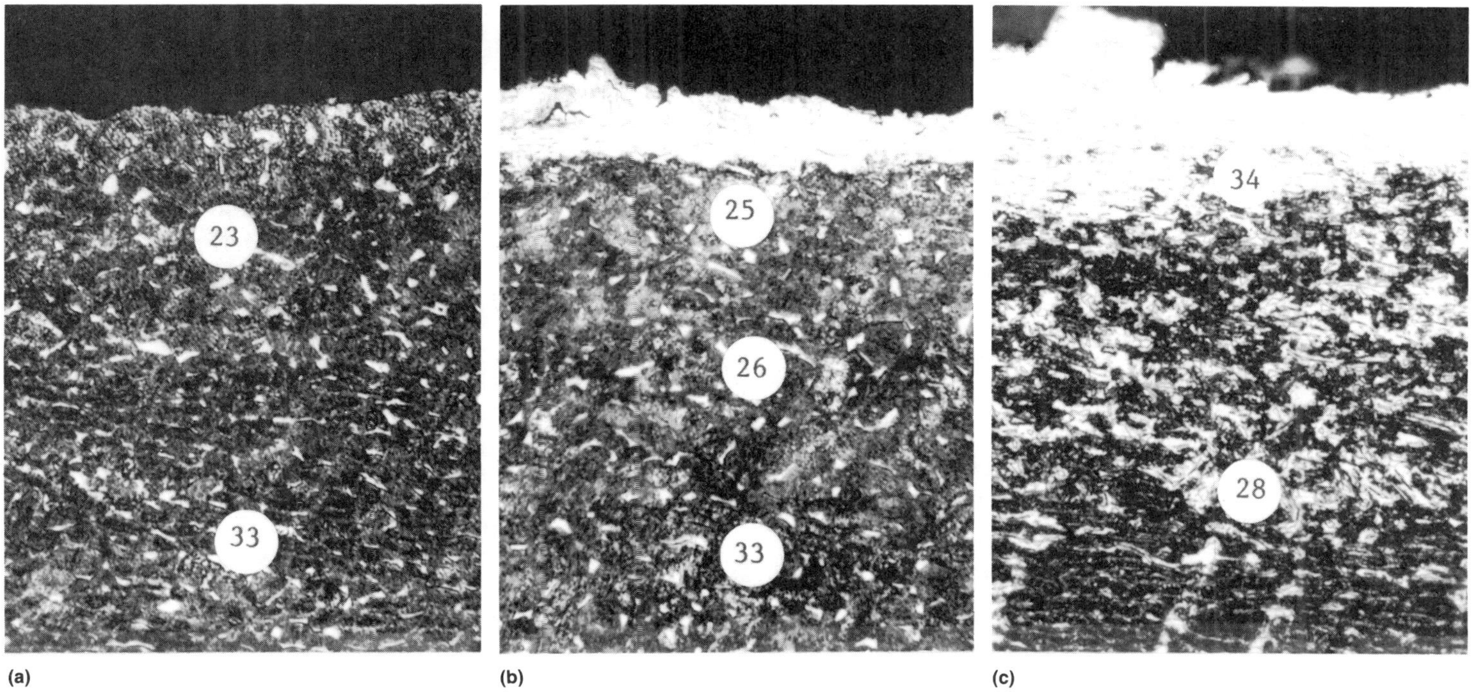


**Fig. 12** Surface characteristics of Ti-6Al-4V (32 to 34 HRC) produced by 3° traverse surface grinding. (a) Gentle grinding. (b) Conventional grinding. (c) Abusive grinding. Numbers inside photomicrographs indicate hardness (HRC values converted from Knoop hardness measurements). Surface finish measurements are averages of readings made on all specimens from each group. Orientation was along the longitudinal grind. Surface sections are perpendicular to the grinding lay. 100×

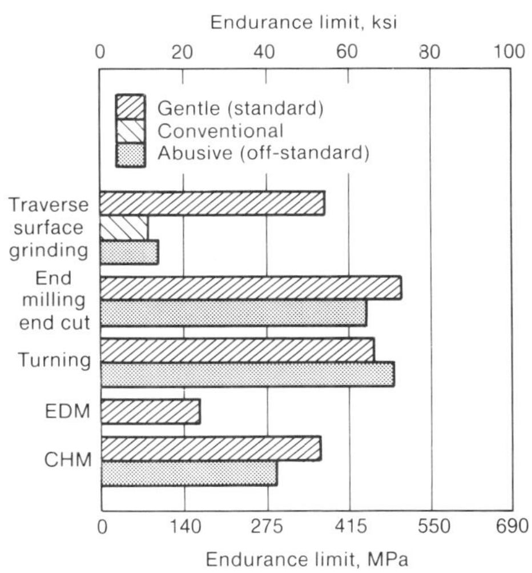


**Fig. 13** Effects of grinding on high-cycle fatigue behavior of annealed Ti-6Al-4V with a hardness of 32 to 34 HRC. EDM, electrical discharge machining; CHM, chemical milling

- A flood coolant system using a water-soluble oil lubricant

Zirconium has a marked tendency to gall and work harden, necessitating higher-than-normal clearance angles on tools to penetrate the previously work-hardened surface and cut a clean, coarse chip.

At the elevated temperatures generated at the tool and chip interface, there is a great tendency for zirconium to alloy with the carbide cutting material. For this reason, it is important to design cutting tool geometry for free cutting and to keep tool surfaces smooth to minimize chip friction. Also, tools need to be sharpened frequently.

Satisfactory results can be obtained with both cemented carbide and high-speed tool steel tools. However, the carbide usually gives better finishes and higher productivity. Polishing or honing cutting edges lengthens tool life. Zirconium alloys machine to an excellent finish, requiring relatively light horsepower in comparison with alloy steel. Tool forces are relatively low.

Advances in insert geometry and the development of a tough carbide grade have made it possible to increase productivity and lower tool costs in the machining of zirconium and hafnium. With the new chip groove configuration (Fig. 14), a typical continuous stringer is directed away from the tool and is kept from balling up or dangerously wrapping around the workpiece. The new carbide grade can withstand the extreme stresses at the toolcutting edge. Grade H-91 has a minimum transverse rupture strength of 2800 MPa (400 ksi). Other grades in the series, in order of increasing wear resistance and decreasing toughness, are H-17 (micrograin), H-21, and HA. Properties and compositions of these carbide grades are listed in Table 12.

The new chip groove (Fig. 14) is designed to direct and control long continuous chips away from the workpiece. The positive action of the groove results in lower tool pressures and increased operator safety. However, caution should be exercised in machining zirconium because of its pyrophoric properties. Clean machines and the use of moderate-to-heavy chip loads minimize danger.

In general, a moderate-to-high positive lead angle should be used whenever possible. This helps to thin chips and reduce cutting pressures.

A coolant of soluble oil and water is recommended. Care should be taken in choosing a coolant because some solvents have an adverse reaction with these materials. Proper lead angles and coolants help reduce or prevent insert notching at the depth-of-cut line.

Precise working equipment, including toolholders, seats, and tool blocks, is necessary to machine reactive alloys. The more rigid the setup, the better the tool performance. Operating parameters for machining with H-91 are set forth in Table 13.

**Turning and Boring.** Zirconium alloys can be turned without difficulty if tools are sharp and a coolant lubricant is used. Typical starting parameters are listed in Table 14. The guidelines that follow are for turning zirconium with carbide tools.

In the hot-worked condition, zirconium alloys have a tight, abrasive scale on them. Rough turning under these conditions can be effective with the use of grade H8 (C-1) tungsten-carbide-cobalt tools. Finish turning calls for a hard wear-resistant grade such as HA (C-2). Grade HF (C-3, C-4) is used on very light cuts.

In roughing and finishing, the following tooling parameters are used:

- *Roughing:* Side rake, 5°; back rake, 0°; side cutting-edge angle, 15 to 30°; nose radius, 0.8 mm (1/32 in.); and relief, 5°

### Table 7 Examples of low-stress grinding operating parameters

| Grinding method | Material | Grinding motion | Grinding wheel: Classification | Grinding wheel: Speed, m/s (sfm) | Grinding wheel: Dressing technique | Lubricant | Workpiece speed, m/min (sfm) | Cross feed, mm/pass (in./pass) | Down feed or infeed, mm/pass (in./pass): Roughing | Finishing steps: First | Finishing steps: Second | Finishing steps: Third | Sparkout passes |
|---|---|---|---|---|---|---|---|---|---|---|---|---|---|
| Surface grinding | Titanium alloys | Traverse or plunge | C60J8V | 12–16 (2400–3100) | Coarse(a) | Sulfochlorinated oil | 18 (60) | 1.3 (0.050) | 0.05 (0.002) | 0.23 at 0.013 (0.009 at 0.0005) | 0.010 at 0.010 (0.0004 at 0.0004) | 0.015 at 0.005 (0.0006 at 0.0002) | 2 |
| | | Notch | Cl80R9B | 25–30 (4800–6000) | Fine(b) | Sulfochlorinated oil | 18 (60) | 0 (0) | 0.025 (0.001) | 0.0025 (0.0001) | . . . | 0.0015 (0.00005) | 0 |
| Cylindrical grinding | Titanium alloys | Traverse | C60J8V | 14–65 (2800–3250) | Coarse | Sulfochlorinated oil | 12–30 (40–100)(c) | 180 (7)(d) | 0.025 (0.001) | 0.23 at 0.013 (0.009 at 0.0005) | 0.010 at 0.010 (0.0004 at 0.0004) | 0.015 at 0.005 (0.0006 at 0.00002) | 2 |
| | | Plunge | C60J8V | 14–65 (2800–3250) | Coarse | Sulfochlorinated oil | 12–30 (40–100)(c) | 0 (0) | 0.005 (0.0002) | 0.0020 (0.00008)(d) | . . . | 0.0005 (0.00002)(e) | 5–20 rev |
| | | Notch | C120R9B | 23–38 (4700–5500) | Fine | Sulfochlorinated oil | 12–30 (40–100)(c) | 0 (0) | 0.005 (0.0002) | 0.0020 (0.00008)(d) | . . . | 0.0005 (0.00002)(e) | 5–20 rev |

(a) Single-point diamond at 25 mm (1 in.) in 7 s. (b) Single-point diamond at 25 mm (1 in.) in 21 s. (c) Use highest possible table speed (workpiece) up to 30 m/min (100 sfm). (d) Units are in mm/min (in./min). (e) Values are approximate; hand feed at a slow, steady rate; units are in mm/rev (in./rev). Source: Metcut Research Associates Inc.

- *Finishing*: Side rake, 5°; back rake, 0°; side cutting-edge angle, 0 to 15°; nose radius, 0.8 mm (1/32 in.); and relief, 5°

**Milling.** Both vertical face and horizontal slab milling give good results. Wherever possible, zirconium should be climb milled to penetrate the work at the maximum approach angle and depth of cut while emerging through the work-hardened area. The faces and edges of milling cutters should be kept very sharp. A set of herringbone cutters will permit positive axial rake angles to be effective at both sides of a recess. Optimum surface finish and tool life are obtained when the tool is ground with a positive 12 to 15° radial rake and when cutting corners are ground to a 3.2 mm (1/8 in.) × 45° chamfer. A high-spiral flute should also be used. The work should be flooded or sprayed with a coolant to wash away all chips from the tool. The feed can range from 0.05 to 0.25 mm/tooth (0.002 to 0.010 in./tooth) at 45 to 75 m/min (150 to 250 sfm). With sharp cutters, the work absorbs about 12% of the cutting energy. Zirconium requires only about 75% of the horsepower required for SAE 1020 cold-rolled steel.

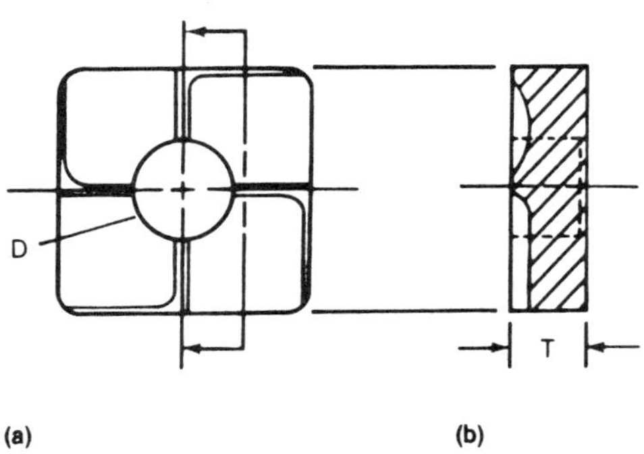


**Fig. 14** New chip groove design developed for machining both reactive and refractory metals. (a) Front view. (b) Side view

**Drilling.** The standard, fresh-ground (118°), thin-webbed drill running at slightly lower than normal speeds (12 to 18 m/min, or 40 to 60 sfm) and a feed of 0.13 to 0.25 mm/rev (0.005 to 0.010 in./rev) with a coolant-lubricant gives good results for zirconium. A firm backing should be used to prevent burrs at the exit. Adequate stock should be left for reaming so that smearing can be minimized. In tapping zirconium, chip-driver or gun-type taps give good tool life, but must be kept sharp.

**Grinding.** Zirconium can be specified for applications for which extremely close dimensional tolerances and high-quality surface finishes are required. The grinding methods used involve standard grinding machine equipment for all functions, such as surface grinding, cylindrical grinding, centerless grinding, and belt grinding. In addition, all standard abrasive equipment such as abrasive wheels, coated abrasives, and lubricants can be used.

The grinding characteristics of zirconium and its alloys are similar, and both wheel and belt grinding can be used. The use of a straight grinding oil or oil coolant not only

### Table 8 Wheel specifications for precision grinding of titanium

| Grinding operation | Conventional speed, 20–30 m/s (4000–6000 sfm) | Low speed, 7.5–10 m/s (1500–2000 sfm) |
|---|---|---|
| Centerless | 37C54-M5B | . . . |
| Cylindrical | 37C80-KVK | 32A60-K5VBE |
| Internal | 39C60-K8VK | 32A60-L8VBE |
| Surface | | |
| Horizontal spindle | 39C80-K8VK | 32A80-L5VBE |
| Vertical spindle | | |
| Cylinder | 37C60-HVK | . . . |
| Segments | 32A24-H12VBEP | . . . |
| Thread | 37C220-T9BH | . . . |

### Table 9 Operational information for abrasive cutting

| Machine setting | Bar diameter(a): ≤760 mm (≤3.00 in.) | Bar diameter(a): >760 mm (>3.00 in.) |
|---|---|---|
| Feed, $mm^2/min$ ($in.^2/min$) | 50–100 (2–4) | 130–150 (5–6) |
| Speed, m/s (sfm) | 35–60 (7000–12 000) | 30–35 (6000–7000) |
| Cutting motion | Oscillating wheel | Oscillating wheel and work rotation |

(a) A 10% water solution of rust inhibitor (nitrite-amine types) and/or 10% water solution of soluble oil may be used as coolants.

### Table 10 Operational data for hacksawing titanium

| Titanium grade | Machine setting: Speed, stroke/min | Feed, mm/stroke (in./stroke), Workpiece size: 100–150 mm (4–6 in.) | 150–205 mm (6–8 in.) | 205–255 mm (8–10 in.) | ≥255 mm (≥10 in.) |
|---|---|---|---|---|---|
| Commercially pure | 90–100 | 0.30 (0.012) | 0.23 (0.009) | 0.15 (0.006) | 0.08 (0.003) |
| All alloy grades | | | | | |
| Annealed | 60–90 | 0.23 (0.009) | 0.15 (0.006) | 0.08 (0.003) | 0.08 (0.003) |
| Heat treated | 30–60 | 0.23 (0.009) | 0.15 (0.006) | 0.08 (0.003) | 0.08 (0.003) |

**Table 11 LBM cutting rates for titanium alloys**

| Type of cut | Work thickness mm | Work thickness in. | Cutting speed(a) mm/min | Cutting speed(a) in./min |
|---|---|---|---|---|
| Contour | 0.5 | 0.020 | 5080 | 200 |
| Contour | 1.6 | 0.062 | 4060 | 160 |
| Contour | 3.1 | 0.125 | 3050 | 120 |
| Straight | 6 | 0.250 | 5080 | 200 |
| Contour | 6 | 0.250 | 1520 | 60 |
| Contour | 13 | 0.50 | 1020 | 40 |
| Contour | 25 | 1.0 | 380 | 15(b) |
| Contour | 50 | 2.0 | 130 | 5(b) |

(a) Data are based on use of a continuous wave $CO_2$ laser with an oxygen assist. (b) 16 kW (21 hp) laser was used. Source: Metcut Research Associates Inc.

produces a better finish and higher yields, but also prevents ignition, which can occur when grinding swarf is fine and dry.

*Wheel Grinding.* Zirconium grinding produces a stream of white sparks. Conventional speeds (15 to 30 m/s, or 3000 to 6000 sfm) and feeds (0.025 to 0.125 mm/pass, or 0.001 to 0.005 in./pass) are satisfactory, and SiC generally gives better results than does $Al_2O_3$. The effect of the grinding fluid on zirconium is the same as for other metals. Straight grinding oils produce higher grinding ratios than water-miscible fluids at all infeeds.

In cutoff work, silicon carbide rubber wheels prove to be the most effective. Operations and wheel recommendations are summarized in Table 15.

*Belt Grinding.* Belt speed and contact wheel selection are two primary considerations. Recommended belt speeds are 10 to 15 m/s (2000 to 3000 sfm) at low grinding pressures with 50-grit and coarser material, and 13 to 18 m/s (2500 to 3500 sfm) with 60-grit and finer belts, with similar working pressure. At high grinding pressures, speeds of 13 to 18 m/s (2500 to 3500 sfm) are recommended with 50-grit and coarser material and 15 to 20 m/s (3000 to 4000 sfm) with 60-grit and finer material.

Soluble-oil coolants either alone or mixed with water and applied in a flood are recommended. Resin abrasive cloth may be used with oil and rubber contact wheels for general polishing operations, while resin industrial cloth type 3 or 6 is recommended for use with oil when grinding pressures are high. Similarly, waterproof cloth SiC for light work and $Al_2O_3$ for heavy work may be effectively employed with soluble oil and water coolants.

**Sawing.** Power hacksawing is done most efficiently with coarse high-speed tool steel blades using a medium stroke rate (90 strokes per minute) and a light feed (cutting rate of about 480 $mm^2$/min, or 0.75 $in.^2$/min).

Band sawing is generally done with a saw having a 16 to 20 saw pitch and running approximately 23 m/min (75 sfm). A 1.1 mm (0.042 in.) rake set and a saw width of 13 mm (½ in.) in conjunction with a flood of soluble lubricant or an air stream to wash the chips away from the cut give satisfactory results. A cutting rate of 840 $mm^2$/min (1.3 $in.^2$/min) gives an excellent finish.

Cold circular sawing is done with a saw having a 15 to 20 saw pitch and running approximately 17 m/min (55 sfm). Teeth should be alternately leveled and square, with the leveled teeth slightly higher than the square ones. The cut and the saw blades should be cooled with soluble coolant. Fine chips should frequently be removed from the area to prevent a buildup of flammable solids.

**Health and Safety Considerations.** Zirconium is nontoxic. Nonetheless, there are serious limitations on its use because of potential health hazards.

Zirconium is pyrophoric because of its heat-producing reaction with oxidizing elements such as oxygen. Large pieces of sheet, plate, bar, tube, and ingot can be heated to high temperatures without excessive oxidation or burning. However, small pieces with a high surface-area-to-mass ratio, such as machine chips and turnings, are easily ignited and burn at extremely high temperatures. Large accumulations of chips and other finely divided materials should be avoided. When storing chips and turnings, care should be taken to place the material in nonflammable containers in isolated areas.

One effective storage method is to keep the material covered with water in containers and, in turn, to use oil on the water to keep it from evaporating. If a fire accidentally starts in zirconium, water or ordinary fire extinguishers should not be used. Instead, dry sand, powdered graphite, or commercially available Metal-X powder (sodium chloride-based) should be used. Large quantities of water can be used to control and extinguish fires in other flammables in the vicinity of a zirconium fire.

## Guidelines for Machining Hafnium

Hafnium is similar to zirconium in machinability. It is readily machined with conventional methods. However, like zirconium, hafnium has a tendency to gall and work harden. This can be minimized by using sharp tools and setting the depth of cut greater than 0.08 mm (0.003 in.) to penetrate the work-hardened area created by previous tool passes. Slow speeds and heavy feeds are recommended, and satisfactory results can be obtained with both carbide and high-speed tool steel tools. Carbide tools generally give better finishes and higher productivity, with tool life being similar to that obtained in the machining of stainless steel.

**Table 12 Properties of carbide grades used in machining zirconium and zirconium and hafnium alloys**

| Proprietary carbide grade | Composition, % WC | Co | TaC | NbC | TiC | Tool material grades General purpose | ISO | Hardness, HRA | Density, g/cm³ | Average transverse rupture strength MPa | ksi |
|---|---|---|---|---|---|---|---|---|---|---|---|
| H-17 | 90.0 | 10.0 | . . . | . . . | . . . | C-1, C-2, C-10 | K20, K30 | 91.4 | 14.85 | 2450 | 355 |
| H-21 | 93.5 | 6.0 | 0.5 | 0.5 | . . . | C-2 | K10, K30 | 92.10 | 14.85 | 2120 | 308 |
| H-91 | 88.5 | 11.0 | 0.5 | 0.5 | 0.5 | C-1, C-10, C-11 | K40 | 89.70 | 12.42 | 2690 | 390 |
| HA | 93.5 | 5.5 | 1.0 | 1.0 | . . . | C-2 | K10, K20 | 92.4 | 14.90 | 1960 | 285 |

**Table 13 Parameters for machining zirconium with grade H-91 carbide inserts**

| Operation | Insert type | Machine settings: Speed m/min | Speed sfm | Feed mm/min | Feed in./min | Depth of cut mm | Depth of cut in. | Tool geometry: Lead angle, degrees | Rake angles |
|---|---|---|---|---|---|---|---|---|---|
| OD turning and facing | | | | | | | | | |
| Roughing | SNMG-642A-3S | 45–75 | 150–250 | 0.50–0.75 | 0.020–0.030 | ≤9.5 | ≤0.375 | +15 to +60 | . . . |
| Finishing | SNMG-642A-2M | 75–120 | 250–400 | 0.13–0.50 | 0.005–0.020 | ≤9.5 | ≤0.375 | +15 to +60 | . . . |
| Planing and shaping | Single-point tooling(a) | 30–45 | 100–150 | 1.0–1.3 | 0.040–0.050 | ≤3 | ≤0.125 | . . . | . . . |
| Milling | | | | | | | | | |
| Roughing | SPG | 45–60 | 150–200 | 75–130 | 3–5 | 1.25–2.5 | 0.050–0.100 | +15 to +30 | Positive axial and positive radial |
| Finishing | SPE | 60–85 | 200–275 | 25–75 | 1–3 | ≤1.25 | ≤0.050 | +15 to +30 | Positive axial and positive radial |

(a) Single-point tooling with high positive rake angles (20 to 25° positive chip groove) is recommended. OD, outside diameter

**Table 14 Typical parameters for turning and boring zirconium alloys**

| Operation | Machine settings(a) | | | | | |
|---|---|---|---|---|---|---|
| | Feed | | Speed | | Depth of cut | |
| | mm/rev | in./rev | m/min | sfm | mm | in. |
| Roughing | 0.51 | 0.020 | 24–30 | 80–100 | 3.18–6.35 | 0.125–0.250 |
| Finishing | 0.13 | 0.005 | 30–46 | 100–150 | 0.25–1.00 | 0.010–0.040 |

(a) Tool geometry: 0.8 mm (1/32 in.) nose radius, 10° end clearance, 4.8 mm (3/16 in.) chip breaker, 15° back rake angle, 10° side rake angle

**Table 15 Wheel selection chart for grinding zirconium**

| Operation | Grain type | Bond type | Grit size | Grade |
|---|---|---|---|---|
| Surface grinding | SiC | Vitrified | 46–60 | J |
| Cylindrical grinding | $Al_2O_3$ | Vitrified | 60–80 | L-M |
| Snagging | SiC | Resinoid | 20 | P |
| Cutoff | SiC | Rubber | 60 | L |

Care should be taken to minimize fine chips, as hafnium is pyrophoric in nature. A flood of water-soluble coolant is recommended. To minimize the fire hazard, chips should not be allowed to accumulate in the bed of the machine and should be continually removed and stored, preferably under water, in isolated areas.

**Turning.** Roughing cuts are the most effective with a feed of about 5 mm/rev (0.20 in./rev) at 25 to 30 m/min (80 to 100 sfm) to a depth of 3.18 to 6.35 mm (0.125 to 0.250 in.). Tool geometry should include a 0.8 mm (1/32 in.) nose radius. Alloyed carbides, such as grade H-91 (see Table 12), are commonly used.

Finishing cuts can be made with the same tool (reground and honed) by reducing feed to 0.13 mm/rev (0.005 in./rev) and depth of cut to 0.25 to 1.0 mm (0.010 to 0.040 in.), and increasing speed to 30 to 45 m/min (100 to 150 sfm). A back rake angle of 15° should be used with clearance angles and side rake of 10°. Sharp tools and a coolant lubricant should always be used.

**Milling.** The vertical face method is recommended, but acceptable results can be obtained with horizontal slab milling. Faces and edges of the milling cutter should be kept very sharp.

Optimum surface finish and tool life are obtained when the tool is ground with a positive 12 to 15° radial rake and insert cutting corners are ground to 3.2 mm (1/8 in.) by 45° chamfer. A high-spiral flute should also be used. Work should be flooded or sprayed with a coolant to wash away all chips from the tool. Feed can range from 0.05 to 0.25 mm/tooth (0.002 to 0.010 in./tooth) at 45 to 75 m/min (150 to 250 sfm).

Information in Table 16 is for machining hafnium with grade H-91 tooling. Machining methods include turning and facing, planing and shaping, and milling.

**Drilling.** Hafnium can be drilled and tapped with standard techniques and equipment. Speeds of 15 m/min (50 sfm) at a feed rate of 0.13 mm/rev (0.005 in./rev) give satisfactory results. Copious coolant should be used to prevent overheating as well as potential tool binding and breakage.

**Health and Safety Considerations.** Like zirconium and titanium, hafnium is nontoxic and does not require the use of complicated or expensive fabrication equipment or procedures because of health hazards. Fire is the only significant hazard.

Hafnium is more pyrophoric than either zirconium or titanium, and special handling precautions are necessary when product forms exhibiting high surface-to-mass ratios are encountered. These include conditions such as thin turnings, chips, and powder. However, hafnium sheet, plate, bar, tube, and ingot can be readily heated to high temperatures without burning or excessive oxidation.

Accumulations of chips, turnings, and finely divided particles are to be avoided. This material should be stored, totally covered with water, in nonflammable, leak-proof containers placed in isolated areas (use of oil on the water to retard evaporation is recommended). Turnings and chips having minimal scrap values should be burned in small quantities in isolated areas by lighting with a fuse or a small flame. Local, state, and federal regulations should be reviewed and observed.

Extreme caution is necessary to prevent ignition and explosion hazards when handling very fine particles that can become airborne. Inert gases such as helium or argon can be used to prevent ignition of powders or sponge.

While water is recommended for storing chips and turnings, once ignited, moist hafnium material will burn more violently than dry hafnium because decomposition of the water produces hydrogen. The use of water to combat hafnium fires is to be avoided. Also, common fire-extinguishing agents such as carbon dioxide, foam, and carbon tetrachloride should never be used on hafnium fires because they will increase the rate of burning. Large quantities of water can be used to control and extinguish fires in other flammables in the vicinity of the hafnium fire. Dry salt (sodium chloride) or commercially available type D fire extinguisher powder is recommended as an extinguishing agent.

## ACKNOWLEDGMENT

The author wishes to express his appreciation to the following individuals for their significant contributions to this article: John F. Kahles, Metcut Research Associates Inc.; Larry Mayer, TIMET; R. Terrence Webster, Teledyne Wah Chang Albany; and Barry Heller, Teledyne Firth Sterling.

## REFERENCES

1. N. Zlatin, M. Field, W.P. Koster, *et al.*, "Final Report on Machinability of Materials," AFML-TR-65-444, Metcut Research Associates Inc., 1969
2. H. Simon, M. Thoma, and K. Maier, Coolants With Chlorine Extreme Pressure Additives, *WT Z. Industr.*, Furtgung 69, 1979, p 79-82 (in German)
3. W.J. Zdeblick and R.E. DeVor, An Ex-

**Table 16 Parameters for machining hafnium with grade H-91 carbide inserts**

| Operation | Insert type | Machine settings | | | | | | Tool geometry | |
|---|---|---|---|---|---|---|---|---|---|
| | | Speed | | Feed | | Depth of cut | | | |
| | | m/min | sfm | mm/min | in./min | mm | in. | Lead angle, degrees | Rake angles |
| OD turning and facing | | | | | | | | | |
| Roughing | SPG | 45–60 | 150–200 | 0.50–0.75 | 0.020–0.030 | ≤4 | ≤0.150 | +15 to (+60) | . . . |
| Finishing | SPU | 60–120 | 200–250 | 0.13–0.25 | 0.005–0.010 | 0.25–0.50 | 0.010–0.020 | +15 to +60 | . . . |
| Planing and shaping | Single-point tooling(a) | 30–45 | 100–150 | 1.0–1.3 | 0.040–0.050 | ≤3 | ≤0.125 | . . . | . . . |
| Milling | | | | | | | | | |
| Roughing | SPG | 45–60 | 150–200 | 75–130 | 3–5 | 1.25–2.4 | 0.050–0.100 | +15 to +30 | Positive axial and positive radial |
| Finishing | SPE | 60–75 | 200–250 | 25–75 | 1–3 | ≤1.25 | ≤0.050 | +15 to +30 | Positive axial and positive radial |

(a) Single-point tooling with high positive rake angle (20 to 25° positive chip groove) is recommended.

perimental Strategy for Designing Tool Life Experiments, *(Trans. ASME) J. Eng. Ind.*, Vol 100 (No. 4), 1978, p 441-450
4. N. Zlatin, M. Field, *et al.*, "Machinability Parameters on New and Selective Aerospace Materials," AFML-TR-69-144, Metcut Research Associates Inc., 1969
5. W.P. Koster, "Surface Integrity of Machined Structural Components," AFML-TR-70-11, Metcut Research Associates Inc., 1970
6. W.P. Koster, "Manufacturing Methods for Surface Integrity of Machined Structural Components," AFML-TR-71-258, Metcut Research Associates Inc., 1974
7. W.P. Koster, "Surface Integrity of Machined Materials," AFML-TR-74-80, Metcut Research Associates Inc., 1974

# Machining of Refractory Metals

Harry E. Chandler, ASM INTERNATIONAL

THE REFRACTORY METALS, niobium, molybdenum, tantalum, and tungsten, differ considerably in machining characteristics. Both tungsten and molybdenum lack ductility at room temperature, but molybdenum is less brittle than unalloyed tungsten. Further, the brittleness of molybdenum varies considerably, depending on its condition. Pressed-and-sintered and arc cast molybdenum are both brittle, but forging greatly reduces their room-temperature brittleness. Niobium and tantalum, on the other hand, are ductile and can be machined by methods similar to those used for other ductile metals.

## Tooling, Cutting Fluids, and Safety

Maximum rigidity of machines and tools is essential in machining tungsten and molybdenum, to obtain satisfactory cutting and to minimize damage to work during processing.

**Niobium.** This metal may be machined by the usual techniques. However, special attention should be paid to tool angles and lubrication because niobium has a tendency to gall.

In lathe turning, for example, these tool angles are recommended: approach angle, 15 to 20°; side rake, 30 to 35°; side and end clearance, 5°; plane relief angle, 15 to 20°; nose radius, 0.50 to 0.75 mm (0.020 to 0.030 in.). Air, soluble oil, and other suitable products are used to cool and lubricate.

**Molybdenum.** Wrought molybdenum, which has a hardness range of 200 to 300 HB, is similar to stainless steel in machining characteristics. Conventional equipment and cutting tools are used. Tools must be kept sharp, and regular tool changes are recommended for best results. To avoid poor surface finish and excessive tool wear, heavy duty machines, rigid tool mounting, and firm workpiece clamping are essential.

**Table 1 Turning wrought, cast, and powder metallurgy (P/M) refractory metals or alloys with single-point and box tools**

| | | | | High-speed steel (HSS) tool | | | Uncoated brazed carbide tool | | |
|---|---|---|---|---|---|---|---|---|---|
| Material | Hardness, HB | Condition | Depth of cut, mm (in.)(a) | Speed, m/min (sfm) | Feed, mm/rev (in./rev) | Tool material grade, ISO(b) (AISI)(c) | Speed, m/min (sfm) | Feed, mm/rev (in./rev) | Tool material grade, ISO(b) (C-grade)(d) |
| Niobium | 170–225 | Stress | 1 (0.040) | 24 (80) | 0.13 (0.005) | S9, S11(e) (T15, M42)(e) | 90 (300) | 0.13 (0.005) | K20, M20 (C-2) |
| C103, C129Y, Nb-1Zr, Cb-752, FS-85, FS-291, WC-3015 | | relieved | 4 (0.150) | 21 (70) | 0.13 (0.005) | S9, S11(e) (T15, M42)(e) | 81 (265) | 0.13 (0.005) | K20, M20 (C-2) |
| Molybdenum | 220–290 | Stress | 1 (0.040) | ... | ... | ... | 100 (325) | 0.13 (0.005) | K20, M20 (C-2) |
| Mo, Mo-50Re, TZC, TZM | | relieved | 4 (0.150) | ... | ... | ... | 88 (290) | 0.25 (0.010) | K20, M20 (C-2) |
| Tantalum | 200–250 | Stress | 1 (0.040) | 18 (60) | 0.13 (0.005) | S9, S11(e) (T15, M42)(e) | 34 (110) | 0.13 (0.005) | K20, M20 (C-2) |
| ASTAR 811C, T-111, T-222, Ta-l0W, Ta-Hf, Ta63 | | relieved | 4 (0.150) | 15 (50) | 0.25 (0.010) | S9, S11(e) (T15, M42)(e) | 26 (85) | 0.25 (0.010) | K20, M20 (C-2) |
| Tungsten, 85% density | 180–220 | Pressed and | 1 (0.040) | ... | ... | ... | 32 (105) | 0.18 (0.007) | K01 (C-4) |
| | | sintered | 4 (0.150) | ... | ... | ... | 26 (85) | 0.25 (0.010) | K01 (C-4) |
| Tungsten, 93% density | 290–320 | Pressed and | 1 (0.040) | ... | ... | ... | 27 (90) | 0.18 (0.007) | K01 (C-4) |
| | | sintered | 4 (0.150) | ... | ... | ... | 24 (80) | 0.25 (0.010) | K01 (C-4) |
| Tungsten, 96 and 100% density | 290–320 | Forged or | 1 (0.040) | ... | ... | ... | 40 (130) | 0.18 (0.007) | K01 (C-4) |
| | | arc cast | 4 (0.150) | ... | ... | ... | 37 (120) | 0.25 (0.010) | K01 (C-4) |
| Tungsten-2 thoria | 260–320 | Pressed and | 1 (0.040) | ... | ... | ... | 49 (160) | 0.18 (0.007) | K01 (C-4) |
| | | sintered | 4 (0.150)(a) | ... | ... | ... | 40 (130) | 0.25 (0.010) | K01 (C-4) |

(a) *Caution:* With the exception of tungsten-2 thoria, check horsepower requirements for depth of cut exceeding 16 mm (0.625 in.); for tungsten-2 thoria, check horsepower requirements when depth of cut is greater than 4 mm (0.15 in.). (b) Designation system of the International Organization for Standardization (ISO). (c) Designation system of the American Iron and Steel Institute (AISI). (d) The C-grade classification system of carbide tools is defined in the article "Cemented Carbides" in this Volume. (e) Any premium high-speed steel (T15, M33, M41-47, or S9, S10, S11, S12). Source: Metcut Research Associates Inc.

**Table 2 Turning niobium and molybdenum alloys (wrought, cast, and P/M) with cutoff and form tools**

| | | | | Feed, mm/rev (in./rev) | | | | | | | | |
|---|---|---|---|---|---|---|---|---|---|---|---|---|
| | | | | Cutoff tool width | | | Form tool width | | | | | |
| Material | Hardness, HB | Condition | Speed, m/min (sfm) | 1.5 mm (0.06 in.) | 3 mm (0.125 in.) | 6 mm (0.250 in.) | 13 mm (0.500 in.) | 19 mm (0.750 in.) | 25 mm (1 in.) | 38 mm (1.5 in.) | 50 mm (2 in.) | Tool material grade, ISO (AISI or C-grade)(a) |
| Niobium | 170–225 | Stress | 17 (55) | 0.025 | 0.025 | 0.038 | 0.038 | 0.038 | 0.025 | 0.018 | 0.018 | S9, S11)(b) |
| C103, C129Y, Nb-1Zr, Cb-752, | | relieved | | (0.001) | (0.001) | (0.0015) | (0.0015) | (0.0015) | (0.001) | (0.0007) | (0.0007) | (T15, M42)(b) |
| FS-85, FS-291, WC-3015 | | | 56 (185) | 0.025 | 0.025 | 0.038 | 0.038 | 0.038 | 0.025 | 0.018 | 0.018 | K40, M40 |
| | | | | (0.001) | (0.001) | (0.0015) | (0.0015) | (0.0015) | (0.001) | (0.0007) | (0.0007) | (C-2) |
| Molybdenum | 220–290 | Stress | 62 (205) | 0.025 | 0.038 | 0.050 | 0.050 | 0.038 | 0.038 | 0.025 | 0.025 | K40, M40 |
| Mo, Mo-50Re, TZC, TZM | | relieved | | (0.001) | (0.0015) | (0.002) | (0.002) | (0.0015) | (0.0015) | (0.001) | (0.001) | (C-2) |

(a) Use submicron carbide when the recommended carbide grade chips excessively. (b) Any premium high-speed steel (T15, M33, M41-47, or S9, S10, S11, S12). Source: Metcut Research Associates Inc.

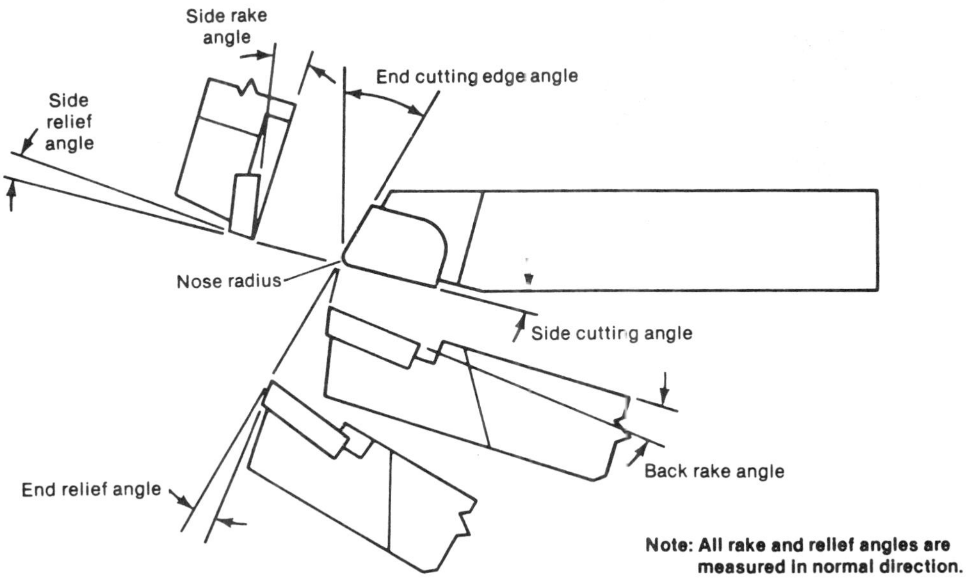


| Material | Hardness, HB | High-speed steel: Back rake angle, degrees | Side rake angle, degrees | End relief angle, degrees | Side relief angle, degrees | Carbide, Brazed: Back rake angle, degrees | Side rake angle, degrees | Relief angles, degrees | Carbide, Indexable: Back rake angle, degrees | Side rake angle, degrees | Relief angles, degrees |
|---|---|---|---|---|---|---|---|---|---|---|---|
| Niobium, molybdenum, and tantalum alloys | 170–290 | 0 | 20 | 5 | 5 | 0 | 20 | 7 | ... | ... | 5 |
| Tungsten alloys | 180–320 | ... | ... | ... | ... | −15 | 0 | 7 | ... | ... | ... |

**Fig. 1** Tool geometry for single point turning and boring of wrought, cast, and P/M refractory metals. Use the largest nose radius and the largest side cutting edge angle or end cutting edge angle that are consistent with part requirements used.

**Tantalum** is readily machined, using high-speed steel or carbide tools. The machining and grinding characteristics of tantalum vary from those of soft copper to those of annealed stainless steel. Fully recrystallized unalloyed tantalum has the machinability of soft copper.

**Tungsten.** Among the refractory metals, turning is a problem only with tungsten. Carbide tools ground with a negative back rake, 15° lead, and 0° side rake are mandatory in this instance, and all turning is at room temperature. In grinding, 60-grit silicon carbide or 46-grit alumina are recommended.

**Nontraditional methods** are also employed in machining refractory metals. Electrical discharge machining (EDM) is used to cut tungsten, niobium, and tantalum. Electrical discharge machining, wire electrical discharge machining, and chemical milling are used to machine both molybdenum and niobium.

**Chucking and Fixturing.** Workpieces of tungsten and molybdenum can be held in standard jaw-type chucks, provided sufficient pressure can be obtained and balanced. Chucking forces can cause cracking, and consequently a four-jaw chuck, and not a three-jaw chuck, is often used for holding parts being machined from these metals.

Except for niobium, the use of copper shims is recommended for chucking or fixturing refractory metal parts, especially parts made of tungsten or molybdenum. Thermoplastic supports are sometimes used for holding fragile parts made of these metals.

**Cutting Tools.** All refractory metals are abrasive (tungsten is the most abrasive). Accordingly, tool wear is more rapid and tool life is shorter than when machining carbon or alloy steel with similar tools.

For machining niobium and tantalum, high-speed steel tools can be used in all operations. However, carbide tools are frequently used, especially for single-point cutting and face milling operations. For a fine surface finish (0.80 μm, or 32 μin. $R_a$), a single-point high-speed steel tool should be used, because it provides a sharper cutting point than a carbide tool.

**Table 3 Boring niobium, molybdenum, tantalum, and tungsten alloys (wrought, cast, and P/M)**

| Material | Hardness, HB | Condition | Depth of cut, mm (in.) | High-speed steel tool: Speed, m/min (sfm) | Feed, mm/rev (in./rev) | Tool material grade, ISO (AISI) | Uncoated carbide tool (brazed): Speed, m/min (sfm) | Feed, mm/rev (in./rev) | Tool material grade, ISO (C-grade) |
|---|---|---|---|---|---|---|---|---|---|
| Niobium | 170–225 | Stress relieved | 0.25 (0.010) | 24 (80) | 0.05 (0.002) | S4, S5 (M2, M3) | 85 (280) | 0.05 (0.002) | K20, M20 (C-2) |
| C103, C129Y, Nb-1Zr, | | | 1.25 (0.050) | 20 (65) | 0.10 (0.004) | S4, S5 (M2, M3) | 69 (225) | 0.10 (0.004) | K20, M20 (C-2) |
| Cb-752, FS-85, FS-291, | | | 2.5 (0.100) | 17 (55) | 0.13 (0.005) | S4, S5 (N2, M3) | 53 (175) | 0.13 (0.005) | K20, M20 (C-2) |
| WC-3015 | | | | | | | | | |
| Molybdenum | 220–290 | Stress relieved | 0.25 (0.010) | ... | ... | ... | 85 (280) | 0.05 (0.002) | K20, M20 (C-2) |
| Mo, Mo-50Re, TZC, TZM | | | 1.25 (0.050) | ... | ... | ... | 69 (225) | 0.10 (0.004) | K20, M20 (C-2) |
| | | | 2.5 (0.100) | ... | ... | ... | 62 (205) | 0.20 (0.008) | K20, M20 (C-2) |
| Tantalum | 200–250 | Stress relieved | 0.25 (0.010) | 18 (60) | 0.05 (0.002) | S4, S5 (M2, M3) | 29 (95) | 0.05 (0.002) | K20, M20 (C-2) |
| ASTAR 811C, T-111, | | | 1.25 (0.050) | 15 (50) | 0.10 (0.004) | S4, S5 (M2, M3) | 23 (75) | 0.10 (0.004) | K20, M20 (C-2) |
| T-222, Ta-l0W, Ta-Hf, Ta63 | | | 2.5 (0.100) | 12 (40) | 0.20 (0.008) | S4, S5 (M2, M3) | 18 (60) | 0.20 (0.008) | K20, M20 (C-2) |
| Tungsten, 85% density | 180–200 | Pressed and sintered | 0.25 (0.010) | ... | ... | ... | 29 (95) | 0.075 (0.003) | K01 (C-4) |
| | | | 1.25 (0.050) | ... | ... | ... | 23 (75) | 0.13 (0.005) | K01 (C-4) |
| Tungsten, 93% density | 290–320 | Pressed and sintered | 0.25 (0.010) | ... | ... | ... | 24 (80) | 0.075 (0.003) | K01 (C-4) |
| | | | 1.25 (0.050) | ... | ... | ... | 20 (65) | 0.13 (0.005) | K01 (C-4) |
| | | | 2.5 (0.100) | ... | ... | ... | 17 (55) | 0.20 (0.008) | K01 (C-4) |
| Tungsten, 96 and 100% density | 290–320 | Forged or arc cast | 0.25 (0.010) | ... | ... | ... | 34 (110) | 0.075 (0.003) | K01 (C-4) |
| | | | 1.25 (0.050) | ... | ... | ... | 27 (90) | 0.13 (0.005) | K01 (C-4) |
| | | | 2.5 (0.100) | ... | ... | ... | 26 (85) | 0.20 (0.008) | K01 (C-4) |
| Tungsten-2 thoria | 260–320 | Pressed and sintered | 0.25 (0.010) | ... | ... | ... | 43 (140) | 0.075 (0.003) | K01 (C-4) |
| | | | 1.25 (0.050) | ... | ... | ... | 34 (110) | 0.13 (0.005) | K01 (C-4) |
| | | | 2.5 (0.100) | ... | ... | ... | 27 (90) | 0.20 (0.008) | K01 (C-4) |
| Tungsten alloys | 260–320 | As cast | 0.25 (0.010) | ... | ... | ... | 43 (140) | 0.075 (0.003) | K01 (C-4) |
| | | | 1.25 (0.050) | ... | ... | ... | 34 (110) | 0.13 (0.005) | K01 (C-4) |
| | | | 2.5 (0.100) | ... | ... | ... | 27 (90) | 0.20 (0.008) | K01 (C-4) |

Source: Metcut Research Associates Inc.

**Table 4 Counterboring and spotfacing niobium, molybdenum, tantalum, and tungsten alloys (wrought, cast, and P/M)**

| Material | Hardness, HB | Condition | Speed, m/min (sfm) | Feed, mm/rev (in./rev)(a) — Nominal hole diameter: 6 mm (¼ in.) | 13 mm (½ in.) | 19 mm (¾ in.) | 25 mm (1 in.) | 38 mm (1½ in.) | 50 mm (2 in.) | 65 mm (2½ in.) | 75 mm (3 in.) | Tool material grade, ISO (AISI or C-grade) |
|---|---|---|---|---|---|---|---|---|---|---|---|---|
| Niobium C103, C129Y, Nb-1Zr, Cb-752, FS-85, FS-291, WC-3015 | 170–225 | Stress relieved | 15 (50) | 0.075 (0.003) | 0.10 (0.004) | 0.10 (0.004) | 0.10 (0.004) | 0.13 (0.005) | 0.13 (0.005) | 0.13 (0.005) | 0.15 (0.006) | S4, S5 (M2, M3) |
| | | | 30 (100) | 0.075 (0.003) | 0.10 (0.004) | 0.10 (0.004) | 0.10 (0.004) | 0.13 (0.005) | 0.13 (0.005) | 0.13 (0.005) | 0.15 (0.006) | K10 (C-2) |
| Molybdenum Mo, Mo-50Re, TZC, TZM | 220–290 | Stress relieved | 30 (100) | 0.075 (0.003) | 0.10 (0.004) | 0.10 (0.004) | 0.10 (0.004) | 0.13 (0.005) | 0.13 (0.005) | 0.13 (0.005) | 0.15 (0.006) | K10 (C-2) |
| Tantalum ASTAR 811C, T-111, T-222, Ta-l0W, Ta-Hf, Ta63 | 200–250 | Stress relieved | 9 (30) | 0.050 (0.002) | 0.075 (0.003) | 0.075 (0.003) | 0.075 (0.003) | 0.10 (0.004) | 0.10 (0.004) | 0.10 (0.004) | 0.13 (0.005) | S9, S11(b) (T15, M42)(b) |
| | | | 21 (70) | 0.050 (0.002) | 0.075 (0.003) | 0.075 (0.003) | 0.075 (0.003) | 0.10 (0.004) | 0.10 (0.004) | 0.10 (0.004) | 0.13 (0.005) | K10 (C-2) |
| Tungsten, 85% density | 180–200 | Pressed and sintered | 9 (30) | 0.050 (0.002) | 0.075 (0.003) | 0.075 (0.003) | 0.075 (0.003) | 0.10 (0.004) | 0.10 (0.004) | 0.10 (0.004) | 0.13 (0.005) | K30, K40 (C-1) |
| Tungsten, 93% density | 290–320 | Pressed and sintered | 8 (25) | 0.050 (0.002) | 0.075 (0.003) | 0.075 (0.003) | 0.075 (0.003) | 0.10 (0.004) | 0.10 (0.004) | 0.10 (0.004) | 0.13 (0.005) | K30, K40 (C-1) |

(a) Feeds are based on three flutes for 6 mm (¼ in.) diam, four flutes for 13 to 65 mm (0.5 to 2.5 in.) diam, and six flutes for 75 mm (3 in.) diam. (b) Any premium high-speed steel (T15, M33, M41-M47, or S9, S10, S11, S12). Source: Metcut Research Associates Inc.

**Table 5 Trepanning niobium, molybdenum, and tantalum alloys (wrought, cast, and P/M)**

| Material | Hardness, HB | Condition | Speed m/min | Speed sfm | Feed(a) mm/rev | Feed(a) in./rev | Tool material grade, ISO (AISI or C-grade) |
|---|---|---|---|---|---|---|---|
| Niobium C103, C129Y, Nb-1Zr, Cb-752, FS-85, FS-291, WC-3015 | 170–225 | Stress relieved | 12 | 40 | 0.13 | 0.005 | S9, S11(b) (T15, M42)(b) |
| | | | 49 | 160 | 0.13 | 0.005 | K20 (C-2) |
| Molybdenum Mo, Mo-50Re, TZC, TZM | 220–290 | Stress relieved | 49 | 160 | 0.13 | 0.005 | K20 (C-2) |
| Tantalum ASTAR 811C, T-111, T-222, Ta-l0W, Ta-Hf, Ta63 | 200–250 | Stress relieved | 9 | 30 | 0.13 | 0.005 | S9, S11(b) (T15, M42)(b) |
| | | | 30 | 100 | 0.15 | 0.006 | K20 (C-2) |

(a) Based on standard 19 mm (¾ in.) cutting edge, single blade. (b) Any premium high-speed steel (T12, M33, M41–M47, or S9, S10, S11, S12). Source: Metcut Research Associates Inc.

**Table 6 Drilling niobium, molybdenum, tantalum, and tungsten alloys (wrought, cast, and P/M)**

| Material | Hardness, HB | Condition | Speed, m/min (sfm)(a) | Feed, mm/rev (in./rev) — Nominal hole diameter: 1.5 mm (1/16 in.) | 3 mm (⅛ in.) | 6 mm (¼ in.) | 13 mm (½ in.) | 19 mm (¾ in.) | 25 mm (1 in.) | 38 mm (1½ in.) | 50 mm (2 in.) | Tool material grade, ISO (AISI or C-grade) |
|---|---|---|---|---|---|---|---|---|---|---|---|---|
| Niobium C103, C129Y, Nb-1Zr, Cb-752, FS-85, FS-291, WC-3015 | 170–225 | Stress relieved | 15 (50) | 0.025 (0.001) | ... | ... | ... | ... | ... | ... | ... | S9, S11(b) (T15, M42)(b) |
| | | | 24 (80) | ... | 0.075 (0.003) | 0.13 (0.005) | 0.18 (0.007) | ... | ... | ... | ... | S9, S11(b) (T15, M42)(b) |
| Molybdenum Mo, Mo-50Re, TZC, TZM | 220–290 | Stress relieved | 23 (75) | 0.025 (0.001) | ... | ... | ... | ... | ... | ... | ... | S9, S11(b) (T15, M42)(b) |
| | | | 34 (110) | ... | 0.075 (0.003) | 0.13 (0.005) | 0.18 (0.007) | ... | ... | ... | ... | S9, S11(b) (T15, M42)(b) |
| Tantalum ASTAR 811C, T-111, T-222, Ta-l0W, Ta-Hf, Ta63 | 200–250 | Stress relieved | 15 (50) | 0.025 (0.001) | ... | ... | ... | ... | ... | ... | ... | S9, S11(b) (T15, M42)(b) |
| | | | 17 (55) | ... | 0.050 (0.002) | 0.050 (0.002) | 0.10 (0.004) | ... | ... | ... | ... | S9, S11(b) (T15, M42)(b) |
| Tungsten, 85% density | 180–200 | Pressed and sintered | 53 (175) | ... | ... | 0.050 (0.002) | 0.050 (0.002) | 0.050 (0.002) | ... | ... | ... | K10 (C-2) |
| Tungsten, 93% density | 290–320 | Pressed and sintered | 46 (150) | ... | ... | 0.050 (0.002) | 0.050 (0.002) | 0.050 (0.002) | ... | ... | ... | K10 (C-2) |
| Tungsten, 96 and 100% density | 290–320 | Forged or arc cast | 46 (150) | ... | ... | 0.050 (0.002) | 0.050 (0.002) | 0.050 (0.002) | ... | ... | ... | K10 (C-2) |

(a) For holes more than two diameters deep, reduce speed and feed. (b) Any premium high-speed steel (T15, M33, M41–M47, or S9, S10, S11, S12). Source: Metcut Research Associates Inc.

For machining molybdenum, carbide is preferred and is used almost exclusively in single-point cutting. On the other hand, high-speed steel is used exclusively for drilling, reaming, and tapping. Both tool materials have been successfully used in milling.

In machining tungsten, carbide tools are used for all operations except tapping. Solid carbide drills are required for drilling. Tapping is generally avoided, but if it is mandatory, high-speed steel taps are used.

Grade C-2 carbide is recommended for machining all refractory metals except tungsten, for which C-4 is recommended. Grade C-1 carbides are also used for refractory metals to reduce chipping caused by built-up edge. As in the machining of other metals, disposable carbide tips are usually more economical than brazed tips.

For machining all refractory metals except tungsten, cutting tools normally have positive rake, although there are exceptions. In some operations zero or negative rake has proved to be better for tungsten.

Sharp cutting tools are essential to obtain the best economy and surface finish and to minimize damage to the workpiece. When tungsten or molybdenum is being machined, the best tool life for any depth of cut is obtained from a combination of speed and feed for which flank wear of the cutting edge proceeds at the same rate as cratering of the tool face.

**Cutting Fluids.** Soluble-oil emulsions are satisfactory for most machining operations on refractory metals; exceptions are drilling, reaming, and tapping. For these operations, cutting oils that contain sulfur or chlorine are preferred. Oils that contain

**Table 7 Oil hole or pressurized-coolant drilling niobium, molybdenum, tantalum, and tungsten alloys (wrought, cast, and P/M)**

| Material | Hardness, HB | Condition | Speed, m/min (sfm) | Feed, mm/rev (in./rev) — Nominal hole diameter: 3 mm (⅛ in.) | 6 mm (¼ in.) | 13 mm (½ in.) | 19 mm (¾ in.) | 25 mm (1 in.) | 38 mm (1½ in.) | 50 mm (2 in.) | Tool material grade, ISO (AISI or C-grade) |
|---|---|---|---|---|---|---|---|---|---|---|---|
| Niobium C103, C129Y, Nb-1Zr, Cb-752, FS-85, FS-291, WC-3015 | 170–225 | Stress relieved | 29 (95) | 0.050 (0.002) | 0.13 (0.005) | 0.15 (0.006) | 0.20 (0.008) | 0.25 (0.010) | 0.30 (0.012) | 0.36 (0.014) | S9, S11(b) (C-2) |
| | | | 53 (175) | 0.050 (0.002) | 0.13 (0.005) | 0.15 (0.006) | 0.20 (0.008) | 0.25 (0.010) | 0.30 (0.012) | 0.36 (0.014) | K10 (C-2) |
| Molybdenum Mo, Mo-50Re, TZC, TZM | 220–290 | Stress relieved | 40 (130) | 0.050 (0.002) | 0.13 (0.005) | 0.15 (0.006) | 0.20 (0.008) | 0.25 (0.010) | 0.30 (0.012) | 0.36 (0.014) | S9, S11(b) (T15, M42)(b) |
| | | | 53 (175) | 0.050 (0.002) | 0.13 (0.005) | 0.15 (0.006) | 0.20 (0.008) | 0.25 (0.010) | 0.30 (0.012) | 0.36 (0.014) | K10 (C-2) |
| Tantalum ASTAR 811C, T-111, T-222, Ta-10W, Ta-Hf, Ta63 | 200–250 | Stress relieved | 20 (65) | 0.025 (0.001) | 0.10 (0.004) | 0.15 (0.006) | 0.20 (0.008) | 0.25 (0.010) | 0.30 (0.012) | 0.36 (0.014) | S9, S11(b) (T15, M42)(b) |
| | | | 46 (150) | 0.025 (0.001) | 0.10 (0.004) | 0.15 (0.006) | 0.20 (0.008) | 0.25 (0.010) | 0.30 (0.012) | 0.36 (0.014) | K10 (C-2) |
| Tungsten, 85% density | 180–200 | Pressed and sintered | 60 (200) | ... | 0.025 (0.001) | 0.050 (0.002) | 0.050 (0.002) | ... | ... | ... | K10 (C-2) |
| Tungsten, 93% density | 290–320 | Pressed and sintered | 53 (175) | ... | 0.025 (0.001) | 0.050 (0.002) | 0.050 (0.002) | ... | ... | ... | K10 (C-2) |
| Tungsten, 96 and 100% density | 290–320 | Forged or arc cast | 53 (175) | ... | 0.025 (0.001) | 0.050 (0.002) | 0.050 (0.002) | ... | ... | ... | K10 (C-2) |

(a) For holes more than five diameters deep, feed must be reduced. (b) Any premium high-speed steel (T15, M33, M41–M47, or S9, S10, S11, S12). Source: Metcut Research Associates Inc.

sulfur or chlorine are also preferred for nearly all machining operations on niobium, with the exception of turning, for which a soluble-oil emulsion is satisfactory.

Specific cutting fluids for niobium alloys are: general machining, PMC 9305; fine finishing, PMC 9318; grinding, PMC 9305; and threading, PMC 9205. When tungsten is heated for machining, no cutting fluid is used. When tungsten is not heated, a chlorinated cutting oil is preferred.

**Precautions.** To avoid chipping, flaking, and damage at breakthrough in machining molybdenum and tungsten, special practice and special tools are often required. The probability of cracking at corners and other surfaces in single-point cutting or milling is reduced by observing certain precautions:

- Machining to an external corner by cutting to the middle of a surface from opposite directions, especially when sharp corners are required, should be avoided. Climb milling, rather than conventional milling, should be used, so that the cutting forces are directed into the work metal. In conventional milling, work metal is likely to break at the point of entrance or exit of the cutter
- Thermoplastic supports should be used when it is necessary to machine to an external corner
- A large chamfer should be machined on the corner before the adjacent surface is machined (this chamfer should be large enough so that some metal is left after the turning cut is completed)
- Tools with high lead angles (about 45°) should be used, especially for roughing

Toxic materials can be generated in working with niobium in abrasive cutoff, surface grinding with a horizontal-spindle and reciprocating-table machine, cylindrical grinding, internal grinding, and centerless grinding. The same precaution that applies to niobium in these processes applies to the internal grinding of tungsten. Appropriate regulations are available from the National Institute for Occupational Safety and Health.

**Table 8 Spade drilling niobium, molybdenum, tantalum, and tungsten alloys (wrought, cast, and P/M)**

| Material | Hardness, HB | Condition | Speed, m/min (sfm) | Feed, mm/rev (in./rev) — Nominal hole diameter: 19–25 mm (¾-1 in.) | 38 mm (1½ in.) | 50 mm (2 in.) | 75 mm (3 in.) | Tool material grade, ISO (AISI or C-grade) |
|---|---|---|---|---|---|---|---|---|
| Niobium C103, C129Y, Nb-1Zr, Cb-752, FS-85, FS-291, WC-3015 | 170–225 | Stress relieved | 20 (65) | 0.15 (0.006) | 0.18 (0.007) | 0.25 (0.010) | 0.30 (0.012) | S9, S11(a) (T15, M42)(a) |
| Molybdenum Mo, Mo-50Re, TZC, TZM | 220–290 | Stress relieved | 24 (80) | 0.20 (0.008) | 0.25 (0.010) | 0.30 (0.012) | 0.40 (0.015) | S9, S11(a) (T15, M42)(a) |
| Tantalum ASTAR 811C, T-111, T-222, Ta-10W, Ta-Hf, Ta63 | 200–250 | Stress relieved | 12 (40) | 0.15 (0.006) | 0.18 (0.007) | 0.25 (0.010) | 0.30 (0.012) | S9, S11(a) (T15, M42)(a) |
| Tungsten, 85% density | 180–200 | Pressed and sintered | 53 (175) | 0.050 (0.002) | 0.075 (0.003) | 0.13 (0.005) | 0.15 (0.006) | K20 (C-2) |
| Tungsten, 93% density | 290–320 | Pressed and sintered | 46 (150) | 0.050 (0.002) | 0.075 (0.003) | 0.13 (0.005) | 0.15 (0.006) | K20 (C-2) |
| Tungsten, 96 and 100% density | 290–320 | Forged or arc cast | 46 (150) | 0.050 (0.002) | 0.075 (0.003) | 0.13 (0.005) | 0.15 (0.006) | K20 (C-2) |
| Tungsten-2 thoria | 260–320 | Pressed and sintered | 53 (175) | 0.050 (0.002) | 0.075 (0.003) | 0.13 (0.005) | 0.15 (0.006) | K20 (C-2) |

(a) Any premium high-speed steel (T15, M33, M41–M47, or S9, S10, S11, S12). Source: Metcut Research Associates Inc.

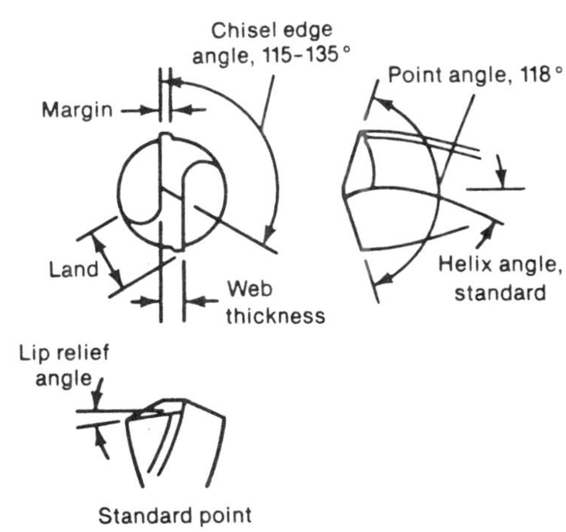


| Lip relief angle, degrees | Drill size |
|---|---|
| 20 | No. 80-No. 61 |
| 18 | No. 60-No. 41 |
| 16 | No. 41-No. 31 |
| 14 | 3.2–6.4 mm (⅛–¼ in.) |
| 12 | 6.4–9.5 mm (¼–⅜ in.) |
| 10 | 9.5–13 mm (⅜–½ in.) |
| 8 | 13–19 mm (½–¾ in.) |
| 7 | ≥25 mm (≥1 in.) |

**Fig. 2** Tool geometry for general-purpose high-speed steel twist drill for tungsten alloys (290 to 320 HB) and niobium, molybdenum, and tantalum alloys (170 to 290 HB)

## Table 9 Reaming niobium, molybdenum, tantalum, and tungsten alloys (wrought, cast, and P/M)

| | | | Roughing | | | | | | | |
|---|---|---|---|---|---|---|---|---|---|---|
| | | | | Feed, mm/rev (in./rev)(a) — Reamer diameter | | | | | | |
| Material | Hardness, HB | Condition | Speed, m/min (sfm) | 3 mm (1/8 in.) | 6 mm (1/4 in.) | 13 mm (1/2 in.) | 25 mm (1 in.) | 38 mm (1 1/2 in.) | 50 mm (2 in.) | Tool material grade, ISO (AISI or C-grade) |
| Niobium; C103, C128Y, Nb-1Zr, Cb-752, FS-85, FS-291, WC-3015 | 170–225 | Stress relieved | 27 (90) | 0.075 (0.003) | 0.10 (0.004) | 0.15 (0.006) | 0.20 (0.008) | 0.23 (0.009) | 0.25 (0.010) | S3, S4, S2 (M1, M2, M7) |
| | | | 37 (120) | 0.10 (0.004) | 0.13 (0.005) | 0.18 (0.007) | 0.23 (0.009) | 0.28 (0.011) | 0.30 (0.012) | K20 (C-2) |
| Molybdenum; Mo, Mo-50Re, TZC, TZM | 220–290 | Stress relieved | 15 (50) | 0.075 (0.003) | 0.15 (0.006) | 0.25 (0.010) | 0.30 (0.012) | 0.36 (0.014) | 0.40 (0.016) | S3, S4, S2 (M1, M2, M7) |
| | | | 37 (120) | 0.075 (0.003) | 0.15 (0.006) | 0.25 (0.010) | 0.30 (0.012) | 0.36 (0.014) | 0.40 (0.016) | K20 (C-2) |
| Tantalum; ASTAR 811C, T-111, T-222, Ta-l0W, Ta-HF, Ta63 | 200–250 | Stress relieved | 21 (70) | 0.075 (0.003) | 0.15 (0.006) | 0.20 (0.008) | 0.25 (0.010) | 0.30 (0.012) | 0.36 (0.014) | S3, S4, S2 (M1, M2, M7) |
| | | | 24 (80) | 0.075 (0.003) | 0.15 (0.006) | 0.20 (0.008) | 0.25 (0.010) | 0.30 (0.012) | 0.36 (0.014) | K20 (C-2) |
| Tungsten, 85% density | 180–200 | Pressed and sintered | 46 (150) | 0.025 (0.001) | 0.050 (0.002) | 0.075 (0.003) | 0.10 (0.004) | 0.13 (0.005) | 0.15 (0.006) | K20 (C-2) |
| Tungsten, 93% density | 290–320 | Pressed and sintered | 38 (125) | 0.025 (0.001) | 0.050 (0.002) | 0.075 (0.003) | 0.10 (0.004) | 0.13 (0.005) | 0.15 (0.006) | K20 (C-2) |
| Tungsten, 96 and 100% density | 290–320 | Forged or arc cast | 38 (125) | 0.025 (0.001) | 0.050 (0.002) | 0.075 (0.003) | 0.10 (0.004) | 0.13 (0.005) | 0.15 (0.006) | K20 (C-2) |

| | | | Finishing | | | | | | | |
|---|---|---|---|---|---|---|---|---|---|---|
| | | | | Feed, mm/rev (in./rev)(a) — Reamer diameter | | | | | | |
| Material | Hardness, HB | Condition | Speed, m/min (sfm) | 3 mm (1/8 in.) | 6 mm (1/4 in.) | 13 mm (1/2 in.) | 25 mm (1 in.) | 38 mm (1 1/2 in.) | 50 mm (2 in.) | Tool material grade, ISO (AISI or C-grade) |
| Niobium; C103, C128Y, Nb-1Zr, Cb-752, FS-85, FS-291, WC-3015 | 170–225 | Stress relieved | 12 (40) | 0.075 (0.003) | 0.10 (0.004) | 0.15 (0.006) | 0.20 (0.008) | 0.23 (0.009) | 0.25 (0.010) | S3, S4, S2 (M1, M2, M7) |
| | | | 18 (60) | 0.10 (0.004) | 0.13 (0.005) | 0.18 (0.007) | 0.23 (0.009) | 0.28 (0.011) | 0.30 (0.012) | K20 (C-2) |
| Molybdenum; Mo, Mo-50Re, TZC, TZM | 220–290 | Stress relieved | 9 (30) | 0.075 (0.003) | 0.15 (0.006) | 0.25 (0.010) | 0.30 (0.012) | 0.36 (0.014) | 0.40 (0.016) | S3, S4, S2 (M1, M2, M7) |
| | | | 12 (40) | 0.075 (0.003) | 0.15 (0.006) | 0.25 (0.010) | 0.30 (0.012) | 0.36 (0.014) | 0.40 (0.016) | K20 (C-2) |
| Tantalum; ASTAR 811C, T-111, T-222, Ta-l0W, Ta-HF, Ta63 | 200–250 | Stress relieved | 11 (35) | 0.075 (0.003) | 0.15 (0.006) | 0.20 (0.008) | 0.25 (0.010) | 0.30 (0.012) | 0.36 (0.014) | S3, S4, S2 (M1, M2, M7) |
| | | | 15 (50) | 0.075 (0.003) | 0.15 (0.006) | 0.20 (0.008) | 0.25 (0.010) | 0.30 (0.012) | 0.36 (0.014) | K20 (C-2) |

(a) Based on four flutes for 3 and 6 mm (1/8 and 1/4 in.) reamers; six flutes for 13 mm (1/2 in.) reamers and eight flutes for 25 mm (1 in.) and larger reamers. Source: Metcut Research Associates Inc.

## Table 10 Tapping niobium and molybdenum alloys (wrought, cast, and P/M)

| | | | Speed, m/min (sfm)(a) — Pitch, threads/1 in. | | | | |
|---|---|---|---|---|---|---|---|
| Material | Hardness, HB | Condition | ≤7 | 8–15 | 16–24 | >24 | High-speed steel tool material grade, ISO (AISI) |
| Niobium; C103, C129Y, Nb-1Zr, Cb-752, FS-85, FS-291, WC-3015 | 170–225 | Stress relieved | 0.9 (3) | 2.1 (7) | 2.7 (9) | 3 (10) | S2, S3 (M10, M7, M1) |
| Molybdenum; Mo, Mo-50Re, TZC, TZM | 220–290 | Stress relieved | 5 (15) | 9 (30) | 14 (45) | 15 (50) | S2, S3 (M10, M7, M1) |

(a) These speeds are for tapping 65 to 75% threads in shallow through holes. Reduce the speed when tapping deep holes, blind holes, or higher percentage of thread. Source: Metcut Research Associates Inc.

## Table 11 Hollow milling niobium, molybdenum, and tantalum alloys (wrought, cast, and P/M)

| Material(a) | Hardness, HB | Condition | Speed, m/min (sfm) | Feed, mm/rev (in./rev) | Tool material grade, ISO (AISI or C-grade) |
|---|---|---|---|---|---|
| Niobium; C103, C129Y, Nb-1Zr, Cb-752, FS-85, FS-291, WC-3015 | 170–225 | Stress relieved | 12 (40) | 0.10 (0.004) | S4, S2 (M2, M7) |
| | | | 34 (110) | 0.20 (0.008) | K20 (C-2) |
| Molybdenum; Mo, Mo-50Re, TZC, TZM | 220–290 | Stress relieved | 34 (110) | 0.20 (0.008) | K20 (C-2) |
| Tantalum; ASTAR 811C, T-111, T-222, Ta-l0W, Ta-Hf, Ta63 | 200–250 | Stress relieved | 9 (30) | 0.075 (0.003) | S9, S11 (T15, M42)(b) |
| | | | 24 (80) | 0.15 (0.006) | K20 (C-2) |

(a) Recommendations based on a stock allowance of 6.5 mm (0.250 in.) on the diameter. (b) Any premium high-speed steel (T15, M33, M41–M47, or S9, S10, S11, S12). Source: Metcut Research Associates Inc.

Both materials are also potential fire hazards in the same operations. Caution is advised in grinding and disposing of swarf.

## Turning, Boring, and Trepanning

Metal removal rates for turning refractory metals range from about 5 to 20% of those for normalized 4130 steel, and about the same as those for nickel-base and cobalt-base heat-resisting alloys. The cost of turning refractory metals runs from 5 to 30 times greater than that for 4130 steel at 200 HB. However, such comparisons tend to be conservative. Experience may suggest ways to significantly increase such factors as depth of cut and metal removal rate.

**Niobium and Tantalum.** High side rake on tools is recommended for turning of both metals. Turning practice varies considerably from shop to shop. Sometimes better results are obtained by significantly lowering speeds and significantly increasing feed rate. Regardless of the technique, smooth finishes are difficult to obtain, and finishing by grinding may be required.

Nominal speeds and feeds for turning niobium and tantalum with single-point and box tools are given in Table 1. Speeds and feeds for turning niobium with cutoff and form tools are given in Table 2. The geometry of a single-point tool for turning refractory metals is shown in Fig. 1.

Data for the boring, counterboring and spotfacing, and trepanning of niobium and tantalum are presented in Tables 3, 4, and 5, respectively.

**Molybdenum.** This metal is extremely abrasive, and all machining operations result in high rates of wear. Carbide tools with high side rake are preferred for turning and boring.

Nominal speeds and feeds for turning molybdenum with single-point and box tools are given in Table 1. The same information for turning with cutoff and form tools is found in Table 2.

Speeds and feeds for boring, for counterboring and spotfacing, and for trepanning are given in Tables 3, 4, and 5, respectively.

**Example.** A shop with considerable experience in turning molybdenum recommends:

| Factor | Roughing | Finishing |
|---|---|---|
| Cutting tool | 94WC-6Co | 94WC-6Co |
| Speed, m/min (sfm) | 30–60 (100–200) | 90–180 (300–600) |
| Feed, mm/rev (in./rev) | 0.15–0.20 (0.006–0.008) | 0.15–0.20 (0.006–0.008) |
| Depth of cut, mm (in.) | 3–6 (⅛–¼) | 0.40–0.50 (0.015–0.020) |
| Tool geometry, degrees | | |
| Back rake | 5–10 | 0 |
| Side rake | 12 | 12 |
| Clearance | 10 | 10 |

This shop recommends a chip curler for both roughing and finishing cuts.

**Tungsten.** Nominal speeds and feeds for turning tungsten with single-point and box tools are given in Table 1. Materials include pressed-and-sintered tungsten of 85 and 93% density, pressed-and-sintered tungsten-2 thoria, and forged or arc cast tungsten of 96 and 100% density. The same information for boring in Table 3 covers the same materials. Parameters for counterboring and spotfacing tungsten with 85 and 93% density are given in Table 4.

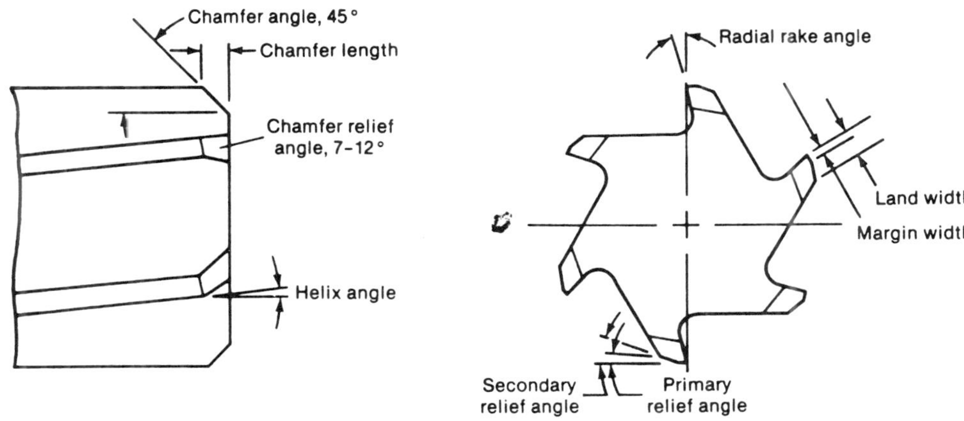


| Diameter, mm | Diameter, in. | Margin width, mm | Margin width, in. | Primary radial relief angle, degrees |
|---|---|---|---|---|
| <3.2 | <⅛ | 0.1–0.15 | 0.004–0.006 | 20–25 |
| 3.2–6.4 | ⅛–¼ | 0.15–0.2 | 0.006–0.008 | 15–20 |
| 6.4–13 | ¼–½ | 0.2–0.25 | 0.008–0.010 | 11–14 |
| 13–19 | ½–¾ | 0.25–0.38 | 0.010–0.015 | 8–12 |
| 19–25 | ¾–1 | 0.30–0.43 | 0.012–0.017 | 7–10 |
| 25–38 | 1–1½ | 0.36–0.46 | 0.014–0.018 | 5–8 |
| 38–50 | 1½–2 | 0.41–0.56 | 0.016–0.022 | 5–8 |
| >50 | >2 | 0.46–0.64 | 0.018–0.025 | 5–8 |

**Fig. 3** Tool geometry for high-speed steel reamer

## Parameters for Drilling the Refractory Metals

Except for tungsten, differences in the drilling properties of refractory alloys and the soft steels are less than those in turning. Niobium, tantalum, and molybdenum are easier to drill than the heat-resisting alloys or steel at a hardness of 50 to 55 HRC. Success in drilling these alloys depends primarily on rigidity of setup and drill design. Problems are more common with portable drills and small holes because of the difficulty of maintaining adequate rigidity and positive feed.

**Niobium.** Accurately ground split-point drills should be used to minimize rubbing and accelerated wear at drill margins. Speeds and feeds for drilling, oil hole or pressurized-coolant drilling, and spade drilling are presented in Tables 6, 7, and 8, respectively. One manufacturer has drilled holes as small as 0.35 mm (0.0135 in.) in a niobium C-103 alloy.

Standard high-speed drills, ground to normal angles, may be used, but peripheral lands wear badly, and care must be taken to see that the drill does not wear undersize.

The high-speed steel twist drill, the geometry of which is shown in Fig. 2, can be used for all four of the refractory metals. Highly active sulfurized or chlorinated cutting oils should be used.

**Molybdenum.** This metal is easier to drill than most other refractory metals. Any of the standard high-speed steels, such as M1, M3, and M10, are suitable for drills, under normal conditions. The standard 118° point

**Table 12 Face milling niobium, molybdenum, tantalum, and tungsten alloys (wrought, cast, and P/M)**

| Materials | Hardness, HB | Condition | Depth of cut, mm (in.)(a) | High-speed steel tool: Speed, m/min (sfm) | High-speed steel tool: Feed, mm/tooth (in./tooth) | High-speed steel tool: Tool material grade, ISO (AISI) | Uncoated carbide tool: Speed, m/min (sfm), Brazed | Uncoated carbide tool: Speed, m/min (sfm), Indexable | Uncoated carbide tool: Feed, mm/tooth (in./tooth) | Tool material grade, ISO (C-grade) |
|---|---|---|---|---|---|---|---|---|---|---|
| Niobium C103, C129Y, Nb-1Zr, Cb-752, FS-85, FS-291, WC-3015 | 170–225 | Stress relieved | 1 (0.040) | 30 (100) | 0.13 (0.005) | S9, S11(b) (T15, M42)(b) | 46 (150) | 50 (165) | 0.15 (0.006) | K10, M20 (C-2) |
| | | | 4 (0.150) | 24 (80) | 0.18 (0.007) | S9, S11(b) (T15, M42)(b) | 43 (140) | 47 (155) | 0.20 (0.008) | K10, M20 (C-2) |
| Molybdenum Mo, Mo-Re, TZC, TZM | 220–290 | Stress relieved | 1 (0.040) | 30 (100) | 0.15 (0.006) | S9, S11(b) (T15, M42)(b) | 78 (255) | 84 (275) | 0.10 (0.004) | K10, M20 (C-2) |
| | | | 4 (0.150) | 18 (60) | 0.20 (0.008) | S9, S11(b) (T15, M42)(b) | 69 (225) | 76 (250) | 0.13 (0.005) | K10, M20 (C-2) |
| Tantalum ASTAR 811C, T-111, T-222, Ta-l0W, Ta-Hf, Ta-63 | 220–250 | Stress relieved | 1 (0.040) | 23 (75) | 0.25 (0.010) | S9, S11(b) (T15, M42)(b) | ... | ... | ... | ... |
| | | | 4 (0.150) | 12 (40) | 0.25 (0.010) | S9, S11(b) (T15, M42)(b) | ... | ... | ... | ... |
| Tungsten, 85% density | 180–200 | Pressed and sintered | 1 (0.040) | ... | ... | ... | 30 (100) | 34 (110) | 0.20 (0.008) | K01 (C-3) |
| | | | 4 (0.150) | ... | ... | ... | 24 (80) | 27 (90) | 0.25 (0.010) | K01 (C-3) |

(a) Depth of cut is measured parallel to axis of cutter. (b) Any premium high-speed steel (T15, M33, M41–47, or S9, S10, S11, S12). Source: Metcut Research Associates Inc.

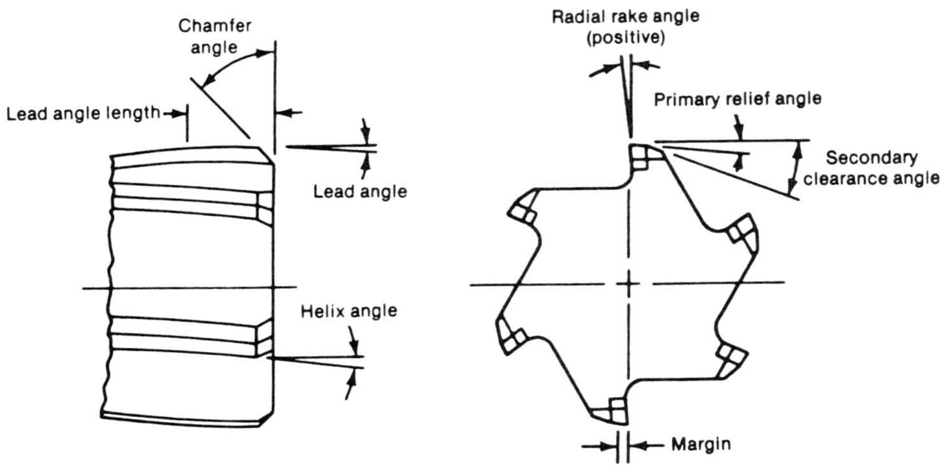


| Diameter, mm (in.) | Margin or circular land, mm (in.) | Primary radial relief angle, degrees | Secondary clearance angle, degrees | Length of chamfer, mm (in.) | Chamfer relief angle, degrees Primary | Secondary |
|---|---|---|---|---|---|---|
| 6.4 (¼) | 0.13–0.175 (0.005–0.007) | 12–15 | 28–32 | 0.5–0.75 (0.020–0.030) | 10–12 | 25–28 |
| 9.5 (⅜) | 0.13–0.20 (0.005–0.008) | 12–15 | 28–32 | 0.5–0.75 (0.020–0.030) | 10–12 | 22–24 |
| 13 (½) | 0.20–0.25 (0.008–0.010) | 11–13 | 26–28 | 0.75–1.0 (0.030–0.040) | 9–11 | 20–22 |
| 16 (⅝) | 0.23–0.30 (0.009–0.012) | 10–12 | 22–26 | 1.0–1.3 (0.040–0.050) | 8–10 | 18–20 |
| 19 (¾) | 0.25–0.38 (0.010–0.015) | 9–10 | 20–22 | 1.0–1.3 (0.040–0.050) | 7–9 | 14–18 |
| 22 (⅞) | 0.25–0.38 (0.010–0.015) | 9–10 | 18–20 | 1.0–1.3 (0.040–0.050) | 7–9 | 14–18 |
| 25 (1) | 0.30–0.38 (0.012–0.015) | 8–10 | 18–20 | 1.3–1.5 (0.050–0.060) | 7–9 | 14–18 |
| 38 (1½) | 0.36–0.41 (0.014–0.016) | 7–8 | 16–18 | 1.3–1.5 (0.050–0.060) | 6–8 | 12–16 |
| 50 (2) | 0.46–0.58 (0.018–0.023) | 7–8 | 16–18 | 1.3–1.5 (0.050–0.060) | 6–8 | 12–16 |

**Fig. 4** Tool geometry for carbide reamer

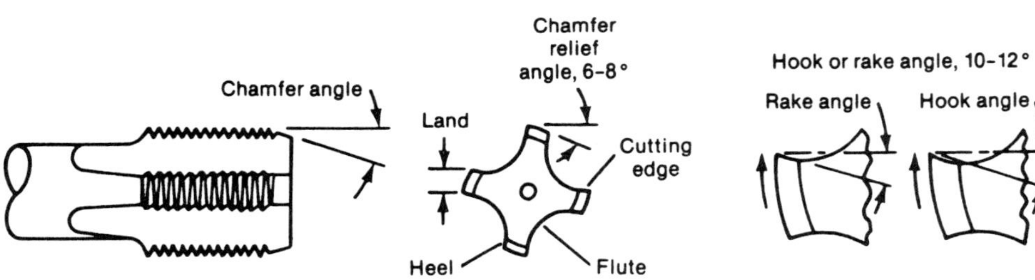


**Fig. 5** Tool geometry of high-speed steel (170 to 290 HB) taps for niobium, molybdenum, and tantalum alloys. A two-flute spiral-point tap is used for a through hole, and a four-flute hand tap is used for a blind hole. Special taps for these alloys are also available.

chisel edge (plain) drill performs as well as drills with special points (except sometimes at breakthrough).

Drill thrust is low in drilling molybdenum—approximately the same as in drilling pearlitic cast iron. Higher feed rates may be used without damaging the drill. However, when expensive metals are drilled, avoidance of damage to the workpiece, rather than maximum tool life, should receive primary consideration.

Chipping of the workpiece during breakthrough of the drill is a difficulty encountered in drilling the more brittle forms of molybdenum. Because 20 to 30% of the total thrust required by the standard drill results from the inefficient chisel edge at the point, damage to workpieces at breakthrough can sometimes be decreased by substituting a split-point drill for the chisel edge drill. Decreasing the feed rate is less efficient in reducing thrust, and it causes poor and erratic drill life. The use of sacrificial workpiece supports and prechamfering of the breakthrough area also help to prevent chipping at breakthrough.

Parameters for drilling, oil hole or pressurized-coolant drilling, and spade drilling of molybdenum are presented in Tables 6, 7, and 8, respectively.

**Tantalum.** Practice and results in drilling tantalum are influenced by the continuous, stringy chips formed. Because chips are likely to clog the holes, the usable range of cutting speeds for drilling small holes is closely related to the depth of the hole. Small-diameter holes up to one diameter deep can be drilled at 12 to 15 m/min (40 to 50 sfm). For holes more than one diameter deep, speed should be reduced to 6 to 12 m/min (20 to 40 sfm).

Hole quality is affected by the drill point and the feed rate. Although drills with crankshaft or split points do not have longer life, they require less thrust force, thus helping to prevent the formation of burrs and extruded edges at the exit side of through holes. Drill feed should not exceed 0.050 mm/rev (0.002 in./rev) for drills up to 9.5 mm (⅜ in.) in diameter. Heavier feeds, although usable from the standpoint of long-

**Table 13 Side and slot milling niobium, molybdenum, and tantalum alloys (wrought, cast, and P/M) with arbor-mounted cutters**

| Material | Hardness, HB | Condition | Depth of cut, mm (in.)(a) | High-speed steel tool: Speed, m/min (sfm) | Feed, mm/tooth (in./tooth) | Tool material grade, ISO (AISI)(b) | Uncoated carbide tool: Speed, m/min (sfm) Brazed | Indexable | Feed, mm/tooth (in./tooth) | Tool material grade, ISO (C-grade) |
|---|---|---|---|---|---|---|---|---|---|---|
| Niobium | 170–225 | Stress relieved | 1 (0.040) | 41 (135) | 0.13 (0.005) | S9, S11 (T15, M42) | 46 (150) | 55 (180) | 0.13 (0.005) | K10, M20 (C-2) |
| C103, C129Y, Nb-1Zr, Cb-752, | | | 4 (0.150) | 38 (125) | 0.15 (0.006) | S9, S11 (T15, M42) | 43 (140) | 52 (170) | 0.15 (0.006) | K10, M20 (C-2) |
| FS-85, FS-291, WC-3015 | | | 8 (0.300) | 35 (115) | 0.18 (0.007) | S9, S11 (T15, M42) | 40 (130) | 49 (160) | 0.18 (0.007) | K10, M20 (C-2) |
| Molybdenum | 220–290 | Stress relieved | 1 (0.040) | 35 (115) | 0.13 (0.005) | S9, S11 (T15, M42) | 60 (200) | 73 (240) | 0.18 (0.007) | K10, M20 (C-2) |
| Mo, Mo-50Re, TZC, TZM | | | 4 (0.150) | 34 (110) | 0.15 (0.006) | S9, S11 (T15, M42) | 56 (185) | 69 (225) | 0.20 (0.008) | K10, M20 (C-2) |
| | | | 8 (0.300) | 30 (100) | 0.18 (0.007) | S9, S11 (T15, M42) | 52 (170) | 62 (205) | 0.23 (0.009) | K10, M20 (C-2) |
| Tantalum | 200–250 | Stress relieved | 1 (0.040) | 23 (75) | 0.10 (0.004) | S9, S11 (T15, M42) | ... | ... | ... | ... |
| ASTAR 811C, T-111, T-222, | | | 4 (0.150) | 20 (65) | 0.13 (0.005) | S9, S11 (T12, M42) | ... | ... | ... | ... |
| Ta-l0W, Ta-Hf, Ta63 | | | 8 (0.300) | 17 (55) | 0.13 (0.005) | S9, S11 (T15, M42) | ... | ... | ... | ... |

(a) Axial depth of cut for side milling; radial depth of cut for slot milling. (b) Any premium high-speed steel (T15, M33, M41–M47, or S9, S10, S11, S12). Source: Metcut Research Associates Inc.

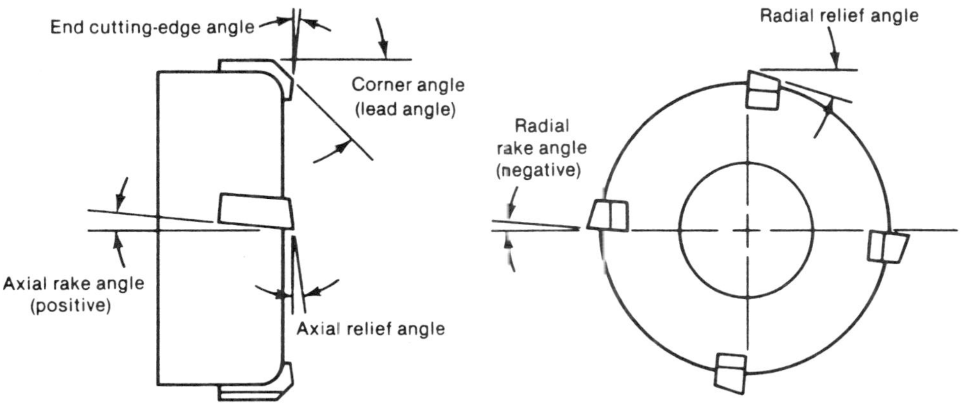


| Material | Hardness, HB | High-speed steel Axial rake angle, degrees | High-speed steel Radial rake angle, degrees | Indexable carbide Axial rake angle, degrees | Indexable carbide Radial rake angle, degrees | Brazed carbide Axial rake angle, degrees | Brazed carbide Radial rake angle, degrees | Corner angle, degrees | End cutting edge angle, degrees | Axial relief angle, degrees | Radial relief angle, degrees |
|---|---|---|---|---|---|---|---|---|---|---|---|
| Niobium alloys, wrought, cast, P/M | 170–225 | 0 | 20 | 5–7 | 0–5 | 0 | 10 | 45 | 5–10 | 10 | 10 |
| Molybdenum alloys, wrought, cast, P/M | 220–290 | 0 | 20 | 5–7 | 0–5 | 0 | 0 | 45 | 5–10 | 10 | 10 |
| Tantalum alloys, wrought, cast, P/M | 200–250 | 0 | 20 | 5–7 | 0–5 | 0 | 0 | 45 | 5–10 | 10 | 10 |
| Tungsten alloys, wrought, cast, P/M | 180–320 | ... | ... | −4 to −8 | −3 to −11 | −5 | 0 | 45 | 5–10 | 15 | 15 |

**Fig. 6** Tool geometry for face mill

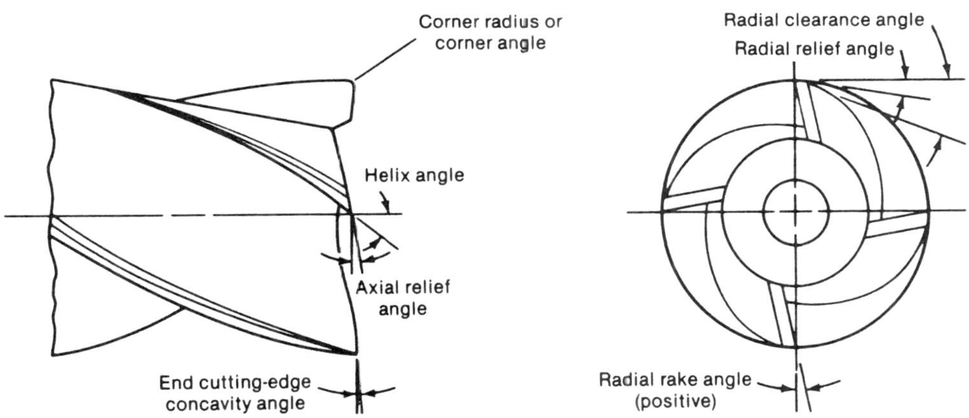


**Fig. 7** Tool geometry for peripheral end mill

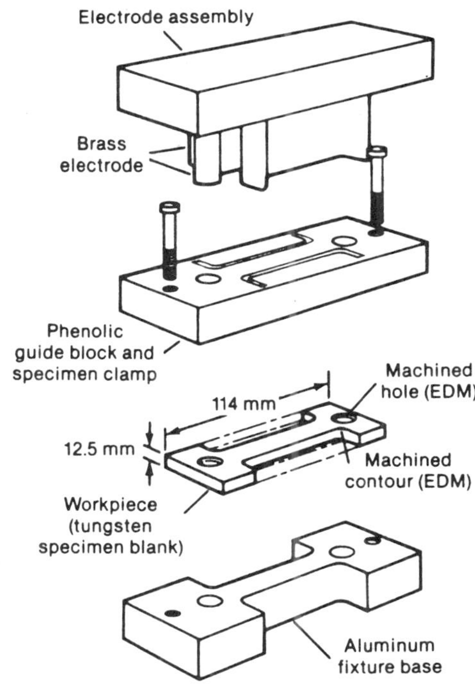


**Operations**

1. Saw 12.5 mm (0.50 in.) thick tungsten plate into blanks, using a diamond-edge band saw
2. Load blank into fixture and close components of assembly
3. Electrical discharge machine contour and two holes, using 30 A gap current (time required, 5 to 7 min/specimen
4. Unload specimen, polish edges, and ream holes

**Fig. 8** Setup for producing a tungsten test specimen by electrical discharge machining

er drill life, will produce less accurate holes with poor surface finish.

**Tungsten.** This metal is more difficult to drill than any of the other refractory metals. Because tungsten begins a brittle-to-ductile transition at about 200 °C (400 °F), the workpiece is often heated to 200 to 425 °C (400 to 800 °F) for machining. However, acceptable holes can be drilled on a production basis at room temperature if the proper procedure is followed and precautions are taken. Parameters for drilling, oil hole or pressurized-coolant drilling, and spade drilling are given in Tables 6, 7, and 8, respectively.

Because of possible damage to the workpiece and shortened drill life, design and manufacturing alternatives to avoid the drilling of holes should be considered. The major problems encountered in conventional drilling of high-density unalloyed tungsten are:

- Chipping at hole edges on the entrance side
- Spalling on the exit side of through holes
- Radial cracking, extending from holes to nearby discontinuities
- Drill fracture, caused by excessive force, nonrigid setup, or chip interference in drill flutes

Recommended procedures for drilling tungsten are:

- Use solid carbide drills that are accurately ground (drill lip runout should be less than 0.025 mm, or 0.001 in.)
- To prevent spalling at breakthrough when drilling holes, use either a tight-fitting backup plate of steel or similar metal clamped over the surface where breakthrough occurs, or drill holes halfway through from each side
- Drill holes no closer than one drill diameter from any edge or other discontinuity
- Direct a heavy stream of cutting fluid into the hole to flush away chips. Retract the drill and flush out chips at least once for holes deeper than two diameters
- Locate and start holes with a carbide center drill. Avoid use of drill bushings because they interfere with chip flow and cause drill fracture
- Replace drills at the first indication of chipping at hole edges or when the wear land at the drill corners reaches 0.40 mm (0.016 in.)

## Reaming and Tapping of Refractory Metals

Nominal speeds and feeds for roughing and finishing reaming of niobium, molybdenum, tantalum, and tungsten are presented in Table 9. Carbide tools are most frequently used, but high-speed steel reamers can

## Table 14 Peripheral end milling niobium, molybdenum, and tantalum alloys (wrought, cast, and P/M)

| Material | Hardness, HB | Condition | Radial depth of cut, mm (in.)(a) | High-speed steel tool: Speed, m/min (sfm) | Feed, mm/tooth (in./tooth), Cutter diameter: 10 mm (3/8 in.) | 13 mm (1/2 in.) | 19 mm (3/4 in.) | 25–50 mm (1–2 in.) | Tool material grade, ISO (AISI) |
|---|---|---|---|---|---|---|---|---|---|
| Niobium | 170–225 | Stress relieved | 0.5 (0.020) | 30 (100) | 0.038 (0.0015) | 0.050 (0.002) | 0.064 (0.0025) | 0.075 (0.003) | S9(b) (T15)(b) |
| C103, C129Y, Nb-1Zr, Cb-752, | | | 1.5 (0.060) | 24 (80) | 0.050 (0.002) | 0.075 (0.003) | 0.075 (0.003) | 0.10 (0.004) | S9(b) (T15)(b) |
| FS-85, FS-291, WC-3015 | | | diam/4 (diam/4) | 21 (70) | 0.038 (0.0015) | 0.050 (0.002) | 0.064 (0.0025) | 0.089 (0.0035) | S9(b) (T15)(b) |
| Molybdenum | 220–290 | Stress relieved | 0.5 (0.020) | 60 (200) | 0.025 (0.001) | 0.025 (0.001) | 0.038 (0.0015) | 0.050 (0.002) | S9(b) (T15)(b) |
| Mo, Mo-50Re, TZC, TZM | | | 1.5 (0.060) | 52 (170) | 0.050 (0.002) | 0.075 (0.003) | 0.10 (0.004) | 0.10 (0.004) | S9(b) (T15)(b) |
| | | | diam/4 (diam/4) | 43 (140) | 0.025 (0.001) | 0.050 (0.002) | 0.050 (0.002) | 0.075 (0.003) | S9(b) (T15)(b) |
| Tantalum | 200–250 | Stress relieved | 0.5 (0.020) | 20 (65) | 0.025 (0.001) | 0.038 (0.0015) | 0.050 (0.002) | 0.064 (0.0025) | S4, S5, S2 (M2, M3, M7) |
| ASTAR 811C, T-111, T-222, Ta-l0W, Ta-Hf, Ta-63 | | | 1.5 (0.060) | 15 (50) | 0.050 (0.002) | 0.050 (0.002) | 0.075 (0.003) | 0.10 (0.004) | S4, S5, S2 (M2, M3, M7) |
| | | | diam/4 (diam/4) | 14 (45) | 0.025 (0.001) | 0.038 (0.0015) | 0.064 (0.0025) | 0.075 (0.003) | S4, S5, S2 (M2, M3, M7) |
| | | | diam/2 (diam/2) | 11 (35) | 0.018 (0.0007) | 0.025 (0.001) | 0.050 (0.002) | 0.064 (0.0025) | S4, S5, S2 (M2, M3, M7) |

| Material | Hardness, HB | Condition | Radial depth of cut, mm (in.)(a) | Carbide tool: Speed, m/min (sfm) | Feed, mm/tooth (in./tooth), Cutter diameter: 10 mm (3/8 in.) | 13 mm (1/2 in.) | 19 mm (3/4 in.) | 25–50 mm (1–2 in.) | Tool material grade, ISO (C-grade) |
|---|---|---|---|---|---|---|---|---|---|
| Niobium | 170–225 | Stress relieved | 0.5 (0.020) | 53 (175) | 0.038 (0.0015) | 0.050 (0.002) | 0.064 (0.0025) | 0.075 (0.003) | K20, M20 (C-2) |
| C103, C129Y, Nb-1Zr, Cb-752, | | | 1.5 (0.060) | 46 (150) | 0.050 (0.002) | 0.075 (0.003) | 0.075 (0.003) | 0.10 (0.004) | K20, M20 (C-2) |
| FS-85, FS-291, WC-3015 | | | diam/4 (diam/4) | | | | | | |
| Molybdenum | 220–290 | Stress relieved | 0.5 (0.020) | 100 (325) | 0.025 (0.001) | 0.025 (0.001) | 0.038 (0.0015) | 0.050 (0.002) | K20, M20 (C-2) |
| Mo, Mo-50Re, TZC, TZM | | | 1.5 (0.060) | 90 (300) | 0.050 (0.002) | 0.075 (0.003) | 0.10 (0.004) | 0.10 (0.004) | K20, M20 (C-2) |
| | | | diam/4 (diam/4) | | | | | | |
| Tantalum | 200–250 | Stress relieved | 0.5 (0.020) | ... | ... | ... | ... | ... | ... |
| ASTAR 811C, T-111, T-222, | | | 1.5 (0.060) | ... | ... | ... | ... | ... | ... |
| Ta-l0W, Ta-Hf, Ta-63 | | | diam/4 (diam/4) | ... | ... | ... | ... | ... | ... |
| | | | diam/2 (diam/2) | ... | ... | ... | ... | ... | ... |

(a) For standard-length end mills, maximum axial depth can be up to 1.5 times the cutter diameter. (b) Any premium high-speed steel (T15, M33, M41–M47, or S9, S10, S11, S12). Source: Metcut Research Associates Inc.

## Table 15 End milling-slotting niobium, molybdenum, tantalum, and tungsten alloys (wrought, cast, and P/M)

| Material | Hardness, HB | Condition | Axial depth of cut, mm (in.) | Speed, m/min (sfm) | Feed, mm/tooth (in./tooth), Width of slot: 10 mm (3/8 in.) | 13 mm (1/2 in.) | 19 mm (3/4 in.) | 25–50 mm (1–2 in.) | HSS tool material except as noted |
|---|---|---|---|---|---|---|---|---|---|
| Niobium | 170–225 | Stress relieved | 0.75 (0.030) | 18 (60) | 0.018 (0.0007) | 0.025 (0.001) | 0.038 (0.0015) | 0.050 (0.002) | S9, S11(a) (T15, M42)(a) |
| C103, C129Y, Nb-1Zr, Cb-752, FS-85, FS-291, WC-3015 | | | 1.5 (0.060) | 17 (55) | 0.013 (0.0005) | 0.038 (0.0015) | 0.050 (0.002) | 0.075 (0.003) | S9, S11(a) (T15, M42)(a) |
| | | | diam/4 (diam/4) | 15 (50) | ... | 0.025 (0.001) | 0.038 (0.0015) | 0.050 (0.002) | S9, S11(a) (T15, M42)(a) |
| Molybdenum | 220–290 | Stress relieved | 0.75 (0.030) | 35 (115) | 0.018 (0.0007) | 0.025 (0.001) | 0.038 (0.0015) | 0.050 (0.002) | S9, S11(a) (T15, M42)(a) |
| Mo, Mo-50Re, TZC, TZM | | | 1.5 (0.060) | 34 (110) | 0.013 (0.0005) | 0.038 (0.0015) | 0.050 (0.002) | 0.075 (0.003) | S9, S11(a) (T15, M42)(a) |
| | | | diam/4 (diam/4) | 32 (105) | ... | 0.025 (0.001) | 0.038 (0.0015) | 0.050 (0.002) | S9, S11(a) (T15, M42)(a) |
| Tantalum | 200–250 | Stress relieved | 0.75 (0.030) | 11 (35) | 0.018 (0.0007) | 0.025 (0.001) | 0.038 (0.0015) | 0.050 (0.002) | S4, S5, S2, (M2, M3, M7) |
| ASTAR 811C, T-111, T-222, Ta-l0W, Ta-Hf, Ta-63 | | | 1.5 (0.060) | 9 (30) | ... | 0.025 (0.001) | 0.050 (0.002) | 0.075 (0.003) | S4, S5, S2 (M2, M3, M7) |
| | | | diam/4 (diam/4) | 8 (25) | ... | ... | 0.038 (0.0015) | 0.050 (0.002) | S4, S5, S2 (M2, M3, M7) |
| Tungsten, 85% density | 180–200 | Pressed and sintered | 0.75 (0.030) | 69 (225) | ... | ... | ... | 0.075 (0.003) | K10 carbide (C-3 carbide) |
| | | | 1.5 (0.060) | 60 (200) | ... | ... | ... | 0.075 (0.003) | K10 carbide (C-3 carbide) |
| Tungsten, 93% density | 290–320 | Pressed and sintered | 0.75 (0.030) | 69 (225) | ... | ... | ... | 0.075 (0.003) | K10 carbide (C-3 carbide) |
| | | | 3 (0.125) | 60 (200) | ... | ... | ... | 0.075 (0.003) | K10 carbide (C-3 carbide) |
| Tungsten, 96 and 100% density | 290–320 | Forged or arc cast | 0.75 (0.030) | 69 (225) | ... | ... | ... | 0.075 (0.003) | K10 carbide (C-3 carbide) |
| | | | 3 (0.125) | 60 (200) | ... | ... | ... | 0.075 (0.003) | K10 carbide (C-3 carbide) |
| Tungsten-2 thoria | 260–320 | Pressed and sintered | 0.75 (0.030) | 30 (100) | ... | ... | ... | 0.050 (0.002) | K10 carbide (C-3 carbide) |
| | | | 3 (0.125) | 27 (90) | ... | ... | ... | 0.050 (0.002) | K10 carbide (C-3 carbide) |

(a) Any premium high-speed steel (T15, M33, M41). Source: Metcut Research Associates Inc.

**Table 16 Thread milling niobium alloys (wrought, cast, and P/M)**

| Material | Hardness, HB | Condition | Speed, m/min (sfm) | Feed, mm/tooth/rev (in./tooth/rev) | HSS tool material grade, ISO (AISI) |
|---|---|---|---|---|---|
| Niobium C103, C129Y, Nb-1Zr, Cb-752, FS-85, FS-291, WC-3015 | 170–225 | Stress relieved | 18 (60) | 0.038 (0.0015) | S4, S2 (M2, M7) |

Source: Metcut Research Associates Inc.

**Table 17 Power hacksawing niobium, molybdenum, and tantalum alloys (wrought, cast, and P/M) with a high-speed steel blade**

| Material | Hardness, HB | Condition | Material thickness, mm (in.) | Pitch, teeth/1 in. | Speed, strokes/min | Feed: Positive, mm/stroke (in./stroke) | Feed: Pressure(a) |
|---|---|---|---|---|---|---|---|
| Niobium | 170–225 | Stress relieved | <6 (<¼) | 10 | 60 | 0.075 (0.003) | L |
| C103, C129Y, Nb-1Zr, | | | 6–19 (¼–¾) | 10 | 60 | 0.075 (0.003) | L |
| Cb-752, FS-85, FS-291, | | | 19–50 (¾–2) | 6 | 55 | 0.15 (0.006) | M |
| WC-3015 | | | >50 (>2) | 4 | 55 | 0.15 (0.006) | M |
| Molybdenum | 220–290 | Stress relieved | <6 (<¼) | 10 | 100 | 0.075 (0.003) | L |
| Mo, Mo-50Re, | | | 6–19 (¼–¾) | 10 | 100 | 0.075 (0.003) | L |
| TZC, TZM | | | 19–50 (¾–2) | 6 | 90 | 0.15 (0.006) | M |
| | | | >50 (>2) | 4 | 90 | 0.15 (0.006) | M |
| Tantalum | 200–250 | Stress relieved | <6 (<¼) | 10 | 55 | 0.075 (0.003) | L |
| ASTAR 811C, | | | 6–19 (¼–¾) | 10 | 55 | 0.075 (0.003) | L |
| T-111, T-222, Ta-l0W, | | | 19–50 (¾–2) | 5 | 50 | 0.15 (0.006) | M |
| Ta-Hf, Ta-63 | | | >50 (>2) | 4 | 50 | 0.15 (0.006) | M |

(a) L, light; M, medium. Source: Metcut Research Associates Inc.

**Table 18 Power band sawing niobium, molybdenum, and tantalum alloys (wrought, cast, and P/M) with a high-speed steel precision tooth form blade**

| Material | Hardness, HB | Condition | Material thickness, mm (in.) | Pitch, teeth/1 in. | Band speed, m/min (sfm) |
|---|---|---|---|---|---|
| Niobium | 170–225 | Stress relieved | <13 (<½) | 14 | 30 (100) |
| C103, C129Y, Nb-1Zr, | | | 13–25 (½–1) | 10 | 27 (90) |
| Cb-752, FS-85, FS-291, | | | 25–75 (1–3) | 8 | 23 (75) |
| WC-3015 | | | >75 (>3) | 6 | 20 (65) |
| Molybdenum | 220–290 | Stress relieved | <13 (<½) | 14 | 37 (120) |
| Mo, Mo-50Re, TZC, TZM | | | 13–25 (½–1) | 10 | 34 (110) |
| | | | 25–75 (1–3) | 8 | 27 (90) |
| | | | >75 (>3) | 6 | 23 (75) |
| Tantalum | 200–250 | Stress relieved | <13 (<½) | 14 | 24 (80) |
| ASTAR 811C, T-111, | | | 13–25 (½–1) | 10 | 21 (70) |
| T-222, Ta-l0W, Ta-Hf, Ta63 | | | 25–75 (1–3) | 8 | 18 (60) |
| | | | >75 (>3) | 6 | 15 (50) |

Source: Metcut Research Associates Inc.

also be used, at slightly lower speeds and feeds. Nominal speeds and feeds for tapping niobium and molybdenum are given in Table 10. Tantalum (Ta-10W) with a hardness of 200 to 250 HB has been tapped with M1, M7, and M10 high-speed steel tools at a speed of 0.9 m/min (3 sfm).

Tapping of tungsten is difficult and should be avoided. If it is tapped, the workpiece should be heated to 200 to 425 °C (400 to 800 °F), and a four-flute stub-type plug tap made of nitrided M10 or T15 high-speed steel should be used. A speed of 1.5 m/min (5 sfm) is recommended.

Tool geometries for high-speed steel and carbide reamers are shown in Fig. 3 and 4, respectively. Tool geometries for tapping niobium, molybdenum, and tantalum alloys are shown in Fig. 5.

## Milling Refractory Metals

Machinability ratings based on metal removal rates in turning are an indication of performance to be expected in milling the refractory metals. Tool geometries for face mills and peripheral end mills are shown in Fig. 6 and 7, respectively.

**Niobium.** This metal is milled with both high-speed steel and carbide cutters. The cutters that have proved to be the best in most applications have 0° axial rake and +40° radial rake. Success in milling depends on the selection of cutting fluid: A highly chlorinated or sulfurized oil is recommended.

Parameters for milling niobium are set forth in Tables 11 to 16. (Table 11, hollow milling; Table 12, face milling; Table 13, side and slot milling; Table 14, peripheral end milling; Table 15, end milling-slotting; Table 16, thread milling).

**Molybdenum.** Normally, carbide-tipped face mills and high-speed steel end mills may be used in machining this metal. In each instance, tool angles should be similar to those for cast iron. Soluble-oil coolant is suggested. Corner and edge breakout experienced with multiple cutters can be minimized by infeeding. Cutter conditions should be checked frequently, and workpiece backup plates should be used, particularly when stock removal is heavy.

For the more brittle grades of molybdenum, cutters with zero axial and radial rakes give good results and minimize chipping. Positive rakes can be used for the more ductile grades.

Although climb milling, rather than conventional milling, is preferred for all refractory metals, it is especially important to use the climb technique for milling the more brittle grades of molybdenum because chipping of the work metal at the entrance and exit of the cutter is minimized.

The use of cutting fluid is mandatory for milling molybdenum. A soluble oil mixed 1:20 with water is satisfactory for most applications. Parameters for milling molybdenum are given in Tables 11 to 15.

**Tantalum.** Either high-speed steel or carbide tools may be used to machine this metal. Feeds are about the same as those for niobium, although tantalum is machined at slower speeds. Cutting fluid should be a soluble-oil emulsion. Parameters for milling tantalum are presented in Tables 11 to 15.

**Tungsten.** The composition and density of tungsten greatly affect the speed used in face milling at a given feed rate. In end milling, composition and density have less effect on speed. The climb technique should always be used.

Carbide cutting edges are used exclusively for milling tungsten. In room-temperature milling, optimum results are obtained with a copious supply of highly chlorinated cutting oil. Some success in warm milling (200 to 425 °C, or 400 to 800 °F), has been reported; under these conditions no cutting fluid is used.

Parameters for machining tungsten are given in Table 12 (face milling) and Table 15 (end milling-slotting).

## Parameters for Sawing the Refractory Metals

Niobium, molybdenum, and tantalum may be sawed with power hacksaws or band saws, using high-speed steel blades. Tungsten may be cut with circular sawing, using high-speed steel blades, and with abrasive cutoff. The latter method is also suitable for cutting the other refractory metals. Circular sawing is confined to use with niobium, tantalum, and tungsten. Parame-

**Table 19 Circular sawing niobium, tantalum, and tungsten alloys (wrought, cast, and P/M) with a high-speed steel blade**

| Material | Hardness, HB | Condition | Solid stock diameter or thickness, mm (in.) | Pitch, mm/tooth (in./tooth) | Cutting speed, m/min (sfm) | Feed, mm/tooth (in./tooth) | High-speed steel tool material grade, ISO (AISI) |
|---|---|---|---|---|---|---|---|
| Niobium | 170–225 | Stress relieved | 6–75 | 5–20 | 21 | 0.13 | S4, S2 |
| C130, C129Y, | | | (1/4–3) | (0.20–0.75) | (70) | (0.005) | (M2–M7) |
| Nb-1Zr, Cb-752, | | | 75–150 | 13–28 | 17 | 0.13 | S4, S2 |
| FS-85, FS-291, | | | (3–6) | (0.50–1.10) | (55) | (0.005) | (M2, M7) |
| WC-3015 | | | 150–250 | 20–33 | 14 | 0.18 | S4, S2 |
| | | | (6–9) | (0.75–1.30) | (45) | (0.007) | (M2, M7) |
| | | | 250–400 | 23–45 | 11 | 0.18 | S4, S2 |
| | | | (9–15) | (0.90–1.75) | (35) | (0.007) | (M2, M7) |
| Tantalum | 200–250 | Stress relieved | 6–75 | 5–20 | 15 | 0.10 | S4, S2 |
| ASTAR 811C, | | | (1/4–3) | (0.20–0.75) | (50) | (0.004) | (M2, M7) |
| T-111, T-222, | | | 75–150 | 13–28 | 12 | 0.10 | S4, S2 |
| Ta-l0W, Ta-Hf, | | | (3–6) | (0.50–1.10) | (40) | (0.004) | (M2, M7) |
| Ta-63 | | | 150–250 | 20–33 | 9 | 0.15 | S4, S2 |
| | | | (6–9) | (0.75–1.30) | (30) | (0.006) | (M2, M7) |
| | | | 250–400 | 23–45 | 8 | 0.15 | S4, S2 |
| | | | (9–15) | (0.90–1.75) | (25) | (0.006) | (M2, M7) |
| Tungsten alloys | 290–320 | Pressed and sintered | 6–75 | 5–18 | 12 | 0.075 | S4, S2 |
| Anviloy 1100 | | | (1/4–3) | (0.20–0.70) | (40) | (0.003) | (M2, M7) |
| Anviloy 1150 | | | 75–150 | 13–25 | 11 | 0.075 | S4, S2 |
| Anviloy 1200 | | | (3–6) | (0.50–1.00) | (35) | (0.003) | (M2, M7) |
| | | | 150–250 | 20–30 | 8 | 0.10 | S4, S2 |
| | | | (6–9) | (0.75–1.20) | (25) | (0.004) | (M2, M7) |
| | | | 250–400 | 23–38 | 6 | 0.10 | S4, S2 |
| | | | (9–15) | (0.90–1.50) | (20) | (0.004) | (M2, M7) |

Source: Metcut Research Associates Inc.

**Table 20 Abrasive cutoff of niobium, molybdenum, tantalum, and tungsten alloys (wrought, cast, and P/M)**

| Material | Hardness, HB | Condition | Wheel identification, wet cutting, ISO (ANSI) |
|---|---|---|---|
| Niobium(a)(b) C102, C129Y, Nb-1Zr, Cb-752, FS-85, FS-281, WC-3015 | 170–225 | Stress relieved | C60-MR (C60-MR) |
| Molybdenum Mo, Mo-50Re, TZC, TZM | 220–290 | Stress relieved | C60-MR (C60-MR) |
| Tantalum ASTAR 811C, T-111, T-222, Ta-l0W, Ta-Hf, Ta-63 | 200–250 | Stress relieved | C60-MR (C60-MR) |
| Tungsten 85, 93, 96, and 100% density | 180–320 | Pressed and sintered, forged, or arc cast | C60-MR (C60-MR) |

(a) *Caution:* toxic material; refer to National Institute for Occupational Safety and Health for precautions. (b) *Caution:* potential fire hazard; exercise caution in grinding and disposing of swarf. Source: Metcut Research Associates Inc.

**Table 21 Surface grinding niobium and molybdenum alloys (wrought, cast, and P/M) with a horizontal spindle and reciprocating table**

| Material | Hardness, HB | Condition | Wheel speed, m/s (ft/s) | Table speed, m/min (sfm) | Down feed, mm/pass (in./pass) Rough | Finish | Cross feed, mm/pass (in./pass) | Wheel identification, ISO or ANSI(a) |
|---|---|---|---|---|---|---|---|---|
| Niobium(b)(c) C103, C129Y, Nb-1Zr, Cb-752, FS-85, FS-291, WC-3015 | 170–225 | Stress relieved | 10–20 (33–66) | 12–18 (40–60) | 0.025 (0.001) | 0.013 (0.0005)(max) | 0.50–5.0 (0.020–0.200) | A46JV |
| Molybdenum Mo, Mo-50Re, TZC, TZM | 220–290 | Stress relieved | 10–20 (33–66) | 6–12 (20–40) | 0.025 (0.001) | 0.013 (0.0005)(max) | 0.50–5.0 (0.020–0.200) | A46KV |

(a) Wheel recommendations are for wet grinding. For dry grinding, use a softer-grade wheel. (b) *Caution:* toxic material; refer to National Institute for Occupational Safety and Health for precautions. (c) *Caution:* potential fire hazard; exercise caution in grinding and disposing of swarf. Source: Metcut Research Associates Inc.

ters for these operations are presented in Tables 17 to 20.

## Nontraditional Machining of Refractory Metals

All of the refractory metals may be machined with the electrical discharge machining (EDM) process. In some applications, metal may be removed selectively from molybdenum sheet, plate, and forgings by chemical machining, chemical milling, or chemical blanking. There are no known production applications of electrochemical machining for niobium and tantalum.

Wire EDM is useful for cutting complex shapes of both niobium and tantalum. For example, settings that cut Inco 718 are also suitable for three of the niobium alloys, C-103, FS-85 and Cb-752. The remarkable factor is that the recast structure is less than 0.013 mm (0.0005 in.) thick and is readily pickled off. Electrochemical machining (ECM) is not selected to machine niobium and tantalum because they tend to form oxide films, which passivate the metal against attack, except with very aggressive electrolytes and with high electrode potential.

Both EDM and ECM work well with molybdenum. With EDM, stock removal rates are as high as $8 \times 10^3$ mm$^3$/min (0.5 in.$^3$/min). Tolerances of ±0.013 mm (±0.0005 in.) have been obtained with EDM. Intricate shapes are made with EDM wire cutting. Electrochemical machining is normally capable of about $16 \times 10^3$ mm$^3$/min (1 in.$^3$/min) stock removal at 10 000 amp.

Because unalloyed tungsten is extremely brittle at room temperature and presents machining problems by conventional methods, electrical discharge machining is sometimes used. The principal advantage is that EDM is essentially force free, and force-induced fracture and chipping of delicate parts is avoided. However, EDM is extremely slow. Metal removal rate for tungsten is approximately 1.6 mm$^3$ (0.0001 in.$^3$) per ampere-minute with brass electrodes and is accompanied by a workpiece-to-electrode wear ratio of 1:2. By comparison, the typical metal removal rate for a hardened-steel workpiece and a brass electrode is 5 mm$^3$ (0.0003 in.$^3$) per ampere-minute with a workpiece-to-electrode wear ratio of 2.5:1.

The tool assembly shown in Fig. 8 was used to machine a contour and two holes in blanks of unalloyed tungsten. With a standard machine and gap current of 30 A, the machining operation was completed in 5 to 7 min.

## Parameters for Grinding Refractory Metals

Even under the best of conditions, refractory metals are difficult and expensive to grind. The grinding ratio is extremely sensitive to grinding fluid.

With the exception of tungsten, refractory metals load grinding wheels rapidly. Frequent dressing of wheels and plentiful quantities of grinding fluid are required. Water-base solutions such as potassium nitrite are effective in minimizing wheel loading. When the residue that the solution leaves on work is objectionable, soluble-oil emulsions or chlorinated grinding fluids may be used.

**Grinding Niobium.** This metal can be ground without surface cracking if the previously mentioned precautions are observed. Parameters for four methods of grinding niobium are found in Tables 21 to 24.

**Molybdenum.** Chatter can be a problem in grinding. In some surface grinding, a

**Table 22 Cylindrical grinding niobium alloys (wrought, cast, and P/M)**

| Material | Hardness, HB | Condition | Wheel speed, m/s (ft/s) | Work speed, m/min (sfm) | Infeed on diameter, mm/pass (in./pass) Rough | Finish | Traverse, wheel width per rev of work Rough | Finish | Wheel identification, ISO or ANSI(a) |
|---|---|---|---|---|---|---|---|---|---|
| Niobium(b)(c) ........ C103, C129Y, Nb-1Zr, Cb-752, FS-85, FS-291, WC-3015 | 170–225 | Stress relieved | 20 (66) | 15–30 (50–100) | 0.025 (0.001) | 0.008 (0.0003) (max) | 1/5 | 1/10 | A60KV |

(a) Wheel recommendations are for wet grinding of 50 to 100 mm (2 to 4 in.) diam work. For dry grinding, use a softer-grade wheel. For larger-diameter work, use a softer-grit and/or coarser-grit wheel. For smaller-diameter work, use a harder-grade wheel. Wheel recommendations also apply to plunge grinding applications. (b) *Caution:* toxic material; refer to National Institute for Occupational Safety and Health for precautions. (c) *Caution:* potential fire hazard; exercise caution in grinding and disposing of swarf. Source: Metcut Research Associates Inc.

**Table 23 Centerless grinding niobium alloys (wrought, cast, and P/M)**

| Material | Hardness, HB | Condition | Wheel speed, m/s (ft/s) | Throughfeed of work, m/min (sfm)(a) | Infeed on diameter, mm/pass (in./pass) Rough | Finish | Wheel identification, ISO or ANSI(b) |
|---|---|---|---|---|---|---|---|
| Niobium(c)(d) ............. C103, C129Y, Nb-1Zr, Cb-752, FS-85, FS-291, WC-3015 | 170–225 | Stress relieved | 20 (66) | 1.3–3.8 (4.3–12.5) | 0.025 (0.001) | 0.013 (0.0005) (max) | A60LV |

(a) Recommended starting conditions; use regulating wheel angle with positive inclination of 3° and a regulating wheel speed of 25 to 40 rev/min. (b) Wheel recommendations are for wet grinding 20 to 50 mm (0.8 to 2 in.) diam work. For larger-diameter work, use softer-grade and/or coarser-grit wheel. For smaller-diameter work, use harder-grade wheel. (c) *Caution:* toxic material; refer to National Institute for Occupational Safety and Health for precautions. (d) *Caution:* Potential fire hazard; use caution in grinding and disposing of swarf. Source: Metcut Research Associates Inc.

harder wheel (higher bond strength) and a slower speed have helped to minimize this condition. However, with a rigid setup and a plentiful supply of grinding fluid (water-base synthetic), molybdenum can usually be ground successfully.

Parameters for surface grinding molybdenum with a horizontal-spindle reciprocating machine are found in Table 21. Those for internal grinding are set forth in Table 24.

**Tantalum.** This metal is sometimes difficult to machine because chatter develops when surface grinding. Speeds as low as 600 m/min (2000 sfm) have been used for this reason. However, with a rigid setup and a flood of grinding fluid (typically, water-base synthetic) chatter usually can be overcome. Wheels with aluminum oxide abrasive and a vitrified bond are used exclusively for grinding this metal.

Parameters for internal grinding of tantalum are presented in Table 24. In one example, the alloy Ta-10W with a hardness of 240 HB was surface ground with an A46J8V wheel, a wheel speed of 600 m/min (2000 sfm), a table speed of 12 m/min (40 sfm), and a down feed of 0.025 mm (0.001 in.). Grinding fluid was a $KNO_2$ solution. Cross feeds were 0.64 mm (0.025 in.) per pass. The alloy had a grinding ratio of 25, which compares with a ratio of 20 for 302 stainless and 4.3 for tungsten. The ratio indicates volume of metal removed in grinding in relation to the volume of wheel wear. The higher the index, the easier a metal is to grind.

**Tungsten.** Surface cracking can be a problem in machining as a result of high thermal stresses. A wheel that is too soft is preferable to one that is too hard. An excessively hard wheel will glaze quicker and is more likely to cause cracking. Aluminum oxide wheels, such as A46N5V operated at 1200 m/min (4000 sfm) with a down feed of 0.015 mm (0.0005 in.) per pass, have been used in surface grinding. Generally, however, silicon carbide wheels perform better than aluminum oxide types in grinding this metal.

Because tungsten is brittle, even the lightest grinding pressure is likely to cause flaking at corners. Grinding toward sharp corners is never recommended. Soluble oil is acceptable as a cutting fluid; heating the fluid to about 40 °C (100 °F) sometimes helps prevent surface cracking. Parameters for internal grinding of tungsten and tungsten-2 thoria are presented in Table 24.


## ACKNOWLEDGMENT

The author wishes to express his appreciation to the following individuals for their valuable assistance in preparing this document: Barry Heller, Teledyne Firth Sterling; John B. Lambert, Fansteel Inc.; Bob McLemore, Marquardt Company; Meinrad J. Ostermann, Teledyne Wah Chang, Huntsville; and R.T. Webster, Teledyne Wah Chang, Albany. Special thanks are also due to Metcut Research Associates Inc. for supplying tables detailing machining data for refractory metals and to Amax Specialty Metals Corporation for literature on the machining of molybdenum.


**Table 24 Internal grinding niobium, molybdenum, tantalum, and tungsten alloys (wrought, cast, and P/M)**

| Material | Hardness, HB | Condition | Wheel speed, m/s (ft/s) | Work speed, m/min (sfm) | Infeed on diameter, mm/pass (in./pass) Rough | Finish | Traverse, wheel width per rev of work Rough | Finish | Wheel identification, ISO or ANSI(a) |
|---|---|---|---|---|---|---|---|---|---|
| Niobium(b)(c) ......................... C103, C129Y, GCb-1Zr, Cb-752, FS-85, FS-291, WC-3015 | 170–225 | Stress relieved | 20 (66) | 15–45 (50–150) | 0.025 (0.001) | 0.013 (0.0005)(max) | 1/3 | 1/6 | A60JV |
| Molybdenum ...................... Mo, Mo-50Re, TZC, TZM | 220–290 | Stress relieved | 20 (66) | 15–45 (50–150) | 0.025 (0.001) | 0.013 (0.0005)(max) | 1/3 | 1/6 | A60JV |
| Tantalum ......................... ASTAR 811C, T-111, T-222, Ta-10W, Ta-Hf, Ta63 | 220–250 | Stress relieved | 20 (66) | 15–45 (50–150) | 0.025 (0.001) | 0.013 (0.0005)(max) | 1/3 | 1/6 | A60JV |
| Tungsten, 85, 93, 96, and 100% density ......................... | 180–320 | Pressed and sintered, forged, or arc cast | 10 (33) | 15–45 (50–150) | 0.013 (0.0005) | 0.005 (0.0002)(max) | 1/3 | 1/6 | C60HV |
| Tungsten-2 thoria(b)(c) ............. | 260–320 | Pressed and sintered | 10 (33) | 15–45 (50–150) | 0.013 (0.0005) | (0.005) (0.0002)(max) | 1/3 | 1/6 | C60HV |

(a) Maximum hole length is 2.5 times hole diameter; maximum wheel width is 1.5 times wheel diameter; wheel recommendations are for wet grinding 20 to 50 mm (0.8 to 2 in.) diam holes. For larger holes, use same or softer-grade wheel. For smaller holes, use harder-grade wheel. (b) *Caution:* Toxic material; refer to National Institute for Occupational Safety and Health for precautions. (c) *Caution:* Potential fire hazard; use caution in grinding and disposing of swarf. Source: Metcut Research Associates Inc.

# Machining of Beryllium*

Douglas V. Gallagher, Rockwell International, and R.E. Hardesty, Electrofusion Corporation

BERYLLIUM metal has properties that require special attention when machining. These properties are as follows:

- Because beryllium is expensive, near-net shape blanks should be used whenever possible to reduce material waste and subsequent recovery costs
- The dust and chips generated in machining are toxic to humans and can cause beryllium disease; therefore, a tested exhaust system and an approved monitoring system are essential
- When beryllium is machined, its abrasiveness will quickly wear tooling and thus degrade accuracy or cause scrap under tight-tolerance conditions
- Beryllium is susceptible to machining damage, which will degrade material properties, particularly under high strain rate conditions; this requires special machining schedules to minimize the damage and subsequent removal of the damage
- Beryllium is a brittle material that will easily chip if handled roughly or if fixture support is inappropriate

## Material Considerations

Because of its high cost, beryllium is typically recovered after machining. Cutting beryllium dry helps facilitate recovery of the metal because oil does not have to be removed from the chips.

Near-net shape technologies such as cold isostatic pressing, sintering, and hot isostatic pressing should be taken advantage of in the purchase of beryllium blanks for parts. Typically, 2.5 mm (0.100 in.) of material or less is all that must be removed. Two benefits of using near-net shape technologies are the reduction in the amount of chips and dust produced and the reduction in cost due to the shorter machining times.

The material used for conducting the tool wear studies and some of the surface damage studies presented in this article was S65 or select S65 beryllium. The compositions of these materials differ only in the few areas indicated in Table 1. Both of the beryllium alloys are powder metallurgy products and are compacted using vacuum hot pressing and/or hot isostatic pressing.

**Table 1 Composition of S65 and select 65 beryllium**

| Beryllium type | Composition, % | | | | | | | | | | Fe/Al ratio |
|---|---|---|---|---|---|---|---|---|---|---|---|
| | Be | BeO | Al | C | Fe | Mg | Si | S | U | Other | |
| S65 | 98.5 | 1.0 | 0.06 | 0.12 | 0.12 | 0.08 | 0.06 | ... | ... | 0.04 max | ... |
| Select 65 | 98.5 | 1.2 | 0.06 | 0.1 | 0.1 | 0.08 | 0.06 | 0.04 | 0.04 | 0.04 max | ~2:1 |

Beryllium is a brittle material and must be handled accordingly. Fixtures should be designed to provide solid support, especially for small or sharp features. Care must be taken in handling so that features are not chipped or broken.

## Health Concerns

Beryllium in solid form is not hazardous and can be handled without special precautions. The hazards to humans from beryllium lie in the dust and chips produced during machining and operations such as polishing that produce aerosols. Therefore, the dust and chips must be controlled, and this can be done by using a tested exhaust system when cutting the material dry or by using a coolant to contain the dust. The principal regulation on airborne beryllium as set by the American Conference of Governmental Industrial Hygienists is that the in-plant atmospheric concentration of beryllium should not exceed 2 $\mu g/m^3$ as an average concentration throughout an 8-h day.

Because of this requirement, housekeeping becomes a very important function in beryllium machining. Work areas and machines must be kept clean of all chips and dust. This involves vacuuming up chips on the key ways, chips pans, cross slide, compound, and so on, of the machine tool being used. An industrial hygienist must be consulted to ensure that personnel and work areas remain free of beryllium hazards. The subject of protection cannot be fully covered in this article and only the highlights of some of the precautions necessary will be discussed. Shop practices must be studied and monitored by an experienced industrial hygienist.

Beryllium can be cut wet, with the advantage that the dust will not easily permeate the air. The disadvantages are that chip recovery is more difficult, the coolant may dry out (which allows the beryllium dust to become airborne), and some coolants are reactive with the metal.

Most beryllium machining is done dry. The dust and chips are controlled with an exhaust system located near the cutting tool where the chips are produced. Generally, the exhaust nozzle is located no farther than 13 to 19 mm (½ to ¾ in.) from the cutting tool. The velocity at the nozzle should be approximately 4390 m/min (14 400 actual ft/min) (Ref 1). For lathe operations, a hood and an exhaust system should be installed around the chuck and the part. The hood has a door that allows loading and unloading of parts. There is a slit in the door that allows the tool to pass through. The velocity at the slit in the door should be approximately 90 m/min (300 actual ft/min) with the door closed. A hood is impractical during milling operations; therefore, an exhaust nozzle is used at the cutting tool.

For operations requiring the polishing of surfaces with emery cloth or a coated nylon abrasive pad (scotch brite), special care must be taken. This is because of the proximity of the machinist to the work and the fine dust produced. In these situations, a respirator should be worn at all times. For polishing, the part should be enclosed in a hood if on a lathe, or in a box provided with an exhaust system. To prevent contamination of the surrounding work areas, small pieces of emery cloth or a coated nylon abrasive pad (scotch brite) should be used and then discarded immediately in an approved waste container.

## Surface Damage

When cutting beryllium, the machining processes introduce an altered layer of material at the part surface. This layer, termed

*The section "Chemical Milling and Photochemical Machining" was prepared by Mr. Hardesty.

**Table 2 Transverse tensile properties of as-machined, annealed, and etched beryllium**

| Grade | Condition | Ultimate tensile strength, MPa | Ultimate tensile strength, ksi | 0.2% yield strength, MPa | 0.2% yield strength, ksi | Yield point, MPa (ksi) | Elongation, % |
|---|---|---|---|---|---|---|---|
| S200E | Before machining | 366 | 53.1 | 227 | 32.9 | . . . | 3.7 |
| | As machined | 263 | 38.2 | 216 | 31.3 | . . . | 0.7 |
| | Annealed | | | | | | |
| | 400 °C (750 °F) | 218 | 31.6 | 193 | 28.0 | . . . | 0.5 |
| | 500 °C (930 °F) | 207 | 30.0 | 199 | 28.8 | . . . | 0.3 |
| | 600 °C (1110 °F) | 282 | 40.9 | 187 | 27.2 | . . . | 2.0 |
| | 700 °C (1290 °F) | 353 | 51.2 | 205 | 29.7 | . . . | 3.6 |
| | 800 °C (1470 °F) | 343 | 49.8 | 230 | 33.3 | . . . | 3.1 |
| | 900 °C (1650 °F) | 347 | 50.4 | 249 | 36.1 | . . . | 3.2 |
| | 1000 °C (1830 °F) | 342 | 49.6 | 239 | 34.7 | . . . | 3.3 |
| | Etched | | | | | | |
| | 25 µm (985 µin.) | 323 | 46.8 | 209 | 30.3 | . . . | 2.8 |
| | 50 µm (1970 µin.) | 337 | 48.9 | 220 | 31.9 | . . . | 2.8 |
| | 75 µm (2950 µin.) | 333 | 48.3 | 216 | 31.3 | . . . | 3.1 |
| 1319A | Before machining | 386 | 56.0 | 273 | 39.6 | . . . | 6.2 |
| | As-machined | 282 | 40.8 | 277 | 40.2 | 286 (41.4) | 0.7 |
| | Annealed | | | | | | |
| | 400 °C (750 °F) | 249 | 36.2 | 245 | 35.6 | 257 (37.3) | 0.4 |
| | 500 °C (930 °F) | 270 | 39.2 | . . . | . . . | . . . | 0.2 |
| | 600 °C (1110 °F) | 242 | 35.1 | . . . | . . . | . . . | 0.2 |
| | 700 °C (1290 °F) | 375 | 54.5 | 257 | 37.2 | 264 (38.2) | 3.1 |
| | 800 °C (1470 °F) | 381 | 55.2 | 235 | 34.1 | . . . | 3.8 |
| | 900 °C (1650 °F) | 388 | 56.2 | 238 | 34.5 | . . . | 4.0 |
| | 1000 °C (1830 °F) | 383 | 55.5 | 231 | 33.5 | . . . | 4.6 |
| | Etched | | | | | | |
| | 25 µm (985 µin.) | 340 | 49.3 | 294 | 42.7 | 311 (45.2) | 1.4 |
| | 50 µm (1970 µin.) | 382 | 55.4 | 289 | 41.9 | 310 (45.0) | 4.0 |
| | 75 µm (2950 µin.) | 372 | 53.9 | 270 | 39.2 | 291 (42.2) | 4.0 |

Source: Ref 5

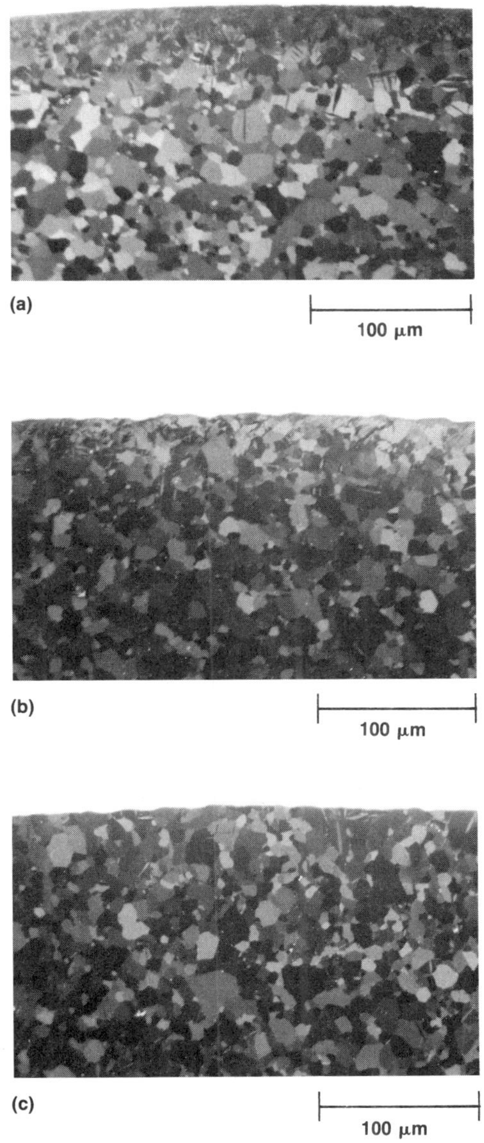


**Fig. 1** Machining-damaged surface of beryllium before etching (a), after 25 µm etched (b) and after 75 µm etched (c). Courtesy of Lawrence Livermore National Laboratory

the surface-damaged or machining-damaged layer, extends from 0.025 to 0.50 mm (0.001 to 0.020 in.) in depth, depending on the method used for machining (Ref 2). Typical damage is in the range of 0.025 to 0.10 mm (0.001 to 0.004 in.) for most machining. The surface-damaged layer consists of a dense network of intersecting twins and some cracking. These twins are a symptom of beryllium machining damage and may contribute to reduced mechanical properties, along with other metallurgical conditions present in the surface layer. It is for this reason that the machining schedules (the number of passes and the depth of each pass) must be controlled (Ref 2-4).

The surface-damaged layer can cause a reduction in the mechanical properties of the part; this appears to be most detrimental in applications in which parts will be exposed to high strain rates (Ref 3, 5, 6). The layer can be minimized by control of the machining schedule but cannot be eliminated by machining alone. Therefore, subsequent methods may be needed to restore the mechanical properties by removing or repairing the surface-damaged layer.

**Etching and heat treating** are two methods that can be used to restore the mechanical properties of machining-damaged beryllium (Table 2). In some rare applications, the machining schedule can control the surface damage to a sufficient degree that no difference is seen in the as-machined and the etched or heat-treated parts. Whether or not the machining-damaged layer will cause problems is dependent on how the part will be used in service. If the part is to be used structurally, then surface damage treatments will be necessary to restore mechanical properties. If the part will not be exposed to severe service, then damage to mechanical properties can be controlled by using a schedule similar to the following:

| Operation | Depth, mm | Depth, in. |
|---|---|---|
| Maximum roughing cut | 1.3 | 0.050 |
| Finishing cuts | 1.3 | 0.050 |
| | 0.5 | 0.020 |
| | 0.25 | 0.010 |
| | 0.25 | 0.010 |
| | 0.13 | 0.005 |
| | 0.05 | 0.002 |
| | 0.05 | 0.002 |
| | 0.025 | 0.001 |
| | 0.025 | 0.001 |

If the parts are to be heat treated or etched to remove machine damage, then the following schedule can be used (Ref 2, 3):

| Operation | Depth, mm | Depth, in. |
|---|---|---|
| Maximum roughing cut | 1.3 | 0.050 |
| Finishing cuts | 1.3 | 0.050 |
| | 0.5 | 0.020 |
| | 0.25 | 0.010 |
| | 0.13 | 0.005 |

These schedules are conservative, and some applications may require fewer cuts. This should be determined on a case-by-case basis.

The chemical milling of beryllium parts removes the machining-damaged layer (Fig. 1) and usually takes off about 0.05 to 0.10 mm (0.002 to 0.004 in.) of material from the surface (Ref 7). This is accomplished by immersing the part in an acid solution for a specified time. An advantage of this process is that the machining-damaged layer is removed. The main disadvantage is that the removal can be uneven, which can result in a loss of precision tolerances.

Vacuum heat treating of the beryllium part at 800 °C (1470 °F) for 2 h will remove most of the machining damage and can restore all or most of the mechanical properties. Heat treating removes the mechanically induced twins near the surface (Fig. 2). However, thermally induced twins will

**Fig. 2** Machining-damaged surface of beryllium before annealing (a), after 600 °C (1100 °F) anneal (b), after 700 °C (1300 °F) anneal (c), and after 800 °C (1470 °F) anneal (d). Courtesy of Lawrence Livermore National Laboratory

be present in the surface and throughout the body of the material after the heat treatment; but they do not have any effect on mechanical properties (Ref 5, 6). The advantage of this method is that dimensional tolerances are not affected except for the distortion from heat treating. If the heat treating is done properly, the potential problem of beryllium sticking to the furnace platen or being deposited inside the furnace can be minimized or eliminated.

## Turning

The machine tools used for beryllium are not different from any other except for the fitting of exhaust systems. The exhaust systems on mills consist of attaching a vacuum nozzle and its hose to the machine. For lathes, a hood and a vacuum nozzle should be fitted around the chuck and the part.

**Cutting oils** are typically not used in the machining of beryllium. The cutting oil must be selected with care because some are reactive with the metal. In particular, cutting oils that contain sulfur or chlorine can discolor or corrode the metal.

**Cutting Tools.** The C4 grades of carbide have been found to perform well in the cutting of beryllium. The exceptions are tools with titanium or tantalum as binders; these tools will fail early in the cutting process (Ref 8). Such C4 grades as Teledyne HF and Carbaloy 999 perform well. Another type of tool material that can be used is cubic boron nitride (CBN). However, speeds and feeds must be reduced with CBN tools to promote tool life and still produce a good surface finish.

If precision machining (profiles ≦0.1 mm, or 0.004 in.) is being performed, the tool geometry is of some importance. Because of its abrasiveness, beryllium will wear tooling quickly. If the primary relief angle of the cutting tool is large, the nose radius of the tool will wear faster than for a smaller angle. If possible, the primary relief angle should be 3 to 6°. The top rake angle is generally held from 0 to 3°. For machining that does not require a high degree of precision, a top rake angle of 3 to 6° and a primary relief angle of 10° are satisfactory.

**Speeds and feeds** in turning beryllium depend on the type of tool material being used. In finishing cuts with the C4 grade carbides, the conditions that yield optimum tool life are a cutting speed of 67 to 73 m/min (220 to 240 sfm) at a feed rate of 0.1 mm/rev (0.004 in./rev). This will produce surface finishes in the range of 1.0 to 1.6 μm (40 to 63 μin.) $R_a$. Finishing with CBN tooling involves speeds of 37 to 46 m/min (120 to 150 sfm) at a feed rate of 0.05 mm/rev (0.002 in./rev). Surface finishes in this case will be in the range of 0.75 to 1.25 μm (30 to 50 μin.) $R_a$.

For rough machining, speeds of 107 to 122 m/min (350 to 400 sfm) can be used for the carbide tooling. Cubic boron nitride is not typically used for rough machining, because of its cost and the slow feeds and speeds necessary to keep the tool from breaking down.

## Chemical Milling and Photochemical Machining

Chemical milling and photochemical machining are commonly used methods of etching beryllium components. This section will discuss both of these processes as applied to beryllium. The articles "Chemical Milling" and "Photochemical Machining" in this Volume describe these processes in more detail.

**Chemical milling** is used to remove damaged surface layers or to create surface pockets for weight reduction. One form of chemical milling consists of completely immersing the part into an etching solution to

**Fig. 3** Beryllium part processed by photochemical machining

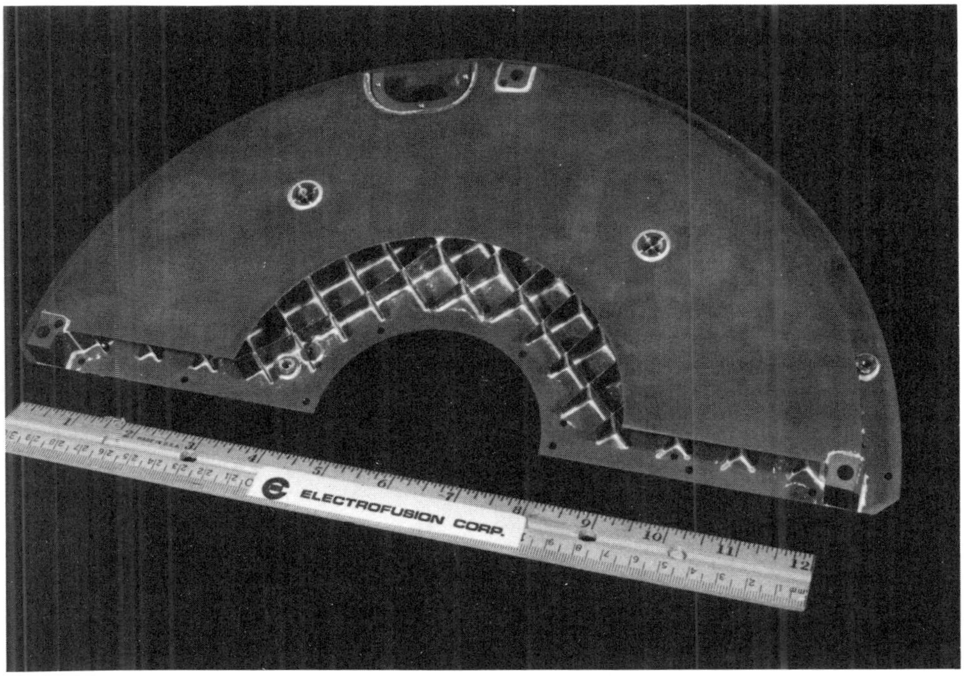


**Fig. 4** Laser base plate made of photochemically machined beryllium

remove the machining-damaged layer. When combined with a thermal stress relief cycle, this approach provides a stress-free and/or damage-free part.

The etchants typically used for the chemical milling of beryllium are sulfuric acid, nitric/hydrofluoric acid solutions, and ammonium bifluoride. Each of these solutions provides a certain capability or advantage, depending on the application.

**The photochemical machining of beryllium** is used to produce detailed precision parts (Fig. 3 and 4). The process uses the same etchants as chemical milling and has the following advantages:

- *Maximum material utilization*: Through proper layout, parts can be produced from the inside of cutouts of other parts, thus maximizing material use
- *Reduced manufacturing costs*: Hundreds of parts can be produced from one sheet of material with one master, one etching process, and no subsequent machining
- *Repeatability*: When a large number of final assemblies are required, the same photochemical machining master can be repeatedly used to manufacture all the required detail parts. Thus, the first set of parts is identical to the last set, and the sets are interchangeable
- *Precision*: Photochemical components offer extremely tight tolerancing—as tight as ±0.025 mm (±0.001 in.) if critically required. Tolerances of ±0.075 mm (±0.003 in.) are easily produced
- *Ductility*: The photochemically machined components exhibit excellent ductility, and no further processing is required to remove any residual stresses or damage typically induced through conventional manufacturing techniques

## REFERENCES


1. T. Lewis, Rockwell International, Rocky Flats Industrial Hygiene, private communication, Sept 1988
2. S. Beitscher, J.F. Capes, W.W. Leslie, J.R. Luckow, and R.L. Riegel, "An Abbreviated Machining Schedule for Fabricating Parts Free of Surface Damage," Report RFP 2838, Rockwell International, Rocky Flats Plant, May 1979
3. S. Beitscher, W.L. Johns, R.R. Corle, J.R. Luckow, S.A. Hoover, and A.G. Grimes, "Further Abbreviation of the Machining Schedule for Producing Surface-Damage-Free Beryllium," Report RFP 3455, Rockwell International, Rocky Flats Plant, March 1983
4. J.E. Hanafee, "Effect of Machining on Properties of Impact-Ground Beryllium," Report CRl-52287, Lawrence Livermore Laboratory, June 1977
5. J.E. Hanafee, Effect of Annealing and Etching on Machining Damage in Structural Beryllium, *J. Appl. Metal Work.*, Vol 1 (No. 3), 1980, p 41-51
6. R.R. Corle, W.W. Leslie, and A.W. Brewer, "The Testing and Heat Treating of Beryllium for Machine Damage Removal," Report RFP 3084, Rockwell International, Rocky Flats Plant, Sept 1981
7. R.L. Riegel, E.L. Childs, and K.R. Souply, "Chemical Milling of Beryllium for Machining Damage Removal," Report RFP 2174, Dow Chemical Company, Rocky Flats Plant, July 1974
8. W.L. Johns and M.H. Jackson, "Right Tool for Machining Beryllium," Report RFP 4180, Rockwell International, Rocky Flats Plant, 1987

# Machining of Uranium and Uranium Alloys

James A. Aris, Rockwell International

URANIUM AND URANIUM ALLOYS are considered easy to machine. Depleted uranium is machined on a large scale at very high rates of production. The problems associated with machining are due to a combination of properties, including toughness and stringiness, abrasiveness, galling, work hardening, pyrophoricity, low modulus, high density, reactivity with coolants, reactivity with tools and grinding wheels, and toxicity. These properties vary with alloy composition and heat treatment.

This article will focus on the basic metallurgy, machining parameters, and health precautions applicable to the machining of uranium and its alloy classes. References made to uranium when machining parameters or metallurgy are discussed will pertain to both depleted and enriched uranium. The machining of uranium is isotopically independent, but health physics concerns vary with the individual isotopes. Only when health physics is concerned will the discussion be limited specifically to depleted uranium.

## Metallurgical Considerations

The three classes of uranium alloys are:

- Dilute alloys, containing less than 0.4 wt% of alloying elements
- Lean alloys, containing between 0.4 and 4 wt% of alloying elements
- Stainless alloys, containing more than 4 wt% of alloying elements (Ref 1)

Metallurgically, uranium consists of an orthorhombic α uranium matrix with carbides and oxides as principal inclusions (Ref 2). Common alloying elements include titanium, which makes uranium hard, and niobium, which makes it corrosion resistant (Ref 1). For example, unalloyed depleted uranium has a yield strength of 240 MPa (35 ksi), 10% elongation, and a hardness of 56.8 HRA in the β quenched condition. Uranium-0.75% titanium alloy (U-0.75Ti) in a solution quenched and aged condition has a yield strength of 760 MPa (110 ksi), 15% elongation, and hardness of 41 to 42 HRC. The yield strength of U-6Nb alloy in the γ quenched condition is 275 MPa (40 ksi), 22% elongation, and a hardness of 54 HRA (Ref 3, 4). (Information may also be found in the article "Heat Treating of Special-Purpose Alloys," in *Heat Treating*, Volume 4, 9th Edition, *Metals Handbook*.) Common uses for depleted uranium and uranium alloys include nuclear weapon components, calorimeter plates, ballistic penetrators (large and small), radiation shielding, gyroscope rotors, and aircraft counterweights (Ref 5).

Some unique characteristics of uranium and uranium alloys that affect their machinability include:

- Shape memory
- A nonisotropic coefficient of thermal expansion
- Their pyrophoric, radioactive, and toxic properties

Shape memory is the tendency of the material to return to its original shape after a temperature change. Shape memory is more prevalent in alloyed than in unalloyed uranium (Ref 1).

A nonisotropic coefficient of thermal expansion causes the material to distort under a thermal influence. Unlike isotropic materials that expand in all three crystallographic directions when heated, unalloyed uranium expands in two directions and contracts in the third direction. To minimize distortion of the part during machining, the cutting tools must be kept sharp and the temperature of the material kept as low as possible (Ref 6, 7).

Uranium can also have a relatively rapid oxidation rate if its surface is left unprotected. To overcome this problem, a finish-machined uranium component might require a coating to protect the workpiece from oxidation.

The pyrophoric, radioactive, and toxic properties of uranium and uranium alloys are discussed in the section "Health Effects and Required Precautions" in this article.

## Machining Parameters

Because uranium and uranium alloys are pyrophoric, roughing operations are performed at a slower cutting speed and faster feed rate than more common materials such as steel or aluminum. This produces as large a chip as possible, thereby minimizing the possibility of a fire (Ref 8, 9). A large coarse chip will not burn as readily as a small fine one. The exception to this practice is the final pass used in finish turning operations.

It is also important to avoid working the metal in a single direction, which imparts a preferred orientation to the crystal structure. This, in turn, increases the likelihood of distortion. This pass can be made at a higher speed, lower feed, and shallower depth of cut to produce a better finish if the amount of chips produced is small and the chip pan contains enough coolant to submerge the chips. These fine chips may ignite, but there is insufficient material to support a large fire.

Uranium and uranium alloys are ground when necessary using water base coolant and a silicon wheel. Grinding of uranium is difficult. Due to the relatively low thermal diffusivity of uranium, the ground surface overheats and oxidizes even at moderate stock removal rates and with a generous supply of coolant. A heavier than normal depth of cut is employed to decrease work hardening. Grinding uranium and uranium alloys presents a serious health and fire hazard, however, and should be avoided whenever possible (Ref 10-12).

Table 1 lists typical speeds and feeds used for cutting uranium and uranium alloys. For most operations, the same parameters apply to both material types.

**Coolants.** Water-base coolants are preferred because of the fire hazard; however, with proper precautions, uranium can be successfully machined in mineral oil. The problem with mineral oil is that if a chip fire starts, it must be extinguished before the flash point of the oil is reached and it too ignites. As chips coming off of the workpiece ignite, immediate submersion in the coolant will lower the temperature sufficiently to stop the burning (Ref 1). The danger of fire is minimized if chips do not accumulate in the chip pan or are kept submerged in the coolant during cutting. Oils having low flash points or aerosol mists

**Table 1 Machining parameters for uranium and uranium alloys**

| | Turning and facing | | Milling | |
|---|---|---|---|---|
| | Roughing | Finishing(a) | Roughing | Finishing(a) |
| Cutting speed, m/min (sfm) | 30–107 (100–350) | 122 (400) | 21–46 (70–150) | 61–76 (200–250) |
| Feed rate | | | | |
| mm/rev (in./rev) | 0.15–0.20 (0.006–0.008) | 0.051–0.102 (0.002–0.004) | ... | ... |
| mm/min (in./min) | ... | ... | 63–76 (2.5–3.0) | 38–50 (1.5–2.0) |
| mm/tooth (in./tooth) | ... | ... | 0.043–0.051 (0.0017–0.002) | 0.025–0.034 (0.001–0.0013) |
| Depth of cut, mm (in.) | 1.27–2.54 (0.050–0.100) | 0.0025–0.13 (0.0001–0.005) | 1.3 (0.050) | 0.13 (0.005) |

(a) For final cuts producing a minimum number of chips because of the fire danger. Source: Ref 13, 14

such as WD-40 must never be used as cutting oils on uranium. (Misting of any coolant as opposed to flooding is discouraged because it is not sufficient to prevent fire (Ref 8, 9).) To prevent fire when machining uranium or uranium alloys, the cutting area must be flooded with coolant, chips must be prevented from accumulating in the chip pan, and chips must be submerged. Machined chips that are being transported should be kept stored under water to prevent self-ignition due to friction.

**Turning and Facing.** Uranium and its alloys are free machining and can be cut on lathes ranging from a jeweler's smallest lathe to the largest computer numerically controlled (CNC) multiple-spindle lathe. Most uranium alloys have the advantage of being neither as gummy nor as pyrophoric as pure uranium (Ref 8, 9, 15).

*Tolerances.* Uranium can generally be machined to within the tolerance band of the lathe. Material properties of uranium and uranium alloys are such that tolerances of the order of 0.008 mm (0.0003 in.) are not uncommon (Ref 8). In order to hold close tolerances, tools must be kept sharp.

*Cold-Working Properties.* The carbide compounds formed in some uranium alloys are harder than the tungsten-carbide cutting tool used to machine the alloys. As these compounds wear the tool, the uranium will cold work (Ref 1), and the heat and stress induced in the part will cause distortion and exaggerate the effect of shape memory.

*Coolant temperature control* is recommended. A constant supply of temperature-moderated coolant helps dissipate heat from the part and thus minimizes distortion (Ref 8, 9).

These practices apply to highly accurate machining of any material but are especially critical in uranium machining operations because of the expansive characteristics of uranium. The nonisotropic coefficient of thermal expansion makes it difficult to predict the manner in which uranium will expand from one part to the next, unless metallurgical controls have been used. Unalloyed, depleted uranium is commonly heated to the β phase 725 °C (1335 °F) and then rapidly cooled to room temperature to randomize the expansion directions and minimize the effect of anisotropy (Ref 16).

To maintain dimensional accuracy, the parts should be rough turned to leave a 0.05 to 0.25 mm (0.002 to 0.010 in.) depth of cut for finish turning, which is done several days after rough turning in order to allow the material to stabilize. As mentioned earlier, care must be taken to prevent fire, particularly when taking light cuts and producing fine chips.

**Threading.** Uranium and uranium alloys lend themselves well to single-point threading (Ref 8, 9). Uranium threads can be very smooth and strong when generated. When threading with a tap or a die, ample chip clearance must be employed to minimize tool breakage. This is further discussed in the "Drilling and Tapping" section of this article.

**Milling** parameters are similar to those for turning. Chips produced should be large, and coolant should flood the workpiece whenever possible to minimize the fire hazard. Typical milling parameters are: 22 to 76 m/min (72 to 250 sfm) cutting speed, 38.1 to 76.2 mm/min (1.5 to 3.0 in./min), or approximately 0.025 to 0.051 mm/tooth (0.001 to 0.002 in./tooth) feed rate, and as deep a cut as possible to prevent fire (Ref 13, 14, 17). As with turning, tools should be kept sharp to minimize the effects of inherent material factors such as shape memory and distortion.

**Drilling and Tapping.** When tap holes of 9.5 mm (⅜ in.) diam or smaller are drilled, the holes should be drilled one size larger than normal (letter, number, or fraction) to allow better chip clearance and less force on the tap. Solid carbide drills covered with a high sulfur content paste lubricant have proved to be superior to carbide-tipped drills when drilling small holes.

To minimize breakage, spiral-pointed taps having a variable rake and a nitride surface finish are commonly used (Ref 8, 9, 17). It is good practice to use quality tapping fluids such as high-pressure fluids with a high-sulfur content, fluids composed of 90% high-pressure $MoS_2$ tapping fluid mixed with 10% Freon, or 90% beeswax mixed with 10% Freon. These fluids work well on all uranium and uranium alloys tested (Ref 8, 9). If a tapping head is used, tap sizes as small as 5-40 can be readily power tapped (Ref 17).

Experience has shown that for drilling large holes in alloyed or unalloyed uranium, a coolant-fed drill having indexable inserts works much better than a standard twist drill. With this type of coolant-fed drill, the cutting time, tool wear, and fire danger are significantly decreased (Ref 8).

## Tool Wear and Types of Tools

From a machining standpoint, uranium and its alloys are generally tough, soft, and gummy (adherent), although some alloys can be quite hard, depending on the metallurgical processing (Ref 18). The carbide inclusions make them abrasive as well. Normal flank wear is typical of the tool wear seen, even when using the proper tool and cutting parameters, when machining uranium. The cutting edge will appear to be mechanically worn, not chipped or notched. Both of the tools in Fig. 1 were used to make a 495 mm (19.5 in.) long cut in unalloyed, depleted uranium, with working parameters of 53.22 m/min (174.6 sfm) cutting speed, 0.25 mm (0.010 in.) depth of cut, and a 0.21 mm/rev (0.0084 in./rev) feed rate. Both tools were TNMP 432 carbide inserts (insert configuration and dimensions are given in Fig. 2), one of plain carbide (94WC-5.5Co-0.5TaC), the other coated with titanium nitride (ISO grade M-20 between C-4 to C-5 industry code, TiC/TiCN/TiN coatings on a special, strengthened substrate). The wear patterns are very similar. As can be seen in Fig. 1, the coated insert wore only slightly less than the plain carbide.

Tools typically giving the best results for turning and facing unalloyed uranium and the softer alloys are uncoated grades of high tantalum-carbide content inserts having nominal compositions, 74WC-20TaC-6Co and 50TaC-44WC-6Co (Ref 13, 18, 19) with a slight positive rake, typically 5 to 12° (Ref 14). Uncoated carbides in the C3 to C7 range generally work better on the harder alloys where abrasion, rather than heat at the tool tip, is the primary wear pattern (Ref 19, 20). Carbide inserts with various coatings and tools made of high-speed steel have been tried with poor results. Coated inserts have either fractured, worn excessively, or the coating has flaked off (Ref 19), and high-speed tools have shown excessive wear (Ref 8, 9).

The high-tantalum carbides are preferred when machining the unalloyed uranium and the lower-strength alloys because they can tolerate the high heat that is generated. However, the higher the tanta-

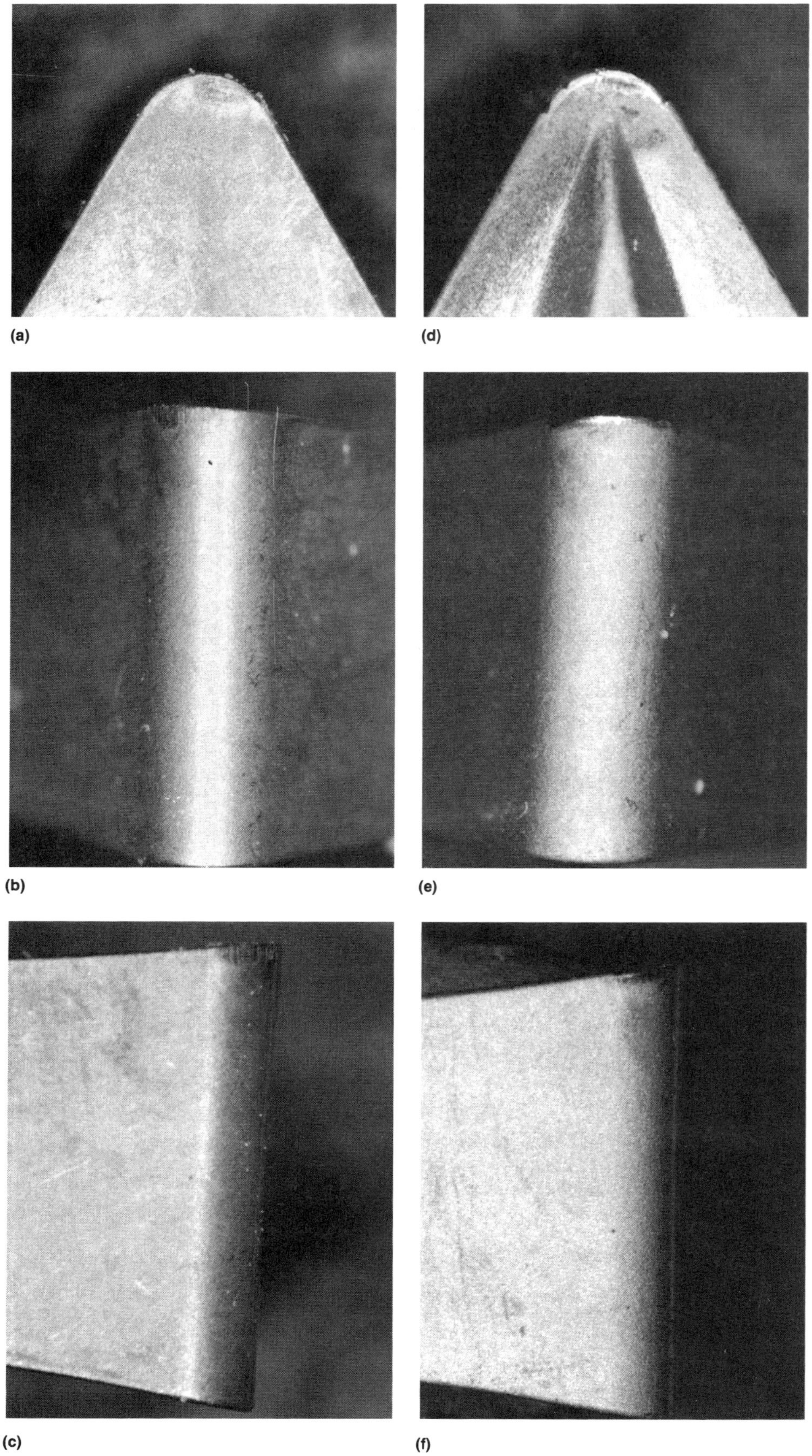


**Fig. 1** Normal flank wear encountered when cutting unalloyed, depleted uranium using two types of TNMP 432 carbide inserts (three views of each insert are shown). (a, b, c) Plain carbide (94WC-5.5Co-0.5TaC). (d, e, f) Titanium-nitride coated (TiC/TiCN/TiN coatings on a special-strengthened substrate) carbide. 15×

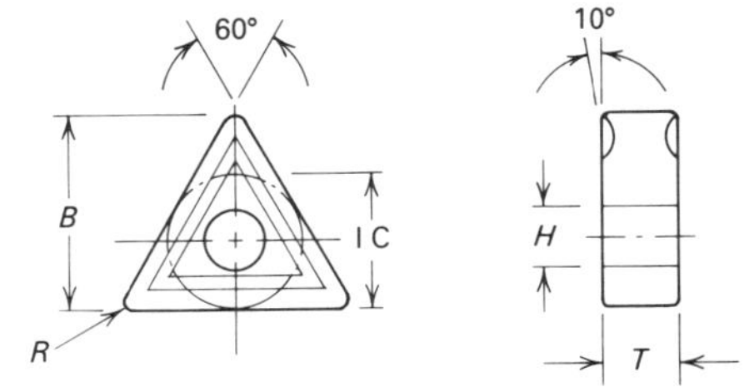


| Specification | mm | in. |
|---|---|---|
| Inscribed circle (IC) | 13 | ½ |
| *T* | 4.8 | 3/16 |
| *R* | 0.8 | 1/32 |
| *H* | 5.16 | 0.203 |
| *B* | 18.26 | 0.7188 |

**Fig. 2** Configuration and dimensions of the TNMP 432 carbide inserts shown in Fig. 1

lum content of the carbide tool, the lower the tool hardness. Therefore, high-tantalum carbides are not acceptable tools for use on most of the higher-hardness uranium alloys because of excessive edge wear on the tool.

## Surface Finish

A trade-off between surface finish and fire control exists when cutting uranium and uranium alloys. The parameters that produce the finest finish, higher cutting speed, and slower feed rate also produce finer chips, which burn easily. This is also the case when light finish cuts are taken to promote a high dimensional accuracy. To minimize the danger of fire, one should cut rough dimensions at low speed and high feed, then clear away the chips and finally take the last, light cut, allowing the chips to be submerged in the coolant.

The surface finish attainable on uranium and its alloys is comparable to that of stainless steel when similar tools, speeds, and feeds are used. An experiment was performed to compare the surface finish of uranium and two of its alloys to that of stainless steel under identical conditions. As stated previously, identical parameters used for other materials, as well as very light cuts and water spray mist, are not recommended for uranium. For strictly experimental purposes, however, this approach was accomplished under carefully controlled conditions. Figure 3 shows the surface finish measurements in microinches (roughness average, $R_a$) of four samples identically machined. The sample materials were type 316 stainless steel, high-purity depleted uranium, U-6Nb, and U-0.75Ti alloy age hardened to 43 HRC. The samples were 13 mm (½ in.) diam coupons faced on a 250 mm (10 in.) toolroom lathe at a spindle speed of 375 rev/min, a feed rate of 0.028 mm/rev (0.0011 in./rev) and a 0.08 mm (0.003 in.) depth of cut. The coolant was water spray mist, the tool was a carbide insert (94WC-5.5Co-0.5TaC nominal composition) having a

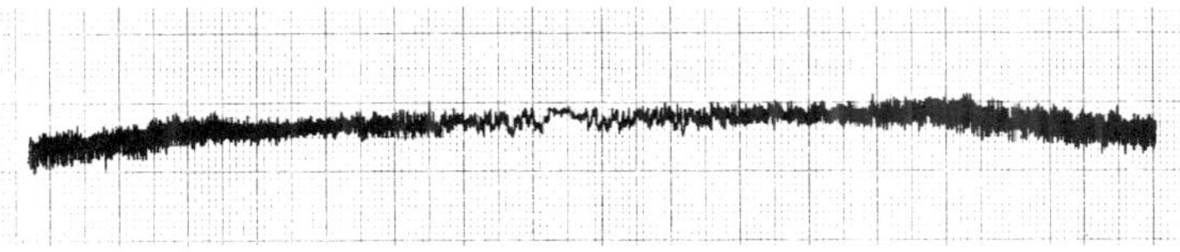

| Parameter | Surface finish: Maximum, µm | Maximum, µin. | Average, µm | Average, µin. | Minimum, µm | Minimum, µin. |
|---|---|---|---|---|---|---|
| $R_a$ | 0.317 | 12.5 | 0.216 | 8.5 | 0.157 | 6.2 |
| $R_{rms}$ | 0.378 | 14.9 | 0.274 | 10.8 | 0.20 | 7.9 |
| $R_{max}$ | 2.00 | 78.7 | 1.44 | 56.8 | 1.08 | 42.7 |
| $R_z$ | 1.70 | 67.1 | 1.28 | 50.7 | 0.978 | 38.5 |

(a)

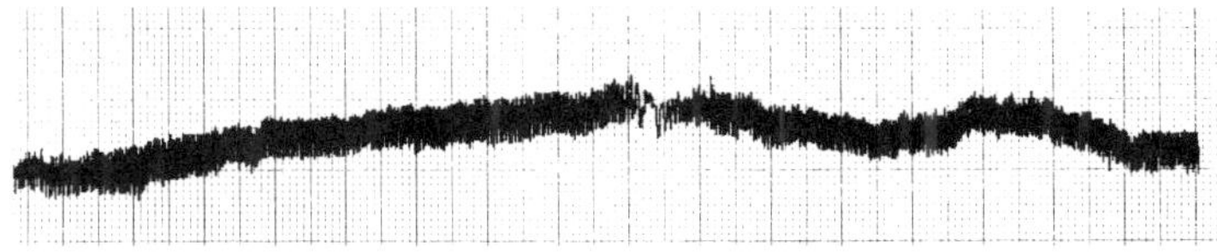

| Parameter | Surface finish: Maximum, µm | Maximum, µin. | Average, µm | Average, µin. | Minimum, µm | Minimum, µin. |
|---|---|---|---|---|---|---|
| $R_a$ | 0.432 | 17.0 | 0.384 | 15.1 | 0.325 | 12.8 |
| $R_{rms}$ | 0.517 | 20.3 | 0.462 | 18.2 | 0.414 | 16.3 |
| $R_{max}$ | 2.53 | 99.5 | 2.05 | 80.6 | 1.91 | 75.1 |
| $R_z$ | 2.02 | 79.4 | 1.86 | 73.2 | 1.77 | 69.6 |

(b)

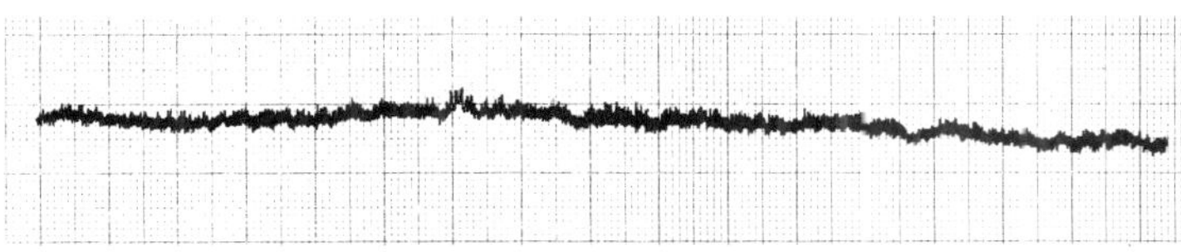

| Parameter | Surface finish: Maximum, µm | Maximum, µin. | Average, µm | Average, µin. | Minimum, µm | Minimum, µin. |
|---|---|---|---|---|---|---|
| $R_a$ | 0.180 | 7.1 | 0.147 | 5.8 | 0.122 | 4.8 |
| $R_{rms}$ | 0.229 | 9.0 | 0.187 | 7.4 | 0.157 | 6.2 |
| $R_{max}$ | 1.47 | 58.0 | 1.17 | 45.2 | 0.960 | 37.8 |
| $R_z$ | 1.15 | 45.2 | 0.960 | 37.8 | 0.869 | 34.2 |

(c)

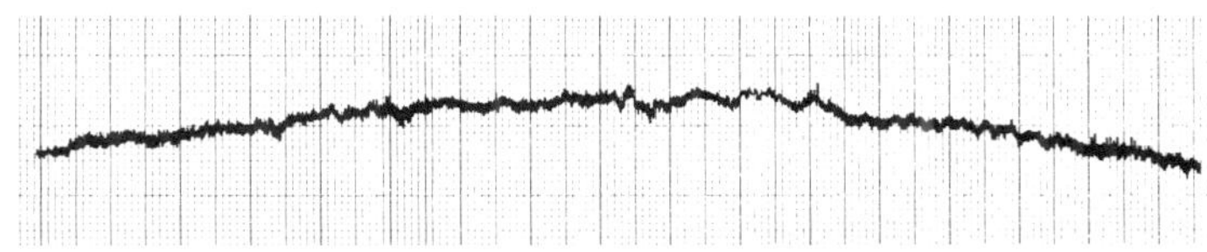

| Parameter | Surface finish: Maximum, µm | Maximum, µin. | Average, µm | Average, µin. | Minimum, µm | Minimum, µin. |
|---|---|---|---|---|---|---|
| $R_a$ | 0.175 | 6.9 | 0.135 | 5.3 | 0.112 | 4.4 |
| $R_{rms}$ | 0.221 | 8.7 | 0.173 | 6.8 | 0.142 | 5.6 |
| $R_{max}$ | 1.27 | 50.1 | 1.01 | 39.7 | 0.805 | 31.7 |
| $R_z$ | 0.945 | 37.2 | 0.838 | 33.0 | 0.729 | 28.7 |

(d)

**Fig. 3** Surface trace and surface finish of four identically machined metals and alloys. (a) Type 316 SS. (b) High-purity depleted uranium. (c) U-6Nb alloy. (d) U-0.75Ti alloy (age-hardened to 43 HRC). Identical test conditions include cutoff (0.8 mm, or 0.03 in.), drive speed (0.25 mm/s, or 0.01 in./s), sample length (25 mm, or 1 in.), and normal polarity. There was a slight variation in transverse length: (a) 8.23 mm (0.324 in.), (b) 8.43 mm (0.332 in.), (c) 8.23 mm (0.324 in.), and (d) 8.36 mm (0.329 in.). Chart scale profile is 0.25 µm/grad (10 µin./grad) in the vertical direction and 0.05 mm/grad (0.002 in./grad) in the horizontal direction. $R_a$, $R_{rms}$, $R_{max}$, and $R_z$ represent arithmetic average, root mean square, maximum peak-to-valley height, and ten-point height, respectively.

15° positive rake and 0.08 mm (0.003 in.) nose radius. Machining the 316 stainless steel produced an $R_a$ of 0.21 µm (8.5 µin.). The $R_a$ obtained on the unalloyed, depleted uranium was 0.38 µm (15.1 µin.), while the roughness average ($R_a$) on the U-Nb and U-Ti alloys was 0.145 and 0.132 µm (5.8 and 5.3 µin.), respectively. It is interesting to note that a smoother finish was obtained on the niobium and titanium alloys of uranium than was obtained on the stainless steel.

Detailed information on surface finish is available in the article "Surface Finish and Surface Integrity" in this Volume.

## Health Effects and Required Precautions

The three major health hazards of uranium are internal and external radiation exposure, biological effects, and the pyrophoric nature of the material. The effects and precautions discussed here pertain only to depleted uranium and its alloys. The hazards of enriched uranium ($U_{235}$) are greater, and more elaborate precautions are necessary.

Three ways for uranium to enter the body are ingestion, inhalation, and through an open wound (Ref 20, 21). Of these three methods, more uranium is inhaled than enters through ingestion or injection (open wound) (Ref 21).

**Internal radiation exposure** is radiation emitted by material that has been taken into the body. Once uranium enters the body, α particles emitted during radioactive decay pose the most serious internal radiation hazard. Because it is not highly penetrating, the α particle travels a short distance into the body tissues, where it is absorbed and its energy is released. This energy is concentrated into a small volume of tissue. This causes localized cellular damage that has been linked to cancer (Ref 21).

**External radiation exposure** is radiation absorbed by the body from an external source. The significant hazard of external radiation is exposure of the skin to β particles when uranium and uranium alloys are handled. Wearing gloves when possible minimizes this hazard (Ref 22). Separating the uranium from the body either by a distance of 1.2 to 1.8 m (4 to 6 ft) or by using shielding materials such as metal or glass is sufficient to minimize the radiation intensity even for long periods of exposure (Ref 21).

**Biological Effects.** Uranium is an extremely toxic substance. Soluble uranium compounds that have entered the body typically damage the kidneys, insoluble uranium compounds damage the lungs, and both soluble and insoluble uranium compounds affect the liver, though not to the degree that they affect either the kidneys or the lungs. The severity of the damage depends on the amount of uranium absorbed and the number of times the body is exposed. In some cases, small amounts of uranium can be expelled gradually, and the damage is minimal or repairable. Absorption of sufficiently large amounts of uranium or prolonged exposure, however, may cause severe and permanent illness (Ref 21).

**Pyrophoric.** Like magnesium, uranium and uranium alloys are extremely pyrophoric. Uranium turnings and fine machining dusts may burn when subjected to heat in the presence of air. A chip fire that is out of control can cause great damage and pose a serious health threat. Severe burns, radiation exposure (both internal and external) (Ref 20), and extensive property damage can occur from fires resulting from chips that are not totally submerged, but merely accumulating in chip pans and waste barrels.

Water thrown on a uranium chip fire that is out of control produces an exothermic reaction involving a disassociation of the hydrogen and oxygen in the water, thus producing a more violent eruption. Out-of-control fires should be smothered with carbon dust.

Uranium turnings should be promptly recycled or otherwise disposed of by approved disposal techniques.

**Precautions Observed in Machining Uranium.** To prevent uranium contamination, eating, drinking, and smoking should be prohibited while personnel are working in the uranium area. Chip fires hot enough to vaporize the metal must be prevented, and an approved respirator should be easily accessible in case of airborne contamination. Protective clothing consisting of coveralls and shoe coverings must be worn in place of street clothes. Upon leaving the uranium working area, personnel should wash hands and faces before eating, drinking, or smoking, and anyone machining or forming uranium should shower before leaving the area. Uranium and materials that come in contact with uranium such as cutting oil, cutting tools, protective clothing, and so on, are considered hazardous materials, the disposal of which is governed by state and federal regulations.

Although the effects of depleted uranium on health are more subdued than those of enriched uranium and plutonium, they are no less real. When one is working with any uranium or uranium alloy, the material should be respected as a hazardous and potentially dangerous material.

When feasible, machining of uranium and uranium alloys should be done on a machine that has negative air relative to the operator/machinist tending the equipment. Air flow should be sufficient to entrain any particulates, and the resulting mixture should be passed through appropriate filters.

## ACKNOWLEDGMENT

The author would like to express his thanks to J.M. Aldrich and S.S. Christiansen for their help in preparing this article.

## REFERENCES

1. R.J. Jackson, Rocky Flats Plant, Rockwell International, private communications, Feb 1988
2. R.J. Jackson, "Reversible Martensitic Transformations Between Transition Phases of Uranium-Base Niobium Alloys," RFP-1535, Rockwell International
3. J.R. Hammetter, "Depleted Uranium Beta Quench Study Task No. 27," submitted for publication
4. R.J. Jackson, "Properties of the Uranium-6.3 wt.% Niobium Alloy," in Rocky Flats Material Datasheets, Uranium-6% Niobium, RFP-1613, Rockwell International, Oct 1971, p 1-19
5. L.J. Frick *et al.*, "Uranium Health Physics Personnel Briefing," Rocky Flats Plant, Rockwell International, unpublished data, March 1986
6. L.T. Lloyd and C.S. Barrett, Thermal Expansion of Alpha Uranium, *J. Nucl. Mater.*, Vol 18, 1966, p 55-59
7. M.C. Mataya and P.A. Steinmeyer, The Effect of Rolling on Thermal Expansion and Crystallographic Texture in Uranium, submitted to *Metall. Trans.*
8. T.L. Gripp, Rocky Flats Plant, Rockwell International, private communications, Jan 1988
9. A.F. Blazon, Rocky Flats Plant, Rockwell International, private communications, Jan 1988
10. W.R. Sharpe, U.S. Department of the Army, Army Research and Development and Engineering Center, private communications, April 1988
11. J.F. Boland and D.J. Sandstrom, Mechanical Fabrication, Heat Treatment, and Machining of Uranium Alloys, in *Physical Metallurgy of Uranium Alloys*, Brook Hill Publishing, 1976
12. E.R. Knight, "Machinability Program for Depleted Uranium Alloy Penetrators," AD-A010 351, Lake City Army Ammunition Plant, Nov 1974, p 9, 18, 20
13. T.O. Morris, "Machining of Uranium and Uranium Alloys," Paper presented in part at the Uranium Technology Seminar, American Society for Metals (Gatlinburg, TN), May 1981
14. J.E. Fuller and J.K. Lynch, "Machining Study of a Uranium-4 Weight Percent Niobium Alloy," RFP-1743, Rocky Flats Plant, Rockwell International, Nov 1971
15. R.L. Clites, Rocky Flats Plant, Rockwell International, private communications, Feb 1988
16. M.C. Mataya, Rocky Flats Plant, Rockwell International, private communications, April 1988
17. R.A. Martin, Rocky Flats Plant, Rockwell International, private communications, Feb 1988
18. G.L. VanArsdale, Battelle, Pacific Northwest Laboratories, private communications, July 1988
19. T.O. Morris, Martin Marietta Corporation, Oak Ridge Y-12 Plant, private communications, April 1988
20. J.M. Aldrich, Rocky Flats Plant, Rockwell International, private communications, Feb 1988
21. "Safety Procedures for Processing Depleted Uranium," U.S. Army Material Development and Readiness Command, Aug 1978

## SELECTED REFERENCES

- S.S. Christiansen, "Shape Memory Effects in Reversible Martensitic Transformations," Colorado School of Mines, unpublished research, May 1988
- J. Conboy and P. Shevchik, "DU Chip Recovery Program, Phase I, A Machining Study for the Production of Contaminant-Free Chips," AD-A131 389/9, South Creek Industries, Inc., July 1983
- A.N. Holden, *Physical Metallurgy of Uranium*, Addison-Wesley, 1958
- C.E. Latham-Brown and F. Porter, "Atmosphere Assisted Machining of Depleted Uranium (DU) Penetrators," AD-A182 138/8/XAB, U.S. Army Armament Research, Development and Engineering Center
- "Radiation Protection for Occupational Workers," Order 5480.11, U.S. Department of Energy, Dec 1987
- R.A. Vandermeer *et al.*, A Phenomenological Study of the Shape Memory Effect in Polycrystalline Uranium-Niobium Alloys, *Metall. Trans. A*, Vol 12A, 1981, p 733-741

# Machining of Powder Metallurgy Materials

Harry E. Chandler, ASM INTERNATIONAL

POWDER METALLURGY (P/M) is a near-net shape process capable of producing complex parts with little or no need for secondary operations such as machining, joining, or assembly. However, the inability to produce certain geometrical figures such as transverse holes, undercuts, and threads frequently necessitates some machining, particularly drilling. Because of the growing use of P/M materials to increase productivity and reduce costs, the study of the machinability of these materials has become increasingly important, especially because powder metals are generally considered to have poor machinability in comparison with wrought or cast metals (Ref 1). In addition, there is a trend toward the design of complex P/M parts that call for machining, in contrast to the past practice to minimize machining (Ref 2).

The machinability of P/M components is dependent upon a number of factors, including workpiece and tool material properties, cutting conditions, machine and cutting tool parameters (Ref 3, 4), as well as some P/M material and production process parameters such as particle size, part geometry, porosity, and compaction and sintering methods.

The cutting characteristics of a workpiece material can be controlled by the alteration of the microstructure by means of changes in chemical composition, the addition of free-machining additives, or various mechanical treatments (Ref 1, 5, 6). These techniques often have limited influence because of interactions among different factors. Optimization techniques include adjustments in cutting conditions and tool geometry and the selection of proper tooling materials. Microstructural changes seldom can be made to improve machinability without introducing some significant negative influence on mechanical and physical properties. Materials prealloyed with manganese sulfide are an exception.

## Optimizing P/M Part Machining

The following are measures that can optimize the machining of P/M materials:

- Taking advantage of the good machining properties of high-density parts (above 92% of theoretical density)
- Machining in the presintered condition
- Using additives that do not require sacrifices in powder processing or part properties
- Following the suggestions of published guidelines for material and machining methods
- Running machinability tests on P/M materials

**High Density.** The machinability of high-density parts is comparable to that of wrought metals. Speeds and feeds are similar, for example (Ref 6).

Forging is one of the methods used to obtain high-density parts. Figure 1 shows a P/M connecting rod for a 1.8 L automotive engine that underwent double disk grinding, deburring, and shot peening. The P/M material was a copper steel composition. The ultimate tensile strength of the finished part was about 825 MPa (120 ksi), the fatigue strength was 55 MPa (8 ksi) at $10^7$ cycles, and the hardness was in the range of 76 to 89 HRG (Ref 7).

**Presintering.** Machining usually follows presintering, which is performed to volatilize and burn off pressing lubricants. The lubricant must be removed because it may cause carburization of ferrous alloys during final sintering; carburization reduces machinability and improves mechanical properties and corrosion resistance of P/M alloys. In terms of machining speed and tool wear, the gains are attractive. However, there can be losses in part tolerances and finishing characteristics (Ref 8).

**Additives.** Free-machining benefits can be obtained by means of small additions to a standard powder composition (Ref 6). Additives for ferrous powders include lead, sulfur, copper, or graphite; for nonferrous powders, lead is used. The advantages of changing composition in this manner can be at least partially offset by side effects. Additions may cause problems such as dimensional changes of parts during sintering, and there could be some deterioration in the properties of parts.

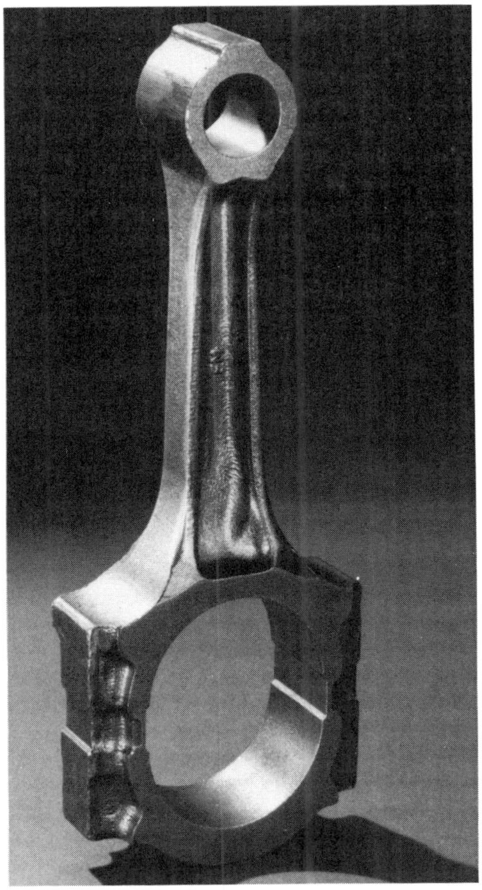

**Fig. 1** Hot forged P/M connecting rod that was double-disk ground. Machinability was the key factor in adopting this part making process. Courtesy of P.K. Johnson, Metal Powder Industries Federation

Prealloyed manganese sulfide powders, such as MP 37, appear to avoid those shortcomings in ferrous alloys (Ref 9). Manganese content is intentionally high to ensure that all sulfur is present in the form of manganese sulfide inclusions. When these inclusions are extensively deformed in the shear plane and in the flow zone adjacent to the tool surface, they contribute to higher cutting speeds, longer tool life, good surface finish on parts, and lower tool forces. In addition, chips are more readily handled

**Table 1 Typical compositions for ferrous P/M structural materials**

| P/M material | MPIF designation | MPIF composition limits, %(a) C | Ni | Mo | Cu | Fe |
|---|---|---|---|---|---|---|
| Iron | F-0000 | 0–0.3 | . . . | . . . | . . . | 97.7–100 |
| Carbon steel | F-0005 | 0.3–0.6 | . . . | . . . | . . . | 97.4–99.7 |
| | F-0008 | 0.6–0.9 | . . . | . . . | . . . | 97.1–99.4 |
| Iron-copper | FC-0200 | 0–0.3 | . . . | . . . | 1.5–3.9 | 93.8–98.5 |
| Copper steel | FC-0205 | 0.3–0.6 | . . . | . . . | 1.5–3.9 | 93.5–98.2 |
| | FC-0208 | 0.6–0.9 | . . . | . . . | 1.5–3.9 | 93.2–97.9 |
| | FC-0505 | 0.3–0.6 | . . . | . . . | 4.0–6.0 | 91.4–95.7 |
| | FC-0508 | 0.6–0.9 | . . . | . . . | 4.0–6.0 | 91.1–95.4 |
| | FC-0808 | 0.6–0.9 | . . . | . . . | 7.0–9.0 | 88.1–90.4 |
| | FC-1000 | 0–0.3 | . . . | . . . | 9.5–10.5 | 87.2–90.5 |
| Iron-nickel | FN-0200 | 0–0.3 | 1.0–3.0 | . . . | 0–2.5 | 92.2–99.0 |
| Nickel steel | FN-0205 | 0.3–0.6 | 1.0–3.0 | . . . | 0–2.5 | 91.9–98.7 |
| | FN-0208 | 0.6–0.9 | 1.0–3.0 | . . . | 0–2.5 | 91.6–98.4 |
| | FN-0405 | 0.3–0.6 | 3.0–5.5 | . . . | 0–2.0 | 89.9–96.7 |
| | FN-0408 | 0.6–0.9 | 3.0–5.5 | . . . | 0–2.0 | 89.6–96.4 |
| Low-alloy steel | FL-4205 | 0.4–0.7 | 0.35–0.55 | 0.50–0.85 | . . . | 95.9–98.7 |
| | FL-4605 | 0.4–0.7 | 1.70–2.00 | 0.40–0.80 | . . . | 94.5–97.5 |
| Copper-infiltrated iron | FX-1000 | 0–0.3 | . . . | . . . | 8.0–14.9 | 82.8–92.0 |
| Copper-infiltrated steel | FX-1005 | 0.3–0.6 | . . . | . . . | 8.0–14.9 | 82.5–91.7 |
| | FX-1008 | 0.6–0.9 | . . . | . . . | 8.0–14.9 | 82.2–91.4 |
| | FX-2000 | 0–0.3 | . . . | . . . | 15.0–25.0 | 72.7–85.0 |
| | FX-2005 | 0.3–0.6 | . . . | . . . | 15.0–25.0 | 72.4–84.7 |
| | FX-2008 | 0.6–0.9 | . . . | . . . | 15.0–25.0 | 72.1–84.4 |

(a) Metal Powder Industries Federation (MPIF) standards require that the total of all other elements be less than or equal to 2.0%. Source: Ref 10

than those produced by conventional P/M materials. More detailed information on MP 37 powder can be found later in this article.

Another additive-type technique, the oil or resin impregnation of porous P/M parts, also improves machinability. The most common application for oil impregnation is for self-lubricating bearings (see the article "P/M Bearings" in Volume 7, 9th Edition of *Metals Handbook*).

**Material Guidelines.** Published information is available to help in the selection of P/M materials for good machinability, in the design of parts that are within the parameters of the P/M process, and in machining P/M materials by proven procedures (Ref 6, 10). Many medium- to high-density materials have been standardized by the Metal Powder Industries Federation (MPIF). The MPIF designations for ferrous P/M materials, described in Ref 10, include a prefix of one of more letters (the first of which is F to indicate an iron-base, or ferrous, material), four numerals, and a two- or three-digit suffix. The second letter in the prefix identifies the principal alloying element (if one is specified); the percentage of that element is indicated by the first two numerals. The third and fourth numerals indicate the amount of combined carbon; the code designation 00 indicates 0.0 to 0.3% C, 05 indicates 0.3 to 0.6% C, and 08 indicates 0.6 to 0.9% C. The suffix number represents the specified minimum strength value (in ksi). For example, FC-0208-60 is a P/M copper steel material containing nominal 2% Cu and 0.8% combined carbon possessing a minimum yield strength of 60 ksi in the as-sintered condition.

For nonferrous materials, the first two numbers in the four-digit series designate the percentage of the major alloying constituent. The last two numbers of the four-digit series designate the percentage of the minor alloying constituent.

For improved machinability, lead is sometimes the third alloying element in a nonferrous alloy system. Lead will then be indicated only by the letter P in the prefix. The percentage of lead, or any other minor alloying element that happens to be excluded from the four-digit nomenclature, is represented in the chemical composition that appears with each standard material.

For example, CNZP-1816-15 is a nickel-silver alloy containing nominal 18% Ni, 16% Zn, and 1 to 2% Pb, with a minimum yield strength of 15 ksi.

Compositions for a variety of P/M structural materials are given in Tables 1 to 3. Hardness and density values for these materials are given in Table 4.

P/M carbon steels are selected primarily for parts with moderate strength and hardness, combined with machinability. Iron-copper and copper steel materials are produced from admixtures of elemental iron powder and elemental copper powder with or without graphite powder (carbon). When secondary machining is required, combined carbon contents of less than 0.5% should be specified. Copper-infiltrated iron and steel materials offer improved machinability because of reductions in interrupted cuts, and machined parts have a smooth surface finish. Among stainless steels, SS-303 is preferred when parts require extensive secondary machining. Brass, bronze, and nickel silver parts usually have good machinability.

**Design.** Certain types of holes, undercuts, and threads are examples of features that cannot be accommodated by the P/M consolidation (pressing) process and therefore require machining. Holes in the direction of pressing, produced with core rods that extend up through the tools, are readily incorporated in parts (Ref 6), but side holes or those not parallel to the direction of pressing cannot be made in the same way and are generally produced by secondary machining.

Undercuts on the horizontal plane (perpendicular to the die centerline) cannot be produced if they prevent the part from ejecting from the die (Fig. 2). Annular grooves around a part are produced by machining or by making the part in an assembly of two pieces. Likewise, a part with a reverse taper (larger on bottom than on top) cannot be ejected from a die. Because threads in holes and on outside diameters prevent a part from being ejected from a die, they cannot be made with conventional P/M methods; machining is required.

**Representative machining methods** are turning, milling, drilling, tapping, and grinding. The basic guideline is to start with cutting conditions for wrought and cast parts and tailor them for the application. The recommendation applies in particular to lower-density parts. High-density parts (>92% of theoretical density) machine like

**Table 2 Typical compositions for P/M stainless steels**

| MPIF designation | MPIF composition limits, %(a) Fe | Cr | Ni | Mn | Si | S | C | P | Mo | N |
|---|---|---|---|---|---|---|---|---|---|---|
| SS-303N1, N2 | rem | 17.0–19.0 | 8.0–13.0 | 0–2.0 | 0–1.0 | 0.15–0.30 | 0–0.15 | 0–0.20 | . . . | 0.2–0.6 |
| SS-303L | rem | 17.0–19.0 | 8.0–13.0 | 0–2.0 | 0–1.0 | 0.15–0.30 | 0–0.03 | 0–0.20 | . . . | . . . |
| SS-304N1, N2 | rem | 18.0–20.0 | 8.0–12.0 | 0–2.0 | 0–1.0 | 0–0.03 | 0–0.08 | 0–0.045 | . . . | 0.2–0.6 |
| SS-304L | rem | 18.0–20.0 | 8.0–12.0 | 0–2.0 | 0–1.0 | 0–0.03 | 0–0.03 | 0–0.045 | . . . | . . . |
| SS-316N1, N2 | rem | 16.0–18.0 | 10.0–14.0 | 0–2.0 | 0–1.0 | 0–0.03 | 0–0.08 | 0–0.045 | 2.0–3.0 | 0.2–0.6 |
| SS-316L | rem | 16.0–18.0 | 10.0–14.0 | 0–2.0 | 0–1.0 | 0–0.03 | 0–0.03 | 0–0.045 | 2.0–3.0 | . . . |
| SS-410 | rem | 11.5–13.0 | . . . | 0–1.0 | 0–1.0 | 0–0.03 | 0–0.25 | 0–0.04 | . . . | 0.2–0.6 |

(a) MPIF standards require that the total of all other elements be less than or equal to 2.0%. rem, remainder. Source: Ref 10

**Table 3 Typical compositions for copper-base P/M structural materials**

| Material | MPIF designation | MPIF composition limits, %(a) | | | | |
|---|---|---|---|---|---|---|
| | | Cu | Zn | Pb | Sn | Ni |
| Bronze | CT-1000 | 87.5–90.5 | rem | 1.0–2.0 | 9.5–10.5 | ... |
| Brass | CZ-1000 | 88.0–91.0 | rem | ... | ... | ... |
| | CZP-1002 | 88.0–91.0 | rem | 1.0–2.0 | ... | ... |
| | CZP-2002 | 77.0–80.0 | rem | 1.0–2.0 | ... | ... |
| | CZ-3000 | 68.5–71.5 | rem | ... | ... | ... |
| | CZP-3002 | 68.5–71.5 | rem | 1.0–2.0 | ... | ... |
| Nickel-silver | CNZ-1818 | 62.5–65.5 | rem | ... | ... | 16.5–19.5 |
| | CNZP-1816 | 62.5–65.5 | rem | 1.0–2.0 | ... | 16.5–19.5 |

(a) MPIF standards require that the total of all other elements be less than or equal to 2.0%. Source: Ref 10

wrought metals. Speeds and feeds, for instance, are similar.

Smearing of self-lubricating porous parts can be a problem. Recommended practice involves the use of sharp tools and light cuts in single-point machining, such as turning or boring.

Coolants are preferred in most machining operations. Coolant pickup can be a problem: The rate of pickup is directly related to amount of porosity. Ideally, all machining except grinding should precede deburring. Retained deburring abrasive can cause excessive tool wear. Ceramic and cubic boron nitride (CBN) inserts are usually run dry; performance is typically better than or at least equal to that obtained with coolants.

Tooling materials, depending upon the application, range from high-speed steels and coated or uncoated carbides to cermets and ceramics.

In drilling, speeds and feeds are 80 to 85% of those for wrought metals of the same composition. For long tool life, nitrided steel, high-speed steels containing cobalt, and carbide-tipped drills are recommended.

Internal threads are produced with conventional tapping methods. Tool life is rated excellent, when using premium-grade high-speed steel or nitrided taps. Roll form tapping is another technique in use, especially for blind holes.

The life of a given tool material used in machining a given P/M material can be determined by simple testing, such as drilling, as described later in this article. For work materials, hardness can be used as an indicator of machinability with this caveat: With P/M an increase in porosity can result in a reduction in hardness and machinability (Ref 11). Tests for both tool and workpiece materials are described in the article "Machinability Test Methods" in this Volume.

Usually, parts with an average hardness of HRB 52 have machining properties similar to those of cast iron. At this hardness level, parts must be held rigid, and weak sections must be supported to prevent distortion. Compressed air is used to cool the tool and maintain swarf clearances. Jets are directed onto tool cutting edges and work surfaces. Liquid coolants cannot be used because parts must be kept dry and clean for subsequent sintering. Carbide tips of ISO designation K10 with a hardness of 92.6 HRA give good results and will accept some interruptions on the cutting surface.

Tools must be held rigid, and cutting edges must be sharp with rake angles of up to 3° positive on top and side and frontal clearances of 3 to 5°. Surface speeds of 105 to 120 m/min (350 to 400 sfm) and feeds of 0.050 to 0.10 mm/rev (0.002 to 0.004 in./rev) are satisfactory for form turning, but surface speeds can be increased to 180 to 210 m/min (600 to 700 sfm) in single-point turning. Feeds can be increased within the boundaries of economic tool life, the standard of accuracy, and surface finish requirements.

In machining fully sintered parts (average hardness, 90 HRB), K10 carbide tips give a satisfactory life with 0° top rake, 7° frontal clearance, and 5° side clearance. Single-point finish turning is used, with a stock removal of 0.125 to 0.20 mm (0.005 to 0.008 in.) of surface depth and 0.050 mm (0.002 in.) of feed per revolution. Surface speeds are 120 to 135 m/min (400 to 450 sfm), and tips require a radius of 0.20 to 0.25 mm (0.008 to 0.010 in.). Form turning is not advisable because of the work-hardening characteristics of the material in this state. When CBN tools are used, surface speeds can be increased from 600 to 1000 m/min (2000 to 3280 sfm) using the same rake and clearances as those for K10 tools. These tools will also accept some degree of interruption on the cutting surface. With K10, a coolant is necessary to keep the tool cool and to maintain swarf clearance. With CBN, the procedure used depends on the workpiece material. Results from tests run dry are usually similar to those run wet. When the wet technique is used, the tool must be flooded with coolant continuously.

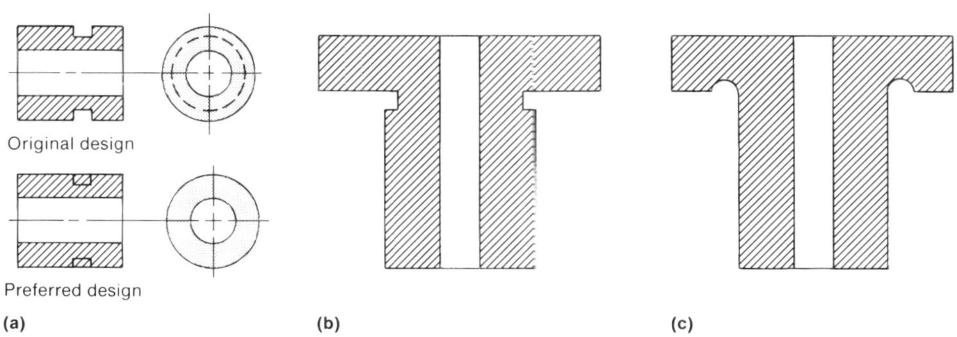


**Fig. 2** P/M part design considerations. (a) Undercuts on horizontal plane cannot be produced in P/M process. Machining is required to obtain such features in parts. (b) Example of undercut in flange that is beyond capability of P/M process. (c) Alternative to part in (b) that can be made without secondary machining

## Factors Affecting P/M Machinability

Researchers have explained the poor machinability of P/M parts on the basis of the interrupted-cutting theory, hard-inclusion theory, and thermal-conductivity theory (Ref 12-16). However, the factors assumed in these theories are hard to isolate, to evaluate independently, and to determine the contribution to machinability. Poor machinability is clearly due to the presence of porosity, which also affects mechanical and physical properties. Both tensile strength and thermal conductivity are reduced dramatically by the presence of porosity. While lower strength often results in better machinability, the disadvantage of lower thermal conductivity more than offsets the advantage of lower strength (Ref 17, 18).

Much of the current research and development on machinability focuses on the influence of porosity, various aspects of the P/M process, chemical compositions, machining variables, and physical properties of the material (Ref 1).

### Effects of Porosity

One of the distinct characteristics of P/M materials—porosity—causes noncontinuous contact between the workpiece and cutting edge of the tool, leading to the postulation of an interrupted-cutting model that attempts to describe the influence of porosity on machinability (Ref 6, 8, 12). In this model, increased tool wear is thought to occur because the cutting tool experiences continuous load variation and is subjected to microshocks and vibration as it passes through the porous material.

It has been proposed that the cutting process includes "microblows" at the cutting edge (Ref 12). Using a scanning electron microscope to examine the tool used to

## Table 4 Hardness and density values for P/M structural materials

| MPIF designation | Density, g/cm³ | Apparent hardness(a) | Matrix hardness(b) |
|---|---|---|---|
| **Irons and steels** | | | |
| F-0000-10 | 6.1 | 40 HRF | ... |
| -15 | 6.7 | 60 HRF | ... |
| -20 | 7.3 | 80 HRF | ... |
| F-0005-15 | 6.1 | 25 HRB | ... |
| -20 | 6.6 | 40 HRB | ... |
| -25 | 6.9 | 55 HRB | ... |
| F-0005-50HT(c) | 6.6 | 20 HRC | 58 HRC |
| -60HT | 6.8 | 22 HRC | 58 HRC |
| -70HT | 7.0 | 25 HRC | 58 HRC |
| F-0008-20 | 5.8 | 35 HRB | ... |
| -25 | 6.2 | 50 HRB | ... |
| -30 | 6.6 | 60 HRB | ... |
| -35 | 7.0 | 70 HRB | ... |
| F-0008-55HT | 6.3 | 22 HRC | 60 HRC |
| -65HT | 6.6 | 28 HRC | 60 HRC |
| -75HT | 6.9 | 32 HRC | 60 HRC |
| -85HT | 7.1 | 35 HRC | 60 HRC |
| **Iron-copper and copper steels** | | | |
| FC-0200-15 | 6.0 | 11 HRB | ... |
| -18 | 6.3 | 18 HRB | ... |
| -21 | 6.6 | 26 HRB | ... |
| -24 | 6.9 | 36 HRB | ... |
| FC-0205-30 | 6.0 | 37 HRB | ... |
| -35 | 6.3 | 48 HRB | ... |
| -40 | 6.7 | 60 HRB | ... |
| -45 | 7.1 | 72 HRB | ... |
| FC-0205-60HT | 6.2 | 19 HRC | 58 HRC |
| -70HT | 6.5 | 25 HRC | 58 HRC |
| -80HT | 6.8 | 31 HRC | 58 HRC |
| -90HT | 7.0 | 36 HRC | 58 HRC |
| FC-0208-30 | 5.8 | 50 HRB | ... |
| -40 | 6.3 | 61 HRB | ... |
| -50 | 6.7 | 73 HRB | ... |
| -60 | 7.2 | 84 HRB | ... |
| FC-0208-50HT | 6.1 | 20 HRC | 60 HRC |
| -65HT | 6.4 | 27 HRC | 60 HRC |
| -80HT | 6.8 | 35 HRC | 60 HRC |
| -95HT | 7.1 | 43 HRC | 60 HRC |
| FC-0505-30 | 5.8 | 51 HRB | ... |
| -40 | 6.3 | 62 HRB | ... |
| -50 | 6.7 | 72 HRB | ... |
| FC-0508-40 | 5.9 | 60 HRB | ... |
| -50 | 6.3 | 68 HRB | ... |
| -60 | 6.8 | 80 HRB | ... |
| FC-0808-45 | 6.0 | 65 HRB | ... |
| FC-1000-20 | 6.0 | 15 HRB | ... |
| **Iron-nickel and nickel steels** | | | |
| FN-0200-15 | 6.6 | ... | ... |
| -20 | 6.7 | 75 HRF | ... |
| -25 | 7.3 | ... | ... |
| FN-0205-20 | 6.6 | 44 HRB | ... |
| -25 | 6.9 | 59 HRB | ... |
| -30 | 7.2 | 69 HRB | ... |
| -35 | 7.4 | 78 HRB | ... |
| FN-0205-80HT | 6.6 | 23 HRC | 55 HRC |
| -105HT | 6.9 | 29 HRC | 55 HRC |
| -130HT | 7.1 | 33 HRC | 55 HRC |
| -155HT | 7.2 | 36 HRC | 55 HRC |
| -180HT | 7.4 | 40 HRC | 55 HRC |
| FN-0208-30 | 6.7 | 63 HRB | ... |
| -35 | 6.9 | 71 HRB | ... |
| -40 | 7.1 | 77 HRB | ... |
| -45 | 7.3 | 83 HRB | ... |
| -50 | 7.4 | 88 HRB | ... |
| FN-0208-80HT | 6.7 | 26 HRC | 57 HRC |
| -105HT | 6.9 | 31 HRC | 57 HRC |
| -130HT | 7.0 | 35 HRC | 57 HRC |
| -155HT | 7.2 | 39 HRC | 57 HRC |
| -180HT | 7.4 | 42 HRC | 57 HRC |
| FN-0405-25 | 6.5 | 49 HRB | ... |
| -35 | 7.0 | 71 HRB | ... |
| -45 | 7.4 | 84 HRB | ... |
| FN-0405-80HT | 6.5 | 19 HRC | 55 HRC |
| -105HT | 6.8 | 25 HRC | 55 HRC |
| -130HT | 7.0 | 31 HRC | 55 HRC |
| -155HT | 7.3 | 37 HRC | 55 HRC |
| -180HT | 7.4 | 40 HRC | 55 HRC |
| FN-0408-35 | 6.5 | 67 HRB | ... |
| -45 | 6.9 | 78 HRB | ... |
| -55 | 7.2 | 87 HRB | ... |
| **Low-alloy steels** | | | |
| FL-4205-80HT | 6.60 | 28 HRC | 60 HRC |
| -100HT | 6.80 | 32 HRC | 60 HRC |
| -120HT | 7.00 | 36 HRC | 60 HRC |
| -140HT | 7.20 | 39 HRC | 60 HRC |
| FL-4605-80HT | 6.55 | 24 HRC | 60 HRC |
| -100HT | 6.75 | 29 HRC | 60 HRC |
| -120HT | 6.95 | 34 HRC | 60 HRC |
| -140HT | 7.20 | 39 HRC | 60 HRC |
| **Copper-infiltrated iron and steel** | | | |
| FX-1000-25 | 7.3 | 65 HRB | ... |
| FX-1005-40 | 7.3 | 82 HRB | ... |
| FX-1005-110HT | 7.3 | 38 HRC | 55 HRC |
| FX-1008-50 | 7.3 | 89 HRB | ... |
| FX-1008-110HT | 7.3 | 43 HRC | 58 HRC |
| FX-2000-25 | 7.3 | 66 HRB | ... |
| FX-2005-45 | 7.3 | 85 HRB | ... |
| FX-2005-90HT | 7.3 | 36 HRC | 55 HRC |
| FX-2008-60 | 7.3 | 90 HRB | ... |
| FX-2008-90HT | 7.3 | 36 HRC | 58 HRC |
| **Stainless steels** | | | |
| SS-303N1-25 | 6.4 | 62 HRB | ... |
| SS-303N2-35 | 6.5 | 63 HRB | ... |
| SS-303L-12 | 6.6 | 21 HRB | ... |
| SS-304N1-30 | 6.4 | 61 HRB | ... |
| SS-304N2-33 | 6.5 | 62 HRB | ... |
| SS-304L-13 | 6.6 | ... | ... |
| SS-316N1-25 | 6.4 | 59 HRB | ... |
| SS-316N2-33 | 6.5 | 62 HRB | ... |
| SS-316L-15 | 6.6 | 20 HRB | ... |
| SS-410-90HT | 6.5 | 23 HRC | 55 HRC |
| **Brass, bronze, and nickel-silver** | | | |
| CZ-1000-9 | 7.60 | 65 HRH | ... |
| -10 | 7.90 | 72 HRH | ... |
| -11 | 8.10 | 80 HRH | ... |
| CZP-2002-11 | 7.60 | 75 HRH | ... |
| -12 | 8.00 | 84 HRH | ... |
| CZ-3000-14 | 7.60 | 84 HRH | ... |
| -16 | 8.00 | 92 HRH | ... |
| CZP-3002-13 | 7.60 | 80 HRH | ... |
| -14 | 8.00 | 88 HRH | ... |
| CNZ-1818-17 | 7.90 | 90 HRH | ... |
| CT-1000-13 (repressed) | 7.20 | 82 HRH | ... |

(a) The hardness value of a P/M part when using a conventional indentation hardness tester is referred to as apparent hardness because it represents a combination of matrix hardness and effect of porosity. Apparent hardness measures the resistance to indentation or brinelling. (b) Matrix hardness is determined by microhardness testers using Knoop (HK) or diamond pyramid hardness (DPH) indentors. It measures the true hardness of the structure by eliminating the effect of porosity, and thus is a measure of resistance to abrasive and adhesive wear. (c) When the letters HT appear after the suffix digits, it is understood that the P/M material specified has been heat treated (quenched and tempered) and that the strength indicated is the ultimate tensile strength in ksi. Source: Ref 10

turn porous iron, it was found that the tool cutting edge was chipped an amount approximately equal to the 1 μm (40 μin.) pore size of the material. Chipping was attributed to microcrumbling, occurring as the tool passed through voids within the porous part. Abrasive tool wear was claimed to be due to the noncontinuous contact between the machined surface and the tool material at the flank face, which allowed abrasive particles to enter this wear junction. The same investigator also found that the grinding ratio decreased with increased porosity while grinding 70 to 88% dense iron material (Ref 19).

In another study it was found that drill flank wear increased with decreasing relative density (from 90 to 70%, as shown in Fig. 3) (Ref 18). At 64% density 304L stainless steel had very poor machinability; the tool failed while drilling the first hole. Higher density materials failed primarily because of flank wear, while those of lower density typically failed because of excessive margin wear. The effect of density on drill life was significant, as shown in Fig. 4. Drill life lengthened with increasing density at two different cutting speeds. Life at the 12 m/min (39 sfm) cutting speed was 50 to 90% lower than that at 8.5 m/min (28 sfm). It is clear that wear data and tool life relationships developed for fully dense materials cannot be expected to be valid for porous materials because of their discontinuous nature.

One major auto maker has found that the machining characteristics of forged P/M parts are similar to those of wrought or cast counterparts (Ref 17). No significant difference was observed in turning, boring, or drilling forged AISI-SAE wrought steel parts or their P/M counterparts made of water-atomized AISI-SAE 1141 powder. Titanium nitride coated carbide tools and ceramics are preferred for machining P/M forgings to reduce abrasive tool wear, which is more apparent than it is with wrought forgings.

**Temperature and Wear.** The porosity of P/M materials also influences their physical properties, such as thermal conductivity and specific heat. An increase in porosity causes a decrease in thermal conductivity (Ref 20-24).

Lower-density materials, having lower absorption and diffusivity coefficients, conduct less heat than do the more dense materials (Ref 17). Higher amounts of heat remain in the cutting zone when low-density materials are machined. This results in higher temperatures in the cutting zone. For this reason, the thermal conductivity of material is important in machining P/M parts.

The conductivity phenomenon can intensify thermal types of tool wear. As a result, the amount of wear can be expected to increase relative to porosity. It is also possible for a temperature increase to have an appreciable effect on oxidation and other chemical reactions on the generated surfaces. This may in turn affect tool wear, depending on reactions at the tool/workpiece and tool/chip interfaces. A high temperature in the cutting zone accelerates chemical reactions in the lubrication films in the contact zones. Many different wear phenomena are influenced by a temperature increase. Several authors have shown that adhesive wear predominates in the machining of P/M steels (Ref 12, 19).

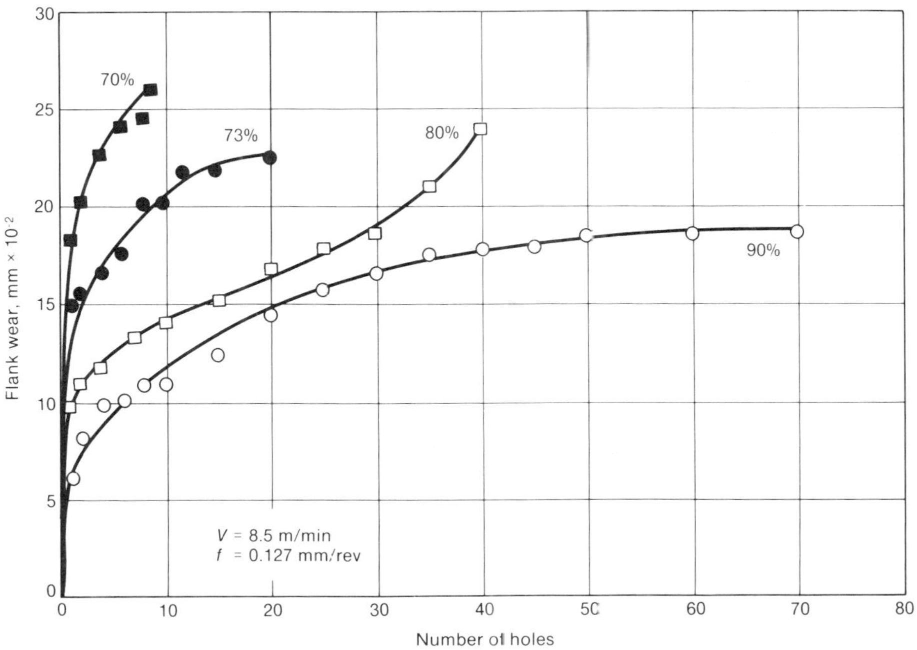


**Fig. 3** Flank wear growth as function of density. P/M material is 304L stainless steel. *V*, speed; *f*, feed

Average drill flank temperature was measured in drilling a P/M version of AISI 304L stainless steel (Ref 1). Variations in temperature with density for new and worn drills at various hole depths and cutting speeds are shown in Fig. 5. Drill temperatures were significantly higher at the higher cutting speeds and also increased with hole depth. In addition, there was a greater difference in temperature between the new and worn drills at greater hole depths.

**Additives.** High temperatures in the cutting zone become an important consideration because additives such as sulfur have low melting points and may melt during machining. Reactions may change the microstructure of the machined material and could lead to the production of parts having undesirable microstructures. Also, cutting temperature becomes very important when polymer materials with low melting temperatures are being used to close porosity and improve surface lubrication. Polymers melt at high cutting temperatures, and poor surface finish results (Ref 1).

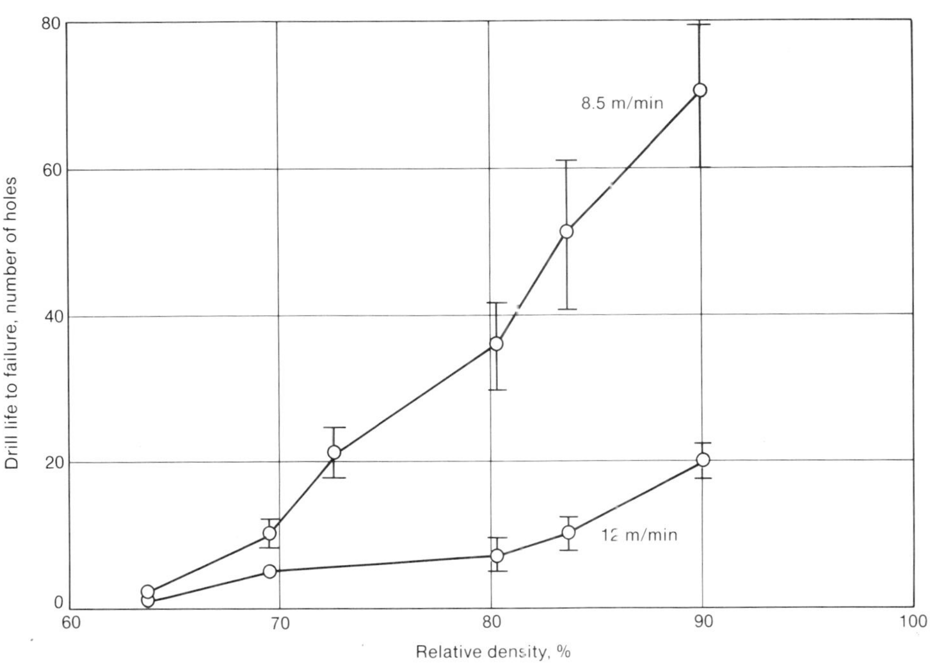


**Fig. 4** Effect of relative density on drill life of 304L P/M material

**Chips.** Porosity affects chip formation and chip continuity. Chips tend to densify, depending on cutting parameters. In general, chips produced in P/M machining are discontinuous, especially at intermediate and low densities. Continuous chips are observed in machining high-density materials. Chips generated when machining forged P/M parts were found to be almost as long as those obtained when machining wrought steel (Ref 13). In general, the higher the material density, the longer the chip. Chips produced in machining low density (porosity >35%) 304L P/M stainless steel are very small—less than 0.2 mm (0.008 in.) in their longest dimension (Ref 17, 25). This could present problems, in vertical drilling or pocket milling for example.

**Surface Finish.** Porosity has a strong influence on the surface finish of machined parts. The surface finish of a porous material can be expected to have characteristics that are different from those of a wrought material, when cutting conditions for both materials are identical. This is because porosity causes variations in tool wear mechanisms, in material work hardening, and in densification at the surface, and, additionally, may cause higher temperatures at the cutting surface.

**Summary.** Because the interrupted-cutting model does not adequately explain phenomena exhibited in drilling P/M 304L, these phenomena are better rationalized in terms of three offsetting parameters: the influence of porosity on thermal conductivity, work hardening, and the bulk strength of the workpiece.

Porosity significantly influences cutting forces, tool wear, thermal conductivity, cutting temperature, chip formation, and generated surface finish. These factors should be considered in evaluating the machinability of P/M parts. In general, higher density improves machinability. As full density is approached, machinability may be close to that of wrought materials.

## Effects of the P/M Process

The production of a P/M part requires processing steps not normally used in the manufacture of wrought materials. For example, the powder production process determines such characteristics as particle size and shape. They in turn greatly influence the further processing, the microstructure, and the properties of a P/M material. The behavior of a powder during compaction and sintering depends on the powder manufacturing method. Methods used to add chemical additives, such as premixing or prealloying with the powder, have a significant effect on final properties and machining characteristics of the part.

Although a rise in tool wear rate accompanies an increase in porosity, the magni-

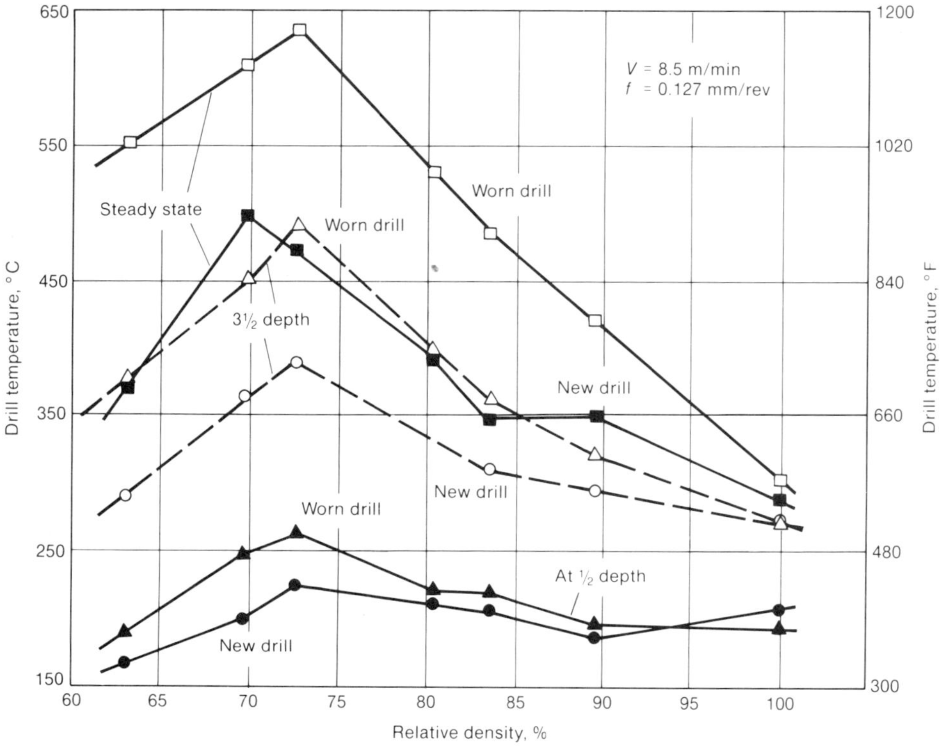


**Fig. 5** Effect of density on drill temperature at different tool life fractions for 304L P/M stainless steel

tude depends on the particular P/M material and the method by which the powder is produced (Ref 26). A sponge iron particle is normally softer than an atomized powder particle, which may explain why sponge iron base materials exhibit a greater tendency to stick to a cutting tool (Ref 27). The size, shape, and distribution of the pores basically depends on compaction pressure and powder characteristics. Green density of a compact determines its green strength and strongly influences sintered density and strength.

Following compaction, green compacts are sintered to boost their strength through neck formation and growth and to develop the desired microstructure and resulting properties. Sintering parameters such as temperature, atmosphere, dew point, and sintering time control the strength level and composition. Oxidation of porous materials may occur during sintering. Pores open to the atmosphere normally have solid oxide films on their surfaces, which may lead to greater abrasive wear of cutting tools (Ref 28). Abrasive wear that occurs in turning porous iron materials when solid oxides are present at the tool/chip interface has been described (Ref 12). Oxide-reduced powders usually contain residual internal oxide inclusions, which, if sufficiently hard, can be detrimental to machinability. Any variation in oxygen, nitrogen, or carbon content during sintering may influence machinability. Increases in the nitrogen and carbon content in type 304L P/M material during sintering have resulted in reduced tool life (Ref 1).

## Effects of Composition

Material composition is an important factor in P/M machining. A lamellar pearlite structure is preferred for sintered steels (Ref 29). Higher carbon has been reported to cause an increase in cutting forces in drilling different types of carbon steels (Ref 30, 31). Carbon content in the range of 0.1 to 0.3% was recommended for good machinability. The addition of graphite lowered cutting forces in turning (Ref 6). In general, forces decrease with additions of up to 0.3% C in an iron-base material.

**Sulfur Additions.** Sulfur is an additive that is commonly used to enhance machinability. Longer tool life and better machinability can be obtained in both turning and drilling operations by adding sulfur to sponge and atomized iron. The presence of sulfur inclusions causes the material matrix to deform more easily, facilitating crack propagation and reducing cutting forces. The addition of sulfur in the range of 0.25 to 0.5% to an iron powder reduced cutting forces in turning (Ref 32). A reduction of 12 to 20% in thrust force and improved surface finish were observed during drilling 90% dense iron when sulfur in the range of 0.25 to 0.5% was added (Ref 7, 32). The addition of 0.15 to 0.25% sulfur into sponge iron powder enhanced machinability (Ref 33). Sulfur in the range of 0.3 to 0.6% was found to improve tool life and surface finish by as much as 25% in the turning of 74 to 80% dense sponge iron that also contained carbon and copper (Ref 25, 26).

**Manganese Sulfide Powder Additions.** Manganese sulfide has been used with success as an addition in the form of a finer powder to promote machinability (Ref 34). Iron powders prealloyed with manganese sulfide, such as MP 37, and MP 36S, are commonly used in free-machining steels. One of the newer grades, MP 37, was suggested by practice in making free-machining low-carbon grades of steel by ingot metallurgy. The wrought steels typically contain 1.2% Mn and have a manganese to sulfur ratio of 3.0:4.0. Until recently, the Mn/S ratios used in making commercial iron powders have been much lower. MP 37 contains primarily iron, and 0.004% C, 0.35% O, 0.236% S, 0.02% P, 0.94% Mn, and 0.06% Si. By comparison, MP 36S contains iron and 0.019% C, 0.15% O, 0.380% S, 0.38% Mn, 0.02% P, and 0.16% Si. The purpose of a higher Mn/S ratio is to ensure that all sulfur present is in the form of MnS inclusions. These inclusions impart significant gains in machinability and the physical properties of P/M parts. In addition, size changes (growth) that take place in parts during sintering are reduced (Table 5).

In one machinability study, the performance of MP 37 was compared with that of an iron-base powder, MP 35 (0.01% C, 0.25% O, 0.02% S, 0.35% Mn, 0.12% Si, with the remainder being iron). Two different mixes were used:

| Constituent | Powder mix A, % | Powder mix B, % |
|---|---|---|
| Powder | 98.0 | 96.0 |
| Lubricant | 1.0 | 1.0 |
| Graphite | 1.0 | 1.0 |
| Copper | 0 | 2.0 |

**Table 5 Properties, sintering performance, and chemical composition of MP 36S and MP 37 transverse rupture bars**

| Mix(a) | Grade(b) | Transverse rupture strength, MPa | Transverse rupture strength, ksi | Hardness, HRB | Dimensional change (length), % | Chemical composition after sintering, % C | Chemical composition after sintering, % S |
|---|---|---|---|---|---|---|---|
| A | MP 37 | 645 | 95 | 60 | 0.04 | 0.678 | 0.214 |
| | MP 36S | 380 | 55 | 55 | 0.37 | 0.696 | 0.361 |
| B | MP 37 | 935 | 135 | 80 | 0.20 | 0.699 | 0.210 |
| | MP 36S | 770 | 112 | 79 | 0.64 | 0.718 | 0.367 |

(a) Mix A contained 98.0% powder, 1.0% zinc stearate, and 1.0% graphite; mix B contained 96.0% powder, 1.0% zinc stearate, 1.0% graphite, and 2.0% Cu. (b) Both MP 36S and MP 37 compositions included 0.02% P.

**Table 6 Time required to drill 25 holes, and the number of holes drilled before drill failure**

| Hole number(a) | Time, s — Mix A: MP 37 | Mix A: MP 35 | Mix B: MP 37 | Mix B: MP 36 |
|---|---|---|---|---|
| 1 | 5.9 | 9.2 | 6.7 | 14.9 |
| 2 | 6.3 | . . . | 8.5 | 16.7 |
| 3 | 7.0 | 10.2 | 8.6 | 13.8 |
| 4 | 7.0 | 11.2 | 8.5 | 18.4 |
| 5 | 6.7 | . . . | 8.8 | 18.7 |
| 6 | 6.6 | 11.3 | 8.6 | 20.1 |
| 7 | 7.6 | 11.9 | 8.7 | 20.2 |
| 8 | 7.8 | 13.0 | 9.4 | 20.7 |
| 9 | 8.2 | 13.6 | 9.8 | 18.1 |
| 10 | 8.1 | 13.7 | 9.7 | 17.3 |
| 11 | 8.4 | 14.2 | 8.7 | 16.1 |
| 12 | 8.5 | 14.7 | 9.3 | 16.5 |
| 13 | 8.5 | 15.7 | 9.5 | 16.2 |
| 14 | 8.5 | 16.4 | 9.8 | 16.9 |
| 15 | 9.0 | 16.6 | 9.9 | 17.6 |
| 16 | 9.2 | 17.0 | 9.4 | 19.1 |
| 17 | 9.2 | 16.2 | 10.1 | 20.2 |
| 18 | 9.3 | 15.3 | 10.8 | Failed |
| 19 | 11.5 | 15.3 | 11.2 | . . . |
| 20 | 8.3 | 15.8 | 10.9 | . . . |
| 21 | 8.1 | 16.3 | 11.5 | . . . |
| 22 | 8.2 | 17.4 | 11.8 | . . . |
| 23 | 7.8 | 16.5 | 12.7 | . . . |
| 24 | 8.3 | 18.0 | 12.3 | . . . |
| 25 | 8.4 | Failed | 12.2 | . . . |
| **Average(s)** | **8.1** | **14.52** | **9.9** | **17.7** |
| **Failure hole number** | **82** | **25** | **32** | **18** |

(a) Holes were 6.35 mm (0.25 in.) in diameter by 25 mm (1 in.) deep.

Machinability specimens were disks 100 mm (4 in.) in diameter and 25 mm (1 in.) thick. They had been pressed to a density of 6.60 g/cm$^3$ ± 0.10 g/cm$^3$. Sintering duration was 20 min in a nitrogen-base atmosphere at a temperature of 1120 °C (2050 °F), using a commercial production line.

**Table 7 Percentage of inclusions by type in MP 36 and MP 37**

| Type | MP 36, % | MP 37, % |
|---|---|---|
| MnS | 34 | 43 |
| MnS + $SiO_2$ | 6 | . . . |
| MnS + $SiO_2$ (weak) | 9 | . . . |
| MnFeO | . . . | 29 |
| MnFeO + S (weak) | . . . | 14 |
| MnO + $SiO_2$ | 12 | . . . |
| MnO + $SiO_2$ + S (weak) | 18 | . . . |
| MnO + $SiO_2$ (weak) | 15 | 14 |
| MnO + S + Si (weak) | 3 | . . . |
| FeS | 3 | . . . |

Performance was determined by the time required to drill holes through the disks. The high-speed drills used were 6.35 mm (0.25 in.) in diameter and were operated at a spindle speed of 2300 rev/min and a constant drill point loading of 70 kgf (154 lbf). Table 6 gives the time required to drill 25 holes and the hole number at which a drill failed for mixes A and B of MP 35 and MP 37. The MnS-containing powders (MP 37) exhibited much better machinability than did the lower sulfur content MP 35 powder.

A scanning electron microscope (SEM) was used to characterize inclusions in samples of MP 37 sintered to a density of 6.8 g/cm$^3$. A specimen of MP 36 (0.01% C, 0.25% O, 0.02% S, 0.35% Mn, 0.12% Si) prepared under the same conditions was also analyzed. Inclusions in MP 36 and MP 37 rupture bars are characterized in Table 7.

MP 36 contained a larger amount of silicate inclusions, which can be explained by the larger silicon content of that grade compared with that of MP 37. Hard silicate inclusions tend to degrade the machinability of MP 36. On the other hand, MP 37 contained mainly MnS and manganese oxide (MnFeO and MnFeO weak inclusions). Manganese sulfide has been known for some time to improve machinability, and a number of mechanisms have been suggested to explain its effect (Ref 35). It has also been suggested that manganese oxide inclusions improve machinability by reducing cutting forces (Ref 36).

Although the volume fraction of inclusions was not determined quantitatively because of the large number of pores near inclusions, the number of inclusions can be qualitatively estimated by comparing the SEM micrographs in Fig. 6.

Clearly, MP 37 contained a much larger number of inclusions than did MP 36. In MP 37, inclusions were mainly MnS and manganese oxides, and their larger numbers could be expected to reduce forces further during machining and favor the formation of discontinuous chips. Finally, MP 37 contained some large aggregates of inclusions (10 μm), which also would contribute to improved machinability (Ref 37).

**Bismuth and Lead Additions.** Bismuth has been shown to improve machinability. Additions of 0.5 to 2% reduce drilling thrust by 25 to 50% in working with 90% dense carbon steel (Ref 13, 30, 31). The machinability of P/M carbon steel was improved by adding up to 2% Pb; additions of 0.5 to 2% Pb reduced drilling thrust by 25 to 50%. However, even such small amounts of lead can lower mechanical properties.

**Selenium and Tellurium Additions.** The addition of 0.15 to 0.25% Se and 0.05 to 0.25% Te into a sponge iron powder enhanced machinability (Ref 33). In comparing sulfur, selenium, and tellurium additions, tellurium was found to be the most effective, while sulfur was the least effective. With additions of selenium of up to 1.5%, thrust and torque decreased in drilling 304L and 303 stainless steels (Ref 38). The presence of these additives resulted in some reduction in mechanical properties and in part dimensions during sintering.

**Copper** in an iron matrix may improve machinability. Metal deformation is improved, and thermal conductivity of the matrix is increased. An addition of 2 to 3% Cu reduced drilling thrust by 13%, while additions of copper up to 1% effectively reduced drill wear (Ref 39).

Generally, additives contribute to lower cutting forces, longer tool life, and better surface finish. Alloying elements form inclusions that reduce friction at the tool/chip interface, reducing tool wear and built-up edge formation during cutting. There is also a tendency for chips to break more readily.

## Machining Variables

Wrought materials have been studied much more extensively than P/M materials in terms of the influence of machining variables on machinability. However, it is known that an increase in feed rate or depth of cut results in greater cutting forces (Ref

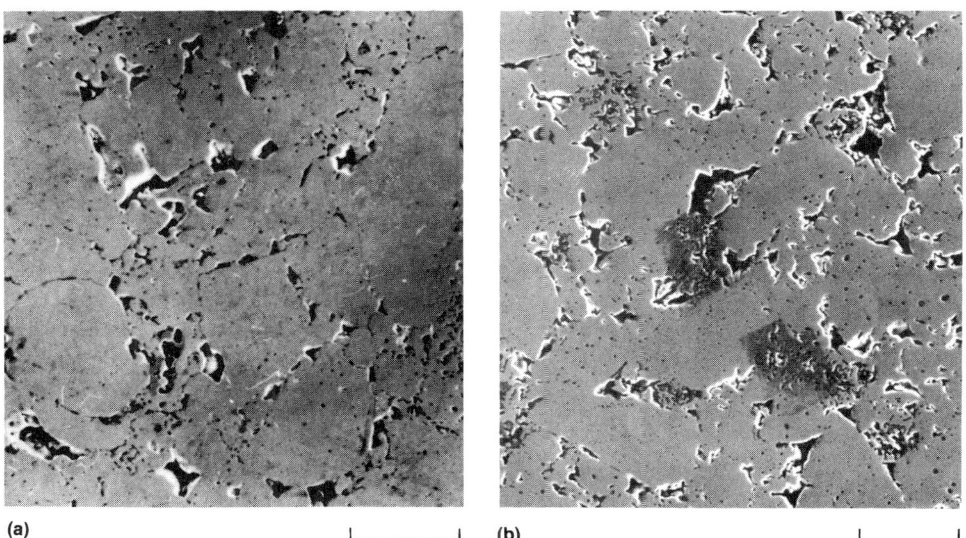

(a) 100 μm (b) 100 μm

**Fig. 6** Comparative number of inclusions in (a) MP 36 and (b) MP 37

12, 19). One study shows that the rate of increase in cutting force decreased at higher relative densities on the order of 80 to 90% (Ref 12). A reduction in cutting force with faster cutting speed was observed for a P/M stainless steel; this is similar to the behavior of wrought materials (Ref 40). Drill life and torque decreased with increasing cutting speeds during the drilling of 304L and 303 stainless steels (Ref 38, 41). A gain in the cutting force was observed in the drilling of 316L P/M materials when feed rates were increased and cutting speeds were decreased (Ref 40).

Cutting tool wear rate is influenced by the cutting speed and feed rate. Wear was observed to be significantly greater with increasing feed rates in the drilling of 77 to 90% dense iron (Ref 27). Surprisingly, a significant decline in wear rate occurred with higher cutting speeds in the turning of 80% dense AISI-SAE 1045 steel (Ref 42). This was explained in terms of greater workpiece material densification in front of the tool. On the other hand, tool life dropped drastically with increasing cutting speed in the turning and drilling of steel and stainless steel P/M materials (Ref 27, 43, 44).

Machining variables strongly influence surface quality. In one study, roughness increased with feed rate while turning 84% dense iron material (Ref 27). The influence of speed and feed on the surface roughness of iron and copper P/M bushings at different density levels has been studied (Ref 45). Surface quality improved with increasing speed, while it deteriorated with higher feed rates. Surface quality of machined P/M materials can be very important when surface porosity must be retained so that there is no loss in capillary action. Other important factors are inclusions at the pore surface and the characteristics of fluids that may be used during cutting. Cutting tools should be in good condition, and a cutting fluid or low pressure air should be used to cool tools and clear chips.

Tool geometry affects machining in a way that differs from that of wrought materials. In one study, drills of slightly different point geometry produced different thrust and torque response curves (Ref 24). Two drills were tested, one with a 148° chisel edge angle, the other with a 128° angle. The life of the latter drill was longer than that of the former. A higher rake angle can result in lower cutting forces in turning. Chips generated in tests were shaped like commas (Ref 27).

Low helix angle drills are not recommended for softer P/M materials because of their poor chip ejection characteristics (Ref 27). Drills with 40° helix angles had twice the tool life of those with 30° helix angles in work performed with soft P/M materials.

In some cases, cutting fluids are not allowed because they are absorbed into pores and could be detrimental to the performance of the part. Ceramic and silicon nitride cutting tools that operate without cutting fluids may improve machinability.

Finally, it is important to optimize cutting conditions, tool geometry, and tool material for specific P/M materials. Available machining data for corresponding wrought materials may be used as a starting point if better data are not available.

**Table 8 Results for selected sintering times of disks**

| Hole number | Drilling time, s: 16 min sinter, OC | 16 min sinter, IC | 31 min sinter, OC | 31 min sinter, IC | Full sinter, MP 37 | Full sinter, MP 35 |
|---|---|---|---|---|---|---|
| 1 | 4.13 | 3.71 | 5.04 | 4.82 | 5.9 | 9.2 |
| 2 | 4.45 | 4.06 | 6.89 | 5.17 | 6.3 | 10.7 |
| 3 | 4.78 | 4.29 | 7.17 | 5.22 | 7.0 | 10.2 |
| 4 | 4.80 | 4.12 | 7.86 | 5.24 | 7.0 | 11.2 |
| 5 | 4.83 | 4.31 | 7.46 | 5.28 | 6.7 | 11.7 |
| 6 | 4.79 | 4.56 | 7.78 | 5.41 | 6.6 | 11.3 |
| 7 | 5.07 | 4.57 | 8.15 | 5.47 | 7.6 | 11.9 |
| 8 | 5.23 | 4.58 | 8.38 | 5.57 | 7.8 | 13.0 |
| 9 | 5.13 | 4.70 | 7.79 | 5.71 | 8.2 | 13.6 |
| 10 | 5.21 | 4.73 | 7.74 | 5.61 | 8.1 | 13.7 |
| 11 | 5.18 | 4.67 | 7.63 | 5.47 | 8.4 | 14.2 |
| 12 | 5.38 | 4.71 | 7.86 | 5.64 | 8.5 | 14.7 |
| 13 | 5.24 | 4.71 | 7.28 | 5.66 | 8.5 | 15.7 |
| 14 | 5.45 | 4.78 | 7.31 | 5.68 | 8.5 | 16.4 |
| 15 | 5.37 | 4.74 | 7.22 | 5.84 | 9.0 | 16.6 |
| 16 | 5.39 | 4.79 | 7.73 | 5.89 | 9.2 | 17.0 |
| 17 | 5.30 | 4.71 | 7.71 | 6.16 | 9.2 | 16.2 |
| 18 | 5.09 | 4.86 | 7.96 | 6.15 | 9.3 | 15.3 |
| 19 | 5.25 | 4.89 | 8.07 | 6.09 | 11.5 | 15.3 |
| 20 | 5.45 | 4.89 | 8.08 | 6.22 | 8.3 | 15.8 |
| 21 | 5.44 | 4.81 | 8.40 | 6.27 | 8.1 | 16.3 |
| 22 | 5.50 | 4.87 | 8.88 | 7.26 | 8.2 | 17.4 |
| 23 | 5.39 | 4.85 | 8.21(a) | 7.26 | 7.8 | 16.5 |
| 24 | 5.26 | 4.48 | 8.25 | 6.26 | 8.3 | 18.0 |
| 25 | 5.42 | 4.90 | 8.29 | 7.36 | 8.4 | 18.4(a) |
| **Average** | **5.141** | **4.612** | **7.726** | **5.868** | **8.10** | **14.41** |
| **Total** | **128.5** | **115.3** | **193.1** | **146.7** | **202.4** | **360.3** |

Note: Underlined values were extrapolated. (a) Drill seized at this hole. OC, outer circle; IC, inner circle

## Properties Versus Machinability

The relationship between properties and machining performance was investigated on specimens sintered at increasing time-temperature settings to obtain a progression of properties on the same base material. As described earlier in this article, MP 37 is a prealloyed MnS iron powder offering improved machinability. A conventional iron powder, MP 35, was used for comparative purposes. Machinability indexes were established by analyzing data obtained in drilling tests. Properties investigated were transverse rupture strength, tensile strength, and dimensional change in sintering. Optical and scanning electron microscope (SEM) techniques were used to analyze sulfur- and manganese-containing inclusions.

A major finding during these tests was that strength and hardness progression during the sintering of a given material is accompanied by increasing difficulty in machining MP 35 and MP 37, in terms of drilling time and tool wear. In evaluating the machinability of a sintered material, strength and hardness levels are essential considerations, especially when a comparison of different material conditions, such as density and chemical composition, is necessary.

Also, in metallographic analysis of inclusions within solid particles, it was found that most of them were of the MnS type. This means that MnS, with its good tool lubrication properties, is found well dispersed in the metal matrix, rather than being concentrated in pores. The result helps to explain the outstanding machinability of MP 37.

**Test Details.** Most of the trials were with MP 37 blended with 0.75% Acrawax and graphite as required to obtain about 0.8% combined carbon. The Mn/S ratio was 4.5: 1.

Three types of specimens were used: disks 100 mm (4 in.) in diameter and 25 mm (1 in.) thick molded to a green density of 6.6 ± 0.05 g/cm$^3$, transverse rupture bars (TRS) 6.35 mm (0.250 in.) thick, and dogbone tensile bars 3.5 mm (0.140 in.) thick.

Disks were sintered at seven different times: 6, 11, 16, 21, 26, 31, and 51 min in the hot zone, with 31 min regarded as representative of production practice, and 51 minutes regarded as longer than average practice. The thinner bars were in the hot zone at different times: TRS, 3, 6, 11, 14, 17, 23, 26, 29, and 60 min; tensile, 6, 14, 23, and 29 min. Results in selected sintering times are shown in Table 8.

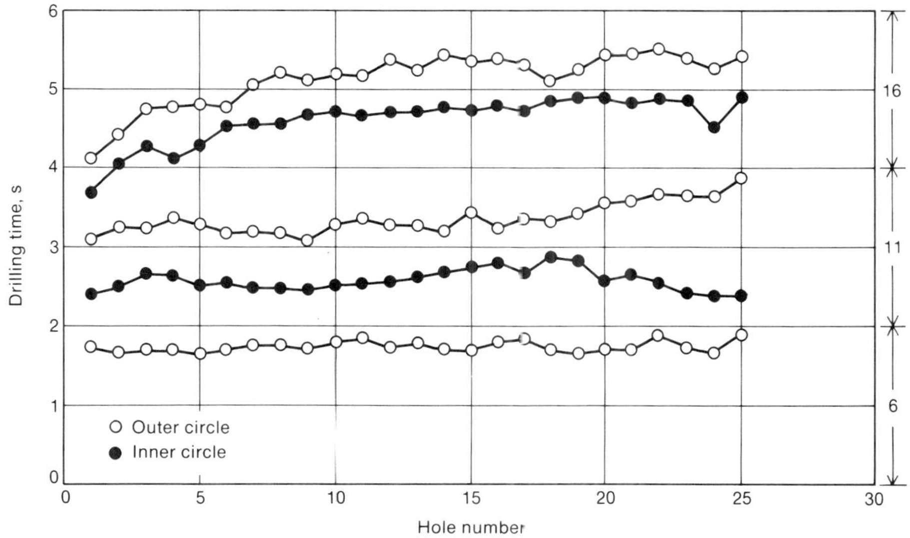


**Fig. 7** Results in drilling P/M materials after sintering at 6, 11, and 16 min. Only the outside circle was drilled in a 6 min sinter.

The drill press used included a simple pulley-and-weight system to obtain a force of 70 kgf (154 lbf), measured at the drill point. Drilling was at 2300 rev/min. A commercial high-speed steel drill 6.35 mm (0.250 in.) in diameter was used. In machining conventional iron powder without sulfide additions, such a drill will fail in 25 or less holes (see Table 6). Drilling times were recorded to the nearest hundredth of a second.

Two series of 25 holes each were drilled around the circumference of each disk, with one series located about 3 to 5 mm (1/8 to 3/16 in.) from the outside diameter (outer-circle values), and the second series (inner-circle values) about 5 mm (3/16 in.) inside the first series. Generally, inside holes were drilled faster and with less drill wear than the outer holes.

**Results.** In Fig. 7, three sets of data are plotted using the same scale. They show drilling time progression for disks sintered at 6, 11, and 16 min. Outer-circle and inner-circle data are plotted separately. In the case of the 5 min sintering, only the outer circle was drilled, on the assumption that the inner part of the slug was not subjected to appreciable sintering action. The following trends are evident in Fig. 7:

- A shift toward longer drilling times with increasing number of holes drilled very evident
- An increase in the incline of each series of plotted data points
- Curves representing inner-circle holes laying below the ones representing outer-circle holes, but the separation narrowing as the sintering progresses

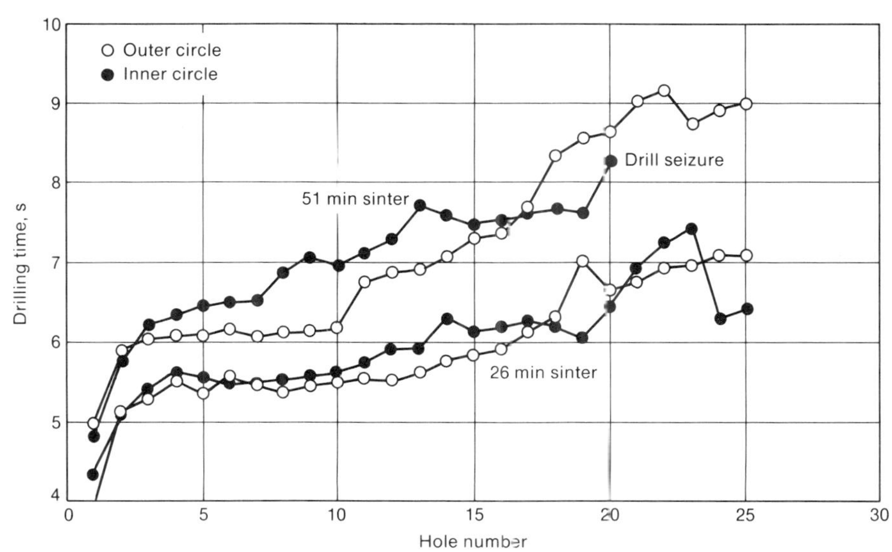


**Fig. 8** Comparison of results in drilling tests for materials sintered for 26 and 51 min (full sinter)

**Table 9 Selected machinability indexes**

| Sintering time, min | Circle(a) | Constant C | x-coefficient A | Total time (25 holes), s |
|---|---|---|---|---|
| 6 | OC | 1.7213 | 0.0026 | 43.9 |
| 11 | IC | 2.5619 | 0.0015 | 64.5 |
| 11 | OC | 3.0871 | 0.0212 | 84.1 |
| 16 | IC | 4.1985 | 0.0318 | 115.3 |
| 16 | OC | 4.6578 | 0.0372 | 128.5 |
| 21 | IC | 4.6558 | 0.0612 | 135.7 |
| 21 | OC | 5.1971 | 0.0485 | 145.0 |
| 26 | IC | 5.0772 | 0.0723 | 149.7 |
| 26 | OC | 4.8222 | 0.0882 | 148.3 |
| 31 | IC | 4.8371 | 0.0796 | 146.7 |
| 31 | OC | 7.3054 | 0.0394 | 193.1 |
| 51 | IC | 5.9026 | 0.1097 | 182.0 |
| 51 | OC | 5.1451 | 0.1603 | 180.4 |
| FS(b) | MP 37 | 6.7630 | 0.1025 | 202.4 |
| FS | MP 35 | 10.0501 | 0.3358 | 360.3 |

(a) OC, outer circle; IC, inner circle. (b) FS, full sinter

It should also be pointed out that a typical diagram comparing tool wear and cutting time shows an initially fast wear stage, followed by an almost stationary wear stage and then by a steep, short, final wear stage, which immediately precedes drill failure (Ref 41).

Results in drilling P/M materials after they have been sintered for 26 and 51 min are shown in Fig. 8. Under the testing conditions used, 51 min in this instance, is representative of a full sinter. The evolution of drilling performance with increasing sintering times is indicated by the numerical values for three leading machinability parameters given in Table 9. The constant C, representing the $y$-axis intercept, was chosen as a measure of what has sometimes been called cuttability of the material. This quantity is free from the influence of drill wear and represents a measure of the inherent resistance of the material to drilling.

The second parameter selected was the $x$-coefficient. This quantity is associated with the ability of the material to wear, or otherwise degrade, the drill bit; the steeper the slope, the faster is the wear rate of the tool. This coefficient represents a measure of wearability.

A more elementary parameter was selected to represent the cumulative effect of both the basic resistance and the wearing capability of each material: total time required to drill the 25 holes in each series.

Calculated values for the indexes are shown in Table 9. A mathematical progression of the indexes with increasing sintering time was not attempted, but the pattern clearly shows parabolic behavior with a fast initial climb and a subsequent flattening with longer sintering times.

The seven disks used for the 25-hole tests were analyzed for basic chemical properties, tested for surface hardness, and measured for outside-diameter (OD) dimensional change.

Carbon, sulfur, and oxygen content were analyzed from drillings of the first 1 mm

## Table 10 Influence of sintering time on hardness of drilled P/M disks

| | Hardness, HRB | | | | | | | | | | | | | |
|---|---|---|---|---|---|---|---|---|---|---|---|---|---|---|
| | 6 min | | 11 min | | 16 min | | 21 min | | 26 min | | 31 min | | 51 min | |
| Specimen | A | B | A | B | A | B | A | B | A | B | A | B | A | B |
| 1 | −16 | −19 | 31 | 32 | 51 | 46 | 50 | 51 | 55 | 54 | 54 | 54 | 60 | 58 |
| 2 | −15 | −18 | 35 | 32 | 50 | 49 | 54 | 52 | 57 | 54 | 58 | 57 | 61 | 59 |
| 3 | −16 | −17 | 35 | 34 | 49 | 50 | 56 | 56 | 57 | 56 | 59 | 58 | 63 | 60 |
| 4 | −15 | −18 | 32 | 30 | 51 | 50 | 56 | 56 | 58 | 59 | 61 | 59 | 64 | 62 |
| 5 | −18 | −17 | 31 | 30 | 54 | 53 | 57 | 58 | 58 | 59 | 63 | 51 | 65 | 63 |
| 6 | −18 | −16 | 32 | 34 | 52 | 52 | 57 | 58 | 59 | 61 | 63 | 62 | 63 | 63 |
| 7 | −18 | −18 | 35 | 35 | 51 | 50 | 55 | 55 | 58 | 59 | 58 | 61 | 62 | 62 |
| 8 | −16 | −18 | 34 | 34 | 52 | 47 | 55 | 55 | 57 | 58 | 59 | 58 | 62 | 61 |
| **Average** | **−16** | **−19** | **33** | **33** | **52** | **47** | **53** | **53** | **56** | **56** | **57** | **56** | **61** | **60** |
| **Center** | **−16** | **−17** | **14** | **21** | **51** | **50** | **54** | **56** | **55** | **58** | **58** | **59** | **63** | **63** |

Note: Face A columns start at top of disk and proceed clockwise at 45° increments. Face B was run counterclockwise so that the points lie opposite those on face A.

## Table 11 Average material properties of P/M disks

See also Fig. 9.

| Sintering time, min | Hardness, HRB | Dimensional change, % | Carbon content, % Core | Carbon content, % Surface | Sulfur content, % Core | Sulfur content, % Surface | Oxygen content, % Core | Oxygen content, % Surface |
|---|---|---|---|---|---|---|---|---|
| 0(a) | ... | 0.26 | 1.007 | 1.007 | 0.204 | 0.204 | 0.40 | 0.40 |
| 6 | −17.0 | 0.38 | 0.949 | 0.948 | 0.191 | 0.174 | 0.48 | 0.46 |
| 11 | 33.0 | 0.54 | 0.928 | 0.889 | 0.185 | 0.175 | ... | ... |
| 16 | 50.5 | 0.55 | 0.888 | 0.832 | 0.181 | 0.180 | ... | ... |
| 21 | 55.0 | 0.55 | 0.832 | 0.770 | 0.178 | 0.178 | 0.32 | 0.34 |
| 26 | 57.0 | 0.44 | 0.823 | 0.746 | 0.178 | 0.176 | ... | ... |
| 31 | 59.0 | 0.40 | 0.789 | 0.746 | 0.181 | 0.174 | ... | ... |
| 51 | 61.5 | 0.33 | 0.734 | 0.728 | 0.178 | 0.174 | 0.38 | 0.20 |

(a) Green (unsintered) condition

(0.04 in.) layer of both circular faces as well as from deeper drillings. The results were considered separately as representative of surface and core values. Hardness was measured by direct Rockwell testing on both circular faces of each disk. Results and comments are reported in Table 10. Dimensional change from OD die size was obtained by measuring minimum and maximum diameters at the two ends of the disk. Green (unsintered) values were considered as zero sintering. Average material properties are summarized in Table 11.

To illustrate the change in properties with increasing sintering time, three graphs were obtained from Table 11. They are reproduced in Fig. 9. Surface hardness averages follow a regular progression (Fig. 9a). Carbon follows an expected decreasing pattern from the initial value of 1.007; the loss was more pronounced on the surface for the intermediate sintering times (Fig. 9b). Sulfur shows a remarkable stability, with a final loss of 15% at the surface, where the atmosphere is active, and 13% at the core (Fig. 9c). The graph also indicates that desulfurization takes place mostly at the initial stage of sintering.

Transverse rupture and tensile bars were also produced from the same batch of powder mix used for the disks. They had a green density of 6.6 g/cm³. Dimensional change from die size was measured, and six hardness HRB readings were taken on both major faces of the transverse rupture bars; then all bars were tested for rupture strength. Ultimate tensile strength, yield strength at 2% offset, and elongation were measured on the dogbone bars. Average properties are given in Table 12. Data show that dimensional change shifted steadily toward smaller values as sintering progressed; the zero time value was measured on green bars. Hardness developed in the early states of sintering, while strength characteristics progressed more slowly.

**Metallography.** Intermetallic bonds made by cold welding during compaction are perfected and enlarged in sintering. Photographs of an unsintered part show thin gray or black lines between particles. As sintering proceeds, original particle boundaries disappear.

Carbon diffusion occurs very rapidly in the base iron used. After 3 min in the hot zone set at 1120 °C (2050 °F), a transverse rupture bar shows no evidence of pearlite, that is, carbon in solution. After 6 min in the hot zone, however, the structure shows all of the carbon in solution, with 100% of the area being pearlite.

The morphology of the MnS inclusion changes only slightly with sintering time. Figure 10 shows the transverse rupture bar sintered for 3 min. Particle boundaries are very clear, and admixed graphite flakes are still residing in the pores. Figure 11 is also from the 3 min sinter of the transverse rupture bar; it shows some of the MnS inclusions present in the spongy region of the iron. After 29 min in the hot zone, which would be at least average or typical for a ferrous sinter, MnS inclusions were still present within particles and also between particles (Fig. 12).

**Correlations.** Data for both hardness and drilling time show a rate of increase in value that tends to be faster at the initial stage of sintering. However, the hardness increases at a higher rate than the drilling time. At 21 min of sintering, for instance, the HRB reading had reached 90% of the final value, while the drilling time had increased to only 77% of the final value. Therefore, the correlation of HRB to drilling time appears to be well defined but not linear (Fig. 13).

Apparent hardness in a P/M material based on iron and graphite mixes compacted to a given density depends primarily on the diffusion of carbon, which takes place in the very early stages of sintering. Conversely, the developing of sintered bonds, which has a less pronounced effect on hardness, appears to have a major influence on drilling

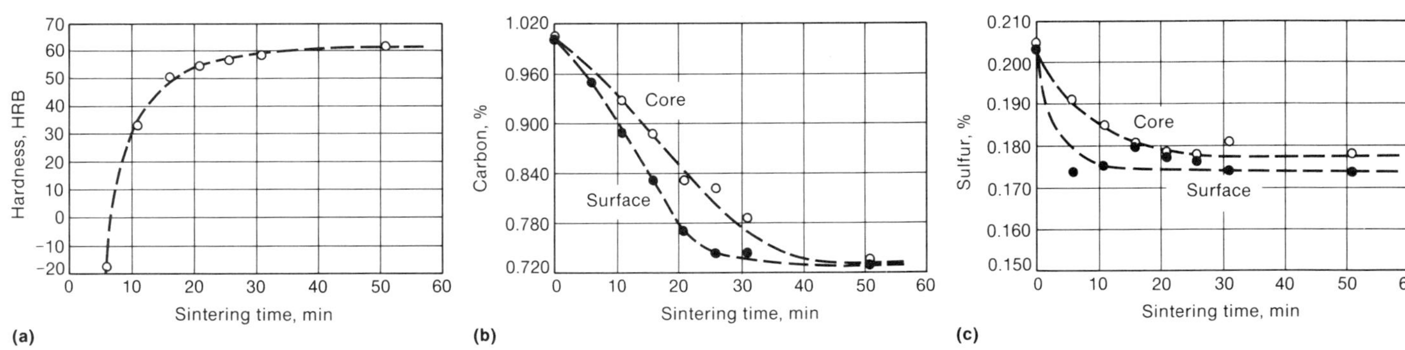


**Fig. 9** Effect of sintering time on (a) disk hardness, (b) carbon content, and (c) sulfur content

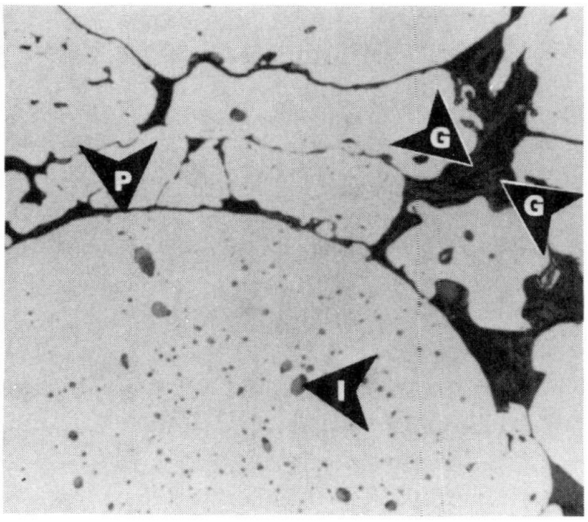


**Fig. 10** Unetched transverse rupture bar sintered for 3 min in the hot zone at 1120 °C (2050 °F). Arrow P shows a particle boundary. Arrow I shows MnS inclusion formed inside iron particle. Arrow G–G shows graphite flakes still present in the pores. 575×

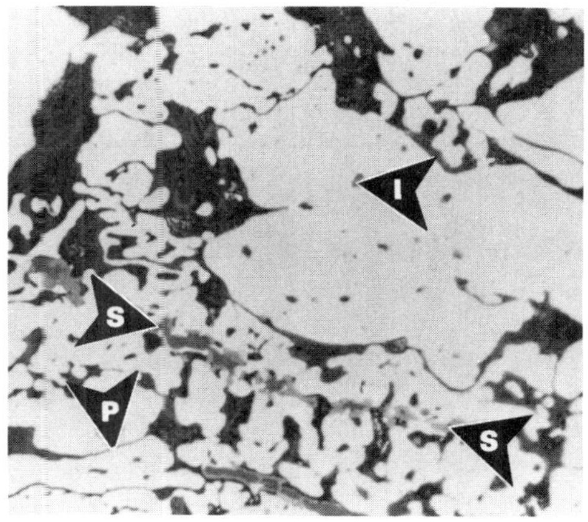


**Fig. 11** Another view of the transverse rupture bar shown in Fig. 10. Arrow P shows a particle boundary. Arrow I shows MnS inclusions formed inside iron particles. Arrow S–S shows some MnS-like inclusion between original particles. 575×

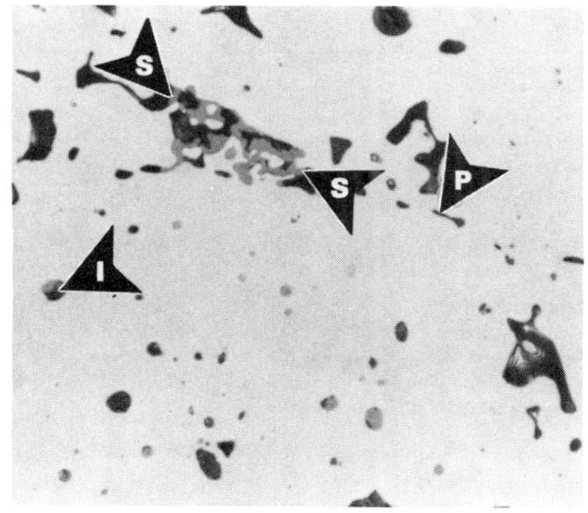


**Fig. 12** Unetched transverse rupture bar sintered for 29 min in the hot zone at 1120 °C (2050 °F). Arrow P shows a particle boundary short segment. Arrow I shows MnS inclusions formed inside iron particles. Arrow S–S shows MnS inclusions in the spongy region between particles. MnS has not changed much in appearance from that in the 3 min sinter (see Fig. 10 and 11).

time. Bond development, as shown by the metallographic analysis, progresses at a slower rate than that of carbon diffusion.

In conclusion, it can be stated that hardness is not the only characteristic, and in many instances not even the leading one, affecting machinability. A numerical correlation between strength and machinability could not be established directly because of the different sintering conditions existing between the massive disks and the much smaller bars. However, by correlating transverse rupture bars to drilling time using a metallographic method for comparing degree of sintering, it was still possible to obtain a valid description of the effect of the increasing strength on machinability. The correlation is illustrated in Table 13. Machinability parameters are compared in Table 14.

## Machining Guidelines

Updated practices for ten methods of machining (turning and boring, milling, drilling, grinding, reaming, burnishing, tapping, and honing and lapping) follow.

**Turning and Boring.** Coated or uncoated carbides, cermets, and ceramics are generally used for turning and boring tools. For interrupted cuts, carbides are preferred to prevent chipping of the insert. In cases in which the preservation of surface porosity is vital, CBN inserts are used, especially with low-porosity materials. Cutting speeds in the range of 250 to 2000 m/min (820 to 6500 sfm), feed rates in the range of 0.050 to 0.075 mm/rev (0.002 to 0.003 in./rev), and depths of cut in the range of 0.13 to 0.40 mm (0.005 to 0.015 in.) may be used with alloy steels with hardnesses in the range of 50 to 400 HB; usually, it is preferable to decrease the cutting speed to below 1000 m/min (3280 sfm) when the hardness is above 250 HB. These conditions are suitable for hot-pressed or cold-pressed ceramic tools. In cases in which uncoated carbide tools are used, cutting speeds should be less than 190 m/min (625 sfm) for machining most ferrous or nonferrous alloys.

**Milling.** Slot and side milling cutters are often used for machining P/M materials. Speeds of 70 to 100 m/min (230 to 330 sfm), feed rates of 0.005 to 0.1 mm (0.0002 to 0.004 in.) per tooth, and depths of cut of 0.13 to 0.4 mm (0.005 to 0.015 in.) are recommended in machining ferrous and nonferrous alloys with uncoated carbides. Higher speeds and feeds should be used in machining aluminum. Aluminum P/M alloys have better chip characteristics than their wrought counterparts. Chips are much smaller and are broken more easily, with little or no stringer buildup (Fig. 14).

In face milling with uncoated carbides, cutting speeds of 90 to 120 m/min (295 to 395 sfm), feed rates of 0.05 to 0.15 mm (0.002 to 0.006 in.) per tooth, and depths of cut of 0.12 to 0.4 mm (0.005 to 0.015 in.) are recommended for carbon and alloy steels and stainless steels; however, nonferrous materials may be cut at speeds up to 170 m/min (560 sfm) and feed rates as high as 0.1 mm (0.004 in.) per tooth.

Speeds of 25 to 50 m/min (80 to 165 sfm) are used in machining P/M iron, steel, stainless steel, copper, and brass with high-speed end mills. On the other hand, cutting speeds in the range of 100 to 200 m/min (330 to 655 sfm) are recommended in machining soft iron, steels, and aluminum with carbide tools. With harder steels, stainless steels, copper, and brass, speeds should be lowered to the range of 60 to 100 m/min (195 to 330 sfm).

**Drilling.** Large amounts of coolant are required in drilling medium-density or low-

**Table 12 Average properties of tensile test bars**

| Sintering time, min | Dimensional change, % | Hardness, HRB | Transverse rupture strength MPa | ksi | Ultimate tensile strength MPa | ksi | Yield strength at 2% offset MPa | ksi | Elongation, % |
|---|---|---|---|---|---|---|---|---|---|
| 3 | ... | −19 | 110 | 16.0 | ... | ... | ... | ... | ... |
| 6 | ... | 69 | 507 | 73.5 | 137 | 19.9 | 134 | 19.4 | 1.5 |
| 11 | 0.27 | 67 | 596 | 86.4 | ... | ... | ... | ... | ... |
| 14 | 0.24 | 68 | 650 | 94.3 | 236 | 34.3 | 211 | 30.7 | 2.3 |
| 17 | 0.24 | 68 | 634 | 92.0 | ... | ... | ... | ... | ... |
| 23 | 0.18 | 69 | 667 | 96.8 | 242 | 35.0 | 209 | 30.4 | 2.0 |
| 26 | 0.16 | 69 | 645 | 93.6 | ... | ... | ... | ... | ... |
| 29 | 0.15 | 70 | 667 | 96.8 | 257 | 37.3 | 220 | 31.9 | 2.0 |
| 60 | 0.06 | 72 | 678 | 98.4 | ... | ... | ... | ... | ... |

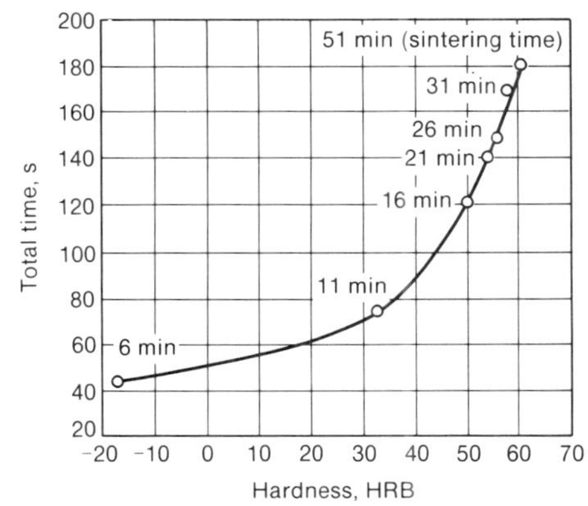


**Fig. 13** Correlation of hardness to drilling time (total time for drilling 25 holes)

**Table 13 Correlation of strength to P/M machinability**

| Degree of sintering(a) | Drilling time OC, s(b) | Transverse rupture strength MPa | ksi | Sintering time, min Disk | Bar |
|---|---|---|---|---|---|
| 1 | (23.8) | 110 | 16.0 | ... | 3 |
| 2 | 43.9 | ... | ... | 6 | ... |
| 3 | (64.0) | 507 | 73.5 | ... | 6 |
| 4 | 84.1 | ... | ... | 11 | ... |
| 5 | ... | ... | ... | ... | ... |
| 6 | 128.5 | ... | ... | 16 | ... |
| 7 | (136.8) | 596 | 86.4 | ... | 11 |
| 8 | 145.0 | ... | ... | 21 | ... |
| 9 | ... | ... | ... | ... | ... |
| 10 | 148.3 | ... | ... | 26 | ... |
| 11 | (149.0) | 650 | 94.3 | ... | 14 |
| 12 | 149.7(c) | ... | ... | 31 | ... |
| 20 | 180.4 | 667 | 96.8 | 51 | 29 |

Note: Parenthetical values are linear interpretations from adjacent values. (a) Arbitrary scale. (b) OC, outside circle of drilled holes. (c) Average value

**Fig. 14** Comparison of machining chips from a P/M aluminum alloy (left) and a conventionally processed, wrought aluminum alloy. Courtesy of Aluminum Company of America

**Table 14 Comparison of P/M machinability parameters**

| Powder grade | Index of machinability Drill time (25 holes), s | Constant C | x-coefficient A | Failure at hole | Density, g/cm³ | Hardness, HRB | Transverse rupture strength MPa | ksi |
|---|---|---|---|---|---|---|---|---|
| MP 35(a) | 360.4 | 10.0501 | 0.3358 | 25 | 6.7 | 71 | 528 | 76.6 |
| MP 37(a) | 202.4 | 6.7630 | 0.1025 | 82 | 6.7 | 70 | 664 | 96.3 |
| MP 37(b) | 181.2 | 5.5238 | 0.1350 | >83(c) | 6.6 | 70 | 667 | 96.8 |

(a) Fully sintered. (b) Average values of inner and outer circles from disk sintered for 51 min. (c) Obtained by averaging 73, 86, and greater than 90

density materials; coolant should operate effectively at the drilling point to reduce abrasive wear due to powder particles at the bottom of the hole. A single-nozzle coolant system does not work properly because small or powdered chips do not easily exit through the drill flutes. A ring design system, however, is effective in eliminating the chip-clogging problem. Oil hole drills are the most effective means for removing chips from the cutting zone.

Cutting speeds of up to 25 m/min (80 sfm) and feed rates up to 0.25 mm/rev (0.010 in./rev) are recommended for high-speed steel drills. Cutting speed and feed rate could be as high as 120 m/min (390 sfm) and 0.5 mm/rev (0.02 in./rev), respectively, when solid-carbide or carbide-tipped drills are used. Carbide indexable drills are efficient because margins are eliminated. Abrasive margin wear and the welding of powder chips are also eliminated.

Holes in planes beyond the capability of the P/M process are best made by drilling when parts are in the presintered (partially sintered) state. In this state, drilling properties are similar to those of cast iron. If it is necessary to qualify the position and size of holes after final sintering, a carbide reamer or carbide-tipped reamer, should be used. Roller burnishing may be used to meet accuracy and surface finish requirements. Any size change in the operation is related to the preburnished surface finish and the size of the hole. A change of 0.019 to 0.025 mm (0.00075 to 0.001 in.) in diameter is representative for a hole about 25 mm (1 in.) in diameter.

A number of studies have been performed to determine ways of enhancing drill life during P/M machining (Ref 43, 45-50). The following example documents the effects of work hardening of P/M parts on drill life.

**Example 1: Influence of Work Hardening on Drilling P/M Types 304L.** The drillability of 304L stainless steel was investigated at different levels of porosity (Ref 43). The major finding was that 70% dense material caused much higher wear rates than the 90% dense material, a conclusion that runs counter to the general notion that higher porosity promotes machinability. However, there is an explanation. By measuring microhardness in the fully dense region at the hole bottom below the tip of the drill, researchers noted that the low-density material hardens to 490 HK, while 90% dense material hardens only to 300 HK. Because the drill is always cutting through the almost fully dense work-hardened material, drill life is determined by the hardness of the workpiece. It appears that the low-density material work hardens more when compressed to full density than does the 90% dense material. There was no evidence of cracking, chipping, or fatiguing of the tip of the drill, counter to the idea that the drill edge fails by fatiguing from interrupted contact inherent to porosity.

**Grinding** of P/M parts can be very complex, especially when materials are low in density because in many cases preservation of surface integrity is essential. Usually, surface porosity decreases during grinding. A large amount of the generated powder chips is forced into pores and many chips are welded due to the high temperature at the wheel/workpiece interface. When grinding is necessary to achieve dimensional functionality of a part, and surface porosity needs to be preserved, special processes such as ultrasonic or solvent cleaning are applied immediately after grinding. For rough applications, a downfeed of 0.025 to 0.075 mm (0.0010 to 0.003 in.) is recommended, while for finish passes, a maximum of 0.013 mm (0.0005 in.) should be used.

Stock removal rates should be either the same as or less than those used in finish turning of cast iron; wheels should be similar. It is important to keep a plentiful supply of coolant (containing an inhibitor) directed onto the wheel and the work to maintain a clean grinding wheel contact.

**Reaming.** To control bore accuracy in powder metallurgy parts, reaming is sometimes used instead of pin sizing, ball sizing, or burnishing. Standard reamers are satisfactory; left-hand spiral reamers have also proved successful. The cutting edges should have the best possible finish to minimize edge buildup, which results in oversize holes. If the surface finish of the hole is not a factor, the drill should leave a reaming allowance, the amount depending on hole size. Guidelines can be used:

| Hole diameter mm | in. | Allowance mm | in. |
|---|---|---|---|
| ≤6.5 | ≤0.25 | 0.050 | 0.002 |
| 6.5–12.5 | 0.25–0.50 | 0.050–0.10 | 0.002–0.004 |
| 12.5–25 | 0.50–1.0 | 0.10–0.15 | 0.004–0.006 |

If the surface finish is critical, reaming allowances should be doubled. When possible, reamers should be used in floating holders and run at 7.5 to 15 m/min (25 to 50 sfm). Recommended feeds are:

| Hole diameter mm | in. | Feed mm/rev | in./rev |
|---|---|---|---|
| ≤6.5 | ≤0.25 | 0.15 | 0.006 |
| 8–12.5 | 0.30–0.50 | 0.18 | 0.007 |
| 14–19 | 0.55–0.75 | 0.25 | 0.01 |

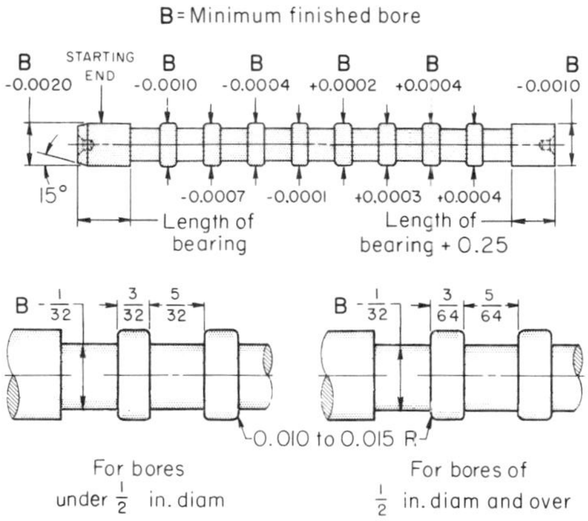


**Fig. 15** Ball broach for burnishing bores in P/M parts. Dimensions given in inches

**Burnishing.** When the clearance between a shaft and a P/M bearing is ±0.012 mm (±0.0005 in.) or less, burnishing the bearing bores after they have been installed in the housing is preferred for correcting the bore size. No more than 0.002 mm/mm (0.002 in./in.) of diameter should be displaced, and the smallest amount of displacement that will produce the true diameter is desirable. The type of burnishing tool recommended for this operation is illustrated in Fig. 15.

Given a finished bore diameter of 38 mm, +0.005 mm/−0.0000 mm (1.500 in., +0.0002 in./−0.0000 in.) (B in Fig. 15), the diameter of the starting end of the burnishing tool then becomes 38.10 mm, −0.050 mm (1.5000 in., −0.0020 in.), or 38.05 mm (1.4980 in.), and bearings would be bored to 38.075 mm, +0.0125 mm/−0.0000 mm (1.4990 in., +0.0005/−0.0000 in.). Thus, there would be a minimum clearance of 0.025 mm (0.001 in.) at the entering end of the tool, and the first land would be a line-to-line fit. The tool then becomes progressively larger, and the bearing is expanded. If there were no springback, the operation would be stopped at the fourth or fifth tool land. However, the bearing would ordinarily be burnished to 0.010 mm (0.0004 in.) oversize to allow for springback.

Roller burnishing is a cold-working operation that compresses metal rather than removes it. The technique is suitable for sintered (not heat-treated) powder metal materials for which maintenance of open surface porosity is not critical. A significant improvement in surface finish can be obtained using a roller burnishing tool. In addition, the tool is adjustable to match individual product specifications as well as to compensate for wear on the rolls and mandrel. Both through holes and blind holes can be roller burnished. Hole size tolerance depends on the input tolerance of the hole; that is, a prepared tolerance of 0.050 mm (0.002 in.) can be reduced to 0.025 mm (0.001 in.), or a ±0.0025 mm (±0.0001 in.) tolerance can be held if the input tolerance is 0.010 mm (0.0004 in.). Surface finishes of 0.25 μm (10 μin.) are common after roller burnishing. A lightweight, low-viscosity lubricating oil is recommended for most P/M materials.

**Tapping.** Conventional tap drill charts should be followed to maintain 65 to 75% depth of thread. Two-flute taps are recommended for diameters up to 8 mm (5/16 in.). Three-flute taps should be used for diameters of 8 to 12.5 mm (5/16 to 1/2 in.). Spiral-point taps are desirable because they throw the chip out instead of driving it into the pores of the workpiece. Some experimenting in tapping P/M parts may be required to determine which tap is best for a specific metal.

**Honing and Lapping.** Holes requiring extreme accuracy can be honed or lapped by normal techniques if retention of porosity is not required. However, size control of holes in P/M parts can usually be obtained more economically by reaming or burnishing.

High-density ferrous metal parts, especially when hardened, have been successfully honed and lapped using conventional procedures.

Diamond- and CBN-plated bore finishing tools are recommended for precise hole size control. These tools can be used on standard drilling or honing machines, as well as on multiple-spindle or numerically controlled machines. The use of an adjustable sleeve attached to a mating tapered mandrel increases tool life. The selection of diamond grit size determines the metal removal rate and the surface achieved. The amount of material to be removed from the hole diameter can be determined by:

$$\frac{\text{Surface finish (start)} - \text{surface finish (after honing)}}{100\,000} = \text{Required stock removal}$$

If the existing finish is 1.25 μm (50 μin.) and the desired finish is 0.25 μm (10 μin.) then 0.010 mm (0.0004 in.) should be removed from the hole diameter.

Honing of infiltrated parts is seldom practical, because the stones become loaded. Neither lapping nor honing is recommended for porous parts, because either of these processes will cause the pores to become filled with abrasive particles. For special applications that require the use of lapping or honing, ultrasonic or solvent cleaning should be performed following grinding.

## ACKNOWLEDGMENT

The author gratefully acknowledges the valuable contributions of John S. Agapiou, General Motors Technical Center; Amedeo deRege, Domfer Metal Powders Ltd.; Robert A. Powell, Hoeganaes Corporation; Frank D. Leone, Pitney Bowes; Howard I. Sanderow, Stanadyne, Inc.; and Peter K. Johnson, Metal Powder Industries Federation.

## REFERENCES

1. J.S. Agapiou and M.F. DeVries, Machinability of Powder Metallurgy Materials, *Int. J. Powder Metall., Powder Technol.*, Vol 34 (No. 1), 1988
2. K.H. Roll, Powder Metallurgy at the Turn of the New Century, *1987 Annual Powder Metallurgy Conference Proceedings*, Metal Powder Industries Federation
3. *Tool and Manufacturing Engineers Handbook*, chapter 1 in Vol 1, *Machining*, Society of Manufacturing Engineers, 1983
4. J.I. ElGomayel, "Fundamentals of the Chip Removal Process," SME Technical Paper MR77-256, Society of Manufacturing Engineers, 1976
5. J.S. Agapiou, General Motors Technical Center, Warren, MI, private communication, 1988
6. *P/M Design Handbook*, Metal Powder Industries Federation, 1983
7. 1987 P/M Part-of-the-Year Competition Winners, *1987 Annual Powder Metallurgy Conference Proceedings*, Metal Powder Industries Federation, 1987, p 920-921
8. L.G. Roy, A.F. deRege, and L.F. Pease, Relationship Between Machinability and Strength in a Prealloyed MnS Sintered Metal, *1988 International Powder Metallurgy Conference Proceedings*, Vol 44, Metal Powder Industries Federation, 1988
9. L.G. Roy, G. L'Esperance, P. Lambert, and L.F. Pease, Prealloyed Mn/S Powders for Improved Machinability in P/M Parts, *1987 Annual Powder Metallurgy Conference Proceedings*, Metal Powder Industries Federation, 1987, p 489
10. "Materials Standards for P/M Structural Parts," MPIF Standard 35, 1987-1988 ed., Metal Powder Industries Federation
11. R.A. Powell, Hoeganaes Corporation, Riverton, NJ, private communication, 1988
12. W.I. Kononenko, Some Laws Governing Tool Wear During Machining of Porous Material, *Powder Metall. Int.*, Vol 2, 1976, p 24-28
13. P.J. Andersen and J.S. Hirschorn, "The Influence of Porosity on the Machinability of Powder Metallurgy Steels," Paper presented at the International Symposium on Influence of Metallurgy on the Machinability of Steel, (Tokyo), 1977, p 333-344
14. A.A. Zakaria and J.I. ElGomayel, On the Reliability of the Cutting Temperature for Monitoring Tool Wear, *Int. J. Mach. Tool Des. Res.*, Vol 15, 1975, p 195-208
15. A.J. Vaccari, A Guide to Machining P/M Parts, *Mater. Eng.*, 1960, p 50-53

16. H.G. Taylor, chapter 20 in *Hoeganaes Iron Handbook: Machining*, Vol 1, Section G, 1960, p 1-8
17. J.S. Agapiou, The Influence of Porosity on the Machinability of Powder Metallurgy 304L Stainless Steel, Using Tool Wear, Tool Life, and Cutting Forces, *Int. J. Powder Metall., Powder Technol.*, in press
18. J.S. Agapiou, G.W. Halldin, and M.F. DeVries, On the Machinability of Powder Metallurgy Austenitic Stainless Steels, *ASME J. Eng. Ind.*, in press
19. A. Artamonov and V.I. Kononenko, Cutter Wear When Machining Porous Materials, *Sov. Powder Metall. Met. Ceram.*, No. 4, 1967, p 316-321
20. A.V. Luikov *et al.*, Thermal Conductivity of Porous Systems, *Int. J. Heat Mass Trans.*, Vol 11, 1968, p 117-140
21. W.D. Kingery and M.C. McQuarrie, Thermal Conductivity: I, Concepts of Measurement and Factors Affecting Thermal Conductivity of Ceramic Materials, *J. Am. Ceram. Soc.*, Vol 37 (No. 2), 1954, p 67-72
22. A.L. Loeb, Thermal Conductivity: VIII, A Theory of Thermal Conductivity of Porous Materials, *J. Am. Ceram. Soc.*, Vol 37 (No. 2), 1954, p 96-99
23 S.A. El-Fekey *et al.*, Mathematical Analysis of the Dependence of Thermal Conductivity on Porosity, *Powder Metall. Int.*, Vol 37, 1978, p 80-81
24. J.S. Agapiou and M.F. DeVries, An Experimental Determination of the Thermal Conductivity of 304L Stainless Steel Powder Metallurgy Materials, *J. Heat Transf. ASME*, to be published
25. L.K. Gillespie, "Machinability as Related to Precision Miniature Parts," MR76-09, SME Technical Paper, Society of Manufacturing Engineers, 1976
26. R.S. Jamison and E. Geijer, "Machinability of Sintered Iron and High Hardness Steel Alloys," Technical Paper EM68-125, American Society of Tool and Manufacturing Engineers, 1968
27. K. Sten-Ake, Turning and Drilling of Some Typical Sintered Steels, *Powder Metall.*, Vol 12 (No. 24), 1969, p 538-565
28. M.V. Kas'yan and G.M. Zaimtsyan, Effect of Porosity on Cutting Forces, *Sov. Powder Metall. Met. Ceram.*, Vol 12 (No. 6), 1973, p 481-484
29. A. Alder, Machinability of Sintered Steels, *Precis. Met.*, June 1971, p 49-52
30. P.J. Andersen and J.S. Hirschorn, The Influence of Some Metallurgical Variables on the Machinability of Sintered Steels, in *Influence of Metallurgy on Machinability*, American Society for Metals, 1975, p 143-158
31. P.J. Andersen and J.S. Hirschorn, Use of Additives to Improve the Machinability of Sintered Steels, *Mod. Dev. Powder Metall.*, Vol 10, 1977, p 477-490
32. O. Svahn and B. Alvelid, The Machinability of Powder-Metallurgy Materials, *Z. Ind. Fertig.*, No. 9, 1963, p 560-564
33. U. Enystrom, Machinability of Sintered Steels, *Progress in Powder Metallurgy, National Powder Metallurgy Conference Proceedings*, 1981, p 417-438
34. K.S. Chopra, Improvement of Machinability in P/M Parts Using Manganese Sulfide, *1987 Annual Powder Metallurgy Conference Proceedings*, Vol 43, Metal Powder Industries Federation, p 501
35. E.M. Trent, *Metal Cutting*, Butterworths, 1977
36. L.M. Riekels, Inclusion Characteristics in Resulfurized AISI 1215 Free Machining Steel With Varying Manganese Contents, in the *Proceedings of the International Conference on the Influence of Metallurgy on the Machinability of Engineering Materials*, 1982
37. G.J. Marston and J.D. Murray, *J. Iron Steel Inst.*, Vol 268, 1970, p 568
38. K. Kusaka *et al.*, Influence of S, Cu, and Sn Additives on the Properties of AISI 304L Type Sintered Stainless Steels, in *Proceedings of the International Powder Metallurgy Conference*, 1984, p 150-163
39. J.M. Capus and C. Fournel, Tool Wear Measurements in the Machining of Sintered Ferrous Alloys, in *Progress in Powder Metallurgy*, Vol 37, 1981, p 165-176
40. A.W. Abduljabbar, "The Effect of Porosity and Metal Cutting Variables on Machinability of Powder Metallurgy—316L Stainless Steel and FC0508 Copper Steel," Master of Science thesis, University of Wisconsin, Madison, 1982
41. J.S. Agapiou, G.W. Halldin, and M.F. DeVries, The Influence of Porosity on the Machinability of Powder Metallurgy 304L Stainless Steel Using Cutting Forces, in *Proceedings of NAMRC-XV*, Society of Manufacturing Engineers, 1987, p 531-537
42. V.A. Danilenko, Effect of Machining on the Wear Resistance of Sintered Materials, *Sov. Powder Metall. Met. Ceram.*, No. 6, 1970, p 509-510
43. J.S. Agapiou, G.W. Halldin, and M.F. DeVries, Drillability of 304 Stainless Steel P/M Material: Tool Wear and Life, *1987 Annual Powder Metallurgy Conference Proceedings*, Vol 43, Metal Powder Industries Federation, p 181
44. Y. Trudel *et al.*, Selected Additives to Improve Machinability of Ferrous P/M Parts, in *Proceedings of the International Powder Metallurgy Conference*, 1984, p 190-201
45. V.V. Podgorkov *et al.*, Finish Machining of Sintered Iron and Copper Base Materials, *Sov. Powder Metall. Met. Ceram.*, Vol 13 (No. 8), 1974, p 674-677
46. S. Suzuki *et al.*, Machinability of 4100 Series Sintered Steel Containing Sulfur, *1987 Annual Powder Metallurgy Conference Proceedings*, Vol 43, Metal Powder Industries Federation, p 511
47. U. Engstrom, Machinability of Sintered Steels, *Progress in Powder Metallurgy 1982*, Vol 38, *1982 National Powder Metallurgy Conference Proceedings*, Metal Powder Industries Federation, p 417
48. J.M. Capus and C. Fournel, Tool Wear Measurements in Machining of Sintered Ferrous Alloys, *Progress in Powder Metallurgy 1981*, Vol 37, *1981 National Powder Metallurgy Conference Proceedings*, Metal Powder Industries Federation, p 165
49. Y. Trudel, C. Ciloglu, and S. Tremblay, Selected Additives to Improve Machinability of Ferrous P/M Parts, Vol 15, *Modern Developments in Powder Metallurgy*, Metal Powder Industries Federation, p 775
50. A. deRege, G. L'Esperance, L.F. Pease, and L. Roy, Prealloyed MnS Powders for Improved Machinability, *Near Net Shape Manufacturing Conference*, P.W. Lee and B.L. Ferguson, Ed., ASM INTERNATIONAL, 1988, p 57-68

# Machining of Metal-Matrix Composites and Honeycomb Structures

Harry E. Chandler, ASM INTERNATIONAL

METAL-MATRIX COMPOSITES (MMCs) consist of a metal base reinforced with one or more constituents (fibers, particulates, whiskers) in continuous or discontinuous form. In the case of the continuous fiber reinforced composites, the fiber is the dominating constituent, and the metal matrix serves as a vehicle for transmitting the load to the reinforcing fiber. Composites that incorporate discontinuous reinforcement are matrix dominated, forming a pseudo dispersion-hardened structure. Optimum properties can be achieved in continuous fiber reinforced composites when the fibers are oriented in one direction, yet this results in a highly anisotropic material. At best, planar isotropy can be achieved with continuous fiber reinforced composites. Composites reinforced with whiskers or particulates tend to be isotropic unless they are given sufficient deformation processing to cause texturing (grain elongation in the metal matrix or whisker alignment). Typical microstructures of MMCs are shown in Fig. 1.

Metal-matrix composites are usually fabricated to near-net shape. They are subsequently trimmed to net size or shape (requiring machining) and attached to an assembly by adhesive bonding, mechanical fasteners (requiring drilling), or welding. Because MMCs are made up of hard, abrasive nonmetallic fibers or particles in a soft metal matrix, they present unusual machining problems. Tool wear, for example, is much more severe than that encountered in machining the matrix metal alone.

Honeycomb structures consist of a core formed into hexagonal-shaped cells (Fig. 2) bonded to facing materials to form a sandwich panel (Fig. 3). These sandwich structures have two types of cores: metallic (aluminum, corrosion-resistant steel, titanium, or nickel-base alloys) and nonmetallic (aramid, fiberglass, or kraft paper). Facing materials are normally aluminum, fiberglass, graphite, or aramid. Detailed information on the fabrication of honeycomb structures can be found in *Composites*, Volume 1 of the *Engineered Materials Handbook*.

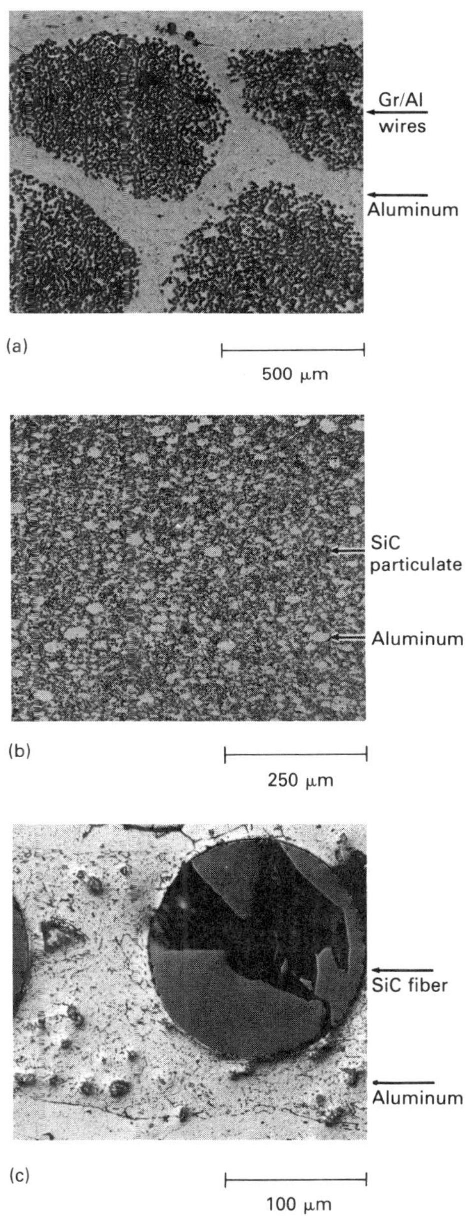


**Fig. 1** Cross sections of typical fiber-reinforced MMCs. (a) Continuous fiber reinforced graphite/aluminum composites. (b) Discontinuous silicon carbide/aluminum composite. (c) Continuous-fiber silicon carbide/aluminum composite

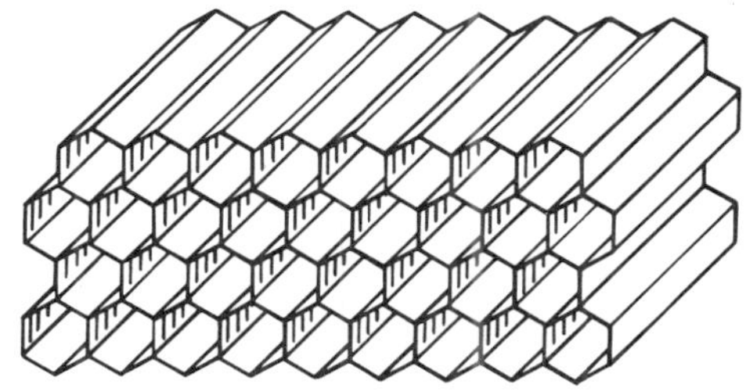

**Fig. 2** Hexagonal cell honeycomb

The machining of honeycombs is usually confined to the core structure. In fact, machining is generally defined as custom shaping of the core to a specification. Shaping methods include perimeter trim, double-relief routing, machining of flat chamfers, and contouring. Band saws are used in cutoff operations.

## General Guidelines for Machining of MMCs

Traditional solid tool methods of machining (with special cutters) and nontraditional methods are in use for MMCs. In both cases, the nature of MMCs, principally their abrasiveness, must be taken into account. Tool materials, cutter shapes, speeds, and feeds are largely determined by the machining properties of the reinforcing material. In addition, precautions must be taken to avoid damage to the workpiece during machining (that is, delamination, fiber fraying, and drill breakthrough) and premature dulling of tools. Dust and excessive noise, caused mainly in cutting reinforcing fibers, are also problems in traditional machining, and machine tool contact can introduce stress into the workpiece.

Nontraditional machining methods, such as abrasive waterjet and laser cutting, provide some relief, but they too have inherent limitations. With the waterjet technique, for example, cutting is provided by a shearing force, which can initiate delamination in the machining of composites containing ceramic fibers. High-pressure water can penetrate the fiber/matrix interface and cause separa-

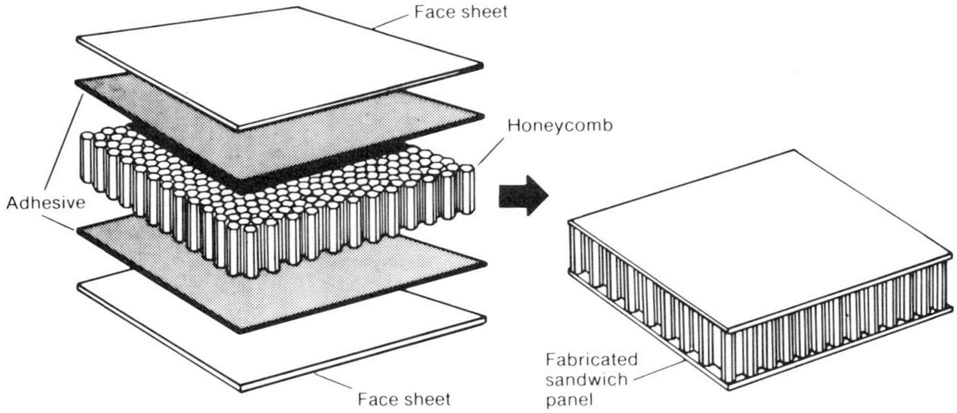


**Fig. 3** Example of a bonded sandwich assembly

**Fig. 4** Carbide router bit with diamond-shaped chisel cut

tion. In addition, the Occupational Safety and Health Administration requires operators of waterjet equipment to wear hearing protectors.

**Traditional Machining.** In solid tool machining and drilling, obtaining acceptable machined edges and drilled holes is a source of trouble. Success in machining and drilling is fiber related and totally independent of the matrix material, including metals.

Trimming is one of the more common operations in the fabrication of composite materials (both MMCs and resin-matrix composites). The cutting tools used include circular saws, router cutters, and abrasive tools. For straight-line cuts, diamond-coated circular saw blades and portable equipment are suggested. Tool life is long.

Good finishes are obtained with carbide router bits that have a diamond-shaped chisel cut (Fig. 4) and with diamond-coated router tools. They can also be used in any trimming operation. A finish of better than 3 μm (125 μin.) in roughness can be obtained with carbide router bits.

Abrasive tools are also used in trimming and final finishing. Examples are sanding drums, disks, belts, and abrasive cloth. Parameters for trimming composites are listed in Table 1.

Coolant may or may not be used in the traditional machining of composites. For example, coolant is not used in the peck drilling of laminated composites. Peck drilling was developed to prevent problems with delamination and shredding; this technique is discussed in the section "Drilling" in this article.

Band sawing with diamond grit blades is preferred for both continuous and discontinuous MMCs (Ref 1). Coolants are usually needed to prevent overheating of the blade and to prevent matrix metal pickup. However, carbide blades can be used for discontinuous MMCs with low fiber content.

Diamond grit grinding bits are suggested for the profile milling, chamfering, drilling, and countersinking of continuous MMCs, using water mist coolants. Conventional cutting tools, such as end mills and twist drills, even the diamond type, cannot shear the fibers and will cause fiber damage and delamination.

Discontinuous metal-matrix composites can be machined by all conventional methods, but speeds and feeds must be adjusted. Drilling, turning, and milling cannot be done at the same rates on aluminum-matrix composites as on conventional aluminum alloys. The more reinforcement in a composite, the faster a tool will wear, and greater penetration forces may be required. The basic approach is to reduce cutting speeds to manage the rate of wear while maintaining high feed rates to improve productivity before the tools wear excessively.

Cutting tools may exhibit different wear patterns. Abrasion, not heat, is the primary problem. Twist drills may wear on margins as well as on cutting lips. End milling cutters may break because of early loss of edge keenness, which greatly boosts side thrust forces. The greater forces may deflect the tool, affecting dimensional accuracy.

**Nontraditional Machining.** Waterjet, abrasive waterjet, and laser cutting are among the nontraditional machining methods used. In the waterjet process, a hydraulically driven intensifier pumps a fluid (typically filtered and conditioned water) at pressures up to 410 MPa (60 ksi). The fluid is expelled through an orifice to form a jet stream. The coherent jet of water is propelled at speeds up to about 850 m/s (2800 ft/s).

An abrasive material, such as garnet in grit sizes ranging from 16 to 150, is used in abrasive waterjet cutting. Grit is introduced into the stream after the primary jet is formed. Cutting is an erosive or shear process. The article "Waterjet/Abrasive Waterjet Machining" in this Volume contains detailed information on the equipment, principles, and application of these techniques.

Many MMCs are cut with high speed and quality using the $CO_2$ laser. In cutting some MMCs, $CO_2$ and YAG lasers are alternatives to diamond tools or grinding. The thermal nature of the process, however, limits its utility in applications in which charring or thermal degradation is not acceptable (Ref 2-4).

Metallic materials, including MMCs, do not exhibit the same behavior as thick, organic materials, because metallics do not break down into gaseous components during laser cutting. The material melts and

**Table 1 Parameters for composite trimming**

| Operation | Equipment | Cutter type | Speed m/s | Speed sfm |
|---|---|---|---|---|
| Straight-line cuts | Pneumatic saw | Diamond-coated circular saw(a) | 60–90 | 12 000–18 000 |
| | Hand router | Diamond router(b) | 10–15 | 2000–3000 |
| | | Carbide router(c) | | |
| Irregular outline | Hand router | Diamond router(b) | 10–15 | 2000–3000 |
| | | Carbide router(c) | | |
| Chamfer, deburr | Hand | Abrasive drum(d) | ... | ... |
| Finish operations | Hand drill motor | Abrasive disk(d) | 5–60 | 1000–12 000 |
| | Hand | Abrasive cloth(d) | ... | ... |

(a) Diamond circular saw, 1.3 mm (0.050 in.) kerf, 36/44-grit. (b) Diamond routers, 36/44-grit roughing, 80/100-grit finishing. (c) Carbide diamond-shaped chisel cut routers. (d) 80-grit (rough). 220-grit finish

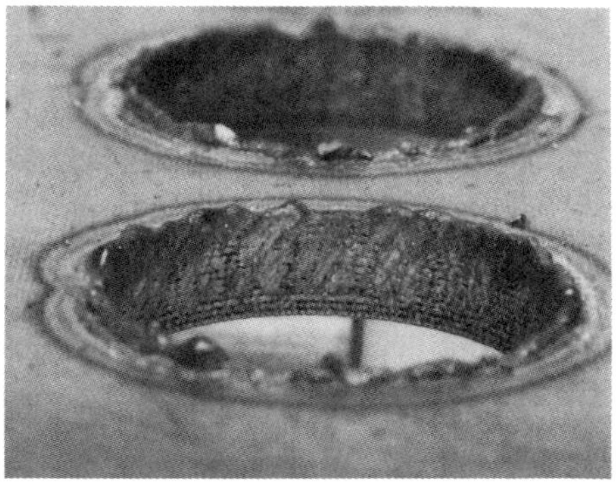

**Fig. 5** Laser-cut 3.2 mm (0.13 in.) titanium-SiC composite in which the titanium matrix material has flowed away from the SiC fibers and formed a recast layer at the exit (top) side of the cut

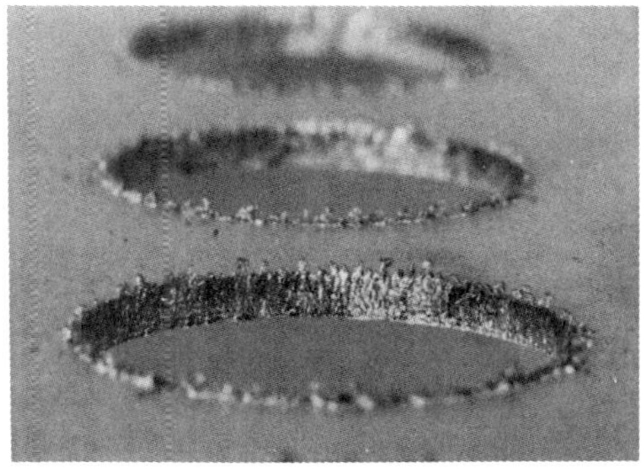

**Fig. 6** Laser-cut 13 mm (0.50 in.) holes cut through 1 mm (0.04 in.) thick aluminum-boron. Recast aluminum forms a burr on the exit (top) side of the cut.

must be physically ejected by the gas-assisted flow. Recast material often adheres to the exit side of the cut (Fig. 5), and its removal may require additional operations. Recast material can also adhere to the walls of a cut, forming a degraded solidified layer. Recast material is minimized by a high-pressure, gas-assisted flow and by using reactive gases if they are metallurgically acceptable.

Failure to penetrate a workpiece completely results in the kerf material being ejected upward toward the laser source. This material generally damages the cut edge, producing a ragged incision surrounded by debris.

A third nontraditional machining method employed for MMCs is wire electrical discharge machining, a spark erosion process in which a small-diameter copper or brass wire serves as the electrode. A dc power supply delivers high-frequency pulses of electricity to the wire and the workpiece. Material is eroded ahead of the traveling wire by spark discharges. The principles of this technique are described in detail in the article "Electrical Discharge Machining" in this Volume.

## Machining Guidelines for Specific MMCs

A wide variety of MMCs have been studied. Matrix materials include aluminum, copper, titanium, magnesium, and steel. Reinforcing materials include boron, silicon carbide (SiC), aluminum oxide ($Al_2O_3$), graphite, and tungsten. This Section will review aluminum-boron, aluminum-SiC, aluminum-$Al_2O_3$, and titanium-SiC MMCs. Additional data on MMCs can be found in *Composites*, Volume 1 of the *Engineered Materials Handbook*. The corrosion properties and casting characteristics of MMCs are discussed in Volumes 13 and 15, respectively, of the 9th Edition of *Metals Handbook*.

### Aluminum-Boron Composites

Aluminum-boron MMCs are machined with both traditional (cutting, milling, drilling) and nontraditional methods (for example, laser cutting). Traditional machining characteristics are similar to those for aluminum-SiC composites (see discussion below). The following example details the use of laser cutting.

**Example 1: Laser Cutting of Aluminum-Boron Composites.** An aluminum-boron composite 1 mm (0.04 in.) thick was cut at a rate of 8 m/min (25 sfm) with a 1500-W $CO_2$ laser. A 63 mm (2.5 in.) focal length lens was used to produce minimum spot size. With air-assisted flow, cut quality was good, with 0.2 mm (0.008 in.) of slag on the exit side of the kerf (Fig. 6). Pulsing provided good cut quality at a lower speed.

A YAG laser provides slightly better cut quality than a $CO_2$ laser in cutting aluminum-boron MMCs because of its high pulsing power. Typical parameters would be 0.3 m/min (1.0 sfm), with an 80 mm (3 in.) focal length lens on a pulsed 400-W unit. Additional information on the laser cutting of composites can be found in Ref 5.

### SiC-Reinforced Composites

**Machining practices for aluminum-SiC** are similar to those for A390 sand cast aluminum (16 to 18% Si), as described in the article "Machining of Aluminum and Aluminum Alloys" in this Volume. Close-tolerance intricate parts can be produced with excellent finish and surface integrity by using polycrystalline diamond cutting tools and slight adaptations of standard metal cutting and finishing techniques developed for abrasive materials (Ref 6).

The extremely hard carbide reinforcement is abrasive. High temperatures between the tool and the workpiece result in rapid tool wear. The following data are based on the use of a group of materials containing SiC particulates or whiskers in volume fractions ranging from 15 to 40%.

These composites produce small, discontinuous (segmented) chips with little or no built-up edge, depending on the matrix material used. For example, in machining a composite with a 6061 aluminum matrix in the annealed condition, moderate amounts of built-up edge may result.

**Power Sawing of Aluminum-SiC MMCs.** In cutoff operations on large stock, a power hacksaw is used, along with a heavy flow of cutting fluid. Effective tooling materials include a high-speed tool steel blade with a standard (precision) tooth configuration having 2½ teeth per 25 mm (1 in.) in a raker set. The blade is 3 mm (0.125 in.) thick, 115 mm (4.5 in.) high, and 900 mm (36 in.) long. A kerf of 6.5 mm (0.25 in.) is produced.

A high rate of blade wear should be expected. Wear is indicated by an absence of chips in the cutting fluid. Cutting action is poor when abrasion wears away the sides of the teeth.

Specific feeds and speeds are usually determined by experimenting with combinations that will not stall the machine. Depending on blade condition, cutting through a 350 mm (13½ in.) diam billet requires 30 to 45 min. Blade replacement is normally required after an average of five cuts through material containing 20 vol% of SiC whiskers. With 40 vol% of SiC, blades last an average of two cuts.

A power hacksaw can be used for cutting off lighter stock and for contour sawing. A horizontal (cutoff type) or vertical bandsaw can be used. A varitooth bimetal blade with raker set high-speed steel teeth is preferred, with an alternating sequence of four large teeth (five teeth per 25 mm, or 1 in., pitch) and five small teeth (eight teeth per 25 mm, or 1 in., pitch). A cutting fluid is not used with this configuration because it impairs the cutting action.

The varitooth blades in use are 0.90 mm (0.035 in.) thick and 25 mm (1 in.) high. The horizontal type is 3.4 m (11 ft, 5 in.) long, and the vertical type is 3.7 m (12 ft, 5 in.) long. Blade speeds are 25 m/min (90 sfm) for the horizontal type and 60 m/min (200 sfm) for the vertical type. Abrasive wear on the sides of the teeth necessitates frequent blade replacement. Typically, a bandsaw blade should be replaced after an average of four cuts through bar stock 150 mm (6 in.) in diameter with 20 vol% of SiC whiskers.

These materials have also been cut with a high-speed abrasive cutoff wheel (nylon-reinforced resinoid with $Al_2O_3$ grit) and a cutting fluid similar to that used with the reciprocating power hacksaw. With a wheel 300 mm (12 in.) in diameter by 6.5 mm (0.25 in.) thick, kerf width is about 7.5 mm (0.3 in.).

**Wire electrical discharge machining (EDM)** is ideally suited to the straight or contour cutting of aluminum-SiC and aluminum-boron MMCs. The rate of removal is influenced by the electrical conductivity of the workpiece, not the hardness or abrasiveness of the workpiece material. Wire EDM leaves virtually no directional machining marks, and the thermally affected zone at the cut surface is insignificant because its depth is less than that of the roughness irregularities. Feed rate depends on the material, wire diameter, and specified surface quality.

The continuously moving wire electrode acts like an electric band saw, with the wire replacing the band and the spark discharge replacing the teeth. Extremely tight corners with almost no radius can be cut. Kerf width is minimal, with the influencing factor being the diameter of the electrode wire. Use of a standard 0.25 mm (0.010 in.) brass wire results in a kerf width of about 0.35 mm (0.0134 in.).

The effectiveness of wire EDM cutting has been demonstrated with a composite containing 20 vol% SiC whiskers. A cut 100 mm (4 in.) deep by 200 mm (8 in.) long was made without difficulty. Other possible cuts with this process include tapers, relief angles, and other complex geometries.

**Abrasive Waterjet Cutting.** In cutting aluminum-SiC MMCs with this process, the kerf is usually 0.025 mm (0.001 in.) greater than the diameter of the nozzle orifice. Excellent cut surfaces have been obtained with a workpiece 12.5 mm (½ in.) thick at a rate of 100 mm (4 in.) per minute. Good cut surfaces (better than those produced with a power hacksaw) have been obtained with a cutting rate of 250 mm/min (10 in./min).

**Drilling and Reaming.** Holes of acceptable quality can be produced with standard high-speed steel drills with 118° points. The procedure is slow and requires close operator attention (drills need to be sharpened frequently, for example). A generous supply of cutting fluid should be used.

A surface cutting speed in the range of 4.5 to 6 m/min (15 to 20 sfm) is preferred. The operator must be careful not to allow the drill to bind (which causes torsion drill failures) or to dwell (which results in extended abrasive contact between the drill point and the chips). Frequent extraction of the drill is necessary to remove congested chips from the flutes.

Tool life is extended by using a carbide spade drill with hand feed and a cutting speed in the range of 90 m/min (300 sfm). Best results have been realized with a straight-flute (0°) helix-angle carbide drill with a split (crankshaft type) point at a 138° angle. Tool life in the range of 100 holes has been obtained, using a water-base emulsion cutting fluid in combination with a speed of 25 m/min (90 sfm) and a feed of 0.25 mm/rev (0.010 in./rev). Best results are obtained by quickly removing the abrasive chips (flushing with coolant) and by moving the tool through the workpiece as rapidly as possible.

Holes of excellent quality have been produced with a rotary ultrasonic drilling machine turning at 2000 to 3000 rev/min with axial vibration. The drill point was coated with 80-grit diamond.

Reaming can be done with standard high-speed steel reamers, using a hand feed and surface cutting speed of 4.5 to 6 m/min (15 to 20 sfm) and a generous supply of cutting fluid. Chips should be frequently removed from the flutes.

**Turning.** Optimum results are obtained with polycrystalline diamond cutting tools having a back rake angle of 0°. Excellent results can be obtained dry, but when metallurgical surface integrity is required, a cutting fluid should be used. The degree of cooling needed calls for any good-quality soluble oil emulsion at a 1:20 ratio with water. Generally, surface cutting speeds are in the range of 300 m/min (1000 sfm), feed rates are in the range of 0.8 to 0.15 mm/rev (0.003 to 0.006 in./rev), and cutting depths are of the order of 1.5 mm (0.06 in.).

Diamond tooling is speed sensitive and wears at higher speeds. Some production-oriented evaluations show that reasonable tool life is possible with cutting speeds in the range of 150 to 240 m/min (500 to 800 sfm). Diamond cutting tools should be considered the standard for common lathe operations such as outside diameter turning, boring, and facing. Operations such as parting and knurling can be done on a prototype basis using standard high-speed steel cutoff and knurling tools.

**Milling.** Most operations call for face milling cutter configurations. Experience with peripheral (slab milling) cutters is limited. Results are excellent with polycrystalline diamond inserts set in a standard FP-5 face mill cutter body. This configuration has a 125 mm (5 in.) diameter and utilizes ten inserts preset for a positive rake of about 7° and a 15° lead angle.

Machining is done dry at a surface cutting speed of 1675 m/min (5500 sfm) and a feed of 0.060 mm/rev (0.0024 in./rev) or 0.006 mm (0.00024 in.) per tooth along with cutting depths up to 1.5 mm (0.06 in.). Surface finish is excellent, but the influence of high cutting speed on tool wear has not been evaluated. On a production basis, optimum tool life will probably be realized with a cutting fluid and surface cutting speeds of no more than 750 m/min (2500 sfm). A special polycrystalline diamond, two-flute side milling cutter is used to machine the edge contour of flat tensile specimens.

**Nontraditional Milling.** Electrical discharge machining is effective in producing complex shapes, particularly those that are internal. For example, a 25 × 50 mm (1 × 2 in.) hole was sunk into an extruded bar of aluminum-SiC using a graphite electrode with negative polarity.

Chemical milling is ideally suited to large, shallow cuts in thin sheets. Burrs are not produced, and workpiece surface stresses are not generated. Techniques developed for aluminum are used. Etch rates in the range of 0.12 to 0.025 mm/min (0.005 to 0.001 in./min) are possible with a sodium hydroxide etchant.

**Threading.** Standard methods are used, including outside and inside diameter, single-point lathe threading, tapping, die threading, and rolled thread forming. In using standard high-speed steel taps, tool wear should be closely monitored to ensure dimensional accuracy.

**Grinding.** The hardness and free-cutting tendency of aluminum-SiC MMCs make these composites easier to grind than unreinforced aluminum alloys. Bright and smooth surface finishes can be obtained with moderate clogging of the wheel. Grinding parameters have not been developed for standard operations such as surface grinding and cylindrical grinding. Other abrasive machining procedures, such as honing, lapping, and free abrasive slurry surface grinding, are easily adapted for aluminum-SiC composites. General guidelines for machining aluminum-SiC MMCs are given in Table 2.

**Titanium-SiC MMCs** can be laser cut in thin layers. An argon-assisted flow should be used to avoid oxidation of the titanium matrix. In one case, a workpiece 0.8 mm (0.03 in.) thick was cut with a 1500-W $CO_2$ laser at 8 m/min (25 sfm). Quality was improved by reducing the cutting speed to 0.6 m/min (2.0 sfm) and using 700 ms pulses of 1000 W at 220 Hz.

The cutting of a 3.2 mm (0.13 in.) titanium-SiC workpiece was done with a $CO_2$ beam of 1.4 ms pulses using 1500 W at 220 Hz at 0.25 m/min (0.82 sfm). The titanium matrix was observed to have melted back from the edge of the cut for a distance of about 0.1 mm (0.004 in.), as shown in Fig. 5. The resulting slag adhered to the exit side of the cut to a height of 0.5 mm (0.02 in.). A YAG laser provides slightly better quality than a $CO_2$ type when used on this material. Typical parameters would be 0.15 m/min (0.50 sfm) with a 75 mm (3 in.) focal length lens on a pulsed 400-W unit.

## Oxide-Reinforced Composites

Alumina fiber reinforced MMCs are difficult to machine. In one case, conventional high-speed steel twist drills had a tool life of less than one hole. The following guidelines for drilling, turning, and milling are based on studies of aluminum-matrix composites reinforced with alumina Fiber FP from E.I. Du Pont de Nemours & Company, Inc. (Ref 7). Alumina fiber concentrations were 55% by volume. In drilling and milling, test plates 12.5 mm (½ in.) thick were used. In turning tests, bars measuring 50 mm (2 in.) in diameter were used. The procedures discussed generally apply to both aluminum- and magnesium-matrix composites of this type.

**Drilling.** All drilling experiments were conducted with 6.35 mm (¼ in.) diam solid-carbide and polycrystalline diamond tools. In a preliminary evaluation, the feed rate boundary condition for solid-carbide drills was established, using a constant speed of 22.5 m/min (75 sfm). A feed rate of 0.36 mm/

**Table 2 Machining guidelines for aluminum-SiC MMCs**

| Process | Required equipment | Tooling | Speed, m/min (sfm) | Feed rate | Comments |
|---|---|---|---|---|---|
| Cutoff sawing | Horizontal band saw with positive hydraulic head feed | Carbide-tip blades | 120 (400) | Maximum | Good cut quality. Maintain moderate pressure to maximize blade life. |
| Sample cutting | Diamond blade wafering saw | Circular diamond blade | ... | ... | Effective for cutting small precision samples |
| Wire EDM | Conventional EDM equipment | Wire electrode | ... | ... | Aluminum-SiC cuts much like unreinforced aluminum alloys containing high silicon contents (~15–20 wt%), although cutting speed may be reduced somewhat. |
| Abrasive waterjet | Conventional abrasive waterjet equipment | ... | ... | 5–125 mm/min (0.2–5.0 in./min) | Cuts thin sections easily. Cutting speed greatly reduced for thick sections |
| Turning | Conventional lathes | Polycrystalline diamond | 180–210 (600–700) | 0.065–0.12 mm/rev (0.0025–0.005 in./rev) | Best surface quality at low feed rates |
| Chemical milling | Conventional etchants used for aluminum | ... | ... | ... | Good surface finish |
| Drilling | Conventional drills with high-speed tool steel drill bits | ... | As high as practical | As high as practical | Drill bit flutes must be kept clear of drill shavings. Bits require frequent resharpening. |
| Tapping | Conventional taps and dies | ... | ... | ... | Same as conventional aluminum |

rev (0.0144 in./rev) caused premature failure of the tool due to chipping of the cutting lips. A value of 0.305 mm/rev (0.012 in./rev) was set as the upper boundary.

A soluble-oil coolant applied to the tool resulted in accelerated wear caused by the creation of an abrasive slurry at the cutting edge. The problem was solved by using compressed air directed at the tool to remove chips.

To check lip wear, new, solid-carbide high-helix drills were run for 120 holes at speeds ranging from 7.5 to 30 m/min (25 to 100 sfm) and feeds ranging from 0.150 to 0.305 mm/rev (0.006 to 0.012 in./rev). Results are shown in Fig. 7. A feed of 0.305 mm/rev (0.012 in./rev) resulted in the least tool wear after 120 holes. This feed, considered optimum for solid-carbide tools, also represents the highest productivity rate of the feeds tested.

Although there was a notable difference in tool wear when feed was altered, this was not true of speed. At the optimum feed of 0.305 mm/rev (0.012 in./rev), there was little difference in tool wear due to speed. Tool wear begins to increase dramatically at speeds over 45 m/min (150 sfm).

In drilling an additional 60 holes, bringing the total to 180, it was determined that the optimum drilling speed with solid-carbide tools was 30 m/min (100 sfm) at a feed of 0.305 mm/rev (0.012 in./rev). This is illustrated in Fig. 8.

Changes in tool geometry did not result in meaningful gains in tool life. Only the solid-carbide high-helix drill with a 135° angle point showed any promise, but the improvement did not outweigh the cost of special tools. Drills tipped with polycrystalline diamond and natural diamond tools are not suitable alternatives to less expensive solid carbides, which outperformed them in terms of wear and productivity.

Parameters for drilling aluminum-$Al_2O_3$ are transferable for drilling magnesium-$Al_2O_3$. However, the machinability of composites with commercially pure (CP) magnesium and ZE41A magnesium matrices differs markedly. Average tool wear in drilling 60 holes in the CP magnesium type was 0.099 mm (0.0039 in.), while that of the other type was 1.067 mm (0.0420 in.).

*Conclusions*. Drilling of Fiber FP aluminum can be performed with solid-carbide high-helix drills at a speed of 30 m/min (100 sfm) and a feed of 0.305 mm/rev (0.012 in./rev).

**Turning** studies were carried out using a C-2 grade uncoated carbide insert with a number one, or 0.397 mm (1/64 in.), corner radius. The insert material was the same as that used to produce the solid-carbide tools used in drilling.

At feeds above 0.320 mm/rev (0.0126 in./rev), the rate at which wear progresses was much less than it was at feed rates below this figure (Fig. 9). In addition, feeds above 0.395 mm/rev (0.0156 in./rev) produced unacceptable surface finishes. As in the drilling study, with speeds over 30 m/min (100 sfm), the rate of wear on the insert in turning increased drastically with changes in speed (Fig. 10). Diamond, in polycrystal-

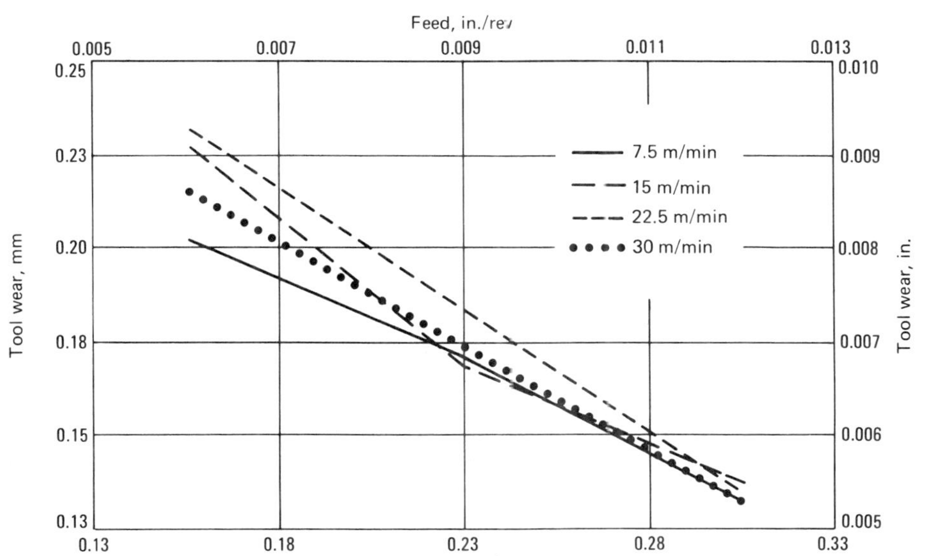


**Fig. 7** Tool wear versus feed rate at four surface speeds in drilling 120 holes in a Fiber FP aluminum MMC using solid carbide tools. Source: Ref 7

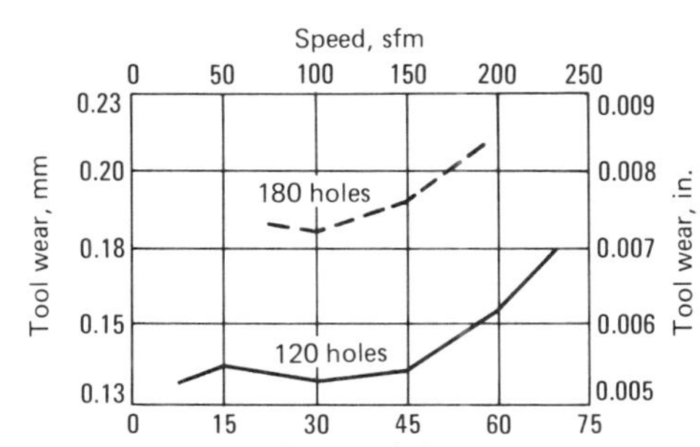


**Fig. 8** Tool wear versus speed after drilling 120 and 180 holes in a Fiber FP aluminum MMC. Tool wear increases after 45 m/min (150 sfm). Source: Ref 7

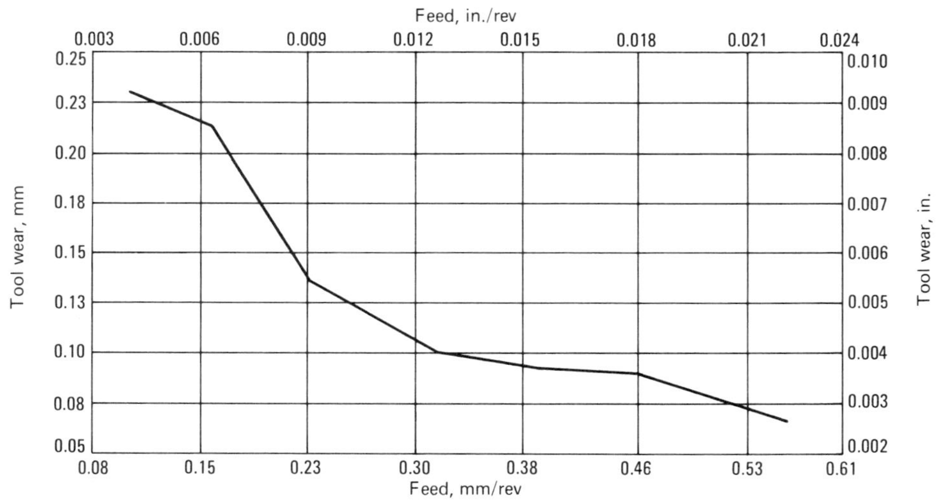


**Fig. 9** Tool wear versus feed rate in turning a Fiber FP aluminum MMC using an uncoated C-2 grade insert. Note how the wear rate progress was significantly less when feed rates reached ≧0.320 mm/rev (0.0126 in./rev). Source: Ref 7

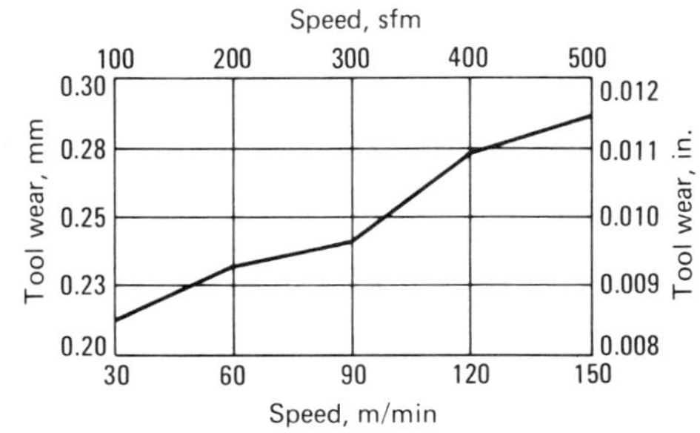


**Fig. 10** Wear of C-2 carbide inserts as a function of surface speed when turning 300 mm (12 in.) of Fiber FP aluminum MMC at 0.40 mm/rev (0.015 in./rev) and taking a 0.50 mm (0.020 in.) depth of cut. Source: Ref 7

line or natural form, is not suitable for turning these composites.

Of the ten tool materials tested, a ceramic-coated carbide showed the most promise as an alternative to uncoated carbide (Table 3). After 305 mm (12 in.) of linear cut, it wore 0.163 mm (0.0064 in.) compared with 0.185 mm (0.0073 in.) for uncoated carbide under identical conditions.

*Conclusions*. Turning of Fiber FP aluminum can be performed with C-2 uncoated carbide or ceramic-coated carbide at a speed of 30 m/min (100 sfm). The highest feed rate that produces an acceptable surface finish should be used to enhance tool life. In turning, large nose radius to round inserts should be used to enhance surface finish at high feed rates.

Using the speed of 30 m/min (100 sfm) selected as optimum for turning and the base knowledge that higher feed rates yield better tool life, milling trials were made with a round insert milling cutter. Feed rates of 0.95 mm/tooth (0.038 in./tooth) and 1.5 mm/tooth (0.062 in./tooth) were used. The depth of cut was 1.25 mm (0.050 in.). The tool was found to exhibit a normal wear pattern.

*Conclusions*. Milling of Fiber FP aluminum can be performed with C-2 uncoated or ceramic-coated carbide tools at 30 m/min (100 sfm). As in turning, the highest feed rate that produces an acceptable surface finish should be used.

## Drilling

**Dissimilar-Material Laminates.** In drilling composites such as a graphite/metal sandwich, peck drilling can be used. This method was invented at Northrop Aircraft Division when problems were encountered in drilling fastener holes in a 38 mm (1.5 in.) thick stackup of titanium/graphite/titanium. A cost-effective, quality hole could not be produced with conventional power feed drilling. Mist coolant caused titanium chips and graphite sludge (mixture of graphite dust and coolant) to impact into the drill flutes. As a result, holes in the graphite were oversize, and the tools eventually broke.

In peck drilling, deep holes are drilled without coolant. The drill advances to an adjustable depth and then withdraws from the hole to clear chips and to dissipate heat. It then reenters the hole to drill deeper and repeats the "woodpecking" action until the desired depth is reached (Fig. 11). Optimum peck drilling parameters for three materials are listed in Table 4. All holes were produced dry, using carbide-tip drills having a standard NAS-907 P3 split point geometry. Data are for hole sizes up to 9.5 mm (0.375 in.) in diameter.

As an alternative, the one-step drilling/countersinking method, which is more cost effective than peck drilling, can be used. However, it is suitable only for drilling structures that are 25 mm (1.000 in.) thick or less and have hard material in the stackup that does not exceed 5 mm (0.200 in.). This procedure uses specially designed equipment that drills, reams, and countersinks in one operation. The drill motor is an air-operated, hydraulically controlled feed rate type that is clamped to the work surface by an expandable collet. Use of a template foot pad with the collet ensures perpendicularity between the cutter axis and the work material. The operator simply inserts the collet into a previously drilled hole while holding the foot pad firmly against the material surface. Clamping and drilling start when the motor is started. Parameters for the one-step drilling/countersinking of graphite/aluminum, graphite/titanium, and graphite/steel composites are given in Table 5. Additional information on producing holes in

**Table 3 Wear developed on inserts of various materials in turning Fiber FP aluminum MMCs**

| Trial No. | Insert grade | Material description | Speed | | Feed | | Depth of cut | | Length of cut | | Flank/ radius wear | |
|---|---|---|---|---|---|---|---|---|---|---|---|---|
| | | | m/min | sfm | mm/rev | in./rev | mm | in. | mm | in. | mm | in. |
| 1 | TC-30 | TiN+TiC | 30 | 100 | 0.40 | 0.0154 | 1.0 | 0.040 | 50 | 2 | 0.23 | 0.0090 |
| 2 | TC-30 | TiN+TiC | 150 | 500 | 0.40 | 0.0154 | 1.0 | 0.040 | 50 | 2 | 0.35 | 0.0137 |
| 3 | K-060 | Cold press ceramic | 150 | 500 | 0.40 | 0.0154 | 1.0 | 0.040 | 50 | 2 | 0.66 | 0.0260 |
| 4 | K-090 | Hot press ceramic | 150 | 500 | 0.40 | 0.0154 | 1.8 | 0.070 | 50 | 2 | 0.84 | 0.0330 |
| 5 | SiAlON | Silicon nitride | 150 | 500 | 0.40 | 0.0154 | 1.8 | 0.070 | 50 | 2 | 0.62 | 0.0243 |
| 6 | KC850 | TiN multi-phase | 30 | 100 | 0.40 | 0.0154 | 1.0 | 0.040 | 250 | 10 | 0.35 | 0.0137 |
| 7 | K68 | C-2 uncoated | 30 | 100 | 0.40 | 0.0154 | 1.0 | 0.040 | 250 | 10 | 0.24 | 0.0094 |
| 8 | 415 | TiN | 30 | 100 | 0.40 | 0.0154 | 0.5 | 0.020 | 250 | 10 | 0.235 | 0.0093 |
| 9 | CBN | Cubic boron nitride | 150 | 500 | 0.40 | 0.0154 | 1.0 | 0.040 | 250 | 10 | 0.43 | 0.0170 |
| 10 | KC910 | Ceramic-coated carbide | 30 | 100 | 0.50 | 0.0200 | 1.0 | 0.040 | 300 | 12 | 0.16 | 0.0064 |
| 11 | K2000 | Silicon nitride | 150 | 500 | 0.40 | 0.0154 | 0.5 | 0.020 | 50 | 2 | 0.18 | 0.0072 |

Source: Ref 7

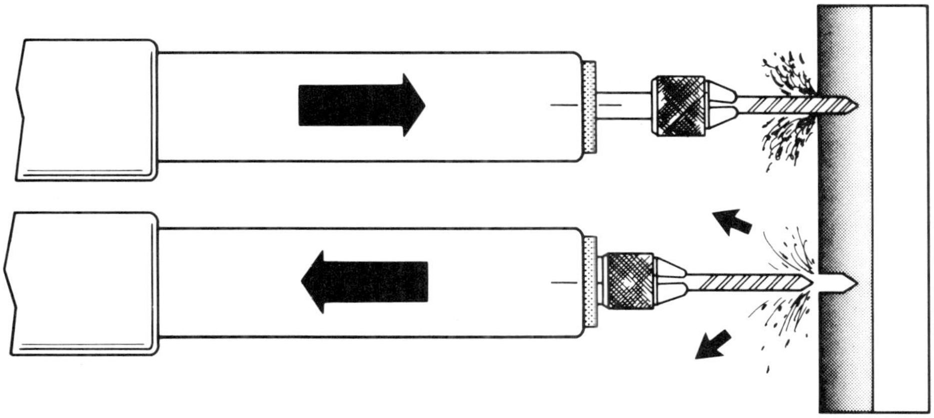

**Fig. 11** Schematic of peck drilling, in which drill advances into material, retracts, cools, and clears chips before repeating the process

dissimilar-material laminates is provided in Ref 8 to 10.

**Drilling Graphite-Epoxy With Aluminum and Titanium Substructures.** The holemaking method and the tool used vary with fiber type, fiber presentation, fiber combinations, fiber-metal combinations, and resin type. Drills that perform well in graphite-only structures, such as the four-flute, tapered drill, do not cut well in metal. A metal structure requires a more conventional, chip-cutting tool. The graphite still imposes a requirement for carbide cutters, and the titanium imposes a maximum cutting speed of about 300 rev/min for most common fastener hole sizes.

In the case of graphite-epoxy composites with a metallic substructure, the metal provides a built-in backup that prevents exit side fiber breakout or delamination in the graphite portion of the structure if it is well clamped. Because it also provides a backing for self-colleting drill units, these tools can be safely used in this type of structure.

The preferred drill configuration for this type of workpiece is the eight-faceted twist drill, and the preferred equipment would be a self-colleting drill motor. Air over oil feed, noncolleting drill motors held in a drill jig or by a robot arm would also be satisfactory. Drill speed should be about 2000 rev/min for aluminum and 300 rev/min for titanium. No cutting fluid is recommended in either case.

Pulling aluminum chips through the graphite structure with the drill should cause no problem, but titanium chips could cause some erosion in the graphite, allowing the hole to become oversize. Erosion can also occur to any cast-in-place plastic shim that may be used at the faying surface.

If erosion proves to be a problem, holes must be drilled undersize and reamed to size with a straight-flute carbide-insert reamer. Peck drilling (described earlier) is an option. Considering the reaming characteristics of titanium, one would expect the final hole to be about 0.025 mm (0.001 in.) smaller in diameter in the titanium than in the graphite; the reverse, however, appears to be true.

When an interference fit fastener is required in the aluminum substructure for fatigue improvement, first the correct-size hole for aluminum should be drilled through both the graphite and the aluminum. The graphite should then be removed and the hole reamed up to its correct clearance size. Recommendations for countersinking in this combination of materials are no different from those for graphite-epoxy unidirectional tape or cloth.

## Machining Guidelines For Honeycomb Structures

In machining honeycomb core (nonmetal or metal) in the expanded condition (Fig. 12), one of the major difficulties is stabilizing the core. A means of holding the workpiece and preventing excessive deflection of the cell walls is needed. If deflections are not reduced, surface finish will be extremely poor.

Stabilization is usually obtained by using temporary or permanent backing materials. Some producers of fiberglass cores supply a one-sided adhesive material that can be attached to the honeycomb structure. The adhesive side is attached to the honeycomb, the smooth side against a vacuum plate. However, temporary backing on aluminum may contaminate the metal, which must be cleaned with extreme care after the backing is removed. The ice chuck is a technique used to avoid the problem.

**Table 4 Parameters for peck drilling**

| Material | Speed, rev/min | Feed rate per revolution mm | in. | Peck cycle/ 25 mm (1 in.)/min |
|---|---|---|---|---|
| Titanium | 550 | 0.050 | 0.002 | 60 |
| Graphite | 550 | 0.10 | 0.004 | 30 |
| Aluminum | 550 | 0.10 | 0.004 | 30 |

Note: All holes were produced dry, using carbide-tip drills. The recommended parameters for hole sizes up to 9.5 mm (0.375 in.) are listed.

With this technique, the honeycomb is partially submerged in a pool of water that is then frozen to secure the core. A special fixture with passages built into it circulates a low-temperature freezing fluid and a high-temperature thawing fluid.

Other proprietary stabilization methods are used by leading aerospace companies. Most of these involve the use of special paper, plastic, or adhesive backing materials.

The stabilization problem can be minimized by designing parts that require little or no machining on one side, allowing the skin to be bonded to one side before the other is machined. Slight steps in the thickness of the material can be produced by the core fabricator, and the material can be supplied ready for bonding on one side. With one side secured by bonding, the part can be vacuum clamped, and the opposite side can then be machined to any contour desired.

**Tools.** Standard aluminum-cutting end mills and valve stem cutters are used to machine aluminum honeycomb. Valve stem cutters are limited to surfaces that are flat or have extremely large radii. Even for this type of cut, they are not as good as standard end mills for the higher-density cores. Solid-carbide tools perform better than high-speed steel in this application.

Machines should be capable of spindle speeds exceeding 10 000 rev/min to produce acceptable surface finishes. Spindle speed should be variable because valve stem cutters have a tendency to develop loading problems that are sensitive to spindle speed. A minimum axial depth of cut of the order of 0.25 mm (0.010 in.) is needed to obtain good surface finishes.

**Table 5 Parameters for one-step drilling/countersinking**

| Materials | Hole diameter (max) mm | in. | Speed, rev/min | Feed rate per revolution mm | in. |
|---|---|---|---|---|---|
| Graphite and aluminum | 3.967 | 0.1562 | 2800 | 0.025–0.040 | 0.0010–0.0015 |
| | 4.763 | 0.1875 | 2800 | 0.025–0.040 | 0.0010–0.0015 |
| | 6.350 | 0.2500 | 2800 | 0.025–0.040 | 0.0010–0.0015 |
| | 7.938 | 0.3125 | 1800 | 0.045–0.055 | 0.0017–0.0022 |
| | 9.525 | 0.3750 | 1800 | 0.045–0.055 | 0.0017–0.0022 |
| Graphite and titanium or steel | 3.967 | 0.1562 | 400 | 0.10–0.15 | 0.0040–0.0050 |
| | 4.763 | 0.1875 | 400 | 0.10–0.15 | 0.0040–0.0050 |
| | 6.350 | 0.2500 | 400 | 0.10–0.15 | 0.0040–0.0050 |
| | 7.938 | 0.3125 | 400 | 0.10–0.15 | 0.0040–0.0050 |
| | 9.525 | 0.3750 | 400 | 0.10–0.15 | 0.0040–0.0050 |

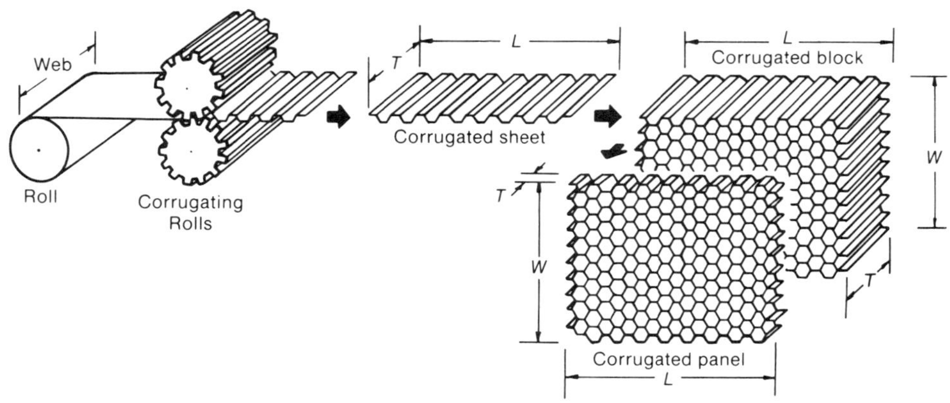


**Fig. 12** Expansion method of honeycomb core fabrication. This method consists of printing an adhesive on ribbon sheets, stacking the sheets, and curing the stack in a press at an elevated temperature to form a block of honeycomb.

A unique machining problem is encountered when two pieces of core are used in the same part and bonded end to end. A strip on either side of the joint is filled with potting compound to strengthen the junction. The machining characteristics of the compound are vastly different from those of the honeycomb core. The material causes a sudden increase in spindle torque that may slow down the spindle or even cause stalling if power is not adequate. In addition, the potting compound is abrasive, which causes accelerated tool wear, and the machining process produces dust and noxious fumes.

Band saws are used to cut flat or tapered slabs from core blocks for later machining or to remove parts that have had their upper surfaces machined. In addition, the periphery of a core block can be trimmed with band saws.

In band sawing honeycomb with paper or fiberglass cores, a blade speed of 360 m/min (1200 sfm) is recommended for all cuts. Blade type for paper cores is 10 pitch, while that for fiberglass may be 10 pitch or diamond grit type, depending on core weight. Feeds also vary with core weight. For example, a 10-pitch blade, a feed of 635 mm/min (25 in./min), and a blade speed of 360 m/min (1200 sfm) are suggested for fiberglass cores weighing 65 or 90 kg/m$^3$ (4.0 or 5.5 lb/ft$^3$).

**Fig. 13** High-speed steel tools used for machining nonmetallic honeycomb structures. Left, circular milling cutter chip shredder to break up chips used for finish machining. Right, chip chaser type with notched bottom used for rough machining. Courtesy of Lockheed Aeronautical Systems Company

A diamond grit blade, a feed of 305 mm/min (12 in./min), and a blade speed of 360 m/min (1200 sfm) are suggested for fiberglass core with weights of 130 and 190 kg/m$^3$ (8.0 and 12.0 lb/ft$^3$). By comparison, a paper core weighing 130 kg/m$^3$ (8.0 lb/ft$^3$) is sawed at a feed of 305 mm/min (12 in./min), using a speed of 360 m/min (1200 sfm) and a 10-pitch blade.

Valve stem and abrasive cutters are used in milling honeycombs with paper or fiberglass cores. Valve stem tools are used primarily to generate flat surfaces, but curved surfaces with sufficiently large radii can be produced. However, any degree of curvature will cause deterioration in surface finish quality. When a part has many or all flat surfaces joined by radiused corners, one practice calls for machining the part with a valve stem cutter and finishing corners with a conventional hand grinder.

In milling paper cores with valve stem cutters, parameters are based on the material and its weight. For example, a speed of 14 000 rev/min and a feed of 12.7 m/min (500 in./min) are used to mill paper with a density of 50 kg/m$^3$ (3 lb/ft$^3$). The valve stem cutter has a diameter of 40 mm (1.5 in.).

Parameters for diamond grit cutters also vary with the material and its weight. For example, in milling fiberglass cores less than 95 kg/m$^3$ (6 lb/ft$^3$) in density, a 50 mm (2 in.) cylinder is used at a speed of 16 000 to 20 000 rev/min and a feed of 7.6 m/min (300 in./min).

The high-speed steel roughing cutter shown on the right-hand side in Fig. 13 is used for machining nonmetallic honeycomb structures (see Table 6). The tool is 63 mm (2.45 in.) in overall diameter. It is a chip chaser type cutter with a notched bottom to aid in cutting. The finish cutter shown on the left-hand side of Fig. 13 is also made from high-speed steel. This circular milling cutter, which has a chip shredder to break up chips, is also 63 mm (2.45 in.) in diameter.

When using roughing cutters, cutter feed should be in the ribbon direction, the direction in which the honeycomb would be pulled apart if delaminated. Circular cutters should cut in the conventional rotational mode. Climb cutting will leave a surface with attached hairs or loose fibers. A good bonding surface must be free of loose fibers. A vortex tube that delivers cooled air has been used as a machining assist. Dust and chips are vacuumed away as-generated by a shop vacuum system mounted on the machine.

**Sandwich Panels.** In some cases, sandwich panels are machined. The core may be nonmetallic or metallic, but the skins are usually aluminum or fiberglass. Routing is the typical machining operation used to cut a smaller pattern from the sheet as a whole and to make holes for attachment points. Solid-carbide router bits perform well in pattern cutting. They are normally run at

**Table 6 Machining parameters for milling nonmetallic honeycomb structures using the roughing and finishing tools shown in Fig. 13**

| Honeycomb core material | Cell size | | Density | | Feed rate | | | |
|---|---|---|---|---|---|---|---|---|
| | | | | | Roughing | | Finishing | |
| | mm | in. | kg/m³ | lb/ft³ | mm/min | in./min | mm/min | in./min |
| HRH-10 aramid phenolic, 0°, 90° fiber orientation | 3 | 1/8 | 30 | 1.8 | 7600 | 300 | 4600–6100 | 180–240 |
| | 3 | 1/8 | 50 | 3.0 | 7600 | 300 | 4600–6100 | 180–240 |
| HRP glass phenolic, 0°, 90° fiber orientation | 5 | 3/16 | 65 | 4.0 | 7600 | 300 | 4600–6100 | 180–240 |
| | 5 | 3/16 | 90 | 5.5 | 7600 | 300 | 4600–6100 | 180–240 |
| | 5 | 3/16 | 130 | 8.0 | 4600 | 180 | 3800–4600 | 150–180 |
| HFT glass phenolic, 45°, 135° fiber orientation | 3 | 1/8 | 50 | 3.0 | 7600 | 300 | 4600–6100 | 180–240 |
| | 3 | 1/8 | 51 | 3.2 | 7600 | 300 | 4600–6100 | 180–240 |
| | 3 | 1/8 | 65 | 4.0 | 7600 | 300 | 4600–6100 | 180–240 |

Note: Chip chasers are preferred for all roughing applications. Circular milling tools should be used for all finish machining with the exception of the 130 kg/m³ (8.0 lb/ft³) HRH-10 aramid phenolic; because of excessive heat buildup on the circular cutter, a chip chaser should be used for roughing and finishing. Rough and finish cutting speeds: 10 000–13 000 rev/min (2000–2600 m/min, or 6550–8515 sfm). Depth of cut: 7.5–12.5 mm (0.300–0.500 in.) for roughing; 0.25–0.50 mm (0.010–0.200 in.) for finishing

about 15 000 rev/min. Feed rates vary widely, being dependent on the material and operations. Diamond abrasive hole saws are preferred in making holes.

## ACKNOWLEDGMENT

The author wishes to thank the following individuals for their valuable assistance during the preparation of this manuscript: Walter R. Mohn, Advanced Composite Materials Corporation; Mark R. van den Bergh, DWA Composite Specialties, Inc.; Thomas Slawson, Ridge Metals Inc.; Octavio L. Mamon, General Dynamics, Space Systems Division; Michael E. Buck, Avco Specialty Materials/subsidiary of Textron Inc.; Dennis Plank, Boeing Company; Ron Wood, Lockheed Aeronautical Systems Company; William P. Koster, Metcut Research Associates, Inc.; and T.B. Gurganus, Aluminum Company of America.

## REFERENCES

1. P.S. Waldrop, "Metal Matrix Composites: Manufacturing Challenges," Report EM85-106, Society of Manufacturing Engineers, 1985
2. J. Darchuk and L. Migliore, Guidelines for Laser Cutting, in *Lasers and Applications*, Springer-Verlag, 1985, p 91-97
3. *Guide for Materials Processing by Lasers*, Section 4, p 5; Section 5; Section 8, p 2, Laser Institute of America, 1977
4. B. Ward, Supersonic Characteristics of Nozzles Used With Lasers for Cutting, in *Proceedings of ICALEO*, Vol 44, Laser Institute of America, 1985, p 94-101
5. W.E. Lawson, Laser Cutting of Composites, in *Composites for Manufacturing*, Society of Manufacturing Engineers, 1986
6. D. Pavluk and W.R. Mohn, *Guide to Machining SXA Engineered Materials*, Advanced Composites Materials Corporation, 1985
7. M.J. McGinty and C.W. Preuss, "Machining of Fiber Metal Matrix Composites," Paper presented at the Congress on High Productivity in Machining, Materials, and Processing (New Orleans, LA), ASM INTERNATIONAL, 1985
8. J.P. Chanani and J.A. Boldt, Manufacturing Methods for Composite Hole Generation, *Trans. SAE*, Vol 90, 1982
9. J.A. Boldt, "Machining and Drilling of Composite and Composite/Metallic Structures," Process Specification MA-113, Northrop Corporation
10. J.P. Chanani, "Optimal Methods for Generating Holes in Advanced Composites and Hybrids," Paper presented at WESTECH, American Society for Metals, 1985

# Metric Conversion Guide

This Section is intended as a guide for expressing weights and measures in the Système International d'Unités (SI). The purpose of SI units, developed and maintained by the General Conference of Weights and Measures, is to provide a basis for worldwide standardization of units and measure. For more information on metric conversions, the reader should consult the following references:

- "Standard for Metric Practice," E 380, *Annual Book of ASTM Standards*, American Society for Testing and Materials, 1916 Race Street, Philadelphia, PA 19103
- "Metric Practice," ANSI/IEEE 268–1982, American National Standards Institute, 1430 Broadway, New York, NY 10018
- *The International System of Units*, SP 330, 1986, National Institute of Standards and Technology. Order from Superintendent of Documents, U.S. Government Printing Office, Washington, DC 20402-9325
- *Metric Editorial Guide*, 4th ed. (revised), 1985, American National Metric Council, 1010 Vermont Avenue NW, Suite 1000, Washington, DC 20005–4960
- *ASME Orientation and Guide for Use of SI (Metric) Units*, ASME Guide SI 1, 9th ed., 1982, The American Society of Mechanical Engineers, 345 East 47th Street, New York, NY 10017

## Base, supplementary, and derived SI units

| Measure | Unit | Symbol |
|---|---|---|
| **Base units** | | |
| Amount of substance | mole | mol |
| Electric current | ampere | A |
| Length | meter | m |
| Luminous intensity | candela | cd |
| Mass | kilogram | kg |
| Thermodynamic temperature | kelvin | K |
| Time | second | s |
| **Supplementary units** | | |
| Plane angle | radian | rad |
| Solid angle | steradian | sr |
| **Derived units** | | |
| Absorbed dose | gray | Gy |
| Acceleration | meter per second squared | m/s$^2$ |
| Activity (of radionuclides) | becquerel | Bq |
| Angular acceleration | radian per second squared | rad/s$^2$ |
| Angular velocity | radian per second | rad/s |
| Area | square meter | m$^2$ |
| Capacitance | farad | F |
| Concentration (of amount of substance) | mole per cubic meter | mol/m$^3$ |
| Conductance | siemens | S |
| Current density | ampere per square meter | A/m$^2$ |
| Density, mass | kilogram per cubic meter | kg/m$^3$ |
| Electric charge density | coulomb per cubic meter | C/m$^3$ |
| Electric field strength | volt per meter | V/m |
| Electric flux density | coulomb per square meter | C/m$^2$ |
| Electric potential, potential difference, electromotive force | volt | V |
| Electric resistance | ohm | Ω |
| Energy, work, quantity of heat | joule | J |
| Energy density | joule per cubic meter | J/m$^3$ |
| Entropy | joule per kelvin | J/K |
| Force | newton | N |
| Frequency | hertz | Hz |
| Heat capacity | joule per kelvin | J/K |
| Heat flux density | watt per square meter | W/m$^2$ |
| Illuminance | lux | lx |
| Inductance | henry | H |
| Irradiance | watt per square meter | W/m$^2$ |
| Luminance | candela per square meter | cd/m$^2$ |
| Luminous flux | lumen | lm |
| Magnetic field strength | ampere per meter | A/m |
| Magnetic flux | weber | Wb |
| Magnetic flux density | tesla | T |
| Molar energy | joule per mole | J/mol |
| Molar entropy | joule per mole kelvin | J/mol · K |
| Molar heat capacity | joule per mole kelvin | J/mol · K |
| Moment of force | newton meter | N · m |
| Permeability | henry per meter | H/m |
| Permittivity | farad per meter | F/m |
| Power, radiant flux | watt | W |
| Pressure, stress | pascal | Pa |
| Quantity of electricity, electric charge | coulomb | C |
| Radiance | watt per square meter steradian | W/m$^2$ · sr |
| Radiant intensity | watt per steradian | W/sr |
| Specific heat capacity | joule per kilogram kelvin | J/kg · K |
| Specific energy | joule per kilogram | J/kg |
| Specific entropy | joule per kilogram kelvin | J/kg · K |
| Specific volume | cubic meter per kilogram | m$^3$/kg |
| Surface tension | newton per meter | N/m |
| Thermal conductivity | watt per meter kelvin | W/m · K |
| Velocity | meter per second | m/s |
| Viscosity, dynamic | pascal second | Pa · s |
| Viscosity, kinematic | square meter per second | m$^2$/s |
| Volume | cubic meter | m$^3$ |
| Wavenumber | 1 per meter | 1/m |

## Conversion factors

| To convert from | to | multiply by |
|---|---|---|
| **Angle** | | |
| degree | rad | 1.745 329 E − 02 |
| **Area** | | |
| in.$^2$ | mm$^2$ | 6.451 600 E + 02 |
| in.$^2$ | cm$^2$ | 6.451 600 E + 00 |
| in.$^2$ | m$^2$ | 6.451 600 E − 04 |
| ft$^2$ | m$^2$ | 9.290 304 E − 02 |
| **Bending moment or torque** | | |
| lbf · in. | N · m | 1.129 848 E − 01 |
| lbf · ft | N · m | 1.355 818 E + 00 |
| kgf · m | N · m | 9.806 650 E + 00 |
| ozf · in. | N · m | 7.061 552 E − 03 |
| **Bending moment or torque per unit length** | | |
| lbf · in./in. | N · m/m | 4.448 222 E + 00 |
| lbf · ft/in. | N · m/m | 5.337 866 E + 01 |
| **Current density** | | |
| A/in.$^2$ | A/cm$^2$ | 1.550 003 E − 01 |
| A/in.$^2$ | A/mm$^2$ | 1.550 003 E − 03 |
| A/ft$^2$ | A/m$^2$ | 1.076 400 E + 01 |
| **Electricity and magnetism** | | |
| gauss | T | 1.000 000 E − 04 |
| maxwell | μWb | 1.000 000 E − 02 |
| mho | S | 1.000 000 E + 00 |
| Oersted | A/m | 7.957 700 E + 01 |
| Ω · cm | Ω · m | 1.000 000 E − 02 |
| Ω circular-mil/ft | μΩ · m | 1.662 426 E − 03 |
| **Energy (impact, other)** | | |
| ft · lbf | J | 1.355 818 E + 00 |
| Btu (thermochemical) | J | 1.054 350 E + 03 |
| cal (thermochemical) | J | 4.184 000 E + 00 |
| kW · h | J | 3.600 000 E + 06 |
| W · h | J | 3.600 000 E + 03 |
| **Flow rate** | | |
| ft$^3$/h | L/min | 4.719 475 E − 01 |
| ft$^3$/min | L/min | 2.831 000 E + 01 |
| gal./h | L/min | 6.309 020 E − 02 |
| gal./min | L/min | 3.785 412 E + 00 |
| **Force** | | |
| lbf | N | 4.448 222 E + 00 |
| kip (1000 lbf) | N | 4.448 222 E + 03 |
| tonf | kN | 8.896 443 E + 00 |
| kgf | N | 9.806 650 E + 00 |
| **Force per unit length** | | |
| lbf/ft | N/m | 1.459 390 E + 01 |
| lbf/in. | N/m | 1.751 268 E + 02 |
| **Fracture toughness** | | |
| ksi $\sqrt{\text{in.}}$ | MPA $\sqrt{\text{m}}$ | 1.098 800 E + 00 |
| **Heat content** | | |
| Btu/lb | kJ/kg | 2.326 000 E + 00 |
| cal/g | kJ/kg | 4.186 800 E + 00 |
| **Heat input** | | |
| J/in. | J/m | 3.937 008 E + 01 |
| kJ/in. | kJ/m | 3.937 008 E + 01 |
| **Length** | | |
| Å | nm | 1.000 000 E − 01 |
| μin. | μm | 2.540 000 E − 02 |
| mil | μm | 2.540 000 E + 01 |
| in. | mm | 2.540 000 E + 01 |
| in. | cm | 2.540 000 E + 00 |
| ft | m | 3.048 000 E − 01 |
| yd | m | 9.144 000 E − 01 |
| mile | km | 1.609 300 E + 00 |
| **Mass** | | |
| oz | kg | 2.834 952 E − 02 |
| lb | kg | 4.535 924 E − 01 |
| ton (short, 2000 lb) | kg | 9.071 847 E + 02 |
| ton (short, 2000 lb) | kg × $10^3$(a) | 9.071 847 E − 01 |
| ton (long, 2240 lb) | kg | 1.016 047 E + 03 |
| **Mass per unit area** | | |
| oz/in.$^2$ | kg/m$^2$ | 4.395 000 E + 01 |
| oz/ft$^2$ | kg/m$^2$ | 3.051 517 E − 01 |
| oz/yd$^2$ | kg/m$^2$ | 3.390 575 E − 02 |
| lb/ft$^2$ | kg/m$^2$ | 4.882 428 E + 00 |
| **Mass per unit length** | | |
| lb/ft | kg/m | 1.488 164 E + 00 |
| lb/in. | kg/m | 1.785 797 E + 01 |
| **Mass per unit time** | | |
| lb/h | kg/s | 1.259 979 E − 04 |
| lb/min | kg/s | 7.559 873 E − 03 |
| lb/s | kg/s | 4.535 924 E − 01 |
| **Mass per unit volume (includes density)** | | |
| g/cm$^3$ | kg/m$^3$ | 1.000 000 E + 03 |
| lb/ft$^3$ | g/cm$^3$ | 1.601 846 E − 02 |
| lb/ft$^3$ | kg/m$^3$ | 1.601 846 E + 01 |
| lb/in.$^3$ | g/cm$^3$ | 2.767 990 E + 01 |
| lb/in.$^3$ | kg/m$^3$ | 2.767 990 E + 04 |
| **Power** | | |
| Btu/s | kW | 1.055 056 E + 00 |
| Btu/min | kW | 1.758 426 E − 02 |
| Btu/h | W | 2.928 751 E − 01 |
| erg/s | W | 1.000 000 E − 07 |
| ft · lbf/s | W | 1.355 818 E + 00 |
| ft · lbf/min | W | 2.259 697 E − 02 |
| ft · lbf/h | W | 3.766 161 E − 04 |
| hp (550 ft · lbf/s) | kW | 7.456 999 E − 01 |
| hp (electric) | kW | 7.460 000 E − 01 |
| **Power density** | | |
| W/in.$^2$ | W/m$^2$ | 1.550 003 E + 03 |
| **Press capacity** | | |
| See Force | | |
| **Pressure (fluid)** | | |
| atm (standard) | Pa | 1.013 250 E + 05 |
| bar | Pa | 1.000 000 E + 05 |
| in. Hg (32 °F) | Pa | 3.386 380 E + 03 |
| in. Hg (60 °F) | Pa | 3.376 850 E + 03 |
| lbf/in.$^2$ (psi) | Pa | 6.894 757 E + 03 |
| torr (mm Hg, 0 °C) | Pa | 1.333 220 E + 02 |
| **Specific heat** | | |
| Btu/lb · °F | J/kg · K | 4.186 800 E + 03 |
| cal/g · °C | J/kg · K | 4.186 800 E + 03 |
| **Stress (force per unit area)** | | |
| tonf/in.$^2$ (tsi) | MPa | 1.378 951 E + 01 |
| kgf/mm$^2$ | MPa | 9.806 650 E + 00 |
| ksi | MPa | 6.894 757 E + 00 |
| lbf/in.$^2$ (psi) | MPa | 6.894 757 E − 03 |
| MN/m$^2$ | MPa | 1.000 000 E + 00 |
| **Temperature** | | |
| °F | °C | 5/9 · (°F − 32) |
| °R | °K | 5/9 |
| **Temperature interval** | | |
| °F | °C | 5/9 |
| **Thermal conductivity** | | |
| Btu · in./s · ft$^2$ · °F | W/m · K | 5.192 204 E + 02 |
| Btu/ft · h · °F | W/m · K | 1.730 735 E + 00 |
| Btu · in./h · ft$^2$ · °F | W/m · K | 1.442 279 E − 01 |
| cal/cm · s · °C | W/m · K | 4.184 000 E + 02 |
| **Thermal expansion** | | |
| in./in. · °C | m/m · K | 1.000 000 E + 00 |
| in./in. · °F | m/m · K | 1.800 000 E + 00 |
| **Velocity** | | |
| ft/h | m/s | 8.466 667 E − 05 |
| ft/min | m/s | 5.080 000 E − 03 |
| ft/s | m/s | 3.048 000 E − 01 |
| in./s | m/s | 2.540 000 E − 02 |
| km/h | m/s | 2.777 778 E − 01 |
| mph | km/h | 1.609 344 E + 00 |
| **Velocity of rotation** | | |
| rev/min (rpm) | rad/s | 1.047 164 E − 01 |
| rev/s | rad/s | 6.283 185 E + 00 |
| **Viscosity** | | |
| poise | Pa · s | 1.000 000 E − 01 |
| stokes | m$^2$/s | 1.000 000 E − 04 |
| ft$^2$/s | m$^2$/s | 9.290 304 E − 02 |
| in.$^2$/s | mm$^2$/s | 6.451 600 E + 02 |
| **Volume** | | |
| in.$^3$ | m$^3$ | 1.638 706 E − 05 |
| ft$^3$ | m$^3$ | 2.831 685 E − 02 |
| fluid oz | m$^3$ | 2.957 353 E − 05 |
| gal. (U.S. liquid) | m$^3$ | 3.785 412 E − 03 |
| **Volume per unit time** | | |
| ft$^3$/min | m$^3$/s | 4.719 474 E − 04 |
| ft$^3$/s | m$^3$/s | 2.831 685 E − 02 |
| in.$^3$/min | m$^3$/s | 2.731 177 E − 07 |
| **Wavelength** | | |
| Å | nm | 1.000 000 E − 01 |

(a) kg × $10^3$ = 1 metric ton or 1 megagram (Mg)

## SI prefixes—names and symbols

| Exponential expression | Multiplication factor | Prefix | Symbol |
|---|---|---|---|
| $10^{18}$ | 1 000 000 000 000 000 000 | exa | E |
| $10^{15}$ | 1 000 000 000 000 000 | peta | P |
| $10^{12}$ | 1 000 000 000 000 | tera | T |
| $10^{9}$ | 1 000 000 000 | giga | G |
| $10^{6}$ | 1 000 000 | mega | M |
| $10^{3}$ | 1 000 | kilo | k |
| $10^{2}$ | 100 | hecto(a) | h |
| $10^{1}$ | 10 | deka(a) | da |
| $10^{0}$ | 1 | BASE UNIT | |
| $10^{-1}$ | 0.1 | deci(a) | d |
| $10^{-2}$ | 0.01 | centi(a) | c |
| $10^{-3}$ | 0.001 | milli | m |
| $10^{-6}$ | 0.000 001 | micro | μ |
| $10^{-9}$ | 0.000 000 001 | nano | n |
| $10^{-12}$ | 0.000 000 000 001 | pico | p |
| $10^{-15}$ | 0.000 000 000 000 001 | femto | f |
| $10^{-18}$ | 0.000 000 000 000 000 001 | atto | a |

(a) Nonpreferred. Prefixes should be selected in steps of $10^3$ so that the resultant number before the prefix is between 0.1 and 1000. These prefixes should not be used for units of linear measurement, but may be used for higher order units. For example, the linear measurement, decimeter, is nonpreferred, but square decimeter is acceptable.

# Abbreviations and Symbols

*a* wheel depth of cut in grinding; crystal lattice length along the *a*-axis
*A* area
$A_c$ area of cut
$A_f$ area of sliding contact on the rake face
$A_s$ shear plane area
**A** ampere
**Å** angstrom
**ac** alternating current
**AC** adaptive control
**ACC** adaptive control with constraints
**ACI** Alloy Casting Institute
**ACO** adaptive control with optimization
**AFM** abrasive flow machining
**AGMA** American Gear Manufacturers Association
**AGV** automated guided vehicle
**AIA** Aerospace Industries Association
**AISI** American Iron and Steel Institute
**AJM** abrasive jet machining
**ANSI** American National Standards Institute
**APT** automatically programmed tool
**ASP** Anti-Segregation Process
**ASTM** American Society for Testing and Materials
**at.%** atomic percent
**atm** atmosphere (pressure)
**AWJ** abrasive waterjet
**AWM** abrasive waterjet machining
**AWS** American Welding Society
*b* crystal lattice length along the *b*-axis
*B* grinding width
**bal** balance or remainder
**bcc** body-centered cubic
**Bé** Baumé (specific-gravity scale)
**BLU** basic length unit
**BUE** built-up edge
*c* specific heat of workpiece material; cost; crystal lattice length along the *c*-axis
**CAD** computer-aided design
**CAM** computer-aided manufacturing
**CAPP** computer-aided process planning
**CBN** cubic boron nitride
**CCPA** Cemented Carbides Producers Association
**CE** carbon equivalent
**CFR** Code of Federal Regulations
**CHM** chemical milling
**CIM** computer integrated manufacturing
**CL** cutter location
**cm** centimeter
**CM** chemical machining (milling)
**CMM** coordinate measuring machine
**CNC** computer numerical control
**cpm** cycles per minute
**CPM** Crucible Particle Metallurgy
**cps** cycles per second
**CPU** central processing unit
**CR** cold rolled
**CRT** cathode ray terminal
**CSD** controlled spray deposition
**CVD** chemical vapor deposition
**CVN** Charpy V-notch (impact test or specimen)
**CW** continuous wave
*d* depth of cut; used in mathematical expressions involving a derivation (denotes rate of change); diameter
*D* diameter; distance
$D_E$ equivalent diameter
$D_S$ wheel diameter
$D_W$ workpiece diameter
**dc** direct current
**DIN** Deutsche Industrie-Normen (German Industrial Standards)
**DNC** direct numerical control
**DOC** depth of cut
*e* natural log base, 2.71828
*E* modulus of elasticity (Young's modulus)
**EB** end-of-block (character); electron beam
**EBM** electron beam machining
**ECEA** end cutting edge angle
**ECM** electrochemical machining
**ECG** electrochemical grinding
**EDM** electrical discharge machining
**EIA** Electronic Industries Association
**ELI** extralow interstitial
**ELP** electropolishing
**EP** extreme pressure
**Eq** equation
***et al.*** and others
**ETP** electrolytic tough pitch
*f* feed rate
$f_t$ feed per tooth
*F* force
$F_c$ primary (horizontal) cutting force
$F_f$ feed force
$F_m$ momentum force
$F_n$ normal force on shear plane
$F_N$ normal force in grinding
$F_r$ radial or thrust force
$F_s$ shear force
$F_t$ tangential force
**fcc** face-centered cubic
**Fig.** figure
**FMS** flexible manufacturing system
**ft** foot
**GAC** geometric adaptive control
**gal.** gallon
**GPa** gigapascal
**h** hour
*h* undeformed chip thickness; height
**HAG** higher-accuracy grinding
**HAZ** heat-affected zone
**HB** Brinell hardness
**HBN** hexagonal boron nitride
**hcp** hexagonal close-packed
**HIP** hot isostatic pressing
**HK** Knoop hardness
**hp** horsepower
**HPG** higher-productivity grinding
**HPSN** hot-pressed silicon nitride
**HR** Rockwell hardness (requires scale designation, such as HRC for Rockwell C hardness)
**HSLA** high-strength low-alloy (steel)
**HSS** high-speed steel
**HV** Vickers hardness
**Hz** hertz
**IBR** integrally bladed rotator
**IC** inscribed circle
**ID** inside diameter
**in.** inch
**INCRA** International Copper Research Association
**ipt** inch per tooth
**ISO** International Organization for Standardization
**J** joule
**JIT** just-in-time

*k* thermal conductivity
**K** Kelvin
*K* stress intensity factor
$K_{Ic}$ plane-strain fracture toughness
**kg** kilogram
**kgf** kilogram force
**km** kilometer
**KB** kilobyte
**kPa** kilopascal
**ksi** kips (1000 lbf) per square inch
**kV** kilovolt
**kW** kilowatt
*ℓ* length
**L** liter
*L* length
**lb** pound
**lbf** pound force
**LBM** laser beam machining
**ln** natural logarithm (base *e*)
**log** common logarithm (base 10)
**LSG** low-stress grinding
**m** meter
**MB** megabyte
**mg** milligram
**Mg** megagram (metric tonne)
**min** minimum; minute
**MHS** material handling system
**mL** milliliter
**mm** millimeter
**MPa** megapascal
**MPIF** Metal Powder Industries Federation
**MRP** materials requirement planning
**MRR** material removal rate
**MRS** minimum residual stress
**ms** millisecond
**MSDS** material safety data sheet
**N** newton
*N* speed of rotation (rev/min); number of cycles to failure
**NASA** National Aeronautics and Space Administration
**NC** numerical control
**NFPA** National Fire Protection Association
**nm** nanometer
**No.** number
**NPS** American National Standard Straight Pipe Thread
**NPSC** American National Standard Straight Pipe Thread for Couplings
**NPT** American National Standard Taper Pipe Thread
**NPTF** Dryseal USA (American) Standard Taper Pipe Thread
**ns** nanoseconds
**OD** outside diameter
**ODS** oxide dispersion-strengthened
**OF** oxygen-free
**OFHC** oxygen-free high-conductivity (copper)
**OSHA** Occupational Safety and Health Administration
**oz** ounce
**p** page
*p* pitch
*P* cutting power
$P_g$ gross power
$P_s$ specific power
**Pa** pascal
**PC** personal computer
**PCBN** polycrystalline CBN (cubic boron nitride)
**PCD** polycrystalline diamond
**PCM** photochemical machining
**pH** negative logarithm of hydrogen-ion activity
**PH** precipitation hardenable
**PLC** programmable logic controller
**P/M** powder metallurgy
**ppm** parts per million
**PSA** pressure-sensitive adhesive
**psi** pounds per square inch
**PTP** point-to-point
**PVD** physical vapor deposition
*r* radius; ratio of uncut chip thickness to chip thickness
*R* stress (load) ratio; radius; gas constant; resultant force
$R_a$ roughness average
$R_{max}$ maximum peak-to-valley roughness height
$R_q$ root-mean-square roughness average
$R_y$ maximum peak-to-valley roughness height
$R_z$ ten-point height (roughness average)
**Ref** reference
**rem** remainder
**rev** revolution
**RHR** roughness height rating
**rms** root mean square
**s** second
**SAE** Society of Automotive Engineers
**SCC** stress-corrosion cracking
**SCEA** side cutting edge angle
**SEM** scanning electron microscopy
**SFFER** silica-flour-filled epoxy resin
**sfm** surface feet per minute
**SFR** smooth face rubber
**SI** Système International d'Unités
**SP** specific power
**SR** serrated rubber
**SUS** Saybolt Universal seconds
*t* depth of cut; feed or uncut chip thickness; time; tool lifetime; thickness
$t_c$ chip thickness
*T* temperature
**TEA** triethanolamine
**TEM** thermal energy method
**TIR** total indicator reading
**TRS** transverse rupture strength
**tsi** tons per square inch
**TSR** thermal shock resistance
*TWR* tool wear rate
*u* specific energy
$u_f$ friction energy per unit volume
$u_m$ kinetic (momentum) energy per unit volume
$u_s$ shear energy per unit volume
**UNS** Unified Numbering System
**USM** ultrasonic machining
**UTS** ultimate tensile strength
**V** volt
*V* cutting velocity
$V_c$ chip velocity
$\overline{V}_F$ slide infeed removal rate (grinding)
$V_s$ shear velocity
$\overline{V}_S$ radial wear rate of grinding wheel
$\overline{V}_W$ radial removal rate of work being ground
**vol** volume
**vol%** volume percent
**W** watt
*w* width; weight
$w_c$ width of chip
*W* chip weight
**WJM** waterjet machining
**WRP** work removal parameter
**wt%** weight percent
**YAG** yttrium-aluminum-garnet
**yr** year
$Z_W$ workpiece total volumetric removal rate (grinding)
$Z'_W$ workpiece unit-width volumetric removal rate (grinding)
° angular measure; degree
°C degree Celsius (centigrade)
°F degree Fahrenheit
↔ direction of reaction
÷ divided by
= equals
≅ approximately equals
≠ not equal to
≡ identical with
> greater than
≫ much greater than
≥ greater than or equal to
∞ infinity
∝ is proportional to; varies as
∫ integral of
< less than
≪ much less than

$\leq$ less than or equal to
$\pm$ maximum deviation
$-$ minus; negative ion charge
$\times$ diameters (magnification); multiplied by
$\cdot$ multiplied by
/ per
% percent
$+$ plus; positive ion charge
$\sqrt{\ }$ square root of
$\sim$ approximately; similar to
$\alpha$ back or rake angle
$\beta$ angle between normal force ($N$) and resultant force ($R$); shape parameter
$\gamma$ shear strain; clearance angle
$\dot{\gamma}$ shear strain rate
$\Delta$ change in quantity; an increment; a range
$\epsilon$ strain
$\dot{\epsilon}$ strain rate
$\eta$ machine efficiency
$\theta$ edge angle; characteristic life parameter; tool temperature
$\mu$ friction coefficient
**μin.** microinch
**μm** micron (micrometer)
**μs** microsecond
$\nu$ Poisson's ratio
$\pi$ pi (3.141592)
$\rho$ density
$\sigma$ stress
$\tau$ shear stress
$\phi$ shear angle
$\Omega$ ohm

## Greek Alphabet

| | | | | | |
|---|---|---|---|---|---|
| Α, α | alpha | Ι, ι | iota | Ρ, ρ | rho |
| Β, β | beta | Κ, κ | kappa | Σ, σ | sigma |
| Γ, γ | gamma | Λ, λ | lambda | Τ, τ | tau |
| Δ, δ | delta | Μ, μ | mu | Υ, υ | upsilon |
| Ε, ε | epsilon | Ν, ν | nu | Φ, φ | phi |
| Ζ, ζ | zeta | Ξ, ξ | xi | Χ, χ | chi |
| Η, η | eta | Ο, ο | omicron | Ψ, ψ | psi |
| Θ, θ | theta | Π, π | pi | Ω, ω | omega |

# Index

# C

## E

## F



## G

## Q



## R

## S

## T